Prefixes and Suffixes in Anatomical and Medical Terminology

Element	Definition and Example
mito-	thread: *mitochondrion*
mono-	alone, one, single: *monocyte*
morph-	form, shape: *morphology*
multi-	many, much: *multinuclear*
myo-	muscle: *myology*
narc-	numbness, stupor: *narcotic*
necro-	corpse, dead: *necrosis*
neo-	new, young: *neonatal*
nephr-	kidney: *nephritis*
neuro-	nerve: *neurolemma*
noto-	back: *notochord*
ob-	against, toward, in front of: *obturator*
oc-	against: *occlusion*
-oid	resembling, likeness: *sigmoid*
oligo-	few, small: *oligodendrocyte*
-oma	tumor: *lymphoma*
oo-	egg: *oocyte*
or-	mouth: *oral*
orchi-	testis: *orchiectomy*
ortho-	straight, normal: *orthopnea*
-ory	pertaining to: *sensory*
-ose	full of: *adipose*
osteo-	bone: *osteoblast*
oto-	ear: *otolith*
ovo-	egg: *ovum*
par-	give birth to, bear: *parturition*
para-	near, beyond, beside: *paranasal*
path-	disease, that which undergoes sickness: *pathology*
-pathy	abnormality, disease: *neuropathy*
ped-	children: *pediatrician*
pen-	need, lack: *penicillin*
-penia	deficiency: *thrombocytopenia*
per-	through: *percutaneous*
peri-	near, around: *pericardium*
phag-	to eat: *phagocyte*
-phil	have an affinity for: *neutrophil*
phleb-	vein: *phlebitis*
-phobia	abnormal fear, dread: *hydrophobia*
-plasty	reconstruction of: *rhinoplasty*
platy-	flat, side: *platysma*
-plegia	stroke, paralysis: *paraplegia*
-pnea	to breathe: *apnea*
pneumo(n)-	lung: *pneumonia*
pod-	foot: *podiatry*
-poiesis	formation of: *hemopoiesis*
poly-	many, much: *polyploid*
post-	after, behind: *postnatal*
pre-	before in time or place: *prenatal*
pro-	before in time or place: *prophase*

Element	Definition and Example
proct-	anus: *proctology*
pseudo-	false: *pseudostratified*
psycho-	mental: *psychology*
pyo-	pus: *pyorrhea*
quad-	fourfold: *quadriceps femoris*
re-	back, again: *repolarization*
rect-	straight: *rectus abdominis*
ren-	kidney: *renal*
rete-	network: *rete testis*
retro-	backward, behind: *retroperitoneal*
rhin-	nose: *rhinitis*
-rrhagia	excessive flow: *menorrhagia*
-rrhea	flow or discharge: *diarrhea*
sanguin-	blood: *sanguineous*
sarc-	flesh: *sarcoma*
-scope	instrument for examining a part: *stethoscope*
-sect	cut: *dissect*
semi-	half: *semilunar*
-sis	process or action: *dialysis*
steno-	narrow: *stenosis*
-stomy	surgical opening: *tracheostomy*
sub-	under, beneath, below: *subcutaneous*
super-	above, beyond, upper: *superficial*
supra-	above, over: *suprarenal*
syn- (sym-)	together, joined, with: *synapse*
tachy-	swift, rapid: *tachycardia*
tele-	far: *telencephalon*
tens-	stretch: *tensor tympani*
tetra-	four: *tetrad*
therm-	heat: *thermogram*
thorac-	chest: *thoracic cavity*
thrombo-	lump, clot: *thrombocyte*
-tomy	cut: *appendectomy*
tox-	poison: *toxemia*
tract-	draw, drag: *traction*
trans-	across, over: *transfuse*
tri-	three: *trigone*
trich-	hair: *trichology*
-trophy	a state relating to nutrition: *hypertrophy*
-tropic	turning toward, changing: *gonadotropic*
ultra-	beyond, excess: *ultrasonic*
uni-	one: *unicellular*
-uria	urine: *polyuria*
uro-	urine, urinary organs or tract: *uroscope*
vas-	vessel: *vasoconstriction*
viscer-	organ: *visceral*
vit-	life: *vitamin*
zoo-	animal: *zoology*
zygo-	union, join: *zygote*

FIFTH EDITION

HUMAN ANATOMY

KENT M. VAN DE GRAAFF

WEBER STATE UNIVERSITY

Boston, Massachusetts Burr Ridge, Illinois Dubuque, Iowa
Madison, Wisconsin New York, New York San Francisco, California St. Louis, Missouri

WCB/McGraw-Hill

A Division of The **McGraw·Hill** *Companies*

HUMAN ANATOMY, FIFTH EDITION

Recycled/acid free paper

 This book is printed on recycled, acid-free paper containing 10% postconsumer waste.

1 2 3 4 5 6 7 8 9 0 QPD/QPD 0 9 8 7

ISBN 0-697-28413-1

Publisher: *Kevin T. Kane/Michael D. Lange*
Sponsoring editor: *Kristine Noel Tibbetts*
Developmental editor: *Kelly A. Drapeau*
Marketing manager: *Keri L. Witman*
Project manager: *Marla K. Irion/Jane C. Morgan*
Production supervisor: *Sandy Ludovissy*
Interior designer: *Katherine Farmer*
Cover designer: *Jamie O'Neal*
Photo research coordinator: *John C. Leland*
Art editor: *Brenda A. Ernzen*
Compositor: *Shepherd, Inc.*
Typeface: *10/12 Goudy*
Printer: *Quebecor Printing Book Group/Dubuque*

The credits section for this book begins on page 797 and is considered an extension of the copyright page.

Library of Congress Cataloging-in-Publication Data

Van De Graaff, Kent M. (Kent Marshall), 1942–
Human anatomy/Kent M. Van De Graaff.—5th ed.
p. cm.
Includes index.
ISBN 0-697-28413-1
1. Human anatomy. I. Title.
[DNLM: 1. Anatomy. QS 4 V225h 1998]
QM23.2. V36 1997
611–dc21
DNLM/DLC
for Library of Congress 97-8453
CIP

When ordering this title, use ISBN 0-07-115625-9

http://www.mhcollege.com

The Dynamic Human

*E*xperience anatomy and physiology in an entirely new dimension. **The Dynamic Human** CD-ROM interactively illustrates the complex relationships between anatomical structures and their functions in the human body. Realistic, three-dimensional visuals are the premier features of this exciting learning tool. After a brief introduction, **The Dynamic Human** covers each body system—demonstrating to the viewer the anatomy, physiology, histology, and clinical applications of each system.

Human Anatomy, by Kent M. Van De Graaff, has been correlated to **The Dynamic Human.** Throughout the text, a "dancing man" icon appears in many figure legends, signaling the reader that information relating to this figure can be found on **The Dynamic Human** CD-ROM. A complete correlation guide is found in the preface of the book.

Windows Version 0-697-37910-8
Macintosh Version 0-697-37909-4

Contents:

Anatomical Orientation
Skeletal System
Muscular System
Nervous System
Endocrine System
Cardiovascular System
Lymphatic System
Digestive System
Respiratory System
Urinary System
Reproductive System

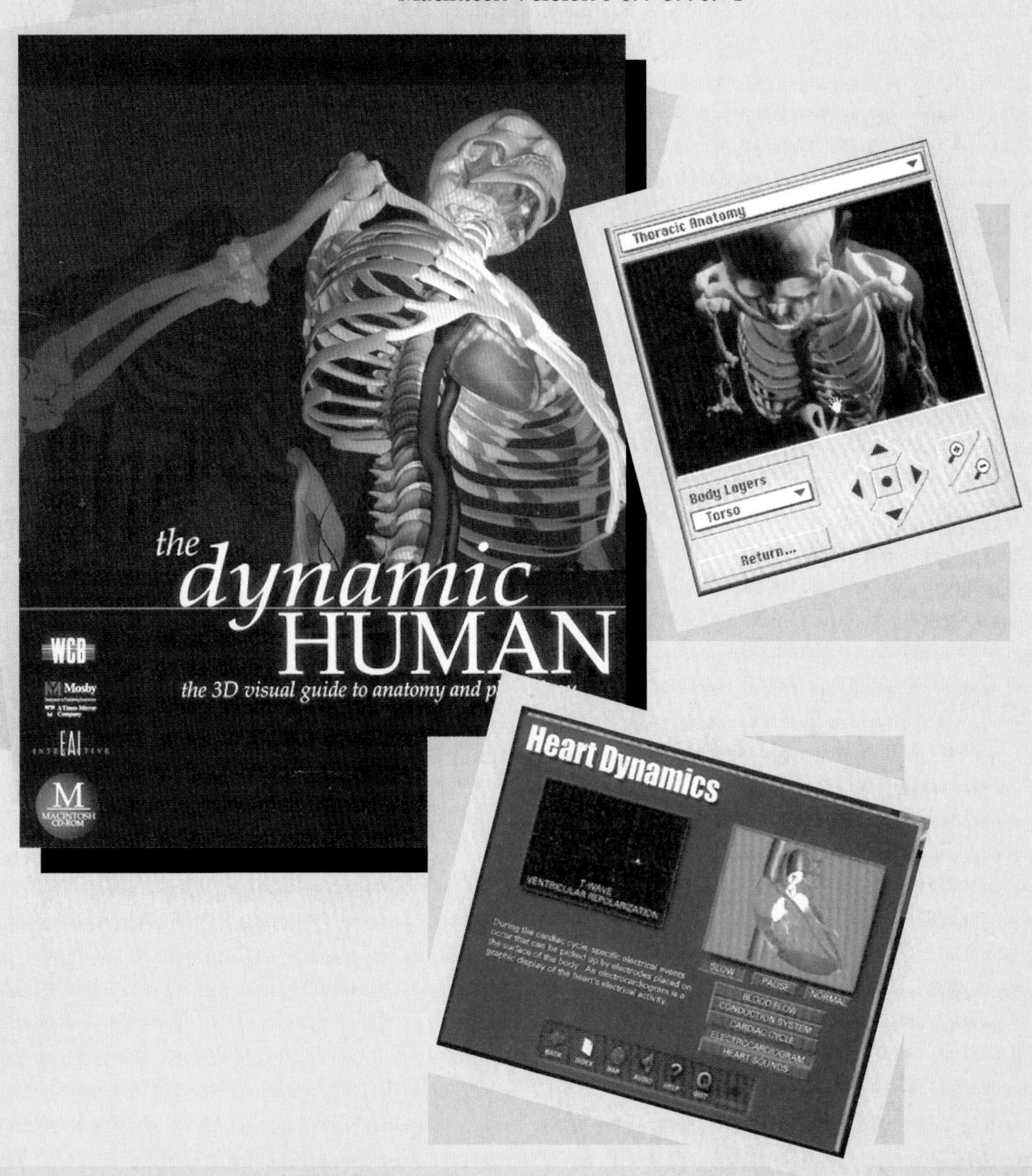

To order any of these products, contact your bookstore manager or call our Customer Service Department at 800-338-3987.

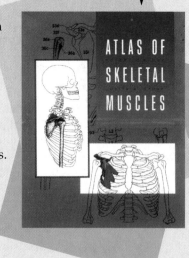

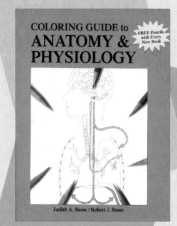

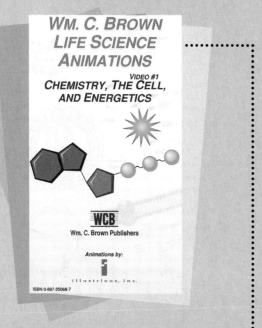

QuickStudy:

Computerized Study Guide for *Human Anatomy*

by Kent M. Van De Graaff

Focus study time only in the areas where you need the most help.

Available on diskette:
IBM: 0-697-34630-7
Macintosh: 0-697-34631-5

How does it work?

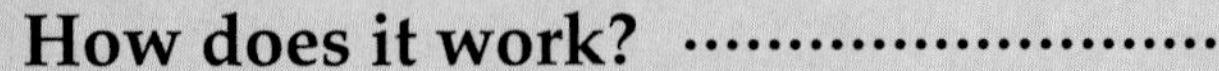

- Take the computerized test for each chapter.
- The program presents you with feedback for each answer.
- A page-referenced study plan is then created for all incorrect answers.
- Look up the answers to the questions you missed.
- This easy-to-use *QuickStudy* is not a duplicate of the printed study guide that accompanies your text. It's full of new study aids.

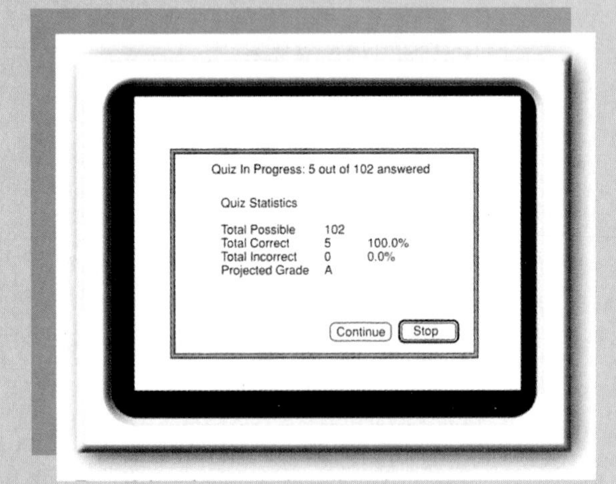

Provides instant feedback on quiz results.

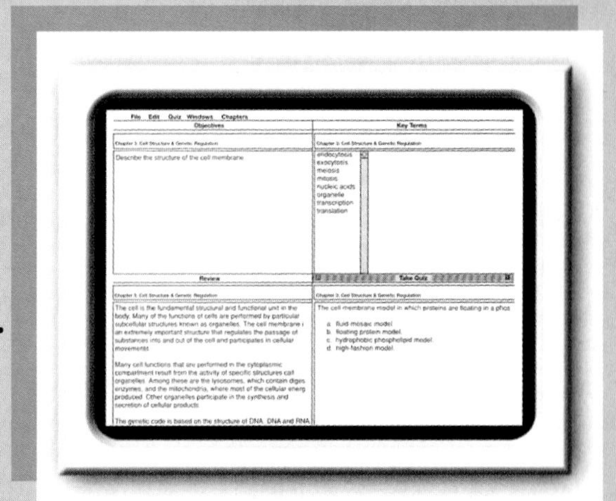

Four different review methods available.

- **Learning Objectives** outline the major points covered in each text chapter.
- **Key Terms** and their definitions are listed for each chapter.
- **Review** presents important concepts and facts in detail.
- **Take Quiz** allows testing on any combination of—or all—chapters.

To order any of these products, contact your bookstore manager or call our Customer Service Department at 800-338-3987.

Brief Contents

Preface xv

UNIT I

Historical Perspective

CHAPTER ONE
History of Anatomy 1

UNIT II

Terminology, Organization, and the Human Organism

CHAPTER TWO
Body Organization and Anatomical Nomenclature 22

UNIT III

Microscopic Structure of the Body

CHAPTER THREE
Cytology 46

CHAPTER FOUR
Histology 72

UNIT IV

Support and Movement

CHAPTER FIVE
Integumentary System 101

CHAPTER SIX
Skeletal System: Introduction and the Axial Skeleton 128

CHAPTER SEVEN
Skeletal System: The Appendicular Skeleton 168

CHAPTER EIGHT
Articulations 191

CHAPTER NINE
Muscular System 226

CHAPTER TEN
Surface and Regional Anatomy 287

UNIT V

Integration and Coordination

CHAPTER ELEVEN
Nervous Tissue and the Central Nervous System 333

CHAPTER TWELVE
Peripheral Nervous System 388

CHAPTER THIRTEEN
Autonomic Nervous System 419

CHAPTER FOURTEEN
Endocrine System 438

CHAPTER FIFTEEN
Sensory Organs 470

UNIT VI

Maintenance of the Body

CHAPTER SIXTEEN
Circulatory System 520

CHAPTER SEVENTEEN
Respiratory System 582

CHAPTER EIGHTEEN
Digestive System 614

CHAPTER NINETEEN
Urinary System 655

UNIT VII

Reproduction and Development

CHAPTER TWENTY
Male Reproductive System 678

CHAPTER TWENTY-ONE
Female Reproductive System 706

CHAPTER TWENTY-TWO
Developmental Anatomy, Postnatal Growth, and Inheritance 734

Appendix A Answers to Objective Questions with Explanations 773

Appendix B Laboratory Demonstrations in Anatomy 778

Appendix C Medical and Pharmacological Abbreviations 778

Appendix D Units of Measurement and Their Equivalents 779

Appendix E Related Web Sites 780

Glossary 784

Credits 797

Index 800

Contents

Preface 000

UNIT I
Historical Perspective

CHAPTER ONE
History of Anatomy 1

- Definition of the Science 2
- Prescientific Period 2
- Scientific Period 5
  - Mesopotamia and Egypt 7
  - China and Japan 8
  - Grecian Period 9
  - Alexandrian Era 10
  - Roman Era 11
  - Middle Ages 12
  - Contributions of Islam 12
  - Renaissance 12
  - Seventeenth and Eighteenth Centuries 14
  - Nineteenth Century 16
  - Twentieth Century 16
- Chapter Summary 20
- Review Activities 20

UNIT II
Terminology, Organization, and the Human Organism

CHAPTER TWO
Body Organization and Anatomical Nomenclature 22

- Clinical Case Study 23
- Classification and Characteristics of Humans 23
  - Phylum Chordata 23
  - Class Mammalia 25
  - Order Primates 25
  - Family Hominidae 27
  - Characteristics of Humans 27
- Body Organization 28
  - Cellular Level 28
  - Tissue Level 28
  - Organ Level 29
  - System Level 29
- Anatomical Nomenclature 33
- Planes of Reference and Descriptive Terminology 33
  - Planes of Reference 33
  - Descriptive Terminology 34
- Body Regions 36
  - Head 37
  - Neck 37
  - Trunk 38
  - Upper Extremity 39
  - Lower Extremity 40
- Body Cavities and Membranes 40
  - Body Cavities 41
  - Body Membranes 41
- Clinical Case Study Answer 43
- Chapter Summary 44
- Review Activities 44

UNIT III
Microscopic Structure of the Body

CHAPTER THREE
Cytology 46

- Introduction to Cytology 47
  - Cells as Functional Units 47
  - Cellular Diversity 47
- Cellular Chemistry 48
  - Elements, Molecules, and Compounds 48
  - Water 49
  - Electrolytes 49
  - Proteins 49
  - Carbohydrates 50
  - Lipids 50
- Cellular Structure 50
  - Cell Membrane 52
  - Cytoplasm and Organelles 54
  - Cell Nucleus 61
- Cell Cycle 62
  - Structure of DNA 63
  - Cell Cycle and Cell Division 64
- Clinical Considerations 67
  - Cellular Adaptations 67
  - Trauma to Cells 67
  - Medical Genetics 68
  - Cancer 68
  - Aging 69
- Chapter Summary 70
- Review Activities 70

CHAPTER FOUR
Histology 72

- Definition and Classification of Tissues 73
- Developmental Exposition: The Tissues 74
- Epithelial Tissue 74
  - Characteristics of Membranous Epithelia 75
  - Simple Epithelia 76
  - Stratified Epithelia 78
  - Body Membranes 80
  - Glandular Epithelia 82
- Connective Tissue 85
  - Characteristics and Classification of Connective Tissue 86
  - Embryonic Connective Tissue 86
  - Connective Tissue Proper 86
  - Cartilage 91
  - Bone Tissue 94
  - Blood (Vascular Tissue) 94
- Muscle Tissue 96
- Nervous Tissue 97
- Clinical Considerations 98
  - Changes in Tissue Composition 98
  - Tissue Analysis 99
  - Tissue Transplantation 99
- Chapter Summary 99
- Review Activities 100

UNIT IV

Support and Movement

Chapter Five

Integumentary System 101

- Clinical Case Study 102
- The Integument as an Organ 102
- Layers of the Integument 103
  - Epidermis 103
  - Dermis 107
- Functions of the Integument 109
  - Physical Protection 109
  - Hydroregulation 109
  - Thermoregulation 109
  - Cutaneous Absorption 109
  - Sensory Reception 110
  - Synthesis 110
  - Communication 110
- Epidermal Derivatives 111
  - Hair 111
  - Nails 113
  - Glands 114
- Developmental Exposition: The Integumentary System 116
- Clinical Considerations 118
  - Inflammatory Conditions (Dermatitis) 118
  - Neoplasms 118
  - Burns 119
  - Frostbite 121
  - Skin Grafts 121
  - Wound Healing 123
  - Aging of the Skin 124
- Clinical Case Study Answer 124
- Internal Affairs 125
- Important Clinical Terminology 126
- Chapter Summary 126
- Review Activities 127

Chapter Six

Skeletal System: Introduction and the Axial Skeleton 128

- Clinical Case Study 129
- Organization of the Skeletal System 129
- Functions of the Skeletal System 131
- Bone Structure 132
  - Shapes of Bones 133
  - Structure of a Typical Long Bone 133
- Bone Tissue 135
  - Bone Cells 135
  - Spongy and Compact Bone Tissues 136
- Bone Growth 137
- Skull 139
  - Cranial Bones 146
  - Facial Bones 149
  - Hyoid Bone 152
  - Auditory Ossicles 152
- Vertebral Column 153
  - General Structure of Vertebrae 154
  - Regional Characteristics of Vertebrae 156
- Rib Cage 159
  - Sternum 159
  - Ribs 159
- Clinical Considerations 161
  - Developmental Disorders 161
  - Nutritional and Hormonal Disorders 161
- Developmental Exposition: The Axial Skeleton 162
  - Neoplasms of Bone 162
  - Aging of the Skeletal System 164
- Clinical Case Study Answer 164
- Internal Affairs 165
- Important Clinical Terminology 166
- Chapter Summary 166
- Review Activities 166

Chapter Seven

Skeletal System: The Appendicular Skeleton 168

- Clinical Case Study 169
- Pectoral Girdle and Upper Extremity 169
  - Pectoral Girdle 169
  - Brachium (Arm) 172
  - Antebrachium (Forearm) 173
  - Manus (Hand) 174
- Pelvic Girdle and Lower Extremity 176
  - Pelvic Girdle 177
  - Thigh 180
  - Leg 180
  - Pes (Foot) 182
- Developmental Exposition: The Appendicular Skeleton 185
- Clinical Considerations 186
  - Developmental Disorders 186
  - Trauma and Injury 187
- Clinical Case Study Answer 188
- Chapter Summary 189
- Review Activities 189

Chapter Eight

Articulations 191

- Clinical Case Study 192
- Classification of Joints 192
- Fibrous Joints 192
  - Sutures 193
  - Syndesmoses 193
  - Gomphoses 194
- Cartilaginous Joints 194
  - Symphyses 194
  - Synchondroses 194
- Synovial Joints 195
  - Structure of a Synovial Joint 196
  - Kinds of Synovial Joints 198
- Movements at Synovial Joints 201
  - Angular Movements 201
  - Circular Movements 201
  - Special Movements 203
  - Biomechanics of Body Movement 203
- Specific Joints of the Body 208
  - Temporomandibular Joint 209
  - Sternoclavicular Joint 209
  - Glenohumeral (Shoulder) Joint 210
  - Elbow Joint 211
  - Metacarpophalangeal and Interphalangeal Joints 212
  - Coxal (Hip) Joint 214
  - Tibiofermoral (Knee) Joint 215
  - Talocrural (Ankle) Joint 217
- Clinical Considerations 218
  - Trauma to Joints 218

Developmental Exposition:
The Synovial Joints 220
Diseases of Joints 221
Treatment of Joint Disorders 222
Clinical Case Study Answer 222
Important Clinical Terminology 224
Chapter Summary 224
Review Activities 224

Chapter Nine

Muscular System 226

Clinical Case Study 227
Introduction to the Muscular System 227
Structure of Skeletal Muscles 229
Muscle Attachments 229
Associated Connective Tissue 229
Muscle Groups 231
Muscle Architecture 231
Blood and Nerve Supply to Skeletal Muscle 231
Skeletal Muscle Fibers and Types of Muscle Contraction 233
Skeletal Muscle Fibers 233
Isotonic and Isometric Contractions 236
Neuromuscular Junction 236
Motor Unit 237
Naming of Muscles 239
Muscles of the Axial Skeleton 241
Muscles of Facial Expression 241
Muscles of Mastication 241
Ocular Muscles 241
Muscles That Move the Tongue 244
Muscles of the Neck 244
Muscles of Respiration 248
Muscles of the Abdominal Wall 248
Muscles of the Pelvic Outlet 250
Muscles of the Vertebral Column 251
Muscles of the Appendicular Skeleton 253
Muscles That Act on the Pectoral Girdle 253
Muscles That Move the Humerus at the Shoulder Joint 254
Muscles That Move the Forearm at the Elbow Joint 257
Muscles of the Forearm That Move the Joints of the Wrist, Hand, and Fingers 258
Muscles of the Hand 263
Muscles That Move the Thigh at the Hip Joint 265
Muscles of the Thigh That Move the Knee Joint 268
Muscles of the Leg That Move the Joints of the Ankle, Foot, and Toes 273
Muscles of the Foot 276
Clinical Considerations 278
Evaluation of Muscle Condition 278
Functional Conditions in Muscles 278
Developmental Exposition:
The Muscular System 280
Diseases of Muscles 282
Aging of Muscles 283
Clinical Case Study Answer 283
Internal Affairs 284
Important Clinical Terminology 285
Chapter Summary 285
Review Activities 285

Chapter Ten

Surface and Regional Anatomy 287

Introduction to Surface Anatomy 288
Surface Anatomy of the Newborn 289
General Appearance 290
Palpable Structures 291
Head 291
Surface Anatomy 292
Internal Anatomy 295
Neck 297
Surface Anatomy 297
Internal Anatomy 300
Trunk 300
Surface Anatomy 301
Internal Anatomy 305
Pelvis and Perineum 310
Shoulder and Upper Extremity 311
Surface Anatomy 311
Internal Anatomy 314
Buttock and Lower Extremity 317
Surface Anatomy 317
Internal Anatomy 321
Clinical Considerations 321
Head and Neck Regions 321
Thoracic Region 326
Abdominal Region 327
Shoulder and Upper Extremity 329
Hip and Lower Extremity 330
Chapter Summary 331
Review Activities 331

UNIT V

Integration and Coordination

Chapter Eleven

Nervous Tissue and the Central Nervous System 333

Clinical Case Study 334
Organization and Functions of the Nervous System 334
Organization of the Nervous System 334
Functions of the Nervous System 334
Developmental Exposition:
The Brain 336
Neurons and Neuroglia 338
Neurons 338
Neuroglia 340
Classification of Neurons and Nerves 342
Transmission of Impulses 346
Nerve Impulse 347
Synapse 347
General Features of the Brain 348
Cerebrum 352
Structure of the Cerebrum 353
Lobes of the Cerebrum 355
Brain Waves 356
White Matter of the Cerebrum 357

Basal Nuclei 358
Language 359
Diencephalon 361
Thalamus 361
Hypothalamus 362
Epithalamus 363
Pituitary Gland 363
Mesencephalon 363
Metencephalon 364
Pons 364
Cerebellum 364
Myelencephalon 366
Medulla Oblongata 366
Reticular Formation 367
Meninges of the Central Nervous System 367
Dura Mater 367
Arachnoid 369
Pia Mater 369
Ventricles and Cerebrospinal Fluid 370
Ventricles of the Brain 370
Cerebrospinal Fluid 372
Blood-Brain Barrier 372
Spinal Cord 372
Structure of the Spinal Cord 373
Spinal Cord Tracts 374
Developmental Exposition: The Spinal Cord 378
Clinical Considerations 379
Neurological Assessment and Drugs 379
Developmental Problems 380
Injuries 380
Disorders of the Nervous System 381
Senescence of the Nervous System 383
Degenerative Diseases of the Nervous System 383
Internal Affairs 384
Clinical Case Study Answer 385
Chapter Summary 386
Review Activities 387

Chapter Twelve

Peripheral Nervous System 388

Clinical Case Study 389
Introduction to the Peripheral Nervous System 389
Cranial Nerves 389
Structure and Function of the Cranial Nerves 389
Neurological Assessment of the Cranial Nerves 398
Spinal Nerves 400
Nerve Plexuses 401
Cervical Plexus 401
Brachial Plexus 401
Lumbar Plexus 404
Sacral Plexus 405
Reflex Arc and Reflexes 410
Components of the Reflex Arc 411
Kinds of Reflexes 411
Clinical Case Study Answer 414
Developmental Exposition: The Peripheral Nervous System 416
Chapter Summary 417
Review Activities 417

Chapter Thirteen

Autonomic Nervous System 419

Clinical Case Study 420
Introduction to the Autonomic Nervous System 420
Organization of the Autonomic Nervous System 420
Visceral Effector Organs 422
Structure of the Autonomic Nervous System 423
Sympathetic (Thoracolumbar) Division 423
Parasympathetic (Craniosacral) Division 425
Functions of the Autonomic Nervous System 429
Neurotransmitters of the Autonomic Nervous System 429
Responses to Adrenergic Stimulation 429
Responses to Cholinergic Stimulation 430
Organs with Dual Innervation 431
Organs without Dual Innervation 432
Control of the Autonomic Nervous System by Higher Brain Centers 433
Medulla Oblongata 433
Hypothalamus 433
Limbic System, Cerebellum, and Cerebrum 434
Clinical Case Study Answer 435
Clinical Considerations 435
Autonomic Dysreflexia 435
Chapter Summary 436
Review Activities 437

Chapter Fourteen

Endocrine System 438

Clinical Case Study 439
Introduction to the Endocrine System 439
Glands of the Endocrine System 440
Hormones and Their Actions 441
Control of Hormone Secretion 443
Pituitary Gland 444
Description of the Pituitary Gland 444
Pituitary Hormones 446
Thyroid and Parathyroid Glands 450
Description of the Thyroid Gland 450
Functions of the Thyroid Gland 451
Parathyroid Glands 452
Pancreas 453
Description of the Pancreas 453
Endocrine Function of the Pancreas 454
Adrenal Glands 455
Description of the Adrenal Glands 455
Functions of the Adrenal Glands 456
Gonads and Other Endocrine Glands 458
Gonads 458
Pineal Gland 458
Thymus 459
Stomach and Small Intestine 459
Placenta 459

Developmental Exposition: The Endocrine System 460
Clinical Considerations 462
Diagnosis of Endocrine Disorders 463
Disorders of the Pituitary Gland 463
Disorders of the Thyroid and Parathyroid Glands 463
Disorders of the Pancreatic Islets 465
Clinical Case Study Answer 465
Disorders of the Adrenal Glands 465
Internal Affairs 466
Chapter Summary 468
Review Activities 468

Chapter Fifteen

Sensory Organs 470

Clinical Case Study 471
Overview of Sensory Perception 471
Classification of the Senses 471
Somatic Senses 472
Tactile and Pressure Receptors 473
Receptors for Heat, Cold, and Pain 474
Proprioceptors 476
Neural Pathways for Somatic Sensations 477
Olfactory Sense 479
Gustatory Sense 480
Visual Sense 481
Accessory Structures of the Eye 483
Structures of the Eyeball 487
Function of the Eyeball 492
Neural Pathways for Vision, Eye Movements, and Processing Visual Information 496
Developmental Exposition: The Eye 498
Senses of Hearing and Balance 499
Outer Ear 499
Middle Ear 501
Inner Ear 502
Sound Waves and Neural Pathways for Hearing 505
Mechanics of Equilibrium 505
Clinical Considerations 510
Diagnosis of Eye and Ear Disorders 511
Developmental Problems of the Eyes and Ears 511
Functional Impairments of the Eye 511
Developmental Exposition: The Ear 512
Infections and Diseases of the Eye 516
Infections, Diseases, and Functional Impairments of the Ear 516
Clinical Case Study Answer 517
Chapter Summary 518
Review Activities 518

UNIT VI

Maintenance of the Body

Chapter Sixteen

Circulatory System 520

Clinical Case Study 521
Functions and Major Components of the Circulatory System 521
Functions of the Circulatory System 521
Major Components of the Circulatory System 521
Blood 523
Formed Elements of Blood 524
Hemopoiesis 528
Blood Plasma 528
Heart 529
Location and General Description 529
Heart Wall 529
Chambers and Valves 529
Circulatory Routes 534
Conduction System of the Heart 535
Electrocardiogram 536
Heart Sounds 537
Blood Vessels 538
Arteries 538
Capillaries 538
Veins 540
Blood Pressure 541
Principal Arteries of the Body 542
Aortic Arch 544
Arteries of the Neck and Head 547
Arteries of the Shoulder and Upper Extremity 547
Branches of the Thoracic Portion of the Aorta 548
Branches of the Abdominal Portion of the Aorta 549
Arteries of the Pelvis and Lower Extremity 550
Principal Veins of the Body 554
Veins Draining the Head and Neck 556
Veins of the Upper Extremity 556
Veins of the Thorax 557
Veins of the Lower Extremity 557
Veins of the Abdominal Region 558
Hepatic Portal System 558
Fetal Circulation 560
Lymphatic System 562
Lymph and Lymphatic Vessels 562
Lymph Nodes 564
Other Lymphoid Organs 565
Clinical Considerations 566
Cardiovascular Assessment 567
Developmental Exposition: The Circulatory System 568
Blood Disorders 572
Heart Diseases 572
Vascular Disorders 574
Disorders of the Lymphatic System 575
Trauma to the Circulatory System 576
Clinical Case Study Answer 577
Internal Affairs 578
Chapter Summary 579
Review Activities 580

Chapter Seventeen

Respiratory System 582

Clinical Case Study 583
Introduction to the Respiratory System 583

Physical Requirements of the Respiratory System 583
Functions of the Respiratory System 583
Basic Structure of the Respiratory System 583
Conducting Passages 585
Nose 585
Paranasal Sinuses 587
Pharynx 587
Larynx 588
Trachea 590
Bronchial Tree 591
Pulmonary Alveoli, Lungs, and Pleurae 592
Pulmonary Alveoli 592
Lungs 593
Pleurae 595
Mechanics of Breathing 598
Inspiration 598
Expiration 599
Respiratory Volumes and Capacities 600
Nonrespiratory Air Movements 601
Regulation of Breathing 603
Developmental Exposition: The Respiratory System 604
Clinical Considerations 606
Developmental Problems of the Respiratory System 606
Trauma or Injury 607
Common Respiratory Disorders 608
Disorders of Respiratory Control 608
Internal Affairs 610
Clinical Case Study Answer 611
Chapter Summary 612
Review Activities 612

Chapter Eighteen

Digestive System 614
Clinical Case Study 615
Introduction to the Digestive System 615
Serous Membranes and Tunics of the Gastrointestinal Tract 616
Serous Membranes 616
Layers of the Gastrointestinal Tract 617
Innervation of the Gastrointestinal Tract 620
Mouth, Pharynx, and Associated Structures 620
Cheeks, Lips, and Palate 620
Tongue 622
Teeth 622
Salivary Glands 625
Pharynx 625
Esophagus and Stomach 628
Esophagus 628
Swallowing Mechanisms 628
Stomach 628
Small Intestine 632
Regions of the Small Intestine 633
Structural Modifications of the Small Intestine 634
Mechanical Activities of the Small Intestine 635
Large Intestine 636
Regions and Structures of the Large Intestine 636
Mechanical Activities of the Large Intestine 638
Liver, Gallbladder, and Pancreas 639
Liver 640
Gallbladder 642
Pancreas 643
Developmental Exposition: The Digestive System 644
Clinical Considerations 648
Developmental Problems of the Digestive System 648
Pathogens and Poisons 648
Clinical Problems of the Teeth and Salivary Glands 649
Disorders of the Liver 649
Disorders of the GI Tract 650
Internal Affairs 651
Clinical Case Study Answer 652
Important Clinical Terminology 652
Chapter Summary 652
Review Activities 653

Chapter Nineteen

Urinary System 655
Clinical Case Study 656
Introduction to the Urinary System 656
Kidneys 657
Position and Appearance of the Kidneys 657
Gross Structure of the Kidney 658
Microscopic Structure of the Kidney 658
Ureters, Urinary Bladder, and Urethra 664
Ureters 664
Urinary Bladder 665
Urethra 666
Micturition 667
Clinical Considerations 669
Developmental Problems of the Urinary Organs 669
Developmental Exposition: The Urinary System 670
Symptoms and Diagnosis of Urinary Disorders 672
Infections of the Urinary Organs 673
Trauma to the Urinary Organs and Functional Impairments 673
Internal Affairs 674
Clinical Case Study Answer 675
Chapter Summary 676
Review Activities 676

UNIT VII

Reproduction and Development

Chapter Twenty

Male Reproductive System 678
Clinical Case Study 679
Introduction to the Male Reproductive System 679
Categories of Reproductive Structures 679

Perineum and Scrotum 681
Perineum 681
Scrotum 681
Testes 683
Structure of the Testes 683
Endocrine Functions of the Testes 684
Spermatogenesis 685
Structure of Spermatozoa 686
Spermatic Ducts, Accessory Reproductive Glands, and the Urethra 688
Spermatic Ducts 688
Accessory Reproductive Glands 689
Urethra 690
Penis 691
Mechanisms of Erection, Emission, and Ejaculation 693
Erection of the Penis 693
Emission and Ejaculation of Semen 693
Clinical Considerations 695
Developmental Problems of the Male Reproductive System 695
Developmental Exposition: The Male Reproductive System 696
Sexual Dysfunction 699
Diseases of the Male Reproductive System 700
Clinical Case Study Answer 702
Internal Affairs 703
Chapter Summary 704
Review Activities 704

Chapter Twenty-One

Female Reproductive System 706

Clinical Case Study 707
Introduction to the Female Reproductive System 707
Structure and Function of the Ovaries 709
Position and Structure of the Ovaries 709
Ovarian Cycle 710
Secondary Sex Organs 714
Uterine Tubes 714
Uterus 715
Vagina 717
Vulva 718
Mechanism of Erection and Orgasm 719
Mammary Glands 719
Structure of the Breast and the Mammary Glands 720
Ovulation and Menstruation 722
Clinical Considerations 723
Diagnostic Procedures 723
Developmental Exposition: The Female Reproductive System 724
Developmental Problems of the Female Reproductive System 726
Problems Involving the Ovaries and Uterine Tubes 726
Problems Involving the Uterus 728
Diseases of the Vagina and Vulva 729
Diseases of the Breasts and Mammary Glands 729
Methods of Contraception 729
Clinical Case Study Answer 732
Chapter Summary 732
Review Activities 733

Chapter Twenty-Two

Developmental Anatomy, Postnatal Growth, and Inheritance 734

Clinical Case Study 735
Fertilization 735
Preembryonic Period 737
Cleavage and Formation of the Blastocyst 737
Implantation 739
Formation of the Germ Layers 740
Embryonic Period 742
Extraembryonic Membranes 742
Placenta 745
Umbilical Cord 746
Structural Changes in the Embryo by Week 748
Fetal Period 752
Structural Changes in the Fetus by Weeks 752
Labor and Parturition 755
Periods of Postnatal Growth 756
Neonatal Period 756
Infancy 757
Childhood 757
Adolescence 759
Adulthood 760
Inheritance 762
Clinical Considerations 765
Abnormal Implantation Sites 765
In Vitro Fertilization and Artificial Implantation 766
Multiple Pregnancy 766
Fetal Monitoring 767
Congenital Defects 769
Clinical Case Study Answer 769
Genetic Disorders of Clinical Importance 770
Chapter Summary 770
Review Activities 772

Appendix A Answers to Objective Questions with Explanations 773
Appendix B Laboratory Demonstrations in Anatomy 778
Appendix C Medical and Pharmacological Abbreviations 778
Appendix D Units of Measurement and Their Equivalents 779
Appendix E Related Web Sites 780
Glossary 784
Credits 797
Index 800

Preface

Human anatomy is an essential foundation course for students pursuing careers in such fields as medicine, paramedicine, dentistry, medical technology, physical therapy, and athletic training. The focus of this text is on presenting practical information that students will be able to apply to real-world situations they might encounter in their chosen field. In addition, numerous examples throughout the text reinforce the principle that learning anatomy helps students to become better acquainted with themselves. The fifth edition of *Human Anatomy* is intended to serve as a basic introduction to human anatomy for students enrolled in medical, allied-health, and physical education programs, or for those majoring in the biological sciences. This text also can be of immense value to any student wanting to know more about the structure, function, and possible dysfunction of the human body.

Objectives

In preparing a text and its ancillaries (study guide, laboratory manual, study cards, instructor's manual, and so forth), it is essential to consider both the needs of the student and the needs of the instructor. A well-written and inviting text is at the heart of an effective educational package. With this in mind, the following objectives were formulated for the fifth edition of *Human Anatomy:*

- To provide an accurate, up-to-date text of practical value that is visually appealing, interesting, and easy to comprehend—a text that will entice readers to study the material and thereby enhance their appreciation of life through a better understanding of the structure, function, and magnificence of their own bodies.
- To clearly state the basic concepts associated with the discipline of human anatomy and to provide a conceptual framework in the form of learning objectives.
- To express the beauty of the body through spectacular art. Anatomy is a visual science where exactness is essential. The numerous high-quality illustrations prepared expressly for this edition augment the acclaimed art program of the previous edition.
- To emphasize the relevance of concepts and to stimulate student interest in the material through a series of thematic commentaries, highlighted by topic icons.
- To provide a systematic, balanced presentation of anatomical concepts at the developmental, cellular, histological, clinical, and gross anatomy levels.
- To build students' technical vocabulary to the point where they feel comfortable with basic medical terminology.
- To encourage proper care of the body in order to enjoy a healthier, more productive life.
- To acquaint students with the history of anatomy, from its primitive beginnings to recent advances in the field. Only with the realization of how long it took to build up knowledge that is now taken for granted—and with what difficulty—can students appreciate the science of anatomy in its proper proportion.

Text Organization

The 22 chapters in this text are grouped into seven units that are identified by *colored tabs* on the outside page margins.

Unit I

In this unit, the stage is set for studying human anatomy by providing a historical perspective on how this science has developed over the centuries. Anatomy is an exciting and dynamic science that remains vital as it continues to broaden its scope. It is hoped that this unit will make the reader feel a part of the heritage of anatomy.

Unit II

In this unit, the characteristics that define humans as a distinct species are described. The various levels of organization of the human body are also described, and the basic terminology necessary for understanding the structure and functioning of the body is introduced.

Unit III

Body organization is considered at the cellular and histological levels in this unit. Cellular chemistry is emphasized as an integral aspect of learning about how the body functions.

Unit IV

Support, protection, and movement of the human body are the themes of this unit. The integumentary system provides the body with external support and protection, and the skeletal system provides internal support and protection for certain organs of the body. Movement is possible at the joints of the skeleton as the associated skeletal muscles are contracted. Surface anatomy and regional anatomy are given detailed coverage in chapter 10 of this unit. Atlas-quality photographs of dissections of human cadavers are included in this chapter.

Unit V

This unit includes chapters on the nervous system, endocrine system, and sensory organs. The concepts identified and discussed in these chapters are concerned with the integration and coordination of body functions and the perception of environmental stimuli.

Unit VI

In this unit, the structure and function of the circulatory, respiratory, digestive, and urinary systems are discussed as they contribute in their individual ways to the overall functioning and general welfare of the organism. All of these systems work together in maintaining a stable internal environment in which the cells of the body can thrive on a day-to-day basis.

Unit VII

The male and female reproductive systems are described in this unit, and the continuance of the human species through sexual reproduction is discussed. Unit VII provides an overview of the entire sequence of human life, including prenatal development and postnatal growth, development, and aging. Basic concepts of genetics and inheritance are also explained.

Learning Aids

Each of the 22 chapters of this text incorporates numerous pedagogical devices that organize and underscore the practicality of the material, clarify important concepts, and stimulate students' natural curiosity about the human body. In short, these aids make the study of human anatomy more effective and enjoyable.

Chapter Introductions

The chapter introductions include an outline of the *chapter contents*, accompanied by a photograph of historical significance. A *clinical investigation* in the form of a *case study in anatomy* also appears at the beginning of most of the chapters. These hypothetical situations are indicative of the type of clinical material that will be presented in the chapters. The solution to the case study is presented at the end of the chapter, following the last major section.

Understanding Terminology

Where each technical term first appears and is defined in the narrative, it is set off in *boldface* or *italic type*, and is often followed by a *phonetic pronunciation* in parentheses. In this fifth edition, many new phonetic pronunciations have been added. In addition, the derivations of many terms are provided in *footnotes*, and the roots of each term can be identified by referring to the *glossary of prefixes and suffixes* on the inside of the front and back covers. If students know how a term was derived, and if they can pronounce the term correctly, they are more likely to retain its meaning. The anatomical terminology used in this edition of *Human Anatomy* is current as presented in *Nomina Anatomica* (NA), sixth edition, Churchill Livingstone, Inc., 1989.

Chapter Sections

Each chapter is divided into several major sections, each of which is prefaced by a *concept statement* and a list of *learning objectives*. A concept statement is a succinct expression of the main idea, or organizing principle, of the information contained in a chapter section. The learning objectives indicate the level of competency needed to understand the concept thoroughly and be able to apply it in practical situations. The narrative that follows discusses the concept in detail, with reference to the objectives. *Review questions* at the end of each chapter section test the student's understanding of the concept and mastery of the learning objectives.

Commentaries and Clinical Information

Set off from the text narrative are short paragraphs highlighted by accompanying topic icons. This interesting information is relevant to the discussion that precedes it, but more importantly, it demonstrates how basic scientific knowledge is applied. New commentaries—some of topical interest—have been added to the fifth edition, and others have been updated. The five icons used are as follows:

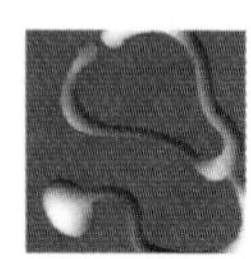

Clinical information is indicated by a stethoscope.

Information relevant to *normal aging* is indicated by an hourglass.

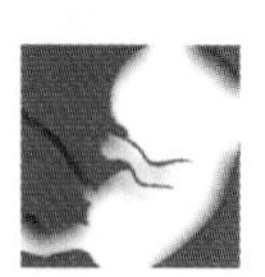

Developmental information of practical importance is indicated by a human embryo.

Information relevant to the body processes that maintain *homeostasis* (a state of dynamic equilibrium) is indicated by a gear mechanism.

Topics relevant to human anatomy that are quite simply of *academic interest* are indicated by a mortarboard.

In addition to the in-text commentaries, selected developmental disorders, clinical procedures, and diseases or dysfunctions of specific organ systems are described in *clinical considerations sections* that appear at the end of most chapters. Photographs of pathological conditions accompany many of these discussions.

Developmental Expositions

In each body system chapter, a discussion of prenatal development is presented just before the clinical considerations section. Each of these discussions includes *exhibits* and *explanations* of the morphogenic events involved in the development of a body system. Placement toward the end of the chapter ensures that the anatomical terminology needed to understand the embryonic structures has been introduced. Occasionally, a developmental exposition will follow the relevant discussion of a particular body part or region; this occurs, for example, in sections on the brain and spinal cord, and on the eye and ear.

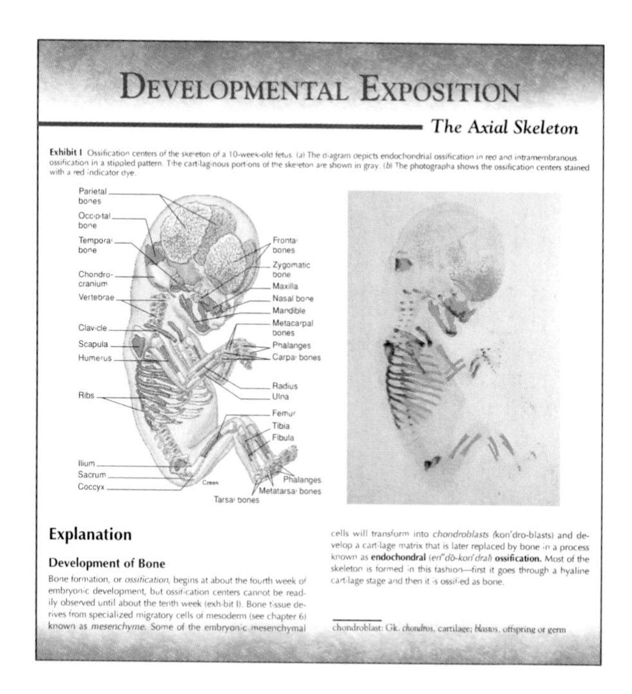

DEVELOPMENTAL EXPOSITION

The Axial Skeleton

Exhibit I Ossification centers of the skeleton of a 10-week-old fetus. (*a*) The diagram depicts endochondrial ossification in red and intramembranous ossification in a stippled pattern. The cartilaginous portions of the skeleton are shown in gray. (*b*) The photographa shows the ossification centers stained with a red indicator dye.

Explanation

Development of Bone

Bone formation, or *ossification,* begins at about the fourth week of embryonic development, but ossification centers cannot be readily observed until about the tenth week (exhibit I). Bone tissue derives from specialized migratory cells of mesoderm (see chapter 6) known as *mesenchyme.* Some of the embryonic mesenchymal cells will transform into *chondroblasts* (kon'dro-blasts) and develop a cartilage matrix that is later replaced by bone in a process known as **endochondral** (*en"dŏ-kon'dral*) **ossification.** Most of the skeleton is formed in this fashion—first it goes through a hyaline cartilage stage and then it is ossified as bone.

chondroblast: Gk. *chondros,* cartilage; *blastos,* offspring or germ

Internal Affairs

The page in each body system chapter entitled "Internal Affairs" ties the functional aspects of one body system to each of the others.

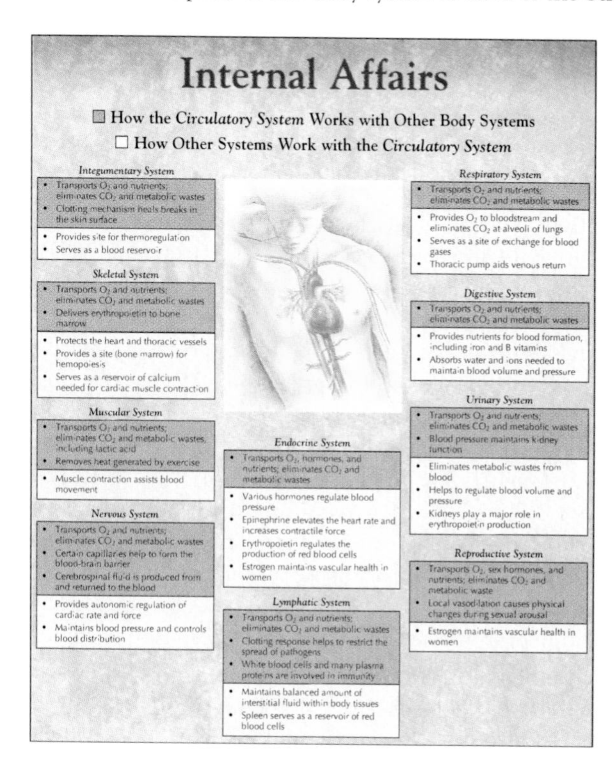

Internal Affairs

■ How the *Circulatory System* Works with Other Body Systems

□ How Other Systems Work with the *Circulatory System*

Integumentary System

- Transports O_2 and nutrients; eliminates CO_2 and metabolic wastes
- Clotting mechanism heals breaks in the skin surface

- Provides site for thermoregulation
- Serves as a blood reservoir

Skeletal System

- Transports O_2 and nutrients; eliminates CO_2 and metabolic wastes
- Delivers erythropoietin to bone marrow

- Protects the heart and thoracic vessels
- Provides a site (bone marrow) for hemopoiesis
- Serves as a reservoir of calcium needed for cardiac muscle contraction

Muscular System

- Transports O_2 and nutrients; eliminates CO_2 and metabolic wastes, including lactic acid
- Removes heat generated by exercise

- Muscle contraction assists blood movement

Nervous System

- Transports O_2 and nutrients; eliminates CO_2 and metabolic wastes
- Certain capillaries help to form the blood-brain barrier
- Cerebrospinal fluid is produced from and returned to the blood

- Provides autonomic regulation of cardiac rate and force
- Maintains blood pressure and controls blood distribution

Endocrine System

- Transports O_2, hormones, and nutrients; eliminates CO_2 and metabolic wastes

- Various hormones regulate blood pressure
- Epinephrine elevates the heart rate and increases contractile force
- Erythropoietin regulates the production of red blood cells
- Estrogen maintains vascular health in women

Lymphatic System

- Transports O_2 and nutrients; eliminates CO_2 and metabolic wastes
- Clotting response helps to restrict the spread of pathogens
- White blood cells and many plasma proteins are involved in immunity

- Maintains balanced amount of interstitial fluid within body tissues
- Spleen serves as a reservoir of red blood cells

Respiratory System

- Transports O_2 and nutrients; eliminates CO_2 and metabolic wastes

- Provides O_2 to bloodstream and eliminates CO_2 at alveoli of lungs
- Serves as a site of exchange for blood gases
- Thoracic pump aids venous return

Digestive System

- Transports O_2 and nutrients; eliminates CO_2 and metabolic wastes

- Provides nutrients for blood formation, including iron and B vitamins
- Absorbs water and ions needed to maintain blood volume and pressure

Urinary System

- Transports O_2 and nutrients; eliminates CO_2 and metabolic wastes
- Blood pressure maintains kidney function

- Eliminates metabolic wastes from blood
- Helps to regulate blood volume and pressure
- Kidneys play a major role in erythropoietin production

Reproductive System

- Transports O_2, sex hormones, and nutrients; eliminates CO_2 and metabolic waste
- Local vasodilation causes physical changes during sexual arousal

- Estrogen maintains vascular health in women

The information contained in these pages has been expanded considerably for the fifth edition, so that the concept of the interrelationships of systems in maintaining homeostasis is emphasized to a greater extent.

Chapter Summaries

A *summary,* in outline form, at the end of each chapter reinforces the learning experience. In addition, these comprehensive summaries can serve as a valuable tool in helping students prepare for examinations.

Review Activities

Following each chapter summary, sets of *objective, essay,* and *critical thinking questions* give students the opportunity to obtain feedback as to the depth of their understanding and learning. The critical thinking questions are a new fifth-edition feature. They challenge students to use the chapter information in novel ways toward the solution of practical problems. The correct responses to the objective questions are provided in Appendix A (page 773). Each answer is also explained, so that students can effectively use the review activities to broaden their understanding of the subject matter.

Illustrations and Tables

Because anatomy is a descriptive science, great care has been taken to provide an outstanding illustration program that maximizes students' learning. In the fifth edition, nearly 25% of the illustrations are new, and many others have been revised. These illustrations represent a collaborative effort between author and illustrator, often involving dissection of cadavers to ensure accuracy. In addition to being aesthetically pleasing, each illustration has been checked and rechecked for conceptual clarity and precision of the linework, labels, and caption. All of the figures are integrated with the text narrative, and most are original *full-color art.* In addition to the anatomical renderings, *color graphics* are used to clarify complex physiological processes. *Light* and *scanning electron micrographs* are also used where appropriate, and carefully selected *photographs* appear throughout the text. *Color-coding* is used in certain art sequences as a technique to aid learning. For example, the bones of the skull in chapter 6 are color-coded so that each bone can be readily identified in the many renderings included in the chapter. In chapters 6, 7, and 8 on the skeletal system and articulations, new *orientation diagrams* have been added to highlight the location of specific bones and joints relative to the body as a whole or to a particular region of the body. The photographs of *dissected human cadavers* in chapter 10 illustrate the complexity of structural relationships that can be fully appreciated only when seen in a human specimen. Elsewhere in the text, photographs of specific organs from cadavers are used to augment the illustrations.

Numerous *tables* throughout the text summarize information and clarify complex data. Several of them are illustrated to communicate information in the most effective manner. Like the figures, all of the tables are referenced in the text narrative and placed as close to the reference as possible to spare students the trouble of flipping through pages.

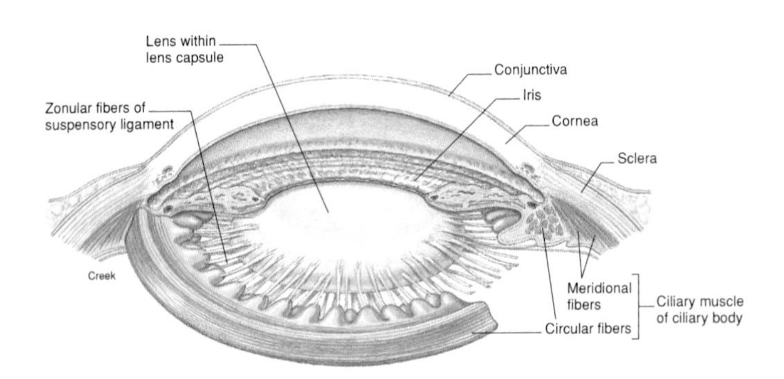

Appendixes, Glossary, and Index

The appendixes, following chapter 22, contain valuable reference information. *Appendix A* (page 773) gives the correct responses to the objective questions that appear at the end of each chapter, along with explanations. *Appendix B* (page 778) offers suggestions for preparing laboratory demonstrations in anatomy. *Appendix C* (page 778) lists the medical and pharmacological abbreviations commonly used in clinical practice. *Appendix D* (page 779) lists commonly used units of measurement and their equivalents, and explains how to convert one unit into another. *Appendix E* (page 780), new to this edition, lists web sites where useful information pertaining to most of the chapters can be found.

The *glossary,* updated and expanded, provides definitions for the important technical terms used in the text. Phonetic pronunciations are included for most of the terms, and an easy-to-use pronunciation guide appears at the beginning of the glossary. Synonyms, including eponymous terms, are indicated, and for some terms antonyms are given as well.

Human Anatomy, Fifth Edition as a Reference Book

Not only is *Human Anatomy*, fifth edition, an effective primary text for a course of study, it is also a valuable reference book. This text can serve as an important resource in any health-oriented course. In addition, it can be used as a family reference book regarding the structure, function, development, senescence (aging), and possible dysfunctions of the human body.

Supplementary Materials

The supplementary materials that accompany the fifth edition of *Human Anatomy* are designed to assist you in your learning activities and to help anatomy instructors plan course work and presentations.

1. An **Instructor's Manual with Test Item File** prepared by Jeffrey and Karianne Prince provides instructional support in the use of the textbook. It also contains a test item file with approximately 70 items for each chapter to help instructors make up examinations (0-697-28415-8).
2. The **Laboratory Manual to accompany Human Anatomy,** fifth edition, by Kent M. Van De Graaff contains cat dissections and selected organ dissections. This laboratory manual emphasizes learning anatomical structures through visual observation, palpation, and knowledge of the functional relationship of one body system to another (0-697-28416-6).
3. An **Instructor's Manual to the Laboratory Manual** (0-697-28417-4) prepared by Jeffrey and Karianne Prince provides the answers to the questions that appear in the Laboratory Manual.
4. A **Student Study Guide to accompany Human Anatomy,** fifth edition, by Kent M. Van De Graaff contains chapter concepts, objective questions, chapter review activities, and detailed explanations for answers to the objective questions (0-697-28418-2).
5. **150 transparencies** of full-color illustrations from the text that have been chosen for their value in reinforcing lecture presentations. Available to instructors who adopt this text (0-697-28419-0).
6. **Visual Resource Library to accompany Human Anatomy** is available on CD-ROM containing all of the line art with an easy-to-use interface program enabling the user to quickly move among the images, show or hide labels, and create a multimedia presentation (0-697-34641-2).
7. **Slides:** 100 histology slides (0-697-26367-3), 60 clinical slides (0-697-26368-1), and 25 radiographic slides (0-697-26366-5) lend support to information given in the text and is free to adopters of this textbook. A narrative describing each slide is provided in the Instructor's Manual.
8. **QuickStudy** is a computerized study guide that contains true/false, multiple-choice, and fill-in-the-blank questions available in IBM 3.5″ (0-697-34630-7) and Macintosh 3.5″ (0-697-34631-5) format.
9. **MicroTest III** is a computerized test generator that allows for quick creation of tests based on questions from the test item file. Requires no programming experience. Available in Windows 3.5″ (0-697-28423-9) and Macintosh 3.5″ (0-697-28424-7) format. It is available free to instructors who adopt this text.
10. **Human Anatomy and Physiology Study Cards,** 3/e, by Kent M. Van De Graaff, R. Ward Rhees, and Christopher H. Creek is a boxed set of 300 3-by-5 inch study cards. It serves as a well-organized and illustrated synopsis of the structure and function of the human body. Clinical information is presented as it applies to specific body organs and systems or physiological processes. The Study Cards offer a quick and effective way for students to review human anatomy and physiology (0-697-26447-5).
11. **The Dynamic Human CD-ROM** illustrates the important relationships between anatomical structures and their functions in the human body. Realistic computer visualization and three-dimensional visualizations are the premier features of this CD-ROM. Available in Macintosh (0-697-37909-4) or Windows (0-697-37910-8) format.
12. **The Dynamic Human Videodisc** contains all of the animations (200+) from the CD-ROM. A barcode directory is also available (0-697-38937-5).
13. **WCB Life Science Animations Videotape Series** consists of five videotapes containing fifty-three animations that cover many key physiological processes as follows: Chemistry, The Cell, and Energetics (0-697-25068-7); Cell Division, Heredity, Genetics, Reproduction, and Development (0-697-25069-5); Animal Biology #1 (0-697-25070-9); Animal Biology #2 (0-697-25071-7);

and Plant Biology, Evolution, and Ecology (0-697-26600-1). Another videotape containing similar animations is also available, entitled **Physiological Concepts of Life Science** (0-697-21512-1).

14. **Explorations in Human Biology CD-ROM** by Johnson consists of sixteen interactive animations that examine issues related to the human side of biology. Available in Macintosh (0-697-22964-5) or Windows (0-697-22963-7) format.
15. **Explorations in Cell Biology and Genetics CD-ROM** by Johnson contains 17 animations that provide a colorful, compelling way to delve into, and understand, often-challenging subject matter (0-697-29214-2).
16. **Life Science Living Lexicon CD-ROM** by Marchuk contains a comprehensive collection of life science terms, including definitions of their roots, prefixes, and suffixes as well as audio pronunciations and illustrations. The Lexicon is student-interactive, featuring quizzing and note taking capabilities (0-697-29266-5).
17. **WCB Anatomy and Physiology Videodisc** contains more than 30 animations of physiological processes, as well as line art and micrographs. A bar code directory is also available (0-697-27716-X).
18. **WCB Anatomy and Physiology Videotape Series:** Introduction to the Human Cadaver and Prosection (0-697-11177-6); Introduction to Cat Dissection: Cat Musculature (0-697-11630-1); Blood Cell Counting, Identification and Grouping (0-697-11629-8); Internal Organs and the Circulatory System of the Cat (0-697-13922-0). These exceptional videotapes are available free to qualified adopters.
19. **Coloring Guide to Anatomy and Physiology,** 1/e, by Stone-Stone emphasizes learning through the process of color association, and provides a thorough review of anatomical and physiological concepts (0-697-17109-4).
20. **Atlas of Skeletal Muscles,** 2/e, by Stone-Stone serves as a guide to the structure and function of human skeletal muscles. The illustrations help students locate muscles and understand their actions (0-697-13790-2).
21. **Laboratory Atlas of Anatomy and Physiology,** 1/e, by Eder et al. is a full-color atlas containing histology, human skeletal anatomy, human muscular anatomy, dissections, and reference tables (0-8151-3077-5).
22. **Survey of Infectious and Parasitic Diseases,** 1/e, by Van De Graaff is a black-and-white booklet that presents the essential information on one hundred of the most common and clinically significant diseases (0-697-27535-3).
23. **Homepage on the Internet** (http://www.mhhe.com/sciencemath/biology/vdghumananatomy/). An attractive site on the World Wide Web features a homepage dedicated to the new edition of **Human Anatomy.** This effective site is designed for use by both instructors and students. It features information about the authors, what's new in the fifth edition, and details about the extensive ancillary package.

Acknowledgments

I am very appreciative of the many colleagues, students, friends, and family members who provided input toward the preparation of the fifth edition of *Human Anatomy*. My sincere gratitude is extended to professor Michael J. Shively who provided valuable suggestions for improvement as he carefully reviewed the entire text manuscript. His constant personal friendship and encouragement are also appreciated. Samuel I. Zeveloff and J. Ronald Galli are colleagues who were especially supportive of my efforts in preparing this edition. Several professors who taught from the previous edition shared suggestions that have been incorporated into this one. Furthermore, many students have reviewed portions of the manuscript and offered suggestions for improvements. Feedback from a student perspective is particularly useful and appreciated.

Several physicians reviewed the clinical information in this edition, ensuring that it is current and accurate. Drs. J. Philip Freestone and Douglas W. Hacking reviewed selected portions of the text. A father's request to three of his sons resulted in additional clinical input. A heartfelt thanks is extended to Drs. Kyle M. Van De Graaff and Eric J. Van De Graaff, as well as soon-to-be physician Ryan J. Van De Graaff, for their generous suggestions and genuine interest in what their dad does.

The visual appeal and accuracy provided by quality photographs and illustrations are essential in an anatomy text. The talented artists who contributed over 50 new illustrations to this edition were Christopher H. Creek and Rictor Lew. Drs. Gary M. Watts and Brent C. Chandler and the Department of Radiology at Utah Valley Regional Medical Center provided many of the radiographs used in the text. The photographs of surface anatomy and some of the photomicrographs were taken by John L. Crawley, Glenn L. Anderson, and Sheril D. Burton, Ph.D. It was a pleasure working with these professionals, and I appreciate their contribution.

Jeffrey Prince and Karianne N. Prince were terrific to work with in the preparation of specific ancillaries that accompany this text. Their involvement is especially rewarding to me, in that they are former students.

Ann Mirels copyedited the manuscript, and her talented and dedicated efforts are evident throughout this edition. I am grateful for her friendship and belief in me as an author. The editorial staff at WCB/McGraw-Hill was superb to work with. Sponsoring Editor Kristine Noel Tibbetts and Developmental Editor Kelly Drapeau worked diligently throughout the rewriting and production. Marla Irion, project manager, Brenda Ernzen, art editor, and John Leland, photo editor, spent countless hours attending to the myriad details that a text such as this involves.

WCB/McGraw-Hill assembled a panel of competent anatomists to review the manuscript and illustrations. These professionals aided my work immeasurably, and I am especially grateful for their frank criticisms, comments, and reassurance.

Jane E. Aloi
Saddleback College
Lindy J. Summers-Bair
Bobby R. Baldridge
Asbury College
Steven Bassett
Southeast Community College
Belinda Beck
University of Oregon
Dianne Y. Bell
Avila College
Raymond L. Bernor
Howard University
Robert H. Blackwell
Utah Valley State College
Leann Blem
Virginia Commonwealth University
Julie Harrill Bowers
East Tennessee State University
Thomas R. Campbell
Los Angeles Pierce College
Diane K. Chaddock
Southwestern Michigan College
Keith Concon
Florida Inernational University
Richard Connett
Monroe Community College
Natalie A. Connors
Saint Louis University School of Medicine
John E. Davis
Alma College
Thomas A. Davis
Loras College
Kamiab Delfanian
University of South Dakota School of Medicine
Alan E. Dietsche
University of Rochester
Shirley Dillaman
Penn State University-Shenango
Maureen A. Donnelly
Florida International University
Douglas Duff
Indiana University South Bend
J. Roger Egan
Adirondack Community College
Charles S. Easley
University of South Carolina-Spartanburg
Leonard Epp
Mount Union College
Gibril O. Fadika
Saint Paul's College
Carl D. Frailey
Johnson County Community College
James B. Gale
Sonoma State University
Paul A. Gardner
Snow College
Anthony J. Gaudin
California State University-Northridge
Robert M. George
Florida International University
Gene Giggleman
Parker College of Chiropractic
Andrew Goliszek
North Carolina A & T State University
Glenn A. Gorelick
Citrus College
Keith E. Graham
Lutheran College of Health Professions
Joseph C. Gregorek
Gannon University
The Pennsylvania State University College of Medicine
William E. Hamilton
Penn State University-New Kensington
Gary B. Hanson
Concordia University-Portland
John P. Harley
Eastern Kentucky University
Donna Price Henry
Florida Gulf Coast University
Susan Herman
Slippery Rock University
Phyllis C. Hirsch
Slippery Rock University
Rita Hoots
Yuba College
Nancy McAllister-Irwin
Lander University
Bert H. Jacobson
Oklahoma State University
James L. Junker
Campbell University School of Pharmacy
Michael H. Kesner
Indiana University of Pennsylvania
Glenn E. Kietzmann
Wayne State College
Ronald L. Koller
Los Angeles Pierce College
Dennis Landin
Louisiana State University
Douglas J. Law
University of Missouri-Kansas City

Victoria L. Lawrence
Penn State University-McKeesport Campus
Community College of Allegheny County-South Campus
Mary Leida
Morningside College
Bruce Maley
University of Kentucky School of Medicine
Corrie Mancinelli
West Virginia University School of Medicine
Murray K. Marks
University of Tennessee
Mary E. Allen-Martin
San Antonio College
Aldred P. McQueen, Sr.
Hampton University
John W. Mills
Clarkson University
Scott Minton
Southern California College
James B. Mitchell
Moravian College
Royce Lee Montgomery
University of North Carolina School of Medicine
David Dee Mowry
Ohio University-Lancaster
Tim R. Mullican
Dakota Wesleyan University
Scott Neiderer
Missouri Valley College
Margaret M. Nowack
Gwynedd Mercy College
Thomas M. O'Connor
Washburn University
Daniel R. Olson
Northern Illinois University
Joyce Jennings-Pineda
Southwest Missouri State University-West Plains Campus
Claude Pyatte
Lees McRae College
Dell Redding
Evergreen Valley College
Tricia A. Reichert
Colby Community College
Roscoe B. Root
Lansing Community College
Zenda Rushing
Francis Marion University
David J. Saxon
Morehead State University
Randall M. Schietzelt
Harper College
Dilip K. Sen
Virginia State University
Blair S. Shean
Morgan Community College
Michael Jay Shively
Utah Valley State College
Deborah Sillman
Penn State University-New Kensington
Danny Sterling
Clinch Valley College
Suzanne Strait
Marshall University
Ruy Tchao
Philadelphia College of Pharmacy and Science
Spencer Jay Turkel
New York Institute of Technology
Sally van Niel
Everett Communitgy College
Robin Vance
Union College
Karin VanMeter
Des Moines Area Community College
James Walker
Purdue University
Barbara Walton
University of Tennessee at Chattanooga
J. S. Waterhouse
Professor Emeritus
State University of New York at Plattsburgh
Ralph E. Werner
Richard Stockton College
M. C. Wicht, Jr.
Limestone College
Mary L. Cochran Wilson
Gordon College
Thomas H. Wilson
Judson College
Paul B. Wissann
Santa Monica College
David M. Wolfrom
Paducah Community College
Michael B. Worrell
Indiana University School of Allied Health Sciences
Don R. Yeltman
Davis & Elkins College
Robert A. Zaccaria
Lycoming College

Once again my wife, Karen, and our six children have endured through another edition. I am especially grateful for their sacrifice, encouragement, and love.

Life Science Animations (LSA)

Figure	Animation
Figure 3.2 LSA 2	Journey into a Cell
Figure 3.19 LSA 15	DNA Replication
Figure 3.21 LSA 12	Mitosis
Figure 9.10 LSA 30	Sliding Filament Model of Muscle Contraction
Figure 11.6 LSA 22	Formation of Myelin Sheath
Figure 12.28 LSA 25	Reflex Arcs
Figure 15.34 LSA 27	The Organ of Corti
Figure 16.48 LSA 39	Common Congenital Defects of the Heart
Figure 20.6 LSA 19	Spermatogenesis
Figure 21.5 LSA 20	Oogenesis
Figure 22.16 LSA 21	Human Embryonic Development
Figure 22.17 LSA 21	Human Embryonic Development
Figure 22.18 LSA 21	Human Embryonic Development

The Dynamic Human CD-ROM

Chapter 1

1.18 Skeletal System/Clinical Concepts/Fractured Femur
Digestive System/Clinical Concepts/Gallstones
1.19 Anatomical Orientation/Visible Human/Head

Chapter 2

2.11 Anatomical Orientation/Planes
2.12 Anatomical Orientation/Planes
Table 2.3 Anatomical Orientation/Directional Terminology

Chapter 3

3.1 Nervous System/Histology/Spinal Neurons
Muscular System/Histology/Skeletal Muscle
3.5 Digestive System/Histology/Duodenum
Nervous System/Histology/Retina
Nervous System/Histology/Organ of Corti
3.15 Reproductive System/Histology/Uterine Tube

Chapter 4

4.3 Urinary System/Histology/Renal Cortex
4.4 Digestive System/Histology/Duodenum
4.5 Reproductive System/Histology/Uterine Tube
4.9 Urinary System/Histology/Bladder
4.20 Skeletal System/Histology/Hyaline Cartilage
4.21 Skeletal System/Histology/Fibrocartilage
4.22 Skeletal System/Histology/Elastic Cartilage
4.24 Skeletal System/Histology/Compact Bone
Skeletal System/Histology/Spongy Bone
4.26 Muscular System/Histology/Cardiac Muscle
Muscular System/Histology/Smooth Muscle
Muscular System/Histology/Skeletal Muscle (cross section)
Muscular System/Histology/Skeletal Muscle (longitudinal)
4.27 Nervous System/Histology/Spinal Neuron

Chapter 6

6.1 Skeletal System/Anatomy/Gross Anatomy
Skeletal System/Explorations/Walking Skeleton
6.3 Skeletal System/Explorations/Cross Section of a Long Bone
6.5 Skeletal System/Explorations/Cross Section of a Long Bone
6.6 Skeletal System/Histology/Compact Bone
6.7 Skeletal System/Histology/Compact Bone
6.8 Skeletal System/Histology/Compact Bone
6.14 Skeletal System/Anatomy/3D Viewer: Cranial Anatomy
6.15 Skeletal System/Anatomy/3D Viewer: Cranial Anatomy
6.16 Skeletal System/Anatomy/3D Viewer: Cranial Anatomy
6.17 Skeletal System/Anatomy/3D Viewer: Cranial Anatomy
6.18 Skeletal System/Anatomy/3D Viewer: Cranial Anatomy
6.19 Skeletal System/Anatomy/3D Viewer: Cranial Anatomy
6.20 Skeletal System/Anatomy/3D Viewer: Cranial Anatomy

Continued

The Dynamic Human CD-ROM—*Continued*

6.32 Skeletal System/Anatomy/3D Viewer: Thoracic Anatomy

6.38 Skeletal System/Anatomy/ 3D Viewer: Thoracic Anatomy

Chapter 7

7.1 Skeletal System/Anatomy/ Gross Anatomy
Skeletal System/Explorations/ Walking Skeleton

7.23 Skeletal System/Clinical Concepts/Fractured Femur

7.24 Skeletal System/Clinical Concepts/Fractured Femur

Chapter 8

8.1 Skeletal System/Explorations/ Types of Joints/Fibrous Joints

8.2 Skeletal System/Explorations/ Types of Joints/Fibrous Joints

8.3 Skeletal System/Explorations/ Types of Joints/Cartilaginous Joints

8.4 Skeletal System/Explorations/ Types of Joints/Cartilaginous Joints

8.7 Skeletal System/Explorations/ Types of Joints/Synovial Joints

8.9 Skeletal System/Explorations/ Types of Joints/Synovial Joints

8.10 Skeletal System/Explorations/ Types of Joints/Synovial Joints

8.11 Skeletal System/Explorations/ Types of Joints/Synovial Joints

8.12 Skeletal System/Explorations/ Types of Joints/Synovial Joints

8.13 Skeletal System/Explorations/ Types of Joints/Synovial Joints

8.14 Skeletal System/Explorations/ Types of Joints/Synovial Joints

8.25 Skeletal System/Explorations/ Types of Joints/Synovial Joints/ Synovial Joint Motion

8.27 Skeletal System/Explorations/ Types of Joints/Synovial Joints/ Synovial Joint Motion

8.29 Skeletal System/Explorations/ Types of Joints/Synovial Joints

8.30 Skeletal System/Explorations/ Types of Joints/Synovial Joints/ Synovial Joint Motion

8.31 Skeletal System/Explorations/ Types of Joints/Synovial Joints/ Synovial Joint Motion
Skeletal System/Explorations/ Types of Joints/Synovial Joints/ Generic Synovial Joint

8.32 Skeletal System/Clinical Concepts/MRI of Knee

8.37 Skeletal System/Clinical Concepts/Arthroscopy

Chapter 9

9.2 Muscular System/Explorations/ Muscle Action Around Joints

9.3 Muscular System/Explorations/ Sliding Filament Theory

9.4 Muscular System/Explorations/ Muscle Action Around Joints

9.6 Muscular System/Explorations/ Sliding Filament Theory
Muscular System/Histology/ Skeletal Muscle (cross section)

9.7 Muscular System/Explorations/ Sliding Filament Theory
Muscular System/Histology/ Skeletal Muscle (longitudinal)

9.8 Muscular System/Explorations/ Sliding Filament Theory
Muscular System/Histology/ Skeletal Muscle (cross section)
Muscular System/Histology/ Skeletal Muscle (longitudinal)

9.9 Muscular System/Histology/ Skeletal Muscle (longitudinal)

9.10 Muscular System/Histology/ Skeletal Muscle (longitudinal)

9.11 Muscular System/Clinical Concepts/ Isometric vs. Isotonic Contraction

9.12 Muscular System/Explorations/ Neuromuscular Junction

Chapter 11

11.2 Nervous System/Histology/ Spinal Neuron

11.3 Nervous System/Histology/ Spinal Neuron

11.15 Muscular System/Explorations/ Neuromuscular Junction

11.16 Nervous System/Anatomy/ Gross Anatomy of the Brain

11.17 Nervous System/Anatomy/ Gross Anatomy of the Brain

11.19 Nervous System/Anatomy/ Gross Anatomy of the Brain

11.22 Nervous System/Explorations/ Motor and Sensory Pathways

11.25 Nervous System/Anatomy/ Gross Anatomy of the Brain

11.35 Nervous System/Anatomy/ Spinal Cord Anatomy
Nervous System/Histology/ Spinal Cord

11.36 Nervous System/Anatomy/ Spinal Cord Anatomy
Nervous System/Histology/ Spinal Cord

11.41 Nervous System/Anatomy/ Spinal Cord Anatomy
Nervous System/Histology/ Spinal Cord

Chapter 12

12.4 Nervous System/Explorations/ Olfaction

12.5 Nervous System/Explorations/Vision

12.9 Nervous System/Explorations/Hearing

12.27 Skeletal System/Clinical Concepts/Herniated Disk

12.28 Nervous System/Explorations/ Reflex Arc

Chapter 13

13.1 Nervous System/Explorations/ Reflex Arc

13.9 Nervous System/Anatomy/ Gross Anatomy of the Brain

13.10 Nervous System/Anatomy/ Gross Anatomy of the Brain

Chapter 14

14.2 Endocrine System/Anatomy/ Gross Anatomy

14.6 Endocrine System/Explorations/ Endocrine Function

14.8 Endocrine System/Anatomy/ Gross Anatomy/Hypothalamus and Pituitary Gland

14.9 Endocrine System/Histology/ Pituitary Gland

14.12 Endocrine System/Gross Anatomy/ Hypothalamus and Pituitary Gland

The Dynamic Human CD-ROM—*Continued*

14.13 Endocrine System/Gross Anatomy/Thyroid Gland
14.14 Endocrine System/Histology/Thyroid Gland
14.15 Endocrine System/Explorations/Hypothalamo-Pituitary Thyroid Axis
14.16 Endocrine System/Gross Anatomy/Thyroid/Parathyroid
14.19 Endocrine System/Gross Anatomy/Pancreas
Endocrine System/Clinical Concepts/Diabetes
14.21 Endocrine System/Gross Anatomy/Adrenal Gland
Endocrine System/Histology/Adrenal Gland
14.22 Endocrine System/Histology/Adrenal Gland

Chapter 15

15.4 Nervous System/Explorations/Motor and Sensory Pathways
15.5 Nervous System/Explorations/Olfaction
15.6 Nervous System/Explorations/Taste
15.7 Nervous System/Explorations/Taste
Nervous System/Explorations/Innervation of Tongue
15.9 Nervous System/Histology/Eye
15.14 Nervous System/Histology/Eye
15.15 Nervous System/Explorations/Vision
15.16 Nervous System/Explorations/Vision
15.17 Nervous System/Explorations/Vision
15.18 Nervous System/Explorations/Vision
15.19 Nervous System/Histology/Retina
Nervous System/Explorations/Vision
15.28 Nervous System/Explorations/Hearing
15.32 Nervous System/Explorations/Dynamic Equilibrium
Nervous System/Explorations/Static Equilibrium
15.33 Nervous System/Explorations/Hearing
15.34 Nervous System/Histology/Organ of Corti
15.35 Nervous System/Histology/Organ of Corti
15.40 Nervous System/Explorations/Static Equilibrium
15.41 Nervous System/Explorations/Dynamic Equilibrium

Chapter 16

16.7 Cardiovascular System/Anatomy/Gross Anatomy
Cardiovascular System/Anatomy/3D Viewer: Thoracic Anatomy
TA16.1 Cardiovascular System/Histology/Purkinje Fiber
16.8 Cardiovascular System/Anatomy/Gross Anatomy
16.9 Cardiovascular System/Anatomy/Gross Anatomy
16.10 Cardiovascular System/Anatomy/Gross Anatomy
16.11 Cardiovascular System/Explorations/Heart Dynamics/ECG
Cardiovascular System/Explorations/Heart Dynamics/Conduction System
16.12 Cardiovascular System/Explorations/Heart Dynamics/Heart Sounds
16.13 Cardiovascular System/Explorations/Generic Vasculature
16.15 Cardiovascular System/Explorations/Generic Vasculature
16.16 Cardiovascular System/Explorations/Generic Vasculature
16.37 Cardiovascular System/Explorations/Generic Portal System
16.41 Immune and Lymphatic Systems/Anatomy/Gross Anatomy
16.42 Immune and Lymphatic Systems/Anatomy/Gross Anatomy
Immune and Lymphatic Systems/Histology/Lymph Node
16.43 Immune and Lymphatic Systems/Anatomy/Gross Anatomy
Immune and Lymphatic Systems/Histology/Spleen
16.44 Immune and Lymphatic Systems/Anatomy/Gross Anatomy

Chapter 17

17.1 Respiratory System/Anatomy/Gross Anatomy
Respiratory System/Anatomy/3D Viewer: Thoracic Anatomy
17.4 Respiratory System/Histology/Alveoli
Respiratory System/Histology/Trachea
Respiratory System/Histology/Nasal Cavity
Respiratory System/Histology/Lung
Respiratory System/Histology/Bronchiole
17.7 Respiratory System/Anatomy/Gross Anatomy
17.9 Respiratory System/Anatomy/Gross Anatomy
17.10 Respiratory System/Histology/Trachea
17.12 Respiratory System/Anatomy/Gross Anatomy
17.14 Respiratory System/Histology/Bronchiole
Respiratory System/Histology/Alveoli
17.15 Respiratory System/Histology/Bronchiole
Respiratory System/Histology/Alveoli
17.16 Respiratory System/Histology/Bronchiole
Respiratory System/Histology/Alveoli
17.17 Respiratory System/Anatomy/Gross Anatomy
Respiratory System/Anatomy/3D Viewer: Thoracic Anatomy
17.18 Respiratory System/Anatomy/Gross Anatomy
17.21 Anatomical Orientation/Visible Human/Thorax
17.22 Respiratory System/Explorations/Mechanics of Breathing
17.24 Respiratory System/Explorations/Mechanics of Breathing
17.26 Respiratory System/Clinical Concepts/Spirometry

Chapter 18

18.1 Digestive System/Anatomy/Gross Anatomy

Continued

The Dynamic Human CD-ROM—*Continued*

18.3 Digestive System/Anatomy/Gross Anatomy
18.4 Digestive System/Histology/Duodenum
18.6 Digestive System/Anatomy/Gross Anatomy
18.7 Nervous System/Histology/Vallate Papillae
18.10 Digestive System/Histology/Tooth
18.12 Digestive System/Histology/Submandibular Gland
18.14 Digestive System/Histology/Esophagus
18.15 Digestive System/Anatomy/Gross Anatomy
Digestive System/Explorations/Peristalsis
18.16 Digestive System/Anatomy/Gross Anatomy
Digestive System/Explorations/Stomach
18.18 Digestive System/Histology/Fundic Stomach
18.19 Digestive System/Histology/Fundic Stomach
18.20 Digestive System/Anatomy/Gross Anatomy
Digestive System/Explorations/Duodenum
Digestive System/Histology/Duodenum
18.21 Digestive System/Anatomy/Gross Anatomy
Digestive System/Anatomy/3D Viewer: Digestive Anatomy
18.22 Digestive System/Histology/Duodenum
18.23 Digestive System/Histology/Duodenal Villi
18.25 Digestive System/Anatomy/Gross Anatomy
18.26 Digestive System/Clinical Concepts/Barium Radiograph
18.30 Anatomical Orientation/Visible Human/Abdomen
18.31 Digestive System/Anatomy/Gross Anatomy
Digestive System/Anatomy/3D Viewer: Digestive Anatomy
18.32 Digestive System/Histology/Liver
18.33 Digestive System/Histology/Liver
18.35 Digestive System/Clinical Concepts/Gallstones
18.36 Digestive System/Anatomy/Gross Anatomy
Digestive System/Anatomy/3D Viewer: Digestive Anatomy
Digestive System/Explorations/Pancreas
18.39 Digestive System/Clinical Concepts/Ulcer

Chapter 19

19.1 Urinary System/Anatomy/Gross Anatomy
19.2 Urinary System/Anatomy/Gross Anatomy
19.3 Anatomical Orientation/Visible Human/Abdomen
19.4 Urinary System/Anatomy/Kidney Anatomy
19.5 Urinary System/Anatomy/Kidney Anatomy
Urinary System/Histology/Renal Cortex
Urinary System/Histology/Renal Medulla
19.7 Urinary System/Anatomy/3D Viewer: Nephron
19.8 Urinary System/Anatomy/Nephron Anatomy
19.14 Urinary System/Anatomy/Gross Anatomy
19.15 Urinary System/Anatomy/Gross Anatomy
Urinary System/Histology/Bladder

Chapter 20

20.1 Reproductive System/Anatomy/Male Reproductive Anatomy
Reproductive System/Anatomy/3D Viewer: Male Reproductive Anatomy
20.3 Reproductive System/Histology/Testis
20.4 Reproductive System/Histology/Testis
20.6 Reproductive System/Explorations/Male/Spermatogenesis
20.7 Reproductive System/Histology/Testis
20.14 Reproductive System/Anatomy/Male Reproductive Anatomy
Reproductive System/Anatomy/3D Viewer: Male Reproductive Anatomy
20.15 Reproductive System/Explorations/Male/Erection
20.19 Reproductive System/Clinical Concepts/Vasectomy

Chapter 21

21.1 Reproductive System/Anatomy/Female Reproductive Anatomy
Reproductive System/Anatomy/3D Viewer: Female Reproductive Anatomy
21.3 Reproductive System/Anatomy/Female Reproductive Anatomy
Reproductive System/Anatomy/3D Viewer: Female Reproductive Anatomy
21.4 Reproductive System/Histology/Ovary
Reproductive System/Histology/Ovarian Follicle
21.5 Reproductive System/Explorations/Female/Oogenesis
21.6 Reproductive System/Histology/Ovary
Reproductive System/Histology/Ovarian Follicle
21.8 Reproductive System/Explorations/Female/Oogenesis
21.9 Reproductive System/Explorations/Female/Ovarian Cycle
21.10 Reproductive System/Histology/Uterine Tube
21.19 Reproductive System/Explorations/Female/Menstrual Cycle
Reproductive System/Explorations/Female/Ovarian Cycle
21.21 Reproductive System/Clinical Concepts/Tubal Ligation

ONE

History of Anatomy

Definition of the Science 2
Prescientific Period 2
Scientific Period 5
- Mesopotamia and Egypt 7
- China and Japan 8
- Grecian Period 9
- Alexandrian Era 10
- Roman Era 11
- Middle Ages 12
- Contributions of Islam 12
- Renaissance 12
- Seventeenth and Eighteenth Centuries 14
- Nineteenth Century 16
- Twentieth Century 16

Chapter Summary 20
Review Activities 20

This woodcut, from a medical book published in 1511, portrays three giants of early medicine—Galen, physician to the gladiators and to the emperor Marcus Aurelius; Avicenna, "the Galen of Islam," whose Canon *was one of the most famous medical books ever written; and Hippocrates, the father of medicine.*

Definition of the Science

The science of human anatomy is concerned with the structural organization of the human body. The descriptive anatomical terminology is principally of Greek and Latin derivation.

Objective 1 Define *anatomy.*

Objective 2 Distinguish between anatomy, physiology, and biology.

Objective 3 Explain why most anatomical terms are derived from Greek and Latin words.

Human anatomy is the science concerned with the structure of the human body. The term **anatomy** is derived from a Greek word meaning "to cut up"; indeed, in ancient times, the word *anatomize* was more commonly used than the word *dissect.* The science of **physiology** is concerned with the function of the body. It is inseparable from anatomy in that structure tends to reflect function. The term *physiology* is derived from another Greek word—this one meaning "the study of nature." The "nature" of an organism is its function. Anatomy and physiology are both subdivisions of the science of **biology,** the study of living organisms. The anatomy of every structure of the body is adapted for performing a function, or perhaps several functions.

The dissection of human **cadavers** (*kă-dav'erz*) has served as the basis for understanding the structure and function of the human body for many centuries. Every beginning anatomy student can discover and learn firsthand as the structures of the body are systematically dissected and examined. The anatomical terms that a student learns while becoming acquainted with a structure represent the work of hundreds of dedicated anatomists of the past, who have dissected, diagrammed, described, and named the multitude of body parts.

Most of the terms that form the language of anatomy are of Greek or Latin derivation. Latin was the language of the Roman Empire, during which time an interest in scientific description was cultivated. With the decline of the Roman Empire, Latin became a "dead language," but it retained its value in nomenclature because it remained unchanged throughout history. As a consequence, if one is familiar with the basic prefixes and suffixes (see the inside front and back covers of this text), many of the terms in the descriptive science of anatomy can be understood. Although the Greeks and Romans made significant contributions to anatomical terminology, it should be noted that many individuals from other cultures have also contributed to the science of human anatomy.

As a scientific field of inquiry, human anatomy has had a rich, long, and frequently troubled heritage. The history of human anatomy parallels that of medicine. In fact, interest in the structure of the body often has been stimulated by the desire of the medical profession to explain a body dysfunction. Various religions, on the other hand, have at one time or another stifled the study of human anatomy through their restrictions on human dissections and their emphasis on nonscientific explanations for diseases and debilitations.

Over the centuries, peoples' innate interest in their own bodies and physical capabilities has found various forms of expression. The Greeks esteemed athletic competition and expressed the beauty of the body in their sculptures. Many of the great masters of the Renaissance portrayed human figures in their art. Indeed, several of these artists were excellent anatomists because their preoccupation with detail demanded it. Such an artistic genius was Michelangelo, who captured the splendor of the human form in sculpture with the *David* (fig. 1.1) and in paintings like those in the Sistine Chapel.

Shakespeare's reverence for the structure of the human body found expression in his writings: "What a piece of work is a man! How noble in reason! how infinite in faculty! In form and moving, how express and admirable! In action how like an angel! In apprehension how like a god! The beauty of the world! The paragon of animals!" (*Hamlet* 2.2.315–319).

In the past, human anatomy was an academic, purely descriptive science, concerned primarily with identifying and naming body structures. Although dissection and description form the basis of anatomy, the importance of human anatomy today is in its functional approach and clinical applications. Human anatomy is a practical, applied science that provides the foundation for understanding physical performance and body health. Studying the history of anatomy helps us to appreciate the relevant science that it is today.

1. What is the derivation and meaning of *anatomy?*
2. Explain the statement, Anatomy is a science based on observation, whereas physiology is based on experimentation and observation.
3. Why does understanding the biology of an organism depend on knowing its anatomy and physiology?
4. Discuss the value of using established Greek or Latin prefixes and suffixes in naming newly described body structures.

Prescientific Period

Evidence indicates that a knowledge of anatomy was of survival value in prehistoric times and that it provided the foundation for medicine.

Objective 4 Explain why an understanding of human anatomy is essential in the science of medicine.

Objective 5 Define *trepanation* and *paleopathology.*

---

anatomy: Gk. *ana,* up; *tome,* a cutting
physiology: Gk. *physis,* nature; *logos,* study
biology: Gk. *bios,* life; *logos,* study
cadaver: L. *cadere,* to fall

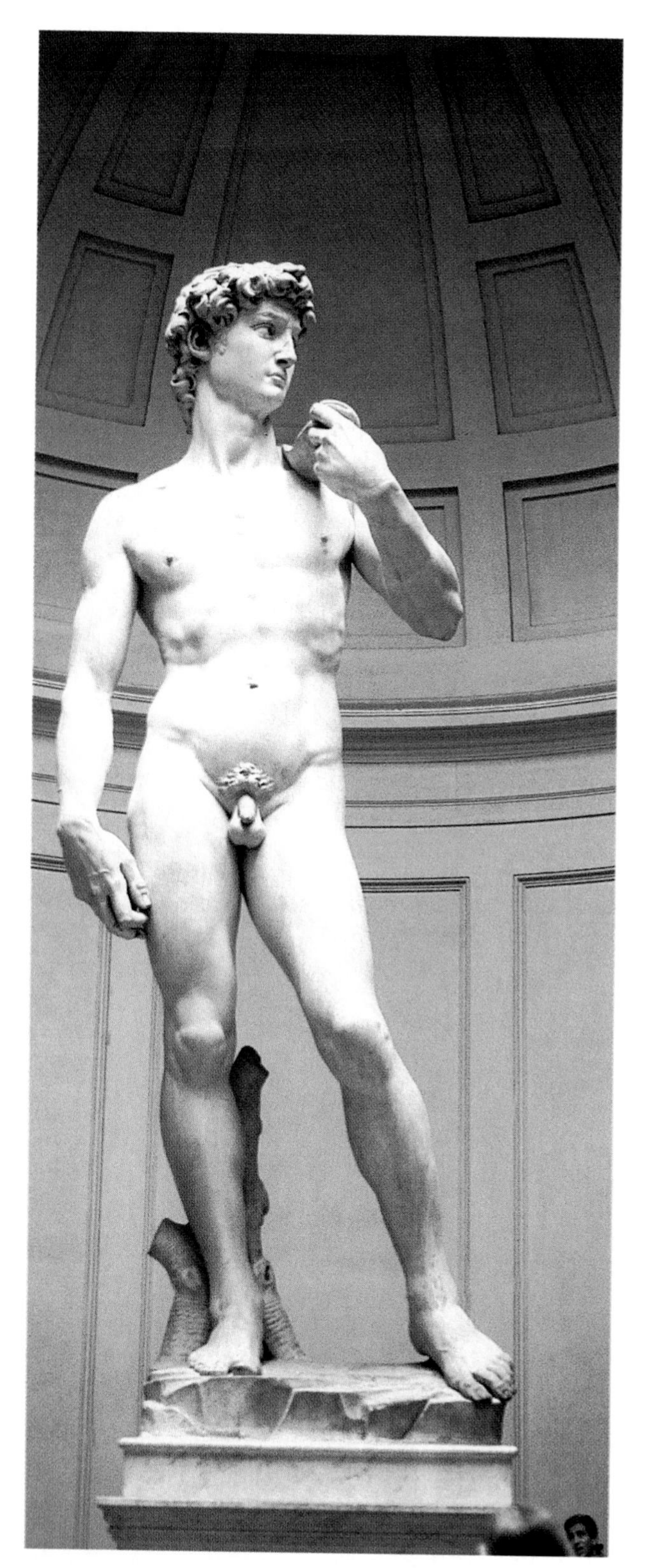

Figure 1.1

Michelangelo completed the seventeen-foot-tall *David* in 1504. Sculptured from a single block of white, unflawed Carrara marble, this masterpiece captures the physical beauty of the human body in an expression of art.

Figure 1.2

Contemporary redrawings of large game mammals that were depicted on the walls of caves occupied by early *Homo sapiens* in western Europe. Presumably the location of the heart is drawn on the mammoth, and vulnerable anatomical sites are shown on the two bison. Prehistoric people needed a practical knowledge of anatomy simply for survival.

It is likely that a type of practical comparative anatomy is the oldest science. Certainly, humans have always been aware of some of their anatomical structures and how they function. Our prehistoric ancestors undoubtedly knew their own functional abilities and limitations as compared to those of other animals. Through the trial and error of hunting, they discovered the "vital organs" of an animal, which, if penetrated with an object, would cause death (fig. 1.2). Likewise, they knew the vulnerable areas of their own bodies.

The butchering of an animal following the kill provided many valuable anatomy lessons for prehistoric people. They knew which parts of an animal's body could be used for food, clothing, or implements. Undoubtedly, they knew that the muscles functioned in locomotion and that they also provided a major source of food. The skin from mammals with its associated fur served as a protective covering for their own sparsely haired skin. Early humans knew that the skeletal system formed a durable framework within their bodies and those of other vertebrates. They used the bones from the animals on which they fed to fashion a variety of tools and weapons. They knew that their own bones could be broken through accidents, and that improper healing would result in permanent disability. They knew that if an animal was wounded, it would bleed, and that excessive loss of blood would cause death. Perhaps they also realized that a severe blow to the head could cause deep sleep and debilitate an animal without killing it. Obviously, they noted anatomical differences between the sexes, even though they could not have understood basic reproductive functions. The knowledge these people had was of the basic, practical type—a knowledge necessary for survival.

Certain surgical skills are also ancient. **Trepanation** (*trep-ă-na'shun*), the drilling of a hole in the skull, or removal of a portion of a cranial bone, seems to have been practiced by several groups of prehistoric people. Trepanation was probably used as a ritualistic procedure to release evil spirits, or on some patients,

trepanation: Gk. *trypanon*, a borer

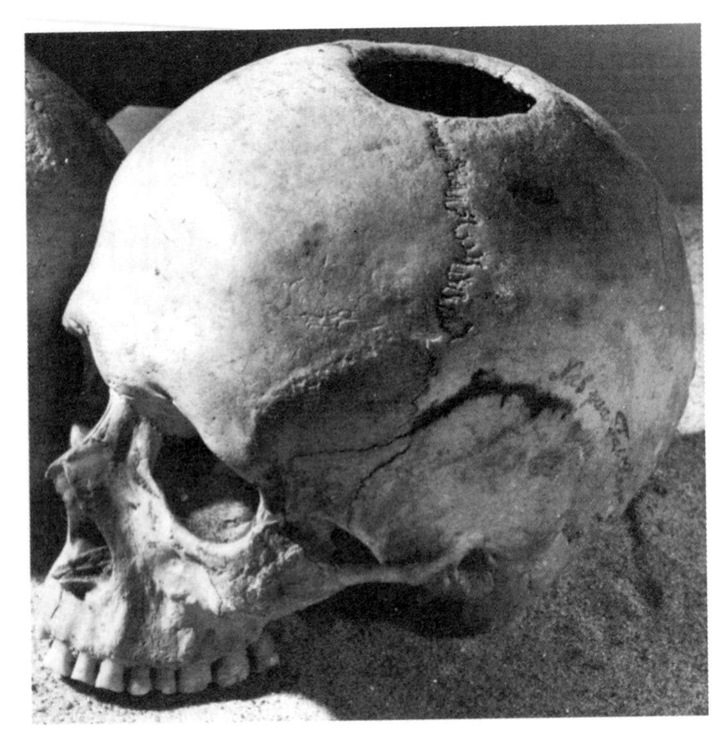

Figure 1.3

The surgical art of trepanation was practiced by several prehistoric cultures. Amazingly, more than a few patients survived this ordeal, as evidenced by ossification around the bony edges of the wound.

perhaps, to relieve cranial pressure resulting from a head wound. Trepanated skulls have been found repeatedly in archaeological sites (fig. 1.3). Judging from the partial reossification in some of these skulls, apparently a fair proportion of the patients survived.

What is known about prehistoric humans is conjectured through information derived from cave drawings, artifacts, and fossils that contain paleopathological information. **Paleopathology** is the science concerned with studying diseases and causes of death in prehistoric humans. A person's approximate age can be determined from skeletal remains, as can the occurrence of certain injuries and diseases, including nutritional deficiencies. Diets and dental conditions, for example, are indicated by fossilized teeth. What cannot be determined, however, is the extent of anatomical information and knowledge that may have been transmitted orally up until the time humans invented symbols to record their thoughts, experiences, and history.

1. Why would it be important to know the anatomy of the skull and brain before performing a surgery such as trepanation?
2. What types of data might a paleopathologist be interested in obtaining from an Egyptian mummy?

paleopathology: Gk. *palaios*, ancient; *pathos*, suffering; *logos*, to study

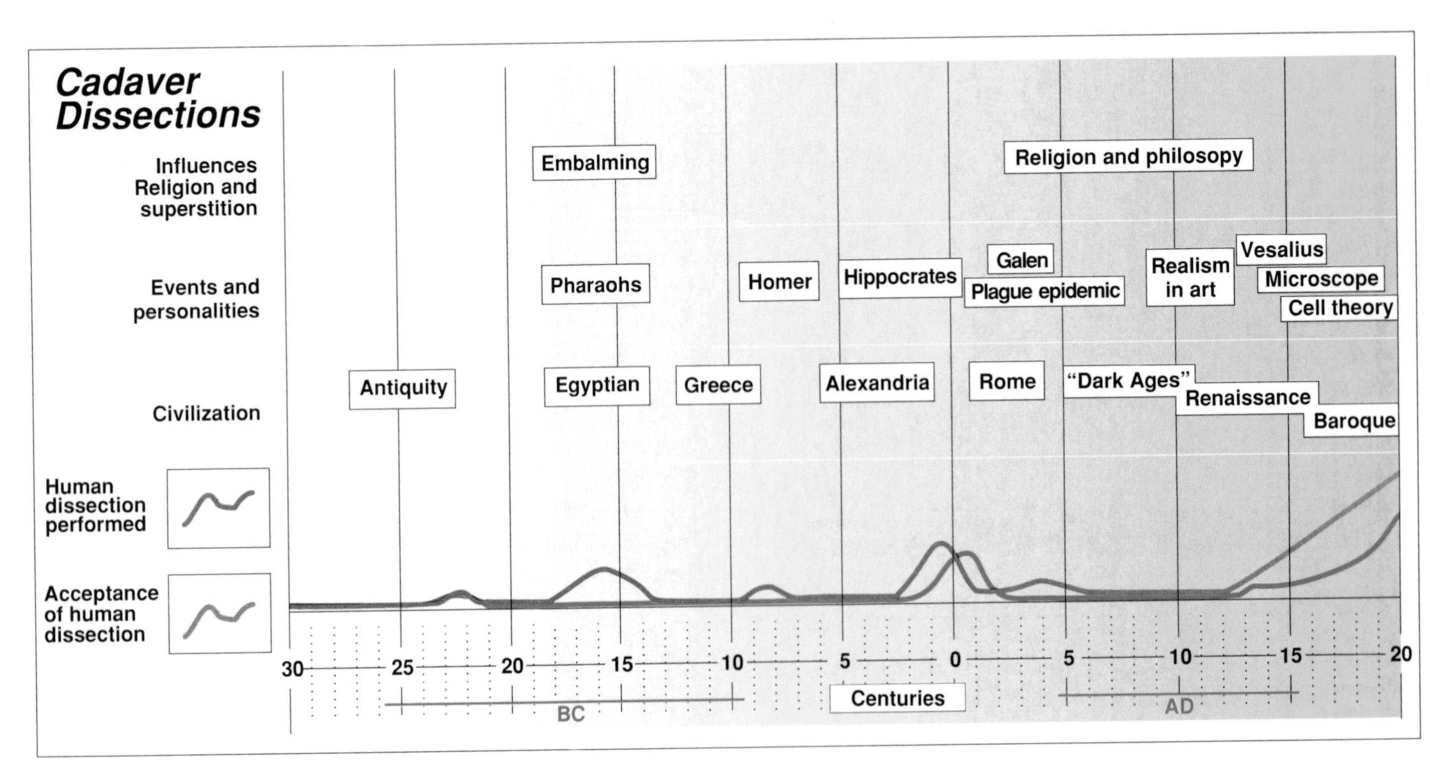

Figure 1.4

A timeline depicting the story of cadaver dissections.

Table 1.1

Survey of Some Important Contributors to the Science of Human Anatomy

Person	Civilization	Lifetime or Date of Contribution	Contribution
Menes	Egyptian	About 3400 B.C.	Wrote the first anatomy manual
Homer	Ancient Greece	About 800 B.C.	Described the anatomy of wounds in the *Iliad*
Hippocrates	Ancient Greece	About 460–377 B.C.	Father of medicine; inspired the Hippocratic oath
Aristotle	Ancient Greece	384–322 B.C.	Founder of comparative anatomy; profoundly influenced Western scientific thinking
Herophilus	Alexandria	About 325 B.C.	Conducted remarkable research on aspects of the nervous system
Erasistratus	Alexandria	About 300 B.C.	Sometimes called father of physiology; attempted to apply physical laws to the study of human function
Celsus	Roman	30 B.C.–A.D. 30	Compiled information from the Alexandrian school; first medical author to be printed (1478) in movable type after Gutenburg's invention
Galen	Greek (lived under Roman domination)	130–201	Probably the most influential medical writer of all time; established principles that went unchallenged for 1,500 years
de' Luzzi	Renaissance	1487	Prepared dissection guide
Leonardo da Vinci	Renaissance	1452–1519	Produced anatomical drawings of unprecedented quality based on human cadaver dissections
Vesalius	Renaissance	1514–64	Refuted past misconceptions about body structure and function by direct observation and experiment; often called father of anatomy
Harvey	Premodern (European)	1578–1657	Demonstrated the function of the circulatory system; applied the experimental method to anatomy
Leeuwenhoek	Premodern (European)	1632–1723	Refined the microscope; described various cells and tissues
Malpighi	Premodern (European)	1628–94	Regarded as father of histology; first to confirm the existence of the capillaries
Sugita	Premodern (Japanese)	1774	Compiled a five-volume treatise on anatomy
Schleiden and Schwann	Modern (European)	1838–39	Formulated the cell theory
Roentgen	Modern (European)	1895	Discovered X rays
Crick and Watson	Modern (English and American)	1953	Determined the structure of DNA

Scientific Period

Human anatomy is a dynamic and growing science with a long, exciting heritage. It continues to provide the foundation for medical, biochemical, developmental, cytogenetic, and biomechanical research.

Objective 6 Discuss some of the key historical events in the science of human anatomy.

Objective 7 List the historical periods in which cadavers were used to study human anatomy.

Objective 8 Explain why an understanding of human anatomy is relevant to all individuals.

Objective 9 Discuss one way of keeping informed about developments in anatomical research and comment on the importance of this endeavor.

The scientific period begins with recorded anatomical observations made in early Mesopotamia on clay tablets in *cuneiform script* over 3,000 years ago and continues to the present day. Obviously, all of the past contributions to the science of anatomy cannot be mentioned; however, certain individuals and cultures had a tremendous impact and will be briefly commented on in this section.

The history of anatomy has an interesting parallel with the history of the dissection of human cadavers, as is depicted in figure 1.4. A few of the individuals who made significant contributions to the field are listed in table 1.1. Some of their contributions were in the form of books (table 1.2) that describe and illustrate the structure of the body and in some cases explain various body functions.

---
cuneiform: L. *cuneus*, wedge; *forma*, shape

Table 1.2

Influential Books and Publications on Anatomy and Related Disciplines

Aristotle. 384–322 B.C. *Historia animalium (History of animals)*, *De partibus animalium (On the parts of animals)*, and *De generatione animalium (On the generation of animals)*. These classic works by the great Greek philosopher profoundly influenced biological thinking for centuries.

Celsus, Cornelius. 30 B.C.–A.D. 30. *De re medicina*. This eight-volume work was primarily a compilation of the medical data that was available from the Alexandrian school.

Galen, Claudius. 130–201. Nearly 500 medical treatises on descriptive anatomy. Although Galen's writings contained numerous errors, his pronouncements on the structure and function of the body held sway until the Renaissance.

de' Luzzi, Mondino. 1487. *Anathomia*. This book was used as a dissection guide for over 225 years, during which time it underwent 40 revisions.

Vesalius, Andreas. 1543. *De humani corporis fabrica (On the structure of the human body)*. The beautifully illustrated *Fabrica* boldly challenged many of the errors that had been perpetuated by Galen. In spite of the controversies it provoked, this book was well accepted and established a new standard of excellence in anatomy texts.

Fabricius ab Aquapendente, Hieronymus. 1600–1621. *De formato foetu (On the formation of the fetus)* and *De formatione ovi et pulli (On the formation of eggs and birds)*. These books marked the beginning of embryological study.

Harvey, William. 1628. *Exercitatio de motu cordis et sanguinis in animalibus (On the motion of the heart and blood in animals)*. Harvey demonstrated that blood must be circulated, and his experimental methods are still regarded as classic examples of scientific methodology.

Descartes, René. 1637. *Discourse on method*. This philosophic thesis stimulated a mechanistic interpretation of biological data.

Linnaeus, Carolus. 1758. *Systema naturae*. The basis for the classification of living organisms is explained in this monumental work. Its anatomical value is in comparative anatomy, where the anatomy of different species is compared.

Haller, Albrecht von. 1760. *Elementa physiologiae (Physiological elements)*. Some basic physiological concepts are presented in this book, including a summary of what was then known of the functioning of the nervous system.

Sugita, Genpaku. 1774. *Kaitai shinsho (A new book of anatomy)*. This book adopted a European conceptualization of body structure and function and ushered in a new era of anatomy for the Japanese.

Cuvier, Georges. 1817. *Le règne animal (The animal kingdom)*. This comprehensive comparative vertebrate anatomy book had enormous influence on contemporary zoological thought.

Baer, Karl Ernst von. 1828–37. *Über entwicklungsgeschichté der thiere (On the development of animals)*. This book helped to pave the way for modern embryology by discussing germ layer formation.

Beaumont, William. 1833. *Experiments and observations on the gastric juice and the physiology of digestion*. Basic digestive functions are accurately described in this classic work.

Müller, Johannes. 1834–40. *Handbuch der physiologie des menschen für vorlesungen (Elements of physiology)*. This book established physiology as a science concerned with the functioning of the body.

Schwann, Theodor. 1839. *Mikroskopische untersuchungen über die übereinstimmung in der struktur und dem wachstum der thiere und pflanzen (Microscopic researches into accordance in the structure and growth of animals and plants)*. The basic theory that all living organisms consist of living cells is presented in this classic study.

Kölliker, Albrecht von. 1852. *Mikroskopische anatomia (Microscopic anatomy)*. This premier textbook in histology laid the foundation for the emerging science.

Gray, Henry. 1858. *Anatomy, descriptive and surgical*. This masterpiece, better known as *Gray's anatomy*, is still in print and contains over 200 of the original illustrations. Thousands of physicians have used it to learn gross human anatomy.

Virchow, Rudolf. 1858. *Die cellularpathologie (Cell pathology)*. Descriptions of normal and diseased tissues are presented in this book.

Darwin, Charles. 1859. *On the origin of species*. The ideas set forth in this classic took many years to be understood and accepted. Its importance to anatomy is that it provided an explanation for the anatomical variation evident among different species.

Mendel, Gregor. 1866. *Versuche über pflanzenhybriden (Experiments with plant hybrids)*. Through observation and experimentation, Mendel demonstrated the basic principles of heredity.

Owen, Richard. 1866. *Anatomy and physiology of the vertebrates*. Some basic concepts of structure and function, such as homologue and analogue, are presented in this book.

Balfour, Francis M. 1880. *Comparative embryology*. This book was considered a primary source of information for the emerging science of experimental embryology.

Weismann, August. 1892. *Das keimplasma (The germplasm)*. Weismann postulated the theory of meiosis, which states that a reduction in the chromosome number is necessary in the gametes of both the male and female for fertilization to occur.

Hertwig, Oskar. 1893. *Zelle and gewebe (Cell and tissue)*. Important distinctions between the sciences of cytology and histology are made in this book.

Wilson, Edmund B. 1896. *The cell in development and heredity*. This book had a profound influence on the development of cytogenetics.

Pavlov, Ivan. 1897. *Le travail des glandes digestives (The work of the digestive glands)*. The physiological functioning of the digestive system is described in this classic experimental work.

Sherrington, Charles. 1906. *The integrative action of the nervous system*. The basic concepts of neurophysiology were first established in this book.

Garrod, Archibald. 1909. *Inborn errors of metabolism*. Genetic defects are discussed in this pioneer book, and are shown to be caused by defective genes.

Bayliss, William M. 1915. *Principles of general physiology*. This book provided the synthesis that was needed for a newly emerging science.

Spemann, Hans. 1938. *Embryonic development and induction*. This masterful book provided the foundation for the science of modern experimental embryology.

Crick, Francis H. C., and James D. Watson. 1953. *Genetic implications of the structure of deoxyribonucleic acid*. This remarkable work explains the process of genetic replication and control of cellular functions.

Steindler, Arthur. 1995. *Kinesiology of the human body under normal and pathological conditions*. This contemporary text stimulated interest in biomechanics and functional anatomy as applied to clinical problems.

Figure 1.5

An inscribed clay model of a sheep's liver from the eighteenth or nineteenth century B.C. The people of ancient Mesopotamia regarded the liver as the seat of human emotions.

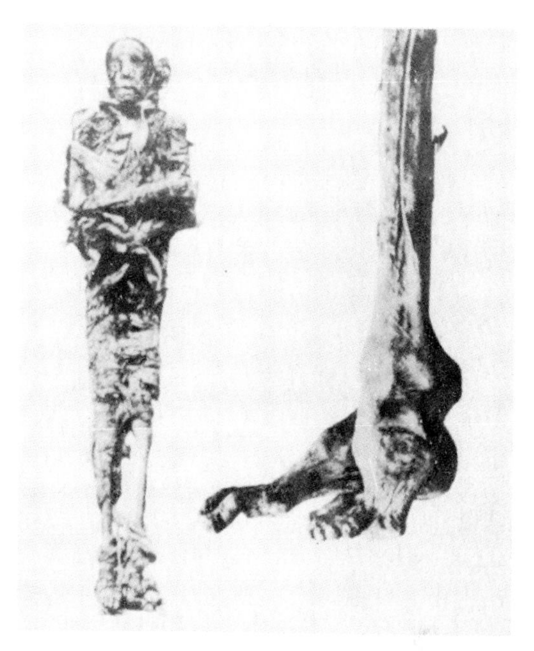

Figure 1.6

Perhaps the greatest contribution of the ancient Egyptian era to anatomy and medicine is the information obtained from the mummies. Certain diseases, injuries, deformities, and occasionally cause of death can be determined from paleopathogenic examination of the mummified specimens. Shown on the right is a congenital clubfoot from a mummy of a person who lived during the Nineteenth Dynasty (about 1300 B.C.).

Mesopotamia and Egypt

Mesopotamia was the name given to the long, narrow wedge of land between the Tigris and Euphrates rivers, which is now a large part of present-day Iraq. Archaeological excavations and ancient records show that this area was settled prior to 4000 B.C. On the basis of recorded information about the culture of the people, Mesopotamia is frequently called the Cradle of Civilization.

Many early investigations of the body represented an attempt to describe basic life forces. For example, people wondered which organ it was that constituted the soul. Some cuneiform writings from ancient Mesopotamia depicted and described body organs that were thought to serve this function. The liver, which was extensively studied in sacrificial animals (fig. 1.5), was thought to be the "guardianship of the soul and of the sentiments that make us men." This was a logical assumption because of the size of the liver and its close association with blood, which was observed to be vital for life. Even today, several European cultures associate the liver with various emotions.

The warm blood and arrangement of blood vessels are obviously a governing system within the body, and this influenced the search for the soul. When excessive blood is lost, the body dies. Therefore, some concluded, blood must contain a vital, life-giving force. The scholars of Mesopotamia were influenced by this idea, as was Aristotle, the Greek scientist who lived centuries later. Aristotle believed that the seat of the soul was the heart and that the brain functioned in cooling the blood that flowed from the heart. The association of the heart, in song and poetry, with the emotions of love and caring has its basis in Aristotelian thought.

The ancient Egyptian culture neighbored Mesopotamia to the west. Here, the sophisticated science of embalming the dead in the form of mummies was perfected (fig. 1.6). No known attempts were made to perform anatomical or pathological studies on the corpses, however, because embalming was strictly a religious ritual. It was reserved for royalty and the wealthy to prepare them for a life after death.

The Egyptian techniques of embalming could have contributed greatly to the science of anatomy had they been recorded. Apparently, however, embalmings were not generally looked on with favor by the general public in ancient Egypt. In fact, embalmers were frequently persecuted and even stoned. Embalming was a mystic art related more to religion than to science, and since it required a certain amount of mutilation of the dead body, it was regarded as demonic. Consequently, embalming techniques that could have provided embalmed cadavers as dissection specimens had to wait until centuries later to be rediscovered.

Several written works concerning anatomy have been discovered from ancient Egypt, but none of these influenced succeeding cultures. Menes, a king-physician during the first Egyptian dynasty of about 3400 B.C. (even before the pyramids were built), wrote what is thought to be the first manual on anatomy. Later writings (2300–1250 B.C.) attempted a systematization of the body, beginning with the head and progressing downward.

embalm: L. *in*, in; *balsamum*, balsam
physician: Gk. *physikos*, natural

Like the people of Mesopotamia and Greece, the ancient Egyptians were concerned with a controlling spirit of the body. In fact, they even had a name for this life force—the Ba spirit—and they believed that it was associated with the bowels and the heart. Food was placed in the tomb of a mummy to feed the Ba spirit during the journey to Osiris, Egyptian god of the underworld.

China and Japan

China

In ancient China, interest in the human body was primarily philosophical. Ideas about anatomy were based on reasoning rather than dissection or direct observation. The Chinese revered the body and abhorred its mutilation. An apparent exception was the practice of binding the feet of young girls and women in an attempt to enhance their beauty. Knowledge of the internal organs came only from wounds and injuries. Only in recent times have dissections of cadavers been permitted in Chinese medical schools.

The ancient Chinese had an abiding belief that everything in the universe depended on the balance of the two opposing cosmic principles of *yin* and *yang*. As for the circulatory system, the blood was the conveyor of the yang, and the heart and vessels represented the yin. Other structures of the body were composed of lesser forces termed *zō* and *fū*.

The Chinese were great herbalists. Writings more than 5,000 years old describe various herbal concoctions and potions to alleviate a wide variety of ailments, including diarrhea, constipation, and menstrual discomfort. Opium was described as an excellent painkiller.

Until recently, the Chinese have been possessive of their beliefs, and for this reason Western cultures were not influenced by Chinese thoughts or writings to an appreciable extent. Perhaps the best known but least understood of the Chinese contributions to human anatomy and medicine is acupuncture.

Acupuncture is an ancient practice that was established to maintain a balance between the yin and the yang. Three hundred sixty-five precise meridian sites, or vital points, corresponding to the number of days in a year, were identified on the body (fig. 1.7). Needles inserted into the various sites were believed to release harmful secretions and rid the tissues of obstructions. Acupuncture is still practiced in China and has gained acceptance with some medical specialists in the United States and other countries as a technique of anesthesia and as a treatment for certain ailments. The painkilling effect of acupuncture has been documented and is more than psychological. Acupuncture sites have been identified on domestic animals and have been used to a limited degree in veterinary medicine. Why acupuncture is effective remains a mystery, although it has recently been correlated with endorphin production within the brain (see chapter 11).

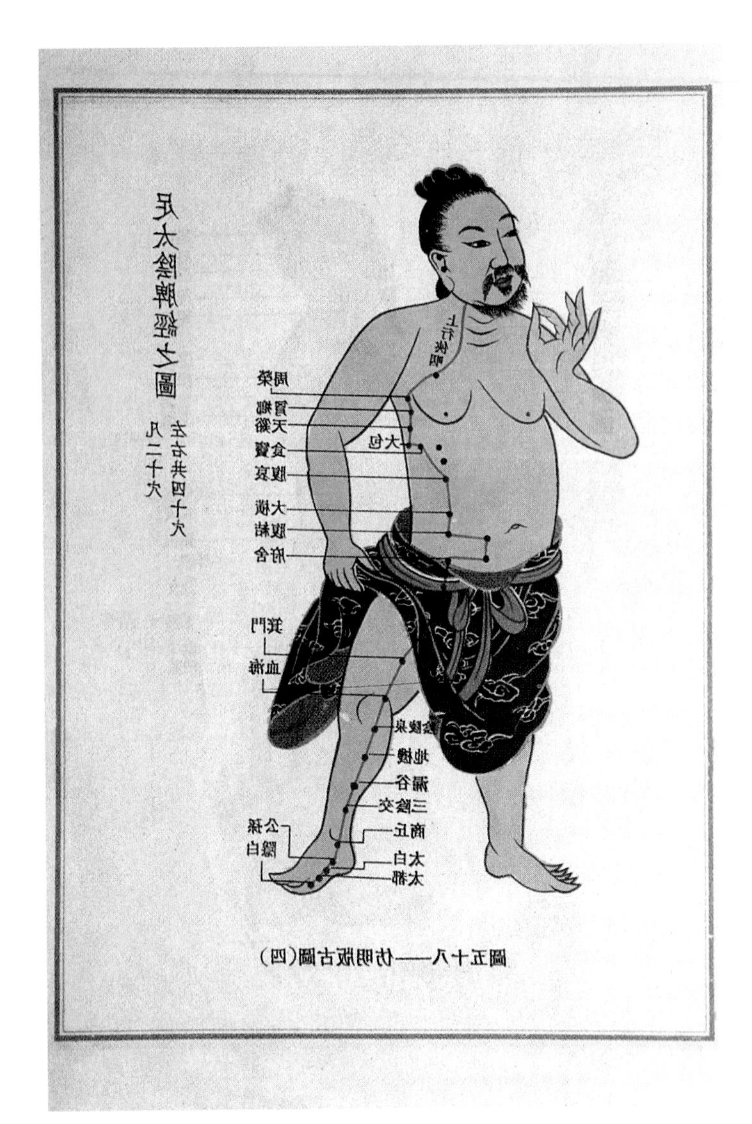

Figure 1.7

An acupuncture chart from the Ming dynasty of ancient China.

Japan

The advancement of anatomy in Japan was strongly influenced by the Chinese and Dutch. The earliest records of anatomical interest in Japan date back to the sixth century. Buddhist monks from Japan were trained in China where they were exposed to Chinese philosophy, and so Chinese beliefs concerning the body became prevalent in Japan as well. By the eighteenth century, Western influences, especially the Dutch, were such that the Japanese sought to determine for themselves which version of anatomy was correct. In 1774, a five-volume work called *Kaitai Shinsho* (*A New Book of Anatomy*), published by a Japanese physician, Genpaku Sugita, totally adopted the Dutch conceptualization of the body. This book marked the beginning of a modern era in anatomy and medicine for the Japanese people.

For several hundred years, Western nations were welcome in Japan. In 1603, however, the Japanese government banned all contact with the Western world because they feared the influences of Christianity on their society. Although this ban was strictly enforced and Japan became isolated, Japanese scholars continued to circulate Western books on anatomy and medicine. These books eventually prompted Japanese physicians to reassess what they had been taught concerning the structure of the body.

Grecian Period

It was in ancient Greece that anatomy first gained wide acceptance as a science. The writings of several Greek philosophers had a tremendous impact on future scientific thinking. During this period, the Greeks were obsessed with the physical beauty of the human body, as reflected by their exquisite sculptures.

The young people of Greece were urged to be athletic and develop their physical abilities, but at about age 18 they were directed more to intellectual pursuits of science, rhetoric, and philosophy. An educated individual was expected to be acquainted with all fields of knowledge, and it was only natural that great strides were made in the sciences.

Perhaps the first written reference to the anatomy of wounds sustained in battle is contained in the *Iliad*, written by Homer in about 800 B.C. Homer's detailed descriptions of the anatomy of wounds were exceedingly accurate. However, he described clean wounds—not the type of traumatic wounds that would likely be suffered on a battlefield. This has led to speculation that human dissections were conducted during this period and that anatomical structure was well understood. Victims of human sacrifice may have served as subjects for anatomical study and demonstration.

Hippocrates

Hippocrates (460–377 B.C.), the most famous of the Greek physicians of his time, is regarded as the father of medicine because of the sound principles of medical practice that his school established (fig. 1.8). His name is memorialized in the Hippocratic oath, which many graduating medical students repeat as a promise of professional stewardship and duty to humankind (table 1.3).

Hippocrates probably had only limited exposure to human dissections, but he was well disciplined in the popular *humoral theory* of body organization. Four body humors were recognized, and each was associated with a particular body organ: blood with the liver; choler, or yellow bile, with the gallbladder; phlegm with the lungs; and melancholy, or black bile, with the spleen. A healthy person was thought to have a balance of the four humors. The concept of humors has long since been discarded, but it dominated medical thought for over 2,000 years.

Perhaps the greatest contribution of Hippocrates was that he attributed diseases to natural causes rather than to the displeasure of the gods. His application of logic and reason to medicine was the beginning of observational medicine.

Figure 1.8

A fourteenth-century painting of the famous Greek physician Hippocrates. Hippocrates is referred to as the father of medicine; his creed is immortalized as the Hippocratic oath.

The four humors are a part of our language and medical practice even today. Melancholy is a term used to describe depression or despondency in a person, while melanous refers to a black or sallow complexion. The prefix melano- means black. Cholera is an infectious intestinal disease that causes diarrhea and vomiting. Phlegm (pronounced flem) within the upper respiratory system is symptomatic of several pulmonary disorders. Sanguine, a term that originally referred to blood, is used to describe a passionate temperament. This term, however, has evolved to refer simply to the cheerfulness and optimism that accompanied a sanguine personality, and no longer refers directly to the humoral theory.

humor: L. *humor*, fluid
melancholy: Gk. *melan*, black; *chole*, bile
cholera: Gk. *chole*, bile
phlegm: Gk. *phlegm*, inflammation
sanguine: L. *sanguis*, bloody

Table 1.3
The Hippocratic Oath

I swear by Apollo Physician and Aesculapius and Hygeia and Panacea and all the gods and goddesses making them my witnesses, that I will fulfill according to my ability and judgment this oath and this covenant:

To hold him who has taught me this art as equal to my parents and to live my life in partnership with him, and if he is in need of money to give him a share of mine, and to regard his offspring as equal to my brothers in male lineage and to teach them this art—if they desire to learn it—without fee and covenant; to give a share of precepts and oral instruction and all the other learning to my sons and to the sons of him who has instructed me and to pupils who have signed the covenant and have taken an oath according to the medical law, but to no one else.

I will apply dietetic measures for the benefit of the sick according to my ability and judgment; I will keep them from harm and injustice.

I will neither give a deadly drug to anybody if asked for it, nor will I make a suggestion to this effect. Similarly I will not give to a woman an abortive remedy. In purity and holiness I will guard my life and my art.

I will not use the knife, not even on sufferers from stone, but will withdraw in favor of such men as are engaged in this work.

Whatever houses I may visit, I will come for the benefit of the sick, remaining free of all intentional injustice, of all mischief, and in particular of sexual relations with both female and male persons, be they free or slaves.

What I may see or hear in the course of the treatment or even outside of the treatment in regard to the life of men, which on no account one must spread abroad, I will keep to myself, holding such things shameful to be spoken about.

If I fulfill this oath and do not violate it, may it be granted to me to enjoy life and art, being honored with fame among all men for all time to come; if I transgress it and swear falsely, may the opposite of all this be my lot.

Figure 1.9

This Roman copy of a Greek sculpture is believed to be of Aristotle, the famous Greek philosopher.

Aristotle

Aristotle (384–322 B.C.), a pupil of Plato, was an accomplished writer, philosopher, and zoologist (fig. 1.9). He was also a renowned teacher and was hired by King Philip of Macedonia to tutor his son, Alexander, who later became known as Alexander the Great.

Aristotle made careful investigations of all kinds of animals, which included references to humans, and he pursued a limited type of scientific method in obtaining data. He wrote the first known account of embryology, in which he described the development of the heart in a chick embryo. He named the aorta and contrasted the arteries and veins. Aristotle's best known zoological works are *History of Animals*, *On the Parts of Animals*, and *On the Generation of Animals* (see table 1.2). These books had a profound influence on the establishment of specialties within anatomy, and they earned Aristotle recognition as founder of comparative anatomy.

In spite of his tremendous accomplishments, Aristotle perpetuated some erroneous theories regarding anatomy. For example, the doctrine of the humors formed the boundaries of his thought. Plato had described the brain as the "seat of feeling and thought," but Aristotle disagreed. He placed the seat of intelligence in the heart and argued that the function of the brain, which was bathed in fluid, was to cool the blood that was pumped from the heart, thereby maintaining body temperature.

Alexandrian Era

Alexander the Great founded Alexandria in 332 B.C. and established it as the capital of Egypt and a center of learning. In addition to a great library, there was also a school of medicine in Alexandria. The study of anatomy flourished because of the acceptance of dissections of human cadavers and human **vivisections** (*viv″ĭ-sek′shunz*) (dissections of living things). This brutal procedure was commonly performed on condemned criminals. People reasoned that to best understand the functions of the body, it should be studied while the subject was alive, and that a condemned man could best repay society through the use of his body for a scientific vivisection.

Aesculapius: Gk. (mythology) son of Apollo and god of medicine
Hygeia: Gk. (mythology) daughter of Aesculapius; personification of health; *hygies*, healthy
Panacea: Gk. (mythology) also a daughter of Aesculapius; assisted in temple rites and tended sacred serpents; *pan*, all, every; *akos*, remedy

vivisection: L. *vivus*, alive; *sectio*, a cutting

Unfortunately, the scholarly contributions and scientific momentum of Alexandria did not endure. Most of the written works were destroyed when the great library was burned by the Romans as they conquered the city in 30 B.C. What is known about Alexandria was obtained from the writings of later scientists, philosophers, and historians, including Pliny, Celsus, Galen, and Tertullian. Two men of Alexandria, Herophilus and Erasistratus, made lasting contributions to the study of anatomy.

Herophilus

Herophilus (about 325 B.C.) was trained in the Hippocratic school but became a great teacher of anatomy in Alexandria. Through vivisections and dissections of human cadavers, he provided excellent descriptions of the skull, eye, various visceral organs and organ relationships, and the functional relationship of the spinal cord to the brain. Two monumental works by Herophilus are entitled *On Anatomy* and *Of the Eyes*. He regarded the brain as the seat of intelligence and described many of its structures, such as the meninges, cerebrum, cerebellum, and fourth ventricle. He was also the first to distinguish nerves as either sensory or motor.

Erasistratus

Erasistratus (about 300 B.C.) was more interested in body functions than structure and is frequently referred to as the father of physiology. In a book on the causes of diseases, he included observations on the heart, vessels, brain, and cranial nerves. Erasistratus noted the toxic effects of snake venom on various visceral organs and described changes in the liver resulting from various diseases. Although some of his writings were scientifically accurate, he also had notions that were primitive and mystical. He thought, for example, that the cranial nerves carried animal spirits and that muscles contracted because of distention by spirits. He also believed that the left ventricle of the heart was filled with a vital air spirit (pneuma) that came in from the lungs, and that the arteries transported this pneuma rather than blood.

Both Herophilus and Erasistratus were greatly criticized later in history for the vivisections they performed. Celsus (about 30 B.C.) and Tertullian (about A.D. 200) were particularly critical of the practice of vivisection. Herophilus was described as a butcher of men who had dissected as many as 600 living human beings, sometimes in public demonstrations.

Roman Era

In many respects, the Roman Empire stifled scientific advancements and set the stage for the Dark Ages. The approach to science shifted from theoretical to practical during this time. Few dissections of cadavers were performed other than in attempts to determine the cause of death in criminal cases. Medicine was not preventive but was limited, almost without exception, to the treatment of soldiers injured in battle. Later in Roman history, laws were established that attested to the influence of the Church on medical practice. According to Roman law, for example, no deceased pregnant woman could be buried without prior removal of the fetus from the womb so that it could be baptized.

The scientific documents that have been preserved from the Roman Empire are mostly compilations of information obtained from the Greek and Egyptian scholars. New anatomical information was scant, and for the most part was derived from dissections of animals other than human. Two important anatomists from the Roman era were Celsus and Galen.

Celsus

Most of what is currently known about the Alexandrian school of medicine is based on the writings of the Roman encyclopedist Cornelius Celsus (30 B.C.–A.D. 30). He compiled this information into an eight-volume work called *De re medicina*. Celsus had only limited influence in his own time, however, probably because of his use of Latin rather than Greek. It was not until the Renaissance that the enormous value of his contribution was recognized.

Galen

Claudius Galen (A.D. 130–201) was perhaps the best physician since Hippocrates. A Greek living under Roman domination, he was certainly the most influential writer of all times on medical subjects. For nearly 1,500 years, the writings of Galen represented the ultimate authority on anatomy and medical treatment. Galen probably dissected no more than two or three human cadavers during his career, of necessity limiting his anatomical descriptions to nonhuman animal dissections. He compiled nearly 500 medical papers (of which 83 have been preserved) from earlier works of others, as well as from his personal studies. He perpetuated the concept of the humors of the body and gave authoritative explanations for nearly all body functions.

Galen's works contain many errors, primarily because of his desire to draw definitive conclusions regarding human body functions on the basis of data obtained largely from animals such as monkeys, pigs, and dogs. He did, however, provide some astute and accurate anatomical details in what are still regarded as classic studies. He proved to be an experimentalist, demonstrating that the heart of a pig would continue to beat when spinal nerves were transected so that nerve impulses could not reach the heart. He showed that the squealing of a pig stopped when the recurrent laryngeal nerve that innervated its vocal cords was cut. Galen also tied off the ureter in a sheep to prove that urine was produced in the kidney, not in the urinary bladder as had been falsely assumed. In addition, he proved that the arteries contained blood rather than pneuma.

Galen compiled a list of many medicinal plants and used medications extensively to treat illnesses. Although he frequently used bloodletting in an effort to balance the four

visceral: L. *viscus*, internal organ
pneuma: Gk. *pneuma*, air

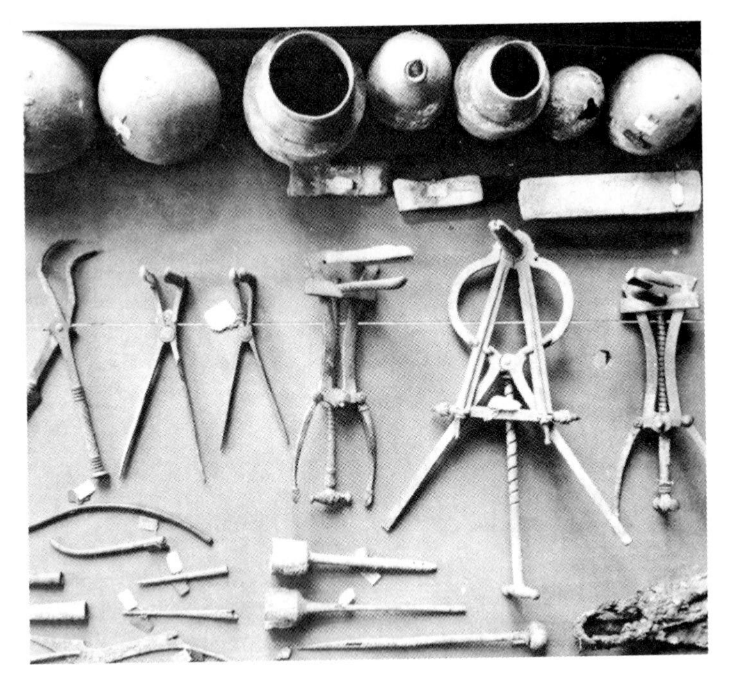

Figure 1.10

These surgical and gynecological instruments were found in the House of the Surgeon (about A.D. 62–79) in Pompeii. They are representative of the medical equipment used throughout the Roman Empire during this time.

humors, he cautioned against removing too much blood. He accumulated a wide variety of medical instruments and suggested their use as forceps, retractors, scissors, and splints (fig. 1.10). He was also a strong advocate of helping nature heal through good hygiene, a proper diet, rest, and exercise.

Middle Ages

The Middle Ages, frequently referred to as the Dark Ages, began with the fall of Rome to the Goths in A.D. 476 and lasted nearly 1,000 years, until Constantinople was conquered by the Turks in 1453. The totalitarian Christian Church suppressed science and medical activity stagnated. Human existence continued to be miserably precarious, and people no longer felt capable of learning from personal observation. Rather, they accepted life "on faith."

Dissections of cadavers were totally prohibited during this period, and molesting a corpse was a criminal act frequently punishable by burning at the stake. If mysterious deaths occurred, examinations by inspection and palpation were acceptable. During the plague epidemic in the sixth century, however, a few *necropsies* (*nek'rop-sēz*) and dissections were performed in the hope of determining the cause of this dreaded disease (see fig. 1.4).

epidemic: Gk. *epi*, upon; *demos*, people
necropsy: Gk. *nekros*, corpse; *opsy*, view

During the Crusades soldiers cooked the bones of their dead comrades so that they could be returned home for proper burial. Even this act, however, was eventually considered sacrilegious and strongly condemned by the Church. One of the ironies of the Middle Ages was that the peasants received far less respect and had fewer rights when they were alive than when they were dead.

Contributions of Islam

The Arabic-speaking people made a profound contribution to the history of anatomy in a most unusual way. It was the Islamic world that saved much of Western scholarship from the ruins of the Roman Empire, the oppression of the Christian Church, and the onset of the Middle Ages. With the expansion of Islam through the Middle East and North Africa during the eighth century, the surviving manuscripts from Alexandria were taken back to the Arab countries, where they were translated from Greek to Arabic.

As the Dark Ages enshrouded Europe, the Christian Church attempted to stifle any scholarship or worldly knowledge that was not acceptable within Christian dogma. The study of the human body was considered heretical, and the Church banned all writings on anatomical subjects. Without the Islamic repository of the writings of Aristotle, Hippocrates, Galen, and others, the progress of centuries in anatomy and medicine would have been lost. It wasn't until the thirteenth century that the Arabic translations were returned to Europe and, in turn, translated to Latin. During the translation process, any Arabic terminology that had been introduced was systematically removed, so that today we find few anatomical terms of Arabic origin.

Renaissance

The period known as the Renaissance was characterized by a rebirth of science. It lasted roughly from the fourteenth through the sixteenth century and was a transitional period from the Middle Ages to the modern age of science.

The Renaissance was ushered in by the great European universities established in Bologna, Salerno, Padua, Montpellier, and Paris. The first recorded human dissections at these newly established centers of learning were the work of the surgeon William of Saliceto (1215–80) from the University of Bologna. The study of anatomy quickly spread to other universities, and by the year 1300 human dissections had become an integral part of the medical curriculum. However, the Galenic dogma that normal human anatomy was sufficiently understood persisted, so interest at this time centered on methods and techniques of dissection rather than on furthering knowledge of the human body.

The development of movable type in about 1450 revolutionized the production of books. Celsus, whose *De re medicina* was "rediscovered" during the Renaissance, was the first medical author to be published in this manner. Among the first anatomy books to be printed in movable type was that of Jacopo Berengario of Carpi, a professor of surgery at Bologna. He described many anatomical structures, including the appendix, thymus, and larynx. The most influential text of the period was written by Mondino de' Luzzi, also of the University of Bologna, in 1316. First published in 1487, it was more a dissection guide than a study of gross anatomy, and in spite of its numerous Galenic errors, 40 editions were published, until the time of Vesalius.

Because of the rapid putrefaction of an unembalmed corpse, the anatomy textbooks of the early Renaissance were organized so that the more perishable portions of the body were considered first. Dissections began with the abdominal cavity, then the chest, followed by the head, and finally the appendages. A dissection was a marathon event, frequently continuing for perhaps 4 days.

With the increased interest in anatomy during the Renaissance, obtaining cadavers for dissection became a serious problem. Medical students regularly practiced grave robbing until finally an official decree was issued that permitted the bodies of executed criminals to be used as specimens.

Corpses were embalmed to prevent deterioration, but this was not especially effective, and the stench from cadavers was apparently a persistent problem. Anatomy professors lectured from a thronelike chair at some distance from the immediate area (fig. 1.11). The phrase, "I wouldn't touch that with a 10-foot pole" probably originated during this time in reference to the smell of a decomposing cadaver.

Figure 1.11

The scene of a cadaver dissection, 1500, from *Fasciculus Medicinae* by Johannes de Ketham. The anatomy professor removed himself from the immediate area to a thronelike chair overlooking the proceedings. The dissections were performed by hired assistants. One of them, the ostensor, pointed to the internal structures with a wand as the professor lectured.

The major advancements in anatomy that occurred during the Renaissance were in large part due to the artistic and scientific abilities of Leonardo da Vinci and Andreas Vesalius. Working in the fifteenth and sixteenth centuries, each produced monumental studies of the human form.

Leonardo

The great Renaissance Italian Leonardo da Vinci (1452–1519) is best known for his artistic works (e.g., *Mona Lisa*) and his scientific contributions. He displayed genius as a painter, sculptor, architect, musician, and anatomist—although his anatomical drawings were not published until the end of the last century. As a young man, Leonardo regularly participated in cadaver dissections and intended to publish a textbook on anatomy with the Pavian professor Marcantonio della Torre. The untimely death of della Torre at the age of 31 halted their plans. When Leonardo died, his notes and sketches were lost and were not discovered for more than 200 years. The advancement of anatomy would have been accelerated by many years if Leonardo's notebooks had been available to the world at the time of his death.

Leonardo's illustrations helped to create a new climate of visual attentiveness to the structure of the human body. He was intent on accuracy, and his sketches are incredibly detailed (fig. 1.12). He experimentally determined the structure of complex body organs such as the brain and the heart. He made wax casts of the ventricles of the brain to study its structure. He constructed models of the heart valves to demonstrate their action.

Vesalius

The contribution of Andreas Vesalius (1514–64) to the science of human anatomy and to modern medicine is immeasurable. Vesalius was born in Brussels into a family of physicians. He received his early medical training at the University of Paris and completed his studies at the University of Padua in Italy, where he began teaching surgery and anatomy immediately after graduation. At Padua, Vesalius participated in human dissections and initiated the use of live models to determine surface landmarks for internal structures (fig. 1.13).

Vesalius apparently had enormous energy and ambition. By the time he was 28 years old, he had already completed the masterpiece of his life, *De Humani Corporis Fabrica,* in which the various body systems and individual organs are beautifully illustrated and described (fig. 1.14). His book was especially important in that it boldly challenged hundreds of Galen's teachings. Vesalius wrote of his surprise upon finding numerous anatomical errors in the works of Galen that were taught as fact, and he refused to accept Galen's explanations on faith. Because he was so outspokenly opposed to Galen, he incurred the wrath of many of the traditional anatomists, including his former teacher Sylvius (Jacques Dubois) of Paris. Sylvius even went so far as to give him

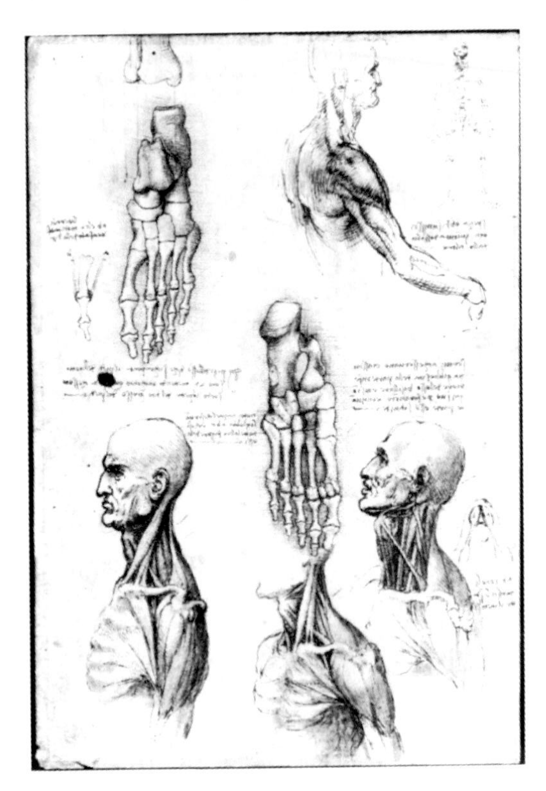

Figure 1.12

Only a master artist could achieve the detail and accuracy of the anatomical sketches of Leonardo da Vinci. "The painter who has acquired a knowledge of the nature of the sinews, muscles and tendons," Leonardo wrote, "will know exactly in the movement of any limb how many and which of the sinews are the cause of it, and which muscle by its swelling is the cause of the sinew's contracting."

Figure 1.13

A painting of the great anatomist Andreas Vesalius, as he dissects a cadaver. From his masterpiece *De Humani Corporis Fabrica*.

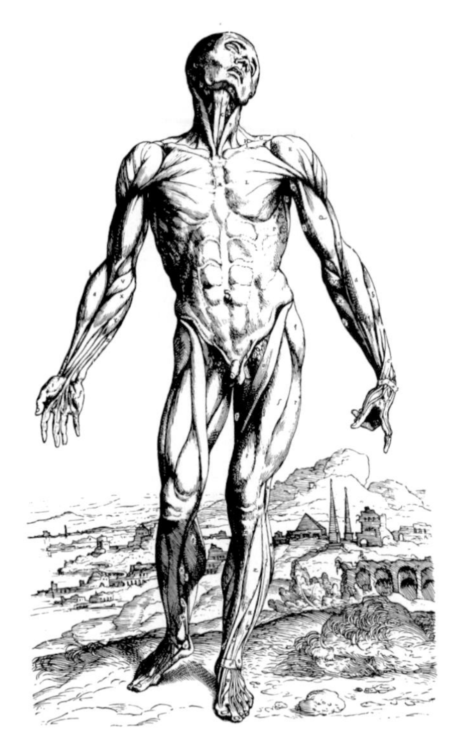

Figure 1.14

A plate from *De Humani Corporis Fabrica*, which Vesalius completed at the age of 28. This book, published in 1543, revolutionized the science of anatomy.

the nickname Vesanus (madman). Vesalius became so unnerved by the relentless attacks that he destroyed much of his unpublished work and ceased his dissections.

Although Vesalius was the greatest anatomist of his epoch, others made significant contributions and to an extent paved the way for Vesalius. Michelangelo pursued anatomy in 1495, being supplied with corpses by the friar of a local monastery. Mondino de' Luzzi and the surgeon Jacopo Berengario of Carpi also corrected many of Galen's errors. Fallopius (1523–62) and Eustachius (1524–74) completed detailed dissections of specific body regions.

Seventeenth and Eighteenth Centuries

During the seventeenth and eighteenth centuries, the science of anatomy attained an unparalleled acceptance. In some of its aspects, it also took on a somewhat theatrical quality. Elaborate amphitheaters were established in various parts of Europe for public demonstrations of human dissections (fig. 1.15). Exorbitantly priced tickets of admission were sold to the wealthy, and the dissections were performed by elegantly robed anatomists who were also splendid orators. The subjects were usually executed prisoners, and the performances were scheduled during cold weather because of the perishable nature of the cadavers.

Fortunately there were also serious, scientific-minded anatomists who made significant contributions during this period.

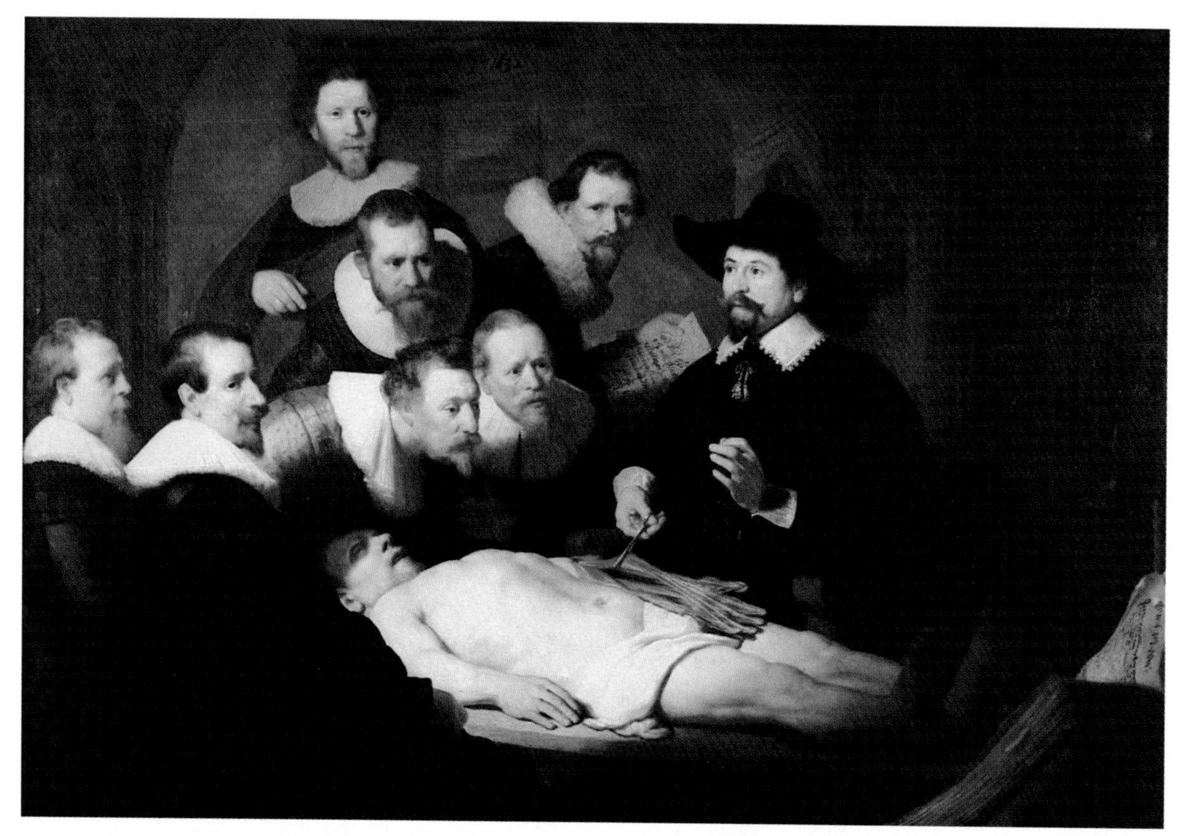

Figure 1.15

The Anatomy Lesson of Dr. Tulp, a famous Rembrandt painting completed in 1632, depicts one of the public anatomies that were popular during this period. Dr. Nicholas Tulp, was a famous Dutch anatomist who described the congenital defect in the spinal column known as spina bifida aperta.

Two of the most important contributions were the explanation of blood flow and the development of the microscope.

Harvey

In 1628, the English physician William Harvey (1578–1657) published his pioneering work *On the Motion of the Heart and Blood in Animals.* Not only did this brilliant research establish proof of the continuous circulation of blood within contained vessels, it also provided a classic example of the scientific method of investigation (fig. 1.16). Like Vesalius, Harvey was severely criticized for his departure from Galenic philosophy. The controversy over circulation of the blood raged for 20 years, until other anatomists finally repeated Harvey's experiments and concurred with his findings.

Leeuwenhoek

Antoni van Leeuwenhoek (*la'ven-hook*) (1632–1723) was a Dutch optician and lens grinder who improved the microscope to the extent that he achieved a magnification of 270 times. His many contributions included developing techniques for examining tissues and describing blood cells, skeletal muscle, and the lens of the eye. Although he was the first to accurately describe sperm cells, Leeuwenhoek did not understand their role in fertilization. Rather, he thought that a spermatozoan contained a miniature human being called a *homunculus.*

Figure 1.16

In the early seventeenth century, the English physician William Harvey demonstrated that blood circulates and does not flow back and forth through the same vessels.

The development of the microscope added an entirely new dimension to anatomy and eventually led to explanations of basic body functions. In addition, the improved microscope was invaluable for understanding the etiologies of many diseases, and thus for discovering cures for many of them. Although Leeuwenhoek improved the microscope, credit for its invention is usually given to the Dutch spectacle maker, Zacharius Janssen. The first scientific investigation using a microscope was performed by Francisco Stelluti in 1625 on the structure of a bee.

Malpighi and Others

Marcello Malpighi (*mal-pe'ge*) (1628–94), an Italian anatomist, is sometimes referred to as the father of histology. He discovered the capillary blood vessels that Harvey had postulated and described the pulmonary alveoli of lungs and the histological structure of the spleen and kidneys.

Many other individuals made significant contributions to anatomy during this 200-year period. In 1672, the Dutch anatomist Regnier de Graaf described the ovaries of the female reproductive system, and in 1775 Lazzaro Spallanzani showed that both ovum and sperm cell were necessary for conception. Francis Glisson (1597–1677) described the liver, stomach, and intestines, and suggested that nerve impulses cause the emptying of the gallbladder. Thomas Wharton (1614–73) and Niels Stensen (1638–86) separately contributed to knowledge of the salivary glands and lymph nodes within the neck and facial regions. In 1664, Thomas Willis published a summary of what was then known about the nervous system.

A number of anatomical structures throughout the body are named in honor of the early anatomists. Thus we have *graafian follicles, Stensen's* and *Wharton's ducts, fallopian tubes, Bartholin's glands,* the *circle of Willis,* and many others. Because these terms have no descriptive basis, they are not particularly useful to a student of anatomy.

Nineteenth Century

The major scientific contribution of the nineteenth century was the formulation of the cell theory. It could be argued that this theory was the most important breakthrough in the history of biology and medicine because all of the body's functions were eventually interpreted as the effects of cellular function.

The term *cell* was coined in 1665 by an English physician, Robert Hooke, as he examined the structure of cork under his microscope in an attempt to explain its buoyancy. What Hooke actually observed were the rigid walls that surrounded the empty cavities of the dead cells. The significance of cellular structure did not become apparent until approximately 150 years after Hooke's work.

With improved microscopes, finer details were observed. In 1809, a French zoologist, Jean Lamarck, observed the jellylike substance within a living cell and speculated that this material was far more important than the outside structure of the cell. Fifteen years later, René H. Dutrochet described the differences between plant and animal cells.

Two German scientists, Matthias Schleiden and Theodor Schwann, are credited with the biological principle referred to as the *cell theory*. Schleiden, a botanist, suggested in 1838 that each plant cell leads a double life—that is, in some respects it behaves as an independent organism, but at the same time it cooperates with the other cells that form the whole plant. A year later, Schwann, a zoologist, concluded that all organisms are composed of cells that are essentially alike. Nineteen years later, the addition of another biological principle seemed to complete the explanation of cells. In 1858, the German pathologist Rudolf Virchow wrote a book entitled *Cell Pathology* in which he proposed that cells can arise only from preexisting cells. The mechanism of cellular replication, however, was not understood for several more decades.

Johannes Müller (1801–58), a comparative anatomist, is noted for applying the sciences of physics, chemistry, and psychology to the study of the human body. When he began his teaching career, science was sufficiently undeveloped to allow him to handle numerous disciplines at once. By the time of his death, however, knowledge had grown so dramatically that several professors were needed to fill the positions he had held alone.

Twentieth Century

Contributions to the science of anatomy during the twentieth century have not been as astounding as they were when little was known about the structure of the body. The study of anatomy grew increasingly specialized, and research became more detailed and complex.

One innovation that gained momentum early in the twentieth century was the simplification and standardization of nomenclature. Because of the proliferation of scientific literature toward the end of the nineteenth century, over 30,000 terms for structures in the human body were on record, many of which were redundant. In 1895, in an attempt to reduce the confusion, the German Anatomical Society compiled a list of approximately 500 terms called the *Basle Nomina Anatomica (BNA)*. The terms on this list were universally approved for use in the classroom and in publications.

Other conferences on nomenclature have been held throughout the century, under the banner of the *International Congress of Anatomists*. At the Seventh International Congress held in New York City in 1960, a resolution was passed to eliminate all eponyms ("tombstone names") from anatomical terminology and instead use descriptive names. Structures like Stensen's duct and Wharton's duct, for example, are now properly referred to as the parotid duct and submandibular duct, respectively. Because eponyms are so entrenched, however, it will be extremely difficult to eliminate all of them from anatomical terminology. But at least there is a trend toward descriptive simplification.

In this text, the preferred descriptive terms are used in both the text narrative and the accompanying illustrations. Where the term first appears in the narrative, however, the preferred form is

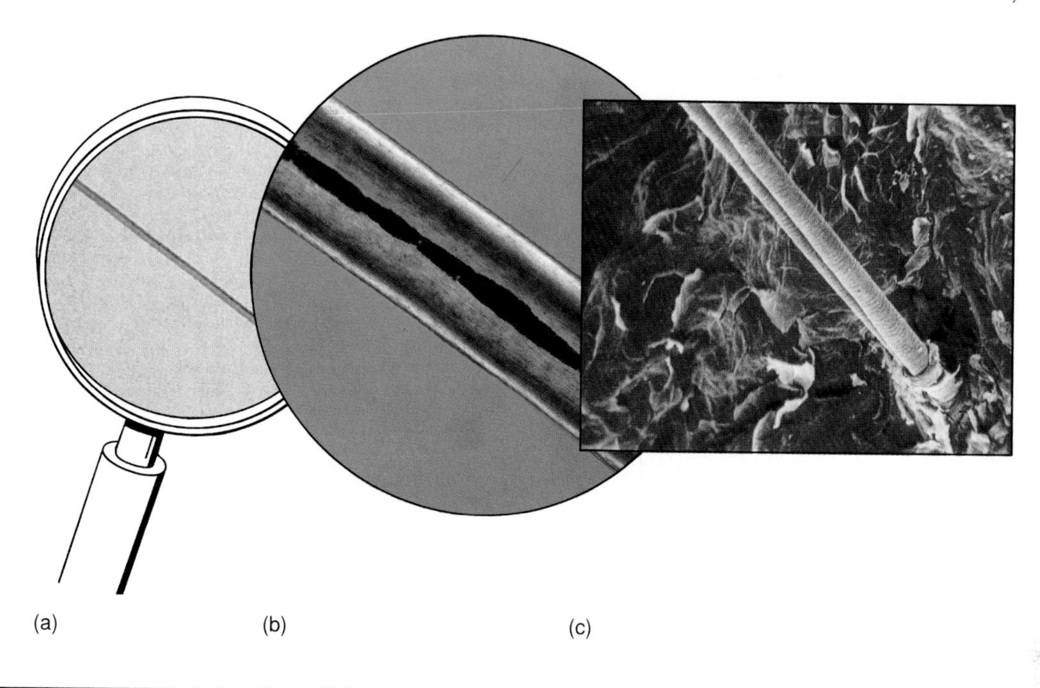

Figure 1.17

Different techniques for viewing microscopic anatomy have greatly enhanced our understanding of the structure and function of the human body. (*a*) The appearance of hair under a simple magnifying glass, (*b*) an observation of a stained section of a hair through a light microscope, and (*c*) a hair emerging from skin as viewed through an electron microscope (340× at 35-mm size).

followed by a parenthetical reference to the traditional name honoring an individual—for example, **uterine** (fallopian) **tube.** The terminology used in this text is in accordance with the official anatomical nomenclature presented in the reference publication, *Nomina Anatomica,* Sixth Edition.

In response to the increased technology and depth of understanding in the twentieth century, new disciplines and specialties have appeared in the science of human anatomy in an attempt to categorize and use the new knowledge. The techniques of such cognate disciplines as chemistry, physics, electronics, mathematics, and computer science have been incorporated into research efforts.

There are several well-established divisions of human anatomy. The oldest, of course, is **gross anatomy,** which is the study of body structures that can be observed with the unaided eye. Stringent courses in gross anatomy in professional schools provide the foundation for a student's entire medical or paramedical training. Gross anatomy also forms the basis for the other specialties within anatomy. **Surface anatomy** (see chapter 10) deals with surface features of the body that can be observed beneath the skin or palpated (examined by touch).

Microscopic Anatomy

Structures smaller than 0.1 mm (100 μm) can be seen only with the aid of a microscope. The sciences of **cytology** (the study of cells), or **cellular biology,** and **histology** (the study of tissues) are specialties of anatomy that have provided additional insight into structure and function of the human body. One can observe greater detail with the *electron microscope* than with the *light microscope* (fig. 1.17). New techniques in staining and histochemistry have aided electron microscopy by revealing the fine details of cells and tissues that are said to compose their *ultrastructure*.

Radiographic Anatomy

Radiographic anatomy, or *radiology*, provides a way of observing structures within the living body. Radiology is based on the principle that substances of different densities absorb different amounts of X rays, resulting in a differential exposure on film. Radiopaque substances such as barium can be ingested (swallowed) or injected into the body to produce even greater contrasts (fig. 1.18). **Angiography** involves making a radiograph after injecting a dye into the bloodstream. In *angiocardiography*, the heart and its associated vessels are x-rayed. *Cineradiography* permits the study of certain body systems through the use of motion picture radiographs. Traditional radiographs have had limitations as diagnostic tools for understanding human anatomy because of the two-dimensional plane that is photographed. Since radiographs compress the body image with an overlap of organs and tissues, diagnosis is often difficult.

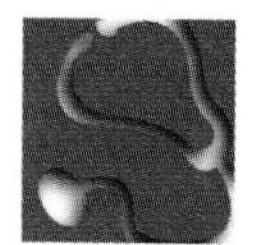

X rays were discovered in 1895 by Wilhelm Konrad Roentgen (*rent'gen*). The radiograph image that is produced on film is frequently referred to as a *roentgenograph.* The recent development of the computerized axial tomography technique has been hailed as the greatest advancement in diagnostic medicine since the discovery of X rays themselves.

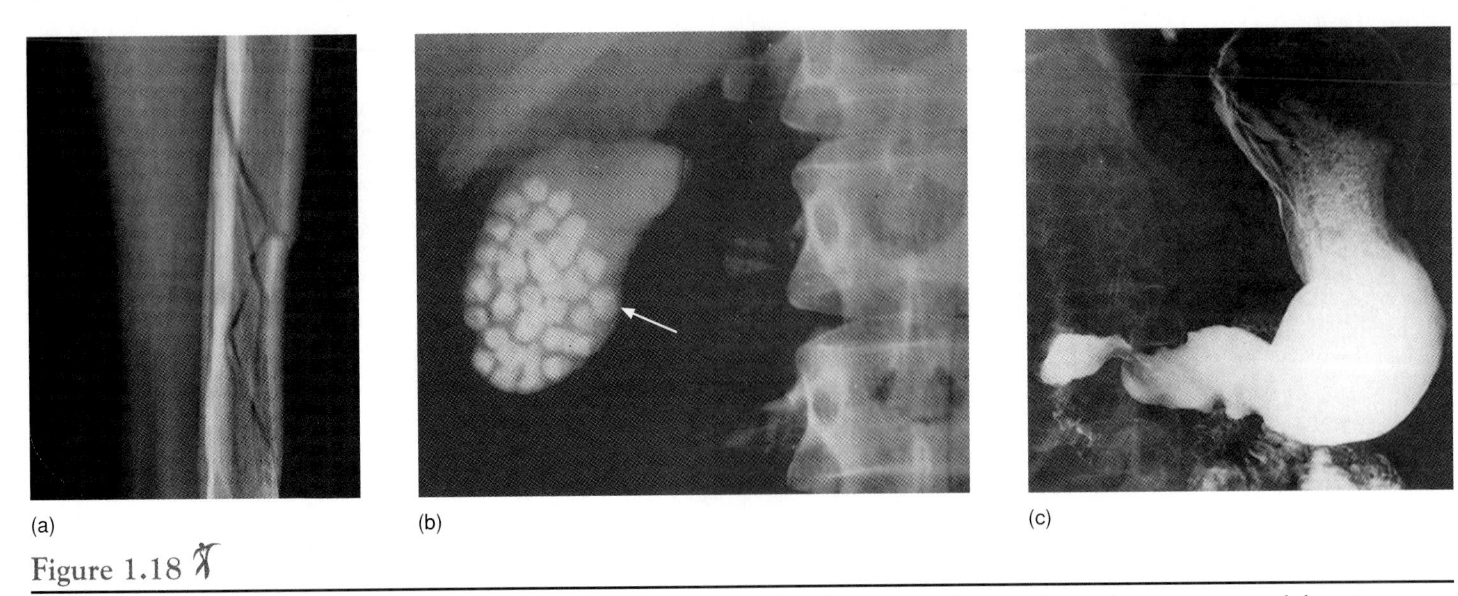

Figure 1.18

The versatility of radiology makes this technique one of the most important tools in diagnostic medicine and provides a unique way of observing specific anatomical structures within the body. (*a*) A radiograph of a healing fracture, (*b*) a radiograph of gallstones within a gallbladder, and (*c*) a radiograph of a stomach filled with a radiopaque contrast medium.

The **computerized axial tomography** technique (**CT,** or **CAT,** scan) has greatly enhanced the versatility of X rays. It uses a computer to display a cross-sectional image similar to that which could only be obtained in an actual section through the body (fig. 1.19*a*).

Another technique of radiographic anatomy is the **dynamic spacial reconstructor (DSR)** scan (fig. 1.19*b*). The DSR functions as an electronic knife that pictorially slices an organ, such as the heart, to provide three-dimensional images. The DSR can be used to observe movements of organs, detect defects, assess the extent of a disease such as cancer, or determine the extent of trauma to tissues after a stroke or heart attack.

Magnetic resonance imaging (MRI), also called **nuclear magnetic resonance (NMR),** provides a new technique for diagnosing diseases and following the response of a disease to chemical treatment (fig. 1.19*c*). An MRI image is created rapidly as hydrogen atoms in tissues, subjected to a strong magnetic field, respond to a pulse of radio waves. MRI has the advantage of being noninvasive—that is, no chemicals are introduced into the body. It is better than a CT scan for distinguishing between soft tissues, such as the gray and white matter of the nervous system.

A **positron emission tomography (PET)** scan is a radiological technique used to observe the metabolic activity in organs (fig. 1.19*d*) following the injection of a radioactive substance, such as treated glucose, into the bloodstream. PET scans are very useful in revealing the extent of damaged heart tissue and in identifying areas where blood flow to the brain is blocked.

Human anatomy will always be a relevant science. Not only does it enhance our personal understanding of body functioning, it is also essential in the clinical diagnosis and treatment of disease. Human anatomy is no longer confined to the observation and description of structures in isolation, but has expanded to include the complexities of how the body functions as an integrated whole. The science of anatomy is dynamic and has remained vital because the two aspects of the body—structure and function—are inseparable.

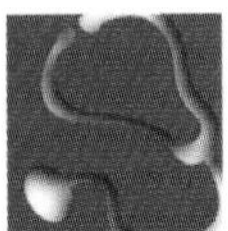

One of the important aspects of human anatomy and medicine is the *autopsy*—a thorough postmortem examination of all of the organs and tissues of a body. Autopsies were routinely performed in the early part of the twentieth century, but their frequency has declined significantly in the last three decades. Currently, only 15% of corpses are autopsied in the United States, which is down from over 50% 30 years ago. Autopsies are of value because (1) they often determine the cause of death, which may confirm or disconfirm preliminary death statements; (2) they frequently reveal diseases or structural defects that may have gone undetected in life; (3) they check the effectiveness of a particular drug therapy for a patient or the success of a particular surgery; and (4) they serve as a means of training medical students.

An interesting piece of information regarding the value of an autopsy in confirming the cause of death was revealed in a study of 2,557 autopsies conducted over a 30-year period to determine the accuracy of physicians' diagnoses of deaths (see "Autopsy," S. A. Geller, *Scientific American,* March 1983). In this study, the causes of death had been improperly or inaccurately recorded in 42% of the cases. This means that because of a decline in the number of autopsies, there may be over a million death certificates filed in the United States each year that are in error. This has important implications for criminal justice and medical genetics, as well as for the insurance industry.

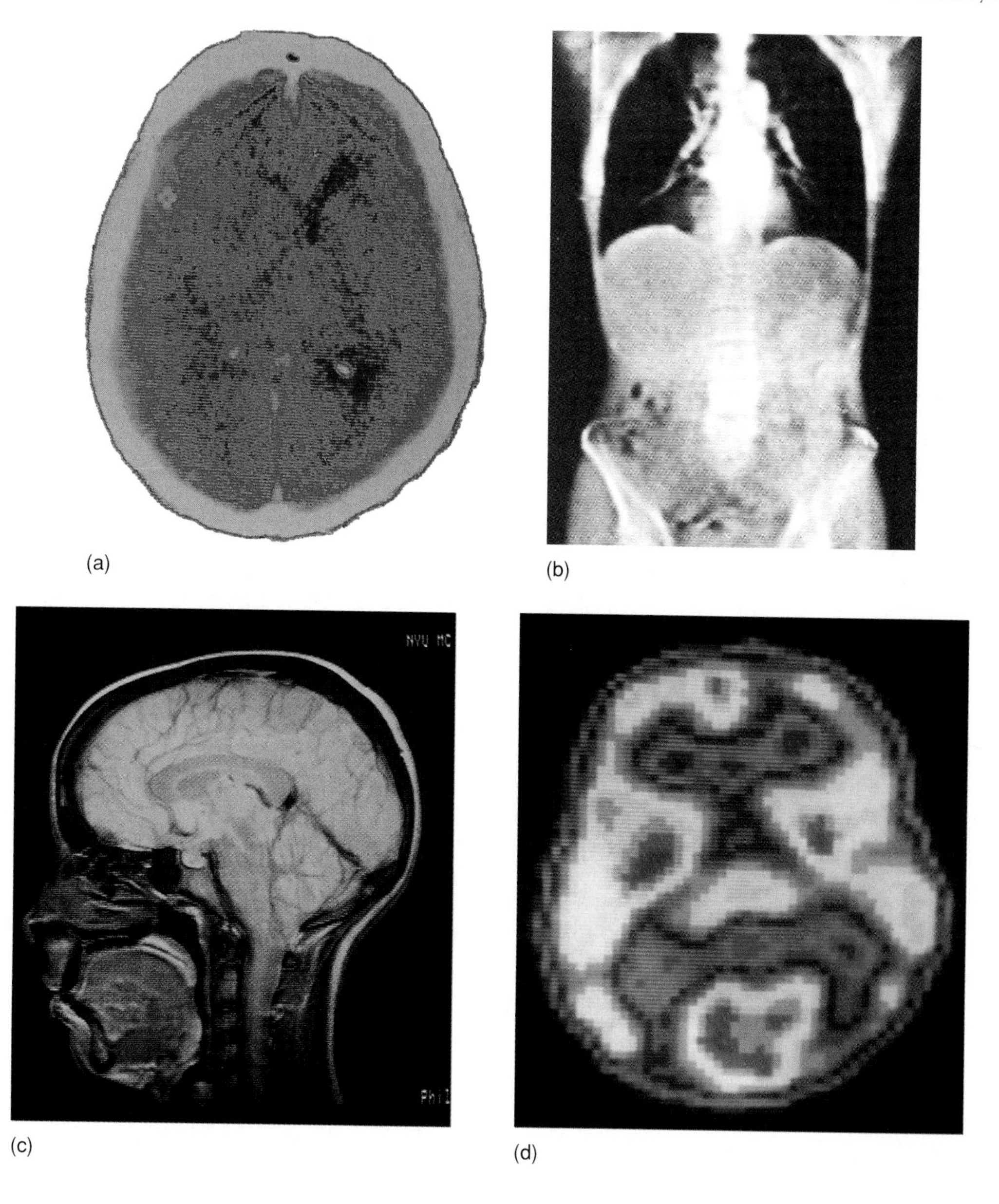

Figure 1.19

Different techniques in radiographic anatomy provide unique views of the human body. (*a*) a CT scan through the head, (*b*) a DSR scan through the trunk (torso), (*c*) an MRI image through the head, and (*d*) a PET scan through the head.

An objective of this text is to enable students to become educated and conversant in anatomy. An excellent way to keep up with anatomy during and after completing the formal course is to subscribe to and read magazines such as *Science*, *Scientific American*, *Discover*, and *Science Digest*. These publications and others include articles on recent scientific findings, many of which pertain to anatomy. If you are to be an educated contributor to society, it is essential to stay informed.

1. Briefly discuss the impact of each of the following on the science of human anatomy: the humoral theory of body organization, vivisections, the Middle Ages, human dissections, movable type, the invention of the microscope, the cell theory, and the development of X-ray techniques.
2. Explain why knowledge of anatomy is personally relevant.
3. Why is it important to stay abreast of new developments in human anatomy? How might this be accomplished?

Chapter Summary

Definition of the Science (p. 2)

1. Human anatomy is the science concerned with the structure of the human body.
2. The terms of anatomy are descriptive and are generally of Greek or Latin derivation.
3. The history of human anatomy parallels that of medicine and has also been greatly influenced by various religions.

Prescientific Period (pp. 2–4)

1. Prehistoric interest in anatomy was undoubtedly limited to practical information necessary for survival.
2. Trepanation was a surgical technique that was practiced by several cultures.
3. Paleopathology is the science concerned with diseases of prehistoric people.

Scientific Period (pp. 5–19)

1. A few anatomical descriptions were inscribed in clay tablets in cuneiform writing by people who lived in Mesopotamia in about 4000 B.C.
2. Egyptians of about 3400 B.C. developed a technique of embalming. It was not recorded, however, and therefore was not of value in furthering the study of anatomy.
3. The belief in a balance between yin and yang was a compelling influence in Chinese philosophy and provided the rationale for the practice of acupuncture.
4. The advancement of anatomy in Japan was largely due to the influence of the Chinese and Dutch.
5. Anatomy first found wide acceptance as a science in ancient Greece.
   (a) Hippocrates is regarded as the father of medicine because of the sound principles of medical practice he established.
   (b) The Greek philosophy of body humors dominated medical thought for over 2,000 years.
   (c) Aristotle pursued a limited type of scientific method in obtaining data; his writings contain some basic anatomy.
6. Alexandria was a center of scientific learning from 300 to 30 B.C.
   (a) Human dissections and vivisections were performed in Alexandria.
   (b) Erasistratus is referred to as the father of physiology because of his interpretations of various body functions.
7. Theoretical data was deemphasized during the Roman era.
   (a) Celsus's eight-volume work was a compilation of medical data from the Alexandrian school.
   (b) Galen was an influential medical writer who made some important advances in anatomy; at the same time he introduced serious errors into the literature that went unchallenged for centuries.
   (c) Science was suppressed for nearly 1,000 years during the Middle Ages, and dissections of human cadavers were prohibited.
   (d) Anatomical writings were taken from Alexandria by Arab armies, and thus saved from destruction during the Dark Ages in Europe.
8. During the Renaissance, many great European universities were established.
   (a) Andreas Vesalius and Leonardo da Vinci were renowned Renaissance men who produced monumental studies of the human form.
   (b) *De Humani Corporis Fabrica*, written by Vesalius, had a tremendous impact on the advancement of human anatomy. Vesalius is regarded as the father of human anatomy.
9. Two major scientific contributions of the seventeenth and eighteenth centuries were the explanation of blood flow and the development of the microscope.
   (a) In 1628, William Harvey correctly described the circulation of blood.
   (b) Shortly after the microscope had been perfected by Antoni van Leeuwenhoek, many investigators added new discoveries to the rapidly changing specialty of microscopic anatomy.
10. The cell theory was formulated during the nineteenth century by Matthias Schleiden and Theodor Schwann, and cellular biology became established as a science separate from anatomy.
11. A trend toward simplification and standardization of anatomical nomenclature began in the twentieth century. In addition, many specialties within anatomy developed, including cytology, histology, embryology, electron microscopy, and radiology.

Review Activities

Objective Questions

1. *Anatomy* is derived from a Greek word meaning
   (a) to cut up. (c) functioning part.
   (b) to analyze. (d) to observe death.
2. The most important contribution of William Harvey was his research on
   (a) the continuous circulation of blood.
   (b) the microscopic structure of spermatozoa.
   (c) the detailed structure of the kidney.
   (d) the striped appearance of skeletal muscle.
3. Which of the following listings is in correct chronological order?
   (a) Galen, Hippocrates, Harvey, Vesalius, Aristotle
   (b) Hippocrates, Galen, Vesalius, Aristotle, Harvey
   (c) Hippocrates, Aristotle, Galen, Vesalius, Harvey
   (d) Aristotle, Hippocrates, Galen, Harvey, Vesalius
4. Anatomy was first widely accepted as a science in ancient
   (a) Rome. (c) China.
   (b) Egypt. (d) Greece.
5. The establishment of sound principles of medical practice earned this man the title of father of medicine.
   (a) Hippocrates (c) Erasistratus
   (b) Aristotle (d) Galen
6. Which of the four body humors was believed by Hippocrates to be associated with the lungs?
   (a) black bile (c) phlegm
   (b) yellow bile (d) blood
7. The anatomical masterpiece *De Humani Corporis Fabrica* was the work of
   (a) Leonardo. (c) Vesalius.
   (b) Harvey. (d) Leeuwenhoek.
8. What event of about 1450 helped to usher in the Renaissance?
   (a) the development of the microscope
   (b) an acceptance of the scientific method
   (c) the development of the cell theory
   (d) the development of movable type
9. The body organ thought by Aristotle to be the seat of intelligence was
   (a) the liver. (c) the brain.
   (b) the heart. (d) the intestine.
10. X rays were discovered during the late nineteenth century by
    (a) Roentgen. (c) Schleiden.
    (b) Hooke. (d) Müller.

Essay Questions

1. Define the terms *anatomize*, *trepanation*, *paleopathology*, *vivisection*, and *cadaver*.
2. Why were the techniques of embalming a corpse, which were perfected in ancient Egypt, not shared with other cultures or recorded for future generations?
3. What is acupuncture? What are some of its uses today?
4. Why is Latin an ideal language from which to derive anatomical terms? What is the current trend regarding the use of proper names (eponyms) in referring to anatomical structures?
5. Why do you suppose the Hippocratic oath has survived for over 2,000 years as a

creed for medical practice? What aspects of the oath are difficult to conform to in today's society?
6. What is meant by the humoral theory of body organization? Which great anatomists were influenced by this theory? When did it cease to be an influence on anatomical investigation and interpretation?
7. Discuss the impact of Galen on the advancement of anatomy and medicine. What circumstances permitted the philosophies of Galen to survive for such a long period?
8. Briefly discuss the establishment of anatomy as a science during the Renaissance.
9. Who invented the microscope? What part did it play in the advancement of anatomy? What specialties of anatomical study have arisen since the introduction of the microscope?
10. Discuss the impact of the work of Andreas Vesalius on the science of anatomy.
11. Give some examples of how culture and religion influenced the science of anatomy.
12. List some techniques currently used to study anatomy and identify the specialties within which these techniques are used.

Critical-Thinking Questions

1. Discuss some of the factors that contributed to a decline in scientific understanding during the Middle Ages. How would you account for the resurgence of interest in the science of anatomy during the Renaissance?
2. Homeostasis is a physiological term that was coined by the American physiologist Walter Cannon in 1932. It refers to the ability of an organism to maintain the stability of its internal environment by adjusting its physiological processes. Discuss the similarities of the humoral theory of body organization and the philosophy of yin and yang as ancient attempts to explain homeostasis.
3. You learned in this chapter that Galen relied on dissections of animals other than human in an attempt to understand human anatomy. Discuss the value and limitations of using mammalian specimens (other than human) in the laboratory portion of a human anatomy course. What advantages are gained by studying human cadavers?
4. Just as geography describes the topography for history, anatomy describes the topography for medicine. Using specific examples, discuss how discoveries in anatomy have resulted in advances in medicine.

TWO

Body Organization and Anatomical Nomenclature

Clinical Case Study 23
Classification and Characteristics of Humans 23
- Phylum Chordata 23
- Class Mammalia 25
- Order Primates 25
- Family Hominidae 27
- Characteristics of Humans 27

Body Organization 28
- Cellular Level 28
- Tissue Level 28
- Organ Level 29
- System Level 29

Anatomical Nomenclature 33
Planes of Reference and Descriptive Terminology 33
- Planes of Reference 33
- Descriptive Terminology 34

Body Regions 36
- Head 37
- Neck 37
- Trunk 38
- Upper Extremity 39
- Lower Extremity 40

Body Cavities and Membranes 40
- Body Cavities 41
- Body Membranes 41

Clinical Case Study Answer 43
Chapter Summary 44
Review Activities 44

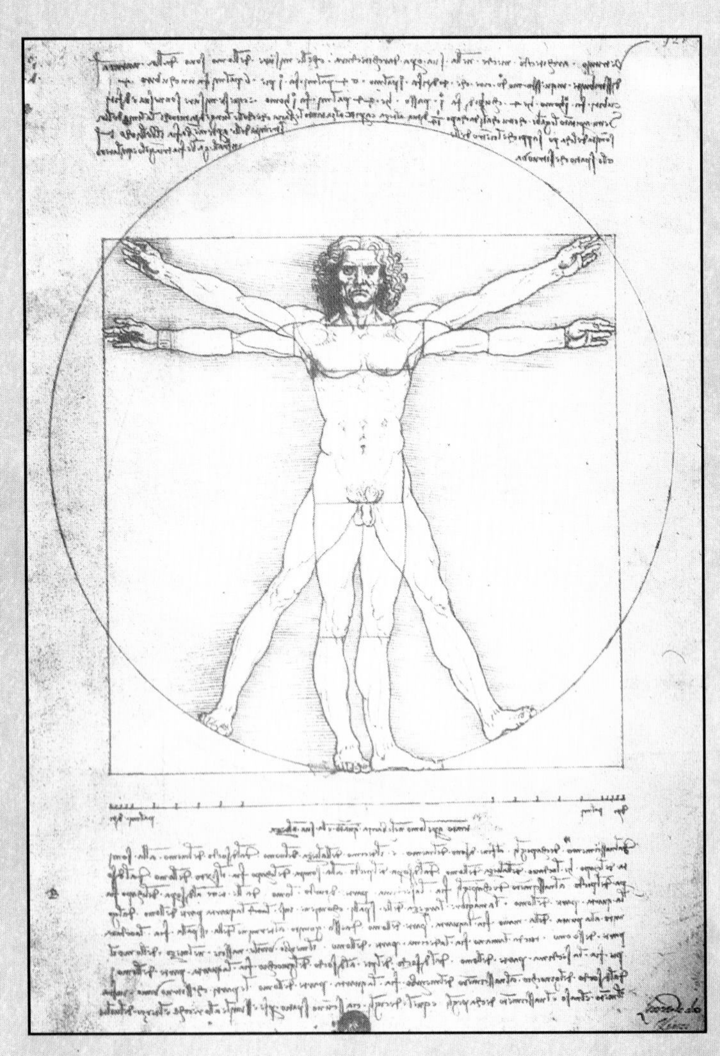

Leonardo da Vinci was the first to draw the parts of the body from different angles and the first to show cross sections of various structures. This famous drawing, annotated with translations from De Architestura *by Vitrubius, illustrates the proportions of the human body.*

Clinical Case Study

A young woman was hit by a car while crossing a street. Upon arrival at the scene, paramedics found the patient to be a bit dazed but reasonably lucid, complaining of pain in her abdomen and the left side of her chest. Otherwise, her vital signs were within normal limits. Initial evaluation in the emergency room revealed a very tender abdomen and left chest. The chest radiograph demonstrated a collapsed left lung resulting from air in the pleural space (pneumothorax). The emergency room physician inserted a drainage tube into the left chest (into the pleural space) to treat the pneumothorax. Attention was then turned to the abdomen. Due to the finding of tenderness, a peritoneal lavage was performed. This procedure involves penetrating the abdominal wall and inserting a tube into the peritoneal cavity. Clear fluid such as sterile water or normal saline is then instilled into the abdomen and siphoned out again. The fluid used in this procedure is called lavage fluid. A return of lavage fluid containing blood, fecal matter, or bile indicates injury to an abdominal organ that requires surgery. The return of lavage fluid from this patient was clear. However, the nurse stated that lavage fluid was draining out of the chest tube.

From what you know about how the various body cavities are organized, do you suppose this phenomenon could be explained based on normal anatomy? What might have caused it to occur in our patient? Does the absence of bile, blood, etc., in the peritoneal lavage fluid guarantee that no organ has been ruptured? If it does not, explain why in terms of the relationship of the various organs to the membranes within the abdomen.

Classification and Characteristics of Humans

Humans are biological organisms belonging to the phylum Chordata within the kingdom Animalia and to the family Hominidae within the class Mammalia and the order Primates.

Objective 1 Classify humans according to the taxonomic system.

Objective 2 List the characteristics that identify humans as chordates and as mammals.

Objective 3 Describe the anatomical characteristics that set humans apart from other primates.

The human organism, or ***Homo sapiens,*** as we have named ourselves, is unique in many ways. Our scientific name translates from the Latin to "man the intelligent," and indeed our intelligence is our most distinguishing feature. It has enabled us to build civilizations, conquer dread diseases, and establish cultures. We have invented a means of communicating through written symbols. We record our own history, as well as that of other organisms, and speculate about our future. We continue to devise ever more ingenious ways for adapting to our changing environment. At the same time, we are so intellectually specialized that we are not self-sufficient. We need one another as much as we need the recorded knowledge of the past.

We are constantly challenged to learn more about ourselves. As we continue to make new discoveries about our structure and function, our close relationship to other living organisms becomes more and more apparent. Often, it is sobering to realize our biological imperfections and limitations.

We share many characteristics with all living animals. As human organisms, we breathe, eat and digest food, excrete bodily wastes, locomote, and reproduce our own kind. We are subject to disease, injury, pain, aging, mutations, and death. Since we are composed of organic materials, we will decompose after death as microorganisms consume our flesh as food. The processes by which our bodies produce, store, and utilize energy are similar to those used by all living organisms. The genetic code that regulates our development is found throughout nature. The fundamental patterns of development of many nonhuman animals also characterize the formation of the human embryo.

In the classification, or taxonomic, system established by biologists to organize the structural and evolutionary relationships of living organisms, each category of classification is referred to as a *taxon*. The highest taxon is the kingdom and the most specific taxon is the species. Humans are species belonging to the **animal kingdom. Phylogeny** (*fi-loj'ĕ-ne*) is the science that studies relatedness on the basis of taxonomy.

Phylum Chordata

Human beings belong to the phylum Chordata (*fī'lum kor-dă'tă*), along with fishes, amphibians, reptiles, birds, and other mammals. All chordates have three structures in common: a **notochord** (*no'to-kord*), a **dorsal hollow nerve cord,** and **pharyngeal** (*fă-rin'je-al*) **pouches** (fig. 2.1). These chordate characteristics are well expressed during the embryonic period of development and, to a certain extent, are present in an adult. The notochord is a flexible rod of tissue that extends the length of the back of an embryo. A portion of the notochord persists in the adult as the **nucleus pulposus,** located within each intervertebral disc (fig. 2.2). The dorsal hollow nerve cord is positioned above the notochord and develops into the **brain** and **spinal cord,** which are supremely functional as the central nervous system in the adult. Pharyngeal pouches form gill openings in fishes and some amphibians. In other chordates, such as humans, embryonic pharyngeal pouches develop, but only one of the pouches persists, becoming the middle-ear cavity. The **auditory** (eustachian) (*yoo-sta'shun*) **tube,** is a persisting connection between the middle-ear cavity and the **pharynx** (*far'ingks*) (throat area).

taxon: Gk. *taxis*, order
phylogeny: L. *phylum*, tribe; Gk. *logos*, study

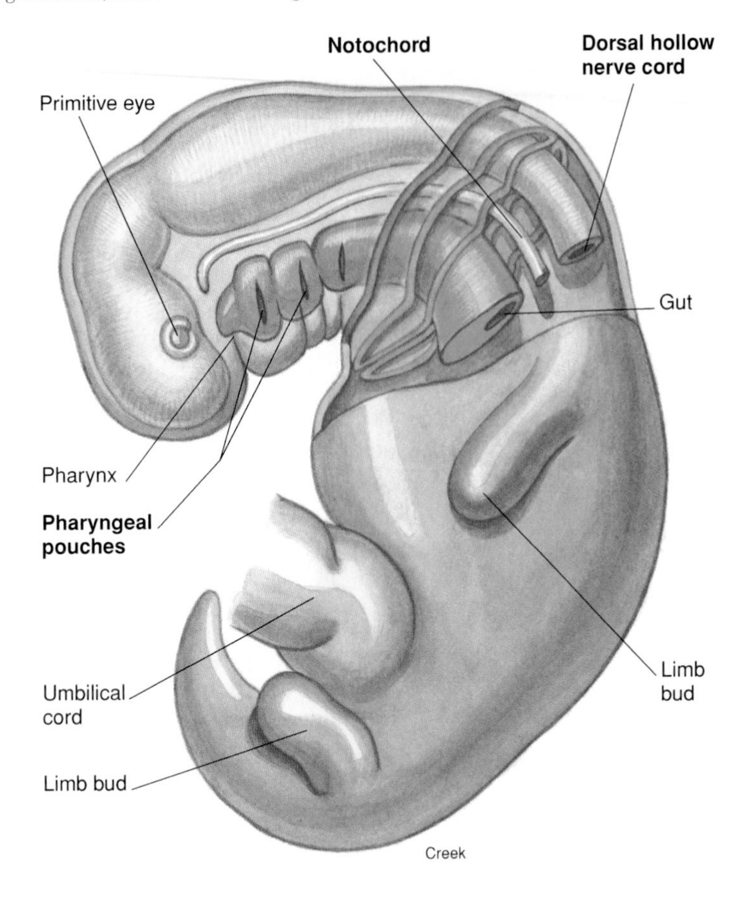

Figure 2.1

A schematic diagram of a chordate embryo. The three diagnostic chordate characteristics are indicated in boldface type.

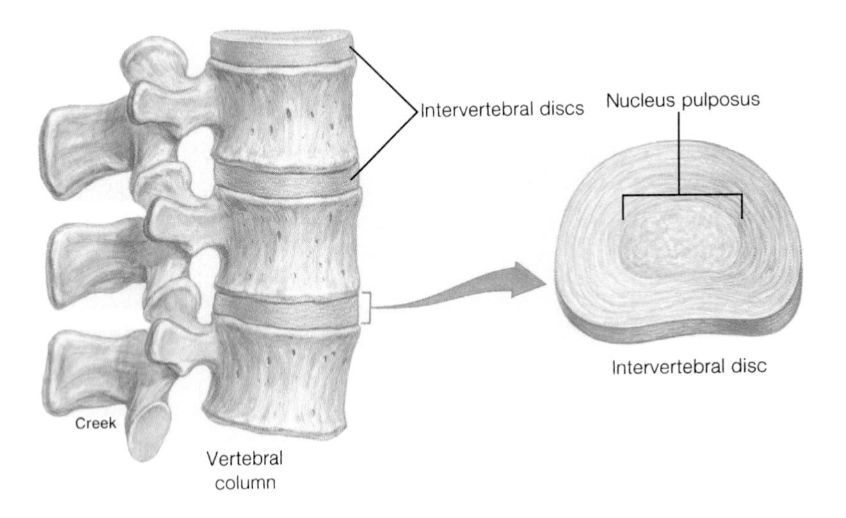

Figure 2.2

A lateral view of the vertebral column showing the intervertebral discs and, to the right, a superior view of an intervertebral disc showing the nucleus pulposus.

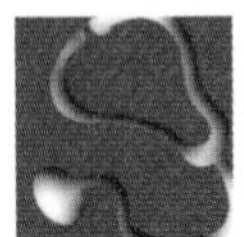

The function of an intervertebral disc and its nucleus pulposus is to allow flexibility between vertebrae for movement of the entire spinal column while preventing compression. Spinal nerves exit between vertebrae, and the discs maintain the spacing to avoid nerve damage. A "slipped disc," resulting from straining the back, is a misnomer. What actually occurs is a herniation, or rupture, because of a weakened wall of the nucleus pulposus. This may cause severe pain as a nerve is compressed.

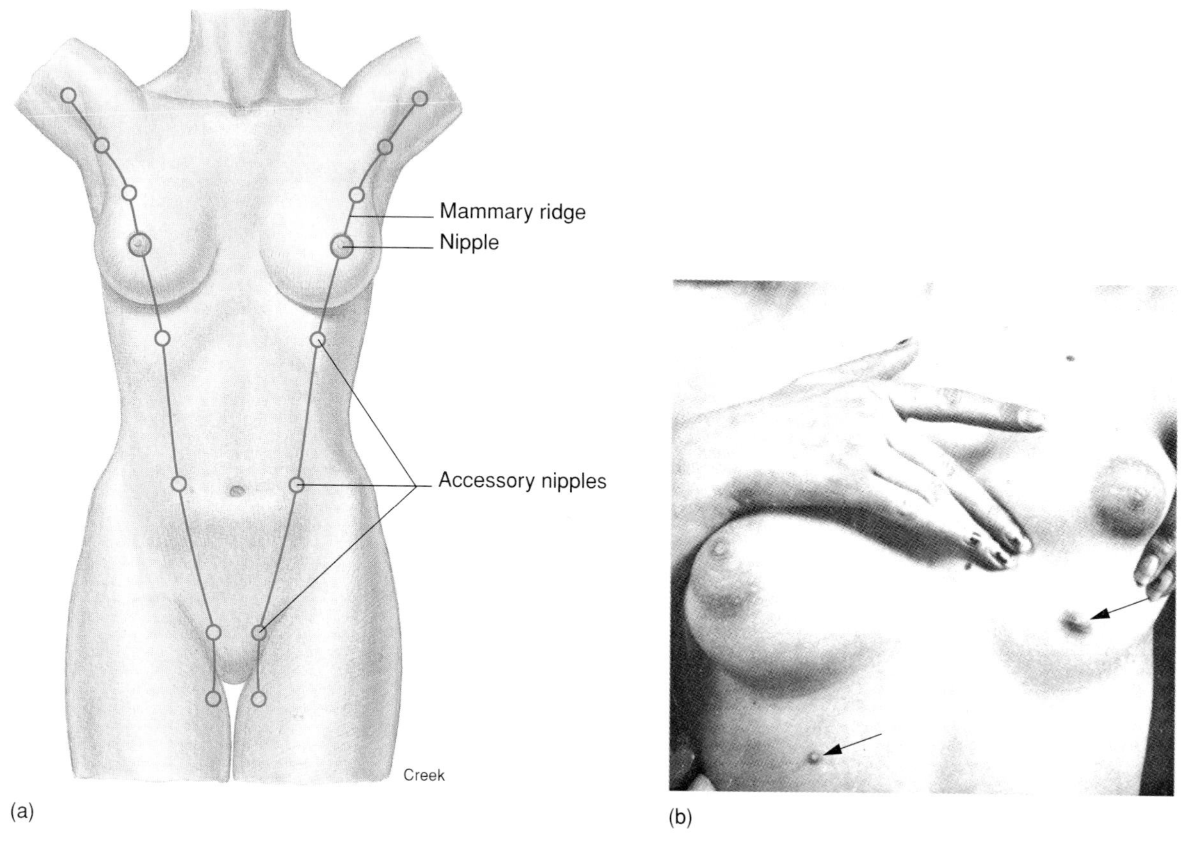

Figure 2.3

The mammary ridge and accessory nipples. (*a*) Mammary glands are positioned along a mammary ridge. (*b*) In humans, additional nipples (polythelia) occasionally develop elsewhere along the mammary ridge (see arrows).

Class Mammalia

Mammals are chordate animals with hair and mammary glands. Hair is a thermoregulatory protective covering for most mammals, and mammary glands serve for suckling the young (fig. 2.3). Other characteristics of mammals include three auditory ossicles (bones), heterodont dentition (teeth of various shapes), squamosal-dentary jaw articulation (a joint between the lower jaw and skull), an attached placenta (*plă-cen'tă*), well-developed facial muscles, a muscular diaphragm, and a four-chambered heart with a left aortic arch (fig. 2.4).

Order Primates

There are several subdivisions of closely related groupings of mammals. These are called orders. Humans, along with lemurs, monkeys, and great apes, belong to the order called Primates. Members of this order have prehensile hands (fig. 2.5), digits modified for grasping, and relatively large, well-developed brains (fig. 2.6).

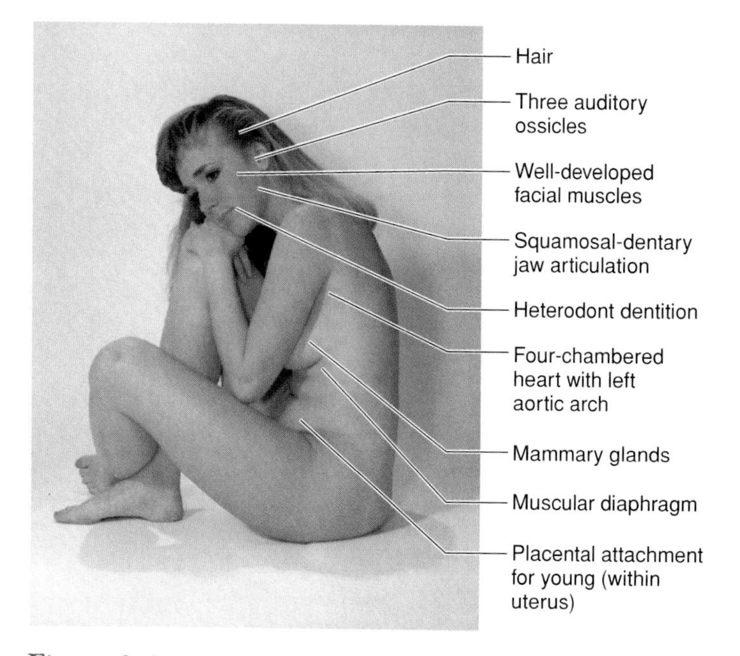

Figure 2.4

Mammals have several distinguishing characteristics; some of these are indicated in the photo with their approximate location within the body.

---

heterodont: Gk. *heteros*, other; *odontos*, tooth
placenta: L. *placenta*, flat cake
Primates: L. *primas*, first
prehensile: L. *prehensus*, to grasp

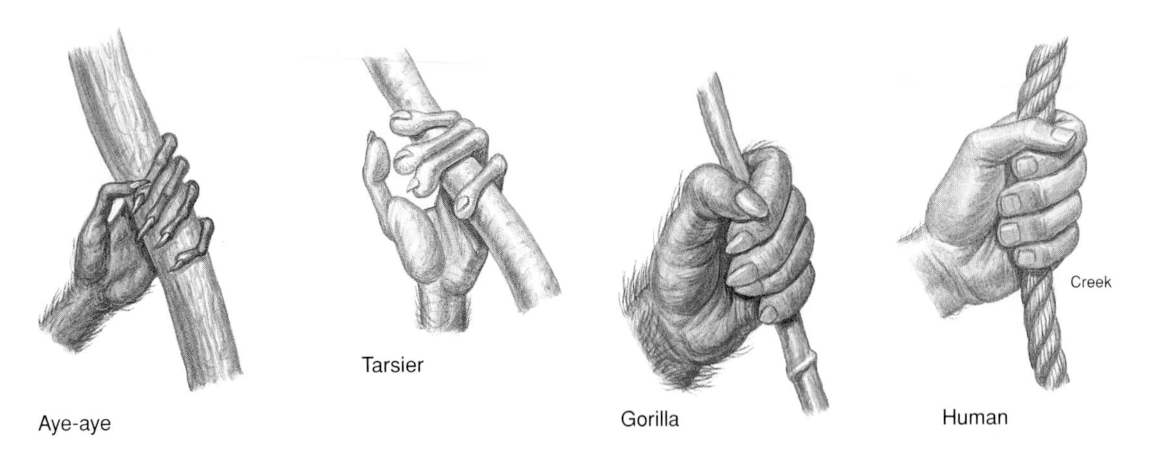

Figure 2.5

An opposable thumb enables a prehensile grip, which is characteristic of primates.

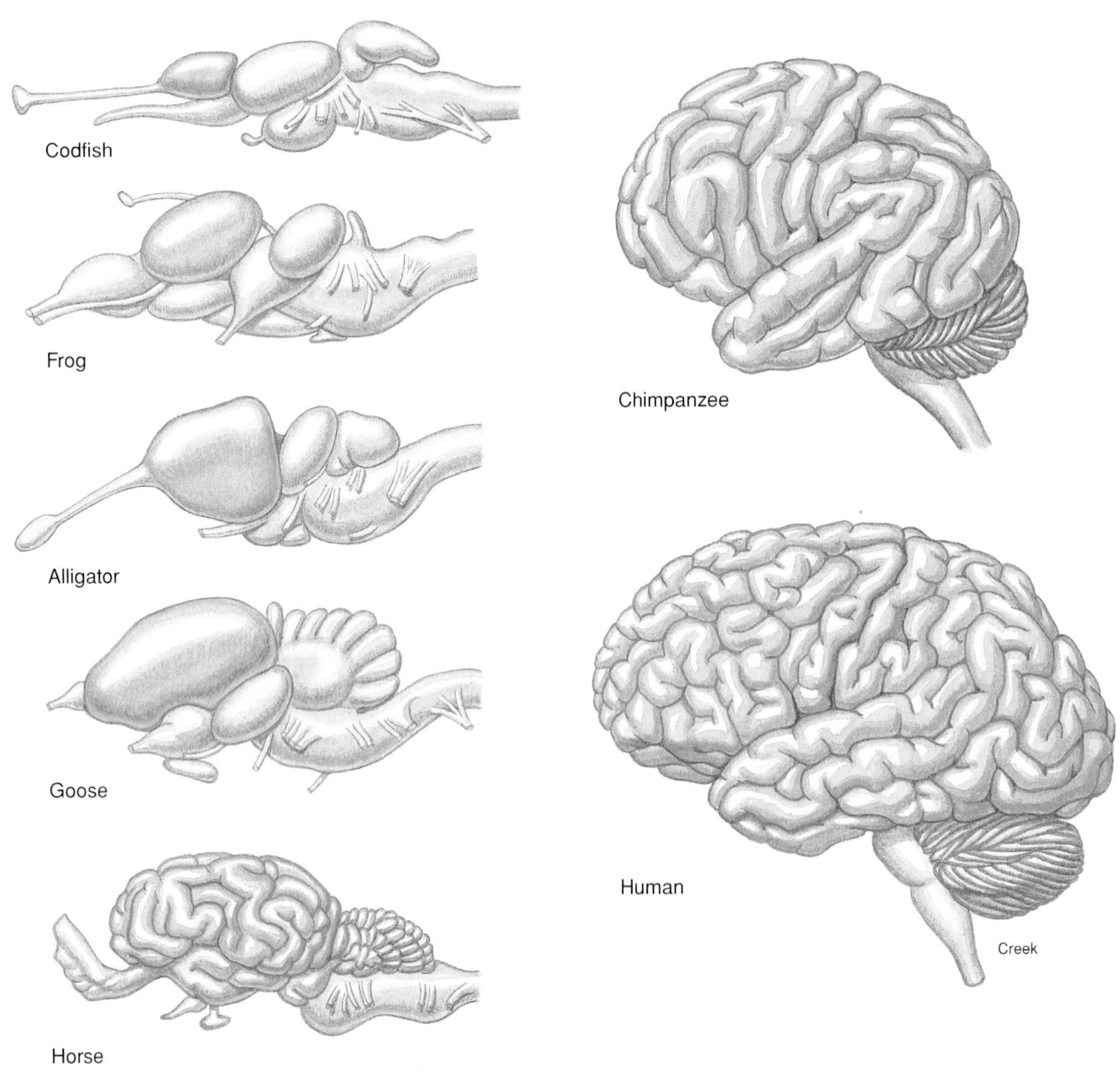

Figure 2.6

The brain of various vertebrates showing the relative size of the cerebrum (shaded) to other structures. (The brains are not drawn to scale. Note that only mammals have a convoluted cerebrum.)

Figure 2.7

The principal races of humans: (*a*) Mongoloid (Thailand), (*b*) Caucasoid (U.S.), (*c*) Negroid (Africa), (*d*) people of Indian subcontinent (Nepal), (*e*) Capoid (Kalahari bushman), and (*f*) Australoid (Ngatatjara man, West Australia).

Family Hominidae

Humans are the sole living members of the family Hominidae. *Homo sapiens* is included within this family, to which all the varieties or ethnic groups of humans belong (fig. 2.7). Each "racial group" has distinguishing features that have been established in isolated populations over thousands of years. Our classification pedigree is presented in table 2.1.

Fostered by greater ease of travel and communication, more frequent contact between diverse cultures has led to a breakdown of some of the traditional barriers to interracial marriage. This may lead to a mixing of the "gene pool," so that distinct ethnic groups become less evident. Perhaps multiple ethnic ties in everyone's pedigree would help to reduce cultural hostility and strife.

Table 2.1 Classification of Human Beings

Taxon	Designated Grouping	Characteristics
Kingdom	Animalia	Eucaryotic cells without walls, plastids, or photosynthetic pigments
Phylum	Chordata	Dorsal hollow nerve cord; notochord; pharyngeal pouches
Subphylum	Vertebrata	Vertebral column
Class	Mammalia	Mammary glands; hair
Order	Primates	Well-developed brain; prehensile hands
Family	Hominidae	Large cerebrum; bipedal locomotion
Genus	*Homo*	Flattened face; prominent chin and nose with inferiorly positioned nostrils
Species	*sapiens*	Largest cerebrum

Characteristics of Humans

As human beings, certain of our anatomical characteristics are so specialized that they are diagnostic in separating us from other animals, and even from other closely related mammals. We also have characteristics that are equally well developed in other animals, but when these function with the human brain, they provide us with remarkable and unique capabilities. Our anatomical characteristics include the following:

1. **A large, well-developed brain.** The adult human brain weighs between 1,350 and 1,400 grams (3 pounds). This gives us a large brain-to-body-weight ratio. But more important is the development of portions of the brain. Certain extremely specialized regions and structures within the brain account for emotion, thought, reasoning, memory, and even precise, coordinated movement.
2. **Bipedal locomotion.** Because humans stand and walk on two appendages, our style of locomotion is said to be *bipedal*. Upright posture imposes other diagnostic structural features, such as the *sigmoid* (S-shaped) *curvature* of the spine, the anatomy of the hips and thighs, and arched feet. Some of these features may cause clinical problems in older individuals.
3. **An opposable thumb.** The human thumb is structurally adapted for tremendous versatility in grasping objects. The saddle joint at the base of the thumb allows a wide range of movement (see fig. 8.13). All primates have opposable thumbs.
4. **Well-developed vocal structures.** Humans, like no other animals, have developed articulated speech. The anatomical structure of our vocal organs (larynx, tongue, and lips), and our well-developed brain have made this possible.
5. **Stereoscopic vision.** Although this characteristic is well developed in several other animals, it is also keen in humans. Our eyes are directed forward so that when we focus on an object, we view it from two angles. Stereoscopic vision gives us depth perception, or a three-dimensional image.

We also differ from other animals in the number and arrangement of our vertebrae (vertebral formula), the kinds and number of our teeth (tooth formula), the degree of development of our facial muscles, and the structural organization of various body organs.

The human characteristics just described account for the splendor of our cultural achievements. As bipedal animals, we have our hands free to grasp and manipulate objects with our opposable thumbs. We can store information in our highly developed brain, make use of it at a later time, and even share our learning through oral or written communication.

bipedal: L. *bi*, two; *pedis*, foot
sigmoid: Gk. *sigma*, shaped like the letter S
stereoscopic: Gk. *stereos*, solid; *skopein*, to view

1. What is a chordate? Why are humans considered members of the phylum Chordata?
2. Why are humans designated as mammals and primates? What characteristics distinguish humans from other primates?
3. Which characteristics of humans are adaptive for social organization?

Body Organization

Structural and functional levels of organization characterize the human body, and each of its parts contributes to the total organism.

Objective 4 Identify the components of a cell, tissue, organ, and system, and explain how these structures relate to one another in constituting an organism.

Objective 5 Describe the general function of each system.

Cellular Level

The **cell** is the basic structural and functional component of life. Humans are multicellular organisms composed of 60 to 100 trillion cells. It is at the microscopic cellular level that such vital functions of life as metabolism, growth, irritability (responsiveness to stimuli), repair, and replication are carried on.

Cells are composed of **atoms**—minute particles that are bound together to form larger particles called **molecules** (fig. 2.8). Certain molecules, in turn, are grouped in specific ways to form small functional structures called **organelles** (*or″gă-nelz′*). Each organelle carries out a specific function within the cell. A cell's nucleus, mitochondria, and endoplasmic reticulum are organelles. The structure of cells and the functions of the organelles will be examined in detail in chapter 3.

The human body contains many distinct kinds of cells, each specialized to perform specific functions. Examples of specialized cells are bone cells, muscle cells, fat cells, blood cells, and nerve cells. The unique structure of each of these cell types is directly related to its function.

Tissue Level

Tissues are layers or groups of similar cells that perform a common function. The entire body is composed of only four principal kinds of tissues: *epithelial*, *connective*, *muscular*, and

cell: L. *cella*, small room
tissue: Fr. *tissu*, woven; from L. *texo*, to weave

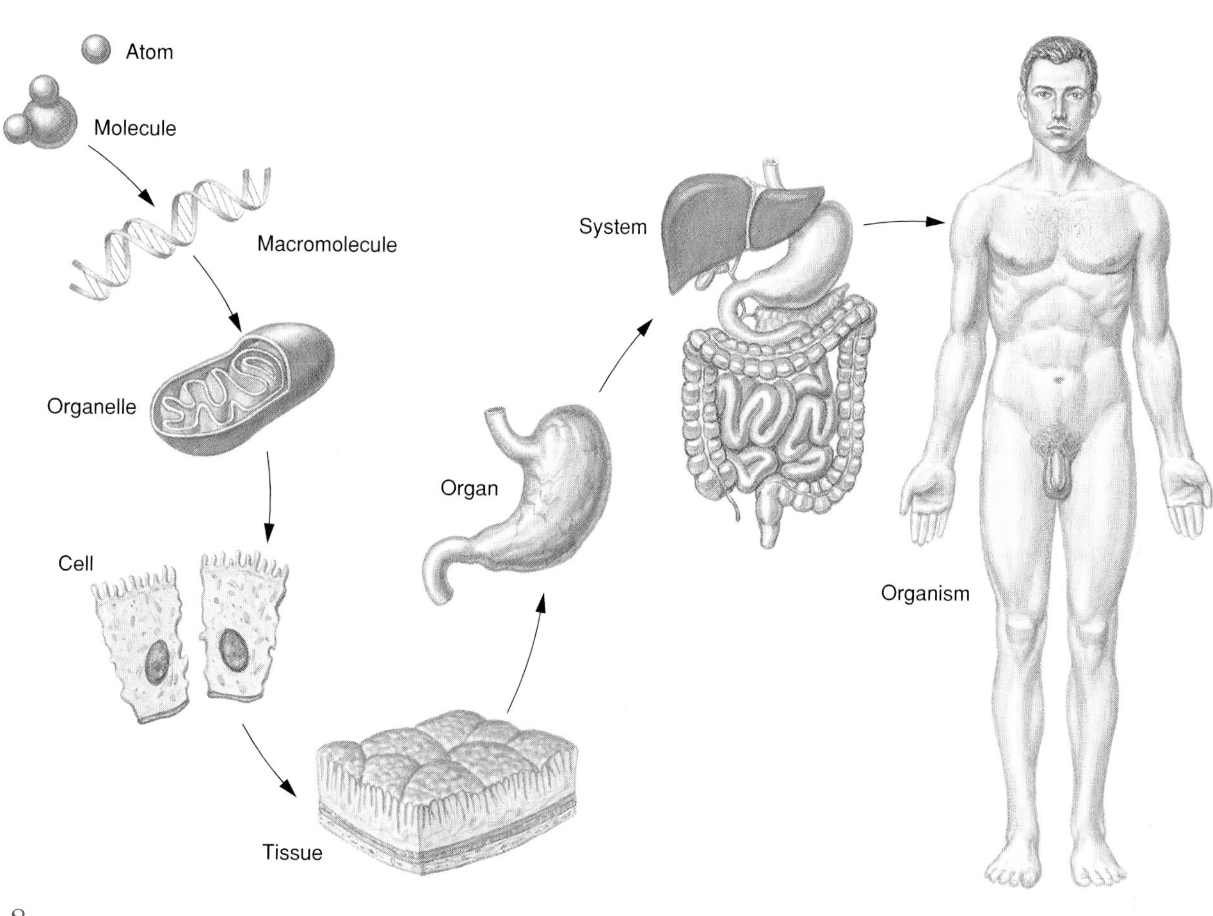

Figure 2.8

The levels of structural organization and complexity within the human body.

nervous tissue. An example of a tissue is the muscle within the heart, whose function it is to pump the blood through the body. The outer layer of skin is a tissue (epithelium) because it is composed of similar cells that together serve as a protective shield for the body. *Histology* is the science concerned with the microscopic study of tissues. The characteristic roles of each tissue type are discussed fully in chapter 4.

Organ Level

An **organ** is an aggregate of two or more tissue types that performs a specific function. Organs occur throughout the body and vary greatly in size and function. Examples of organs are the heart, spleen, pancreas, ovary, skin, and even any of the bones within the body. Each organ usually has one or more primary tissues and several secondary tissues. In the stomach, for example, the inside epithelial lining is considered the primary tissue because the basic functions of secretion and absorption occur within this layer. Secondary tissues of the stomach are the supporting connective tissue and vascular, nervous, and muscle tissues.

System Level

The **systems** of the body constitute the next level of structural organization. A body system consists of various organs that have similar or related functions. Examples of systems are the circulatory system, nervous system, digestive system, and endocrine system. Certain organs may serve two systems. For example, the pancreas functions with both the endocrine and digestive systems and the pharynx serves both the respiratory and digestive systems. All the systems of the body are interrelated and function together, making up the *organism*.

organ: Gk. *organon*, instrument

system: Gk. *systema*, being together

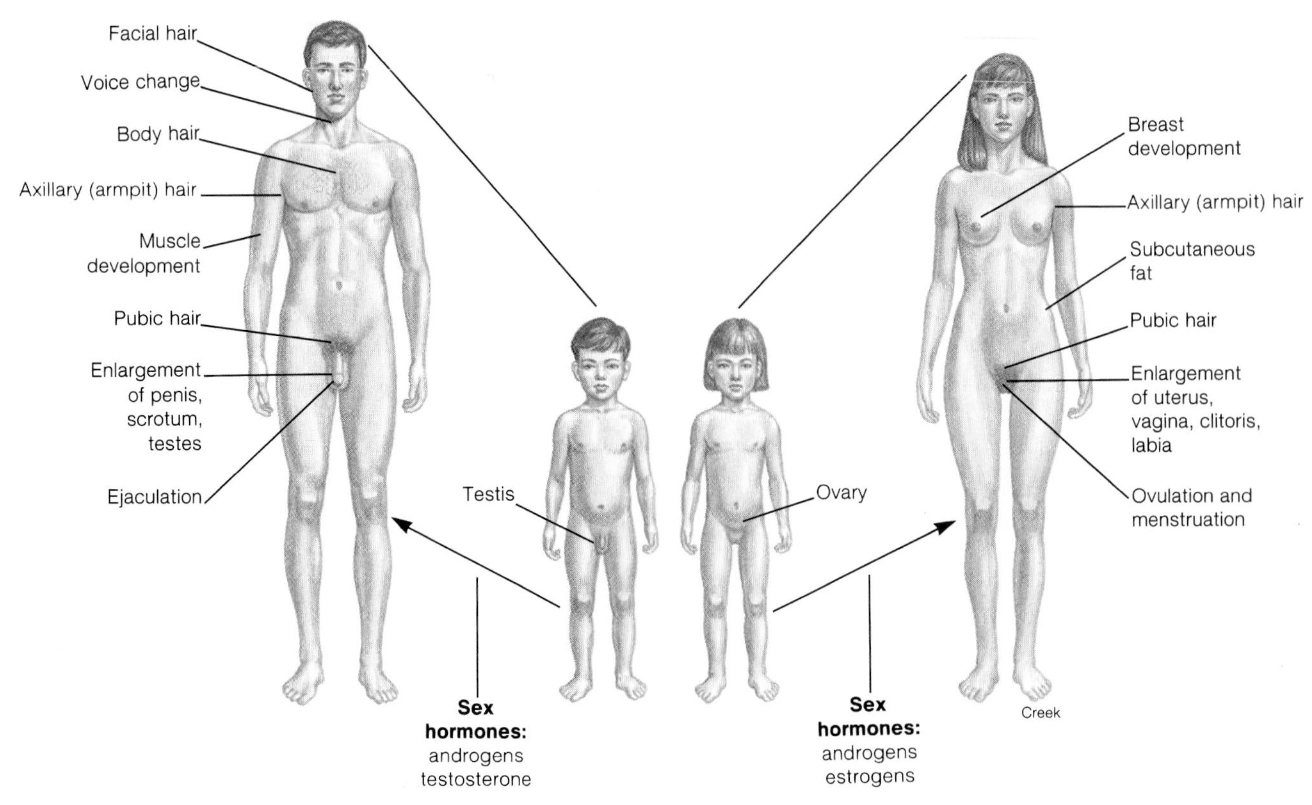

Figure 2.9

Developmental changes from childhood to adulthood. Controlled by hormones, puberty is the period of body growth when sexual characteristics are expressed and the sexual organs become functional.

Growth is a normal process by which an organism increases in size as a result of the accretion of cells and tissues similar to those already present within organs. Growth is an integral part of development that continues until adulthood. Normal growth depends not only on proper nutrition but on the concerted effect of several hormones (chemicals produced by endocrine glands), including insulin, growth hormone, and (during adolescence) the sex hormones. It is through the growth process that each of the body systems eventually matures (fig. 2.9). *Puberty* is the developmental transition during the growth process when sexual features become expressed in several of the body systems and the reproductive organs become functional.

A *systematic* (*systemic*) *approach* to studying anatomy emphasizes the purposes of various organs within a system. For example, the functional role of the digestive system can be best understood if all of the organs within that system are studied together. In a *regional approach*, all of the organs and structures in one particular region are examined at the same time. The regional approach has merit in graduate professional schools (medical, dental, etc.) because the structural relationships of portions of several systems can be observed simultaneously. Dissections of cadavers are usually conducted on a regional basis. Trauma or injury usually affects a region of the body, whereas a disease that affects a region may also involve an entire system.

This text uses a systematic approach to anatomy. In the chapters that follow, you will become acquainted, system by system, with the functional anatomy of the entire body. An overview of the structure and function of each of the body systems is presented in figure 2.10.

1. Construct a diagram to illustrate the levels of structural organization that characterize the body. Which of these levels are microscopic?
2. Why is the skin considered an organ?
3. Which body systems control the functioning of the others? Which are supportive of the organism? Which serve a transportive role?

Integumentary system
Function: external support and protection of body

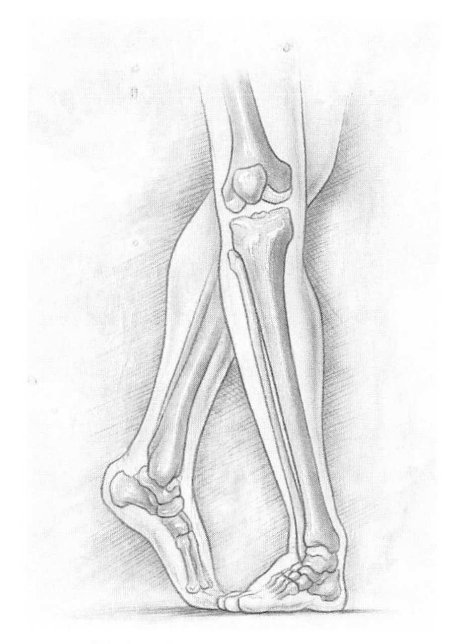

Skeletal system
Function: internal support and flexible framework for body movement; production of blood cells

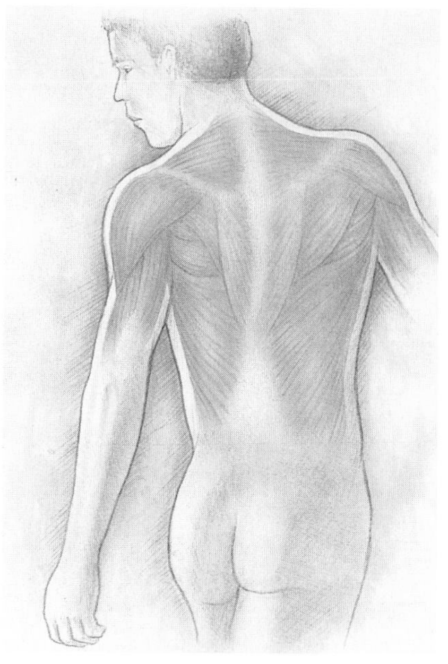

Muscular system
Function: body movement; production of body heat

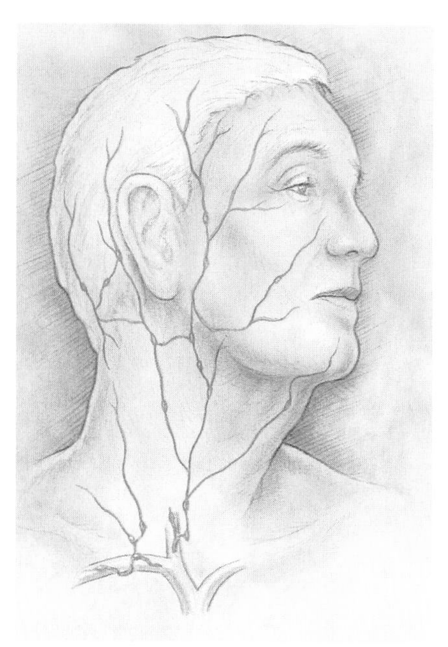

Lymphatic system
Function: body immunity; absorption of fats; drainage of tissue fluid

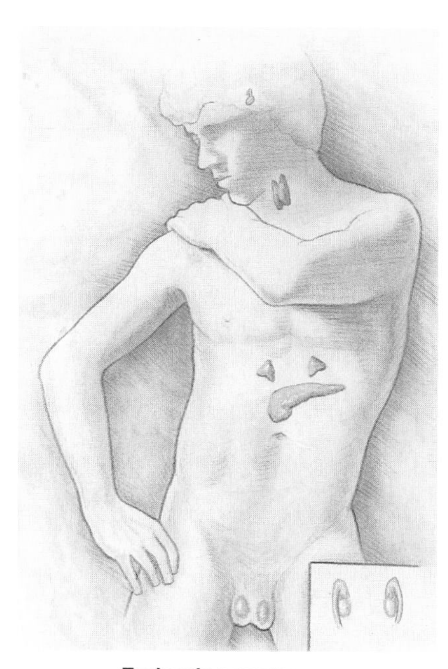

Endocrine system
Function: secretion of hormones for chemical regulation

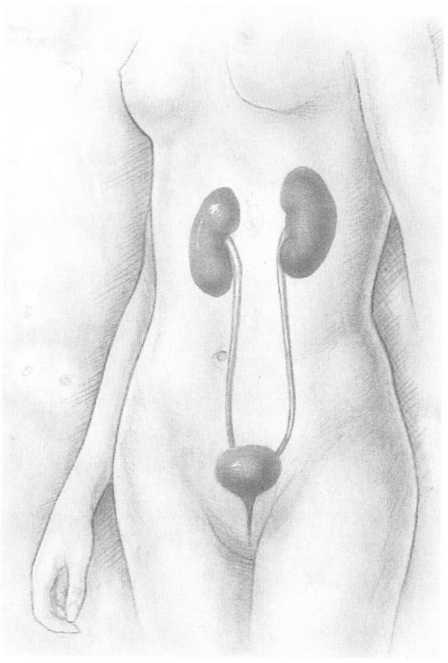

Urinary system
Function: filtration of blood; maintenance of volume and chemical composition of blood; removal of metabolic wastes from body

Figure 2.10

The body systems.

Continued

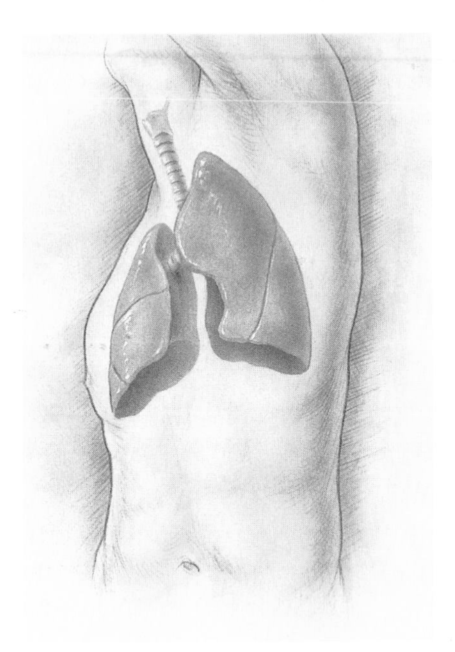

Respiratory system
Function: gaseous exchange between external environment and blood

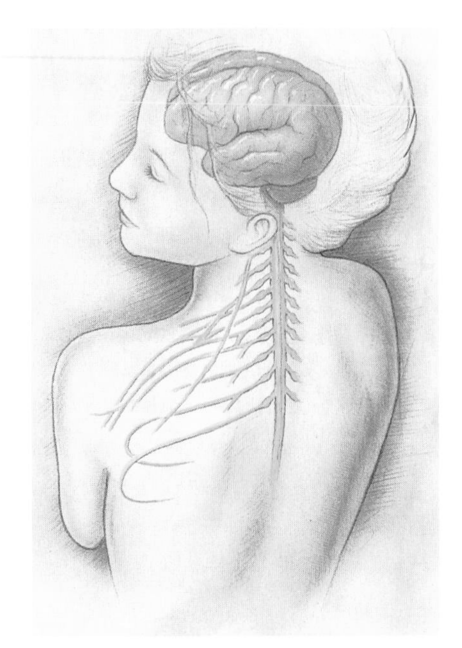

Nervous system
Function: control and regulation of all other systems of the body

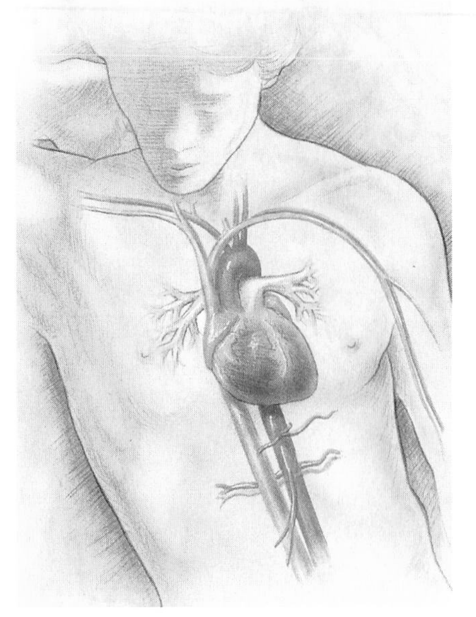

Circulatory system
Function: transport of life-sustaining materials to body cells; removal of metabolic wastes from cells

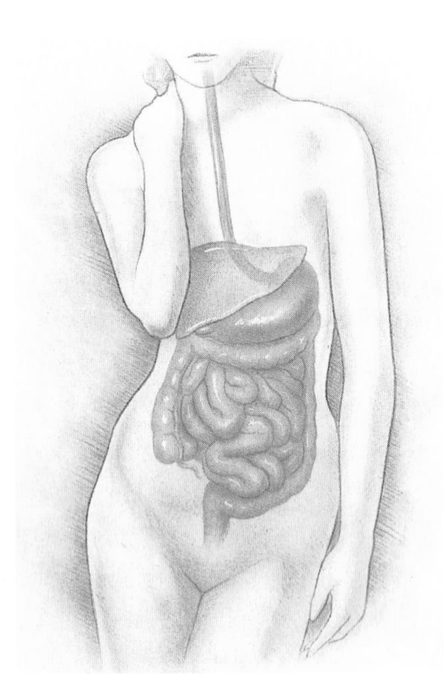

Digestive system
Function: breakdown and absorption of food materials

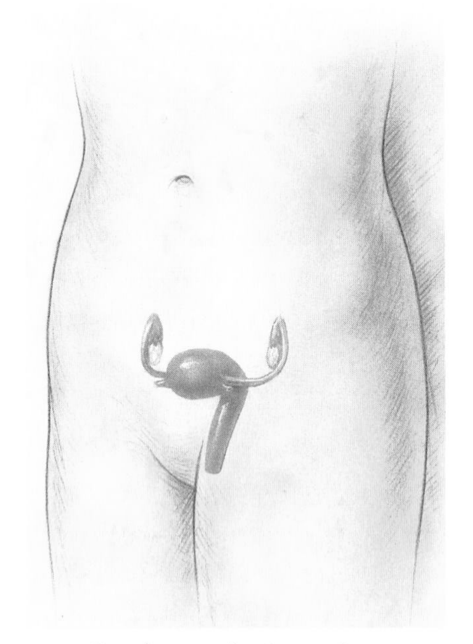

Female reproductive system
Function: production of female sex cells (ova); receptacle for sperm from male; site for fertilization of ovum, implantation, and development of embryo and fetus; delivery of fetus

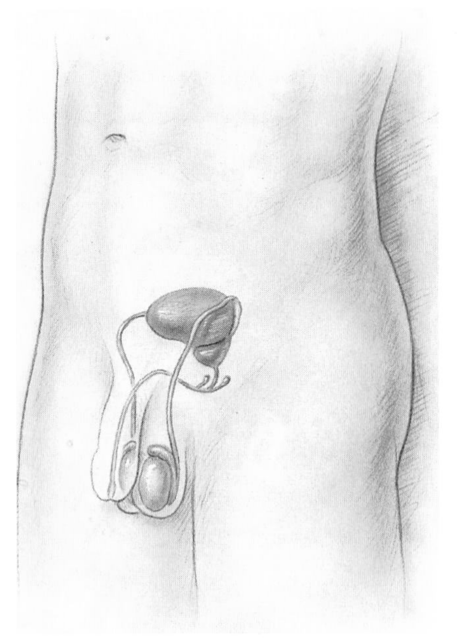

Male reproductive system
Function: production of male sex cells (sperm); transfer of sperm to reproductive system of female

Figure 2.10—*Continued*

Anatomical Nomenclature

In order to understand the science of anatomy, students must master its descriptive terminology.

Objective 6 Explain how anatomical terms are derived.

Objective 7 Describe what is meant by prefixes and suffixes.

Anatomy is a descriptive science. Analyzing anatomical terminology can be a rewarding experience in that one learns something of the character of antiquity in the process. However, understanding the roots of words is not only of academic interest. Familiarity with technical terms reinforces the learning process. Most anatomical terms are derived from Greek or Latin, but some of the more recent terms are of German and French origin. As mentioned in chapter 1, some anatomical structures bear the names of people who discovered or described them. Such terms are totally nondescriptive; unfortunately, they have little meaning in and of themselves.

Many Greek and Latin terms were coined more than 2,000 years ago. Deciphering the meanings of these terms affords a glimpse into our medical heritage. Many terms referred to common plants or animals. Thus, the term *vermis* means worm; *cochlea* (*kok'le-ă*), snail shell; *cancer*, crab; and *uvula*, little grape. Even the term *muscle* comes from the Latin *musculus*, which means mouse. Other terms suggest the warlike environment of ancient Greece and Rome. *Thyroid*, for example, means shield; *xiphos* (*zi'fos*) means sword; and *thorax*, breastplate. *Sella* means saddle and *stapes* (*sta'pēz*) means stirrup. Various tools or instruments were referred to in early anatomy. The malleus and anvil resemble miniatures of a blacksmith's implements, and tympanum refers to a drum.

You will encounter many new terms throughout your study of anatomy. You can learn these terms more easily if you know the meaning of their prefixes and suffixes. Use the glossary of prefixes and suffixes (on the inside front and back covers) as an aid in learning new terms. Pronouncing these terms as you learn them will also help you remember them. A guide to the singular and plural forms of words is presented in table 2.2.

The material presented in the remainder of this chapter provides a basic foundation for anatomy, as well as for all medical and paramedical fields. Anatomy is a very precise science because of its universally accepted reference language for describing body parts and locations.

1. Explain the statement, Anatomy is a descriptive science.
2. Refer to the glossary of prefixes and suffixes on the inside front and back covers to decipher the terms *blastocoel*, *hypodermic*, *dermatitis*, and *orchiectomy*.

Table 2.2
Some Examples of Singular and Plural Word Endings

Singular Ending	Plural Ending	Examples
-a	-ae	Axilla, axillae
-ax	-aces	Thorax, thoraces
-en	-ina	Lumen, lumina
-ex	-ices	Cortex, cortices
-is	-es	Diagnosis, diagnoses
-is	-ides	Epididymis, epididymides
-ix	-ices	Appendix, appendices
-ma	-mata	Carcinoma, carcinomata
-on	-a	Mitochondrion, mitochondria
-um	-a	Cilium, cilia
-us	-i	Tarsus, tarsi
-us	-ora	Corpus, corpora
-us	-era	Viscus, viscera
-x	-ges	Pharynx, pharynges
-y	-ies	Ovary, ovaries

Source: Dr. Kenneth S. Saladin, Georgia College.

Planes of Reference and Descriptive Terminology

All of the descriptive planes of reference and terms of direction used in anatomy are standardized because of their reference to the body in anatomical position.

Objective 8 Identify the planes of reference used to locate structures within the body.

Objective 9 Describe the anatomical position.

Objective 10 Define and be able to properly use the descriptive and directional terms that refer to the body.

Planes of Reference

In order to visualize and study the structural arrangements of various organs, the body may be sectioned (cut) and diagrammed according to three fundamental planes of reference: a **sagittal** (*saj'ĭ-tal*) **plane,** a **coronal plane,** and a **transverse plane**

coronal: L. *corona*, crown

(figs. 2.11 and 2.12). A sagittal plane extends vertically through the body dividing it into right and left portions. A *midsagittal plane* is a sagittal plane that passes lengthwise through the midplane of the body, dividing it equally into right and left halves. Coronal, or *frontal*, planes also pass lengthwise and divide the body into anterior (front) and posterior (back) portions. Transverse planes, also called *horizontal*, or *cross-sectional, planes*, divide the body into superior (upper) and inferior (lower) portions.

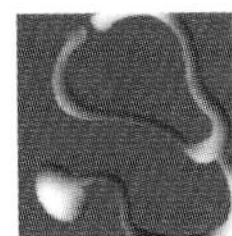

The value of the computerized tomographic X-ray (CT) scan (see fig. 1.19*a*) is that it displays an image along a transverse plane similar to that which could otherwise be obtained only in an actual section through the body. Prior to the development of this technique, the vertical plane of conventional radiographs made it difficult, if not impossible, to assess the extent of body irregularities.

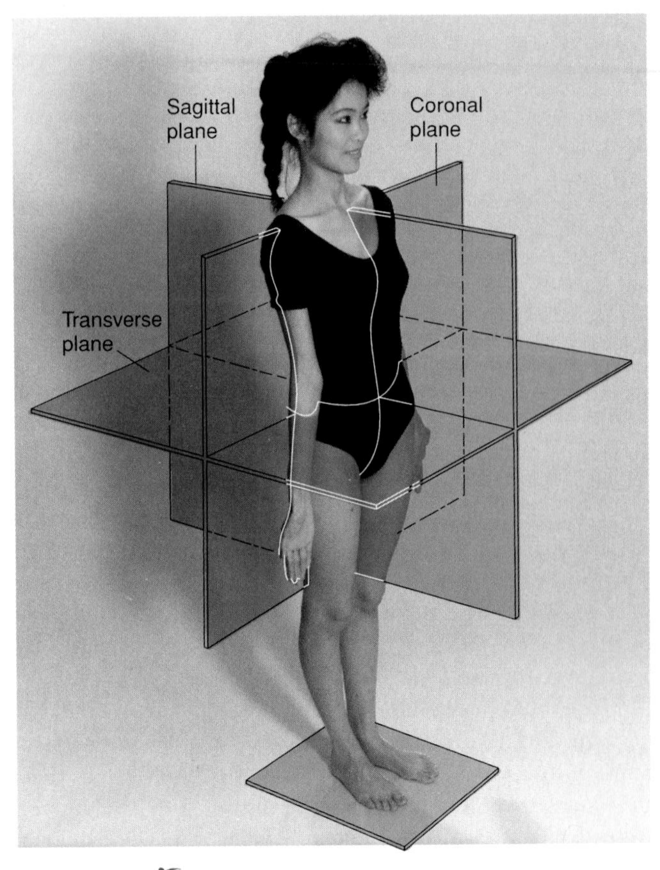

Figure 2.11

Planes of reference through the body.

Descriptive Terminology

Anatomical Position

All terms of direction that describe the relationship of one body part to another are made in reference to the **anatomical position.** In the anatomical position, the body is erect, the feet are parallel to each other and flat on the floor, the eyes are directed forward, and the arms are at the sides of the body with the palms of the hands turned forward and the fingers pointed straight down (fig. 2.13).

Directional Terms

Directional terms are used to locate structures and regions of the body relative to the anatomical position. A summary of directional terms is presented in table 2.3.

Clinical Procedures

Certain clinical procedures are important in determining anatomical structure and function in a living individual. The most common of these are as follows:

1. **Inspection.** Visually observing the body to note any clinical symptoms, such as abnormal skin color, swelling, or rashes. Other observations may include needle marks on the skin, irregular breathing rates, or abnormal behavior.
2. **Palpation.** Applying the fingers with firm pressure to the surface of the body to feel surface landmarks, lumps, tender spots, or pulsations.
3. **Percussion.** Tapping sharply on various locations on the thorax or abdomen to detect resonating vibrations as an aid in locating excess fluids or organ abnormalities.
4. **Auscultation.** Listening to the sounds that various organs make (breathing, heartbeat, digestive sounds, and so forth).
5. **Reflex testing.** Observing a person's automatic response to a stimulus. One test of a reflex mechanism involves tapping a predetermined tendon with a reflex hammer and noting the response.

1. Explain why a transverse plane through an organ is more important in studying specific systems than a transverse plane through the body.
2. What do we mean when we say that directional terms are relative and must be used in reference to a body structure or a body in anatomical position?
3. Write a list of statements, similar to the examples in table 2.3, that correctly express the directional terms used to describe the relative positions of various body structures.

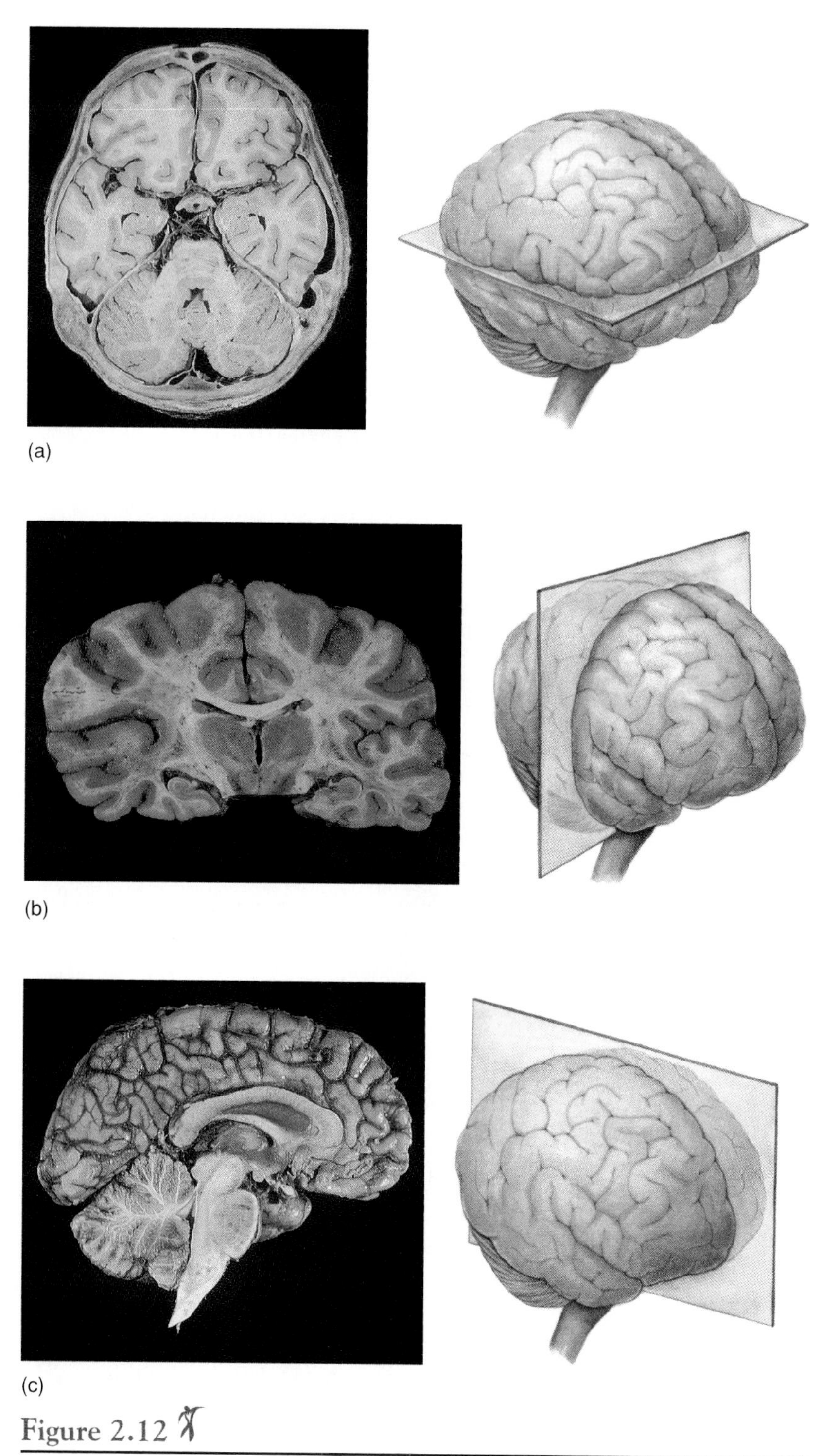

Figure 2.12

The human brain sectioned along (*a*) a transverse plane, (*b*) a coronal plane, and (*c*) a sagittal plane.

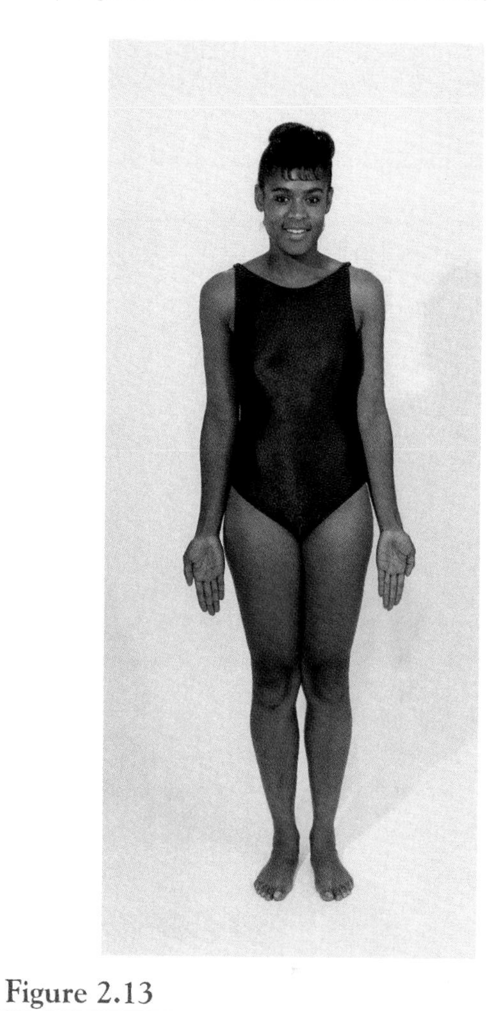

Figure 2.13

In the anatomical position, the body is erect, the feet parallel, the eyes directed forward, the arms to the sides with the palms directed forward, and the fingers pointed straight down.

Table 2.3
Directional Terms for the Human Body

Term	Definition	Example
Superior (cranial, cephalic)	Toward the head; toward the top	The thorax is superior to the abdomen.
Inferior (caudal)	Away from the head; toward the bottom	The legs are inferior to the trunk.
Anterior (ventral)	Toward the front	The navel is on the anterior side of the body.
Posterior (dorsal)	Toward the back	The kidneys are posterior to the intestine.
Medial	Toward the midline of the body	The heart is medial to the lungs.
Lateral	Away from the midline of the body	The ears are lateral to the nose.
Internal (deep)	Away from the surface of the body	The brain is internal to the cranium.
External (superficial)	Toward the surface of the body	The skin is external to the muscles.
Proximal	Toward the trunk of the body	The knee is proximal to the foot.
Distal	Away from the trunk of the body	The hand is distal to the elbow.

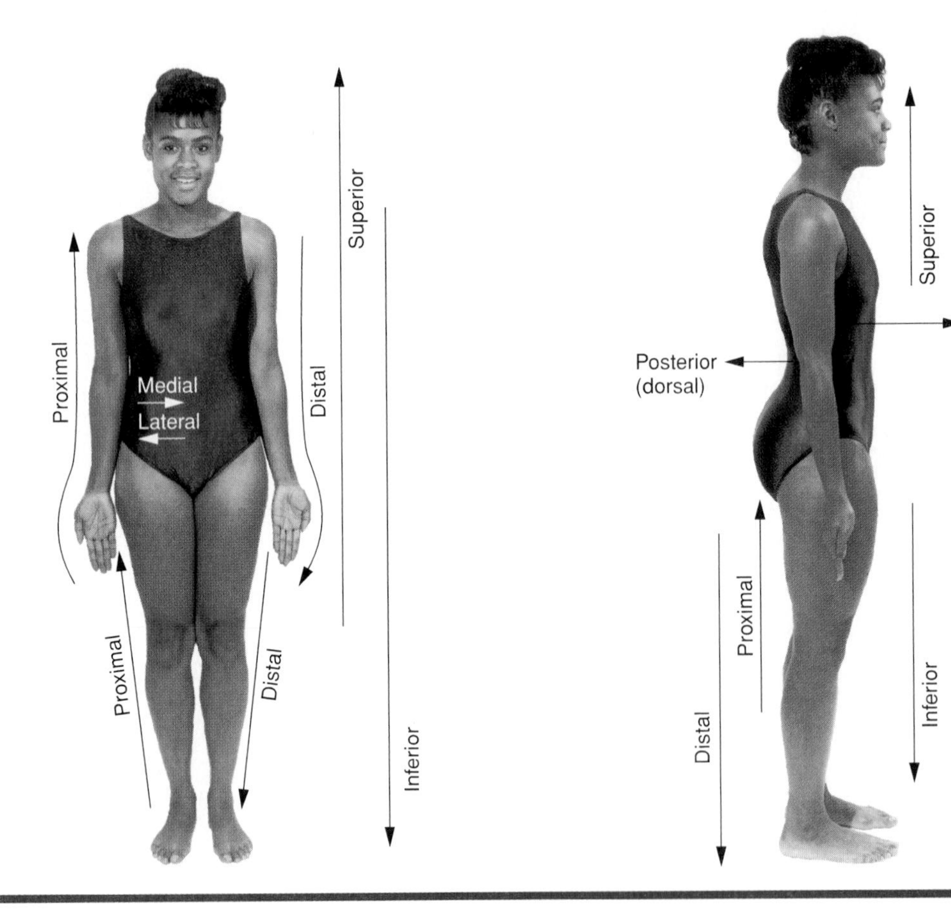

Body Regions

The human body is divided into regions and specific local areas that can be identified on the surface. Each region contains internal organs, the locations of which are anatomically and clinically important.

Objective 11 List the regions of the body and the principal local areas that make up each region.

Objective 12 Explain why it is important to be able to describe the body areas and regions in which major internal organs are located.

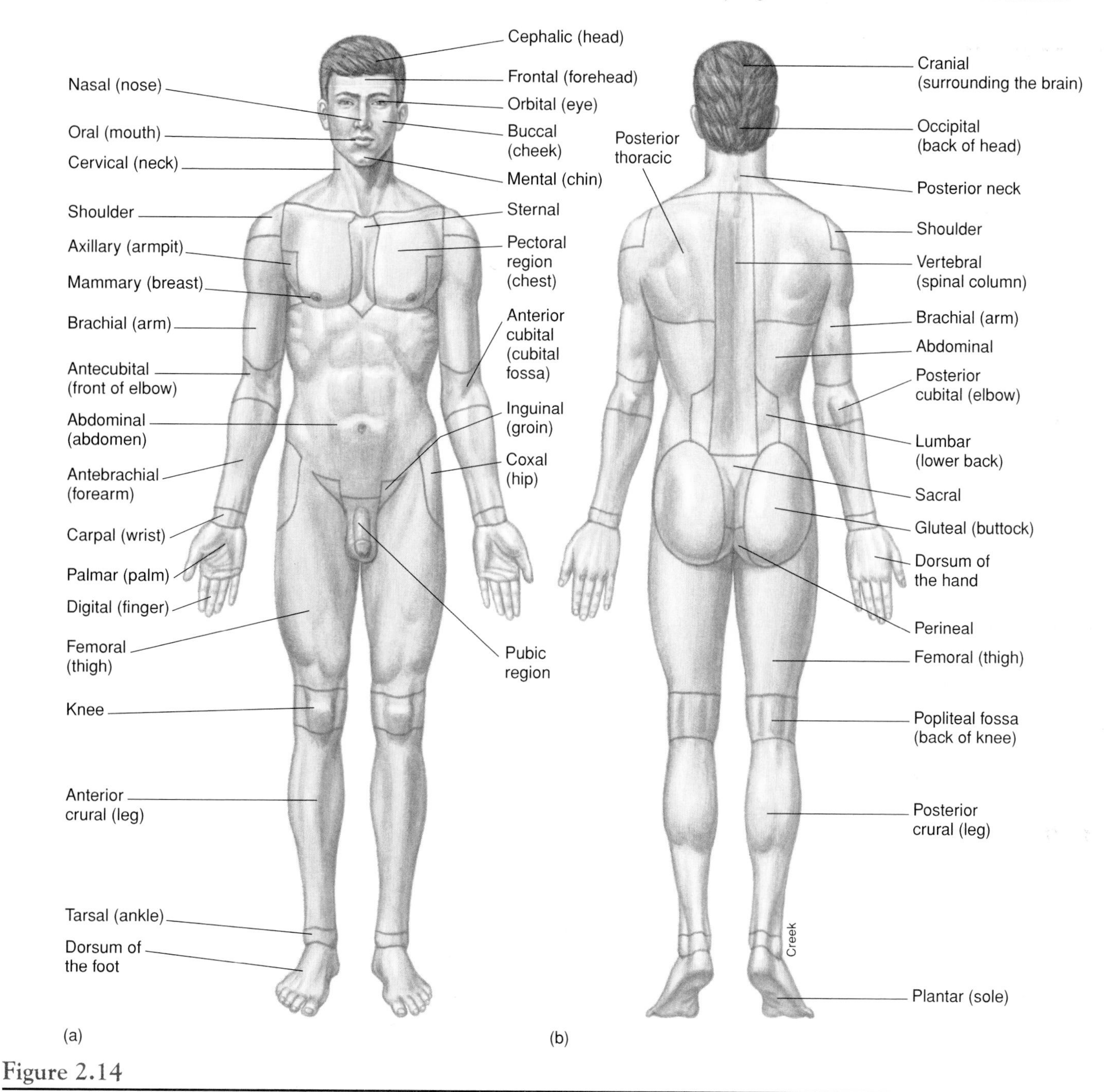

Figure 2.14

Body regions. (*a*) An anterior view and (*b*) a posterior view.

The human body is divided into several regions that can be identified on the surface of the body. Learning the terms that refer to these regions now will make it easier to learn the names of underlying structures later. The major body regions are the *head, neck, trunk, upper extremity,* and *lower extremity* (fig. 2.14). The trunk is frequently divided into the thorax, abdomen, and pelvis.

Head

The **head** is divided into a **facial region,** which includes the eyes, nose, and mouth, and a **cranial region,** or **cranium** (*kra'ne-um*), which covers and supports the brain. The identifying names for specific surface regions of the face are based on associated organs—for example, the orbital (eye), nasal (nose), oral (mouth), and auricular (ear) regions—or underlying bones—for example, the frontal, temporal parietal, zygomatic, and occipital regions.

Neck

The **neck,** referred to as the **cervical region,** supports the head and permits it to move. As with the head, detailed subdivisions

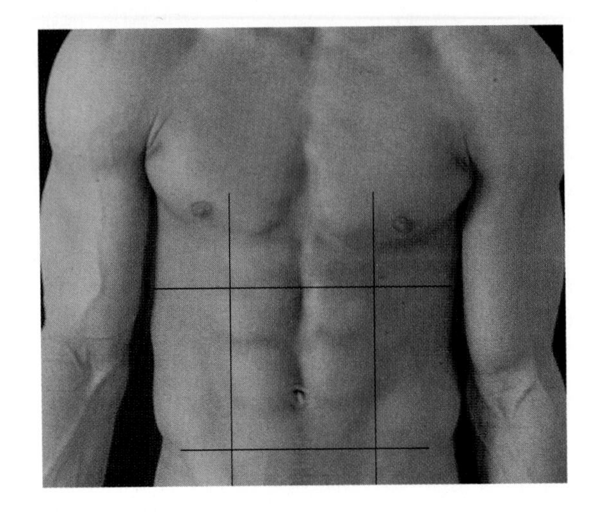

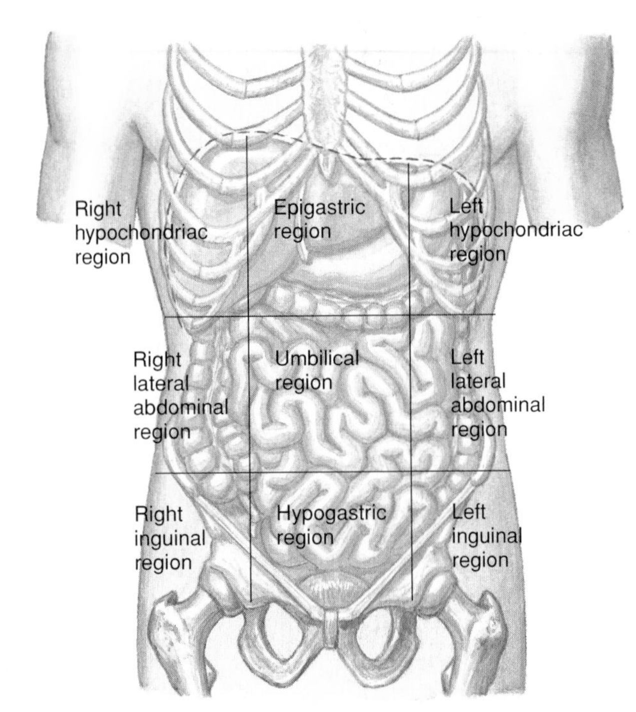

Figure 2.15

The abdomen is frequently subdivided into nine regions. The upper vertical planes are positioned lateral to the rectus abdominis muscles, the upper horizontal plane is positioned at the level of the rib cage, and the lower horizontal plane is even with the upper border of the hipbones.

of the neck can be identified. Additional information concerning the neck region can be found in chapter 10.

Trunk

The **trunk**, or *torso*, is the portion of the body to which the neck and upper and lower extremities attach. It includes the thorax, abdomen, and pelvic region.

Thorax

The thorax (*thor'aks*), or **thoracic** (*thŏ-ras'ik*) **region,** is commonly referred to as the chest. The **mammary region** of the thorax surrounds the nipple and in sexually mature females is enlarged as the breast. Between the mammary regions is the **sternal region.** The armpit is called the **axillary fossa,** or simply the **axilla,** and the surrounding area, the **axillary region.** The **vertebral region** extends the length of the back, following the vertebral column.

thorax: L. *thorax*, chest
mammary: L. *mamma*, breast
axillary: L. *axilla*, armpit

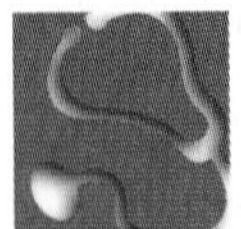

The heart and lungs are located within the thoracic cavity. Easily identified surface landmarks are helpful in assessing the condition of these organs. A physician must know, for example, where the valves of the heart can best be detected and where to listen for respiratory sounds. The axilla becomes important when examining for infected lymph nodes. When fitting a patient for crutches, a physician will instruct the patient to avoid supporting the weight of the body on the axillary region because of the possibility of damaging the underlying nerves and vessels.

Abdomen

The abdomen (*ab'dŏ-men*) is located below the thorax. Centered on the front of the abdomen, the **umbilicus (navel)** is an obvious landmark. The abdomen has been divided into nine regions to describe the location of internal organs. The subdivisions of the abdomen are diagrammed in figure 2.15, and the internal organs located within these regions are identified in table 2.4. Subdividing the abdomen into four quadrants (fig. 2.16) is a common clinical practice for locating the sites of pains, tumors, or other abnormalities.

Pelvic Region

The pelvic region forms the lower portion of the trunk. Within the pelvic region is the **pubic area,** which is covered with pubic hair in

Table 2.4

Regions of the Abdomen and Pelvis

Region	Location	Internal Organs
Right hypochondriac	Right, upper one-third of abdomen	Gallbladder; portions of liver and right kidney
Epigastric	Upper, central one-third of abdomen	Portions of liver, stomach, pancreas, and duodenum
Left hypochondriac	Left, upper one-third of abdomen	Spleen; splenic flexure of colon; portions of left kidney and small intestine
Right lateral	Right, lateral one-third of abdomen	Cecum; ascending colon; hepatic flexure; portions of right kidney and small intestine
Umbilical	Center of abdomen	Jejunum; ileum; portions of duodenum, colon, kidneys, and major abdominal vessels
Left lateral	Left, lateral one-third of abdomen	Descending colon; portions of left kidney and small intestine
Right inguinal	Right, lower one-third of abdomen	Appendix; portions of cecum and small intestine
Pubic (hypogastric)	Lower, center one-third of abdomen	Urinary bladder; portions of small intestine and sigmoid colon
Left inguinal	Left, lower one-third of abdomen	Portions of small intestine, descending colon, and sigmoid colon

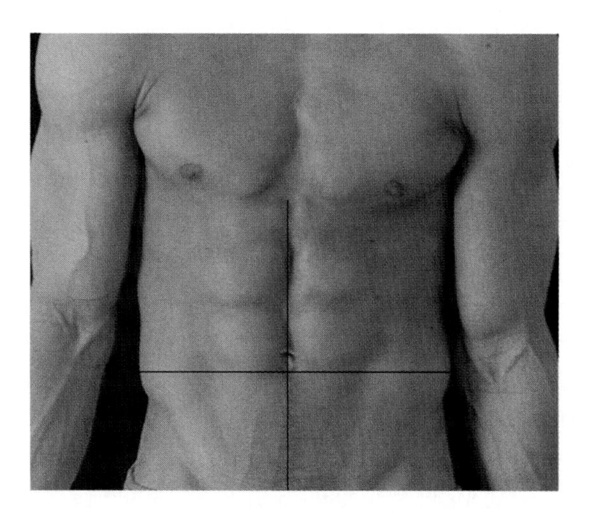

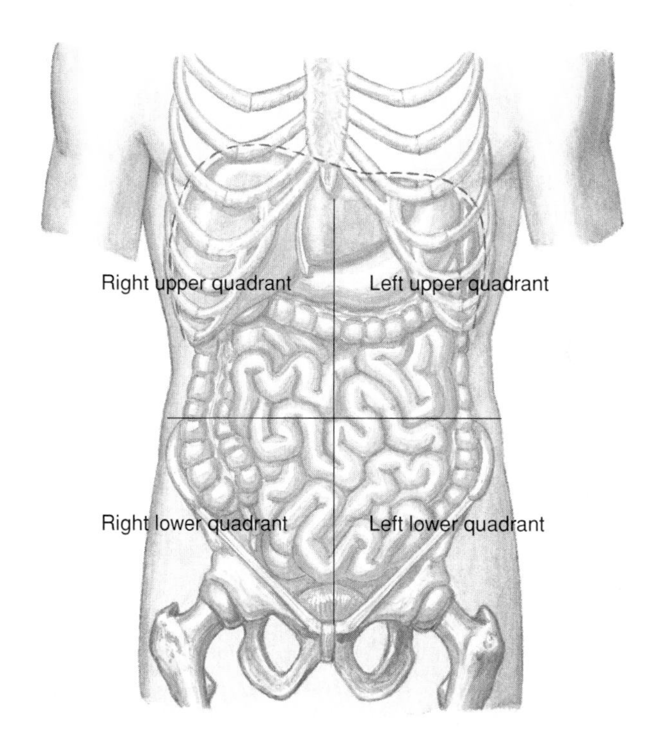

Figure 2.16

A clinical subdivision of the abdomen into four quadrants by a median plane and a transverse plane through the umbilicus.

sexually mature individuals. The **perineum** (*per″ĭ-ne′um*) (fig. 2.17) is the region containing the external sex organs and the anal opening. The center of the back side of the abdomen, commonly called the small of the back, is the **lumbar region.** The **sacral region** is located further down, at the point where the vertebral column terminates. The large hip muscles form the **buttock,** or **gluteal region.** This region is a common injection site for hypodermic needles.

Upper Extremity

The **upper extremity** is anatomically divided into the **shoulder, brachium** (*bra′ke-um*) (arm), **antebrachium** (forearm), and **manus** (hand) (see fig. 2.14). The shoulder is the region between the pectoral girdle and the brachium that contains the shoulder joint. The shoulder is also referred to as the **omos,** or **deltoid region.** The

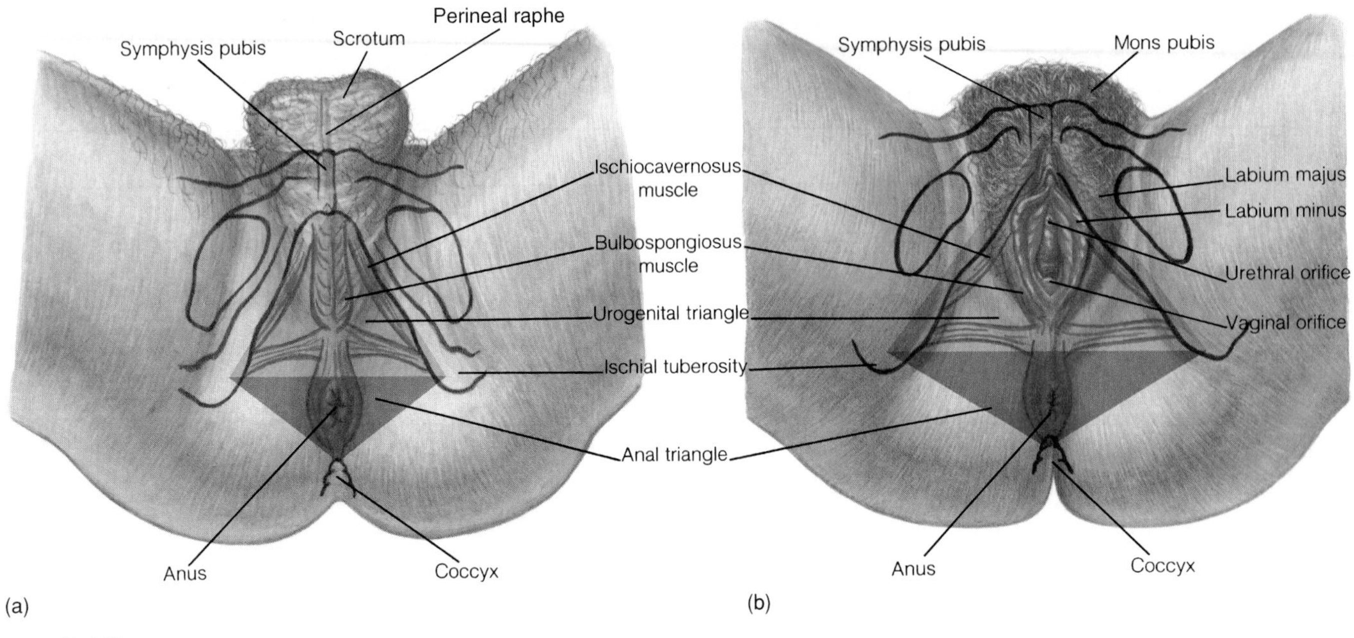

Figure 2.17

A superficial view of the perineum of (*a*) a male and (*b*) a female. The perineal region can be divided into a urogenital triangle (anteriorly) and an anal triangle (posteriorly).

cubital region is the area between the arm and forearm that contains the elbow joint. The **cubital fossa** is the depressed anterior portion of the cubital region. It is an important site for intravenous injections or the withdrawal of blood.

The manus has three principal divisions: the **carpus**, containing the carpal bones (see fig. 7.8); the **metacarpus**, containing the metacarpal bones; and the five **digits** (commonly called fingers), containing the phalanges. The front of the hand is referred to as the **palmar region (palm)** and the back of the hand is called the **dorsum of the hand.**

Lower Extremity

The **lower extremity** consists of the **hip, thigh, knee, leg,** and **pes** (foot). The thigh is commonly called the **upper leg,** or **femoral region.** The knee has two surfaces: the front surface is the **patellar region,** or kneecap; the back of the knee is called the **popliteal** (*pop"lĭ-te'al*) **fossa.** The leg has anterior and posterior **crural regions** (see fig. 2.14). The *shin* is a prominent bony ridge extending longitudinally along the anterior crural region, and the *calf* is the thickened muscular mass of the posterior crural region.

The pes has three principal divisions: the **tarsus**, containing the tarsal bones (see fig. 7.19); the **metatarsus**, containing the metatarsal bones; and the five **digits** (commonly called toes), containing the phalanges. The **ankle** is the junction between the leg and the foot. The **heel** is the back of the foot, and the **sole** of the foot is referred to as the **plantar surface.** The **dorsum of the foot** is the top surface.

cubital: L. *cubitis*, elbow
popliteal: L. *poples*, ham (hamstring muscles) of the knee

1. Using yourself as a model, identify the various body regions depicted in figure 2.14. Which of these regions have surface landmarks that help distinguish their boundaries?
2. In which region of the body are intravenous injections given?
3. Distinguish the pubic area and perineum within the pelvic region.
4. Identify the joint between the following regions: the brachium and antebrachium, the pectoral girdle and brachium, the leg and foot, the antebrachium and hand, and the thigh and leg.
5. Explain how knowledge of the body regions is applied in a clinical setting.

Body Cavities and Membranes

For functional and protective purposes, the viscera are compartmentalized and supported in specific body cavities by connective and epithelial membranes.

Objective 13 Identify the various body cavities and the organs found in each.

Objective 14 Discuss the types and functions of the various body membranes.

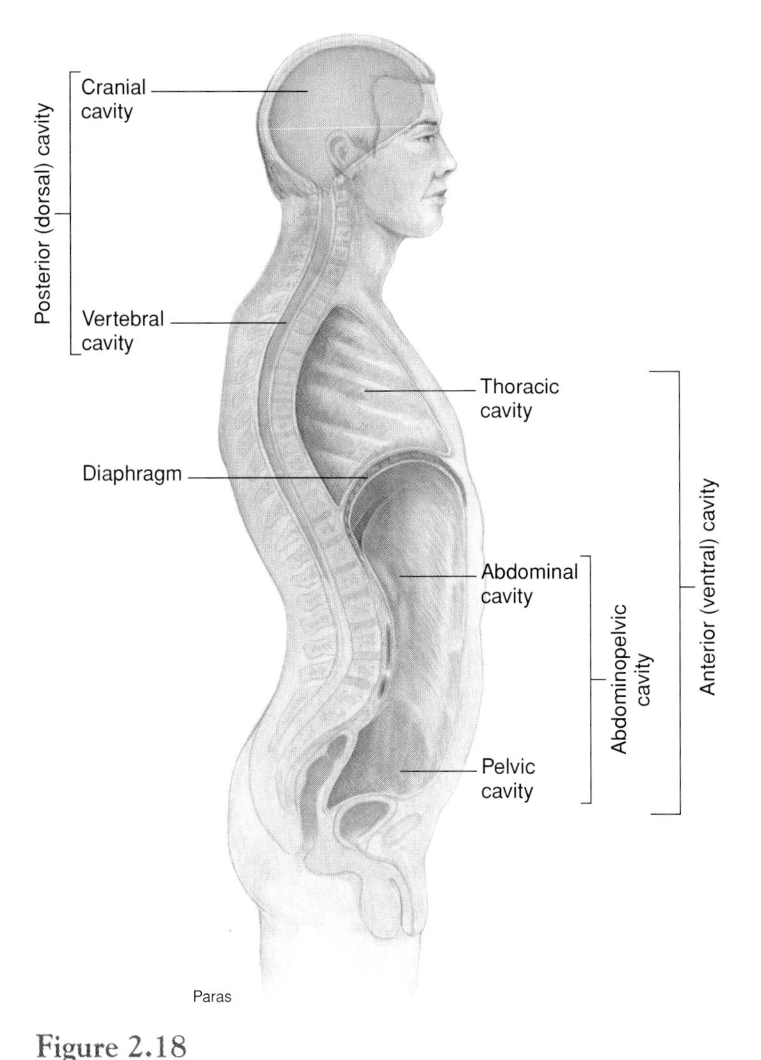

Figure 2.18

A midsagittal (median) section showing the body cavities.

Body Cavities

Body cavities are confined spaces within the body. They contain organs that are protected, compartmentalized, and supported by associated membranes. There are two principal body cavities: the **posterior (dorsal) body cavity** and the larger **anterior (ventral) body cavity.** The posterior body cavity contains the brain and the spinal cord. During development, the anterior cavity forms from a cavity within the trunk called the **coelom** (*se'lom*). The coelom is lined with a membrane that secretes a lubricating fluid. As development progresses, the coelom is partitioned by the muscular **diaphragm** into an upper **thoracic cavity,** or chest cavity, and a lower **abdominopelvic cavity** (figs. 2.18 and 2.19). Organs within the coelom are collectively called **viscera** (*vis'er-ă*), or **visceral organs** (fig. 2.20). Within the thoracic cavity are two **pleural** (*ploor'al*) **cavities** surrounding the right and left lungs and a **pericardial** (*per"ĭ-kar'de-al*) **cavity** surrounding the heart (fig. 2.21). The area between the two lungs is known as the **mediastinum** (*me"de-ă-sti'num*).

The abdominopelvic cavity consists of an upper **abdominal cavity** and a lower **pelvic cavity.** The abdominal cavity contains the stomach, small intestine, large intestine, liver, gallbladder, pancreas, spleen, and kidneys. The pelvic cavity is occupied by the terminal portion of the large intestine, the urinary bladder, and certain reproductive organs (uterus, uterine tubes, and ovaries in the female; seminal vesicles and prostate in the male).

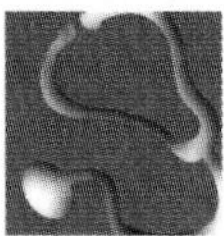

Body cavities serve to confine organs and systems that have related functions. The major portion of the nervous system occupies the posterior cavity; the principal organs of the respiratory and circulatory systems are in the thoracic cavity; the primary organs of digestion are in the abdominal cavity; and the reproductive organs are in the pelvic cavity. Not only do these cavities house and support various body organs, they also effectively compartmentalize them so that infections and diseases cannot spread from one compartment to another. For example, pleurisy of one lung membrane does not usually spread to the other, and an injury to the thoracic cavity will usually result in the collapse of only one lung rather than both.

In addition to the large anterior and posterior cavities, there are several smaller cavities within the head. The **oral cavity** functions primarily in digestion and secondarily in respiration. It contains the teeth and tongue. The **nasal cavity,** which is part of the respiratory system, has two chambers created by a nasal septum. There are two **orbits,** each of which houses an eyeball and its associated muscles, vessels, and nerves. Likewise, there are two **middle-ear cavities** containing the auditory ossicles (ear bones). The location of the cavities within the head is shown in figure 2.22.

Body Membranes

Body membranes are composed of thin layers of connective and epithelial tissue that cover, separate, and support visceral organs and line body cavities. There are two basic types of body membranes: **mucous** (*myoo'kus*) **membranes** and **serous** (*se'rus*) **membranes.**

Mucous membranes secrete a thick, sticky fluid called *mucus*. Mucus generally lubricates or protects the associated organs where it is secreted. Mucous membranes line various cavities and tubes that enter or exit the body, such as the oral and nasal cavities and the tubes of the respiratory, reproductive, urinary, and digestive systems.

Serous membranes line the thoracic and abdominopelvic cavities and cover visceral organs, secreting a watery lubricant called *serous fluid*. **Pleurae** are serous membranes associated with the lungs. Each pleura (pleura of right lung and pleura of left

coelom: Gk. *koiloma*, a cavity

orbital: L. *orbis*, circle

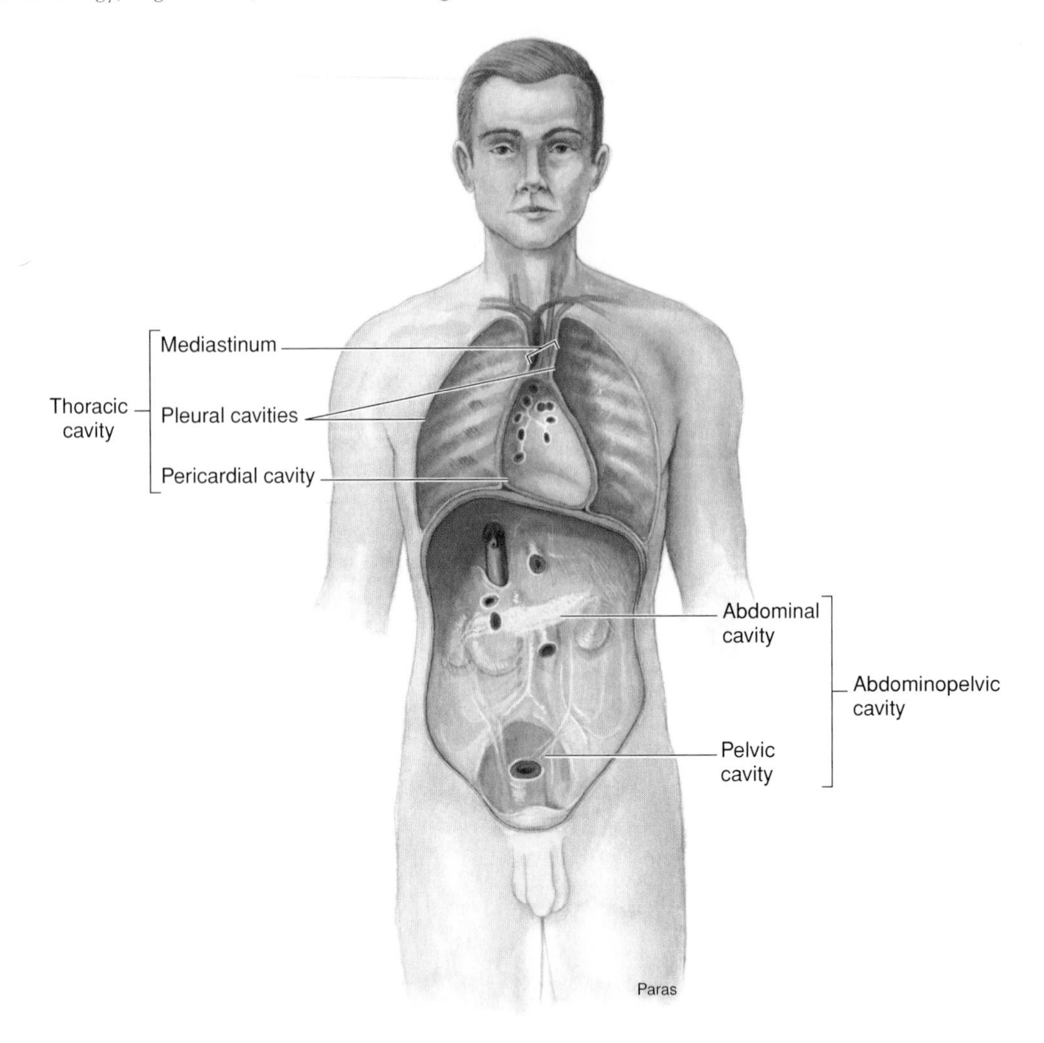

Figure 2.19

An anterior view showing the cavities of the trunk.

lung) has two parts. The **visceral pleura** adheres to the outer surface of the lung, while the **parietal** (*pă-ri'ĕ-tal*) **pleura** lines the thoracic walls and the thoracic surface of the diaphragm. The moistened space between the two pleurae is known as the **pleural cavity** (fig. 2.21).

Pericardial membranes are the serous membranes covering the heart. A thin **visceral pericardium** covers the surface of the heart, and a thicker **parietal pericardium** surrounds the heart. The space between these two membranes is called the **pericardial cavity.**

Serous membranes of the abdominal cavity are called **peritoneal** (*per"ĭ-tŏ-ne'al*) **membranes.** The **parietal peritoneum** lines the abdominal wall, and the **visceral peritoneum** covers the visceral organs. The **peritoneal cavity** is the potential space within the abdominopelvic cavity between the parietal and visceral peritoneal membranes. The *lesser omentum* and the *greater omentum* are folds of the peritoneum that extend from the stomach. They store fat and cushion and protect visceral organs of the abdominal cavity. Certain organs, such as the kidneys, adrenal glands, and a portion of the pancreas, which are within the abdominopelvic cavity, are positioned behind the parietal peritoneum, and are therefore said to be *retroperitoneal*. **Mesenteries** (*mes'en-ter"ēz*) are double folds of peritoneum that connect the parietal peritoneum to the visceral peritoneum (see figs. 2.20 and 18.3).

peritoneum: Gk. *peritonaion*, stretched over

1. Describe the divisions and boundaries of the anterior body cavity and list the major organs contained within each division.
2. Distinguish between mucous and serous membranes and list the specific serous membranes of the thoracic and abdominopelvic cavities.
3. Explain the importance of separate and distinct body cavities.

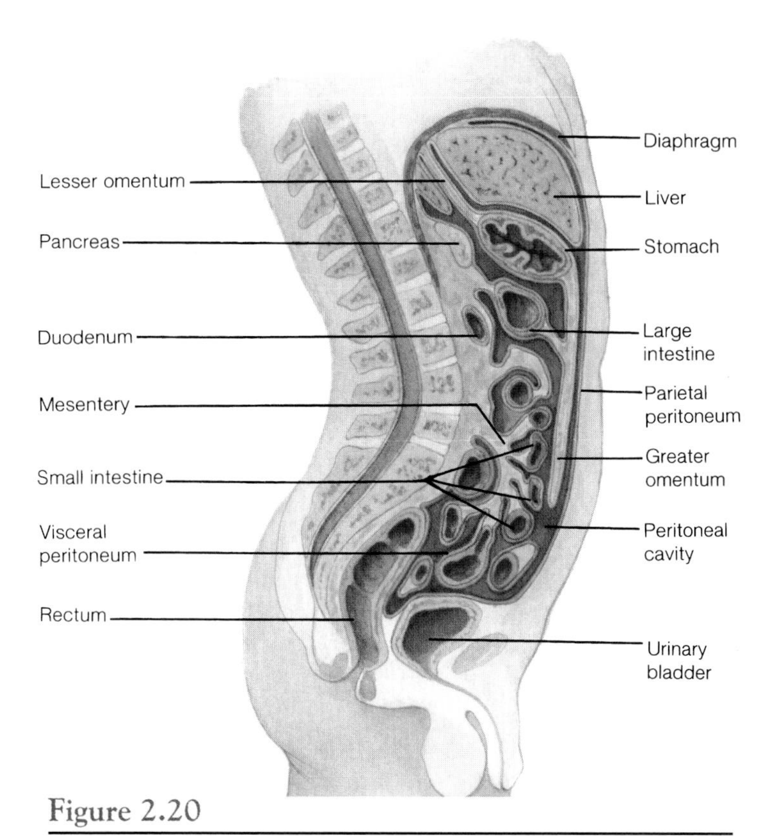

Figure 2.20

Visceral organs of the abdominopelvic cavity and the supporting serous membranes.

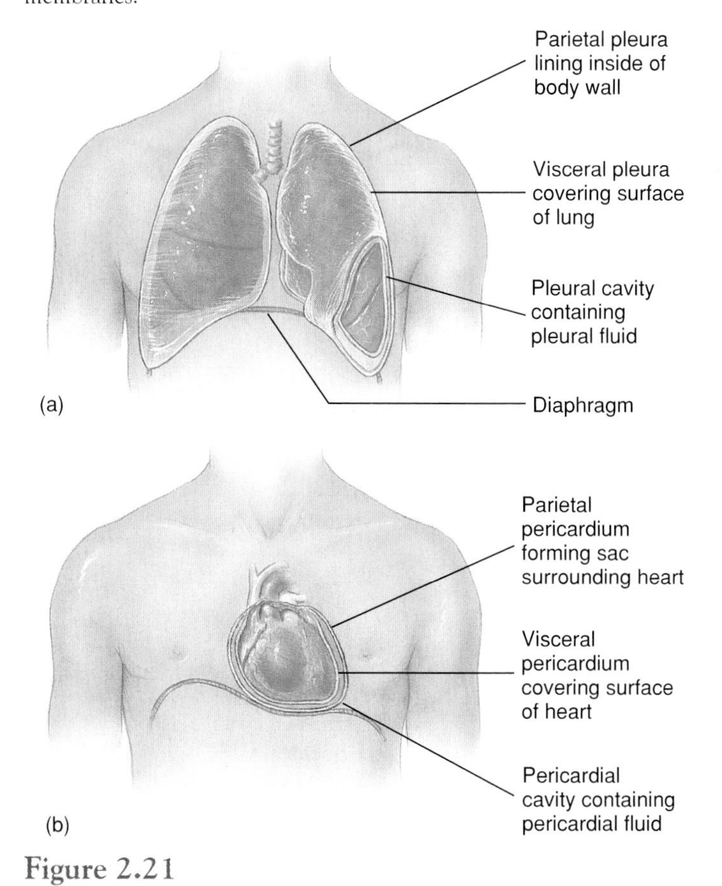

Figure 2.21

Serous membranes of the thorax (*a*) surrounding the lungs and (*b*) surrounding the heart.

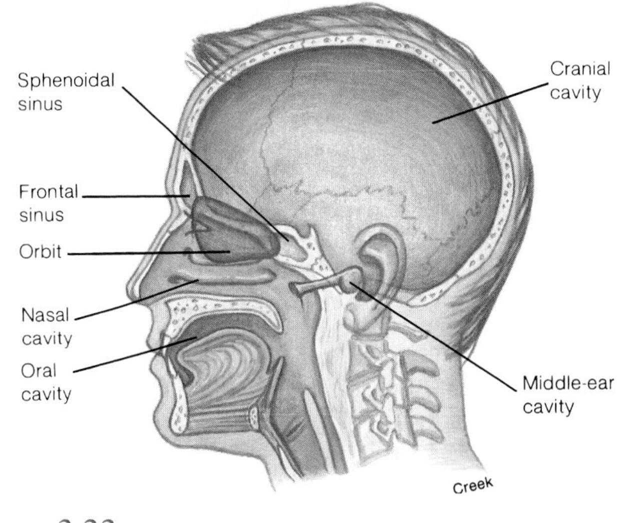

Figure 2.22

Cavities within the head.

Clinical Case Study Answer

Normally, the thoracic cavity is effectively separated from the abdominopelvic cavity by the diaphragm, peritoneum, and pleura. The phenomenon of peritoneal lavage fluid draining out of a tube properly placed in the chest can be explained only by the presence of a defect in the diaphragm. The defect in our patient is likely a traumatic rupture or laceration of the diaphragm that was caused by a sharp blow to the abdomen. The blow would have produced sudden upward pressure against the diaphragm, causing it to rupture. The absence of bile, blood, etc., in peritoneal lavage fluid does not guarantee the absence of trauma to organs such as the duodenum and pancreas. These organs are not located within the peritoneal cavity; rather they are retroperitoneal, or fully posterior to the peritoneal membrane. This keeps blood or leaked enzymes from these organs out of the peritoneal space and away from lavage fluid. Other signs must therefore be relied upon, and a high index of suspicion maintained, so as not to overlook injury to these organs. Evidence of injury to any of the intraabdominal organs, or the presence of diaphragmatic rupture, calls for emergency laparotomy to repair the structures involved.

Chapter Summary

Classification and Characteristics of Humans (pp. 23–28)

1. Our scientific name, *Homo sapiens*, means "man the intelligent," and our intelligence is our most distinguishing feature.
2. Humans belong to the phylum Chordata because of the presence of a notochord, a dorsal hollow nerve cord, and pharyngeal pouches during the embryonic stage of human development.
3. Humans are mammals, and as such have mammalian characteristics. These include hair, mammary glands, three auditory ossicles, heterodontia, a placenta, a muscular diaphragm, and a four-chambered heart with a left aortic arch.
4. Humans are also classified within the order Primates. Primates have prehensile hands, digits modified for grasping, and well-developed brains.
5. Humans are the sole members of the family Hominidae.
6. Some of the characteristics of humans are a large, well-developed brain; bipedal locomotion; an opposable thumb; well-developed vocal structures; and stereoscopic vision.

Body Organization (pp. 28–32)

1. Cells are the fundamental structural and functional components of life.
2. Tissues are aggregations of similar cells that perform specific functions.
3. An organ is a structure consisting of two or more tissues that performs a specific function.
4. A body system is composed of a group of organs that function together.

Anatomical Nomenclature (p. 33)

1. Most anatomical terms are derived from Greek or Latin words that provide clues to the meaning of the terms.
2. Familiarity with the basic prefixes and suffixes facilitates learning and remembering anatomical terminology.
3. Anatomy is a foundation science for all of the medical and paramedical fields.

Planes of Reference and Descriptive Terminology (pp. 33–35)

1. The body or organs of the body may be sectioned according to planes of reference. These include a midsagittal plane that runs vertically through a structure, dividing it into right and left halves; a sagittal plane that runs vertically through a structure, dividing it into right and left portions; a coronal (frontal) plane that runs vertically through a structure, dividing it into anterior (front) and posterior (back) portions; and a transverse (cross-sectional) plane that runs horizontally through a structure, dividing it into upper and lower portions.
2. In the anatomical position, the subject is standing with feet parallel, eyes directed forward, and arms at the sides of the body with palms turned forward and fingers pointing downward.
3. Directional terms are used to describe the location of one body part with respect to another part in anatomical position.
4. Clinical procedures include observation (visual inspection), palpation (feeling with firm pressure), percussion (detecting resonating vibrations), auscultation (listening to organ sounds), and reflex-response testing (determining involuntary movements).

Body Regions (pp. 36–40)

1. The head is divided into a facial region, which includes the eyes, nose, and mouth, and a cranial region, which covers and supports the brain.
2. The neck is called the cervical region and functions to support the head and permit movement.
3. The front of the thorax is subdivided into two mammary regions and one sternal region.
4. On either side of the thorax is an axillary fossa and a lateral pectoral region.
5. The abdomen may be divided into nine anatomical regions or four quadrants.
6. Regional names pertaining to the upper extremity include the shoulder, brachium, antebrachium, and manus.
7. Regional names pertaining to the lower extremity include the hip, thigh, leg, and foot.

Body Cavities and Membranes (pp. 40–43)

1. The posterior cavity, which encompasses the cranial and spinal cavities, encloses and protects the brain and spinal cord—the central nervous system.
2. The anterior cavity, which encompasses the thoracic and abdominopelvic cavities, contains the visceral organs.
3. Other body cavities include the oral, nasal, and middle-ear cavities.
4. The body has two principal types of membranes: mucous membranes, which secrete protective mucus, and serous membranes, which line the ventral cavities and cover visceral organs. Serous membranes secrete a lubricating serous fluid.
5. Serous membranes may be categorized as pleural membranes (associated with the lungs), pericardial membranes (associated with the heart), or peritoneal membranes (associated with the abdominal viscera).

Review Activities

Objective Questions

1. Which of the following is *not* a principal chordate characteristic?
   (a) a dorsal hollow nerve cord
   (b) a distinct head, thorax, and abdomen
   (c) a notochord
   (d) pharyngeal pouches
2. Prehensile hands, digits modified for grasping, and large, well-developed brains are structural characteristics of the grouping of animals referred to as
   (a) primates. (c) mammals.
   (b) vertebrates. (d) chordates.
3. Layers or aggregations of similar cells that perform specific functions are called
   (a) organelles. (c) organs.
   (b) tissues. (d) glands.
4. Filtration and maintenance of the volume and the chemical composition of the blood are functions of
   (a) the urinary system.
   (b) the lymphatic system.
   (c) the circulatory system.
   (d) the endocrine system.
5. The cubital fossa is located in
   (a) the thorax.
   (b) the upper extremity.
   (c) the abdomen.
   (d) the lower extremity.
6. Because it is composed of more than one tissue type, the skin is considered
   (a) a composite tissue.
   (b) a system.
   (c) an organ.
   (d) an organism.
7. Which of the following is *not* a reference plane?
   (a) coronal (c) vertical
   (b) transverse (d) sagittal
8. The external genitalia (reproductive organs) are located in
   (a) the popliteal fossa.
   (b) the perineum.
   (c) the hypogastric region.
   (d) the epigastric region.

9. The region of the thoracic cavity between the two pleural cavities is called
   (a) the midventral space.
   (b) the mediastinum.
   (c) the ventral cavity.
   (d) the median cavity.
10. The abdominal region superior to the umbilical region that contains most of the stomach is
   (a) the hypochondriac region.
   (b) the epigastric region.
   (c) the diaphragmatic region.
   (d) the inguinal region.
11. Regarding serous membranes, which of the following word pairs is *incorrect?*
   (a) visceral pleura/lung
   (b) parietal peritoneum/body wall
   (c) mesentery/heart
   (d) parietal pleura/body wall
   (e) visceral peritoneum/intestines
12. The plane of reference that divides the body into anterior and posterior portions is
   (a) sagittal. (c) coronal.
   (b) transverse. (d) cross-sectional.
13. In the anatomical position,
   (a) the arms are extended away from the body.
   (b) the palms of the hands face posteriorly.
   (c) the body is erect and the palms face anteriorly.
   (d) the body is in a fetal position.
14. Listening to sounds that functioning visceral organs make is called
   (a) percussion. (c) audiotation.
   (b) palpation. (d) auscultation.

Essay Questions

1. Discuss the characteristics an animal must possess to be classified as a chordate; a mammal; a human.
2. Describe the relationship of the notochord to the vertebral column, the pharyngeal pouches to the ear, and the dorsal hollow nerve cord to the central nervous system.
3. Which grouping of animals are we most closely related to? What anatomical characteristics do we have in common with them?
4. Identify the levels of complexity that characterize the human body.
5. Outline the systems of the body and identify the major organs that compose each system.
6. What is meant by anatomical position? Why is the anatomical position important in studying anatomy?
7. Define the terms *palpation*, *percussion*, and *auscultation*.
8. Which major region of the body contains each of the following structures or minor regions:
   (a) scapular region,
   (b) brachium,
   (c) popliteal fossa,
   (d) lumbar region,
   (e) cubital fossa,
   (f) hypochondriac region,
   (g) perineum,
   (h) axillary fossa.
9. Diagram the thoracic cavity showing the relative position of the pericardial cavity, pleural cavities, and mediastinum.
10. What is a serous membrane? Explain how the names of the serous membranes differ in accordance with each body cavity.

Critical-Thinking Questions

1. Smooth muscle is a tissue type. Using examples, discuss the level of body organization "below" smooth muscle tissue and the level of body organization "above" smooth muscle tissue. Which of these levels would be studied as microscopic anatomy and which would be studied as gross anatomy?
2. Arteries transport blood away from the heart and veins transport blood toward the heart. Using the terms *anterior*, *proximal*, *distal*, *medial*, and *lateral*, describe blood flow from the heart to the palm of the hand and right thumb and back to the heart. In answering this question, keep the anatomical position in mind. Also, be aware that adding an "ly" to these directional terms changes them from adjectives to adverbs.
3. A 25-year-old man sustained trauma to the left lateral side of his rib cage following a rock-climbing accident. Upon arrival at the hospital emergency room, the ER doctor indicated that the extent of the injury could be determined only through the techniques of inspection, palpation, percussion, and auscultation. Describe how each of these techniques may be used to assess the condition of the thorax and thoracic organs.

THREE

Cytology

Introduction to Cytology 47
- Cells as Functional Units 47
- Cellular Diversity 47

Cellular Chemistry 48
- Elements, Molecules, and Compounds 48
- Water 49
- Electrolytes 49
- Proteins 49
- Carbohydrates 50
- Lipids 50

Cellular Structure 50
- Cell Membrane 52
- Cytoplasm and Organelles 54
- Cell Nucleus 61

Pictured here are James Watson (left) and Francis Crick with their demonstration model of the molecular structure of DNA—the famous double helix. Watson and Crick drew heavily on the research findings of Maurice Wilkins and Rosalind Franklin in solving the DNA puzzle. Their discovery, which ushered in an era of molecular genetics, has come to rank as one of the major scientific breakthroughs of this century.

Cell Cycle 62
- Structure of DNA 63
- Cell Cycle and Cell Division 64

Clinical Considerations 67
- Cellular Adaptations 67
- Trauma to Cells 67
- Medical Genetics 68
- Cancer 68
- Aging 69

Chapter Summary 70

Review Activities 70

Introduction to Cytology

The cell is the fundamental structural and functional unit of the body. Although cells vary widely in size and shape, they have basic structural similarities, and all cells metabolize to stay alive.

Objective 1 Define the terms *cell, metabolism,* and *cytology.*

Objective 2 Using examples, explain how cells differ from one another and how the structure of a cell determines its function.

Cells as Functional Units

Human anatomy is concerned with the structure of the human body and the relationship of its parts. The body is a masterpiece of organization for which the **cell** provides the basis. For this reason, the cell is called the functional unit. As discussed in chapter 2, cellular organization forms tissues, whose organization in turn forms the organs, which in turn form systems. If the organs and systems are to function properly, cells must function properly. Cellular function is referred to as **metabolism.** In order for cells to remain alive and metabolize, certain requirements must be met. Each cell must have access to nutrients and oxygen and be able to eliminate wastes. In addition, a constant, protective environment must be maintained. All of these requirements are achieved through organization.

Cells were first observed more than 300 years ago by the English scientist Robert Hooke. Using his crude microscope to examine a thin slice of cork, he saw a network of cell walls and boxlike cavities. He called them "little boxes or cells," after the barren cubicles of a monastery. As better microscopes were developed, the intriguing architectural details of cellular structure were gradually revealed. The improved lenses resulted in a series of developments that culminated in the formulation of the **cell theory** in 1938 and 1939 by two German biologists, Matthias Schleiden and Theodor Schwann. This theory states that all living organisms are composed of one or more cells and that the cell is the basic unit of structure for all organisms. The work of Schleiden and Schwann laid the groundwork for a new science called **cytology,** which is concerned with the structure and function of cells.

A knowledge of the cellular level of organization is important for understanding the basic body processes of cellular respiration, protein synthesis, mitosis, and meiosis. An understanding of cellular structure gives meaning to the concept of tissue, organ, and system levels of functional body organization. Furthermore, many dysfunctions and diseases of the body originate in the cells. Although cellular structure and function have been investigated for many years, we still have much to learn about cells. The *etiologies*, or causes, of a number of complex diseases are as yet unknown. Scientists are seeking why and how the body ages. The answers will come only through a better understanding of cellular structure and function.

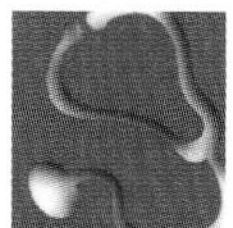

Advancements in microscopy have revolutionized the science of cytology. In a new process called *microtomography,* the capabilities of electron microscopy are combined with those of CT scanning to produce high-magnification, three-dimensional, microtomographic images of living cells. With this technology, living cells can be observed as they move, grow, and divide. The clinical applications are immense, as scientists can observe the response of diseased cells (including cancer cells) to various drug treatments.

Cellular Diversity

It is amazing that from a single cell, the fertilized egg, hundreds of kinds of cells arise, producing the estimated 60 trillion to 100 trillion cells that make up an adult human. Cells vary greatly in size and shape. The smallest cells are visible only through a high-powered microscope. Even the largest, an egg cell (ovum), is barely visible to the unaided eye. The sizes of cells are measured in micrometers (μm)—one micrometer equals 1/1,000th of a millimeter. Using this basis of comparison, an ovum is about 140 μm in diameter and a red blood cell is about 7.5 μm in diameter. The most common type of white blood cell varies in size from 10 to 12 μm in diameter. Although still microscopic, some cells can be extremely long. A nerve cell (neuron), for example, may extend the entire length of a limb and be over a meter long.

Although a typical diagram of a cell depicts it as round or cube-shaped, the shapes of cells are actually highly variable. They can be flat, oval, elongate, stellate, columnar, and so on (fig. 3.1). The shape of a cell is frequently an indication of its function. A disc-shaped red blood cell is adapted to transport oxygen. Thin, flattened cells may be bound together to form selectively permeable membranes. An irregularly shaped cell, such as a neuron, has a tremendous ratio of surface area to volume, which is ideal for receiving and transmitting stimuli.

The surfaces of some cells are smooth, so that substances pass over them easily. Other cells have distinct depressions and elevations on their cell membranes to facilitate absorption. Some cell surfaces support such structures as cilia, flagella, and gelatinous coats, which assist movement and provide adhesion. Regardless of the sizes and shapes of cells, they all have structural modifications that serve functional purposes.

1. Why is the cell considered the basic structural and functional unit of the body?
2. What conditions are necessary for metabolism to occur?
3. Give some examples of structural modifications that allow cells to perform specific functions.

---

metabolism: Gk. *metabole*, change
cytology: L. *cella*, small room; Gk. *logos*, study of
etiology: L. *aitia*, cause; Gk. *logos*, study of

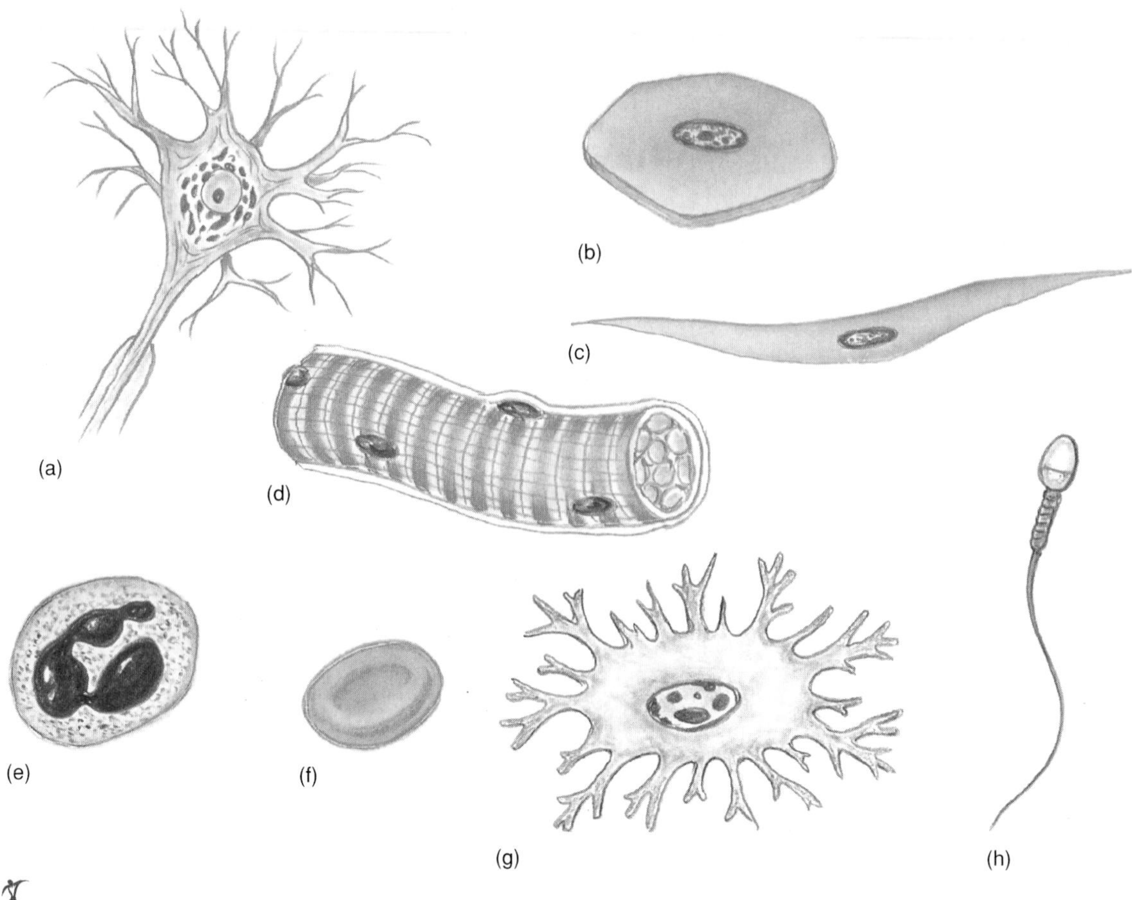

Figure 3.1

Examples of the various shapes of cells within the body. (The cells are not drawn to scale.) (*a*) A neuron (nerve cell) showing the cell body surrounded by numerous dendritic extensions and a portion of the axon extending below, (*b*) a squamous epithelial cell from the lining of a blood vessel, (*c*) a smooth muscle cell from the intestinal wall, (*d*) a skeletal muscle cell, (*e*) a leukocyte (white blood cell), (*f*) an erythrocyte (red blood cell), (*g*) an osteocyte (bone cell), and (*h*) a spermatozoon (sperm cell).

Cellular Chemistry

All tissues and organs are composed of cellular structures that have basically the same chemical components. The most important inorganic substances in the body include water, acids, bases, and salts. The most important organic substances in the body include proteins, carbohydrates, and lipids.

Objective 3 List the common chemical elements found within cells.

Objective 4 Differentiate between inorganic and organic compounds and give examples of each.

Objective 5 Explain the importance of water in maintaining body homeostasis.

Objective 6 Differentiate between proteins, carbohydrates, and lipids.

To understand cellular structure and function, one must have a knowledge of basic cellular and general body chemistry. All of the processes that occur in the body comply with principles of chemistry. Furthermore, many of the dysfunctions of the body have a chemical basis.

Elements, Molecules, and Compounds

Elements are the simplest chemical substances. Four elements compose over 95% of the body's mass. These elements and their percentages of body weight are oxygen (O) 65%, carbon (C) 18%, hydrogen (H) 10%, and nitrogen (N) 3%. Additional common elements found in the body include calcium (Ca), potassium (K), sodium (Na), phosphorus (P), magnesium (Mg), and sulfur (S).

A few elements exist separately in the body, but most are chemically bound to others to form **molecules**. Some molecules are composed of like elements—an oxygen molecule (O_2), for example. Others, such as water (H_2O), are composed of different kinds of elements. **Compounds** are molecules composed of two or more different elements. Thus, the chemical structure of water may be referred to as both a molecule and a compound.

Table 3.1

Compounds Found in Adult Males and Females (Expressed as Percentage of Body Weight)

Substance	Male	Female
Water	62	59
Protein	18	15
Lipid	14	20
Carbohydrates	1	1
Other (electrolytes, nucleic acids)	5	5

Organic compounds are those that are composed of carbon, hydrogen, and oxygen. They include common body substances such as proteins, carbohydrates, and lipids. **Inorganic compounds** generally lack carbon and include common body substances such as water and *electrolytes* (acids, bases, and salts). The percentages of organic and inorganic compounds found in adult males and females are compared in table 3.1.

Water

Water is by far the most abundant compound found within cells and in the extracellular environment. Water generally occurs within the body as a homogeneous mixture of two or more compounds called a *solution*. In this condition, the water is the *solvent*, or the liquid portion of the solution, and the *solutes* are substances dissolved in the solution. Water is an almost universal solvent, meaning that almost all chemical compounds dissolve in it. In addition, it is also used to transport many solutes through the cell membrane of a cell or from one part of the cell to another. Water is also important in maintaining a constant cellular temperature, and thus a constant body temperature, because it absorbs and releases heat slowly. Evaporative cooling (sweating) through the skin also involves water. Another function of water is as a *reactant* in the breakdown (*hydrolysis*) of food material in digestion.

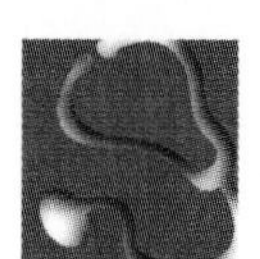

Dehydration is a condition in which fluid loss exceeds fluid intake, with a resultant decrease in the volume of intracellular and extracellular fluids. Rapid dehydration through vomiting, diarrhea, or excessive sweating can lead to serious medical problems by impairing cellular function. Infants are especially vulnerable because their fluid volume is so small. They can die from dehydration resulting from diarrhea within a matter of hours.

solution: L. *solvere*, loosen or dissolve
hydrolysis: Gk. *hydor*, water; *lysis*, a loosening

Table 3.2

Kinds of Electrolytes

	Characteristic	Examples
Acid	Ionizes to release hydrogen ions (H^+)	Carbonic acid, hydrochloric acid, acetic acid, phosphoric acid
Base	Ionizes to release hydroxyl ions (OH^-) that combine with hydrogen ions	Sodium hydroxide, potassium hydroxide, magnesium hydroxide, aluminum hydroxide
Salt	Substance formed by the reaction between an acid and a base	Sodium chloride, aluminum chloride, magnesium sulfate

Electrolytes

Electrolytes are inorganic compounds that break down into ions when dissolved in water, forming a solution capable of conducting electricity. An electrolyte is classified according to the ions it yields when dissolved in water. The three classes of electrolytes are *acids*, *bases*, and *salts*, all of which are important for normal cellular function. The functions of ions include the control of water movement through cells and the maintenance of normal acid-base (*pH*) balance. Ions are also essential for nerve and muscle function, and some ions serve as cofactors that are needed for optimal activity of enzymes. Symptoms of electrolyte imbalances range from muscle cramps and brittle bones to coma and cardiac arrest. The three kinds of electrolytes are summarized in table 3.2.

Proteins

Proteins are nitrogen-containing organic compounds composed of amino acid subunits. An *amino acid* is an organic compound that contains an amino group ($—NH_2$) and a carboxyl group ($—COOH$). There are 20 different types of amino acids that can contribute to a given protein. This variety allows each type of protein to be constructed to function in very specific ways.

Proteins are the most abundant of the organic compounds. They may exist by themselves or be *conjugated* (joined) with other compounds; for example, with nucleic acids (RNA or DNA) to form nucleoproteins, with carbohydrates to form glycoproteins, or with lipids to form lipoproteins.

Proteins may be categorized according to their role in the body as structural or functional. **Structural proteins** contribute significantly to the structure of different tissues. Examples include *collagen* in connective tissue and *keratin* in the epidermis of the skin. **Functional proteins** assume a more active role in

electrolyte: L. *electrum*, amber; Gk. *lysis*, a loosening
acid: L. *acidus*, sour
protein: Gk. *proteios*, of the first quality

Table 3.3

Chemical Substances of Cells: Location and Function

Substance	Location in Cell	Functions
Water	Throughout	Dissolves, suspends, and ionizes materials; helps to regulate temperature
Electrolytes	Throughout	Establish osmotic gradients, pH, and membrane potentials
Proteins	Membranes, cytoskeleton, ribosomes, enzymes	Provide structure, strength, and contractility; catalyze; buffer
Lipids	Membranes, Golgi complex, inclusions	Provide reserve energy source; shape, protect, and insulate
Carbohydrates	Inclusions	Preferred fuel for metabolic activity
Nucleic acids		
DNA	Nucleus, in chromosomes and genes	Controls cell activity
RNA	Nucleolus, cytoplasm	Transmits genetic information; transports amino acids
Trace materials		
Vitamins	Cytoplasm, nucleus	Work with enzymes in metabolism
Minerals	Cytoplasm, nucleus	Essential for normal metabolism; involved in osmotic balance; add strength; buffer

the body, exerting some form of control of metabolism. Examples include *enzymes* and *antibodies*. Many *hormones* belong to a specialized group of messenger and regulator proteins produced by endocrine glands. Cellular growth, repair, and division depend on the availability of functional proteins. Proteins, under certain conditions, may even be metabolized to supply cellular energy.

Carbohydrates

Carbohydrates are organic compounds that contain carbon, hydrogen, and oxygen, with a 2:1 ratio of hydrogen to oxygen. Carbohydrates include *monosaccharides*, or simple sugars, *disaccharides*, or double sugars, and *polysaccharides*, or long-chained sugars. Carbohydrates are the body's most readily available energy source and also may be used as a fuel reserve. Excessive carbohydrate intake is converted to *glycogen* (animal starch) or to *fat* for storage in adipose tissue.

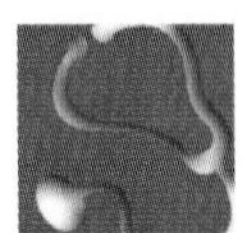

If a person is deprived of food, the body uses the glycogen and fat reserves first and then metabolizes the protein within the cells. The gradual destruction of cellular protein accounts for the lethargy, extreme emaciation, and ultimate death of starvation victims.

Lipids

Lipids are a third group of important organic compounds found in cells. They are insoluble in water and include both fats and fat-related substances, such as phospholipids and cholesterol. Fats are important in building cell parts and supplying metabolic energy. They also protect and insulate various parts of the body. Phospholipids and protein molecules make up the cell membrane and play an important role in regulating which substances enter or leave a cell.

Lipids, like carbohydrates, are composed of carbon, hydrogen, and oxygen. Lipids, however, contain a smaller proportion of oxygen than do carbohydrates.

The locations and functions of inorganic and organic substances within cells are summarized in table 3.3.

1. List the four most abundant elements in the body and state their relative percentages of body weight.
2. Define *molecule* and *compound*. What are the two kinds of compounds that exist in the body? On what basis are they distinguished?
3. List some of the functions of water relative to cells and define *solvent* and *solute*.
4. Discuss the importance of electrolytes in maintaining homeostasis within cells.
5. Define *protein* and describe how proteins function within cells. Explain how proteins differ from carbohydrates and lipids.

Cellular Structure

The cell membrane separates the interior of a cell from the extracellular environment. The passage of substances into and out of the cell is regulated by the cell membrane. Most of the metabolic activities of a cell occur within the cytoplasmic organelles. The nucleus functions in protein synthesis and cell reproduction.

hormone: Gk. *hormon*, setting in motion
lipid: Gk. *lipos*, fat

Table 3.4
Cellular Components: Structure and Function

Component	Structure	Functions
Cell (plasma) membrane	Membrane composed of a double layer of phospholipids in which proteins are embedded	Gives form to cell and controls passage of materials in and out of cell
Cytoplasm	Fluid, jellylike substance between the cell membrane and the nucleus in which organelles are suspended	Serves as matrix substance in which chemical reactions occur
Endoplasmic reticulum	System of interconnected membrane-forming canals and tubules	Provides supporting framework within cytoplasm; transports materials and provides attachment for ribosomes
Ribosomes	Granular particles composed of protein and RNA	Synthesize proteins
Golgi complex	Cluster of flattened membranous sacs	Synthesizes carbohydrates and packages molecules for secretion; secretes lipids and glycoproteins
Mitochondria	Double-walled membranous sacs with folded inner partitions	Release energy from food molecules and transform energy into usable ATP
Lysosomes	Single-walled membranous sacs	Digest foreign molecules and worn and damaged cells
Peroxisomes	Spherical membranous vesicles	Contain enzymes that detoxify harmful molecules and break down hydrogen peroxide
Centrosome	Nonmembranous mass of two rodlike centrioles	Helps to organize spindle fibers and distribute chromosomes during mitosis
Vacuoles	Membranous sacs	Store and release various substances within the cytoplasm
Fibrils and microtubules	Thin, hollow tubes	Support cytoplasm and transport materials within the cytoplasm
Cilia and flagella	Minute cytoplasmic projections that extend from the cell surface	Move particles along cell surface or move the cell
Nuclear membrane (envelope)	Double-walled membrane composed of protein and lipid molecules that surrounds the nucleus	Supports nucleus and controls passage of materials between nucleus and cytoplasm
Nucleolus	Dense nonmembranous mass composed of protein and RNA molecules	Forms ribosomes
Chromatin	Fibrous strands composed of protein and DNA molecules	Contains genetic code that determines which proteins (especially enzymes) will be manufactured by the cell

Objective 7 Describe the components of a cell.

Objective 8 Describe the composition and structure of the cell membrane and relate its structure to the functions it performs.

Objective 9 Distinguish between passive and active transport and describe the different ways in which each is accomplished.

Objective 10 Describe the structure and function of the endoplasmic reticulum, ribosomes, Golgi complex, lysosomes, and mitochondria.

Objective 11 Describe the structure and function of the nucleus.

As the basic functional unit of the body, the cell is a highly organized molecular factory. As previously discussed, cells come in a great variety of shapes and sizes. This variation, which is also apparent in subcellular structures (organelles), reflects the diversity of function of different cells in the body. All cells, however, have certain features in common—a cell membrane, for example, and most of the other structures listed in table 3.4. Thus, although no one cell can be considered "typical," the general structure of cells can be indicated by a single illustration (fig. 3.2).

For descriptive purposes, a cell can be divided into three principal parts:

1. **Cell (plasma) membrane.** The selectively permeable cell membrane gives form to the cell. It controls the passage of molecules into and out of the cell and separates the cell's internal structures from the extracellular environment.
2. **Cytoplasm** and **organelles.** The cytoplasm (*si'tŏ-plaz"em*) is the cellular material between the nucleus and the cell membrane. Organelles (*or"gă-nelz'*) are the specialized structures within the cell that perform specific functions.
3. **Nucleus.** The nucleus (*noo'kle-us*) is the large spheroid or oval body usually located near the center of the cell. It contains the DNA, or genetic material, that directs the

plasma: Gk. *plasma*, to form or mold
nucleus: L. *nucleus*, kernel or nut

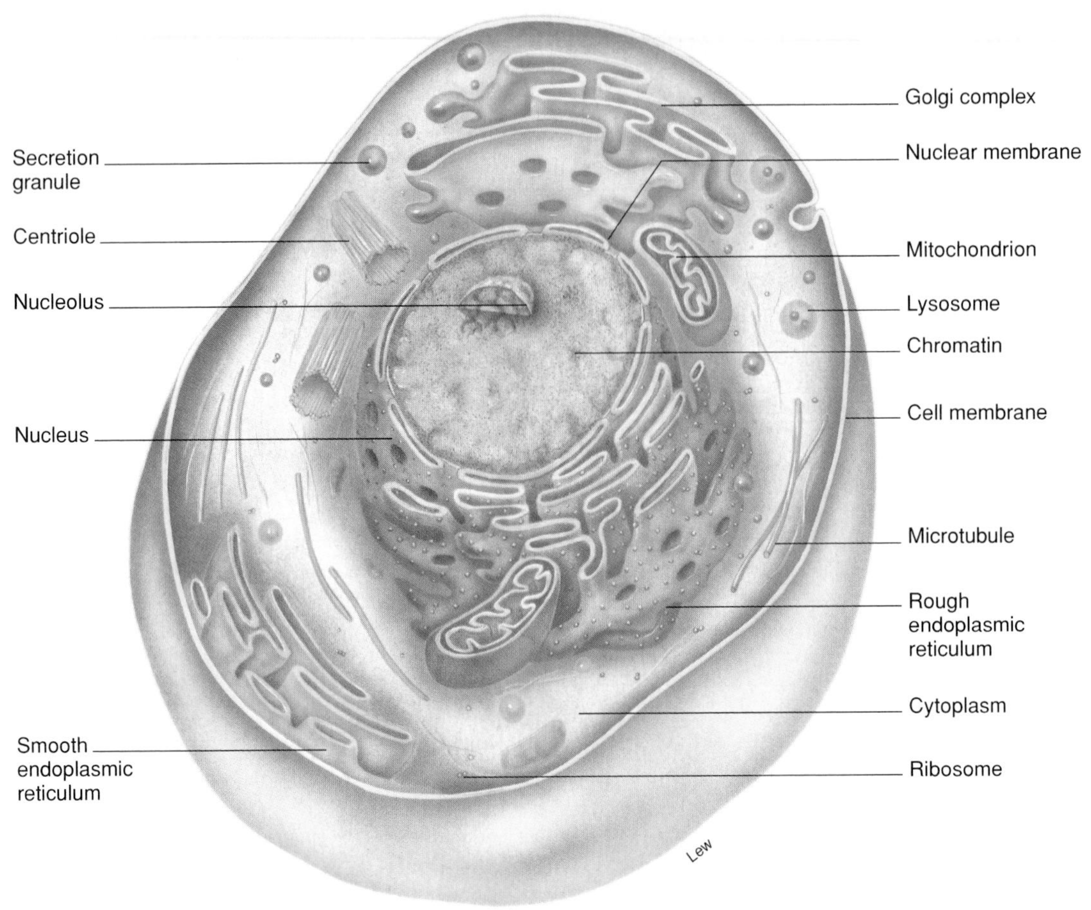

Figure 3.2

Structural features of a generalized cell.

activities of the cell. Within the nucleus, one or more dense bodies called **nucleoli** (singular, *nucleolus*) may be seen. The nucleolus contains the subunits for ribosomes, the structures that serve as sites for protein synthesis.

Cell Membrane

The extremely thin **cell (plasma) membrane** is composed primarily of phospholipid and protein molecules. Its thickness ranges from 65 to 100 angstroms (Å); that is, it is less than a millionth of an inch thick. The structure of the cell membrane is not fully understood, but most cytologists believe that it consists of a double layer of phospholipids in which larger globular proteins are embedded (fig. 3.3). The proteins are free to move within the membrane. As a result, they are not uniformly distributed, but rather form a constantly changing mosaic. Minute openings, or pores, ranging between 7 and 10 Å in diameter extend through the membrane.

The two most important functions of the cell membrane are to enclose the components of the cell and to regulate the passage of substances into and out of the cell. A highly selective exchange of substances occurs across the membrane boundary, involving several types of passive and active processes. The various kinds of movement across a cell membrane are summarized in table 3.5 and illustrated in figure 3.4.

The permeability of the cell membrane depends on the following factors:

1. **Structure of the cell membrane.** Although cell membranes of all cells are composed of phospholipids, there is evidence that their thickness and structural arrangement—both of which could affect permeability—vary considerably.
2. **Size of the molecules.** Macromolecules, such as certain proteins, are not allowed into the cell. Water and amino acids are small molecules and can readily pass through the cell membrane.
3. **Ionic charge.** The protein portion of the cell membrane carries a positive or negative ionic charge. Ions with an opposite charge are attracted to and readily pass through the membrane, whereas those with a similar charge are repelled.
4. **Lipid solubility.** Substances that are easily dissolved in lipids pass into the cell with no problem, since a portion of the cell membrane is composed of lipid material.
5. **Presence of carrier molecules.** Specialized carrier molecules within the cell membrane are capable of attracting and transporting substances across the membrane, regardless of size, ionic charge, or lipid solubility.

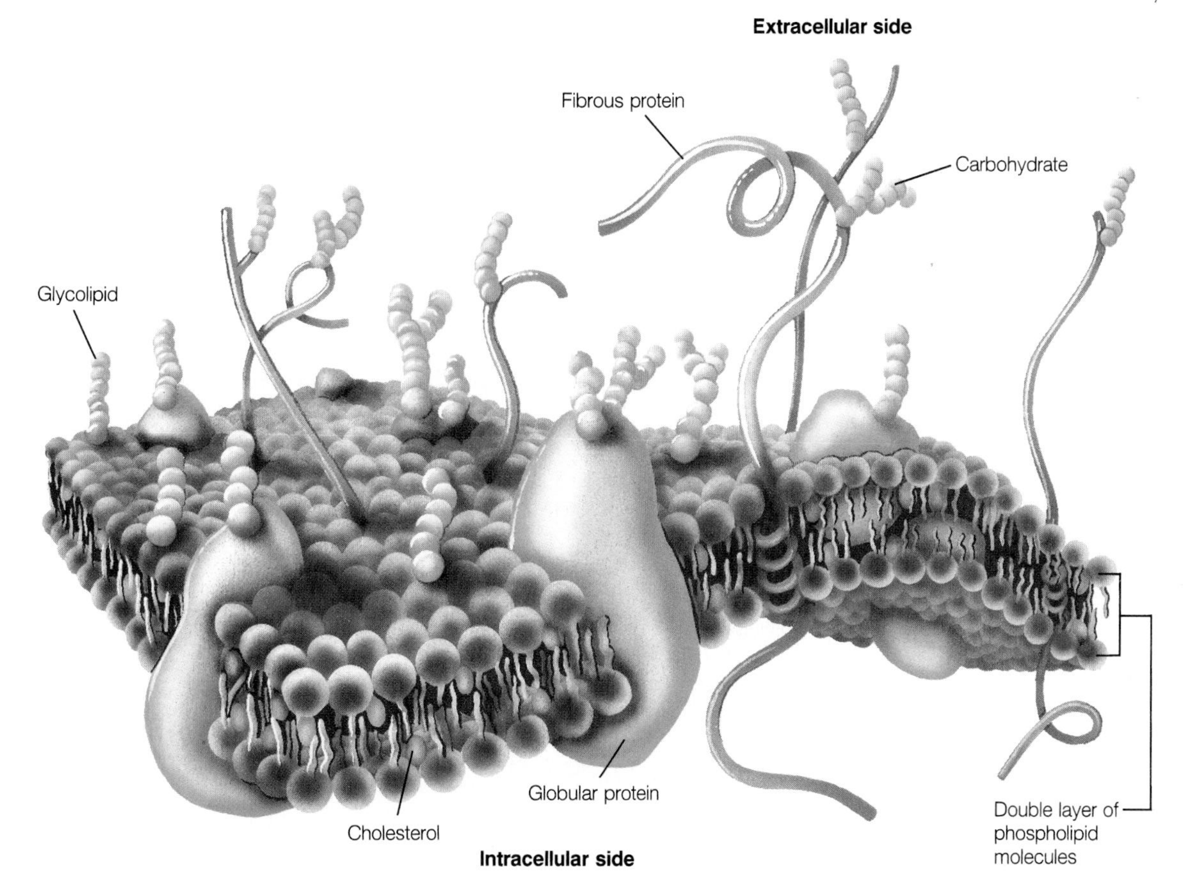

Figure 3.3

The cell membrane consists of a double layer of phospholipids, with the phosphates (shown by spheres) oriented outward and the hydrophobic hydrocarbons (wavy lines) oriented toward the center. Proteins may completely or partially span the membrane. Carbohydrates are attached to the outer surface.

Table 3.5
Movement through Cell Membranes

Processes	Characteristics	Energy Source	Example
Simple diffusion	Tendency of molecules to move from regions of high concentration to regions of lower concentration	Molecular motion	Respiratory gases are exchanged in lungs
Facilitated diffusion	Diffusion of molecules through semipermeable membrane with the aid of membrane carriers	Carrier energy and molecular motion	Glucose enters cell attached to carrier protein
Osmosis	Passive movement of water molecules through semipermeable membrane from regions of high water concentration to regions of lower water concentration	Molecular motion	Water moves through cell wall to maintain constant turgidity of cell
Filtration	Movement of molecules from regions of high pressure to regions of lower pressure as a result of hydrostatic pressure	Blood pressure	Wastes are removed within kidneys
Active transport	Carrier-mediated transport of solutes from regions of their low concentration to regions of their higher concentration (against their concentration gradient)	Cellular energy (ATP)	Glucose and amino acids move through membranes
Endocytosis			
Pinocytosis	Process in which membrane engulfs minute droplets of fluid from surroundings	Cellular energy	Membrane forms vacuoles containing solute and solvent
Phagocytosis	Process in which membrane engulfs solid particles from surroundings	Cellular energy	White blood cell membrane engulfs bacterial cell
Exocytosis	Release of molecules from cell as vesicles rupture	Cellular energy	Hormones and mucus are secreted out of cell; neurotransmitter is released at synapse

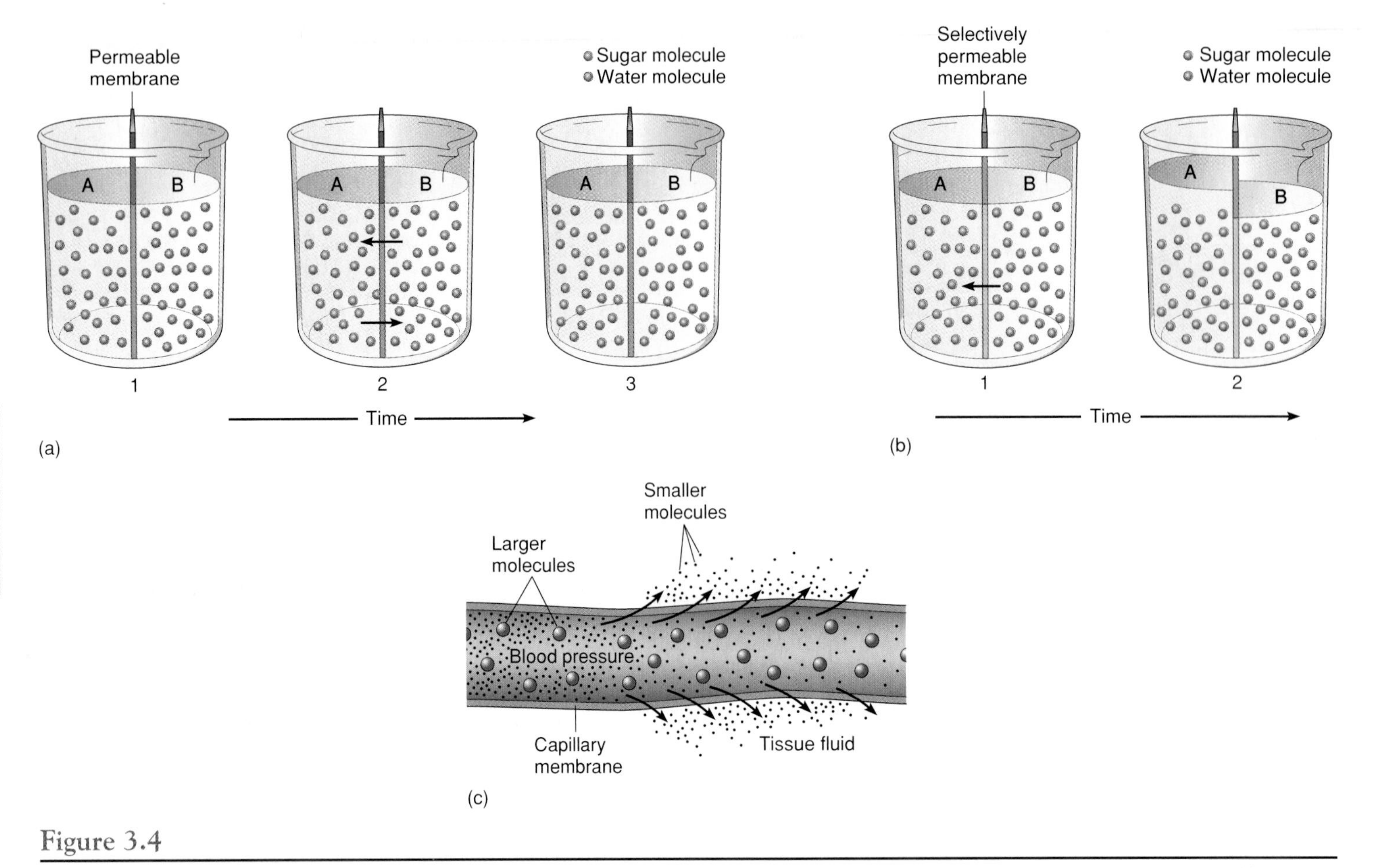

Figure 3.4

Examples of various kinds of movements through membranes. (*a*) Sugar molecules diffuse from compartment A to compartment B until equilibrium is achieved in 3. (*b*) Osmosis occurs as a selectively permeable membrane allows only water to diffuse through the membrane between compartments A and B, causing the level of the liquid to rise in A. (*c*) Filtration occurs as small molecules are forced through a membrane by blood pressure, leaving the larger molecules behind.

6. **Pressure differences.** The pressure difference on the two sides of a cell membrane may greatly aid movement of molecules either into or out of a cell.

Cell membranes of certain cells are highly specialized to facilitate specific functions (fig. 3.5). The columnar cells lining the lumen (hollow portion) of the intestinal tract, for example, have numerous fine projections, or **microvilli** *(mi″kro-vil′i)*, that aid in the absorptive process of digestion. A single columnar cell may have as many as 3,000 microvilli on the exposed portion of the cell membrane, and a square millimeter of surface area may contain over 200 million microvilli.

Certain sensory organs contain cells that have specialized cell membranes. The photoreceptors, or light-responding rods and cones of the eye, have double-layered, disc-shaped membranes called **sacs.** These structures contain pigments associated with vision. Within the spiral organ (organ of Corti) in the inner ear are **hair cells.** These tactile (touch) receptors are stimulated through mechanical vibration. Hair cells are so named because of the fine hairlike processes that extend from their cell membranes.

microvilli: Gk. *mikros*, small; *villus*, tuft of hair

Cytoplasm and Organelles

Cytoplasm refers to the material located within the cell membrane but outside of the nucleus. The material within the nucleus is frequently called the **nucleoplasm.** The term **protoplasm** is sometimes used to refer to the cytoplasm and nucleoplasm collectively.

When observed through an electron microscope (fig. 3.6), distinct cellular components called **organelles** can be seen in the highly structured cytoplasm. The matrix of the cytoplasm is a jellylike substance that is 80% to 90% water. The organelles and inorganic *colloid substances* (suspended particles) are dispersed throughout the cytoplasm. Colloid substances have similar ionic charges that space them uniformly.

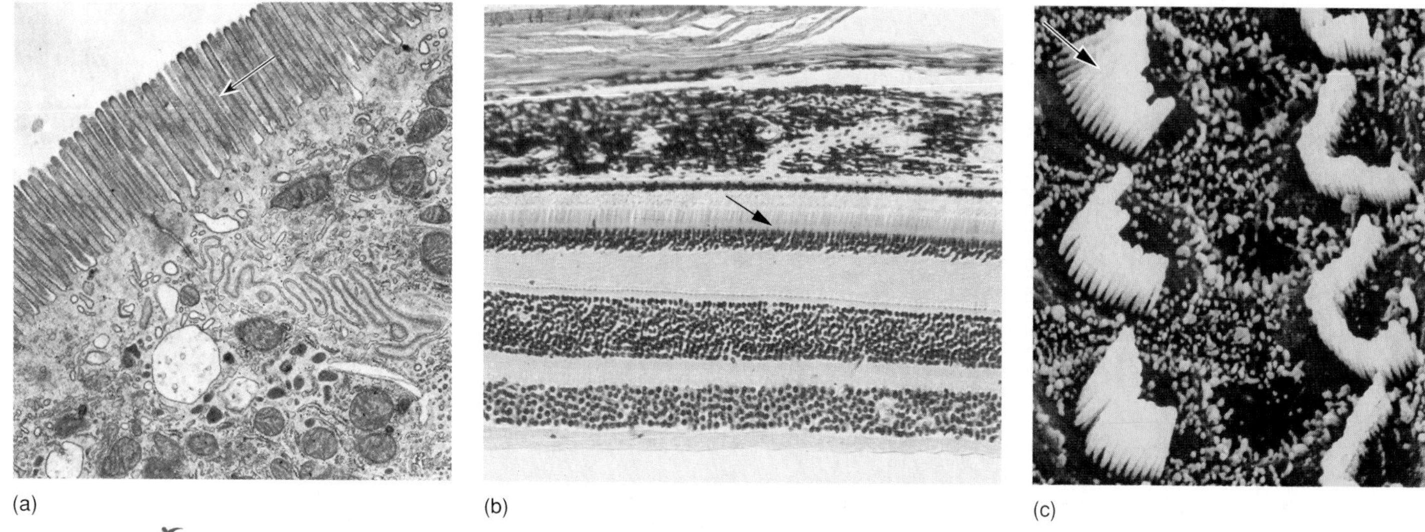

Figure 3.5

Specialized features of cell membranes, as indicated by arrow. (*a*) Microvilli of the cells that line the lumen (hollow portion) of the small intestine for absorption (640×), (*b*) photoreceptors within the eye for sight (160×), and (*c*) hair cells within the spiral organ (organ of Corti) of the inner ear for hearing.

Metabolic activity occurs within the organelles of the cytoplasm. Specific roles such as heat production, cellular maintenance, repair, storage, and protein synthesis are carried out within the organelles.

The structure and functions of each of the major organelles are discussed in the following paragraphs and summarized in table 3.4.

Endoplasmic Reticulum

Often abbreviated **ER,** the endoplasmic reticulum (*en″do-plaz′mik rĕ-tik′yŭ-lum*) is widely distributed throughout the cytoplasm as a complex network of interconnected membranes (fig. 3.7). Although the name sounds complicated, *endoplasmic* simply means "within the plasm" (cytoplasm of the cell) and *reticulum* means "network." Between the interconnected membranes are minute spaces, or **cisterna,** that are connected at one end to the cell membranes. The tubules may also be connected to other organelles or to the outer nuclear envelope.

The ER provides a pathway for transportation of substances within the cell and a storage area for synthesized molecules. There are two distinct varieties, either of which may predominate in a given cell:

1. a **rough,** or **granular, endoplasmic reticulum** (*rough ER*), characterized by numerous small granules called **ribosomes** that are attached to the outer surface of the membranous wall; and
2. a **smooth endoplasmic reticulum** (*smooth ER*) that lacks ribosomes.

The membranous wall of rough ER provides a site for protein synthesis within ribosomes. Smooth ER manufactures certain

Figure 3.6

A transmission electron microscope (TEM) like this one is used to observe and photograph organelles within the cytoplasm of a cell.

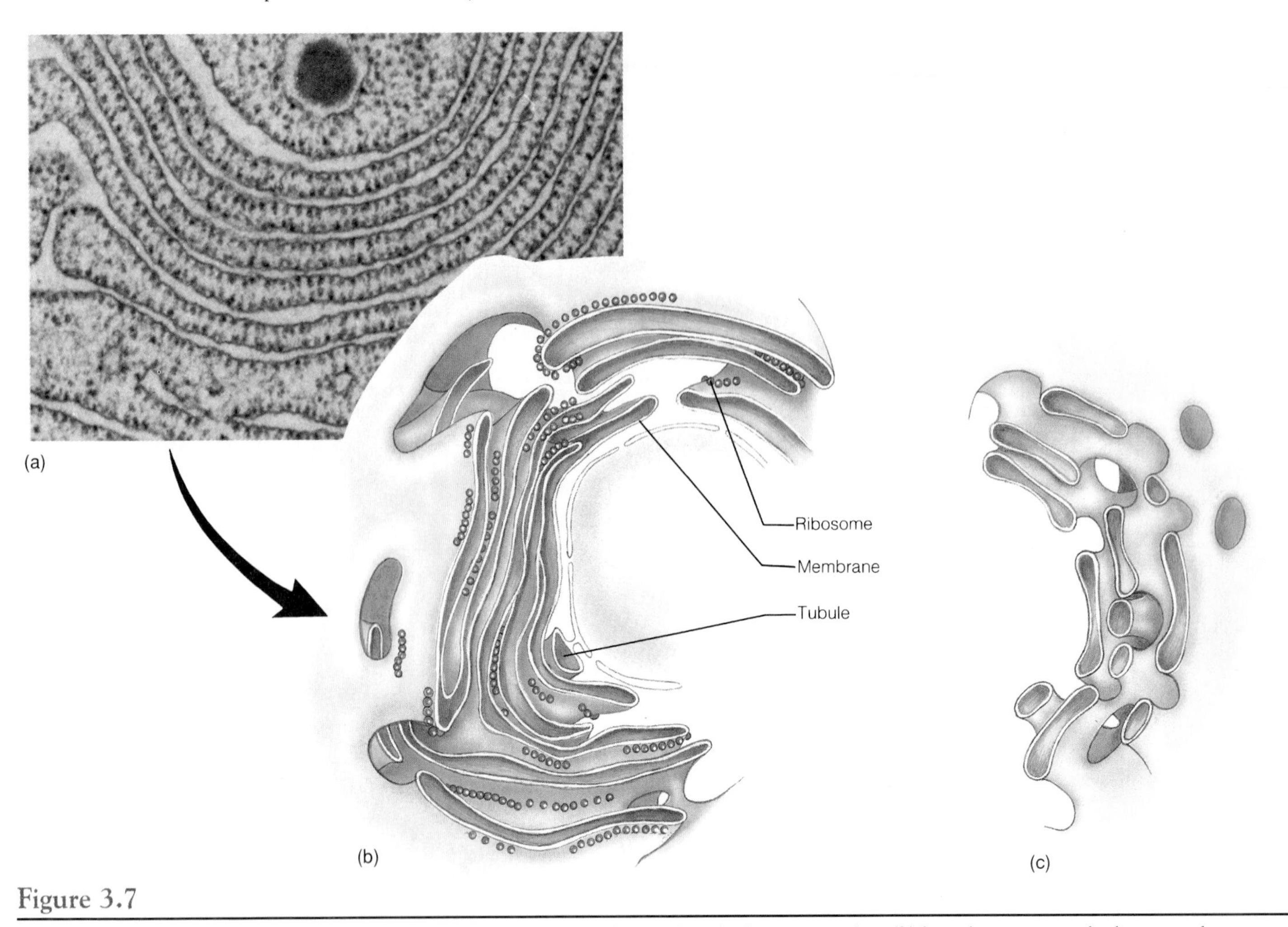

Figure 3.7

(*a*) An electron micrograph of endoplasmic reticulum (about 100,000×). Rough endoplasmic reticulum (*b*) has ribosomes attached to its surface, whereas smooth endoplasmic reticulum (*c*) lacks ribosomes.

lipid molecules. Also, enzymes within the smooth ER of liver cells inactivate or detoxify a variety of chemicals.

A person who repeatedly uses certain drugs, such as alcohol or phenobarbital, develops a tolerance to them, so that greater quantities are required to achieve the effect they had originally. The cytological explanation for this is that repeated use causes the smooth endoplasmic reticulum to proliferate in an effort to detoxify these drugs and protect the cell. With increased amounts of smooth endoplasmic reticulum, cells can handle an increased concentration of drugs.

Ribosomes

Ribosomes (*ri'bŏ-somz*) may occur as free particles suspended within the cytoplasm, or they may be attached to the membranous wall of the rough endoplasmic reticulum. Ribosomes are small, granular organelles (fig. 3.7) composed of protein and RNA molecules. They synthesize protein molecules that may be used to build cell structures or to function as enzymes. Some of the proteins synthesized by ribosomes are secreted by the cell to be used elsewhere in the body.

Golgi Complex

The Golgi (*gol'je*) complex (*Golgi apparatus*) consists of several tiny membranous sacs located near the nucleus (fig. 3.8*a*).

The Golgi complex is involved in the synthesis of carbohydrates and cellular secretions. As large carbohydrate molecules are synthesized, they combine with proteins to form compounds called *glycoproteins* that accumulate in the channels of the Golgi complex. When a critical volume is reached, the vesicles break off from the complex and are carried to the cell membrane and released as a secretion (fig. 3.8*b*). Once the vesicle has fused with the cell membrane, it ruptures to release its contents, thus completing the process known as *exocytosis*.

The Golgi complex is prominent in cells of certain secretory organs of the digestive system, including the pancreas and the salivary glands. Pancreatic cells, for example, produce digestive enzymes that are packaged in the Golgi complex and secreted as droplets that flow into the pancreatic duct and are transported to the gastrointestinal (GI) tract.

Golgi complex: from Camillo Golgi, Italian histologist, 1843–1926

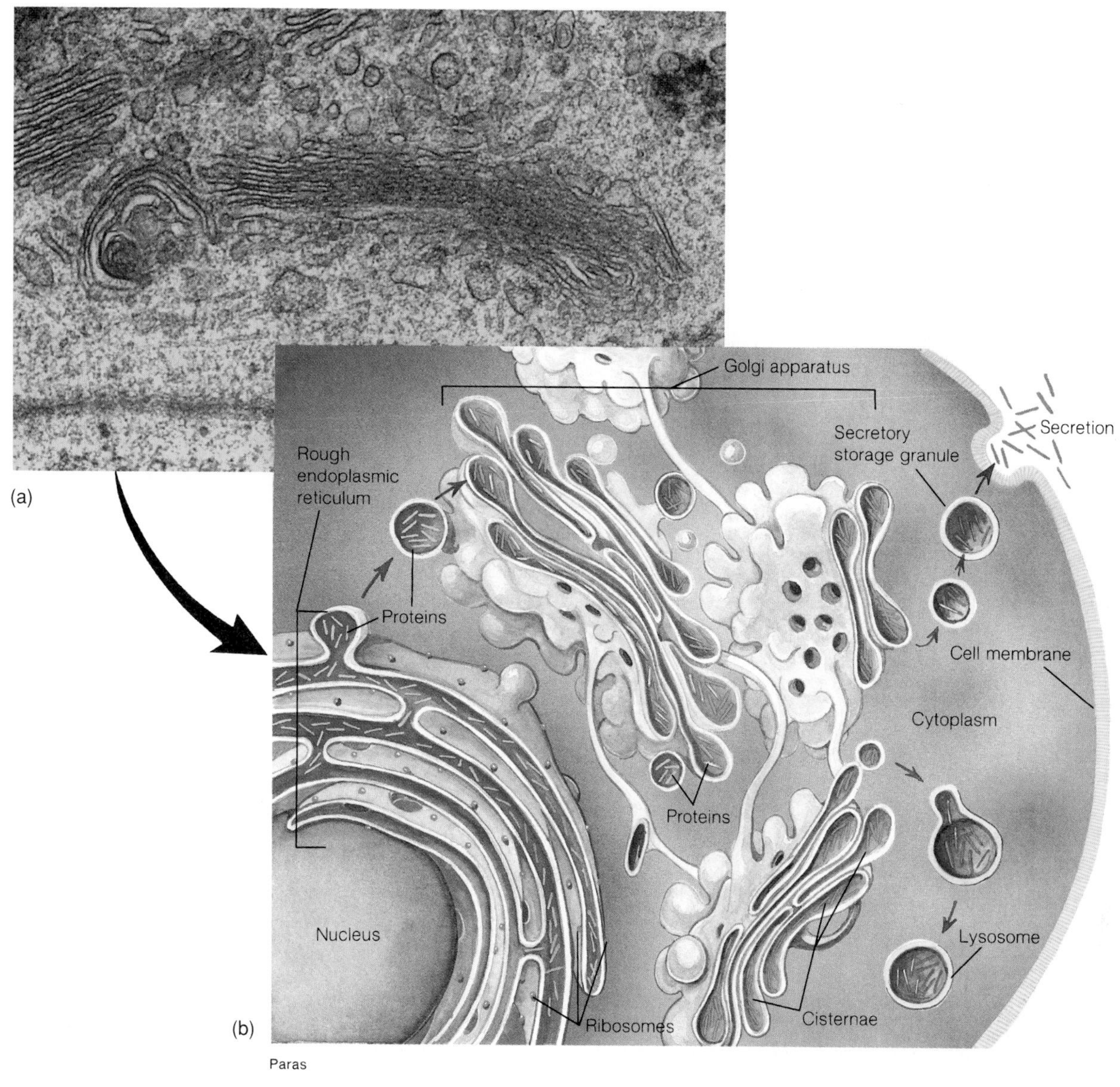

Figure 3.8

(*a*) An electron micrograph of a Golgi complex (apparatus). Notice the formation of vesicles at the ends of some of the flattened sacs. (*b*) An illustration of the processing of proteins by the rough endoplasmic reticulum and Golgi complex.

Mitochondria

Mitochondria (*mi″tŏ-kon′dre-ă*) are double-membraned saclike organelles. They are found in all cells in the body, with the exception of mature red blood cells. The outer mitochondrial membrane is smooth, whereas the inner membrane is arranged in intricate folds called **cristae** (*kris′te*) (fig. 3.9). The cristae create a enormous surface area for chemical reactions.

Mitochondria vary in size and shape. They can migrate through the cytoplasm and can reproduce themselves by budding or cleavage. They are often called the "powerhouses" of cells because of their role in producing metabolic energy. Enzymes connected to the cristae control the chemical reactions that form ATP. Metabolically active cells, such as muscle cells, liver cells, and kidney cells, have a large number of mitochondria because of their high energy requirements.

The darker color of some cuts of meat (a chicken thigh, for example, as compared to a breast) is due to larger amounts of *myoglobin,* a pigmented compound in muscle tissue that acts to store oxygen. Mitochondria are likewise more abundant in red meat. Both mitochondria and myoglobin are important for the high level of metabolic activity in red muscle tissue.

mitochondrion: Gk. *mitos*, a thread; *chondros*, lump, grain
cristae: L. *crista*, crest

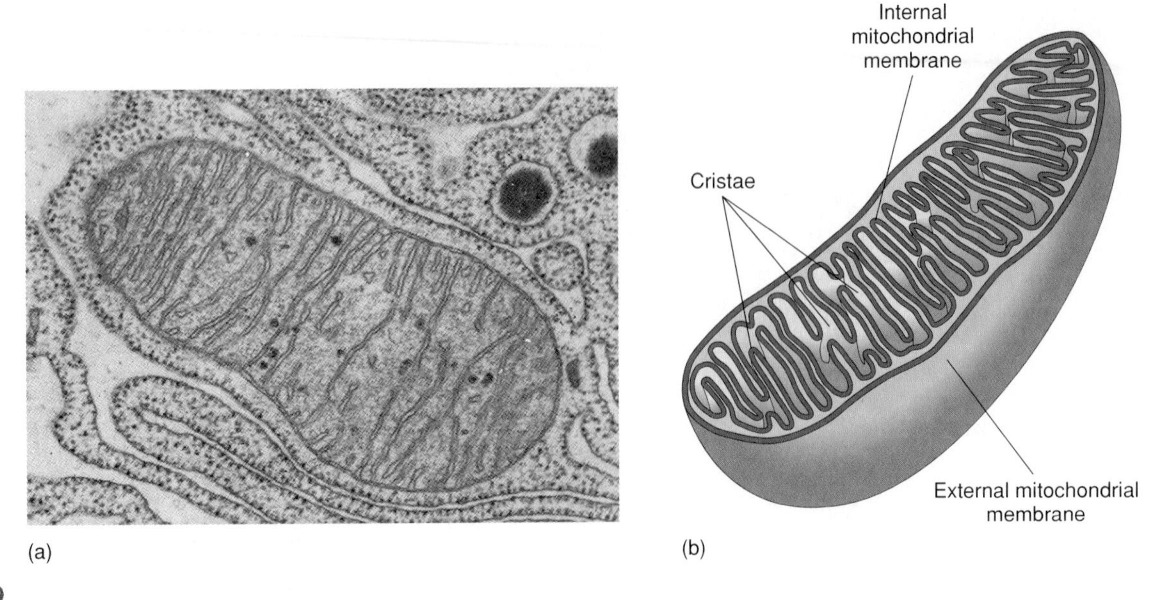

Figure 3.9

(*a*) An electron micrograph of a mitochondrion (about 40,000×). The external mitochondrial membrane and the infoldings (cristae) of the internal mitochondrial membrane are clearly seen. (*b*) A diagram of a mitochondrion.

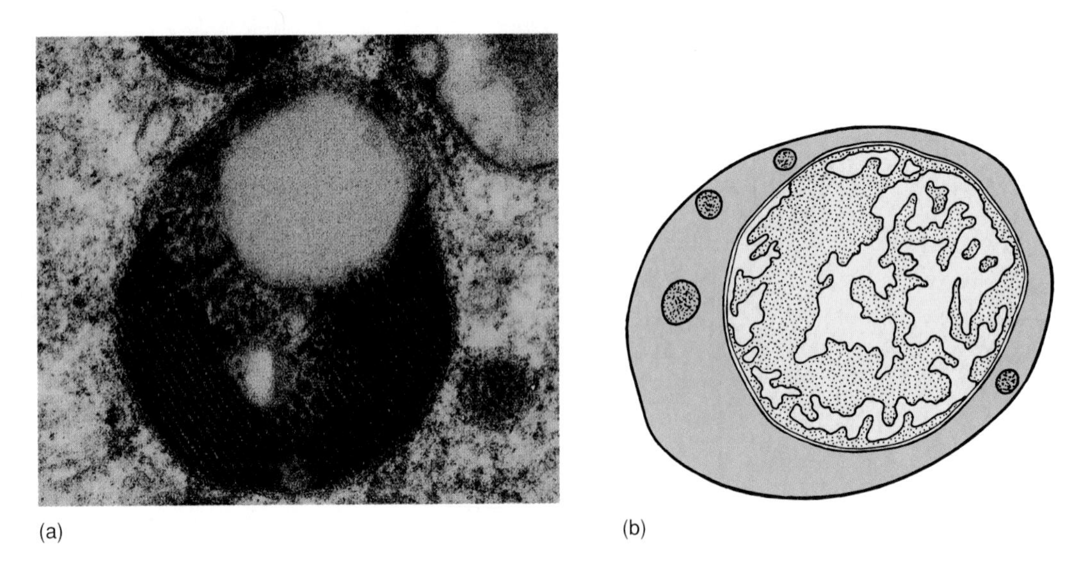

Figure 3.10

(*a*) An electron micrograph of a lysosome (about 30,000×). (*b*) A diagram of a lysosome.

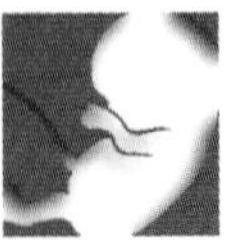

Because mitochondria are contained within ova (egg cells) but not within the heads of sperm cells, all of the mitochondria in a fertilized egg are derived from the mother. As cells divide during the developmental process, the mitochondria likewise replicate themselves; thus, all of the mitochondria in a fetus are genetically identical to those in the original ovum. This accounts for a unique form of inheritance that is passed only from mother to child. A rare cause of blindness—*Leber's hereditary optic neuropathy*—and perhaps some genetically based neuromuscular disorders, are believed to be inherited in this manner.

Lysosomes

Lysosomes (*li'sŏ-sōmz*) vary in appearance from granular bodies to small vesicles to membranous spheres (fig. 3.10). They are scattered throughout the cytoplasm. Powerful digestive enzymes enclosed within lysosomes are capable of breaking down protein and carbohydrate molecules. White blood cells contain large numbers of lysosomes and are said to be *phagocytic*, meaning that they will ingest, kill, and digest bacteria through the enzymatic activity of their lysosomes.

Leber's hereditary optic neuropathy: from Theodor Leber, German ophthalmologist, 1840–1917

lysosome: Gk. *lysis*, a loosening; *somo*, body
phagocytic: Gk. *phagein*, to eat; *kytos*, a cell

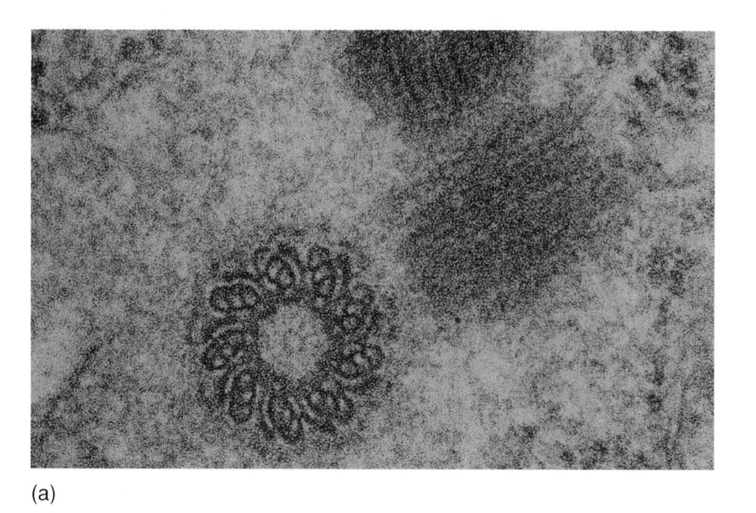
(a)

(b)

Figure 3.11

(*a*) An electron micrograph of centrioles in a centrosome (about 14,200×). (*b*) A diagram showing that the centrioles are positioned at right angles to each other.

The normal atrophy, or decrease in size, of the uterus following the birth of a baby is due to lysosomal digestive activity. Likewise, the secretions of lysosomes are responsible for the regression of the mammary tissue of the breasts after the weaning of an infant.

Lysosomes also digest worn-out cell parts, and if their membranes are ruptured they destroy the entire cell within which they reside. For this reason, lysosomes are frequently called "suicide packets."

Several diseases arise from abnormalities in lysosome function. The painful inflammation of *rheumatoid arthritis,* for example, occurs when enzymes from lysosomes are released into the joint capsule and initiate digestion of the surrounding tissue.

Lysosomes were not discovered until the early 1950s, but their existence and functions had been predicted before these organelles were actually observed in cells. Such was not the case with other organelles, whose structures generally were observed and described before their functional roles in the cell were understood.

Peroxisomes

Peroxisomes *(pĕ-roks'ĭ-sōmz)* are membranous sacs that resemble lysosomes structurally and they too contain enzymes. Peroxisomes occur in most cells but are particularly abundant in the kidney and liver. Some of the enzymes in peroxisomes promote the breakdown of fats, producing hydrogen peroxide—a highly toxic substance—as a by-product. Hydrogen peroxide is an important compound in white blood cells, which phagocytize diseased or worn-out cells. Peroxisomes also contain the enzyme *catalase*, which breaks down excess hydrogen peroxide into water and oxygen so that there is no toxic effect on other organelles within the cytoplasm.

Centrosome and Centrioles

The **centrosome** (central body) is a nonmembranous spherical mass positioned near the nucleus. Within the centrosome, a pair of rodlike structures called **centrioles** *(sen'tre-ōlz)* (fig. 3.11) are positioned at right angles to each other. The wall of each centriole is composed of nine evenly spaced bundles, and each bundle contains three microtubules.

Centrosomes are found only in those cells that can divide. During the mitotic (replication) process, the centrioles move away from each other and take positions on either side of the nucleus. They are then involved in the distribution of the chromosomes during cellular reproduction. Mature muscle and nerve cells lack centrosomes, and thus cannot divide.

Vacuoles

Vacuoles *(vak'yoo-ōlz)* are membranous sacs of various sizes that usually function as storage chambers. They are formed when a portion of the cell membrane invaginates and pinches off during endocytosis. Vacuolation is initiated either by *pinocytosis (pin"ŏ-si-to'sis)*, in which cells take in minute droplets of liquid through the cell membrane, or by *phagocytosis (fag"ŏ-si-to'sis)*, in which the cell membrane engulfs solid particles (fig. 3.12). Vacuoles may contain liquid or solid materials that were previously outside the cell.

Fibrils and Microtubules

Both fibrils and microtubules are found throughout the cytoplasm. The fibrils are minute rodlike structures, whereas the microtubules are fine, threadlike tubular structures of varying lengths (fig. 3.13). Both provide the cell with support by forming a type of cytoskeleton. Specialized fibrils called myofilaments are

vacuole: L. *vacuus*, empty

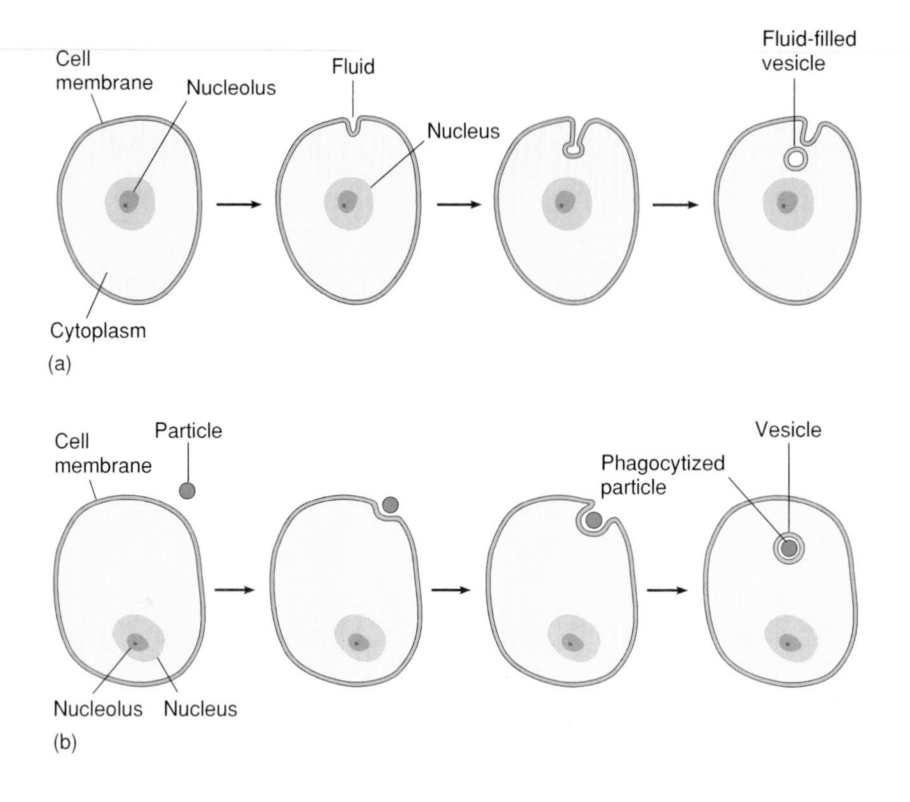

Figure 3.12

Pinocytosis and phagocytosis compared. (*a*) During pinocytosis, the cell takes in a minute droplet of fluid from its surroundings. (*b*) During phagocytosis, a solid particle is engulfed and ingested through the cell membrane.

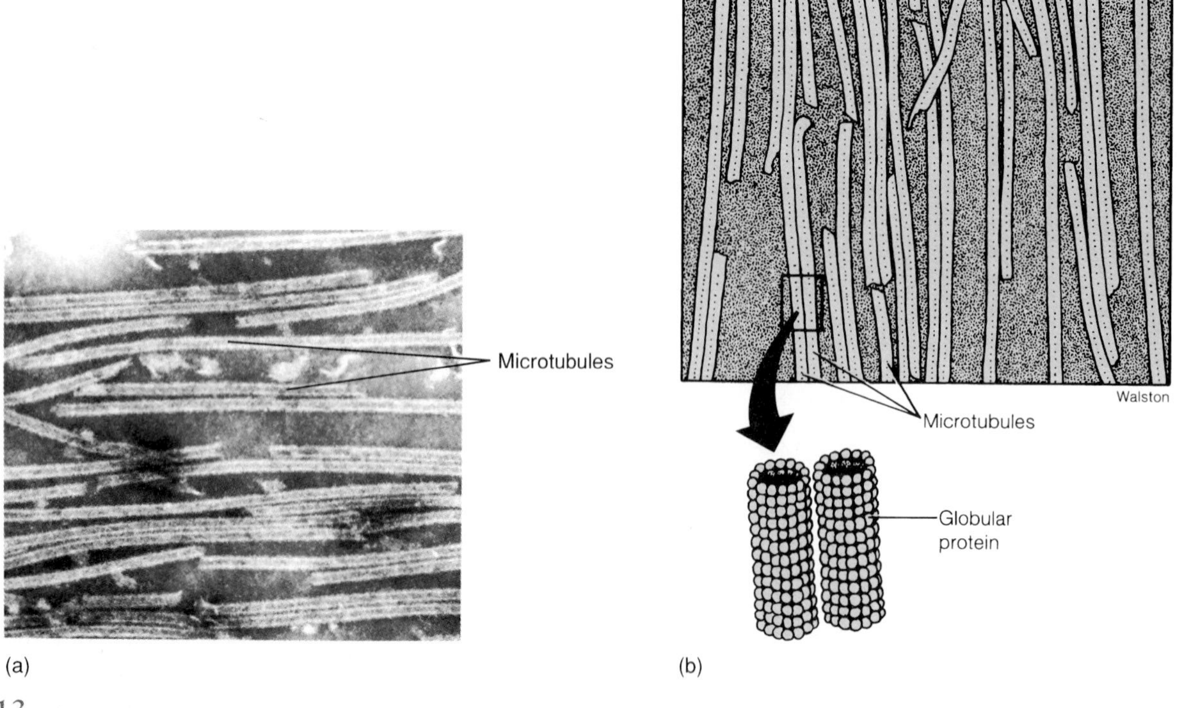

Figure 3.13

(*a*) An electron micrograph showing microtubules forming a type of cytoskeleton (about 30,000×). (*b*) A diagram of a microtubule showing the precisely arranged globular proteins of which they are composed.

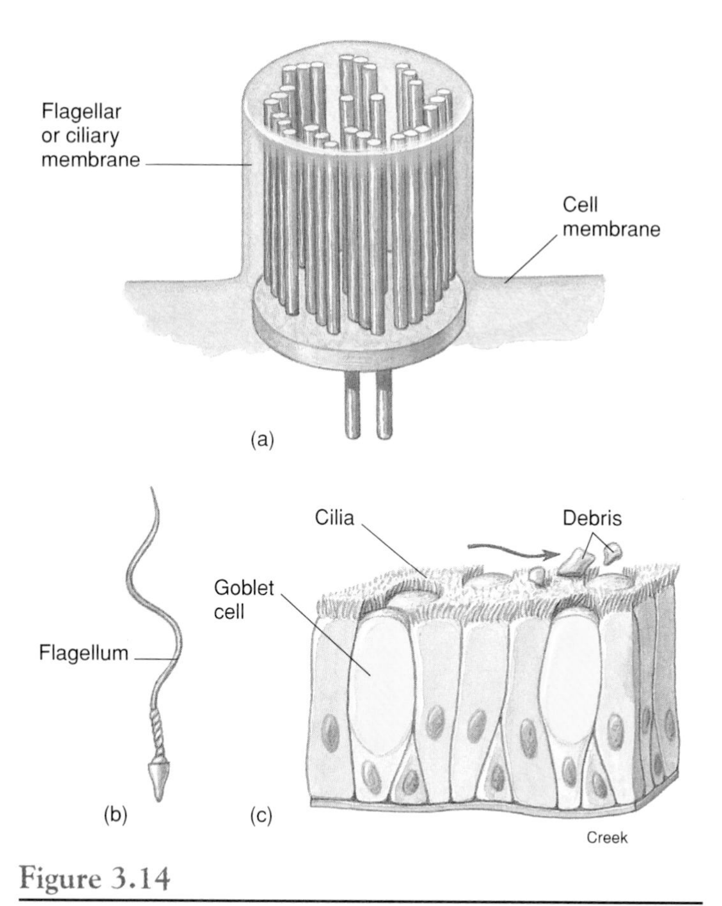

Figure 3.14

(*a*) Cilia and flagella are similar in the structural arrangement of their microtubules. (*b*) A sperm cell (spermatozoon) has a single flagellum for propulsion. (*c*) Cilia produce a wavelike motion to move particles toward the outside of the body.

Figure 3.15

An electron micrograph of ciliated cells that line the lumen of the uterine tube (640×).

particularly abundant in muscle cells, where they aid in the contraction of these cells. Microtubules are also involved in the transportation of macromolecules throughout the cytoplasm. They are especially abundant in the cells of endocrine organs, where they aid the movement of hormones to be secreted into the blood. Microtubules in certain cells provide flexible support for cilia and flagella.

Cilia and Flagella

Although cilia and flagella appear to be extensions of the cell membrane, they are actually cytoplasmic projections from the interior of the cell. These projections contain cytoplasm and supportive microtubules bounded by the cell membrane (fig. 3.14). Cilia and flagella should not be confused with microvilli or with stereocilia, both of which are specializations of cell membranes.

Cilia (sil'e-ă) are numerous short projections from the exposed border of certain cells (fig. 3.15). Ciliated cells are interspersed with mucus-secreting **goblet cells.** There is always a film of mucus on the free surface of ciliated cells. Ciliated cells line the lumina (hollow portions) of sections of the respiratory and reproductive tracts. The function of the cilia is to move the mucus and any adherent material toward the exterior of the body.

Flagella (flă-jel'ă) are similar to cilia in basic microtubular structure (see fig. 3.14), but they are somewhat longer than cilia. The only example of a flagellated cell in humans is the sperm cell, which uses the single structure for locomotion.

Cell Nucleus

The spherical **nucleus** is usually located near the center of the cell (fig. 3.16). It is the largest structure of the cell and contains the genetic material that determines cellular structure and controls cellular activity.

Most cells contain a single nucleus. Certain cells, however, such as skeletal muscle cells, are multinucleated. The long skeletal muscle fibers contain so much cytoplasm that several governing centers are necessary. Other cells, such as mature red blood cells, lack nuclei. These cells are limited to certain types of chemical activities and are not capable of cell division.

The nucleus is enclosed by a bilayered **nuclear membrane** *(nuclear envelope)* (fig. 3.16). The narrow space between the inner and outer layers of the nuclear membrane is called the **nucleolemma cisterna** *(sis-ter'na)*. Minute **nuclear pores** are located along the nuclear membrane. These openings are lined with proteins that act as selective gates, allowing certain

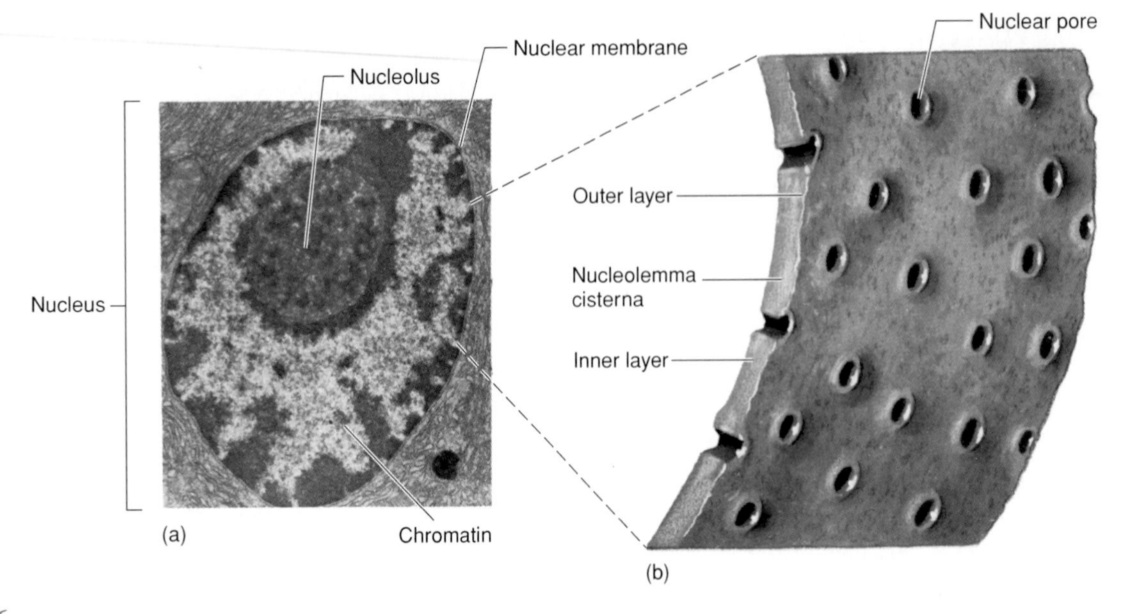

Figure 3.16

(*a*) An electron micrograph of the cell nucleus (about 20,000×). The nucleus contains a nucleolus and masses of chromatin. (*b*) The double-layered nuclear membrane has pores that permit substances to pass between nucleus and cytoplasm.

molecules, such as proteins, RNA, and protein-RNA complexes, to move between the nucleoplasm and the cytoplasm.

Two important structures within the nucleoplasm of the nucleus determine what a cell will look like and what functions it will perform:

1. **Nucleoli.** Nucleoli *(noo-kle'ŏ-li)* are small, nonmembranous spherical bodies composed largely of protein and RNA. It is thought that they function in the production of ribosomes. As ribosomes are formed, they migrate through the nuclear membrane into the cytoplasm.
2. **Chromatin.** Chromatin *(kro'mă-tin)* is a coiled, threadlike mass. It is the genetic material of the cell and consists principally of protein and DNA molecules. When a cell begins to divide, the chromatin shortens and thickens into rod-shaped structures called *chromosomes (kro'mŏ-sōmz)* (fig. 3.17). Each chromosome carries thousands of genes that determine the structure and function of a cell.

1. Describe the composition and specializations of the cell membrane. Discuss the importance of the selective permeability of the cell membrane.
2. Describe the various kinds of movements across the cell membrane. Which are passive and which are active?
3. Describe the structure and function of the following cytoplasmic organelles: rough endoplasmic reticulum, Golgi complex, lysosomes, and mitochondria.
4. Distinguish between the nucleus and nucleoli.
5. Distinguish between chromatin and chromosomes.

Cell Cycle

A cell cycle consists of growth, synthesis, and mitosis. Growth is the increase in cellular mass resulting from metabolism. Synthesis is the production of DNA and RNA to regulate cellular activity. Mitosis is the division of the nucleus and cytoplasm of a cell that results in the formation of two daughter cells.

Objective 12 Describe the structure of a DNA molecule.

Objective 13 List the stages of mitosis and discuss the events of each stage.

Objective 14 Discuss the significance of mitosis.

Cellular replication is one of the principal concepts of biology. Through the process of cellular division called **mitosis** *(mi-to'sis)*, a multicellular organism can develop and be maintained. Mitosis enables body growth and the replacement of damaged, diseased, or worn-out cells. The process ensures that each daughter cell will have the same number and kind of chromosomes as the original parent cell.

In an average healthy adult, over 100 billion cells will die and be mitotically replaced during a 24-hour period. This represents a replacement of about 2% of the body mass each day. Some of the most mitotically active sites are the outer layer of skin, the internal lining of the digestive tract, and the liver.

Before a cell can divide, it must first duplicate its chromosomes so that the genetic traits can be passed to the succeeding

mitosis: Gk. *mitos*, thread

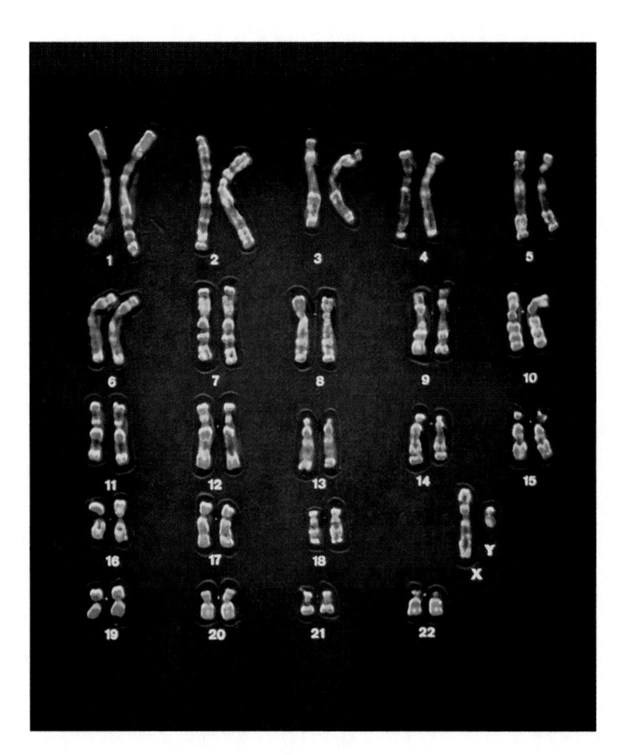

Figure 3.17

A color-enhanced light micrograph showing the full complement of male chromosomes arranged in numbered homologous pairs.

generations of cells. A chromosome consists of a coiled **deoxyribonucleic acid (DNA) molecule** that is complexed with protein. As mentioned previously, chromosomes are formed by the shortening and thickening of the chromatin within the nucleus when the cell begins to divide, at which time they are clearly visible under the compound microscope. There are 23 pairs of chromosomes in each human body (somatic) cell and approximately 20,000 genes are positioned on each chromosome.

Chromosomes are of varied lengths and shapes—some twisted, some rodlike. During mitosis, they shorten and condense, each pair assuming a characteristic shape (fig. 3.17). On the chromosome is a small, buttonlike body called a **centromere** to which are attached the **spindle fibers** that direct the chromosome toward the pole of the cell during mitosis.

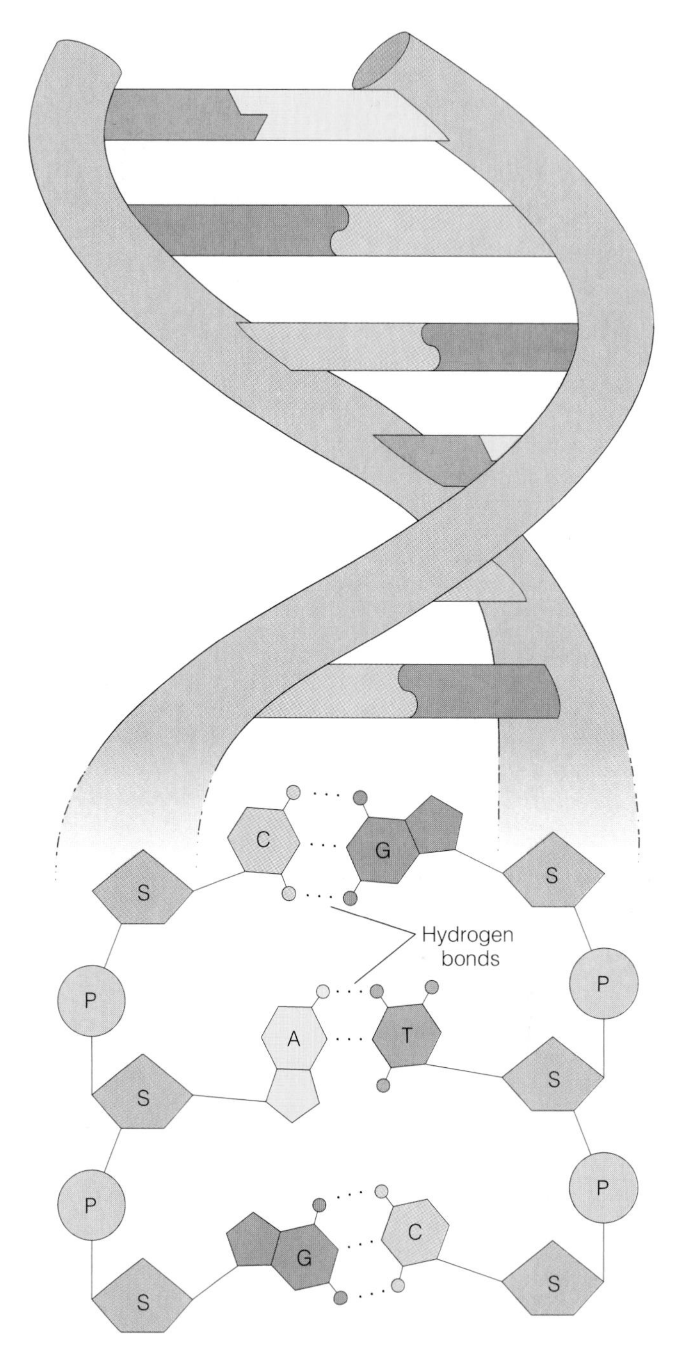

Figure 3.18

The double-helix structure of DNA. Each strand of the helix contains only four kinds of organic bases (A, T, C, and G).

Structure of DNA

The DNA molecule is frequently called a *double helix* because of its resemblance to a spiral ladder (fig. 3.18). The sides of the DNA molecule are formed by alternating units of the sugar *deoxyribose* and phosphoric acid called the *phosphate group*. The rungs of the molecule are composed of pairs of *nitrogenous bases*. The ends of each nitrogenous base are attached to the deoxyribose-phosphate units. There are only four types of nitrogenous bases in a DNA molecule: adenine (A), thymine (T), cytosine (C), and guanine (G).

The basic structural units of the DNA molecule are called **nucleotides.** Each nucleotide consists of a molecule of deoxyribose, a phosphate group, and one of the four nitrogenous bases. Thus, there is a nucleotide type for each of the four bases.

The pairing of the nitrogenous bases of the nucleotides is highly specific. The molecular configuration of each base is such that adenine always pairs with thymine and cytosine always pairs with guanine. The hydrogen bonds between these bases are relatively weak and can be easily split during cellular division (fig. 3.19). During division, the sequence of bases along the sides of the DNA molecule serves as a template that determines the sequence along each new strand.

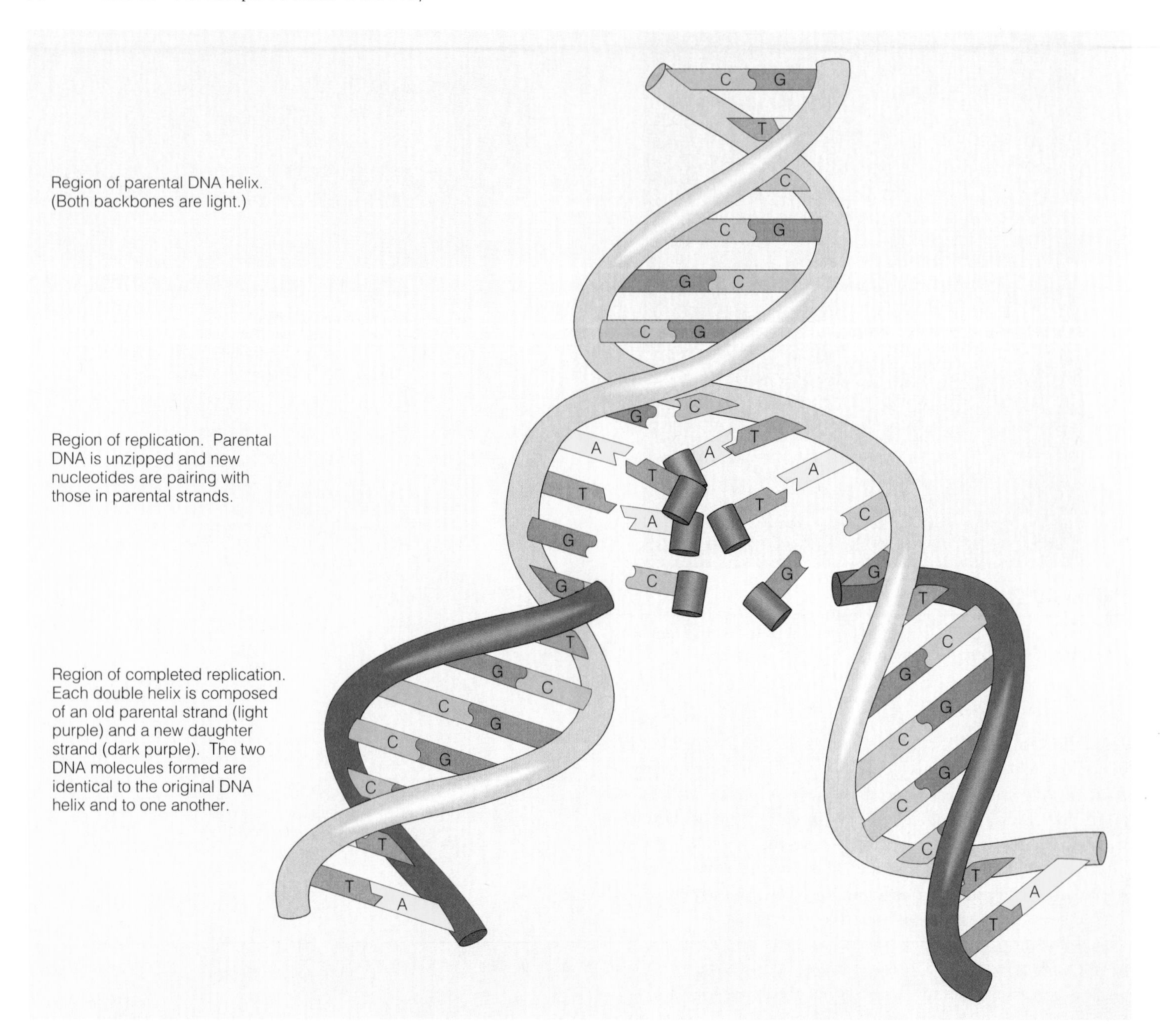

Region of parental DNA helix. (Both backbones are light.)
Region of replication. Parental DNA is unzipped and new nucleotides are pairing with those in parental strands.
Region of completed replication. Each double helix is composed of an old parental strand (light purple) and a new daughter strand (dark purple). The two DNA molecules formed are identical to the original DNA helix and to one another.

James Watson and Francis Crick, who devised the double-helix model (see chapter opening photograph), first described their vision of DNA in 1953, in the journal *Nature* (see table 1.2). The closing sentence of their brief article (a mere 900 words) is a marvel of humility and restraint: "It has not escaped our notice that the specific pairing we have postulated . . . immediately suggests a possible copying mechanism for the genetic material."

Cell Cycle and Cell Division

A **cell cycle** is the series of changes that a cell undergoes from the time it is formed until it has completed a division and reproduced itself. **Interphase** is the first period of the cycle, from cell formation to the start of cell division (fig. 3.20). During interphase, the cell grows, carries on metabolic activities, and prepares itself for division.

Interphase is divided into **G1, S,** and **G2 phases.** During the G1 (first growth) phase, the cell grows rapidly and is meta-

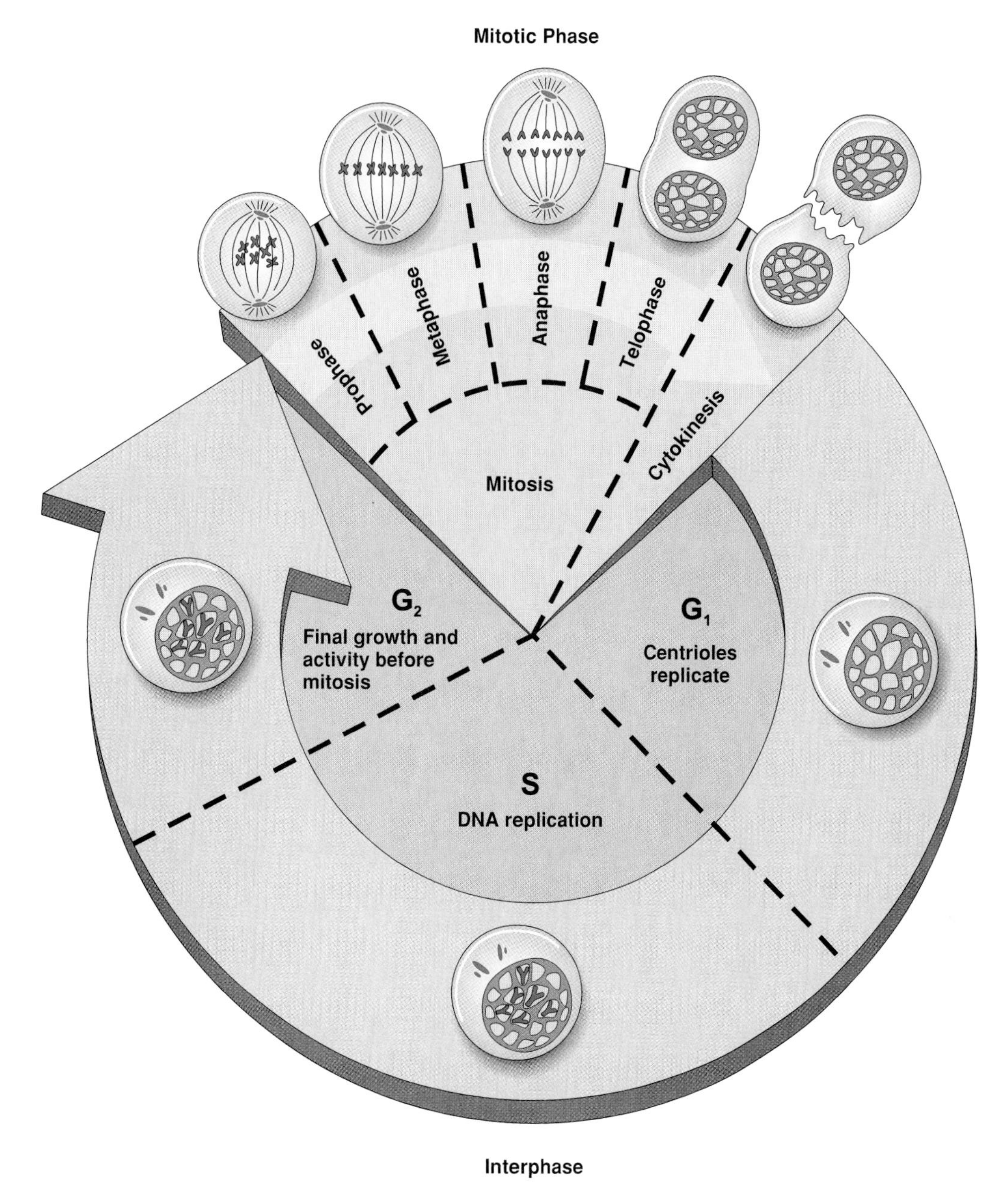

Figure 3.20

Interphase and the mitotic phase are the two principal divisions of the cell cycle. During the mitotic phase, nuclear division is followed by cytoplasmic division and the formation of two daughter cells.

bolically active. The duration of G1 varies considerably in different types of cells. It may last only hours in cells that have rapid division rates, or it may be a matter of days or even years for other cells. At the end of G1, the centrioles replicate in preparation for their role in cell division. During the S (synthetic) phase, the DNA in the nucleus of the cell replicates, so that the two future cells will receive identical copies of the genetic material. During the G2 (second growth) phase, the enzymes and other proteins needed for the division process are synthesized, and the cell continues to grow.

The actual division of a cell is referred to as the **mitotic phase,** or simply **M phase** (fig. 3.20). The mitotic phase is further divided into mitosis and cytokinesis. **Mitosis** is the period of a cell cycle during which there is nuclear division and the duplicated chromosomes separate to form two genetically identical daughter nuclei. The process of mitosis takes place in four successive stages, each stage passing into the next without sharp structural distinctions. These stages are *prophase*, *metaphase*, *anaphase*, and *telophase* (fig. 3.21). **Cytokinesis** *(si″to-kĭ-ne′sis)* is division of the cytoplasm, which takes place during telophase.

---

cytokinesis: Gk. *kytos*, a hollow; *kinesis*, movement

(a) Interphase

- The chromosomes are in an extended form and seen as chromatin in the electron microscope.
- The nucleus is visible.

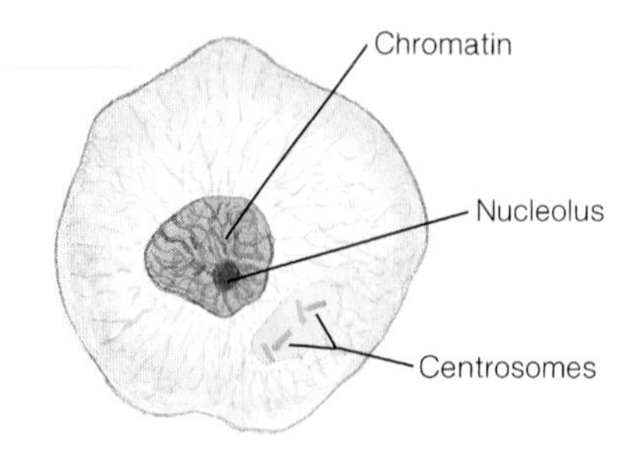

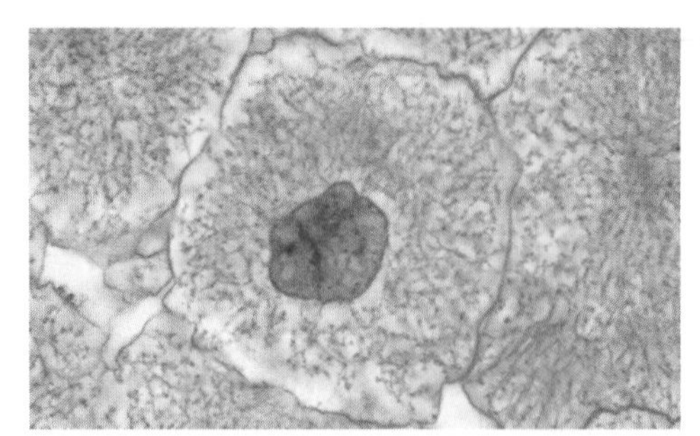

(b) Prophase

- The chromosomes are seen to consist of two chromatids joined by a centromere.
- The centrioles move apart toward opposite poles of the cell.
- Spindle fibers are produced and extended from each centrosome.
- The nuclear membrane starts to disappear.
- The nucleolus is no longer visible.

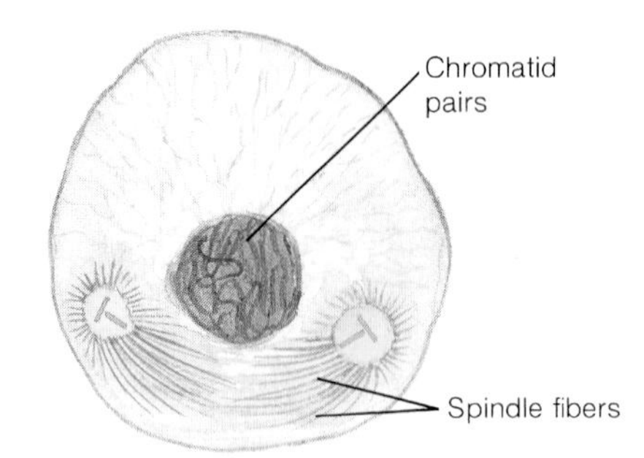

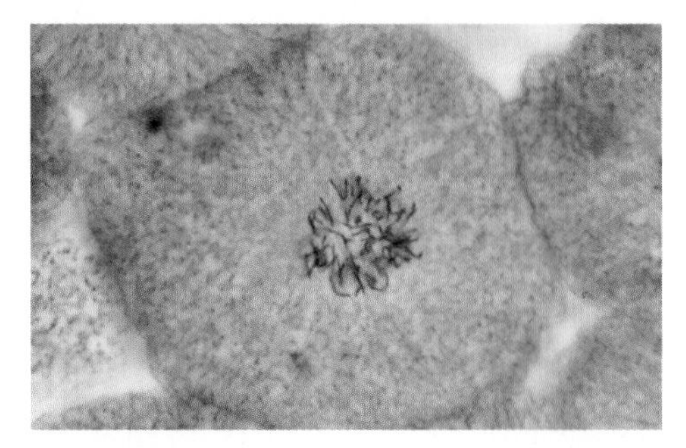

(c) Metaphase

- The chromosomes are lined up at the equator of the cell.
- The spindle fibers from each centriole are attached to the centromeres of the chromosomes.
- The nuclear membrane has disappeared.

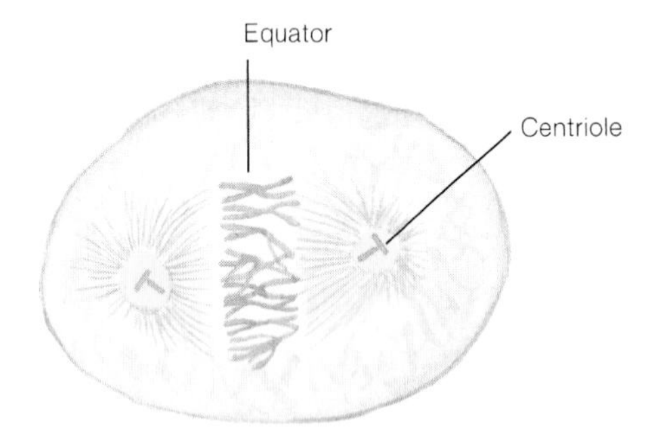

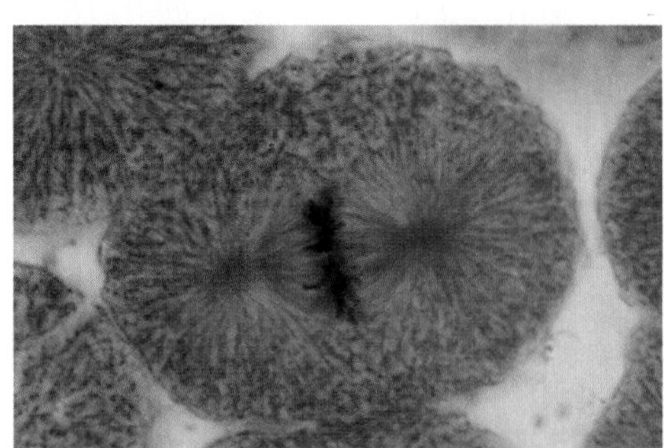

(d) Anaphase

- The centromeres split, and the sister chromatids separate as each is pulled to an opposite pole.

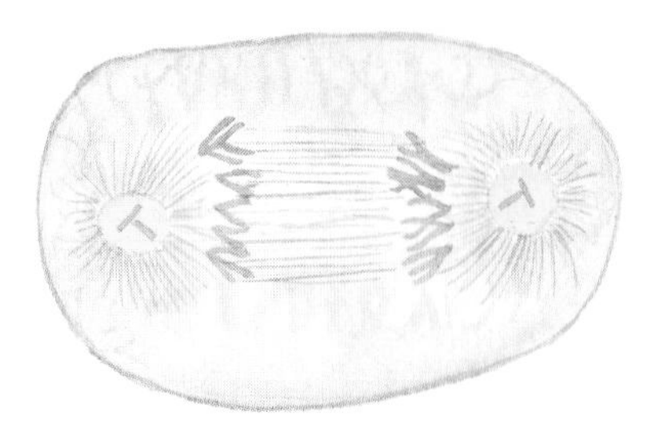

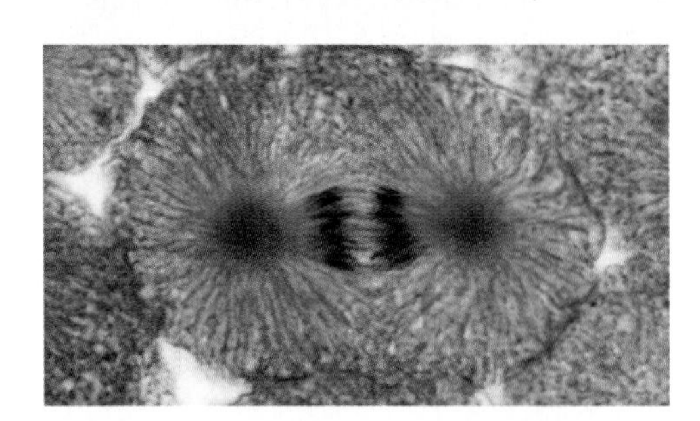

(e) Telophase

- The chromosomes become longer, thinner and less distance.
- New nuclear membranes form.
- The nucleolus reappears.
- Cell division is nearly complete.

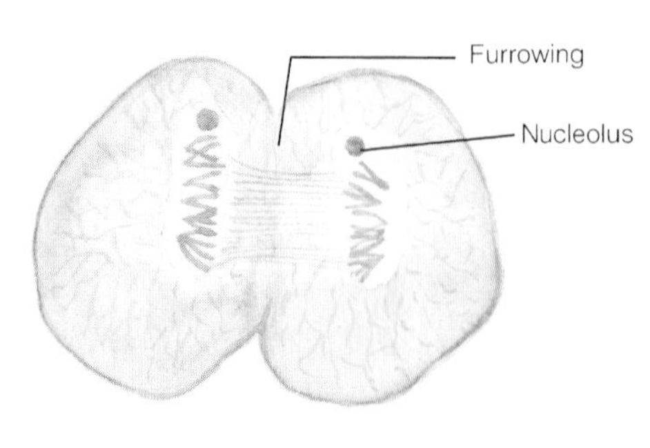

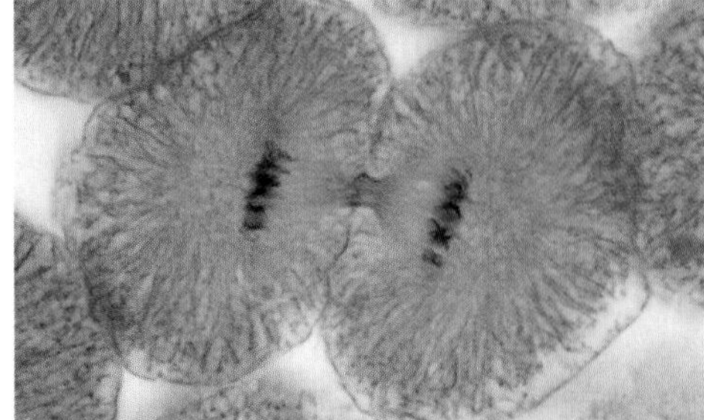

Figure 3.21

The stages of mitosis.

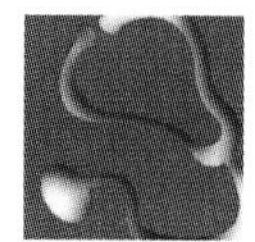

Highly specialized cells, such as muscle and nerve cells, do not replicate after a person is born. If these cells die, as the result of disease, injury, or even disuse, they are not replaced and scar tissue may form. Nerve cells are especially vulnerable to damage from oxygen deprivation, alcohol, and various other drugs.

1. Explain why the DNA molecule is described as a double helix.
2. List the phases in the life cycle of a cell and describe the principal events that occur during each phase.
3. Explain why mitosis is such an important biological process.

Clinical Considerations

Cellular Adaptations

Apparently included within cellular specialization of structure and function is mitotic potential. Certain cells do not require further division once the organ to which they contribute becomes functional. Others, as part of their specialization, require continuous mitosis to keep an organ healthy. Thus, in the adult, it is found that some cells divide continually, some occasionally, and some not at all. For example, epidermal cells, hemopoietic cells, and cells that line the lumen of the GI tract divide continually throughout life. Cells within specialized organs, such as the liver or kidneys, divide as the need becomes apparent. Naturally occurring cellular death, disease, or trauma from surgery or injury may necessitate mitosis in these organs. Still other cells, such as muscle or nerve cells, lose their mitotic ability as they become differentiated. Trauma to these cells frequently causes a permanent loss of function.

Although the factors regulating mitosis are unclear, evidence suggests that mitotic ability is genetically controlled and, for those cells that do divide, even the number of divisions is predetermined. If this is true, it may account for the aging process. Physical stress, nutrition, and hormones definitely have an effect on mitotic activity. It is thought that the replication activity of cells might be controlled through a feedback mechanism involving the release of a *growth-inhibiting substance*. Such a substance might slow or inhibit the cell divisions and growth of particular organs once they had amassed a certain number of cells or had reached a certain size.

Except for cells on exposed surfaces, most cells of the body are located in a fairly homogeneous environment, where continual adaptation to change is not necessary for survival. However, cells do have remarkable adaptability and resilience, enabling them to withstand conditions that might otherwise be lethal. Prolonged exposure to sunlight, for example, stimulates the synthesis of melanin and tanning of the skin. Likewise, mechanical friction to the skin stimulates mitotic activity and the synthesis of a fibrous protein, *keratin*, which results in the formation of a protective callus.

Cells adapt to potentially injurious stimuli by several specific mechanisms. **Hypertrophy** *(hi″per′trŏ-fe)* refers to an increase in the size of cells resulting from increased synthesis of protein, nucleic acids, and lipids. Cellular hypertrophy can be either compensatory or hormonal. **Compensatory hypertrophy** occurs when increased metabolic demands on particular cells result in an increase in cellular mass. Examples of compensatory hypertrophy include the enlargement of skeletal muscle fibers as a result of exercise and cardiac (heart) muscle fibers or kidney cells because of an increased work demand. Hypertension (high blood pressure) causes cardiac cells to hypertrophy because they must pump blood against raised pressures. After the removal of a diseased kidney, there is a compensatory increase in the size of the cells of the remaining kidney so that its normal weight is approximately doubled. Examples of **hormonal hypertrophy** are the increased size of the breasts and smooth muscles of the uterus in a pregnant woman.

Hyperplasia *(hi″per-pla′ze-ă)* refers to an increase in the number of cells formed as a result of increased mitotic activity. The removal of a portion of the liver, for example, leads to regeneration, or hyperplasia, of the remaining liver cells to restore the loss. But the triggering mechanism for hyperplasia is not known. In women, a type of hormonally induced hyperplasia occurs in cells of the endometrium of the uterus after menstruation, which restores this layer to a suitable state for possible implantation of an embryo.

Atrophy *(at′rŏ-fe)* refers to a decrease in the size of cells and a corresponding decrease in the size of the affected organ. Atrophy can occur in the cells of any organ and may be classified as **disuse atrophy, disease atrophy,** or **aging (senile) atrophy.**

Metaplasia *(met″ă-pla′ze-ă)* is a specialized cellular change in which one type of cell transforms into another. Generally, it involves the change of highly specialized cells into more generalized, protective cells. For example, excessive exposure to inhaled smoke causes the specialized ciliated columnar epithelial cells lining the bronchial airways to change into stratified squamous epithelium, which is more resistant to injury from smoke.

Trauma to Cells

As adaptable as cells are to environmental changes, they are subject to damage from aging and disease. If a trauma causes extensive cellular death, the condition may become life threatening. A person dies when a vital organ can no longer perform its metabolic role in sustaining the body.

hypertrophy: Gk. *hyper*, over; *trophe*, nourishment
hyperplasia: Gk. *hyper*, over; *plasis*, a molding
atrophy: Gk. *a*, without; *trophe*, nourishment
metaplasia: Gk. *meta*, between; *plasis*, a molding

Energy deficit means that more energy is required by a cell than is available. Cells can tolerate certain mild deficits because of various reserves stored within the cytoplasm, but a severe or prolonged deficit will cause cells to die. An energy deficit occurs when the cells do not have enough glucose or oxygen to allow for glucose combustion. Examples of energy deficits are low levels of blood sugar (hypoglycemia) and the impermeability of the cell membrane to glucose (as in diabetes mellitus). Malnutrition also may result in an energy deficit. Few cells can tolerate an interruption in oxygen supply. Cells of the brain and the heart have tremendous oxygen demands, and an interruption of the supply to these organs can cause death in a matter of minutes.

Physical injury to cells, another type of trauma, occurs in a variety of ways. High temperature **(hyperthermia)** is generally less tolerable to cells than low temperature **(hypothermia).** Respiratory rate, heart rate, and metabolism accelerate with hyperthermia. Continued hyperthermia causes protein coagulation within cells, and eventually cellular death. In frostbite, rapid or prolonged chilling causes cellular injury. In severe frostbite, ice crystals form and cause the cells to burst.

Burns are particularly significant if they cause damage to the deeper skin layers, which interferes with the mitotic activity of cells (see chapter 5). Of immediate concern with burns, however, is the devastating effect of fluid loss and infection through traumatized cell membranes.

Accidental poisoning and suicide through drug overdose account for large numbers of deaths in the United States and elsewhere. **Drugs** and **poisons** can cause cellular dysfunction by disrupting DNA replication, RNA transcription, enzyme systems, or cell membrane activity.

Radiation causes a type of cell trauma that is cumulative in effect. When X rays are administered for therapeutic purposes (radiotherapy), small doses are focused on a tumorous area over a course of many days to prevent widespread cellular injury. Some cells are more sensitive to radiation than others. Immature or mitotically active cells are highly sensitive, whereas cells that are no longer growing, such as neurons and muscle cells, are not as vulnerable to radiation injury.

Infectious agents, or **pathogens,** also cause cellular dysfunction. Viruses and bacteria are the most common pathogens. Viruses usually invade and destroy cells as they reproduce themselves. Bacteria, on the other hand, do not usually invade cells but will frequently poison cells with their toxic metabolic wastes.

Medical Genetics

Medical genetics is a branch of medicine concerned with diseases that have a genetic origin. Genetic factors include abnormalities in chromosome number or structure and mutant genes. Genetic diseases are a diverse group of disorders, including malformed blood cells (sickle-cell anemia), defective blood clotting (hemophilia), and mental retardation (Down syndrome).

Chromosomal abnormalities occur in approximately 0.6% of live-birth infants. The majority (70%) are subtle, cause no problems, and usually go undetected. Structural changes in the DNA that are passed from parent to offspring by means of sex cells are called **mutations** *(myoo-ta'shunz)*. Mutations either occur naturally or are environmentally induced through chemicals or radiation. Natural mutations are not well understood. About 12% of all congenital malformations are caused by mutations and probably come about through an interaction of genetic and environmental factors. Many of these problems can be predicted by knowing the genetic pedigree of prospective parents and prevented through genetic counseling. **Teratology** *(ter-ă-tol'ŏ-je)* is the science concerned with developmental defects and the diagnosis, treatment, and prevention of malformations.

Genetic problems are occasionally caused by having too few or too many chromosomes. The absence of an entire chromosome is termed **monosomy** *(mon'o-som"me)*. Embryos with monosomy usually die. People with **Turner's syndrome** have only one X chromosome and have a better chance of survival than those who are missing one of the other chromosomes. **Trisomy** *(tri'so-me)*, a genetic condition in which an extra chromosome is present, occurs more frequently than monosomy. The best known among the trisomies is **Down syndrome.**

In an attempt to better understand medical genetics, the *Human Genome Project* was launched by Congress in 1988 with the ambitious goal of completely mapping the human genome by September 30, 2005. That allowed just 17 years to determine the exact sequences of bases with which the 3 billion base pairs are arranged to form the 50,000 to 100,000 genes in the haploid human genome of a sperm cell or ovum. Knowing this information will provide the ultimate reference for diagnosis and treatment of the 4,000 genetic diseases that are known to be directly caused by particular abnormal genes.

Cancer

Cancer refers to a complex group of diseases characterized by uncontrolled cell replication. The rapid proliferation of cells results in the formation of a **neoplasm,** or new cellular mass. Neoplasms, frequently called *tumors*, are classified as benign or malignant based on their cytological and histological features. **Benign neoplasms** usually grow slowly and are confined to a particular area. These types are usually not life threatening unless they grow to large sizes in vital organs like the brain. **Malignant neoplasms** (fig. 3.22) grow rapidly and **metastasize** *(mĕ-tas'tă-sīz)* (fragment and spread) easily through lymphatic or blood vessels. The original malignant neoplasm is called the *primary growth* and the new tumors, or metastatic tumors, are called *secondary growths*.

Cancer cells resemble undifferentiated or primordial cell types. Generally they do not mature before they divide and are not capable of maintaining normal cell function. Cancer causes death when a vital organ regresses because of competition from

---

mutation: L. *mutare*, to change

teratology: Gk. *teras*, monster; *logos*, study

Turner's syndrome: from Henry H. Turner, American endocrinologist, 1892–1970

Down syndrome: from John L. H. Down, English physician, 1828–96

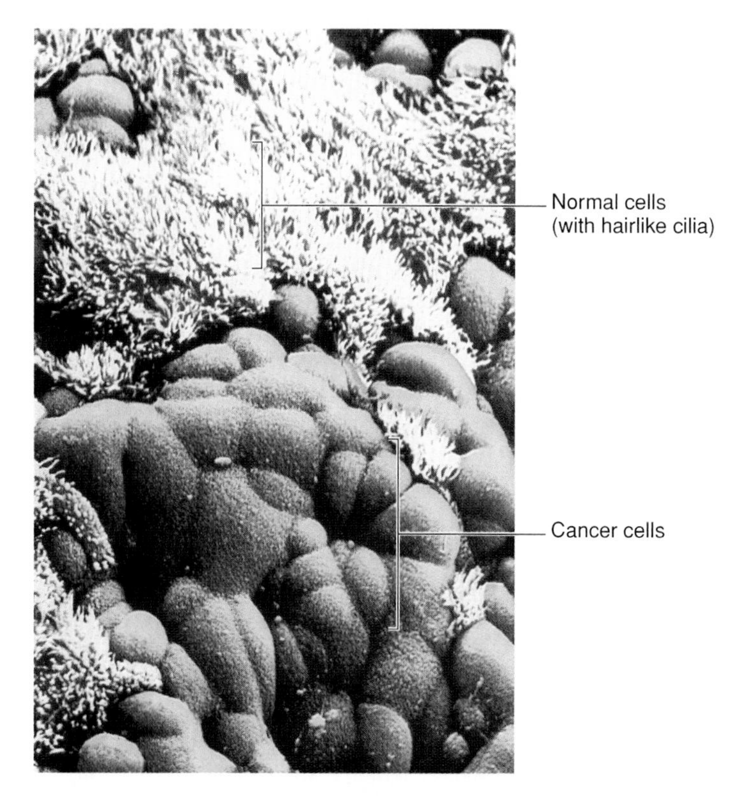

Figure 3.22

An electron micrograph of cancer cells from the respiratory tract (59,800×).

cancer cells for space and nutrients. The pain associated with cancer develops when the growing neoplasm affects sensory neurons.

The various types of cancers are classified on the basis of the tissue in which they develop. Lymphoma, for example, is a cancer of lymphoid tissue; osteogenic cancer is a type of bone cancer; myeloma is cancer of the bone marrow; and sarcoma is a general term for any cancer arising from cells of connective tissue.

The etiology (cause) of cancers is largely unknown. However, initiating factors, or **carcinogens** *(kar-sin'ŏ-jenz),* such as viruses, chemicals, or irradiation, may provoke cancer to develop. Cigarette smoking, for example, causes various respiratory cancers to develop. The tendency to develop other types of cancers has a genetic basis. Some researchers even think that physiological stress can promote certain types of cancerous activity. Because the causes of cancer are not well understood, emphasis is placed on early detection with prompt treatment.

Aging

Although there are obvious external indicators of aging—graying and loss of hair, wrinkling of skin, loss of teeth, and decreased muscle mass—changes within cells as a result of aging are not as apparent and are not well understood. Certain organelles alter with age. The mitochondria, for example, may change in structure and number, and the Golgi complex may fragment. Also, lipid vacuoles tend to accumulate in the cytoplasm, and the cytoplasmic food stores that contain glycogen decrease.

The chromatin and chromosomes within the nucleus show changes with aging, such as clumping, shrinking, or fragmenting with repeated mitotic divisions. There is strong evidence that certain cell types have a predetermined number of mitotic divisions that are genetically controlled, thus determining the overall vitality and longevity of an organ. If this is true, identifying and genetically manipulating the "aging gene" might be possible.

Extracellular substances also change with age. Protein strands of *collagen* and *elastin* change in quality and number in aged tissues. Elastin plays an important role in the walls of arteries, and its deterioration is thought to be associated with such vascular diseases as arteriosclerosis in aged persons.

carcinogen: Gk, *karkinos,* cancer

Chapter Summary

Introduction to Cytology (p. 47)

1. Cells are the structural and functional units of the body. Cellular function is referred to as metabolism and the study of cells is referred to as cytology.
2. Cellular function depends on the specific membranes and organelles characteristic of each type of cell.
3. All cells have structural modifications that serve functional purposes.

Cellular Chemistry (pp. 48–50)

1. Four elements (oxygen, carbon, hydrogen, and nitrogen) compose over 95% of the body's mass and are linked together to form inorganic and organic compounds.
2. Water is the most abundant inorganic compound in cells and is an excellent solvent.
   (a) Water is important in temperature control and hydrolysis.
   (b) Dehydration, a condition in which fluid loss exceeds fluid intake, may be a serious problem—especially in infants.
3. Electrolytes are inorganic compounds that form ions when dissolved in water.
   (a) The three classes of electrolytes are acids, bases, and salts.
   (b) Electrolytes are important in maintaining pH, in conducting electrical currents, and in regulating the activity of enzymes.
4. Proteins are organic compounds that may exist by themselves or be conjugated with other compounds.
   (a) Proteins are important structural components of the body and are necessary for cellular growth, repair, and division.
   (b) Enzymes and hormones are examples of specialized proteins.
5. Carbohydrates are organic compounds containing carbon, hydrogen, and oxygen, with a 2:1 ratio of hydrogen to oxygen.
   (a) The carbohydrate group includes the starches and sugars.
   (b) Carbohydrates are the most abundant source of cellular energy.
6. Lipids are organic fats and fat-related substances.
   (a) Lipids are composed primarily of carbon, hydrogen, and oxygen.
   (b) Lipids serve as an important source of energy, form parts of membranes, and protect and insulate various parts of the body.

Cellular Structure (pp. 50–62)

1. A cell is composed of a cell membrane, cytoplasm and organelles, and a nucleus.
2. The cell membrane, composed of phospholipid and protein molecules, encloses the contents of the cell and regulates the passage of substances into and out of the cell.
   (a) The permeability of the cell membrane depends on its structure, the size of the molecules, ionic charge, lipid solubility, and the presence of carrier molecules.
   (b) Cell membranes may be specialized with such structures as microvilli, sacs, and hair cells.
3. Cytoplasm refers to the material between the cell membrane and the nucleus. Nucleoplasm is the material within the nucleus. Protoplasm is a collective term for both the cytoplasm and nucleoplasm.
4. Organelles are specialized components within the cytoplasm of cells.
   (a) Endoplasmic reticulum provides a framework within the cytoplasm and forms a site for the attachment of ribosomes. It functions in the synthesis of lipids and proteins and in cellular transport.
   (b) Ribosomes are particles of protein and RNA that function in protein synthesis. The protein particles may be used within the cell or secreted.
   (c) The Golgi complex consists of membranous vesicles that synthesize glycoproteins and secrete lipids. The Golgi complex is extensive in secretory cells, such as those of the pancreas and salivary glands.
   (d) Mitochondria are membranous sacs that consist of outer and inner mitochondrial layers and folded membranous extensions of the inner layer called cristae. The mitochondria produce ATP and are called the "powerhouses" of a cell.
   (e) Lysosomes are spherical bodies that contain digestive enzymes. They are abundant in the phagocytic white blood cells.
   (f) Peroxisomes are enzyme-containing membranous sacs that are abundant in the kidneys and liver. Some of the enzymes in peroxisomes generate hydrogen peroxide, and one of them, catalase, breaks down excess hydrogen peroxide.
   (g) The centrosome is the dense area of cytoplasm near the nucleus that contains the centrioles. The paired centrioles play an important role in cell division.
   (h) Vacuoles are membranous sacs that function as storage chambers.
   (i) Fibrils and microtubules provide support in the form of a cytoskeleton.
   (j) Cilia and flagella are projections of the cell that have the same basic structure and that function in producing movement.
5. The cell nucleus is enclosed in a nuclear membrane that controls the movement of substances between the nucleoplasm and the cytoplasm.
   (a) The nucleoli are small bodies of protein and RNA within the nucleus that produce ribosomes.
   (b) Chromatin is a coiled fiber of protein and DNA that shortens to form chromosomes during cell reproduction.

Cell Cycle (pp. 62–67)

1. The cell cycle consists of growth, synthesis, and mitosis.
   (a) Growth is the increase in cellular mass that results from metabolism. Synthesis is the production of DNA and RNA to regulate cellular activity. Mitosis is the splitting of the cell's nucleus and cytoplasm that results in the formation of two diploid cells.
   (b) Mitosis permits an increase in the number of cells (body growth) and allows for the replacement of damaged, diseased, or worn-out cells.
2. A DNA molecule is in the shape of a double helix. The structural unit of the molecule is a nucleotide, which consists of deoxyribose (sugar), phosphate, and a nitrogenous base.
3. Cell division consists of a division of the chromosomes (mitosis) and a division of the cytoplasm (cytokinesis). The stages of mitosis include prophase, metaphase, anaphase, and telophase.

Review Activities

Objective Questions

1. Inorganic compounds that form ions when dissociated in water are
   (a) hydrolites.
   (b) metabolites.
   (c) electrolytes.
   (d) ionizers.
   (e) nucleic acids.
2. The four elements that compose over 95% of the body are
   (a) oxygen, potassium, hydrogen, carbon.
   (b) carbon, sodium, nitrogen, oxygen.
   (c) potassium, sodium, magnesium, oxygen.
   (d) carbon, oxygen, nitrogen, hydrogen.
   (e) oxygen, carbon, hydrogen, sulfur.

3. Which organelle contains strong hydrolytic enzymes?
   (a) the lysosome
   (b) the Golgi complex
   (c) the ribosome
   (d) the vacuole
   (e) the mitochondrion
4. Ciliated cells occur in
   (a) the trachea.
   (b) the ductus deferens.
   (c) the bronchioles.
   (d) the uterine tubes.
   (e) all of the above.
5. Osmosis deals with the movement of
   (a) gases.
   (b) water only.
   (c) oxygen only.
   (d) both a and c.
6. The phase of mitosis in which the chromosomes line up at the equator (equatorial plane) of the cell is called
   (a) interphase.
   (b) prophase.
   (c) metaphase.
   (d) anaphase.
   (e) telophase.
7. The phase of mitosis in which the chromatids separate is called
   (a) interphase.
   (b) prophase.
   (c) metaphase.
   (d) anaphase.
   (e) telophase.
8. The organelle that combines protein with carbohydrates and packages them within vesicles for secretion is
   (a) the Golgi complex.
   (b) the rough endoplasmic reticulum.
   (c) the smooth endoplasmic reticulum.
   (d) the ribosome.
9. The enlarged skeletal muscle fibers that result from an increased work demand serve to illustrate
   (a) disuse atrophy.
   (b) compensatory hypertrophy.
   (c) metaplasia.
   (d) inertia.
10. Regeneration of liver cells is an example of
   (a) compensatory hypertrophy.
   (b) hyperplasia.
   (c) metaplasia.
   (d) hypertrophy.

Essay Questions

1. Explain why a knowledge of cellular anatomy is necessary for understanding tissue and organ function within the body.
2. Why is water a good fluid medium of the cell?
3. How are proteins, carbohydrates, and lipids similar? How are they different? What are enzymes and hormones?
4. Describe the cell membrane. List the various kinds of movement through the cell membrane and give an example of each.
5. Describe, diagram, and list the functions of the following:
   (a) endoplasmic reticulum,
   (b) ribosome,
   (c) mitochondrion,
   (d) Golgi complex,
   (e) centrioles, and
   (f) cilia.
6. Define *inorganic compound* and *organic compound* and give examples of each.
7. Define the terms *protoplasm*, *cytoplasm*, and *nucleoplasm*. Describe the position of the membranes associated with each of these substances.
8. Describe the structure of the nucleus and the functions of its parts.
9. What is a nucleotide? How does it relate to the overall structure of a DNA molecule?
10. Explain the relationship between DNA, chromosomes, chromatids, and genes.
11. Distinguish between mitosis and cytokinesis. Describe the major events of mitosis and discuss the significance of the mitotic process.
12. Give examples of factors that contribute cellular hypertrophy, hyperplasia, atrophy, and metaplasia.
13. Explain how cells respond to
   (a) energy deficit,
   (b) hyperthermia,
   (c) burns,
   (d) radiation, and
   (e) pathogens.
14. Define the following genetic terms: *teratology*, *monosomy*, *trisomy*, and *mutation*.
15. In what ways do cells of a neoplasm differ from normal cells. How may a malignant neoplasm cause death?

Critical-Thinking Questions

1. How is the structural organization of its individual cells essential to a multicellular organism?
2. Construct a table comparing the structure and function of several kinds of cells. Indicate which organelles would be of particular importance to each kind of cell
3. Define *medical genetics* and give some examples of genetic diseases. Is spending the billions of dollars required to complete the Human Genome Project justified? Why or why not?
4. The brain is protected to some extent by the blood-brain barrier—a membrane between circulating blood and the brain that keeps certain damaging substances from reaching brain tissue. However, the brain is still subject to trauma that can cause it to swell, much like an ankle swells with a sprain. Since the cranium is a cavity of fixed size, brain edema (swelling) can rapidly lead to coma and death. Knowing what you do about movement of water across a membrane, can you explain why *mannitol*, a type of sugar that does not cross the blood-brain barrier, is commonly used to treat patients who have suffered head trauma?
5. Your friend knows that you have just reviewed cellular chemistry, and so he asks for your opinion about his new diet. In an attempt to eliminate the lipid content in adipose tissue and thus lose weight, he has completely eliminated fat from his diet. He feels that he is now free to eat as much food as he likes, provided it consists only of carbohydrates and protein. Is your friend's logic flawed? Would you advise him to stick with this diet?

FOUR

Histology

Definition and Classification of Tissues 73
Developmental Exposition: The Tissues 74
Epithelial Tissue 74
- Characteristics of Membranous Epithelia 75
- Simple Epithelia 76
- Stratified Epithelia 78
- Body Membranes 80
- Glandular Epithelia 82

Connective Tissue 85
- Characteristics and Classification of Connective Tissue 86
- Embryonic Connective Tissue 86
- Connective Tissue Proper 86
- Cartilage 91
- Bone Tissue 94
- Blood (Vascular Tissue) 94

Muscle Tissue 96
Nervous Tissue 97
Clinical Considerations 98
- Changes in Tissue Composition 98
- Tissue Analysis 99
- Tissue Transplantation 99

Chapter Summary 99
Review Activities 100

It was through this microscope that Robert Hooke first observed the rigid walls of cork that he called "cells." Although it followed not too long after Leeuwenhoek's microscope, the one used by Hooke is considerably more elaborate.

Definition and Classification of Tissues

Histology is the specialty of anatomy that involves study of the microscopic structure of tissues. Tissues are assigned to four basic categories on the basis of their cellular composition and histological appearance.

Objective 1 Define *tissue* and discuss the importance of histology.

Objective 2 Describe the functional relationship between cells and tissues.

Objective 3 List the four principal tissue types and briefly describe the functions of each type.

Although cells are the structural and functional units of the body, the cells of a complex multicellular organism are so specialized that they do not function independently. *Tissues* are aggregations of similar cells and cell products that perform specific functions. The various types of tissues are established during early embryonic development. As the embryo grows, organs form from specific arrangements of tissues. Many adult organs, including the heart and muscles, contain the original cells and tissues that were formed prenatally, although some functional changes occur in the tissues as they are acted upon by hormones or as their effectiveness diminishes with age.

The study of tissues is referred to as **histology.** It provides a foundation for understanding the structure and functions of the organs discussed in the chapters that follow. Many diseases profoundly alter the tissues within an affected organ; therefore, by knowing the normal tissue structure, a physician can recognize the abnormal. In medical schools a course in histology is usually followed by a course in *pathology*, the study of abnormal tissues in diseased organs.

Although histologists employ many different techniques for preparing, staining, and sectioning tissues, only two basic kinds of microscopes are used to view the prepared tissues. The *light microscope* is used to observe overall tissue structure (fig. 4.1), and the *electron microscope* to observe the fine details of tissue and cellular structure. Most of the histological photomicrographs in this text are at the light microscopic level. However, where fine structural detail is needed to understand a particular function, electron micrographs are used.

Tissue cells are surrounded and bound together by a nonliving intercellular **matrix** (*ma'triks*) that the cells secrete. Matrix varies in composition from one tissue to another and may take the form of a liquid, semisolid, or solid. Blood, for example, has a liquid matrix, permitting this tissue to flow through vessels. By contrast, bone cells are separated by a solid matrix, permitting this tissue to support the body.

The tissues of the body are assigned to four principal types on the basis of structure and function: (1) *epithelial* (*ep"ĭ-the'le-al*) *tissue* covers body surfaces, lines body cavities and ducts, and forms glands; (2) *connective tissue* binds, supports, and protects body parts; (3) *muscle tissue* contracts to produce movement; and (4) *nervous tissue* initiates and transmits nerve impulses from one body part to another.

histology: Gk. *histos*, web (tissue); *logos*, study
pathology: Gk. *pathos*, suffering, disease; *logos*, study
matrix: L. *matris*, mother

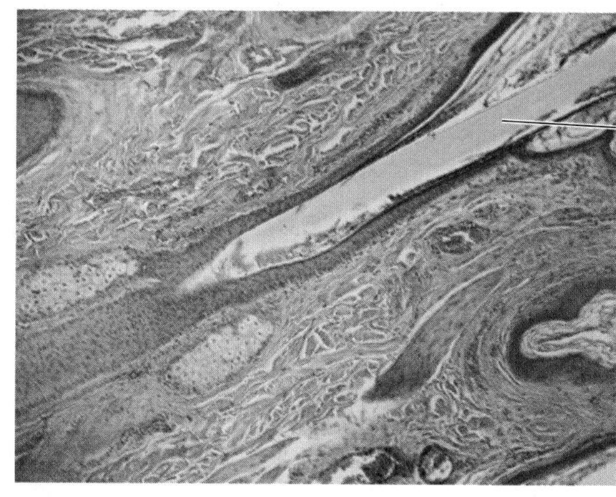

Figure 4.1

The appearance of skin (*a*) magnified 25 times, as seen through a compound light microscope, and (*b*) magnified 280 times, as seen through a scanning electron microscope (SEM).

1. Define *tissue* and explain why histology is important to the study of anatomy, physiology, and medicine.
2. Cells are the functional units of the body. Explain how the matrix permits specific kinds of cells to be even more effective and functional as tissues.
3. What are the four principal kinds of body tissues? What are the basic functions of each type?

Developmental Exposition

The Tissues

Exhibit I The early stages of embryonic development. (*a*) Fertilization and the formation of the zygote, (*b*) the morula at about the third day, (*c*) the early blastocyst at the time of implantation between the fifth and seventh day, (*d*) a blastocyst at 2 weeks, and (*e*) a blastocyst at 3 weeks showing the three primary germ layers that constitute the embryonic disc.

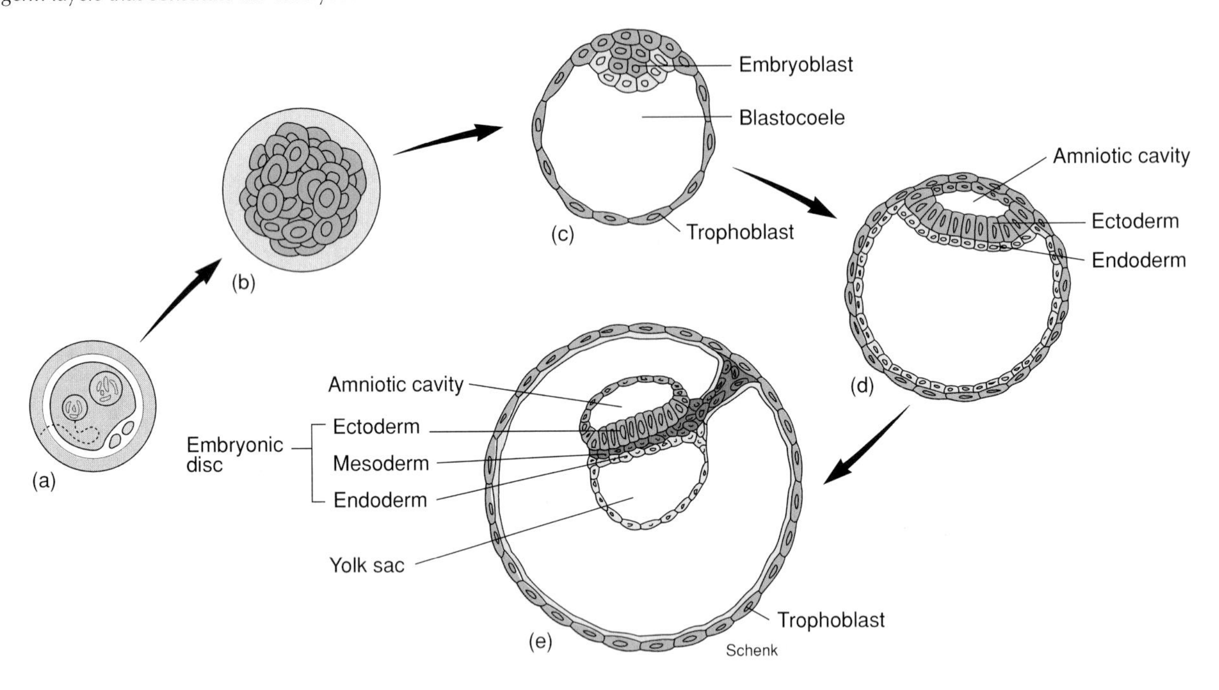

Explanation

Human prenatal development is initiated by the fertilization of an ovulated ovum (egg) from a female by a sperm cell from a male. The chromosomes within the nucleus of a **zygote** (*zi′ gōt*) (fertilized egg) contain all the genetic information necessary for the differentiation and development of all body structures.

Within 30 hours after fertilization, the zygote undergoes a mitotic division as it moves through the uterine tube toward the uterus (see chapter 22). After several more cellular divisions, the embryonic mass consists of 16 or more cells and is called a **morula** (*mor′yoo-lă*), as shown in exhibit I. Three or 4 days after conception, the morula enters the uterine cavity where it remains unattached for about 3 days. During this time, the center of the morula fills with fluid absorbed from the uterine cavity. As the fluid-filled

zygote: Gk. *zygotos*, yolked

morula: Gk. *morus*, mulberry

Epithelial Tissue

There are two major categories of epithelia: membranous and glandular. Membranous epithelia are located throughout the body and form such structures as the outer layer of the skin; the inner lining of body cavities, tubes, and ducts; and the covering of visceral organs. Glandular epithelia are specialized tissues that form the secretory portion of glands.

Objective 4 Compare and contrast the various types of membranous epithelia.

Objective 5 Discuss the functions of the membranous epithelia in different locations in the body.

Objective 6 Define *gland* and compare and contrast the various types of glands in the body.

epithelium: Gk. *epi*, upon; *thelium*, to cover

Table 4A

Derivatives of the Germ Layers

Ectoderm	Mesoderm	Endoderm
Epidermis of skin and epidermal derivatives: hair, nails, glands of the skin; linings of oral, nasal, anal, and vaginal cavities	Muscle: smooth, cardiac, skeletal	Epithelium of pharynx, auditory canal, tonsils, thyroid, parathyroid, thymus, larynx, trachea, lungs, GI tract, urinary bladder and urethra, and vagina
Nervous tissue; sense organs	Connective tissue: embryonic, connective tissue proper, cartilage, bone, blood	Liver and pancreas
Lens of eye; enamel of teeth	Dermis of skin; dentin of teeth	
Pituitary gland	Epithelium (endothelium) of blood vessels, lymphatic vessels, body cavities, joint cavities	
Adrenal medulla	Internal reproductive organs	
	Kidneys and ureters	
	Adrenal cortex	

space develops inside the morula, two distinct groups of cells form. The single layer of cells forming the outer wall is known as the **trophoblast,** and the inner aggregation of cells is known as the **embryoblast.** With further development, the trophoblast differentiates into a structure that will later form part of the placenta; the embryoblast will eventually become the embryo. With the establishment of these two groups of cells, the morula becomes known as a **blastocyst** (*blas'tŏ-sist*). Implantation of the blastocyst in the uterine wall begins between the fifth and seventh day (see chapter 22).

As the blastocyst completes implantation during the second week of development, the embryoblast undergoes marked differentiation. A slitlike space called the **amniotic** (*am'ne-ot-ic*) **cavity** forms within the embryoblast, adjacent to the trophoblast. The embryoblast now consists of two layers: an upper **ectoderm,** which is closer to the amniotic cavity, and a lower **endoderm,** which borders the **blastocoel** (blastocyst cavity). A short time later, a third layer called the **mesoderm** forms between the endoderm and ectoderm. These three layers constitute the **primary germ layers.**

The primary germ layers are of great significance because all the cells and tissues of the body are derived from them (see fig. 22.10). Ectodermal cells form the nervous system; the outer layer of skin (epidermis), including hair, nails, and skin glands; and portions of the sensory organs. Mesodermal cells form the skeleton, muscles, blood, reproductive organs, dermis of the skin, and connective tissue. Endodermal cells produce the lining of the GI tract, the digestive organs, the respiratory tract and lungs, and the urinary bladder and urethra. The derivatives of the primary germ layers are summarized in table 4A.

trophoblast: Gk. *trophe*, nourishment; *blastos*, germ
embryoblast: Gk. *embryon*, to be full, swell; *blastos*, germ
ectoderm: Gk. *ecto*, outside; *derm*, skin

endoderm: Gk. *endo*, within; *derm*, skin
mesoderm: Gk. *meso*, middle; *derm*, skin

Characteristics of Membranous Epithelia

Membranous epithelia always have one free surface exposed to a body cavity, a lumen (hollow portion of a body tube), or to the skin surface. Some membranous epithelia are derived from ectoderm, such as the outer layer of the skin; some from mesoderm, such as the inside lining of blood vessels; and others from endoderm, such as the inside lining of the digestive tract (gastrointestinal, or GI, tract).

Membranous epithelia may be one or several cell layers thick. The upper surface may be exposed to gases, as in the case of epithelium in the integumentary and respiratory systems; to liquids, as in the circulatory and urinary systems; or to semisolids, as in the GI tract. The deep surface of most membranous epithelia is bound to underlying supportive tissue by a **basement membrane,** that consists of glycoprotein from the epithelial cells and a meshwork of collagenous and reticular fibers from the underlying connective tissue. With few exceptions, membranous epithelia are avascular (without blood vessels) and must be nourished by diffusion from underlying connective tissues. Cells that make up membranous epithelia are tightly packed together, with little intercellular matrix between them.

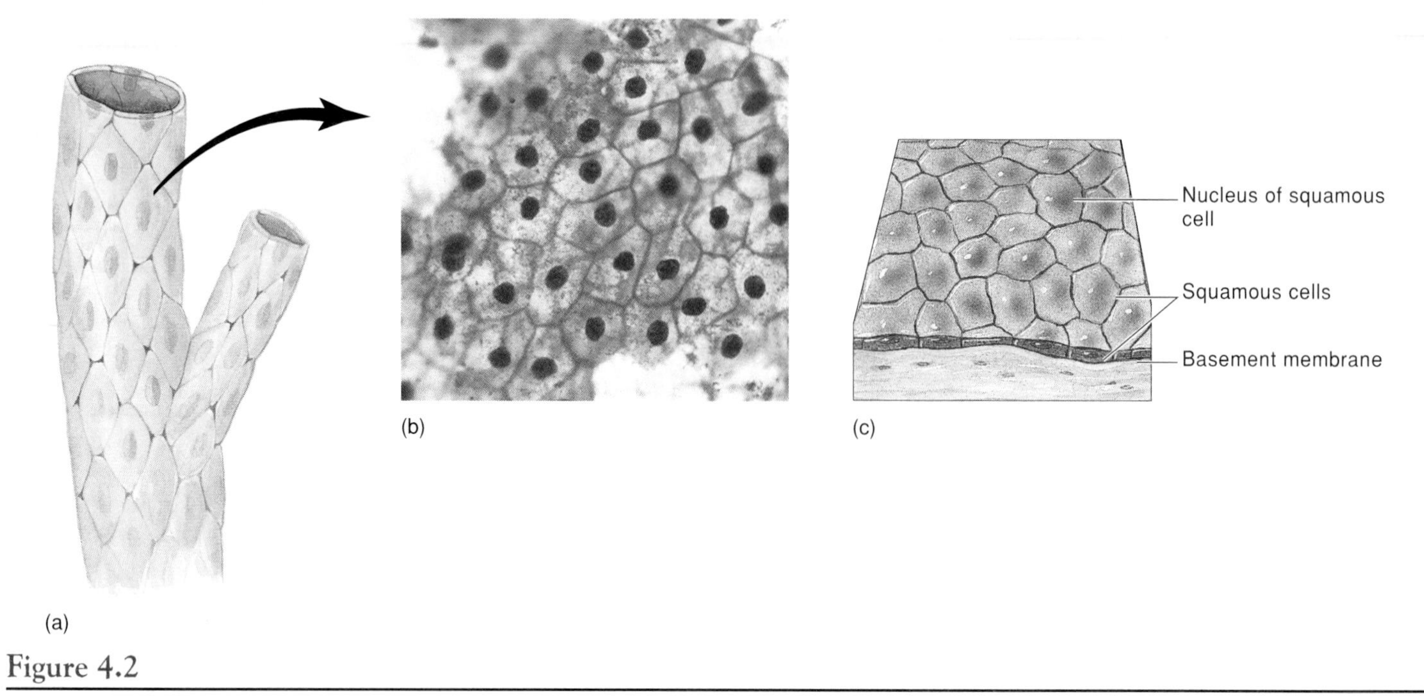

Figure 4.2

(*a*) Simple squamous epithelium lines the lumina of vessels, where it permits diffusion. (*b*) A photomicrograph of this tissue and (*c*) a labeled diagram.

Some of the functions of membranous epithelia are quite specific, but certain generalities can be made. Epithelia that cover or line surfaces provide protection from pathogens, physical injury, toxins, and desiccation. Epithelia lining the GI tract function in absorption. The epithelium of the kidneys provides filtration, while that within the pulmonary alveoli (air sacs) of the lungs allows for diffusion. Highly specialized *neuroepithelium* in the taste buds and in the nasal region has a chemoreceptor function.

Many membranous epithelia are exposed to friction or harmful substances from the outside environment. For this reason, epithelial tissues have remarkable regenerative abilities. The mitotic replacement of the outer layer of skin and the lining of the GI tract, for example, is a continuous process.

Membranous epithelia are histologically classified by the number of layers of cells and the shape of the cells along the exposed surface. Epithelial tissues that are composed of a single layer of cells are called *simple;* those that are layered are said to be *stratified. Squamous* cells are flattened; *cuboidal* cells are cube-shaped; and *columnar* cells are taller than they are wide.

Simple Epithelia

Simple epithelial tissue is a single cell layer thick and is located where diffusion, absorption, filtration, and secretion are principal functions. The cells of simple epithelial tissue range from thin, flattened cells to tall, columnar cells. Some of these cells have cilia that create currents for the movement of materials across cell surfaces. Others have microvilli that increase the surface area for absorption.

Simple Squamous Epithelium

Simple squamous (*skwa'mus*) epithelium is composed of flattened, irregularly shaped cells that are tightly bound together in a mosaiclike pattern (fig. 4.2). Each cell contains an oval or spherical central nucleus. This epithelium is adapted for diffusion and filtration. It occurs in the pulmonary alveoli within the lungs (where gaseous exchange occurs), in portions of the kidney (where blood is filtered), on the inside walls of blood vessels, in the lining of body cavities, and in the covering of the viscera. The simple squamous epithelium lining the inner walls of blood and lymphatic vessels is termed **endothelium** (*en"do-the'le-um*) (fig. 4.2*c*). That which covers visceral organs and lines body cavities is called **mesothelium** (*mes"ŏ-the'le-um*).

Simple Cuboidal Epithelium

Simple cuboidal epithelium is composed of a single layer of tightly fitted cube-shaped cells (fig. 4.3). This type of epithelium is found lining small ducts and tubules that have excretory, secretory, or absorptive functions. It occurs on the surface of the ovaries, forms a portion of the tubules within the kidney, and lines the ducts of the salivary glands and pancreas.

Simple Columnar Epithelium

Simple columnar epithelium is composed of tall, columnar cells (fig. 4.4). The height of the cells varies, depending on the site

squamous: L. *squamosus*, scaly
endothelium: Gk. *endon*, within; *thelium*, to cover
mesothelium: Gk. *meso*, middle; *thelium*, to cover

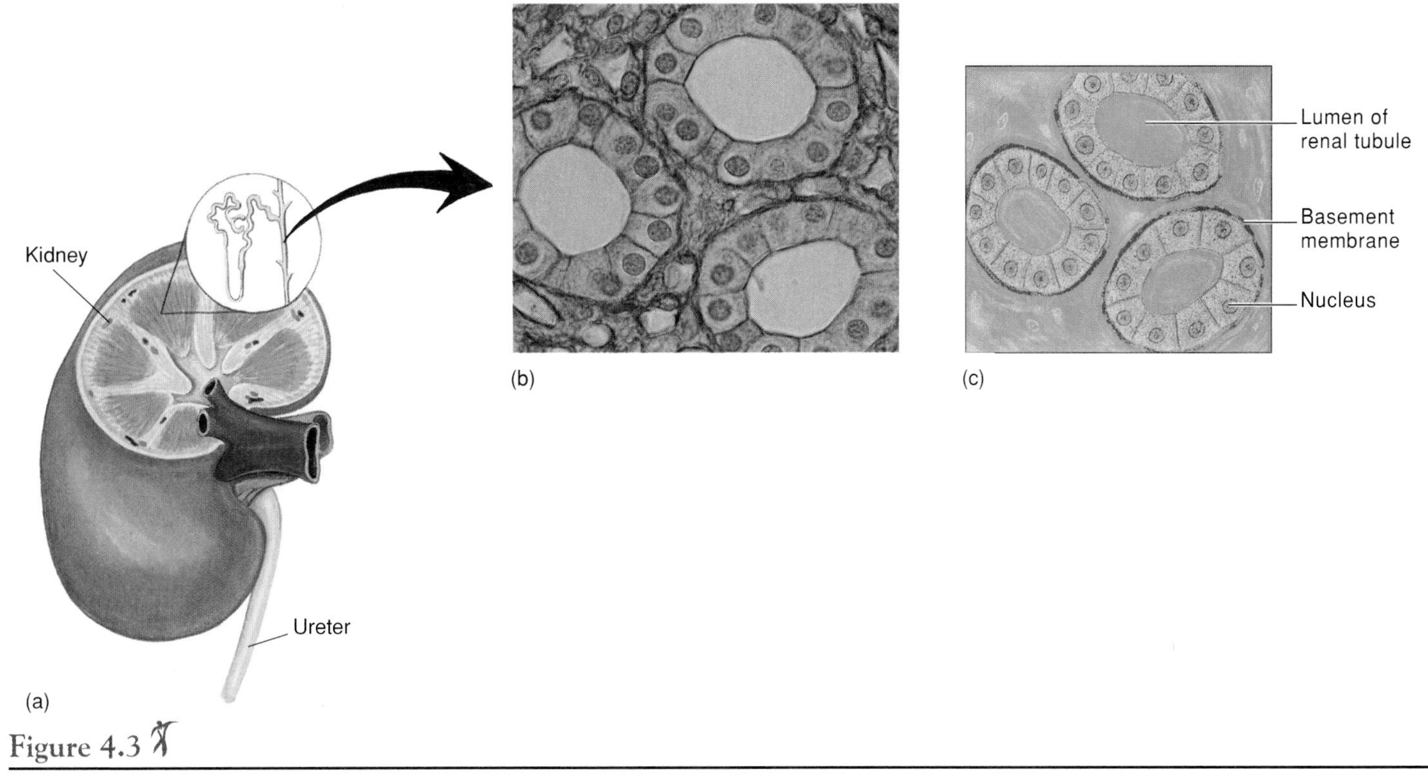

Figure 4.3

(*a*) Simple cuboidal epithelium lines the lumina of ducts; for example, in the kidneys, where it permits movement of fluids and ions. (*b*) A photomicrograph of this tissue and (*c*) a labeled diagram.

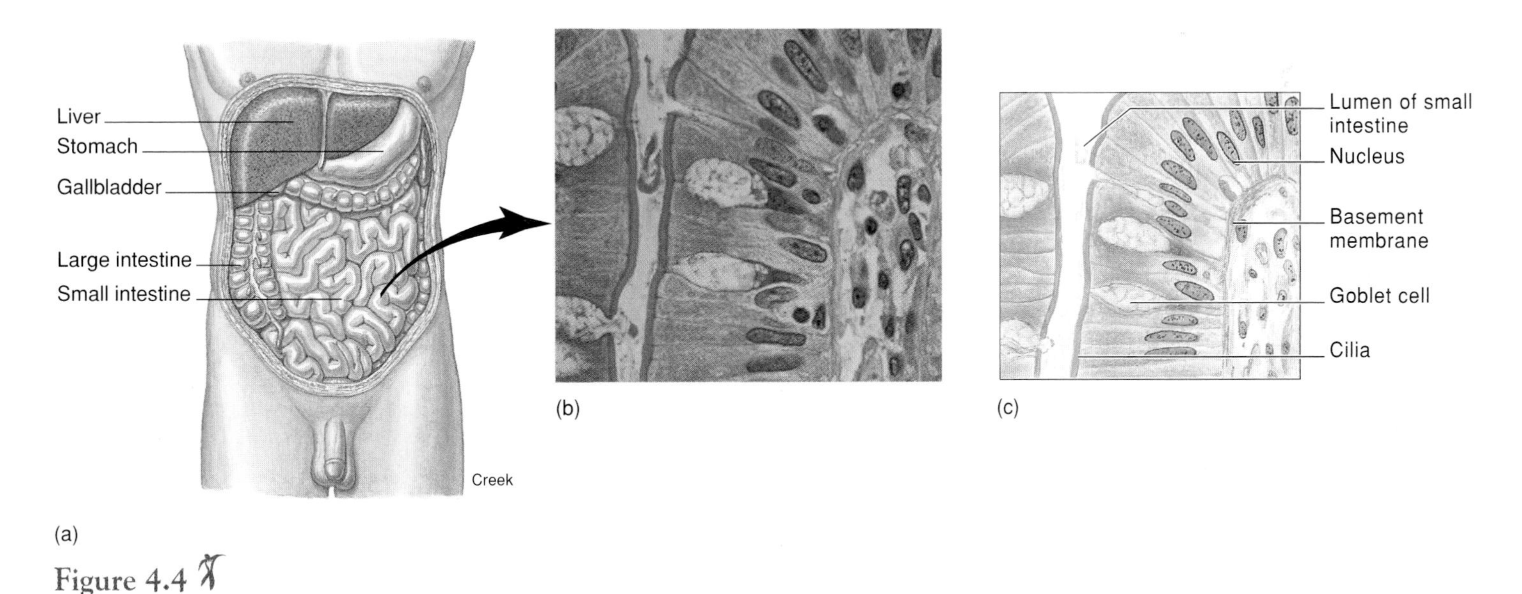

Figure 4.4

(*a*) Simple columnar epithelium lines the lumen of the digestive tract, where it permits secretion and absorption. (*b*) A photomicrograph of this tissue and (*c*) a labeled diagram.

and function of the tissue. Each cell contains a single nucleus which is usually located near the basement membrane. Specialized unicellular glands called **goblet cells** are scattered through this tissue at most locations. Goblet cells secrete a lubricative and protective mucus along the free surfaces of the cells. Simple columnar epithelium is found lining the inside walls of the stomach and intestine. In the digestive system, it forms a highly absorptive surface and also secretes certain digestive chemicals. Within the stomach, simple columnar epithelium has a tremendous mitotic rate. It replaces itself every 2 to 3 days.

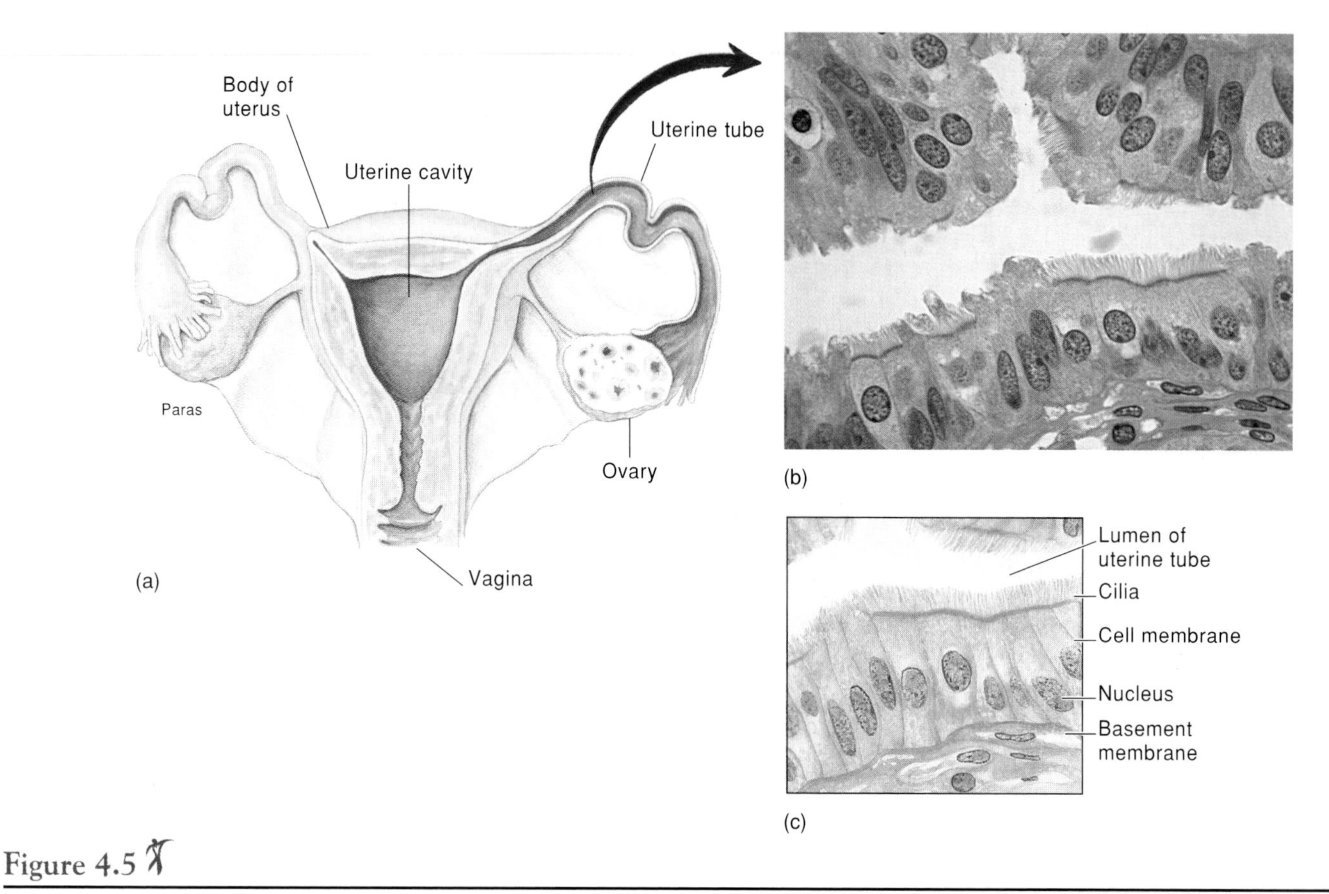

Figure 4.5

(*a*) Simple ciliated columnar epithelium lines the lumen of the uterine tube, where currents generated by the cilia propel the egg cell toward the uterus. (*b*) A photomicrograph of this tissue and (*c*) a labeled diagram.

Simple Ciliated Columnar Epithelium

Simple ciliated columnar epithelium is characterized by the presence of cilia along its free surface (fig. 4.5). By contrast, the simple columnar type is unciliated. Cilia produce wavelike movements that transport materials through tubes or passageways. This type of epithelium occurs in the female uterine tubes to move the ovum (egg cell) toward the uterus.

Not only do the cilia function to propel the ovum, but recent evidence indicates that sperm introduced into the female vagina during sexual intercourse may be moved along the return currents, or eddies, generated by ciliary movement. This greatly enhances the likelihood of fertilization.

Pseudostratified Ciliated Columnar Epithelium

As the name implies, this type of epithelium has a layered appearance (*strata* = layers). Actually, it is not multilayered (*pseudo* = false), since each cell is in contact with the basement membrane. Not all cells are exposed to the surface, however (fig. 4.6). The tissue appears to be stratified because the nuclei of the cells are located at different levels. Numerous goblet cells and a ciliated exposed surface are characteristic of this epithelium. It is found lining the inside walls of the trachea and the bronchial tubes; hence, it is frequently called *respiratory epithelium*. Its function is to remove foreign dust and bacteria entrapped in mucus from the lower respiratory system.

Coughing and sneezing, or simply "clearing the throat," are protective reflex mechanisms for clearing the respiratory passages of obstruction or of inhaled particles that have been trapped in the mucus along the ciliated lining. The material that is coughed up consists of the mucus-entrapped particles.

Stratified Epithelia

Stratified epithelia have two or more layers of cells. In contrast to the single-layered simple epithelia, they are poorly suited for absorption and secretion. Stratified epithelia have a primarily protective function that is enhanced by rapid cell divisions. They are classified according to the shape of the surface layer of cells, since the layer in contact with the basement membrane is always cuboidal or columnar in shape.

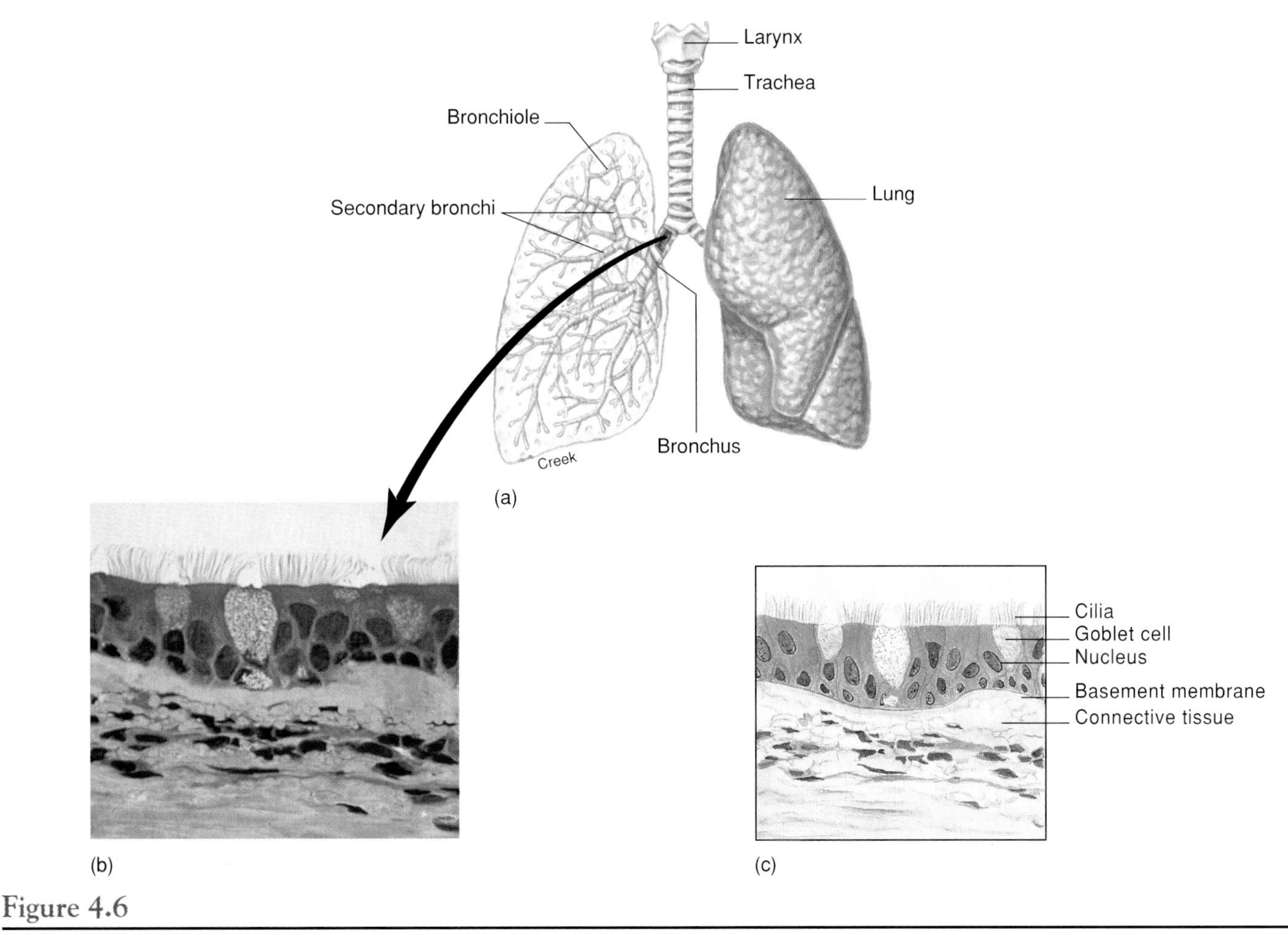

Figure 4.6

(*a*) Pseudostratified ciliated columnar epithelium lines the lumen of the respiratory tract, where it traps foreign material and moves it away from the pulmonary alveoli of the lungs. (*b*) A photomicrograph of this tissue and (*c*) a labeled diagram.

Stratified Squamous Epithelium

Stratified squamous epithelium is composed of a variable number of cell layers that are flattest at the surface (fig. 4.7). Mitosis occurs only at the deepest layers (see table 5.2). The mitotic rate approximates the rate at which cells are sloughed off at the surface. As the newly produced cells grow in size, they are pushed toward the surface, where they replace the cells that are sloughed off. Movement of the epithelial cells away from the supportive basement membrane is accompanied by the production of *keratin* (*ker'ă-tin*) (described below), progressive dehydration, and flattening.

There are two types of stratified squamous epithelial tissues: *keratinized* and *nonkeratinized.*

1. **Keratinized stratified squamous epithelium** contains *keratin*, a protein that strengthens the tissue. Keratin makes the epidermis (outer layer) of the skin somewhat waterproof and protects it from bacterial invasion. The outer layers of the skin are dead, but glandular secretions keep them soft (see chapter 5).
2. **Nonkeratinized stratified squamous epithelium** lines the oral cavity and pharynx, nasal cavity, vagina, and anal canal. This type of epithelium, called *mucosa* (*myoo-ko'să*), is well adapted to withstand moderate abrasion but not fluid loss. The cells on the exposed surface are alive and are always moistened.

Stratified squamous epithelium is the first line of defense against the entry of living organisms into the body. Stratification, rapid mitotic activity, and keratinization within the epidermis of the skin are important protective features. An acidic pH along the surfaces of this tissue also helps to prevent disease. The pH of the skin is between 4.0 and 6.8. The pH in the oral cavity ranges from 5.8 to 7.1, which tends to retard the growth of microorganisms. The pH of the anal region is about 6, and the pH along the vaginal lining is 4 or lower.

keratin: Gk. *keras*, horn

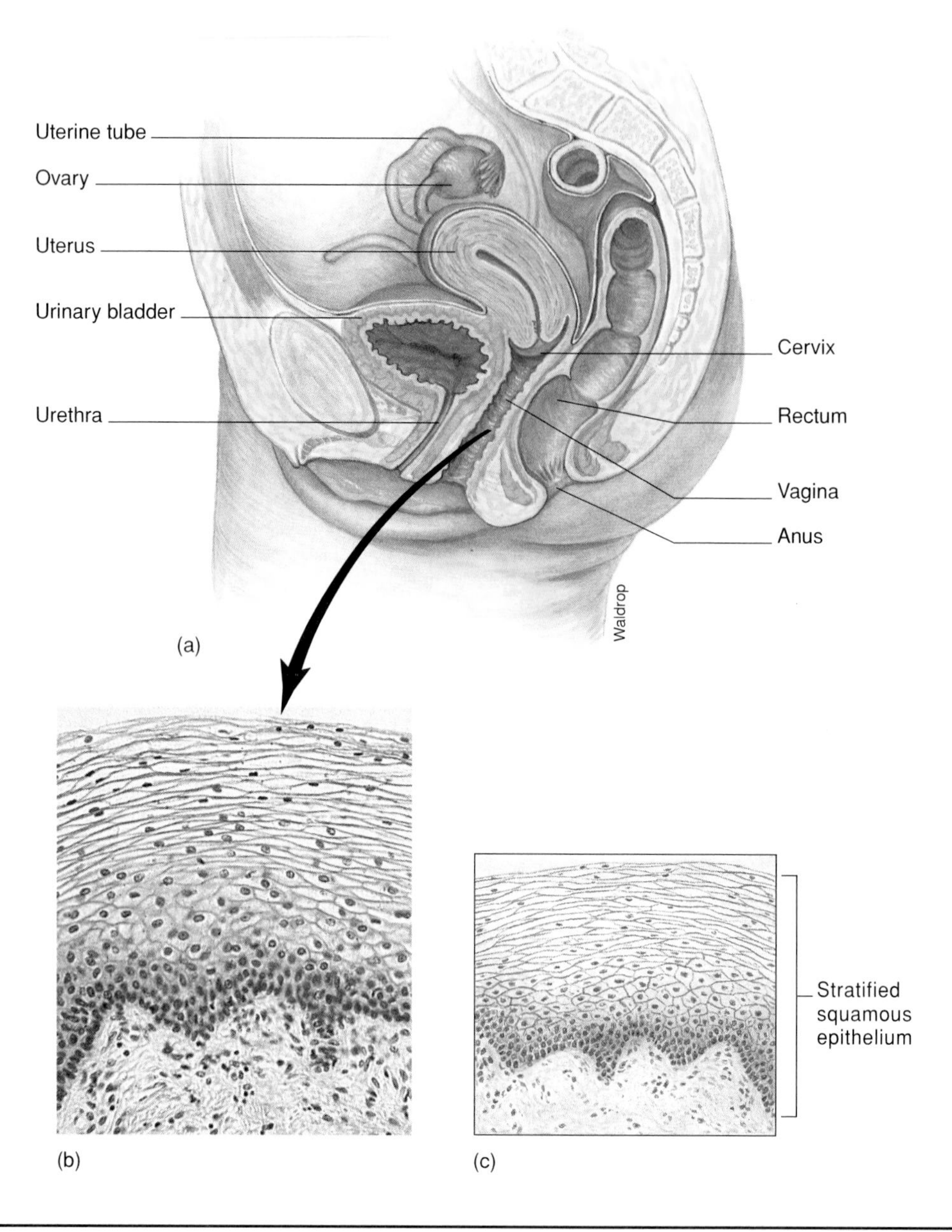

Figure 4.7

Stratified squamous epithelium forms the outer layer of skin and the lining of body openings. In the moistened areas, such as in the vagina (*a*), it is nonkeratinized, whereas in the epidermis of the skin it is keratinized. (*b*) A photomicrograph of this tissue and (*c*) a labeled diagram.

Stratified Cuboidal Epithelium

Stratified cuboidal epithelium usually consists of only two or three layers of cuboidal cells (fig. 4.8). This type of epithelium is confined to the linings of the large ducts of sweat glands, salivary glands, and the pancreas, where its stratification probably provides a more robust lining than would simple epithelium.

Transitional Epithelium

Transitional epithelium is similar to nonkeratinized stratified squamous epithelium except that the surface cells of the former are large and round rather than flat, and some may have two nuclei (fig. 4.9). Transitional epithelium is found only in the urinary system, particularly lining the cavity of the urinary bladder and lining the lumina of the ureters. This tissue is specialized to permit distension (stretching) of the urinary bladder as it fills with urine.

A summary of membranous epithelial tissue is presented in table 4.1.

Body Membranes

Body membranes are composed of thin layers of epithelial tissue and, in certain locations, epithelial tissue coupled with supporting connective tissue. Body membranes cover, separate,

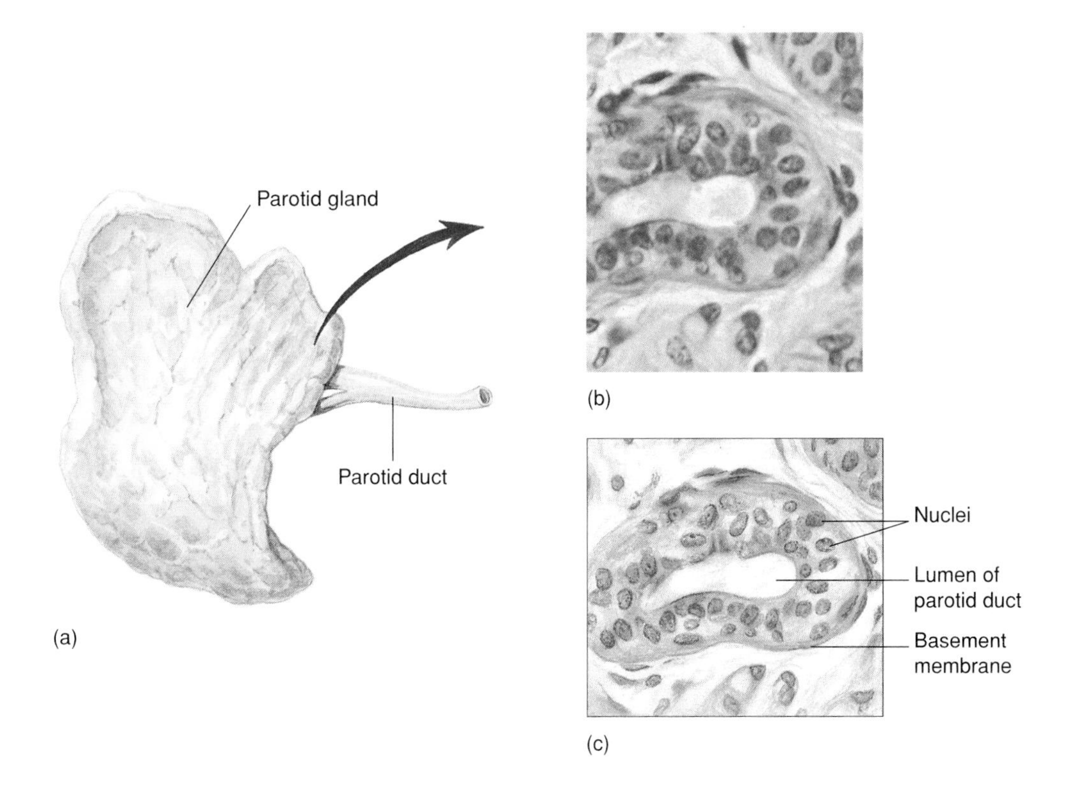

Figure 4.8

(*a*) Stratified cuboidal epithelium lines the lumina of large ducts like the parotid duct, which drains saliva from the parotid gland. (*b*) A photomicrograph of this tissue and (*c*) a labeled diagram.

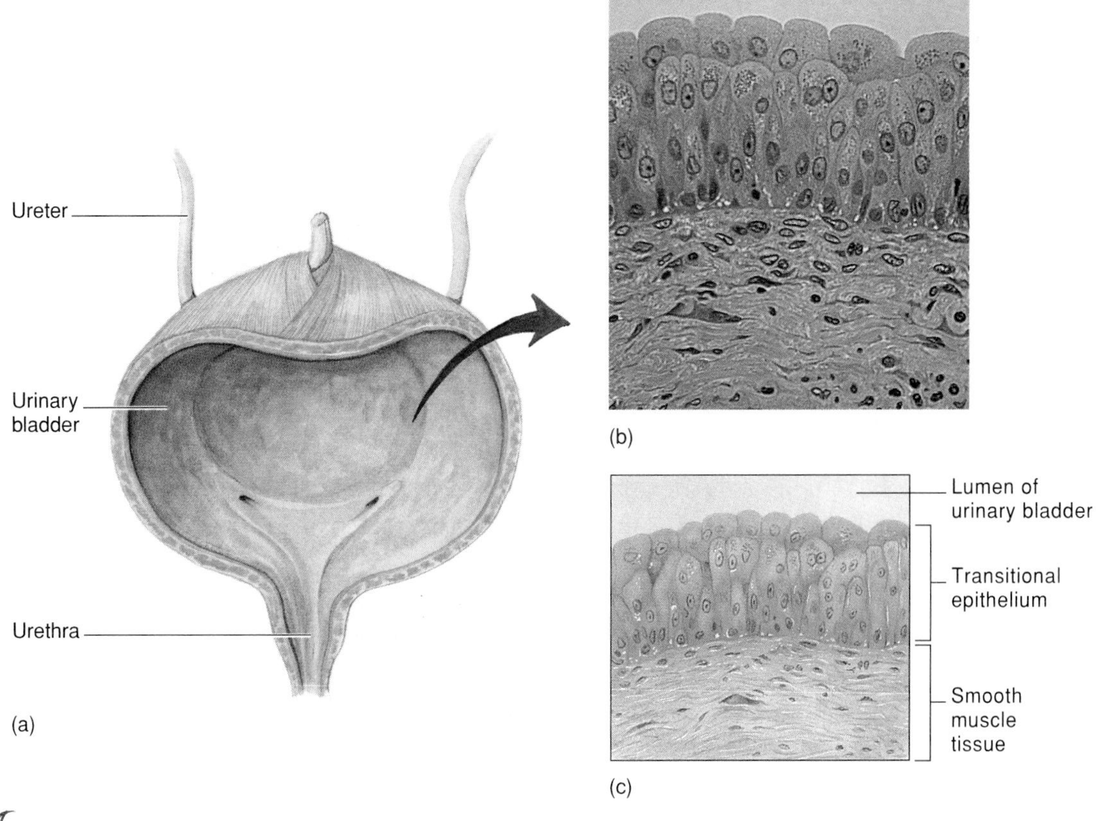

Figure 4.9

(*a*) Transitional epithelium lines the lumina of the ureters and the cavity of the urinary bladder, where it permits distention. (*b*) A photomicrograph of this tissue and (*c*) a labeled diagram.

Table 4.1
Summary of Membranous Epithelial Tissue

Type	Structure and Function	Location
Simple Epithelia	Single layer of cells; function varies with type	Covering visceral organs; linings of body cavities, tubes, and ducts
Simple squamous epithelium	Single layer of flattened, tightly bound cells; diffusion and filtration	Capillary walls; pulmonary alveoli of lungs; covering visceral organs; linings of body cavities
Simple cuboidal epithelium	Single layer of cube-shaped cells; excretion, secretion, or absorption	Surface of ovaries; linings of renal tubules, salivary ducts, and pancreatic ducts
Simple columnar epithelium	Single layer of nonciliated, tall, column-shaped cells; protection, secretion, and absorption	Lining of most of GI tract
Simple ciliated columnar epithelium	Single layer of ciliated, column-shaped cells; transportive role through ciliary motion	Lining of uterine tubes
Pseudostratified ciliated columnar epithelium	Single layer of ciliated, irregularly shaped cells; many goblet cells; protection, secretion, ciliary movement	Lining of respiratory passageways
Stratified Epithelia	Two or more layers of cells; function varies with type	Epidermal layer of skin; linings of body openings, ducts, and urinary bladder
Stratified squamous epithelium (keratinized)	Numerous layers containing keratin, with outer layers flattened and dead; protection	Epidermis of skin
Stratified squamous epithelium (nonkeratinized)	Numerous layers lacking keratin, with outer layers moistened and alive; protection and pliability	Linings of oral and nasal cavities, vagina, and anal canal
Stratified cuboidal epithelium	Usually two layers of cube-shaped cells; strengthening of luminal walls	Large ducts of sweat glands, salivary glands, and pancreas
Transitional epithelium	Numerous layers of rounded, nonkeratinized cells; distension	Walls of ureters, part of urethra, and urinary bladder

and support visceral organs and line body cavities. The two basic types of body membranes, **mucous membranes** and **serous membranes,** are described in detail in chapter 2 under the heading "Body Cavities and Membranes" (see p. 40).

Glandular Epithelia

As tissues develop in the embryo, tiny invaginations (infoldings) of membranous epithelia give rise to specialized secretory structures called **exocrine** (*ek'sŏ-krin*) **glands.** These glands remain connected to the epithelium by ducts, and their secretions pass through the ducts onto body surfaces or into body cavities. Exocrine glands should not be confused with endocrine glands, which are ductless, and which secrete their products (hormones) into the blood or surrounding extracellular fluid. Exocrine glands within the skin include oil (sebaceous) glands, sweat glands, and mammary glands. Exocrine glands within the digestive system include the salivary and pancreatic glands.

Exocrine glands are classified according to their structure and how they discharge their products. Classified according to structure, there are two types of exocrine glands, *unicellular* and *multicellular glands*.

1. **Unicellular glands** are single-celled glands, such as *goblet cells* (fig. 4.10). They are modified columnar cells that occur within most epithelial tissues. Goblet cells are found in the epithelial linings of the respiratory and digestive systems. The mucus secretion of these cells lubricates and protects the surface linings.
2. **Multicellular glands,** as their name implies, are composed of both secretory cells and cells that form the walls of the ducts. Multicellular glands are classified as *simple* or *compound glands*. The ducts of the simple glands do not branch, whereas those of the compound type do (fig. 4.11). Multicellular glands are also classified according to the shape of their secretory portion. They are identified as *tubular glands* if the secretory portion resembles a tube and as acinar glands if the secretory portion resembles a flask. Multicellular glands with a secretory portion that resembles both a tube and a flask are termed *tubuloacinar glands*.

Multicellular glands are also classified according to the means by which they release their product (fig. 4.12).

1. **Merocrine** (*mer'ŏ-krin*) **glands** are those that secrete a watery substance through the cell membrane of the

exocrine: Gk. *exo*, outside; *krinein*, to separate

merocrine: Gk. *meros*, part; *krinein*, to separate

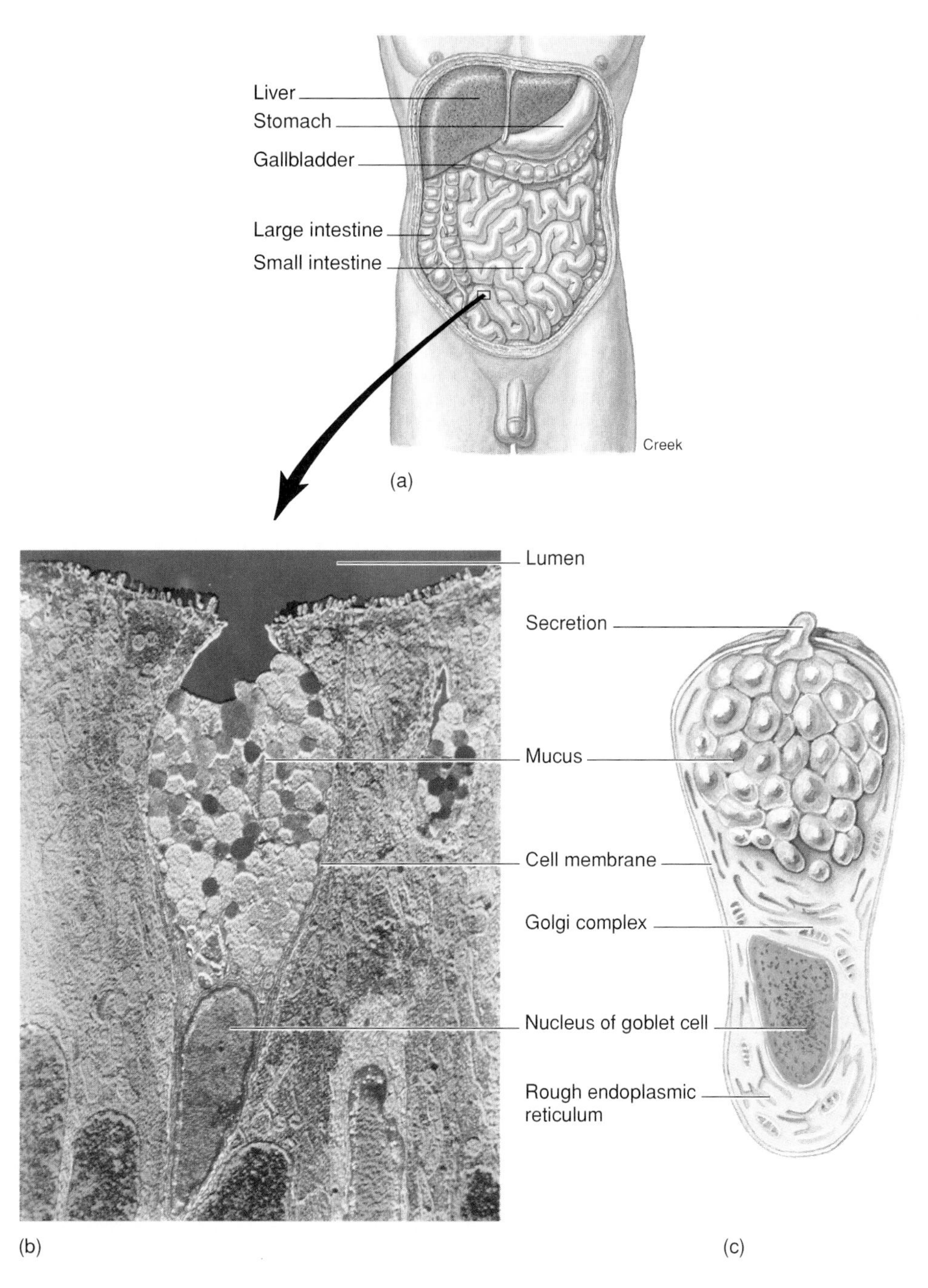

Figure 4.10

A goblet cell is a unicellular gland that secretes mucus, which lubricates and protects surface linings. (*a*) Goblet cells are abundant in the columnar epithelium lining the lumen of the small intestine. (*b*) A photomicrograph of a goblet cell and (*c*) a labeled diagram.

secretory cells. Salivary glands, pancreatic glands, and certain sweat glands are of this type.

2. **Apocrine** (*ap'ŏ-krin*) **glands** are those in which the secretion accumulates on the surface of the secretory cell; then, a portion of the cell, along with the secretion, is pinched off to be discharged. Mammary glands are of this type.
3. **Holocrine** (*hol'ŏ-krin*) **glands** are those in which the entire secretory cell is discharged, along with the secretory product. An example of a holocrine gland is an oil-secreting (sebaceous) gland of the skin (see chapter 5).

A summary of glandular epithelial tissue is presented in table 4.2.

apocrine: Gk. *apo*, off; *krinein*, to separate

holocrine: Gk. *holos*, whole; *krinein*, to separate

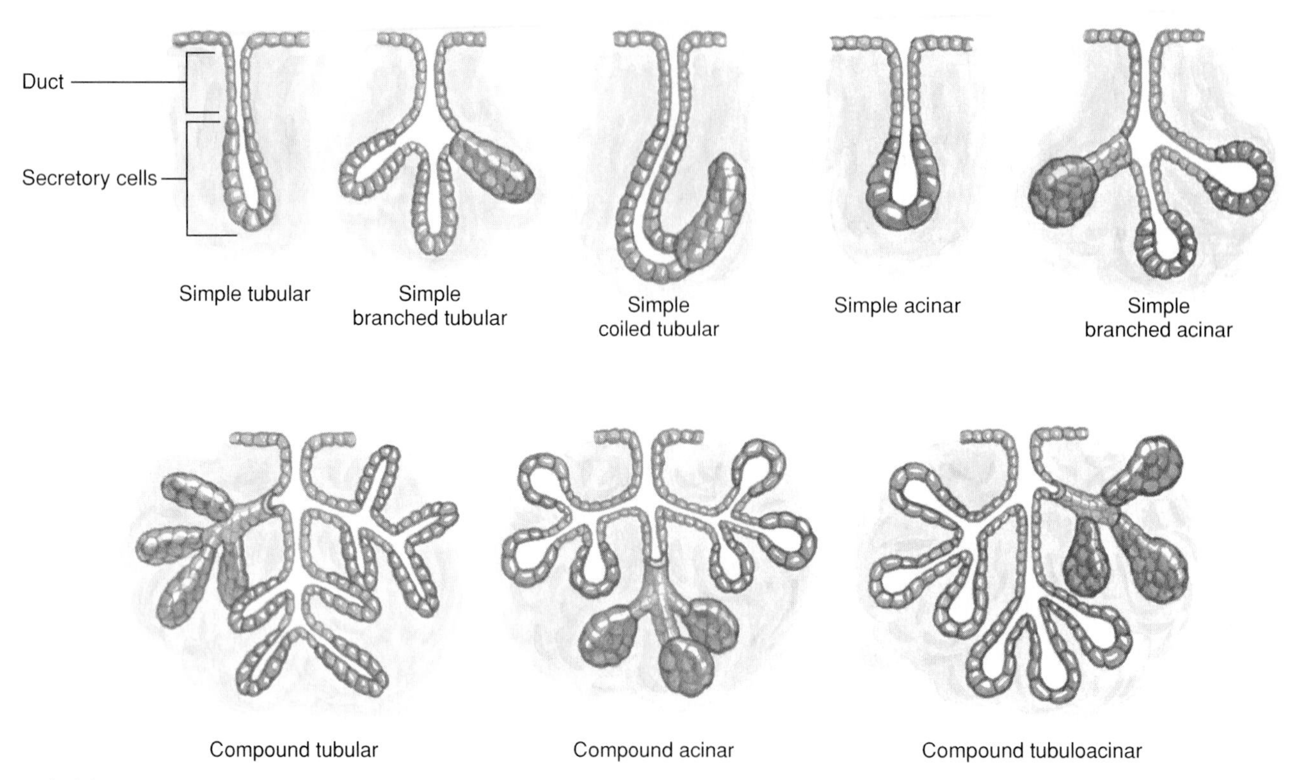

Figure 4.11

Structural classification of multicellular exocrine glands. The ducts of the simple glands either do not branch or have few branches, whereas those of the compound glands have multiple branches.

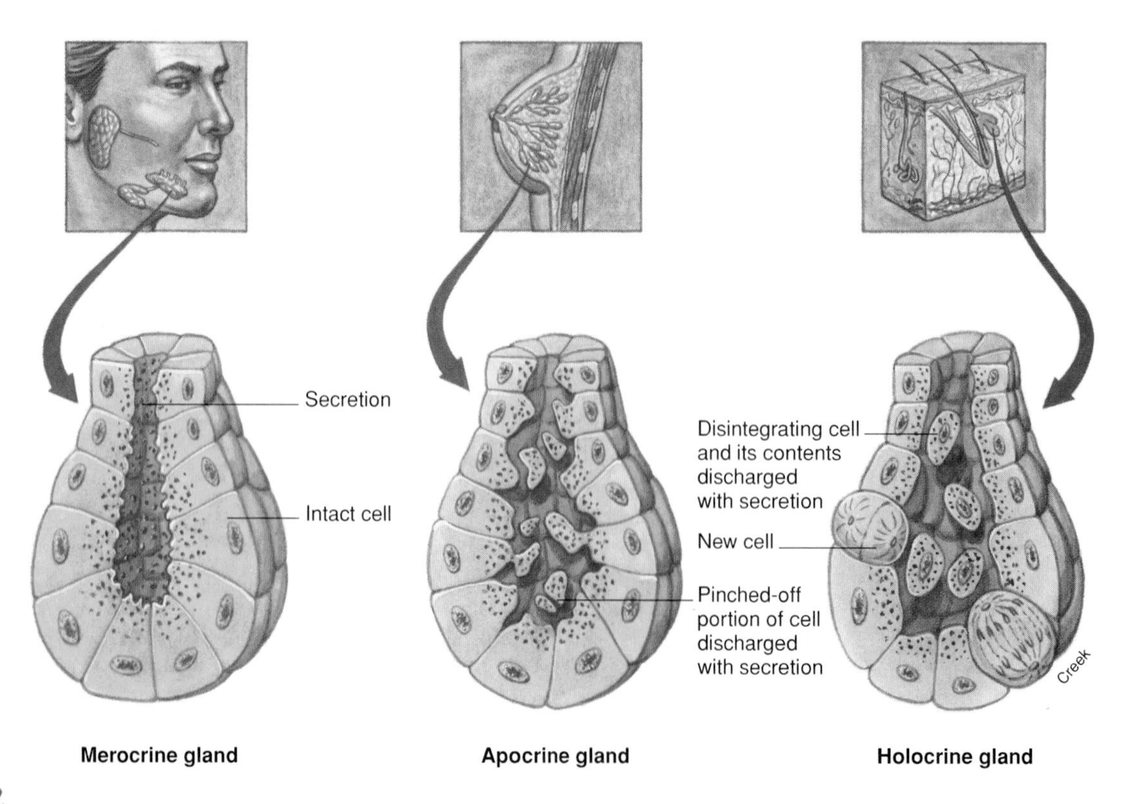

Figure 4.12

Examples of multicellular exocrine glands.

Table 4.2
Summary of Glandular Epithelial Tissue

Classification of Exocrine Glands by Structure

Type	Function	Example
I. Unicellular	Lubricate and protect	Goblet cells of digestive, respiratory, urinary, and reproductive systems
II. Multicellular	Protect, cool body, lubricate, aid in digestion, maintain body homeostasis	Sweat glands, digestive glands, liver, mammary glands, sebaceous glands
A. Simple		
1. Tubular	Aid in digestion	Intestinal glands
2. Branched tubular	Protect, aid in digestion	Uterine glands, gastric glands
3. Coiled tubular	Regulate temperature	Certain sweat glands
4. Acinar	Provide additive for spermatozoa	Seminal vesicles of male reproductive system
5. Branched acinar	Condition skin	Sebaceous glands of the skin
B. Compound		
1. Tubular	Lubricate urethra of male, assist body digestion	Bulbourethral glands of male reproductive system, liver
2. Acinar	Provide nourishment for infant, aid in digestion	Mammary glands, salivary glands (sublingual and submandibular)
3. Tubuloacinar	Aid in digestion	Salivary gland (parotid), pancreas

Classification of Exocrine Glands by Mode of Secretion

Type	Description of Secretion	Example
Merocrine glands	Watery secretion for regulating temperature or enzymes that promote digestion	Salivary and pancreatic glands, certain sweat glands
Apocrine glands	Portion of secretory cell and secretion are discharged; provides nourishment for infant, assists in regulating temperature	Mammary glands, certain sweat glands
Holocrine glands	Entire secretory cell with enclosed secretion is discharged; conditions skin	Sebaceous glands of the skin

1. List the functions of simple squamous epithelia.
2. What are the three types of columnar epithelia? What do they have in common? How are they different?
3. What are the two types of stratified squamous epithelia and how do they differ?
4. Distinguish between unicellular and multicellular glands. Explain how multicellular glands are classified according to their mechanism of secretion.
5. In what ways are mammary glands and certain sweat glands similar?

Connective Tissue

Connective tissue is divided into subtypes according to the matrix that binds the cells. Connective tissue provides structural and metabolic support for other tissues and organs of the body.

Objective 7 Describe the general characteristics, locations, and functions of connective tissue.

Objective 8 Explain the functional relationship between embryonic and adult connective tissue.

Objective 9 List the various ground substances, fiber types, and cells that constitute connective tissue and explain their functions.

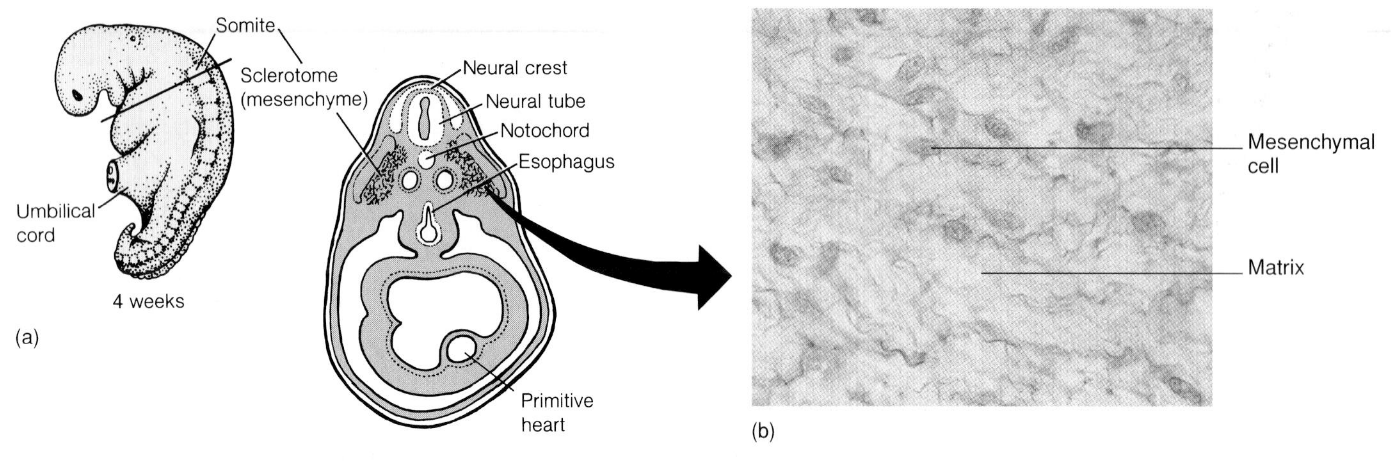

Figure 4.13

Mesenchyme is a type of embryonic connective tissue that can migrate and give rise to all other kinds of connective tissue. (*a*) It is found within an early developing embryo and (*b*) consists of irregularly shaped cells lying in a jellylike homogeneous matrix.

Characteristics and Classification of Connective Tissue

Connective tissue is the most abundant tissue in the body. It supports other tissues or binds them together and provides for the metabolic needs of all body organs. Certain types of connective tissue store nutritional substances; other types manufacture protective and regulatory materials.

Although connective tissue varies widely in structure and function, all types of connective tissue have similarities. With the exception of mature cartilage, connective tissue is highly vascular and well nourished. It is able to replicate and, by so doing, is responsible for the repair of body organs. Unlike epithelial tissue, which is composed of tightly fitted cells, connective tissue contains considerably more matrix (intercellular material) than cells. Connective tissue does not occur on free surfaces of body cavities or on the surface of the body, as does epithelial tissue. Furthermore, connective tissue is embryonically derived from mesoderm, whereas epithelial tissue derives from ectoderm, mesoderm, and endoderm.

The classification of connective tissue is not exact, and several schemes have been devised. In general, however, the various types are named according to the kind and arrangement of the matrix. The following are the basic kinds of connective tissues:

A. Embryonic connective tissue
B. Connective tissue proper
   1. Loose (areolar) connective tissue
   2. Dense regular connective tissue
   3. Dense irregular connective tissue
   4. Elastic connective tissue
   5. Reticular connective tissue
   6. Adipose tissue
C. Cartilage
   1. Hyaline cartilage
   2. Fibrocartilage
   3. Elastic cartilage
D. Bone tissue
E. Blood (vascular tissue)

Embryonic Connective Tissue

The embryonic period of development, which lasts 6 weeks (from the start of the third to the end of the eighth week), is characterized by extensive tissue differentiation and organ formation. At the beginning of the embryonic period, all connective tissue looks alike and is referred to as **mesenchyme** (*mez'en-kīm*). Mesenchyme is undifferentiated embryonic connective tissue that is derived from mesoderm. It consists of irregularly shaped cells surrounded by large amounts of a homogeneous, jellylike matrix (fig. 4.13). In certain periods of development, mesenchyme migrates to predisposed sites where it interacts with other tissues to form organs. Once mesenchyme has completed its embryonic migration to a predetermined destination, it differentiates into all other kinds of connective tissue.

Some mesenchymal-like tissue persists past the embryonic period in certain sites within the body. Good examples are the undifferentiated cells that surround blood vessels and form fibroblasts if the vessels are traumatized. Fibroblasts assist in healing wounds (see chapter 5).

Another kind of prenatal connective tissue exists only in the fetus (the fetal period is from 9 weeks to birth) and is called *mucous connective tissue* or *Wharton's jelly*. It gives a turgid consistency to the umbilical cord.

Connective Tissue Proper

Connective tissue proper has a loose, flexible matrix, frequently called *ground substance*. The most common cell within connective tissue proper is called a **fibroblast** (*fi'bro-blast*). Fibroblasts are large, star-shaped cells that produce **collagenous** (*kŏ-laj'ĕ-nus*), **elastic,** and **reticular** (*rĕ-tik'yoo-lar*) **fibers. Collagenous fibers** are composed of a protein called *collagen* (*kol'ă-jen*); they

Wharton's jelly: from Thomas Wharton, English anatomist, 1614–73
collagen: Gk. *kolla,* glue

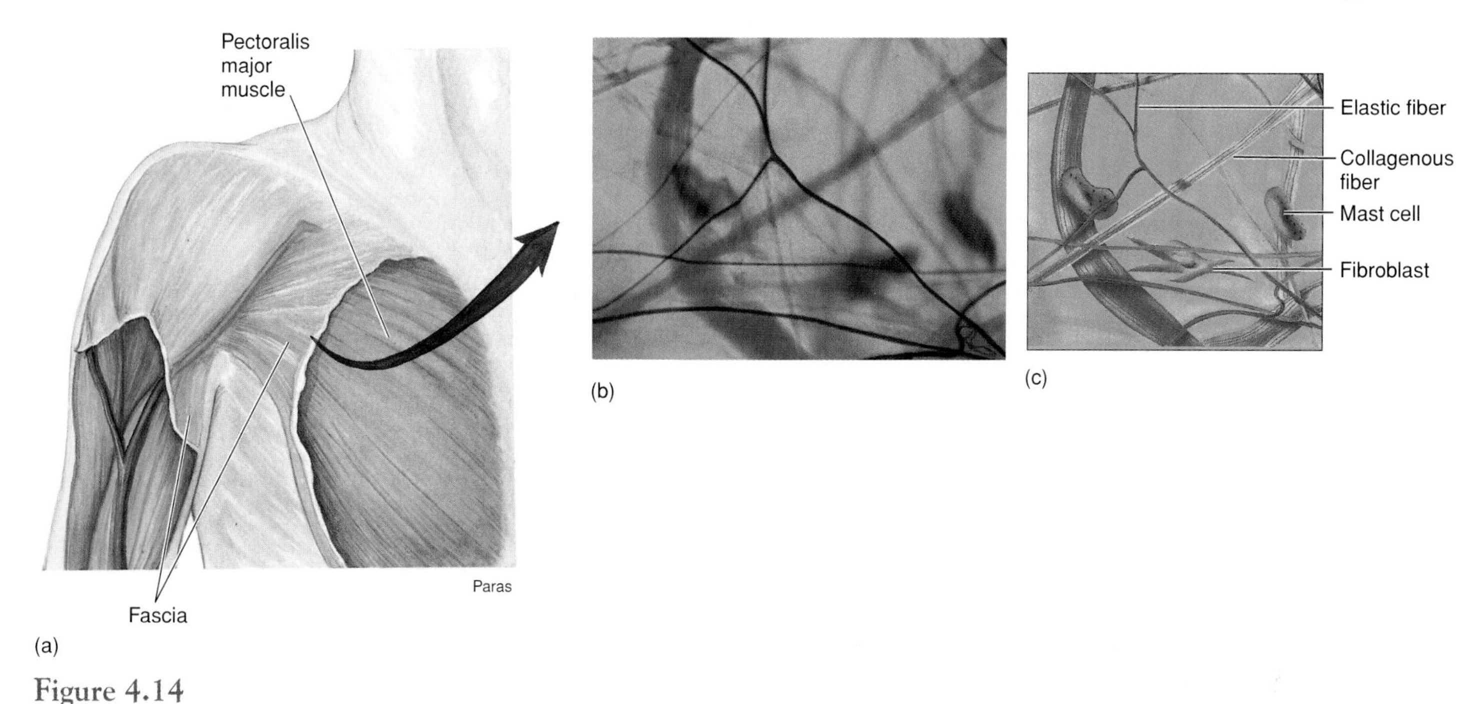

Figure 4.14

Loose connective tissue is packing and binding tissue that surrounds muscles (*a*), nerves, and vessels and binds the skin to the underlying muscles. (*b*) A photomicrograph of the tissue and (*c*) a labeled diagram.

are flexible, yet they have tremendous strength. **Elastic fibers** are composed of a protein called *elastin*, which provides certain tissues with elasticity. Collagenous and elastic fibers may be either sparse and irregularly arranged, as in loose connective tissue, or tightly packed, as in dense connective tissue. Tissues with loosely arranged fibers generally form packing material that cushions and protects various organs, whereas those that are tightly arranged form the binding and supportive connective tissues of the body.

Resilience in tissues that contain elastic fibers is extremely important for several physical functions of the body. Consider, for example, that elastic fibers are found in the walls of arteries and in the walls of the lower respiratory passageways. As these walls are expanded by blood moving through vessels or by inspired air, the elastic fibers must first stretch and then recoil. This maintains the pressures of the fluid or air moving through the lumina, thus ensuring adequate flow rates and rates of diffusion through capillary and lung surfaces.

Reticular fibers reinforce by branching and joining to form a delicate lattice or reticulum. Reticular fibers are common in lymphatic glands, where they form a meshlike center called the *stroma*.

Six basic types of connective tissue proper are generally recognized. These tissues are distinguished by the consistency of the ground substance and the type and arrangement of the reinforcement fibers.

elastin: Gk. *elasticus*, to drive
reticular: L. *rete*, net or netlike
stroma: Gk. *stroma*, a couch or bed

Loose Connective (Areolar) Tissue

Loose connective tissue is distributed throughout the body as a binding and packing material. It binds the skin to the underlying muscles and is highly vascular, providing nutrients to the skin. Loose connective tissue that binds skin to underlying muscles is known as **fascia** (*fash'e-ă*). It also surrounds blood vessels and nerves, where it provides both protection and nourishment. Specialized cells called **mast cells** are dispersed throughout the loose connective tissue surrounding blood vessels. Mast cells produce *heparin* (*hep'ă-rin*), an anticoagulant that prevents blood from clotting within the vessels. They also produce *histamine*, which is released during inflammation. Histamine acts as a powerful vasodilator.

The cells of loose connective tissue are predominantly fibroblasts, with collagenous and elastic fibers dispersed throughout the ground substance (fig. 4.14). The irregular arrangement of this tissue provides flexibility, yet strength, in any direction. It is this tissue layer, for example, that permits the skin to move when a part of the body is rubbed.

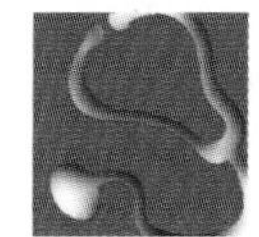

Much of the fluid of the body is found within loose connective tissue and is called *interstitial fluid* (tissue fluid). Sometimes excessive tissue fluid accumulates, causing a swollen condition called *edema* (*ĕ-de'mă*). Edema is a symptom of numerous dysfunctions or disease processes.

fascia: L. *fascia*, a band or girdle
heparin: Gk. *hepatos*, the liver

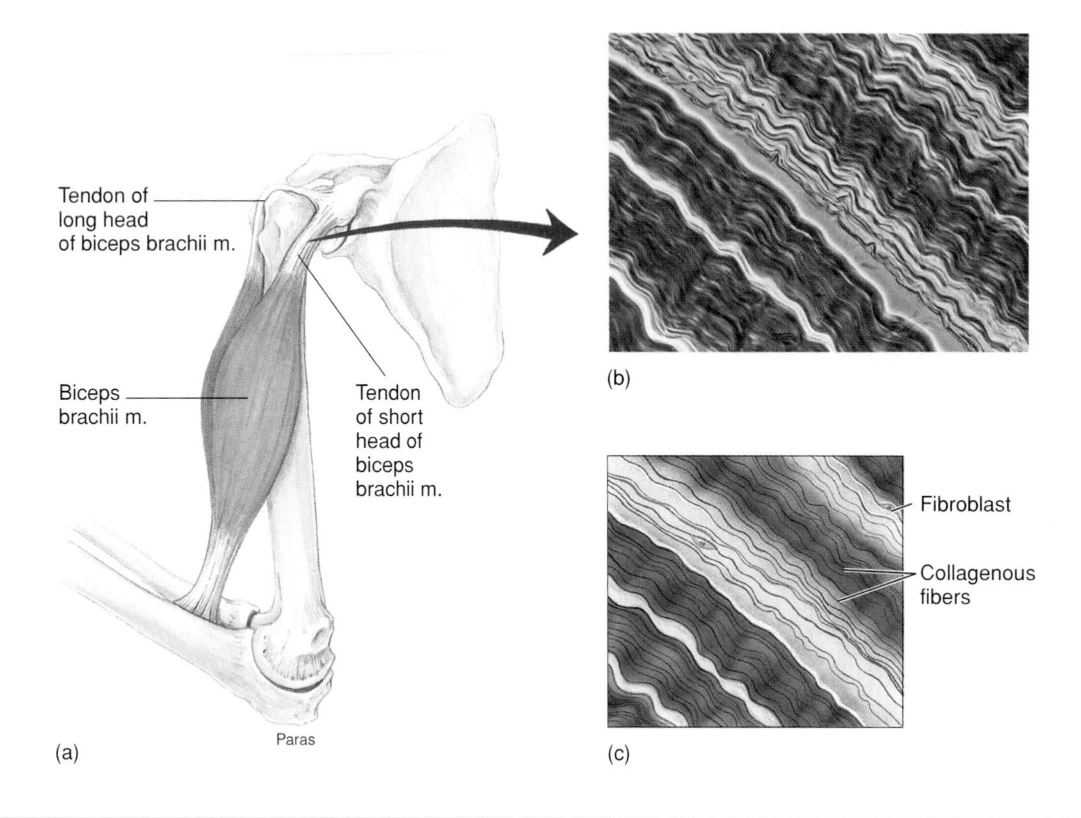

Figure 4.15

Dense regular connective tissue forms the strong and highly flexible tendons (*a*) and ligaments. (*b*) A photomicrograph of the tissue and (*c*) a labeled diagram.

Dense Regular Connective Tissue

Dense regular connective tissue is characterized by large amounts of densely packed collagenous fibers that run parallel to the direction of force placed on the tissue during body movement. Because this tissue is silvery white in appearance, it is sometimes called white fibrous connective tissue.

Dense regular connective tissue occurs where strong, flexible support is needed (fig. 4.15). **Tendons,** which attach muscles to bones and transfer the forces of muscle contractions, and **ligaments,** which connect bone to bone across articulations, are composed of this type of tissue.

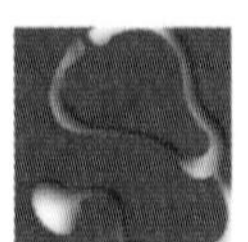

Trauma to ligaments, tendons, and muscles are common sports-related injuries. A *strain* is an excessive stretch of the tissue composing the tendon or muscle, with no serious damage. A *sprain* is a tearing of the tissue of a ligament and may be slight, moderate, or complete. A complete tear of a major ligament is especially painful and disabling. Ligamentous tissue does not heal well because it has a poor blood supply. Surgical reconstruction is generally needed for the treatment of a severed ligament.

tendon: L. *tendere*, to stretch
ligament: L. *ligare*, bind

Dense Irregular Connective Tissue

Dense irregular connective tissue is characterized by large amounts of densely packed collagenous fibers that are interwoven to provide tensile strength in any direction. This tissue is found in the dermis of the skin and the submucosa of the GI tract. It also forms the fibrous capsules of organs and joints (fig. 4.16).

Elastic Connective Tissue

Elastic connective tissue is composed primarily of elastic fibers that are irregularly arranged and yellowish in color (fig. 4.17). They can be stretched to one and a half times their original lengths and will snap back to their former size. Elastic connective tissue is found in the walls of large arteries, in portions of the larynx, and in the trachea and bronchial tubes of the lungs. It is also present between the arches of the vertebrae that make up the vertebral column.

Reticular Connective Tissue

Reticular connective tissue is characterized by a network of reticular fibers woven through a jellylike matrix (fig. 4.18). Certain specialized cells within reticular tissue are *phagocytic* (*fag"ŏ-sit'ik*), and therefore can ingest foreign materials. The liver, spleen, lymph nodes, and bone marrow contain reticular connective tissue.

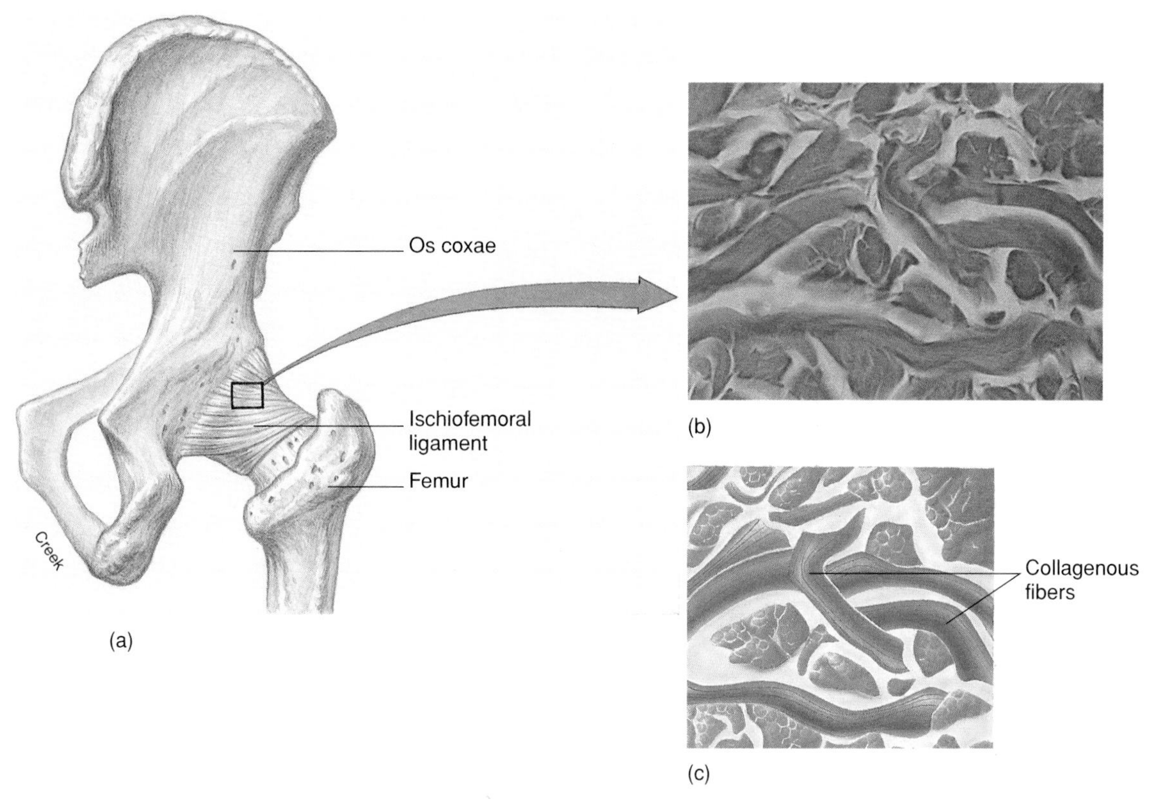

Figure 4.16

Dense irregular connective tissue forms joint capsules (*a*) that contain synovial fluid for lubricating movable joints. (*b*) A photomicrograph of the tissue and (*c*) a labeled diagram.

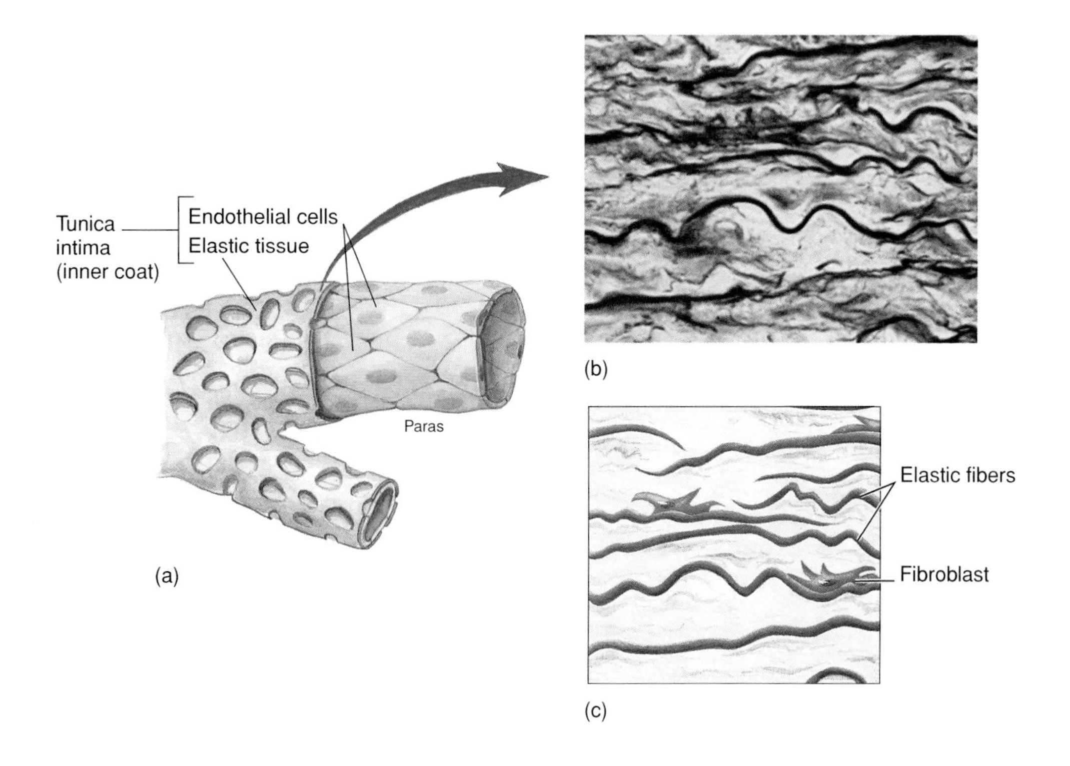

Figure 4.17

Elastic connective tissue permits stretching of a large artery (*a*) as blood flows through. (*b*) A photomicrograph of the tissue and (*c*) a labeled diagram.

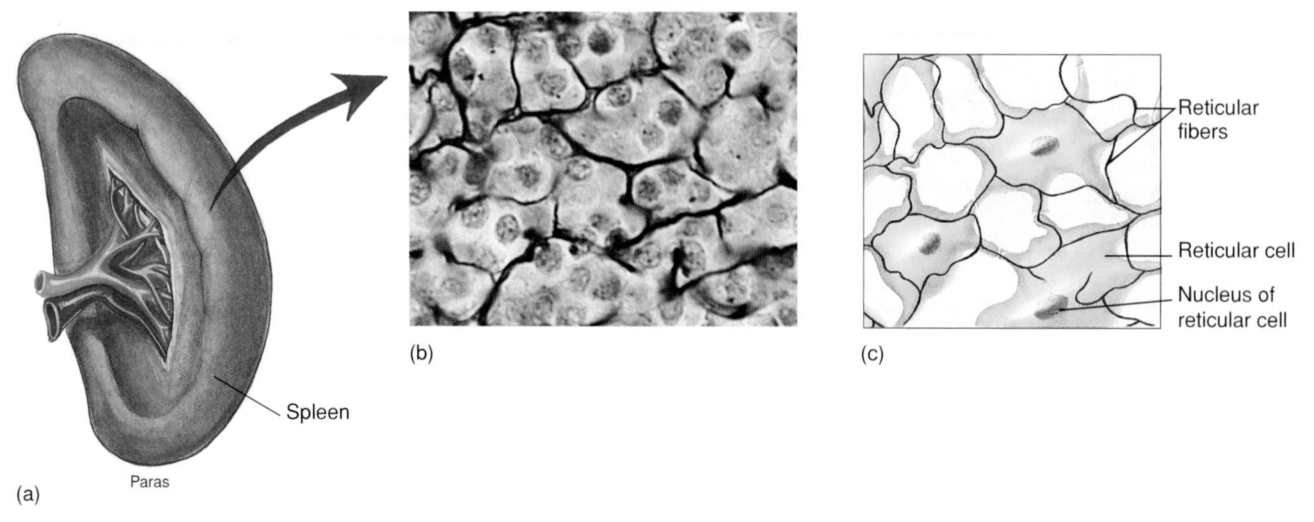

Figure 4.18

Reticular connective tissue forms the stroma, or framework, of such organs as the spleen (*a*), liver, thymus, and lymph nodes. (*b*) A photomicrograph of this tissue and (*c*) a labeled diagram.

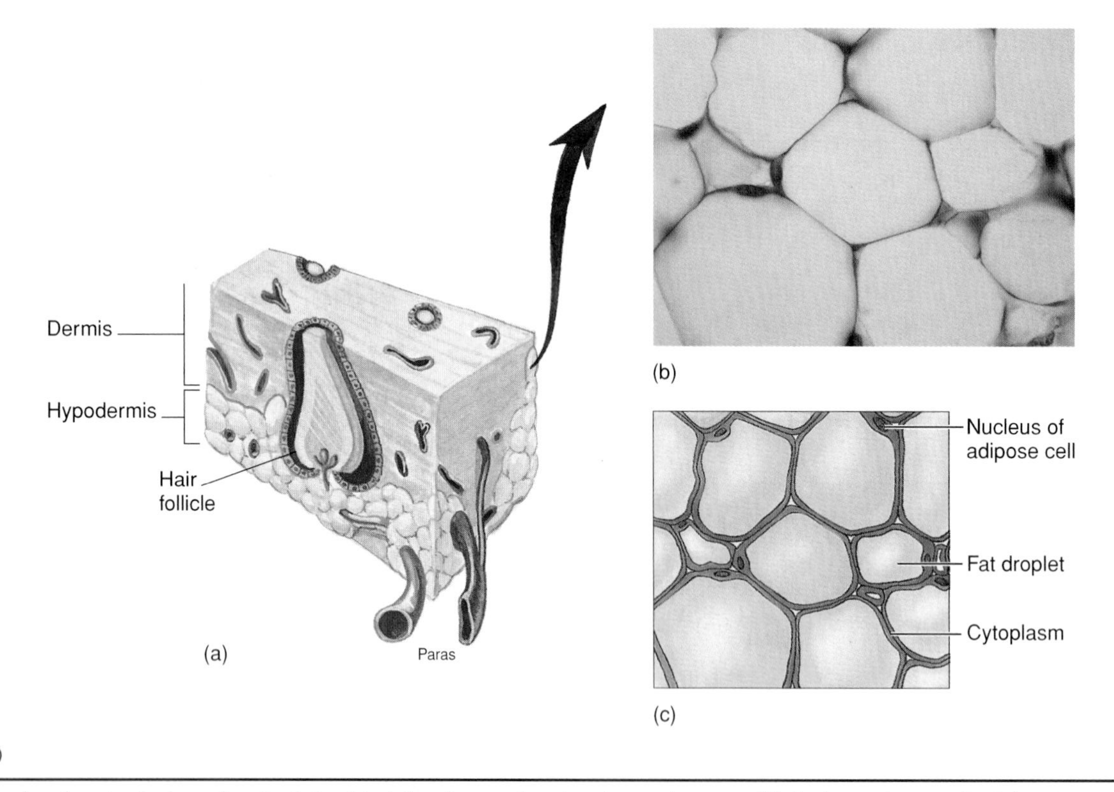

Figure 4.19

Adipose tissue is abundant in the hypodermis of the skin (*a*) and around various internal organs. (*b*) A photomicrograph of the tissue and (*c*) a labeled diagram.

Adipose Tissue

Adipose tissue is a specialized type of loose fibrous connective tissue that contains large quantities of **adipose cells,** or **adipocytes** (*ad′ĭ-po-sīts*). Adipose cells form from mesenchyme and, for the most part, are formed prenatally and during the first year of life. Adipose cells store droplets of fat within their cytoplasm, causing them to swell and forcing their nuclei to one side (fig. 4.19).

Adipose tissue is found throughout the body but is concentrated around the kidneys, in the hypodermis of the skin, on the

adipose: L. *adiposus,* fat

Table 4.3
Summary of Connective Tissue Proper

Type	Structure and Function	Location
Loose connective (areolar) tissue	Predominantly fibroblast cells with lesser amounts of collagen and elastin proteins; binds organs, holds tissue fluids	Surrounding nerves and vessels, between muscles, beneath the skin
Dense regular connective tissue	Densely packed collagenous fibers that run parallel to the direction of force; provides strong, flexible support	Tendons, ligaments
Dense irregular connective tissue	Densely packed collagenous fibers arranged in a tight interwoven pattern; provides tensile strength in any direction	Dermis of skin, fibrous capsules of organs and joints, periosteum of bone
Elastic connective tissue	Predominantly irregularly arranged elastic fibers; supports, provides framework	Large arteries, lower respiratory tract, between the arches of vertebrae
Reticular connective tissue	Reticular fibers that form a supportive network; stores, performs phagocytic function	Lymph nodes, liver, spleen, thymus, bone marrow
Adipose tissue	Adipose cells; protects, stores fat, insulates	Hypodermis of skin, surface of heart, omentum, around kidneys, back of eyeball, surrounding joints

surface of the heart, surrounding joints, and in the breasts of sexually mature females. Fat functions not only as a food reserve, but also supports and protects various organs. It is a good insulator against cold because it is a poor conductor of heat.

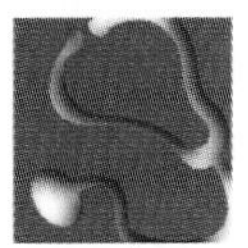

Excessive fat can be unhealthy by placing a strain on the heart and perhaps causing early death. For these reasons, good exercise programs and sensible diets are extremely important. Adipose tissue can also retain environmental pollutants that are ingested or absorbed through the skin. Dieting eliminates the fat stored within adipose tissue but not the tissue itself.

The surgical procedure of *suction lipectomy* may be used to remove small amounts of adipose tissue from localized body areas such as the breasts, abdomen, buttocks, and thighs. Suction lipectomy is used for cosmetic purposes rather than as a treatment for obesity, and the risks for potentially detrimental side-effects need to be seriously considered. Potential candidates should be between 30 and 40 years old and only about 15 to 20 pounds overweight. They should also have good skin elasticity.

The characteristics, functions, and locations of connective tissue proper are summarized in table 4.3.

Cartilage

Cartilage (*kar′tĭ-lij*) consists of cartilage cells, or **chondrocytes** (*kon′dro-sīts*), and a semisolid matrix that imparts marked elastic properties to the tissue. It is a supportive and protective connective tissue that is frequently associated with bone. Cartilage forms a precursor to one type of bone and persists at the articular surfaces on the bones of all movable joints.

The chondrocytes within cartilage may occur singly but are frequently clustered. Chondrocytes occupy cavities, called **lacunae** (*lă-kyoo′ne*—singular *lacuna*), within the matrix. Most cartilage is surrounded by a dense irregular connective tissue called **perichondrium** (*per″ĭ-kon′dre-um*). Cartilage at the articular surfaces of bones (articular cartilage) lacks a perichondrum. Because mature cartilage is avascular, it must receive nutrients through diffusion from the perichondrium and the surrounding tissue. For this reason, cartilaginous tissue has a slow rate of mitotic activity; if damaged, it heals with difficulty.

There are three kinds of cartilage: *hyaline* (*hi′-ă-līn*) *cartilage*, *fibrocartilage*, and *elastic cartilage*. They are distinguished by the type and amount of fibers embedded within the matrix.

Hyaline Cartilage

Hyaline cartilage, commonly called "gristle," has a homogeneous, bluish-staining matrix in which the collagenous fibers are so fine that they can be observed only with an electron microscope. When viewed through a light microscope, hyaline cartilage has a clear, glassy appearance (fig. 4.20).

Hyaline cartilage is the most abundant cartilage within the body. It covers the articular surfaces of bones, supports the tubular trachea and bronchi of the respiratory system, reinforces the nose, and forms the flexible bridge, called **costal cartilage,** between the ventral end of each of the first 10 ribs and the sternum. Most of the bones of the body form first as hyaline cartilage and later become bone in a process called *endochondral ossification*.

Fibrocartilage

Fibrocartilage has a matrix that is reinforced with numerous collagenous fibers (fig. 4.21). It is a durable tissue adapted to

lacuna: L. *lacuna*, hole or pit
hyaline: Gk. *hyalos*, glass

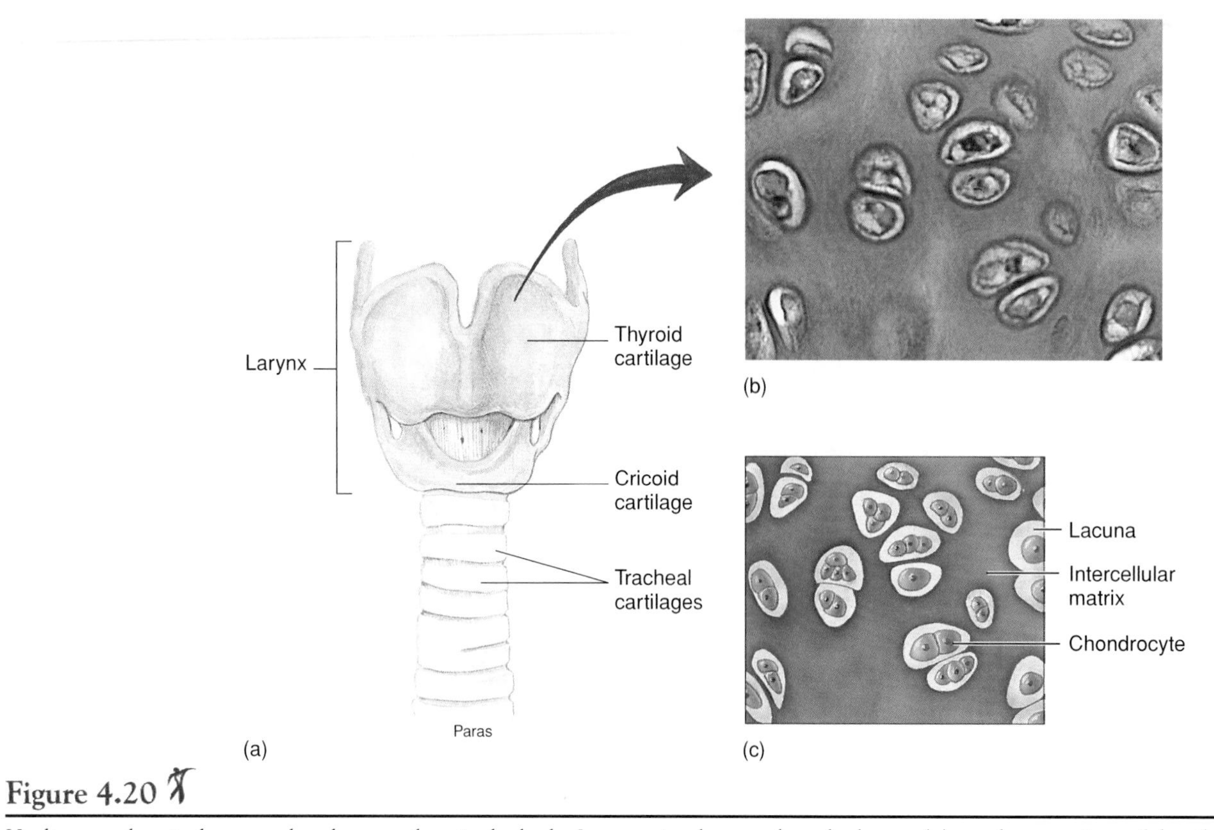

Figure 4.20

Hyaline cartilage is the most abundant cartilage in the body. It occurs in places such as the larynx (*a*), trachea, portions of the rib cage, and embryonic skeleton. (*b*) A photomicrograph of the tissue and (*c*) a labeled diagram.

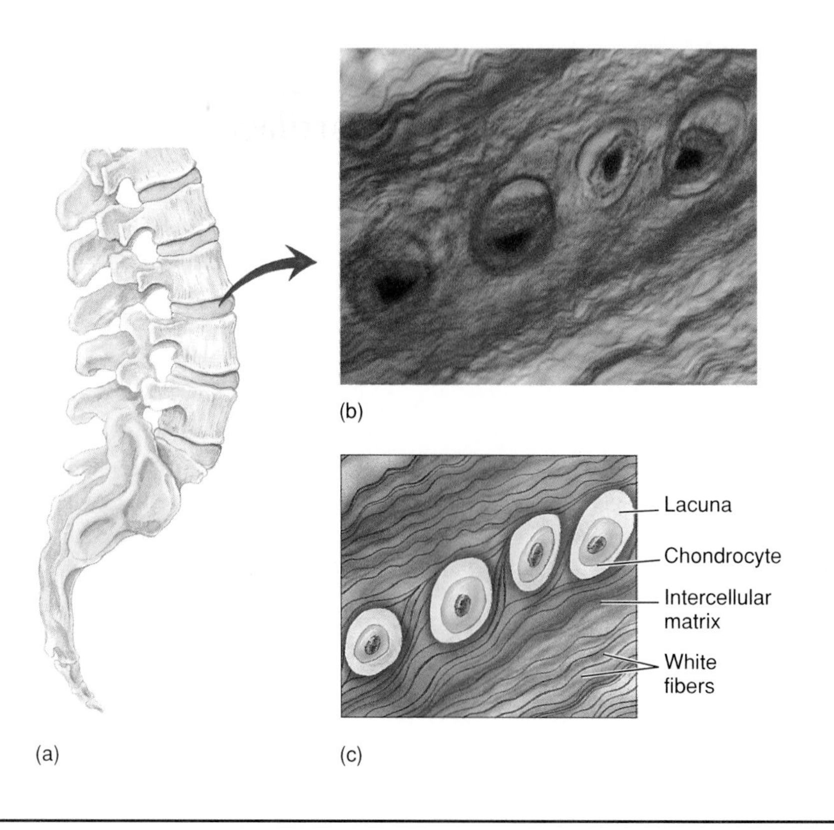

Figure 4.21

Fibrocartilage is located at the symphysis pubis, within the knee joints, and between the vertebrae as the intervertebral discs (*a*). A photomicrograph of the tissue is shown in (*b*) and a labeled diagram in (*c*).

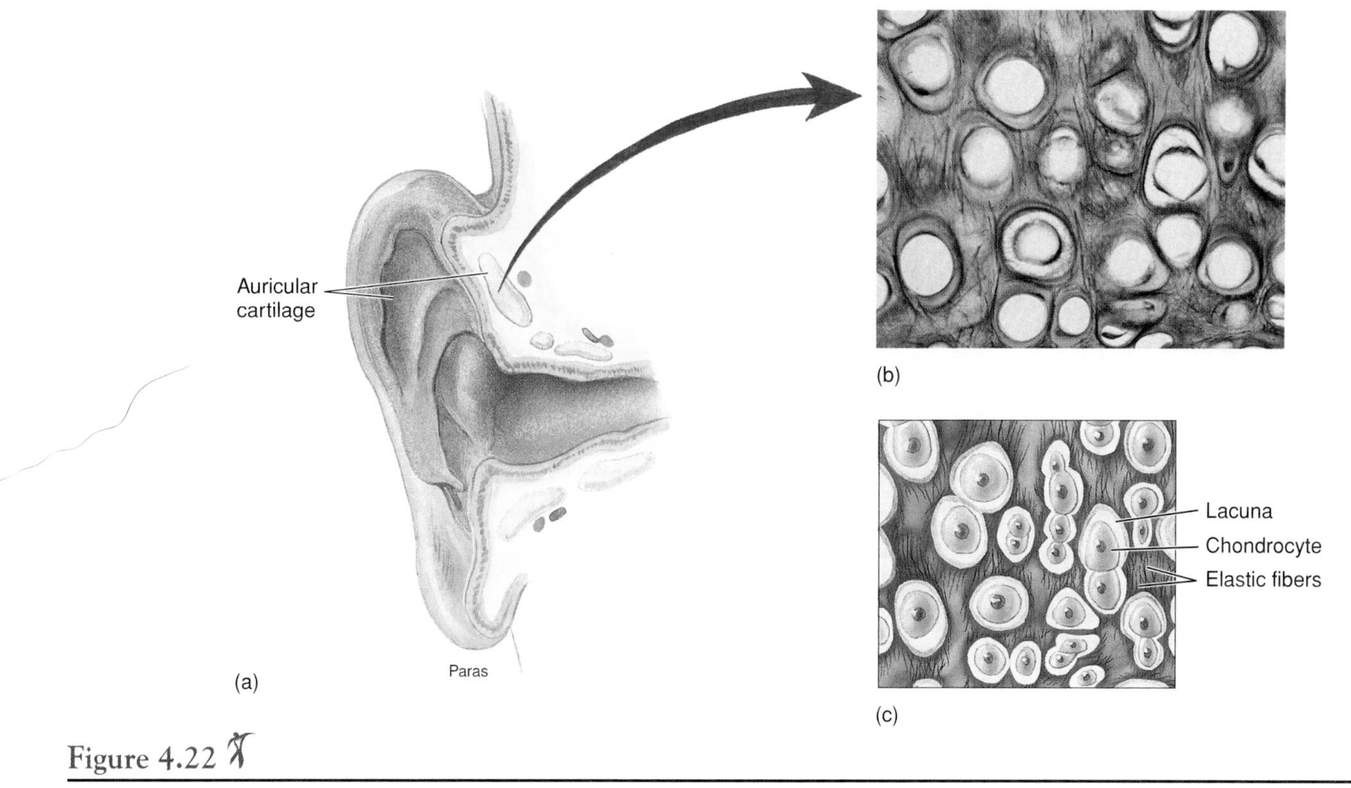

Figure 4.22

Elastic cartilage gives support to the outer ear (*a*), auditory canal, and parts of the larynx. A photomicrograph of the tissue is shown in (*b*) and a labeled diagram in (*c*).

Table 4.4
Summary of Cartilage

Type	Structure and Function	Location
Hyaline cartilage	Homogeneous matrix with extremely fine collagenous fibers; provides flexible support, protects, is precursor to bone	Articular surfaces of bones, nose, walls of respiratory passages, fetal skeleton
Fibrocartilage	Abundant collagenous fibers within matrix; supports, withstands compression	Symphysis pubis, intervertebral discs, knee joint
Elastic cartilage	Abundant elastic fibers within matrix; supports, provides flexibility	Framework of outer ear, auditory canal, portions of larynx

Fibrocartilage

withstand tension and compression. It is found at the symphysis pubis, where the two pelvic bones articulate, and between the vertebrae as intervertebral discs. It also forms the cartilaginous wedges within the knee joint, called *menisci* (see chapter 8).

By the end of the day, the intervertebral discs of the vertebral column are somewhat compacted. So a person is actually slightly shorter in the evening than in the morning, following a recuperative rest. Aging, however, brings with it a gradual compression of the intervertebral discs that is irreversible.

Elastic Cartilage

Elastic cartilage is similar to hyaline cartilage except for the presence of abundant elastic fibers that make elastic cartilage very flexible without compromising its strength (fig. 4.22). The numerous elastic fibers also give it a yellowish appearance. This tissue is found in the outer ear, portions of the larynx, and in the auditory canal.

The three types of cartilage are summarized in table 4.4.

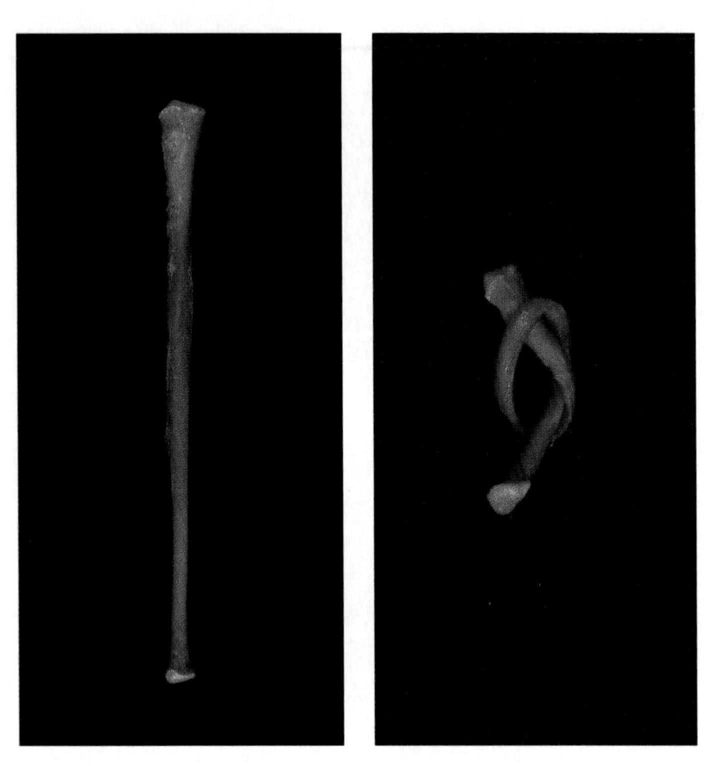

Figure 4.23

A bone soaked in a weak acid, such as the acidic acid in vinegar, demineralizes and becomes flexible.

Bone Tissue

Bone is the most rigid of all the connective tissues. Unlike cartilage, bone has a rich vascular supply and is the site of considerable metabolic activity. The hardness of bone is largely due to the calcium phosphate (calcium hydroxyapatite) deposited within the intercellular matrix. Numerous collagenous fibers, also embedded within the matrix, give bone some flexibility.

When bone is placed in a weak acid, the calcium salts dissolve away and the bone becomes pliable. It retains its basic shape but can be easily bent and twisted (fig. 4.23). In calcium deficiency diseases, such as *rickets,* the bone tissue becomes pliable and bends under the weight of the body (see fig. 5.10).

Based on porosity, bone tissue is classified as either compact or spongy, and most bones have both types (fig. 4.24). **Compact** (*dense*) **bone tissue** constitutes the hard outer portion of a bone, and **spongy** (*cancellous*) **bone tissue** constitutes the porous, highly vascular inner portion. The outer surface of a bone is covered by a connective tissue layer called the *periosteum* that serves as a site of attachment for ligaments and tendons, provides protection, and gives durable strength to the bone. Spongy bone tissue makes the bone lighter and provides a space for red bone marrow, where blood cells are produced.

In compact bone tissue, mature bone cells, called **osteocytes,** are arranged in concentric layers around a **central** (haversian) **canal,** which contains a vascular and nerve supply. Each osteocyte occupies a cavity called a **lacuna.** Radiating from each lacuna are numerous minute canals, or **canaliculi,** which traverse the dense matrix of the bone tissue to adjacent lacunae. Nutrients diffuse through the canaliculi to reach each osteocyte. The matrix is deposited in concentric layers called **lamellae.** Bone tissue is described in detail in chapter 6.

Blood (Vascular Tissue)

Blood, or **vascular tissue,** is a highly specialized fluid connective tissue that plays a vital role in maintaining homeostasis. The cells, or **formed elements,** of blood are suspended in a liquid matrix called **blood plasma** (fig. 4.25). The three types of formed elements are **erythrocytes** (red blood cells), **leukocytes** (white blood cells), and **thrombocytes** (platelets). Blood is discussed fully in chapter 16.

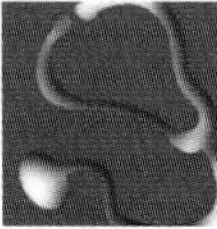

An injury to a portion of the body may stimulate tissue repair activity, usually involving connective tissue. A minor scrape or cut results in platelet and plasma activity of the exposed blood and the formation of a scab. The epidermis of the skin regenerates beneath the scab. A severe open wound heals through connective tissue granulation. In this process, collagenous fibers form from surrounding fibroblasts to strengthen the traumatized area. The healed area is known as a *scar.*

1. List the basic types of connective tissue and describe the structure, function, and location of each.
2. Which of the previously discussed connective tissues function to protect body organs? Which type is phagocytic? Which types bind and support various structures? Which types are associated in some way with the skin?
3. What is the developmental significance of mesenchyme and how does it differ functionally from adult connective tissue?
4. Briefly describe reticular fibers, fibroblasts, collagenous fibers, elastic fibers, and mast cells.

haversian canal: from Clopton Havers, English anatomist, 1650–1702
erythrocyte: Gk. *erythros,* red; *kytos,* hollow (cell)
leukocyte: Gk. *leukos,* white; *kytos,* hollow (cell)
thrombocyte: Gk. *thrombos,* a clot; *kytos,* hollow (cell)

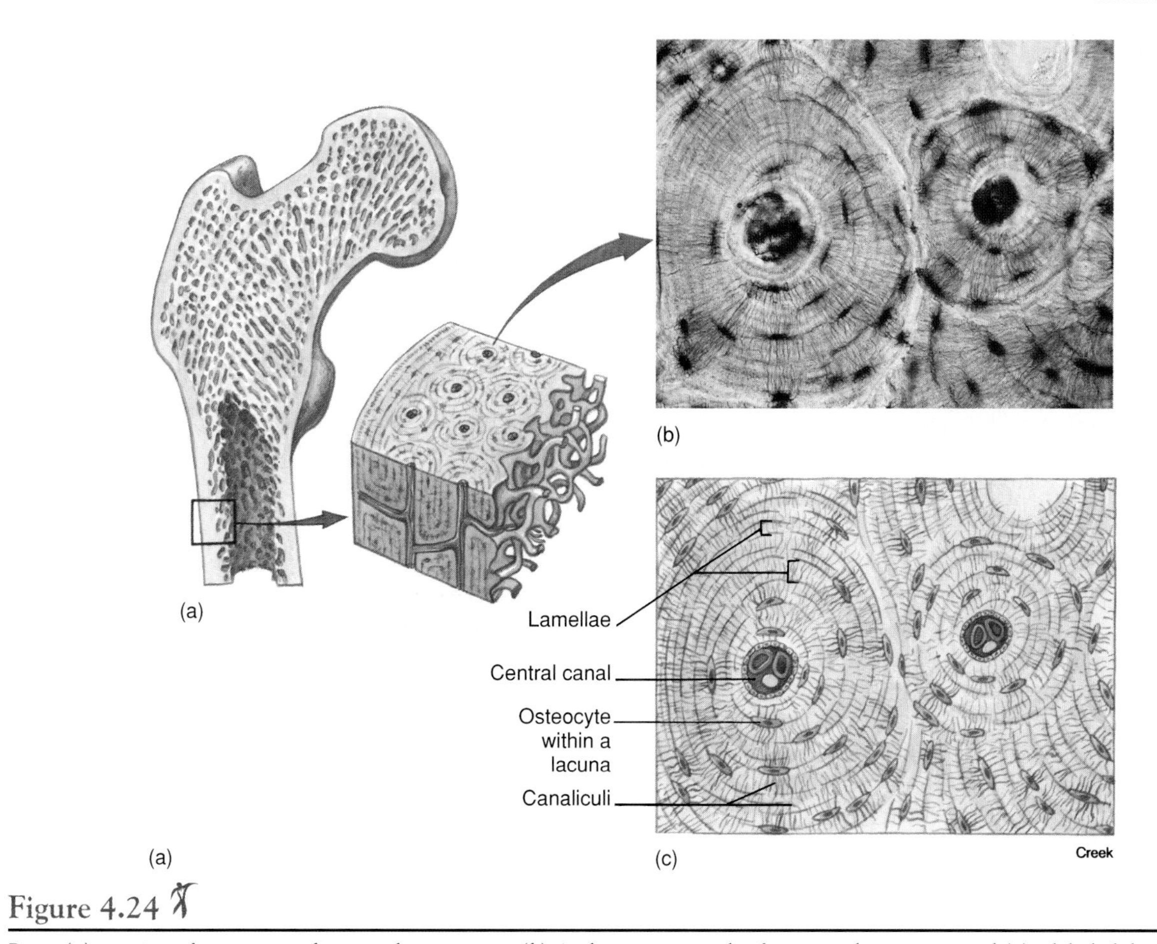

Figure 4.24

Bone (*a*) consists of compact and spongy bone tissues. (*b*) A photomicrograph of compact bone tissue and (*c*) a labeled diagram.

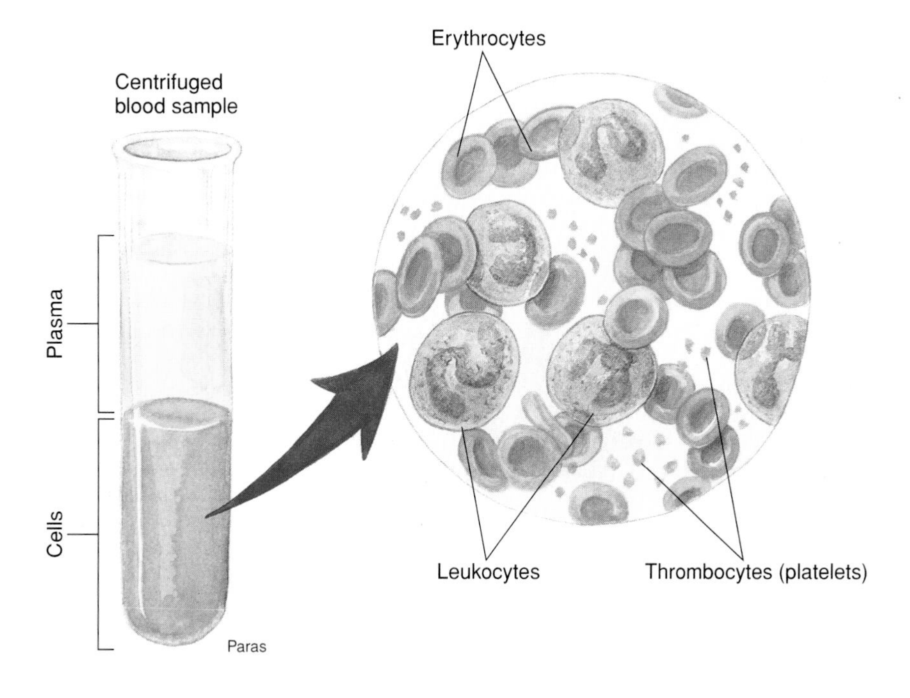

Figure 4.25

Blood consists of formed elements—erythrocytes (red blood cells), leukocytes (white blood cells), and thrombocytes (platelets)—suspended in a liquid plasma matrix.

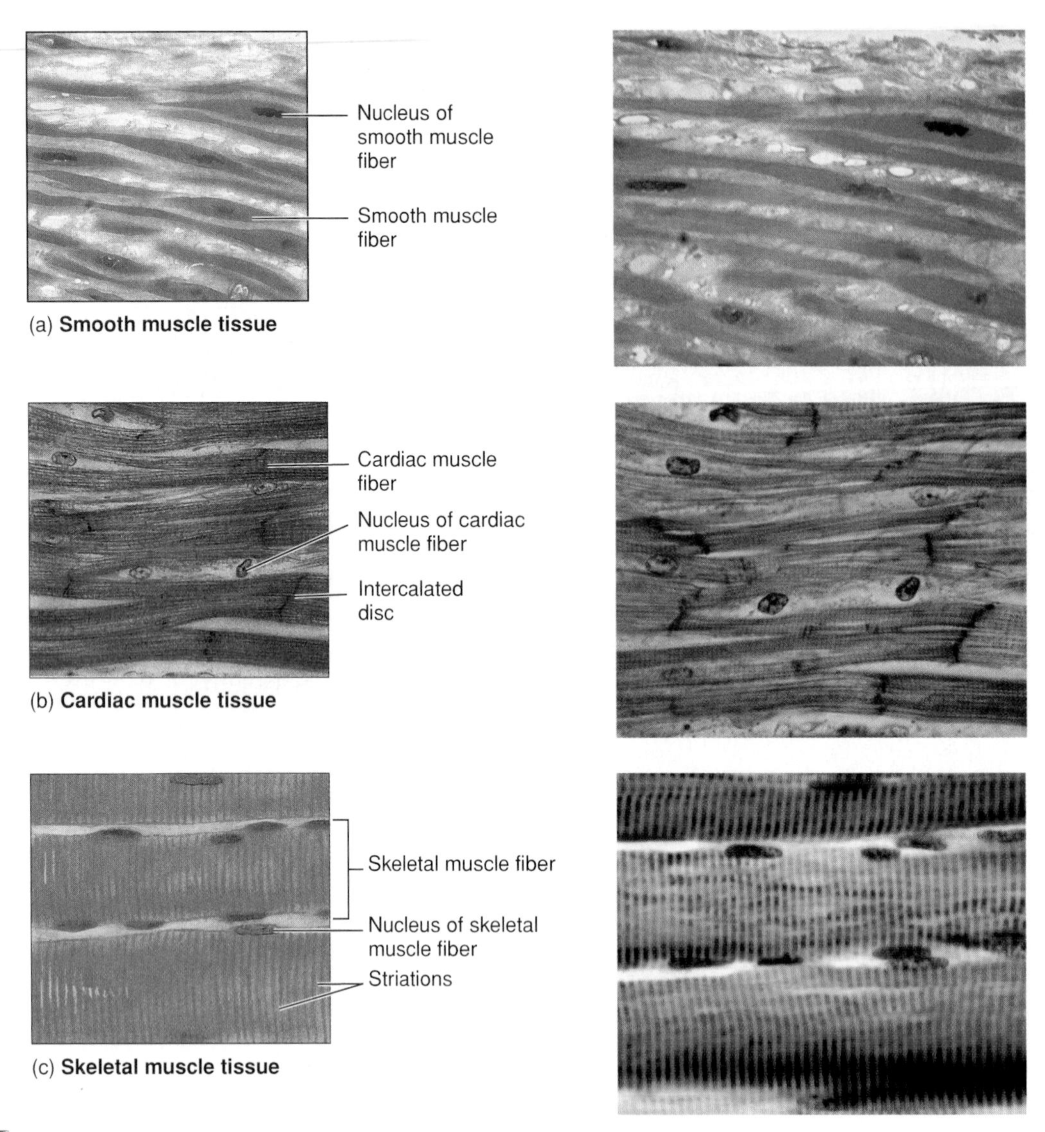

Figure 4.26 X

Muscle tissue: (*a*) smooth, (*b*) cardiac, and (*c*) skeletal.

Muscle Tissue

Muscle tissue is responsible for the movement of materials through the body, the movement of one part of the body with respect to another, and for locomotion. Fibers in the three kinds of muscle tissue are adapted to contract in response to stimuli.

Objective 10 Describe the structure, location, and function of the three types of muscle tissue.

Muscle tissue is unique in its ability to contract, and thus make movement possible. The muscle cells, or *fibers*, are elongated in the direction of contraction, and movement is accomplished through the shortening of the fibers in response to a stimulus. Muscle tissue is derived from mesoderm. There are three types of muscle tissue in the body: *smooth*, *cardiac*, and *skeletal muscle tissue* (fig. 4.26).

Smooth Muscle

ANS

Smooth muscle tissue is common throughout the body, occurring in many of the systems. For example, in the wall of the GI tract it provides the contractile force for the peristaltic movements involved in the mechanical digestion of food. Smooth muscle is also found in the walls of arteries, the walls of respiratory passages, and in the urinary and reproductive ducts. The contraction of smooth muscle is under autonomic (involuntary) nervous control, and is discussed in more detail in chapter 13.

Smooth muscle fibers are long, spindle-shaped cells. They contain a single nucleus and lack striations. These cells

Table 4.5
Summary of Muscle Tissue

Type	Structure and Function	Location
Smooth	Elongated, spindle-shaped fiber with single nucleus; involuntary movements of internal organs	Walls of hollow internal organs
Cardiac	Branched, striated fiber with single nucleus and intercalated discs; involuntary rhythmic contraction	Heart wall
Skeletal	Multinucleated, striated, cylindrical fiber that occurs in fasciculi; voluntary movement of skeletal parts	Associated with skeleton; spans joints of skeleton via tendons

are usually grouped together in flattened sheets, forming the muscular portion of a wall around a lumen.

Cardiac Muscle

Cardiac muscle tissue makes up most of the wall of the heart. This tissue is characterized by bifurcating (branching) fibers, each with a single, centrally positioned nucleus, and by transversely positioned **intercalated** (*in-ter'kă-lāt-ed*) **discs.** Intercalated discs help to hold adjacent cells together and transmit the force of contraction from cell to cell. Like skeletal muscle, cardiac muscle is striated, but unlike skeletal muscle it experiences rhythmic involuntary contractions. Cardiac muscle is further discussed in chapter 16.

Skeletal Muscle

Skeletal muscle tissue attaches to the skeleton and is responsible for voluntary body movements. Each elongated, multinucleated fiber has distinct transverse striations. Fibers of this muscle tissue are grouped into parallel fasciculi (bundles) that can be seen without a microscope in fresh muscle. Both cardiac and skeletal muscle fibers cannot replicate once tissue formation has been completed shortly after birth. Skeletal muscle tissue is further discussed in chapter 9. The three types of muscle tissue are summarized in table 4.5.

1. Describe the general characteristics of muscle tissue. What is meant by *voluntary* and *involuntary* as applied to muscle tissue?
2. Distinguish between smooth, cardiac, and skeletal muscle tissue on the bases of structure, location, and function.

Nervous Tissue

Nervous tissue is composed of neurons, which respond to stimuli and conduct impulses to and from all body organs, and neuroglia, which functionally support and physically bind neurons.

Objective 11 Describe the basic characteristics and functions of nervous tissue.

Objective 12 Distinguish between neurons and neuroglia.

Neurons

Although there are several kinds of neurons (*noor'onz*) in nervous tissue, they all have three principal components: (1) a cell body, or perikaryon; (2) dendrites; and (3) an axon (fig. 4.27). **Dendrites** are branched processes that receive stimuli and conduct impulses toward the cell body. The **cell body,** or **perikaryon** (*per"ĭ-kar'e-on*), contains the nucleus and specialized organelles and microtubules. The **axon** is a cytoplasmic extension that conducts impulses away from the cell body. The term **nerve fiber** refers to any process extending from the cell body of a neuron and the myelin sheath that surrounds it (see fig. 11.5).

Neurons derive from ectoderm and are the basic structural and functional units of the nervous system. They are specialized to respond to physical and chemical stimuli, convert stimuli into nerve impulses, and conduct these impulses to other neurons, muscle fibers, or glands. Of all the body's cells, neurons are probably the most specialized. As with muscle cells, the number of neurons is established shortly after birth; thereafter, they lack the ability to undergo mitosis, although under certain circumstances a severed portion can regenerate.

Neuroglia

In addition to neurons, nervous tissue contains neuroglia (*noo-rog'le-ă*) (fig. 4.27). Neuroglial cells, sometimes called *glial cells*, are about five times as abundant as neurons and have limited mitotic abilities. They do not transmit impulses but support and bind neurons together. Certain neuroglial cells are phagocytic; others assist in providing sustenance to the neurons.

Neurons and neuroglia are discussed in detail in chapter 11.

1. Compare and contrast neurons and neuroglia in terms of structure, function, and location.
2. List the structures of a neuron following the sequence of a nerve impulse passing through the cell.

neuron: Gk. *neuron*, sinew or nerve
perikaryon: Gk. *peri*, around; *karyon*, nut or kernel
neuroglia: Gk. *neuron*, nerve; *glia*, glue

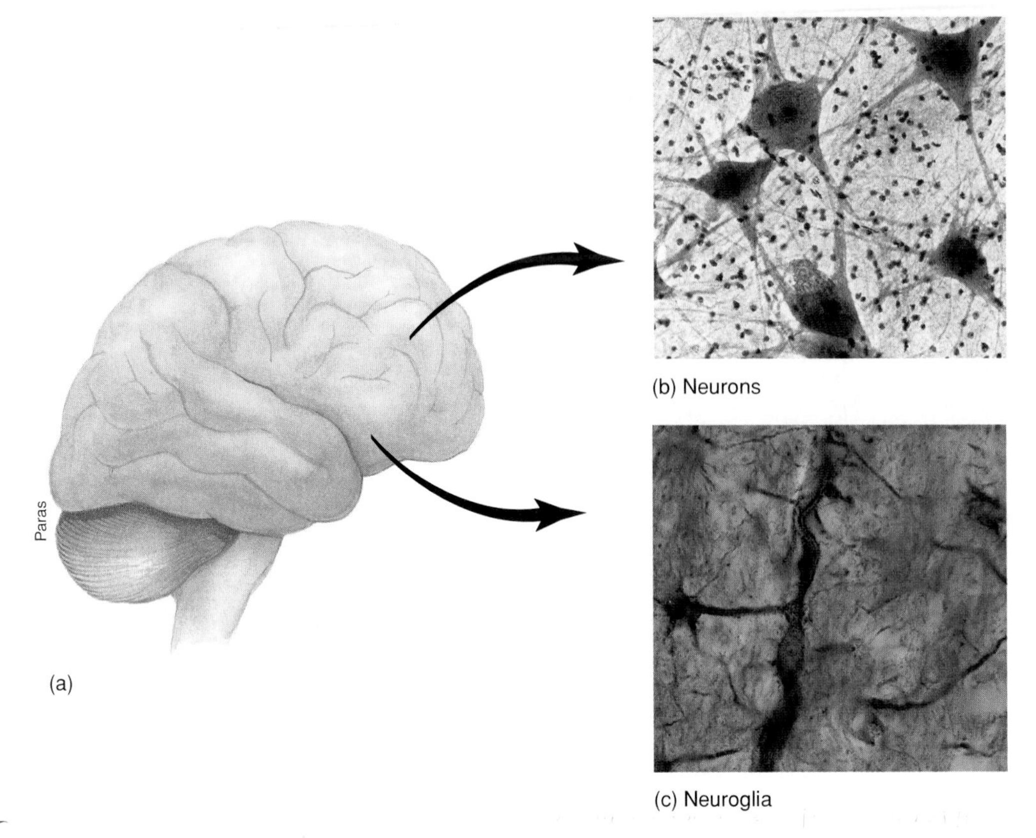

Figure 4.27

Nervous tissue is found in the brain (*a*), spinal cord, nerves, and ganglia. It consists of two principal kinds of cells: (*b*) neurons and (*c*) neuroglia.

Clinical Considerations

As stated at the beginning of this chapter, the study of tissues is extremely important in understanding the structure and function of organs and body systems. Histology has immense clinical importance as well. Many diseases are diagnosed through microscopic examination of tissue sections. Even in performing an autopsy, an examination of various tissues is vital in establishing the cause of death.

Several sciences are concerned with specific aspects of tissues. *Histopathology* is the study of diseased tissues. *Histochemistry* is concerned with the physiology of tissues as they maintain homeostasis. *Histotechnology* explores the ways in which tissues can be better stained and observed. In all of these disciplines, a thorough understanding of normal, or healthy, tissues is imperative for recognizing altered, or abnormal, tissues.

Changes in Tissue Composition

Most diseases alter tissue structure *locally*, where the disease is prevalent. Some diseases, however, called *general conditions*, cause changes that are far removed from the locus of the disease. **Atrophy** (wasting of body tissue), for example, may be limited to a particular organ where the disease interferes with the metabolism of that organ, but it may also involve an entire limb if nourishment or nerve impulses are impaired. *Muscle atrophy* can be caused by a disease of the nervous system like polio, or it can be the result of a diminished blood supply to a muscle. *Senescence* (*sĕ-nĕ'sens*) *atrophy*, or simply *senescence*, is the natural aging of tissues and organs within the body. *Disuse atrophy* is a local atrophy that results from the inactivity of a tissue or organ. Muscular dystrophy causes a disuse atrophy that decreases muscle size and strength due to the loss of sarcoplasm within the muscle.

Necrosis (*nĕ-kro'sis*) is death of cells or tissues within the living body. It can be recognized by physical changes in the dead tissues. Necrosis can be caused by severe injury; physical agents (trauma, heat, radiant energy, chemical poisons); or poor nutrition of tissues. When histologically examined, the necrotic tissue usually appears opaque, with a whitish or yellowish cast. **Gangrene** is a massive necrosis of tissue accompanied by an invasion of microorganisms that live on decaying flesh.

atrophy: Gk. *a*, without; *trophe*, nourishment
necrosis: Gk. *nekros*, corpse
gangrene: Gk. *gangraina*, gnaw or eat

Somatic death is the death of the body as a whole. Following somatic death, tissues undergo irreversible changes, such as **rigor mortis** (muscular rigidity), clotting of the blood, and cooling of the body. Postmortem (after death) changes occur under varying conditions at predictable rates, which is useful in estimating the approximate time of death.

Tissue Analysis

In diagnosing a disease, it is frequently important to examine tissues from a living person histologically. When this is necessary, a **biopsy** (*bi'op-se*) (removal of a section of living tissue) is performed. There are several techniques for biopsies. *Surgical removal* is usually done on large masses or tumors. *Curettage* (*kyoo" rĕ-tazh'*) involves cutting and scraping tissue, as may be done in examining for uterine cancer. In a *percutaneous needle biopsy*, a biopsy needle is inserted through a small skin incision and tissue samples are withdrawn. Both normal and diseased tissues are removed for purposes of comparison.

Preparing tissues for examination is a multistep process. **Fixation** is fundamental for all histological preparation. It is the rapid killing, hardening, and preservation of tissue to maintain its existing structure. **Embedding** the tissue in a supporting medium such as paraffin wax usually follows fixation. The next step, **sectioning** the tissue into extremely thin slices, is followed by **mounting** the specimen on a slide. Some tissues are fixed by rapid freezing and then sectioned while frozen, making embedding unnecessary. Frozen sections enable the pathologist to make a quick diagnosis during a surgical operation. These are done frequently, for example, in cases of suspected breast cancer. **Staining** is the next step. Hematoxylin and eosin (H & E) stains are routinely used on all tissue specimens. They give a differential blue and red color to the basic and acidic structures within the tissue. Other dyes may be needed to stain for specific structures.

Examination is first done with the unaided eye and then with a microscope. Practically all histological conditions can be diagnosed with low magnification (40×). Higher magnification is used to clarify specific details. Further examination may be performed with an electron microscope, which reveals the intricacy of cellular structure. Histological observation provides the foundation for subsequent diagnosis, prognosis, treatment, and reevaluation.

Tissue Transplantation

In the last two decades, medical science has made tremendous advancements in tissue transplants. Tissue transplants are necessary for replacing nonfunctional, damaged, or lost body parts. The most successful transplant is one where tissue is taken from one place on a person's body and moved to another place, such as a skin graft from the thigh to replace burned tissue of the hand. Transfer of one's own tissue is termed an **autograft. Isografts** are transplants between genetically identical individuals, the only example being identical twins. These transplants also have a high success rate. **Homotransplants,** or **allografts,** are grafts between individuals of the same species but of different genotype, and **heterografts,** or **xenografts,** are grafts between individuals of different species. Both allografts and xenografts present the problem of a possible *tissue-rejection reaction*. When this occurs, the recipient's immune mechanisms are triggered, and the donor's tissue is identified as foreign and is destroyed. The reaction can be minimized by "matching" recipient and donor tissue. Immunosuppressive drugs also may lessen the rejection rate. These drugs act by interfering with the recipient's immune mechanisms. Unfortunately, immunosuppressive drugs may also lower the recipient's resistance to infections. New techniques involving blood transfusions from donor to recipient before a transplant are proving successful. In any event, tissue transplants are an important aspect of medical research, and significant breakthroughs are on the horizon.

Chapter Summary

Definition and Classification of Tissues (p. 73)

1. Tissues are aggregations of similar cells that perform specific functions. The study of tissues is called histology.
2. Cells are surrounded and bound together by an intercellular matrix, the composition of which varies from solid to liquid.
3. The four principal types of tissues are epithelial tissue, connective tissue, muscle tissue, and nervous tissue.

Epithelial Tissue (pp. 74–85)

1. Epithelia are derived from all three germ layers and may be one or several layers thick. The lower surface of most membranous epithelia is supported by a basement membrane.
2. Simple epithelium consists of a single cell layer that varies in shape and surface characteristics. It is located where diffusion, filtration, and secretion occur.
3. Stratified epithelium consists of two or more layers of cells and is adapted for protection.
4. Transitional epithelium lines the urinary bladder, ureters, and parts of the urethra. The cells of transitional epithelium permit distension.
5. Body membranes are composed of thin layers of epithelial tissue that may be coupled with supporting connective tissue. The two basic types are mucous membranes and serous membranes.
6. Glandular epithelia are derived from developing epithelial tissue and function as secretory exocrine glands.

Connective Tissue (pp. 85–95)

1. Connective tissues are derived from mesoderm and, with the exception of cartilage, are highly vascular.
2. Connective tissue proper contains fibroblasts, collagenous fibers, and elastic fibers within a flexible ground substance.
3. Cartilage provides a flexible framework for many organs. It consists of a semisolid matrix of chondrocytes and various fibers.
4. Bone tissue consists of osteocytes, collagenous fibers, and a durable matrix of mineral salts.
5. Blood consists of formed elements (erythrocytes, leukocytes, and thrombocytes) suspended in a fluid plasma matrix.

Muscle Tissue (pp. 96–97)

1. Muscle tissues (smooth, cardiac, and skeletal) are responsible for the movement of materials through the body, the movement of one part of the body with respect to another, and for locomotion.
2. Fibers in muscle tissue are adapted to contract in response to stimuli.

Nervous Tissue (p. 97)

1. Neurons are the functional units of the nervous system. They respond to stimuli and conduct impulses to and from all body organs.
2. Neuroglia support and bind neurons. Some are phagocytic; others provide sustenance to neurons.

Review Activities

Objective Questions

1. Which of the following is *not* a principal type of body tissue?
   (a) nervous (d) muscular
   (b) integumentary (e) epithelial
   (c) connective
2. Which statement regarding tissues is *false?*
   (a) They are aggregations of similar kinds of cells that perform specific functions.
   (b) All of them are microscopic and are studied within the science of histology.
   (c) All of them are stationary within the body at the location of their developmental origin.
   (d) An organ is composed of two or more tissue types.
3. Connective tissue, muscle, and the dermis of the skin derive from embryonic
   (a) mesoderm. (c) ectoderm.
   (b) endoderm.
4. Which statement regarding epithelia is *false?*
   (a) They are derived from mesoderm, ectoderm, and endoderm.
   (b) They are strengthened by elastic and collagenous fibers.
   (c) One side is exposed to a lumen, a body cavity, or to the external environment.
   (d) They have very little intercellular matrix.
5. A gastric ulcer of the stomach would involve
   (a) simple cuboidal epithelium.
   (b) transitional epithelium.
   (c) simple ciliated columnar epithelium.
   (d) simple columnar epithelium.
6. Which structural and secretory designation describes mammary glands?
   (a) acinar, apocrine
   (b) tubular, holocrine
   (c) tubular, merocrine
   (d) acinar, holocrine
7. Dense regular connective tissue is found in
   (a) blood vessels. (c) tendons.
   (b) the spleen. (d) the wall of the uterus.
8. The phagocytic connective tissue found in the lymph nodes, liver, spleen, and bone marrow is
   (a) reticular. (c) mesenchyme.
   (b) loose fibrous. (d) elastic.
9. Cartilage is slow in healing following an injury because
   (a) it is located in body areas that are under constant physical strain.
   (b) it is avascular.
   (c) its chondrocytes cannot reproduce.
   (d) it has a semisolid matrix.
10. Cardiac muscle tissue has
    (a) striations.
    (b) intercalated discs.
    (c) rhythmic involuntary contractions.
    (d) all of the above.

Essay Questions

1. Define *tissue*. What are the differences between cells, tissues, glands, and organs?
2. What physiological functions are epithelial tissues adapted to perform?
3. Identify the epithelial tissue
   (a) in the pulmonary alveoli of the lungs,
   (b) lining the lumen of the GI tract,
   (c) in the outer layer of skin,
   (d) lining the cavity of the urinary bladder,
   (e) lining the uterine tube, and
   (f) lining the trachea and bronchial tubes.
   Describe the function of the tissue in each case.
4. Why are both keratinized and nonkeratinized epithelia found within the body?
5. Describe how epithelial glands are classified according to structural complexity and secretory function.
6. Identify the connective tissue
   (a) on the surface of the heart and surrounding the kidneys,
   (b) within the wall of the aorta,
   (c) forming the symphysis pubis,
   (d) supporting the outer ear,
   (e) forming the lymph nodes, and
   (f) forming the tendo calcaneus.
   Describe the function of the tissue in each case.
7. Compare and contrast the structure and location of the following: reticular fibers, collagenous fibers, elastin, fibroblasts, and mast cells.
8. What is the relationship between adipose cells and fat? Discuss the function of fat and explain the potential danger of excessive fat.
9. Discuss the mitotic abilities of each of the four principal tissue types.
10. Define the following terms: *atrophy*, *necrosis*, *gangrene*, and *somatic death*.

Critical-Thinking Questions

1. The function of a tissue is actually a function of its cells. And the function of a cell is a function of its organelles. Knowing this, what type of organelles would be particularly abundant in cardiac muscle tissue that requires a lot of energy; in reticular tissue within the liver, where cellular debris and toxins are ingested; and in dense regular connective tissue that consists of tough protein strands?
2. Your aunt was recently diagnosed as having brain cancer. In talking with your aunt's physician, she indicated that the cancer was actually a neuroglioma, and went on to say that cancer of neurons and muscle cells is a rare occurrence. Explain why neuroglial cells are much more susceptible to cancer than are neurons or muscle cells.
3. Compare the vascular supply of bones and ligaments and discuss how this may be relevant to the clinical course of an ankle sprain and an ankle fracture.
4. The connective tissue diseases are a group of disorders most likely caused by an abnormal immune response to a person's own connective tissue. The best known of these is rheumatoid arthritis, in which small joints of the body become inflamed and the articulating surfaces erode away. Knowing where connective tissue is found in the body, can you predict which organs might be involved in other connective tissue diseases?

FIVE

Integumentary System

Clinical Case Study 102

The Integument as an Organ 102

Layers of the Integument 103
- Epidermis 103
- Dermis 107

Functions of the Integument 109
- Physical Protection 109
- Hydroregulation 109
- Thermoregulation 109
- Cutaneous Absorption 110
- Synthesis 110
- Sensory Reception 110
- Communication 110

Epidermal Derivatives 111
- Hair 111
- Nails 113
- Glands 114

Developmental Exposition: The Integumentary System 116

Clinical Considerations 118
- Inflammatory Conditions (Dermatitis) 118
- Neoplasms 118
- Burns 119
- Frostbite 121
- Skin Grafts 121
- Wound Healing 123
- Aging of the Skin 124

Clinical Case Study Answer 124

Internal Affairs 125

Important Clinical Terminology 126

Chapter Summary 126

Review Activities 127

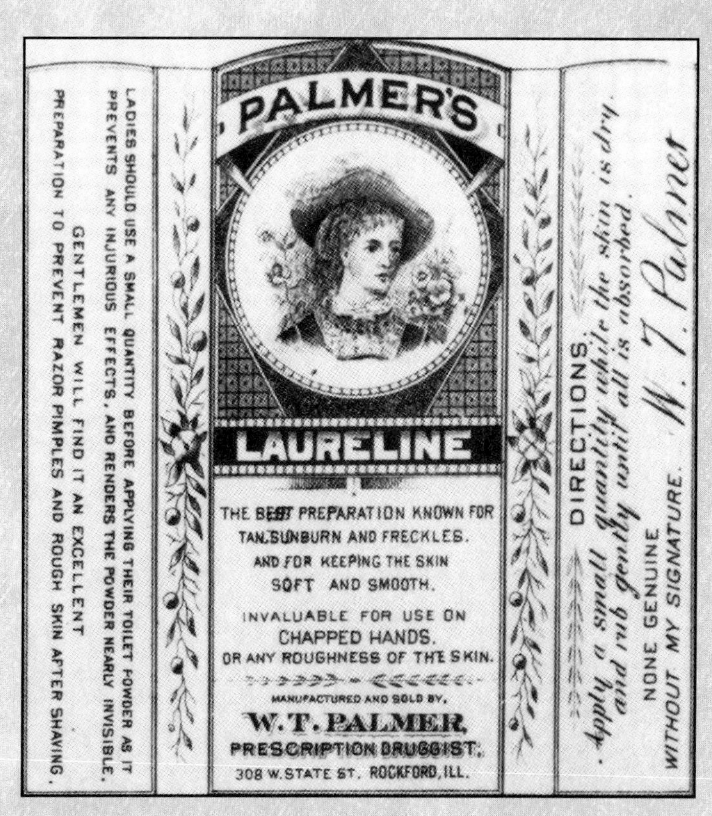

Americans spend billions of dollars each year on a seemingly infinite variety of preparations that promise to banish wrinkles and dry skin and camouflage the effects of aging. Early advertisements for skin creams, such as this one issued by the J. F. Lawrence Printing Company of Chicago, were geared toward the "gentlemen," as well as the "ladies."

Clinical Case Study

A 27-year-old male was involved in a gasoline explosion and sustained burns to his face, neck, chest, and arms. Upon arrival at the emergency room, he complained of intense pain in his face and neck, both of which exhibited extensive blistering and erythema (redness). These findings were all curiously absent on the burned chest and arms, which had a pale, waxy appearance.

Examination revealed the skin on the patient's chest and arms to be leathery and lacking sensation. The emergency room physician commented to an observing medical student that third-degree burns were present on the skin of these regions and that excision of the burn eschar (traumatized tissue) with subsequent skin grafting would be required.

Why would the areas that sustained second-degree burns be red, blistered, and painful, while the third-degree burns were pale and insensate (without sensation, including pain)? Why would the chest and arms require skin grafting, but probably not the face and neck? *Hints:* Think in terms of functions of the skin and survival of the germinal cells in functioning skin. Carefully examine figures 5.1 and 5.19.

The Integument as an Organ

The integument (skin) is the largest organ of the body, and together with its accessory organs (hair, glands, and nails), it constitutes the integumentary system. In certain areas of the body, it has adaptive modifications that accommodate protective or metabolic functions. In its role as a dynamic interface between the continually changing external environment and the body's internal environment, the skin helps to maintain homeostasis.

Objective 1 Explain why the integument is considered an organ and a component of the integumentary system.

Objective 2 Describe some common clinical conditions of the integument that result from nutritional deficiencies or body dysfunctions.

We are more aware of and concerned with our integumentary system than perhaps any other system of our body. One of the first things we do in the morning is to look in a mirror and see what we have to do to make our skin and hair presentable. Periodically, we examine our skin for wrinkles and our scalp for gray hairs as signs of aging. We recognize other people to a large extent by features of their skin.

The appearance of our skin frequently determines the initial impression we make on others. Unfortunately, it may also determine whether or not we succeed in gaining social acceptance. For example, social rejection as a teenager, imagined or real, can be directly associated with skin problems such as acne. A person's self-image and consequent social behavior may be closely associated with his or her physical appearance.

Even clothing styles are somewhat determined by how much skin we, or the designers, want to expose. But our skin is much more than a showpiece. It helps to regulate certain body functions and protect certain body structures.

The skin, or *integument* (*in-teg'yoo-ment*), and its accessory structures (hair, glands, and nails) constitute the integumentary system. Included in this system are the millions of sensory receptors of the skin and its extensive vascular network. The skin is a dynamic interface between the body and the external environment. It protects the body from the environment even as it allows for communication with the environment.

The skin is an organ, since it consists of several kinds of tissues that are structurally arranged to function together. It is the largest organ of the body, covering over 7,600 sq cm (3,000 sq in.) in the average adult, and accounts for approximately 7% of a person's body weight. The skin is of variable thickness, averaging 1.5 mm. It is thickest on the parts of the body exposed to wear and abrasion, such as the soles of the feet and palms of the hand. In these areas, it is about 6 mm thick. It is thinnest on the eyelids, external genitalia, and tympanic membrane (eardrum), where it is approximately 0.5 mm thick. Even its appearance and texture varies from the rough, callous skin covering the elbows and knuckles to the soft, sensitive areas of the eyelids, nipples, and genitalia.

The general appearance of the skin is clinically important because it provides clues to certain body dysfunctions. Pale skin may indicate shock, whereas red, flushed, overwarm skin may indicate fever and infection. A rash may indicate allergies or local infections. Abnormal textures of the skin may be the result of glandular or nutritional problems (table 5.1). Even chewed fingernails may be a clue to emotional problems.

1. Explain why the skin is considered an organ and why the skin, together with the integumentary derivatives, is considered a system.
2. Which vitamins and minerals are important for healthy skin? (See table 5.1)
3. Describe the appearance of the skin that may accompany each of the following conditions: allergy; shock; infection; dry, stiff hair; hyperpigmentation; and general dermatitis.

integument: L. *integumentum*, a covering

Table 5.1

Conditions of the Skin and Associated Structures Indicating Nutritional Deficiencies or Body Dysfunctions

Condition	Deficiency	Comments
General dermatitis	Zinc	Redness and itching
Scrotal or vulval dermatitis	Riboflavin	Inflammation in genital region
Hyperpigmentation	Vitamin B_{12}, folic acid, or starvation	Dark pigmentation on backs of hands and feet
Dry, stiff, brittle hair	Protein, calories, and other nutrients	Usually occurs in young children or infants
Follicular hyperkeratosis	Vitamin A, unsaturated fatty acids	Rough skin caused by keratotic plugs from hair follicles
Pellagrous dermatitis	Niacin and tryptophan	Lesions on areas exposed to sun
Thickened skin at pressure points	Niacin	Noted at belt area at the hips
Spoon nails	Iron	Thin nails that are concave or spoon-shaped
Dry skin	Water or thyroid hormone	Dehydration, hypothyroidism, rough skin
Oily skin (acne)		Hyperactivity of sebaceous glands

Layers of the Integument

The integument consists of two principal layers. The outer epidermis is stratified into four or five structural layers, and the thick and deeper dermis consists of two layers. The hypodermis (subcutaneous tissue) connects the skin to underlying organs.

Objective 3 Describe the histological characteristics of each layer of the integument.

Objective 4 Summarize the transitional events that occur within each of the epidermal layers.

Epidermis

The **epidermis** (*ep″ĭ-der′mis*) is the superficial protective layer of the skin. It is composed of stratified squamous epithelium that varies in thickness from 0.007 to 0.12 mm. All but the deepest layers are composed of dead cells. Either four or five layers may be present, depending on where the epidermis is located (figs. 5.1 and 5.2). The epidermis of the palms and soles has five layers because these areas are exposed to the most friction. In all other areas of the body, the epidermis has only four layers. The names and characteristics of the epidermal layers are as follows.

1. **Stratum basale** (basal layer). The stratum basale (*stra′tum bă-sal′e*) consists of a single layer of cells in contact with the dermis. Four types of cells compose the stratum basale: *keratinocytes* (*ker″ă-tin′o-sīts*), *melanocytes* (*mel′ă-no-sīts*), *tactile cells* (Merkel cells), and *nonpigmented granular dendrocytes* (*Langerhans cells*). With the exception of tactile cells, these cells are constantly dividing mitotically and moving outward to renew the epidermis. It usually takes between 6 to 8 weeks for the cells to move from the stratum basale to the surface of the skin.

   Keratinocytes are specialized cells that produce the protein **keratin** (*ker′ă-tin*), which toughens and waterproofs the skin. As keratinocytes are pushed away from the vascular nutrient and oxygen supply of the dermis, their nuclei degenerate, their cellular content becomes dominated by keratin, and the process of *keratinization* is completed. By the time keratinocytes reach the surface of the skin, they resemble flat dead scales. They are completely filled with keratin enclosed in loose cell membranes. **Melanocytes** are specialized epithelial cells that synthesize the pigment **melanin** (*mel′ă-nin*) which provides a protective barrier to the ultraviolet radiation in sunlight. **Tactile cells** are sparse compared to keratinocytes and melanocytes. These sensory receptor cells aid in tactile (touch) reception. **Nonpigmented granular dendrocytes** are scattered throughout the stratum basale. They are protective *macrophagic cells* that ingest bacteria and other foreign debris.
2. **Stratum spinosum** (spiny layer). The stratum spinosum (*spi-no′sum*) contains several stratified layers of cells. The

stratum: L. *stratum*, something spread out
basale: Gk. *basis*, base

keratinocyte: Gk. *keras*, hornlike; *kytos*, cell
melanocyte: Gk. *melas*, black; *kytos*, cell
Merkel cells: from F. S. Merkel, German anatomist, 1845–1919
Langerhans cells: from Paul Langerhans, German anatomist, 1847–1888
macrophagic: Gk. *makros*, large; *phagein*, to eat
spinosum: L. *spina*, thorn

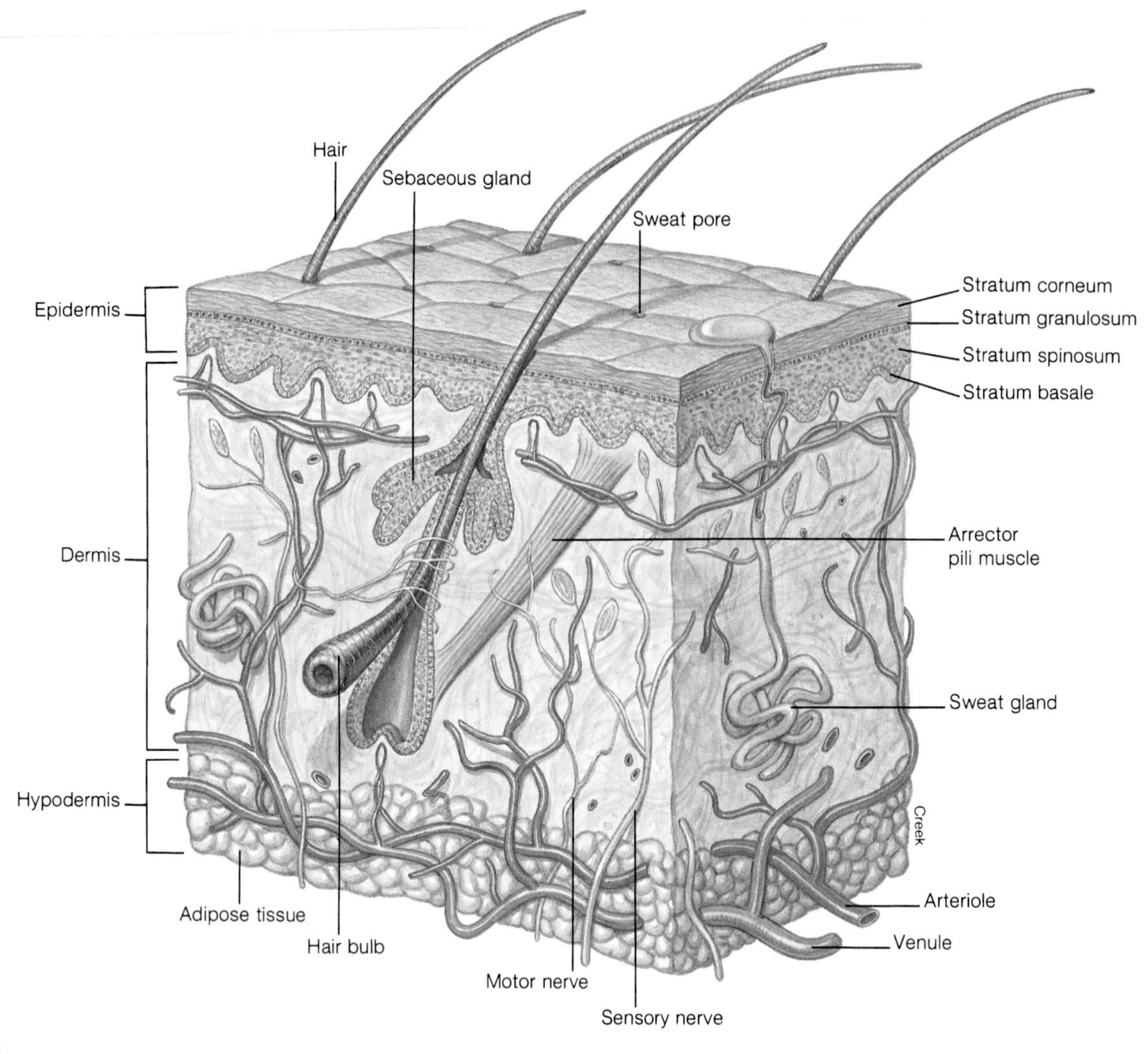

Figure 5.1

A diagram of the skin.

spiny appearance of this layer is due to the spinelike extensions that arise from the keratinocytes when the tissue is fixed for microscopic examination. Since there is limited mitosis in the stratum spinosum, this layer and the stratum basale are collectively referred to as the **stratum germinativum** (*jer-mĭ″nă-ti′vum*).

3. **Stratum granulosum** (granular layer). The stratum granulosum (*gran″yoo-lo′sum*) consists of only three or four flattened rows of cells. These cells contain granules that are filled with *keratohyalin*, a chemical precursor to keratin.
4. **Stratum lucidum** (clear layer). The nuclei, organelles, and cell membranes are no longer visible in the cells of the stratum lucidum (*loo′sĭ-dum*), and so histologically this layer appears clear. It exists only in the lips and in the thickened skin of the soles and palms.
5. **Stratum corneum** (hornlike layer). The stratum corneum (*kor′ne-um*) is composed of 25 to 30 layers of flattened, scalelike cells. Thousands of these dead cells shed from the skin surface each day, only to be replaced by new ones from deeper layers. This surface layer is cornified; it is the layer that actually protects the skin (fig. 5.3). *Cornification*, brought on by keratinization, is the drying and flattening of the stratum corneum and is an important protective adaptation of the skin. Friction at the surface of the skin stimulates additional mitotic activity in the stratum basale and stratum spinosum, which may result in the formation of a *callus* for additional protection.

The specific characteristics of each epidermal layer are described in table 5.2.

germinativum: L. *germinare*, sprout or growth
granulosum: L. *granum*, grain
lucidum: L. *lucidus*, light

corneum: L. *corneus*, hornlike

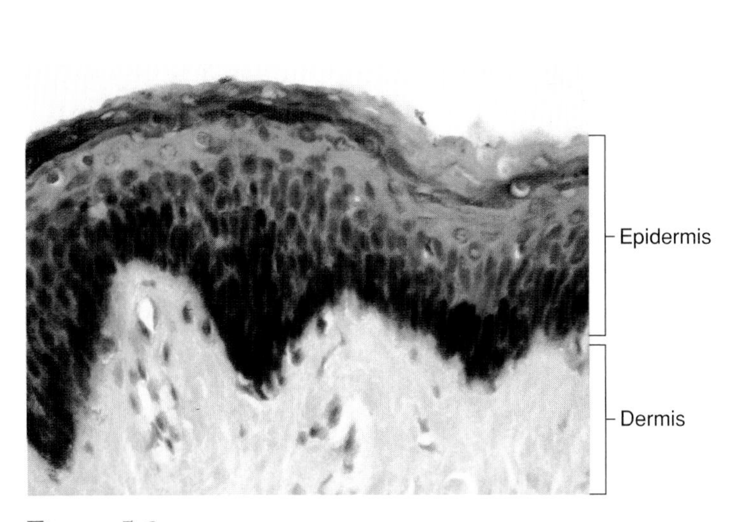
Epidermis
Dermis

Figure 5.2

A photomicrograph of the epidermis (250×).

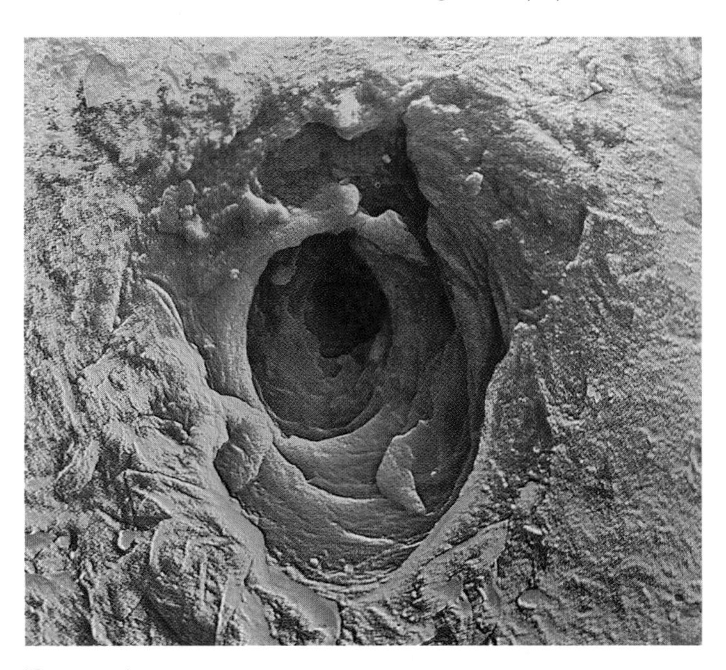

Figure 5.3

A scanning electron micrograph of the surface of the skin showing the opening of a sweat gland.

Table 5.2
Layers of the Epidermis

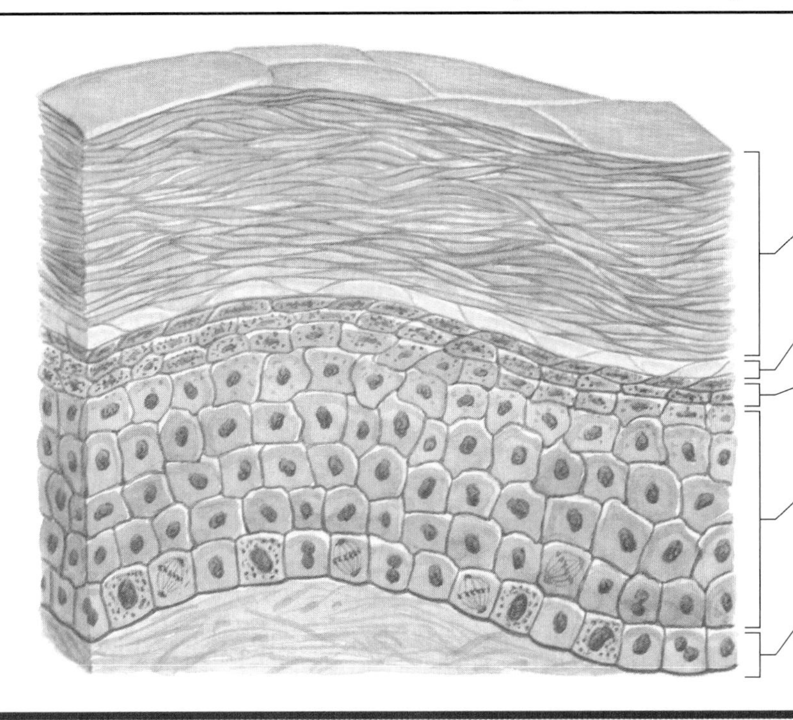
Stratum corneum
Consists of many layers of keratinized dead cells that are flattened and nonnucleated; cornified
Stratum lucidum
A thin, clear layer found only in the epidermis of the lips, palms, and soles
Stratum granulosum
Composed of one or more layers of granular cells that contain fibers of keratin and shriveled nuclei
Stratum spinosum
Composed of several layers of cells with centrally located, large, oval nuclei and spinelike processes; limited mitosis
Stratum basale
Consists of a single layer of cuboidal cells in contact with the basement membrane that undergo mitosis; contains pigment-producing melanocytes

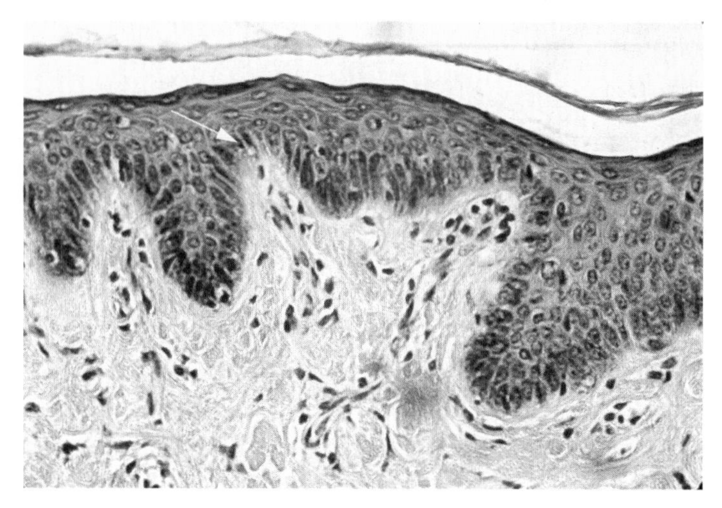

Figure 5.4

Melanocytes throughout the stratum basale (see arrow) produce melanin.

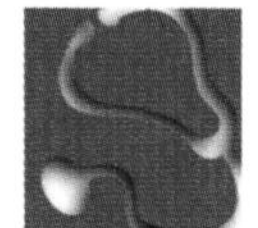

Tattooing colors the skin permanently because dyes are injected below the mitotic basal layer of the epidermis into the underlying dermis. In nonsterile conditions, infectious organisms may be introduced along with the dye. Small tattoos can be removed by skin grafting; for large tattoos, mechanical abrasion of the skin is preferred.

Coloration of the Skin

Normal skin color is the expression of a combination of three pigments: *melanin*, *carotene*, and *hemoglobin*. **Melanin** is a brown-black pigment produced in the melanocytes of the stratum basale (fig. 5.4). All individuals of similar size have approximately the same number of melanocytes, but the amount of melanin produced and the distribution of the melanin determine racial variations in skin color, such as black, brown, yellow, and white. Melanin protects the basal layer against the damaging effect of the ultraviolet (UV) rays of the sun. A gradual exposure to the sunlight promotes the increased production of melanin within the melanocytes, and hence tanning of the skin. The skin of a person with *albinism* (*al'bĭ-niz-em*) has the normal number of melanocytes in the epidermis but lacks the enzyme *tyrosinase* that converts the amino acid *tyrosine* to melanin. Albinism is a hereditary condition.

Other genetic expressions of melanocytes are more common than albinism. *Freckles*, for example, are caused by aggregated patches of melanin. A lack of melanocytes in localized areas of the skin causes distinct white spots in the condition called *vitiligo* (*vit-ĭ-li'go*). After the age of 50, brown plaquelike growths, called *seborrheic* (*seb"ŏ-re'ik*) *hyperkeratoses*, may appear on the skin, particularly on exposed portions. Commonly called "liver spots," these pigmented patches are benign growths of pigment-producing melanocytes. Usually no treatment is required, unless for cosmetic purposes.

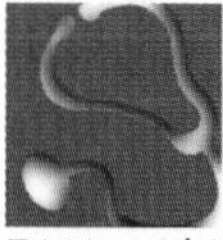

Excessive exposure to sunlight can cause skin cancer (see Clinical Considerations and fig. 5.18). In sunlight, the skin absorbs two wavelengths of ultraviolet rays known as *UVA* and *UVB*. The DNA within the basal skin cells may be damaged as the sun's more dangerous UVB rays penetrate the skin. Although it was once believed that UVA rays were harmless, findings now indicate that excessive exposure to these rays may inhibit the DNA repair process that follows exposure to UVB. Therefore, individuals who are exposed solely to UVA rays in tanning salons are still in danger of basal cell carcinoma, since they will later be exposed to UVB rays of sunlight when they go outdoors.

Carotene (*kar'ŏ-tēn*) is a yellowish pigment found in certain plant products, such as carrots, that tends to accumulate in cells of the stratum corneum and fatty parts of the dermis. It was once thought to account for the yellow-tan skin of people of Asian descent, but this coloration is now known to be caused by variations in melanin.

Hemoglobin (*he"mo-glo'bin*) is not a pigment of the skin; rather, it is the oxygen-binding pigment found in red blood cells. Oxygenated blood flowing through the dermis gives the skin its pinkish tones.

Certain physical conditions or diseases cause symptomatic discoloration of the skin. *Cyanosis* (*si-ă-no'sis*) is a bluish discoloration of the skin that appears in people with certain cardiovascular or respiratory diseases. People also become cyanotic during an interruption of breathing. In *jaundice,* the skin appears yellowish because of an excess of bile pigment in the bloodstream. Jaundice is usually symptomatic of liver dysfunction and sometimes of liver immaturity, as in a jaundiced newborn. *Erythema* (*er"ĭ-the'mă*) is a redness of the skin generally due to vascular trauma, such as from a sunburn.

Surface Patterns

The exposed surface of the skin has recognizable patterns that are either present at birth or develop later. **Fingerprints** (*friction ridges*) are congenital patterns that are present on the finger and toe pads, as well as on the palms and soles. The designs formed by these lines have basic similarities but are not identical in any two individuals (fig. 5.5). They are formed by the pull of elastic fibers within the dermis and are well established prenatally. The ridges of fingerprints function to prevent slippage when grasping objects. Because they are precise and easy to reproduce, fingerprints are customarily used for identifying individuals.

vitiligo: L. *vitiatio*, blemish

carotene: L. *carota*, carrot (referring to orange coloration)
hemoglobin: Gk. *haima*, blood; *globus*, globe
cyanosis: Gk. *kyanosis*, dark blue color
jaundice: L. *galbus*, yellow
erythema: Gk. *erythros*, red; *haima*, blood

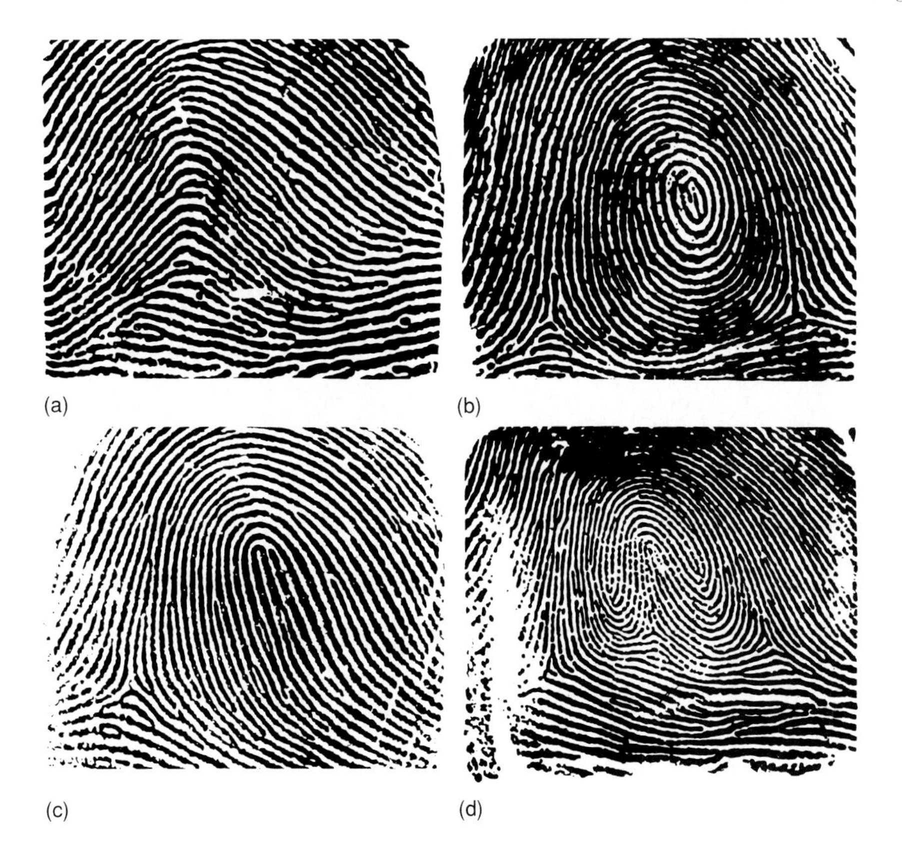

Figure 5.5

The four basic fingerprint patterns: (*a*) arch, (*b*) whorl, (*c*) loop, and (*d*) combination.

Acquired lines include the deep **flexion creases** on the palms and the shallow **flexion lines** that can be seen on the knuckles and on the surface of other joints. Furrows on the forehead and face are acquired from continual contraction of facial muscles, such as from smiling or squinting in bright light or against the wind. Facial lines become more strongly delineated as a person ages.

The science known as *dermatoglyphics* is concerned with the classification and identification of fingerprints. Every individual's prints are unique, including those of identical twins. Fingerprints, however, are not exclusive to humans. All other primates have fingerprints, and even dogs have a characteristic "nose print" that is used for identification in the military canine corps and in certain dog kennels.

Dermis

The **dermis** is deeper and thicker than the epidermis (see fig. 5.1). Elastic and collagenous fibers within the dermis are arranged in definite patterns, producing *lines of tension* in the skin and providing skin tone (fig. 5.6). There are many more elastic fibers in the dermis of a young person than in an elderly one, and a decreasing number of elastic fibers is apparently associated with aging. The extensive network of blood vessels in the dermis provides nourishment to the living portion of the epidermis.

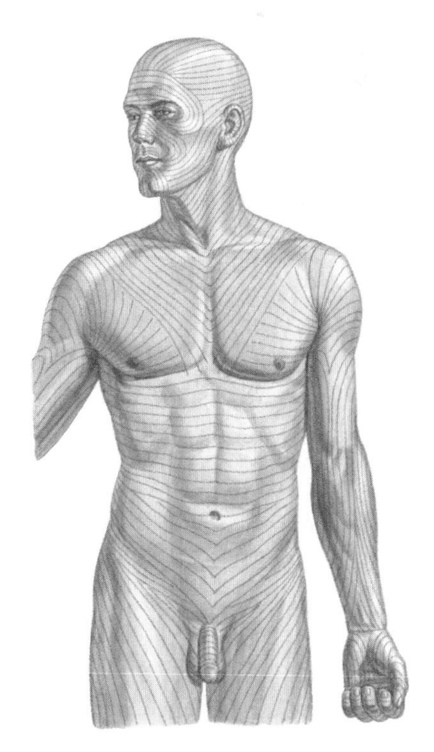

Figure 5.6

Lines of tension are caused by the pull of elastic collagenous fibers within the dermis of the skin. Surgical incisions made parallel to the lines of tension heal more rapidly and create less scar tissue than those made across the lines of tension.

The dermis also contains many sweat glands, oil-secreting glands, nerve endings, and hair follicles.

Layers of the Dermis

The dermis is composed of two layers. The upper layer, called the **stratum papillarosum** (papillary layer), is in contact with the epidermis and accounts for about one-fifth of the entire dermis. Numerous projections, called *papillae* (*pă-pil'e*), extend from the upper portion of the dermis into the epidermis. Papillae form the base for the friction ridges on the fingers and toes.

The deeper and thicker layer of the dermis is called the **stratum reticularosum** (reticular layer). Fibers within this layer are more dense and regularly arranged to form a tough, flexible meshwork. It is quite distensible, as is evident in pregnant women or obese individuals, but it can be stretched too far, causing "tearing" of the dermis. The repair of a strained dermal area leaves a white streak called a stretch mark, or **linea albicans** (*lin'e-ă al'bĭ-kanz*). Lineae albicantes are frequently found on the buttocks, thighs, abdomen, and breasts (fig. 5.7).

It is the strong, resilient reticular layer of domestic mammals that is used in making leather and suede. In the tanning process, the hide of an animal is treated with various chemicals that cause the epidermis with its hair and the papillary layer of the dermis to separate from the underlying reticular layer. The reticular layer is then softened and treated with protective chemicals before being cut and assembled into consumer goods.

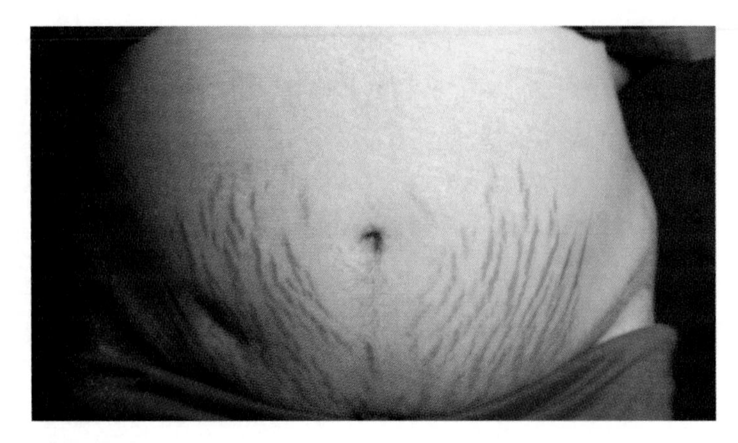

Figure 5.7

Stretch marks (lineae albicantes) on the abdomen of a pregnant woman. Stretch marks generally fade with time but may leave permanent markings.

Innervation of the Skin

The dermis of the skin has extensive innervation. Specialized integumentary *effectors* consist of muscles or glands within the dermis that respond to motor impulses transmitted from the central nervous system to the skin by autonomic nerve fibers.

Several types of **sensory receptors** respond to various tactile (touch), pressure, temperature, tickle, or pain sensations. Some are free nerve endings, some form a network around hair follicles, and some extend into the papillae of the dermis. Certain areas of the body, such as the palms, soles, lips, and external genitalia, have a greater concentration of sensory receptors and are therefore more sensitive to touch. Chapter 15 includes a detailed discussion of the structure and function of the various sensory receptors.

Vascular Supply of the Skin

Blood vessels within the dermis supply nutrients to the mitotically active stratum basale of the epidermis and to the cellular structures of the dermis, such as glands and hair follicles. Dermal blood vessels play an important role in regulating body temperature and blood pressure. Autonomic vasoconstriction or vasodilation responses can either shunt the blood away from the superficial dermal arterioles or permit it to flow freely throughout dermal vessels. Fever or shock can be detected by the color and temperature of the skin. Blushing is the result of involuntary vasodilation of dermal blood vessels.

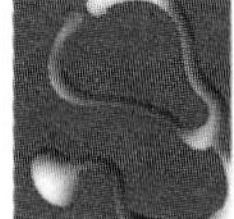

It is important to maintain good blood circulation in people who are bedridden to prevent bedsores, or *decubitus* (*de-kyoo'bĭ-tus*) *ulcers*. When a person lies in one position for an extended period, the dermal blood flow is restricted where the body presses against the bed. As a consequence, cells die and open wounds may develop (fig. 5.8). Changing the position of the patient frequently and periodically massaging the skin to stimulate blood flow are good preventive measures against decubitus ulcers.

Hypodermis

The **hypodermis,** or *subcutaneous tissue*, is not actually a part of the skin, but it binds the dermis to underlying organs. The hypodermis is composed primarily of loose connective tissue and adipose cells interlaced with blood vessels (see fig. 5.1). Collagenous and elastic fibers reinforce the hypodermis—particularly on the palms and soles, where the skin is firmly attached to underlying structures. The amount of adipose tissue in the hypodermis varies with the region of the body and the sex, age, and nutritional state of the individual. Females generally have

papilla: L. *papula*, swelling or pimple

decubitus: L. *decumbere*, lie down
ulcer: L. *ulcus*, sore
hypodermis: Gk. *hypo*, under; *derma*, skin

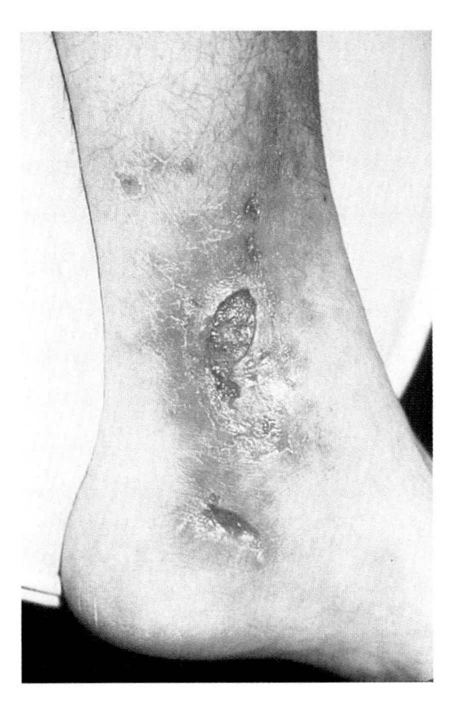

Figure 5.8

A bedsore (decubitus ulcer) on the medial surface of the ankle. Bedsores are most common on skin overlying a bony projection, such as at the hip, ankle, heel, shoulder, or elbow.

about an 8% thicker hypodermis than males. This layer functions to store lipids, insulate and cushion the body, and regulate temperature.

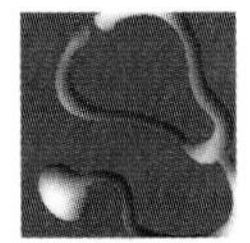

The hypodermis is the site for subcutaneous injections. Using a hypodermic needle, medicine can be administered to patients who are unconscious or uncooperative, and when oral medications are not practical. Subcutaneous devices to administer slow-release, low-dosage medications are now available. For example, insulin may be administered in this way to treat some forms of diabetes. Even a subcutaneous birth-control device (Norplant) is currently being marketed (see fig. 21.26).

1. List the layers of the epidermis and dermis and explain how they differ in structure and function.
2. Describe the sequence of cellular replacement within the epidermis and the processes of keratinization and cornification.
3. How do both the dermis and hypodermis function in thermoregulation?
4. What two basic types of innervation are found within the dermis?

Functions of the Integument

The integument not only protects the body from pathogens and external injury, it is a highly dynamic organ that plays a key role in maintaining body homeostasis.

Objective 5 Discuss the role of the integument in the protection of the body from disease and external injury, the regulation of body fluids and temperature, absorption, synthesis, sensory reception, and communication.

Physical Protection

The skin is a barrier to microorganisms, water, and excessive sunlight (UV light). Oily secretions onto the surface of the skin form an acidic protective film (pH 4.0–6.8) that waterproofs the body and retards the growth of most pathogens. The protein keratin in the epidermis also waterproofs the skin, and the cornified outer layer (stratum corneum) resists scraping and keeps out microorganisms. As mentioned previously, exposure to UV light stimulates the melanocytes in the stratum basale to synthesize melanin, which absorbs and disperses sunlight. In addition, surface friction causes the epidermis to thicken by increasing the rate of mitosis in the cells of the stratum basale and stratum spinosum, resulting in the formation of a protective *callus*.

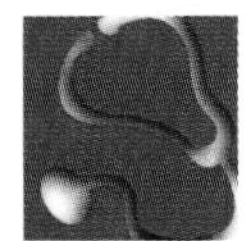

Regardless of skin pigmentation, everyone is susceptible to skin cancer if his or her exposure to sunlight is sufficiently intense. There are an estimated 800,000 new cases of skin cancer yearly in the United States, and approximately 9,300 of these are diagnosed as the potentially life-threatening *melanoma* (*mel-ă-no'mă*) (cancer of melanocytes). Melanomas (see fig. 5.18) are usually termed malignant, as they may spread rapidly. Sunscreens are advised for people who must be in direct sunlight for long periods of time.

Hydroregulation

The thickened, keratinized, and cornified epidermis of the skin is adapted for continuous exposure to the air. In addition, the outer layers are dead and scalelike, and a protein-polysaccharide basement membrane adheres the stratum basale to the dermis. Human skin is virtually waterproof, protecting the body from desiccation (dehydration) on dry land, and even from water absorption when immersed in water.

Thermoregulation

The skin plays a crucial role in the regulation of body temperature. Body heat comes from cellular metabolism, particularly in

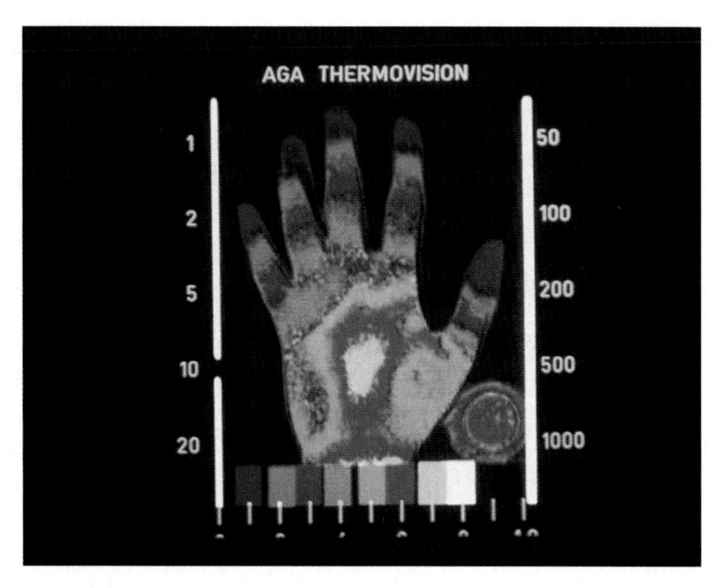

Figure 5.9

A thermogram of the hand showing differential heat radiation. Hair and body fat are good insulators. Red and yellow indicate the warmest parts of the body. Blue, green, and white indicate the coolest.

muscle cells as they maintain tone or a degree of tension. A normal body temperature of 37°C (98.6°F) is maintained in three ways, all involving the skin:

1. through radiant heat loss from dilated blood vessels,
2. through evaporation of perspiration, and
3. through retention of heat from constricted blood vessels (fig. 5.9).

The volume of perspiration produced is largely a function of how much the body is overheated. This volume increases approximately 100 to 150 ml/day for each 1°C elevation in body temperature. For each hour of hard physical work out-of-doors in the summertime, a person may produce 1 to 10 L of perspiration.

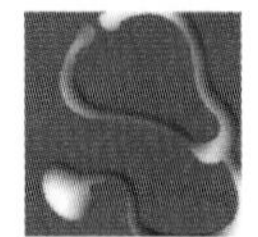

A serious danger of continued exposure to heat and excessive water and salt loss is *heat exhaustion,* characterized by nausea, weakness, dizziness, headache, and a decreased blood pressure. *Heat stroke* is similar to heat exhaustion, except that in heat stroke sweating is prevented (for reasons that are not clear) and body temperature rises. Convulsions, brain damage, and death may follow.

Excessive heat loss triggers a shivering response in muscles, which increases cellular metabolism. Not only do skeletal muscles contract, but tiny smooth muscles called **arrectores pilorum** (*ă″rek-to′rēz pil-o′rum*—singular, *arrector pili*), which are attached to hair follicles, are also contracted involuntarily and cause goose bumps.

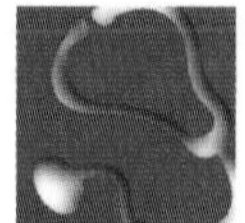

When the body's heat-producing mechanisms cannot keep pace with heat loss, *hypothermia* results. A lengthy exposure to temperatures below 20°C (68°F) and dampness may lead to this condition. This is why it is so important that a hiker, for example, dress appropriately for the weather conditions, especially on cool, rainy spring or fall days. The initial symptoms of hypothermia are numbness, paleness, delirium, and uncontrolled shivering. If the core temperature falls below 32°C (90°F), the heart loses its ability to pump blood and will go into fibrillation (erratic contractions). If the victim is not warmed, extreme drowsiness, coma, and death follow.

Cutaneous Absorption

Because of the effective protective barriers of the integument already described, cutaneous absorption (absorption through the skin) is limited. Some gases, such as oxygen and carbon dioxide, may pass through the skin and enter the blood. Small amounts of UV light, necessary for synthesis of vitamin D, are absorbed readily. Of clinical consideration is the fact that certain chemicals such as lipid-soluble toxins and pesticides can easily enter the body through the skin.

Synthesis

The integumentary system synthesizes melanin and keratin, which remain in the skin synthesis of vitamin D, which is used elsewhere in the body, begins in the skin with activation of a precursor molecule by UV light. The molecule is modified in the liver and kidneys to produce *calcitriol* (*kal-sĭ-tre′ol*), the most active form of vitamin D. Only small amounts of UV light are necessary for vitamin D synthesis, but these amounts are very important to a growing child. Active vitamin D enters the blood and helps to regulate the metabolism of calcium and phosphorus, which are important in the development of strong and healthy bones. *Rickets* is a disease caused by vitamin D deficiency (fig. 5.10).

Sensory Reception

Highly specialized sensory receptors (see chapter 15) that respond to the precise stimuli of heat, cold, pressure, touch, vibration, and pain are located throughout the dermis. Called **cutaneous receptors,** these sensory nerve cells are especially abundant in the skin of the face and palms, the fingers, the soles of the feet, and the genitalia. They are less abundant along the back and on the back of the neck and are sparse in the skin over joints, especially the elbow. Generally speaking, the thinner the skin, the greater the sensitivity.

Communication

Humans are highly social animals, and the integument plays an important role in communication. Various emotions, such as

(a)

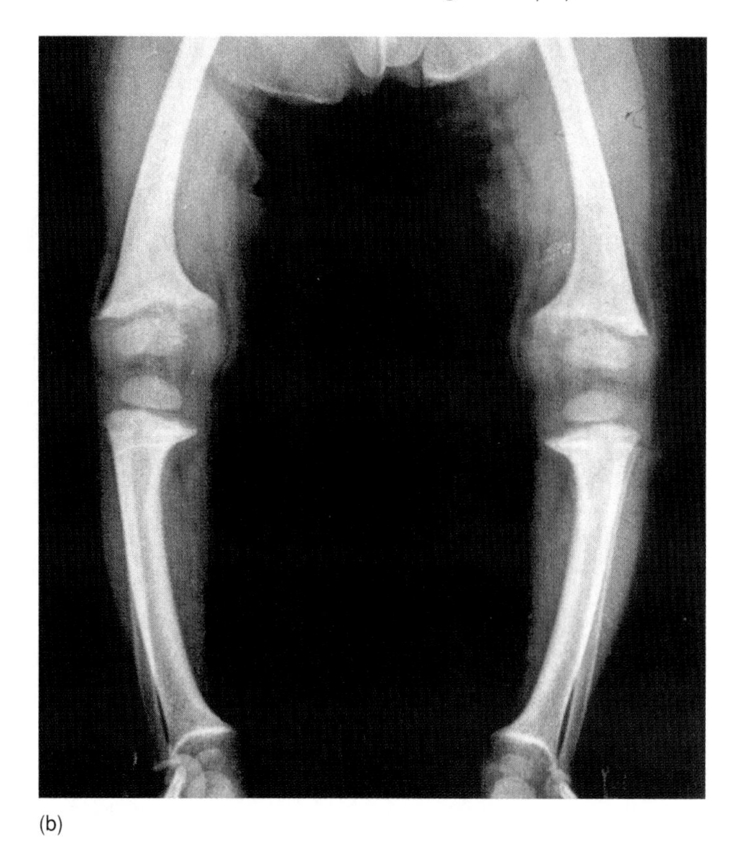
(b)

Figure 5.10

(*a*) Rickets in a child from a Nepalese village, whose inhabitants live in windowless huts. During the rainy season, which may last 5 to 6 months, the children are kept indoors. (*b*) A radiograph of a 10-month-old child with rickets. Rickets develops from an improper diet and also from lack of the sunlight needed to synthesize vitamin D.

anger or embarrassment, may be reflected in changes of skin color. The contraction of specific facial muscles produces facial expressions that convey an array of emotions, including love, surprise, happiness, sadness, and despair. Secretions from certain integumentary glands have odors that frequently elicit subconscious responses from others who detect them.

1. List five modifications of the integument that are structurally or functionally protective.
2. Explain how the integument functions to regulate body fluids and temperature.
3. What substances are synthesized in the integument?

Epidermal Derivatives

Hair, nails, and integumentary glands form from the epidermal layer, and are therefore of ectodermal derivation. Hair and nails are structural features of the integument and have a limited functional role. By contrast, integumentary glands are extremely important in body defense and maintenance of homeostasis.

Objective 6 Describe the structure of hair and list the three principal types.

Objective 7 Discuss the structure and function of nails.

Objective 8 Compare and contrast the structure and function of the three principal kinds of integumentary glands.

Hair

The presence of **hair** on the body is one of the distinguishing features of mammals, but its distribution, function, density, and texture varies across mammalian species. Humans are relatively hairless, with only the scalp, face, pubis, and axillae being densely haired. Men and women have about the same density of hair on their bodies, but hair is generally more obvious on men (fig. 5.11) as a result of male hormones. Certain structures and regions of the body are hairless, such as the palms, soles, lips, nipples, penis, and parts of the female genitalia.

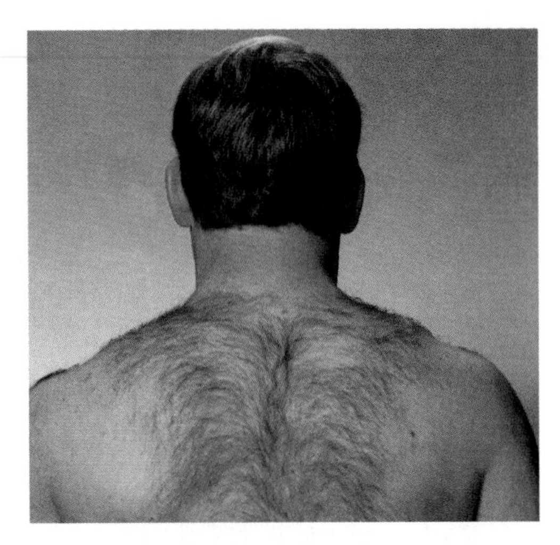

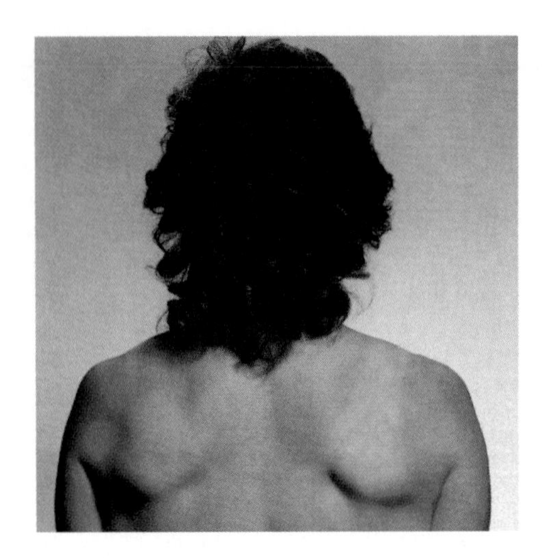

Figure 5.11

A comparison of the expression of body hair in males and females.

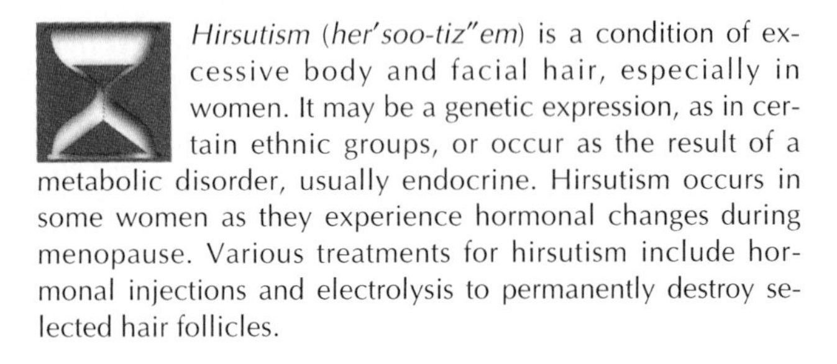

Hirsutism (*her'soo-tiz"em*) is a condition of excessive body and facial hair, especially in women. It may be a genetic expression, as in certain ethnic groups, or occur as the result of a metabolic disorder, usually endocrine. Hirsutism occurs in some women as they experience hormonal changes during menopause. Various treatments for hirsutism include hormonal injections and electrolysis to permanently destroy selected hair follicles.

The primary function of hair is protection, even though its effectiveness is limited. Hair on the scalp and eyebrows protect against sunlight. The eyelashes and the hair in the nostrils protect against airborne particles. Hair on the scalp may also protect against mechanical injury. Some secondary functions of hair are to distinguish individuals and to serve as a sexual attractant.

Each hair consists of a diagonally positioned **shaft, root,** and **bulb** (fig. 5.12). The shaft is the visible, but dead, portion of the hair projecting above the surface of the skin. The bulb is the enlarged base of the root within the **hair follicle.** Each hair develops from stratum basale cells within the bulb of the hair, where nutrients are received from dermal blood vessels. As the cells divide, they are pushed away from the nutrient supply toward the surface, and cellular death and keratinization occur. In a healthy person, hair grows at the rate of approximately 1 mm every 3 days. As the hair becomes longer, however, it enters a resting period, during which there is minimal growth.

The life span of a hair varies from 3 to 4 months for an eyelash to 3 to 4 years for a scalp hair. Each hair lost is replaced by a new hair that grows from the base of the follicle and pushes the old hair out. Between 10 and 100 hairs are lost daily. Baldness results when hair is lost and not replaced. This condition may be disease-related, but it is generally inherited and most frequently occurs in males because of genetic influences combined with the action of the male sex hormone *testosterone* (*tes-tos'tĕ-rōn*). No treatment is effective in reversing genetic baldness; however, flaps or plugs of skin containing healthy follicles from hairy parts of the body can be grafted onto hairless regions.

hirsutism: L. *hirsutus*, shaggy

Three layers can be observed in hair that is cut in cross section. The inner **medulla** (*mĕ-dul'ă*) is composed of loosely arranged cells separated by numerous air cells. The thick **cortex** surrounding the medulla consists of hardened, tightly packed cells. A **cuticle** covers the cortex and forms the toughened outer layer of the hair. Cells of the cuticle have serrated edges that give a hair a scaly appearance when observed under a dissecting scope.

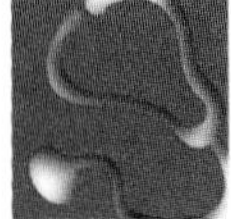

People exposed to heavy metals, such as lead, mercury, arsenic, or cadmium, will have concentrations of these metals in their hair that are 10 times as great as those found in their blood or urine. Because of this, hair samples can be extremely important in certain diagnostic tests.

Even evidence of certain metabolic diseases or nutritional deficiencies may be detected in hair samples. For example, the hair of children with cystic fibrosis will be deficient in calcium and display excessive sodium. There is a deficiency of zinc in the hair of malnourished individuals.

Hair color is determined by the type and amount of pigment produced in the stratum basale at the base of the hair follicle. Varying amounts of melanin produce hair ranging in color from blond to brunette to black. The more abundant the melanin, the darker the hair. A pigment with an iron base (*trichosiderin*) produces red hair. Gray or white hair is the result of a lack of pigment production and air spaces within the layers of the shaft of the hair. The texture of hair is determined by the cross-sectional shape: straight hair is round in cross section, wavy hair is oval, and kinky hair is flat.

Sebaceous glands and arrector pili muscles (described previously) are attached to the hair follicle (fig. 5.12c). The arrector

medulla: L. *medulla*, marrow
cortex: L. *cortex*, bark
cuticle: L. *cuticula*, small skin

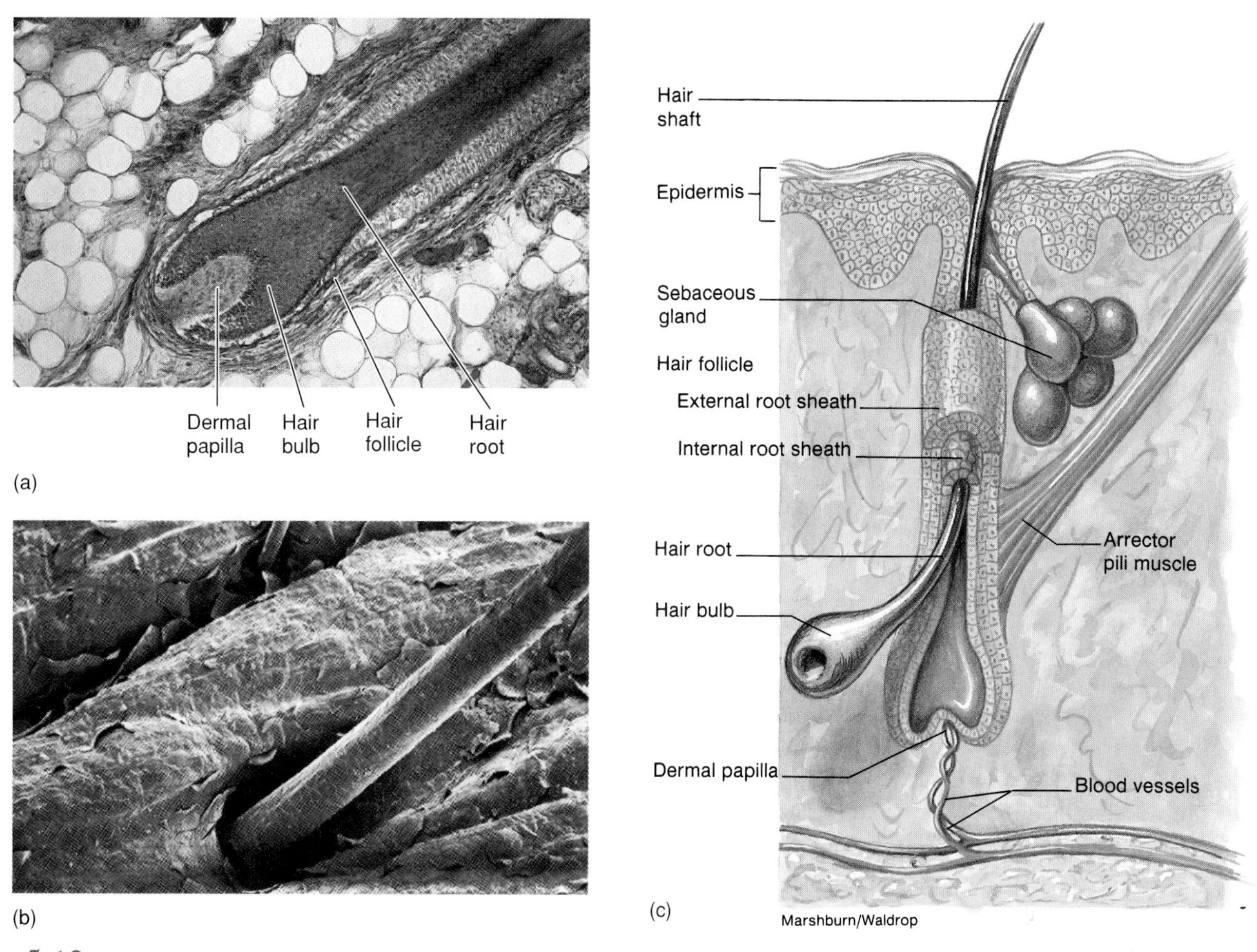

Figure 5.12

The structure of hair and the hair follicle. (*a*) A photomicrograph (63×) of the bulb and root of a hair within a hair follicle. (*b*) A scanning electron micrograph (280×) of a hair as it extends from a follicle. (*c*) A diagram of hair, a hair follicle, and sebaceous gland, and an arrector pili muscle.

pili muscles are involuntary, responding to thermal or psychological stimuli. When they contract, the hair is pulled into a more vertical position, causing goose bumps.

Humans have three distinct kinds of hair:

1. **Lanugo.** Lanugo (*lă-noo'go*) is a fine, silky fetal hair that appears during the last trimester of development. It is usually seen only on premature infants.
2. **Angora.** Angora hair grows continuously. It is found on the scalp and on the faces of mature males.
3. **Definitive.** Definitive hair grows to a certain length and then stops. It is the most common type of hair. Eyelashes, eyebrows, pubic, and axillary hair are examples.

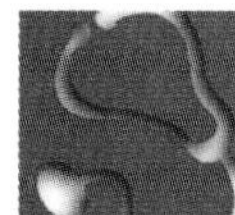

Anthropologists have referred to humans as the naked apes because of our relative hairlessness. The clothing that we wear over the exposed surface areas of our bodies functions to insulate and protect us, just as hair or fur does in other mammals. However, the nakedness of our skin does lead to some problems. *Skin cancer* occurs frequently in humans, particularly in regions of the skin exposed to the sun. *Acne,* another problem unique to humans, is partly related to the fact that hair is not present to dissipate the oily secretion from the sebaceous glands.

Nails

The **nails** on the ends of the fingers and toes are formed from the compressed outer layer (stratum corneum) of the epidermis. The hardness of the nail is due to the dense keratin fibrils running parallel between the cells. Both fingernails and toenails protect the digits, and fingernails also aid in grasping and picking up small objects.

Each nail consists of a **body, free border,** and **hidden border** (fig. 5.13). The platelike body of the nail rests on a **nail bed,** which is actually the stratum spinosum of the epidermis. The body and nail bed appear pinkish because of the underlying vascular tissue. The sides of the nail body are protected by a **nail fold,** and the furrow between the sides and body is the **nail groove.** The free border of the nail extends over a thickened region of the stratum

lanugo: L. *lana,* wool

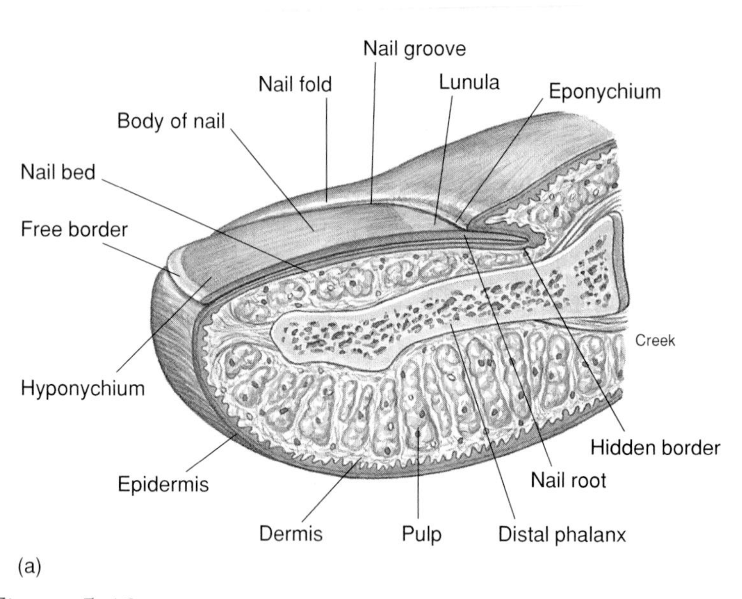

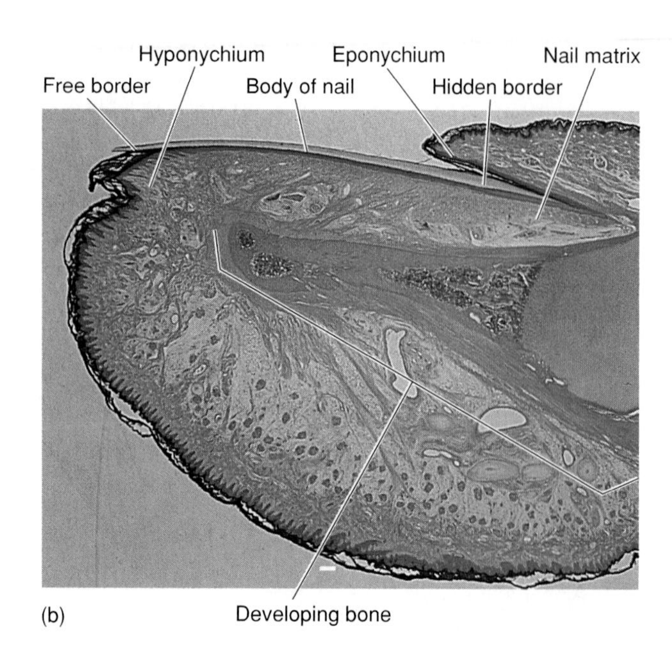

Figure 5.13

The fingertip and the associated structures of the nail. (*a*) A diagram of a dissected nail, and (*b*) a photomicrograph of a nail from a fetus (3.5×).

corneum called the **hyponychium** (*hi″pŏ-nik′e-um*) (quick). The root of the nail is attached at the base.

An **eponychium** (cuticle) covers the hidden border of the nail. The eponychium frequently splits, causing a hangnail. The growth area of the nail is the **nail matrix.** A small part of the nail matrix, the **lunula** (*loo′nyoo-lă*), can be seen as a half-moon-shaped area near the eponychium of the nail.

The nail grows by the transformation of the superficial cells of the nail matrix into nail cells. These harder, transparent cells are then pushed forward over the strata basale and spinosum of the nail bed. Fingernails grow at the rate of approximately 1 mm each week. The growth rate of toenails is somewhat slower.

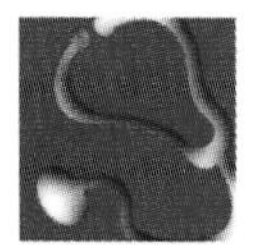

The condition of nails may be indicative of a person's general health and well-being. Nails should appear pinkish, showing the rich vascular capillaries beneath the translucent nail. A yellowish hue may indicate certain glandular dysfunctions or nutritional deficiencies. Split nails may also be caused by nutritional deficiencies. A prominent bluish tint may indicate improper oxygenation of the blood. Spoon nails (concave body) may be the result of iron-deficiency anemia, and clubbing at the base of the nail may be caused by lung cancer. Dirty or ragged nails may indicate poor personal hygiene, and chewed nails may suggest emotional problems.

Glands

Although they originate in the epidermal layer, all of the glands of the skin are located in the dermis, where they are physically supported and receive nutrients. Glands of the skin are referred to as *exocrine*, because they are externally secreting glands that either release their secretions directly or through ducts. The glands of the skin are of three basic types: *sebaceous* (*sĕ-ba′shus*), *sudoriferous* (*soo″dor-if′er-us*), and *ceruminous* (*sĕ-roo′mĭ-nus*).

Sebaceous Glands

Commonly called oil glands, sebaceous glands are associated with hair follicles, since they develop from the follicular epithelium of the hair. They are holocrine glands (see chapter 4) that secrete **sebum** (*se′bum*) onto the shaft of the hair (fig. 5.12). Sebum, which consists mainly of lipids, is dispersed along the shaft of the hair to the surface of the skin, where it lubricates and waterproofs the stratum corneum and also prevents the hair from becoming brittle. If the ducts of sebaceous glands become blocked for some reason, the glands may become infected, resulting in acne. Sex hormones regulate the production and secretion of sebum, and hyperactivity of sebaceous glands can result in serious acne problems, particularly during teenage years.

Sudoriferous Glands

Commonly called sweat glands, **sudoriferous glands** excrete perspiration, or sweat, onto the surface of the skin. Perspiration is composed of water, salts, urea, and uric acid. It serves not only for evaporative cooling, but also for the excretion of certain wastes. Sweat glands are most numerous on the palms, soles, axillary and pubic regions, and on the forehead. They are coiled and

hyponychium: Gk. *hypo*, under; *onyx*, nail
lunula: L. *lunula*, small moon

sebum: L. *sebum*, tallow or grease
sudoriferous: L. *sudorifer*, sweat; *ferre*, to bear

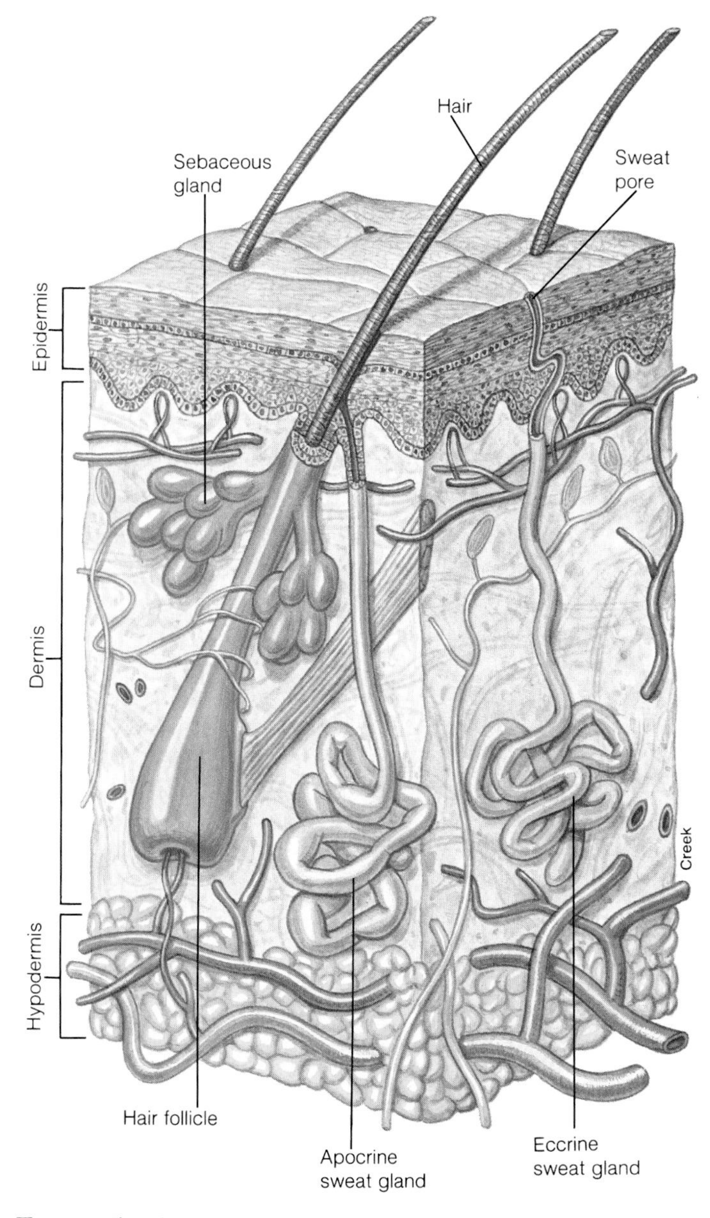

Figure 5.14

Types of skin glands.

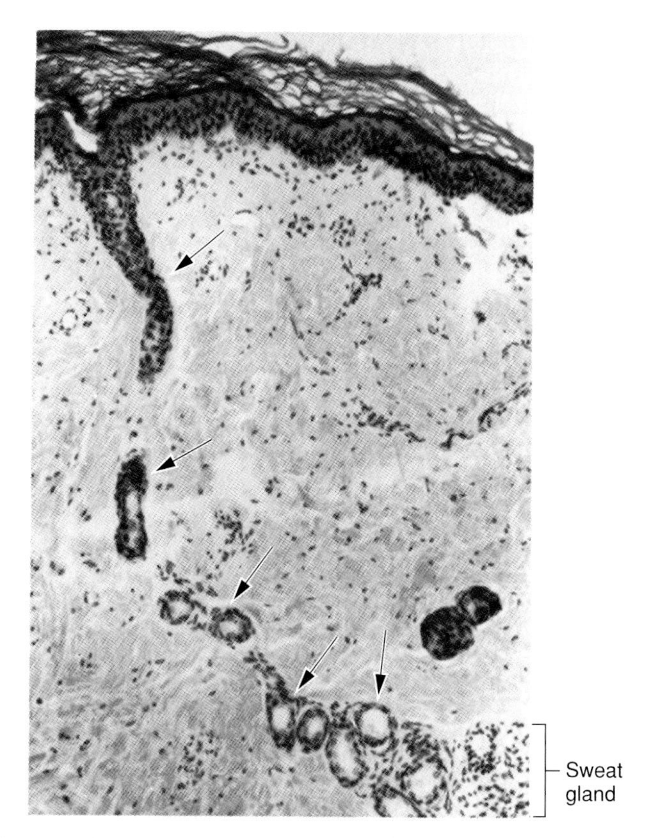

Figure 5.15

A photomicrograph of an eccrine sweat gland (100×). The coiled structure of the ductule portion of the gland (see arrows) accounts for its discontinuous appearance.

tubular (fig. 5.14) and are of two types: *eccrine* (*ek′rin*) and *apocrine* (*ap′ŏ-krin*) *sweat glands.*

1. **Eccrine sweat glands** are widely distributed over the body, especially on the forehead, back, palms, and soles. These glands are formed before birth and function in evaporative cooling (figs. 5.14 and 5.15).
2. **Apocrine sweat glands** are much larger than the eccrine glands. They are found in the axillary and pubic regions, where they secrete into hair follicles. Apocrine glands are not functional until puberty, and their odoriferous secretion is thought to act as a sexual attractant.

Mammary glands, found within the breasts, are specialized sudoriferous glands that secrete milk during lactation (fig. 5.16).

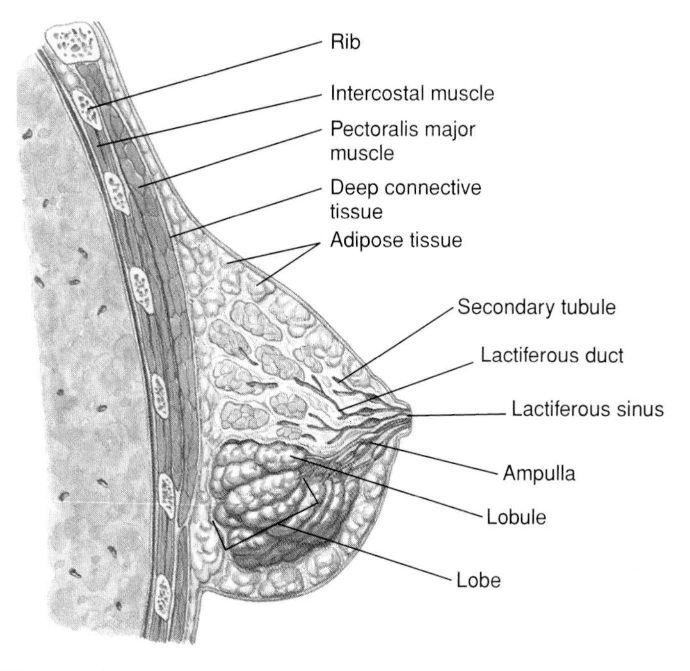

Figure 5.16

A sagittal section of a mammary gland within the human breast.

Developmental Exposition

The Integumentary System

Explanation

Both the ectodermal and mesodermal germ layers (see chapter 4) function in the formation of the structures of the integumentary system. The epidermis and the hair, glands, and nails of the skin develop from the ectodermal germ layer (exhibits, I, II, and III). The dermis develops from a thickened layer of undifferentiated mesoderm called *mesenchyme* (*mez′en-kīm*).

By 6 weeks, the ectodermal layer has differentiated into an outer flattened *periderm* and an inner cuboidal *germinal* (*basal*) *layer* in contact with the mesenchyme. The periderm eventually sloughs off, forming the *vernix caseosa* (*ka″se-o′să*), a cheeselike protective coat that covers the skin of the fetus.

By 11 weeks, the mesenchymal cells below the germinal cells have differentiated into the distinct collagenous and elastic connective tissue fibers of the dermis. The tensile properties of these fibers cause a buckling of the epidermis and the formation of dermal papillae. During the early fetal period (about 10 weeks), specialized neural crest cells called *melanoblasts* migrate into the developing dermis and differentiate into *melanocytes*. The melanocytes soon migrate to the germinal layer of the epidermis, where they produce the pigment *melanin* that colors the epidermis.

Before hair can form, a *hair follicle* must be present. Each hair follicle begins to develop at about 12 weeks (exhibit II), as a mass of germinal cells called a *hair bud* proliferates into the underlying mesenchyme. As the hair bud becomes club-shaped, it is referred to as a *hair bulb*. The hair follicle, which physically supports and provides nourishment to the hair, is derived from specialized mesenchyme called the *hair papilla,* which is localized around the hair bulb, and from the epithelial cells of the hair bulb called the *hair matrix*. Continuous mitotic activity in the epithelial cells of the hair bulb results in the growth of the hair.

Sebaceous glands and sweat glands are the two principal types of integumentary glands. Both develop from the germinal layer of the epidermis (exhibit II). Sebaceous glands develop as proliferations from the sides of the developing hair follicle. Sweat glands become coiled as the secretory portion of the developing gland proliferates into the dermal mesenchyme. Mammary glands (exhibit III) are modified sweat glands that develop in the skin of the anterior thoracic region.

Exhibit I The development of the skin.

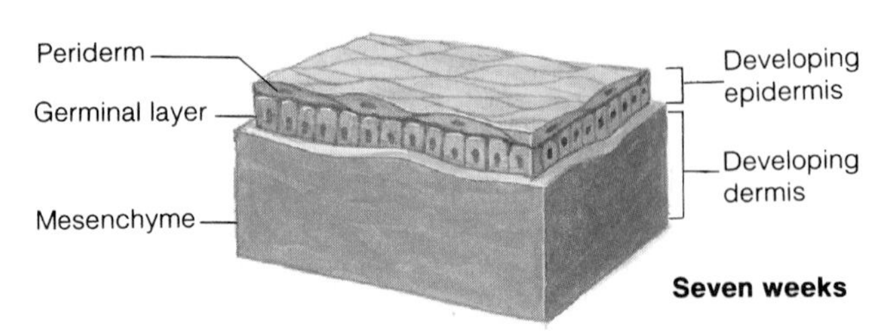

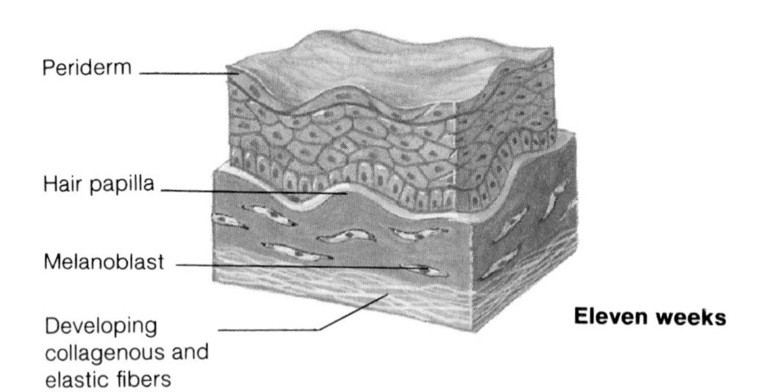

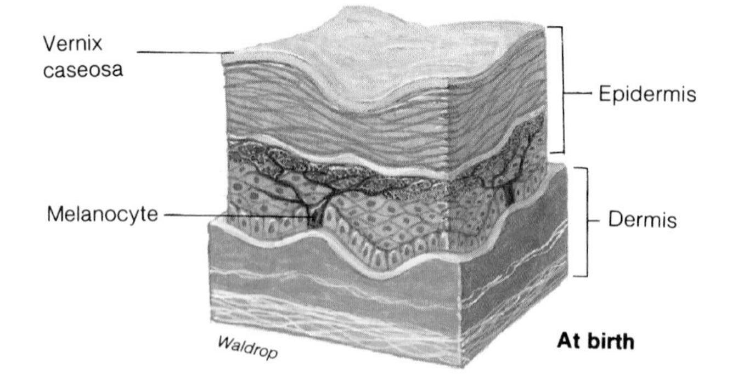

Exhibit II The development of hair and glands.

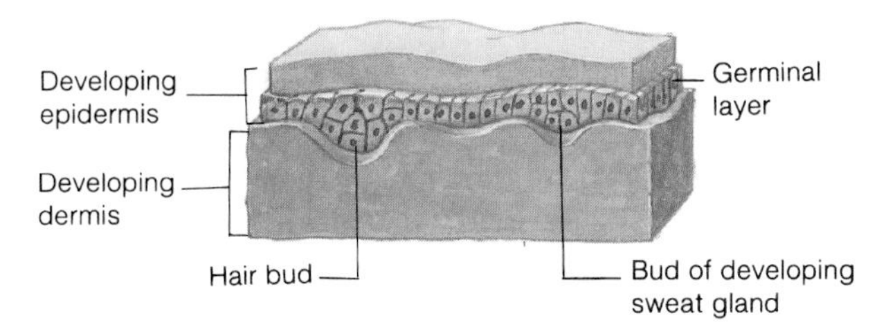

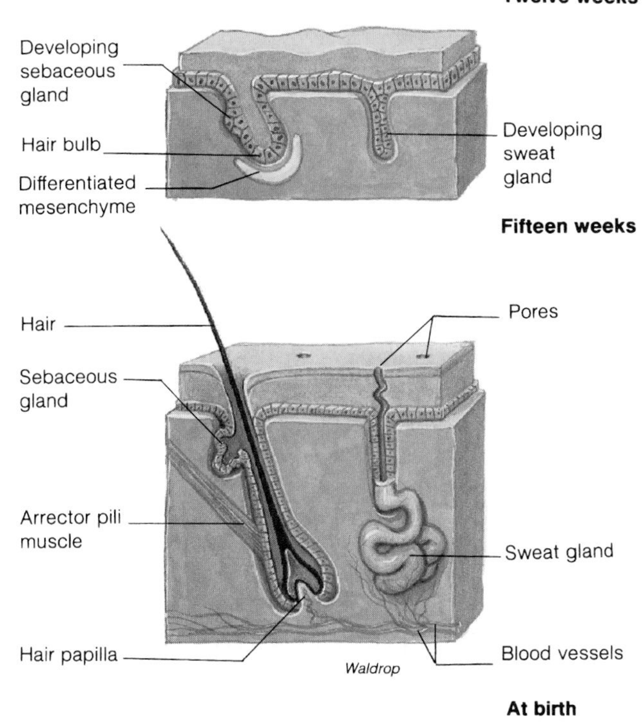

Exhibit III The development of mammary glands. (*a*) Twelve weeks, (*b*) sixteen weeks, and (*c*) about twenty-eight weeks.

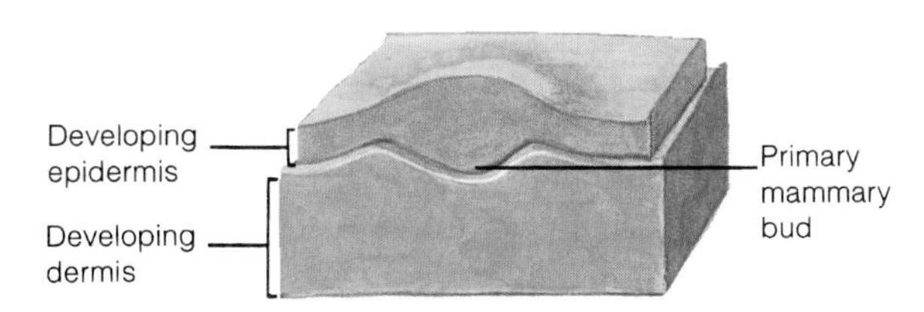

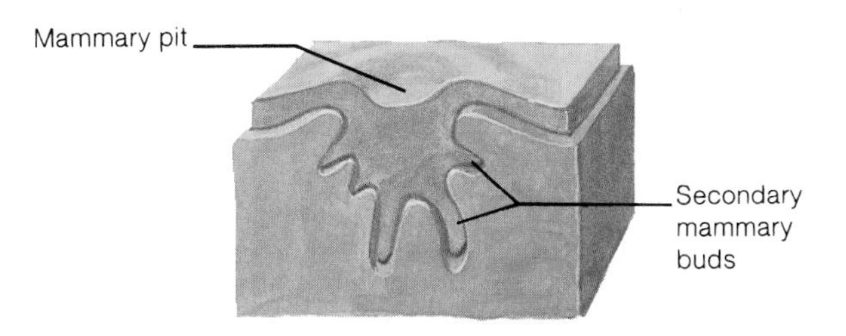

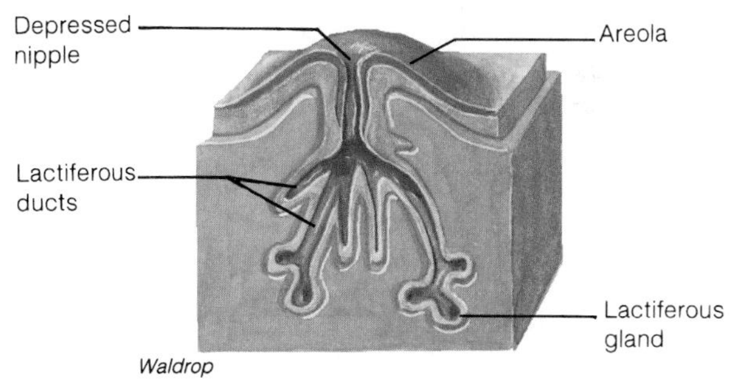

The breasts of the female reach their greatest development during the childbearing years, under the stimulus of pituitary and ovarian hormones.

Good routine hygiene is very important for health and social reasons. Washing away the dried residue of perspiration and sebum eliminates dirt. Excessive bathing, however, can wash off the natural sebum and dry the skin, causing it to itch or crack. The commercial lotions used for dry skin are, for the most part, refined and perfumed lanolin, which is sebum from sheep.

Ceruminous Glands

These specialized glands are found only in the external auditory canal (ear canal) where they secrete **cerumen** (*sĕ-roo′men*), or earwax. Cerumen is a water and insect repellent, and also keeps the tympanic membrane (eardrum) pliable. Excessive amounts of cerumen may interfere with hearing.

1. Draw and label a hair. Indicate which portion is alive and discuss what causes the cells in a hair to die.
2. Describe the structure and function of nails.
3. List the three types of integumentary glands and describe the structure and function of each.
4. Are skin glands mesodermal or ectodermal in derivation? Are they epidermal or dermal in functional position?

Clinical Considerations

The skin is a buffer against the external environment and is therefore subject to a variety of disease-causing microorganisms and physical assaults. A few of the many diseases and disorders of the integumentary system are briefly discussed here.

Inflammatory Conditions (Dermatitis)

Inflammatory skin disorders are caused by immunologic hypersensitivity, infectious agents, poor circulation, or exposure to environmental assaults such as wind, sunlight, or chemicals. Some people are allergic to certain foreign proteins and, because of this inherited predisposition, experience such hypersensitive reactions as asthma, hay fever, hives, drug and food allergies, and eczema. **Lesions,** as applied to inflammatory conditions, are defined as more or less circumscribed pathologic changes in the tissue. Some of the more common inflammatory skin disorders and their usual sites are illustrated in figure 5.17.

There are also a number of *infectious diseases* of the skin, which is not surprising considering the highly social and communal animals we are. Most of these diseases can now be prevented, but too frequently people fail to take appropriate precautionary measures. Infectious diseases of the skin include childhood viral infections (measles and chicken pox); bacteria, such as staphylococcus (impetigo); sexually transmitted diseases; leprosy; fungi (ringworm, athlete's foot, candida); and mites (scabies).

Neoplasms

Both benign and malignant neoplastic conditions or diseases are common in the skin. *Pigmented moles* (nevi), for example, are a type of benign neoplastic growth of melanocytes. *Dermal cysts* and *benign viral infections* are also common. *Warts* are virally caused abnormal growths of tissue that occur frequently on the hands and feet. These warts are usually treated effectively with liquid nitrogen or acid. A different type of wart, called a *venereal wart*, occurs in the anogenital region of affected sexual partners. Risk factors for cervical cancer may be linked to venereal warts, so they are treated aggressively with chemicals, cryosurgery, cautery, or laser therapy.

Skin cancer is the most common malignancy in the United States. As shown in figure 5.18, there are three frequently encountered types. **Basal cell carcinoma**, the most common skin cancer, accounts for about 70% of total cases. It usually occurs where exposure to sunlight is the greatest—on the face and arms. This type of cancer arises from cells in the stratum basale. It appears first on the surface of the skin as a small, shiny bump. As the bump enlarges, it often develops a central crater that erodes, crusts, and bleeds. Fortunately, there is little danger that it will spread (metastasize) to other body areas. These carcinomas are usually treated by excision (surgical removal).

Squamous cell carcinoma arises from cells immediately superficial to the stratum basale. Normally, these cells undergo very little division, but in squamous cell carcinoma they continue to divide as they produce keratin. The result is usually a firm, red keratinized tumor, confined to the epidermis. If untreated, however, it may invade the dermis and metastasize. Treatment usually consists of excision and radiation therapy.

Malignant melanoma, the most life-threatening form of skin cancer, arises from the melanocytes located in the stratum basale. Often, it begins as a small molelike growth, which enlarges, changes color, becomes ulcerated, and bleeds easily. Metastasis occurs quickly, and unless treated early—usually by widespread excision and radiation therapy—this cancer is often fatal.

cerumen: L. *cera*, wax

neoplasm: Gk. *neo*, new; *plasma*, something formed
benign: L. *benignus*, good-natured
malignant: L. *malignus*, acting from malice

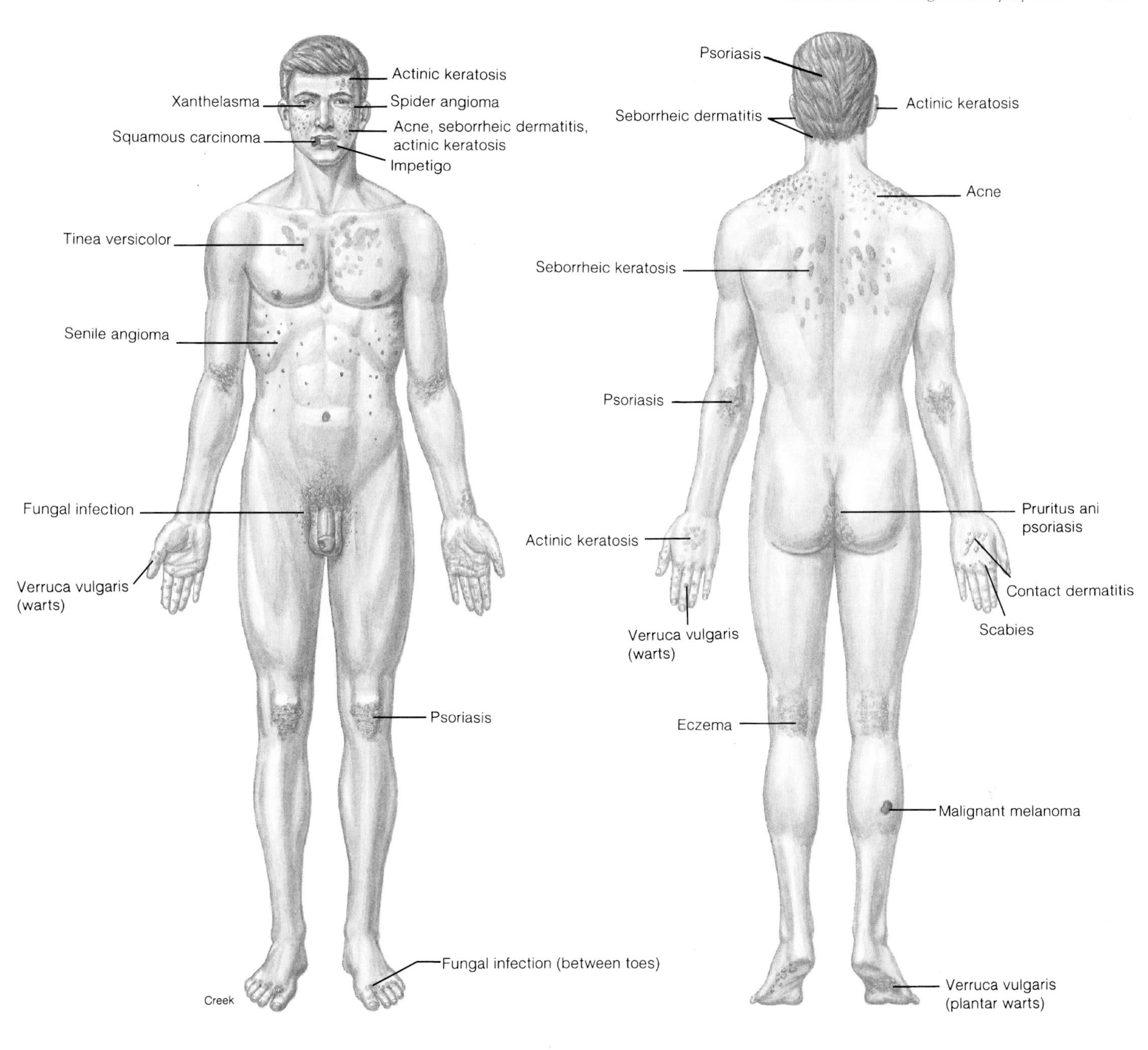

Figure 5.17

Common inflammatory skin disorders and their usual sites of occurrence.

Burns

A burn is an epithelial injury caused by contact with thermal, radioactive, chemical, or electrical agent. Burns generally occur on the skin, but they can involve the linings of the respiratory and GI tracts. The extent and location of a burn is frequently less important than the degree to which it disrupts body homeostasis. Burns that have a **local effect** (local tissue destruction) are not as serious as those that have a **systemic effect.** Systemic effects directly or indirectly involve the entire body and are a threat to life. Possible systemic effects include body dehydration, shock, reduced circulation and urine production, and bacterial infections.

Burns are classified as first degree, second degree, or third degree, based on their severity (fig. 5.19). In **first-degree burns,** the epidermal layers of the skin are damaged and symptoms are restricted to local effects such as redness, pain, and edema (swelling). A shedding of the surface layers (desquamation) generally follows in a few days. A sunburn is an example. **Second-degree burns** involve both the epidermis and dermis. Blisters appear and recovery is usually complete, although slow. **Third-degree burns** destroy the entire thickness of the skin and

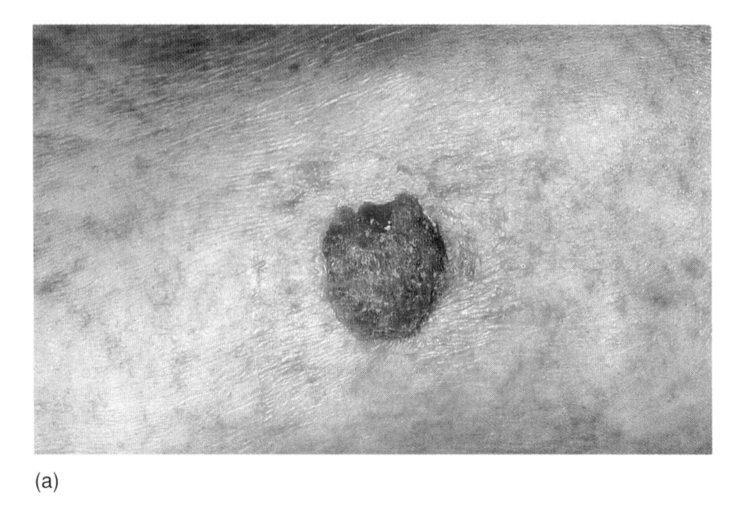

(a)

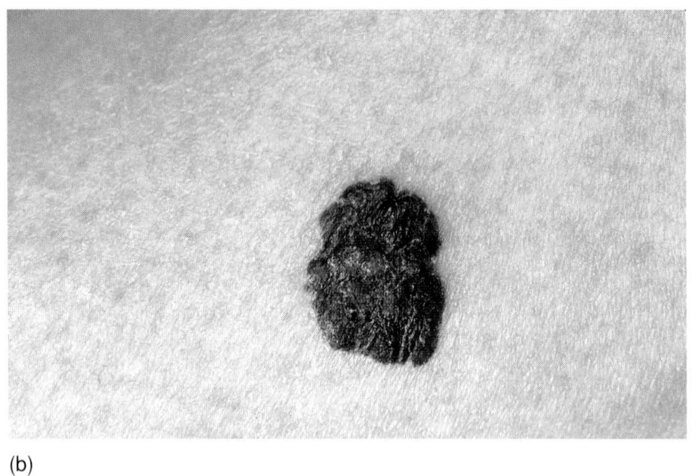

(b)

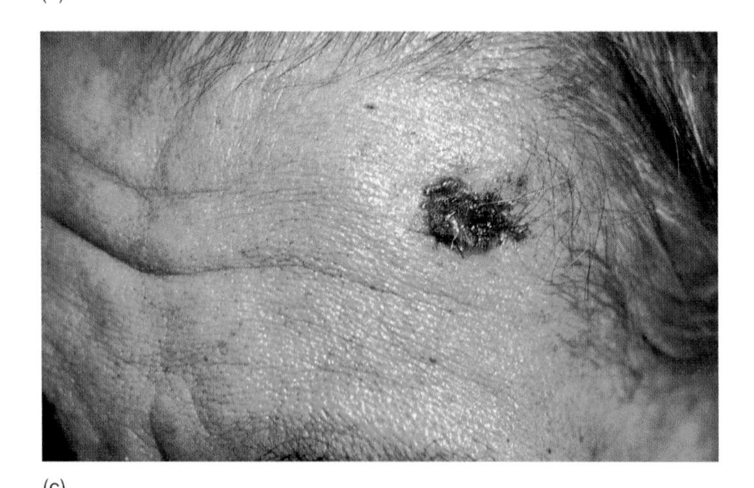

(c)

Figure 5.18

Types of skin cancer. (*a*) Squamous cell carcinoma, (*b*) malignant melanoma, and (*c*) basal cell carcinoma.

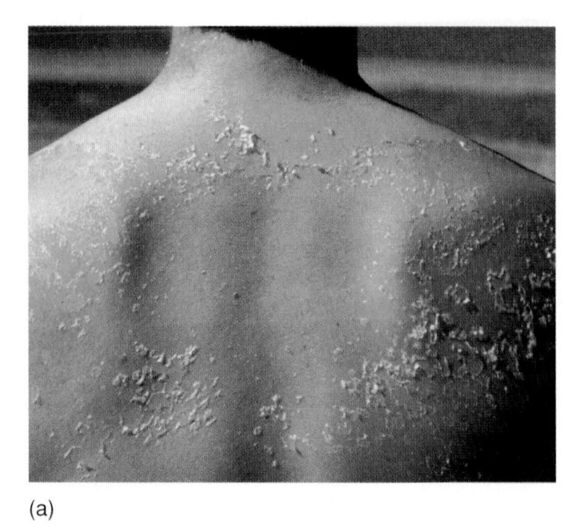

(a)

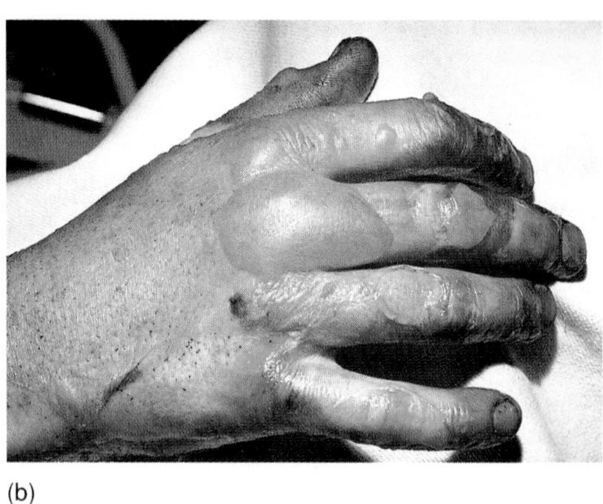

(b)

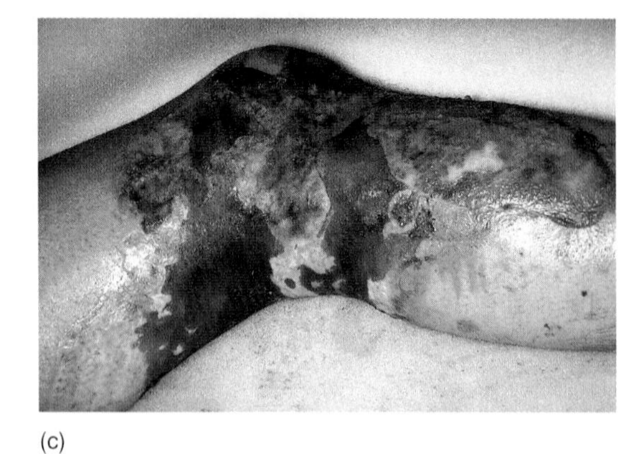

(c)

Figure 5.19

The classification of burns. (*a*) First-degree burns involve the epidermis and are characterized by redness, pain, and edema—such as with a sunburn; (*b*) second-degree burns involve the epidermis and dermis and are characterized by intense pain, redness, and blistering; and (*c*) third-degree burns destroy the entire skin and frequently expose the underlying organs. The skin is charred and numb and does not protect against fluid loss.

frequently some of the underlying muscle. The skin appears waxy or charred and is insensitive to touch. As a result, ulcerating wounds develop, and the body attempts to heal itself by forming scar tissue. Skin grafts are frequently used to assist recovery.

As a way of estimating the extent of damaged skin suffered in burned patients, the *rule of nines* (fig. 5.20) is often applied. The surface area of the body is divided into regions, each of which accounts for about 9% (or a multiple of 9%) of the total skin surface. An estimation of the percentage of surface area damaged is important in treating with intravenous fluid, which replaces the fluids lost from tissue damage.

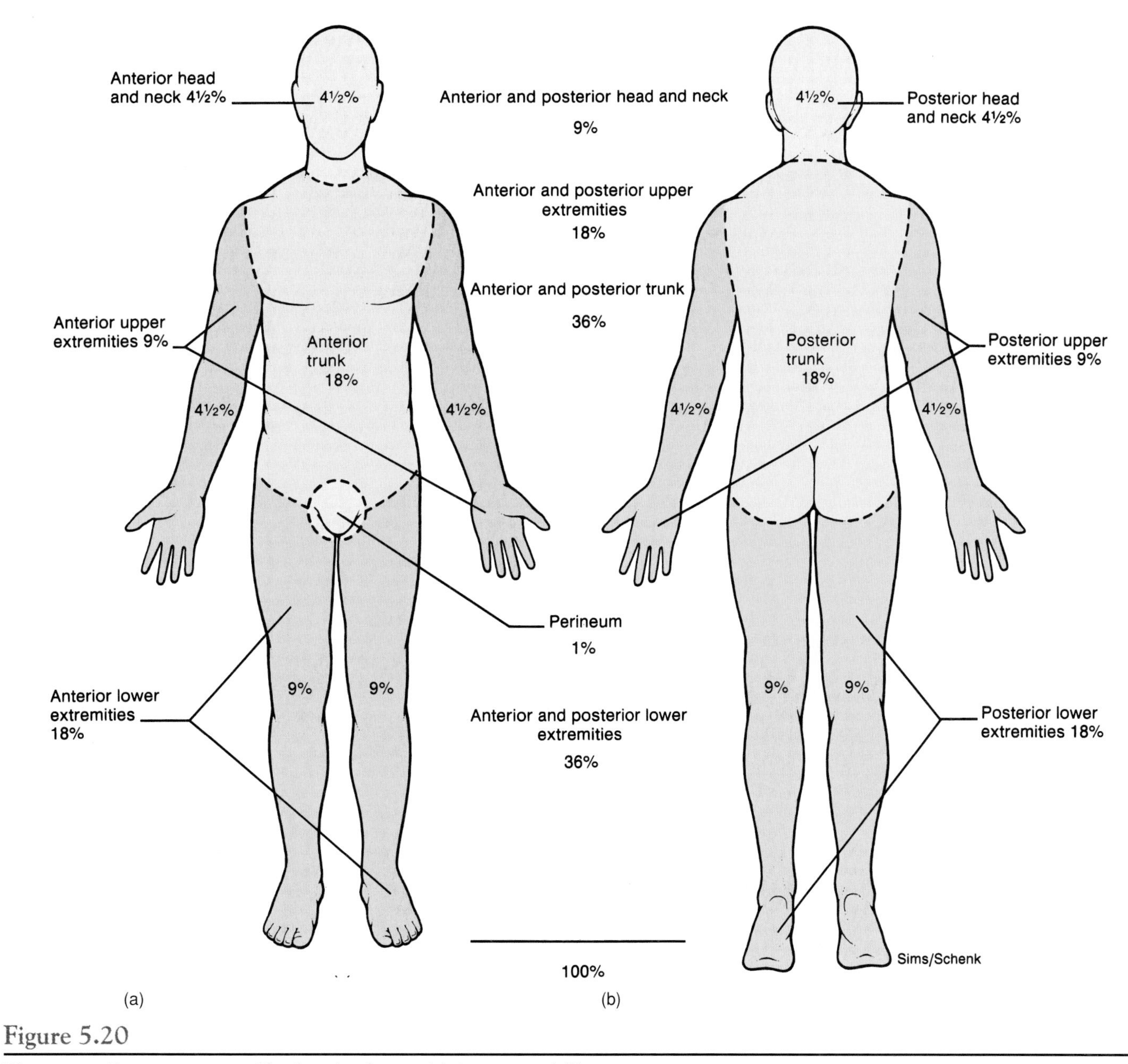

Figure 5.20

The extent of burns, as estimated by the rule of nines. (*a*) Anterior and (*b*) posterior.

Frostbite

Frostbite is a local destruction of the skin resulting from freezing. Like burns, frostbite is classified by its degree of severity: first degree, second degree, and third degree. In **first-degree frostbite,** the skin will appear cyanotic (bluish) and swollen. Vesicle formation and hyperemia (engorgement with blood) are symptoms of **second-degree frostbite.** As the affected area is warmed, there will be further swelling, and the skin will redden and blister. In **third-degree frostbite,** there will be severe edema, some bleeding, and numbness followed by intense throbbing pain and necrosis of the affected tissue. Gangrene will follow untreated third-degree frostbite.

Skin Grafts

If extensive areas of the stratum basale of the epidermis are destroyed in second-degree or third-degree burns or frostbite, new skin cannot grow back. In order for this type of wound to heal, a skin graft must be performed.

A **skin graft** is a segment of skin that has been excised from a *donor site* and transplanted to the *recipient site,* or *graft bed.* As stated in chapter 4, an *autograft* is the most successful type of tissue transplant. It involves taking a thin sheet of healthy epidermis from a donor site of the burn or frostbite patient and moving it to the recipient site (fig. 5.21). A *heterotransplant*

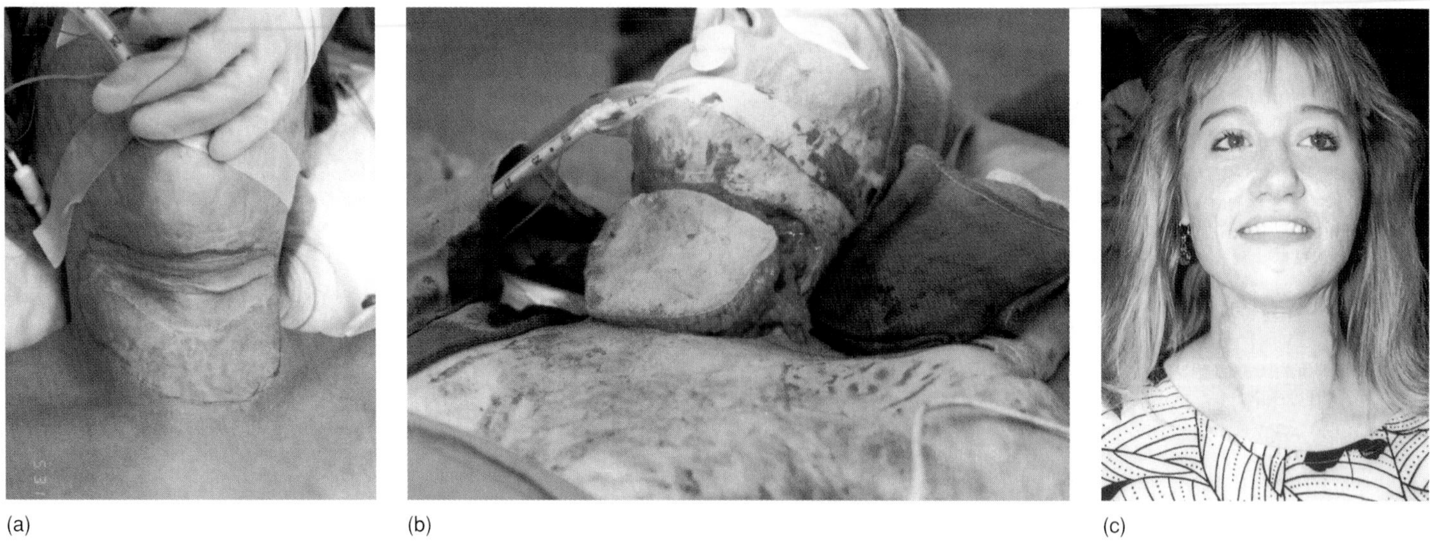

Figure 5.21

A skin graft to the neck. (*a*) Traumatized skin is prepared for excision; (*b*) healthy skin from another body location is transplanted to the graft site; and (*c*) 1 year following the successful transplant, healing is complete.

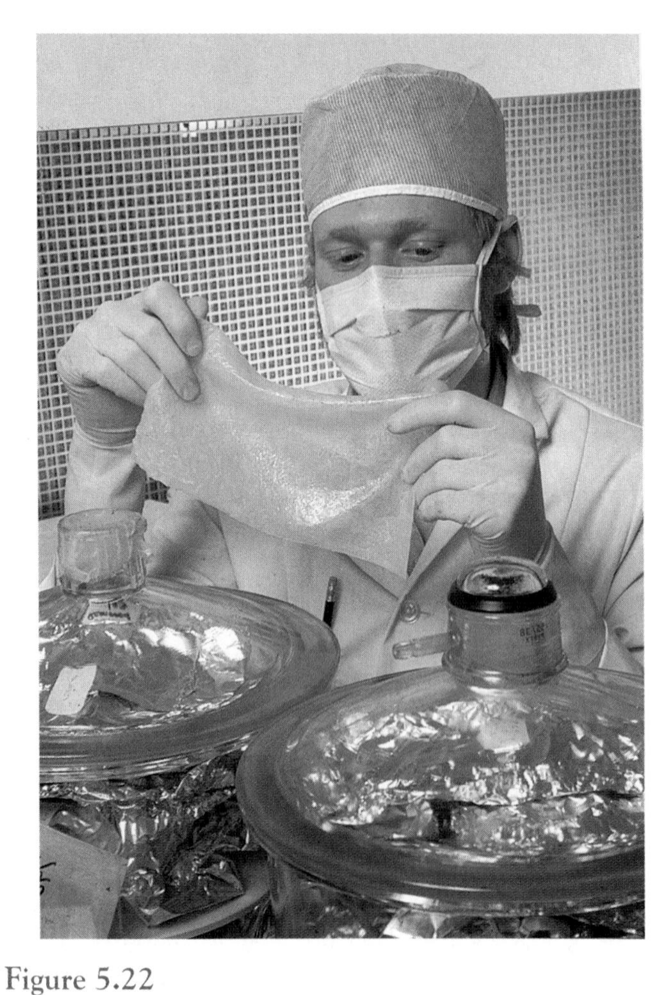

Figure 5.22

Synthetic skin used in grafting.

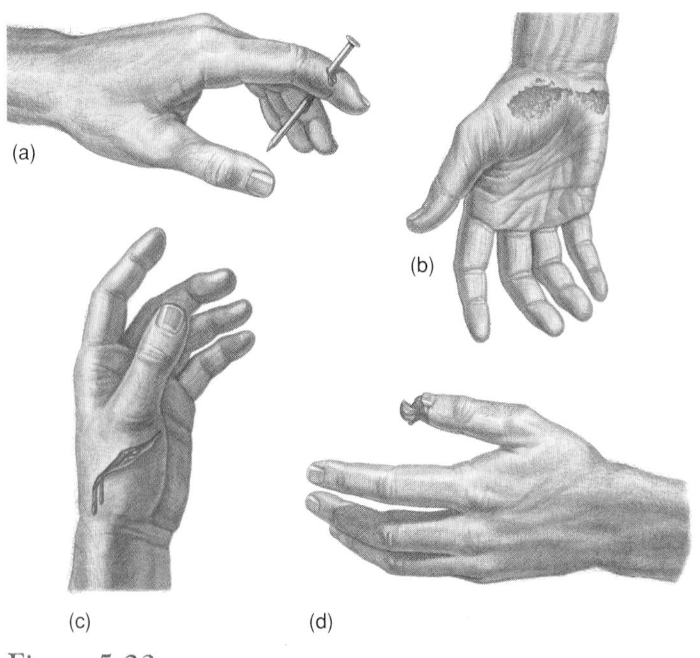

Figure 5.23

Various kinds of wounds: (*a*) puncture, (*b*) abrasion, (*c*) laceration, and (*d*) avulsion.

(*xenograph*—between two different species) can serve as a temporary treatment to prevent infection and fluid loss.

Synthetic skin fabricated from animal tissue bonded to a silicone film (fig. 5.22) may be used on a patient who is extensively burned. The process includes seeding the synthetic skin with basal skin cells obtained from healthy locations on the patient. This treatment eliminates some of the problems of skin grafting—for example, additional trauma, widespread scarring, and rejection, as in the case of skin obtained from a cadaver.

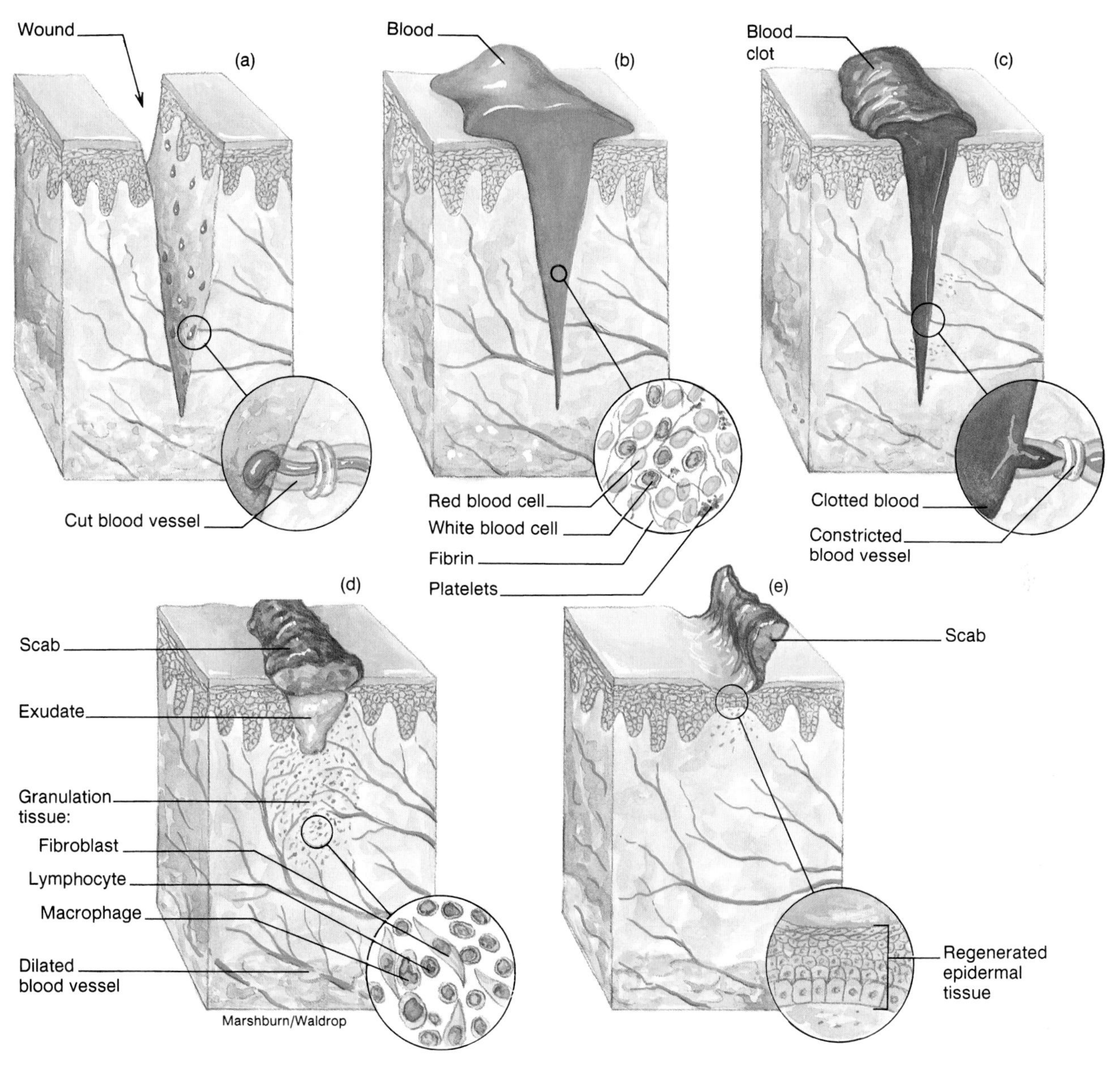

Figure 5.24

The process of wound healing. (*a*) A penetrating wound into the dermis ruptures blood vessels. (*b*) Blood cells, fibrinogen, and fibrin flow out of the wound. (*c*) Vessels constrict and a clot blocks the flow of blood. (*d*) A protective scab is formed from the clot, and granulation occurs within the site of the wound. (*e*) The scab sloughs off as the epidermal layers are regenerated.

Wound Healing

The skin effectively protects against many abrasions, but if a wound does occur (fig. 5.23) a sequential chain of events promotes rapid healing. The process of wound healing depends on the extent and severity of the injury. Trauma to the epidermal layers stimulates increased mitotic activity in the stratum basale, whereas injuries that extend to the dermis or subcutaneous layer elicit activity throughout the body, not just within the wound area. General body responses include a temporary elevation of temperature and pulse rate.

In an open wound (fig. 5.24), blood vessels are broken and bleeding occurs. Through the action of **blood platelets** and protein molecules called **fibrinogen** (*fi-brin'ŏ-jen*), a clot forms and soon blocks the flow of blood. The scab that forms from the clot covers and protects the damaged area. Mechanisms are activated to destroy bacteria, dispose of dead or injured cells, and isolate the injured area. These responses are collectively referred to as *inflammation* and are characterized by redness, heat, edema, and pain. Inflammation is a response that confines the injury and promotes healing.

The next step in healing is the differentiation of binding **fibroblasts** from connective tissue at the wound margins. Together with new branches from surrounding blood vessels, **granulation tissue** is formed. Phagocytic cells migrate into the wound and ingest dead cells and foreign debris. Eventually, the damaged area is repaired and the protective scab is sloughed off.

If the wound is severe enough, the granulation tissue may develop into **scar tissue** (fig. 5.25). The collagenous fibers of scar

Figure 5.25

Scars for body adornment on the face of this Buduma man from the islands of Lake Chad are created by instruments that make crescent-shaped incisions into the skin in beadlike patterns. Special ointments are applied to the cuts to retard healing and promote scar formation.

Figure 5.26

Aging of the skin results in a loss of elasticity and the appearance of wrinkles.

tissue, are more dense than those of normal tissue, and scar tissue has no stratified squamous or epidermal layer. Scar tissue also has fewer blood vessels than normal skin, and may lack hair, glands, and sensory receptors. The closer the edges of a wound, the less granulation tissue develops and the less obvious a scar. This is one reason for suturing a large break in the skin.

Aging of the Skin

As the skin ages, it becomes thin and dry, and begins to lose its elasticity. Collagenous fibers in the dermis become thicker and stiffer, and the amount of adipose tissue in the hypodermis diminishes, making it thinner. Skinfold measurements indicate that the diminution of the hypodermis begins at about the age of 45. With a loss of elasticity and a reduction in the thickness of the hypodermis, wrinkling, or permanent infolding of the skin, becomes apparent (fig. 5.26).

During the aging of the skin, the number of active hair follicles, sweat glands, and sebaceous glands also declines. Consequently, there is a marked thinning of scalp hair and hair on the extremities, reduced sweating, and decreased sebum production. Since elderly people cannot perspire as freely as they once did, they are more likely to complain of heat and are at greater risk for heat exhaustion. They also become more sensitive to cold because of the loss of insulating adipose tissue and diminished circulation. A decrease in the production of sebum causes the skin to dry and crack frequently.

The integument of an elderly person is not as well protected from the sun because of thinning, and melanocytes that produce melanin gradually atrophy. The loss of melanocytes accounts for graying of the hair and pallor of the skin.

Clinical Case Study Answer

The blistering and erythema characteristic of second-degree burns is a manifestation of intact and functioning blood vessels, which exist in abundance within the spared dermis. In third-degree burns, the entire dermis and its vasculature is destroyed, thus explaining the absence of these findings. In addition, nerve endings and other nerve end organs that reside in the dermis are destroyed in third-degree burns, resulting in a desensitized area. By contrast, significant numbers of these structures are spared and functional in second-degree burns, thus preserving sensation—including pain. The third-degree burn areas will all require skin grafting in order to prevent infection, one of the skin's most vital functions. In second-degree burns, the spared dermis serves somewhat of a barrier to bacteria. Consequently, skin grafting is usually unnecessary, especially if sufficient numbers of skin adnexa (hair follicles, sweat glands, and so forth), which generally lie deep within the dermis, are spared. These structures serve as starting points for regeneration of surface epithelium and skin organs.

Internal Affairs

■ How the Integumentary System Works with Other Body Systems
□ How Other Systems Work with the Integumentary System

Skeletal System

■
- Covers and protects the skeletal system
- Role in vitamin D synthesis promotes calcium absorption needed for bone growth and maintenance

□
- Supports the skin
- Stores minerals needed by skin

Muscular System

■
- Covers and protects the muscles
- Role in vitamin D synthesis promotes calcium absorption needed for muscle contraction
- Permits radiant heat loss (sweating) during muscle contractions

□
- Generates body heat to warm the skin
- Skeletal muscle contractions pull on skin, producing facial expressions

Nervous System

■
- Houses cutaneous (tactile) receptors that convey sensory sensations to the brain

□
- Provides autonomic motor impulses to cutaneous vessels and glands

Endocrine System

■
- Covers and protects certain endocrine glands
- Converts some steroid hormones to their active forms

□
- Sex hormones cause integumentary features to change—pubic, axillary, and facial hair and oily skin

Circulatory System

■
- Protects the circulatory system
- Maintains constant body temperature
- Prevents fluid loss (formation of scabs)

□
- Transports O_2 and CO_2, nutrients, and fluids to and from the skin

Lymphatic System

■
- Protects against pathogen invasion
- Prevents edema (retention of interstitial fluid)

□
- Maintains a balanced amount of interstitial fluid within the skin
- Immune cells protect skin from infection and promote tissue repair

Respiratory System

■
- Protects organs of the upper respiratory tract

□
- Provides O_2 and eliminates CO_2

Digestive System

■
- Role in vitamin D synthesis enables intestinal absorption of calcium

□
- Provides nutrients for growth, maintenance, and repair of skin

Urinary System

■
- Excretes salts and some nitrogenous wastes

□
- Eliminates metabolic wastes
- Activates vitamin D

Reproductive System

■
- Major component of scrotum that covers and protects testes
- Cutaneous receptors respond to erotic stimuli
- Mammary glands produce milk

□
- Gonads produce sex hormones that promote growth, maturation, and maintenance of the skin

Figure 5.27

The individual on the left has melanocytes within his skin, but as a result of a mutant gene he is affected with albinism—an inability to synthesize melanin.

Important Clinical Terminology

acne An inflammatory condition of sebaceous glands. Acne is effected by gonadal hormones, and is therefore common during puberty and adolescence. Pimples and blackheads on the face, chest, and back are expressions of this condition.

albinism (*al'bĭ-niz"em*) A congenital condition in which the pigment of the skin, hair, and eyes is deficient as a result of a metabolic block in the synthesis of melanin (fig. 5.27).

alopecia (*al"ŏ-pe'she-ă*) Loss of hair; baldness. Male pattern baldness is genetically determined and irreversible. Other types of hair loss may respond to treatment.

athlete's foot (tinea pedis) A fungus disease of the skin of the foot.

blister A collection of fluid between the epidermis and dermis resulting from excessive friction or a burn.

boil (furuncle) A localized bacterial infection originating in a hair follicle or skin gland.

carbuncle A bacterial infection similar to a boil, except that a carbuncle infects the subcutaneous tissues.

cold sore (fever blister) A lesion on the lip or oral mucous membrane caused by type I herpes simplex virus (HSV) and transmitted by oral or respiratory exposure.

comedo (*kom'e-do*) A plug of sebum and epithelial debris in the hair follicle and excretory duct of the sebaceous gland; also called a *blackhead* or *whitehead*.

corn A type of callus localized on the foot, usually over toe joints.

dandruff Common dandruff is the continual shedding of epidermal cells of the scalp; it can be removed by normal washing and brushing of the hair. Abnormal dandruff may be caused by certain skin diseases, such as seborrhea or psoriasis.

decubitus (*de-kyoo'bĭ-tus*) ulcer A bedsore—an exposed ulcer caused by a continual pressure that restricts dermal blood flow to a localized portion of the skin (see fig. 5.8).

dermabrasion A procedure for removing tattoos or acne scars by high-speed sanding or scrubbing.

dermatitis An inflammation of the skin.

dermatology A specialty of medicine concerned with the study of the skin—its anatomy, physiology, histopathology, and the relationship of cutaneous lesions to systemic disease.

eczema (*ek'ze-mă*) A noncontagious inflammatory condition of the skin producing itchy, red vesicular lesions that may be crusty or scaly.

erythema (*er"ĭ-the-mă*) Redness of the skin, generally is a result of vascular trauma.

furuncle A boil—a localized abscess resulting from an infected hair follicle.

gangrene Necrosis of tissue resulting from the obstruction of blood flow. It may be localized or extensive and may be infected secondarily with anaerobic microorganisms.

hives (urticaria) (*ur"tĭ-ka're-ă*) A skin eruption of reddish wheals usually accompanied by extreme itching. It may be caused by drugs, food, insect bites, inhalants, emotional stress, or exposure to heat or cold.

impetigo (*im-pĕ-ti'go*) A contagious skin infection that results in lesions followed by scaly patches. It generally occurs on the face and is caused by staphylococci or streptococci.

keratosis Any abnormal growth and hardening of the stratum corneum of the skin.

melanoma (*mel-ă-no'mă*) A cancerous tumor originating from proliferating melanocytes within the epidermis of the skin.

nevus (*ne'vus*) A mole or birthmark—a congenital pigmentation of a limited area of the skin.

papilloma (*pap-ĭ-lo'mă*) A benign epithelial neoplasm, such as a wart or corn.

papule A small inflamed elevation of the skin, such as a pimple.

pruritus (*proo-ri'tus*) Itching. It may be symptomatic of systemic disorders but is generally due to dry skin.

psoriasis (*so-ri'ă-sis*) An inherited inflammatory skin disease, usually expressed as circular scaly patches of skin.

pustule A small, localized pus-filled elevation of the skin.

seborrhea (*seb-ŏ-re'ă*) A disease characterized by an excessive activity of the sebaceous glands and accompanied by oily skin and dandruff. It is known as "cradle cap" in infants.

wart A roughened projection of epidermal cells caused by a virus.

Chapter Summary

The Integument as an Organ (p. 102)

1. The skin is considered an organ because it consists of several kinds of tissues.
2. The appearance of the skin is clinically important because it provides clues to certain body conditions or dysfunctions.

Layers of the Integument (pp. 103–109)

1. The stratified squamous epithelium of the epidermis is composed of five structural and functional layers: the stratum basale, stratum spinosum, stratum granulosum, stratum lucidum, and stratum corneum.
   (a) Normal skin color is the result of a combination of melanin and carotene in the epidermis and hemoglobin in the blood of the dermis and hypodermis.
   (b) Fingerprints on the surface of the epidermis are congenital patterns, unique to each individual; flexion creases and flexion lines are acquired.
2. The thick dermis of the skin is composed of fibrous connective tissue interlaced with elastic fibers. The two layers of the dermis are the papillary layer and the deeper reticular layer.
3. The hypodermis, composed of adipose and loose connective tissue, binds the dermis to underlying organs.

Functions of the Integument (pp. 109–111)

1. Structural features of the skin protect the body from disease and external injury.
   (a) Keratin and acidic oily secretions on the surface of the skin protect it from water and microorganisms.
   (b) Cornification of the skin protects against abrasion.
   (c) Melanin is a barrier to UV light.

2. The skin regulates body fluids and temperatures.
   (a) Fluid loss is minimal as a result of keratinization and cornification.
   (b) Temperature regulation is maintained by radiation, convection, and the antagonistic effects of sweating and shivering.
3. The skin permits the absorption of UV light, respiratory gases, steroids, fat-soluble vitamins, and certain toxins and pesticides.
4. The integument synthesizes melanin and keratin, which remain in the skin, and has a role in the synthesis of vitamin D, which is used elsewhere in the body.
5. Sensory reception in the skin is provided through cutaneous receptors throughout the dermis and hypodermis. Cutaneous receptors respond to precise sensory stimuli and are more sensitive in thin skin.
6. Certain emotions are reflected in changes in the skin.

Epidermal Derivatives (pp. 111–118)

1. Hair is characteristic of all mammals, but its distribution, function, density, and texture varies across mammalian species.
   (a) Each hair consists of a shaft, root, and bulb. The bulb is the enlarged base of the root within the hair follicle.
   (b) The three layers of a hair shaft are the medulla, cortex, and cuticle.
   (c) Lanugo, angora, and definitive are the three distinct kinds of human hair.
2. Hardened, keratinized nails are found on the distal dorsum of each digit, where they protect the digits; fingernails aid in grasping and picking up small objects.
   (a) Each nail consists of a body, free border, and hidden border.
   (b) The hyponychium, eponychium, and nail fold support the nail on the nail bed.
3. Integumentary glands are exocrine, since they either secrete or excrete substances through ducts.
   (a) Sebaceous glands secrete sebum onto the shaft of the hair.
   (b) The two types of sudoriferous (sweat) glands are eccrine and apocrine.
   (c) Mammary glands are specialized sudoriferous glands that secrete milk during lactation.
   (d) Ceruminous glands secrete cerumen (earwax).

Review Activities

Objective Questions

1. Hair, nails, integumentary glands, and the epidermis of the skin are derived from embryonic
   (a) ectoderm. (c) endoderm.
   (b) mesoderm. (d) mesenchyme.
2. Spoon-shaped nails may be the result of a dietary deficiency of
   (a) zinc. (c) niacin.
   (b) iron. (d) vitamin B_{12}.
3. The epidermal layer *not* present in the thin skin of the face is the stratum
   (a) granulosum. (c) spinosum.
   (b) lucidum. (d) corneum.
4. Which of the following does *not* contribute to skin color?
   (a) dermal papillae (c) carotene
   (b) melanin (d) hemoglobin
5. Which of the following is *not* true of the epidermis?
   (a) It is composed of stratified squamous epithelium.
   (b) As the epidermal cells die, they undergo keratinization and cornification.
   (c) Rapid mitotic activity (cell division) within the stratum corneum accounts for the thickness of this epidermal layer.
   (d) In most areas of the body, the epidermis lacks blood vessels and nerves.
6. Integumentary glands that empty their secretions into hair follicles are
   (a) sebaceous glands.
   (b) endocrine glands.
   (c) eccrine glands.
   (d) ceruminous glands.
7. Fetal hair that is present during the last trimester of development is referred to as
   (a) angora. (c) lanugo.
   (b) definitive. (d) replacement.
8. Which of these conditions is potentially life threatening?
   (a) acne (c) eczema
   (b) melanoma (d) seborrhea
9. The skin of a burn victim has been severely damaged through the epidermis and into the dermis. Integumentary regeneration will be slow with some scarring, but it will be complete. Which kind of burn is this?
   (a) first degree (c) third degree
   (b) second degree
10. The technical name for a blackhead or whitehead is
   (a) a carbuncle. (c) a nevus.
   (b) a melanoma. (d) a comedo.

Essay Questions

1. Discuss the development of the skin and associated hair, glands, and nails. What role do the ectoderm and mesoderm play in integumentary development?
2. List the functions of the skin. Which of these occur(s) passively as a result of the structure of the skin? Which occur(s) dynamically as a result of physiological processes?
3. What are types of tissues found in each of the three layers of skin?
4. Discuss the growth process and regeneration of the epidermis.
5. What are some physical and chemical features of the skin that make it an effective protective organ?
6. Of what practical value is it for the outer layers of the epidermis and hair to be composed of dead cells?
7. Define the following: *lines of tension*, *friction ridges*, and *flexion lines*. What causes each of these to develop?
8. Distinguish between a hair follicle and a hair. Aside from hair and hair follicles, what are the other epidermal derivatives?
9. Compare and contrast the structure and function of sebaceous, sudoriferous, mammary, and ceruminous glands.
10. Discuss what is meant by an inflammatory lesion. What are some frequent causes of skin lesions?

Critical-Thinking Questions

1. Why is it important that the epidermis serve as a barrier against UV rays, yet not block them out completely?
2. Review the structure and function of the skin by explaining (a) the mechanisms involved in thermoregulation; (b) variations in skin color; (c) abnormal coloration of the skin (for example, cyanosis, jaundice, and pallor); and (d) the occurrence of acne.
3. Do you think that humans derive any important benefit from contraction of the arrector pili muscles? Justify your answer.
4. The relative hairlessness of humans is unusual among mammals. Why should it be that we have any hair at all?
5. Compounds such as lead, zinc, and arsenic may accumulate in the hair and nails. Chemical toxins from pesticides and pollutants may accumulate in the adipose tissue (subcutaneous fat) of the hypodermis. Discuss some of the possible clinical situations where this knowledge would be of importance.

Six

Skeletal System: Introduction and the Axial Skeleton

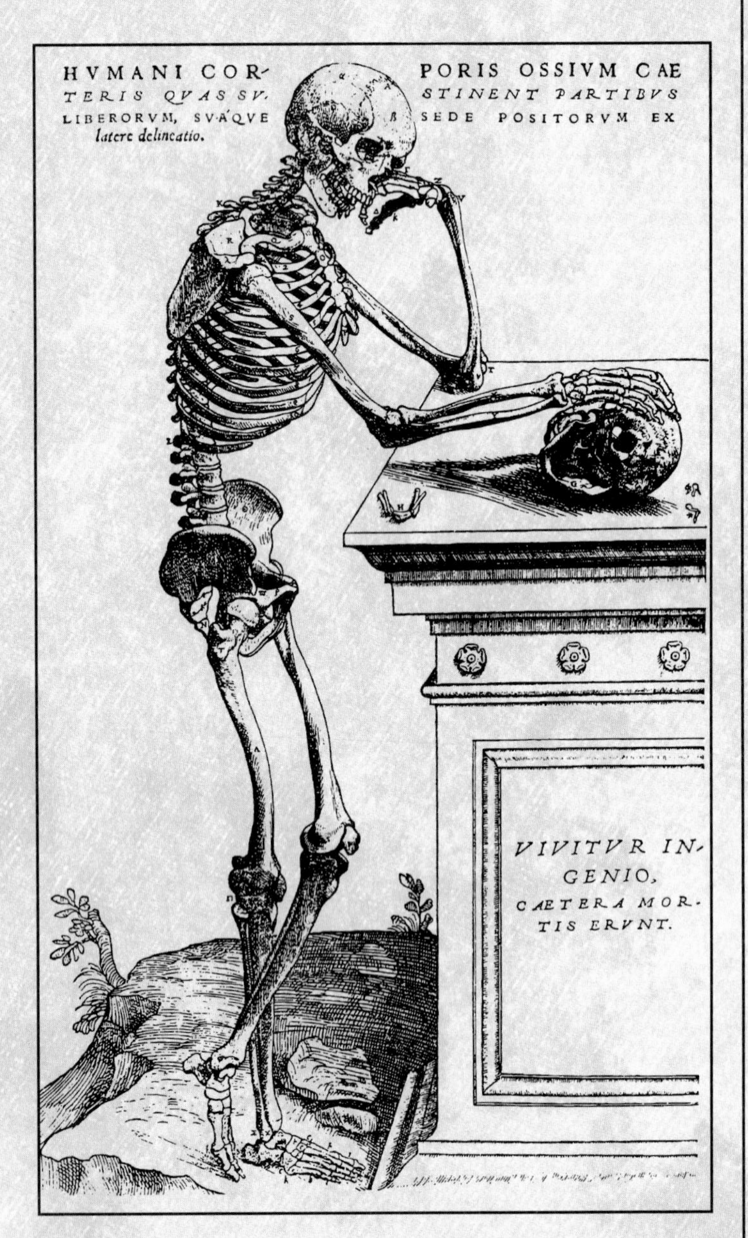

In this woodcut from the first volume of Vesalius's De Humani Corporis Fabrica, *a skeleton contemplates a skull. The hyoid bone, mistakenly that of a dog, lies below the thinker's elbow. The malleus and incus of the middle ear, barely visible at the far right, are often mistaken for the artist's signature.*

Clinical Case Study 129

Organization of the Skeletal System 129

Functions of the Skeletal System 131

Bone Structure 132
- Shapes of Bones 133
- Structure of a Typical Long Bone 133

Bone Tissue 135
- Bone Cells 135
- Spongy and Compact Bone Tissues 136

Bone Growth 137

Skull 139
- Cranial Bones 146
- Facial Bones 149
- Hyoid Bone 152
- Auditory Ossicles 152

Vertebral Column 153
- General Structure of Vertebrae 154
- Regional Characteristics of Vertebrae 156

Rib Cage 159
- Sternum 159
- Ribs 159

Clinical Considerations 161
- Developmental Disorders 161
- Nutritional and Hormonal Disorders 161

Developmental Exposition: The Axial Skeleton 162
- Neoplasms of Bone 162
- Aging of the Skeletal System 164

Clinical Case Study Answer 164

Internal Affairs 165

Important Clinical Terminology 166

Chapter Summary 166

Review Activities 166

Clinical Case Study

A 68-year-old man visited his family doctor for his first physical examination in 30 years. Upon sensing a disgruntled patient, the doctor gently tried to determine the reason. In response to the doctor's inquiry, the patient blurted out, "The nurse who measured my height is incompetent! I know for a fact I used to be six feet even when I was in the Navy, but she tells me I'm 5′10″!" The doctor then performed the measurement himself, noting that although the patient's posture was excellent, he was indeed 5′10″, just as the nurse had said. He explained to the patient that the spine contains some nonbony tissue, which shrivels up a bit over the years. The patient interrupted, stating indignantly that he knew anatomic terms and principles and would like a detailed explanation. How would you explain the anatomy of the vertebral column and the changes it undergoes during the aging process?

Hints: The patient's normal posture and the fact that he had no complaints of pain indicated good health for his age. Examine figure 6.32 and carefully read the accompanying caption. Also see the Clinical Considerations section at the end of the chapter.

Organization of the Skeletal System

The axial and appendicular components of the skeletal system of an adult human consist of 206 individual bones arranged to form a strong, flexible body framework.

Objective 1 Describe the division of the skeletal system into axial and appendicular components.

The adult skeletal system consists of approximately 206 bones. The exact number of bones differs from person to person depending on age and genetic factors. At birth, the skeleton consists of about 270 bones. As further bone development (ossification) occurs during infancy, the number increases. During adolescence, however, the number of bones decreases, as separate bones gradually fuse. Each bone is actually an organ that plays a part in the total functioning of the skeletal system. The science concerned with the study of bones is called *osteology*.

Some adults have extra bones within the sutures (joints) of the skull called **sutural** (*wormian*) **bones.** Additional bones may develop in tendons in response to stress as the tendons repeatedly move across a joint. Bones formed this way are called **sesamoid** (*ses′ă-moid*) **bones.** Sesamoid bones, like the sutural bones, vary in number. The patellae ("kneecaps") are two sesamoid bones all people have.

For convenience of study, the skeleton is divided into *axial* and *appendicular portions*, as shown in figure 6.1 and summarized in table 6.1. The **axial skeleton** consists of the bones that form the axis of the body and support and protect the organs of the head, neck, and trunk. The components of the axial skeleton are as follows:

1. **Skull.** The skull consists of two sets of bones: the cranial bones that form the cranium, or braincase, and the facial bones that support the eyes and nose and form the bony framework of the oral cavity.
2. **Auditory ossicles.** Three auditory ossicles ("ear bones") are present in the middle-ear chamber of each ear and serve to transmit sound impulses.
3. **Hyoid bone.** The hyoid bone is located above the larynx ("voice box") and below the mandible ("jawbone"). It supports the tongue and assists in swallowing.
4. **Vertebral column.** The vertebral column ("backbone") consists of 26 individual bones separated by cartilaginous intervertebral discs. In the pelvic region, several vertebrae are fused to form the *sacrum*, which is the attachment portion of the pelvic girdle. A few terminal vertebrae are fused to form the *coccyx* ("tailbone").
5. **Rib cage.** The rib cage forms the bony and cartilaginous framework of the thorax. It articulates posteriorly with the thoracic vertebrae and includes the 12 pairs of *ribs*, the flattened *sternum*, and the *costal cartilages* that connect the ribs to the sternum.

The **appendicular** (*ap″en-dik′yoo-lar*) **skeleton** is composed of the bones of the upper and lower extremities and the bony girdles that anchor the appendages to the axial skeleton. The components of the appendicular skeleton are as follows:

1. **Pectoral girdle.** The paired *scapulae* ("shoulder blades") and *clavicles* ("collarbones") are the appendicular components of the pectoral girdle, and the *sternum* ("breastbone") is the axial component. The primary function of the pectoral girdle is to provide attachment for the muscles that move the brachium (arm) and antebrachium (forearm).
2. **Upper extremities.** Each upper extremity contains a proximal *humerus* within the brachium, an *ulna* and *radius* within the antebrachium, the *carpal bones*, the *metacarpal bones*, and the *phalanges* ("finger bones") of the hand.
3. **Pelvic girdle.** The two *ossa coxae* ("hipbones") are the appendicular components of the pelvic girdle, and the sacrum is the axial component. The ossae coxae are united anteriorly by the *symphysis* (*sim′fĭ-sis*) *pubis* and posteriorly by the sacrum. The pelvic girdle supports the weight of the body through the vertebral column and protects the viscera within the pelvic cavity.

---

wormian bone: from Ole Worm, Danish physician, 1588–1654
sesamoid: Gk. *sesamon*, like a sesame seed

ossicle: L. *ossiculum*, little bone

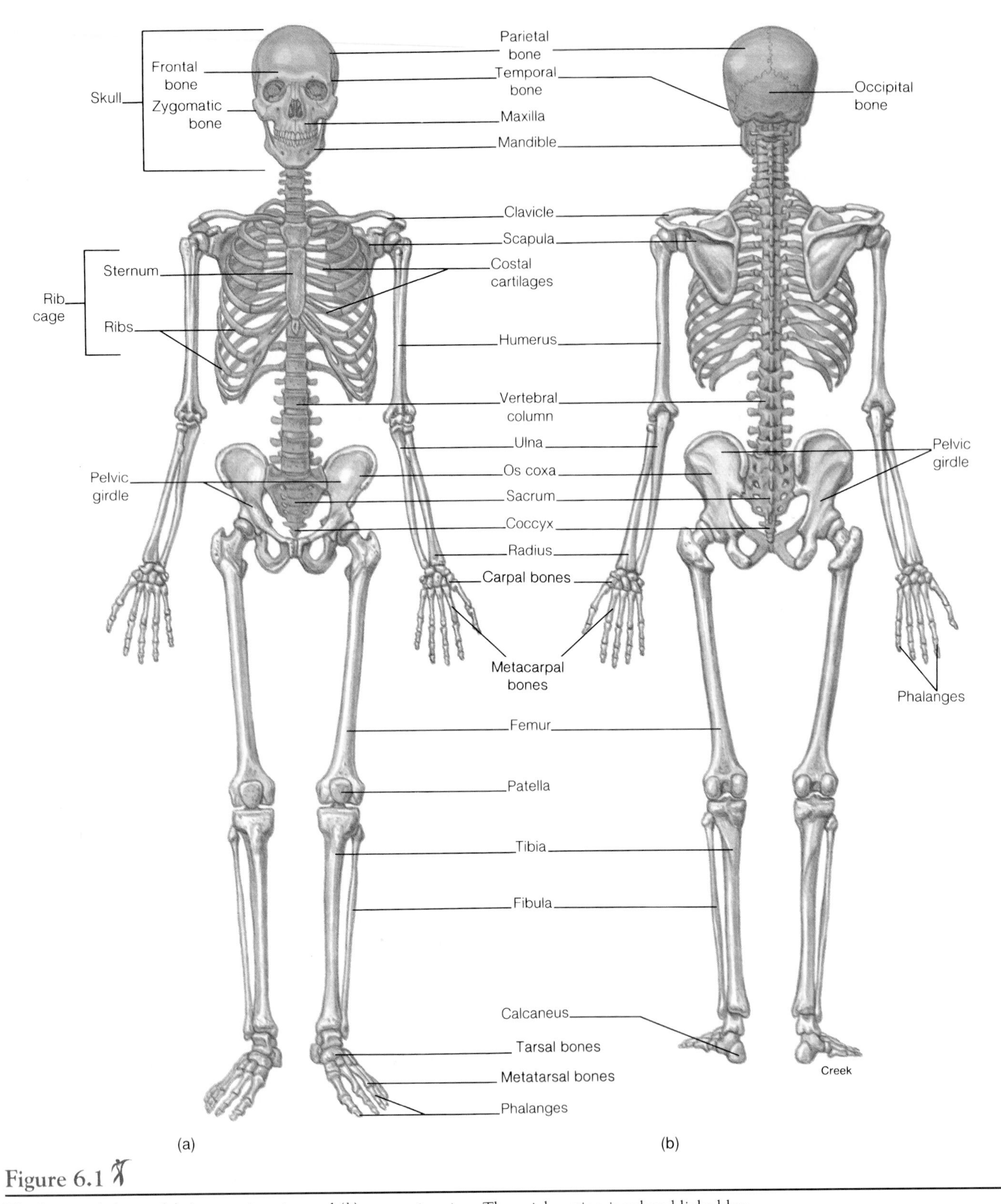

Figure 6.1

The human skeleton. (*a*) An anterior view and (*b*) a posterior view. The axial portion is colored light blue.

Table 6.1
Bones of the Adult Skeleton

Axial Skeleton			Appendicular Skeleton	
Skull—22 Bones		**Auditory Ossicles**—6 Bones	**Pectoral Girdle**—5 Bones	
14 Facial Bones	*8 Cranial Bones*	malleus (2)	sternum* (1)	
maxilla (2)	frontal (1)	incus (2)	scapula (2)	
palatine (2)	parietal (2)	stapes (2)	clavicle (2)	
zygomatic (2)	occipital (1)	**Hyoid**—1 bone	**Upper Extremities**—60 Bones	
lacrimal (2)	temporal (2)	**Vertebral Column**—26 Bones	humerus (2)	carpal bones (16)
nasal (2)	sphenoid (1)	cervical vertebra (7)	radius (2)	metacarpal bones (10)
vomer (1)	ethmoid (1)	thoracic vertebra (12)	ulna (2)	phalanges (28)
inferior nasal concha (2)		lumbar vertebra (5)	**Pelvic Girdle**—3 Bones	
mandible (1)		sacrum (1) (4 or 5 fused bones)	sacrum* (1)	
		coccyx (1) (3–5 fused bones)	os coxae (2) (each contains 3 fused bones)	
		Rib Cage—25 Bones	**Lower Extremities**—60 Bones	
		rib (24)	femur (2)	tarsal bones (14)
		sternum (1)	tibia (2)	metatarsal bones (10)
			fibula (2)	phalanges (28)
			patella (2)	

*Although the sternum and sacrum are bones of the axial skeleton, technically speaking they are also considered bones of the pectoral and pelvic girdles, respectively.

4. **Lower extremities.** Each lower extremity contains a proximal *femur* ("thighbone") within the thigh, a *tibia* ("shinbone") and *fibula* within the leg, the *tarsal bones*, the *metatarsal bones*, and the *phalanges* ("toe bones") of the foot. In addition, the *patella* (*pă-tel′ă;* "kneecap") is located on the anterior surface of the knee joint, between the thigh and leg.

1. List the bones of the body that you can palpate. Indicate which are bones of the axial skeleton and which are bones of the appendicular skeleton.
2. What are sesamoid bones and where are they found?
3. Describe the locations and functions of the pectoral and pelvic girdles.

Functions of the Skeletal System

The bones of the skeleton perform the mechanical functions of support, protection, and leverage for body movement and the metabolic functions of hemopoiesis and mineral storage.

Objective 2 Discuss the principal functions of the skeletal system and identify the body systems served by these functions.

The strength of bone comes from its inorganic components, of such durability that they resist decomposition even after death. Much of what we know of prehistoric animals, including humans, has been determined from preserved skeletal remains. When we think of bone, we frequently think of a hard, dry structure. In fact, the term *skeleton* comes from a Greek word meaning "dried up." Living bone, however, is not inert material; it is dynamic and adaptable. It performs many body functions, including support, protection, leverage for body movement, hemopoiesis, and mineral storage.

1. **Support.** The skeleton forms a rigid framework to which the softer tissues and organs of the body are attached. It is of interest that the skeleton's 206 bones support a mass of muscles and organs that may weigh five times as much as the bones themselves.
2. **Protection.** The skull and vertebral column enclose the brain and spinal cord; the rib cage protects the heart, lungs, great vessels, liver, and spleen; and the pelvic cavity supports and protects the pelvic viscera. Even the site where red blood cells are produced is protected within the central hollow portion of certain bones.
3. **Body movement.** Bones serve as anchoring attachments for most skeletal muscles. In this capacity, the bones act as levers (with the joints functioning as pivots) when muscles contract and cause body movement.
4. **Hemopoiesis.** The process of blood cell formation is called hemopoiesis (*hem″ ŏ-poi-e′sis*). It takes place in tissue called

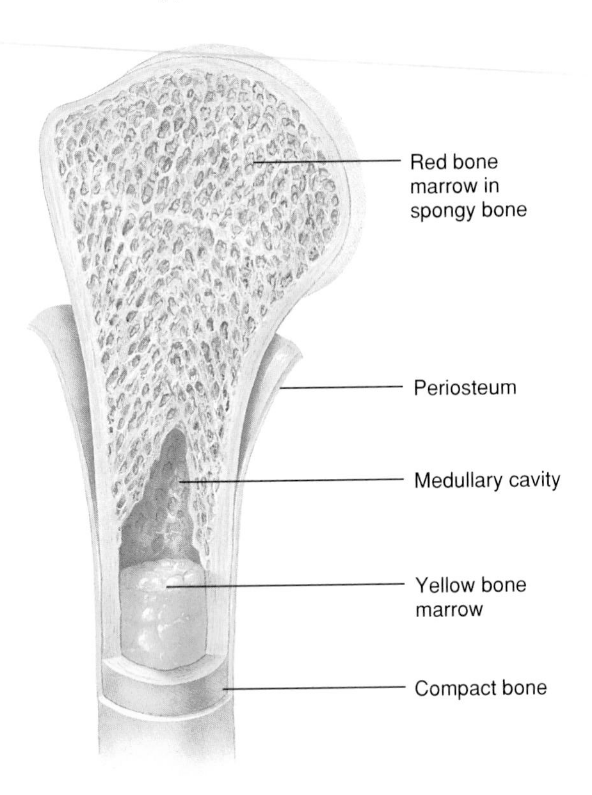

Figure 6.2

Hemopoiesis is the process by which blood cells are formed. In an adult, blood cells are formed in the red bone marrow.

red bone marrow located internally in some bones (fig. 6.2). In an infant, the spleen and liver produce red blood cells, but as the bones mature, the bone marrow takes over this formidable task. It is estimated that an average of 2.5 million red blood cells are produced every second by the bone marrow to replace those that are worn out and destroyed by the liver.

5. **Mineral storage.** The inorganic matrix of bone is composed primarily of the minerals calcium and phosphorus. These minerals which account for approximately two-thirds of the weight of bone, give bone its firmness and strength. About 95% of the calcium and 90% of the phosphorus within the body are deposited in the bones and teeth. Although the concentration of these inorganic salts within the blood is kept within narrow limits, both are essential for other body functions. Calcium is necessary for muscle contraction, blood clotting, and the movement of ions and nutrients across cell membranes. Phosphorus is required for the activities of the nucleic acids DNA and RNA, as well as for ATP utilization. If mineral salts are not present in the diet in sufficient amounts, they may be withdrawn from the bones until they are replenished through proper nutrition. In addition to calcium and phosphorus, lesser amounts of magnesium, sodium, fluorine, and strontium are stored in bone tissue.

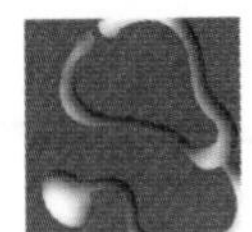

Vitamin D assists in the absorption of calcium and phosphorus from the small intestine into the blood. As bones develop in a child, it is extremely important that the child's diet contain an adequate amount of these two minerals and vitamin D. If the diet is deficient in these essentials, the blood level falls below that necessary for calcification, and a condition known as *rickets* develops (see fig. 5.10). Rickets is characterized by soft bones that may result in bowlegs and malformation of the head, chest, and pelvic girdle.

In summary, the skeletal system is not an isolated body system. It is associated with the muscle system in storing calcium needed for muscular contraction and providing attachments for muscles as they span the movable joints. The skeletal system serves the circulatory system by producing blood cells in protected sites. Directly or indirectly, the skeletal system supports and protects all of the systems of the body (see Internal Affairs, p. 165).

1. List the functions of the skeletal system.
2. Discuss two ways in which the skeletal system serves the circulatory system in the production of blood. What are two ways in which it serves the muscular system?

Bone Structure

Each bone has a characteristic shape and diagnostic surface features that indicate its functional relationship to other bones, muscles, and to the body structure as a whole.

Objective 3 Classify bones according to their shapes and give an example of each type.

Objective 4 Describe the various markings on the surfaces of bones.

Objective 5 Describe the gross features of a typical long bone and list the functions of each surface feature.

The shape and surface features of each bone indicate its functional role in the skeleton (table 6.2). Bones that are long, for example, provide body support and function as levers during body movement. Bones that support the body are massive and have large articular surfaces and processes for muscle attachment. Roughened areas on these bones may serve for the attachment of ligaments, tendons, or muscles. A flattened surface provides an attachment site for a large muscle or may provide protection. Grooves around an articular end of a bone indicate where a tendon or nerve passes, and openings through a bone permit the passage of nerves or blood vessels.

Table 6.2 Surface Features of Bone

Structure	Description and Example
Articulating Surfaces	
condyle (*kon′dil*)	A large, rounded articulating knob (the occipital condyle of the occipital bone)
facet	A flattened or shallow articulating surface (the costal facet of a thoracic vertebra)
head	A prominent, rounded articulating end of a bone (the head of the femur)
Depressions and Openings	
alveolus (*al-ve′ŏ-lus*)	A deep pit or socket (the dental alveoli [tooth sockets] in the maxilla and mandible)
fissure	A narrow, slitlike opening (the superior orbital fissure of the sphenoid bone)
foramen (*fŏ-ra′men*—plural, *foramina*)	A rounded opening through a bone (the foramen magnum of the occipital bone)
fossa (*fos′ă*)	A flattened or shallow surface (the mandibular fossa of the temporal bone)
sinus	A cavity or hollow space in a bone (the frontal sinus of the frontal bone)
sulcus	A groove that accommodates a vessel, nerve, or tendon (the intertubercular sulcus of the humerus)
Nonarticulating Prominences	
crest	A narrow, ridgelike projection (the iliac crest of the os coxae)
epicondyle	A projection adjacent to a condyle (the medial epicondyle of the femur)
process	Any marked bony prominence (the mastoid process of the temporal bone)
spine	A sharp, slender process (the spine of the scapula)
trochanter	A massive process found only on the femur (the greater trochanter of the femur)
tubercle (*too′ber-k′l*)	A small, rounded process (the greater tubercle of the humerus)
tuberosity	A large, roughened process (the radial tuberosity of the radius)

Shapes of Bones

The bones of the skeleton are grouped on the basis of shape into four principal categories: *long bones*, *short bones*, *flat bones*, and *irregular bones* (fig. 6.3).

1. **Long bones.** Long bones are longer than they are wide and function as levers. Most of the bones of the upper and lower extremities are of this type (e.g., the humerus, radius, ulna, metacarpal bones, femur, tibia, fibula, metatarsal bones, and phalanges).

facet: Fr. *facette*, little face
trochanter: Gk. *trochanter*, runner
tuberosity: L. *tuberosus*, lump

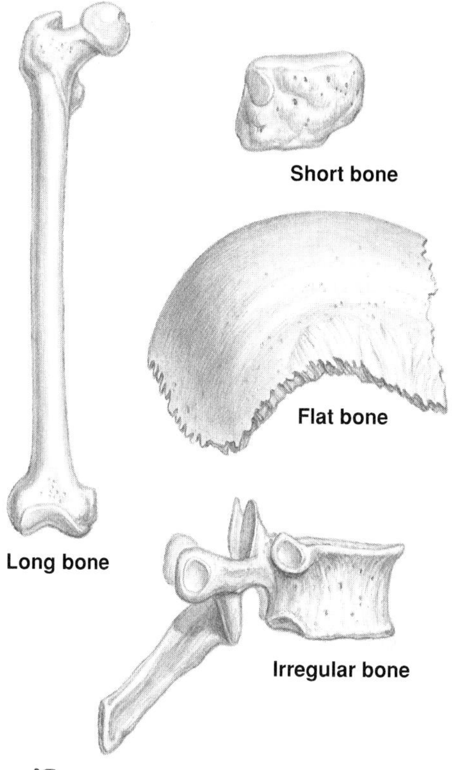

Figure 6.3

Examples of bone types, as classified by shape.

2. **Short bones.** Short bones are somewhat cube-shaped and are found in the wrist and ankle where they transfer forces of movement.
3. **Flat bones.** Flat bones have a broad surface for muscle attachment or protection of underlying organs (e.g., the cranial bones, ribs, and bones of the shoulder girdle).
4. **Irregular bones.** Irregular bones have varied shapes and many surface features for muscle attachment or articulation (e.g., the vertebrae and certain bones of the skull).

Structure of a Typical Long Bone

Bone tissue is organized as *compact* (*dense*) *bone* or *spongy* (*cancellous*) *bone*, and most bones have both types. Compact bone is hard and dense, and is the protective exterior portion of all bones. The spongy bone, when it occurs, is deep to the compact bone and is quite porous. The microscopic structure of spongy and compact bone will be considered shortly.

In a flat bone of the skull, the spongy bone is sandwiched between the compact bone and is called a *diploe* (*dip′lo-e*) (fig. 6.4). Because of this protective layering of bone tissue, a blow to the head may fracture the outer compact bone layer without harming the inner compact bone layer and the brain.

The long bones of the skeleton have a descriptive terminology all their own. In a long bone from an appendage, the bone

diploe: Gk. *diplous*, double

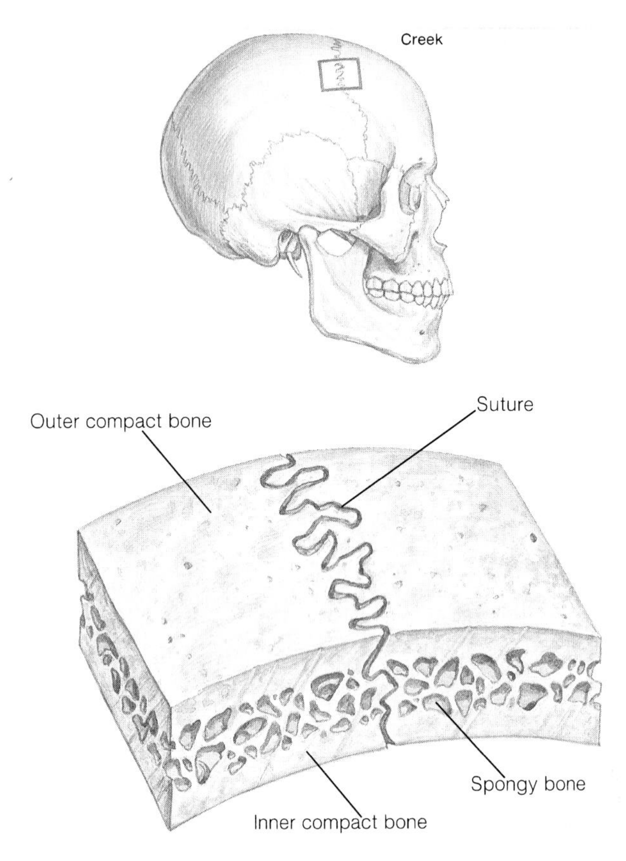

Figure 6.4

A section through the skull showing diploe. Diploe is a layer of spongy bone sandwiched between two surface layers of compact bone. It is extremely strong yet light in weight.

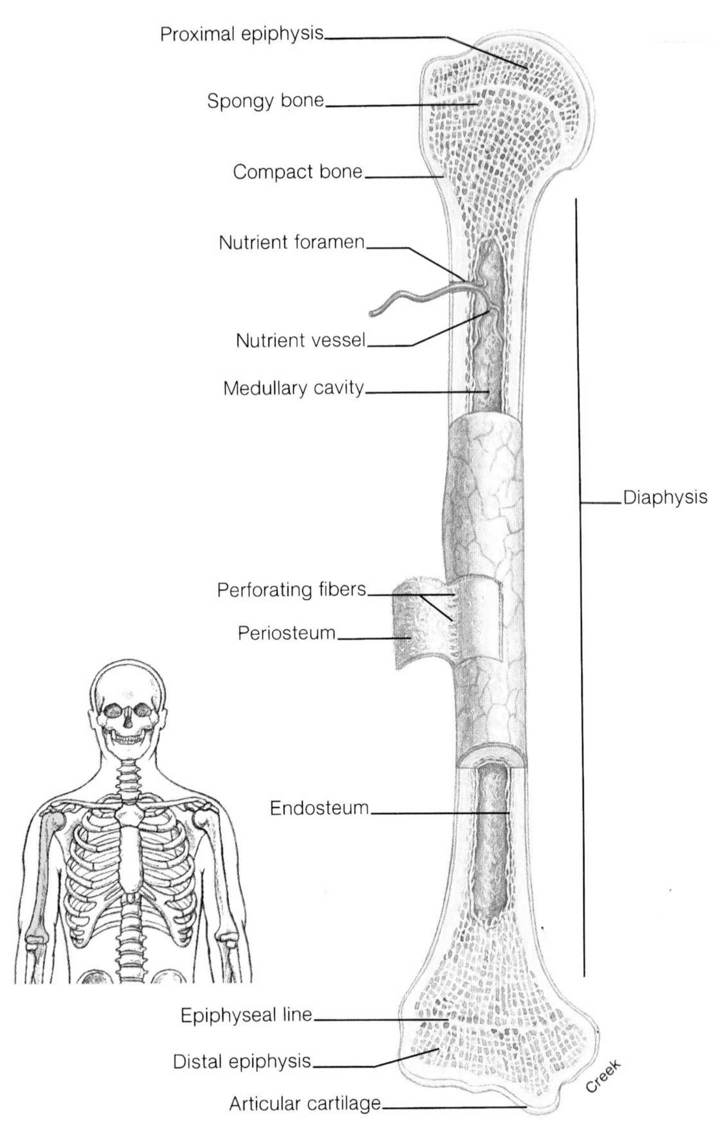

Figure 6.5

A diagram of a long bone (the humerus) shown in a partial longitudinal section.

shaft, or **diaphysis** (*di-af'ĭ-sis*), consists of a cylinder of compact bone surrounding a central cavity called the **medullary** (*med'yoo-lar-e*) **cavity** (fig. 6.5). The medullary cavity is lined with a thin layer of connective tissue called the **endosteum** (*en-dos'te-um*). In an adult, the cavity contains **yellow bone marrow,** so named because it contains large amounts of yellow fat. On each end of the diaphysis is an **epiphysis** (*ĕ-pif'ĕ-sis*), consisting of spongy bone surrounded by a layer of compact bone. **Red bone marrow** is found within the porous chambers of spongy bone. In an adult, erythropoiesis (the production of red blood cells; see chapter 16) occurs in the red bone marrow, especially that of the sternum, vertebrae, portions of the ossa coxae, and the proximal epiphyses of the femora and humeri. The red bone marrow is also responsible for the formation of white blood cells and platelets and for the phagocytosis of worn-out red blood cells. **Articular cartilage,** which is composed of thin hyaline cartilage, caps each epiphysis and facilitates joint movement. Along the diaphysis are **nutrient foramina**—small openings into the bone that allow nutrient vessels to pass into the bone for nourishment of the living tissue.

Between the diaphysis and epiphysis is a cartilaginous **epiphyseal** (*ep"ĭ-fiz'e-al*) **plate**—a region of mitotic activity that is responsible for *linear bone growth*. As bone growth is completed, an **epiphyseal line** replaces the plate and final ossification occurs between the epiphysis and the diaphysis. A **periosteum** (*per"e-os'te-um*) of dense regular connective tissue covers the surface of the bone, except over the articular cartilage. This highly vascular layer serves as a place for a tendon-muscle attachment and is responsible for *appositional bone growth* (increase in width). The periosteum is secured to the bone by **perforating** (Sharpey's) **fibers** (fig. 6.5), composed of bundles of collagenous fibers.

diaphysis: Gk. *dia*, throughout; *physis*, growth
epiphysis: Gk. *epi*, upon; *physis*, growth

periosteum: Gk. *peri*, around; *osteon*, bone
Sharpey's fibers: from William Sharpey, Scottish physiologist and histologist, 1802–80

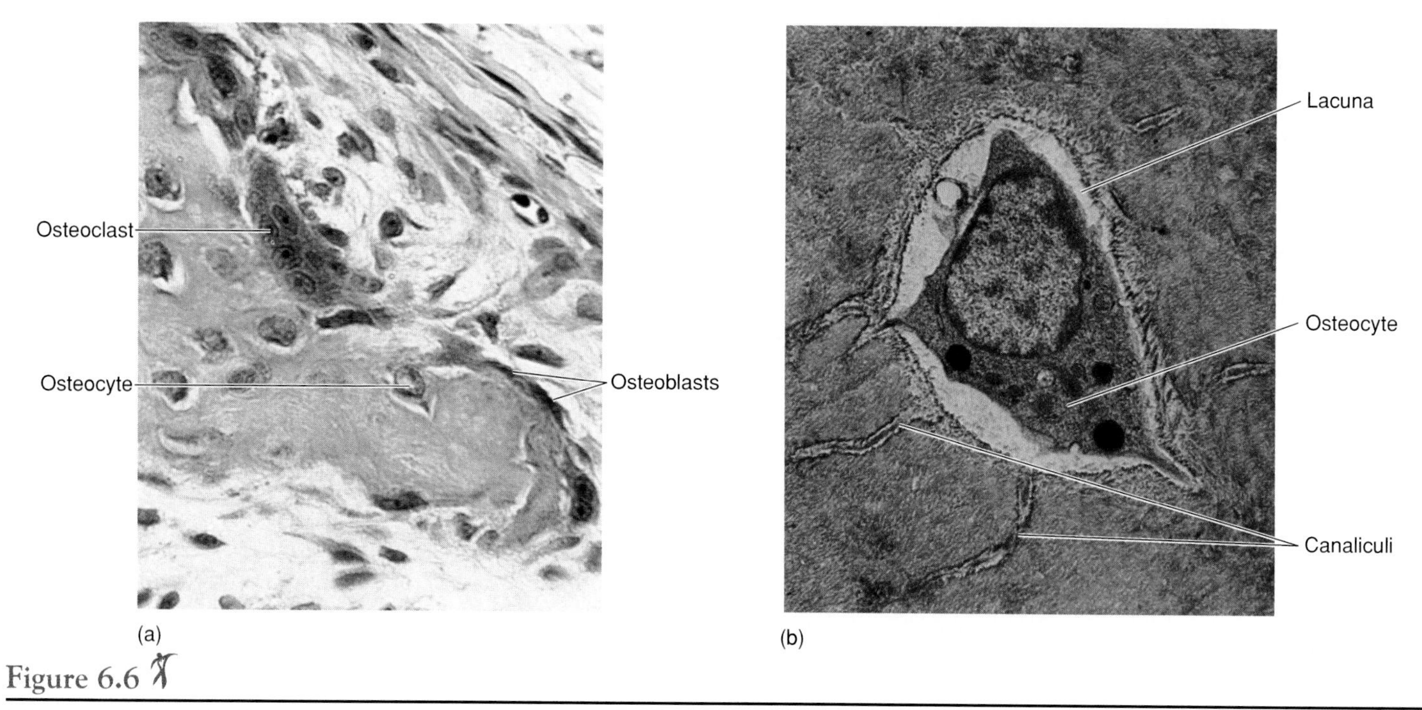

Figure 6.6

(*a*) Types of bone cells. (*b*) An osteocyte within a lacuna.

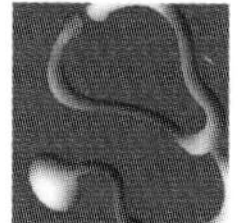

Fracture of a long bone in a young person may be especially serious if it damages an epiphyseal plate. If such an injury goes untreated, or is not treated properly, longitudinal growth of the bone may be arrested or slowed, resulting in permanent shortening of the affected limb.

1. Using examples, discuss the function of each of the four kinds of bones as determined by shape.
2. Define each of the following surface markings on bones: *condyle*, *head*, *facet*, *process*, *crest*, *epicondyle*, *fossa*, *alveolus*, *foramen*, and *sinus*.
3. Diagram a sagittal view of a typical long bone and label the diaphysis, medullary cavity, epiphyses, articular cartilages, nutrient foramen, periosteum, and epiphyseal plates. Explain the function of each of these structures.

Bone Tissue

Bone tissue is composed of several types of bone cells embedded in a matrix of ground substance, inorganic salts (calcium and phosphorus), and collagenous fibers. Bone cells and ground substance give bone flexibility and strength; the inorganic salts give it hardness.

Objective 6 Identify the five types of bone cells and list the functions of each.

Objective 7 Distinguish between spongy and compact bone tissues.

Bone Cells

There are five principal types of bone cells contained within bone tissue. **Osteogenic** (*os″te-ŏ-jen′ik*) **cells** are found in the bone tissues in contact with the endosteum and the periosteum. These cells respond to trauma, such as a fracture, by giving rise to bone-forming cells (*osteoblasts*) and bone-destroying cells (*osteoclasts*). **Osteoblasts** (*os′te-ŏ-blasts*) are bone-forming cells (fig. 6.6) that synthesize and secrete unmineralized ground substance. They are abundant in areas of high metabolism within bone, such as under the periosteum and bordering the medullary cavity. **Osteocytes** (*os′te-ŏ-sīts*) are mature bone cells (figs. 6.6 and 6.7) derived from osteoblasts that have secreted bone tissue around themselves. Osteocytes maintain healthy bone tissue by secreting enzymes and influencing bone mineral content. They also regulate the calcium release from bone tissue to blood. **Osteoclasts** (*os′te-ŏ-klasts*) are large multinuclear cells (fig. 6.6) that enzymatically break down bone tissue. These cells are important in bone growth, remodeling, and healing. **Bone-lining cells** are derived from osteoblasts along the surface of most bones in the adult skeleton. These cells are thought to regulate the movement of calcium and phosphate into and out of bone matrix.

osteoblast: Gk. *osteon*, bone; *blastos*, offspring or germ
osteoclast: Gk. *osteon*, bone; *klastos*, broken

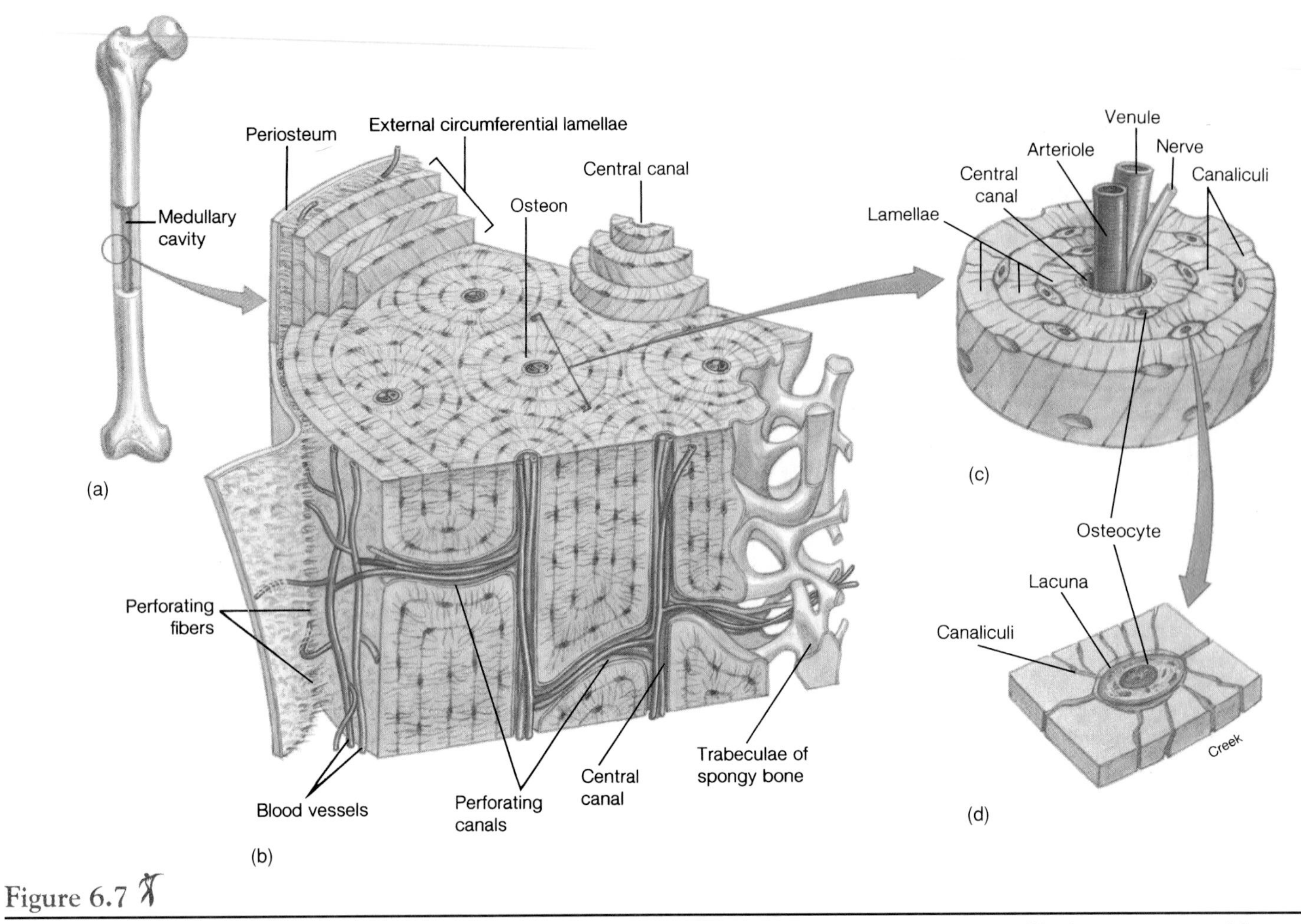

Figure 6.7

Compact bone tissue. (*a*) A diagram of the femur showing a cut through the compact bone into the medullary cavity. (*b*) The arrangement of the osteons within the diaphysis of the bone. (*c*) An enlarged view of an osteon showing the osteocytes within lacunae and the concentric lamellae. (*d*) An osteocyte within a lacuna.

Spongy and Compact Bone Tissues

As mentioned earlier, most bones contain both spongy and compact bone tissues (fig. 6.7). **Spongy bone tissue** is located deep to the compact bone tissue, and is quite porous. Minute spikes of bone tissue, called **trabeculae** (*tră-bek'yŭ-le*), give spongy bone a latticelike appearance. Spongy bone is highly vascular and provides great strength to bone with minimal weight.

Compact bone tissue forms the external portion of a bone and is very hard and dense. It consists of precise arrangements of microscopic cylindrical structures oriented parallel to the long axis of the bone (fig. 6.7). These columnlike structures are the **osteons** (*os'te-onz*), or **haversian systems,** of the bone tissue. The matrix of an osteon is laid down in concentric rings, called **lamellae** (*lă-mel'e*), that surround a **central** (haversian) **canal** (fig. 6.8). The central canal contains minute nutrient vessels and a nerve. Osteocytes within spaces called **lacunae** (*lă-kyoo'ne*) are regularly arranged between the lamellae. The lacunae are connected by tiny channels called **canaliculi** (*kan"ă-lik'yŭ-li*), through which nutrients diffuse. Metabolic activity within bone tissue occurs at the osteon level. Between osteons there are incomplete remnants of osteons, called **interstitial systems. Perforating** (Volkmann's) **canals** penetrate compact bone, connecting osteons with blood vessels and nerves.

1. Construct a simple table listing the location and function of each type of cell found within bone tissue.
2. Define *osteon* and sketch the arrangement of osteons within compact bone tissue.

haversian system: from Clopton Havers, English anatomist, 1650–1702

Volkmann's canal: from Alfred Volkmann, German physiologist, 1800–1877

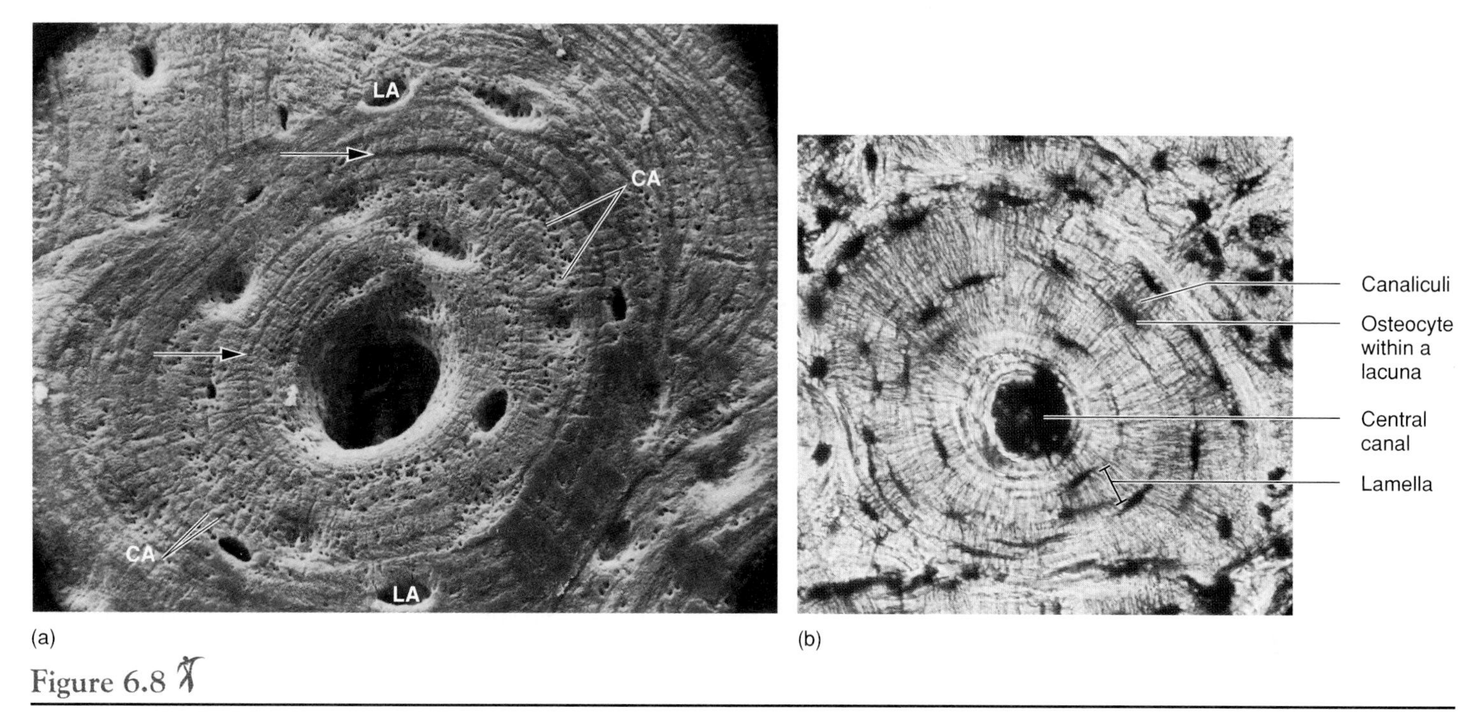

Figure 6.8

Bone tissue as seen in (*a*) a scanning electron micrograph and (*b*) a photomicrograph. The lacunae (La) provides spaces for the osteocytes, which are connected to one another by canaliculi (Ca). Note the divisions between the lamellae (*arrows*).

Bone Growth

The development of bone from embryo to adult depends on the orderly processes of cell division, growth, and ongoing remodeling. Bone growth is influenced by genetics, hormones, and nutrition.

Objective 8 Describe the process of endochondral ossification as related to bone growth.

In most bone development, a cartilaginous model is gradually replaced by bone tissue during *endochondral bone formation*. As the cartilage model grows, the *chondrocytes* (cartilage cells) in the center of the shaft hypertrophy, and minerals are deposited within the matrix in a process called *calcification* (fig. 6.9). Calcification restricts the passage of nutrients to the chondrocytes, causing them to die. At the same time, some cells of the perichondrium (dense regular connective tissue surrounding cartilage) differentiate into *osteoblasts*. These cells secrete **osteoid** (*os′te-oid*), the hardened organic component of bone. As the perichondrium calcifies, it gives rise to a thin plate of compact bone called the **periosteal bone collar.** The periosteal bone collar is surrounded by the periosteum.

A **periosteal bud,** consisting of osteoblasts and blood vessels, invades the disintegrating center of the cartilage model from the periosteum. Once in the center, the osteoblasts secrete osteoid, and a **primary ossification center** is established. Ossification then expands into the deteriorating cartilage. This process is repeated in both the proximal and distal epiphyses, forming **secondary ossification centers** where spongy bone develops.

Once the secondary ossification centers have been formed, bone tissue totally replaces cartilage tissue, except at the articular ends of the bone and at the epiphyseal plates. An **epiphyseal plate** contains five histological zones (fig. 6.10). The *reserve zone* borders the epiphysis and consists of small chondrocytes irregularly dispersed throughout the intercellular matrix. The chondrocytes in this zone anchor the epiphyseal plate to the bony epiphysis. The *proliferation zone* consists of larger, regularly arranged chondrocytes that are constantly dividing. The *hypertrophic zone* consists of very large chondrocytes that are arranged in columns. The linear growth of long bones is due to the cellular proliferation at the proliferation zone and the growth and maturation of these new cells within the hypertrophic zone. The *resorption zone* is the area where a change in mineral content is occurring. The *ossification zone* is a region of transformation from cartilage tissue to bone tissue. The chondrocytes within this zone die because the intercellular matrix surrounding them becomes calcified. Osteoclasts then break down the calcified matrix and the area is invaded by osteoblasts and capillaries from the bone tissue of the diaphysis. As the osteoblasts mature, osteoid is secreted and bone tissue is formed. The result of this process is a gradual increase in the length of the bone at the epiphyseal plates.

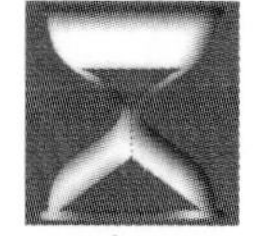

The time at which epiphyseal plates ossify varies greatly from bone to bone, but it usually occurs between the ages of 18 and 20 within the long bones (table 6.3). Because ossification of the epiphyseal cartilages within each bone occurs at predictable times, radiologists can determine the ages of people who are still growing by examining radiographs of their bones (fig. 6.11). Large discrepancies between bone age and chronological age may indicate a genetic or endocrine abnormality.

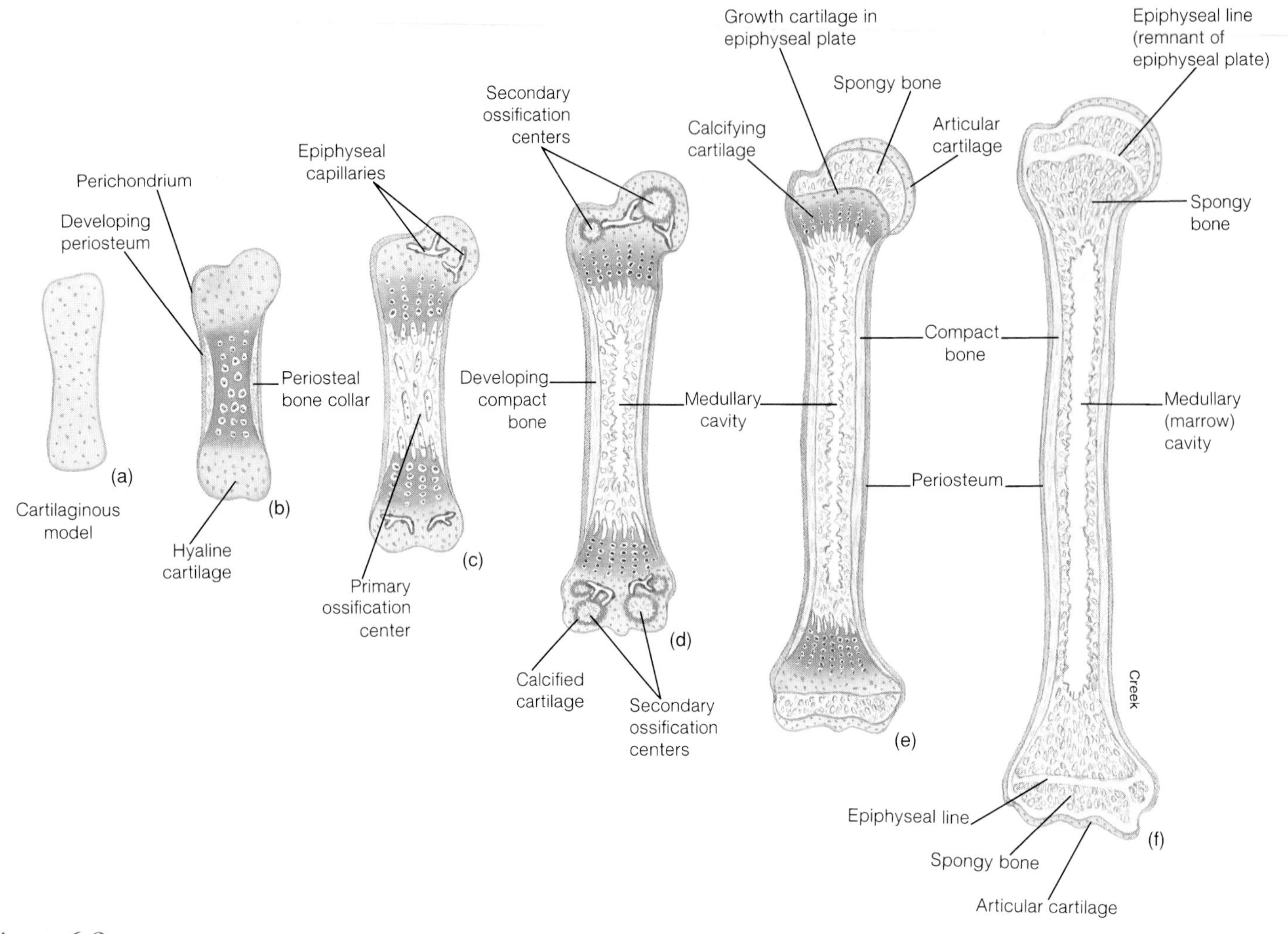

Figure 6.9

The growth process of a long bone, beginning with (*a*) the cartilaginous model as it occurs in an embryo at 6 weeks. The bone develops (*b–e*) through intermediate stages to (*f*) adult bone.

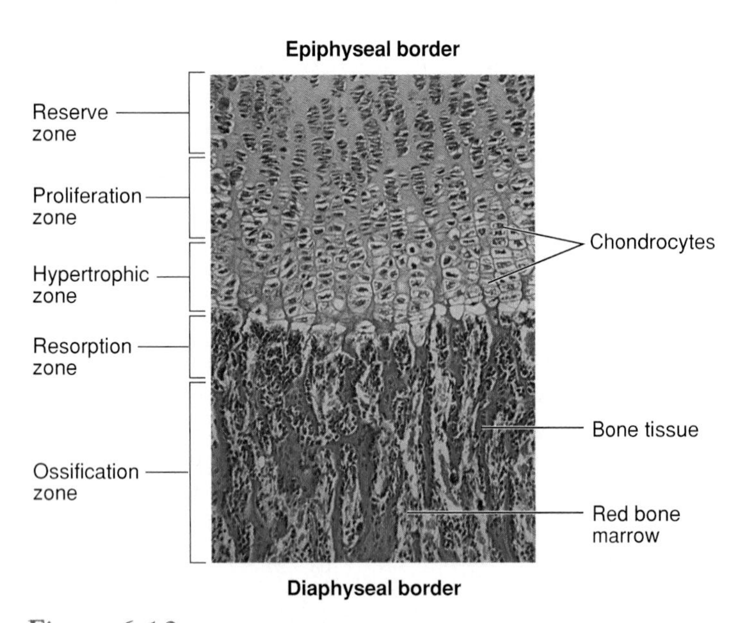

Figure 6.10

A photomicrograph from an epiphyseal plate (63×).

Table 6.3

Average Age of Completion of Bone Ossification

Bone	Chronological Age of Fusion
Scapula	18–20
Clavicle	23–31
Bones of upper extremity	17–20
Os coxae	18–23
Bones of lower extremity	18–22
Vertebra	25
Sacrum	23–25
Sternum (body)	23
Sternum (manubrium, xiphoid)	30+

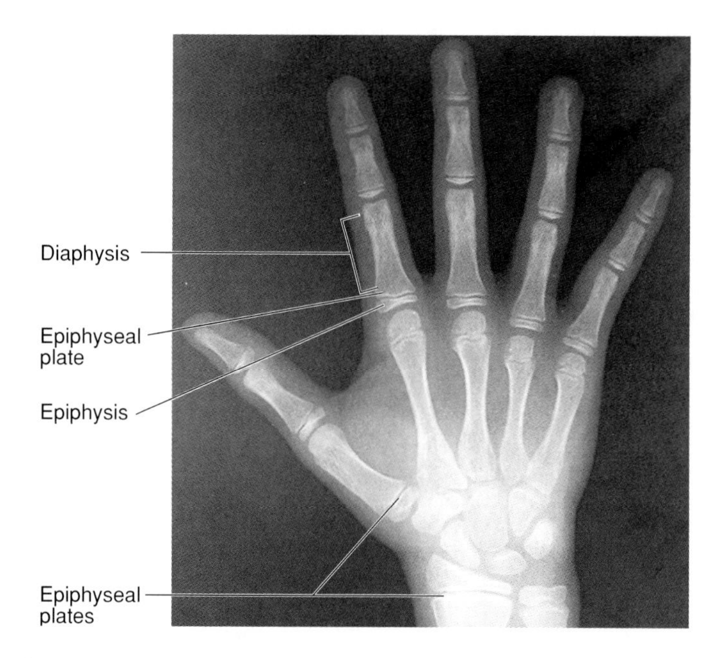

Figure 6.11

The presence of epiphyseal plates as seen in a radiograph of a child's hand. The plates indicate that the bones are still growing in length.

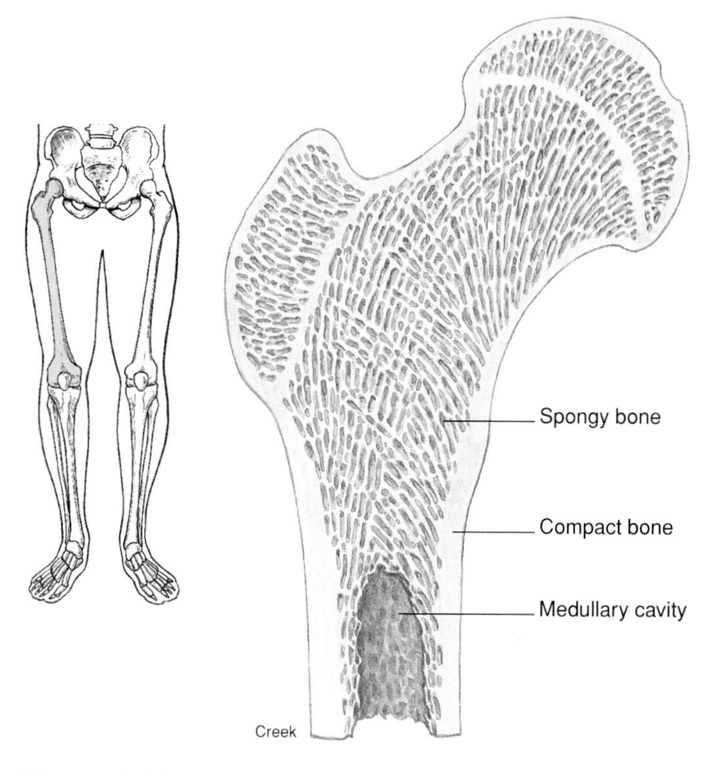

Figure 6.12

A longitudinal section of the proximal end of a femur showing stress lines within spongy bone.

Bone is continually being remodeled over the course of a person's life. Bony prominences develop as stress is applied to the periosteum, causing the osteoblasts to secrete osteoid and form new bone tissue. The greater trochanter of the femur, for example, develops in response to forces of stress applied to the periosteum where the tendons of muscles attach (fig. 6.12). Even though a person has stopped growing in height, bony processes may continue to enlarge somewhat if he or she remains physically active.

As new bone layers are deposited on the outside surface of the bone, osteoclasts dissolve bone tissue adjacent to the medullary cavity. In this way, the size of the cavity keeps pace with the increased growth of the bone.

Even the absence of stress causes a remodeling of bones. This effect can best be seen in the bones of bedridden or paralyzed individuals. Radiographs of their bones reveal a marked loss of bone tissue. The absence of gravity that accompanies space flight may result in mineral loss from bones if an exercise regimen is not followed.

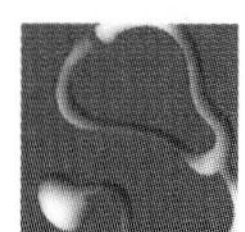

The movement of teeth in orthodontics involves bone remodeling. The dental alveoli (tooth sockets) are reshaped through the activity of osteoclast and osteoblast cells as stress is applied with braces. The use of traction in treating certain skeletal disorders has a similar effect.

1. List the zones of an epiphyseal plate and briefly describe the characteristics of each.
2. Explain the function of osteoblasts and osteoclasts in endochondral ossification and bone growth.

Skull

The human skull, consisting of 8 cranial and 14 facial bones, contains several cavities that house the brain and sensory organs. Each bone of the skull articulates with the adjacent bones and has diagnostic and functional processes, surface features, and foramina.

Objective 9 List the fontanels and discuss their functions.

Objective 10 Identify the cranial and facial bones of the skull and describe their structural characteristics.

Objective 11 Describe the location of each of the bones of the skull and identify the articulations that affix one to the other.

The skull consists of *cranial bones* and *facial bones*. The 8 bones of the cranium articulate firmly with one another to enclose and protect the brain and sensory organs. The 14 facial bones form the framework for the facial region and support the teeth. Variation in size, shape, and density of the facial bones is a major contributor to the individuality of each human face. The facial bones, with the exception of the mandible ("jawbone"), are also firmly interlocked with one another and the cranial bones.

The skull has several cavities. The **cranial cavity** is the largest, with an approximate capacity of 1,300 to 1,350 cc. The **nasal cavity** is formed by both cranial and facial bones and is

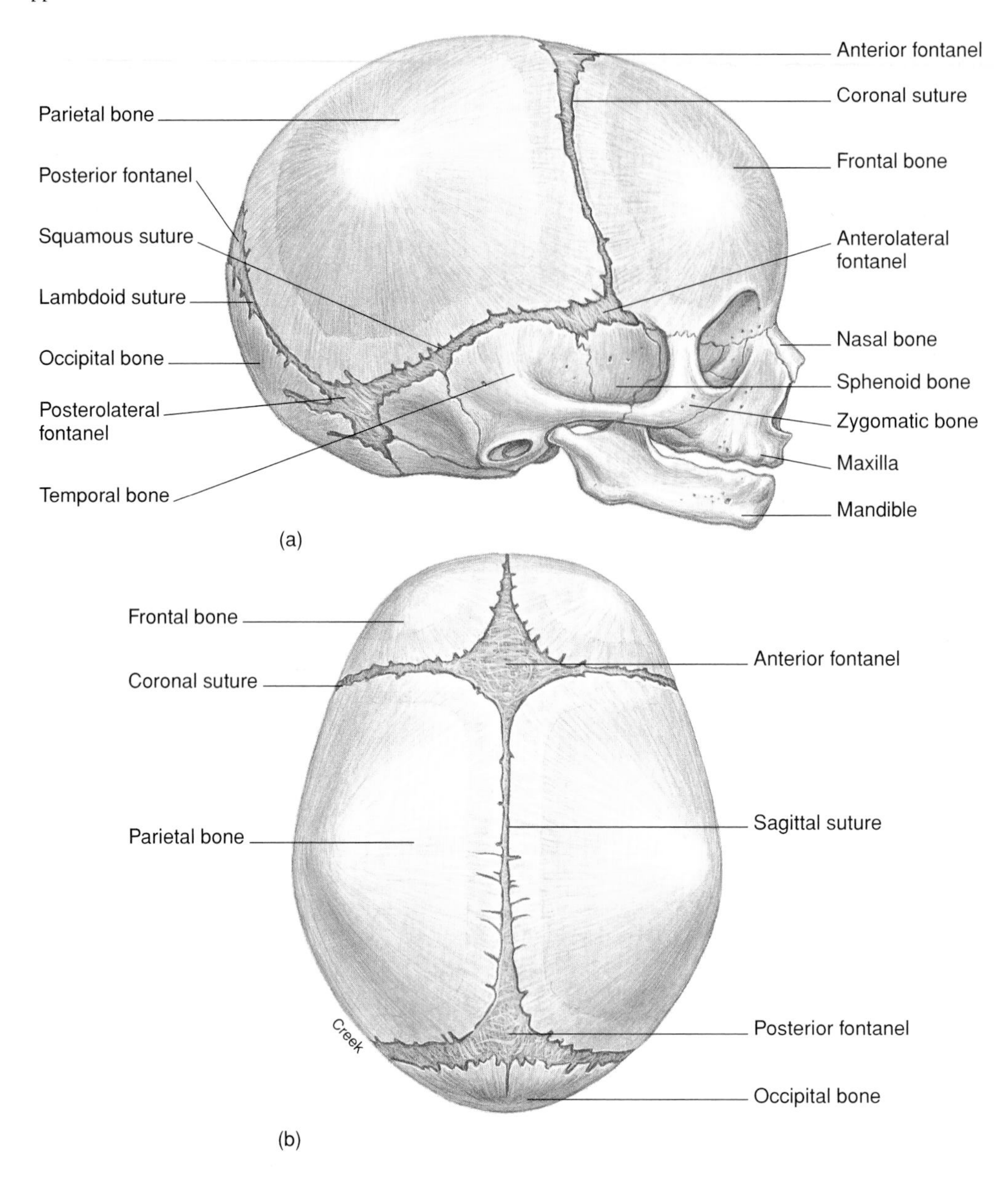

Figure 6.13

The fetal skull showing the six fontanels and the sutures. (*a*) A right lateral view and (*b*) a superior view.

partitioned into two chambers, or **nasal fossae,** by a **nasal septum** of bone and cartilage. Four sets of **paranasal sinuses,** located within the bones surrounding the nasal area, communicate via ducts into the nasal cavity. **Middle-** and **inner-ear cavities** are positioned inferior to the cranial cavity and house the organs of hearing and balance. The two **orbits** for the eyeballs are formed by facial and cranial bones. The **oral,** or **buccal** (*buk'al*) **cavity** (mouth), which is only partially formed by bone, is completely within the facial region (see fig. 2.22).

During fetal development and infancy, the bones of the cranium are separated by fibrous unions. There are also six large areas of connective tissue membrane that cover the gaps between the developing bones. These membranous sheets are called **fontanels** (*fon"tă-nelz'*), meaning "little fountains." The name derives from the fact that a baby's pulse can be felt surging in these "soft spots" on the skull. The fontanels permit the skull to undergo changes in shape, called *molding*, during parturition (childbirth), and they accommodate the rapid growth of the brain during infancy. Ossification of the fontanels is normally complete by 20 to 24 months of age. The fontanels are illustrated in figure 6.13 and briefly described below.

1. **Anterior** (frontal) **fontanel.** The anterior fontanel is diamond-shaped and is the most prominent. It is located on the anteromedian portion of the skull.
2. **Posterior** (occipital) **fontanel.** The posterior fontanel is positioned at the back of the skull on the median line. It is also diamond-shaped, but smaller than the anterior fontanel.

fontanel: Fr. *fontaine*, little fountain

Table 6.4
Major Foramina of the Skull

Foramen	Location	Structures Transmitted
Carotid canal	Petrous part of temporal bone	Internal carotid artery and sympathetic nerves
Greater palatine foramen	Palatine bone of hard palate	Greater palatine nerve and descending palatine vessels
Hypoglossal canal	Anterolateral edge of occipital condyle	Hypoglossal nerve and branch of ascending pharyngeal artery
Incisive foramen	Anterior region of hard palate, posterior to incisors	Branches of descending palatine vessels and nasopalatine nerve
Inferior orbital fissure	Between maxilla and greater wing of sphenoid bone	Maxillary nerve of trigeminal cranial nerve, zygomatic nerve, and infraorbital vessels
Infraorbital foramen	Inferior to orbit in maxilla	Infraorbital nerve and artery
Jugular foramen	Between petrous portion of temporal and occipital bones, posterior to carotid canal	Internal jugular vein; vagus, glossopharyngeal, and accessory nerves
Foramen lacerum	Between petrous portion of temporal and sphenoid bones	Branch of ascending pharyngeal artery and internal carotid artery
Lesser palatine foramen	Posterior to greater palatine foramen in hard palate	Lesser palatine nerves
Foramen magnum	Occipital bone	Union of medulla oblongata and spinal cord, meningeal membranes, and accessory nerves; vertebral and spinal arteries
Mandibular foramen	Medial surface of ramus of mandible	Inferior alveolar nerve and vessels
Mental foramen	Below second premolar on lateral side of mandible	Mental nerve and vessels
Nasolacrimal canal	Lacrimal bone	Nasolacrimal (tear) duct
Cribriform foramina	Cribriform plate of ethmoid bone	Olfactory nerves
Optic foramen	Back of orbit in lesser wing of sphenoid bone	Optic nerve and ophthalmic artery
Foramen ovale	Greater wing of sphenoid bone	Mandibular nerve of trigeminal cranial nerve
Foramen rotundum	Within body of sphenoid bone	Maxillary nerve of trigeminal cranial nerve
Foramen spinosum	Posterior angle of sphenoid bone	Middle meningeal vessels
Stylomastoid foramen	Between styloid and mastoid processes of temporal bone	Facial nerve and stylomastoid artery
Superior orbital fissure	Between greater and lesser wings of sphenoid bone	Four cranial nerves (oculomotor, trochlear, ophthalmic nerve of trigeminal, and abducens)
Supraorbital foramen	Supraorbital ridge of orbit	Supraorbital nerve and artery
Zygomaticofacial foramen	Anterolateral surface of zygomatic bone	Zygomaticofacial nerve and vessels

3. **Anterolateral** (sphenoid) **fontanels.** The paired anterolateral fontanels are found on both sides of the skull, directly lateral to the anterior fontanel. They are relatively small and irregularly shaped.
4. **Posterolateral** (mastoid) **fontanels.** The paired posterolateral fontanels, also irregularly shaped, are located on the posterolateral sides of the skull.

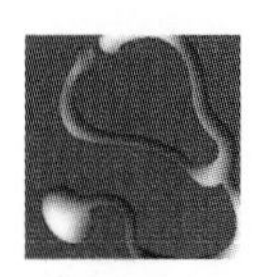

During normal childbirth, the fetal skull comes under tremendous pressure. Bones may even shift, altering the shape of the skull. A common occurrence during molding of the fetal skull is for the occipital bone to be repositioned under the two parietal bones. In addition, one parietal bone may shift so as to overlap the other. This makes delivery easier for the mother. If a baby is born breech (buttocks first), these shifts do not occur. Delivery becomes much more difficult, often requiring the use of forceps.

A prominent **sagittal suture** extends the anteroposterior median length of the skull between the anterior and posterior fontanels. A **coronal suture** extends from the anterior fontanel to the anterolateral fontanel. A **lambdoid suture** extends from the posterior fontanel to the posterolateral fontanel. A **squamous suture** connects the posterolateral fontanel to the anterolateral fontanel.

The bones of the skull contain numerous foramina (see table 6.2) to accommodate nerves, vessels, and other structures. The foramina of the skull are summarized in table 6.4. Various views of the skull are shown in figures 6.14 through 6.21; radiographs are shown in figure 6.22.

Although the hyoid bone and the three paired auditory ossicles are not considered part of the skull, they are associated with it. These bones are described in this section, immediately following the discussion of the facial bones.

lambdoid: Gk. *lambda*, letter λ in Greek alphabet

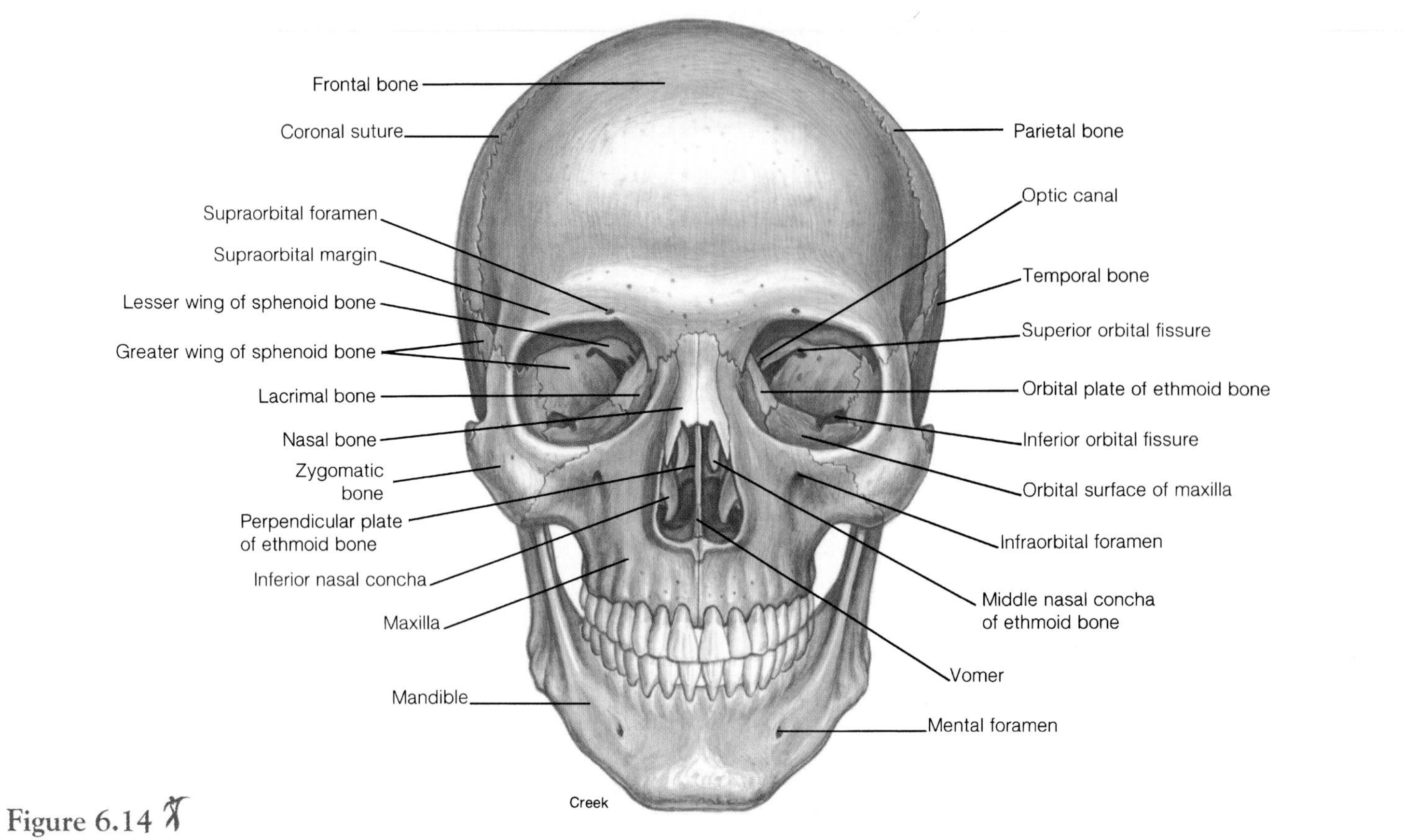
Frontal bone
Coronal suture
Supraorbital foramen
Supraorbital margin
Lesser wing of sphenoid bone
Greater wing of sphenoid bone
Lacrimal bone
Nasal bone
Zygomatic bone
Perpendicular plate of ethmoid bone
Inferior nasal concha
Maxilla
Mandible
Parietal bone
Optic canal
Temporal bone
Superior orbital fissure
Orbital plate of ethmoid bone
Inferior orbital fissure
Orbital surface of maxilla
Infraorbital foramen
Middle nasal concha of ethmoid bone
Vomer
Mental foramen
Creek

Figure 6.14

An anterior view of the skull.

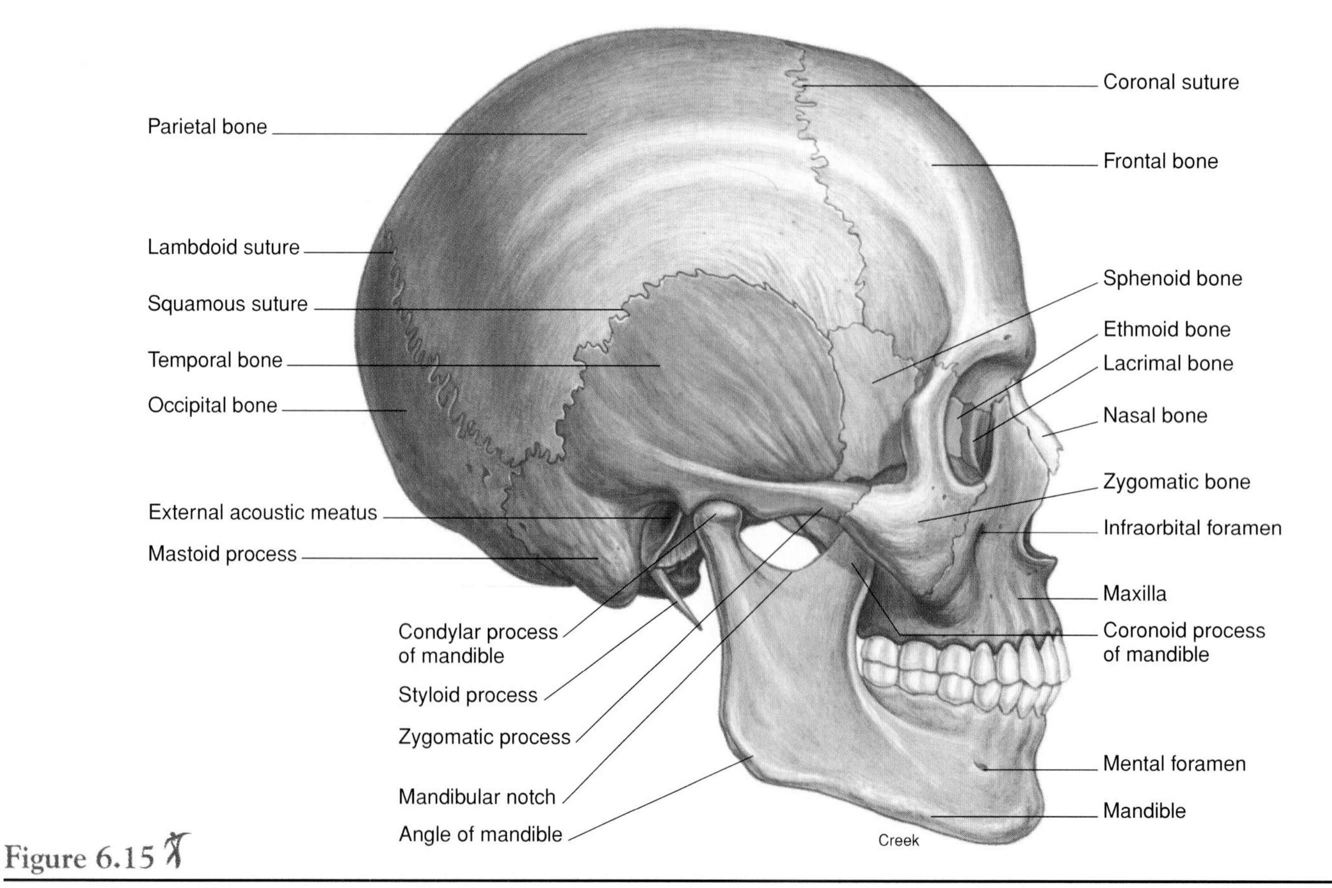
Parietal bone
Lambdoid suture
Squamous suture
Temporal bone
Occipital bone
External acoustic meatus
Mastoid process
Condylar process of mandible
Styloid process
Zygomatic process
Mandibular notch
Angle of mandible
Coronal suture
Frontal bone
Sphenoid bone
Ethmoid bone
Lacrimal bone
Nasal bone
Zygomatic bone
Infraorbital foramen
Maxilla
Coronoid process of mandible
Mental foramen
Mandible
Creek

Figure 6.15

A lateral view of the skull.

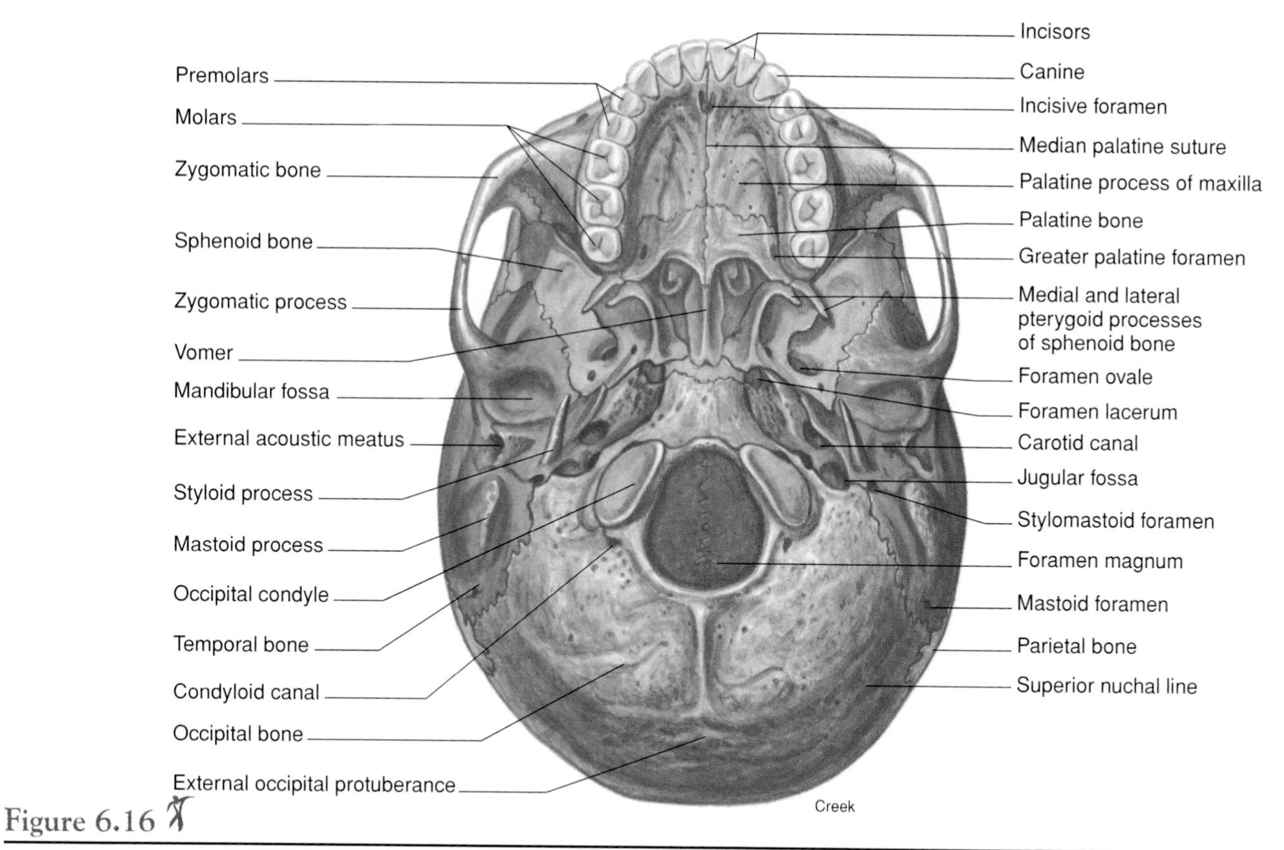
Incisors
Canine
Incisive foramen
Median palatine suture
Palatine process of maxilla
Palatine bone
Greater palatine foramen
Medial and lateral pterygoid processes of sphenoid bone
Foramen ovale
Foramen lacerum
Carotid canal
Jugular fossa
Stylomastoid foramen
Foramen magnum
Mastoid foramen
Parietal bone
Superior nuchal line
Premolars
Molars
Zygomatic bone
Sphenoid bone
Zygomatic process
Vomer
Mandibular fossa
External acoustic meatus
Styloid process
Mastoid process
Occipital condyle
Temporal bone
Condyloid canal
Occipital bone
External occipital protuberance
Creek

Figure 6.16

An inferior view of the skull.

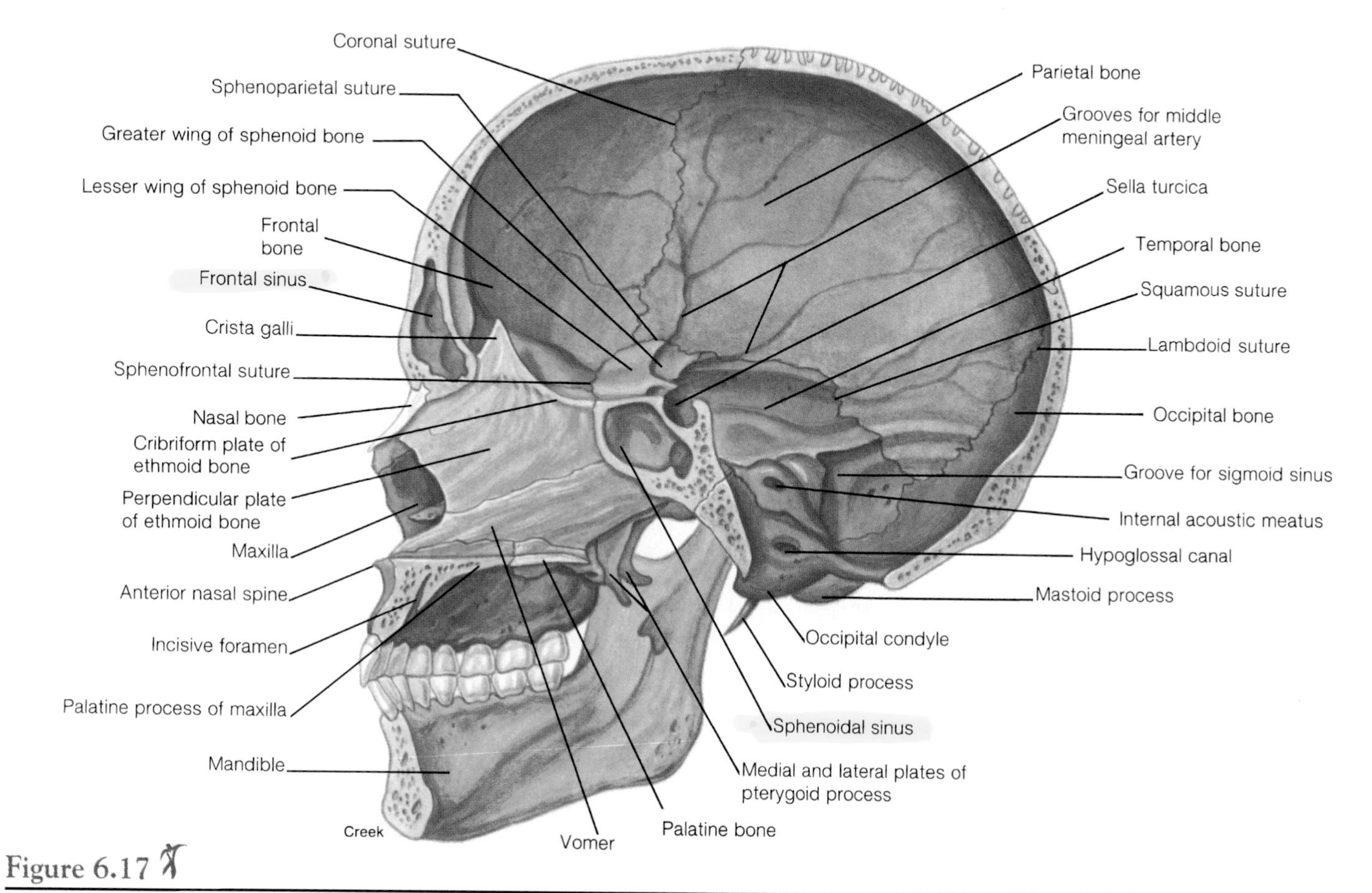
Coronal suture
Sphenoparietal suture
Greater wing of sphenoid bone
Lesser wing of sphenoid bone
Frontal bone
Frontal sinus
Crista galli
Sphenofrontal suture
Nasal bone
Cribriform plate of ethmoid bone
Perpendicular plate of ethmoid bone
Maxilla
Anterior nasal spine
Incisive foramen
Palatine process of maxilla
Mandible
Parietal bone
Grooves for middle meningeal artery
Sella turcica
Temporal bone
Squamous suture
Lambdoid suture
Occipital bone
Groove for sigmoid sinus
Internal acoustic meatus
Hypoglossal canal
Mastoid process
Occipital condyle
Styloid process
Sphenoidal sinus
Medial and lateral plates of pterygoid process
Palatine bone
Vomer
Creek

Figure 6.17

A sagittal view of the skull.

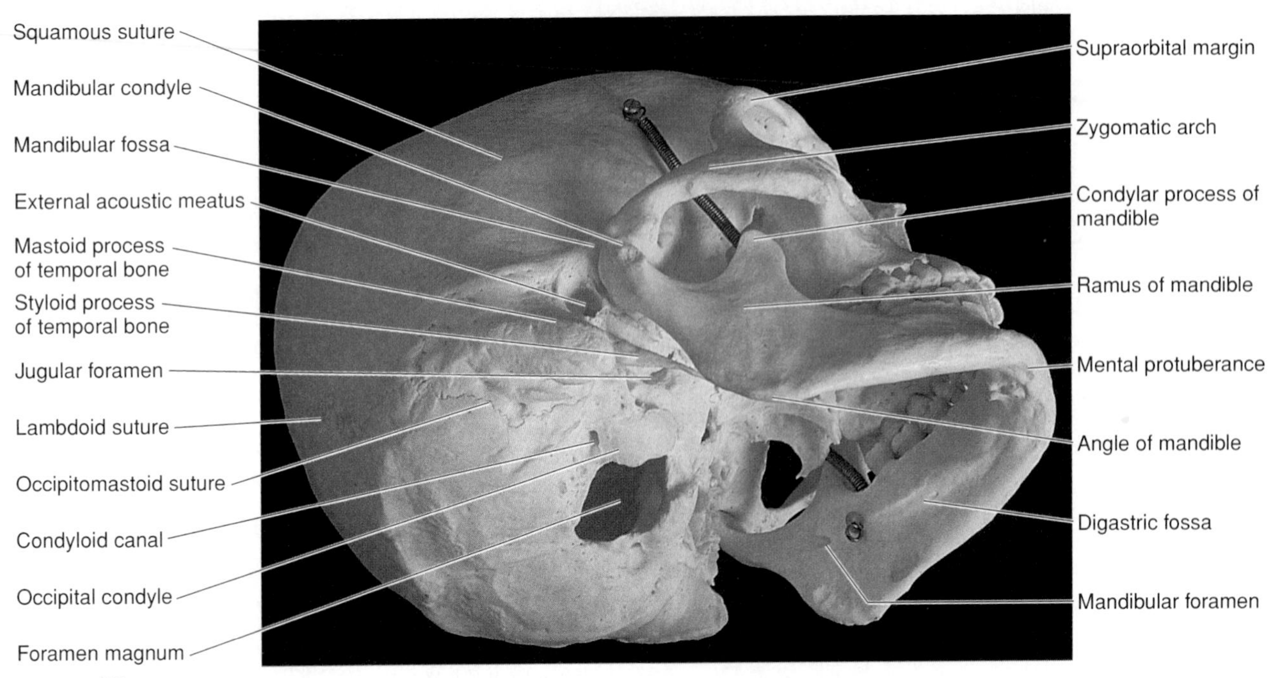
Squamous suture
Mandibular condyle
Mandibular fossa
External acoustic meatus
Mastoid process of temporal bone
Styloid process of temporal bone
Jugular foramen
Lambdoid suture
Occipitomastoid suture
Condyloid canal
Occipital condyle
Foramen magnum
Supraorbital margin
Zygomatic arch
Condylar process of mandible
Ramus of mandible
Mental protuberance
Angle of mandible
Digastric fossa
Mandibular foramen

Figure 6.18

An inferolateral view of the skull.

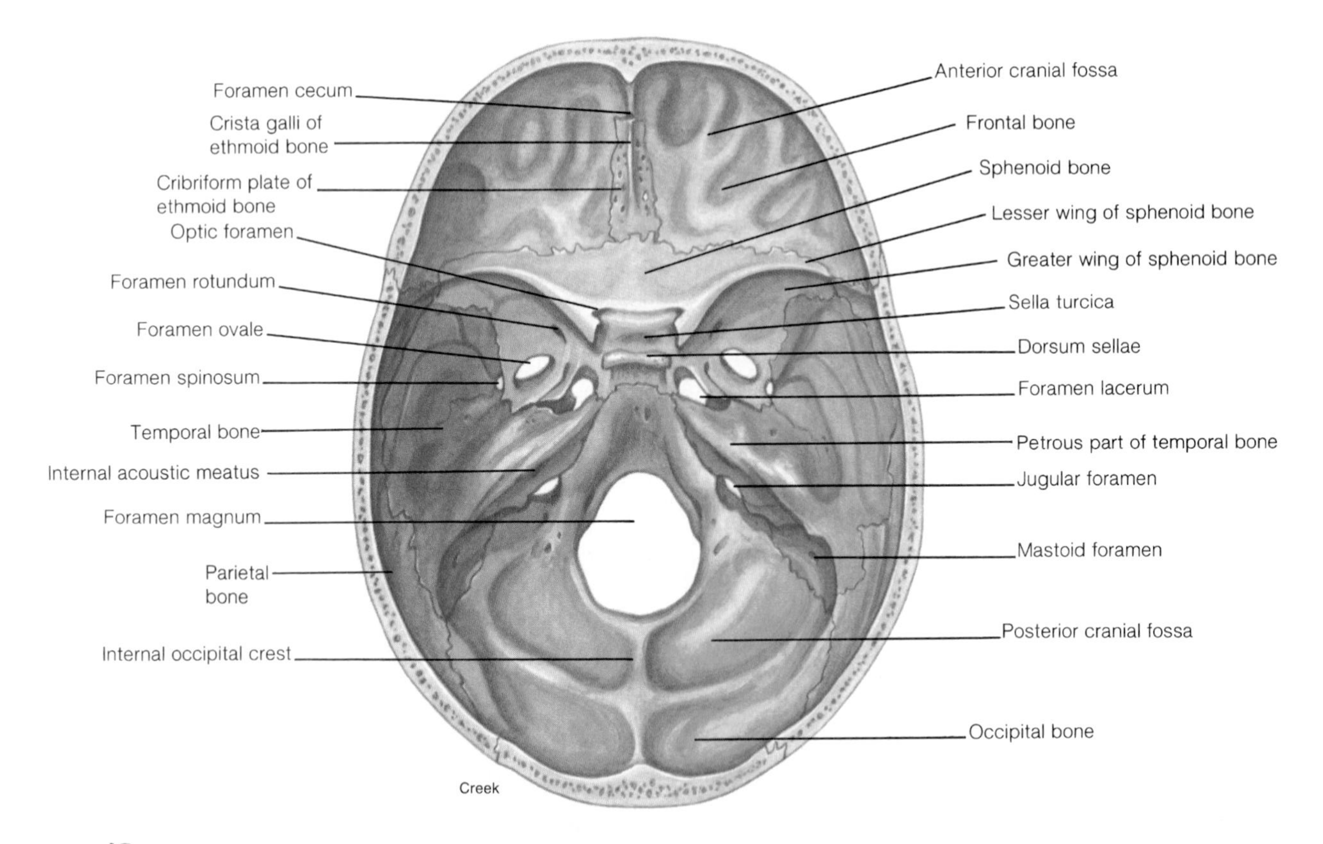
Foramen cecum
Crista galli of ethmoid bone
Cribriform plate of ethmoid bone
Optic foramen
Foramen rotundum
Foramen ovale
Foramen spinosum
Temporal bone
Internal acoustic meatus
Foramen magnum
Parietal bone
Internal occipital crest
Anterior cranial fossa
Frontal bone
Sphenoid bone
Lesser wing of sphenoid bone
Greater wing of sphenoid bone
Sella turcica
Dorsum sellae
Foramen lacerum
Petrous part of temporal bone
Jugular foramen
Mastoid foramen
Posterior cranial fossa
Occipital bone
Creek

Figure 6.19

The floor of the cranial cavity.

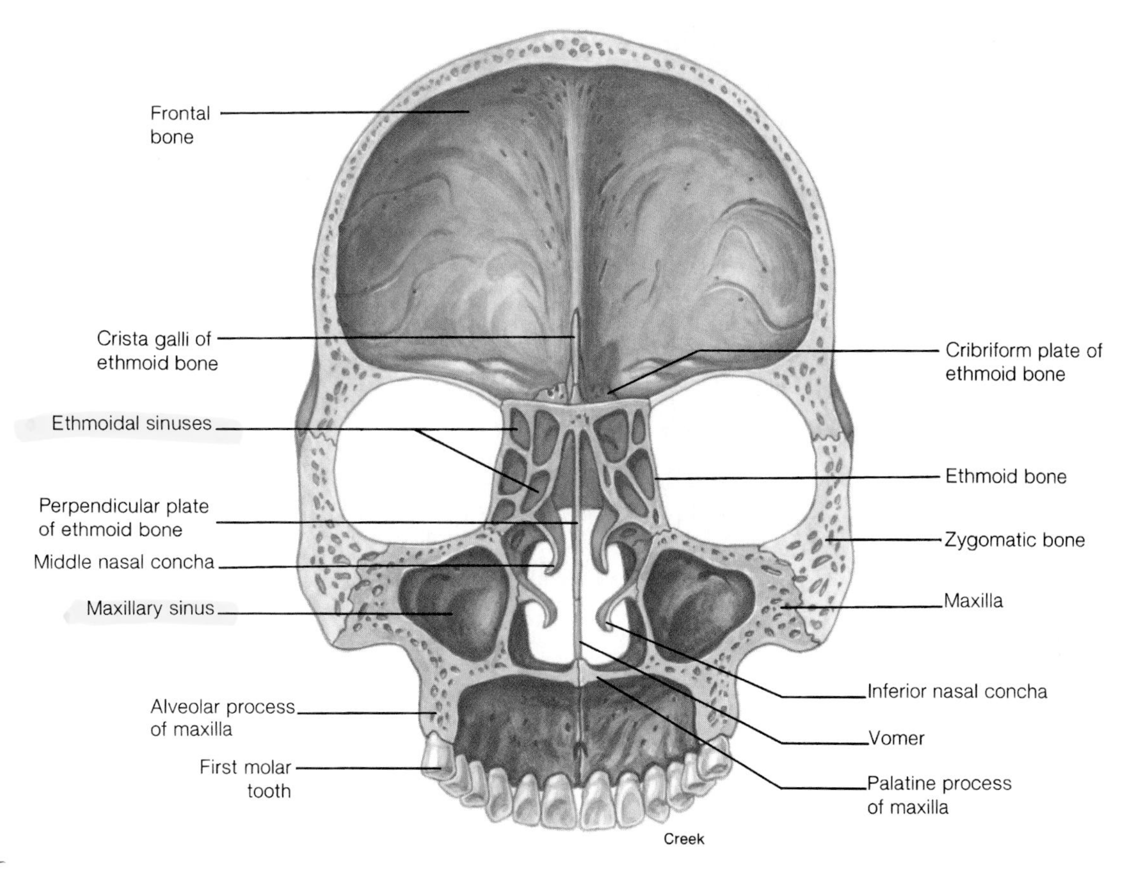

Figure 6.20

A posterior view of a frontal (coronal) section of the skull.

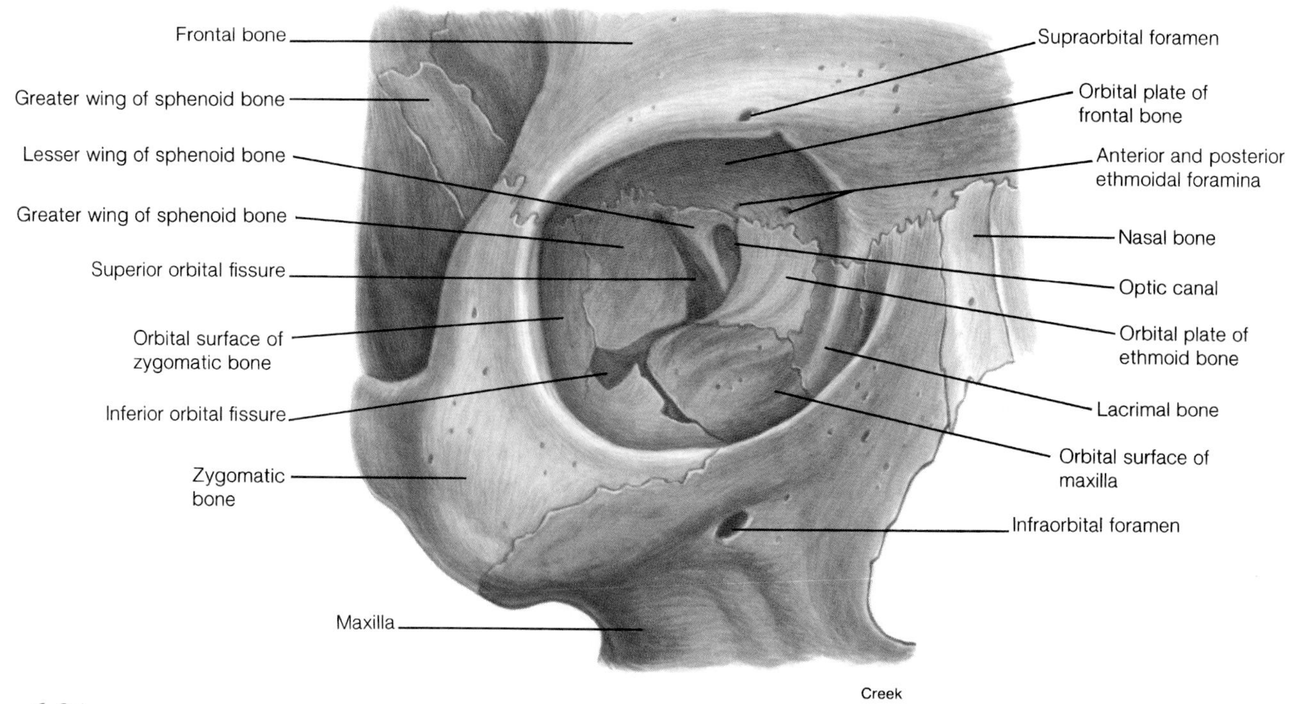

Figure 6.21

Bones of the orbit.

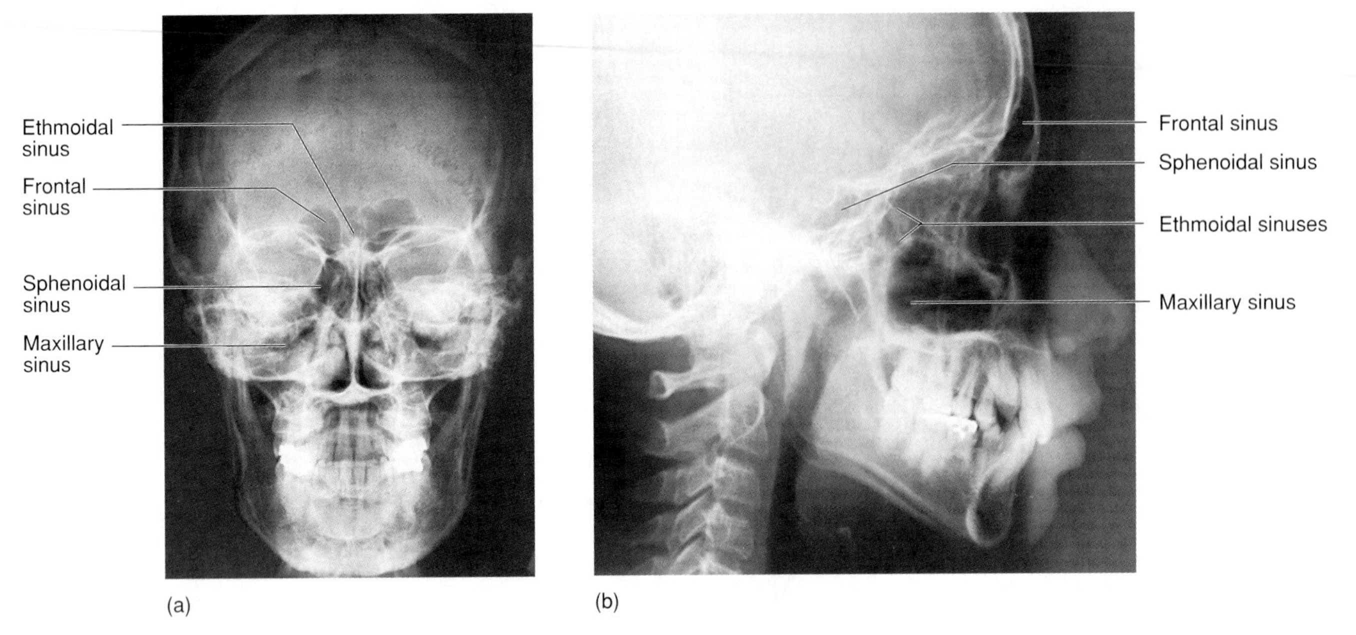

Figure 6.22

Radiographs of the skull showing the paranasal sinuses. (*a*) An anteroposterior view and (*b*) a right lateral view.

Cranial Bones

The cranial bones enclose and protect the brain and associated sensory organs. They consist of one *frontal*, two *parietals*, two *temporals*, one *occipital*, one *sphenoid*, and one *ethmoid*.

Frontal Bone

The frontal bone forms the anterior roof of the cranium, the forehead, the roof of the nasal cavity, and the superior arch of the *orbits*, which contain the eyeballs. The bones of the orbit are summarized in table 6.5. The frontal bone develops in two halves that grow together. Generally, they are completely fused by age 5 or 6. A suture sometimes persists between these two portions beyond age 6 and is referred to as a *metopic* (*mĕ-top′ik*) *suture*. The **supraorbital margin** is a prominent bony ridge over the orbit. Slightly medial to its midpoint is an opening called the **supraorbital foramen**, which provides passage for a nerve, artery, and veins.

The frontal bone also contains **frontal sinuses,** which are connected to the nasal cavity (fig 6.22). These sinuses, along with the other paranasal sinuses, lessen the weight of the skull and act as resonance chambers for voice production.

Parietal Bone

The two parietal bones form the upper sides and roof of the cranium (figs. 6.15 and 6.17). The **coronal suture** separates the frontal bone from the parietal bones, and the **sagittal suture** along the superior midline separates the right and left parietals from each other. The inner concave surface of each parietal bone, as well as the inner concave surfaces of other cranial bones, is marked by shallow impressions from convolutions of the brain and vessels serving the brain.

Table 6.5

Bones Forming the Orbit

Region of the Orbit	Contributing Bones
Roof (superior)	Frontal bone; lesser wing of sphenoid bone
Floor (inferior)	Maxilla; zygomatic bone; palatine bone
Lateral wall	Zygomatic bone
Posterior wall	Greater wing of sphenoid bone
Medial wall	Maxilla; lacrimal bone; ethmoid bone
Superior margin	Frontal bone
Lateral margin	Zygomatic bone
Medial margin	Maxilla

Temporal Bone

The two temporal bones form the lower sides of the cranium (figs. 6.15, 6.16, 6.17, and 6.23). Each temporal bone is joined to its adjacent parietal bone by the **squamous suture.** Structurally, each temporal bone has four parts.

1. **Squamous part.** The squamous part is the flattened plate of bone at the sides of the skull. Projecting forward is a **zygomatic** (*zi″go-mat′ik*) **process** that forms the posterior portion of the **zygomatic arch.** On the inferior surface of

cranium: Gk. *kranion*, skull
metopic suture: Gk. *metopon*, forehead; L. *sutura*, sew

zygomatic: Gk. *zygoma*, yolk

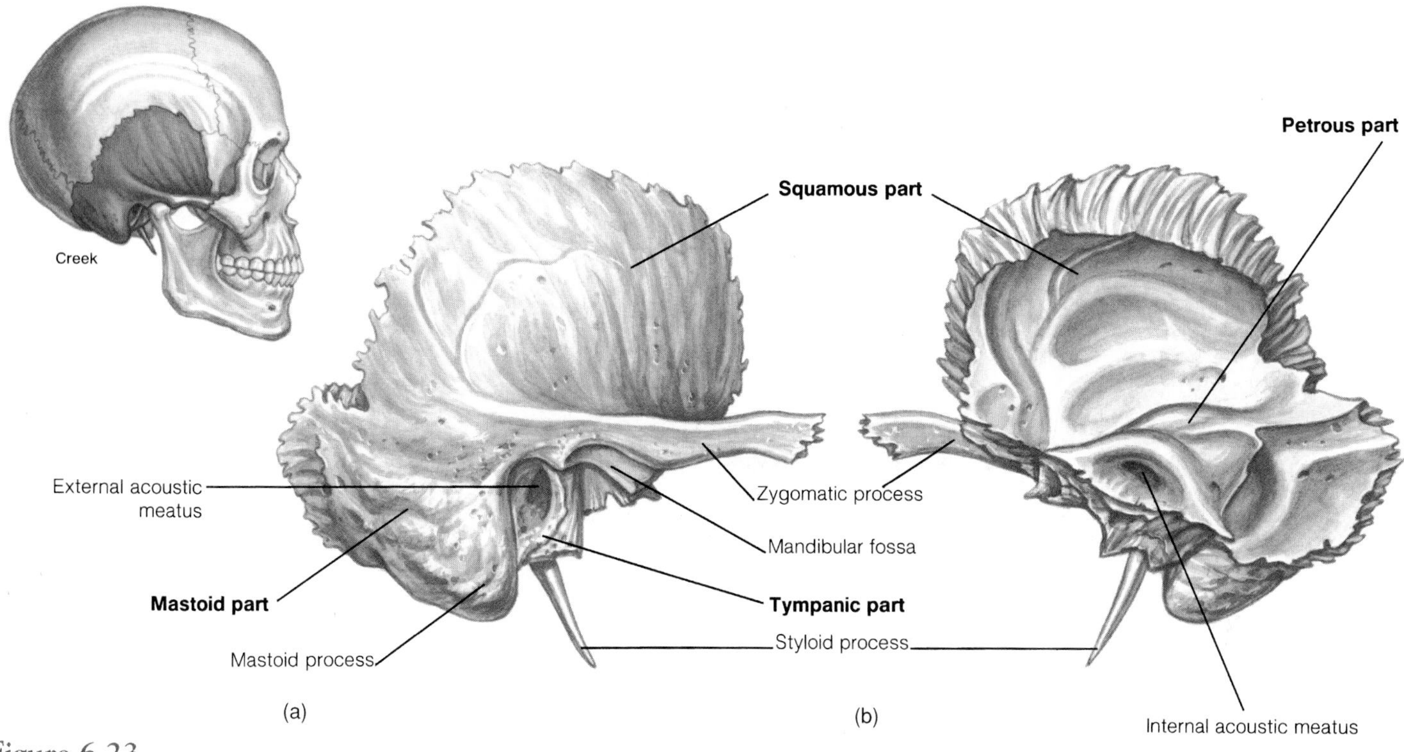

Figure 6.23

The temporal bone. (*a*) A lateral view and (*b*) a medial view.

the squamous part is the cuplike **mandibular fossa,** which forms a joint with the condyle of the mandible. This articulation is the *temporomandibular joint.*

2. **Tympanic part.** The tympanic part of the temporal bone contains the **external acoustic meatus** (*me-a'tus*), or ear canal, which is posterior to the mandibular fossa. A thin, pointed **styloid process** (figs. 6.16, 6.17, and 6.18) projects inferiorly from the tympanic part.
3. **Mastoid part.** The **mastoid process,** a rounded projection posterior to the external acoustic meatus, accounts for the mass of the mastoid part. The **mastoid foramen** (fig. 6.16) is directly posterior to the mastoid process. The **stylomastoid foramen,** located between the mastoid and styloid processes (fig. 6.16), provides the passage for part of the facial nerve.
4. **Petrous part.** The petrous (*pet'rus*) part can be seen in the floor of the cranium (figs. 6.19 and 6.23). The structures of the middle ear and inner ear are housed in this dense part of the temporal bone. The **carotid** (*kă-rot'id*) **canal** and the **jugular foramen** border on the medial side of the petrous part at the junction of the temporal and occipital bones. The carotid canal allows blood into the brain via the internal carotid artery, and the jugular foramen lets blood drain from the brain via the internal jugular vein. Three cranial nerves also pass through the jugular foramen.

styloid: Gk. *stylos*, pillar
mastoid: Gk. *mastos*, breast
petrous: Gk. *petra*, rock

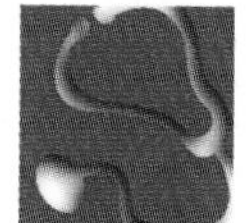

The mastoid process of the temporal bone can be easily palpated as a bony knob immediately behind the earlobe. This process contains a number of small air-filled spaces called *mastoid cells* that can become infected in *mastoiditis,* as a result, for example, of a prolonged middle-ear infection.

Occipital Bone

The occipital bone forms the posterior and most of the base of the skull. It articulates with the parietal bones at the **lambdoid suture.** The **foramen magnum** is the large hole in the occipital bone through which the spinal cord passes to attach to the brain stem. On each side of the foramen magnum are the **occipital condyles** (fig. 6.16), which articulate with the first vertebra (the atlas) of the vertebral column. At the anterolateral edge of the occipital condyle is the **hypoglossal canal** (fig. 6.17), through which the hypoglossal nerve passes. A **condyloid** (*kon'dĭ-loid*) **canal** lies posterior to the occipital condyle (fig. 6.16). The **external occipital protuberance** is a prominent posterior projection on the occipital bone that can be felt as a definite bump just under the skin. The **superior nuchal** (*noo'kal*) **line** is a ridge of bone extending laterally from the occipital protuberance to the mastoid portion of the temporal bone. **Sutural bones** are small clusters of irregularly shaped bones that frequently occur along the lambdoid suture.

magnum: L. *magnum*, great
nuchal: Fr. *nuque*, nape of neck

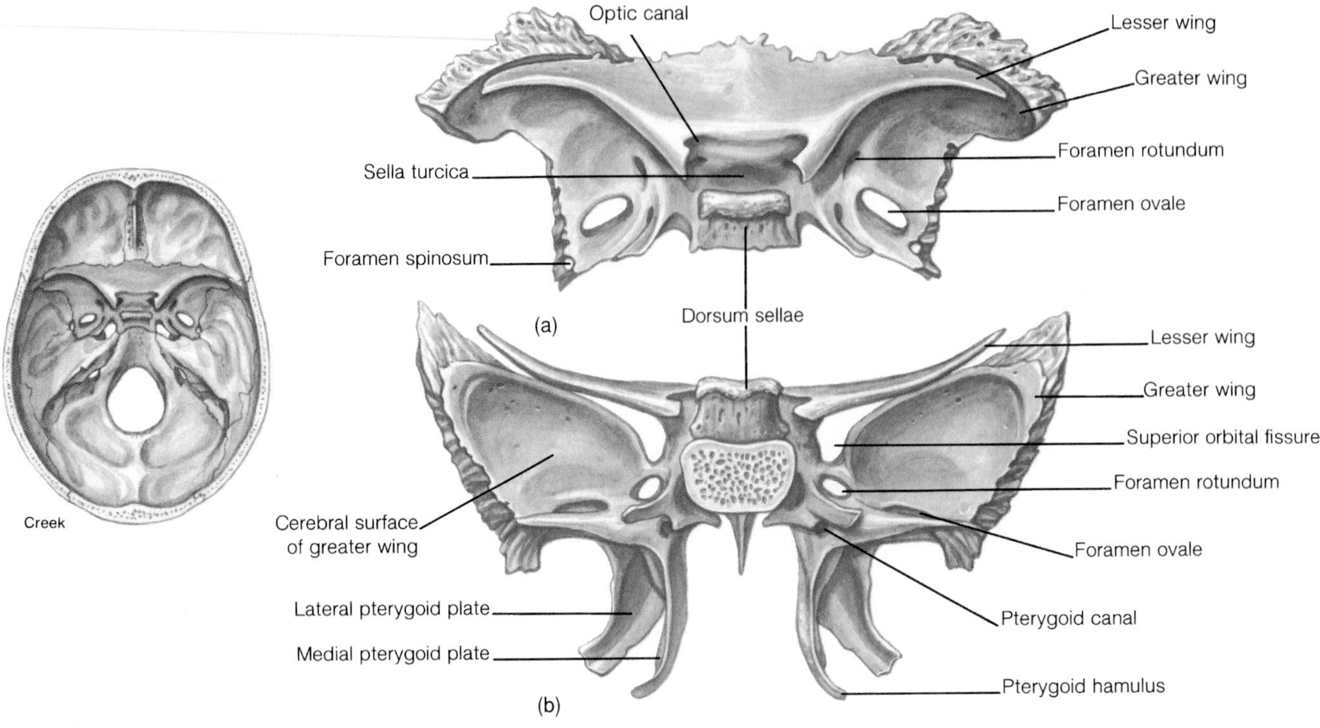

Figure 6.24

The sphenoid bone. (*a*) A superior view and (*b*) a posterior view.

Sphenoid Bone

The sphenoid (*sfe'noid*) bone forms part of the anterior base of the cranium and can be viewed laterally and inferiorly (figs. 6.15 and 6.16). This bone has a somewhat mothlike shape (fig. 6.24). It consists of a **body** and laterally projecting **greater** and **lesser wings** that form part of the orbit. The wedgelike body contains the **sphenoidal sinuses** and a prominent saddlelike depression, the **sella turcica** (*sel'ă tur'sĭ-kă*). Commonly called "Turk's saddle," the sella turcica houses the pituitary gland. A pair of **pterygoid** (*ter'ĭ-goid*) **processes** project inferiorly from the sphenoid bone and help form the lateral walls of the nasal cavity.

Several foramina (figs. 6.16, 6.19, and 6.24) are associated with the sphenoid bone.

1. The **optic canal** is a large opening through the lesser wing into the back of the orbit that provides passage for the optic nerve and the ophthalmic artery.
2. The **superior orbital fissure** is a triangular opening between the wings of the sphenoid bone that provides passage for the ophthalmic nerve, a branch of the trigeminal cranial nerve and for the oculomotor, trochlear, and abducens cranial nerves.
3. The **foramen ovale** is an opening at the base of the lateral pterygoid plate, through which the mandibular nerve passes.
4. The **foramen spinosum** is a small opening at the posterior angle of the sphenoid bone that provides passage for the middle meningeal vessels.
5. The **foramen lacerum** (*las'er-um*) is an opening between the sphenoid and the petrous part of the temporal bone, through which the internal carotid artery and the meningeal branch of the ascending pharyngeal artery pass.
6. The **foramen rotundum** is an opening just posterior to the superior orbital fissure, at the junction of the anterior and medial portions of the sphenoid bone. The maxillary nerve passes through this foramen.

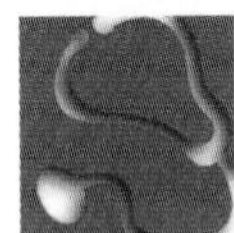

Located on the inferior side of the cranium, the sphenoid bone would seem to be well protected from trauma. Actually, just the opposite is true—and in fact the sphenoid is the most frequently fractured bone of the cranium. It has several broad, thin, platelike extensions that are perforated by numerous foramina. A blow to almost any portion of the skull causes the buoyed, fluid-filled brain to rebound against the vulnerable sphenoid bone, often causing it to fracture.

Ethmoid Bone

The ethmoid bone is located in the anterior portion of the floor of the cranium between the orbits, where it forms the roof of the nasal cavity (figs. 6.17, 6.20, and 6.25). An inferior projection of the ethmoid bone, called the **perpendicular plate,** forms the superior part of the nasal septum that separates the nasal cavity into two chambers. Each chamber of the nasal cavity is

sphenoid: Gk. *sphenoeides*, wedgelike

ethmoid: Gk. *ethmos*, sieve

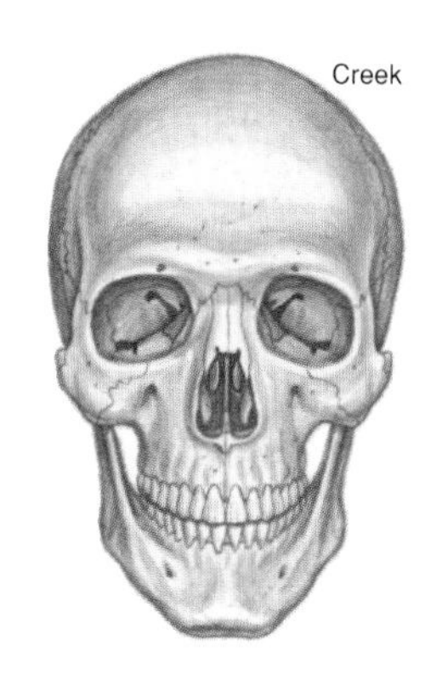

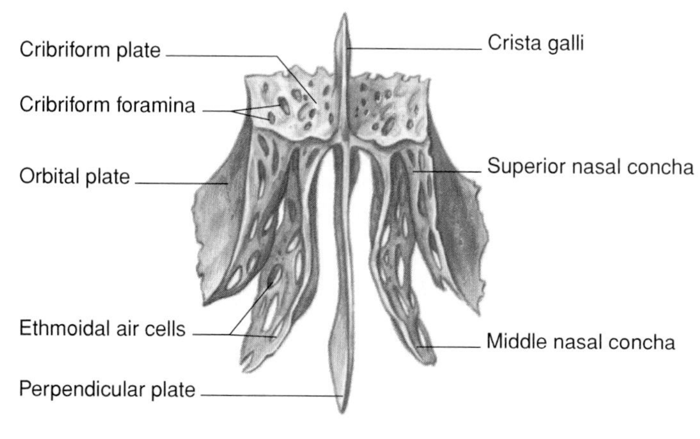

Figure 6.25

An anterior view of the ethmoid bone.

referred to as a **nasal fossa.** Flanking the perpendicular plate on each side is a large but delicate mass of bone riddled with ethmoidal air cells, collectively constituting the **ethmoid sinus.** A spine of the perpendicular plate, the **crista galli** (*kris'tă gal'e*), projects superiorly into the cranial cavity and serves as an attachment for the meninges covering the brain. On both lateral walls of the nasal cavity are two scroll-shaped plates of the ethmoid bone, the **superior** and **middle nasal conchae** (*kong'ke*—singular, *concha*) (fig. 6.26), also known as *turbinates.* At right angles to the perpendicular plate, within the floor of the cranium, is the **cribriform** (*krib'rĭ-form*) **plate,** which has numerous **cribriform foramina** for the passage of olfactory nerves from the nasal cavity. The bones of the nasal cavity are summarized in table 6.6.

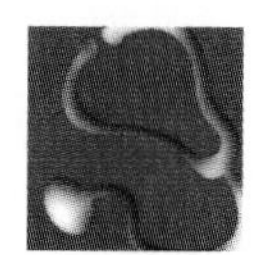

The moist, warm vascular lining within the nasal cavity is susceptible to infections, particularly if a person is not in good health. Infections of the nasal cavity can spread to several surrounding areas. The paranasal sinuses connect to the nasal cavity and are especially prone to infection. The eyes may become reddened and swollen during a nasal infection because of the connection of the nasolacrimal duct, through which tears drain from the orbit to the nasal cavity. Organisms may spread via the auditory tube from the nasopharynx to the middle ear. With prolonged nasal infections, organisms may even ascend to the meninges covering the brain via the sheaths of the olfactory nerves and pass through the cribriform plate to cause *meningitis.*

crista galli: L. *crista,* crest; *galli,* cock's comb
conchae: L. *conchae,* shells
cribriform: L. *cribrum,* sieve; *forma,* like

Facial Bones

The 14 bones of the skull not in contact with the brain are called **facial bones.** These bones, together with certain cranial bones (frontal bone and portions of the ethmoid and temporal bones), give shape and individuality to the face. Facial bones also support the teeth and provide attachments for various muscles that move the jaw and cause facial expressions. With the exceptions of the vomer and mandible, all of the facial bones are paired. The articulated facial bones are illustrated in figures 6.14 through 6.21.

Maxilla

The two maxillae (*mak-sil'e*) unite at the midline to form the upper jaw, which supports the upper teeth. **Incisors, canines** (cuspids), **premolars,** and **molars** are anchored in **dental alveoli,** (tooth sockets), within the **alveolar** (*al-ve'ŏ-lar*) **process** of the maxilla (fig. 6.27). The **palatine** (*pal'ă-tīn*) **process,** a horizontal plate of the maxilla, forms the greater portion of the **hard palate** (*pal'it*), or roof of the mouth. The **incisive foramen** (fig. 6.16) is located in the anterior region of the hard palate, behind the incisors. An **infraorbital foramen** is located under each orbit and serves as a passageway for the infraorbital nerve and artery to the nose (figs. 6.14, 6.15, 6.21, and 6.27). A final opening within the maxilla is the **inferior orbital fissure.** It is located between the maxilla and the greater wing of the sphenoid (fig. 6.14) and is the external opening for the maxillary nerve of the trigeminal nerve and infraorbital vessels. The large **maxillary sinus** located within the maxilla is one of the four paranasal sinuses (figs. 6.20 and 6.22).

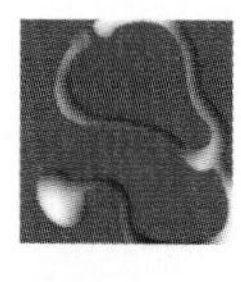

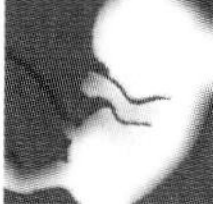

If the two palatine processes fail to join during early prenatal development (about 12 weeks), a *cleft palate* results. A cleft palate may be accompanied by a *cleft lip* lateral to the midline. These conditions can be surgically treated with excellent cosmetic results. An immediate problem, however, is that a baby with a cleft palate may have a difficult time nursing because it is unable to create the necessary suction within the oral cavity to swallow effectively.

incisor: L. *incidere,* to cut
canine: L. *canis,* dog
molar: L. *mola,* millstone
alveolus: L. *alveus,* little cavity

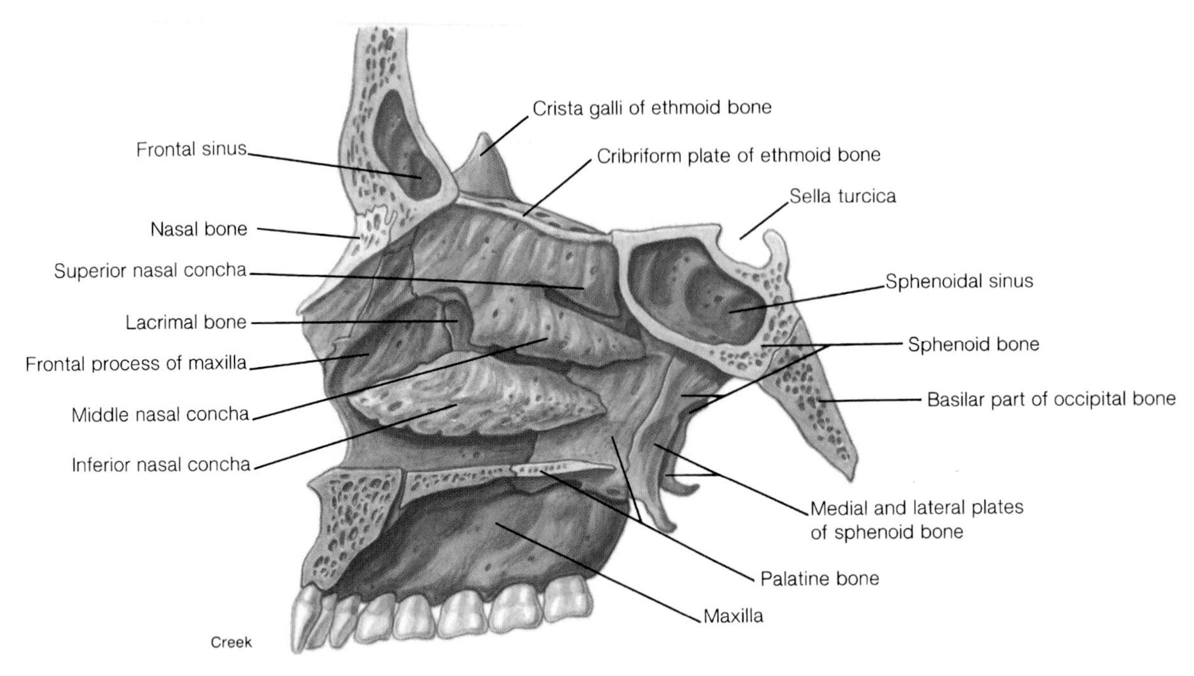

Figure 6.26

The lateral wall of the nasal cavity.

Table 6.6
Bones That Enclose the Nasal Cavity

Region of Nasal Cavity	Contributing Bones
Roof (superior)	Ethmoid bone (cribriform plate); frontal bone
Floor (inferior)	Maxilla; palatine bone
Lateral wall	Maxilla; palatine bone
Nasal septum (medial)	Ethmoid bone (perpendicular plate); vomer; nasal bone
Bridge	Nasal bone
Conchae	Ethmoid bone (superior and middle conchae); inferior nasal concha

Palatine Bone

The L-shaped palatine bones form the posterior third of the hard palate, a part of the orbits, and a part of the nasal cavity. The **horizontal plates** of the palatines contribute to the formation of the hard palate (fig. 6.28). At the posterior angle of the hard palate is the large **greater palatine foramen** that provides passage for the greater palatine nerve and descending palatine vessels (fig. 6.16). Two or more smaller **lesser palatine foramina** are positioned posterior to the greater palatine foramen. Branches of the lesser palatine nerve pass through these openings.

Zygomatic Bone

The two zygomatic bones form the cheekbones of the face. A posteriorly extending *temporal process* of this bone unites with the *zygomatic process* of the temporal bone to form the **zygomatic arch** (fig. 6.16). The zygomatic bone also forms the lateral margin of the orbit. A small **zygomaticofacial** (*zi″gŏ-mat″ĭ-kŏ-fa′shal*) **foramen,** located on the anterolateral surface of this bone, allows passage of the zygomatic nerves and vessels.

Lacrimal Bone

The thin lacrimal bones form the anterior part of the medial wall of each orbit (fig. 6.21). These are the smallest of the facial bones. Each one has a **lacrimal sulcus**—a groove that helps to form the **nasolacrimal canal.** This opening permits the tears of the eye to drain into the nasal cavity.

Nasal Bone

The small, rectangular nasal bones (fig. 6.14) join at the midline to form the bridge of the nose. The nasal bones support the flexible cartilaginous plates, which are a part of the framework of the nose. Fractures of the nasal bones or fragmentation of the associated cartilages are common facial injuries.

Inferior Nasal Concha

The two inferior nasal conchae are fragile, scroll-like bones that project horizontally and medially from the lateral walls of the nasal cavity (figs. 6.14 and 6.20). They extend into the nasal cavity just below the superior and middle nasal conchae, which

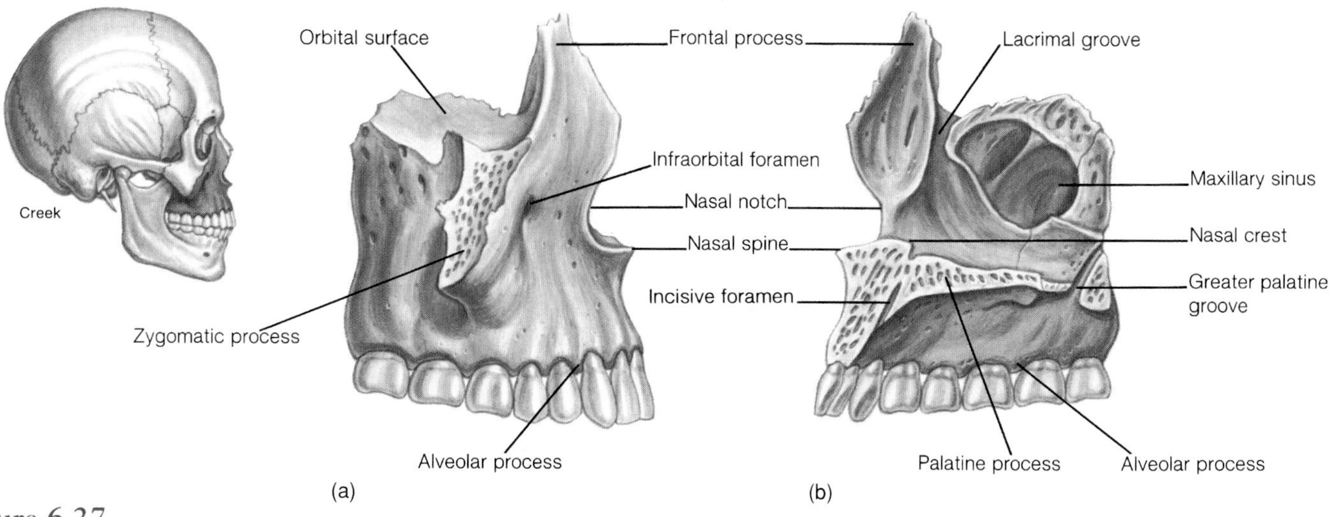

Figure 6.27

The maxilla. (*a*) A lateral view and (*b*) a medial view.

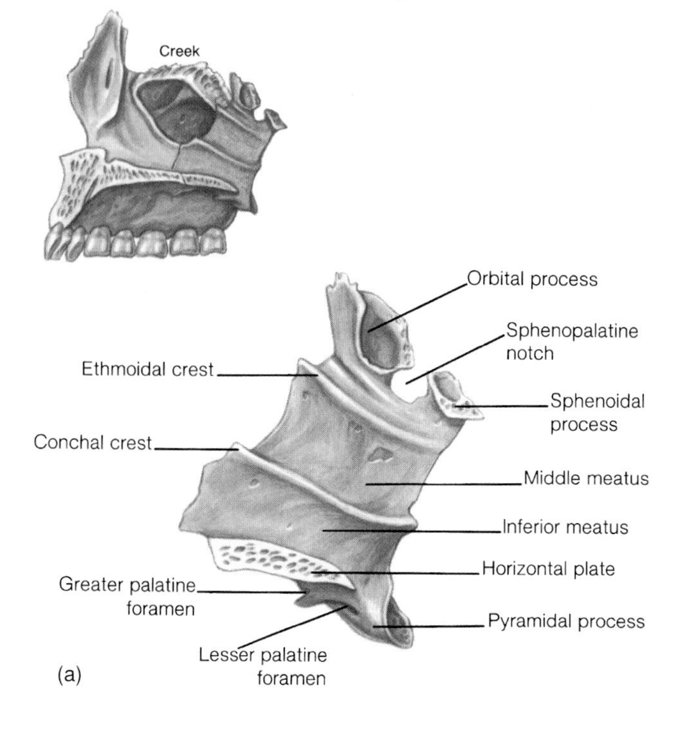

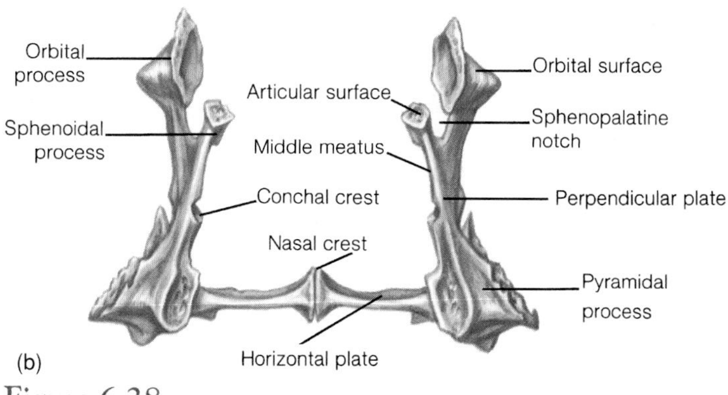

Figure 6.28

The palatine bone. (*a*) A medial view and (*b*) the two palatine bones viewed posteriorly. The two palatine bones form the posterior portion of the hard palate.

are part of the ethmoid bone (see fig. 6.25). The inferior nasal conchae are the largest of the three paired conchae, and, like the other two, are covered with a mucous membrane to warm, moisten, and cleanse inhaled air.

Vomer

The vomer (*vo'mer*) is a thin, flattened bone that forms the lower part of the nasal septum (figs. 6.16, 6.17, and 6.20). Along with the perpendicular plate of the ethmoid bone, it supports the layer of septal cartilage that forms most of the anterior part of the nasal septum.

Mandible

The mandible ("jawbone") is the largest, strongest bone in the face. It is attached to the skull by a temporomandibular joint, and is the only movable bone of the skull. The horseshoe-shaped front and horizontal lateral sides of the mandible are referred to as the **body** (fig. 6.29). Extending vertically from the posterior part of the body are two **rami** (*ra'mi*—singular, *ramus*). At the superior margin of each ramus is a knoblike **condylar process,** which articulates with the mandibular fossa of the temporal bone, and a pointed **coronoid process** for the attachment of the temporalis muscle. The depressed area between these two processes is called the **mandibular notch.** The angle of the mandible is where the horizontal body and vertical ramus meet at the corner of the jaw.

Two sets of foramina are associated with the mandible: the **mental foramen,** on the anterolateral aspect of the body of the mandible below the first molar, and the **mandibular foramen,** on

---

vomer: L. *vomer*, plowshare
mandible: L. *mandere*, to chew
ramus: L. *ramus*, branch
condylar: L. *condylus*, knucklelike
coronoid: Gk. *korone*, like a crow's beak

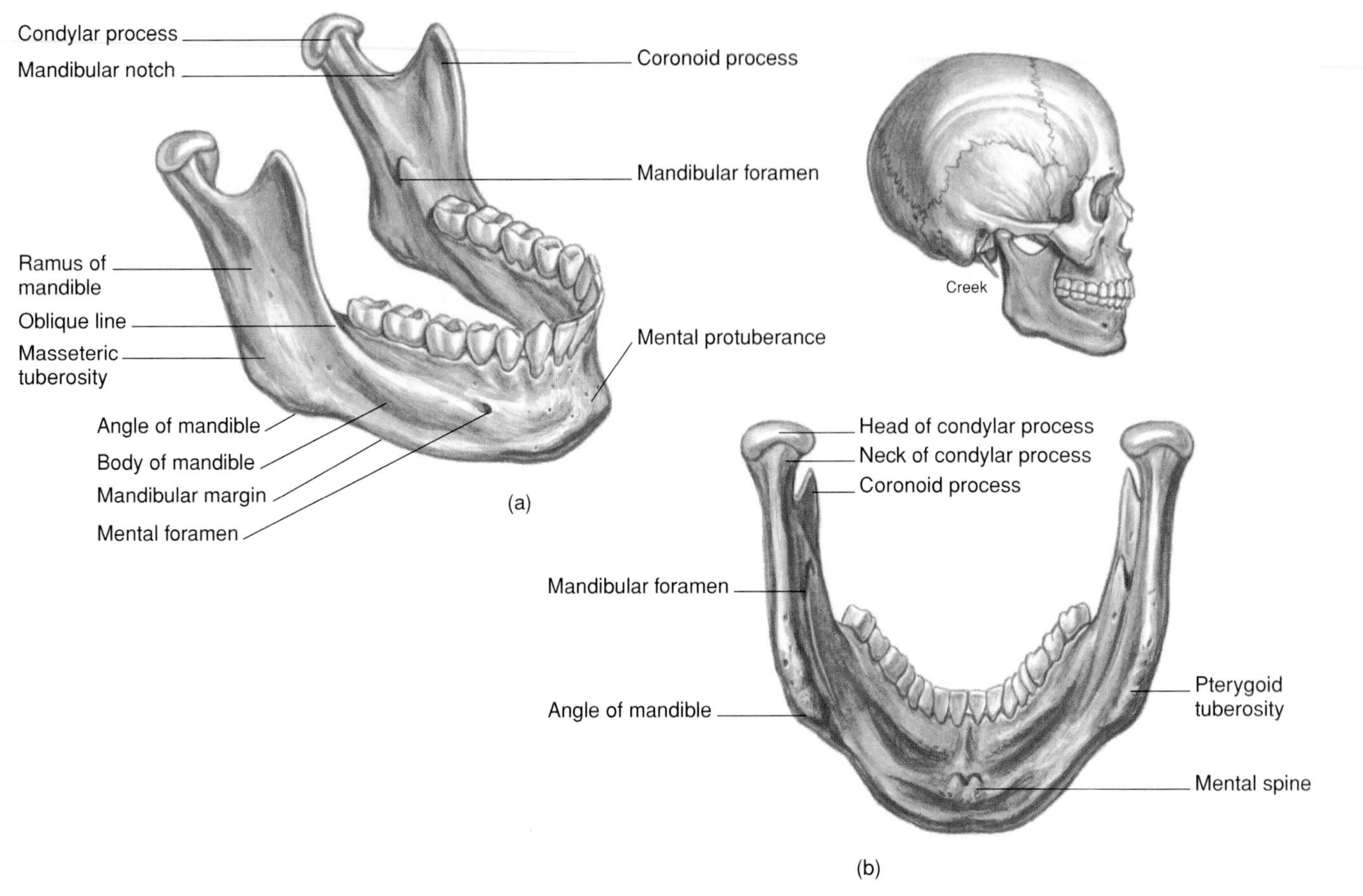

Figure 6.29

The mandible. (*a*) A lateral view and (*b*) a posterior view.

the medial surface of the ramus. The mental nerve and vessels pass through the mental foramen, and the inferior alveolar nerve and vessels are transmitted through the mandibular foramen. Several muscles that close the jaw extend from the skull to the mandible (see chapter 9). The mandible of an adult supports 16 teeth within dental alveoli, which occlude with the teeth of the maxilla.

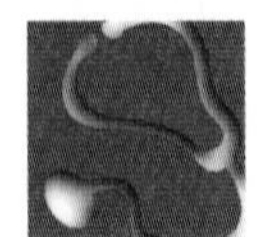

Dentists use bony landmarks of the facial region to locate the nerves that traverse the foramina in order to inject anesthetics. For example, the trigeminal nerve is composed of three large nerves, the lower two of which convey sensations from the teeth, gums, and jaws. The mandibular teeth can be desensitized by an injection near the mandibular foramen called a *third-division,* or *lower, nerve block.* An injection near the foramen rotundum of the skull, called a *second-division nerve block,* desensitizes all of the upper teeth on one side of the maxilla.

Hyoid Bone

The single **hyoid bone** is a unique part of the skeleton in that it does not attach directly to any other bone. It is located in the neck region, below the mandible, where it is suspended from the styloid process of the temporal bone by the stylohyoid muscles and ligaments. The hyoid bone has a **body,** two **lesser cornua** (*kor'nyoo-ă*—singular, *cornu*) extending anteriorly, and two **greater cornua** (fig. 6.30), which project posteriorly to the stylohyoid ligaments.

The hyoid bone supports the tongue and provides attachment for some of its muscles (see fig. 9.18). It may be palpated by placing a thumb and a finger on either side of the upper neck under the lateral portions of the mandible and firmly squeezing medially. This bone is carefully examined in an autopsy when strangulation is suspected, since during strangulation it is frequently fractured.

Auditory Ossicles

Three small paired bones, called **auditory ossicles,** are located within the middle-ear cavities in the petrous part of the temporal bones (fig. 6.31). From outer to inner, these bones are the **malleus** ("hammer"), **incus** ("anvil"), and **stapes** ("stirrup"). As described in chapter 15, their movements transmit sound impulses through the middle-ear cavity (see p. 501).

cornu: L. *cornu,* horn
malleus: L. *malleus,* hammer
incus: L. *incus,* anvil
stapes: L. *stapes,* stirrup

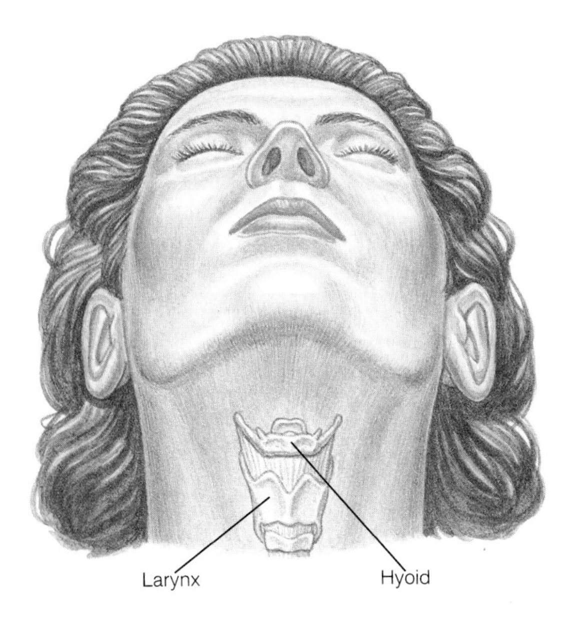

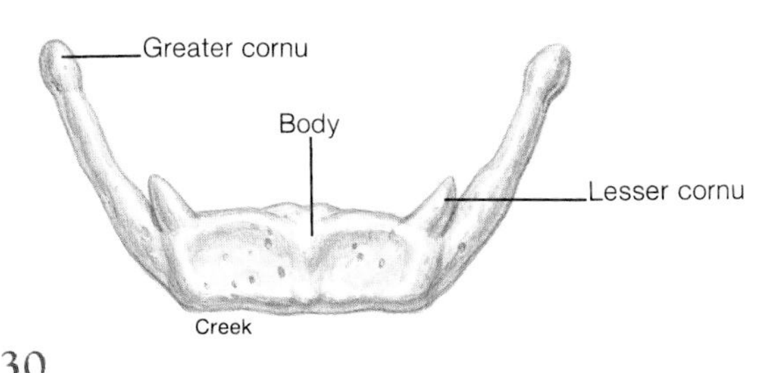

Figure 6.30

An anterior view of the hyoid bone.

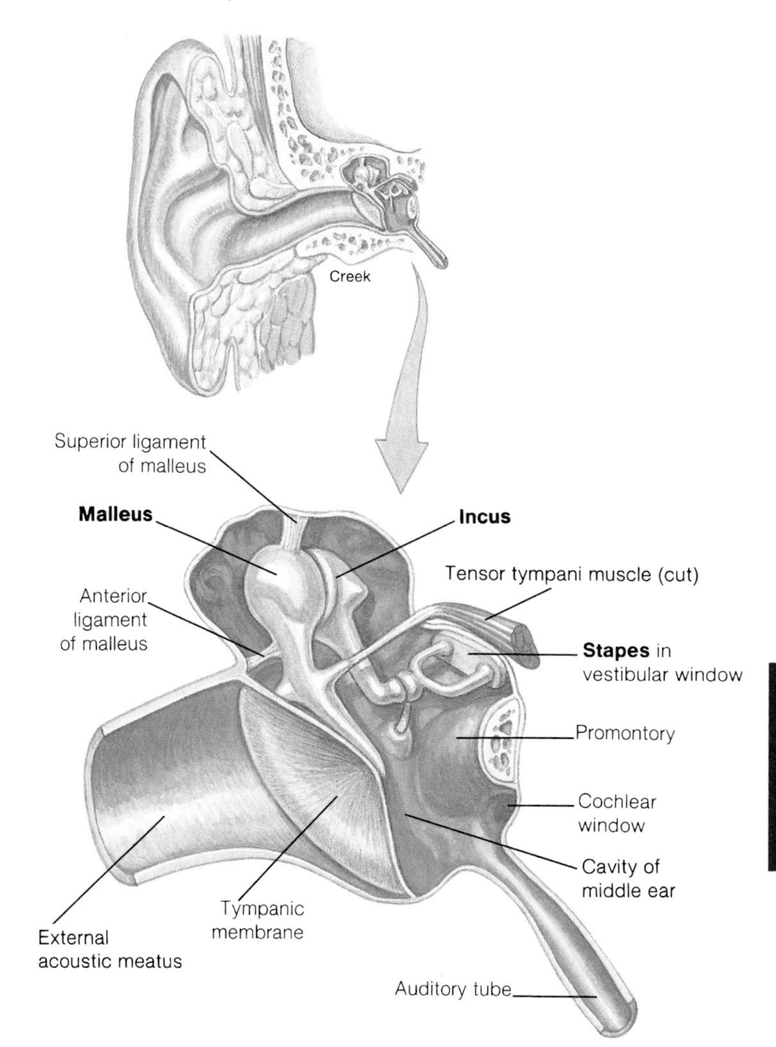

Figure 6.31

The three auditory ossicles within the middle-ear cavity.

1. State which facial and cranial bones of the skull are paired and which are unpaired. Also, indicate at least two structural features associated with each bone of the skull.
2. Describe the location of each bone of the skull and indicate the sutures that join these bones.
3. What is the function of each of the following: sella turcica, foramen magnum, petrous part of the temporal bone, crista galli, and nasal conchae?
4. Which facial bones support the teeth?

Vertebral Column

The vertebral column consists of a series of irregular bones called vertebrae, separated from each other by fibrocartilaginous intervertebral discs. Vertebrae enclose and protect the spinal cord, support the skull and allow for its movement, articulate with the rib cage, and provide for the attachment of trunk muscles. The intervertebral discs lend flexibility to the vertebral column and absorb vertical shock.

Objective 12 Identify the bones of the five regions of the vertebral column and describe the characteristic curves of each region.

Objective 13 Describe the structure of a typical vertebra.

The **vertebral column** ("backbone") and the *spinal cord* of the nervous system constitute the *spinal column*. The vertebral column has three functions:

1. to support the head and upper extremities while permitting freedom of movement;
2. to provide attachment for various muscles, ribs, and visceral organs; and
3. to protect the spinal cord and permit passage of the spinal nerves.

The vertebral column is typically composed of 33 individual vertebrae, some of which are fused. There are seven **cervical,** twelve **thoracic,** five **lumbar,** three to five fused **sacral,** and four or five fused **coccygeal** (*kok-sij'e-al*) **vertebrae;** thus, the adult vertebral column is composed of a total of 26 movable parts. Vertebrae are separated by fibrocartilaginous intervertebral discs and are secured to each other by interlocking processes and binding ligaments. This structural arrangement permits only limited movement between adjacent vertebrae but extensive movement for the vertebral column as a whole. Between the vertebrae are openings called **intervertebral foramina** that allow passage of spinal nerves.

When viewed from the side, four curvatures of the vertebral column can be identified (fig. 6.32). The **cervical, thoracic,** and **lumbar curves** are identified by the type of vertebrae they include. The **pelvic curve** (sacral curve) is formed by the shape of the sacrum and coccyx (*kok'siks*). The curves of the vertebral column play an important functional role in increasing the strength and maintaining the balance of the upper part of the body; they also make possible a bipedal stance.

The four vertebral curves are not present in an infant. The cervical curve begins to develop at about 3 months as the baby begins holding up its head, and it becomes more pronounced as the baby learns to sit up (fig. 6.33). The lumbar curve develops as a child begins to walk. The thoracic and pelvic curves are called *primary curves* because they retain the shape of the fetus. The cervical and lumbar curves are called *secondary curves* because they are modifications of the fetal shape.

General Structure of Vertebrae

Vertebrae are similar in their general structure from one region to another. A typical vertebra consists of an anterior drum-shaped **body,** which is in contact with intervertebral discs above and below (fig. 6.34). The **vertebral arch** is attached to the posterior surface of the body and is composed of two supporting **pedicles** (*ped'ĭ-kulz*) and two arched **laminae** (*lam' ĭ-ne*). The space formed by the vertebral arch and body is the **vertebral foramen,** through which the spinal cord passes. Between the pedicles of adjacent vertebrae are the **intervertebral foramina,** through which spinal nerves emerge as they branch off the spinal cord.

Seven processes arise from the vertebral arch of a typical vertebrae: the **spinous process,** two **transverse processes,** two **superior articular processes,** and two **inferior articular processes** (fig. 6.35). The spinous process and transverse processes serve for muscle attachment and the superior and inferior articular processes limit twisting of the vertebral column. The spinous process protrudes posteriorly and inferiorly from the vertebral arch. The transverse processes extend laterally from each side of a vertebra at the point where the lamina and pedicle join. The superior articular processes of a vertebra interlock with the inferior articular processes of the bone above.

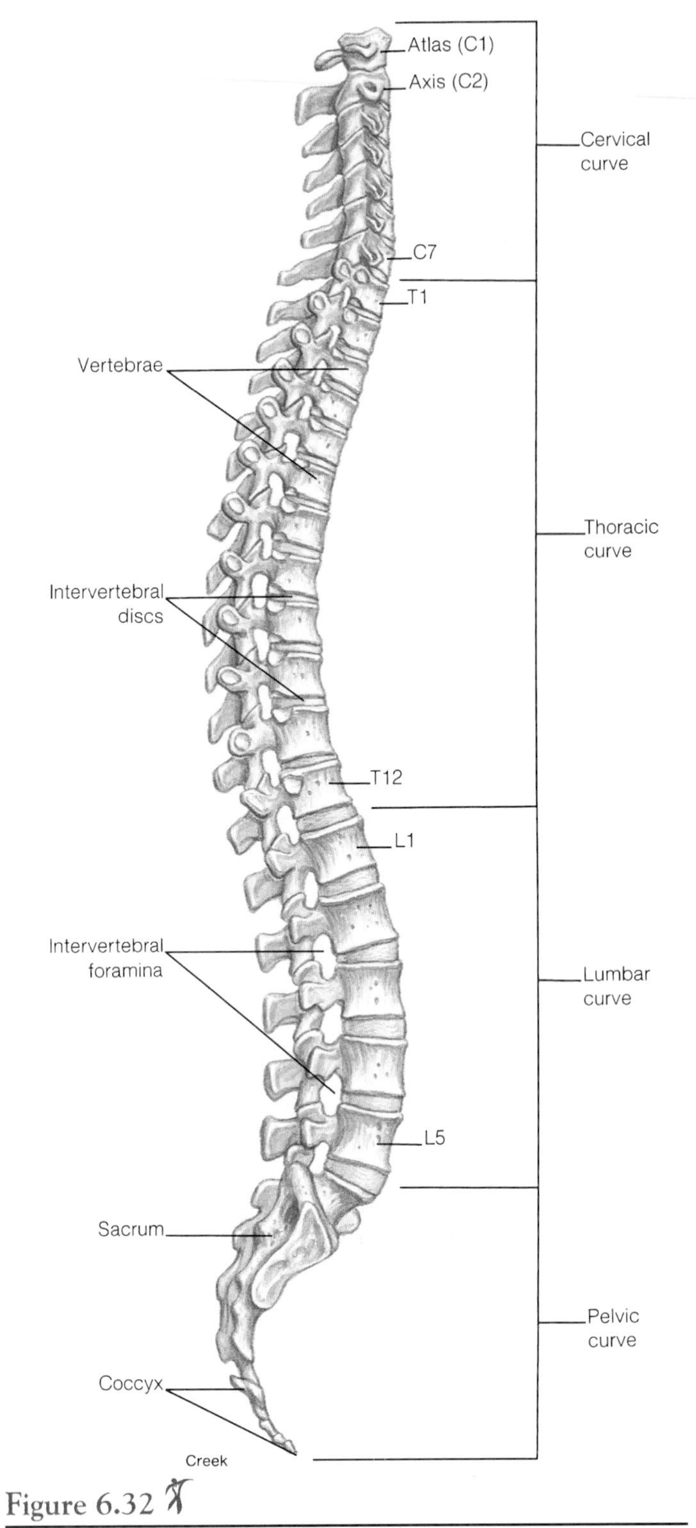

Figure 6.32

The vertebral column of an adult has four curves named according to the region in which they occur. The bodies of the vertebrae are separated by intervertebral discs, which allow flexibility.

pedicle: L. *pediculus*, small foot
lamina: L. *lamina*, thin layer

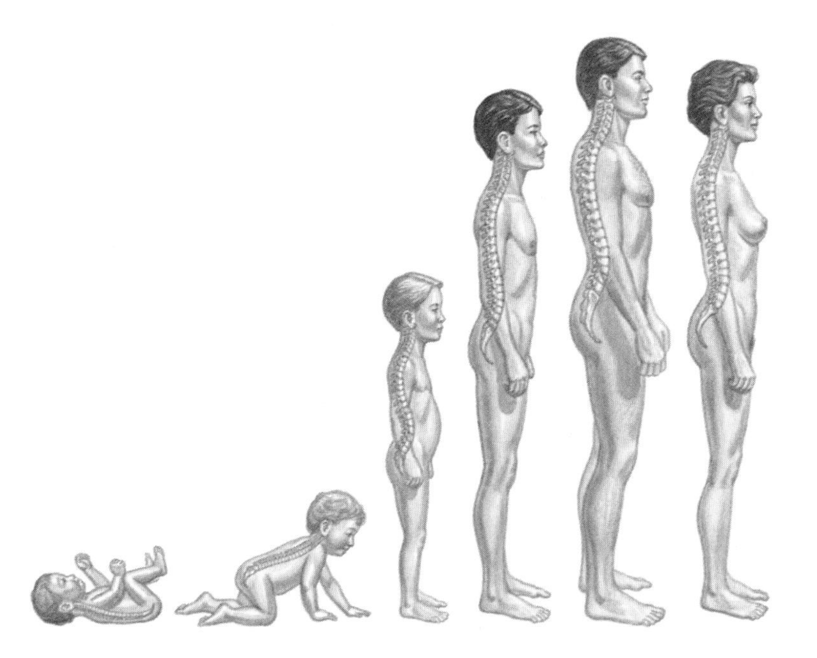

Figure 6.33

The development of the vertebral curves. An infant is born with the two primary curves but does not develop the secondary curves until it begins sitting upright and walking. (Note the differences in the curves between the sexes.)

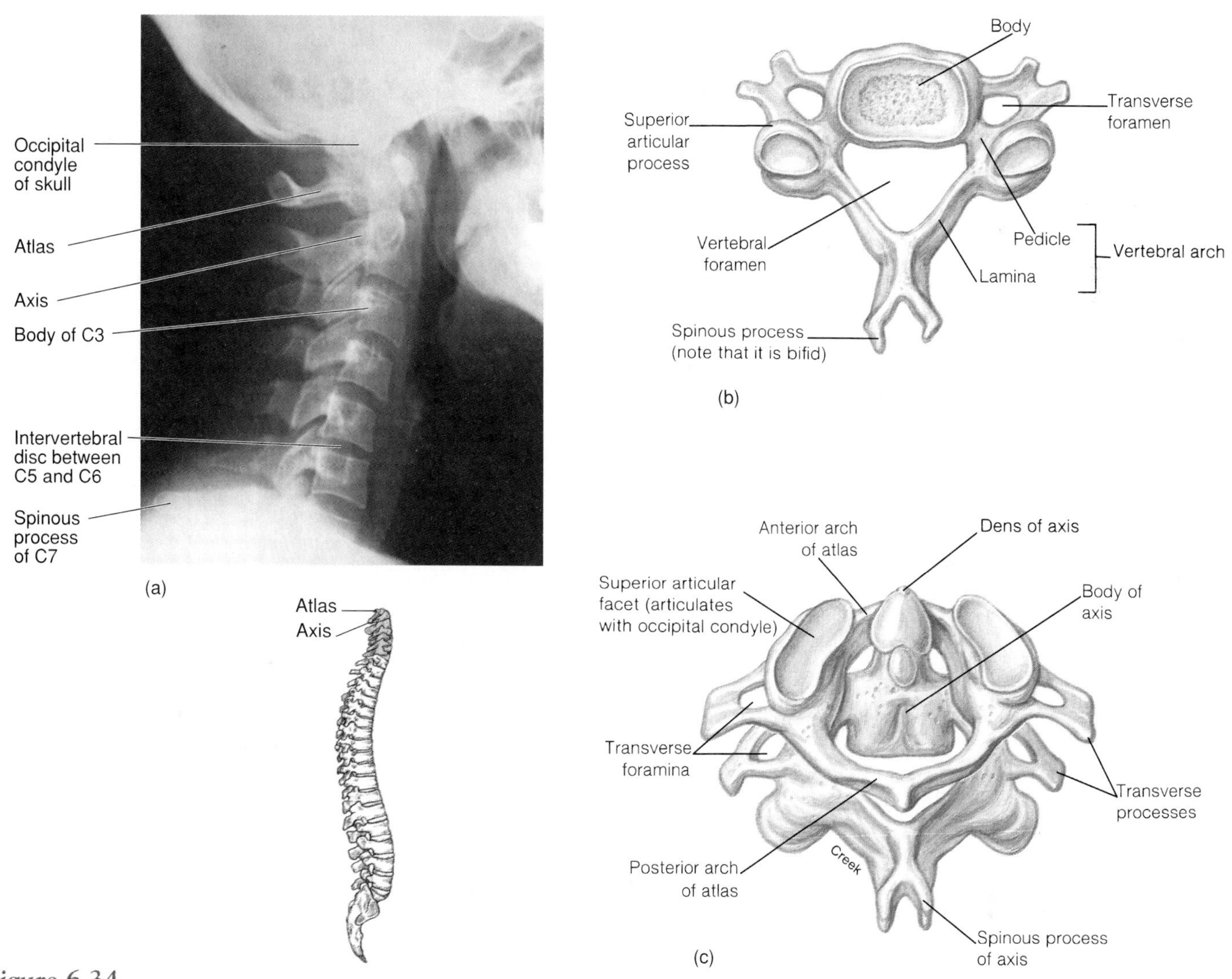

Figure 6.34

Cervical vertebrae. (*a*) A radiograph of the cervical region, (*b*) a superior view of a typical cervical vertebra, and (*c*) the articulated atlas and axis.

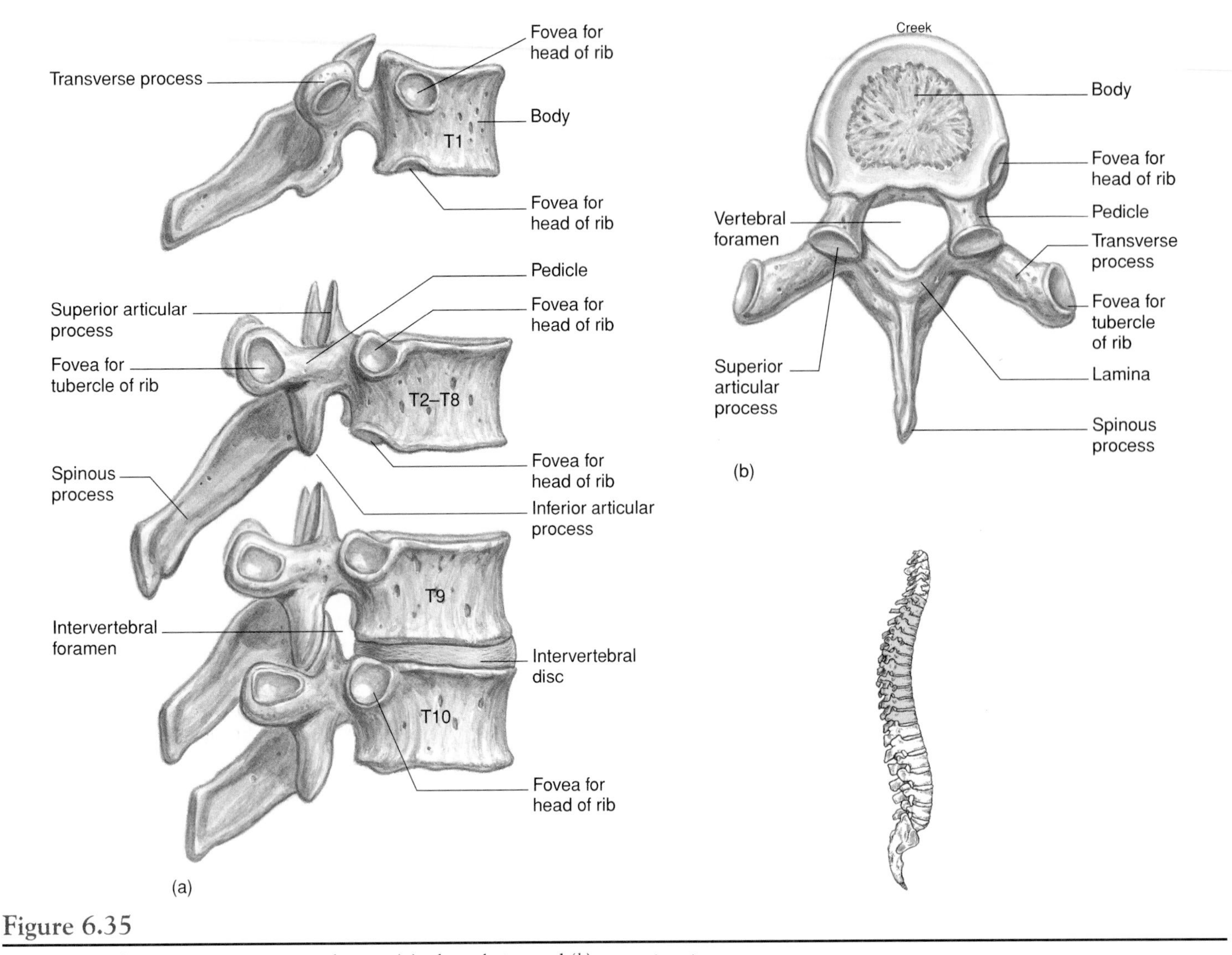

Figure 6.35

Thoracic vertebrae. Representative vertebrae in (*a*) a lateral view and (*b*) a superior view.

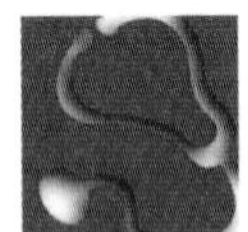

A *laminectomy* is the surgical removal of the spinous processes and their supporting vertebral laminae in a particular region of the vertebral column. A laminectomy may be performed to relieve pressure on the spinal cord or nerve root caused by a blood clot, a tumor, or a herniated (ruptured) disc. It may also be performed on a cadaver to expose the spinal cord and its surrounding meninges.

Regional Characteristics of Vertebrae

Cervical Vertebrae (7)

The seven cervical vertebrae form a flexible framework for the neck and support the head. The bone tissue of cervical vertebrae is more dense than that found in the other vertebral regions, and, except for those in the coccygeal region, the cervical vertebrae are smallest. Cervical vertebrae are distinguished by the presence of a **transverse foramen** in each transverse process (fig. 6.34). The vertebral arteries and veins pass through this opening as they contribute to the blood flow associated with the brain. Cervical vertebrae C2–C6 generally have a *bifid*, or notched, spinous process. The bifid spinous processes increase the surface area for attachment of the strong *nuchal ligament* that attaches to the back of the skull. The first cervical vertebra has no spinous process, and the process of C7 is not bifid and is larger than those of the other cervical vertebrae.

The **atlas** is the first cervical vertebra (sometimes called cervical 1 or C1). The atlas lacks a body, but it does have a short, rounded spinous process called the **posterior tubercle.** It also has cupped **superior articular surfaces** that articulate with the oval occipital condyles of the skull. This *atlanto-occipital joint* supports the skull and permits the nodding of the head in a "yes" movement.

atlas: from Gk. mythology, Atlas—the Titan who supported the heavens

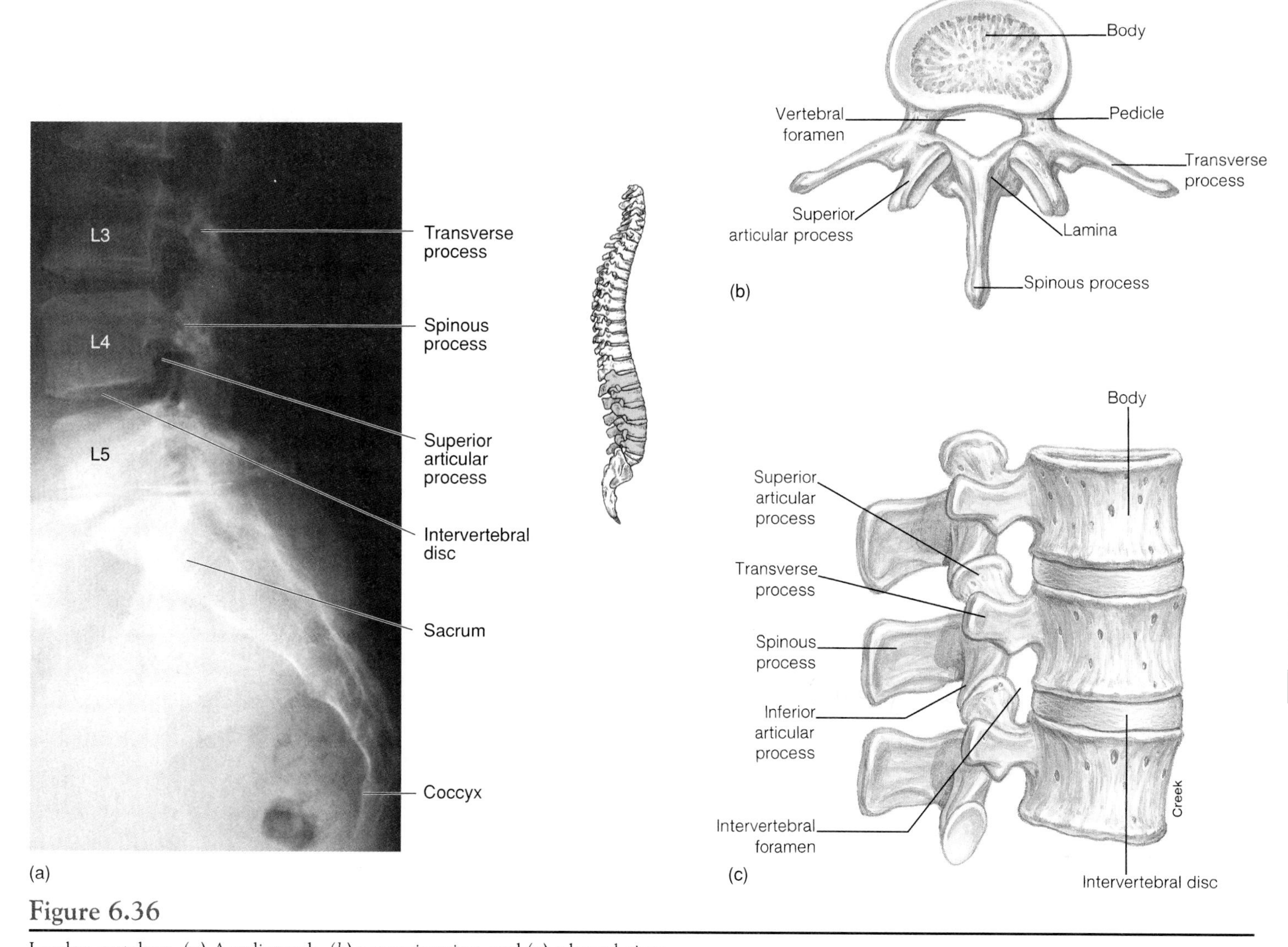

Figure 6.36

Lumbar vertebrae. (*a*) A radiograph, (*b*) a superior view, and (*c*) a lateral view.

The **axis** is the second cervical vertebra (C2). It has a peglike **dens** (*odontoid process*) for rotation with the atlas in turning the head from side to side, as in a "no" movement.

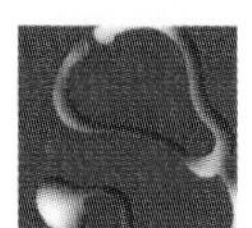

Whiplash is a common term for any injury to the neck. Muscle, bone, or ligament injury in this portion of the spinal column is relatively common in individuals involved in automobile accidents and sports injuries. Joint dislocation occurs commonly between the fourth and fifth or fifth and sixth cervical vertebrae, where neck movement is greatest. Bilateral dislocations are particularly dangerous because of the probability of spinal cord injury. Compression fractures of the first three cervical vertebrae are common and follow abrupt forced flexion of the neck. Fractures of this type may be extremely painful because of pinched spinal nerves.

Thoracic Vertebrae (12)

Twelve thoracic vertebrae articulate with the ribs to form the posterior anchor of the rib cage. Thoracic vertebrae are larger than cervical vertebrae and increase in size from superior (T1) to inferior (T12). Each thoracic vertebra has a long spinous process, which slopes obliquely downward, and **foveae** (*facets*) for articulation with the ribs (fig. 6.35).

Lumbar Vertebrae (5)

The five lumbar vertebrae are easily identified by their heavy bodies and thick, blunt spinous processes (fig. 6.36) for attachment of powerful back muscles. They are the largest vertebrae of the vertebral column. Their articular processes are also distinctive in that the facets of the superior pair are directed medially instead of posteriorly and the facets of the inferior pair are directed laterally instead of anteriorly.

---

axis: L. *axis*, axle
odontoid: Gk. *odontos*, tooth

lumbar: L. *lumbus*, loin

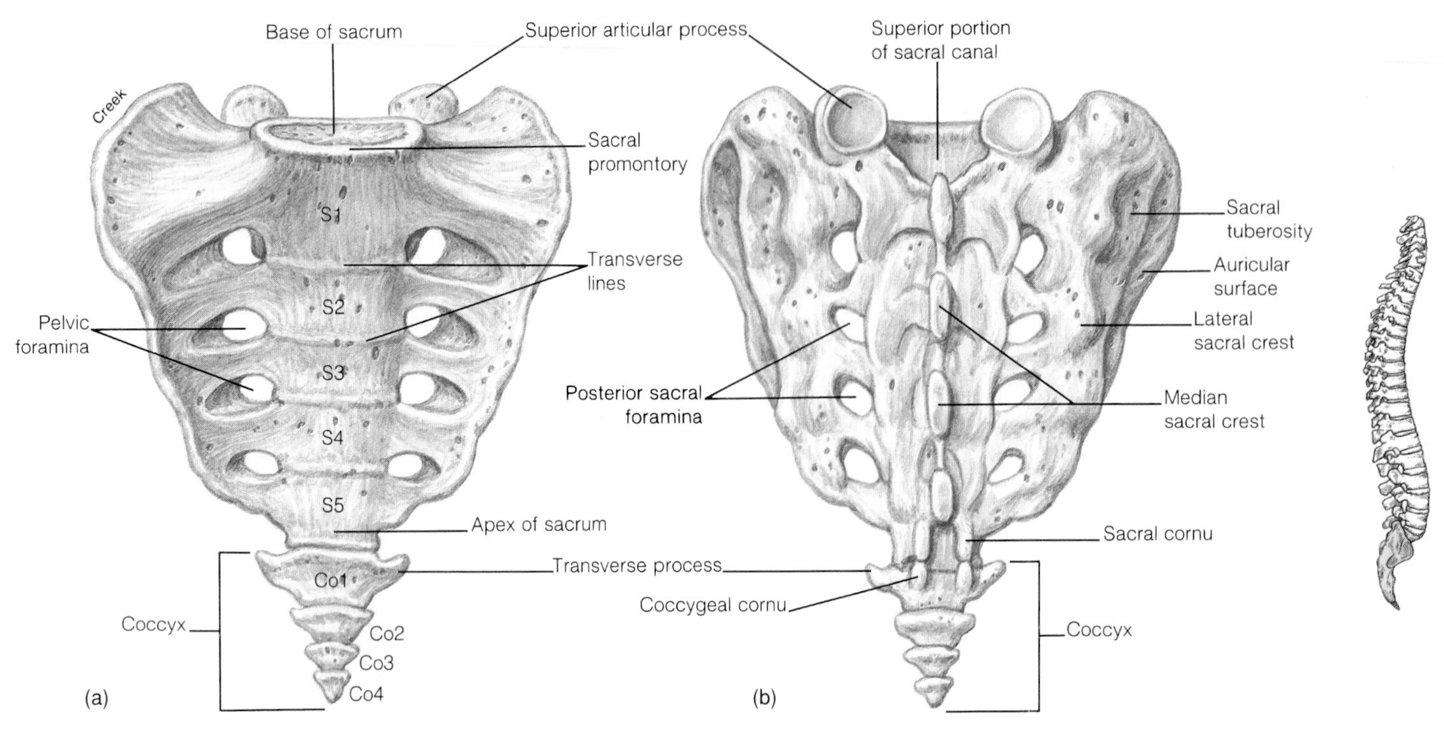

Figure 6.37

The sacrum and coccyx. (*a*) An anterior view and (*b*) a posterior view.

Sacrum

The wedge-shaped sacrum provides a strong foundation for the pelvic girdle. It consists of four or five sacral vertebrae (fig. 6.37) that become fused after age 26. The sacrum has an extensive **auricular surface** on each lateral side for the formation of a slightly movable **sacroiliac** (*sak"ro-il'e-ak*) **joint** with the ilium of the hip. A **median sacral crest** is formed along the posterior surface by the fusion of the spinous processes. **Posterior sacral foramina** on either side of the crest allow for the passage of nerves from the spinal cord. The **sacral canal** is the tubular cavity within the sacrum that is continuous with the vertebral canal. Paired **superior articular processes,** which articulate with the fifth lumbar vertebra, arise from the roughened **sacral tuberosity** along the posterior surface.

The smooth anterior surface of the sacrum forms the posterior surface of the pelvic cavity. It has four **transverse lines** denoting the fusion of the vertebral bodies. At the ends of these lines are the paired **pelvic foramina (anterior sacral foramina).** The superior border of the anterior surface of the sacrum, called the **sacral promontory** (*prom'on-tor"e*), is an important obstetric landmark for pelvic measurements.

Coccyx

The triangular coccyx ("tailbone") is composed of three to five fused coccygeal vertebrae. The first vertebra of the fused coccyx has two long **coccygeal cornua,** which are attached by ligaments to the sacrum (fig. 6.37). Lateral to the cornua are the transverse processes.

---

sacrum: L. *sacris*, sacred
coccyx: Gk. *kokkyx*, like a cuckoo's beak

Distinct losses in height occur during middle and old age. Between the ages of 50 and 55, there is a decrease of 0.5 to 2.0 cm (0.25 to 0.75 in.) because of compression and shrinkage of the intervertebral discs. Elderly individuals may suffer a further loss of height because of osteoporosis (see Clinical Considerations at the end of this chapter).

The regions of the vertebral column are summarized in table 6.7.

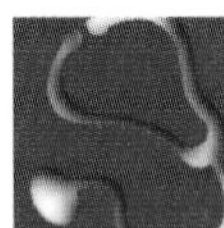

When a person sits, the coccyx flexes anteriorly, acting as a shock absorber. An abrupt fall on the coccyx, however, may cause a painful subperiosteal bruising, fracture, or fracture-dislocation of the sacrococcygeal joint. An especially difficult childbirth can even injure the coccyx of the mother. Coccygeal trauma is painful and may require months to heal.

1. Which are the primary curves of the vertebral column and which are the secondary curves? Describe the characteristic curves of each region.
2. What is the function of the transverse foramina of the cervical vertebrae?
3. Describe the diagnostic differences between a thoracic and a lumbar vertebra. Which structures are similar and could therefore be characteristic of a typical vertebra?

Table 6.7
Regions of the Vertebral Column

Region	Number of Bones	Diagnostic Features
Cervical	7	Transverse foramina; superior facets of atlas articulate with occipital condyle; dens of axis; spinous processes of third through sixth vertebrae are generally bifid
Thoracic	12	Long spinous processes that slope obliquely downward; fovea for articulation with ribs
Lumbar	5	Large bodies; prominent transverse processes; short, thick spinous processes
Sacrum	4 or 5 fused vertebrae	Extensive auricular surface; median sacral crest; posterior sacral foramina; sacral promontory; sacral canal
Coccyx	3 to 5 fused vertebrae	Small and triangular; coccygeal cornua

Rib Cage

The cone-shaped, flexible rib cage consists of the thoracic vertebrae, 12 paired ribs, costal cartilages, and the sternum. It encloses and protects the thoracic viscera and is directly involved in the mechanics of breathing.

Objective 14 Identify the parts of the rib cage and compare and contrast the various types of ribs.

The *sternum* ("breastbone"), *ribs*, *costal cartilages*, and the previously described thoracic vertebrae form the **rib cage** (fig. 6.38). The rib cage is anteroposteriorly compressed and more narrow superiorly than inferiorly. It supports the pectoral girdle and upper extremities, protects and supports the thoracic and upper abdominal viscera, and plays a major role in breathing (see fig. 9.21). Certain bones of the rib cage contain active sites in the bone marrow for the production of red blood cells.

Sternum

The **sternum** is an elongated, flattened bony plate consisting of three separate bones: the upper **manubrium,** (*mă-noo′bre-um*), the central **body,** and the lower **xiphoid** (*zif′oid; zi′foid*) **process.** On the lateral sides of the sternum are **costal notches** where the costal cartilages attach. A **jugular notch** is formed at the superior end of the manubrium, and a **clavicular notch** for articulation with the clavicle is present on both sides of the sternal notch. The manubrium articulates with the costal cartilages of the first and second ribs. The body of the sternum attaches to the costal cartilages of the second through the tenth ribs. The xiphoid process does not attach to ribs but is an attachment for abdominal muscles. The costal cartilages of the eighth, ninth, and tenth ribs fuse to form the **costal margin** of the rib cage. A **costal angle** is formed where the two costal margins come together at the xiphoid process. The **sternal angle** (angle of Louis) may be palpated as an elevation between the manubrium and body of the sternum at the level of the second rib (fig. 6.38). The costal angle, costal margins, and sternal angle are important surface landmarks of the thorax and abdomen (see figs. 10.19 and 10.20).

Ribs

Embedded in the muscles of the body wall are twelve pairs of **ribs,** each pair attached posteriorly to a thoracic vertebra. Anteriorly, the first seven pairs are anchored to the sternum by individual *costal cartilages;* these ribs are called **true ribs.** The remaining five pairs (ribs 8, 9, 10, 11, and 12) are termed **false ribs.** Because the last two pairs of false ribs do not attach to the sternum at all, they are referred to as **floating ribs.**

Although the ribs vary structurally, each of the first ten pairs has a **head** and a **tubercle** for articulation with a vertebra. The last two have a head but no tubercle. In addition, each of the twelve pairs has a **neck, angle,** and **body** (fig. 6.39). The head projects posteriorly and articulates with the body of a thoracic vertebra (fig. 6.40). The tubercle is a knoblike process, just lateral to the head. It articulates with the fovea on the transverse process of a thoracic vertebra. The neck is the constricted area between the head and the tubercle. The body is the curved main part of the rib. Along the inner surface of the body is a depressed canal called the **costal groove** that protects the costal vessels and nerve. Spaces between the ribs are called **intercostal spaces** and are occupied by the intercostal muscles.

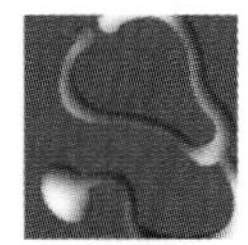

Fractures of the ribs are relatively common, and most frequently occur between ribs 3 and 10. The first two pairs of ribs are protected by the clavicles; the last two pairs move freely and will give with an impact. Little can be done to assist the healing of broken ribs other than binding them tightly to limit movement.

1. Describe the rib cage and list its functions. What determines whether a rib is true, false, or floating?
2. Explain how the costal margin and costal angle are formed.

sternum: Gk. *sternon*, chest
manubrium: L. *manubrium*, a handle
xiphoid: Gk. *xiphos*, sword
costal: L. *costa*, rib

angle of Louis: from Pierre C. A. Lewis, French physician, 1787–1872

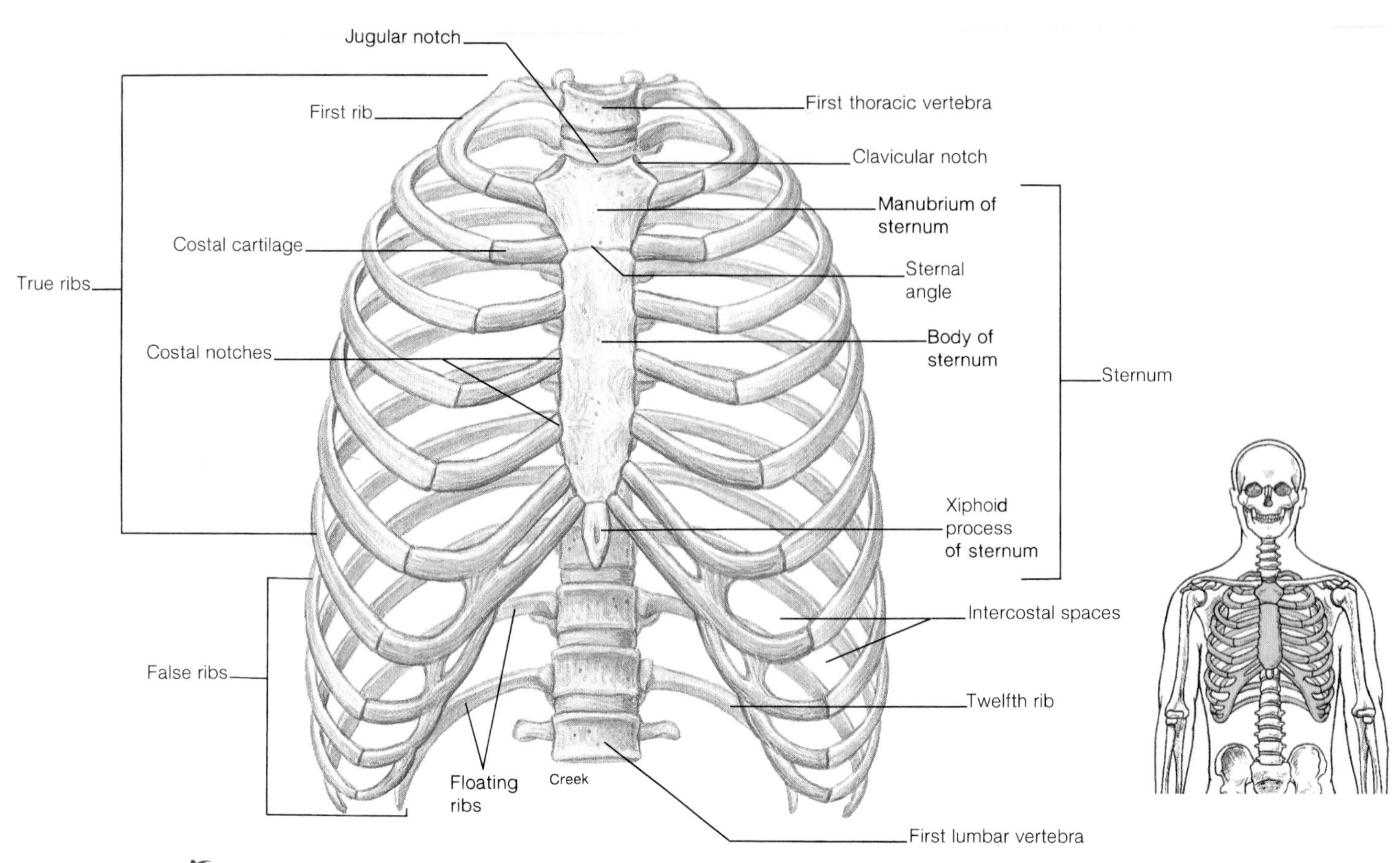

Figure 6.38

The rib cage.

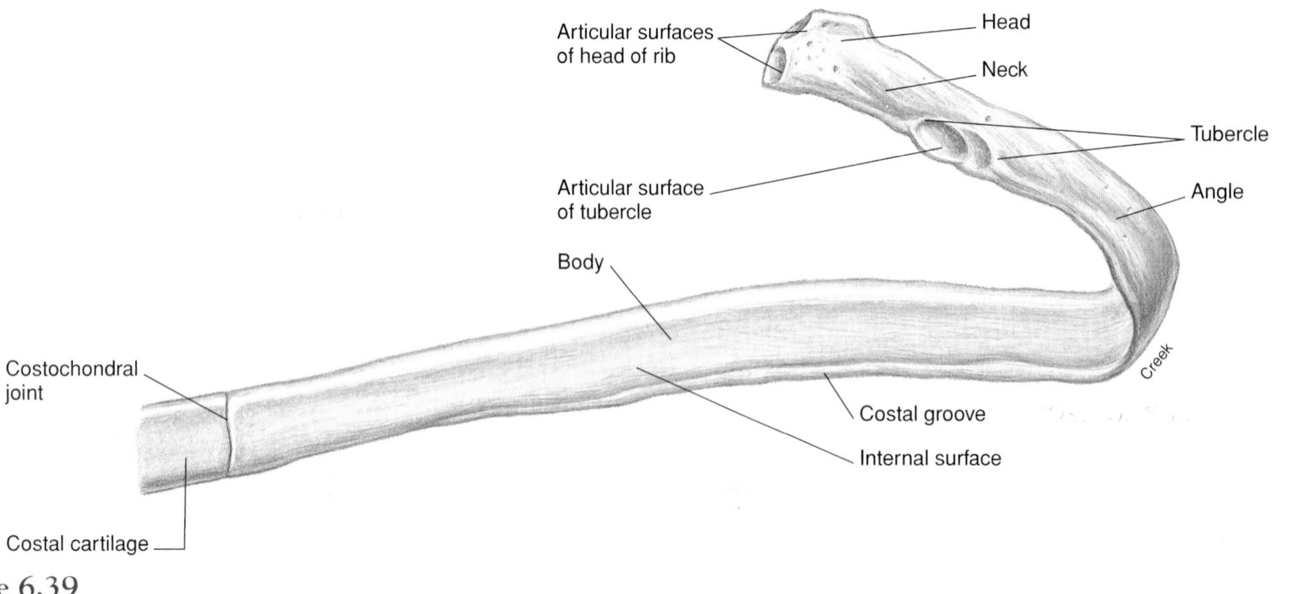

Figure 6.39

The structure of a rib.

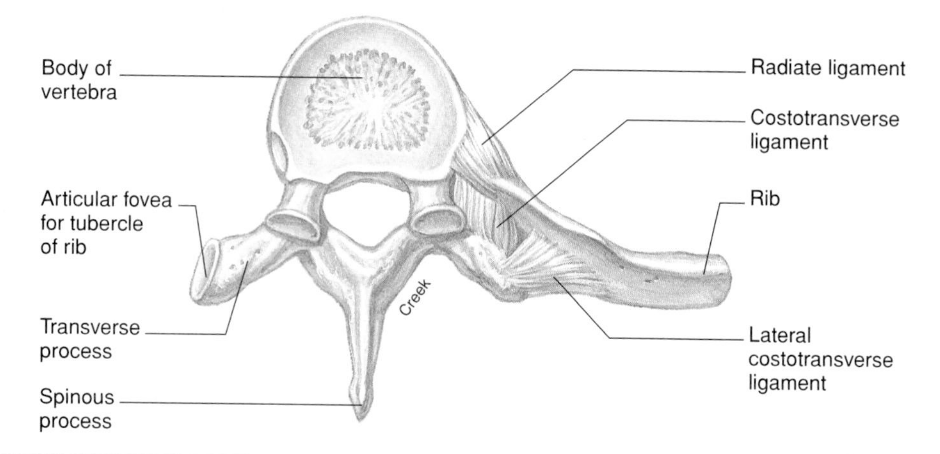

Figure 6.40

Articulation of a rib with a thoracic vertebra as seen in a superior view.

Clinical Considerations

Each bone is a dynamic living organ that is influenced by hormones, diet, aging, and disease. Since the development of bone is genetically controlled, congenital abnormalities may occur. The hardness of bones gives them strength, yet they lack the resiliency to avoid fracture when they undergo severe trauma. (Fractures are discussed in chapter 7 and joint injuries are discussed in chapter 8.) All of these aspects of bone make for some important and interesting clinical considerations.

Developmental Disorders

Congenital malformations account for several types of skeletal deformities. Certain bones may fail to form during osteogenesis, or they may form abnormally. **Cleft palate** and **cleft lip** are malformations of the palate and face. They vary in severity and seem to involve both genetic and environmental factors. **Spina bifida** (*spī′nă bif′ĭ-dă*) is a congenital defect of the vertebral column resulting from a failure of the laminae of the vertebrae to fuse, leaving the spinal cord exposed (fig. 6.41). The lumbar area is most likely to be affected, and frequently only a single vertebra is involved.

Nutritional and Hormonal Disorders

Several bone disorders result from nutritional deficiencies or from excessive or deficient amounts of the hormones that regulate bone development and growth. Vitamin D has a tremendous influence on bone structure and function. When there is a deficiency of this vitamin, the body is unable to metabolize calcium and phosphorus. Vitamin D deficiency in children causes **rickets.** The bones of a child with rickets remain soft and structurally weak, and bend under the weight of the body (see fig. 5.10).

A vitamin D deficiency in the adult causes the bones to give up stored calcium and phosphorus. This demineralization results in a condition called **osteomalacia** (*os″te-o-mă-la′sha*).

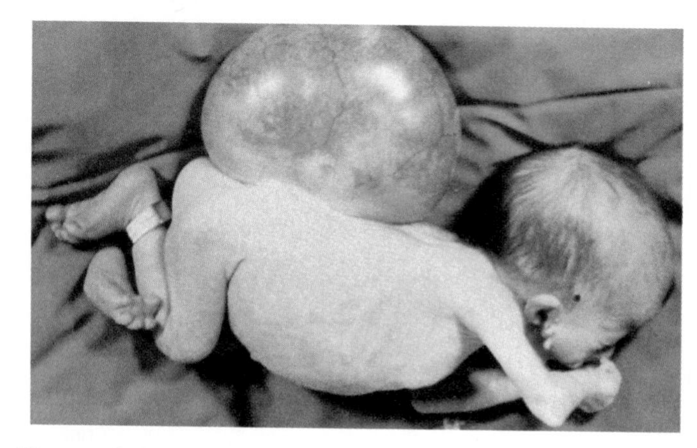

Figure 6.41

In spina bifida, failure of the vertebral arches to fuse permits a herniation of the meninges that cover the spinal cord through the vertebral column. This results in a condition called meningomyelocele.

Osteomalacia occurs most often in malnourished women who have repeated pregnancies and who experience relatively little exposure to sunlight. It is marked by increasing softness of the bones, so that they become flexible and thus cause deformities.

The consequences of endocrine disorders are described in chapter 14. Since hormones exert a strong influence on bone development, however, a few endocrine disorders will be briefly mentioned here. Hypersecretion of growth hormone from the pituitary gland leads to **gigantism** in young people if it begins before ossification of their epiphyseal plates. In adults, it leads to **acromegaly** (*ak″ro-meg′ă-le*), which is characterized by hypertrophy of the bones of the face, hands, and feet. In a child, growth hormone deficiency results in slowed bone growth—a condition called **dwarfism.**

Paget's disease, a bone disorder that affects mainly older adults, occurs more frequently in males than in females. It is characterized by disorganized metabolic processes within bone tissue. The activity of osteoblasts and osteoclasts becomes irregular,

Paget's disease: from Sir James Paget, English surgeon, 1814–99

Developmental Exposition

The Axial Skeleton

Exhibit I Ossification centers of the skeleton of a 10-week-old fetus. (*a*) The diagram depicts endochondrial ossification in red and intramembranous ossification in a stippled pattern. Tlhe cartilaginous portions of the skeleton are shown in gray. (*b*) The photograph shows the ossification centers stained with a red indicator dye.

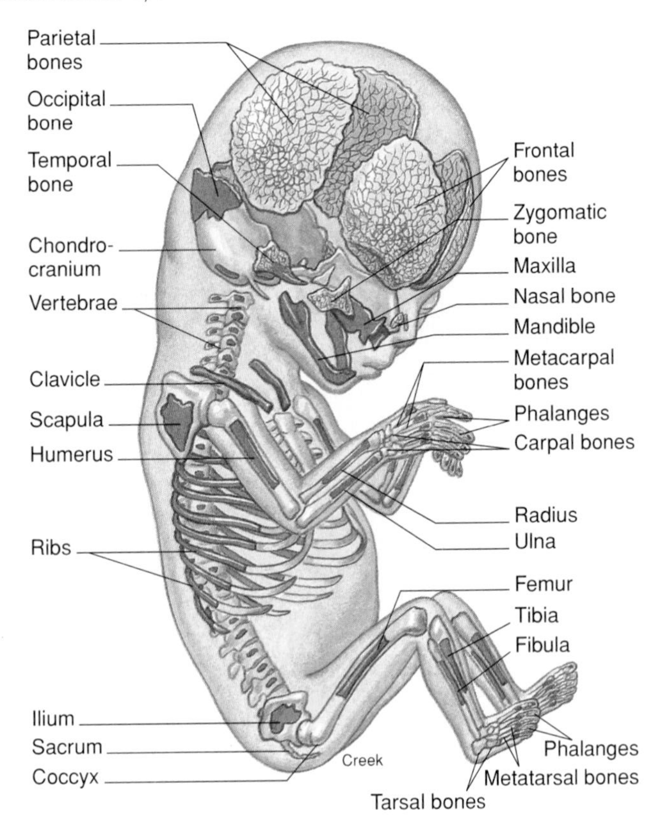

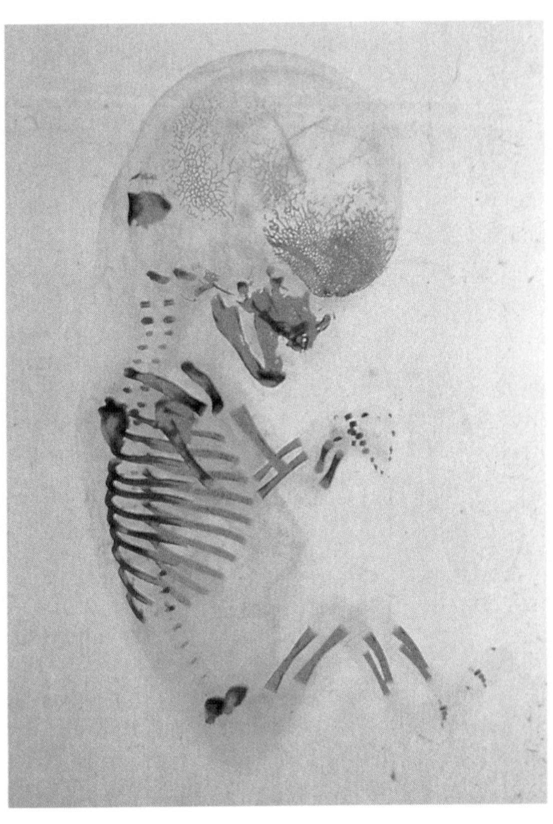

Explanation

Development of Bone

Bone formation, or *ossification,* begins at about the fourth week of embryonic development, but ossification centers cannot be readily observed until about the tenth week (exhibit I). Bone tissue derives from specialized migratory cells of mesoderm (see chapter 6) known as *mesenchyme.* Some of the embryonic mesenchymal cells will transform into *chondroblasts* (*kon′dro-blasts*) and develop a cartilage matrix that is later replaced by bone in a process known as **endochondral** (*en″dŏ-kon′dral*) **ossification.** Most of the skeleton is formed in this fashion—first it goes through a hyaline cartilage stage and then it is ossified as bone.

chondroblast: Gk. *chondros,* cartilage; *blastos,* offspring or germ

resulting in thick bony deposits in some areas of the skeleton and fragile, thin bones in other areas. The vertebral column, pelvis, femur, and skull are most often involved, and become increasingly painful and deformed. Bowed leg bones, abnormal curvature of the spine, and enlargement of the skull may develop. The cause of Paget's disease is currently not known.

Neoplasms of Bone

Malignant bone tumors are three times more common than benign tumors. Pain is the usual symptom of either type of osseous neoplasm, although benign tumors may not have accompanying pain.

Exhibit II The embryonic skull at 12 weeks is composed of bony elements from three developmental sources: the chondrocranium (colored blue-gray), the neurocranium (colored light yellow), and the viscerocranium (colored salmon).

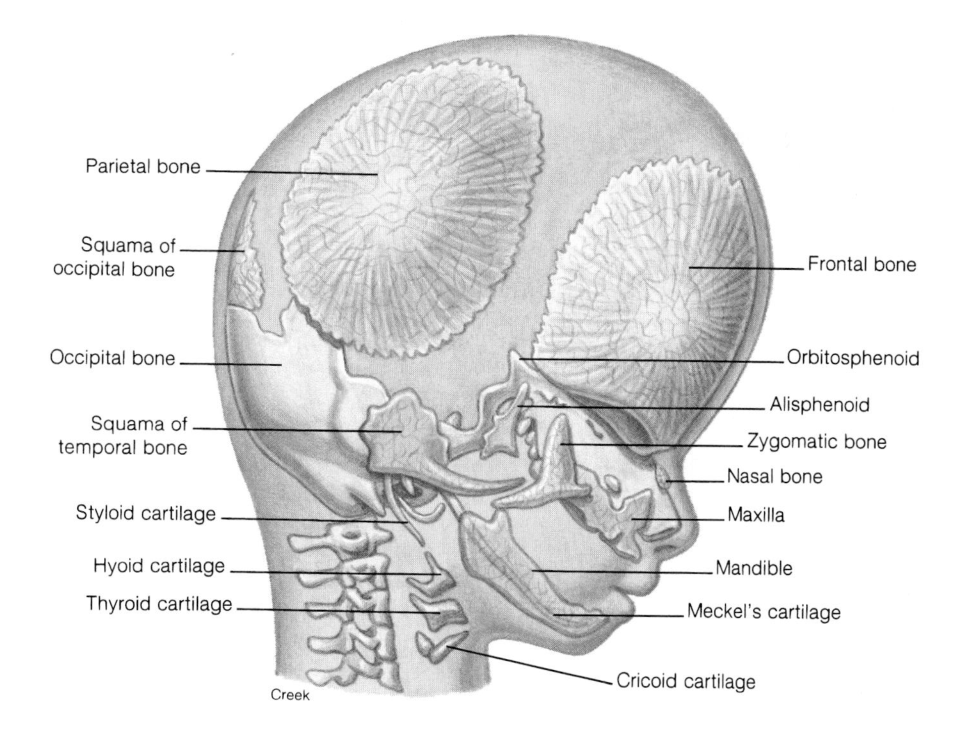

A smaller number of mesenchymal cells develop into bone directly, without first going through a cartilage stage. This type of bone-formation process is referred to as **intramembranous** (*in″tră-mem′bră-nus*) **ossification.** The clavicles, facial bones, and certain bones of the cranium are formed this way. *Sesamoid bones* are specialized intramembranous bones that develop in tendons. The patella is an example of a sesamoid bone.

Development of the Skull

The formation of the skull is a complex process that begins during the fourth week of embryonic development and continues well beyond the birth of the baby. Three aspects of the embryonic skull are involved in this process: the chondrocranium, the neurocranium, and the viscerocranium (exhibit II). The **chondrocranium** is the portion of the skull that undergoes endochondral ossification to form the bones supporting the brain. The **neurocranium** is the portion of the skull that develops through membranous ossification to form the bones covering the brain and facial region. The **viscerocranium** (splanchnocranium) is the portion that develops from the embryonic visceral arches to form the mandible, auditory ossicles, the hyoid bone, and specific processes of the skull.

chondrocranium: Gk. *chondros*, cartilage; *kranion*, skull
viscerocranium: L. *viscera*, soft parts; Gk. *kranion*, skull

Two types of benign bone tumors are **osteomas,** which are the more frequent and which often involve the skull, and **osteoid osteomas,** which are painful neoplasms of the long bones, usually in children.

Osteogenic sarcoma (*sar-ko′mă*) is the most virulent type of bone cancer. It frequently metastasizes through the blood to the lungs. This disease usually originates in the long bones and is accompanied by aching and persistent pain.

A **bone scan** (fig. 6.42) is a diagnostic procedure frequently done on a person who has had a malignancy elsewhere in the body that may have metastasized to the bone. The patient receiving a bone scan may be injected with a radioactive substance

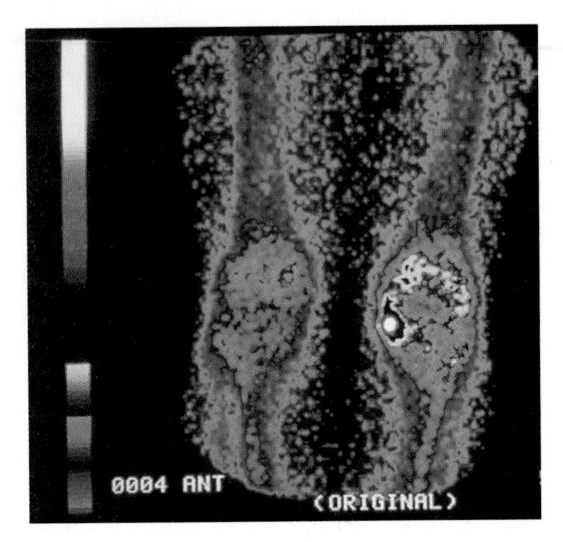

Figure 6.42

A bone scan of the legs of a patient suffering from arthritis in the left knee joint. In a bone scan, an image of an arthritic joint shows up lighter than most of a normal joint.

that accumulates more rapidly in malignant tissue than in normal tissue. Entire body radiographs show malignant bone areas as intensely dark dots.

Aging of the Skeletal System

Senescence affects the skeletal system by decreasing skeletal mass and density and increasing porosity and erosion (fig. 6.43). Bones become more brittle and susceptible to fracture. Articulating surfaces also deteriorate, contributing to arthritic conditions. Arthritic diseases are second to heart disease as the most common debilitation in the elderly.

Osteoporosis (*os″te-o-pŏ-ro′sis*) is a weakening of the bones, primarily as a result of calcium loss. The causes of osteoporosis include aging, inactivity, poor diet, and an imbalance in hormones or other chemicals in the blood. It is most common in older women because low levels of estrogens after menopause lead to increased bone resorption, and the formation of new bone is not sufficient to keep pace. People with osteoporosis are prone to bone fracture, particularly at the pelvic girdle and vertebrae, as the bones become too brittle to support the weight of the body. Complications of hip fractures often lead to permanent disability, and vertebral compression fractures may produce a permanent curved deformity of the spine.

Although there is no known cure for osteoporosis, good eating habits and a regular program of exercise, established at an early age and continued throughout adulthood, can minimize its effects. Treatment in women through dietary calcium, exercise, and estrogens has had limited positive results. In addition, a drug called *alendronate* (Fosamax), approved by the FDA in 1995, has been shown to be effective in managing osteoporosis. This drug works without hormones to block osteoclast activity, making it useful for women who choose not to be treated with estrogen replacement therapy.

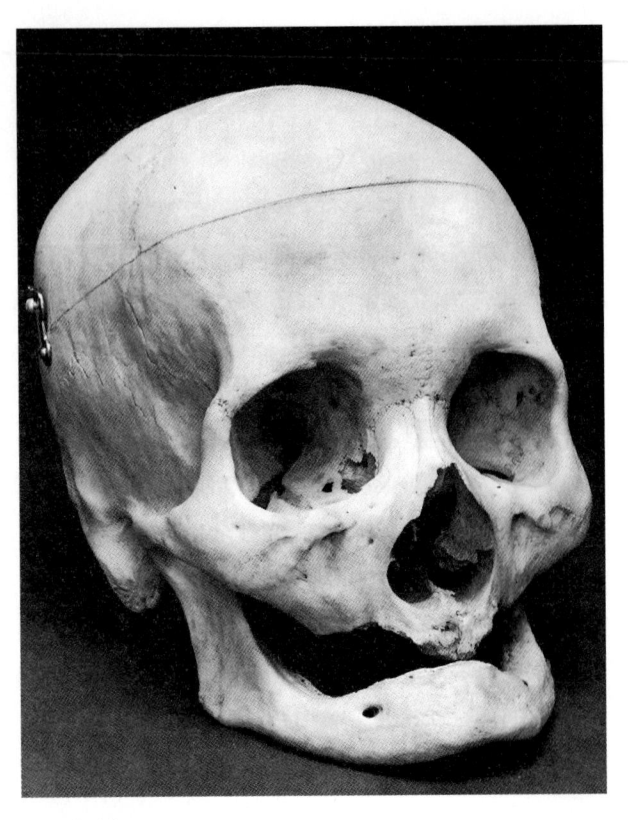

Figure 6.43

A geriatric skull. Note the loss of teeth and the degeneration of bone, particularly in the facial region.

Clinical Case Study Answer

The height (overall length) of the vertebral column is equal to the sum of the thicknesses of the vertebrae plus the sum of the thicknesses of the intervertebral discs. The body of a vertebra consists of outer compact bone and inner spongy bone. An intervertebral disc consists of a fibrocartilage sheath called the *anulus fibrosus* and a mucoid center portion called the *nucleus pulposus*. The intervertebral discs generally change their anatomical configuration as one ages. In early adulthood, the nucleus pulposus is spongy and moist. With advanced age, however, it desiccates, resulting in a flattening of the intervertebral disc. Collectively, the intervertebral discs account for 25% of the height of the vertebral column. As they flatten with age, there is a gradual decrease in a person's overall height. Height loss may also result from undetectable compression fractures of the vertebral bodies, which are common in elderly people. This phenomenon, however, is considered pathological and is not an aspect of the normal aging process. In a person with osteoporosis, there is often a marked decrease in height and perhaps more serious clinical problems as well, such as compression of spinal nerves.

Internal Affairs

How the Skeletal System Works with Other Body Systems
How Other Systems Work with the Skeletal System

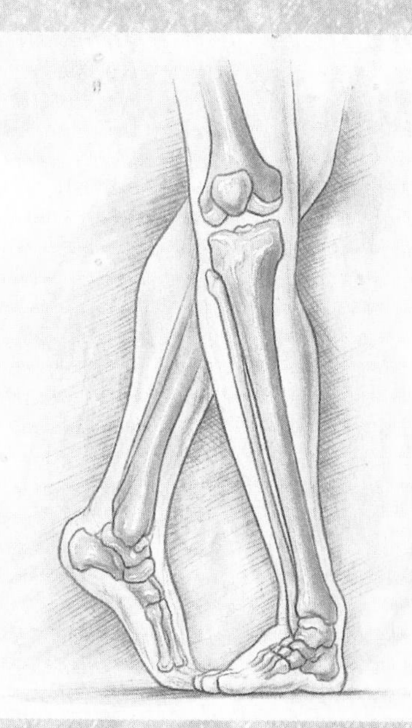

Integumentary System
- Provides skin with physical support
- Initiates synthesis of vitamin D needed for absorption of calcium and phosphorus

Muscular System
- Provides attachment sites for muscles
- Source of calcium for muscle contraction
- Causes bones to move at joints
- Partially responsible for bone shape and strength

Nervous System
- Protects central nervous system with bony encasement
- Source of calcium for neural function
- Sensory receptors provide sensations of body position and pain from bones and joints

Endocrine System
- Protects endocrine glands in head and pelvis
- Source of calcium for production of certain hormones
- Hormonally controls bone growth and maintenance

Circulatory System
- Bone marrow produces blood cells
- Source of calcium for cardiac muscle contraction
- Transports O_2 and CO_2, nutrients, and hormones to and from bone tissue

Lymphatic System
- Bone marrow produces and stores lymphocytes and other cells of immune system
- Maintains balanced amount of interstitial fluid within bone tissue
- Lymphocytes protect bone tissue following trauma

Respiratory System
- Forms respiratory passageway through nasal cavity
- Protects lungs and aids in ventilation
- Provides O_2 and eliminates CO_2

Digestive System
- Provides organs of GI tract with physical support and protection
- Stores minerals
- Provides nutrients for growth, maintenance, and repair of bone tissue

Urinary System
- Provides organs of urinary system with physical support and protection
- Eliminates metabolic wastes
- Activates vitamin D

Reproductive System
- Provides organs of reproductive system with physical support and protection
- Gonads produce sex hormones that promote growth and development, and maintain bone tissue

Important Clinical Terminology

achondroplasia (*ă-kon″dro-pla′ze-ă*) A genetic defect that inhibits formation of cartilaginous bone during fetal development.
craniotomy Surgical cutting into the cranium to provide access to the brain.
epiphysiolysis (*ep″ĭ-fiz″e-ol′ĭ-sis*) A separation of the epiphysis from the diaphysis of a growing long bone.
laminectomy The surgical removal of the posterior arch of a vertebra, usually to repair a herniated intervertebral disc.
orthopedics The branch of medicine concerned with the diagnosis and treatment of trauma, diseases, and abnormalities involving the skeletal and muscular systems.
osteitis (*os-te-i′tis*) An inflammation of bone tissue.
osteoblastoma (*os″te-o-blas-to′mă*) A benign tumor produced from bone-forming cells, most frequently in the vertebrae of young children.
osteochondritis (*os″te-o-kon-dri′tis*) An inflammation of bone and cartilage tissues.
osteomyelitis (*os″te-o-mi″ĕ-li′tis*) An inflammation of bone marrow caused by bacteria or fungi.
osteonecrosis (*os″te-o-nĕ-kro′sis*) The death of bone tissue, usually caused by obstructed arteries.
osteopathology The study of bone diseases.
osteosarcoma A malignant tumor of bone tissue.
osteotomy (*os″te-ot′ŏ-me*) The cutting of a bone, usually by means of a saw or a chisel.

Chapter Summary

Organization of the Skeletal System (pp. 129–131)

1. The axial skeleton consists of the skull, auditory ossicles, hyoid bone, vertebral column, and rib cage.
2. The appendicular skeleton consists of the bones within the pectoral girdle, upper extremities, pelvic girdle, and lower extremities.

Functions of the Skeletal System (pp. 131–132)

1. The mechanical functions of bones include the support and protection of softer body tissues and organs. In addition, certain bones function as levers during body movement.
2. The metabolic functions of bones include hemopoiesis and mineral storage.

Bone Structure (pp. 132–135)

1. Bone structure includes the shape and surface features of each bone, along with gross internal components.
2. Bones may be structurally classified as long, short, flat, or irregular.
3. The surface features of bones are classified as articulating surfaces, nonarticulating prominences, and depressions and openings.
4. A typical long bone has a diaphysis, or shaft, filled with marrow in the medullary cavity; epiphyses; epiphyseal plates for linear growth; and a covering of periosteum for appositional growth and the attachments of ligaments and tendons.

Bone Tissue (pp. 135–136)

1. Compact bone is the dense outer portion; spongy bone is the porous, vascular inner portion.
2. The five types of bone cells are osteogenic cells, in contact with the endosteum and periosteum; osteoblasts (bone-forming cells); osteocytes (mature bone cells); osteoclasts (bone-destroying cells); and bone-lining cells, along the surface of most bones.
3. In compact bone, the lamellae of osteons are the layers of inorganic matrix surrounding a central canal. Osteocytes are mature bone cells, located within capsules called lacunae.

Bone Growth (pp. 137–139)

1. Bone growth is an orderly process determined by genetics, diet, and hormones.
2. Most bones develop through endochondral ossification.
3. Bone remodeling is a continual process that involves osteoclasts in bone resorption and osteoblasts in the formation of new bone tissue.

Skull (pp. 139–153)

1. The eight cranial bones include the frontal (1), parietals (2), temporals (2), occipital (1), sphenoid (1), and ethmoid (1).
   (a) The cranium encloses and protects the brain and provides for the attachment of muscles.
   (b) Sutures are fibrous joints between cranial bones.
2. The 14 facial bones include the nasals (2), maxillae (2), zygomatics (2), mandible (1), lacrimals (2), palatines (2), inferior nasal conchae (2), and vomer (1).
   (a) The facial bones form the basic shape of the face, support the teeth, and provide for the attachment of the facial muscles.
   (b) The hyoid bone is located in the neck, between the mandible and the larynx.
   (c) The auditory ossicles (malleus, incus, and stapes) are located within each middle-ear chamber of the petrous part of the temporal bone.

Vertebral Column (pp. 153–158)

1. The vertebral column consists of 7 cervical, 12 thoracic, 5 lumbar, 4 or 5 fused sacral, and 3 to 5 fused coccygeal vertebrae.
2. Cervical vertebrae have transverse foramina; thoracic vertebrae have fovea for articulation with ribs; lumbar vertebrae have large bodies; sacral vertebrae are triangularly fused and contribute to the pelvic girdle; and the coccygeal vertebrae form a small triangular bone.

Rib Cage (pp. 159–160)

1. The sternum consists of a manubrium, body, and xiphoid process.
2. There are seven pairs of true ribs and five pairs of false ribs. The inferior two pairs of false ribs (pairs 11 and 12) are called floating ribs.

Review Activities

Objective Questions

1. A bone is considered to be
   (a) a tissue. (c) an organ.
   (b) a cell. (d) a system.
2. Which of the following statements is *false?*
   (a) Bones are important in the synthesis of vitamin D.
   (b) Bones and teeth contain about 99% of the body's calcium.
   (c) Red bone marrow is the primary site for hemopoiesis.
   (d) Most bones develop through endochondral ossification.

3. Match each of the following foramina with the bone in which it occurs.

1. foramen rotundum	(a) ethmoid bone
2. mental foramen	(b) occipital bone
3. carotid canal	(c) sphenoid bone
4. cribriform foramina	(d) mandible
5. foramen magnum	(e) temporal bone

4. With respect to the hard palate, which of the following statements is *false?*
   (a) It is composed of two maxillae and two palatine bones.
   (b) It separates the oral cavity from the nasal cavity.
   (c) The mandible articulates with the posterolateral angles of the hard palate.
   (d) The median palatine suture, incisive foramen, and greater palatine foramina are three of its structural features.
5. The location of the sella turcica is immediately
   (a) superior to the sphenoidal sinus.
   (b) inferior to the frontal sinus.
   (c) medial to the petrous parts of the temporal bones.
   (d) superior to the perpendicular plate of the ethmoid bone.
6. Specialized bone cells that enzymatically reabsorb bone tissue are
   (a) osteoblasts. (c) osteons.
   (b) osteocytes. (d) osteoclasts.
7. The mandibular fossa is located in which structural part of the temporal bone?
   (a) the squamous part
   (b) the tympanic part
   (c) the mastoid part
   (d) the petrous part
8. The crista galli is a structural feature of which bone?
   (a) the sphenoid bone
   (b) the ethmoid bone
   (c) the palatine bone
   (d) the temporal bone
9. Transverse foramina are characteristic of
   (a) lumbar vertebrae.
   (b) sacral vertebrae.
   (c) thoracic vertebrae.
   (d) cervical vertebrae.
10. The bone disorder that frequently develops in elderly people, particularly if they experience prolonged inactivity, malnutrition, or a hormone imbalance is
   (a) osteitis. (c) osteoporosis.
   (b) osteonecrosis. (d) osteomalacia.

Essay Questions

1. What are the functions of the skeletal system? Do the individual bones of the skeleton carry out these functions equally? Explain.
2. Explain why there are approximately 270 bones in an infant but 206 bones in a mature adult.
3. List the bones of the skull that are paired. Which are unpaired? Identify the bones of the skull that can be palpated.
4. Describe the development of the skull. What are the fontanels, where are they located, and what are their functions?
5. Which facial bones contain foramina? What structures pass through these openings?
6. Distinguish between the axial and appendicular skeletons. Describe where these two components articulate.
7. List four types of bones based on shape and give an example of each type.
8. Diagram a typical long bone. Label the epiphyses, diaphysis, epiphyseal plates, medullary cavity, nutrient foramina, periosteum, and articular cartilages.
9. List the bones that form the cranial cavity, the orbit, and the nasal cavity. Describe the location of the paranasal sinuses, the mastoidal sinus, and the inner-ear cavity.
10. Describe how bones grow in length and width. How are these processes similar, and how do they differ? Explain how radiographs can be used to determine normal bone growth.
11. Explain the process of endochondral ossification of a long bone. Why is it important that a balance be maintained between osteoblast activity and osteoclast activity?
12. Describe the curvature of the vertebral column. What do the terms *primary curves* and *secondary curves* refer to?
13. List two or more characteristics by which vertebrae from each of the five regions of the vertebral column can be identified.
14. Identify the bones that form the rib cage. What functional role do the bones and the costal cartilages have in respiration?

Critical-Thinking Questions

1. Many people think that the bones in our bodies are dead—understandable considering that we associate bones with graveyards, Halloween, and left-over turkey from a Thanksgiving dinner. Your kid brother is convinced of this. What information could you use to try to get him to change his mind?
2. The sensory organs involved with sight, smell, and hearing are protected by bone. Describe the locations of each of these sensory organs and list the associated bones that provide protection.
3. Explain why a proper balance of vitamins, hormones, and minerals is essential in maintaining healthy bone tissue. Give examples of diseases or skeletal conditions that may occur in the event of an imbalance of any of these three essential substances.
4. The most common surgical approach to a pituitary gland tumor is through the nasal cavity. With the knowledge that the pituitary gland is supported by the sella turcica of the brain case, list the bones that would be involved in the removal of the tumor.
5. The contour of a child's head is distinctly different from that of an adult. Which skull bones exhibit the greatest amount of change as a child grows to adulthood?

SEVEN

Skeletal System: The Appendicular Skeleton

Clinical Case Study 169
Pectoral Girdle and Upper Extremity 169
- Pectoral Girdle 169
- Brachium (Arm) 172
- Antebrachium (Forearm) 173
- Manus (Hand) 174

Pelvic Girdle and Lower Extremity 176
- Pelvic Girdle 177
- Thigh 180
- Leg 180
- Pes (Foot) 182

Developmental Exposition:
The Appendicular Skeleton 185
Clinical Considerations 186
- Developmental Disorders 186
- Trauma and Injury 187

Clinical Case Study Answer 188
Chapter Summary 189
Review Activities 189

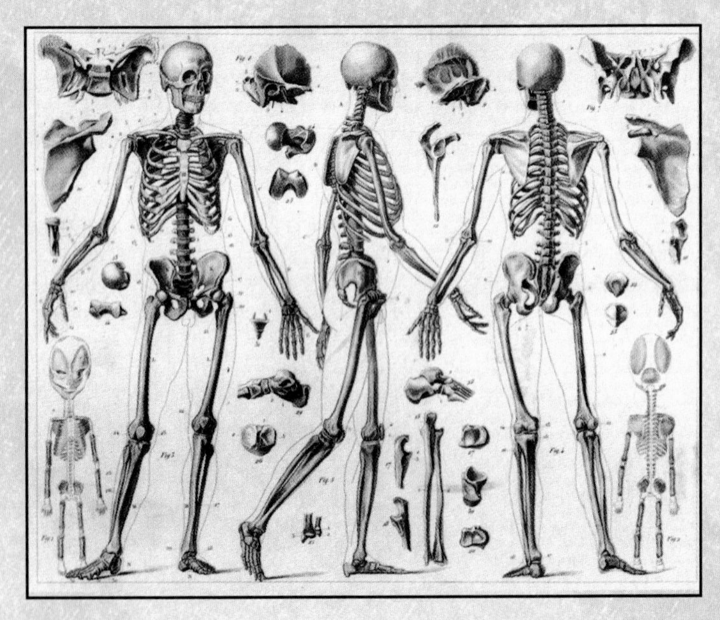

Many early anatomical renderings attempted to portray the skeleton in lifelike poses, with meticulous attention to detail. The growing emphasis on accuracy in the universities of Europe in the twelfth and thirteenth centuries paved the way for the tremendous progress in the study of the human form that occurred during the Renaissance.

Clinical Case Study

A 12-year-old boy was hit by a car while crossing a street. He was brought to the emergency room in stable condition, complaining of severe pain in his right leg. Radiographs revealed a 4-inch fracture extending inferiorly from the surface of the tibial plateau into the anterior body of the tibia. The fragment of bone created by the fracture was moderately displaced. With the radiographs in hand, the orthopedic surgeon went into the waiting room and conferred with the boy's parents. He told them that this kind of injury was more serious in children and growing adolescents than in adults. He went on to say that future growth of the bone might be jeopardized and that surgery, although recommended, could not guarantee normal growth. The parents asked, "What is it about this particular fracture that threatens future growth?"

If you were the surgeon, how would you respond?

Hints: Review the section on bone growth in chapter 6. Carefully examine figures 6.5 and 6.9 in chapter 6 and figures 7.17 and 7.23 in this chapter.

Pectoral Girdle and Upper Extremity

The structure of the pectoral girdle and upper extremities is adaptive for freedom of movement and extensive muscle attachment.

Objective 1 Describe the bones of the pectoral girdle and the articulations between them.

Objective 2 Identify the bones of the upper extremity and list the distinguishing features of each.

Pectoral Girdle

Two *scapulae* and two *clavicles* make up the **pectoral** (*shoulder*) **girdle** (fig. 7.1). It is not a complete girdle, having only an anterior attachment to the axial skeleton at the sternum. As an axial bone, the sternum was described in chapter 6 (see fig. 6.38). Lacking a posterior attachment to the axial skeleton, the pectoral girdle has a wide range of movement. Because it is not weight-bearing, it is structurally more delicate than the pelvic girdle. The primary function of the pectoral girdle is to provide attachment areas for the numerous muscles that move the shoulder and elbow joints.

Clavicle

The slender S-shaped clavicle (*klav'ĭ-kul*; "collarbone") connects the upper extremity to the axial skeleton and holds the shoulder joint away from the trunk to permit freedom of movement. The articulation of the medial **sternal extremity** (fig. 7.2) of the clavicle to the manubrium of the sternum is referred to as the *sternoclavicular joint.* The lateral **acromial** (*a-kro'me-al*) **extremity** of the clavicle articulates with the acromion of the scapula (fig. 7.3). This articulation is referred to as the *acromioclavicular joint.* A **conoid tubercle** is present on the acromial extremity of the clavicle, and a **costal tuberosity** is present on the inferior surface of the sternal extremity. Both processes serve as attachments for ligaments.

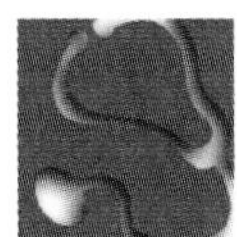

The long, delicate clavicle is the most commonly broken bone in the body. When a person receives a blow to the shoulder, or attempts to break a fall with an outstretched hand, the force is transmitted to the clavicle, possibly causing it to fracture. The most vulnerable part of this bone is through its center, immediately proximal to the conoid tubercle. Because the clavicle is directly beneath the skin and is not covered with muscle, a fracture can easily be palpated, and frequently seen.

Scapula

The scapula (*skap'yoo-lă*; "shoulder blade") is a large, triangular flat bone on the posterior side of the rib cage, overlying ribs 2 through 7. The **spine** of the scapula is a prominent diagonal bony ridge seen on the posterior surface (fig. 7.3). The spine strengthens the scapula, making it more resistant to bending. Above the spine is the **supraspinous fossa,** and below the spine is the **infraspinous fossa.** The spine broadens toward the shoulder as the **acromion** (figs. 7.3 and 7.4). This process serves for the attachment of several muscles, as well as for articulation with the clavicle. Inferior to the acromion is a shallow depression, the **glenoid** (*gle'*noid) **cavity,** into which the head of the humerus fits. The **coracoid** (*kor'ă-koid*) **process** is a thick upward projection lying superior and anterior to the glenoid cavity. On the anterior surface of the scapula is a slightly concave area known as the **subscapular fossa.**

The scapula has three borders delimited by three angles. The superior edge is called the **superior border.** The **medial border** is nearest to the vertebral column, and the **lateral border** is directed toward the arm. The **superior angle** is located between the superior and medial borders; the **inferior angle,** at the junction of the medial and lateral borders; and the **lateral angle,** at the junction of the superior and lateral borders. It is at the lateral angle that the scapula articulates with the head of the humerus. Along the superior border, a distinct depression called the **scapular notch** is a passageway for the suprascapular nerve.

---

clavicle: L. *clavicula,* a small key
acromial: Gk. *akros,* peak; *omos,* shoulder
conoid tubercle: Gk. *konos,* cone; L. *tuberculum,* a small swelling
costal tuberosity: L. *costa,* rib; *tuberous,* a knob
scapula: L. *scapula,* shoulder
glenoid: Gk. *glenoeides,* shallow form
coracoid: Gk. *korakodes,* like a crow's beak

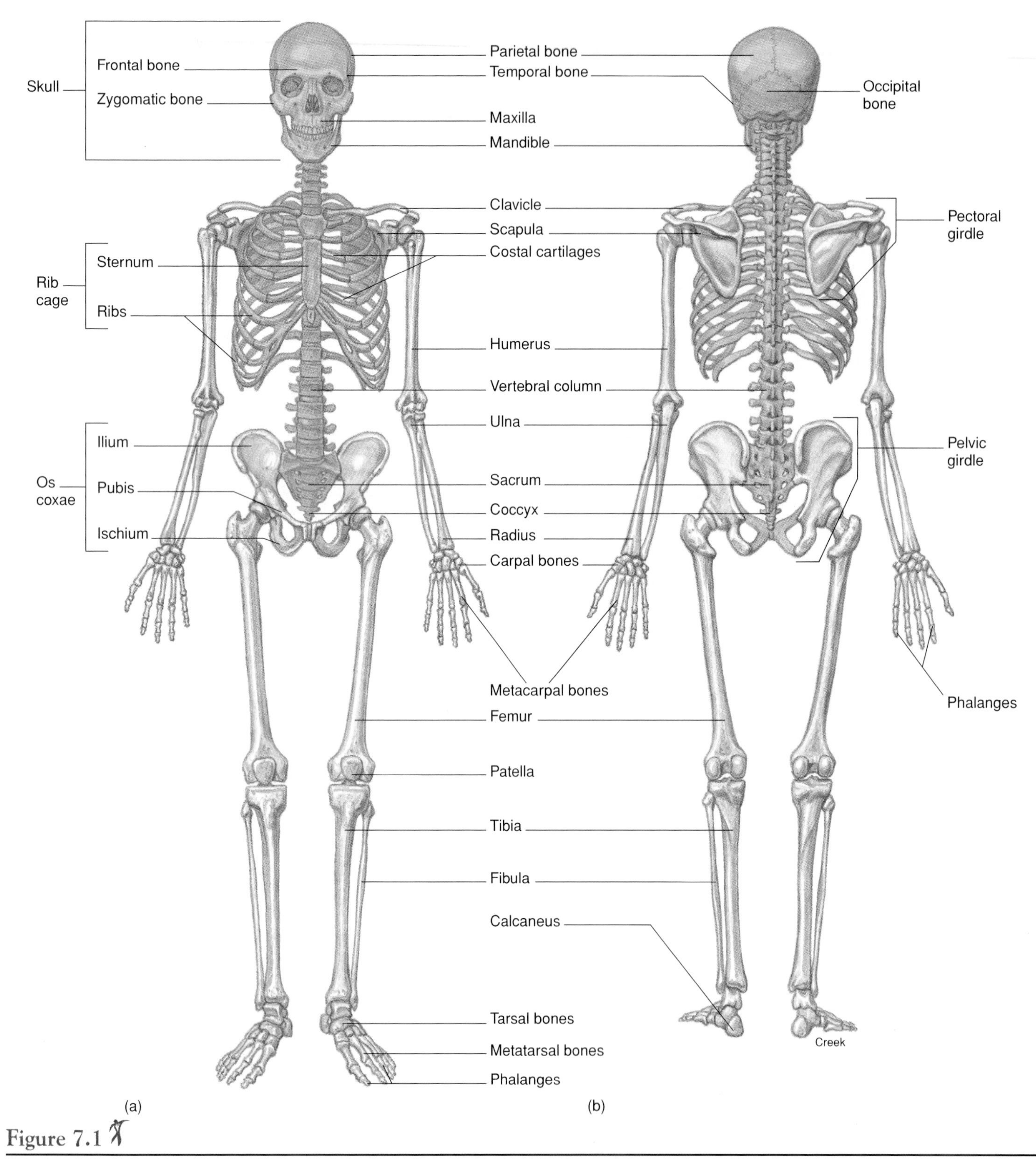

Figure 7.1

The human skeleton. (*a*) An anterior view and (*b*) a posterior view. The axial portion is colored light blue.

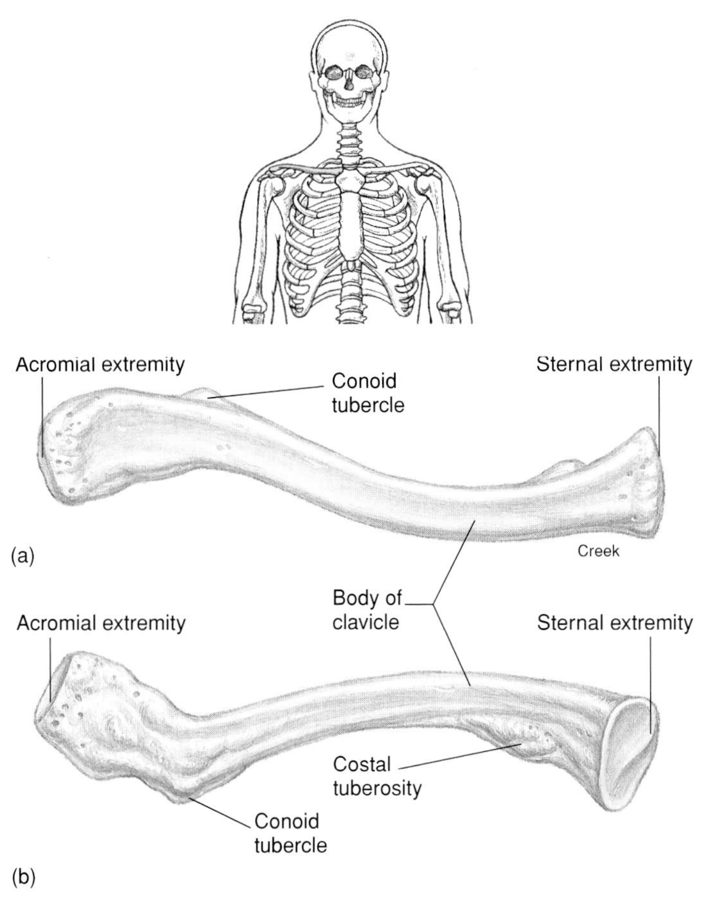

Figure 7.2

The right clavicle. (*a*) A superior view and (*b*) an inferior view.

Acromion of scapula
Clavicle
Coracoid process of scapula
Head of humerus
Scapula
Greater tubercle of humerus
Body of humerus

Figure 7.3

This radiograph of the right shoulder shows the positions of the clavicle, scapula, and humerus.

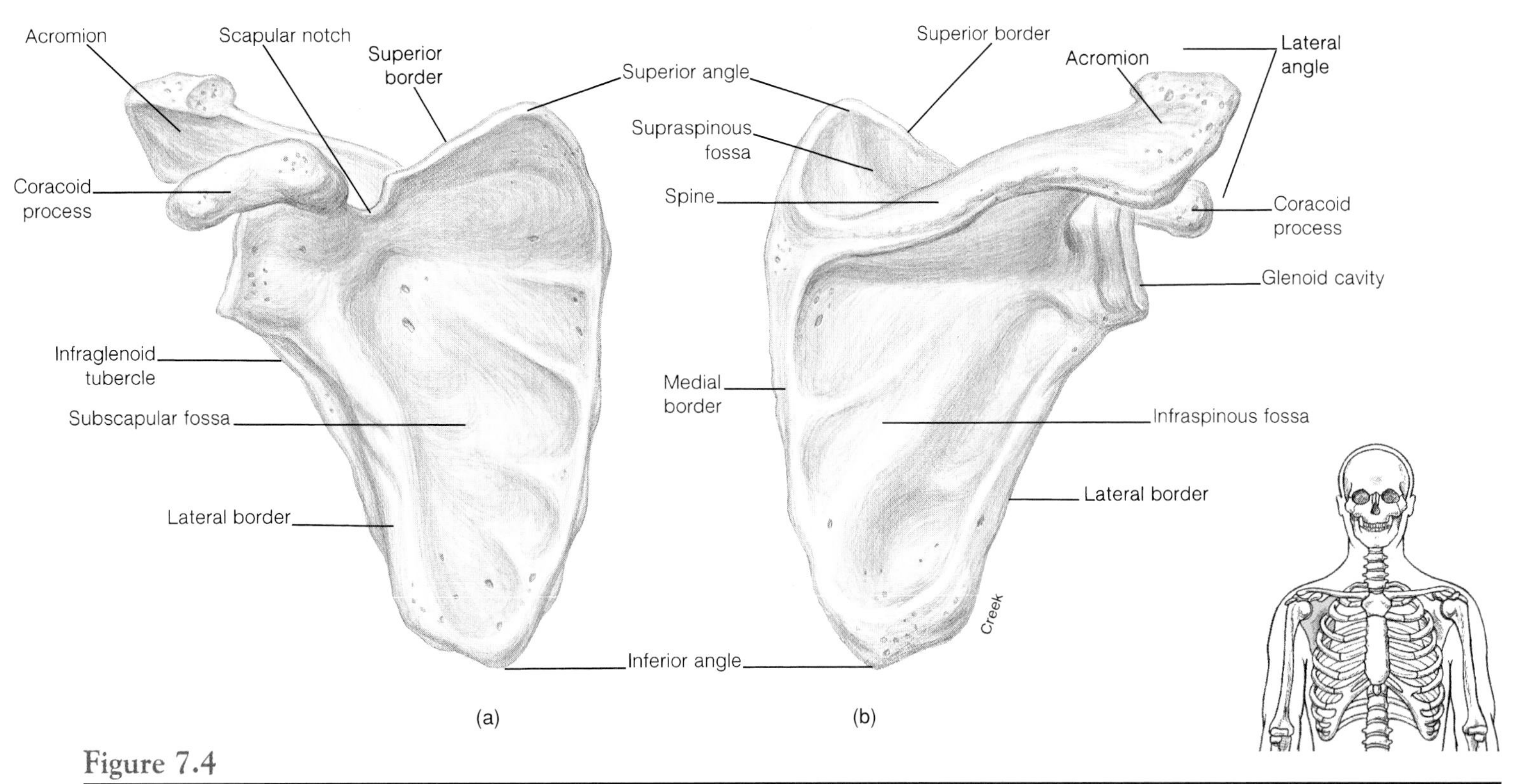

Figure 7.4

The right scapula. (*a*) An anterior view and (*b*) a posterior view.

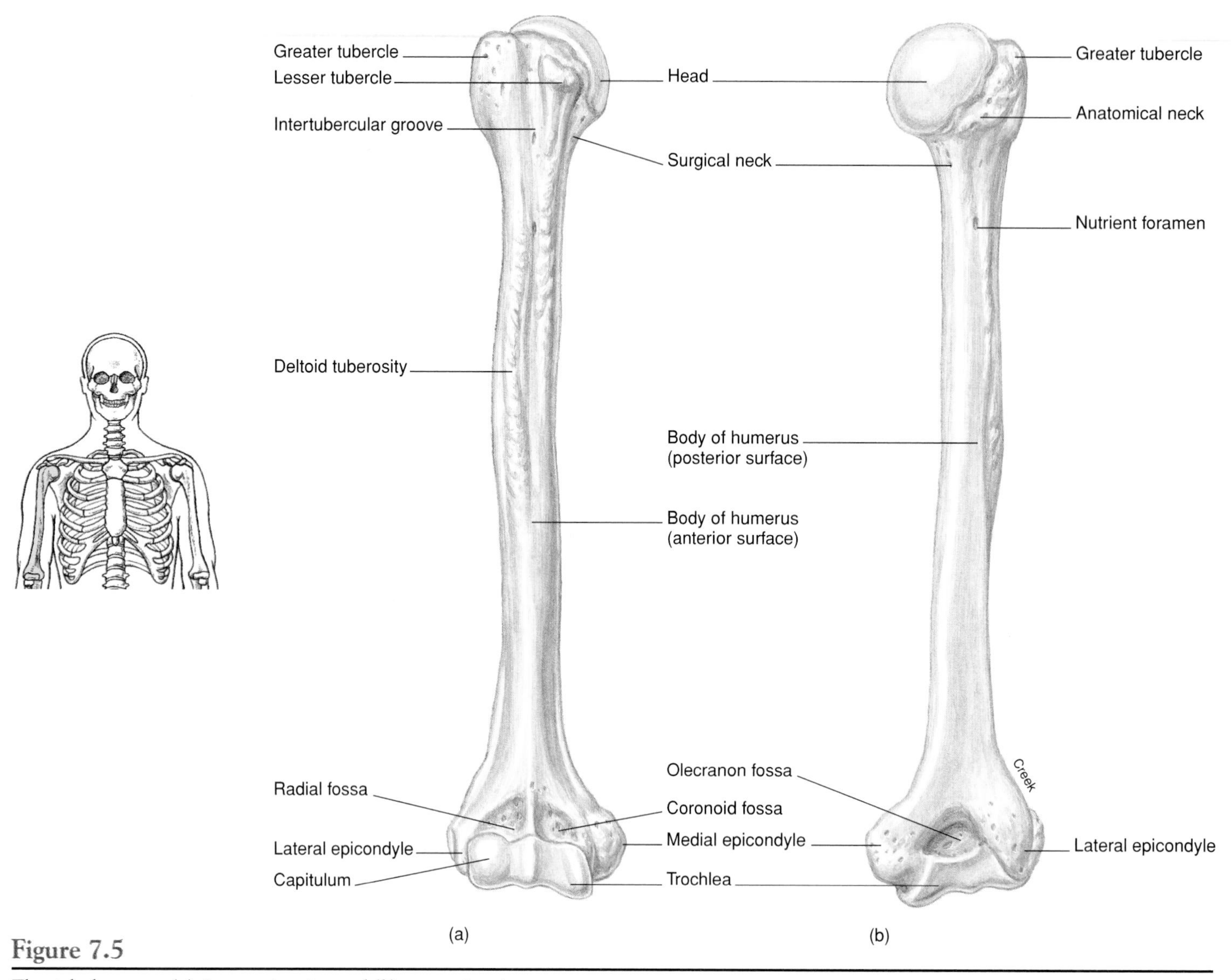

Figure 7.5

The right humerus. (*a*) An anterior view and (*b*) a posterior view.

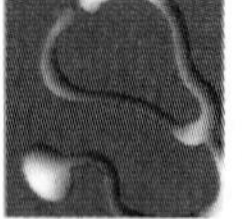

The scapula has numerous surface features because 15 muscles attach to it. Clinically, the pectoral girdle is significant because the clavicle and acromion of the scapula are frequently broken in trying to break a fall. The acromion is used as a landmark for identifying the site for an injection in the arm. This site is chosen because the musculature of the shoulder is quite thick and contains few nerves.

Brachium (Arm)

The brachium (*bra´ke-um*) extends from the shoulder to the elbow. In strict anatomical usage, *arm* refers only to this portion of the upper limb. The brachium contains a single bone—the *humerus*.

Humerus

The humerus (fig. 7.5) is the longest bone of the upper extremity. It consists of a proximal **head,** which articulates with the glenoid cavity of the scapula; a **body** ("shaft"); and a distal end, which is modified to articulate with the two bones of the forearm. Surrounding the margin of the head is a slightly indented groove denoting the **anatomical neck.** The **surgical neck,** the constriction just below the head, is a frequent fracture site. The **greater tubercle** is a large knob on the lateral proximal portion of the humerus. The **lesser tubercle** is slightly anterior to the greater tubercle and is separated from the greater by an **intertubercular groove.** The tendon of the biceps brachii muscle passes through this groove. Along the lateral midregion of the body of the humerus is a roughened area, the **deltoid tuberosity,** for the attachment of the deltoid muscle. Small openings in the body are called **nutrient foramina.**

The **humeral condyle** on the distal end of the humerus has two articular surfaces. The **capitulum** (*kă-pit´yoo-lum*) is the

deltoid: Gk. *deltoeides*, shaped like the letter Δ
capitulum: L. *caput*, little head

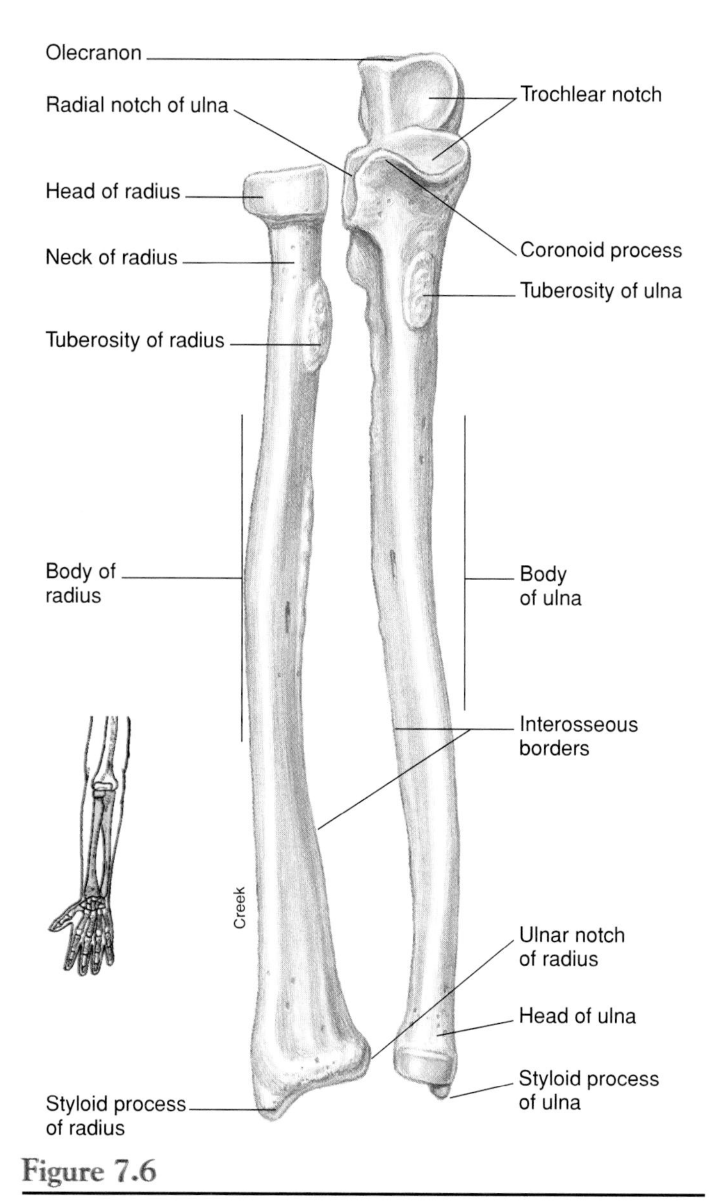

Figure 7.6

An anterior view of the right radius and ulna.

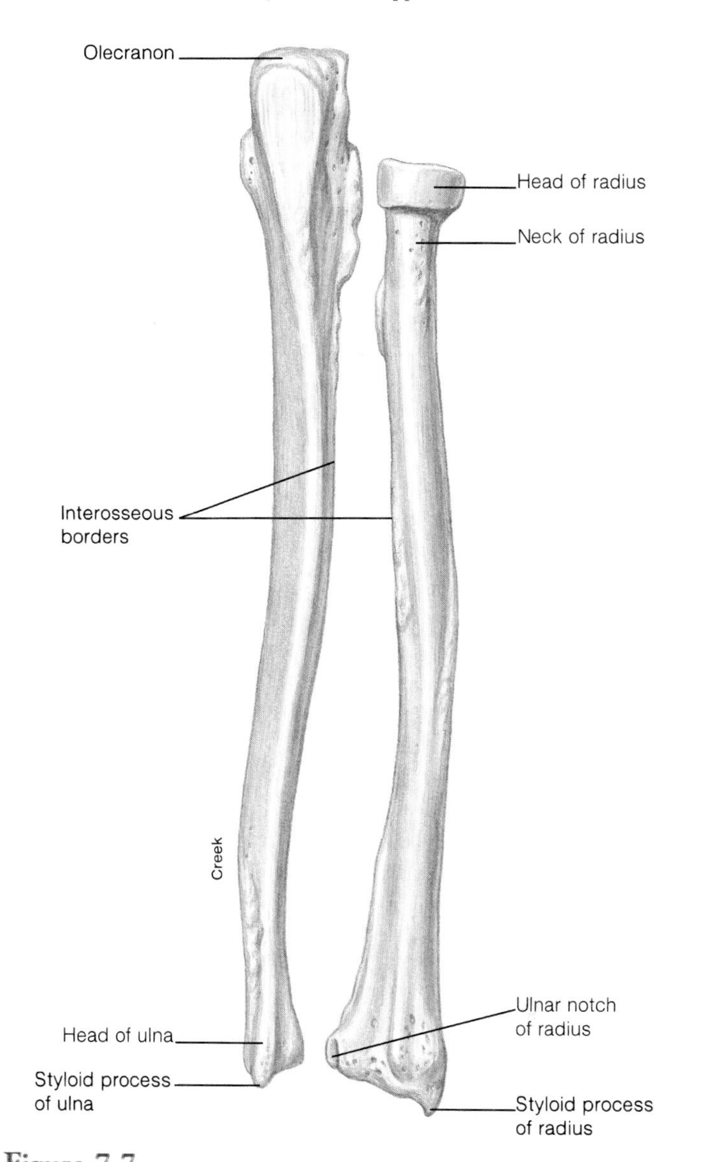

Figure 7.7

A posterior view of the right radius and ulna.

lateral rounded part that articulates with the radius. The **trochlea** (*trok' le-ă*) is the pulleylike medial part that articulates with the ulna. On either side above the condyle are the **lateral** and **medial epicondyles.** The large medial epicondyle protects the ulnar nerve that passes posteriorly through the ulnar sulcus (see fig. 10.34). It is popularly known as the "funny bone" because striking the elbow on the edge of a table, for example, stimulates the ulnar nerve and produces a tingling sensation. The **coronoid fossa** is a depression above the trochlea on the anterior surface. The **olecranon** (*o-lek'ră-non*) **fossa** is a depression on the distal posterior surface. Both fossae are adapted to work with the ulna during movement of the forearm.

trochlea: Gk. *trochilia*, a pulley
olecranon: Gk. *olene*, ulna; *kranion*, head

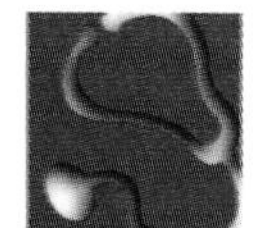

The medical term for tennis elbow is *lateral epicondylitis,* which means inflammation of the tissues surrounding the lateral epicondyle of the humerus. At least six muscles that control backward (extension) movement of the wrist and finger joints originate on the lateral epicondyle. Repeated strenuous contractions of these muscles, as in stroking with a tennis racket, may strain the periosteum and muscle attachments, resulting in swelling, tenderness, and pain around the epicondyle. Binding usually eases the pain, but only rest can eliminate the causative factor, and recovery generally follows.

Antebrachium (Forearm)

The skeletal structures of the antebrachium are the *ulna* on the medial side and the *radius* on the lateral (thumb) side (figs. 7.6 and 7.7). The ulna is more firmly connected to the humerus than the radius, and it is longer than the radius. The radius, however, contributes more significantly to the articulation at the wrist joint than does the ulna.

Ulna

The proximal end of the ulna articulates with the humerus and radius. A distinct depression, the **trochlear notch,** articulates with the trochlea of the humerus. The **coronoid process** forms the anterior lip of the trochlear notch, and the **olecranon** forms the posterior portion. Lateral and inferior to the coronoid process is the **radial notch,** which accommodates the head of the radius.

On the tapered distal end of the ulna is a knobbed portion, the **head,** and a knoblike projection, the **styloid process.** The ulna articulates at both ends with the radius.

Radius

The radius consists of a **body** with a small proximal end and a large distal end. A proximal disc-shaped **head** articulates with the capitulum of the humerus and the radial notch of the ulna. The prominent **tuberosity of radius** (radial tuberosity), for attachment of the biceps brachii muscle, is located on the medial side of the body, just below the head. On the distal end of the radius is a double-faceted surface for articulation with the proximal carpal bones. The distal end of the radius also has a **styloid process** on the lateral tip and an **ulnar notch** on the medial side that receives the distal end of the ulna. The styloid processes on the ulna and radius provide lateral and medial stability for articulation at the wrist.

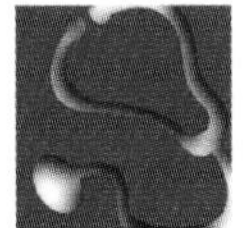

When a person falls, the natural tendency is to extend the hand to break the fall. This reflexive movement frequently results in fractured bones. Common fractures of the radius include a fracture of the head, as it is driven forcefully against the capitulum; a fracture of the neck; or a fracture of the distal end (*Colles' fracture*), caused by landing on an outstretched hand.

When falling, it is less traumatic to the body to withdraw the appendages, bend the knees, and let the entire body hit the surface. Athletes learn that this is the safe way to fall.

Manus (Hand)

The hand contains 27 bones, grouped into the *carpus, metacarpus,* and *phalanges* (figs. 7.8, 7.9, and 7.10).

Carpus

The carpus, or wrist, contains eight carpal bones arranged in two transverse rows of four bones each. The proximal row, naming from the lateral (thumb) to the medial side, consists of the **scaphoid** (navicular), **lunate, triquetral** (*tri-kwe'* tral), and **pisiform.** The pisiform forms in a tendon as a sesamoid bone. The distal row, from lateral to medial, consists of the **trapezium** (greater multangular), **trapezoid** (lesser multangular), **capitate,** and **hamate.** The scaphoid and lunate of the proximal row articulate with the distal end of the radius.

Metacarpus

The metacarpus, or palm of the hand, contains five metacarpal bones. Each metacarpal bone consists of a proximal **base,** a **body,** and a distal **head** that is rounded for articulation with the base of each proximal phalanx. The heads of the metacarpal bones are distally located and form the knuckles of a clenched fist.

Phalanges

The 14 phalanges are the bones of the digits. A single finger bone is called a **phalanx** (*fa'langks*). The phalanges of the fingers are arranged in a proximal row, a middle row, and a distal row. The thumb, or *pollex* (adjective, *pollicis*), lacks a middle phalanx. The digits are sequentially numbered I to V starting with the thumb—the lateral side, in reference to anatomical position.

A summary of the bones of the upper extremities is presented in table 7.1.

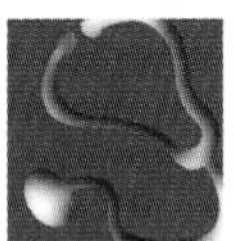

The hand is a marvel of structural complexity that can withstand considerable abuse. Other than sprained ligaments of the fingers and joint dislocations, the most common bone injury is a fracture to the scaphoid—a wrist bone that accounts for about 70% of carpal fractures. When immobilizing the wrist joint, the wrist is positioned in the plane of relaxed function. This is the position in which the hand is about to grasp an object between the thumb and index finger.

1. Describe the structure of the pectoral girdle. Why is the pectoral girdle considered an incomplete girdle?
2. Identify the fossae and processes of the scapula.
3. Describe each of the long bones of the upper extremity.
4. Where are the styloid processes of the wrist area? What are their functions?
5. Name the bones in the proximal row of the carpus. Which of these bones articulate(s) with the radius?

styloid: Gk. *stylos,* pillar
carpus: Gk. *karpos,* wrist

navicular: L. *navicula,* small ship
lunate: L. *lunare,* crescent or moon-shaped
triquetral: L. *triquetrous,* three-cornered
pisiform: Gk. *pisos,* pea
trapezium: Gk. *trapesion,* small table
capitate: L. *capitatus,* head
hamate: L. *hamatus,* hook
phalanx: Gk. *phalanx,* finger bone or toe bone

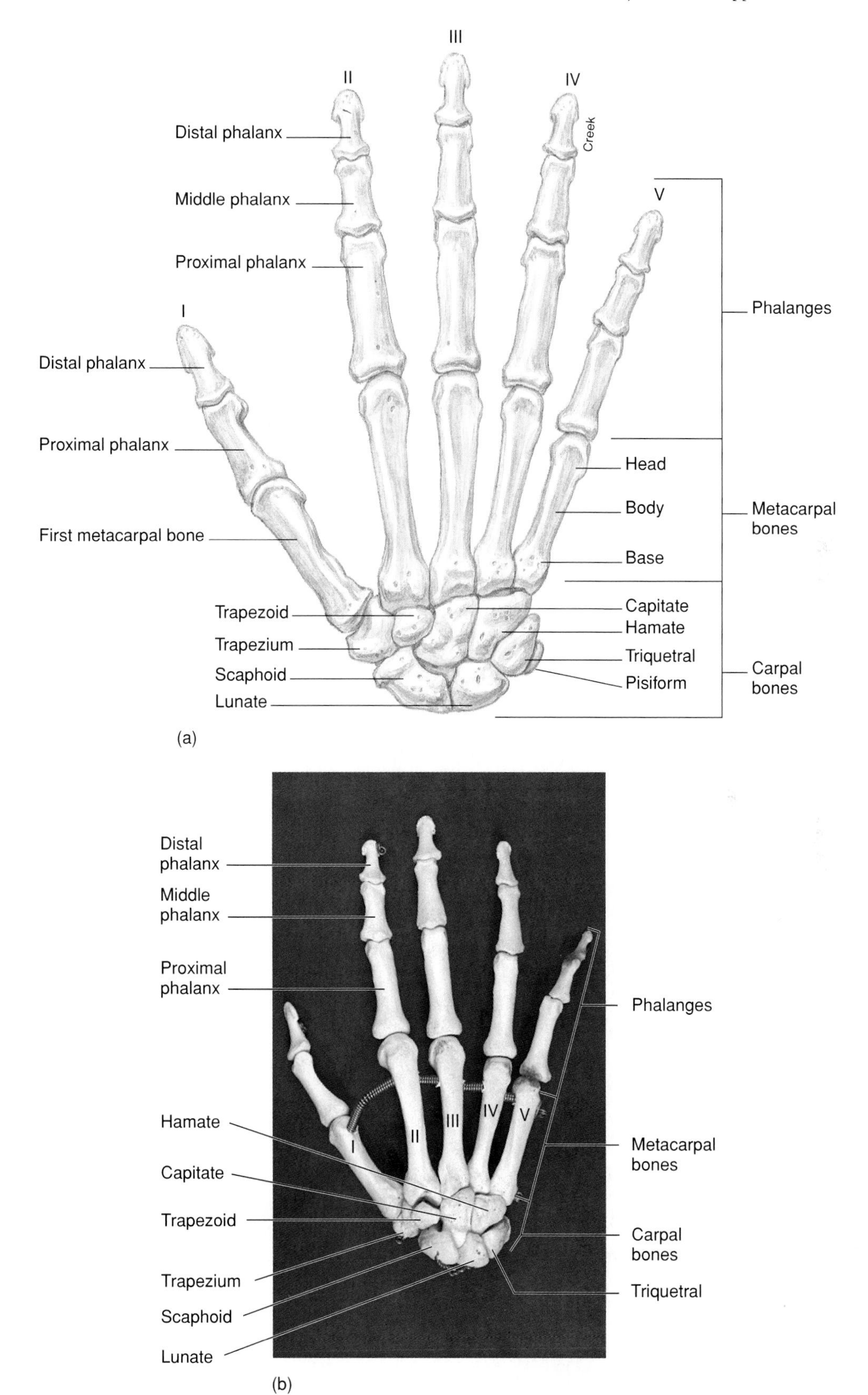

Figure 7.8

A posterior view of the bones of the right hand as shown in (*a*) a drawing and (*b*) a photograph. Each digit (finger) is indicated by a Roman numeral, the first digit, or thumb, being Roman numeral I.

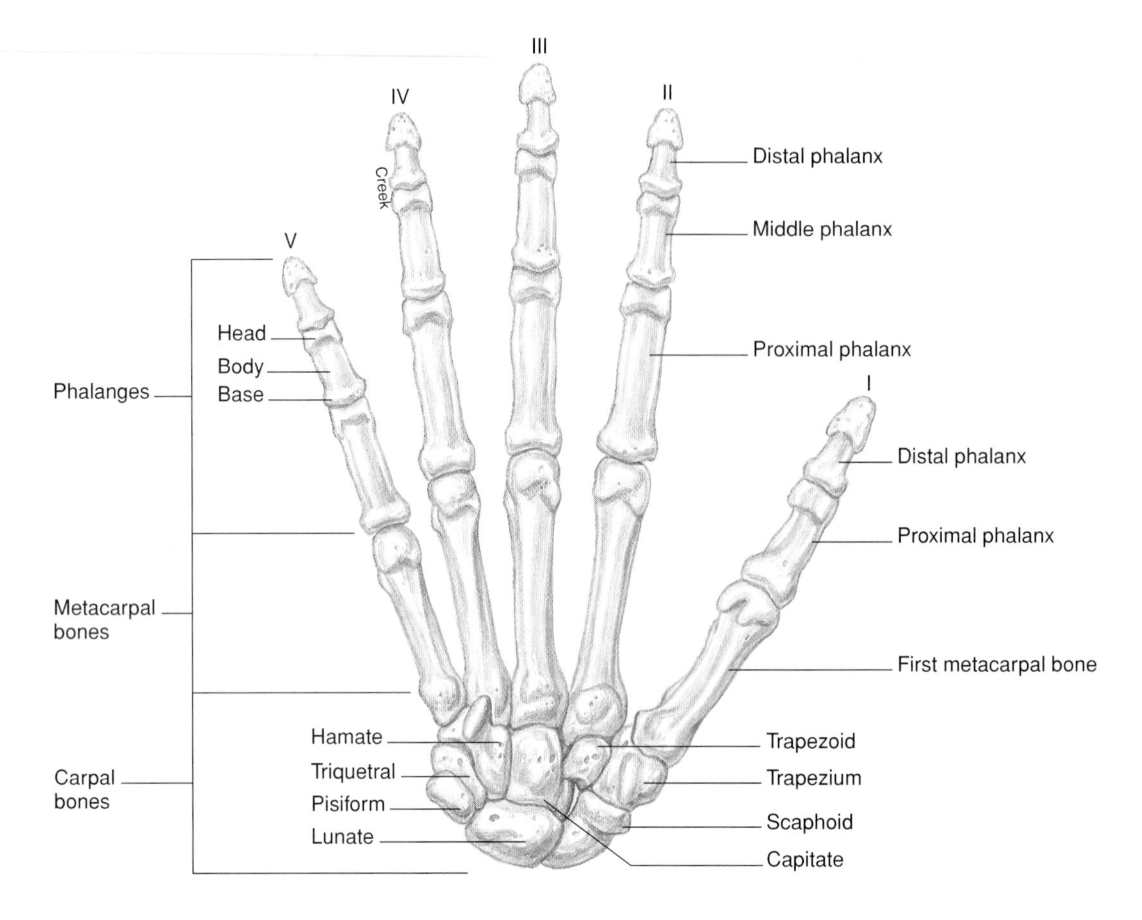

Figure 7.9

An anterior view of the bones of the right hand.

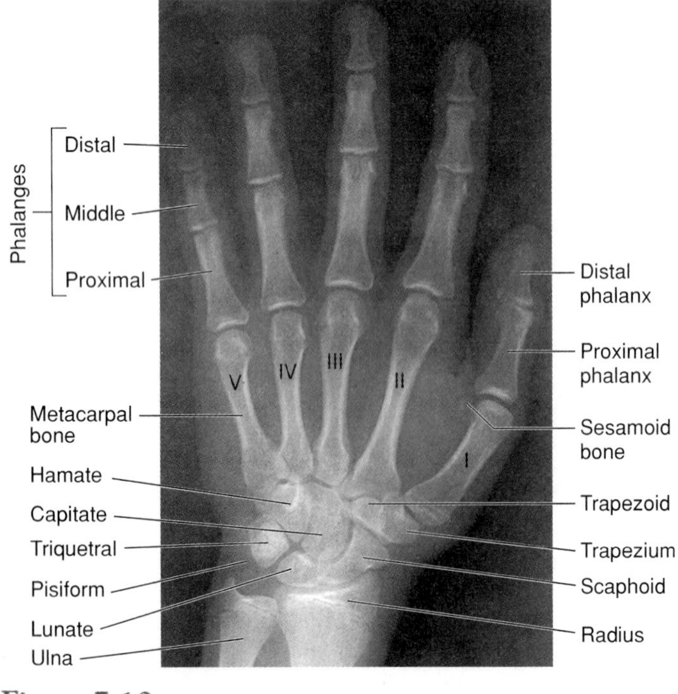

Figure 7.10

A radiograph of the right hand shown in an anteroposterior projection. (Note the presence of a sesamoid bone at the thumb joint.)

Pelvic Girdle and Lower Extremity

The structure of the pelvic girdle and lower extremities is adaptive for support and locomotion. Extensive processes and surface features on certain bones of the pelvic girdle and lower extremities accommodate massive muscles used in body movement and in maintaining posture.

Objective 3 Describe the structure of the pelvic girdle and list its functions.

Objective 4 Describe the structural differences in the male and female pelves.

Objective 5 Identify the bones of the lower extremity and list the distinguishing features of each.

Objective 6 Describe the structural features and functions of the arches of the foot.

Table 7.1

Bones of the Pectoral Girdle and Upper Extremities

Name and Number	Location	Major Distinguishing Features
Clavicle (2)	Anterior base of neck, between sternum and scapula	S-shaped; sternal and acromial extremities; conoid tubercle; costal tuberosity
Scapula (2)	Upper back forming part of the shoulder	Triangular; spine; subscapular, supraspinous, and infraspinous fossae; glenoid cavity; coracoid process; acromion
Humerus (2)	Brachium, between scapula and elbow	Longest bone of upper extremity; greater and lesser tubercles; intertubercular groove; surgical neck; deltoid tuberosity; capitulum; trochlea; lateral and medial epicondyles; coronoid and olecranon fossae
Ulna (2)	Medial side of forearm	Trochlear notch; olecranon; coronoid and styloid processes; radial notch
Radius (2)	Lateral side of forearm	Head; radial tuberosity; styloid process; ulnar notch
Carpal bone (16)	Wrist	Short bones arranged in two rows of four bones each
Metacarpal bone (10)	Palm of hand	Long bones, each aligned with a digit
Phalanx (28)	Digits	Three in each digit, except two in thumb

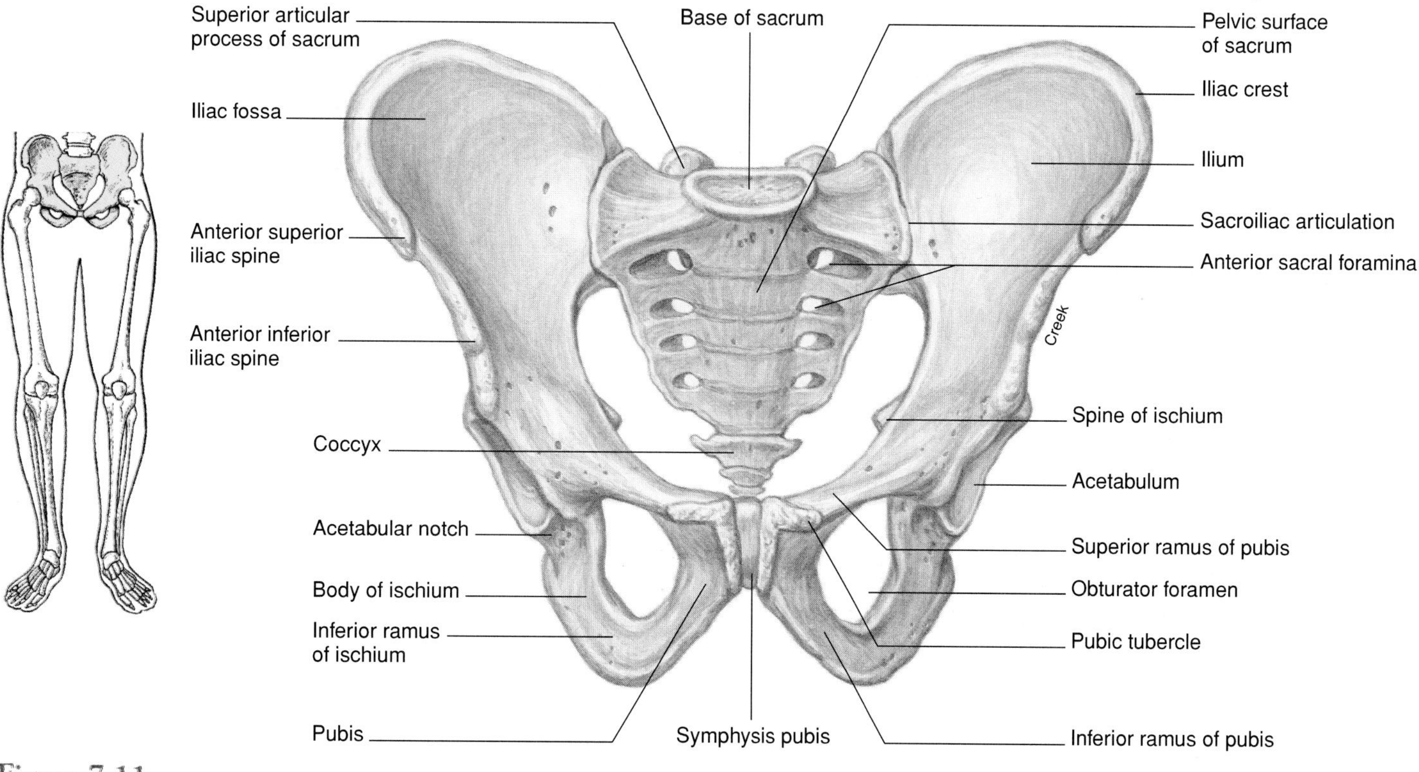

Figure 7.11

An anterior view of the pelvic girdle.

Pelvic Girdle

The **pelvic girdle** is formed by two *ossa coxae* (*os'ă kuk'se*; "hipbones"), united anteriorly at the *symphysis pubis* (figs. 7.11 and 7.12). It is attached posteriorly to the sacrum of the vertebral column. The sacrum, a bone of the axial skeleton, was described in chapter 6 (see fig. 6.37). The deep, basinlike structure formed by the ossa coxae, together with the sacrum and coccyx, is called the **pelvis** (plural, *pelves* or *pelvises*). The pelvic girdle and its associated ligaments support the weight of the body from the vertebral column. The pelvic girdle also supports and protects the lower viscera, including the urinary bladder, the reproductive organs, and in a pregnant woman, the developing fetus.

The pelvis is divided into a **greater** (false) **pelvis** and a **lesser** (true) **pelvis** (see fig. 7.15). These two components are divided by

coxae: L. *coxae*, hips

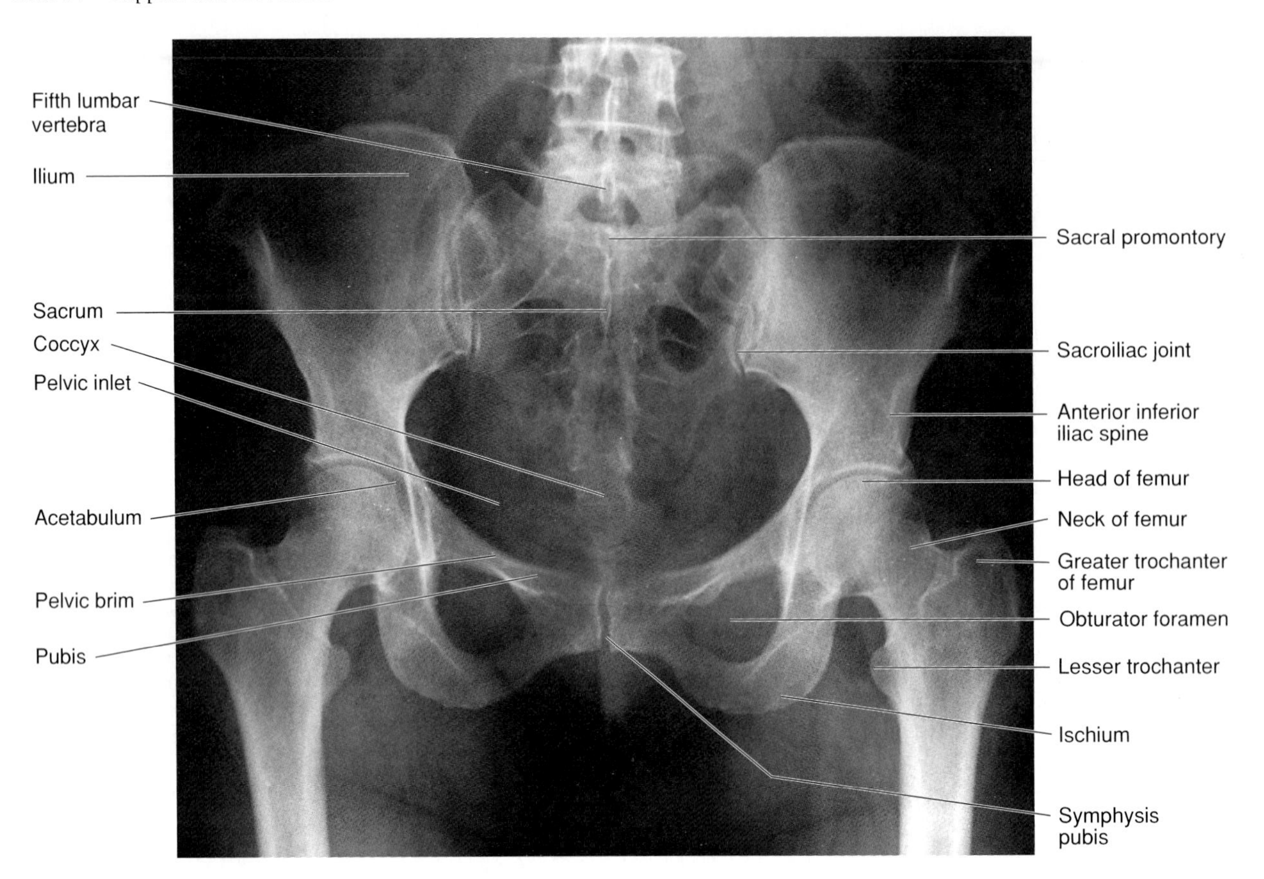

Figure 7.12

A radiograph of the pelvic girdle and the articulating femurs.

the **pelvic brim,** a curved bony rim passing inferiorly from the sacral promontory to the upper margin of the symphysis pubis. The greater pelvis is the expanded portion of the pelvis, superior to the pelvic brim. The pelvic brim not only divides the two portions but surrounds the **pelvic inlet** of the lesser pelvis. The lower circumference of the lesser pelvis bounds the pelvic outlet.

Each os coxae ("hipbone") actually consists of three separate bones: the *ilium*, the *ischium*, and the *pubis* (figs. 7.13 and 7.14). These bones are fused together in the adult. On the lateral surface of the os coxae, where the three bones ossify, is a large circular depression, the **acetabulum** (*as″ĕ-tab′yŭ-lum*), which receives the head of the femur. Although both ossa coxae are single bones in the adult, the three components of each one are considered separately for descriptive purposes.

Ilium

The ilium is the uppermost and largest of the three pelvic bones. It has a crest and four angles, or spines—important surface landmarks that serve for muscle attachment. The **iliac crest** forms the

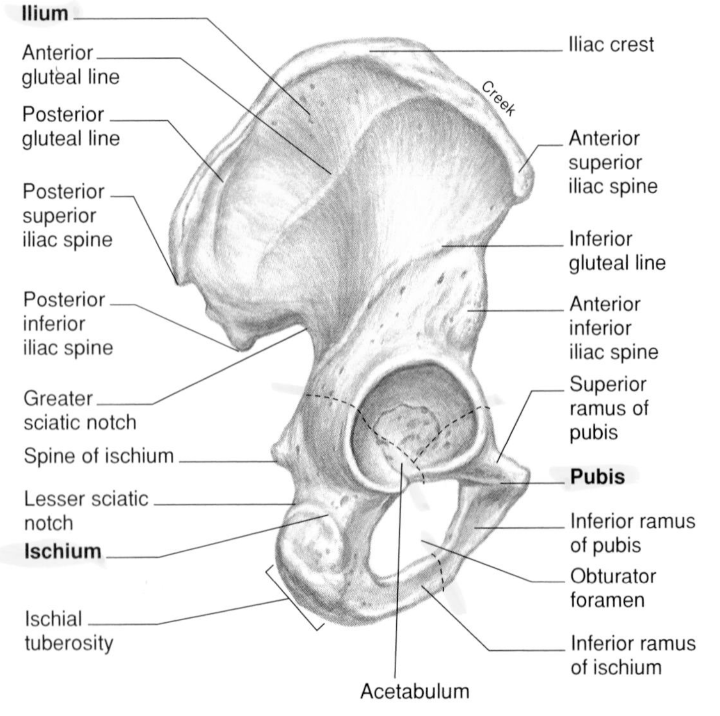

Figure 7.13

The lateral aspect of the right os coxae. (The three bones comprising the os coxae are labeled in boldface type.)

ilium: L. *ilia*, loin
ischium: Gk. *ischion*, hip joint
pubis: L. *pubis*, genital area
acetabulum: L. *acetabulum*, vinegar cup

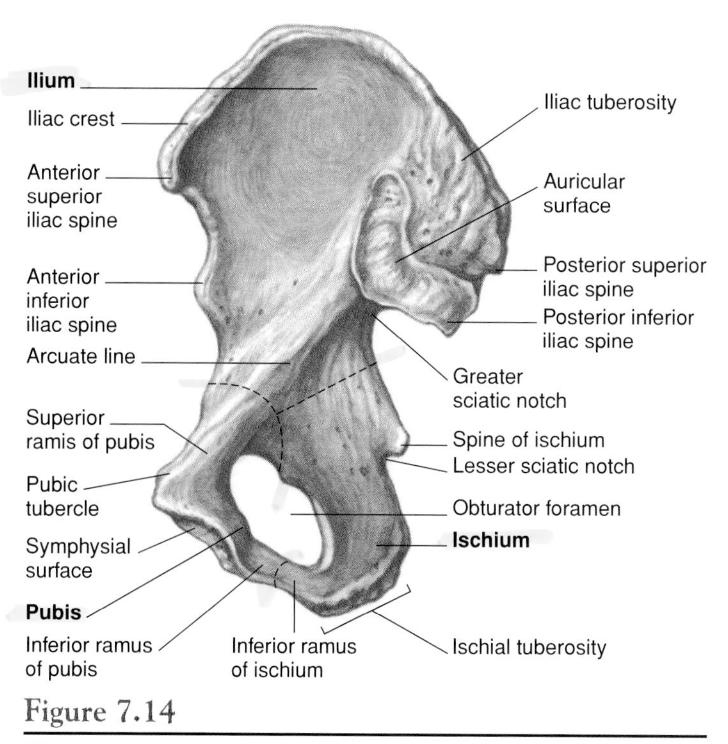

Figure 7.14

The medial aspect of the right os coxae. (The three bones comprising the os coxae are labeled in boldface type.)

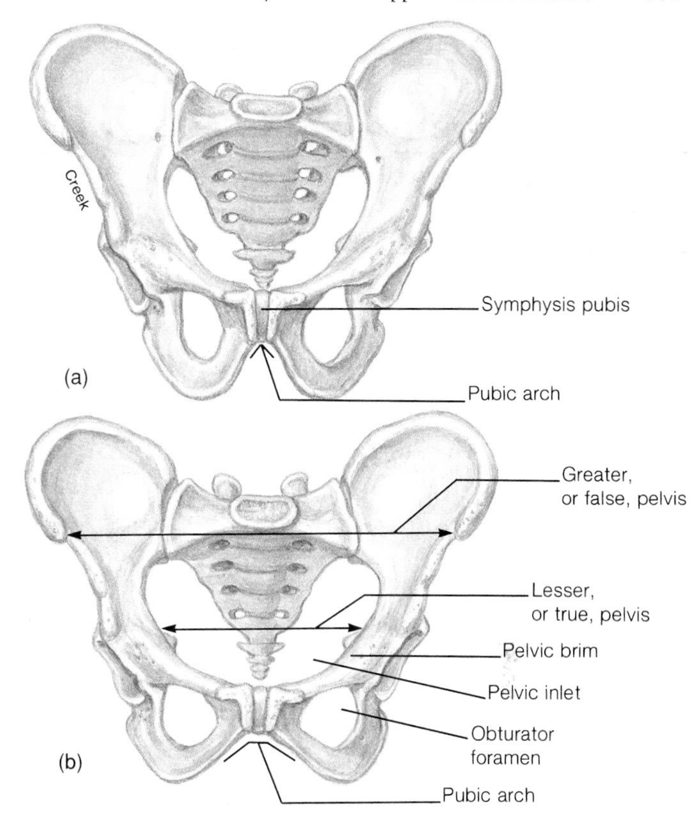

Figure 7.15

A comparison of (*a*) the male and (*b*) the female pelvic girdle.

prominence of the hip. This crest terminates anteriorly as the **anterior superior iliac spine.** Just below this spine is the **anterior inferior iliac spine.** The posterior termination of the iliac crest is the **posterior superior iliac spine,** and just below this is the **posterior inferior iliac spine.**

Below the posterior inferior iliac spine is the **greater sciatic notch,** through which the sciatic nerve passes. On the medial surface of the ilium is the roughened **auricular surface,** which articulates with the sacrum. The **iliac fossa** is the smooth, concave surface on the anterior portion of the ilium. The iliacus muscle originates from this fossa. The **iliac tuberosity,** for the attachment of the sacroiliac ligament, is positioned posterior to the iliac fossa. Three roughened ridges are present on the **gluteal surface** of the posterior aspect of the ilium. These ridges, which serve to attach the gluteal muscles, are the **inferior, anterior,** and **posterior gluteal lines.**

Ischium

The ischium (*is′ke-um*) is the posteroinferior bone of the os coxae. This bone has several distinguishing features. The **spine of the ischium** is the projection immediately posterior and inferior to the greater sciatic notch of the ilium. Inferior to this spine is the **lesser sciatic notch** of the ischium. The **ischial tuberosity** is the bony projection that supports the weight of the body in the sitting position. A deep **acetabular** (*as″ĕ-tab′yŭ-lar*) **notch** is present on the inferior portion of the acetabulum. The large **obturator** (*ob′tŭ-ra″tor*) **foramen** is formed by the **ramus** of the ischium, together with the pubis. The obturator foramen is covered by the obturator membrane, to which several muscles attach.

Pubis

The pubis is the anterior bone of the os coxae. It consists of a **superior ramus** and an **inferior ramus** that support the **body** of the pubis. The body contributes to the formation of the symphysis pubis—the joint between the two ossa coxae. At the lateral end of the anterior border of the body is the **pubic tubercle,** one of the attachments for the inguinal ligament.

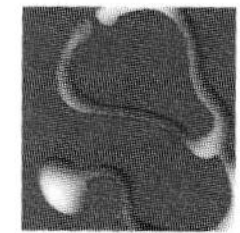

The structure of the human pelvis, in its attachment to the vertebral column, permits an upright posture and locomotion on two appendages (bipedal locomotion). An upright posture may cause problems, however. The sacroiliac joint may weaken with age, causing lower back pains. The weight of the viscera may weaken the walls of the lower abdominal area and cause hernias. Some of the problems of childbirth are related to the structure of the mother's pelvis. Finally, the hip joint tends to deteriorate with age, so that many elderly people suffer from degenerative arthritis (*osteoarthrosis*).

Sex-Related Differences in the Pelvis

Structural differences between the pelvis of an adult male and that of an adult female (fig. 7.15 and table 7.2) reflect the female's role in pregnancy and parturition. In a vaginal delivery, a baby must pass through its mother's lesser pelvis. *Pelvimetry* (*pel-vim′ĕ-tre*) is the measurement of the dimensions of the pelvis—especially of

Table 7.2

Comparison of the Male and Female Pelves

Characteristics	Male Pelvis	Female Pelvis
General structure	More massive; prominent processes	More delicate; processes not so prominent
Pelvic inlet	Heart-shaped	Round or oval
Pelvic outlet	Narrower	Wider
Anterior superior iliac spines	Not as wide apart	Wider apart
Obturator foramen	Oval	Triangular
Acetabulum	Faces laterally	Faces more anteriorly
Symphysis pubis	Deeper, longer	Shallower, shorter
Pubic arch	Angle less than 90°	Angle greater than 90°

the adult female pelvis—to determine whether a cesarean section might be necessary. Diameters may be determined by vaginal palpation or by sonographic images.

Thigh

The *femur* is the only bone of the thigh. In the following discussion, however, the *patella* will also be discussed.

Femur

The femur (*fe′mur*; "thighbone") is the longest, heaviest, strongest bone in the body (fig. 7.16). The proximal rounded **head** of the femur articulates with the acetabulum of the os coxae. A roughened shallow pit, the **fovea capitis femoris,** is present in the lower center of the head of the femur. The fovea capitis femoris provides the point of attachment for the ligamentum teres, which helps to support the head of the femur against the acetabulum. The constricted region supporting the head is called the **neck** and is a common site for fractures in the elderly.

The **body** of the femur has a slight medial curve to bring the knee joint in line with the body's plane of gravity. The degree of curvature is greater in the female because of the wider pelvis. The body of the femur has several distinguishing features for muscle attachment. On the proximolateral side of the body is the **greater trochanter,** and on the medial side is the **lesser trochanter.** On the anterior side, between the trochanters, is the **intertrochanteric** (*in″ter-tro″kan-ter′ik*) **line.** On the posterior side, between the trochanters, is the **intertrochanteric crest.** The **linea aspera** (*lin′e-ă as′per-ă*) is a roughened vertical ridge on the posterior surface of the body of the femur.

The distal end of the femur is expanded for articulation with the tibia. The **medial** and **lateral condyles** are the articular processes for this joint. The shallow depression between the condyles on the posterior aspect is called the **intercondylar fossa.** The **patellar surface** is located between the condyles on the anterior side. Above the condyles on the lateral and medial sides are the **epicondyles,** which serve for ligament and tendon attachment.

Patella

The patella (*pă-tel′ă*; "kneecap") is a large, triangular sesamoid bone positioned on the anterior side of the distal femur (figs. 7.17 and 7.18). It develops in response to strain in the tendon of the quadriceps femoris muscle. It has a broad **base** and an inferiorly pointed **apex.** Articular facets on the **articular surface** of the patella articulate with the medial and lateral condyles of the femur.

The functions of the patella are to protect the knee joint and to strengthen the quadriceps tendon. It also increases the leverage of the quadriceps femoris muscle as it extends (straightens) the knee joint.

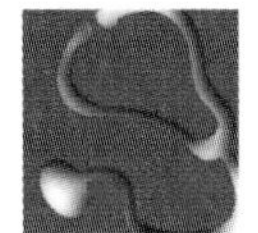

The patella can be fractured by a direct blow. It usually does not fragment, however, because it is confined within the tendon. Dislocations of the patella may result from injury or from underdevelopment of the lateral condyle of the femur.

Leg

Technically speaking, *leg* refers only to that portion of the lower limb between the knee and foot. The *tibia* and *fibula* are the bones of the leg. The tibia is the larger and more medial of the two.

Tibia

The tibia (*tib′e-ă*; "shinbone") articulates proximally with the femur at the knee joint and distally with the talus of the ankle. It also articulates both proximally and distally with the fibula. Two slightly concave surfaces on the proximal end of the tibia, the **medial** and **lateral condyles** (fig. 7.18) articulate with the condyles of the femur. The condyles are separated by a slight upward projection called the **intercondylar eminence.** The **tibial tuberosity,** for attachment of the patellar ligament, is located on the proximoanterior part of the body of the tibia. The **anterior crest** is a sharp ridge along the anterior surface of the body.

The **medial malleolus** (*mă-le′o-lus*) is a prominent medial knob of bone located on the distomedial end of the tibia. A **fibular notch,** for articulation with the fibula, is located on the

femur: L. *femur*, thigh
linea aspera: L. *linea*, line; *asperare*, rough

patella: L. *patina*, small plate
tibia: L. *tibia*, shinbone, pipe, flute

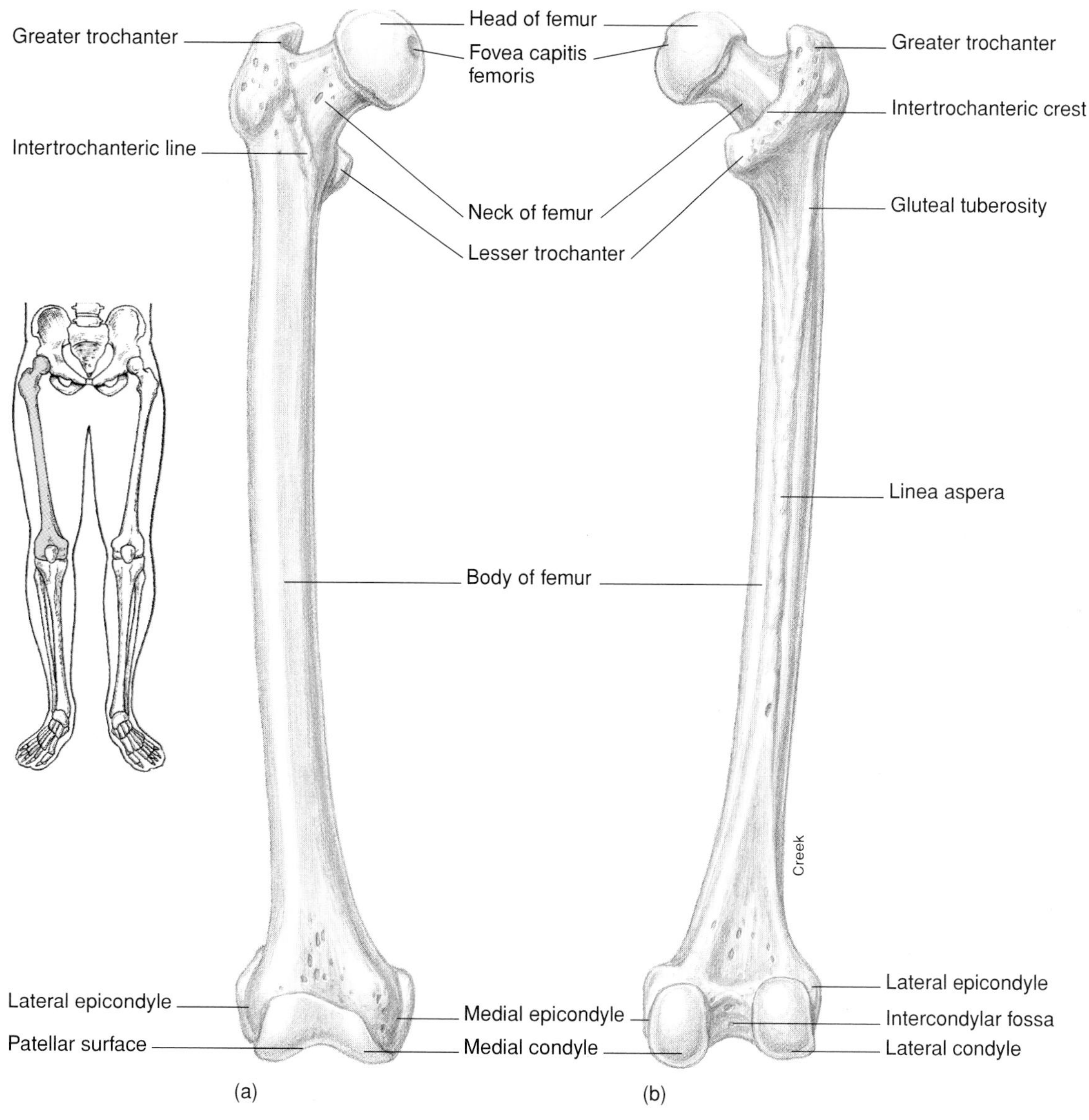

Figure 7.16

The right femur. (*a*) An anterior view and (*b*) a posterior view.

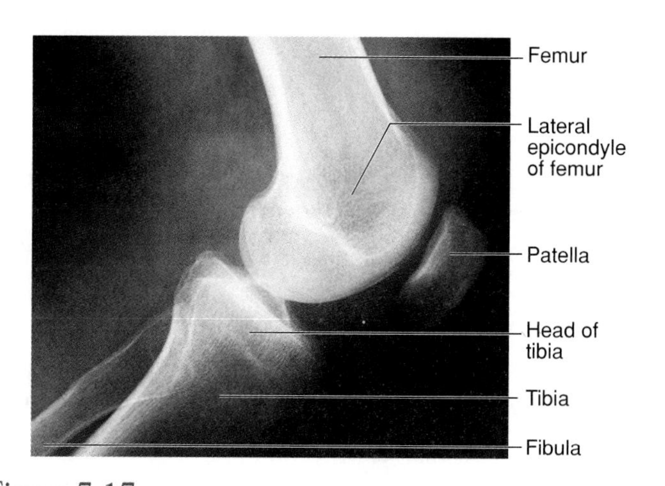

Figure 7.17

A radiograph of the right knee.

distolateral end. The **medial** and **lateral epicondyles** are located above the condyles, on either side. In that the tibia is the weight-bearing bone of the leg, it is much larger than the fibula.

Fibula

The fibula (*fib'yŭ-lă*) is a long, slender bone that is more important for muscle attachment than for support. The **head** of the fibula articulates with the proximolateral end of the tibia. The distal end has a prominent knob called the **lateral malleolus.**

---

fibula: L. *fibula*, clasp or brooch
malleolus: L. *malleolus*, small hammer

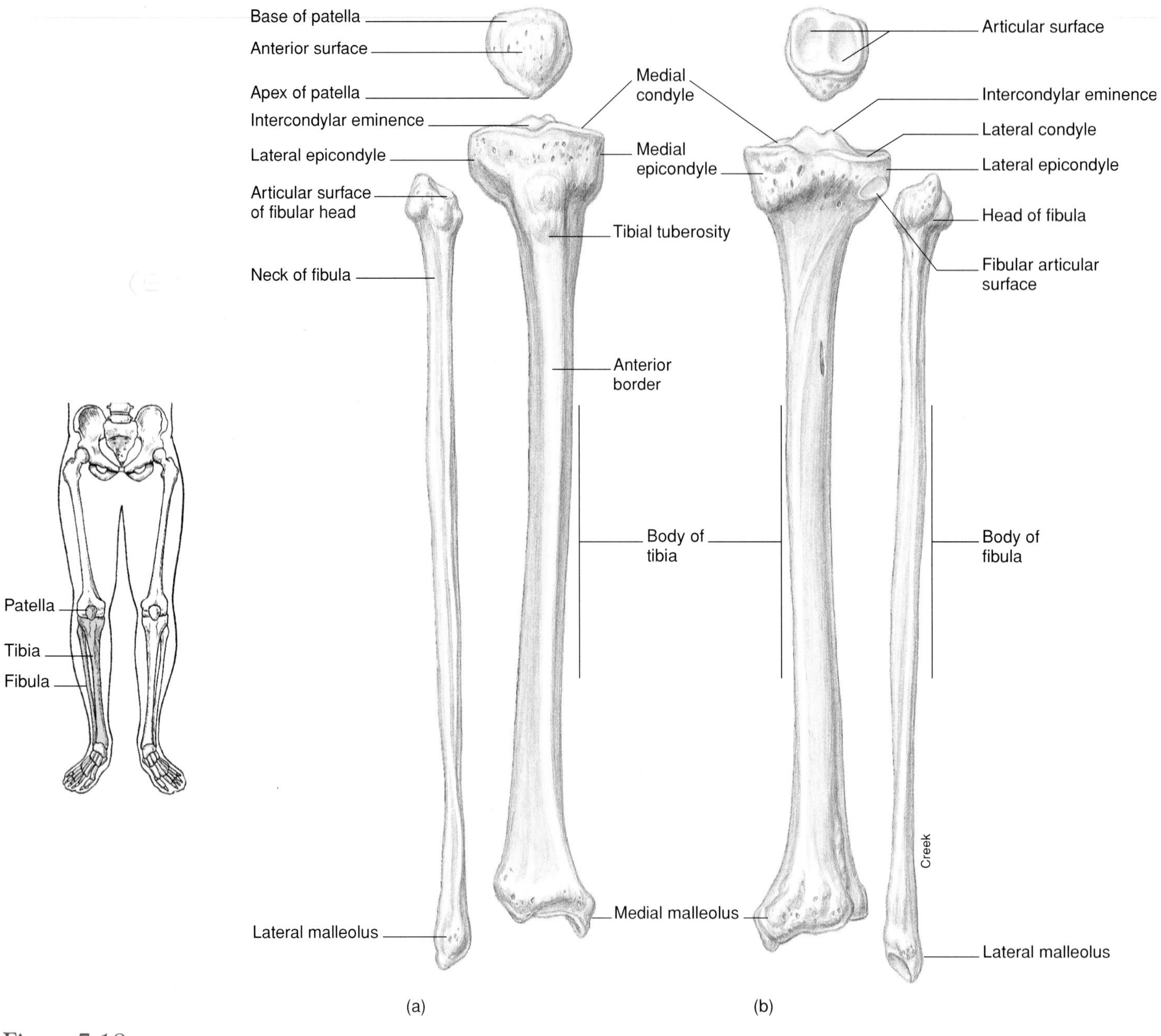

Figure 7.18

The right tibia, fibula, and patella. (*a*) An anterior view and (*b*) a posterior view.

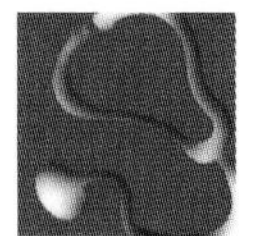

The lateral and medial malleoli are positioned on either side of the talus and help to stabilize the ankle joint. Both processes can be seen as prominent surface features and are easily palpated. Fractures to the fibula above the lateral malleolus are common in skiers. Clinically referred to as *Pott's fracture,* it is caused by a shearing force acting at a vulnerable spot on the leg.

Pes (Foot)

The foot contains 26 bones, grouped into the *tarsus*, *metatarsus*, and *phalanges* (fig. 7.19). Although similar to the bones of the hand, the bones of the foot have distinct structural differences in order to support the weight of the body and provide leverage and mobility during walking.

Tarsus

There are seven tarsal bones. The most superior in position is the **talus,** which articulates with the tibia and fibula to form the

tarsus: Gk. *tarsos*, flat of the foot
talus: L. *talus*, ankle

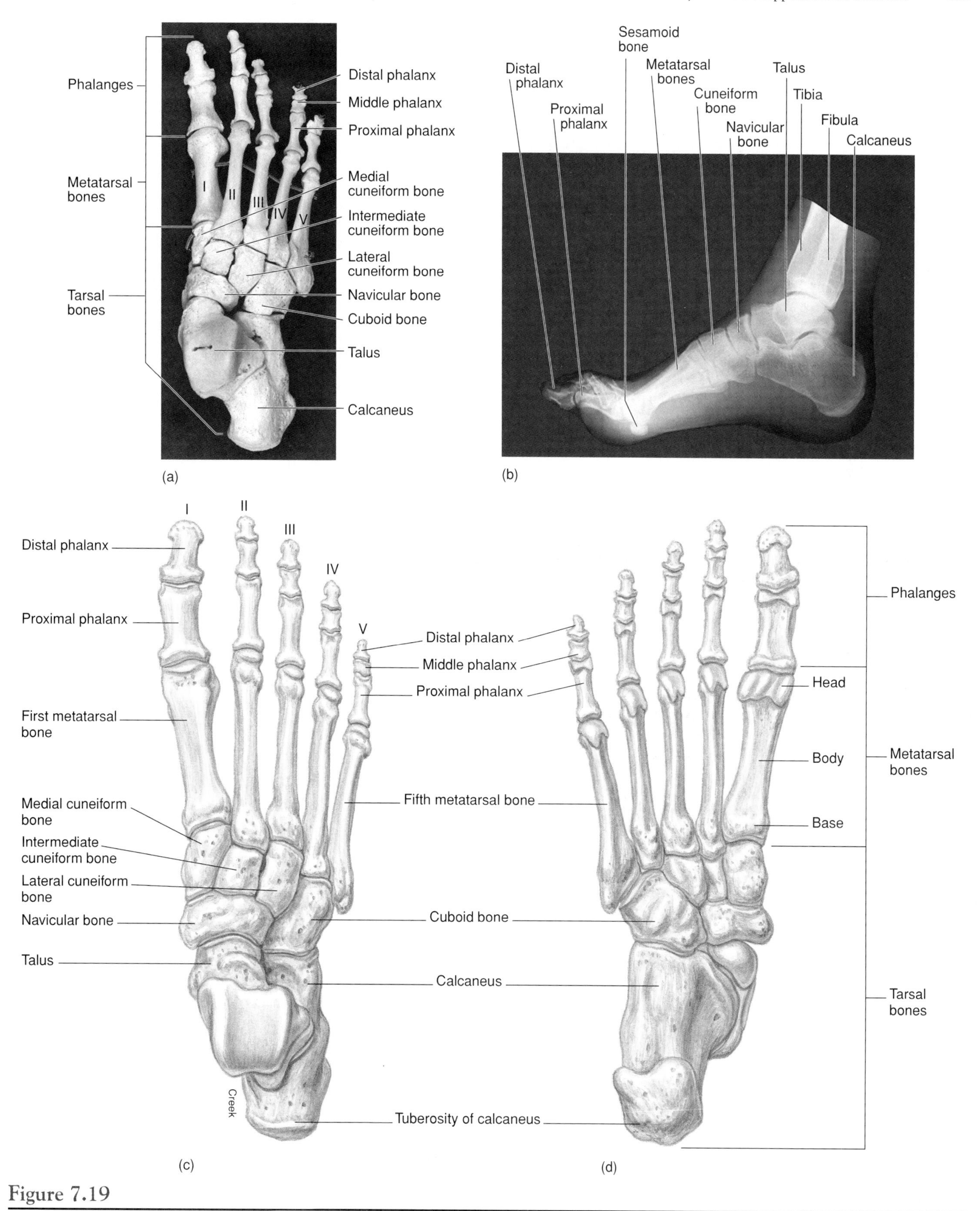

Figure 7.19

The bones of the right foot. (*a*) A photograph of a superior view, (*b*) a radiograph of a medial view, (*c*) a superior view, and (*d*) an inferior view. Each digit (toe) is indicated by a Roman numeral, the first digit, or great toe, being Roman numeral I.

ankle joint. The **calcaneus** (*kal-ka'ne-us*) is the largest of the tarsal bones and provides skeletal support for the heel of the foot. It has a large posterior extension, called the **tuberosity of the calcaneus,** for the attachment of the calf muscles. Anterior to the talus is the block-shaped **navicular bone.** The remaining four tarsal bones form a distal series that articulate with the metatarsal bones. They are, from the medial to the lateral side, the **medial, intermediate,** and **lateral cuneiform** (*kyoo-ne' ĭ-form*) **bones** and the **cuboid bone.**

Metatarsus

The metatarsal bones and phalanges are similar in name and number to the metacarpals and phalanges of the hand. They differ in shape, however, because of their load-bearing role. The metatarsal bones are numbered I to V, starting with the medial (great toe) side of the foot. The first metatarsal bone is larger than the others because of its major role in supporting body weight.

The metatarsal bones each have a **base, body,** and **head.** The proximal bases of the first, second, and third metatarsals articulate proximally with the cuneiform bones. The heads of the metatarsals articulate distally with the proximal phalanges. The proximal joints are called *tarsometatarsal joints*, and the distal joints are called *metatarsophalangeal (met" ă-tar" so-fă-lan' je-al) joints*. The ball of the foot is formed by the heads of the first two metatarsal bones.

Phalanges

The 14 phalanges are the skeletal elements of the toes. As with the fingers of the hand, the phalanges of the toes are arranged in a proximal row, a middle row, and a distal row. The great toe, or *hallux* (adjective, *hallucis*) has only a proximal and a distal phalanx.

Arches of the Foot

The foot has two arches. They are formed by the structure and arrangement of the bones and maintained by ligaments and tendons (fig. 7.20). The arches are not rigid; they "give" when weight is placed on the foot, and they spring back as the weight is lifted.

The **longitudinal arch** is divided into medial and lateral parts. The medial part is the more elevated of the two. The talus is keystone of the medial part, which originates at the calcaneus, rises at the talus, and descends to the first three metatarsal bones. The shallower lateral part consists of the calcaneus, cuboid, and fourth and fifth metatarsal bones. The cuboid is the keystone bone of this arch.

The **transverse arch** extends across the width of the foot and is formed by the calcaneus, navicular, and cuboid bones posteriorly and the bases of all five metatarsal bones anteriorly.

calcaneus: L. *calcis*, heel

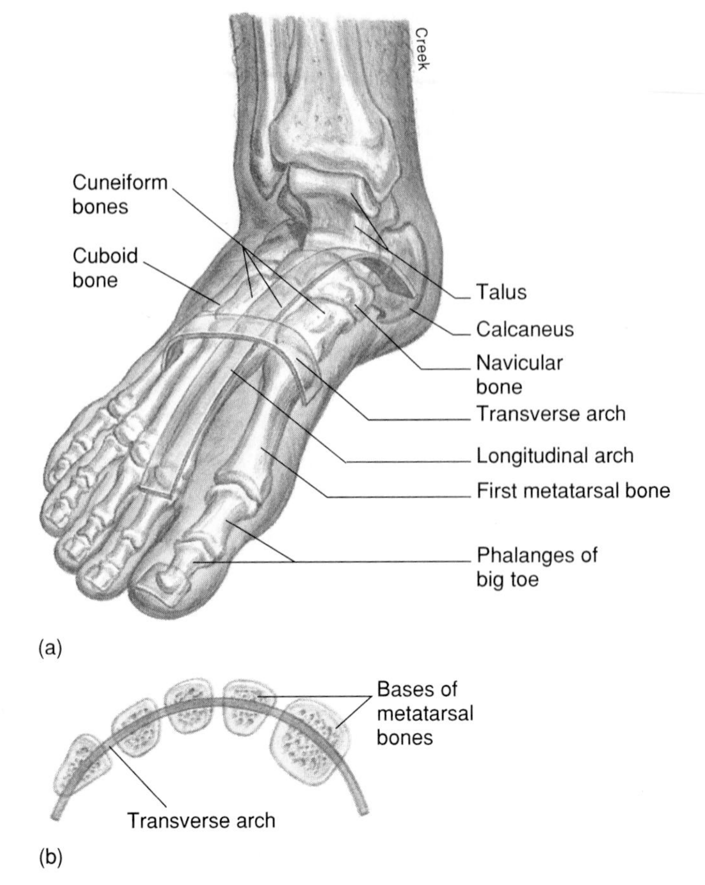

Figure 7.20

The arches of the foot. (*a*) A medial view of the right foot showing both arches and (*b*) a transverse view through the bases of the metatarsal bones showing a portion of the transverse arch.

A weakening of the ligaments and tendons of the foot may cause the arches to "fall"—a condition known as *pes planus*, or, more commonly, "flatfoot."

The bones of the lower extremities are summarized in table 7.3.

1. Describe the structure and functions of the pelvic girdle. How does its structure reflect its weight-bearing role?
2. How can female and male pelves be distinguished? Why is the lesser pelvis clinically significant in females?
3. Describe the structure of each of the long bones of the lower extremity and the position of each of the tarsal bones.
4. Which bones of the foot contribute to the formation of the arches? What are the functions of the arches?

DEVELOPMENTAL EXPOSITION

The Appendicular Skeleton

Exhibit I The development of the appendicular skeleton. (*a*) Limb buds are apparent in an embryo at 28 days and (*b*) an ectodermal ridge is the precursor of the skeletal and muscular structures. (*c*) Mesenchymal primordial cells are present at 33 days. (*d*) Hyaline cartilaginous models of individual bones develop early in the sixth week. (*e*) Later in the sixth week, the cartilaginous skeleton of the upper extremity is well formed.

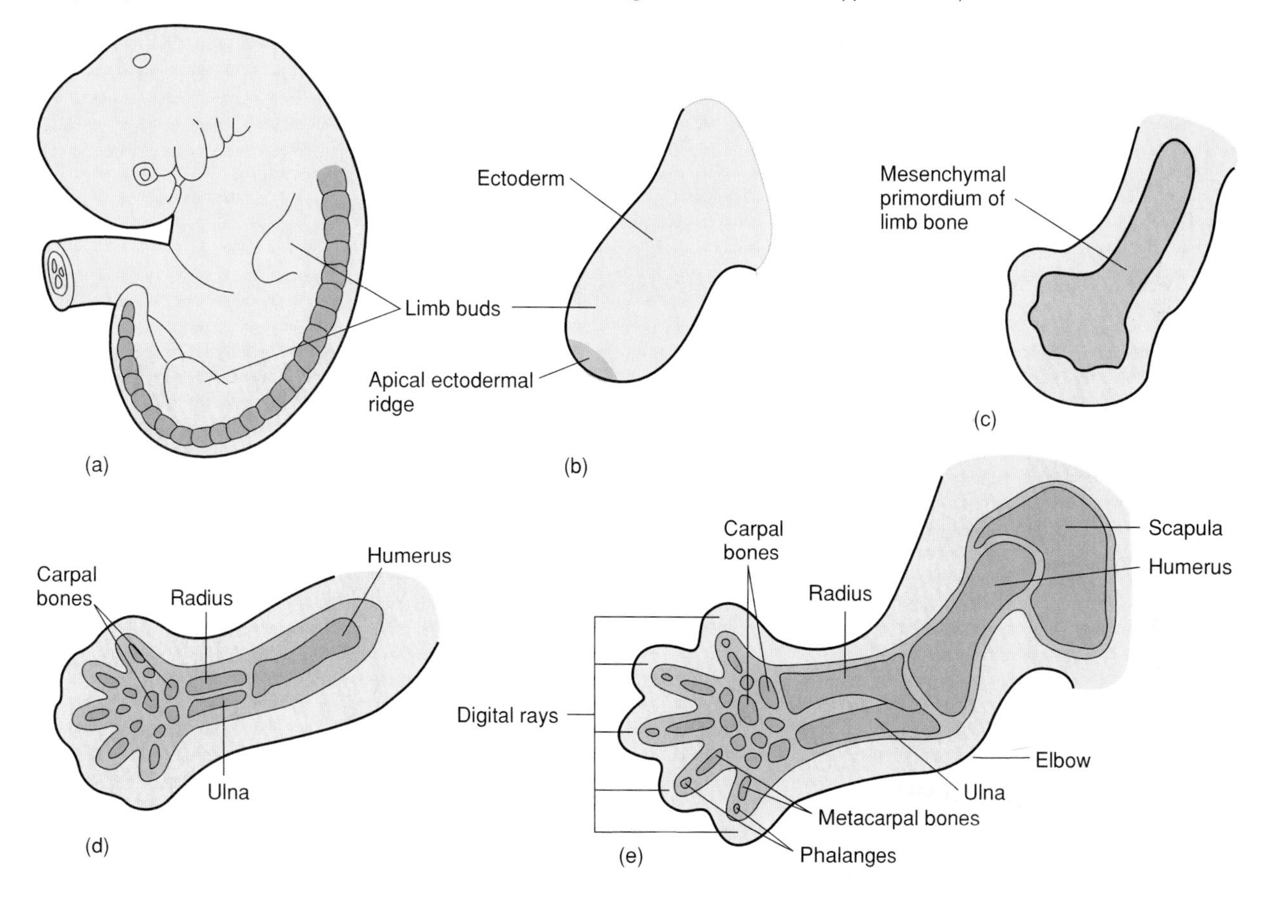

Explanation

The development of the upper and lower extremities is initiated toward the end of the fourth week with the appearance of four small elevations called **limb buds** (exhibit I). The superior pair are the arm buds, whose development precedes that of the inferior pair of leg buds by a few days. Each limb bud consists of a mass of undifferentiated mesoderm partially covered with a layer of ectoderm. This **apical** (*a'pĭ-kal*) **ectodermal ridge** promotes bone and muscle development.

As the limb buds elongate, migrating mesenchymal tissues differentiate into specific cartilaginous bones. Primary ossification centers soon form in each bone, and the hyaline cartilage tissue is gradually replaced by bony tissue in the process of *endochondral ossification* (see chapter 6).

Initially, the developing limbs are directed caudally, but later there is a lateral rotation in the upper extremity and a medial rotation in the lower extremity. As a result, the elbows are directed backward and the knees directed forward.

Digital rays that will form the hands and feet are apparent by the fifth week, and the individual digits separate by the end of the sixth week.

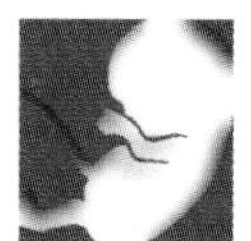

A large number of limb deformities occurred in children born between 1957 and 1962. During this period, the sedative thalidomide was used by large numbers of pregnant women to relieve "morning sickness." It is estimated that 7,000 infants suffered severe limb malformations as a result of exposure to this drug in their early intrauterine life. The malformations ranged from *micromelia* (short limbs) to *amelia* (absence of limbs).

micromelia: Gk. *mikros*, small; *melos*, limb
amelia: Gk. *a*, without; *melos*, limb

Table 7.3

Bones of the Pelvic Girdle and Lower Extremities

Name and Number	Location	Major Distinguishing Features
Os coxae (2)	Hip, part of the pelvic girdle; composed of the fused ilium, ischium, and pubis	Iliac crest; acetabulum; anterior superior iliac spine; greater sciatic notch of the ilium; ischial tuberosity; lesser sciatic notch of the ischium; obturator foramen; pubic tubercle
Femur (2)	Bone of the thigh, between hip and knee	Head; fovea capitis femoris; neck; greater and lesser trochanters; linea aspera; lateral and medial condyles; lateral and medial epicondyles
Patella (2)	Anterior surface of distal femur	Triangular sesamoid bone
Tibia (2)	Medial side of leg, between knee and ankle	Medial and lateral condyles; intercondylar eminence; tibial tuberosity; anterior crest; medial malleolus; fibular notch; medial and lateral epicondyles
Fibula (2)	Lateral side of leg, between knee and ankle	Head; lateral malleolus
Tarsal bones (14)	Ankle	Large talus and calcaneus to receive weight of leg; five other wedge-shaped bones to help form arches of foot
Metatarsal bones (10)	Sole of foot	Long bones, each in line with a digit
Phalanx (28)	Digits	Three in each digit except two in great toe

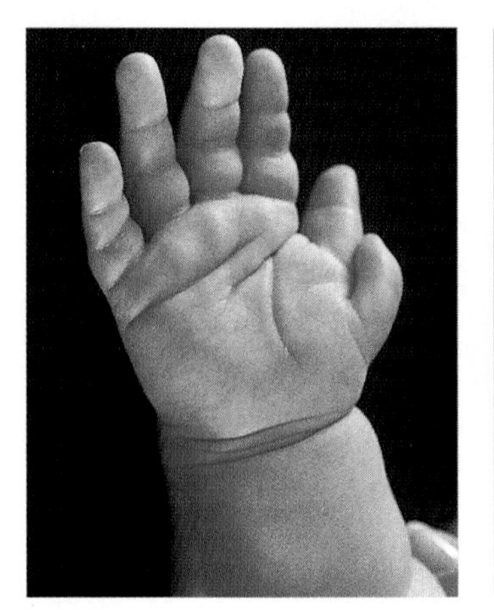

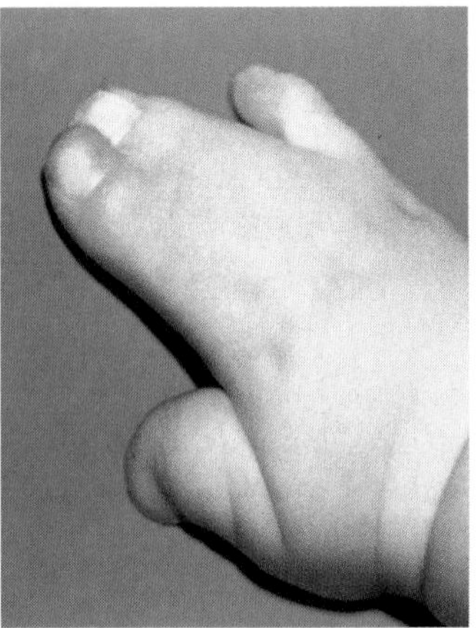

Figure 7.21

Polydactyly is the condition in which there are extra digits. It is the most common congenital deformity of the foot, although it also occurs in the hand. Syndactyly is the condition in which two or more digits are webbed together. It is a common congenital deformity of the hand, although it also occurs in the foot. Both conditions can be surgically corrected.

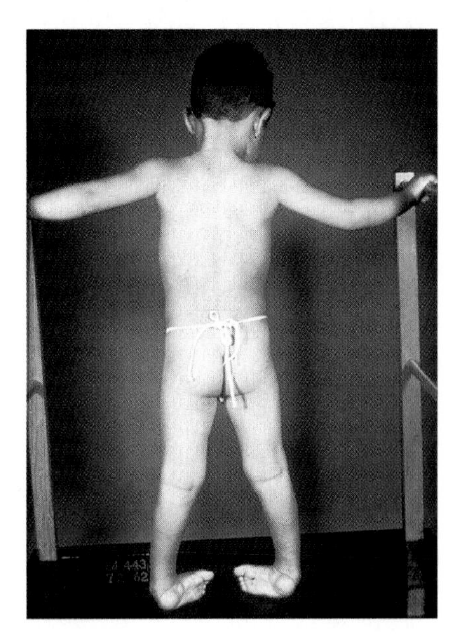

Figure 7.22

Talipes, or clubfoot, is a congenital malformation of a foot or both feet. The condition can be effectively treated surgically if the procedure is done at an early age.

Clinical Considerations

Developmental Disorders

Minor defects of the extremities are relatively common malformations. Extra digits, a condition called **polydactyly** (*pol″e-dak′tĭ-le*) (fig. 7.21), is the most common limb deformity. Usually an extra digit is incompletely formed and does not function. **Syndactyly** (*sin-dak′tĭ-le*), or webbed digits, is also common. Polydactyly is inherited as a dominant trait, whereas syndactyly is a recessive trait.

Talipes (*tal′ĭ-pēz*), or clubfoot (fig. 7.22), is a congenital malformation in which the sole of the foot is twisted medially. It is not certain whether it is abnormal positioning or restricted movement in utero that causes this condition, but both genetics and environmental conditions are involved in most cases.

polydactyly: Gk. *polys*, many; *daktylos*, finger
syndactyly: Gk. *syn*, together; *daktylos*, finger

talipes: L. *talus*, heel; *pes*, foot

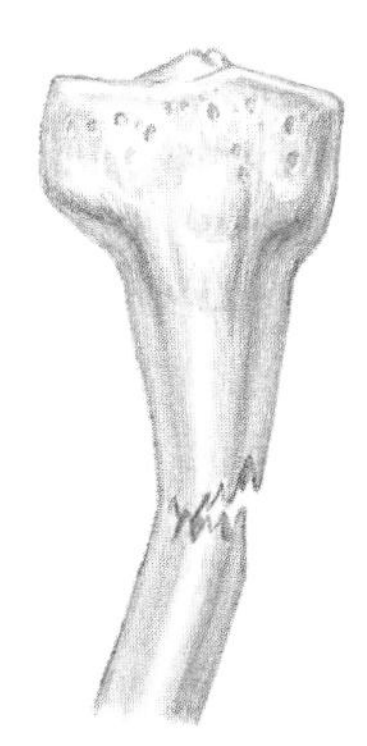

A *greenstick* fracture is incomplete, and the break occurs on the convex surface of the bend in the bone.

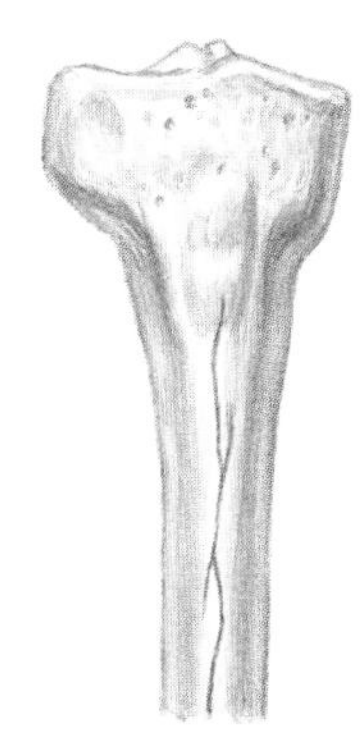

A *partial* (*fissured*) fracture involves an incomplete break.

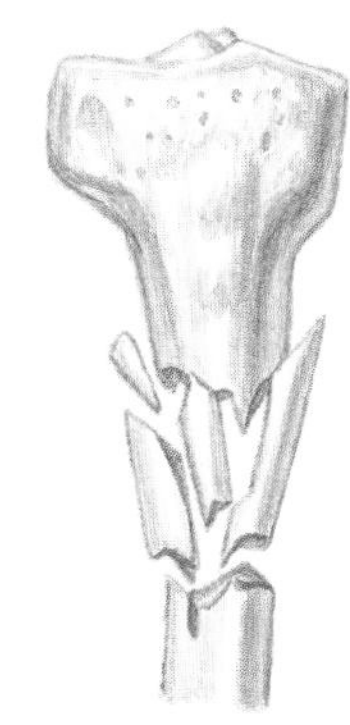

A *comminuted* fracture is complete and results in several bony fragments.

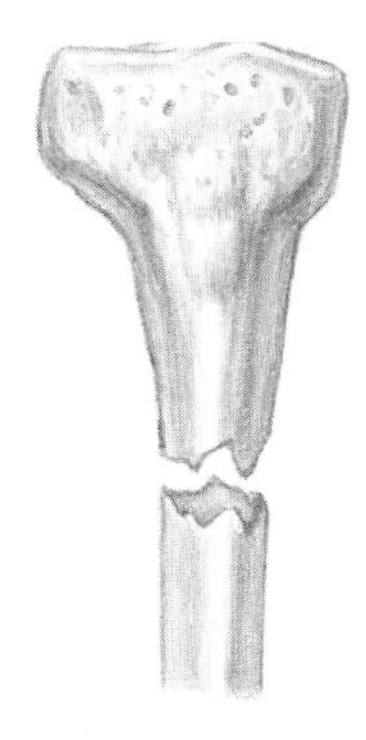

A *transverse* fracture is complete, and the fracture line is horizontal.

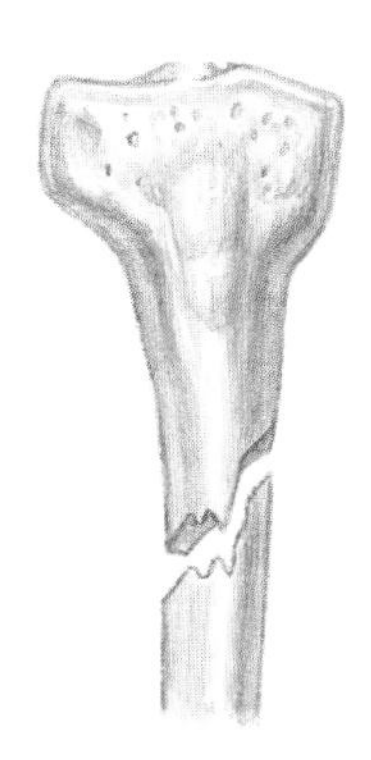

An *oblique* fracture is complete, and the fracture line is at an angle to the long axis of the bone.

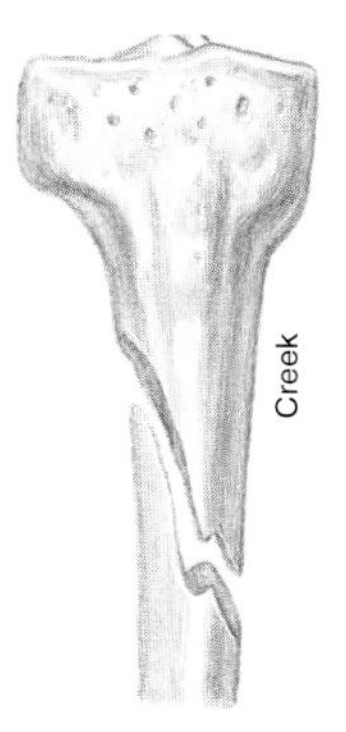

A *spiral* fracture is caused by twisting a bone excessively.

Figure 7.23

Examples of fractures.

Trauma and Injury

The most common type of bone injury is a **fracture**—the cracking or breaking of a bone. Radiographs are often used to diagnose the precise location and extent of a fracture. Fractures may be classified in several ways, and the type and severity of the fracture is often related to the age and general health of the individual. **Pathologic fractures,** for example, result from diseases that weaken the bones. Most fractures, however, are called **traumatic fractures** because they are caused by injuries. The following are descriptions of several kinds of traumatic fractures (fig. 7.23).

1. **Simple,** or **closed.** The fractured bone does not break through the skin.
2. **Compound,** or **open.** The fractured bone is exposed to the outside through an opening in the skin.
3. **Partial (fissured).** The bone is incompletely broken.
4. **Complete.** The fracture has separated the bone into two pieces.
5. **Comminuted** (*kom' ĭ-noot"ed*). The bone is splintered into several fragments.
6. **Spiral.** The fracture line is twisted as it is broken.
7. **Greenstick.** An incomplete break, in which one side of the bone is broken, and the other side is bowed.
8. **Impacted.** One end of a broken bone is driven into the other.
9. **Transverse.** The fracture occurs across the bone at a right angle to the long axis.
10. **Oblique.** The fracture occurs across the bone at an oblique angle to the long axis.
11. **Colles'.** A fracture of the distal portion of the radius.
12. **Pott's.** A fracture of either or both of the distal ends of the tibia and fibula at the level of the malleoli.
13. **Avulsion.** A portion of a bone is torn off.
14. **Depressed.** The broken portion of the bone is driven inward, as in certain skull fractures.
15. **Displaced.** A fracture in which the bone fragments are not in anatomical alignment.
16. **Nondisplaced.** A fracture in which the bone fragments remain in anatomical alignment.

Colles' fracture: from Abraham Colles, Irish surgeon, 1773–1843
Pott's fracture: from Percivall Pott, British surgeon, 1713–1788

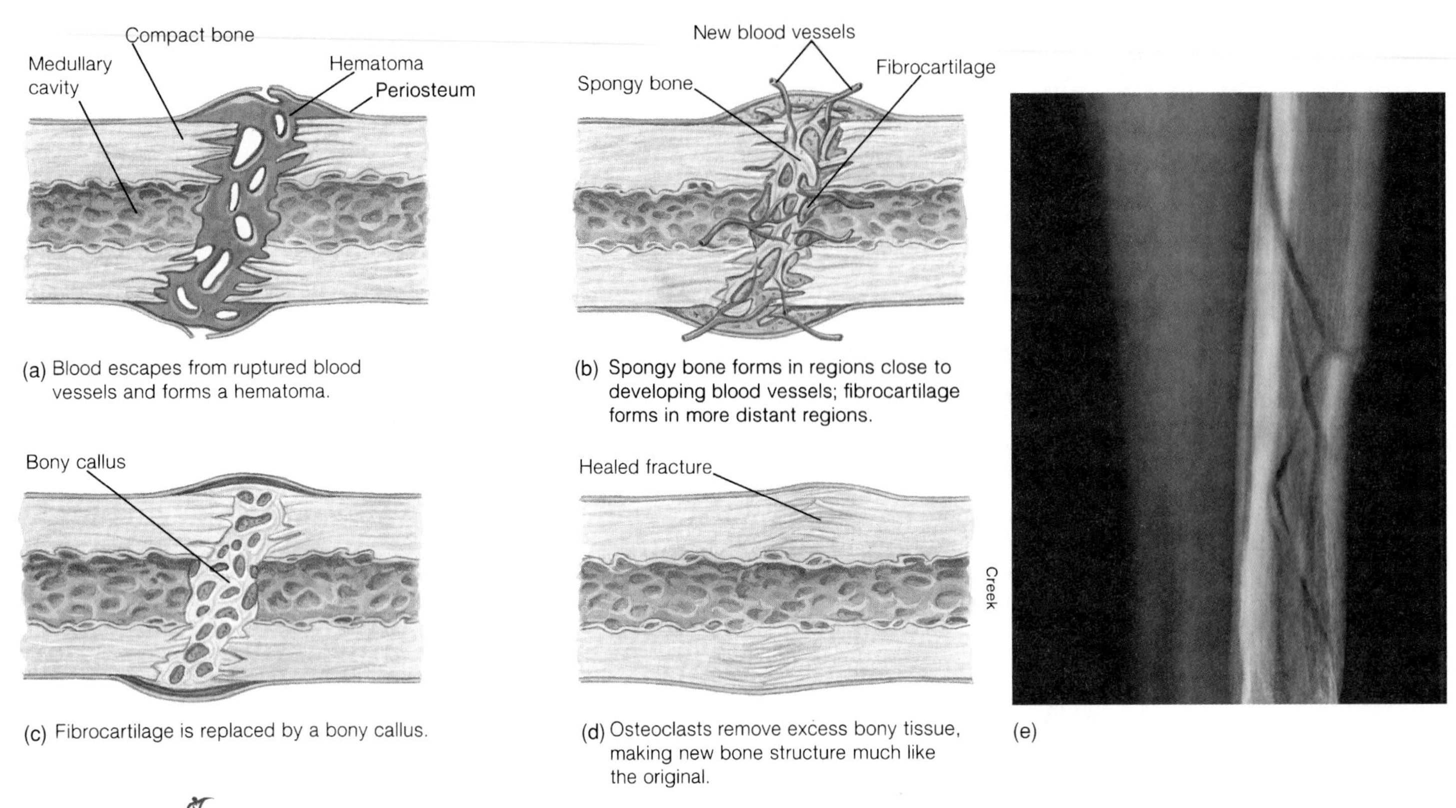

(a) Blood escapes from ruptured blood vessels and forms a hematoma.

(b) Spongy bone forms in regions close to developing blood vessels; fibrocartilage forms in more distant regions.

(c) Fibrocartilage is replaced by a bony callus.

(d) Osteoclasts remove excess bony tissue, making new bone structure much like the original.

(e)

Figure 7.24

Stages (*a–d*) of the repair of a fracture. (*e*) A radiograph of a healing fracture.

When a bone fractures, medical treatment involves realigning the broken ends and then immobilizing them until new bone tissue has formed and the fracture has healed. The site and severity of the fracture and the age of the patient determines the type of immobilization. The methods of immobilization include tape, splints, casts, straps, wires, screws, plates, and steel pins. Certain fractures seem to resist healing, however, even with this array of treatment options. New techniques for treating fractures include applying weak electrical currents to fractured bones. This method has shown promise in promoting healing and significantly reducing the time of immobilization.

Physicians can realign and immobilize a fracture, but the ultimate repair of the bone occurs naturally within the bone itself. Several steps are involved in this process (fig. 7.24).

1. When a bone is fractured, the surrounding periosteum is usually torn and blood vessels in both tissues are ruptured. A blood clot called a **fracture hematoma** (*hēm″ă-tō′mă*) soon forms throughout the damaged area. A disrupted blood supply to osteocytes and periosteal cells at the fracture site causes localized cellular death. This is followed by swelling and inflammation.
2. The traumatized area is "cleaned up" by the activity of phagocytic cells within the blood and osteoclasts that resorb bone fragments. As the debris is removed, fibrocartilage fills the gap within the fragmented bone, and a cartilaginous mass called a **bony callus** is formed. The bony callus becomes the precursor of bone formation in much the same way that hyaline cartilage serves as the precursor of developing bone.
3. The remodeling of the bony callus is the final step in the healing process. The cartilaginous callus is broken down, a new vascular supply is established, and compact bone develops around the periphery of the fracture. A healed fracture line is frequently undetectable in a radiograph, except that for a period of time the bone in this area may be slightly thicker.

Clinical Case Study Answer

The injury involves the cartilaginous epiphyseal growth plate, which is the site of linear growth in long bones. At cessation of growth, this plate disappears as the epiphysis and diaphysis fuse. Until this occurrence, however, disruption of the growth plate can adversely affect growth of the bone.

hematoma: Gk. *hema*, blood; *oma*, tumor

callus: L. *callosus*, hard

Chapter Summary

Pectoral Girdle and Upper Extremity (pp. 169–176)

1. The pectoral girdle consists of the paired scapular and clavicles. Anteriorly, each clavicle articulates with the sternum at the sternoclavicular joint.
   (a) Distinguishing features of the clavicle include the acromial and sternal extremities, conoid tubercle, and costal tuberosity.
   (b) Distinguishing features of the scapula include the spine, acromion, and coracoid process; the supraspinous, infraspinous, and subscapular fossae; the glenoid cavity; the coracoid process; superior, medial, and lateral borders; and superior, inferior, and lateral angles.
2. The brachium contains the humerus, which extends from the shoulder to the elbow.
   (a) Proximally, distinguishing features of the humerus include a rounded head, greater and lesser tubercles, an anatomical neck, and an intertubercular groove. Distally, they include medial and lateral epicondyles, coronoid and olecranon fossae, a capitulum, and a trochlea.
   (b) The head of the humerus articulates proximally with the glenoid cavity of the scapula; distally, the trochlea and capitulum articulate with the ulna and radius, respectively.
3. The antebrachium contains the ulna (medially) and the radius (laterally).
   (a) Proximally, distinguishing features of the ulna include the olecranon and coronoid processes, the trochlear notch, and the radial notch. Distally, they include the styloid process and head of ulna.
   (b) Proximally, distinguishing features of the radius include the head and neck of radius and the tuberosity of radius. Distally, they include the styloid process and ulnar notch.
4. The hand contains 27 bones including 8 carpal bones, 5 metacarpal bones, and 14 phalanges.

Pelvic Girdle and Lower Extremity (pp. 176–186)

1. The pelvic girdle is formed by two ossa coxae, united anteriorly at the symphysis pubis. It is attached posteriorly to the sacrum—a bone of the axial skeleton.
2. The pelvis is divided into a greater pelvis, which helps to support the pelvic viscera, and a lesser pelvis, which forms the walls of the birth canal.
3. Each os coxae consists of an ilium, ischium, and pubis. Distinguishing features of the os coxae include an obturator foramen and an acetabulum, the latter of which is the socket for articulation with the head of the femur.
   (a) Distinguishing features of the ilium include an iliac crest, iliac fossa, anterior superior iliac spine, anterior inferior iliac spine, and greater sciatic notch.
   (b) Distinguishing features of the ischium include the body, ramus, ischial tuberosity, and lesser sciatic notch.
   (c) Distinguishing features of the pubis include the ramus and pubic tubercle. The two pubic bones articulate at the symphysis pubis.
4. The thigh contains the femur, which extends from the hip to the knee, where it articulates with the tibia and the patella.
   (a) Proximally, distinguishing features of the femur include the head, fovea capitus femoris, neck, and greater and lesser trochanters. Distally, they include the lateral and medial epicondyles, the lateral and medial condyles, and the patellar surface. The linea aspera is a roughened ridge positioned vertically along the posterior aspect of the body of the femur.
   (b) The head of the femur articulates proximally with the acetabulum of the os coxae and distally with the condyles of the tibia and the articular facets of the patella.
5. The leg contains the tibia medially and the fibula laterally.
   (a) Proximally, distinguishing features of the tibia include the medial and lateral condyles, medial and lateral epicondyles, intercondylar eminence, and tibial tuberosity. Distally, they include the medial malleolus and fibular notch. The anterior crest is a sharp ridge extending the anterior length of the tibia.
   (b) Distinguishing features of the fibula include the head proximally and the lateral malleolus distally.
6. The foot contains 26 bones including 7 tarsal bones, 5 metatarsal bones, and 14 phalanges.

Review Activities

Objective Questions

1. In anatomical position, the subscapular fossa of the scapula faces
   (a) anteriorly.
   (b) medially.
   (c) posteriorly.
   (d) laterally.
2. The clavicle articulates with
   (a) the scapula and the humerus.
   (b) the humerus and the manubrium.
   (c) the manubrium and the scapula.
   (d) the manubrium, the scapula, and the humerus.
3. Which of the following bones has a conoid tubercle?
   (a) the scapula
   (b) the humerus
   (c) the radius
   (d) the clavicle
   (e) the ulna
4. The proximal process of the ulna is
   (a) the lateral epicondyle.
   (b) the olecranon.
   (c) the coronoid process.
   (d) the styloid process.
   (e) the medial epicondyle.
5. Which of the following statements concerning the carpus is *false?*
   (a) It consists of eight carpal bones arranged in two transverse rows of four bones each.
   (b) All of the carpal bones are considered sesamoid bones.
   (c) The navicular and the lunate articulate with the radius.
   (d) The trapezium, trapezoid, capitate, and hamate articulate with the metacarpals.
6. Pelvimetry is a measurement of
   (a) the os coxae.
   (b) the symphysis pubis.
   (c) the pelvic brim.
   (d) the lesser pelvis.
7. Which of the following is *not* a structural feature of the os coxae?
   (a) the obturator foramen
   (b) the acetabulum
   (c) the auricular surface
   (d) the greater sciatic notch
   (e) the linea aspera
8. A fracture across the intertrochanteric line would involve
   (a) the ilium.
   (b) the femur.
   (c) the tibia.
   (d) the fibula.
   (e) the patella.

9. Relative to the male pelvis, the female pelvis
   (a) is more massive.
   (b) is narrower at the pelvic outlet.
   (c) is tilted backward.
   (d) has a shallower symphysis pubis.
10. Clubfoot is a congenital foot deformity that is clinically referred to as
    (a) talipes.
    (b) syndactyly.
    (c) pes planus.
    (d) polydactyly.

Essay Questions

1. Contrast the structure of the pectoral and pelvic girdles. How do the structural differences relate to differences in function?
2. Explain why the clavicle is more frequently fractured than the scapula.
3. List the processes of the bones of the upper and lower extremities that can be palpated. Why are these bony landmarks important to know?
4. The bones of the hands are similar to those of the feet, but there are some important differences in structure and arrangement. Compare and contrast the anatomy of these appendages, taking into account their functional roles.
5. Define *bipedal locomotion* and discuss the adaptations of the pelvic girdle and lower extremities that allow for this type of movement.
6. Explain how the female pelvis is adapted to the needs of pregnancy and childbirth.
7. Explain the significance of the limb buds, apical ectodermal ridges, and digital rays in limb development. When does limb development begin and when is it complete?
8. What is meant by a congenital skeletal malformation? Give two examples of such abnormalities that occur within the appendicular skeleton.
9. What are the differences between pathological and traumatic fractures? Give some examples of traumatic fractures.
10. How does a fractured bone repair itself? Why is it important that the fracture be immobilized?

Critical-Thinking Questions

1. James Smithson, benefactor of the Smithsonian Institution, died in 1829 at the age of 64. Although his body has been buried in Italy, it was reinterred in 1904 near the front entry of the Smithsonian in Washington, D.C. Before the reburial, scientists at the Smithsonian carefully examined Smithson's skeleton to learn more about him. From the bones they concluded that Smithson was rather slightly built but athletic—he had a large chest and powerful arms and hands. His teeth were worn on the left side from chewing a pipe. The scientists also reported that "certain peculiarities of the right little finger suggest that he may have played the harpsichord, piano, or a stringed instrument such as a violin." Preserved bones can serve as a storehouse of information. Considering current technology, what other types of information might be gleaned from examination of a preserved skeleton?
2. Which would you say has been more important in human evolution—adaptation of the hand or adaptation of the foot? Explain your reasoning.
3. Speculate as to why a single bone is present in both the brachium and the thigh, whereas the antebrachium and leg each have two bones.
4. Compare the tibia and fibula with respect to structure and function. Which would be more debilitating, a compound fracture of the tibia or a compound fracture of the fibula?

EIGHT

Articulations

Clinical Case Study 192
Classification of Joints 192
Fibrous Joints 192
  Sutures 193
  Syndesmoses 193
  Gomphoses 194
Cartilaginous Joints 194
  Symphyses 194
  Synchondroses 194
Synovial Joints 195
  Structure of a Synovial Joint 196
  Kinds of Synovial Joints 198
Movements at Synovial Joints 201
  Angular Movements 201
  Circular Movements 201
  Special Movements 203
  Biomechanics of Body Movement 203
Specific Joints of the Body 208
  Temporomandibular Joint 209
  Sternoclavicular Joint 209
  Glenohumeral (Shoulder) Joint 210
  Elbow Joint 211
  Metacarpophalangeal and Interphalangeal Joints 212
  Coxal (Hip) Joint 214
  Tibiofemoral (Knee) Joint 215
  Talocrural (Ankle) Joint 217
Clinical Considerations 218
  Trauma to Joints 218
Developmental Exposition: The Synovial Joints 220
  Diseases of Joints 221
  Treatment of Joint Disorders 222
Clinical Case Study Answer 222
Important Clinical Terminology 224
Chapter Summary 224
Review Activities 224

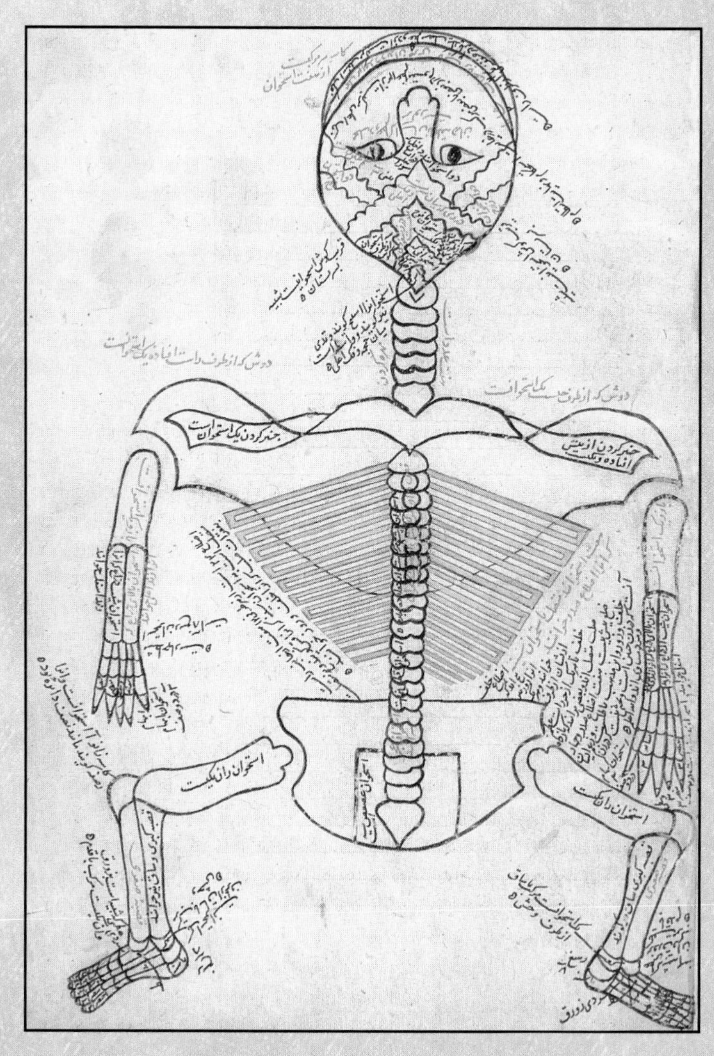

This illustration from a fifteenth-century Persian manuscript emphasizes the articulations of the body—the junctions between bones that hold the skeleton together and permit it to move.

Clinical Case Study

A 20-year-old college football player sustained injury to his right knee during the opening game of the season. Because of rapid swelling and intense pain, he was taken to the emergency room of the local hospital. When the attending physician asked him to describe how the injury occurred, the athlete responded, "I was carrying the ball on an end run left on third down and two. As I planted my right foot just before I was going to make my cut, I was hit in the knee from the side. I felt my knee give way, and then I felt a stabbing pain on the inside of my knee."

Close examination by the physician revealed marked swelling on the medial part of the knee. The doctor determined that *valgus stress* (an inward bowing stress on the knee) caused the medial aspect of the joint to "open." Which stabilizing structure is most likely injured? Which cartilaginous structure is frequently injured in association with the previously mentioned structure? Is there an anatomical explanation? What are some other stabilizing structures within the knee that are frequently injured in sports?

Hint: An impact to one side of the knee generally results in greater trauma to the other side. Carefully read the sections in this chapter on the tibiofemoral (knee) joint and trauma to joints. In addition, examine figures 8.31 and 8.32.

Classification of Joints

On the basis of anatomical structure, the articulations between the bones of the skeleton are classified as fibrous joints, cartilaginous joints, or synovial joints. Fibrous joints firmly bind skeletal elements together with fibrous connective tissue. Cartilaginous joints firmly unite skeletal elements with cartilage. Synovial joints are freely movable joints; they are enclosed by joint capsules that contain synovial fluid.

Objective 1 Define *arthrology* and *kinesiology*.

Objective 2 Compare and contrast the three principal kinds of joints.

One of the functions of the skeletal system is to permit body movement. It is not the bones themselves that allow movement, but rather the unions between the bones, called **articulations** or **joints.** Although the joints of the body are actually part of the skeletal system, this chapter is devoted entirely to them.

The structure of a joint determines the direction range of movement it permits. Not all joints are flexible, however, and as one part of the body moves, other joints remain rigid to stabilize the body and maintain balance. The coordinated activity of the joints permits the sinuous, elegant movements of a gymnast or ballet dancer, just as it permits all of the commonplace actions associated with walking, eating, writing, and speaking.

Arthrology is the science concerned with the study of joints. Generally speaking, an arthrologist is interested in the structure, classification, and function of joints, including any dysfunctions that may develop. *Kinesiology* (*kĭ-ne″se-ol′ŏ-je*), a more applied and dynamic science, is concerned with the mechanics of human motion—the functional relationship of the bones, muscles, and joints as they work together to produce coordinated movement. Kinesiology is a subdiscipline of *biomechanics*, which deals with a broad range of mechanical processes, including the forces that govern blood circulation and respiration.

In studying the joints, a kinetic approach allows for the greatest understanding. The student should be able to demonstrate the various movements permitted at each of the movable joints. Additionally, he or she should be able to explain the adaptive advantage, as well as the limitations, of each type of movement.

The articulations of the body are grouped by their structure into three principal categories.

1. **Fibrous joints.** In fibrous joints, the articulating bones are held together by fibrous connective tissue. These joints lack joint cavities.
2. **Cartilaginous joints.** In cartilaginous joints, the articulating bones are held together by cartilage. These joints also lack joint cavities.
3. **Synovial joints.** In synovial (*sĭ-no′ve-al*) joints, the articulating bones are capped with cartilage, and ligaments frequently help to support them. These joints are distinguished by fluid-filled joint cavities.

1. Explain the statement that kinesiology is applied arthrology.
2. List the three types of joints and speculate as to which would be most supportive and which most vulnerable to trauma.

Fibrous Joints

As the name suggests, the articulating bones in fibrous joints are tightly bound by fibrous connective tissue. Fibrous joints range from rigid and relatively immovable joints to those that are slightly movable. The three kinds of fibrous joints are sutures, syndesmoses, and gomphoses.

---

arthrology: Gk. *arthron*, joint; *logos*, study
kinesiology: Gk. *kinesis*, movement; *logos*, study

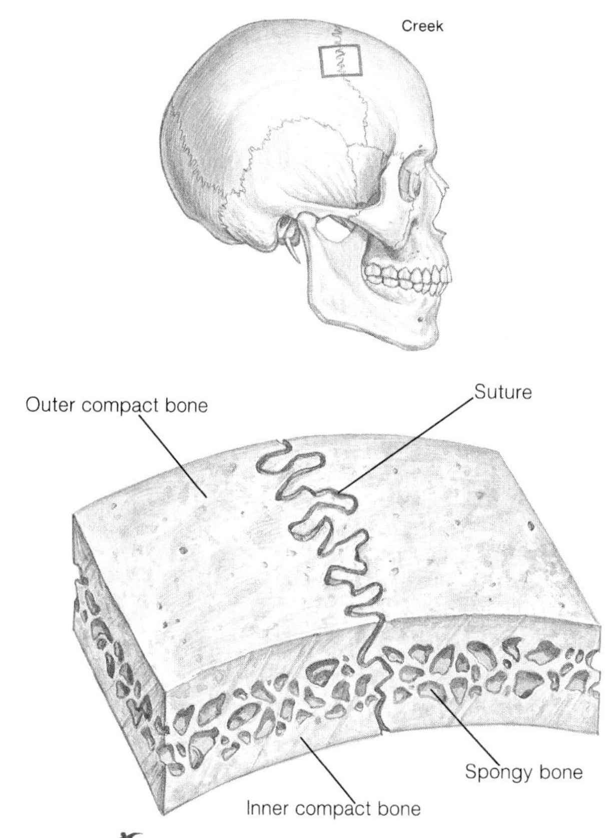

Figure 8.1

A section of the skull showing a suture.

Objective 3 Describe the structure of a suture and indicate where sutures are located.

Objective 4 Describe the structure of a syndesmosis and indicate where syndesmoses are located.

Objective 5 Describe the structure of gomphoses and note their location. Also, discuss the importance of these joints in dentistry.

Sutures

Sutures are found only within the skull. They are characterized by a thin layer of dense irregular connective tissue that binds the articulating bones (fig. 8.1). Sutures form at about 18 months of age and replace the pliable fontanels of an infant's skull (see fig. 6.13).

Different types of sutures can be distinguished by the appearance of the articulating edge of bone. A **serrate suture** is characterized by interlocking sawlike articulations. This is the most common type of suture, an example of which is the sagittal suture between the two parietal bones. In a **squamous** (*lap*) **suture,** the edge of one bone overlaps that of the articulating bone. The squamous suture formed between the temporal and parietal bones is an example. In a **plane** (*butt*) **suture,** the edges of the articulating bones are fairly smooth and do not overlap. An example is the median palatine suture, where the paired maxillary and palatine bones articulate to form the hard palate (see fig. 6.16).

suture: L. *sutura*, sew

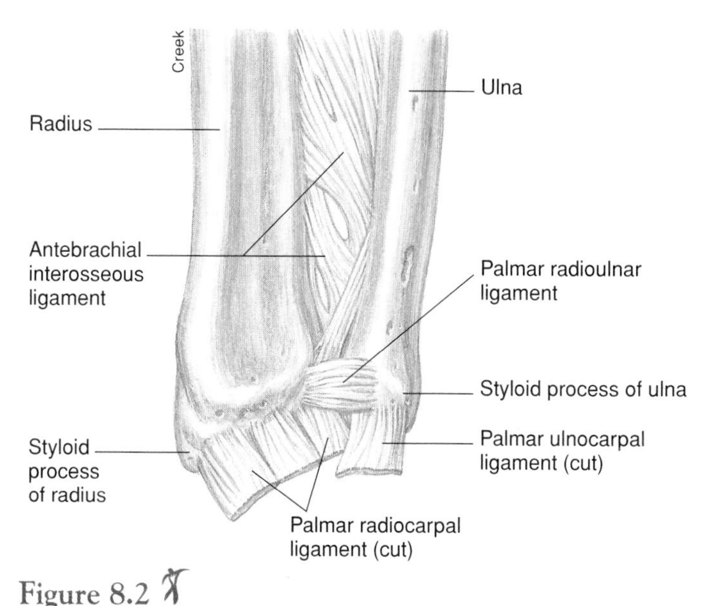

Figure 8.2

The side-to-side articulation of the ulna and radius forms a syndesmotic joint. An interosseous ligament tightly binds these bones and permits only slight movement between them.

The nomenclature in human anatomy is extensive and precise. There are over 30 named sutures in the skull, even though just a few of them are mentioned by name in figures 6.15, 6.16, and 6.17. Review these illustrations and make note of the bones that articulate to form the sutures identified.

A *synostosis* (sin"os-to'sis) is a sutural joint in that it is present during growth of the skull, but in the adult it becomes totally ossified. For example, the frontal bone forms as two separate components, but the suture becomes obscured in most individuals as the skull completes its growth.

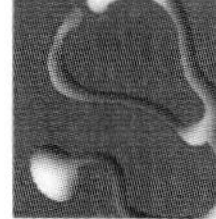

Fractures of the skull are fairly common in an adult but much less so in a child. The skull of a child is resilient to blows because of the nature of the bone and the layer of fibrous connective tissue within the sutures. The skull of an adult is much like an eggshell in its lack of resilience. It will frequently splinter on impact.

Syndesmoses

Syndesmoses (*sin"des-mo'sēz*) are fibrous joints held together by collagenous fibers or sheets of fibrous tissue called *interosseous ligaments*. The tympanostapedial joint in the middle-ear chamber is a syndesmosis. This type of joint also occurs in the antebrachium (forearm) between the distal parts of the radius and ulna (fig. 8.2)

syndesmosis: Gk. *syndesmos*, binding together

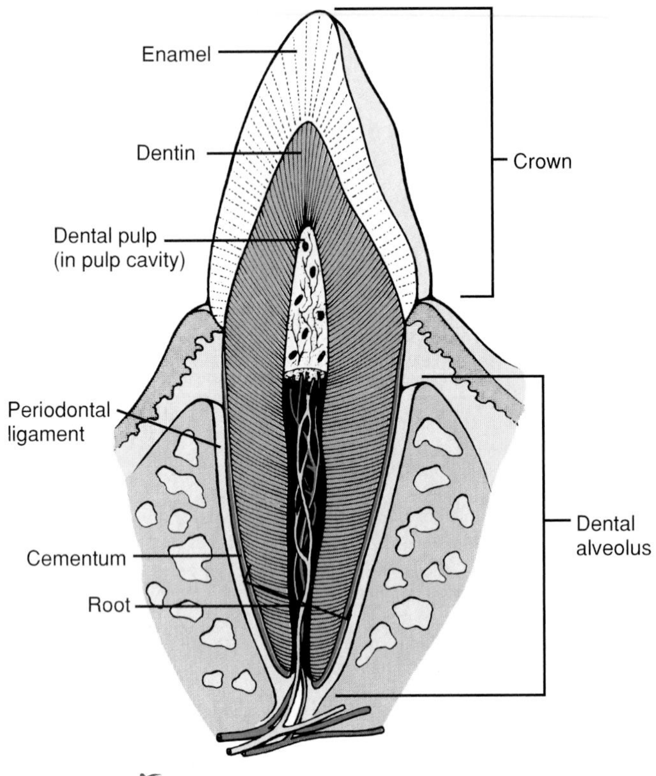

Figure 8.3

A gomphosis is a fibrous joint in which a tooth is held in its socket.

and in the leg between the distal parts of the tibia and fibula. Slight movement is permitted at these joints as the antebrachium or leg is rotated.

1. Compare and contrast the three kinds of sutures. Give an example of each and note its location.
2. In what way does the structure of a syndesmosis permit it to move slightly?
3. A gomphosis is called a peg-and-socket joint. What do the "peg" and "socket" represent?

Gomphoses

Gomphoses (*gom-fo'sēz*) are fibrous joints that occur between the teeth and the supporting bones of the jaws. More specifically, a gomphosis, or *dentoalveolar joint*, is where the root of a tooth is attached to the periodontal ligament of the dental alveolus (tooth socket) of the bone (fig. 8.3).

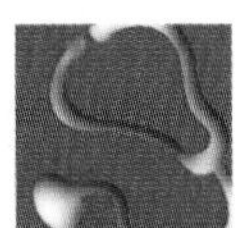

Periodontal disease occurs at gomphoses. It refers to the inflammation and degeneration of the gum, periodontal ligaments, and alveolar bone tissue. With this condition, the teeth become loose and plaque accumulates on the roots. Periodontal disease may be caused by poor oral hygiene, compacted teeth (poor alignment), or local irritants, such as impacted food, chewing tobacco, or cigarette smoke.

gomphosis: Gk. *gompho*, nail or bolt

Cartilaginous Joints

Cartilaginous joints allow limited movement in response to twisting or compression. The two types of cartilaginous joints are symphyses and synchondroses.

Objective 6 Describe the structure of a symphysis and indicate where symphyses occur.

Objective 7 Describe the structure of a synchondrosis and indicate where synchondroses occur.

Symphyses

The adjoining bones of a **symphysis** (*sim'fĭ-sis*) are covered with hyaline cartilage, which becomes infiltrated with collagenous fibers to form an intervening pad of fibrocartilage. This pad cushions the joint and allows limited movement. The symphysis pubis and the intervertebral joints formed by the intervertebral discs (fig. 8.4) are examples of symphyses. Although only limited motion is possible at each intervertebral joint, the combined movement of all of the joints of the vertebral column results in extensive spinal action.

Synchondroses

Synchondroses (*sin" kon-dro'sēz*) are cartilaginous joints that have hyaline cartilage between the articulating bones. Some of these joints are temporary, forming the epiphyseal plates (growth plates) between the diaphyses and epiphyses in the long bones of children (fig. 8.5). When growth is complete, these synchondrotic joints ossify. A totally ossified synchondrosis may also be referred to as a *synostosis*.

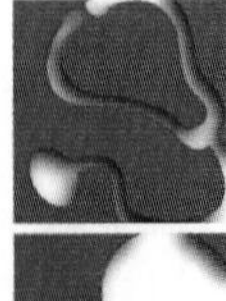
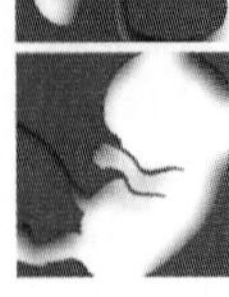

A fracture of a long bone in a child may be extremely serious if it involves the mitotically active epiphyseal plate of a synchondrotic joint. If such an injury is left untreated, bone growth is usually retarded or arrested, so that the appendage will be shorter than normal.

symphysis: Gk. *symphysis*, growing together
synchondrosis: Gk. *syn*, together; *chondros*, cartilage
synostosis: Gk. *syn*, together; *osteon*, bone

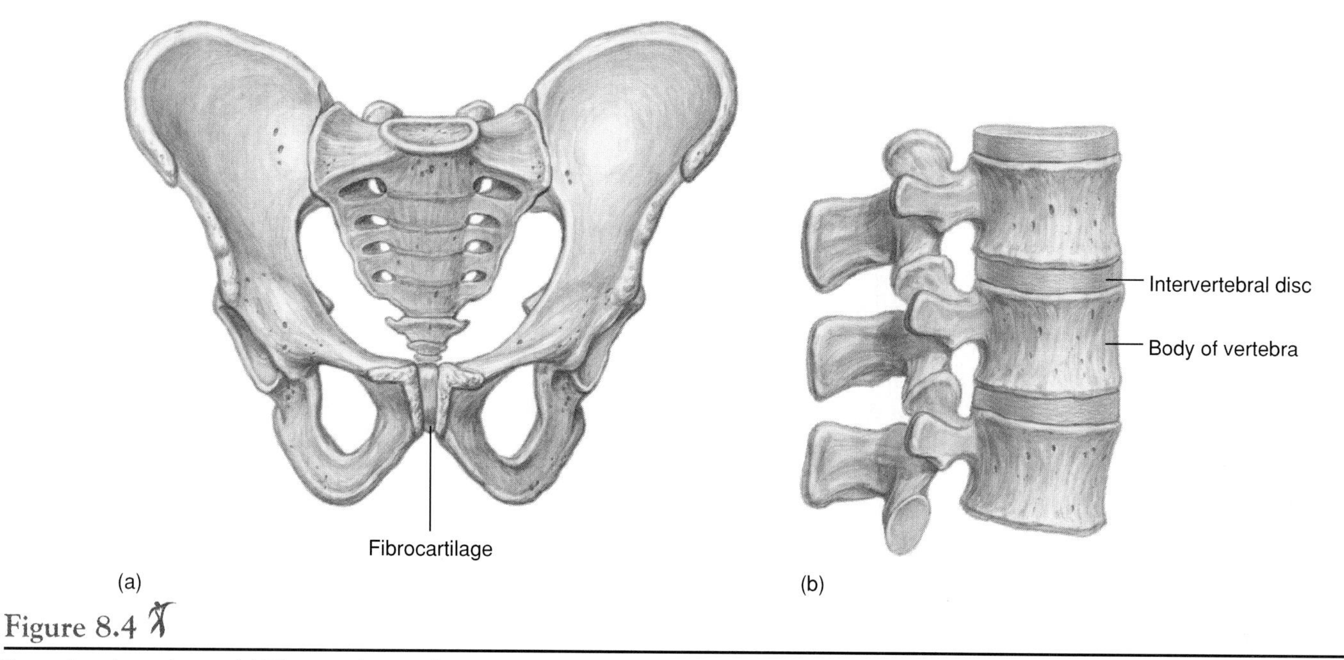

Figure 8.4

Examples of symphyses. (*a*) The symphysis pubis and (*b*) the intervertebral joints between vertebral bodies.

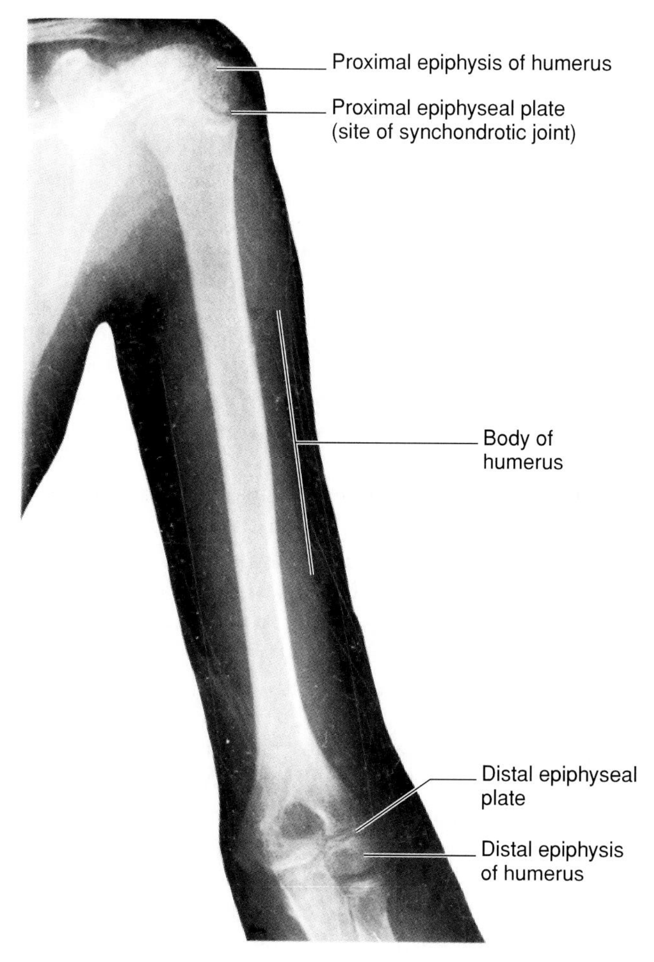

Figure 8.5

A radiograph of the left humerus of a 10-year-old child showing a synchondrotic joint. In a long bone, this type of joint occurs at both the proximal and distal epiphyseal plates. The mitotic activity at synchondrotic joints is responsible for bone growth in length.

Synchondroses that do not ossify as a person ages are those between the occipital, sphenoid, temporal, and ethmoid bones of the skull. In addition, the costochondral articulations between the ends of the ribs and the costal cartilages that attach to the sternum are examples of synchondroses. Elderly people often exhibit some ossification of the costal cartilages of the rib cage. This may restrict movement of the rib cage and obscure an image of the lungs in a thoracic radiograph.

1. Discuss the function of the pad of fibrocartilage in a symphysis and give two examples of symphyses.
2. What structural feature is characteristic of all synchondroses? Give two examples of synchondroses.

Synovial Joints

The freely movable synovial joints are enclosed by joint capsules containing synovial fluid. Based on the shape of the articular surfaces and the kinds of motion they permit, synovial joints are categorized as gliding, hinge, pivot, condyloid, saddle, or ball-and-socket.

Objective 8 Describe the structure of a synovial joint.

Objective 9 Discuss the various kinds of synovial joints, noting where they occur and the movements they permit.

The most obvious type of articulation in the body is the freely movable synovial joint. The function of synovial joints is to

Figure 8.6

Although joint flexibility is structurally determined and limited, some individuals can achieve an extraordinary range of movement through extensive training.

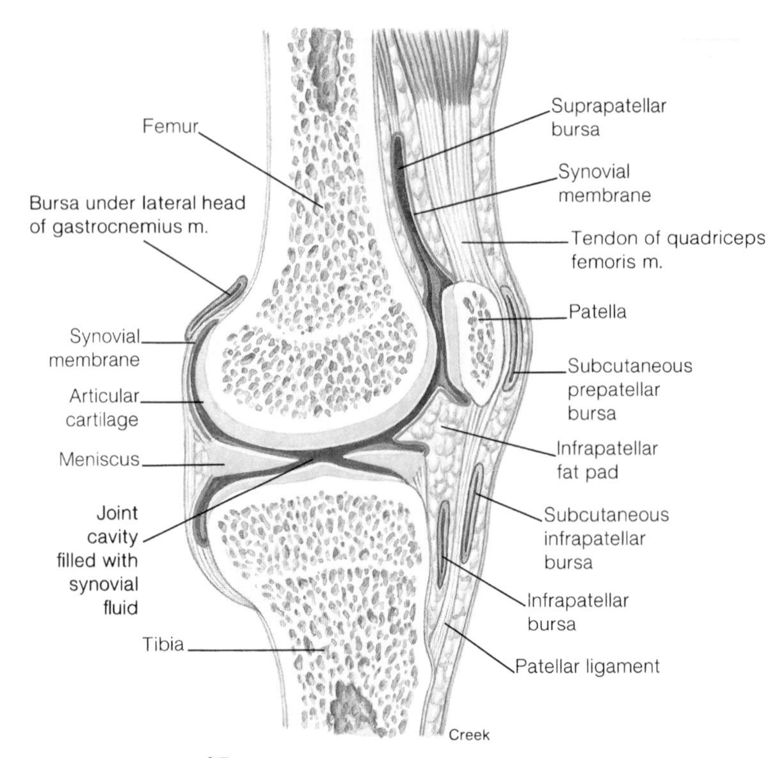

Figure 8.7

A synovial joint is represented by the knee joint, shown here in a sagittal view.

provide a wide range of precise, smooth movements, at the same time maintaining stability, strength, and, in certain aspects, rigidity in the body.

Synovial joints are the most complex and varied of the three major types of joints. A synovial joint's range of motion is determined by three factors:

1. the structure of the bones involved in the articulation (for example, the olecranon of the ulna prevents hyperextension of the elbow joint);
2. the strength of the joint capsule and the strength and tautness of the associated ligaments and tendons; and
3. the size, arrangement, and action of the muscles that span the joint. Range of motion at synovial joints is characterized by tremendous individual variation, most of which is related to body conditioning (fig. 8.6). Although some people can perform remarkable contortions and are said to be "double-jointed," they have no extra joints that help them do this. Rather, through conditioning, they are able to stretch the ligaments that normally inhibit movement.

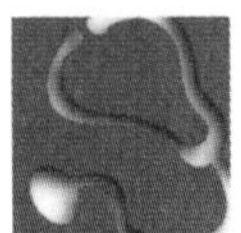

Arthroplasty is the surgical repair or replacement of joints. Advancements in this field continue as new devices are developed to restore lost joint function and permit movement that is free of pain. A recent advancement in the repair of soft tissues involves the use of *artificial ligaments.* A material consisting of carbon fibers coated with a plastic called polylactic acid is sewn in and around torn ligaments and tendons. This reinforces the traumatized structures and provides a scaffolding on which the body's collagenous fibers can grow. As healing progresses, the polylactic acid is absorbed and the carbon fibers break down.

Structure of a Synovial Joint

Synovial joints are enclosed by a fibroelastic **joint capsule** (articular capsule), which is filled with lubricating **synovial fluid** (fig. 8.7). The term *synovial* is derived from a Greek word meaning "egg white," which this fluid resembles. It is secreted by a thin **synovial membrane** that lines the inside of the joint capsule. Synovial fluid is similar to interstitial fluid (fluid between the cells). It is rich in hyaluronic acid and albumin, and also contains phagocytic cells that clean up tissue debris resulting from wear on the joint cartilages. The bones that articulate in a synovial joint are capped with a smooth layer of hyaline cartilage called the **articular cartilage.** Articular cartilage is only about 2 mm thick. Because articular cartilage lacks blood vessels, it has to be nourished by the movement of synovial fluid during joint activity.

Ligaments are flexible connective tissue cords that connect from bone to bone as they help bind synovial joints. Ligaments may be located within the joint cavity or on the outside of the capsule. Tough, fibrous cartilaginous pads called **menisci** (*mĕ-nis′ ki*—singular, *meniscus*) are unique to the knee joint, where they cushion and guide the articulating bones. A few other synovial joints, such as the temporomandibular joint (see fig. 8.23), have a fibrocartilaginous pad called an **articular disc** that provides functions similar to menisci.

arthroplasty: Gk. *arthron*, joint; *plasso*, to form

meniscus: Gk. *meniskos*, small moon

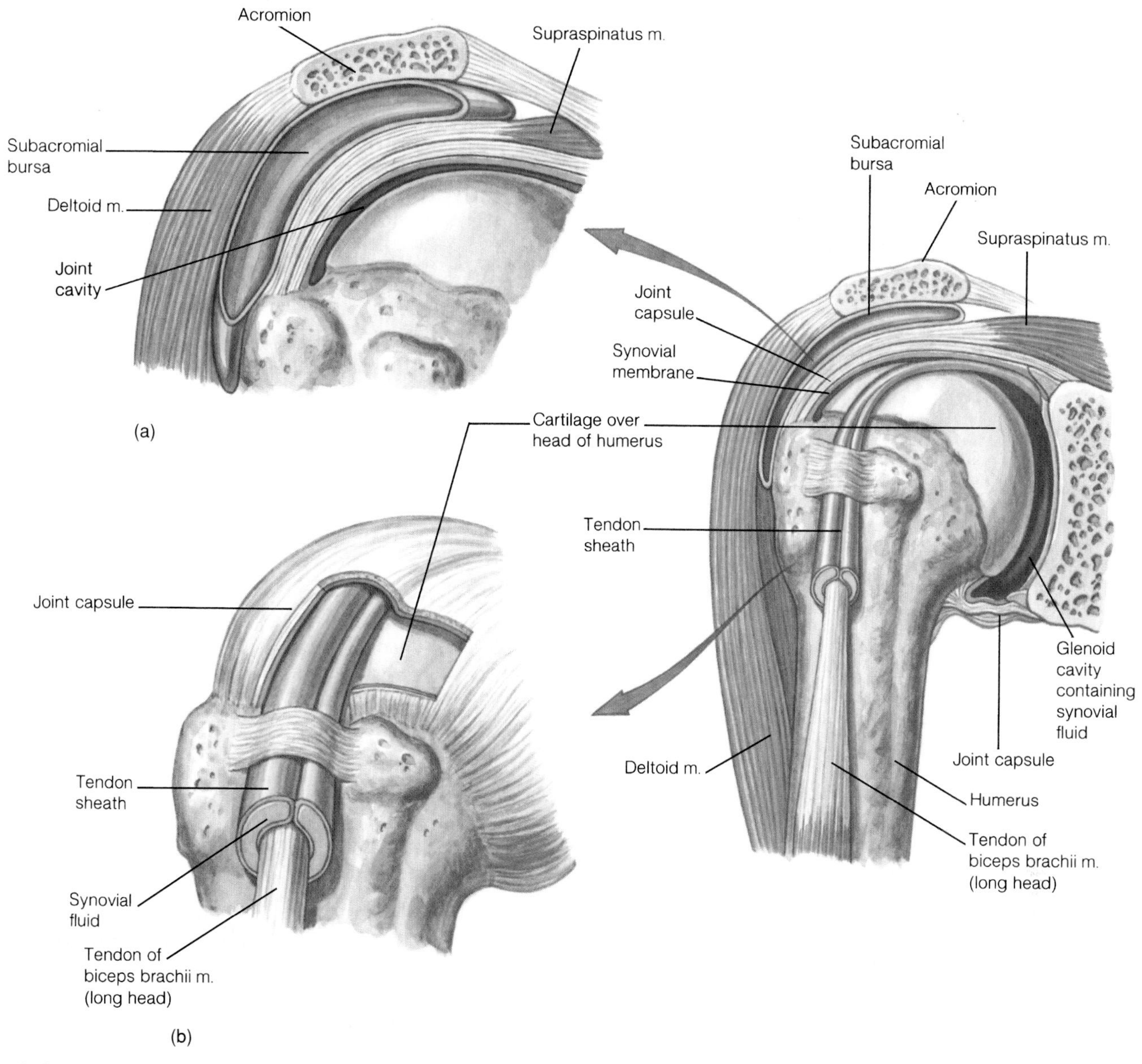

Figure 8.8

Bursae and tendon sheaths are friction-reducing structures found in conjunction with synovial joints. (*a*) A bursa is a closed sac filled with synovial fluid. Bursae are commonly located between muscles or between tendons and joint capsules. (*b*) A tendon sheath is a double-layered sac of synovial fluid that completely envelops a tendon.

Many people are concerned about the cracking sounds they hear as joints move, or the popping sounds that result from "popping" or "cracking" the knuckles by forcefully pulling on the fingers. These sounds are actually quite normal. When a synovial joint is pulled upon, its volume is suddenly expanded and the pressure of the joint fluid is lowered, causing a partial vacuum within the joint. As the joint fluid is displaced and hits against the articular cartilage, air bubbles burst and a popping or cracking sound is heard. Similarly, displaced water in a sealed vacuum tube makes this sound as it hits against the glass wall. Popping your knuckles does not cause arthritis, but it can lower your social standing.

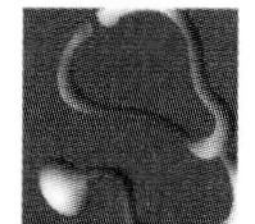

The articular cartilage that caps the articular surface of each bone and the synovial fluid that circulates through the joint during movement are protective features of synovial joints. They serve to minimize friction and cushion the articulating bones. Should trauma or disease render either of them nonfunctional, the two articulating bones will come in contact. Bony deposits will then form, and a type of arthritis will develop within the joint.

Closely associated with some synovial joints are flattened, pouch-like sacs called **bursae** (*bur'se*—singular *bursa*) that are filled with synovial fluid (fig. 8.8*a*). These closed sacs are commonly

bursa: Gk. *byrsa*, bag or purse

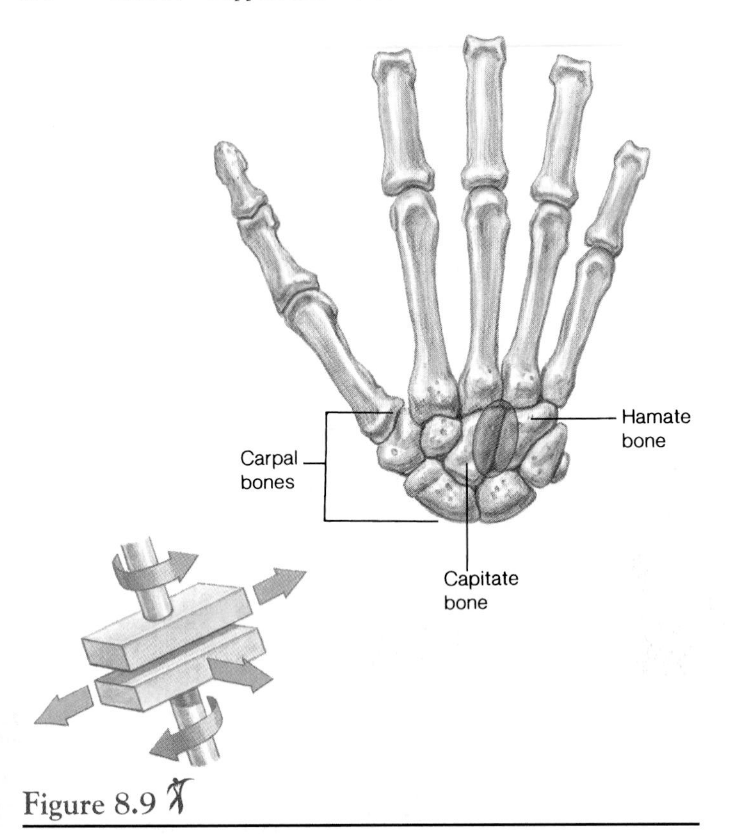

Figure 8.9

The intercarpal articulations in the wrist are examples of gliding joints in which the articulating surfaces of the adjacent bones are flattened or slightly curved. Note the diagrammatic representation showing the direction of possible movement.

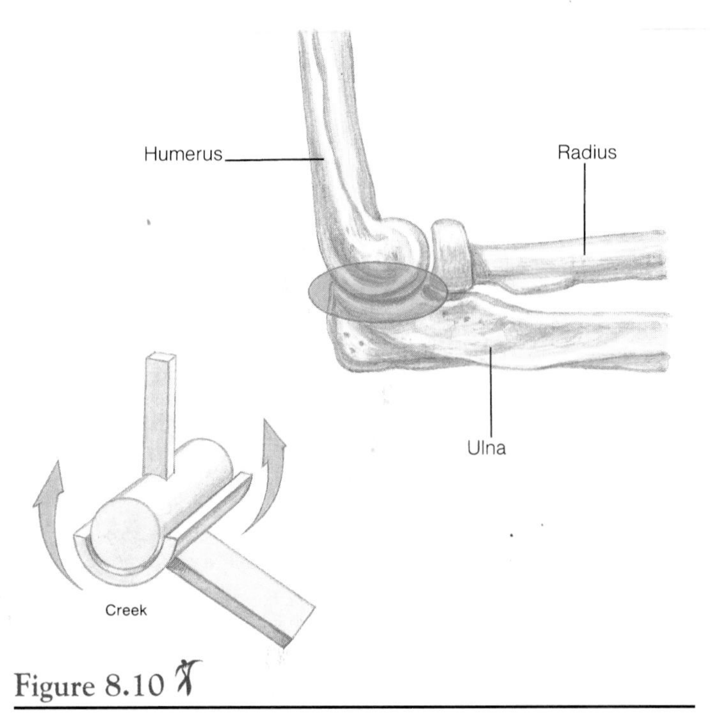

Figure 8.10

A hinge joint permits only a bending movement (flexion and extension). The hinge joint of the elbow involves the articulation of the distal end of the humerus with the proximal end of the ulna. Note the diagrammatic representation showing the direction of possible movement.

located between muscles, or in areas where a tendon passes over a bone. They function to cushion certain muscles and assist the movement of tendons or muscles over bony or ligamentous surfaces. A **tendon sheath** (fig. 8.8*b*) is a modified bursa that surrounds and lubricates the tendons of certain muscles, particularly those that cross the wrist and ankle joints.

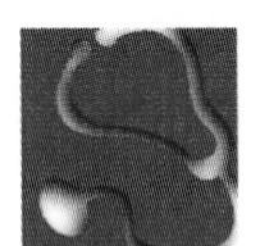

Improperly fitted shoes or inappropriate shoes can cause joint-related problems. People who perpetually wear high-heeled shoes often have backaches and leg aches because their posture has to counteract the forward tilt of their bodies when standing or walking. Their knees are excessively flexed, and their spine is thrust forward at the lumbar curvature in order to maintain balance. Tightly fitted shoes, especially those with pointed toes, may result in the development of *hallux valgus*—a lateral deviation of the hallux (great toe) in the direction of the other toes. Hallux valgus is generally accompanied by the formation of a *bunion* at the medial base of the proximal phalanx of the hallux. A bunion is an inflammation and accompanying callus that develops in response to pressure and rubbing of a shoe.

Kinds of Synovial Joints

Synovial joints are classified into six main categories on the basis of their structure and the motion they permit. The six categories are *gliding*, *hinge*, *pivot*, *condyloid*, *saddle*, and *ball-and-socket*.

Gliding

Gliding joints allow only side-to-side and back-and-forth movements, with some slight rotation. This is the simplest type of joint movement. The articulating surfaces are nearly flat, or one may be slightly concave and the other slightly convex (fig. 8.9). The intercarpal and intertarsal joints, the sternoclavicular joint, and the joint between the articular processes of adjacent vertebrae are examples.

Hinge

Hinge joints are *monaxial*—like the hinge of a door, they permit movement in only one plane. In this type of articulation, the surface of one bone is always concave, and the other convex (fig. 8.10). Hinge joints are the most common type of synovial joints. Examples include the knee, the humeroulnar articulation within the elbow, and the joints between the phalanges.

Pivot

The movement at a pivot joint is limited to rotation about a central axis. In this type of articulation, the articular surface on one bone is conical or rounded and fits into a depression on another bone (fig. 8.11). Examples are the proximal articulation of the radius and ulna for rotation of the forearm, as in turning a doorknob, and the articulation between the atlas and axis that allows rotational movement of the head.

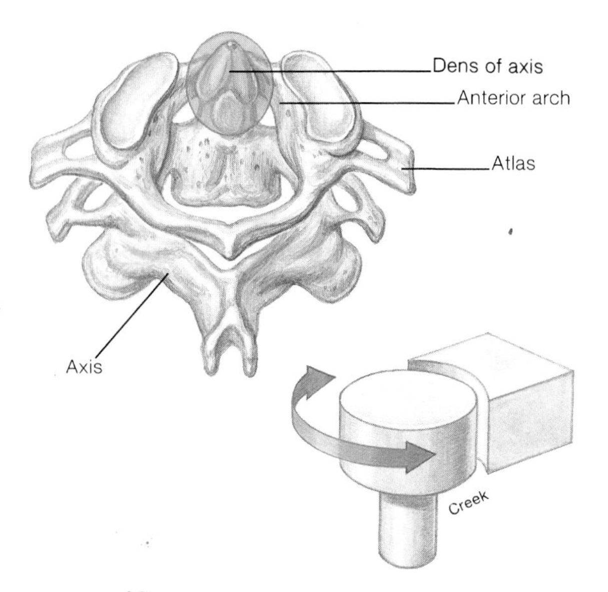

Figure 8.11

The articulation of the atlas with the axis forms a pivot joint that permits a rotation. Note the diagrammatic representation showing the direction of possible movement. (Refer to figure 8.10 and determine which articulating bones of the elbow region form a pivot joint.)

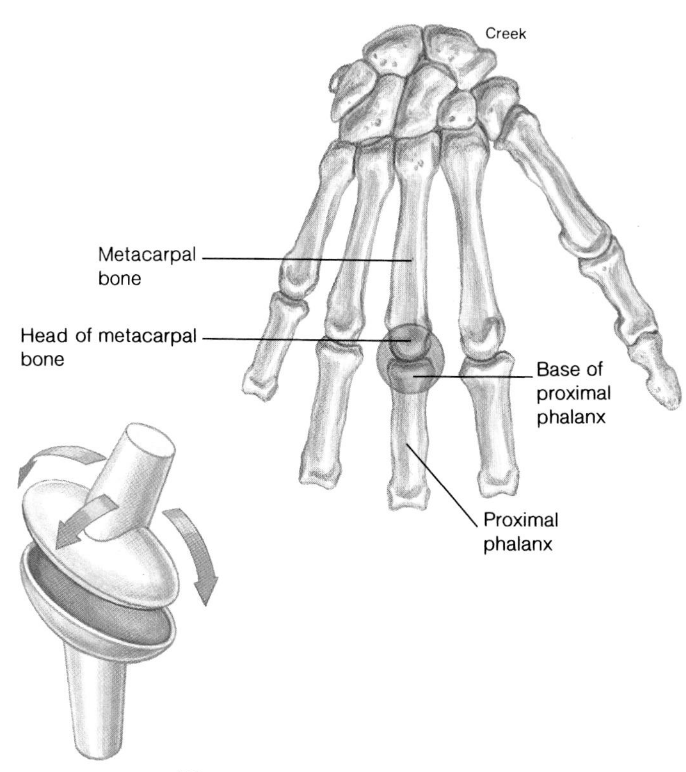

Figure 8.12

The metacarpophalangeal articulations of the hand are examples of condyloid joints in which the oval condyle of one bone articulates with the cavity of another. Note the diagrammatic representation showing the direction of possible movement.

Condyloid

A condyloid articulation is structured so that an oval, convex articular surface of one bone fits into a concave depression on another bone (fig. 8.12). This permits angular movement in two directions, as in up-and-down and side-to-side motions. Condyloid joints are therefore said to be *biaxial* joints. The radiocarpal joint of the wrist and the metacarpophalangeal joints are examples.

Saddle

Each articular process of a saddle joint has a concave surface in one direction and a convex surface in another. This articulation is a modified condyloid joint that allows a wide range of movement. There are two places in the body where a saddle joint occurs. One is at the articulation of the trapezium of the carpus with the first metacarpal bone (fig. 8.13). This carpometacarpal joint is the one responsible for the opposable thumb—a hallmark of primate anatomy. The other is at the articulation between the malleus and incus, two of the auditory ossicles of the middle ear (see fig. 6.31).

Ball-and-Socket

Ball-and-socket joints are formed by the articulation of a rounded convex surface with a cuplike cavity (fig. 8.14). This *multiaxial* type of articulation provides the greatest range of movement of all the synovial joints. Examples are the hip and shoulder joints.

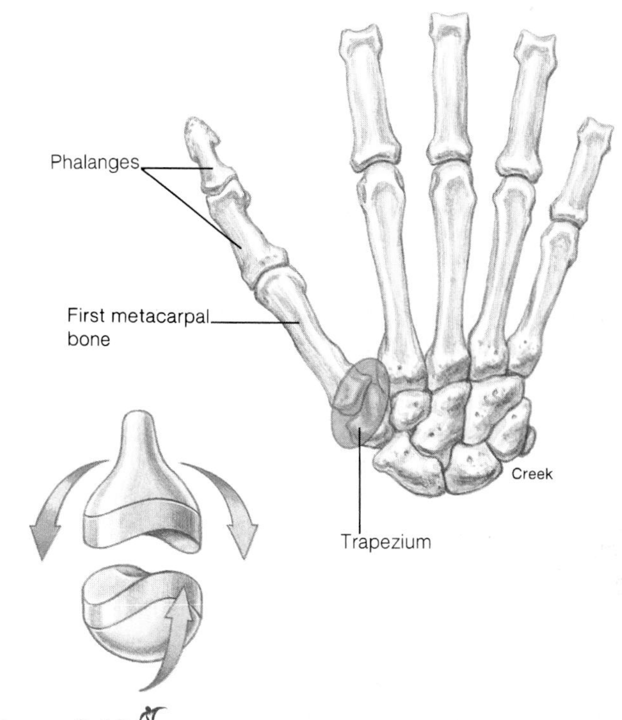

Figure 8.13

A saddle joint is formed as the trapezium articulates with the base of the first metacarpal bone. Note the diagrammatic representation showing the direction of possible movement.

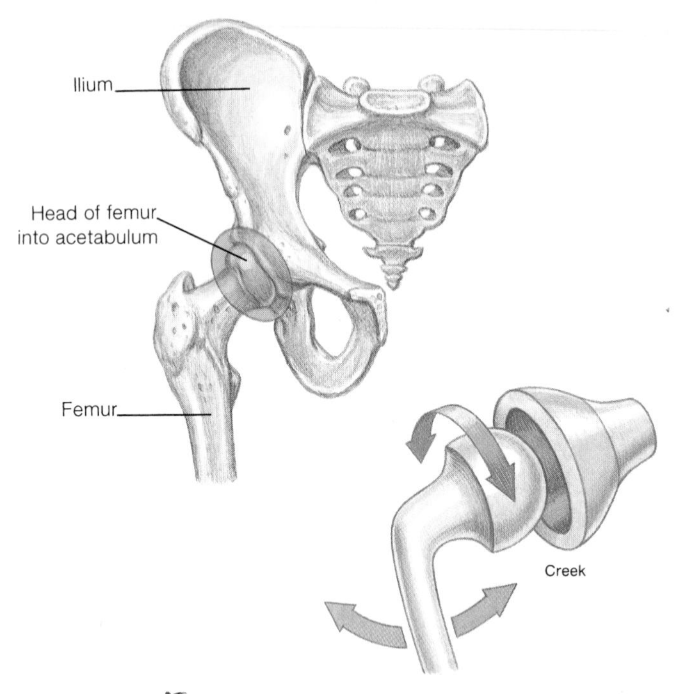

Figure 8.14

A ball-and-socket articulation illustrated by the hip joint. Note the diagrammatic representation showing the direction of possible movement.

A summary of the various types of joints is presented in table 8.1.

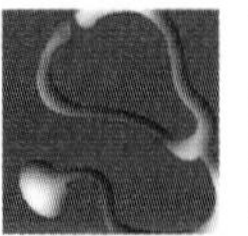

Trauma to a synovial joint causes the excessive production of synovial fluid in an attempt to cushion and immobilize the joint. This leads to swelling of the joint and discomfort. The most frequent type of joint injury is a *sprain,* in which the supporting ligaments or the joint capsule are damaged to varying degrees.

1. List the structures of a synovial joint and explain the function of each.
2. What three factors limit the range of movement in synovial joints?
3. Give an example of each type of synovial joint and describe the range of movement allowed by each.

Table 8.1
Types of Articulations

Type	Structure	Movements	Example
Fibrous Joints	Skeletal elements joined by fibrous connective tissue		
1. Suture	Edges of articulating bones frequently jagged; separated by thin layer of fibrous tissue	None	Sutures between bones of the skull
2. Syndesmoses	Articulating bones bound by interosseous ligament	Slightly movable	Joints between tibia-fibula and radius-ulna
3. Gomphoses	Teeth bound into dental alveoli of bone by periodontal ligament	Slightly movable	Dentoalveolar joints (teeth secured in dental alveoli)
Cartilaginous Joints	Skeletal elements joined by fibrocartilage or hyaline cartilage		
1. Symphyses	Articulating bones separated by pad of fibrocartilage	Slightly movable	Intervertebral joints; symphysis pubis
2. Synchondroses	Mitotically active hyaline cartilage located between skeletal elements	None	Epiphyseal plates within long bones; costal cartilages of rib cage
Synovial Joints	Joint capsule containing synovial membrane and synovial fluid		
1. Gliding	Flattened or slightly curved articulating surfaces	Sliding	Intercarpal and intertarsal joints
2. Hinge	Concave surface of one bone articulates with convex surface of another	Bending motion in one plane	Knee; elbow; joints of phalanges
3. Pivot	Conical surface of one bone articulates with depression of another	Rotation about a central axis	Atlantoaxial joint; proximal radioulnar joint
4. Condyloid	Oval condyle of one bone articulates with elliptical cavity of another	Movement in two planes	Radiocarpal joint; metacarpophalangeal joint
5. Saddle	Concave and convex surface on each articulating bone	Wide range of movements	Carpometacarpal joint of thumb
6. Ball-and-socket	Rounded convex surface of one bone articulates with cuplike socket of another	Movement in all planes and rotation	Shoulder and hip joints

Movements at Synovial Joints

Movements at synovial joints are produced by the contraction of skeletal muscles that span the joints and attach to or near the bones forming the articulation. In these actions, the bones act as levers, the muscles provide the force, and the joints are the fulcra, or pivots.

Objective 10 List and discuss the various kinds of movements that are possible at synovial joints.

Objective 11 Describe the components of a lever and explain the role of synovial joints in lever systems.

Objective 12 Compare the structures of first-, second-, and third-class levers.

As previously mentioned, the range of movement at a synovial joint is determined by the structure of the individual joint and the arrangement of the associated muscle and bone. The movement at a hinge joint, for example, occurs in only one plane, whereas the structure of a ball-and-socket joint permits movement around many axes. Joint movements are broadly classified as *angular* and *circular*. Each of these categories includes specific types of movements, and certain special movements may involve several of the specific types. The description of joint movements are in reference to anatomical position (see fig. 2.13).

Angular Movements

Angular movements increase or decrease the joint angle produced by the articulating bones. The four types of angular movements are *flexion*, *extension*, *abduction*, and *adduction*.

Flexion

Flexion is movement that decreases the joint angle on an anteroposterior plane (fig. 8.15*a*). Examples of flexion are the bending of the elbow or knee. Flexion of the elbow joint is a forward movement, whereas flexion of the knee is a backward movement. Flexion of the ankle and shoulder joints is a bit more complicated. In the ankle joint, flexion occurs as the top surface (dorsum) of the foot is elevated. This movement is frequently called **dorsiflexion** (fig. 8.15*b*). Pressing the foot downward (as in rising on the toes) is called **plantar flexion.** Flexion of the shoulder joint consists of raising the arm anteriorly from anatomical position, as if to point forward.

flexion: L. *flectere*, to bend

Extension

In extension, which is the reverse of flexion, the joint angle is increased (fig. 8.15*a*). Extension returns a body part to anatomical position. In an extended joint, the angle between the articulating bones is 180°. An exception is the ankle joint, in which there is a 90° angle between the foot and leg in anatomical position. Examples of extension are straightening of the elbow or knee joints from flexion positions. **Hyperextension** occurs when a part of the body is extended beyond the anatomical position so that the joint angle is greater than 180°. An example of hyperextension is bending the neck to tilt the head backward, as in looking at the sky.

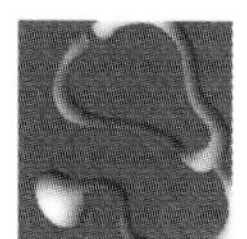

A common injury in runners is *patellofemoral stress syndrome,* commonly called "runner's knee." This condition is characterized by tenderness and aching pain around or under the patella. During normal knee movement, the patella glides up and down the patellar groove between the femoral condyles. In patellofemoral stress syndrome, the patella rubs laterally, causing irritation to the membranes and articular cartilage within the knee joint. Joggers frequently experience this condition from prolonged running on the slope of a road near the curb.

Abduction

Abduction is movement of a body part away from the main axis of the body, or away from the midsagittal plane, in a lateral direction (fig. 8.15c). This term usually applies to the arm or leg but can also apply to the fingers or toes, in which case the line of reference is the longitudinal axis of the limb. An example of abduction is moving the arms sideward, away from the body. Spreading the fingers apart is another example.

Adduction

Adduction, the opposite of abduction, is movement of a body part toward the main axis of the body (fig. 8.15c). In anatomical position, the arms and legs have been adducted toward the midplane of the body.

Circular Movements

In joints that permit **circular movement,** a bone with a rounded or oval surface articulates with a corresponding depression on another bone. The two basic types of circular movements are *rotation* and *circumduction*.

extension: L. *ex*, out, away from; *tendere*, stretch
abduction: L. *abducere*, lead away
adduction: L. *adductus*, bring to

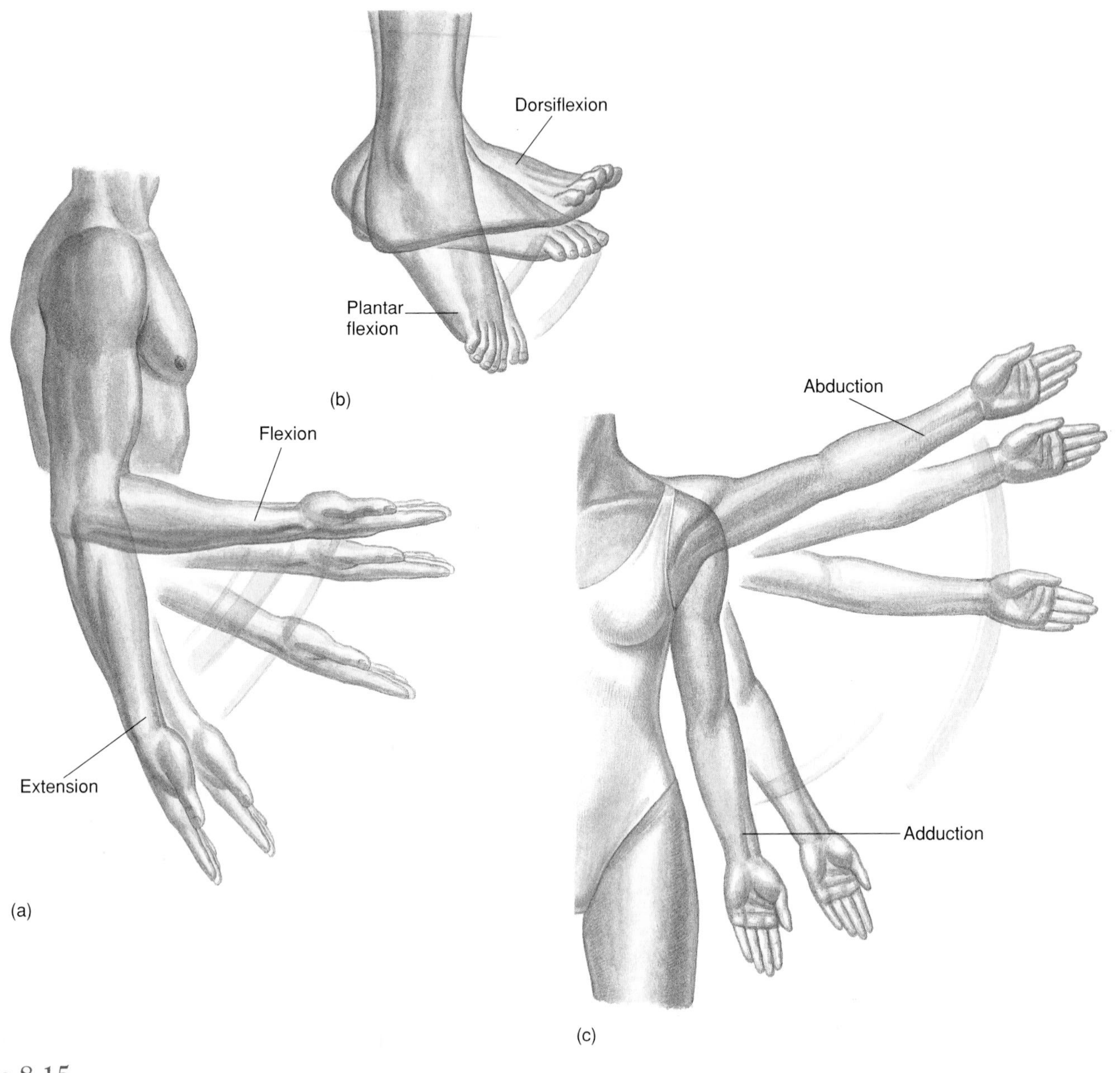

Figure 8.15

Angular movements within synovial joints include (*a*) flexion and extension, (*b*) dorsiflexion and plantar flexion, and (*c*) abduction and adduction.

Rotation

Rotation is movement of a body part around its own axis (see figs. 8.11 and 8.16*a*). There is no lateral displacement during this movement. Examples are turning the head from side to side, as if gesturing "no," and twisting at the waist.

Supination (*soo″pĭ-na′shun*) is a specialized rotation of the forearm so that the palm of the hand faces forward (anteriorly) or upward (superiorly). In anatomical position, the forearm is already supine. **Pronation** (*pro-na′shun*) is the opposite of supination. It is a rotational movement of the forearm so that the palm is directed to the rear (posteriorly) or downward (inferiorly).

Applied to the foot, the term *pronation* describes a combination of eversion (see p. 203) and abduction movements that result in a lowering of the longitudinal arch.

rotation: L. *rotare*, a wheel

Circumduction

Circumduction is the circular movement of a body part so that a cone-shaped airspace is traced. The distal extremity performs the circular movement and the proximal attachment serves as

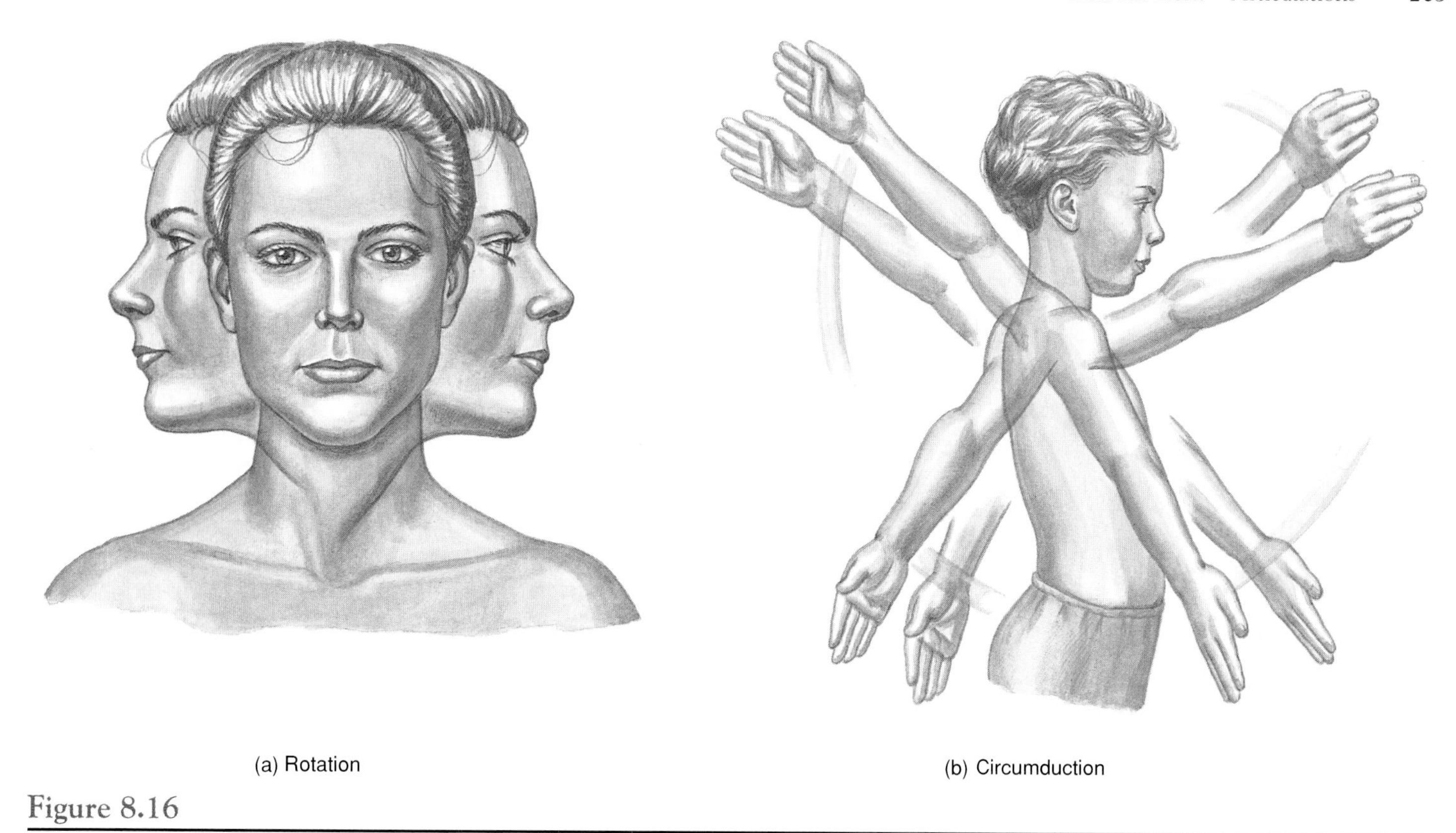

Figure 8.16

Circular movements within synovial joints include (*a*) rotation and (*b*) circumduction.

the pivot (fig. 8.16*b*). This type of motion is possible at the trunk, shoulder, wrist, metacarpophalangeal, hip, ankle, and metatarsophalangeal joints.

Special Movements

Because the terms used to describe generalized movements around axes do not apply to movement at certain joints or areas of the body, other terms must be used. **Inversion** is movement of the sole of the foot inward or medially (fig. 8.17*a*). **Eversion,** the opposite of inversion, is movement of the sole of the foot outward or laterally. The pivot axes for these movements are at the ankle and intertarsal joints. Both inversion and eversion are clinical terms that are usually used to describe developmental abnormalities.

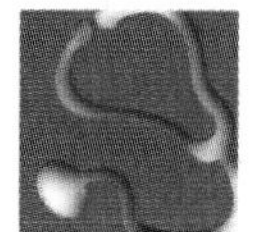

The condition of the heels of your shoes can tell you whether you invert or evert your foot as you walk. If the heel is worn down on the outer side, you tend to invert your foot as you walk. If the heel is worn down on the inside, you tend to evert your foot.

Protraction is movement of part of the body forward, on a plane parallel to the ground. The thrusting out of the lower jaw (fig. 8.17*b*) and the movement of the shoulder and upper extremity forward are examples. **Retraction,** the opposite of protraction, is the pulling back of a protracted part of the body on a plane parallel to the ground. Retraction of the mandible brings the lower jaw back in alignment with the upper jaw, so that the teeth occlude.

Elevation is movement that raises a body part. Examples include elevating the mandible to close the mouth and lifting the shoulders to shrug (fig. 8.17*c*). **Depression** is the opposite of elevation. Both the mandible and shoulders are depressed when moved downward.

Many of the movements permitted at synovial joints are visually summarized in figures 8.18 through 8.20.

Biomechanics of Body Movement

A lever is any rigid structure that turns about a fulcrum when force is applied. Levers are generally associated with machines but can also apply to other mechanical structures, such as the human body. There are four basic elements in the function of a lever: (1) the lever itself—a rigid bar or other such structure; (2) a pivot or fulcrum; (3) an object or resistance to be moved; and (4) a force that is applied to one portion of the rigid structure. In the body, synovial joints usually serve as the fulcra (*F*), the muscles provide the force, or effort (*E*), and the bones act as the rigid lever arms that move the resisting object (*R*).

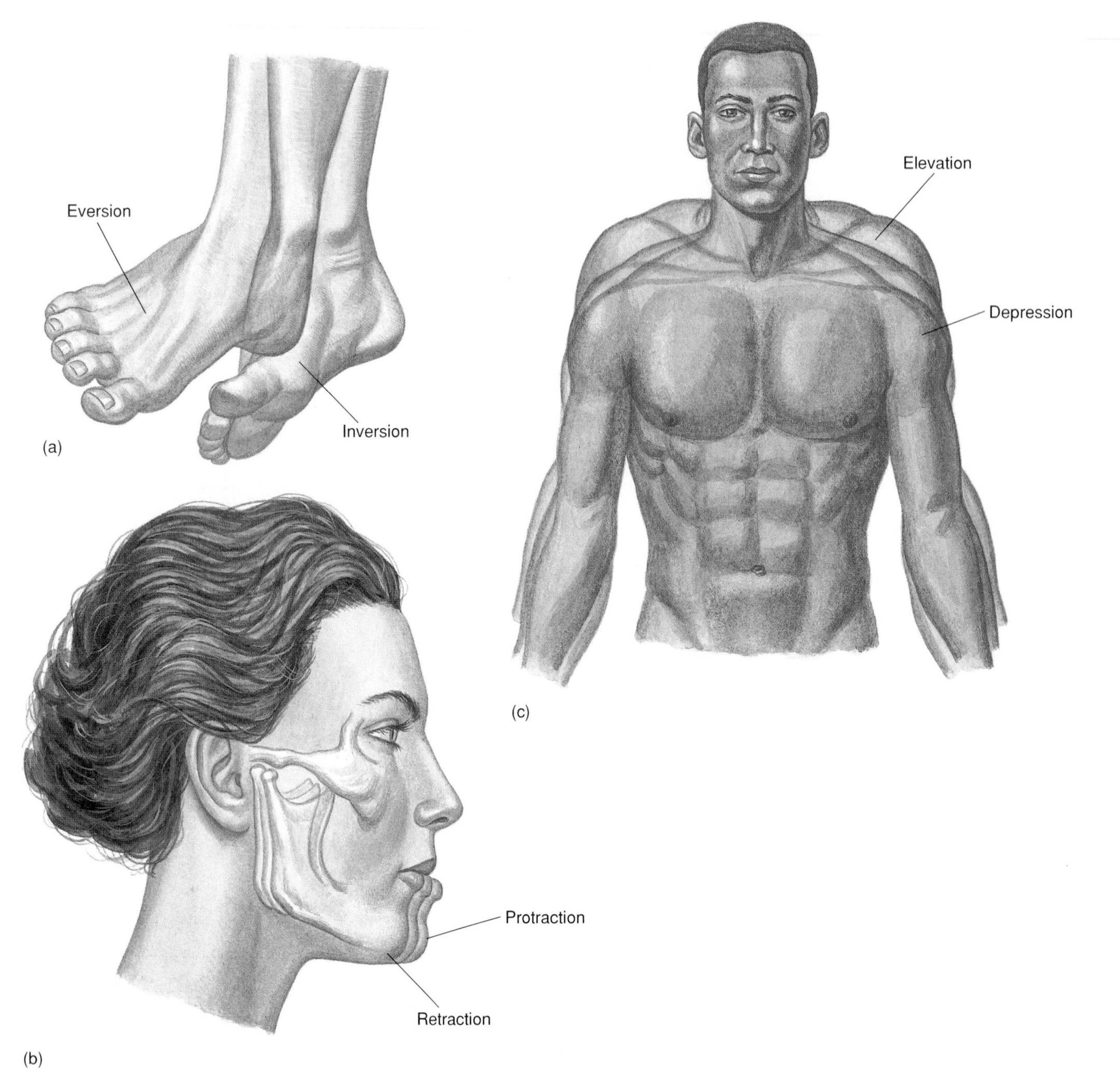

Figure 8.17

Special movements within synovial joints include (*a*) inversion and eversion, (*b*) protraction and retraction, and (*c*) elevation and depression.

There are three kinds of levers, determined by the arrangement of their parts (fig. 8.21).

1. In a **first-class lever,** the fulcrum is positioned between the effort and the resistance. The sequence of elements in a first-class lever is much like that of a seesaw—a sequence of resistance-pivot-effort. Scissors and hemostats are mechanical examples of first-class levers. In the body, the head at the atlanto-occipital (*at-lan'to-ok-sip'ĭ-tal*) joint is a first-class lever. The weight of the skull and facial portion of the head is the resistance, and the posterior neck muscles that contract to oppose the tendency of the head to tip forward provide the effort.
2. In a **second-class lever,** the resistance is positioned between the fulcrum and the effort. The sequence of elements is pivot-resistance-effort, as in a wheelbarrow or the action of a crowbar when one end is placed under a rock and the other end lifted. Contraction of the calf muscles (E) to elevate the body (R) on the toes, with the ball of the foot acting as the fulcrum, is another example.

Figure 8.18

A photographic summary of joint movements. (*a*) Adduction of shoulder, hip, and carpophalangeal joints; (*b*) abduction of shoulder, hip, and carpophalangeal joints; (*c*) rotation of vertebral column; (*d*) lateral flexion of vertebral column; (*e*) flexion of vertebral column; (*f*) hyperextension of vertebral column; (*g*) flexion of shoulder, hip, and knee joints of right side of body and extension of elbow and wrist joints; (*h*) hyperextension of shoulder and hip joints on right side of body and plantar flexion of right ankle joint.

3. In a **third-class lever,** the effort lies between the fulcrum and the resistance. The sequence of elements is pivot-effort-resistance, as in the action of a pair of forceps in grasping an object. The third-class lever is the most common type in the body. The flexion of the elbow is an example. The effort occurs as the biceps brachii muscle is contracted to move the resistance of the forearm, with the joint between the ulna and humerus forming the fulcrum.

Each bone-muscle interaction at a synovial joint represents some kind of lever system, and each lever system confers an advantage. Certain joints are adapted for force at the expense of speed, whereas most are clearly adapted for speed. The specific attachment of muscles that span a joint plays an extremely important role in determining the mechanical advantage (fig. 8.22). The position of the insertion of a muscle relative to the joint is an important factor in the biomechanics of the contraction. An insertion close to the

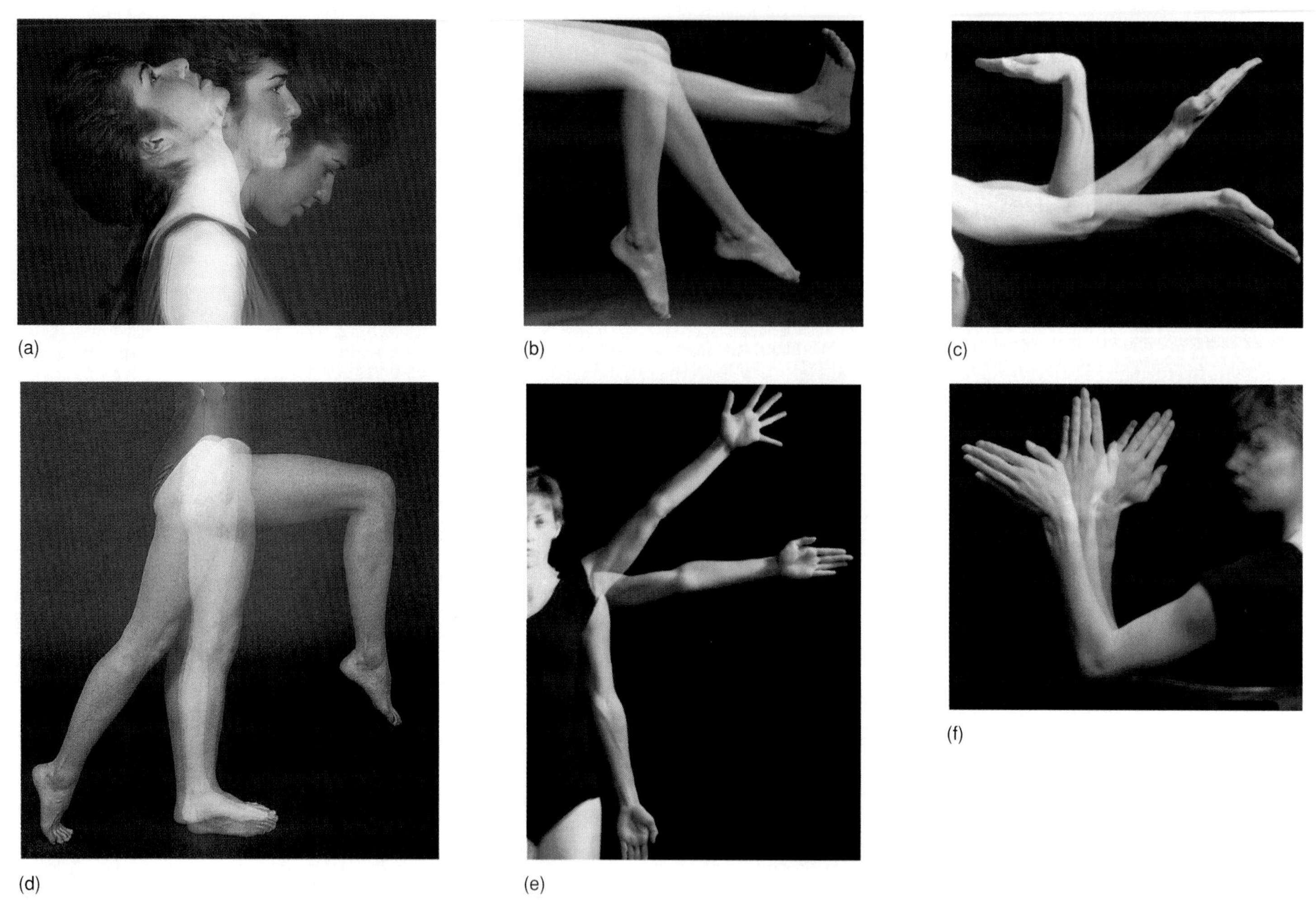

Figure 8.19

A photographic summary of some angular movements at synovial joints. (*a*) Flexion, extension, and hyperextension in the cervical region; (*b*) flexion and extension at the knee joint, and plantar flexion and dorsiflexion at the ankle joint; (*c*) flexion and extension at the elbow joint, and flexion, extension, and hyperextension at the wrist joint; (*d*) flexion, extension, and hyperextension at the hip joint, and flexion and extension at the knee joint; (*e*) adduction and abduction of the arm and fingers; (*f*) abduction and adduction of the wrist joint (posterior view). Note that the range of abduction at the wrist joint is less extensive than the range of adduction as a result of the length of the styloid process of the radius.

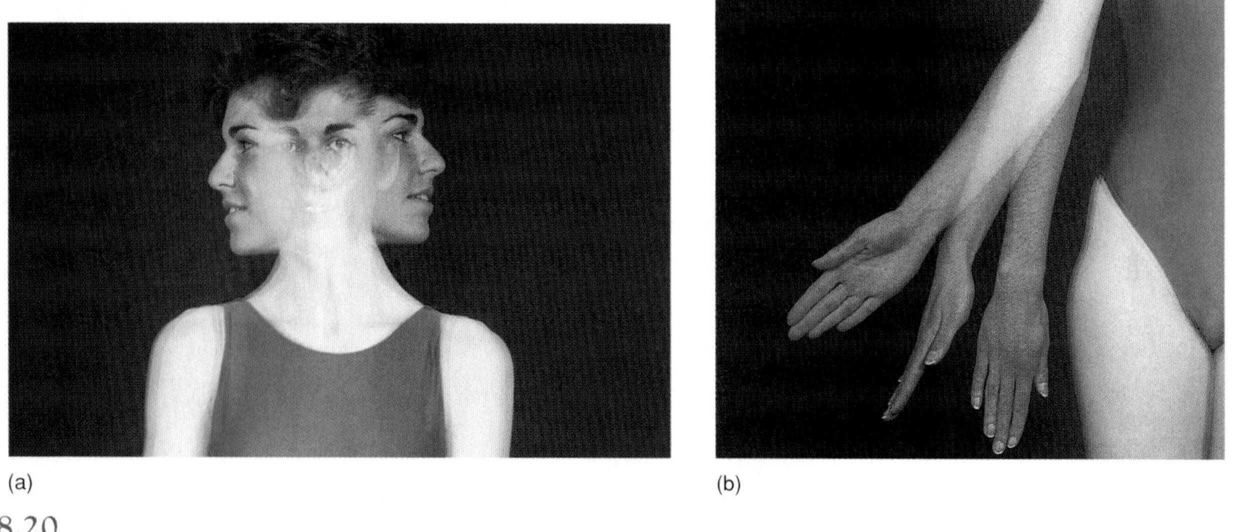

Figure 8.20

A photographic summary of some rotational movements at synovial joints. (*a*) Rotation of the head at the cervical vertebrae, especially at the atlantoaxial joint, and (*b*) rotation of the forearm (antebrachium) at the proximal radioulnar joint.

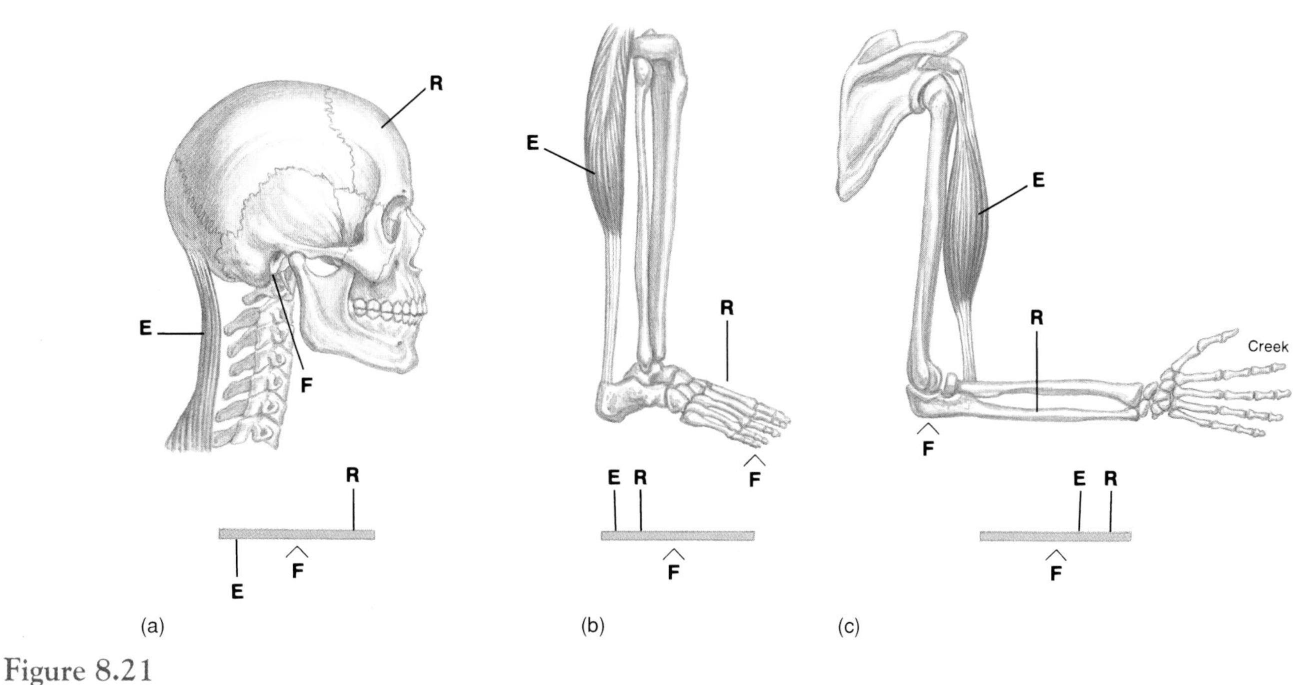

Figure 8.21

The three classes of levers. (*a*) In a first-class lever, the fulcrum (**F**) is positioned between the resistance (**R**) and the effort (**E**). (*b*) In a second-class lever, the resistance is between the fulcrum and the effort. (*c*) In a third-class lever, the effort is between the fulcrum and the resistance.

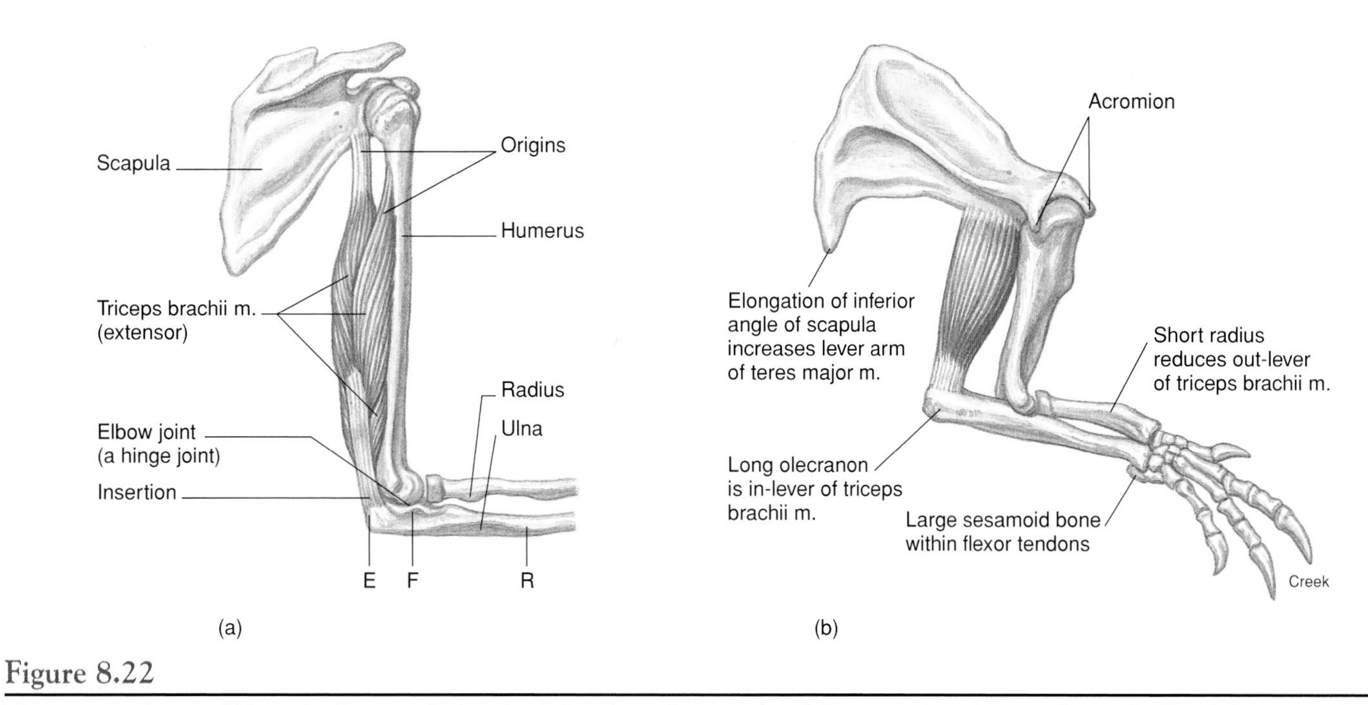

Figure 8.22

The position of a joint (fulcrum) relative to the length of a long bone (lever arm) and the point of attachment of a muscle (force) determines the mechanical advantage when movement occurs. (*a*) The elbow joint and extensor muscles of a human and (*b*) the elbow joint and extensor muscles of an armadillo.

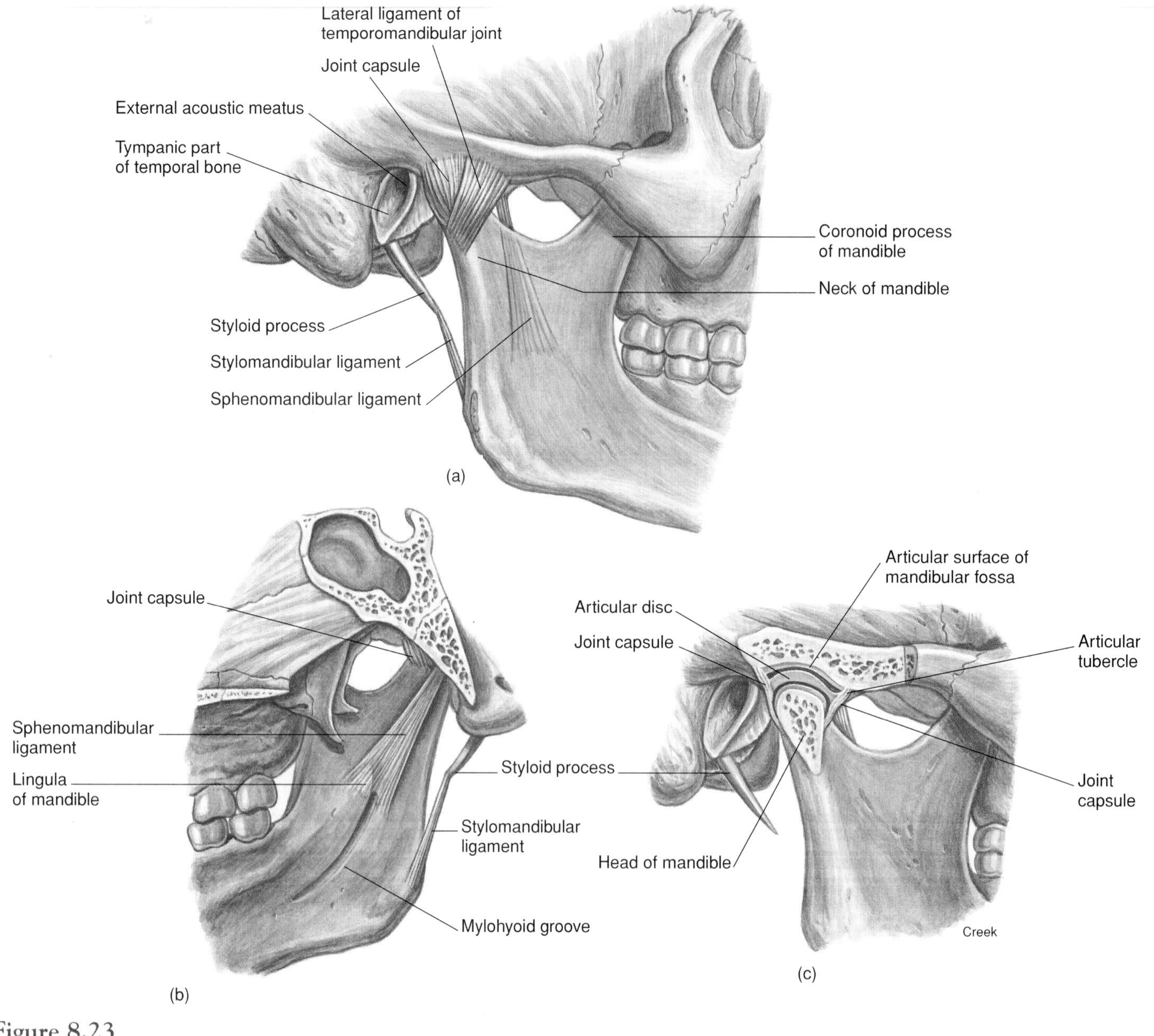

Figure 8.23

The temporomandibular joint. (*a*) A lateral view, (*b*) a medial view, and (*c*) a sagittal view.

joint (fulcrum), for example, will produce a faster movement and greater range of movement than an insertion that is more distant from the joint. An attachment far from the joint capitalizes on the length of the lever arm (bone), and increases force at the sacrifice of speed and range of movement.

1. Describe the structure of a joint that permits rotational movement.
2. What types of joints are involved in the body's lever systems?
3. Which is the most common type of lever in the body?

Specific Joints of the Body

Of the numerous joints in the body, some have special structural features that enable them to perform particular functions. These joints are also somewhat vulnerable to trauma and are therefore clinically important.

Objective 13 Describe the structure, function, and possible clinical importance of the following joints: temporomandibular, sternoclavicular, glenohumeral, elbow, metacarpophalangeal, interphalangeal, coxal, tibiofemoral, and talocrural.

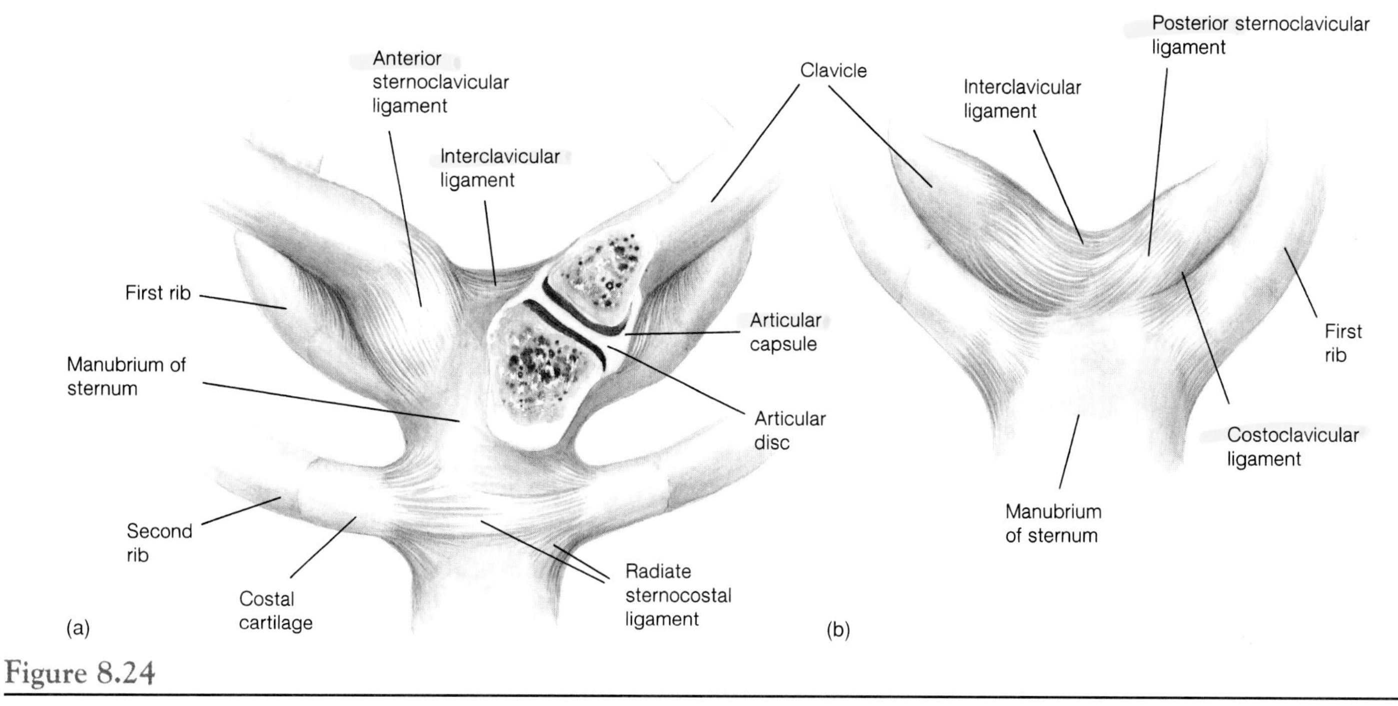

Figure 8.24

The sternoclavicular joint and associate ligaments. (*a*) An anterior view showing a coronal (frontal) section and (*b*) a posterior view.

Temporomandibular Joint

The **temporomandibular joint** represents a unique combination of a hinge joint and a gliding joint (fig. 8.23). It is formed by the condylar process of the mandible and the mandibular fossa and articular tubercle of the temporal bone. An *articular disc* separates the joint cavity into superior and inferior compartments.

Three major ligaments support and reinforce the temporomandibular joint. The **lateral ligament** of the temporomandibular joint is positioned on the lateral side of the joint capsule and is covered by the parotid gland. This ligament prevents the head of the mandible from being displaced posteriorly and fracturing the tympanic plate when the chin suffers a severe blow. The **stylomandibular ligament** is not directly associated with the joint but extends inferiorly and anteriorly from the styloid process to the posterior border of the ramus of the mandible. On the medial side by the joint, a **sphenomandibular** (*sfe″no-man-dib′yŭ-lar*) **ligament** extends from the spine of the sphenoid bone to the ramus of the mandible.

The movements of the temporomandibular joint include depression and elevation of the mandible as a hinge joint, protraction and retraction of the mandible as a gliding joint, and lateral rotatory movements. The lateral motion is made possible by the articular disc.

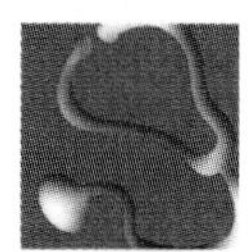

The temporomandibular joint can be easily palpated by applying firm pressure to the area in front of your ear and opening and closing your mouth. This joint is most vulnerable to dislocation when the mandible is completely depressed, as in yawning. Relocating the jaw is usually a simple task, however, and is accomplished by pressing down on the molars while pushing the jaw backward.

Temporomandibular joint (TMJ) syndrome is a recently recognized ailment that may afflict an estimated 75 million Americans. The apparent cause of TMJ syndrome is a malalignment of one or both temporomandibular joints. The symptoms of the condition range from moderate and intermittent facial pain to intense and continuous pain in the head, neck, shoulders, or back. Clicking sounds in the jaw and limitation of jaw movement are common symptoms. Some vertigo (dizziness) and tinnitus (ringing in the ears) may also occur.

Sternoclavicular Joint

The **sternoclavicular** (*ster″no-klă-vik′yŭ-lar*) **joint** is formed by the sternal extremity of the clavicle and the manubrium of the sternum (fig. 8.24). Although a gliding joint, the sternoclavicular joint has a relatively wide range of movement because of the presence of an articular disc within the joint capsule.

Four ligaments support the sternoclavicular joint and provide flexibility. An **anterior sternoclavicular ligament** covers the anterior surface of the joint, and a **posterior sternoclavicular ligament** covers the posterior surface. Both ligaments extend from the sternal end of the clavicle to the manubrium. An **interclavicular ligament** extends between the sternal ends of both clavicles, binding them together. The **costoclavicular ligament** extends from the costal cartilage of the first rib to the costal tuberosity of the clavicle.

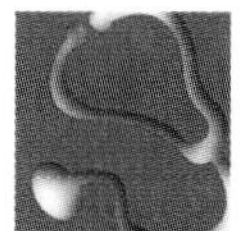

Of all the joints associated with the rib cage, the sternoclavicular joint is the one most frequently dislocated. Excessive force along the long axis of the clavicle may displace the clavicle forward and downward. Injury to the costal cartilages is painful and is caused most frequently by a forceful, direct blow to the costal cartilages.

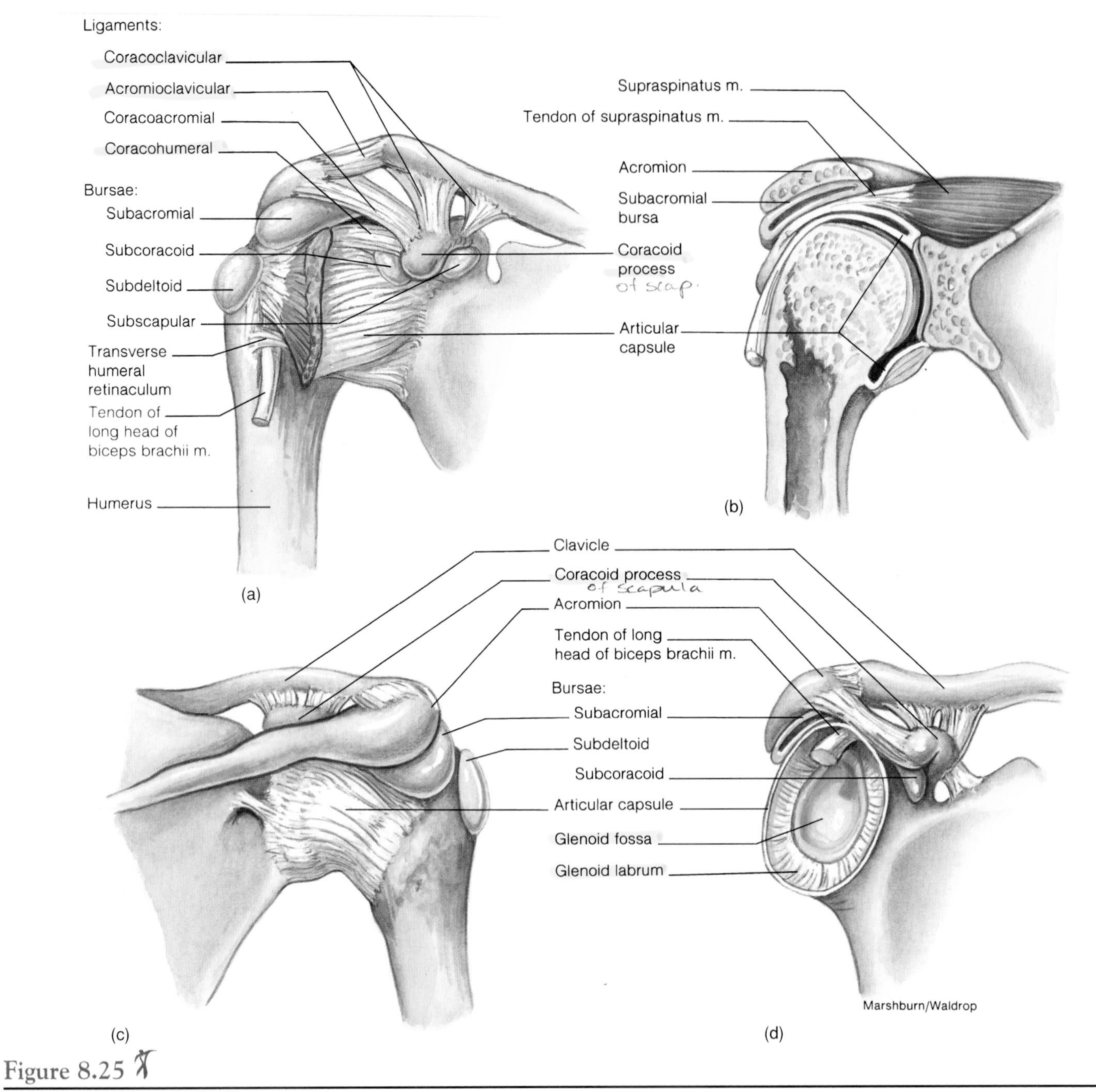

Figure 8.25

The humeral (shoulder) joint. (*a*) An anterior view, (*b*) a coronally sectioned anterior view, (*c*) a posterior view, and (*d*) a lateral view with the humerus removed.

Glenohumeral (Shoulder) Joint

The **shoulder joint** is formed by the head of the humerus and the glenoid fossa of the scapula (fig. 8.25). It is a ball-and-socket joint and the most freely movable joint in the body. A circular band of fibrocartilage called the **glenoid labrum** passes around the rim of the shoulder joint and deepens the concavity of the glenoid fossa (figs. 8.25 and 8.26). The shoulder joint is protected from above by an arch formed by the acromion and coracoid process of the scapula and by the clavicle.

Although two ligaments and one retinaculum surround and support the shoulder joint, most of the stability of this joint depends on the powerful muscles and tendons that cross over it. Thus, it is an extremely mobile joint in which stability has been sacrificed for mobility. The **coracohumeral** (*kor″ă-ko-hyoo′mer-al*) **ligament** extends from the coracoid process of the scapula to the greater tubercle of the humerus. The joint capsule is reinforced with three ligamentous bands called the **glenohumeral ligaments.** The final support of the shoulder joint is the **transverse humeral**

labrum: L. *labrum*, lip

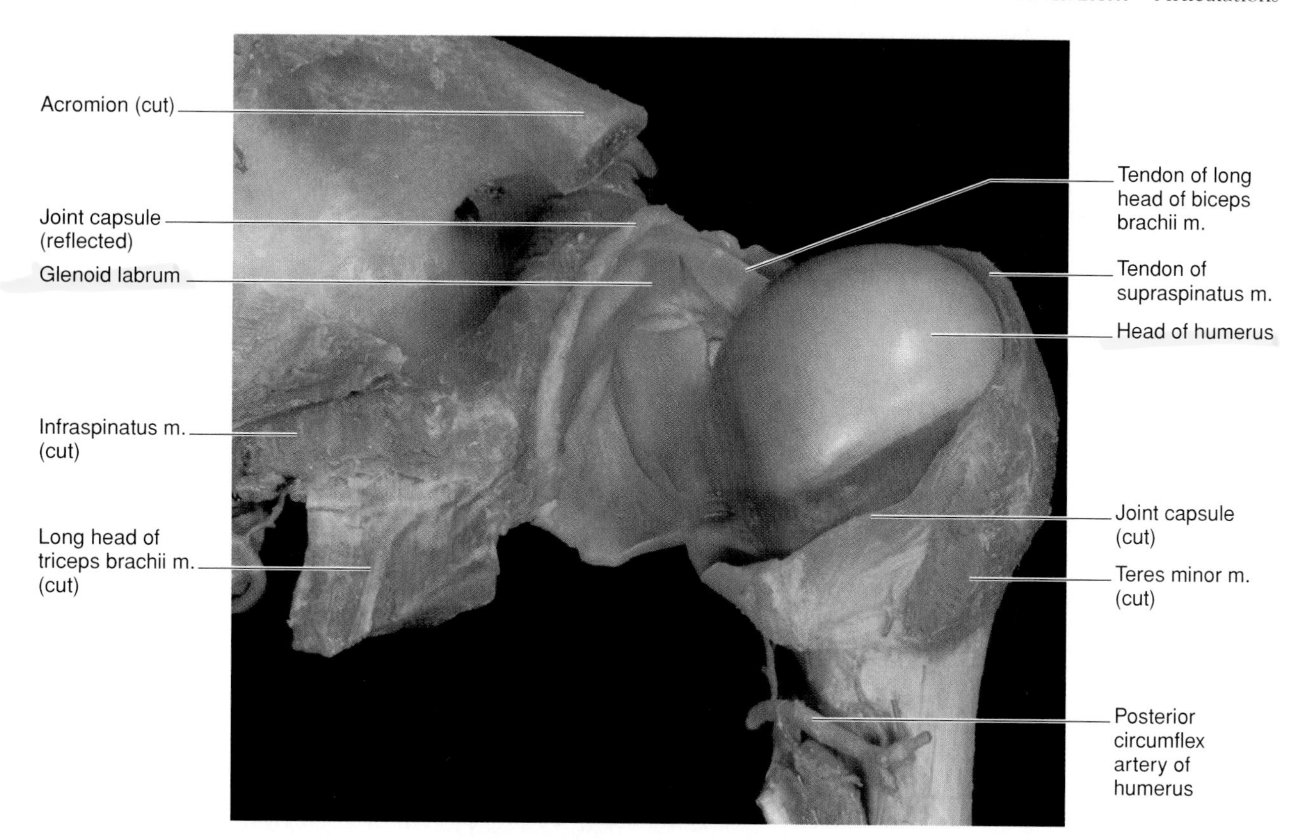

Figure 8.26

A posterior view of a dissected humeral joint. An incision has been made into the joint capsule and the humerus has been retracted laterally and rotated posteriorly.

retinaculum, a thin band that extends from the greater tubercle to the lesser tubercle of the humerus.

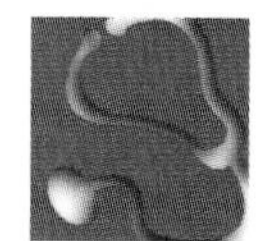

The stability of the shoulder joint is provided mainly by the tendons of the subscapularis, supraspinatus, infraspinatus, and teres minor muscles, which together form the *musculotendinous* (*rotator*) *cuff.* The cuff is fused to the underlying capsule, except in its inferior aspect. Because of the lack of inferior stability, most dislocations (subluxations) occur in this direction. The shoulder is most vulnerable to trauma when the arm is fully abducted and then receives a blow from above—as for example, when the outstretched arm is struck by heavy objects falling from a shelf. Degenerative changes in the musculotendinous cuff produce an inflamed, painful condition known as *pericapsulitis.*

Two major and two minor bursae are associated with the shoulder joint. The larger bursae are the **subdeltoid bursa,** located between the deltoid muscle and the joint capsule, and the **subacromial bursa,** located between the acromion and joint capsule. The **subcoracoid bursa,** which lies between the coracoid process and the joint capsule, is frequently considered an extension of the subacromial bursa. A small **subscapular bursa** is located between the tendon of the subscapularis muscle and the joint capsule.

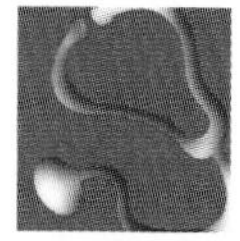

The shoulder joint is vulnerable to dislocations from sudden jerks of the arm, especially in children before strong shoulder muscles have developed. Because of the weakness of this joint in children, parents should be careful not to force a child to follow by yanking on the arm. Dislocation of the shoulder is extremely painful and may cause permanent damage or perhaps muscle atrophy as a result of disuse.

Elbow Joint

The **elbow joint** is a hinge joint composed of two articulations—the **humeroulnar joint,** formed by the trochlea of the humerus and the trochlear notch of the ulna, and the **humeroradial joint,** formed by the capitulum of the humerus and the head of the radius (figs. 8.27 and 8.28). Both of these articulations are enclosed in a single joint capsule. On the posterior side of the elbow, there is a large **olecranon bursa** to lubricate the area. A **radial (lateral) collateral ligament** reinforces the elbow joint on the lateral side and an **ulnar (medial) collateral ligament** strengthens the medial side.

A third joint occurs in the elbow region—the **proximal radioulnar joint**—but it is not part of the hinge. At this joint, the head of the radius fits into the radial notch of the ulna and is held in place by the **annular ligament.**

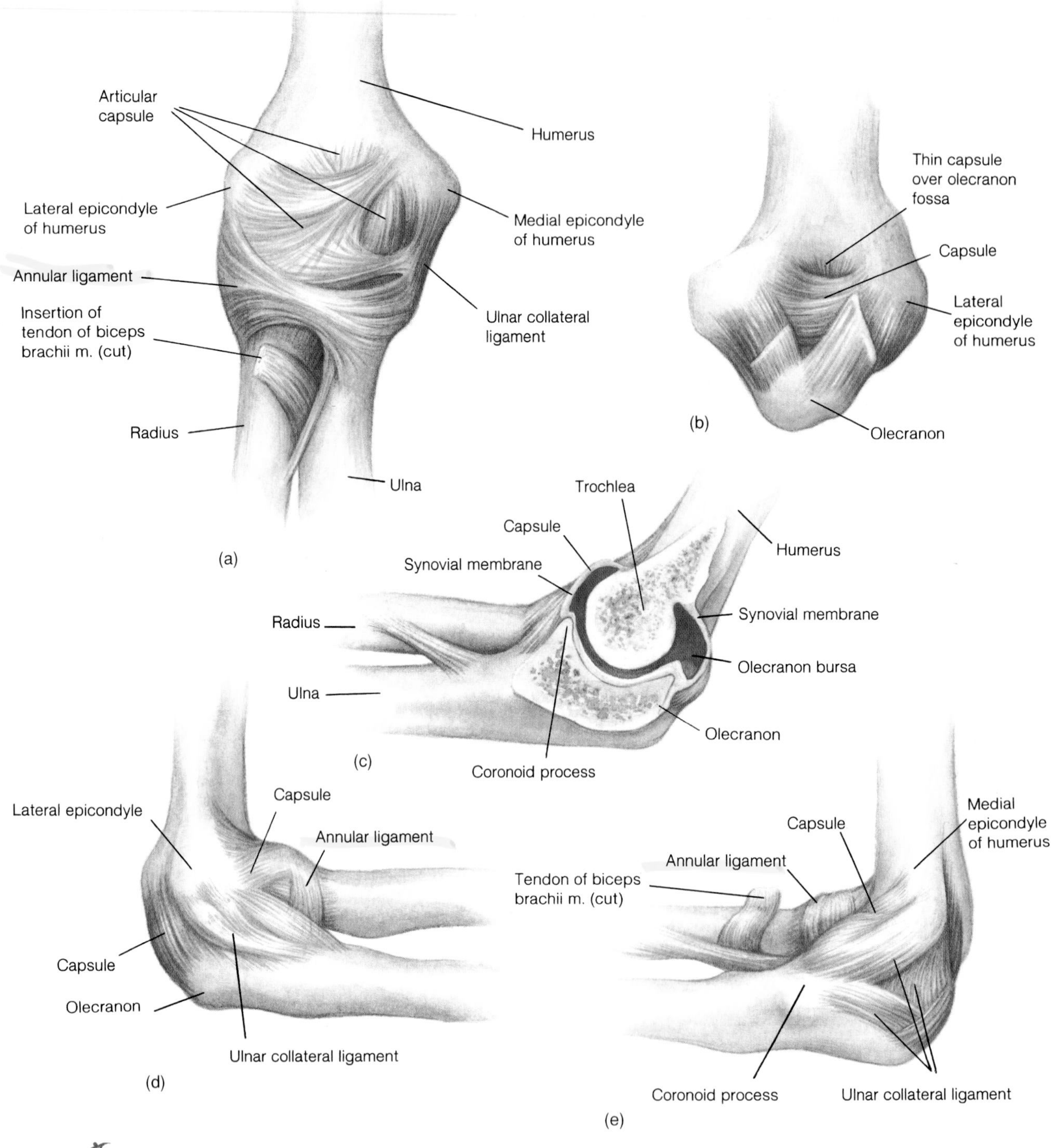

Figure 8.27

The right elbow region. (*a*) An anterior view, (*b*) a posterior view, (*c*) a sagittal view, (*d*) a lateral view, and (*e*) a medial view.

Because so many muscles originate or insert near the elbow, it is a common site of localized tenderness, inflammation, and pain. *Tennis elbow* is a general term for musculotendinous soreness in this area. The structures most generally strained are the tendons attached to the lateral epicondyle of the humerus. The strain is caused by repeated extension of the wrist against some force, as occurs during the backhand stroke in tennis.

Metacarpophalangeal and Interphalangeal Joints

The **metacarpophalangeal joints** are condyloid joints, and the **interphalangeal joints** are hinge joints. The articulating bones of the former are the metacarpal bones and the proximal phalanges; those of the latter are adjacent phalanges (fig. 8.29). Each joint in both joint types has three ligaments. A **palmar ligament** spans each joint on the palmar, or anterior, side of the joint capsule.

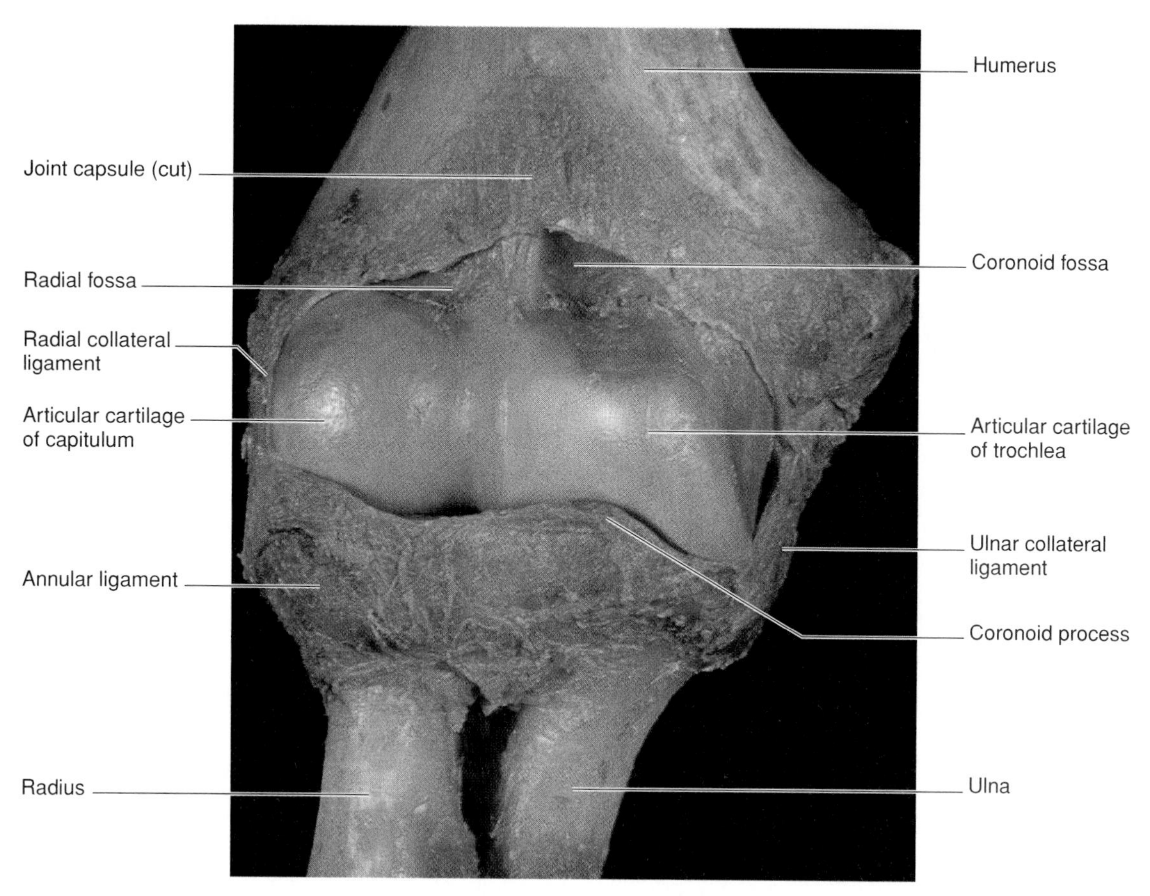

Figure 8.28

A posterior view of a dissected elbow joint. A portion of the joint capsule has been removed to show the articular surface of the humerus.

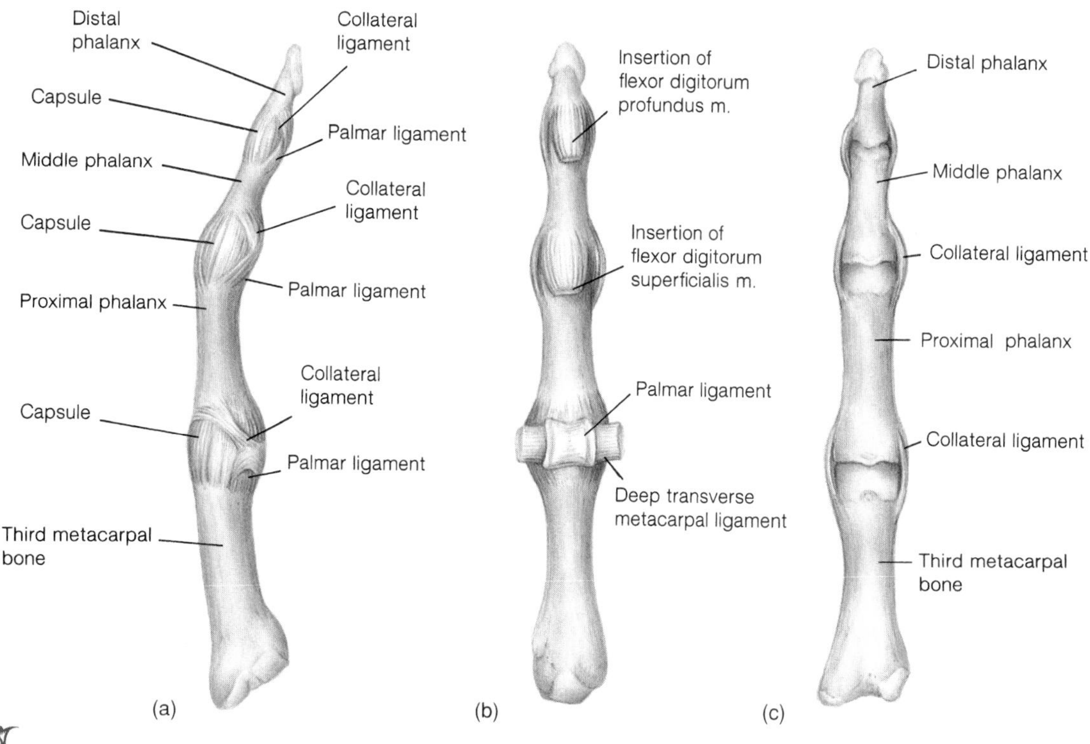

Figure 8.29

Metacarpophalangeal and interphalangeal joints. (*a*) A lateral view, (*b*) an anterior (palmar) view, and (*c*) a posterior view.

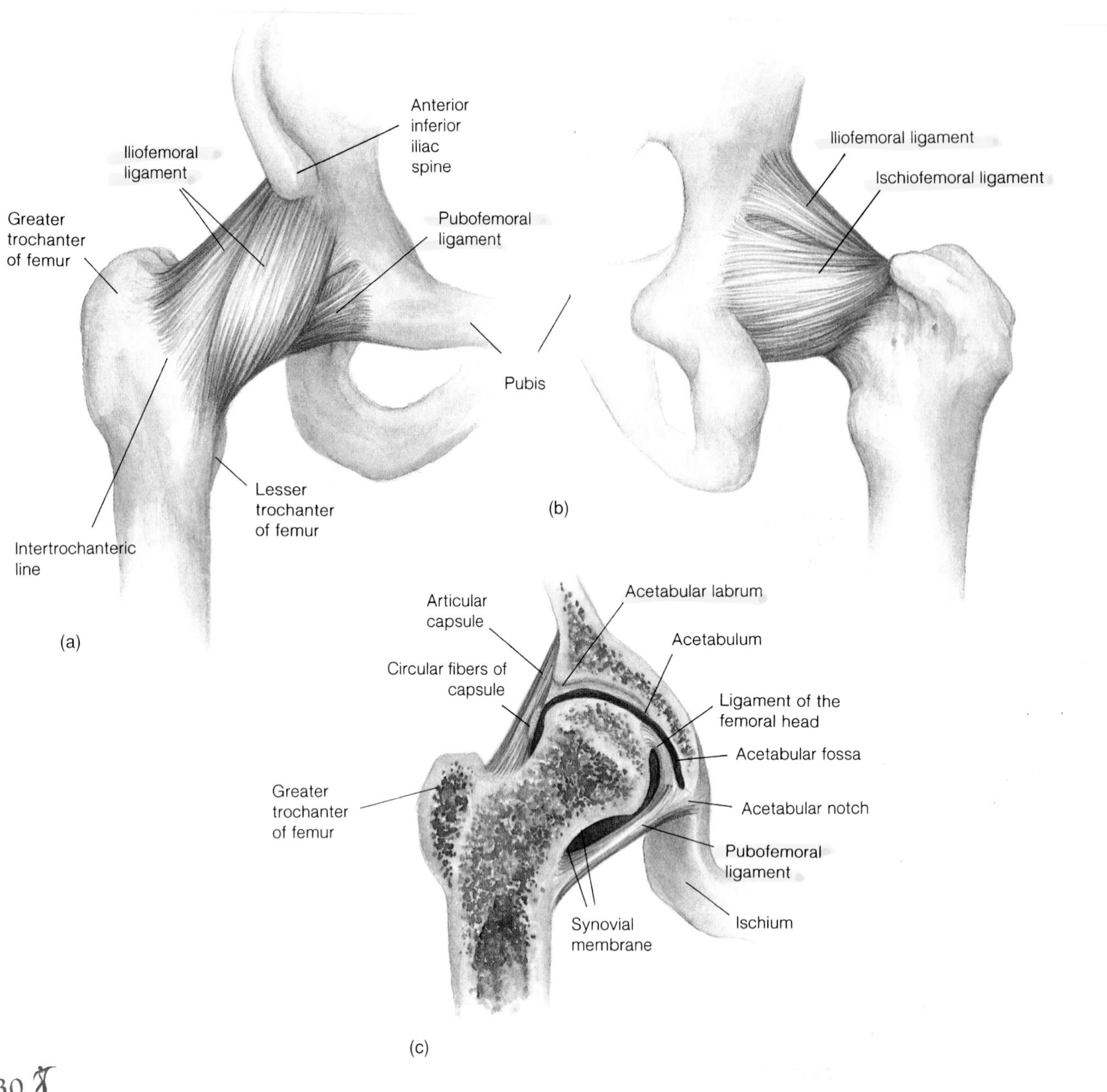

Figure 8.30

The right coxal (hip) joint. (*a*) An anterior view, (*b*) a posterior view, and (*c*) a coronal view.

Each joint also has two **collateral ligaments,** one on the lateral side and one on the medial side, to further reinforce the joint capsule. There are no supporting ligaments on the posterior side.

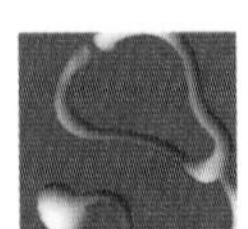

Athletes frequently jam a finger. It occurs when a ball forcefully strikes a distal phalanx as the fingers are extended, causing a sharp flexion at the joint between the middle and distal phalanges. No ligaments support the joint on the posterior side, but there is a tendon from the digital extensor muscles of the forearm. It is this tendon that is damaged when the finger is jammed. Treatment involves splinting the finger for a period of time. If splinting is not effective, surgery is generally performed to avoid a permanent crook in the finger.

Coxal (Hip) Joint

The ball-and-socket **hip joint** is formed by the head of the femur and the acetabulum of the os coxae (fig. 8.30). It bears the weight of the body and is therefore much stronger and more stable than the shoulder joint. The hip joint is secured by a strong fibrous joint capsule, several ligaments, and a number of powerful muscles.

The primary ligaments of the hip joint are the anterior **iliofemoral** (*il″e-o-fem′or-al*) and **pubofemoral ligaments** and the posterior **ischiofemoral** (*is″ke-o-fem′or-al*) **ligament.** The **ligamentum capitis femoris** is located within the articular capsule and attaches the head of the femur to the acetabulum. This is a relatively slack ligament, and does not play a significant role in holding the femur in its socket. However, it does contain a small

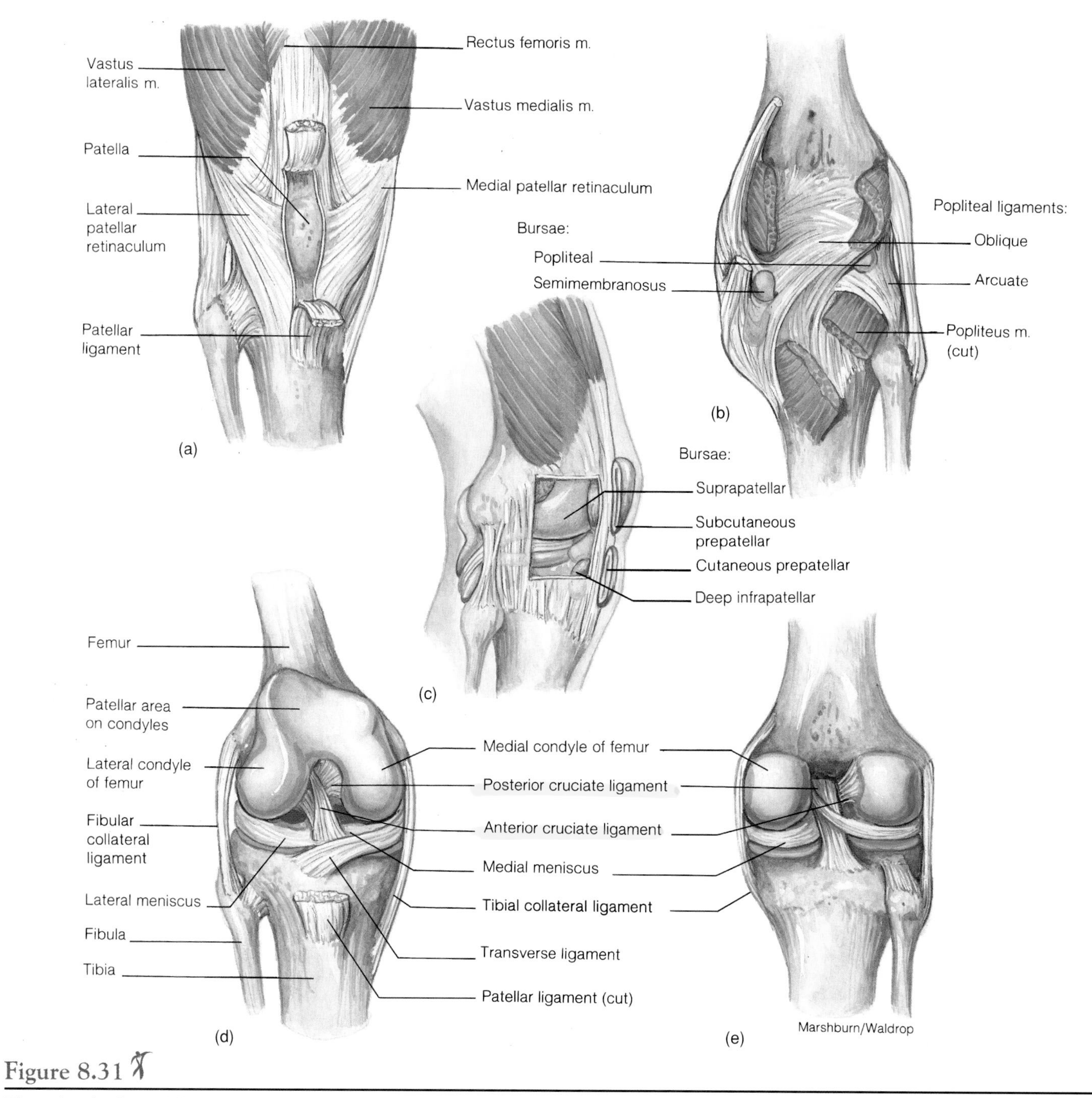

Figure 8.31

The right tibiofemoral (knee) joint. (*a*) An anterior view, (*b*) a superficial posterior view, (*c*) a lateral view showing the bursae, (*d*) an anterior view with the knee slightly flexed and the patella removed, and (*e*) a deep posterior view.

artery that supplies blood to the head of the femur. The **transverse acetabular** (*as″ĕ-tab′yŭ-lar*) **ligament** crosses the acetabular notch and connects to the joint capsule and the ligamentum capitis femoris. The **acetabular labrum,** a fibrocartilaginous rim that rings the head of the femur as it articulates with the acetabulum, is attached to the margin of the acetabulum.

Tibiofemoral (Knee) Joint

The **knee joint,** located between the femur and tibia, is the largest, most complex, and probably the most vulnerable joint in the body. It is a complex hinge joint that permits limited rolling and gliding movements in addition to flexion and extension. On the anterior side, the knee joint is stabilized and protected by the patella and the **patellar ligament,** forming a gliding **patellofemoral joint.**

Because of the complexity of the knee joint, only the relative positions of the ligaments, menisci, and bursae will be covered here. Although the attachments will not be discussed in detail, the locations of these structures can be seen in figures 8.31 and 8.32.

In addition to the patella and the patellar ligament on the anterior surface, the tendinous insertion of the quadriceps

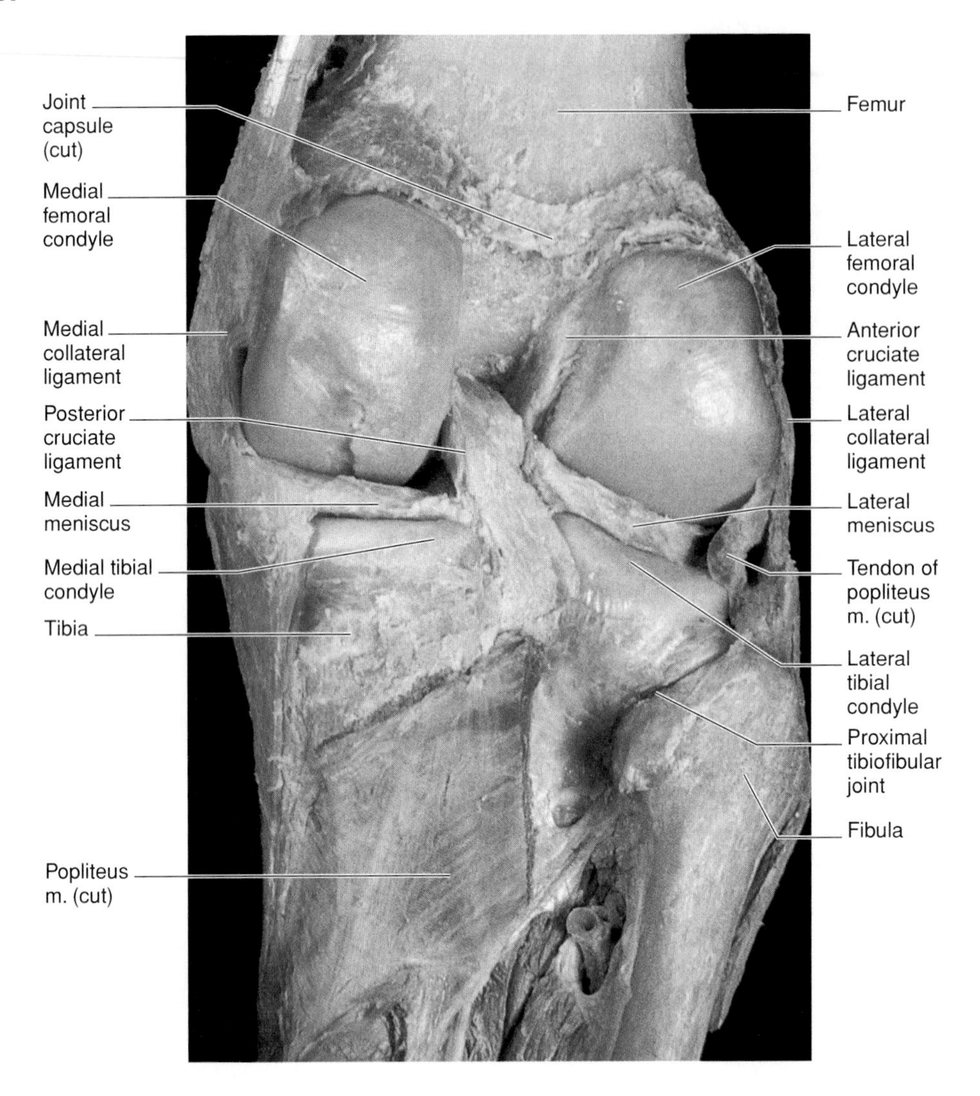

Figure 8.32

A posterior view of a dissected tibiofemoral joint. The joint capsule has been removed to expose the cruciate ligaments and the menisci.

femoris muscle forms two supportive bands called the **lateral** and **medial patellar retinacula** (*ret″ĭ-nak-yŭ-lă*). Four bursae are associated with the anterior aspect of the knee: the **subcutaneous prepatellar bursa,** the **suprapatellar bursa,** the **cutaneous prepatellar bursa,** and the **deep infrapatellar bursa.**

The posterior aspect of the knee is referred to as the **popliteal** (*pop″lĭ-te′al*) **fossa.** The broad **oblique popliteal ligament** and the **arcuate** (*ar′kyoo-āt*) **popliteal ligament** are superficial in position, whereas the **anterior** and **posterior cruciate** (*kroo′she-āt*) **ligaments** lie deep within the joint. The **popliteal bursa** and the **semimembranosus bursa** are the two bursae associated with the back of the knee.

Strong **collateral ligaments** support both the medial and lateral sides of the knee joint. Two fibrocartilaginous discs called the **lateral** and **medial menisci** are located within the knee joint interposed between the distal femoral and proximal tibial condyles. The two menisci are connected by a **transverse ligament.** In addition to the four bursae on the anterior side and the two on the posterior side, there are seven bursae on the lateral and medial sides, for a total of thirteen.

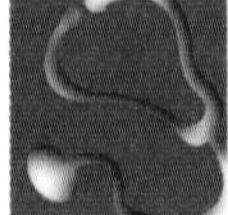

During normal walking and running, and in the support of the body, the knee joint functions superbly. It can tolerate considerable stress without tissue damage. However, the knee lacks bony support to withstand sudden forceful stresses, which frequently occur in athletic competition. Knee injuries often require surgery, and they heal with difficulty because of the avascularity of the cartilaginous tissue. Knowledge of the anatomy of the knee provides insight as to its limitations. The three C's—the anterior cruciate ligament, the collateral ligaments, and the cartilage—are the most likely sites of crippling injury.

cruciate: L. *crucis*, cross

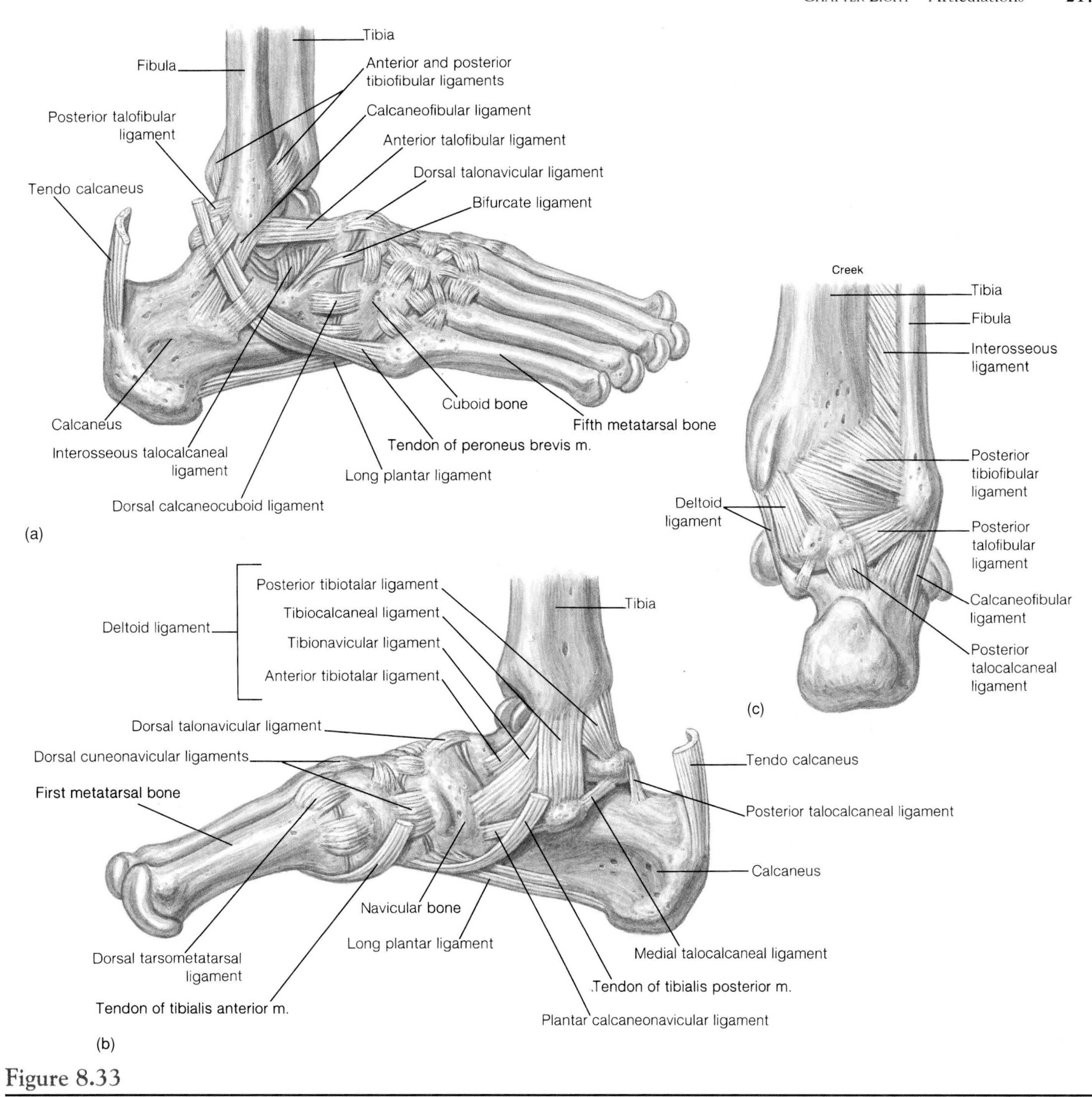

Figure 8.33

The right talocrural (ankle) joint. (*a*) A lateral view, (*b*) a medial view, and (*c*) a posterior view.

Talocrural (Ankle) Joint

There are actually two principal articulations within the **ankle joint,** both of which are hinge joints (figs. 8.33 and 8.34). One is formed as the distal end of the tibia and its medial malleolus articulate with the talus; the other is formed as the lateral malleolus of the fibula articulates with the talus.

One joint capsule surrounds the articulations of the three bones, and four ligaments support the ankle joint on the outside of the capsule. The strong **deltoid ligament** is associated with the tibia, whereas the **lateral collateral ligaments, anterior talofibular** (*ta-lo-fib-yoo'lar*) **ligament, posterior talofibular ligament,** and **calcaneofibular** (*kal-ka"ne-o-fib'yoo-lar*) **ligament** are associated with the fibula.

The malleoli form a cap over the upper surface of the talus that prohibits side-to-side movement at the ankle joint. Unlike the condyloid joint at the wrist, the movements of the ankle are limited to flexion and extension. Dorsiflexion of the ankle is checked primarily by the tendo calcaneus, whereas plantar flexion, or ankle extension, is checked by the tension of the extensor tendons on the front of the joint and the anterior portion of the joint capsule.

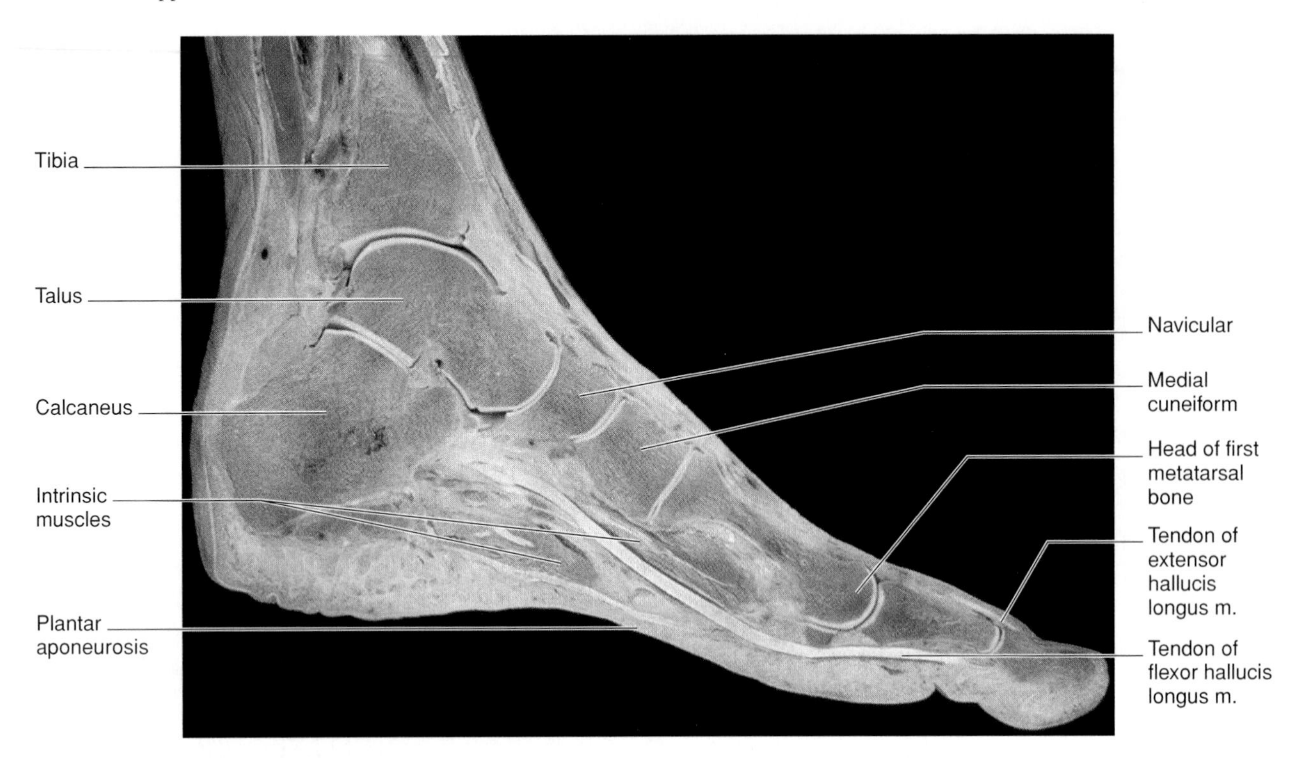

Figure 8.34

A sagittal section of the foot from a cadaver.

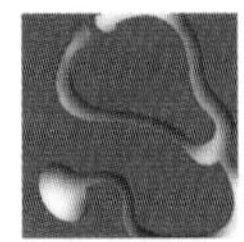

Ankle sprains are a common type of locomotor injury. They vary widely in seriousness but tend to occur in certain locations. The most common cause of ankle sprain is excessive inversion of the foot, resulting in partial tearing of the anterior talofibular ligament and the calcaneofibular ligament. Less commonly, the deltoid ligament is injured by excessive eversion of the foot. Torn ligaments are extremely painful and are accompanied by immediate local swelling. Reducing the swelling and immobilizing the joint are about the only treatments for moderate sprains. Extreme sprains may require surgery and casting of the joint to facilitate healing.

A summary of the principal joints of the body and their movement is presented in table 8.2.

1. What are the only joints that have menisci?
2. What two types of joints are found in the shoulder region? Why is the shoulder joint so vulnerable?
3. Which joints are reinforced with muscles that span the joint?
4. Describe the structure of the knee joint and indicate which structures protect and reinforce its anterior surface.

Clinical Considerations

A synovial joint is a remarkable biologic system. Its self-lubricating action provides a shock-absorbing cushion between articulating bones and enables almost frictionless movement under tremendous loads and impacts. Under normal circumstances and in most people, the many joints of the body perform without problems throughout life. Joints are not indestructible, however, and are subject to various forms of trauma and disease. Although not all of the diseases of joints are fully understood, medical science has made remarkable progress in the treatment of arthrological problems.

Trauma to Joints

Joints are well adapted to withstand compression and tension forces. Torsion or sudden impact to the side of a joint, however, can be devastating. These types of injuries frequently occur in athletes.

In a **strained joint,** unusual or excessive exertion stretches the tendons or muscles surrounding a joint. The damage is not serious. Strains are frequently caused by not "warming up" the muscles and not "stretching" the joints prior to exercise. A **sprain** is a tearing of the ligaments or tendons surrounding a joint. There are various grades of sprains, and the severity determines the treatment. Severe sprains damage articular cartilages

Table 8.2
Principal Articulations

Joint	Type	Movement
Most skull joints	Fibrous (suture)	Immovable
Temporomandibular	Synovial (hinge; gliding)	Elevation, depression; protraction, retraction
Atlanto-occipital	Synovial (condyloid)	Flexion, extension, circumduction
Atlantoaxial	Synovial (pivot)	Rotation
Intervertebral		
bodies of vertebrae	Cartilaginous (symphysis)	Slight movement
articular processes	Synovial (gliding)	Flexion, extension, slight rotation
Sacroiliac	Cartilaginous (gliding)	Slight gliding movement; may fuse in adults
Costovertebral	Synovial (gliding)	Slight movement during breathing
Sternocostal	Synovial (gliding)	Slight movement during breathing
Sternoclavicular	Synovial (gliding)	Slight movement when shrugging shoulders
Sternal	Cartilaginous (symphysis)	Slight movement during breathing
Acromioclavicular	Synovial (gliding)	Protraction, retraction; elevation, depression
Glenohumeral (shoulder)	Synovial (ball-and-socket)	Flexion, extension; adduction, abduction; rotation; circumduction
Elbow	Synovial (hinge)	Flexion, extension
Proximal radioulnar	Synovial (pivot)	Rotation
Distal radioulnar	Fibrous (syndesmosis)	Slight side-to-side movement
Radiocarpal (wrist)	Synovial (condyloid)	Flexion, extension; adduction, abduction; circumduction
Intercarpal	Synovial (gliding)	Slight movement
Carpometacarpal		
fingers	Synovial (condyloid)	Flexion, extension; adduction, abduction
thumb	Synovial (saddle)	Flexion, extension; adduction, abduction
Metacarpophalangeal	Synovial (condyloid)	Flexion, extension; adduction, abduction
Interphalangeal	Synovial (hinge)	Flexion, extension
Symphysis pubis	Fibrous (symphysis)	Slight movement
Coxal (hip)	Synovial (ball-and-socket)	Flexion, extension; adduction, abduction; rotation; circumduction
Tibiofemoral (knee)	Synovial (hinge)	Flexion, extension; slight rotation when flexed
Proximal tibiofibular	Synovial (gliding)	Slight movement
Distal tibiofibular	Fibrous (syndesmosis)	Slight movement
Talocrural (ankle)	Synovial (hinge)	Dorsiflexion, plantar flexion; slight circumduction; inversion, eversion
Intertarsal	Synovial (gliding)	Inversion, eversion
Tarsometatarsal	Synovial (gliding)	Flexion, extension; adduction, abduction

and may require surgery. Sprains are usually accompanied by **synovitis,** an inflammation of the joint capsule.

Luxation, or **joint dislocation,** is a derangement of the articulating bones that compose the joint. Joint dislocation is more serious than a sprain and is usually accompanied by sprains. The shoulder and knee joints are the most vulnerable to dislocation. Self-healing of a dislocated joint may be incomplete, leaving the person with a "trick knee," for example, that may unexpectedly give way.

Subluxation is partial dislocation of a joint. Subluxation of the hip joint is a common type of birth defect that can be treated by bracing or casting to promote suitable bone development.

Bursitis (*bur-si'tis*) is an inflammation of the bursa associated with a joint. Because the bursa is close to the joint, the joint capsule may be affected as well. Bursitis may be caused by excessive stress on the bursa from overexertion, or it may be a local or systemic inflammatory process. As the bursa swells, the surrounding muscles become sore and stiff. **Tendonitis** involves inflammation of a tendon; it usually comes about in the same way as bursitis.

The flexible vertebral column is a marvel of mechanical engineering. Not only do the individual vertebrae articulate one

luxation: L. *luxus,* out of place

Developmental Exposition

The Synovial Joints

Exhibit I Development of synovial joints. (*a*) At 6 weeks, different densities of mesenchyme denote where the bones and joints will form. (*b*) At 9 weeks, a basic synovial model is present. At 12 weeks, the synovial joints are formed and have either (*c*) a free joint cavity (e.g., interphalangeal joint); (*d*) a cavity containing menisci (e.g., knee joint); or (*e*) a cavity with a complete articular disc (e.g., sternoclavicular joint).

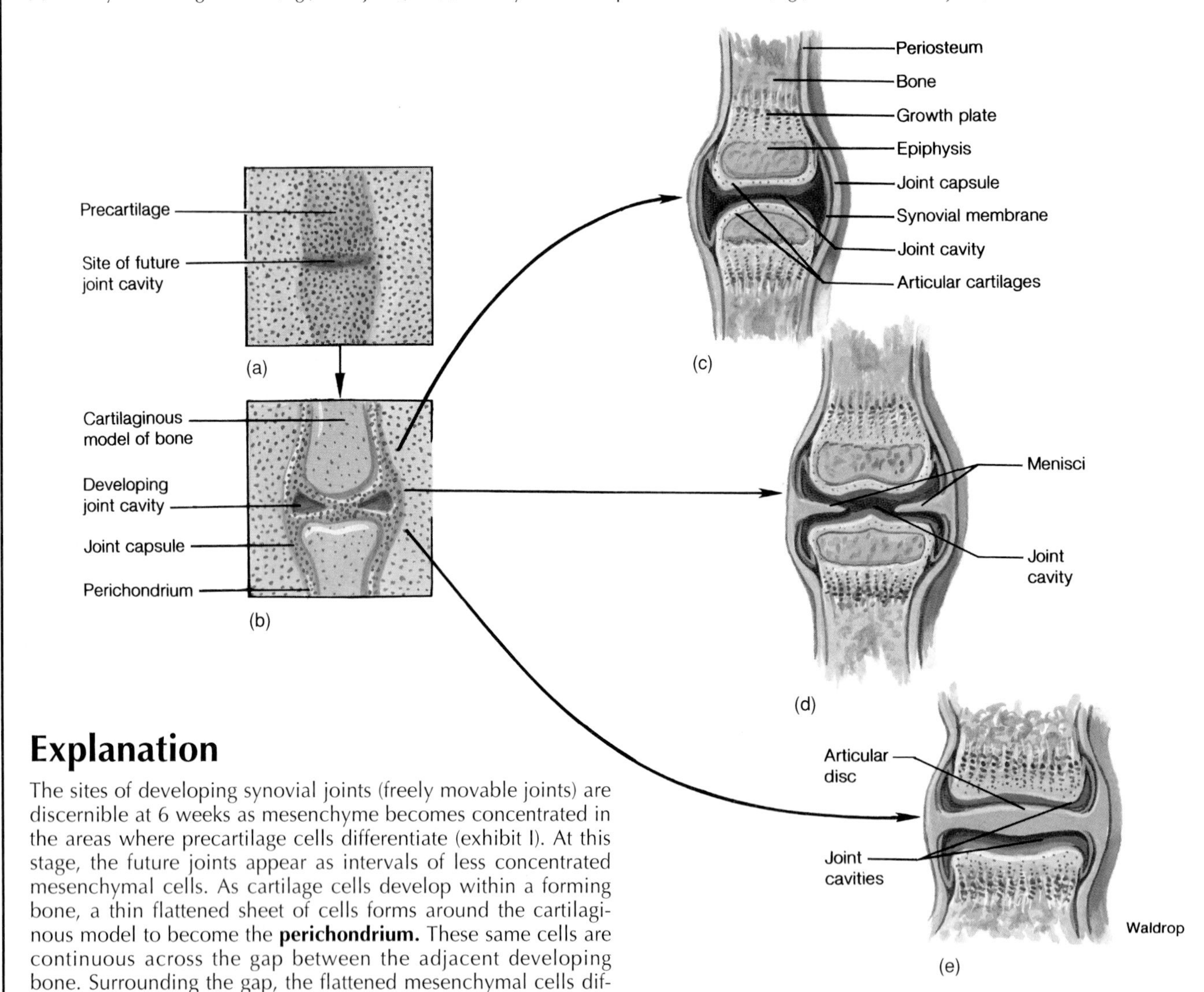

Explanation

The sites of developing synovial joints (freely movable joints) are discernible at 6 weeks as mesenchyme becomes concentrated in the areas where precartilage cells differentiate (exhibit I). At this stage, the future joints appear as intervals of less concentrated mesenchymal cells. As cartilage cells develop within a forming bone, a thin flattened sheet of cells forms around the cartilaginous model to become the **perichondrium.** These same cells are continuous across the gap between the adjacent developing bone. Surrounding the gap, the flattened mesenchymal cells differentiate to become the **joint capsule.**

During the early part of the third month of development, the mesenchymal cells still remaining within the joint capsule begin to migrate toward the epiphyses of the adjacent developing bones. The cleft eventually enlarges to become the **joint cavity.** Thin pads of hyaline cartilage develop on the surfaces of the epiphyses that contact the joint cavity. These pads become the **articular cartilages** of the functional joint. As the joint continues to develop, a highly vascular **synovial membrane** forms on the inside of the joint capsule and begins secreting a watery *synovial fluid* into the joint cavity.

In certain developing synovial joints, the mesenchymal cells do not migrate away from the center of the joint cavity. Rather, they give rise to cartilaginous wedges called **menisci,** as in the knee joint, or to complete cartilaginous pads called **articular discs,** as in the sternoclavicular joint.

Most synovial joints have formed completely by the end of the third month. Shortly thereafter, fetal muscle contractions, known as *quickening,* cause movement at these joints. Joint movement enhances the nutrition of the articular cartilage and prevents the fusion of connective tissues within the joint.

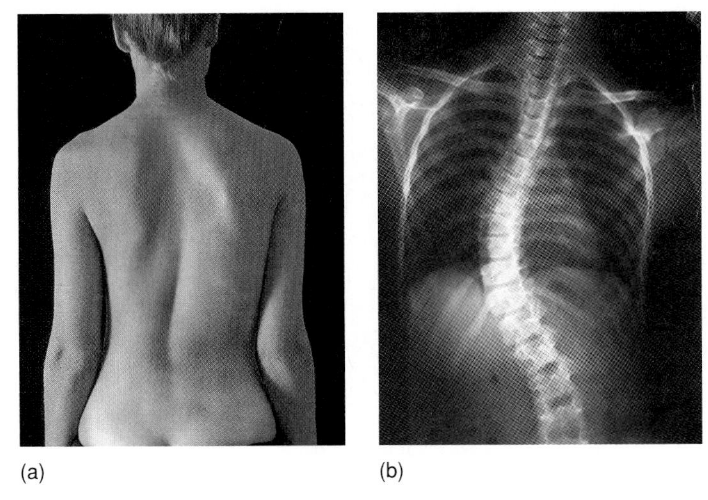

(a) (b)

Figure 8.35

Scoliosis is a lateral curvature of the spine, usually in the thoracic region. It may be congenital, disease-related, or idiopathic (of unknown cause). (*a*) A posterior view of a 19-year-old woman and (*b*) a radiograph.

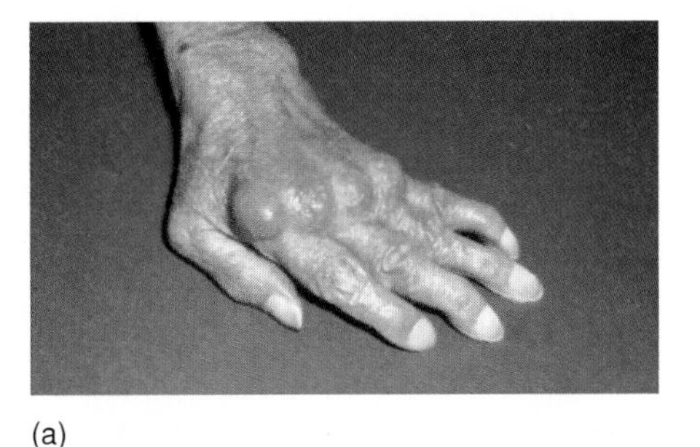

(a)

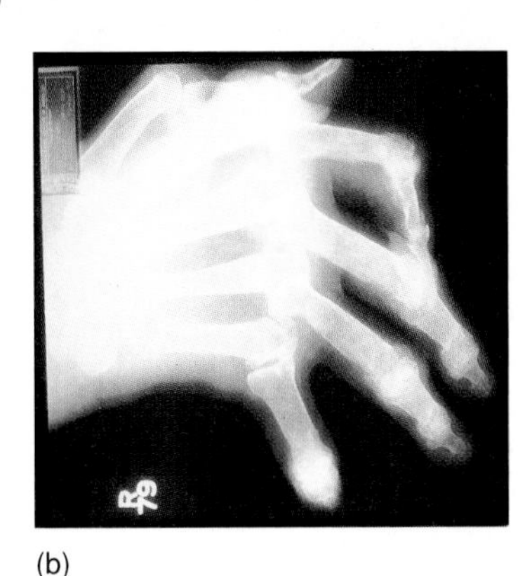

(b)

Figure 8.36

Rheumatoid arthritis may eventually cause joint ossification and debilitation as seen in (*a*) a photograph of a patient's hand and (*b*) a radiograph.

with another, but together they form the portion of the axial skeleton with which the head, ribs, and ossa coxae articulate. The vertebral column also encloses the spinal cord and provides exits for 31 pairs of spinal nerves. Considering all the articulations in the vertebral column and the physical abuse it takes, it is no wonder that back ailments are second only to headaches as the most common physical complaint. Our way of life causes many of the problems associated with the vertebral column. Improper shoes, athletic exertion, sudden stops in vehicles, or improper lifting can all cause the back to go awry. Body weight, age, and general physical condition influence a person's susceptibility to back problems.

The most common cause of back pain is *strained muscles*, generally as a result of overexertion. The second most frequent back ailment is a *herniated disc*. The dislodged nucleus pulposus of a disc may push against a spinal nerve and cause excruciating pain. The third most frequent back problem is a *dislocated articular facet* between two vertebrae, caused by sudden twisting of the vertebral column. The treatment of back ailments varies from bed rest to spinal manipulation to extensive surgery.

Curvature disorders are another problem of the vertebral column. **Kyphosis** (*ki-fo'sis*) (hunchback) is an exaggeration of the thoracic curve. **Lordosis** (swayback) is an abnormal anterior convexity of the lumbar curve. **Scoliosis** (*sko-le-o'sis*) (crookedness) is an abnormal lateral curvature of the vertebral column (fig. 8.35). It may be caused by abnormal vertebral structure, unequal length of the legs, or uneven muscular development on the two sides of the vertebral column.

Diseases of Joints

Arthritis is a generalized designation for over 50 different joint diseases (fig. 8.36), all of which have the symptoms of edema, inflammation, and pain. The causes are unknown, but certain types follow joint trauma or bacterial infection. Some types are genetic and others result from hormonal or metabolic disorders. The most common forms are *rheumatoid arthritis*, *osteoarthritis*, and *gouty arthritis*.

Rheumatoid (*roo'mă-toid*) **arthritis** results from an autoimmune attack against the joint tissues. The synovial membrane thickens and becomes tender, and synovial fluid accumulates. This is generally followed by deterioration of the articular cartilage, which eventually exposes bone tissue. When bone tissue is unprotected, joint ossification produces the crippling effect of this disease. Females are affected more often than males, and the disease usually begins between the ages of 30 and 50. Rheumatoid arthritis tends to occur bilaterally. If the right wrist or hip develops the disease, so does the left.

Osteoarthritis is a degenerative joint disease that results from aging and irritation of the joints. Although osteoarthritis is far more common than rheumatoid arthritis, it is usually less damaging. Osteoarthritis is a progressive disease in which the articular cartilages gradually soften and disintegrate. The affected joints seldom swell, and the synovial membrane is rarely damaged. As the articular cartilage deteriorates, ossified spurs

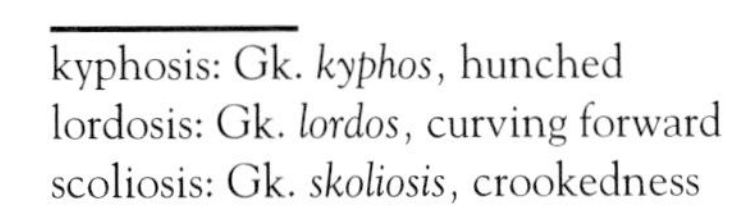

kyphosis: Gk. *kyphos*, hunched
lordosis: Gk. *lordos*, curving forward
scoliosis: Gk. *skoliosis*, crookedness

rheumatoid: Gk. *rheuma*, a flowing

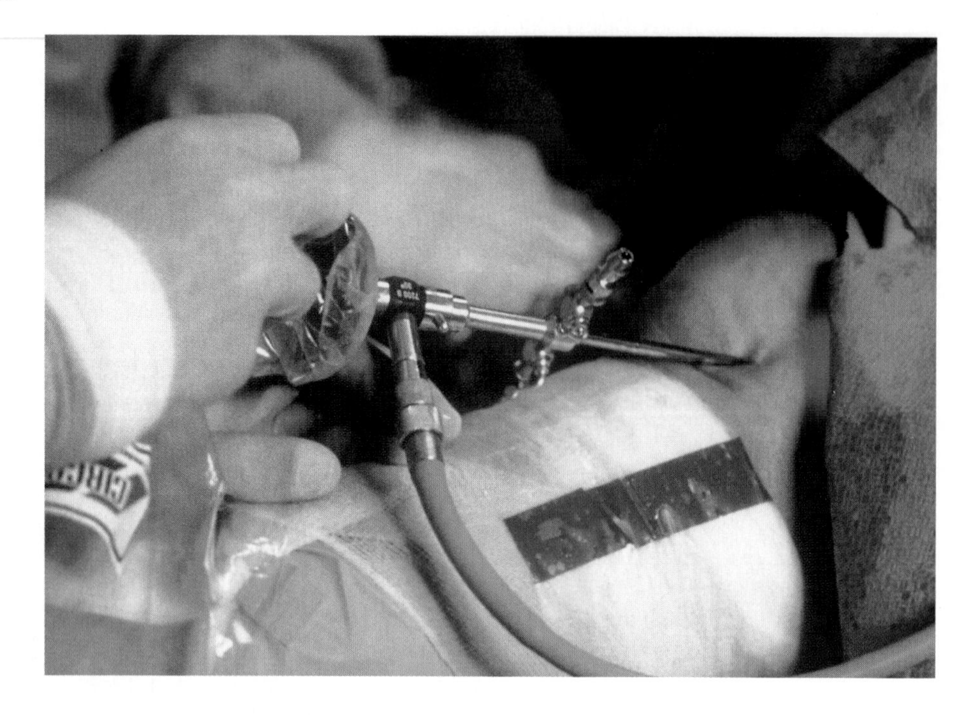

Figure 8.37

Arthroscopy. In this technique, a needlelike viewing arthroscope is threaded into the joint capsule through a tiny incision. The arthroscope has a fiberoptic light source that illuminates the interior of the joint. Thus, the position of the surgical instruments that may be inserted through other small incisions can be seen.

are deposited on the exposed bone, causing pain and restricting the movement of articulating bones. Osteoarthritis most frequently affects the knee, hip, and intervertebral joints.

Gouty arthritis results from a metabolic disorder in which an abnormal amount of uric acid is retained in the blood and sodium urate crystals are deposited in the joints. The salt crystals irritate the articular cartilage and synovial membrane, causing swelling, tissue deterioration, and pain. If gout is not treated, the affected joint fuses. Males have a greater incidence of gout than females, and apparently the disease is genetically determined. About 85% of gout cases affect the joints of the foot and legs. The most common joint affected is the metatarsophalangeal joint of the hallux (great toe).

Treatment of Joint Disorders

Arthroscopy (*ar-thros'kŏ-pe*) is widely used in diagnosing and, to a limited extent, treating joint disorders (fig. 8.37). Arthroscopic inspection involves making a small incision into the joint capsule and inserting a tubelike instrument called an *arthroscope*. In arthroscopy of the knee, the articular cartilage, synovial membrane, menisci, and cruciate ligaments can be observed. Samples can be extracted, and pictures taken for further evaluation.

Remarkable advancements have been made in the last 15 years in the development of **joint prostheses** (*pros-the'sēz*) (fig. 8.38). These artificial articulations do not take the place of normal, healthy joints, but they are a valuable option for chronically disabled arthritis patients. They are now available for finger, shoulder, and elbow joints, as well as for hip and knee joints.

Clinical Case Study Answer

The way in which the knee was injured and the location of the pain, taken together with the exam findings, indicate a complete or near-complete tear of the medial collateral ligament. Because the medial meniscus is attached to this ligament, it is frequently torn as well in an injury of this sort. Other ligaments susceptible to athletic injury are the anterior cruciate ligament (most common) and the lateral collateral and posterior cruciate ligaments. Complete tears of these ligaments usually require surgical repair for acceptable results. Incomplete tears can often be managed by nonsurgical means.

gout: L. *gutta*, a drop (thought to be caused by "drops of viscous humors")

prosthesis: Gk. *pros*, in addition to; *thesis*, a setting down

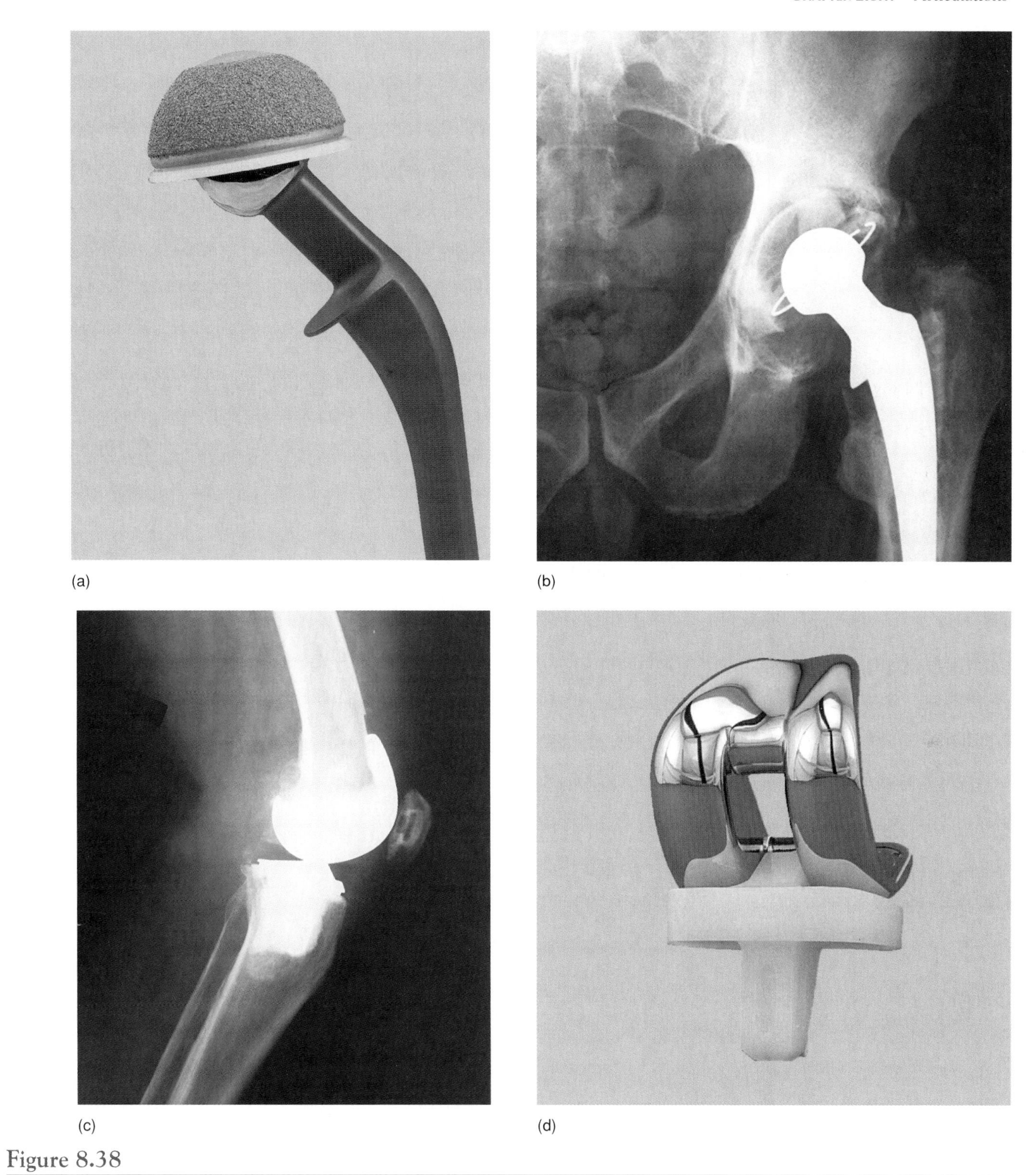

Figure 8.38

Two examples of joint prostheses. (*a, b*) The coxal (hip) joint and (*c, d*) the tibiofemoral (knee) joint.

Important Clinical Terminology

ankylosis (***ang″kĭ-lo′sis***) Stiffening of a joint, resulting in severe or complete loss of movement.
arthralgia (***ar-thral′je-ă***) Severe pain in a joint.
arthrolith (***ar′thro-lith***) A gouty deposit in a joint, also called *arthrdynia*.
arthrometry (***ar-throm′ĕ-tre***) Measurement of the range of movement in a joint.
arthroncus (***ar-thron′kus***) Swelling of a joint as a result of trauma or disease.
arthropathy (***ar-throp′ă-the***) Any disease affecting a joint.
arthroplasty (***ar′thro-plas″te***) Surgical repair of a joint.
arthrosis (***ar-thro′sis***) A joint or an articulation; also, a degenerative condition of a joint.
arthrosteitis (***ar″thros-te-i′tis***) Inflammation of the bony structure of a joint.
chondritis (***kon-dri′tis***) Inflammation of the articular cartilage of a joint.
coxarthrosis (***koks″ar-thro′sis***) A degenerative condition of the hip joint.
hemarthrosis (***hem-ar-thro′sis***) An accumulation of blood in a joint cavity.
rheumatology (***roo″mă-tol′ŏ-je***) The medical specialty concerned with the diagnosis and treatment of rheumatic diseases.
spondylitis (***spon-dil-i′tis***) Inflammation of one or more vertebrae.
synovitis (***sin″o-vi′tis***) Inflammation of the synovial membrane lining the inside of a joint capsule.

Chapter Summary

Classification of Joints (p. 192)

1. Joints are formed as adjacent bones articulate. Arthrology is the science concerned with the study of joints; kinesiology is the study of movements involving certain joints.
2. Joints are classified as fibrous, cartilaginous, or synovial.

Fibrous Joints (pp. 192–194)

1. Articulating bones in fibrous joints are tightly bound by fibrous connective tissue. Fibrous joints are of three types: sutures, syndesmoses, and gomphoses.
2. Sutures are found only in the skull; they are classified as serrate, lap, or plane.
3. Syndesmoses are found in the vertebral column, middle ear, antebrachium, and leg. The articulating bones of syndesmoses are held together by interosseous ligaments, which permit slight movement.
4. Gomphoses are found only in the skull, where the teeth are bound into their sockets by the periodontal ligaments.

Cartilaginous Joints (pp. 194–195)

1. The fibrocartilage or hyaline cartilage of cartilaginous joints allows limited motion in response to twisting or compression. The two types of cartilaginous joints are symphyses and synchondroses.
2. The symphysis pubis and the joints formed by the intervertebral discs are examples of symphyses.
3. Some synchondroses are temporary joints formed in the growth plates between the diaphyses and epiphyses in the long bones of children. Other synchondroses are permanent; for example, the joints between the ribs and the costal cartilages of the rib cage.

Synovial Joints (pp. 195–200)

1. The freely movable synovial joints are enclosed by joint capsules that contain synovial fluid. Synovial joints include gliding, hinge, pivot, condyloid, saddle, and ball-and-socket types.
2. Synovial joints contain a joint cavity, articular cartilages, and synovial membranes that produce the synovial fluid. Some also contain articular discs, accessory ligaments, and associated bursae.
3. The movement of a synovial joint is determined by the structure of the articulating bones, the strength and tautness of associated ligaments and tendons, and the arrangement and tension of the muscles that act on the joint.

Movements at Synovial Joints (pp. 201–208)

1. Movements at synovial joints are produced by the contraction of the skeletal muscles that span the joints and attach to or near the bones forming the articulations. In these actions, the bones act as levers, the muscles provide the force, and the joints are the fulcra, or pivots.
2. Angular movements increase or decrease the joint angle produced by the articulating bones. Flexion decreases the joint angle on an anterior-posterior plane; extension increases the same joint angle. Abduction is the movement of a body part away from the main axis of the body; adduction is the movement of a body part toward the main axis of the body.
3. Circular movements can occur only where the rounded surface of one bone articulates with a corresponding depression on another bone. Rotation is the movement of a bone around its own axis. Circumduction is a conelike movement of a body part.
4. Special joint movements include inversion and eversion, protraction and retraction, and elevation and depression.
5. Synovial joints and their associated bones and muscles can be classified as first-, second-, or third-class levers. In a first-class lever, the fulcrum is positioned between the effort and the resistance. In a second-class lever, the resistance lies between the fulcrum and the effort. In a third-class lever, the effort is applied between the fulcrum and the resistance.

Specific Joints of the Body (pp. 208–218)

1. The temporomandibular joint, a combined hinge and gliding joint, is of clinical importance because of temporomandibular joint (TMJ) syndrome.
2. The glenohumeral (shoulder) joint, a ball-and-socket joint, is vulnerable to dislocations from sudden jerks of the arm, especially in children before strong shoulder muscles have developed.
3. There are two sets of articulations at the elbow joint as the distal end of the humerus articulates with the proximal ends of the ulna and radius. It is a hinge joint that is subject to strain during certain sports.
4. The metacarpophalangeal joints (knuckles) are condyloid joints, and the interphalangeal joints (between adjacent phalanges) are hinge joints.
5. The ball-and-socket coxal (hip) joint is adapted for weight bearing. Its capsule is extremely strong and is reinforced by several ligaments.
6. The hinged tibiofemoral (knee) joint is the largest, most vulnerable joint in the body.
7. There are two hinged articulations within the talocrural (ankle) joint. Sprains are frequently associated with this joint.

Review Activities

Objective Questions

1. Which statement regarding joints is *false?*
   (a) They are places where two or more bones articulate.
   (b) All joints are movable.
   (c) Arthrology is the study of joints; kinesiology is the study of the biomechanics of joint movement.
2. Synchondroses are a type of
   (a) fibrous joint.
   (b) synovial joint.
   (c) cartilaginous joint.

3. An interosseous ligament is characteristic of
   (a) a suture. (c) a symphysis.
   (b) a synchondrosis. (d) a syndesmosis.
4. Which of the following joint type-function word pairs is *incorrect?*
   (a) synchondrosis/growth at the epiphyseal plate
   (b) symphysis/movement at the intervertebral joint between vertebral bodies
   (c) suture/strength and stability in the skull
   (d) syndesmosis/movement of the jaw
5. Which of the following is a *false* statement?
   (a) Synchondroses occur in the long bones of children and young adults.
   (b) Sutures occur only in the skull.
   (c) Saddle joints occur in the thumb and in the neck, where rotational movement is possible.
   (d) Syndesmoses occur in the antebrachium and leg.
6. Which of the following is *not* characteristic of all synovial joints?
   (a) articular cartilage
   (b) synovial fluid
   (c) a joint capsule
   (d) menisci
7. The atlantoaxial and the proximal radioulnar synovial joints are specifically classified as
   (a) hinge. (c) pivotal.
   (b) gliding. (d) condyloid.
8. Which of the following joints can be readily and comfortably hyperextended?
   (a) an interphalangeal joint
   (b) a coxal joint
   (c) a tibiofemoral joint
   (d) a sternocostal joint
9. Which of the following is most vulnerable to luxation?
   (a) the elbow joint
   (b) the glenohumeral joint
   (c) the coxal joint
   (d) the tibiofemoral joint
10. A thickening and tenderness of the synovial membrane and the accumulation of synovial fluid are signs of the development of
    (a) arthroscopitis.
    (b) gouty arthritis.
    (c) osteoarthritis.
    (d) rheumatoid arthritis.

Essay Questions

1. What are the three structural classes of joints? Describe the characteristics of each.
2. Why is anatomical position so important in explaining the movements that are possible at joints?
3. What are the structural elements of a synovial joint that determine its range of movement?
4. What are the advantages of a hinge joint over a ball-and-socket type? If ball-and-socket joints allow a greater range of movement, why aren't all the synovial joints of this type?
5. What is synovial fluid? Where is it produced and what are its functions?
6. Describe a bursa and discuss its function. What is bursitis?
7. Identify four types of synovial joints found in the wrist and hand regions, and state the types of movement permitted by each.
8. Discuss the articulations between the pectoral and pelvic regions and the axial skeleton with regard to range of movement, ligamentous attachments, and potential clinical problems.
9. What is meant by a sprained ankle? How does a sprain differ from a strain or a luxation?
10. What occurs within the joint capsule in rheumatoid arthritis? How does rheumatoid arthritis differ from osteoarthritis?

Figure 8.39

Which joints of the body are being flexed as this person assumes a fetal position?

Critical-Thinking Questions

1. Refer to figure 8.39 and identify the joints being flexed. In the upper and lower extremities of your own body, which are larger and stronger, the flexor muscles or the extensor muscles? Why?
2. Considering the type of synovial joint at the hip and the location of the gluteal muscles of the buttock, explain why this type of lever system is adapted for rapid, wide-ranging movements.
3. The star runningback of a local high school football team was taken to the emergency room of the local hospital following a knee injury during the championship game. The injury resulted from a hard blow ("clipping") to the back of his right knee as it was supporting the weight of his body. Suspecting a rupture of the anterior cruciate ligament, the ER physician informed the football player that this diagnosis could be confirmed by pulling the tibia forward as the knee was flexed. He explained that if the tibia slipped forward at the knee ("bureau drawer sign"), it could be assumed that the anterior cruciate ligament was ruptured.

   In terms of the anatomy of the knee joint, explain the occurrence of bureau drawer sign. What structure most likely would be traumatized if the tibia could be displaced backward?
4. In what ways do the anatomical differences between the jaw, shoulder, elbow, hip, knee, and ankle joints relate to their differences in function?

NINE

Muscular System

Clinical Case Study 227
Introduction to the Muscular System 227
Structure of Skeletal Muscles 229
- Muscle Attachments 229
- Associated Connective Tissue 229
- Muscle Groups 231
- Muscle Architecture 231
- Blood and Nerve Supply to Skeletal Muscle 231

This engraving by the artist Jan Wandelaer appeared in a work by Bernard Albinus published in the eighteenth century. Albinus, a German anatomist and professor at the University of Leiden for 50 years, was the first to classify and arrange the muscles in a proper manner.

Skeletal Muscle Fibers and Types of Muscle Contraction 233
- Skeletal Muscle Fibers 233
- Isotonic and Isometric Contractions 236
- Neuromuscular Junction 236
- Motor Unit 237

Naming of Muscles 239

Muscles of the Axial Skeleton 241
- Muscles of Facial Expression 241
- Muscles of Mastication 241
- Ocular Muscles 241
- Muscles That Move the Tongue 244
- Muscles of the Neck 244
- Muscles of Respiration 248
- Muscles of the Abdominal Wall 248
- Muscles of the Pelvic Outlet 250
- Muscles of the Vertebral Column 251

Muscles of the Appendicular Skeleton 253
- Muscles That Act on the Pectoral Girdle 253
- Muscles That Move the Humerus at the Shoulder Joint 254
- Muscles That Move the Forearm at the Elbow Joint 257
- Muscles of the Forearm That Move the Joints of the Wrist, Hand, and Fingers 258
- Muscles of the Hand 263
- Muscles That Move the Thigh at the Hip Joint 265
- Muscles of the Thigh That Move the Knee Joint 268
- Muscles of the Leg That Move the Joints of the Ankle, Foot, and Toes 273
- Muscles of the Foot 276

Clinical Considerations 278
- Evaluation of Muscle Condition 278
- Functional Conditions in Muscles 278

Developmental Exposition: The Muscular System 280
- Diseases of Muscles 282
- Aging of Muscles 283

Clinical Case Study Answer 283

Internal Affairs 284

Important Clinical Terminology 285

Chapter Summary 285

Review Activities 285

Clinical Case Study

A 66-year-old man went to a doctor for a routine physical exam. The man's medical history revealed that he had been treated surgically for cancer of the oropharynx 6 years earlier. The patient stated that the cancer had spread to the lymph nodes in the left side of his neck. He pointed to the involved area, explaining that lymph nodes, a vein, and a muscle, among other things, had been removed. On the right side, only lymph nodes had been removed, and they were found to be benign. The patient then stated that he had difficulty turning his head to the right. Obviously perplexed, he commented, "It seems to me Doc, that if they took the muscle out of the left side of my neck, I would be able to turn my head only to the right."

Does the patient have a valid point? If not, how would you explain the reason for his disability in terms of neck musculature?

Hints: The action of a muscle can always be explained on the basis of its points of attachment and the joint or joints it spans. Carefully examine the muscles shown in figure 9.20 and described in table 9.6.

Introduction to the Muscular System

Skeletal muscles are adapted to contract in order to carry out the functions of generating body movement, producing heat, and supporting the body and maintaining posture.

Objective 1 Define the term *myology* and describe the three principal functions of muscles.

Objective 2 Explain how muscles are described according to their anatomical location and cooperative function.

Myology is the study of muscles. More than 600 skeletal muscles make up the muscular system, and technically each one is an organ—it is composed of skeletal muscle tissue, connective tissue, and nervous tissue. Each muscle also has a particular function, such as moving a finger or blinking an eyelid. Collectively, the skeletal muscles account for approximately 40% of the body weight.

Muscle cells (fibers) contract when stimulated by nerve impulses. The stimulation of just a few fibers is not enough to cause a noticeable effect, but isolated fiber contractions are important and occur continuously within a muscle. When a sufficient number of skeletal muscle fibers are activated, the muscle contracts and causes body movement.

Muscles perform three principal functions: (1) movement, (2) heat production, and (3) body support and maintenance of posture.

1. **Movement.** The most obvious function performed by skeletal muscles is to move the body or parts of the body, as in walking, running, writing, chewing, and swallowing. Even the eyeball and the auditory ossicles have associated skeletal muscles that are responsible for various movements. The contraction of skeletal muscle is equally important in breathing and in moving internal body fluids. The stimulation of individual skeletal muscle fibers maintains a state of muscle contraction called *tonus*, which is important in the movement of blood and lymph.

   The involuntary contraction of smooth muscle tissue is also essential for movement of materials through the body. Likewise, the involuntary contraction of cardiac muscle tissue continuously pumps blood throughout the body.
2. **Heat production.** Body temperature is held remarkably constant. Metabolism within the cells releases heat as an end product. Since muscles constitute approximately 40% of body weight and are in a continuous state of fiber activity, they are the primary source of body heat. The rate of heat production increases greatly during strenuous exercise.
3. **Posture and body support.** The skeletal system provides a framework for the body, but skeletal muscles maintain posture, stabilize the flexible joints, and support the viscera. Certain muscles are active postural muscles whose primary function is to work in opposition to gravity. Some postural muscles are working even when you think you are relaxed. As you are sitting, for example, the weight of your head is balanced at the atlanto-occipital joint through the efforts of the muscles located at the back of the neck. If you start to get sleepy, your head will suddenly nod forward as the postural muscles relax and the weight (resistance) overcomes the effort.

Muscle tissue in the body is of three types: *smooth*, *cardiac*, and *skeletal* (see fig. 4.26). Although these three types differ in structure and function, and the muscular system refers to the *skeletal* muscle system, the following basic properties characterize all muscle tissue:

1. **Irritability.** Muscle tissue is sensitive to stimuli from nerve impulses.
2. **Contractility.** Muscle tissue responds to stimuli by contracting lengthwise, or shortening.
3. **Extensibility.** Once a stimulus has subsided and the fibers within muscle tissue are relaxed, they may be stretched even beyond their resting length by the contraction of an opposing muscle. The fibers are then prepared for another contraction.
4. **Elasticity.** Muscle fibers, after being stretched, have a tendency to recoil to their original resting length.

---

myology: Gk. *myos*, muscle; *logos*, study of
muscle: L. *mus*, mouse (from the appearance of certain muscles)

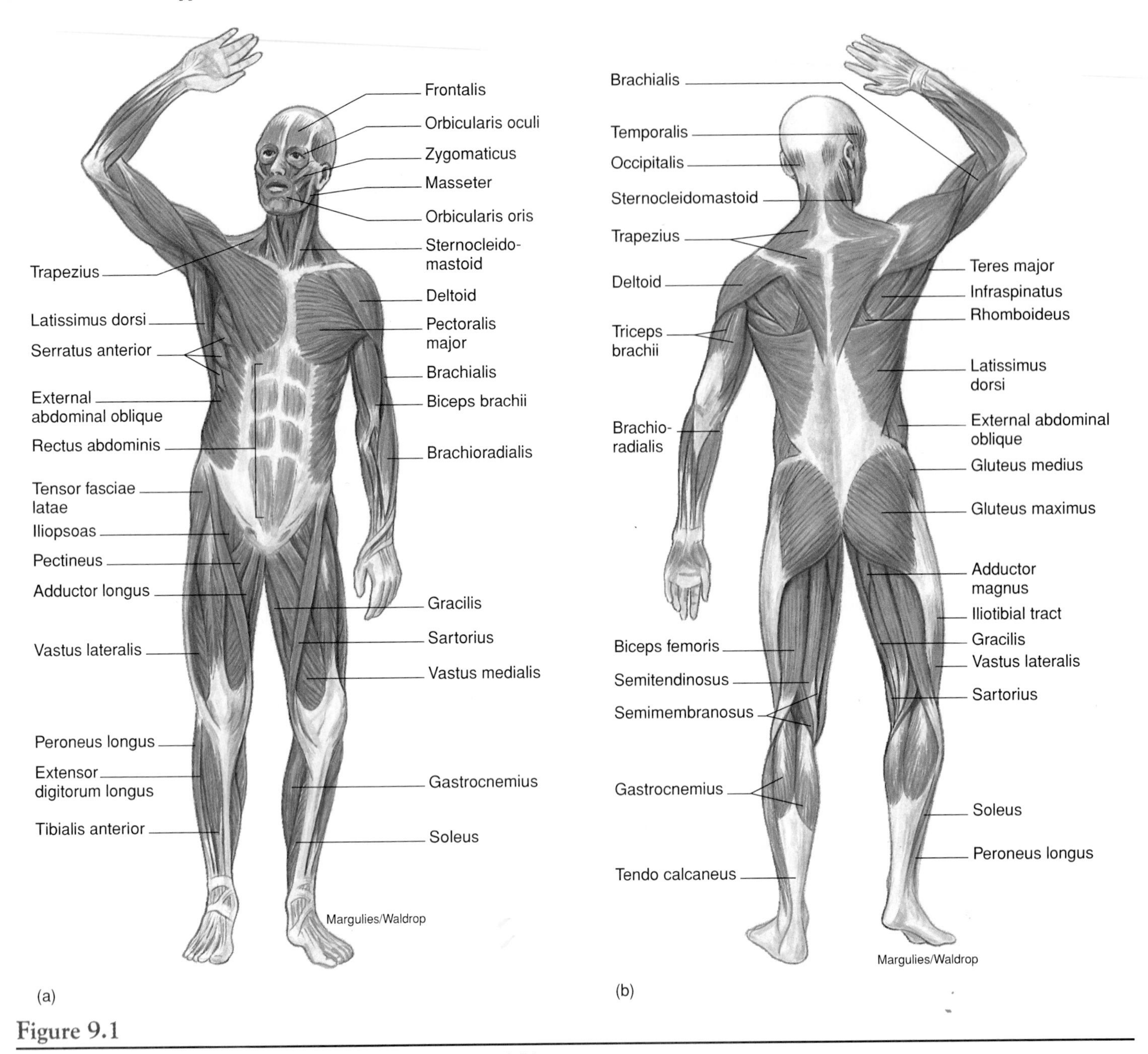

Figure 9.1

The principal superficial skeletal muscles. (*a*) An anterior view and (*b*) a posterior view.

A histological description of each of the three muscle types was presented in chapter 4 and should be reviewed at this time. Cardiac muscle is involuntary and is discussed further in chapter 13 in the autonomic nervous system and in chapter 16, in connection with the heart. Smooth muscle is widespread throughout the body and is also involuntary. It is discussed in chapter 13 and, when appropriate, in connection with the organs in which it occurs. The remaining information presented in this chapter pertains only to skeletal muscle and the skeletal muscle system of the body.

Muscles are usually described in groups according to anatomical location and cooperative function. The *muscles of the axial skeleton* include the facial muscles, neck muscles, and anterior and posterior trunk muscles. The *muscles of the appendicular skeleton* include those that act on the pectoral and pelvic girdles and those that move limb joints. The principal superficial muscles are shown in figure 9.1.

1. How do the functions of muscles help to maintain body homeostasis?
2. What is meant by a postural muscle?
3. Distinguish between the axial and the appendicular muscles.

Structure of Skeletal Muscles

Skeletal muscle tissue and its binding connective tissue are arranged in a highly organized pattern that unites the forces of the contracting muscle fibers and directs them onto the structure being moved.

Objective 3 Compare and contrast the various binding connective tissues associated with skeletal muscles.

Objective 4 Distinguish between isometric and isotonic contractions.

Objective 5 Describe the various types of muscle fiber architecture and discuss the biomechanical advantage of each type.

Muscle Attachments

Skeletal muscles are attached to a bone on each end by **tendons** (fig. 9.2). A tendon is composed of dense regular connective tissue and binds a muscle to the periosteum of a bone. When a muscle contracts, it shortens, and this places tension on its tendons and attached bones. The muscle tension causes movement of the bones at a synovial joint (see figs. 8.7 and 8.8), where one of the attached bones generally moves more than the other. The more movable bony attachment of the muscle, known as the **insertion,** is pulled toward its less movable attachment, the **origin.** In muscles associated with the girdles and appendages, the origin is the proximal attachment and the insertion is the distal attachment. The fleshy, thickened portion of a muscle is referred to as its **belly (gaster).**

Flattened, sheetlike tendons are called **aponeuroses** (*ap″ŏ-noo-ro′sēz*). An example is the galea aponeurotica, which is found on the top and sides of the skull (see fig. 9.14). In certain places, especially in the wrist and ankle, the tendons are not only enclosed by protective **tendon sheaths** (see fig. 8.8), but also the entire group of tendons is covered by a thin but strong band of connective tissue called a **retinaculum** (*ret″ĭ-nak′yoo-lum*) (see, for example, the extensor retinaculum in fig. 9.43). Attached to articulating bones, retinacula anchor groups of tendons and keep them from bowing during muscle contraction.

Associated Connective Tissue

Contracting muscle fibers would not be effective if they worked as isolated units. Each fiber is bound to adjacent fibers to form bundles, and the bundles in turn are bound to other bundles.

aponeurosis: Gk. *aponeurosis*, change into a tendon
retinaculum: L. *retinere*, to hold back (retain)

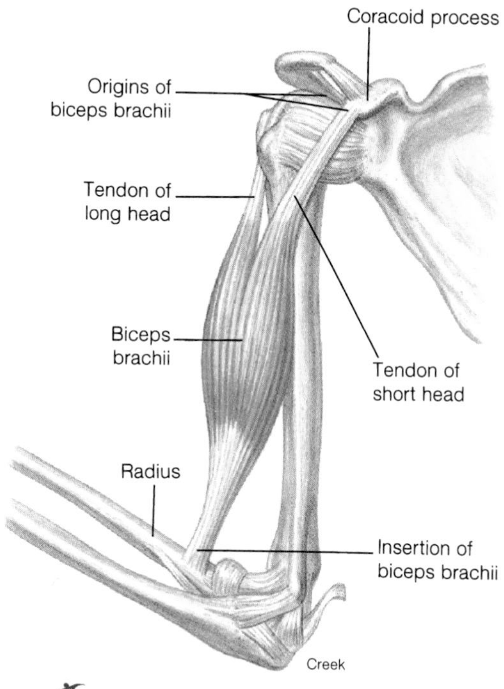

Figure 9.2

The skeletomuscular relationship. The more proximal, fixed point of muscle attachment is the origin; the distal, maneuverable point of attachment is the insertion. The contraction of muscle fibers causes one bone to move relative to another around a joint.

With this arrangement, the contraction in one area of a muscle works in conjunction with contracting fibers elsewhere in the muscle. The binding substance within muscles is the associated loose connective tissue.

Connective tissue is structurally arranged within muscle to protect, strengthen, and bind muscle fibers into bundles and bind the bundles together (fig. 9.3). The individual fibers of skeletal muscles are surrounded by a fine sheath of connective tissue called **endomysium** (*en″do-mis′e-um*). The endomysium binds adjacent fibers together and supports capillaries and nerve endings serving the muscle. Another connective tissue, the **perimysium,** binds groups of muscle fibers together into bundles called **fasciculi** (*fă-sik′yŭ-li*—singular, *fasciculus*). The perimysium supports blood vessels and nerve fibers serving the various fasciculi. The entire muscle is covered by the **epimysium,** which in turn is continuous with a tendon.

Fascia (*fash′e-ă*) is a fibrous connective tissue of varying thickness that covers muscle and attaches to the skin. *Superficial fascia* secures the skin to the underlying structures. The superficial fascia over the buttocks and abdominal wall is thick and

endomysium: Gk. *endon*, within; *myos*, muscle
perimysium: Gk. *peri*, around; *myos*, muscle
fasciculus: L. *fascis*, bundle
epimysium: Gk. *epi*, upon; *myos*, muscle
fascia: L. *fascia*, a band or girdle

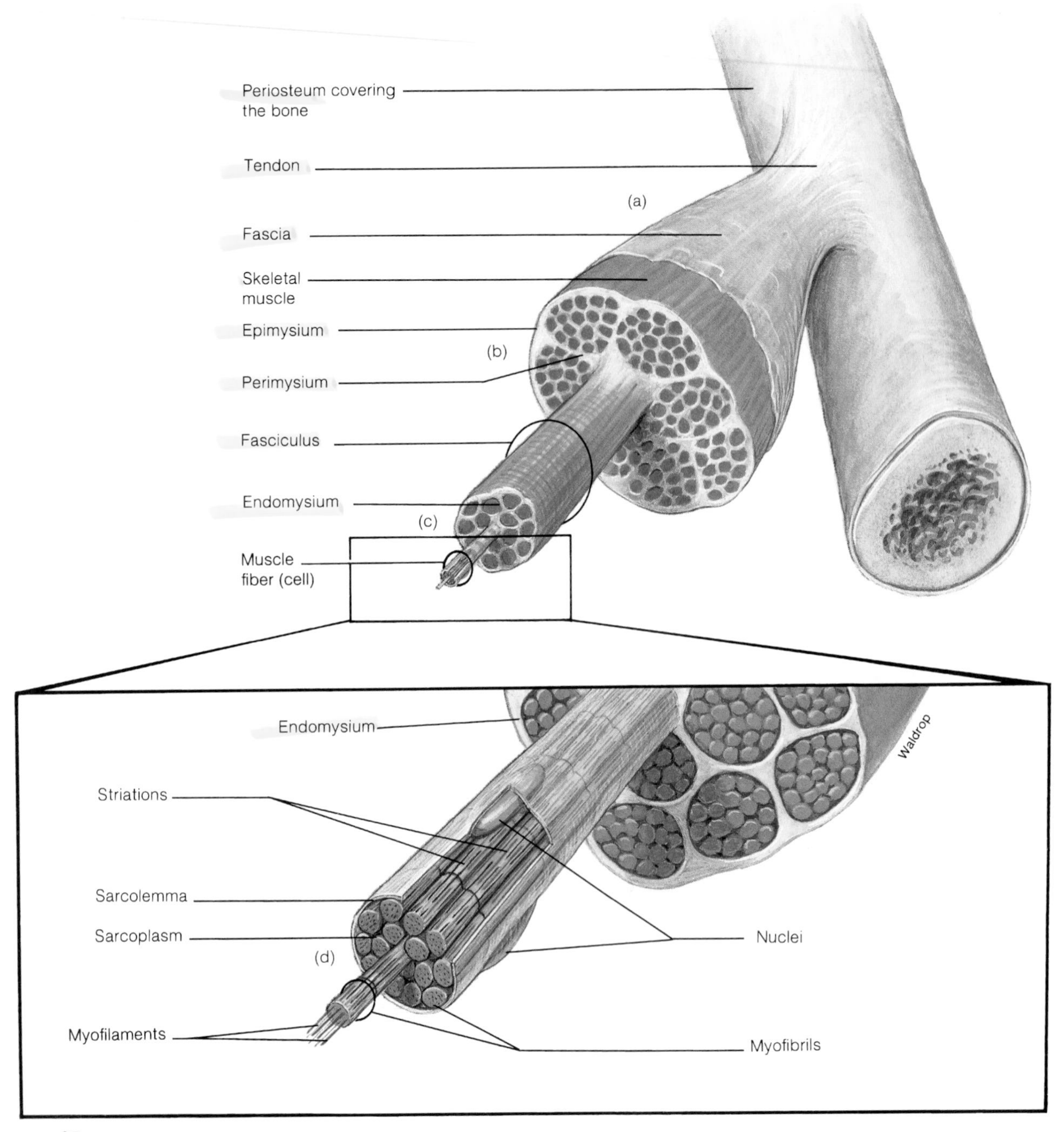

Figure 9.3

The relationship between skeletal muscle tissue and its associated connective tissue. (*a*) The fascia and tendon attaches a muscle to the periosteum of a bone. (*b*) The epimysium surrounds the entire muscle, and the perimysium separates and binds the fasciculi (muscle bundles). (*c*) The endomysium surrounds and binds individual muscle fibers. (*d*) An individual muscle fiber contains myofibrils (specialized contractile organelles) composed of thin (actin) and thick (myosin) myofilaments.

laced with adipose tissue. By contrast, the superficial fascia under the skin of the back of the hand, elbow, and facial region is thin. *Deep fascia* is an inward extension of the superficial fascia. It lacks adipose tissue and blends with the epimysium of muscle. Deep fascia surrounds adjacent muscles, compartmentalizing and binding them into functional groups.

The tenderness of meat is due in part to the amount of connective tissue present in a particular cut. A slice of meat from the ends of a muscle contains much more connective tissue than a cut through the belly of the muscle. Fibrous meat is difficult to chew and may present a social problem in trying to extract it discreetly from between the teeth.

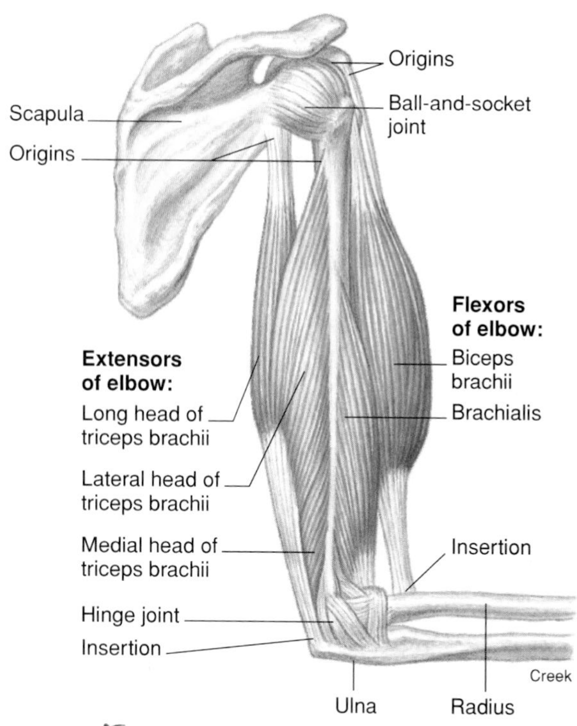

Figure 9.4

Examples of synergistic and antagonistic muscles. The two heads of the biceps brachii and the brachialis muscle are synergistic to each other, as are the three heads of the triceps brachii. The biceps brachii and the brachialis are antagonistic to the triceps brachii, and the triceps brachii is antagonistic to the biceps brachii and the brachialis muscle. When one antagonistic group contracts, the other one must relax; otherwise, movement does not occur.

Muscle Groups

Just as individual muscle fibers seldom contract independently, muscles generally do not contract separately but work as functional groups. Muscles that contract together in accomplishing a particular movement are said to be *synergistic* (*sin″er-jis′tik*) (fig. 9.4). *Antagonistic* muscles perform opposite functions and are generally located on the opposite sides of the joint. For example, the two heads of the biceps brachia muscle, together with the brachialis muscle, contract to *flex* the elbow joint. The triceps brachii muscle, the antagonist to the biceps brachii and brachialis muscles, *extends* the elbow as it is contracted.

Seldom does the action of a single muscle cause a movement at a joint. Utilization of several synergistic muscles rather than one massive muscle allows for a division of labor. One muscle may be an important postural muscle, for example, whereas another may be adapted for rapid, powerful contraction.

synergistic: Gk. *synergein*, cooperate
antagonistic: Gk. *antagonistes*, struggle against

Table 9.1
Muscle Architecture

Type and Description	Appearance
Parallel—straplike; long excursion (contract over a great distance); good endurance; not especially strong; e.g., sartorius and rectus abdominis muscles	
Convergent—fan-shaped; force of contraction focused onto a single point of attachment; stronger than parallel type; e.g., deltoid and pectoralis major	
Sphincteral—fibers concentrically arranged around a body opening (*orifice*); act as a sphincter when contracted; e.g., orbicularis oculi and orbicularis oris	
Pennate—many fibers per unit area; strong muscles; short excursions; highly dexterous; tire quickly; three types: (a) unipennate, (b) bipennate, and (c) multipennate	(a) (b) (c)

orifice: L. *orificium*, mouth; facere, to make
pennate: L. *pennatus*, feather

Muscle Architecture

Skeletal muscles may be classified on the basis of fiber arrangement as *parallel*, *convergent*, *sphincteral* (circular), or *pennate* (table 9.1). Each type of fiber arrangement provides the muscle with distinct capabilities.

Muscle fiber architecture can be observed on a cadaver or other dissection specimen. If you have the opportunity to learn the muscles of the body from a cadaver, observe the fiber architecture of specific muscles and try to determine the advantages afforded to each muscle by its location and action.

Blood and Nerve Supply to Skeletal Muscle

Muscle cells have a high rate of metabolic activity and therefore require extensive vascularity to receive nutrients and oxygen and

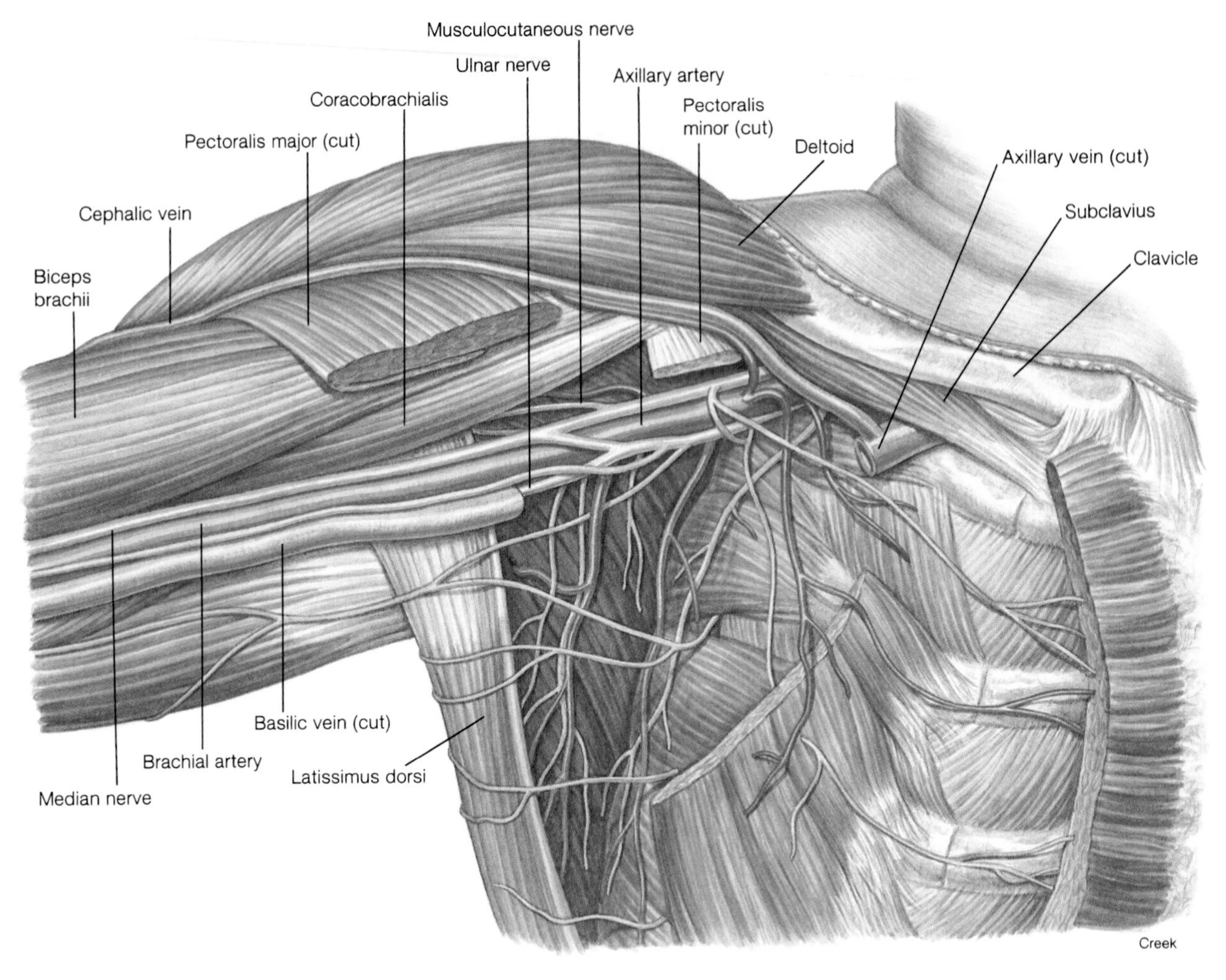

Figure 9.5

The relationship of blood vessels and nerves to skeletal muscles of the axillary region. (Note the close proximity of the nerves and vessels as they pass between muscle masses.)

to eliminate waste products. Smaller muscles generally have a single artery supplying blood and perhaps two veins returning blood (fig. 9.5). Large muscles may have several arteries and veins. The microscopic capillary exchange between arteries and veins occurs throughout the endomysium that surrounds individual fibers.

A skeletal muscle fiber cannot contract unless it is stimulated by a nerve impulse. This means that there must be extensive *innervation* (served with neurons) to a muscle to ensure the connection of each muscle fiber to a nerve cell. Actually there are two nerve pathways for each muscle. A **motor (efferent) neuron** is a nerve cell that conducts nerve impulses *to the muscle fiber*, stimulating it to contract. A **sensory (afferent) neuron** conducts nerve impulses *away from the muscle fiber* to the central nervous system, which responds to the activity of the muscle fiber. Muscle fibers will atrophy if they are not periodically stimulated to contract.

For years it was believed that muscle soreness was simply caused by a buildup of lactic acid within the muscle fibers during exercise. Although lactic acid accumulation probably is a factor related to soreness, recent research has shown that there is also damage to the contractile proteins within the muscle. If a muscle is used to exert an excessive force (for example, to lift a heavy object or to run a distance farther than it is conditioned to), some of the actin and myosin filaments become torn apart. This microscopic damage causes an inflammatory response that results in swelling and pain. If enough proteins are torn, use of the entire muscle may be compromised. Staying in good physical condition guards against muscle soreness following exercise. Conditioning the body not only improves vascularity but enlarges muscle fibers and allows them to work more efficiently over a longer duration.

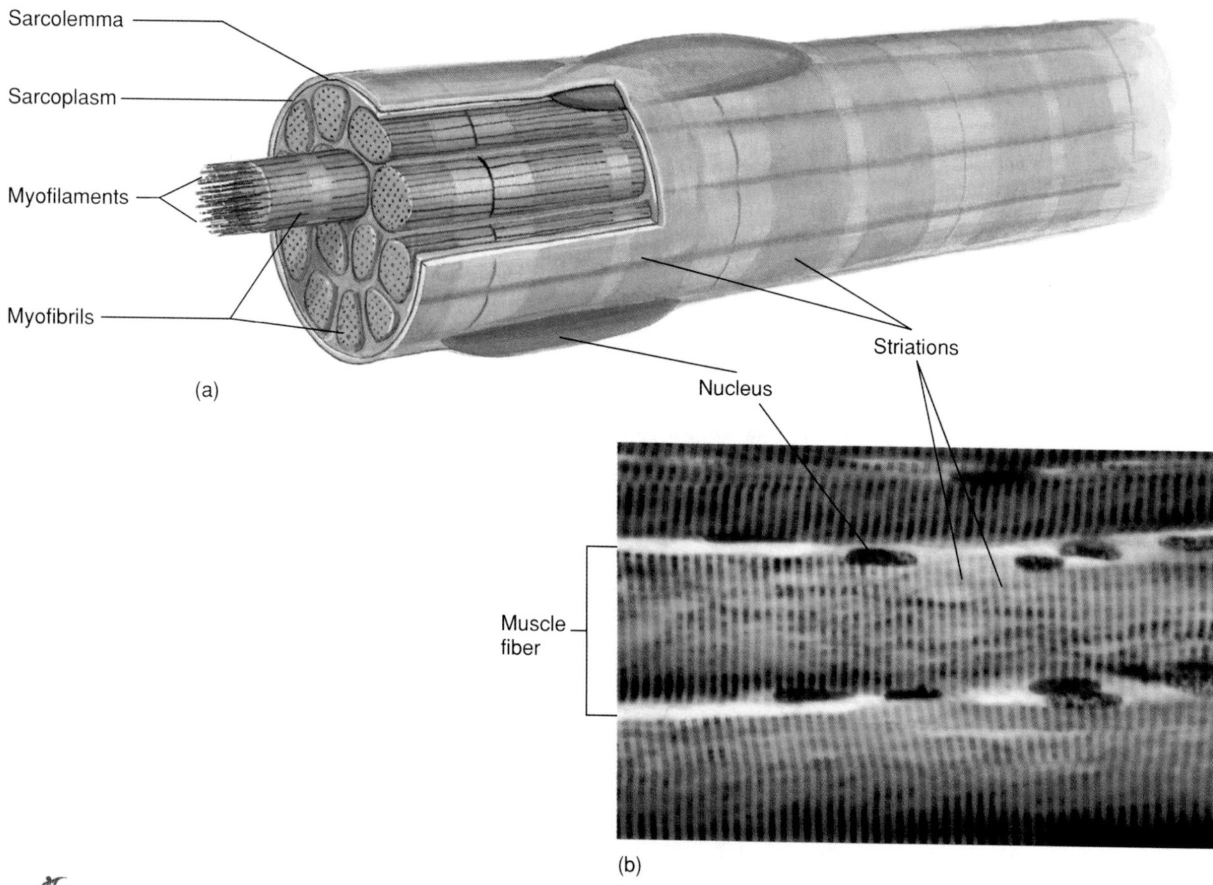

Figure 9.6

(*a*) A skeletal muscle fiber contains numerous organelles called myofibrils composed of the thick and thin myofilaments of actin and myosin. A skeletal muscle fiber is striated and multinucleated. (*b*) A light micrograph of skeletal muscle fibers showing the striations and the peripheral location of the nuclei.

1. Contrast the following terms: *endomysium* and *epimysium*; *fascia* and *tendon*; *aponeurosis* and *retinaculum*.
2. Discuss the biomechanical advantage of having synergistic muscles. Give some examples of synergistic muscles and state which muscles are antagonistic.
3. Which type of muscle architecture provides dexterity and strength?

Skeletal Muscle Fibers and Types of Muscle Contraction

Muscle fiber contraction in response to a motor impulse results from a sliding movement within the myofibrils in which the length of the sarcomeres is reduced.

Objective 6 Identify the major components of a muscle fiber and discuss the function of each part.

Objective 7 Distinguish between isotonic and isometric contractions.

Objective 8 Define *motor unit* and discuss the role of motor units in muscular contraction.

Skeletal Muscle Fibers

Despite their unusual elongated shape, muscle cells have the same organelles as other cells: mitochondria, intracellular membranes, glycogen granules, and so forth. Unlike most other cells in the body, however, skeletal muscle fibers are multinucleated and striated (fig. 9.6). In addition, some skeletal muscle fibers may reach lengths of 30 cm (12 in.) and have diameters of 10 to 100 μm.

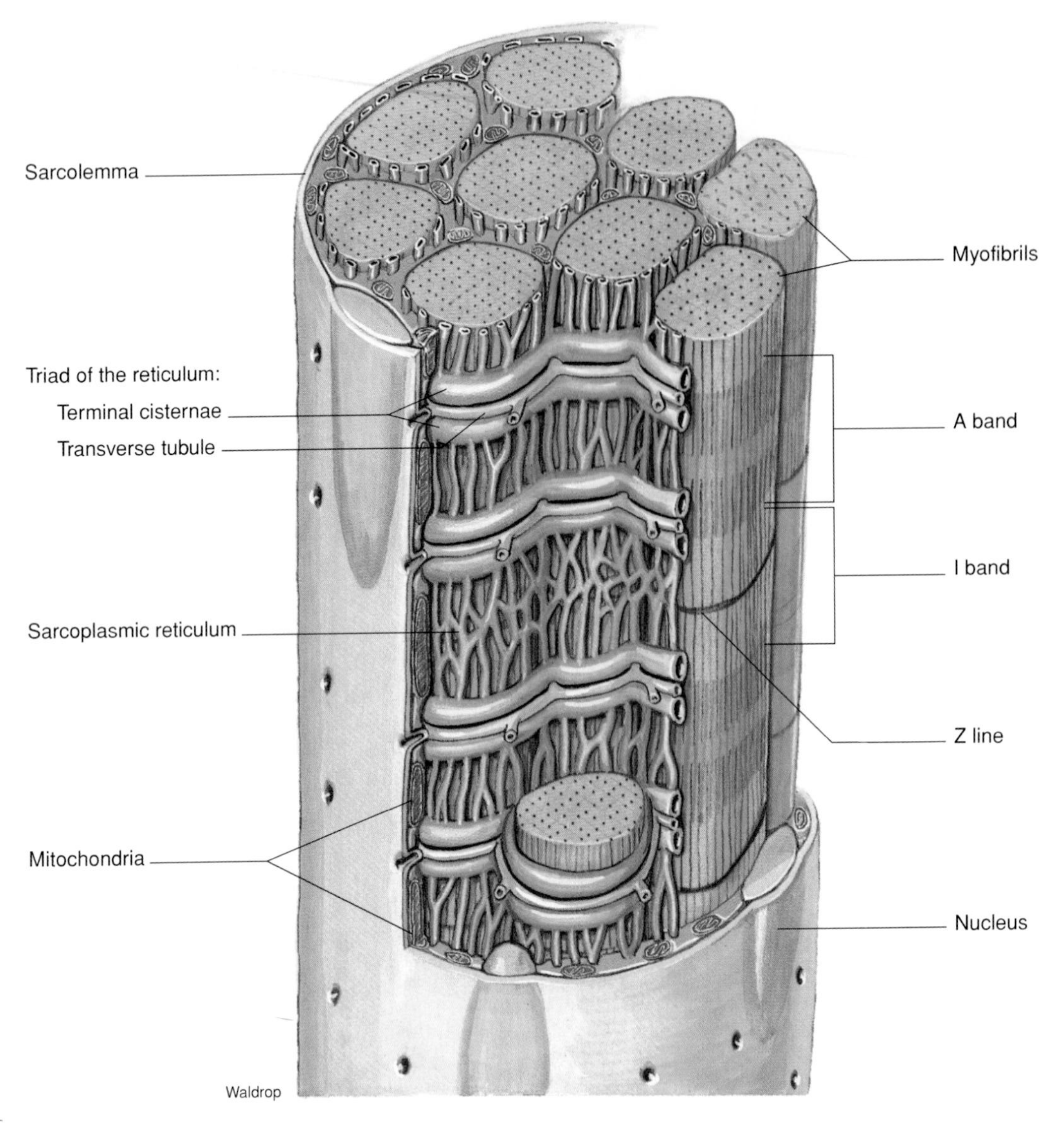

Figure 9.7

The structural relationship of the myofibrils of a muscle fiber to the sarcolemma, transverse tubules, and sarcoplasmic reticulum. (Note the position of the mitochondria.)

Each muscle fiber is surrounded by a cell membrane called the **sarcolemma** (*sar″ kŏ-lem′ ă*). A network of membranous channels, the **sarcoplasmic reticulum,** extends throughout the cytoplasm of the fiber, which is called **sarcoplasm** (fig. 9.7). A system of **transverse tubules** (T tubules) runs perpendicular to the sarcoplasmic reticulum and opens to the outside through the sarcolemma. Also embedded in the sarcolemma are many threadlike structures called **myofibrils** (fig. 9.8). These myofibrils are approximately one micrometer (1μm) in diameter and extend in parallel from one end of the muscle fiber to the other. They are so densely packed that other organelles—such as mitochondria and intracellular membranes—are restricted to the narrow spaces in the sarcoplasm that remain between adjacent myofibrils. Each myofibril is composed of even smaller protein filaments, or **myofilaments.** *Thin filaments* are about 6 nm in diameter and are composed of the protein **actin.** *Thick filaments* are about 16 nm in diameter and are composed of the protein **myosin**.

The characteristic dark and light striations of skeletal muscle myofibrils are due to the arrangement of these myofilaments. The dark bands are called *A bands*, and the light bands are called *I bands*. At high magnification, thin dark lines can be seen in the middle of the I bands. These are called *Z lines*. The arrangement of thick and thin filaments between a pair of Z lines forms a repeating pattern that serves as the basic subunit of skeletal muscle contraction. These subunits, from Z line to Z line, are known as **sarcomeres** (fig. 9.8). A longitudinal section of a myofibril thus presents a side view of successive sarcomeres (fig. 9.9 a, b).

---

actin: L. *actus*, motion, doing
myosin: L. *myosin*, within muscle

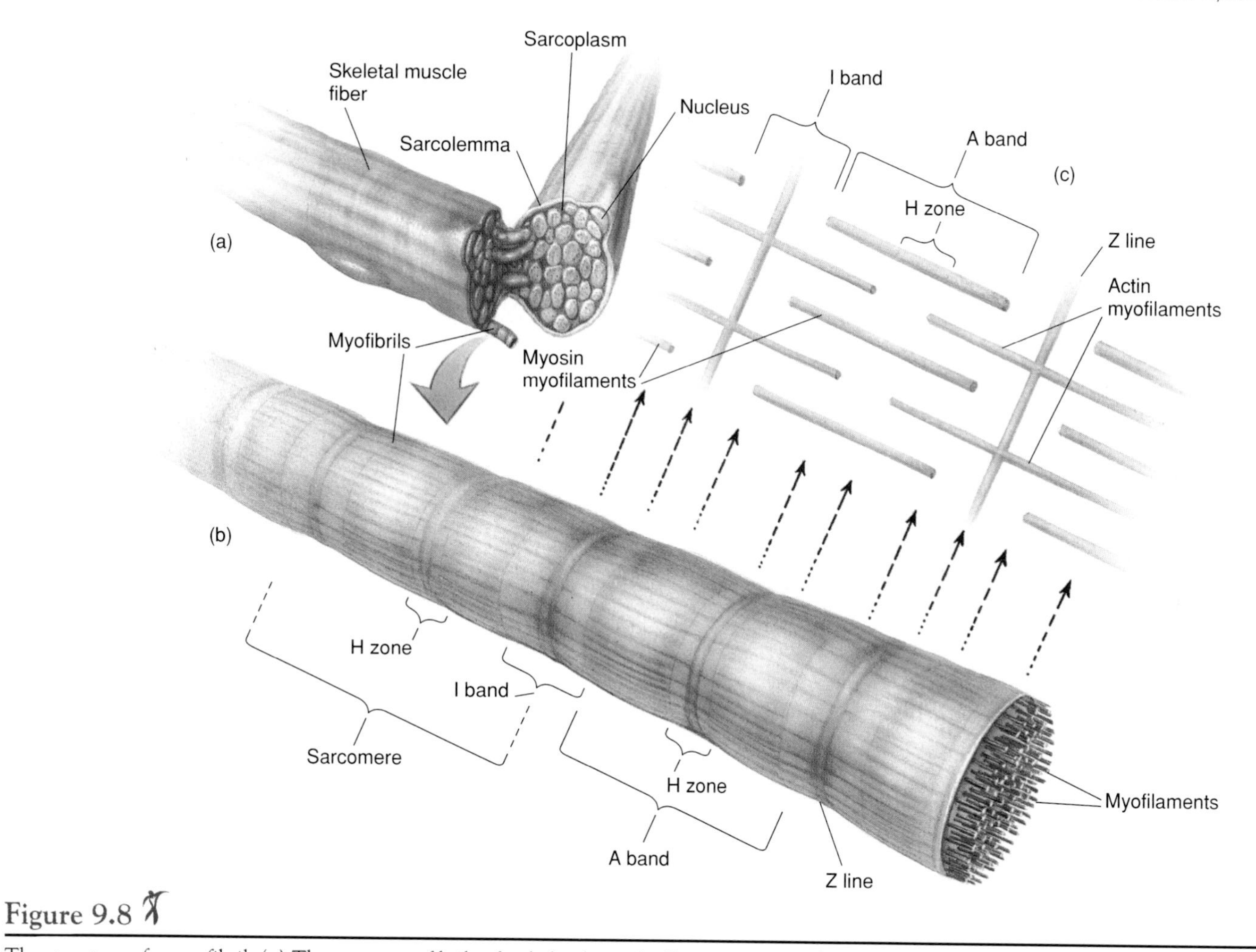

Figure 9.8

The structure of a myofibril. (*a*) The many myofibrils of a skeletal muscle fiber are arranged into compartments (*b*) called sarcomeres. (*c*) The characteristic striations of a sarcomere are due to the arrangement of thin and thick myofilaments, composed of actin and myosin, respectively.

The I bands within a myofibril are the lighter areas that extend from the edge of one stack of thick myosin filaments to the edge of the next stack of thick filaments. They are light in appearance because they contain only thin filaments. The thin filaments, however, do not end at the edges of the I bands. Instead, each thin filament extends part way into the A bands on each side. Since thick and thin filaments overlap at the edges of each A band, the edges of the A band are darker in appearance than the central region. The central lighter regions of the A bands are called *H zones* (for *helle*, a German word meaning "bright"). The central H zones thus contain only thick filaments that are not overlapped by thin filaments.

The side view of successive sarcomeres in figure 9.9*b* is, in a sense, misleading. There are numerous sarcomeres within each myofibril that are out of the plane of the section (and out of the picture). A better appreciation of the three-dimensional structure of a myofibril can be obtained by viewing the myofibril in transverse section. In this view, shown in figure 9.9*c*, it can be seen that the Z lines are actually disc-shaped (Z stands for *Zwischenscheibe*, a German word meaning "between disc"), and that the thin filaments that penetrate these Z discs surround the thick filaments in a hexagonal arrangement. If one concentrates on a single row of dark thick myofilaments in this transverse section, the alternating pattern of thick and thin filaments seen in longitudinal section becomes apparent.

When a muscle is stimulated to contract, it decreases in length as a result of the shortening of its individual fibers. Shortening of the muscle fibers, in turn, is produced by shortening of their myofibrils, which occurs as a result of the shortening of the distance from Z line to Z line (fig. 9.10). As the sarcomeres shorten in length, however, the A bands do *not* shorten but instead appear closer together. The I bands—which represent the distance between A bands of successive myomeres—decrease in length.

The thin actin filaments composing the I band do not shorten, however. Close examination reveals that the length of the thick and thin myofilaments remains constant during muscle contraction. Shortening of the sarcomeres is produced not by shortening of the myofilaments, but rather by the *sliding* of thin filaments over and between thick ones. In the process of contraction, the thin filaments on either side of each A band extend deeper and deeper toward the center, thereby increasing the amount of overlap with the thick filaments. The central H bands thus get shorter and shorter during contraction.

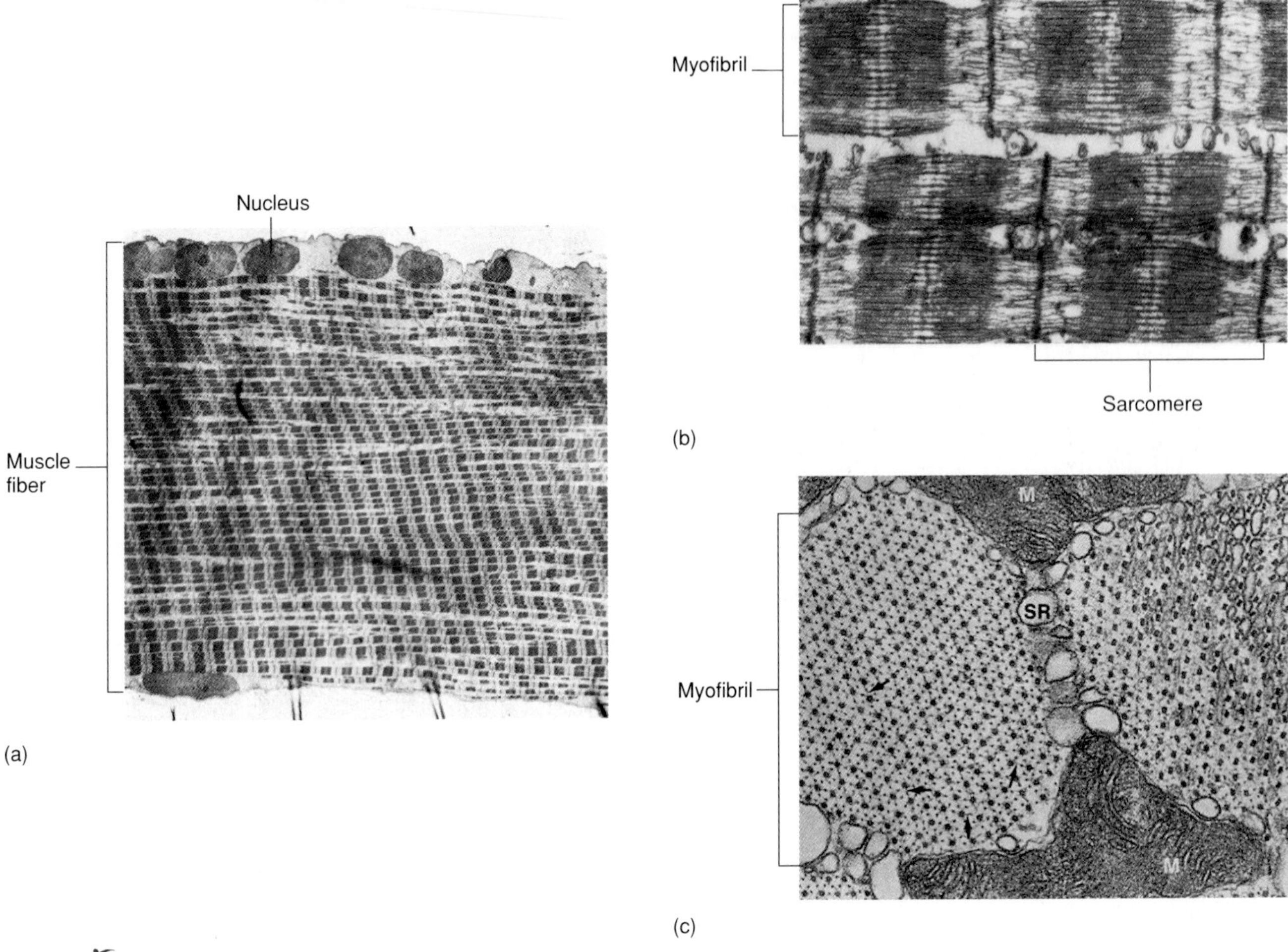

Figure 9.9

Electron micrographs of myofibrils of a muscle fiber. (*a*) At a low power (1,600×), a single muscle fiber containing numerous myofibrils. (*b*) At high power (53,000×), myofibrils in longitudinal section. (Note the sarcomeres and overlapping thick and thin myofilaments.) (*c*) The hexagonal arrangement of thick and thin filaments as seen in transverse section (arrows point to cross bridges; SR = sarcoplasmic reticulum).

(From R. G. Kessel and R. H. Kardon. Tissues and Organs: A Text-Atlas of Scanning Electron Microscopy *© 1979 W. H. Freeman and Company.)*

Isotonic and Isometric Contractions

In order for muscle fibers to shorten when they contract, they must generate a force that is greater than the opposing forces that act to prevent movement of the muscle's insertion. Flexion of the elbow, for example, occurs against the force of gravity and the weight of the objects being lifted. The tension produced by the contraction of each muscle fiber separately is insufficient to overcome these opposing forces, but the combined contractions of large numbers of muscle fibers may be sufficient to overcome them and flex the elbow as the muscle fibers shorten.

Contraction that results in visible muscle shortening is called *isotonic contraction* because the force of contraction remains relatively constant throughout the shortening process (fig. 9.11). If the opposing forces are too great or if the number of muscle fibers activated is too few to shorten the muscle, however, an *isometric contraction* is produced, and movement does not occur.

Neuromuscular Junction

A nerve serving a muscle is composed of both motor and sensory neurons. Each motor neuron has a threadlike **axon** that extends from the CNS to a group of skeletal muscle fibers. Close to these skeletal muscle fibers, the axon divides into numerous branches called **axon terminals.** The axon terminals contact the sarcolemma of the muscle fibers by means of **motor end**

isotonic: Gk. *isos*, equal; *tonos*, tension

isometric: Gk. *isos*, equal; *metron*, measure
axon: Gk. *axon*, axis

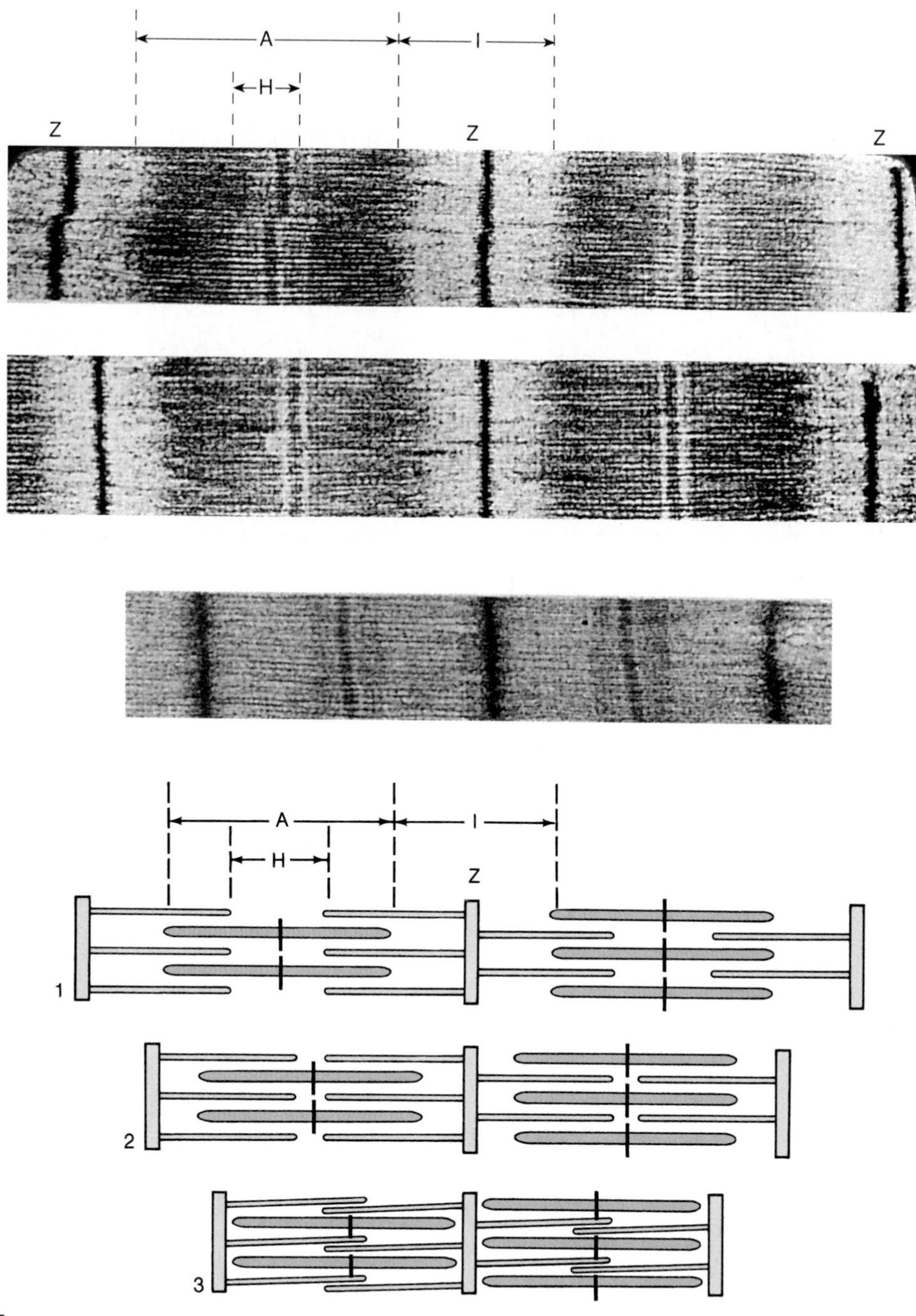

Figure 9.10

The sliding filament model of contraction. As the myofilaments slide, the Z lines are brought closer together. The A bands remain the same length during contraction, but the I and H bands narrow progressively and eventually may be obliterated. (1) Relaxed muscle, (2) partially contracted muscle, and (3) fully contracted muscle.

plates (fig. 9.12). The area consisting of the motor end plate and the cell membrane of a muscle fiber is known as the **neuromuscular** (myoneural) **junction.**

Acetylcholine (*ă-sēt″ l-kō′ lēn*) is a neurotransmitter chemical stored in **synaptic vesicles** at the axon terminals. A nerve impulse reaching the axon terminal causes the release of acetylcholine into the **neuromuscular cleft** of the neuromuscular junction. As this chemical mediator contacts the receptor sites of the sarcolemma, it initiates physiological activity within the muscle fiber, resulting in contraction.

Motor Unit

A **motor unit** consists of a single motor neuron and the aggregation of muscle fibers innervated by the motor neuron (fig. 9.12*b*).

(a)

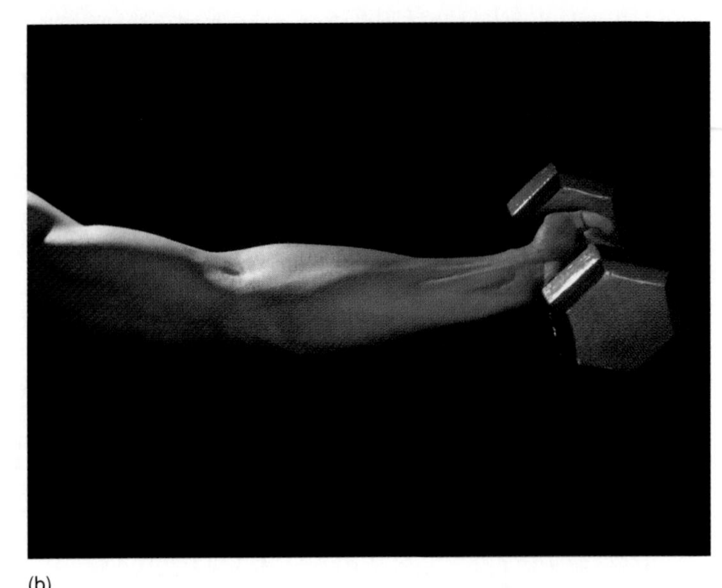

(b)

Figure 9.11

(*a*) Isometric and (*b*) isotonic contraction.

When a nerve impulse travels through a motor unit, all of the fibers served by it contract simultaneously to their maximum. Most muscles have an innervation ratio of 1 motor neuron for each 100 to 150 muscle fibers. Muscles capable of precise, dexterous movements, such as an eye muscle, may have an innervation ratio of 1:10. Massive muscles that are responsible for gross body movements, such as those of the thigh, may have innervation ratios exceeding 1:500.

All of the motor units controlling a particular muscle, however, are not the same size. Innervation ratios in a large thigh muscle may vary from 1:100 to 1:2,000. Neurons that innervate smaller numbers of muscle fibers have smaller cell bodies and axon diameters than neurons that have larger innervation ratios. The smaller neurons also are stimulated by lower levels of excitatory input. The small motor units, as a result, are the ones that are used most often. The larger motor units are activated only when very forceful contractions are required.

Skeletal muscles are voluntary in that they can be consciously contracted. The magnitude of the task determines the number of motor units that are activated. Performing a light task, such as lifting a book, requires few motor units, whereas lifting a table requires many. Muscles with pennate architecture have many motor units and are strong and dexterous; however, they generally fatigue more readily than muscles with fewer motor units. Being mentally "psyched up" to accomplish an athletic feat involves voluntary activation of more motor units within the muscles. Although a person seldom utilizes all of the motor units within a muscle, the secretion of *epinephrine* (*ep"ĭ-nef'rin*) from the adrenal gland does promote an increase in the force that can be produced when a given number of motor units are activated.

Steroids are hormones produced by the adrenal glands, testes, and ovaries. Because they are soluble in lipids, they readily pass through cell membranes and enter the cytoplasm, where they combine with proteins to form steroid-protein complexes that are necessary for the syntheses of specific kinds of messenger RNA molecules. Synthetic steroids were originally developed to promote weight gain in cancer and anorexic patients. It soon became apparent, however, that steroids taken by bodybuilders and athletes could provide them with increased muscle mass, strength, and aggressiveness. The use of steroids is now considered illegal by most athletic associations. Not only do they confer unfair advantages in physical competition, they also can have serious side effects. These include gonadal atrophy, hypertension, induction of malignant tumors of the liver, and overly aggressive behavior, to name just a few.

1. Draw three successive sarcomeres in a myofibril of a resting muscle fiber. Label the myofibril, sarcomeres, A bands, I bands, H bands, and Z lines.
2. Why do the A bands appear darker than the I bands?
3. Draw three successive sarcomeres in a myofibril of a contracted fiber. Indicate which bands get shorter during contraction and explain how this occurs.
4. Describe how the antagonistic muscles in the brachium can be exercised through both isotonic and isometric contractions.
5. Explain why motor units are considered the basic functional units of muscle contraction.

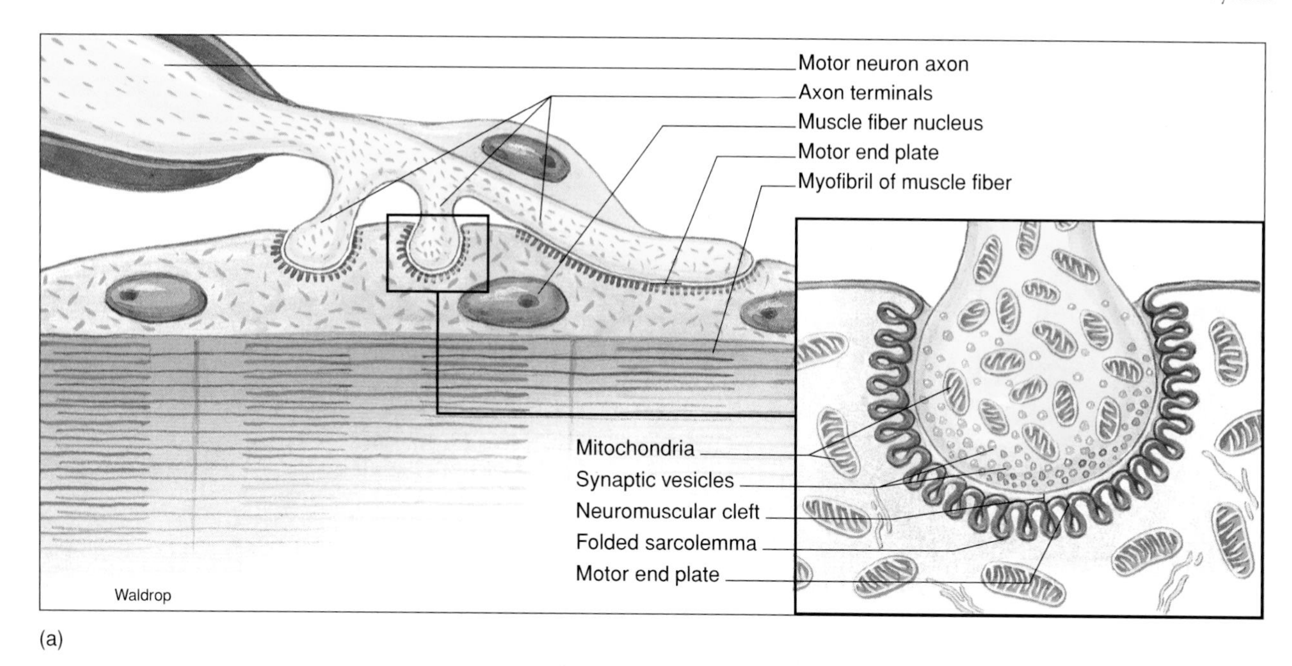

(a)

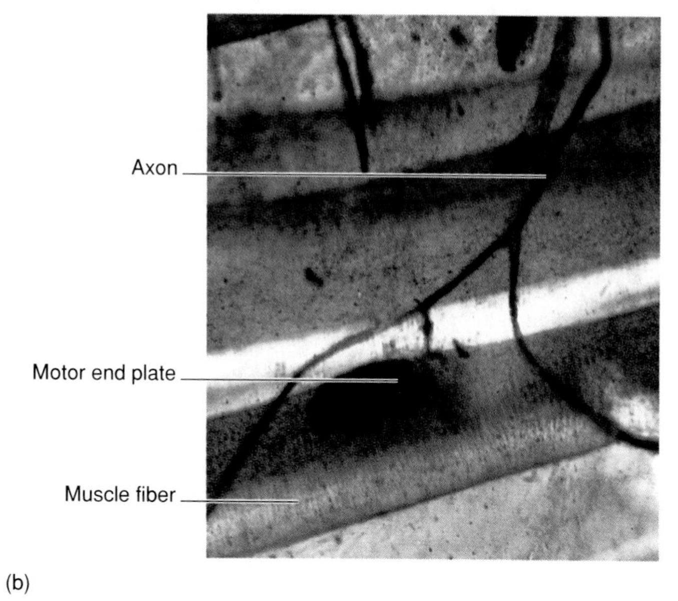

(b)

Figure 9.12

A motor end plate at the neuromuscular junction. (*a*) A neuromuscular junction is the site where the nerve fiber and muscle fiber meet. The motor end plate is the specialized portion of the sarcolemma of a muscle fiber surrounding the terminal end of the axon. (Note the slight gap between the membrane of the axon and that of the muscle fiber.) (*b*) A photomicrograph of muscle fibers and motor end plates. A motor neuron and the skeletal muscle fibers it innervates constitute a motor unit.

Naming of Muscles

Skeletal muscles are named on the basis of shape, location, attachment, orientation of fibers, relative position, or function.

Objective 9 Use examples to describe the various ways in which muscles are named.

One of your tasks as a student of anatomy is to learn the names of the principal muscles of the body. Although this may seem overwhelming, keep in mind that most of the muscles are paired; that is, the right side is the mirror image of the left. To help you further, most muscles have names that are descriptive.

As you study the muscles of the body, consider how each was named. Identify the muscle on the figure referenced in the text narrative and locate it on your own body as well. Use your

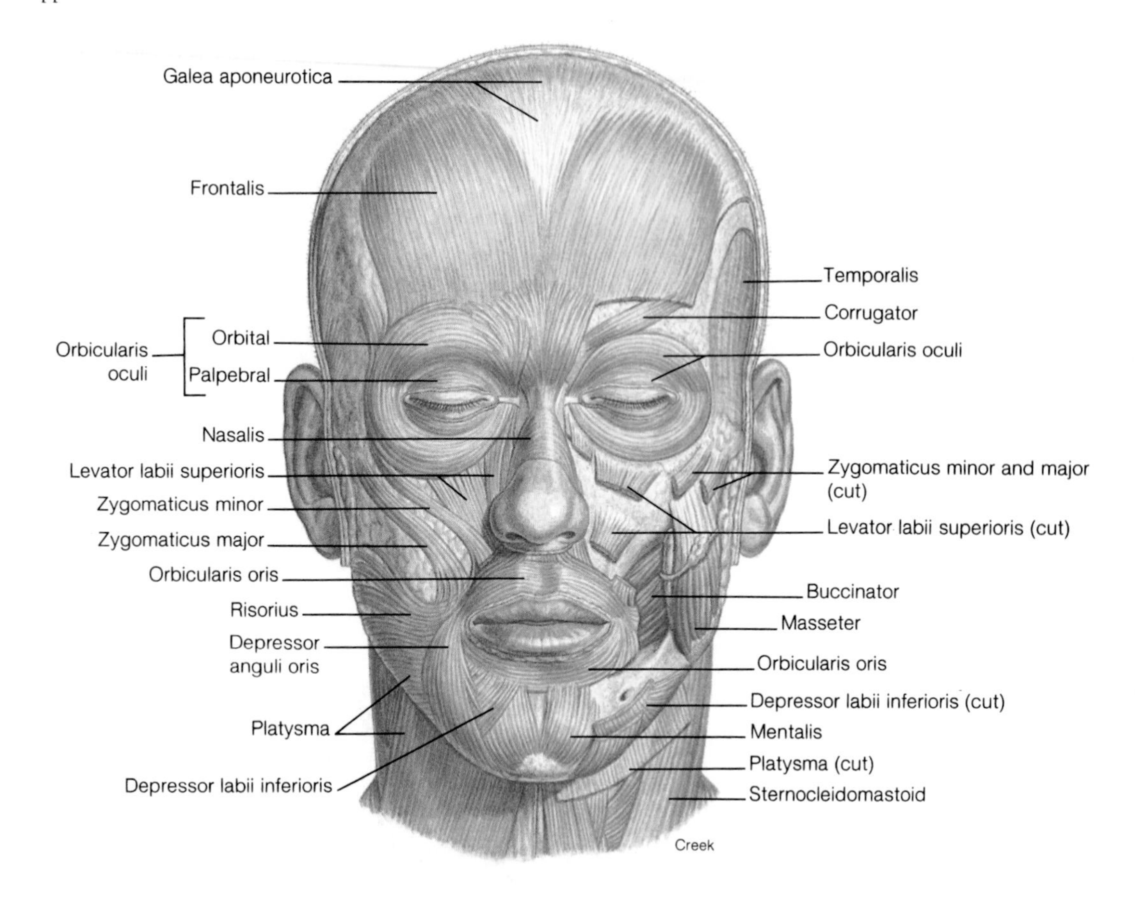

Figure 9.13

An anterior view of the superficial facial muscles involved in facial expression.

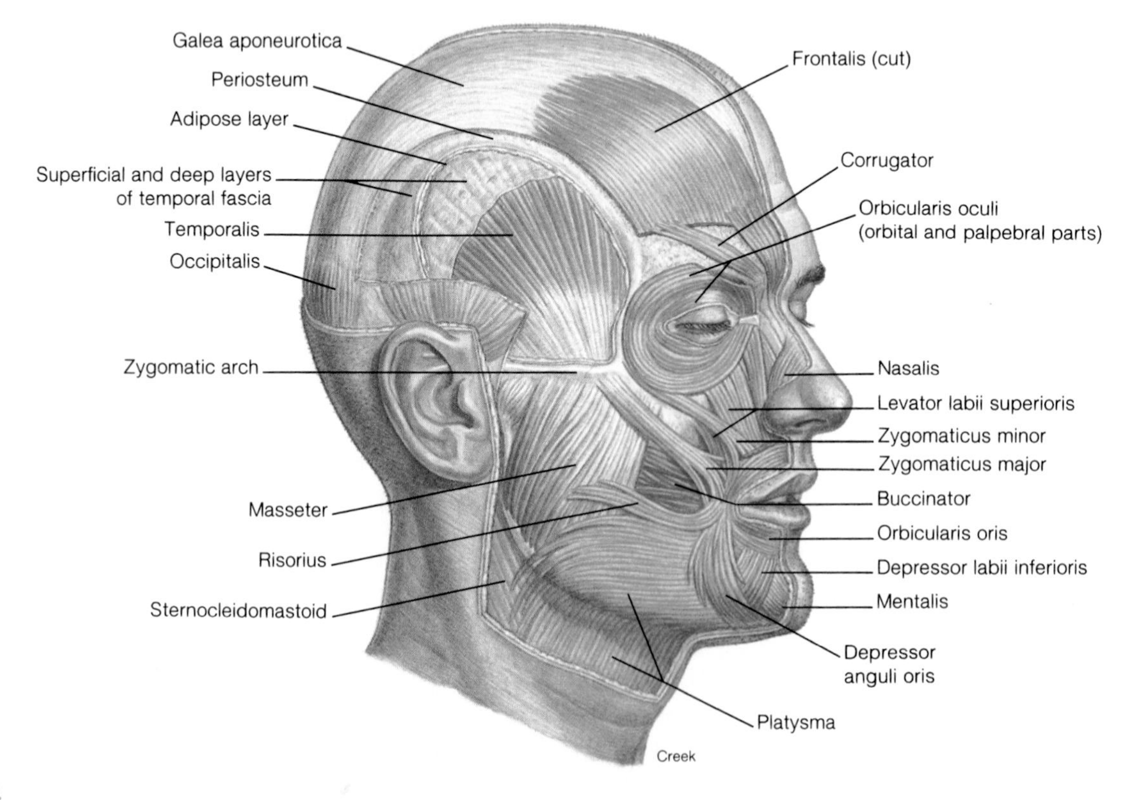

Figure 9.14

A lateral view of the superficial facial muscles involved in facial expression.

body to act out its movement. Feel it contracting beneath your skin and note the movement that occurs at the joint. Learning the muscles in this way will simplify the task and make it more meaningful.

The following are some criteria by which the names of muscles have been logically derived:

1. **Shape:** rhomboideus (like a rhomboid); trapezius (like a trapezoid); or denoting the number of heads of origin: triceps (three heads), biceps (two heads)
2. **Location:** pectoralis (in the chest, or pectus); intercostal (between ribs); brachia (arm)
3. **Attachment:** many facial muscles (zygomaticus, temporalis, nasalis); sternocleidomastoid (sternum, clavicle, and mastoid process of the temporal bone)
4. **Size:** maximus (larger, largest); minimus (smaller, smallest); longus (long); brevis (short)
5. **Orientation of fibers:** rectus (straight); transverse (across); oblique (in a slanting or sloping direction)
6. **Relative position:** lateral, medial, internal, and external
7. **Function:** adductor, flexor, extensor, pronator, and levator (lifter)

1. Refer to chapter 2 (fig. 2.14) and review the location of the following body regions: cervical, pectoral, abdominal, gluteal, perineal, brachial, antebrachial, inguinal, thigh, and popliteal.
2. Refer to chapter 8 and review the movements permitted at synovial joints.

Muscles of the Axial Skeleton

Muscles of the axial skeleton include those responsible for facial expression, mastication, eye movement, tongue movement, neck movement, and respiration, and those of the abdominal wall, the pelvic outlet, and the vertebral column.

Objective 10 Locate the major muscles of the axial skeleton. Identify synergistic and antagonistic muscles and describe the action of each one.

Muscles of Facial Expression

Humans have a well-developed facial musculature (figs. 9.13 and 9.14) that allows for complex facial expression as a means of social communication. Very often we let our feelings be known without a word spoken.

The muscles of facial expression are located in a superficial position on the scalp, face, and neck. Although highly variable in size and strength, these muscles all originate on the bones of the skull or in the fascia and insert into the skin (table 9.2). They are all innervated by the facial nerves (see fig. 12.8). The locations and points of attachments of most of the facial muscles are such that, when contracted, they cause movements around the eyes, nostrils, or mouth (fig. 9.15).

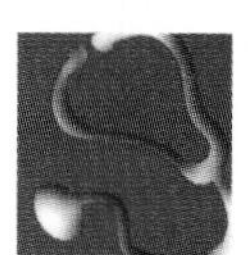

The muscles of facial expression are of clinical concern for several reasons, all of which involve the facial nerve. Located right under the skin, the many branches of the facial nerve are vulnerable to trauma. Facial lacerations and fractures of the skull frequently damage branches of this nerve. The extensive pattern of motor innervation becomes apparent in stroke victims and persons suffering from *Bell's palsy.* The facial muscles on one side of the face are affected in these people, and that side of the face appears to sag.

Muscles of Mastication

The large **temporalis** and **masseter** (*mă-se´ter*) muscles (fig. 9.16) are powerful elevators of the mandible in conjunction with the **medial pterygoid** (*ter´ĭ-goid*) muscle. The primary function of the medial and lateral pterygoid muscles is to provide grinding movements of the teeth. The **lateral pterygoid** also protracts the mandible (table 9.3).

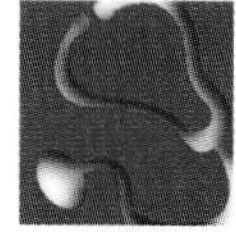

Tetanus is a bacterial disease caused by the introduction of anaerobic *Clostridium tetani* into the body, usually from a puncture wound. The bacteria produce a neurotoxin that is carried to the spinal cord by sensory nerves. The motor impulses relayed back cause certain muscles to contract continuously (tetany). The muscles that move the mandible are affected first, which is why the disease is commonly known as *lockjaw.*

Ocular Muscles

The movements of the eyeball are controlled by six extrinsic ocular (eye) muscles (fig. 9.17 and table 9.4). Five of these muscles arise from the margin of the optic foramen at the back of the orbital cavity and insert on the outer layer (sclera) of the eyeball. Four **rectus muscles** maneuver the eyeball in the direction indicated by their names **(superior, inferior, lateral,** and **medial),** and two **oblique muscles (superior** and **inferior)** rotate the eyeball on its axis. The medial rectus on one side contracts with the medial rectus of the opposite eye when focusing on close objects. When looking to the side, the lateral rectus of one eyeball works with the medial rectus of the opposite eyeball to keep both eyes functioning together. The superior oblique muscle passes through a pulleylike cartilaginous loop, the *trochlea,* before attaching to the eyeball.

Table 9.2

Muscles of Facial Expression

Muscle	Origin	Insertion	Action
Epicranius	Galea aponeurotica and occipital bone	Skin of eyebrow and galea aponeurotica	Wrinkles forehead and moves scalp
Frontalis	Galea aponeurotica	Skin of eyebrow	Wrinkles forehead and elevates eyebrow
Occipitalis	Occipital bone and mastoid process	Galea aponeurotica	Moves scalp backward
Corrugator	Fascia above eyebrow	Root of nose	Draws eyebrow toward midline
Orbicularis oculi	Bones of medial orbit	Tissue of eyelid	Closes eye
Nasalis	Maxilla and nasal cartilage	Aponeurosis of nose	One part widens nostrils; another part depresses nasal cartilages and compresses nostrils
Orbicularis oris	Fascia surrounding lips	Mucosa of lips	Closes and purses lips
Levator labii superioris	Upper maxilla and zygomatic bone	Orbicularis oris and skin above lips	Elevates upper lip
Zygomaticus	Zygomatic bone	Superior corner of orbicularis oris	Elevates corner of mouth
Risorius	Fascia of cheek	Orbicularis oris at corner of mouth	Draws angle of mouth laterally
Depressor anguli oris	Mandible	Inferior corner of orbicularis oris	Depresses corner of mouth
Depressor labii inferioris	Mandible	Orbicularis oris and skin of lower lip	Depresses lower lip
Mentalis	Mandible (chin)	Orbicularis oris	Elevates and protrudes lower lip
Platysma	Fascia of neck and chest	Inferior border of mandible	Depresses mandible and lower lip
Buccinator	Maxilla and mandible	Orbicularis oris	Compresses cheek

corrugator: L. *corrugo*, a wrinkle
risorius: L. *risor*, a laughter
mentalis: L. *mentum*, chin
platysma: Gk. *platys*, broad
buccinator: L. *bucca*, cheek

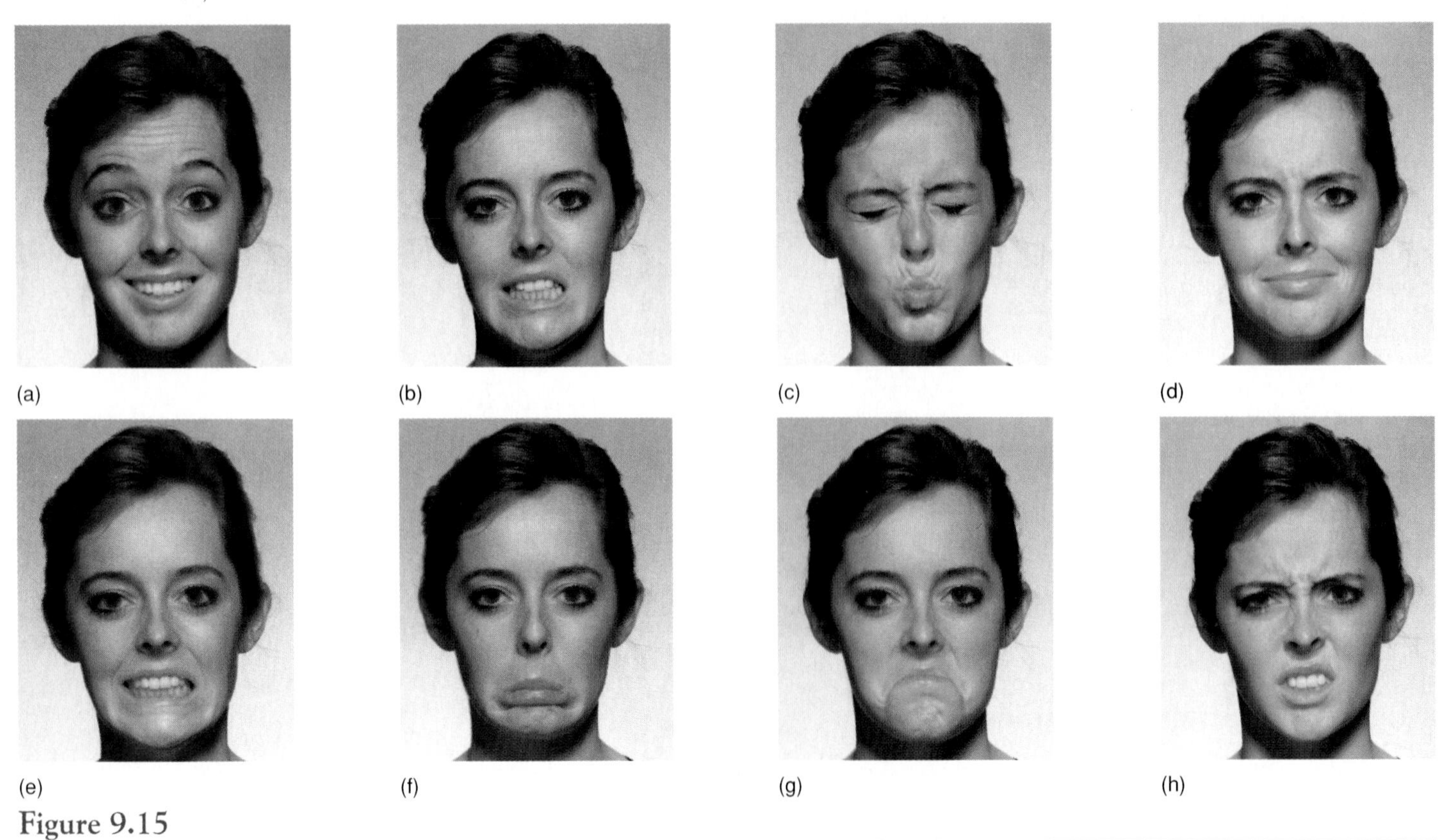

Figure 9.15

Expressions produced by contractions of facial muscles. In each of these photographs, identify the muscles that are being contracted.

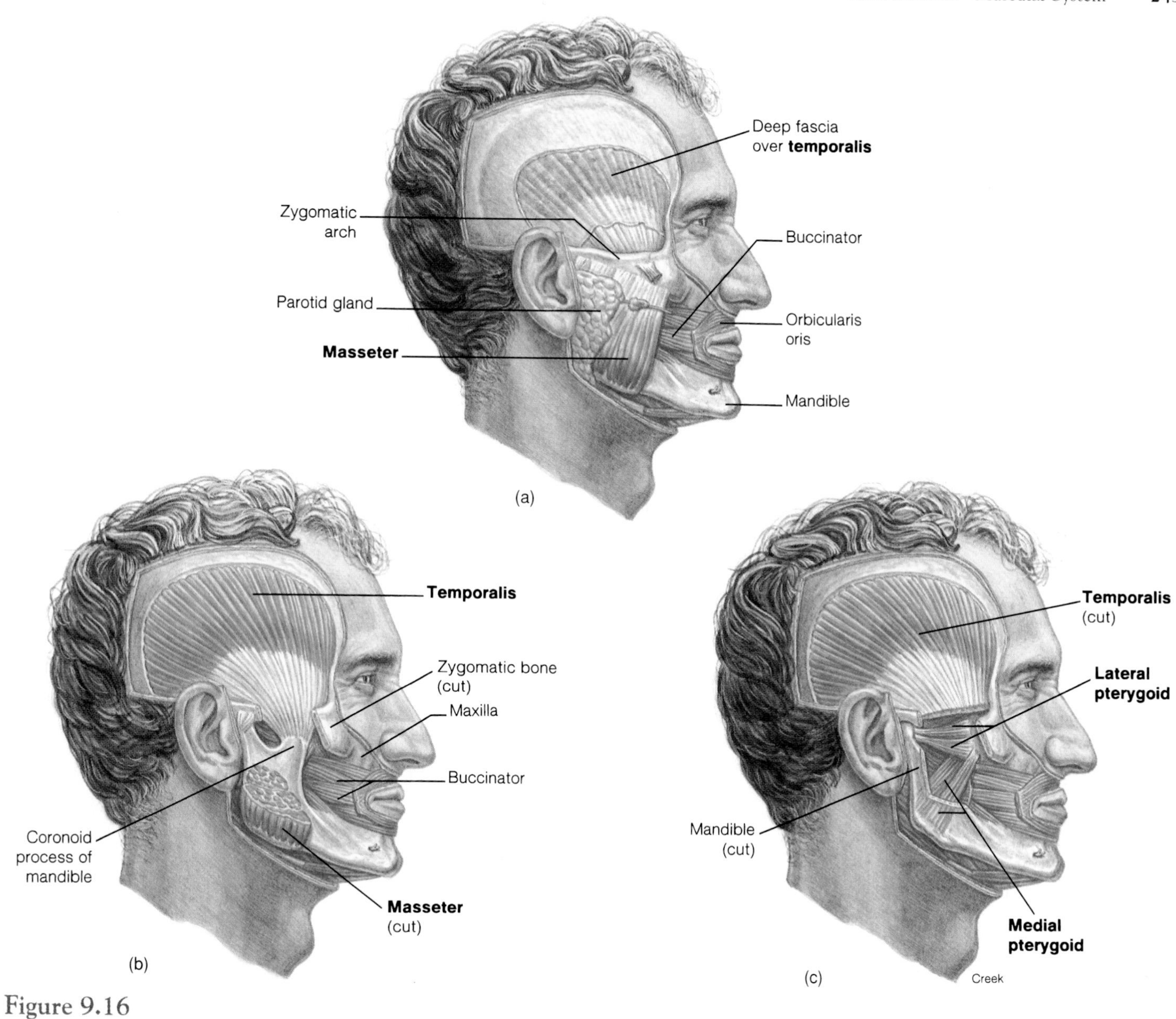

Figure 9.16

Muscles of mastication. (*a*) A superficial view, (*b*) a deep view, and (*c*) the deepest view, showing the pterygoid muscles. (The muscles of mastication are labeled in boldface type.)

Table 9.3
Muscles of Mastication

Muscle	Origin	Insertion	Action
Temporalis	Temporal fossa	Coronoid process of mandible	Elevates mandible
Masseter	Zygomatic arch	Lateral part of ramus of mandible	Elevates mandible
Medial pterygoid	Sphenoid bone	Medial aspect of mandible	Elevates mandible and moves mandible laterally
Lateral pterygoid	Sphenoid bone	Anterior side of mandibular condyle	Protracts mandible

masseter: Gk. *maseter*, chew
pterygoid: Gk. *pteron*, wing

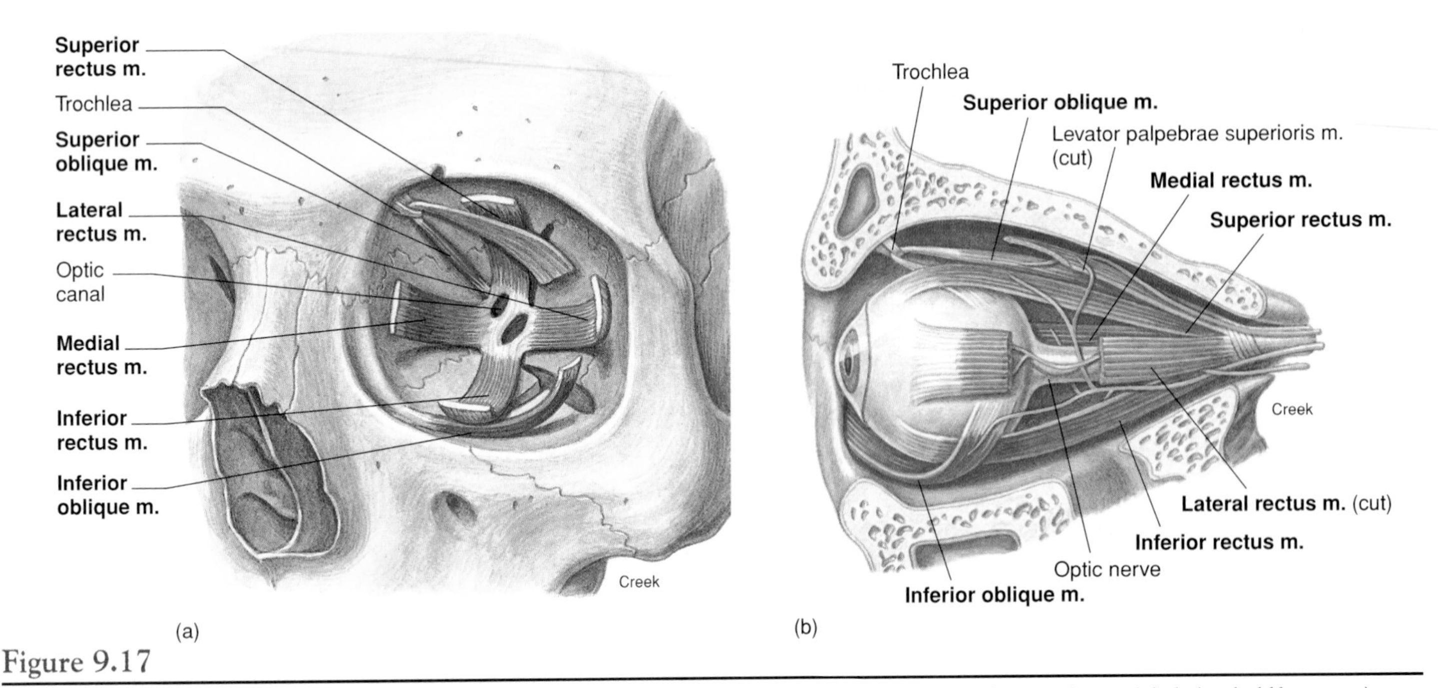

Figure 9.17

Extrinsic ocular muscles of the left eyeball. (*a*) An anterior view and (*b*) a lateral view. (The extrinsic ocular muscles are labeled in boldface type.)

Table 9.4
Ocular Muscles

Muscle	Cranial Nerve Innervation	Movement of Eyeball
Lateral rectus	Abducens	Lateral
Medial rectus	Oculomotor	Medial
Superior rectus	Oculomotor	Superior and medial
Inferior rectus	Oculomotor	Inferior and medial
Inferior oblique	Oculomotor	Superior and lateral
Superior oblique	Trochlear	Inferior and lateral

Another muscle, the **levator palpebrae** (*le-va'tor pal'pe-bre*) **superioris** (fig. 9.17*b*), is located in the ocular region but is not attached to the eyeball. It extends into the upper eyelid and raises the eyelid when contracted.

Muscles That Move the Tongue

The tongue is a highly specialized muscular organ that functions in speaking, manipulating food, cleansing the teeth, and swallowing. The *intrinsic tongue muscles* are located within the tongue and are responsible for its mobility and changes of shape. The *extrinsic tongue muscles* are those that originate on structures other than the tongue and insert onto it to cause gross tongue movement (fig. 9.18 and table 9.5). The four paired extrinsic muscles are the **genioglossus** (*je-ne"o-glos'us*) **styloglossus, hyoglossus,** and **palatoglossus.** When the anterior portion of the genioglossus muscle is contracted, the tongue is depressed and thrust forward. If both genioglossus muscles are contracted together along their entire lengths, the superior surface of the tongue becomes transversely concave. This muscle is extremely important to nursing infants; the tongue is positioned around the nipple with a concave groove channeled toward the pharynx.

Muscles of the Neck

Muscles of the neck either support and move the head or are attached to structures within the neck region, such as the hyoid bone and larynx. Only the more obvious neck muscles will be considered in this chapter.

You can observe many of the muscles in this section and those that follow on your own body. Refer to chapter 10 to determine which muscles form important surface landmarks. These muscles are illustrated in figures 9.19 and 9.20 and are summarized in table 9.6.

Posterior Muscles

The posterior muscles include the *sternocleidomastoid* (originates anteriorly), *trapezius*, *splenius capitis*, *semispinalis capitis*, and *longissimus capitis*.

As the name implies, the **sternocleidomastoid** (*ster"no-kli"do-mas'toid*) muscle originates on the sternum and clavicle and inserts on the mastoid process of the temporal bone (fig. 9.20). When contracted on one side, it turns the head sideways in the direction opposite the side on which the muscle is located. If both

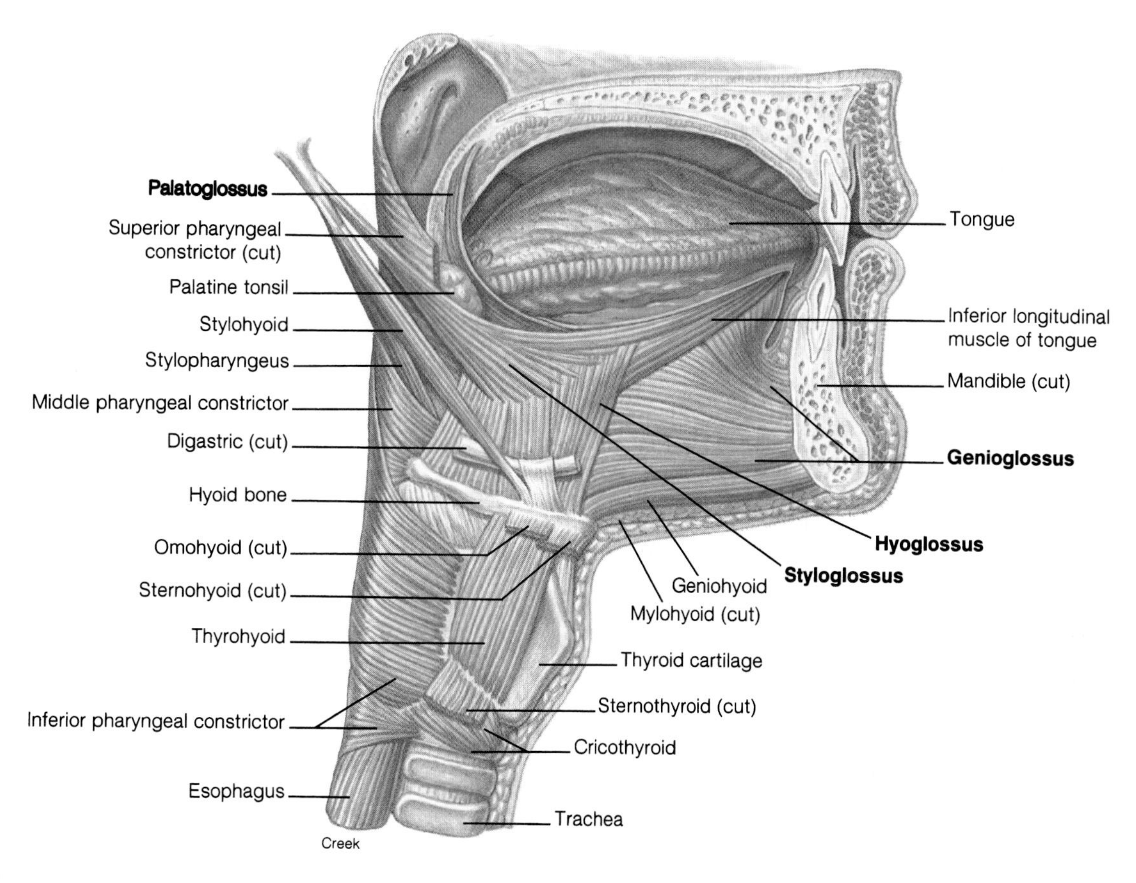

Figure 9.18

Extrinsic muscles of the tongue and deep structures of the neck. (The extrinsic muscles of the tongue are labeled in boldface type.)

Table 9.5
Extrinsic Tongue Muscles

Muscle	Origin	Insertion	Action
Genioglossus	Mental spine of mandible	Undersurface of tongue	Depresses and protracts tongue
Styloglossus	Styloid process of temporal bone	Lateral side and undersurface of tongue	Elevates and retracts tongue
Hyoglossus	Body of hyoid bone	Side of tongue	Depresses sides of tongue
Palatoglossus	Soft palate	Side of tongue	Elevates posterior tongue; constricts fauces (opening from oral cavity to pharynx)

genioglossus: L. *geneion*, chin; *glossus*, tongue

sternocleidomastoid muscles are contracted, the head is pulled forward and down. The sternocleidomastoid muscle is covered by the platysma muscle (see fig. 9.14 and table 9.6).

Although a portion of the **trapezius** muscle extends over the posterior neck region, it is primarily a superficial muscle of the back and will be described later.

The **splenius capitis** (*sple′ne-us kap′ĭ-tis*) is a broad muscle, positioned deep to the trapezius (fig. 9.19). It originates on the ligamentum nuchae and the spinous processes of the seventh cervical and first three thoracic vertebrae. It inserts on the back of the skull below the superior nuchal line and on the mastoid process of the temporal bone. When the splenius capitis contracts on one side, the head rotates and extends to one side. Contracted together, the splenius capitis muscles extend the head at the neck. Further contraction causes hyperextension of the neck and head.

The broad, sheetlike **semispinalis capitis** muscle extends upward from the seventh cervical and first six thoracic vertebrae to insert on the occipital bone (fig. 9.19). When the two semispinalis capitis muscles contract together, they extend the head

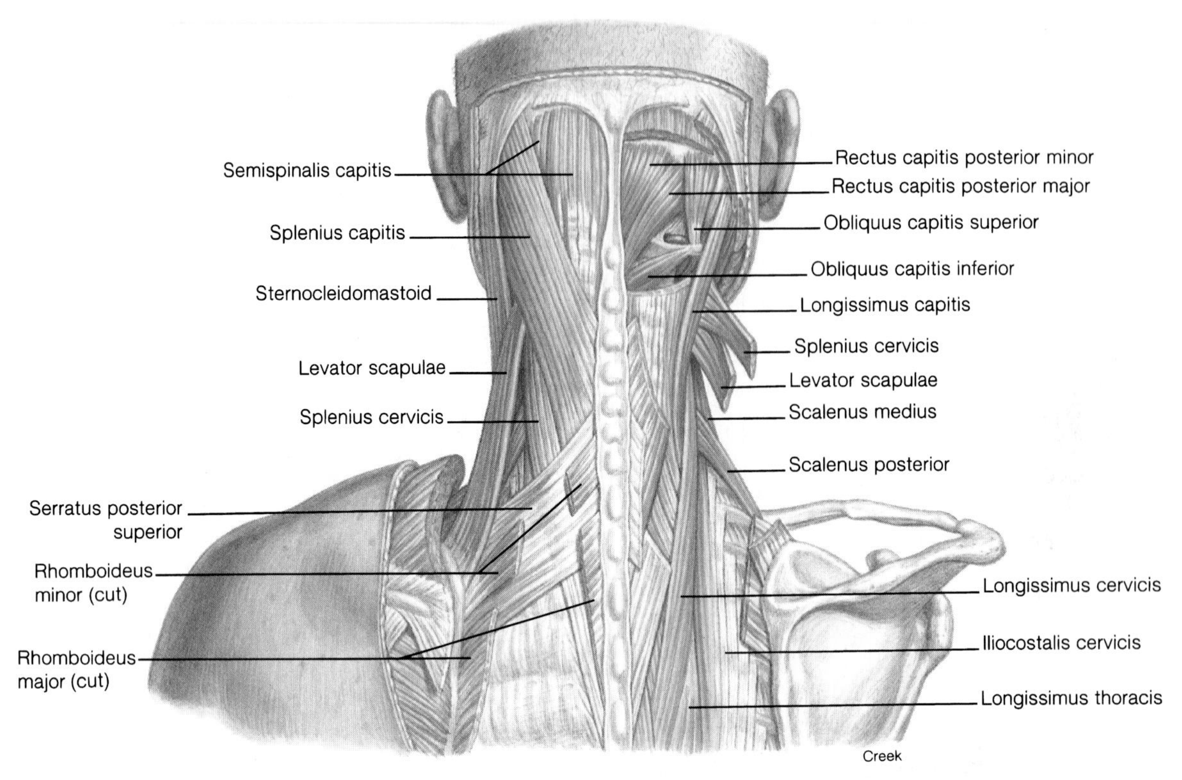

Figure 9.19

Deep muscles of the posterior neck and upper back regions.

at the neck, along with the splenius capitis muscle. If one of the muscles acts alone, the head is rotated to the side.

The narrow, straplike **longissimus** (*lon-jis' ĭ-mus*) **capitis** muscle ascends from processes of the lower four cervical and upper five thoracic vertebrae and inserts on the mastoid process of the temporal bone (fig. 9.19). This muscle extends the head at the neck, bends it to one side, or rotates it slightly.

Suprahyoid Muscles

The group of suprahyoid muscles located above the hyoid bone includes the *digastric*, *mylohyoid*, and *stylohyoid* muscles (fig. 9.20).

The **digastric** is a two-bellied muscle of double origin that inserts on the hyoid bone. The anterior origin is on the mandible at the point of the chin, and the posterior origin is near the mastoid process of the temporal bone. The digastric muscle can open the mouth or elevate the hyoid bone.

The **mylohyoid** muscle forms the floor of the mouth. It originates on the inferior border of the mandible and inserts on the median raphe and body of the hyoid bone. As this muscle contracts, the floor of the mouth is elevated. It aids swallowing by forcing food toward the back of the mouth.

The slender **stylohyoid** muscle extends from the styloid process of the skull to the hyoid bone, which it elevates as it contracts. The secondary effect of this muscle on tongue movement has already been described.

Infrahyoid Muscles

The thin, straplike infrahyoid muscles are located below the hyoid bone. They are individually named on the basis of their origin and insertion and include the *sternohyoid*, *sternothyroid*, *thyrohyoid*, and *omohyoid* muscles (fig. 9.20).

The **sternohyoid** muscle originates on the manubrium of the sternum and inserts on the hyoid bone. It depresses the hyoid bone as it contracts.

The **sternothyroid** muscle also originates on the manubrium but inserts on the thyroid cartilage of the larynx. When this muscle contracts, the larynx is pulled downward.

The short **thyrohyoid** muscle extends from the thyroid cartilage to the hyoid bone. It elevates the larynx and lowers the hyoid bone.

The long, thin **omohyoid** muscle originates on the superior border of the scapula and inserts on the hyoid bone. It acts to depress the hyoid bone.

The coordinated movements of the hyoid bone and the larynx are impressive. The hyoid bone does not articulate with any other bone, yet it has eight paired muscles attached to it. Two involve tongue movement, one lowers the jaw, one elevates the floor of the mouth, and four depress the hyoid bone or elevate the thyroid cartilage of the larynx.

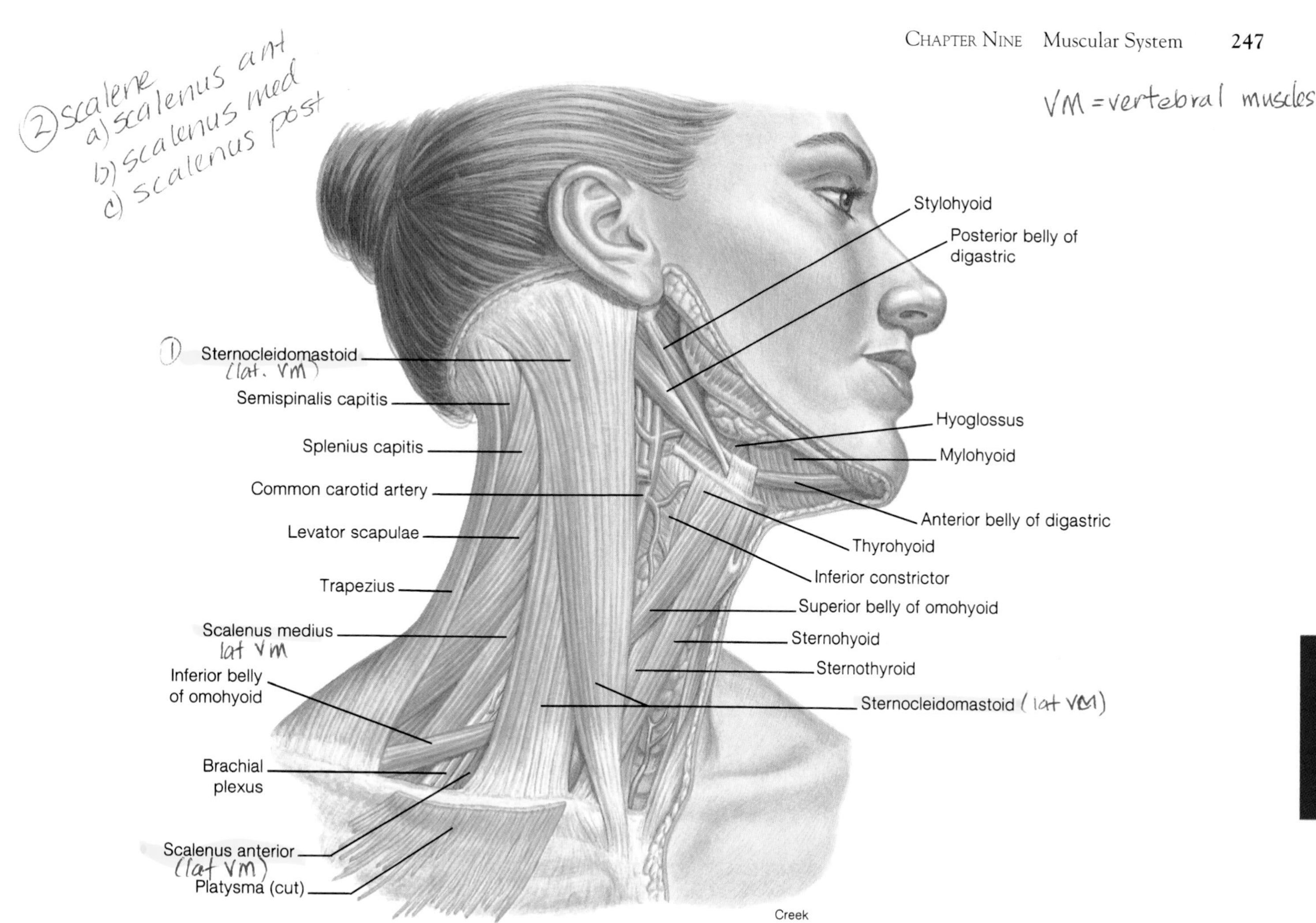

Figure 9.20

Muscles of the anterior and lateral neck regions.

Table 9.6
Muscles of the Neck

Muscle	Origin	Insertion	Action	Innervation
Sternocleidomastoid	Sternum and clavicle	Mastoid process of temporal bone	Turns head to side; flexes neck	Accessory n.
Digastric	Inferior border of mandible and mastoid process of temporal bone	Hyoid bone	Opens mouth; elevates hyoid bone	Trigeminal n. (ant. belly); facial n. (post. belly)
Mylohyoid	Inferior border of mandible	Body of hyoid bone and median raphe	Elevates hyoid bone and floor of mouth	Trigeminal n.
Geniohyoid	Medial surface of mandible at chin	Body of hyoid bone	Elevates hyoid bone	Spinal n. (C1)
Stylohyoid	Styloid process of temporal bone	Body of hyoid bone	Elevates and retracts tongue	Facial n.
Sternohyoid	Manubrium	Body of hyoid bone	Depresses hyoid bone	Spinal nn. (C1–C3)
Sternothyroid	Manubrium	Thyroid cartilage	Depresses thyroid cartilage	Spinal nn. (C1–C3)
Thyrohyoid	Thyroid cartilage	Great cornu of hyoid bone	Depresses hyoid bone; elevates thyroid	Spinal nn. (C1–C3)
Omohyoid	Superior border of scapula	Body of hyoid bone	Depresses hyoid bone	Spinal nn. (C1–C3)

digastric: L. *di*, two; Gk. *gaster*, belly
mylohyoid: Gk. *mylos*, akin to; *hyoeides*, pertaining to hyoid bone
omohyoid: Gk. *omos*, shoulder

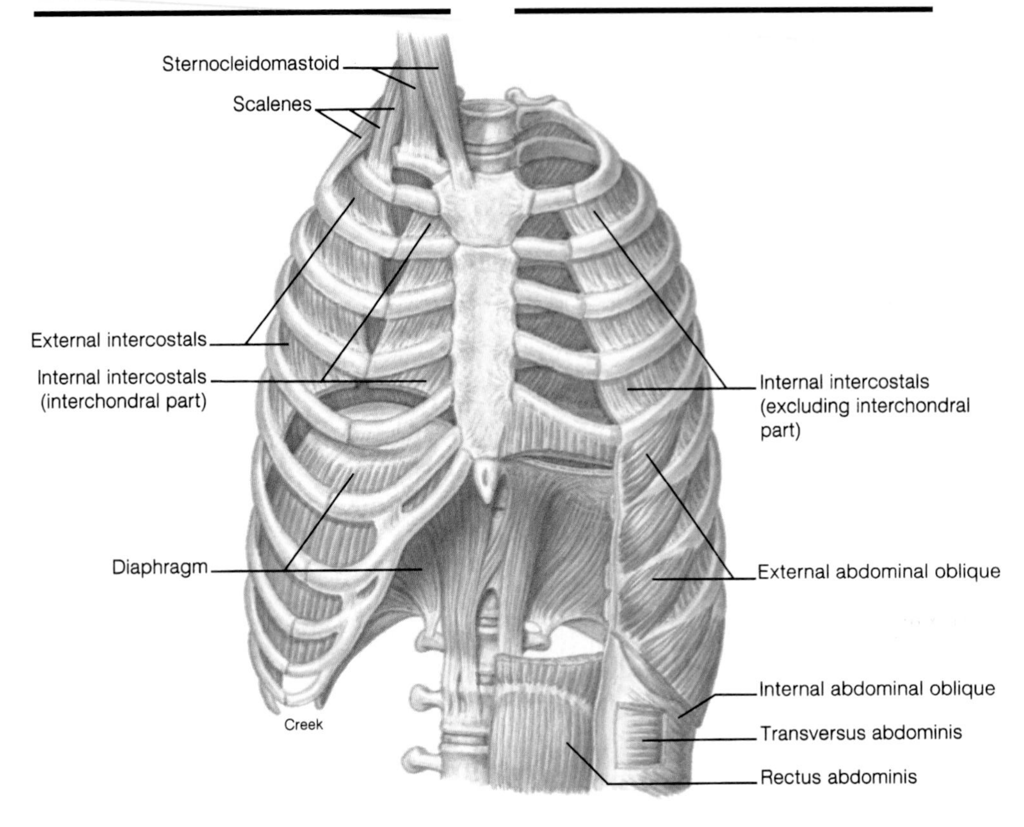

Figure 9.21

Muscles of respiration.

Muscles of Respiration

The muscles of respiration are skeletal muscles that continually contract rhythmically, usually involuntarily. Breathing, or *pulmonary ventilation*, is divided into two phases: *inspiration (inhalation)* and *expiration (exhalation)*.

During normal, relaxed inspiration, the contracting muscles are the **diaphragm,** the **external intercostal** muscles, and the interchondral portion of the **internal intercostal** muscles (fig. 9.21). A downward contraction of the dome-shaped diaphragm causes a vertical increase in thoracic dimension. A simultaneous contraction of the external intercostals and the interchondral portion of the internal intercostals produces an increase in the lateral dimension of the thorax. In addition, the **sternocleidomastoid** and **scalene** (*ska'lēn*) muscles may assist in inspiration through elevation of the first and second ribs, respectively. The intercostal muscles are innervated by the intercostal nerves, and the diaphragm receives its stimuli through the phrenic nerves.

Expiration is primarily a passive process, occurring as the muscles of inspiration are relaxed and the rib cage recoils to its original position. During forced expiration, the interosseous portion of the **internal intercostals** contracts, causing the rib cage to be depressed. This portion of the internal intercostals lies under the external intercostals, and its fibers are directed downward and backward. The *abdominal muscles* may also contract during forced expiration, which increases pressure within the abdominal cavity and forces the diaphragm superiorly, squeezing additional air out of the lungs.

Muscles of the Abdominal Wall

The anterolateral abdominal wall is composed of four pairs of flat, sheetlike muscles: the *external abdominal oblique, internal abdominal oblique, transversus abdominis,* and *rectus abdominis* muscles (fig. 9.22). These muscles support and protect the organs of the abdominal cavity and aid in breathing. When they contract, the pressure in the abdominal cavity increases, which can aid in defecation and in stabilizing the spine during heavy lifting.

The **external abdominal oblique** is the strongest and most superficial of the three layered muscles of the lateral abdominal wall (figs. 9.22 and 9.23). Its fibers are directed inferiorly and medially. The **internal abdominal oblique** lies deep to the external abdominal oblique, and its fibers are directed at right angles to those of the external abdominal oblique. The **transversus abdominis** is the deepest of the abdominal muscles; its fibers run horizontally across the abdomen. The long, straplike **rectus abdominis** muscle is entirely enclosed in a fibrous sheath formed from the aponeuroses of the other three abdominal muscles. The *linea alba* is a band of connective tissue on the midline of the abdomen that separates the two rectus

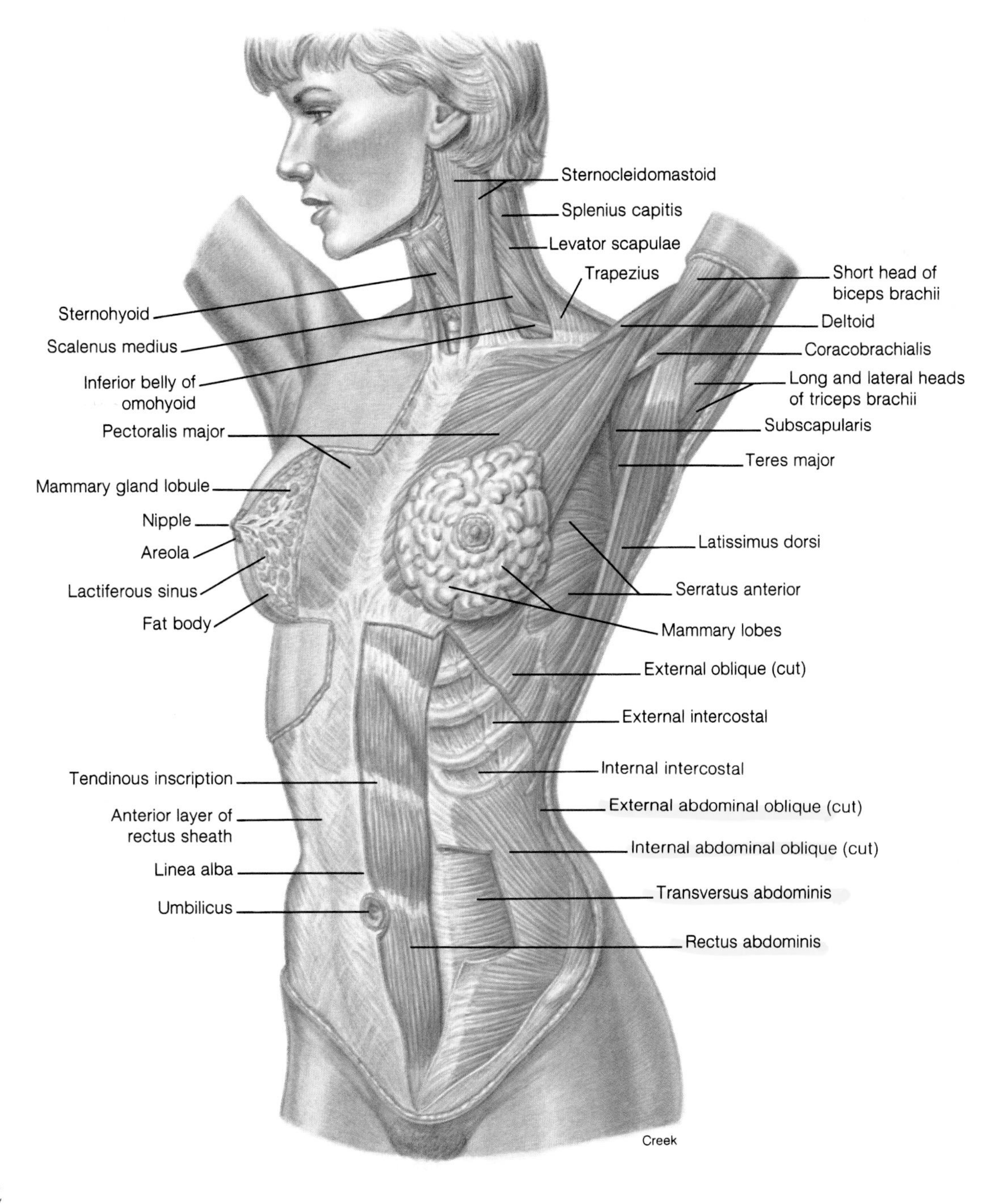

Figure 9.22

Muscles of the anterolateral neck, shoulder, and trunk regions. The mammary gland is an integumentary structure positioned over the pectoralis major muscle.

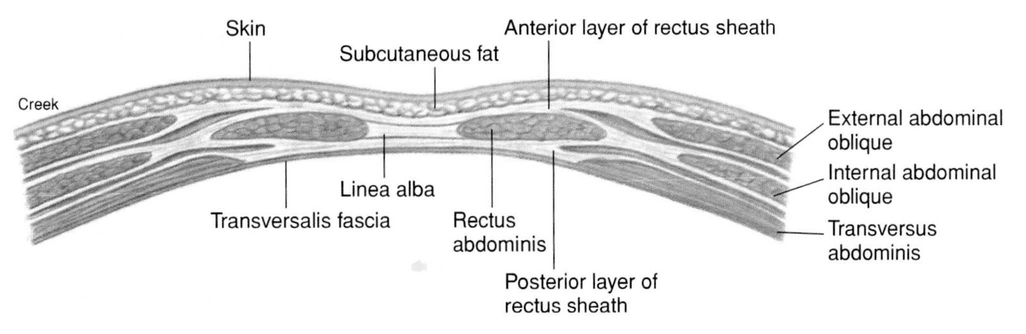

Figure 9.23

Muscles of the anterior abdominal wall shown in a transverse view.

Table 9.7
Muscles of the Abdominal Wall

Muscle	Origin	Insertion	Action
External abdominal oblique	Lower eight ribs	Iliac crest and linea alba	Compresses abdomen; lateral rotation; draws thorax downward
Internal abdominal oblique	Iliac crest, inguinal ligament, and lumbodorsal fascia	Linea alba and costal cartilage of last three or four ribs	Compresses abdomen; lateral rotation; draws thorax downward
Transversus abdominis	Iliac crest, inguinal ligament, lumbar fascia, and costal cartilage of last six ribs	Xiphoid process, linea alba, and pubis	Compresses abdomen
Rectus abdominis	Pubic crest and symphysis pubis	Costal cartilage of fifth to seventh ribs and xiphoid process of sternum	Flexes vertebral column (contraction)

rectus: L. *rectus*, straplike; *abdomino*, belly

Table 9.8
Muscles of the Pelvic Outlet

Muscle	Origin	Insertion	Action
Levator ani	Spine of ischium and pubic bone	Coccyx	Supports pelvic viscera; aids in defecation
Coccygeus	Ischial spine	Sacrum and coccyx	Supports pelvic viscera; aids in defecation
Transversus perinei	Ischial tuberosity	Central tendon	Supports pelvic viscera
Bulbospongiosus	Central tendon	Males: base of penis; females: root of clitoris	Constricts urethral canal; constricts vagina
Ischiocavernosus	Ischial tuberosity	Males: pubic arch and crus of the penis; females: pubic arch and crus of the clitoris	Aids erection of penis and clitoris

abdominis muscles. *Tendinous inscriptions* transect the rectus abdominis muscles at several points, causing the abdominal region of a well-muscled person with low body fat to appear segmented.

Refer to table 9.7 for a summary of the muscles of the abdominal wall.

Muscles of the Pelvic Outlet

Any sheet that separates cavities may be termed a diaphragm. The **pelvic outlet**—the entire muscular wall at the bottom of the pelvic cavity—contains two: the pelvic diaphragm and the urogenital diaphragm. The **urogenital diaphragm** lies immediately deep to the external genitalia; the **pelvic diaphragm** is situated closer to the internal viscera. Together, these sheets of muscle provide support for pelvic viscera and help to regulate the passage of urine and feces.

The pelvic diaphragm consists of the levator ani and the coccygeus muscles (table 9.8). The **levator ani** (*le-va'tor a'ni*) (fig. 9.24) is a thin sheet of muscle that helps to support the pelvic viscera and constrict the lower part of the rectum, pulling it forward and aiding defecation. The deeper, fan-shaped **coccygeus** (*kok-sij'e-us*) aids the levator ani in its functions.

An *episiotomy* is a surgical incision, for obstetrical purposes, of the vaginal orifice and a portion of the levator ani muscle of the perineum. Following a pudendal nerve block, an episiotomy may be done during childbirth to accommodate the head of an emerging fetus with minimal tearing of the tissues. After delivery, the cut is sutured.

The urogenital diaphragm consists of the deep, sheetlike **transversus perinei** (*per-ĭ-ne'i*) muscle, and the associated **external anal sphincter** muscle. The external anal sphincter is a funnel-shaped constrictor muscle that surrounds the anal canal.

Inferior to the pelvic diaphragm are the *perineal muscles*, which provide the skeletal muscular support to the genitalia. They include the *bulbocavernosus*, *ischiocavernosus*, and the *superficial transversus perinei* muscles (fig. 9.24). The muscles of the pelvic diaphragm and the urogenital diaphragm are similar in the male and female, but the perineal muscles exhibit marked sex-based differences.

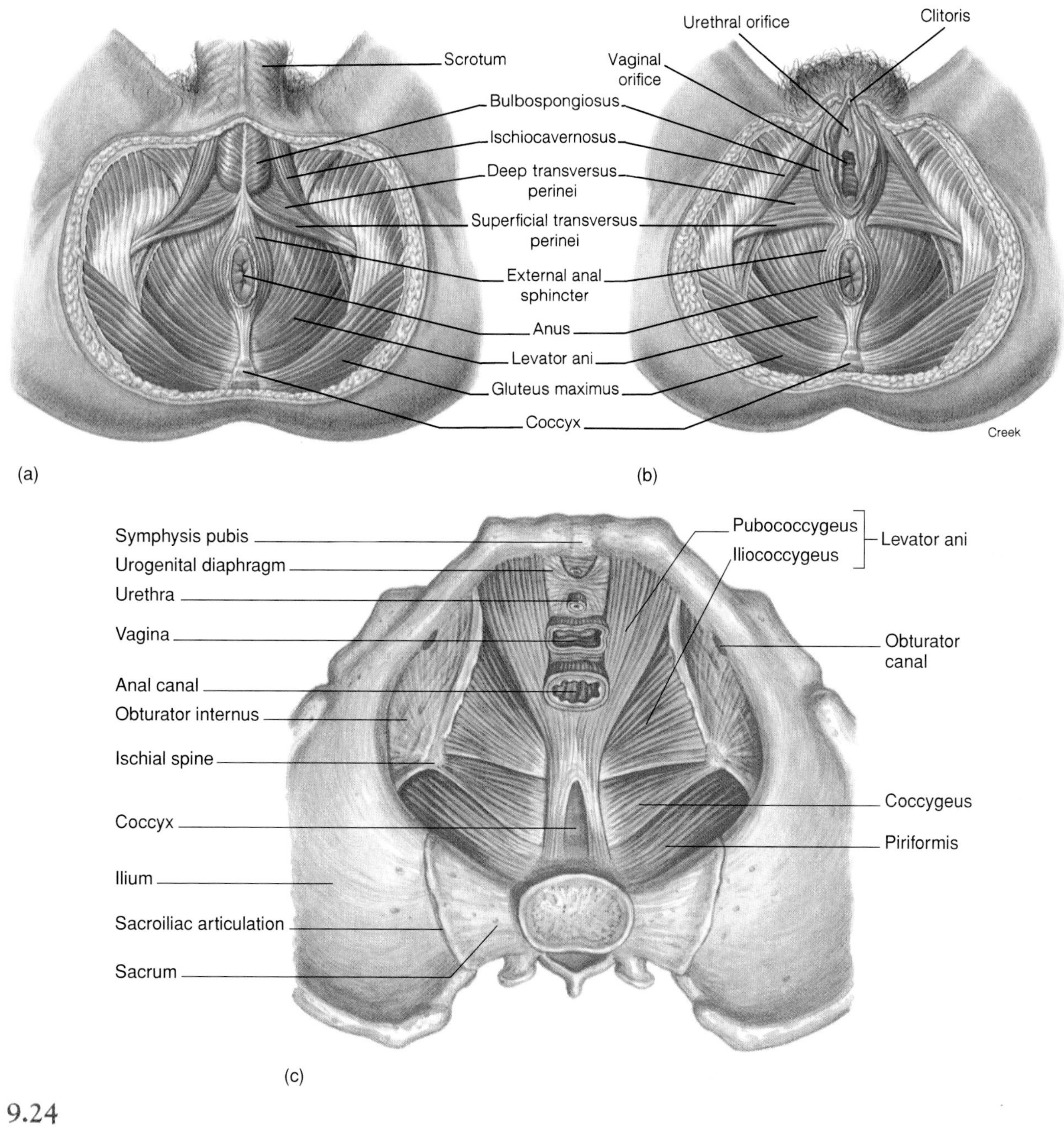

Figure 9.24

Muscles of the pelvic outlet: (*a*) male and (*b*) female. (*c*) A superior view of the internal muscles of the female pelvic outlet.

In males, the **bulbospongiosus** (*bul"bo-spon"je-o-sus*) of one side unites with that of the opposite side to form a muscular constriction surrounding the base of the penis. When contracted, the two muscles constrict the urethral canal and assist in emptying the urethra. In females, these muscles are separated by the vaginal orifice, which they constrict as they contract. The **ischiocavernosus** (*is"ke-o-kă"ver-no-sus*) muscle inserts onto the pubic arch and crus of the penis in the male and the pubic arch and crus of the clitoris of the female. This muscle assists the erection of the penis and clitoris during sexual arousal.

Muscles of the Vertebral Column

The strong, complex muscles of the vertebral column are adapted to provide support and movement in resistance to the effect of gravity.

The vertebral column can be flexed, extended, hyperextended, rotated, and laterally flexed (right or left). The muscle that flexes the vertebral column, the rectus abdominis, has already been described as a long, straplike muscle of the anterior abdominal wall. The extensor muscles located on the posterior side of the vertebral column have to be stronger than the flexors

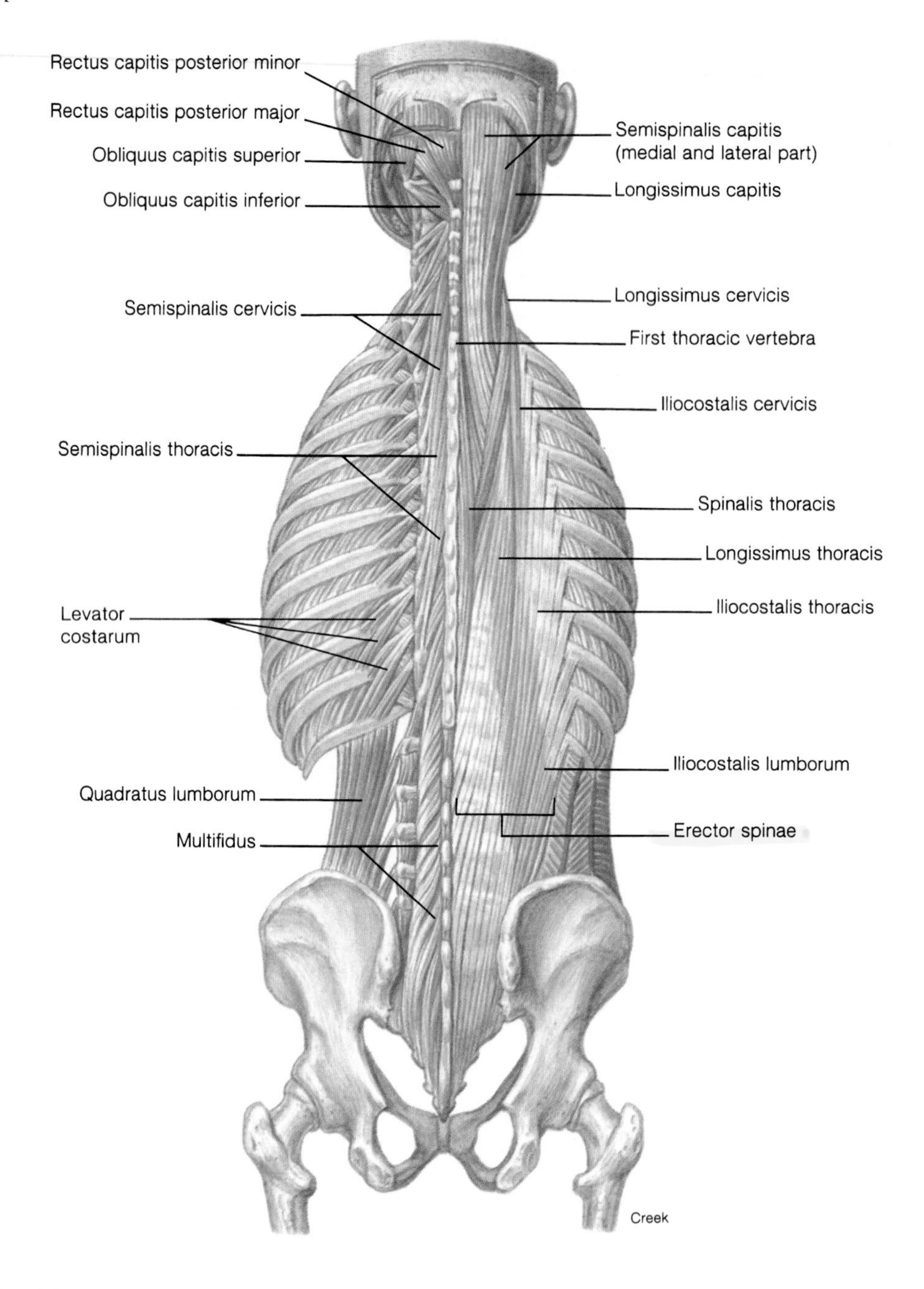

Figure 9.25

Muscles of the vertebral column. The superficial neck muscles and erector spinae group of muscles are illustrated on the right, and the deep neck and back muscles are illustrated on the left.

because extension (such as lifting an object) is in opposition to gravity. The extensor muscles consist of a superficial group and a deep group. Only some of the muscles of the vertebral column will be described here.

The **erector spinae** (*spi'ne*) muscles constitute a massive superficial muscle group that extends from the sacrum to the skull. It actually consists of three groups of muscles: the **iliocostalis, longissimus,** and **spinalis** muscles (fig. 9.25 and table 9.9). Each of these groups, in turn, consists of overlapping slips of muscle. The iliocostalis is the most lateral group, the longissimus is intermediate in position, and the spinalis, in the medial position, comes in contact with the spinous processes of the vertebrae.

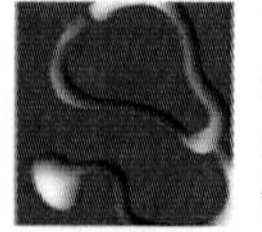

The erector spinae muscles are frequently strained through improper lifting. A heavy object should not be lifted with the vertebral column flexed; instead, the hip and knee joints should be flexed so that the pelvic and leg muscles can aid in the task.

Pregnancy may also put a strain on the erector spinae muscles. Pregnant women will try to counterbalance the effect of a protruding abdomen by hyperextending the vertebral column. This results in an exaggerated lumbar curvature, strained muscles, and a peculiar gait.

The deep **quadratus lumborum** (*kwod-ra'tus lum-bor'um*) muscle originates on the iliac crest and the lower three lumbar

Table 9.9
Muscles of the Vertebral Column

Muscle	Origin	Insertion	Action	Innervation
Quadratus lumborum	Iliac crest and lower three lumbar vertebrae	Twelfth rib and upper four lumbar vertebrae	Extends lumbar region; laterally flexes vertebral column	Intercostal nerve T12 and lumbar nerves L2–L4
Erector spinae	Consists of three groups of muscles: iliocostalis, longissimus, and spinalis. The iliocostalis and longissimus are further subdivided into three groups each on the basis of location along the vertebral column.			
Iliocostalis lumborum	Crest of ilium	Lower six ribs	Extends lumbar region	Posterior rami of lumbar nerves
Iliocostalis thoracis	Lower six ribs	Upper six ribs	Extends thoracic region	Posterior rami of thoracic nerves
Iliocostalis cervicis	Angles of third to sixth rib	Transverse processes of fourth to sixth cervical vertebrae	Extends cervical region	Posterior rami of cervical nerves
Longissimus thoracis	Transverse processes of lumbar vertebrae	Transverse processes of all the thoracic vertebrae and lower nine ribs	Extends thoracic region	Posterior rami of spinal nerves
Longissimus cervicis	Transverse processes of upper four or five thoracic vertebrae	Transverse processes of second to sixth cervical vertebrae	Extends and laterally flexes cervical region	Posterior rami of spinal nerves
Longissimus capitis	Transverse processes of upper five thoracic vertebrae and articular processes of lower three cervical vertebrae	Posterior margin of cranium and mastoid process of temporal bone	Extends head; acting separately, turns face toward that side	Posterior rami of middle and lower cervical nerves
Spinalis thoracis	Spinous processes of upper lumbar and lower thoracic vertebrae	Spinous processes of upper thoracic vertebrae	Extends vertebral column	Posterior rami of spinal nerves
Semispinalis thoracis	Transverse processes of T6–T10	Spinous processes of C6–T4	Extends vertebral column	Posterior rami of spinal nerves
Semispinalis cervicis	Transverse processes of T1–T6	Spinous processes of C2–C5	Extends vertebral column	Posterior rami of spinal nerves
Semispinalis capitis	Transverse processes of C7–T7	Nuchal line of occipital bone	Extends head	Posterior rami of spinal nerves

vertebrae. It inserts on the transverse processes of the first four lumbar vertebrae and the inferior margin of the twelfth rib. When the right and left quadratus lumborum contract together, the vertebral column in the lumbar region extends. Separate contraction causes lateral flexion of the spine.

1. Identify the facial muscles responsible for (a) wrinkling the forehead, (b) pursing the lips, (c) protruding the lower lip, (d) smiling, (e) frowning, (f) winking, and (g) elevating the upper lip to show the teeth.
2. Describe the actions of the extrinsic muscles that move the tongue.
3. Which muscles of the neck either originate from or insert on the hyoid bone?
4. Describe the actions of the muscles of inspiration. Which muscles participate in forced expiration?
5. Which muscles of the pelvic outlet support the floor of the pelvic cavity? Which are associated with the genitalia?
6. List the subgroups of the erector spinae group of muscles and describe their locations?

Muscles of the Appendicular Skeleton

The muscles of the appendicular skeleton include those of the pectoral girdle, arm, forearm, wrist, hand, and fingers, and those of the pelvic girdle, thigh, leg, ankle, foot, and toes.

Objective 11 Locate the major muscles of the appendicular skeleton. Identify synergistic and antagonistic muscles and describe the action of each one.

Muscles That Act on the Pectoral Girdle

The shoulder is attached to the axial skeleton only at the sternoclavicular joint; therefore, strong, straplike muscles are necessary in this region. Furthermore, muscles that move the brachium originate on the scapula, and during brachial movement the scapula has to be held stationary. The muscles that act on the pectoral girdle originate on the axial skeleton and can be divided into anterioı and posterior groups.

The anterior group of muscles that act on the pectoral girdle includes the **serratus** (*ser-a'tus*) **anterior, pectoralis** (*pek"to-ra'lis*) **minor,** and **subclavius** (*sub-kla've-us*) muscles (fig. 9.26). The posterior group includes the **trapezius, levator scapulae**

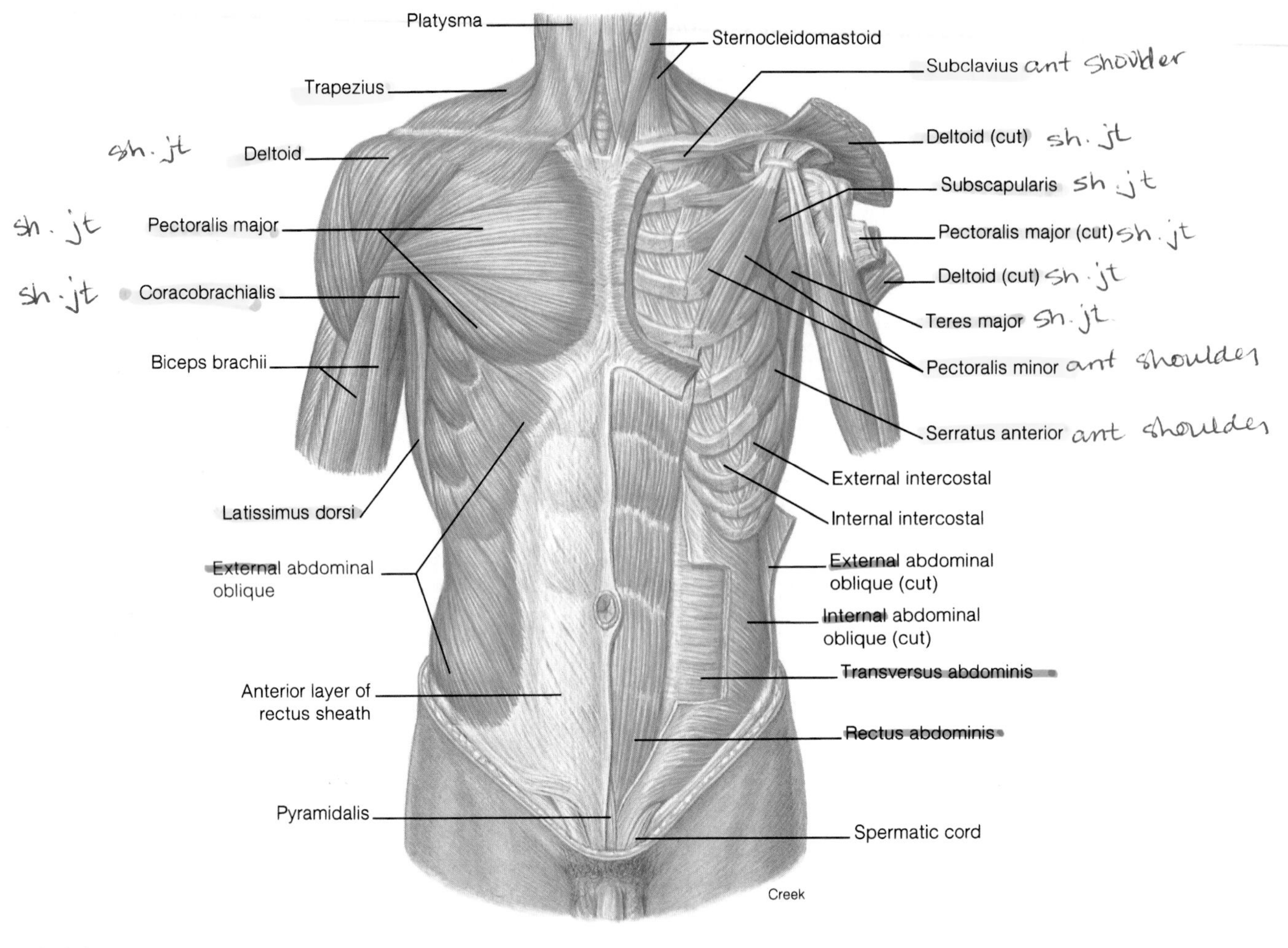

Figure 9.26

Muscles of the anterior trunk and shoulder regions. The superficial muscles are illustrated on the right, and the deep muscles are illustrated on the left.

(*skap-yŭ'le*) and **rhomboideus** (*rom-boid'e-us*) muscles (fig. 9.27). These muscles are positioned so that one of them does not cause an action on its own. Rather, several muscles contract synergistically to result in any movement of the girdle.

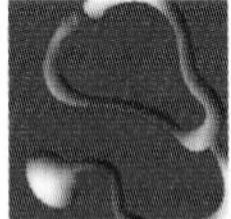

Treatment of advanced stages of *breast cancer* requires the surgical removal of both pectoralis major and pectoralis minor muscles in a procedure called a *radical mastectomy.* Postoperative physical therapy is primarily geared toward strengthening the synergistic muscles of this area. As the muscles that act on the brachium are learned, determine which are synergists with the pectoralis major.

Muscles That Move the Humerus at the Shoulder Joint

Of the nine muscles that span the shoulder joint to insert on the humerus, only two—the *pectoralis major* and *latissimus dorsi*—do not originate on the scapula (table 9.10). These two are designated as axial muscles, whereas the remaining seven are scapular muscles. The muscles of this region are shown in figures 9.26 and 9.27, and the attachments of all the muscles that either originate or insert on the scapula are shown in figure 9.28.

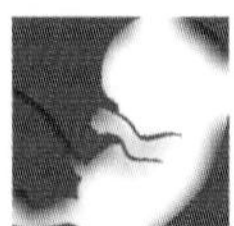

In terms of their development, the pectoralis major and the latissimus dorsi muscles are not axial muscles at all. They develop in the forelimb and extend to the trunk secondarily. They are considered axial muscles only because their origins are on the axial skeleton.

Axial Muscles

The **pectoralis major** is a large, fan-shaped chest muscle (see fig. 9.26) that binds the humerus to the pectoral girdle. It is the principal flexor muscle of the shoulder joint. The large, flat, triangular **latissimus dorsi** (*la-tis'ĭ-mus dor'si*) muscle covers the inferior half of the thoracic region of the back (see fig. 9.27) and is the antagonist to the pectoralis major muscle. The latissimus dorsi is frequently called the "swimmer's muscle" because

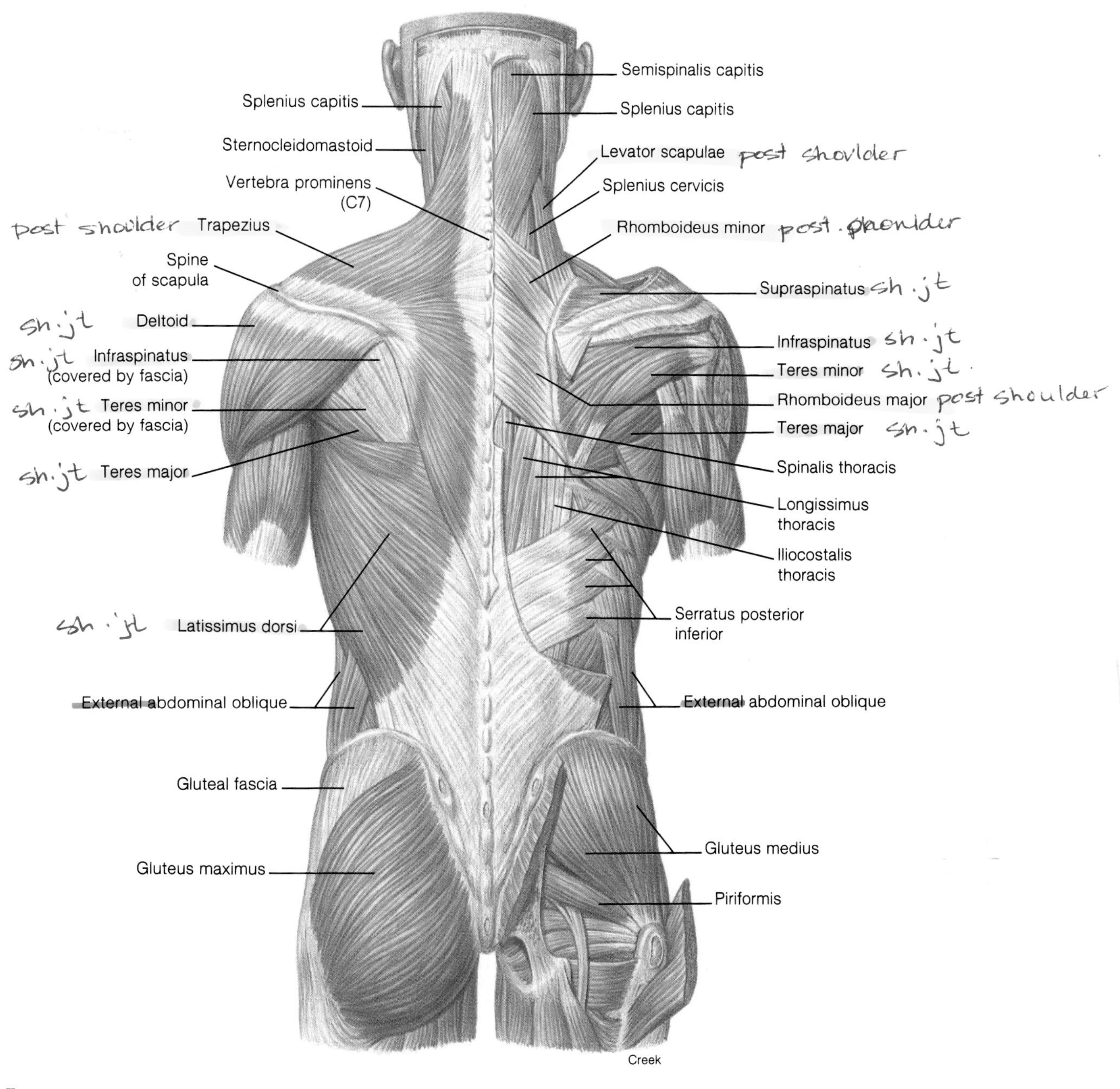

Figure 9.27

Muscles of the posterior neck, shoulder, trunk, and gluteal regions. The superficial muscles are illustrated on the left, and the deep muscles are illustrated on the right.

it powerfully extends the shoulder joint, drawing the arm downward and backward while it rotates medially. Extension of the shoulder joint is in reference to anatomical position and is therefore a backward, retracting (increasing the shoulder joint angle) movement of the arm.

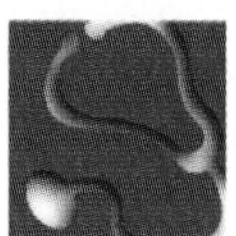

A latissimus dorsi muscle, conditioned with pulsated electrical impulses, will in time come to resemble cardiac muscle tissue in that it will be indefatigable, using oxygen at a steady rate. Following conditioning, the muscle may be used in an autotransplant to repair a surgically removed portion of a patient's

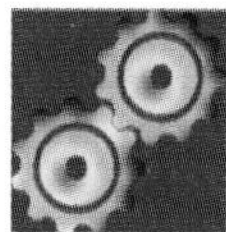

diseased heart. The procedure involves detaching the latissimus dorsi muscle from its vertebral origin, leaving the blood supply and innervation intact, and slipping it into the pericardial cavity where it is wrapped around the heart like a towel. A pacemaker is required to provide the continuous rhythmic contractions.

Scapular Muscles

The nonaxial scapular muscles include the *deltoid*, *supraspinatus*, *infraspinatus*, *teres major*, *teres minor*, *subscapularis*, and *coracobrachialis* muscles.

Table 9.10

Muscles That Act on the Pectoral Girdle and That Move the Shoulder Joint

Muscle	Origin	Insertion	Action	Innervation
Serratus anterior	Upper eight or nine ribs	Anterior vertebral border of scapula	Pulls scapula forward and downward	Long thoracic n.
Pectoralis minor	Sternal ends of third, fourth, and fifth ribs	Coracoid process of scapula	Pulls scapula forward and downward	Medial and lateral pectoral nn.
Subclavius	First rib	Subclavian groove of clavicle	Draws clavicle downward	Spinal nerves C5, C6
Trapezius	Occipital bone and spines of seventh cervical and all thoracic vertebrae	Clavicle, spine of scapula, and acromion	Elevates, depresses, and adducts scapula; hyperextends neck; braces shoulder	Accessory nerve
Levator scapulae	First to fourth cervical vertebrae	Medial border of scapula	Elevates scapula	Dorsal scapular n.
Rhomboideus major	Spines of second to fifth thoracic vertebrae	Medial border of scapula	Elevates and adducts scapula	Dorsal scapular n.
Rhomboideus minor	Seventh cervical and first thoracic vertebrae	Medial border of scapula	Elevates and adducts scapula	Dorsal scapular n.
Pectoralis major	Clavicle, sternum, and costal cartilages of second to sixth rib; rectus sheath	Crest of greater tubercle of humerus	Flexes, adducts, and rotates shoulder joint medially	Medial and lateral pectoral nn.
Latissimus dorsi	Spines of sacral, lumbar, and lower thoracic vertebrae; iliac crest and lower four ribs	Intertubercular groove of humerus	Extends, adducts, and rotates shoulder joint medially; adducts shoulder joint	Thoracodorsal n.
Deltoid	Clavicle, acromion and spine of scapula	Deltoid tuberosity of humerus	Abducts, extends, or flexes shoulder joint	Axillary n.
Supraspinatus	Fossa—superior to spine of scapula	Greater tubercle of humerus	Abducts and laterally rotates shoulder joint	Suprascapular n.
Infraspinatus	Fossa—inferior to spine of scapula	Greater tubercle of humerus	Rotates shoulder joint laterally	Suprascapular n.
Teres major	Inferior angle and lateral border of scapula	Crest of lesser tubercle of humerus	Extends shoulder joint, or adducts and rotates shoulder joint medially	Lower subscapular n.
Teres minor	Axillary border of scapula	Greater tubercle and groove of humerus	Rotates shoulder joint laterally	Axillary n.
Subscapularis	Subscapular fossa	Lesser tubercle of humerus	Rotates shoulder joint medially	Subscapular nn.
Coracobrachialis	Coracoid process of scapula	Body of humerus	Flexes and adducts shoulder joint	Musculocutaneous n.

serratus: L. *serratus*, saw-shaped
trapezius: Gk. *trapezoeides*, trapezoid-shaped
rhomboideus: Gk. *rhomboides*, rhomboid-shaped
pectoralis: L. *pectus*, chest
latissimus: L. *latissimus*, widest
deltoid: Gk. *delta*, triangular
teres: L. *teres*, rounded

The **deltoid** is a thick, powerful muscle that caps the shoulder joint (figs. 9.29 and 9.30). Although it has several functions (table 9.10), its principal action is abduction of the shoulder joint. Functioning together, the pectoralis major and the latissimus dorsi muscles are antagonists to the deltoid muscle in that they cause adduction of the shoulder joint. The deltoid muscle is a common site for intramuscular injections.

The remaining six scapular muscles also help to stabilize the shoulder and have specific actions at the shoulder joint (table 9.10). The **supraspinatus** (*soo″pra-spi-na′tus*) muscle laterally rotates the arm and is synergistic with the deltoid muscle in abducting the shoulder joint. The **infraspinatus** muscle rotates the arm laterally. The action of the **teres** (*tĕ′rēz*) **major** muscle is similar to that of the latissimus dorsi, adducting and medially rotating the shoulder joint. The **teres minor** muscle works with the infraspinatus muscle in laterally rotating the arm at shoulder joint. The **subscapularis** muscle is a strong stabilizer of the shoulder and also aids in medially rotating the arm at the shoulder joint. The **coracobrachialis** (*kor″ă-ko-bra″ke-al′ is*) muscle is a synergist to the pectoralis major in flexing and adducting the shoulder joint.

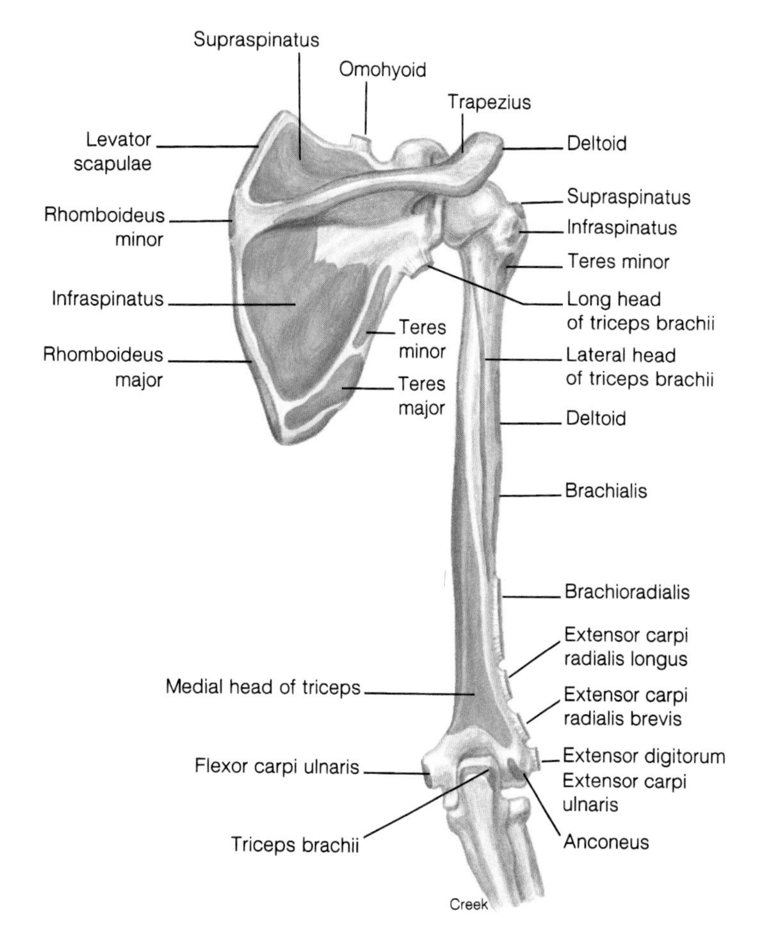

Figure 9.28

A posterior view of the scapula and humerus showing the areas of attachment of the associated muscles. (Points of origin are color coded red, and points of insertion are color coded blue.)

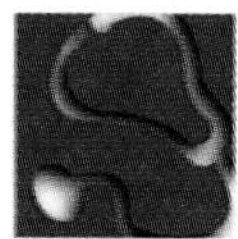

Four of the nine muscles that cross the shoulder joint, the supraspinatus, infraspinatus, teres minor, and subscapularis, are commonly called the *musculotendinous cuff,* or *rotator cuff.* Their distal tendons blend with and reinforce the fibrous capsule of the shoulder joint en route to their points of insertion on the humerus. This structural arrangement plays a major role in stabilizing the shoulder joint. Musculotendinous cuff injuries are common among baseball players. When throwing a baseball, an abduction of the shoulder is followed by a rapid and forceful rotation and flexion of the shoulder joint, which may strain the musculotendinous cuff.

Muscles That Move the Forearm at the Elbow Joint

The powerful muscles of the brachium are responsible for flexion and extension of the elbow joint. These muscles are the *biceps brachii, brachialis, brachioradialis,* and *triceps brachii* (figs. 9.29 and 9.30). In addition, a short triangular muscle, the *anconeus,* is positioned over the distal end of the triceps brachii muscle, near the elbow. A transverse section through the brachium in figure 9.31 provides a different perspective of the brachial region.

The powerful **biceps brachii** (*bi'ceps bra' ke-i*) muscle, positioned on the anterior surface of the humerus, is the most familiar muscle of the arm, yet it has no attachments on the humerus. This muscle has a dual origin: a medial tendinous head, the **short head,** arises from the coracoid process of the scapula, and the **long head** originates on the superior tuberosity of the glenoid cavity, passes through the shoulder joint, and descends in the intertubercular groove on the humerus (see fig. 8.8). Both heads of the biceps brachii muscle insert on the radial tuberosity. The **brachialis** (*bra" ke-al' is*) muscle is located on the distal anterior half of the humerus, deep to the biceps brachii. It is synergistic to the biceps brachii in flexing the elbow joint.

The **brachioradialis** (*bra" ke-o-ra" de-ă' lis*) is the prominent muscle positioned along the lateral (radial) surface of the forearm. It, too, flexes the elbow joint.

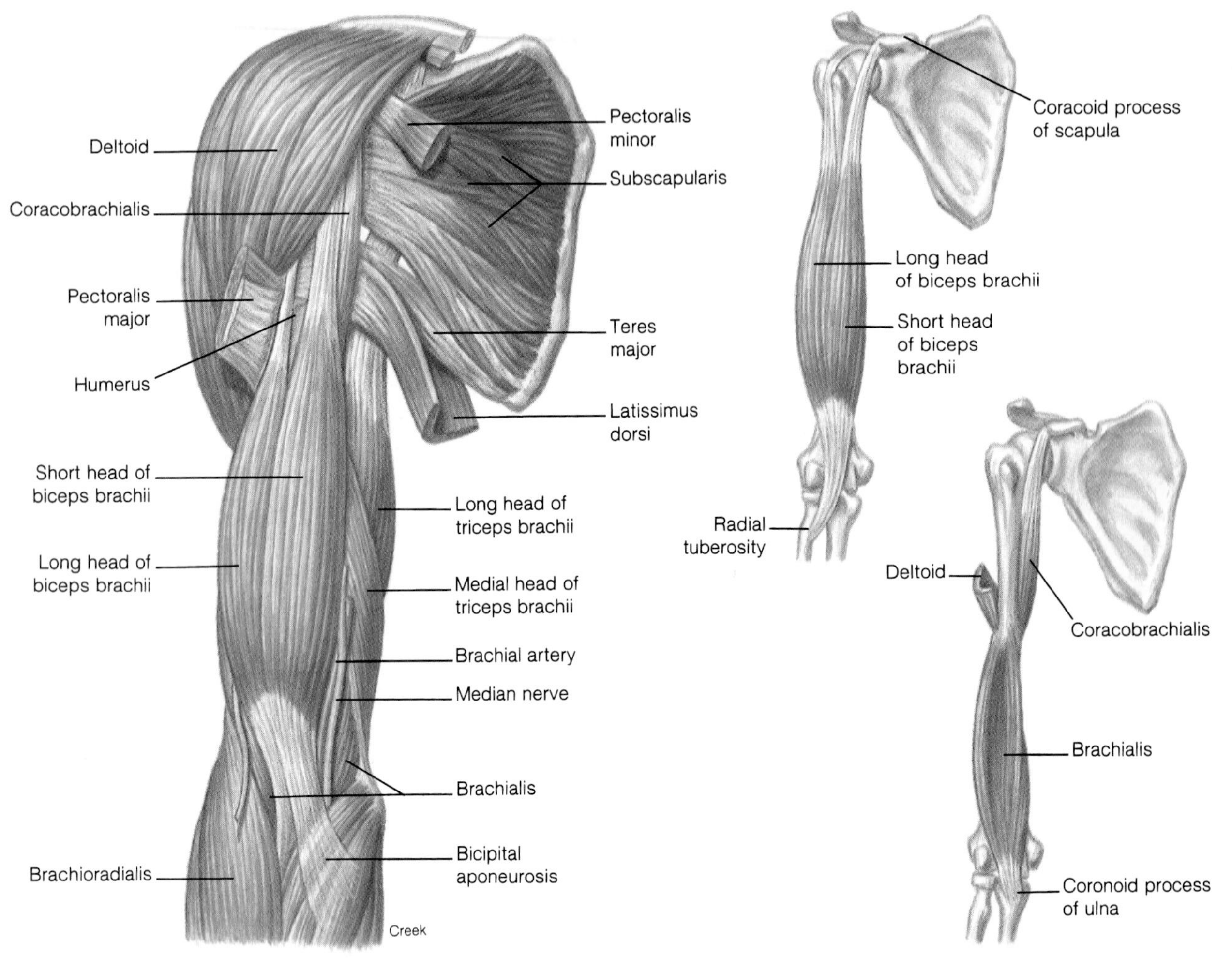

Figure 9.29

Muscles of the right anterior shoulder and brachium.

The **triceps brachii** muscle, located on the posterior surface of the brachium, extends the forearm at the elbow joint, in opposition to the action of the biceps brachii. Thus, these two muscles are antagonists. The triceps brachii has three heads, or origins. Two of the three, the **lateral head** and **medial head,** arise from the humerus, whereas the **long head** arises from the infraglenoid tuberosity of the scapula. A common tendinous insertion attaches the triceps brachii muscle to the olecranon of the ulna. The small **anconeus** (*an-ko'ne-us*) muscle is a synergist of the triceps brachii in elbow extension.

Refer to table 9.11 for a summary of the muscles that act on the forearm at the elbow joint.

Muscles of the Forearm That Move the Joints of the Wrist, Hand, and Fingers

The muscles that cause most of the movements in the joints of the wrist, hand, and fingers are positioned along the forearm (figs. 9.32 and 9.33). Several of these muscles act on two joints—the elbow and wrist. Others act on the joints of the wrist, hand, and digits. Still others produce rotational movement at the radioulnar joint. The four primary movements typically effected at the hand and digits are: supination, pronation, flexion, and extension. Other movements of the hand include adduction and abduction.

Supination and Pronation of the Hand

The **supinator** (*soo"pĭ-na'tor*) muscle wraps around the upper posterior portion of the radius (fig. 9.33), where it works synergistically with the biceps brachii muscle to supinate the hand. Two muscles are responsible for pronating the hand—the pronator teres and pronator quadratus. The **pronator teres** muscle is located on the upper medial side of the forearm, whereas the deep, anteriorly positioned **pronator quadratus** muscle extends between the ulna and radius on the distal fourth of the forearm. These two muscles work synergistically to rotate the palm of the hand posteriorly and position the thumb medially.

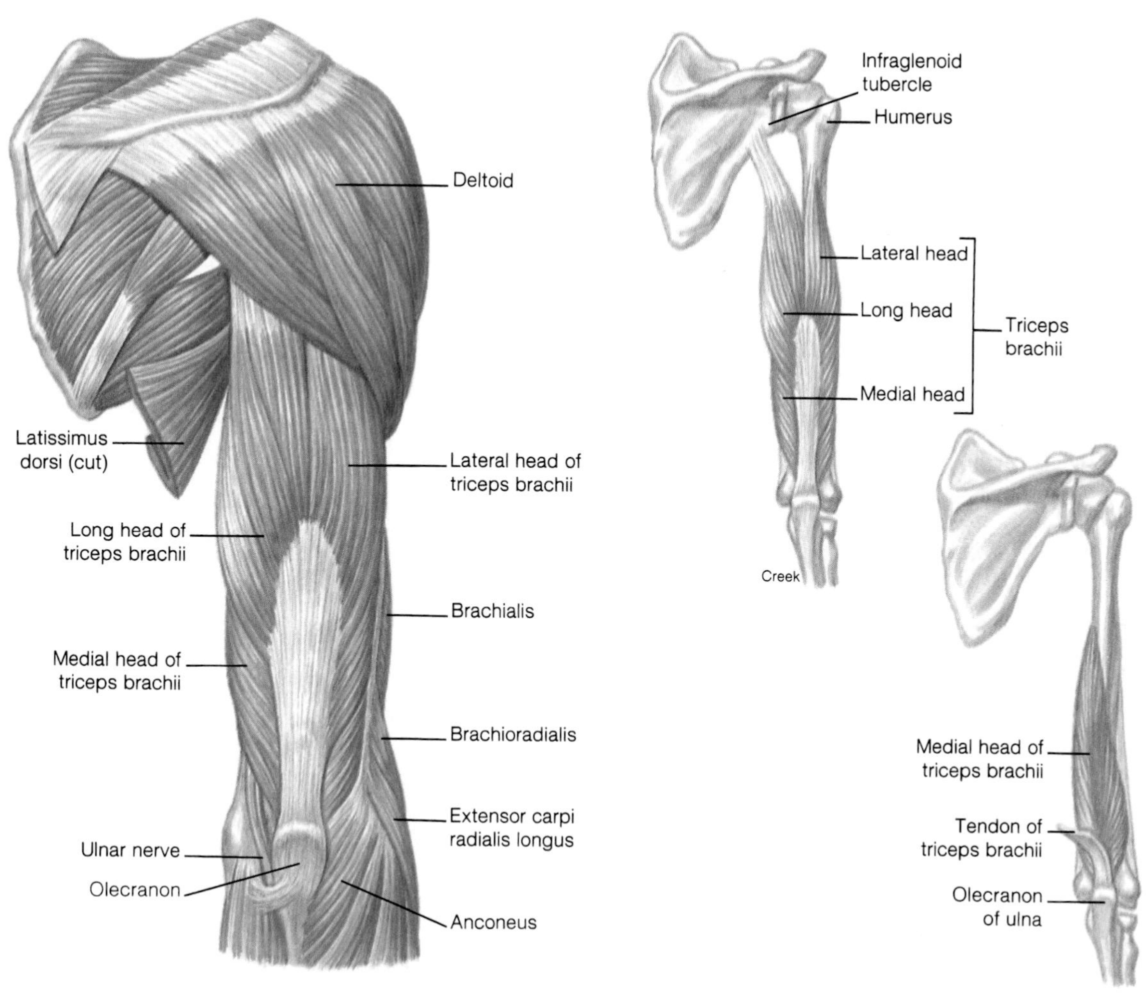

Figure 9.30

Muscles of the right posterior shoulder and brachium.

Table 9.11
Muscles That Act on the Forearm at the Elbow Joint

Muscle	Origin	Insertion	Action	Innervation
Biceps brachii	Coracoid process and tuberosity above glenoid cavity of scapula	Radial tuberosity	Flexes elbow joint; supinates forearm and hand at radioulnar joint	Musculocutaneous n.
Brachialis	Anterior body of humerus	Coronoid process of ulna	Flexes elbow joint	Musculocutaneous n.
Brachioradialis	Lateral supracondylar ridge of humerus	Proximal to styloid process of radius	Flexes elbow joint	Radial n.
Triceps brachii	Tuberosity below glenoid cavity; lateral and medial surfaces of humerus	Olecranon of ulna	Flexes elbow joint	Radial n.
Anconeus	Lateral epicondyle of humerus	Olecranon of ulna	Flexes elbow joint	Radial n.

biceps: L. *biceps*, two heads
triceps: L. *triceps*, three heads
anconeus: Gk. *ancon*, elbow

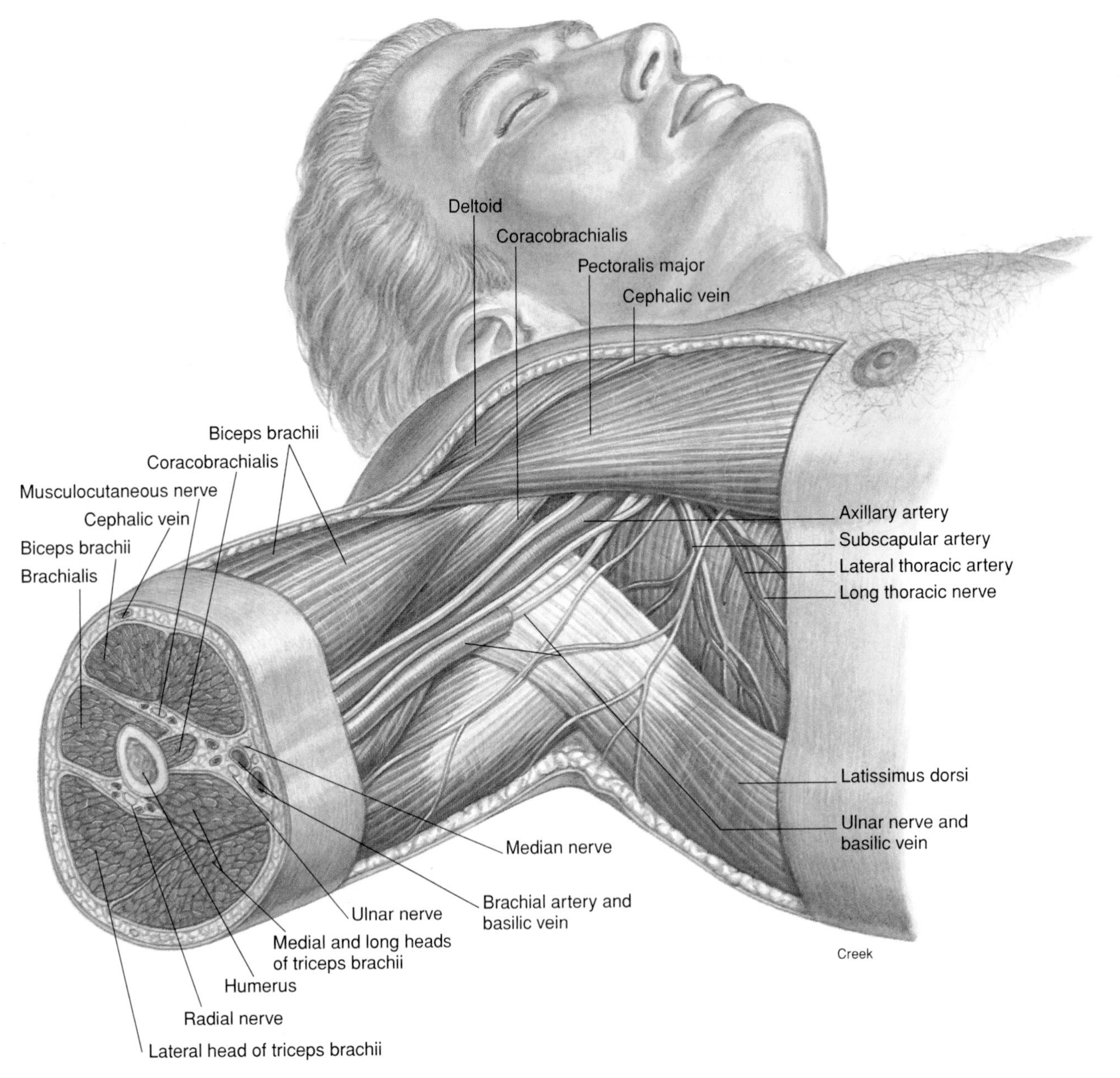

Figure 9.31

The axillary region and a transverse section through the brachium.

Flexion of the Wrist, Hand, and Fingers

Six of the muscles that flex the joints of the wrist, hand, and fingers will be described from lateral to medial and from superficial to deep (figs. 9.32 and 9.33). Although four of the six arise from the medial epicondyle of the humerus (see table 9.12), their actions on the elbow joint are minimal. The brachioradialis, already described, is an obvious reference muscle for locating the muscles of the forearm that flex the joints of the hand.

Lateral side ① The **flexor carpi radialis** muscle extends diagonally across the anterior surface of the forearm, and its distal cordlike tendon crosses the wrist under the *flexor retinaculum*. This muscle is an important landmark for locating the radial artery, where the pulse is usually taken.

The narrow **palmaris longus** muscle is superficial in position on the anterior surface of the forearm. It has a long, slender tendon that attaches to the *palmar aponeurosis*, where it assists in flexing the wrist joints.

The palmaris longus is the most variable muscle in the body. It is totally absent in approximately 8% of all people, and in 4% it is absent in one or the other forearm. Furthermore, it is absent more often in females than males, and on the left side in both sexes. Because of the superficial position of the palmaris longus muscle, you can readily determine whether it is present in your own forearm by flexing the wrist while touching the thumb and little finger, and then examining for its tendon just proximal to the wrist (see figs. 10.37 and 10.41*b*).

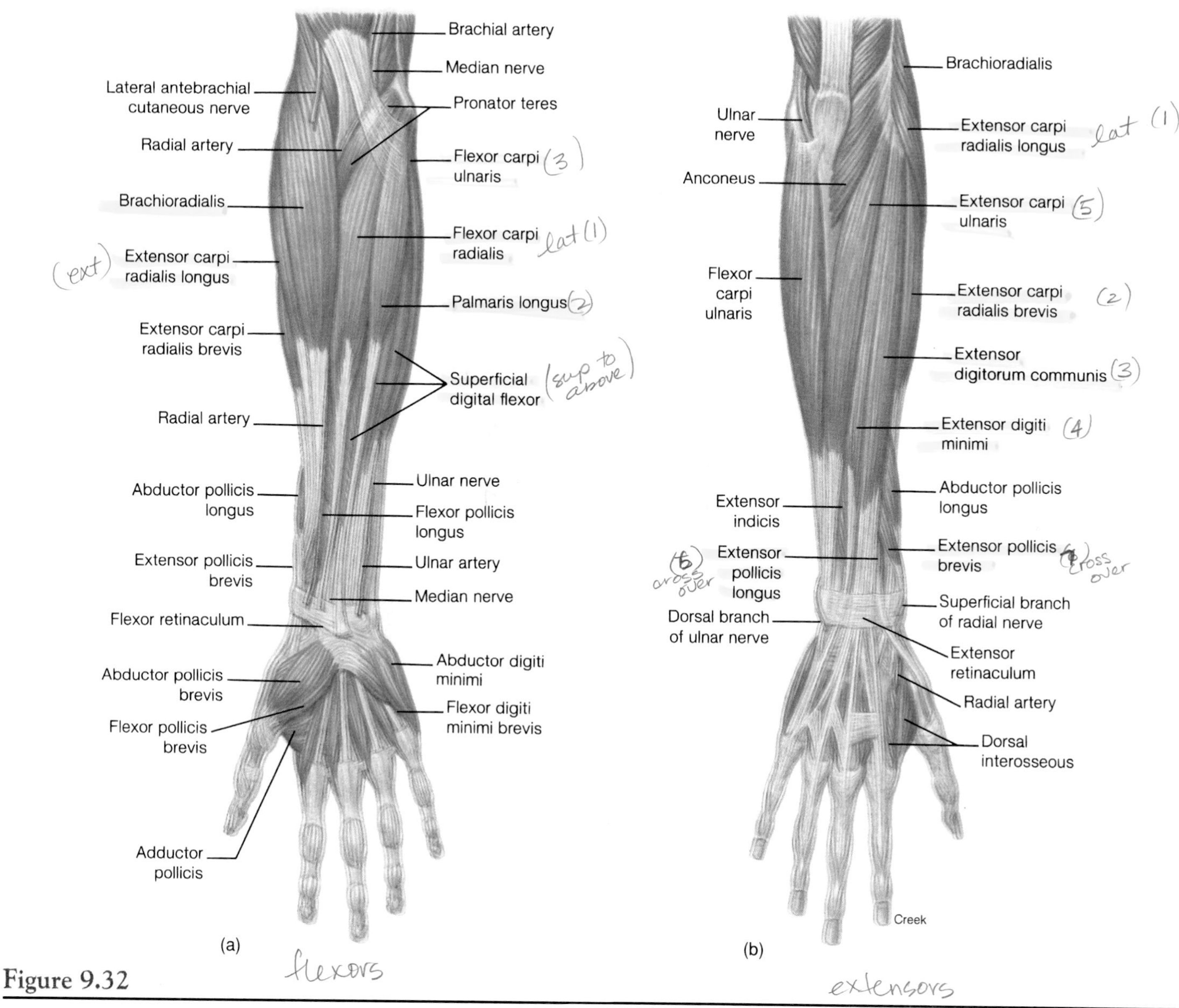

Figure 9.32

Superficial muscles of the right forearm. (*a*) An anterior view and (*b*) a posterior view.

The **flexor carpi ulnaris** muscle is positioned on the medial anterior side of the forearm, where it assists in flexing the wrist joints and adducting the hand.

The broad **superficial digital flexor** muscle lies directly beneath the three flexor muscles just described (figs. 9.32 and 9.33). It has an extensive origin, involving the humerus, ulna, and radius (see table 9.12). The tendon at the distal end of this muscle is united across the wrist joint but then splits to attach to the middle phalanx of digits II through V.

The **deep digital flexor** muscle lies deep to the superficial digital flexor. It inserts on the distal phalanges two (II) through five (V). These two muscles flex the joints of the wrist, hand, and the second, third, fourth, and fifth digits.

The **flexor pollicis longus** muscle is a deep lateral muscle of the forearm. It flexes the joints of the thumb, assisting the grasping mechanism of the hand.

The tendons of the muscles that flex the joints of the hand can be seen on the wrist as a fist is made. These tendons are securely positioned by the flexor retinaculum (fig. 9.32*a*), which crosses the wrist area transversely.

Extension of the Hand

The muscles that extend the joints of the hand are located on the posterior side of the forearm. Most of the primary extensor muscles can be seen superficially in figure 9.32*b* and will be discussed from lateral to medial.

The long, tapered **extensor carpi radialis longus** muscle is medial to the brachioradialis muscle. It extends the carpal joint and abducts the hand at the wrist. Immediately medial to

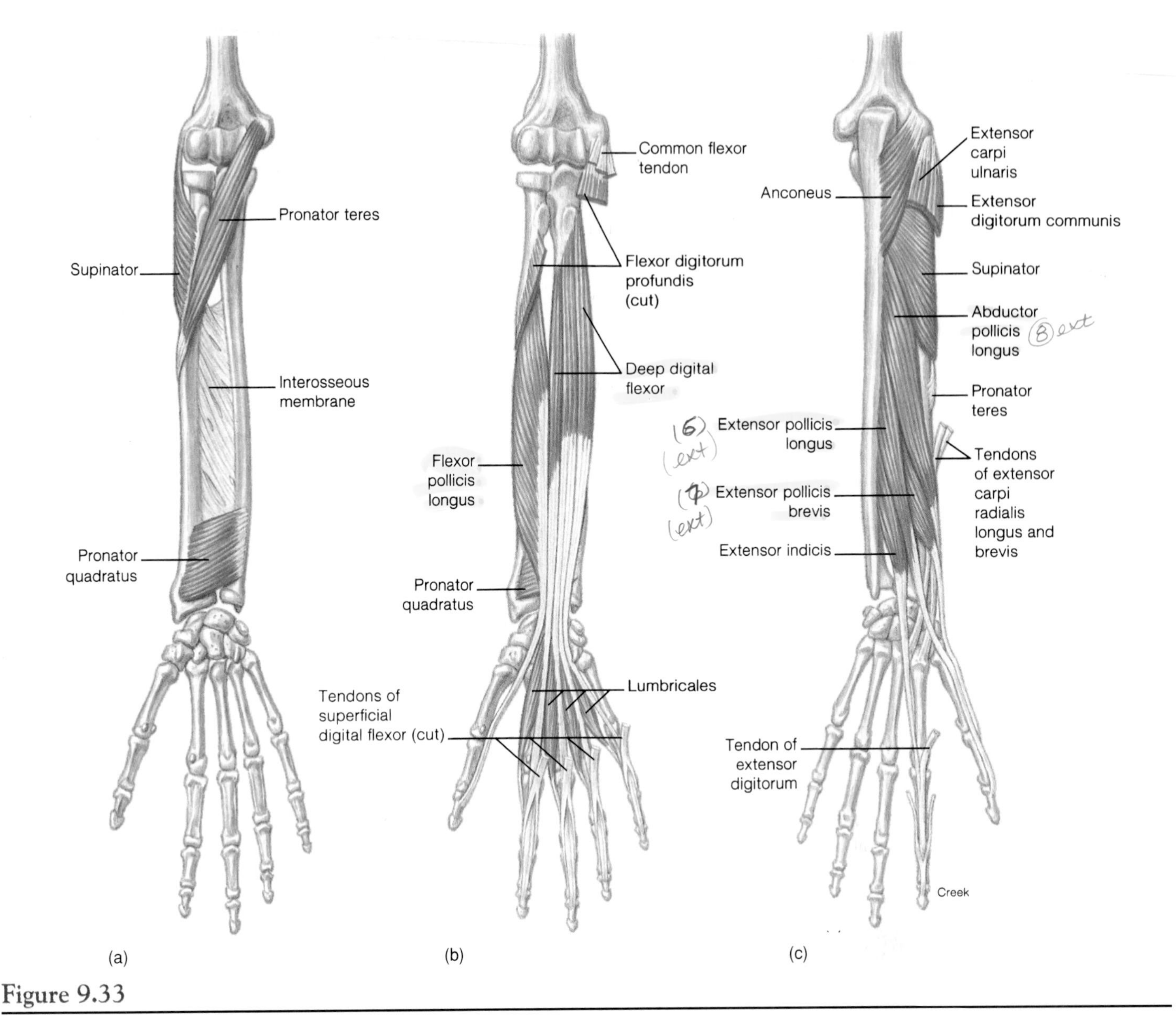

Figure 9.33

Deep muscles of the right forearm. (*a*) rotators, (*b*) flexors, and (*c*) extensors.

the extensor carpi radialis longus is the **extensor carpi radialis brevis,** which performs approximately the same functions. The origin and insertion of the latter muscle are different, however (see table 9.12).

The **extensor digitorum communis** muscle is positioned in the center of the forearm, along the posterior surface. It originates on the lateral epicondyle of the humerus. Its tendon of insertion divides at the wrist, beneath the extensor retinaculum, into four tendons that attach to the distal tip of the medial phalanges of digits II through V.

The **extensor digiti minimi** is a long, narrow muscle located on the ulnar side of the extensor digitorum communis muscle. Its tendinous insertion fuses with the tendon of the extensor digitorum communis going to the fifth digit.

The **extensor carpi ulnaris** is the most medial muscle on the posterior surface of the forearm. It inserts on the base of the fifth metacarpal bone, where it functions to extend and adduct the joints of the hand.

The **extensor pollicis longus** muscle arises from the midulnar region, crosses the lower two-thirds of the forearm, and inserts on the base of the distal phalanx of the thumb (fig. 9.33). It extends the joints of the thumb and abducts the hand. The **extensor pollicis brevis** muscle arises from the lower midportion of the radius and inserts on the base of the proximal phalanx of the thumb (fig. 9.33). The action of this muscle is similar to that of the extensor pollicis longus.

As its name implies, the **abductor pollicis longus** muscle abducts the joints of the thumb and hand. It originates on the interosseous ligament, between the ulna and radius, and inserts on the base of the first metacarpal bone.

The muscles that act on the wrist, hand, and digits are summarized in table 9.12.

Table 9.12
Muscles of the Forearm That Move the Joints of the Wrist, Hand, and Digits

Muscle	Origin	Insertion	Action	Innervation
Supinator	Lateral epicondyle of humerus and crest of ulna	Lateral surface of radius	Supinates forearm and hand	Radial n.
Pronator teres	Medial epicondyle of humerus	Lateral surface of radius	Pronates forearm and hand	Median n.
Pronator quadratus	Distal fourth of ulna	Distal fourth of radius	Pronates forearm and hand	Median n.
Flexor carpi radialis	Medial epicondyle of humerus	Base of second and third metacarpal bones	Flexes and abducts hand at wrist	Median n.
Palmaris longus	Medial epicondyle of humerus	Palmar aponeurosis	Flexes wrist	Median n.
Flexor carpi ulnaris	Medial epicondyle and olecranon	Carpal and metacarpal bones	Flexes and adducts wrist	Ulnar n.
Flexor digitorum superficialis	Medial epicondyle, coronoid process, and anterior border of radius	Middle phalanges of digits II–V	Flexes wrist and digits at metacarpophalangeal and interphalangeal joints	Median n.
Flexor digitorum profundus	Proximal two-thirds of ulna and interosseous ligament	Distal phalanges of digits III–V	Flexes wrist and digits at metacarpophalangeal and interphalangeal joints	Median and ulnar nn.
Flexor pollicis longus	Body of radius, interosseous membrane, and coronoid process of ulna	Distal phalanx of thumb	Flexes joints of thumb	Median n.
Extensor carpi radialis longus	Lateral supracondylar ridge of humerus	Second metacarpal bone	Extends and abducts wrist	Radial n.
Extensor carpi radialis brevis	Lateral epicondyle of humerus	Third metacarpal bone	Extends and abducts wrist	Radial n.
Extensor digitorum communis	Lateral epicondyle of humerus	Posterior surfaces of digits II–V	Extends wrist and phalanges at joints of carpophalangeal and interphalangeal joints	Radial n.
Extensor digiti minimi	Lateral epicondyle of humerus	Extensor aponeurosis of fifth digit	Extends joints of fifth digit and wrist	Radial n.
Extensor carpi ulnaris	Lateral epicondyle of humerus and olecranon	Base of fifth metacarpal bone	Extends and adducts wrist	Radial n.
Extensor pollicis longus	Middle of body of ulna, lateral side	Base of distal phalanx of thumb	Extends joints of thumb; abducts joints of hand	Radial n.
Extensor pollicis brevis	Distal body of radius and interosseous ligament	Base of first phalanx of thumb	Extends joints of thumb; abducts joints of hand	Radial n.
Abductor pollicis longus	Distal radius and ulna and interosseous ligament	Base of first metacarpal bone	Abducts joints of thumb and joints of hand	Radial n.

supinator: L. *supin*, bend back
pronator: L. *pron*, bend forward
palmaris: L. *palma*, flat of hand
profundus: L. *profundus*, deep
pollicis: L. *pollex*, thumb

Notice that the joints of your hand are partially flexed even when the hand is relaxed. The muscles that extend the hand are not as strong as the muscles that flex it. This is why people who receive strong electrical shocks through the arms will tightly flex their hands and cling to a cord or wire. All the muscles of the arm are stimulated to contract, but the flexors, being stronger, cause the hands to close tightly.

Muscles of the Hand

The hand is a marvelously complex structure, adapted to permit an array of intricate movements. Flexion and extension movements of the hand and phalanges are accomplished by the muscles of the forearm just described. Precise finger movements that require coordinating abduction and adduction with flexion and extension are the function of the small intrinsic muscles of the hand. These muscles and associated structures of the hand are depicted in figure 9.34. The position and actions of the muscles of the hand are listed in table 9.13.

The muscles of the hand are divided into **thenar** (*the'nar*), **hypothenar** (*hi-poth'ĕ-nar*), and **intermediate** groups. The *thenar eminence* is the fleshy base of the thumb and is formed by three muscles: the **abductor pollicis brevis,** the **flexor pollicis brevis,** and the **opponens pollicis.** The most important of the thenar muscles is the opponens pollicis, which opposes the thumb to the palm of the hand.

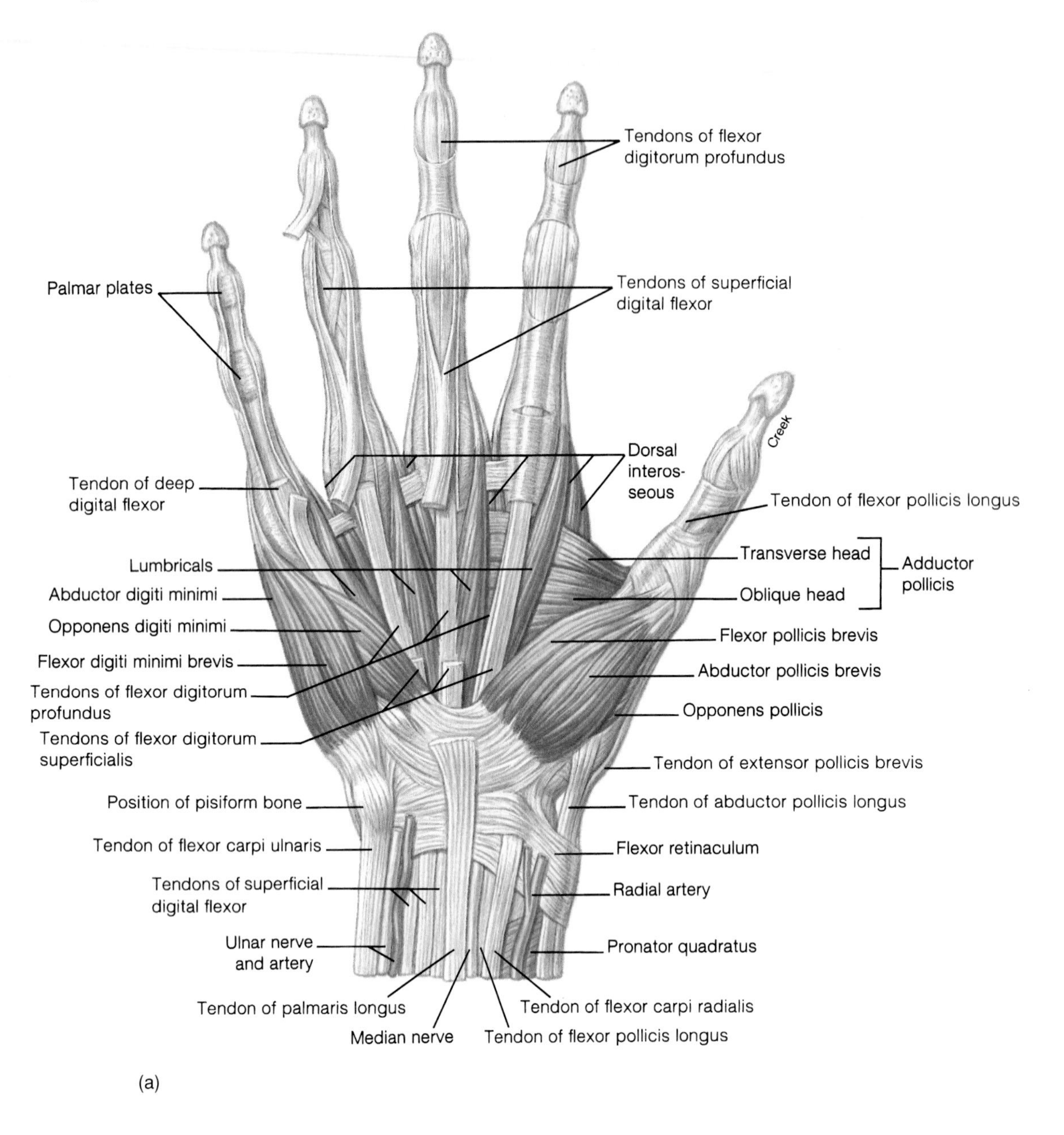

Tendons of flexor digitorum profundus
Palmar plates
Tendons of superficial digital flexor
Creek
Dorsal interosseous
Tendon of deep digital flexor
Tendon of flexor pollicis longus
Transverse head
Adductor pollicis
Lumbricals
Oblique head
Abductor digiti minimi
Opponens digiti minimi
Flexor pollicis brevis
Flexor digiti minimi brevis
Abductor pollicis brevis
Tendons of flexor digitorum profundus
Opponens pollicis
Tendons of flexor digitorum superficialis
Tendon of extensor pollicis brevis
Position of pisiform bone
Tendon of abductor pollicis longus
Tendon of flexor carpi ulnaris
Flexor retinaculum
Tendons of superficial digital flexor
Radial artery
Ulnar nerve and artery
Pronator quadratus
Tendon of palmaris longus
Tendon of flexor carpi radialis
Median nerve
Tendon of flexor pollicis longus
(a)

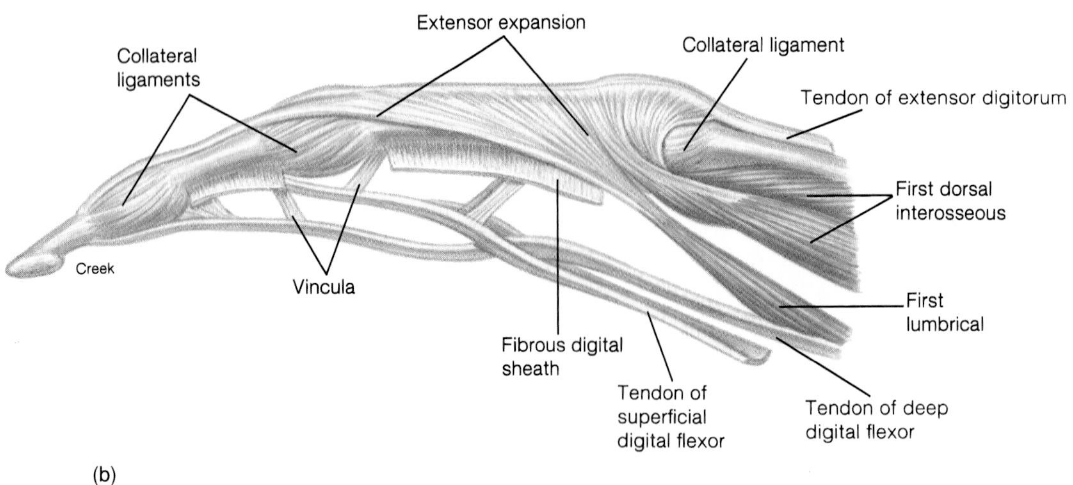

Extensor expansion
Collateral ligaments
Collateral ligament
Tendon of extensor digitorum
First dorsal interosseous
Creek
Vincula
First lumbrical
Fibrous digital sheath
Tendon of superficial digital flexor
Tendon of deep digital flexor
(b)

Figure 9.34

Muscles of the hand. (*a*) An anterior view and (*b*) a lateral view of the second digit (index finger).

Table 9.13
Intrinsic Muscles of the Hand

Muscle	Origin	Insertion	Action	Innervation
Thenar Muscles				
Abductor pollicis brevis	Flexor retinaculum, scaphoid, and trapezium	Proximal phalanx of thumb	Abducts joints of thumb	Median n.
Flexor pollicis brevis	Flexor retinaculum and trapezium	Proximal phalanx of thumb	Flexes joints of thumb	Median n.
Opponens pollicis	Trapezium and flexor retinaculum	First metacarpal bone	Opposes joints of thumb	Median n.
Intermediate Muscles				
Adductor pollicis (oblique and transverse heads)	Oblique head, capitate; transverse head, second and third metacarpal bones	Proximal phalanx of thumb	Adducts joints of thumb	Ulnar n.
Lumbricales (4)	Tendons of flexor digitorum profundus	Extensor expansions of digits II–V	Flexes digits at metacarpophalangeal joints; extends digits at interphalangeal joints	Median and ulnar nn.
Palmar interossei (3)	Medial side of second metacarpal bone; lateral sides of fourth and fifth metacarpal bones	Proximal phalanges of index, ring, and little fingers and extensor digitorum communis	Adducts fingers toward middle finger at metacarpophalangeal joints	Ulnar n.
Dorsal interossei (4)	Adjacent sides of metacarpal bones	Proximal phalanges of index and middle fingers (lateral sides) plus proximal phalanges of middle and ring fingers (medial sides) and extensor digitorum communis	Abducts fingers away from middle finger at metacarpophalangeal joints	Ulnar n.
Hypothenar Muscles				
Abductor digiti minimi	Pisiform and tendon of flexor carpi ulnaris	Proximal phalanx of digit V	Abducts joints of digit V	Ulnar n.
Flexor digiti minimi	Flexor retinaculum and hook of hamate	Proximal phalanx of digit V	Flexes joints of digit V	Ulnar n.
Opponens digiti minimi	Flexor retinaculum and hook of hamate	Fifth metacarpal bone	Opposes joints of digit V	Ulnar n.

opponens: L. *opponens*, against

The *hypothenar eminence* is the elongated, fleshy bulge at the base of the little finger. It also is formed by three muscles: the **abductor digiti minimi** muscle, the **flexor digiti minimi** muscle, and the **opponens digiti minimi** muscle.

Muscles of the intermediate group are positioned between the metacarpal bones in the region of the palm. This group includes the **adductor pollicis** muscle, the **lumbricales** (*lum′brĭ-ka′lēz*), and the **palmar** and **dorsal interossei** (*in″ter-os′e-i*) muscles.

Muscles That Move the Thigh at the Hip Joint

The muscles that move the thigh at the hip joint originate from the pelvic girdle and the vertebral column and insert on various places on the femur. These muscles stabilize a highly movable hip joint and provide support for the body during bipedal stance and locomotion. The most massive muscles of the body are found in this region, along with some extremely small muscles. The muscles that move the thigh at the hip joint are divided into anterior, posterior, and medial groups.

Anterior Muscles

The anterior muscles that move the thigh at the hip joint are the *iliacus* and *psoas major* (figs. 9.35 and 9.36).

The triangular **iliacus** (*il″e-ak′us; ĭ-li′ă-kus*) muscle arises from the iliac fossa and inserts on the lesser trochanter of the femur.

The long, thick **psoas** (*so′as*) **major** muscle originates on the bodies and transverse processes of the lumbar vertebrae; it inserts, along with the iliacus muscle, on the lesser trochanter (fig. 9.36). The psoas major and the iliacus work synergistically in flexing and rotating the hip joint and flexing the vertebral column. These two muscles are collectively called the **iliopsoas** (*il″e-o-so′as*) muscle.

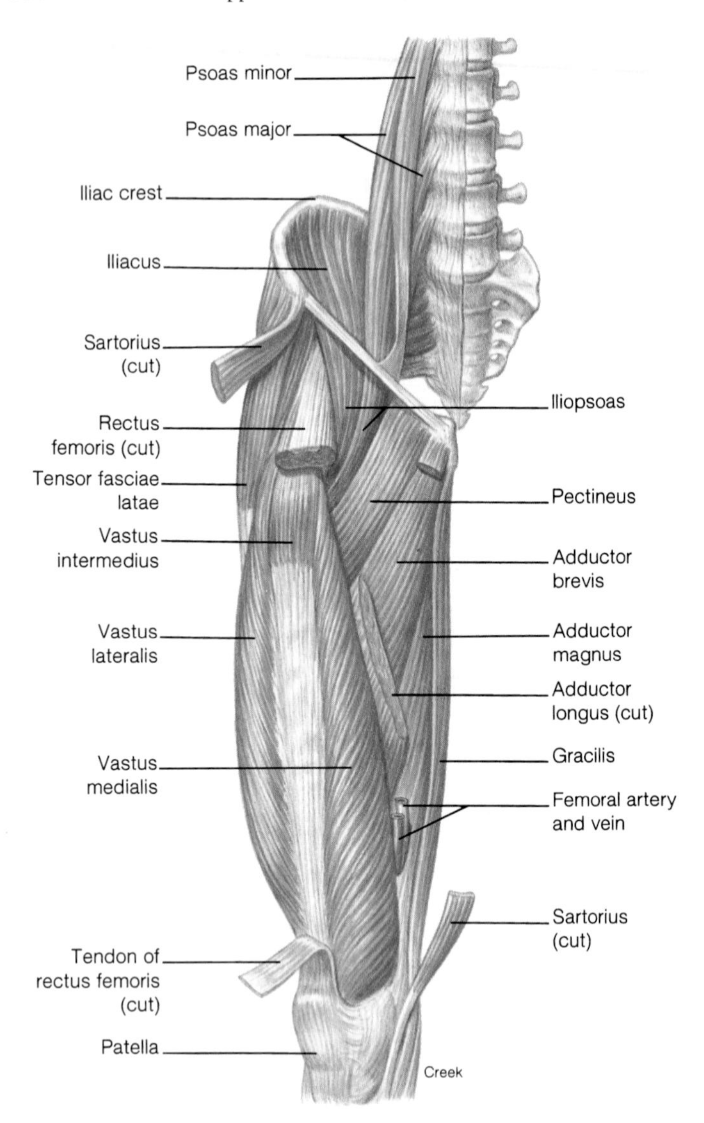

Figure 9.35

Muscles of the right anterior pelvic and thigh regions.

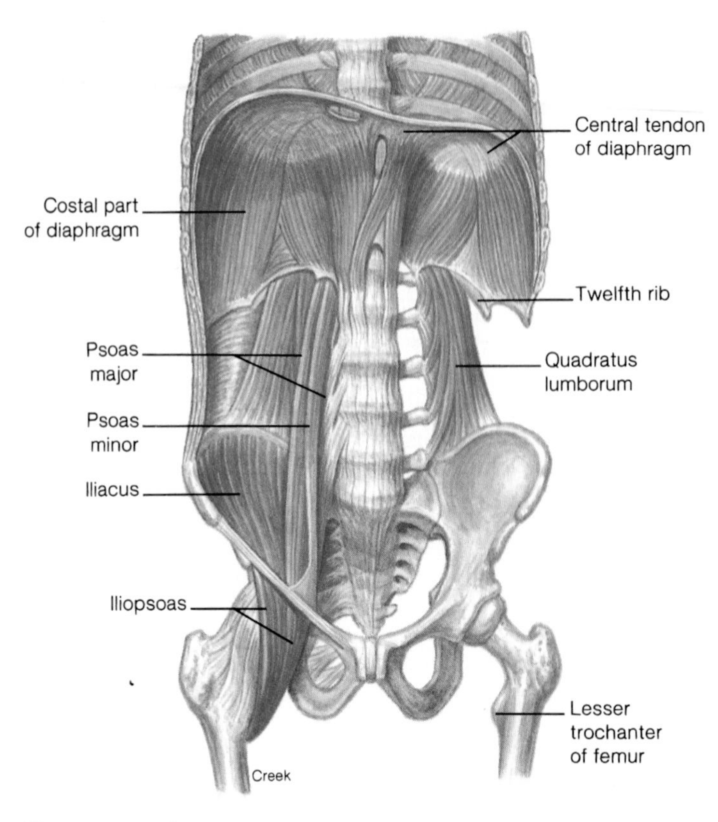

Figure 9.36

Anterior pelvic muscles that move the hip.

Posterior and Lateral (Buttock) Muscles

The posterior muscles that move the thigh at the hip joint include the *gluteus maximus*, *gluteus medius*, *gluteus minimis*, and *tensor fasciae latae*.

The large **gluteus** (*gloo'te-us*) **maximus** muscle forms much of the prominence of the buttock (figs. 9.37 and 9.40). It is a powerful extensor muscle of the hip joint and is very important for bipedal stance and locomotion. The gluteus maximus originates on the ilium, sacrum, coccyx, and aponeurosis of the lumbar region. It inserts on the gluteal tuberosity of the femur and the *iliotibial tract*, a thickened tendinous region of the fascia lata extending down the thigh (see fig. 9.39).

The **gluteus medius** muscle is located immediately deep to the gluteus maximus (fig. 9.37). It originates on the lateral surface of the ilium and inserts on the greater trochanter of the femur. The gluteus medius abducts and medially rotates the hip joint. The mass of this muscle is of clinical significance as a site for intramuscular injections.

The **gluteus minimus** muscle is the smallest and deepest of the gluteal muscles (figs. 9.37 and 9.40). It also arises from the lateral surface of the ilium, and it inserts on the lateral surface of the greater trochanter, where it acts synergistically with the gluteus medius muscle to abduct and medially rotate the hip joint.

The quadrangular **tensor fasciae latae** (*fash'e-e la'te*) muscle is positioned superficially on the lateral surface of the hip (see fig. 9.39). It originates on the iliac crest and inserts on a broad lateral fascia of the thigh called the *iliotibial tract*. The tensor fasciae latae muscle and the gluteus medius muscle are synergistic abductor muscles of the hip joint.

A deep group of six lateral rotator muscles of the hip joint is positioned directly over the posterior aspect of the hip. These muscles are not discussed here but are identified in figure 9.37 from superior to inferior as the **piriformis,** (*pĭ-rĭ-for'mis*), **superior gemellus** (*je-mel'us*), **obturator internus, inferior gemellus, obturator externus,** and **quadratus femoris** muscles.

The anterior and posterior group of muscles that move the hip joint are summarized in table 9.14.

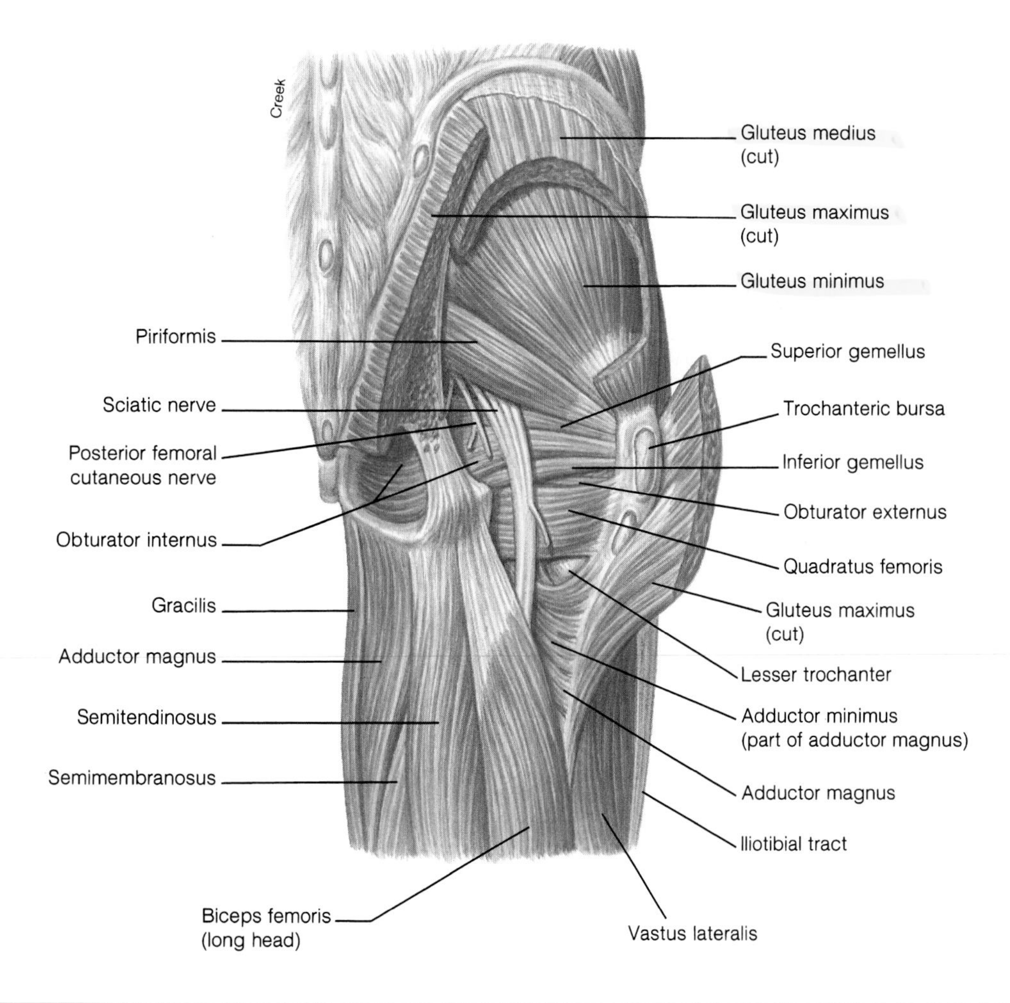

Figure 9.37

Deep gluteal muscles.

Table 9.14

Anterior and Posterior Muscles That Move the Thigh at the Hip Joint

Muscle	Origin	Insertion	Action	Innervation
Iliacus	Iliac fossa	Lesser trochanter of femur, along with psoas major	Flexes and rotates thigh laterally at the hip joint; flexes joints of vertebral column	Femoral n.
Psoas major	Transverse processes of all lumbar vertebrae	Lesser trochanter, along with iliacus	Flexes and rotates thigh laterally at the hip joint; flexes joints of vertebral column	Spinal nerves L2, L3
Gluteus maximus	Iliac crest, sacrum, coccyx, and aponeurosis of the lumbar region	Gluteal tuberosity and iliotibial tract	Extends and rotates thigh laterally at the hip joint	Inferior gluteal n.
Gluteus medius	Lateral surface of ilium	Greater trochanter	Abducts and rotates thigh medially at the hip joint	Superior gluteal n.
Gluteus minimus	Lateral surface of lower half of ilium	Greater trochanter	Abducts and rotates thigh medially at the hip joint	Superior gluteal n.
Tensor fasciae latae	Anterior border of ilium and iliac crest	Iliotibial tract	Abducts thigh at the hip joint	Superior gluteal n.

psoas: Gk. *psoa*, loin
gluteus: Gk. *gloutos*, rump

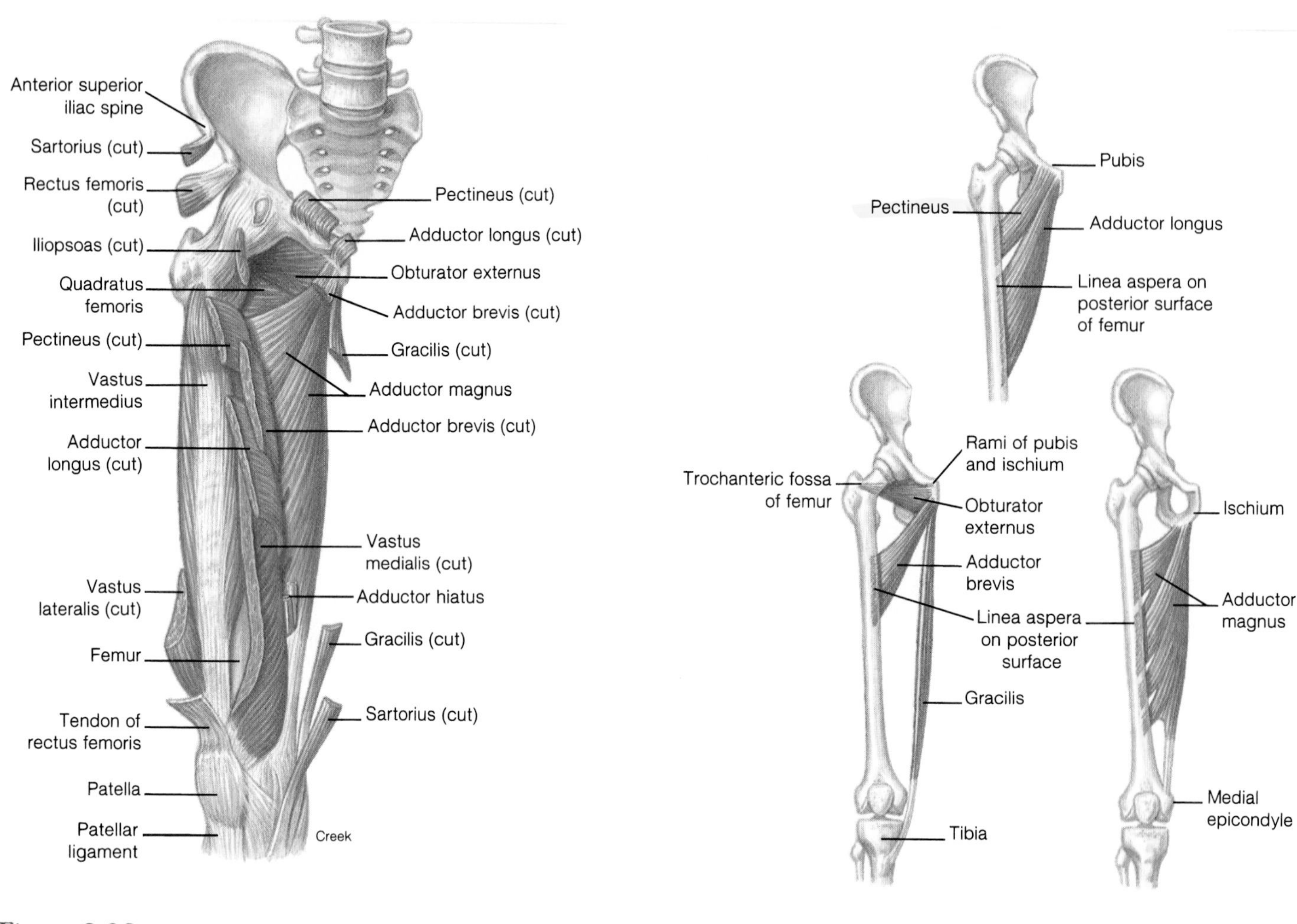

Figure 9.38

Adductor muscles of the right thigh.

Medial, or Adductor, Muscles

The medial muscles that move the hip joint include the *gracilis pectineus*, *adductor longus*, *adductor brevis*, and *adductor magnus* muscles (figs. 9.38, 9.39, 9.40, and 9.41).

The long, thin **gracilis** (*gras′ĭ-lis*) muscle is the most superficial of the medial thigh muscles. It is a two-joint muscle and can adduct the hip joint or flex the knee.

The **pectineus** (*pek-tin′e-us*) muscle is the uppermost of the medial muscles that move the hip joint. It is a flat, quadrangular muscle that flexes and adducts the hip.

The **adductor longus** muscle is located immediately lateral to the gracilis on the upper third of the thigh; it is the most anterior of the adductor muscles. The **adductor brevis** muscle is a triangular muscle located deep to the adductor longus and pectineus muscles, which largely conceal it. The **adductor magnus** muscle is a large, thick muscle, somewhat triangular in shape. It is located deep to the other two adductor muscles. The adductor longus, adductor brevis, and the adductor magnus are synergistic in adducting, flexing, and laterally rotating the hip joint.

The muscles that adduct the hip joint are summarized in table 9.15.

Muscles of the Thigh That Move the Knee Joint

The muscles that move the knee originate on the pelvic girdle or thigh. They are surrounded and compartmentalized by tough fascial sheets, which are a continuation of the fascia lata and iliotibial tract. These muscles are divided according to function and position into two groups: anterior extensors and posterior flexors.

Anterior, or Extensor, Muscles

The anterior muscles that move the knee joint are the *sartorius* and *quadriceps femoris* muscles (fig. 9.38).

The long, straplike **sartorius** (*sar′to′re-us*) muscle obliquely crosses the anterior aspect of the thigh. It can act on both the

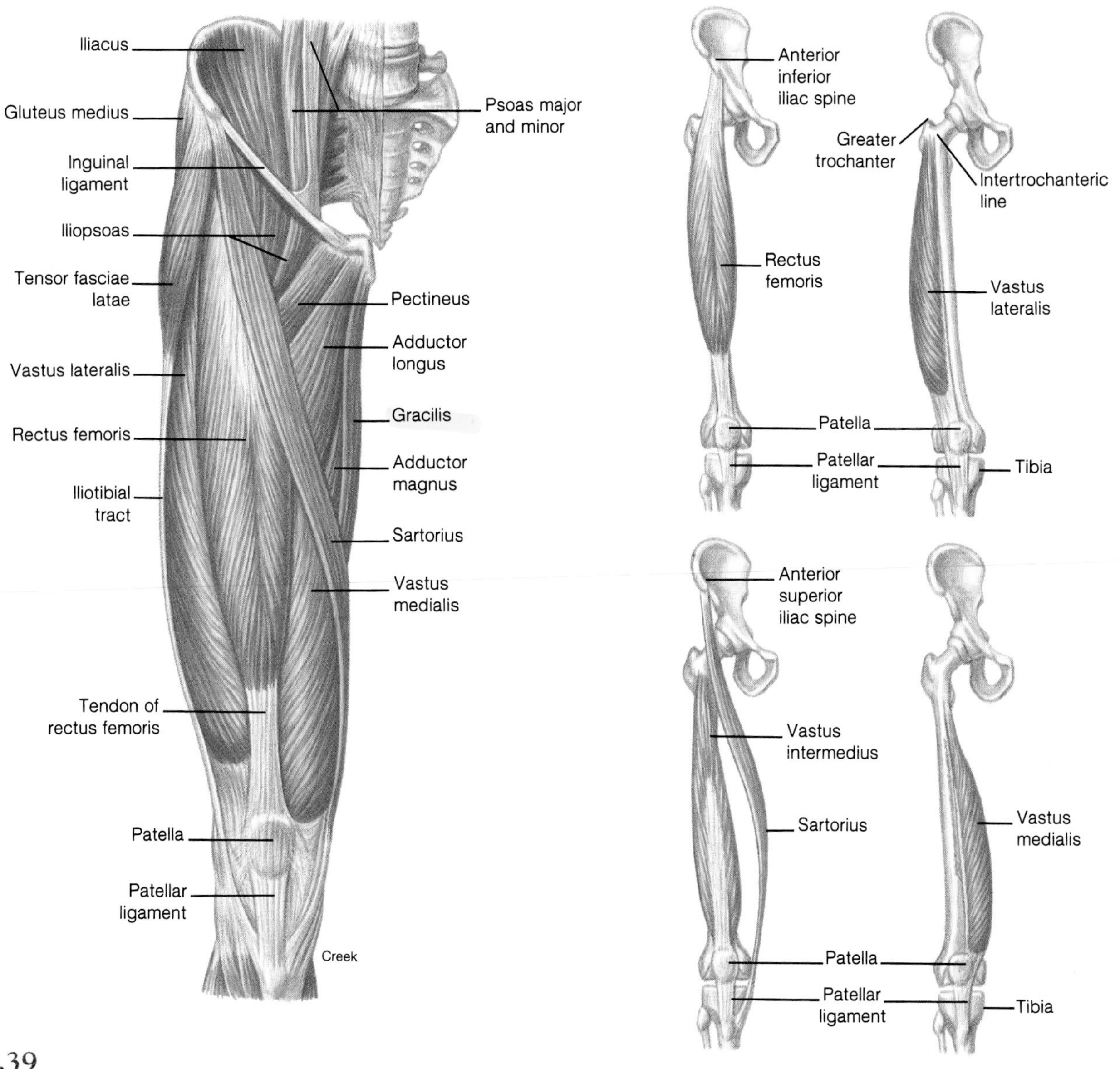

Figure 9.39

Muscles of the right anterior thigh.

Table 9.15

Medial Muscles That Move the Thigh at the Hip Joint

Muscle	Origin	Insertion	Action	Innervation
Gracilis	Inferior edge of symphysis pubis	Proximal medial surface of tibia	Adducts thigh at hip joint; flexes and rotates leg at knee joint	Obturator n.
Pectineus	Pectineal line of pubis	Distal to lesser trochanter of femur	Adducts and flexes thigh at hip joint	Femoral n.
Adductor longus	Pubis—below pubic crest	Linea aspera of femur	Adducts, flexes, and laterally rotates thigh at hip joint	Obturator n.
Adductor brevis	Inferior ramus of pubis	Linea aspera of femur	Adducts, flexes, and laterally rotates thigh at hip joint	Obturator n.
Adductor magnus	Inferior ramus of ischium and pubis	Linea aspera and medial epicondyle of femur	Adducts, flexes, and laterally rotates thigh at hip joint	Obturator and tibial nn.

gracilis: Gk. *gracilis*, slender

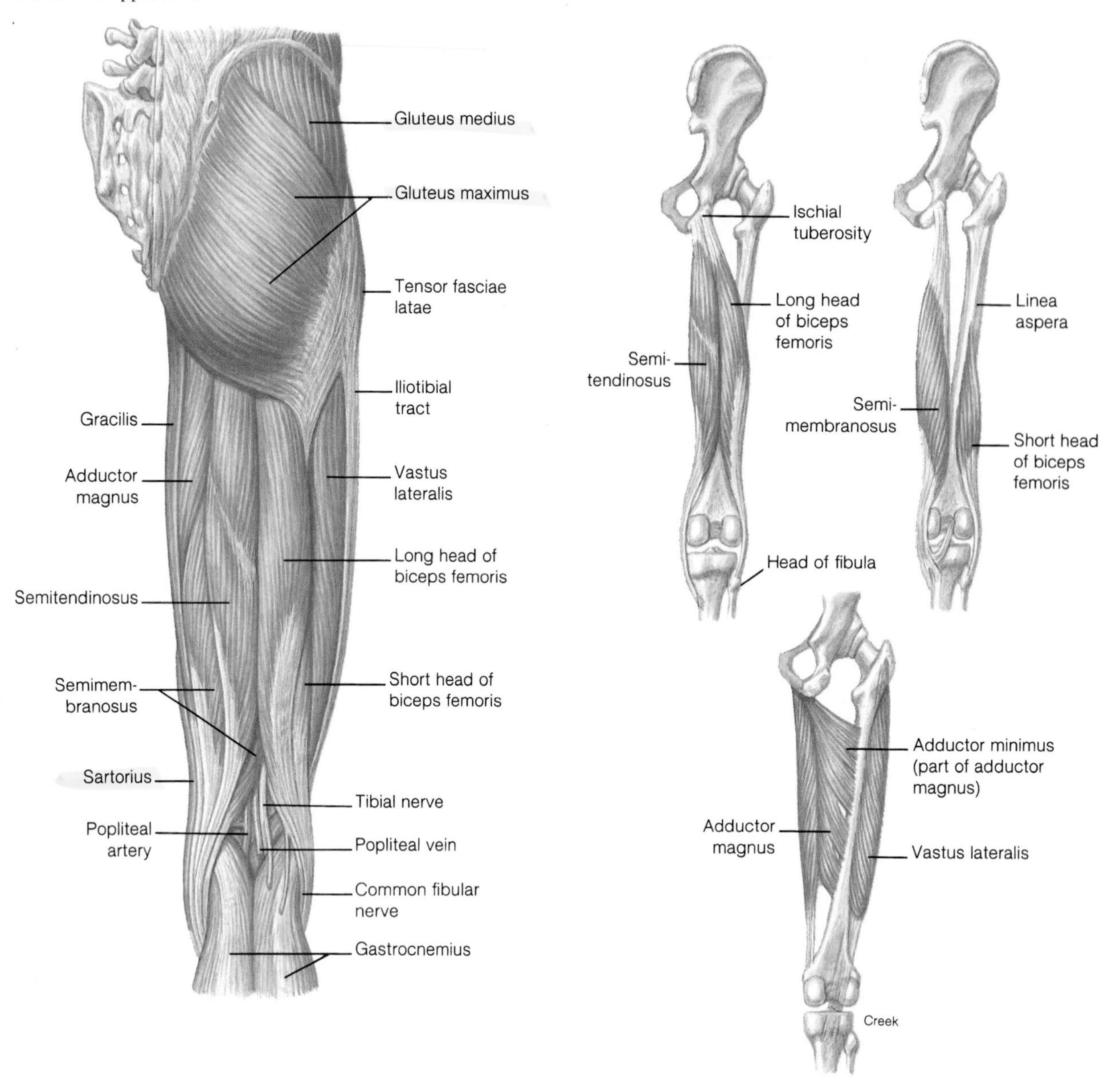

Figure 9.40

Muscles of the right posterior thigh.

hip and knee joints to flex and rotate the hip laterally, and also to assist in flexing the knee joint and rotating it medially. The sartorius is the longest muscle of the body. It is frequently called the "tailor's muscle" because it helps to effect the cross-legged sitting position in which tailors are often depicted.

The **quadriceps femoris** muscle is actually a composite of four distinct muscles that have separate origins but a common insertion on the patella via the tendon of the rectus femoris muscle. The tendon of the rectus femoris is continuous over the patella and becomes the *patellar ligament* as it attaches to the tibial tuberosity (fig. 9.39). These muscles function synergistically to extend the knee, as in kicking a football. The four muscles of the quadriceps femoris muscle are the *rectus femoris*, *vastus lateralis*, *vastus medialis*, and *vastus intermedius*.

The **rectus femoris** muscle occupies a superficial position and is the only one of the four quadriceps that functions in both the hip and knee joints. The laterally positioned **vastus lateralis** is the largest muscle of the quadriceps femoris. It is a common intramuscular injection site in infants who have small, underdeveloped buttock and shoulder muscles. The **vastus medialis** muscle occupies a medial position along the thigh. The **vastus intermedius** muscle lies deep to the rectus femoris.

The anterior thigh muscles that move the knee joint are summarized in table 9.16.

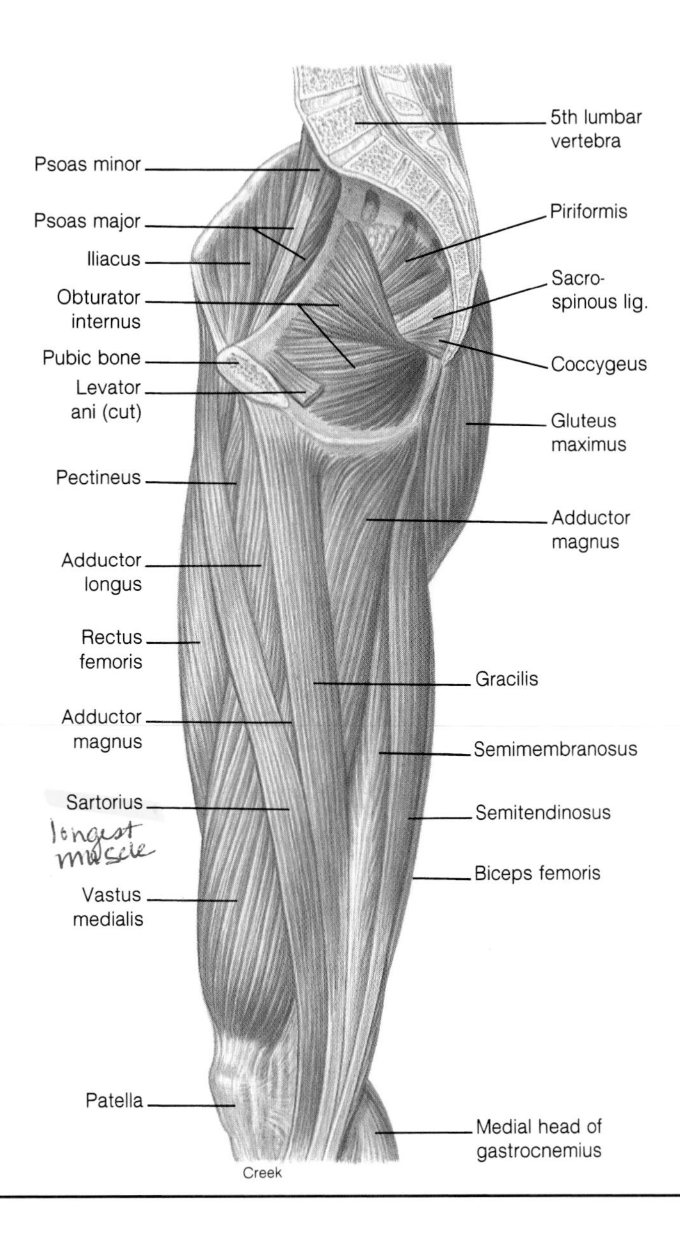

Figure 9.41

Muscles of the right medial thigh.

Table 9.16

Anterior Thigh Muscles That Move the Knee Joint

Muscle	Origin	Insertion	Action	Innervation
Sartorius	Anterior superior iliac spine	Medial surface of tibia	Flexes knee and hip joints, abducts hip joint, rotates thigh laterally at hip joint, and rotates thigh medially at hip joint	Femoral n.
Quadriceps femoris		Patella by common tendon, which continues as patellar ligament to tibial tuberosity	Extends leg at knee joint	Femoral n.
Rectus femoris	Anterior superior iliac spine and lip of acetabulum			
Vastus lateralis	Greater trochanter and linea aspera of femur			
Vastus medialis	Medial surface and linea aspera of femur			
Vastus intermedius	Anterior and lateral surfaces of femur			

sartorius: L. *sartor*, a tailor (muscle used to cross legs in a tailor's position)

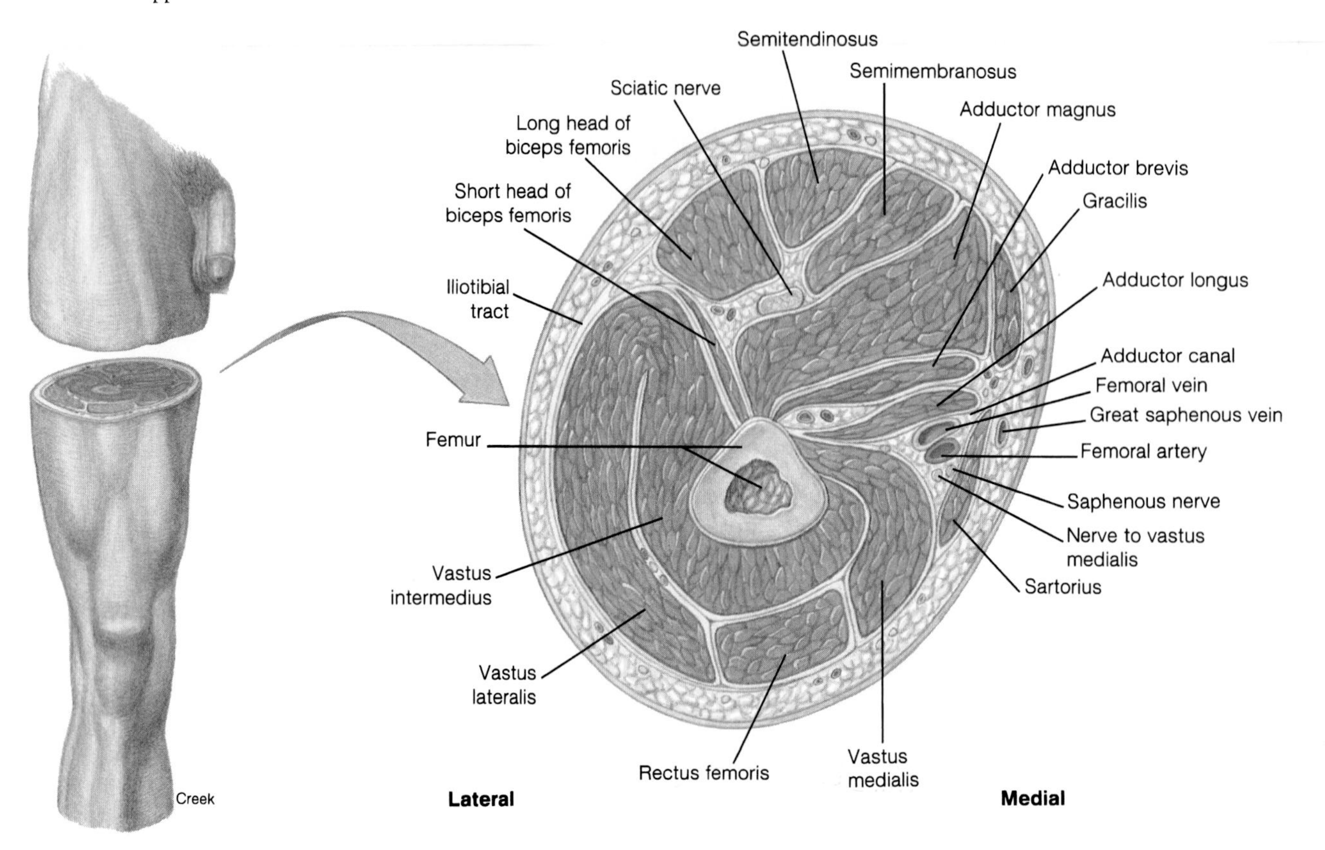

Figure 9.42

A transverse section of the right thigh as seen from above. (Note the position of the vessels and nerves.)

Table 9.17

Posterior Thigh Muscles That Move the Knee Joint

Muscle	Origin	Insertion	Action
Biceps femoris	Long head—ischial tuberosity; short head—linea aspera of femur	Head of fibula and lateral epicondyle of tibia	Flexes knee joint; extends and laterally rotates thigh at hip joint
Semitendinosus	Ischial tuberosity	Proximal portion of medial surface of body of tibia	Flexes knee joint; extends and medially rotates thigh at hip joint
Semimembranosus	Ischial tuberosity	Medial epicondyle of tibia	Flexes knee joint; extends and medially rotates thigh at hip joint

Posterior, or Flexor, Muscles

There are three posterior thigh muscles, which are antagonistic to the quadriceps femoris muscles in flexing the knee joint. These muscles are known as the **hamstrings** (fig. 9.40). The name derives from the butchers' practice of using the tendons of these muscles at the knee of a hog to hang a ham for curing.

The **biceps femoris** muscle occupies the posterior lateral aspect of the thigh. It has a superficial long head and a deep short head, and causes movement at both the hip and knee joints. The superficial **semitendinosus** muscle is fusiform and is located on the posterior medial aspect of the thigh. It also works over two joints. The flat **semimembranosus** muscle lies deep to the semitendinosus on the posterior medial aspect of the thigh.

The posterior thigh muscles that move the leg at the knee joint are summarized in table 9.17. The relative positions of the muscles of the thigh are illustrated in figure 9.42.

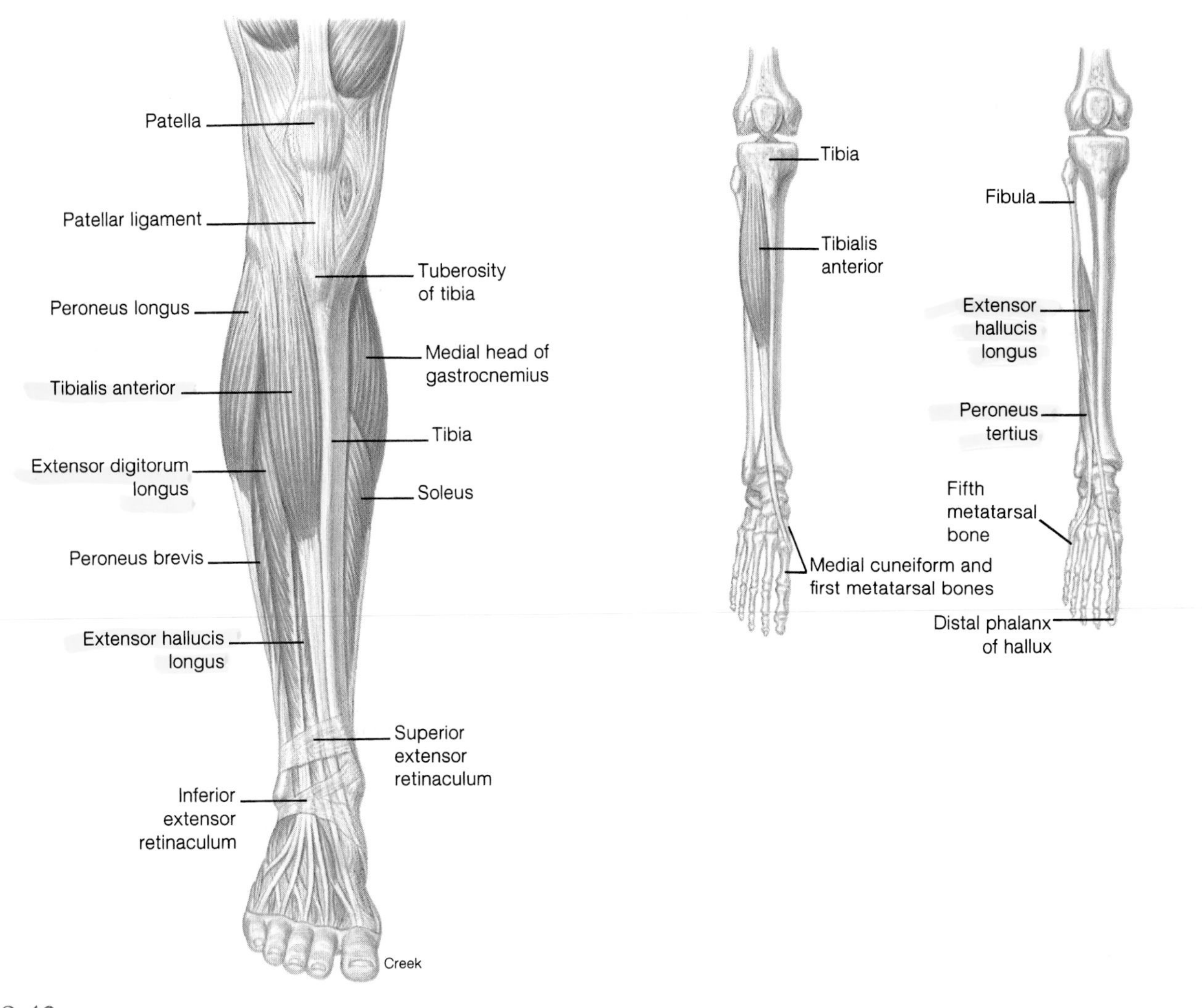

Figure 9.43

Anterior crural muscles.

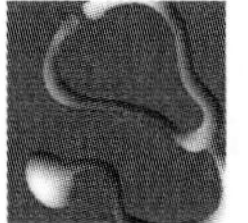

Hamstring injuries are a common occurrence in some sports. The injury usually occurs when sudden lateral or medial stress to the knee joint tears the muscles or tendons. Because of its structure and the stress applied to it in competition, the knee joint is highly susceptible to injury. Altering the rules in contact sports could reduce the incidence of knee injury. At the least, additional support and protection should be provided for this vulnerable joint.

Muscles of the Leg That Move the Joints of the Ankle, Foot, and Toes

The muscles of the leg, the **crural** muscles, are responsible for the movements of the foot. There are three groups of crural muscles: anterior, lateral, and posterior. The anteromedial aspect of the leg along the body of the tibia lacks muscle attachment.

Anterior Crural Muscles

The anterior crural muscles include the *tibialis anterior*, *extensor digitorum longus*, *extensor hallucis longus*, and *peroneus tertius* muscles (figs. 9.43, 9.44, and 9.45).

The large, superficial **tibialis anterior** muscle can be easily palpated on the anterior lateral portion of the tibia (fig. 9.43). It parallels the prominent anterior crest of the tibia. The **extensor digitorum longus** muscle is positioned lateral to the tibialis anterior on the anterolateral surface of the leg. The **extensor hallucis** (*hă-loo'sis*) **longus** muscle is positioned deep between the tibialis anterior muscle and the extensor digitorum longus muscle. The

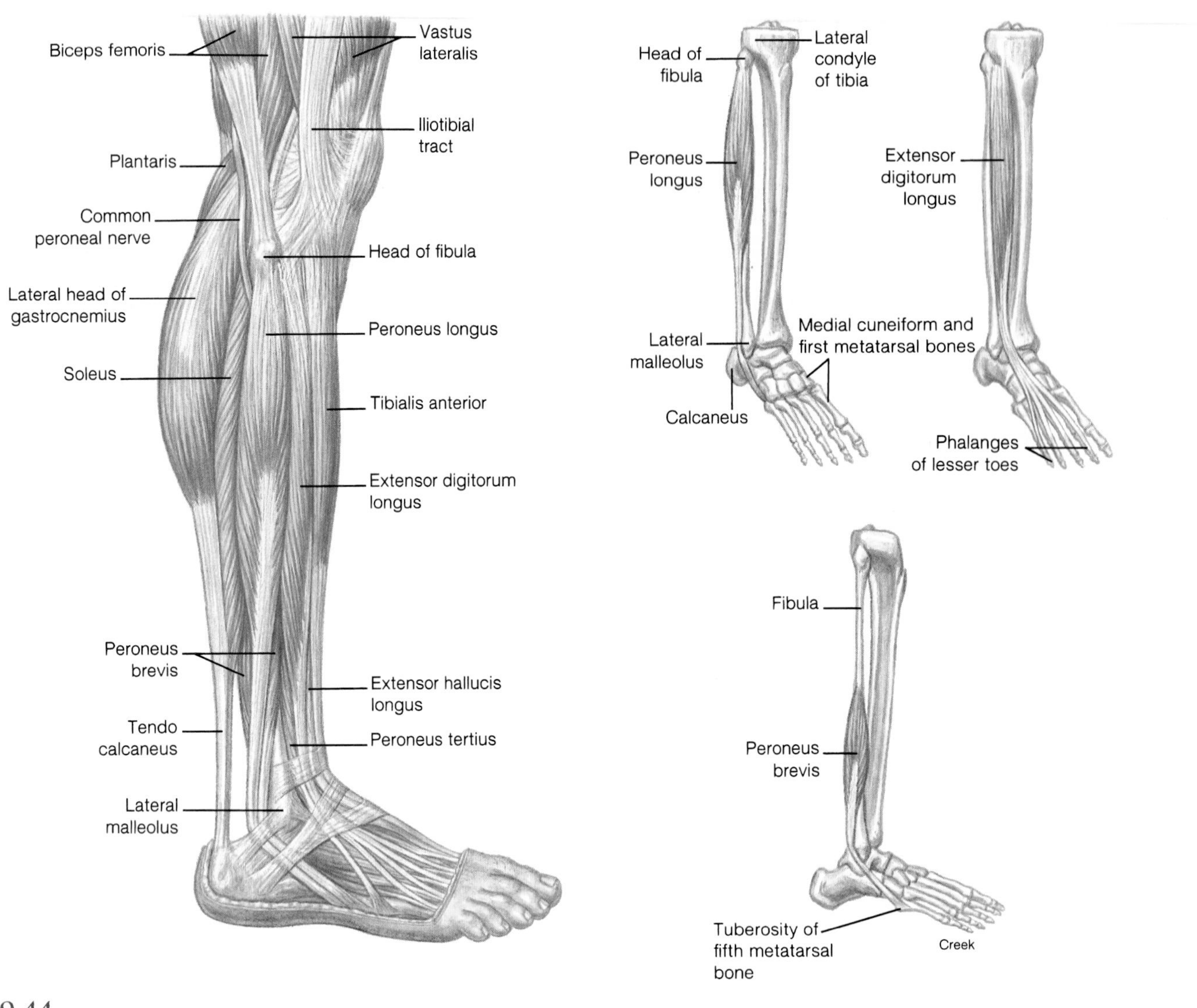

Figure 9.44

Lateral crural muscles.

small **peroneus tertius** muscle is continuous with the distal portion of the extensor digitorum longus muscle.

Lateral Crural Muscles

The lateral crural muscles are the *peroneus longus* and *peroneus brevis* (figs. 9.43 and 9.44).

The long, flat **peroneus longus** muscle is a superficial lateral muscle that overlies the fibula. The **peroneus brevis** muscle lies deep to the peroneus longus and is positioned closer to the foot. These two muscles are synergistic in flexing the ankle joint and everting the foot (see table 9.18).

Posterior Crural Muscles

The seven posterior crural muscles can be grouped into a superficial and a deep group. The superficial group is composed of the *gastrocnemius*, *soleus*, and *plantaris muscles* (fig. 9.46). The four deep posterior crural muscles are the *popliteus*, *flexor hallucis longus*, *flexor digitorum longus*, and *tibialis posterior muscles* (fig. 9.47).

The **gastrocnemius** (*gas″trok-ne′me-us*) muscle is a large superficial muscle that forms the major portion of the calf of the leg. It consists of two distinct heads that arise from the posterior surfaces of the medial and lateral epicondyles of the femur. This muscle and the deeper soleus muscle insert onto the calcaneus via the common *tendo calcaneus (tendon of Achilles)*. This is the strongest tendon in the body, but it is frequently ruptured from sudden stress during athletic competition. The gastrocnemius acts over two joints to cause flexion of the knee joint and plantar flexion of the foot at the ankle joint.

The **soleus** muscle lies deep to the gastrocnemius. These two muscles are frequently referred to as a single muscle, the **triceps surae** (*sur′e*). The soleus and gastrocnemius muscles have a

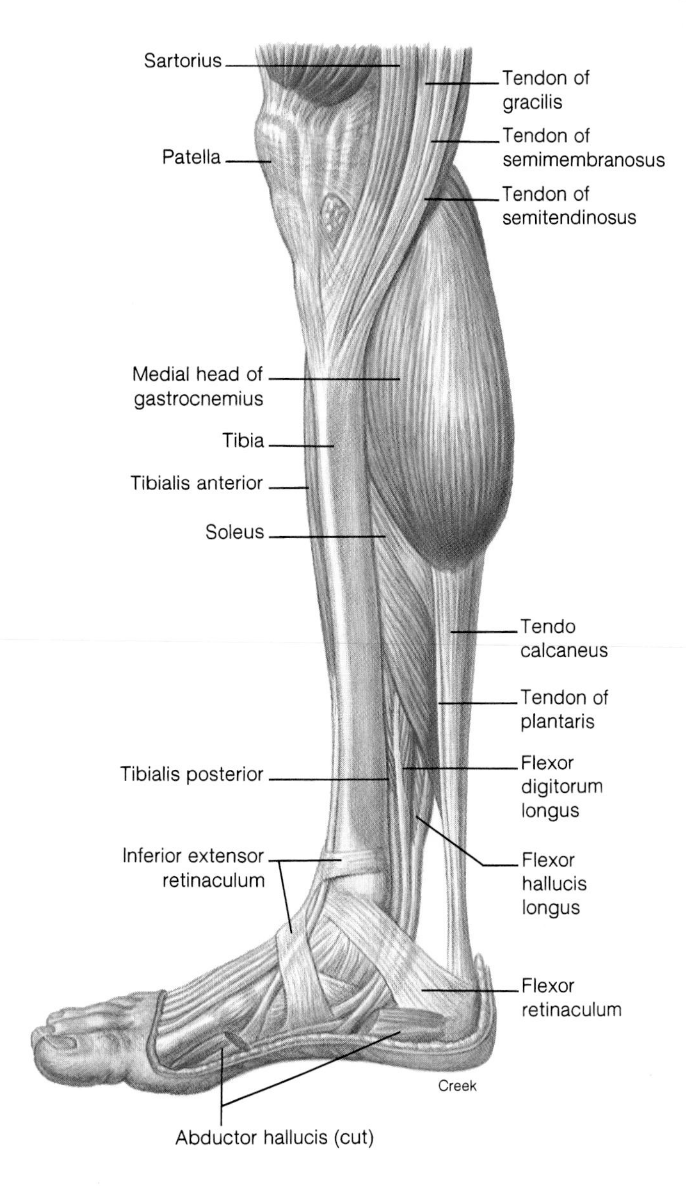

Figure 9.45

A medial view of the crural muscles.

common insertion, but the soleus acts on only the ankle joint, in plantar flexing the foot.

The small **plantaris** muscle arises just superior to the origin of the lateral head of the gastrocnemius muscle on the lateral supracondylar ridge of the femur. It has a very long, slender tendon of insertion onto the calcaneus. The tendon of this muscle is frequently mistaken for a nerve by those dissecting it for the first time. The plantaris is a weak muscle, with limited ability to flex the knee and plantar flex the ankle joint.

The thin, triangular **popliteus** (*pop-lit'e-us*) muscle is situated deep to the heads of the gastrocnemius muscle, where it forms part of the floor of the *popliteal fossa*—the depression on the back side of the knee joint (fig. 9.48). The popliteus muscle is a medial rotator of the tibia on the femur. The bipennate **flexor hallucis longus** muscle lies deep to the soleus muscle on the posterolateral side of the leg. It flexes the joints of the great toe (hallux) and assists in plantar flexing ankle joint and inverting the foot.

The **flexor digitorum longus** muscle also lies deep to the soleus, and it parallels the flexor hallucis longus muscle on the medial side of the leg. Its distal tendon passes posterior to the medial malleolus and continues along the plantar surface of the foot, where it branches into four tendinous slips that attach to the bases of the distal phalanges of the second, third, fourth, and fifth digits (fig. 9.49). The flexor digitorum longus works over several joints, flexing the joints in four of the digits and assisting in plantar flexing the ankle joint and inverting the foot.

The **tibialis posterior** muscle is located deep to the soleus muscle, between the posterior flexors. Its distal tendon passes behind the medial malleolus and inserts on the plantar surfaces of the navicular, cuneiform, and cuboid bones, and

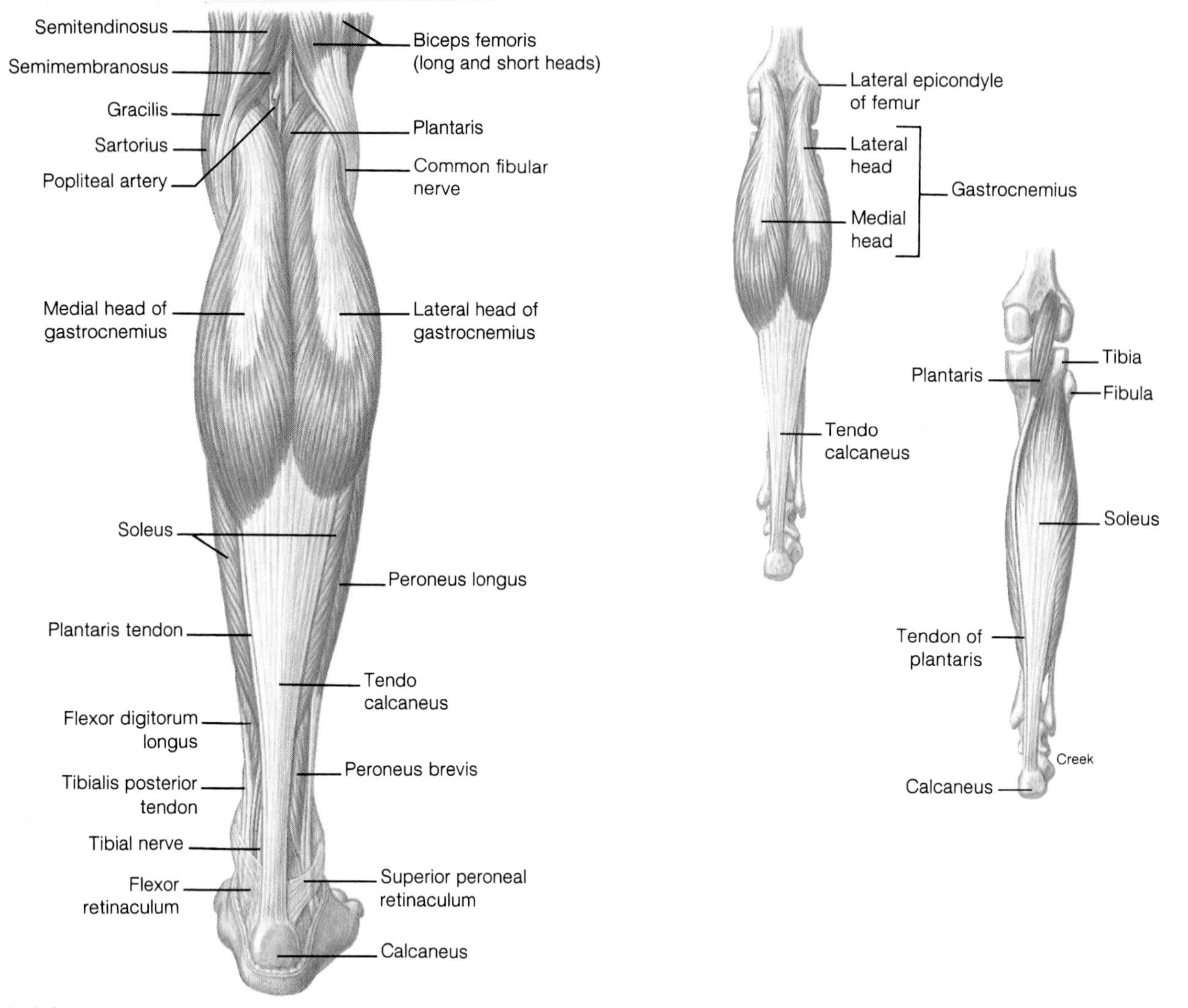

Figure 9.46

Posterior crural muscles and the popliteal region.

the second, third, and fourth metatarsal bones (fig. 9.49). The tibialis posterior plantar flexes the ankle joint, inverts the foot, and lends support to the arches of the foot.

The crural muscles are summarized in table 9.18.

Muscles of the Foot

With the exception of one additional intrinsic muscle, the **extensor digitorum brevis,** the muscles of the foot are similar in name and number to those of the hand. The functions of the muscles of the foot are different, however, because the foot is adapted to provide support while bearing body weight rather than to grasp objects.

The muscles of the foot can be grouped into four layers (fig. 9.49), but these are difficult to dissociate, even in dissection. The muscles function either to move the toes or to support the arches of the foot through their contraction. Because of their complexity, the muscles of the foot will be presented only in illustrations (see figs. 9.49 and 9.50).

1. List all the muscles that either originate from or insert on the scapula.
2. On the basis of function, categorize the muscles of the upper extremity as flexors, extensors, abductors, adductors, or rotators. (Each muscle may fit into two or more categories.)
3. Which muscles of the lower extremity span two joints, and therefore have two different actions?

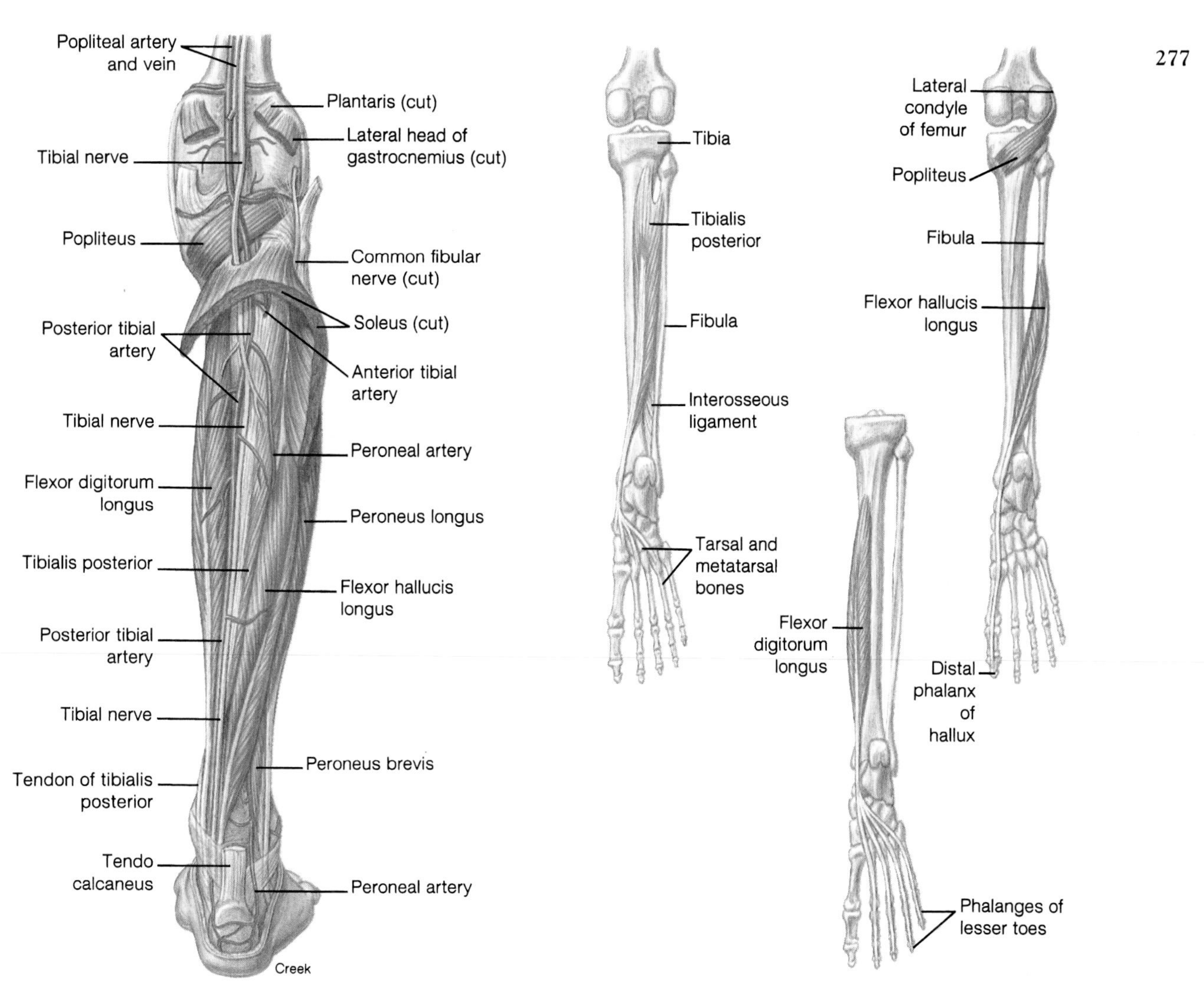
Popliteal artery and vein
Plantaris (cut)
Lateral head of gastrocnemius (cut)
Tibial nerve
Popliteus
Common fibular nerve (cut)
Soleus (cut)
Posterior tibial artery
Anterior tibial artery
Tibial nerve
Peroneal artery
Flexor digitorum longus
Peroneus longus
Tibialis posterior
Flexor hallucis longus
Posterior tibial artery
Tibial nerve
Peroneus brevis
Tendon of tibialis posterior
Tendo calcaneus
Peroneal artery
Creek
Tibia
Tibialis posterior
Fibula
Interosseous ligament
Tarsal and metatarsal bones
Lateral condyle of femur
Popliteus
Fibula
Flexor hallucis longus
Flexor digitorum longus
Distal phalanx of hallux
Phalanges of lesser toes

Figure 9.47

Deep posterior crural muscles.

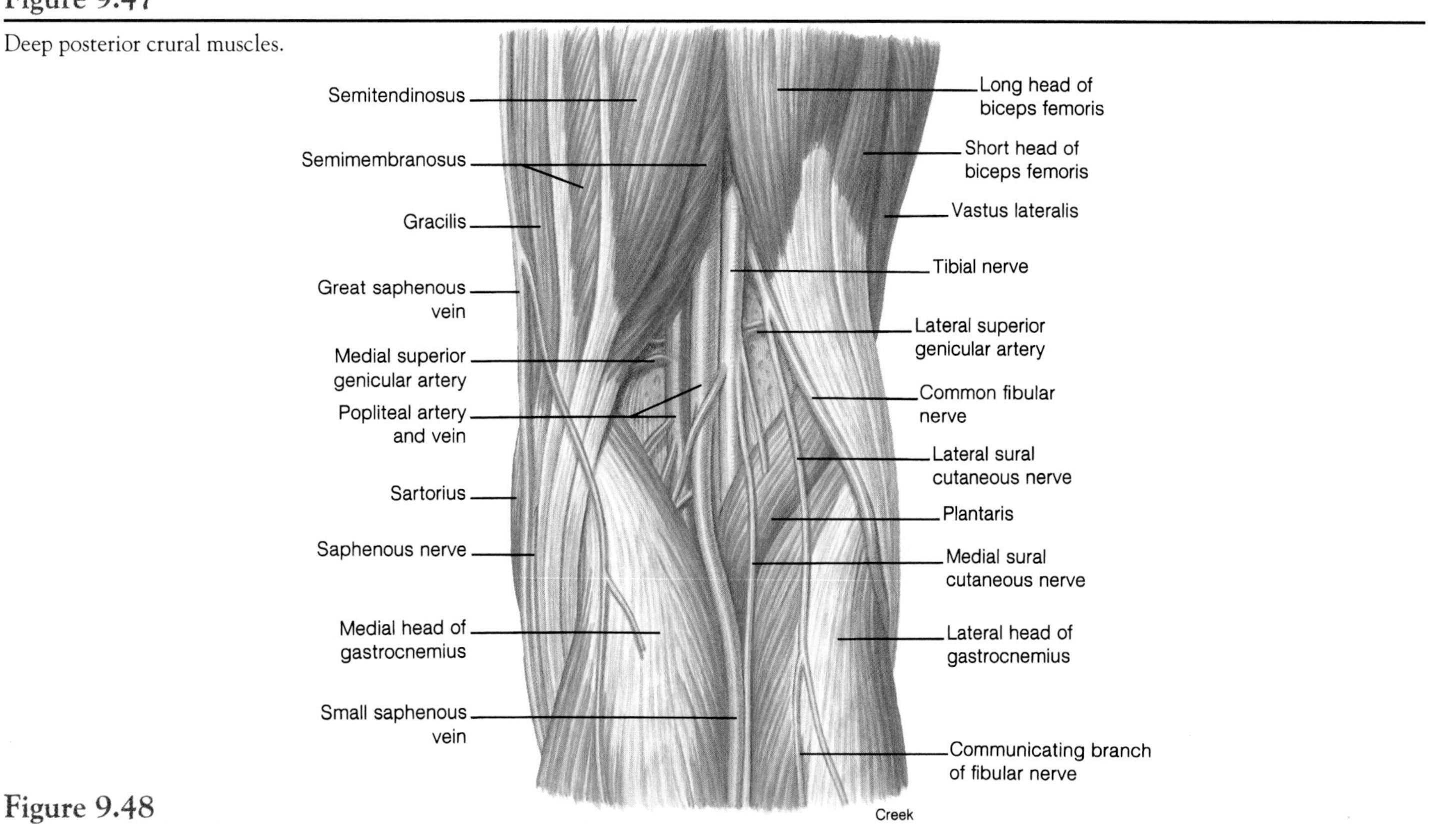
Semitendinosus
Long head of biceps femoris
Semimembranosus
Short head of biceps femoris
Gracilis
Vastus lateralis
Tibial nerve
Great saphenous vein
Lateral superior genicular artery
Medial superior genicular artery
Common fibular nerve
Popliteal artery and vein
Lateral sural cutaneous nerve
Sartorius
Plantaris
Saphenous nerve
Medial sural cutaneous nerve
Medial head of gastrocnemius
Lateral head of gastrocnemius
Small saphenous vein
Communicating branch of fibular nerve
Creek

Figure 9.48

Muscles that surround the popliteal fossa.

Table 9.18

Muscles of the Leg That Move the Joints of the Ankle, Foot, and Toes

Muscle	Origin	Insertion	Action	Innervation
Tibialis anterior	Lateral epicondyle and body of tibia	First metatarsal bone and first cuneiform	Dorsiflexes ankle and inverts foot at ankle	Deep fibular n.
Extensor digitorum longus	Lateral epicondyle of tibia and anterior surface of fibula	Extensor expansions of digits II–V	Extends joints of digits II–V and dorsiflexes foot at ankle	Deep fibular n.
Extensor hallucis longus	Anterior surface of fibula and interosseous membrane	Distal phalanx of digit I	Extends joints of great toe and assists dorsiflexion of foot at ankle	Deep fibular n.
Peroneus tertius	Anterior surface of fibula and interosseous membrane	Dorsal surface of fifth metatarsal bone	Dorsiflexes and everts foot at ankle	Deep fibular n.
Peroneus longus	Lateral epicondyle of tibia and head and body of fibula	First cuneiform and metatarsal bone I	Plantar flexes and everts foot at ankle	Superficial fibular n.
Peroneus brevis	Lower aspect of fibula	Metatarsal bone V	Plantar flexes and everts foot at ankle	Superficial fibular n.
Gastrocnemius	Lateral and medial epicondyle of femur	Posterior surface of calcaneus	Plantar flexes foot at ankle; flexes knee joint	Tibial n.
Soleus	Posterior aspect of fibula and tibia	Calcaneus	Plantar flexes foot at ankle	Tibial n.
Plantaris	Lateral supracondylar ridge of femur	Calcaneus	Plantar flexes foot at ankle	Tibial n.
Popliteus	Lateral condyle of femur	Upper posterior aspect of tibia	Flexes and medially rotates leg at knee joint	Tibial n.
Flexor hallucis longus	Posterior aspect of fibula	Distal phalanx of great toe	Flexes joint of distal phalanx of great toe	Tibial n.
Flexor joints of digitorum longus	Posterior surface of tibia	Distal phalanges of digits II–V	Flexes joints of distal phalanges of digits II–V	Tibial n.
Tibialis posterior	Tibia and fibula and interosseous membrane	Navicular, cuneiform, cuboid, and metatarsal bones II–IV	Plantar flexes and inverts foot at ankle; supports arches	Tibial n.

hallucis: L. *hallus*, great toe
peroneus tertius: Gk. *perone*, fibula; *tertius*, third
gastrocnemius: Gk. *gaster*, belly; *kneme*, leg
soleus: L. *soleus*, sole of foot
popliteus: L. *poples*, ham of the knee

Clinical Considerations

Compared to the other systems of the body, the muscular system is extremely durable. If properly conditioned, the muscles of the body can adequately serve a person for a lifetime. Muscles are capable of doing incredible amounts of work; through exercise, they can become even stronger.

Clinical considerations include evaluation of muscle condition, functional conditions in muscles, diseases of muscles, and aging of muscles.

Evaluation of Muscle Condition

The clinical symptoms of muscle diseases include weakness, loss of muscle mass (atrophy), and pain. The most obvious diagnostic procedure is a clinical examination of the patient. Following this, it may be necessary to test muscle function using **electromyography (EMG)** to measure conduction rates and motor unit activity within a muscle. Laboratory tests may include serum enzyme assays or muscle biopsies. A biopsy is perhaps the most definitive diagnostic tool. Progressive atrophy, polymyositis, and metabolic diseases of muscles can be determined through a biopsy.

Functional Conditions in Muscles

Muscles depend on systematic periodic contraction to maintain optimal health. Obviously, overuse or disease will cause a change in muscle tissue. The immediate effect of overexertion on muscle tissue is the accumulation of lactic acid, which results in fatigue and soreness. Excessive contraction of a muscle can also damage the fibers or associated connective tissue, resulting in a **strained muscle.**

A **cramp** within a muscle is an involuntary, painful, prolonged contraction. Cramps can occur while muscles are in use or

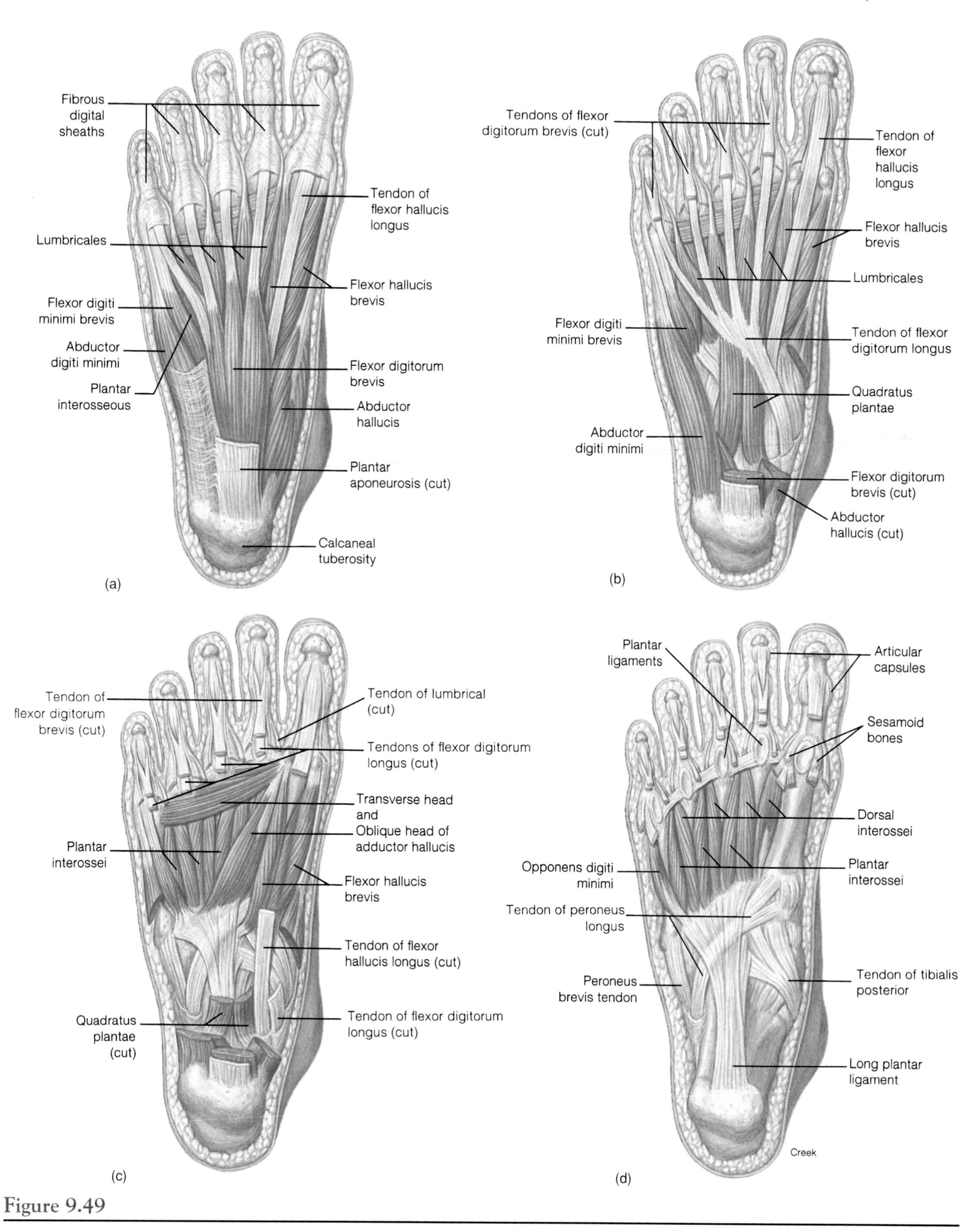

Figure 9.49

The four musculotendinous layers of the plantar aspect of the foot: (*a*) superficial layer, (*b*) second layer, (*c*) third layer, and (*d*) deep layer.

DEVELOPMENTAL EXPOSITION

The Muscular System

Explanation

The formation of skeletal muscle tissue begins during the fourth week of embryonic development as specialized mesodermal cells called *myoblasts* begin rapid mitotic division (exhibit I). The proliferation of new cells continues while the myoblasts migrate and fuse together into *syncytial myotubes*. (A *syncytium* [*sin-sish'e-um*] is a multinucleated protoplasmic mass formed by the union of originally separate cells.) At 9 weeks, primitive myofilaments course through the myotubes, and the nuclei of the contributing myoblasts are centrally located. Growth in length continues through the addition of myoblasts.

The process of muscle fiber development occurs within specialized mesodermal masses called *myotomes* in the embryonic trunk area and from loosely organized masses of mesoderm in the head and appendage areas. At 6 weeks, the trunk of an embryo is segmented into distinct myotomes (exhibit II) that are associated dorsally with specific *sclerotomes*—paired masses of mesenchymal tissue that give rise to vertebrae and ribs. As will be explained in chapter 12, spinal nerves arise from the spinal cord and exit between vertebrae to innervate developing muscles in the adjacent myotomes. As myotomes develop, additional myoblasts migrate ventrally, toward the midline of the body, or distally, into the developing limbs. The muscles of the entire muscular system have been differentiated and correctly positioned by the eighth week. The orientation of the developing muscles is preceded and influenced by cartilaginous models of bones.

It is not certain when skeletal muscle is sufficiently developed to sustain contractions, but by week 17 the fetal movements known as *quickening* are strong enough to be recognized by the mother. The individual muscle fibers have now thickened, the nuclei have moved peripherally, and the filaments (myofilaments) can be recognized as alternating dark and light bands. Growth in length still continues through addition of myoblasts. Shortly before a baby is born, the formation of myoblast cells ceases, and all of the muscle cells have been determined. Differences in strength, endurance, and coordination are somewhat genetically determined but are primarily the result of individual body conditioning. Muscle coordination is an ongoing process of achieving a fine neural control of muscle fibers. Mastery of muscle movement is comparatively slow in humans. It is several months before a human infant has the coordination to crawl, and about a year before it can stand or walk. By contrast, most mammals can walk and run within a few hours after they are born.

myoblast: Gk. *myos*, muscle; *blastos*, germ
syncytial: Gk. *syn*, with; *cyto*, cell

Exhibit I The development of skeletal muscle fibers. (*a*) At 5 weeks, the myotube is formed as individual cell membranes are broken down. Myotubes grow in length by incorporating additional myoblasts; each adds an additional nucleus. (*b*) Muscle fibers are distinct at 9 weeks, but the nuclei are still centrally located, and growth in length continues through the addition of myoblasts. (*c*) At 5 months, thin (actin) and thick (myosin) myofilaments are present and moderate growth in length still continues. (*d*) By birth, the striated myofilaments have aggregated into bundles, the fiber has thickened, and the nuclei have shifted to the periphery. Myoblast activity ceases and all the muscle fibers a person will have are formed. (*e*) The appearance of a mature muscle fiber.

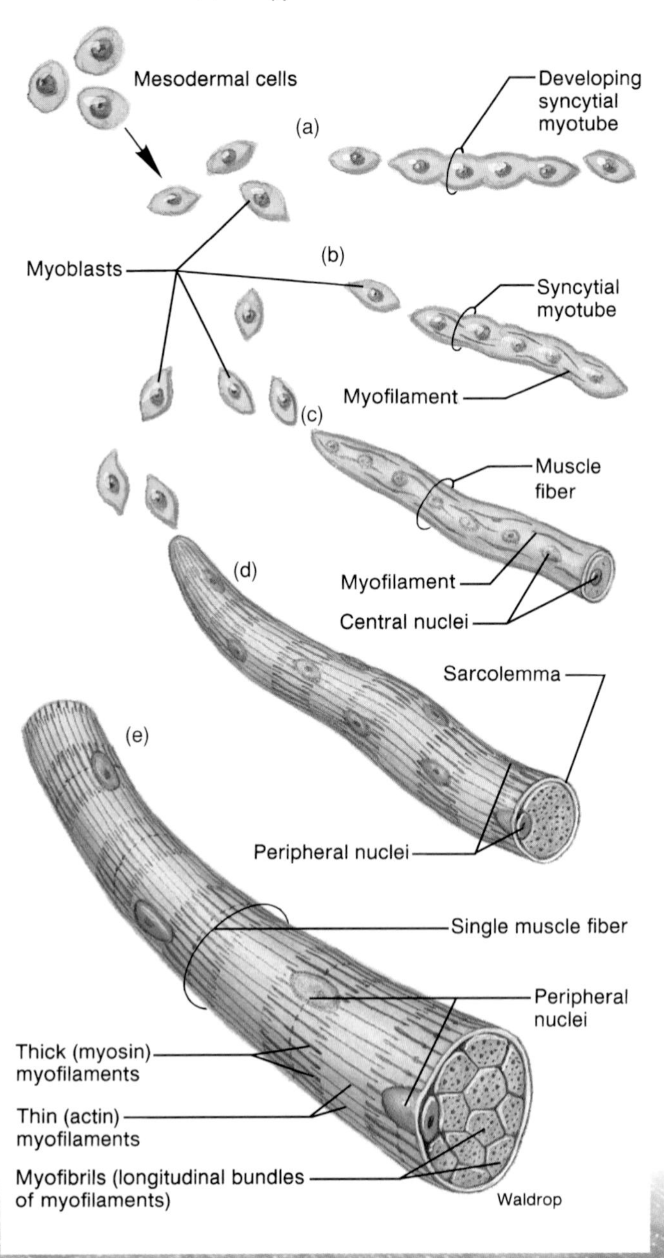

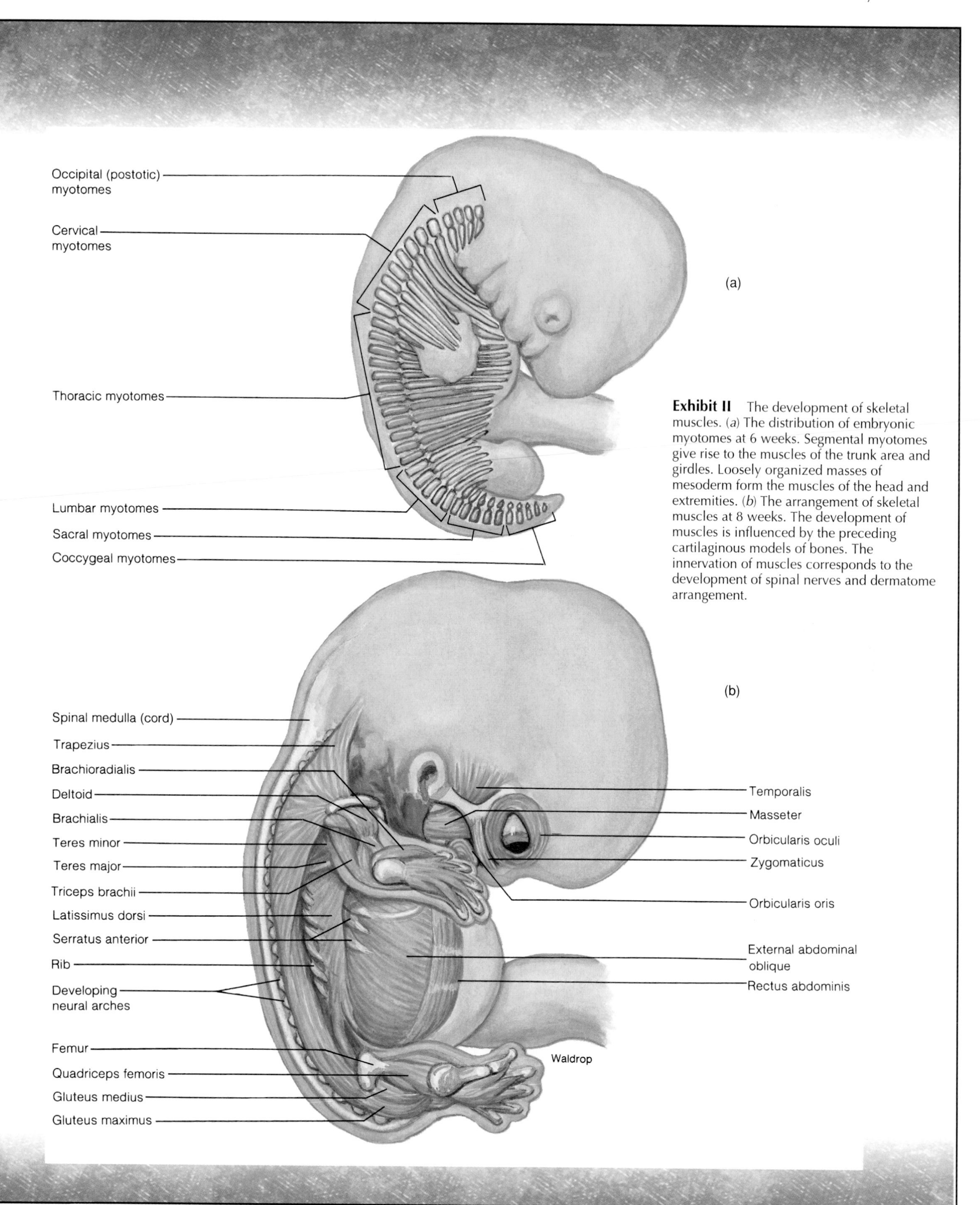

Exhibit II The development of skeletal muscles. (*a*) The distribution of embryonic myotomes at 6 weeks. Segmental myotomes give rise to the muscles of the trunk area and girdles. Loosely organized masses of mesoderm form the muscles of the head and extremities. (*b*) The arrangement of skeletal muscles at 8 weeks. The development of muscles is influenced by the preceding cartilaginous models of bones. The innervation of muscles corresponds to the development of spinal nerves and dermatome arrangement.

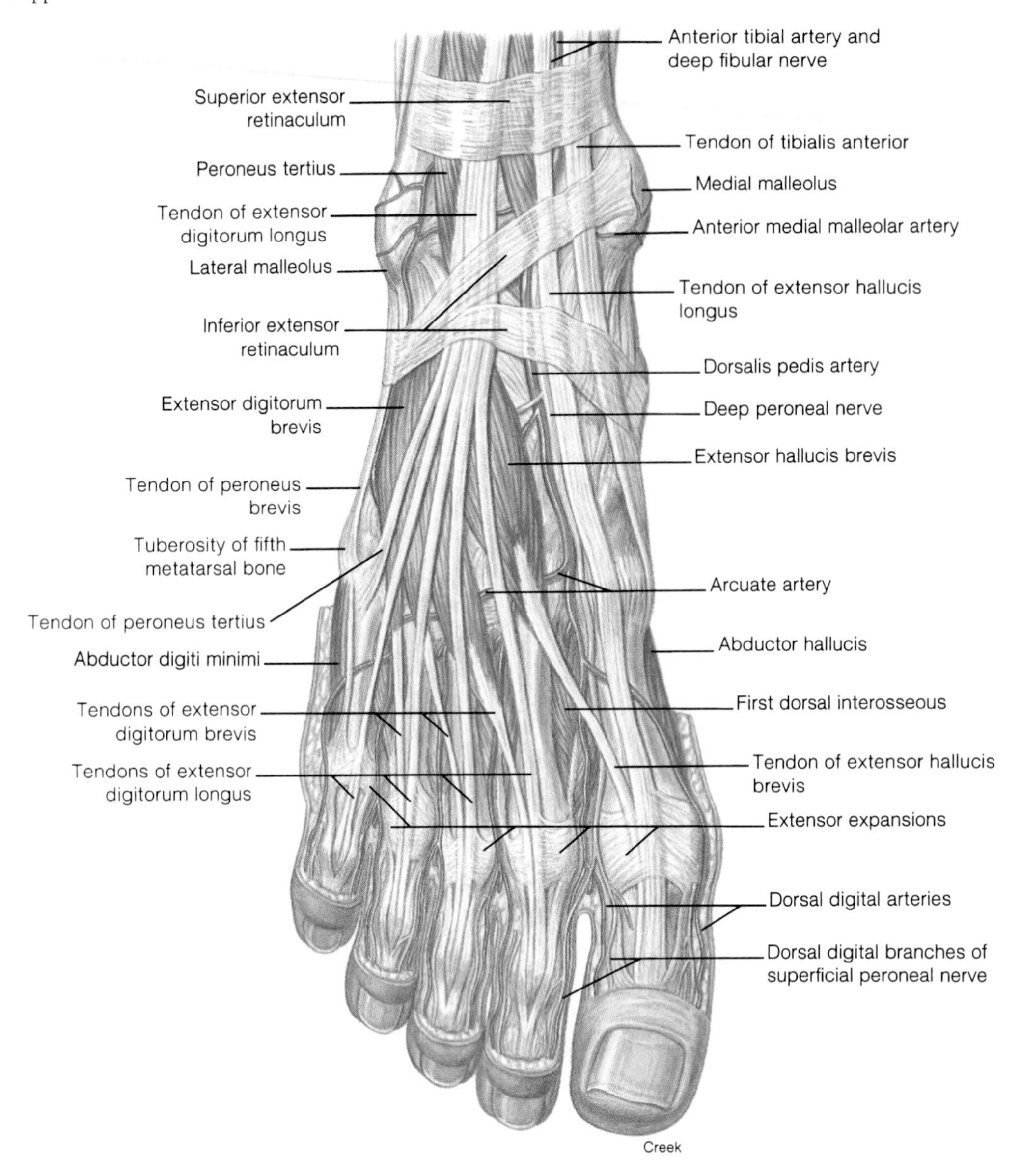

Figure 9.50

An anterior view of the dorsum of the foot.

at rest. The precise cause of cramps is unknown, but evidence indicates that they may be related to conditions within the muscle. They may result from general dehydration, deficiencies of calcium or oxygen, or from excessive stimulation of the motor neurons.

Torticollis (*tor" tĭ-kol' is*), or **wryneck,** is an abnormal condition in which the head is inclined to one side as a result of a contracted state of muscles on that side of the neck. This disorder may be either inborn or acquired.

A condition called **rigor mortis** (rigidity of death) affects skeletal muscle tissue several hours after death, as depletion of ATP within the fibers causes stiffness of the joints. This is similar to *physiological contracture*, in which muscles become incapable of either contracting or relaxing as a result of a lack of ATP.

When skeletal muscles are not contracted, either because the motor nerve supply is blocked or because the limb is immobilized (as when a broken bone is in a cast), the muscle fibers **atrophy** (*at'rŏ-fe*), or diminish in size. Atrophy is reversible if exercise is resumed, as after a healed fracture, but tissue death is inevitable if the nerves cannot be stimulated.

The fibers in healthy muscle tissue increase in size, or **hypertrophy,** if a muscle is systematically exercised. This increase in muscle size and strength is not due to an increase in the number of muscle cells, but rather to the increased production of myofibrils, accompanied by a strengthening of the associated connective tissue.

Diseases of Muscles

Fibromyositis (*fi"bro-mi"ŏ-si' tis*) is an inflammation of both skeletal muscular tissue and the associated connective tissue. Its causes are not fully understood. Pain and tenderness frequently occur in the extensor muscles of the lumbar region of the spinal

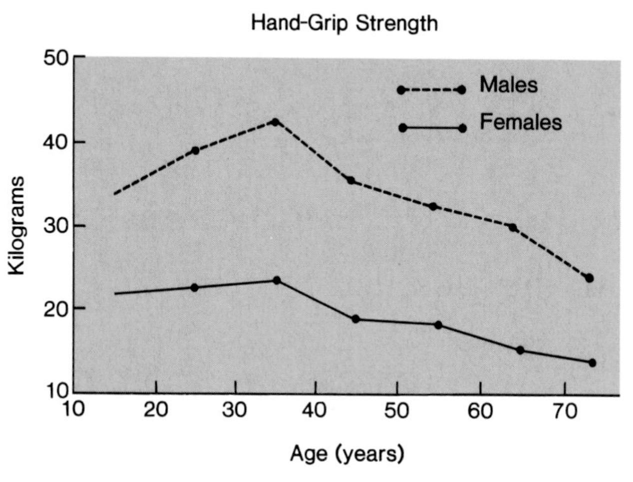

Figure 9.51

A gradual diminishing of muscle strength occurs after the age of 35, as shown with a graph of hand-grip strength.

Figure 9.52

Competitive runners in late adulthood, recognizing the benefits of exercise.

column, where there are extensive aponeuroses. Fibromyositis of this region is called **lumbago** (*lum-ba′go*), or **rheumatism.**

Muscular dystrophy is a genetic disease characterized by a gradual atrophy and weakening of muscle tissue. There are several kinds of muscular dystrophy, none of whose etiology is completely understood. The most frequent type affects children and is sex-linked to the male child. As muscular dystrophy progresses, the muscle fibers atrophy and are replaced by adipose tissue. Most children who have muscular dystrophy die before the age of 20.

The disease **myasthenia gravis** (*mi″as-the′ne-ă grav′is*) is characterized by extreme muscle weakness and low endurance. It results from a defect in the transmission of impulses at the neuromuscular junction. Myasthenia gravis is believed to be an autoimmune disease, and it typically affects women between the ages of 20 and 40.

Poliomyelitis (polio) is actually a viral disease of the nervous system that causes muscle paralysis. The viruses are usually localized in the anterior (ventral) horn of the spinal cord, where they affect the motor nerve impulses to skeletal muscles.

Neoplasms (abnormal growths of new tissue) are rare in muscles, but when they do occur, they are usually malignant. **Rhabdomyosarcoma** (*rab″do-mi″ŏ-sar-ko′mă*) is a malignant tumor of skeletal muscle. It can arise in any skeletal muscle, and most often afflicts young children and elderly people.

Aging of Muscles

Although elderly people experience a general decrease in the strength and fatigue-resistance of skeletal muscle (fig. 9.51), the extent of senescence varies considerably among individuals. Apparently, the muscular system is one of the body systems in which a person may actively slow degenerative changes. A decrease in muscle mass is partly due to changes in connective and circulatory tissues. Atrophy of the muscles of the appendages causes the arms and legs to appear thin and bony. Degenerative changes in the nervous system decrease the effectiveness of motor activity. Muscles are slower to respond to stimulation, causing a marked reduction in physical capabilities. The diminished strength of the respiratory muscles may limit the ability of the lungs to ventilate.

Exercise is important at all stages of life but is especially beneficial as one approaches old age (fig. 9.52). Exercise not only strengthens bones and muscles, but it also contributes to a healthy circulatory system and thus ensures an adequate blood supply to all body tissues. If an older person does not maintain muscular strength through exercise, he or she will be more prone to accidents. Loss of strength is a major contributor to falls and fractures. It often results in dependence on others to perform even the routine tasks of daily living.

Clinical Case Study Answer

When cancer of the head or neck involves lymph nodes in the neck, a number of structures on the affected side are removed surgically. This procedure usually includes the sternocleidomastoid muscle. This muscle, which originates on the sternum and clavicle, inserts on the mastoid process of the temporal bone. When it contracts, the mastoid process, which is located posteriorly at the base of the skull, is pulled forward, causing the chin to rotate away from the contracting muscle. This explains why the patient who had his left sternocleidomastoid muscle removed would have difficulty turning his head to the right.

---

lumbago: L. *lumbus*, loin
myasthenia: Gk. *myos*, muscle; *astheneia*, weakness
poliomyelitis: Gk. *polios*, gray; *myolos*, marrow
rhabdomyosarcoma: Gk. *rhabdos*, rod; *myos*, muscle; *oma*, a growth

Internal Affairs

How the Muscular System Works with Other Body Systems

How Other Systems Work with the Muscular System

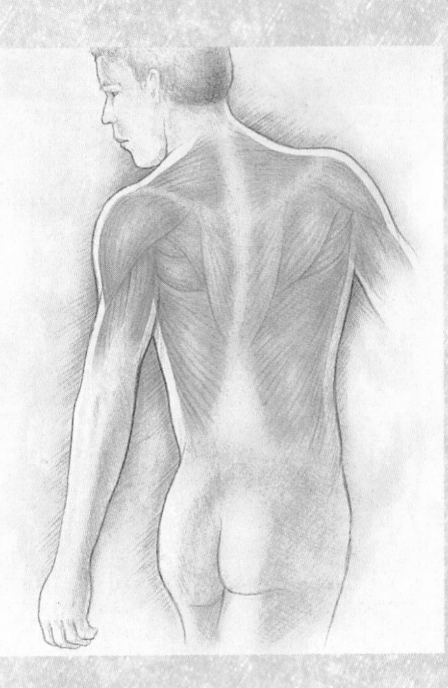

Integumentary System

- Facial musculature attached to skin produces facial expression when contracted

- Covers and protects body musculature
- Removes excessive body heat
- Initiates synthesis of vitamin D needed for muscle contraction

Skeletal System

- Enables body movement and stabilizes joints
- Muscle contractions maintain the health and strength of bone

- Source of calcium and phosphate
- Provides attachment sites for muscles
- Joints of skeleton provide levers for movement

Nervous System

- Sensory receptors monitor body position via autonomic nervous system
- Muscles give expression to thoughts, emotions, and motor commands that arise in central nervous system

- Coordinates muscle contraction
- Increases cardiac output and respiratory rates during periods of muscle activity

Endocrine System

- Provides protection to certain endocrine glands
- Exercise stimulates secretion of stress hormones

- Sex hormones promote muscle development and maintenance
- Specific hormones regulate calcium and phosphate concentrations
- Epinephrine and norepinephrine stimulate muscle contraction

Circulatory System

- Tonus and voluntary muscle contractions assist blood movement, particularly within veins

- Transports O_2 and CO_2, nutrients, and fluids to and from muscles; removes lactic acid and heat

Lymphatic System

- Supports and protects superficial lymph nodes
- Muscle contractions assist lymph movement
- Exercise elevates levels of immune cells and antibodies

- Maintains balanced amount of interstitial fluid within muscle tissue
- Lymphocytes provide defense against infection

Respiratory System

- Respiratory muscles enable ventilation of lungs; sound production
- CO_2 generated by exercise stimulates respiratory rate and depth

- Provides O_2 and eliminates CO_2 to and from muscles

Digestive System

- Enables chewing and swallowing
- Supports and protects organs of GI tract

- Provides nutrients for growth, maintenance, and repair of muscles
- Liver regulates blood glucose levels

Urinary System

- Muscles of urinary tract surround urinary bladder and form urethral sphincter
- Muscles of pelvic floor support urinary bladder

- Eliminates metabolic wastes from muscles
- Assists regulation of calcium and phosphate concentrations

Reproductive System

- Supports pelvic viscera
- Contributes to erection; pelvic thrust during coitus; abdominal and pelvic muscles aid childbirth

- Gonads produce sex hormones that promote muscle development and maintenance

Important Clinical Terminology

convulsion An involuntary, spasmodic contraction of skeletal muscle.

fibrillation *(fib-rĭ-la'shun)* Rapid randomized involuntary contractions of individual motor units within a muscle.

hernia The rupture or protrusion of a portion of the underlying viscera through muscle tissue. Most hernias occur in the normally weak places of the abdominal wall. There are four common hernia types:

1. **femoral**—viscera descending through the femoral ring;
2. **hiatal**—the superior portion of the stomach protruding into the thoracic cavity through the esophageal opening of the diaphragm;
3. **inguinal**—viscera protruding through the inguinal ring into the inguinal canal; and
4. **umbilical**—a hernia occurring at the navel.

intramuscular injection A hypodermic injection into a heavily muscled area to avoid damaging nerves. The most common site is the buttock.

myalgia *(mi-al'je-ă)* Pain in a muscle.

myokymia *(mi-o-ki'me-ă)* Twitching of isolated segments of muscle; also called *kymatism*.

myoma *(mi-o'mă)* A tumor of muscle tissue.

myopathy *(mi-op'ă-the)* Any muscular disease.

myotomy *(mi-ot'ŏ-me)* Surgical cutting or anatomical dissection of muscle tissue.

myotonia *(mi"ŏ-to'ne-ă)* A prolonged muscular spasm.

paralysis The loss of nervous control of a muscle.

shinsplints Tenderness and pain on the anterior surface of the leg generally caused by straining the tibialis anterior or extensor digitorum longus muscle.

Chapter Summary

Introduction to the Muscular System (pp. 227–228)

1. The contraction of skeletal muscle fibers results in body motion, heat production, and the maintenance of posture and body support.
2. The four basic properties characteristic of all muscle tissue are irritability, contractility, extensibility, and elasticity.
3. Axial muscles include facial muscles, neck muscles, and trunk muscles. Appendicular muscles include those that act on the girdles and those that move the segments of the appendages.

Structure of Skeletal Muscles (pp. 229–233)

1. The origin of a muscle is the more stationary attachment. The insertion is the more movable attachment.
2. Individual muscle fibers are covered by endomysium. Muscle bundles, called fasciculi, are covered by perimysium. The entire muscle is covered by epimysium.
3. Synergistic muscles work together to promote a particular movement. Muscles that oppose or reverse the actions of other muscles are antagonists.
4. Muscles may be classified according to fiber arrangement as parallel, convergent, pennate, or sphincteral.
5. Motor neurons conduct nerve impulses to the muscle fiber, stimulating it to contract. Sensory neurons conduct nerve impulses away from the muscle fiber to the central nervous system.

Skeletal Muscle Fibers and Types of Muscle Contraction (pp. 233–238)

1. Each skeletal muscle fiber is a multinucleated, striated cell. It contains a large number of long, threadlike myofibrils and is enclosed by a cell membrane called a sarcolemma.
   (a) Myofibrils have alternating A and I bands. Each I band is bisected by a Z line, and the subunit between two Z lines is called the sarcomere.
   (b) Extending through the sarcoplasm are a network of membranous channels called the sarcoplasmic reticulum and a system of transverse tubules (T tubules).
2. During muscle contraction, shortening of the sarcomeres is produced by sliding of the thin (actin) myofilaments over and between the thick (myosin) myofilaments.
   (a) The actin on each side of the A bands is pulled toward the center.
   (b) The H bands thus appear to be shorter as more actin overlaps the myosin.
   (c) The I bands also appear to be shorter as adjacent A bands are pulled closer together.
   (d) The A bands stay the same length because the myofilaments (both thick and thin) do not shorten during muscle contraction.
3. When a muscle exerts tension without shortening, the contraction is termed isometric; when shortening does occur, the contraction is isotonic.
4. The neuromuscular junction is the area consisting of the motor end plate and the sarcolemma of a muscle fiber. In response to a nerve impulse, the synaptic vesicles of the axon terminal secrete a neurotransmitter, which diffuses across the neuromuscular cleft of the neuromuscular junction and stimulates the muscle fiber.
5. A motor unit consists of a motor neuron and the muscle fibers it innervates.
   (a) Where fine control is needed, each motor neuron innervates relatively few muscle fibers. Where strength is more important than precision, each motor unit innervates a large number of muscle fibers.
   (b) The neurons of small motor units have relatively small cell bodies and tend to be easily excited. Those of large motor units have larger cell bodies and are less easily excited.

Naming of Muscles (pp. 239–241)

1. Skeletal muscles are named on the basis of shape, location, attachment, orientation of fibers, relative position, and function.
2. Most muscles are paired; that is, the right side of the body is a mirror image of the left.

Muscles of the Axial Skeleton (pp. 241–253)

The muscles of the axial skeleton include those responsible for facial expression, mastication, eye movement, tongue movement, neck movement, and respiration, and those of the abdominal wall, the pelvic outlet, and the vertebral column. They are summarized in tables 9.2 through 9.9.

Muscles of the Appendicular Skeleton (pp. 253–278)

The muscles of the appendicular skeleton include those of the pectoral girdle, humerus, forearm, wrist, hand, and fingers, and those of the pelvic girdle, thigh, leg, ankle, foot, and toes. They are summarized in tables 9.10 through 9.18.

Review Activities

Objective Questions

1. The site at which a nerve impulse is transmitted from the motor nerve ending to the skeletal muscle cell membrane is
   (a) the sarcomere.
   (b) the neuromuscular junction.
   (c) the myofilament.
   (d) the Z line.

2. Muscles capable of highly dexterous movements contain
   (a) one motor unit per muscle fiber.
   (b) many muscle fibers per motor unit.
   (c) few muscle fibers per motor unit.
   (d) many motor units per muscle fiber.
3. Which of the following is *not* used as a means of naming muscles?
   (a) location
   (b) action
   (c) shape
   (d) attachment
   (e) strength of contraction
4. Neurotransmitters are stored in synaptic vesicles within
   (a) the sarcolemma.
   (b) the motor units.
   (c) the myofibrils.
   (d) the axon terminals.
5. Which of the following muscles have motor units with the lowest innervation ratio?
   (a) brachial muscles
   (b) muscles of the forearm
   (c) thigh muscles
   (d) abdominal muscles
6. An eyebrow is drawn toward the midline of the face through contraction of which muscle?
   (a) the corrugator (c) the nasalis
   (b) the risorius (d) the frontalis
7. A flexor of the shoulder joint is
   (a) the pectoralis major.
   (b) the supraspinatus.
   (c) the teres major.
   (d) the trapezius.
   (e) the latissimus dorsi.
8. Which of the following muscles does *not* have either an origin or insertion on the humerus?
   (a) the teres minor
   (b) the biceps brachii
   (c) the supraspinatus
   (d) the brachialis
   (e) the pectoralis major
9. Which muscle of the four that compose the quadriceps femoris muscle may act on the hip and knee joints?
   (a) the vastus medialis
   (b) the vastus intermedius
   (c) the rectus femoris
   (d) the vastus lateralis
10. Which of the following muscles plantar flexes the ankle joint and inverts the foot as it supports the arches?
   (a) the flexor digitorum longus
   (b) the tibialis posterior
   (c) the flexor hallucis longus
   (d) the gastrocnemius

Essay Questions

1. Describe how muscle fibers are formed and explain why the fibers are multinucleated.
2. Describe the special characteristics of muscle tissue that are essential for muscle contraction.
3. Define *fascia*, *aponeurosis*, and *retinaculum*.
4. Describe the structural arrangement of the muscle fibers and fasciculi within muscle.
5. What are the advantages and disadvantages of pennate-fibered muscles?
6. List the major components of a skeletal muscle fiber and describe the function of each part.
7. What is a motor unit, and what is its role in muscle contraction?
8. Give three examples of synergistic muscle groups within the upper extremity and identify the antagonistic muscle group for each.
9. Attempt to contract, one at a time, each of the neck muscles depicted in figure 9.20.
10. List all the muscles that either originate or insert on the scapula.
11. Give three examples of synergistic muscle groups within the lower extremity and identify the antagonistic muscle group for each.
12. Describe the flexor and extensor compartments of the muscles of the forearm.
13. List the muscles that border the popliteal fossa. Describe the structures that are located in this region.
14. Firmly press your fingers on the front, sides, and back of your ankle as you move your foot. The tendons of which muscles can be palpated anteriorly, laterally, and posteriorly?

Critical-Thinking Questions

1. In the sixteenth century, Andreas Vesalius demonstrated that cutting a muscle along its length has very little effect on its function; on the other hand, a transverse cut puts a muscle out of action. How would you explain Vesalius's findings?
2. As a result of a severe head trauma sustained in an automobile accident, a 17-year-old male lost function of his right oculomotor nerve. Explain what will happen to the function of the affected eye.
3. Discuss the position of flexor and extensor muscles relative to the shoulder, elbow, and wrist joints.
4. Based on function, describe exercises that would strengthen the following muscles: (a) the pectoralis major, (b) the deltoid, (c) the triceps, (d) the pronator teres, (e) the rhomboideus major, (f) the trapezius, (g) the serratus anterior, and (h) the latissimus dorsi.
5. Why is it necessary to have dual (sensory and motor) innervation to a muscle? Give an example of a disease that results in loss of motor innervation to specific skeletal muscles, and describe the effects of this denervation.
6. Compare muscular dystrophy and myasthenia gravis as to causes, symptoms, and the effect they have on muscle tissue.

TEN

Surface and Regional Anatomy

Introduction to Surface Anatomy 288
Surface Anatomy of the Newborn 289
- General Appearance 290
- Palpable Structures 291

Head 291
- Surface Anatomy 292
- Internal Anatomy 295

Neck 297
- Surface Anatomy 297
- Internal Anatomy 300

Trunk 300
- Surface Anatomy 301
- Internal Anatomy 305

Pelvis and Perineum 310
Shoulder and Upper Extremity 311
- Surface Anatomy 311
- Internal Anatomy 314

Buttock and Lower Extremity 317
- Surface Anatomy 317
- Internal Anatomy 321

Clinical Considerations 321
- Head and Neck Regions 321
- Thoracic Region 326
- Abdominal Region 327
- Shoulder and Upper Extremity 329
- Hip and Lower Extremity 330

Chapter Summary 331
Review Activities 331

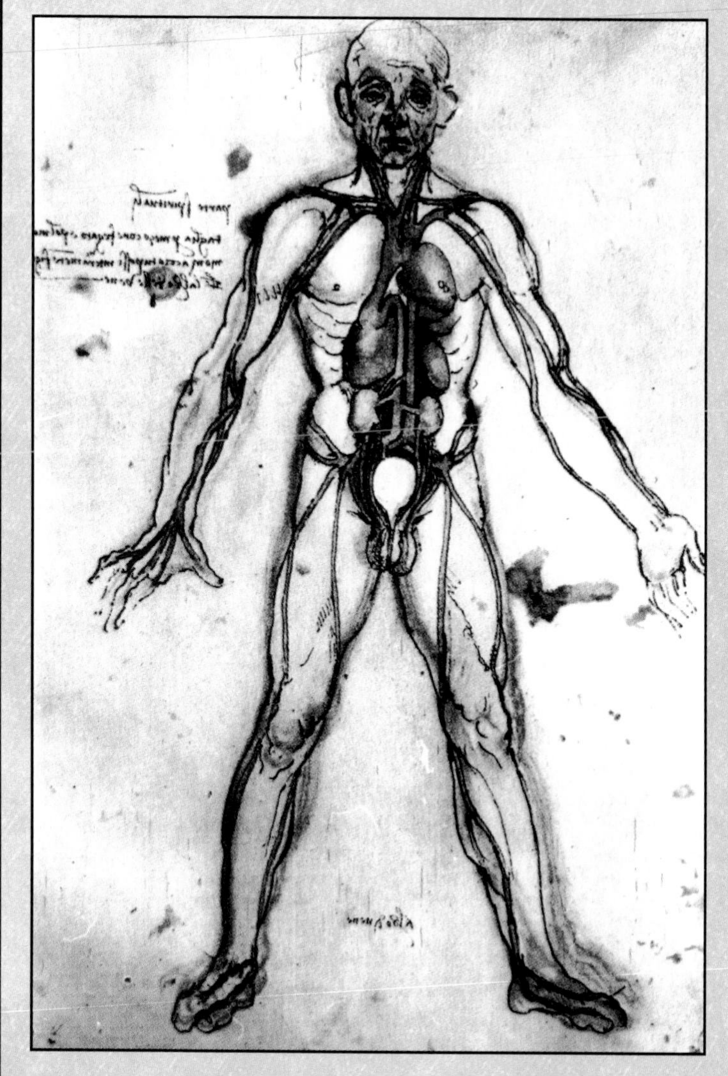

Surface anatomy was largely ignored throughout the Dark Ages, until the artists of the fifteenth century began to take intense interest in the human form. Leonardo da Vinci was an anatomist as well as an artist. Many of his drawings, this one included, were based on dissections he himself performed.

Introduction to Surface Anatomy

Surface anatomy, a branch of gross anatomy, is the study of the form and markings of the surface of the body as they relate to deeper structures. Knowledge of surface anatomy is essential in performing a physical examination, treating diseases or dysfunctions of the body, and maintaining physical fitness.

Objective 1 Discuss the value of surface anatomy in learning about internal anatomical structures.

Objective 2 Explain why surface anatomy is important in the diagnosis and treatment of diseases or dysfunctions of the body.

Objective 3 Explain why individual differences in body physique may have a bearing on the effectiveness of observation and palpation.

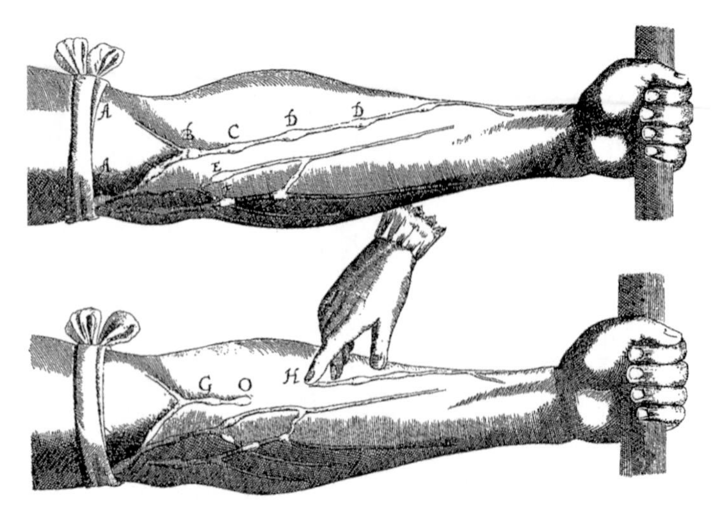

Figure 10.1

A demonstration of the presence and function of valves within the veins of the forearm, conducted by the great English anatomist William Harvey. (After William Harvey, *On the Motion of the Heart and Blood in Animals*, 1628.) In order to understand the concept of a closed circulatory system (e.g., blood contained within vessels), a knowledge of surface anatomy was essential.

It is amazing how much anatomical information you can acquire by examining the surface anatomy of your own body. Surface anatomy is the study of the structure and markings of the surface of the body through visual inspection or palpation. Surface features can be readily identified through *visual inspection*, and anatomical features beneath the skin can be located by *palpation* (feeling with firm pressure or perceiving by the sense of touch). Knowledge of surface anatomy is clinically important in locating precise sites for *percussion* (tapping sharply to detect resonating vibrations) and *auscultation* (listening to sounds emitted from organs).

With the exception of certain cranial bones, the bones of the entire skeleton can be palpated. Once the position, shape, and processes of these bones are identified, these skeletal features can serve as landmarks for locating other anatomical structures. Many of the skeletal muscles and their tendinous attachments are clearly visible as they are contracted and caused to move. The location and range of movement of the joints of the body can be determined as the articulating bones are moved by muscle contractions. On some individuals, the positions of superficial veins can be located and their courses traced. Even the location and function of the valves within the veins can be demonstrated on the surface of the skin (fig. 10.1). Some of the arteries also can be seen as they pulsate beneath the skin. Knowing where the arterial pressure points are is an important clinical aspect of surface anatomy (see fig. 16.31). Other structures can be identified from the body surface, including certain nerves, lymph nodes, glands, and other internal organs.

Surface anatomy is an essential aspect of the study of anatomy. Knowing where muscles and muscle groups are located can be extremely important in maintaining physical fitness. In many medical and paramedical professions, the surface anatomy of a patient is of immeasurable value in diagnosis and treatment. Knowing where to record a pulse, insert needles and tubes, listen to the functioning of internal organs, take radiographs, and perform physical therapy requires a knowledge of surface landmarks.

The effectiveness of observation and palpation in studying a person's surface anatomy is related to the amount of subcutaneous adipose tissue present (fig. 10.2). In examining an obese person, it may be extremely difficult to observe or palpate certain internal structures that are readily discernible in a thin person. The hypodermis of females is normally thicker than that of males (fig. 10.3). This tends to smooth the surface contours of females and obscure the muscles, veins, and bony prominences that are apparent in males.

This chapter will be of great value in reviewing the bones articulations, and muscles you have already studied. In the photographs of dissected cadavers, you will be able to see the relationships between the various body organs and systems in specific regions. Refer back to this chapter as you study the anatomy of the remaining body systems. Reviewing in this way will broaden your perspective in locating various organs and structures.

If you use yourself as a model from which to learn and review, anatomy as a science will take on new meaning. As you learn about a bone or a process on a bone, palpate that part of your body. Contract the muscles you are studying so that you better understand their locations, attachments, and actions. In this way, you will become better acquainted with your body, and anatomy will become more enjoyable and easier to learn. Your body is one crib sheet you can take with you to exams.

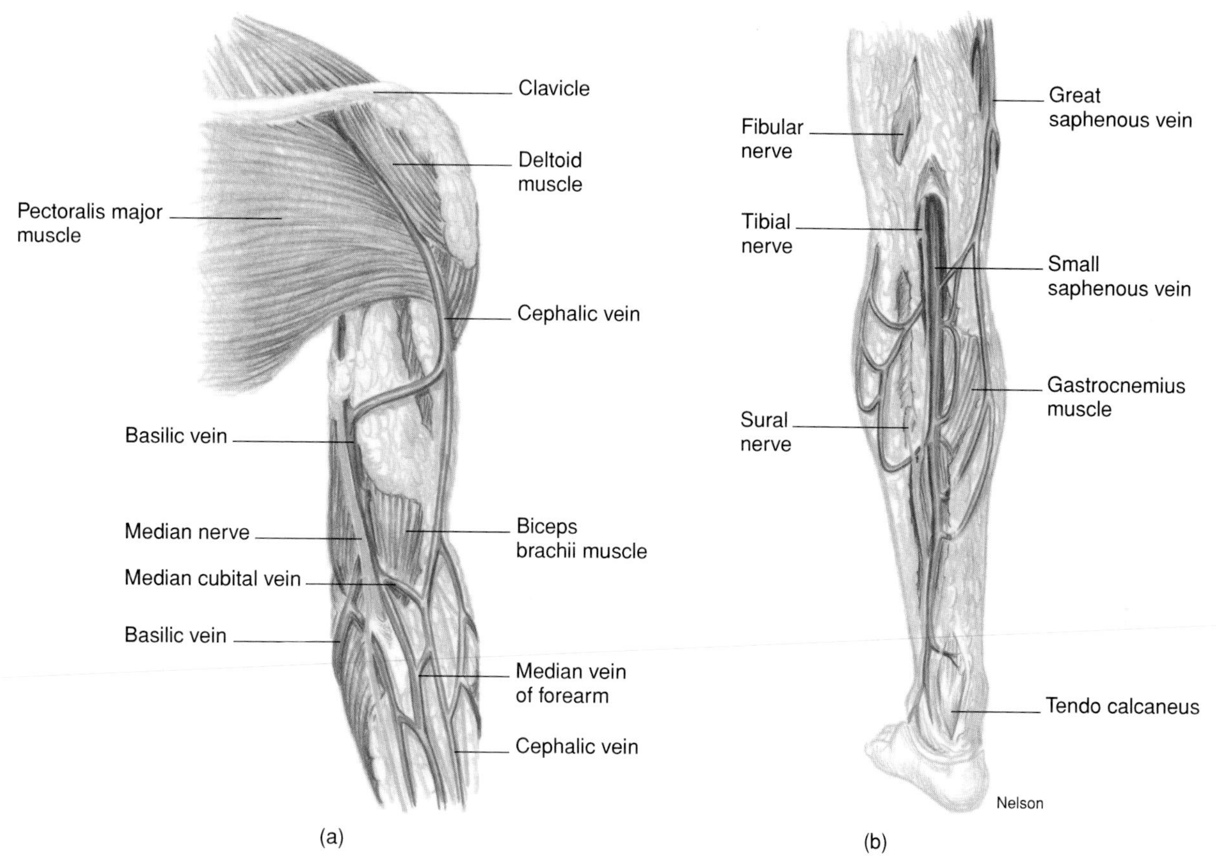

Figure 10.2

Subcutaneous adipose tissue. (*a*) An anterior view of the left brachial region and (*b*) a posterior view of the left leg.

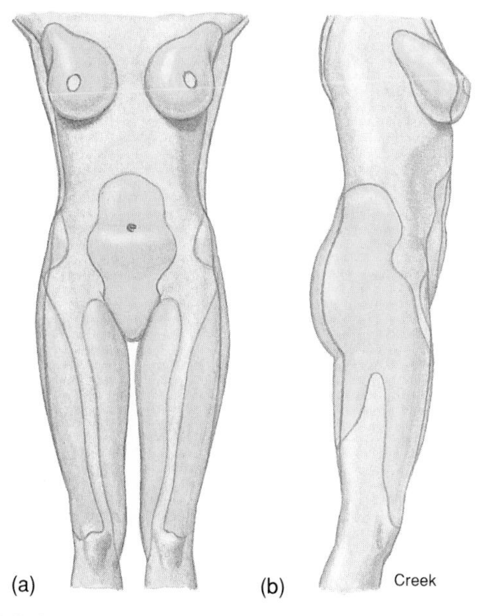

Figure 10.3

Principal areas of adipose deposition of a female. (*a*) An anterior view and (*b*) a lateral view. The outline of a male is superposed in both views. There is significantly more adipose tissue interlaced in the fascia covering the muscles, vessels, and nerves in a female than in a male. The hypodermis layer of the skin is also approximately 8% thicker in a female than in a male.

1. Explain what is meant by visual inspection and palpation, and discuss the value of surface anatomy in locating internal structures.
2. Why is a knowledge of surface anatomy important in a clinical setting?
3. How does the hypodermis of the skin differ in males and females? What are the clinical implications of this difference?

Surface Anatomy of the Newborn

The surface anatomy of a newborn infant represents an early stage of human development; therefore, it differs from that of an adult. Certain aspects of the surface anatomy of a neonate are of clinical importance in ascertaining the degree of physical development, general health, and possible congenital abnormalities.

---

neonate: Gk. *neos*, new; L. *natalis*, birth

Table 10.1

Surface Anatomy of the Neonate

Body Structure	Normal Conditions	Common Variations
General posture	Joints of vertebral column and extremities flexed	Extended legs and neck; abducted and rotated thighs (breech birth)
Skin	Red or pink, with vernix caseosa and lanugo; edematous face, extremities, and genitalia	Neonatal jaundice; integumentary blisters; mongolian spots
Skull	Fontanels large, flat, and firm, but soft to the touch	Molded skull, bulging fontanels; cephalhematoma
Eyes	Lids edematous; color—gray, dark blue, or brown; absence of tears; corneal, pupillary, and blink reflexes	Conjunctivitis, subconjunctival hemorrhage
Ears	Auricle flexible, with cartilage present; top of auricle positioned on horizontal line with outer canthus of eye	Auricle flat against head
Neck	Short and thick, surrounded by neck folds	Torticollis
Chest	Equal anteroposterior and lateral dimensions; xiphoid process evident; breast enlargement	Funnel or pigeon chest; additional nipples (polythelia); secretions from breast (witch's milk)
Abdomen	Cylindric in shape; liver and kidneys palpable	Umbilical hernia
Genitalia	(♂ and ♀) Edematous and darkly pigmented; (♂) testes palpable in scrotum; periodic erection of penis	(♀) Blood-tinged discharge (pseudomenstruation); hymenal tag; (♂) testes palpable in inguinal canal; inability to retract prepuce; inguinal hernia
Extremities	Symmetrical; 10 fingers and toes; soles flat with moderate to deep creases	Partial syndactyly; asymmetric length of toes

Objective 4 Describe the surface anatomy of a normal, full-term neonate

Objective 5 List some of the internal structures that can be palpated in a neonate.

The birth of a baby is the dramatic culmination of a 9-month gestation, during which the miraculous development of the fetus prepares it for extrauterine life. Although the normal, full-term neonate is physiologically prepared for life, it is totally dependent on the care of others. The physical assessment of the neonate is extremely important to ensure its survival. Much of the assessment is performed through inspection and palpation of its surface anatomy. The surface anatomy of a neonate obviously differs from that of an adult because of the transitional stage of development from fetus to infant.

Although the surface anatomy of a neonate is discussed at this point in the text, prenatal development and body growth with its accompanying physiological changes are discussed in detail in chapter 22. A summary of the surface anatomy of the neonate is presented in table 10.1.

General Appearance

As a result of in utero position, the posture of the full-term neonate is one of flexion (fig. 10.4). The neonate born vertex (head first) keeps the neck and vertebral column flexed, with the chin resting on the upper chest. The hands are clenched into fists, the elbow joints are flexed, and the arms are held to the

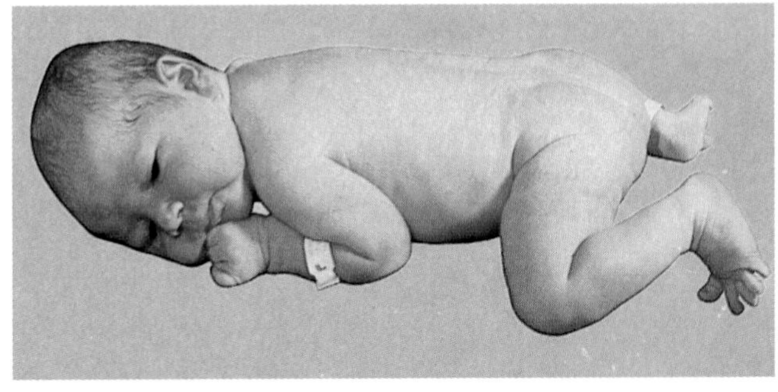

(a)

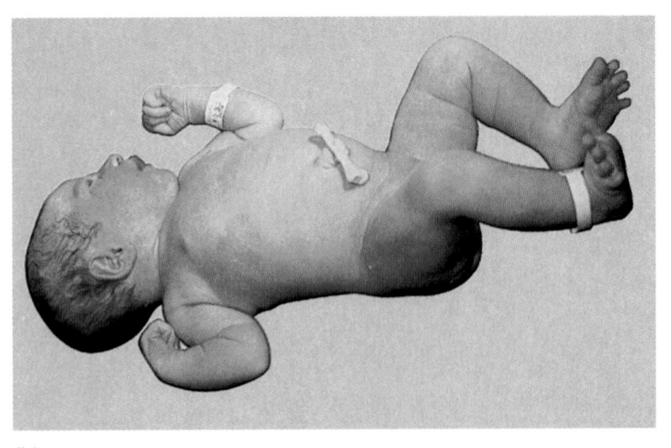

(b)

Figure 10.4

The flexion position of a neonate (*a, b*) is an indication of a healthy gestation and a normal delivery.

chest. The knee and hip joints are flexed, drawing the thighs toward the abdomen. The ankle joints are dorsiflexed.

The skin is the one organ of the neonate that is completely visible and is therefore a source of considerable information concerning its state of development and clinical condition. At birth, the skin is covered with a grayish, cheeselike substance called **vernix caseosa** (*ver'niks ka"se-o'să*). If it is not washed away during bathing, the vernix will dry and disappear within a couple of days. Fine, silklike hair called **lanugo** (*lă-noo'go*) may be present on the forehead, cheeks, shoulders, and back. Distended sebaceous glands called **milia** (*mil'e-ă*) may appear as tiny white papules on the nose, cheeks, and chin. Skin color depends on genetic background, although certain areas, such as the genitalia, areola, and linea alba may appear darker than the rest of skin because of a response to maternal and placental hormones that enter the fetal circulation. **Mongolian spots** occur in about 90% of newborn Blacks, Asians, and American Indians. These blue-gray pigmented areas vary in size and are usually located in the lumbosacral region. Mongolian spots generally fade within the first year or two.

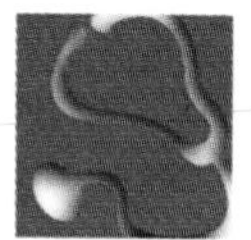

Abnormal skin color is clinically important in the physical assessment of the neonate. *Cyanosis* (*si"ă-no'sis*) (bluish discoloration) is usually due to a pulmonary disease (for example, atelectasis or pneumonia) or to congenital heart disease. Although *jaundice* (yellowish discoloration) is common in infants and is usually of no concern, it may indicate liver or bone marrow problems. *Pallor* (paleness) may indicate anemia, edema, or shock.

The appearance of the nails and nail beds is especially valuable in determining body dysfunctions, certain genetic conditions, and even normal gestations. Cyanosis, pallor, and capillary pulsations are best observed at the nails. Jaundice is common in postmature neonates and can be visibly detected by yellow nails.

Local edema (swelling) is not uncommon in the neonate, particularly in the skin of the face, legs, hands, feet, and genitalia. Creases on the palms of the hands and soles of the feet should be prominent; the absence of creases accompanies prematurity (fig. 10.5). The nose is usually flattened after birth and there may be bruises there, or on other areas of the face. The auricle of the ear is flexible, with the top edge positioned on a horizontal line with the outer canthus (corner) of the eye.

The neck of a neonate is short, thick, and surrounded by neck folds. The chest is rounded in cross section, and the abdomen is cylindrical. The abdomen may bulge in the upper right quadrant because of the large liver.

If the newborn is thin, peristaltic intestinal waves may be observed. At birth, the **umbilical cord** appears bluish white and moist. After clamping, it begins to dry and appears yellowish brown. It progressively shrivels and becomes greenish black prior to its falling off by the second week.

The genitalia of both sexes may appear darkly pigmented due to maternal hormonal influences. In a female neonate, a **hymenal** (*hi'men-al*) **tag** is frequently present and is visible at the back of the vaginal opening. It is composed of tissue from the hymen and labia minora. The hymenal tag usually disappears by the end of the first month.

vernix caseosa: L. *vernix*, varnish; *caseus*, cheese
milia: L. *miliarius*, relating to millet

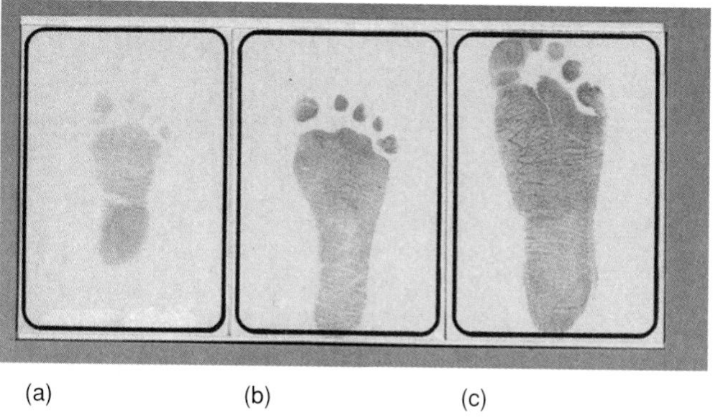

Figure 10.5

Sole creases at different ages of gestation as seen from footprints of two premature babies (*a, b*) and a full-term baby (*c*). (*a*) At 26 weeks of gestation, only an anterior transverse crease is present. (*b*) By 33 weeks, creases have developed along the medial instep. (*c*) The entire sole has developed creases by 38 weeks.

Palpable Structures

The six **fontanels** (see fig. 6.13) can be lightly palpated as the "soft spots" on the infant's head. The liver is palpable 2–3 cm (1 in.) below the right costal margin. During a physical examination of a neonate, the physician will palpate both kidneys soon after delivery, before the intestines fill with air. The suprapubic area is also palpated for an abnormally distended urinary bladder. The newborn should void urine within the first 24 hours after birth.

The testes of the male neonate should always be palpated in the scrotum. If the neonate is small or premature, the testes may be palpable in the inguinal canals. An examination for inguinal hernias is facilitated by the crying of an infant, which creates abdominal pressure.

1. Describe the appearance of each of the following in a normal neonate: skin, head, thorax, abdomen, genitalia, and extremities. What is meant by the normal flexion position of a neonate?
2. Which internal body organs are palpable in a neonate?

Head

The head is the most highly integrated region of the body, since it communicates with and controls all of the body systems. The head is of clinical concern because it contains important sense

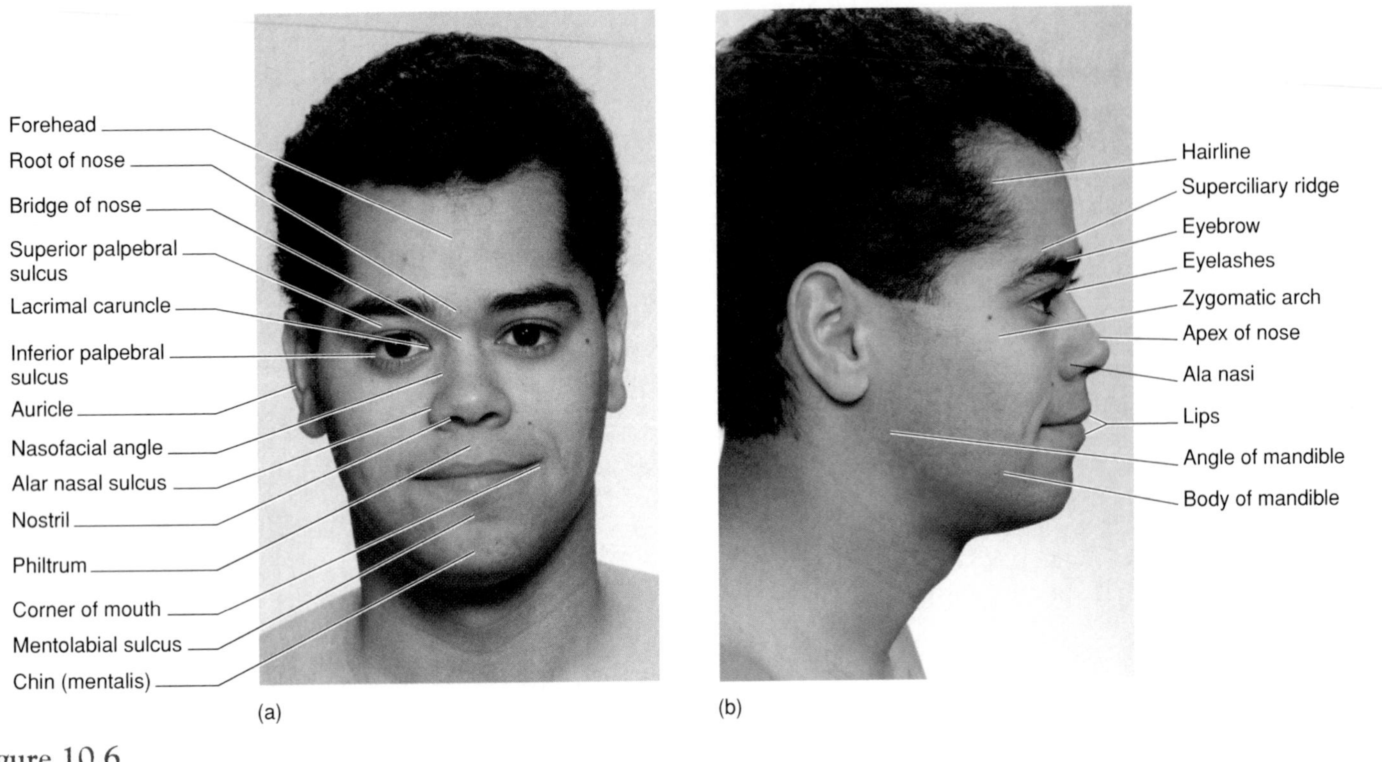

Figure 10.6

The surface anatomy of the facial region. (*a*) An anterior view and (*b*) a lateral view.

organs and provides openings into the respiratory and digestive systems. Of social importance is the aesthetics (pleasing appearance) of the head, which in some cases is also of clinical concern.

Objective 6 Identify various surface features of the cranial and facial regions by observation or palpation.

Objective 7 Describe the basic internal anatomy of the head.

Surface Anatomy

The head contains the brain and the special sense organs—the eyes, ears, nose, and taste buds. It also provides openings into the respiratory and digestive systems. The head is structurally and developmentally divided into the cranium and the face.

Cranium

The **cranium,** also known as the **braincase,** is covered by the *scalp*. The scalp is attached anteriorly, at the level of the eyebrows, to the **supraorbital ridges.** It extends posteriorly over the area commonly called the forehead and across the crown (vertex) of the head to the **superior nuchal** *(noo'kal)* **line,** a ridge on the back of the skull. Both the supraorbital ridge above the **orbit,** or socket of the eye, and the superior nuchal line at the back of the skull can be easily palpated. Laterally, the scalp covers the **temporal region** and terminates at the fleshy portion of the ear called the **auricle** (*or'ĭ-kul*), or **pinna.** The temporal region is the attachment for the **temporalis** muscle, which can be palpated when the jaw is repeatedly clenched. This region is clinically important because it is a point of entrance to the cranial cavity in many surgical procedures.

Only a portion of the scalp is covered with hair, and the variable **hairline** is genetically determined. The scalp is clinically important because of the dense connective tissue layer that supports nerves and vessels beneath the skin. When the scalp is cut, the wound is held together by the connective tissue, but at the same time the vessels are held open, resulting in profuse bleeding.

Face

The **face** (fig. 10.6) is divided into four regions: the **ocular region,** which includes the eye and associated structures; the **auricular region,** which includes the ear; the **nasal region,** which includes the external and internal structures of the nose; and the **oral region,** which includes the mouth and associated structures.

The skin of the face is relatively thin and contains many sensory receptors, particularly in the oral region. Certain facial

cranium: Gk. *kranion*, skull

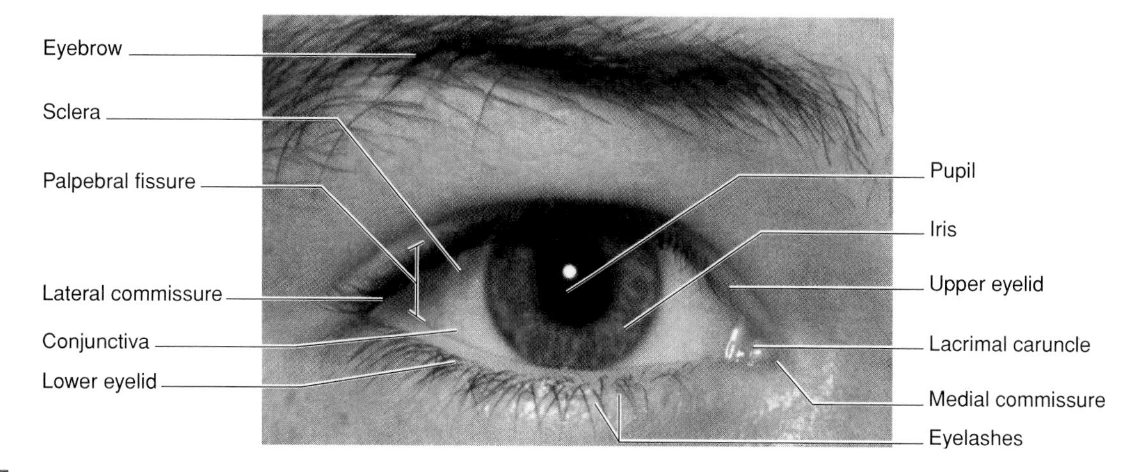

Figure 10.7

The surface anatomy of the ocular region.

regions also have numerous **sweat glands** and **sebaceous** (*sĕ-ba'shus*) (oil-secreting) **glands.** Facial acne is a serious dermatological problem for many teenagers. Facial hair appears over most of the facial region in males after they go through puberty; unwanted facial hair may occur sparsely on some females and can be a social problem.

The muscles of facial expression are important in their effect on surface features. As they are contracted, various emotions are conveyed. These muscles originate on the facial bones and insert into the dermis (second major layer) of the skin. Repeated contraction of these muscles may eventually cause permanent crease lines in the skin.

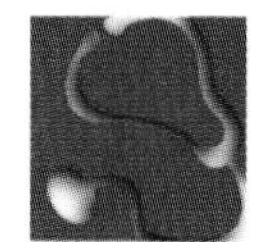

Because the organs of the facial region are so complex and specialized, there are professional fields of specialty associated with the various regions. *Optometry* and *ophthalmology* are concerned with the structure and function of the eye. *Dentistry* is entirely devoted to the health and functional and cosmetic problems of the oral region, particularly the teeth. An *otorhinolaryngologist* (*o"to-ri"no-lar"ing-gol'ŏ-jist*) is an ear, nose, and throat specialist.

The *ocular region* includes the eyeball and associated structures. Most of the surface features of the ocular region protect the eye. **Eyebrows** protect against potentially damaging sunlight and mechanical blows; **eyelids** reflexly close to protect against objects moving toward the eye or visual stimuli; **eyelashes** prevent airborne particles from contacting the eyeball; and *lacrimal* (*lak'rĭ-mal*) *secretions* (tears) wash away chemicals or foreign materials and prevent the surface of the eyeball from drying. Many of the surface features of the ocular region are shown in figure 10.7 and described in table 10.2.

The *auricular region* includes the visible surface structures and internal organs that function in hearing and maintaining equilibrium. The fleshy **auricle** and the tubular opening into the middle ear, called the **external acoustic canal,** are the only observable surface features of the auricular region. The rim of the auricle, shaped and supported by elastic cartilage, is called the **helix;** the inferior portion is referred to as the **earlobe.** The earlobe is composed primarily of connective and fatty tissue, and therefore can be easily pierced. For this reason, it is sometimes used when obtaining blood for a blood count. The **tragus** (*tra' gus*) is a small, posteriorly directed projection partially covering and protecting the external acoustic canal. Further protection is provided by the many fine hairs that surround the opening into this canal. The condyle of the mandible can be palpated at the opening of the external acoustic canal by placing the little finger in the opening, and then vigorously moving the jaw. Refer to figure 10.8 and table 10.3 for an illustration and description of other surface features of the auricular region.

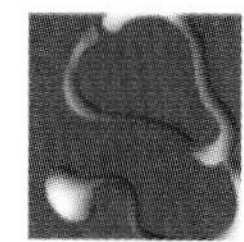

The inspection of some of the internal structures of the ear is part of a routine physical examination and is performed using an otoscope. *Cerumen* (*sĕ-roo'men*) (earwax) may accumulate in the canal, but this is a protective substance. It waterproofs the tympanic membrane (eardrum) and because of its bitter taste is thought to be an insect repellent. In some cases, it may become impacted and require physical removal.

A few structural features of the *nasal region* are apparent from its surface anatomy (fig. 10.9 and table 10.4). The principal function of the nose is associated with the respiratory system, and the need for a permanent body opening to permit gaseous ventilation accounts for its surface features. The **root** (nasion) of the nose is the point in the skull where the nasal and frontal bones unite. It is located at about the level of the eyebrows. The firm, narrow part between the eyes is the **bridge** (dorsum nasi) of the nose and is formed by the union of the nasal bones. The nose below this level has a pliable cartilaginous framework that maintains an opening. The tip of the nose

Table 10.2
Surface Anatomy of the Ocular Region

Structure	Comments
Eyebrow	Ridge of hair that superiorly arches the eye. It protects the eye against sunlight and is important in facial expression.
Eyelids	Movable folds of skin and muscle that cover the eyeball anteriorly. They assist in lubricating the anterior surface of the eyeball and reflexly close to protect the eyeball.
Eyelashes	Rows of hairs on the margins of the eyelids. They prevent airborne particles from contacting the eyeball.
Conjunctiva	Thin mucous membrane that covers the anterior surface of the eyeball and lines the undersurface of the eyelids. It aids in reducing friction during blinking.
Sclera	Outer fibrous layer of the eyeball; the "white" of the eye that gives form to the eyeball
Cornea	Transparent anterior portion of the eyeball. It is slightly convex to refract incoming light waves.
Iris	Circular, colored, muscular portion of the eyeball that surrounds the pupil. It reflexly regulates the amount of incoming light.
Pupil	Opening in the center of the iris through which light enters the eyeball
Palpebral fissure	Space between the eyelids when they are open
Subtarsal sulcus	Groove beneath the eyelid that parallels the margin of the lid. It traps small foreign particles that contact the conjunctiva.
Medial commissure	Medial junction of the upper and lower eyelids
Lateral commissure	Lateral junction of the upper and lower eyelids
Lacrimal caruncle	Fleshy, pinkish elevation at the medial commissure. It contains sebaceous and sweat glands.

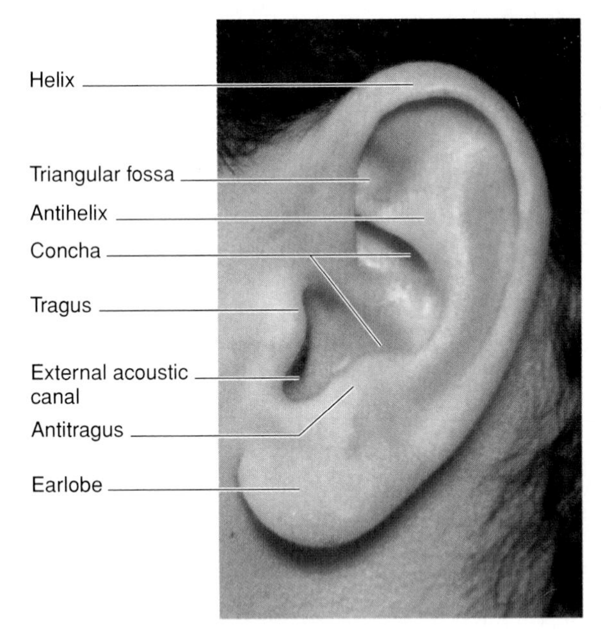

Figure 10.8

The surface anatomy of the auricular region.

Table 10.3
Surface Anatomy of the Auricular Region

Structure	Comments
Auricle (pinna)	Expanded portion of the ear projecting from the side of the head. It funnels sound waves into the external acoustic canal.
Helix	Outer rim of the auricle. It gives form and shape to the auricle.
Earlobe	Fleshy inferior portion of the auricle
Tragus	Small projection of the auricle, just anterior to the external acoustic canal.
Antitragus	Small, cartilaginous anterior projection opposite the tragus
Antihelix	Semicircular ridge anterior to the greater portion of the helix
Concha	Depressed hollow of the auricle. It funnels sound waves.
External acoustic canal	Slightly S-shaped tube extending inward to the tympanic membrane. It contains glands that secrete earwax for protection.
Triangular fossa	Triangular depression in the superior part of the antihelix

is called the **apex.** The **nostrils,** or **nares,** (*na'rēz*) are the paired openings into the nose. The **wing** (ala) of the nose forms the flaired outer margin of each nostril.

Structures of the *oral region* that are important in surface anatomy include the fleshy upper and lower **lips** (labia), the **chin** (mentum), and the structures of the *oral cavity* that can be observed when the mouth is open. The lips and chin are shown in figure 10.9 and the structures of the oral cavity, in figure 10.10.

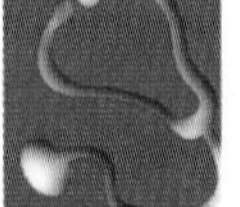

The color of the lips and other mucous membranes of the oral cavity are diagnostic of certain body dysfunctions. The lips may appear pale in people with severe anemia, or bluish in those with abnormal amounts of reduced hemoglobin in the blood. A lemon yellow tint to the lips may indicate pernicious anemia or jaundice.

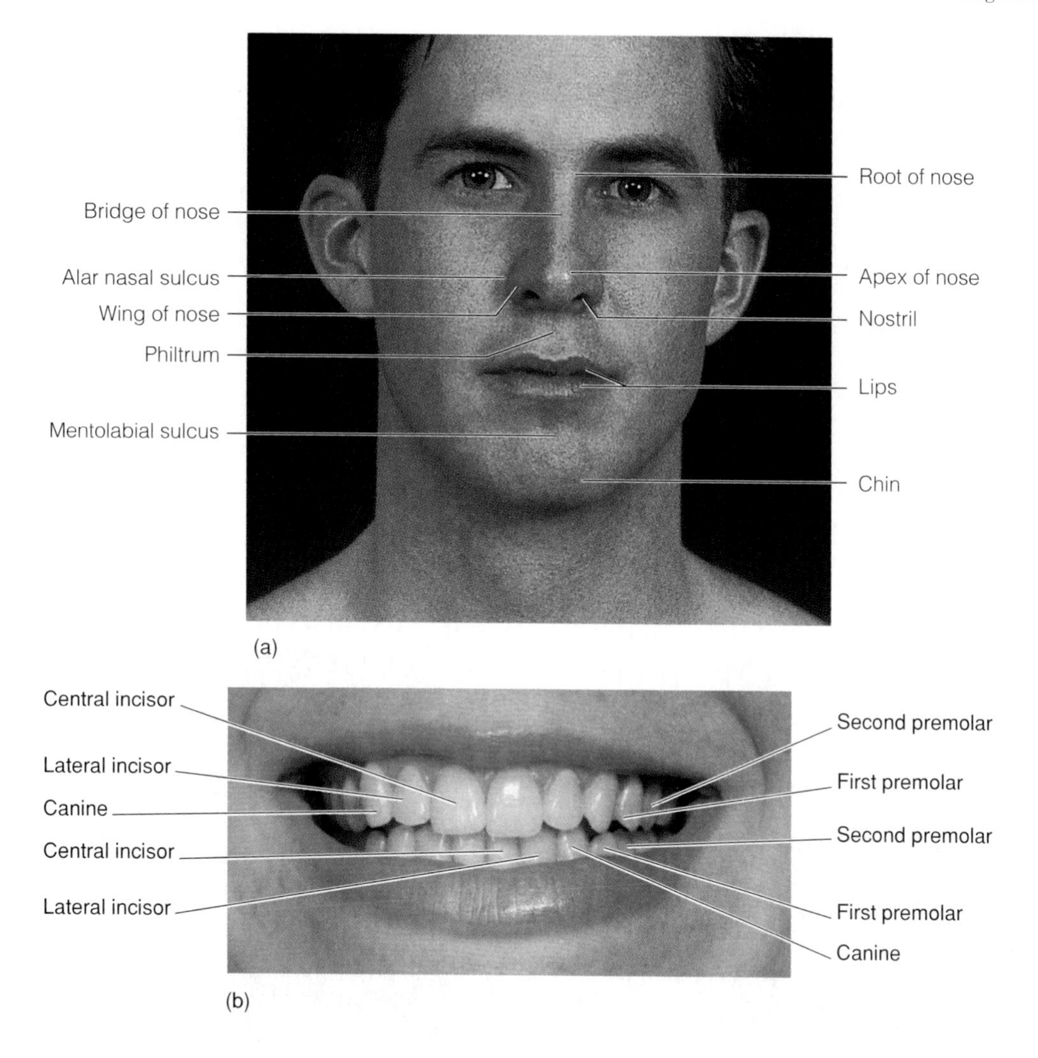

Figure 10.9

The surface anatomy of the nasal and oral regions. (*a*) The nose and lips and (*b*) the teeth.

Table 10.4
Surface Anatomy of the Nasal and Oral Regions

Structure	Comments
Root of nose (nasion)	Superior attachment of the nose to the cranium
Bridge of nose (dorsum nasi)	Bony upper framework of the nose formed by the union of nasal bones
Alar nasal sulcus	Lateral depression where the ala of the nose contacts the tissues of the face
Apex of nose	Tip of the nose
Nostril (naris)	External opening into the nasal cavity
Wing of nose (ala)	Laterally expanded border of the nostril
Philtrum	Vertical depression in the medial part of the upper lip
Lip (labium)	Upper and lower anterior borders of the mouth
Chin (mentum)	Anterior portion of the lower jaw

Internal Anatomy

The internal anatomy of the head from cadaver dissections is shown in figures 10.11 through 10.13. Figure 10.13 depicts the brain in sagittal section within the cranium. A detailed discussion of the brain, with accompanying illustrations, is presented in chapter 11. The sensory organs of the head (eyes, ears, taste buds, and olfactory receptors) are discussed and illustrated in chapter 15.

1. What are the boundaries of the cranial region and why is this region clinically important?
2. Why do scalp wounds bleed so freely? How might this relate to infections?
3. What are the subdivisions of the facial region?

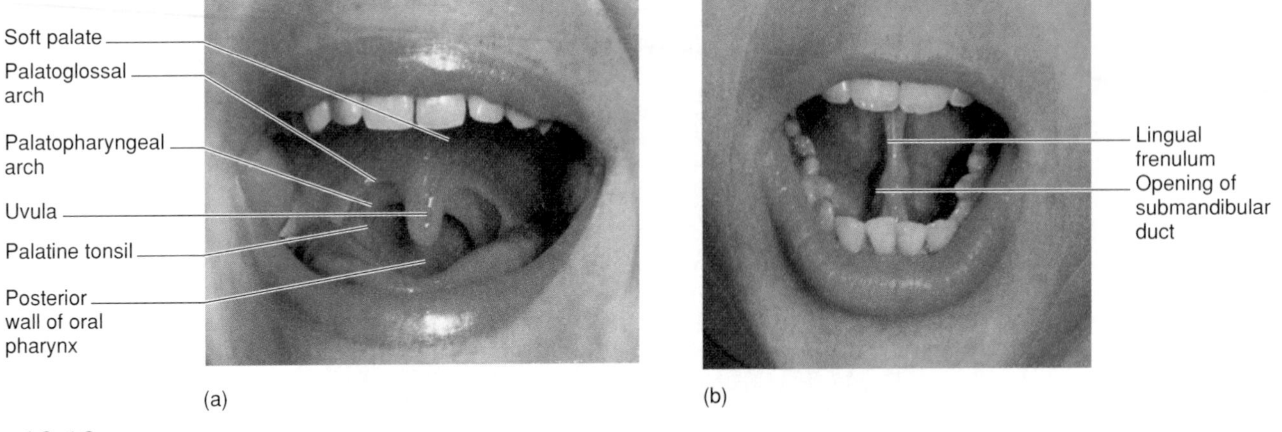

Figure 10.10

Surface structures of the oral cavity (*a*) with the mouth open and (*b*) with the mouth open and the tongue elevated.

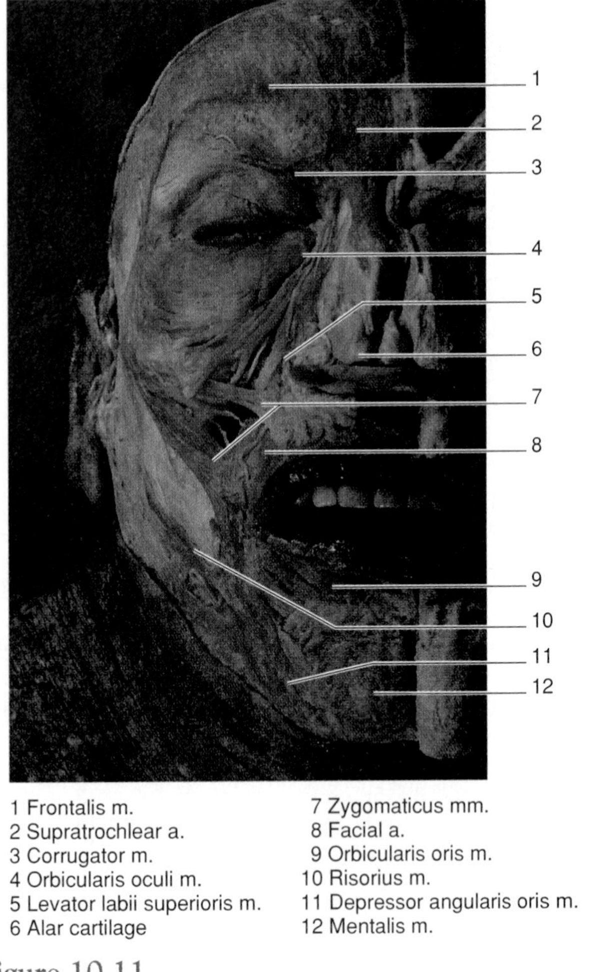

1 Frontalis m.
2 Supratrochlear a.
3 Corrugator m.
4 Orbicularis oculi m.
5 Levator labii superioris m.
6 Alar cartilage
7 Zygomaticus mm.
8 Facial a.
9 Orbicularis oris m.
10 Risorius m.
11 Depressor angularis oris m.
12 Mentalis m.

Figure 10.11

An anterior view of the muscles of the head (m. = muscle, mm. = muscles, a. = artery).

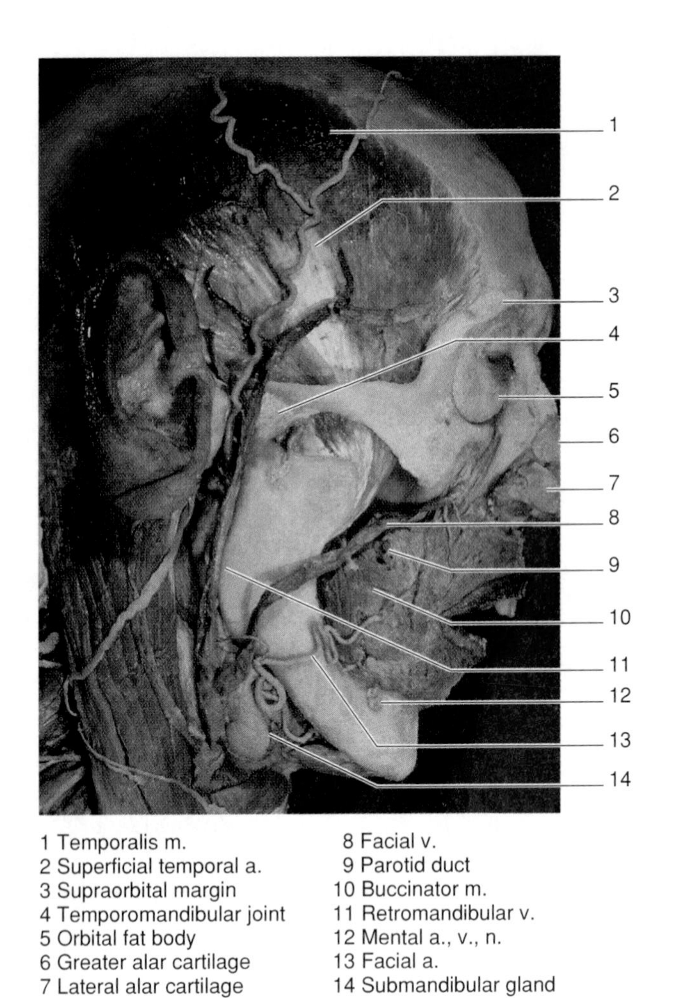

1 Temporalis m.
2 Superficial temporal a.
3 Supraorbital margin
4 Temporomandibular joint
5 Orbital fat body
6 Greater alar cartilage
7 Lateral alar cartilage
8 Facial v.
9 Parotid duct
10 Buccinator m.
11 Retromandibular v.
12 Mental a., v., n.
13 Facial a.
14 Submandibular gland

Figure 10.12

A lateral view of the deep muscles of the head (m. = muscle, a. = artery, v. = vein, n. = nerve).

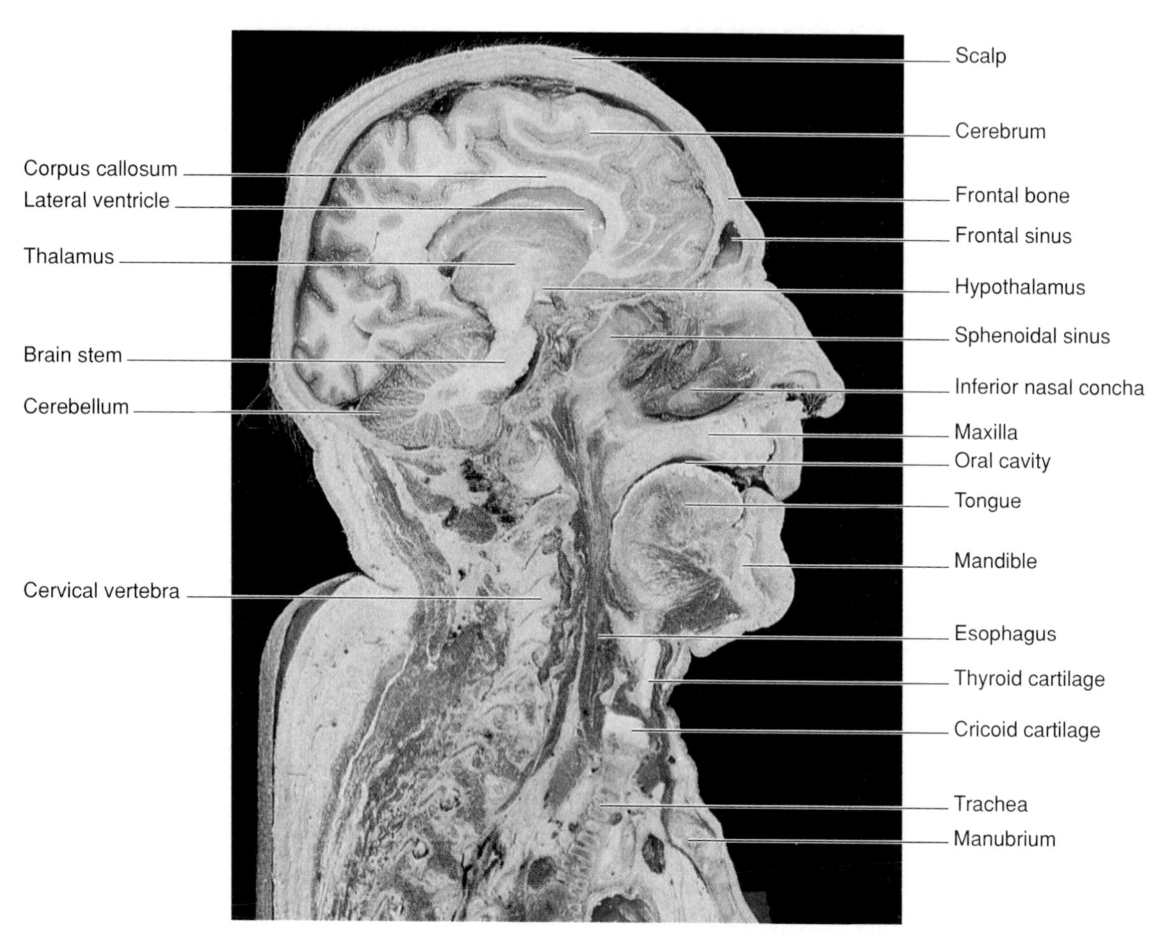

Figure 10.13

A sagittal section of the head and neck.

Neck

The flexible neck has a number of important external features. In addition, several major organs are contained within the neck, and other vital structures pass through it.

Objective 8 Discuss the functions of the neck.

Objective 9 Name and locate the triangles of the neck and list the structures contained within these triangles.

The **neck** is a complex region of the body that connects the head to the thorax. The spinal cord, trachea, esophagus, and major vessels traverse this highly flexible area. In addition, other organs are contained entirely within the neck, as are several important glands. Remarkable musculature in the neck produces an array of movements. Because of this complexity, the neck is a clinically important area. Its surface features provide landmarks for locating internal structures.

Surface Anatomy

The neck is divided into four regions: (1) an *anterior region* called the **cervix** (*ser'viks*) that contains portions of the digestive and respiratory tracts, the **larynx** (*lar'ingks*) (voice box), vessels passing to and from the head, nerves, and the **thyroid** and **parathyroid glands;** (2) right and (3) left *lateral regions*, each composed of major neck muscles and **cervical lymph nodes;** and (4) a *posterior region*, referred to as the **nucha** (*noo'ka*) which includes the spinal cord, cervical vertebrae, and associated structures.

The most prominent structure of the cervix of the neck is the **thyroid cartilage** of the larynx (fig. 10.14). The **laryngeal prominence** of the thyroid cartilage, commonly called the "Adam's apple," can be palpated on the midline of the neck. The thyroid cartilage supports the vocal folds. It is larger in males than in females because male sex hormones stimulate its growth during puberty. The **hyoid bone** can be palpated just above the larynx.

cervix: L. *cervix*, neck
larynx: Gk. *larynx*, upper windpipe
hyoid: Gk. *hyoeides*, U-shaped

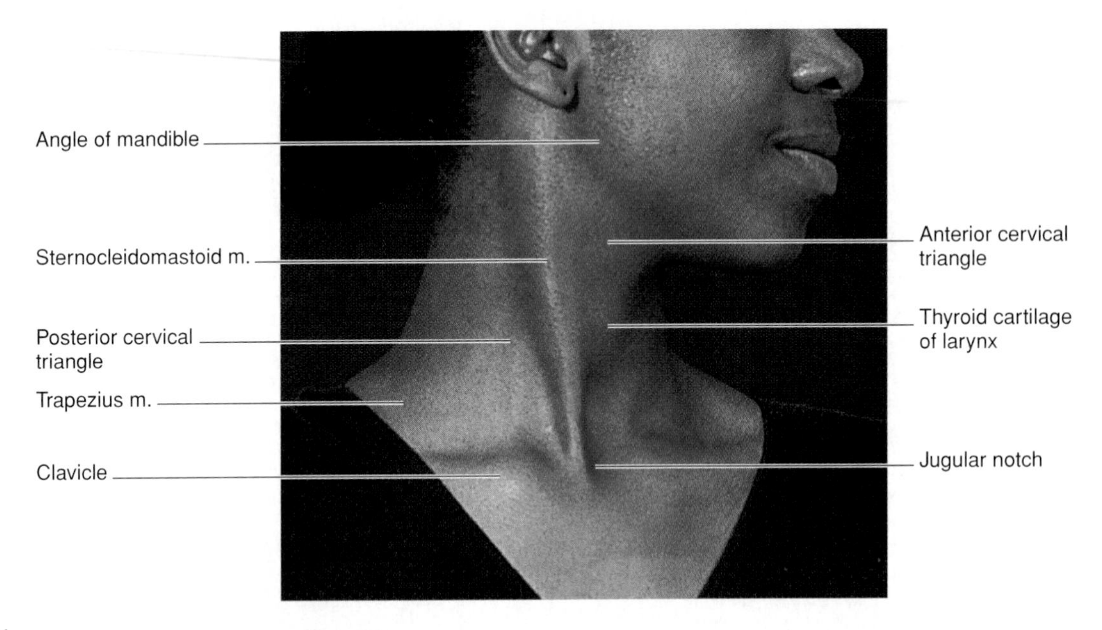

Figure 10.14

An anterolateral view of the neck.

Both of these structures are elevated during swallowing, which is one of the actions that directs food and fluid into the esophagus. Note this action on yourself by gently cupping your fingers on the larynx, and then swallowing. Directly below the thyroid cartilage is the **cricoid** (*kri'coid*) **cartilage,** followed by the **trachea** (*tra'ke-ă*) (windpipe). Both of these structures can be palpated. The cricoid cartilage serves as a landmark for locating the rings of cartilage of the trachea when creating an emergency airway *(tracheostomy)*. The thyroid gland can be palpated on either side of the neck, just below the level of the larynx. In addition, pulsations of the **common carotid** (*kă-rot'id*) **artery** can be observed on either side of the neck, just lateral and a bit superior to the level of the larynx.

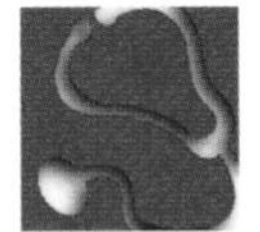
The arteries of the head and neck are rarely damaged because of their elasticity. In a severe lateral blow to the head, however, the internal carotid artery may rupture, resulting in the perception of a roaring sound as blood rushes into the cavernous sinuses of the temporal bone. Containment of carotid hemorrhage within the sinuses may actually be lifesaving.

The **jugular notch** is a V-shaped groove in the manubrium of the sternum, which creates a depression on the inferior midline of the cervix. The two **clavicles** are obvious in all people because they lie just under the skin.

The **sternocleidomastoid** (*ster"no-kli"do-mas'toid*) and **trapezius** muscles are the prominent structures of each lateral region (figs. 10.14 and 10.15). The sternocleidomastoid muscle can be palpated along its entire length when the head is turned to the side. The tendon of this muscle is especially prominent to the side of the jugular notch. The trapezius muscle can be felt when the shoulders are shrugged. An inflammation of the trapezius causes a "stiff neck." If a person is angry or if a shirt collar is too tight, the **external jugular vein** can be seen as it courses obliquely across the sternocleidomastoid muscle. **Cervical lymph nodes** of the lateral neck region may become swollen and painful from infectious diseases of the oral or pharyngeal regions.

Most of the structures of the nucha are too deep to be of importance in surface anatomy. The spines of the lower cervical vertebrae (especially C7), however, can be observed and palpated when the neck is flexed. In this same position, the **ligamentum nuchae** (*noo'ke*) (not shown) is raised, forming a firm ridge that extends superiorly from vertebra C7 to the external occipital protuberance of the skull. Clinically, the ligamentum nuchae is extremely important because of the debilitating damage it sustains from whiplash injury or a broken neck.

Triangles of the Neck

The triangles of the neck, created by the arrangement of specific muscles and bones, are clinically important because of the specific structures included in each. The structures of the neck that are important in surface anatomy have already been described, however. Thus, the two major and six minor triangles are depicted in figure 10.15 and presented in table 10.5 in summary form. The sternocleidomastoid muscle obliquely transects the neck, dividing it into an **anterior cervical triangle** and a **posterior cervical triangle.** The apex of the anterior cervical triangle is directed inferiorly. The median line of the neck forms the anterior boundary of the anterior cervical triangle; the inferior border of the mandible forms its superior boundary; and the sternocleidomastoid muscle forms its posterior boundary. The posterior cervical triangle is formed by the sternocleidomastoid

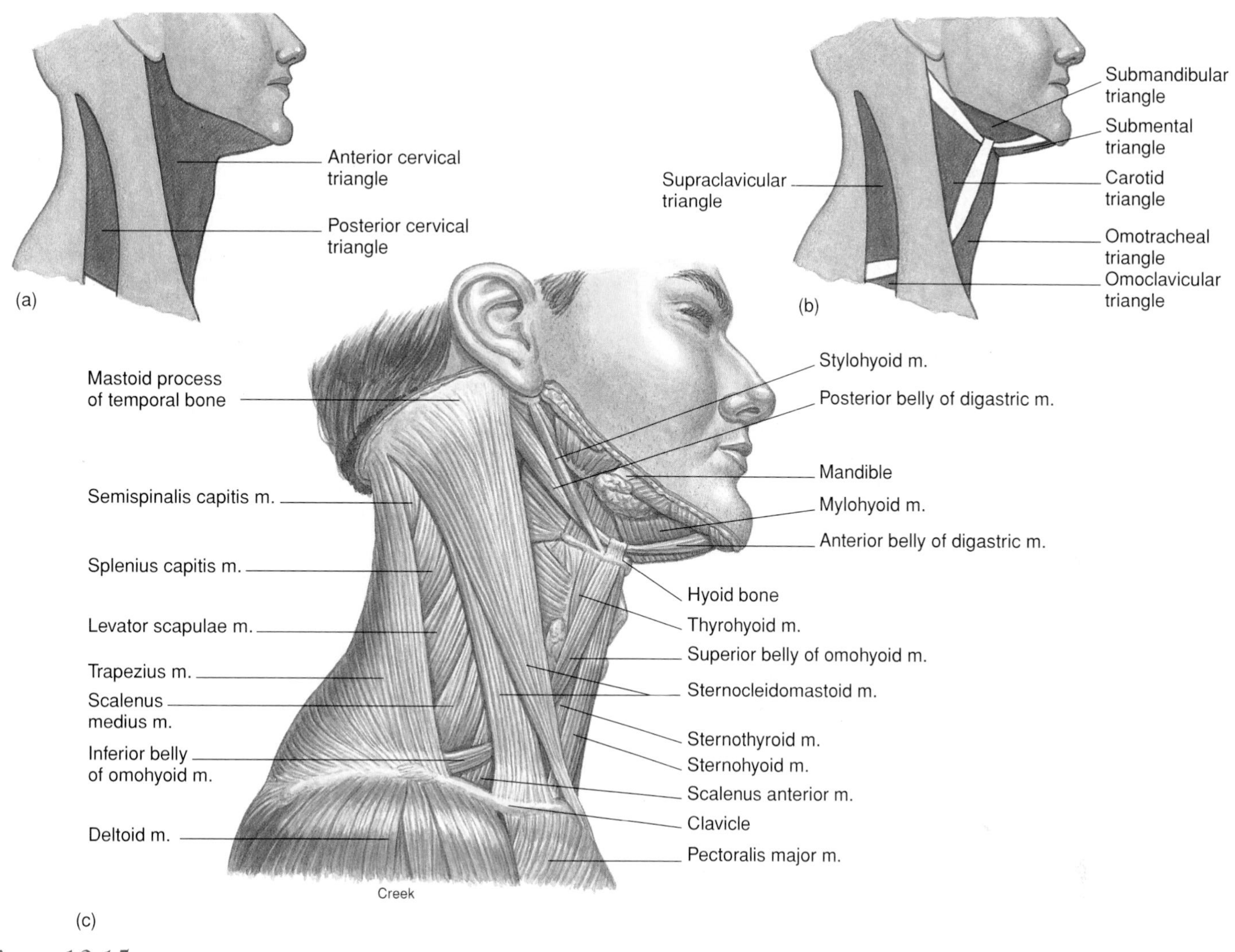

Figure 10.15

Triangles of the neck. (*a*) The two large triangular divisions, (*b*) the six lesser triangular subdivisions, and (*c*) the detailed muscular anatomy of the neck.

Table 10.5

Boundaries and Internal Structures of the Triangles of the Neck

Triangle	Boundaries	Internal Structures
Anterior cervical	Sternocleidomastoid muscle; median line of neck; inferior border of mandible	Four lesser triangles contain salivary glands, larynx, trachea, thyroid glands, and various vessels and nerves
Carotid	Sternocleidomastoid, posterior digastric, and omohyoid muscles	Carotid arteries, internal jugular vein, and vagus nerve
Submandibular	Digastric muscle (both heads); inferior border of mandible	Salivary glands
Submental	Digastric muscle; hyoid bone (This is the only unpaired triangle of the neck.)	Muscles of the floor of the mouth and salivary glands and ducts
Omotracheal	Sternocleidomastoid and omohyoid muscles; midline of neck	Larynx, trachea, thyroid gland, and carotid sheath
Posterior cervical	Sternocleidomastoid and trapezius muscles; clavicle	Nerves and vessels
Supraclavicular	Sternocleidomastoid, trapezius, and omohyoid muscles	Cervical plexus and accessory nerve
Omoclavicular	Sternocleidomastoid and omohyoid muscles; clavicle	Brachial plexus and subclavian artery

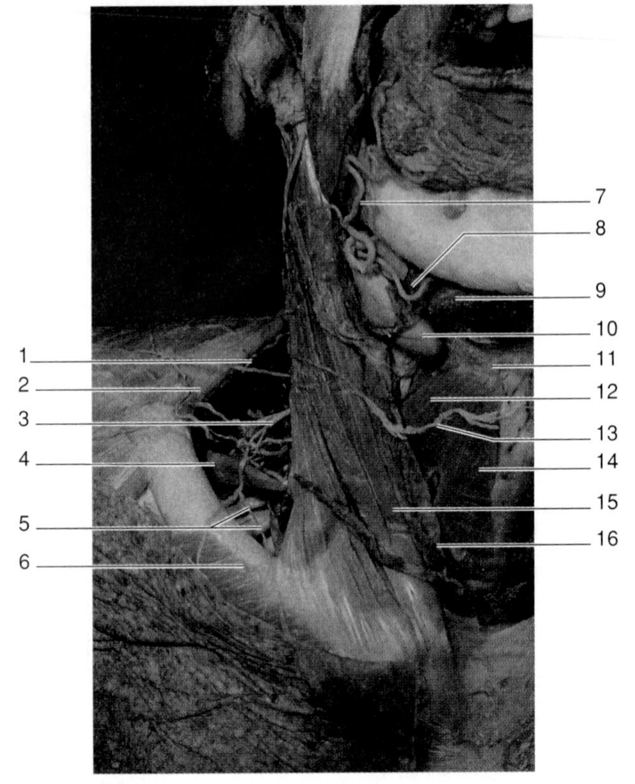

1 Accessory n.
2 Trapezius m.
3 Supraclavicular n.
4 Omohyoid m.
5 Brachial plexus
6 Clavicle
7 Facial a.
8 Mylohyoid m.
9 Digastric m.
10 Submandibular gland
11 Hyoid bone
12 Omohyoid m.
13 Transverse cervical n.
14 Sternohyoid m.
15 Sternocleidomastoid m.
16 External jugular v.

Figure 10.16

An anterior view of the right cervical region.

muscle anteriorly and the trapezius muscle posteriorly; the clavicle forms its base inferiorly.

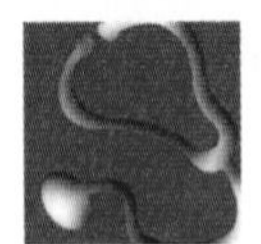

Three structures traversing the neck are extremely important and potentially vulnerable. These structures are the common carotid artery, which carries blood to the head; the internal jugular vein, which drains blood from the head; and the vagus nerve, which conducts nerve impulses to visceral organs. These structures are protected in the neck by their deep position behind the sternocleidomastoid muscle and by their enclosure in a tough connective tissue called the *carotid sheath.*

Internal Anatomy

The internal anatomy of the neck from cadaver dissections is shown in figures 10.16 and 10.17. The organs of the neck are highly integrated and packed into a relatively small area. The neck has to support the head, at the same time permitting flexibility.

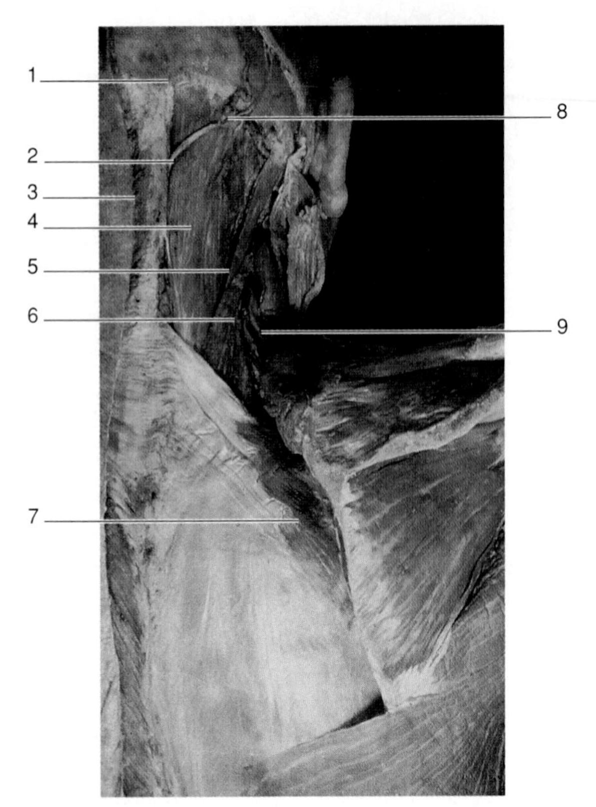

1 Occipital bone
2 Greater occipital n.
3 Ligamentum nuchae
4 Semispinalis capitis m.
5 Longissimus capitis m.
6 Longissimus cervicis m.
7 Serratus posterior m.
8 Occipital a.
9 Levator scapulae m.

Figure 10.17

A posterior view of the deep cervical muscles.

1. List four functions of the neck. Which body systems are located, in part, within the neck.
2. What are the structural regions of the neck? Identify the structures included in each region?
3. With reference to the triangles of the neck, where would you palpate to feel (a) a pulse, (b) the trachea, (c) cervical lymph nodes, and (d) the thyroid gland?

Trunk

The locations of vital visceral organs in the cavities of the trunk make the surface anatomy of this body region especially important.

Objective 10 Identify various surface features of the trunk by observation or palpation.

Objective 11 List the auscultation sites of the thorax and abdomen.

Surface Anatomy

The **trunk,** or *torso,* is divided into the **back, thorax** (chest), **abdomen** (venter), and **pelvis.** A region called the *perineum* forms the floor of the pelvis and includes the external genitalia. The pelvis and perineum are discussed in the following section.

The surface anatomy of the trunk is particularly important in determining the location and condition of the visceral organs. Some of the surface features may be obscured, however, because of age, sex, or body weight.

Back

No matter how obese a person may be, a **median furrow** can be seen on the back, along with some of the spinous process of the vertebrae (fig. 10.18). The entire series of vertebral spines can be observed when the vertebral column is flexed. This position is important in determining defects of the vertebral column (see Clinical Considerations in chapters 8 and 11). The back of the **scapula** (*skap'yoo-lă*) presents other important surface landmarks. The base of the spine of the scapula is level with the third thoracic vertebra, and the inferior angle of the scapula is even with the seventh thoracic vertebra. Several muscles of the scapula can be observed on a lean, muscular person and are identified in figure 10.18. Many of the ribs and muscles that attach to the ribs can be seen in a lateral view (fig. 10.19).

Two pairs of triangles on the back are clinically important. The **triangle of auscultation** (see fig. 10.18) is bounded by the trapezius muscle, the latissimus dorsi muscle, and the medial border of the scapula. Because there is a space between the superficial back muscles in this area, heart and respiratory sounds are not muffled by the muscles when a stethoscope is placed here.

Thorax

The leading causes of death in the United States are associated with disease or dysfunction of the thoracic organs. With the exception of the breasts and surrounding lymph nodes, the organs of the thorax are located within the rib cage. The paired clavicles and the jugular notch have already been identified as important surface features of the neck, with regard to the thoracic region (fig. 10.20), these structures serve as reference points for counting the ribs. Many of the ribs can be seen on a thin person. All but the first, and at times the twelfth, can be palpated. The sternum is composed of three separate bones (manubrium, body, and xiphoid process), each of which can be palpated. The **sternal angle** is felt as an elevation between the manubrium and body of the sternum. The sternal angle is important because it is located at the level of the second rib. The articulation between the body of the sternum and the xiphoid process, called the **xiphisternal** (*zif"ĭ-ster'nal*) **joint,** is positioned over the lower border of the heart and the diaphragm. The **costal arch** of the rib cage is the lower oblique boundary and can be easily identified when a person inhales and holds his or her breath.

The nipples in the male (fig. 10.20) are located at the fourth intercostal spaces (the area between the fourth and fifth ribs), about 10 cm (4 in.) from the midline. In sexually mature women, their position varies according to age and the size and pendulousness of the breasts (fig. 10.21). The position of the left nipple in males is an important landmark for knowing where to listen to various heart sounds and for determining whether the heart is enlarged. For diagnostic purposes, an imaginary line, the **midclavicular line,** can be extended vertically from the middle of the clavicle through the nipple. Several superficial chest muscles can be observed or palpated and are therefore important surface features. These muscles are depicted in figures 10.20 and 10.21.

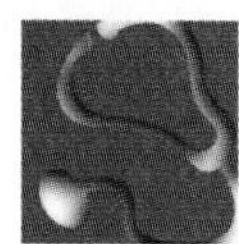

In addition to helping one know where to listen with a stethoscope to heart sounds, surface features of the thorax are important for auscultations of the lungs, radiographs, tissue biopsies, sternal taps for bone marrow studies, and thoracic surgery. Although the anatomical features of the rib cage are quite consistent, some people exhibit slight deformities and asymmetries. They are generally not disabling and require no treatment. Most of the abnormalities are congenital and include such conditions as a projecting sternum (pigeon breast) or a receding sternum (funnel chest).

Abdomen

The abdomen is the portion of the body between the diaphragm and the pelvis. Because it does not have a bony framework like that of the thorax the surface anatomy is not as well defined. Bony landmarks of both the thorax and pelvis are used when referring to abdominal structures (fig. 10.22). The **right costal arch** of the rib cage is located over the liver on the right side, and the **left costal arch** is positioned over the stomach on the left. The xiphoid process is important because from this point a tendinous, midventral raphe, the **linea alba** (*lin'e'ă al'bă*), extends the length of the abdomen to attach to the **symphysis pubis.** The symphysis pubis can be palpated at the anterior union of the two halves of the pelvic girdle. The **navel,** or **umbilicus,** is the site of attachment of the fetal umbilical cord and is located along the linea alba. The linea alba separates the paired, straplike **rectus abdominis** muscles, which can be seen when a person flexes the abdomen (as when doing sit-ups).

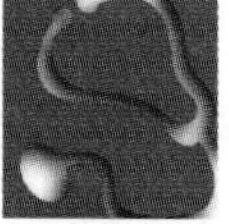

Clinically, the linea alba is a favored site for abdominal surgery because an incision made along this line severs no muscles and few vessels or nerves. Moreover, the linea alba heals readily. It has been said that only a zipper would provide a more convenient entry to the abdominal cavity.

---

linea alba: L. *linea,* line; *alba,* white
navel: O. E. *nafela,* umbilicus
umbilicus: L. *umbilicus,* navel

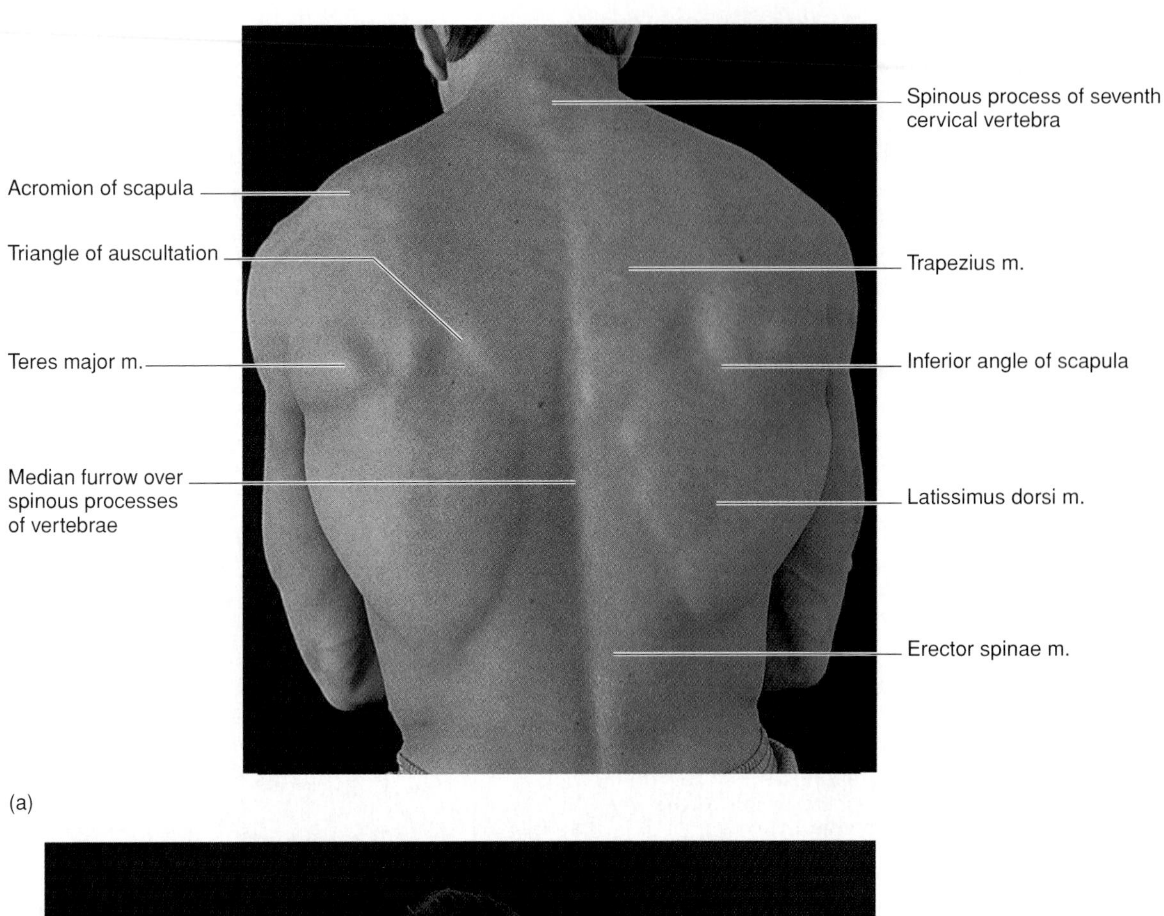

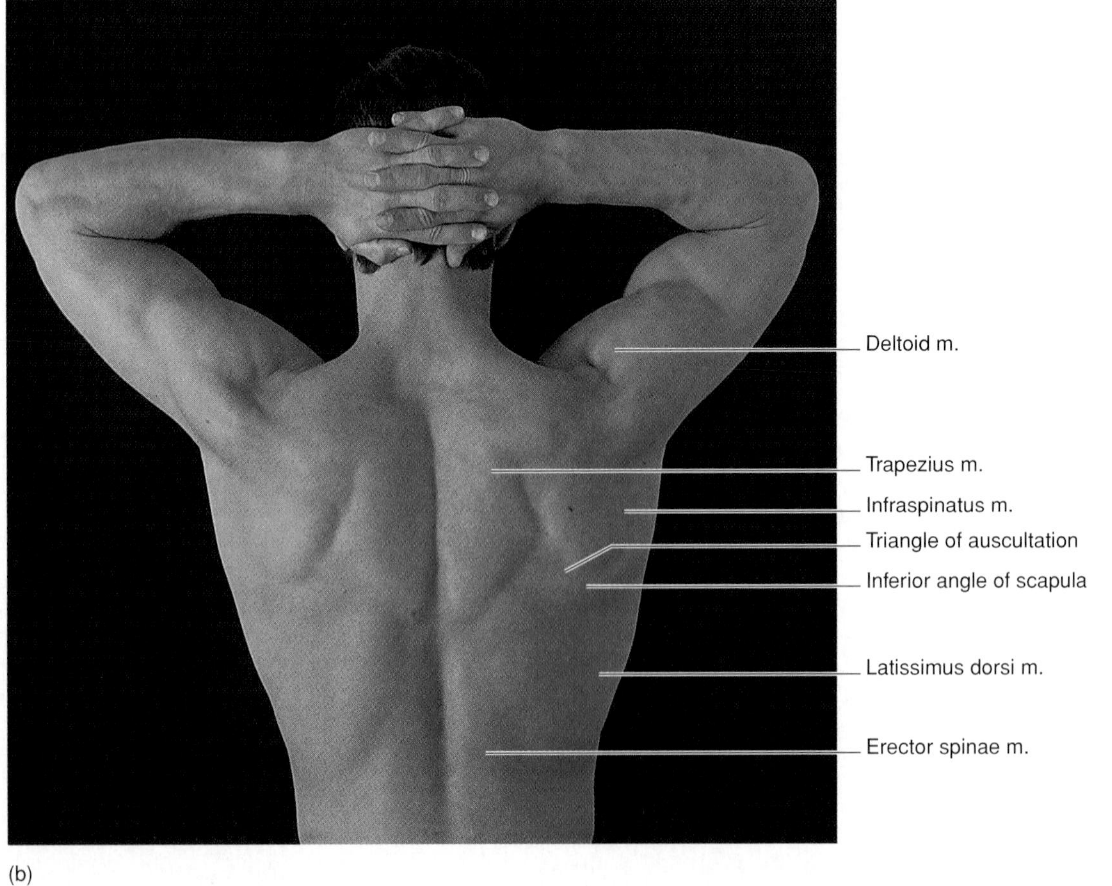

Figure 10.18

The surface anatomy of the back. (*a*) Adduction of the limbs at the shoulder joints and (*b*) abduction of the shoulder joints and flexion of the elbow joints.

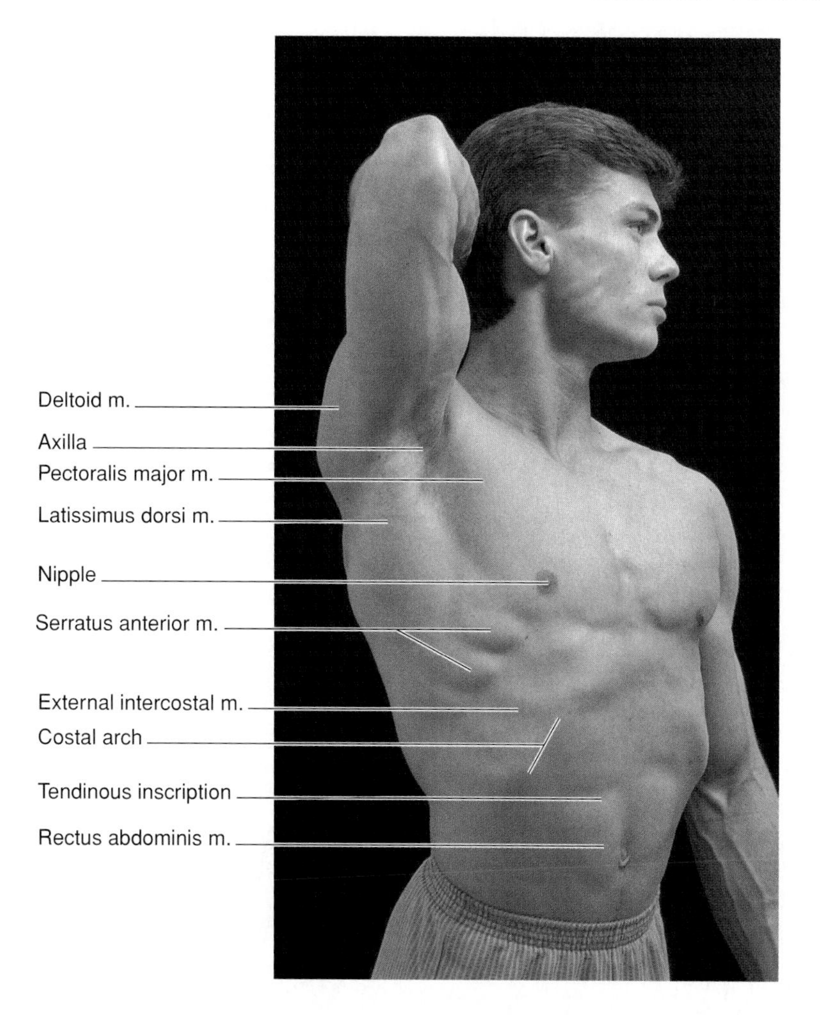

Figure 10.19

An anterolateral view of the trunk and axilla.

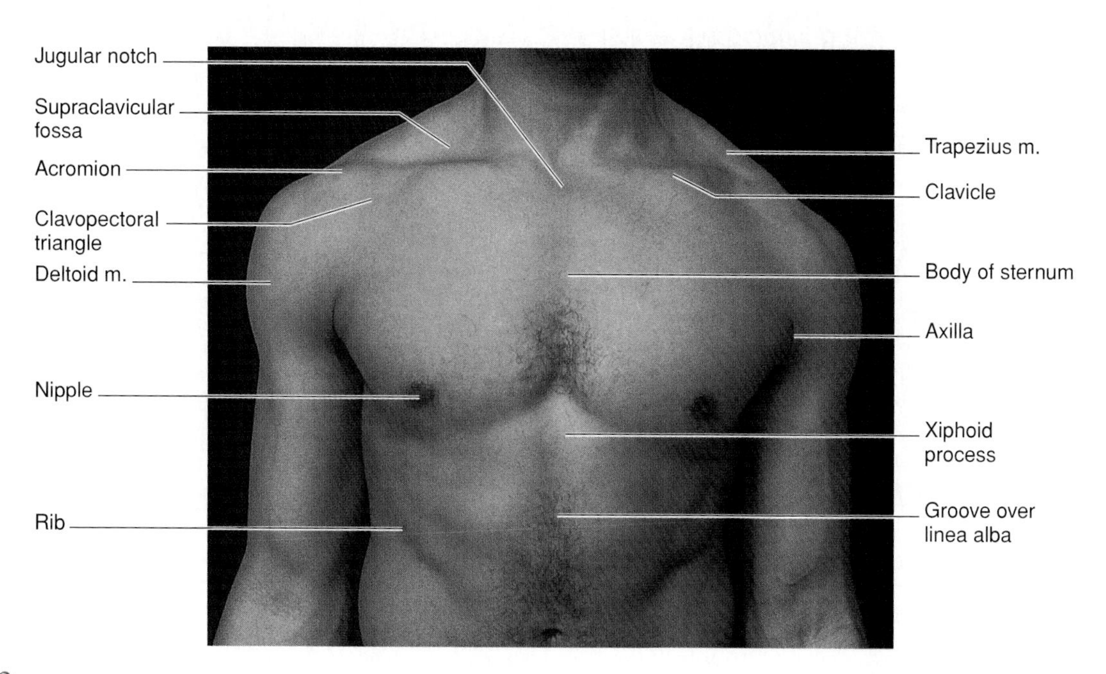

Figure 10.20

The surface anatomy of the anterior thoracic region of the male.

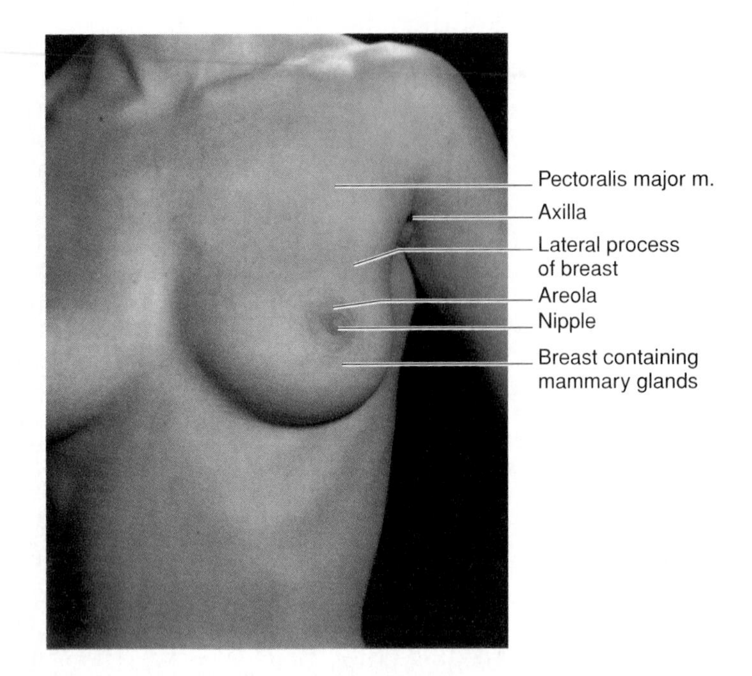

Figure 10.21

The surface anatomy of the female breast.

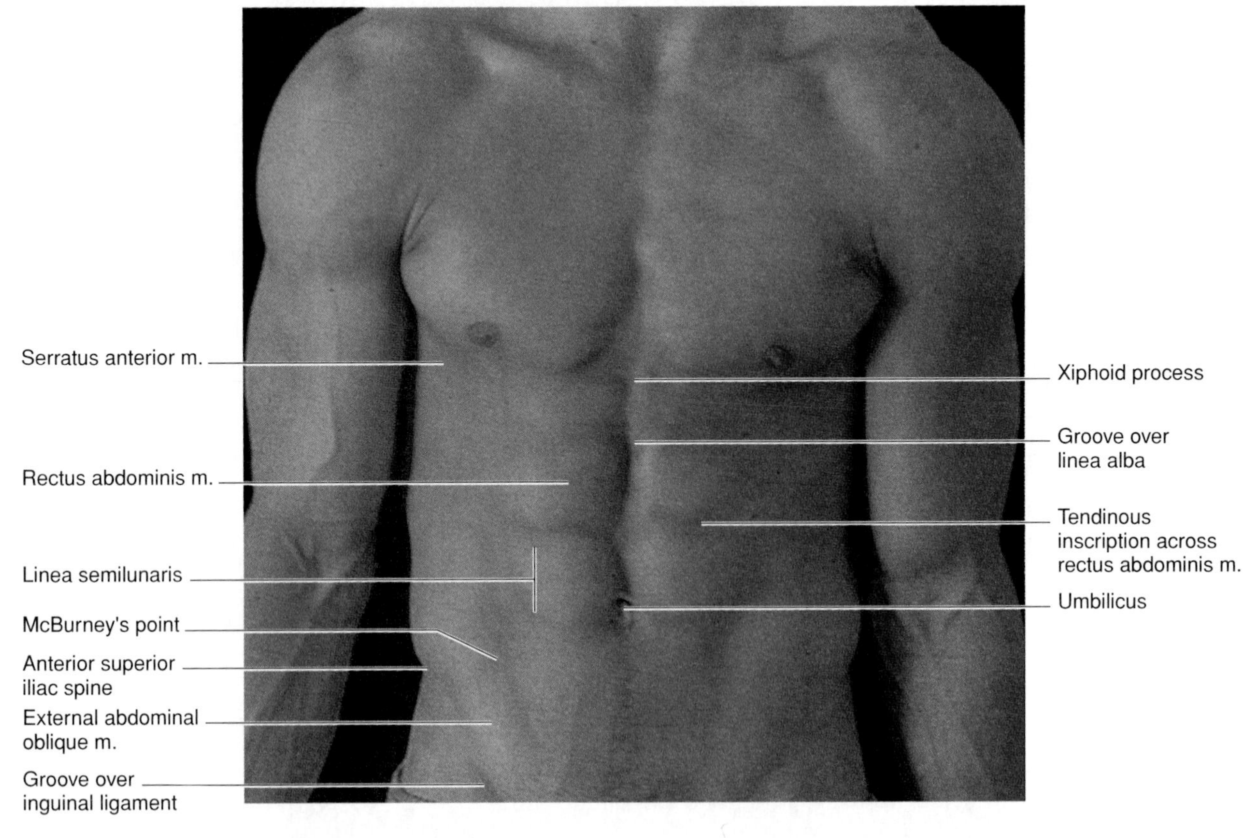

Figure 10.22

The surface anatomy of the anterior abdominal region.

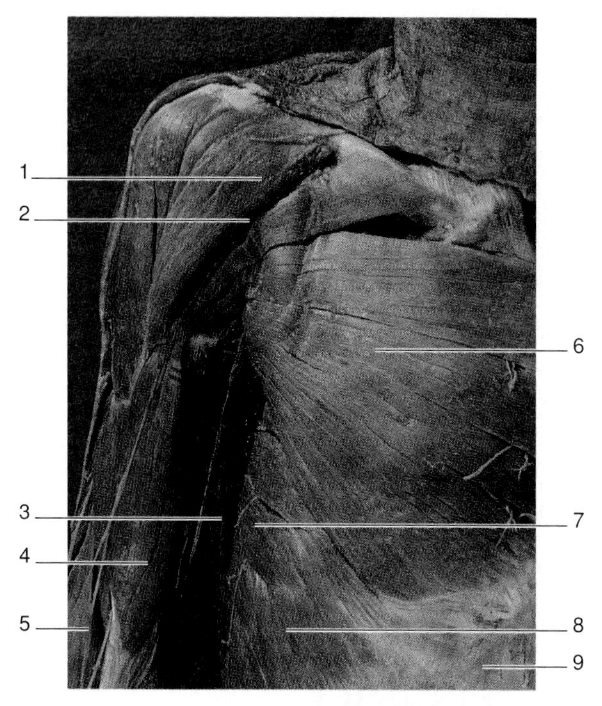

1 Deltoid m.
2 Cephalic v.
3 Latissimus dorsi m.
4 Biceps brachii m.
5 Brachioradialis m.
6 Pectoralis major m.
7 Serratus anterior m.
8 External abdominal oblique m.
9 Sheath of rectus abdominis m.

Figure 10.23

An anterior view of the superficial muscles of the right thorax, shoulder, and brachium.

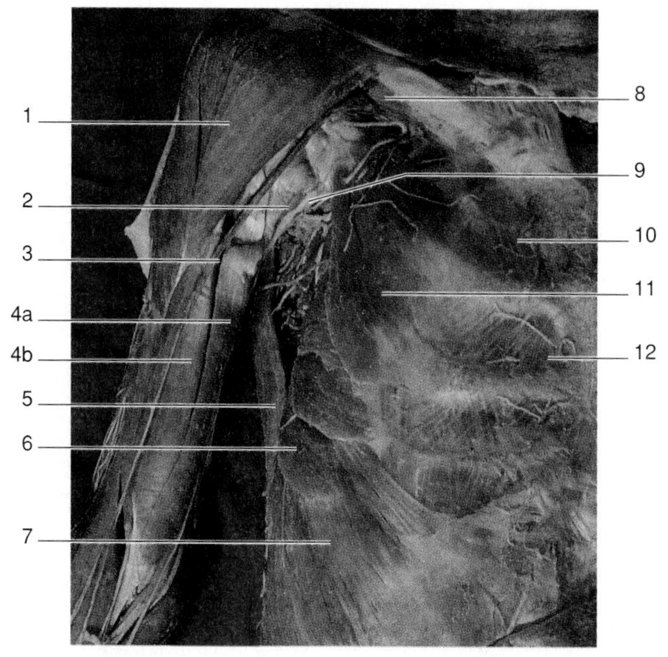

1 Deltoid m.
2 Coracobrachialis m.
3 Cephalic vein
4 Biceps brachii m.:
 4a Short head
 4b Long head
5 Latissimus dorsi m.
6 Serratus anterior m.
7 External abdominal oblique m.
8 Subclavius m.
9 Brachial plexus
10 Internal intercostal m.
11 Pectoralis minor m.
12 External intercostal m.

Figure 10.24

An anterior view of the deep muscles of the right thorax, shoulder, and brachium.

The lateral margin of the rectus abdominis muscle can be observed on some individuals, and the surface line it produces is called the **linea semilunaris.** The **external abdominal oblique** muscle is the superficial layer of the muscular abdominal wall. The **iliac crest** is subcutaneous and can be palpated along its entire length. The highest point of the crest lies opposite the body of the fourth lumbar vertebra, an important level in spinal anesthesia. Another important landmark is **McBurney's point,** located about one-third of the distance from the right anterior superior iliac spine on a line between that spine and the navel (fig. 10.22). This point overlies the appendix of the GI tract. In surgical removal of the appendix (*appendectomy*), an oblique incision is made through McBurney's point.

The abdominal region is frequently divided into nine regions or four quadrants in order to describe the location of internal organs and to clinically identify the sites of various pains or conditions. These regions have been adequately described in chapter 2 (see figs. 2.15 and 2.16).

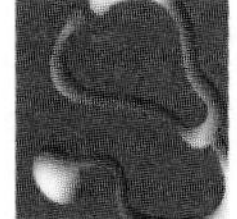

Although the position of the umbilicus is relatively consistent in all people, its shape and health is not. For example, there may be an opening to the outside, called a *fistula,* or herniation of some of the abdominal contents. Acquired umbilical hernias may develop in children who have a weak abdominal wall in this area, or they may develop in pregnant women because of the extra pressure exerted at this time.

The umbilicus is a common site for an incision into the abdominal cavity in a procedure called *laparotomy.* Laparotomy is frequently done to examine or perform surgery on the internal female reproductive organs. A depressed umbilicus on an obese person is difficult to keep clean, and so various types of infections may occur there.

Internal Anatomy

Thorax

Included in the internal anatomy of the thorax (figs. 10.23 through 10.29) are the rib cage and its contents, the thoracic

McBurney's point: from Charles McBurney, American surgeon, 1845–1914

laparotomy: Gk. *lapara,* flank; *tome,* incision

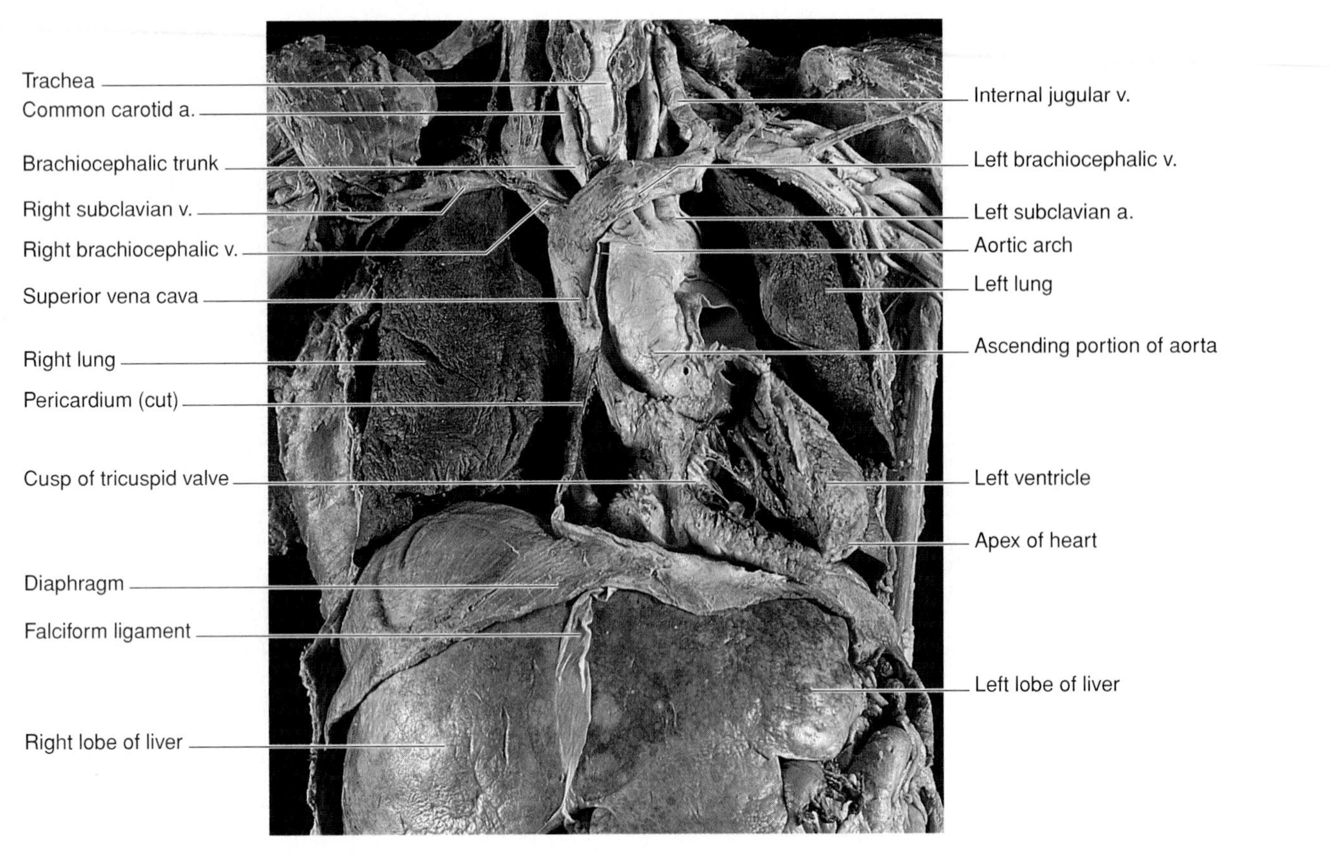

Figure 10.25

Viscera of the thorax. The heart has been coronally sectioned to expose the chambers.

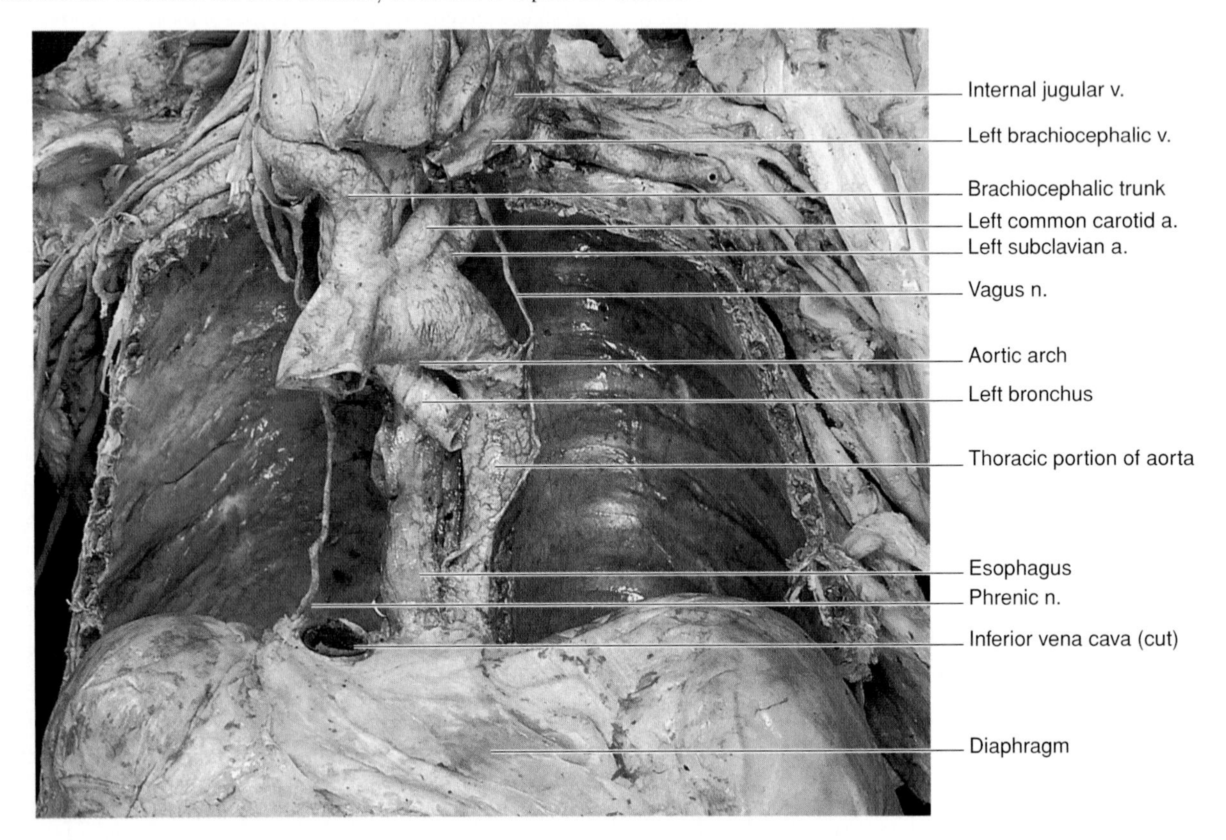

Figure 10.26

The thoracic cavity with the heart and lungs removed.

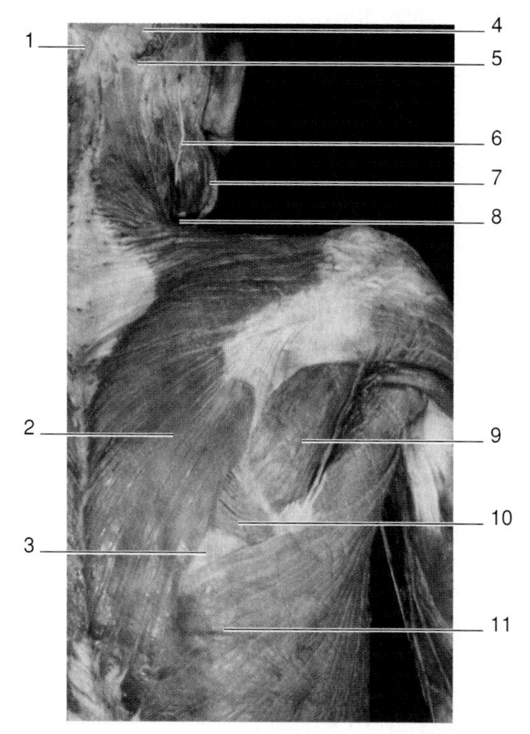

1 External occipital protuberance
2 Trapezius m.
3 Triangle of auscultation
4 Occipital a.
5 Greater occipital n.
6 Lesser occipital n.
7 Sternocleidomastoid m.
8 Great auricular n.
9 Infraspinatus m.
10 Rhomboideus major m.
11 Latissimus dorsi m.

Figure 10.27

A posterior view of the superficial muscles of the right thorax and neck.

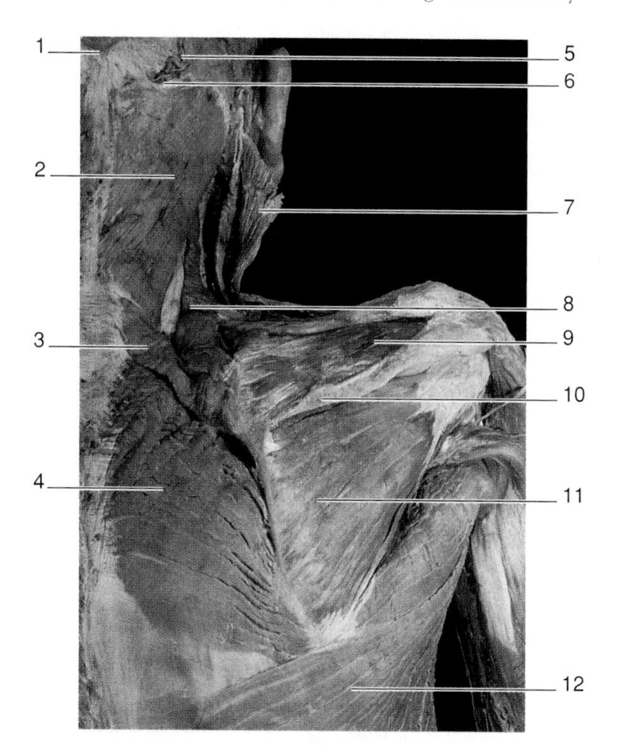

1 External occipital protuberance
2 Splenius capitis m.
3 Rhomboideus minor m.
4 Rhomboideus major m.
5 Occipital a.
6 Great occipital n.
7 Sternocleidomastoid m. (cut)
8 Levator scapulae m.
9 Supraspinatus m.
10 Spine of scapula
11 Infraspinatus m.
12 Latissimus dorsi m.

Figure 10.28

A posterior view of the deep structures of the right thorax and neck.

musculature, and the mammary glands and breasts of a female. The rib cage is formed by the sternum, the costal cartilages, and the ribs attached to the thoracic vertebrae. It protects the lungs, several large vessels, and the heart. It also affords a site of attachment for the muscles of the thorax, upper extremities, back, and diaphragm. The principal organs of the respiratory and circulatory systems are located within the thorax, and the esophagus of the digestive system passes through the thorax. Because the viscera of the thoracic cavity are vital organs, the thorax is of immense clinical importance.

Abdomen

The cavity of the abdomen contains the stomach and intestines, the liver and gallbladder, the kidneys and adrenal glands, the spleen, the internal genitalia, and major vessels and nerves. Because of the domed shape of the diaphragm, some of the abdominal viscera are protected by the rib cage. The abdominal region is shown in photographs of cadavers in figures 10.30, 10.31, and 10.32.

1. Which structures of the trunk can be readily observed? Which can be palpated?
2. Where are the common auscultation sites of the trunk located?
3. Why are the linea alba, costal margins, linea semilunaris, and McBurney's point important landmarks?

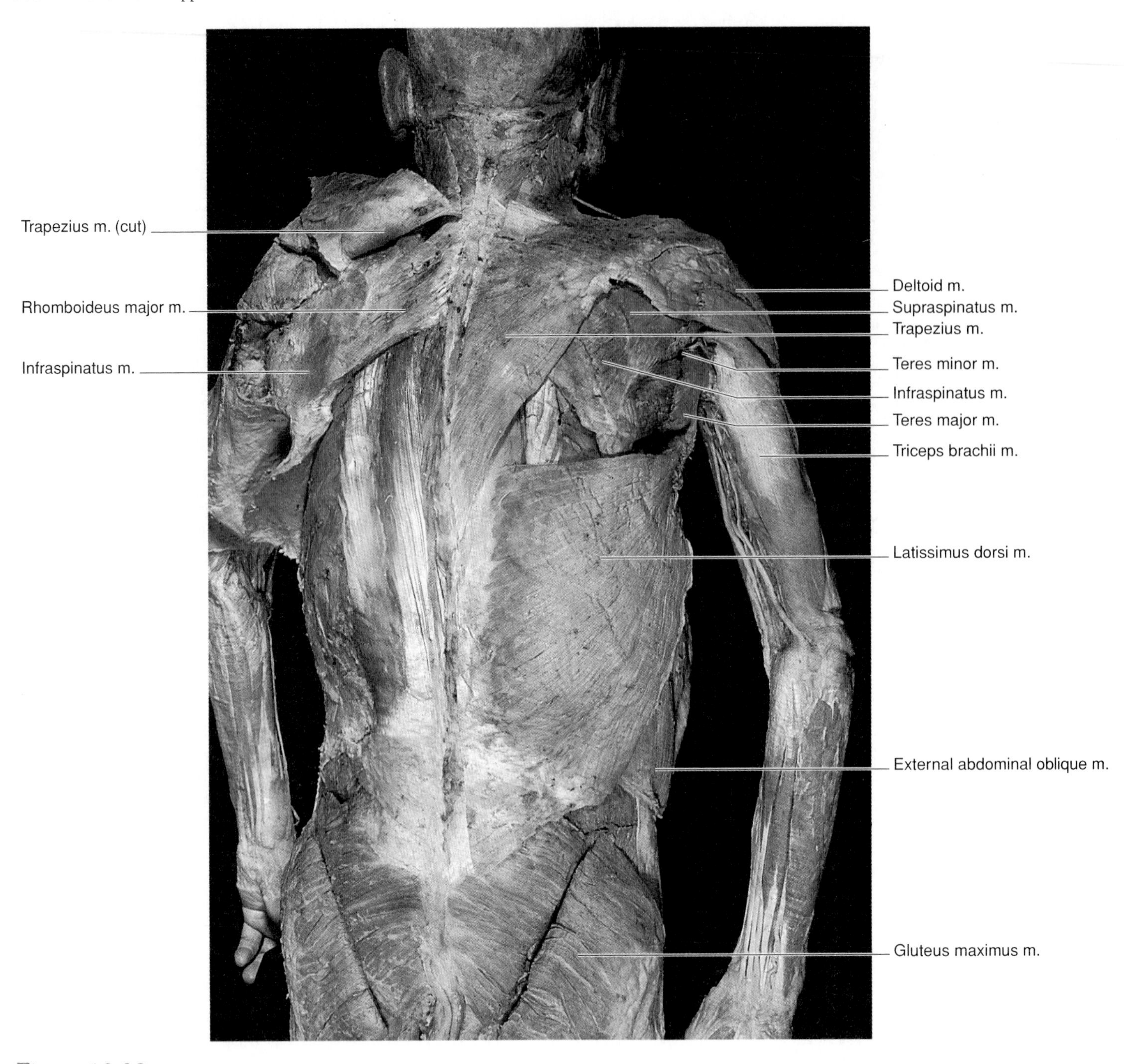

Figure 10.29

A posterior view of the trunk with deep muscles exposed on the left.

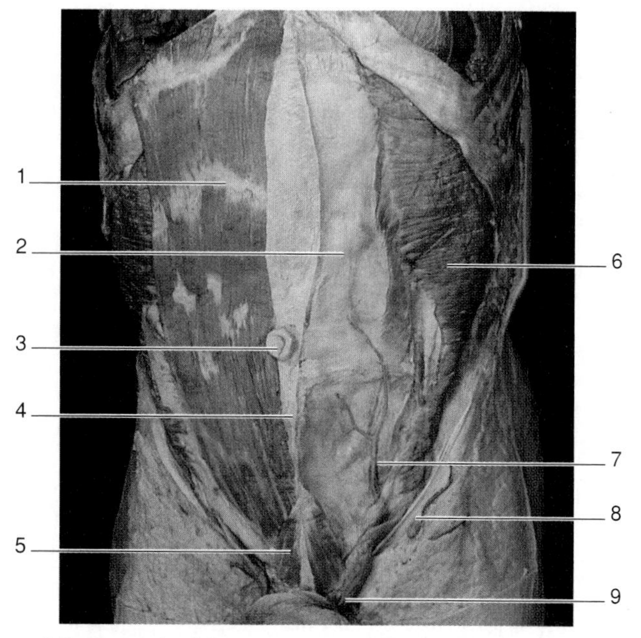

1 Rectus abdominis m.
2 Sheath of rectus abdominis m.
3 Umbilicus
4 Linea alba
5 Pyramidalis m.
6 Transverse abdominis m.
7 Inferior epigastric a.
8 Inguinal ligament
9 Spermatic cord

Figure 10.30

An anterior view of the structures of the abdominal wall.

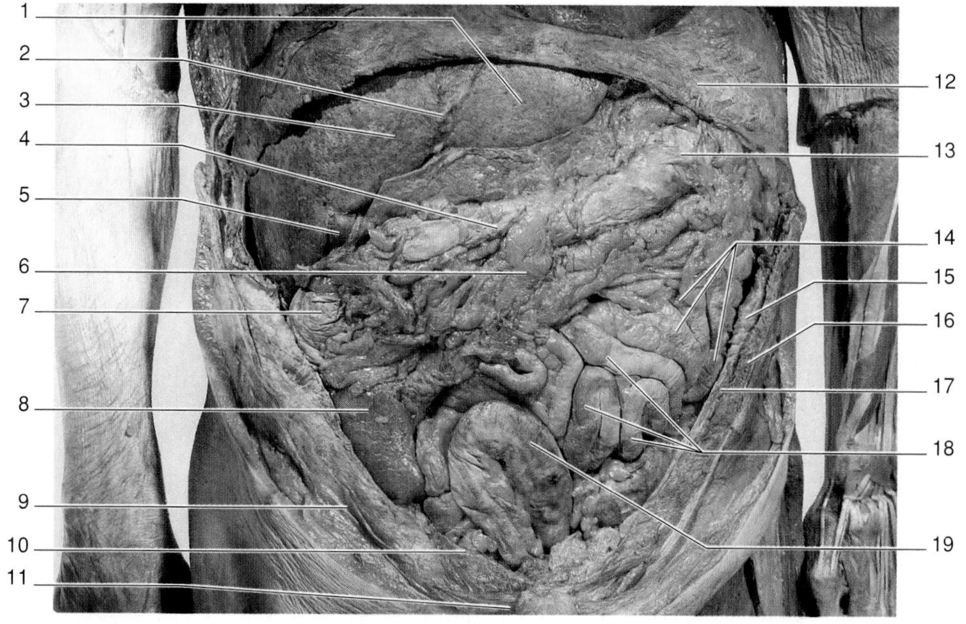

1 Left lobe of liver
2 Falciform ligament
3 Right lobe of liver
4 Transverse colon
5 Gallbladder
6 Greater omentum
7 Hepatic flexure of colon
8 Fat deposit within greater omentum
9 Aponeurosis of internal abdominal oblique m.
10 Rectus abdominis m. (cut)
11 Sheath of rectus abdominis m. (cut)
12 Diaphragm
13 Splenic flexure of colon
14 Jejunum
15 Transversus abdominis m. (cut)
16 Internal and external abdominal oblique mm. (cut)
17 Parietal peritoneum (cut)
18 Ileum
19 Sigmoid colon

Figure 10.31

An anterior view of the abdominal viscera.

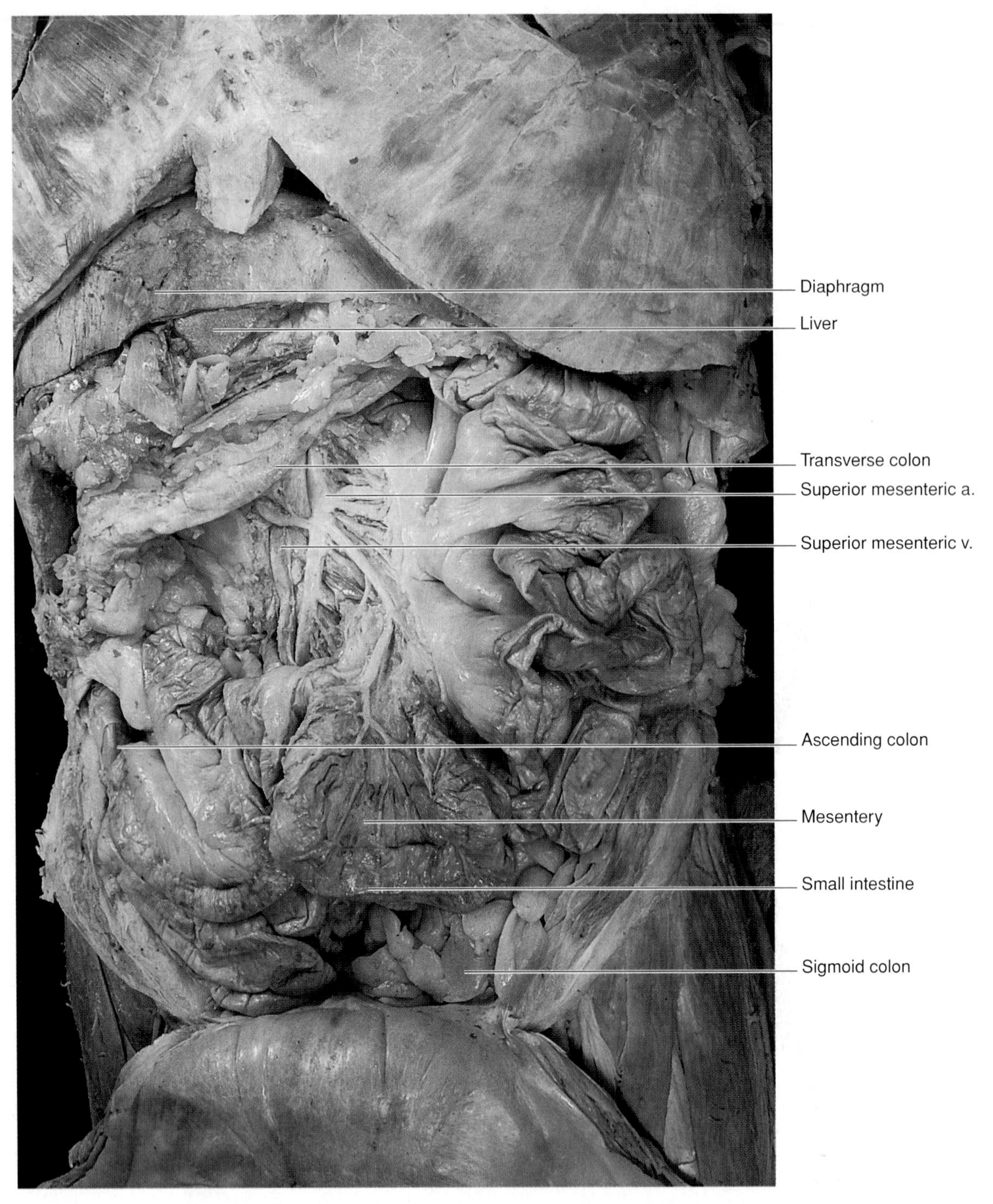

Figure 10.32

An anterior view of the abdominal viscera with the greater omentum removed and the small intestine displaced to the left.

Pelvis and Perineum

The surface features of the pelvic region are important primarily to identify reproductive organs and clinical problems of these organs.

Objective 12 Describe the location of the perineum and list the organs of the pelvic and perineal regions.

The important bony structures of the **pelvis** include the crest of the ilium and the symphysis pubis, located anteriorly, and the ischium and coccyx, which are palpable posteriorly. An **inguinal** (*ing'wĭ-nal*) **ligament** extends from the anterior superior iliac spine to the symphysis pubis and is clinically important because

inguinal: L. *inguinalis*, groin

hernias occur along it. Although the inguinal ligament cannot be seen, an oblique groove overlying the ligament is an apparent surface feature.

The **perineum** (*per″ĭ-ne′um*) (see fig. 2.17) is the region that contains the external sex organs and the anal opening. The surface features of this region are further discussed in chapters 20 and 21. The surface anatomy of the perineum of a female becomes particularly important during parturition.

1. Define the term *perineum*. What structures are located within the perineum?
2. List three body systems that have openings within the pelvic region.

Shoulder and Upper Extremity

The anatomy of the shoulder and upper extremity is of clinical importance because of frequent trauma to these body regions. In addition, vessels of the upper extremity are used as pressure points and as sites for venipuncture for drawing blood, providing nutrients and fluids, and administering medicine.

Objective 13 Identify various surface features of the shoulder and upper extremity by observation or palpation.

Objective 14 Discuss the clinical importance of the axilla, cubital fossa, and wrist.

Surface Anatomy

Shoulder

The scapula, clavicle, and proximal portion of the humerus are the bones of shoulder, and portions of each of them are important surface landmarks in this region. Posteriorly, the spine of the scapula and acromion are subcutaneous and easily located.

The acromion and clavicle, as well as several large shoulder muscles, can be seen anteriorly (fig. 10.33). The rounded curve of the shoulder is formed by the thick deltoid muscle that covers the greater tubercle of the humerus. The deltoid muscle frequently serves as a site for intramuscular injections. The large pectoralis major muscle is prominent as it crosses the shoulder joint and attaches to the humerus. A small depression, the **clavipectoral triangle** (fig. 10.33), is situated below the clavicle and is bounded on either side by the deltoid and pectoralis major muscles.

acromion: Gk. *akros*, extreme, tip; *omion*, small shoulder

Figure 10.33

An anterior view of the right shoulder region.

Axilla

The axilla is commonly called the armpit. This depressed region of the shoulder supports axillary hair in sexually mature individuals. The axilla is clinically important because of the subcutaneous position of vessels, nerves, and lymph nodes in this region. Two muscles form the anterior and posterior borders (fig. 10.34). The anterior axillary fold is formed by the pectoralis major muscle, and the posterior axillary fold consists primarily of the latissimus dorsi muscle as it extends from the lumbar vertebrae to the humerus. Axillary lymph nodes are palpable in some individuals.

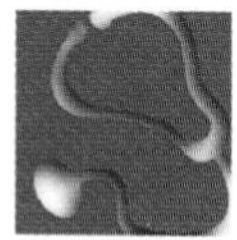

In sexually mature females, the lateral process of the mammary gland, which is positioned on the pectoralis major muscle (see figs. 9.22 and 10.21), extends partially into the axilla. In doing a *breast self-examination* (see fig. 21.22), a woman should palpate the axillary area as well as the entire breast because the lymphatic drainage pathway is toward the axilla (see fig. 21.18).

Brachium

Several muscles are clearly visible in the brachium (figs. 10.34 and 10.35). The belly of the biceps brachii muscle becomes prominent when the elbow is flexed. While the arm is in this position, the

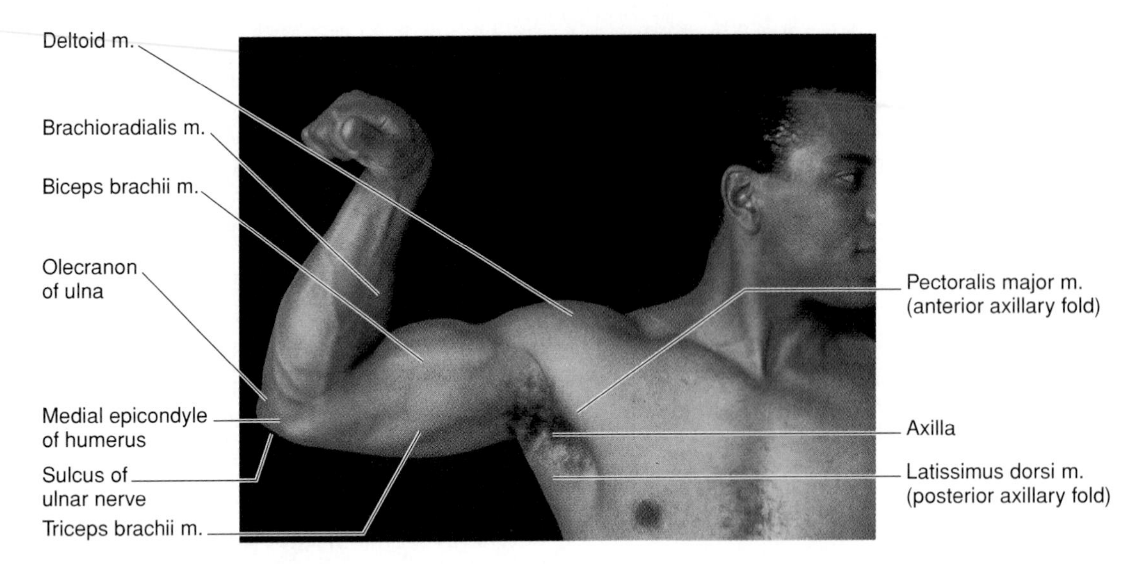

Figure 10.34

An anterior view of the right shoulder region and upper extremity.

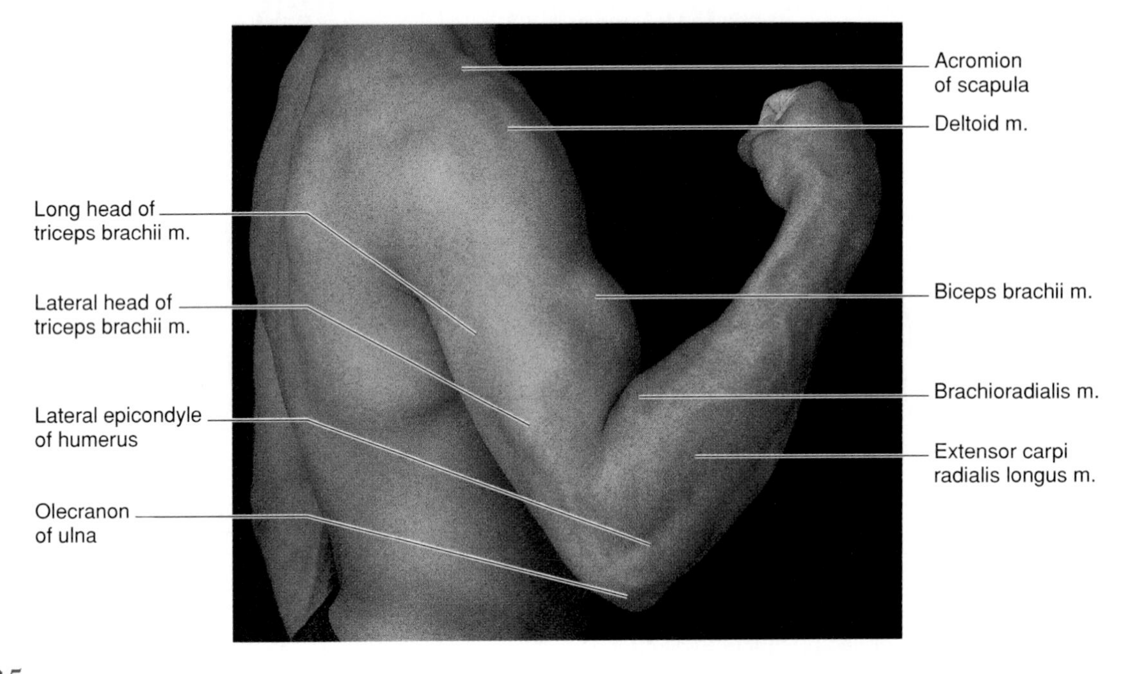

Figure 10.35

A lateral view of the upper extremity.

deltoid muscle can be traced as it inserts on the humerus. The triceps brachii muscle forms the bulk of the posterior surface of the brachium. A groove forms on the medial side of the brachium between the biceps brachii and triceps brachii muscles, where pulsations of the brachial artery may be felt as it carries blood toward the forearm (see fig. 10.37). This region is clinically important because it is where arterial blood pressure is taken with a sphygmomanometer. It is also the place to apply pressure in case of severe arterial hemorrhage in the forearm or hand.

Three bony prominences can be located in the region of the elbow (fig. 10.36). The medial and lateral epicondyles are processes on the humerus, whereas the olecranon is a proximal process of the ulna. When the elbow is extended, these prominences lie on the same transverse plane; when the elbow is flexed, they form a triangle. The ulnar nerve can be palpated in the sulcus (groove) posterior to the medial epicondyle. This sulcus overlying the ulnar nerve is commonly known as the "funny bone," or "crazy bone."

The **cubital fossa** is the depression on the anterior surface of the elbow region, where the median cubital vein links the cephalic and basilic veins. These veins are subcutaneous and become more conspicuous when a proximal compression is applied. For this reason, they are an important location (particularly the median cubital) for the removal of venous blood for analyses and transfusions or for intravenous therapy (fig. 10.37).

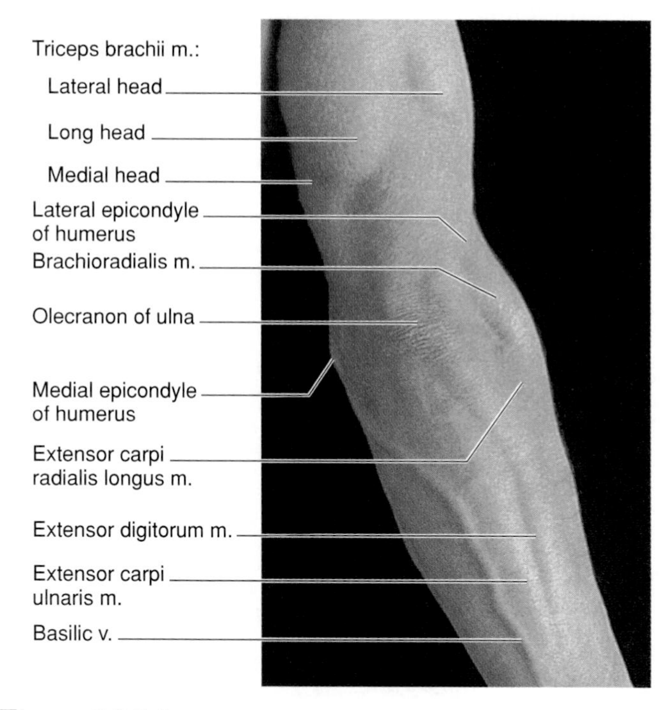

Figure 10.36

A posterior view of the elbow.

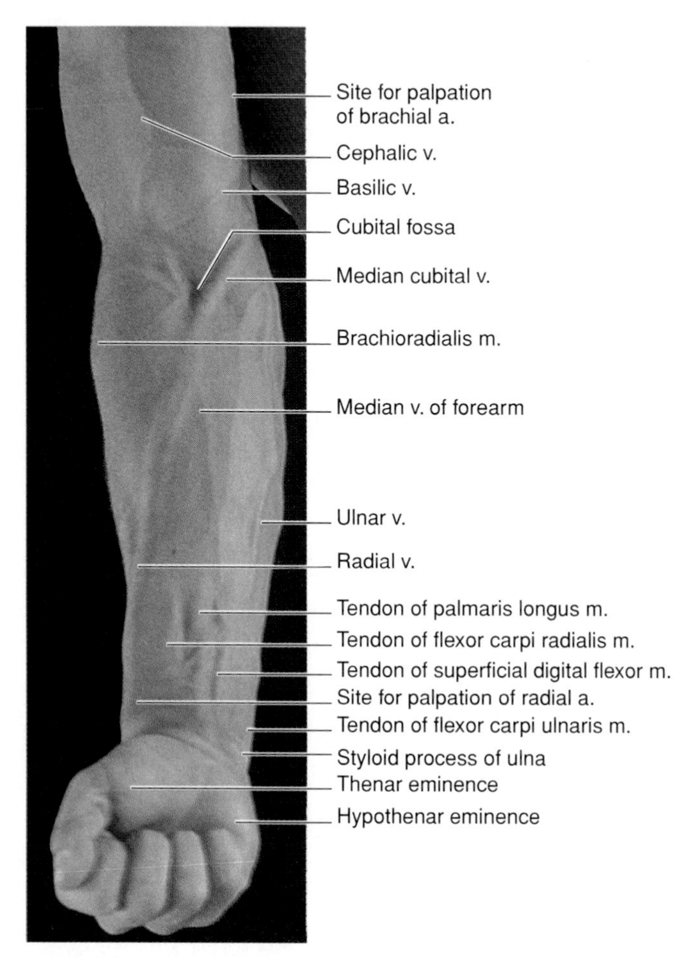

Figure 10.37

An anterior view of the forearm and hand.

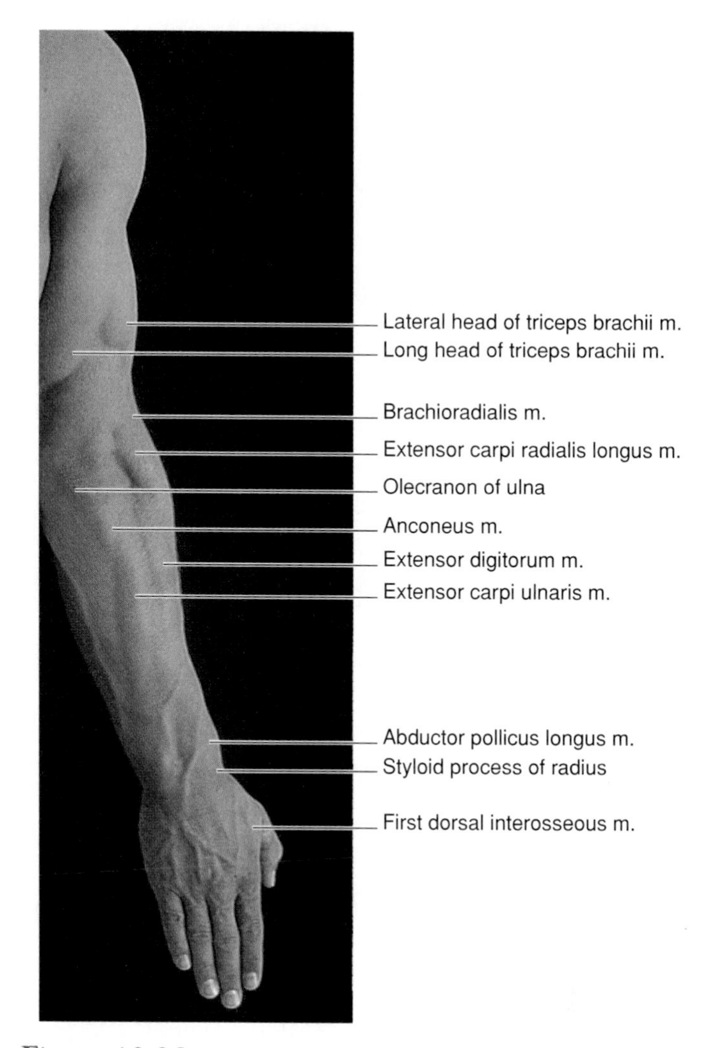

Figure 10.38

A posterior view of the forearm and hand.

Antebrachium

Contained within the antebrachium (forearm) are two parallel bones (the ulna and radius) and the muscles that control the movements of the hand. The muscles of the forearm taper distally over the wrist, where their tendons attach to various bones of the hand. Several muscles of the forearm can be identified as surface features and are depicted in figures 10.37 and 10.38.

Because of the frequency of fractures involving the forearm, bony landmarks are clinically important when setting broken bones. The ulna can be palpated along its entire length from the olecranon to the styloid process. The distal half of the radius is palpable as the forearm is rotated, and its styloid process can be located.

Nerves, tendons, and vessels are close to the surface at the wrist, making cuts to this area potentially dangerous. Tendons from four flexor muscles can be observed as surface features if the anterior forearm muscles are strongly contracted while making a fist. The tendons that can be observed along this surface, from lateral to medial, are from the following muscles: flexor

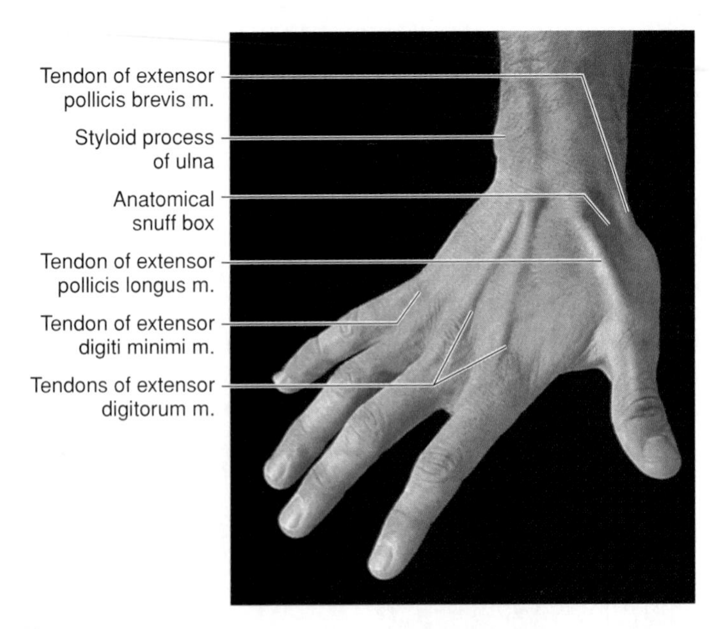

Figure 10.39

An posteromedial view of the right hand showing the anatomical snuffbox.

carpi radialis, palmaris longus, superficial digital flexor, and flexor carpi ulnaris. The median nerve going to the hand is located under the tendon of the palmaris longus muscle (see fig. 10.37), and the ulnar nerve is lateral to the tendon of the flexor carpi ulnaris muscle. The radial artery lies along the surface of the radius, immediately lateral to the tendon of the flexor carpi radialis muscle. This is the artery commonly used when monitoring the pulse. By careful palpation, pulsations can also be detected in the ulnar artery, lateral to the tendon of the flexor carpi ulnaris.

Two tendons that attach to the thumb can be seen on the posterior surface of the wrist as the thumb is extended backward. The tendon of the extensor pollicis brevis muscle is positioned anterolaterally along the thumb, and the tendon of the extensor pollicis longus muscle lies posteromedially (fig. 10.39). The depression created between these two tendons as they are pulled taut is referred to as the *anatomical snuffbox*. Pulsations of the radial artery can be detected in this depression.

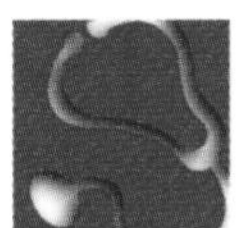

The median nerve, which serves the thumb, is the nerve most commonly injured by stab wounds or the penetration of glass into the wrist or hand. Severing of this nerve paralyzes the adductor muscles of the thumb; they waste away, resulting in an inability to oppose the thumb in grasping.

Hand

Much of the surface anatomy of the hand, such as flexion creases, fingerprints, and fingernails, involves features of the skin discussed in chapter 5. Other surface features are the extensor tendons from the extensor digitorum muscle, which can be seen going to each of the fingers on the back side of the hand as the digital joints are extended (fig. 10.40). The knuckles of the hand

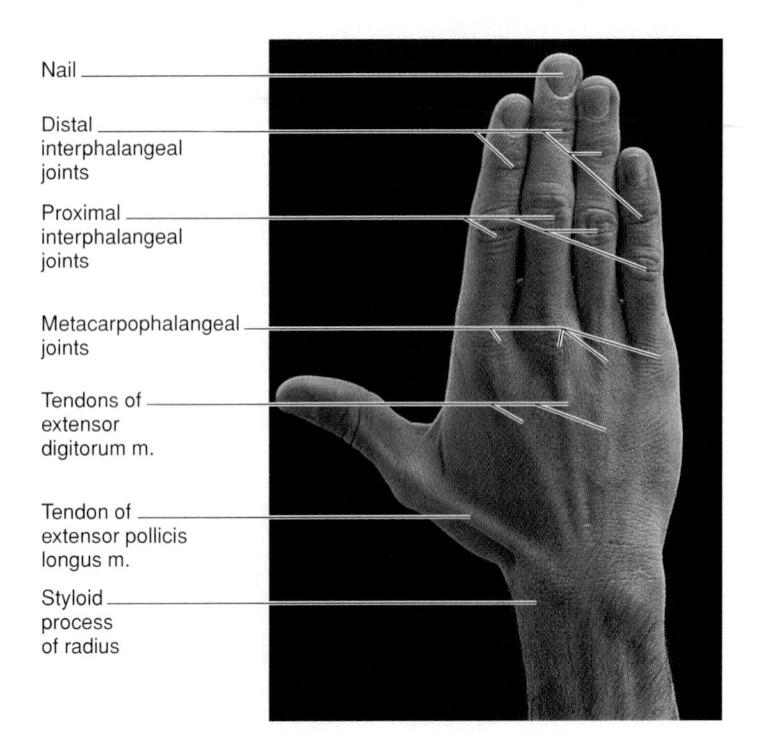

Figure 10.40

A posterior view of the hand.

are the distal ends of the second through the fifth metacarpal bones. Each of the joints of the fingers and the individual phalanges can be palpated. The **thenar** (*the'nar*) **eminence** is the thickened, muscular portion of the hand that forms the base of the thumb (fig. 10.41).

Internal Anatomy

The internal anatomy of the shoulder and upper extremity includes the structures of the shoulder, brachium, cubitus (elbow), antebrachium, and hand. The principal structures of these regions are shown in the cadaver dissections in figures 10.42 through 10.46.

1. List the clinically important structures that can be observed or palpated in the shoulder and upper extremity.
2. Describe the locations of the axilla, brachium, cubital fossa, and wrist.
3. Bumping the ulnar nerve causes a tingling sensation along the medial part of the forearm and into the little finger of the hand. What does this tell you about its distribution?
4. Which of the two bones of the forearm is the more stationary as the arm is rotated?

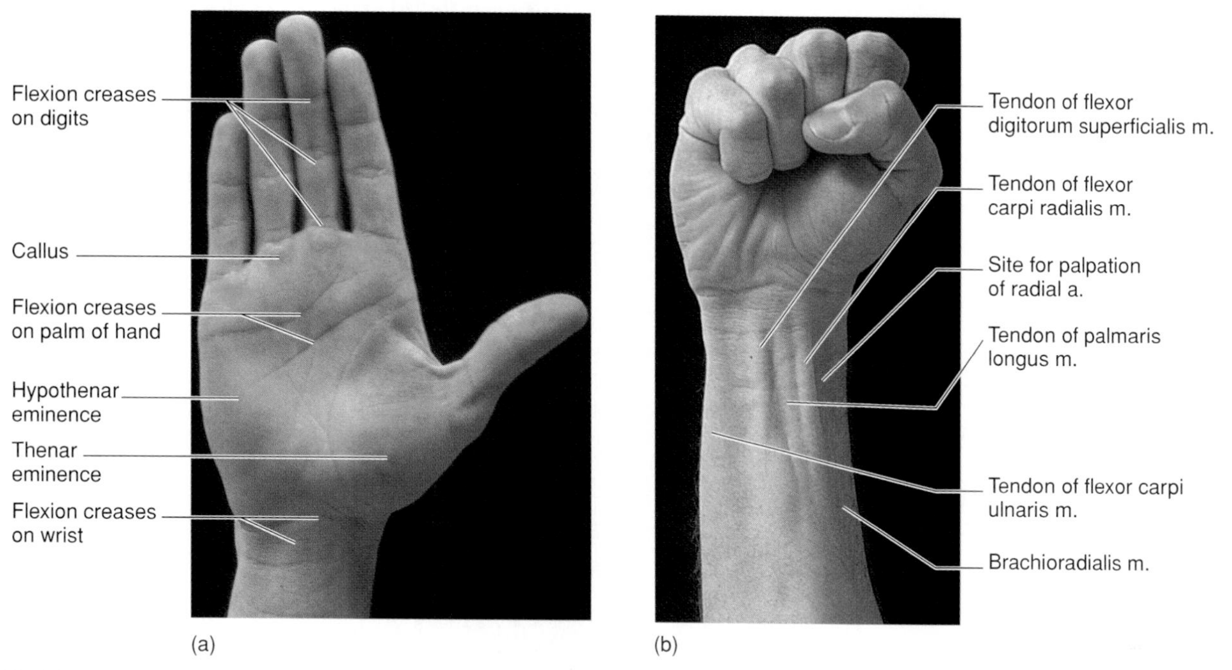

Figure 10.41

An anterior view of the wrist and hand (*a*) with the hand open and (*b*) with a clenched fist.

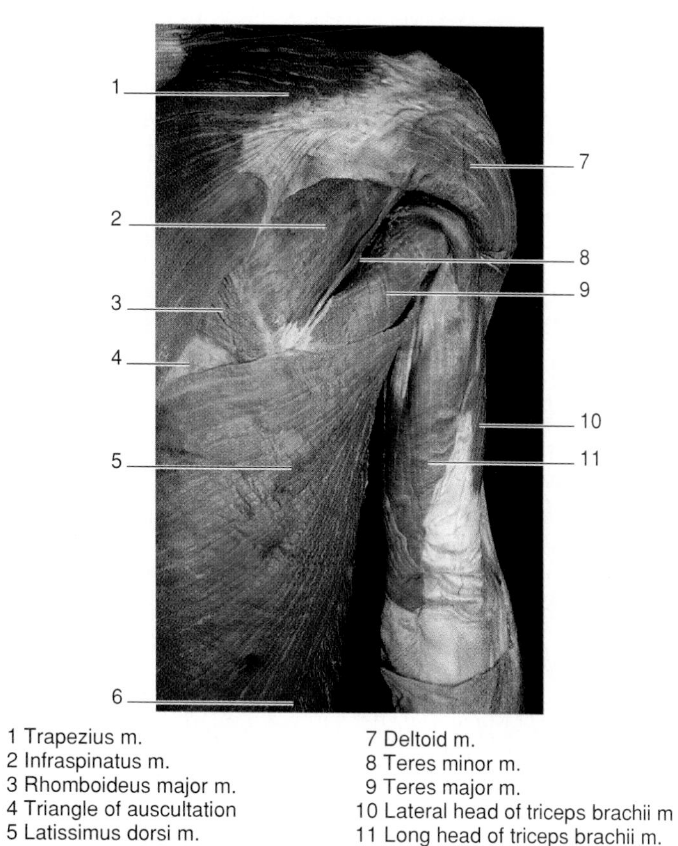

Figure 10.42

A posterior view of the superficial muscles of the right shoulder and brachium.

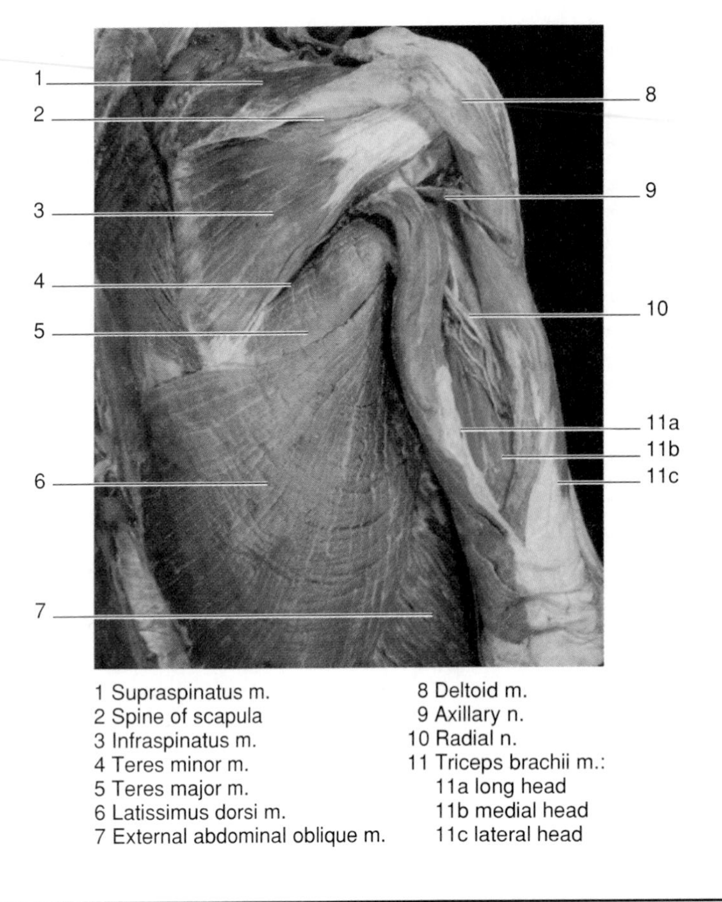

Figure 10.43

A posterior view of the deep muscles of the right shoulder and brachium.

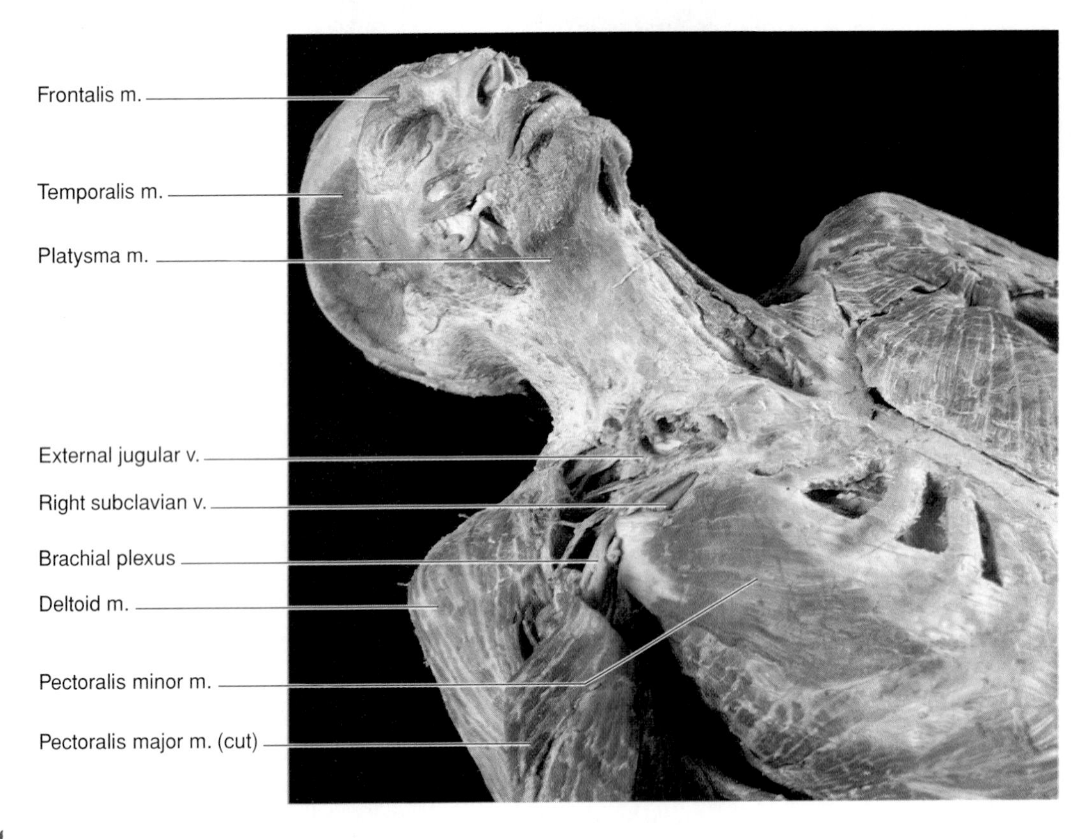

Figure 10.44

An anterolateral view of the head, neck, and thorax.

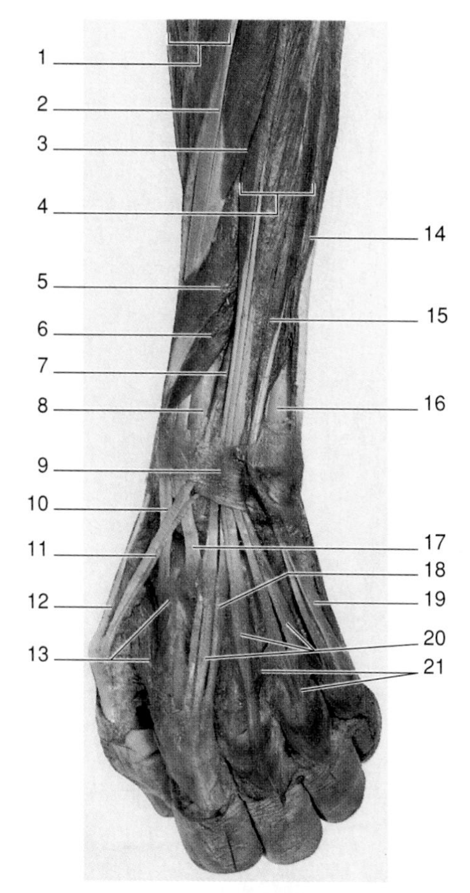

1 Brachioradialis m.
2 Tendon of extensor carpi radialis longus m.
3 Extensor carpi radialis brevis m.
4 Extensor digitorum communis m.
5 Abductor pollicis longus m.
6 Extensor pollicis brevis m.
7 Extensor pollicis longus m.
8 Radius
9 Carpal extensor retinaculum
10 Tendon of extensor carpi radialis longus m.
11 Tendon of extensor pollicis longus m.
12 Tendon of extensor pollicis brevis m.
13 First dorsal interosseous m.
14 Extensor carpi ulnaris m.
15 Extensor digiti minimi m.
16 Ulna
17 Tendon of extensor carpi radialis brevis m.
18 Tendon of extensor indicis m.
19 Tendon of extensor digiti minimi m.
20 Tendons of extensor digitorum m.
21 Intertendinous connections

Figure 10.45

A posterior view of the left forearm and hand.

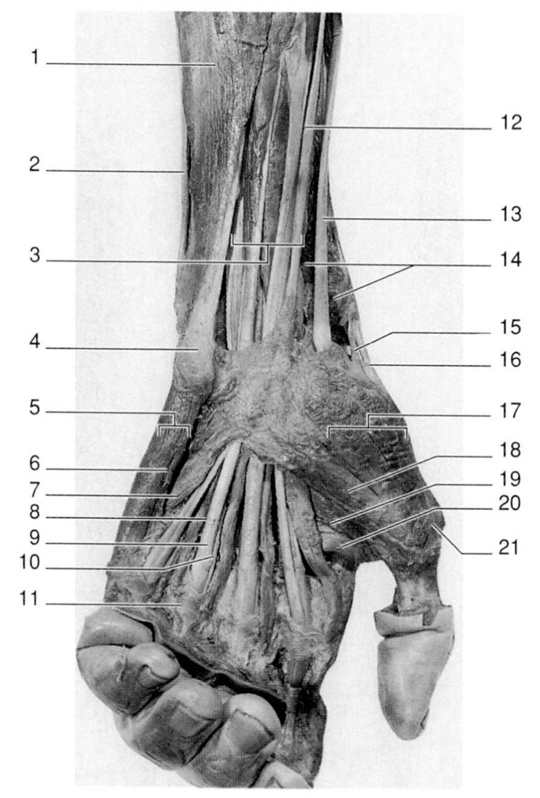

1 Flexor carpi ulnaris m.
2 Extensor carpi ulnaris m.
3 Flexor digitorum superficialis m.
4 Pisiform bone
5 Abductor digiti minimi m.
6 Flexor digiti minimi m.
7 Opponens digiti minimi m.
8 Lumbrical m.
9 Tendon of superficial digital flexor m.
10 Tendon of deep digital flexor m.
11 Fibrous digital sheath
12 Tendon of palmaris longus m.
13 Tendon of flexor carpi radialis m.
14 Pronator quadratus m.
15 Tendon of extensor pollicis brevis m.
16 Tendon of extensor pollicis longus m.
17 Abductor pollicis brevis m.
18 Flexor pollicis brevis m.
19 Adductor pollicis m. (oblique head)
20 Adductor pollicis m. (transverse head)
21 Opponens pollicis m.

Figure 10.46

An anterior view of the left forearm and hand.

Buttock and Lower Extremity

The massive bones and muscles of the buttock and lower extremity are important as weight-bearers and locomotors. Many of the surface features of these regions are important with respect to locomotion or locomotor dysfunction.

Objective 15 Identify various surface features of the buttock and lower extremity by observation or palpation.

Objective 16 Discuss the clinical importance of the buttock, femoral triangle, popliteal fossa, ankle, and arches of the foot.

Surface Anatomy

Buttock

The superior borders of the buttocks, or *gluteal region*, are formed by the iliac crests (fig. 10.47). Each crest can be palpated medially to the level of the second sacral vertebra. From this point, the **natal cleft** extends vertically to separate the buttocks into two prominences, each formed by pads of fat and by the massive gluteal muscles. An ischial tuberosity can be palpated in the lower portion of each buttock. In a sitting position, the ischial tuberosities support the weight of the body. When standing, these processes are covered by the gluteal

buttock: O. E. *buttuc*, end or rump

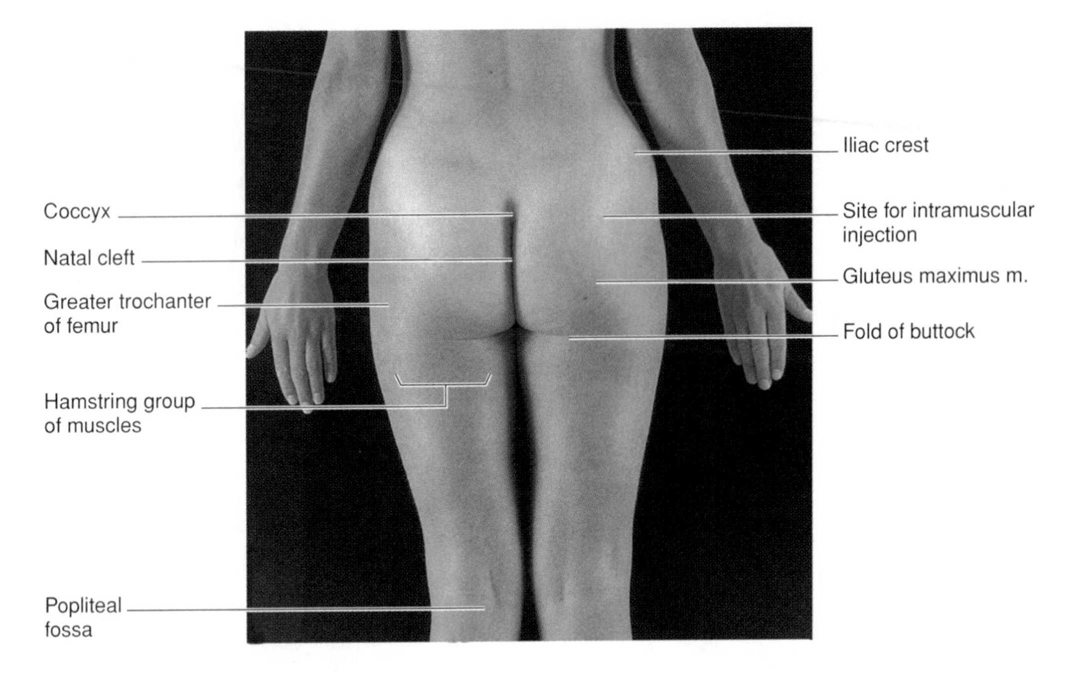

Figure 10.47

The buttocks and the posterior aspect of the thigh. (Note the relation of the angle of the elbow joint to the pelvic region, which is characteristic of females.)

muscles. The sciatic nerve, which is the major nerve to the lower extremity, lies deep in the gluteus maximus muscle. The inferior border of the gluteus maximus muscle forms the **fold of the buttock.**

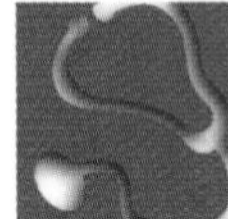

Because of the thickness of the gluteal muscles and the rich blood supply, the buttock is a preferred site for intramuscular injections. Care must be taken, however, not to inject into the sciatic nerve. For this reason, the surface landmark of the iliac crest is important. The injection is usually administered 5–7 cm (2–3 in.) below the iliac crest, in what is known as the upper lateral quadrant of the buttock.

Thigh

The femur is the only bone of the thigh, but there are three groups of thigh muscles. The anterior group of muscles, referred to as the *quadriceps femoris*, extends the knee joint when it is contracted (fig. 10.48). The medial muscles are the adductors, and when contracted they draw the thigh medially. The "hamstrings" are positioned on the posterior aspect of the thigh (see fig. 10.47) and serve to extend the hip joint, as well as flex the knee joint, when they are contracted. The tendinous attachments of the hamstrings can be palpated along the posterior aspect of the knee joint when it is flexed. The hamstrings or their attachments are often injured in athletic competition.

The **femoral** (*fem'or-al*) **triangle** is an extremely important element of the surface anatomy of the thigh. It can be seen as a depression inferior to the location of the inguinal ligament on the anteromedial surface in the upper part of the thigh (see fig. 16.30). The major vessels of the lower extremity, as well as

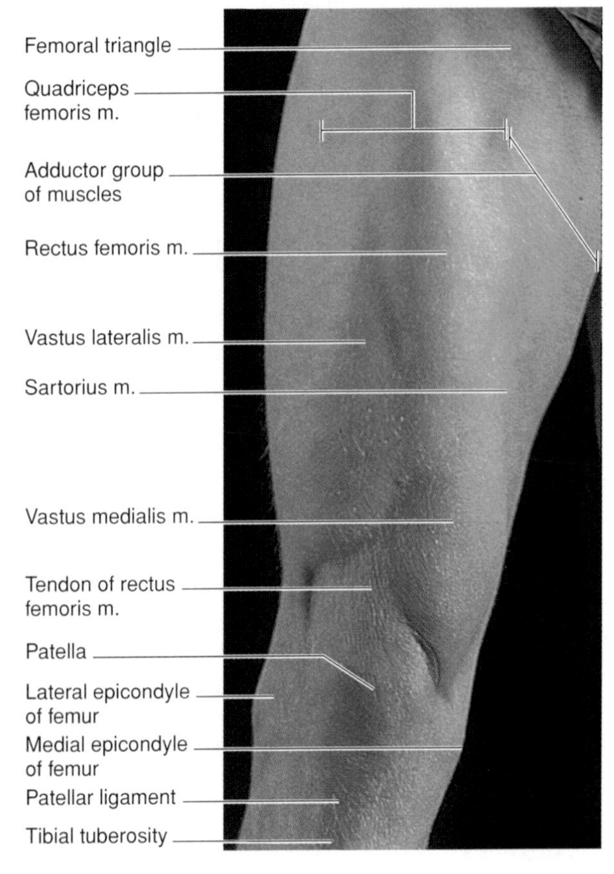

Figure 10.48

An anterior view of the right thigh and knee.

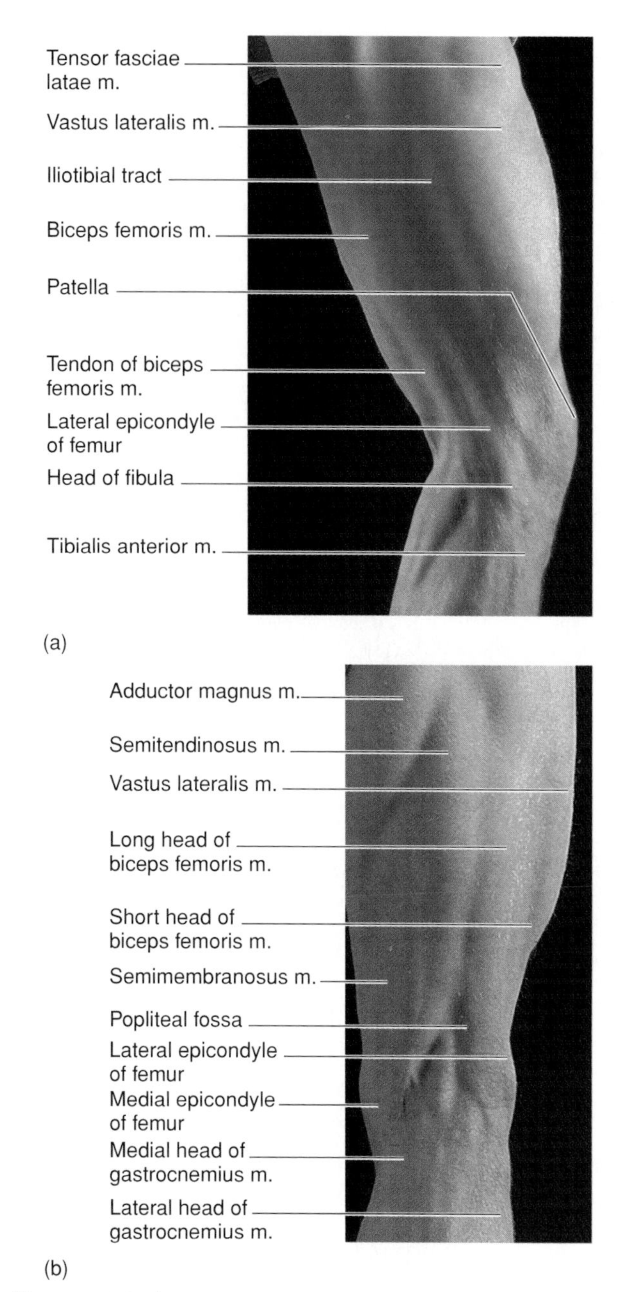

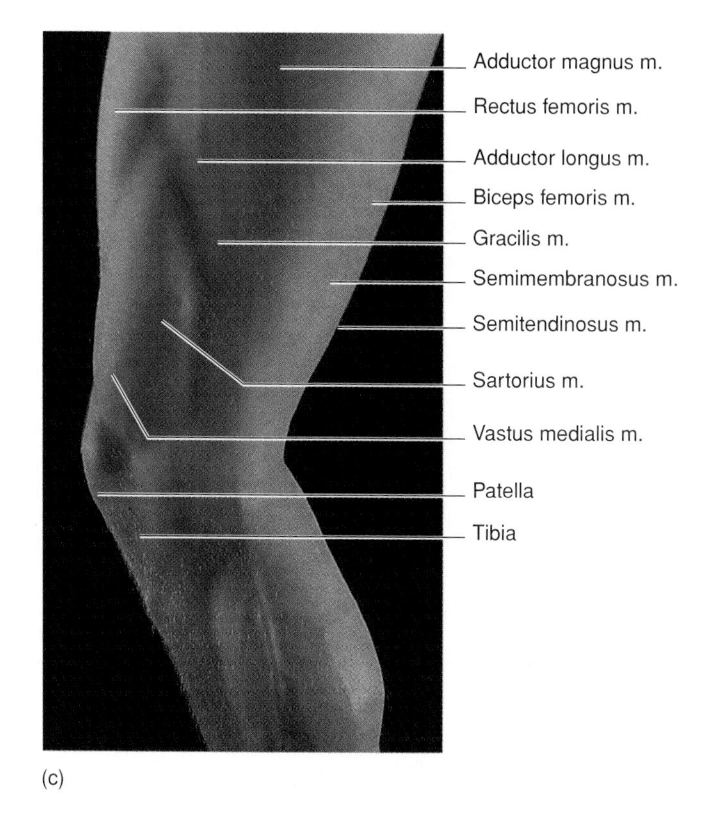

Figure 10.49

The right thigh and knee. (*a*) A lateral view, (*b*) posterior view, and (*c*) a medial view.

the femoral nerve, traverse this region. Hernias are frequent in this area. More importantly, the femoral triangle serves as an arterial pressure point (see fig. 16.31) in the case of uncontrolled hemorrhage of the lower extremity.

The greater trochanter of the femur can be palpated on the upper lateral surface of the thigh (see fig. 10.47). At the knee, the lateral and medial condyles of the femur and tibia can be identified (fig. 10.48). The patella ("kneecap") can be easily located within the patellar tendon, anterior to the knee joint. Stress or injury to this joint may cause swelling, commonly called "water on the knee."

The depression on the posterior aspect of the knee joint is referred to as the **popliteal** (*pop″lĭ-te′al*) **fossa** (fig. 10.49). This area becomes clinically important in elderly people who suffer degenerative conditions. Aneurysms of the popliteal artery are common, as are popliteal abscesses resulting from infected lymph nodes. The small saphenous vein as it traverses the popliteal fossa may become varicose in the elderly.

Leg

Portions of the tibia and fibula, the bones of the leg, can be observed as surface features. The medial surface and anterior border (commonly called "shin") of the tibia are subcutaneous and are palpable throughout their length. At the ankle, the medial malleolus of the tibia and the lateral malleolus of the fibula are

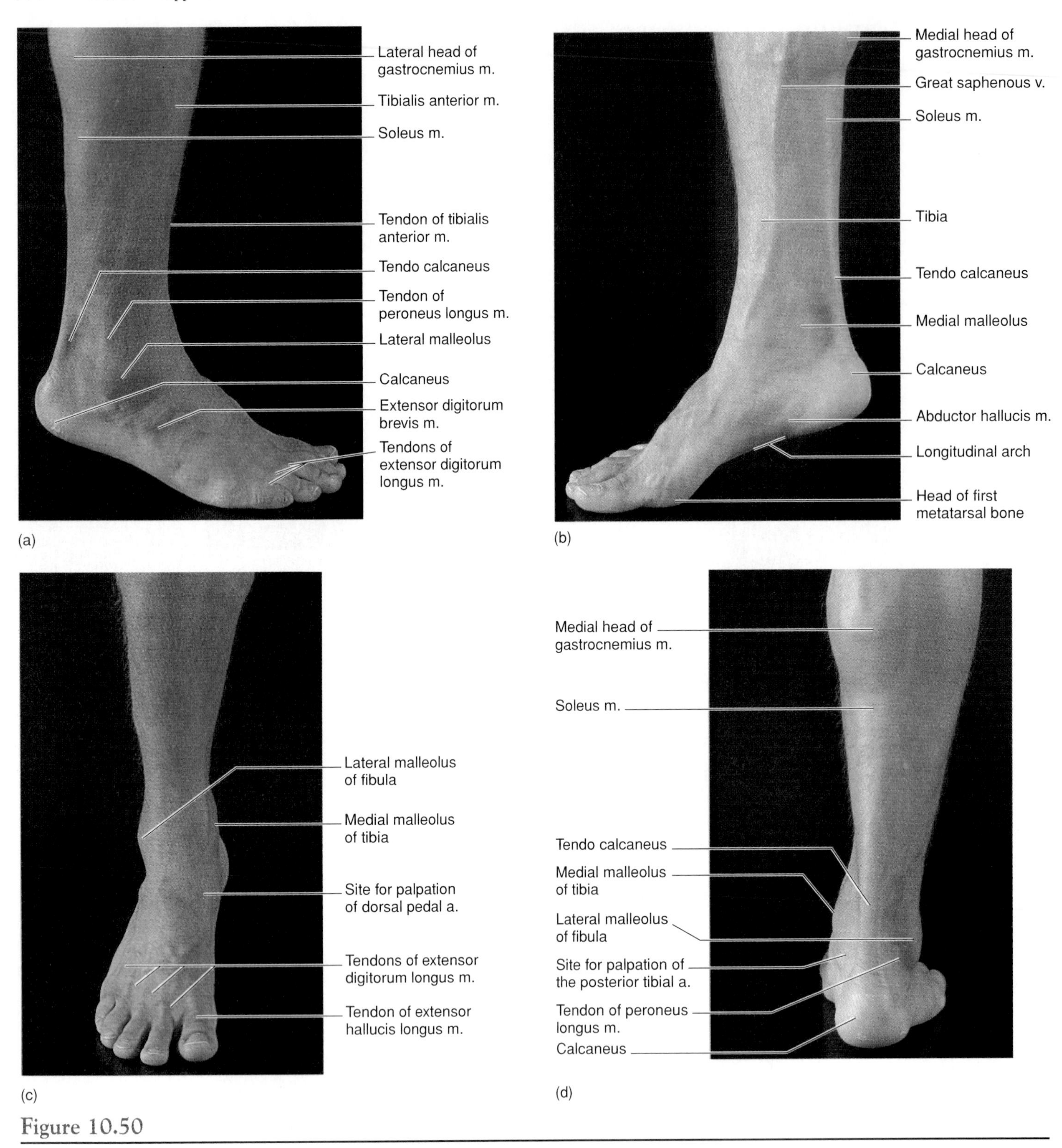

Figure 10.50

The right leg and foot. (*a*) A lateral view, (*b*) a medial view, (*c*) an anterior view, and (*d*) a posterior view.

easy to observe as prominent eminences (fig. 10.50). Of clinical importance in setting fractures of the leg is knowing that the top of the medial malleolus lies about 1.3 cm (0.6 in.) proximal to the level of the tip of the lateral malleolus.

The heel is not part of the leg; rather, it is the posterior part of the calcaneus. It warrants mention with the leg, however, because of its functional relationship to it. The **tendo calcaneus** (tendon of Achilles) is the strong, cordlike tendon that attaches to the calcaneus from the calf of the leg. The muscles forming the belly of the calf are the gastrocnemius and soleus. Pulsations from the posterior tibial artery can be detected by palpating between the medial malleolus and the calcaneus.

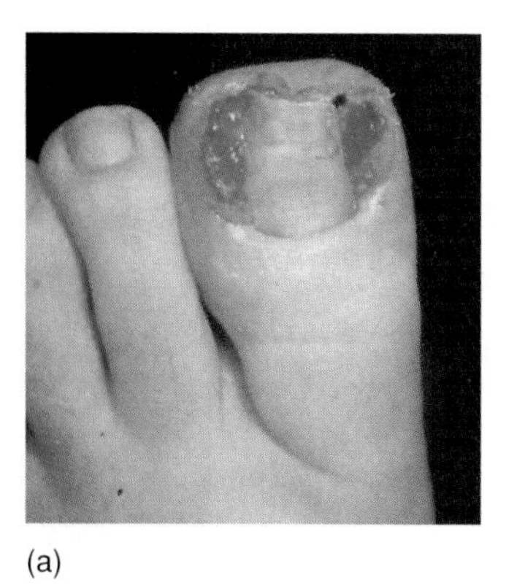
(a)

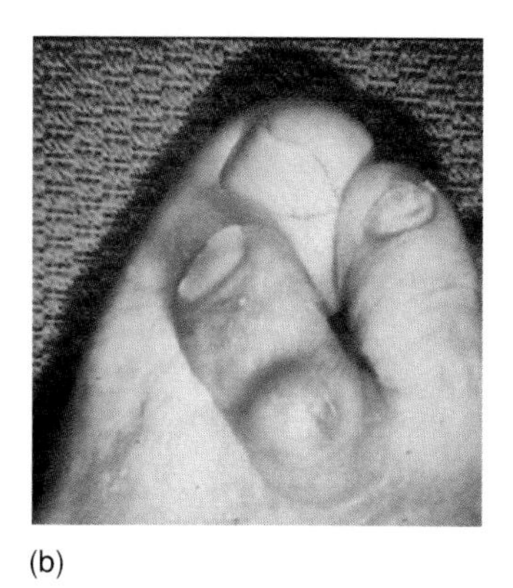
(b)

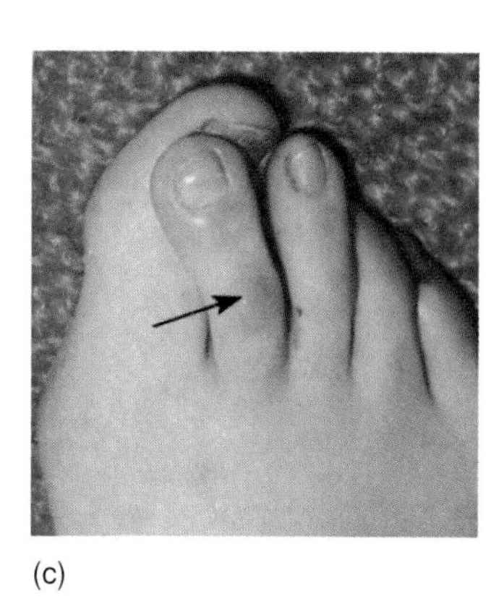
(c)

Figure 10.51

Common clinical conditions of the foot and toes. (*a*) Ingrown toenail, (*b*) hammertoe, and (*c*) corn.

The superficial veins of the leg can be observed on many individuals (see fig. 10.2). The great saphenous vein can be seen subcutaneously along the medial aspect of the leg. The less conspicuous small saphenous vein drains the lateral surface of the leg. If these veins become excessively enlarged, they are called *varicose veins*.

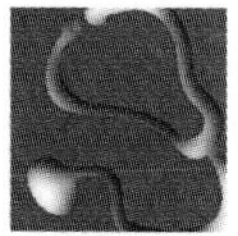

Leg injuries are common among athletes. *Shin-splints,* probably the result of a stress fracture or periosteum damage of the tibia, is a common condition in runners. A fracture of one or both malleoli is caused by a severe twisting of the ankle region. Skiing fractures are generally caused by strong torsion forces on the body of the tibia or fibula.

Foot

The feet are adapted to support the weight of the body, to maintain balance, and to function mechanically during locomotion. The structural features and surface anatomy of the foot are indicative of these functions. The **longitudinal arch** of the foot, located on the medial portion of the plantar surface (see figs. 7.20 and 10.50c), provides a spring effect when locomoting. The head of the first metatarsal bone forms the medial ball of the foot, just proximal to the hallux (great toe).

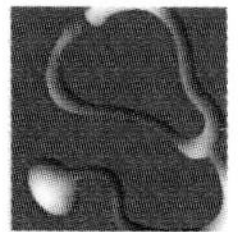

The feet and toes are adapted to endure tremendous compression forces during locomotion. Although appropriate shoes help to minimize trauma to the feet and toes, there is still an array of common clinical conditions (fig. 10.51) that may impede walking or running. An *ingrown toenail* occurs as the sharp edge of a toenail becomes embedded in the skin fold, causing inflammation and pain. *Hammertoe* is a condition resulting from a forceful hyperextension at the metatarsophalangeal joint with flexion at the proximal interphalangeal joint. A *corn* is a cone-shaped horny mass of thickened skin resulting from recurrent pressure on the skin over a bony prominence. Most often it occurs on the outside of the little toe or the upper surfaces of the other toes. *Soft corns* occur between the toes and are kept soft by moisture.

The fifth metatarsal bone forms much of the lateral border of the plantar surface of the foot. The tendons of the extensor digitorum longus muscle can be seen along the superior surface of the foot, especially if the toes are elevated. Pulsations of the dorsal pedal artery can be palpated on the superior surface of the foot between the first and second metatarsal bones. The individual phalanges of the toes, the joints between these bones, and the toenails are obvious surface landmarks.

Internal Anatomy

The internal anatomy of the buttock and lower extremity include the structures of the hip, thigh, knee, leg, and foot. The principal structures of these regions are shown in the cadaver dissections in figures 10.52 through 10.58.

1. What are the surface features that form the boundaries of a buttock?
2. List the clinically important structures that can be observed or palpated in the buttock and lower extremity.

Clinical Considerations

Head and Neck Regions

The highly specialized head and neck regions are extremely vulnerable to trauma and disease. Furthermore, because of the incredible complexity of these body regions, they are susceptible to numerous congenital conditions that occur during prenatal development.

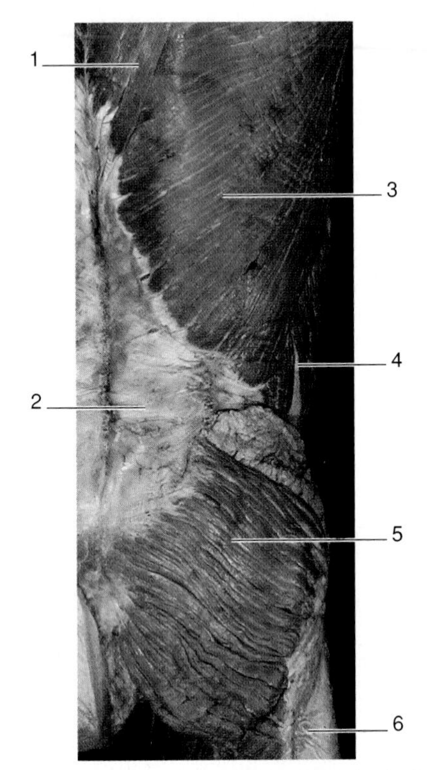

Figure 10.52

A posterior view of the superficial muscles of the right abdominal and gluteal regions.

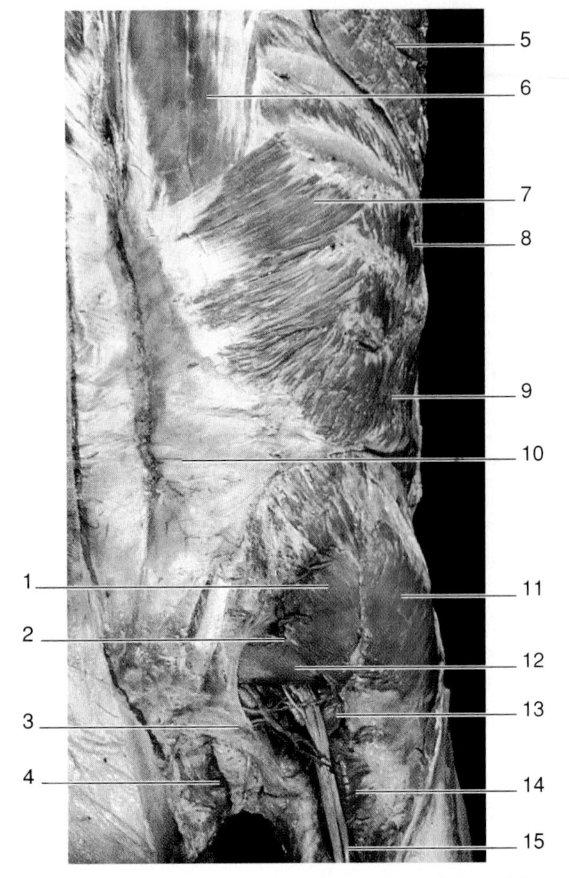

Figure 10.53

A posterior view of the deep muscles of the right abdominal and gluteal regions.

The aggressive nature of humans, reflected in part by a penchant for contact sports and fast-moving vehicles, puts the human head and neck in constant danger of injury. Pathogens readily gain access to the internal structures of the head and neck through the several openings into the head. Also, the risk for contracting certain diseases is greatly increased by the social nature of humans.

Developmental Conditions

Congenital malformations of the head and neck regions result from genetic or environmental causes and are generally very serious. The less severe malformations may result in functional disability, whereas the more severe malformations usually make life impossible.

Anencephaly (*an″en-sef′ă-le*), a severe underdevelopment of the brain and surrounding cranial bones, is always fatal. The cause of anencephaly is unknown, but genetic and geographic factors are believed to be involved. South Wales, for example, reports incidences of anencephaly as high as 1 in every 105 births. It occurs more frequently in females than males, and the damage to the developing embryo occurs between day 16 and day 26 following conception.

Altered cranial bones and sutures, resulting in pressure on the brain, accompany several kinds of congenital conditions. **Microcephaly** is characterized by premature closure of the sutures of the skull. If the child is untreated, underdevelopment of the brain and mental retardation will result. **Cranial encephalocele** (*en-sef′ă-lo-sēl*) is a condition in which the skull does not develop properly, and portions of the brain often protrude through it. In **hydrocephalus** (*hi″dro-sef′ă-lus*), an excessive accumulation of cerebrospinal fluid dilates the ventricles of the brain and causes a separation of the cranial bones.

A **cleft palate** and **cleft lip** is a common congenital condition of varying degrees of severity. A vertical split on one side, where the maxillary and median nasal processes fail to unite, is referred to as a unilateral cleft. A bilateral, or double, cleft occurs when the maxillary and median nasal process on both sides fail to unite.

By the age of 30, the sutures of the skull normally synostose and cranial bone growth ceases. **Premature synostosis**

synostose: Gk. *syn*, together; *osteon*, bone

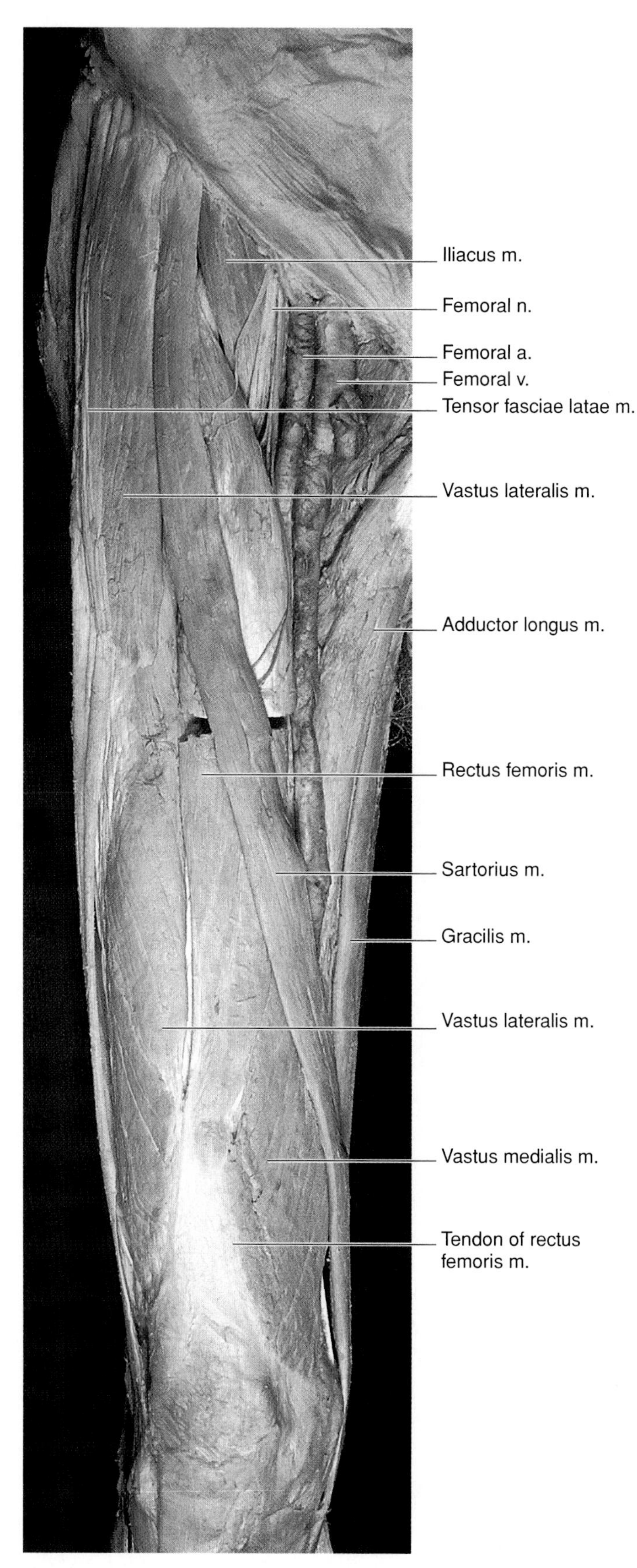

Figure 10.54

An anterior view of the superficial muscles of the right thigh.

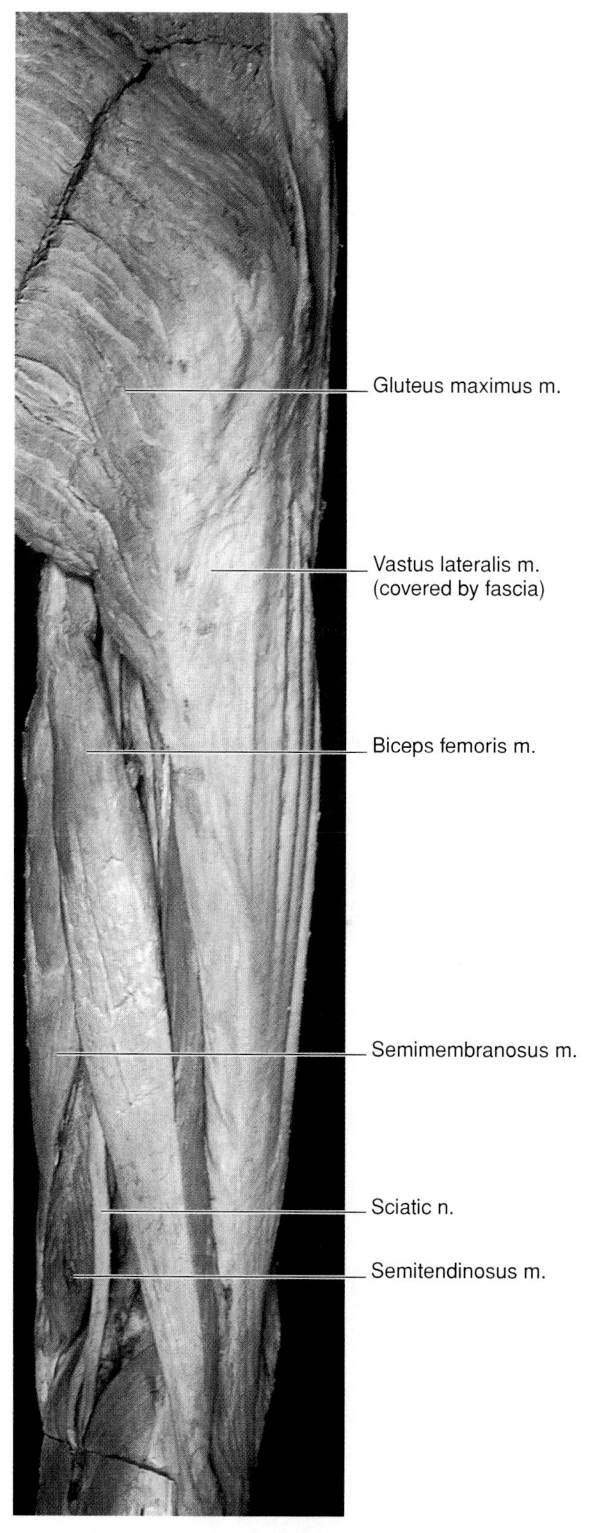

Figure 10.55

A posterior view of the superficial muscles of the right hip and thigh.

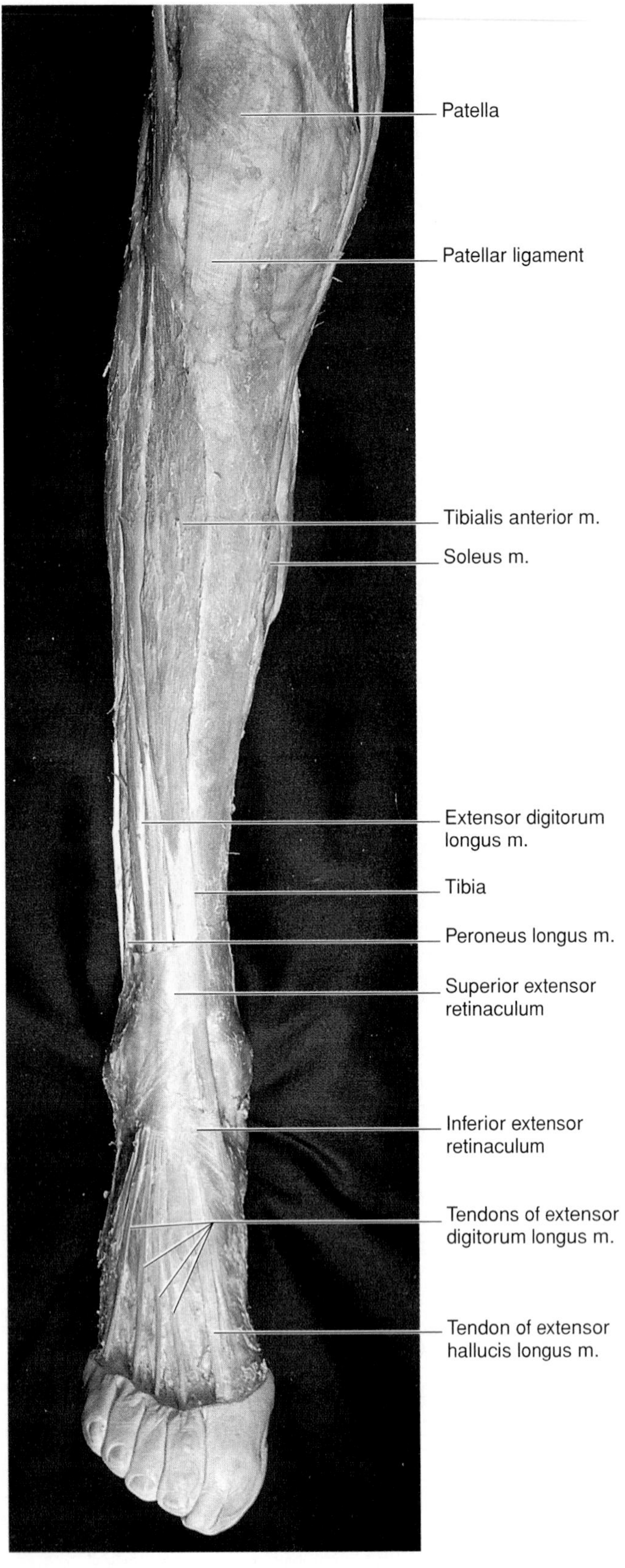

Figure 10.56

An anterior view of the superficial muscles of the right leg.

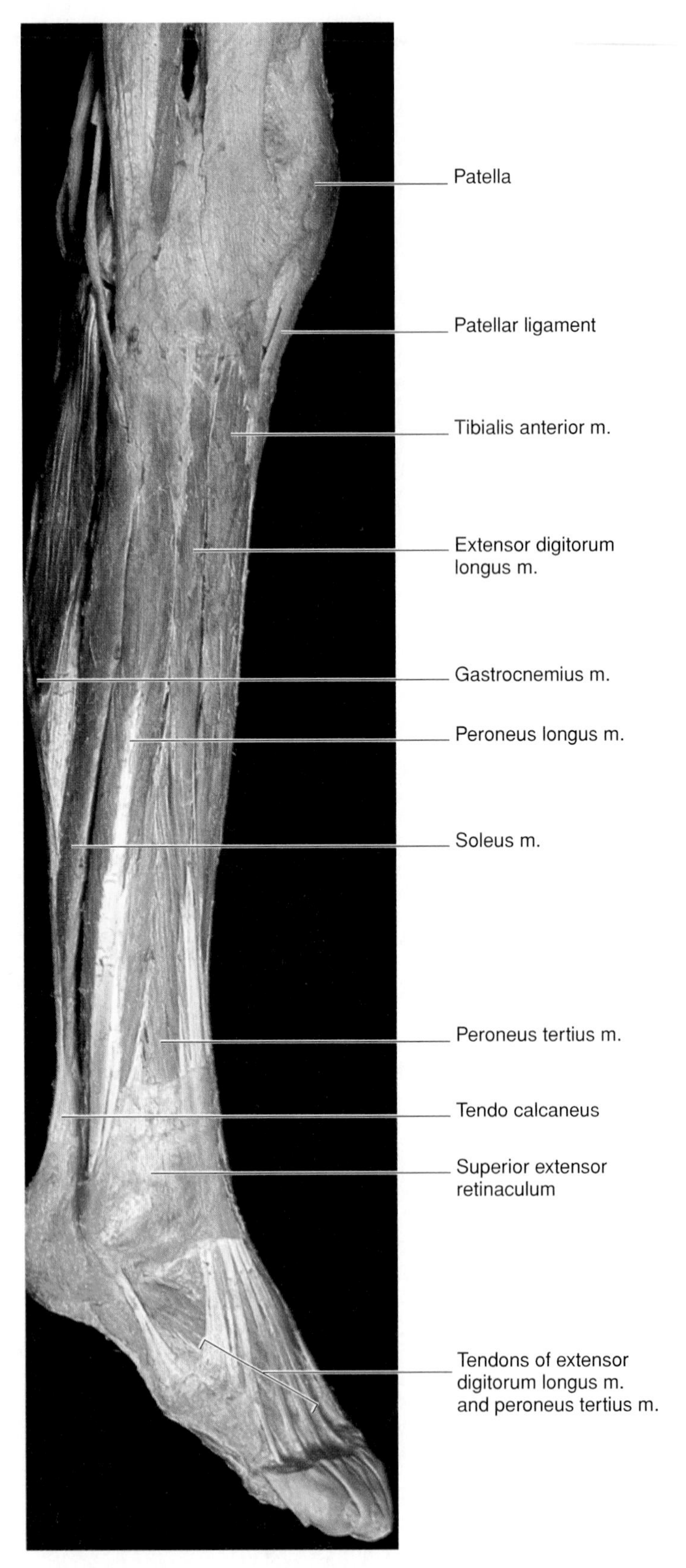

Figure 10.57

A lateral view of the superficial muscles of the right leg.

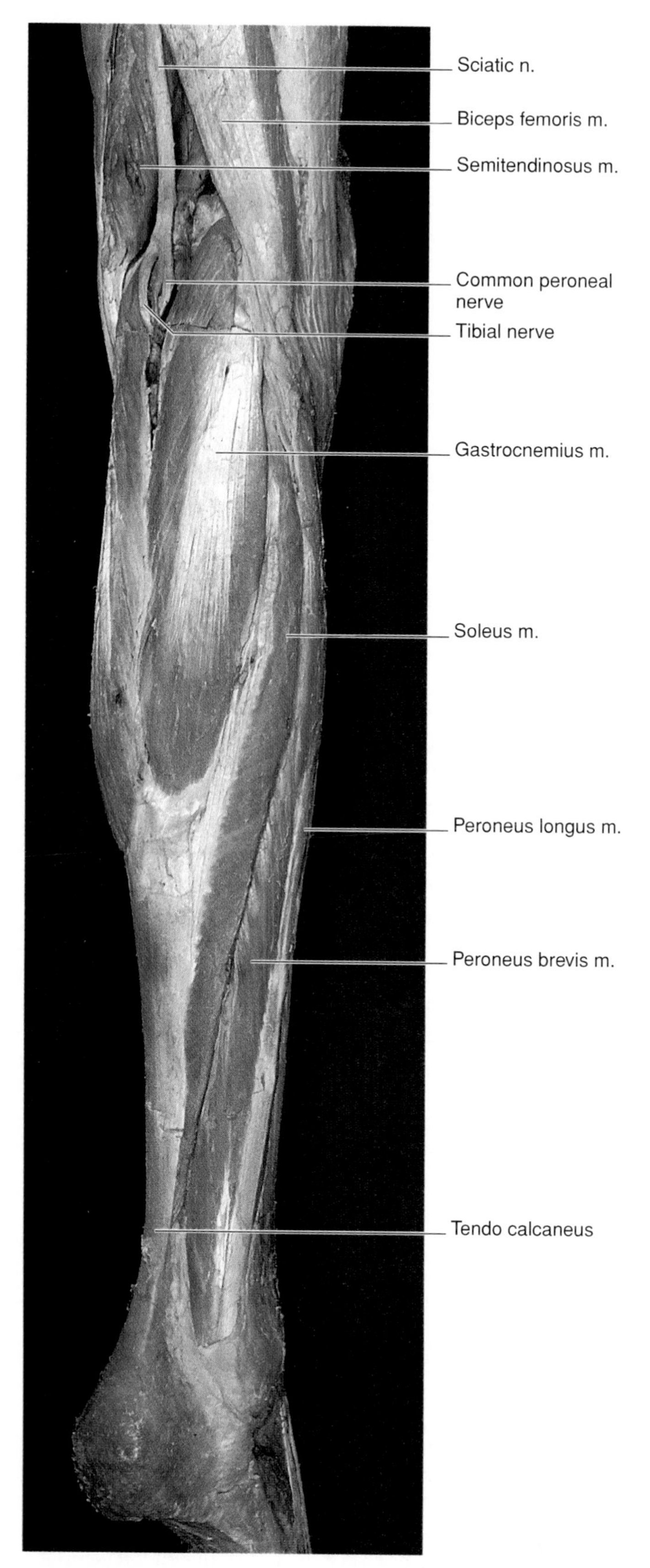

Figure 10.58

A posterior view of the superficial muscles of the right leg.

(microcephaly) is an early union of the cranial sutures before the brain has reached its normal size. **Scaphocephaly** is a malformation in which the sagittal suture prematurely closes. The skull will be noticeably crooked in a condition called **plagiocephaly** (*pla″je-o-sef′ă-le*).

Trauma to the Head and Neck

The head and neck are extremely susceptible to trauma and blows, which are frequently physically debilitating if not fatal. Striking the head from the front or back often causes **subdural hemorrhage,** resulting from the tearing of the superior cerebral veins at their points of entrance to the superior sagittal sinuses. Blows to the side of the head tend to be less severe because the falx cerebelli and tentorium cerebelli (see table 11.5) restrict the displacement of the brain sideways. With a sudden violent lateral movement of the head, such as in a serious automobile accident, the serrated edge of the lesser wing of the sphenoid bone may severely damage the brain and tear cranial nerves.

The arteries of the head and neck are rarely damaged because of their elasticity. In a severe lateral blow to the head, however, the internal carotid artery may rupture, and a roaring sound will be perceived by the injured person as blood quickly fills the cavernous sinuses of the temporal bone.

Skull fractures are fairly common in adults but much less common in children. The cranial bones of a child are resilient, and sutures are not yet ossified. The cranium of an adult, however, has limited resilience and tends to splinter. A hard blow to the head frequently breaks the bone on the opposite side of the skull in what is called a **contrecoup fracture.** The sphenoid bone, with its numerous foramina, is the weakest bone of the cranium. It frequently sustains a contrecoup fracture as a result of a hard blow to the top of the head.

The most frequently fractured bones of the face are the nasal bones and the mandible. Trauma to these bones generally results in a simple fracture, which is not usually serious. If the nasal septum or cribriform plate of the ethmoid bone is fractured, however, careful treatment is required. If the cribriform plate is severely fractured, a tear in the meninges may result, causing a sudden loss of cerebrospinal fluid and death.

Whiplash is a common injury to the neck due to a sudden and forceful displacement of the head (see fig. 11.47). The muscle, bone, or ligaments may be injured, in addition to the spinal cord and cervical nerves. A whiplash is usually extremely painful and difficult to treat because of the difficulty in diagnosing the extent of the injury.

The sensory organs within the head are also very prone to trauma. The eyes may be injured by sudden bright light, and loud noise can rupture the tympanic membrane of the middle ear. A nonpenetrating blow to the eye may result in a herniation of the orbital contents through a fracture created in the floor of the orbit. The nerves that control the eye may also be damaged.

---

scaphocephalus: Gk. *skaphe*, boat; *kephale*, head
plagiocephaly: Gk. *plagios*, oblique; *kephale*, head

Diseases of the Head and Neck

The head and neck are extremely susceptible to infection, especially along the mucous membranes lining body openings. Sinusitis, tonsillitis, laryngitis, pharyngitis, esophagitis, and colds are common, periodically recurring ailments of the mucous-lined digestive and respiratory tracts of the head and neck.

The cutaneous area of the head most susceptible to infections extends from the upper lip to the midportion of the scalp. An infection of the scalp may spread via the circulatory system to the bones of the skull, causing **osteomyelitis** (*os″te-o-mi″ĕ-li′tis*). The infection may even spread into the sagittal venous sinus, causing **venous sinus thrombosis.** A **boil** in the facial region may secondarily cause thrombosis of the facial vein or the spread of the infection to the sinuses of the skull. Before antibiotics, such sinus infections had a mortality rate of 90%.

Close observation of the head by a physician can be helpful in diagnosing several diseases and body conditions. The nose becomes greatly enlarged in a person with **acromegaly** (*ak″ro-meg′ă-le*) and very wide in a person with **hypothyroidism.** The bridge of the nose is depressed in a person with **congenital syphilis.**

The color of the mucous membranes of the mouth may be important in diagnosing illness. Pale lips generally indicate *anemia*, yellow lips indicate *pernicious anemia*, and blue lips are characteristic of *cyanosis*, or cardiovascular problems. In *Addison's disease*, the normally pinkish mucous membranes of the cheeks have brownish areas of pigmentation.

Thoracic Region

Developmental Conditions

When serious deformities of the chest do occur, they are almost always due to an overgrowth of the ribs. In **pigeon breast** (*pectus carinatum*), the sternum is pushed forward and downward like the keel of a boat. In **funnel chest** (*pectus excavatum*), the sternum is pushed posteriorly, causing an anterior concavity in the thorax. Rarely, there may be a congenital absence of a pair, or pairs, of ribs. The absence of ribs is due to incomplete development of the thoracic vertebrae, a condition termed **hemivertebrae,** and may result in impaired respiratory function. There is a 0.5% occurrence of **cervical rib,** and half the time it is bilateral. A cervical rib is attached to the transverse process of the seventh cervical vertebra, and it either has a free anterior portion or is attached to the first (thoracic) rib. Pressure of a cervical rib on the brachial plexus may produce a burning, prickling sensation (*paresthesia*) along the ulnar border of the forearm and atrophy of the medial (hypothenar) muscles of the hand.

The pectoralis major muscle may be congenitally absent, either partially or wholly. A person with this anomaly appear to have a sunken chest and must rely on the contractions of muscles that are synergistic to the pectoralis major for flexion at the shoulder joint.

The rapid and complex development of the heart and major thoracic vessels accounts for the numerous congenital abnormalities that may affect these organs (see chapter 16). Congenital heart problems occur in approximately 3 of every 100 births and account for about 50% of early childhood deaths. Cardiac malformations usually arise from developmental defects in the heart valves, septa (atrial and/or ventricular), or both. A **patent foramen ovale** (*fŏ-ra′men o-val′e*) is an example of a septal defect. Such a malformation may permit venous blood from the right atrium to mix freely with the oxygenated blood in the left atrium. A **ventricular septal defect** usually occurs in the upper portion of the interventricular septum and is generally more serious than an atrial septal defect because of the greater fluid pressures in the ventricles and the greater chance of heart failure.

Congenital valvular problems are classified as either an **incompetence** (leakage) or a **stenosis** (constriction) of valves. Improper closure of a valve permits some backflow of blood, causing an abnormal sound referred to as a **murmur.** Murmurs are common and generally have no adverse affect on a person's health.

The **tetralogy of Fallot** is a combination of four defects within the heart of a newborn: (1) a ventricular septal defect, (2) an overriding aorta, (3) pulmonary stenosis, and (4) right ventricular hypertrophy. It immediately causes a cyanotic condition (blue baby). Although tetralogy of Fallot is one of the most common cardiac defects, it is also one of the simplest to correct surgically.

Abnormal development of the primitive aortic arches occasionally results in both a left and a right aortic arch. In this case, there are generally anomalies of other vessels as well.

Trauma to the Thorax

Because of its resilience, the rib cage generally provides considerable protection for the thoracic viscera. The ribs of children are highly elastic and fractures are rare. By contrast, the ribs of adults, are frequently fractured by direct trauma or, indirectly, by crushing injuries. Ribs 3 through 8 are the ones most commonly fractured. The first and second ribs are somewhat protected by the clavicle and the last four ribs are more flexible to blows. The costal cartilages in elderly people may undergo some ossification, reducing the flexibility of the rib cage and causing some confusion when examining a chest radiograph.

A possible complication of a rib fracture is a puncture of the lung or the protrusion of a bone fragment through the skin. In either case, pleural membranes will likely be ruptured resulting in a **pneumothorax** (*noo″mo-thor′aks*) (accumulation of air in the pleural cavity) or a **hemothorax** (blood in the pleural cavity).

Any puncture wound to the thorax—from a bullet or a knife, for example—may cause a pneumothorax. **Atelectasis** (*at′l-ek′tă-sis*), the collapse of a lung or part of it, generally results from a pneumothorax, and makes respiration extremely difficult. Because each lung is surrounded by its own pleural cavity, trauma to one lung does not usually directly affect the other lung.

The heart, the ascending aorta, and the pulmonary trunk are enclosed by the fibrous pericardium. A severe blow to the chest, such as hitting the steering wheel in an automobile accident, may

---

tetralogy of Fallot: from Etienne L. A. Fallot, French physician, 1850–1911

cause a sudden surge of blood from the ventricles sufficient to rupture the ascending aorta. Such an injury will flood the pericardial sac with blood, causing **cardiac tamponade** (fluid compression) and almost immediate death.

Diseases of the Thorax

The leading causes of death in the United States are due to disease or dysfunction of the thoracic organs. Consequently, the surface features of the thorax are extremely important to a physician as reference locations for *palpation* (feeling with firm pressure, *percussion* tapping with the fingertips), and *auscultation* (listening with a stethoscope). As mentioned earlier, many of the ribs are evident on a thin person, and all but the first, and at times the twelfth, can be palpated. The sternum, clavicles, and scapulae also provide important bony landmarks in conducting a physical examination. The nipples in the male and prepubescent female are located at the fourth intercostal spaces. The position of the left nipple in males provides a guide for where to listen for various heart sounds.

The clinical importance of the female breasts lies in their periodic phases of activity (during pregnancy and lactation) and their susceptibility to neoplastic change. The breasts and mammary glands are highly susceptible to infections, cysts, and tumors. The superficial position of the breasts allows for effective treatment by way of surgery and radiotherapy if tumors are detected early. The importance of *breast self-examination (BSE)* (see chapter 21) cannot be overemphasized. Breast cancer, or **carcinoma of the breast,** is surpassed only by lung cancer as the most common malignancy in women. Untreated, breast cancer is eventually fatal. One in nine women will develop breast cancer, and one-third of these will die from the disease.

The lungs can be examined through percussion, auscultation, or observation. *Bronchoscopy* enables a physician to examine the trachea, carina, primary bronchi, and secondary bronchi. A bronchoscope also enables a physician to remove foreign objects from these passageways. Swallowed objects that are aspirated beyond the glottis lodge in the right primary bronchus 90% of the time because of its near vertical alignment with the trachea.

The lungs are a common site for cancer. Fortunately, each lung is divided into distinct lobes, which allows a surgeon to remove a diseased portion and leave the rest of the lung intact. Pneumonia, tuberculosis, asthma, pleurisy, and emphysema are other common diseases that directly or indirectly afflict the lungs.

Cardiovascular diseases are the leading cause of death in the United States. Included among these diseases are heart attacks, which are caused by an insufficient blood supply to the myocardium (**myocardial ischemia** [*is-ke'me-ă*]). Poor cardiac circulation is due to an accumulation of atherosclerotic plaques or the presence of a thrombus (clot). A heart attack is generally accompanied by severe chest pains (**angina pectoris**) and usually by referred pain, perceived as arising from the left arm and shoulder. If the coronary deficiency is continuous, local tissue necrosis results, causing a permanent loss of cardiac muscle fibers (**myocardial infarction**). Extensive cardiac necrosis results in cardiac arrest.

Ventricular fibrillation is random, disorganized electrical activity within the ventricular wall of the heart. The consequent loss of coordinated ventricular contraction impairs coronary circulation, resulting in low blood pressure, or **hypotension.** If untreated, continuous ventricular fibrillation results in death.

Various other heart diseases include infections of the serous membrane (**pericarditis**), infection of the lining of the heart chambers (**endocarditis**), infection of the valves (*bacterial endocarditis*), and immune-mediated damage to the valves, as occurs in *rheumatic fever*. Valvular disease may cause the cusps to function poorly and may result in an enlarged heart.

Abdominal Region

Developmental Conditions

The development of the abdominal viscera from endoderm and mesoderm is a highly integrated, complex, and rapidly occurring process; the viscera are therefore susceptible to a wide range of congenital malformations.

The diaphragm develops in four directions simultaneously as skeletal muscle tissues coalesce toward the posterior center. Failure of the muscle tissues to fuse results in a **congenital diaphragmatic (hiatal) hernia.** This abnormal opening in the diaphragm may permit certain abdominal viscera to project into the thoracic cavity.

The role of the umbilicus in the development of the fetal urinary, circulatory, and digestive systems may present some interesting congenital defects. A **patent urachus** (*yoor'ă-kus*) is an opening from the urinary bladder to the outside through the umbilicus. For a short time during development, this opening is normal. Closure of the urachus occurs in most fetuses with progressive development of the urinary system. A patent urachus will generally go undetected unless there is extreme difficulty with urination, such as may be caused by an enlarged prostate. In such a case, some urine may be forced through the patent urachus and out the umbilicus.

Meckel's diverticulum is the most common anomaly of the small intestine. It is the result of failure of the embryonic yolk sac to atrophy completely. Present in about 3% of the population, a Meckel's diverticulum consists of a pouch, approximately 6.5 cm (2.5 in.) long, that resembles the appendix. It arises near the center of the ileum, and may terminate freely or be attached to the anterior abdominal wall near the umbilicus. Like the appendix, a Meckel's diverticulum is prone to infections; it may become inflamed, producing symptoms similar to appendicitis. For this reason, it is usually removed as a precautionary measure when discovered during abdominal surgery.

The connection from the ileum of the small intestine to the outside sometimes is patent at the time of birth; this condition is called a *fecal fistula*. It permits the passage of fecal material through the umbilicus and must be surgically corrected in a newborn.

Other parts of the GI tract are also common sites for congenital problems. In **pyloric stenosis,** there is a narrowing of the pyloric orifice of the stomach resulting from hypertrophy of the muscular layer of the pyloric sphincter. This condition is more

---

Meckel's diverticulum: from Johann Friedrich Meckel, German anatomist, 1724–74

common in males than in females, and the symptoms usually appear early in infancy. The constricted opening interferes with the passage of food into the duodenum and therefore causes dilation of the stomach, vomiting, and weight loss. Treatment involves a surgical incision of the pyloric sphincter.

Congenital megacolon (Hirschsprung's disease) is a condition in which ganglia fail to develop in the submucosal and myenteric plexuses in a portion of the colon. The absence of these ganglia results in enlargement of the affected portion of the colon because of lack of innervation and muscle tone. In the absence of peristalsis, there is severe constipation. Treatment involves surgical resection of the affected portion of the colon.

Congenital malformations may occur in any of the abdominal viscera, but most of these conditions are inconsequential. Accessory spleens, for example, occur in about 10% of the population. Located near the hilum of the spleen, these anomalous organs are small (about 1 cm. in diameter), number from two to five, and are only moderately functional. They usually atrophy within a few years after birth.

Tremendous variation can occur in the formation of the kidneys. They are frequently multilobed, fused, or malpositioned (see fig. 19.17). There also may be more than the normal two. In the case of an anomalous kidney, there is usually an accompanying variation in the vascular supply. It is common to have multiple renal arteries serving a kidney. Most renal anomalies do not pose serious problems.

An abnormal pattern of sex hormone production in the embryo may result in considerable malformation of the developing genitalia. These anomalies may be cosmetic concern only, or they may render the organ nonfunctional. Some may be so severe as to preclude determination of an individual's sex based on external appearance. Most of these conditions can be surgically corrected. Also of clinical concern and treatable are the various problems that may occur during descent of the testes into the scrotum. In the normal development of the male fetus, the testes will be in scrotal position by the twenty-eighth week of gestation. If they are undescended at birth, a condition called **cryptorchidism** (*krip-tor'kĭ-diz"em*), medical intervention may be necessary.

Trauma to the Abdomen

The rib cage, the omentum (see fig. 18.3), and the pendant support of the abdominal viscera offer some protection from trauma. However, puncture wounds, compression, and severe blows to the abdomen may result in serious abdominal injury.

The large and dense liver, located in the upper right quadrant of the abdomen, is quite vulnerable to traumatic blows, stab wounds, or puncture wounds from fractured ribs. A lacerated liver is extremely serious because of the possibility of internal hemorrhage from such a vascular organ.

The spleen is another highly vascular organ that is frequently injured, especially from blunt abdominal trauma. A ruptured spleen causes severe internal hemorrhage and shock. Its prompt removal (*splenectomy*) is necessary to keep the patient from bleeding to death. The spleen may also rupture spontaneously due to infectious diseases that cause it to hypertrophy.

Rupture of the pancreas is not nearly as common as rupture of the spleen, but it could occur if a strong compression of the upper abdomen were to force the pancreas against the vertebral column. The danger of a ruptured pancreas is the flow of pancreatic juice into the peritoneal cavity, the subsequent digestive action, and peritonitis.

The kidneys are vulnerable to trauma in the lumbar region, such as from a traumatic blow. Because the kidney is fluid-filled, a blow to one side propagates through the kidney and may possibly rupture the renal pelvis or the proximal portion of the ureter. Blood in the urine is symptomatic of kidney damage. Medical treatment of a traumatized kidney varies with the severity of the injury.

Trauma to the external genitalia of both males and females is a relatively common occurrence. The pendant position of the penis and scrotum makes them vulnerable to compression forces. For example, if a construction worker were to slip and land astride a steel beam, his external genitalia would be compressed between the beam and his pubic bone. In this type of accident, the penis (including the urethra) might split open, and one or both testes might be crushed.

Trauma to the female genitalia usually results from sexual abuse. Vaginal tearing and a displaced uterus are common in molested girls. The physical and mental consequences are generally severe.

Diseases of the Abdomen

Any of the abdominal organs may be afflicted by an array of diseases. It is beyond the scope of this text to cover all of these diseases; instead, an overview of some general conditions will be presented.

Knowledge of the clinical regions of the abdomen (see figs. 2.15 and 2.16) and the organs within these regions (see table 2.3) is fundamental to the physician in performing a physical examination. Also important are the locations of the linea alba, extending from the xiphoid process to the symphysis pubis; the umbilicus; the inguinal ligament; the bones and processes that can be palpated on the rib cage; and the pelvic girdle.

Peritonitis is of major clinical concern. The peritoneum is the serous membrane of the abdominal cavity. It lines the abdominal wall as the parietal peritoneum, and it covers the visceral organs as the visceral peritoneum. The peritoneal cavity is the moistened space between the parietal and visceral portions of the peritoneum. Peritonitis results from any type of contamination of the peritoneal cavity, such as from a puncture wound, blood-borne diseases, or a ruptured visceral organ. In females, peritonitis is frequently a complication of infections of the reproductive tract that have entered the peritoneal cavity via the uterine tubes. Without medical treatment, peritonitis is generally fatal.

Ulcers may occur throughout the GI tract. *Peptic ulcers*—erosions of the mucous membranes of the stomach or duodenum—are produced by the action of hydrochloric acid (HCl) contained in gastric juice. Agents that weaken the mucosal lining of the stomach, including alcohol and aspirin, and hypersecretion

Hirschsprung's disease: from Harold Hirschsprung, Danish physician, 1830–1916

of gastric juice, as may accompany chronic stress, increase the likelihood of developing peptic ulcers. The bacterium *H. pylori,* which may be present in the GI tract, also may contribute to the weakening of mucosal barriers.

Enteritis, or inflammation of the intestinal mucosa, is frequently referred to as intestinal flu. Causes of enteritis include bacterial or viral infections, irritating foods or fluids, and emotional stress. The symptoms are abdominal pain, nausea, and diarrhea. *Diarrhea* is symptomatic of inflammation, stress, and other body dysfunctions. In children, it is of immense clinical importance because of the rapid loss of body fluids.

Shoulder and Upper Extremity

Developmental Conditions

Twenty-eight days after conception, a limb bud appears on the upper lateral side of the embryo, which eventually becomes a shoulder and an upper extremity. Three weeks later (7 weeks after conception) the shoulder and upper extremity are present in the form of mesenchymal primordium of bone and muscle. It is during this crucial 3 weeks of development that malformations of the extremities can occur.

If a pregnant woman uses certain teratogenic drugs or is exposed to certain diseases (*Rubella* virus, for example) during development of the embryo, there is a strong likelihood that the appendage will be incompletely developed. A large number of limb deformities occurred between 1957 and 1962 as a result of women ingesting the sedative thalidomide during early pregnancy to relieve morning sickness. It is estimated that 7,000 infants were malformed by this drug. The malformations ranged from *micromelia* (short limbs) to *amelia* (the absence of limbs).

Although genetic deformities of the shoulder and upper extremity are numerous, only a few are relatively common. **Sprengel's deformity** affects the development of one or both scapulae. In this condition, the scapula is smaller than normal and is positioned at an elevated level. As a result, abduction of the arm is not possible beyond a right angle to the plane of the body.

Minor defects of the extremities are relatively common malformations. Extra digits, a condition called **polydactyly** (*pol″ e-dak′ tĭ-le*) is the most common limb deformity. Usually an extra digit is incompletely formed and nonfunctional. **Syndactyly,** or webbed digits, is likewise a relatively common limb malformation. Polydactyly is inherited as a dominant trait, whereas syndactyly is a recessive trait.

Trauma to the Shoulder and Upper Extremity

The wide variety of injuries to the shoulder and upper extremity range from damaged bones and surrounding muscles, tendons, vessels, and nerves to damaged joints in the form of sprains or dislocations.

It is not uncommon to traumatize the shoulder and upper extremity of a newborn during a difficult delivery. **Upper arm birth palsy** (*Erb–Duchenne palsy*) is the most common type of birthing injury, caused by a forcible widening of the angle between the head and shoulder. Using forceps to rotate the fetus in utero, or pulling on the head during delivery, may cause this injury. The site of injury is at the junction of vertebrae C5 and C6 (*Erb's point*), as they form the upper trunk of the brachial plexus. The expression of the injury is paralysis of the abductors and lateral rotators of the shoulder and the flexors of the elbow. The arm will permanently hang at the side in medial rotation.

The stability of the shoulder is largely dependent on the support of the clavicle and acromion of the scapula superiorly and the tendons forming the rotator cuff anteriorly. Because support is weak along the inferior aspect of the shoulder joint, dislocations are frequent in this direction. Injuries of this sort are common in athletes engaging in contact sports. Sudden jerks of the arm are also likely to dislocate the shoulder, especially in children who have weak muscles spanning this area.

Fractures are common in any location of the shoulder and upper extremity. Many fractures result from extending the arm to break a fall. The clavicle is the most frequently broken bone in the body. Also common are fractures of the humerus, which are often serious because of injury to the nerves and vessels that parallel the bone. The surgical neck of the humerus is a common fracture site. At this point, the axillary nerve is often damaged, thus limiting abduction of the arm. A fracture in the middle third of the humerus may damage the radial nerve, causing paralysis of the extensor muscles of the hand (wristdrop). A fracture of the olecranon of the ulna often damages the ulnar nerve, resulting in paralysis of the flexor muscles of the hand and the adductor muscles of the thumb. The distal part of the radius is frequently fractured (*Colles' fracture*) by falling on an outstretched arm. In this fracture, the hand is displaced backward and upward.

Sports injuries frequently involve the upper extremity. Repeated extension of the wrist against a force, such as occurs during the backhand stroke in tennis, may cause **lateral epicondylitis** (*ep″ ĭ-kon″dĭ-lī′tis*) (tennis elbow). Wearing an elbow brace or a compression band may help to reduce the pain, but only if the cause is eliminated will the area be allowed to heal.

Athletes frequently jam a finger when a ball forcefully strikes a distal phalanx as the fingers are extended, causing a sharp flexion at the joint between the middle and distal phalanges. Splinting the finger for a period of time may be curative; however, surgery may be required to avoid a permanent crook in the finger.

Diseases of the Shoulder and Upper Extremity

Inflammations in specific locations of the shoulder or upper extremity are the only common clinical conditions endemic to

Sprengel's deformity: from Otto G. K. Sprengel, German surgeon, 1852–1915

Erb–Duchenne palsy: from Wilhelm H. Erb, German neurologist, 1840–1921, and Guillaume G. A. Duchenne, French neurologist, 1806–75

Colles' fracture: from Abraham Colles, Irish surgeon, 1773–1843

these regions. **Bursitis,** for example, may specifically afflict any of the numerous bursae of the shoulder, elbow, or wrist joints. There are several types of **arthritis,** but generally they involve synovial joints throughout the body rather than just those in the hands and fingers.

Carpal tunnel syndrome is caused by compression of the median nerve by the carpal flexor retinaculum that forms the palmar aspect of the carpal tunnel. The nerve compression results in a painful burning sensation or numbness of the first three fingers and some muscle atrophy. The compression is due to an inflammation of the transverse carpal ligament, which may be eased through surgery.

Tenosynovitis (*ten"o-sin"o-vi'tis*) is an inflammation of the synovial tendon sheath in the wrist or hand. Digital sheath infections are quite common following a puncture wound in which pathogens enter the closed synovial sheath. The increased pressure from the swollen, infected sheath may cause severe pain and eventually result in necrosis of the flexor tendons. The loss of hand function can be prevented by draining the synovial sheath and providing antibiotic treatment.

Hip and Lower Extremity

Developmental Conditions

The embryonic development of the hip and lower extremity follows the developmental pattern of the shoulder and upper extremity: the appearance of the limb bud is followed by the formation of the mesenchymal primordium of bone and muscle in the shape of an appendage. Development of the lower extremity, however, lags behind that of the upper extremity by 3 or 4 days.

The likelihood of congenital deformities of the hips and lower extremities in a newborn is slim if the pregnant mother has been healthy and well nourished, especially prior to and during embryonic development. The few congenital malformations that occur generally have a genetic basis.

In **congenital dislocation of the hip,** the acetabulum fails to develop adequately, and the head of the femur slides out of the acetabulum onto the gluteal surface of the ilium. If this condition goes untreated, the infant will never be able to walk normally.

Polydactyly and syndactyly occur in the feet as well as in the hands. Treatment of the feet is the same as treatment of the hands.

Talipes, (*tal'ĭ-pēz*) or **clubfoot,** is a congenital malformation in which the sole of the foot is twisted medially. It is uncertain whether abnormal positioning or restricted movement in utero causes talipes, but both genetic and environmental factors are involved in most cases.

Trauma to the Hip and Lower Extremity

As with the shoulder and upper extremity, a variety of traumatic conditions are associated with the hip and lower extremity. These range from injury to the bones and surrounding muscles, tendons, vessels, and nerves to damage of the joints in the form of sprains or dislocations.

Dislocation of the hip is a common and severe result of an automobile accident when a seat belt is not worn. When the hip is in the flexed position, as in sitting in a seat, a sudden force applied at the distal end of the femur will drive the head of the femur out of the acetabular socket, fracturing the posterior acetabular hip. In this kind of injury, there is usually damage to the sciatic nerve.

Trauma to the nerve roots that form the sciatic nerve may also occur from a herniated disc or pressure from the uterus during pregnancy. An improperly administered injection into the buttock may damage the sciatic nerve itself. Sciatic nerve damage is usually very painful and is expressed throughout the posterior length of the lower extremity.

Fractures are common in any location of the hip and lower extremity. Athletes (such as skiers) and elderly people seem to be most vulnerable. *Osteoporosis* markedly weakens the bones of the hip and thigh regions, making them vulnerable to fracture. A common fracture site, especially in elderly women, is across the femoral neck. A fracture of this kind may be complicated by vascular and nerve interruption. A direct blow to the knee will frequently fracture the patella. A potentially more serious knee trauma, however, is a clipping injury, caused by a blow to the lateral side. In this type of injury, there is generally damage to the cruciate ligament and menisci. Serious complications arise if the common peroneal nerve, traversing the popliteal fossa, is damaged. Damage to this nerve results in paralysis of the ankle and foot extensors (*footdrop*) and inversion of the foot.

Stress fractures of the long bones of the lower extremity are common to athletes. *Shinsplints,* a painful condition of the anterior muscles of the leg and their periosteal attachments, are often accompanied by tibial stress fractures. Stress fractures of the metatarsal bones may be very painful, even though they may not show up on radiographs. Frequently, the only way to heal stress fractures is to abstain from exercise.

Sprains are common in the joints of the lower extremity. Ligaments and tendons are torn to varying degrees in sprains. Sprains are usually accompanied by *synovitis,* an inflammation of the joint capsule.

Diseases of the Hip and Lower Extremity

As in the shoulder and upper extremity, infections in the hip and lower extremity—such as bursitis and tendinitis—can be localized in any part of the hip or lower extremity. Likewise, several types of arthritis may affect joints in these regions.

A variety of skin diseases afflict the foot, including athlete's foot, plantar warts, and dyshidrosis. Most of the diseases of the feet can be prevented, or if they do occur, they can be treated effectively.

Because arterial occlusive disease is common in elderly people, palpation of the posterior tibial artery is clinically important in general physical assessment. This can be accomplished by gently palpating between the medial malleolus and the tendo calcaneus.

Many neuromuscular diseases have a direct effect on the functional capabilities of the lower extremities. *Muscular dystrophy* and *poliomyelitis* are both serious immobilizing diseases because of muscle paralysis.

Chapter Summary

Introduction to Surface Anatomy (pp. 288–289)

1. Surface anatomy is concerned with identifying body structures through visual inspection and palpation. It has tremendous application in the maintenance of physical fitness and in medical diagnosis and treatment.
2. Most of the bones of the skeleton are palpable and provide landmarks for locating other anatomical structures.
3. The effectiveness of visual inspection and palpation in studying a person's surface anatomy is influenced by the thickness of the hypodermis, which varies in accordance with the amount of subcutaneous adipose tissue present.

Surface Anatomy of the Newborn (pp. 289–291)

1. Certain aspects of the surface anatomy of a neonate are of clinical importance in ascertaining the degree of physical development, general health, and possible congenital abnormalities.
2. The posture of a full-term, normal neonate is one of flexion.
3. Portions of the skin and subcutaneous tissues of a neonate are edematous. Vernix caseosa covers the body, and lanugo may be present on the head, neck, and back.
4. The fontanels, liver, and kidneys should be palpable, as well as the testes of a male.

Head (pp. 291–295)

1. Surface features of the cranium include the forehead, crown, temporalis muscles, and the hair and hairline.
2. The face is composed of the ocular region that surrounds the eye, the auricular region of the ear, the nasal region serving the respiratory system, and the oral region serving the digestive and respiratory systems.

Neck (pp. 296–300)

1. Major organs are located within the flexible neck, and structures that are essential for body sustenance pass through the neck to the trunk.
2. The neck consists of an anterior cervix, right and left lateral regions, and a posterior nucha.
3. Two major and six minor triangles, each of which contains specific structures, are located on each side of the neck.
   (a) The anterior cervical triangle encompasses the carotid, submandibular, submental, and muscular triangles.
   (b) The posterior cervical triangle encompasses the supraclavicular and omoclavicular triangles.

Trunk (pp. 300–309)

1. Vital visceral organs in the trunk make the surface anatomy of this region especially important.
2. The median furrow is visible, and the vertebral spines and scapulae are palpable on the back.
3. Palpable structures of the thorax include the sternum, the ribs, and the costal arch.
4. The important surface anatomy features of the abdomen include the linea alba, umbilicus, costal arch, iliac crests, and the pubis.

Pelvis and Perineum (pp. 310–311)

1. The crest of the ilium, the symphysis pubis, and the inguinal ligament are important pelvic landmarks.
2. The perineum is the region that contains the external genitalia and the anal opening.

Shoulder and Upper Extremity (pp. 311–317)

1. The surface anatomy of the shoulder and upper extremity is important because of frequent trauma to these regions. Vessels of the upper extremity are also used as pressure points and for intravenous injections or blood withdrawal.
2. The scapula, clavicle, and humerus are palpable in the shoulder.
3. The axilla is clinically important because of the vessels, nerves, and lymph nodes located there.
4. The brachial artery is an important pressure point in the brachium. The median cubital vein is important for the removal of blood or for intravenous therapy.
5. The ulna, radius, and their processes are palpable landmarks of the forearm.
6. The knuckles, fingernails, and tendons for the extensor muscles of the forearm can be observed on the posterior aspect of the hand.
7. Flexion creases and the thenar eminence are important features on the anterior surface of the hand.

Buttock and Lower Extremity (pp. 317–321)

1. The massive bones and muscles in the buttock and lower extremity serve as weight-bearers and locomotors. Many of the surface features of these regions are important with respect to locomotion or locomotor dysfunctions.
2. The prominences of the buttocks are formed by the gluteal muscles and are separated by the natal cleft.
3. The thigh has three muscle groups: anterior (quadriceps), medial (adductors), and posterior (hamstrings).
4. The femoral triangle and popliteal fossa are clinically important surface landmarks.
5. The structures of the leg include the tibia and fibula, the muscles of the calf, and the saphenous veins.
6. The surface anatomy of the foot includes structures adapted to support the weight of the body, maintain balance, and function during locomotion.

Review Activities

Objective Questions

1. Eyebrows are located on
   (a) the palpebral fissure.
   (b) the subtarsal sulcus.
   (c) the scalp.
   (d) the supraorbital ridges.
   (e) both c and d.
2. Which of the following structures is *not* part of the auricle (pinna) of the ear?
   (a) the tragus (c) the earlobe
   (b) the ala (d) the helix
3. Which of the following clinical-structural word pairs is *incorrectly* matched?
   (a) cleft lip/philtrum
   (b) broken nose/nasion
   (c) pierced ear/earlobe
   (d) black eye/concha
4. The conjunctiva
   (a) covers the entire eyeball.
   (b) is a thick nonmucous membrane.
   (c) secretes tears.
   (d) none of the above apply.

5. Which of the following could *not* be palpated within the cervix of the neck?
   (a) the larynx
   (b) the hyoid bone
   (c) the trachea
   (d) the cervical vertebrae
6. Palpation of an arterial pulse in the neck is best accomplished at
   (a) the carotid triangle.
   (b) the supraclavicular triangle.
   (c) the submental triangle.
   (d) the submandibular triangle.
   (e) the omotracheal triangle.
7. Which nerve lies posterior to the medial epicondyle of the humerus?
   (a) the ulnar nerve
   (b) the median nerve
   (c) the radial nerve
   (d) the brachial nerve
   (e) the cephalic nerve
8. Which of the following surface features could *not* be observed on obese people?
   (a) the jugular notch
   (b) scapular muscles
   (c) clavicles
   (d) vertebral spines
   (e) the natal cleft
9. Which pair of muscles forms the anterior and posterior borders of the axilla?
   (a) deltoid/pectoralis minor
   (b) biceps brachii/triceps brachii
   (c) latissimus dorsi/pectoralis major
   (d) triceps brachii/pectoralis major
   (e) latissimus dorsi/deltoid
10. Varicose veins occur when which of the following become(s) excessively enlarged?
   (a) saphenous veins
   (b) tibial veins
   (c) the external iliac vein
   (d) the popliteal vein
   (e) all of the above

Essay Questions

1. List four surface features of the cranium and explain how the cranium relates to the scalp.
2. Identify the four regions of the face and list at least two surface features of each region.
3. Which surface features can be observed on the trunk of any person, regardless of how obese that person might be?
4. Identify the two major triangles of the neck and list the associated six minor triangles. Discuss the importance of knowing the boundaries of these triangles and the specific structures included in each.
5. Name four structures that are palpable along the anterior midline of the neck.
6. What three bones are found in the shoulder region? List the surface features that can be either observed or palpated on each of these bones.
7. Identify the tendons or vessels that can be observed or palpated along the anterior surface of the wrist. Which nerves pass through this region?
8. Describe the locations of the arteries that can be palpated as they pulsate in the following regions: (a) neck, (b) brachium, (c) antebrachium, (d) thigh, and (e) ankle. Which of these are considered clinical pressure points?
9. Describe the locations of the following regions: (a) cubital fossa, (b) femoral triangle, (c) axilla, (d) perineum, and (e) popliteal fossa. Comment on the clinical importance of each of these regions.
10. Describe the anatomical location where each of the following could be observed or palpated: (a) the distal tendinous attachments of the (a) hamstring muscles; (b) the greater trochanter; (c) the great and small saphenous veins; (d) the femoral, posterior tibial, and dorsal pedal arteries; and (e) the medial malleolus

Critical-Thinking Questions

1. A Saturday afternoon athlete crashed while mountain biking without a helmet. He sustained deep cuts across the front of his knee, across the back of his elbow, horizontally through his scalp, and across the length of the cheek on his face. Which areas would be the most difficult to hold together with sutures, thus requiring more time to heal? Why the disparity in the various regions?
2. Knowledge of surface anatomy is crucial to the intensive care physician when vascular access to the large veins of the neck is required for the rapid administration of fluids and medications. Cannulation of the internal jugular and subclavian veins is frequently employed. Can you associate the position of these veins with surface landmarks on the neck? Refer to figures 16.32 and 16.33 for the location of these veins. What might be a possible complication of using the subclavian vein?
3. It is often necessary in the critical care setting for a physician to obtain vascular access to the femoral artery or vein. How do these structures lie in relation to each other and what other structures are in the vicinity? (Refer to fig. 10.54.)

ELEVEN

Nervous Tissue and the Central Nervous System

Clinical Case Study 334
Organization and Functions of the Nervous System 334
- Organization of the Nervous System 334
- Functions of the Nervous System 334

Developmental Exposition: The Brain 336
Neurons and Neuroglia 338
- Neurons 338
- Neuroglia 340
- Classification of Neurons and Nerves 342

Transmission of Impulses 346
- Nerve Impulse 347
- Synapse 347

General Features of the Brain 348
Cerebrum 352
- Structure of the Cerebrum 353
- Lobes of the Cerebrum 355
- Brain Waves 356
- White Matter of the Cerebrum 357
- Basal Nuclei 358
- Language 359

Diencephalon 361
- Thalamus 361
- Hypothalamus 362
- Epithalamus 363
- Pituitary Gland 363

Mesencephalon 363
Metencephalon 364
- Pons 364
- Cerebellum 364

Myelencephalon 366
- Medulla Oblongata 366
- Reticular Formation 367

Meninges of the Central Nervous System 367
- Dura Mater 367
- Arachnoid 369
- Pia Mater 369

Ventricles and Cerebrospinal Fluid 370
- Ventricles of the Brain 370
- Cerebrospinal Fluid 372
- Blood-Brain Barrier 372

Spinal Cord 372
- Structure of the Spinal Cord 373
- Spinal Cord Tracts 374

Developmental Exposition: The Spinal Cord 378
Clinical Considerations 379
- Neurological Assessment and Drugs 379
- Developmental Problems 380
- Injuries 380
- Disorders of the Nervous System 381
- Senescence of the Nervous System 383
- Degenerative Diseases of the Nervous System 383

Internal Affairs 384
Clinical Case Study Answer 385
Chapter Summary 386
Review Activities 387

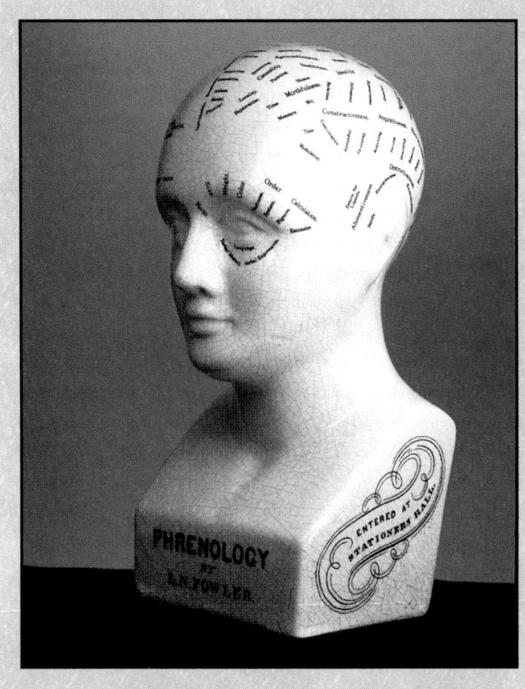

Phrenology, a popular pseudoscience of the nineteenth century, was based on the now discredited belief that the shape and bumps of the skull reveal character and mental capacity. In this bust by L. N. Fowler, a wide range of personality traits are mapped on the cranium surrounding the brain.

Clinical Case Study

A 56-year-old woman visited her family doctor for evaluation of a headache that had persisted for nearly a month. Upon questioning the patient, the doctor learned that her left arm, as she put it, "was a bit unwieldy, hard to control, and weak." Through examination, the doctor determined that the entire left upper extremity was generally weak. He also found weakness, although less significant, of the left lower extremity. Sensation in the limbs seemed to be normal, although mild rigidity and hyperactive reflexes were present. Expressing concern, the doctor told the patient that she needed a CT scan of her head, and explained that there could be a problem within the brain, possibly a tumor or other lesion. The doctor then picked up the phone and contacted a radiologist. After explaining the patient's case, the doctor remarked parenthetically that he believed he knew where the problem was located.

Why did the doctor suggest to the patient that there might be a problem within her brain when the symptoms were weakness of the extremities, and then just on one side of her body? Also, how would he know the location of the suspected brain tumor? In which side of the brain and in which lobe would it be? Explain the muscle weakness in terms of neuronal pathways from the brain to the periphery.

Hints: Remember the controlling and integrating function of the brain. Carefully study the information and accompanying figures concerning the structures and functions of the brain and the neuronal tracts.

Organization and Functions of the Nervous System

The central nervous system and the peripheral nervous system are structural components of the nervous system, whereas the autonomic nervous system is a functional component. Together they orient the body, coordinate body activities, permit the assimilation of experiences, and program instinctual behavior.

Objective 1 Describe the divisions of the nervous system.

Objective 2 Define *neurology*; define *neuron*.

Objective 3 List the functions of the nervous system.

The immensely complex brain and its myriad of connecting pathways constitute the nervous system. The nervous system, along with the endocrine system, regulates the functions of the other body systems. The brain, however, does much more than that—and its potential is perhaps greatly underestimated. It is incomprehensible that one's personality, thoughts, and aspirations result from the functioning of a body organ. Plato referred to the brain as "the divinest part of us." The thought processes of this organ have devised the technology for launching rockets into space, curing diseases, and splitting atoms. But with all of these achievements, the brain still remains largely ignorant of its own workings.

Neurology, the study of the nervous system, has been referred to as the last frontier of functional anatomy. Basic questions concerning the functioning of the nervous system remain unanswered: How do nerve cells store and retrieve memory? What are the roles of the many chemical compounds within the brain? What causes mental illness or senility? Scientists are still developing the technology and skills necessary to understand the functional complexity of the nervous system. The next few decades will undoubtedly witness major progress toward the achievement of research goals in this field of study.

Organization of the Nervous System

The nervous system is divided into the **central nervous system (CNS),** which includes the brain and spinal cord, and the **peripheral nervous system (PNS),** which includes the *cranial nerves* arising from the brain and the *spinal nerves* arising from the spinal cord (fig. 11.1 and table 11.1).

The **autonomic nervous system (ANS)** is a functional subdivision of the nervous system. The controlling centers of the ANS are located within the brain and are considered part of the CNS; the peripheral portions of the ANS are subdivided into the **sympathetic** and **parasympathetic divisions.**

Functions of the Nervous System

The nervous system is specialized for perceiving and responding to events in our internal and external environments. An awareness of one's environment is made possible by **neurons** (nerve cells), which are highly specialized with respect to excitability and conductivity. The nervous system functions throughout the body in conjunction with the endocrine system (see chapter 14) to closely coordinate the activities of the other body systems. In addition to integrating body activities, the nervous system has the ability to store experiences (*memory*) and to establish patterns of response based on prior experiences (*learning*).

The functions of the nervous system include

1. orientation of the body to internal and external environments;
2. coordination and control of body activities;
3. assimilation of experiences requisite to memory, learning, and intelligence; and
4. programming of instinctual behavior (apparently more important in vertebrates other than humans).

neurology: Gk. *neuron*, nerve; L *logos*, study of

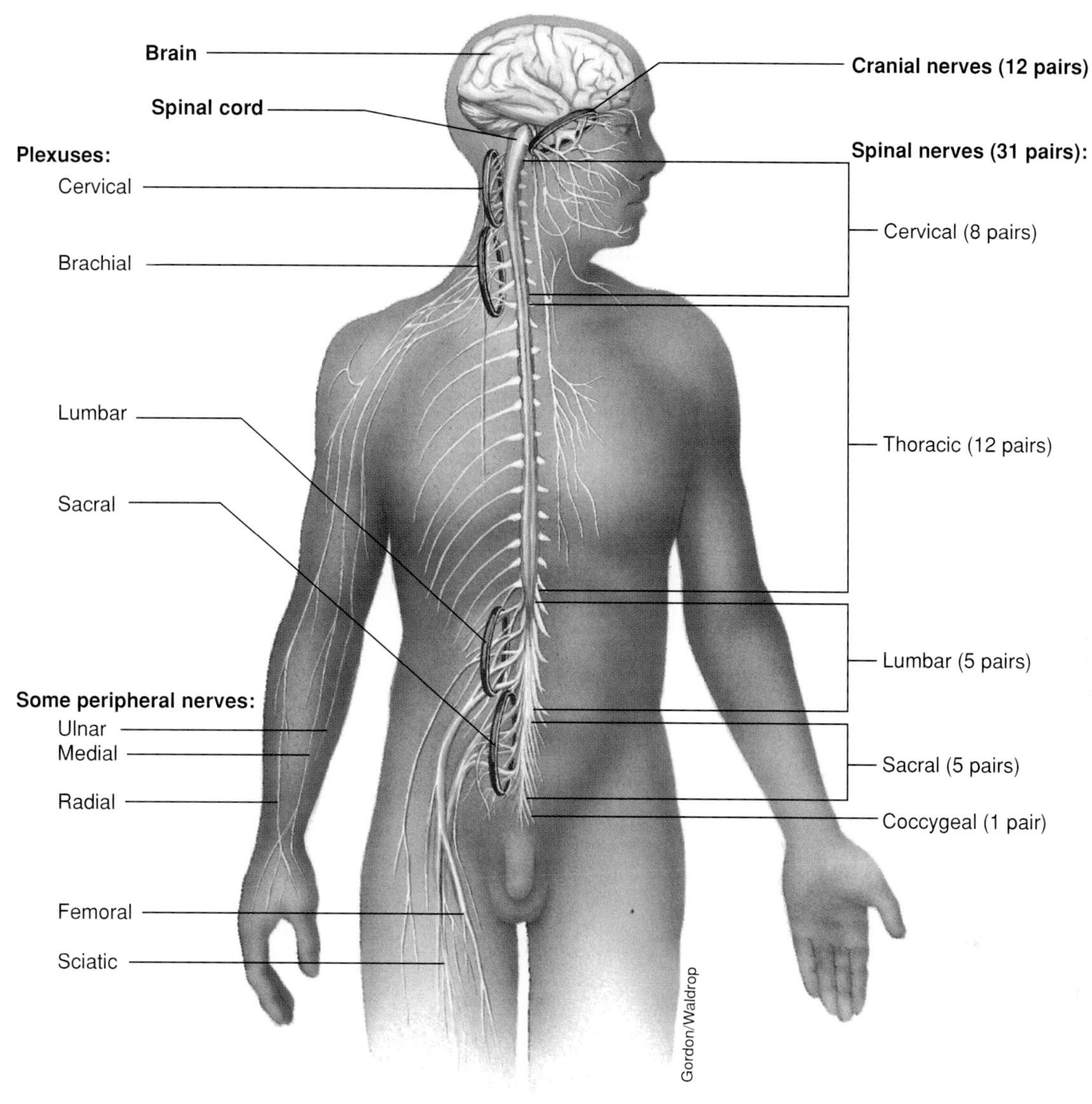

Figure 11.1

The nervous system. The central nervous system (CNS) consists of the brain and spinal cord. The peripheral nervous system (PNS) consists of cranial nerves and spinal nerves. Also part of the PNS are the plexuses and additional nerves that arise from the cranial and spinal nerves. The autonomic nervous system (ANS) is a functional subdivision of the nervous system.

Table 11.1
Terminology Pertaining to the Nervous System

Term	Definition
Central nervous system (CNS)	Brain and spinal cord
Peripheral nervous system (PNS)	Nerves, ganglia, and nerve plexuses
Sensory nerve fiber (afferent fiber)	Neuron that transmits impulses from a sensory receptor to the CNS
Motor nerve fiber (efferent fiber)	Neuron that transmits impulses from the CNS to an effector organ; for example, a muscle or gland
Nerve	Cablelike collection of nerve fibers; may be "mixed" (contain both sensory and motor fibers)
Plexus	Network of interlaced nerves
Somatic motor nerve	Nerve that stimulates contraction of skeletal muscles
Autonomic motor nerve	Nerve that stimulates contraction (or inhibits contraction) of smooth muscle and cardiac muscle and that stimulates secretion of glands
Ganglion	Grouping of neuron cell bodies located outside the CNS
Nucleus	Grouping of neuron cell bodies within the CNS
Tract	Grouping of nerve fibers that interconnect regions of the CNS

DEVELOPMENTAL EXPOSITION

The Brain

Exhibit I The early development of the nervous system from embryonic ectoderm. (*a*) A dorsal view of an 18-day-old embryo showing the formation of the neural plate and the position of a transverse cut indicated in (a_1). (*b*) A dorsal view of a 22-day-old embryo showing cranial and caudal neuropores and the positions of three transverse cuts indicated in (b_1-b_3). (Note the amount of fusion of the neural tube at the various levels of the 22-day-old embryo. Note also the relationship of the notochord to the neural tube.)

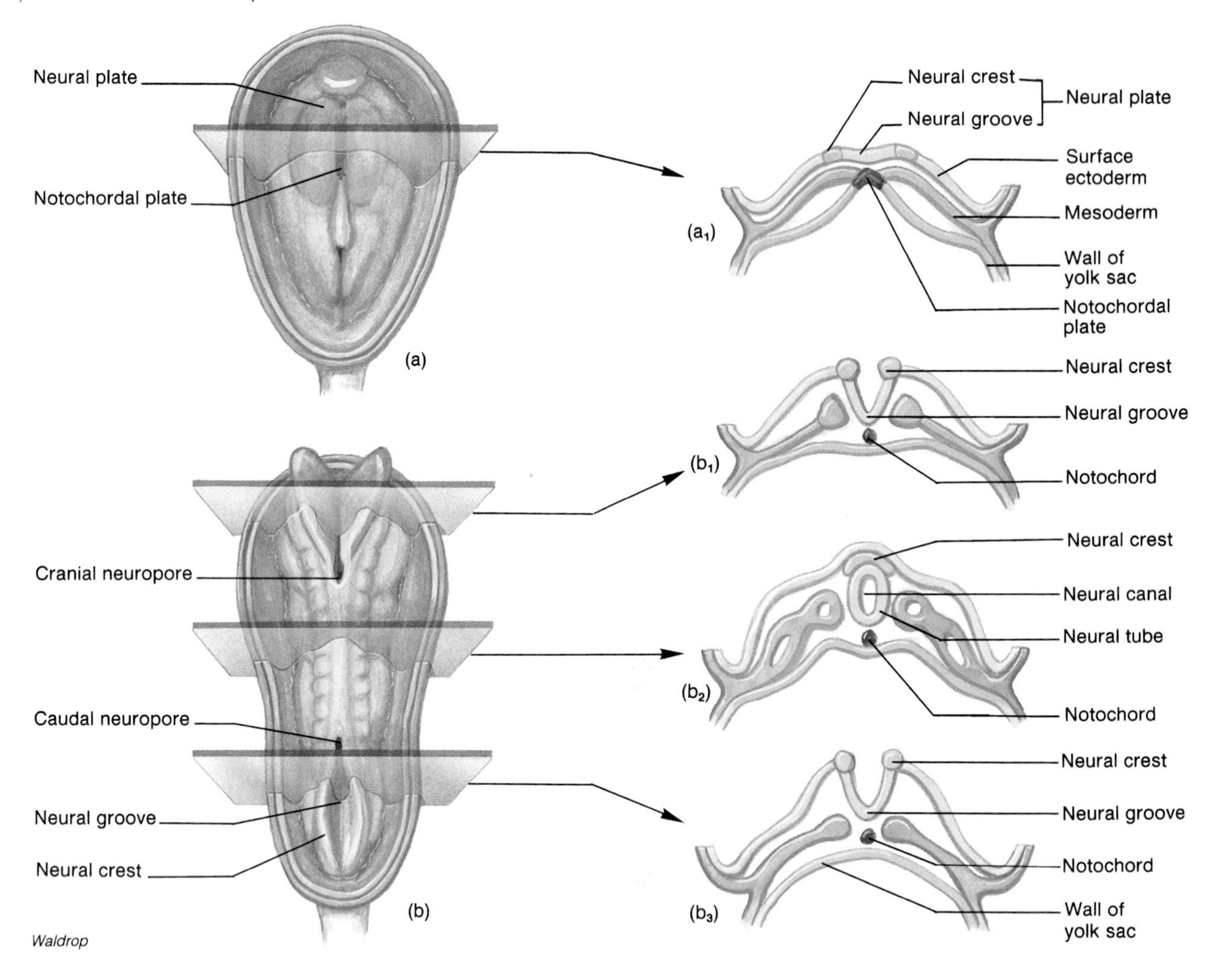

Exhibit II The developmental sequence of the brain. During the fourth week, the three principal regions of the brain are formed. During the fifth week, a five-regioned brain develops and specific structures begin to form.

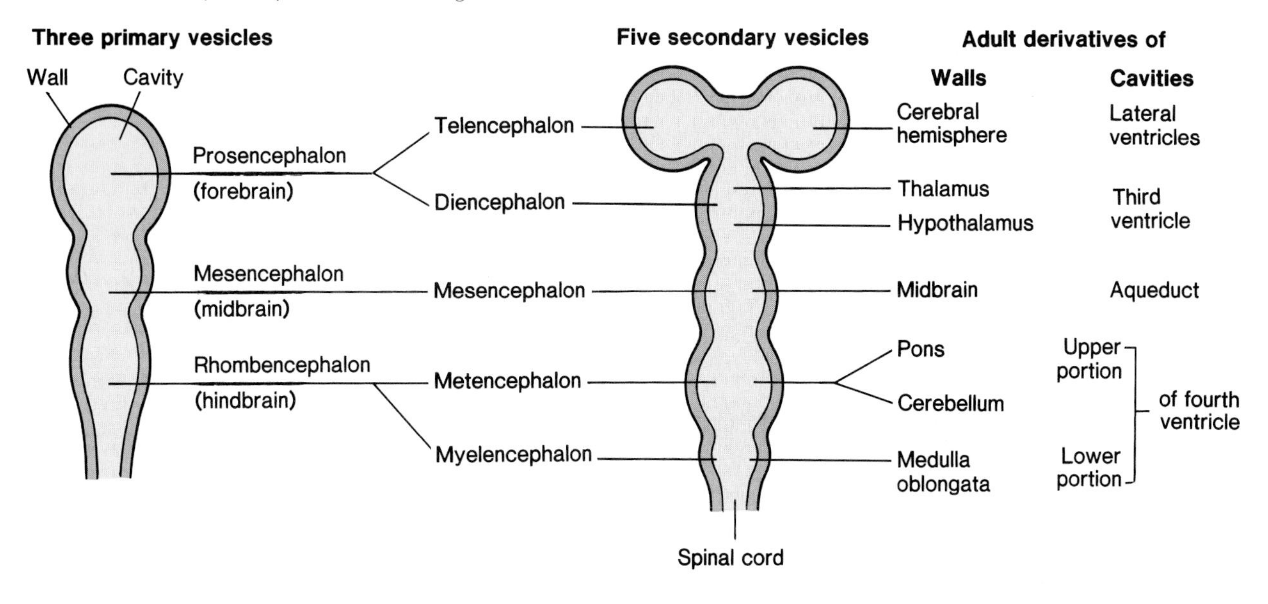

Explanation

The first indication of nervous tissue development occurs about 17 days following conception, when a thickening appears along the entire dorsal length of the embryo. This thickening, called the **neural plate** (exhibit I), differentiates and eventually gives rise to all of the **neurons** and to most of the **neuroglia** that support the neurons. As development progresses, the midline of the neural plate invaginates to become the **neural groove.** At the same time, there is a proliferation of cells along the lateral margins of the neural plate, which become the thickened **neural folds.** The neural groove continues to deepen as the neural folds elevate. By day 20, the neural folds have met and fused at the midline, and the neural groove has become a **neural tube.** For a short time, the neural tube is open both cranially and caudally. These openings, called **neuropores,** close during the fourth week. Once formed, the neural tube separates from the surface ectoderm and eventually develops into the central nervous system (brain and spinal cord). The **neural crest** forms from the neural folds as they fuse longitudinally along the dorsal midline. Most of the peripheral nervous system (cranial and spinal nerves) forms from the neural crest. Some neural crest cells break away from the main tissue mass and migrate to other locations, where they differentiate into motor nerve cells of the sympathetic nervous system or into *neurolemocytes* (Schwann cells), which are a type of glial cell important in the peripheral nervous system.

The brain begins its embryonic development as the cephalic end of the neural tube starts to grow rapidly and differentiate (exhibit II). By the middle of the fourth week, three distinct swellings are evident: the **prosencephalon** (*pros"en-sef'ă-lon*) (forebrain), the **mesencephalon** (midbrain), and the **rhombencephalon** (hindbrain). Further development during the fifth week results in the formation of five specific regions. The **telencephalon** and the **diencephalon** (*di"en-sef'-ă-lon*) derive from the forebrain, the mesencephalon remains unchanged, and the **metencephalon** and **myelencephalon** form from the hindbrain. The caudal portion of the myelencephalon is continuous with and resembles the spinal cord.

These four functions depend on the ability of the nervous system to monitor changes, or *stimuli*, from both inside and outside the body; to interpret the changes in a process called *integration;* and to effect responses by activating muscles or glands. Thus, broadly speaking, the nervous system has *sensory*, *integrative*, and *motor functions*, all of which work together to maintain the internal constancy, or homeostasis, of the body.

An instinct also may be called a *fixed action pattern;* typically, it is genetically specified with little environmental modification. It is triggered only by a specific stimulus. some of the basic instincts in humans include survival, feeding, drinking, voiding, and specific vocalization. some ethologists (scientists who study animal behavior) believe that reproduction becomes an instinctive behavior following puberty.

1. Distinguish between the CNS and PNS. What are the subdivisions of the peripheral portions of the ANS?
2. Explain why neurology is considered a dynamic science.

Neurons and Neuroglia

Neurons come in many forms, but all contain dendrites for reception and an axon for the conduction of nerve impulses. The various types of neurons may be classified on the basis of structure or function. Different types of neuroglia support the neurons, both structurally and functionally.

Objective 4 Describe the microscopic structure of a neuron.

Objective 5 Describe how a neurilemmal sheath and a myelin sheath are formed.

Objective 6 List the types of neuroglia and describe their functions.

Objective 7 Describe the functions and locations of sensory and motor nerve fibers.

The highly specialized and complex nervous system is composed of only two principal categories of cells—neurons and neuroglia. **Neurons** are the basic structural and functional units of the nervous system. They are specialized to respond to physical and chemical stimuli, conduct impulses, and release specific chemical regulators. Through these activities, neurons perform such functions as storing memory, thinking, and regulating other organs and glands. Neurons cannot divide mitotically, although some neurons can regenerate a severed portion or sprout small new branches under certain conditions.

Mitotic activity of neurons is completed during prenatal development. Thus, a person is born with all the neurons he or she is capable of producing. However, neurons continue to grow and to specialize after a person is born, particularly in the first several years of postnatal life.

Neuroglia, or **glial cells,** are supportive cells in the nervous system that aid the function of neurons. Neuroglia are about five times as abundant as neurons and have limited mitotic abilities.

Neurons

Although **neurons** vary considerably in size and shape, they all have three principal components: (1) a cell body, (2) dendrites, and (3) an axon (figs. 11.2 and 11.3).

The **cell body** is the enlarged portion of the neuron that more closely resembles other cells. It contains a nucleus with a prominent nucleolus and the bulk of the cytoplasm. Besides containing organelles typically found in cells, the cytoplasm of neurons is characterized by the presence of **chromatophilic substances** (Nissl bodies) and filamentous strands of protein called **neurofibrils** (*noor″ŏ-fi′ brilz*). Chromatophilic substances are specialized layers of granular (rough) endoplasmic reticulum, whose function is protein synthesis, and minute **microtubules,** which appear to be involved in transporting material within the cell. The cell bodies within the CNS are frequently clustered into regions called **nuclei** (not to be confused with the nucleus of a cell). Cell bodies in the PNS generally occur in clusters called **ganglia** (*gang′gle-ă*).

Dendrites (*den′drīts*) are branched processes that extend from the cytoplasm of the cell body. Dendrites function to receive stimuli and conduct impulses to the cell body. Some dendrites are covered with minute **dendritic spinules** that greatly increase their surface area and provide contact points for other neurons. The area occupied by dendrites is referred to as the **dendritic zone** of a neuron.

The **axon** (*ak′son*) is the second type of cytoplasmic extension from the cell body. The term *nerve fiber* is commonly used in reference to either an axon or an elongated dendrite. An axon is a relatively long, cylindrical process that conducts impulses away from the cell body. Axons vary in length from a few millimeters in the CNS to over a meter between the distal portions of the extremities and the spinal cord. Side branches called **collateral branches** extend a short distance from the axon. The cytoplasm of an axon contains many mitochondria, microtubules, and neurofibrils.

Nissl body: from Franz Nissl, German neuroanatomist, 1860–1919
ganglion: Gk. *ganglion*, swelling
dendrite: Gk. *dendron*, tree branch
axon: Gk. *axon*, axis

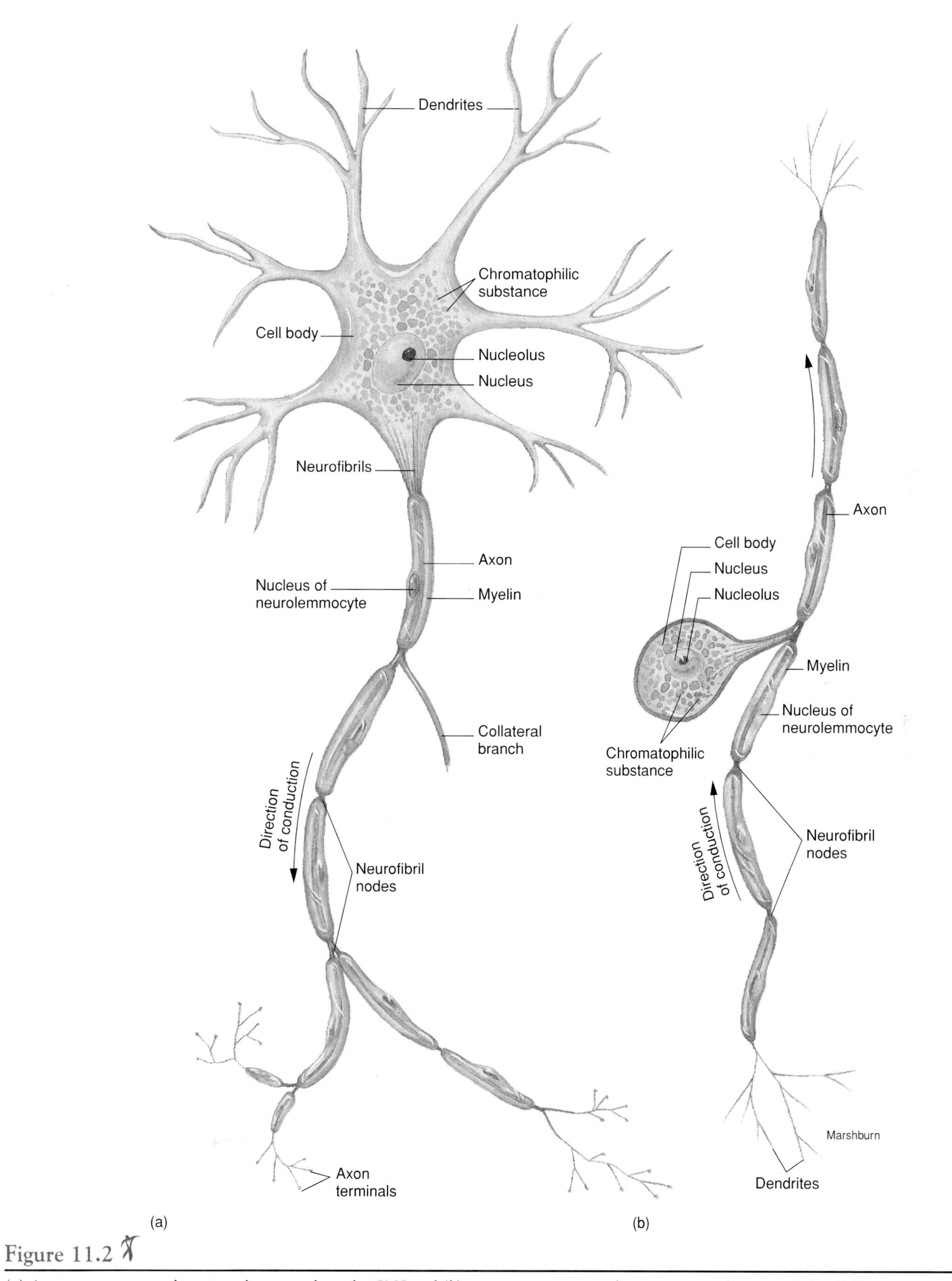

Figure 11.2

(*a*) A motor neuron conducts impulses away from the CNS and (*b*) a sensory neuron conducts nerve impulses toward the CNS.

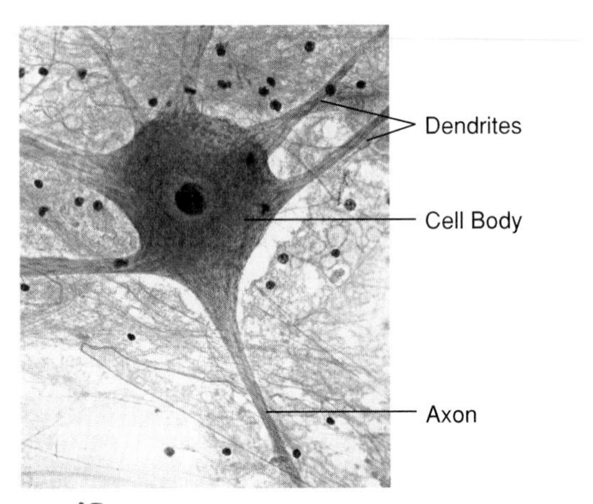

Figure 11.3

The neuron as seen in a photomicrograph of nervous tissue.

Proteins and other molecules are transported rapidly through the axon by two different mechanisms: axoplasmic flow and axonal transport. *Axoplasmic flow,* the slower of the two, results from rhythmic waves of contraction that push cytoplasmic contents from the axon hillock, where the axon originates, to the nerve fiber endings. *Axonal transport,* which is more rapid and more selective, may occur in a retrograde as well as a forward direction. Indeed, such retrograde transport may be responsible for the movement of herpes virus, rabies virus, and tetanus toxin from nerve terminals into cell bodies.

Neuroglia

Unlike other organs that are packaged in connective tissue derived from mesoderm, all but one type of the supporting **neuroglia (glial cells)** (fig. 11.4) are derived from the same ectoderm that produces neurons. There are six categories of neuroglia: (1) **neurolemmocytes** (Schwann cells), which form myelin layers around axons in the PNS; (2) **oligodendrocytes** (*ol″ĭ-go-den′drŏ-sīts*), which form myelin layers around axons in the CNS; (3) **microglia** (*mi-krog′le-ă*), which are derived from mesoderm and migrate through the CNS, removing foreign and degenerated material; (4) **astrocytes,** which help to regulate the passage of molecules from the blood to the brain; (5) **ependymal** (*ĕ-pen′dĭ-mal*) **cells,** which line the ventricles of the brain and the central canal of the spinal cord; and (6) **ganglionic gliocytes,** which support neuron cell bodies within the ganglia of the PNS. The six types of neuroglia are summarized in table 11.2.

Myelination

Neurons are either *myelinated* or *unmyelinated.* **Myelination** (*mi″ĕ-lĭ-na′shun*) is the process in which a neurolemmocyte or oligodendrocyte surrounds a portion of the axon or dendrite to provide support and aid in the conduction of impulses (figs. 11.5 and 11.6). The neuroglia that participate in the myelination process contain a white lipid-protein substance called **myelin** (*mi′ĕ-lin*). As several neuroglia are positioned in sequence, a **myelin layer** that encloses the axon or dendrite is formed. Myelinated neurons occur both in the CNS and the PNS. Myelin is responsible for the color of the *white matter* of the brain and spinal cord and the white coloration of nerves.

Myelination in the PNS occurs as neurolemmocytes grow and wrap around an axon or dendrite (figs. 11.2, 11.5, and 11.6). The outer surface of the myelin sheath is encased in a glycoprotein **neurolemmal sheath** that promotes neuron regeneration in the event that the neuron is injured. Each neurolemmocyte wraps only about 1 mm of axon, leaving gaps of exposed axon between adjacent neurolemmocytes. These gaps in the myelin sheath and neurolemmal sheath are known as the **neurofibril nodes** (nodes of Ranvier [ron′ve-a; ran′vēr]). It is at the neurofibril nodes that a nerve impulse is propagated along a neuron.

The myelin sheaths of the CNS are formed by **oligodendrocytes.** Unlike a neurolemmocyte, which forms a myelin sheath around only one axon, each oligodendrocyte has extensions that form myelin sheaths around several axons.

Regeneration of a Cut Axon

When an axon in a peripheral nerve is cut, the distal region of the axon that was severed from the cell body degenerates and is phagocytosed by neurolemmocytes. The neurolemmocytes then form a *regeneration tube* (fig. 11.7), as the part of the axon that is connected to the cell body begins to grow and exhibit ameboid movement. The neurolemmocytes of the regeneration tube are believed to secrete chemicals that attract the growing axon tip, and the tube helps to guide the regenerating axon to its proper destination. Even a severed major nerve may be surgically reconnected and the function of the nerve reestablished if the surgery is performed before tissue death. The CNS lacks neurolemmocytes, and central axons are generally believed to have a much more limited ability to regenerate than peripheral axons.

---

neuroglia: Gk. *neuron,* nerve; *glia,* glue
Schwann cell: from Theodor Schwann, German histologist, 1810–82
oligodendrocyte: Gk. *oligos,* few; L. *dens,* tooth; Gk. *kytos,* hollow (cell)
microglia: Gk. *mikros,* small; *glia,* glue
astrocyte: Gk. *aster,* star; *kytos,* hollow (cell)
ependyma: Gk. *ependyma,* upper garment

myelin: Gk. *myelos,* marrow
nodes of Ranvier: From Louis A. Ranvier, French pathologist, 1835–1922

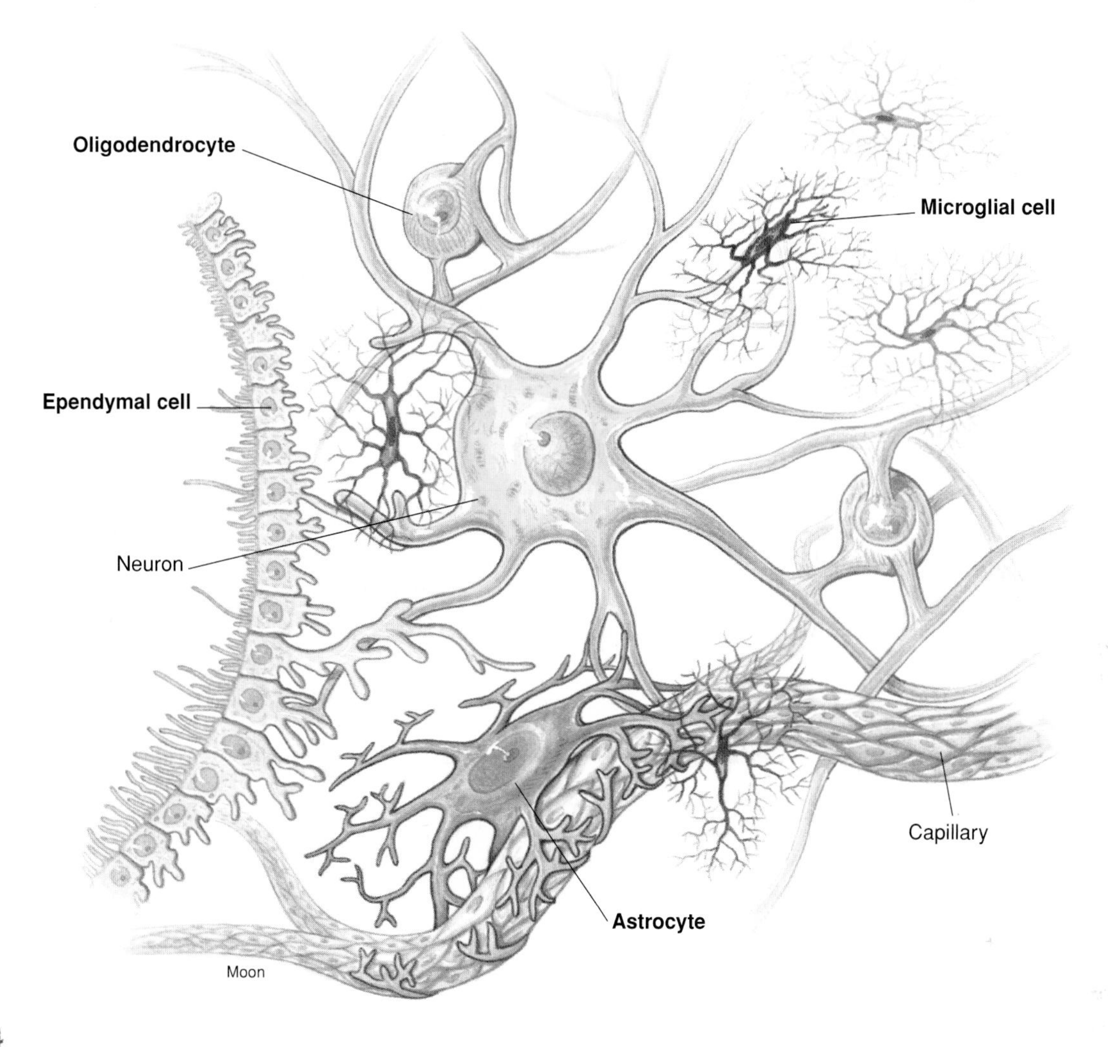

Figure 11.4

The four types of neuroglia found within the central nervous system.

Table 11.2
Neuroglia

Type	Structure	Function/Location	Type	Structure	Function/Location
Astrocytes	Stellate with numerous processes	Form structural support between capillaries and neurons of the CNS; contribute to blood-brain barrier	Ependymal cells	Columnar cells, some of which have ciliated free surfaces	Line ventricles and the central canal of the CNS, where cerebrospinal fluid is circulated by ciliary motion
Oligodendrocytes	Similar to astrocytes, but with shorter and fewer processes	Form myelin in the CNS; guide development of neurons in the CNS	Ganglionic gliocytes	Small, flattened cells	Support ganglia in the PNS
Microglia	Minute cells with few short processes	Phagocytize pathogens and cellular debris within the CNS	Neurolemmocytes (Schwann cells)	Flattened cells arranged in series around axons or dendrites	Form myelin in the PNS

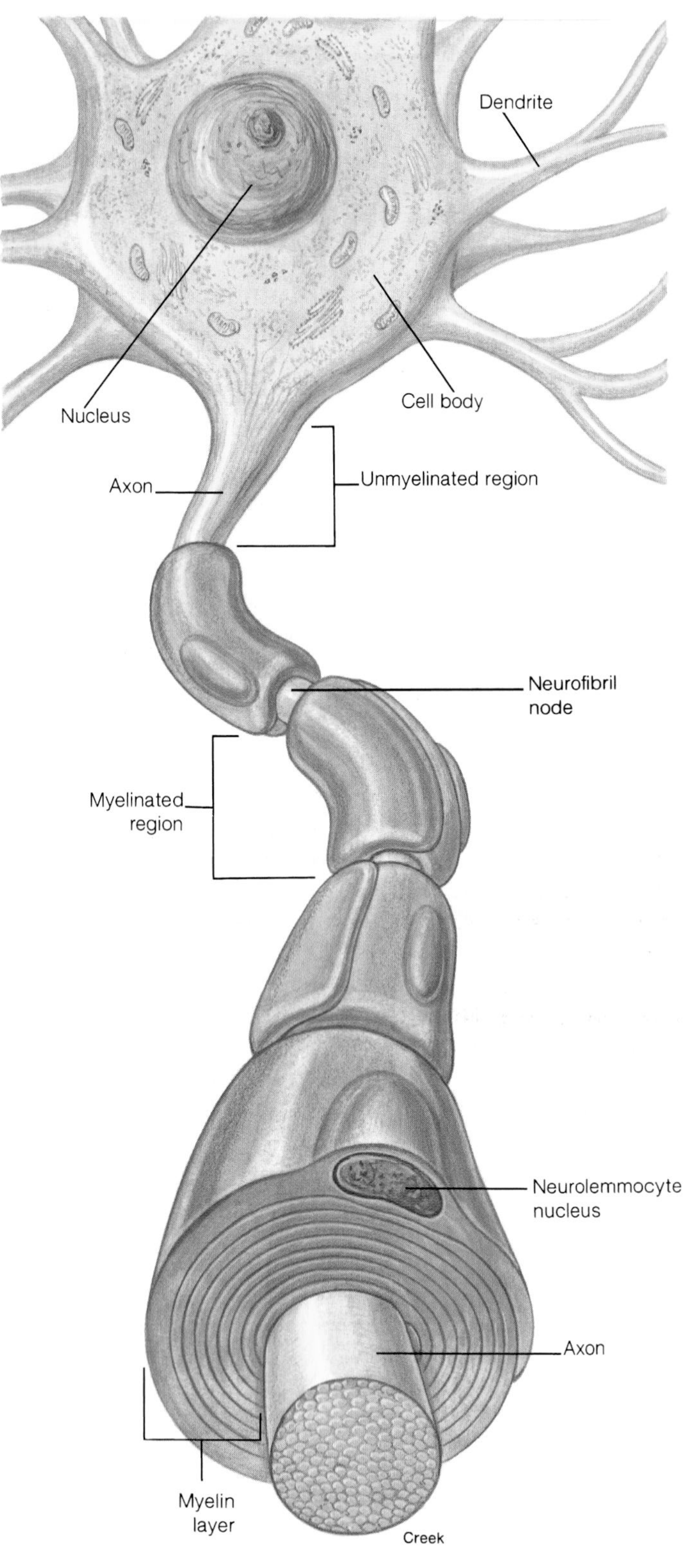

Figure 11.5

A myelinated neuron. A myelin layer is formed by the wrapping of neurolemmocytes around the axon of a neuron.

Astrocytes and the Blood-Brain Barrier

Astrocytes are large, stellate cells with numerous cytoplasmic processes that radiate outward. These are the most abundant neuroglia in the CNS, and in some locations in the brain they constitute 90% of the nervous tissue.

Capillaries in the brain, unlike those of most other organs, do not have pores between adjacent endothelial cells. Molecules within these capillaries must thus be moved through the endothelial cells by active transport, endocytosis, and exocytosis. Astrocytes within the brain have numerous extensions called **vascular processes** that surround most of the outer surface of the brain capillaries (fig. 11.8). Before molecules in the blood can enter neurons in the CNS, they may have to pass through both the endothelial cells and the astrocytes. Astrocytes, therefore, contribute to the **blood-brain barrier,** which is highly selective; some molecules are permitted to pass, whereas closely related molecules may not be allowed to cross the barrier.

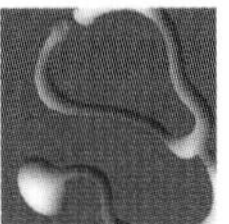

The blood-brain barrier presents difficulties in the chemotherapy of brain diseases because drugs that could enter other organs may not be able to enter the brain. In the treatment of *Parkinson's disease,* for example, patients who need a chemical called dopamine in the brain must be given a precursor molecule called levodopa (L-dopa). This is because dopamine cannot cross the blood-brain barrier, whereas L-dopa can enter neurons and be converted to dopamine in the brain.

Classification of Neurons and Nerves

Neurons may be classified according to structure or function. The functional classification is based on the direction of conducted impulses. Sensory impulses originate in sensory receptors and are conducted by **sensory,** or **afferent, neurons** to the CNS. Motor impulses originate in the CNS and are conducted by **motor,** or **efferent, neurons** to a muscle or gland (fig. 11.9). Motor neurons may be *somatic* (nonvisceral) or *autonomic* (visceral). **Association neurons,** or **interneurons,** are located between sensory and motor neurons and are found within the spinal cord and brain.

The structural classification of neurons is based on the number of processes that extend from the cell body of the neuron (fig. 11.10). The spindle-shaped **bipolar neuron** has a process at both ends; this type occurs in the retina of the eye. **Pseudounipolar neurons** have a single process that divides into two. They are called pseudounipolar (*pseudo* = false) because they originate as bipolar neurons, but then their two processes converge and partially fuse during early embryonic development. Most sensory neurons are pseudounipolar and have their cell bodies located in sensory ganglia of spinal and cranial nerves. **Multipolar neurons** are the most common type and are characterized by several dendrites and one axon extending from the cell body. Motor neurons are a good example of this type.

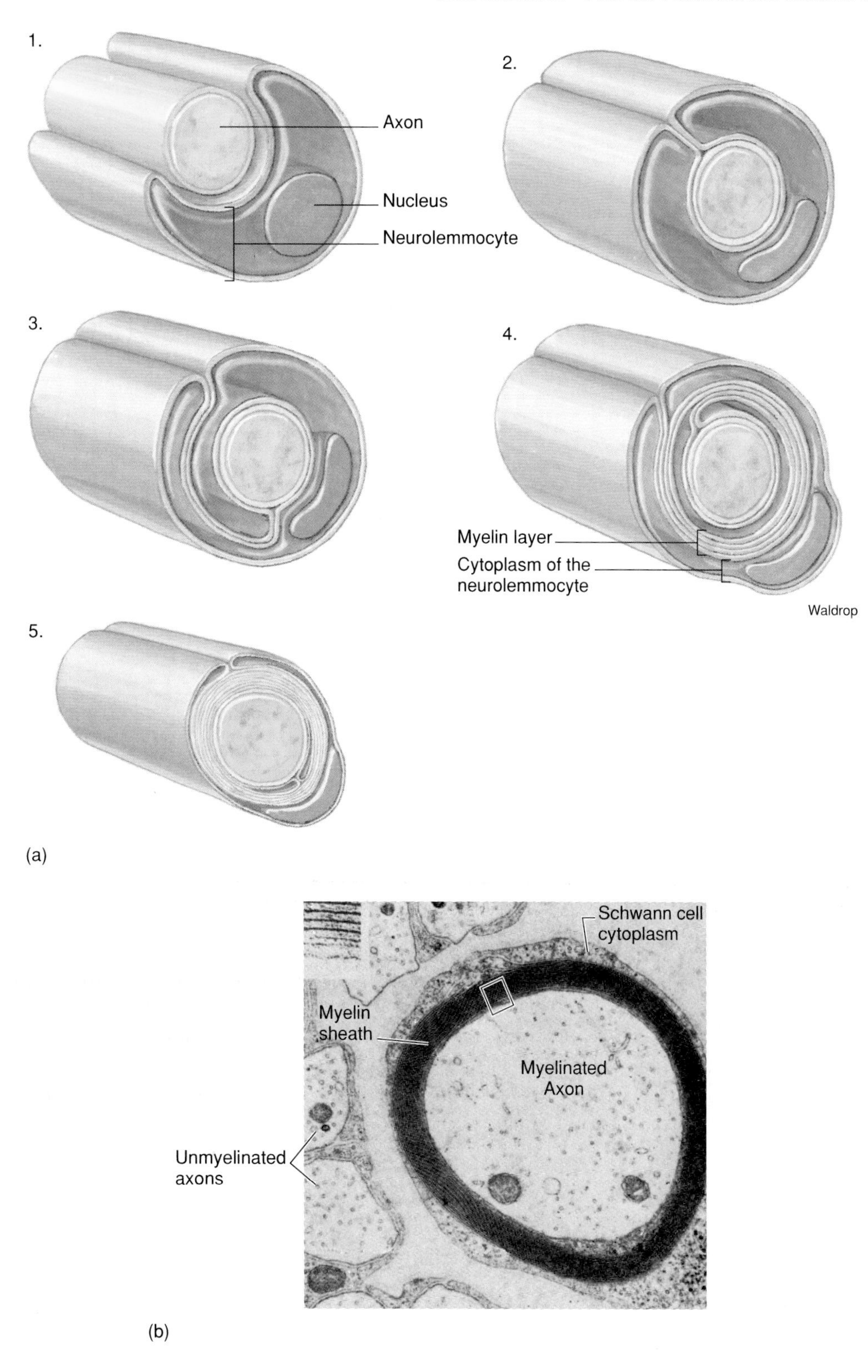

Figure 11.6

The process of myelination. (*a*) An axon of a peripheral neuron becomes myelinated as neurolemmocytes wrap around it to form a myelin layer. (*b*) A photomicrograph of a myelinated axon in cross section.

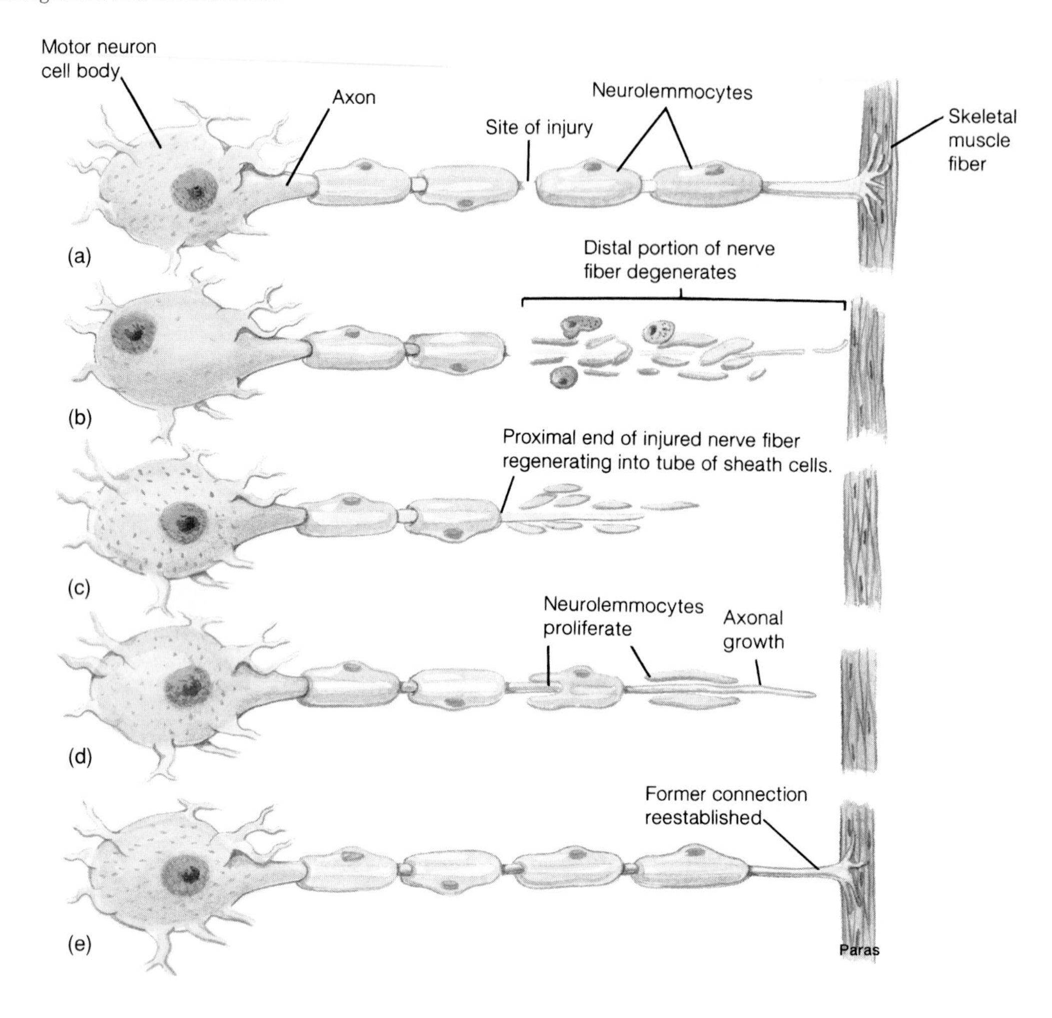

Figure 11.7

Neuron regeneration. (*a*) If a neuron is severed through a myelinated axon, the proximal portion may survive but (*b*) the distal portion will degenerate through phagocytosis. (*c–d*) The myelin layer provides a pathway for regeneration of the axon, and (*e*) innervation is restored.

A **nerve** is a collection of nerve fibers outside the CNS. Fibers within a nerve are held together and strengthened by loose connective tissue (fig. 11.11). Each individual nerve fiber is enclosed in a connective tissue sheath called the **endoneurium** (*en"do-nyoo're-um*). A group of fibers, called a **fasciculus** (*fă-sik' yŭ-lus*), is surrounded by a connective tissue sheath called a **perineurium.** The entire nerve is surrounded and supported by connective tissue called the **epineurium,** which contains tiny blood vessels and, often, adipose cells. Perhaps less than a quarter of the bulk of a nerve consists of nerve fibers. More than half is associated connective tissue, and approximately a quarter is the myelin that surrounds the nerve fibers.

Most nerves are composed of both motor and sensory fibers, and thus are called **mixed nerves.** Some of the cranial nerves, however, are composed either of sensory neurons only **(sensory nerves)** or of motor neurons only **(motor nerves).** Sensory nerves serve the special senses, such as taste, smell, sight, and hearing. Motor nerves conduct impulses to muscles, causing them to contract, or to glands, causing them to secrete.

fasciculus: L. diminutive of *fascis*, bundle

Neurons and their fibers within nerves may be classified according to the area of innervation into the following scheme (fig. 11.12):

1. **Somatic sensory.** Sensory receptors within the skin, bones, muscles, and joints receive stimuli and convey nerve impulses through somatic sensory (afferent) fibers to the CNS for interpretation. Sensory receptors within the eyes and ears are also somatic sensory.
2. **Somatic motor.** Impulses from the CNS travel through somatic motor (efferent) fibers and cause the contraction of skeletal muscles.
3. **Visceral sensory.** Visceral sensory (afferent) fibers convey impulses from visceral organs and blood vessels to the CNS for interpretation. Sensory receptors within the tongue (taste) and nasal epithelium (smell) are also visceral sensory.
4. **Visceral motor.** Visceral motor (efferent) fibers, also called **autonomic motor fibers,** are part of the autonomic nervous system. They originate in the CNS and innervate cardiac muscle, glands, and smooth muscle within the visceral organs.

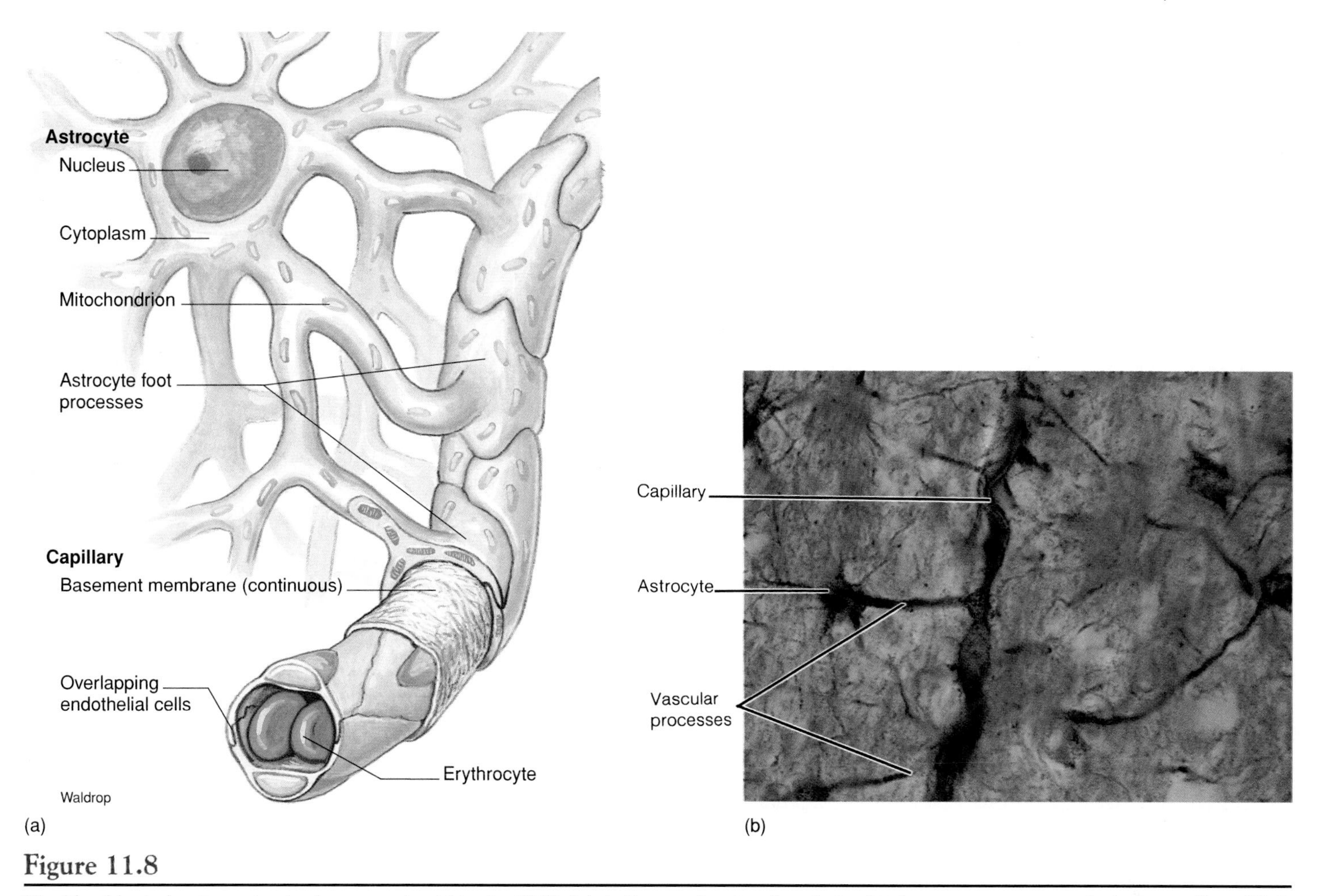

Figure 11.8

The blood-brain barrier maintains homeostasis within the CNS. (*a*) A diagram showing the relationship between astrocytes and a brain capillary. (*b*) A photomicrograph showing the vascular processes of astrocytes.

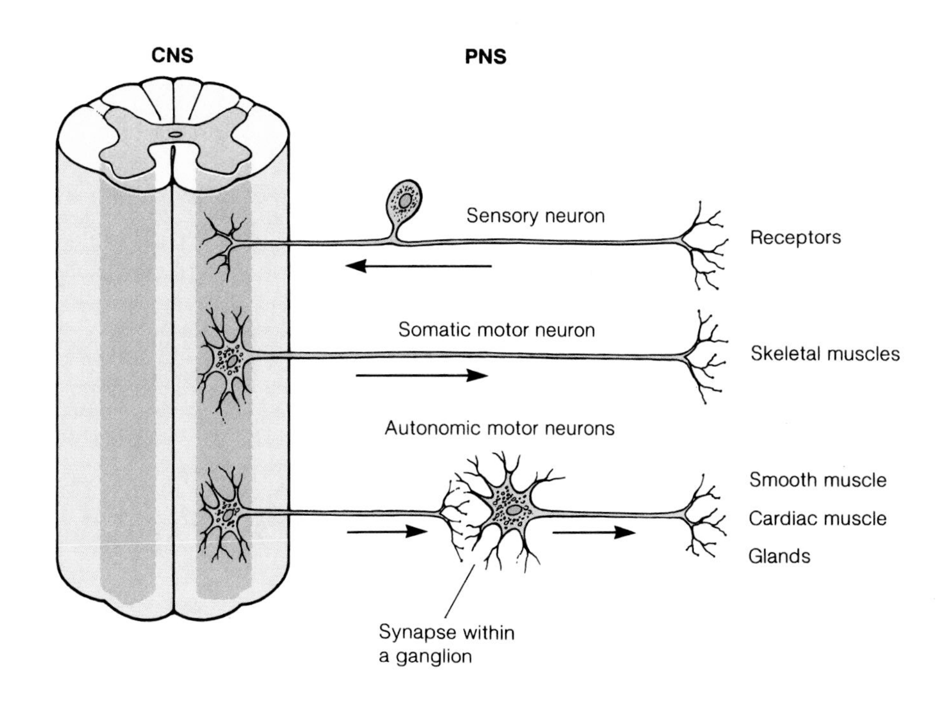

Figure 11.9

The relationship between sensory and motor fibers of the peripheral nervous system (PNS) and the central nervous system (CNS).

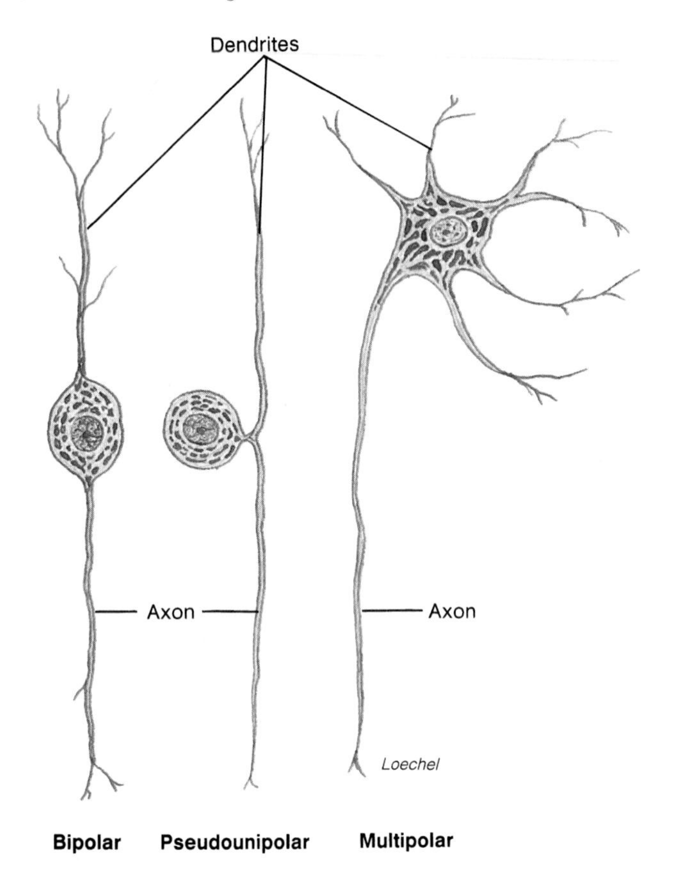

Figure 11.10

The three structural classes of neurons.

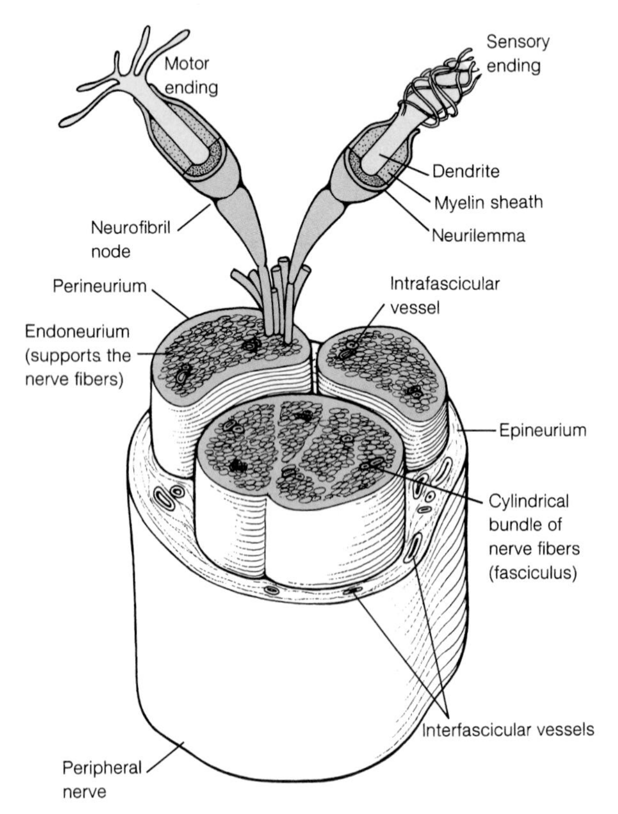

Figure 11.11

The structure of a nerve.

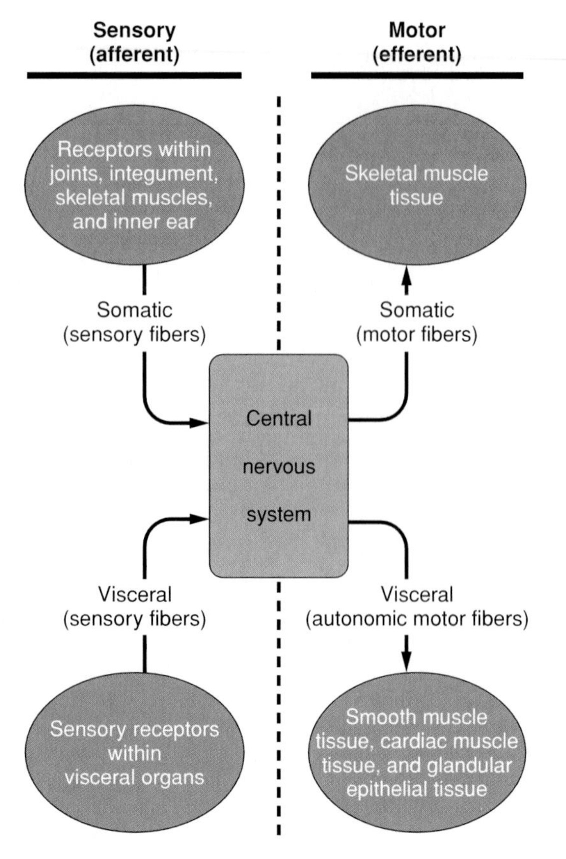

Figure 11.12

The classification of nerve fibers by origin and function.

1. Briefly describe the functions of dendrites, cell bodies, and axons and distinguish between bipolar, pseudounipolar, and multipolar neurons.
2. Distinguish between myelinated and unmyelinated axons.
3. Discuss specific ways in which neuroglia aid neurons.
4. Explain the nature of the blood-brain barrier and describe its structure.
5. Contrast neurons, nerve fibers, and nerves, and describe how nerves are classified.

Transmission of Impulses

Movements of sodium and potassium ions across the axon membrane trigger impulses that travel through the neuron toward a synapse. Synaptic transmission is facilitated by the secretion of a neurotransmitter chemical.

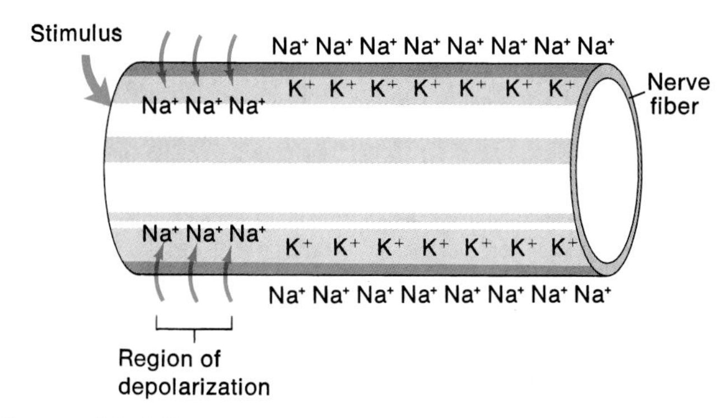

Figure 11.13

A nerve impulse is initiated if a sufficient stimulus occurs at the receptor site, causing movement of ions across the membrane of the nerve fiber.

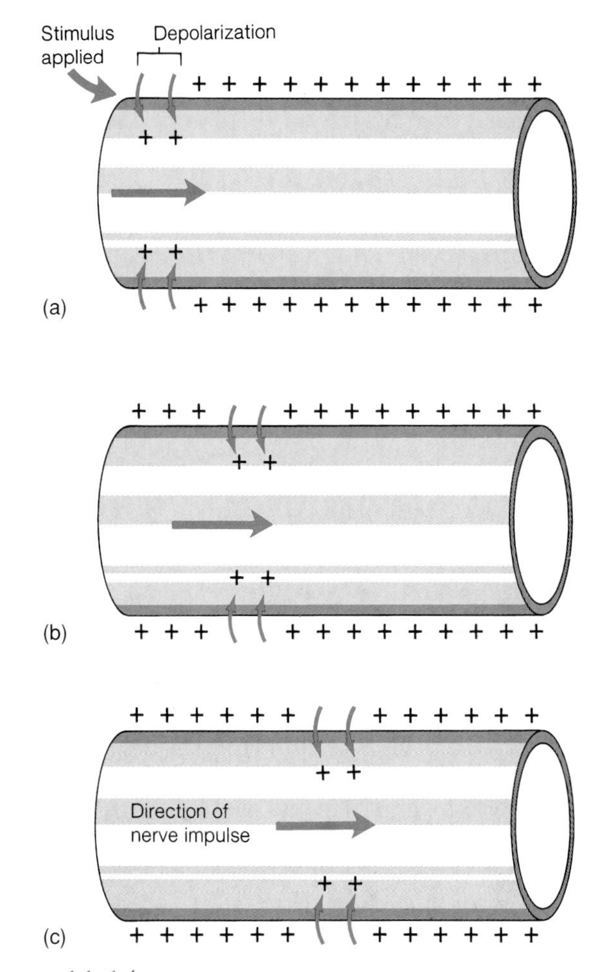

Figure 11.14

(*a–c*) Depolarization occurs in a sequential pattern along a nerve fiber as sodium ions diffuse inward and a wave of action potentials, or a nerve impulse, moves in one direction along the fiber.

Objective 8 Explain how a nerve fiber first becomes depolarized and then repolarized.

Objective 9 Describe the structure of a presynaptic nerve fiber ending and explain how neurotransmitters are released.

Nerve Impulse

Two functional properties of neurons are irritability and conductivity, both of which are involved in the transmission of a nerve impulse. *Irritability* is the ability of dendrites and cell bodies to respond to a stimulus and convert it into an impulse. *Conductivity* is the transmission of an impulse along the axon or a dendrite of a neuron. A *nerve impulse* is the actual movement, or exchange, of sodium (Na^+) and potassium (K^+) ions along the length of a nerve fiber, resulting in the creation of a stimulus that activates another neuron or another tissue.

Before a nerve fiber can respond to a stimulus, it must be *polarized*. A polarized nerve fiber has an abundance of sodium ions on the outside of the axon membrane, which produces a difference in electrical charge called the *resting potential*. When a stimulus of sufficient strength arrives at the receptor portion of the neuron, the polarized nerve fiber becomes *depolarized*, and a nerve impulse is initiated (fig. 11.13). Once depolarization has started, a sequence of ionic exchange occurs along the axon, and the nerve impulse is transmitted (fig. 11.14). After the axon membrane has reached maximum depolarization, the original concentrations of sodium and potassium ions are reestablished in a process called *repolarization*. This restores the reusing potential, and the nerve fiber is now ready to send another impulse.

A nerve impulse travels in one direction only and is an *all-or-none* response. This means that if an impulse is initiated, it will invariably travel the length of the nerve fiber and proceed without a loss in voltage. The speed of a nerve impulse is determined by the diameter of the nerve fiber, its type (myelinated or unmyelinated), and the general physiological condition of the neuron. For example, unmyelinated nerve fibers with small diameters conduct impulses at the rate of about 0.5 m/sec; myelinated neurons conduct impulses at speeds up to 130 m/sec.

Synapse

A **synapse** (*sin'aps*) is the functional connection between the axon terminal of a **presynaptic neuron** and a dendrite of a **postsynaptic neuron** (fig. 11.15). The **axon terminal** (synaptic knob), is the distal portion of the presynaptic neuron at the end of the axon; it is characterized by the presence of numerous mitochondria and **synaptic vesicles.** Synaptic vesicles contain a neurotransmitter chemical, the most common of which is *acetylcholine* (*ă-set"l-ko'lēn*). When a nerve impulse reaches the axon terminal, some of the vesicles respond by releasing their neurotransmitter into the **synaptic cleft,** a tiny gap separating the presynaptic and postsynaptic membranes. If a sufficient number

synapse Gk. *syn*, together; *haptein*, to fasten

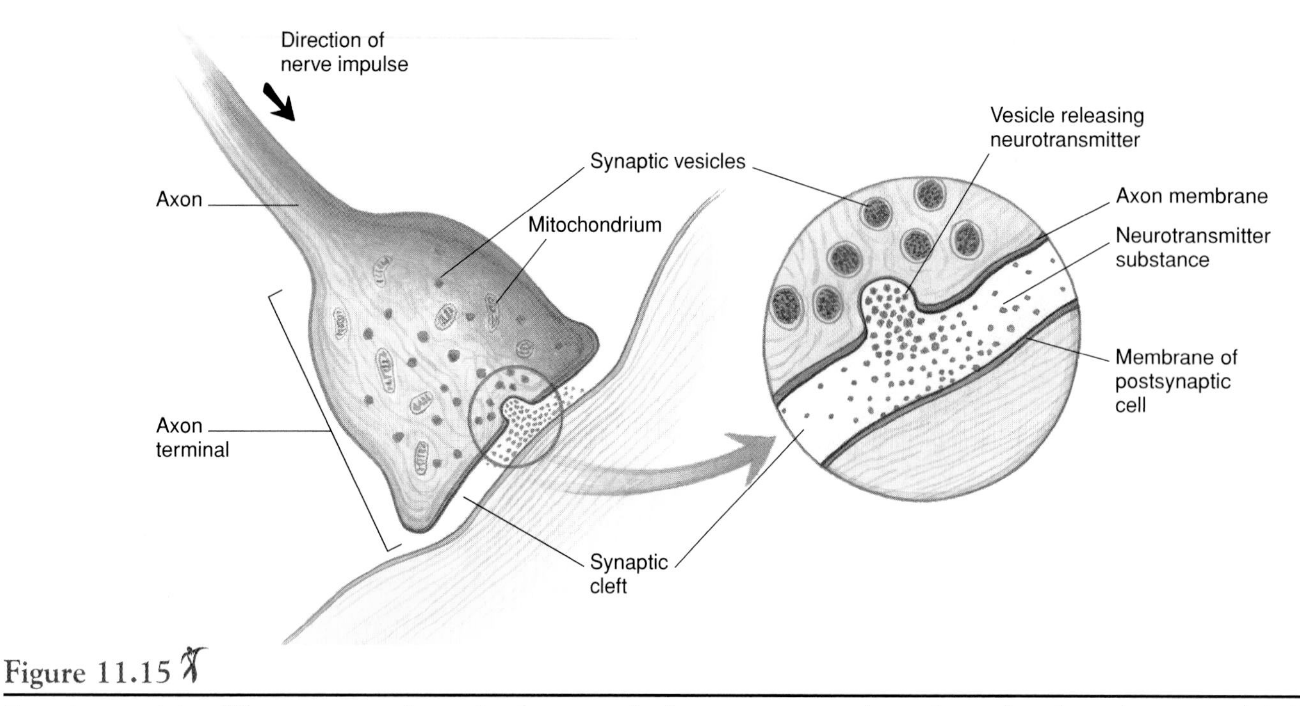

Figure 11.15

Synaptic transmission. When a nerve impulse reaches the synaptic knob, neurotransmitter chemicals are released into the synaptic cleft. Synaptic transmission occurs if sufficient amounts of the neurotransmitter chemicals are released.

of nerve impulses occur in a short time interval, enough neurotransmitter will accumulate within the synaptic cleft to stimulate the postsynaptic neuron to depolarize.

Neurotransmitters are decomposed by enzymes present in the synaptic cleft. *Cholinesterase* (*ko″lĭ-nes′tĭ-rās*) is the enzyme that decomposes acetylcholine and prepares the synapse to receive additional neurotransmitter with the next impulse.

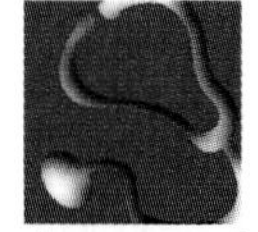

Synaptic transmission may be affected by various drugs or diseases. *Caffeine* is a stimulant that increases the rate of transmission across the synapse. *Aspirin* causes a moderate decrease in the transmission rate. The drug *strychnine* profoundly reduces synaptic transmission; it affects breathing and paralyzes respiratory structures, causing death. In *Parkinson's disease,* there is a deterioration of neurons within the brain that synthesize the neurotransmitter dopamine.

1. Define the terms *depolarization* and *repolarization* and illustrate these processes graphically.
2. Describe the mechanism by which a neurotransmitter affects impulse conduction across a synapse?

General Features of the Brain

The brain is enclosed by the cranium and meninges and is bathed in cerebrospinal fluid. The tremendous metabolic rate of the brain makes it highly susceptible to oxygen deprivation.

Objective 10 Describe the general features of the brain.

Objective 11 Comment on the importance of neuropeptides.

Objective 12 List the metabolic needs of the brain.

The entire delicate CNS is protected by a bony encasement—the cranium (figs. 11.16 and 11.17*d*) surrounding the brain and the vertebral column surrounding the spinal cord. The **meninges** (*mĕ-nin′jēz*) are connective tissue encasements that form a protective membrane between the bone and the soft tissue of the CNS. The CNS is bathed in **cerebrospinal** (*ser″ĕ-bro-spi′nal*) **fluid** that circulates within the hollow **ventricles** of the brain, the **central canal** of the spinal cord, and the **subarachnoid** (*sub″ă-rak′noid*) **space** surrounding the entire CNS (see figs. 11.34 and 11.35).

The CNS is composed of gray and white matter. **Gray matter** consists of either nerve cell bodies and dendrites or bundles of unmyelinated axons and neuroglia. The gray matter of the brain exists as the outer convoluted **cortex layer** of the cerebrum and cerebellum. In addition, specialized gray matter clusters of nerve cells called **nuclei** are found deep within the white matter. **White matter** forms the tracts within the CNS. It consists of aggregations of dendrites and myelinated axons, along with associated neuroglia.

The brain of an adult weighs nearly 1.5 kg (3–3.5 lb) and is composed of an estimated 100 billion (10^{11}) neurons. As described in the previous section, neurons communicate with one another by means of innumerable synapses between the axons and dendrites within the brain. Neurotransmission within the brain is regulated by specialized neurotransmitter chemicals

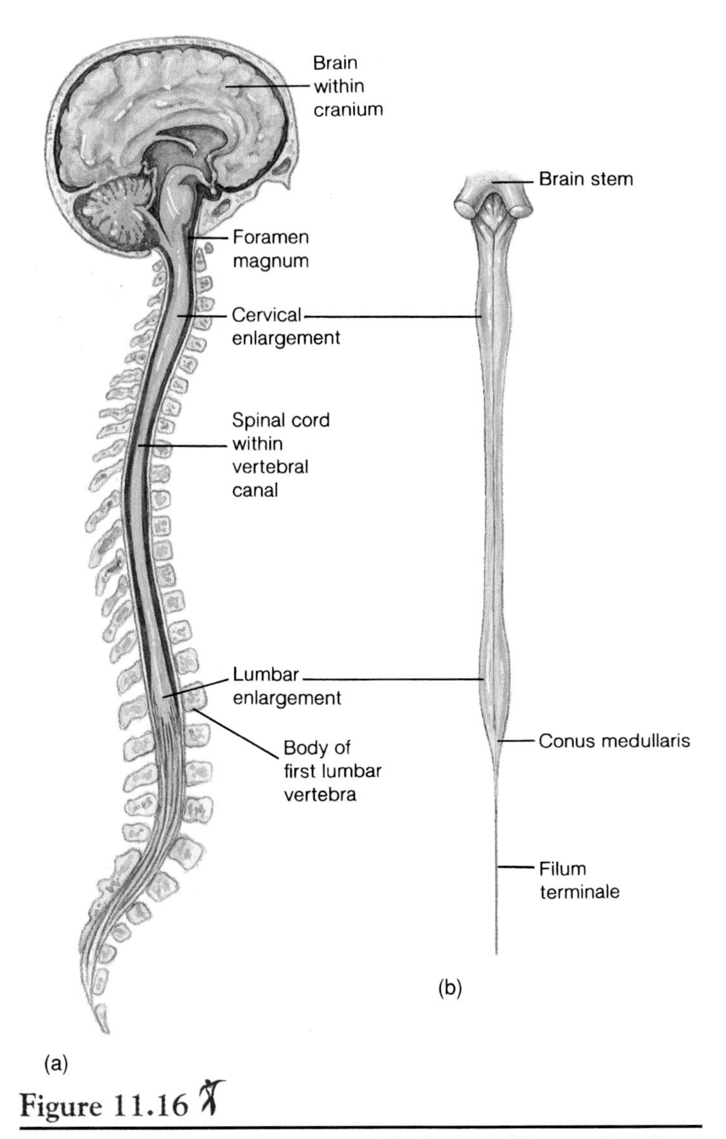

Figure 11.16

The central nervous system consists of the brain and the spinal cord, both of which are covered with meninges and bathed in cerebrospinal fluid. (*a*) A sagittal section showing the brain within the cranium of the skull and the spinal cord within the vertebral canal. (*b*) The spinal cord, shown in a posterior view, extends from the level of the foramen magnum to the first lumbar vertebra (L1).

called *neuropeptides*. These specialized chemical protein messengers are thought to account for specific mental functions.

Over 200 neuropeptides have been identified within the brain, yet most of their functions are not well understood. Two groups of neuropeptides that have received considerable attention are *enkephalins* and *endorphins*. These classes of substances numb the brain to pain, functioning in a manner similar to morphine. They are released in response to stress or pain in a traumatized person.

The brain has a tremendous metabolic rate and therefore needs a continuous supply of oxygen and nutrients. Although it accounts for only 2% of a person's body weight, the brain receives approximately 20% of the total resting cardiac output. This amounts to a flow of about 750 ml of blood per minute. The volume remains relatively constant even with changes in physical or mental activity. This continuous flow is so crucial that a failure of cerebral circulation for as short an interval as 10 seconds causes unconsciousness.

The brain is composed of perhaps the most sensitive tissue of the body. Because of its high metabolic rate, it not only requires continuous oxygen, but also a continuous nutrient supply and the rapid removal of wastes. It is also very sensitive to certain toxins and drugs. The cerebrospinal fluid aids the metabolic needs of the brain by serving as a medium for exchange of nutrients and waste products between the blood and nervous tissue. Cerebrospinal fluid also maintains a protective homeostatic environment within the brain. The blood-brain barrier and the secretory activities of neural tissue also help to maintain homeostasis. The brain has an extensive vascular supply through the paired internal carotid and vertebral arteries that unite at the cerebral arterial circle (circle of Willis) (see chapter 16 and fig. 16.23).

The brain of a newborn is especially sensitive to oxygen deprivation or to excessive oxygen. If complications arise during childbirth and the oxygen supply from the mother's blood to the baby is interrupted while it is still in the birth canal, the infant may be stillborn or suffer brain damage that can result in cerebral palsy, epilepsy, paralysis, or mental retardation. Excessive oxygen administered to a newborn may cause blindness.

Measurable increases in regional blood flow and in glucose and oxygen metabolism within the brain accompany mental functions, including perception and emotion. These metabolic changes can be assessed through the use of *positron emission tomography (PET)*. The technique of a PET scan (fig. 11.18) is based on injecting radioactive tracer molecules labeled with carbon-11, fluorine-18, and oxygen-15 into the bloodstream and photographing the gamma rays that are subsequently emitted from the patient's brain through the skull. PET scans are of value in studying neurotransmitters and neuroreceptors, as well as the substrate metabolism of the brain.

The development of the five basic regions of the brain, was described earlier in this chapter. From each of these regions, distinct functional structures are formed. These structures are summarized in table 11.3 and will be discussed in greater detail in the following sections.

1. What characteristics do the brain and spinal cord have in common? Describe the general features of each.
2. Explain how the study of neuropeptides can enhance understanding of brain function.
3. Using specific examples, describe the metabolic requirements of the brain.

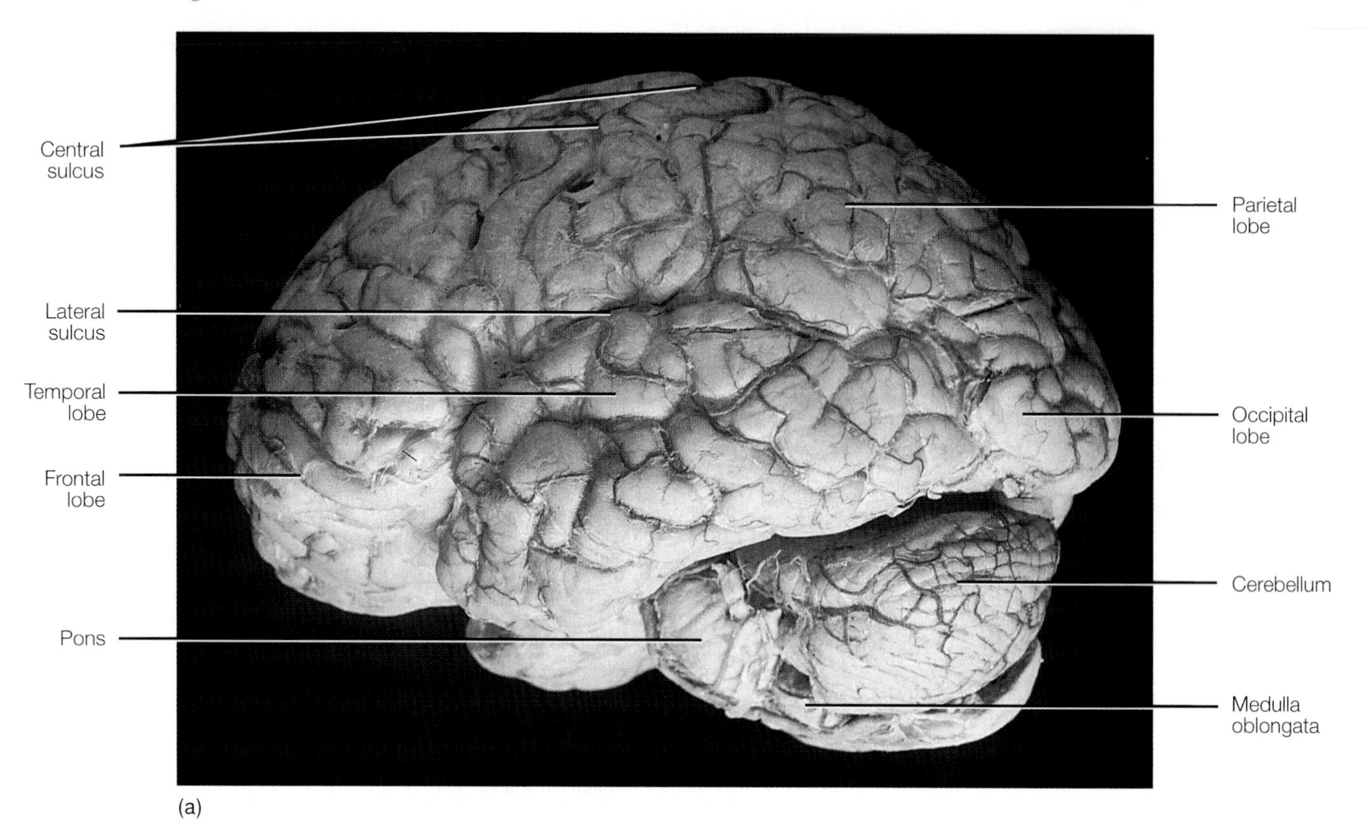

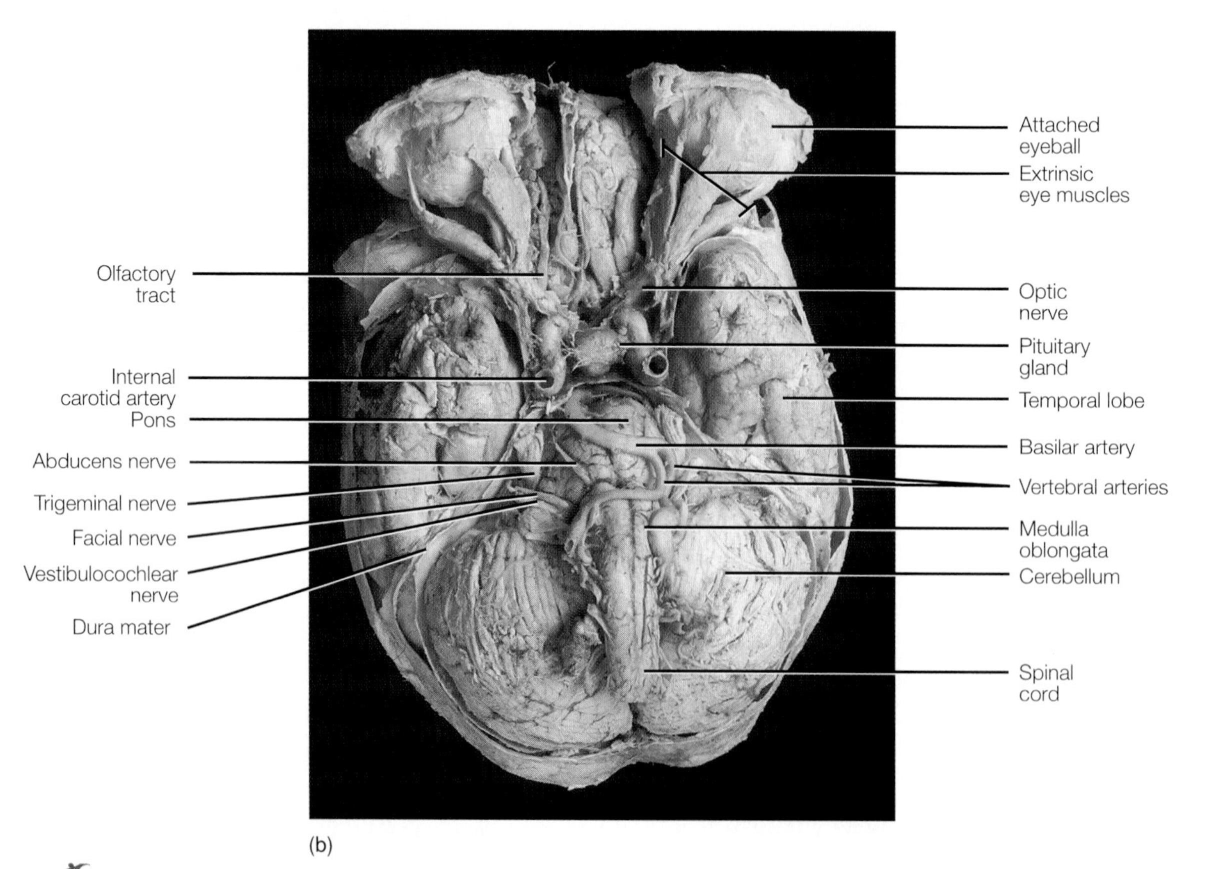

Figure 11.17

The brain: (*a*) a lateral view, (*b*) an inferior view, and (*c*) a sagittal view. (*d*) A left/right magnetic resonance image (MRI) of the skull, brain, and cervical portion of the spinal cord.

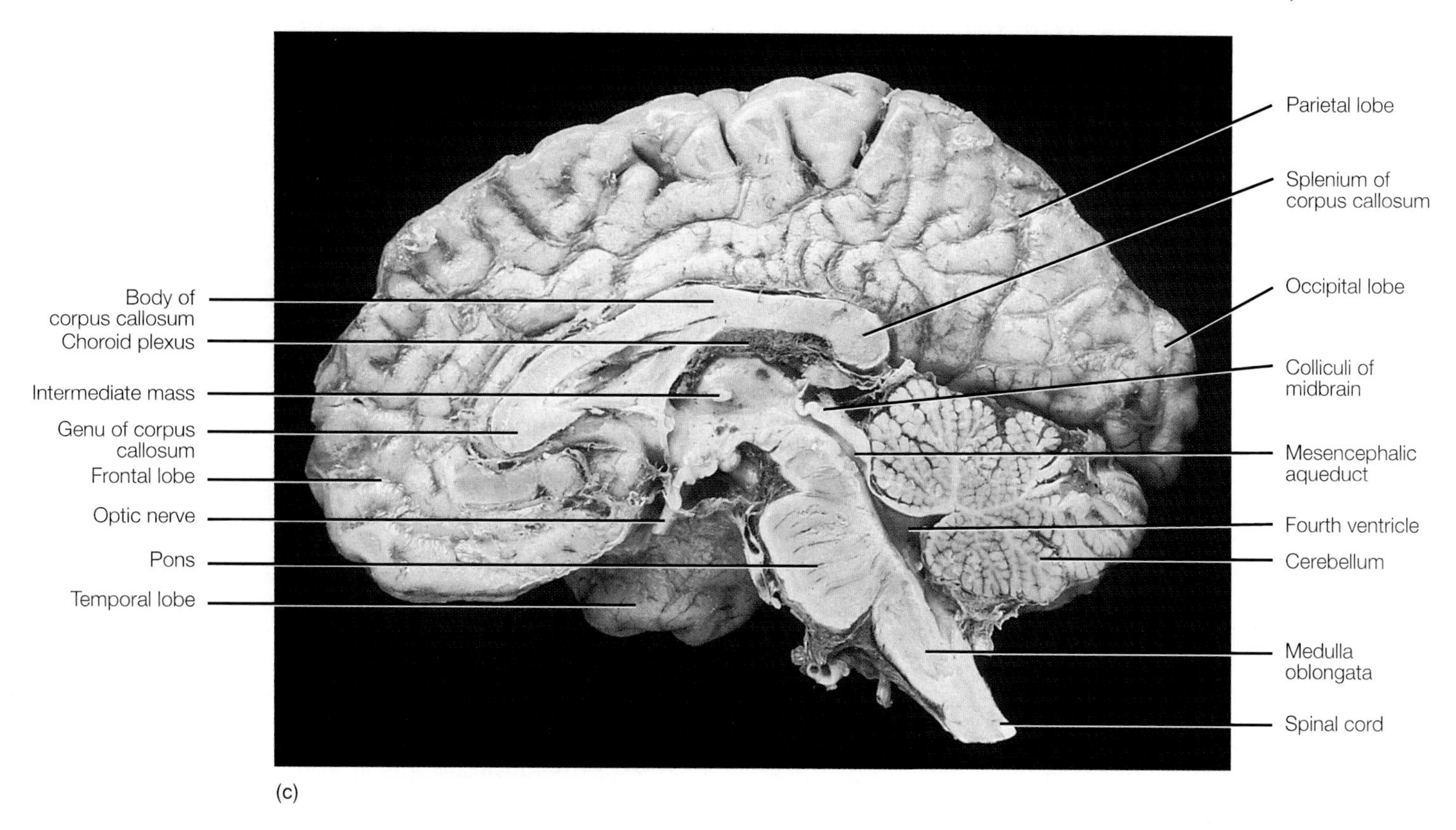
Parietal lobe
Splenium of corpus callosum
Occipital lobe
Colliculi of midbrain
Mesencephalic aqueduct
Fourth ventricle
Cerebellum
Medulla oblongata
Spinal cord
Body of corpus callosum
Choroid plexus
Intermediate mass
Genu of corpus callosum
Frontal lobe
Optic nerve
Pons
Temporal lobe
(c)

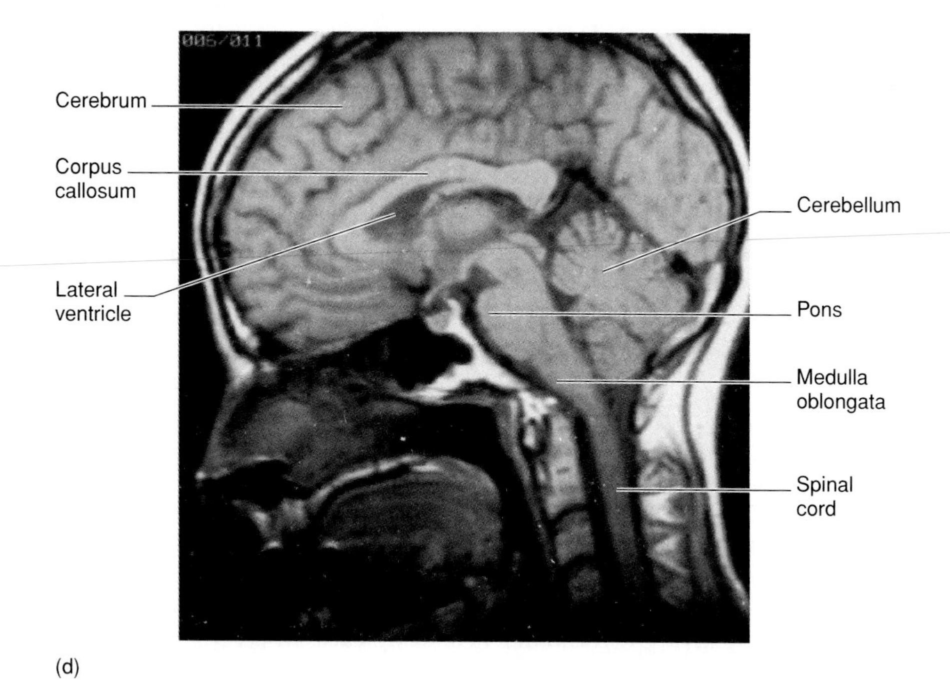
Cerebrum
Corpus callosum
Lateral ventricle
Cerebellum
Pons
Medulla oblongata
Spinal cord
(d)

Figure 11.17

Continued.

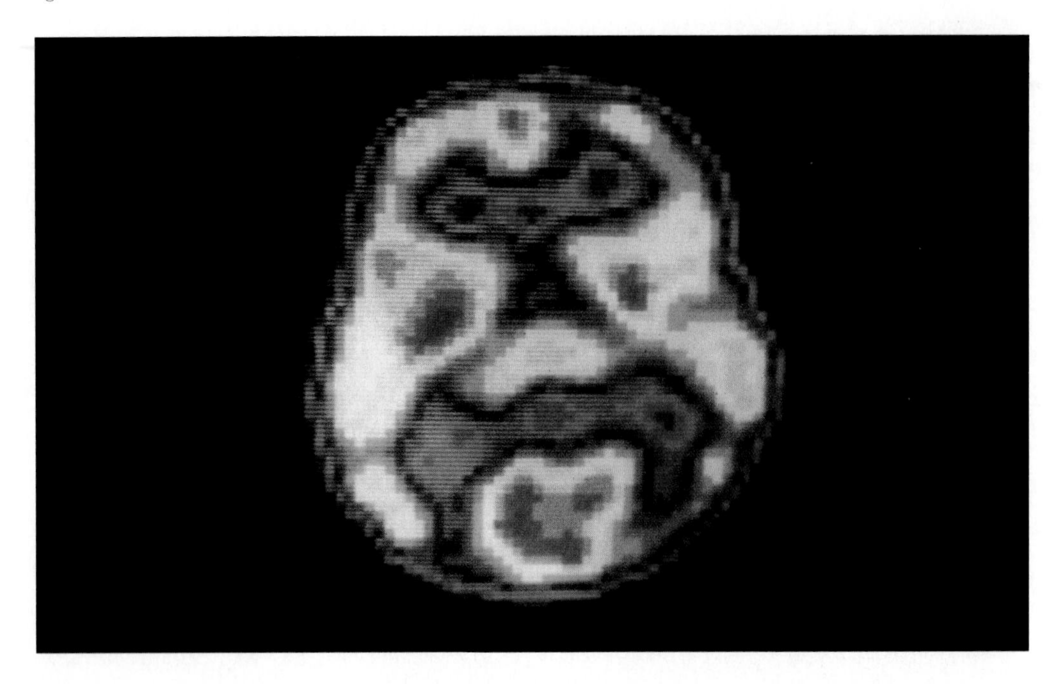

Figure 11.18

A positron emission tomographic (PET) scan of the brain (in transverse section) of an unmedicated patient with schizophrenia. Red areas indicate high glucose use (uptake of 18-F-deoxyglucose). The scan shows highest glucose uptake in the posterior region, where the brain's visual center is located.

Table 11.3

Derivation and Functions of the Major Brain Structures

	Region	Structure	Function
Prosencephalon (forebrain)	Telencephalon	Cerebrum	Control of most sensory and motor activities; reasoning, memory, intelligence, etc.; instinctual and limbic functions
	Diencephalon	Thalamus	Relay center; all impulses (except olfactory) going into the cerebrum synapse here; some sensory interpretation; initial autonomic response to pain
		Hypothalamus	Regulation of food and water intake, body temperature, heartbeat, etc.; control of secretory activity in anterior pituitary gland; instinctual and limbic functions
		Pituitary gland	Regulation of other endocrine glands
Mesencephalon (midbrain)	Mesencephalon	Superior colliculi	Visual reflexes (eye-hand coordination)
		Inferior colliculi	Auditory reflexes
		Cerebral peduncles	Reflex coordination; contain many motor fibers
Rhombencephalon (hindbrain)	Metencephalon	Cerebellum	Balance and motor coordination
		Pons	Relay center; contains nuclei (pontine nuclei)
	Myelencephalon	Medulla oblongata	Relay center; contains many nuclei; visceral autonomic center (e.g., respiration, heart rate, vasoconstriction)

Cerebrum

Objective 13 Describe the structure of the cerebrum and list the functions of the cerebral lobes.

Objective 14 Define the term *electroencephalogram* and discuss the clinical importance of the EEG.

Objective 15 Describe the fiber tracts within the cerebrum.

The cerebrum, consisting of five paired lobes within two convoluted hemispheres, is concerned with higher brain functions, including the perception of sensory impulses, the instigation of voluntary movement, the storage of memory, thought processes, and reasoning ability. The cerebrum is also concerned with instinctual and limbic (emotional) functions.

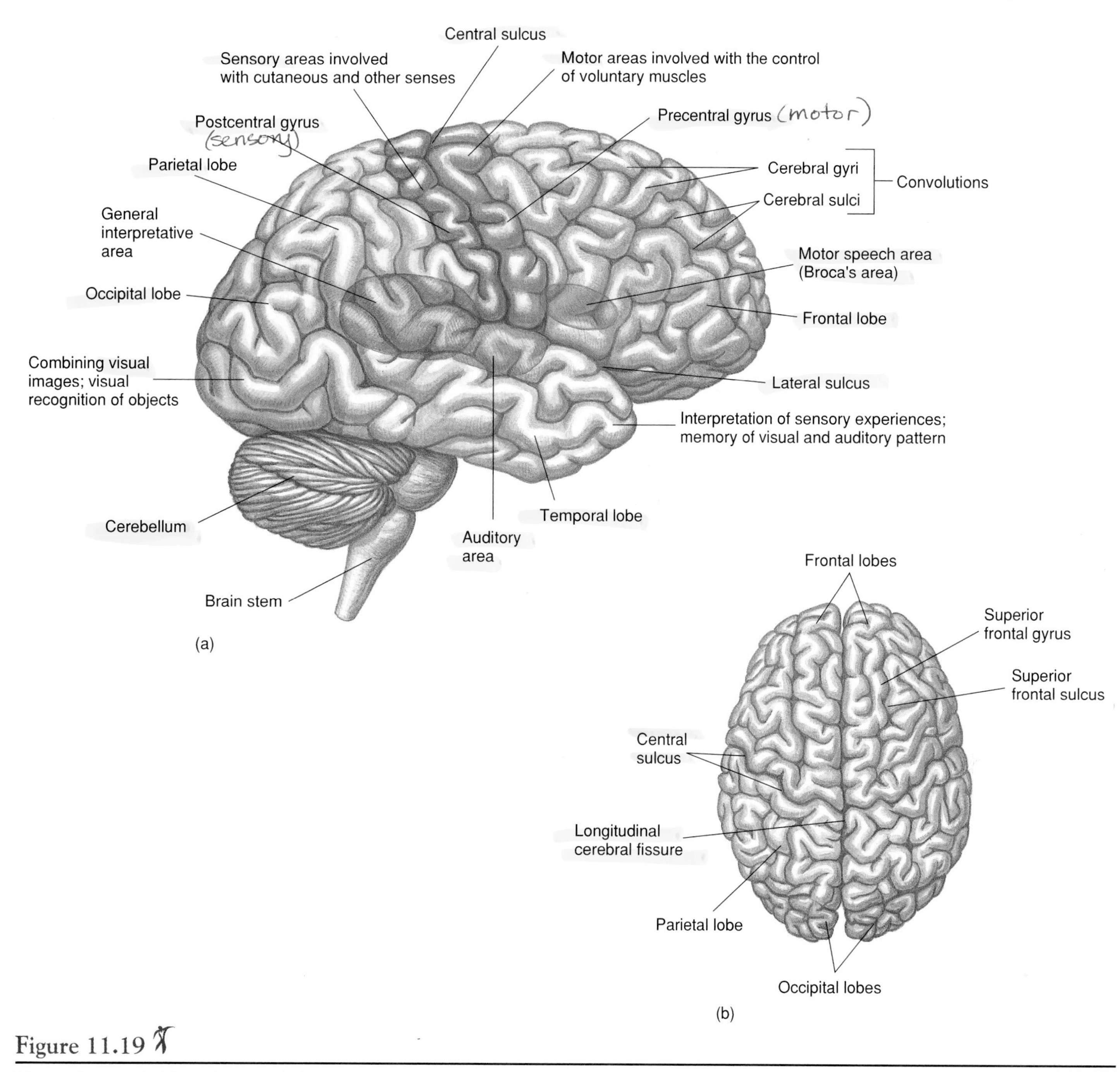

Figure 11.19

The cerebrum. (*a*) A lateral view and (*b*) a superior view.

Structure of the Cerebrum

The **cerebrum** (*ser'ĕ-brum*), located in the region of the telencephalon, is the largest and most obvious portion of the brain. It accounts for about 80% of the mass of the brain and is responsible for the higher mental functions, including memory and reason. The cerebrum consists of the **right** and **left hemispheres,** which are incompletely separated by a **longitudinal cerebral fissure** (fig. 11.19). Portions of the two hemispheres are connected internally by the **corpus callosum** (*kă-lo'sum*), a large tract of white matter (see fig. 11.17*c,d*). A portion of the meninges called the **falx** (*falks*) **cerebri** extends into the longitudinal fissure. Each cerebral hemisphere contains a central cavity, the **lateral ventricle** (fig. 11.20), which is lined with ependymal cells and filled with cerebrospinal fluid.

cerebrum: L. *cerebrum*, brain

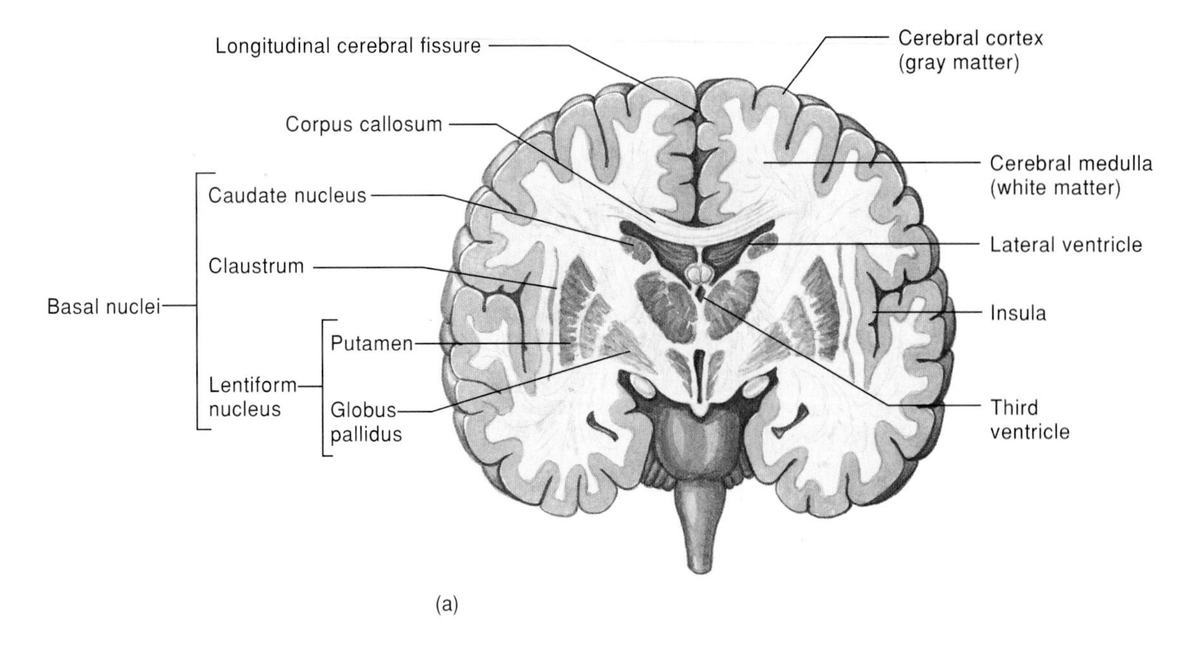

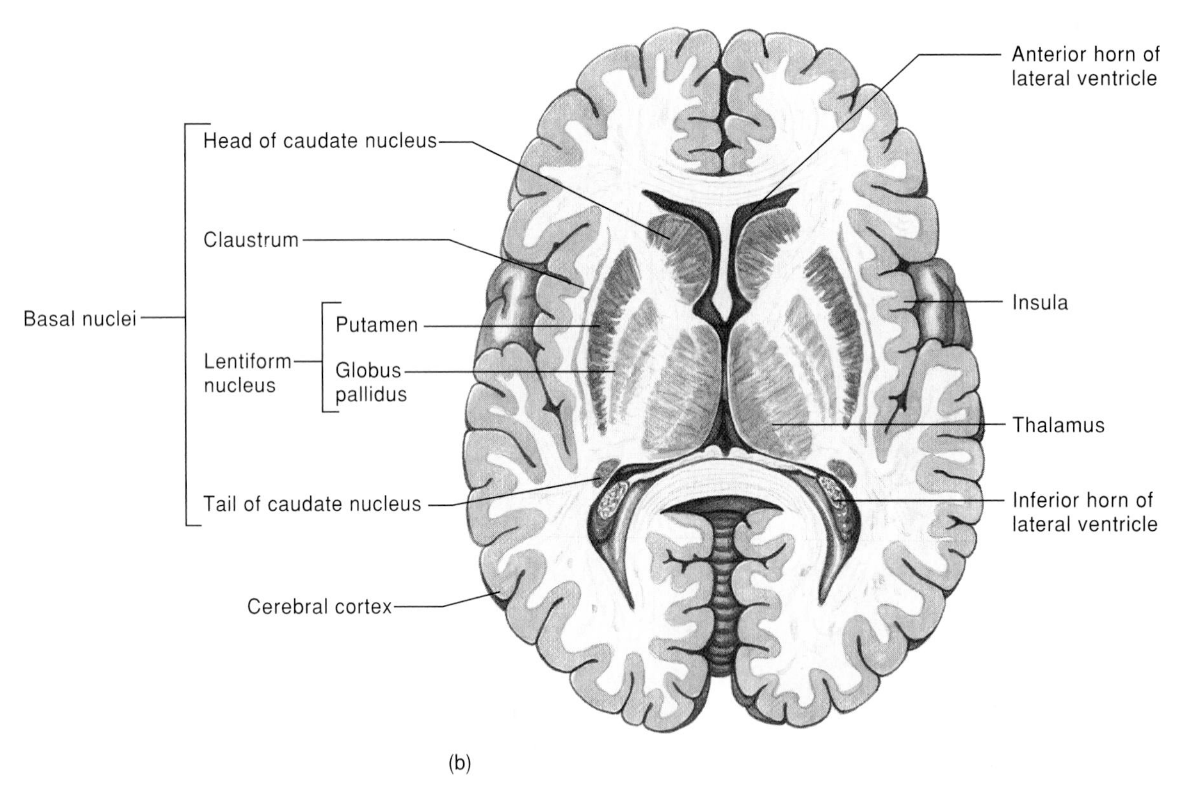

Figure 11.20

Sections through the cerebrum and diencephalon. (*a*) A coronal view and (*b*) a cross section.

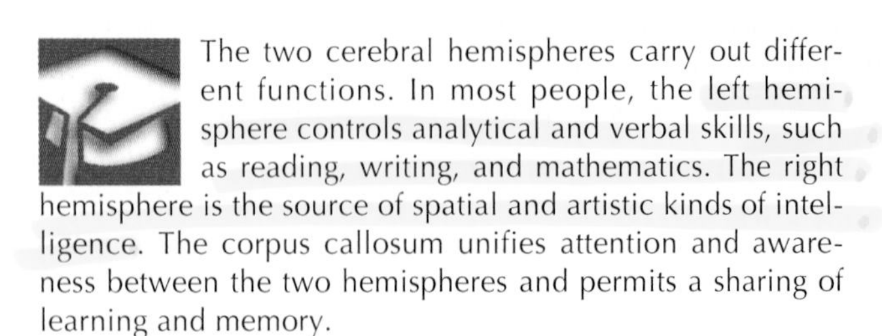

The two cerebral hemispheres carry out different functions. In most people, the left hemisphere controls analytical and verbal skills, such as reading, writing, and mathematics. The right hemisphere is the source of spatial and artistic kinds of intelligence. The corpus callosum unifies attention and awareness between the two hemispheres and permits a sharing of learning and memory.

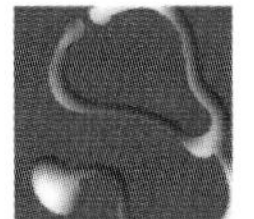

Severing the corpus callosum is a radical treatment for controlling severe epileptic seizures. Although this surgery has proven successful, it results in the cerebral hemispheres functioning as separate structures, each with its own information, competing for control. A more recent and effective technique of controlling epileptic seizures is a precise laser treatment of the corpus callosum.

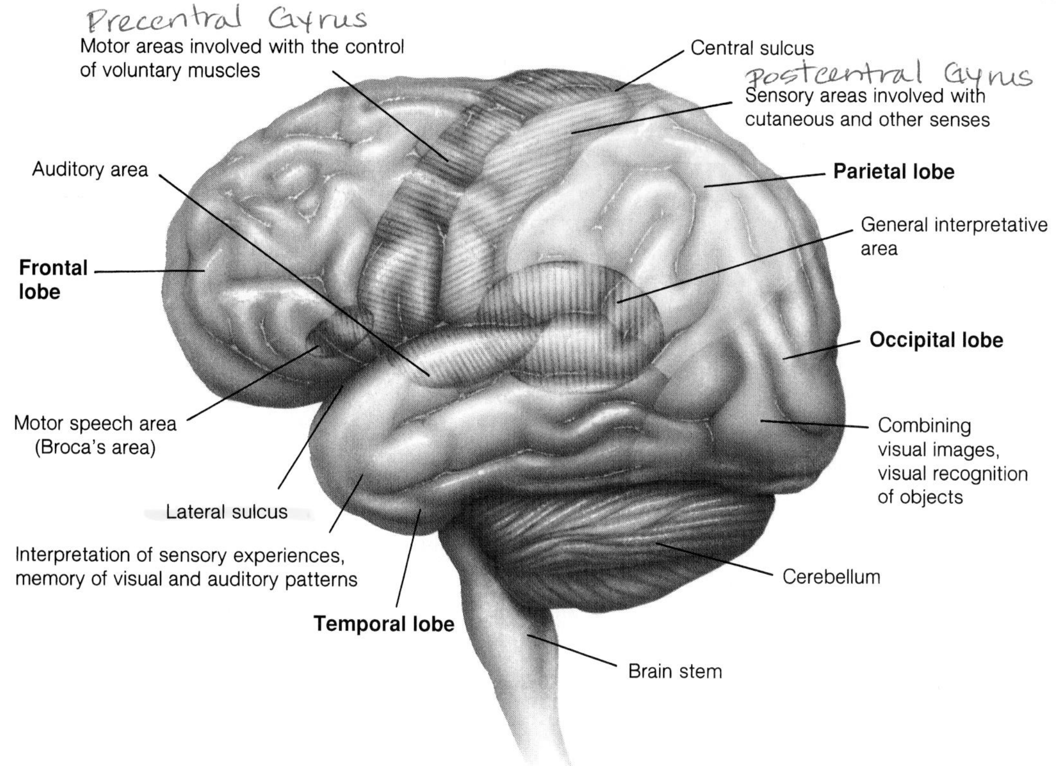

Figure 11.21

The lobes of the left cerebral hemisphere showing the principal motor and sensory areas of the cerebral cortex.

The cerebrum consists of two layers. The surface layer, referred to as the **cerebral cortex,** is composed of gray matter that is 2–4 mm (0.08–0.16 in.) thick (fig. 11.20). Beneath the cerebral cortex is the thick **white matter** of the cerebrum, which constitutes the second layer. The cerebral cortex is characterized by numerous folds and grooves called **convolutions.** Convolutions form during early fetal development, when brain size increases rapidly and the cortex enlarges out of proportion to the underlying white matter. The elevated folds of the convolutions are the **cerebral gyri** (*jī'ri'*singular, *gyrus*), and the depressed grooves are the **cerebral sulci** (*sul'si*—singular, *sulcus*). The convolutions effectively triple the area of the gray matter, which is composed of nerve cell bodies.

Recent studies indicate that increased learning is accompanied by an increase in the number of synapses between neurons within the cerebrum. Although the number of neurons is established during prenatal development, the number of synapses is variable depending upon the learning process. The number of cytoplasmic extensions from the cell body of a neuron determines the extent of nerve impulse conduction and the associations that can be made to cerebral areas already containing stored information.

gyrus: Gk. *gyros*, circle
sulcus: L. *sulcus*, a furrow or ditch

Lobes of the Cerebrum

Each cerebral hemisphere is subdivided into five lobes by especially deep sulci called fissures. Four of these lobes appear on the surface of the cerebrum and are named according to the overlying cranial bones (fig. 11.21). The reasons for the separate cerebral lobes, as well as two cerebral hemispheres, have to do with specificity of function (table 11.4).

Frontal Lobe

The frontal lobe forms the anterior portion of each cerebral hemisphere (fig. 11.21). A prominent deep furrow called the **central sulcus** (fissure of Rolando) separates the frontal lobe from the parietal lobe. The central sulcus extends at right angles from the longitudinal fissure to the lateral sulcus. The **lateral sulcus** (fissure of Sylvius) extends laterally from the inferior surface of the cerebrum to separate the frontal and temporal lobes. The **precentral gyrus** (see figs. 11.19 and 11.21), an important motor area, is positioned immediately in front of the central sulcus. The frontal lobe's functions include initiating voluntary motor impulses for the movement of skeletal muscles, analyzing sensory experiences, and providing responses relating to personality. The frontal lobes

fissure of Rolando: from Luigi Rolando, Italian anatomist, 1773–1831
fissure of Sylvius: from Franciscus Sylvius del la Boë, Dutch anatomist, 1614–72

Table 11.4
Functions of the Cerebral Lobes

Lobe	Functions	Lobe	Functions
Frontal	Voluntary motor control of skeletal muscles; personality; higher intellectual processes (e.g., concentration, planning, and decision making); verbal communication	Temporal	Interpretation of auditory sensations; storage (memory) of auditory and visual experiences
Parietal	Somatesthetic interpretation (e.g., cutaneous and muscular sensations); understanding speech and formulating words to express thoughts and emotions; interpretation of textures and shapes	Occipital	Integration of movements in focusing the eye; correlation of visual images with previous visual experiences and other sensory stimuli; conscious perception of vision
		Insula	Memory; integration of other cerebral activities

also mediate responses related to memory, emotions, reasoning, judgment, planning, and verbal communication.

Parietal Lobe

The parietal lobe lies posterior to the central sulcus of the frontal lobe. An important sensory area called the **postcentral gyrus** (see figs. 11.19 and 11.21) is positioned immediately behind the central sulcus. The postcentral gyrus is designated as a somatesthetic area because it responds to stimuli from cutaneous and muscular receptors throughout the body.

Portions of the precentral gyrus responsible for motor movement and portions of the postcentral gyrus that respond to sensory stimuli do not correspond in size to the part of the body being served, but rather to the number of motor units activated or to the density of receptors (fig. 11.22). For example, because the hand has many motor units and sensory receptors, larger portions of the precentral and postcentral gyri serve it than serve the thorax, even though the thorax is much larger.

In addition to responding to somatesthetic stimuli, the parietal lobe functions in understanding speech and in articulating thoughts and emotions. The parietal lobe also interprets the textures and shapes of objects as they are handled.

Temporal Lobe

The temporal lobe is located below the parietal lobe and the posterior portion of the frontal lobe. It is separated from both by the lateral sulcus (see fig. 11.21). The temporal lobe contains auditory centers that receive sensory fibers from the cochlea of the ear. This lobe also interprets some sensory experiences and stores memories of both auditory and visual experiences.

Occipital Lobe

The occipital lobe forms the posterior portion of the cerebrum and is not distinctly separated from the temporal and parietal lobes (see fig. 11.21). It lies superior to the cerebellum and is separated from it by an infolding of the meningeal layer called the **tentorium cerebelli** (*ten-to're-um ser"ĕ-bel'i*). The principal functions of the occipital lobe concern vision. It integrates eye movements by directing and focusing the eye. It is also responsible for visual association—correlating visual images with previous visual experiences and other sensory stimuli.

Insula

The insula is a deep lobe of the cerebrum that cannot be viewed on the surface (see fig. 11.20). It lies deep to the lateral sulcus and is covered by portions of the frontal, parietal, and temporal lobes. Little is known of the function of the insula except that it integrates other cerebral activities. It is also thought to have some function in memory.

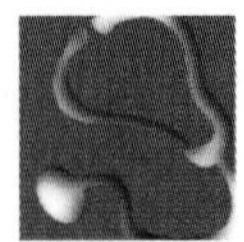

Because of its size and position, portions of the cerebrum frequently suffer brain trauma. A concussion to the brain may cause a temporary or permanent impairment of cerebral functions. Much of what is known about cerebral function comes from observing body dysfunctions when specific regions of the cerebrum are traumatized.

Brain Waves

Neurons within the cerebral cortex continuously generate electrical activity. This activity can be recorded by electrodes attached to precise locations on the scalp, producing an **electroencephalogram** (*ĕ-lek"tro-en-sef'ă-lŏ-gram*) (**EEG**). An EEG pattern, commonly called *brain waves,* is the collective expression of millions of action potentials from cerebral neurons.

Brain waves are first emitted from a developing brain during early fetal development and continue throughout a person's life. The cessation of brain-wave patterns (a "flat EEG") may be a decisive factor in the legal determination of death.

Certain distinct EEG patterns signify healthy mental functions. Deviations from these patterns are of clinical significance in diagnosing trauma, mental depression, hematomas, tumors, infections, and epileptic lesions. Normally, there are four kinds of EEG patterns (fig. 11.23).

1. **Alpha waves** are best recorded from the parietal and occipital regions while a person is awake and relaxed but

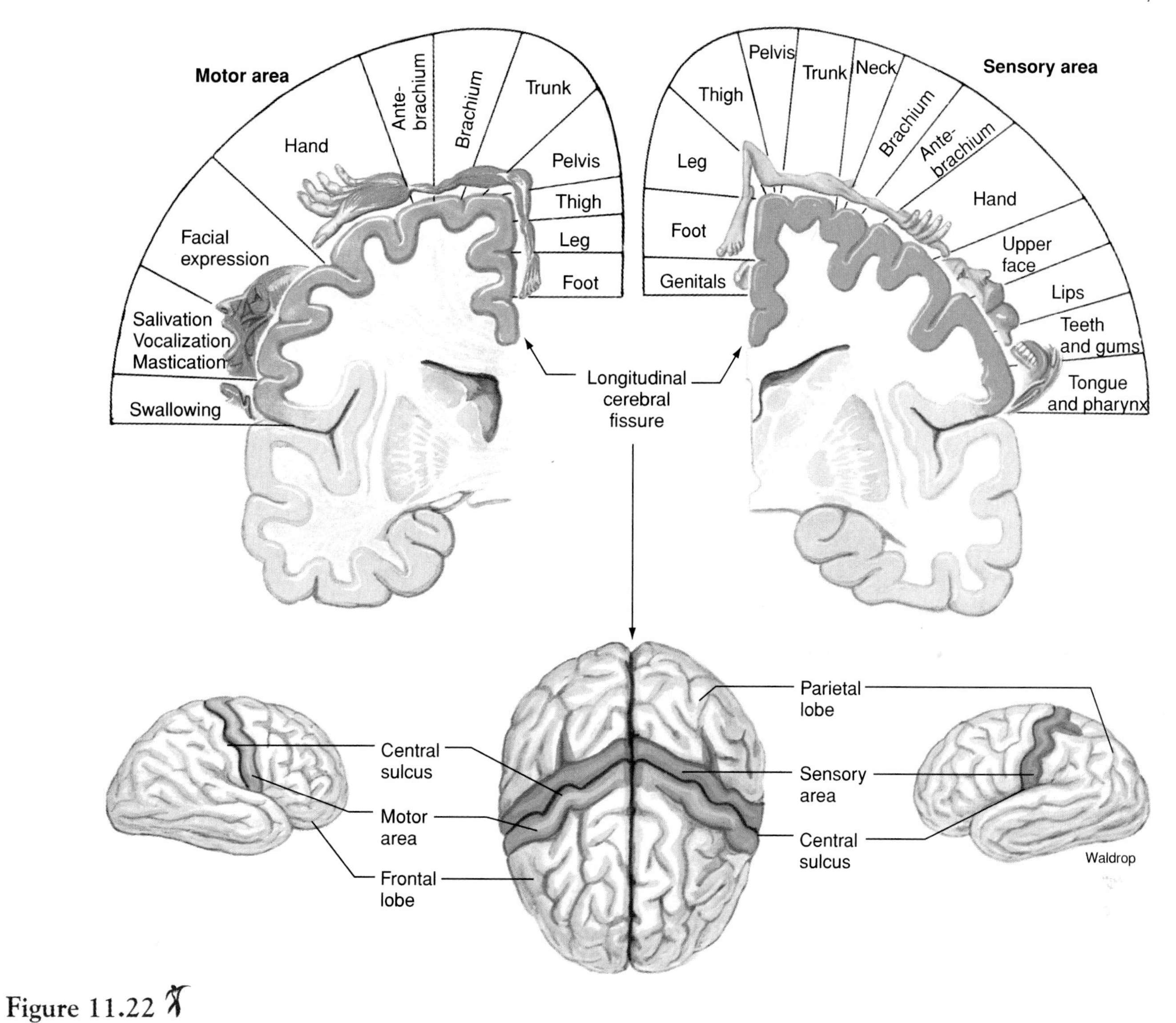

Figure 11.22

Motor and sensory areas of the cerebral cortex. Motor areas control skeletal muscles and sensory areas receive somatesthetic sensations.

with the eyes closed. These waves are rhythmic oscillations of about 10 to 12 cycles/second. The alpha rhythm of a child under the age of 8 occurs at a slightly lower frequency of 4 to 7 cycles/second.

2. **Beta waves** are strongest from the frontal lobes, especially the area near the precentral gyrus. These waves are sensory-evoked and respond to visual and mental activity. Because they respond to stimuli from receptors and are superimposed on the continuous activity patterns of the alpha waves, they constitute *evoked activity*. Beta waves have a frequency of 13 to 25 cycles/second.
3. **Theta waves** are emitted from the temporal and occipital lobes. They have a frequency of 5 to 8 cycles/second and are common in newborn infants. The recording of theta waves in adults generally indicates severe emotional stress and can be a forewarning of a nervous breakdown.
4. **Delta waves** are seemingly emitted in a general pattern from the cerebral cortex. These waves have a frequency of 1 to 5 cycles/second and are common during sleep and in an awake infant. The presence of delta waves in an awake adult indicates brain damage.

White Matter of the Cerebrum

The thick white matter of the cerebrum is deep to the cerebral cortex (see fig. 11.20) and consists of dendrites, myelinated axons, and associated neuroglia. These fibers form the billions of connections within the brain by which information is transmitted to the appropriate places in the form of electrical impulses. The three types of fiber tracts within the white matter are named according to location and the direction in which they conduct impulses (fig. 11.24).

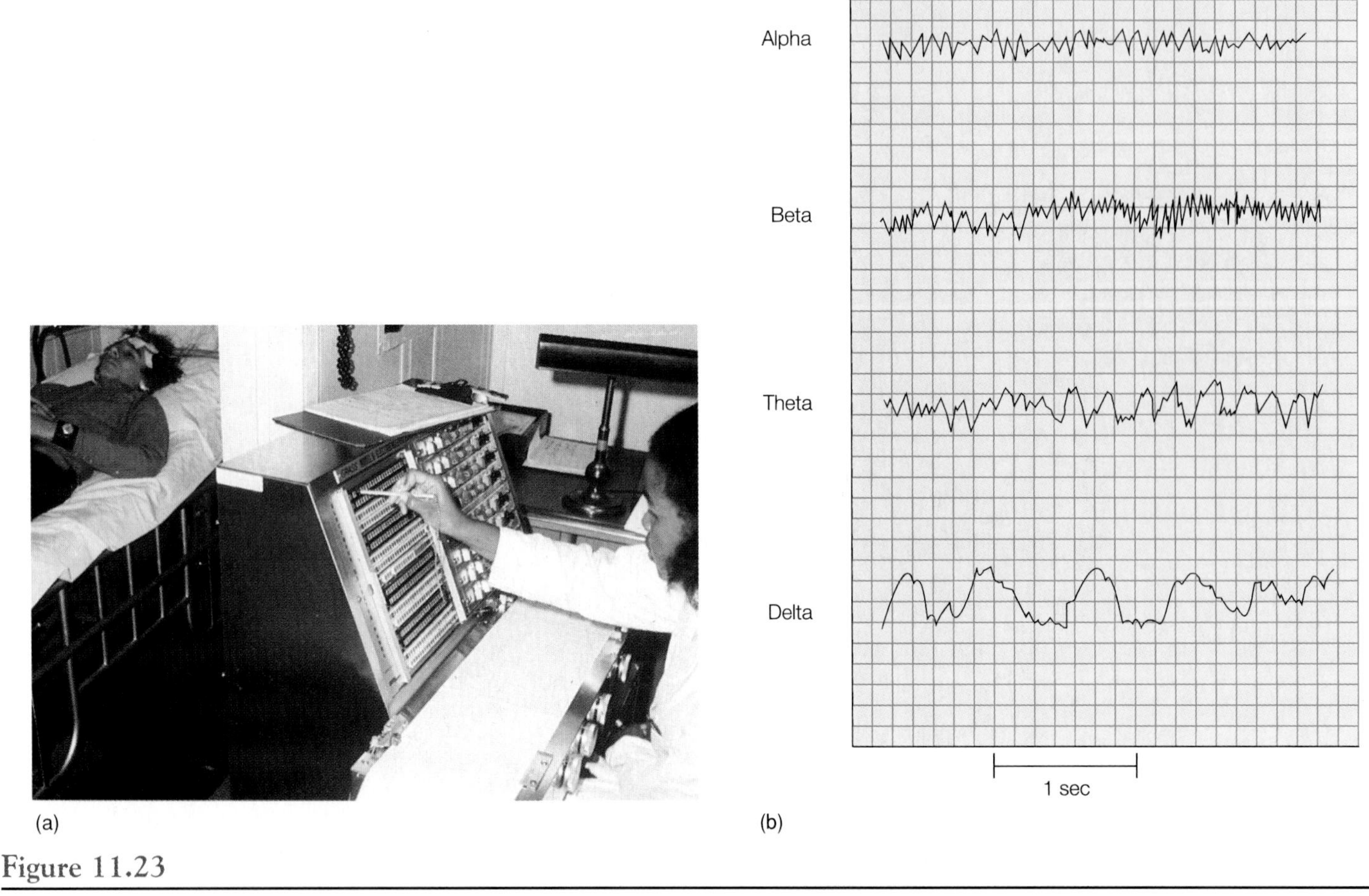

Figure 11.23

(*a*) A technician using an electroencephalograph to take the EEG of a patient. (*b*) Types of EEG patterns.

1. **Association fibers** are confined to a given cerebral hemisphere and conduct impulses between neurons within that hemisphere.
2. **Commissural** (*kă-mĭ-shur′al*) **fibers** connect the neurons and gyri of one hemisphere with those of the other. The **corpus callosum** and **anterior commissure** (fig. 11.25) are composed of commissural fibers.
3. **Projection fibers** form the ascending and descending tracts that transmit impulses from the cerebrum to other parts of the brain and spinal cord and from the spinal cord and other parts of the brain to the cerebrum.

Basal Nuclei

The **basal nuclei** are specialized paired masses of gray matter located deep within the white matter of the cerebrum (figs. 11.20 and 11.26). The most prominent of the basal nuclei is the **corpus striatum** (*stri-a′tum*), so named because of its striped appearance. The corpus striatum is composed of several masses of nuclei. The **caudate nucleus** is the upper mass. A thick band of white matter lies between the caudate nucleus and the next two masses underneath, collectively called the **lentiform nucleus.** The lentiform nucleus consists of a lateral portion, called the **putamen** (*pyoo-ta′ men*), and a medial portion, called the **globus pallidus** (fig. 11.26). The **claustrum** (*klos′trum*) is another portion of the basal nuclei. It is a thin layer of gray matter, lying just deep to the cerebral cortex of the insula.

The basal nuclei are associated with other structures of the brain, particularly within the mesencephalon. The caudate nucleus and putamen of the basal nuclei control unconscious contractions of certain skeletal muscles, such as those of the upper extremities involved in involuntary arm movements during walking. The globus pallidus regulates the muscle tone necessary for specific intentional body movements. Neural diseases or physical trauma to the basal nuclei generally cause a variety of motor movement dysfunctions, including rigidity, tremor, and rapid and aimless movements.

corpus striatum: L. *corpus*, body; *striare*, striped

lentiform: L. *lentis*, elongated
putamen: L. *putare*, to cut, prune
globus pallidus: L. *globus*, sphere; *pallidus*, pale

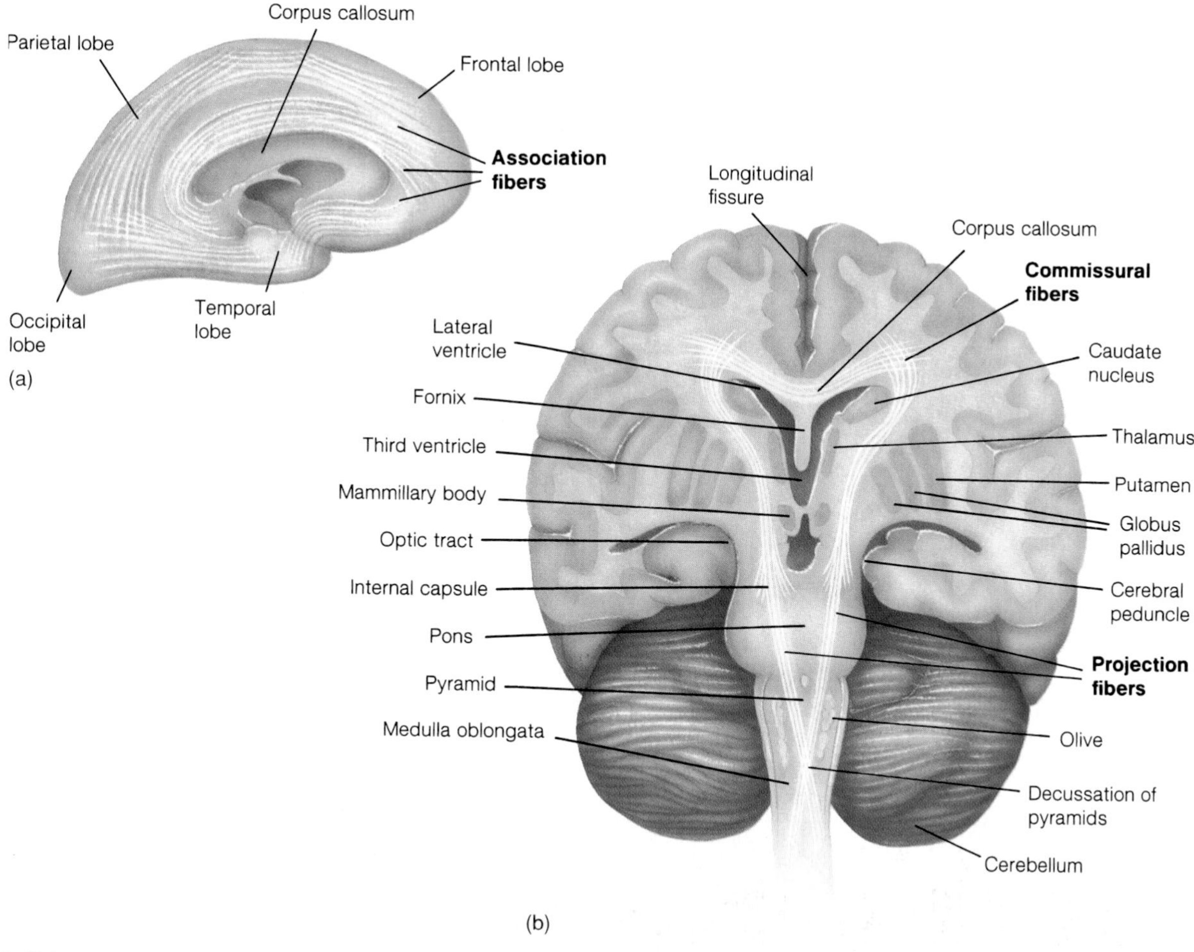

Figure 11.24

Types of fiber tracts within the white matter associated with the cerebrum. (*a*) Association fibers of a given hemisphere and (*b*) commissural fibers connecting the hemispheres and projection fibers connecting the cerebral hemispheres with other structures of the CNS. (Note the decussation [crossing-over] of projection fibers within the medulla oblongata.)

Language

Knowledge of the brain regions involved in language has been gained primarily by the study of *aphasias*—speech and language disorders caused by damage to specific language areas of the brain. These areas (fig. 11.27) are generally located in the cerebral cortex of the left hemisphere in both right-handed and left-handed people.

The **motor speech area** (Broca's area) is located in the left inferior gyrus of the frontal lobe. Neural activity in the motor speech area causes selective stimulation of motor impulses in motor centers elsewhere in the frontal lobe, which in turn causes coordinated skeletal muscle movement in the pharynx and larynx. At the same time, motor impulses are sent to the respiratory muscles to regulate air movement across the vocal folds. The combined muscular stimulation translates thought patterns into speech.

Wernicke's area (*ver'nĭ-kēz*) is located in the superior gyrus of the temporal lobe and is directly connected to the motor speech area by a fiber tract called the **arcuate fasciculus.** People with *Wernicke's aphasia* produce speech that has been described as a "word salad." The words used may be real words that are randomly mixed together, or they may be made-up words. Language comprehension has been destroyed in people with Wernicke's aphasia; they cannot understand either spoken or written language.

It appears that the concept of words to be spoken originates in Wernicke's area and is then communicated to the motor speech area through the arcuate fasciculus. Damage to the arcuate fasciculus produces *conduction aphasia,* which is fluent but nonsensical speech as in Wernicke's aphasia, even though both the motor speech area and Wernicke's area are intact.

The **angular gyrus,** located at the junction of the parietal, temporal, and occipital lobes, is believed to be a center for the integration of auditory, visual, and somatesthetic information. Damage to the angular gyrus produces aphasias, which suggests that this

aphasia: L. *a,* without; Gk. *phasis,* speech
Broca's area: from Pierre P. Broca, French neurologist, 1824–80

Wernicke's area: from Karl Wernicke, German neurologist, 1848–1905

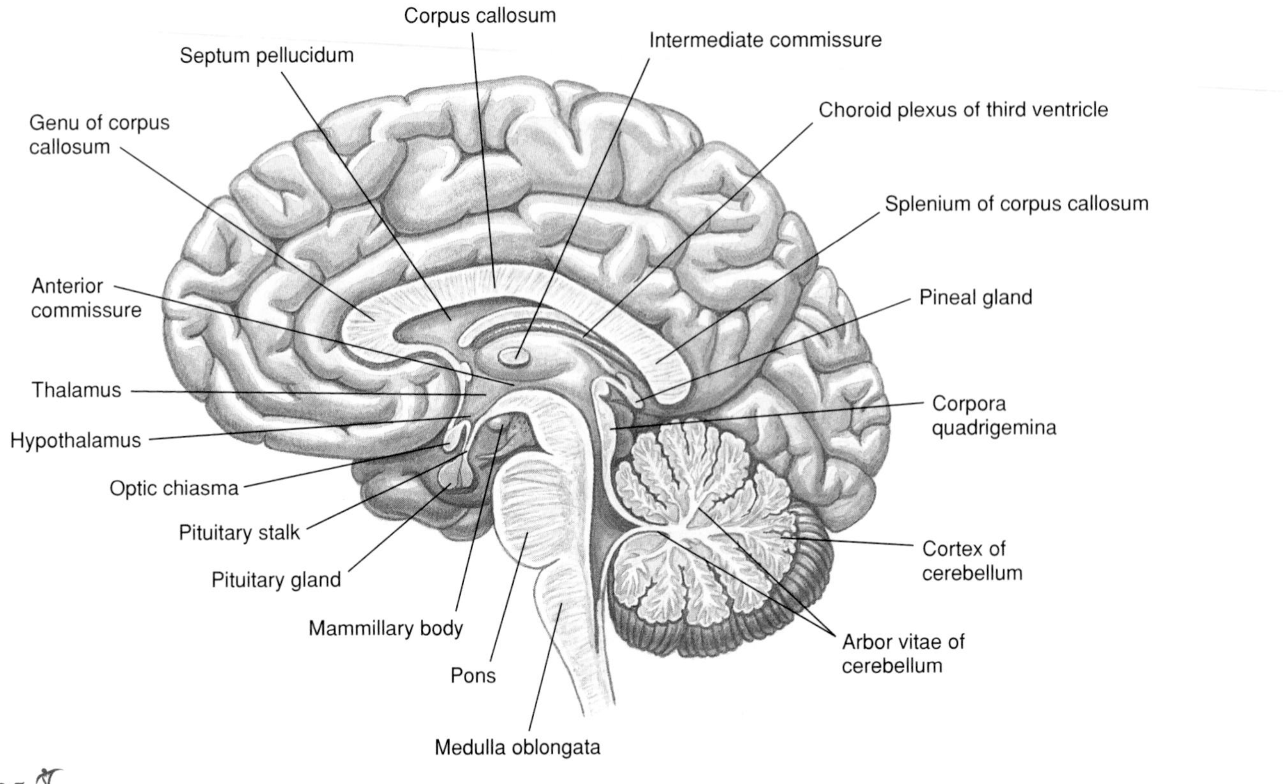

Figure 11.25

A midsagittal section through the brain.

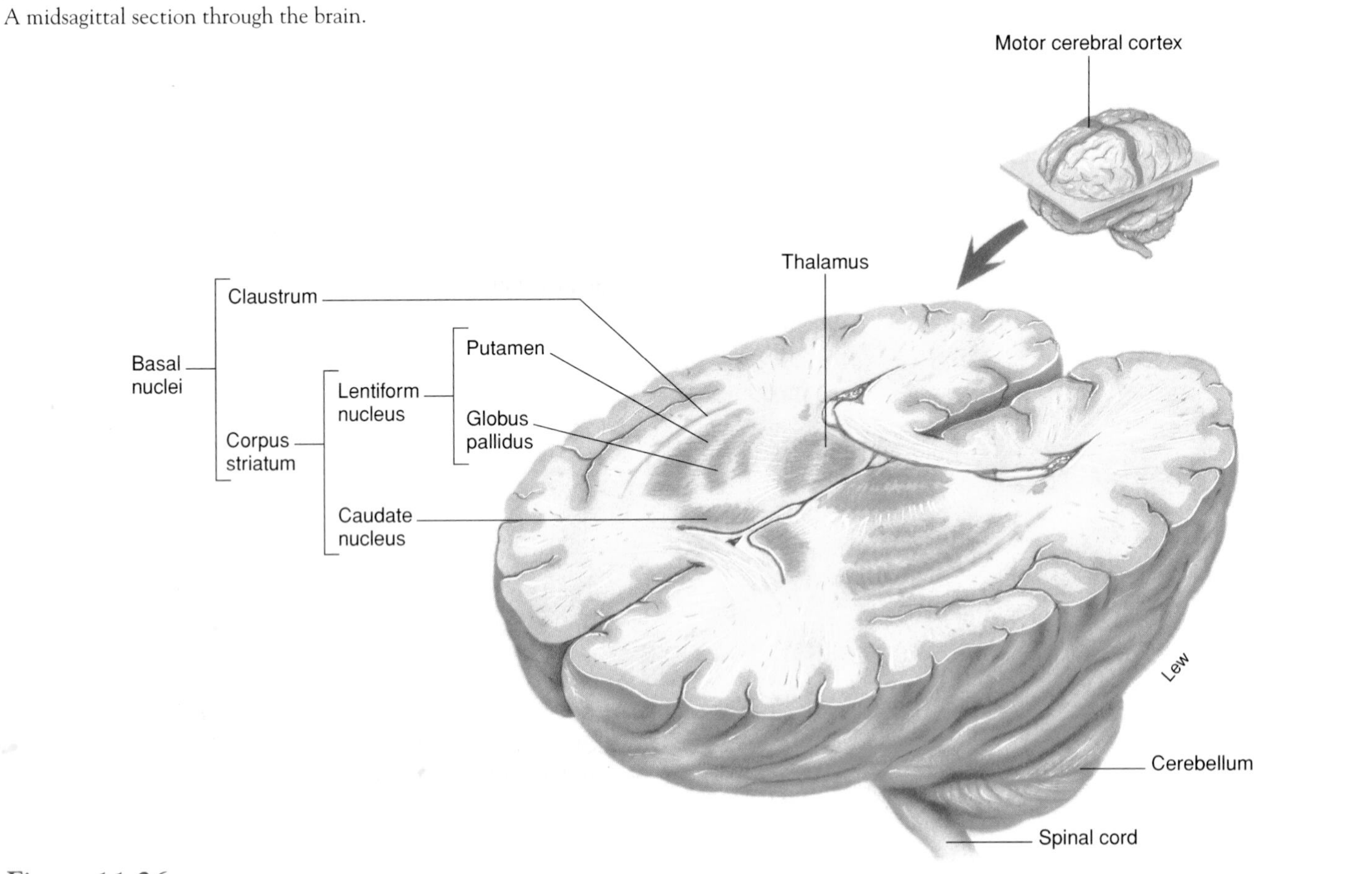

Figure 11.26

Structures of the cerebrum containing nuclei involved in the control of skeletal muscles. The thalamus is a relay center between the motor cerebral cortex and the other brain areas.

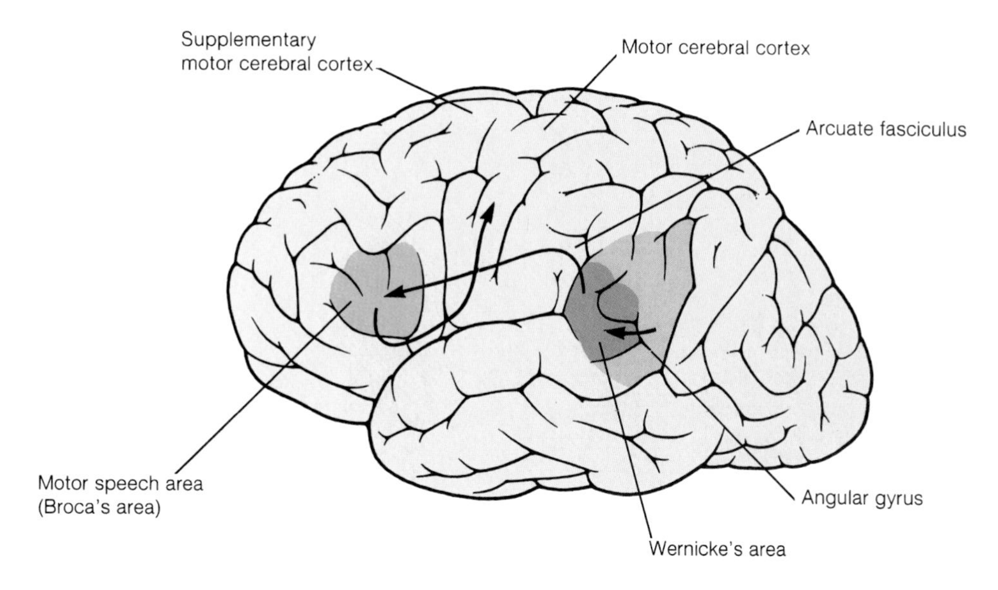

Figure 11.27

Brain areas involved in the control of speech. Arrows indicate the direct communication between these areas.

area projects to Wernicke's area. Some patients with damage to the left angular gyrus can speak and understand spoken language but cannot read or write. Other patients can write a sentence but cannot read it, presumably due to damage to the projections from the occipital lobe (involved in vision) to the angular gyrus.

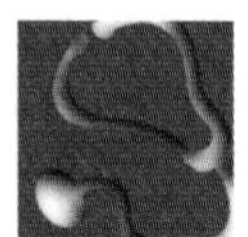

Recovery of language ability, by transfer to the right hemisphere after damage to the left hemisphere, is very good in children but decreases after adolescence. Recovery is reported to be faster in left-handed people, possibly because language ability is more evenly divided between the two hemispheres in left-handed people. Some recovery usually occurs after damage to the motor speech area, but damage to Wernicke's area produces more severe and permanent aphasias.

1. Diagram a lateral view of the cerebrum and label the four superficial lobes and the fissures that separate them.
2. List the functions of each of the paired cerebral lobes.
3. What is a brain-wave pattern? How are these patterns monitored clinically?
4. Describe the arrangement of the fiber tracts within the cerebrum.
5. Name and describe the basal nuclei and list their functions.
6. Describe the aphasias that result from damage to the motor speech area and Wernicke's area from damage to the arcuate fasciculus and from damage to the angular gyrus. Explain how these areas may interact in the production of speech.

Diencephalon

The diencephalon is a major autonomic region of the brain that consists of such vital structures as the thalamus, hypothalamus, epithalamus, and pituitary gland.

Objective 16 List the autonomic functions of the thalamus and the hypothalamus.

Objective 17 Describe the location and structure of the pituitary gland.

The **diencephalon** (*di″en-sef′ă-lon*) is the second subdivision of the forebrain and is almost completely surrounded by the cerebral hemispheres of the telencephalon. The third ventricle (see fig. 11.36) is a narrow midline cavity within the diencephalon. The most important structures of the diencephalon are the *thalamus*, *hypothalamus*, *epithalamus*, and *pituitary gland*.

Thalamus

The **thalamus** (*thal′ă-mus*) is a large oval mass of gray matter, constituting nearly four-fifths of the diencephalon. It is actually a paired organ, with each portion positioned immediately below the lateral ventricle of its respective cerebral hemisphere (see figs. 11.25 and 11.26). The principal function of the thalamus is to act as a relay center for all sensory impulses, except smell, to the cerebral cortex. Specialized masses of nuclei relay the incoming impulses to precise locations within the cerebral lobes for interpretation.

thalamus: L. *thalamus*, inner room

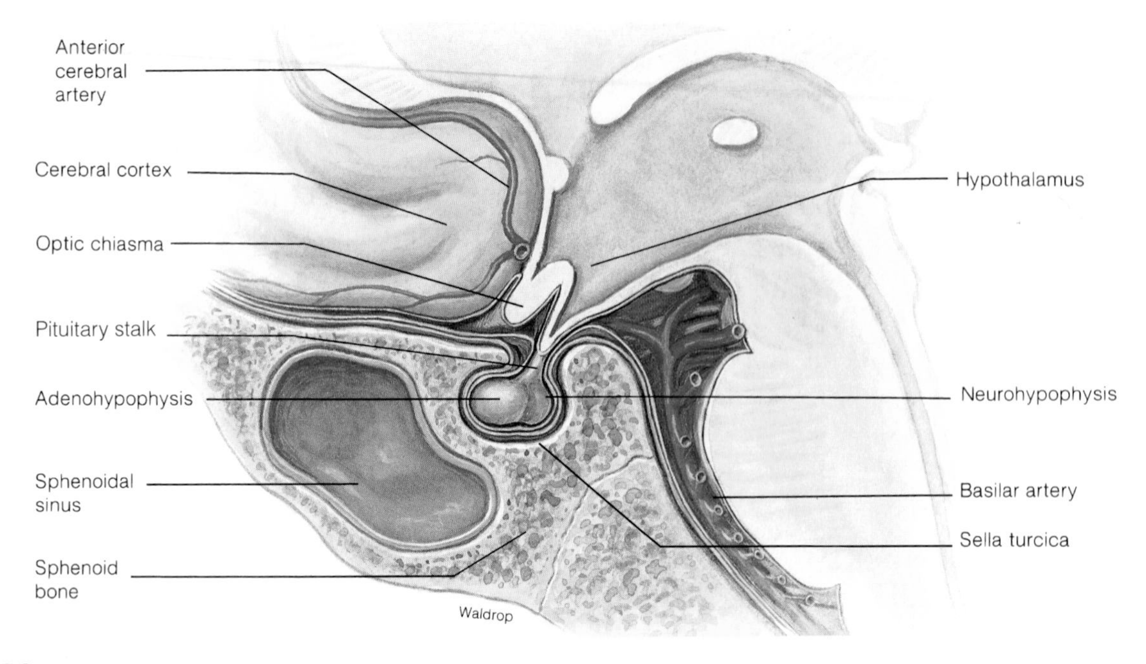

Figure 11.28

The pituitary gland is positioned within the sella turcica of the sphenoid bone and is attached to the brain by the stalk of the pituitary.

The thalamus also performs some sensory interpretation. The cerebral cortex discriminates pain and other tactile stimuli, but the thalamus responds to general sensory stimuli and provides crude awareness. The thalamus probably plays a role in the initial autonomic response of the body to intense pain, and is therefore partially responsible for the physiological shock that frequently follows serious trauma.

Hypothalamus

The **hypothalamus** (*hi″po-thal′ă-mus*), named for its position below the thalamus, is the most inferior portion of the diencephalon. It forms the floor and part of the lateral walls of the third ventricle (figs. 11.25 and 11.28) and contains several masses of nuclei that are interconnected with other parts of the nervous system. Despite its small size, the hypothalamus performs numerous vital functions, most of which relate directly or indirectly to the regulation of visceral activities. It also has emotional and instinctual functions (see also fig. 14.12).

The hypothalamus acts as an autonomic nervous center in accelerating or decelerating certain body functions. It secretes eight hormones, including two released from the posterior pituitary. These hormones and their functions are discussed in chapter 14. The principal autonomic and limbic (emotional) functions of the hypothalamus are as follows:

1. **Cardiovascular regulation.** Although the heart has an innate pattern of contraction, impulses from the hypothalamus cause autonomic acceleration or deceleration of the heart rate. Impulses from the posterior hypothalamus produce a rise in arterial blood pressure and an increase of the heart rate. Impulses from the anterior portion have the opposite effect. Rather than traveling directly to the heart, impulses from these regions pass first to the cardiovascular centers of the medulla oblongata.
2. **Body-temperature regulation.** Specialized nuclei within the anterior portion of the hypothalamus are sensitive to changes in body temperature. If the arterial blood flowing through this portion of the hypothalamus is above normal temperature, the hypothalamus initiates impulses that cause heat loss through sweating and vasodilation of cutaneous vessels of the skin. A below-normal blood temperature causes the hypothalamus to relay impulses that result in heat production and retention through shivering, contraction of cutaneous blood vessels, and cessation of sweating.
3. **Regulation of water and electrolyte balance.** Specialized *osmoreceptors* in the hypothalamus continuously monitor the osmotic concentration of the blood. An increased osmotic concentration resulting from lack of water causes the production of antidiuretic hormone (ADH) by the hypothalamus and its release from the posterior pituitary. At the same time, a *thirst center* within the hypothalamus produces feelings of thirst.
4. **Regulation of hunger and control of gastrointestinal activity.** The *feeding center* is a specialized portion of the lateral hypothalamus that monitors blood glucose, fatty acid, and amino acid levels. Low levels of these substances in the blood are partially responsible for a sensation of hunger elicited from the hypothalamus. When enough food has been eaten, the *satiety* (*să-ti′ĭ-te*) *center* in the midportion of the hypothalamus inhibits the feeding center. The hypothalamus also receives sensory impulses from the abdominal viscera and regulates glandular secretions and the peristaltic movements of the GI tract.

5. **Regulation of sleeping and wakefulness.** The hypothalamus has both a *sleep center* and a *wakefulness center* that function with other parts of the brain to determine the level of conscious alertness.
6. **Sexual response.** Specialized *sexual center* nuclei within the dorsal portion of the hypothalamus respond to sexual stimulation of the tactile receptors within the genital organs. The experience of orgasm involves neural activity within the sexual center of the hypothalamus.
7. **Emotions.** A number of nuclei within the hypothalamus are associated with specific emotional responses, including anger, fear, pain, and pleasure (see fig. 14.12).
8. **Control of endocrine functions.** The hypothalamus produces neurosecretory chemicals that stimulate the anterior and posterior pituitary to release various hormones.

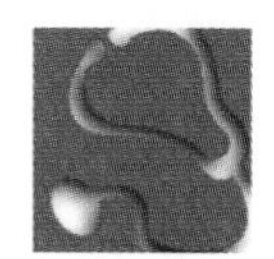

The hypothalamus is a vital structure in maintaining overall body homeostasis. Dysfunction of the hypothalamus may seriously affect autonomic, somatic, or psychic body functions. Not surprisingly, this organ is implicated as a principal factor in *psychosomatic illness.* Insomnia, peptic ulcers, palpitation of the heart, diarrhea, and constipation are a few symptoms of psychophysiologic disorders.

Epithalamus

The **epithalamus** is the posterior portion of the diencephalon that forms a thin roof over the third ventricle. The inside lining of the roof consists of a vascular **choroid plexus,** where cerebrospinal fluid is produced (see fig. 11.25). A small mass of tissue called the **pineal** (*pin'e-al*) **gland,** named for its resemblance to a pine cone, extends outward from the posterior end of the epithalamus (see fig. 11.25). It is thought to have a neuroendocrine function. The **posterior commissure,** located inferior to the pineal gland, is a tract of commissural fibers that connects the right and left superior colliculi of the midbrain (see fig. 11.31).

Pituitary Gland

The rounded, pea-shaped **pituitary gland,** or **cerebral hypophysis** (*hi-pof'ĭ-sis*), is positioned on the inferior aspect of the diencephalon. It is attached to the hypothalamus by the funnel-shaped **infundibulum** (*in"fun-dib'yŭ-lum*) and is supported by the sella turcica of the sphenoid bone (fig. 11.28). The pituitary, which has an endocrine function, is structurally and functionally divided into an anterior portion, called the **adenohypophysis** (*ad"ĕ-no-hi-pof'ĭ-sis*), and a posterior portion, called the **neurohypophysis** (see chapter 14).

pineal: L. *pinea*, pine cone
pituitary: L. *pituita*, phlegm (this gland was originally thought to secrete mucus into nasal cavity)
infundibulum: L. *infundibulum*, funnel

1. What is meant by the statement that the thalamus is the pain center of the brain?
2. List the body systems and functions over which the hypothalamus has some control.
3. What are the components of the epithalamus?
4. Describe the location of the pituitary gland relative to the rest of the diencephalon and the rest of the brain. What are the two portions of the pituitary gland?

Mesencephalon (Midbrain)

The mesencephalon contains the corpora quadrigemina, concerned with visual and auditory reflexes, and the cerebral peduncles, composed of fiber tracts. It also contains specialized nuclei that help to control posture and movement.

Objective 18 List the structures of the mesencephalon and explain their functions.

The *brain stem* contains nuclei for autonomic functions of the body and their connecting tracts. It is that portion of the brain that attaches to the spinal cord and includes the midbrain, pons, and medulla oblongata. The **mesencephalon** (*mes"en-sef'ă-lon*), or midbrain, is the short section of the brain stem between the diencephalon and the pons (see fig. 11.31). Within the midbrain is the **mesencephalic aqueduct** (aqueduct of Sylvius) (see fig. 11.36), which connects the third and fourth ventricles. The midbrain also contains the corpora quadrigemina (see fig. 11.25), the cerebral peduncles (see fig. 11.24), the *red nucleus*, and the *substantia nigra*.

The **corpora quadrigemina** (*kwad"rĭ-jem-ĭ-nă*) are the four rounded elevations on the posterior portion of the midbrain. The two upper eminences, the **superior colliculi,** (*ko-lik'yŭ-li*) are concerned with visual reflexes. The two posterior eminences, the **inferior colliculi,** are responsible for auditory reflexes. The **cerebral peduncles** (*pĕ-dung'kulz*) are a pair of cylindrical structures composed of ascending and descending projection fiber tracts that support and connect the cerebrum to the other regions of the brain.

The **red nucleus** lies deep within the midbrain between the cerebral peduncle and the cerebral aqueduct. It connects the cerebral hemispheres and the cerebellum and functions in reflexes concerned with motor coordination and maintenance of posture. Its reddish color is due to its rich blood supply and an iron-containing pigment in the cell bodies of its neurons. Another nucleus, the **substantia nigra** (*ni'gră*), lies inferior to the red nucleus. The substantia nigra is thought to inhibit forced involuntary movements. Its dark color reflects its high content of melanin pigment.

aqueduct of Sylvius: from Jacobus Sylvius, French anatomist, 1478–1555
corpora quadrigeminal: L. *corpus*, body; *quadri*, four; *geminus*, twin
colliculus: L. *colliculus*, small mound

1. Describe symptoms that might indicate a tumor in the midbrain.
2. Which structures of the midbrain function with the cerebellum in controlling posture and movement?

Metencephalon (Hindbrain)

The metencephalon contains the pons, which relays impulses, and the cerebellum, which coordinates skeletal muscle contractions.

Objective 19 Describe the location and structure of the pons and cerebellum and list their functions.

The **metencephalon** (*met″en-sef′ă-lon*) is the most superior portion of the hindbrain. Two vital structures of the metencephalon are the *pons* and *cerebellum*. The mesencephalic aqueduct of the mesencephalon enlarges to become the **fourth ventricle** (see fig. 11.37) within the metencephalon and myelencephalon.

Pons

The **pons** can be observed as a rounded bulge on the inferior surface of the brain, between the midbrain and the medulla oblongata (fig. 11.29). It consists of white fiber tracts that course in two principal directions. The surface fibers extend transversely to connect with the cerebellum through the middle cerebellar peduncles. The deeper longitudinal fibers are part of the motor and sensory tracts that connect the medulla oblongata with the tracts of the midbrain.

Scattered throughout the pons are several nuclei associated with specific cranial nerves. The cranial nerves that have nuclei within the pons include the trigeminal (V), which transmits impulses for chewing and sensory sensations from the head; the abducens (VI), which controls certain movements of the eyeball; the facial (VII), which transmits impulses for facial movements and sensory sensations from the taste buds; and the vestibular branches of the vestibulocochlear (VIII), which maintain equilibrium.

Other nuclei of the pons function with nuclei of the medulla oblongata to regulate the rate and depth of breathing. The two respiratory centers of the pons are called the **apneustic** and **pneumotaxic areas** (fig. 11.29).

Cerebellum

The **cerebellum** (*ser′ĕ-bel′um*) is the second largest structure of the brain. It is located in the metencephalon and occupies the inferior and posterior aspect of the cranial cavity. The cerebellum is separated from the overlying cerebrum by a **transverse fissure.**

pons: L. *pons*, bridge
cerebellum: L. *cerebellum*, diminutive of *cerebrum*, brain

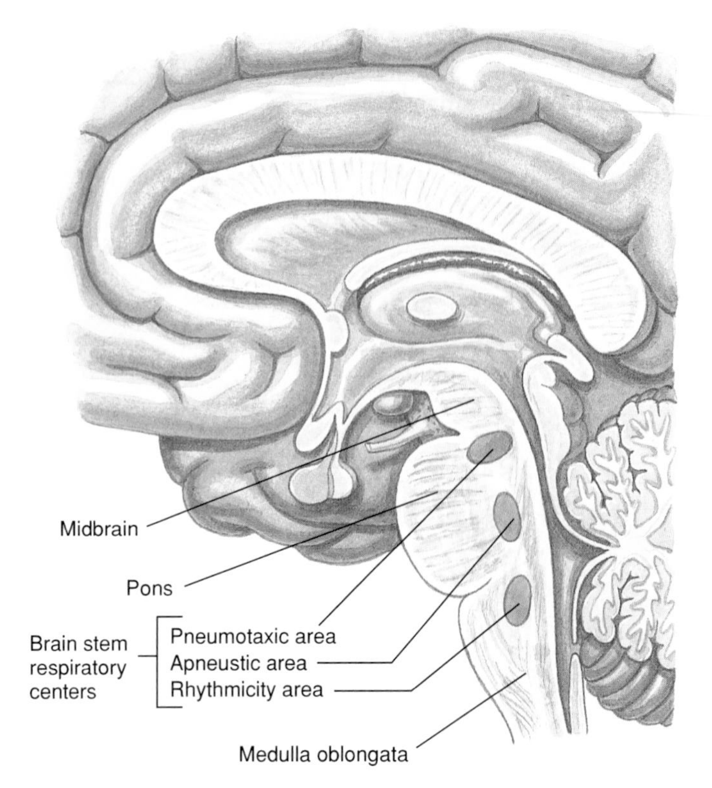

Figure 11.29

Nuclei within the pons and medulla oblongata that constitute the respiratory center.

A portion of the meninges called the **tentorium cerebelli** extends into the transverse fissure. The cerebellum consists of two **hemispheres** and a central constricted area called the **vermis** (fig. 11.30). The **falx cerebelli** is the portion of the meninges that partially extends between the hemispheres.

Like the cerebrum, the cerebellum has a thin outer layer of gray matter, the **cerebellar cortex,** and a thick, deeper layer of white matter. The cerebellum is convoluted into a series of parallel slender **folia.** The tracts of white matter within the cerebellum have a distinctive branching pattern called the **arbor vitae** (*vi′te*) that can be seen in a sagittal view (fig. 11.30).

Three paired bundles of nerve fibers called **cerebellar peduncles** support the cerebellum and provide it with tracts for communicating with the rest of the brain (fig. 11.31). Following is a description of the cerebellar peduncles.

1. **Superior cerebellar peduncles** connect the cerebellum with the midbrain. The fibers within these peduncles originate primarily from specialized **dentate nuclei** within the cerebellum and pass through the red nucleus to the thalamus, and then to the motor areas of the cerebral cortex. Impulses through the fibers of these peduncles provide feedback to the cerebrum.

vermis: L. *vermis*, worm
arbor vitae: L. *arbor*, tree; *vitae*, life
peduncle: L. *peduncle*, diminutive of *pes*, foot

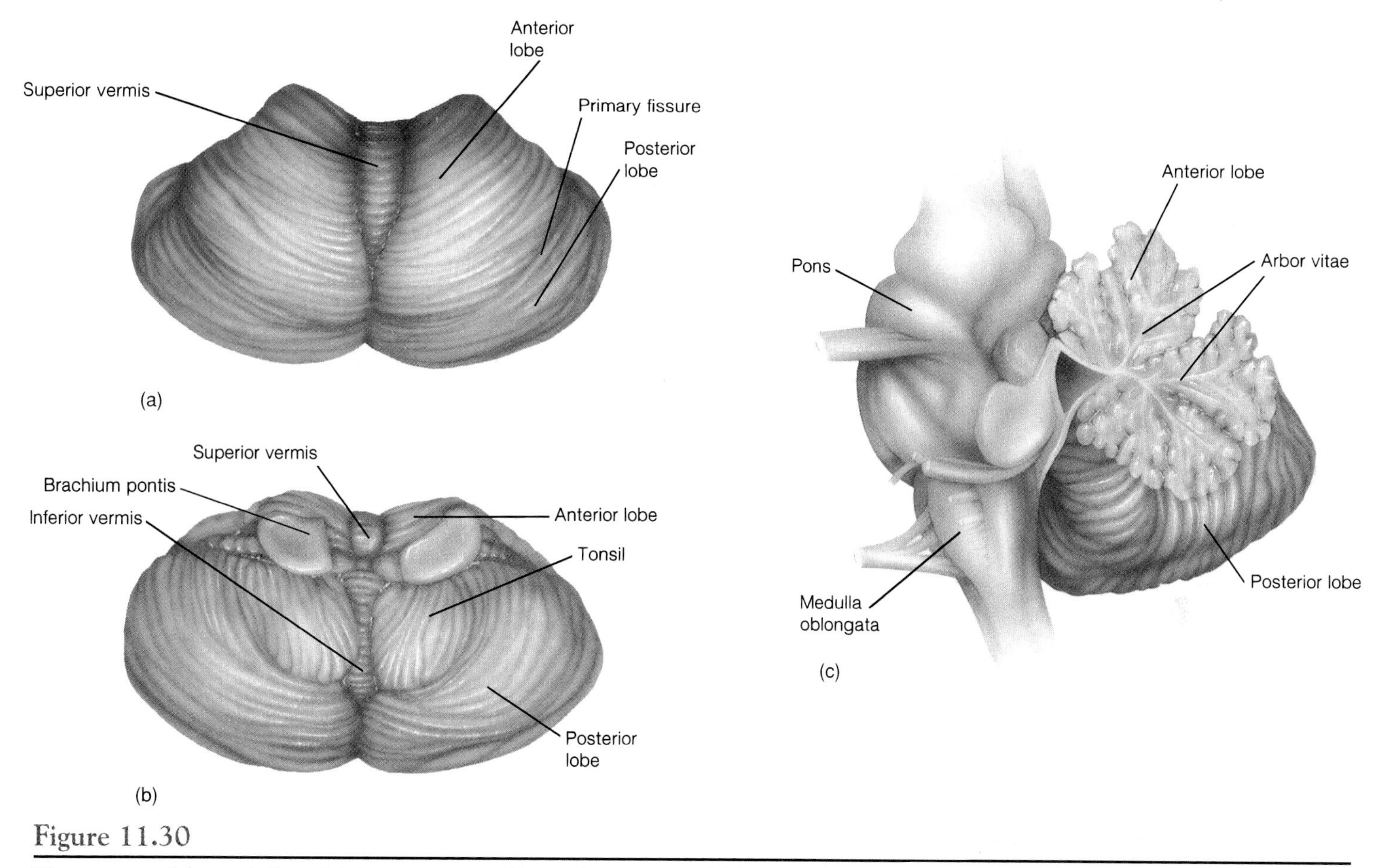

Figure 11.30

The structure of the cerebellum. (*a*) A superior view, (*b*) an inferior view, and (*c*) a sagittal view.

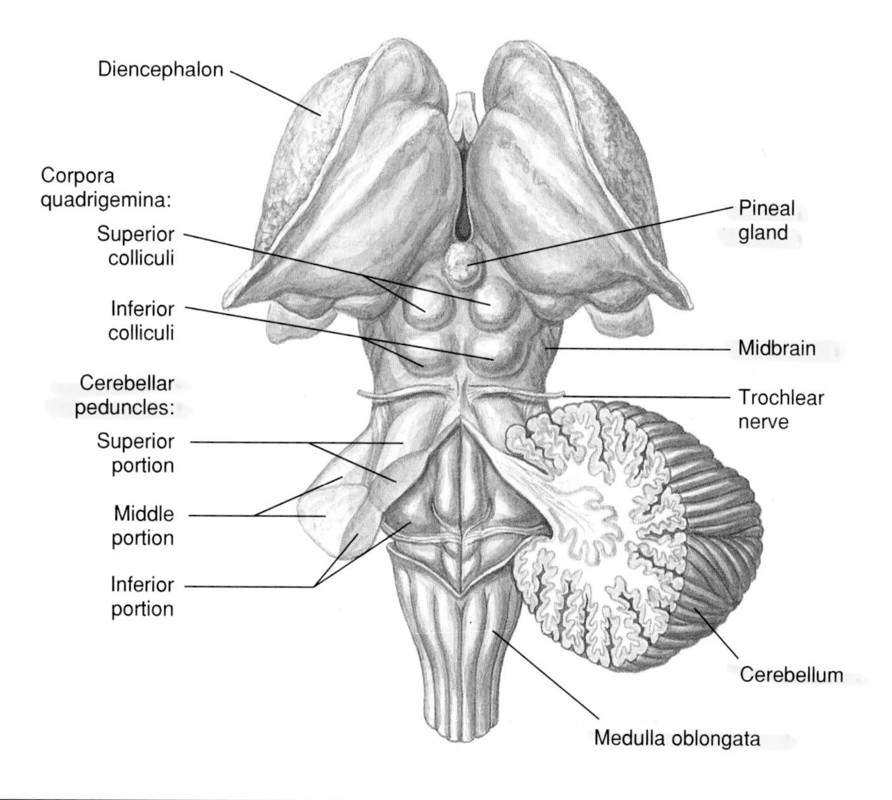

Figure 11.31

The cerebellar peduncles can be seen when the cerebellar hemisphere has been removed from its attachment to the brain stem.

2. **Middle cerebellar peduncles** convey impulses of voluntary movement from the cerebrum through the pons and to the cerebellum.
3. **Inferior cerebellar peduncles** connect the cerebellum with the medulla oblongata and the spinal cord. They contain both incoming vestibular and proprioceptive fibers and outgoing motor fibers.

The principal function of the cerebellum is coordinating skeletal muscle contractions by recruiting precise motor units within the muscles. Impulses for voluntary muscular movement originate in the cerebral cortex and are coordinated by the cerebellum. The cerebellum constantly initiates impulses to selective motor units for maintaining posture and muscle tone. The cerebellum also processes incoming impulses from **proprioceptors** (*pro″pre-o-sep′torz*) within muscles, tendons, joints, and special sense organs to refine learned movement patterns. A proprioceptor is a sensory nerve ending that is sensitive to changes in the tension of a muscle or tendon.

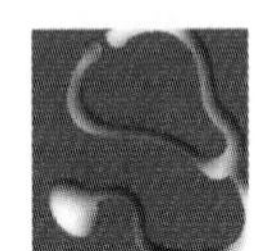

Trauma or diseases of the cerebellum, such as a *stroke* or *cerebral palsy,* frequently cause an impairment of skeletal muscle function. Movements become jerky and uncoordinated in a condition known as *ataxia.* There is also a loss of equilibrium, resulting in a disturbance of gait. *Alcohol intoxication* causes similar uncoordinated body movements.

1. Describe the locations and relative sizes of the pons and the cerebellum.
2. Which cranial nerves have nuclei located within the pons?
3. Define the terms *tentorium cerebelli, vermis, arbor vitae,* and *cerebellar peduncles.*
4. List the functions of the pons and the cerebellum.

Myelencephalon

The medulla oblongata, contained within the myelencephalon, connects to the spinal cord and contains nuclei for the cranial nerves and vital autonomic functions.

Objective 20 Describe the location, structure, and functions of the medulla oblongata.

Objective 21 Describe the location and primary functions of the reticular formation.

Medulla Oblongata

The **medulla oblongata** (*mĕ-dul′ă ob″long-gă′tă*), is a bulbous structure about 3 cm (1 in.) long, is the most inferior structure of the brain stem. It is continuous with the pons anteriorly and the spinal cord posteriorly at the level of the foramen magnum (see figs. 11.24 and 11.25). Externally, the medulla oblongata resembles the spinal cord, except for two triangular elevations called **pyramids** on the inferior side and an oval enlargement called the **olive** (see fig. 11.24) on each lateral surface. The **fourth ventricle,** the space within the medulla oblongata, is continuous posteriorly with the central canal of the spinal cord and anteriorly with the mesencephalic aqueduct (see fig. 11.36).

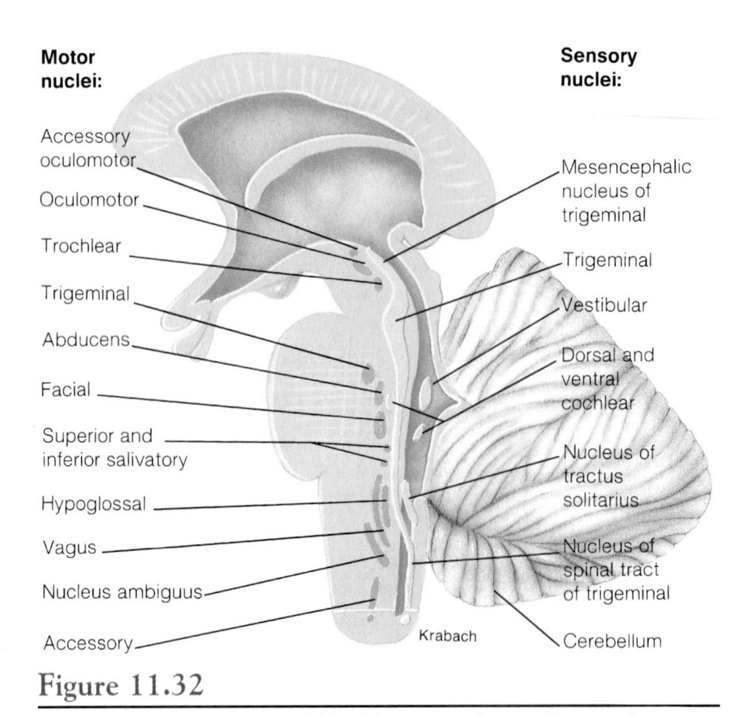

Figure 11.32

A sagittal section of the medulla oblongata and pons showing the cranial nuclei of gray matter.

The medulla oblongata is composed of vital nuclei and white matter that form all of the descending and ascending tracts communicating between the spinal cord and various parts of the brain. Most of the fibers within these tracts cross over to the opposite side through the pyramidal region of the medulla oblongata, permitting one side of the brain to receive information from and send information to the opposite side of the body (see fig. 11.24).

The gray matter of the medulla oblongata consists of several important nuclei for the cranial nerves and sensory relay (fig. 11.32). The **nucleus ambiguus** (*am-big′yoo-us*) and the **hypoglossal nucleus** are the centers from which arise the vestibulocochlear (VIII), glossopharyngeal (IX), accessory (XI), and hypoglossal (XII) nerves. The vagus nerves (X) arise from **vagus nuclei,** one on each lateral side of the medulla oblongata, adjacent to the fourth ventricle. The **nucleus gracilis** (*gras′ĭ-lis*) and the **nucleus cuneatus** (*kyoo-ne-a′tus*) relay sensory information to the thalamus, and the impulses are then relayed to the cerebral cortex via the thalamic nuclei (not illustrated). The **inferior olivary nuclei** and the **accessory olivary nuclei** of the olive mediate impulses passing from the forebrain and midbrain through the inferior cerebellar peduncles to the cerebellum.

Three other nuclei within the medulla oblongata function as autonomic centers for controlling vital visceral functions.

1. **Cardiac center.** Both *inhibitory* and *accelerator fibers* arise from nuclei of the cardiac center. Inhibitory impulses constantly travel through the vagus nerves to slow the heartbeat. Accelerator impulses travel through the spinal cord and eventually innervate the heart through fibers within spinal nerves T1–T5.
2. **Vasomotor center.** Nuclei of the vasomotor center send impulses via the spinal cord and spinal nerves to the smooth muscles of arteriole walls, causing them to constrict and elevate arterial blood pressure.
3. **Respiratory center.** The respiratory center of the medulla oblongata controls the rate and depth of breathing and functions in conjunction with the respiratory nuclei of the pons (see fig. 11.29) to produce rhythmic breathing.

Other nuclei of the medulla oblongata function as centers for reflexes involved in sneezing, coughing, swallowing, and vomiting. Some of these activities (swallowing, for example) may be initiated voluntarily, but once they progress to a certain point they become involuntary and cannot be stopped.

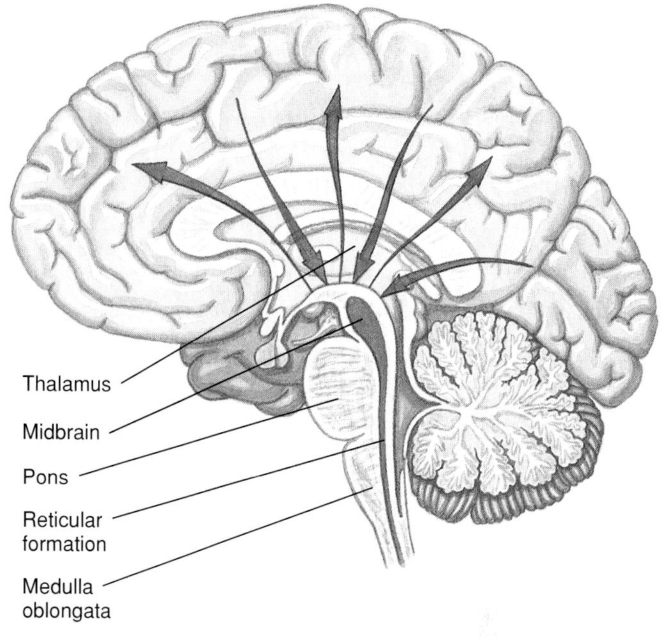

Figure 11.33

The reticular activating system. The arrows indicate the direction of impulses along nerve pathways that connect with the RAS.

Reticular Formation

The **reticular formation** is a complex network of nuclei and nerve fibers within the brain stem that functions as the *reticular activating system* (RAS) in arousing the cerebrum. Portions of the reticular formation are located in the spinal cord, pons, midbrain, and parts of the thalamus and hypothalamus (fig. 11.33). The reticular formation contains ascending and descending fibers from most of the structures within the brain.

Nuclei within the reticular formation generate a continuous flow of impulses unless they are inhibited by other parts of the brain. The principal functions of the RAS are to keep the cerebrum in a state of alert consciousness and to selectively monitor the sensory impulses perceived by the cerebrum. The RAS also helps the cerebellum activate selected motor units to maintain muscle tonus and produce smooth, coordinated contractions of skeletal muscles.

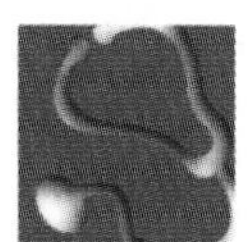

The RAS is sensitive to changes in and trauma to the brain. The sleep response is thought to occur because of a decrease in activity within the RAS, perhaps due to the secretion of specific neurotransmitters. A blow to the head or certain drugs and diseases may damage the RAS, causing unconsciousness. A *coma* is a state of unconsciousness and inactivity of the RAS that even the most powerful external stimuli cannot disturb.

1. Describe the major reflex centers of the medulla oblongata that regulate autonomic functions.
2. Which cranial nerves arise from the medulla oblongata?
3. Explain the statement that the RAS is the brain's chief watchguard.

Meninges of the Central Nervous System

The CNS is covered by protective meninges; namely, the dura mater, the arachnoid, and the pia mater.

Objective 22 Describe the position of the meninges as they protect the CNS.

As mentioned previously in the discussion of the brain's general features, the entire delicate CNS is protected by a bony encasement—the cranium surrounding the brain and the vertebral column surrounding the spinal cord. It is also protected by three membranous connective tissue coverings called the **meninges** (figs. 11.34 and 11.35). Individually, from the outside in, they are known as the *dura mater*, the *arachnoid*, and the *pia mater*.

Dura Mater

The **dura mater** (*door'ă ma'ter*) is in contact with the bone and is composed primarily of dense connective tissue. The **cranial dura mater** is a double-layered structure. The outer **periosteal** (*per"e-os'te-al*) **layer** adheres lightly to the cranium, where it constitutes the periosteum (fig. 11.34). The inner **meningeal layer,** which is thinner, follows the general contour of the brain. The

meninges: L. plural form of *meninx*, membrane
dura mater: L. *dura*, hard; *mater*, mother

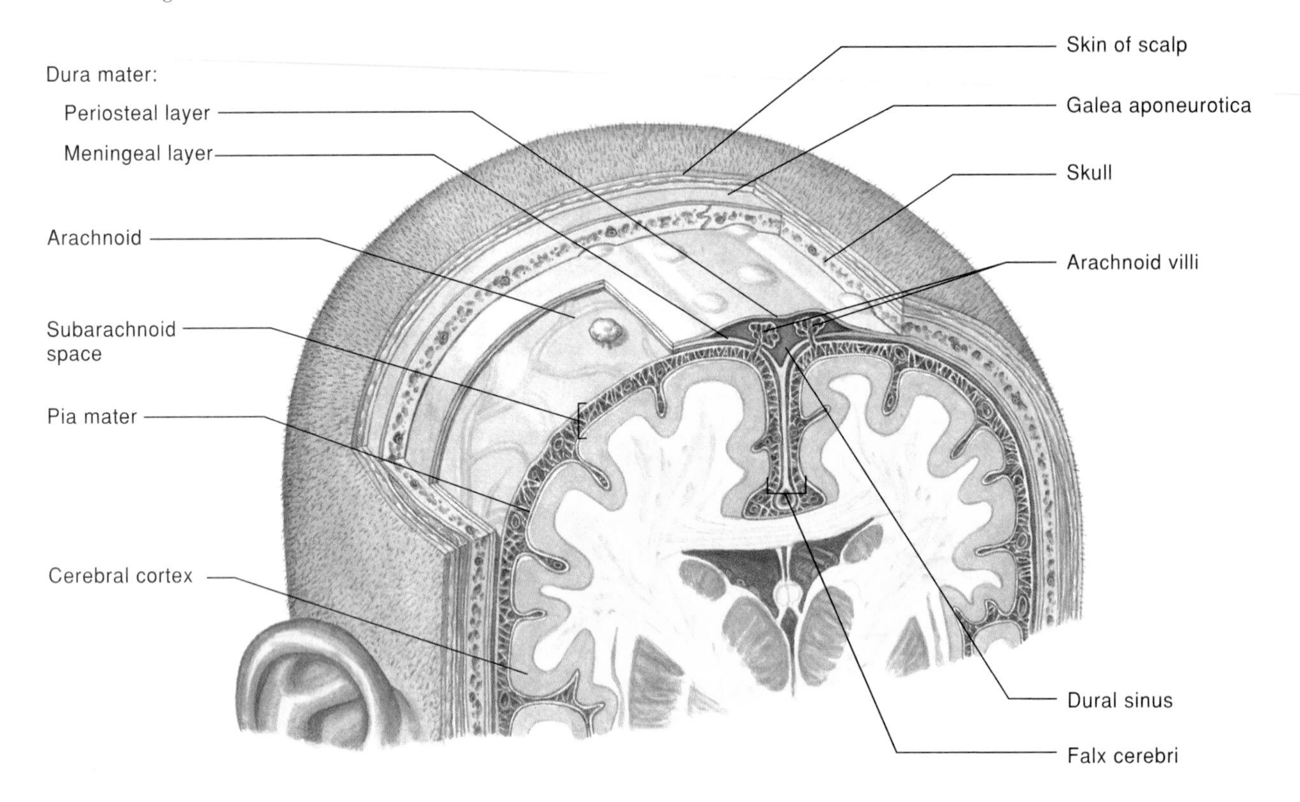

Figure 11.34

Meninges and associated structures surrounding the brain.

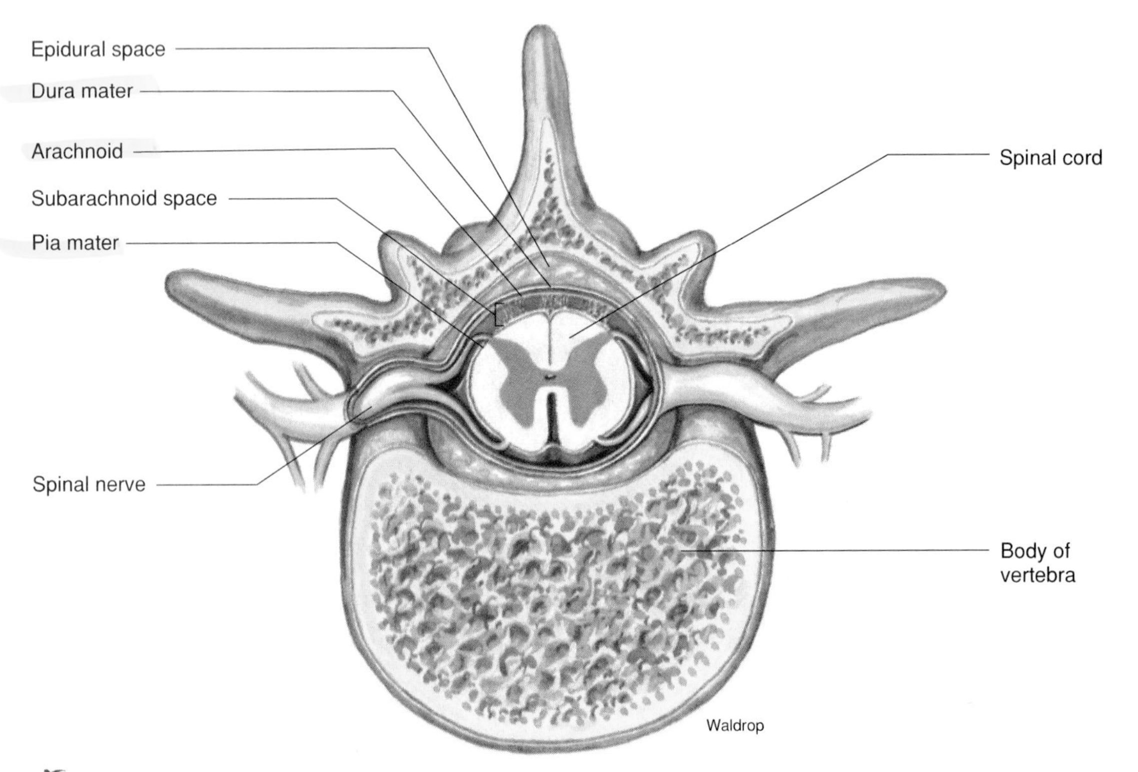

Figure 11.35

Meninges and associated structures surrounding the spinal cord. The epidural space in the lower lumbar region is of clinical importance as a site for an epidural block that may be administered to facilitate childbirth.

Table 11.5

Septa of the Cranial Dura Mater

Septa	Location
Falx cerebri	Extends downward into the longitudinal fissure to partition the right and left cerebral hemispheres; anchored anteriorly to the crista galli of the ethmoid bone and posteriorly to the tentorium
Tentorium cerebelli	Separates the occipital and temporal lobes of the cerebrum from the cerebellum; anchored to the tentorium, petrous parts of the temporal bones, and occipital bone
Falx cerebelli	Partitions the right and left cerebellar hemispheres; anchored to the occipital crest
Diaphragma sellae	Forms the roof of the sella turcica

spinal dura mater is not double layered. It is similar to the meningeal layer of the cranial dura mater.

The two layers of the cranial dura mater are generally fused and cover most of the brain. In certain regions, however, the layers separate, enclosing **dural sinuses** (see fig. 11.34) that collect venous blood and drain it to the internal jugular veins of the neck.

In four locations, the meningeal layer of the cranial dura mater forms distinct septa to partition major structures on the surface of the brain and anchor the brain to the inside of the cranial case. These septa were identified earlier and are reviewed in table 11.5.

The spinal dura mater (figs. 11.35 and 11.36) forms a tough, tubular **dural sheath** that continues into the vertebral canal and surrounds the spinal cord. There is no connection between the dural sheath and the vertebrae forming the vertebral canal, but instead there is a potential cavity called the **epidural space** (see fig. 11.35). The epidural space is highly vascular and contains loose fibrous and adipose connective tissues that form a protective pad around the spinal cord.

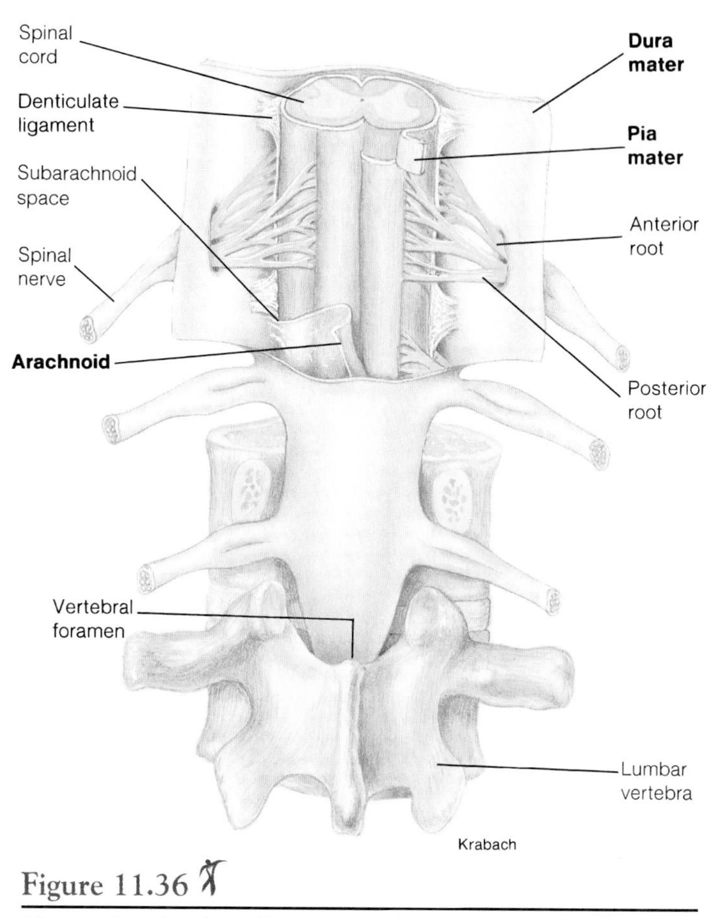

Figure 11.36

The spinal cord and spinal meninges. (The meninges are labeled in boldface type.)

Arachnoid

The **arachnoid** (*ă-rak'noid*) is the middle of the three meninges. This delicate, netlike membrane spreads over the CNS but generally does not extend into the sulci or fissures of the brain. The **subarachnoid space,** located between the arachnoid mater and the deepest meninx, the pia mater, contains cerebrospinal fluid. The subarachnoid space is maintained by weblike strands that connect the arachnoid and pia mater (see fig. 11.34).

Pia Mater

The thin **pia** (*pi'ă*) **mater,** which is tightly bound to the convolutions of the brain and the irregular contours of the spinal cord, is composed of modified loose connective tissue. It is highly vascular and functions to support the vessels that nourish the underlying cells of the brain and spinal cord. The pia mater is specialized over the roofs of the ventricles, where it contributes to the formation of the choroid plexuses along with the arachnoid. Lateral extensions of the pia mater along the spinal cord form the **ligamentum denticulatum,** which attaches the spinal cord to the dura mater (fig. 11.36).

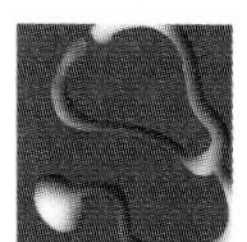

Meningitis, an inflammation of the meninges, is usually caused by bacteria or viruses. The arachnoid and the pia mater are the two meninges most frequently affected. Meningitis is accompanied by high fever and severe headache. Complications may cause sensory impairment, paralysis, or mental retardation. Untreated meningitis generally results in coma and death.

1. Contrast the cranial dura mater and the spinal dura mater.
2. Explain how the meninges support and protect the CNS.
3. Describe the location of the dural sinuses and the epidural space.

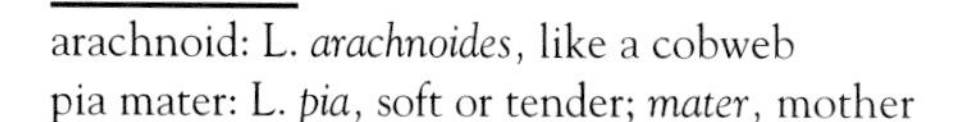

arachnoid: L. *arachnoides*, like a cobweb
pia mater: L. *pia*, soft or tender; *mater*, mother

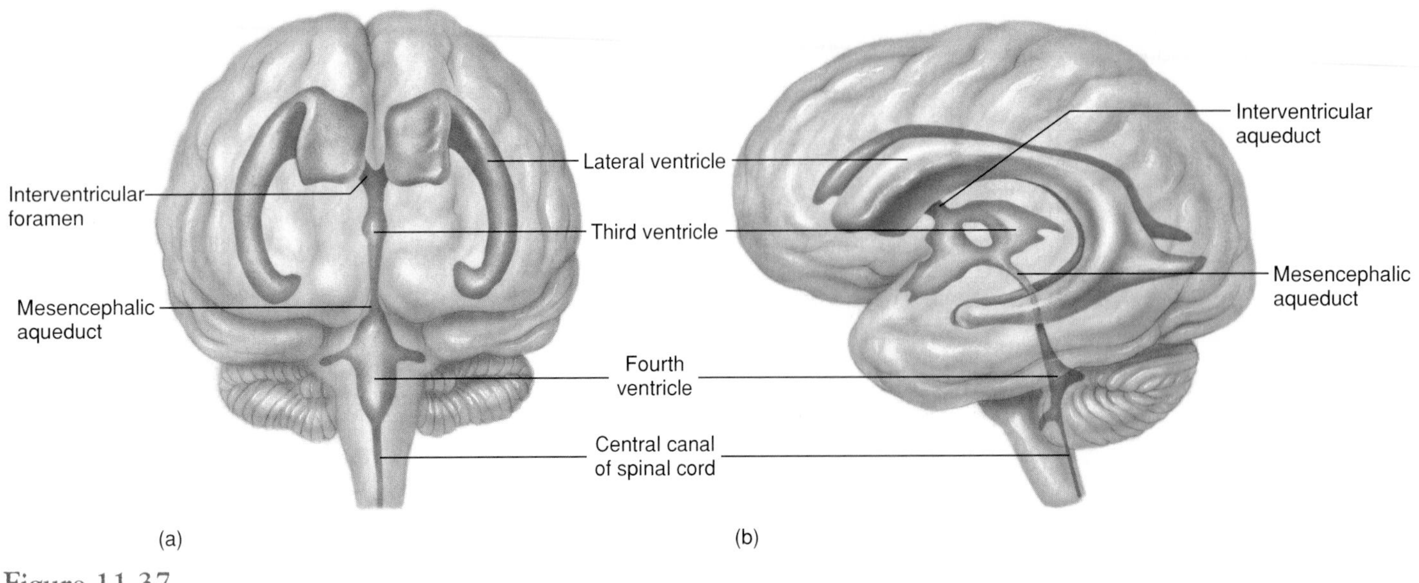

Figure 11.37

The ventricles of the brain. (*a*) An anterior view and (*b*) a lateral view.

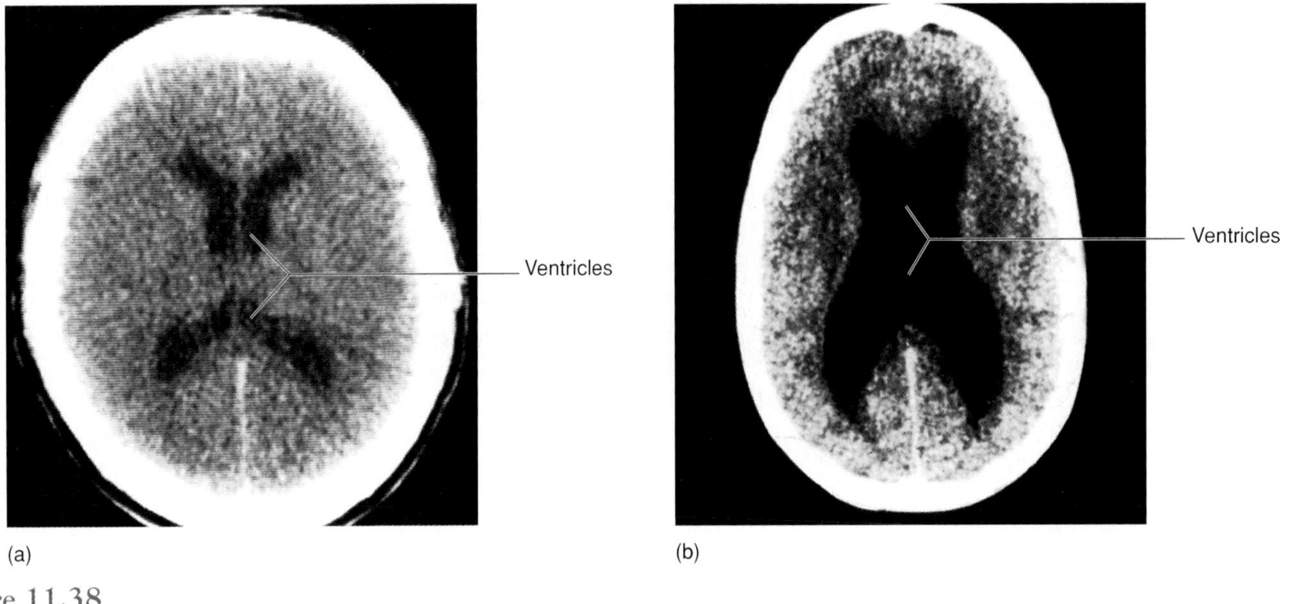

Figure 11.38

CT scans of the brain showing (*a*) a normal configuration of the ventricles and (*b*) an abnormal configuration of the ventricles as a result of hydrocephalism.

Ventricles and Cerebrospinal Fluid

The ventricles, central canal, and subarachnoid space contain cerebrospinal fluid, formed by the active transport of substances from blood plasma in the choroid plexuses.

Objective 23 Discuss the formation, function, and flow of cerebrospinal fluid.

Objective 24 Explain the importance of the blood-brain barrier in maintaining homeostasis within the brain.

Cerebrospinal fluid (CSF) is a clear, lymphlike fluid that forms a protective cushion around and within the CNS. The fluid also buoys the brain. CSF circulates through the various **ventricles** of the brain, the **central canal** of the spinal cord, and the **subarachnoid space** around the entire CNS. The cerebrospinal fluid returns to the circulatory system by draining through the walls of the **arachnoid villi,** which are venous capillaries.

Ventricles of the Brain

The ventricles of the brain are connected to one another and to the central canal of the spinal cord (figs. 11.37 and 11.38). Each of the two **lateral ventricles** (first and second ventricles) is located in

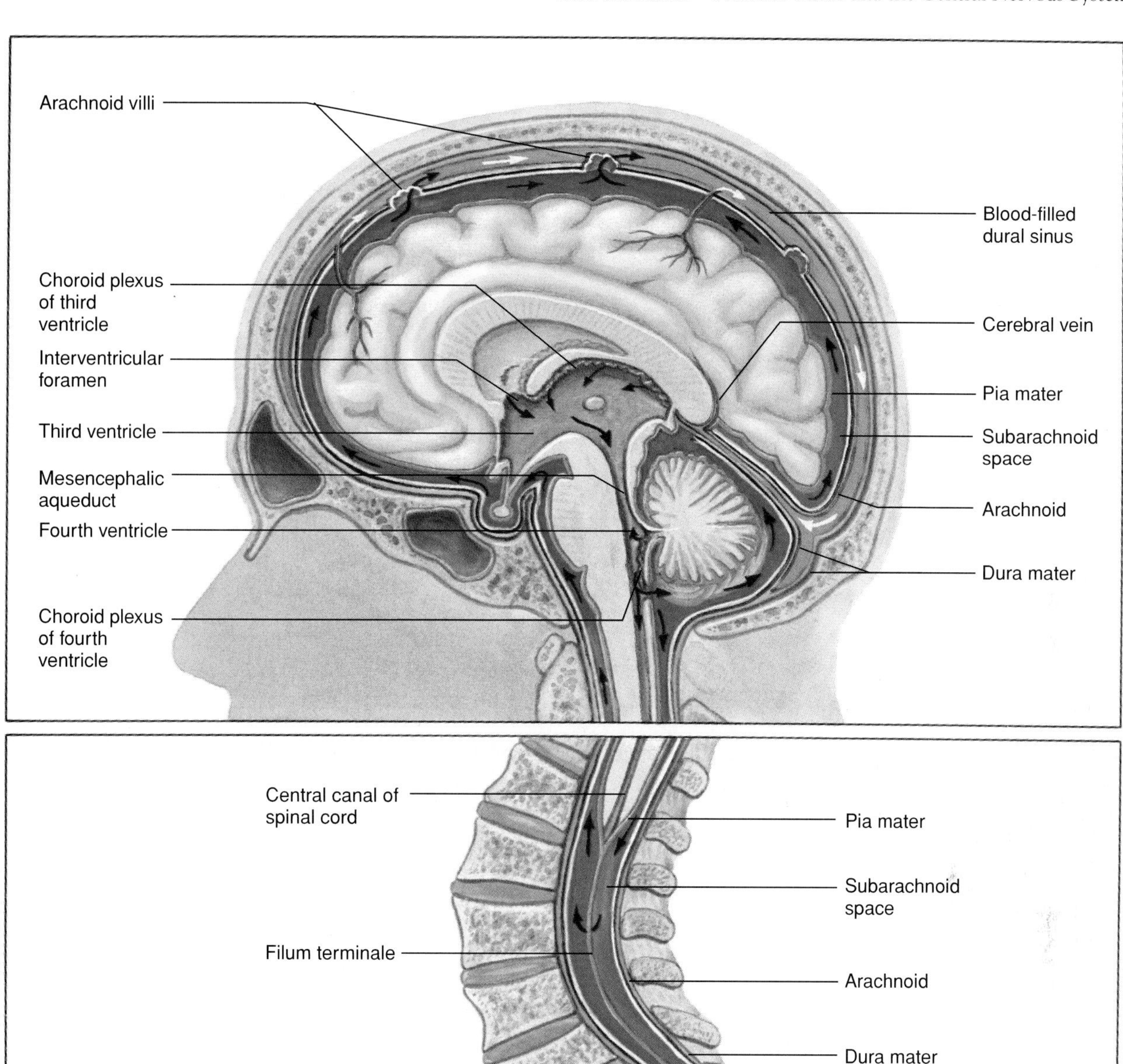

Figure 11.39

The flow of cerebrospinal fluid. Cerebrospinal fluid is secreted by choroid plexuses in the ventricular walls. The fluid circulates through the ventricles and central canal, enters the subarachnoid space, and is reabsorbed into the blood of the dural sinuses through the arachnoid villi.

one of the hemispheres of the cerebrum, inferior to the corpus callosum. The **third ventricle** is located in the diencephalon, between the thalami. Each lateral ventricle is connected to the third ventricle by a narrow, oval opening called the **interventricular foramen** (foramen of Monro). The **fourth ventricle** is located in the brain stem, within the pons, cerebellum, and medulla oblongata. The **mesencephalic aqueduct** (cerebral aqueduct) passes through the midbrain to link the third and fourth ventricles. The fourth ventricle also communicates posteriorly with the central canal of the spinal cord. Cerebrospinal fluid exits from the fourth ventricle into the subarachnoid space (fig. 11.39) through three foramina: the **median aperture** (foramen of Magendie), a medial opening, and

foramen of Monro: from Alexander Monro Jr., Scottish anatomist, 1733–1817

foramen of Magendie: from François Magendie, French physiologist, 1783–1855

two **lateral apertures** (foramina of Luschka) (not illustrated). Cerebrospinal fluid returns to the venous blood through the arachnoid villi.

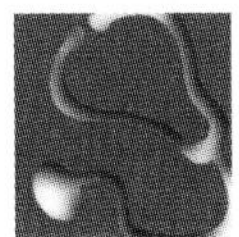

Internal hydrocephalus (*hi″dro-sef′ă-lus*) is a condition in which cerebrospinal fluid builds up within the ventricles of the brain (fig. 11.38*b*). It is more common in infants, whose cranial sutures have not yet strengthened or ossified, than in older individuals. If the pressure is excessive, the condition may have to be treated surgically.

External hydrocephalus, an accumulation of fluid within the subarachnoid space, usually results from an obstruction of drainage at the arachnoid villi. The external pressure compresses neural tissue and is likely to cause brain damage.

Cerebrospinal Fluid

Cerebrospinal fluid buoys the CNS and protects it from mechanical injury. The brain weights about 1,500 grams, but suspended in CSF its buoyed weight is about 50 grams. This means that the brain has a near neutral buoyancy; at a true neutral buoyancy, an object does not float or sink but is suspended in its fluid environment.

In addition to buoying the CNS, CSF reduces the damaging effect of an impact to the head by spreading the force over a larger area. It also helps to remove metabolic wastes from nervous tissue. Since the CNS lacks lymphatic circulation, the CSF moves cellular wastes into the venous return at its places of drainage.

The clear, watery CSF is continuously produced by the filtration of blood plasma through masses of specialized capillaries called **choroid plexuses** and, to a lesser extent, by secretions of ependymal cells. The ciliated ependymal cells cover the choroid plexuses, as well as line the central canal, and presumably aid the movement of the CSF. The tight junctions between the ependymal cells also help to form a *blood–cerebrospinal fluid barrier* that prohibits certain potentially harmful substances in the blood from entering the CSF.

CSF is similar in composition to the blood plasma from which it is formed. Like blood plasma, it contains proteins, glucose, urea, and white blood cells. Comparing electrolytes, CSF contains more sodium, chloride, magnesium and hydrogen, and fewer calcium and potassium ions than does blood plasma.

Up to 800 ml of cerebrospinal fluid are produced each day, although only 140–200 ml are bathing the CNS at any given moment. A person lying in a horizontal position has a slow but continuous circulation of cerebrospinal fluid, with a fluid pressure of about 10 mmHg.

The homeostatic consistency of the CSF composition is critical, and a chemical imbalance may have marked effects on CNS functions. An increase in amino acid glycine concentration, for example, produces hypothermia and hypotension as temperature and blood pressure regulatory mechanisms are disrupted. A slight change in pH may affect the respiratory rate and depth.

foramen of Luschka: from Hubert Luschka, German anatomist, 1820–75

Blood-Brain Barrier

The **blood-brain barrier (BBB)** is a structural arrangement of capillaries, surrounding connective tissue, and specialized neuroglia called astrocytes (see figs. 11.4 and 11.8) that selectively determine which substances can move from the plasma of the blood to the extracellular fluid of the brain. Certain substances such as water, oxygen, carbon dioxide, glucose, and lipid-soluble compounds (alcohol, for example) pass readily through the BBB. Certain inorganic ions (Ca^{+} and K^{+}) pass more slowly, so that the concentrations of these ions in the brain differ from those in the blood plasma. Other substances, such as proteins, lipids, creatine, urea, inulin, certain toxins, and most antibiotics, are restricted in passage. The BBB must be taken into account when planning drug therapy for neurological disorders.

While the BBB is an important protective device, it is essential that the brain be able to monitor and respond to fluctuations in blood glucose, pH, salinity, osmolarity, and pressure. For this reason, the BBB is absent in limited brain areas, including portions of the hypothalamus.

1. Describe the location of the ventricles within the brain.
2. What are the functions of cerebrospinal fluid?
3. Where is cerebrospinal fluid produced and where does it drain?
4. Name some structures to which the BBB is highly permeable and some to which it is slightly permeable. What substances are prevented from crossing?

Spinal Cord

The spinal cord consists of centrally located gray matter, involved in reflexes, and peripherally located ascending and descending tracts of white matter, that conduct impulses to and from the brain.

Objective 25 Describe the structure of the spinal cord.

Objective 26 Describe the arrangement of ascending and descending tracts within the spinal cord.

The **spinal cord** is the portion of the CNS that extends through the vertebral canal of the vertebral column (fig. 11.40). It is continuous with the brain through the foramen magnum of the skull. The spinal cord has two principal functions.

1. **Impulse conduction.** It provides a means of neural communication to and from the brain through tracts of white matter. **Ascending tracts** conduct impulses from the

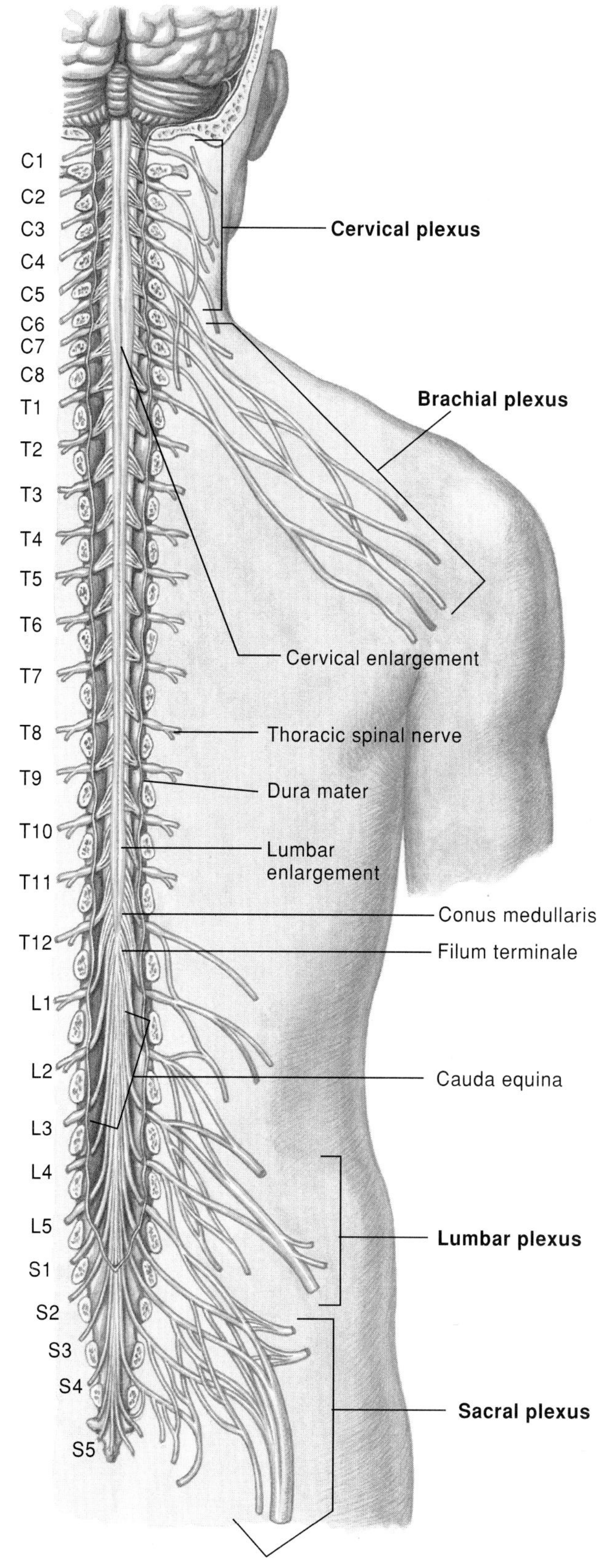

Figure 11.40

The spinal cord and plexuses. (The plexuses are indicated in boldface type).

peripheral sensory receptors of the body to the brain. **Descending tracts** conduct motor impulses from the brain to the muscles and glands.

2. **Reflex integration.** It serves as a center for spinal reflexes. Specific nerve pathways allow for reflexive movements rather than those initiated voluntarily by the brain. Movements of this type are not confined to skeletal muscles; reflexive movements of cardiac and smooth muscles control heart rate, breathing rate, blood pressure, and digestive activities. Spinal nerve pathways are also involved in swallowing, coughing, sneezing, and vomiting.

Structure of the Spinal Cord

The spinal cord extends interiorly from the position of the foramen magnum of the occipital bone to the level of the first lumbar vertebra (L1). It is somewhat flattened posteroventrally, making it oval in cross section. Two prominent enlargements can be seen in a posterior view (fig. 11.40). The **cervical enlargement** is located between the third cervical and the second thoracic vertebrae. Nerves emerging from this region serve the upper extremities. The **lumbar enlargement** lies between the ninth and twelfth thoracic vertebrae. Nerves from the lumbar enlargement supply the lower extremities.

The embryonic spinal cord develops more slowly than the associated vertebral column; thus, in the adult, the cord does not extend beyond L1. The tapering, terminal portion of the spinal cord is called the **conus medullaris** (*med-yoo-lar'is*). The **filum terminale** (*fi'lum ter-mĭ-nal'e*), a fibrous strand composed mostly of pia mater, extends inferiorly from the conus medullaris at the level of L1 to the coccyx. Nerve roots also radiate inferiorly from the conus medullaris through the vertebral canal. These nerve roots are collectively referred to as the **cauda equina** (*kaw'dă e-qui'nă*) because they resemble a horse's tail.

The spinal cord develops as 31 segments, each of which gives rise to a pair of **spinal nerves** that emerge from the spinal cord through the intervertebral foramina. Two grooves, an **anterior median fissure** and a **posterior median sulcus,** extend the length of the spinal cord and partially divide it into right and left portions. Like the brain, the spinal cord is protected by three distinct meninges and is cushioned by cerebrospinal fluid. The pia mater contains an extensive vascular network.

The **gray matter** of the spinal cord is centrally located and surrounded by white matter. It is composed of nerve cell bodies, neuroglia, and unmyelinated association neurons (interneurons). The **white matter** consists of bundles, or tracts, of myelinated fibers of sensory and motor neurons.

The relative size and shape of the gray and white matter varies throughout the spinal cord. The amount of white matter increases toward the brain as the nerve tracts become thicker. More

filum terminale: L. *filum*, filament; *terminus*, end
cauda equina: L. *cauda*, tail; *equus*, horse

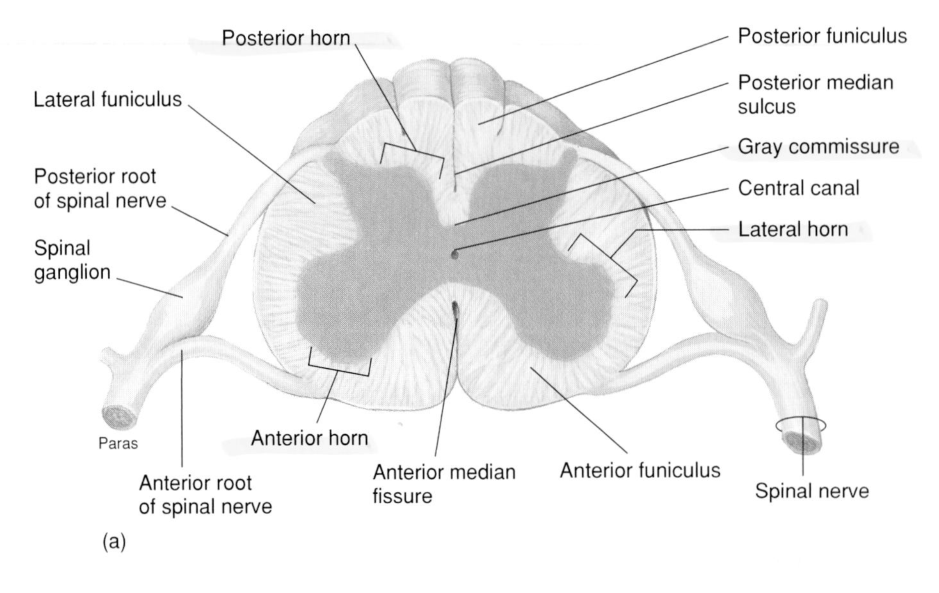

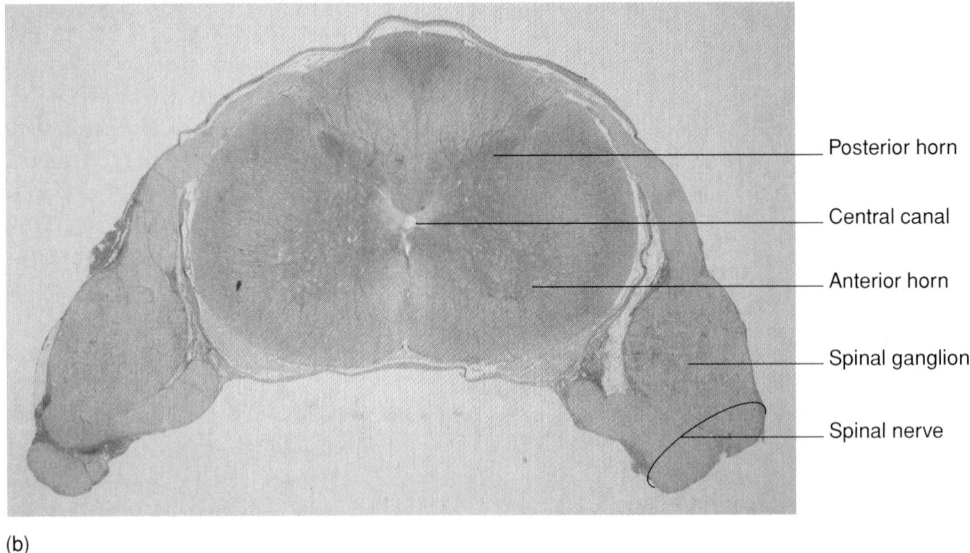

Figure 11.41

The spinal cord in cross section. (*a*) A diagram and (*b*) a photomicrograph.

gray matter is found in the cervical and lumbar enlargements where innervations from the upper and lower extremities, respectively, make connections.

The core of gray matter roughly resembles the letter H (fig. 11.41). Projections of the gray matter within the spinal cord are called *horns*, and are named according to the direction in which they project. The paired **posterior horns** extend posteriorly and the paired **anterior horns** project anteriorly. Between the posterior and anterior horns, the short paired **lateral horns** extend to the sides. Lateral horns are prominent only in the thoracic and upper lumbar regions. The transverse bar of gray matter that connects the paired horns across the center of the spinal cord is called the **gray commissure.** Within the gray commissure is the **central canal.** It is continuous with the ventricles of the brain and is filled with cerebrospinal fluid.

Spinal Cord Tracts

Impulses are conducted through the ascending and descending tracts of the spinal cord within the columns of white matter. The spinal cord has six columns of white matter called **funiculi** (*fyoo-nik'yŭ-li*), which are named according to their relative position within the spinal cord. The two **anterior funiculi** are located between the two anterior horns of gray matter, to either side of the anterior median fissure (fig. 11.41). The two

commissure: L. *commissura*, a joining

funiculus: L. diminutive of *funis*, cord or rope

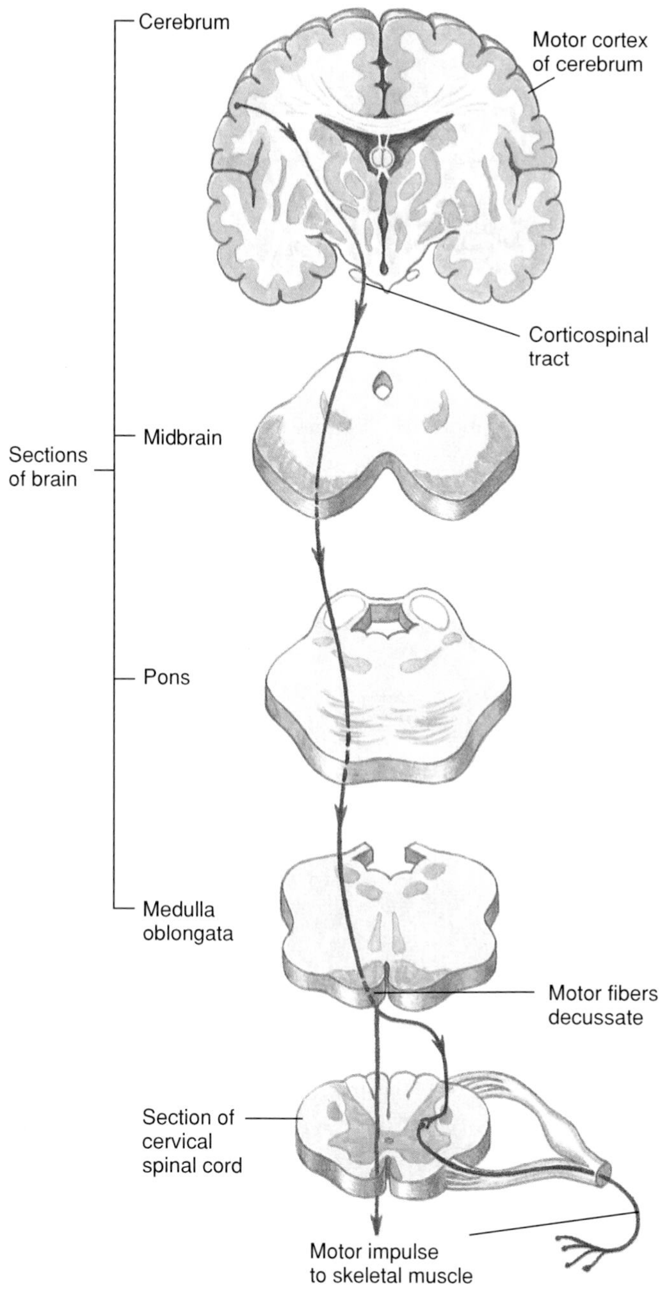

Figure 11.42

A descending corticospinal tract composed of motor fibers that decussate in the medulla oblongata of the brain stem.

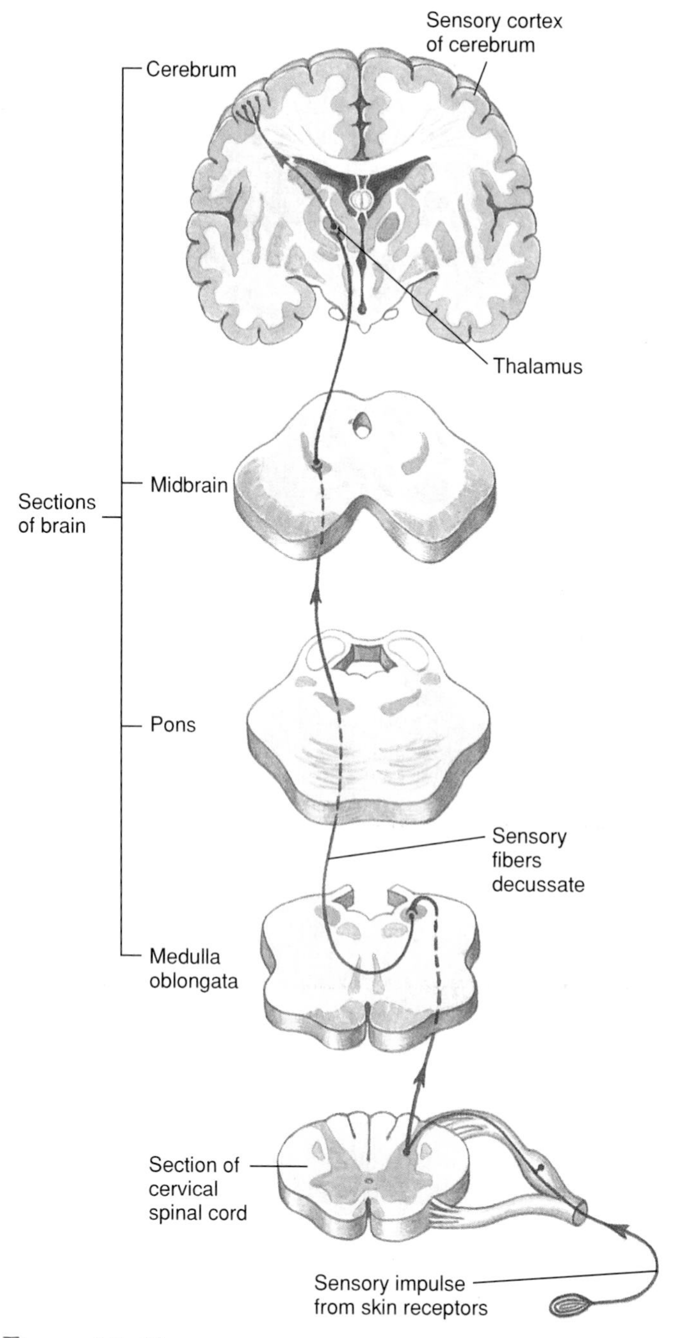

Figure 11.43

An ascending tract composed of sensory fibers that decussate (cross over) in the medulla oblongata of the brain stem.

posterior funiculi are located between the two posterior horns of gray matter, to either side of the posterior median sulcus. Two **lateral funiculi** are located between the anterior and posterior horns of gray matter.

Each funiculus consists of both ascending and descending tracts. The nerve fibers within the tracts are generally myelinated and are named according to their origin and termination. The fibers of the tracts either remain on the same side of the brain and spinal cord or cross over within the medulla oblongata or the spinal cord. The crossing over of nerve tracts is referred to as *decussation* (*de″kus-a′shun*). Illustrated in figures 11.42 and 11.43 are descending and ascending tracts, respectively, that decussate within the medulla oblongata.

The principal ascending and descending tracts within the funiculi are summarized in table 11.6 and illustrated in figure 11.44.

Descending tracts are grouped according to place of origin as either corticospinal or extrapyramidal. **Corticospinal (pyramidal) tracts** descend directly, without synaptic interruption, from the cerebral cortex to the lower motor neurons. The cell bodies of the neurons that contribute fibers to these tracts are located

Table 11.6

Principal Ascending and Descending Tracts of the Spinal Cord

Tract	Funiculus	Origin	Termination	Function
Ascending Tracts				
Anterior spinothalamic	Anterior	Posterior horn on one side of spinal cord; crosses to opposite side	Thalamus, then cerebral cortex	Conducts sensory impulses for crude touch and pressure
Lateral spinothalamic	Lateral	Posterior horn on one side of spinal cord; crosses to opposite side	Thalamus, then cerebral cortex	Conducts pain and temperature impulses that are interpreted within cerebral cortex
Fasciculus gracilis and fasciculus cuneatus	Posterior	Peripheral sensory neurons; does not cross over	Nucleus gracilis and nucleus cuneatus of medulla oblongata; crosses to opposite side; eventually thalamus, then cerebral cortex	Conducts sensory impulses from skin, muscles, tendons, and joints, which are interpreted as sensations of fine touch, precise pressures, and body movements
Posterior spinocerebellar	Lateral	Posterior horn; does not cross over	Cerebellum	Conducts sensory impulses from one side of body to same side of cerebellum for subconscious proprioception required for coordinated muscular contractions
Anterior spinocerebellar	Lateral	Posterior horn; some fibers cross, others do not	Cerebellum	Conducts sensory impulses from both sides of body to cerebellum for subconscious proprioception required for coordinated muscular contractions
Descending Tracts				
Anterior corticospinal	Anterior	Cerebral cortex on one side of brain; crosses to opposite side of spinal cord	Anterior horn	Conducts motor impulses from cerebrum to spinal nerves, and outward to cells of anterior horns for coordinated, precise voluntary movements of skeletal muscle
Lateral corticospinal	Lateral	Cerebral cortex on one side of brain; crosses in base of medulla oblongata to opposite side of spinal cord	Anterior horn	Conducts motor impulses from cerebrum to spinal nerves, and outward to cells of anterior horns for coordinated, precise voluntary movements
Tectospinal	Anterior	Mesencephalon; crosses to opposite side of spinal cord	Anterior horn	Conducts motor impulses to cells of anterior horns, and eventually to muscles that move the head in response to visual, auditory, or cutaneous stimuli
Rubrospinal	Lateral	Mesencephalon (red nucleus); crosses to opposite side of spinal cord	Anterior horn	Conducts motor impulses concerned with muscle tone and posture
Vestibulospinal	Anterior	Medulla oblongata; does not cross over	Anterior horn	Conducts motor impulses that regulate body tone and posture (equilibrium) in response to movements of head
Anterior and medial reticulospinal	Anterior	Reticular formation of brain stem; does not cross over	Anterior horn	Conducts motor impulses that control muscle tone and sweat gland activity
Bulboreticulospinal	Lateral	Reticular formation of brain stem; does not cross over	Anterior horn	Conducts motor impulses that control muscle tone and sweat gland activity

primarily in the precentral gyrus of the frontal lobe. Most (about 85%) of the corticospinal fibers decussate in the pyramids of the medulla oblongata (see fig. 11.24). The remaining 15% do not cross from one side to the other. The fibers that cross compose the **lateral corticospinal tracts,** and the remaining uncrossed fibers compose the **anterior corticospinal tracts.** Because of the crossing of fibers from higher motor neurons in the pyramids, the right hemisphere primarily controls the musculature on the left side of the body, whereas the left hemisphere controls the right musculature.

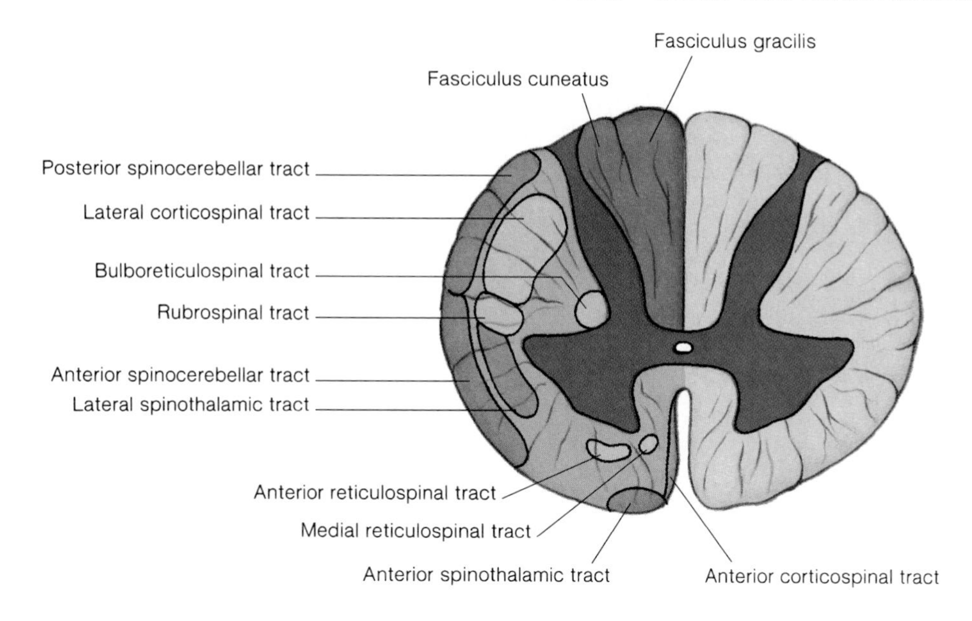

Figure 11.44

A cross section showing the principal ascending and descending tracts within the spinal cord.

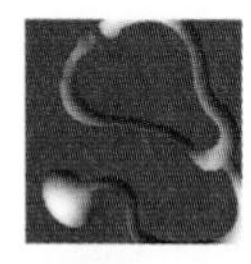

The corticospinal tracts appear to be particularly important in voluntary movements that require complex interactions between the motor cortex and sensory input. Speech, for example, is impaired when the corticospinal tracts are damaged in the thoracic region of the spinal cord, whereas involuntary breathing continues. Damage to the pyramidal motor system can be detected clinically by the presence of *Babinski's reflex,* in which stimulation of the sole of the foot causes extension (upward movement) of the great toe and fanning out of the other toes. Babinski's reflex is normally present in infants because neural control is not yet fully developed.

The remaining descending tracts are **extrapyramidal tracts** that originate in the brain stem region. Electrical stimulation of the cerebral cortex, the cerebellum, and the basal nuclei indirectly evokes movements because of synaptic connections within extrapyramidal tracts.

The **reticulospinal** (*re-tik"yŭ-lo-spi'nal*) **tracts** are the major descending pathways of the extrapyramidal system. These tracts originate in the reticular formation of the brain stem. Neurostimulation of the reticular formation by the cerebrum or cerebellum either facilitates or inhibits the activity of lower motor neurons (depending on the area stimulated) (fig. 11.45).

There are no descending tracts from the cerebellum. The cerebellum can influence motor activity only indirectly, through the vestibular nuclei, red nucleus, and basal nuclei. These structures, in turn, affect lower motor neurons via the **vestibulospinal tracts, rubrospinal tracts,** and **reticulospinal tracts.** Damage to the cerebellum disrupts the coordination of movements with spatial judgment. Underreaching or overreaching for an object may occur, followed by *intention tremor,* in which the limb moves back and forth in a pendulum-like motion.

The basal nuclei, acting through synapses in the reticular formation in particular, appear normally to exert an inhibitory influence on the activity of lower motor neurons. Damage to the basal nuclei thus results in decreased muscle tone. People with such damage display *akinesia* (*a"kĭ-ne'ze-ă*) (complete or partial loss of muscle movement) and *chorea* (*ko-re'ă*) (sudden and uncontrolled random movements).

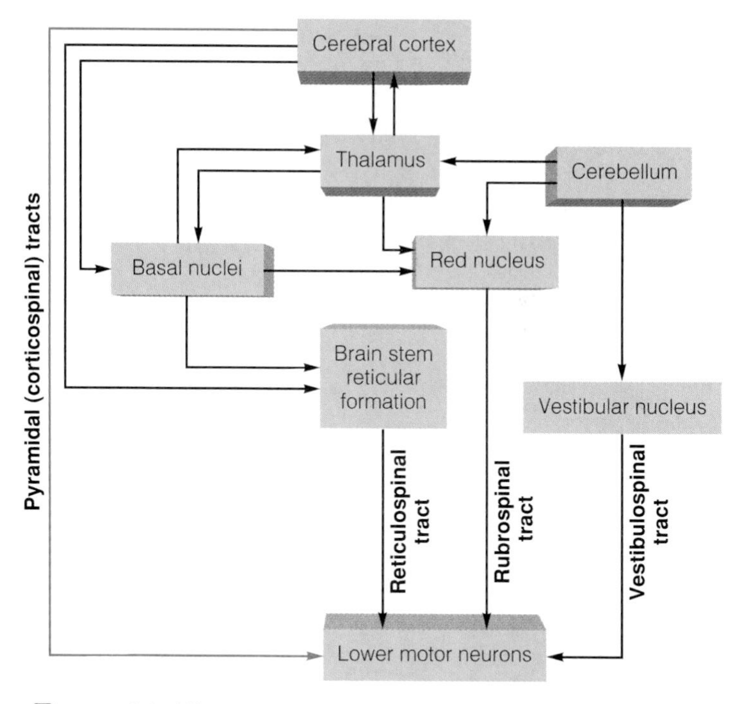

Figure 11.45

Pathways involved in the higher motor neuron control of skeletal muscles. (The pyramidal [corticospinal] tracts are shown in red and the extrapyramidal tracts are shown in black.)

akinesia: Gk. *a*, without; *kinesis*, movement
chorea: Fr. *choros*, a dance

DEVELOPMENTAL EXPOSITION

The Spinal Cord

Exhibit III The development of the spinal cord. (*a*) A dorsal view of an embryo at 23 days with the position of a transverse cut indicated in (*b*). (*c*) The formation of the alar and basal plates is evident in a transverse section through the spinal cord at 6 weeks. (*d*) The size of the central canal has decreased, and functional posterior and anterior horns have formed at 9 weeks.

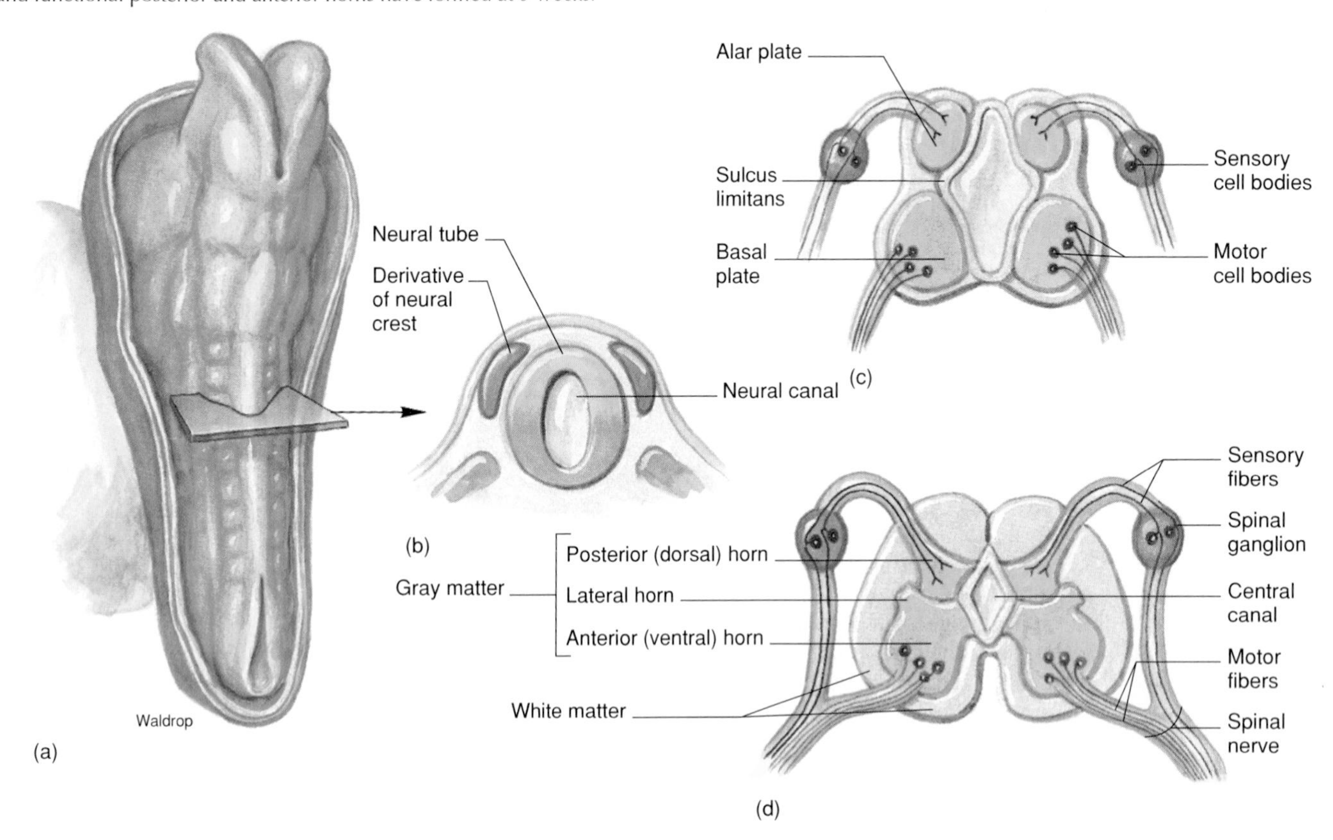

Explanation

The spinal cord, like the brain, develops as the neural tube undergoes differentiation and specialization. Throughout the developmental process, the hollow central canal persists while the specialized white and gray matter forms (exhibit III). Changes in the neural tube become apparent during the sixth week as the lateral walls thicken to form a groove called the **sulcus limitans** along each lateral wall of the central canal. A pair of **alar plates** forms dorsal to the sulcus limitans, and a pair of **basal plates** forms ventrally. By the ninth week, the alar plates have specialized to become the **posterior horns,** containing fibers of the sensory cell bodies, and the basal plates have specialized to form the **anterior** and **lateral horns,** containing motor cell bodies. Sensory neurons of spinal nerves conduct impulses toward the spinal cord, whereas motor neurons conduct impulses away from the spinal cord.

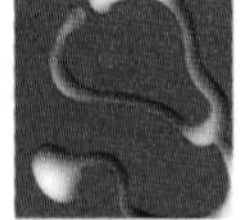

Paralysis agitans, better known as *Parkinson's disease,* is a disorder of the basal nuclei involving the degeneration of fibers from the substantia nigra. These fibers, which use dopamine as a neurotransmitter, are required to antagonize the effects of other fibers that use acetylcholine (ACh) as a transmitter. The relative deficiency of dopamine compared to ACh is believed to produce the symptoms of Parkinson's disease, including *resting tremor.* This shaking of the limbs tends to disappear during voluntary movements and then reappear when the limb is again at rest.

Parkinson's disease is treated with drugs that block the effects of ACh and by the administration of L-dopa, which can be converted to dopamine in the brain. (Dopamine cannot be given directly because it does not cross the blood-brain barrier.)

1. Diagram a cross section of the spinal cord and label the structures of the gray matter and the white matter. Describe the location of the spinal cord.
2. List the structures and function of the corticospinal tracts. Make a similar list for the extrapyramidal tracts.
3. Explain why damage to the right side of the brain primarily affects motor activities on the left side of the body.

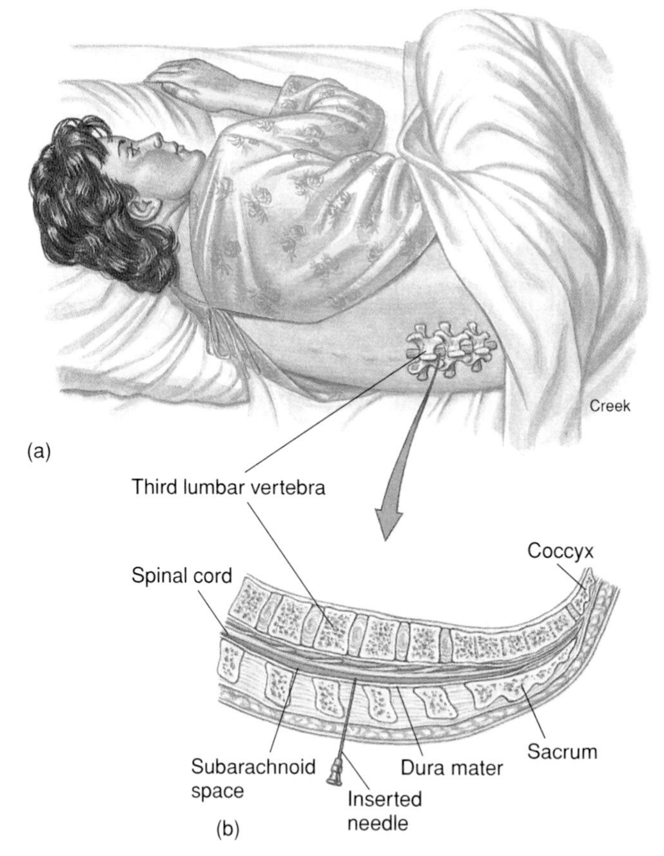

Figure 11.46

(*a*) A lumbar puncture is performed by inserting a needle between the third and fourth lumbar vertebrae (L3–L4) and (*b*) withdrawing cerebrospinal fluid from the subarachnoid space.

Clinical Considerations

The clinical aspects of the central nervous system are extensive and usually complex. Numerous diseases and developmental problems directly involve the nervous system, and the nervous system is indirectly involved with most of the diseases that afflict the body because of the location and activity of sensory pain receptors. Pain receptors are free nerve endings that are present throughout living tissue. The pain sensations elicited by disease or trauma are important in localizing and diagnosing specific diseases or dysfunctions.

Only a few of the many clinical considerations of the central nervous system will be discussed here. These include neurological assessment and drugs, developmental problems, injuries, infections and diseases, and degenerative disorders.

Neurological Assessment and Drugs

Neurological assessment has become exceedingly sophisticated and accurate in the past few years. In a basic physical examination, only the reflexes and sensory functions are assessed. But if the physician suspects abnormalities involving the nervous system, further neurological tests may be done, employing the following techniques.

A **lumbar puncture** is performed by inserting a fine needle between the third and fourth lumbar vertebrae and withdrawing a sample of CSF from the subarachnoid space (fig. 11.46). A **cisternal puncture** is similar to a lumbar puncture except that the CSF is withdrawn from a cisterna at the base of the skull, near the foramen magnum. The pressure of the CSF, which is normally about 10 mmHg, is measured with a *manometer*. Samples of CSF may also be examined for abnormal constituents. In addition, excessive fluid, accumulated as a result of disease or trauma, may be drained.

The condition of the arteries of the brain can be determined through a **cerebral angiogram** (*an'je-ŏ-gram*). In this technique, a radiopaque substance is injected into the common carotid arteries and allowed to disperse through the cerebral vessels. Aneurysms and vascular constrictions or displacements by tumors may then be revealed on radiographs.

The development of the **CT scanner,** or **computerized axial tomographic scanner,** has revolutionized the diagnosis of brain disorders. The CT scanner projects a sharply focused, detailed tomogram, or cross section, of a patient's brain onto a television screen. The versatile CT scanner allows quick and accurate diagnoses of tumors, aneurysms, blood clots, and hemorrhage. The CT scanner may also be used to detect certain types of birth defects, brain damage, scar tissue, and evidence of old or recent strokes.

A machine with even greater potential than the CT scanner is the **DSR,** or **dynamic spatial reconstructor.** Like the CT scanner, the DSR is computerized to transform radiographs into composite video images. However, with the DSR, a three-dimensional view is obtained, and the image is produced much faster than with the CT scanner. The DSR can produce 75,000 cross-sectional images in 5 seconds, whereas the CT scanner can produce only one. With that speed, body functions as well as structures may be studied. Blood flow through vessels of the brain can be observed. This type of data is important in detecting early symptoms of a stroke or other disorders.

Certain disorders of the brain may be diagnosed more simply by examining brain-wave patterns using an **electroencephalogram** (see fig. 11.23). Sensitive electrodes placed on the scalp record particular EEG patterns being emitted from evoked cerebral activity. EEG recordings are used to monitor epileptic patients to predict seizures and to determine proper drug therapy, and also to monitor comatose patients.

The fact that the nervous system is extremely sensitive to various drugs is fortunate; at the same time, this sensitivity has potential for disaster. *Drug abuse* is a major clinical concern because of the addictive and devastating effect that certain drugs have on the nervous system. Much has been written on drug abuse, and it is beyond the scope of this text to elaborate on the effects of drugs. A positive aspect of drugs is their administration in medicine to temporarily interrupt the passage or perception of sensory impulses. Injecting an anesthetic drug near a nerve, as in dentistry, desensitizes a specific area and causes a *nerve block*. Nerve blocks of a limited extent occur if an appendage is cooled or if a nerve is compressed for a period of time. Before the discovery of pharmacological drugs, physicians would frequently cool an affected appendage with ice or snow before performing surgery. **General anesthetics** affect the brain and render a person unconscious. A **local anesthetic** causes a nerve block by desensitizing a specific area.

Developmental Problems

Congenital malformations of the CNS are common and frequently involve overlying bone, muscle, and connective tissue. The more severe abnormalities make life impossible, and the less severe malformations frequently result in functional disability. Neurological malformations usually have a genetic basis, but they also may result from environmental factors such as anoxia, infectious agents, drugs, and ionizing radiation. Some of these malformations were briefly described in the previous chapter.

Spina bifida (*spi'nă bif'ĭ-dă*) is a defective fusion of the vertebral elements and may or may not involve the spinal cord. **Spina bifida occulta** is the most common and least serious type of spina bifida. This defect usually involves few vertebrae, is not externally apparent except for perhaps a pigmented spot with a tuft of hair, and usually does not cause neurological disturbances. **Spina bifida cystica,** a severe type of spina bifida, is a saclike protrusion of skin and underlying meninges that may contain portions of the spinal cord and nerve roots. It is most common in the lower thoracic, lumbar, and sacral regions (see fig. 6.41). The position and extent of the defect determines the degree of neurological impairment.

Anencephaly (*an"en-sef'ă-le*) is a markedly defective development of the brain and surrounding cranial bones. Anencephaly occurs once per thousand births and makes sustained extrauterine life impossible. This congenital defect apparently results from the failure of the neural folds at the cranial portion of the neural plate to fuse and form the prosencephalon.

Microcephaly is an uncommon condition in which brain development is not completed. If enough neurological tissue is present, the infant will survive but will be severely mentally retarded.

Defective skull development frequently causes **cranial encephalocele** (*en-sef'ă-lo-sēl*). This condition occurs approximately once per two thousand births. It is characterized by protrusion of the brain and meninges through a cranial fissure, usually in the occipital region. Occasionally the herniation involves fluid with the meninges, and not brain tissue. In this case, it is referred to as a **cranial meningocele** (*mĕ-ning'go-sēl*).

Hydrocephalus (*hi"dro-sef'ă-lus*) is the abnormal accumulation of cerebrospinal fluid in the ventricles and subarachnoid or subdural space. Hydrocephalus may be caused by the excessive production or blocked flow of cerebrospinal fluid. It may also be associated with other congenital problems, such as spina bifida cystica or encephalocele. Hydrocephalus frequently causes the cranial bones to thin and the cerebral cortex to atrophy.

Many congenital disorders cause an impairment of intellectual function known as **mental retardation.** Chromosomal abnormalities, maternal and fetal infections such as syphilis and German measles, and excessive irradiation of the fetus are all commonly associated with mental retardation.

Injuries

Although the brain and spinal cord seem to be well protected within a bony encasement, they are sensitive organs, highly susceptible to injury.

Certain symptomatic terms are used when determining possible trauma within the CNS. **Headaches** are the most common ailment of the CNS. Most headaches are due to dilated blood vessels within the meninges of the brain. Headaches are generally symptomatic of brain disorders; rather, they tend to be associated with physiological stress, eyestrain, or fatigue. Persistent and intense headaches may indicate a more serious problem, such as a brain tumor. A **migraine** is a specific type of headache that is commonly preceded or accompanied by visual impairments and GI unrest. It is not known why only 5%–10% of the population periodically suffer from migraines or why they are more common in women. Fatigue, allergy, and emotional stress tend to trigger migraines.

Fainting is a brief loss of consciousness that may result from a rapid pooling of blood in the lower extremities. It may occur when a person rapidly arises from a reclined position, receives a blow to the head, or experiences an intense psychologic stimulus, such as viewing a cadaver for the first time. Fainting is of more concern when it is symptomatic of a particular disease.

A **concussion** is an injury resulting from a violent jarring of the brain, usually by a forceful blow to the head. Bones of the skull may or may not be fractured. A concussion usually results in a brief period of unconsciousness, followed by mild **delirium** in which the patient is in a state of confusion. **Amnesia** is a more intense disorientation in which the patient suffers varying degrees of memory loss.

A person who survives a severe head injury may be **comatose** for a short or an extended period of time. A coma is a state of unconsciousness from which the patient cannot be aroused, even by the most intense external stimuli. Severe injury

amnesia: L. *amnesia*, forgetfulness
comatose: Gk. *koma*, deep sleep

to the reticular activating system is likely to result in irreversible coma. Although a head injury is the most common cause of coma, chemical imbalances associated with certain diseases (e.g., diabetes) or the ingestion of drugs or poisons may also be responsible.

The flexibility of the vertebral column is essential for body movements, but because of this flexibility the spinal cord and spinal nerves are somewhat vulnerable to trauma. Falls or severe blows to the back are a common cause of injury. A skeletal injury, such as a fracture, dislocation, or compression of the vertebrae, usually traumatizes nervous tissue as well. Other frequent causes of trauma to the spinal cord include gunshot wounds, stabbings, herniated discs, and birth injuries. The consequences of the trauma depend on the location and severity of the injury and the medical treatment the patient receives. If nerve fibers of the spinal cord are severed, motor or sensory functions will be permanently lost.

Paralysis is a permanent loss of motor control, usually resulting from disease or a lesion of the spinal cord or specific nerves. Paralysis of both lower extremities is called **paraplegia.** Paralysis of both the upper and lower extremity on the same side is called **hemiplegia,** and paralysis of all four extremities is **quadriplegia.** Paralysis may be flaccid or spastic. **Flaccid** (*flak'sid*) **paralysis** generally results from a lesion of the anterior horn cells and is characterized by noncontractile muscles that atrophy. **Spastic paralysis** results from lesions of the corticospinal tracts of the spinal cord and is characterized by hypertonicity of the skeletal muscles.

Whiplash is a sudden hyperextension and flexion of the cervical vertebrae (fig. 11.47) such as may occur during a rear-end automobile collision. Recovery of a minor whiplash (muscle and ligament strains) is generally complete, albeit slow. Severe whiplash (spinal cord compression) may cause permanent paralysis to the structures below the level of injury.

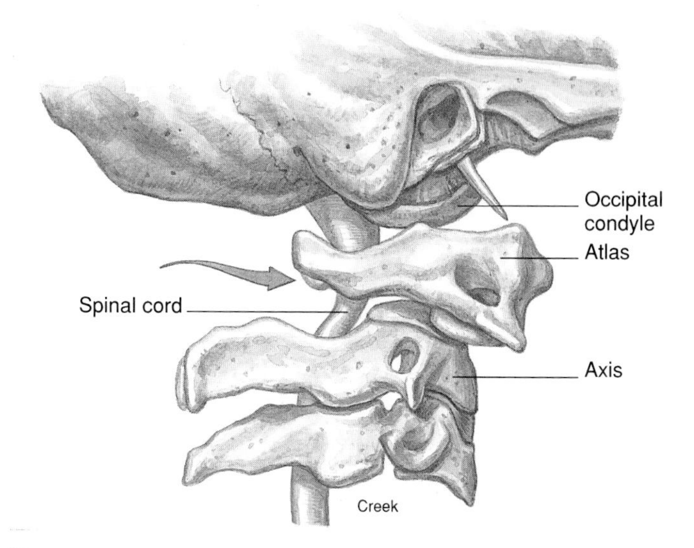

Figure 11.47

Whiplash varies in severity from muscle and ligament strains to dislocation of the vertebrae and compression of the spinal cord. Injuries such as this may cause permanent loss of some or all of the spinal cord functions.

Disorders of the Nervous System

Mental Illness

Mental illness is a major clinical consideration of the nervous system and is perhaps the least understood. Traditionally, mental disorders have been grouped into two broad categories: neurosis and psychosis. In **neurosis,** a maladjustment to certain aspects of life interferes with normal functioning, but contact with reality is maintained. An irrational fear (*phobia*) is an example of neurosis. Neurosis frequently causes intense anxiety or abnormal distress that brings about increased sympathetic stimulation. **Psychosis,** a more serious mental condition, is typified by personality distintegration and a loss of contact with reality. The more common forms of psychosis include *schizophrenia,* in which a person withdraws into a world of fantasy; *paranoia,* in which a person has systematized delusions, often of a persecutory nature; and *manic-depressive psychosis,* in which a person's moods swing widely from intense elation to deepest despair.

Epilepsy

Epilepsy is a relatively common brain disorder with a strong hereditary basis, but it also can be caused by head injuries, tumors, or childhood infectious diseases. It is sometimes idiopathic (without demonstrable cause). A person with epilepsy may periodically experience an *epileptic seizure,* which has various symptoms depending on the type of epilepsy.

The most common kinds of epilepsy are petit mal, psychomotor epilepsy, and grand mal. **Petit mal** (*pet'e-mal'*) occurs almost exclusively in children between the ages of 3 and 12. A child experiencing a petit mal seizure loses contact with reality for 5 to 30 seconds but does not lose consciousness or display convulsions. There may, however, be slight uncontrollable facial gestures or eye movements, and the child will stare, as if in a daydream. During a petit mal seizure, the thalamus and hypothalamus produce an extremely slow EEG pattern of 3 waves per second. Children with petit mal usually outgrow the condition by age 9 or 10 and generally require no medication.

Psychomotor epilepsy is often confused with mental illness because of the symptoms characteristic of the seizure. During such a seizure, EEG activity accelerates in the temporal lobes, causing a person to become disoriented and lose contact with reality. Occasionally during a seizure, specific cerebral motor areas will cause involuntary lip smacking or hand clapping. If motor areas in the brain are not stimulated, a person having a psychomotor epileptic seizure may wander aimlessly until the seizure subsides.

Grand mal is a more serious form of epilepsy characterized by periodic convulsive seizures that generally render a person unconscious. Grand mal epileptic seizures are accompanied by rapid

paralysis: Gk. *paralysis,* loosening
paraplegia: Gk. *para,* beside; *plessein,* to strike

epilepsy: Gk. *epi,* upon; *lepsis,* seize
petit mal: L. *pitinnus,* small child; *malus,* bad

EEG patterns of 25 to 30 waves per second. This sudden increase from the norm of about 10 waves per second may cause extensive stimulation of motor units and, therefore, uncontrollable urinary muscle activity. During a grand mal seizure, a person loses consciousness, convulses, and may lose bladder and bowel control. After a few minutes, the muscles relax and the person awakes but he or she remains disoriented for a short time.

Epilepsy almost never affects intelligence and can be effectively treated with drugs in about 85% of the patients.

Cerebral Palsy

Cerebral palsy is a motor nerve disorder characterized by paresis (partial paralysis) and lack of muscular coordination. It is caused by damage to the motor areas of the brain during prenatal development, birth, or infancy. During neural development within an embryo, radiation or bacterial toxins (such as from German measles) transferred through the placenta of the mother may cause cerebral palsy. Oxygen deprivation resulting from complications at birth and hydrocephalus in a newborn may also cause cerebral palsy. The three areas of the brain most severely affected by this disease are the cerebral cortex, the basal nuclei, and the cerebellum. The type of cerebral palsy is determined by the particular region of the brain that is affected.

Some degree of mental retardation occurs in 60% to 70% of cerebral palsy victims. Partial blindness, deafness, and speech problems frequently accompany this disease. Cerebral palsy is nonprogressive; that is, the impairments do not worsen as a person ages. However, neither are there physical improvements.

Neoplasms of the CNS

Neoplasms of the CNS are either **intracranial tumors,** which affect brain cells or cells associated with the brain, or they are **intravertebral** (intraspinal) **tumors,** which affect cells within or near the spinal cord. **Primary neoplasms** develop within the CNS. Approximately half of these are benign, but they may become lethal because of the pressure they exert on vital centers as they grow. Patients with **secondary,** or **metastatic, neoplasms** within the brain have a poor prognosis because the cancer has already established itself in another body organ—frequently the liver, lung, or breast—and has only secondarily spread to the brain. The symptoms of a brain tumor include headache, convulsions, pain, paralysis, or a change in behavior.

Neoplasms of the CNS are classified according to the tissues in which the cancer occurs. Tumors arising in neuroglia are called **gliomas** (*gli-o'maz*) and account for about half of all primary neoplasms within the brain. Gliomas are frequently spread throughout cerebral tissue, develop rapidly, and usually cause death within a year after diagnosis. **Astrocytomas** (*as"tro-si-to'maz*), **oligodendrogliomas** (*ol"ĭ-go-den"drog-le-o'maz*), and **ependymomas** (*ĕ-pen"dĭ-mo'maz*) are common types of gliomas.

Meningiomas arise from meningeal coverings of the brain and account for about 15% of primary intracranial tumors. Meningiomas are usually harmless if they are treated readily.

Intravertebral tumors are classified as **extramedullary** when they develop on the outside of the spinal cord and as **intramedullary** when they develop within the substance of the spinal cord. Extramedullary neoplasms may cause pain and numbness in body structures distant from the tumor as the growing tumor compresses the spinal cord. An intramedullary neoplasm causes a gradual loss of function below the spinal-segmental level of the affliction.

Methods of detecting and treating cancers within the CNS have greatly improved in the last few years. Early detection and competent treatment have lessened the likelihood of death from these cancers and have reduced the probability of physical impairment.

Dyslexia

Dyslexia is a defect in the language center within the brain. In dyslexia, otherwise intelligent people reverse the order of letters in syllables, of syllables in words, and of words in sentences. The sentence: "The man saw a red dog," for example might be read by the dyslexic as "A red god was the man." Dyslexia is believed to result from the failure of one cerebral hemisphere to respond to written language, perhaps due to structural defects. Dyslexia can usually be overcome by intense remedial instruction in reading and writing.

Meningitis

The nervous system is vulnerable to a variety of organisms and viruses that may cause abscesses or infections. Meningitis is an infection of the meninges. It may be confined to the spinal cord, in which case it is referred to as **spinal meningitis,** or it may involve the brain and associated meninges, in which case it is known as **encephalitis.** When both the brain and spinal cord are involved, the correct term is encephalomyelitis (*en-sef"ă-lo-mi" ĕ-li'tis*). The microorganisms that most commonly cause meningitis are meningococci, streptococci, pneumococci, and tubercle bacilli. Viral meningitis is more serious than bacterial meningitis; nearly 20% of viral encephalitides are fatal. The organisms that cause meningitis probably enter the body through respiratory passageways.

Poliomyelitis

Poliomyelitis, or infantile paralysis, is primarily a childhood disease caused by a virus that destroys nerve cell bodies within the anterior horn of the spinal cord, especially those within the cervical and lumbar enlargements. This degenerative disease is characterized by fever, severe headache, stiffness and pain in the head and back, and the loss of certain somatic reflexes. Muscle paralysis follows within several weeks, and eventually the muscles atrophy. Death results if the virus invades the vasomotor and respiratory nuclei within the medulla oblongata or anterior horn cells controlling respiratory muscles. Poliomyelitis has been effectively controlled with immunization.

Syphilis

Syphilis is a sexually transmitted disease that, if untreated, progressively destroys body organs. When syphilis causes organ degeneration, it is said to be in the *tertiary stage* (10 to 20 years after the primary infection). The organs of the nervous system are frequently infected, causing a condition called **neurosyphilis.** Neurosyphilis is classified according to the tissue involved, and the symptoms vary correspondingly. If the meninges are infected, the condition is termed **chronic meningitis. Tabes** (*ta'bēz*) **dorsalis** is a form of neurosyphilis in which there is a progressive degeneration of the posterior funiculi of the spinal cord and posterior roots of spinal nerves. Motor control is gradually lost and patients eventually become bedridden, unable even to feed themselves.

Senescence of the Nervous System

With respect to the nervous system, little is known about the extent of changes related to normal aging. It has been estimated that perhaps 100,000 neurons die each day of our adult life. Empirical studies, however, show that such claims are unfounded. It is believed that relatively few neural cells are lost during the normal aging process. Neurons are extremely sensitive, however. They are susceptible to various drugs or interruptions of vascular supply such as caused by strokes or other cardiovascular diseases.

There is evidence that senescence alters neurotransmitters. Depression and other age-related conditions, or specific diseases such as Parkinson's disease, may be caused by an imbalance of neurotransmitter chemicals. Changes in sleeping patterns in elderly people also probably result from neurotransmitter problems.

The slowing of the nervous system with age is most apparent in tests of reaction time. It is not certain whether this is a result of a slowing in the transmission of impulses along neurons or in neurotransmitter relays at the synapses.

Although the nervous system does deteriorate with age (in ways that are not well understood), in most people it functions effectively throughout life. Brain dysfunction is not a common characteristic of senescence.

Degenerative Diseases of the Nervous System

Degenerative diseases of the CNS are characterized by a progressive, symmetrical deterioration of vital structures of the brain or spinal cord. The etiologies of these diseases are poorly understood, but it is thought that most of them are genetic.

Cerebrovascular Accident (CVA)

Cerebrovascular accident is the most common disease of the nervous system. It is the third most frequent cause of death in the United States, and perhaps the number one cause of disability. The term **stroke** is frequently used as a synonym for CVA, but actually a stroke refers to the sudden and dramatic appearance of a neurological defect. *Cerebral thrombosis*, in which a thrombus, or clot, forms in an artery of the brain, is the most common cause of CVA. Other causes of CVA include intracerebral hemorrhages, aneurysms, atherosclerosis, and arteriosclerosis of the cerebral arteries.

Patients who recover from CVA frequently suffer partial paralysis and mental disorders, such as loss of language skills. The dysfunction depends upon the severity of the CVA and the regions of the brain that were injured. Patients surviving a CVA can often be rehabilitated, but approximately two-thirds die within 3 years of the initial damage.

Multiple Sclerosis

Multiple sclerosis (MS) is a relatively common neurological disease in people between the ages of 20 and 40. MS is a chronic, degenerating, remitting, and relapsing disease that progressively destroys the myelin layers of neurons in multiple areas of the CNS. Initially, lesions form on the myelin layers and soon develop into hardened *scleroses* or scars (hence the name). The destruction of myelin layers prohibits the normal conduction of impulses, resulting in a progressive loss of functions. Because myelin degeneration is widely distributed, MS has a wider variety of symptoms than any other neurologic disease. This characteristic, coupled with remission, frequently causes the disease to be misdiagnosed.

During the early stages of MS, many patients are believed to be neurotic because of the variability and temporary nature of their symptoms. As the disease progresses, the symptoms may include double vision (diplopia), spots in the visual field, blindness, tremor, numbness of appendages, and locomotor difficulty. Eventually the patient is bedridden, and death may occur anytime from 7 to 30 years after the first symptoms appear.

Syringomyelia

Syringomyelia (*sĭ-ring"go-mi-e'le-ă*) is a relatively uncommon condition characterized by the appearance of cystlike cavities, called *syringes*, within the gray matter of the spinal cord. These syringes progressively destroy the cord from the inside out. As the spinal cord deteriorates, the patient experiences muscular weakness and atrophy and sensory loss, particularly of the senses of pain and temperature. The cause of syringomyelia is unknown.

Tay–Sachs Disease

In Tay-Sachs disease, the myelin sheaths are destroyed by the excessive accumulation of one of the lipid components of the myelin. This results from an enzyme defect caused by the inheritance of genes carried by the parents in a recessive state. The disease is inherited primarily by individuals of Eastern European Jewish descent and appears before the infant is a year old. It causes blindness, loss of mental and motor ability, and ultimately death by the age of 3. Potential parents can tell if they are carriers for this condition by the use of a special blood test for the defective enzyme.

---

multiple sclerosis: L. *multiplus*, many parts; Gk. *skleros*, hardened

Tay–Sachs disease: from Warren Tay, English physician, 1843–1927, and Bernard Sachs, American neurologist, 1858–1944

Internal Affairs

■ How the *Nervous System* Works with Other Body Systems
□ How Other Body Systems Work with the *Nervous System*

Integumentary System

- Influences secretions from integumentary glands and contractions of arrector pili muscles
- Controls continuous blood flow to regulate heat loss

- Supports and protects peripheral receptors
- Provides sensations of heat, cold, pressure, pain, and vibration

Skeletal System

- Innervates bones and monitors movements within joints
- Generates muscle tension needed for bone growth and maintenance

- Supports and protects the brain and spinal cord
- Stores calcium needed for neural function

Muscular System

- Innervates muscles for autonomic and voluntary muscle contractions

- Generates body heat to maintain constant temperature for neural function
- Proprioceptors transmit impulses from muscles to the brain
- Facial muscles express emotional state

Endocrine System

- Innervates endocrine glands causing rapid, autonomic secretion of hormones
- Hypothalamus controls the pituitary gland: sympathetic nervous system stimulates the adrenal medulla

- Hormones augment and sustain autonomic stimuli to body organs

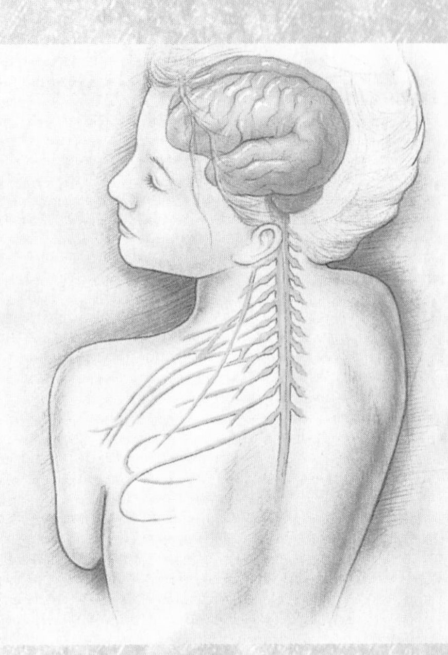

Circulatory System

- Innervates the heart and blood vessels to modify heart rate, blood vessel diameters, blood pressure, and routing of blood

- Transports O_2 and CO_2, nutrients, and fluids to and from the brain and spinal cord
- Cerebrospinal fluid is produced from and returned to the blood

Lymphatic System

- Innervates lymphoid organs
- Plays a role in regulating the immune response

- Protects against infections within the brain and spinal cord

Respiratory System

- Respiratory centers within the brain stem control respiratory rates and depth of respiration

- Provides O_2 and eliminates CO_2

Digestive System

- Innervates digestive organs and autonomically regulates GI tract movements and secretions
- Regulates feeding behavior and defecation

- Provides nutrients for growth, maintenance, and repair of the nervous system

Urinary System

- Innervates organs of urinary system to control urination
- Modifies renal blood pressure

- Eliminates metabolic wastes
- Regulates pH, body fluids, and electrolyte concentrations

Reproductive System

- Innervates reproductive organs to control sexual function

- Gonads produce sex hormones that influence brain development and sexual behavior

Diseases Involving Neurotransmitters

Parkinson's disease, or **paralysis agitans,** is a major cause of neurological disability in people over 60 years of age. It is a progressive degenerative disease of unknown cause. Nerve cells within the substantia nigra, an area within the basal nuclei of the brain, are destroyed. This causes muscle tremors, muscular rigidity, speech defects, and other severe problems. The symptoms of this disease can be partially treated by altering the neurotransmitter status of the brain. Patients are given L-dopa to increase the production of dopamine in the brain, and may also be given anticholinergic drugs to decrease the production of acetylcholine.

Alzheimer's disease is the most common cause of dementia, often beginning in middle age and producing progressive mental deterioration. The cause of Alzheimer's disease is unknown, but evidence suggests that it is associated with the decreased ability of the brain to produce acetylcholine. Attempts to increase ACh production by increased ingestion of precursor molecules (choline or lecithin) have thus far not been successful. Drugs that block acetylcholinesterase, an enzyme that inactivates ACh, offer promise but are so far still in the experimental stage.

At least some psychiatric disorders may be produced by dysfunction of neurotransmitters. There is evidence that **schizophrenia** is associated with hyperactivity of the neurons that use dopamine as a neurotransmitter. Drugs that are effective in the treatment of schizophrenia (e.g., chlorpromazine) act by blocking dopamine receptor proteins. **Depression** is associated with decreased activity of the neurons that use monoamines—norepinephrine and serotonin—as neurotransmitters. Antidepressant drugs enhance the action of these neurotransmitters. In a similar way, barbiturates and benzodiazepine (e.g., Valium), which decrease **anxiety,** act by enhancing the action of the neurotransmitter GABA in the CNS.

Parkinson's disease: From James Parkinson, British physician, 1755–1824
Alzheimer's disease: from Alois Alzheimer, German neurologist, 1864–1915

Clinical Case Study Answer

The affected upper motor neurons have cell bodies that reside in the right cerebral hemisphere. They give rise to fibers that, as they course downward, cross in the medulla oblongata to the left side of the brain stem and spinal cord. They continue on the left side, eventually synapsing at the appropriate level with lower motor neurons in the anterior horn of the spinal cord. The lower motor neurons then give rise to fibers that travel to the periphery, where they innervate end organs; namely, muscle cells in the left side of the body. Since the patient's neurological deficits are all motor as opposed to sensory, the tumor is most likely located in the right frontal lobe. The parietal lobe contains sensory neurons. The persistent headache is due to the pressure of the tumorous mass on the meninges, which are heavily sensory innervated.

Chapter Summary

Organization and Functions of the Nervous System (pp. 334–338)

1. The central nervous system (CNS) consists of the brain and spinal cord and contains gray and white matter. It is covered with meninges and bathed in cerebrospinal fluid.
2. The functions of the nervous system include orientation, coordination, assimilation, and programming of instinctual behavior.

Neurons and Neuroglia (pp. 338–346)

1. Neurons are the basic structural and functional units of the nervous system. Specialized cells called neuroglia provide structural and functional support for the activities of neurons.
2. A neuron contains dendrites, a cell body, and an axon.
   (a) The cell body contains the nucleus, chromatophilic substances, neurofibrils, and other organelles.
   (b) Dendrites receive stimuli and the axon conducts nerve impulses away from the cell body.
3. Neuroglia are of six types: neurolemmocytes form myelin layers around axons in the PNS; oligodendrocytes form myelin layers around axons in the CNS; microglia perform a phagocytic function in the CNS; astrocytes regulate passage of substances from the blood to the CNS; ependymal cells assist the movement of cerebrospinal fluid in the CNS; and ganglionic gliocytes support neuron cell bodies in the PNS.
   (a) The neuroglia that surround an axon form a covering called a myelin layer.
   (b) Myelinated neurons have limited capabilities for regeneration following trauma.
4. A nerve is a collection of dendrites and axons in the PNS.
   (a) Sensory (afferent) neurons are pseudounipolar.
   (b) Motor (efferent) neurons are multipolar.
   (c) Association (interneurons) are located entirely within the CNS.
   (d) Somatic motor nerves innervate skeletal muscle; visceral motor (autonomic) nerves innervate smooth muscle, cardiac muscle, and glands.

Transmission of Impulses (pp. 346–348)

1. Irritability and conductivity are properties of neurons that permit nerve impulse transmission.
2. Neurotransmitters facilitate synaptic impulse transmission.

General Features of the Brain (pp. 348–352)

1. The brain, composed of gray matter and white matter, is protected by meninges and is bathed in cerebrospinal fluid.
2. About 750 ml of blood flows to the brain each minute.

Cerebrum (pp. 352–361)

1. The cerebrum, consisting of two convoluted hemispheres, is concerned with higher brain functions, such as the perception of sensory impulses, the instigation of voluntary movement, the storage of memory, thought processes, and reasoning ability.
2. The cerebral cortex is convoluted with gyri and sulci.
3. Each cerebral hemisphere contains frontal, parietal, temporal, and occipital lobes. The insula lies deep within the cerebrum and cannot be seen in an external view.
4. Brain waves generated by the cerebral cortex are recorded as an electroencephalogram and may provide valuable diagnostic information.
5. The white matter of the cerebrum consists of association, commissural, and projection fibers.
6. Basal nuclei are specialized masses of gray matter located within the white matter of the cerebrum.

Diencephalon (pp. 361–363)

1. The diencephalon is a major autonomic region of the brain.
2. The thalamus is an ovoid mass of gray matter that functions as a relay center for sensory impulses and responds to pain.
3. The hypothalamus is an aggregation of specialized nuclei that regulate many visceral activities. It also performs emotional and instinctual functions.
4. The epithalamus contains the pineal gland and the vascular choroid plexus over the roof of the third ventricle.

Mesencephalon (pp. 363–364)

1. The mesencephalon contains the corpora quadrigemina, the cerebral peduncles, and specialized nuclei that help to control posture and movement.
2. The superior colliculi of the corpora quadrigemina are concerned with visual reflexes and the inferior colliculi are concerned with auditory reflexes.
3. The red nucleus and the substantia nigra are concerned with motor activities.

Metencephalon (pp. 364–366)

1. The pons consists of fiber tracts connecting the cerebellum and medulla oblongata to other structures of the brain. The pons also contains nuclei for certain cranial nerves and the regulation of respiration.
2. The cerebellum consists of two hemispheres connected by the vermis and supported by three paired cerebellar peduncles.
   (a) The cerebellum is composed of a white matter tract called the arbor vitae, surrounded by a thin convoluted cortex of gray matter.
   (b) The cerebellum is concerned with coordinated contractions of skeletal muscle.

Myelencephalon (pp. 366–367)

1. The medulla oblongata is composed of the ascending and descending tracts of the spinal cord and contains nuclei for several autonomic functions.
2. The reticular formation functions as the reticular activating system in arousing the cerebrum.

Meninges of the Central Nervous System (pp. 367–369)

1. The cranial dura mater consists of an outer periosteal layer and an inner meningeal layer. The spinal dura mater is a single layer surrounded by the vascular epidural space.
2. The arachnoid is a netlike meninx surrounding the subarachnoid space, which contains cerebrospinal fluid.
3. The thin pia mater adheres to the contours of the CNS.

Ventricles and Cerebrospinal Fluid (pp. 370–372)

1. The lateral (first and second), third, and fourth ventricles are interconnected chambers within the brain that are continuous with the central canal of the spinal cord.
2. These chambers are filled with cerebrospinal fluid, which also flows throughout the subarachnoid space.
3. Cerebrospinal fluid is continuously formed by the choroid plexuses from blood plasma and is returned to the blood at the arachnoid villi.
4. The blood-brain barrier determines which substances within blood plasma can enter the extracellular fluid of the brain.

Spinal Cord (pp. 372–379)

1. The spinal cord is composed of 31 segments, each of which gives rise to a pair of spinal nerves.

(a) It is characterized by a cervical enlargement, a lumbar enlargement, and two longitudinal grooves that partially divide it into right and left halves.
(b) The conus medullaris is the terminal portion of the spinal cord, and the cauda equina are nerve roots that radiate interiorly from that point.

2. Ascending and descending spinal cord tracts are referred to as funiculi.
(a) Descending tracts are grouped as either corticospinal (pyramidal) or extrapyramidal.
(b) Many of the fibers in the funiculi decussate (cross over) in the spinal cord or in the medulla oblongata of the brain stem.

Review Activities

Objective Questions

Match the following structures of the brain to the region in which they are located:

1. cerebellum	(a) telencephalon
2. cerebral cortex	(b) diencephalon
3. medulla oblongata	(c) mesencephalon
	(d) metencephalon
	(e) myelencephalon

4. The neuroglial cells that form myelin sheaths in the peripheral nervous system are
(a) oligodendrocytes.
(b) ganglionic gliocytes.
(c) neurolemmocytes.
(d) astrocytes.
(e) microglia.
5. A collection of neuron cell bodies located outside the CNS is called
(a) a tract. (c) a nucleus.
(b) a nerve. (d) a ganglion.
6. Which of the following types of neurons are pseudounipolar?
(a) sensory neurons
(b) somatic motor neurons
(c) neurons in the retina
(d) autonomic motor neurons
7. Depolarization of an axon is produced by the movement of
(a) Na^+ into the axon and K^+ out of the axon.
(b) Na^+ into the axon to bond with K^+.
(c) K^+ into the axon and Na^+ out of the axon.
(d) Na^+ and K^+ within the axon toward the axon terminal.
8. The principal connection between the cerebral hemispheres is
(a) the corpus callosum.
(b) the pons.
(c) the intermediate mass.
(d) the vermis.
(e) the precentral gyrus.
9. The structure of the brain that is most directly involved in the autonomic response to pain is
(a) the pons.
(b) the hypothalamus.
(c) the medulla oblongata.
(d) the thalamus.
10. Which statement is *false* concerning the basal nuclei?
(a) They are located within the cerebrum.
(b) They regulate the basal metabolic rate.
(c) They consist of the caudate nucleus, lentiform nucleus, putamen, and globus pallidus.
(d) They indirectly exert an inhibitory influence on lower motor neurons.
11. The corpora quadrigemina, red nucleus, and substantia nigra are structures of
(a) the diencephalon.
(b) the metencephalon.
(c) the mesencephalon.
(d) the myelencephalon.
12. The fourth ventricle is contained within
(a) the cerebrum.
(b) the cerebellum.
(c) the midbrain.
(d) the metencephalon.

Essay Questions

1. Describe the formation of the neural crest and explain how derivatives of neural crest tissue can be located in different parts of the body.
2. Diagram and label a neuron. Beside each label, list the function of the identified structure. Why are neurons considered the basic functional units of the nervous system?
3. What is meant by a myelinated neuron? Describe how an injured nerve fiber may regenerate.
4. List the six principal types of neuroglia and discuss the location, structure, and function of each.
5. What is a nerve impulse and how is it generated? Why is it called an all-or-none response?
6. What is a synapse and what is the role of a neurotransmitter in relaying a nerve impulse?
7. List the types of brain waves recorded on an electroencephalogram and explain the diagnostic value of each.
8. List the functions of the hypothalamus. Why is the hypothalamus considered a major part of the autonomic nervous system?
9. What structures are located within the midbrain? List the nuclei located in the midbrain and state the function of each.
10. Describe the location and structure of the medulla oblongata. List the nuclei contained within it. What are the functions of the medulla oblongata?
11. What is cerebrospinal fluid? Explain how it is produced and describe its path of circulation?
12. What do EEG, ANS, CSF, PNS, RAS, CT scan, MS, DSR, and CVA stand for?
13. What are some of the psychological terms used to describe mental illness? Define these terms.
14. What is epilepsy? What causes it and how is it controlled?
15. What do meningitis, poliomyelitis and neurosyphilis have in common? How do these conditions differ?

Critical-Thinking Questions

1. In a patient's case study, any observed or suspected structural/functional abnormalities of body structures are reported. Prepare a brief case study of a patient who has suffered severe trauma to the medulla oblongata from a blow to the back of the skull.
2. A young man develops weakness in his legs over the course of several days, which worsens until he can no longer walk. He also loses urinary bladder control and complains of loss of sensation from the umbilicus down. Locate the probable site of neurological impairment.
3. Electrical stimulation of the cerebellum or the basal nuclei can produce skeletal movements. How would damage to these two regions of the brain affect skeletal muscle function differently?
4. If an entire cerebral hemisphere is destroyed, a person can still survive. Yet, damage to the medulla oblongata, a much smaller mass of tissue, can be fatal. Explain this difference.
5. A seizure occurs when abnormal electrical activity overwhelms the brain's normal function. A *focal seizure* originates from a specific irritable tissue, such as a tumorous mass, and causes isolated muscle jerking or sensory abnormalities. Immediately before a focal seizure, the sufferer frequently perceives a fleeting sensation, called an aura, that suggests the origin of the electrical burst. Using your knowledge of cerebral lobe function, predict the origin of a seizure that was preceded by the perception of a foul odor? A flash of light? A painful hand? A familiar song?
6. Explain how meningitis contracted through the meninges over the roof of the nose may be detected from a spinal tap performed in the lumbar region.

TWELVE

Peripheral Nervous System

Clinical Case Study 389
Introduction to the Peripheral Nervous System 389
Cranial Nerves 389
- Structure and Function of Cranial Nerves 389
- Neurological Assessment of the Cranial Nerves 398

Spinal Nerves 400
Nerve Plexuses 401
- Cervical Plexus 401
- Brachial Plexus 401
- Lumbar Plexus 404
- Sacral Plexus 405

Reflex Arc and Reflexes 410
- Components of the Reflex Arc 411
- Kinds of Reflexes 411

Clinical Case Study Answer 414
Developmental Exposition: The Peripheral Nervous System 416
Chapter Summary 417
Review Activities 417

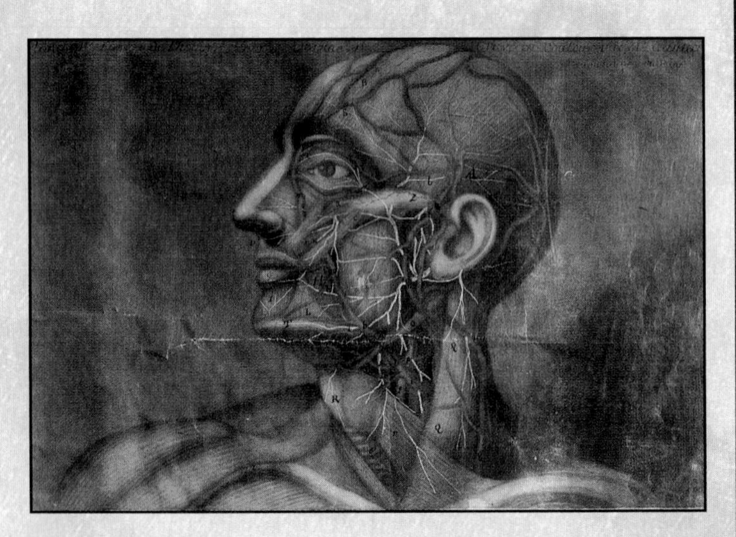

This old medical illustration (undated) shows part of the intricate web of cranial nerves (light yellow) supplying the left side of the head. Some of the principal arteries and veins are also shown. Galen's pronouncement that there were only seven pairs of cranial nerves was accepted throughout the Middle Ages. Actually, there are twelve pairs.

Clinical Case Study

Following an auto accident, a 23-year-old male was brought to the emergency room for treatment of a fractured right humerus. Although the skin was not broken, there was an obvious deformity caused by an angulated fracture at the midshaft. While conducting an examination on the patient's injured arm, the attending orthopedist noticed that the patient was unable to extend the joints of his hand.

What structure could be injured in the brachial region of this patient that would account for his inability to extend his hand? List the muscles that would be affected and describe the movements that would be diminished. Do you think there might be other neurological defects? Explain.

Hints: Because the nervous system functions to coordinate body movement, nerve trauma may be expressed in structures far removed from the site of injury. Carefully read the section dealing with the brachial plexus.

Introduction to the Peripheral Nervous System

The peripheral nervous system consists of all of the nervous tissue outside the central nervous system, including sensory receptors, nerves and their associated ganglia, and nerve plexuses. It provides a communication pathway for impulses traveling between the CNS and the rest of the body.

Objective 1 Define *peripheral nervous system* and distinguish between sensory and mixed nerves.

The **peripheral nervous system (PNS)** is that portion of the nervous system outside the central nervous system. The PNS conveys impulses to and from the brain and spinal cord. Sensory receptors within the sensory organs, nerves, ganglia, and plexuses are all part of the PNS, which serves virtually every part of the body (fig. 12.1). The sensory receptors are discussed in chapter 15.

The nerves of the PNS are classified as cranial nerves or spinal nerves depending on whether they arise from the brain or the spinal cord. A cross section of a spinal nerve is shown in figure 12.2. The terms *sensory nerve*, *motor nerve*, and *mixed nerve* relate to the direction in which the nerve impulses are being conducted. **Sensory nerves** consists of sensory (afferent) neurons that convey impulses toward the CNS. **Motor nerves** consist primarily of motor (efferent) neurons that convey impulses away from the CNS. (Technically speaking, there are no nerves that are motor only; all motor nerves contain some proprioceptor fibers that convey sensory information to the CNS.) **Mixed nerves** are composed of both sensory and motor neurons in about equal numbers, and they convey impulses both to and from the CNS.

1. The tongue responds to tastes and pain and moves to manipulate food. Make a quick sketch of the brain and the tongue to depict the relationship between the CNS and the PNS. Use lines and arrows to indicate the sensory and motor innervation of the tongue. Define *mixed nerve*. What kinds of sensory stimulation arise from the tongue? What type of response is caused by motor stimulation to the tongue?
2. List the structures of the nervous system that are considered part of the PNS.

Cranial Nerves

Twelve pairs of cranial nerves emerge from the inferior surface of the brain and pass through the foramina of the skull to innervate structures in the head, neck, and visceral organs of the trunk.

Objective 2 List the 12 pairs of cranial nerves and describe the location and function of each.

Objective 3 Describe the clinical methods for determining cranial nerve dysfunction.

Structure and Function of the Cranial Nerves

Of the 12 pairs of cranial nerves, 2 pairs arise from the forebrain and 10 pairs arise from the midbrain and brain stem (fig. 12.3). The cranial nerves are designated by Roman numerals and names. The Roman numerals refer to the order in which the nerves are positioned from the front of the brain to the back. The names indicate the structures innervated or the principal functions of the nerves. A summary of the cranial nerves is presented in table 12.1.

Although most cranial nerves are mixed, some are associated with special senses and consist of sensory neurons only. The cell bodies of sensory neurons are located in ganglia outside the brain.

Generations of anatomy students have used a mnemonic device to help them remember the order in which the cranial nerves emerge from the brain: "On old Olympus's towering top, a Finn and German viewed a hop." The initial letter of each word in this jingle corresponds to the initial letter of each pair of cranial nerves. A problem with this classic verse is that the eighth cranial nerve represented by *and* in the jingle, which used to be referred to as auditory, is currently recognized as the vestibulocochlear nerve. Hence, the following topical mnemonic: "On old Olympus's towering top, a fat vicious goat vandalized a hat."

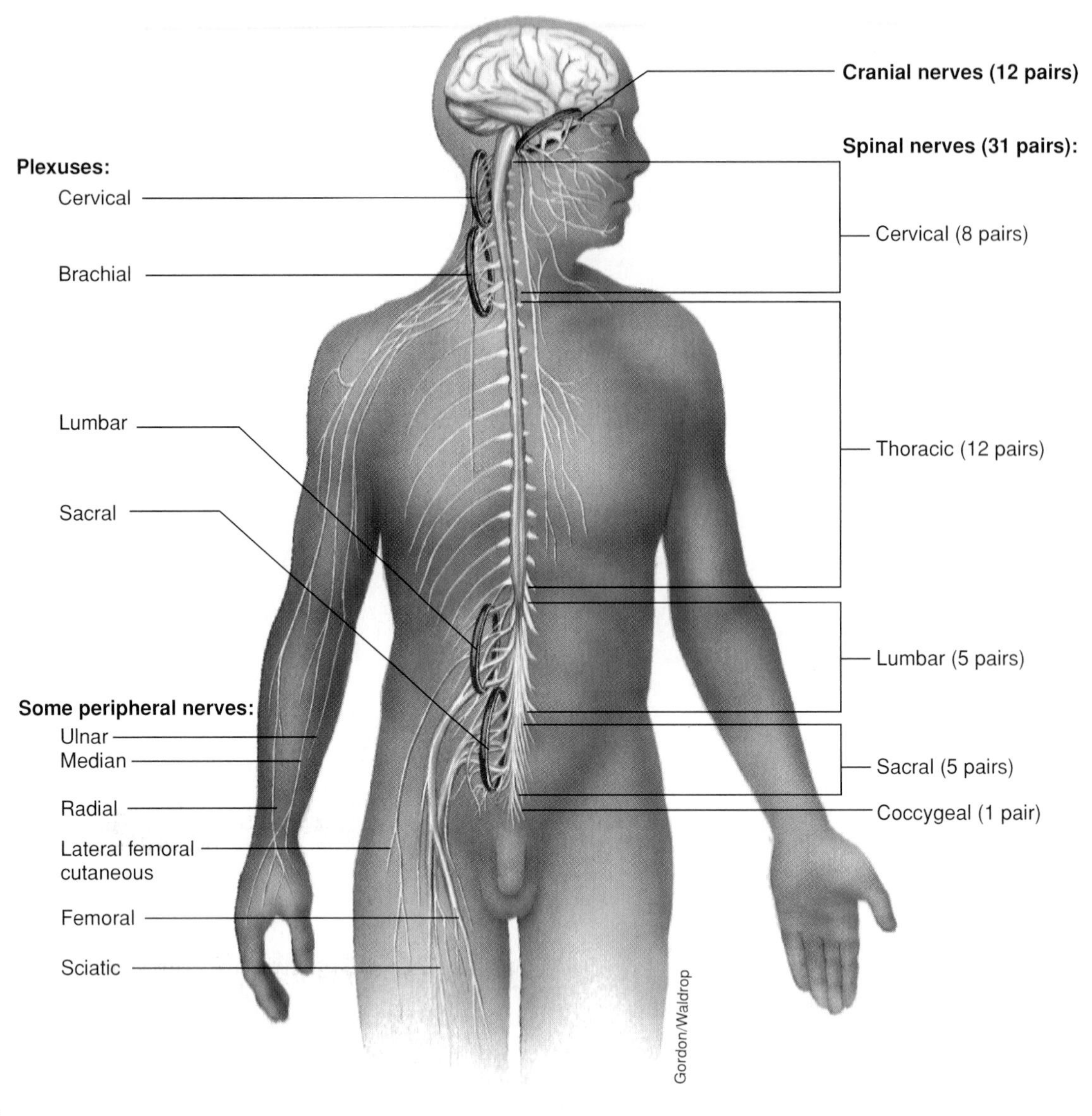

Figure 12.1

The peripheral nervous system includes cranial nerves and spinal nerves, and the nerves that arise from them. Plexuses and ganglia (not shown) are also part of the peripheral nervous system.

I Olfactory Nerve

Actually, numerous olfactory nerves relay sensory impulses of smell from the mucous membranes of the nasal cavity (fig. 12.4). Olfactory nerves are composed of bipolar neurons that function as *chemoreceptors,* responding to volatile chemical particles breathed into the nasal cavity. The dendrites and cell bodies of olfactory neurons are positioned within the mucosa, primarily that which covers the superior nasal conchae and adjacent nasal septum. The axons of these neurons pass through the cribriform plate of the ethmoid bone to the **olfactory bulb** where synapses are made, and the sensory impulses travel through the **olfactory tract** to the primary olfactory area in the cerebral cortex.

olfactory: L. *olfacere,* smell out

II Optic Nerve

The optic nerve, another sensory nerve, conducts impulses from the *photoreceptors* (rods and cones) in the retina of the eye. Each optic nerve is composed of an estimated 1.25 million nerve fibers that converge at the back of the eyeball and enter the cranial cavity through the optic canal. The two optic nerves unite on the floor of the diencephalon to form the **optic chiasma** (*ki-as'mă*) (fig. 12.5). Nerve fibers that arise from the medial half of each retina cross at the optic chiasma to the opposite side of the brain, whereas fibers arising from the lateral half remain on the same side of the brain. The optic nerve fibers pass posteriorly from the optic chiasma to the thalamus via the **optic tracts.** In the thalamus, a majority of the fibers

optic: L. *optica,* see
chiasma: Gk. *chiasma,* an X-shaped arrangement

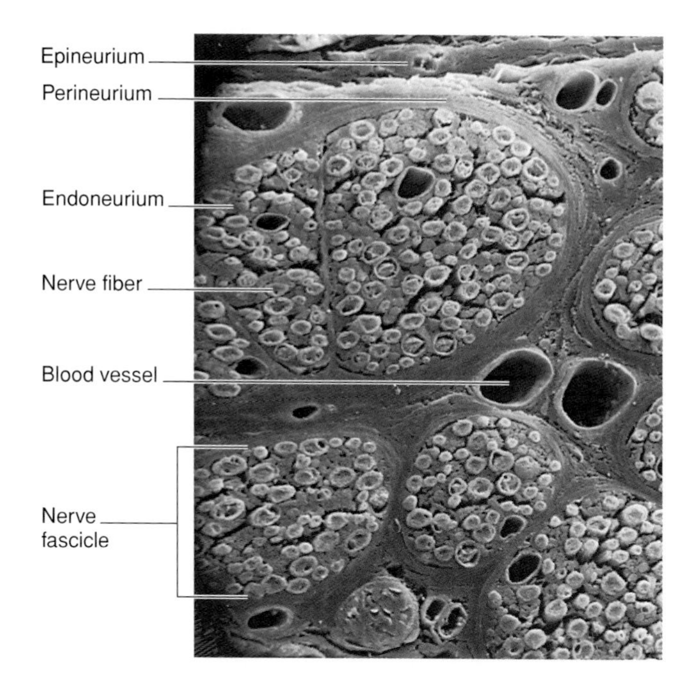

Figure 12.2

A scanning electron micrograph of a spinal nerve seen in cross section (about 1,000×).
(*From: R. G. Kessel and R. H. Kardon,* Tissues and Organs: A Text-Atlas of Scanning Electron Microscopy, *© 1979 W. H. Freeman and Company.*)

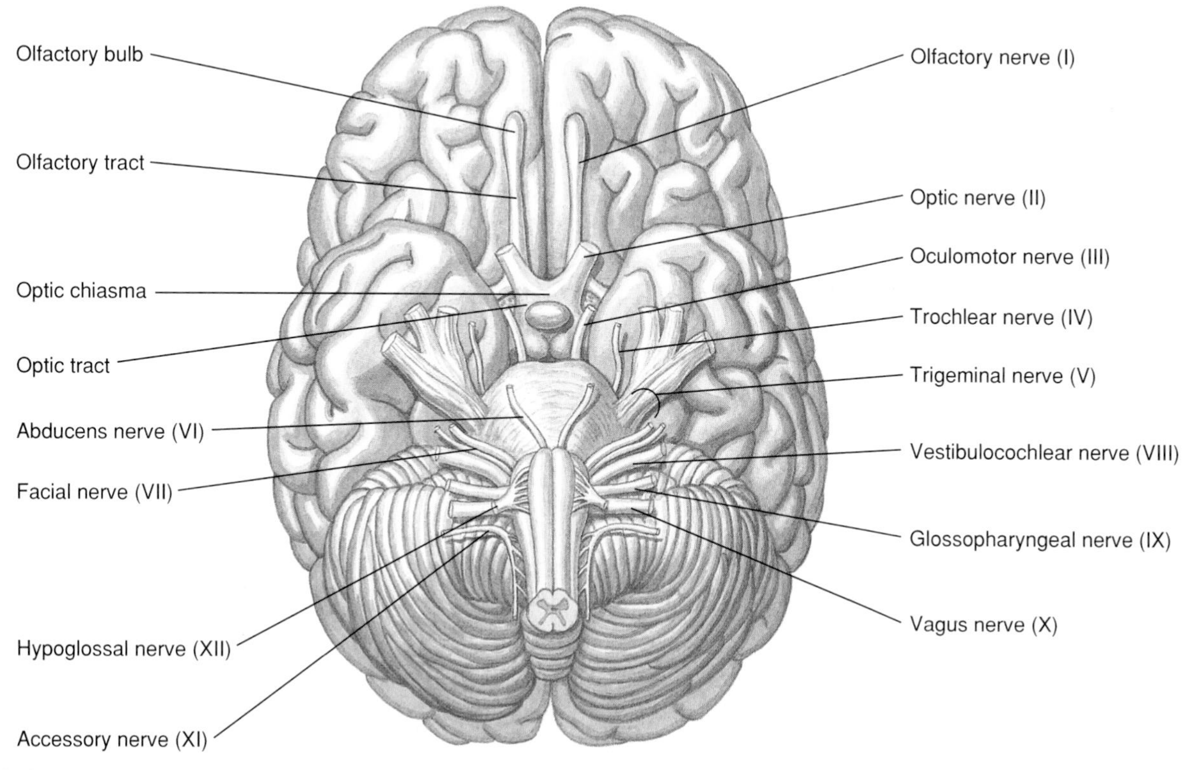

Figure 12.3

The cranial nerves. With the exception of the olfactory nerves, each cranial nerve is composed of a bundle of nerve fibers. The olfactory nerves are minute and diffuse strands of nerve fibers that attach to the olfactory bulb (see fig. 12.4).

Table 12.1

Summary of Cranial Nerves

Number and Name	Foramen Transmitting	Composition	Location of Cell Bodies	Function
I Olfactory	Foramina in cribriform plate of ethmoid bone	Sensory	Bipolar cells in nasal mucosa	Olfaction
II Optic	Optic canal	Sensory	Ganglion cells of retina	Vision
III Oculomotor	Superior orbital fissure	Somatic motor	Oculomotor nucleus	Motor impulses to levator palpebrae superioris and extrinsic eye muscles, except superior oblique and lateral rectus
		Motor: parasympathetic		Innervation to muscles that regulate amount of light entering eye and that focus the lens
		Sensory: proprioception		Proprioception from muscles innervated with motor fibers
IV Trochlear	Superior orbital fissure	Somatic motor	Trochlear nucleus	Motor impulses to superior oblique muscle of eyeball
		Sensory: proprioception		Proprioception from superior oblique muscle of eyeball
V Trigeminal				
Ophthalmic nerve	Superior orbital fissure	Sensory	Trigeminal ganglion	Sensory impulses from cornea, skin of nose, forehead, and scalp
Maxillary nerve	Foramen rotundum	Sensory	Trigeminal ganglion	Sensory impulses from nasal mucosa, upper teeth and gums, palate, upper lip, and skin of cheek
Mandibular nerve	Foramen ovale	Sensory	Trigeminal ganglion	Sensory impulses from temporal region, tongue, lower teeth and gums, and skin of chin and lower jaw
		Sensory: proprioception		Proprioception from muscles of mastication
		Somatic motor	Motor trigeminal nucleus	Motor impulses to muscles of mastication and muscle that tenses the tympanum
VI Abducens	Superior orbital fissure	Somatic motor	Abducens nucleus	Motor impulses to lateral rectus muscle of eyeball
		Sensory: proprioception		Proprioception from lateral rectus muscle of eyeball
VII Facial	Stylomastoid foramen	Somatic motor	Motor facial nucleus	Motor impulses to muscles of facial expression and muscle that tenses the stapes
		Motor: parasympathetic	Superior salivatory nucleus	Secretion of tears from lacrimal gland and salivation from sublingual and submandibular glands
		Sensory	Geniculate ganglion	Sensory impulses from taste buds on anterior two-thirds of tongue; nasal and palatal sensation
		Sensory: proprioception		Proprioception from muscles of facial expression
VIII Vestibulocochlear	Internal acoustic meatus	Sensory	Vestibular ganglion	Sensory impulses associated with equilibrium
			Spiral ganglion	Sensory impulses associated with hearing
IX Glossopharyngeal	Jugular foramen	Somatic motor	Nucleus ambiguus	Motor impulses to muscles of pharynx used in swallowing
		Sensory: proprioception	Petrosal ganglion	Proprioception from muscles of pharynx
		Sensory	Petrosal ganglion	Sensory impulses from taste buds on posterior one-third of tongue, pharynx, middle-ear cavity, and carotid sinus
		Parasympathetic	Inferior salivatory nucleus	Salivation from parotid gland
X Vagus	Jugular foramen	Somatic motor	Nucleus ambiguus	Contraction of muscles of pharynx (swallowing) and larynx (phonation)
		Sensory: proprioception		Proprioception from visceral muscles
		Sensory	Nodose ganglion	Sensory impulses from taste buds on rear of tongue; sensations from auricle of ear; general visceral sensations
		Motor: parasympathetic	Dorsal motor nucleus	Motor impulses to visceral muscles

On Old Olympus' towering top, a fat vicious goat vandalized a hat

Table 12.1
Continued

Number and Name	Foramen Transmitting	Composition	Location of Cell Bodies	Function
XI Accessory	Jugular foramen	Somatic motor	Nucleus ambiguus	Laryngeal movement; soft palate
			Accessory nucleus	Motor impulses to trapezius and sternocleidomastoid muscles for movement of head, neck, and shoulders
		Sensory: proprioception		Proprioception from muscles that move head, neck, and shoulders
XII Hypoglossal	Hypoglossal canal	Somatic motor	Hypoglossal nucleus	Motor impulses to intrinsic and extrinsic muscles of tongue and infrahyoid muscles
		Sensory: proprioception		Proprioception from muscles of tongue

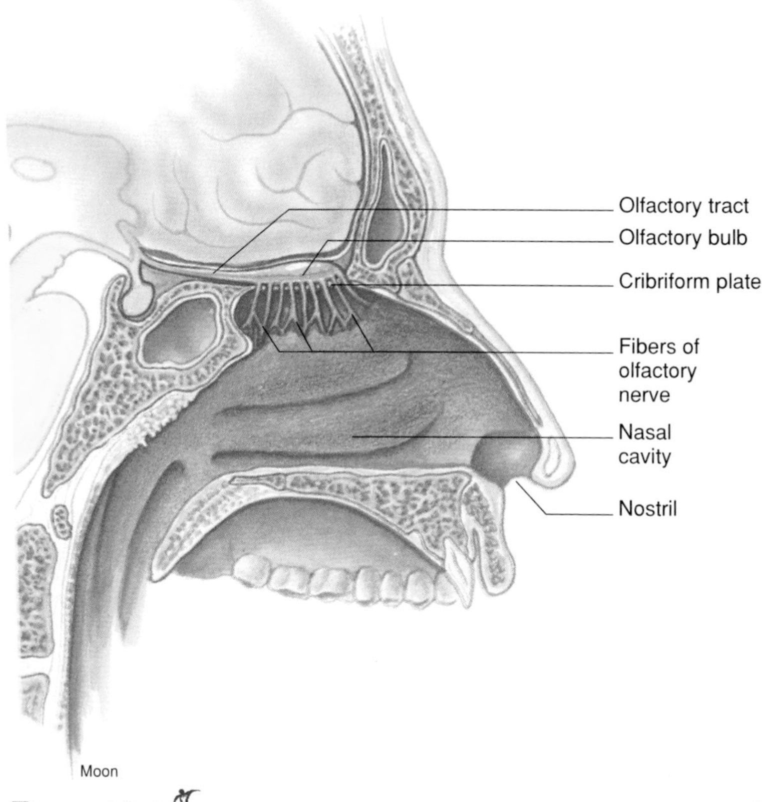

Figure 12.4

The olfactory nerve.

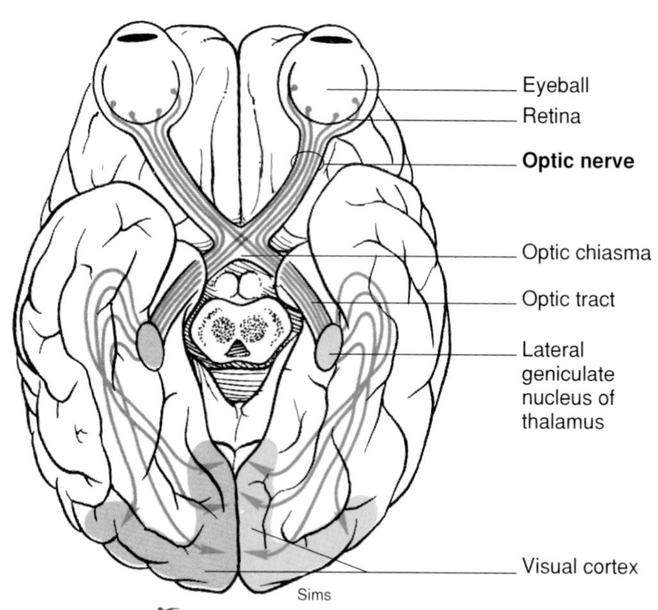

Figure 12.5

The optic nerve and visual pathways.

terminate within certain thalamic nuclei. A few of the ganglion-cell axons that reach the thalamic nuclei have collaterals that convey impulses to the superior colliculi. Synapses within the thalamic nuclei, however, permit impulses to pass through neurons to the **visual cortex** within the occipital lobes. Other synapses permit impulses to reach the nuclei for the oculomotor, trochlear, and abducens nerves, which regulate intrinsic (internal) and extrinsic (from orbit to eyeball) eye muscles. The visual pathway into the eyeball functions reflexively to produce motor responses to light stimuli. If an optic nerve is damaged, the eyeball served by that nerve is blinded.

III Oculomotor

Nerve impulses through the oculomotor nerve produce certain extrinsic and intrinsic movements of the eyeball. The oculomotor is primarily a motor nerve that arises from nuclei within the midbrain. It divides into superior and inferior branches as it passes

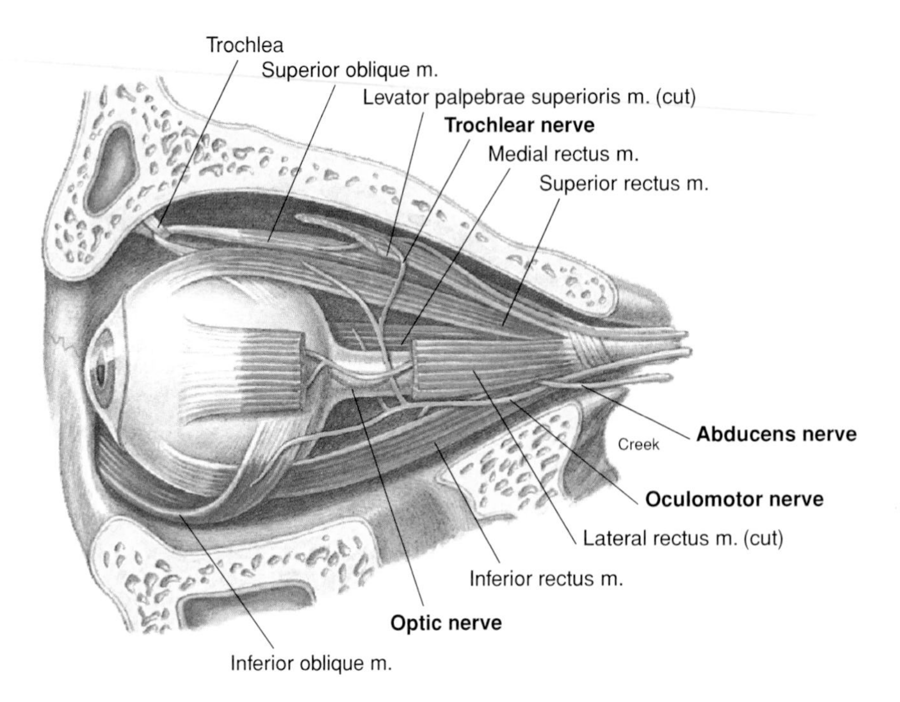

Figure 12.6

The optic nerve, which provides sensory innervation to the eye, and the oculomotor, trochlear, and abducens nerves, which provide motor innervation to the extrinsic eye muscles.

through the superior orbital fissure in the orbit (fig. 12.6). The superior branch innervates the **superior rectus** muscle, which moves the eyeball superiorly, and the **levator palpebrae** (*le-va'tor pal'pĕ-bre*) **superioris** muscle, which raises the upper eyelid. The inferior branch innervates the **medial rectus, inferior rectus,** and **inferior oblique** eye muscles for medial, inferior, and superior and lateral movement of the eyeball, respectively. In addition, fibers from the inferior branch of the oculomotor nerve enter the eyeball to supply autonomic motor innervation to the intrinsic smooth muscles of the iris for pupil constriction and to the muscles within the ciliary body for lens accommodation.

A few sensory fibers of the oculomotor nerve originate from proprioceptors within the intrinsic muscles of the eyeball. These fibers convey impulses that affect the position and activity of the muscles they serve. A person whose oculomotor nerve is damaged may have a drooping upper eyelid or dilated pupil, or be unable to move the eyeball in the directions permitted by the four extrinsic muscles innervated by this nerve.

IV Trochlear

The trochlear (*trok'le-ar*) nerve is a very small mixed nerve that emerges from a nucleus within the midbrain and passes from the cranium through the superior orbital fissure of the orbit. The trochlear nerve innervates the **superior oblique** muscle of the eyeball with both motor and sensory fibers (fig. 12.6). Motor impulses to the superior oblique cause the eyeball to rotate downward and away from the midline. Sensory impulses originate in proprioceptors of the superior oblique muscle and provide information about its position and activity. Damage to the trochlear nerve impairs movement in the direction permitted by the superior oblique eye muscle.

V Trigeminal

The large trigeminal (*tri-jem'in-al*) nerve is a mixed nerve with motor functions originating from the nuclei within the pons and sensory functions terminating in nuclei within the midbrain, pons, and medulla oblongata. Two roots of the trigeminal nerve are apparent as they emerge from the anterolateral side of the pons (see fig. 12.3). The larger **sensory root** immediately enlarges into a swelling called the **trigeminal** (semilunar) **ganglion,** located in a bony depression on the inner surface of the petrous part of the temporal bone. Three large nerves arise from the trigeminal ganglion (fig. 12.7): the **ophthalmic nerve** enters the orbit through the superior orbital fissure, the **maxillary nerve** extends through the foramen rotundum, and the **mandibular nerve** passes through the foramen ovale. The smaller **motor root** consists of motor fibers of the trigeminal nerve that accompany the mandibular nerve through the foramen ovale and innervate the muscles of mastication and certain muscles in the floor of the mouth. Impulses through the motor portion of the mandibular nerve of the trigeminal ganglion stimulate contraction of the muscles involved in chewing, including the **medial** and **lateral pterygoids, masseter, temporalis, mylohyoid,** and the anterior belly of the **digastric** muscle.

trochlear: Gk. *trochos*, a wheel

trigeminal: L. *trigeminus*, three born together
ophthalmic: L. *opthalmia*, region of the eye

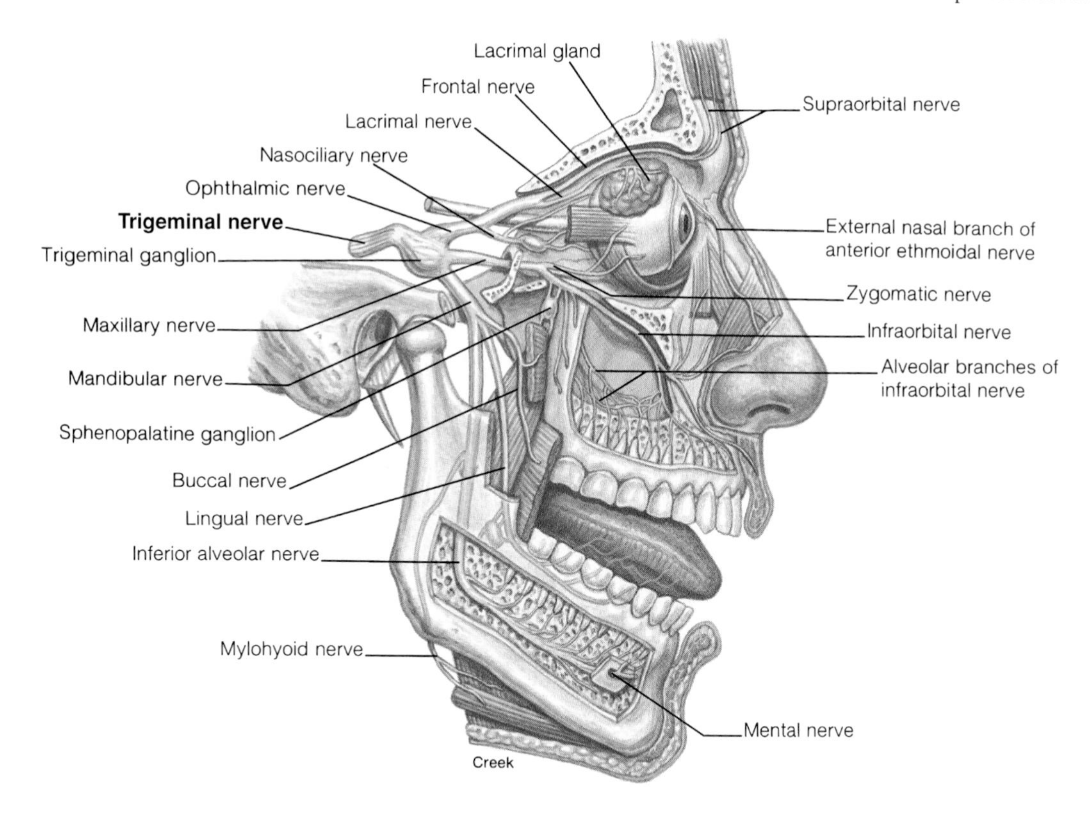

Figure 12.7

The trigeminal nerve and its distribution.

Although the trigeminal is a mixed nerve, its sensory functions are much more extensive than its motor functions. The three sensory nerves of the trigeminal ganglion respond to touch, temperature, and pain sensations from the face. More specifically, the ophthalmic nerve consists of sensory fibers from the anterior half of the scalp, skin of the forehead, upper eyelid, surface of the eyeball, lacrimal (tear) gland, side of the nose, and upper mucosa of the nasal cavity. The maxillary nerve is composed of sensory fibers from the lower eyelid, lateral and inferior mucosa of the nasal cavity, palate and portions of the pharynx, teeth and gums of the upper jaw, upper lip, and skin of the cheek. Sensory fibers of the mandibular nerve transmit impulses from the teeth and gums of the lower jaw, anterior two-thirds of the tongue (not taste), mucosa of the mouth, auricle of the ear, and lower part of the face. Trauma to the trigeminal nerve results in a lack of sensation from specific facial structures. Damage to the mandibular nerve impairs chewing.

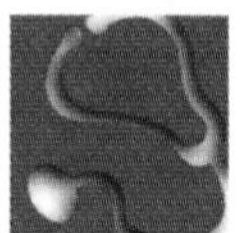

The trigeminal nerve is the principal nerve relating to the practice of dentistry. Before teeth are filled or extracted, anesthetic is injected near the appropriate nerve to block sensation. A *maxillary,* or *second-division, nerve block,* performed by injecting near the sphenopalatine ganglion (see fig. 12.7), desensitizes the teeth in the upper jaw. A *mandibular,* or *third-division, nerve block* desensitizes the lower teeth. This is performed by injecting anesthetic near the inferior alveolar nerve, which branches off the mandibular nerve as it enters the mandible through the mandibular foramen.

VI Abducens

The small abducens (*ab-doo'senz*) nerve originates from a nucleus within the pons and emerges from the lower portion of the pons and the anterior border of the medulla oblongata. It is a mixed nerve that traverses the superior orbital fissure of the orbit to innervate the **lateral rectus** eye muscle (see fig. 12.6). Impulses through the motor fibers of the abducens nerve cause the lateral rectus eye muscle to contract and the eyeball to move away from the midline laterally. Sensory impulses through the abducens nerve originate in proprioceptors in the lateral rectus muscle and are conveyed to the pons, where muscle contraction is mediated. If the abducens nerve is damaged, not only will the patient be unable to move the eyeball laterally, but because of the lack of muscle tonus to the lateral rectus muscle, the eyeball will be pulled medially.

VII Facial

The facial nerve arises from nuclei within the lower portion of the pons, traverses the petrous part of the temporal bone (see fig. 12.9), and emerges on the side of the face near the parotid (salivary) gland. The facial nerve is mixed. Impulses through the motor fibers cause contraction of the posterior belly of the digastric muscle and the muscles of facial expression, including the scalp and platysma muscles (fig. 12.8). The submandibular and sublingual (salivary) glands also receive some autonomic motor innervation from the facial nerve, as does the lacrimal gland.

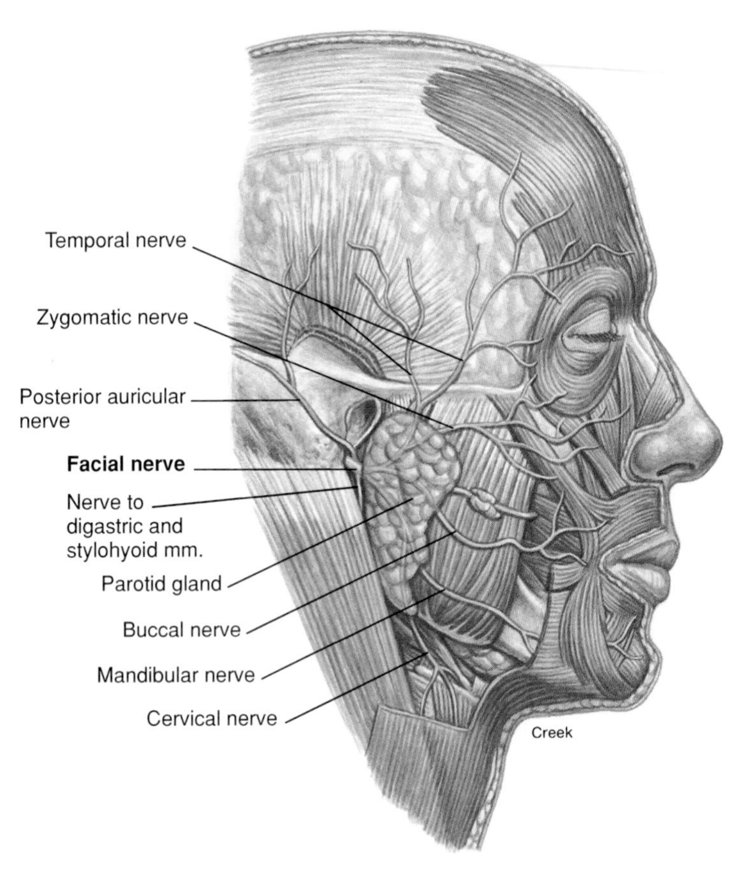

Figure 12.8

The facial nerve and its distribution to superficial structures.

Sensory fibers of the facial nerve arise from taste buds on the anterior two-thirds of the tongue. Taste buds function as *chemoreceptors* because they respond to specific chemical stimuli.

The **geniculate** (*jĕ-nik'yoo-lāt*) **ganglion** is the enlargement of the facial nerve just before the entrance of the sensory portion into the pons. Sensations of taste are conveyed to nuclei within the medulla oblongata, travel through the thalamus, and ultimately to the gustatory (taste) area in the parietal lobe of the cerebral cortex.

Trauma to the facial nerve results in inability to contract facial muscles on the affected side of the face and distorts taste perception, particularly of sweets. The affected side of the face tends to sag because muscle tonus is lost. *Bell's palsy* is a functional disorder (probably of viral origin) of the facial nerve.

VIII Vestibulocochlear

The vestibulocochlear (*ves-tib"yŭ-lo-kok'le-ar*) nerve, also referred to as the **auditory, acoustic,** or **statoacoustic nerve,** serves structures contained within the skull. It is the only cranial nerve that does not exit the cranium through a foramen. A purely sensory nerve, the vestibulocochlear is composed of two nerves that arise within the inner ear (fig. 12.9). The **vestibular nerve** arises from the **vestibular organs** associated with equilibrium and balance. Bipolar neurons from the vestibular organs (saccule, utricle, and semicircular ducts) extend to the **vestibular ganglion,** where cell bodies are contained. From there, fibers convey impulses to the **vestibular nuclei** within the pons and medulla oblongata. Fibers from there extend to the thalamus and the cerebellum.

The **cochlear nerve** arises from the **spiral organ** (organ of Corti) within the cochlea and is associated with hearing. The cochlear nerve is composed of bipolar neurons that convey impulses through the **spiral ganglion** to the **cochlear nuclei** within the medulla oblongata. From there, fibers extend to the thalamus and synapse with neurons that convey the impulses to the auditory areas of the cerebral cortex.

Injury to the cochlear nerve results in perception deafness, whereas damage to the vestibular nerve causes dizziness and loss of balance.

IX Glossopharyngeal

The glossopharyngeal (*glos"o-fă-rin'je-al*) nerve is a mixed nerve that innervates part of the tongue and pharynx (fig. 12.10). The motor fibers of this nerve originate in a nucleus within the medulla oblongata and pass through the jugular foramen. The motor fibers innervate the muscles of the pharynx and the parotid gland to stimulate the swallowing reflex and the secretion of saliva.

The sensory fibers of the glossopharyngeal nerve arise from the pharyngeal region, the parotid gland, the middle-ear cavity, and the taste buds on the posterior one-third of the tongue. These taste buds, like those innervated by the facial nerve, are *chemoreceptors*. Some sensory fibers also arise from sensory receptors within the carotid sinus of the neck and help to regulate blood pressure. Impulses from the glossopharyngeal nerve travel through the medulla oblongata and into the thalamus, where they synapse with fibers that convey the impulses to the gustatory area of the cerebral cortex.

Damage to the glossopharyngeal nerve results in the loss of perception of bitter and sour taste from taste buds on the posterior portion of the tongue. If the motor portion of this nerve is damaged, swallowing becomes difficult.

X Vagus

The vagus (*va'gus*) nerve has motor and sensory fibers that innervate visceral organs of the thoracic and abdominal cavities (fig. 12.11). The motor portion arises from the **nucleus ambiguus** and **dorsal motor nucleus** of the vagus within the medulla oblongata and passes through the jugular foramen. The

Bell's palsy: from Sir Charles Bell, Scottish physician, 1774–1842
vestibulocochlear: L. *vestibulum*, chamber; *cochlea*, snail shell

organ of Corti: from Alfonso Corti, Italian anatomist, 1822–88
glossopharyngeal: L. *glossa*, tongue; Gk. *pharynx*, throat
vagus: L. *vagus*, wandering

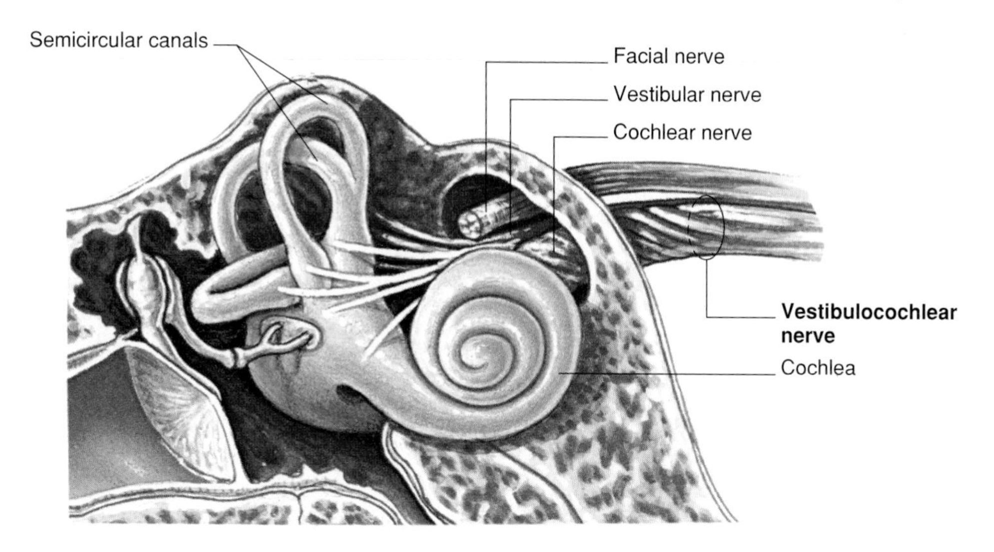

Figure 12.9

The vestibulocochlear nerve. The structures of the inner ear are served by this nerve. The semicircular ducts, concerned with balance and equilibrium, form the membranous labyrinth within the semicircular canals. The cochlea contains the structures concerned with hearing.

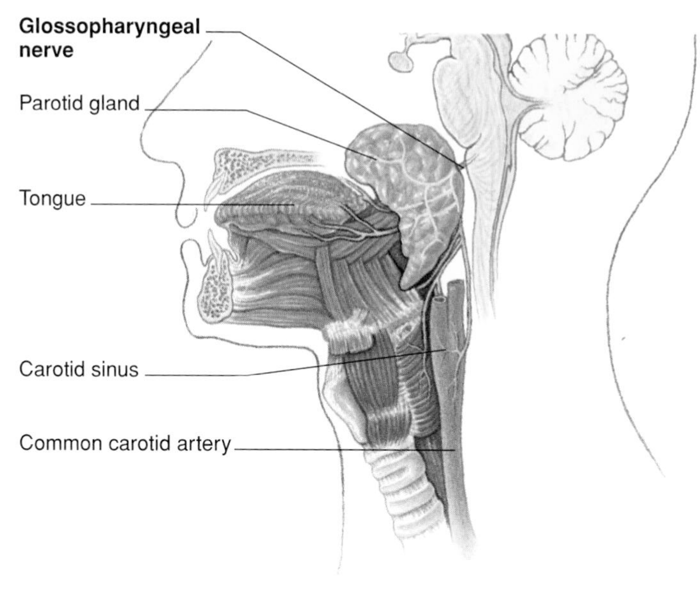

Figure 12.10

The glossopharyngeal nerve.

vagus is the longest of the cranial nerves, and through various branches it innervates the muscles of the pharynx, larynx, respiratory tract, lungs, heart, and esophagus, and those of the abdominal viscera, with the exception of the lower portion of the large intestine. One motor branch of the vagus nerve, the **recurrent laryngeal nerve,** innervates the larynx, enabling speech.

Sensory fibers of the vagus nerve convey impulses from essentially the same organs served by motor fibers. Impulses through the sensory fibers relay specific sensations, such as hunger pangs, distension, intestinal discomfort, or laryngeal movements. Sensory fibers also arise from proprioceptors in the muscles innervated by the motor fibers of this nerve.

If both vagus nerves are seriously damaged, death ensues rapidly because vital autonomic functions stop. The injury of one nerve causes vocal impairment, difficulty in swallowing, or other visceral disturbances.

XI Accessory

The accessory nerve is principally a motor nerve, but it does contain some sensory fibers from proprioceptors within the muscles it innervates. The accessory nerve is unique in that it arises from both the brain and the spinal cord (fig. 12.12). The **cranial root** arises from nuclei within the medulla oblongata (ambiguus and accessory), passes through the jugular foramen with the vagus nerve, and innervates the skeletal muscles of the soft palate, pharynx, and larynx, which contract reflexively during swallowing. The **spinal root** arises from the first five segments of the cervical portion of the spinal cord, passes cranially through the foramen magnum to join with the cranial root, and then passes through the jugular foramen. The spinal root of the accessory nerve innervates the sternocleidomastoid and the trapezius muscles that move the head, neck, and shoulders. Damage to an accessory nerve makes it difficult to move the head or shrug the shoulders.

XII Hypoglossal

The hypoglossal nerve is a mixed nerve. The motor fibers arise from the hypoglossal nucleus within the medulla oblongata

hypoglossal: Gk. *hypo,* under; L. *glossa,* tongue

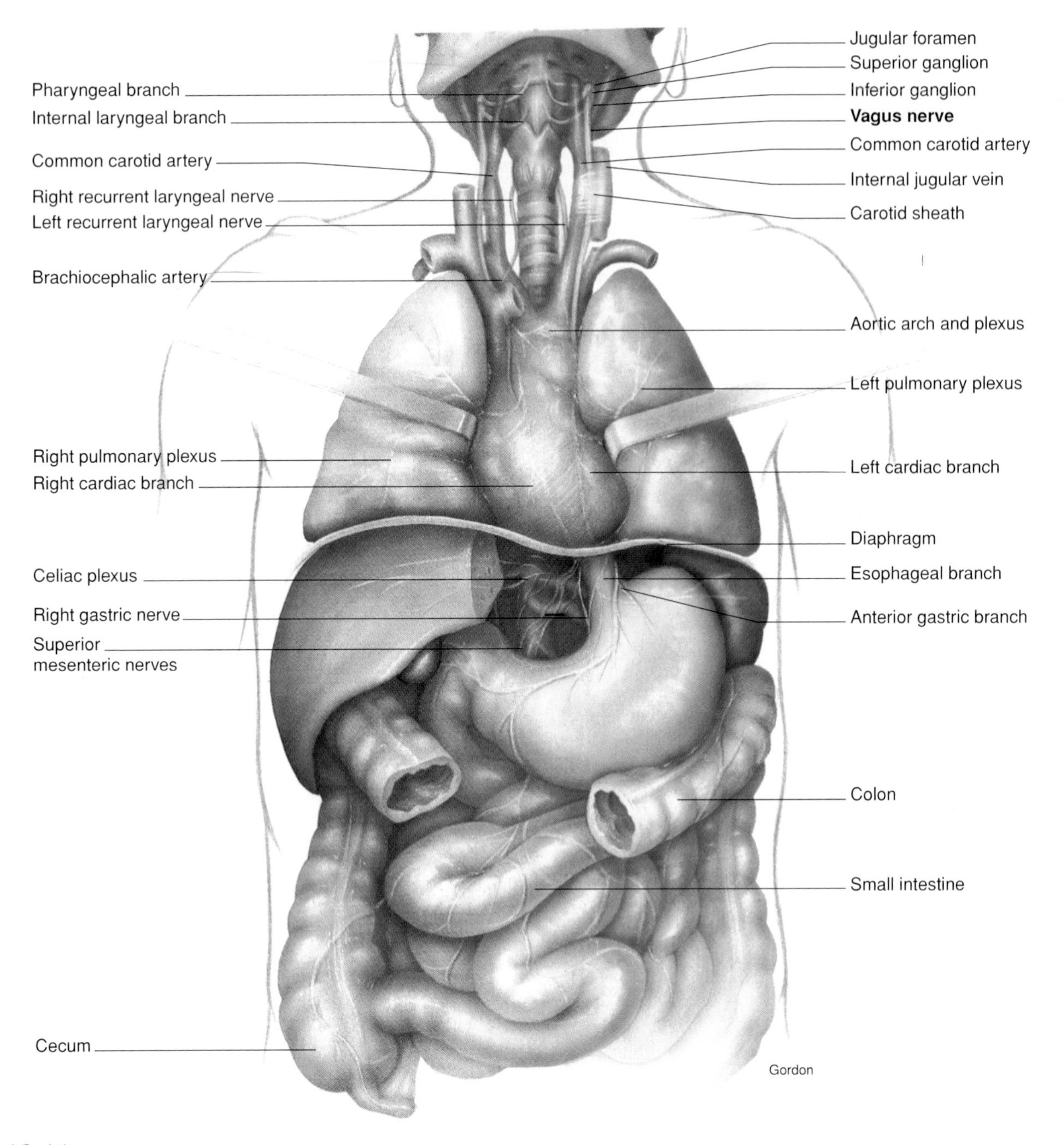

Figure 12.11

Distribution of the vagus nerves.

and pass through the hypoglossal canal of the skull to innervate both the extrinsic and intrinsic muscles of the tongue (fig. 12.12). Motor impulses along these fibers account for the coordinated contraction of the tongue muscles that is needed for such activities as food manipulation, swallowing, and speech.

The sensory portion of the hypoglossal nerve arises from proprioceptors within the same tongue muscles and conveys impulses to the medulla oblongata regarding the position and function of the muscles.

If a hypoglossal nerve is damaged, a person will have difficulty in speaking, swallowing, and protruding the tongue.

Neurological Assessment of the Cranial Nerves

Head injuries and brain concussions are common occurrences in automobile accidents. The cranial nerves would seem to be well protected on the inferior side of the brain. But the brain, immersed in and filled with cerebrospinal fluid, is like a water-sodden log; a blow to the top of the head can cause a serious rebound of the brain from the floor of the cranium. Routine neurological examinations involve testing for cranial nerve dysfunction.

Commonly used clinical methods for determining cranial nerve dysfunction are presented in table 12.2.

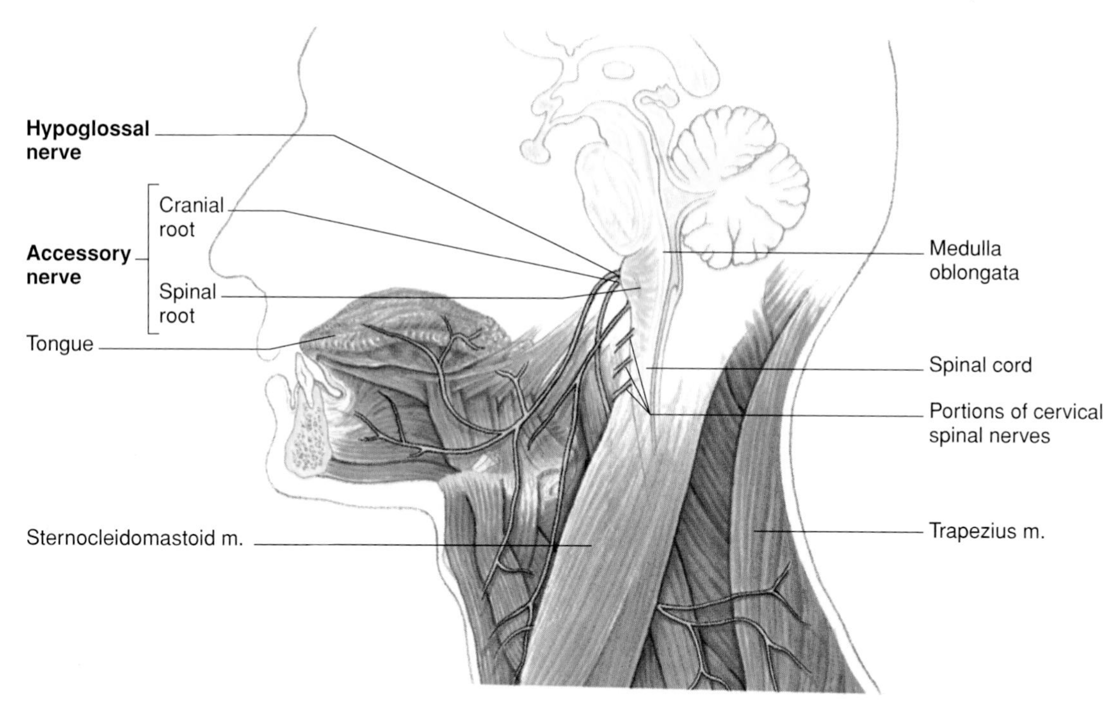

Figure 12.12

The accessory and hypoglossal nerves.

Table 12.2

Methods of Determining Cranial Nerve Dysfunction

Nerve	Techniques of Examination	Comments
Olfactory	Patient asked to differentiate odors (tobacco, coffee, soap, etc.) with eyes closed.	Nasal passages must be patent and tested separately by occluding the opposite side.
Optic	Retina examined with ophthalmoscope; visual acuity tested with eye charts.	Visual acuity must be determined with lenses on, if patient wears them.
Oculomotor	Patient follows examiner's finger movement with eyes—especially movement that causes eyes to cross; pupillary change observed by shining light into each eye separately.	Examiner should note rate of pupillary change and coordinated constriction of pupils. Light in one eye should cause a similar pupillary change in other eye, but to a lesser degree.
Trochlear	Patient follows examiner's finger movement with eyes—especially lateral and downward movement.	
Trigeminal	Motor portion: Temporalis and masseter muscles palpated as patient clenches teeth; patient asked to open mouth against resistance applied by examiner.	Muscles of both sides of the jaw should show equal contractile strength.
	Sensory portion: Tactile and pain receptors tested by lightly touching patient's entire face with cotton and then with pin stimulus.	Patient's eyes should be closed and innervation areas for all three nerves branching from the trigeminal nerve should be tested.
Abducens	Patient follows examiner's finger movement—especially lateral movement.	Motor functioning of cranial nerves III, IV, and VI may be tested simultaneously through selective movements of eyeball.
Facial	Motor portion: Patient asked to raise eyebrows, frown, tightly constrict eyelids, smile, puff out cheeks, and whistle.	Examiner should note lack of tonus expressed by sagging regions of face.
	Sensory portion: Sugar placed on each side of tip of patient's tongue.	Not reliable test for specific facial-nerve dysfunction because of tendency to stimulate taste buds on both sides of tip of tongue.
Vestibulocochlear	Vestibular portion: Patient asked to walk a straight line.	Not usually tested unless patient complains of dizziness or balance problems.
	Cochlear portion: Tested with tuning fork.	Examiner should note ability to discriminate sounds.
Glossopharyngeal and vagus	Motor: Examiner notes disturbances in swallowing, talking, and movement of soft palate; gag reflex tested.	Visceral innervation of vagus cannot be examined, except for innervation to larynx, which is also served by glossopharyngeal.
Accessory	Patient asked to shrug shoulders against resistance of examiner's hand and to rotate head against resistance.	Sides should show uniformity of strength.
Hypoglossal	Patient asked to protrude tongue; tongue thrust may be resisted with tongue blade.	Tongue should protrude straight out; deviation to side indicates ipsilateral-nerve dysfunction; asymmetry, atrophy, or lack of strength should be noted.

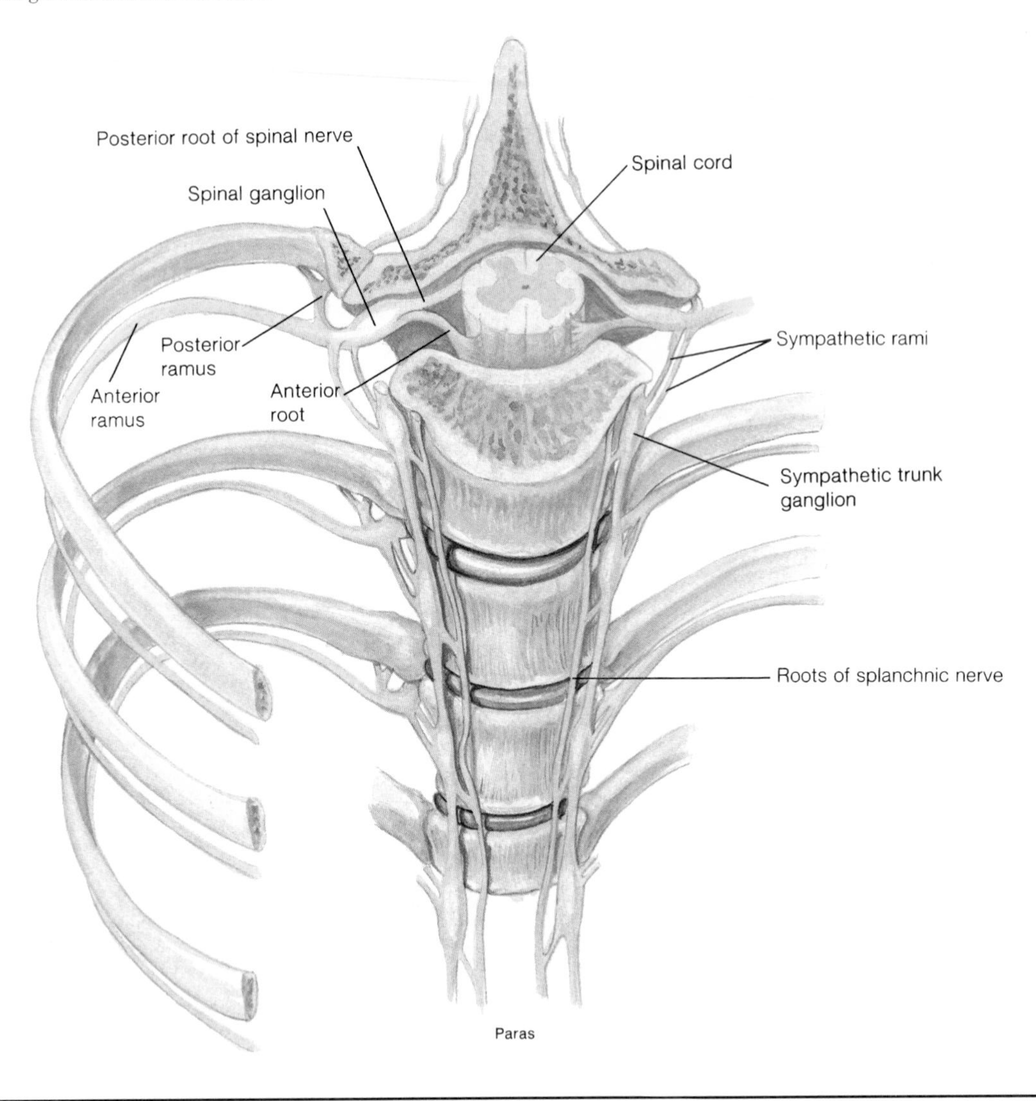

Figure 12.13

A section of the spinal cord and thoracic spinal nerves.

1. Which cranial nerves consist of sensory fibers only?
2. Which cranial nerves pass through the superior orbital fissure? Through the jugular foramen?
3. Which cranial nerves are involved in tasting, chewing and manipulating food, and swallowing?
4. Which cranial nerves have to do with the structure, function, or movement of the eyeball?
5. List the cranial nerves and indicate how each would be tested (both motor and sensory fibers) for possible dysfunction.

Spinal Nerves

Each of the 31 pairs of spinal nerves is formed by the union of a posterior and an anterior spinal root that emerges from the spinal cord through an intervertebral foramen to innervate a body dermatome.

Objective 4 Discuss how the spinal nerves are grouped.

Objective 5 Describe the general distribution of a spinal nerve.

The 31 pairs of **spinal nerves** (see fig. 12.1) are grouped as follows: 8 cervical, 12 thoracic, 5 lumbar, 5 sacral, and 1 coccygeal. With the exception of the first cervical nerve, the spinal nerves leave the spinal cord and vertebral canal through intervertebral foramina. The first pair of cervical nerves emerges between the occipital bone of the skull and the atlas. The second through the seventh pairs of cervical nerves emerge above the vertebrae for which they are named, whereas the eighth pair of cervical nerves passes between the seventh cervical and first thoracic vertebrae. The remaining pairs of spinal nerves emerge below the vertebrae for which they are named.

A spinal nerve is a mixed nerve attached to the spinal cord by a **posterior** (dorsal) **root,** composed of sensory fibers, and an **anterior** (ventral) **root,** composed of motor fibers (fig. 12.13).

The posterior root contains an enlargement called the **spinal** (sensory) **ganglion,** where the cell bodies of sensory neurons are located. The axons of sensory neurons convey sensory impulses through the posterior root into the spinal cord, where synapses occur with dendrites of other neurons. The anterior root consists of axons of motor neurons, which convey motor impulses away from the CNS. A spinal nerve is formed as the fibers from the posterior and anterior roots converge and emerge through an intervertebral foramen.

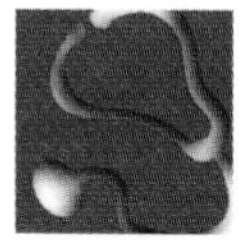

The disease *herpes zoster,* also known as *shingles,* is a viral infection of the spinal ganglia. Herpes zoster causes painful, often unilateral, clusters of fluid-filled vesicles in the skin along the paths of the affected peripheral sensory neurons. The disease develops in adults who were first exposed to the virus as children, and is usually self-limited. Treatment may involve large doses of the antiviral drug acyclovir (Zorivax).

A spinal nerve divides into several branches immediately after it emerges through the intervertebral foramen. The small **meningeal branch** reenters the vertebral canal to innervate the meninges, vertebrae, and vertebral ligaments. A larger branch, called the **posterior ramus,** innervates the muscles, joints, and skin of the back along the vertebral column (fig. 12.13). An **anterior ramus** of a spinal nerve innervates the muscles and skin on the lateral and anterior side of the trunk. Combinations of anterior rami innervate the limbs.

The **rami communicantes** are two branches from each spinal nerve that connect to a **sympathetic trunk ganglion,** which is part of the autonomic nervous system. The rami communicantes are composed of a **gray ramus,** containing unmyelinated fibers, and a **white ramus,** containing myelinated fibers. This arrangement is described in more detail in chapter 13.

1. List the number of nerves in each of the five regions of the vertebral column.
2. What are the four principal branches, or rami, from a spinal nerve, and what structures does each innervate?

Nerve Plexuses

Except in thoracic nerves T2 through T12, the anterior rami of the spinal nerves combine and then split again as networks of nerve fibers referred to as nerve plexuses. There are four plexuses of spinal nerves: the cervical, the brachial, the lumbar, and the sacral. Nerves emerging from the plexuses are named according to the structures they innervate or the general course they take.

Objective 6 List the spinal nerve composition of each of the plexuses arising from the spinal cord.

Objective 7 List the principal nerves that emerge from the plexuses and describe their general innervation.

Cervical Plexus

The **cervical plexus** (*plek'sus*) is positioned deep on the side of the neck, lateral to the first four cervical vertebrae (fig. 12.14). It is formed by the anterior rami of the first four cervical nerves (C1–C4) and a portion of C5. Branches of the cervical plexus innervate the skin and muscles of the neck and portions of the head and shoulders. Some fibers of the cervical plexus also combine with the accessory and hypoglossal cranial nerves to supply dual innervation to some specific neck and pharyngeal muscles. Fibers from the third, fourth, and fifth cervical nerves unite to become the **phrenic** (*fren'ik*) **nerve,** which innervates the diaphragm. Motor impulses through the paired phrenic nerves cause the diaphragm to contract, moving air into the lungs.

The nerves of the cervical plexus are summarized in table 12.3.

Brachial Plexus

The **brachial plexus** is positioned to the side of the last four cervical vertebrae and the first thoracic vertebra. It is formed by the anterior rami of C5 through T1, with occasional contributions from C4 and T2. From its emergence, the brachial plexus extends downward and laterally, passes over the first rib behind the clavicle, and enters the axilla. Each brachial plexus innervates the entire upper extremity of one side, as well as a number of shoulder and neck muscles.

Structurally, the brachial plexus is divided into *roots, trunks, divisions,* and *cords* (fig. 12.15). The roots of the brachial plexus are simply continuations of the anterior rami of the cervical nerves. The anterior rami of C5 and C6 converge to become the **superior trunk,** the C7 ramus becomes the **middle trunk,** and the ventral rami of C8 and T1 converge to become the **inferior trunk.** Each of the three trunks immediately divides into an **anterior division** and a **posterior division.** The divisions then converge to form three cords. The **posterior cord** is formed by the convergence of the posterior divisions of the upper, middle, and lower trunks; hence, it contains fibers from C5 through C8. The **medial cord** is a continuation of the anterior division of the lower trunk and primarily contains fibers from C8 and T1. The **lateral cord** is formed by the convergence of the anterior division of the upper and middle trunk and consists of fibers from C5 through C7.

In summary, the brachial plexus is composed of nerve fibers from the anterior branches of spinal nerves C5 through T1 and a few fibers from C4 and T2. Roots are continuations of the anterior rami. The roots converge to form trunks, and the trunks branch into divisions. The divisions in turn form cords, and the nerves of the upper extremity arise from the cords.

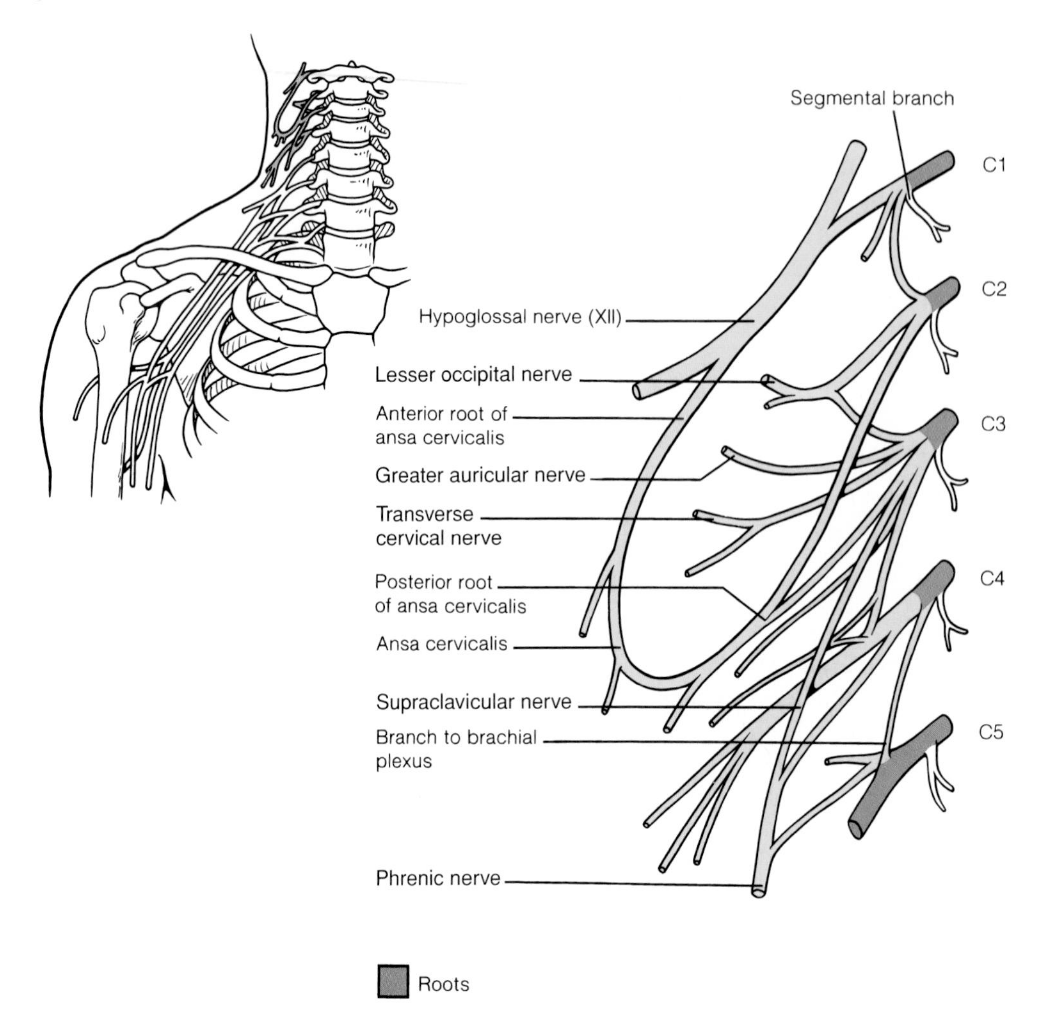

Figure 12.14

The cervical plexus.

Table 12.3
Branches of the Cervical Plexus

Nerve	Spinal Component	Innervation
Superficial Cutaneous Branches		
Lesser occipital	C2,C3	Skin of scalp above and behind ear
Greater auricular	C2,C3	Skin in front of, above, and below ear
Transverse cervical	C2,C3	Skin of anterior aspect of neck
Supraclavicular	C3,C4	Skin of upper portion of chest and shoulder
Deep Motor Branches		
Ansa cervicalis		
Anterior root	C1,C2	Geniohyoid, thyrohyoid, and infrahyoid muscles of neck
Posterior root	C3,C4	Omohyoid, sternohyoid, and sternothyroid muscles of neck
Phrenic	C3–C5	Diaphragm
Segmental branches	C1–C5	Deep muscles of neck (levator scapulae ventralis, trapezius, scalenus, and sternocleidomastoid)

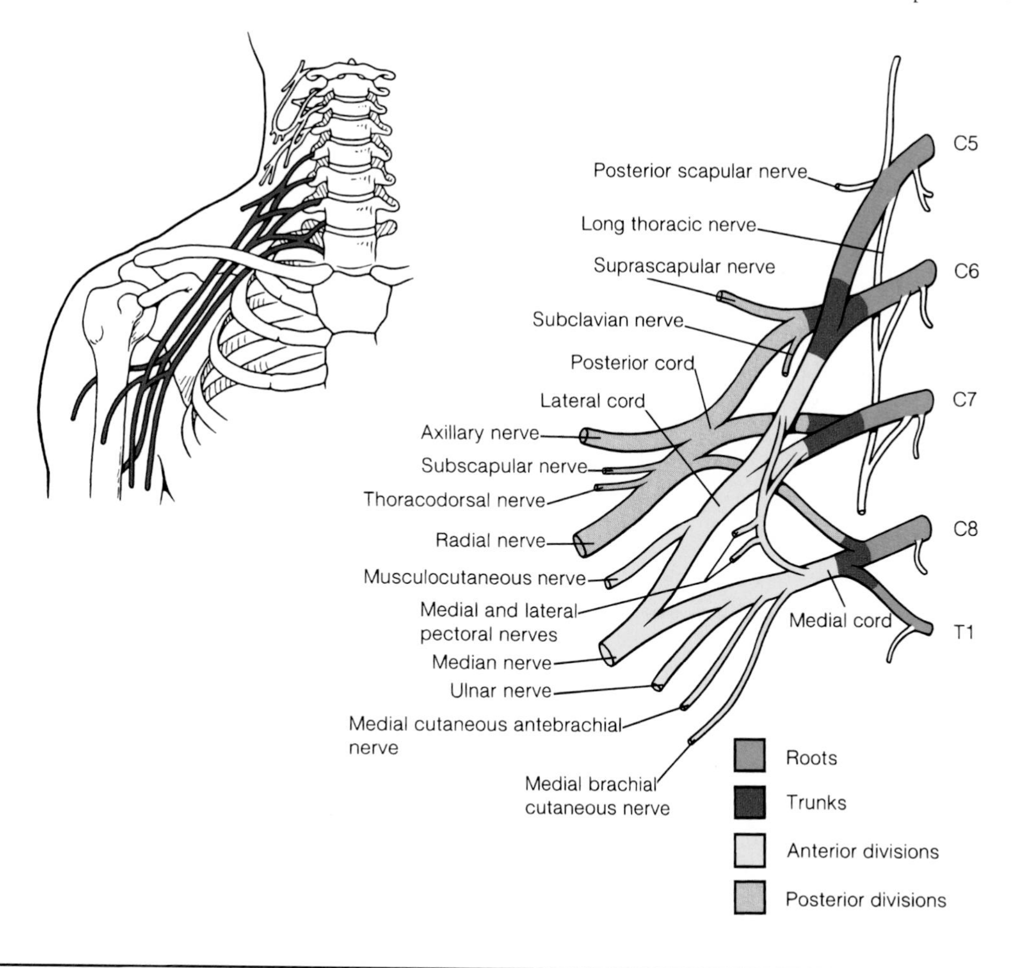

Figure 12.15

The brachial plexus.

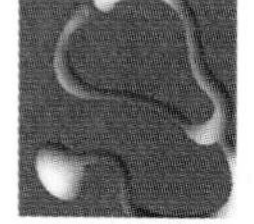

The brachial plexus may suffer trauma, especially if the clavicle, upper ribs, or lower cervical vertebrae are seriously fractured. Occasionally, the brachial plexus of a newborn is severely strained during a difficult delivery when the baby is pulled through the birth canal. In such cases, the arm of the injured side is paralyzed and eventually withers as the muscles atrophy in relation to the extent of the injury.

The entire upper extremity can be anesthetized in a procedure called a *brachial block* or *brachial anesthesia.* The site for injection of the anesthetic is located midway between the base of the neck and the shoulder, posterior to the clavicle. At this point, the anesthetic can be injected close to the brachial plexus.

Five major nerves—the axillary, radial, musculocutaneous, ulnar, and median—arise from the three cords of the brachial plexus to supply cutaneous and muscular innervation to the upper extremity (table 12.4). The **axillary nerve** arises from the posterior cord. It provides sensory innervation to the skin of the shoulder and shoulder joint, and motor innervation to the deltoid and teres minor muscles (fig. 12.16). The **radial nerve** arises from the posterior cord and extends along the posterior aspect of the brachial region to the radial side of the forearm. It provides sensory innervation to the skin of the posterior lateral surface of the upper extremity, including the posterior surface of the hand (fig. 12.17), and motor innervation to all of the extensor muscles of the upper extremity, the supinator muscle, and two muscles that flex the elbow joint.

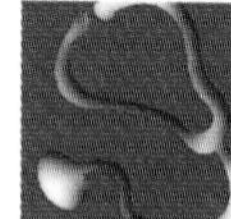

The radial nerve is vulnerable to several types of trauma. *Crutch paralysis* may result when a person improperly supports the weight of the body for an extended period of time with a crutch pushed tightly into the axilla. Compression of the radial nerve between the top of the crutch and the humerus may result in radial nerve damage. Likewise, dislocation of the shoulder frequently traumatizes the radial nerve. Children are particularly at risk as adults yank on their arms. A fracture to the body of the humerus may damage the radial nerve, which parallels the bone at this point. The principal symptom of radial nerve damage is *wristdrop,* in which the extensor muscles of the fingers and wrist fail to function. As a result, the joints of the fingers, wrist, and elbow are in a constant state of flexion.

Table 12.4
Selected Branches of the Brachial Plexus

Nerve	Cord and Spinal Components	Innervation
Axillary	Posterior cord (C5,C6)	Skin of shoulder; shoulder joint, deltoid and teres minor muscles
Radial	Posterior cord (C5–C8,T1)	Skin of posterior lateral surface of arm, forearm, and hand; posterior muscles of brachium and antebrachium (triceps brachii, supinator, anconeus, brachioradialis, extensor carpi radialis brevis, extensor carpi radialis longus, extensor carpi ulnaris)
Musculocutaneous	Lateral cord (C5–C7)	Skin of lateral surface of forearm; anterior muscles of brachium (coracobrachialis, biceps brachii, brachialis)
Ulnar	Medial cord (C8,T1)	Skin of medial third of hand; flexor muscles of anterior forearm (flexor carpi ulnaris, flexor digitorum), medial palm, and intrinsic flexor muscles of hand (profundus, third and fourth lumbricales)
Median	Medial cord (C6–C8,T1)	Skin of lateral two-thirds of hand; flexor muscles of anterior forearm, lateral palm, and first and second lumbricales

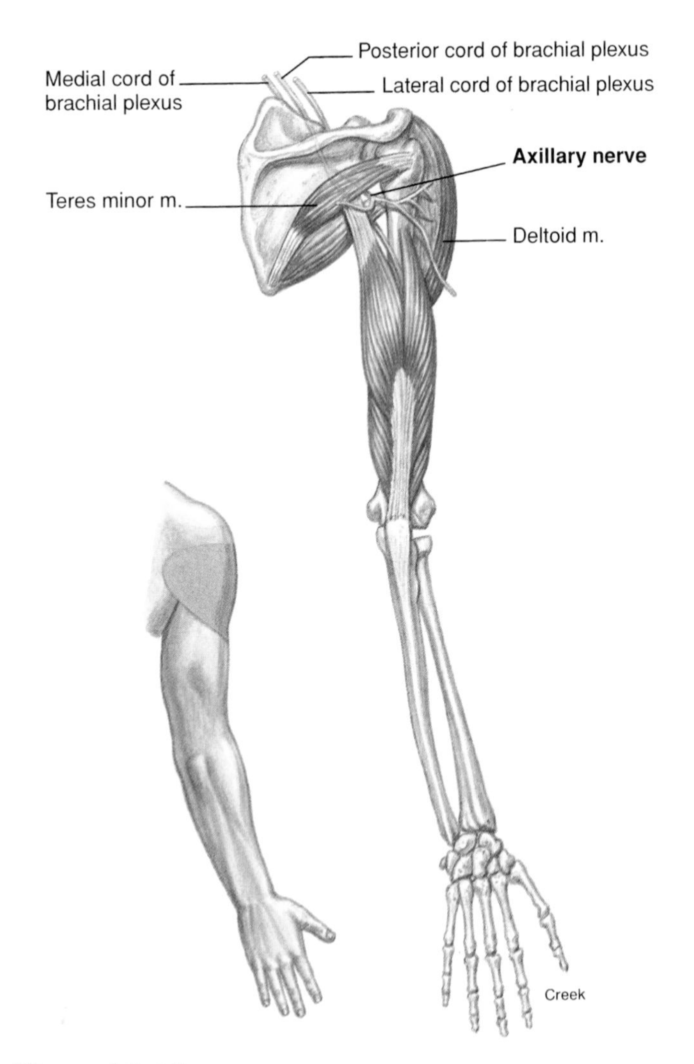

Figure 12.16

Muscular and cutaneous distribution of the axillary nerve.

The **musculocutaneous nerve** arises from the lateral cord. It provides sensory innervation to the skin of the posterior lateral surface of the arm and motor innervation to the anterior muscles of the brachium (fig. 12.18). The **ulnar nerve** arises from the medial cord and provides sensory innervation to the skin on the medial (ulnar side) third of the hand (fig. 12.19). The motor innervation of the ulnar nerve is to two muscles of the forearm and the intrinsic muscles of the hand (except some that serve the thumb).

The ulnar nerve can be palpated in the groove between the medial epicondyle of the humerus and the olecranon of the ulna. This area is commonly known as the "funny bone" or "crazy bone." Damage to the ulnar nerve may occur as the medial side of the elbow is banged against a hard object. The immediate perception of this trauma is a painful tingling that extends down the ulnar side of the forearm and into the hand and medial two digits. Although common, ulnar nerve damage is generally not serious.

The **median nerve** arises from the medial cord. It provides sensory innervation to the skin on the radial portion of the palm of the hand (fig. 12.20) and motor innervation to all but one of the flexor muscles of the forearm and to most of the hand muscles of the thumb (thenar muscles).

Lumbar Plexus

The **lumbar plexus** is positioned to the side of the first four lumbar vertebrae. It is formed by the anterior rami of spinal nerves L1 through L4 and some fibers from T12 (fig. 12.21). The nerves that arise from the lumbar plexus innervate structures of the lower abdomen and anterior and medial portions of the lower extremity. The lumbar plexus is not as complex as the brachial

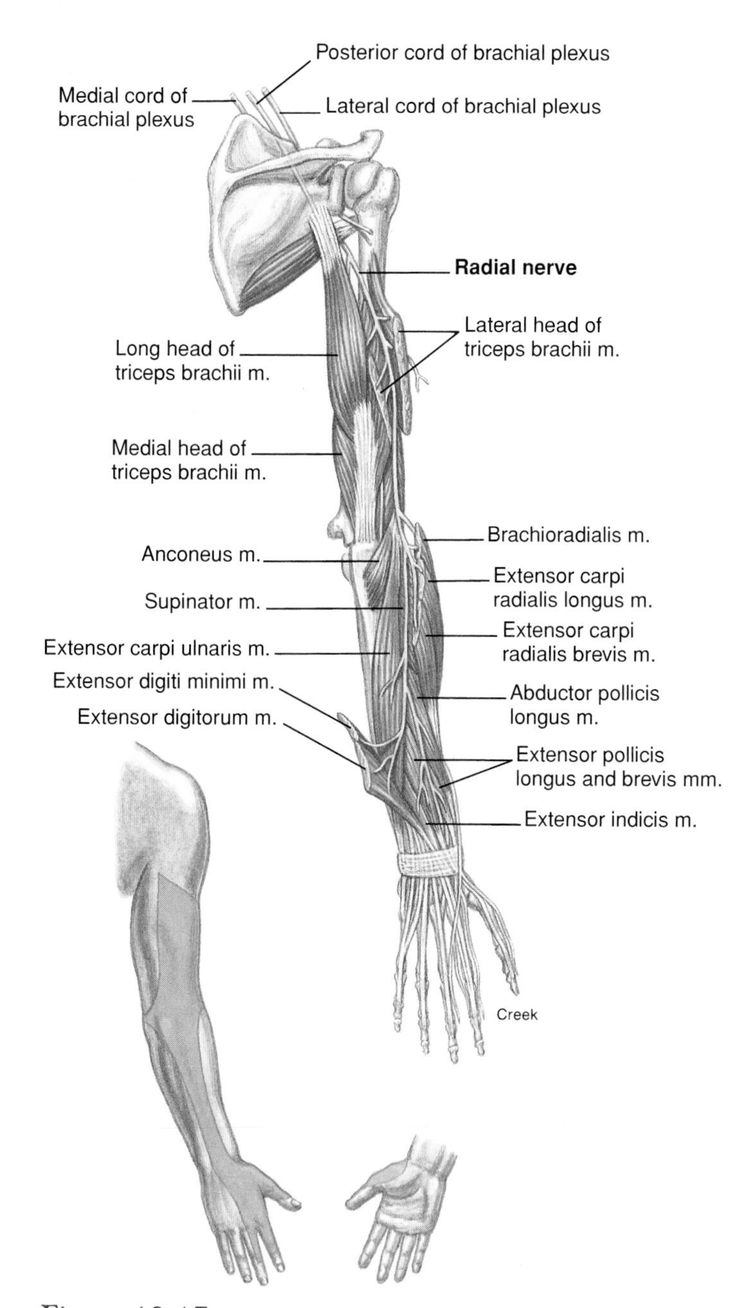

Figure 12.17

Muscular and cutaneous distribution of the radial nerve.

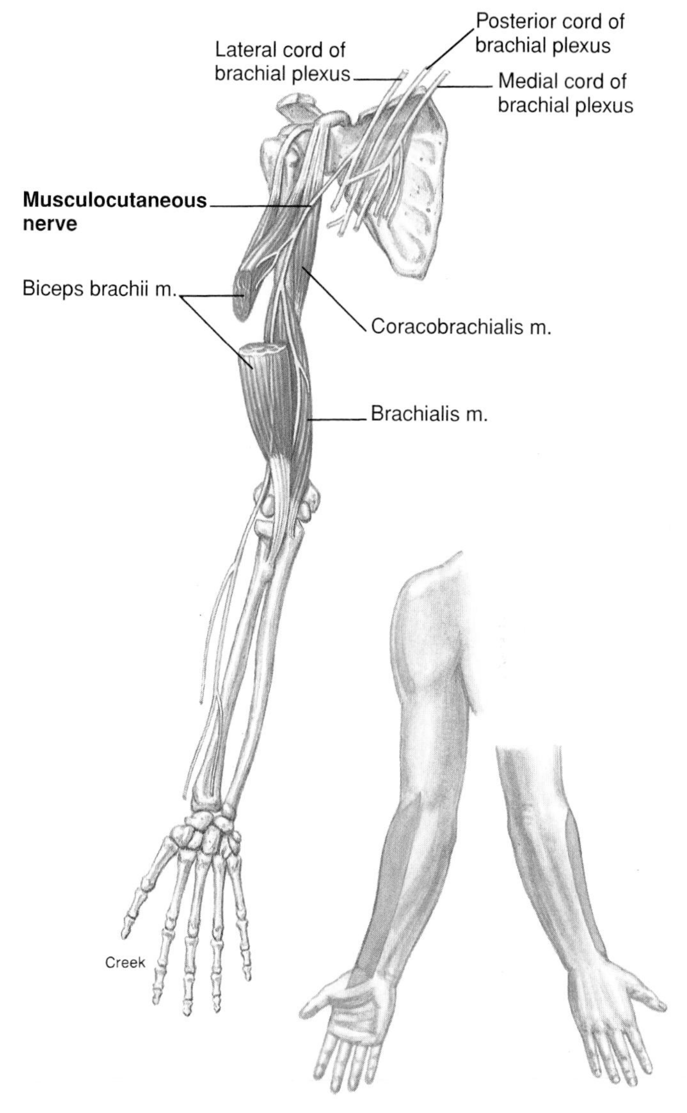

Figure 12.18

Muscular and cutaneous distribution of the musculocutaneous nerve.

plexus, having only roots and divisions rather than the roots, trunks, divisions, and cords of the brachial plexus.

Structurally, the **posterior division** of the lumbar plexus passes obliquely outward, deep to the psoas major muscle, whereas the **anterior division** is superficial to the quadratus lumborum muscle. The nerves that arise from the lumbar plexus are summarized in table 12.5. Because of their extensive innervation, the femoral nerve and the obturator nerve are illustrated in figures 12.22 and 12.23, respectively.

The **femoral** (*fem'or-al*) **nerve** arises from the posterior division of the lumbar plexus and provides cutaneous innervation to the anterior and lateral thigh and the medial leg and foot (fig. 12.22). The motor innervation of the femoral nerve is to the anterior muscles of the thigh, including the iliopsoas and sartorius muscles and the quadriceps femoris group.

The **obturator** (*ob'too-ra"tor*) **nerve** arises from the anterior division of the lumbar plexus. It provides cutaneous innervation to the medial thigh and motor innervation to the adductor muscles of the thigh (fig. 12.23).

Sacral Plexus

The **sacral plexus** lies immediately inferior to the lumbar plexus. It is formed by the anterior rami of spinal nerves L4,

sacral: L. *sacris*, sacred

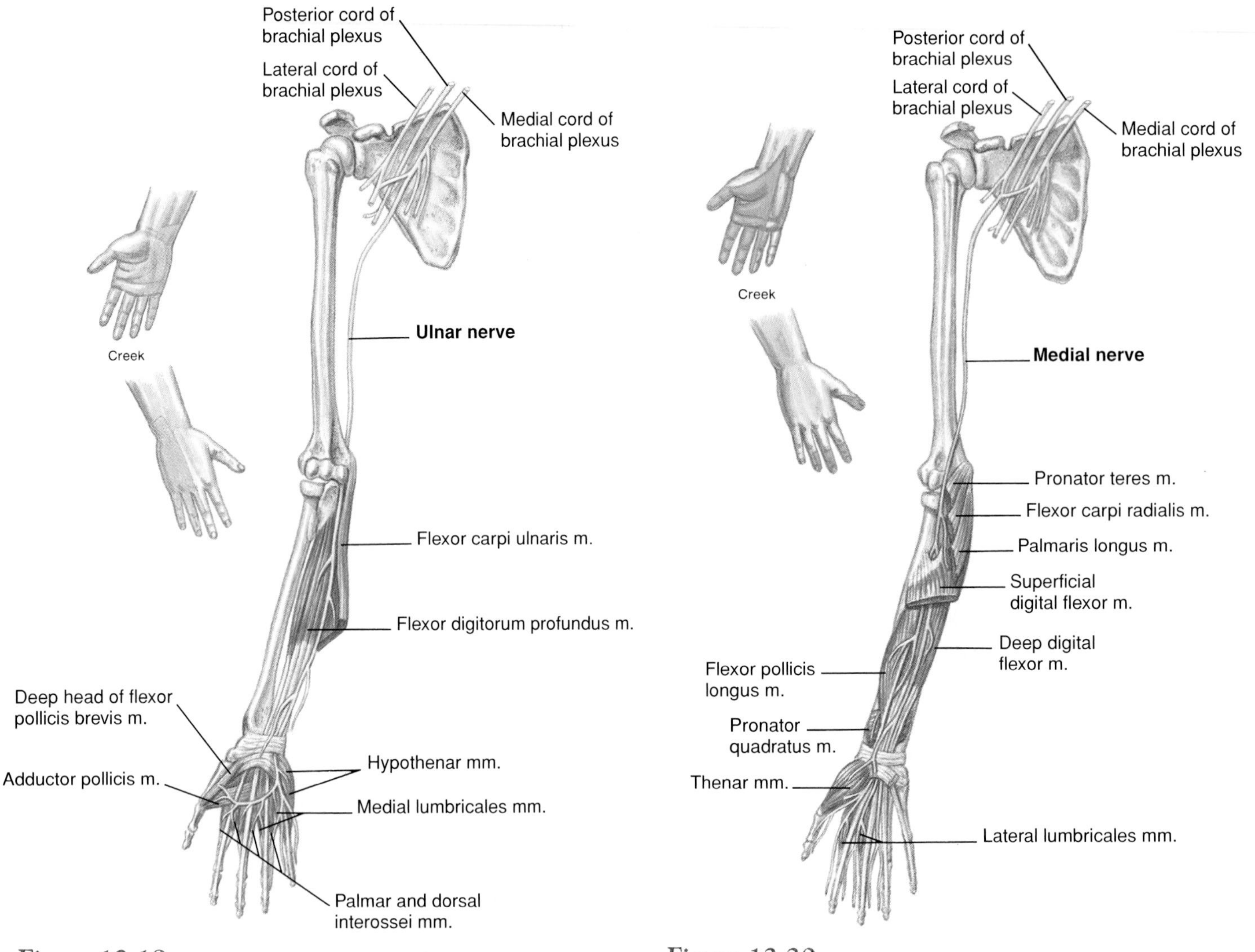

Figure 12.19

Muscular and cutaneous distribution of the ulnar nerve.

Figure 12.20

Muscular and cutaneous distribution of the median nerve.

Table 12.5
Branches of the Lumbar Plexus

Nerve	Spinal Components	Innervation
Iliohypogastric	T12, L1	Skin of lower abdomen and buttock; muscles of anterolateral abdominal wall (external abdominal oblique, internal abdominal oblique, transversus abdominis)
Ilioinguinal	L1	Skin of upper median thigh, scrotum and root of penis in male, and labia majora in female; muscles of anterolateral abdominal wall with iliohypogastric nerve
Genitofemoral	L1, L2	Skin of middle anterior surface of thigh, scrotum in male, and labia majora in female; cremaster muscle in male
Lateral cutaneous femoral	L2, L3	Skin of anterior, lateral, and posterior aspects of thigh
Femoral	L2–L4	Skin of anterior and medial aspect of thigh and medial aspect of leg and foot; anterior muscles of thigh (iliacus, psoas major, pectineus, rectus femoris, sartorius) and extensor muscles of leg (rectus femoris, vastus lateralis, vastus medialis, vastus intermedius)
Obturator	L2–L4	Skin of medial aspect of thigh; adductor muscles of lower extremity (external obturator, pectineus, adductor longus, adductor brevis, adductor magnus, gracilis)
Saphenous	L2–L4	Skin of medial aspect of lower extremity

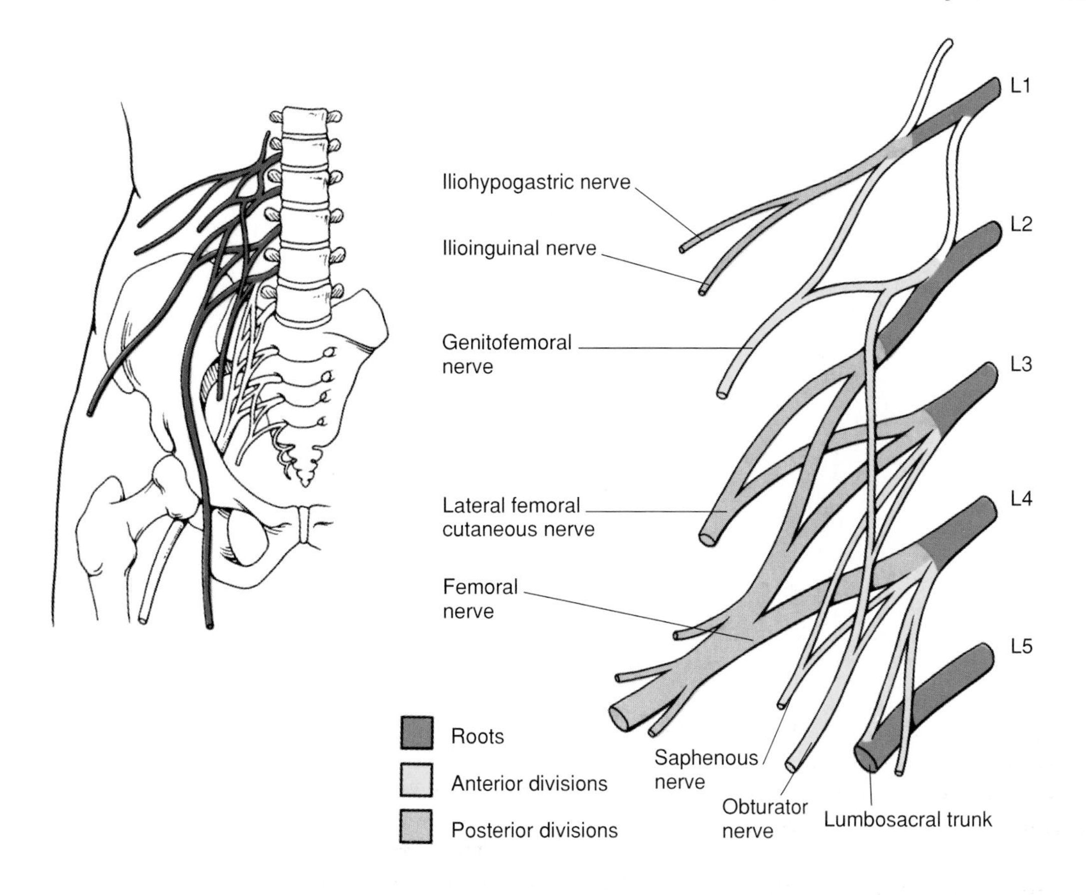

Figure 12.21

The lumbar plexus.

L5, and S1 through S4 (fig. 12.24). The nerves arising from the sacral plexus innervate the lower back, pelvis, perineum, posterior surface of the thigh and leg, and dorsal and plantar surfaces of the foot (table 12.6). Like the lumbar plexus, the sacral plexus consists of **roots** and **anterior** and **posterior divisions,** from which nerves arise. Because some of the nerves of the sacral plexus also contain fibers from the nerves of the lumbar plexus through the **lumbosacral trunk,** these two plexuses are frequently described collectively as the **lumbosacral plexus.**

The **sciatic** (*si-at'ik*) **nerve** is the largest nerve arising from the sacral plexus and is the largest nerve in the body. The sciatic nerve passes from the pelvis through the greater sciatic notch of the os coxae and extends down the posterior aspect of the thigh. It is actually composed of two nerves—the tibial and common fibular nerves—wrapped in a connective tissue sheath.

The **tibial nerve** arises from the anterior division of the sacral plexus, extends through the posterior regions of the thigh and leg, and branches in the foot to form the **medial** and **lateral plantar nerves** (fig. 12.25). The cutaneous innervation of the tibial nerve is to the calf of the leg and the plantar surface of the foot. The motor innervation of the tibial nerve is to most of the posterior thigh and leg muscles and to many of the intrinsic muscles of the foot.

The **common fibular nerve** (peroneal nerve) arises from the posterior division of the sacral plexus, extends through the posterior region of the thigh, and branches in the upper portion of the leg into the **deep** and **superficial fibular nerves** (fig. 12.26). The cutaneous innervation of the common fibular nerve and its branches is to the anterior and lateral leg and to the dorsum of the foot. The motor innervation is to the anterior and lateral muscles of the leg and foot.

sciatic: L. *sciaticus*, hip joint

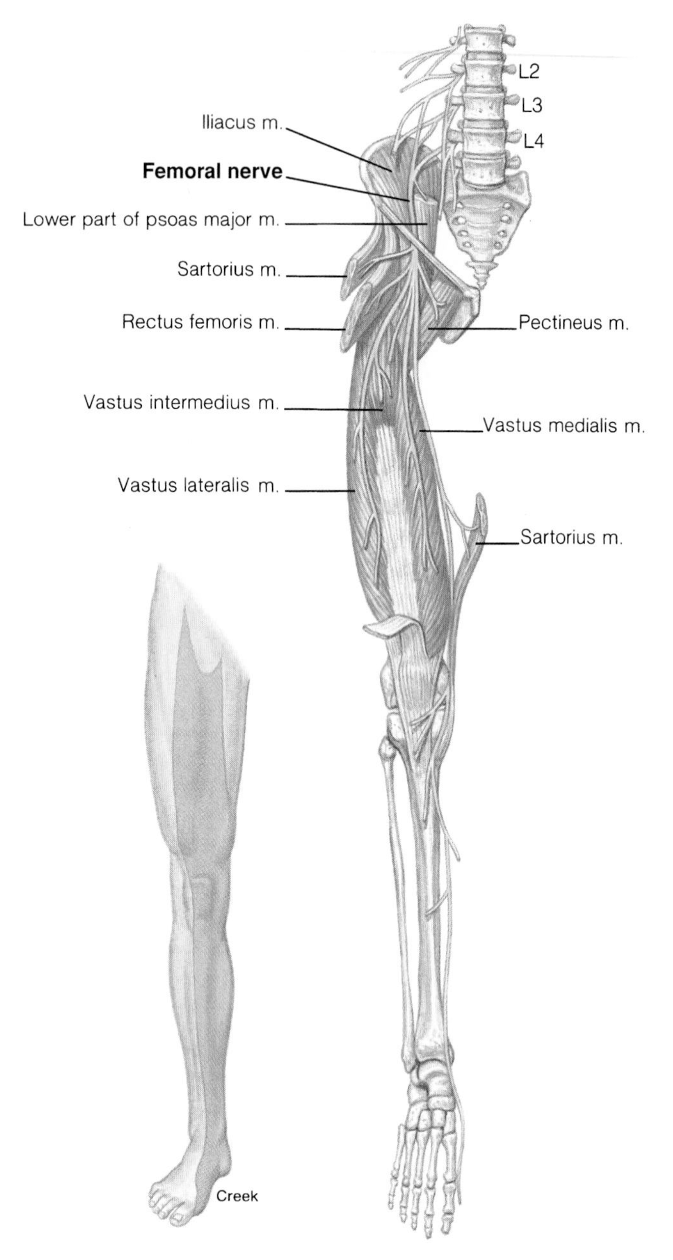

Figure 12.22

Muscular and cutaneous distribution of the femoral nerve.

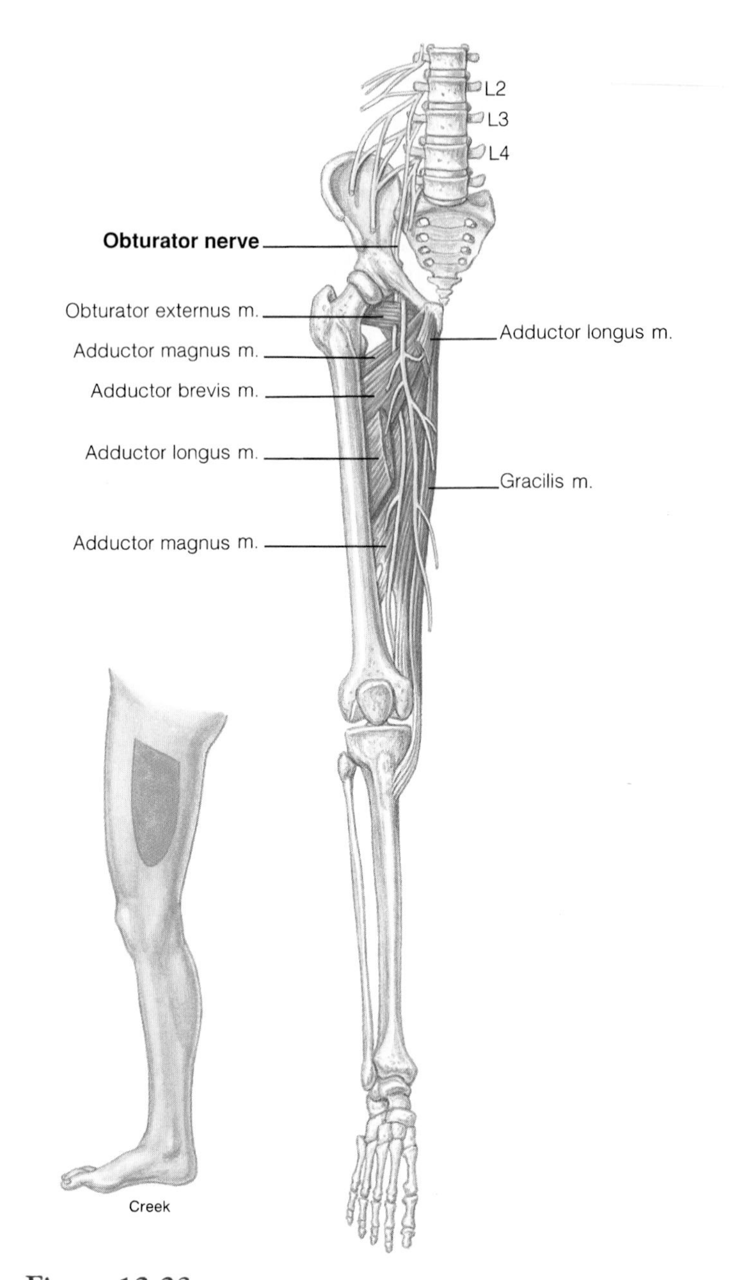

Figure 12.23

Muscular and cutaneous distribution of the obturator nerve.

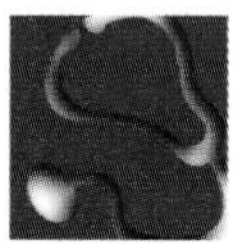

The sciatic nerve in the buttock lies deep to the gluteus maximus muscle, midway between the greater trochanter and the ischial tuberosity. Because of its position, the sciatic nerve is of tremendous clinical importance. A posterior dislocation of the hip joint will generally injure the sciatic nerve. A *herniated disc* (fig. 12.27) or pressure from the uterus during pregnancy may damage the nerve roots, resulting in a condition called *sciatica* (*si-at′ĭ-kă*). Sciatica is characterized by sharp pain in the gluteal region that extends down the posterior side of the thigh. An improperly administered injection into the buttock may injure the sciatic nerve itself. Even a temporary compression of the sciatic nerve as a person sits on a hard surface for a period of time may result in the perception of tingling throughout the limb as the person stands up. The limb is said to have "gone to sleep."

1. Define *nerve plexus*. What are the four spinal nerve plexuses and which spinal nerves contribute to each?
2. Which spinal nerves are not involved in a plexus?
3. Distinguish between a posterior ramus and an anterior ramus. Which is involved in the formation of plexuses?
4. Construct a table that lists the plexus of origin and the general region of innervation for the following nerves: (a) pudendal, (b) phrenic, (c) femoral, (d) ulnar, (e) median, (f) sciatic, (g) saphenous, (h) axillary, (i) radial, (j) ansa cervicalis, (k) tibial, and (l) common fibular.

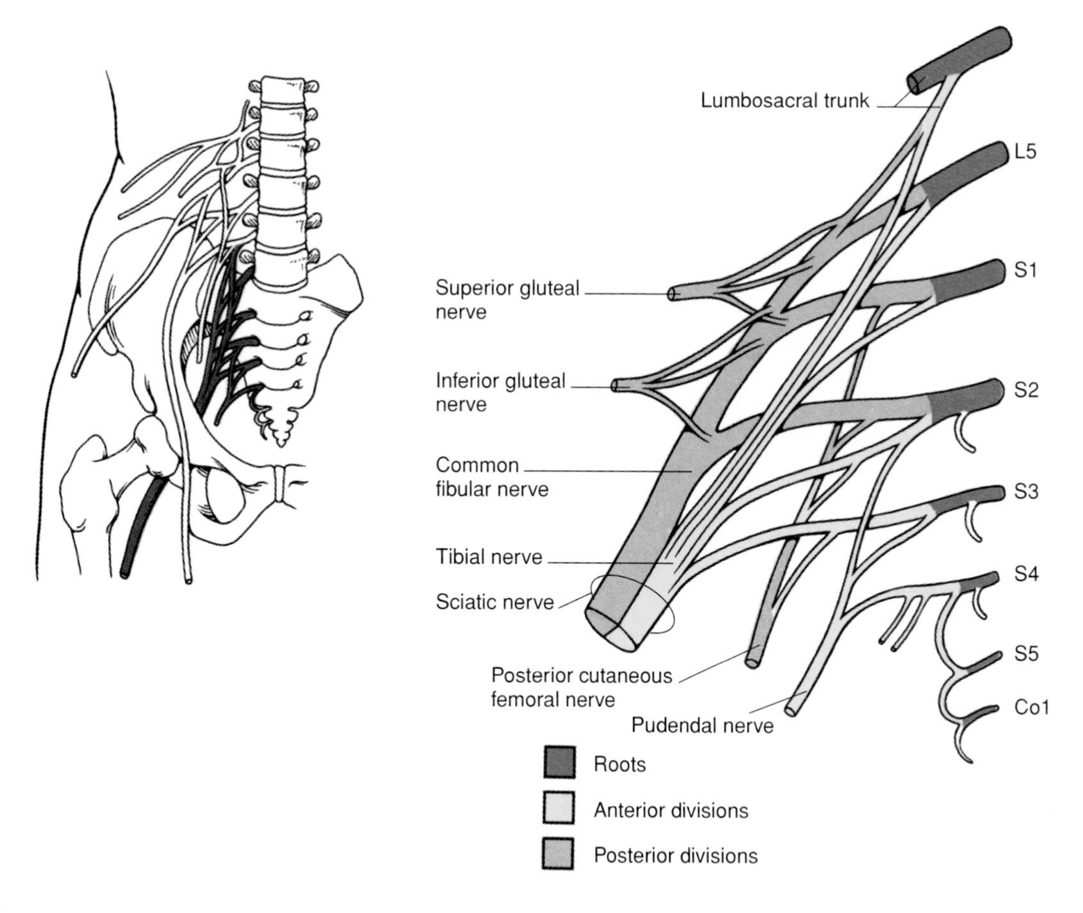

Figure 12.24

The sacral plexus.

Table 12.6

Branches of the Sacral Plexus

Nerve	Spinal Components	Innervation
Superior gluteal	L4, L5, S1	Abductor muscles of thigh (gluteus minimus, gluteus medius, tensor fasciae latae)
Inferior gluteal	L5–S2	Extensor muscle of hip joint (gluteus maximus)
Nerve to piriformis	S1, S2	Abductor and rotator of thigh (piriformis)
Nerve to quadratus femoris	L4, L5, S1	Rotators of thigh (gemellus inferior, quadratus femoris)
Nerve to internal obturator	L5–S2	Rotators of thigh (gemellus superior, internal obturator)
Perforating cutaneous	S2, S3	Skin over lower medial surface of buttock
Posterior cutaneous femoral	S1–S3	Skin over lower lateral surface of buttock, anal region, upper posterior surface of thigh, upper aspect of calf, scrotum in male and labia majora in female
Sciatic	L4–S3	Composed of two nerves (tibial and common fibular); splits into two portions at popliteal fossa; branches from sciatic in thigh region to "hamstring muscles" (biceps femoris, semitendinosus, semimembranosus) and adductor magnus muscle
Tibial (sural, medial and lateral plantar)	L4–S3	Skin of posterior surface of leg and sole of foot; muscle innervation includes gastrocnemius, soleus, flexor digitorum longus, flexor hallucis longus, tibialis posterior, popliteus, and intrinsic foot muscles
Common fibular (superficial and deep fibular)	L4–S2	Skin of anterior surface of the leg and dorsum of foot; muscle innervation includes peroneus tertius, peroneus brevis, peroneus longus, tibialis anterior, extensor hallucis longus, extensor digitorum longus, extensor digitorum brevis
Pudendal	S2–S4	Skin of penis and scrotum in male and skin of clitoris, labia majora, labia minora, and lower vagina in female; muscles of perineum

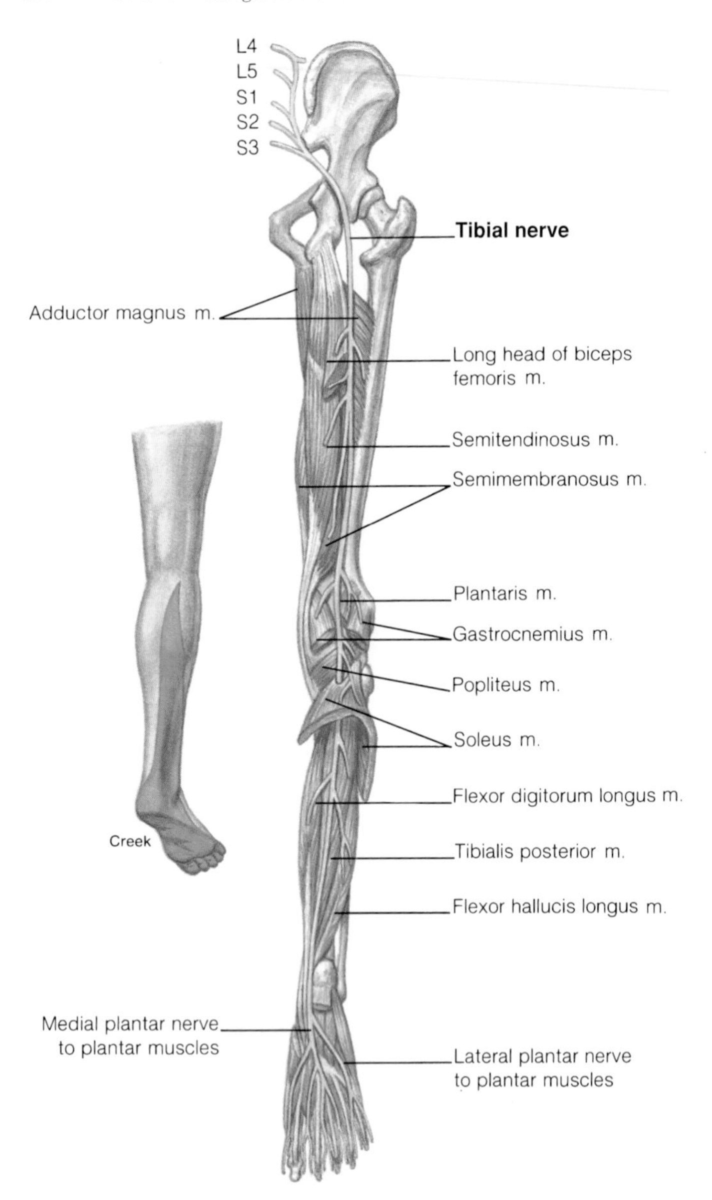

Figure 12.25

Muscular and cutaneous distribution of the tibial nerve.

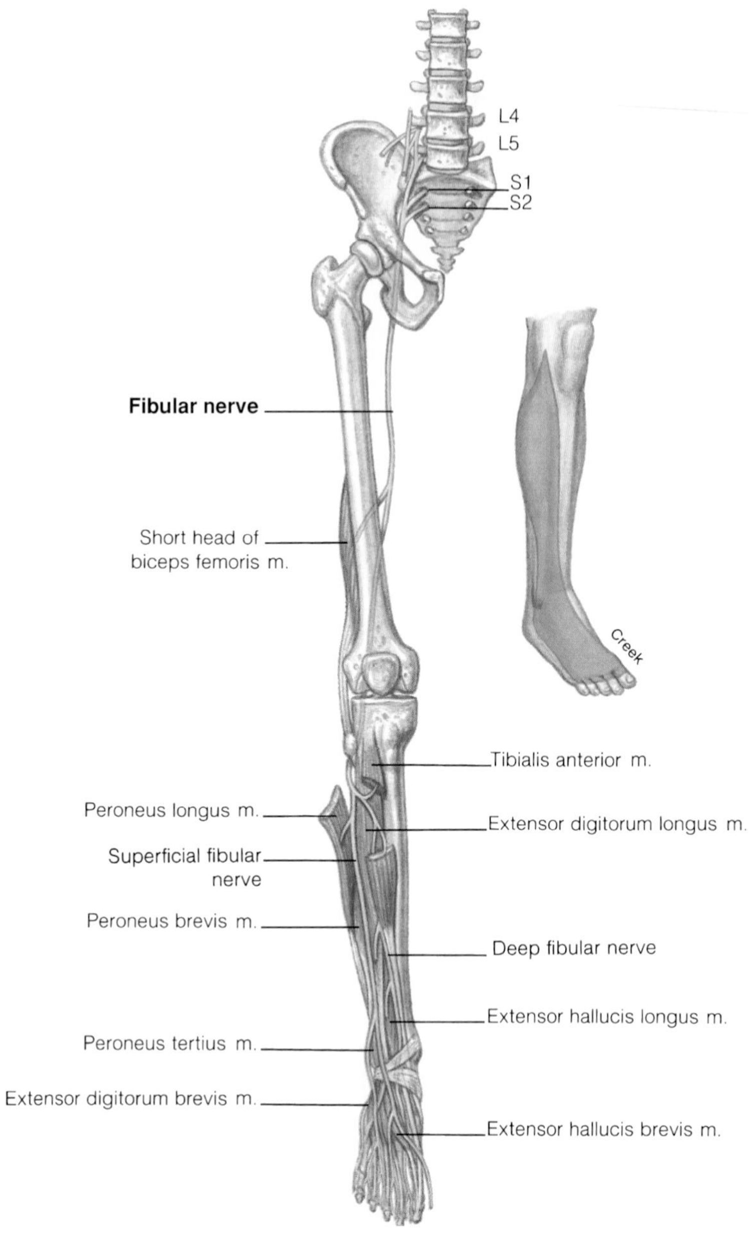

Figure 12.26

Muscular and cutaneous distribution of the fibular nerve.

Reflex Arc and Reflexes

The conduction pathway of a reflex arc consists of a receptor, a sensory neuron, a motor neuron and its innervation in the PNS, and an association neuron in the CNS. The reflex arc provides the mechanism for a rapid, automatic response to a potentially threatening stimulus.

Objective 8 Define *reflex arc* and list its five components.

Objective 9 Distinguish between the various kinds of reflexes.

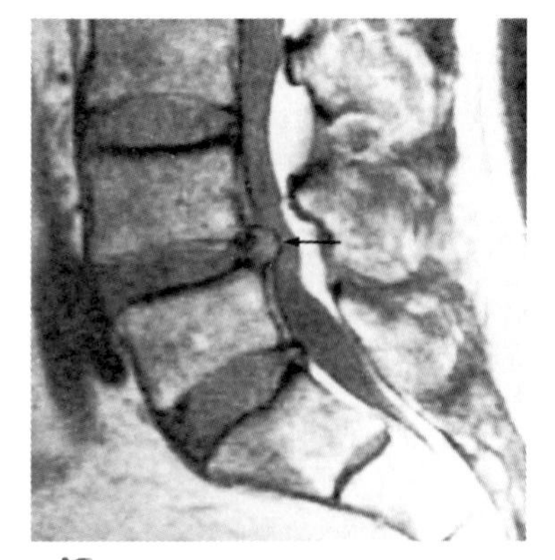

Figure 12.27

An MR scan of a herniated disc (arrow) in the lumbar region.

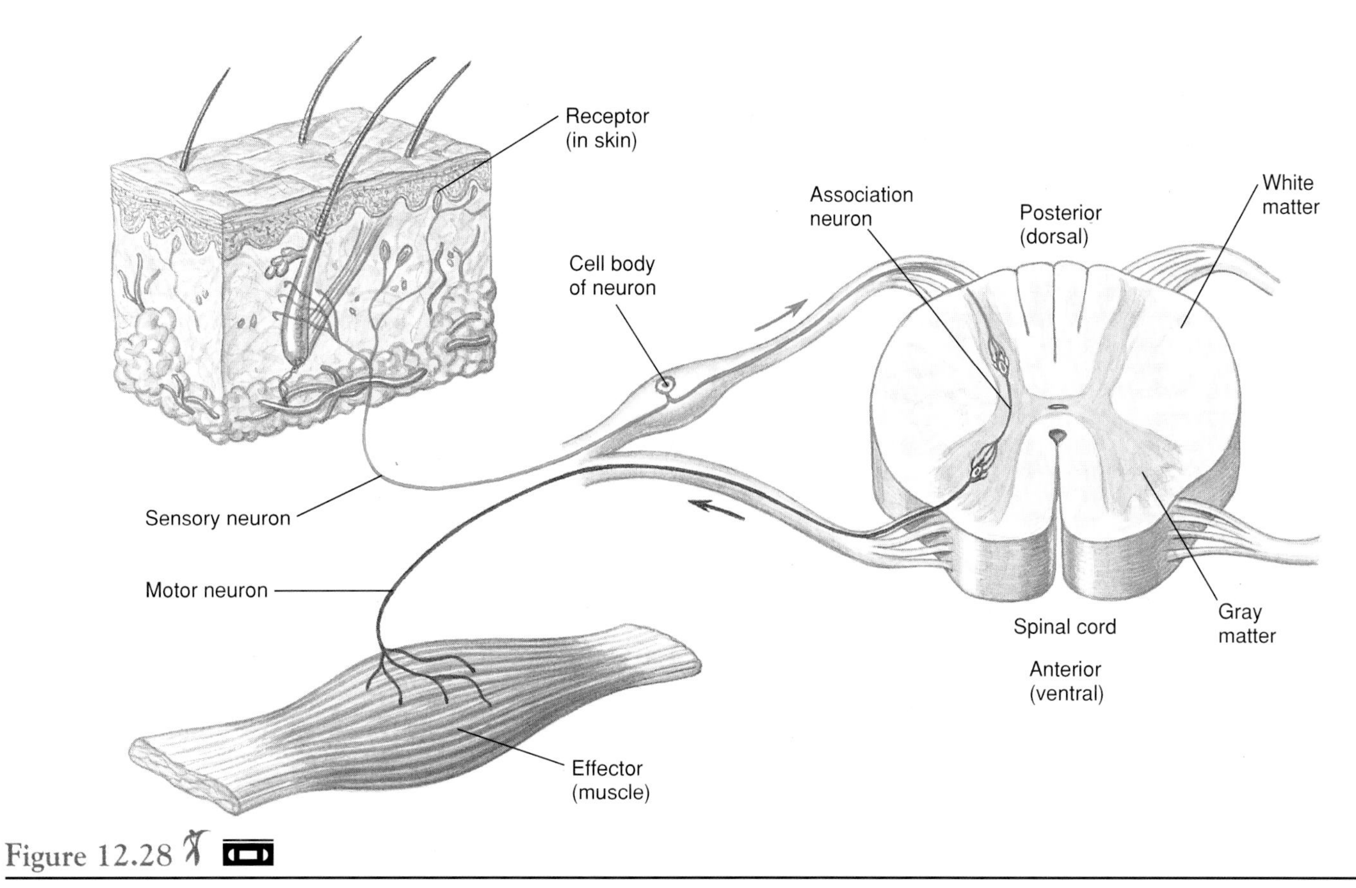

Figure 12.28

A reflex arc.

Specific **nerve pathways** provide routes by which impulses travel through the nervous system. Frequently, a nerve pathway begins with the conduction of impulses to the CNS through sensory receptors and sensory neurons of the PNS. Once within the CNS, impulses may immediately travel back through motor portions of the PNS to activate specific skeletal muscles, glands, or smooth muscles. Impulses may also be sent simultaneously to other parts of the CNS through ascending tracts within the spinal cord.

Components of the Reflex Arc

The simplest type of nerve pathway is a **reflex arc** (fig. 12.28). A reflex arc implies an automatic, unconscious, protective response to a situation in an attempt to maintain body homeostasis. Impulses are conducted over a short route from sensory to motor neurons, and only two or three neurons are involved. The five components of a reflex arc are the receptor, sensory neuron, center, motor neuron, and effector. The **receptor** includes the dendrite of a sensory neuron and the place where the electrical impulse is initiated. The **sensory neuron** relays the impulse through the posterior root to the CNS. The **center** is located within the CNS and usually involves an association neuron (interneuron). It is here that the arc is made and other impulses are sent through synapses to other parts of the body. The **motor neuron** conducts the impulse to an **effector organ** (generally a skeletal muscle). The response of the effector is called a *reflex action* or, simply, a *reflex*.

Kinds of Reflexes

Visceral Reflexes

Reflexes that cause smooth or cardiac muscle to contract or glands to secrete are **visceral** (autonomic) **reflexes.** Visceral reflexes help control the body's many involuntary processes such as heart rate, respiratory rate, blood flow, and digestion. Swallowing, sneezing, coughing, and vomiting may also be reflexive, although they involve the involuntary action of skeletal muscles.

Somatic Reflexes

Somatic reflexes are those that result in the contraction of skeletal muscles. The three principal kinds of somatic reflexes are named according to the response they produce.

The **stretch reflex** involves only two neurons and one synapse in the pathway; it is therefore called a *monosynaptic reflex arc*. Slight stretching of the neuromuscular spindle receptors (described in chapter 15) within a muscle initiates an impulse along a sensory neuron to the spinal cord. A synapse with a motor neuron occurs in the anterior gray column, and activation of a motor unit causes specific muscle fibers to contract. Since the receptor and effector organs of the stretch reflex involve structures on the same side of the spinal cord, the reflex arc is an *ipsilateral reflex arc*. The knee-jerk reflex is an ipsilateral reflex (fig. 12.29), as are all monosynaptic reflex arcs.

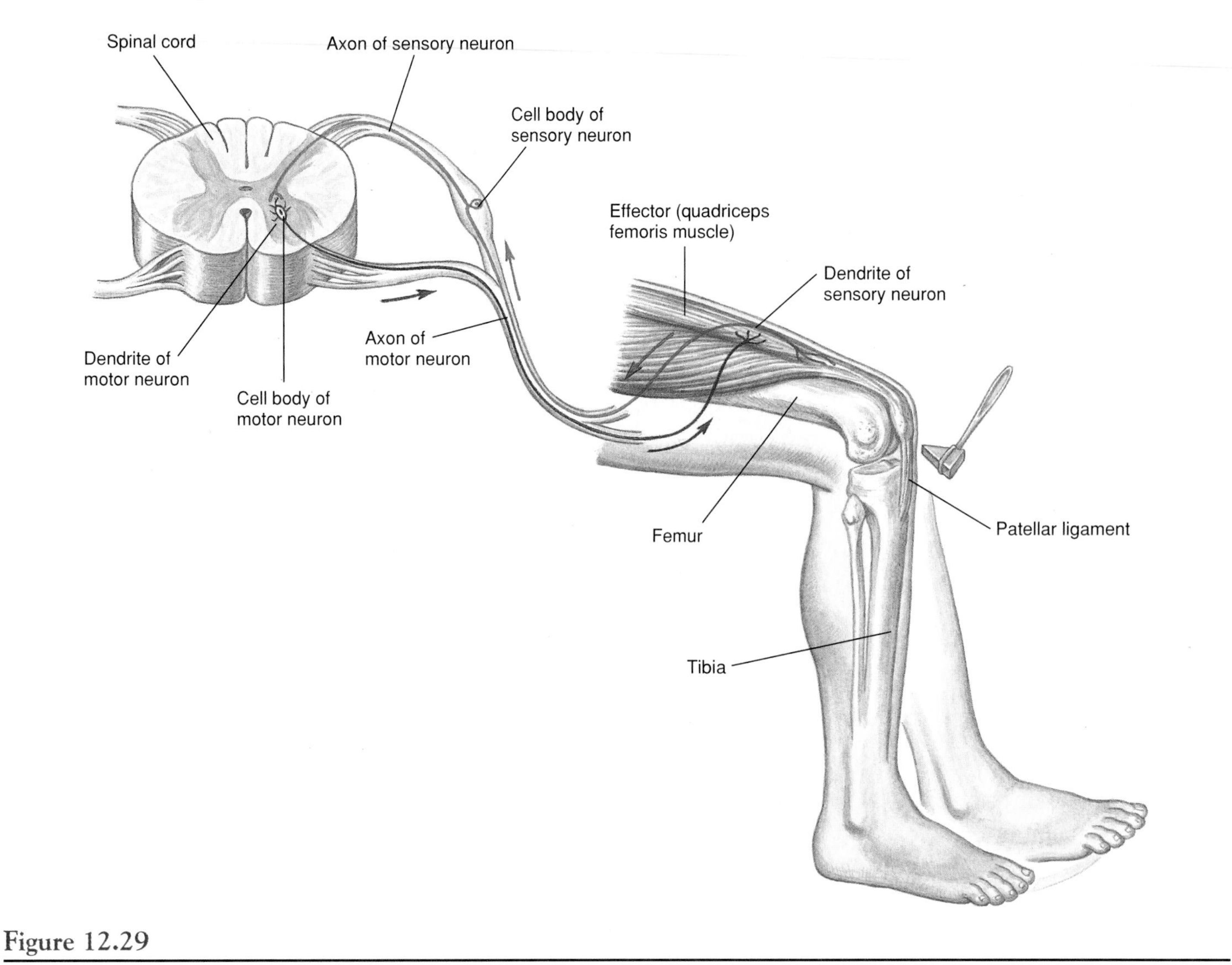

Figure 12.29

The knee-jerk reflex is an ipsilateral reflex. The receptor and effector organs are on the same side of the spinal cord. The knee-jerk reflex is also a monosynaptic reflex because it involves only two neurons and one synapse.

A **flexor reflex,** or **withdrawal reflex,** involves a *polysynaptic reflex arc* (fig. 12.30). Flexor reflexes involve association neurons in addition to the sensory and motor neurons. A flexor reflex is initiated as a person encounters a painful stimulus, such as a hot or sharp object. As a receptor organ is stimulated, sensory neurons transmit the impulse to the spinal cord, where association neurons are activated. Here, the impulses are directed through motor neurons to flexor muscles, which contract in response. Simultaneously, antagonistic muscles are inhibited (relaxed) so that the traumatized extremity can be quickly withdrawn from the harmful source of stimulation.

Several additional reflexes may be activated while a flexor reflex is in progress. In an *intersegmental reflex arc,* motor units from several segments of the spinal cord are activated by impulses coming in from the receptor organ. In an intersegmental reflex arc, more than one effector organ is stimulated. Frequently, sensory impulses from a receptor organ cross over through the spinal cord to activate effector organs in the opposite (*contralateral*) limb. This type of reflex is called a **crossed extensor reflex** (fig. 12.31) and is important for maintaining body balance while a flexor reflex is in progress. For example, withdrawal of one leg, after stepping on broken glass, requires extension of the other in order to keep from falling. The reflexive inhibition of certain muscles to contract, called *reciprocal inhibition*, also helps to maintain balance while either flexor or crossed extensor reflexes are in progress.

Certain reflexes are important for physiological functions, while others are important for avoiding injury. Some of the more common reflexes are described in table 12.7 and illustrated in figure 12.32.

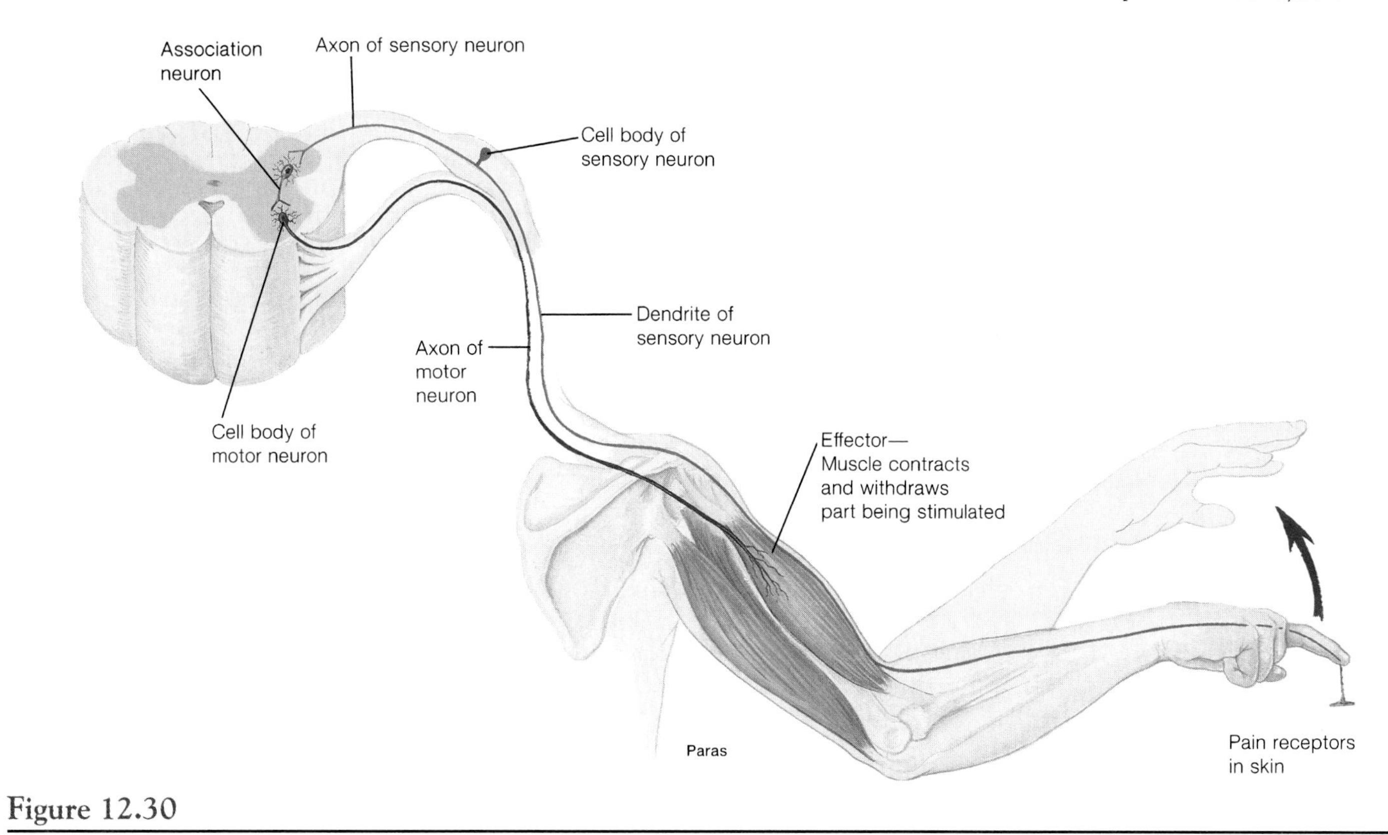

Figure 12.30

The flexor, or withdrawal, reflex is a disynaptic (two-synapse) reflex and involves association neurons in addition to sensory and motor neurons.

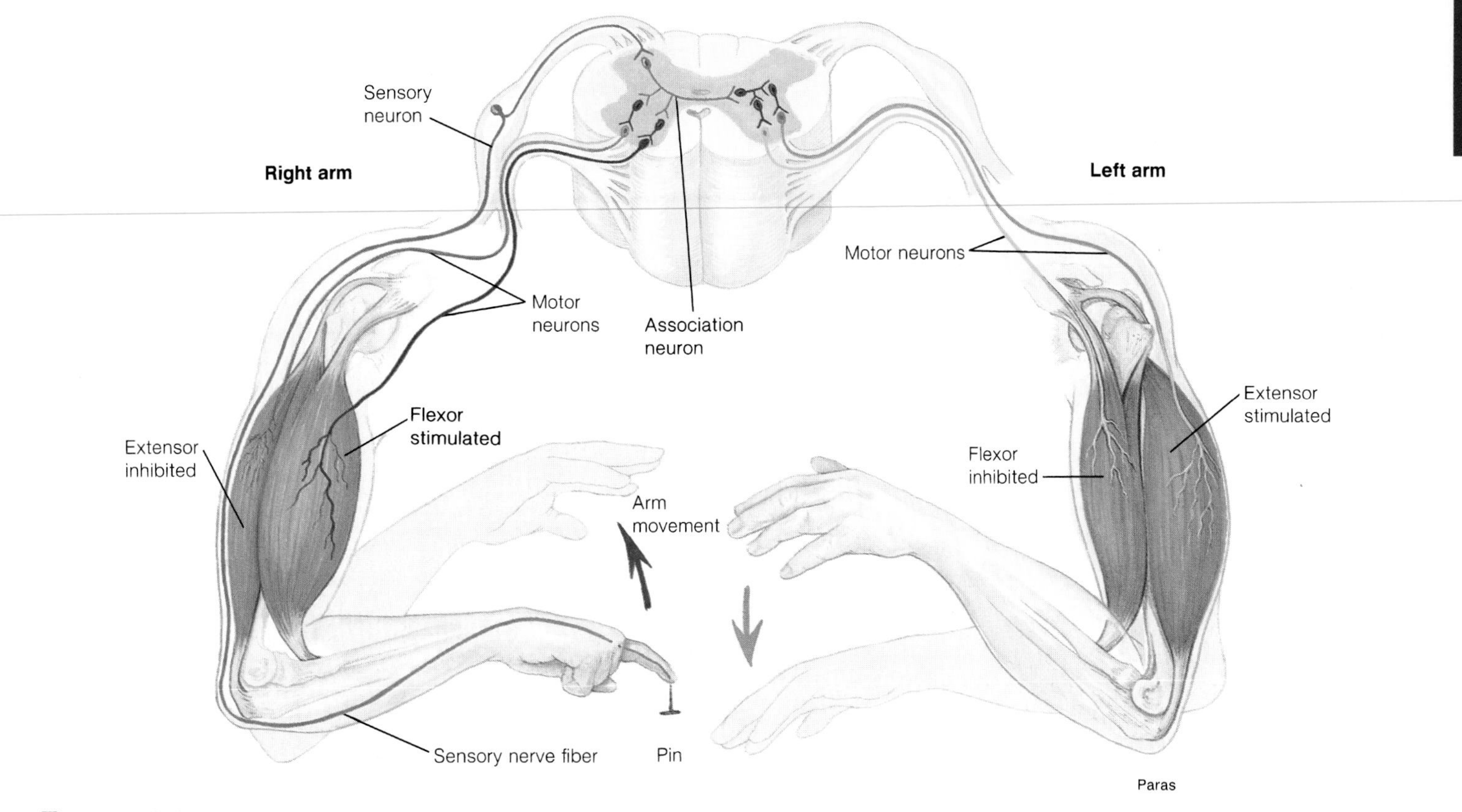

Figure 12.31

A crossed extensor reflex causes a reciprocal inhibition of muscles within the opposite appendage. This type of reflex inhibition is important in maintaining balance.

Table 12.7
Selected Reflexes of Clinical Importance

Reflex	Spinal Segment	Site of Receptor Stimulation	Effector Action
Biceps reflex	C5, C6	Tendon of biceps brachii muscle, near attachment on radial tuberosity	Contracts biceps brachii muscle to flex elbow
Triceps reflex	C7, C8	Tendon of triceps brachii muscle, near attachment on olecranon	Contracts triceps brachii muscle to extend elbow
Supinator or brachioradialis reflex	C5, C6	Radial attachment of supinator and brachioradialis muscles	Supinates forearm and hand
Knee-jerk	L2–L4	Patellar ligament, just below patella	Contracts quadriceps muscle to extend the knee
Ankle reflex	S1, S2	Tendo calcaneus, near attachment on calcaneus	Plantar flexes ankle
Plantar reflex	L4, L5, S1, S2	Lateral aspect of sole, from heel to ball of foot	Plantar flexes foot and flexes toes
Babinski's reflex*	L4, L5, S1, S2	Lateral aspect of sole, from heel to ball of foot	Extends great toe and fans other toes
Abdominal reflexes	T8–T10 above umbilicus and T10–T12 below umbilicus	Sides of abdomen, above and below level of umbilicus	Contract abdominal muscles and deviate umbilicus toward stimulus
Cremasteric reflex	L1, L2	Upper inside of thigh in males	Contracts cremasteric muscle and elevates testis on same side of stimulation

*If Babinski's reflex rather than the plantar reflex occurs as the sole of the foot is stimulated, damage to the corticospinal tract within the spinal cord may be indicated. However, Babinski's reflex is present in infants up to 12 months of age because of the immaturity of their corticospinal tracts.

Babinski's reflex: from Joseph F. Babinski, French neurologist, 1857–1932

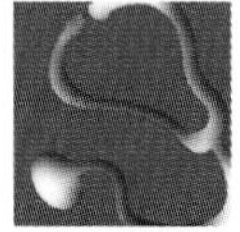

Part of a routine physical examination involves testing a person's reflexes. Several reflexes are used to assess certain neurological conditions, including functioning of the synapses. If some portion of the nervous system has been injured, the testing of certain reflexes may indicate the location and extent of the injury. Also, an anesthesiologist may try to initiate a reflex to ascertain the effect of an anesthetic.

1. How are reflexes important in maintaining body homeostasis?
2. List the five components of a reflex arc.
3. Define the following terms: *visceral reflex, somatic reflex, stretch reflex, flexor reflex, crossed extensor reflex, ipsilateral reflex,* and *contralateral reflex.*

Clinical Case Study Answer

The radial nerve lies in the radial groove of the humerus as it extends through the brachial region toward the arm and hand. In this position, the radial nerve is susceptible to injury in the case of a fracture to the midshaft of the humerus. The muscles innervated by the radial nerve include those of wrist extension (extensor carpi radialis longus and brevis, extensor carpi ulnaris); wrist adduction (extensor carpi ulnaris); wrist abduction (extensor carpi radialis brevis and longus); supination (supinator); and finger extension (finger extensors). A detectable weakness in flexion of the elbow could also result from impairment of the brachioradialis muscle. The triceps brachii muscle would not be affected, however, because of the more proximal branching point of its motor nerve supply. The radial nerve is a mixed nerve, carrying both motor and sensory fibers; therefore, a sensory deficit would be present. Decreased sensation would be detectable on the posterolateral aspect of the hand (see fig. 12.17).

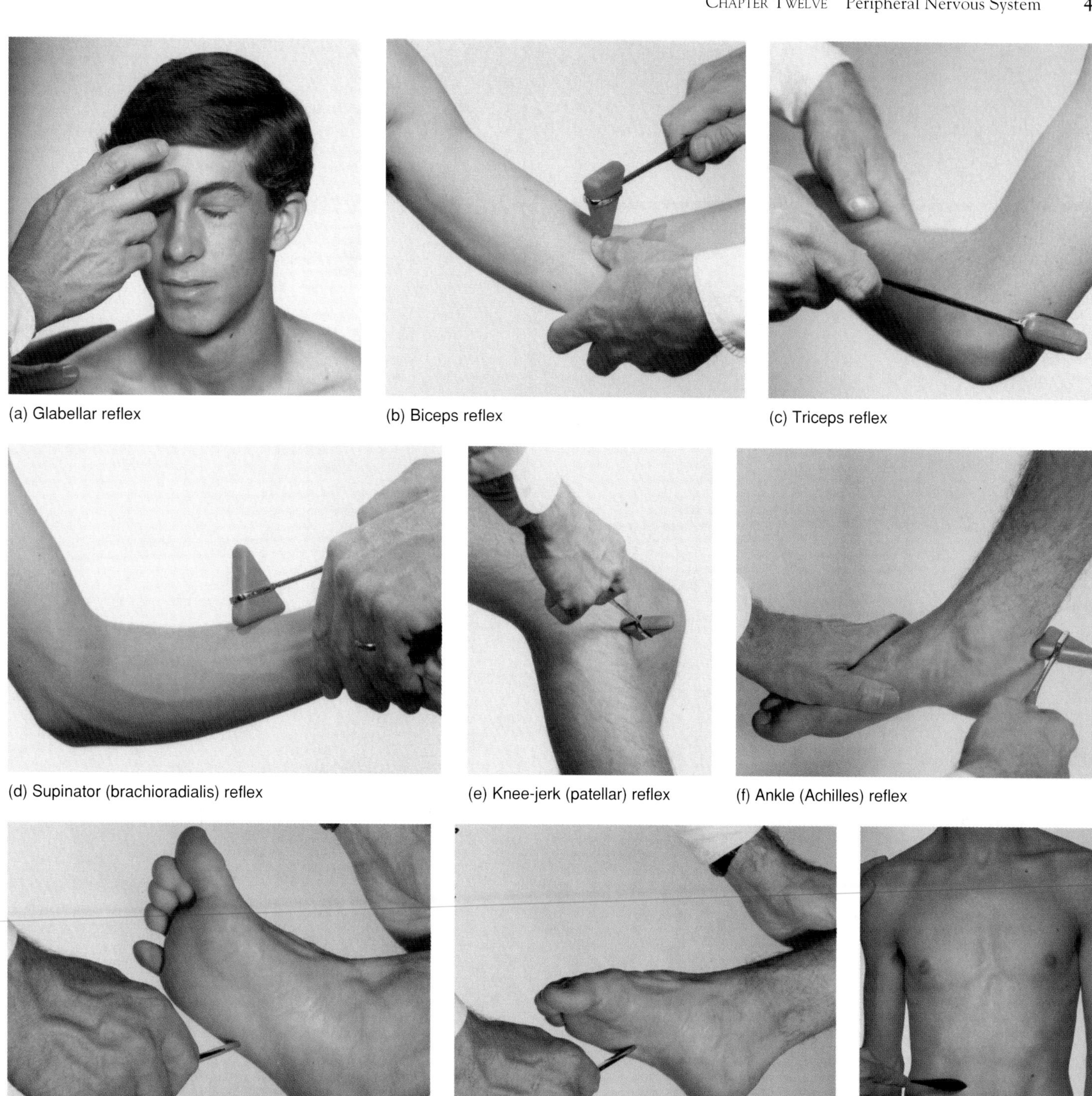
(a) Glabellar reflex
(b) Biceps reflex
(c) Triceps reflex
(d) Supinator (brachioradialis) reflex
(e) Knee-jerk (patellar) reflex
(f) Ankle (Achilles) reflex
(g) Babinski's reflex
(h) Plantar reflex
(i) Abdominal reflex

Figure 12.32

Some reflexes of clinical importance.

Developmental Exposition

The Peripheral Nervous System

Exhibit I The pattern of dermatomes within the body and the peripheral distribution of spinal nerves. (*a*) An anterior view and (*b*) a posterior view.

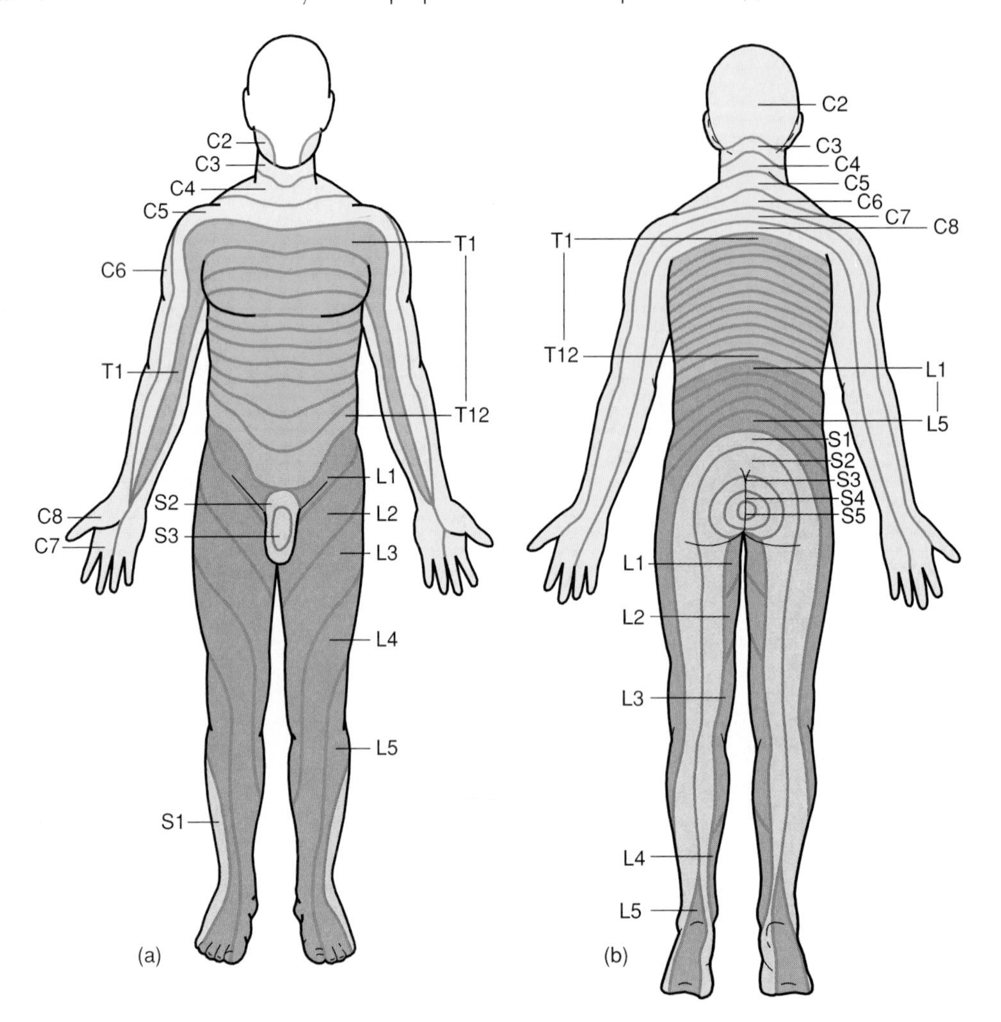

Explanation

Development of the peripheral nervous system produces the pattern of dermatomes within the body (exhibit I). A **dermatome** (*der′mă-tōm*) is an area of skin innervated by all the cutaneous neurons of a single spinal nerve or cranial nerve V. Most of the scalp and face is innervated by sensory neurons from the trigeminal (fifth cranial) nerve. With the exception of the first cervical nerve (C1), all of the spinal nerves are associated with specific dermatomes. Dermatomes are consecutive in the neck and trunk regions. In the appendages, however, adjacent dermatome innervations overlap. The apparently uneven dermatome arrangement in the appendages is due to the uneven rate of nerve growth into the limb buds. Actually, the limbs are segmented, and dermatomes overlap only slightly.

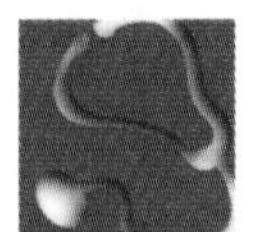

The pattern of dermatome innervation is of clinical importance when a physician wants to anesthetize a particular portion of the body. Because adjacent dermatomes overlap in the appendages, at least three spinal nerves must be blocked to produce complete anesthesia in these regions. Abnormally functioning dermatomes provide clues about injury to the spinal cord or specific spinal nerves. If a dermatome is stimulated but no sensation is perceived, the physician can infer that the injury involves the innervation to that dermatome.

dermatome: Gk. *derma*, skin; *tomia*, a cutting

Chapter Summary

Introduction to the Peripheral Nervous System (p. 389)

1. The peripheral nervous system consists of sensory receptors and the nerves that convey impulses to and from the central nervous system. Ganglia and nerve plexuses are also part of the PNS.
2. The cranial nerves arise from the brain and the spinal nerves arise from the spinal cord.
3. Sensory (afferent) nerves convey impulses toward the CNS, whereas motor (efferent) nerves convey impulses away from the CNS. Mixed nerves are composed of both sensory and motor fibers.

Cranial Nerves (pp. 389–400)

1. Twelve pairs of cranial nerves emerge from the inferior surface of the brain and, with the exception of the vestibulocochlear nerve, pass through foramina of the skull to innervate structures in the head, neck, and visceral organs of the trunk.
2. The names of the cranial nerves indicate their primary function or the general distribution of their fibers.
3. The olfactory, optic, and vestibulocochlear cranial nerves are sensory only; the trigeminal, glossopharyngeal, and vagus are mixed; and the others are primarily motor, with a few proprioceptive sensory fibers.
4. Some of the cranial nerve fibers are somatic; others are visceral.
5. Tests for cranial-nerve dysfunction are clinically important in a neurological examination.

Spinal Nerves (pp. 400–401)

1. Each of the 31 pairs of spinal nerves is formed by the union of an anterior (ventral) and posterior (dorsal) spinal root that emerges from the spinal cord through an intervertebral foramen to innervate a body dermatome.
2. The spinal nerves are grouped according to the levels of the spinal column from which they arise, and they are numbered in sequence.
3. Each spinal nerve is a mixed nerve consisting of a posterior root of sensory fibers and an anterior root of motor fibers.
4. Just beyond its intervertebral foramen, each spinal nerve divides into several branches.

Nerve Plexuses (pp. 401–410)

1. Except in thoracic nerves T2 through T12, the anterior rami of the spinal nerves combine and then split again as networks of nerves called plexuses.
   (a) There are four plexuses of spinal nerves: the cervical, the brachial, the lumbar, and the sacral.
   (b) Nerves that emerge from the plexuses are named according to the structures they innervate or the general course they take.
2. The cervical plexus is formed by the anterior rami of C1 through C4 and a portion of C5.
3. The brachial plexus is formed by the anterior rami of C5 through T1, and occasionally by some fibers from C4 and T2.
   (a) The brachial plexus is divided into roots, trunks, divisions, and cords.
   (b) The axillary, radial, musculocutaneous, ulnar, and median are the five largest nerves arising from the brachial plexus.
4. The lumbar plexus is formed by the anterior rami of L1 through L4 and by some fibers from T12.
   (a) The lumbar plexus is divided into roots and divisions.
   (b) The femoral and obturator are two important nerves arising from the lumbar plexus.
5. The sacral plexus is formed by the anterior rami of L4, L5, and S1 through S4.
   (a) The sacral plexus is divided into roots and divisions.
   (b) The sciatic nerve, composed of the common fibular and tibial nerves, arises from the sacral plexus.
   (c) The lumbar plexus and the sacral plexus are collectively referred to as the lumbosacral plexus.

Reflex Arc and Reflexes (pp. 410–414)

1. The conduction pathway of a reflex arc consists of a receptor, a sensory neuron, a motor neuron and its innervation in the PNS, and a center containing an association neuron in the CNS. The reflex arc enables a rapid, automatic response to a potentially threatening stimulus.
2. A reflex arc is the simplest type of nerve pathway.
3. Visceral reflexes cause smooth or cardiac muscle to contract or glands to secrete.
4. Somatic reflexes cause skeletal muscles to contract.
   (a) The stretch reflex is a monosynaptic reflex arc.
   (b) The flexor reflex is a polysynaptic reflex arc.

Review Activities

Objective Questions

1. Which of the following is a *false* statement concerning the peripheral nervous system?
   (a) It consists of cranial and spinal nerves only.
   (b) It contains components of the autonomic nervous system.
   (c) Sensory receptors, nerves, ganglia, and plexuses are all part of the PNS.
2. An inability to cross the eyes would most likely indicate a problem with which cranial nerve?
   (a) the optic nerve
   (b) the oculomotor nerve
   (c) the abducens nerve
   (d) the facial nerve
3. Which cranial nerve innervates the muscle that raises the upper eyelid?
   (a) the trochlear nerve
   (b) the oculomotor nerve
   (c) the abducens nerve
   (d) the facial nerve
4. The inability to walk a straight line may indicate damage to which cranial nerve?
   (a) the trigeminal nerve
   (b) the facial nerve
   (c) the vestibulocochlear nerve
   (d) the vagus nerve
5. Which cranial nerve passes through the stylomastoid foramen?
   (a) the facial nerve
   (b) the glossopharyngeal nerve
   (c) the vagus nerve
   (d) the hypoglossal nerve
6. Which of the following cranial nerves does not contain parasympathetic fibers?
   (a) the oculomotor nerve
   (b) the accessory nerve
   (c) the vagus nerve
   (d) the facial nerve
7. Which of the following is not a spinal nerve plexus?
   (a) the cervical plexus
   (b) the brachial plexus
   (c) the thoracic plexus
   (d) the lumbar plexus
   (e) the sacral plexus

8. Roots, trunks, divisions, and cords are characteristic of
   (a) the sacral plexus.
   (b) the thoracic plexus.
   (c) the lumbar plexus.
   (d) the brachial plexus.
9. Which of the following nerve-plexus associations is incorrect?
   (a) median/sacral
   (b) phrenic/cervical
   (c) axillary/brachial
   (d) femoral/lumbar
10. Extending the knee joint when the patellar ligament is tapped is an example of
   (a) a visceral reflex.
   (b) a flexor reflex.
   (c) an ipsilateral reflex.
   (d) a crossed extensor reflex.

Essay Questions

1. Explain the structural and functional relationships between the central nervous system, the autonomic nervous system, and the peripheral nervous system.
2. List the cranial nerves and describe the major function(s) of each. How is each cranial nerve tested for dysfunction?
3. Describe the structure of a spinal nerve.
4. List the roots of each of the spinal plexuses. Describe where each plexus is located and state the nerves that originate from it.
5. What is a reflex arc? Explain how reflexes are important in maintaining body homeostasis.
6. Distinguish between monosynaptic, polysynaptic, ipsilateral, stretch, and flexor reflexes.

Critical-Thinking Questions

1. A person with quadriplegia from a spinal cord injury at the level of C5 can speak, digest food, breath, and regulate his or her heartbeat, yet the person cannot move muscles from the shoulder down. Explain why.
2. The doctor taps your patellar ligament with a reflex hammer and can't elicit a kick. What's the matter with you? If it's good to have your leg jump a little, is it better to have your leg jump a lot?
3. A 63-year-old truck driver made an appointment with the company doctor because of pain and numbness in his left leg. Following a routine physical exam, the physician scheduled the man for a magnetic resonance imaging of his lumbar region. The MR scan indicated a herniated disc in the lumbar region. Explain how such a condition could develop and account for the man's symptoms. Discuss the possible treatment of this condition. (You may have to refer to medical textbooks in your library to find the answer.)
4. After being revived from a knockout punch during the world featherweight championship fight, the ringside physician determined that the boxer had sustained oculomotor and facial nerve damage when his right zygomatic bone was shattered from a hard left hook. What symptoms might the boxer have displayed that would cause the physician to come to this conclusion?

THIRTEEN

Autonomic Nervous System

Clinical Case Study 420
Introduction to the Autonomic Nervous System 420
Organization of the Autonomic Nervous System 420
Visceral Effector Organs 422
Structure of the Autonomic Nervous System 423
Sympathetic (Thoracolumbar) Division 423
Parasympathetic (Craniosacral) Division 425
Functions of the Autonomic Nervous System 429
Neurotransmitters of the Autonomic Nervous System 429
Responses to Adrenergic Stimulation 429
Responses to Cholinergic Stimulation 430
Organs with Dual Innervation 431
Organs without Dual Innervation 432
Control of the Autonomic Nervous System by Higher Brain Centers 433
Medulla Oblongata 433
Hypothalamus 433
Limbic System, Cerebellum, and Cerebrum 434
Clinical Case Study Answer 435
Clinical Considerations 435
Autonomic Dysreflexia 435
Chapter Summary 436
Review Activities 437

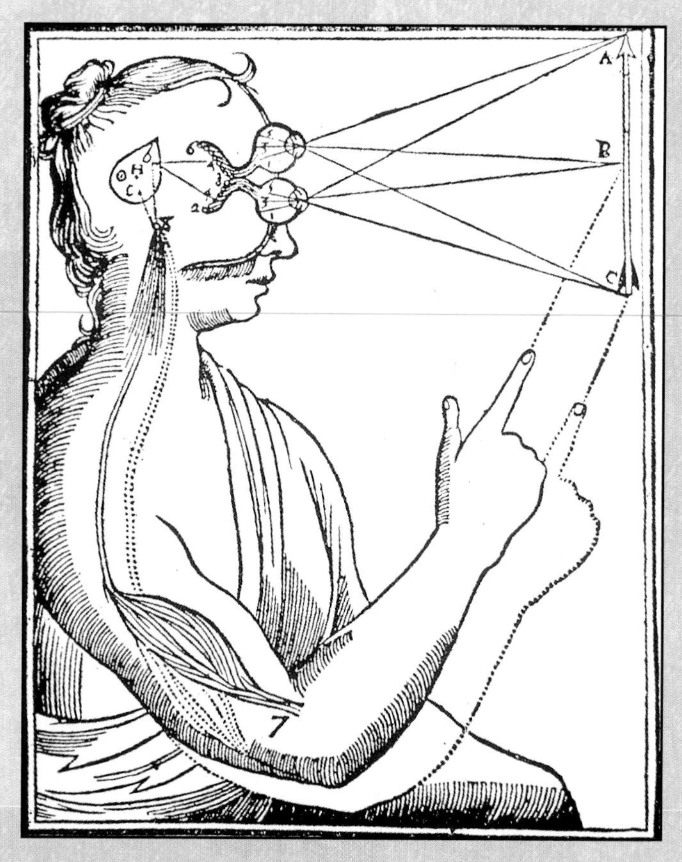

This illustration from The Treatise of Man *by seventeenth-century French philosopher René Descartes shows the supposed relationship between the sensory perception of an image and muscular action. Descartes was partly correct in his description of the visual process and about the reflex arc, but his belief that animal spirits passed through the optic nerve to activate the pineal gland was pure Galen.*

Clinical Case Study

A 67-year-old female with unresectable lung cancer went to a tumor clinic for a routine checkup. The only trouble she had experienced in the 6 months following her diagnosis was occasional coughing, slight shortness of breath, and a mild but nagging pain in the right side of her chest. She explained to her doctor that lately the right side of her face had been "feeling funny." Her symptoms, upon closer questioning, included lack of perspiration on the right side of her face during her regular morning walks. Examination of the patient revealed a right pupil that was much smaller than the left, along with a drooping right eyelid. Recognizing the patient's abnormal face and eye findings as Horner's syndrome, the doctor turned to the chest CT scan in her file to look for the cause. He knowingly nodded as he examined the copy of the scan, noting that the lung tumor, located in the medial aspect of the right upper lobe, had invaded the mediastinum and the ascending nerve tracts into the base of the neck.

What area or general structure(s) within the autonomic nervous system has the tumor invaded and compromised? Using autonomic nervous system terminology, explain the reason for each of the three findings (lack of facial sweating on the right side, pupillary constriction, and eyelid droop) and include a description of neural pathways and effector organs.

Introduction to the Autonomic Nervous System

The action of effectors (muscle tissue and glandular epithelium) is controlled to a large extent by motor neuron impulses. Skeletal muscles, which are the voluntary effectors, are regulated by somatic motor impulses. The involuntary effectors (smooth muscle tissue, cardiac muscle tissue, and glandular epithelium) are regulated by autonomic motor impulses through the autonomic nervous system.

Objective 1 Define the terms *preganglionic neuron* and *postganglionic neuron* and explain how the motor pathways of the somatic motor and autonomic motor systems differ.

Objective 2 Explain how the autonomic innervation of involuntary effectors differs from the innervation of skeletal muscle.

Objective 3 Compare single-unit smooth muscle tissue and multiunit smooth muscle tissue in terms of structure and regulation by autonomic nerve impulses.

Organization of the Autonomic Nervous System

The autonomic portion of the nervous system is concerned with maintaining homeostasis within the body by increasing or decreasing the activity of various organs in response to changing physiological conditions. Although the **autonomic nervous system (ANS)** is composed of portions of both the central nervous system and peripheral nervous system, it functions independently and without a person's conscious control.

Autonomic motor nerves innervate organs whose functions are not usually under voluntary control. The effectors that respond to autonomic regulation include **cardiac muscle tissue** (within the heart), **smooth muscle tissue** (within the viscera), and **glandular epithelium.** These effectors are part of the organs of the *viscera*, of blood vessels, and of specialized structures within other organs. The involuntary effects of autonomic innervation contrast with the voluntary control of skeletal muscles by way of somatic motor innervation.

The traditional distinction between the somatic system and the autonomic nervous system is based on the fact that the former is under conscious control whereas the latter is not. Recently, however, it has been discovered that we have the remarkable ability to consciously influence autonomic activity using techniques such as biofeedback and meditation. This "discovery" comes as old news to Indian yogis, who have been exploiting this ability for generations.

Unlike the somatic motor system, in which impulses are conducted along a single axon from the spinal cord to the neuromuscular junction, the autonomic motor pathway involves two neurons in the motor transmission of impulses (table 13.1). The first of these autonomic motor neurons has its cell body in the gray matter of the brain or spinal cord. Rather than directly innervating the effector organ, the axon of this neuron synapses with a second neuron within an *autonomic ganglion*. (A ganglion is a collection of neuron cell bodies outside the CNS.) The first neuron is thus called a **preganglionic,** or presynaptic, **neuron.** The second neuron in this pathway, called a **postganglionic,** or postsynaptic, **neuron,** has an axon that extends from the autonomic ganglion and synapses with the cells of an effector organ (fig. 13.1).

Preganglionic autonomic neurons originate in the midbrain and hindbrain and from the upper thoracic to the fourth sacral portions of the spinal cord, with the exception of the area between L3 and S1. Autonomic ganglia are located in the head, neck, and abdomen. Chains of autonomic ganglia also parallel the spinal cord along each side. The origin of the preganglionic neurons and the location of the autonomic ganglia help to differentiate the **sympathetic** and **parasympathetic divisions** of the autonomic system, discussed in later sections of this chapter.

viscera: L. *viscera*, internal organs
autonomic: Gk. *auto*, self; *nomos*, law
ganglion: Gk. *ganglion*, a swelling or knot

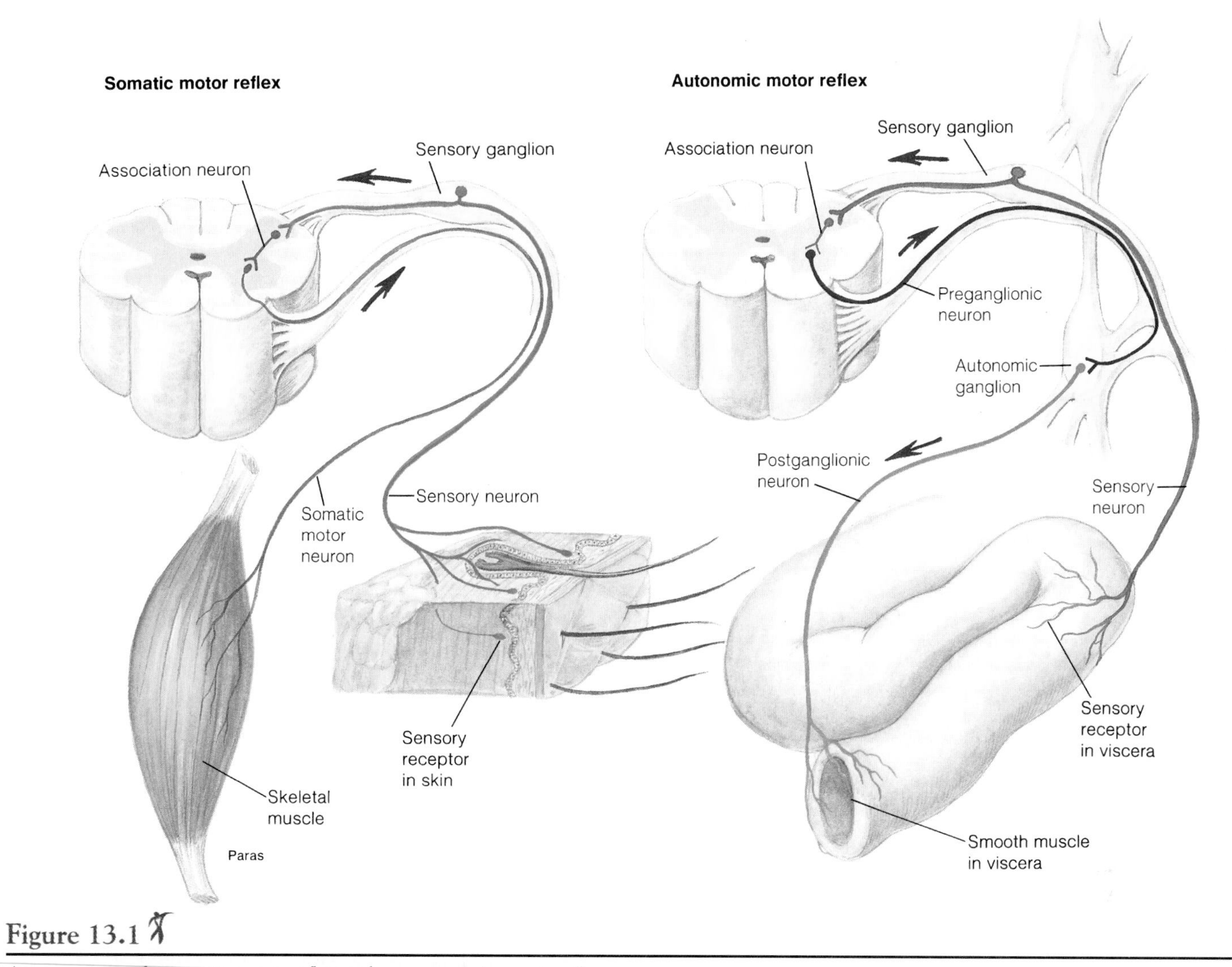

Figure 13.1

A comparison of a somatic motor reflex and an autonomic motor reflex.

Table 13.1

Comparison of the Somatic Motor and Autonomic Motor Systems

Feature	Somatic Motor	Autonomic Motor
Effector organs	Skeletal muscle tissue	Cardiac muscle tissue, smooth muscle tissue, and glandular epithelium
Presence of ganglia	No ganglia	Cell bodies of postganglionic autonomic fibers located in paravertebral, prevertebral (collateral), and terminal ganglia
Number of neurons from CNS to effector	One	Two
Type of neuromuscular junction	Specialized motor end plate	No specialization of postsynaptic membrane; all areas of smooth muscle cells contain receptor proteins for neurotransmitters
Effect of nerve impulse on muscle	Excitatory only	Either excitatory or inhibitory
Type of nerve fibers	Fast-conducting, thick (9–13 μm), and myelinated	Slow-conducting; preganglionic fibers lightly myelinated but thin (3 μm); postganglionic fibers unmyelinated and very thin (about 1.0 μm)
Effect of denervation	Flaccid paralysis and atrophy	Minimal effect on muscle tone and function; target cells show denervation hypersensitivity

Visceral Effector Organs

Unlike skeletal muscles, which enter a state of flaccid paralysis when their motor nerves are severed, the involuntary effectors are somewhat independent of their innervation. Smooth muscles maintain a resting tone (tension) in the absence of nerve stimulation. Damage to an autonomic nerve, in fact, makes its target muscle more sensitive than normal to stimulating agents.

In addition to their intrinsic (built-in) muscle tone, cardiac muscle and many smooth muscles contract rhythmically, even in the absence of nerve stimulation, in response to electrical waves of depolarization initiated by the muscles themselves. Autonomic nerves also maintain a resting tone in the sense that they maintain a baseline firing rate that can be either increased or decreased. Changes in tonic neural activity produce changes in the intrinsic activity of the effector organ. A decrease in the excitatory input to the heart, for example, will slow its rate of beat.

Cardiac Muscle

Like skeletal muscle fibers, cardiac muscle fibers are striated. The long, fibrous skeletal muscle fibers, however, are structurally and functionally separated from each other, whereas the cardiac fibers are short, branched, and interconnected by **intercalated** (*in-ter'kă-lāt-ed*) **discs.**

Electrical impulses that originate at any point in the mass of cardiac fibers called the **myocardium** (*mi"o-kar'de-um*) can spread to all cells in the mass that are joined by intercalated discs. Because all of the cells in the myocardium are electrically joined, the myocardium behaves as a single functional unit, or a *functional syncytium* (*sin-sish'e-um*). Unlike skeletal muscles, which can produce graded contractions with a strength that depends on the number of cells stimulated, the heart contracts with an *all-or-none contraction*.

Furthermore, whereas skeletal muscles require external stimulation by somatic motor nerves before they can produce action potentials and contract, cardiac muscle is able to produce action potentials automatically. Cardiac action potentials normally originate in a specialized group of cells called the *pacemaker* (see fig. 16.11). However, the rate of this spontaneous depolarization, and thus the rate of the heartbeat, is regulated by autonomic innervation.

Smooth Muscles

Smooth (visceral) muscle tissue is arranged in circular layers around the walls of blood vessels, bronchioles (small air passages in the lungs), and in the sphincter muscles of the GI tract. However, both circular and longitudinal smooth muscle layers are found in the tubular GI tract, the ureters (which transport urine), the ductus deferentia (which transport sperm), and the uterine tubes (which transport ova). The alternate contraction of circular and longitudinal smooth muscle layers produces **peristaltic waves** that propel the contents of these tubes in one direction.

Smooth muscle fibers do not contain sarcomeres (which produce striations in skeletal and cardiac muscle). Smooth muscle fibers do, however, contain a great deal of actin and some myosin, which produces a ratio of thin-to-thick myofilaments of about 16:1 (in striated muscles the ratio is 2:1).

The long length of myosin myofilaments and the fact that they are not organized into sarcomeres helps the smooth muscles to function optimally. Smooth muscles must be able to exert tension even when greatly stretched—in the urinary bladder, for example, the smooth muscle cells may be stretched up to two and a half times their resting length. Skeletal muscles, by contrast, lose their ability to contract when the sarcomeres are stretched to the point where actin and myosin no longer overlap.

Single-Unit and Multiunit Smooth Muscles

Smooth muscles are often grouped into two functional categories: **single-unit** and **multiunit.** Single-unit smooth muscles have numerous gap junctions (electrical synapses) between adjacent cells that weld them together electrically; thus, they behave as a single unit. Multiunit smooth muscles have few, if any, gap junctions; thus, the individual cells must be stimulated separately by nerve fibers. This is similar to the control of skeletal muscles, in which numerous motor units are activated.

Single-unit smooth muscles display *pacemaker activity*, in which certain cells stimulate others in the mass. Single-unit smooth muscles also display intrinsic, or *myogenic* (*mi"ŏ-jen'ik*), electrical activity and contraction in response to stretch. For example, the stretch induced by an increase in the luminal contents of a small artery or a section of the GI tract can stimulate myogenic contraction. Such contraction does not require stimulation by autonomic nerves. By contrast, contraction of multiunit smooth muscles requires nerve stimulation. Single-unit and multiunit smooth muscles are compared in table 13.2.

Autonomic Innervation of Smooth Muscles

The neural control of skeletal muscles and that of smooth muscles differ markedly. A skeletal muscle fiber has only one junction with a somatic nerve fiber, and the receptors for the neurotransmitter are localized at the neuromuscular junction in the membrane of the skeletal muscle fiber. By contrast, the entire surface of smooth muscle fibers contains neurotransmitter receptor proteins. Neurotransmitter molecules are released along a stretch of an autonomic nerve fiber that is located some distance from the smooth muscle fibers. The regions of the autonomic fiber that release transmitters appear as bulges, or *varicosities*, and the neurotransmitters released from these varicosities stimulate a number of smooth muscle fibers.

---

myogenic: Gk. *mys*, muscle; *genesis*, origin

Table 13.2

Comparison of Single-Unit and Multiunit Smooth Muscles

Feature	Single-Unit Muscle	Multiunit Muscle
Location	Gastrointestinal tract; uterus; ureter; small arteries (arterioles)	Arrector pili muscles of hair follicles; ciliary muscle (controls shape of lens); iris; ductus deferens; large arteries
Origin of electrical activity	Spontaneous activity by pacemakers (myogenic)	Not spontaneously active; potentials are neurogenic
Type of potentials	Action potentials	Graded depolarizations
Response to stretch	By contraction; not dependent on nerve stimulation	No inherent response
Presence of gap junctions	Numerous gap junctions join all cells together electrically	Few (if any) gap junctions
Type of contraction	Slow and sustained	Slow and sustained

1. How does the neural regulation of cardiac and smooth muscle fibers differ from that of skeletal muscle fibers? How are these three types of muscle tissue affected by the experimental removal of their innervation?
2. Define the terms *preganglionic* and *postganglionic neurons* in the ANS and use a diagram to illustrate how motor innervation differs in somatic and autonomic nerves.
3. Distinguish between single-unit and multiunit smooth muscles. Explain how the two categories are regulated differently by autonomic nerves.

Structure of the Autonomic Nervous System

The sympathetic and parasympathetic divisions of the autonomic nervous system both consist of preganglionic neurons with cell bodies located in the CNS and postganglionic neurons with cell bodies located outside of the CNS in ganglia. However, the specific origin of the preganglionic neurons and the location of the ganglia differ in the two subdivisions of the autonomic nervous system.

Objective 4 Describe the origin of preganglionic sympathetic neurons and the location of sympathetic ganglia.

Objective 5 Explain the relationship between the sympathetic division and the adrenal medulla.

Objective 6 Describe the origin of the preganglionic parasympathetic neurons and the location of the parasympathetic ganglia.

Objective 7 Describe the distribution of the vagus nerve and comment on its significance within the parasympathetic division.

Sympathetic (Thoracolumbar) Division

The **sympathetic division** is also called the *thoracolumbar division* of the ANS because its preganglionic neurons exit the vertebral column from the first thoracic (T1) to the second lumbar (L2) levels. Most sympathetic neurons, however, separate from the somatic motor neurons and synapse with postganglionic neurons within chains of sympathetic trunk ganglia located on either side of the vertebral column (fig. 13.2).

Since the preganglionic sympathetic neurons are myelinated and thus appear white, the "side roads" to the sympathetic ganglia are called **white rami communicantes** (*ra'mi kŏ"myoo-nĭ-kan'tēz*—singular, *ramus communicans*) (fig. 13.3). Some of these preganglionic sympathetic neurons synapse with postganglionic neurons located at their same level in the chain of sympathetic ganglia. Other preganglionic neurons travel up or down within the sympathetic chain before synapsing with postganglionic neurons. Since the postganglionic sympathetic neurons are unmyelinated and thus appear gray, they form the **gray rami communicantes.** Postganglionic axons in the gray rami extend directly back to the anterior roots of the spinal nerves and travel distally within the spinal nerves to innervate their effector organs.

Within the sympathetic trunk ganglia, *divergence* is apparent as preganglionic neurons branch to synapse with numerous postganglionic neurons located at different levels in the chain. *Convergence* is apparent also when a postganglionic neuron receives synaptic input from a large number of preganglionic neurons. The divergence of impulses from the spinal cord to the ganglia and the convergence of impulses within the ganglia usually results in the *mass activation* of almost all of the postganglionic neurons. This explains why the sympathetic division is usually activated as a unit and affects all of its effector organs at the same time.

Many preganglionic neurons that exit the spinal cord in the upper thoracic level travel through the sympathetic chain

ramus: L. *ramus*, a branch

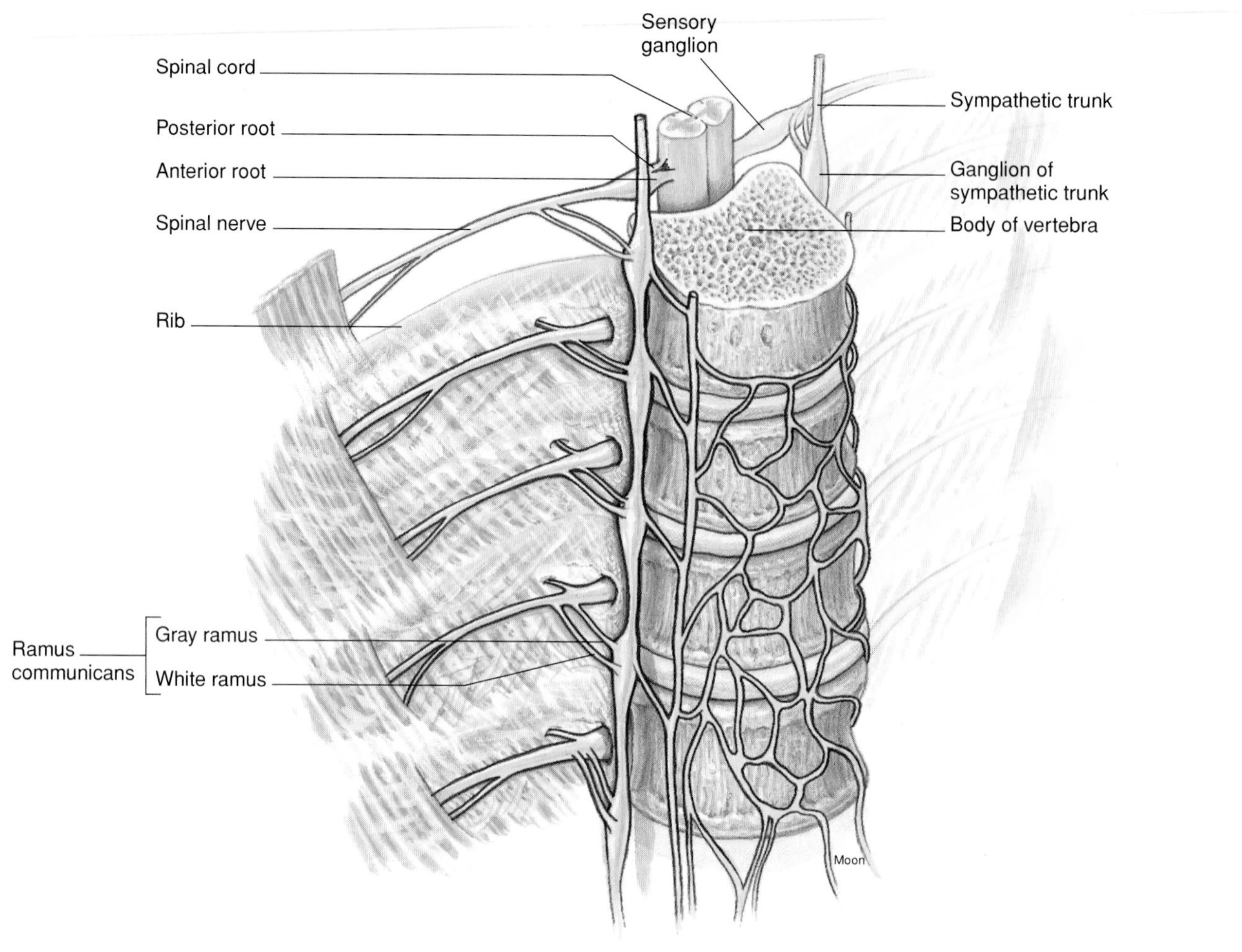

Figure 13.2

The sympathetic trunk of paravertebral ganglia showing its relationship to the vertebral column and spinal cord.

into the neck, where they synapse in cervical sympathetic ganglia (fig. 13.4). Postganglionic neurons from here innervate the smooth muscles and glands of the head and neck.

Peripheral Ganglia

Many preganglionic neurons that exit the spinal cord below the level of the diaphragm pass through the sympathetic trunk without synapsing. Beyond the sympathetic trunk, these preganglionic neurons form **splanchnic** (*splank'nik*) **nerves** (fig. 13.3). Preganglionic neurons in the splanchnic nerves synapse in peripheral ganglia, which include the **celiac** (*se'le-ak*), **superior mesenteric** (*mes"en-ter'ik*), and **inferior mesenteric ganglia** (figs. 13.5 and 13.6).

The *greater splanchnic nerve* arises from preganglionic sympathetic neurons T4–T9 and synapses in the celiac ganglion. These neurons contribute to the *celiac (solar) plexus*. Postganglionic neurons from the celiac ganglion innervate the stomach, spleen, pancreas, liver, small intestine, and kidneys. The *lesser splanchnic nerve* terminates in the superior mesenteric ganglion. Postganglionic neurons from here innervate the small intestine and colon. The *lumbar splanchnic nerve* synapses in the inferior mesenteric ganglion, and the postganglionic neurons innervate the distal colon and rectum, urinary bladder, and genital organs.

Adrenal Glands

The paired adrenal glands are located above each kidney (see fig. 13.5). Each adrenal is composed of two parts: an outer **adrenal cortex** and an inner **adrenal medulla.** These two parts are actually two functionally different glands with different embryonic origins, different hormones, and different regulatory mechanisms (see chap. 14). The adrenal cortex secretes steroid hormones; the adrenal medulla secretes the hormone **epineph-**

splanchnic: Gk. *splanchno*, relating to viscera

adrenal: L. *ad*, to; *renes*, kidney
cortex: L. *cortex*, bark
medulla: L. *medulla*, marrow

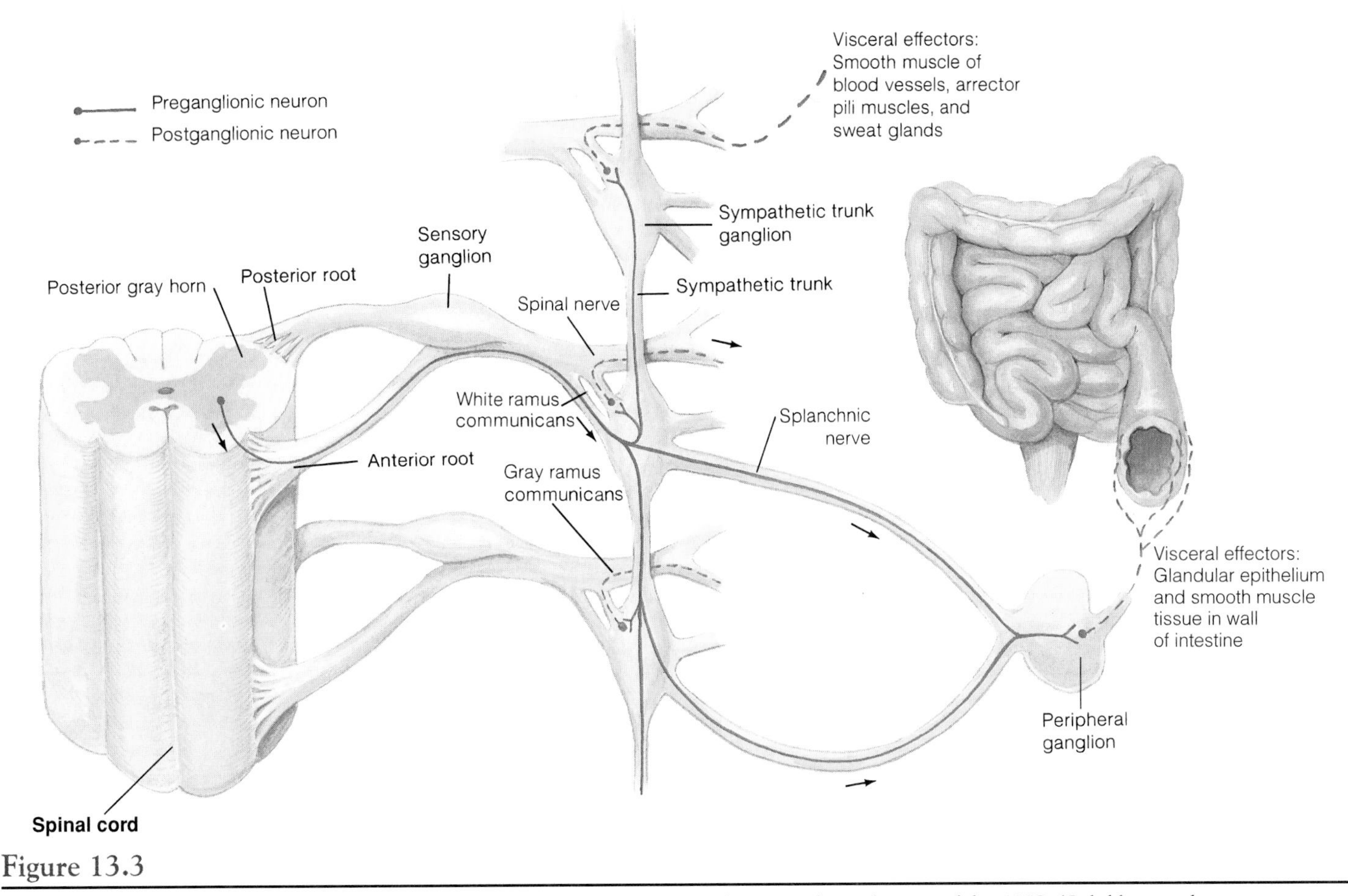

Figure 13.3

Sympathetic trunk ganglia, the sympathetic trunk, and rami communicantes of the sympathetic division of the ANS. (Solid lines indicate preganglionic neurons and dashed lines indicate postganglionic neurons.)

rine (*ep"ĭ-nef'rin*) (adrenaline) and, to a lesser degree, **norepinephrine** when it is stimulated by the sympathetic system.

The adrenal medulla is a modified sympathetic ganglion whose cells are derived from postganglionic sympathetic neurons. The cells of the adrenal medulla are innervated by preganglionic sympathetic neurons originating in the thoracic level of the spinal cord; they secrete epinephrine into the blood in response to sympathetic stimulation. The effects of epinephrine are complementary to those of the neurotransmitter norepinephrine, which is released from postganglionic sympathetic nerve endings.

Parasympathetic (Craniosacral) Division

The **parasympathetic division** is also known as the *craniosacral division* of the autonomic system. This is because its preganglionic neurons originate in the brain (specifically, the midbrain, pons, and medulla oblongata of the brain stem) and in the second through fourth sacral segments of the spinal cord. These preganglionic parasympathetic neurons synapse in ganglia that are located next to (or actually within) the organs innervated. These parasympathetic ganglia, which are called **terminal ganglia,** supply the postganglionic neurons that synapse with the effector cells.

Tables 13.3 and 13.4 show the comparative structures of the sympathetic and parasympathetic divisions. It should be noted that, unlike sympathetic neurons, most parasympathetic neurons do not travel within spinal nerves. Cutaneous effectors (blood vessels, sweat glands, and arrector pili muscles) and blood vessels in skeletal muscles thus receive sympathetic but not parasympathetic innervation.

Four of the 12 pairs of cranial nerves contain preganglionic parasympathetic neurons. These are the oculomotor (III), facial (VII), glossopharyngeal (IX), and vagus (X) nerves. Parasympathetic neurons within the first three of these cranial nerves synapse in ganglia located in the head; neurons in the vagus nerve synapse in terminal ganglia located in many regions of the body.

The oculomotor nerve contains somatic motor and parasympathetic neurons that originate in the oculomotor nuclei of the midbrain. These parasympathetic neurons synapse in the **ciliary ganglion,** whose postganglionic neurons innervate the ciliary muscle and constrictor muscles in the iris of the eye. Preganglionic neurons that originate in the pons travel in the facial nerve to the **pterygopalatine** (*ter"ĭ-go-pal'ă-tēn*) **ganglion,** which sends postganglionic neurons to the nasal mucosa, pharynx, palate, and lacrimal glands. Another group of neurons in the facial nerve terminate in the **submandibular ganglion,** which

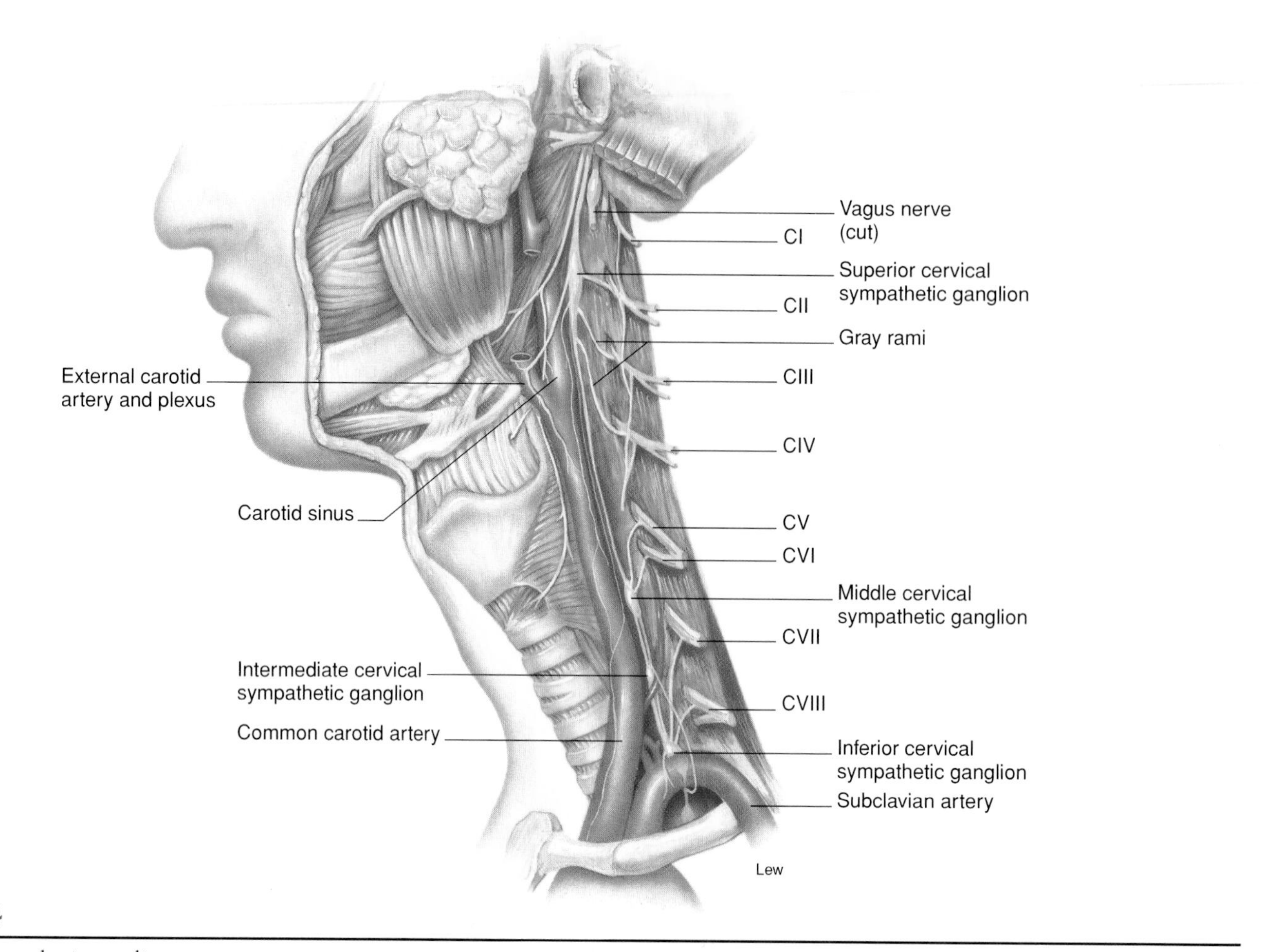

Figure 13.4

The cervical sympathetic ganglia.

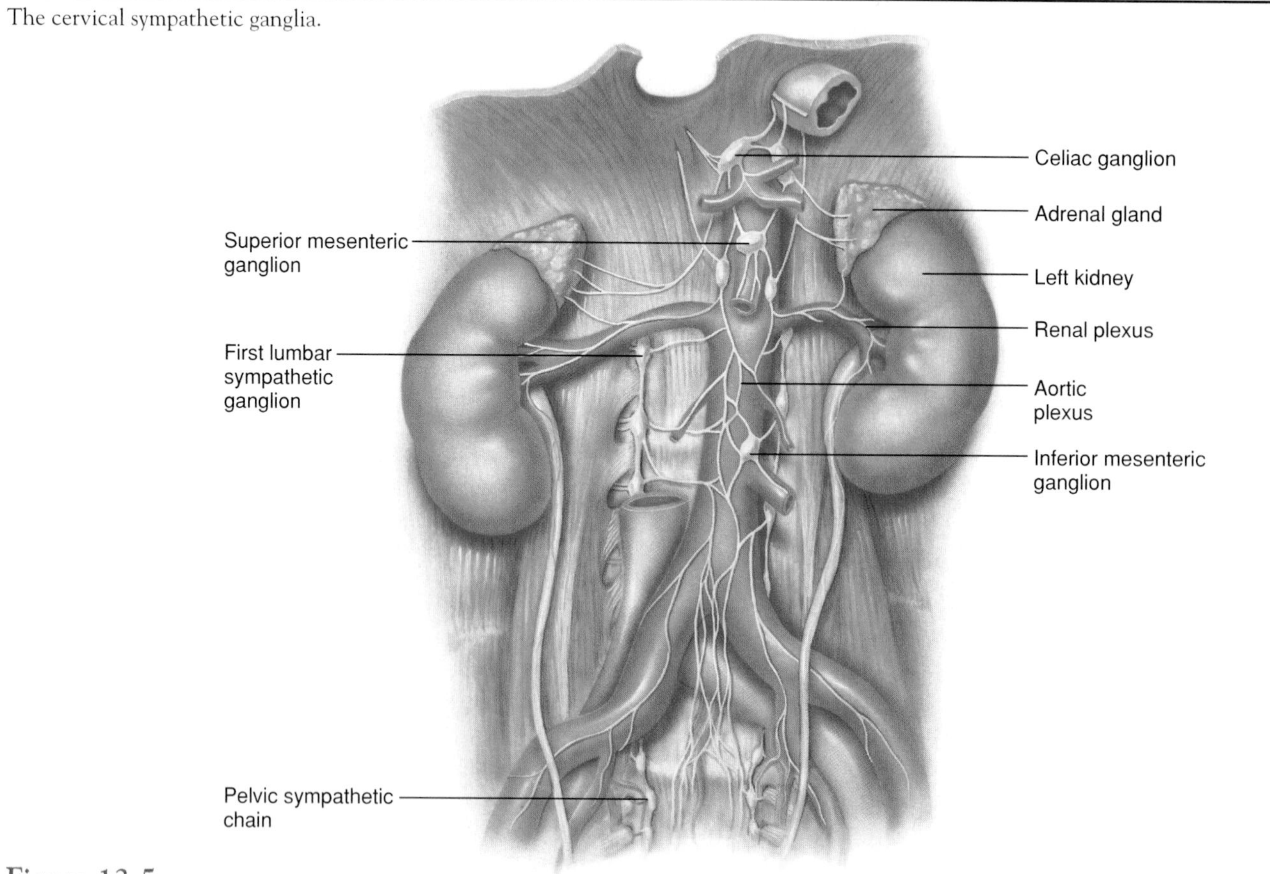

Figure 13.5

Peripheral sympathetic plexuses and ganglia of the abdomen.

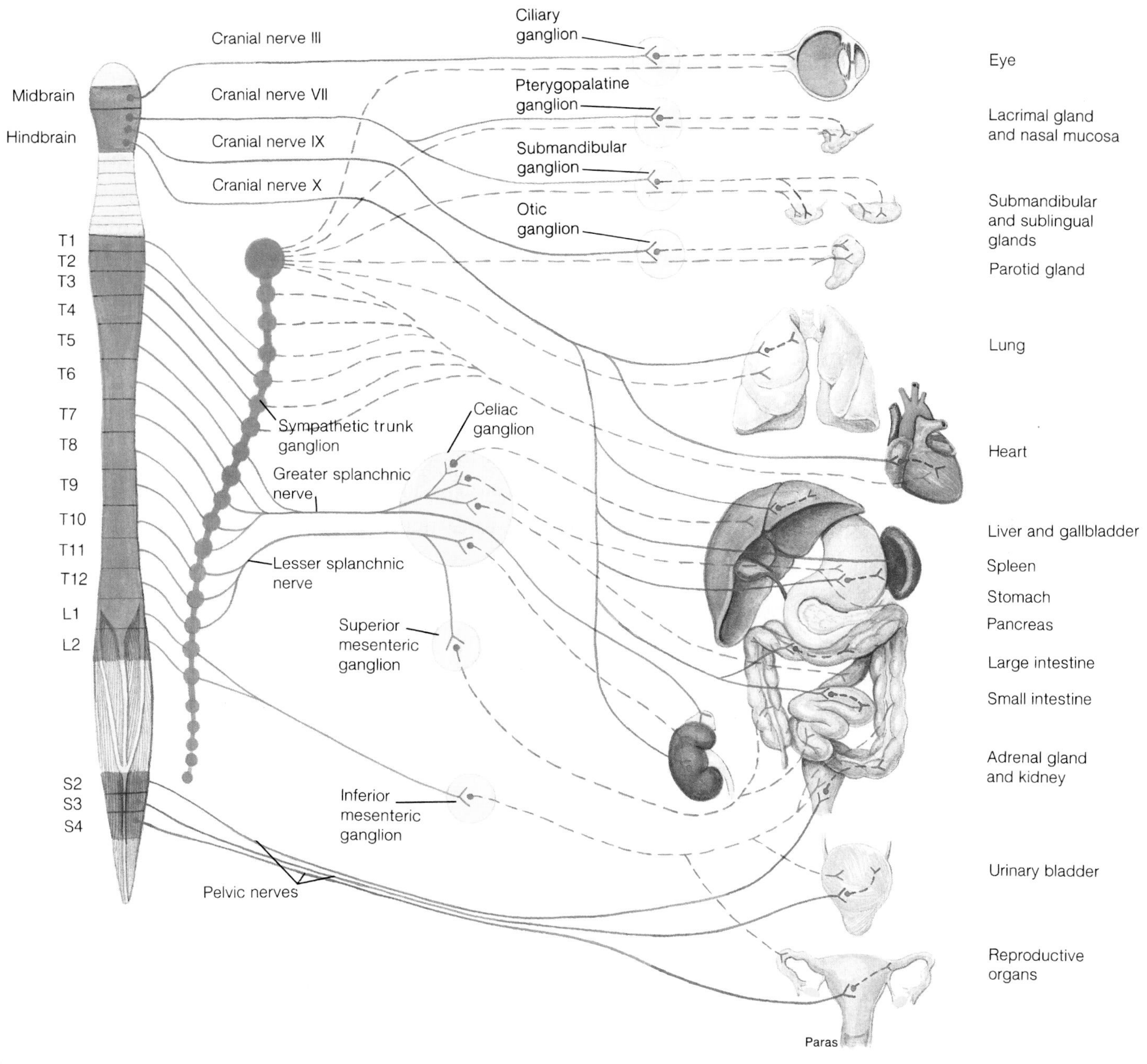

Figure 13.6

The autonomic nervous system. The sympathetic division is shown in red; the parasympathetic, in blue. Solid lines indicate preganglionic neurons and dashed lines indicate postganglionic neurons.

sends postganglionic neurons to the submandibular and sublingual glands. Preganglionic neurons of the glossopharyngeal nerve synapse in the **otic ganglion,** which sends postganglionic neurons to innervate the parotid gland.

Nuclei in the medulla oblongata contribute preganglionic neurons to the very long vagus nerves, which provide the most extensive parasympathetic innervation in the body. As the paired vagus nerves pass through the thorax, they contribute to the *pulmonary plexuses* within the mediastinum. Branches of the pulmonary plexuses accompany blood vessels and bronchi into the lungs. Below the pulmonary plexuses, branches of the vagus nerves merge to form the *esophageal plexuses*.

At the lower end of the esophagus, vagal neurons collect to form an **anterior** and **posterior vagal trunk,** each composed of neurons from both vagus nerves. The vagal trunks enter the abdominal cavity through the esophageal hiatus (opening) in the diaphragm. Neurons from the vagal trunks innervate the stomach

---

vagus: L. *vagus*, wandering

Table 13.3
The Sympathetic (Thoracolumbar) Division

Parts of Body Innervated	Spinal Origin of Preganglionic Neurons	Origin of Postganglionic Neurons
Eye	C8, T1	Cervical ganglia
Head and neck	T1–T4	Cervical ganglia
Heart and lungs	T1–T5	Upper thoracic ganglia
Upper extremities	T2–T9	Lower cervical and upper thoracic ganglia
Upper abdominal viscera	T4–T9	Celiac and superior mesenteric ganglia
Adrenal glands	T10, T11	Adrenal medulla
Urinary and reproductive systems	T12–L2	Celiac and inferior mesenteric ganglia
Lower extremities	T9–L2	Lumbar and upper sacral ganglia

Table 13.4
The Parasympathetic (Craniosacral) Division

Nerve	Origin of Preganglionic Neurons	Location of Terminal Ganglia	Effector Organs
Oculomotor nerve	Midbrain (cranial)	Ciliary ganglion	Eye (smooth muscle in iris and ciliary body)
Facial nerve	Pons (cranial)	Pterygopalatine and submandibular ganglia	Lacrimal, mucous, and salivary glands in head
Glossopharyngeal nerve	Medulla oblongata (cranial)	Otic ganglion	Parotid gland
Vagus nerve	Medulla oblongata (cranial)	Terminal ganglia in or near organ	Heart, lungs, GI tract, liver, pancreas
Through pelvic spinal nerves	S2–S4 (sacral)	Terminal ganglia near organs	Lower half of large intestine, urinary bladder, and reproductive organs

on the anterior and posterior sides. Branches of the vagus nerves within the abdominal cavity also contribute to the *celiac plexus* and *plexuses of the abdominal aorta*.

The preganglionic neurons in the vagus synapse with postganglionic neurons that are actually located *within* the innervated organs. These preganglionic neurons are thus quite long. They provide parasympathetic innervation to the heart, lungs, esophagus, stomach, pancreas, liver, small intestine, and upper half of the large intestine. Postganglionic parasympathetic neurons arise from terminal ganglia within these organs and innervate the smooth muscle tissue and glandular epithelium of these same organs.

Preganglionic neurons from the sacral levels of the spinal cord provide parasympathetic innervation to the lower half of the large intestine, the rectum, and to the urinary and reproductive systems. These neurons, like those of the vagus, synapse with terminal ganglia located within the effector organs. Parasympathetic nerves to the visceral organs thus consist of preganglionic neurons, whereas sympathetic nerves to these organs contain postganglionic neurons.

A composite view of the sympathetic and parasympathetic divisions is provided in figure 13.6, and the comparisons are summarized in table 13.5.

1. Compare the origins of preganglionic sympathetic and parasympathetic neurons and the locations of sympathetic and parasympathetic ganglia.
2. Using a simple line drawing, illustrate the sympathetic pathway from the spinal cord to the heart. Label the preganglionic neuron, postganglionic neuron, and the ganglion.
3. Use a simple diagram to show the parasympathetic innervation of the heart. Label the preganglionic and postganglionic neurons, the nerve involved, and the terminal ganglion.
4. Describe the distribution of the vagus nerve and discuss the functional significance of this distribution.
5. Define the terms *white rami* and *gray rami* and explain why blood vessels in the skin and skeletal muscles receive sympathetic but not parasympathetic innervation.
6. Describe the structure of the adrenal gland and explain its relationship to the sympathetic division of the ANS.

Table 13.5

Comparison of the Structural Features of the Sympathetic and Parasympathetic Divisions

Feature	Sympathetic	Parasympathetic
Location of cell bodies of preganglionic neurons	Thoracolumbar portion of spinal cord	Midbrain, hindbrain, and sacral portion of spinal cord
Location of ganglia	Chain of paravertebral ganglia and prevertebral ganglia	Terminal ganglia in or near effector organs
Distribution of postganglionic neurons	Throughout the body	Mainly limited to the head and viscera
Divergence of impulses from pre- to postganglionic neurons	Great divergence (one preganglionic may activate 20 postganglionic neurons)	Little divergence (one preganglionic only activates a few postganglionic neurons)
Mass discharge of system as a whole	Usually	Not normally

Functions of the Autonomic Nervous System

The actions of the autonomic nervous system, together with the effects of hormones, help to maintain a state of dynamic constancy in the internal environment. The sympathetic division gears the body for action through adrenergic effects; the parasympathetic division conserves and restores the body's energy through cholinergic effects. Homeostasis thus depends, in large part, on the complementary and often antagonistic effects of sympathetic and parasympathetic innervation.

Objective 8 List the neurotransmitters of the preganglionic and postganglionic neurons of the sympathetic and parasympathetic divisions.

Objective 9 Describe the effects of acetylcholine released by postganglionic parasympathetic neurons.

Objective 10 Explain the antagonistic, complementary, and cooperative effects of sympathetic and parasympathetic innervation.

The sympathetic and parasympathetic divisions of the ANS (fig. 13.6) affect the visceral organs in different ways. Mass activation of the sympathetic division prepares the body for intense physical activity in emergencies; the heart rate increases, blood glucose rises, and blood is diverted to the skeletal muscles (away from the visceral organs and skin). These and other effects are listed in table 13.6. The theme of the sympathetic division is aptly summarized in the phrase **fight or flight.**

The effects of parasympathetic nerve stimulation are in many ways opposite to the effects of sympathetic stimulation. The parasympathetic division, however, is not normally activated as a whole. Stimulation of separate parasympathetic nerves can result in slowing of the heart, dilation of visceral blood vessels, and increased activity of the GI tract (table 13.6). The different responses of visceral organs to sympathetic and parasympathetic nerve activity is due to the fact that the postganglionic neurons of these two divisions release different neurotransmitters.

Neurotransmitters of the Autonomic Nervous System

The neurotransmitter released by most postganglionic sympathetic neurons is **norepinephrine** (noradrenaline). Transmission at these synapses is thus said to be **adrenergic** (*ad″rĕ-ner′jik*). There are a few exceptions to this rule: some sympathetic neurons that innervate blood vessels in skeletal muscles, as well as sympathetic neurons to sweat glands, release acetylcholine (are cholinergic).

Acetylcholine (*ă-sēt″l-ko′lēn*) (**ACh**) is the neurotransmitter of all preganglionic neurons (both sympathetic and parasympathetic). Acetylcholine is also the transmitter released by all parasympathetic postganglionic neurons at their synapses with effector cells (fig. 13.7). Transmission at the autonomic ganglia and at synapses of postganglionic neurons is thus said to be **cholinergic** (*ko″lĭ-ner′jik*). In other words, a *cholinergic fiber* is a neuron that secretes ACh at the terminal end of its axon.

Responses to Adrenergic Stimulation

Adrenergic stimulation—by epinephrine in the blood and by norepinephrine released from sympathetic nerve endings—has both excitatory and inhibitory effects. The heart, dilatory muscles of the iris, and the smooth muscles of many blood vessels are stimulated to contract. The smooth muscles of the bronchioles and of some blood vessels, however, are inhibited from contracting; adrenergic chemicals, therefore, cause these structures to dilate.

cholinergic: Gk. *chole*, bile; *ergon*, work

Table 13.6

Effects of Autonomic Nerve Stimulation on Various Visceral Effector Organs

Effector Organ	Sympathetic Effect	Parasympathetic Effect
Eye		
Iris (pupillary dilator muscle)	Dilation of pupil	—
Iris (pupillary constrictor muscle)	—	Constriction of pupil
Ciliary muscle	Relaxation (for far vision)	Contraction (for near vision)
Glands		
Lacrimal (tear)	—	Stimulation of secretion
Sweat	Stimulation of secretion	—
Salivary	Decreased secretion; saliva becomes thick	Increased secretion; saliva becomes thin
Stomach	—	Stimulation of secretion
Intestine	—	Stimulation of secretion
Adrenal medulla	Stimulation of hormone secretion	—
Heart		
Rate	Increased	Decreased
Conduction	Increased rate	Decreased rate
Strength	Increased	—
Blood vessels	Mostly constriction; affects all organs	Dilation in a few organs (e.g., penis)
Lungs		
Bronchioles	Dilation	Constriction
Mucous glands	Inhibition of secretion	Stimulation of secretion
GI tract		
Motility	Inhibition of movement	Stimulation of movement
Sphincters	Closing stimulated	Closing inhibited
Liver	Stimulation of glycogen hydrolysis	—
Adipocytes (fat cells)	Stimulation of fat hydrolysis	—
Pancreas	Inhibition of exocrine secretions	Stimulation of exocrine secretions
Spleen	Stimulation of contraction	—
Urinary bladder	Muscle tone aided	Stimulation of contraction
Arrector pili muscles	Stimulation of hair erection, causing goose bumps	—
Uterus	If pregnant, contraction; if not pregnant, relaxation	—
Penis	Ejaculation	Erection (due to vasodilation)

Responses to Cholinergic Stimulation

Somatic motor neurons, postganglionic parasympathetic neurons, and all preganglionic autonomic neurons are cholinergic—they use acetylcholine as a neurotransmitter. The cholinergic effects of somatic motor neurons and preganglionic autonomic neurons are always excitatory. The cholinergic effects of postganglionic parasympathetic neurons are usually excitatory, with some notable exceptions; the parasympathetic neurons innervating the heart, for example, cause slowing of the heart rate.

The drug *muscarine* (*mus′kă-rēn*), a poison derived from certain mushrooms, mimics the cholinergic effects of parasympathetic nerves in the heart, smooth muscles, and glands by stimulating the acetylcholine receptors located in these organs. This drug, however, does not affect the cholinergic receptors of skeletal muscle or those of autonomic ganglia. The acetylcholine receptors of visceral organs are therefore said to be *muscarinic*.

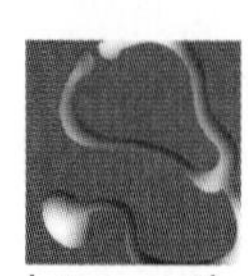

The muscarinic effects of ACh are specifically inhibited by the drug *atropine,* derived from the deadly nightshade plant (*Atropa belladonna*). Indeed, extracts of this plant were used by women during the Middle Ages to dilate their pupils (atropine inhibits parasympathetic stimulation of the iris). This was thought to enhance their beauty (in Italian, *bella* = beautiful, *donna* = woman). Atropine is used clinically today to dilate pupils during eye examinations, to reduce secretions of the respiratory tract prior to general anesthesia, and to inhibit spasmodic contractions of the lower GI tract.

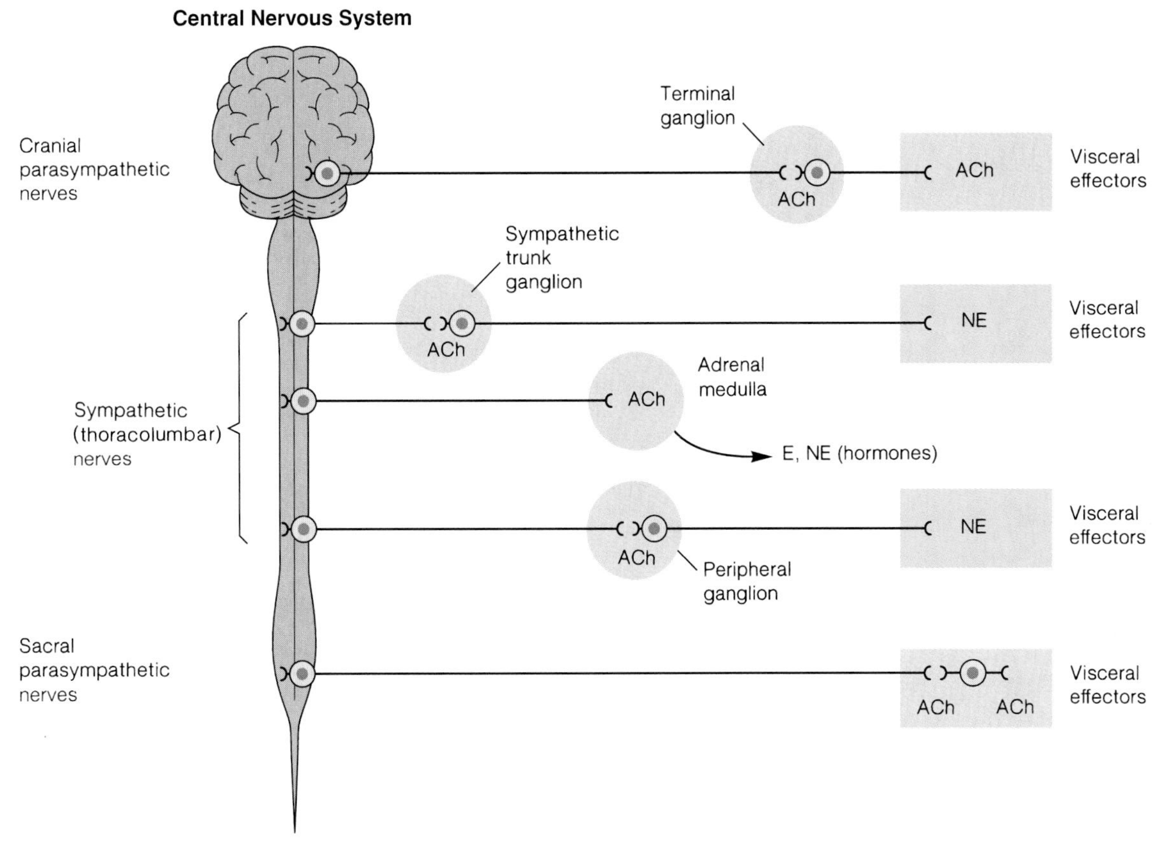

Figure 13.7

Neurotransmitters of the autonomic motor system (ACh = acetylcholine, NE = norepinephrine, E = epinephrine). Those nerves that release ACh are called cholinergic; those that release NE are called adrenergic. The adrenal medulla secretes both epinephrine (85%) and norepinephrine (15%) as hormones into the blood.

Organs with Dual Innervation

Many organs receive dual innervation—they are innervated by both sympathetic and parasympathetic neurons. When this occurs, the effects of these two divisions may be antagonistic, complementary, or cooperative.

Antagonistic Effects

The effects of sympathetic and parasympathetic innervation on the pacemaker region of the heart is the best example of the antagonism of these two systems. In this case, sympathetic and parasympathetic neurons innervate the same cells. Adrenergic stimulation from sympathetic neurons increases the heart rate, whereas cholinergic stimulation from parasympathetic neurons inhibits the pacemaker cells, and thus decreases the heart rate. Antagonism is also seen in the GI tract, where sympathetic nerves inhibit and parasympathetic nerves stimulate intestinal movements and secretions.

The effects of sympathetic and parasympathetic stimulation on the diameter of the pupil of the eye are analogous to the reciprocal innervation of flexor and extensor skeletal muscles by somatic motor neurons. This is because the iris contains antagonistic muscle layers. Contraction of the pupillary dilator muscle, which is stimulated by impulses through sympathetic nerve endings, causes dilation; contraction of the pupillary constrictor muscle, which is innervated by parasympathetic nerve endings, causes constriction of the pupil (fig. 13.8).

Complementary Effects

The effects of sympathetic and parasympathetic stimulation on salivary gland secretion are complementary. The secretion of watery saliva is stimulated through parasympathetic nerves, which also stimulate the secretion of other exocrine glands in the GI tract. Impulses through sympathetic nerves stimulate the constriction of blood vessels throughout the GI tract. The resultant decrease in blood flow to the salivary glands causes the production of a thicker, more viscous saliva.

Cooperative Effects

The effects of sympathetic and parasympathetic stimulation on the urinary and reproductive systems are cooperative. Erection of the penis, for example, is due to vasodilation resulting from parasympathetic nerve stimulation; ejaculation is due to stimulation through sympathetic nerves. Although the contraction of the urinary bladder is myogenic (independent of nerve stimulation), it is

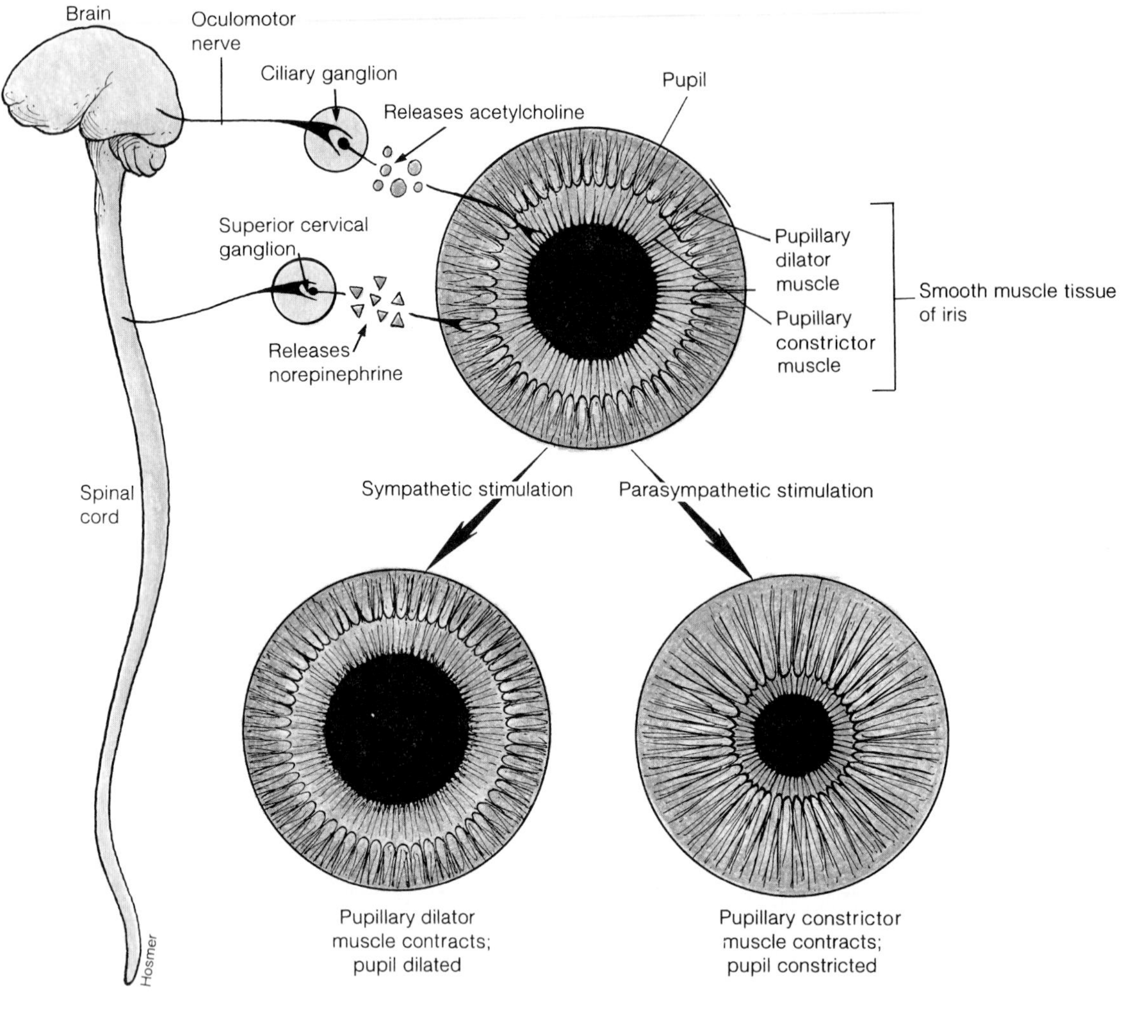

Figure 13.8

Reciprocal sympathetic and parasympathetic innervation of smooth muscle tissue within the iris of the eye. Stimulation through sympathetic nerves causes the dilator muscles to contract, which dilates (enlarges) the size of the pupil. Stimulation through the parasympathetic nerves causes the constrictor muscles to contract, which constricts (decreases) the size of the pupil.

promoted in part by the action of parasympathetic nerves. This *micturition* (*mik″tŭ-rish′un*), or urination, urge and reflex is also enhanced by sympathetic nerve activity, which increases the tone of the bladder muscles. Emotional states that are accompanied by high sympathetic nerve activity may thus result in reflex urination at bladder volumes that are normally too low to trigger this reflex.

Organs without Dual Innervation

Although most organs are innervated by both sympathetic and parasympathetic nerves, some—including the adrenal medulla, arrector pili muscles, sweat glands, and most blood vessels—receive only sympathetic innervation. In these cases, regulation is achieved by increases or decreases in the "tone" (firing rate) of the sympathetic neurons. Constriction of blood vessels, for example, is produced by increased sympathetic activity, which stimulates adrenergic receptors, and vasodilation results from decreased sympathetic nerve stimulation.

Sympathetic activity is required for proper thermoregulatory responses to heat. In a hot room, for example, decreased sympathetic stimulation produces dilation of the blood vessels in the surface of the skin, which increases cutaneous blood flow and provides better heat radiation. During exercise, on the other hand, there is increased sympathetic activity, which causes constriction of the blood vessels in the skin of the limbs and stimulation of sweat glands in the trunk.

The eccrine sweat glands in the trunk secrete a watery fluid in response to sympathetic stimulation. Evaporation of this dilute sweat helps to cool the body. The eccrine sweat glands also secrete a chemical called **bradykinin** (*brad″ĭ-ki′nin*) in response to sympathetic stimulation. Bradykinin stimulates dilation of the surface blood vessels near the sweat glands, helping to radiate heat. At the conclusion of exercise, sympathetic stimulation is reduced and blood flow to the surface of the limbs is increased,

Table 13.7

Some Reflexes Stimulated by Input from Sensory Neurons in the Vagus Nerves That Is Transmitted to Nuclei in the Medulla Oblongata

Organs	Type of Receptors	Reflex Effects
Lungs	Stretch receptors	Further inhalation inhibited; increase in cardiac rate and vasodilation stimulated
	Type J receptors	Stimulated by pulmonary congestion—produces feelings of breathlessness and causes a reflex fall in cardiac rate and blood pressure
Aorta	Chemoreceptors	Stimulated by rise in CO_2 and fall in O_2—produces increased rate of breathing, rise in heart rate, and vasoconstriction
Aorta (cont.)	Baroreceptors	Stimulated by increased blood pressure—produces a reflex decrease in heart rate
Heart	Atrial stretch receptors	Antidiuretic hormone secretion inhibited, thus increasing the volume of urine excreted
	Stretch receptors in ventricles	Produces a reflex decrease in heart rate and vasodilation
GI tract	Stretch receptors	Feelings of satiety, discomfort, and pain

which aids in the elimination of metabolic heat. Notice that all of these thermoregulatory responses are achieved without the direct involvement of the parasympathetic division.

1. Define the terms *adrenergic* and *cholinergic* and use these terms to describe the neurotransmitters of different autonomic neurons.
2. Describe the effects of the drug *atropine* and explain these effects in terms of the actions of the parasympathetic division.
3. Explain how the sympathetic and parasympathetic divisions can have antagonistic, cooperative, and complementary effects. Give an example of each of these effects.

Control of the Autonomic Nervous System by Higher Brain Centers

Visceral functions are largely regulated by autonomic reflexes. In most autonomic reflexes, sensory input is directed to brain centers, which in turn regulate the activity of descending pathways to preganglionic autonomic neurons. The neural centers that directly control the activity of autonomic nerves are influenced by higher brain areas, as well as by sensory input.

Objective 11 Describe the area of the brain that most directly controls the activity of autonomic nerves. Also describe the higher brain areas that influence autonomic activity.

Objective 12 Explain how the activity of the autonomic system and the activity of the endocrine system can be coordinated.

Objective 13 Explain how autonomic functions can be affected by emotions.

Medulla Oblongata

The **medulla oblongata** of the brain stem is the structure that most directly controls the activity of the ANS. Almost all autonomic responses can be elicited by experimental stimulation of the medulla oblongata, which contains centers for the control of the circulatory, respiratory, urinary, reproductive, and digestive systems. Much of the sensory input to these centers travels through the sensory neurons of the vagus nerves. The reflexes that result are listed in table 13.7.

Hypothalamus

The **hypothalamus** (fig. 13.9), located just above the pituitary gland, is the overall control and integration center of the ANS. By means of motor fibers to the brain stem and posterior pituitary, and also by means of hormones that regulate the anterior pituitary, the hypothalamus serves to orchestrate somatic, autonomic, and endocrine responses during various behavioral states.

Experimental stimulation of different areas of the hypothalamus can evoke the autonomic responses characteristic of aggression, sexual behavior, eating, or satiety. Chronic stimulation of the lateral hypothalamus, for example, can make an animal eat and become obese, whereas stimulation of the medial hypothalamus inhibits eating. Other areas contain osmoreceptors that stimulate thirst and the secretion of antidiuretic hormone (ADH) from the posterior pituitary.

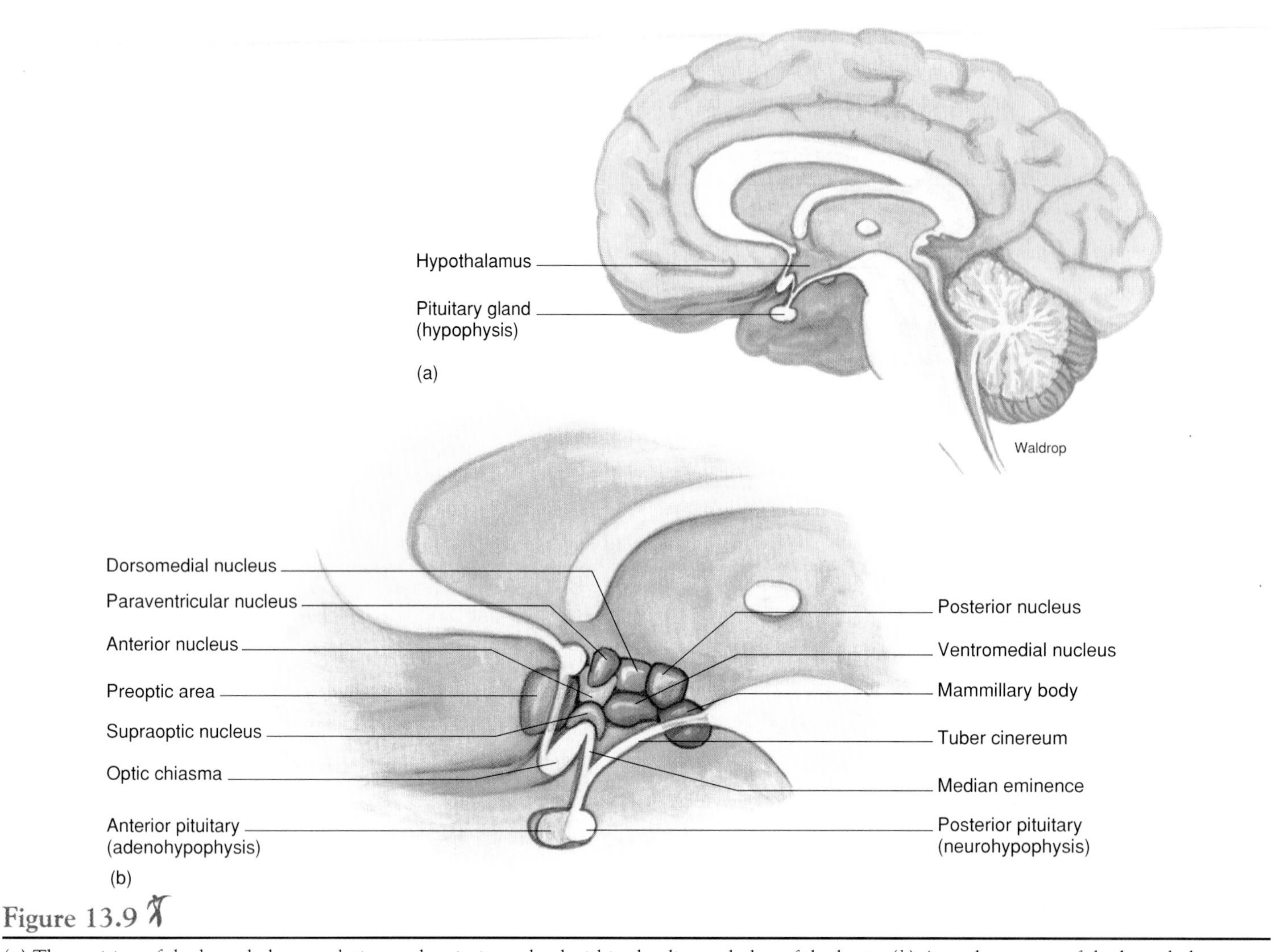

Figure 13.9

(*a*) The position of the hypothalamus relative to the pituitary gland within the diencephalon of the brain. (*b*) An enlargement of the hypothalamus to diagrammatically show the hypothalamic nuclei and the anterior and posterior parts of the pituitary gland.

The hypothalamus is also where the body's thermostat is located. Experimental cooling of the preoptic-anterior hypothalamus causes shivering (a somatic response) and nonshivering thermogenesis (a sympathetic response). Experimental heating of this hypothalamic area results in hyperventilation (stimulated by somatic motor nerves), vasodilation, salivation, and sweat gland secretion (stimulated by autonomic nerves).

The coordination of sympathetic and parasympathetic reflexes by the medulla oblongata is thus integrated with the control of somatic and endocrine responses by the hypothalamus. The activities of the hypothalamus are in turn influenced by higher brain centers.

Limbic System, Cerebellum, and Cerebrum

The **limbic system** is a group of fiber tracts and nuclei that form a ring (limbus) around the brain stem. It includes the cingulate gyrus of the cerebral cortex, the hypothalamus, the fornix (a fiber tract), the hippocampus, and the amygdaloid nucleus (fig. 13.10). These structures, which were derived early in the course of vertebrate evolution, were once called the *rhinencephalon* (*ri″nen-sef′ă-lon*), or "smell brain," because of their importance in the central processing of olfactory information.

In higher vertebrates, these structures are now recognized as centers involved in such basic emotional drives as anger, fear, sex, and hunger, and in short-term memory. Complex circuits between the hypothalamus and other parts of the limbic system (illustrated in fig. 13.10) contribute visceral responses to emotions, including blushing, pallor, fainting, and "butterflies in the stomach."

The autonomic correlates of motion sickness—nausea, sweating, and cardiovascular changes—are eliminated by cutting the motor tracts of the cerebellum. This demonstrates that impulses from the cerebellum to the medulla oblongata influence activity of the ANS. Experimental and clinical observations have also demonstrated that the frontal and temporal lobes of the cerebral cortex influence lower brain areas as part of their involvement in emotion and personality.

limbic: L. *limbus*, edge or border

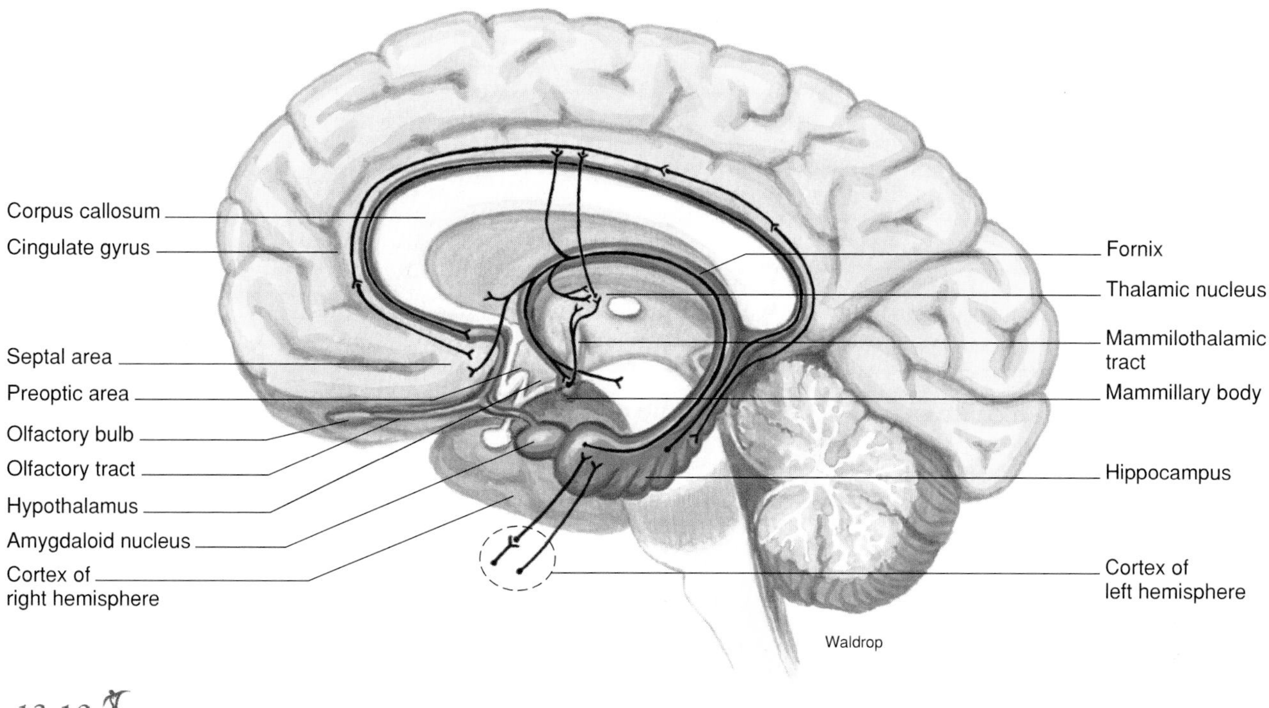

Figure 13.10

The limbic system and the pathways that interconnect the structures of the limbic system. (Note that the left temporal lobe of the cerebral cortex has been removed.)

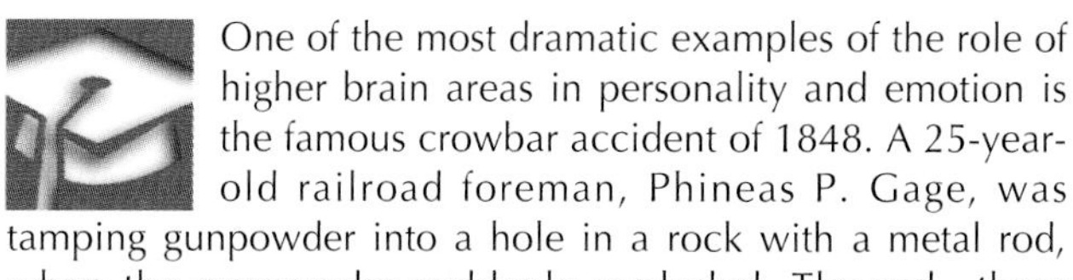

One of the most dramatic examples of the role of higher brain areas in personality and emotion is the famous crowbar accident of 1848. A 25-year-old railroad foreman, Phineas P. Gage, was tamping gunpowder into a hole in a rock with a metal rod, when the gunpowder suddenly exploded. The rod—three feet, seven inches long and one and one-fourth inches thick—was driven through his left eye and through his brain, finally emerging through the back of his skull.

After a few minutes of convulsions, Gage got up, rode a horse three-quarters of a mile into town, and walked up a long flight of stairs to see a doctor. He recovered well, with no noticeable sensory or motor deficits. His associates, however, noted striking personality changes. Before the accident Gage was a responsible, capable, and financially prudent man. Afterward, he was much less inhibited socially, engaging, for example, in gross profanity (which he had never done previously). He also seemed to be tossed about by chance whims. Eventually, Gage was fired from his job, and his old friends remarked that he was "no longer Gage."

Clinical Case Study Answer

The tumor has invaded the sympathetic ganglia and connecting pathways within the base of the neck, known as the cervical ganglia and sympathetic trunk. Preganglionic sympathetic neurons originating in the upper thoracic spinal cord travel upward through the sympathetic trunk, where they synapse on a postganglionic neuron in one of the cervical ganglia. From that point, postganglionic neurons travel through various routes (mostly paralleling arteries) to innervate the effector organs involved. In the case of our patient, these include the pupillary dilator muscle of the iris, the smooth muscle portion of the levator palpebrae superioris, and sweat glands on the right side of the face.

1. Describe the role of the medulla oblongata in the regulation of the ANS.
2. Describe the role of the hypothalamus in the regulation of the autonomic and endocrine systems.
3. What mechanisms are involved when a person blushes? What structures are involved in this response?

Clinical Considerations

Autonomic Dysreflexia

Autonomic dysreflexia, a serious condition producing rapid elevations in blood pressure that can lead to stroke (cerebrovascular accident), occurs in 85% of people with quadriplegia and others with spinal cord lesions above the sixth thoracic level. Lesions to the spinal cord first produce the symptoms of spinal shock, characterized by the loss of both skeletal muscle and autonomic reflexes. After a period of time, both types of reflexes return in an exaggerated state; the skeletal muscles may become spastic due to the absence of higher inhibitory influences, and the visceral organs experience denervation hypersensitivity. Patients in this state have difficulty emptying their urinary bladders and must often be catheterized.

Noxious stimuli, such as overdistension of the urinary bladder, can result in reflex activation of the sympathetic nerves below the spinal cord lesion. This produces goose bumps, cold skin, and vasoconstriction in the regions served by the spinal cord below the level of the lesion. The rise in blood pressure resulting from this vasoconstriction activates pressure receptors that transmit impulses along sensory neurons to the medulla oblongata. In response to this sensory input, the medulla oblongata directs a reflex slowing of the heart and vasodilation. Since descending impulses are blocked by the spinal lesion, however, the skin above the lesion is warm and moist (due to vasodilation and sweat gland secretion), whereas it is cold below the level of spinal cord damage.

Chapter Summary

Introduction to the Autonomic Nervous System (pp. 420–423)

1. The autonomic nervous system (ANS) is a functional division of the nervous system; it is composed of portions of the central nervous system (CNS) and portions of the peripheral nervous system (PNS).
2. Preganglionic autonomic neurons originate in the brain or spinal cord; postganglionic neurons originate in ganglia outside the CNS.
3. Smooth muscle, cardiac muscle, and glands receive autonomic innervation.
   (a) The involuntary effectors are somewhat independent of their innervation and become hypersensitive when their innervation is removed.
   (b) Myocardial cells are interconnected by electrical synapses, or gap junctions, to form a functional syncytium with independent pacemaker activity.
   (c) Single-unit smooth muscles are characterized by gap junctions and pacemaker activity; multiunit smooth muscles have few, if any, gap junctions, and thus their individual cells must be stimulated separately by neurons.

Structure of the Autonomic Nervous System (pp. 423–428)

1. Preganglionic neurons of the sympathetic (thoracolumbar) division originate in the spinal cord (T1–L2).
   (a) Many of these neurons synapse with postganglionic neurons, whose cell bodies are located in a trunk of sympathetic ganglia outside the spinal cord.
   (b) Some preganglionic neurons synapse in peripheral ganglia; included in these are the celiac, superior mesenteric, and the inferior mesenteric ganglia.
   (c) Some preganglionic neurons innervate the adrenal medulla, which secretes epinephrine (and some norepinephrine) into the blood in response to this stimulation.
2. Preganglionic parasympathetic neurons originate in the brain and in the sacral levels of the spinal cord.
   (a) Preganglionic parasympathetic neurons contribute to the oculomotor, facial, glossopharyngeal, and vagus cranial nerves.
   (b) Preganglionic neurons of the vagus nerve are very long and synapse in terminal ganglia located next to or within the innervated organ; short postganglionic neurons then innervate the effector cells.
   (c) The vagus nerves provide parasympathetic innervation to the heart, lungs, esophagus, stomach, liver, small intestine, and upper half of the large intestine.
   (d) Parasympathetic outflow from the sacral levels of the spinal cord innervates terminal ganglia in the lower half of the large intestine, the rectum, and the urinary and reproductive systems.

Functions of the Autonomic Nervous System (pp. 429–433)

1. The effects of sympathetic and parasympathetic activity, together with those of hormones, help to maintain homeostasis. The sympathetic division activates the body to "fight or flight" through adrenergic effects; the parasympathetic division conserves and restores the body's energy through cholinergic effects.
2. All preganglionic autonomic neurons are cholinergic (use acetylcholine as a neurotransmitter).
   (a) All postganglionic parasympathetic neurons are cholinergic.
   (b) Most postganglionic sympathetic neurons are adrenergic (use norepinephrine at their synapses).
   (c) Sympathetic neurons that innervate sweat glands and those that innervate blood vessels in skeletal muscles are cholinergic.
3. Adrenergic effects include stimulation of the heart, vasoconstriction in the viscera and skin, bronchodilation, and glycogenolysis in the liver.
4. Cholinergic effects of parasympathetic nerves are promoted by the drug muscarine and inhibited by atropine.
5. In organs with dual innervation, the effects of the sympathetic and parasympathetic divisions can be antagonistic, complementary, or cooperative.
   (a) The effects are antagonistic in the heart and pupils.
   (b) The effects are complementary in the regulation of salivary gland secretion; they are cooperative in the regulation of the reproductive and urinary systems.
6. In organs without dual innervation (such as most blood vessels), regulation is achieved by increases or decreases in sympathetic nerve activity.

Control of the Autonomic Nervous System by Higher Brain Centers (pp. 433–435)

1. Visceral sensory input to the brain may result in the activity of the descending pathways to the preganglionic autonomic neurons. The centers in the brain that control autonomic activity are influenced by higher brain areas, as well as by sensory input.
2. The medulla oblongata of the brain stem is the structure that most directly controls the activity of the ANS.
   (a) The medulla oblongata is in turn influenced by sensory input and by input from the hypothalamus.
   (b) The hypothalamus orchestrates somatic, autonomic, and endocrine responses during various behavioral states.
3. The activity of the hypothalamus is influenced by input from the limbic system, cerebellum, and cerebrum; these interconnections provide an autonomic component to changes in body position, emotion, and various expressions of personality.

Review Activities

Objective Questions

1. Which of the following statements about the superior mesenteric ganglion is *true?*
   (a) It is a parasympathetic ganglion.
   (b) It is a paravertebral sympathetic ganglion.
   (c) It is located in the head.
   (d) It contains postganglionic sympathetic neurons.
2. The pterygopalatine, ciliary, submandibular, and otic ganglia are
   (a) collateral sympathetic ganglia.
   (b) cervical sympathetic ganglia.
   (c) parasympathetic ganglia that receive neurons from the vagus.
   (d) parasympathetic ganglia that receive neurons from the third, seventh, and ninth cranial nerves.
3. Parasympathetic ganglia are located
   (a) in a trunk parallel to the spinal cord.
   (b) in the posterior roots of spinal nerves.
   (c) next to or within the organs innervated.
   (d) in the brain.
4. The neurotransmitter of preganglionic sympathetic neurons is
   (a) norepinephrine.
   (b) epinephrine.
   (c) acetylcholine.
   (d) dopamine.
5. The preganglionic neurons of the sympathetic division of the autonomic nervous system originate in
   (a) the midbrain and the medulla oblongata.
   (b) the entire spinal nerve complex.
   (c) the first cervical (C1) to the first lumbar (L1) vertebrae.
   (d) the first thoracic (T1) to the second lumbar (L2) vertebrae.
6. Which of the following neurons release norepinephrine?
   (a) preganglionic parasympathetic neurons
   (b) postganglionic parasympathetic neurons
   (c) postganglionic sympathetic neurons in the heart
   (d) postganglionic parasympathetic neurons in sweat glands
   (e) all of the above
7. The actions of sympathetic and parasympathetic neurons are cooperative in
   (a) the heart.
   (b) the reproductive system.
   (c) the digestive system.
   (d) the eyes.
8. Which of the following is *not* a result of parasympathetic nerve stimulation?
   (a) increased movement of the GI tract
   (b) increased mucus secretion
   (c) constriction of the pupils
   (d) constriction of visceral blood vessels
9. Atropine blocks parasympathetic nerve effects. It would therefore result in
   (a) dilation of the pupils.
   (b) a decrease in mucus secretion.
   (c) a decrease in GI tract movement.
   (d) an increase in heart rate.
   (e) all of the above.
10. The area of the brain that is most directly involved in the reflex control of the autonomic system is
    (a) the hypothalamus.
    (b) the cerebral cortex.
    (c) the medulla oblongata.
    (d) the cerebellum.

Essay Questions

1. Compare the sympathetic and parasympathetic divisions in terms of ganglia location and nerve distribution.
2. Explain the structural and functional relationship between the sympathetic nervous system and the adrenal glands.
3. Compare the effects of adrenergic and cholinergic stimulation on the cardiovascular and digestive systems.
4. Explain how effectors that receive only sympathetic innervation are regulated by the ANS.
5. Explain why a person may sweat more profusely immediately after exercise than during exercise.

Critical-Thinking Questions

1. Shock is the medical condition that occurs when body tissues do not receive enough oxygen-carrying blood. It is characterized by low blood flow to the brain, leading to decreased levels of consciousness. Why would a patient with a cervical spinal cord injury be at risk of going into shock?
2. Imagine yourself at the starting block of the 100-meter dash of the Olympics. The gun is about to go off in the biggest race of your life. What is your autonomic nervous system doing at this point? How are your organs reacting?
3. Suppose you lift the wrist of a man who has fainted to feel for a pulse. How does his skin feel? How would you characterize his pulse? What specific role would the autonomic nervous system have in producing these effects?
4. Why would someone be given a prescription for atropine if they had gastritis? Why would the person's mouth feel dry after taking this drug?
5. Most agents used in chemical warfare affect the autonomic nervous system. Nerve gas, for example, stimulates activity of the parasympathetic nervous system to such an extent that it causes rapid death. Based on your knowledge of the autonomic nervous system, can you predict the type of symptoms a nerve-gas victim might suffer?
6. Give evidence for the argument that the autonomic nervous system is somewhat misnamed.

FOURTEEN

Endocrine System

Clinical Case Study 439
Introduction to the Endocrine System 439
Glands of the Endocrine System 440
Hormones and Their Actions 441
Control of Hormone Secretion 443
Pituitary Gland 444
Description of the Pituitary Gland 444
Pituitary Hormones 446
Thyroid and Parathyroid Glands 450
Description of the Thyroid Gland 450
Functions of the Thyroid Gland 451
Parathyroid Glands 452
Pancreas 453
Description of the Pancreas 453
Endocrine Function of the Pancreas 454
Adrenal Glands 455
Description of the Adrenal Glands 455
Functions of the Adrenal Glands 456
Gonads and Other Endocrine Glands 458
Gonads 458
Pineal Gland 458
Thymus 459
Stomach and Small Intestine 459
Placenta 459
Developmental Exposition: The Endocrine System 460
Clinical Considerations 462
Diagnosis of Endocrine Disorders 463
Disorders of the Pituitary Gland 463
Disorders of the Thyroid and Parathyroid Glands 463
Disorders of the Pancreatic Islets 465
Disorders of the Adrenal Glands 465
Internal Affairs 466
Clinical Case Study Answer 467
Chapter Summary 468
Review Activities 468

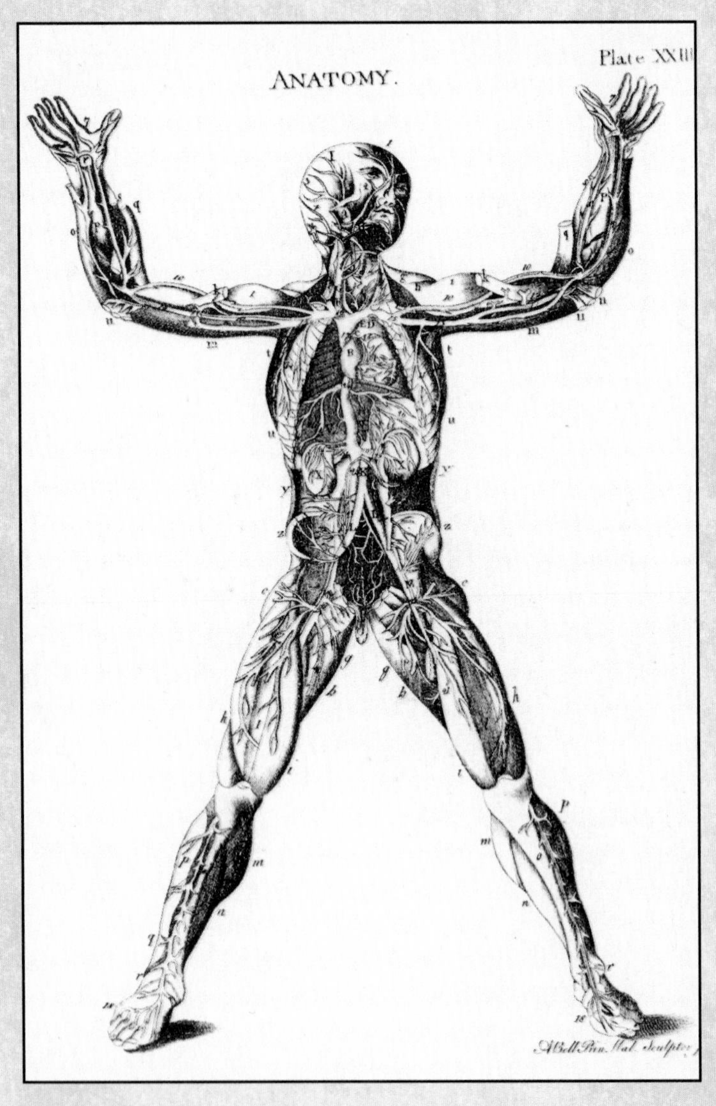

Although this undated medical drawing does not illustrate the endocrine system in particular, it could be used to point out that the endocrine organs are widely scattered throughout the body, with no anatomical continuity. In its pose, the figure is reminiscent of those that appear in Vesalius's Fabrica.

Clinical Case Study

A 38-year-old woman visited her family doctor because she had been experiencing chronic fatigue and weakness, especially in her legs. Upon greeting the patient, the doctor noted that although she was mildly obese, her face seemed unusually round. During questioning, he learned that at her recent 20-year high school reunion, nobody recognized her because her face had changed so much. He also learned that she suffered a chronic unrelenting headache. Physical examination yielded, in addition to the facial findings, an unusual fat distribution that included a hump on the upper back and marked truncal obesity. Additional findings included hypertension and blindness in the lateral visual fields of both eyes (bitemporal hemianopsia). Laboratory findings were remarkable for elevated blood glucose and other evidence that glucocorticoid levels were high.

Can a correlation be drawn between the patient's visual findings and the rest of her clinical picture? Explain. In which endocrine organ does the primary disease process reside? Does that same organ produce the hormones that directly cause the patient's signs and symptoms? Explain. Do you suppose that a tumor elsewhere in the endocrine system could produce similar findings?

Hints: Study figure 14.8 and note the position of the endocrine glands with respect to the optic nerve and optic chiasma. Also, review the actions of hormones in table 14.4.

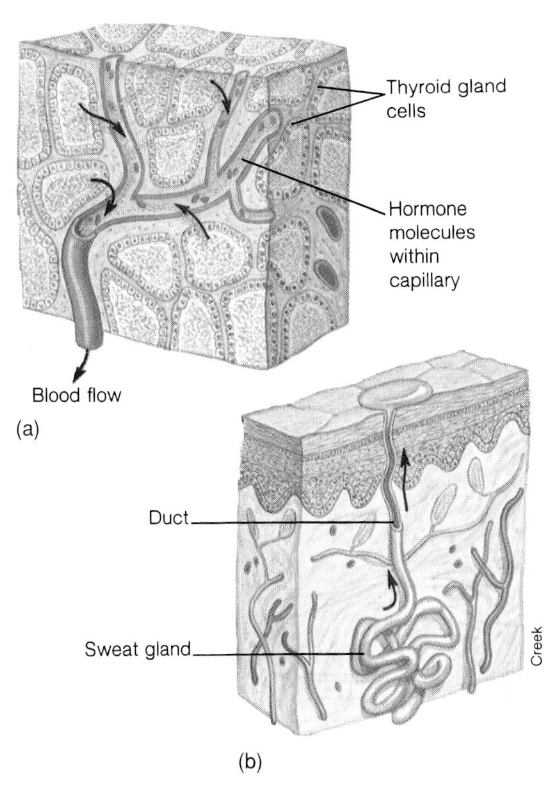

Figure 14.1

A comparison of (*a*) an endocrine gland and (*b*) an exocrine gland. An endocrine gland, such as the thyroid gland shown here, is a ductless gland that releases hormones into the blood or surrounding interstitial fluid. By contrast, exocrine glands, such as sweat glands in the skin, secrete their products directly onto body surfaces or into ducts that lead to body surfaces.

Introduction to the Endocrine System

Hormones are regulatory chemicals secreted by the endocrine glands into the blood, which transports them to their target cells. Feedback mechanisms in the target cells control the secretion (production) of the hormones.

Objective 1 Distinguish between endocrine and exocrine glands.

Objective 2 Compare and contrast the nervous and endocrine systems as to how they regulate body functions and maintain homeostasis.

Objective 3 Define *mixed gland* and identify the endocrine glands that are mixed.

Objective 4 Describe the action of a hormone on its target cell and explain how negative feedback regulates hormonal secretion.

Objective 5 Differentiate between the three principal kinds of hormones.

The numerous glands of the body can be classified as either of two types based on structure and function: *exocrine* or *endocrine*. Exocrine glands, such as sweat, salivary, and mucous glands, produce secretions that are transported through ducts to their respective destinations. Each of the exocrine glands functions within a particular system of the body. The endocrine glands constitute a system of their own, the endocrine system. In contrast to exocrine glands, endocrine glands are ductless; they secrete specific chemicals called *hormones* directly into the blood or surrounding interstitial fluid (fig. 14.1). The blood then transports these hormones to specific sites called *target cells*, where they perform precise functions.

The endocrine system functions closely with the nervous system in regulating and integrating body processes and maintaining homeostasis. The nervous system regulates body activities through the action of electrochemical impulses that are transmitted by means of neurons, resulting in rapid, but usually brief, responses. By contrast, the glands of the endocrine system secrete chemical regulators that travel through the bloodstream or interstitial fluid to their intended sites; their action is relatively slow,

exocrine: Gk. *exo*, outside; *krinein*, to separate
endocrine: Gk. *endon*, within; *krinein*, to separate
hormone: Gk. *hormon*, to set in motion

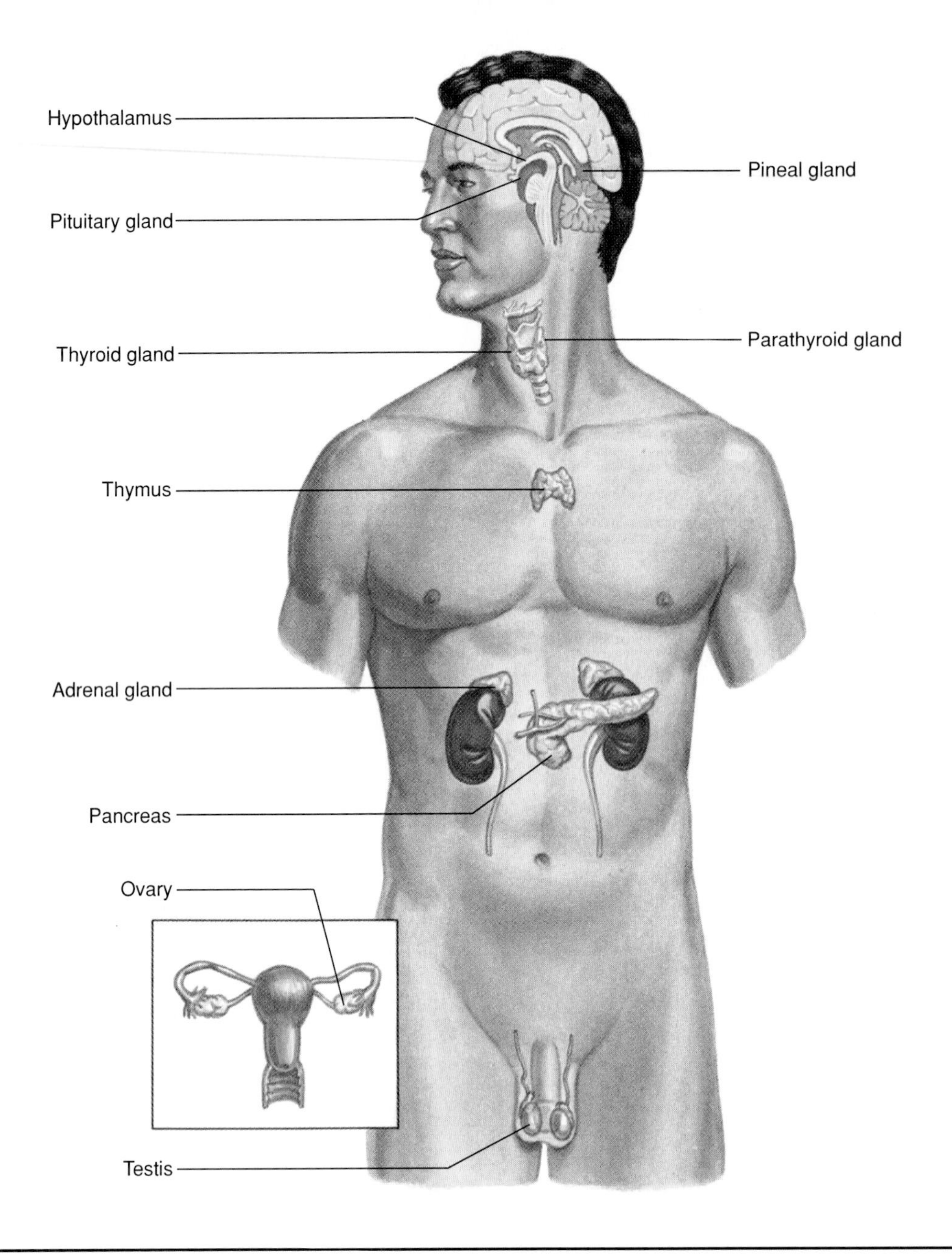

Figure 14.2

The location of the major endocrine organs of the body.

Table 14.1

Comparison of the Endocrine and Nervous Systems

Endocrine System	Nervous System
Secretes hormones that are transported to target cells via the blood or by surrounding interstitial fluid	Transmits neurochemical impulses via nerve fibers
Causes changes in the metabolic activities in specific cells	Causes muscles to contract or glands to secrete
Action is relatively slow (seconds or even days)	Action is very rapid (milliseconds)
Effects are relatively prolonged	Effects are relatively brief

but their effects are prolonged. Neurological responses are measured in milliseconds, but hormonal action requires seconds or days to elicit a response. Some hormones may have an effect that lasts for minutes; for others, the effect may last for weeks or months.

The nervous and endocrine systems are closely coordinated in autonomically controlling the functions of the body. Three endocrine glands are located within the cranial cavity, where certain structures of the brain routinely stimulate or inhibit the release of hormones. Likewise, certain hormones may stimulate or inhibit the activities of the nervous system. The functions of the two body control systems are compared in table 14.1.

Glands of the Endocrine System

The endocrine glands are distributed throughout the body (fig. 14.2). The **pituitary gland,** the **hypothalamus,** and the

Table 14.2
The Principal Endocrine Glands

Endocrine Gland	Major Hormones	Primary Target Organs	Primary Effects
Adrenal cortex	Glucocorticoids Aldosterone	Liver and muscles Kidneys	Glucocorticoids influence glucose metabolism; aldosterone promotes Na^+ retention, K^+ excretion
Adrenal medulla	Epinephrine	Heart, bronchioles, and blood vessels	Causes adrenergic stimulation
Hypothalamus	Releasing and inhibiting hormones	Anterior pituitary	Regulate secretion of anterior pituitary hormones
Small intestine	Secretin and cholecystokinin	Stomach, liver, and pancreas	Inhibit gastric motility and stimulate bile and pancreatic juice secretion
Pancreatic islets	Insulin Glucagon	Many organs Liver and adipose tissue	Insulin promotes cellular uptake of glucose and formation of glycogen and fat; glucagon stimulates hydrolysis of glycogen and fat
Ovaries	Estradiol-17β and progesterone	Female reproductive tract and mammary glands	Maintain structure of reproductive tract and promote secondary sex characteristics
Parathyroid glands	Parathyroid hormone	Bone, intestine, and kidneys	Increases Ca^{++} concentration in blood
Pineal gland	Melatonin	Hypothalamus and anterior pituitary	Affects secretion of gonadotrophic hormones
Pituitary, anterior	Trophic hormones	Endocrine glands and other organs	Stimulate growth and development of target organs; stimulate secretion of other hormones
Pituitary, posterior	Antidiuretic hormone Oxytocin	Kidneys and blood vessels Uterus and mammary glands	Antidiuretic hormone promotes water retention and vasoconstriction; oxytocin stimulates contraction of uterus and mammary secretory units
Stomach	Gastrin	Stomach	Stimulates acid secretion
Testes	Testosterone	Prostate, seminal vesicles, and other organs	Stimulates secondary sexual development
Thymus	Thymosin	Lymph nodes	Stimulates white blood cell production
Thyroid gland	Thyroxine (T_4), triiodothyronine, and (T_3); calcitonin	Blood and most organs	Thyroxine and triiodothyronine promote growth and development and stimulate basal rate of cell respiration (basal metabolic rate or BMR); calcitonin regulates Ca^{++} levels within blood by inhibiting bone decalcification
Kidneys	Erythropoietin	Bone marrow	Stimulates red blood cell production

pineal gland are associated with the brain within the cranial cavity. The **thyroid gland** and **parathyroid glands** are located in the neck. The **adrenal glands** and **pancreas** are located within the abdominal region. The **gonads** (*ovaries*) of the female are located within the pelvic cavity, whereas the **gonads** (*testes*) of the male are located in the scrotum. The pancreas and gonads are frequently classified as *mixed glands* because they have exocrine as well as endocrine functions.

The endocrine system is unique in that its glands are widely scattered throughout the body, with no anatomical continuity. By contrast, the organs of the other body systems are physically linked together in some fashion.

In addition to the glands just mentioned, several others may be considered part of the endocrine system because they have endocrine functions. These include the *thymus*, located in the lower median neck region; the *stomach;* the *kidneys;* the *mucosal cells of the duodenum;* and the *placenta*, associated with the fetus. The principal endocrine glands of the body, their hormones, and the effects of these hormones are listed in table 14.2.

Hormones and Their Actions

Hormones are specific organic substances that act as the chemical messengers of the endocrine system. The three basic kinds of hormones (proteins, steroids, and amines) are derived either from amino acids or cholesterol (fig. 14.3).

Proteins are composed of amino acids bound together in peptide chains (fig. 14.3*a*). Most of the hormones of the body are proteins, including calcitonin from the thyroid gland and hormones secreted by the pituitary gland, the pancreas, and the parathyroid glands. Protein hormones cannot be administered orally because the peptide bonds would be split during the hydrolytic reaction of digestion; thus, they must be injected intravenously, intramuscularly, or subcutaneously.

A *steroid* is a lipid synthesized from cholesterol. Steroids exist as complex rings of carbon and hydrogen atoms (fig. 14.3*b*).

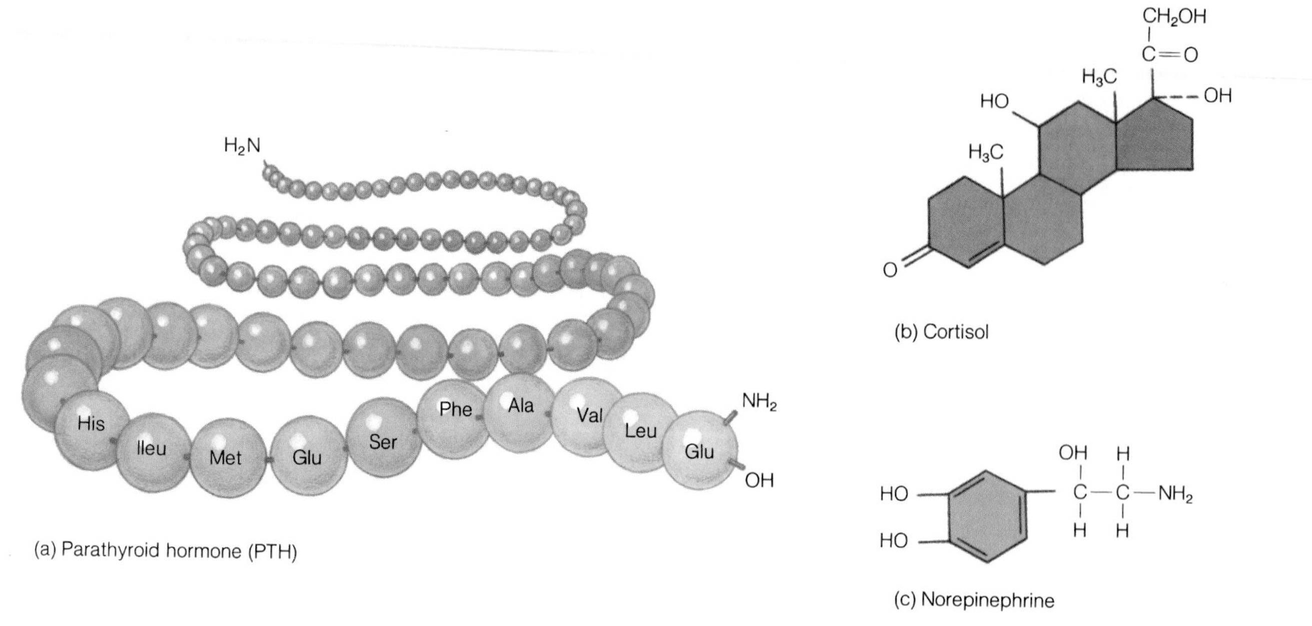

Figure 14.3

Chemical structures exemplifying the three basic kinds of hormones: (*a*) a protein, (*b*) a steroid, and (*c*) an amine.

Table 14.3
Classes of Hormones

Type of Hormone	Composition/Structure	Examples	Method of Administration
Protein	Amino acids bonded by peptide chains	Pituitary hormones, pancreatic hormones, parathyroid hormones, and calcitonin from thyroid gland	Intravenously, intramuscularly, or subcutaneously
Steroid	Lipid with a cholesterol-type nucleus	Sex hormones and hormones from adrenal cortex	Orally, intravenously, or intramuscularly
Amine	Amino acids with no peptide bonds; molecule contains $—NH_2$ group	Thyroxine from thyroid gland, adrenaline from adrenal gland, and melatonin from pineal gland	Orally, subcutaneously, intravenously, or as an inhalant

The types of atoms attached to the rings determine the specific kind of steroid. There are more than 20 steroid hormones in the body, including such common ones as *cortisol, cortisone, estrogen, progesterone*, and *testosterone*. The sex hormones produced by the gonads and the hormones produced by the adrenal cortex are all steroids. Steroids can be taken orally or intravenously to regulate body activity if glandular dysfunction prevents the natural production of normal amounts.

Amines are produced from amino acids but do not contain peptide bonds. Amine molecules contain atoms of carbon, hydrogen, and nitrogen and always have an associated amine group ($—NH_2$) (fig. 14.3c). Thyroxine produced in the thyroid gland, epinephrine (adrenaline) and norepinephrine produced in the adrenal gland, and melatonin produced in the pineal gland are examples of amines.

Thyroxine is usually administered orally, whereas epinephrine is administered intravenously to produce a fast response. Epinephrine can also be administered as an inhalant when it is necessary to enlarge the air-conducting passageways within the lungs.

A summary of the basic types of hormones is presented in table 14.3.

The usual action of hormones is to speed up or slow down metabolism in the target cells. Hormones are extremely specific as to which cells they affect and the cellular changes they elicit. The ability of a hormone to affect a particular cell depends on the presence of receptor molecules in the cell or specific receptor sites on its cell membrane.

Steroid hormones are soluble in lipids and can readily pass through a cell membrane. Once inside a cell, the steroid

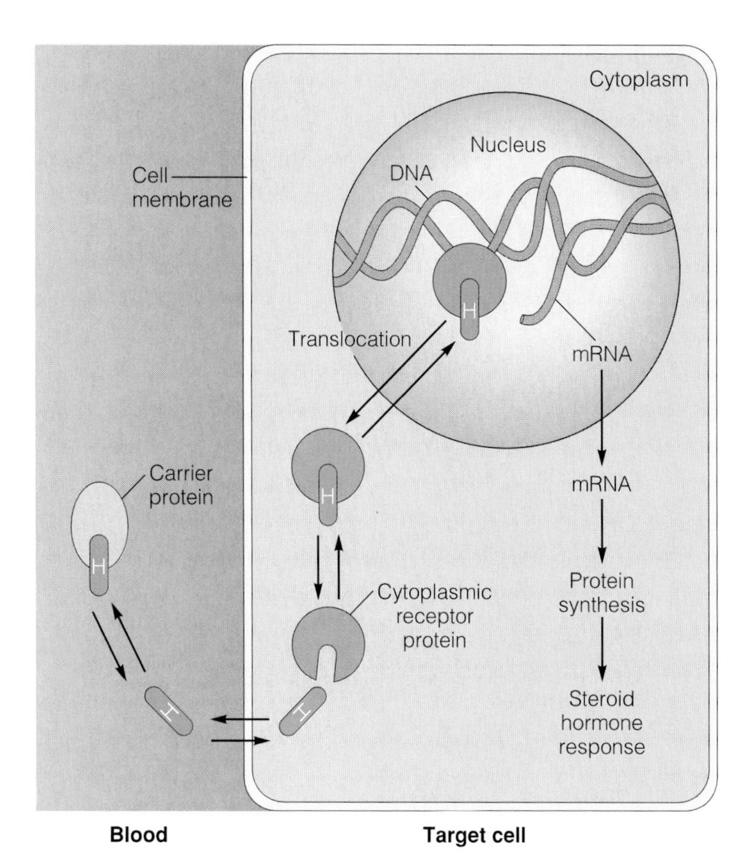

Figure 14.4

The mechanism of the action of a steroid hormone (H) on the target cells. A steroid hormone passes through the cell membrane and unites with a receptor protein in the cytoplasm. The steroid-protein complex then enters the cell nucleus and activates the synthesis of messenger RNA. The messenger RNA then disperses into the cytoplasm where it activates the synthesis of proteins by means of ribosomes.

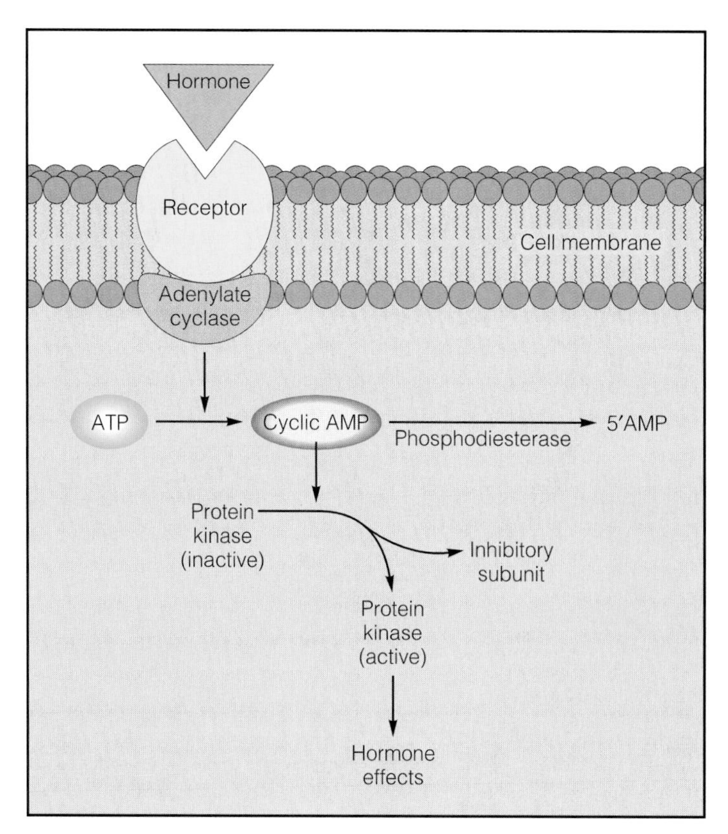

Figure 14.5

The relationship between a protein hormone, the target cells, and cyclic AMP. Hormones arrive at target cells through body fluids and attach to specific receptor sites on the cell membrane. This causes molecules of adenylate cyclase to diffuse into the cytoplasm and promote the change of ATP into cyclic AMP. Cyclic AMP in turn causes changes in cellular processes.

combines with a protein to form a steroid-protein complex (fig. 14.4). This complex is necessary for the synthesis of specific kinds of messenger RNA molecules.

Protein and amine hormones are insoluble in lipids and therefore must attach to specific receptor sites on the cell membrane. These hormones do not enter the cell, but their attachment increases the activity of an enzyme in the cell membrane called *adenylate cyclase (ă-den′l-it si′klās)*. In the presence of adenylate cyclase, ATP molecules in the cell are converted to *cyclic AMP* (adenosine monophosphate), which in turn disperses throughout the cell to cause changes in cellular processes (fig. 14.5). These changes may include increasing protein synthesis, altering membrane permeability, or activating certain cellular enzymes.

Control of Hormone Secretion

The rate of secretion of a particular hormone and the rate of usage by the target cells are closely balanced. The stability of hormone levels is maintained by a negative feedback system and autonomic neural impulses.

Negative feedback (fig. 14.6) is a homeostatic mechanism that maintains the status quo of supply and demand between hormone normal levels and the needs of the target cells. An endocrine gland will continue to secrete hormones that affect target cells until messages come back from the cells reporting that sufficient amounts of these hormones are present. These messages are generally in the form of hormones secreted by the target cells. This chemical feedback information signals the endocrine gland to inhibit secretion (fig. 14.6). The primary endocrine gland will resume secretions when the blood levels of inhibiting chemicals become low again.

Neural impulses through the autonomic nervous system cause certain endocrine glands, such as the adrenal medulla, to secrete hormones. One specialized kind of neural impulse involves the hypothalamus and the pituitary gland (fig. 14.7). In this system, chemical secretions from neurosecretory cells in the hypothalamus called *releasing factors* influence specific target cells in the pituitary gland. Stimulation of the pituitary by a releasing factor causes the secretion of specific hormones. The hypothalamus continues to secrete the releasing factor until a given level of hormones is present in the body fluids and is detected by the hypothalamus through a negative feedback mechanism.

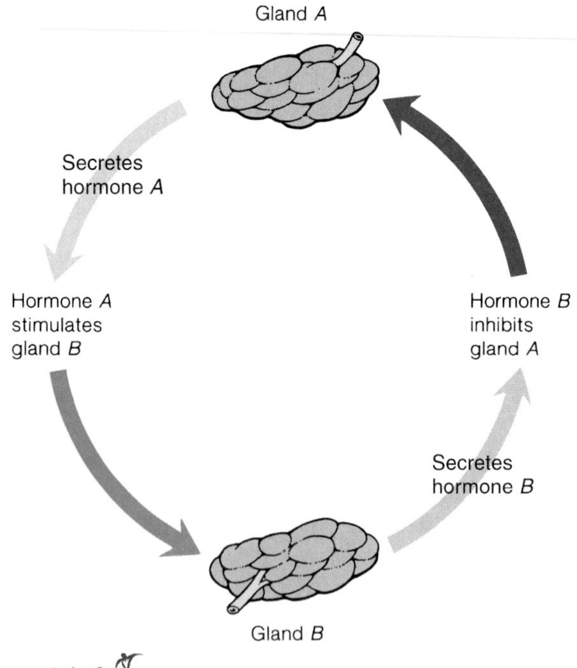

Figure 14.6

In a negative feedback system, gland A secretes hormone A (first step) that stimulates gland B to release its hormone; the hormone from gland B (second step) inhibits the secretion of hormone A by gland A.

1. What are some of the ways in which endocrine and exocrine glands differ structurally?
2. How are the nervous and endocrine systems functionally related?
3. List the body organs that are exclusively endocrine in function. Also list the organs that serve other body functions, in addition to secreting hormones.
4. Using diagrams, describe the mechanism of steroid hormone action and the mechanism of protein hormone action within cells.
5. List the three kinds of hormones, give examples of each, and describe their chemical compositions.

Pituitary Gland

The neurohypophysis of the pituitary gland releases hormones that are produced by the hypothalamus, whereas the adenohypophysis of the pituitary gland secretes its own hormones in response to regulation from hypothalamic hormones. The secretions of the pituitary gland are thus controlled by the hypothalamus, as well as by negative feedback influences from the target glands.

Objective 6 List the hormones secreted by the adenohypophysis and neurohypophysis.

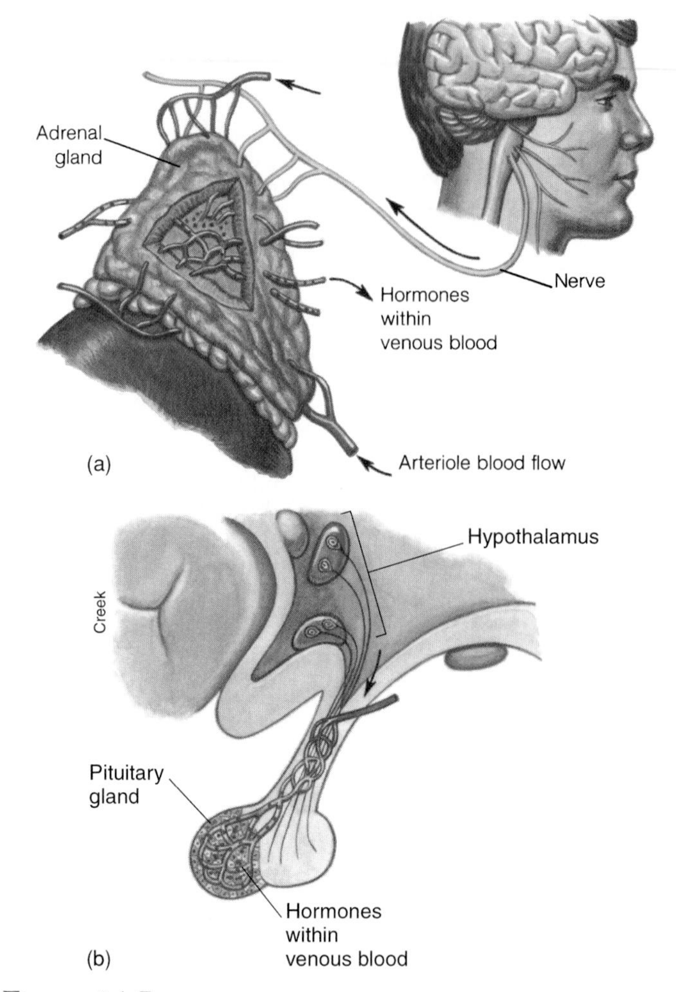

Figure 14.7

Neural control of endocrine secretion. (*a*) The adrenal medulla secretes hormones in response to autonomic stimulation. (*b*) The hypothalamus has a neurosecretory role in stimulating target cells in the pituitary gland to release hormones.

Objective 7 Describe, in a general way, the actions of anterior pituitary hormones.

Objective 8 Explain how the secretions of anterior and posterior pituitary hormones are controlled by the hypothalamus.

Objective 9 Explain how the secretion of anterior pituitary hormones is regulated by negative feedback.

Description of the Pituitary Gland

The **pituitary** (*pĭ-too'ĭ-ter-e*) **gland,** or **cerebral hypophysis** (*hi-pof'ĭ-sis*)is located on the inferior aspect of the brain in the region of the

pituitary: L. *pituita*, phlegm (this gland was originally thought to secrete mucus into the nasal cavity)

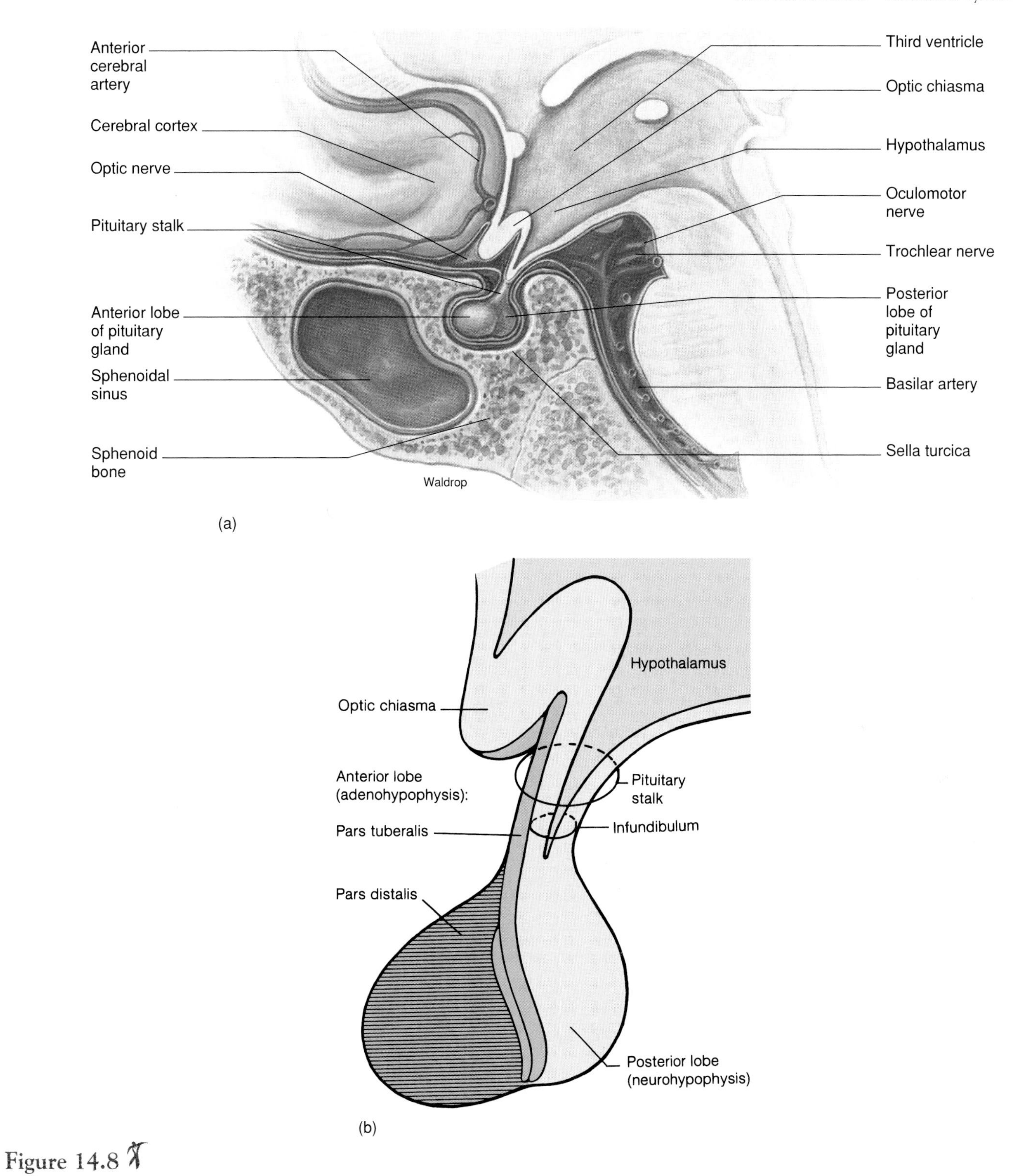

Figure 14.8

The pituitary gland. (*a*) Attached by the pituitary stalk to the hypothalamus, the pituitary gland lies in the sella turcica of the sphenoid bone. (*b*) A diagram of the pituitary gland showing the various portions.

diencephalon and is attached to the brain by a structure called the *pituitary stalk* (fig. 14.8). The pituitary is a rounded, pea-shaped gland measuring about 1.3 cm (0.5 in.) in diameter. It is covered by the dura mater and is supported by the sella turcica of the sphenoid bone. The cerebral arterial circle surrounds the highly vascular pituitary gland, providing it with a rich blood supply (see fig. 16.23).

The pituitary gland is structurally and functionally divided into an anterior lobe, or **adenohypophysis** (*ad″n-o-hi-pof′ĭ-sis*), and a posterior lobe called the **neurohypophysis**

adenohypophysis: Gk. *adeno*, gland; *hypo*, under; *physis*, a growing

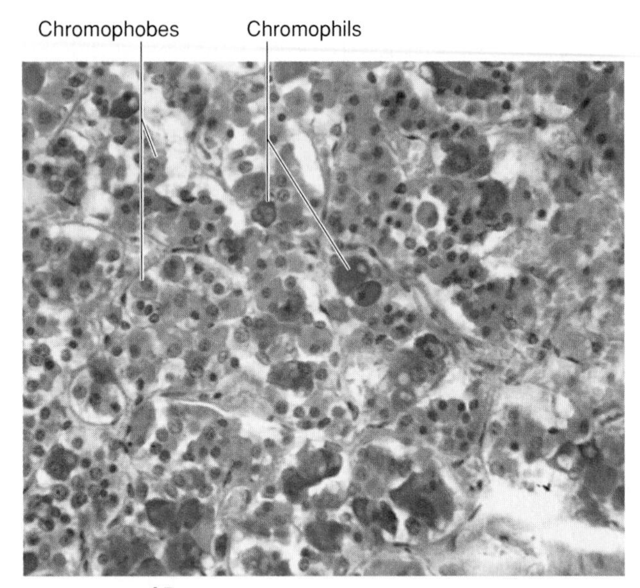

Figure 14.9

The histology of the pars distalis of the anterior pituitary (adenohypophysis).

(*noor″o-hi-pof′ĭ-sis*). The adenohypophysis consists of two parts in adults: (1) the **pars distalis** (*anterior pituitary*) is the bulbar portion, and (2) the **pars tuberalis** is the thin extension in contact with the infundibulum (fig. 14.8). A **pars intermedia,** a strip of tissue between the anterior and posterior lobes, exists in the fetus. During fetal development, its cells mingle with those of the anterior lobe, and in adults they no longer constitute a separate structure. The histology of the adenohypophysis is shown in figure 14.9.

The neurohypophysis is the neural part of the pituitary gland. It consists of the bulbar **lobus nervosa** (*posterior pituitary*), which is in contact with the adenohypophysis, and the funnel-shaped **infundibulum,** a stalk of tissue that connects the pituitary to the base of the hypothalamus. Nerve fibers extend through the infundibulum, along with minute neuroglia-like cells called **pituicytes** (*pĭ-too′ĭ-sīts*).

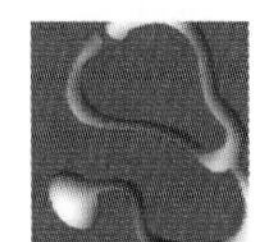

The pituitary gland is the structure of the brain perhaps most subject to neoplasms. A tumor of the pituitary is generally detected easily as it begins to grow and interfere with hormonal activity. If the tumor grows superiorly, it may exert sufficient pressure on the optic chiasma to cause *bitemporal hemianopia,* which is blindness in the temporal field of vision of both eyes. Surgical removal of a neoplasm of the pituitary gland *(hypophysectomy)* may be performed transcranially through the frontal bone or through the nasal cavity and sphenoidal sinus.

Pituitary Hormones

The pituitary gland releases nine important hormones. The first seven in the following list are secreted by the anterior pituitary. The seventh is secreted by certain cells that are remnants of the pars intermedia. The last two are produced in the hypothalamus, transported through axons in the infundibulum to the posterior pituitary, and released from the posterior pituitary into the blood.

The hormones secreted by the pars distalis are called **trophic** (*trof′ik*) **hormones.** The term *trophic* means "food"; it is used because high amounts of the anterior pituitary hormones cause their target glands to hypertrophy, while low levels cause their target glands to atrophy.

1. **Growth hormone (GH).** Growth hormone, or *somatotropin* (*so″măt-o-trōp′in*) regulates the rate of growth of all body cells and promotes mitotic activity. The secretion of GH is regulated by growth hormone–releasing hormone (GH–RH) and growth hormone–inhibiting hormone (GH–IH), or *somatostatin,* from the hypothalamus. The precise mechanism of GH is not understood, but it seems to promote the movement of amino acids through cell membranes and the utilization of these substances in protein synthesis. Pathological hyposecretion of GH during adolescence limits body growth, causing a type of *dwarfism.* Hypersecretion of GH during adolescence may result in *gigantism.* In both of these conditions, body proportions are greatly distorted (see fig. 14.25*a,b*). Hypersecretion of GH in an adult, after the epiphyses (see chapter 6) are fused, causes *acromegaly* (*ak″ro-meg′ă-le*). In this condition, the soft tissues rapidly proliferate and certain body features, such as the hands, feet, nose, jaw, and tongue, become greatly distorted (see fig. 14.25*c*).
2. **Thyroid-stimulating hormone (TSH).** TSH, frequently called *thyrotropin,* regulates the hormonal activity of the thyroid gland. The secretion of TSH, however, is partly regulated by the hypothalamus through the secretion of *thyrotropin-releasing hormone (TRH)*. External factors may influence the release of TSH as well. Exposure to cold, certain illnesses, and emotional stress may trigger an increased output of TSH.
3. **Adrenocorticotropic hormone (ACTH).** ACTH promotes normal functioning of the adrenal cortex. It also acts on all body cells by assisting in the breakdown of fats. The release of ACTH is controlled by a *corticotropin-releasing hormone (CRH)* produced in the hypothalamus. As with TSH, stress further influences the release of ACTH.
4. **Follicle-stimulating hormone (FSH).** In males, FSH stimulates the testes to produce sperm. In females, FSH regulates the monthly development of the follicle and egg. It also stimulates the secretion of the female sex hormone *estrogen.*
5. **Luteinizing hormone (LH).** LH works with FSH, and together they are referred to as *gonadotrophins,* which means their target cells are located within the gonads or reproductive organs. In females, LH works with FSH in bringing about ovulation. It also stimulates the formation of the corpus luteum and the production of another female sex hormone, *progesterone* (see chapter 21).

infundibulum: L. *infundibulum,* a funnel

In males, the luteinizing hormone is called *interstitial cell–stimulating hormone (ICSH)* and stimulates the interstitial cells of the testes to develop and secrete the male sex hormone testosterone (see chapter 20).

The mechanism that controls the production and release of gonadotrophic hormones is not well understood. It is known, however, that following puberty (see chapters 20 and 21) the hypothalamus releases a hormone called *gonadotrophin-releasing hormone (GnRH)* that regulates the secretion of both LH and FSH.

6. **Prolactin.** Prolactin is secreted in both males and females, but it functions primarily in females after parturition. Prolactin assists other hormones in initiating and sustaining milk production by the mammary glands. The hypothalamus plays an important role in the release of this hormone through the production of *prolactin-inhibiting hormone (PIH)*, now known to be dopamine. When PIH is secreted, the secretion of prolactin is inhibited; when PIH is not secreted, prolactin is released.
7. **Melanocyte-stimulating hormone (MSH).** The exact action of MSH in humans is unknown, but it can cause darkening of the skin by stimulating the dispersion of melanin granules within melanocytes. Secretion of MSH is stimulated by *corticotropin-releasing hormone* and inhibited by *dopamine*, both of which come from the hypothalamus.
8. **Oxytocin.** Oxytocin is produced by specialized cells in the hypothalamus. It then travels through axons in the infundibulum to the lobus nervosa, where it is stored and released in response to neural impulses from the hypothalamus. Oxytocin influences physiological activity in the female reproductive system. It is released near the end of gestation and causes uterine contractions during labor. It also stimulates contraction of the mammary gland alveoli and ducts, producing the milk-ejection reflex during lactation (see chapter 21). In males, a rise in oxytocin at the time of ejaculation has been measured, but the physiological significance of this hormone in males has yet to be demonstrated.

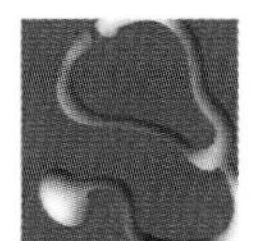

Injections of oxytocin may be given to a woman to induce labor if the pregnancy is prolonged or if the fetal membranes have ruptured and there is a danger of infection. Oxytocin administration after parturition causes the uterus to regress in size and squeezes the blood vessels, thus minimizing the danger of hemorrhage.

9. **Antidiuretic hormone (ADH).** ADH is similar to oxytocin in its site of production and release. Like oxytocin, it is a polypeptide. The major function of ADH is to inhibit the formation of urine in the kidneys, or more specifically, to reduce the amount of water excreted from the kidneys. This hormone is also called *vasopressin* because it causes vasoconstriction at high concentrations.

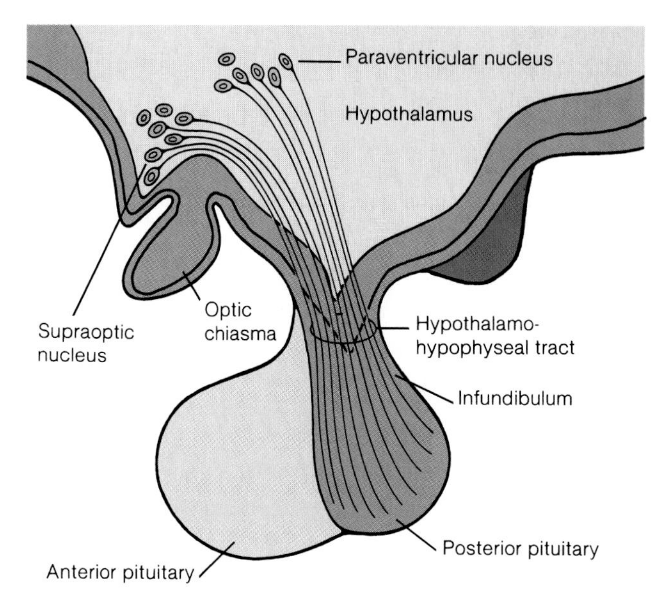

Figure 14.10

The posterior pituitary (neurohypophysis) stores and secretes oxytocin and antidiuretic hormone produced in neuron cell bodies within the supraoptic and paraventricular nuclei of the hypothalamus. These hormones are transported to the posterior pituitary by nerve fibers of the hypothalamo-hypophyseal tract.

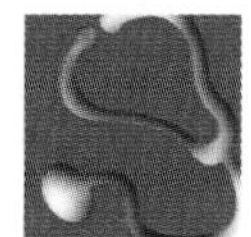

Diabetes insipidus results from a marked decrease in ADH output caused by trauma or disease to the hypothalamus or neurohypophysis. The symptoms of this disease are polyuria (voiding excessive dilute urine), concentrated body fluids with dehydration, and a particularly heightened sensation of thirst.

Oxytocin and antidiuretic hormone are released by the posterior pituitary (pars nervosa of the neurohypophysis). These two hormones, however, are actually produced in neuron cell bodies of the **supraoptic nuclei** and **paraventricular nuclei** of the hypothalamus. These nuclei within the hypothalamus are thus endocrine glands; the hormones they produce are transported along axons of the **hypothalamo-hypophyseal** *(hi-pof'ĭ-se'al)* **tract** (fig. 14.10) to the posterior pituitary, which stores and later releases them. The posterior pituitary is thus more of a storage organ than a true gland.

The secretion of oxytocin and ADH from the posterior pituitary is controlled by **neuroendocrine reflexes.** In nursing mothers, for example, the stimulus of sucking acts via sensory nerve impulses to the hypothalamus to stimulate the reflex secretion of oxytocin. The secretion of ADH is stimulated by osmoreceptor neurons in the hypothalamus in response to a rise in blood osmotic pressure; its secretion is inhibited by sensory impulses from stretch receptors in the left atrium of the heart in response to a rise in blood volume.

At one time, the anterior pituitary was called the "master gland" because it secretes hormones that regulate some other endocrine glands (see table 14.4 and fig. 14.11). Adrenocorticotrophic hormone (ACTH), thyroid-stimulating hormone

Table 14.4
Hormones Released by the Pituitary Gland

Hormone	Action	Regulation of Secretion
Adenohypophysis		
Growth hormone (GH), or somatotropin	Regulates mitotic activity and growth of body cells; promotes movement of amino acids through cell membranes	Stimulated by growth hormone–releasing hormone (GH–RH) from the hypothalamus; inhibited by growth hormone–inhibiting hormone (somatostatin) from the hypothalamus
Thyroid-stimulating hormone (TSH), or thyrotropin	Regulates hormonal activity of thyroid gland	Stimulated by thyrotropin-releasing hormone (TRH) from the hypothalamus; inhibited by thyroid hormones
Adrenocorticotropic hormone (ACTH)	Promotes release of glucocorticoids and mineralocorticoids from the adrenal cortex; assists in breakdown of fats	Stimulated by corticotropin-releasing hormone (CRH) from the hypothalamus; inhibited by glucocorticoids from the adrenal cortex
Follicle-stimulating hormone (FSH)	In males, stimulates production of sperm cells; in females, regulates follicle development in ovary and stimulates secretion of estrogen	Stimulated by gonadotrophin-releasing hormone (GnRH) from the hypothalamus; inhibited by sex steroids from the gonads
Luteinizing hormone (LH), or ICSH in males	Promotes secretion of sex hormones; in females, plays role in release of ovum and stimulates formation of corpus luteum and production of progesterone; in males, stimulates testosterone secretion	Stimulated by gonadotrophin-releasing hormone (GnRH) from the hypothalamus
Prolactin	Promotes secretion of milk from mammary glands (lactation)	Inhibited by prolactin-inhibiting hormone (PIH) from the hypothalamus
Melanocyte-stimulating hormone (MSH)	Can cause darkening of the skin	Stimulated by corticotropin-releasing hormone (CRH) from the hypothalamus; inhibited by dopamine, also from the hypothalamus
Neurohypophysis		
Oxytocin	Stimulates contractions of muscles in uterine wall; causes contraction of muscles in mammary glands	Hypothalamus, in response to stretch in uterine walls and stimulation of breasts
Antidiuretic hormone (ADH)	Reduces water loss from kidneys; elevates blood pressure	Hypothalamus, in response to changes in water concentration in the blood

(TSH), and the gonadotrophic hormones (FSH and LH) stimulate the adrenal cortex, thyroid, and gonads, respectively, to secrete their hormones. The anterior pituitary hormones also have a trophic effect on their target glands in that the health of these glands depends on adequate stimulation by anterior pituitary hormones. The anterior pituitary, however, is not really the master gland, since secretion of its hormones is in turn controlled by hormones secreted by the hypothalamus.

Releasing and inhibiting hormones from the hypothalamus travel through the **hypothalamo-hypophyseal portal system** to control the secretion of hormones from the anterior pituitary. Neurons in the hypothalamus secrete hormones into the region of the **median eminence,** where they enter a network of primary capillaries (fig. 14.12). Venous drainage through the pituitary stalk transports the hypothalamic hormones to a network of secondary capillaries within the anterior pituitary. This system is considered a *portal system* because there are two sets of capillaries in a series (fig. 14.12). (This is analogous to the hepatic portal system that delivers venous blood from the pancreas, spleen, and GI tract to the liver, as described in chapter 16.)

1. List the hormones secreted by the anterior pituitary and explain how the hypothalamus controls the secretion of each hormone.
2. Which hormone secreted by the anterior pituitary does not affect some other endocrine gland?
3. List the hormones released by the posterior pituitary. State the origin of these hormones and the mechanisms by which their secretions are regulated.
4. Which hormone secreted by the pituitary gland affects only females?

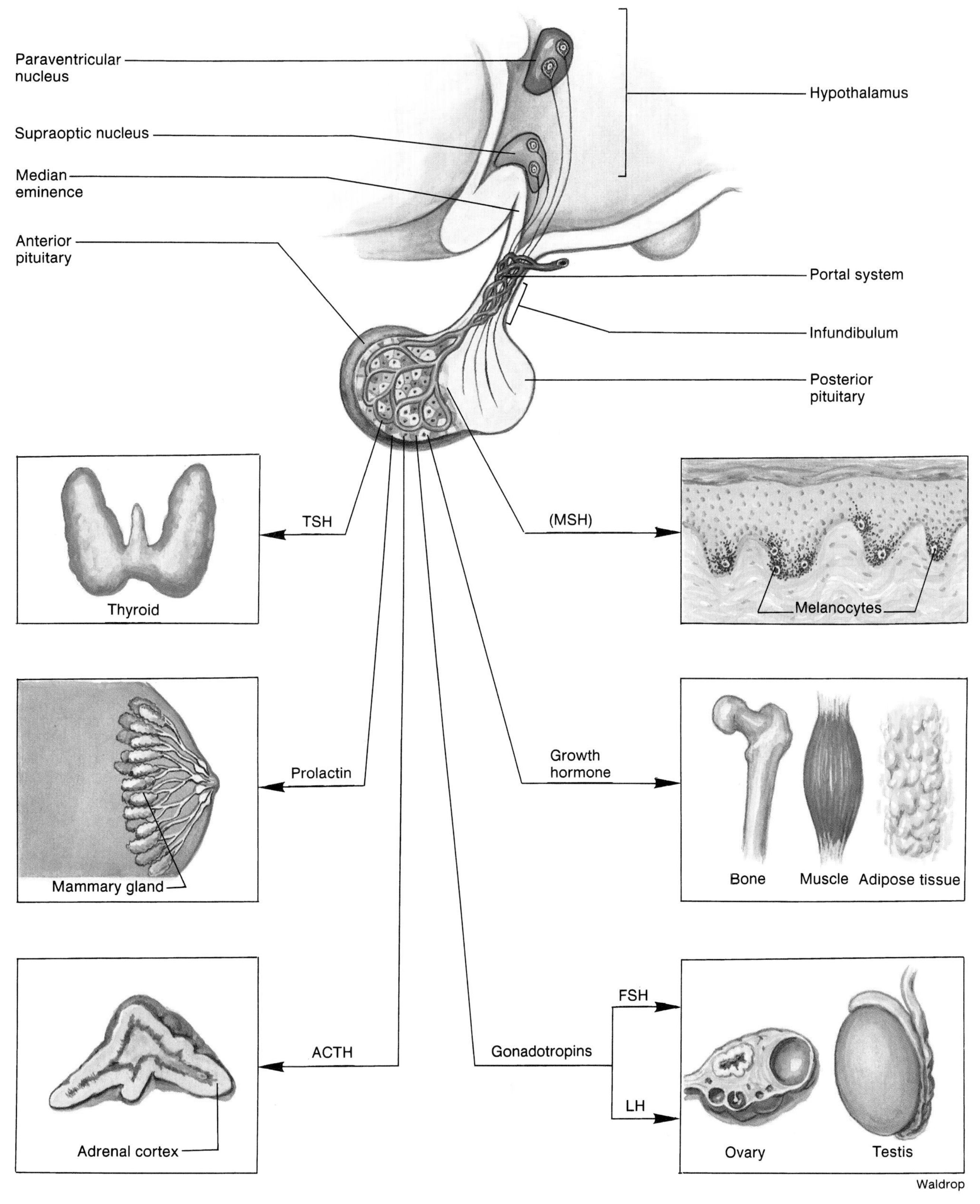

Figure 14.11

The hormones secreted by the anterior pituitary and the target organs for those hormones.

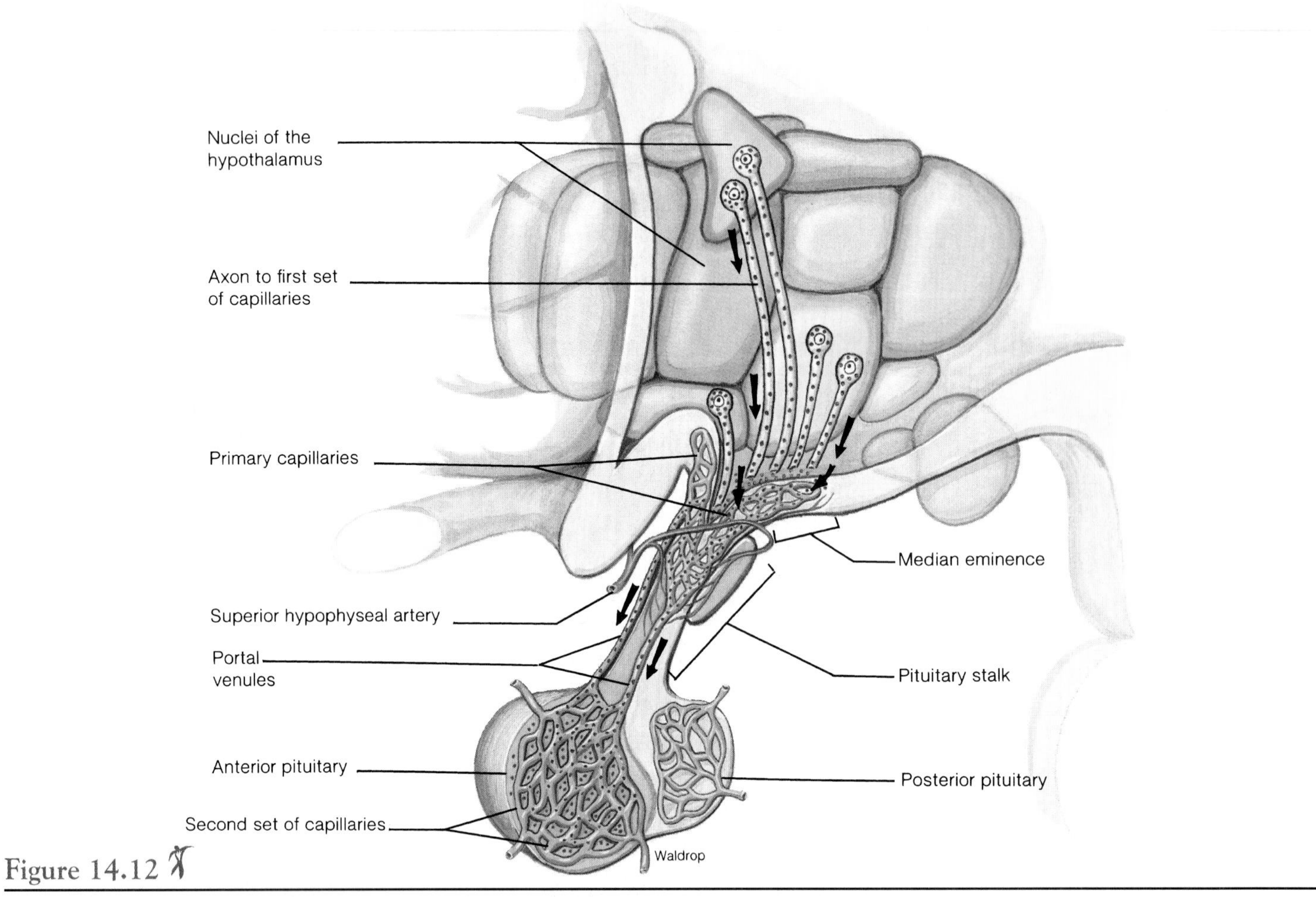

Figure 14.12

The hypothalamo-hypophyseal portal system. Hypothalamic hormones (shown as dots) enter this system in the first set of capillaries of the median eminence and are transported through the portal veins of the pituitary stalk to a second set of capillaries of the anterior pituitary.

Thyroid and Parathyroid Glands

The thyroid gland secretes thyroxine and triiodothyronine, which function in the regulation of energy metabolism. These hormones are critically important for proper growth and development. The thyroid also secretes calcitonin, which may antagonize the action of parathyroid hormone in the regulation of calcium and phosphate balance.

Objective 10 Describe the location and structure of the thyroid gland and list the actions of the thyroid hormones.

Objective 11 Describe the location and structure of the parathyroid glands and list the actions of parathyroid hormone.

Description of the Thyroid Gland

The **thyroid gland** is located in the neck, just below the larynx (fig. 14.13). Its two lobes, each about 5 cm (2 in.) long, are positioned on either lateral side of the trachea and connected anteriorly by a bridge of tissue called the **isthmus.** The thyroid is the largest of the endocrine glands, weighing between 20 and 25 g. It receives an abundant blood supply (80–120 ml/min) through the paired superior thyroid branches of the external carotid arteries and the paired inferior thyroid branches of the subclavian arteries. The venous return is through the paired superior and middle thyroid veins that pass into the internal jugular veins and through the inferior thyroid veins that empty into the brachiocephalic veins.

On a microscopic level, the thyroid gland consists of numerous spherical hollow sacs called **thyroid follicles** (fig. 14.14). These follicles are lined with a simple cuboidal epithelium composed of **follicular cells.** The follicular cells synthesize the two principal thyroid hormones (see table 14.5). The interior of the follicles contains **colloid** (*kol'oid*), a protein-rich fluid. Between the follicles are epithelial cells called **parafollicular cells** that produce a hormone called *calcitonin (kal"sĭ-to'nin)*, or *thyrocalcitonin*.

thyroid: Gk. *thyreos*, oblong shield

isthmus: L. *isthmus*, narrow portion
colloid: Gk. *kolla*, glue

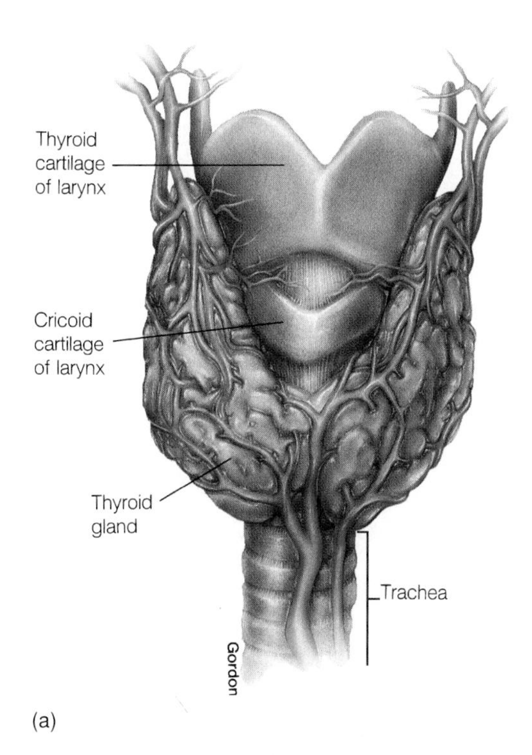

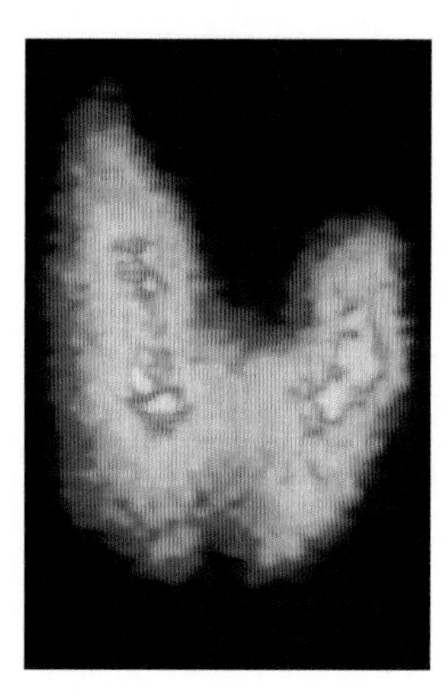

Figure 14.13

The thyroid gland. (*a*) Its relationship to the larynx and trachea and (*b*) a scan of the thyroid gland 24 hours after the intake of radioactive iodine.

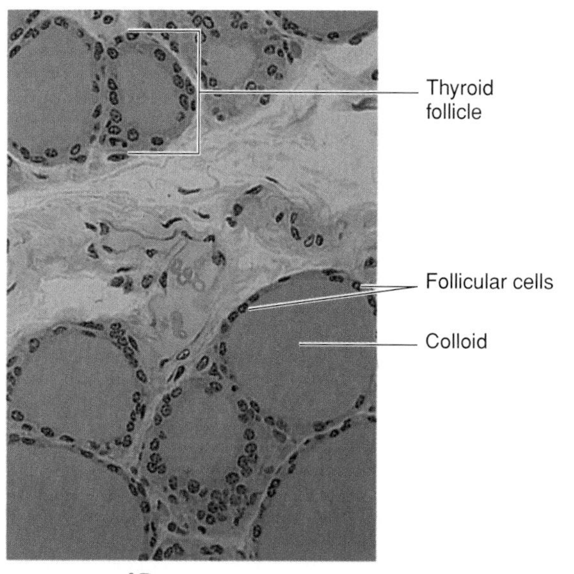

Figure 14.14

The histology of the thyroid gland showing numerous thyroid follicles. Each follicle consists of follicular cells surrounding the fluid known as colloid.

Table 14.5

Hormones of the Thyroid Gland

Hormone	Action	Source of Regulation
Thyroxine (T_4)	Increases rate of protein synthesis and rate of energy release from carbohydrates; regulates rate of growth; stimulates maturity of nervous system	Hypothalamus and release of TSH from adenohypophysis of pituitary gland
Triiodothyronine (T_3)	Same as above	Same as above
Calcitonin (thyrocalcitonin)	Lowers blood calcium by inhibiting the release of calcium from bone tissue	Calcium levels in blood

The thyroid is innervated by postganglionic neurons from the superior and middle cervical sympathetic ganglia and preganglionic neurons from ganglia derived from the second through the seventh thoracic segment of the spinal cord.

Functions of the Thyroid Gland

The thyroid gland produces two major hormones, *thyroxine (thi-rok'sin)* (T_4) and *triiodothyronine (tri''i-ō''dō-thi'ro-nēn)* (T_3), and the minor hormone *calcitonin (thyrocalcitonin)*. The release of thyroxine and triiodothyronine is controlled by the hypothalamus and by the TSH secreted from the adenohypophysis of the pituitary gland. Thyroxine and triiodothyronine are stored in the thyroid follicles and released as needed to control the metabolic rate of the body. More specifically, they act to increase the rate of protein synthesis and the rate of energy release from carbohydrates. They also regulate the growth rate in young people and are associated with sexual maturity and early maturation of the nervous system.

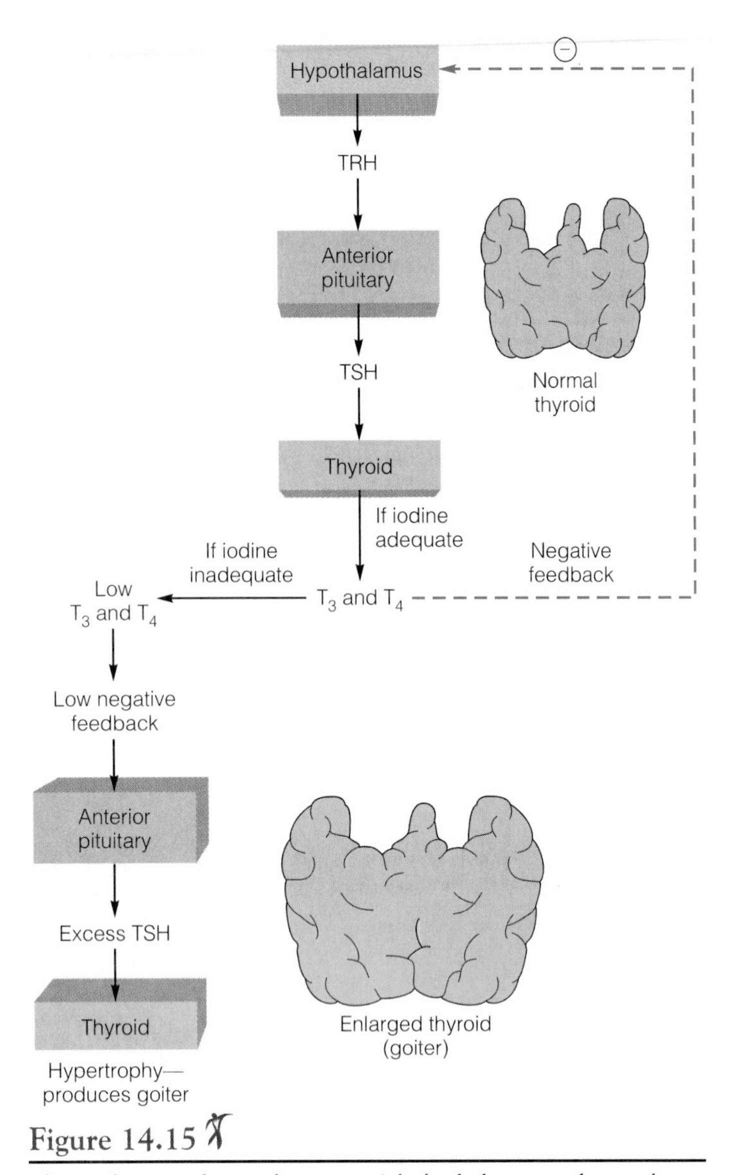

Figure 14.15

The mechanism of goiter formation. A lack of adequate iodine in the diet interferes with the negative feedback control of TSH secretion, resulting in abnormal enlargement of the thyroid gland.

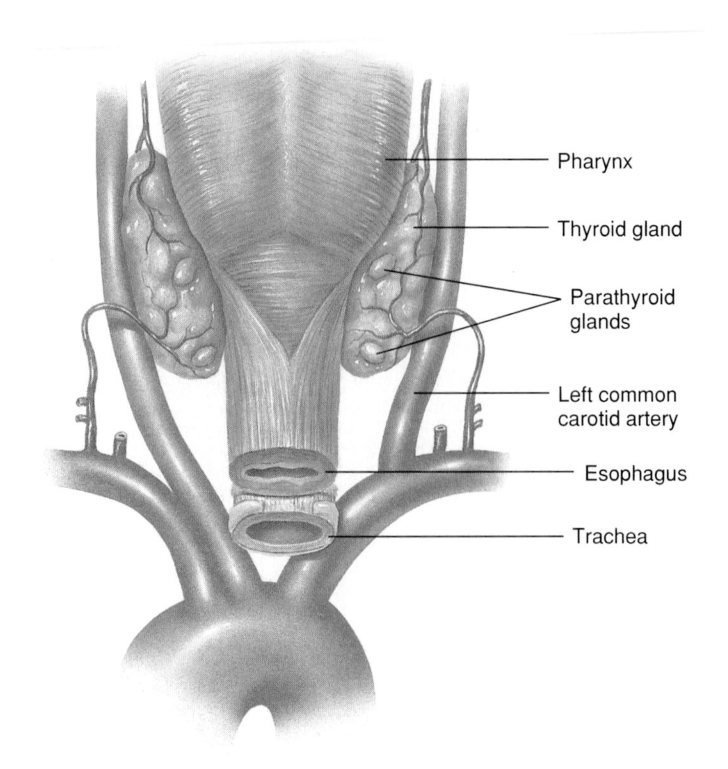

Figure 14.16

A posterior view of the thyroid and parathyroid glands.

Iodine is the most common component of thyroxine and triiodothyronine, and a continual intake of iodine is essential for normal thyroid function. Seafood contains adequate amounts of iodine, and commercial salt generally has iodine as an additive. Absorbed iodine is transported through the blood to the thyroid gland, where an active transport mechanism called an iodine pump moves the iodides into the follicle cells. Here, they combine with amino acids in the synthesis of thyroid hormones.

Most of the thyroxine in blood is attached to carrier proteins. Only the very small percentage of thyroxine that is free in blood plasma can enter the target cells. In the target cells, it is converted to triiodothyronine and attached to nuclear receptor proteins. Through the activation of genes, thyroid hormones stimulate protein synthesis, promote maturation of the nervous system, and increase the rate of energy utilization by the body.

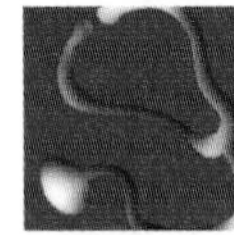

Thyroid-stimulating hormone (TSH) from the anterior pituitary stimulates the thyroid to secrete thyroxine and exerts a trophic effect on the thyroid gland. This trophic effect shows up dramatically in people who develop an *iodine-deficiency (endemic) goiter.* In the absence of sufficient dietary iodine, the thyroid cannot produce adequate amounts of T_4 and T_3. The resulting lack of negative feedback inhibition causes abnormally high levels of TSH secretion, which in turn stimulate the abnormal growth of the thyroid (a goiter). These events are summarized in figure 14.15.

Calcitonin is a polypeptide hormone produced by the parafollicular cells. It works in concert with parathyroid hormone (discussed below) to regulate calcium levels in the blood. Calcitonin inhibits the breakdown of bone tissue and stimulates the excretion of calcium by the kidneys. Both actions result in the lowering of blood calcium levels.

Parathyroid Glands

The small, flattened **parathyroid glands** are embedded in the posterior surfaces of the lateral lobes of the thyroid gland (fig. 14.16). There are usually four parathyroid glands: a *superior* and an *inferior pair*. Each parathyroid gland is a small yellow-brown body 3–8 mm (0.1–0.3 in.) long, 2–5 mm (0.07–0.2 in.) wide, and about 1.5 mm (0.05 in.) deep.

On a microscopic level, the parathyroid glands are composed of two types of epithelial cells (fig. 14.17). The cells that

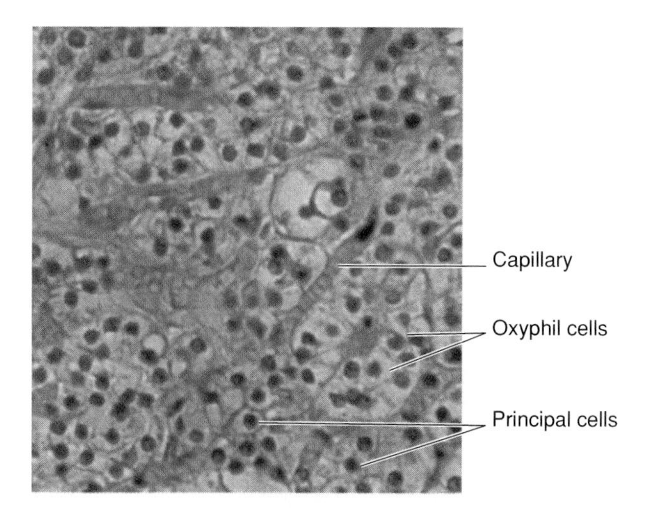

Figure 14.17

The histology of the parathyroid gland.

synthesize parathyroid hormone are called **principal cells** and are scattered among **oxyphil** *(ok'sĕ-fil)* **cells.** Oxyphil cells support the principal cells and are believed to produce reserve quantities of parathyroid hormone.

The blood supply and drainage of the parathyroid glands is similar to that of the thyroid gland. The innervation of the parathyroids, however, is a bit different. The parathyroids receive neurons from the pharyngeal branches of the vagus nerves in addition to neurons arising from the cervical sympathetic ganglia.

The parathyroid glands secrete one hormone called **parathyroid hormone (PTH).** This hormone promotes a rise in blood calcium levels by acting on the bones, kidneys, and small intestine (fig. 14.18); thus, it opposes the effects of calcitonin, released by the thyroid gland.

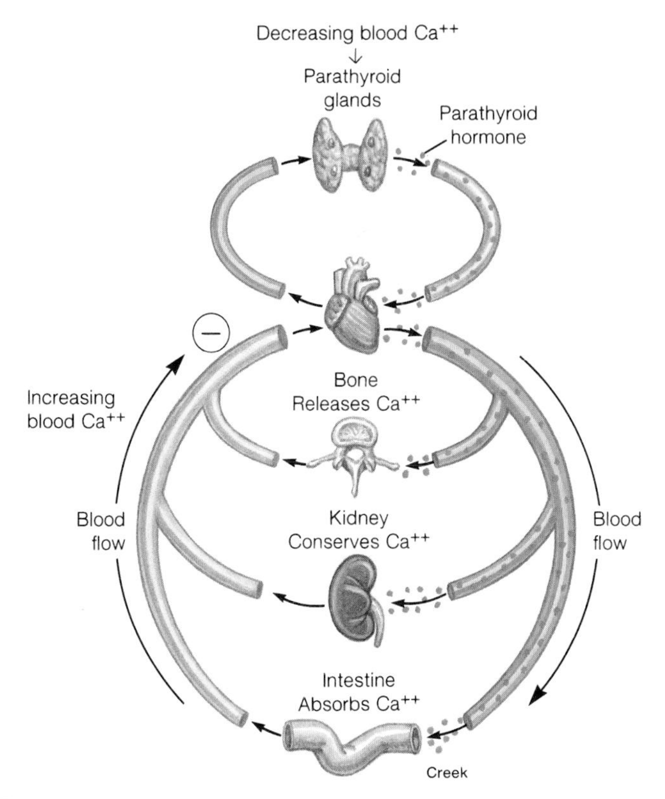

Figure 14.18

Actions of parathyroid hormone. An increased level of parathyroid hormone causes the bones to release calcium, the kidneys to conserve calcium that would otherwise be lost through the urine, and the small intestine to absorb calcium. Negative feedback of increased calcium levels in the blood inhibits the secretion of this hormone.

1. Describe the location and structure of the thyroid gland and list the effects of thyroid hormones.
2. What are some possible metabolic consequences of an overactive thyroid gland?
3. Why is a continual supply of iodine important for body metabolism?
4. Describe the location and structure of the parathyroid glands and identify the target organs of parathyroid hormone.

Pancreas

The pancreatic islets in the pancreas secrete two hormones, insulin and glucagon, which are critically involved in the regulation of blood sugar levels in the body.

Objective 12 Describe the structure of the endocrine portion of the pancreas and the origin of insulin and glucagon.

Objective 13 Describe the actions of insulin and glucagon.

Description of the Pancreas

The **pancreas** *(pan'kre-us)* is both an endocrine and an exocrine gland. The gross structure of this gland and its exocrine functions in digestion are described in chapter 18. The endocrine portion of the pancreas consists of scattered clusters of cells called **pancreatic islets** (islets of Langerhans [*i'lets of lang'er-hanz*]). These endocrine structures are most common in the body and tail of the pancreas (fig. 14.19) and mainly consist of two types of secretory cells called **alpha cells** and **beta cells** (fig. 14.20).

pancreas: Gk. *pan*, all; *kreas*, flesh
islets of Langerhans: from Paul Langerhans, German anatomist, 1847–88

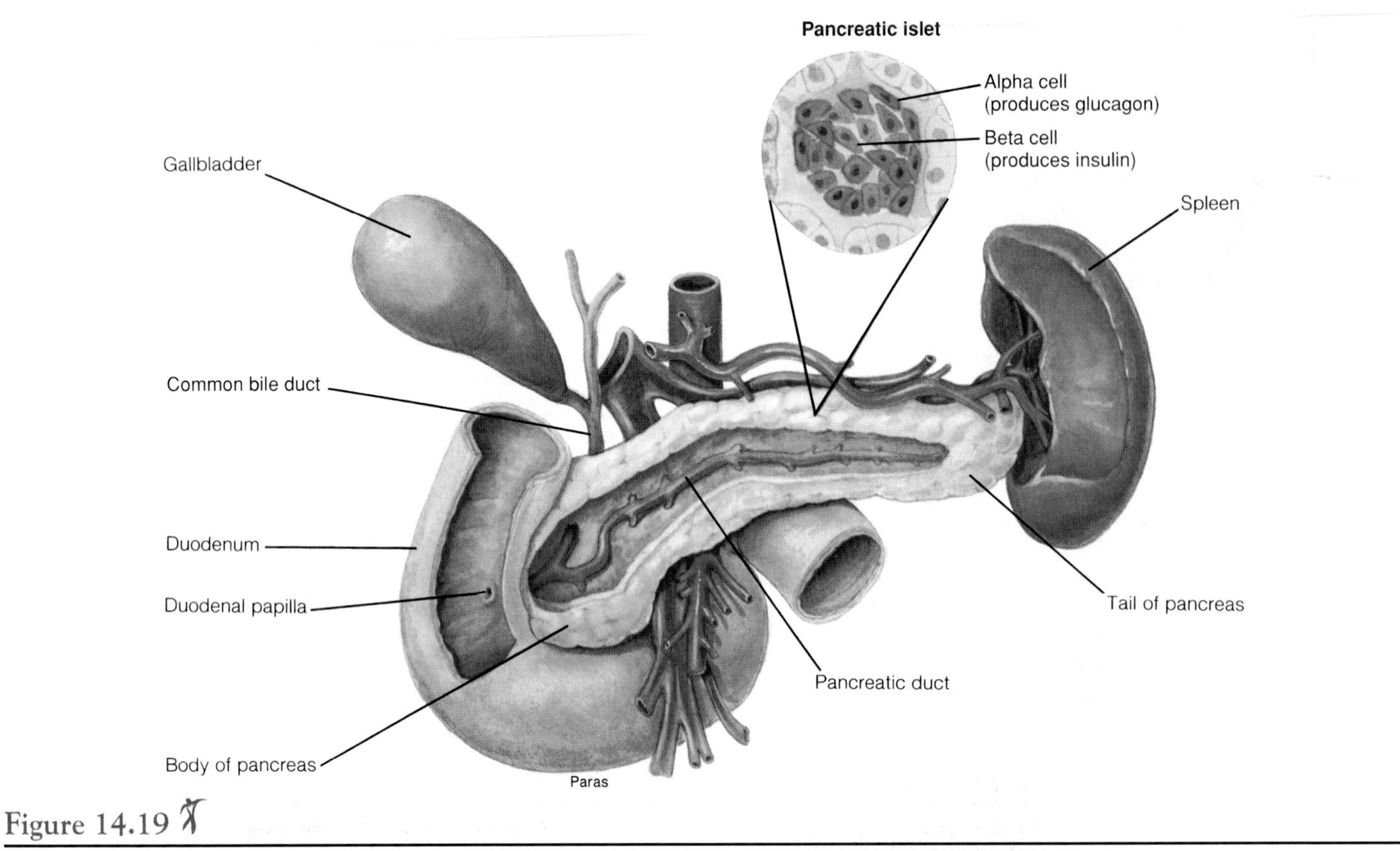

Figure 14.19

The pancreas and associated pancreatic islets.

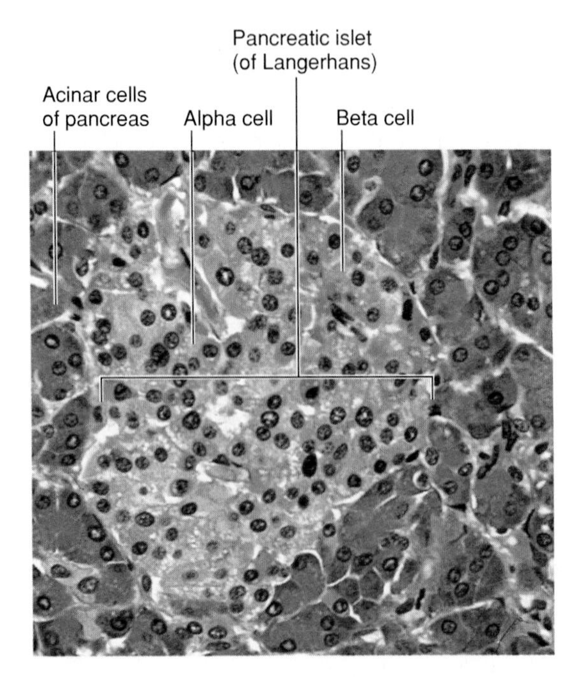

Figure 14.20

The histology of the pancreatic islets.

Endocrine Function of the Pancreas

The endocrine function of the pancreas is to produce and secrete the hormones *glucagon (gloo'că-gon)*and *insulin*. The alpha cells of the pancreatic islets secrete glucagon and the beta cells secrete insulin.

Glucagon stimulates the liver to convert glycogen into glucose, which causes the blood glucose level to rise. Apparently, the alpha cells themselves regulate the glucagon output by monitoring the blood glucose level through a feedback mechanism. If for some reason alpha cells secrete glucagon continuously, high blood sugar, or *hyperglycemia* (*hi"per-gli-se'me-ă*), may result.

Insulin has a physiological function opposite to that of glucagon: it decreases the level of blood sugar. Insulin promotes the movement of glucose through cell membranes, especially in muscle and adipose cells. As the glucose enters the cells, the sugar level of the blood decreases. Other functions of insulin include stimulating muscle and liver cells to convert glucose to glycogen, helping amino acids to enter cells, and assisting the synthesis of proteins and fats. Failure of beta cells to produce insulin causes the common hereditary disease *diabetes mellitus (mĕ-li'tus)*.

The action of the hormones from the pancreatic islets is summarized in table 14.6.

---

insulin: L. *insula*, island

Table 14.6

Hormones of the Pancreas

Hormone	Action	Source of Regulation
Glucagon	Stimulates the liver to convert glycogen to glucose, causing the blood glucose level to rise	Blood glucose level through negative feedback in pancreas
Insulin	Promotes movement of glucose through cell membranes; stimulates the liver to convert glucose to glycogen; promotes the transport of amino acids to cells; assists in synthesis of proteins and fats	Blood glucose level through negative feedback in pancreas

1. Describe the specific sites of glucagon and insulin production.
2. What is the function of glucagon? Of insulin?
3. What disease is caused by the insufficient production of insulin?

Adrenal Glands

The adrenal cortex and adrenal medulla are structurally and functionally different. The adrenal medulla secretes catecholamine hormones that complement the action of the sympathetic division of the ANS. The adrenal cortex secretes corticosteroids that function in the regulation of mineral balance, energy balance, and reproductive activity.

Objective 14 Describe the location of the adrenal glands and distinguish between the adrenal cortex and the adrenal medulla.

Objective 15 List the hormones secreted by the adrenal glands and discuss their effects.

Description of the Adrenal Glands

The **adrenal** (*ă-dre'nal*) **glands** (suprarenal glands) are paired organs that cap the superior borders of the kidneys (fig. 14.21). The adrenal glands, along with the kidneys, are retroperitoneal and are embedded against the muscles of the back in a protective pad of fat.

adrenal: L. *ad*, to; *renes*, kidney

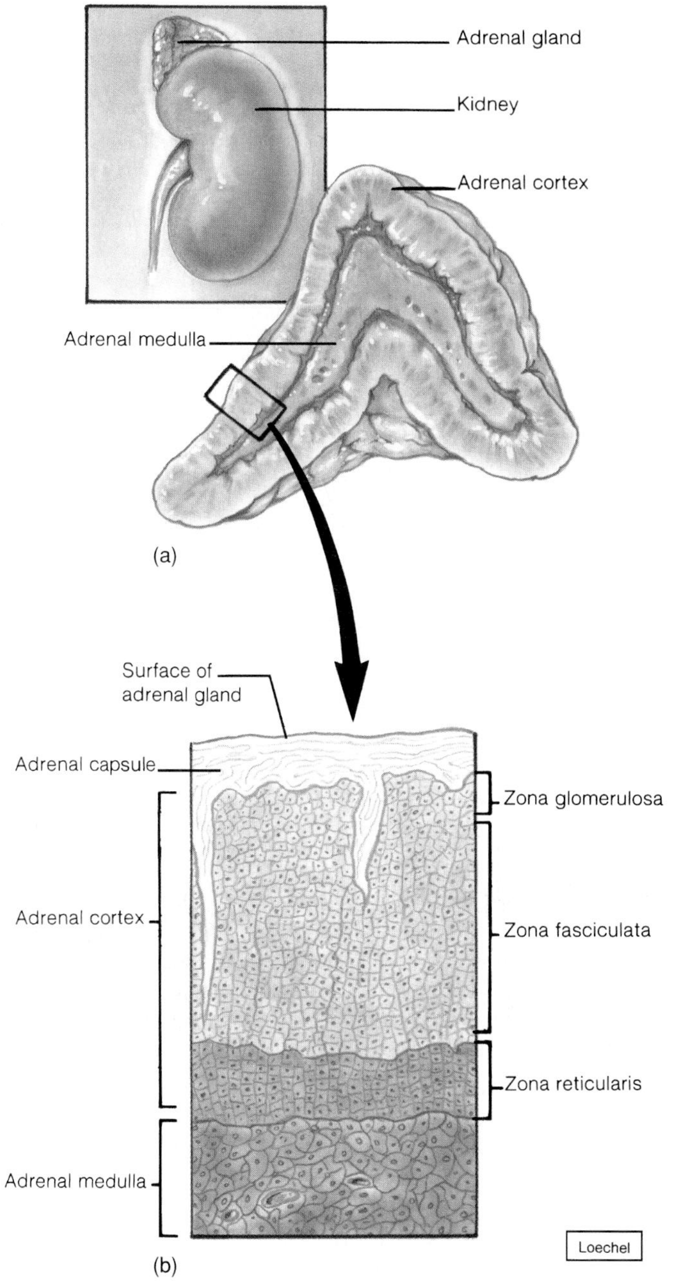

Adrenal gland
Kidney
Adrenal cortex
Adrenal medulla
(a)
Surface of adrenal gland
Adrenal capsule
Adrenal cortex
Adrenal medulla
Zona glomerulosa
Zona fasciculata
Zona reticularis
Loechel
(b)

Figure 14.21

The structure of the adrenal gland. (*a*) The gross structure and (*b*) the histological structure showing the three zones of the adrenal cortex.

Each of the pyramid-shaped adrenal glands is about 50 mm (2 in.) long, 30 mm (1.1 in.) wide, and 10 mm (0.4 in.) deep. Each consists of an outer adrenal cortex and inner adrenal medulla (figs. 14.21 and 14.22) that function as separate glands.

The **adrenal cortex** makes up the bulk of the gland and is histologically subdivided into three zones: an outer **zona glomerulosa** (*glo-mer"yoo-lo'să*), an intermediate **zona fasciculata** (*fă-sik"yoo-lă'tă*), and an inner **zona reticularis.** The **adrenal medulla** is composed of tightly packed clusters of **chromaffin** (*kro-maf'in*) **cells,** which are arranged around blood vessels. Each cluster of chromaffin cells receives direct autonomic innervation.

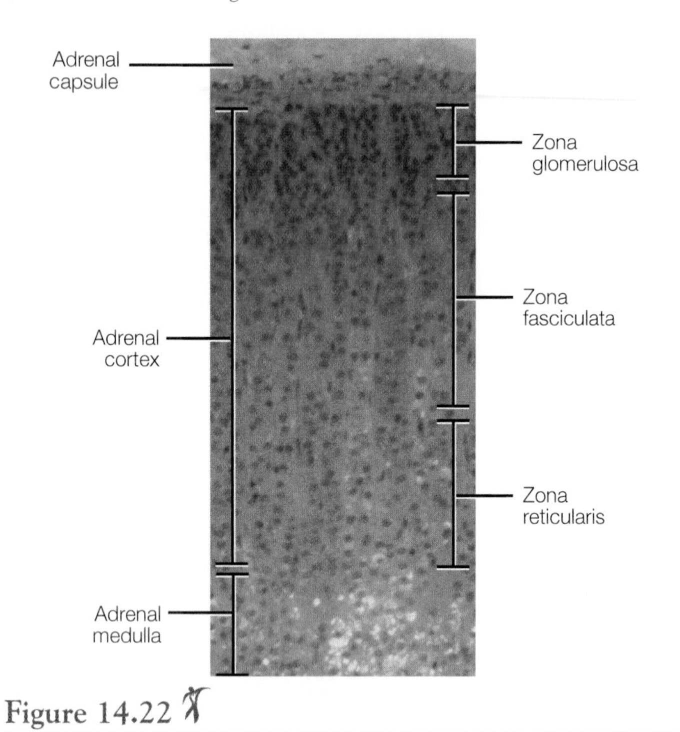

Figure 14.22

The histology of the adrenal gland.

Like other endocrine glands, the adrenal glands are highly vascular. Three separate suprarenal arteries supply blood to each adrenal gland. One arises from the inferior phrenic artery, another from the aorta, and a third is a branch of the renal artery. The venous drainage passes through a suprarenal vein into the inferior vena cava for the right adrenal gland and through the suprarenal vein into the left renal vein for the left adrenal gland.

The adrenal glands are innervated by preganglionic neurons of the splanchnic nerves and by fibers of the celiac and associated sympathetic plexuses.

Functions of the Adrenal Glands

Over 30 hormones have been identified as being produced by the adrenal cortex. These hormones are called **corticosteroids** *(kor"tĭ-ko-ster'oidz)*, or **corticoids,** for short. The adrenal corticoids are grouped into three functional categories: mineralocorticoids, glucocorticoids, and gonadocorticoids.

The *mineralocorticoids* are produced by the zona glomerulosa of the adrenal cortex and regulate the concentrations of extracellular electrolytes. Of the three hormones secreted by this layer, *aldosterone (al-dos'ter-ōn)* is the most important. Aldosterone affects the kidneys, causing them to reabsorb sodium and increase potassium excretion. At the same time, it promotes water reabsorption and reduces urine output.

The *glucocorticoids* are produced primarily by the zona fasciculata of the adrenal gland and influence the metabolism of carbohydrates, proteins, and fats. The glucocorticoids also promote vasoconstriction, act as antiinflammatory compounds, and help the body resist stress. The most abundant and physiologically important glucocorticoid is *cortisol (kor'tĭ-sol) (hydrocortisone)*.

Table 14.7
Hormones of the Adrenal Cortex

Hormone	Action	Source of Regulation
Mineralocorticoids	Regulate the concentration of extracellular electrolytes, especially sodium and potassium	Electrolyte concentration in blood
Glucocorticoids	Influence the metabolism of carbohydrates, proteins, and fats; promote vasoconstriction; act as anti-inflammatories	ACTH from the adenohypophysis of the pituitary gland in response to stress
Gonadocorticoids	Supplement the sex hormones from the gonads	

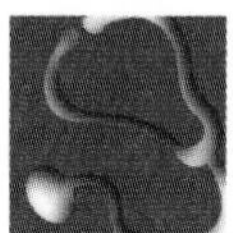

Cortisol and related anti-inflammatory compounds are commonly used to treat patients suffering from arthritis and various allergies. They are also frequently used to treat traumatized joints and to suppress the immune rejection of transplanted tissues. Unfortunately, cortisol inhibits the regeneration of connective tissues, and therefore should be used sparingly and only when necessary.

Gonadocorticoids are the sex hormones that are secreted by the zona reticularis of the adrenal cortex. The majority of these hormones are adrenal *androgens*, but small quantities of adrenal *estrogens* and *progesterones* are produced as well. It is thought that these hormones supplement the sex hormones produced in the gonads. There is also evidence that androgen concentrations in both males and females play a role in determining the sex drive.

A summary of the hormones secreted from the adrenal cortex is presented in table 14.7.

The chromaffin cells of the adrenal medulla produce two closely related hormones: *epinephrine* and *norepinephrine*. Both of these hormones are classified as amines—more specifically, as catecholamines *(kat"ĕ-kol'ă-mēz)*—because they contain amine groups (see fig. 14.3). The effects of these hormones are similar to those caused by stimulation of the sympathetic division of the ANS, except that the hormonal effects are about 10 times longer lasting. The hormones from the adrenal medulla increase cardiac output and heart rate, dilate coronary blood vessels, increase mental alertness, increase the respiratory rate, and elevate metabolic rate. The effects of epinephrine and norepinephrine are compared in table 14.8.

The adrenal medulla is innervated by sympathetic neurons. The impulses are initiated from the hypothalamus via the spinal cord in response to stress. Stress therefore activates the adrenal medulla, as well as the adrenal cortex. Activation of the adrenal medulla prepares the body for greater physical performance—the *fight-or-flight* response (fig. 14.23).

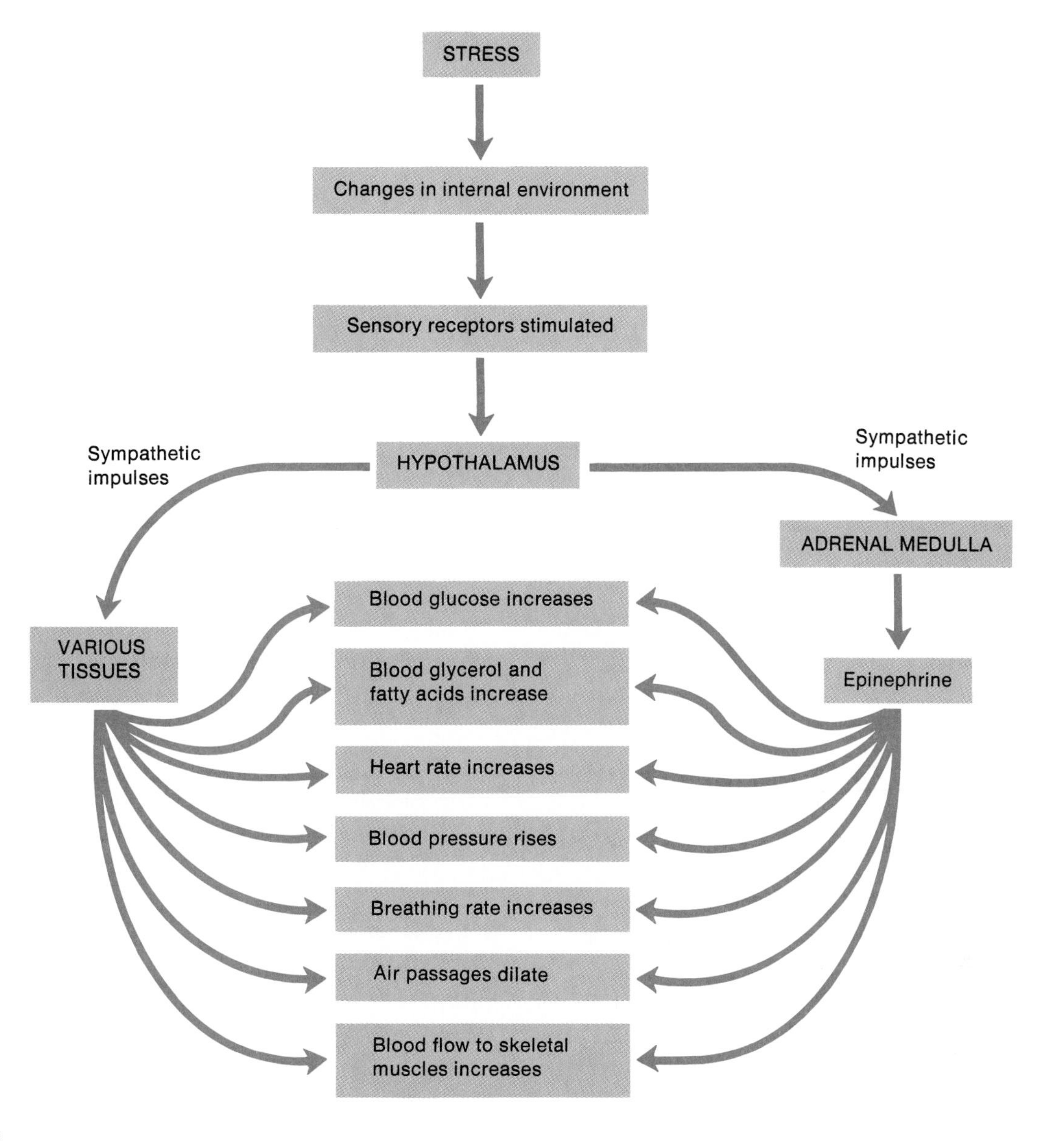

Figure 14.23

Stress is the physiological change in the body that prepares it for fight or flight. These changes are elicited by neural impulses from the hypothalamus that directly influence body tissues and by impulses to the adrenal medulla that secondarily influence the same body tissues through the release of epinephrine.

Table 14.8

Comparison of Adrenal Medullary Hormones

Epinephrine	Norepinephrine
Elevates blood pressure because of increased cardiac output and peripheral vasoconstriction	Elevates blood pressure because of generalized vasoconstriction
Accelerates respiratory rate and dilates respiratory passageways	Similar effect, but to a lesser degree
Increases efficiency of muscular contraction	Similar effect, but to a lesser degree
Increases rate of glycogen breakdown into glucose, so that level of blood glucose rises	Similar effect, but to a lesser degree
Increases rate of fatty acid released from fat, so that level of blood fatty acids rises	Similar effect, but to a lesser degree
Increases release of ACTH and TSH from the adenohypophysis of the pituitary gland	No effect

Excessive stimulation of the adrenal medulla can result in depletion of the body's energy reserves, and high levels of corticosteroid secretion from the adrenal cortex can significantly impair the immune system. It is reasonable to expect, therefore, that prolonged stress can result in increased susceptibility to disease. Indeed, many studies show that prolonged stress results in an increased incidence of cancer and other diseases.

1. Describe the location and appearance of the adrenal gland. Diagram and label the layers of the adrenal gland that can be seen in a sagittal section as, for example, in a histological slide.
2. List the hormones of the adrenal medulla and describe their effects.
3. List the categories of corticosteroids and identify the zones of the adrenal cortex that secrete these hormones.

Gonads and Other Endocrine Glands

The gonads produce sex hormones that control the development and function of the male and female reproductive systems. Additionally, many other organs secrete hormones that help regulate digestion, metabolism, growth, and immunity.

Objective 16 Discuss the endocrine functions of the gonads.

Objective 17 Describe the structure and location of the pineal and thymus glands and their endocrine functions.

Gonads

The **gonads** are the male and female primary sex organs. The male gonads are called **testes** and the female gonads are called **ovaries.** The gonads are mixed glands in that they produce both sex hormones and sex cells, or **gametes** (see chapters 20 and 21).

Testes

The **interstitial cells** of the testes produce and secrete the male sex hormone *testosterone*. Testosterone controls the development and function of the male secondary sex organs—the penis, accessory glands, and ducts. It also promotes the male secondary sex characteristics (see chapter 20) and somewhat determines the sex drive.

Ovaries

The endocrine function of the ovaries is the production of the female sex hormones, *estrogens* and *progesterone*. Estrogens are produced in the **ovarian** (graafian) **follicles** and **corpus luteum** of the ovaries. They are also produced in the placenta, adrenal cortex, and even in the testes of the male. Estrogens are responsible for (1) development and function of the secondary sex organs, (2) menstrual changes of the uterus, (3) development of the female secondary sex characteristics (see chapter 21), and (4) regulation of the sex drive.

Progesterone is produced by the corpus luteum and is primarily associated with pregnancy in preparing the uterus for implantation and preventing abortion of the fetus.

Most cultures of the world practice birth control, or contraception, in one form or another. It has a long history, dating back to the ancient Egyptians who used various substances to inhibit sperm survival and motility. In the age of hormonal biochemistry, birth-control techniques have become increasingly sophisticated. The female, rather than the male, has been the target of hormonal birth-control techniques for the following reasons: (1) ovulation is cyclic; (2) the genetic structure of each ovum is established by the time of the female's birth, whereas sperm production is a continuous process, and therefore more vulnerable to genetic damage; (3) the female system has more potential sites for hormonal interference than does the male system; and (4) the female is usually more conscientious about practicing birth control because she has far more invested in pregnancy than does the male.

Pineal Gland

The small, cone-shaped **pineal** (*pin'e-al*) **gland** (pineal body) (see fig. 14.2), is located in the roof of the third ventricle, near the corpora quadrigemina, where it is encapsulated by the meninges covering the brain. In a child, the pineal gland weighs about 0.2 g and is 5–8 mm (0.2–0.3 in.) long and 9 mm wide. It begins to regress in size at about the age of 7, and in the adult it appears as a thickened strand of fibrous tissue. Histologically, the pineal gland consists of specialized parenchymal and neuroglial cells. Although it lacks direct nervous connection to the rest of the brain, the pineal gland is highly innervated by the sympathetic division of the ANS from the superior cervical ganglion.

The function of the pineal gland in some vertebrates is well known but is not well understood in humans. Secretion of its principal hormone, *melatonin*, follows a circadian (daily) rhythm tied to daily and seasonal changes in light. Melatonin is thought to affect the hypothalamus by stimulating the secretion of certain releasing factors. These factors in turn affect the secretion of gonadotrophin and the ACTH from the adenohypophysis of the pituitary gland. Excessive melatonin secretion in humans is associated with a delay in the onset of puberty; however the role of melatonin in sexual maturation is still highly controversial.

pineal: L. *pinea*, pine cone

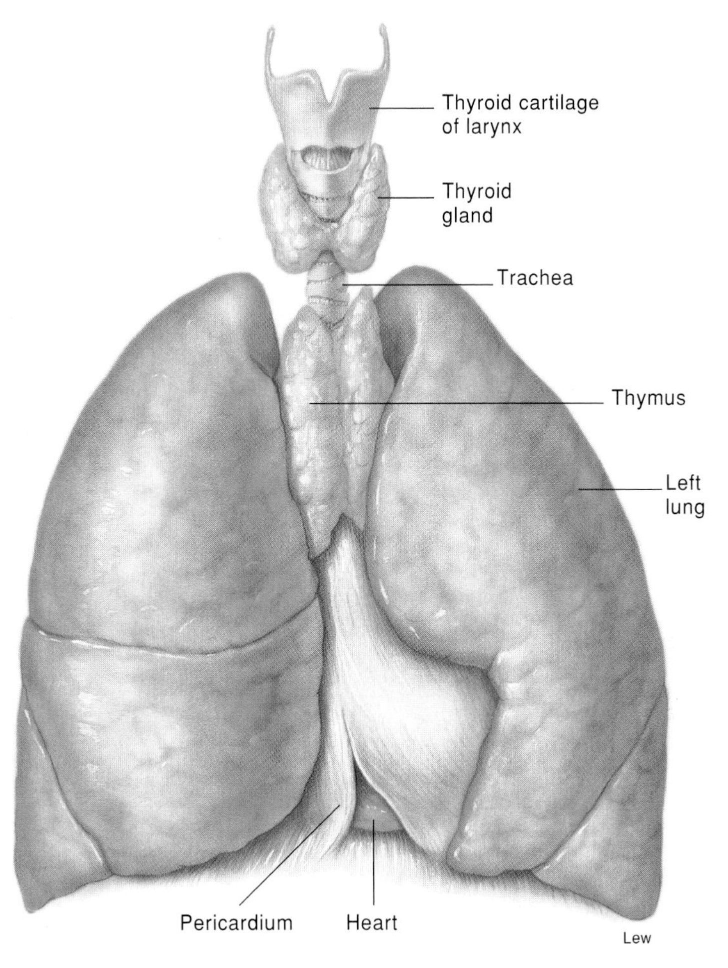

Figure 14.24

The thymus is a bilobed organ located deep in the sternum in the mediastinum of the thorax.

Thymus

The **thymus** *(thi'mus)* is a bilobed organ positioned in the upper mediastinum, in front of the aorta and behind the manubrium of the sternum (fig. 14.24). Although the size of the thymus varies considerably from person to person, it is relatively large in newborns and children and then sharply regresses in size after puberty. Besides decreasing in size, the thymus of adults becomes infiltrated with strands of fibrous and fatty connective tissue.

The principal function of the thymus is associated with the lymphatic system (see chapter 16) in maintaining body immunity through the maturation and discharge of a specialized group of lymphocytes called *T cells (thymus-dependent cells)*. The thymus also secretes a hormone called *thymosin*, which is believed to stimulate the T cells after they leave the thymus.

Stomach and Small Intestine

Certain cells of the mucosal linings of the stomach and small intestine secrete hormones that promote digestive activities (see chapter 18). The effects of these hormones, summarized in table 14.9, coordinate the activities of different regions of the GI tract and the secretions of pancreatic juice and bile in conjunction with regulation by the autonomic nervous system.

Placenta

The **placenta** *(plă-sen'tă)* is the organ responsible for nutrient and waste exchange between the fetus and the mother (see chapter 22).

thymus: Gk. *thymos*, a warty excrescence
placenta: L. *placenta*, flat cake

Table 14.9

Summary of the Physiological Effects of Gastrointestinal Hormones

Hormone	Secreted By	Effects
Gastrin	Stomach	Stimulates parietal cells to secrete HCl Stimulates chief cells to secrete pepsinogen Maintains structure of gastric mucosa
Secretin	Small intestine	Stimulates water and bicarbonate secretion in pancreatic juice Potentiates actions of cholecystokinin on pancreas
Cholecystokinin (CCK)	Small intestine	Stimulates contraction of the gallbladder Stimulates secretion of pancreatic juice enzymes Potentiates action of secretin on pancreas Maintains structure of exocrine pancreas (acini)
Gastric inhibitory peptide (GIP)	Small intestine	Inhibits gastric emptying Inhibits gastric acid secretion Stimulates secretion of insulin from pancreatic islets

DEVELOPMENTAL EXPOSITION

The Endocrine System

Exhibit I The development of the pituitary gland. (*a*) The head end of an embryo at 4 weeks showing the position of a midsagittal cut seen in the developmental sequence (*b–e*). The pituitary gland arises from a specific portion of the neuroectoderm, called the neurohypophyseal bud, which evaginates downward during the fourth and fifth weeks respectively in (*b*) and (*c*), and from a specific portion of the oral ectoderm, called the hypophyseal (Rathke's) pouch, which evaginates upward from a specific portion of the primitive oral cavity. At 8 weeks (*d*), the hypophyseal pouch is no longer connected to the pharyngeal roof of the oral cavity. During the fetal stage (*e*), the development of the pituitary gland is completed.

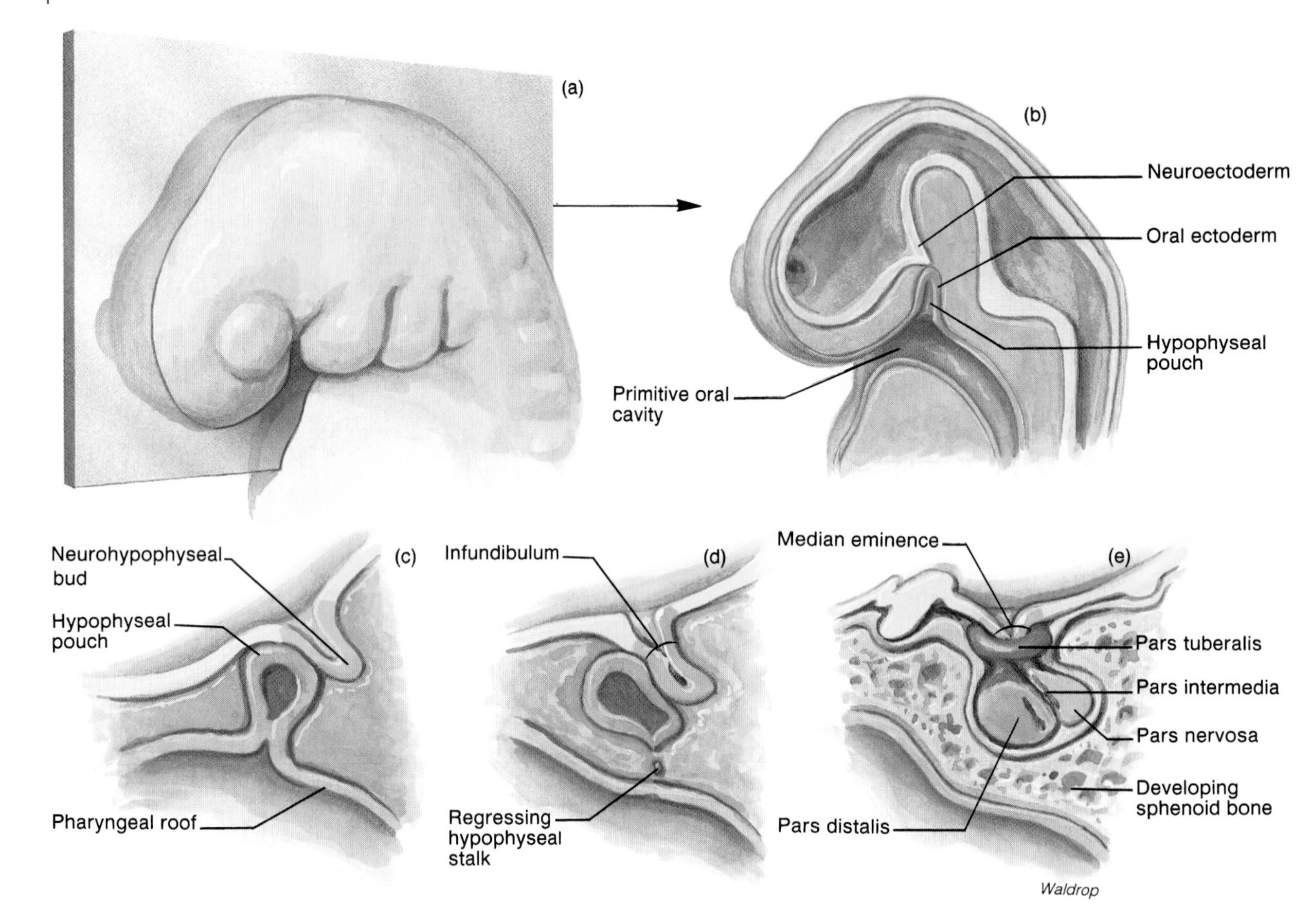

Explanation

The endocrine system is the only anatomical body system whose organs are not structurally connected. Because these organs are isolated from each other and are distributed throughout the body, each endocrine organ has a separate and independent development. All three embryonic germ layers (endoderm, mesoderm, and ectoderm) contribute to the development of the endocrine system.

The following sections describe, in turn, the development of the pituitary, thyroid, pancreas, and adrenal glands. The development of the testes and ovaries is discussed in chapters 20 and 21, respectively.

Pituitary Gland

Although the **pituitary gland** is a single organ, it is actually composed of two distinct types of tissues that have different embryonic origins. These two types of tissues release different hormones and are under different control systems. The anterior portion of the hypophysis, called the **adenohypophysis,** develops from ectoderm that lines the primitive oral cavity. The posterior portion, called the **neurohypophysis,** develops from neuroectoderm of the developing brain (exhibit I).

The adenohypophysis begins to develop during the third week as a diverticulum, a pouchlike extension, called the

Exhibit II The embryonic development of the thyroid gland. (*a*) At 4 weeks, a thyroid diverticulum begins to form in the floor of the pharynx at the level of the second brachial arch. (*b,c*) The thyroid diverticulum extends downward during the fifth and sixth weeks. (*d*) A sagittal section through the head and neck of an adult shows the path of development and final position of the thyroid gland.

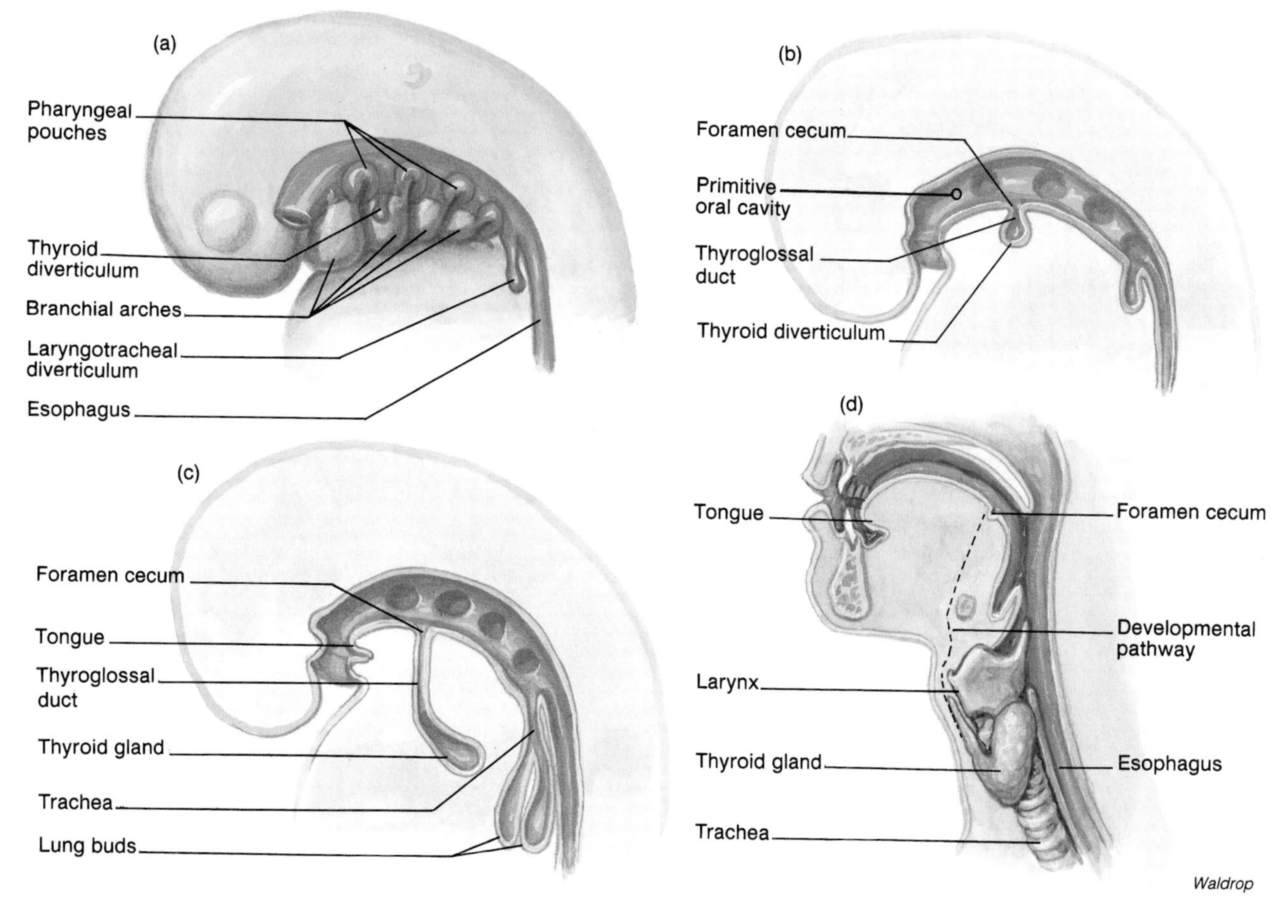

hypophyseal (Rathke's) **pouch.** It arises from the roof of the primitive oral cavity and grows toward the brain. At the same time, another diverticulum called the **infundibulum** forms from the diencephalon on the inferior aspect of the brain. As the two diverticula come in contact, the hypophyseal pouch loses its connection with the oral cavity, and the primordial tissue of the adenohypophysis is formed. The fully developed adenohypophysis includes the **pars distalis** and the **pars tuberalis.** Cells of the **pars intermedia** become mingled with those of the pars distalis, so that the pars intermedia no longer constitutes a separate structure in the adult.

The neurohypophysis develops as the infundibulum extends interiorly from the diencephalon to come in contact with the developing adenohypophysis. The fully formed neurohypophysis consists of the infundibulum and the **lobus nervosa.** Specialized nerve fibers that connect the hypothalamus with the lobus nervosa develop within the infundibulum.

Notice that the neurohypophysis is essentially an extension of the brain; indeed, the entire pituitary gland, like the brain, is surrounded by the meninges. The adenohypophysis, by contrast, is derived from nonneural tissue (the same embryonic tissue that will form the epithelium over the roof of the mouth). These different embryologies have important consequences with regard to the functions of the two parts of the pituitary gland.

Thyroid Gland

The **thyroid gland** is derived from endoderm and begins its development during the third week as a thickening in the floor of the primitive pharynx. The thickening soon evaginates (outpouches) downward as the **thyroid diverticulum** (exhibit II). A

Rathke's pouch: from Martin H. Rathke, German anatomist, 1793–1860

Continued

Developmental Exposition

Continued

narrow **thyroglossal duct** connects the descending primordial thyroid tissue to the pharynx. As the descent continues, the tongue starts to develop, and the opening into the thyroglossal duct, called the **foramen cecum** *(se'kum)* pierces through the base of the tongue. By the seventh week, the thyroid gland occupies a position immediately inferior to the larynx, surrounding the front and lateral sides of the trachea. At this time, the thyroglossal duct disappears and the foramen cecum regresses in size to a vestigial pit that persists throughout life.

Pancreas

The **pancreas** begins development during the fifth week, as dorsal and ventral pancreatic buds of endoderm arise from the caudal portion of the foregut. These primordial pancreatic tissues continue to grow independently until the ventral bud is carried dorsally and fuses with the dorsal bud as the duodenum rotates to the right. The fusion of the two portions of the pancreas occurs during the seventh week.

The pancreas develops from and maintains connections with the small intestine (via the pancreatic duct). The exocrine products of the pancreas, contained in *pancreatic juice,* are channeled through the pancreatic duct to the small intestine. The endocrine parts of the pancreas—the **pancreatic islets**—however, secrete their products (insulin and glucagon) into the blood. The precise origin of the endocrine cells of the pancreas remains uncertain. It is believed that some of the endocrine cells develop as buds from pancreatic ductules and that others arise from neuroectodermal cells that migrate to the pancreas to form these endocrine cells and the autonomic innervation of the pancreas.

Adrenal Glands

The **adrenal glands** begin development during the fifth week from two different germ layers. Each adrenal gland has an outer part, or **adrenal cortex,** which develops from mesoderm, and an inner part, or **adrenal medulla,** which develops from neuroectoderm. The mesodermal ridge that forms the adrenal cortex is in the same region from which the gonads develop.

The neuroectodermal cells that form the adrenal medulla are derived from the neural crest of the neural tube. The developing adrenal medulla is gradually encapsulated by the adrenal cortex, a process that continues into the fetal stage. The formation of the adrenal gland is not completed until the end of the third year of age.

Notice that the adrenal gland, like the pituitary, has a dual origin; part is neural and part is not. Like the pituitary, the adrenal cortex and adrenal medulla are in fact two different endocrine tissues. Although they are located in the same organ, they secrete different hormones and are regulated by different control systems.

foramen cecum: L. *foramen*, open passage; *caecum*, blind tube

The placenta is also an endocrine gland; it secretes large amounts of estrogens and progesterone, as well as a number of polypeptide and protein hormones that are similar to some hormones secreted by the anterior pituitary. These latter hormones include *human chorionic gonadotrophin* (hCG), which is similar to LH, and *somatomammotrophin*, which is similar in action to growth hormone and prolactin. Detection of hCG in urine is an indication of pregnancy and is the basis of home pregnancy tests.

1. Where is testosterone produced? What are its functions?
2. What are the functions of estrogens and progesterone? Where are they produced?
3. Describe the location of the pineal gland and the action of melatonin.
4. Describe the location and function of the thymus.

Clinical Considerations

Endocrinology *(en"dŏ-crĭ-nol'ŏ-je)* is the science concerned with the structure and function of the glands of the endocrine system and the consequences of glandular dysfunctions. The most frequent endocrine disorders result from an unusual increase (hypersecretion) or decrease (hyposecretion) of hormones. Hypersecretion of an endocrine gland is generally caused by hyperplasia (increase in size) of the gland, whereas hyposecretion of hormones is the consequence of a damaged or atrophied gland.

The diagnosis and treatment of endocrine problems can be difficult because of three complex physiological effects of hormones. (1) Hormones are extremely specific, and their target tissues are sensitive to minute changes in hormonal levels. (2) The responses of hormones usually represent a type of chain reaction; that is, the hormones secreted from one gland cause the secretion of hormones by another gland. Thus, the clinical symptoms obscure the source of the problem. (3) Endocrine problems frequently manifest as metabolic problems because the primary action of hormones is to regulate metabolism.

Common diagnostic methods will be discussed in this section, along with the more important endocrine disorders that affect the major endocrine glands.

Diagnosis of Endocrine Disorders

Certain endocrine disorders affect the patient's physical appearance and behavior; therefore, observation is very important in diagnosis. The patient's clinical history is also important in evaluating the rate of progress and stage of development of an endocrine disorder.

The confirmation of endocrine disorders requires laboratory tests, particularly of blood and urine samples. These samples are important because hormones are distributed via the blood, and urine is produced from the metabolic wastes filtered from the blood. A **radioimmunoassay (RIA)** is a laboratory test to determine the concentration of hormones in blood and urine. Other blood tests include the **protein-bound iodine (PBI) test** to determine the level of iodine in the blood, and hence thyroid problems; measurement of blood cholesterol content for thyroid problems; measurement of sodium-potassium ratios to detect Addison's disease; and blood sugar tests, including testing glucose tolerance in a fasting patient to examine for diabetes mellitus.

A *urinalysis* can be important in the diagnosis of several endocrine disorders. A high level of glucose in a fasting patient indicates diabetes mellitus. A patient who has diabetes insipidus will produce a large volume (5–10 L per day) of dilute urine of low specific gravity. Certain diseases of the adrenal glands can also be detected by examining for changes in urine samples collected over a 24-hour period.

Basal metabolism rate (BMR) and **thyroid scans** (see fig. 14.13b) are tests for thyroid disorders. A visual-field examination may be important in detecting a tumor of the pituitary gland. Radiographs and electrocardiograms also may be helpful in diagnosing endocrine disorders.

Disorders of the Pituitary Gland

The pituitary is a remarkable gland. It simultaneously carries out several functions, yet more than 90% of the gland must be destroyed before pituitary function becomes severely impaired.

Panhypopituitarism

A reduction in the activity of the pituitary gland is called **hypopituitarism** (*hi″po-pĭ-too′ĭ-tă-rism*). It can result from intracranial hemorrhage, a blood clot, prolonged steroid treatments, or a tumor. Total pituitary impairment, termed **panhypopituitarism,** brings about a progressive and general loss of hormonal activity. For example, the gonads stop functioning and the person suffers from amenorrhea (lack of menstruation) or aspermia (no sperm production) and loss of pubic and axillary hair. The thyroid and adrenals also eventually stop functioning. People with this condition and those who have had their pituitary surgically removed—a procedure called **hypophysectomy** (*hi-pof″ĭ-sek′to-me*)—receive thyroxine, cortisone, growth hormone, and gonadal hormones throughout life to maintain normal body function.

Abnormal Growth Hormone Secretion

Inadequate growth hormone secretion during childhood causes **pituitary dwarfism** (fig. 14.25*a*). Hyposecretion of growth hormone in an adult produces a rare condition called **pituitary cachexia** (*kă-kek′se-a*) (*Simmonds' disease*). One of the symptoms of this disease is premature aging caused by tissue atrophy. By contrast, oversecretion of growth hormone during childhood causes **gigantism** (fig. 14.25*b*), an abnormal increase in the length of long bones. Excessive growth hormone secretion in an adult does not cause further growth in bone length because the epiphyseal plates have already ossified. Hypersecretion of growth hormone in an adult causes **acromegaly** (*ak″ro-meg′ă-le*) (fig. 14.25*c*), in which the person's appearance gradually changes as a result of thickening of bones and growth of soft tissues, particularly in the face, hands, and feet.

Inadequate ADH Secretion

A dysfunction of the neurohypophysis results in a deficiency in ADH secretion, causing a condition called **diabetes insipidus.** Symptoms of this disease include polyuria (excessive urination), polydipsia (consumption of large amounts of water), and severe ionic imbalances. Diabetes insipidus is treated by injections of ADH.

Disorders of the Thyroid and Parathyroid Glands

Hypothyroidism

The infantile form of hypothyroidism is known as **cretinism** (*kre′tĭ-nizm*). An affected child usually appears normal at birth because thyroxine is received from the mother through the placenta. The clinical symptoms of cretinism are stunted growth, thickened facial features, abnormal bone development, mental retardation, low body temperature, and general lethargy. If cretinism is diagnosed early, it can be successfully treated by administering thyroxine.

Myxedema

Hypothyroidism in an adult causes myxedema (*mik″sĕ-de′mă*). This disorder affects body fluids, causing edema and increasing

Simmonds' disease: from Morris Simmonds, German physician, 1855–1925
acromegaly: Gk. *akron*, extremity; *megas*, large
diabetes: Gk. *diabetes*, to pass through a siphon
myxedema: Gk. *myxa*, mucus; *oidema*, swelling

(a)

(b)

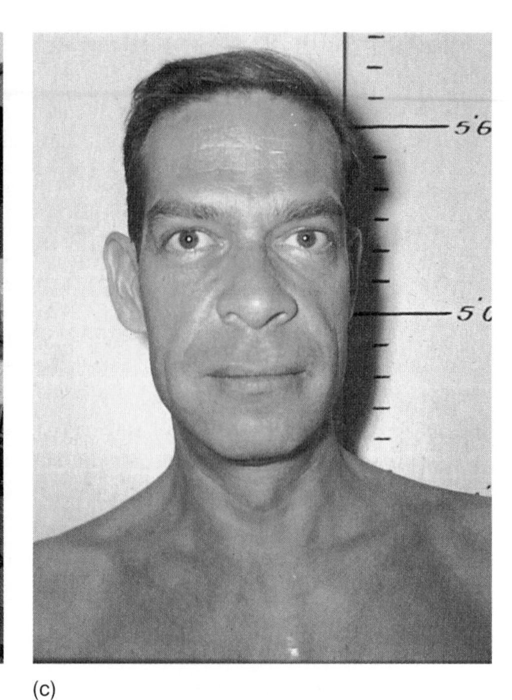

(c)

Figure 14.25

Both hyposecretion and hypersecretion of growth hormone may result in abnormal physical features. (*a*) Pituitary dwarfism is caused by inadequate GH secretion during childhood; (*b*) gigantism is caused by excessive GH secretion during childhood; and (*c*) acromegaly is caused by excessive GH secretion during adulthood.

blood volume, hence increasing blood pressure. A person with myxedema has a low metabolic rate, lethargy, sensitivity to cold, and a tendency to gain weight. This condition is treated with thyroxine or triiodothyronine, both of which are taken orally.

Endemic Goiter

A goiter is an abnormal growth of the thyroid gland. When this is a result of inadequate dietary intake of iodine, the condition is called endemic goiter (fig. 14.26). In this case, growth of the thyroid is due to excessive TSH secretion, which results from low levels of thyroxine secretion. Endemic goiter is thus associated with hypothyroidism.

Graves' Disease

Graves' disease, also called **toxic goiter,** involves growth of the thyroid associated with hypersecretion of thyroxine. This hyperthyroidism is produced by antibodies that act like TSH and stimulate the thyroid; it is an autoimmune disease. As a consequence of high levels of thyroxine secretion, the metabolic rate and heart rate increase, there is loss of weight, and the autonomic nervous system induces excessive sweating. In about half of the cases, **exophthalmos** (*ek"sof-thal'mos*), or bulging of the eyes, also develops (fig. 14.27) because of edema in the tissues of the eye sockets and swelling of the extrinsic eye muscles.

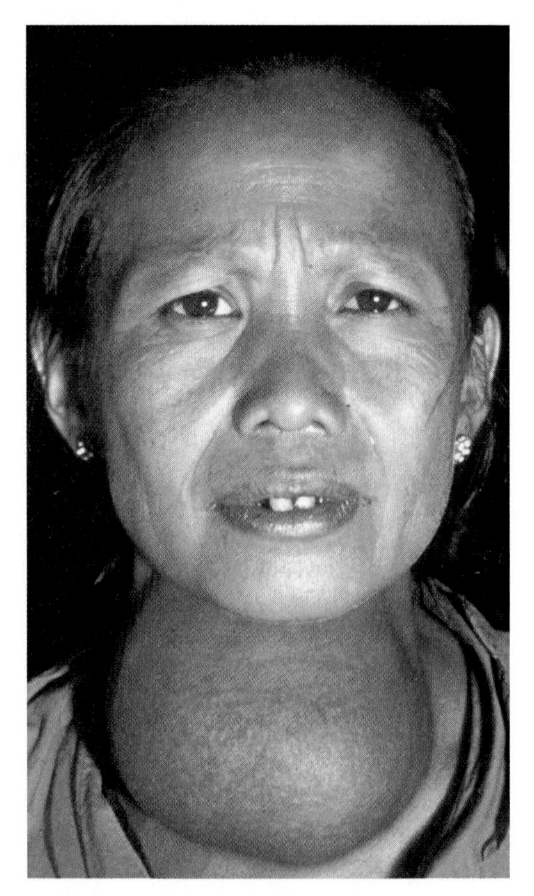

Figure 14.26

An endemic goiter is caused by insufficient iodine in the diet.

Graves' disease: from Robert James Graves, Irish physician, 1796–1853
exophthalmos: Gk. *ex,* out; *opthalmos,* eyeball

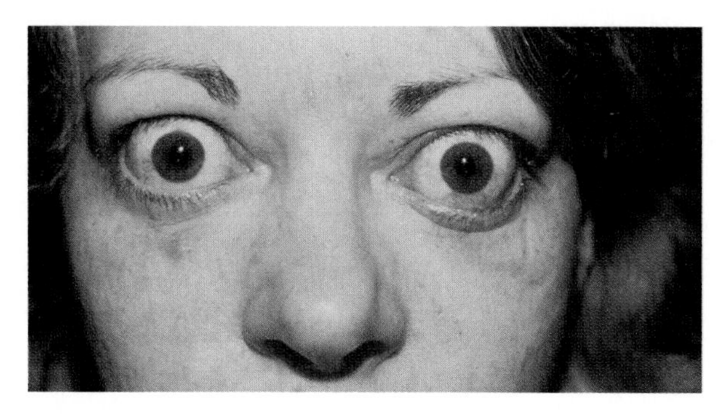

Figure 14.27

Hyperthyroidism is characterized by an increased metabolic rate, weight loss, muscular weakness, and nervousness. Protrusion of the eyeballs (exophthalmos) may also occur.

Disorders of the Parathyroid Glands

Surgical removal of the parathyroid glands sometimes unintentionally occurs when the thyroid is removed because of a tumor or the presence of Graves' disease. The resulting fall in parathyroid hormone (PTH) causes a decrease in plasma calcium levels, which can lead to severe muscle tetany. Hyperparathyroidism is usually caused by a tumor that secretes excessive amounts of PTH. This stimulates demineralization of bone, which makes the bones soft and raises the blood levels of calcium and phosphate. As a result of these changes, the bones are subject to deformity and fracture, and stones composed of calcium phosphate are likely to develop in the urinary tract.

Disorders of the Pancreatic Islets

Diabetes Mellitus

Diabetes mellitus is characterized by fasting hyperglycemia and the presence of glucose in the urine. There are two forms of this disease. **Type I,** or **insulin-dependent diabetes mellitus,** is caused by destruction of the beta cells and the resulting lack of insulin secretion. **Type II,** or **non-insulin-dependent diabetes mellitus** (which is the more common form), is caused by decreased tissue sensitivity to the effects of insulin, so that increasingly large amounts of insulin are required to produce a normal effect. Both types of diabetes mellitus are also associated with abnormally high levels of glucagon secretion.

Treatment of diabetes mellitus requires administering the necessary amounts of insulin to maintain a balanced carbohydrate metabolism. If excessive amounts of insulin are given, the person will experience extreme nervousness and tremors, perhaps followed by convulsion and loss of consciousness. This condition, commonly called *insulin shock*, can be treated by administering glucose intravenously.

Reactive Hypoglycemia

People who have a genetic predisposition for type II diabetes mellitus often first develop reactive hypoglycemia. In this condition, the rise in blood glucose that follows the ingestion of carbohydrates stimulates an excessive secretion of insulin, which in turn causes the blood glucose levels to fall below the normal range. This can result in weakness, changes in personality, and mental disorientation.

Disorders of the Adrenal Glands

Tumors of the Adrenal Medulla

Tumors of the chromaffin cells of the adrenal medulla are referred to as **pheochromocytomas** (*fe-o-kro″mo-si-to′maz*). These tumors cause hypersecretion of epinephrine and norepinephrine whose effects are similar to those of continuous sympathetic nervous stimulation. The symptoms of this condition are hypertension, elevated metabolism, hyperglycemia and sugar in the urine, nervousness, digestive problems, and sweating. It does not take long for the body to become totally fatigued under these conditions, making the patient susceptible to other diseases.

Addison's Disease

This disease is caused by an inadequate secretion of both glucocorticoids and mineralocorticoids, which results in hypoglycemia, sodium and potassium imbalance, dehydration, hypotension, rapid weight loss, and generalized weakness. A person with this condition who is not treated with corticosteroids will die within a few days because of severe electrolyte imbalance and dehydration. Another symptom of this disease is darkening of the skin. This is caused by high secretion of ACTH and possibly MSH (because MSH is derived from the same parent molecule as ACTH), which is a result of lack of negative feedback inhibition of the pituitary by corticosteroids.

Cushing's Syndrome

Hypersecretion of corticosteroids results in Cushing's syndrome. This condition is generally caused by a tumor of the adrenal cortex or by oversecretion of ACTH from the adenohypophysis. Cushing's syndrome is characterized by changes in carbohydrate and protein metabolism, hyperglycemia, hypertension, and muscular weakness. Metabolic problems give the body a puffy appearance and can cause structural changes characterized as "buffalo hump" and "moon face." Similar effects are also seen when people with chronic inflammatory diseases receive prolonged treatment with corticosteroids, which are given to reduce inflammation and inhibit the immune response.

---

pheochromocytomas: Gk. *phaios*, dusky; *chroma*, color; *oma*, tumor
Addison's disease: from Thomas Addison, English physician, 1793–1860
Cushing's syndrome: from Harvey Cushing, American physician, 1869–1939

Internal Affairs

How the *Endocrine System* Works with Other Body Systems
How Other Systems Work with the *Endocrine System*

Integumentary System

- Sex hormones affect skin pigmentation, development of body hair and sweat glands, thickening of the skin in males, and deposition of subcutaneous fat

- Covers and protects the body
- Initiates synthesis of vitamin D, which serves as a hormone

Skeletal System

- Many hormones affect bone growth and development
- PTH and calcitonin regulate calcium blood levels
- Sex hormones help to maintain bone mass in adults

- Provides protection and support for certain endocrine glands in the cranium, thorax, and pelvic cavity
- Stores calcium needed for endocrine function

Muscular System

- Growth hormone and sex hormones stimulate muscular growth and development
- Other hormones affect the electrolyte balance needed for muscle contraction

- Skeletal muscles provide protection for certain endocrine organs

Nervous System

- Several hormones affect neural metabolism
- Sex hormones affect brain development and behaviors
- Hormones affect the electrolyte balance needed for neural function

- Autonomic nervous system works in conjunction with certain adrenal hormones
- Hypothalamic hormones directly control hormonal release from the pituitary

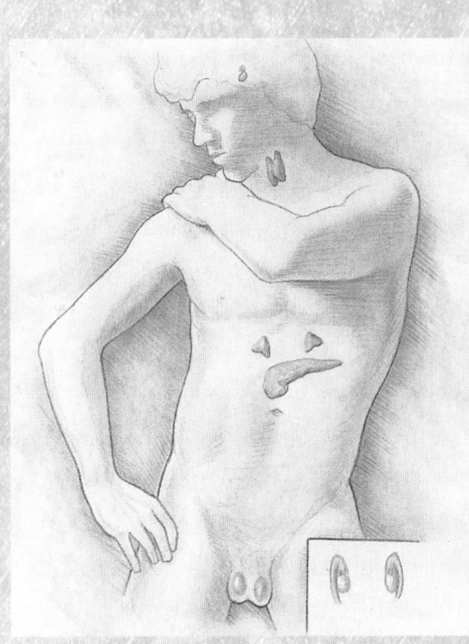

Circulatory System

- Several hormones influence blood volume and pressure
- Epinephrine elevates the heart rate
- Erythropoietin regulates red blood cell production

- Transports O_2 and CO_2, nutrients, and fluids to and from endocrine glands
- Transports hormones to target cells

Lymphatic System

- Several hormones affect immune function
- Glucocorticoids have anti-inflammatory effects

- Maintains balanced amount of interstitial fluid within endocrine organs
- Lymphocytes provide defense against infection

Respiratory System

- Epinephrine and norepinephrine stimulate respiratory activity and dilate respiratory passageways

- Provides O_2 and eliminates CO_2 to and from endocrine organs

Digestive System

- GI tract hormones coordinate secretory activities that affect digestion
- Activated vitamin D enabled intestinal absorption of calcium

- Provides nutrients for growth, maintenance, and repair of endocrine glands

Urinary System

- ADH and aldosterone regulate rates of fluid and electrolyte reabsorption in the kidneys

- Eliminates metabolic wastes
- Kidney cells release renin and erythropoietin when local blood pressure declines

Reproductive System

- Hormones play a major role in regulating sexual development and function, gamete formation, the menstrual cycle, fetal development, parturition, and lactation

- Gonads produce sex hormones that suppress secretory activities in the hypothalamus and pituitary gland

Adrenogenital Syndrome

Usually associated with Cushing's syndrome, this condition is caused by hypersecretion of adrenal sex hormones, particularly the androgens. Adrenogenital syndrome in young children causes premature puberty and enlarged genitals, especially the penis in males and the clitoris in females. An increase in body hair and a deeper voice are other characteristics. This condition in a mature woman can cause growth of a beard.

As with most other endocrine disorders, problems of the adrenal cortex are treated either through surgery (if there is a tumor causing hypersecretion) or by the administration of the necessary levels of appropriate hormones (in the case of tissue dysfunction with hyposecretion).

Clinical Case Study Answer

The visual findings are probably the result of a pituitary tumor (adenoma) eroding and pushing upward on the optic chiasma. The optic chiasma contains nerve tracts that emanate from the medial portions of each retina and receive light from the lateral (temporal) visual fields. The pituitary tumor is secreting either increased or unregulated amounts of ACTH (adrenocorticotropic hormone), which in turn are causing overproduction of adrenocortical hormones. These include glucocorticoids, mineralocorticoids, and various androgens, which in combination are producing the patient's constellation of findings.

This specific condition, hypersecretion of adrenocortical hormones secondary to an ACTH-producing pituitary adenoma, is known as *Cushing's disease*. It is usually treated by transsphenoidal excision of the tumor. When conditions are not right for surgery, radiation or drugs may be used. Elevated levels of adrenocortical hormones can result from endocrine tumors other than pituitary. An example would be an adrenal tumor, either cancerous or benign, that was producing adrenocortical hormones. The physical effects of this condition and all others that involve high levels of adrenocortical hormones are known collectively as *Cushing's syndrome*, which is not to be confused with *Cushing's disease*. The latter term is specifically reserved for hyperadrenalism caused by a pituitary adenoma.

Chapter Summary

Introduction to the Endocrine System (pp. 439–444)

1. Hormones are regulatory molecules released into the blood by endocrine glands. The action of a hormone on target cells is dependent on its concentration and the specific receptor sites on cell membranes.
2. Hormones are classified chemically as steroids, proteins, and amines.
3. Negative feedback occurs when information concerning an imbalance in hormone concentration is fed back to an organ that acts to correct the imbalance.

Pituitary Gland (pp. 444–449)

1. The pituitary gland (hypophysis) is divided into an anterior adenohypophysis and a posterior neurohypophysis.
   (a) In adults the adenohypophysis consists of a glandular pars distalis and a thin proximal extension called the pars tuberalis. A pars intermedia is present in the fetus but does not constitute a separate structure in the adult.
   (b) The neurohypophysis consists of the lobus nervosa and the infundibulum.
2. The anterior pituitary produces and secretes growth hormone, thyroid-stimulating hormone, adrenocorticotropic hormone, follicle-stimulating hormone, luteinizing hormone, prolactin, and melanocyte-stimulating hormone.
3. The posterior pituitary releases oxytocin and antidiuretic hormone.
4. Secretions of the anterior pituitary are controlled by hypothalamic hormones and regulated by the feedback of hormones from the target cells. Release of hormones from the posterior pituitary are controlled by the hypothalamo-hypophyseal nerve tract.

Thyroid and Parathyroid Glands (pp. 450–453)

1. The bilobed thyroid gland is located in the neck, just below the larynx. Four small parathyroid glands are embedded in its posterior surface.
2. Thyroid follicles secrete thyroxine and triiodothyronine, which increase the rate of protein synthesis and the rate of energy release from carbohydrates. They also regulate the rate of growth and the rate of maturation of the nervous system.
3. Parafollicular cells of the thyroid secrete the hormone calcitonin, which lowers blood calcium by inhibiting the release of calcium from bone tissue and stimulating the excretion of calcium by the kidneys.
4. Parathyroid hormone causes an increase in blood calcium and a decrease in blood phosphate levels. It acts on the large intestine, kidneys, and bones.

Pancreas (pp. 453–455)

1. The pancreas is a mixed endocrine and exocrine gland, located in the abdominal cavity.
2. The pancreatic islets contain beta cells that secrete insulin and alpha cells that secrete glucagon.
   (a) Insulin lowers blood glucose and stimulates the production of glycogen, fat, and protein.
   (b) Glucagon raises blood glucose by stimulating the breakdown of liver glycogen.

Adrenal Glands (pp. 455–458)

1. Each adrenal gland consists of an adrenal cortex and an adrenal medulla and is positioned along the superior border of a kidney.
2. Hormones of the adrenal cortex include mineralocorticoids, which regulate sodium reabsorption and potassium excretion; glucocorticoids, which influence metabolism by promoting vasoconstriction and resistance to stress; and gonadocorticoids, which supplement gonadal hormones.
3. Epinephrine and norepinephrine, secreted from the adrenal medulla, produce effects similar to those of the sympathetic division of the ANS.

Gonads and Other Endocrine Glands (pp. 458–462)

1. Testes are the male gonads that produce the male sex hormone testosterone within the interstitial cells.
2. Ovaries are the female gonads that produce estrogens within the ovarian follicles and corpus luteum.
3. The pineal gland, located in the roof of the third ventricle of the brain, secretes melatonin, which seems to have an effect on the hypothalamus in the release of gonadotrophin.
4. The thymus is positioned behind the sternum within the mediastinum. It produces T cells that are important in maintaining body immunity.
5. Certain gastrointestinal cells secrete hormones that aid digestion.
6. The maternal placenta secretes hormones that maintain pregnancy.

Review Activities

Objective Questions

Match the gland to its embryonic origin.

1. adenohypophysis
2. neurohypophysis
3. adrenal medulla
4. pancreas
5. thyroid gland

(a) endoderm of pharynx
(b) diverticulum from brain
(c) endoderm of foregut
(d) neural crest ectoderm
(e) hypophyseal pouch

6. The sella turcica that supports the pituitary gland is located in which bone?
   (a) the ethmoid bone
   (b) the frontal bone
   (c) the sphenoid bone
   (d) the occipital bone
7. The hormone primarily responsible for setting the basal metabolic rate and for promoting the maturation of the brain is
   (a) cortisol. (c) TSH.
   (b) ACTH. (d) thyroxine.
8. Which of the following statements about the adrenal medulla is *true?*
   (a) It develops from mesoderm.
   (b) It secretes some androgens.
   (c) Its secretion prepares the body for the fight-or-flight response.
   (d) The zona fasciculata is stimulated by ACTH.
   (e) All of the above are true.
9. Which of the following statements about the hormone insulin is *true?*
   (a) It is secreted by alpha cells in the pancreatic islets.
   (b) It is secreted in response to a rise in blood glucose.
   (c) It stimulates the production of glycogen and fat.
   (d) Both a and b are true.
   (e) Both b and c are true.

Match the hormone with the primary agent that stimulates its secretion.

10. epinephrine
11. thyroxine
12. corticosteroids
13. ACTH

(a) TSH
(b) ACTH
(c) growth hormone
(d) sympathetic nerve impulse
(e) CRF

14. Steroid hormones are secreted by
   (a) the adrenal cortex.
   (b) the gonads.
   (c) the thyroid.
   (d) both a and b.
   (e) both b and c.

15. The secretion of which of the following hormones would be *increased* in a person with endemic goiter?
    (a) TSH (c) triiodothyronine
    (b) thyroxine (d) all of the above

Essay Questions

1. Explain how the nervous and endocrine systems differ in maintaining body homeostasis.
2. List the glands of the endocrine system and describe their general locations.
3. Give examples of steroids, proteins, and amines and describe the various ways in which they are administered.
4. Define *negative feedback*. Why is this a reliable mechanism for controlling hormonal secretion?
5. Which endocrine glands, or portions of endocrine glands, develop through the process of germ layer invagination? Which develop as an outgrowth or budding of germ layer tissue?
6. Why is the pituitary frequently considered two separate glands?
7. List the hormones secreted by the adenohypophysis and describe the general functions of each.
8. Describe the location and gross structure of the neurohypophysis. What hormones are released from this portion of the pituitary gland and how is their release regulated?
9. List the hormones secreted by the thyroid gland and discuss the general function of each.
10. Describe the location of the parathyroid glands. What specialized cells constitute these glands and what hormone do they secrete?
11. Describe the location and structure of the pancreatic islets, list the hormones they produce, and discuss the general function of each hormone.
12. Distinguish between the adrenal cortex and adrenal medulla in structure, function, and value to the body.
13. Define *chromaffin cells*, *interstitial cells*, *alpha* and *beta cells*, *oxyphil cells*, *follicular cells*, and *pituicytes*.
14. Explain the causes of each of the following endocrine diseases: gigantism, acromegaly, pituitary cachexia, and pituitary dwarfism.
15. Distinguish between cretinism, myxedema, and Graves' disease.

Critical-Thinking Questions

1. The anterior pituitary has often been characterized as the "master gland." In what sense might this description be correct? In what sense is it misleading?
2. A 38-year-old woman experienced a milky discharge through the nipples of her breasts that occurred periodically over a 6-month period. She was not pregnant and her youngest child was 7 years old. Her physician referred to this as *galactorrhea*—the medical term for a whitish or greenish discharge from one or both breasts—and scheduled her for a blood test to determine possible hormonal dysfunction that might be linked to a benign growth on the pituitary gland. How could a tumor of the pituitary result in this woman's symptoms?
3. Brenda, your roommate, has been having an awful time lately. She can't even muster enough energy to go out on a date. She's been putting on weight, she's always cold, and every time she pops in the "Buns of Steel" work-out video, she complains of weakness. When she finally goes to her doctor, he finds her to have a slow pulse and low blood pressure. Further tests indicate thyroid hormone deficiency—a condition termed *hypothyroidism*. This endocrine disorder is very common and occurs eight times more often in females than males. Why would Brenda's symptoms be typical of this disorder? What type of treatment will the doctor prescribe?
4. You decide it's time that your college basketball team make it to the final four. You figure your pal Bubba has the competitive spirit needed to step in as the new star center—if only he weren't five foot eight. Confronting the problem head-on, you start slipping growth hormone into Bubba's morning orange juice—a clever strategy, you think.

   After a time, however, you begin to realize that Bubba isn't growing any taller. In fact, he hasn't grown an inch. Instead, his jaw and forehead seem disproportionately large, and his hands and feet are swollen. To make matters worse, he has developed a horrendous body odor.

   Explain why giving growth hormones to an adult does not change the person's height. Explain the physical changes that can occur when an adult is administered growth hormones.
5. Suppose a person with a normal thyroid gland took thyroid pills. Would he or she lose weight? Why or why not? What endocrine changes would be produced by those pills? If, after a few months, the person stopped taking the pills, what would happen to his or her body weight? Explain.

FIFTEEN

Sensory Organs

Clinical Case Study 471

Overview of Sensory Perception 471

Classification of the Senses 471

Somatic Senses 472
- Tactile and Pressure Receptors 473
- Receptors for Heat, Cold, and Pain 474
- Proprioceptors 476
- Neural Pathways for Somatic Sensation 477

Olfactory Sense 479

Gustatory Sense 480

Visual Sense 481
- Accessory Structures of the Eye 483
- Structure of the Eyeball 487
- Function of the Eyeball 492
- Neural Pathways for Vision, Eye Movements, and Processing Visual Information 496

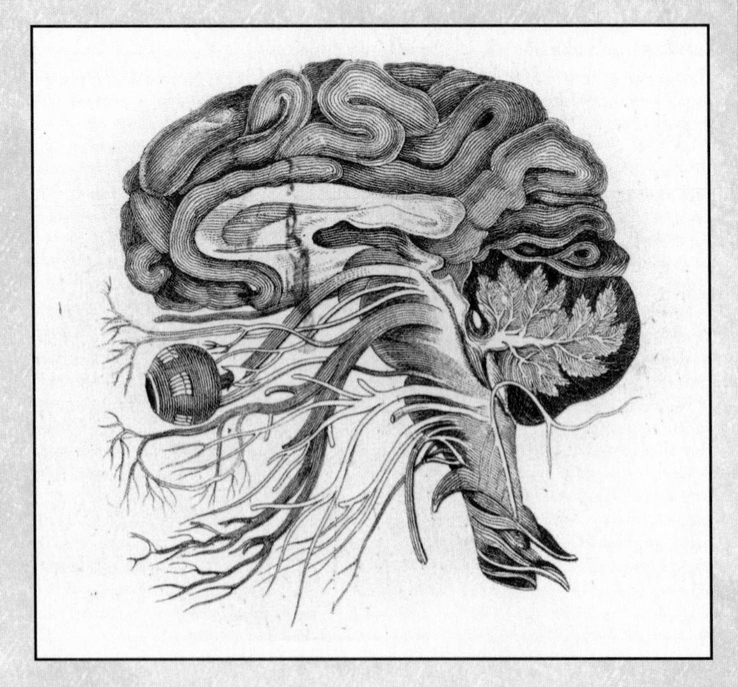

In this delicate, somewhat fanciful, rendering of the brain, the nearly spherical eyeball, attached to the optic nerve, appears to float in space. Neural pathways from the retina of the eyeball to specific areas of the brain permit visual perception. (Across from the eyeball, note the distinctive treelike branching of the white matter of the cerebellum.)

Developmental Exposition: The Eye 498

Senses of Hearing and Balance 499
- Outer Ear 499
- Middle Ear 501
- Inner Ear 502
- Sound Waves and Neural Pathways for Hearing 505
- Mechanics of Equilibrium 505

Clinical Considerations 510
- Diagnosis of Eye and Ear Disorders 511
- Developmental Problems of the Eyes and Ears 511
- Functional Impairments of the Eye 511

Developmental Exposition: The Ear 512
- Infections and Diseases of the Eye 516
- Infections, Diseases, and Functional Impairments of the Ear 516

Clinical Case Study Answer 517

Chapter Summary 518

Review Activities 518

Clinical Case Study

A 50-year-old man complained to his family doctor of progressive hearing loss in his right ear. In order to rule out visible abnormalities or wax buildup, the physician performed an otoscopic examination, which revealed no abnormalities. The physician then struck a tuning fork and placed it on the mastoid process of the patient's right temporal bone. The patient immediately exclaimed, "I can hear that really well, even in my bad ear!" After a moment or so the patient noted that the tone had nearly died out. The doctor then moved the instrument 2 centimeters away from the same ear. At that location, the patient was unable to hear anything. The doctor explained that while someone with normal hearing will hear a vibrating fork when it is held against the mastoid process, the person will hear it better when it is held just outside the external acoustic canal.

What components of the patient's hearing mechanism were bypassed when the handle of the tuning fork was placed on his mastoid process? Describe the type of hearing problem this patient has. Is it a conduction problem or a perception problem?

Hints: The hearing organs of the inner ear can receive and effectively process sound waves directly from the bone in which they are encased. List in order the structures bypassed by sound waves being processed through the mastoid process. Carefully read the section on the structure of the ear and the one on sound waves and neural pathways for hearing. Also review the Clinical Considerations section on functional impairments of the ear.

Overview of Sensory Perception

Sensory organs are highly specialized extensions of the nervous system. They contain sensory neurons adapted to respond to specific stimuli and conduct nerve impulses to the brain.

Objective 1 State the conditions necessary for perceiving a sensation.

Objective 2 Discuss the selectivity of sensory receptors for specific stimuli.

The sense organs are actually extensions of the nervous system that respond to changes in the internal and external environment and transmit nerve impulses to the brain. It is through the sense organs that we achieve awareness of the environment, and for this reason they have been described as "windows for the brain." A stimulus must first be received before the sensation can be interpreted in the brain and the necessary body adjustments made. Not only do we depend on our sense organs to experience pleasure, they also ensure our very survival. For example, they enable us to hear warning sounds, see dangers, avoid toxic substances, and perceive sensations of pain, hunger, and thirst.

A *sensation* is a feeling or awareness of a bodily state or condition that occurs whenever a sensory impulse is transmitted to the brain. The interpretation of a sensation is referred to as *perception*. Perceptions are the creations of our brain; in other words, we see, hear, taste, and smell with our brain. In order to perceive a sensation, the following conditions are necessary.

1. A *stimulus* sufficient to initiate a response in the nervous system must be present.
2. A *receptor* must convert the stimulus to a nerve impulse. A receptor is a specialized peripheral dendritic ending of a sensory nerve fiber or the specialized receptor cells associated with it.
3. The *conduction of the nerve impulse* must occur from the receptor to the brain along a nervous pathway.
4. The *interpretation of the impulse* in the form of a perception must occur within a specific portion of the brain.

Only those impulses that reach the cerebral cortex of the brain are consciously interpreted. If impulses terminate in the spinal cord or brain stem, they may initiate a reflexive motor response but not a conscious awareness. Impulses reaching the cerebral cortex travel through nerve fibers composing sensory, or ascending, tracts. Clusters of neuron cell bodies, called *nuclei,* are synaptic sites along sensory tracts within the CNS. The nuclei that sensory impulses pass through before reaching the cerebral cortex are located in the spinal cord, medulla oblongata, pons, and thalamus.

Through the use of scientific instruments, it is known that the senses act as energy filters that allow perception of only a narrow range of energy. Vision, for example, is limited to light rays in the visible spectrum. Other types of rays of the same type of energy as visible light, such as X rays, radio waves, and ultraviolet and infrared light, cannot normally excite the sensory receptors in the eyes. Although filtered and distorted by the limitations of sensory function, our perceptions allow us to interact effectively with our environment and are of obvious survival value.

1. Distinguish between sensation and perception.
2. List the four conditions necessary for perception and identify which of the four must always involve consciousness in order for perception to occur.
3. Use examples to explain the statement that each of the senses acts as a filter.

Classification of the Senses

The senses are classified as general or special according to the degree of complexity of the receptors and neural pathways. They are also classified as somatic or visceral according to the location of the receptors.

Objective 3 Compare and contrast somatic, visceral, and special senses.

Objective 4 Describe the three basic kinds of receptors and give examples of each.

Structurally, the sensory receptor can be the dendrites of sensory neurons, which are either free (such as those in the skin that respond to pain and temperature) or encapsuled within nonneural structures (such as lamellated corpuscles or pressure receptors in the skin) (see table 15.1). Other receptors form from epithelial cells that synapse with sensory dendrites. These include taste buds on the tongue, photoreceptors in the eyes, and hair cells in the inner ears.

Although we usually speak of five senses, in reality we possess many more. The senses of the body can be classified as general or special according to the degree of complexity of their receptors and sensory pathways. **General senses** are widespread through the body and are structurally simple. Examples are touch, pressure, cold-heat, and pain. **Special senses** are localized in complex receptor organs and have extensive neural pathways (tracts) in the brain. The special senses are taste, smell, sight, hearing, and balance.

The senses can also be classified as somatic or visceral according to the location of the receptors. **Somatic senses** are those in which the receptors are localized within the body wall. These include the cutaneous (skin) receptors and those within muscles, tendons, and joints. **Visceral senses** are those in which the receptors are located within visceral organs. Both classification schemes may be used in describing some senses; for example, hearing (a special somatic sense) or pain from the gastrointestinal tract (a general visceral sense).

Senses are also classified according to the location of the receptors and the types of stimuli to which they respond. There are three basic kinds of receptors: *exteroceptors*, *visceroceptors (enteroceptors)*, and *proprioceptors*.

Exteroceptors

Exteroceptors *(ek″ster-o-sep′torz)* are located near the surface of the body, where they respond to stimuli from the external environment. They include the following:

1. rod and cone cells in the retina of the eye—*photoreceptors;*
2. hair cells in the spiral organ (organ of Corti) within the inner ear—*mechanoreceptors;*
3. olfactory receptors in the nasal epithelium of the nasal cavity—*chemoreceptors;*
4. taste receptors on the tongue—*chemoreceptors;* and
5. skin receptors within the dermis—*tactile receptors* for touch, *mechanoreceptors* for pressure, *thermoreceptors* for temperature, and *nociceptors (no″sĭ-sep′torz)* for pain.

Pain receptors are stimulated by chemicals released from damaged tissue cells, and thus are a type of chemoreceptor. Although there are specific pain receptors, nearly all types of receptors transmit impulses that are perceived as pain if they are stimulated excessively. Pain receptors are located throughout the body, but only those located within the skin are classified as exteroceptors.

Visceroceptors

As the name implies, visceroceptors *(vis″er-o-sep′torz)* are sensory nerve cells that produce sensations arising from the viscera, such as internal pain, hunger, thirst, fatigue, or nausea. Specialized visceroceptors located within the circulatory system are sensitive to changes in blood pressure; these are called *baroreceptors*.

Proprioceptors

Proprioceptors are sensory nerve cells that relay information about body position, equilibrium, and movement. They are located in the inner ear, in and around joints, and between tendons and muscles.

Receptors may also be classified on the basis of sensory adaptation (accommodation). Some receptors respond with a burst of activity when a stimulus is first applied, but then quickly decrease their firing rate—adapt to the stimulus—when the stimulus is maintained. Receptors with this response pattern are called **phasic receptors.** Receptors that produce a relatively constant rate of firing as long as the stimulus is maintained are known as **tonic receptors.**

Phasic receptors alert us to changes in sensory stimuli and are in part responsible for the fact that we can cease paying attention to constant stimuli. This ability is called *sensory adaptation.* Odor and touch, for example, adapt rapidly; bathwater feels hotter when we first enter it. Sensations of pain, by contrast, adapt little if at all.

1. Using examples, explain how sensory receptors can be classified according to complexity, location, structure, and the type of stimuli to which they respond.
2. Distinguish between phasic and tonic receptors.

Somatic Senses

The somatic senses arise in cutaneous receptors and proprioceptors. The perception of somatic sensations is determined by the density of the receptors in the stimulated receptive field and the intensity of the sensation.

somatic: Gk. *somatikos*, body
visceral: L. *viscera*, body organs
nociceptor: L. *nocco*, to injure; *ceptus*, taken

proprioceptor: L. *proprius*, one's own; *ceptus*, taken

Table 15.1
Cutaneous Receptors

Type	Location	Function
Corpuscles of touch (Meissner's corpuscles) (mechanoreceptors)	Papillae of dermis; numerous in hairless portions of body (eyelids, fingertips, lips, nipples, external genitalia)	Detect light motion against surface of skin
Free nerve endings (tactile receptors; thermoreceptors; pain receptors)	Lower layers of epidermis	Detect touch and pressure, changes in temperature, and tissue damage
Root hair plexuses (tactile receptors)	Surrounding hair follicles	Detect movement of hair
Lamellated (pacinian) corpuscles (mechanoreceptors)	Hypodermis; synovial membranes; perimysium; certain visceral organs	Detect deep pressure and high-frequency vibration
Organs of Ruffini (mechanoreceptors)	Lower layers of dermis	Detect deep pressure and stretch
Bulbs of Krause (mechanoreceptors)	Dermis; lips; mouth; conjunctiva of eye	Detect light pressure and low-frequency vibration

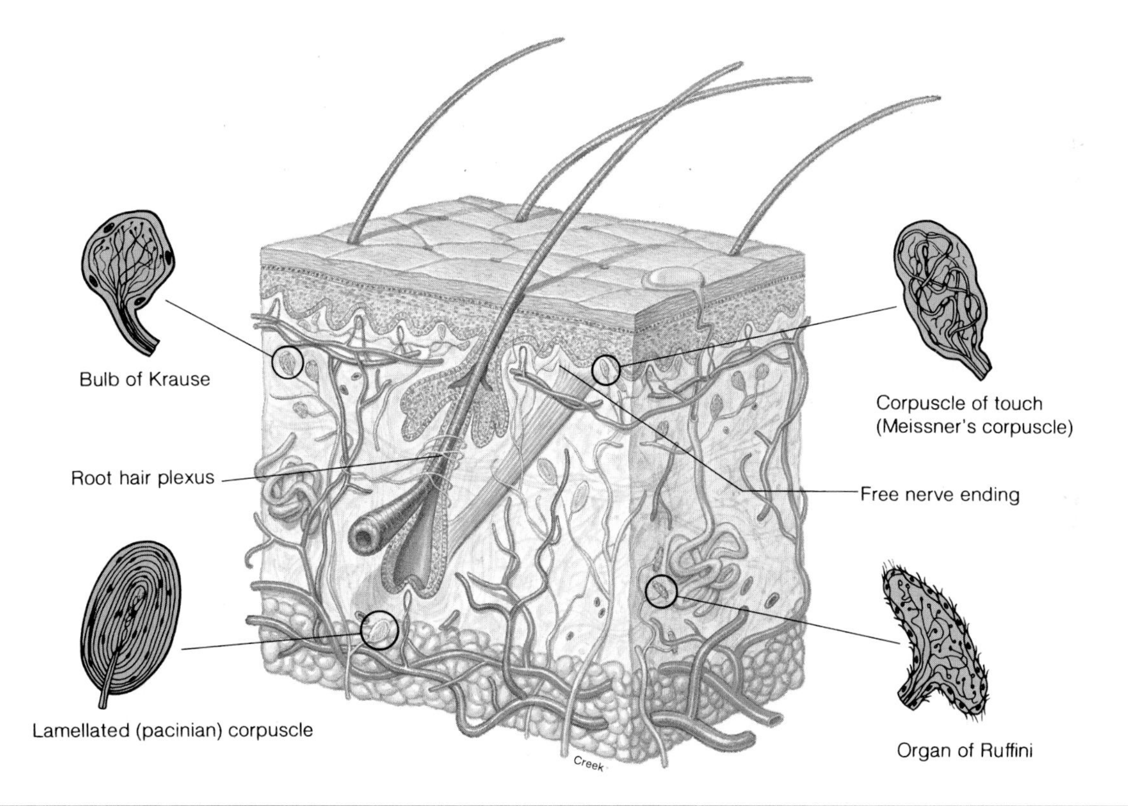

Objective 5 Describe the structure, function, and location of the various tactile and pressure receptors.

Objective 6 Explain the purpose of pain and describe the receptors that respond to pain and the neural pathways for pain sensation.

Objective 7 Explain what is meant by referred pain and phantom pain and give examples of each.

The **somatic senses,** or **somesthetic senses,** arise in cutaneous receptors and proprioceptors. Cutaneous sensations include touch, tickle, pressure, cold, heat, and pain. The proprioceptors located in the inner ears, joints, tendons, and muscles relay information about body position, equilibrium, and movement.

Tactile and Pressure Receptors

Both tactile receptors and pressure receptors are sensitive to mechanical forces that distort or displace the tissue in which they are located. **Tactile receptors** respond to fine, or light, touch and are located primarily in the dermis and hypodermis of the skin. **Pressure receptors** respond to pressure, vibration, and stretch and are commonly found in the hypodermis of the skin and in the tendons and ligaments of joints. The tactile and pressure receptors are summarized in table 15.1.

Corpuscles of Touch

A corpuscle of touch *(Meissner's corpuscle)* is an oval receptor composed of a mass of dendritic endings from two or three nerve fibers enclosed by connective tissue sheaths. These corpuscles are numerous in the hairless portions of the body, such as the eyelids, lips, tip of the tongue, fingertips, palms of the hands, soles of the feet, nipples, and external genitalia. Corpuscles of touch lie within the papillary layer of the dermis, where they are especially sensitive to the movement of objects that barely contact the skin (see chapter 5). Sensations of fine or light touch are perceived as these receptors are stimulated. They also function when a person touches an object to determine its texture.

The highly sensitive fingertips are used in reading braille. Braille symbols consist of dots that are raised 1 mm from the surface of the page and separated from each other by 2.5 mm. Experienced braille readers can scan words at about the same speed that a sighted person can read aloud—a rate of about 100 words per minute.

Free Nerve Endings

Free nerve endings are the least modified and the most superficial of the tactile receptors. These receptors extend into the lower layers of the epidermis, where they end as knobs between the epithelial cells. Free nerve endings respond chiefly to pain and temperature (discussed shortly), but they also detect touch and pressure, for example from clothing. Some free nerve endings are particularly sensitive to tickle and itch.

Root Hair Plexuses

Root hair plexuses are a specialized type of free nerve ending. They are coiled around hair follicles, where they respond to movement of the hair.

Lamellated Corpuscles

Lamellated *(pacinian)* corpuscles are large, onion-shaped receptors composed of the dendritic endings of several sensory nerve fibers enclosed by connective tissue layers. They are commonly found within the synovial membranes of synovial joints, in the perimysium of skeletal muscle tissue, and in certain visceral organs. Lamellated corpuscles are also abundant in the skin of the palms and fingers of the hand, soles of the feet, external genitalia, and breasts. They respond to heavy pressures, generally those that are constantly applied. They can also detect deep vibrations in tissues and organs.

Organs of Ruffini

The organs of Ruffini are encapsulated nerve endings that are found in the deep layers of the dermis and in subcutaneous tissue, where they respond to deep continuous pressure and to stretch. They are also present in joint capsules and function in the detection of joint movement.

Bulbs of Krause

The bulbs of Krause are thought to be a variation of Meissner's corpuscles. They are most abundant in the mucous membranes, and therefore are sometimes called *mucocutaneous corpuscles*.

Historically, both the organs of Ruffini and the bulbs of Krause have been considered to be thermoreceptors—the former heat receptors and the latter cold receptors. However, both are actually mechanoreceptors. The bulbs of Krause respond to light pressure and low-frequency vibration.

Receptors for Heat, Cold, and Pain

The principal receptors for heat and cold (thermoreceptors) and for pain (nociceptors) are the free nerve endings. Several million of them are distributed throughout the skin and internal tissues. The free nerve endings responsible for cold sensations are closer to the surface of the skin and are 10 to 15 times more abundant in any given area of skin than those responsible for sensations of heat.

Pain receptors respond to damage to tissues and are activated by all types of stimuli. They are sparse in most visceral organs and absent entirely within the nervous tissue of the brain. Although the free nerve endings are specialized to respond to tissue damage, all of the cutaneous receptors will relay impulses that are interpreted as pain if stimulated excessively.

The protective value of pain receptors is obvious. Unlike other cutaneous receptors, free nerve endings exhibit little accommodation, so impulses are relayed continuously to the CNS as long as the irritating stimulus is present. Pain receptors are particularly sensitive to chemical stimulation. Muscle spasms, muscle fatigue, or an inadequate supply of blood to an organ may also cause pain.

Impulses for pain are conducted to the spinal cord through sensory neurons. The pain sensations are then conducted to the thalamus along the *lateral spinothalamic tract* of the spinal cord, and from there to the somatesthetic area of the cerebral cortex. Although an awareness of pain occurs in the thalamus, the type and intensity of pain is interpreted in specialized areas of the cerebral cortex.

The sensation of pain can be clinically classified as **somatic pain** or **visceral pain.** Stimulation of the cutaneous pain

---

corpuscle: L. *corpusculum*, diminutive of *corpus*, body
Meissner's corpuscle: from George Meissner, German histologist, 1829–1905
braille: from Louis Braille, French teacher of the blind, 1809–52
pacinian corpuscle: from Filippo Pacini, Italian anatomist, 1812–83

organs of Ruffini: from Angelo Ruffini, Italian anatomist, 1864–1929
bulbs of Krause: from Wilhelm J. F. Krause, German anatomist, 1833–1910

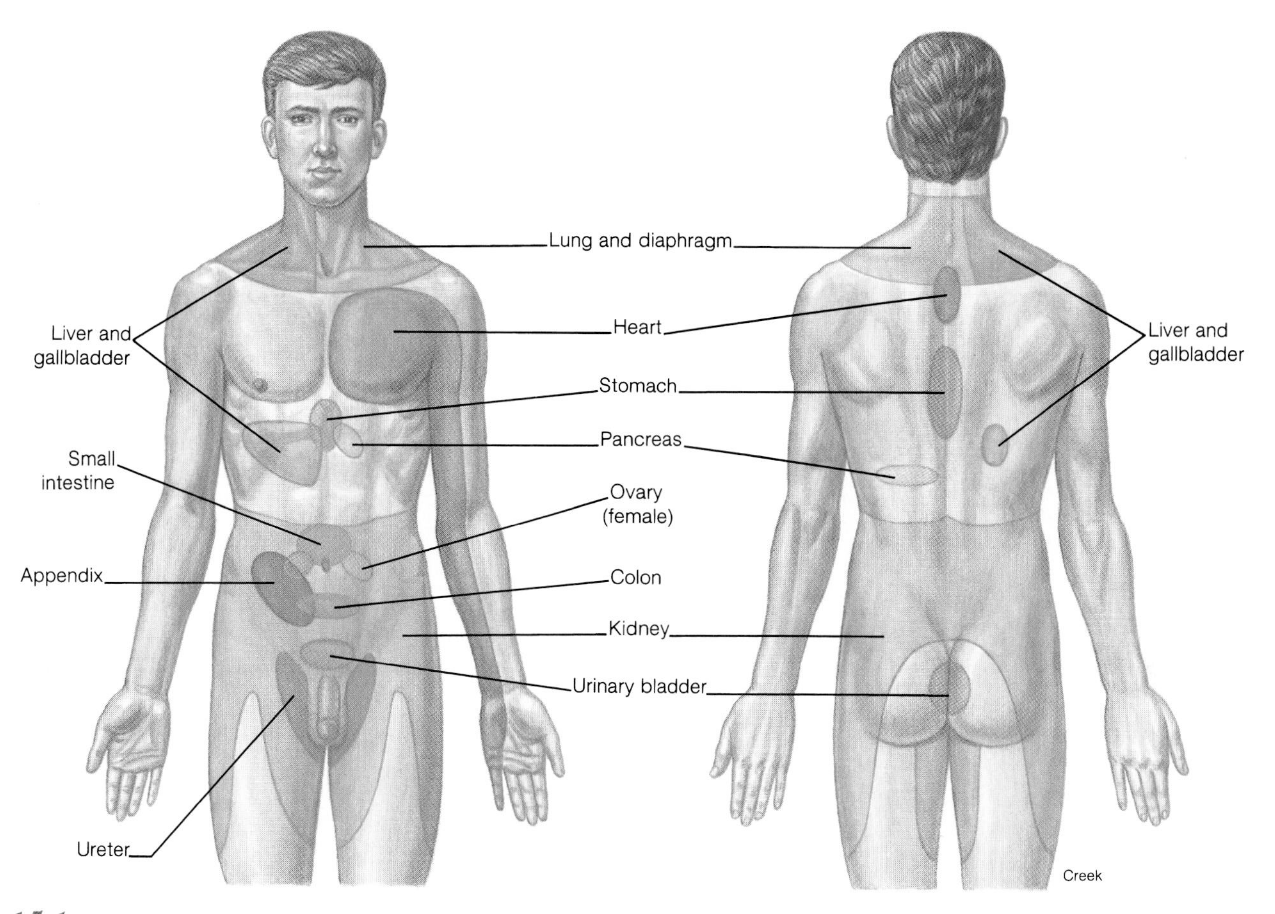

Figure 15.1

Sites of referred pain are perceived cutaneously but actually originate from specific visceral organs.

receptors results in the perception of superficial somatic pain. Deep somatic pain comes from stimulation of receptors in skeletal muscles, joints, and tendons.

Stimulation of the receptors within the viscera causes the perception of visceral pain. Through precise neural pathways, the brain is able to perceive the area of stimulation and project the pain sensation back to that area. The sensation of pain from certain visceral organs, however, may not be perceived as arising from those organs but from other somatic locations. This phenomenon is known as **referred pain** (fig. 15.1). The sensation of referred pain is consistent from one person to another and is clinically important in diagnosing organ dysfunctions. The pain of a heart attack, for example, may be perceived subcutaneously over the heart and down the medial side of the left arm. Ulcers of the stomach may cause pain that is perceived as coming from the upper central (epigastric) region of the trunk. Pain from problems of the liver or gallbladder may be perceived as localized visceral pain or as referred pain arising from the right neck and shoulder regions.

Referred pain is not totally understood but seems to be related to the development of the tracts within the spinal cord. There are thought to be some *common nerve pathways* that are used by sensory impulses coming from both the cutaneous areas and from visceral organs (fig. 15.2). Consequently, impulses along these pathways may be incorrectly interpreted as arising cutaneously rather than from within a visceral organ.

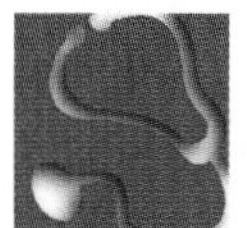

The perception of pain is of survival value because it alerts the body to an injury, disease, or organ dysfunction. *Acute pain* is sudden, usually short-term, and can generally be endured and attributed to a known cause. *Chronic pain,* however, is long-term and tends to weaken a person as it interferes with the ability to function effectively. Certain diseases, such as arthritis, are characterized by chronic pain. In these patients, relief of pain is of paramount concern. Treatment of chronic pain often requires the use of moderate pain-reducing drugs (analgesics) or intense narcotic drugs. Treatment in severely tormented chronic pain patients may include severing sensory nerves or implanting stimulating electrodes in appropriate nerve tracts.

Phantom pain is frequently experienced by an amputee who continues to feel pain from the body part that was amputated, as if it were still there. After amputation, the severed sensory neurons heal and function in the remaining portion of the appendage. Although it is not known why impulses that are interpreted as pain are sent periodically through these neurons, the

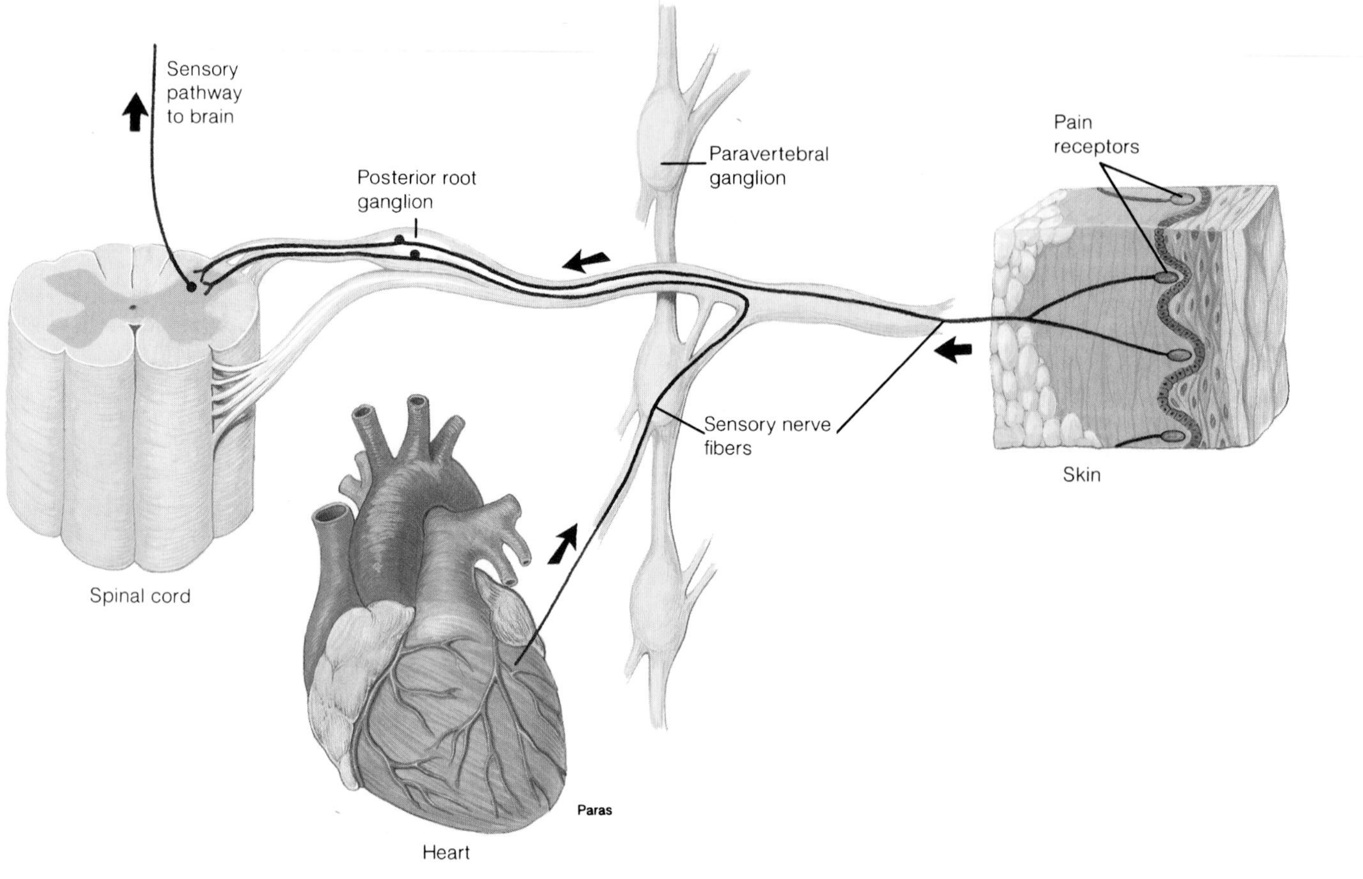

Figure 15.2

An explanation of referred pain. Pain originating from the myocardium of the heart may be perceived as coming from the skin of the left arm because sensory impulses from these two organs are conducted through common nerve pathways to the brain.

sensations evoked in the brain are projected to the region of the amputation, resulting in phantom pain.

Proprioceptors

Proprioceptors advise the brain of our own movements (*proprius* means "one's own") by responding to changes in stretch and tension. Proprioceptor information is used to adjust the strength and timing of muscle contractions to produce coordinated movements. Some of the sensory impulses from proprioceptors reach the level of consciousness as the **kinesthetic sense,** by which the position of the body parts is perceived. With the kinesthetic sense, the position and movement of the limbs can be determined without visual sensations, such as when dressing or walking in the dark. The kinesthetic sense, along with hearing, becomes keenly developed in a blind person.

High-speed transmission is a vital characteristic of the kinesthetic sense because rapid feedback to various body parts is essential for quick, smooth, coordinated body movements.

Proprioceptors are located in and around synovial joints, in skeletal muscle, between tendons and muscles, and in the inner ear. They are of four types: joint kinesthetic receptors, neuromuscular spindles, neurotendinous receptors, and sensory hair cells.

1. **Joint kinesthetic receptors** are located in the connective tissue capsule in synovial joints, where they are stimulated by changes in body position as the joints are moved.
2. **Neuromuscular spindles** are located in skeletal muscle, particularly in the muscles of the limbs. They consist of the endings of sensory neurons that are spiraled around specialized individual muscle fibers (fig. 15.3). Neuromuscular spindles are stimulated by an increase in muscle tension caused by the lengthening or stretching of the individual fibers, and thus provide information about the length of the muscle and the speed of muscle contraction.
3. **Neurotendinous receptors** (*Golgi tendon organs*) are located where a muscle attaches to a tendon (fig. 15.3). They are stimulated by the tension produced in a tendon when the attached muscle is either stretched or contracted.

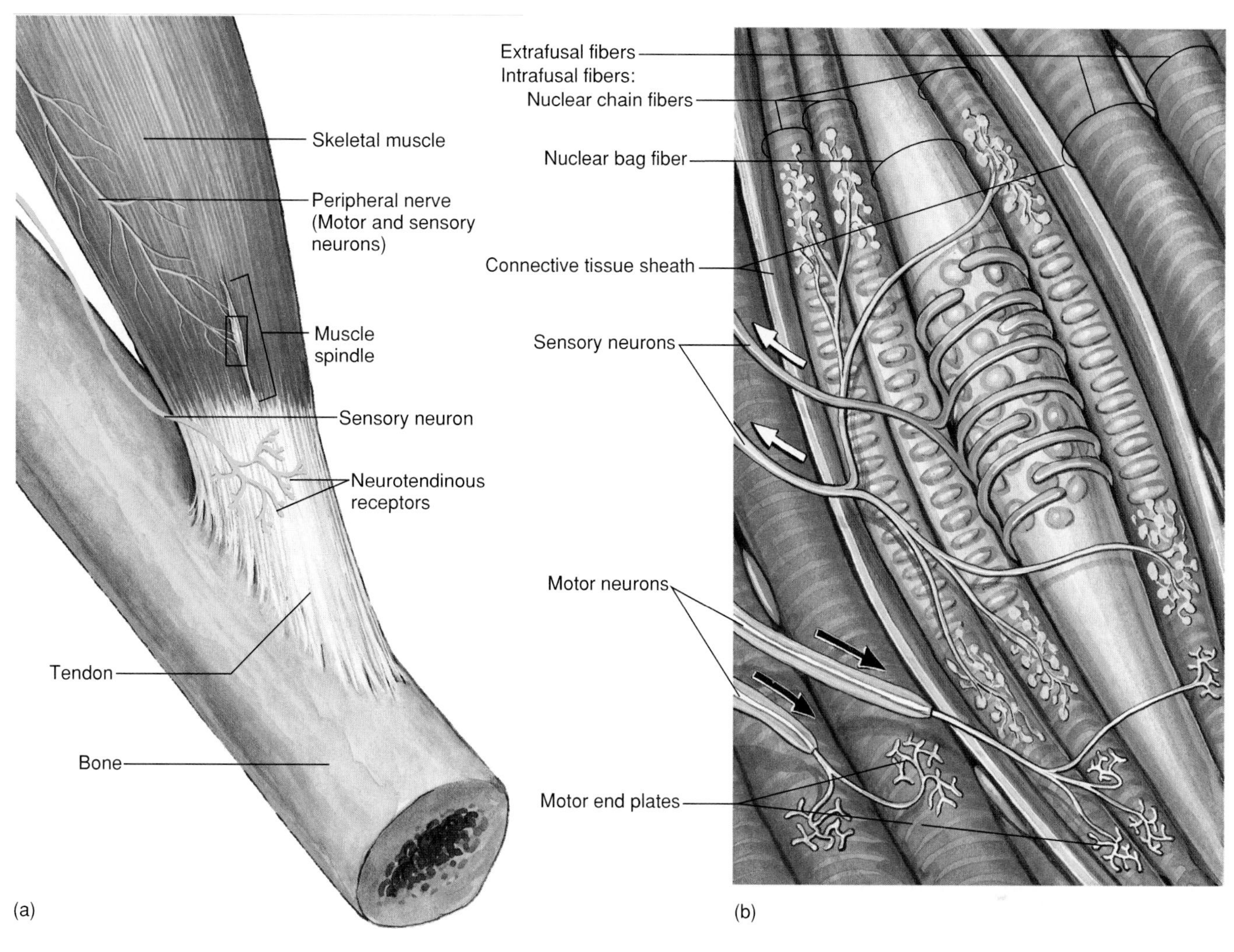

Figure 15.3

Proprioceptors are located within skeletal muscle tissue, tendons, and joint membranes. (*a*) The location of muscle spindles and neurotendinous receptors. (*b*) A magnification of the structure and innervation of muscle spindles.

4. **Sensory hair cells** of the inner ear are located in a fluid-filled, ductule structure called the membranous labyrinth. Their function in equilibrium is discussed later in this chapter in connection with the mechanics of equilibrium.

Neural Pathways for Somatic Sensation

The conduction pathways for the somatic senses are shown in figure 15.4. Sensations of proprioception and of touch and pressure are carried by large, myelinated nerve fibers that ascend in the posterior columns of the spinal cord on the ipsilateral (same) side. These fibers do not synapse until they reach the medulla oblongata of the brain stem; hence, fibers that carry these sensations from the feet are incredibly long. After synapsing in the medulla oblongata with second-order sensory neurons, information in the latter neurons crosses over to the contralateral (opposite) side as it ascends via a fiber tract called the **medial lemniscus** (*lem-nis'kus*) to the thalamus. Third-order sensory neurons in the thalamus that receive this input in turn project to the **postcentral gyrus** in the cerebral cortex.

Sensations of heat, cold, and pain are carried by thin, unmyelinated sensory neurons into the spinal cord. These synapse within the spinal cord with second-order interneurons that cross over to the contralateral side and ascend to the brain in the **lateral spinothalamic tract.** Fibers that mediate touch and pressure ascend in the **ventral spinothalamic tract.** Fibers of both spinothalamic tracts synapse in the thalamus with third-order neurons, which in turn project to the postcentral gyrus. Note that, in all cases, somatic information is carried to the postcentral gyrus in third-order neurons. Also, because of crossing-over, somatic information from each side

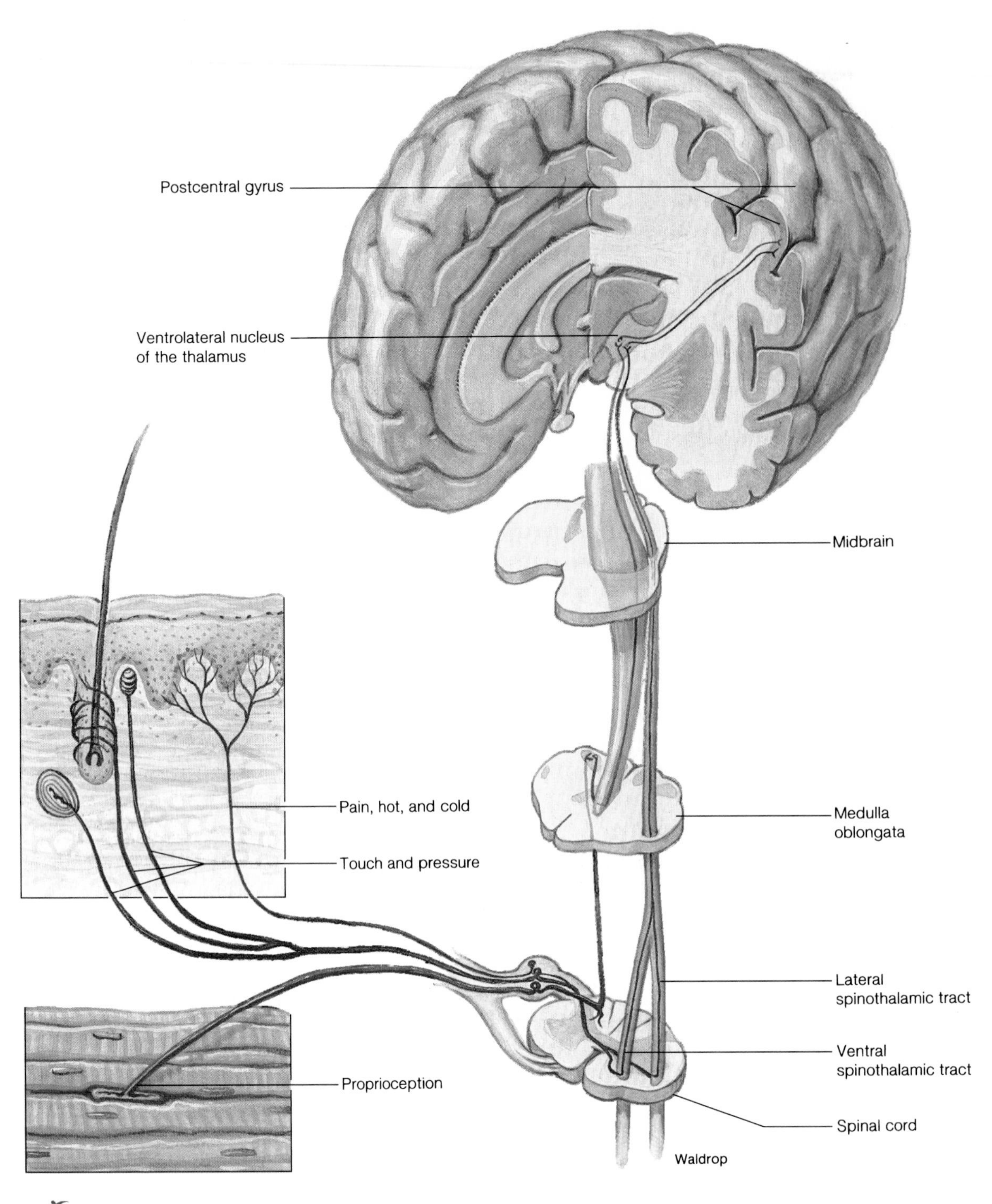

Figure 15.4

Pathways that lead from the cutaneous receptors and proprioreceptors into the postcentral gyrus in the cerebral cortex.

of the body is projected to the postcentral gyrus of the contralateral cerebral hemisphere.

All somatic information from the same area of the body projects to the same area of the postcentral gyrus. It is therefore possible to map out areas of the postcentral gyrus that receive sensory information from different parts of the body (see fig. 11.22). Such a map is greatly distorted, however, because it shows larger areas of cerebral cortex devoted to sensation in the face and hands than in other areas of the body. The disproportionately large areas of the caricature-like *sensory homunculus (ho-mung'kyoo-lus)* drawn on the gyrus reflect the fact that there is a higher density of sensory receptors in the face and hands than in other parts of the body.

---

homunculus: L. *homunculus*, diminutive of *homo*, man ("little man")

1. List the different types of cutaneous receptors and state where they are located. What portion of the brain interprets tactile sensations?
2. Discuss the importance of pain. List the receptors that respond to pain and the structures of the brain that are particularly important in the perception of pain sensation.
3. Using examples, distinguish between referred pain and phantom pain. Discuss why it is important for a physician to know the referred pain sites.
4. Using a flowchart, describe the neural pathways leading from cutaneous pain and pressure receptors to the postcentral gyrus. Indicate where crossing-over occurs.

Olfactory Sense

Olfactory receptors are the dendritic endings of the olfactory nerve (I) that respond to chemical stimuli and transmit the sensation of olfaction directly to the olfactory portion of the cerebral cortex.

Objective 8 Describe the sensory pathway for olfaction.

Olfactory reception in humans is not highly developed compared to that of certain other vertebrates. Because we do not rely on smell for communicating or for finding food, the olfactory sense is probably the least important of our senses. It is more important in detecting the presence of an odor rather than its intensity. Accommodation occurs relatively rapidly with this sense. Olfaction functions closely with gustation (taste) in that the receptors for both are *chemoreceptors*, which require dissolved substances for stimuli.

Olfactory receptor cells are located in the nasal epithelium within the roof of the nasal cavity on both sides of the nasal septum (fig. 15.5). Olfactory cells are moistened by the surrounding glandular goblet cells. The cell bodies of the bipolar olfactory cells lie between the supporting columnar cells. The free end of each olfactory cell contains several dendritic endings, called **olfactory hairs** that constitute the sensitive portion of the receptor cell. These hairs respond to airborne molecules that enter the nasal cavity.

The sensory pathway for olfaction consists of several neural segments. The unmyelinated axons of the olfactory cells unite to form the **olfactory nerves,** which traverse the foramina of the cribriform plate and terminate in the paired masses of gray and white matter called the **olfactory bulbs.** The olfactory bulbs lie on both sides of the crista galli of the ethmoid bone, beneath the frontal lobes of the cerebrum. Within the olfactory bulb, neurons of the olfactory nerves synapse with dendrites of neurons forming the **olfactory tract.** Sensory impulses are conveyed along the olfactory tract and into the olfactory portion of the cerebral cortex, where they are interpreted as odor and cause the perception of smell.

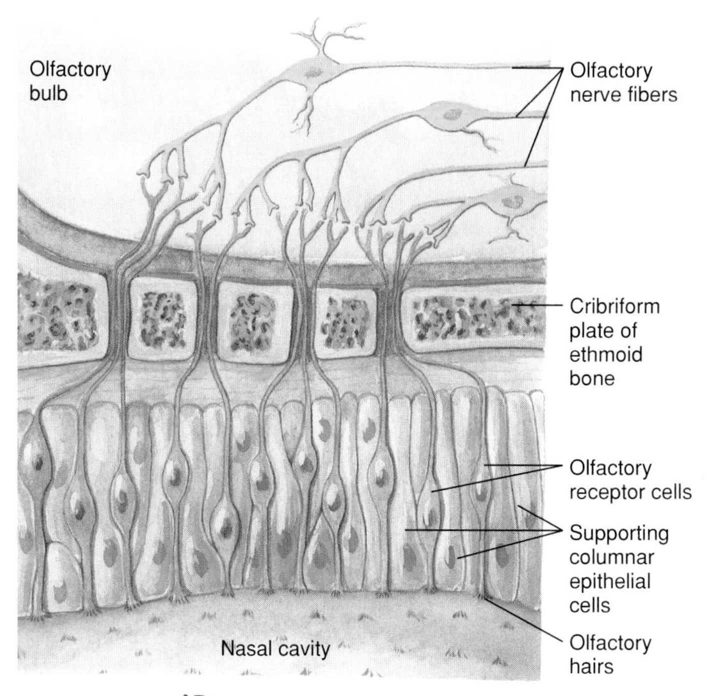

Figure 15.5

The olfactory receptor area within the roof of the nasal cavity.

Unlike taste, which is divisible into only four modalities, thousands of distinct odors can be distinguished by people who are trained in this capacity (as in the perfume industry). The molecular basis of olfaction is not understood, but it is known that a single odorant molecule is sufficient to excite an olfactory receptor.

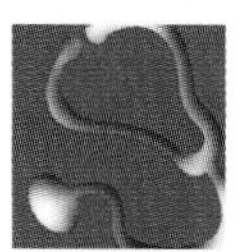

Only about 2% of inhaled air comes in contact with the olfactory receptors, which are positioned above the mainstream of airflow. Olfactory sensitivity can be increased by forceful sniffing, which draws the air into contact with the receptors.

Certain chemicals activate the trigeminal nerves (V) as well as the olfactory nerves (I) and cause reactions. Pepper, for example, may cause sneezing; onions cause the eyes to water; and smelling salts (ammonium salts) initiate respiratory reflexes and are used to revive unconscious persons.

1. What are olfactory hairs? Where are they located?
2. Trace the pathway of an olfactory stimulus from the olfactory hairs to the cerebral cortex, where interpretation occurs.

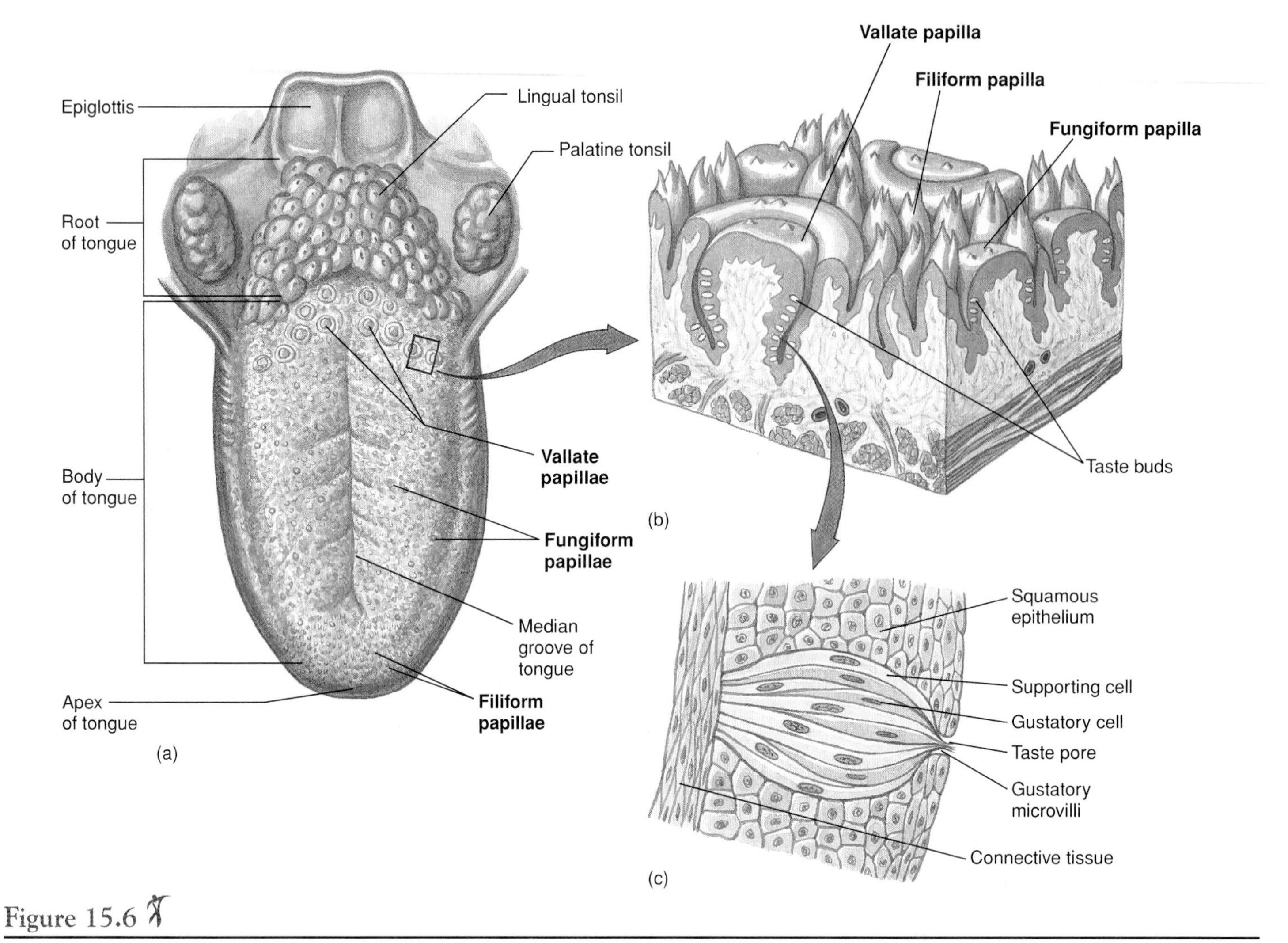

Figure 15.6

Papillae of the tongue and associated taste buds. (*a*) The surface of the tongue. (*b*)Numerous taste buds are positioned within the vallate and fungiform papillae. (*c*)Each gustatory cell and its associated gustatory microvillus are encapsuled by supporting cells.

Gustatory Sense

Taste receptors are specialized epithelial cells, clustered together in taste buds, that respond to chemical stimuli and transmit the sense of taste through the glossopharyngeal nerve (IX) or the facial nerve (VII) to the taste area in the parietal lobe of the cerebral cortex for interpretation.

Objective 9 List the three principal types of papillae and explain how they function in the perception of taste.

Objective 10 Identify the cranial nerves and the sensory pathways of gustation.

The *gustatory* (taste) *receptors* are located in the **taste buds.** Taste buds are specialized sensory organs that are most numerous on the surface of the tongue, but they are also present on the soft palate and on the walls of the oropharynx. The cylindrical taste bud is composed of numerous sensory **gustatory cells** that are encapsulated by **supporting cells** (fig. 15.6). Each gustatory cell contains a dendritic ending called a **gustatory microvillus** that projects to the surface through an opening in the taste bud called the **taste pore.** The gustatory microvilli are the sensitive portion of the receptor cells. Saliva provides the moistened environment necessary for a chemical stimulus to activate the gustatory microvilli.

Taste buds are elevated by surrounding connective tissue and epithelium to form *papillae (pă-pil'e)* (fig. 15.6). Three principal types of papillae can be identified:

1. **Vallate papillae.** The largest but least numerous are the vallate (*val'āt*) papillae, which are arranged in an inverted V-shape pattern on the back of the tongue.
2. **Fungiform papillae.** Knoblike fungiform (*fun'gĭ-form*) papillae are present on the tip and sides of the tongue.

gustatory: L. *gustare*, to taste

papillae: L. *papilla*, nipple

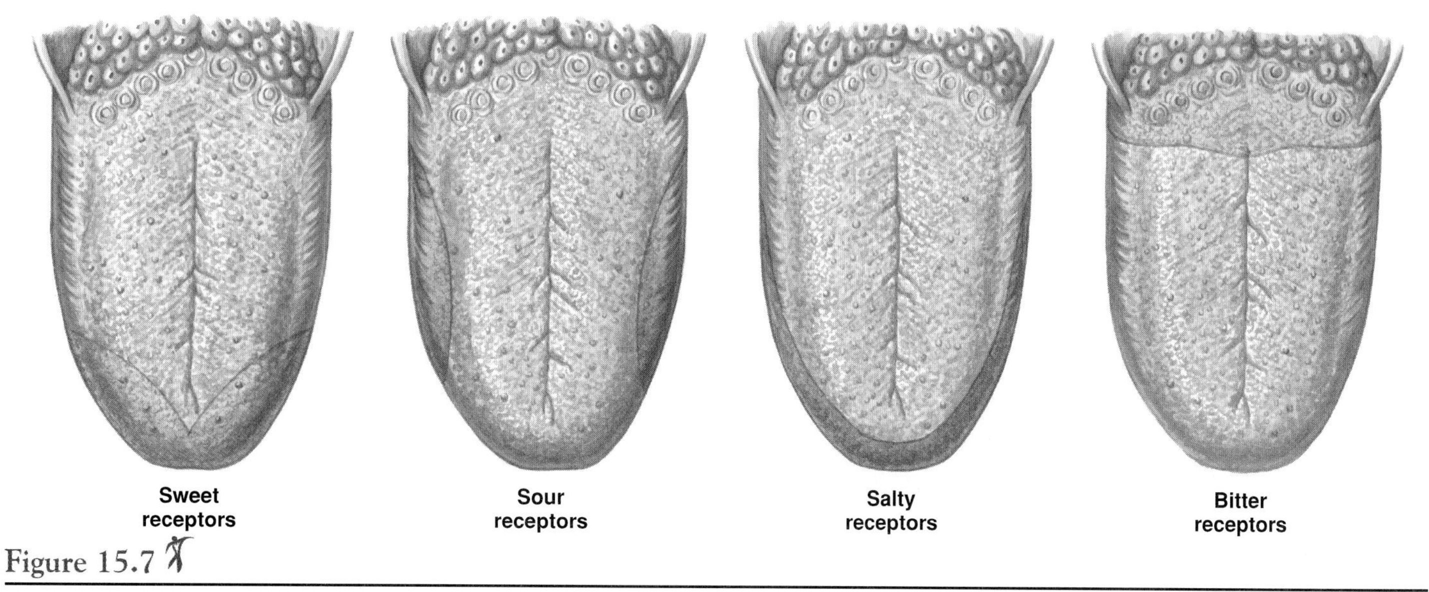

Figure 15.7

Patterns of taste receptor distribution on the surface of the tongue.

3. **Filiform papillae.** Short, thickened, threadlike filiform (*fil'ĭ-form*) papillae are located on the anterior two-thirds of the tongue.

Taste buds are found only in the vallate and fungiform papillae. The filiform papillae, although the most numerous of the human tongue papillae, are not involved in the perception of taste. Their outer cell layers are continuously converted into scalelike projections, which give the tongue surface its somewhat abrasive feel.

There are only four basic tastes, which are sensed most acutely on particular parts of the tongue (fig. 15.7). These are *sweet* (tip of tongue), *sour* (sides of tongue), *bitter* (back of tongue), and *salty* (over most of the tongue, but concentrated on the sides).

Sour taste is produced by hydrogen ions (H^+); all acids therefore taste sour. Most organic molecules, particularly sugars, taste sweet to varying degrees. Only pure table salt (NaCl) has a pure salty taste. Other salts, such as KCl (commonly used in place of NaCl by people with hypertension), taste salty but have bitter overtones. Bitter taste is evoked by quinine and seemingly unrelated molecules.

The sensory pathway that relays taste sensations to the brain mainly involves two cranial nerves (fig. 15.8). Taste buds on the posterior third of the tongue have a sensory pathway through the *glossopharyngeal nerve*, whereas the anterior two-thirds of the tongue is served by the *chorda tympani branch of the facial nerve*. Taste sensations passing through the nerves just mentioned are conveyed through the medulla oblongata and thalamus to the parietal lobe of the cerebral cortex, where they are interpreted.

Because taste and smell are both chemoreceptors, they complement each other. We often confuse a substance's smell with its taste; and if we have a head cold or hold our nose while eating, food seems to lose its flavor.

1. Distinguish between papillae, taste buds, and gustatory microvilli. Discuss the function of each as it relates to taste.
2. Describe the three principal types of papillae.
3. Which cranial nerves have sensory innervation associated with taste? What are the sensory pathways to the brain where the perception of taste occurs?

Visual Sense

Rod and cone cells are the photoreceptors within the eyeball that are sensitive to light energy. They are stimulated to transmit nerve impulses through the optic nerve and optic tract to the visual cortex of the occipital lobes, where the interpretation of vision occurs. Formation of the sensory components of the eye is complete at 20 weeks, and the accessory structures have been formed by 32 weeks.

Objective 11 Describe the accessory structures of the eye and the structure of the eyeball.

Objective 12 Trace the path of light rays through the eye and explain how they are focused on distant and near objects.

Objective 13 Describe the neural pathway of a visual impulse and discuss the neural processing of visual information.

The eyes are organs that refract (bend) and focus incoming light rays onto the sensitive photoreceptors at the back of each eye. Nerve impulses from the stimulated photoreceptors are conveyed through visual pathways within the brain to the occipital lobes of

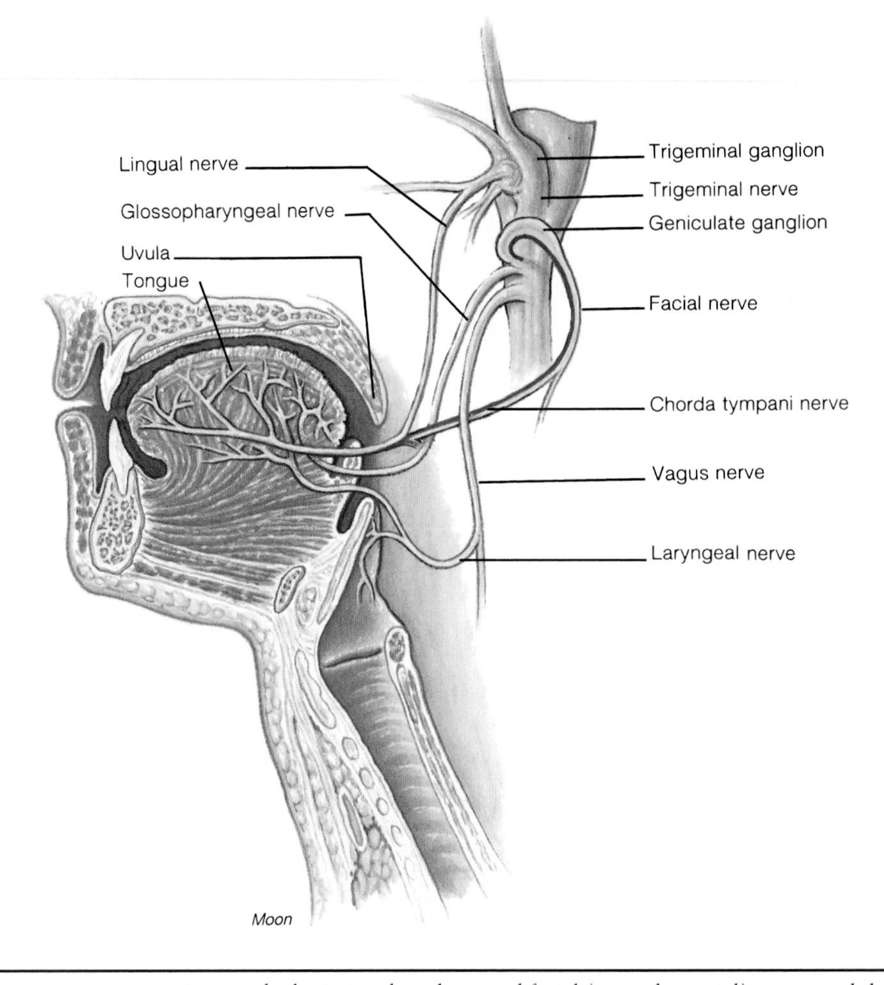

Figure 15.8

The gustatory pathway that conveys taste sensations to the brain involves the paired facial (seventh cranial) nerves and the glossopharyngeal (ninth cranial) nerves. The chorda tympani nerve is the sensory branch of the facial nerve innervating the tongue. Branches from the paired vagus (tenth cranial) nerves and the trigeminal (fifth cranial) nerves also provide some sensory innervation. The hypoglossal (twelfth cranial) nerve (not shown) provides motor innervation to the tongue. The lingual nerve transmits general sensory information from the tongue (hot, cold, pressure, and pain).

the cerebrum, where the sense of vision is perceived. The specialized photoreceptor cells can respond to an incredible 1 billion different stimuli each second. Further, these cells are sensitive to about 10 million gradations of light intensity and 7 million different shades of color.

The eyes are anteriorly positioned on the skull and set just far enough apart to achieve *binocular (stereoscopic) vision* when focusing on an object. This three-dimensional perspective allows a person to assess depth. Often likened to a camera (table 15.2), the eyes are responsible for approximately 80% of all knowledge that is assimilated.

The eyes of other vertebrates are basically similar to ours. Certain species, however, have adaptive modifications. Consider, for example, the extremely keen eyesight of a hawk, which soars high in the sky searching for food, or the eyesight of the owl, which feeds only at night. Note how the location of the eyes on the head corresponds to behavior. Predatory species, such as cats, have eyes that are directed forward, allowing depth perception. Prey species, such as deer, have eyes positioned high on the sides of their heads, allowing panoramic vision to detect distant threatening movements, even while grazing.

Table 15.2

Eye Structures and Their Camera Analogues

Eye Structure	Camera Structure
Cornea and lens	Lens system
Iris and pupil	Variable aperture system
Eyelid	Lens cap
Sclera	Camera frame
Pigment epithelium and choroid	Black interior of camera
Retina*	Film

*Since neural processing begins in the retina, this eye structure may also be considered analogous (in part) to the photographer.

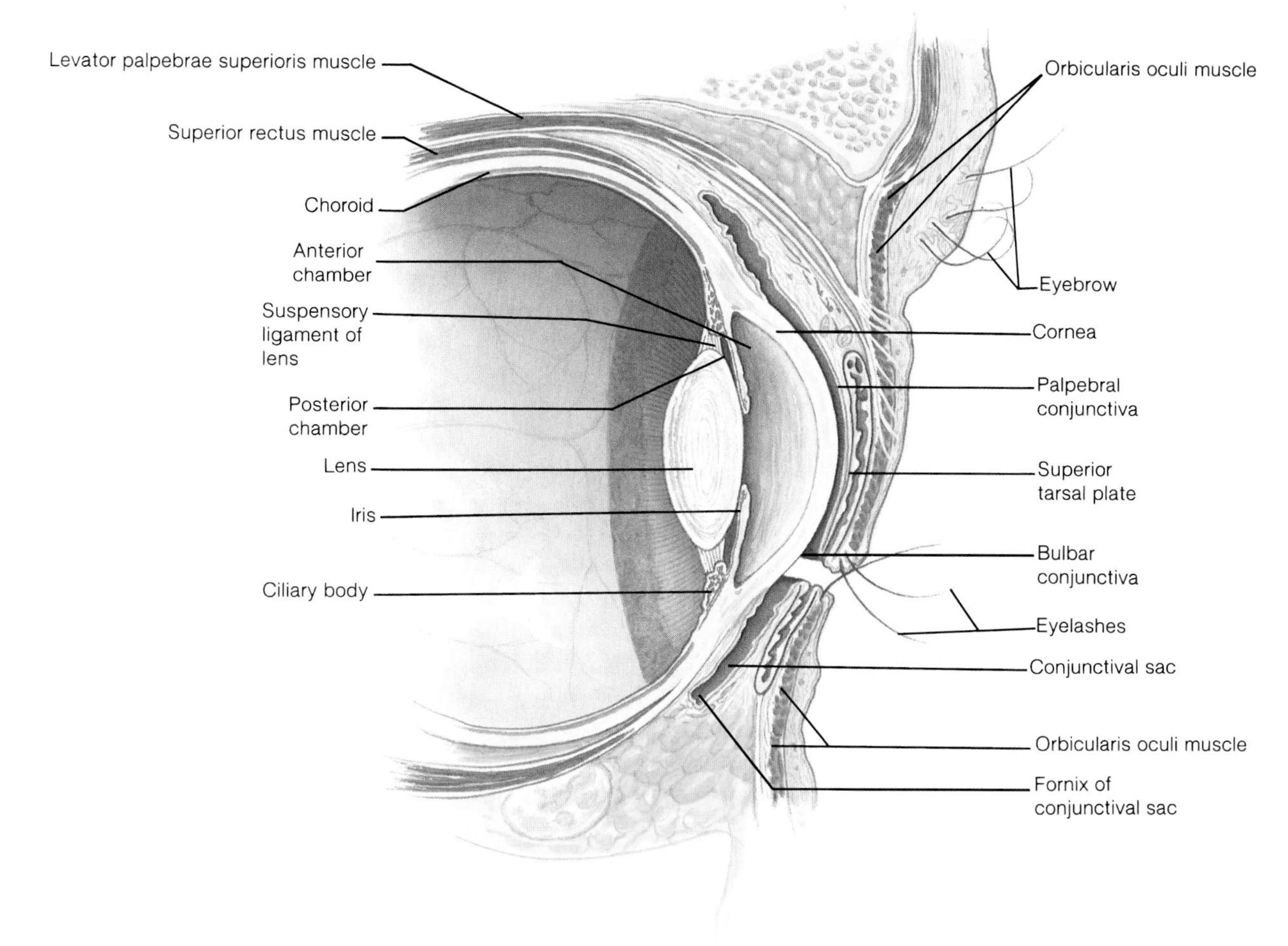

Figure 15.9

The eyeball and associated structures in sagittal section.

Accessory Structures of the Eye

Accessory structures of the eye either protect the eyeball or enable eye movement. Protective structures include the bony orbit, eyebrow, facial muscles, eyelids, eyelashes, conjunctiva, and the lacrimal apparatus that produces tears. Eyeball movements depend on the actions of the extrinsic ocular eye muscles that arise from the orbit and insert on the outer layer of the eyeball.

Orbit

Each eyeball is positioned in a bony depression in the skull called the orbit (see fig. 6.21 and table 6.5). Seven bones of the skull (frontal, lacrimal, ethmoid, zygomatic, maxilla, sphenoid, and palatine) form the walls of the orbit that support and protect the eye.

Eyebrows

Eyebrows consist of short, thick hairs positioned transversely above both eyes along the superior orbital ridges (figs. 15.9 and 15.10). Eyebrows shade the eyes from the sun and prevent perspiration or falling particles from getting into the eyes. Underneath the skin of each eyebrow is the orbital portion of the orbicularis oculi muscle and a portion of the corrugator muscle (see fig. 9.13). Contraction of either of these muscles causes the eyebrow to move, often reflexively, to protect the eye.

Eyelids and Eyelashes

Eyelids, or **palpebrae** (*pal'pĕ-bre*), develop as reinforced folds of skin with attached skeletal muscle that make them movable. In addition to the orbicularis oculi muscle attached to the skin that surrounds the front of the eye, the levator palpebrae superioris muscle attaches along the upper eyelid and provides it with greater movability than the lower eyelid. Contraction of the orbicularis oculi muscle closes the eyelids over the eye, and contraction of the levator palpebrae superioris muscle elevates the upper eyelid to expose the eye.

The eyelids protect the eyeball from desiccation by reflexively blinking about every 7 seconds and moving fluid across the anterior surface of the eyeball. Reflexively blinking as a moving object approaches the eye is obviously of great protective value. To avoid a blurred image, the eyelid will generally blink when the eyeball moves to a new position of fixation.

palpebra: L. *palpebra*, eyelid (related to *palpare*, to pat gently)

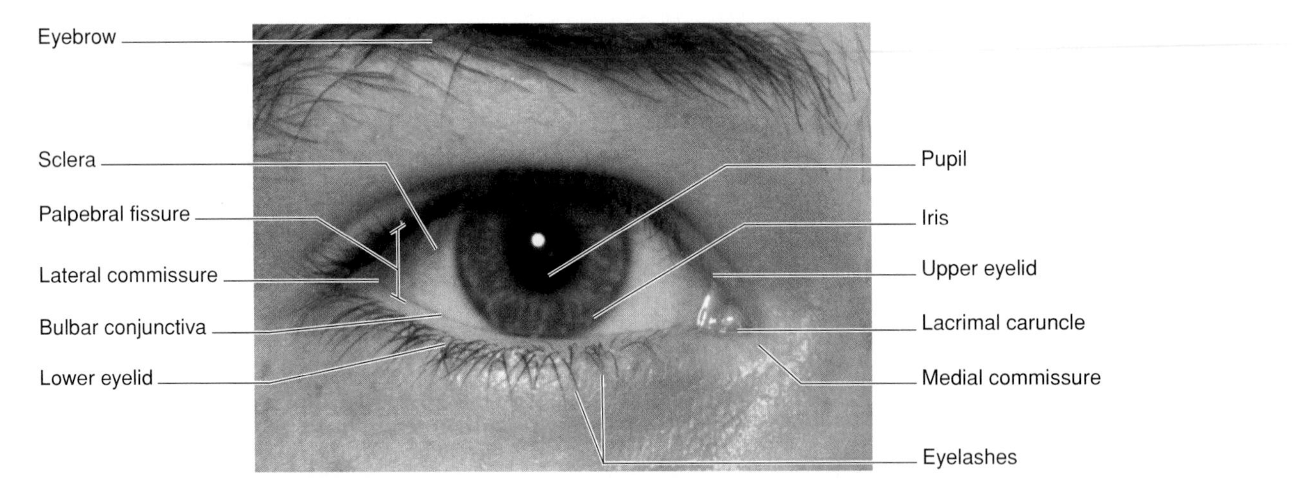

Figure 15.10

The surface anatomy of the eye.

The **palpebral fissure** (fig. 15.10) is the space between the upper and lower eyelids. The shape of the palpebral fissure is elliptical when the eyes are open. The **commissures** (*canthi*) of the eye are the medial and lateral angles where the eyelids come together. The **medial commissure,** which is broader than the **lateral commissure,** is characterized by a small, reddish, fleshy elevation called the **lacrimal caruncle** (*kar'ung-kul*) (fig. 15.11). The lacrimal caruncle contains sebaceous and sudoriferous glands; it produces the whitish secretion, commonly called "sleep dust" that sometimes collects during sleep.

In people of Asian descent, a fold of skin of the upper eyelid, called the *epicanthic fold,* may normally cover part of the medial commissure. An epicanthic fold may also be present in some infants with Down syndrome.

Each eyelid supports a row of **eyelashes** that protect the eye from airborne particles. The shaft of each eyelash is surrounded by a root hair plexus that makes the hair sensitive enough to cause a reflexive closure of the lids. Eyelashes of the upper lid are long and turn upward, whereas those of the lower lid are short and turn downward.

In addition to the layers of the skin and the underlying connective tissue and orbicularis oculi muscle fibers, each eyelid contains a tarsal plate, tarsal glands, and conjunctiva. The **tarsal plates,** composed of dense regular connective tissue, are important in maintaining the shape of the eyelids (fig. 15.9). Specialized sebaceous glands called **tarsal glands** are embedded in the tarsal plates along the exposed inner surfaces of the eyelids. The ducts of the tarsal glands open onto the edges of the eyelids, and their oily secretions help to keep the eyelids from sticking to each other. Modified sweat glands called **ciliary glands** are also located within the eyelids, along with additional sebaceous glands at the bases of the hair follicles of the eyelashes. An infection of these sebaceous glands is referred to as a *sty* (also spelled *stye*).

commissure: L. *commissura,* a joining
caruncle: L. *caruncula,* diminutive of *caro,* flesh
tarsal: Gk. *tarsos,* flat basket

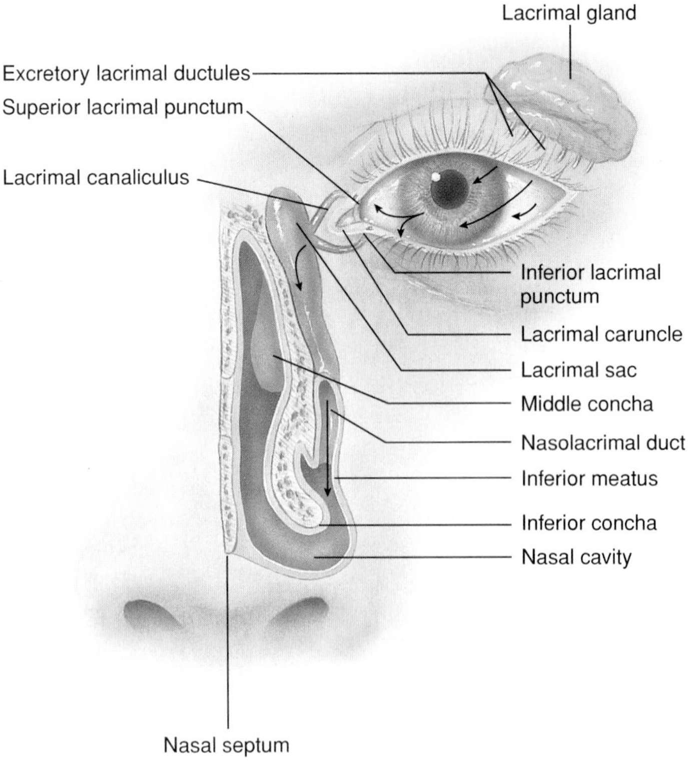

Figure 15.11

The lacrimal apparatus consists of the lacrimal gland, which produces lacrimal fluid (tears), and a series of ducts through which the lacrimal fluid drains into the nasal cavity. Lacrimal fluid moistens and cleanses the conjunctiva that lines the interior surface of the eyelids and covers the exposed anterior surface of the eyeball.

Conjunctiva

The conjunctiva (*con"jungk-ti'vă*) is a thin mucus-secreting epithelial membrane that lines the interior surface of each eyelid and exposed anterior surface of the eyeball (see fig. 15.9). It consists of stratified squamous epithelium that varies in thickness in different regions. The **palpebral conjunctiva** is thick and adheres

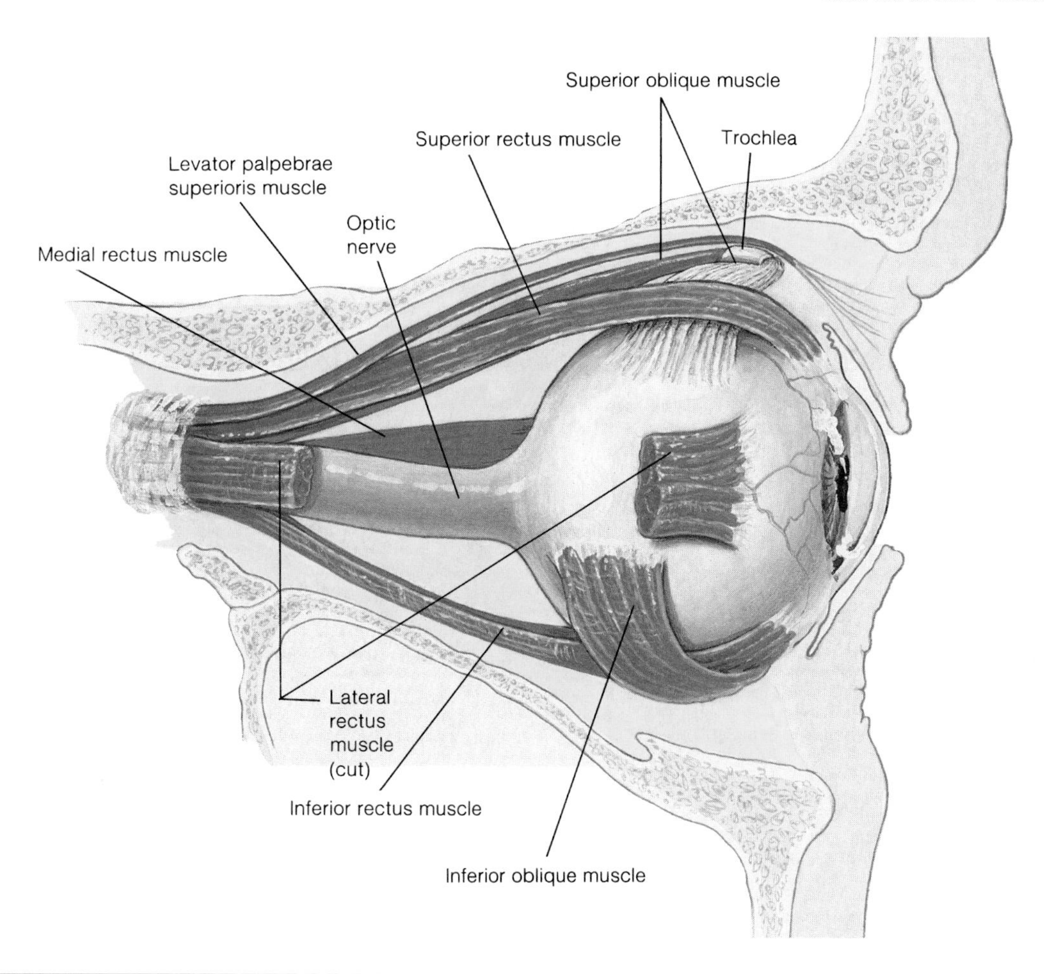

Figure 15.12

The extrinsic ocular muscles of the right eyeball.

to the tarsal plates of the eyelids. Where the conjunctiva reflects onto the anterior surface of the eyeball, it is known as the **bulbar conjunctiva.** This portion is transparent and especially thin where it covers the cornea. Because the conjunctiva is continuous from the eyelids to the anterior surface of the eyeball, a space called the **conjunctival sac** is present when the eyelids are closed. The conjunctival sac protects the eyeball by preventing foreign objects from passing beyond the confines of the sac. The conjunctiva heals rapidly if scratched.

Lacrimal Apparatus

The lacrimal apparatus consists of the lacrimal gland, which secretes the *lacrimal fluid* (tears), and a series of ducts that drain the lacrimal fluid into the nasal cavity (fig. 15.11). The **lacrimal gland,** which is about the size and shape of an almond, is located in the superolateral portion of the orbit. It is a compound tubuloacinar gland that secretes lacrimal fluid through several excretory lacrimal ductules into the conjunctival sac of the upper eyelid. With each blink of the eyelids, lacrimal fluid is spread over the surface of the eye—much like windshield wipers spread windshield washing fluid. Lacrimal fluid drains into two small openings, called **lacrimal puncta,** on both sides of the lacrimal caruncle. From here, the lacrimal fluid drains through the **superior** and **inferior lacrimal canaliculi** (*kan"ă-lik'yŭ-li*) into the **lacrimal sac** and continues through the **nasolacrimal duct** to the inferior meatus of the nasal cavity (fig. 15.10).

Lacrimal fluid is a lubricating mucus secretion that contains a bactericidal substance called *lysozyme*. Lysozyme reduces the likelihood of infections. Normally, about 1 milliliter of lacrimal fluid is produced each day by the lacrimal gland of each eye. If irritating substances, such as particles of sand or chemicals from onions, make contact with the conjunctiva, the lacrimal glands secrete greater volumes. The extra lacrimal fluid protects the eye by diluting and washing away the irritating substance.

Humans are the only animals known to weep in response to emotional stress. While crying, the volume of lacrimal secretion is so great that the tears may spill over the edges of the eyelids and the nasal cavity fill with fluid. The crying response is an effective means of communicating one's emotions and results from stimulation of the lacrimal glands by parasympathetic motor neurons of the facial nerve.

Extrinsic Eye Muscles

The movements of the eyeball are controlled by six extrinsic eye muscles called the **extrinsic ocular muscles** (figs. 15.12 and 15.13).

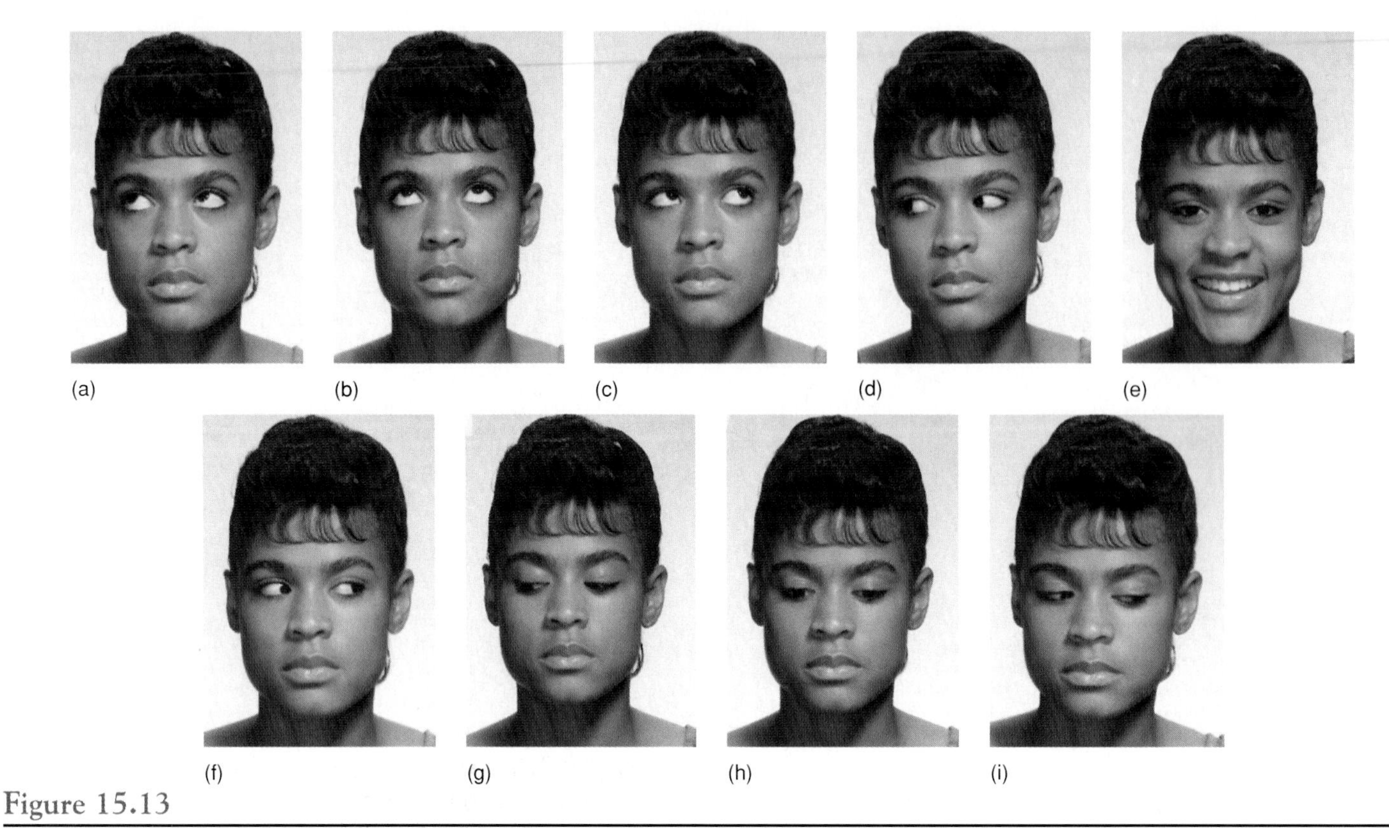

Figure 15.13

The positions of the eyes as the ocular muscles are contracted. (*a*) Right eye, inferior oblique muscle; left eye, superior and medial recti muscles. (*b*) Both eyes, superior recti and inferior oblique muscles. (*c*) Right eye, superior and medial recti muscles; left eye, inferior oblique muscle. (*d*) Right eye, lateral rectus muscle; left eye, medial rectus muscle. (*e*) Primary position with the eyes fixed on a distant fixation point. (*f*) Right eye, medial rectus muscle; left eye, lateral rectus muscle. (*g*) Right eye, superior oblique muscle; left eye, inferior and medial recti muscles. (*h*) Both eyes, inferior recti and superior oblique muscles. (*i*) Right eye, inferior and medial recti muscles; left eye, superior oblique muscle.

Each extrinsic ocular muscle originates from the bony orbit and inserts by a tendinous attachment to the tough outer tunic of the eyeball. Four *recti muscles* (singular, *rectus*) maneuver the eyeball in the direction indicated by their names **(superior, inferior, lateral,** and **medial),** and two **oblique muscles (superior** and **inferior)** rotate the eyeball on its axis (see also fig. 9.17). One of the extrinsic ocular muscles, the superior oblique, passes through a pulleylike cartilaginous loop called the **trochlea** (*trok'le-ă*) before attaching to the eyeball. Although stimulation of each muscle causes a precise movement of the eyeball, most of the movements involve the combined contraction of usually two or more muscles.

The motor units of the extrinsic ocular muscles are the smallest in the body. This means that a single motor neuron serves about 10 muscle fibers, resulting in precise movements. The eyes move in synchrony by contracting synergistic muscles while relaxing antagonistic muscles.

The extrinsic ocular muscles are innervated by three cranial nerves (table 15.3). Innervation of the other skeletal and smooth muscles that serve the eye is also indicated in table 15.3.

trochlea: Gk. *trochos*, a wheel

Table 15.3
Muscles of the Eye

Extrinsic Muscles (Skeletal Muscles)

Muscle	Nerve	Action
Superior rectus	Oculomotor nerve (III)	Rotates eye upward and toward midline
Inferior rectus	Oculomotor nerve (III)	Rotates eye downward and toward midline
Medial rectus	Oculomotor nerve (III)	Rotates eye toward midline
Lateral rectus	Abducens nerve (VI)	Rotates eye away from midline
Superior oblique	Trochlear nerve (IV)	Rotates eye downward and away from midline
Inferior oblique	Oculomotor nerve (III)	Rotates eye upward and away from midline

Intrinsic Muscles (Smooth Muscles)

Muscle	Nerve	Action
Ciliary muscle	Oculomotor nerve (III) parasympathetic fibers	Causes suspensory ligament to relax
Pupillary constrictor muscle	Oculomotor nerve (III) parasympathetic fibers	Causes pupil to constrict
Pupillary dilator muscle	Sympathetic fibers	Causes pupil to dilate

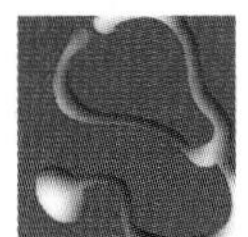

A physical examination may include an eye movement test. As the patient's eyes follow the movement of a physician's finger, the physician can assess weaknesses in specific muscles or dysfunctions of specific cranial nerves. The patient experiencing *double vision (diplopia)* when moving his eyes may be suffering from muscle weakness. Looking laterally tests the abducens nerve; looking inferiorly and laterally tests the trochlear nerve; and crossing the eyes tests the oculomotor nerves of both eyes.

Structure of the Eyeball

The eyeball of an adult is essentially spherical, approximately 25 mm (1 in.) in diameter. About four-fifths of the eyeball lies within the orbit of the skull. The eyeball consists of three basic layers: the *fibrous tunic*, the *vascular tunic*, and the *internal tunic* (fig. 15.14).

Fibrous Tunic

The fibrous tunic is the outer layer of the eyeball. It is divided into two regions: the posterior five-sixths is the opaque *sclera* and the anterior one-sixth is the transparent *cornea* (fig. 15.15).

The toughened **sclera** (*skler'ă*) is the white of the eye. It is composed of tightly bound elastic and collagenous fibers that give shape to the eyeball and protect its inner structures. The sclera is avascular but does contain sensory receptors for pain. The large **optic nerve** exits through the sclera at the back of the eyeball.

The transparent **cornea** is convex, so that it refracts (bends in a converging pattern) incoming light rays. The transparency of the cornea is due to tightly packed, avascular dense connective tissue. Also, the relatively few cells that are present in the cornea are arranged in unusually regular patterns. The circumferential edge of the cornea is continuous structurally with the sclera. The outer surface of the cornea is covered with a thin, nonkeratinized stratified squamous epithelial layer called the **anterior corneal epithelium,** which is actually a continuation of the bulbar conjunctiva of the sclera (see fig. 15.9).

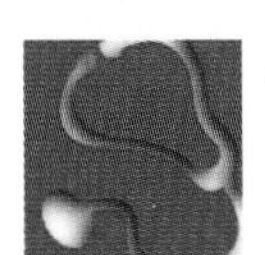

A defective cornea can be replaced with a donor cornea in a surgical procedure called a *corneal transplant (keratoplasty)*. A defective cornea is one that does not transmit or refract light effectively because of its shape, scars, or disease. During a corneal transplant, the defective cornea is excised and replaced with a transplanted cornea that is sutured into place. It is considered to be the most successful type of homotransplant (between individuals of the same species).

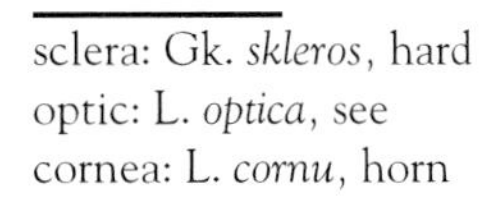

sclera: Gk. *skleros*, hard
optic: L. *optica*, see
cornea: L. *cornu*, horn

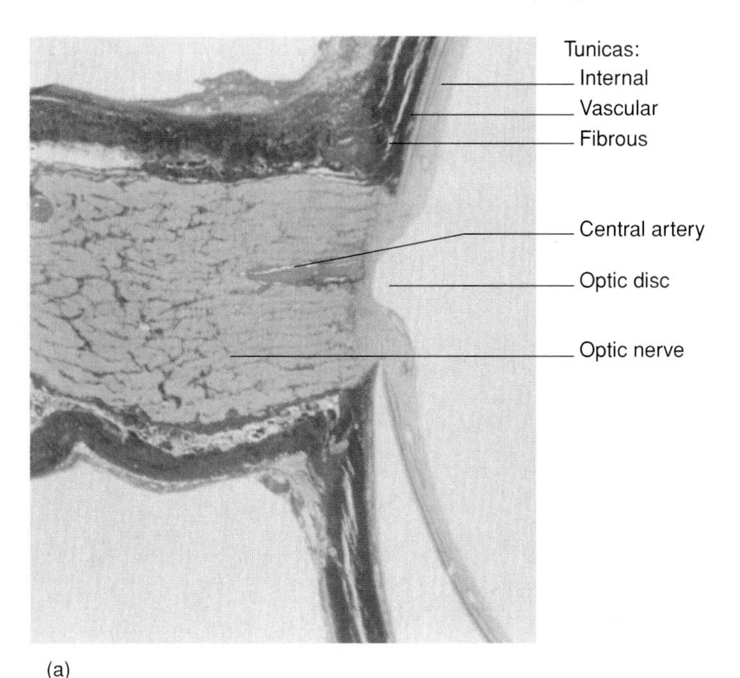

(a)

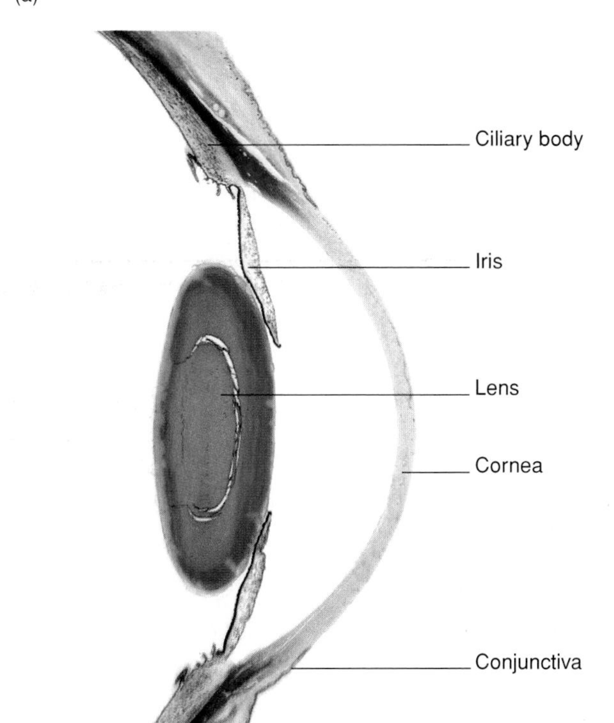

(b)

Figure 15.14

Photomicrographs of the eyeball. (*a*) A posterior portion showing the tunics of the eye, the optic disc, and the optic nerve (7×) and (*b*) an anterior portion showing the cornea ciliary body, iris, and the lens (7×).

Vascular Tunic

The vascular tunic, or **uvea** (*yoo've-ă*) of the eyeball, consists of the *choroid*, the *ciliary body*, and the *iris* (fig. 15.15).

uvea: L. *uva*, grape

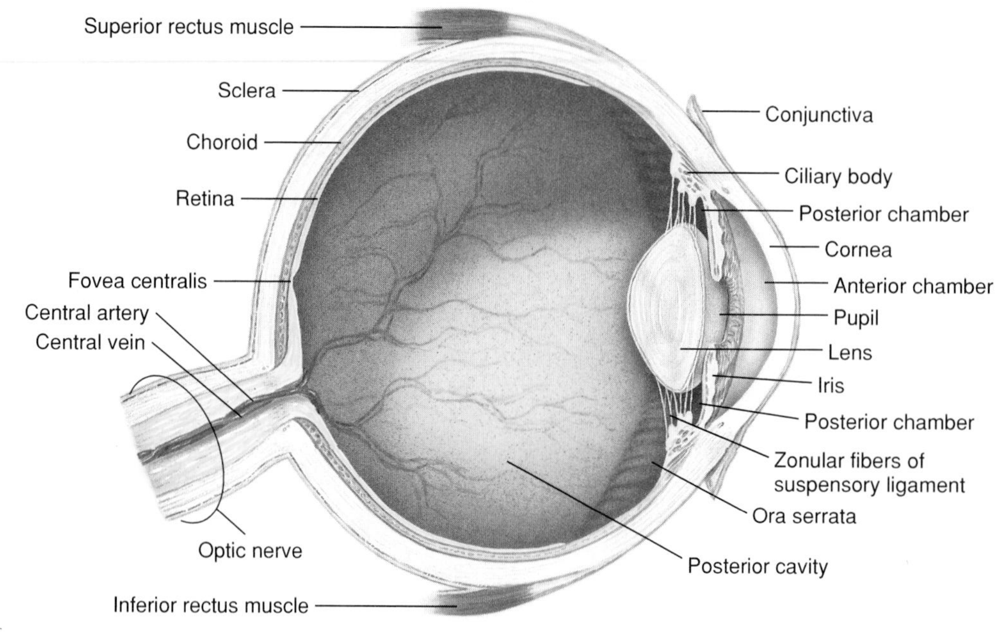

Figure 15.15

A sagittal section of the eyeball.

The **choroid** (*kor'oid*) is a thin, highly vascular layer that lines most of the internal surface of the sclera. The choroid contains numerous pigment-producing melanocytes, which give it a brownish color that prevents light rays from being reflected out of the eyeball. There is an opening in the choroid at the back of the eyeball where the optic nerve is located.

The **ciliary body** is the thickened anterior portion of the vascular tunic that forms an internal muscular ring toward the front of the eyeball (fig. 15.16). Bands of smooth muscle fibers, collectively called the **ciliary muscles** are found within the ciliary body. Numerous extensions of the ciliary body called **ciliary processes** attach to the **zonular fibers,** which in turn attach to the *lens capsule*. Collectively, the zonular fibers constitute the **suspensory ligament.** The transparent **lens** consists of tight layers of protein fibers arranged like the layers of an onion. A thin, clear **lens capsule** encloses the lens and provides attachment for the suspensory ligament (see figs. 15.16 and 15.24).

The shape of the lens determines the degree to which the light rays that pass through will be refracted. Constant tension of the suspensory ligament, when the ciliary muscles are relaxed, flattens the lens somewhat (fig. 15.17). Contraction of the ciliary muscles relaxes the suspensory ligament and makes the lens more spherical. The constant tension within the lens capsule causes the surface of the lens to become more convex when the suspensory ligament is not taut. A flattened lens permits viewing of a distant object, whereas a rounded lens permits viewing of a close object.

The **iris** is the anterior portion of the vascular tunic and is continuous with the choroid. The iris is viewed from the outside as the colored portion of the eyeball (figs. 15.15 and 15.16). It consists of smooth muscle fibers arranged in a circular and a radial pattern. The contraction of the smooth muscle fibers regulates the diameter of the **pupil** (table 15.4 and fig. 15.18), an opening in the center of the iris. Contraction of the pupillary constrictor muscle of the iris, stimulated by bright light, constricts the pupil and diminishes the amount of light entering the eyeball (see fig. 13.8). Contraction of the pupillary dilator muscle, in response to dim light, enlarges the pupil and permits more light to enter.

The amount of dark pigment, melanin, in the iris is what determines its color. In newborns, melanin is concentrated in the folds of the iris, so that all newborn babies have blue eyes. After a few months, the melanin moves to the surface of the iris and gives the baby his or her permanent eye color, ranging from steel blue to dark brown.

Internal Tunic (Retina)

The **retina** (*ret'ĭ-nă*) covers the choroid as the innermost layer of the eye (fig. 15.15). It consists of an outer **pigmented layer,** in contact with the choroid, and an inner **nervous layer,** or *visual portion*. The thick nervous layer of the retina terminates in a jagged margin near the ciliary body called the **ora serrata**

choroid: Gk. *chorion*, membrane
zonular: L. *zona*, a girdle

iris: Gk. *irid*, rainbow
ora serrata: L. *ora*, margin; *serra*, saw

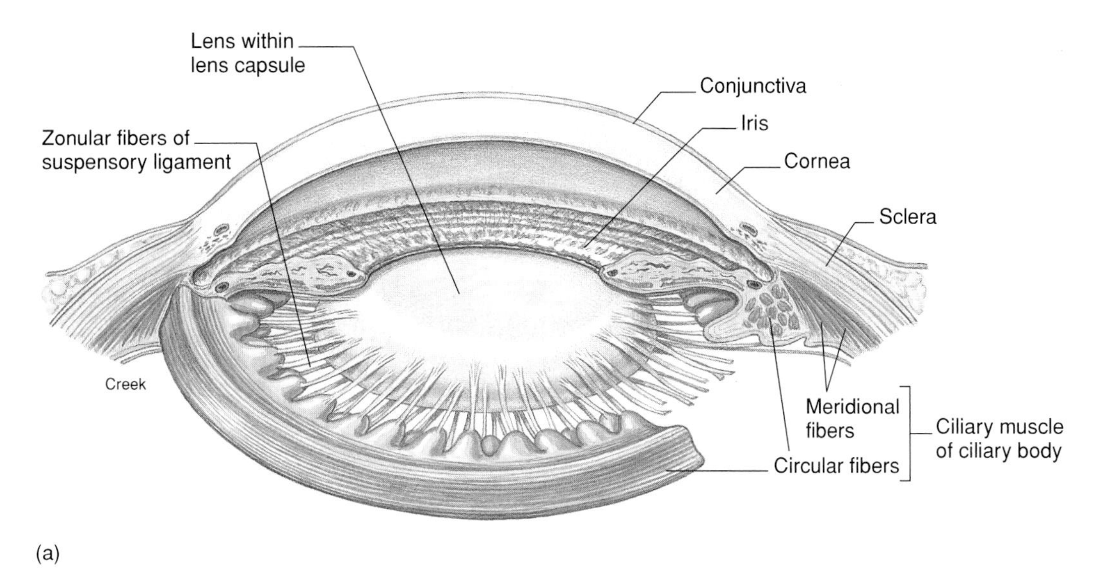

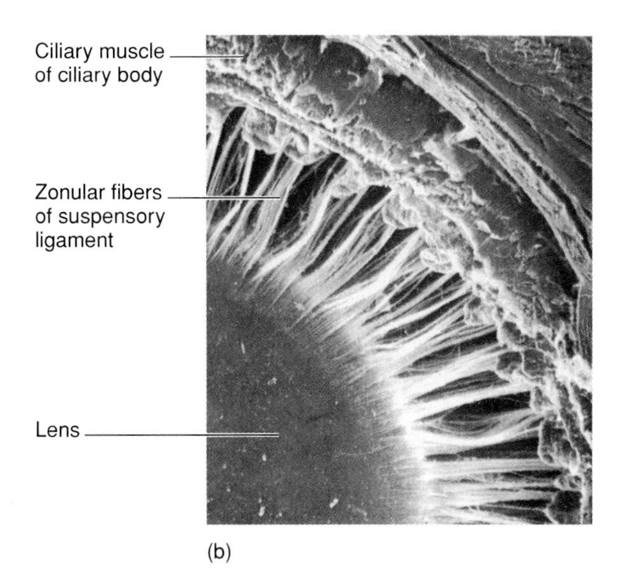

Figure 15.16

The structure of the anterior portion of the eyeball. (*a*) The relationship between the lens, zonular fibers, and ciliary muscles of the eye and (*b*) a scanning electron micrograph in anterior view showing that same relationship.

(*o'ră ser-ra'tă*). The thin pigmented layer extends anteriorly over the back of the ciliary body and iris.

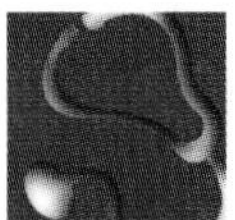

The pigmented layer and nervous layer of the retina are not attached to each other, except where they surround the optic nerve and at the ora serrata. Because of this loose connection, the two layers may become separated as a *detached retina*. Such a separation can be corrected by fusing the layers with a laser.

The nervous layer of the retina is composed of three principal layers of neurons. Listing them in the order in which they conduct impulses, they are the *rod* and *cone cells*, *bipolar neurons*, and *ganglion neurons* (fig. 15.19). In terms of the passage of light, however, the order is reversed. Light must first pass through the layer of ganglion cells and then the layer of bipolar cells before reaching and stimulating the rod and cone cells.

Rod and **cone cells** are photoreceptors. Rod cells number over 100 million per eye and are more slender and elongated than cone cells (fig. 15.20). Rod cells are positioned on the peripheral parts of the retina, where they respond to dim light for black-and-white vision. They also respond to form and movement but provide poor visual acuity. Cone cells, which number about 7 million per eye, provide daylight color vision and greater visual acuity. The photoreceptors synapse with **bipolar neurons,**

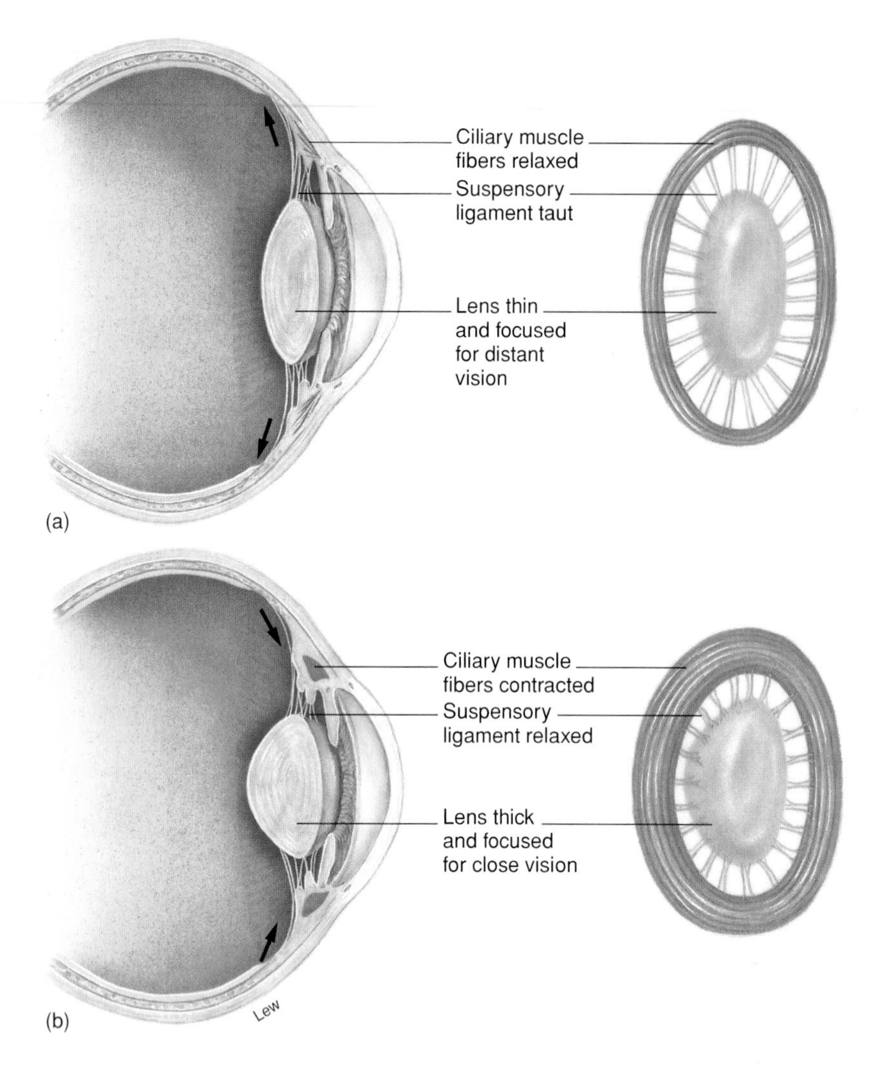

Figure 15.17

Changes in the shape of the lens to bring light rays into sharp focus on the retina. (*a*) The lens is flattened for distant vision when the ciliary muscle fibers are relaxed and the suspensory ligament is taut. (*b*) The lens is more spherical for close-up vision when the ciliary muscle fibers are contracted and the suspensory ligament is relaxed.

which in turn synapse with the **ganglionic neurons.** The axons of ganglionic neurons leave the eye as the optic nerve.

Cone cells are concentrated in a depression near the center of the retina called the **fovea centralis,** which is the area of keenest vision (figs. 15.15 and 15.21). Surrounding the fovea centralis is the yellowish **macula lutea** (*mak'yŭ-lă loo'te-ă*), which also has an abundance of cone cells (fig. 15.21). There are no photoreceptors in the area where the optic nerve is attached to the eyeball. This area is a *blind spot* and is referred to as the **optic disc** (figs. 15.21 and 15.22). A person is normally unaware of the blind spot because (1) the eyes continually move about, (2) an object is viewed from a different angle with each eye, and (3) the image of an object that falls on the blind spot of one retina will fall on receptors of the other retina. The blind spot can easily be demonstrated as described in figure 15.23.

fovea: L. *fovea*, small pit
macula lutea: L. *macula*, spot; *luteus*, yellow

Blood Supply to the Eyeball

Both the choroid and the retina are richly supplied with blood. Two **ciliary arteries** pierce the sclera at the posterior aspect of the eyeball and traverse the choroid to the ciliary body and base of the iris. Although the ciliary arteries enter the eyeball independently, they anastomose (connect) extensively throughout the choroid.

The **central artery** (central retinal artery) branches from the ophthalmic artery and enters the eyeball in contact with the optic nerve. As the central artery passes through the optic disc, it divides into superior and inferior branches, each of which then divides into temporal and nasal branches to serve the inner layers of the retina (see fig. 15.15). The **central vein** drains blood from the eyeball through the optic disc. The

Table 15.4
Summary of Structures Associated with the Eyeball

Tunic and Structure	Location	Composition	Function
Fibrous Tunic	Outer layer of eyeball	Avascular connective tissue	Gives shape to eyeball
Sclera	Posterior outer layer; white of eye	Tightly bound elastic and collagen fibers	Supports and protects eyeball
Cornea	Anterior surface of eyeball	Tightly packed dense connective tissue—transparent and convex	Transmits and refracts light
Vascular Tunic (Uvea)	Middle layer of eyeball	Highly vascular pigmented tissue	Supplies blood; prevents reflection
Choroid	Middle layer in posterior portion of eyeball	Vascular layer	Supplies blood to eyeball
Ciliary body	Anterior portion of vascular tunic	Smooth muscle fibers and glandular epithelium	Supports the lens through suspensory ligament and determines its thickness; secretes aqueous humor
Iris	Anterior portion of vascular tunic, continuous with ciliary body	Pigment cells and smooth muscle fibers	Regulates the diameter of pupil, and hence the amount of light entering the posterior cavity
Internal Tunic	Inner layer of eyeball	Tightly packed photoreceptors, neurons, blood vessels, and connective tissue	Provides location and support for rod and cone cells
Retina	Principal portion of internal tunica	Photoreceptor neurons (rod and cone cells), bipolar neurons, and ganglion neurons	Photoreception; transmits impulses
Lens (not part of any tunic)	Between anterior and posterior chambers; supported by suspensory ligament of ciliary body	Tightly arranged protein fibers; transparent	Refracts light and focuses onto fovea centralis

branches of the central artery can be observed within the eyeball through an ophthalmoscope (fig. 15.21).

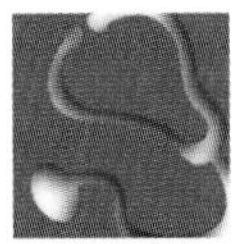

An examination of the internal eyeball with an ophthalmoscope is frequently part of a routine physical examination. Arterioles can be seen within the eyeball. If they appear abnormal (for example, constricted, dilated, or hemorrhaged), they may be symptomatic of certain diseases or body dysfunctions. Diseases such as arteriosclerosis, diabetes, cataracts, or glaucoma can be detected by examining the internal eyeball.

Cavities and Chambers of the Eyeball

The interior of the eyeball is separated by the lens into an *anterior cavity* and a *posterior cavity* (see fig. 15.15). The anterior cavity is subdivided by the iris into an **anterior chamber** and a **posterior chamber** (see fig. 15.15). The anterior chamber is located between the cornea and the iris. The posterior chamber is located between the iris and the suspensory ligament and lens. The anterior and posterior chambers connect through the pupil and are filled with a watery fluid called **aqueous humor.** The constant production of aqueous humor maintains an *intraocular pressure* of about 12 mmHg within the anterior and posterior chambers. Aqueous humor also provides nutrients and oxygen to the avascular lens and cornea. An estimated 5.5 ml of aqueous humor is secreted each day from the vascular epithelium of the ciliary body (fig. 15.24). From its site of secretion within the posterior chamber, the aqueous humor passes through the pupil into the anterior chamber. From here, it drains from the eyeball through the **scleral venous sinus** (canal of Schlemm) into the bloodstream. The scleral venous sinus is located at the junction of the cornea and iris.

The large posterior cavity is filled with a transparent jellylike **vitreous humor.** Vitreous humor contributes to the intraocular pressure that maintains the shape of the eyeball and holds the retina against the choroid. Unlike aqueous humor, vitreous humor is not continuously produced; rather, it is formed prenatally. Additional vitreous humor forms as a person's eyes become larger through normal body growth.

The structures within the eyeball are summarized in table 15.4.

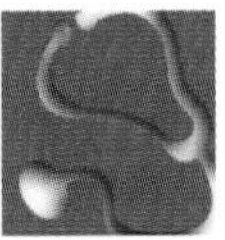

Puncture wounds to the eyeball are especially dangerous and frequently cause blindness. Protective equipment such as goggles, shields, and shatterproof lenses should be used in hazardous occupations and certain sports. If the eye is punctured, the main thing to remember is to *leave the object in place* if it is still impaling the eyeball. Removal may allow the fluids to drain from the eyeball, causing loss of intraocular pressure, a detached retina, and possibly blindness.

canal of Schlemm: from Friedrich S. Schlemm, German anatomist, 1795–1858
vitreous: L. *vitreus*, glassy

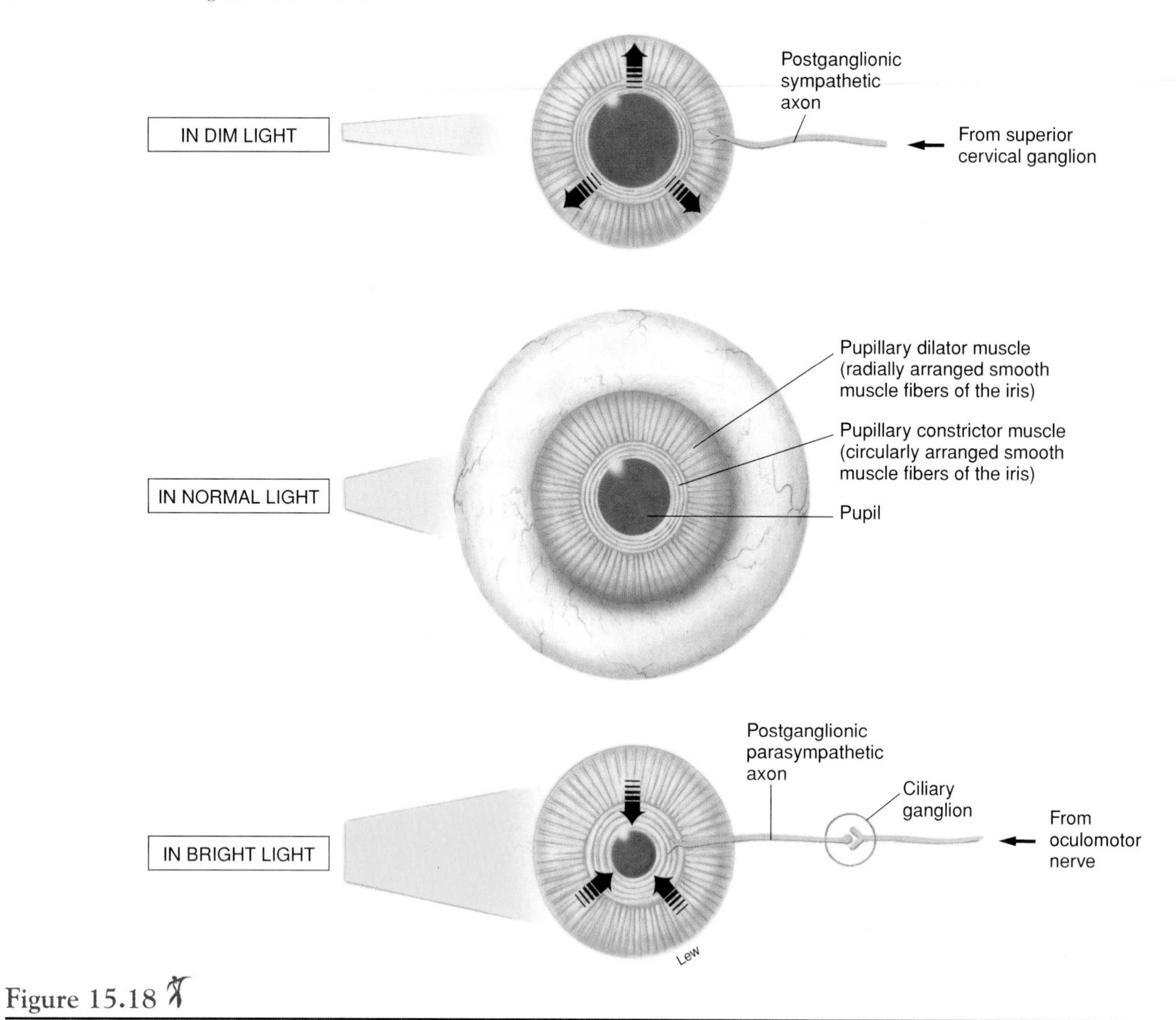

Figure 15.18

Dilation and constriction of the pupil. In dim light, the radially arranged smooth muscle fibers are stimulated to contract by sympathetic stimulation, dilating the pupil. In bright light, the circularly arranged smooth muscle fibers are stimulated to contract by parasympathetic stimulation, constricting the pupil.

Function of the Eyeball

The focusing of light rays and stimulation of photoreceptors of the retina require five basic processes:

1. *transmission of light rays* through transparent media of the eyeball;
2. *refraction of light rays* through media of different densities;
3. *accommodation of the lens* to focus the light rays;
4. *constriction of the pupil* by the iris to regulate the amount of light entering the vitreous chamber; and
5. *convergence of the eyeballs*, so that visual acuity is maintained.

Visual impairment may result if one or more of these processes does not function properly (see Clinical Considerations).

Transmission of Light Rays

Light rays entering the eyeball pass through four transparent media before they stimulate the photoreceptors. In sequence, the media through which light rays pass are the cornea, aqueous humor, lens, and vitreous humor. The cornea and lens are solid media composed of tightly packed, avascular protein fibers. An additional thin, transparent membranous continuation of the conjunctiva covers the outer surface of the cornea. The aqueous humor is a low-viscosity fluid, whereas the vitreous humor is jellylike in consistency.

Refraction of Light Rays

Refraction is the bending of light rays. Refraction occurs as light rays pass at an oblique angle from a medium of one optical density to a medium of a different optical density. The convex cornea is

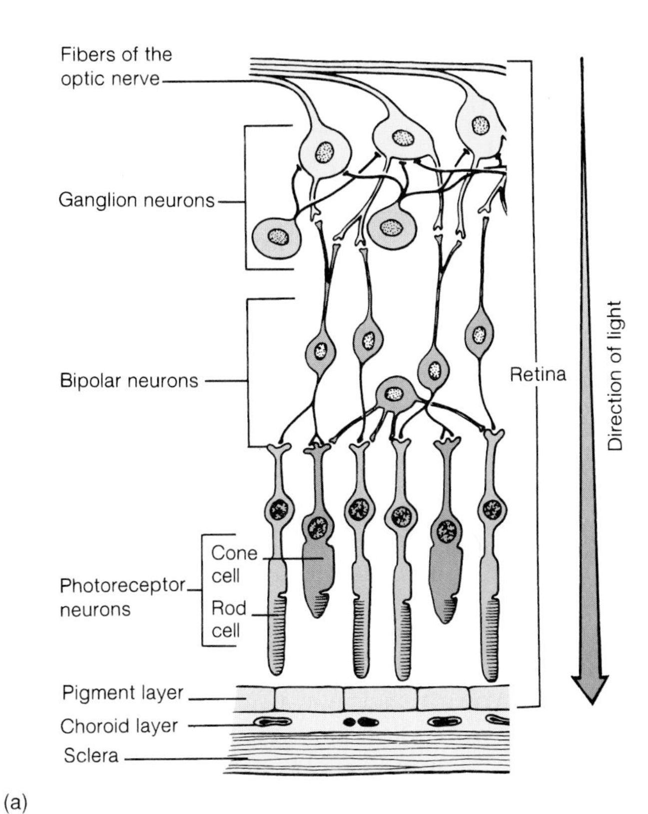

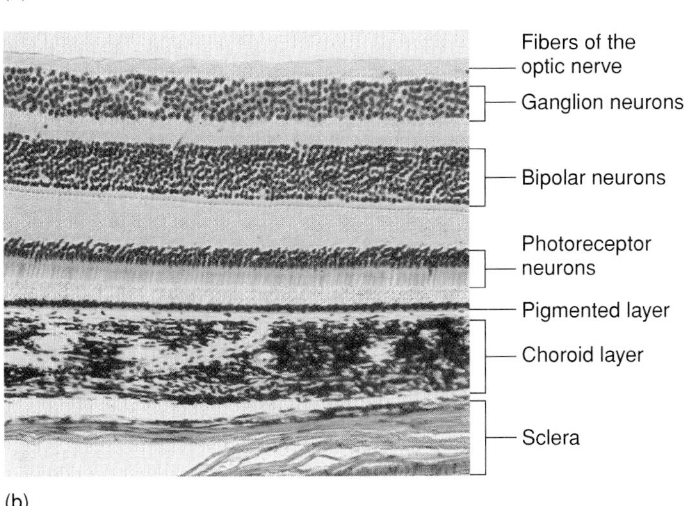

Figure 15.19

The layers of the retina. The retina is inverted, so that light must pass through various layers of nerve cells before reaching the photoreceptors (rod cells and cone cells). (*a*) A schematic diagram and (*b*) a light micrograph.

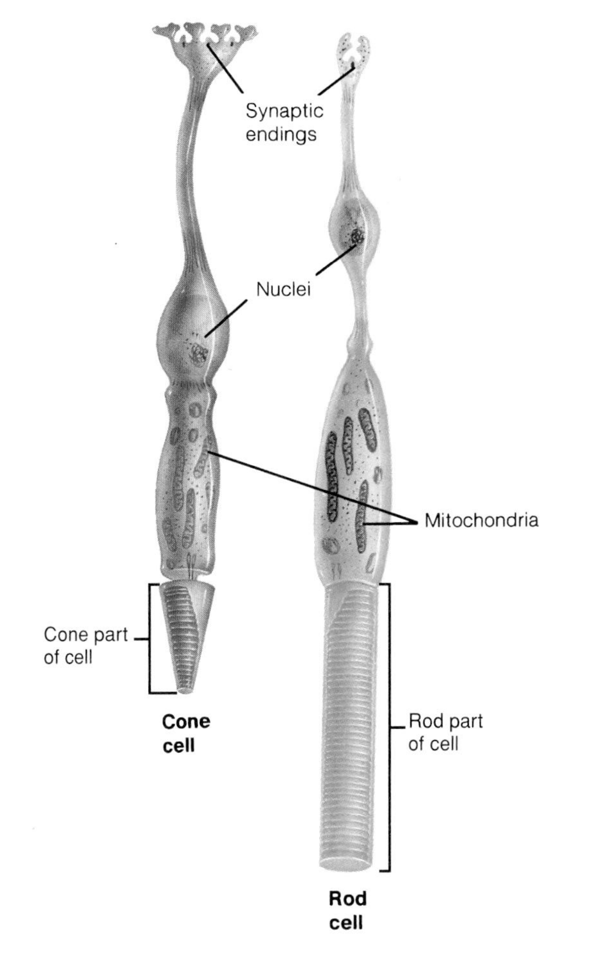

Figure 15.20

Photoreceptor cells of the retina.

the principal refractive medium; the fluids within the various chambers produce minimal refraction. The lens is particularly important for refining and altering refraction. Of the refractive media, only the lens can be altered in shape to achieve precise refraction.

The refraction of light rays is so extensive that the visual image is formed upside down on the retina (fig. 15.25). Nerve impulses of the image in this position are relayed to the visual cortex of the occipital lobe, where the inverted image is interpreted as right side up.

Accommodation of the Lens

Accommodation is the automatic adjustment of the curvature of the lens by contraction of ciliary muscles to bring light rays into sharp focus on the retina. The lens of the eyeball is biconvex. When an object is viewed from a distance of less than about 20 feet, the lens must make an adjustment, or accommodation, for clear focus on the retina. Contraction of the smooth muscle fibers of the ciliary body causes the suspensory ligament to relax and the lens to become thicker (see fig. 15.17). A thicker, more convex lens causes the greater refraction of light required for viewing close objects.

Constriction of the Pupil

Constriction of the pupil occurs through parasympathetic stimulation that causes the pupillary constrictor muscles of the iris to contract (see fig. 15.18). Pupillary constriction is important for two reasons. One is that it reduces the amount of light that enters the posterior cavity. A reflexive constriction of the pupil protects the retina from sudden or intense bright light. More important, a reduced pupil diameter prevents light rays from entering the posterior cavity through the periphery of the lens.

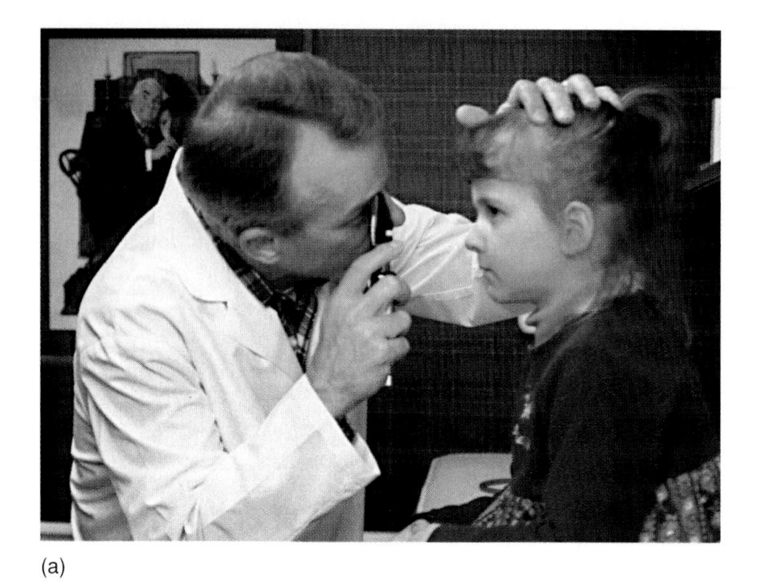

(a)

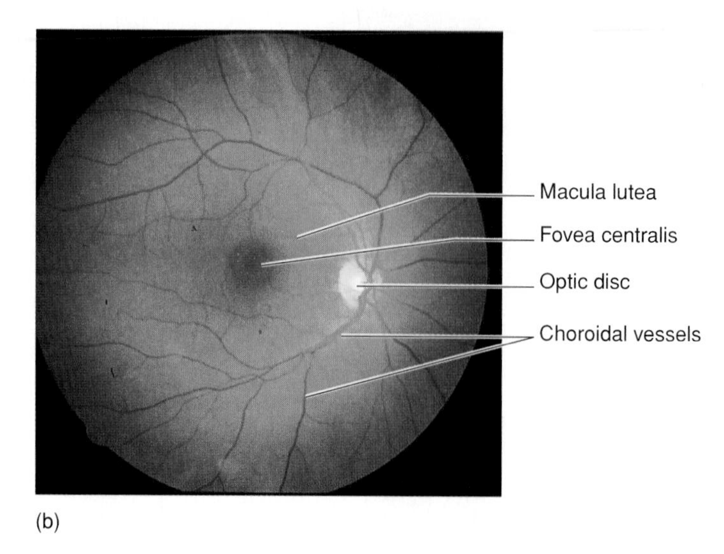

(b)

Figure 15.21

(*a*) A physician viewing the internal anatomy of the eyeball. (*b*) The appearance of the retina as viewed with an ophthalmoscope. Optic nerve fibers leave the eyeball at the optic disc to form the optic nerve. (Note the blood vessels that can be seen entering the eyeball at the optic disc.)

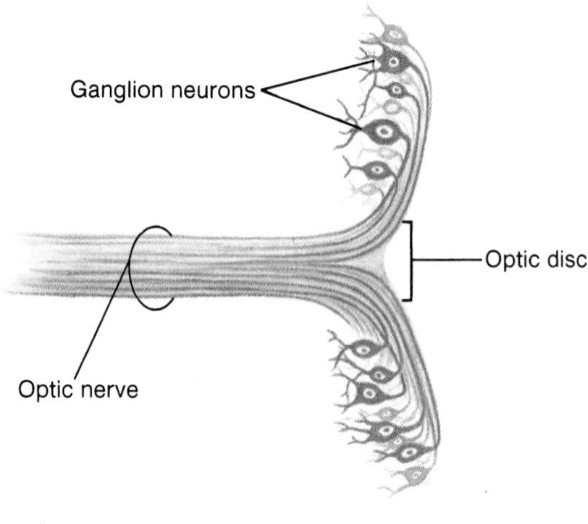

Figure 15.22

The optic disc is a small area of the retina where the fibers of the ganglion neurons emerge to form the optic nerve. The optic disc is frequently called the blind spot because it is devoid of rod and cone cells.

Light rays refracted from the periphery would not be brought into focus on the retina and would cause blurred vision. Autonomic constriction of the pupil and accommodation of the lens occur simultaneously.

Convergence of the Eyeballs

Convergence refers to the medial rotation of the eyeballs when fixating on a close object. In fact, focusing on an object close to the tip of the nose causes a person to appear cross-eyed. The eyeballs must converge when viewing close objects because only then can the light rays focus on the same portions in both retinas.

Figure 15.23

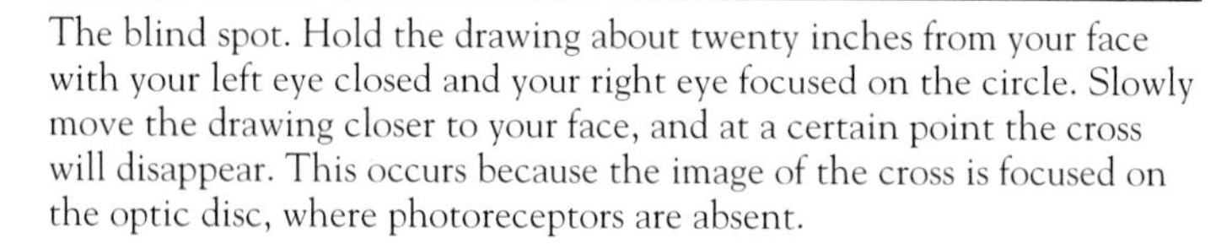

The blind spot. Hold the drawing about twenty inches from your face with your left eye closed and your right eye focused on the circle. Slowly move the drawing closer to your face, and at a certain point the cross will disappear. This occurs because the image of the cross is focused on the optic disc, where photoreceptors are absent.

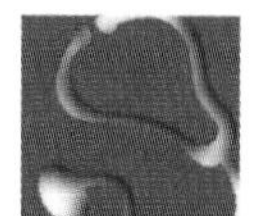

Amblyopia (am″ble-o′pe-ă) ex anopsia, commonly called "lazy eye," is a condition of ocular muscle weakness. This causes a deviation of one eye, so that there is not a concurrent convergence of both eyeballs. With this condition, two images are received by the optic portion of the cerebral cortex—one of which is suppressed to avoid *diplopia (double vision),* or images of unequal clarity. A person who has amblyopia will experience dimness of vision and partial loss of sight. Amblyopia is frequently tested for in young children; if it is not treated before the age of 6, little can be done to strengthen the afflicted muscle.

Visual Spectrum

The eyes transduce the energy of the *electromagnetic spectrum* (fig. 15.26) into nerve impulses. Only a limited part of this spectrum can excite the photoreceptors. Electromagnetic energy with wavelengths between 400 and 700 nanometers (nm) constitute *visible light*. Light of longer wavelengths, which are in the infrared regions of the spectrum, does not have sufficient energy to excite photoreceptors but is perceived as heat. Ultraviolet light, which has shorter wavelengths and more energy than visible light, is filtered out by the yellow color of the eye's lens. Certain insects, such as honeybees, and people who have had their lenses removed, can see light in the ultraviolet range.

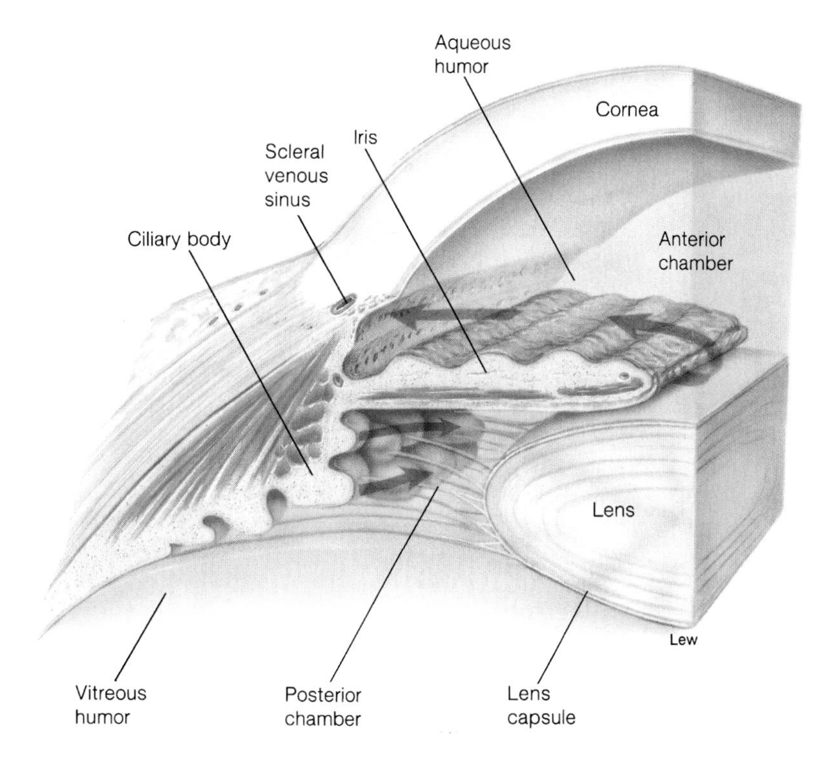

Figure 15.24

Aqueous humor maintains the intraocular pressure within the anterior and posterior chambers of the anterior cavity of the eyeball. It is secreted into the posterior chamber, flows through the pupil into the anterior chamber, and drains from the eyeball through the scleral venous sinus.

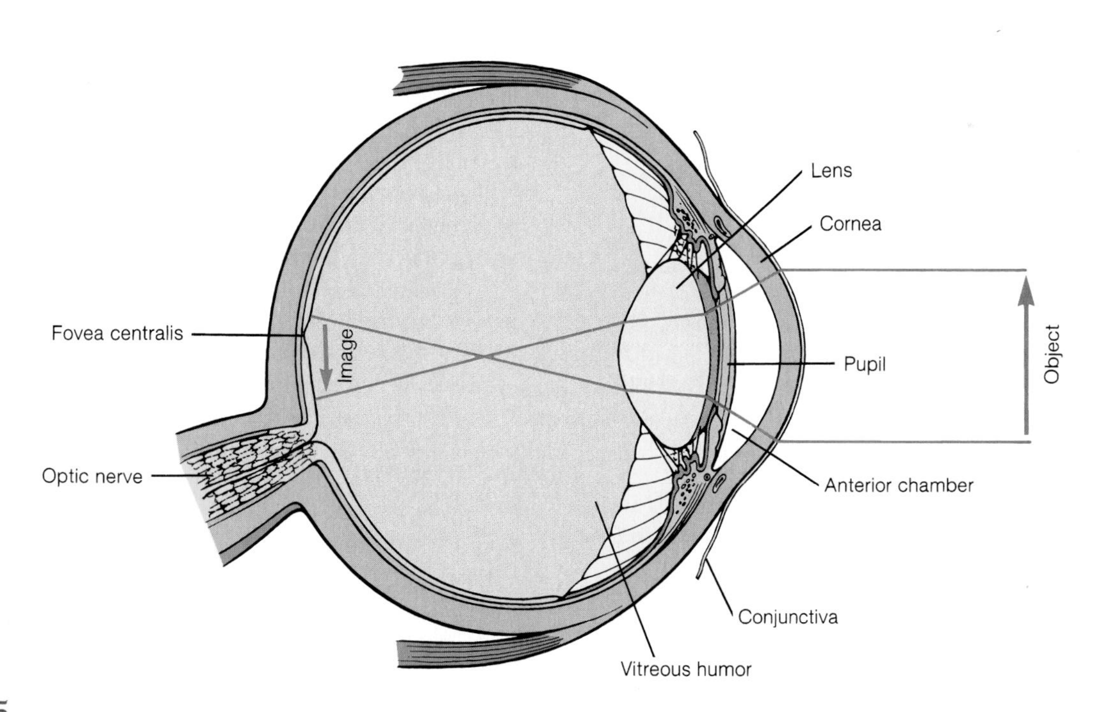

Figure 15.25

The refraction of light waves within the eyeball causes the image of an object to be inverted on the retina.

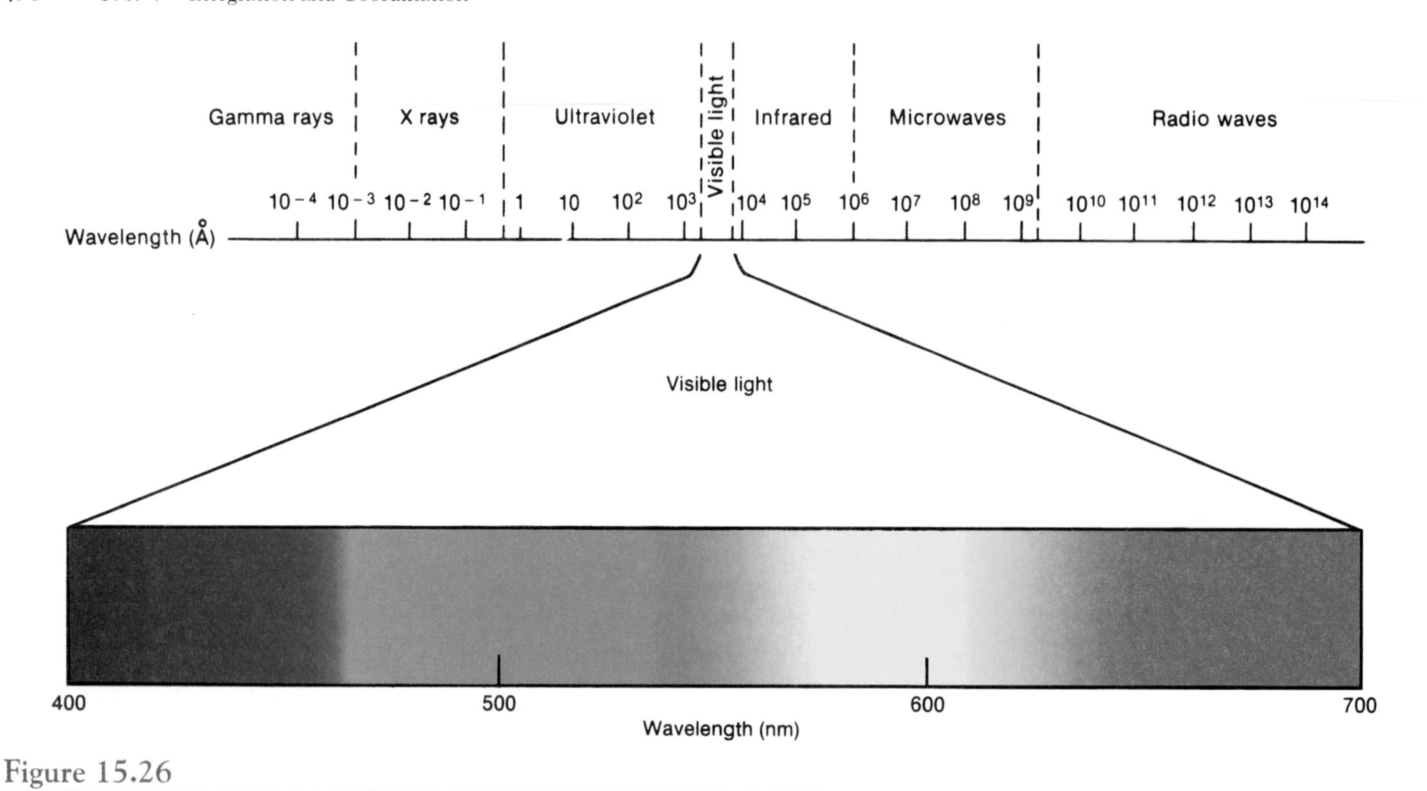

Figure 15.26

The electromagnetic spectrum (*top*) is shown in Angstrom units (1 Å = 10^{-10} meter). The visible spectrum (*bottom*) constitutes only a small range of this spectrum and is shown in nanometer units (1 nm = 10^{-9} meter).

Three distinct and specialized types of cone cells within the retina permit color vision (fig. 15.26). Different photosensitive pigments enable each type to absorb light rays primarily in the blue, green, or red portion of the color spectrum. Blue cone cells are stimulated by wavelengths between 400 and 550 nm; green cone cells are stimulated by wavelengths between 450 and 550 nm; and red cone cells are stimulated by wavelengths between 500 and 700 nm. *Color blindness* is the inability to distinguish colors, particularly reds and greens. It affects about 5% of the U.S. population. Color blindness for the majority of these people is a misnomer, however, since in nearly all cases there is a deficiency in the number of specific cone cells, not a total lack of them. Hence, there is some ability to distinguish color.

A driver with red-green color blindness—the most common type—is able to distinguish traffic signals without difficulty because yellow has been added to the red light, and blue has been added to the green.

Neural Pathways for Vision, Eye Movements, and Processing Visual Information

The photoreceptor neurons, rod and cone cells, are the functional units of sight in that they respond to light rays and produce nerve impulses. Nerve impulses from the rod and cone cells pass through bipolar neurons to ganglion neurons (see fig. 15.19). The optic nerve consists of axons of aggregated ganglion neurons that emerge through the posterior aspect of the eyeball. The two optic nerves (one from each eyeball) converge at the **optic chiasma** (*ki-as'mă*) (fig. 15.27). At this point, all of the fibers arising from the medial (nasal) half of each retina cross to the opposite side. Those fibers of the optic nerve that arise from the lateral (temporal) half of the retina do not cross, however. The **optic tract** is a continuation of optic nerve fibers from the optic chiasma. It is composed of fibers arising from the retinas of both eyeballs.

As the optic tracts enter the brain, some of the fibers in the tracts terminate in the **superior colliculi** (*ko-lik'yoo-li*). These fibers and the motor pathways they activate constitute the **tectal system,** which is responsible for body-eye coordination.

Approximately 70%–80% of the fibers in the optic tract pass to the **lateral geniculate** (*jĕ-nik'yŭ-lit*) **body** of the thalamus (fig. 15.27). Here the fibers synapse with neurons whose axons constitute a pathway called the **optic radiation.** The optic radiation transmits impulses to the **striate cortex** area of the occipital lobe of the cerebrum. This arrangement of visual fibers, known as the **geniculostriate system,** is responsible for perception of the visual field.

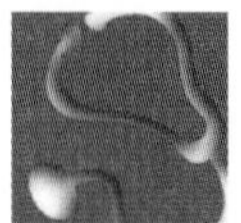

The nerve fibers that cross at the optic chiasma arise from the retinas in the medial portions of both eyes. The photoreceptors of these fibers are stimulated by light entering the eyeball from the periphery. If the optic chiasma is cut longitudinally, peripheral vision will be lost, leaving only "tunnel vision." If an optic tract is cut, both eyes will be partially blind, the lateral field of vision will be lost for one eye, and the medial field of vision lost for the other.

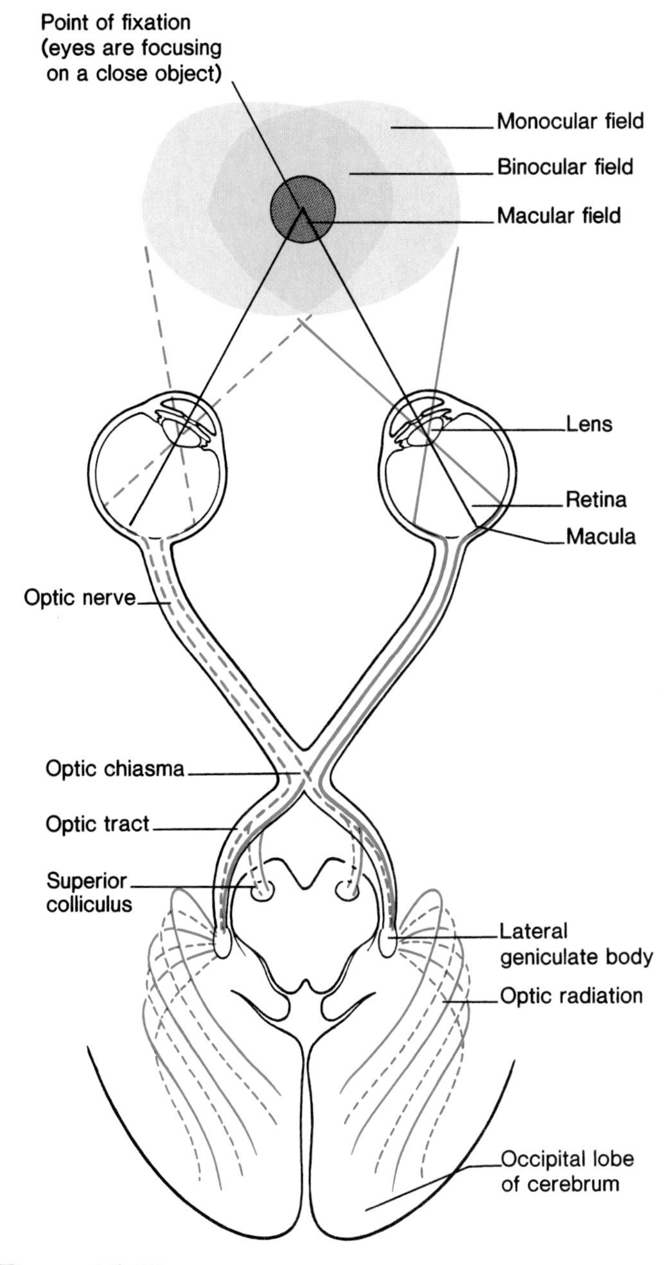

Figure 15.27

Visual fields of the eyes and neural pathways for vision. An overlapping of the visual field of each eye provides binocular vision—the ability to perceive depth.

Superior Colliculi and Eye Movements

Neural pathways from the superior colliculi to motor neurons in the spinal cord help to mediate the startle response to the sight, for example, of an unexpected intruder. Other nerve fibers from the superior colliculi stimulate the **extrinsic eye muscles** (see table 15.3), which are the skeletal muscles that move the eyes.

Two types of eye movements are coordinated by the superior colliculi. *Smooth pursuit movements* track moving objects and keep the image focused on the fovea centralis. *Saccadic (să-kad'ik) eye movements* are quick (lasting 20–50 msec), jerky movements that occur while the eyes appear to be still. These saccadic movements are believed to be important in maintaining visual acuity.

The tectal system is also involved in the control of the **intrinsic eye muscles**—the iris and the muscles of the ciliary body. Shining a light into one eye stimulates the *pupillary reflex* in which both pupils constrict. This is caused by activation of parasympathetic neurons by fibers from the superior colliculi. Postganglionic neurons in the ciliary ganglia behind the eyes, in turn, stimulate constrictor fibers in the iris. Contraction of the ciliary body during *accommodation* also involves stimulation of the superior colliculi.

Processing of Visual Information

For visual information to have meaning, it must be associated with past experience and integrated with information from other senses. Some of this higher processing occurs in the inferior temporal lobes of the cerebral cortex. Experimental removal of these areas from monkeys impairs their ability to remember visual tasks that they previously learned and hinders their ability to associate visual images with the significance of the objects viewed. Monkeys with their inferior temporal lobes removed, for example, will fearlessly handle a snake. The symptoms produced by loss of the inferior temporal lobes are known as *Klüver–Bucy syndrome*.

In an attempt to reduce the symptoms of severe epilepsy, surgeons at one time would cut the corpus callosum in some patients. This fiber tract, as previously described, transmits impulses between the right and left cerebral hemispheres. The right cerebral hemisphere of patients with such *split brains* would therefore, receive sensory information only from the left half of the external world. The left hemisphere, similarly cut off from communication with the right hemisphere, would receive sensory information only from the right half of the external world. In some situations, these patients would behave as if they had two separate minds.

Experiments with split-brain patients have revealed that the two hemispheres have separate abilities. This is true even though each hemisphere normally receives input from both halves of the external world through the corpus callosum. If the sensory image of an object, such as a key, is delivered only to the left hemisphere (by showing it only to the right visual field), the object can be named. If the object is presented to the right cerebral cortex, the person knows what the object is but cannot name it. Experiments such as this suggest that (in right-handed people) the left hemisphere is needed for language and the right hemisphere is responsible for pattern recognition.

Klüver–Bucy syndrome: from Heinrich Klüver, German neurologist, 1897–1979 and Paul C. Bucy, American neurologist, b. 1904

DEVELOPMENTAL EXPOSITION

The Eye

Exhibit I The development of the eye. (*a*) An anterior view of the developing head of a 22-day-old embryo and the formation of the optic vesicle from the neuroectoderm of the prosencephalon (forebrain). (*b*) The development of the optic cup. The lens vesicle is formed (*c*) as the ectodermal lens placode invaginates during the fourth week. The hyaloid vessels become enclosed within the optic nerve (c_1 and e_1) as there is fusion of the optic fissure. (*d*) The basic shape of the eyeball and the position of its internal structures are established during the fifth week. The successive development of the eye is shown at 6 weeks (*e*) and at 20 weeks (*f*), respectively. (*g*) The eye of the newborn.

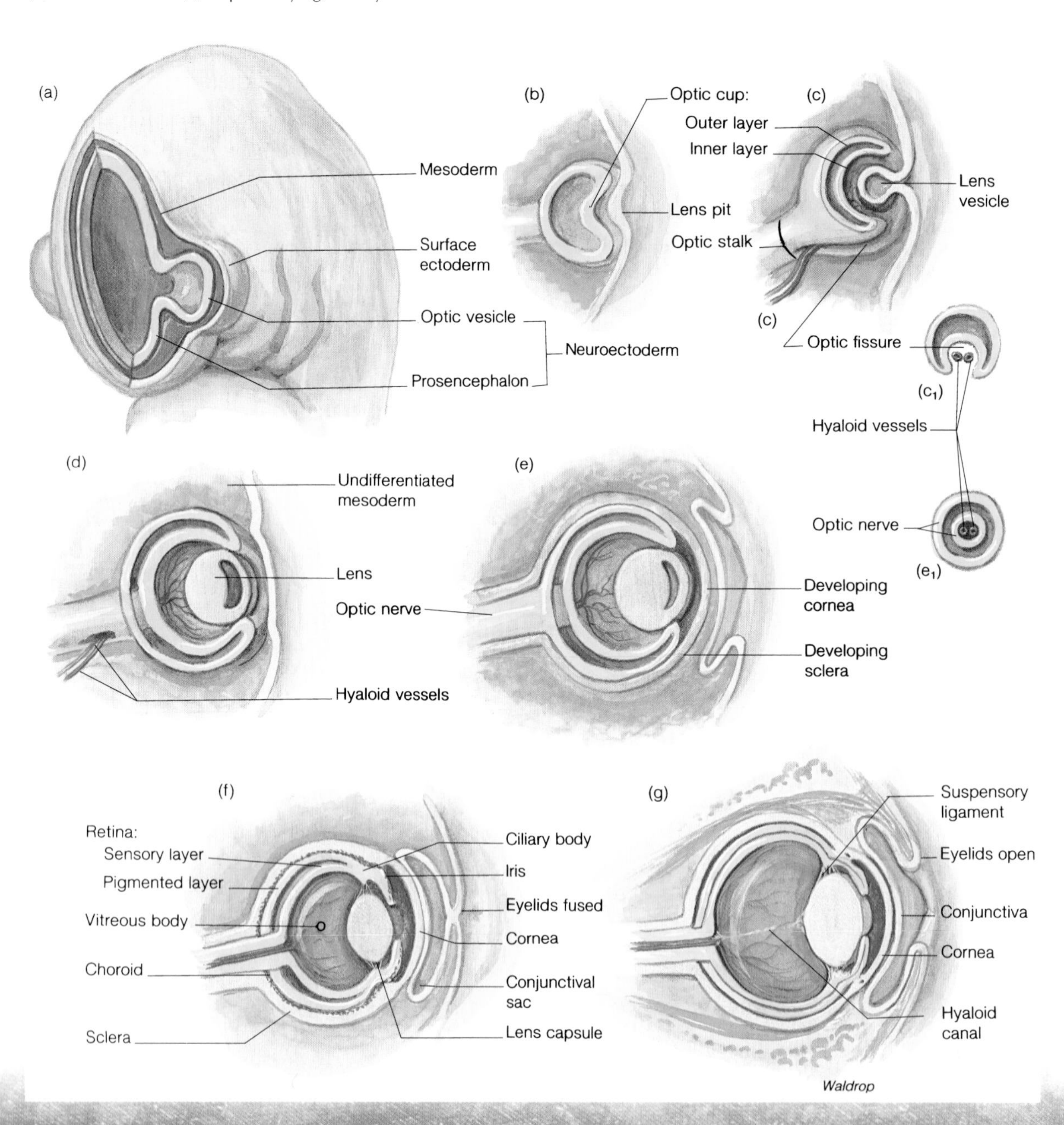

Explanation

The development of the eye is a complex process that involves the precise interaction of neuroectoderm, surface ectoderm, and mesoderm. It begins early in the fourth week, as the neuroectoderm forms a lateral diverticulum on each side of the prosencephalon (forebrain). As the diverticulum increases in size, the distal portion dilates to become the **optic vesicle** and the proximal portion constricts to become the **optic stalk** (exhibit I). Once the optic vesicle has formed, the overlying surface ectoderm thickens and invaginates. The thickened portion is the **lens placode** (*plak'ōd*) and the invagination is the **lens fovea.**

During the fifth week, the lens placode is depressed and eventually cut off from the surface ectoderm, causing the formation of the **lens vesicle.** Simultaneously, the optic vesicle invaginates and differentiates into the two-layered **optic cup.** Along the inferior surface of the optic cup, a groove called the **optic fissure** allows for passage of the **hyaloid artery** and **hyaloid vein** that serve the developing eyeball. The walls of the optic fissure eventually close, so that the hyaloid vessels are within the tissue of the optic stalk. They become the **central vessels of the retina** of the mature eye. The optic stalk eventually becomes the **optic nerve,** composed of sensory axons from the retina.

By the early part of the seventh week, the optic cup has differentiated into two sheets of epithelial tissue that become the sensory and pigmented layers of the **retina.** Both of these layers also line the entire vascular coat, including the **ciliary body, iris,** and the **choroid.** A proliferation of cells in the lens vesicle leads to the formation of the lens. The **lens capsule** forms from the mesoderm surrounding the lens, as does the **vitreous body.** Mesoderm surrounding the optic cup differentiates into two distinct layers of the developing eyeball. The inner layer of mesoderm becomes the vascular **choroid,** and the outer layer becomes the toughened **sclera** posteriorly and the transparent **cornea** anteriorly. Once the cornea has formed, additional surface ectoderm gives rise to the thin **conjunctiva** covering the anterior surface of the eyeball. Epithelium of the **eyelids** and the **lacrimal glands** and **duct** develop from surface ectoderm, whereas the **extrinsic ocular muscles** and all connective tissues associated with the eye develop from mesoderm. These accessory structures of the eye develop gradually during the embryonic period and continue to develop into the fetal period as late as the fifth month.

hyaloid: Gk. *hyalos*, glass; *eiodos*, form

1. List the accessory structures of the eye that either cause the eye to move or protect it within the orbit.
2. Diagram the structure of the eye and label the following: sclera, cornea, choroid, retina, fovea centralis, iris, pupil, lens, and ciliary body. What are the principal cells or tissues in each of the three layers of the eye?
3. Trace the path of light through the three chambers of the eye and explain the mechanism of light refraction. Describe how the eye is focused for viewing distant and near objects.
4. List the different layers of the retina and describe the path of light and of nerve activity through these layers. Continue tracing the path of a visual impulse to the cerebral cortex, and list in order the structures traversed.

Senses of Hearing and Balance

Structures of the outer, middle, and inner ear are involved in the sense of hearing. The inner ear also contains structures that provide a sense of balance, or equilibrium. The development of the ear begins during the fourth week and is complete by week 32.

Objective 14 List the structures of the ear that relate to hearing and describe their locations and functions.

Objective 15 Trace the path of sound waves as they travel through the ear and explain how they are transmitted and converted to nerve impulses.

Objective 16 Explain the mechanisms by which equilibrium is maintained.

The ear is the organ of hearing and equilibrium. It contains receptors that convert sound waves into nerve impulses and receptors that respond to movements of the head. Impulses from both receptor types are transmitted through the vestibulocochlear nerve (VIII) to the brain for interpretation. The ear consists of three principal regions: the *outer ear*, the *middle ear*, and the *inner ear* (fig. 15.28).

Outer Ear

The **outer ear** consists of the auricle, or pinna, and the external acoustic canal. The external acoustic canal is the fleshy tube that is fitted into the bony tube called the *external acoustic meatus (me-a'tus)*. The **auricle** *(or'ĭ-kul)* is the visible fleshy appendage attached to the side of the head. It consists of a cartilaginous

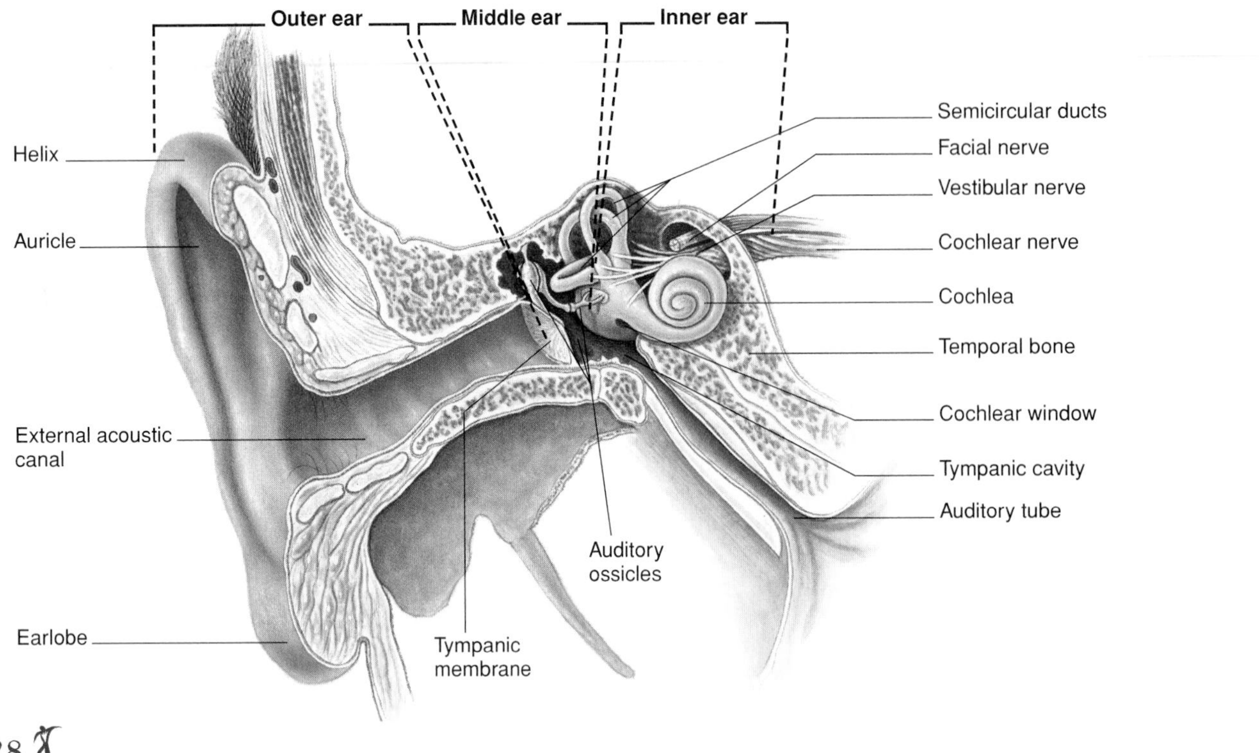

Figure 15.28

The ear. (Note the outer, middle, and inner regions indicated by dashed lines.)

framework of elastic connective tissue covered with skin. The rim of the auricle is the **helix,** and the inferior fleshy portion is the **earlobe** (fig. 15.29). The earlobe is the only portion of the auricle that is not supported with cartilage. The auricle has a ligamentous attachment to the skull and poorly developed auricular muscles that insert within it anteriorly, superiorly, and posteriorly. The blood supply to the auricle is from the posterior auricular and occipital arteries, which branch from the external carotid and superficial temporal arteries, respectively. The structure of the auricle directs sound waves into the external acoustic canal.

The **external acoustic canal** is a slightly S-shaped tube, about 2.5 cm (1 in.) long, that extends slightly upward from the auricle to the *tympanic membrane* (fig. 15.28). The skin that lines the canal contains fine hairs and sebaceous glands near the entrance. Deep within the canal, the skin contains specialized wax-secreting glands, called **ceruminous** *(sĕ-roo′mĭ-nus)* **glands.** The *cerumen* (earwax) secreted from these glands keeps the tympanic membrane soft and waterproof. The cerumen and hairs also help to prevent small foreign objects from reaching the tympanic membrane. The bitter cerumen is probably an insect repellent as well.

The **tympanic** *(tim-pan′ik)* **membrane** ("eardrum") is a thin, double-layered, epithelial partition, approximately 1 cm in diameter, between the external acoustic canal and the middle ear. It is composed of an outer concave layer of stratified squamous epithelium and an inner convex layer of low columnar epithelium. Between the epithelial layers is a layer of connective tissue. The tympanic membrane is extremely sensitive to pain and is innervated by the auriculotemporal nerve (a branch of the mandibular nerve of the trigeminal nerve) and the auricular nerve (a branch of the vagus nerve).

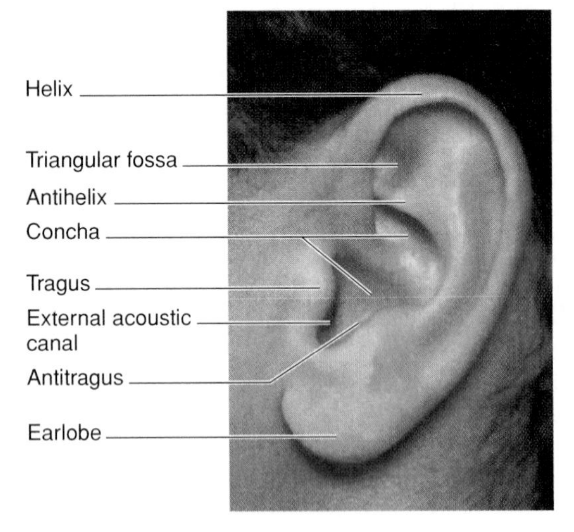

Figure 15.29

The surface anatomy of the auricle of the ear.

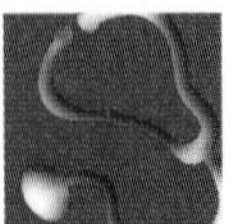

Inspecting the tympanic membrane with an otoscope during a physical examination provides clinical information about the condition of the middle ear (fig. 15.30). The color of the membrane, its curvature, the presence of lesions, and the position of the malleus of the middle ear are features of particular importance. If ruptured, the tympanic membrane will generally regenerate, healing itself within days.

otoscope: Gk. *otikos*, ear; *skopein*, to examine

(a)

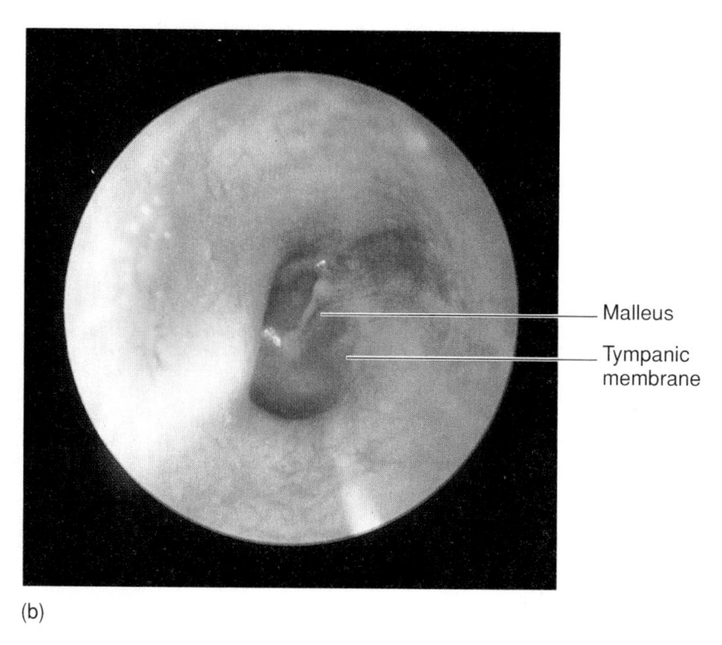

(b)

Figure 15.30

(*a*) A physician viewing the anatomy of the outer ear. (*b*) The appearance of the tympanic membrane as viewed with an otoscope.

Middle Ear

The laterally compressed **middle ear** is an air-filled chamber called the **tympanic cavity** in the petrous part of the temporal bone (see figs. 15.28 and 15.31). The tympanic membrane separates the middle ear from the external acoustic canal of the outer ear. A bony partition containing the **vestibular window** (*oval window*) and the **cochlear window** (*round window*) separates the middle ear from the inner ear.

There are two openings into the tympanic cavity. The **epitympanic recess** in the posterior wall connects the tympanic cavity to the **mastoidal air cells** within the mastoid process of the temporal bone. The **auditory** (eustachian) **tube** connects the tympanic cavity anteriorly with the nasopharynx and equalizes air pressure on both sides of the tympanic membrane.

Three **auditory ossicles** extend across the tympanic cavity from the tympanic membrane to the vestibular window (fig. 15.31). These tiny bones (the smallest in the body), from outer to inner, are the **malleus** *(mal'e-us)* (hammer), **incus** *(ing'kus)* (anvil), and **stapes** *(sta'pēz)* (stirrup). The auditory ossicles are attached to the wall of the tympanic cavity by ligaments. Vibrations of the tympanic membrane cause the auditory ossicles to move and transmit sound waves across the tympanic cavity to the vestibular window. Vibration of the vestibular window moves a fluid within the inner ear and stimulates the receptors of hearing.

eustachian tube: from Bartolommeo E. Eustachio, Italian anatomist, 1520–74
malleus: L. *malleus*, hammer
incus: L. *incus*, anvil
stapes: L. *stapes*, stirrup

As the auditory ossicles transmit vibrations from the tympanic membrane, they act as a lever system to amplify sound waves. In addition, the sound waves are intensified as they are transmitted from the relatively large surface of the tympanic membrane to the smaller surface area of the vestibular window. The combined effect increases sound amplification about 20 times.

Two small skeletal muscles, the **tensor tympani** muscle and the **stapedius** muscle (fig. 15.31), attach to the malleus and stapes, respectively, and contract reflexively to protect the inner ear against loud noises. When contracted, the tensor tympani muscle pulls the malleus inward, and the stapedius muscle pulls the stapes outward. This combined action reduces the force of vibration of the auditory ossicles.

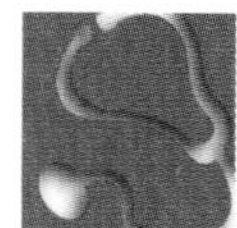

The mucous membranes that line the tympanic cavity, the mastoidal air cells, and the auditory tube are continuous with those of the nasopharynx. For this reason, infections that spread to the nose or throat may spread to the tympanic cavity and cause a middle-ear infection. They may also spread to the mastoidal air cells and cause *mastoiditis.* Forcefully blowing the nose advances the spread of the infection.

An equalization of air pressure on both sides of the tympanic membrane is important in hearing. When atmospheric pressure is reduced, as occurs when traveling to higher altitudes, the tympanic membrane bulges outward in response to the greater air pressure within the tympanic cavity. The bulging is painful and may impair hearing by reducing flexibility. The auditory tube, which is collapsed most of the time in adults, opens during swallowing or yawning and allows the air pressure on the two sides of the tympanic membrane to equalize.

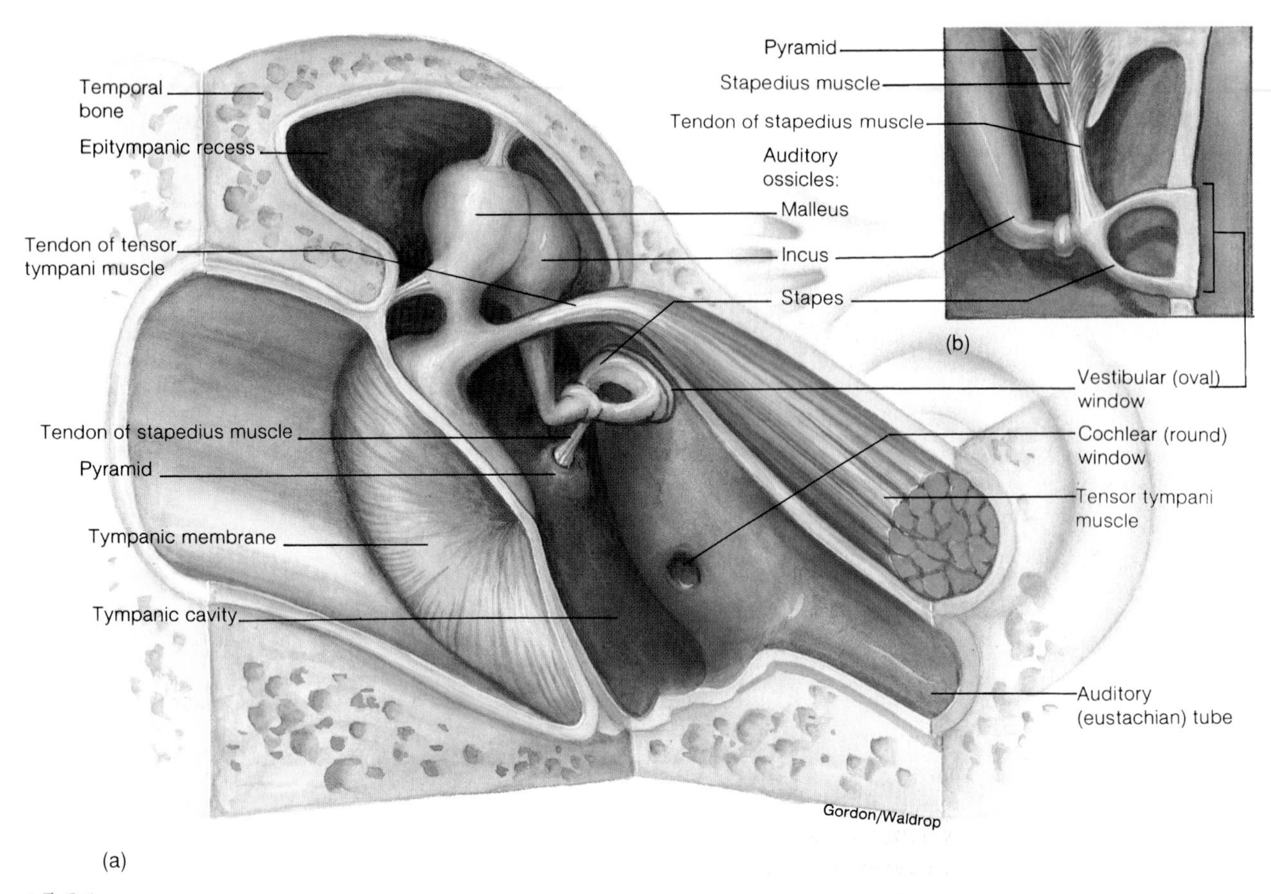

Figure 15.31

(*a*) The auditory ossicles and associated structures within the tympanic cavity. (*b*) A rotated view of the stapes positioned against the vestibular window.

Inner Ear

The entire structure of the **inner ear** is referred to as the **labyrinth** (*lab'ĭ-rinth*). The labyrinth consists of an outer shell of dense bone called the *bony labyrinth* that surrounds and protects a *membranous labyrinth* (fig. 15.32). The space between the bony labyrinth and the membranous labyrinth is filled with a fluid called *perilymph*, which is secreted by cells lining the bony canals. Within the tubular chambers of the membranous labyrinth is yet another fluid called *endolymph*. These two fluids provide a liquid-conducting medium for the vibrations involved in hearing and the maintenance of equilibrium.

The bony labyrinth is structurally and functionally divided into three areas: the *vestibule*, *semicircular canals*, and *cochlea*. The functional organs for hearing and equilibrium are located in these areas.

Vestibule

The vestibule is the central portion of the bony labyrinth. It contains the *vestibular* (oval) *window*, into which the stapes fits, and the *cochlear* (round) *window* on the opposite end (fig. 15.32).

The membranous labyrinth within the vestibule consists of two connected sacs called the **utricle** (*yoo'trĭ-kul*) and the **saccule** (*sak'yool*). The utricle is larger than the saccule and lies in the upper back portion of the vestibule. Both the utricle and saccule contain receptors that are sensitive to gravity and linear movement (acceleration) of the head.

Semicircular Canals

Posterior to the vestibule are the three bony semicircular canals, positioned at nearly right angles to each other. The thinner **semicircular ducts** form the membranous labyrinth within the semicircular canals (fig. 15.32). Each of the three semicircular ducts has a **membranous ampulla** (*am-pool'ă*) at one end and connects with the upper back part of the utricle. Receptors within the semicircular ducts are sensitive to angular acceleration and deceleration of the head, as in rotational movement.

Cochlea

The snail-shaped cochlea is coiled two and a half times around a central core of bone (fig. 15.33). There are three chambers in the cochlea (fig. 15.34). The upper chamber, the **scala**

cochlea: L. *cochlea*, snail shell

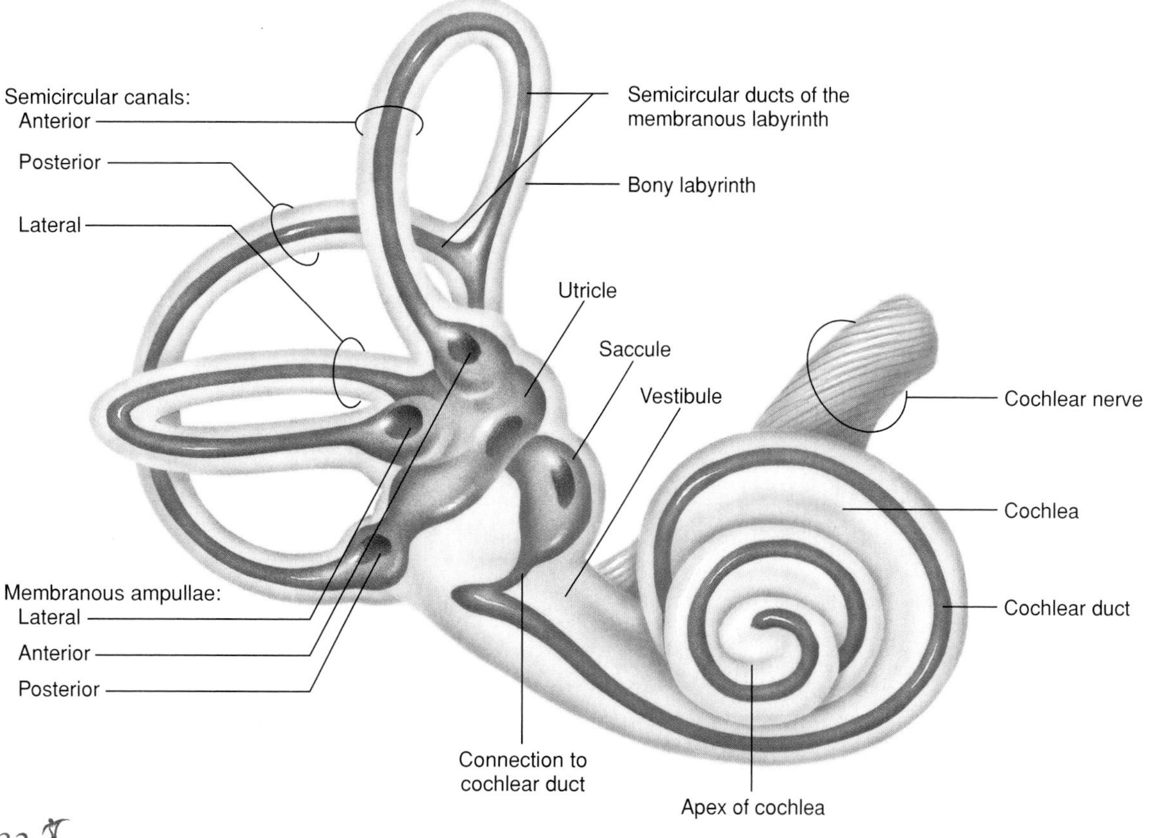

Figure 15.32

The labyrinths of the inner ear. The membranous labyrinth (darker color) is contained within the bony labyrinth. The principal structures of the inner ear are the cochlea for hearing and the vestibular organs (semicircular canals, utricle, and saccule) for balance and equilibrium.

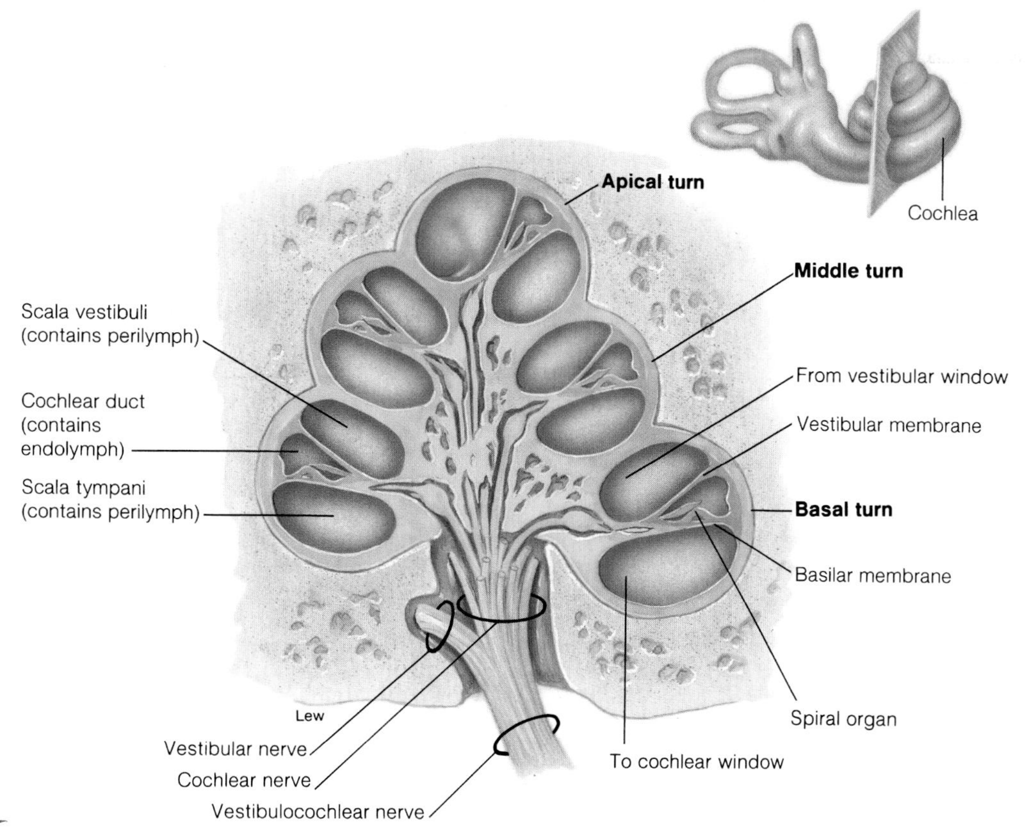

Figure 15.33

A coronal section of the cochlea showing its three turns (indicated in boldface type) and its three compartments—the scala vestibuli, cochlear duct, and scala tympani.

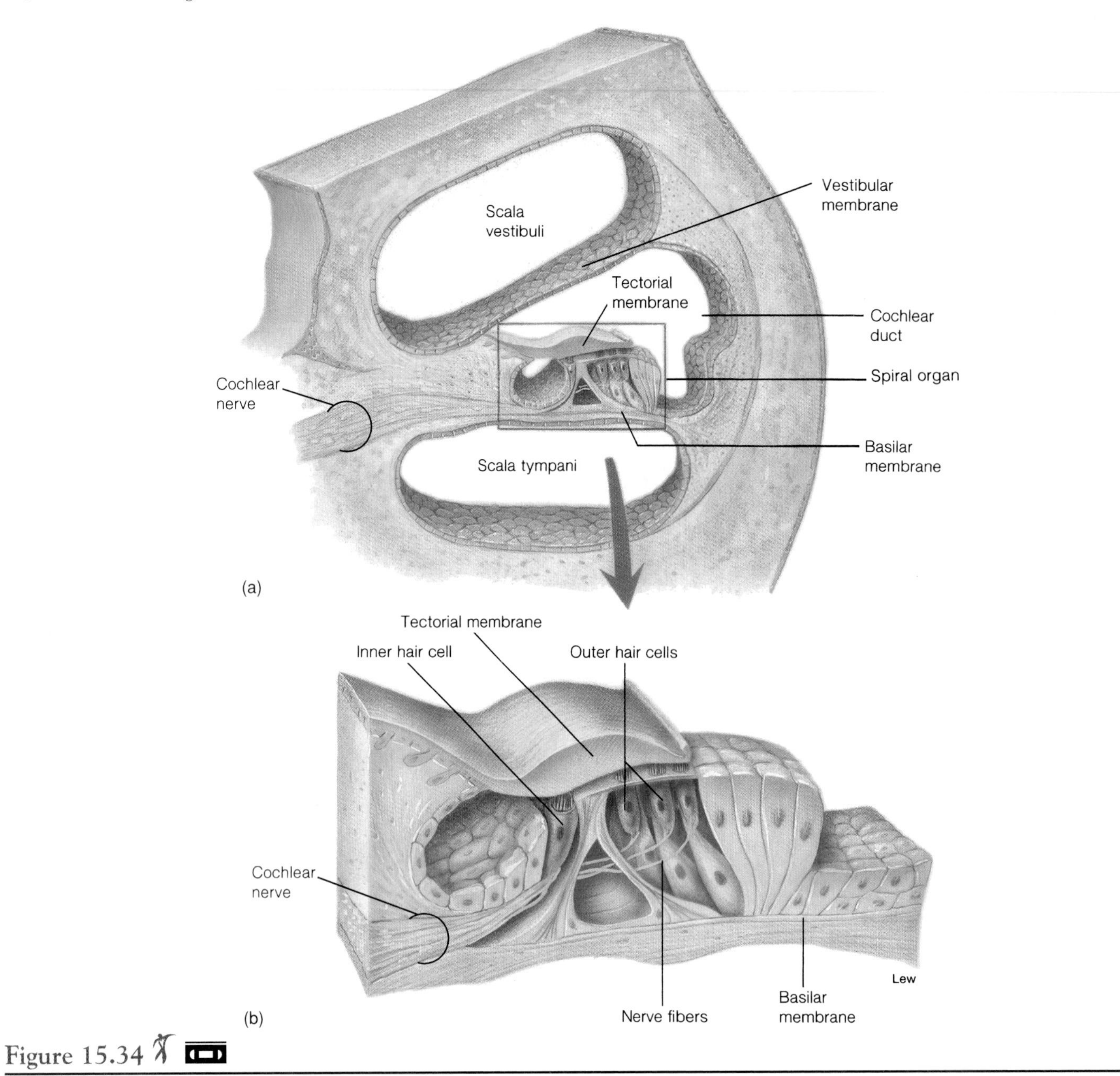

Figure 15.34

A section of the cochlea showing (*a*) the spiral organ within the cochlear duct and (*b*) the spiral organ in greater detail.

(*ska'lă*) **vestibuli,** begins at the vestibular window and extends to the apex (end) of the coiled cochlea. The lower chamber, the **scala tympani,** begins at the apex and terminates at the cochlear window. Both the scala vestibuli and the scala tympani are filled with perilymph. They are completely separated, except at the narrow apex of the cochlea, called the **helicotrema** (*hel"ĭ-kŏ-tre'mă*), where they are continuous (see fig. 15.36). Between the scala vestibuli and the scala tympani is the **cochlear duct,** the triangular middle chamber of the cochlea. The roof of the cochlear duct is called the **vestibular membrane,** and the floor is called the **basilar membrane.** The cochlear duct, which is filled with endolymph, ends at the helicotrema.

Within the cochlear duct is a specialized structure called the **spiral organ** (organ of Corti). The sound receptors that transform mechanical vibrations into nerve impulses are located along the basilar membrane of this structure, making it the functional unit of hearing. The epithelium of the spiral organ consists of supporting cells and hair cells (figs. 15.34 and 15.35). The

scala: Gk. *scala*, staircase
helicotrema: Gk. *helix*, a spiral; *trema*, a hole

organ of Corti: from Alfonso Corti, Italian anatomist, 1822–88

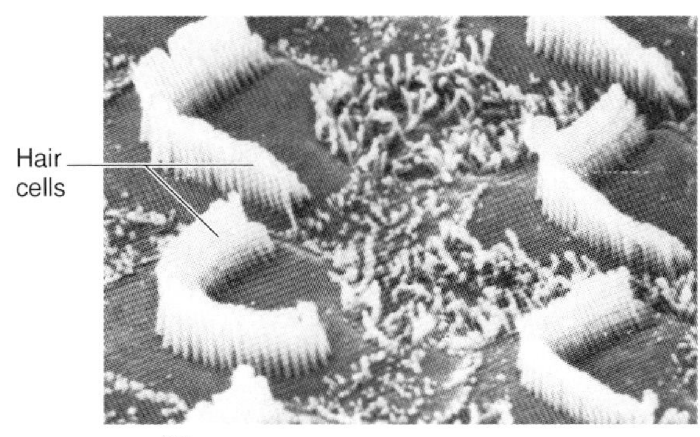

Figure 15.35

A scanning electron micrograph of the hair cells of the spiral organ.

bases of the hair cells are anchored in the basilar membrane, and their tips are embedded in the **tectorial membrane,** which forms a gelatinous canopy over them.

Sound Waves and Neural Pathways for Hearing

Sound Waves

Sound waves travel in all directions from their source, like ripples in a pond after a stone is dropped. These waves of energy are characterized by their frequency and their intensity. The *frequency*, or number of waves that pass a given point in a given time, is measured in *hertz* (*Hz*). The *pitch* of a sound is directly related to its frequency—the higher the frequency of a sound, the higher its pitch. For example, striking the high C on a piano produces a high frequency of sound that has a high pitch.

The *intensity*, or loudness of a sound, is directly related to the amplitude of the sound waves. Sound intensity is measured in units known as *decibels* (*dB*). A sound that is barely audible—at the threshold of hearing—has an intensity of zero decibels. Every 10 decibels indicates a tenfold increase in sound intensity: a sound is 10 times higher than threshold at 10 dB, 100 times higher at 20 dB, a million times higher at 60 dB, and 10 billion times higher at 100 dB. The healthy human ear can detect very small differences in sound intensity—from 0.1 to 0.5 dB.

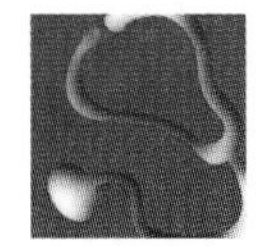

A snore can be as loud as 70 dB, as compared with 105 dB for a power mower. Frequent or prolonged exposure to sounds with intensities over 90 dB (including amplified rock music) can result in hearing loss.

Sound waves funneled through the external acoustic canal produce extremely small vibrations of the tympanic membrane. Movements of the tympanum during ordinary speech (with an average intensity of 60 dB) are estimated to be equal to the diameter of a molecule of hydrogen.

Sound waves passing through the solid medium of the auditory ossicles are amplified about 20 times as they reach the footplate of the stapes, which is seated within the vestibular window. As the vestibular window is displaced, pressure waves pass through the fluid medium of the scala vestibuli (fig. 15.36) and pass around the helicotrema to the scala tympani. Movements of perilymph within the scala tympani, in turn, displace the cochlear window into the tympanic cavity.

When the sound frequency (pitch) is sufficiently low, there is adequate time for the pressure waves of perilymph within the scala vestibuli to travel around the helicotrema to the scala tympani. As the sound frequency increases, however, these pressure waves do not have time to travel all the way to the apex of the cochlea. Instead, they are transmitted through the vestibular membrane, which separates the scala vestibuli from the cochlear duct, and through the basilar membrane, which separates the cochlear duct from the scala tympani, to the perilymph of the scala tympani. The distance that these pressure waves travel, therefore, decreases as the sound frequency increases.

Sounds of low pitch (with frequencies below about 50 Hz) cause movements of the entire length of the basilar membrane—from the base to the apex. Higher sound frequencies result in maximum displacement of the basilar membrane closer to its base, as illustrated in figure 15.36.

Displacement of the basilar membrane and hair cells by movements of perilymph causes the hair cell microvilli that are embedded in the tectorial membrane to bend. This stimulation excites the sensory cells, which causes the release of an unknown neurotransmitter that excites sensory endings of the cochlear nerve.

Neural Pathways for Hearing

Cochlear sensory neurons in the *vestibulocochlear nerve* (VIII) synapse with neurons in the medulla oblongata, which project to the inferior colliculi of the midbrain (fig. 15.37). Neurons in this area in turn project to the thalamus, which sends axons to the auditory cortex of the temporal lobe, where the auditory sensations (nerve impulses) are perceived as sound.

Mechanics of Equilibrium

Maintaining equilibrium is a complex process that depends on continuous input from sensory neurons in the vestibular organs of both inner ears. Although the vestibular organs are the principal source of sensory information for equilibrium, the photoreceptors of the eyes, tactile receptors within the skin, and proprioceptors of tendons, muscles, and joints also provide sensory input that is needed to maintain equilibrium (fig. 15.38).

The vestibular organs provide the CNS with two kinds of receptor information. One kind is provided by receptors within the *saccule* and *utricle*, which are sensitive to gravity and to linear acceleration and deceleration of the head, as occur when riding in a car. The other is provided by receptors within the *semicircular ducts*, which are sensitive to rotational movements, as occur when turning the head, spinning, or tumbling.

The receptor hair cells of the vestibular organs contain 20 to 50 microvilli and one cilium, called a **kinocilium** (*ki″no-sil′e-um*)

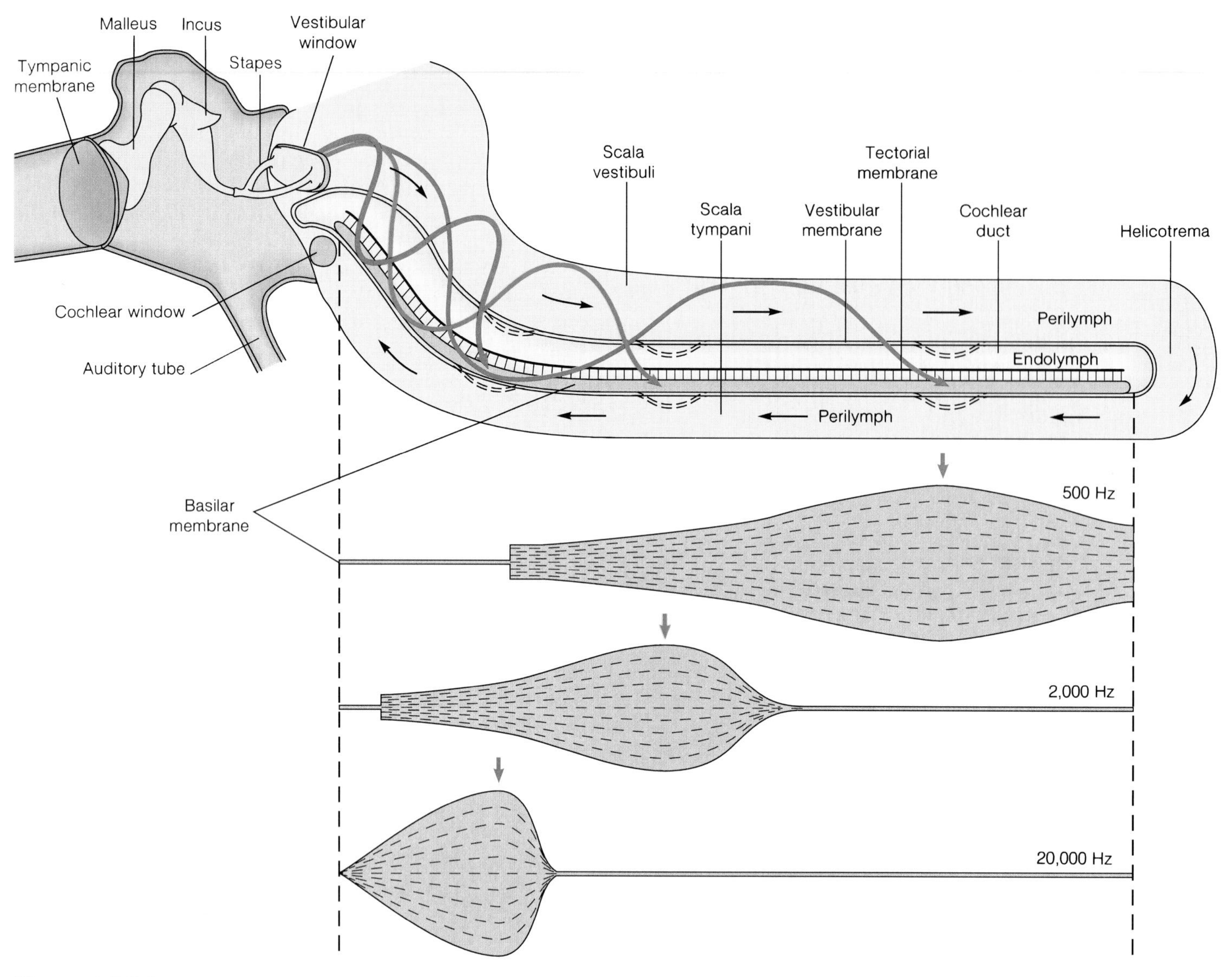

Figure 15.36

An illustration of the cochlea, straightened to show the mechanism of sound-wave generation at specific sites along the basilar membrane. The scala vestibuli and the scala tympani, which contain perilymph, are continuous at the helicotrema. The cochlear duct, which contains endolymph, separates the scala vestibuli and the scala tympani. Sounds of low frequency (blue arrow) cause pressure waves of perilymph to pass through the helicotrema and displace the basilar membrane near its apex. Sounds of medium frequency (green arrow) cause pressure waves to displace the basilar membrane near its center. Sounds of high frequency (red arrow) cause pressure waves to displace the basilar membrane near its base. (The frequency of sound waves is measured in hertz [Hz].)

(fig. 15.39). When the hair cells are displaced in the direction of the kinocilium, the cell membrane is depressed and becomes depolarized. When the hair cells are displaced in the opposite direction, the membrane becomes hyperpolarized.

Saccule and Utricle

Receptor hair cells of the saccule and utricle are located in a small, thickened area of the walls of these organs called the **macula** (*mak'yoo-lă*) (fig. 15.40). Cytoplasmic extensions of the hair cells project into a gelatinous mass, called the **statoconial** (otolithic) **membrane,** that supports microscopic crystals of calcium carbonate called **statoconia** (otoliths). The statoconia increase the weight of the statoconial membrane, which results in a higher *inertia* (resistance to change in movement).

When a person is upright, the hairs of the utricle project vertically into the statoconial membrane, whereas those of the saccule project horizontally. During forward acceleration, the statoconial membrane lags behind the hair cells, so the hair cells of the utricle are bent backward. This is similar to the backward thrust of the body when a car accelerates rapidly forward. The

macula: L. *macula*, spot

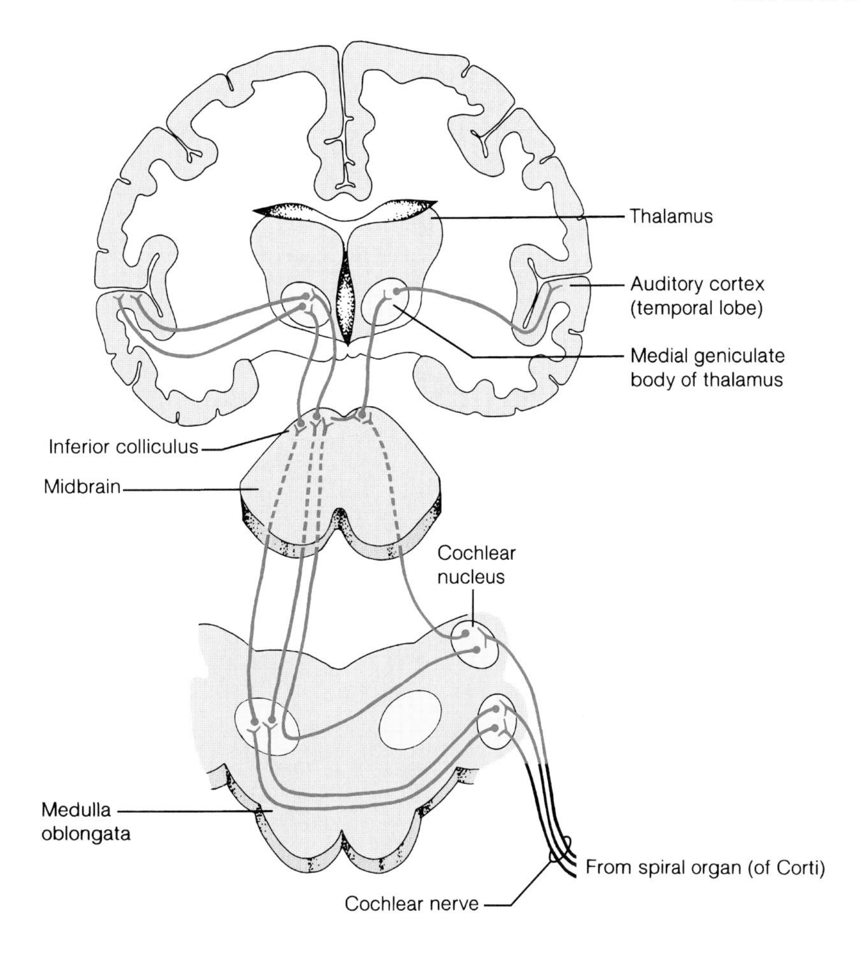

Figure 15.37

Neural pathways for the sense of hearing.

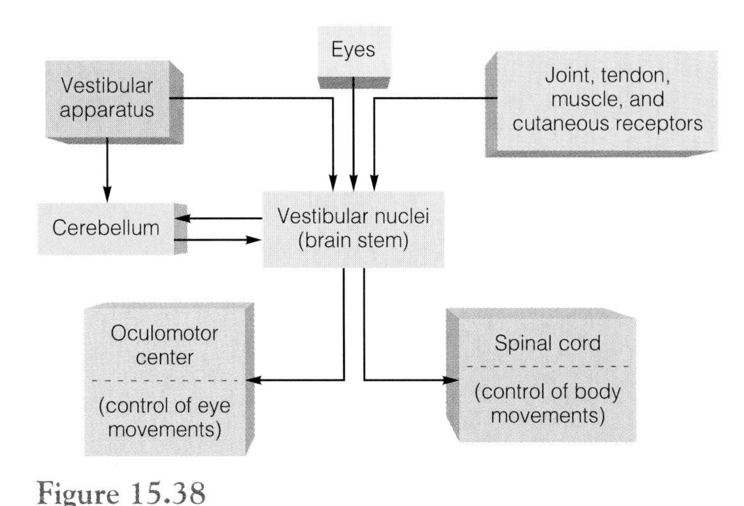

Figure 15.38

The neural processing involved in maintaining equilibrium and balance.

inertia of the statoconial membrane similarly causes the hair cells of the saccule to be pushed upward when a person jumps from a raised platform. Thus, because of the orientation of their hair cell processes, the utricle is more sensitive to horizontal acceleration, and the saccule is more sensitive to vertical acceleration. The changed pattern of action potentials in sensory nerve fibers that results from stimulation of the hair cells allows us to maintain our equilibrium with respect to gravity during linear acceleration.

Sensory impulses from the vestibular organs are conveyed to the brain by way of the vestibular portion of the vestibulocochlear nerve.

Semicircular Canals

Receptors of the semicircular canals are contained within the ampulla at the base of each semicircular duct. An elevated area of the ampulla called the **crista ampullaris** *(am″poo-lar′is)* contains numerous hair cells and supporting cells (fig. 15.41). Like the saccule and utricle, the hair cells have cytoplasmic extensions that project into a dome-shaped gelatinous mass called the **cupula** *(kyoop′loo-lă)*. When the hair cells within the cupula are bent by rapid displacement of the fluid within the semicircular ducts, as in spinning around, sensory impulses travel to the brain by way of the vestibular portion of the vestibulocochlear nerve.

---

cupula: L. *cupula*, cup-shaped

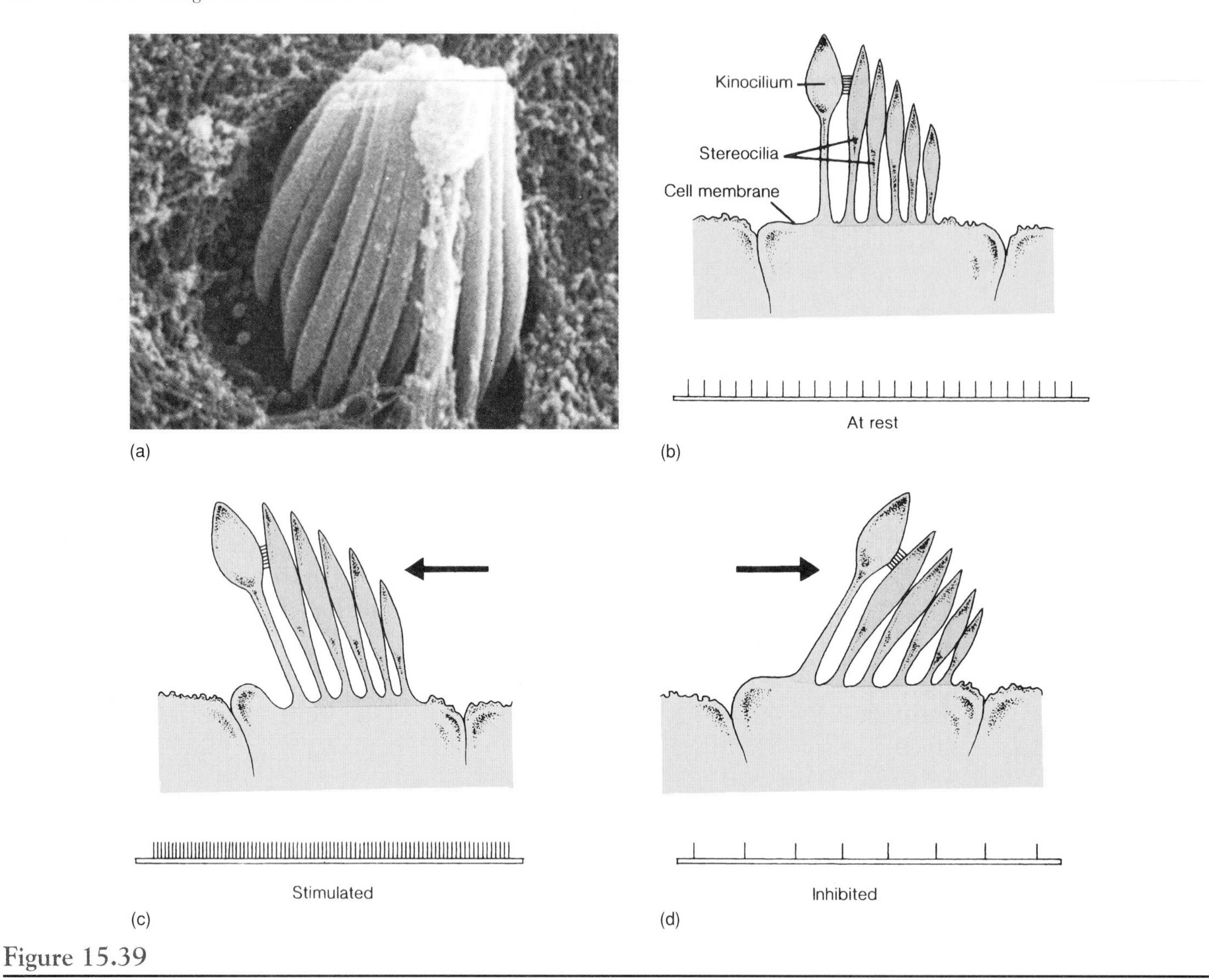

Figure 15.39

Sensory hair cells of a vestibular organ. (*a*) A scanning electron photomicrograph of a kinocilium. (*b*) Sensory hairs (microvilli) and one kinocilium. (*c*) When hair cells are bent in the direction of the kinocilium, the cell membrane is depressed (see arrow) and the sensory neuron innervating the hair cell is stimulated. (*d*) When the hairs are bent in the opposite direction, the sensory neuron is inhibited.

Neural Pathways

Stimulation of the hair cells in the vestibular apparatus activates the sensory neurons of the *vestibulocochlear nerve*. These fibers transmit impulses to the cerebellum and to the vestibular nuclei of the medulla oblongata. The vestibular nuclei, in turn, send fibers to the oculomotor center of the brain stem and to the spinal cord. Neurons in the oculomotor center control eye movements, and neurons in the spinal cord stimulate movements of the head, neck, and limbs. Movements of the eyes and body produced by these pathways serve to maintain balance and track the visual field during rotation.

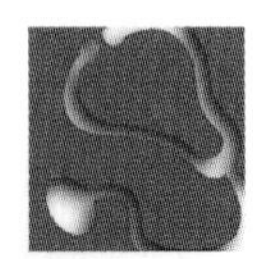

The dizziness and nausea that some people experience when they spin rapidly is explained by the activity occurring within the vestibular organs. When a person first begins to spin, the inertia of the endolymph within the semicircular ducts causes the cupula to bend in the opposite direction. As the spin continues, however, the endolymph and the cupula will eventually be moving in the same direction and at the same speed. If movement is suddenly stopped, the greater inertia of the endolymph causes it to continue moving in the direction of spin and to bend the cupula in that direction.

Bending of the cupula after movement has stopped affects muscular control of the eyes and body. The eyes slowly drift in the direction of the previous spin, and then are rapidly jerked back to the midline position, producing involuntary movements called *postrotatory vestibular nystagmus (ni-stag'mus)*. People experiencing this effect may feel that they are spinning, or that the room is. The loss of equilibrium that results is called *vertigo*. If the vertigo is sufficiently severe, or if the person particularly susceptible, the autonomic nervous system may become involved. This can produce dizziness, pallor, sweating, and nausea.

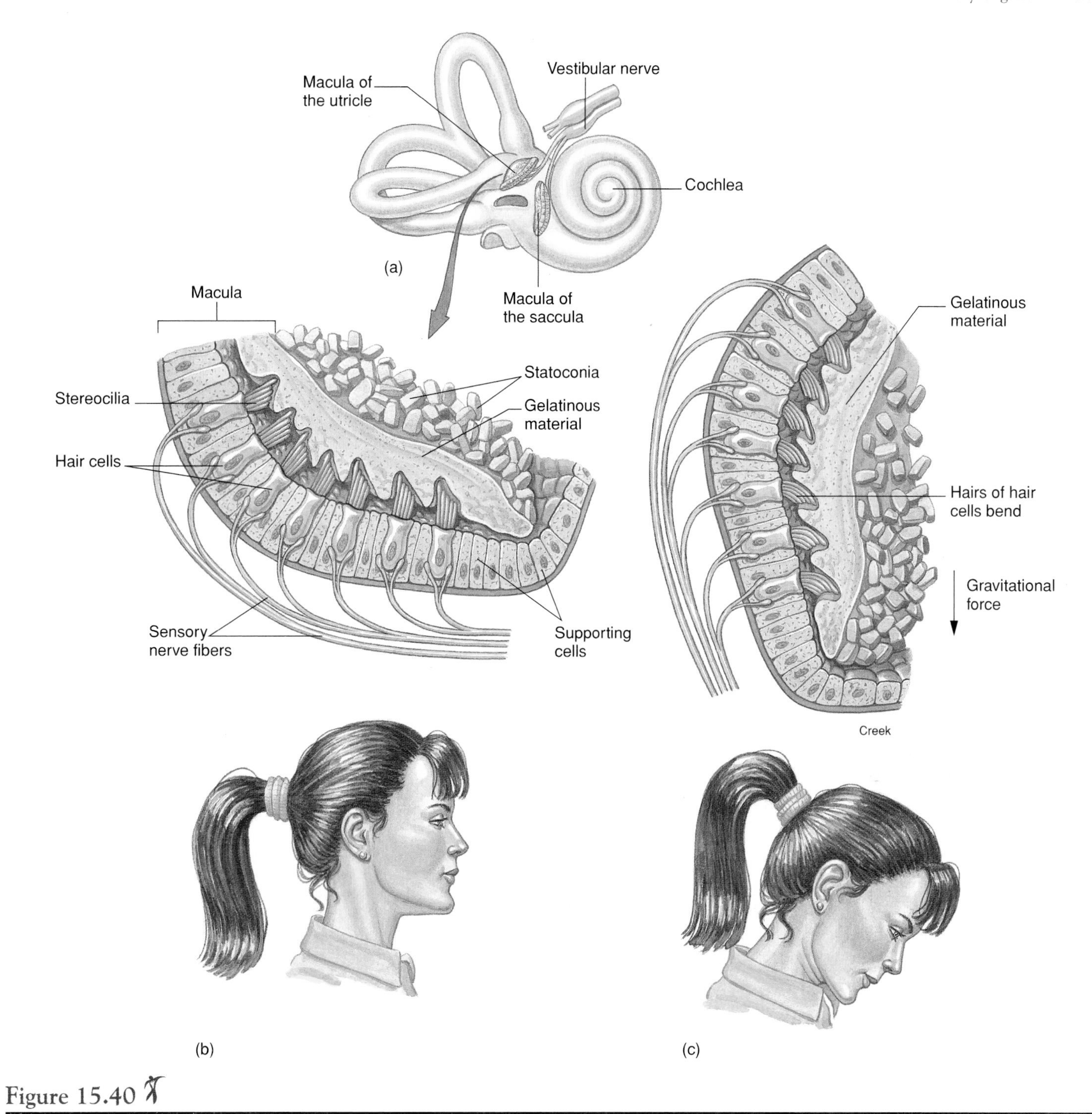

Figure 15.40

(*a*) Maculae of the inner ear. (*b*) When the head is upright, the weight of the statoconia applies direct pressure to the sensitive cytoplasmic extensions of the hair cells. (*c*) As the head is tilted forward, the extensions of the hair cells bend in response to gravitational force and cause the hair cells to be stimulated.

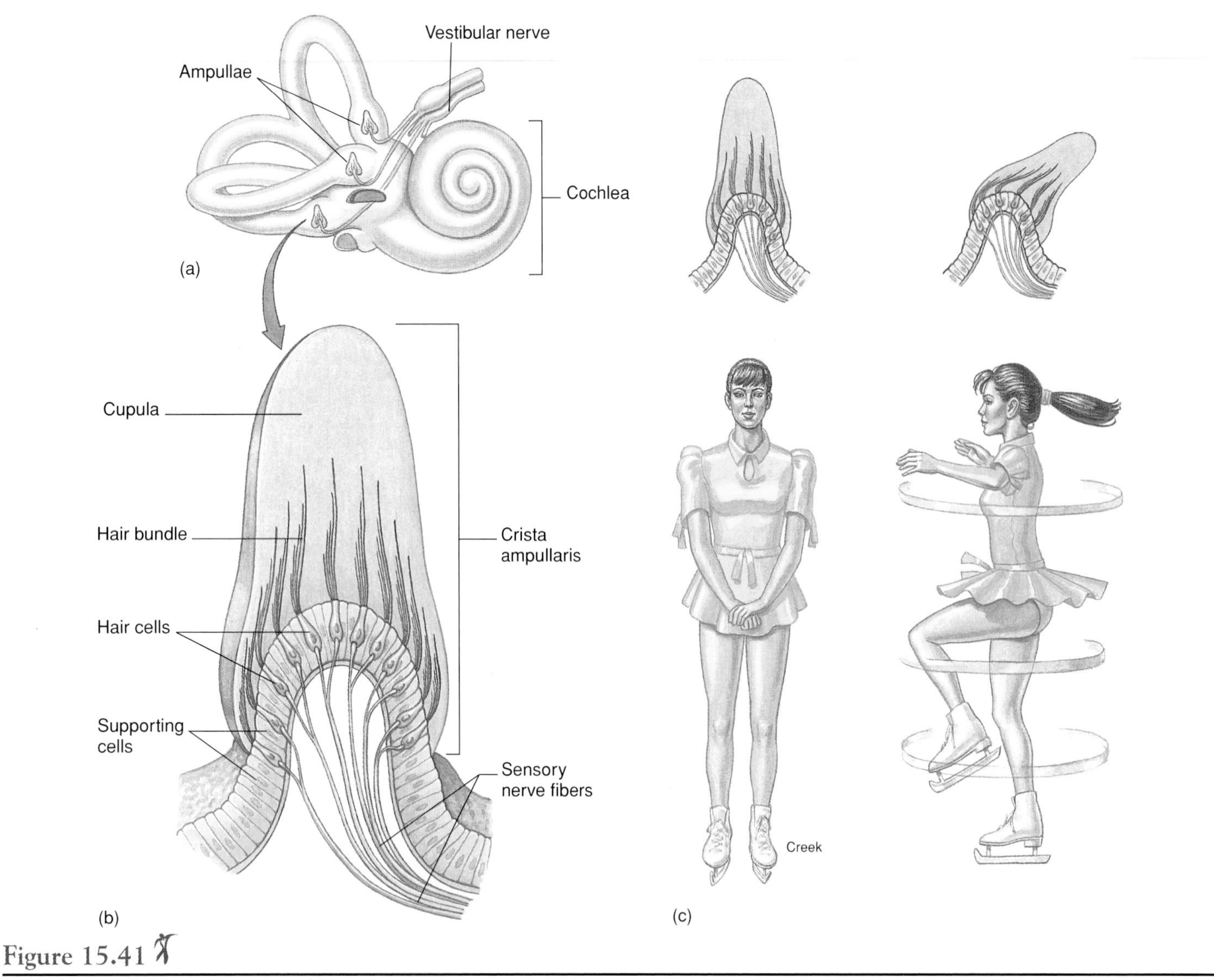

Figure 15.41

(*a*) Ampullae within the inner ear. (*b*) A crista ampullaris within an ampulla. (*c*) Movement of the endolymph during rotation causes the cupula to displace, thus stimulating the hair cells.

1. List the structures of the outer ear, middle ear, and inner ear and explain the function of each as related to hearing.
2. Use a flowchart to describe how sound waves in air within the external acoustic canal are transduced into movements of the basilar membrane.
3. Explain how movements of the basilar membrane can code for different sound frequencies (pitches).
4. Explain how the vestibular organs maintain a sense of balance and equilibrium.

Clinical Considerations

Numerous disorders and diseases afflict the sensory organs. Some of these occur during the sensitive period of prenatal development; others, some of which are avoidable, can occur at any time of life. Still other sensory impairments are the result of changes associated with the natural aging process. The loss of a sense frequently involves a traumatic adjustment. Fortunately, however, when a sensory function is impaired or lost, the other senses seem to become keener to lessen the extent of the handicap. A blind person, for example, compensates somewhat for the loss of sight by developing a remarkable hearing ability.

Entire specialties of medicine are devoted to specific sensory organs. It is beyond the scope of this text to attempt a comprehensive discussion of the numerous diseases and dysfunctions of these organs. Some general comments will be made, however, on the diagnosis of sensory disorders and on developmental problems that can affect the eyes and ears. In addition, the more common diseases and dysfunctions of the eyes and ears are noted.

Diagnosis of Eye and Ear Disorders

Eye

There are two distinct professional specialties concerned with the structure and function of the eye. *Optometry* is the paramedical profession concerned with assessing vision and treating visual problems. An *optometrist* prescribes corrective lenses or visual training but is not a medical doctor and does not treat eye diseases. *Ophthalmology (of"thal-mol'ŏ-je)* is the specialty of medicine concerned with diagnosing and treating eye diseases.

Although the eyeball is an extremely complex organ, it is quite accessible to examination. The following devices are frequently employed: (1) a *cycloplegic drug,* which is instilled into the eyes to dilate the pupils and temporarily inactivate the ciliary muscles; (2) a *Snellen's chart,* which is used to determine the visual acuity of a person standing 20 feet from the chart (a reading of 20/20 is considered normal for the test); (3) an *ophthalmoscope,* which contains a light, mirrors, and lenses to illuminate and magnify the interior of the eyeball so that the structures within may be examined; and (4) a *tonometer,* which is used to measure ocular tension, important in detecting glaucoma.

Ear

Otorhinolaryngology (o"to-ri"no-lar"in-gol'ŏ-je) is the specialty of medicine dealing with the diagnosis and treatment of diseases or conditions of the ear, nose, and throat. *Audiology* is the study of hearing, particularly assessment of the ear and its functioning.

There are three common instruments or techniques used to examine the ears to determine auditory function: (1) an *otoscope* is an instrument used to examine the tympanic membrane of the ear (abnormalities of this membrane are informative when diagnosing specific auditory problems, including middle-ear infections); (2) *tuning fork tests* are useful in determining hearing acuity and especially for discriminating the various kinds of hearing loss; and (3) *audiometry* is a functional examination for hearing sensitivity and speech discrimination.

Snellen's chart: from Herman Snellen, Dutch ophthalmologist, 1834–1908

Developmental Problems of the Eyes and Ears

Although there are many congenital abnormalities of the eyes and ears, most of them are rare. For these organs, the sensitive period of development is between 24 and 45 days after conception. Indeed, 85% of newborns suffer anomalies if infected during this interval. Most congenital disorders of the eyes and ears are caused by genetic factors or intrauterine infections such as *rubella virus.*

If a pregnant woman contracts rubella (German measles), there is a 90% probability that the embryo or fetus will contract it also. An embryo afflicted with rubella is 30% more likely to be aborted, stillborn, or congenitally deformed than one that is not afflicted. Rubella interferes with the mitotic process, and thus causes underdeveloped organs. An embryo with rubella may suffer from a number of physical deformities, including *cataracts* and *glaucoma,* which are common deformities of the eye.

Eye

Most **congenital cataracts** are hereditary, but they may also be caused by maternal rubella infection during the critical fourth to sixth week of eye development. In this condition, the lens is opaque and frequently appears grayish white.

Cyclopia *(si-klo'pe-ă)* is a rare condition in which the eyes are partially fused into a median eye enclosed by a single orbit. Other severe malformations, which are incompatible with life, are generally expressed with this condition.

Ear

Congenital deafness is generally caused by an autosomal recessive gene but may also be caused by a maternal rubella infection. The actual functional impairment is generally either a defective set of auditory ossicles or improper development of the neurosensory structures of the inner ear.

Although the shape of the auricle varies widely, **auricular abnormalities** are not uncommon, especially in infants with chromosomal syndromes causing mental deficiencies. In addition, the external acoustic canal frequently does not develop in these children, producing a condition called **atresia** *(ă-tre'se-ă)* of the external acoustic canal.

Functional Impairments of the Eye

Few people have perfect vision. Slight variations in the shape of the eyeball or curvature of the cornea or lens cause an imperfect focal point of light rays onto the retina. Most variations are slight, however, and the error of refraction goes unnoticed. Severe deviations that are not corrected may cause blurred vision, fatigue, chronic headaches, and depression.

The primary clinical considerations associated with defects in the refractory structures or general shape of the eyeball

DEVELOPMENTAL EXPOSITION

The Ear

Exhibit II The development of the inner ear. (*a*) A lateral view of a 22-day-old embryo showing the position of a transverse cut through the otic placode. (a_1) The otic placode of surface ectoderm begins to invaginate at 22 days. (a_2) By 24 days, a distinct otic fovea has formed and the neural ectoderm is positioned to give rise to the brain. (*b*) A lateral view of a 28-day-old embryo showing the position of a transverse cut through the otocyst. (b_1) By 28 days, the otic fovea has become a distinct otocyst. (b_2) The otocyst is in position in the 30-day-old embryo, where it gives rise to the structures of the inner ear. (*c–e*) Lateral views of the differentiating otocyst into the cochlea and semicircular canals from the fifth to the eighth week.

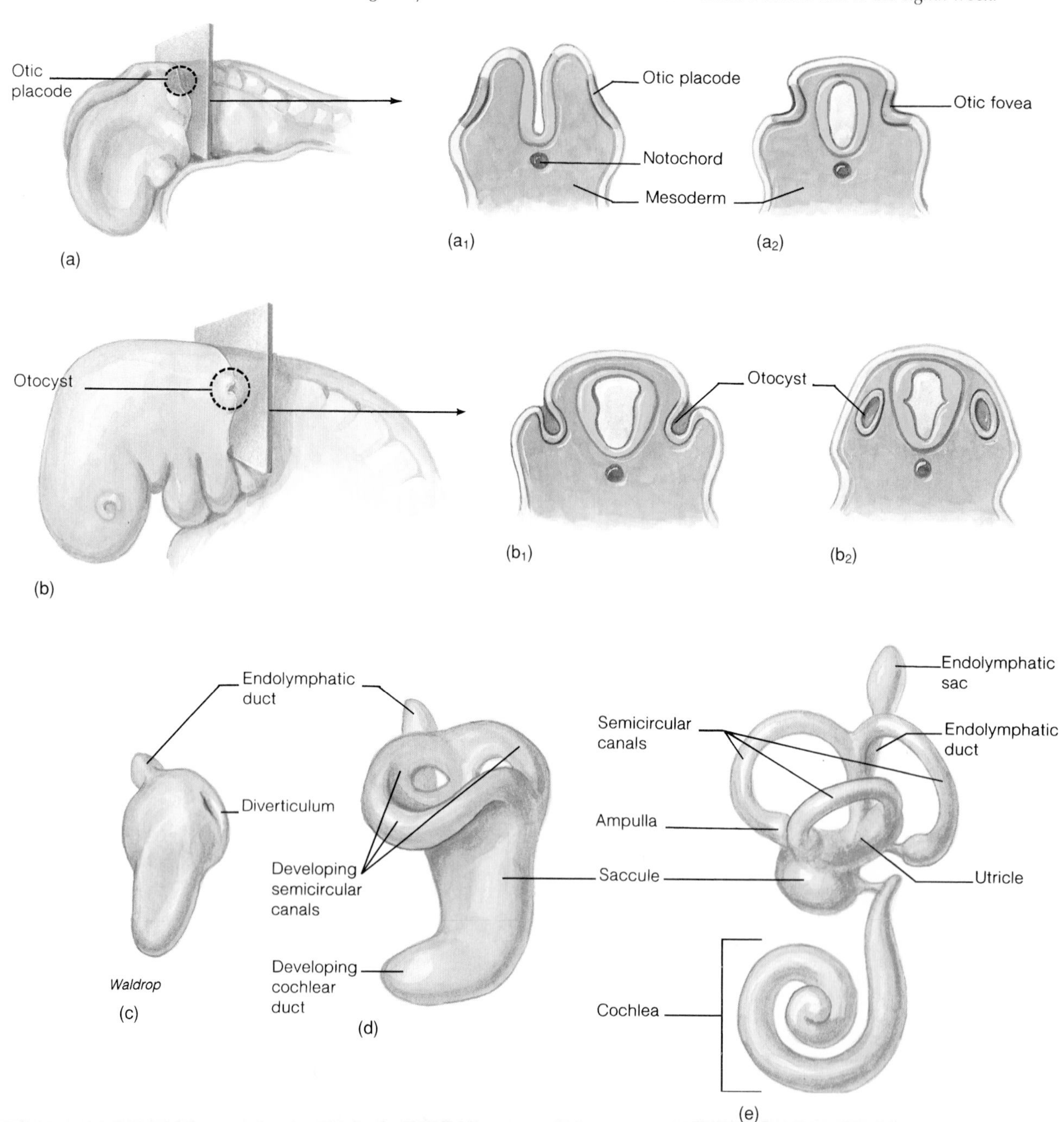

Explanation

The ear begins to develop at the same time as the eye, early during the fourth week. All three embryonic germ layers—ectoderm, mesoderm, and endoderm—are involved in the formation of the ear. Both types of ectoderm (neuroectoderm and surface ectoderm) play a role.

The ear of an adult is structurally and functionally divided into an **outer ear,** a **middle ear,** and an **inner ear,** each of which has a separate embryonic origin. The inner ear does not develop from deep embryonic tissue as one might expect, but rather begins to form early in the fourth week when a plate of surface ectoderm called the **otic** (*o'tik*) **placode** appears lateral to the developing hindbrain (exhibit II). The otic placode soon invaginates and forms an **otic fovea.** Toward the end of the fourth week, the outer edges of the invaginated otic fovea come together and fuse to form an **otocyst.** The otocyst soon pinches off and separates from the surface ectoderm. The otocyst further differentiates to form a dorsal **utricular portion** and a ventral **saccular portion.** Three separate diverticula extend outward from the utricular portion and develop into the **semicircular canals,** which later function in balance and equilibrium. A tubular diverticulum called the **cochlear duct** extends in a coiled fashion from the saccular portion and forms the membranous portion of the **cochlea** of the ear (exhibit II). The **spiral organ,** which is the functional portion of the cochlea, differentiates from cells along the wall of the cochlear duct (exhibit III). The sensory nerves that innervate the inner ear are derived from neuroectoderm from the developing brain.

The differentiating otocyst is surrounded by mesodermal tissue that soon forms a cartilaginous **otic capsule** (exhibit III). As the otocyst and surrounding otic capsule grow, vacuoles containing the fluid **perilymph** form within the otic capsule. The vacuoles soon enlarge and coalesce to form the **perilymphatic space,** which divides into the **scala tympani** and the **scala vestibuli.** Eventually, the cartilaginous otic capsule ossifies to form the **bony** (osseous) **labyrinth** of the inner ear. The middle-ear chamber is referred to as the **tympanic cavity** and derives from the first pharyngeal pouch (exhibit IV). The **auditory ossicles,** which amplify incoming sound waves, derive from the first and second pharyngeal arch cartilages. As the tympanic cavity enlarges, it surrounds and encloses the developing ossicles (exhibit IV). The connection of the tympanic cavity to the pharynx gradually elongates to develop into the **auditory** (eustachian) **tube,** which remains patent throughout life and is important in maintaining an equilibrium of air pressure between the pharyngeal and tympanic cavities.

Exhibit III The formation of the cochlea and the spiral organ from the otic capsule. (*a–d*) Successive stages of development of the perilymphatic space and the spiral organ from the eighth to the twentieth week.

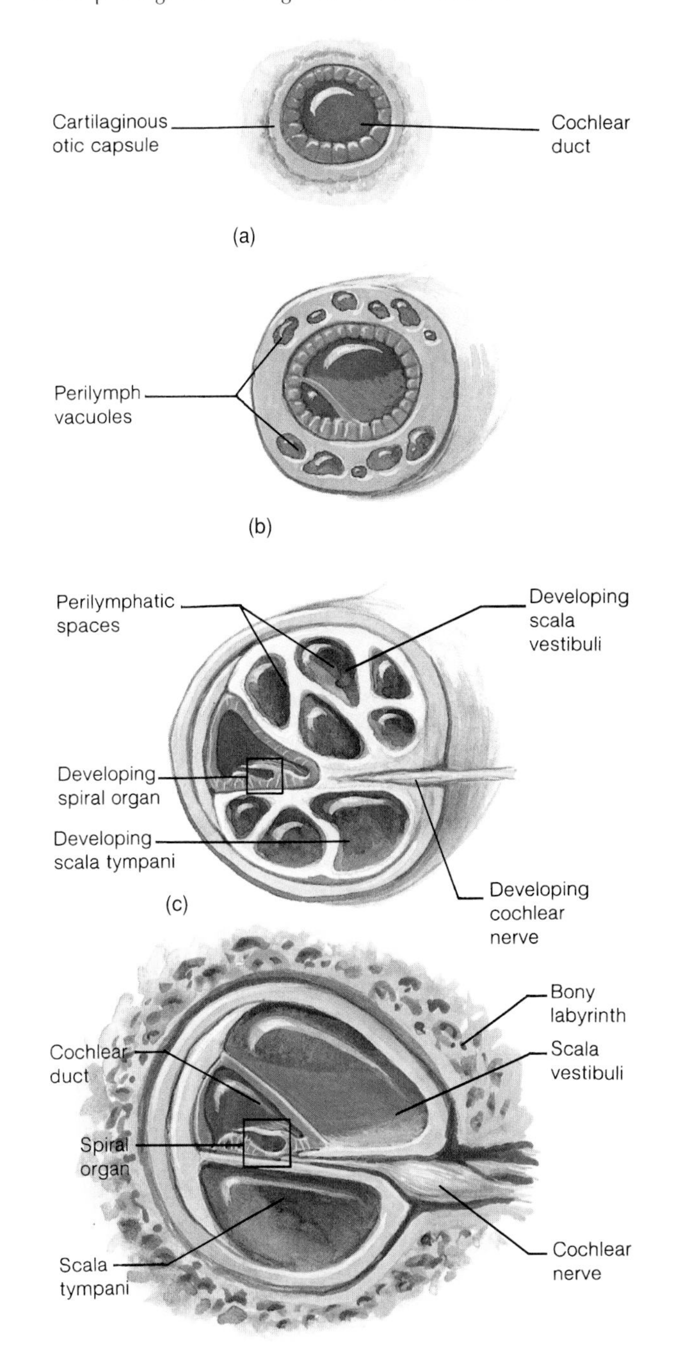

Continued

Developmental Exposition

Continued

Exhibit IV The development of the outer- and middle-ear regions and the auditory ossicles. (*a*) A lateral view of a 4-week-old embryo showing the position of the cut depicted in the sequential development (*b–e*). (*b*) The embryo at 4 weeks illustrating the invagination of the surface ectoderm and the evagination of the endoderm at the level of the first pharyngeal pouch. (*c*) During the fifth week, mesenchymal condensations are apparent, from which the auditory ossicles will be derived. (*d*) Further invagination and evagination at 6 weeks correctly position the structures of the outer- and middle-ear regions. (*e*) By the end of the eighth week, the auditory ossicles, tympanic membrane, auditory tube, and external auditory canal have formed.

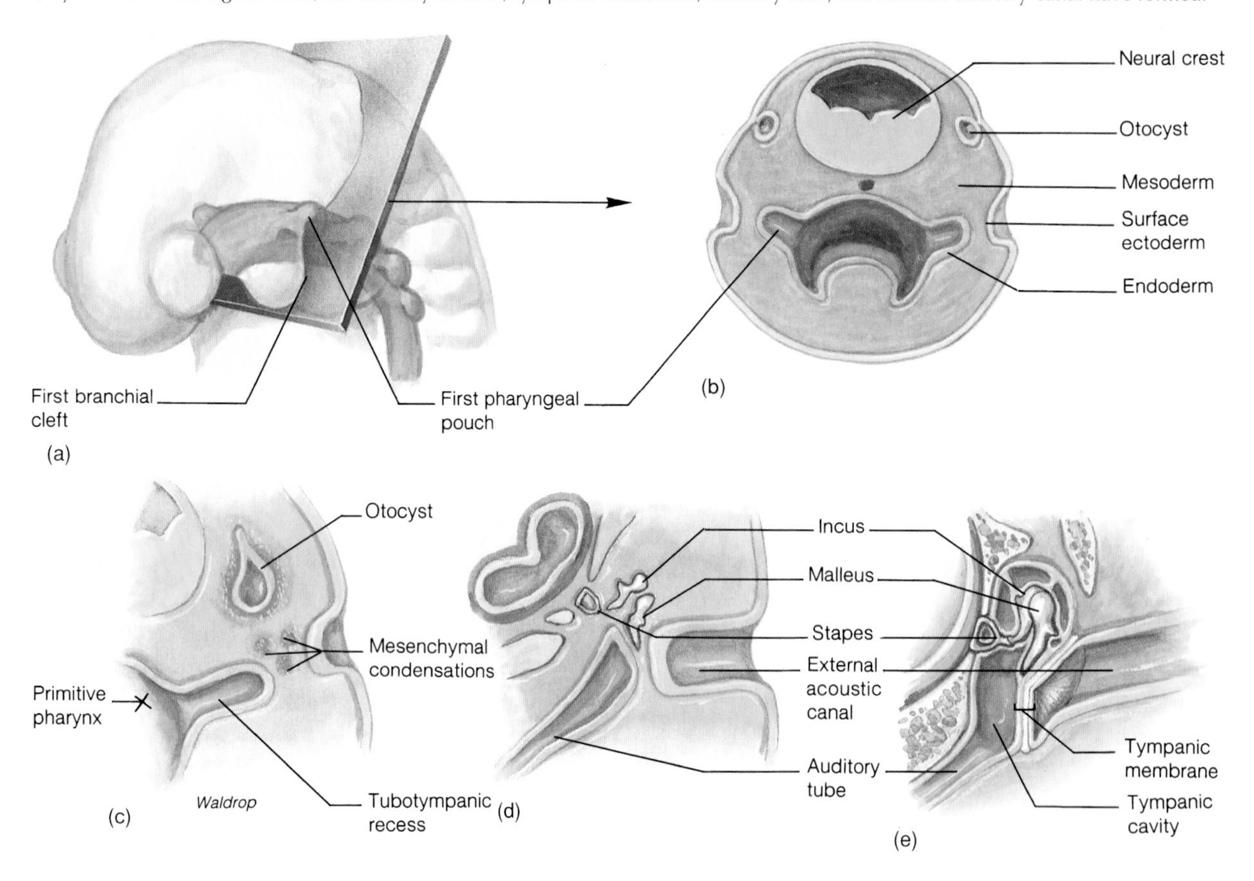

The outer ear includes the fleshy **auricle** attached to the side of the head and the tubular **external acoustic meatus** that extends into the temporal bone of the skull. The external acoustic meatus forms from the surface ectoderm that covers the dorsal end of the first branchial groove (exhibit IV). A solid epithelial plate called the **meatal plug** soon develops at the bottom of the funnel-shaped branchial groove. The meatal plug is involved in the formation of the inner wall of the external acoustic meatus and contributes to the **tympanic membrane** (eardrum). The tympanic membrane has a dual origin from surface ectoderm and the endoderm lining the first pharyngeal pouch (exhibit IV).

are myopia, hyperopia, presbyopia, and astigmatism. **Myopia** (nearsightedness) is an elongation of the eyeball. As a result, light rays focus at a point in the vitreous humor in front of the retina (fig. 15.42). Only light rays from close objects can be focused clearly on the retina; distant objects appear blurred, hence the common term nearsightedness. **Hyperopia** (farsightedness) is a condition in which the eyeball is too short, which causes light rays to be brought to a focal point behind the

myopia: Gk. *myein*, to shut; *ops*, eye

hyperopia: Gk. *hyper*, over; *ops*, eye

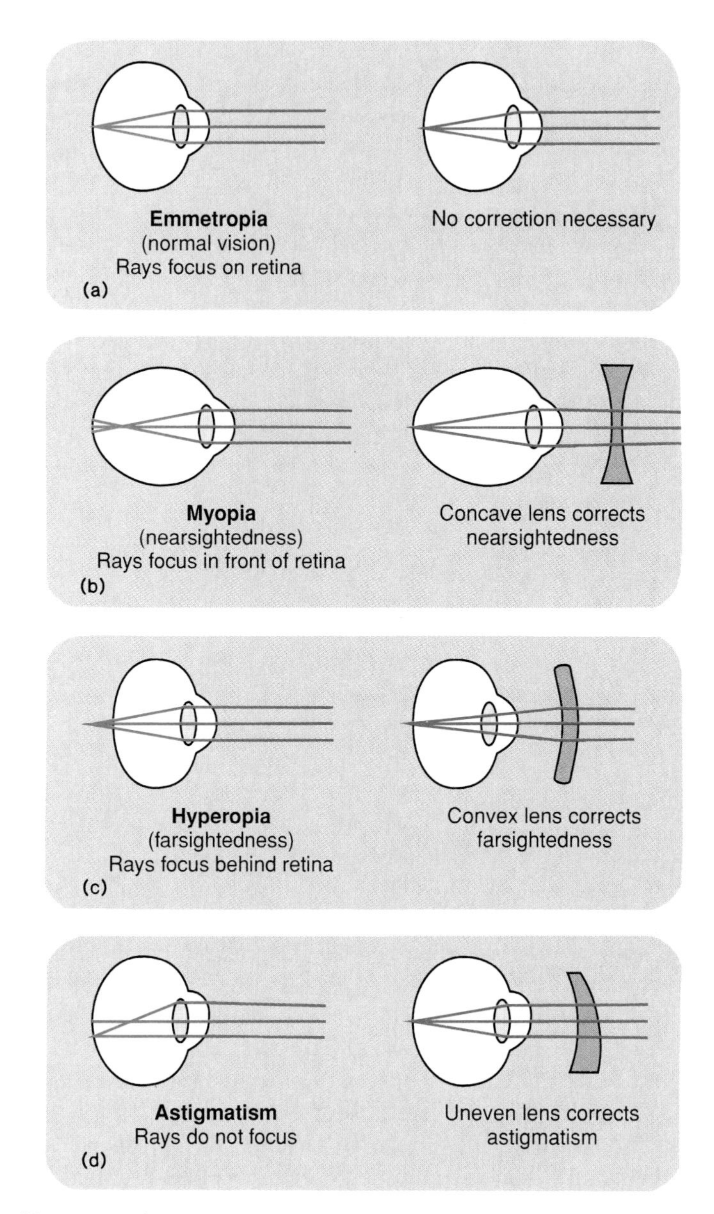

Figure 15.42

In a normal eye (*a*), parallel rays of light are brought to a focus on the retina by refraction in the cornea and lens. If the eye is too long, as in myopia (*b*), the focus is in front of the retina. This can be corrected by a concave lens. If the eye is too short, as in hyperopia (*c*), the focus is behind the retina. This is corrected by a convex lens. In astigmatism (*d*), light refraction is uneven due to an abnormal shape of the cornea or lens.

retina. Although visual accommodation aids a hyperopic person, it generally does not help enough for the person to clearly see very close or distant objects. **Presbyopia** *(prez-be-o'pe-ă)* is a condition in which the lens tends to lose its elasticity and ability to accommodate. It is relatively common in people over 50 years of age. In order to read print on a page, a person with presbyopia must hold the page farther from the eyes than the normal reading distance. **Astigmatism** is a condition in which an irregular curvature of the cornea or lens of the eye distorts the refraction of light rays. If a person with astigmatism views a circle, the image will not appear clear in all 360 degrees; the part of the circle that appears blurred can be used to map the astigmatism.

Various glass or plastic lenses are generally prescribed for people with the visual impairments just described. Myopia may be corrected with a biconcave lens; hyperopia with a biconvex lens; and presbyopia with bifocals, or a combination of two lenses adjusted for near and distant vision. Correction for astigmatism requires a careful assessment of the irregularities and a prescription of a specially ground corrective lens.

As an alternative to a concave lens, a surgical procedure called **radial keratotomy** *(ker-ă-tot'ŏ-me)* is sometimes used to treat moderate myopia. In this technique, 8 to 16 microscopic slashes, like the spokes of a wheel, are made in the cornea from the center to the edge. The ocular pressure inside the eyeball bulges the weakened cornea and flattens its center, changing the focal length of the eyeball. In a relatively new laser surgery called **photorefractive keratectomy** *(ker-ă-tek'tŏ-me)* the cornea is flattened by vaporizing microscopic slivers from its surface.

Cataracts

A cataract is a clouding of the lens that leads to a gradual blurring of vision and the eventual loss of sight. A cataract is not a growth within or upon the eye, but rather a chemical change in the protein of the lens. It is caused by injury, poisons, infections, or age degeneration. Recent evidence indicates that even excessive UV light may cause cataracts.

Cataracts are the leading cause of blindness. A cataract can be removed surgically, however, and vision restored by implanting a tiny intraocular lens that either clips to the iris or is secured into the vacant lens capsule. Special contact lenses or thick lenses for glasses are other options.

Detachment of the Retina

Retinal detachment is a separation of the nervous or visual layer of the retina from the underlying pigment epithelium. It generally begins as a minute tear in the retina that gradually extends as vitreous fluid accumulates between the layers. Retinal detachment may result from hemorrhage, a tumor, degeneration, or trauma from a violent blow to the eye. A detached retina may be repaired by using laser beams, cryoprobes, or intense heat to destroy the tissue beneath the tear and rejoin the layers.

Macular Degeneration

Another retinal disorder, macular degeneration, is common after age 70, but its cause is not fully known. It occurs when the blood supply to the macular area of the retina is reduced, often eventually resulting in hemorrhages, or when other fluid collects in this area, reducing central vision sharpness. People with

presbyopia: Gk. *presbys*, old man; *ops*, eye
astigmatism: Gk. *a*, without; *stigma*, point

cataract: Gk. *katarrhegnynai*, to break down

macular degeneration retain peripheral vision, but have difficulty focusing on an object directly in front of them. A simple home test called the *Amsler grid* can be used to detect this disorder in its early stages.

Strabismus

Strabismus is a condition in which both eyes do not focus on the same axis of vision. This prevents stereoscopic vision, and results in varied visual impairments. Strabismus is usually caused by a weakened extrinsic eye muscle.

Strabismus is assessed while the patient attempts to look straight ahead. If the afflicted eye is turned toward the nose, the condition is called convergent strabismus **(esotropia).** If the eye is turned outward, it is called divergent strabismus **(exotropia).** Disuse of the afflicted eye causes a visual impairment called **amblyopia.** Visual input from the normal eye and the eye with strabismus results in **diplopia** *(dip-lo'pe-ă)*, or double vision. A normal, healthy person who has overindulged in alcoholic beverages may experience diplopia.

Infections and Diseases of the Eye

Infections

Infections and inflammation can occur in any of the accessory structures of the eye or in structures within or on the eyeball itself. The causes of infections are usually microorganisms, mechanical irritation, or sensitivity to particular substances.

Conjunctivitis (inflammation of the conjunctiva) may result from sensitivity to light, allergens, or an infection caused by viruses or bacteria. Bacterial conjunctivitis is commonly called "pinkeye."

Keratitis (inflammation of the cornea) may develop secondarily from conjunctivitis or be caused by such diseases as tuberculosis, syphilis, mumps, or measles. Keratitis is painful and may cause blindness if untreated.

A **chalazion** (*kă-la'ze-on*) is a tumor or cyst on the eyelid that results from infection of the tarsal glands and a subsequent blockage of the ducts of these glands.

A **sty** (*hordeola*) is a relatively common mild infection of the follicle of an eyelash or the sebaceous gland of the follicle. A sty may easily spread from one eyelash to another if untreated. Poor hygiene and the excessive use of cosmetics may contribute to its development.

Diseases

Trachoma *(tră-ko'mă)* is a highly contagious bacterial disease of the conjunctiva and cornea. Although rare in the United States, it is estimated that over 500 million people are afflicted with this disease. Trachoma responds well to treatment with sulfonamides and some antibiotics, but if untreated it will spread progressively until it covers the cornea. At this stage, vision is lost and the eye undergoes degenerative changes.

Glaucoma *(glau-ko'mă)* is the second leading cause of blindness and is particularly common in underdeveloped countries. Although it can afflict individuals of any age, 95% of the cases involve people over the age of 40. Glaucoma is characterized by an abnormal increase in the intraocular pressure of the eyeball. Aqueous humor does not drain through the scleral venous sinuses as quickly as it is produced. Accumulation of fluid causes compression of the blood vessels in the eyeball and compression of the optic nerve. Retinal cells die and the optic nerve may atrophy, producing blindness.

Infections, Diseases, and Functional Impairments of the Ear

Disorders of the ear are common and may affect both hearing and the vestibular functions. The ear is subject to numerous infections and diseases—some of which can be prevented.

Infections and Diseases

External otitis *(o-ti'tis)* is a general term for infections of the outer ear. The causes of external otitis range from dermatitis to fungal and bacterial infections.

Acute purulent otitis media is a middle-ear infection. Pathogens that cause this disease usually enter through the auditory tube, often following a cold or tonsillitis. Children frequently have middle-ear infections because of their susceptibility to infections and their short and straight auditory tubes. As a middle-ear infection progresses to the inflammatory stage, the auditory tube closes and drainage is blocked. An intense earache is a common symptom of a middle-ear infection. The pressure from the inflammation may eventually rupture the tympanic membrane to permit drainage.

Repeated middle-ear infections, particularly in children, usually call for an incision of the tympanic membrane known as a **myringotomy** *(mir"in-got'ŏ-me)* and the implantation of a tiny tube within the tympanic membrane (fig. 15.43). The tube, which is eventually sloughed out of the ear, permits the infection to heal and helps prohibit further infections by keeping the auditory tube open.

Perforation of the tympanic membrane may occur as the result of infections or trauma. The membrane might be ruptured, for example, by a sudden, intense noise. Spontaneous perforation of the membrane usually heals rapidly, but scar tissue may form and lessen sensitivity to sound vibrations.

Otosclerosis *(o"to-sklĕ-ro'sis)* is a progressive deterioration of the normal bone in the bony labyrinth and its replacement with vascular spongy bone. This frequently causes hearing loss as

Amsler grid: from Marc Amsler, Swiss opthalmologist, 1891–1968
amblyopia: Gk. *amblys*, dull; *ops*, vision

glaucoma: Gk. *glaukos*, gray

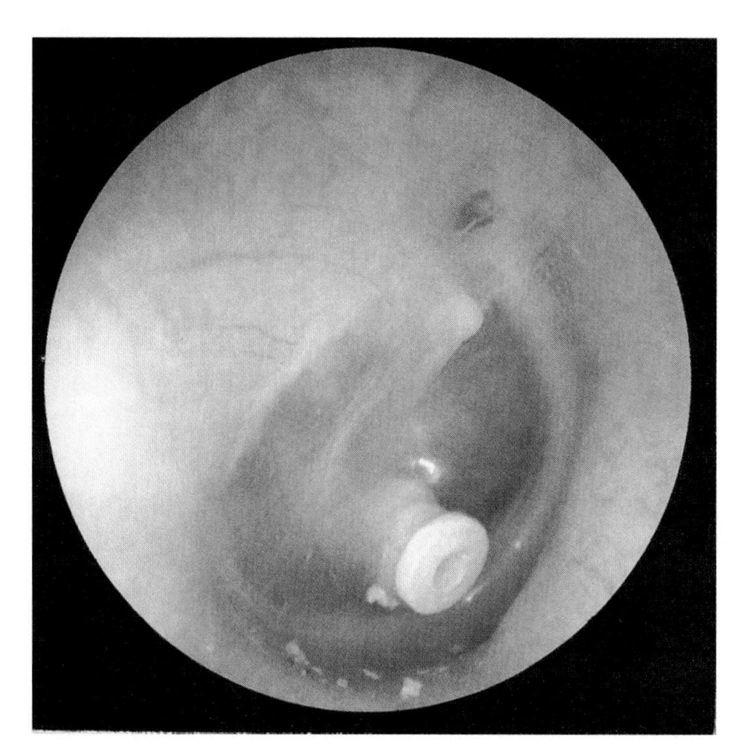

Figure 15.43

An implanted ventilation tube in the tympanic membrane following a myringotomy.

the auditory ossicles are immobilized. Surgical scraping of the bone growth and replacing the stapes with a prosthesis usually restores hearing.

Ménière's *(mān-e-ārz')* **disease** afflicts the inner ear and may cause hearing loss as well as equilibrium disturbance. The causes of Ménière's disease are not completely understood, but they are thought to be related to a dysfunction of the autonomic nervous system that causes a vasoconstriction within the inner ear. The disease is characterized by recurrent periods of **vertigo** (dizziness and a sensation of rotation), **tinnitus** (*tĭ-ni'tus*) (ringing in the ear), and progressive deafness in the affected ear. Ménière's disease is chronic and affects both sexes equally. It is more common in elderly people.

Ménière's disease: from Prosper Ménière, French physician, 1799–1862
vertigo: L. *vertigo*, dizziness
tinnitus: L. *tinnitus*, ring or tingle

Auditory Impairment

Loss of hearing results from disease, trauma, or developmental problems involving any portion of the auditory apparatus, cochlear nerve and auditory pathway, or areas of auditory perception within the brain. Hearing impairment varies from slight disablement, which may or may not worsen, to total deafness. Some types of hearing impairment, including deafness, can be mitigated through hearing aids or surgery.

Based on the structures involved, there are two types of deafness. **Conduction deafness** is caused by an interference with the sound waves through the outer or middle ear. Conduction problems include impacted cerumen (wax), a ruptured tympanic membrane, a severe middle-ear infection, and adhesions (tissue growths) of one or more auditory ossicles *(otosclerosis)*. Medical treatment usually improves the hearing loss from conductive deafness.

Perceptive (sensorineural) deafness results from disorders that affect the inner ear, the cochlear nerve or nerve pathway, or auditory centers within the brain. Perceptive impairment ranges in severity from the inability to hear certain frequencies to total deafness. Perceptive deafness may be caused by diseases, trauma, or genetic or developmental problems. Elderly people frequently experience some perceptive deafness. The ability to perceive high-frequency sounds is generally lost first. Hearing aids may help some patients with perceptive deafness. This type of deafness is permanent, however, because it involves destruction of sensory structures that cannot regenerate.

Clinical Case Study Answer

When the handle of the vibrating tuning fork was placed on the mastoid process, the temporal bone transmitted sound waves directly to the inner ear. This bypassed the conductive components of the middle ear, which are (beginning with the first to receive sound waves) the tympanic membrane, malleus, incus, and stapes. Since the patient could hear well when the sound waves were transmitted through the temporal bone, it could be assumed that the inner-ear organs, as well as the neural pathways, were functioning. It was therefore concluded that the problem was in the conductive components. This is often the result of otosclerosis (a spongy proliferation of bone) or other conditions affecting the auditory ossicles and/or tympanic membrane, many of which are treatable surgically.

Chapter Summary

Overview of Sensory Perception (p. 471)

1. Sensory organs are specialized extensions of the nervous system that respond to specific stimuli and conduct nerve impulses.
2. A stimulus to a receptor that conducts an impulse to the brain is necessary for perception.
3. Sensory organs act as energy filters that permit perception of only a narrow range of energy.

Classification of the Senses (pp. 471–472)

1. The senses are classified according to structure or location of the receptors, or on the basis of the stimuli to which the receptors respond.
2. The receptor cells for the general senses are widespread throughout the body and are simple in structure. The receptor cells for the special sensory organs are localized in complex receptor organs and have extensive neural pathways.
3. The somatic senses arise in cutaneous receptors and proprioceptors; visceral senses arise in receptors located within the visceral organs.
4. Phasic receptors respond quickly to a stimulus but then adapt and decrease their firing rate. Tonic receptors produce a constant rate of firing.

Somatic Senses (pp. 472–479)

1. Corpuscles of touch, free nerve endings, and root hair plexuses are tactile receptors, responding to light touch.
2. Lamellated corpuscles are pressure receptors located in the deep dermis or hypodermis. They are also associated with synovial joints.
3. The organs of Ruffini and bulbs of Krause are both mechanoreceptors; they respond to deep and light pressure, respectively.
4. Free nerve endings respond to light touch and are the principal pain receptors. They also serve as thermoreceptors, responding to changes in temperature.
5. Joint kinesthetic receptors, neuromuscular spindles, and neurotendinous receptors are proprioceptors that are sensitive to changes in stretch and tension.

Olfactory Sense (p. 479)

1. Olfactory receptors of the olfactory nerve respond to chemical stimuli and transmit the sensation of olfaction (smell) to the cerebral cortex.
2. Olfaction functions closely with gustation (taste) in that the receptors of both are chemoreceptors, requiring dissolved substances for stimuli.

Gustatory Sense (pp. 480–481)

1. Taste receptors in taste buds are chemoreceptors and transmit the sensation of gustation to the cerebral cortex.
2. Taste buds are found in the vallate and fungiform papillae of the tongue. Filiform papillae are not involved in taste perception; they give the tongue an abrasive feel.
3. The kinds of taste sensation are sweet, salty, sour, and bitter.

Visual Sense (pp. 481–499)

1. Protective structures of the eye include the eyebrow, eyelids, eyelashes, conjunctiva, and lacrimal gland.
2. Six extrinsic ocular muscles control the movement of the eyeball.
3. The eyeball consists of the fibrous tunic, which is divided into the sclera and cornea; the vascular tunic, which consists of the choroid, the ciliary body, and the iris; and the internal tunic, or retina. The retina has an outer pigmented layer and an inner nervous layer. The transparent lens is not part of any tunic.
4. Rod and cone cells, which are the photoreceptors in the nervous layer of the retina, respond to dim and colored light, respectively. Cone cells are concentrated in the fovea centralis, the area of keenest vision.
5. Rod and cone cells contain specific pigments that provide sensitivity to different light rays.
6. The visual process includes the transmission and refraction of light rays, accommodation of the lens, constriction of the pupil, and convergence of the eyes.
   (a) Refraction is achieved as incoming light rays pass through the cornea, aqueous humor, lens, and vitreous humor.
   (b) A sharp focus is accomplished as the curvature of the lens is changed by autonomic contraction of smooth muscles within the ciliary body.
7. Neural pathways from the retina to the superior colliculus help to regulate eye and body movements. Most fibers from the retina project to the lateral geniculate body, and from there to the striate cortex.
8. The sensory components of the eye have been formed by 20 weeks; the accessory structures have been formed by 32 weeks.

Senses of Hearing and Balance (pp. 499–510)

1. The outer ear consists of the auricle and the external acoustic canal.
2. The middle ear (tympanic cavity), bounded by the tympanic membrane and the vestibular and cochlear windows, contains the auditory ossicles (malleus, incus, and stapes) and the auditory muscles (tensor tympani and stapedius).
3. The middle-ear cavity connects to the pharynx through the auditory tube.
4. The inner ear contains the spiral organ for hearing. It also contains the semicircular canals, saccule, and utricle (located in the vestibule) for maintaining balance and equilibrium.
5. The development of the ear begins during the fourth week and is complete by the thirty-second week.

Review Activities

Objective Questions

1. Which of the following conditions is (are) necessary for the perception of a sensation to take place?
   (a) presence of a stimulus
   (b) nerve impulse conduction
   (c) activation of a receptor
   (d) all of the above
2. The cutaneous receptor sensitized to detect deep pressure is
   (a) a root hair plexus.
   (b) a lamellated corpuscle.
   (c) a bulb of Krause.
   (d) a free nerve ending.
3. Proprioceptors that are located within the connective tissue capsule in synovial joints are called
   (a) neuromuscular spindles.
   (b) Golgi tendon organs.
   (c) neurotendinous receptors.
   (d) joint kinesthetic receptors.
4. The sensation of visceral pain perceived as arising from another somatic location is known as
   (a) related pain. (c) referred pain.
   (b) phantom pain. (d) parietal pain.
5. When a person with normal vision views an object from a distance of at least 20 feet,
   (a) the ciliary muscles are relaxed.
   (b) the suspensory ligament is taut.
   (c) the lens is flat, having the least convex shape.
   (d) all of the above apply.
6. Which of the following is an avascular ocular tissue?
   (a) the sclera (c) the ciliary body
   (b) the choroid layer (d) the iris

7. Additional light may enter the eyeball in response to contraction of
   (a) ciliary muscles.
   (b) pupillary dilator muscles.
   (c) pupillary constrictor muscles.
   (d) orbicularis oculi muscles.
8. The stimulation of hair cells in the semicircular ducts results from the movement of
   (a) endolymph.
   (b) perilymph.
   (c) the statoconial membrane.
9. The middle ear is separated from the inner ear by
   (a) the cochlear window.
   (b) the tympanic membrane.
   (c) the vestibular window.
   (d) both a and c.
10. Glasses with concave lenses help to correct
    (a) presbyopia.
    (b) myopia.
    (c) hyperopia.
    (d) astigmatism.

Essay Questions

1. What four events are necessary for perception of a sensation? Explain the statement that perception is the step beyond sensation in the taking in of environmental information.
2. List the senses of the body and differentiate between the special and somatic senses. How are these two classes of senses similar?
3. List the functions of proprioceptors and differentiate between the various types. What role do proprioceptors play in the kinesthetic sense?
4. Compare and contrast olfaction and gustation. Identify the cranial nerves that serve both of these senses and trace the sensory pathways of each.
5. Describe the accessory structures of the eye and list their functions.
6. Diagram the structure of the eyeball and label the sclera, cornea, choroid, macula lutea, ciliary body, suspensory ligament, lens, iris, pupil, retina, optic disc, and fovea centralis.
7. Outline and explain the process of focusing light rays onto the fovea centralis.
8. Diagram the ear and label the structures of the outer, middle, and inner ear.
9. Trace a sound wave through the structures of the ear and explain the mechanism of hearing.
10. Explain the mechanism by which equilibrium is maintained and the role played by the two kinds of receptor information.
11. Outline the major events in the development of the eye and ear. When are congenital deformities most likely to occur?
12. List some techniques used to examine the eye and the ear. Give examples of congenital abnormalities of the eye and ear.

Critical-Thinking Questions

1. Explain the phenomenon called sensory adaptation. What advantages does it confer? Can you think of any disadvantages?
2. You know your contact lens is somewhere in your eye, but you can't seem to find it. It's causing you great discomfort, and you're desperate to get it out. You're also worried that it might be displaced into the orbit. Considering the anatomy of the eye, do you have cause for concern? Why or why not?
3. How would you account for the high success rate of corneal transplants as compared to other types of tissue transplantations from one person to another?
4. Nearsighted adults may find that as they grow older they can read without glasses. Explain.
5. People with conduction deafness often speak quietly. By contrast, people with perceptive deafness tend to speak in tones that are louder than normal. Explain the difference in anatomical terms.
6. Describe the procedure called myringotomy. Why is it usually successful in treating children who suffer recurring middle-ear infections? Following this procedure, would a child still experience the discomfort of changing air pressure in the ear brought on by a rapid drop in elevation? (To help you answer these questions, consider the advantage of punching two holes in an oil can—one across from the other—as opposed to punching a single hole.)

SIXTEEN

Circulatory System

Clinical Case Study 521

Functions and Major Components of the Circulatory System 521
- Functions of the Circulatory System 521
- Major Components of the Circulatory System 521

Blood 523
- Formed Elements of Blood 524
- Hemopoiesis 528
- Blood Plasma 528

Heart 529
- Location and General Description 529
- Heart Wall 529
- Chambers and Valves 529
- Circulatory Routes 534
- Conduction System of the Heart 535
- Electrocardiogram 536
- Heart Sounds 537

Blood Vessels 538
- Arteries 538
- Capillaries 538
- Veins 540
- Blood Pressure 541

Principal Arteries of the Body 542
- Aortic Arch 542
- Arteries of the Neck and Head 544
- Arteries of the Shoulder and Upper Extremity 547
- Branches of the Thoracic Portion of the Aorta 548
- Branches of the Abdominal Portion of the Aorta 549
- Arteries of the Pelvis and Lower Extremity 550

Principal Veins of the Body 554
- Veins Draining the Head and Neck 556
- Veins of the Upper Extremity 556
- Veins of the Thorax 557
- Veins of the Lower Extremity 557
- Veins of the Abdominal Region 558
- Hepatic Portal System 558

Fetal Circulation 560

Lymphatic System 562
- Lymph and Lymphatic Vessels 562
- Lymph Nodes 564
- Other Lymphoid Organs 565

Clinical Considerations 566
- Cardiovascular Assessment 567

Developmental Exposition: The Circulatory System 568
- Blood Disorders 572
- Heart Diseases 572
- Vascular Disorders 574
- Disorders of the Lymphatic System 575
- Trauma to the Circulatory System 576

Clinical Case Study Answer 577

Internal Affairs 578

Chapter Summary 579

Review Activities 580

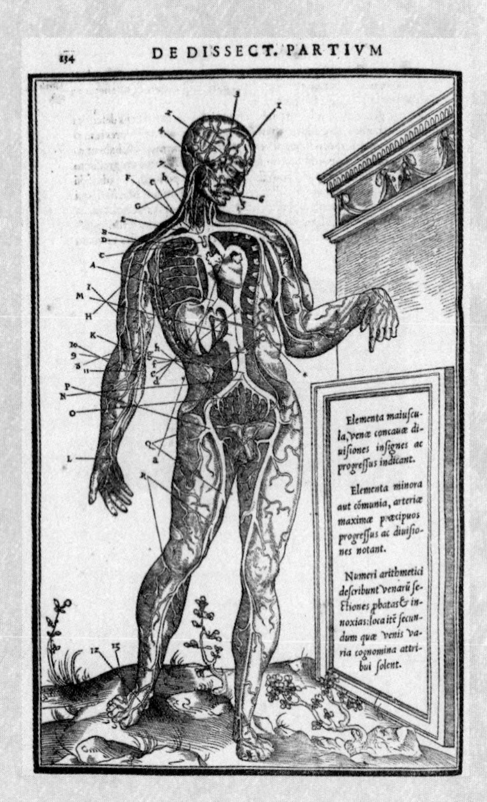

Charles Estienne, French physician of the sixteenth century, was born into a family of scholars, printers, and bookdealers. His text on anatomy, De Dissectione Partium Corporis Humani, *contained a wealth of detailed illustrations, such as this one of the blood vessels. Estienne was the first to remark on the presence of valves in the veins, crucial to the later understanding of blood circulation.*

Clinical Case Study

A 65-year-old woman who had been discharged from the hospital following a myocardial infarction (heart attack) returned to the emergency room several weeks later because of the sudden onset of pain in her right lower extremity. The attending physician noticed that the patient's leg was also pale and cool from the knee down. Furthermore, he was unable to detect a pulse over the dorsal pedal and popliteal arteries. A good femoral pulse, however, was palpable in the inguinal region. Upon questioning, the patient stated that she was sent home with blood-thinning medication because tests had revealed that the heart attack had caused a blood clot to form within her heart. The doctors were worried, she added, that "a piece of the clot could break off and go to other parts of the body."

Explain how the patient's heart attack could have led to her current leg complaints. In which side of her heart (right or left) did the previously diagnosed blood clot most likely reside? Explain anatomically where the cause of her new problem is located and describe how it came to arrive there.

Functions and Major Components of the Circulatory System

An efficient circulatory system is necessary for maintaining the life of complex multicellular organisms.

Objective 1 Describe the functions of the circulatory system.

Objective 2 Describe the major components of the circulatory system.

A unicellular organism can provide for its own maintenance and continuity by performing the wide variety of functions needed for life. By contrast, the complex human body is composed of trillions of specialized cells that demonstrate a division of labor. Cells of a multicellular organism depend on one another for the very basics of their existence. The majority of the cells of the body are firmly implanted in tissues and are incapable of procuring food and oxygen on their own, or even moving away from their own wastes. Therefore, a highly specialized and effective means of transporting materials within the body is needed.

The blood contained within vessels serves this transportation function. An estimated 60,000 miles of vessels throughout the body of an adult ensure that continued sustenance reaches each of the trillions of living cells. However, the blood can also transport disease-causing viruses, bacteria, and their toxins. As a safeguard, the circulatory system has defense mechanisms—the white blood cells and lymphatic system. In order to perform its various functions, the circulatory system works together with the respiratory, urinary, digestive, endocrine, and integumentary systems in maintaining homeostasis (fig. 16.1; see also Internal Affairs, p. 578).

Functions of the Circulatory System

The many functions of the circulatory system can be grouped into two broad areas: transportation and protection.

1. **Transportation.** All of the substances involved in cellular metabolism are transported by the circulatory system. These substances can be categorized as follows:
   a. *Respiratory*. Red blood cells called **erythrocytes** (*ĕ-rith′rŏ-sīts*) transport oxygen to the tissue cells. In the lungs, oxygen from the inhaled air attaches to hemoglobin molecules within the erythrocytes and is transported to the cells for aerobic respiration. Carbon dioxide produced by cellular respiration is carried by the blood to the lungs for elimination in the exhaled air (fig. 16.2).
   b. *Nutritive*. The digestive system is responsible for the mechanical and chemical breakdown of food to forms that can be absorbed through the intestinal wall into the blood and lymph vessels. The blood then carries these absorbed products of digestion through the liver to the cells of the body.
   c. *Excretory*. Metabolic wastes, excess water and ions, as well as other molecules in plasma (the fluid portion of blood), are filtered through the capillaries of the kidneys into kidney tubules and excreted in urine.
   d. *Regulatory*. The blood carries hormones and other regulatory molecules from their site of origin to distant target tissues.
2. **Protection.** The circulatory system protects against injury and foreign microbes or toxins introduced into the body. The clotting mechanism protects against blood loss when vessels are damaged, and white blood cells called **leukocytes** (*loo′kŏ-sīts*) render the body immune to many disease-causing agents. Leukocytes may also protect the body through phagocytosis (see fig. 3.12).

Major Components of the Circulatory System

The circulatory system is frequently divided into the **cardiovascular system,** which consists of the heart, blood vessels, and blood, and the **lymphatic system,** which consists of lymphatic vessels and lymphoid tissues within the spleen, thymus, tonsils, and lymph nodes.

The **heart** is a four-chambered double pump. Its pumping action creates the pressure needed to push blood in the vessels to the lungs and body cells. At rest, the heart of an adult pumps

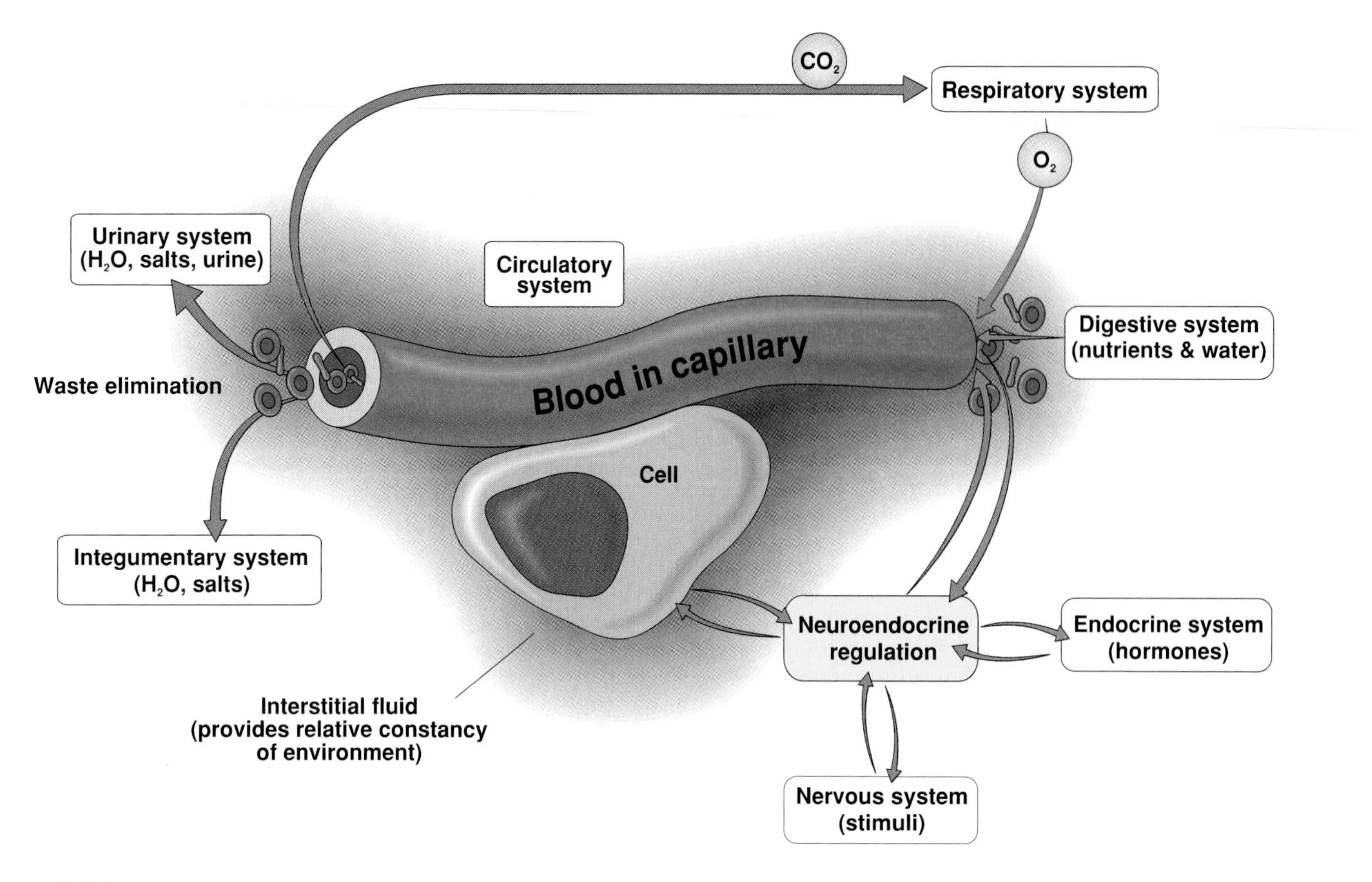

Figure 16.1

The relationship of the circulatory system to the other body systems in maintaining homeostasis.

about 5 liters of blood per minute. It takes only about a minute for blood to be circulated to the most distal extremity and back to the heart.

Blood vessels form a tubular network that permits blood to flow from the heart to all living cells of the body and then back to the heart. *Arteries* carry blood away from the heart, while *veins* return blood to the heart. Arteries and veins are continuous with each other through smaller blood vessels.

Arteries branch extensively to form a network of progressively smaller vessels. Those that are microscopic are called *arterioles* (*ar-te″re-o′lēz*). Conversely, microscopic-sized veins called *venules* (*ven′yoolz*) deliver blood to progressively larger vessels that empty into the large veins. Blood passes from the arterial to the venous system in *capillaries*, which are the thinnest and most numerous blood vessels. All exchanges of fluid, nutrients, and wastes between the blood and tissue cells occur across the walls of capillaries.

Fluid derived from plasma passes out of capillary walls into the surrounding tissues, where it is called *interstitial fluid* or *tissue fluid*. Some of this fluid returns directly to capillaries and some enters into **lymphatic vessels** located in the connective tissues around the blood vessels. Fluid in lymphatic vessels is called *lymph*. This fluid is returned to the venous blood at particular sites. **Lymph nodes,** positioned along the way, cleanse the lymph prior to its return to the venous blood.

Endothermic (warm-blooded) animals, including humans, need an efficient circulatory system to transport oxygen-rich blood rapidly to all parts of the body. Of the vertebrates, only birds and mammals with their consistently warm body temperatures are considered endothermic, and only birds, mammals, and a few reptiles (crocodiles and alligators) have a four-chambered heart.

1. Name the components of the circulatory system that function in oxygen transport, in the transport of nutrients from the digestive system, and in protection.
2. Define the terms *artery*, *vein*, and *capillary* and describe the function of each of these vessels.
3. Define the terms *interstitial fluid* and *lymph*. How do these fluids relate to blood plasma?

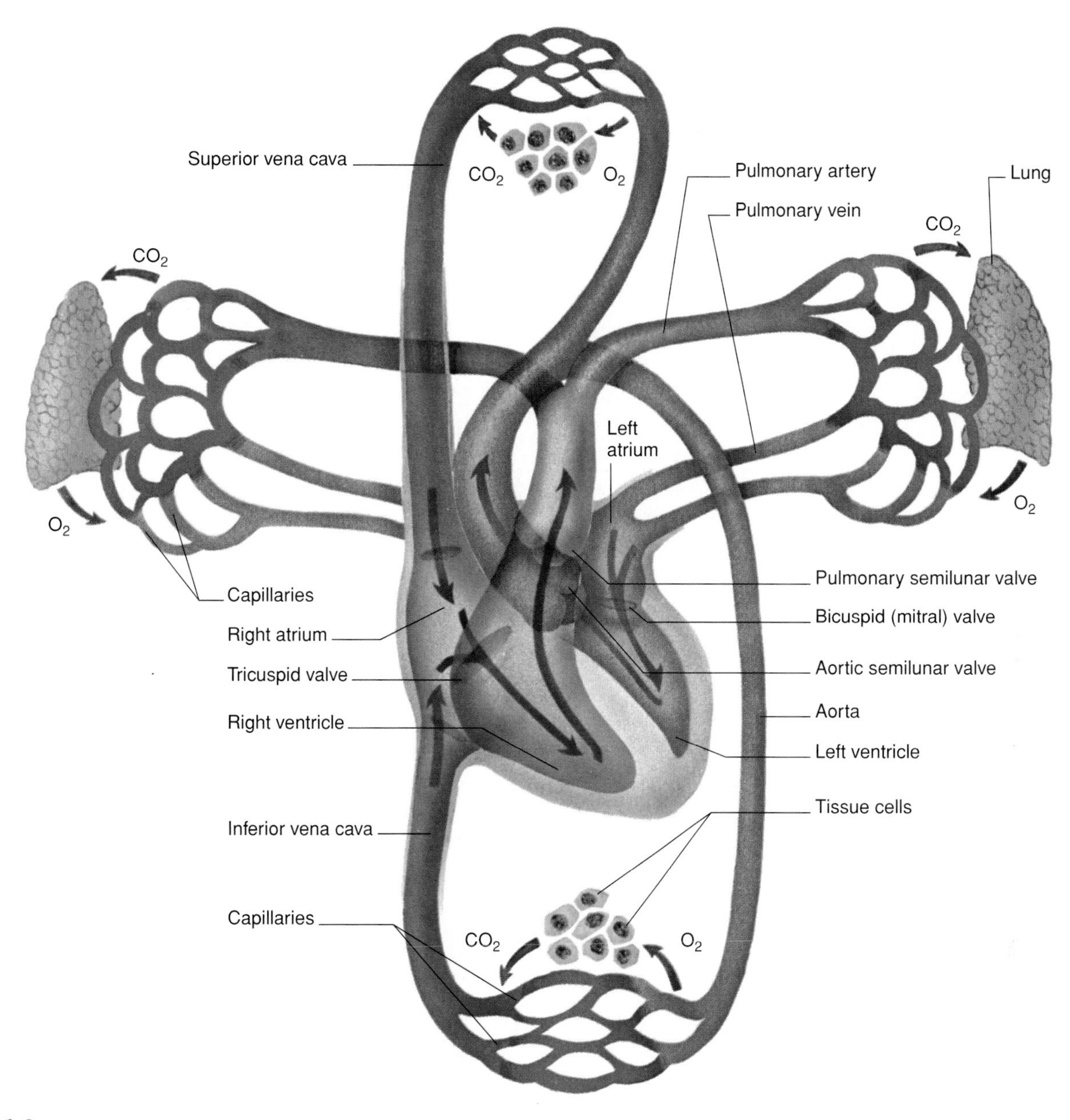

Figure 16.2

A schematic diagram of the circulatory system.

Blood

Blood, a highly specialized connective tissue, consists of formed elements—erythrocytes, leukocytes, and platelets (thrombocytes)—that are suspended and carried in the blood plasma. The constituents of blood function in transport, immunity, and blood-clotting mechanisms.

Objective 3 List the different types of formed elements of blood and describe their appearance.

Objective 4 Describe the origin of erythrocytes, leukocytes, and platelets.

Objective 5 List the different types of substances found in blood plasma.

Objective 6 Describe the origin and function of the different categories of blood plasma proteins.

The total blood volume in the average-sized adult is about 5 liters, constituting about 8% of the total body weight. Blood leaving the heart is referred to as *arterial blood*. Arterial blood, with the exception of that going to the lungs, is bright red in color because of the high concentration of oxyhemoglobin (the combination of oxygen and hemoglobin) in the erythrocytes. *Venous blood* is blood returning to the heart. Except for the venous

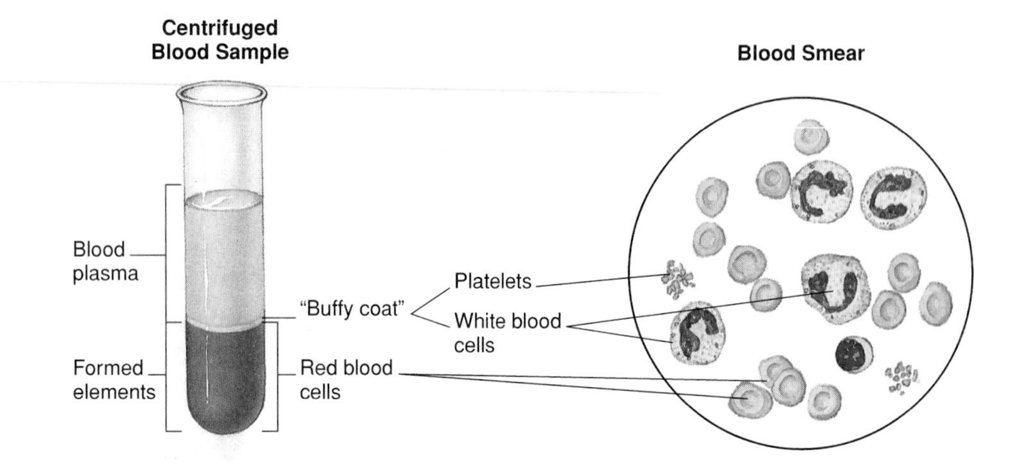

Figure 16.3

The formed elements (blood cells) become packed at the bottom of the test tube when whole blood is centrifuged, leaving the fluid blood plasma at the top of the tube. Red blood cells are the most abundant of the blood cells—white blood cells and platelets form only a thin, light-colored buffy coat at the interface of the packed red blood cells and the blood plasma.

blood from the lungs, it contains less oxygen and is, therefore, a darker red than the oxygen-rich arterial blood. Blood has a viscosity that ranges between 4.5 and 5.5. This means that it is thicker than water, which has a viscosity of 1.0. Blood has a pH range of 7.35 to 7.45 and a temperature within the thorax of the body of about 38° C (100.4° F). When you donate blood, a "unit" (half a liter) is drained. This represents approximately a tenth of your total blood volume.

Blood is composed of a cellular portion, called **formed elements,** and a fluid portion, called **blood plasma.** When a blood sample is centrifuged, the heavier formed elements are packed into the bottom of the tube, leaving blood plasma at the top (fig. 16.3). The formed elements constitute approximately 45% of the total blood volume, a percentage known as the *hematocrit*. The blood plasma accounts for the remaining 55%. The hematocrit closely approximates the percentage of red blood cells per given volume of blood and is an important indicator of the oxygen-carrying capacity of blood.

Formed Elements of Blood

The formed elements of blood include **erythrocytes** (red blood cells, or RBCs); **leukocytes** (white blood cells, or WBCs); and **platelets** (thrombocytes). Erythrocytes are by far the most numerous of these three types. A cubic millimeter of blood contains 5.1 million to 5.8 million erythrocytes in males and 4.3 million to 5.2 million erythrocytes in females. By contrast, the same volume of blood contains only 5,000 to 10,000 leukocytes and 250,000 to 450,000 platelets.

plasma: Gk. *plasma*, to form or mold
hematocrit: Gk. *haima*, blood; *krino*, to separate
erythrocytes: Gk. *erythros*, red; *kytos*, hollow (cell)
leukocytes: Gk. *leukos*, white; *kytos*, hollow (cell)

Erythrocytes

Erythrocytes are biconcave discs—flattened, with a depressed center—about 7.5 µm in diameter and 2.5 µm thick. Their unique shape relates to their function of transporting oxygen; it provides an increased surface area through which gas can diffuse (fig. 16.4). Mature erythrocytes lack a nucleus and mitochondria (they obtain energy from anaerobic respiration). They have a circulating life span of only about 120 days, after which they are destroyed by phagocytic cells in the liver and spleen.

Oxygen molecules attached to *hemoglobin molecules* within erythrocytes give blood its red color. A hemoglobin molecule consists of four protein chains called *globins,* each of which is bound to one *heme,* a red-pigmented molecule. Each heme contains an atom of iron that can combine with one molecule of oxygen. Thus, the hemoglobin molecule as a whole can transport up to four molecules of oxygen. Considering that each erythrocyte contains approximately 280 million hemoglobin molecules, a single erythrocyte can transport over a billion molecules of oxygen. It is within the lungs that the oxygen molecules contained in inhaled air attach to the hemoglobin molecules and are transported via erythrocytes to the trillions of body cells.

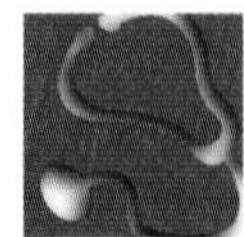

Anemia refers to any condition in which there is an abnormally low hemoglobin concentration and/or erythrocyte count. The most common type is *iron-deficiency anemia,* which is due to a deficient intake or absorption of iron, or to excessive iron loss. Iron is an essential component of the hemoglobin molecule. In *pernicious anemia,* the production of red blood cells is insufficient because of lack of a substance needed for the absorption of vitamin B_{12} by intestinal cells. *Aplastic anemia* is anemia due to destruction of the red bone marrow, which may be caused by chemicals (including benzene and arsenic) or by radiation.

hemoglobin: Gk. *haima*, blood; *globus*, globe

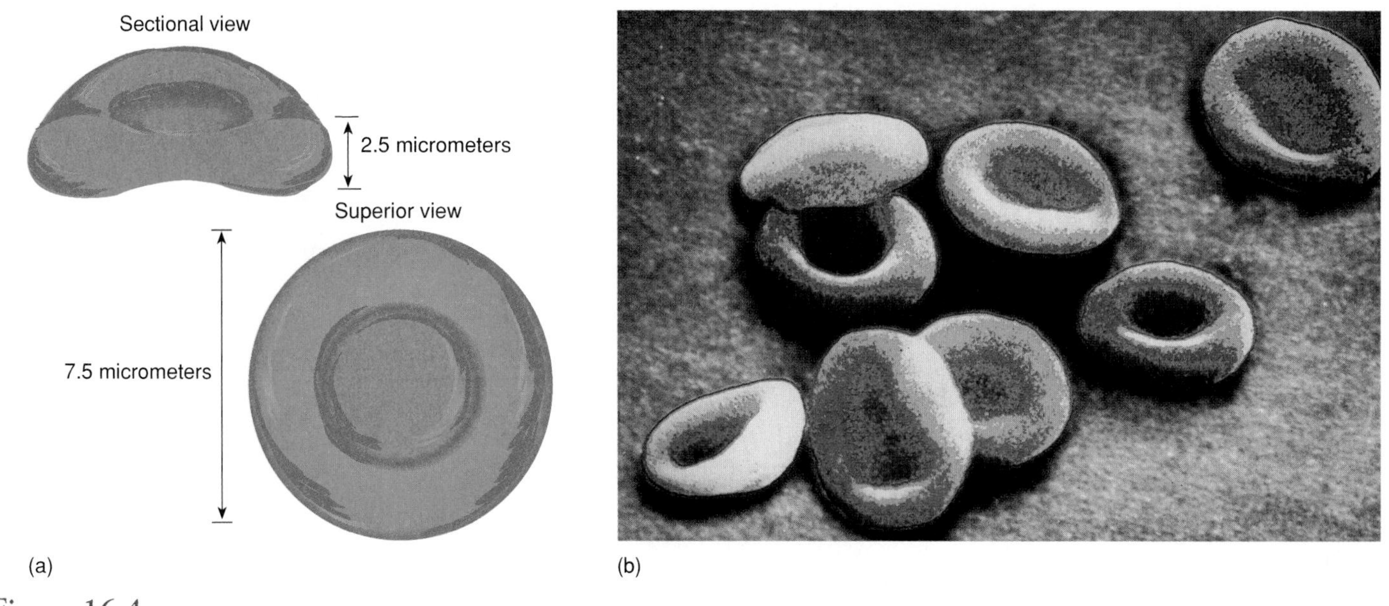

Figure 16.4

Erythrocytes. (*a*) A diagram and (*b*) a scanning electron micrograph.

Leukocytes

Leukocytes are larger than erythrocytes and are different in other ways as well. Leukocytes contain nuclei and mitochondria and can move in an ameboid fashion (erythrocytes are not able to move independently). Because of their ameboid ability, leukocytes can squeeze through pores in capillary walls and move to an extravascular site of infection, whereas erythrocytes usually remain confined within blood vessels. The movement of leukocytes through capillary walls is called *diapedesis* (*di″ă-pĕ-de′sis*).

Leukocytes are almost invisible under the microscope unless they are stained; therefore, they are classified on the basis of their stained appearance. Those leukocytes that have granules in their cytoplasm are called **granular leukocytes.** Those with granules so small that they cannot be seen easily with the light microscope are called **agranular** (or nongranular) **leukocytes.** The granular leukocytes are also identified by their oddly shaped nuclei, which in some cases are contorted into lobes attached by thin strands. The granular leukocytes are therefore known as **polymorphonuclear** (*pol″e-mor″fo-noo′kle-ar*) **(PMN) leukocytes.**

The stain used to identify leukocytes is usually a mixture of a pink-to-red stain called *eosin* (*e′o-sin*) and a blue-to-purple (hematoxylin) stain called a "basic stain." Granular leukocytes with pink-staining granules are therefore called **eosinophils** (*e″ŏ-sin′ŏ-filz*) (fig. 16.5), and those with blue-staining granules are called **basophils.** Those with granules that have little affinity for either stain are **neutrophils.** Neutrophils are the most abundant type of leukocyte, constituting 54%–62% of the leukocytes in the blood.

The leukocytes of the agranular type include **monocytes** and **lymphocytes.** Monocytes are the largest cells found in the blood, and their large nuclei may vary considerably in shape. Lymphocytes have large nuclei surrounded by a relatively thin layer of cytoplasm.

diapedesis: Gk. *dia*, through; *pedester*, on foot

Platelets

Platelets are the smallest of the formed elements and are actually fragments of large cells called **megakaryocytes** (*meg″ă-kar′e-ŏ-sīts*) found in red bone marrow. (This is why the term *formed elements* is used rather than *blood cells* to describe erythrocytes, leukocytes, and platelets.) The fragments that enter the circulation as platelets lack nuclei but, like leukocytes, are capable of ameboid movement. The platelet count per cubic millimeter of blood is 130,000 to 360,000. Platelets survive for about 5 to 9 days and then are destroyed by the spleen and liver.

Platelets play an important role in blood clotting. They constitute the major portion of the mass of the clot, and phospholipids in their cell membranes activate the clotting factors in blood plasma that result in threads of fibrin, which reinforce the platelet plug. Platelets that attach together in a blood clot also release a chemical called *serotonin*, which stimulates constriction of blood vessels, reducing the flow of blood to the injured area.

The appearance of the formed elements of the blood is shown in figures 16.5 and 16.6, and their characteristics are summarized in table 16.1.

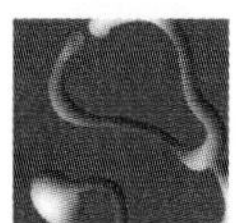

Blood cell counts are an important source of information in assessing the health of a person. An abnormal increase in erythrocytes, for example, is termed *polycythemia* (*pol″e-si-the′me-ă*) and is indicative of several dysfunctions. As previously mentioned, an abnormally low red blood cell count is termed *anemia.* An elevated leukocyte count, called *leukocytosis,* is often associated with localized infection. A large number of immature leukocytes within a blood sample is diagnostic of the disease *leukemia.*

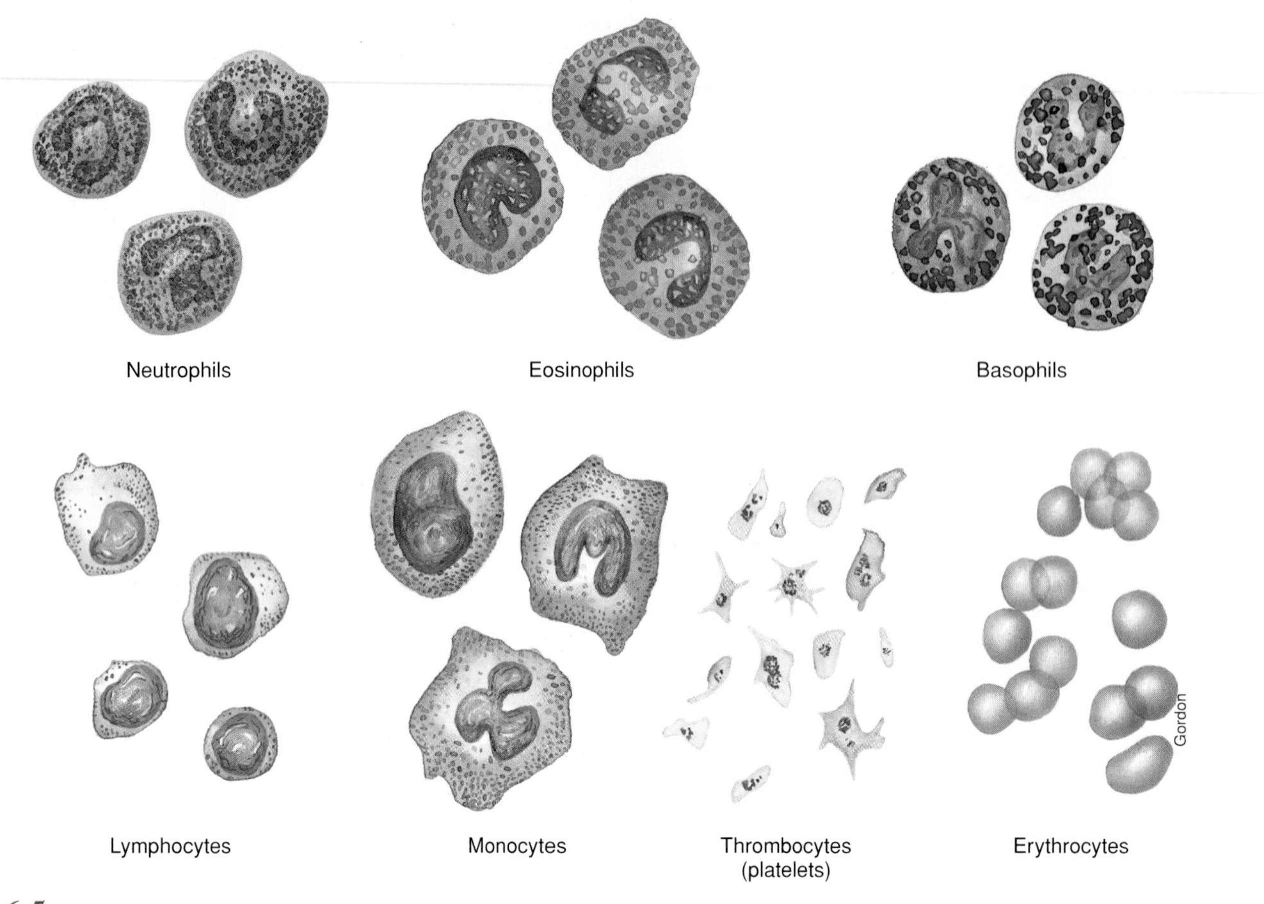

Figure 16.5

Types of formed elements in blood.

Table 16.1
Formed Elements of Blood

Component	Description	Number Present	Function
Erythrocyte (red blood cell)	Biconcave disc without nucleus; contains hemoglobin; survives 100 to 120 days	4,000,000 to 6,000,000/mm^3	Transports oxygen and carbon dioxide
Leukocytes (white blood cells)		5,000 to 10,000/mm^3	Aid in defense against infections by microorganisms
Granulocytes	About twice the size of red blood cells; large cytoplasmic granules; survive 12 hours to 3 days		
1. Neutrophil	Nucleus with 2 to 5 lobes; cytoplasmic granules stain slightly pink	54% to 62% of white cells present	Phagocytic
2. Eosinophil	Nucleus bilobed; cytoplasmic granules stain red in eosin stain	1% to 3% of white cells present	Helps to detoxify foreign substances; secretes enzymes that break down clots
3. Basophil	Nucleus lobed; cytoplasmic granules stain blue in hematoxylin stain	Less than 1% of white cells present	Releases anticoagulant heparin
Agranulocytes	Cytoplasmic granules not visible; survive 100 to 300 days (some much longer)		
1. Monocyte	2 to 3 times larger than red blood cell; nuclear shape varies from round to lobed	3% to 9% of white cells present	Phagocytic
2. Lymphocyte	Only slightly larger than red blood cell; nucleus nearly fills cell	25% to 33% of white cells present	Provides specific immune response (including antibodies)
Platelet (thrombocyte)	Cytoplasmic fragment; survives 5 to 9 days	250,000 to 450,000/mm^3	Enables clotting; releases serotonin, which causes vasoconstriction

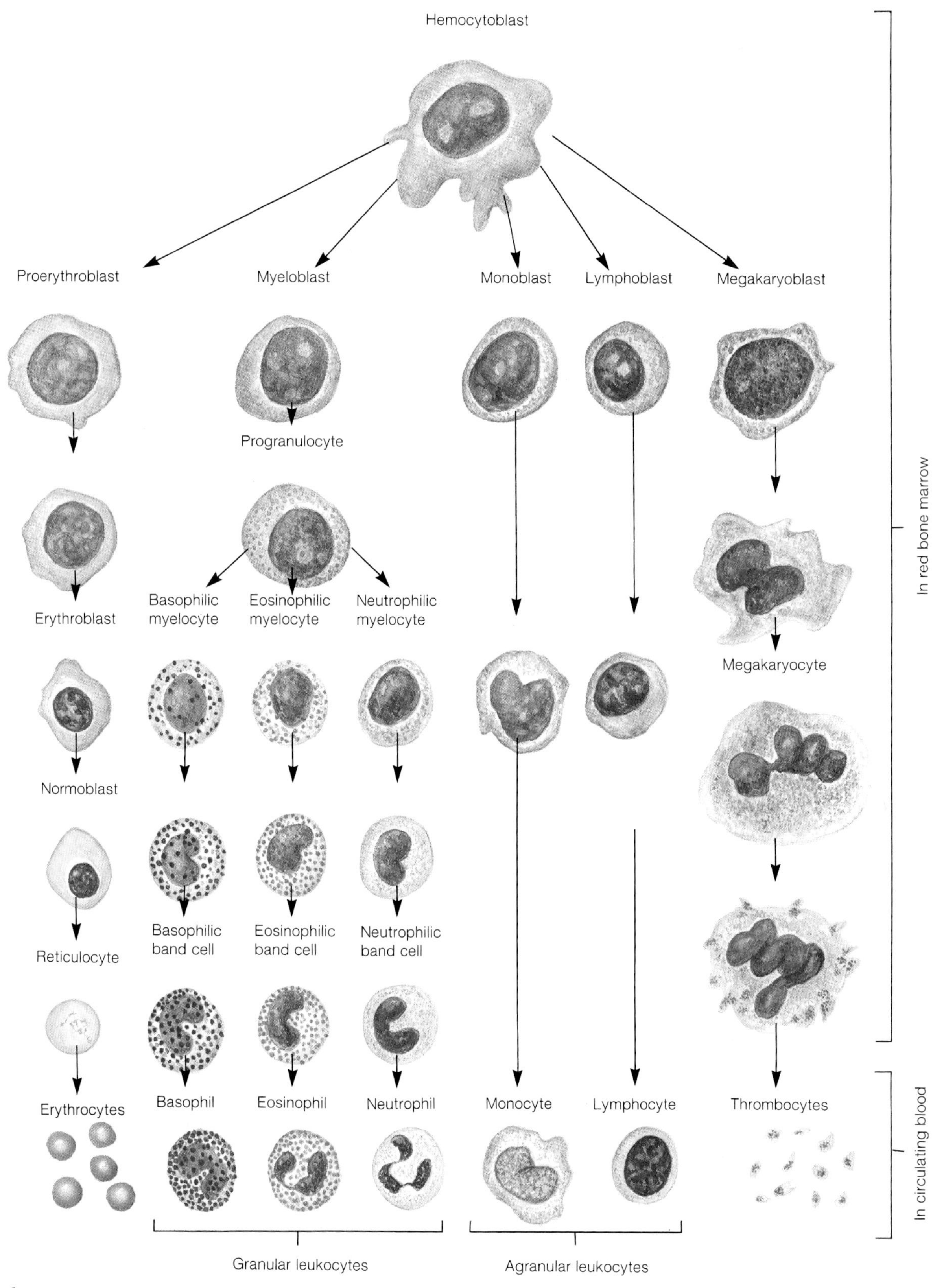

Figure 16.6

The processes of hemopoiesis. Formed elements begin as hemocytoblasts and differentiate into the various kinds of blood cells, depending on the needs of the body.

Hemopoiesis

Blood cells are constantly formed through a process called *hemopoiesis* (fig. 16.6). The term **erythropoiesis** (*ĕ-rith″ro-poi-e′sis*) refers to the formation of erythrocytes; **leukopoiesis** refers to the formation of leukocytes. These processes occur in two classes of tissues. **Myeloid tissue** is the red bone marrow of the humeri, femora, ribs, sternum, pelvis, and portions of the skull that produces erythrocytes, granular leukocytes, and platelets.

Lymphoid tissue—including the lymph nodes, tonsils, spleen, and thymus—produces the agranular leukocytes (monocytes and lymphocytes). During embryonic and fetal development, the hemopoietic centers are located in the yolk sac, liver, and spleen. After birth, the liver and spleen become the sites of blood cell destruction.

Erythropoiesis is an extremely active process. It is estimated that about 2.5 million erythrocytes are produced every second in order to replace the number that are continuously destroyed by the liver and spleen. (Recall that the life span of an erythrocyte is approximately 120 days.) During the destruction of erythrocytes, iron is salvaged and returned to the red bone marrow where it is used again in the formation of erythrocytes. Agranular leukocytes remain functional for 100 to 300 days under normal body conditions. Granular leukocytes, by contrast, have an extremely short life span of 12 hours to 3 days.

Hemopoiesis begins the same way in both myeloid and lymphoid tissues (fig. 16.6). Undifferentiated mesenchymal-like cells develop into stem cells called **hemocytoblasts.** These stem cells are able to divide rapidly. Some of the daughter cells become new stem cells (hence, the stem cell population is never depleted), whereas other daughter cells become specialized along different paths of blood cell formation. Hemocytoblasts, for example, may develop into **proerythroblasts** (*pro″ĕ-rith′rŏ-blasts*), which form erythrocytes; **myeloblasts,** which form granular leukocytes (neutrophils, eosinophils, and basophils); **lymphoblasts,** which form lymphocytes; **monoblasts,** which form monocytes; or **megakaryoblasts,** which form platelets.

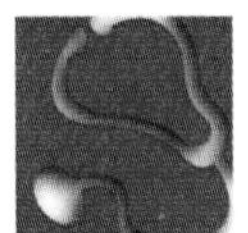

The major purpose of a bone marrow transplant is to provide competent hemopoietic hemocytoblasts to the recipient. If the bone marrow is returned to the same person, the procedure is called an *autotransplant.* If the donor and recipient are different people, it is termed an *allogenic transplant.*

Blood Plasma

Blood plasma is the fluid portion, or *matrix,* of the blood. It can best be visualized when the formed elements are removed. Blood plasma constitutes approximately 55% of a given volume of blood. It is a straw-colored liquid, about 90% water. The remainder of the blood plasma consists of proteins, inorganic salts, carbohydrates, lipids, amino acids, vitamins, and hormones. The functions of blood plasma include transporting nutrients, gases, and vitamins; regulating electrolyte and fluid balances; and maintaining a consistent blood pH.

Plasma proteins constitute 7%–9% of the blood plasma. These proteins remain within the blood and interstitial fluid and assist in maintaining body homeostasis. The three types of plasma proteins are albumins, globulins, and fibrinogen. **Albumins** (*al-byoo′minz*) account for about 60% of the plasma proteins and are the smallest of the three types. They are produced by the liver and provide the blood with the viscosity needed to maintain and regulate blood pressure. **Globulins** (*glob′yoo-linz*) make up about 36% of the plasma proteins. The three types of globulins are **alpha globulins, beta globulins,** and **gamma globulins.** The alpha and beta globulins are synthesized in the liver and function in transporting lipids and fat-soluble vitamins. Gamma globulins are produced by lymphoid tissues and are antibodies of immunity. The third type of plasma protein, **fibrinogen** (*fi-brin′ŏ-jen*), accounts for only about 4% of the protein content. Fibrinogens are large molecules that are synthesized in the liver and, together with platelets, play an important role in clotting the blood. When fibrinogen is removed from blood plasma, the remaining fluid is referred to as *serum*.

The plasma proteins of blood are summarized in table 16.2.

Table 16.2
Plasma Proteins

Protein	Percentage of Total	Origin	Function
Albumin	60%	Liver	Gives viscosity to blood and assists in maintaining blood osmotic pressure
Globulin	36%		
Alpha globulins		Liver	Transport lipids and fat-soluble vitamins
Beta globulins		Liver	Transport lipids and fat-soluble vitamins
Gamma globulins		Lymphoid tissue	Constitute antibodies of immunity
Fibrinogen	4%	Liver	Assists platelets in the formation of clots

hemopoiesis: Gk. *haima,* blood; *poiesis,* production
erythropoiesis: Gk. *erythros,* red; *poiesis,* production
leukopoiesis: Gk. *leukos,* white; *poiesis,* production
megakaryoblasts: Gk. *megas,* great; *karyon,* nut; *blastos,* germ

albumin: L. *albumen,* white
globulin: L. *globulus,* small globe
fibrinogen: L. *fibra,* fibrous

1. Describe erythropoiesis in terms of where it occurs, its rate, and the specific steps involved. What does a mature erythrocyte look like?
2. Describe the different types of leukocytes and explain where and how these different types are produced.
3. Explain how platelets are produced. How do they differ from other formed elements of blood?
4. List the different classes of plasma proteins and describe their origin and functions.
5. Explain how leukocytes get to the site of an infection.

Heart

The structure of the heart enables it to serve as a transport system pump that keeps blood continuously circulating through the blood vessels of the body.

Objective 7 Describe the location of the heart in relation to other organs of the thoracic cavity and their associated serous membranes.

Objective 8 Describe the structure and functions of the three layers of the heart wall.

Objective 9 Describe the chambers and valves of the heart and identify the grooves on its surface.

Objective 10 Trace the flow of blood through the heart and distinguish between the pulmonary and systemic circulations.

Objective 11 Describe the location of the conduction system components of the heart and trace the path of impulse conduction.

Location and General Description

The hollow, four-chambered, muscular **heart** is roughly the size of a clenched fist. It averages 255 grams in adult females and 310 grams in adult males. The heart contracts an estimated 42 million times a year, pumping 700,000 gallons of blood.

The heart is located in the thoracic cavity between the lungs in the mediastinum (fig. 16.7). About two-thirds of the heart is located left of the midline, with its *apex*, or cone-shaped end, pointing downward and resting on the diaphragm. The *base* of the heart is the broad superior end, where the large vessels attach.

The **parietal pericardium** (*per″ĭ-kar′de-um*) is a loose-fitting serous sac of dense fibrous connective tissue that encloses and protects the heart (fig. 16.7). It separates the heart from the other thoracic organs and forms the wall of the *pericardial cavity* (table 16.3), which contains a watery, lubricating *pericardial fluid*. The parietal pericardium is actually composed of an outer *fibrous pericardium* and an inner *serous pericardium*. It is the serous pericardium that produces the lubricating pericardial fluid that allows the heart to beat in a kind of frictionless bath.

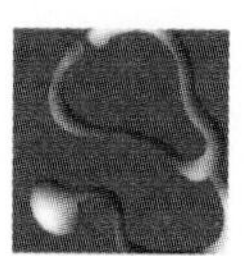

Pericarditis is an inflammation of the parietal pericardium that results in an increased secretion of fluid into the pericardial cavity. Because the tough, fibrous portion of the parietal pericardium is inelastic, an increase in fluid pressure impairs the movement of blood into and out of the chambers of the heart. Some of the pericardial fluid may be withdrawn for analysis by injecting a needle to the left of the xiphoid process to pierce the parietal pericardium.

Heart Wall

The wall of the heart is composed of three distinct layers (table 16.3). The outer layer is the **epicardium,** also called the *visceral pericardium*. The space between this layer and the parietal pericardium is the pericardial cavity, just described. The thick middle layer of the heart wall is called the **myocardium.** It is composed of cardiac muscle tissue (see chapter 4) and arranged in such a way that the contraction of the muscle bundles results in squeezing or wringing of the heart chambers. The thickness of the myocardium varies in accordance with the force needed to eject blood from the particular chamber. Thus, the thickest portion of the myocardium surrounds the left ventricle and the atrial walls are relatively thin. The inner layer of the wall, called the **endocardium,** is continuous with the endothelium of blood vessels. The endocardium also covers the valves of the heart. Inflammation of the endocardium is called *endocarditis*.

Chambers and Valves

The interior of the heart is divided into four chambers: upper right and left **atria** (*a′tre-ă*—singular *atrium*) and lower right and left **ventricles.** The atria contract and empty simultaneously into the ventricles (fig. 16.8), which also contract in unison. Each atrium has an ear-shaped, expandable appendage called an **auricle** (*or′ĭ-kul*). The atria are separated from each other by the thin, muscular **interatrial septum;** the ventricles are separated from each other by the thick, muscular **interventricular septum. Atrioventricular valves** (*AV valves*) lie between the atria and ventricles, and **semilunar valves** are located

pericardium: Gk. *peri*, around; *kardia*, heart

atrium: L. *atrium*, chamber
ventricle: L. *ventriculus*, diminutive of *venter*, belly
auricle: L. *auricula*, a little ear

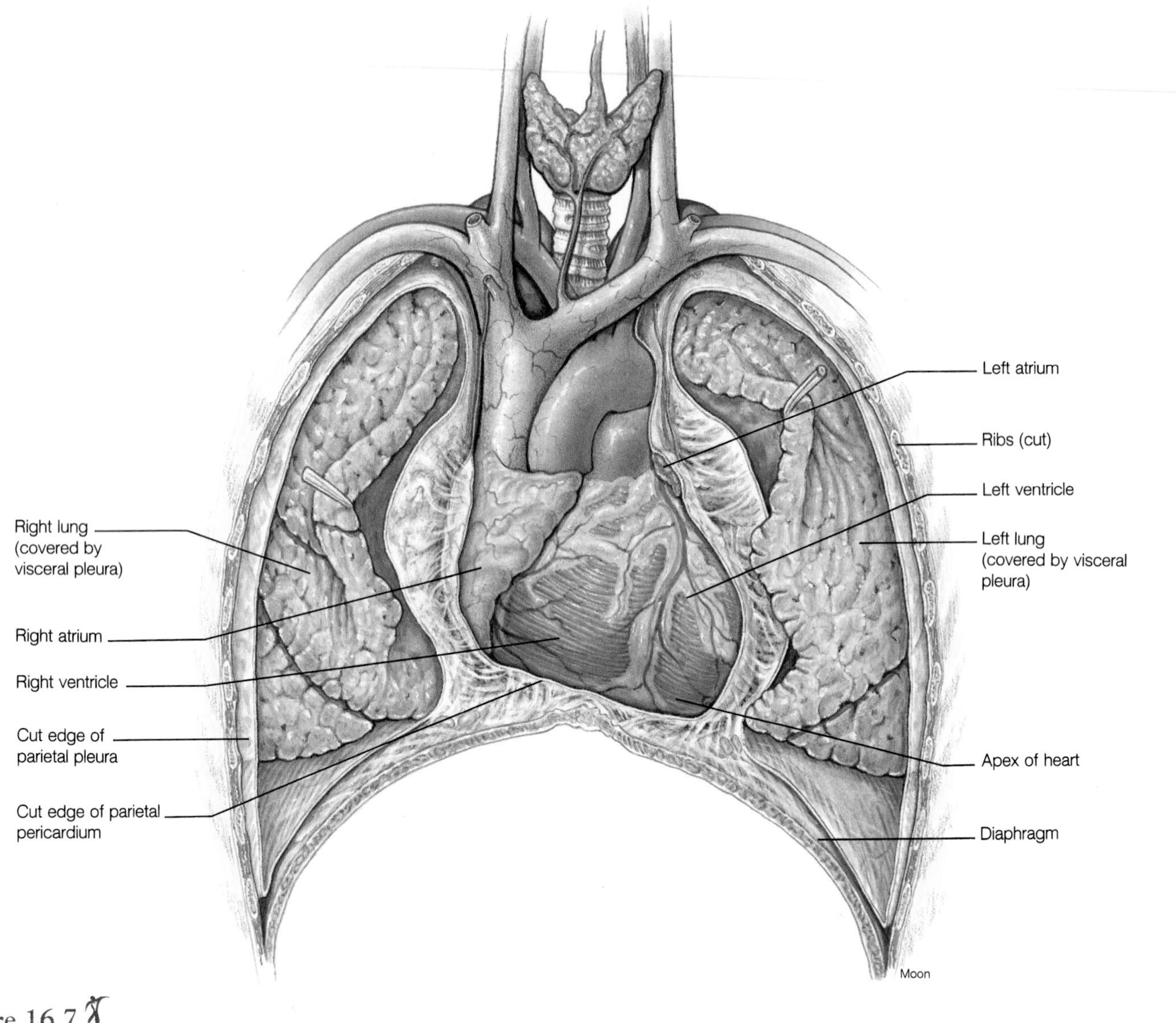

Figure 16.7

The position of the heart and associated serous membranes within the thoracic cavity.

at the bases of the two large vessels leaving the heart. Heart valves (see table 16.4) maintain a one-way flow of blood.

Grooved depressions on the surface of the heart indicate the partitions between the chambers and also contain **cardiac vessels** that supply blood to the muscular wall of the heart. The most prominent groove is the *coronary sulcus* that encircles the heart and marks the division between the atria and ventricles. The partition between the right and left ventricles is denoted by two (anterior and posterior) *interventricular sulci*.

The following discussion follows the sequence in which blood flows through the atria, ventricles, and valves. It is important to keep in mind that the right side of the heart (right atrium and right ventricle) receives deoxygenated blood (blood low in oxygen) and pumps it to the lungs. The left side of the heart (left atrium and left ventricle) receives oxygenated blood (blood rich in oxygen) from the lungs and pumps it throughout the body.

Right Atrium

The right atrium receives systemic venous blood from the **superior vena cava,** which drains the upper portion of the body, and from the **inferior vena cava,** which drains the lower portion (fig. 16.8). The *coronary sinus* is an additional opening into the right atrium that receives venous blood from the myocardium of the heart itself.

Right Ventricle

Blood from the right atrium passes through the **right atrioventricular (AV) valve** (also called the *tricuspid valve*) to fill the right ventricle. The right AV valve is characterized by three

vena cava: L. *vena,* vein; *cava,* empty

Table 16.3
Layers of the Heart Wall

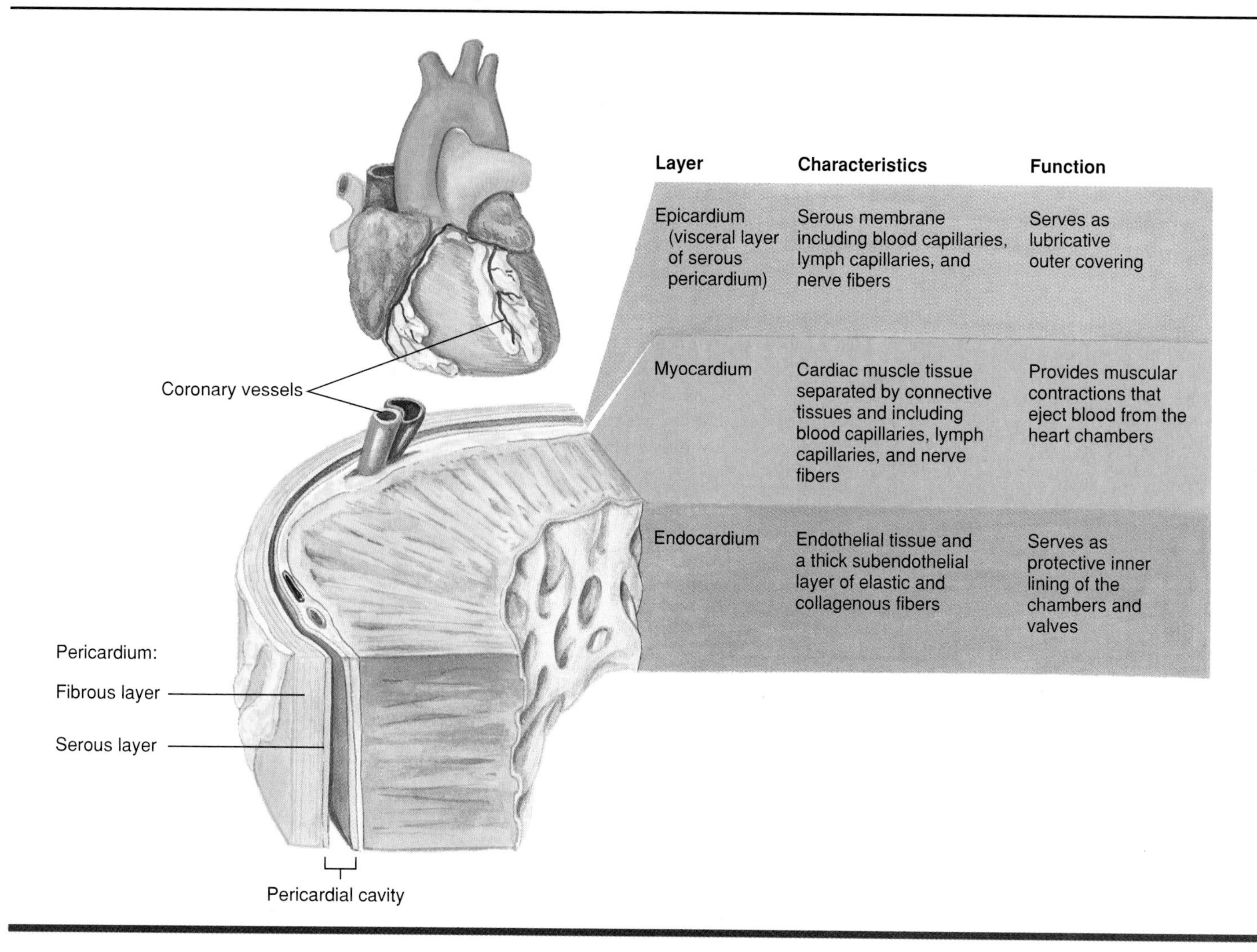

Layer	Characteristics	Function
Epicardium (visceral layer of serous pericardium)	Serous membrane including blood capillaries, lymph capillaries, and nerve fibers	Serves as lubricative outer covering
Myocardium	Cardiac muscle tissue separated by connective tissues and including blood capillaries, lymph capillaries, and nerve fibers	Provides muscular contractions that eject blood from the heart chambers
Endocardium	Endothelial tissue and a thick subendothelial layer of elastic and collagenous fibers	Serves as protective inner lining of the chambers and valves

valve leaflets, or *cusps*. Each cusp is held in position by strong tendinous cords called **chordae tendineae** (*kor'de ten-din'e-e*). The chordae tendineae are secured to the ventricular wall by cone-shaped **papillary muscles.** These structures prevent the valves from everting, like an umbrella in a strong wind, when the ventricles contract and the ventricular pressure increases.

Ventricular contraction causes the right AV valve to close and the blood to leave the right ventricle through the **pulmonary trunk** and to enter the capillaries of the lungs via the **right** and **left pulmonary arteries.** The **pulmonary valve** (also called the *pulmonary semilunar valve*) lies at the base of the pulmonary trunk, where it prevents the backflow of ejected blood into the right ventricle.

Left Atrium

After gas exchange has occurred within the capillaries of the lungs, oxygenated blood is transported to the left atrium through two right and two left **pulmonary veins.**

Left Ventricle

The left ventricle receives blood from the left atrium. These two chambers are separated by the **left atrioventricular (AV) valve** (also called the *bicuspid valve* or *mitral valve*). When the left ventricle is relaxed, the valve is open, allowing blood to flow from the atrium into the ventricle; when the left ventricle contracts,

---

chordae tendineae: L. *chorda*, string; *tendere*, to stretch

bicuspid: L. *bi*, two; *cuspis*, tooth point or spike
mitral: L. *mitra*, like a bishop's mitre

Table 16.4
Valves of the Heart

Valve	Location	Comments
Right atrioventricular valve	Between right atrium and right ventricle	Composed of three cusps that prevent a backflow of blood from right ventricle into right atrium during ventricular contraction
Pulmonary valve	Entrance to pulmonary trunk	Composed of three half-moon-shaped flaps that prevent a backflow of blood from pulmonary trunk into right ventricle during ventricular relaxation
Left atrioventricular (mitral) valve	Between left atrium and left ventricle	Composed of two cusps that prevent a backflow of blood from left ventricle to left atrium during ventricular contraction
Aortic valve	Entrance to ascending aorta	Composed of three half-moon-shaped flaps that prevent a backflow of blood from aorta into left ventricle during ventricular relaxation

Pulmonary valve
Aortic valve
Right AV valve
Left AV valve
Fibrous connective tissue

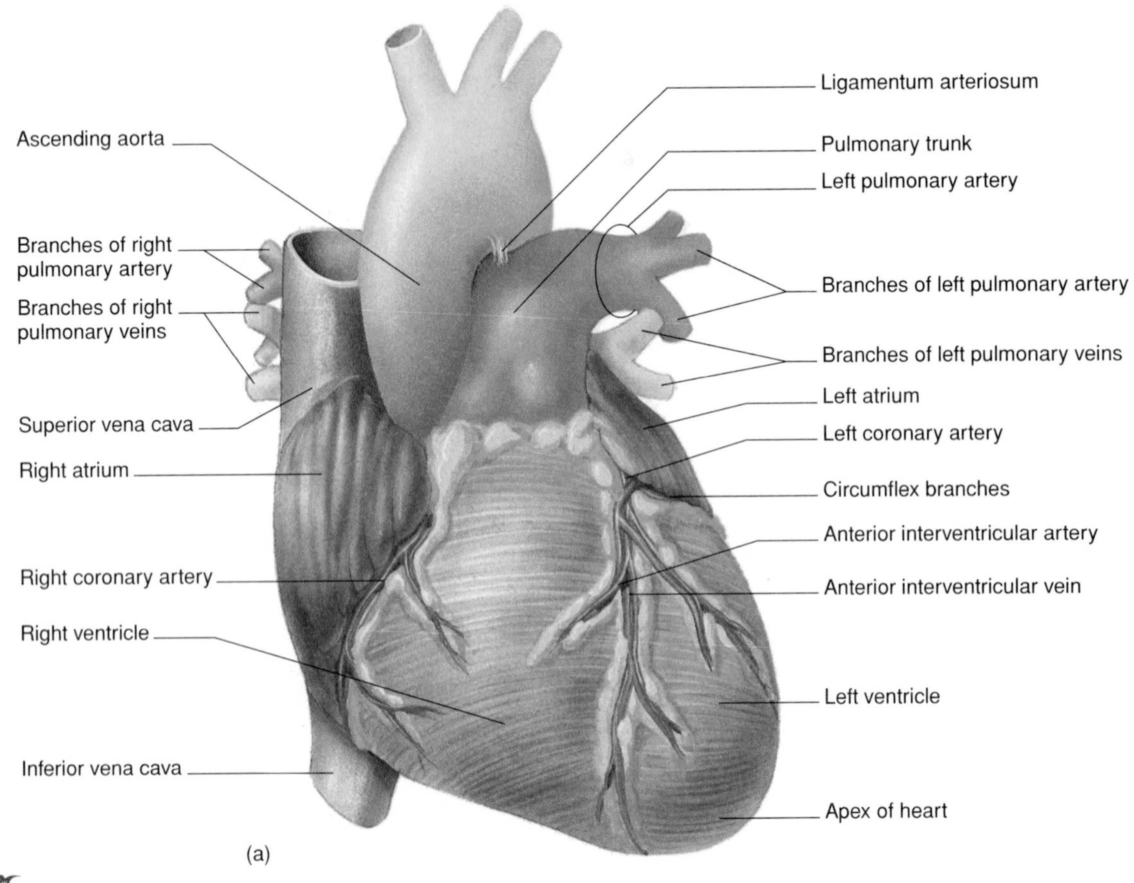

Figure 16.8

The structure of the heart. (*a*) An anterior view, (*b*) a posterior view, and (*c*) an internal view.

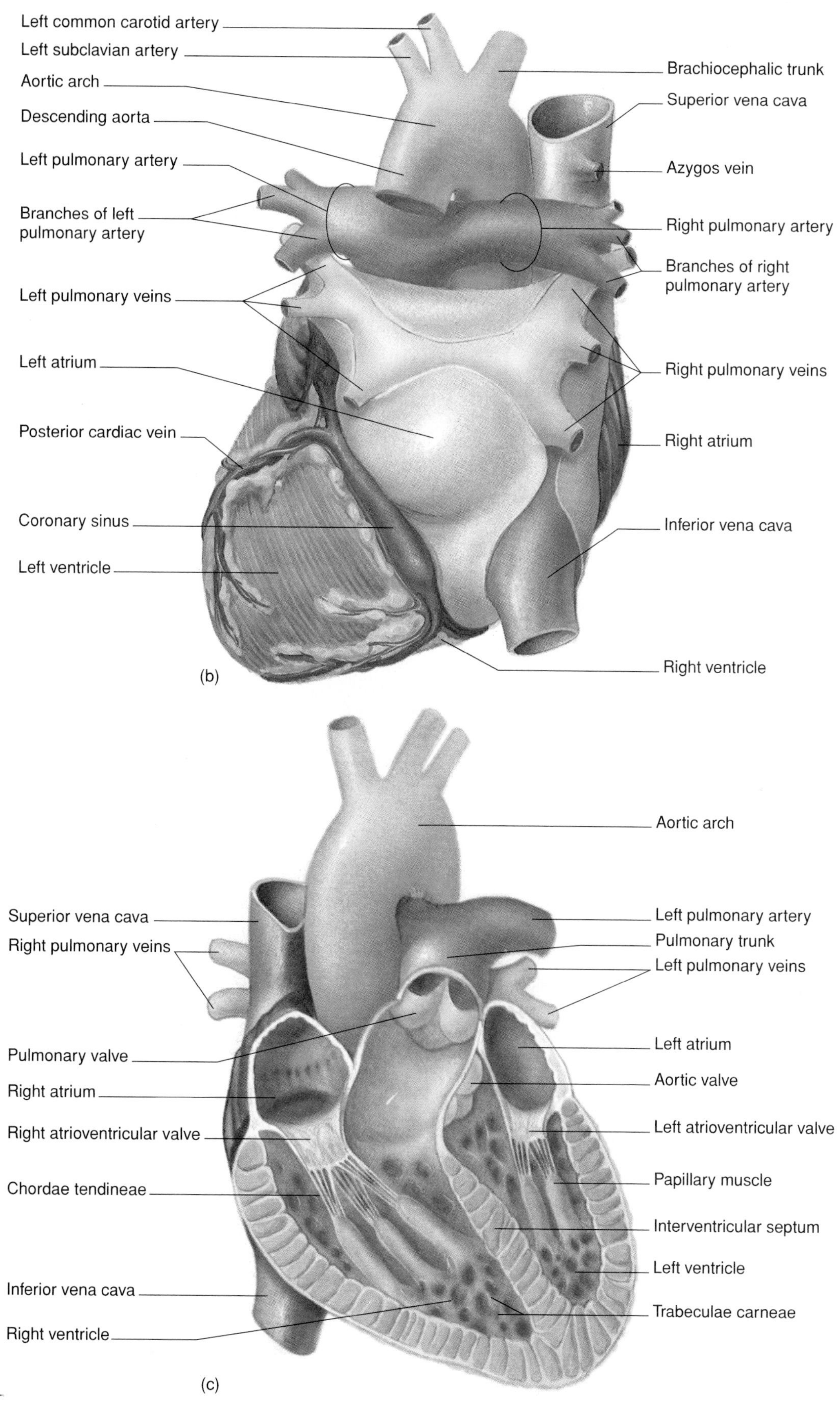

Left common carotid artery
Left subclavian artery
Aortic arch
Descending aorta
Left pulmonary artery
Branches of left pulmonary artery
Left pulmonary veins
Left atrium
Posterior cardiac vein
Coronary sinus
Left ventricle
Brachiocephalic trunk
Superior vena cava
Azygos vein
Right pulmonary artery
Branches of right pulmonary artery
Right pulmonary veins
Right atrium
Inferior vena cava
Right ventricle
(b)
Aortic arch
Superior vena cava
Right pulmonary veins
Pulmonary valve
Right atrium
Right atrioventricular valve
Chordae tendineae
Inferior vena cava
Right ventricle
Left pulmonary artery
Pulmonary trunk
Left pulmonary veins
Left atrium
Aortic valve
Left atrioventricular valve
Papillary muscle
Interventricular septum
Left ventricle
Trabeculae carneae
(c)

Figure 16.8

Continued.

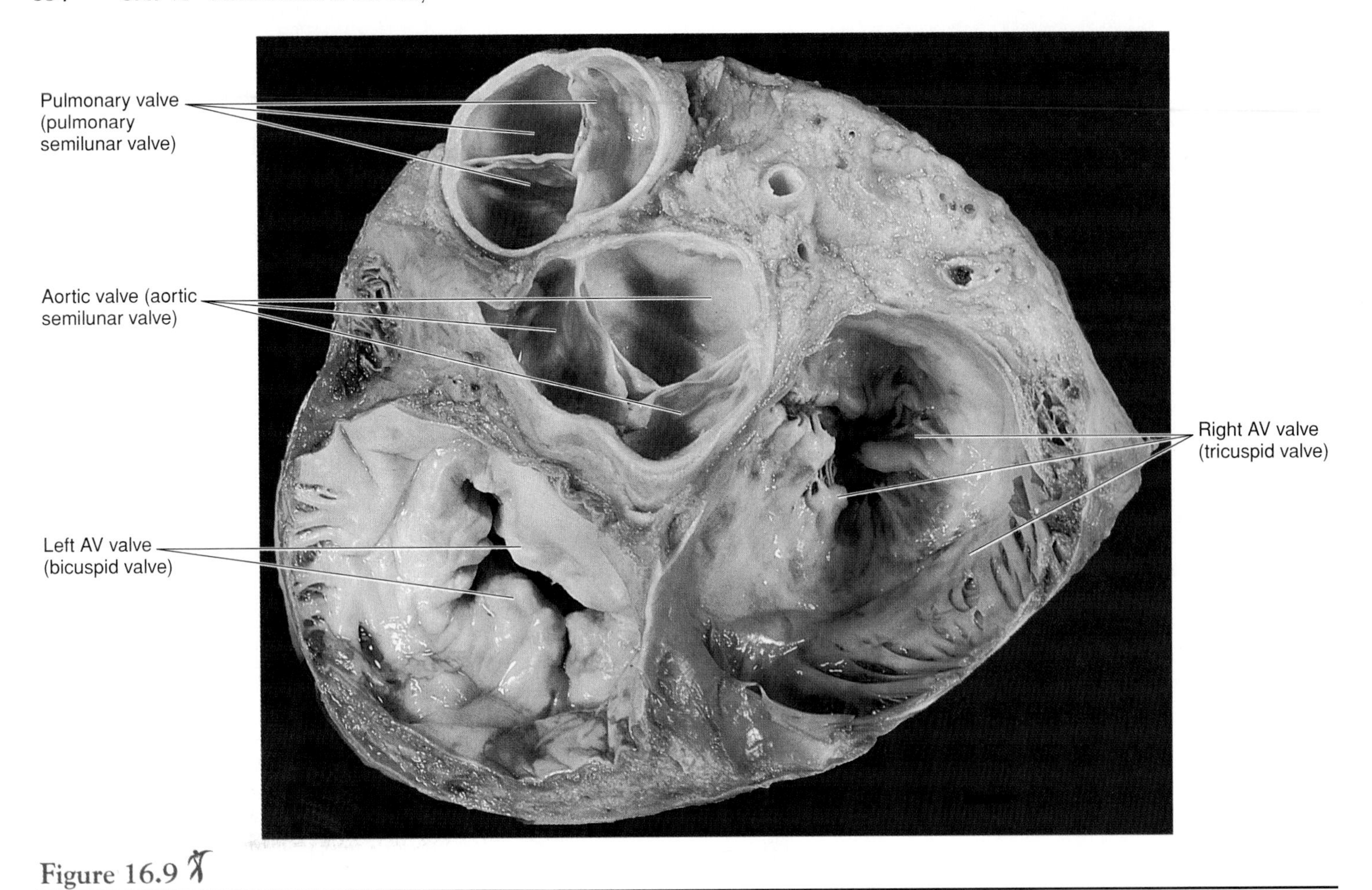

Figure 16.9

A superior view of the heart valves seen with the atria removed.

the valve closes. Closing of the valve during ventricular contraction prevents the backflow of blood into the atrium.

The walls of the left ventricle are thicker than those of the right ventricle because the left ventricle bears a greater workload, pumping blood through the entire body. The endocardium of both ventricles is characterized by distinct ridges called **trabeculae carneae** (*tră-bek'yŭ-le kar'ne-e*) (fig. 16.8c). Oxygenated blood leaves the left ventricle through the **ascending portion of the aorta.** The **aortic valve** (also called the *aortic semilunar valve*), located at the base of the ascending portion of the aorta, closes as a result of the pressure of the blood when the left ventricle relaxes, and thus prevents the backflow of blood into the relaxed ventricle.

The valves are shown in figure 16.9, and their actions are summarized in table 16.4.

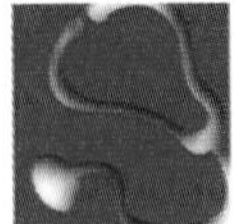

Cardiac catheterization is a procedure used in the diagnosis of certain heart disorders. A tube, or catheter, is inserted into a vein in the leg or the arm and threaded through the lumen of the vessel until it enters the heart. The location of the catheter can be monitored by means of a fluoroscope. When the catheter reaches the desired location within the heart, a radiopaque dye is released. The patient may then be radiographed, blood samples removed for analyses, or blood pressures monitored.

trabeculae carneae: L. *trabecula*, small beams; *carneus*, flesh

Circulatory Routes

The circulatory routes of the blood are illustrated in figure 16.2. The principal divisions of the circulatory blood flow are the *pulmonary* and *systemic circulations*.

The *pulmonary circulation* includes blood vessels that transport blood to the lungs for gas exchange and then back to the heart. It consists of the right ventricle that ejects the blood, the pulmonary trunk with its pulmonary valve, the pulmonary arteries that transport deoxygenated blood to the lungs, the pulmonary capillaries within each lung, the pulmonary veins that transport oxygenated blood back to the heart, and the left atrium that receives the blood from the pulmonary veins.

The **systemic circulation** involves all of the vessels of the body that are not part of the pulmonary circulation. It includes the right atrium, the left ventricle, the aorta with its aortic valve, all of the branches of the aorta, all capillaries other than those in the lungs involved with gas exchange, and all veins other than the pulmonary veins. The right atrium receives all of the venous return of oxygen-depleted blood from the systemic veins.

Coronary Circulation

The wall of the heart has its own supply of blood vessels to meet its vital needs. The myocardium is supplied with blood by the **right** and **left coronary arteries** (fig. 16.10). These two vessels arise from the ascending part of the aorta, at the location of the aortic (semilunar) valve. The coronary arteries encircle the heart within the atrioventricular sulcus, the depression between the atria and ventricles. Two branches arise from both the right and left coronary arteries to serve the atrial and ventricular walls. The left coronary artery gives rise to the **anterior interventricular artery,** which courses within the anterior interventricular sulcus to serve both ventricles, and the **circumflex artery,** which supplies oxygenated blood to the walls of the left atrium and left ventricle. The right coronary artery gives rise to the **right marginal artery,** which serves the walls of the right atrium and right ventricle, and the **posterior interventricular artery,** which courses through the posterior interventricular sulcus to serve the two ventricles. The main trunks of the right and left coronaries anastomose (join together) on the posterior surface of the heart.

From the capillaries in the myocardium, the blood enters the **cardiac veins.** The course of these vessels parallels that of the coronary arteries. The cardiac veins, however, have thinner walls and are more superficial than the arteries. The two principal cardiac veins are the **anterior interventricular vein,** which returns blood from the anterior aspect of the heart, and the **posterior cardiac vein,** which drains the posterior aspect of the heart. These cardiac veins converge to form the **coronary sinus** channel on the posterior surface of the heart (figs. 16.8 and 16.10). The coronary venous blood then enters the heart through an opening into the right atrium.

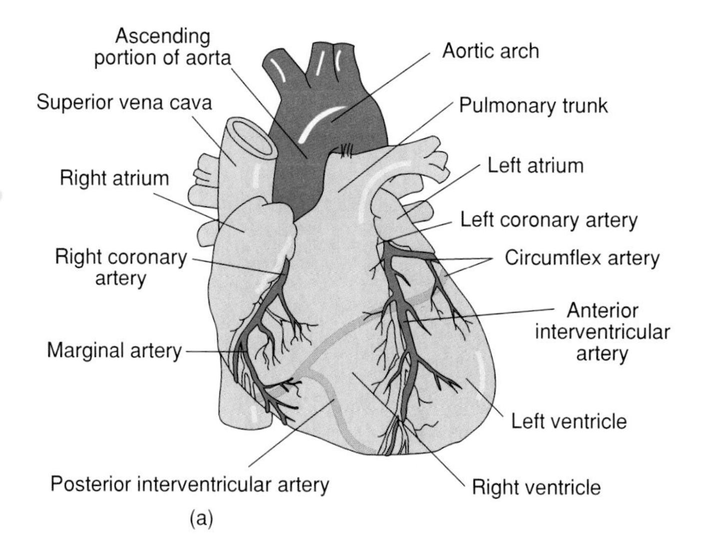

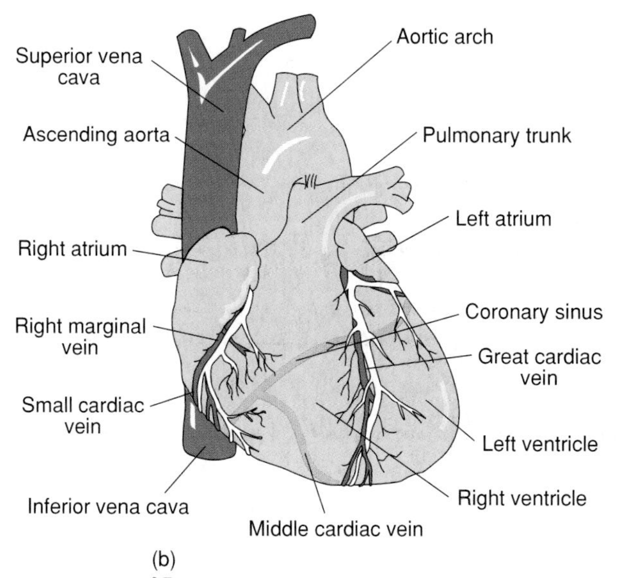

Figure 16.10

Coronary circulation. (*a*) An anterior view of the arterial supply to the heart and (*b*) an anterior view of the venous drainage.

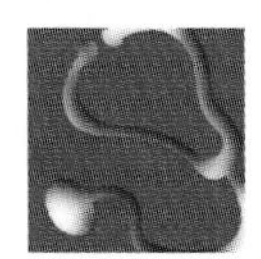

Heart attacks are the leading cause of death in the United States. The most common type of heart attack involves an occlusion of a coronary artery, which reduces the delivery of oxygen to the myocardium. Several strategies are recommended to minimize the risk of heart attack: (1) reduce excess weight; (2) avoid hypertension through proper diet, stress reduction, or medication if necessary; (3) avoid excessive saturated fats that may raise plasma cholesterol levels; (4) refrain from smoking; and (5) exercise regularly.

Conduction System of the Heart

Cardiac muscle has an intrinsic rhythmicity that allows the heartbeat to originate in and be conducted through the heart without extrinsic stimulation. Specialized strands of interconnecting cardiac muscle tissue that coordinate cardiac contraction constitute the **conduction system.** The conduction system enables the **cardiac cycle,** which refers to the events surrounding the filling and emptying of the chambers of the heart. The conduction system consists of specialized tissues that generate and distribute electrical impulses through the heart. The components of the conduction system are the **sinoatrial node** (SA node), **atrioventricular node** (AV node), **atrioventricular bundle** (bundle of His [pronounced "hiss"]), and **conduction myofibers** (Purkinje [pur-kin´jē] fibers). None of these structures are macroscopic, but their locations can be noted (fig. 16.11*a*). The SA node, or *pacemaker*, is located in the posterior wall of the right atrium. The SA node initiates the cardiac cycle by

circumflex: L. *circum*, around; *flectere*, to bend

bundle of His: from Wilhelm His Jr., Swiss physician, 1863–1934
Purkinje fibers: from Johannes E. von Purkinje, Bohemian anatomist, 1787–1869

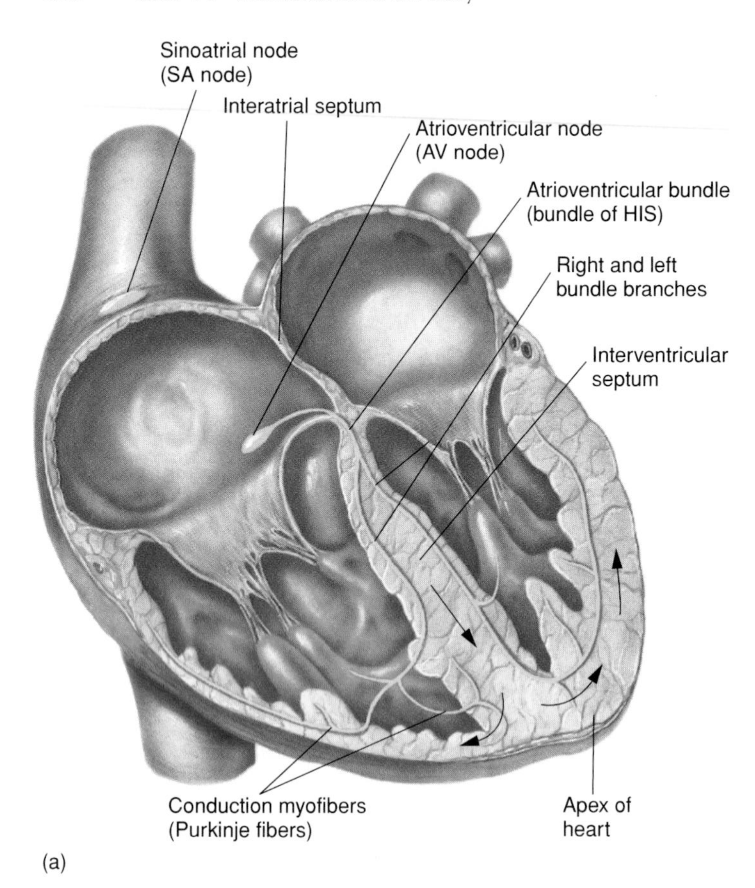

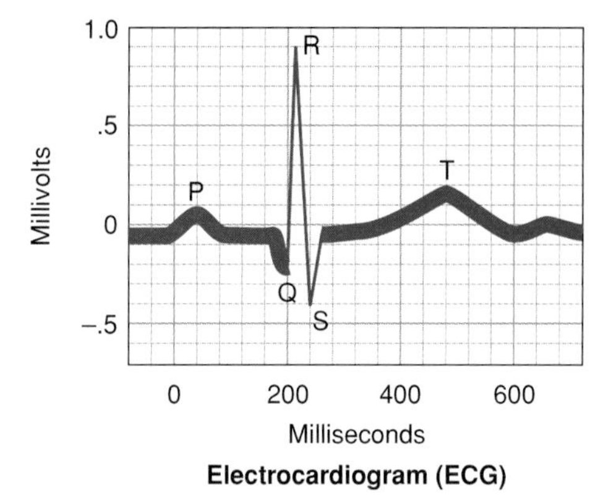

Figure 16.11

(*a*) The conduction system of the heart and (*b*) the electrocardiogram (ECG).

producing an electrical impulse that spreads over both atria, causing them to contract simultaneously and force blood into the ventricles. The basic depolarization rate of the SA node is 70–80 times per minute. The impulse then passes to the AV node, located in the inferior portion of the interatrial septum. From here, the impulse continues through the atrioventricular bundle, located at the top of the interventricular septum. The atrioventricular bundle divides into right and left bundle branches, which are continuous with the conduction myofibers within the ventricular walls. Stimulation of these fibers causes the ventricles to contract simultaneously.

Contraction of the ventricles is referred to as **systole** (*sis'tŏ-le*). Systole, together with the tension of the elastic fibers and contraction of smooth muscles within the systemic arteries, accounts for the systolic pressure within arteries. Ventricular relaxation is called **diastole** (*di-as'tŏ-le*). During diastole, the diastolic pressure within arteries can be recorded.

Variation in functioning within vital organs should be regarded as normal for a healthy adult. The heartbeat can vary by as much as 20 or 30 beats per minute in 24 hours, but the heart maintains an average of about 70 beats per minute during a day. Blood pressure recorded at 120 over 80 in the morning can rise to 140 over 100 by evening. The normal body temperature of 98.6° F will fluctuate between 97° and 99° F over a 24-hour period.

Although the heart does have an innate contraction pattern, it is also innervated by the autonomic nervous system in order to respond to the ever-changing physiological needs of the body. The SA and AV nodes have both sympathetic and parasympathetic innervation. Sympathetic stimulation accelerates the heart rate and dilates the coronary arteries, enabling the heart to meet its own increased metabolic demands as well as those of the rest of the body. Parasympathetic stimulation has the opposite effect. Sympathetic innervation is through fibers from the cervical and upper thoracic ganglia. Parasympathetic innervation is through branches of the vagus nerves. Branches from the right vagus innervate the SA node, and branches from the left vagus innervate the AV node.

Cardiac output is the volume of blood ejected by the heart into the systemic circulation each minute. It is determined by multiplying the stroke rate, or heart rate, by the stroke volume. The stroke volume is the amount of blood pumped from the heart into systemic circulation with each ventricular contraction, which amounts to about 70 ml of blood. A normal resting cardiac output ranges between 4.2 and 5.6 liters per minute. Exercise increases the heart rate, as do the hormones epinephrine and thyroxine. Atropine, caffeine, and camphor are drugs that have a stimulatory effect. Blood pressure and body temperature also have a profound effect on the heart rate.

Electrocardiogram

The electrical impulses that pass through the conduction system of the heart during the cardiac cycle can be recorded as an *electrocardiogram* (ECG or EKG). The electrical changes result from depolarization and repolarization of cardiac muscle fibers and can be detected on the surface of the skin using an instrument called the *electrocardiograph*.

The principal aspects of an ECG are shown in figure 16.11*b*. The wave deflections, designated P, QRS, and T, are produced as specific events of the cardiac cycle occur. Any heart

systole: Gk. *systole*, contraction
diastole: Gk. *diastole*, prolonged or expansion

disease that disturbs the electrical activity will produce characteristic changes in one or more of these waves, so understanding the normal wave-deflection patterns is clinically important.

P Wave

Depolarization of the atrial fibers of the SA node produces the P wave. The actual contraction of the atria follows the P wave by a fraction of a second. The ventricles of the heart are in diastole during the expression of the P wave. A missing or abnormal P wave may indicate a dysfunction of the SA node.

P-R Interval

On the ECG recording, the P-R interval is the period of time from the start of the P wave to the beginning of the QRS complex. This interval indicates the amount of time required for the SA depolarization to reach the ventricles. A prolonged P-R interval suggests a conduction problem at or below the AV node.

QRS Complex

The QRS complex begins as a short downward deflection (Q), continues as a sharp upward spike (R), and ends as a downward deflection (S). The QRS complex indicates the depolarization of the ventricles. During this interval, the ventricles are in systole and blood is being ejected from the heart. It is also during this interval that the atria repolarize, but this event is obscured by the greater depolarization occurring in the ventricles. An abnormal QRS complex generally indicates cardiac problems of the ventricles. An enlarged R spike, for example, generally indicates enlarged ventricles.

S-T Segment

The time duration known as the S-T segment represents the period between the completion of ventricular depolarization and initiation of repolarization. The S-T segment is depressed when the heart receives insufficient oxygen; in acute myocardial infarction, it is elevated.

T Wave

The T wave is produced by ventricular repolarization. An arteriosclerotic heart will produce altered T waves, as will various other heart diseases.

Heart Sounds

Closing of the AV and semilunar valves produces sounds that can be heard at the surface of the chest with a stethoscope. These sounds are often verbalized as "lub-dub." The "lub," or **first sound,** is produced by the closing of the AV valves. The "dub," or **second sound,** is produced by the closing of the aortic and pulmonary semilunar valves. The first sound is thus heard when the ventricles contract at systole, and the second sound is heard when the ventricles relax at the beginning of diastole.

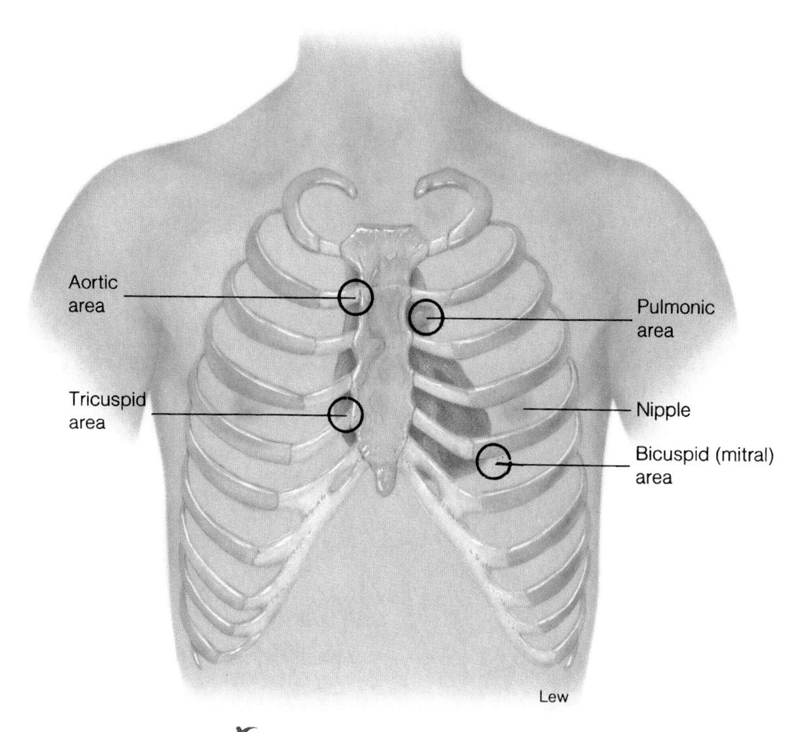

Figure 16.12

The valvular auscultatory areas are the standard stethoscope positions for listening to the heart sounds.

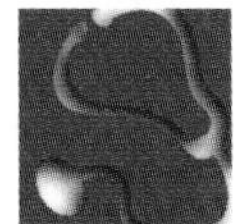

Heart sounds are of clinical importance because they provide information about the condition of the heart valves and may indicate heart problems. Abnormal sounds are referred to as *heart murmurs* and are caused by valvular leakage or restriction and the subsequent turbulence of the blood as it passes through the heart. In general, three basic conditions cause murmurs: (1) *valvular insufficiency,* in which the cusps of the valves do not form a tight seal; (2) *stenosis,* in which the walls surrounding a valve are roughened or constricted; and (3) a *functional murmur,* which is frequent in children and is caused by turbulence of the blood moving through the heart during heavy exercise. Functional murmurs are not pathological and are considered normal.

The valves of the heart are positioned directly deep to the sternum, which tends to obscure and dissipate valvular sounds. For this reason, a physician listening for heart sounds will place a stethoscope at locations designated as **valvular auscultatory areas.** These areas are named according to the valve that can be detected (fig. 16.12). The **aortic area** is immediately to the right of the sternum, at the second intercostal space. The **pulmonic area** is to the left of the sternum, almost directly across from the aortic area. The **tricuspid** and **bicuspid areas** are both near the fifth intercostal space, with the bicuspid (mitral) area more laterally placed. Surface landmarks are extremely important in identifying auscultatory areas.

stenosis: Gk. *stenosis*, a narrowing

1. Distinguish between the pericardial and the thoracic cavities. Describe the serous membranes of the heart.
2. List the three layers of the heart wall and describe the structures associated with each layer.
3. List the valves that aid blood flow through the heart and describe their locations and functions.
4. Describe the flow of electrical impulses through the cardiac conduction system.

Blood Vessels

The structure of arteries and veins allows them to transport blood from the heart to the capillaries and from the capillaries back to the heart. The structure of capillaries permits the exchange of blood plasma and dissolved molecules between the blood and surrounding tissues.

Objective 12 Describe the structure and function of arteries, capillaries, and veins.

Objective 13 Explain why capillaries are considered the functional units of the circulatory system.

Blood vessels form a closed tubular network that permits blood to flow from the heart to all the living cells of the body and then back to the heart. Blood leaving the heart passes through vessels of progressively smaller diameters referred to as **arteries, arterioles,** and **capillaries.** Capillaries are microscopic vessels that join the arterial flow to the venous flow. Blood returning to the heart from the capillaries passes through vessels of progressively larger diameters called **venules** and **veins.**

The walls of arteries and veins are composed of three layers, or tunics, as shown in figure 16.13.

1. The **tunica externa,** or *adventitia,* the outermost layer, is composed of loose connective tissue.
2. The **tunica media,** the middle layer, is composed of smooth muscle. The tunica media of arteries has variable amounts of elastic fibers.
3. The **tunica interna,** the innermost layer, is composed of simple squamous epithelium and elastic fibers composed of *elastin.*

The layer of simple squamous epithelium is referred to as the *endothelium.* This layer lines the inner wall of all blood vessels. Capillaries consist of endothelium only, supported by a basement membrane.

Although arteries and veins have the same basic structure, there are some important differences between the two types of vessels. Arteries have more muscle in proportion to their diameter than do comparably sized veins. Also, arteries appear rounder than veins in cross section. Veins are usually partially collapsed because they are not usually filled to capacity. They can stretch when they receive more blood, and thus function as reservoirs or capacitance vessels. In addition, many veins have valves, which are absent in arteries.

tunic: L. *tunica,* covering or coat

Arteries

In the tunica media of large arteries, there are numerous layers of elastic fibers between the smooth muscle cells. Thus, the large arteries expand when the pressure of the blood rises as a result of the heart's contraction; they recoil, like a stretched rubber band, when blood pressure falls during relaxation of the heart. This elastic recoil helps to produce a smoother, less pulsatile flow of blood through the smaller arteries and arterioles.

Small arteries and arterioles are less elastic than the larger arteries and have a thicker layer of smooth muscle in proportion to their diameter. Unlike the larger *elastic arteries,* therefore, the smaller *muscular arteries* retain a relatively constant diameter as the pressure of the blood rises and falls during the heart's pumping activity. Since small muscular arteries and arterioles have narrow lumina, they provide the greatest resistance to blood flow through the arterial system.

Small muscular arteries that are 100 μm or less in diameter branch to form smaller arterioles (20–30 μm in diameter). In some tissues, blood from the arterioles enters the venules directly through **thoroughfare channels** (metarterioles) that form *vascular shunts* (fig. 16.14). In most cases, however, blood from arterioles passes into capillaries. Capillaries are the narrowest of blood vessels (7–10 μm in diameter), and serve as the functional units of the circulatory system. It is across their walls that exchanges of gases (O_2 and CO_2), nutrients, and wastes between the blood and the tissues take place.

Capillaries

The arterial system branches extensively to deliver blood to over 40 billion capillaries in the body. So extensive is this branching, that scarcely any cell in the body is more than a fraction of a millimeter away from any capillary; moreover, the tiny capillaries provide a total surface area of 1,000 square miles for exchanges between blood and tissue fluid.

Despite their large number, at any given time capillaries contain only about 250 ml of blood out of a total blood volume of about 5,000 ml (most blood is contained in the venous system). The amount of blood flowing through a particular capillary bed is determined in part by the action of the **precapillary sphincter muscles** (fig. 16.14). These muscles allow only 5% to 10% of the capillary beds in skeletal muscles, for example, to be open at rest. Blood flow to an organ is regulated by the action of these precapillary

lumen: L. *lumen,* opening

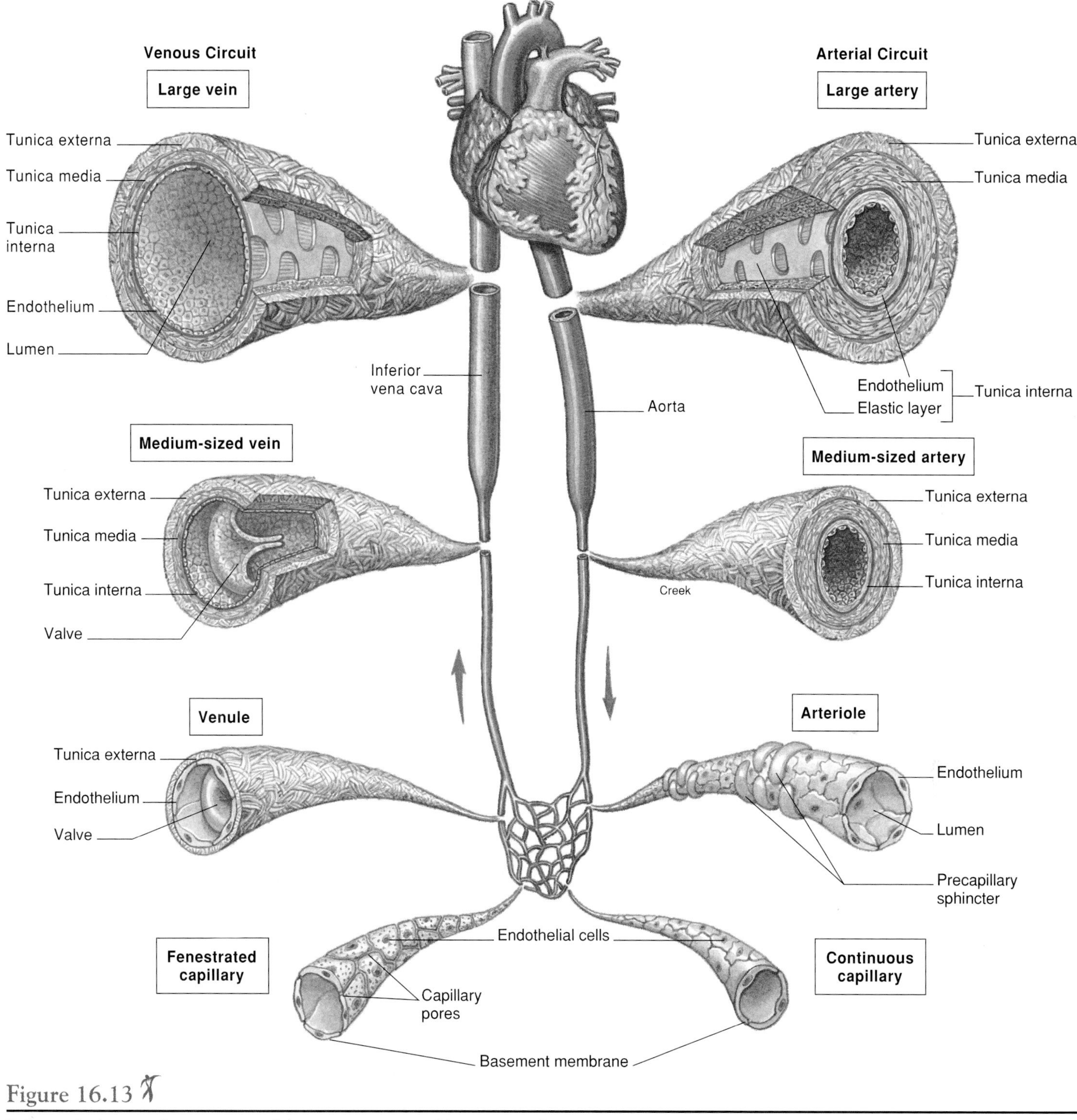

Figure 16.13

Relative thickness and composition of the tunics in comparable arteries and veins.

sphincters and by the degree of resistance to blood flow provided by the small arteries and arterioles in the organ.

Unlike the vessels of the arterial and venous systems, the walls of capillaries are composed of just one cell layer—a simple squamous epithelium, or endothelium. The absence of smooth muscle and connective tissue layers allows for a more rapid exchange of materials between the blood and the tissues.

Types of Capillaries

There are several different types of capillaries, distinguished by significant differences in structure. In terms of their endothelial lining, these capillary types include those that are *continuous*, those that are *discontinuous*, and those that are *fenestrated* (fig. 16.15).

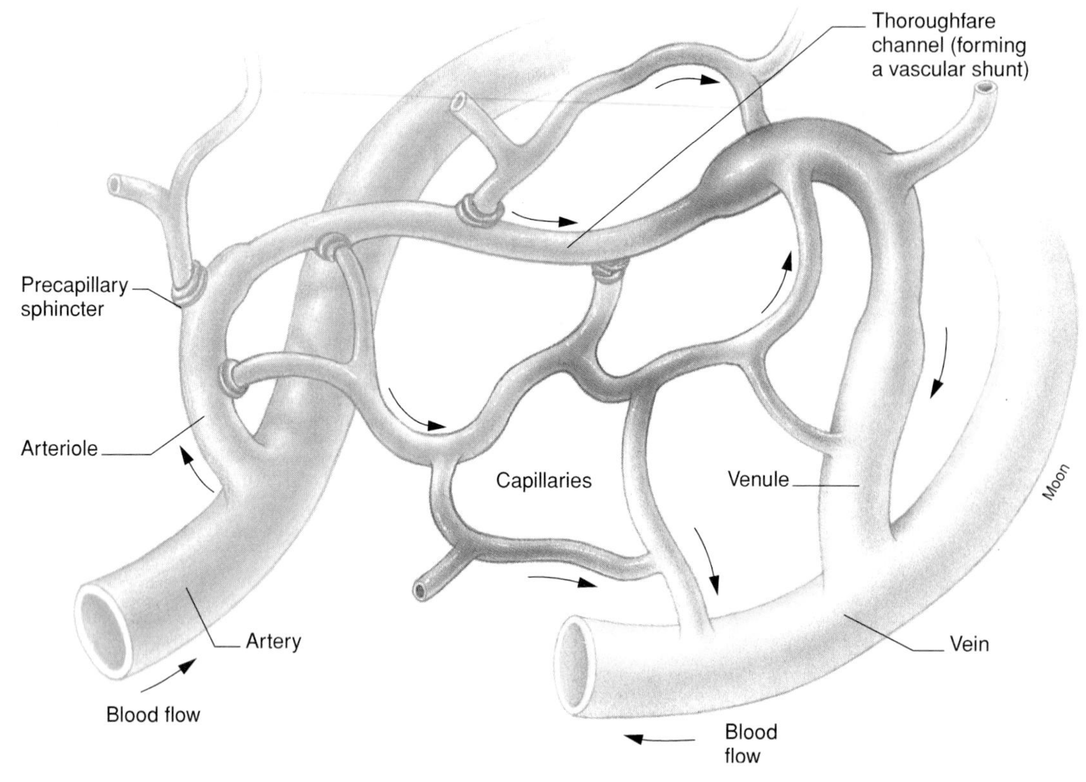

Figure 16.14

Microcirculation at the capillary level. Thoroughfare channels form vascular shunts, providing paths of least resistance between arterioles and venules. Precapillary sphincter muscles regulate the flow of blood through the capillaries.

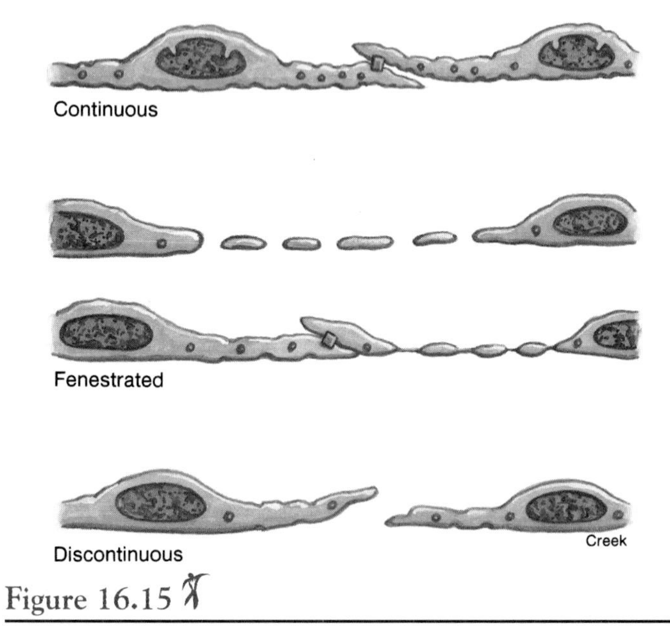

Figure 16.15

Diagrams of continuous, fenestrated, and discontinuous capillaries as they appear in the electron microscope. This classification is based on the continuity of the endothelial layer. (Dark circles in the cytoplasm indicate pinocytotic vesicles.)

Continuous capillaries are those in which adjacent endothelial cells are tightly joined together. These are found in muscles, lungs, adipose tissue, and in the central nervous system. The fact that continuous capillaries in the CNS lack intercellular channels contributes to the blood-brain barrier. Continuous capillaries in other organs have narrow intercellular channels (about 40–45 Å wide) that permit the passage of molecules other than protein between the capillary blood and tissue fluid.

The examination of endothelial cells with an electron microscope has revealed the presence of pinocytotic vesicles, which suggests that the intracellular transport of material may occur across the capillary walls. This type of transport appears to be the only available mechanism of capillary exchange within the central nervous system and may account, in part, for the selective nature of the blood-brain barrier (see fig. 11.8).

Fenestrated (*fen'es-tra-tid*) **capillaries** occur in the kidneys, endocrine glands, and intestines. These capillaries are characterized by wide intercellular pores (800–1,000 Å) that are covered by a layer of mucoprotein, which may serve as a diaphragm. **Discontinuous capillaries** are found in the bone marrow, liver, and spleen. The space between endothelial cells is so great that these capillaries look like little cavities (*sinusoids*) in the organ.

Veins

Veins are vessels that carry blood from capillaries back to the heart. The blood is delivered from microscopic vessels called

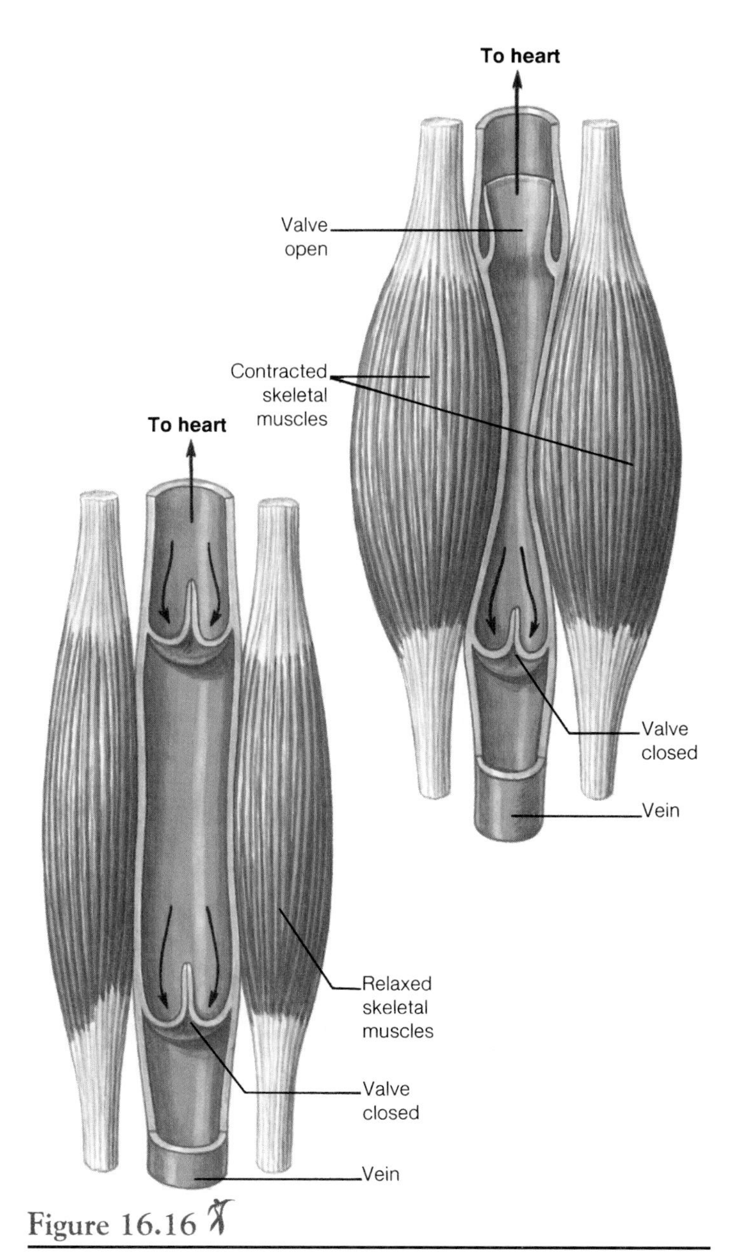

Figure 16.16

The action of the one-way venous valves. The contraction of skeletal muscles helps to pump blood toward the heart, but the flow of blood away from the heart is prevented by closure of the venous valves.

venules into progressively larger vessels that empty into the large veins. The average pressure in the veins is only 2 mmHg, compared to a much higher average arterial pressure of about 100 mmHg. These pressures represent the hydrostatic pressure that the blood exerts on the walls of the vessels.

The low venous pressure is insufficient to return blood to the heart, particularly from the lower limbs. Veins, however, pass between skeletal muscle groups that provide a massaging action as they contract (fig. 16.16). As the veins are squeezed by contracting skeletal muscles, a one-way flow of blood to the heart is ensured by the presence of **venous valves.**

The effect of the massaging action of skeletal muscles on venous blood flow is often described as the *skeletal muscle pump*. The rate of venous return to the heart is dependent, in large part, on the action of skeletal muscle pumps. When these pumps are less active—for example, when a person stands still or is bedridden—blood accumulates in the veins and causes them to bulge. When a person is more active, blood returns to the heart at a faster rate and less is left in the venous system.

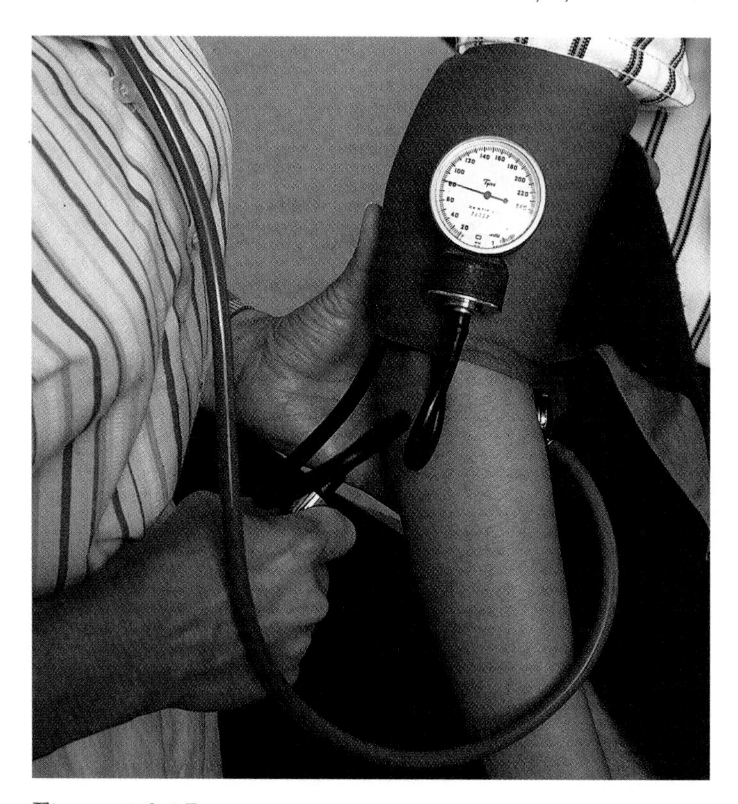

Figure 16.17

The use of a pressure cuff and a sphygmomanometer to measure blood pressure.

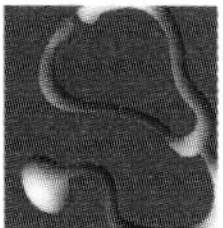

The accumulation of blood in the veins of the legs over a long period of time, as may occur in people with occupations that require standing still all day, can cause the veins to stretch to the point where the venous valves are no longer efficient. This can produce *varicose veins*. During walking, the movements of the foot activate the soleus muscle pump. This effect can be produced in bedridden people by extending and flexing the ankle joints.

Blood Pressure

Blood pressure is the force exerted by the blood against the inner walls of the vessels through which it flows. It plays its primary role in the arteries and arterioles, where the pressure is by far the highest. In capillaries and venules, blood pressure is considerably lower, and so the movement of blood is slower. In the veins, blood pressure plays only a minor role, since the action of the valves and skeletal muscle pumps provides most of the force needed to move blood to the heart. Along with measurements of pulse rate, rate of breathing, and body temperature, blood pressure is usually considered a vital sign.

Arterial blood pressure can be measured with a device called a *sphygmomanometer* (*sfig″mo-mă-nom′ĭ-ter*) (fig. 16.17).

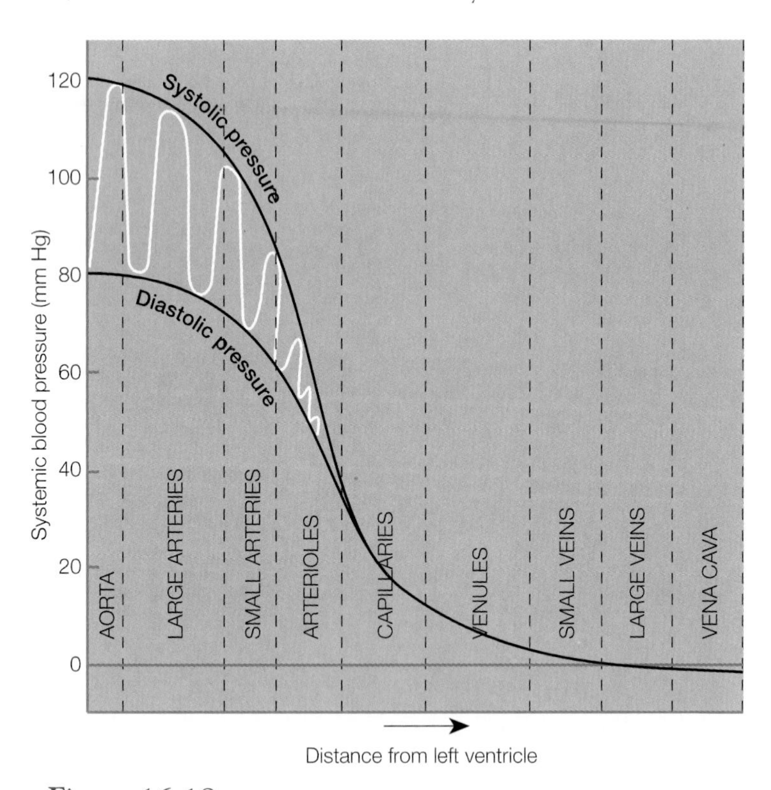

Figure 16.18

Blood pressure as recorded at various vascular sites.

This instrument has an inflatable cuff that is used to constrict an artery at a pressure point (see fig. 16.31). Most commonly, the cuff is wrapped around the upper arm and a stethoscope is applied over the brachial artery, which is compressed against the humerus. As the blood pulsates through the constricted brachial artery, the pressure produced is indicated by the sphygmomanometer.

The normal blood pressure of an adult is about 120/80. This is an expression of the **systolic pressure** (120 mmHg) over the **diastolic pressure** (80 mmHg). The systolic pressure is created as blood is ejected from the heart during ventricular systole. When the ventricles relax during ventricular diastole, the arterial pressure drops and the diastolic pressure is recorded. For the most part, diastolic pressure results from contraction of the smooth muscles within the arterial walls. The difference between the systolic and diastolic pressure is called the **pulse pressure** and is generally about 40 mmHg.

Several factors influence the arterial blood pressure. Blood pressure decreases as the distance from the heart increases (fig. 16.18). An increased cardiac output (stroke volume × heart rate) increases blood pressure. Blood volume and blood viscosity both influence blood pressure, as do various drugs. Changes in the diameters of the vascular lumina through autonomic vasoconstriction or vasodilation have a direct bearing on blood pressure. Blood pressures are also greatly influenced by the general health of the cardiovascular system. Elevated blood pressure is referred to as *hypertension* and is potentially dangerous because of the added strain it puts on the heart.

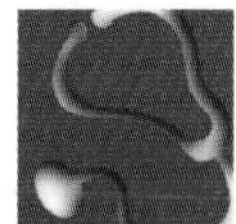

One of the harmful effects of nicotine inhaled in cigarette smoke is vasoconstriction of arterioles. Nicotine is also a heart stimulant and increases cardiac output. Both of these responses raise the blood pressure and increase the strain on the heart.

1. Describe the basic structural pattern of arteries and veins. Describe how arteries and veins differ in structure and how these differences contribute to the resistance function of arteries and the capacitance function of veins.
2. Describe the functional significance of the skeletal muscle pump and explain the action of venous valves.
3. Discuss the functions of capillaries and describe the structural differences between capillaries in different organs.
4. Describe the cardiovascular events that determine systolic and diastolic blood pressures.

Principal Arteries of the Body

The aorta ascends from the left ventricle to a position just above the heart, where it arches to the left and then descends through the thorax and abdomen. Branches of the aorta carry oxygenated blood to all of the cells of the body.

Objective 14 In the form of a flowchart, list the arterial branches of the ascending aorta and aortic arch.

Objective 15 Describe the arterial supply to the brain.

Objective 16 Describe the arterial pathways that supply the upper extremity.

Objective 17 Describe the major arteries serving the thorax, abdomen, and lower extremity.

Contraction of the left ventricle forces oxygenated blood into the arteries of the systemic circulation. The principal arteries of the body are shown in figure 16.19. They will be described by region and identified in order from largest to smallest, or as the blood flows through the system. The major systemic artery is the **aorta** (*a-or'tă*), from which all of the primary systemic arteries arise.

Aortic Arch

The systemic vessel that ascends from the left ventricle of the heart is called the **ascending portion of the aorta.** The **right** and **left coronary arteries,** which serve the myocardium of the heart with blood, are the only branches that arise from the ascending aorta. The aorta arches to the left and posteriorly over the pulmonary

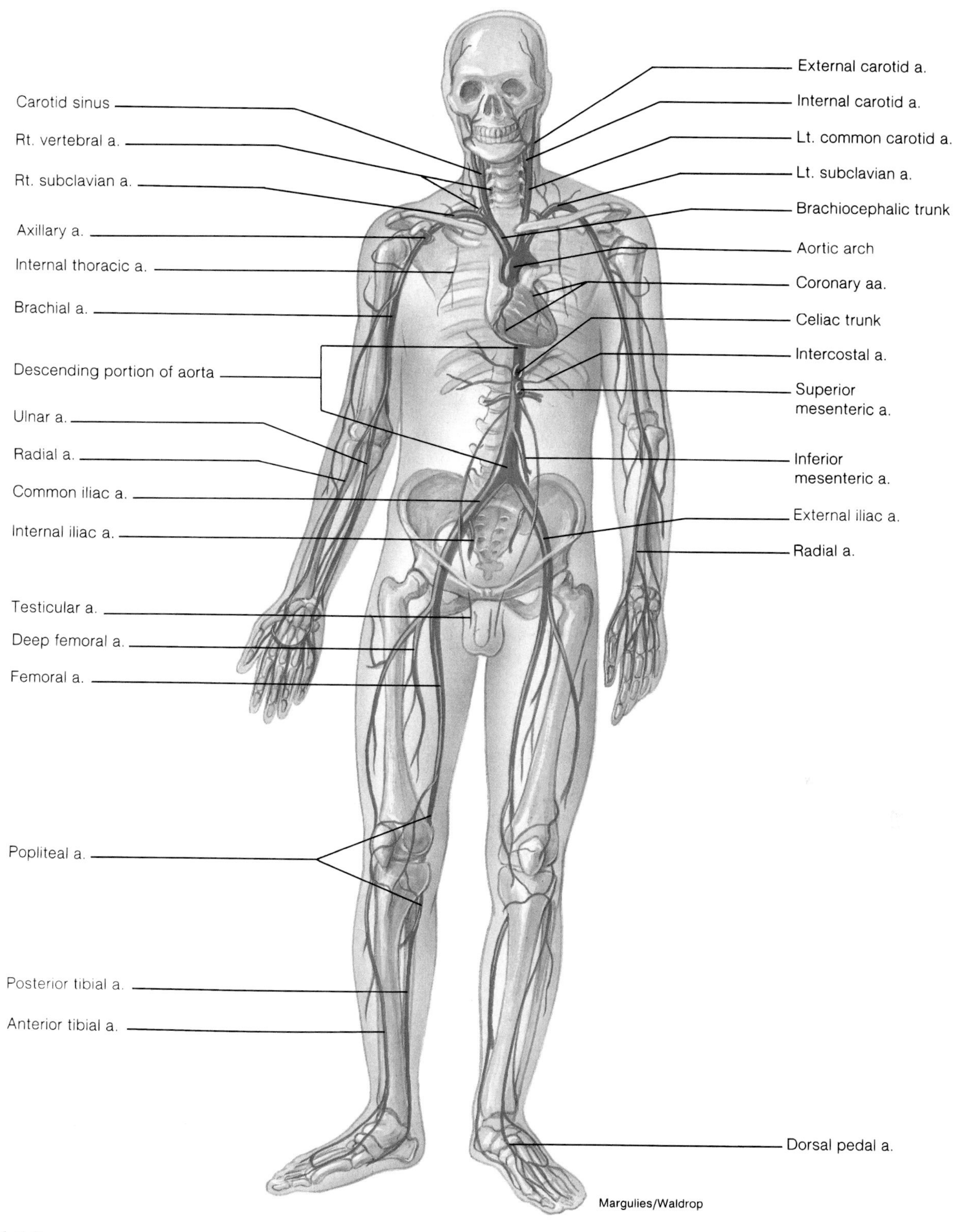

Figure 16.19

Principal arteries of the body (a. = artery; aa. = arteries).

arteries as the **aortic arch** (fig. 16.20). Three vessels arise from the aortic arch: the **brachiocephalic** (*bra″ke-o-sĕ-fal′ik*) **trunk,** the **left common carotid** (*kă-rot′id*) **artery,** and the **left subclavian artery.**

The brachiocephalic trunk is the first vessel to branch from the aortic arch and, as its name suggests, supplies blood to the structures of the shoulder, upper extremity, and head on the right side of the body. It is a short vessel, rising superiorly through the mediastinum to a point near the junction of the sternum and the right clavicle. There it branches into the **right common carotid artery,** which extends to the right side of the neck and head, and

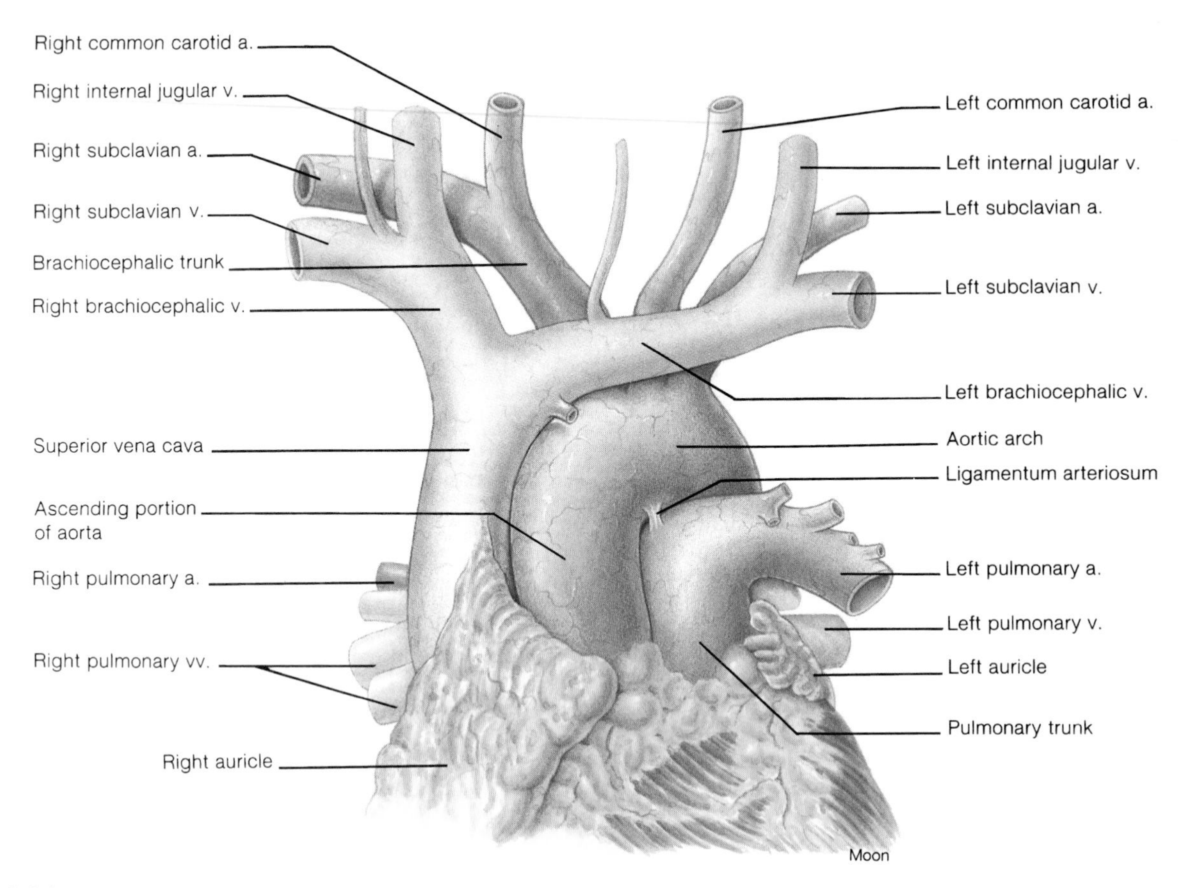

Figure 16.20

The structural relationship between the major arteries and veins to and from the heart (v. = vein; vv. = veins).

the **right subclavian artery,** which carries blood to the right shoulder and upper extremity.

The remaining two branches from the aortic arch are the left common carotid and the left subclavian arteries. The left common carotid artery transports blood to the left side of the neck and head, while the left subclavian artery supplies the left shoulder and upper extremity.

Arteries of the Neck and Head

The common carotid arteries course upward in the neck along the lateral sides of the trachea (fig. 16.21). Each common carotid artery branches into the **internal** and **external carotid arteries** slightly below the angle of the mandible. By pressing gently in this area, a pulse can be detected (see fig. 16.31). At the base of the internal carotid artery is a slight dilation called the **carotid sinus.** The carotid sinus contains *baroreceptors*, which monitor blood pressure. Surrounding the carotid sinus are the **carotid bodies,** small neurovascular organs that contain *chemoreceptors*, which respond to chemical changes in the blood.

Blood Supply to the Brain

The brain is supplied with arterial blood that arrives through four vessels. These vessels eventually unite on the inferior surface of the brain in the area surrounding the pituitary gland (fig. 16.22). The four vessels are the paired *internal carotid arteries* and the paired *vertebral arteries*. The value of four separate vessels coming together at one location is that if one becomes occluded, the three alternate routes may still provide an adequate blood supply to the brain.

The **vertebral arteries** arise from the subclavian arteries at the base of the neck (fig. 16.21). They pass superiorly through the transverse foramina of the cervical vertebrae and enter the skull through the foramen magnum. Within the cranium, the two vertebral arteries unite to form the **basilar artery** at the level of the pons. The basilar artery ascends along the inferior surface of the brain stem and terminates by forming two **posterior cerebral arteries** that supply the posterior portion of the cerebrum (figs. 16.22 and 16.23). The **posterior communicating arteries** are branches that arise from the posterior cerebral arteries and participate in forming the **cerebral arterial circle** (*circle of Willis*) surrounding the pituitary gland.

Each **internal carotid artery** arises from the common carotid artery and ascends in the neck until it reaches the base of the skull, where it enters the carotid canal of the temporal bone. Several branches arise from the internal carotid artery once it is

circle of Willis: from Thomas Willis, English physician, 1621–75

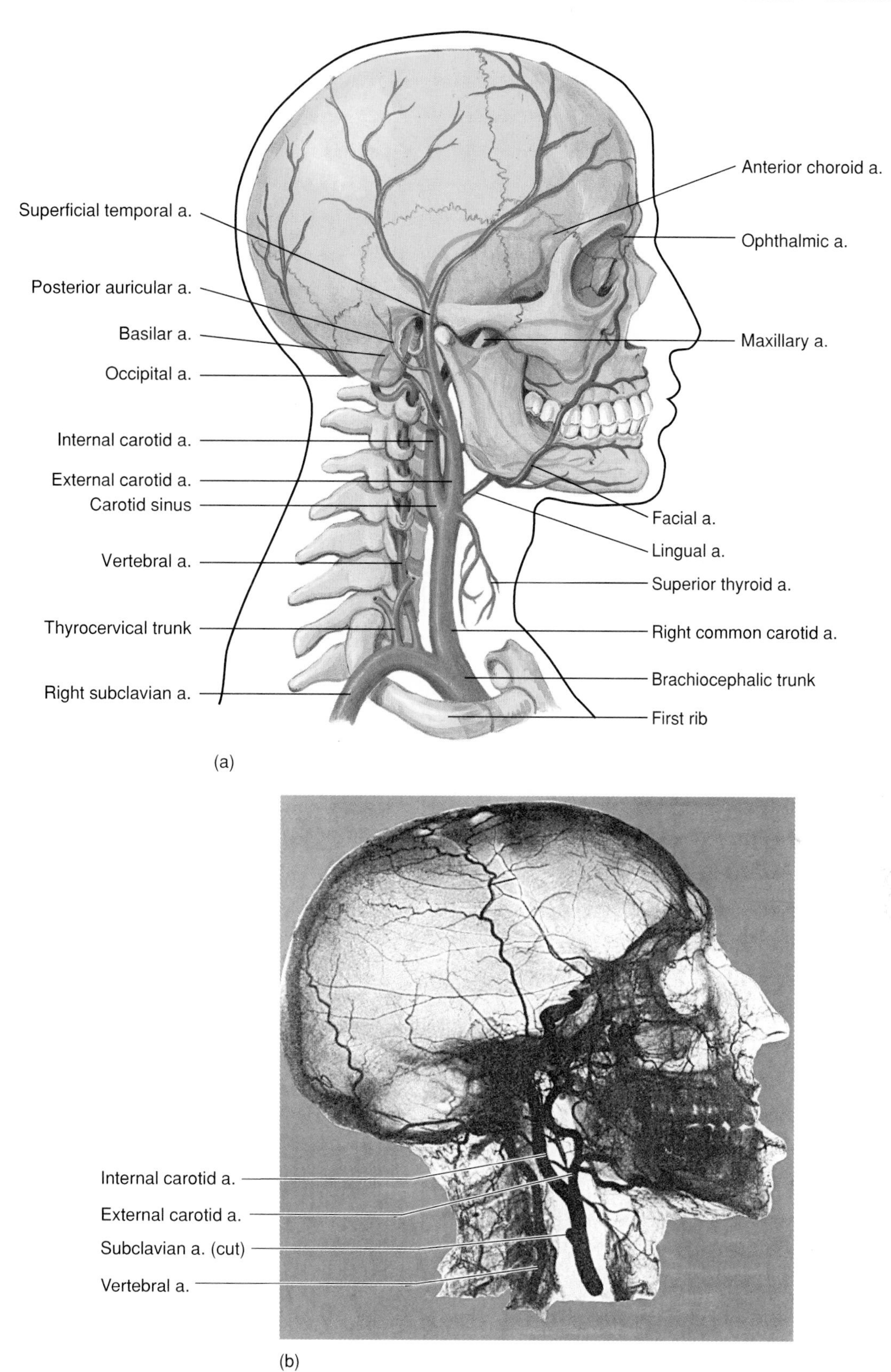

Figure 16.21

Arteries of the neck and head. (*a*) Major branches of the right common carotid and right subclavian arteries. (*b*) A radiograph of the head following a radiopaque injection of the arteries.

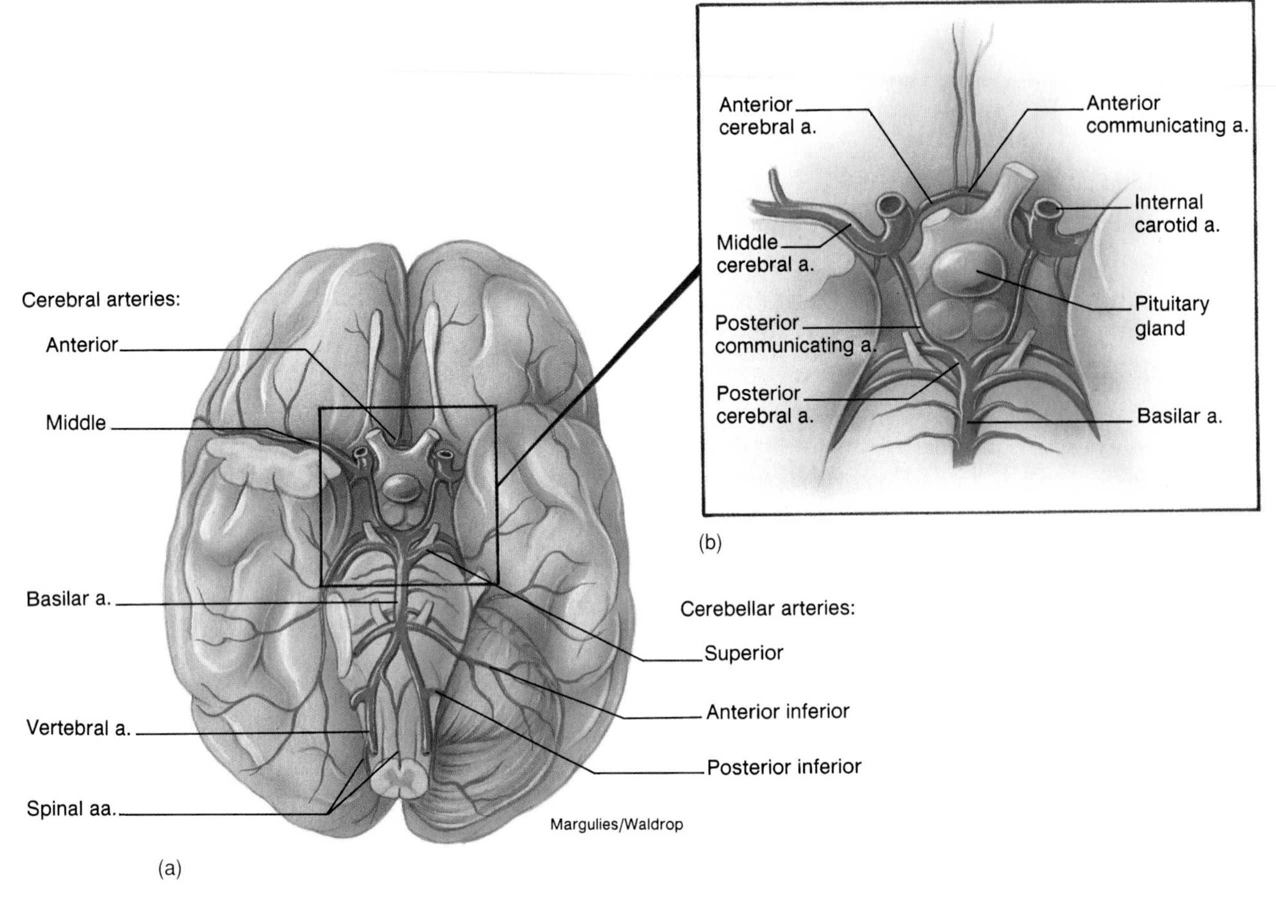

Figure 16.22

Arteries that supply blood to the brain. (*a*) An inferior view of the brain and (*b*) a close-up view of the region of the pituitary gland. The cerebral arterial circle (circle of Willis) consists of the arteries that ring the pituitary gland.

on the inferior surface of the brain. Three of the more important ones are the **ophthalmic artery** (see fig. 16.21*a*), which supplies the eye and associated structures, and the **anterior** and **middle cerebral arteries,** which provide blood to the cerebrum. The internal carotid arteries are connected to the posterior cerebral arteries at the cerebral arterial circle.

Capillaries within the pituitary gland receive both arterial and venous blood. The venous blood arrives from venules immediately superior to the pituitary gland, which drain capillaries in the hypothalamus of the brain. This arrangement of two capillary beds in series—whereby the second capillary bed receives venous blood from the first—is called a *portal system.* The venous blood that travels from the hypothalamus to the pituitary contains hormones from the hypothalamus that help regulate pituitary hormone secretion.

External Carotid Artery

The external carotid artery gives off several branches as it extends upward along the side of the neck and head (see fig. 16.21*a*). The names of these branches are determined by the areas or structures they serve. The principal vessels that arise from the external carotid artery are the following:

1. the **superior thyroid artery,** which serves the muscles of the hyoid region, the larynx and vocal folds, and the thyroid gland;
2. the **ascending pharyngeal artery** (not shown), which serves the pharyngeal area and various lymph nodes;
3. the **lingual artery,** which provides extensive vascularization to the tongue and sublingual gland;
4. the **facial artery,** which traverses a notch on the inferior margin of the mandible to serve the pharyngeal area, palate, chin, lips, and nasal region—an important point in controlling bleeding from the face;
5. the **occipital artery,** which serves the posterior portion of the scalp, the meninges over the brain, the mastoid process, and certain posterior neck muscles; and
6. the **posterior auricular artery,** which serves the auricle of the ear and the scalp over the auricle.

The external carotid artery terminates at a level near the mandibular condyle by dividing into **maxillary** and **superficial**

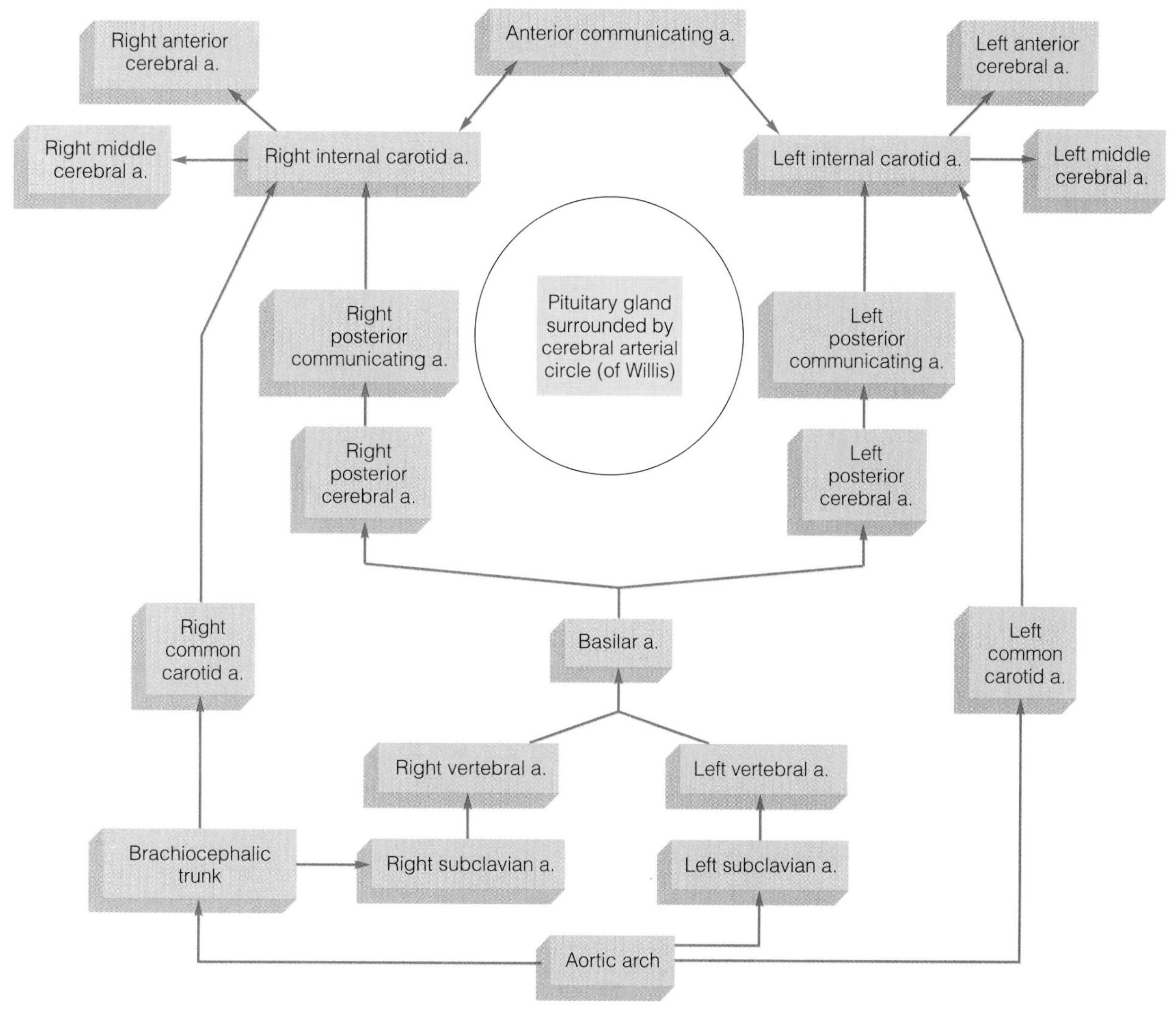

Figure 16.23

The path of arterial blood flow to the brain. (Note the ring of vessels surrounding the pituitary gland to form the cerebral arterial circle.)

temporal arteries. The maxillary artery gives off branches to the teeth and gums, the muscles of mastication, the nasal cavity, the eyelids, and the meninges. The superficial temporal artery supplies blood to the parotid gland and to the superficial structures on the side of the head. Pulsations through the temporal artery can be easily detected by placing the fingertips immediately in front of the ear at the level of the eye (see fig. 16.31). This vessel is frequently used by anesthesiologists to check a patient's pulse rate during surgery.

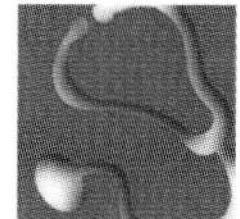

Headaches are usually caused by vascular pressure on the sensitive meninges covering the brain. The two principal vessels serving the meninges are the meningeal branches of the occipital and maxillary arteries. Vasodilation of these vessels creates excessive pressure on the sensory receptors within the meninges, resulting in a headache.

Arteries of the Shoulder and Upper Extremity

As mentioned earlier, the **right subclavian artery** arises from the brachiocephalic trunk, and the **left subclavian artery** arises directly from the aortic arch (see fig. 16.20). Each subclavian artery passes laterally deep to the clavicle, carrying blood toward the arm (fig. 16.24). The pulsations of the subclavian artery can be detected by pressing firmly on the tissue just above the medial portion of the clavicle (see fig. 16.31). From each subclavian artery arises a **vertebral artery** that carries blood to the brain (already described); a short **thyrocervical trunk** that serves the thyroid gland, trachea, and larynx; and an **internal thoracic artery** that descends into the thorax to serve the thoracic wall, thymus, and pericardium. The **costocervical trunk** branches to serve the upper intercostal muscles, posterior neck muscles, and the spinal cord and its meninges. A branch of the internal thoracic artery supplies blood to the muscles and tissues (mammary glands) of the anterior thorax.

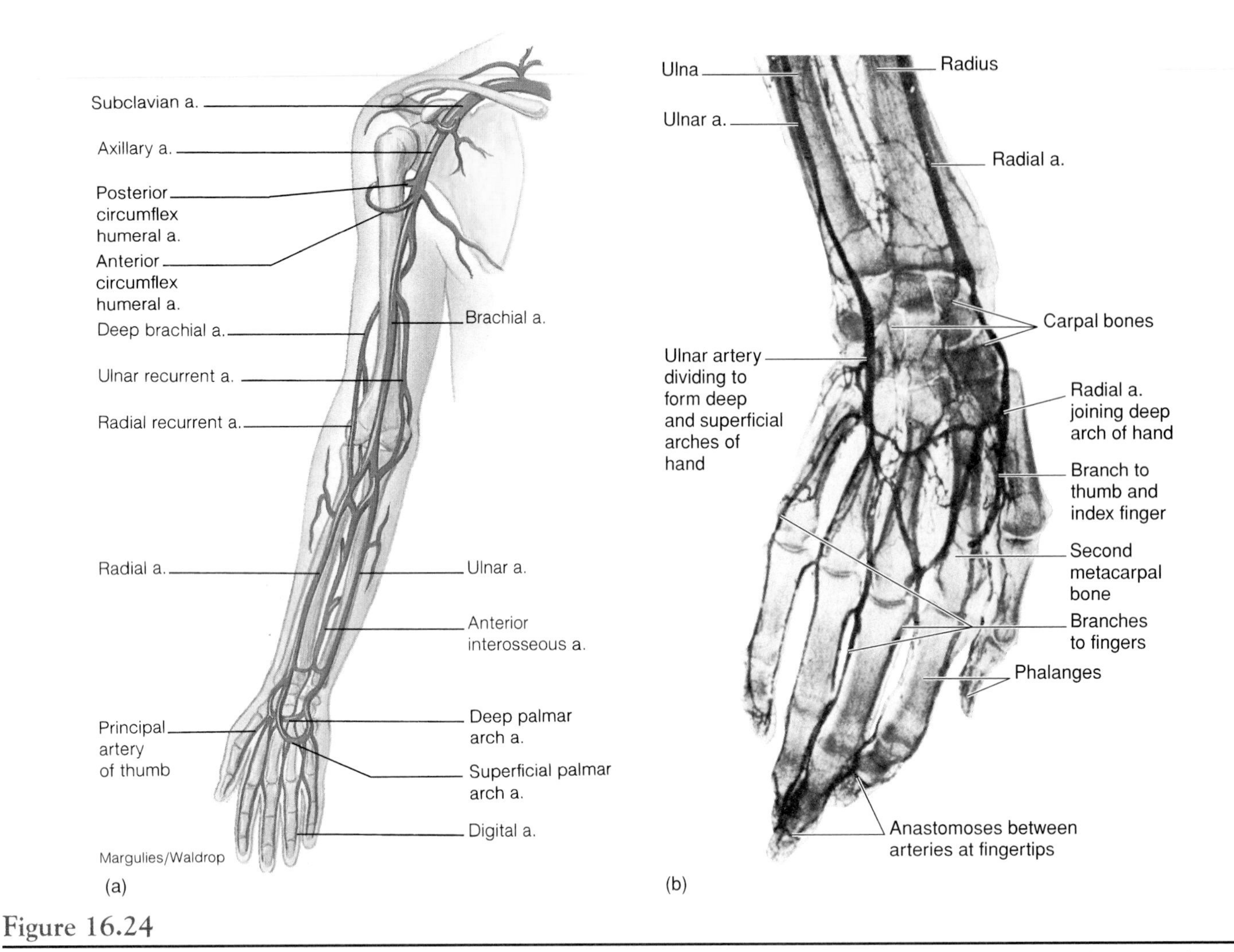

Figure 16.24

Arteries of the right upper extremity. (*a*) An anterior view of the major arteries, and (*b*) a radiograph of the forearm and hand following a radiopaque injection of the arteries.

The **axillary** (*ak'sĭ-lar"e*) **artery** is the continuation of the subclavian artery as it passes into the axillary region (figs. 16.24 and 16.25). The axillary artery is that portion of the major artery of the upper extremity between the outer border of the first rib and the lower border of the teres major muscle. Several small branches arise from the axillary artery and supply blood to the tissues of the upper thorax and shoulder region.

The **brachial** (*bra'ke-al*) **artery** is the continuation of the axillary artery through the brachial region. The brachial artery courses on the medial side of the humerus, where it is a major pressure point and the most common site for determining blood pressure. A **deep brachial artery** branches from the brachial artery and curves posteriorly near the radial nerve to supply the triceps brachii muscle. Two additional branches from the brachial, the **anterior** and **posterior humeral circumflex arteries,** form a continuous ring of vessels around the proximal portion of the humerus.

Just proximal to the cubital fossa, the brachial artery branches into the **radial** and **ulnar arteries,** which supply blood to the forearm and a portion of the hand and digits. The radial artery courses down the lateral, or radial, side of the arm, where it sends numerous small branches to the muscles of the forearm. The **radial recurrent artery** serves the region of the elbow and is the first and largest branch of the radial artery. The radial artery is important as a site for recording the pulse near the wrist.

The ulnar artery extends down the medial, or ulnar, side of the forearm and gives off many small branches to the muscles on that side. It, too, has an initial large branch, the **ulnar recurrent artery,** which arises from the proximal portion near the elbow. At the wrist, the ulnar and radial arteries anastomose to form the **superficial** and **deep palmar arches.** The **metacarpal arteries** of the hand (not shown) arise from the deep palmar arch, and the **digital arteries** of the fingers arise from the superficial palmar arch.

Branches of the Thoracic Portion of the Aorta

The **thoracic portion of the aorta** is a continuation of the aortic arch as it descends through the thoracic cavity to the diaphragm.

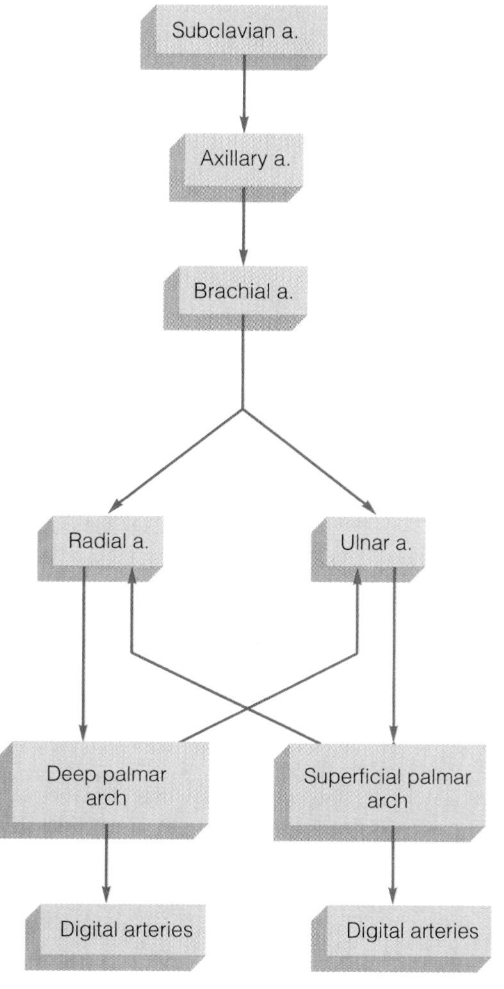

Figure 16.25

The path of arterial blood flow from the subclavian artery to the digital arteries of the fingers.

This large vessel gives off branches to the organs and muscles of the thoracic region. These branches include **pericardial arteries,** going to the pericardium of the heart; **bronchial arteries** for systemic circulation to the lungs; **esophageal** (*ĕ-sof″ă-je′al*) **arteries,** going to the esophagus as it passes through the mediastinum; segmental **posterior intercostal arteries,** serving the intercostal muscles and structures of the wall of the thorax (fig. 16.26); and **superior phrenic** (*fren′ik*) **arteries,** supplying blood to the diaphragm. These vessels are summarized according to their location and function in table 16.5.

Branches of the Abdominal Portion of the Aorta

The **abdominal portion of the aorta** is the segment of the aorta between the diaphragm and the level of the fourth lumbar vertebra, where it divides into the **right** and **left common iliac arteries.** Small **inferior phrenic arteries** that serve the diaphragm are the first vessels to arise from the abdominal portion of the aorta. The next vessel is the short and thick **celiac** (*se′le-ak*) **trunk,** an unpaired vessel that divides immediately into three arteries: the **splenic,** going to the spleen; the **left gastric,** going to the stomach; and the **common hepatic,** going to the liver (fig. 16.27).

The **superior mesenteric artery** is another unpaired vessel. It arises anteriorly from the abdominal portion of the aorta, just below the celiac trunk. The superior mesenteric artery supplies blood to the small intestine (except for a portion of the

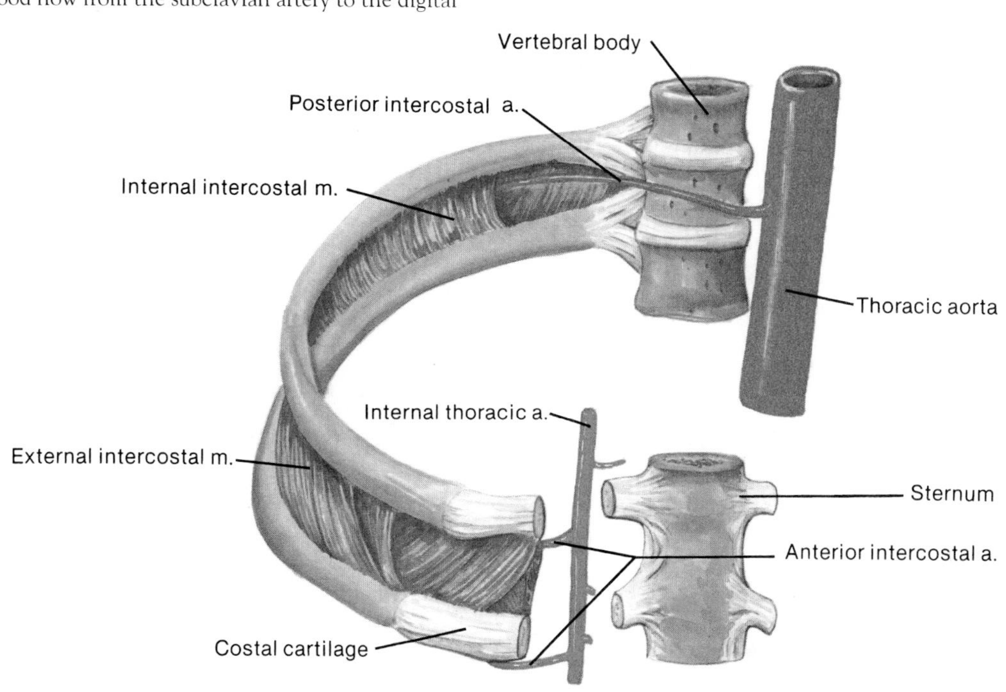

Figure 16.26

Arteries that serve the thoracic wall.

Table 16.5
Segments and Branches of the Aorta

Segment of Aorta	Arterial Branch	General Region or Organ Served
Ascending portion of aorta	Right and left coronary aa.	Heart
Aortic arch	Brachiocephalic trunk	
	Right common carotid a.	Right side of head and neck
	Right subclavian a.	Right shoulder and right upper extremity
	Left common carotid a.	Left side of head and neck
	Left subclavian a.	Left shoulder and left upper extremity
Thoracic portion of aorta	Pericardial aa.	Pericardium of heart
	Posterior intercostal aa.	Intercostal and thoracic muscles, and pleurae
	Bronchial aa.	Bronchi of lungs
	Superior phrenic aa.	Superior surface of diaphragm
	Esophageal aa.	Esophagus
Abdominal portion of aorta	Inferior phrenic aa.	Inferior surface of diaphragm
	Celiac trunk	
	Common hepatic a.	Liver, upper pancreas, and duodenum
	Left gastric a.	Stomach and esophagus
	Splenic a.	Spleen, pancreas and stomach
	Superior mesenteric a.	Small intestine, pancreas, cecum, appendix, ascending colon, and transverse colon
	Suprarenal aa.	Adrenal (suprarenal) glands
	Lumbar aa.	Muscles and spinal cord of lumbar region
	Renal aa.	Kidneys
	Gonadal aa.	
	Testicular aa.	Testes
	Ovarian aa.	Ovaries
	Inferior mesenteric a.	Transverse colon, descending colon, sigmoid colon, and rectum
	Common iliac aa.	
	External iliac aa.	Lower extremities
	Internal iliac aa.	Genital organs and gluteal muscles

duodenum), the cecum, the appendix, the ascending colon, and the proximal two-thirds of the transverse colon.

The next major vessels to arise from the abdominal portion of the aorta are the paired **renal arteries** that carry blood to the kidneys. Smaller **suprarenal arteries,** located just above the renal arteries, serve the adrenal (suprarenal) glands. The **testicular arteries** in the male and the **ovarian arteries** in the female are small paired vessels that arise from the abdominal portion of the aorta, just below the renal arteries. These vessels serve the gonads.

The **inferior mesenteric artery** is the last major branch of the abdominal portion of the aorta. It is an unpaired anterior vessel that arises just before the iliac bifurcation. The inferior mesenteric supplies blood to the distal one-third of the transverse colon, the descending colon, the sigmoid colon, and the rectum.

Several **lumbar arteries** branch posteriorly from the abdominal portion of the aorta throughout its length and serve the muscles and the spinal cord in the lumbar region. In addition, an unpaired **middle sacral artery** (fig. 16.28) arises from the posterior terminal portion of the abdominal portion of the aorta to supply the sacrum and coccyx.

Arteries of the Pelvis and Lower Extremity

The abdominal portion of the aorta terminates in the posterior pelvic area as it bifurcates into the right and left common iliac arteries. These vessels pass downward approximately 5 cm on their respective sides and terminate by dividing into the *internal* and *external iliac arteries*.

The *internal iliac artery* has extensive branches to supply arterial blood to the gluteal muscles and the organs of the pelvic region (fig. 16.28). The wall of the pelvis is served by the **iliolumbar** and **lateral sacral arteries.** The internal visceral organs of the pelvis are served by the **middle rectal** and the **superior, middle,** and **inferior vesicular arteries** to the urinary bladder. In addition, **uterine** and **vaginal arteries** branch from the internal iliac arteries to serve the reproductive organs of the female. The muscles of the buttock are served by the **superior** and **inferior**

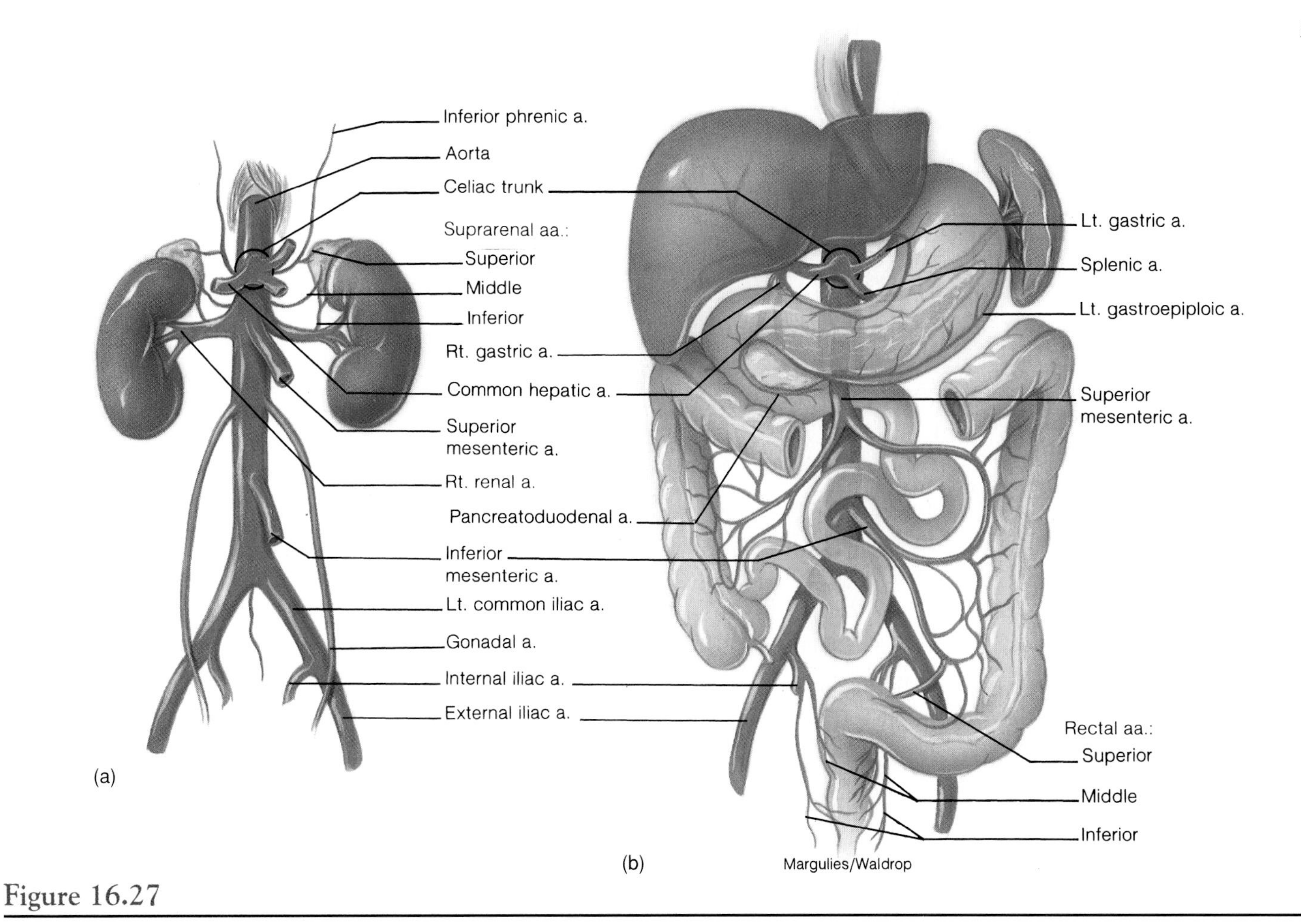

Figure 16.27

An anterior view of the abdominal aorta and its principal branches. In (*a*) the abdominal viscera have been removed; in (*b*) they are intact.

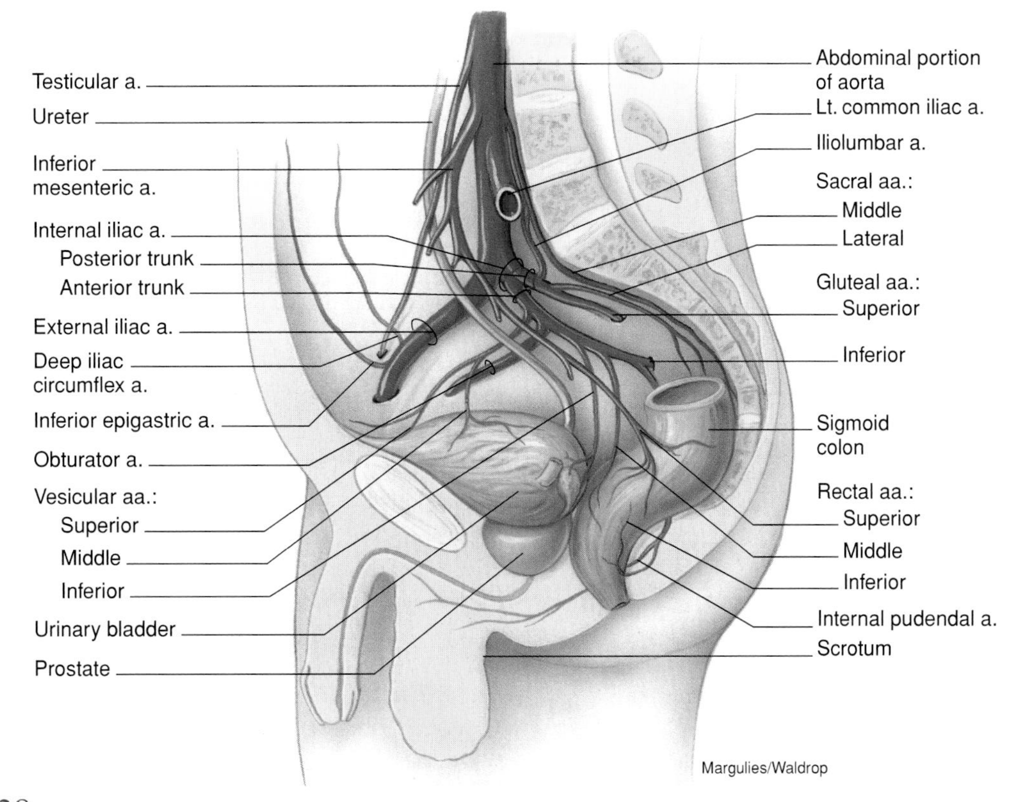

Figure 16.28

Arteries of the pelvic region.

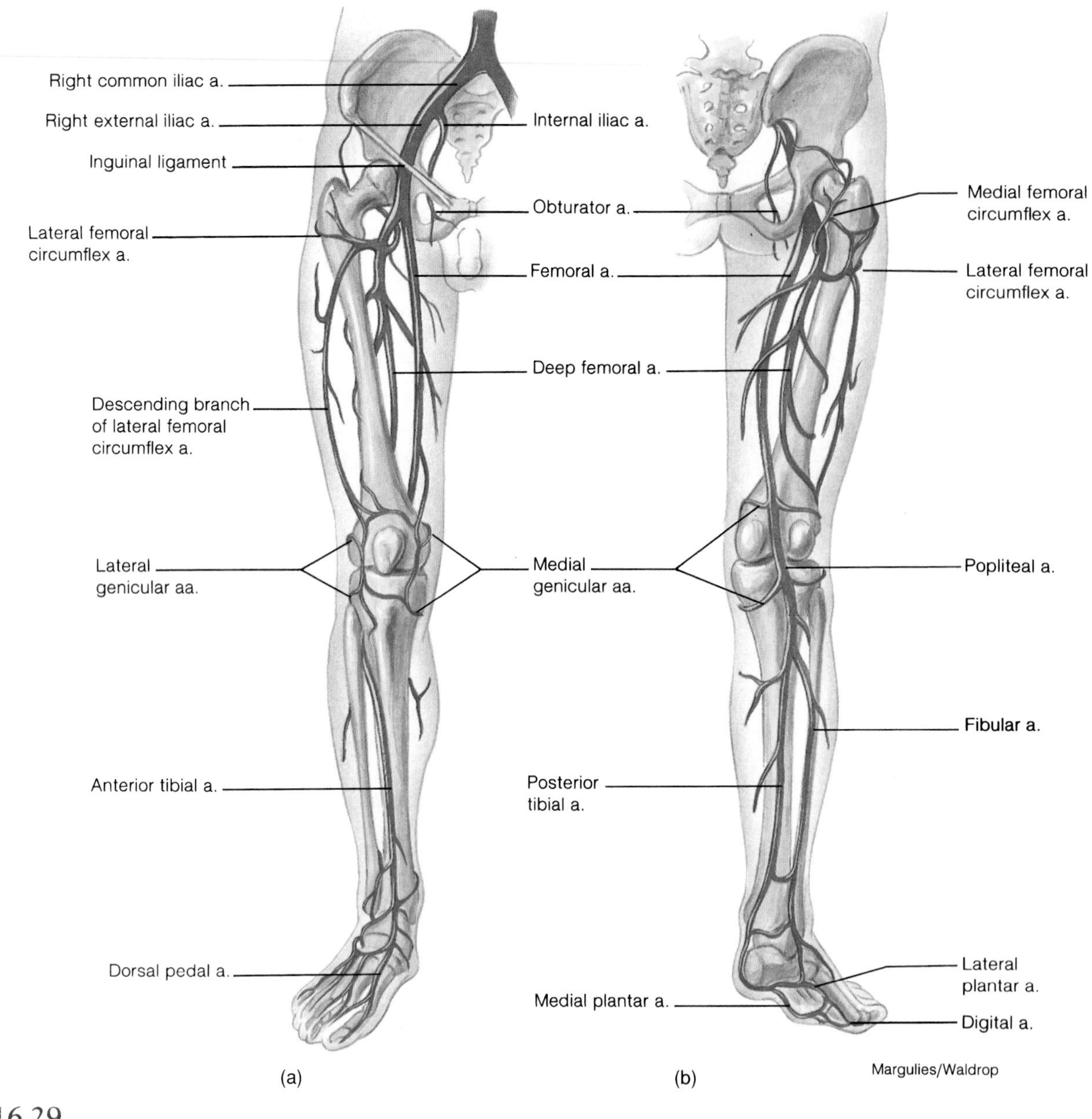

Figure 16.29

Arteries of the right lower extremity. (*a*) An anterior view and (*b*) a posterior view.

gluteal arteries. Some of the upper medial thigh muscles are supplied with blood from the **obturator artery.** The **internal pudendal artery** of the internal iliac artery serves the musculature of the perineum and the external genitalia. During sexual arousal it supplies the blood for vascular engorgement of the penis in the male and clitoris in the female.

The external iliac artery passes out of the pelvic cavity deep to the inguinal ligament (fig. 16.29) and becomes the **femoral** (*fem′or-al*) **artery.** Two branches arise from the external iliac artery, however, before it passes beneath the inguinal ligament. An **inferior epigastric artery** branches from the external iliac artery and passes superiorly to supply the skin and muscles of the abdominal wall. The **deep circumflex iliac artery** is a small branch that extends laterally to supply the muscles attached to the iliac fossa.

The femoral artery passes through an area called the **femoral triangle** on the upper medial portion of the thigh (figs. 16.29 and 16.30). At this point, it is close to the surface and its pulse can be palpated. Several vessels arise from the femoral artery to serve the thigh region. The largest of these, the **deep femoral artery,** passes posteriorly to serve the hamstring muscles. The **lateral** and **medial femoral circumflex arteries** encircle the proximal end of the femur and serve muscles in this region. The femoral artery becomes the **popliteal** (*pop″lĭ-te′al*) **artery** as it passes across the posterior aspect of the knee.

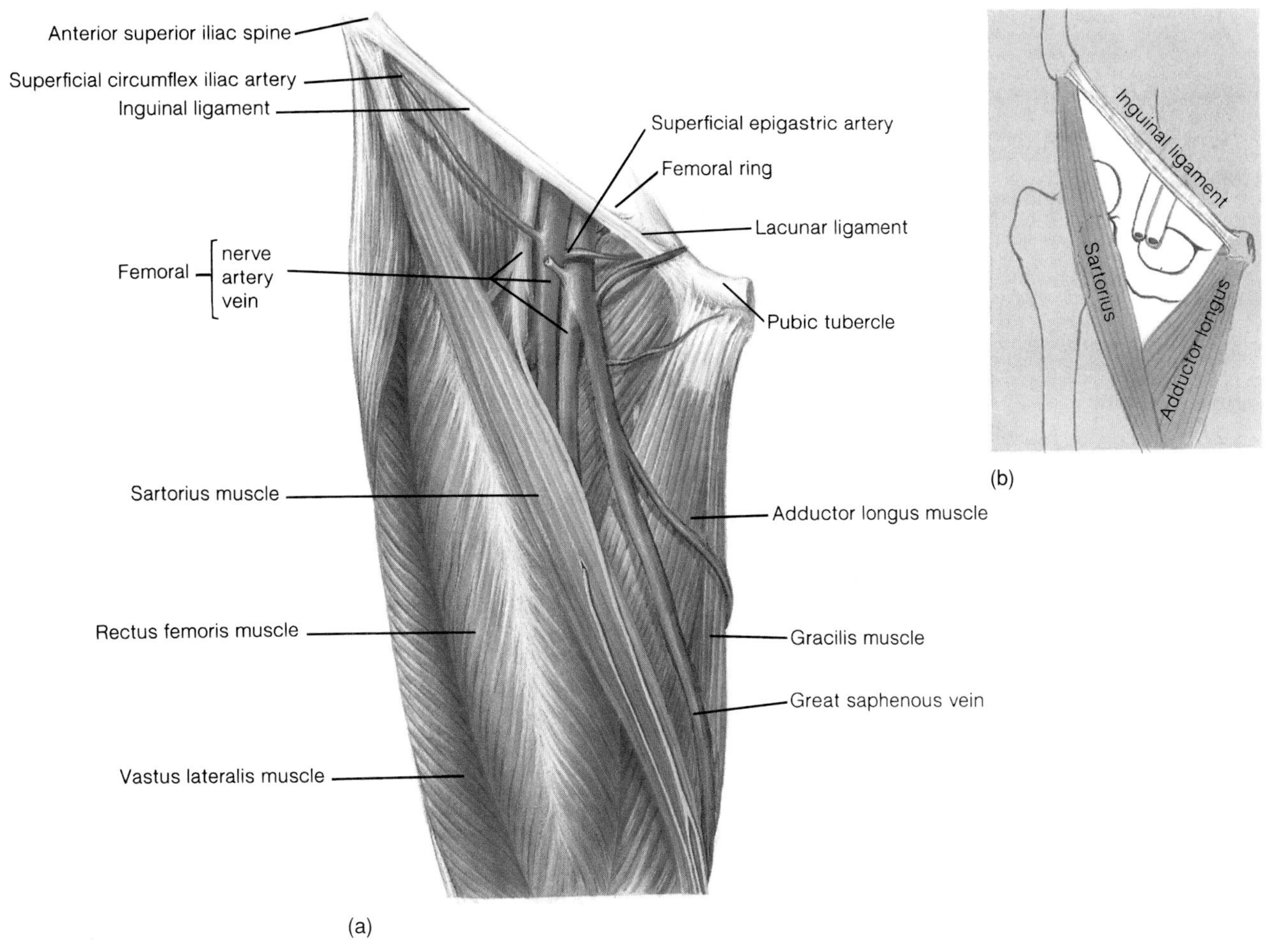

Figure 16.30

The femoral triangle. The structures within the femoral triangle are shown in (*a*); the boundaries of the triangle are shown in (*b*).

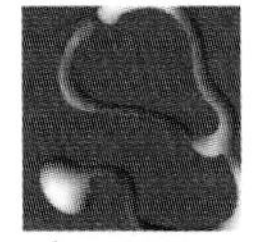

Hemorrhage can be a serious problem in many accidents. To prevent a victim from bleeding to death, it is important to know where to apply pressure to curtail the flow of blood (fig. 16.31). The pressure points for the appendages are the brachial artery on the medial side of the arm and the femoral artery in the groin. Firmly applied pressure to these regions greatly diminishes the flow of blood to traumatized areas below. A tourniquet may have to be applied if bleeding is severe enough to endanger life.

The popliteal artery supplies small branches to the knee joint, and then divides into an **anterior tibial artery** and a **posterior tibial artery** (fig. 16.29). These vessels traverse the anterior and posterior aspects of the leg, respectively, providing blood to the muscles of these regions and to the foot.

At the ankle, the anterior tibial artery becomes the **dorsal pedal artery** that serves the ankle and dorsum of the foot and then contributes to the formation of the **plantar arch** of the foot. Clinically, palpation of the dorsal pedal artery can provide information about circulation to the foot; more important, it can provide information about the circulation in general because its pulse is taken at the most distal portion of the body.

The posterior tibial artery gives off a large **fibular,** or **peroneal, artery** to serve the peroneal muscles of the leg. At the ankle, the posterior tibial bifurcates into the **lateral** and **medial plantar arteries** that supply the sole of the foot. The lateral plantar artery anastomoses with the dorsal pedal artery to form the plantar arch, similar to the arterial arrangement in the hand. **Digital arteries** arise from the plantar arch to supply the toes with blood.

1. Describe the blood supply to the brain. Where is the cerebral arterial circle located and how is it formed?
2. Describe the clinical significance of the brachial and radial arteries.
3. Describe the arterial pathway from the subclavian artery to the digital arteries.
4. List the arteries that supply blood to the lower abdominal wall, the external genitalia, the hamstring muscles, the knee joint, and the dorsum of the foot.

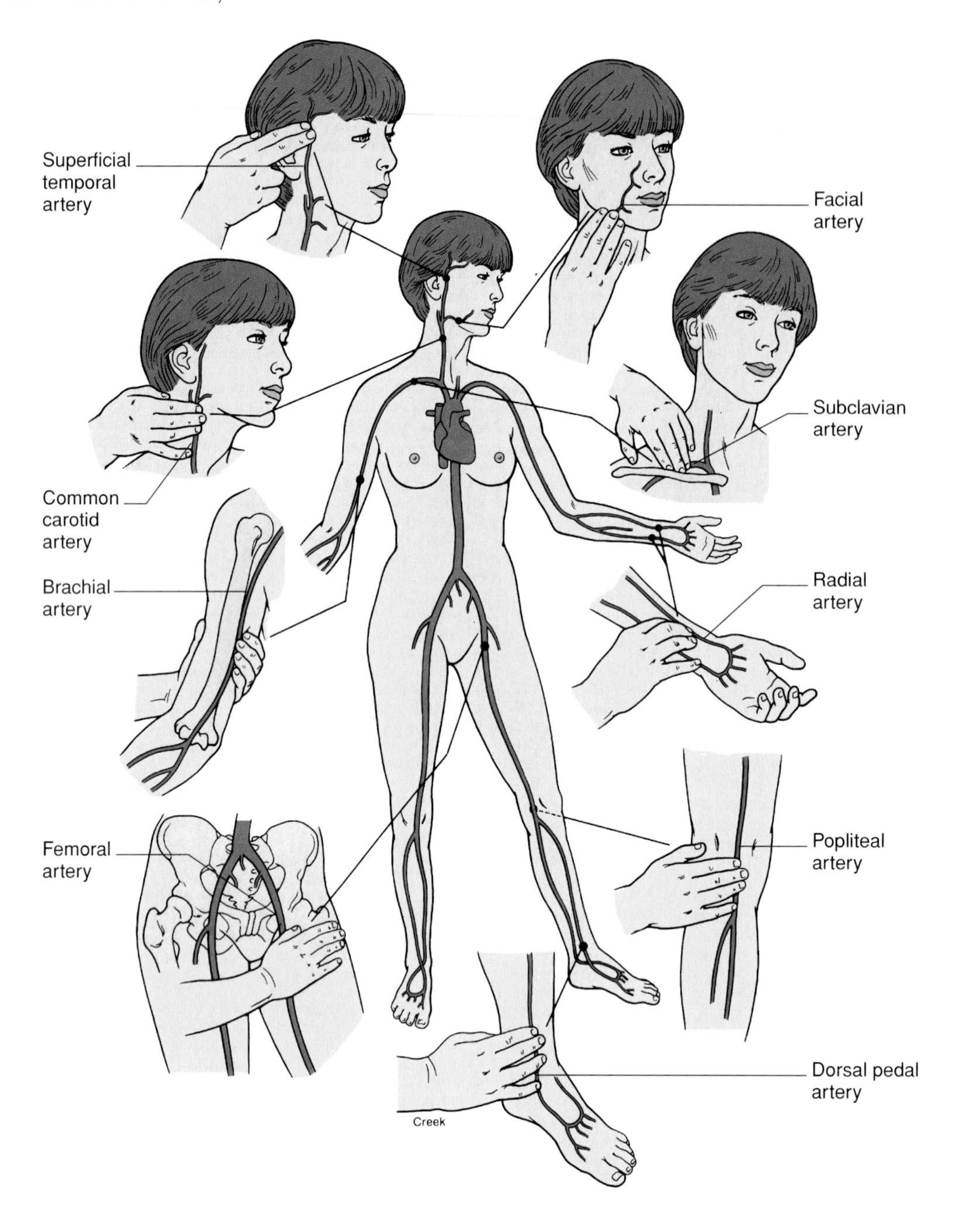

Figure 16.31

Important arterial pressure points and the locations at which arterial pulsations can best be detected.

Principal Veins of the Body

After systemic blood has passed through the tissue, this oxygenated-poor blood is returned through veins of progressively larger diameters to the right atrium of the heart.

Objective 18 Describe the venous drainage of the head, neck, and upper extremity.

Objective 19 Describe the venous drainage of the thorax, lower extremities, and abdominal region.

Objective 20 Describe the vessels involved in the hepatic portal system.

In the venous portion of the systemic circulation, blood flows from smaller vessels into larger ones, so that a vein receives smaller tributaries instead of giving off branches as an artery does. The veins from all parts of the body (except the lungs) converge into two major vessels that empty into the right atrium: the **superior vena cava** (*ve'nă ka'vă*) and the **inferior vena cava** (fig. 16.32). Veins are more numerous than arteries and are both superficial and deep. Superficial veins generally can be seen just beneath the skin and are clinically important in drawing blood and giving injections.

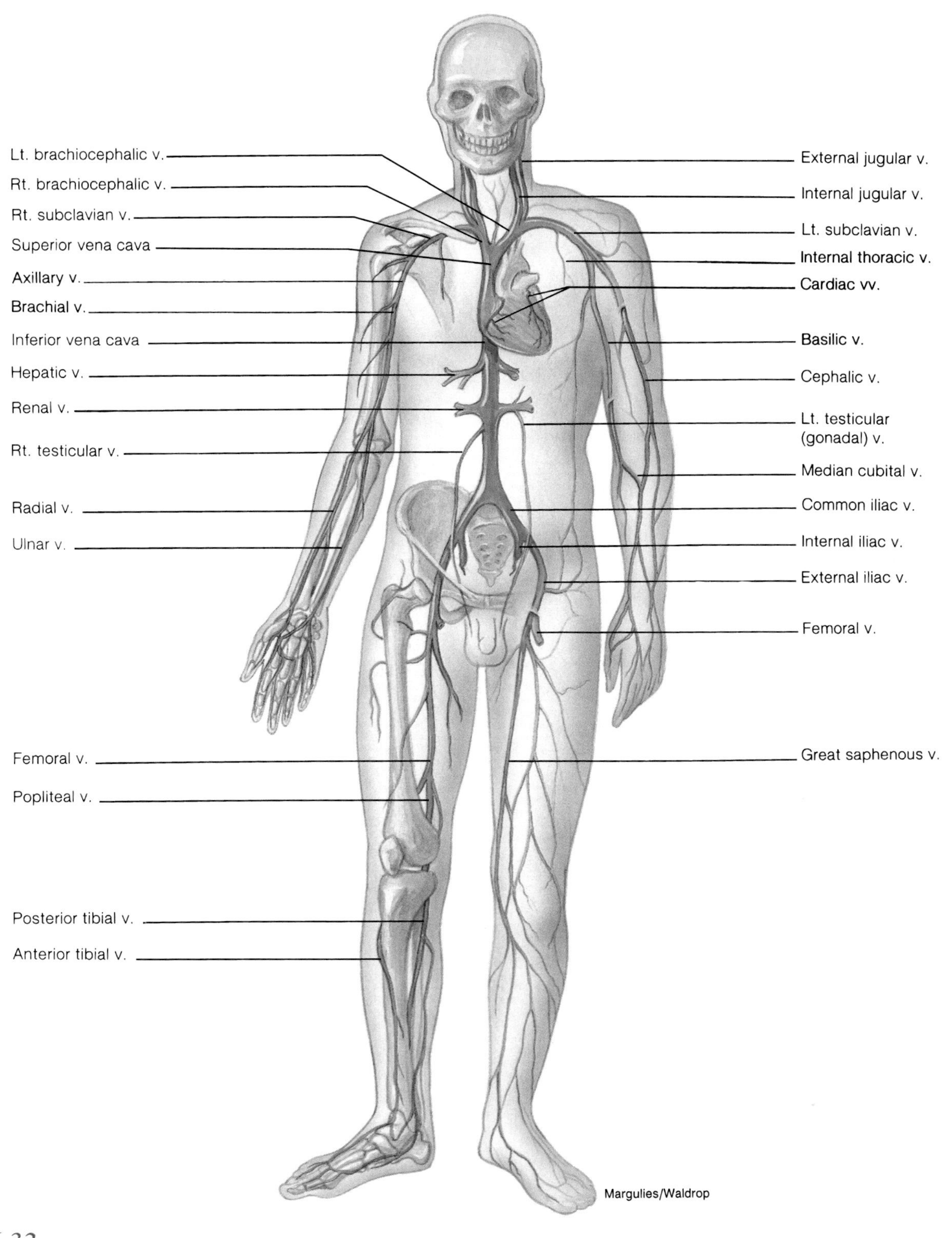

Figure 16.32

Principal veins of the body. Superficial veins are depicted in the left extremities and deep veins in the right extremities.

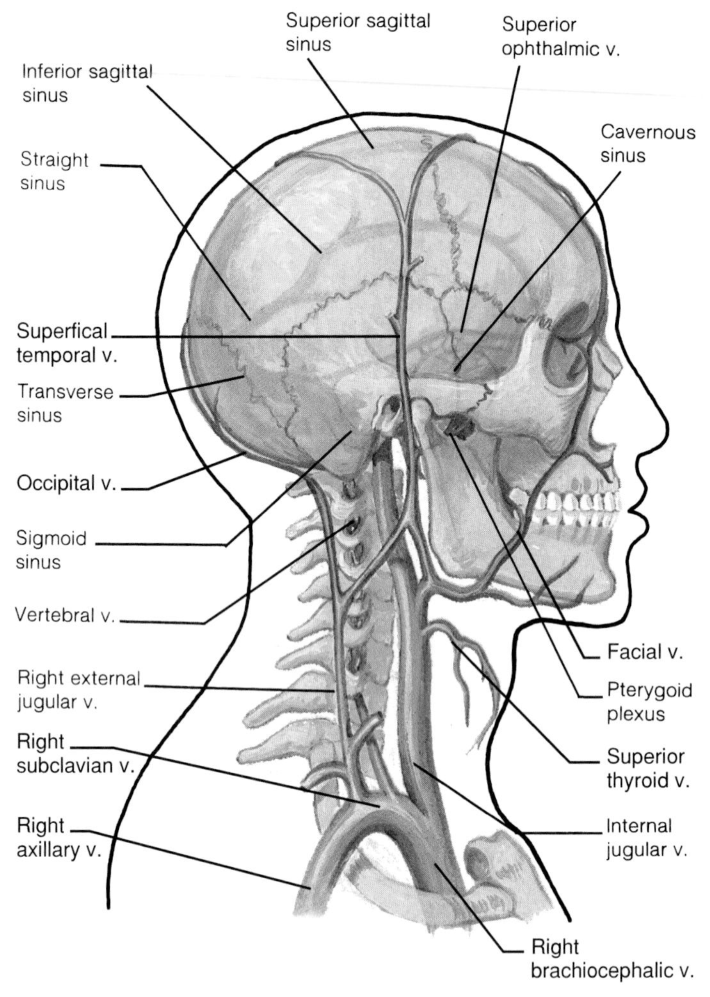

Figure 16.33

Veins that drain the head and neck. (Note the cranial venous sinuses that drain blood from the brain into the internal jugular vein.)

Deep veins are close to the principal arteries and are usually similarly named. As with arteries, veins are named according to the region in which they are found or the organ that they serve. (Note that when a vein serves an organ, it drains blood *away* from it.)

Veins Draining the Head and Neck

Blood from the scalp, portions of the face, and the superficial neck regions is drained by the **external jugular veins** (fig. 16.33). These vessels descend on the lateral sides of the neck, superficial to the sternocleidomastoid muscle and deep to the platysma muscle. They empty into the **right** and **left subclavian veins,** located just behind the clavicles.

The paired **internal jugular veins** drain blood from the brain, meninges, and deep regions of the face and neck. The internal jugular veins are larger and deeper than the external jugular veins. They arise from numerous cranial **venous sinuses** (fig. 16.33), a series of both paired and unpaired channels positioned between the two layers of dura mater. The venous sinuses, in turn, receive venous blood from the **cerebral, cerebellar, ophthalmic** and **meningeal veins.**

The internal jugular vein passes inferiorly down the neck adjacent to the common carotid artery and the vagus nerve. All three of these structures are surrounded by the protective *carotid sheath* and are positioned beneath the sternocleidomastoid muscle. The convergence of the internal jugular vein with the subclavian vein forms a large **brachiocephalic** (*bra″ke-o-sĕ-fal′ik*) **vein** on each side. The two brachiocephalic veins then merge to form the *superior vena cava,* which drains into the right atrium of the heart (see fig. 16.32).

The external jugular vein is the vessel that is seen on the side of the neck when a person is angry or wearing a tight collar. You can voluntarily distend this vein by performing *Valsalva's maneuver.* To do this, take a deep breath and hold it while you forcibly contract your abdominal muscles as in a forced exhalation. This procedure is automatically performed when lifting a heavy object or defecating. The increased thoracic pressure that results compresses the vena cavae and interferes with the return of blood to the right atrium. Thus, Valsalva's maneuver can be dangerous if performed by an individual with cardiovascular disease.

Veins of the Upper Extremity

The upper extremity has both superficial and deep venous drainage (fig. 16.34). The superficial veins are highly variable and form an extensive network just below the skin. The deep veins accompany the arteries of the same region and bear similar names. The deep veins of the upper extremity will be described first.

Both the **radial vein** on the lateral side of the forearm and the **ulnar vein** on the medial side drain blood from the **deep** and **superficial palmar arches** of the hand. The radial and ulnar veins join in the cubital fossa to form the **brachial vein,** which continues up the medial side of the brachium.

The main superficial vessels of the upper extremity are the **basilic vein** and the **cephalic vein.** The basilic vein passes on the ulnar side of the forearm and the medial side of the arm. Near the head of the humerus, the basilic vein merges with the brachial vein to form the **axillary vein.**

The cephalic vein drains the superficial portion of the hand and forearm on the radial side, and then continues up the lateral side of the arm. In the shoulder region, the cephalic vein pierces the fascia and joins the axillary vein. The axillary vein then passes the first rib to form the subclavian vein, which unites with the internal jugular to form the brachiocephalic vein of that side.

Superficially, in the cubital fossa of the elbow, the **median cubital vein** ascends from the cephalic vein on the lateral side

jugular: L. *jugulum,* throat or neck

Valsalva's maneuver: from Antonio Valsalva, Italian anatomist, 1666–1723

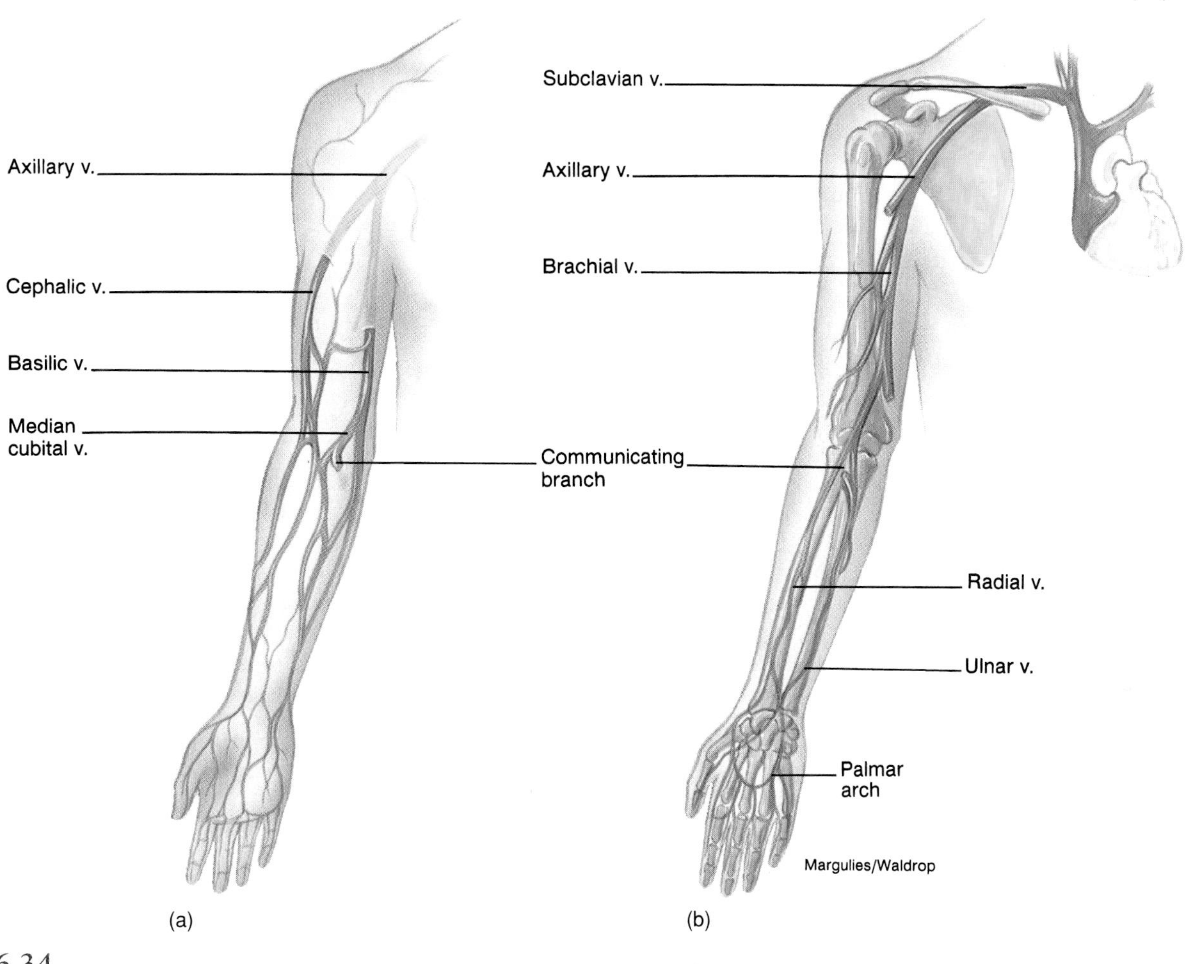

Figure 16.34

An anterior view of the veins that drain the upper right extremity. (*a*) Superficial veins and (*b*) deep veins.

to connect with the basilic vein on the medial side. The median cubital vein is a frequent site for venipuncture to remove a sample of blood or add fluids to the blood.

Veins of the Thorax

The superior vena cava, formed by the union of the two brachiocephalic veins, empties venous blood from the head, neck, and upper extremities directly into the right atrium of the heart. These large vessels lack the valves that are characteristic of most other veins in the body.

In addition to receiving blood from the brachiocephalic veins, the superior vena cava collects blood from the azygos system of veins arising from the posterior thoracic wall (fig. 16.35). The **azygos** (*az'ĭ-gos*) **vein** extends superiorly along the posterior abdominal and thoracic walls on the right side of the vertebral column. It ascends through the mediastinum to join the superior vena cava at the level of the fourth thoracic vertebra. Tributaries of the azygos include the **ascending lumbar veins,** which drain the lumbar and sacral regions; **intercostal veins,** draining from the intercostal muscles; and the **accessory hemiazygos** (*hĕ'me-az-ĭ-gos*) and **hemiazygos veins,** which form the major tributaries to the left of the vertebral column.

Veins of the Lower Extremity

The lower extremities, like the upper extremities, have both a deep and a superficial group of veins (fig. 16.36). The deep veins accompany corresponding arteries and have more valves than do the superficial veins. The deep veins will be described first.

The **posterior** and **anterior tibial veins** originate in the foot and course upward behind and in front of the tibia to the back of the knee, where they merge to form the **popliteal vein.** The popliteal vein receives blood from the knee region. Just above the knee, this vessel becomes the **femoral vein.** The femoral vein in turn continues up the thigh and receives blood from the **deep femoral vein** near the groin. Just above this point, the femoral vein receives blood from the **great saphenous** (*să-fe'nus*) **vein** and then becomes the **external iliac vein** as it passes under the inguinal ligament. The external iliac curves upward to

azygos: Gk. *a*, without; *zygon*, yoke

saphenous: L. *saphena*, the hidden one

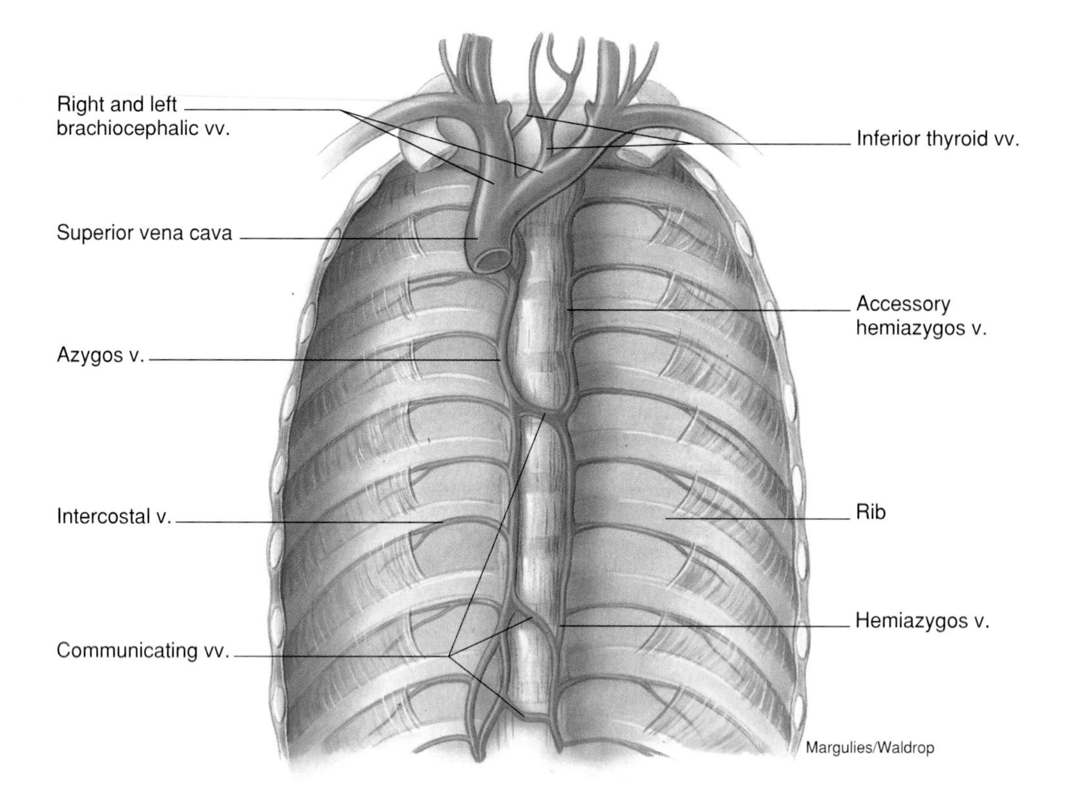

Figure 16.35

Veins of the thoracic region. (The lungs and heart have been removed.)

the level of the sacroiliac joint, where it merges with the **internal iliac vein** at the pelvic and genital regions to form the **common iliac vein.** At the level of the fifth lumbar vertebra, the right and left common iliacs unite to form the large inferior vena cava (fig. 16.36).

The superficial veins of the lower extremity are the **small** and **great saphenous veins.** The small saphenous vein arises from the lateral side of the foot and ascends deep to the skin along the posterior aspect of the leg. It empties into the popliteal vein, posterior to the knee. The great saphenous vein is the longest vessel in the body. It originates from the medial side of the foot and ascends along the medial aspect of the leg and thigh before emptying into the femoral vein. The great saphenous vein is frequently excised and used as a coronary bypass vessel.

Veins of the Abdominal Region

The **inferior vena cava** parallels the abdominal aorta on the right side as it ascends through the abdominal cavity to penetrate the diaphragm and enter the right atrium (see fig. 16.32). It is the largest in diameter of the vessels in the body and is formed by the union of the two common iliac veins that drain the lower extremities. As the inferior vena cava ascends through the abdominal cavity, it receives tributaries from veins that correspond in name and position to arteries previously described.

Four paired **lumbar veins** (not shown) drain the posterior abdominal wall, the vertebral column, and the spinal cord. The **renal veins** drain blood from the kidneys and ureters into the inferior vena cava. The **right testicular vein** in males (or the **right ovarian vein in females**) drains the corresponding gonads, and the **right suprarenal vein** drains the right adrenal gland. These veins empty into the inferior vena cava. The **left testicular vein** (or **left ovarian vein**) and the **left suprarenal vein,** by contrast, empty into the left renal vein. The **inferior phrenic veins** receive blood from the inferior side of the diaphragm and empty into the inferior vena cava. **Right** and **left hepatic veins** originate from the capillary sinusoids of the liver and empty into the inferior vena cava immediately below the diaphragm.

Note that the inferior vena cava does not receive blood directly from the GI tract, pancreas, or spleen. Instead, the venous outflow from these organs first passes through capillaries in the liver.

Hepatic Portal System

A *portal system* is a pattern of circulation in which the vessels that drain one group of capillaries deliver blood to a second group of capillaries, which in turn are drained by more usual systemic veins that carry blood to the vena cavae and the right atrium of the heart. There are thus two capillary beds in series. The **hepatic** (*hĕ-pat′ik*) **portal system** is composed of veins that drain blood from capillaries in the intestines, pancreas, spleen, stomach, and gallbladder into capillaries in the liver (called *sinusoids*) and of the **right** and **left hepatic veins** that drain the

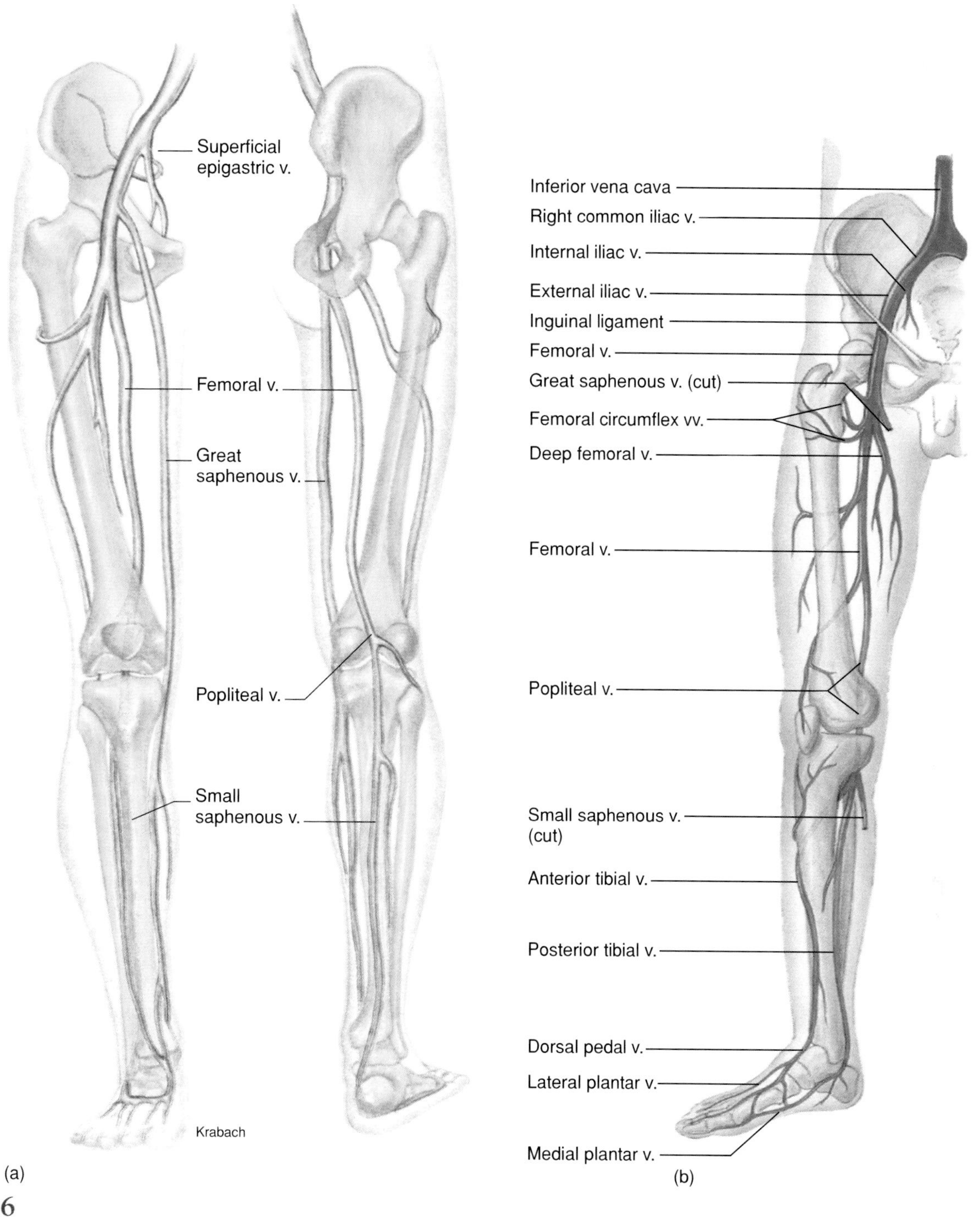

Figure 16.36

Veins of the lower extremity. (*a*) Superficial veins, medial and posterior aspects, and (*b*) deep veins, a medial view.

liver and empty into the inferior vena cava (fig. 16.37). As a consequence of the hepatic portal system, the absorbed products of digestion must first pass through the liver before entering the general circulation.

The **hepatic portal vein** is the large vessel that receives blood from the digestive organs. It is formed by a union of the **superior mesenteric vein,** which drains nutrient-rich blood from the small intestine, and the **splenic vein,** which drains the spleen. The splenic vein is enlarged because of a convergence of the following three tributaries: (1) the **inferior mesenteric vein** from the large intestine, (2) the **pancreatic vein** from the pancreas, and (3) the **left gastroepiploic** (*gas″tro-ep″ĭ-plo′ik*) **vein** from the stomach. The **right gastroepiploic vein,** also from the stomach, drains directly into the superior mesenteric vein.

gastroepiploic: Gk. *gastros,* stomach; *epiplein,* to float on (referring to greater omentum)

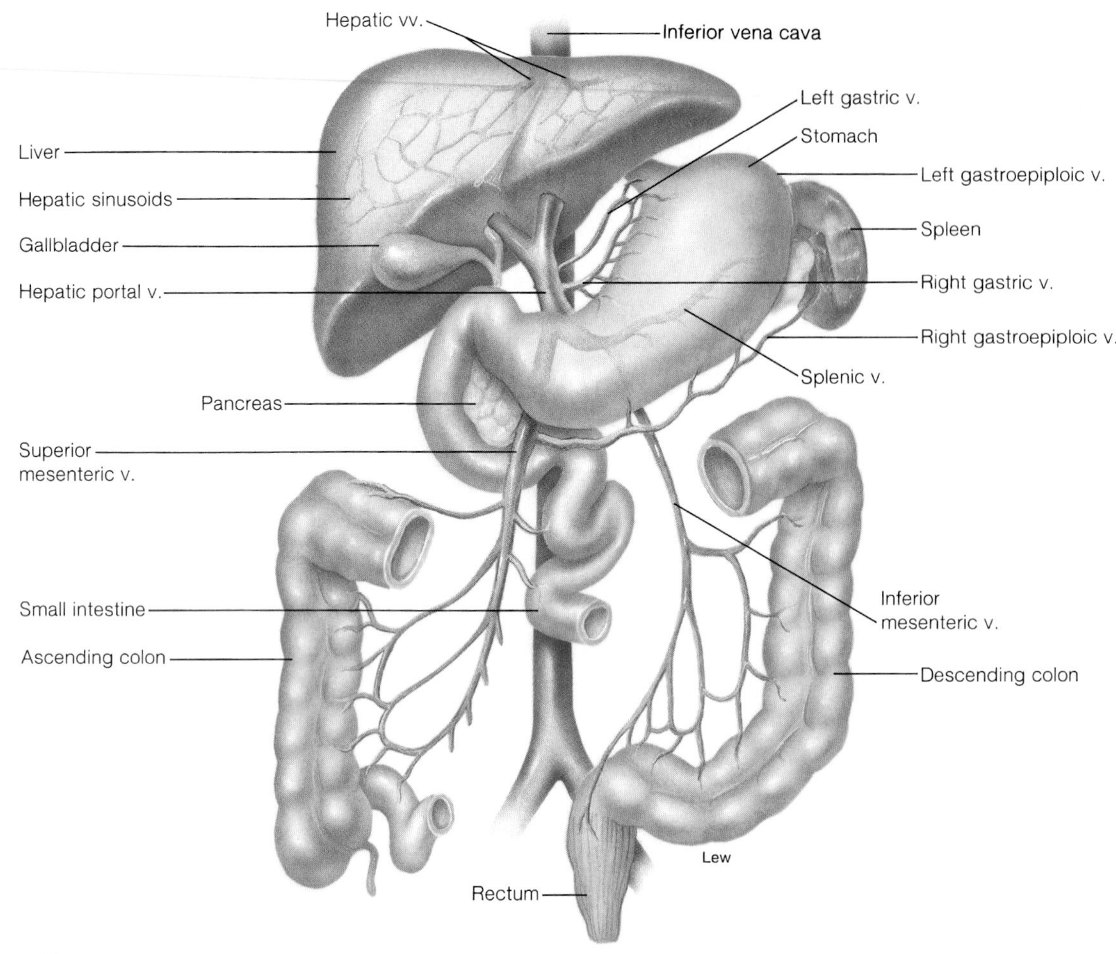

Figure 16.37

The hepatic portal system.

Three additional veins empty into the hepatic portal vein. The **right** and **left gastric veins** drain the lesser curvature of the stomach and the **cystic vein** drains blood from the gallbladder.

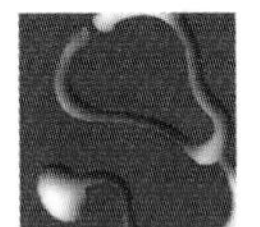

One of the functions of the liver is to detoxify harmful substances, such as alcohol, that are absorbed into the blood from the small intestine. However, excessive quantities of alcohol cannot be processed during a single pass through the liver, and so a person becomes intoxicated. Eventually, the liver is able to process the alcohol as the circulating blood is repeatedly exposed to the liver sinusoids via the hepatic artery. Alcoholics may eventually suffer from *cirrhosis* of the liver as the normal liver tissue is destroyed.

In summary, it is important to note that the sinusoids of the liver receive blood from two sources. The hepatic artery supplies oxygen-rich blood to the liver, whereas the hepatic portal vein transports nutrient-rich blood from the small intestine for processing. These two blood sources become mixed in the liver sinusoids. Liver cells exposed to this blood obtain nourishment from it and are uniquely qualified (because of their anatomical position and enzymatic ability) to modify the chemical nature of the venous blood that enters the general circulation from the GI tract.

1. Using a flowchart, indicate the venous drainage from the head and neck to the superior vena cava. Point out the vein that may bulge in the side of the neck when a person performs Valsalva's maneuver and the vein that is commonly used as a site for venipuncture.
2. Describe the positions, sources, and drainages of the small and great saphenous veins.
3. Describe the hepatic portal system and comment on the functional importance of this system.

Fetal Circulation

All of the respiratory, excretory, and nutritional needs of the fetus are provided for by diffusion across the placenta instead of by the fetal lungs, kidneys, and gastrointestinal tract. Fetal circulation is adaptive to these conditions.

Objective 21 Describe the fetal circulation to and from the placenta.

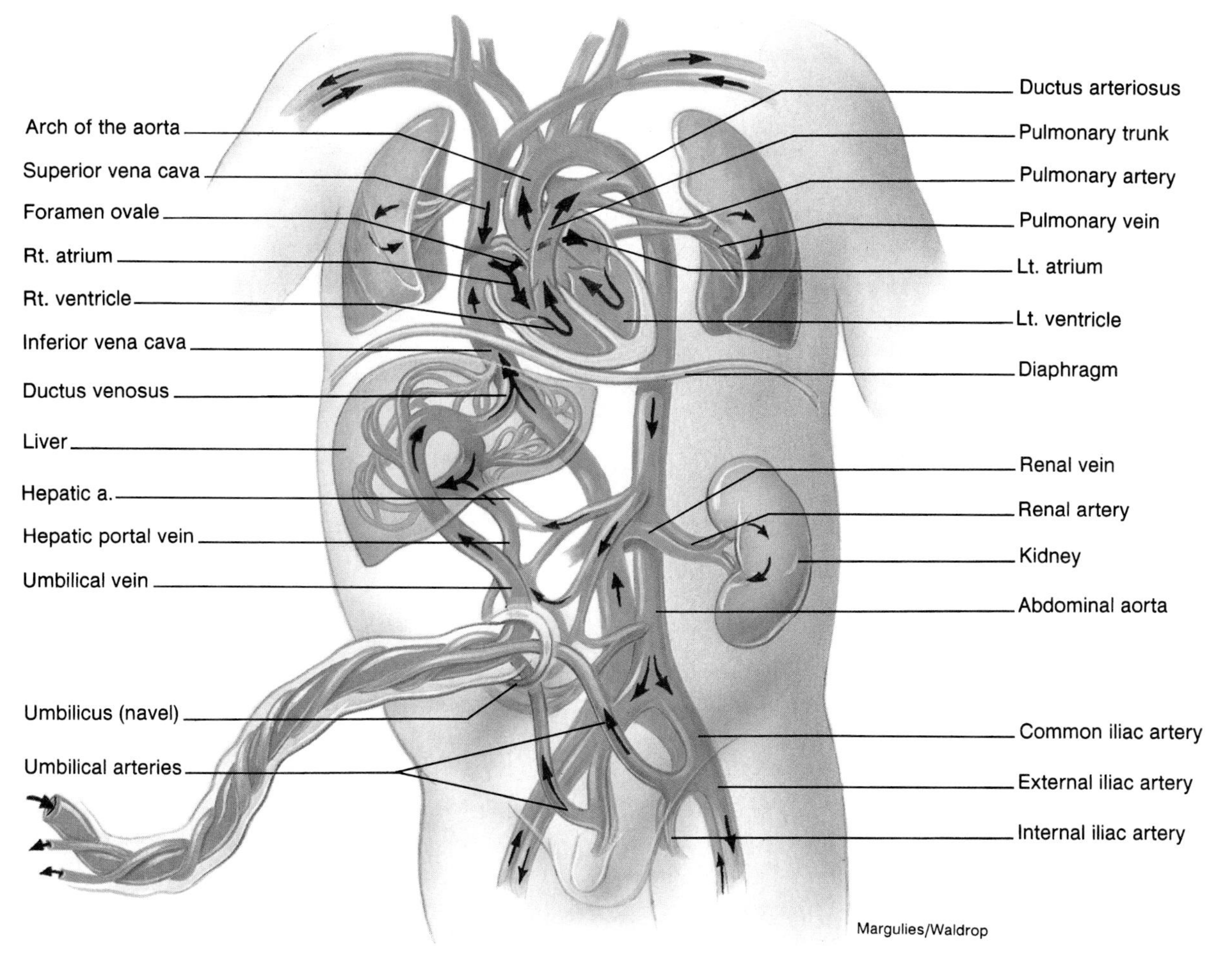

Figure 16.38

Fetal circulation. (Arrows indicate the direction of blood flow.)

Objective 22 Describe the structure and function of the foramen ovale, ductus venosus, and ductus arteriosus.

The circulation of blood through a fetus is by necessity different from blood circulation in a newborn (fig. 16.38). Respiration, the procurement of nutrients, and the elimination of metabolic wastes occur through the maternal blood instead of through the organs of the fetus. The capillary exchange between the maternal and fetal circulation occurs within the **placenta** (*plă-sen′tă*) (see fig. 22.12). This remarkable structure, which includes maternal and fetal capillary beds, is discharged following delivery as the afterbirth.

The **umbilical cord** is the connection between the placenta and the fetal umbilicus. It includes one **umbilical vein** and two **umbilical arteries,** surrounded by a gelatinous substance. Oxygenated and nutrient-rich blood flows through the umbilical vein toward the inferior surface of the liver. At this point, the umbilical vein divides into two branches. One branch merges with the portal vein, while the other branch, called the **ductus venosus** (*ve-no′sus*), enters the inferior vena cava. Thus, oxygenated blood is mixed with venous blood returning from the lower extremities of the fetus before it enters the heart. The umbilical vein is the only vessel of the fetus that carries fully oxygenated blood.

The inferior vena cava empties into the right atrium of the fetal heart. Most of the blood passes from the right atrium into the left atrium through the **foramen ovale** (*fŏ-ra′men o-val′e*), an opening between the two atria. Here, it mixes with a small quantity of blood returning through the pulmonary circulation. The blood then passes into the left ventricle, from which it is pumped into the aorta and through the body of the fetus. Some blood entering the right atrium passes into the right ventricle and out of the heart via the pulmonary trunk. Since the lungs of the fetus are not functional, only a small portion of blood continues through the pulmonary circulation (the resistance to blood flow is very high in the collapsed fetal lungs). Most of the blood in the pulmonary trunk passes through the **ductus arteriosus** (*ar-te″re-o′sus*) into the aortic arch, where it mixes with blood coming from the left ventricle. Blood is returned to the placenta by the two umbilical arteries that arise from the internal iliac arteries.

Notice that, in the fetus, oxygen-rich blood is transported by the inferior vena cava to the heart, and via the foramen ovale and ductus arteriosus to the systemic circulation.

Table 16.6

Cardiovascular Structures of the Fetus and Changes in the Neonate

Structure	Location	Function	Neonate Transformation
Umbilical vein	Connects the placenta to the liver; forms a major portion of the umbilical cord	Transports nutrient-rich oxygenated blood from the placenta to the fetus	Forms the round ligament of the liver
Ductus venosus	Venous shunt within the liver that connects the umbilical vein and the inferior vena cava	Transports oxygenated blood directly into the inferior vena cava	Forms the ligamentum venosum, a fibrous cord in the liver
Foramen ovale	Opening between the right and left atria	Acts as a shunt to bypass the pulmonary circulation	Closes at birth and becomes the fossa ovalis, a depression in the interatrial septum
Ductus arteriosus	Connects the pulmonary trunk and the aortic arch	Acts as a shunt to bypass the pulmonary circulation	Closes shortly after birth, atrophies, and becomes the ligamentum arteriosum
Umbilical arteries	Arise from internal iliac arteries and form a portion of the umbilical cord	Transport blood from the fetus to the placenta	Atrophies to become the lateral umbilical ligaments

Important changes occur in the cardiovascular system at birth. The foramen ovale, ductus arteriosus, ductus venosus, and the umbilical vessels are no longer necessary. The foramen ovale abruptly closes with the first breath of air because the reduced pressure in the right side of the heart causes a flap to cover the opening. This reduction in pressure occurs because the vascular resistance to blood flow in the pulmonary circulation falls far below that of the systemic circulation when the lungs fill with air. The pressure in the inferior vena cava and right atrium falls as a result of the loss of the placental circulation.

The constriction of the ductus arteriosus occurs gradually over a period of about 6 weeks after birth as the vascular smooth muscle fibers constrict in response to the higher oxygen concentration in the postnatal blood. The remaining structure of the ductus arteriosus gradually atrophies and becomes nonfunctional as a blood vessel. Transformation of the unique fetal cardiovascular system is summarized in table 16.6.

1. Trace the path of blood from the fetal heart through the placenta and back to the fetal heart.
2. Trace the path of oxygenated blood through the fetal circulation.
3. Describe the foramen ovale and ductus arteriosus and explain why blood flows the way it does through these structures in the fetal circulation.

Lymphatic System

The lymphatic system, consisting of lymphatic vessels and various lymphoid tissues and organs, helps to maintain fluid balance in tissues and to absorb fats from the gastrointestinal tract. It also is part of the body's defense system against disease.

Objective 23 Describe the pattern of lymph flow from the lymphatic capillaries to the venous system.

Objective 24 Describe the general location, histological structure, and functions of the lymph nodes.

Objective 25 Describe the location and functions of the other lymphoid organs of the body.

The lymphatic system is closely related to the circulatory system, both structurally and functionally (fig. 16.39). A network of lymphatic vessels drains excess interstitial fluid (the approximate 15% that has not been returned directly to the capillaries) and returns it to the bloodstream in a one-way flow that moves slowly toward the subclavian veins. Additionally, the lymphatic system functions in fat absorption and in the body's defense against microorganisms and other foreign substances.

In short, the lymphatic system has three principal functions.

1. It transports excess interstitial (tissue) fluid, which was initially formed as a blood filtrate, back to the blood.
2. It serves as the route by which absorbed fats and some vitamins are transported from the small intestine to the blood.
3. Its cells (called *lymphocytes*), located in lymphatic tissues, help to provide immunological defenses against disease-causing agents.

Lymph and Lymphatic Vessels

The lymphatic network of vessels begins with the microscopic **lymphatic capillaries.** Lymphatic capillaries are closed-ended tubes that form vast networks in the intercellular spaces within most tissues. Within the villi of the small intestine, for example, lymphatic capillaries called *lacteals* (*lak'te-alz*), transport the

lacteal: L. *lacteus*, milk

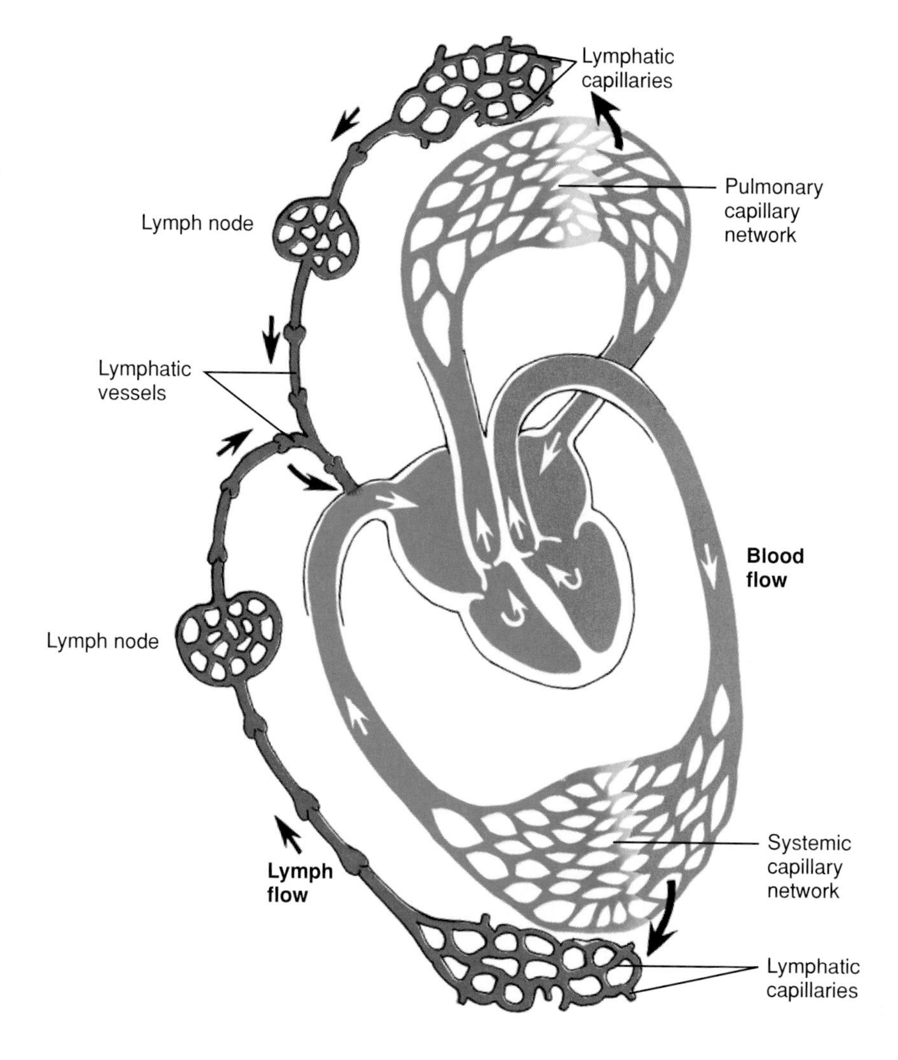

Figure 16.39

The schematic relationship between the circulatory and lymphatic systems. Interstitial (tissue) fluid is an extract of blood plasma formed at the pulmonary and systemic capillary networks. Lymph is the interstitial fluid that enters the lymphatic capillaries to be transported by lymphatic vessels to the venous bloodstream.

products of fat absorption away from the GI tract. Because the walls of lymphatic capillaries are composed of endothelial cells with porous junctions, interstitial fluid, proteins, microorganisms, and absorbed fats (in the small intestine) can easily enter. Once fluid enters the lymphatic capillaries, it is referred to as **lymph** (*limf*). Adequate lymphatic drainage is needed to prevent the accumulation of interstitial fluid, a condition called *edema* (ĕ-de′mă).

From merging lymphatic capillaries, the lymph is carried into larger lymphatic vessels called **lymph ducts.** The walls of lymph ducts are much like those of veins. They have the same three layers and also contain valves to prevent backflow (fig. 16.40). The pressure that keeps the lymph moving comes from the massaging action produced by skeletal muscle contractions and intestinal movements, and from peristaltic contractions of some lymphatic vessels. The valves keep the lymph moving in one direction.

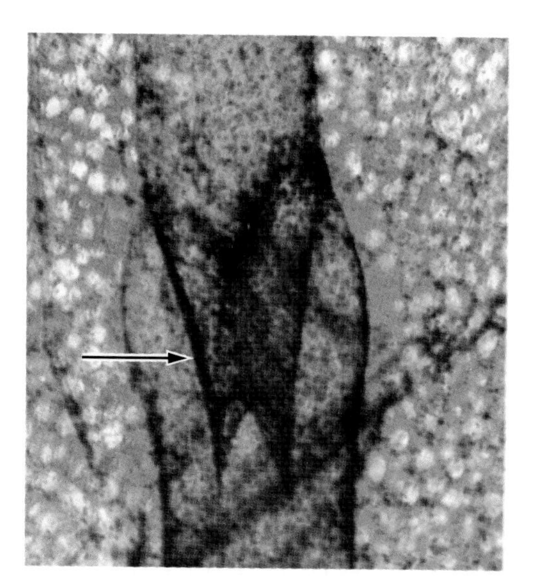

Figure 16.40

A photomicrograph of a valve (arrow) within a lymph vessel.

lymph: L. *lympha*, clear water
edema: Gk. *edema*, a swelling

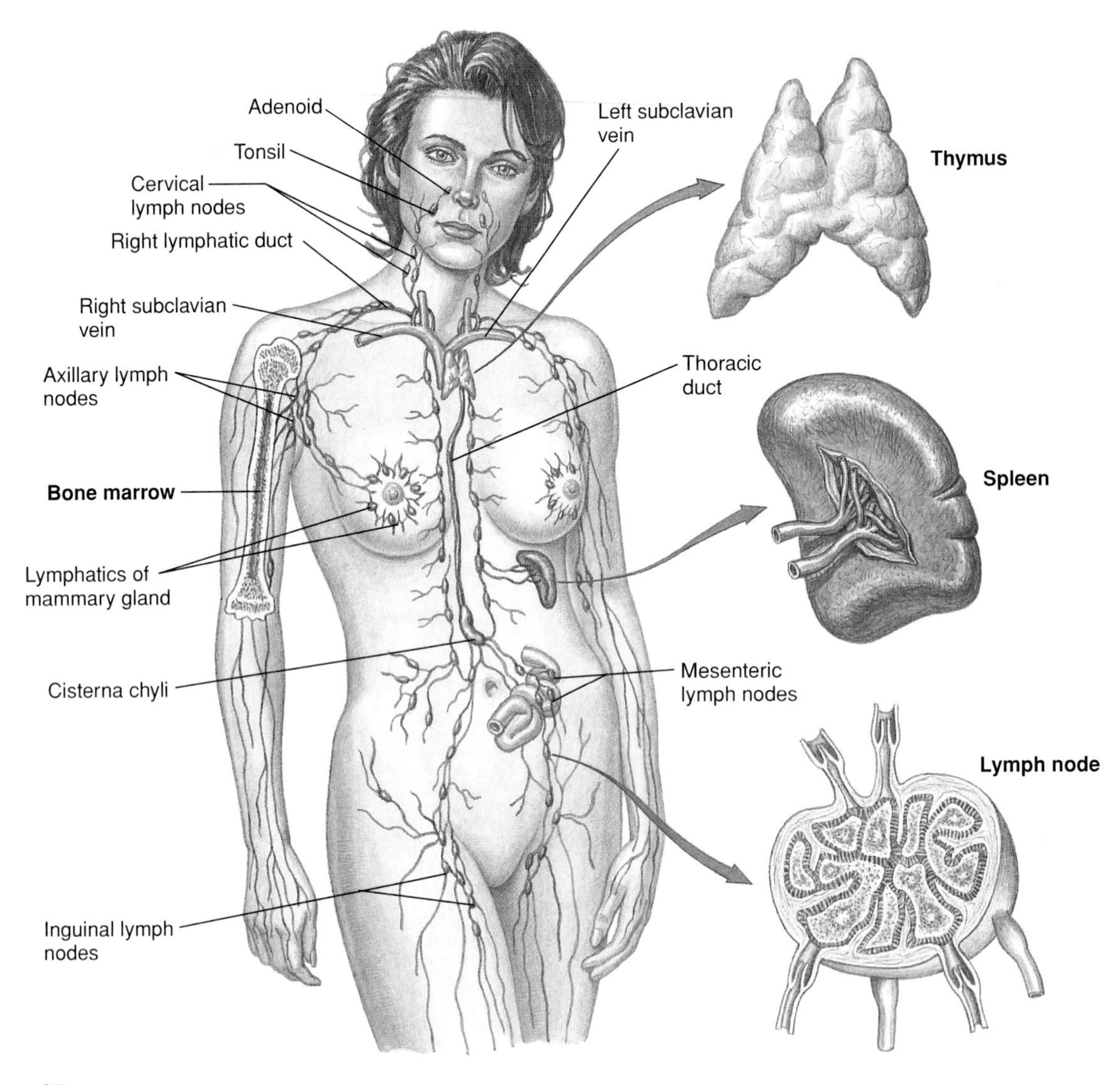

Figure 16.41

The lymphatic system showing the principal lymph nodes. Lymph from the upper right extremity, the right side of the head and neck, and the right thoracic region drains through the right lymphatic duct into the right subclavian vein. Lymph from the remainder of the body drains through the thoracic duct into the left subclavian vein.

Interconnecting lymph ducts eventually empty into one of the two principal vessels: the **thoracic duct** and the **right lymphatic duct** (fig. 16.41). The larger thoracic duct drains lymph from the lower extremities, abdomen, left thoracic region, left upper extremity, and left side of the head and neck. The main trunk of this vessel ascends along the spinal column and drains into the left subclavian vein. In the abdominal area, there is a saclike enlargement of the thoracic duct called the **cisterna chyli** (*sis-ter'nă ki'le*), which collects lymph from the lower extremities and the intestinal region. The smaller right lymphatic duct drains lymphatic vessels from the right upper extremity, right thoracic region, and right side of the head and neck. The right lymphatic duct empties into the right subclavian vein near the internal jugular vein.

cisterna chyli: L. *cisterna*, box; Gk. *chylos*, juice

Lymph Nodes

Lymph filters through the reticular tissue of hundreds of **lymph nodes** clustered along the lymphatic vessels (fig. 16.42). The reticular tissue contains phagocytic cells that help purify the fluid. Lymph nodes are small bean-shaped bodies enclosed within fibrous connective tissue *capsules*. Specialized connective tissue bands called *trabeculae* (*tră-bek'yŭ-le*) divide the node. *Afferent lymphatic vessels* carry lymph into the node, where it is circulated through the *sinuses*, a series of irregular channels. Lymph leaves the node through the *efferent lymphatic vessel*, which emerges from the *hilum*—a depression on the concave side of the node. **Lymphatic nodules** within the node are the sites of lymphocyte production, and are thus important in the development of an immune response.

Lymph nodes usually occur in clusters in specific regions of the body (see fig. 16.41). Some of the principal groups of lymph

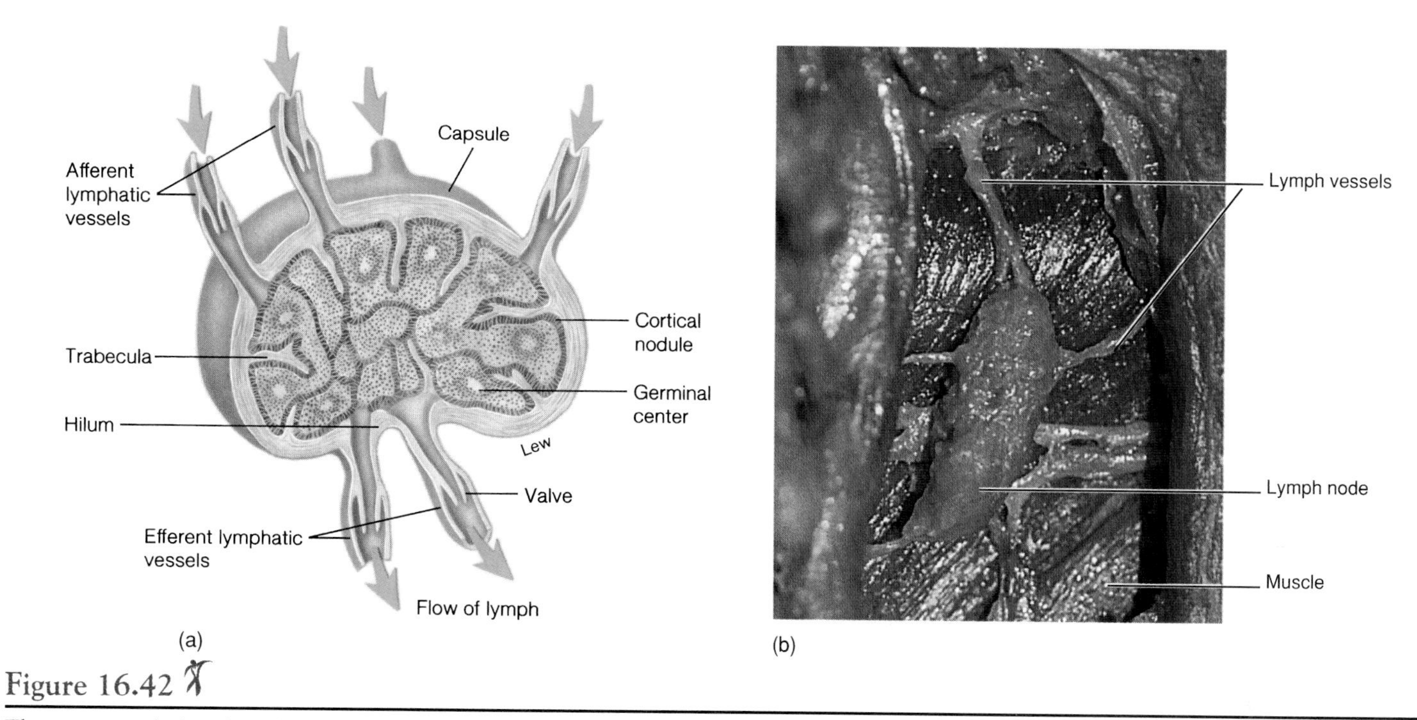

Figure 16.42

The structure of a lymph node. (*a*) A diagram of a sectioned lymph node and associated vessels, and (*b*) a photograph of a lymph node positioned near a blood vessel.

nodes are the **popliteal** (not illustrated) and **inguinal nodes** of the lower extremity, the **lumbar nodes** of the pelvic region, the **cubital** and **axillary nodes** of the upper extremity, the **thoracic nodes** of the chest, and the **cervical nodes** of the neck. The submucosa of the small intestine contains numerous scattered lymphocytes and lymphatic nodules, and larger clusters of lymphatic tissue called **mesenteric** (Peyer's) **patches.**

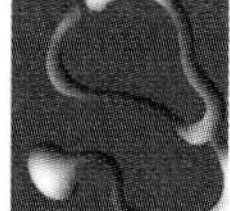

Migrating cancer cells (metastases) are especially dangerous if they enter the lymphatic system, which can disperse them widely. On entering the lymph nodes, the cancer cells can multiply and establish secondary tumors in organs that are along the lymphatic drainage of the site of the primary tumor.

Other Lymphoid Organs

In addition to the lymph nodes just described, the *tonsils*, *spleen*, and *thymus* are lymphoid organs. The **tonsils** form a protective ring of lymphatic tissue around the openings between the nasal and oral cavities and the pharynx (see fig. 17.3).

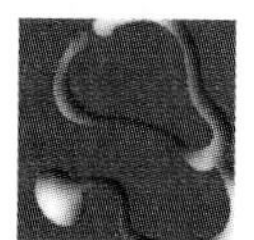

The tonsils, of which there are three pairs, combat infection of the ear, nose, and throat regions. Because of the persistent infections that some children suffer, the tonsils may become so overrun with infections that they, themselves, become a source of infections that spread to other parts of the body. A *tonsillectomy* may then have to be performed. This operation is not as common as it was in the past because of the availability of powerful antibiotics and because the functional value of the tonsils is appreciated to a greater extent.

The **spleen** (fig. 16.43) is located on the left side of the abdominal cavity, to the left of the stomach from which it is suspended. The spleen is not a vital organ in an adult, but it does assist other body organs in producing lymphocytes, filtering the blood, and destroying old erythrocytes. In an infant, the spleen is an important site for the production of erythrocytes. In an adult, it contains *red pulp*, which serves to destroy old erythrocytes, and *white pulp*, which contains germinal centers for the production of lymphocytes.

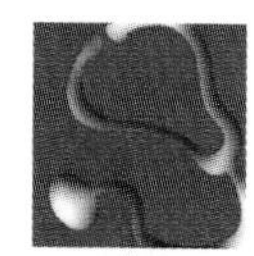

Of all of the abdominal organs, the spleen is the one most easily and frequently injured. Because it is highly vascular, extensive—sometimes massive—hemorrhage occurs when the spleen is ruptured. To prevent death from loss of blood, a *splenectomy* (removal of the spleen) is performed. Without immediate surgery for a ruptured spleen, the mortality rate is 90%.

The **thymus** is located in the anterior thorax, deep to the manubrium of the sternum (fig. 16.44). Because it regresses in size during puberty, it is much larger in a fetus and child than in an adult. The thymus plays a key role in the immune system.

The lymphoid organs are summarized in table 16.7.

Peyer's patches: from Johann K. Peyer, Swiss anatomist, 1653–1712

spleen: L. *splen*, low spirits (thought to cause melancholy)
thymus: Gk. *thymos*, thyme (compared to the flower of this plant by Galen)

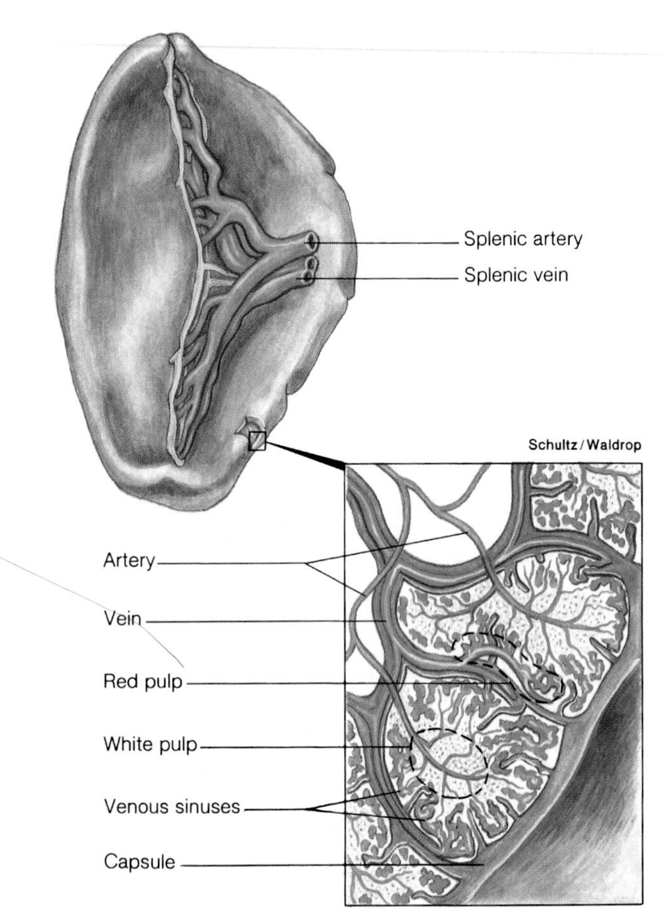

Figure 16.43

The structure of the spleen.

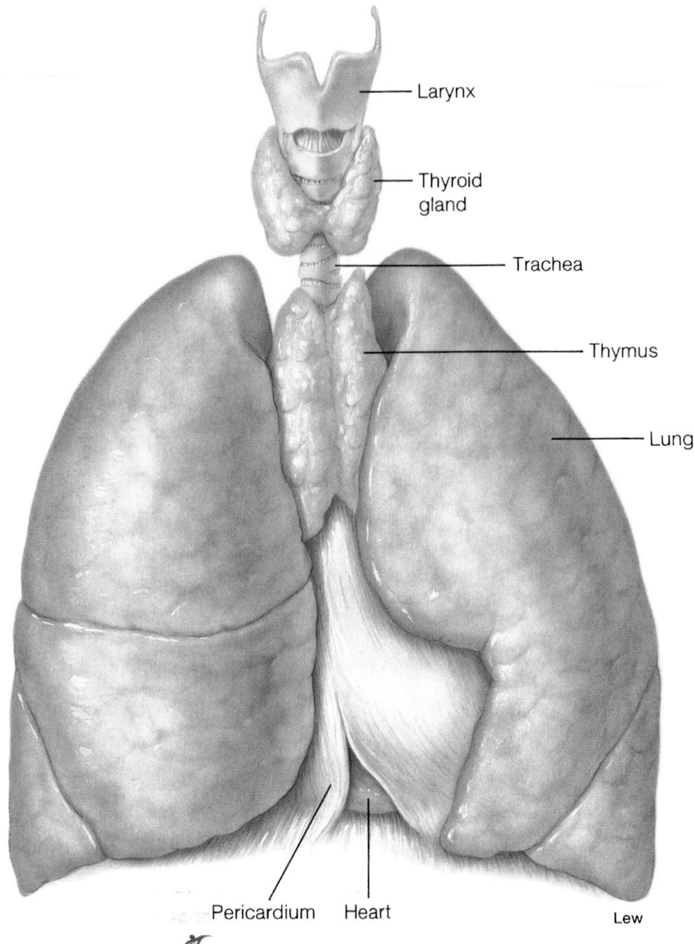

Figure 16.44

The thymus is located in the mediastinum, medial to the lungs.

Table 16.7
Lymphoid Organs

Organ	Location	Function
Lymph nodes	In clusters or chains along the paths of larger lymphatic vessels	Sites of lymphocyte production; house T lymphocytes and B lymphocytes that are responsible for immunity; phagocytes filter foreign particles and cellular debris from lymph
Tonsils	In a ring at the junction of the oral cavity and pharynx	Protect against invasion of foreign substances that are ingested or inhaled
Spleen	In upper left portion of abdominal cavity, beneath the diaphragm and suspended from the stomach	Serves as blood reservoir; phagocytes filter foreign particles, cellular debris, and worn erythrocytes from the blood; houses lymphocytes
Thymus	Within the mediastinum, behind the manubrium	Important site of immunity in a child; houses lymphocytes; changes undifferentiated lymphocytes into T lymphocytes

1. Describe the relationship between lymph and blood in terms of their origin, composition, and fate.
2. List the two major lymph vessels of the body and describe their relationships to the vascular system.
3. Describe the structure, location, and function of the lymph nodes and identify the other lymphoid organs.

Clinical Considerations

It has been said that as the circulatory system goes, so goes the rest of the body. The circulatory system is the lifeline to the other organs of the body. If the vital organs are deprived of the essentials for life, they will falter in their functions, ultimately resulting in body death. Some gerontologists believe that the first system of the body to show signs of aging is the circulatory system, and that this accelerates the aging process in the other systems.

The circulatory system is extremely important clinically because of the frequent irregularities and diseases that afflict it and the effect of its dysfunction on the maintenance of homeostasis. Given the confines of this text, it would not be practical to attempt a comprehensive discussion of the numerous clinical aspects of the cardiovascular system. However, some general comments will be made on cardiovascular assessment, on dysfunctions of the cardiovascular and lymphatic systems, and on the prevention and treatment of heart problems. In addition, first-aid treatment for victims of hemorrhage and shock will be discussed.

Cardiovascular Assessment

Cardiovascular assessment is an extremely important aspect of a physical examination. Several techniques are routinely used in gathering information. Auscultation of heart sounds with a **stethoscope** is not only important for detecting murmurs, but may help the physician evaluate general cardiac functioning or the progress of a patient with a cardiovascular problem.

An **electrocardiogram (ECG)** provides a "picture" of the electrical activity of the heart. The most important diagnostic use of the resting ECG is identifying abnormal cardiac rhythms. Injury to cardiac muscle, as in myocardial infarction, elevates or depresses the S-T segment while inverting and shortening the T wave.

Several diagnostic techniques, including **radiograms, CT scans,** and **ultrasound,** are employed for determining the size and position of the heart, the condition of the pericardial sac, and even the internal structure and condition of the vessels, chambers, and valves. The accumulation of fluids can be easily detected using these techniques. **Catheterization** of the heart in conjunction with a **fluoroscopic analysis** permits observation of the vessels, chambers, and valves of the heart. In addition, internal blood pressures can be recorded and samples of blood taken. Selective **arteriography** (*ar″te-re-og′ră-fe*) enables a physician to study the wall of a particular vessel. An **angiocardiogram** (*an″je-o-kar′de-ŏ-gram*) is a radiograph of the chambers of the heart and its vessels following an intravenous injection of a radiopaque dye.

Arterial blood pressure can be measured using a **sphygmomanometer** (see fig. 16.17). Blood pressure is indicative of the general health of the cardiovascular system. **Hypertension** is sustained high blood pressure and can result in mechanical damage to the cardiovascular system. Hypertension can be caused by resistance in the vessels from physiological actions (vasoconstriction) or by various vascular diseases (arteriosclerosis). Certain characteristics of the blood itself may cause hypertension.

Cardiac Arrhythmia

Arrhythmias, or abnormal heart rhythms, can be detected and described by the abnormal ECG patterns they produce. Since a heartbeat occurs whenever a normal QRS complex (see fig. 16.11*b*) is

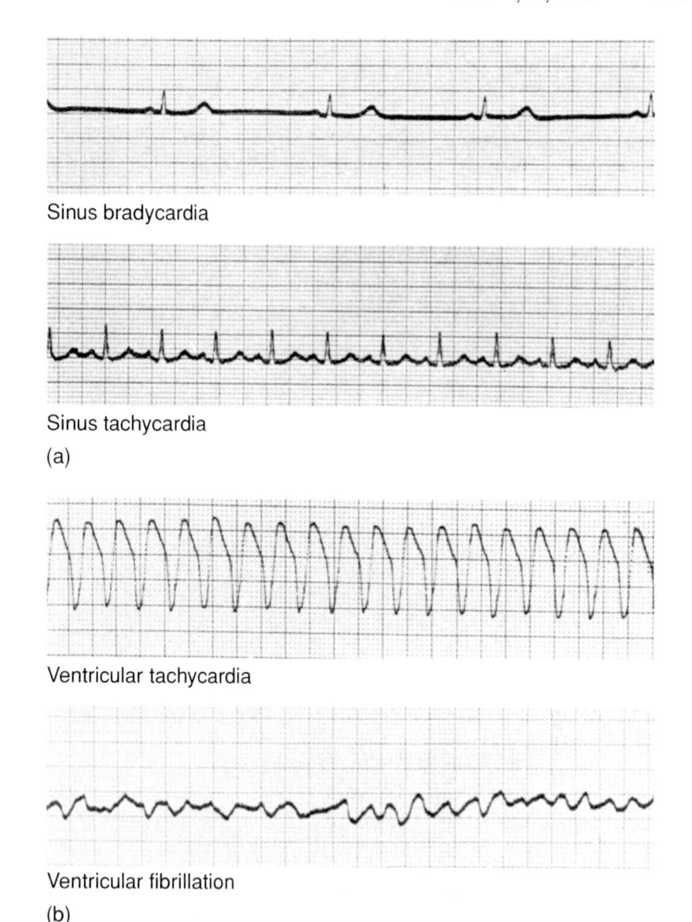

Figure 16.45

In (*a*) the heartbeat is paced by the normal pacemaker—the SA node (hence the name *sinus rhythm*). This can be abnormally slow (bradycardia—46 beats per minute in this example) or fast (tachycardia—136 beats per minute in this example). Compare the pattern of tachycardia in (*a*) with the tachycardia in (*b*). Ventricular tachycardia is produced by an ectopic pacemaker in the ventricles. This dangerous condition can quickly lead to ventricular fibrillation, also shown in (*b*).

seen, and since the ECG chart paper moves at a known speed, its x-axis indicates time, and the cardiac rate (beats per minute) can easily be obtained from the ECG recording. A cardiac rate slower than 60 beats per minute indicates **bradycardia** (*brad″ĭ-kar′de-ă*); a rate faster than 100 beats per minute is described as **tachycardia** (*tak″ĭ-kar′de-ă*).

Both bradycardia and tachycardia can occur normally (fig. 16.45*a*). Endurance-trained athletes, for example, commonly have a slower heart rate than the general population. This *athlete's bradycardia* occurs as a result of higher levels of parasympathetic inhibition of the SA node and is a beneficial adaptation. Activation of the sympathetic division of the ANS during exercise or emergencies causes a normal tachycardia to occur.

An abnormal bradycardia may be caused by a heart block, various drugs, shock, or increased intracranial pressure.

stethoscope: Gk. *stethos*, breast; *skopos*, watch, look at
sphygmomanometer: Gk. *sphygmos*, throbbing; *manos*, at intervals; L. *metrum*, measure

bradycardia: Gk. *bradys*, slow; *kardia*, heart
tachycardia: Gk. *tachys*, rapid; *kardia*, heart

DEVELOPMENTAL EXPOSITION

The Circulatory System

Exhibit I The early development of the heart from embryonic mesoderm. (*a*) A dorsal view of an embryo at day 20 showing the position of a transverse cut depicted in (a_1), (*b*), and (*c*). (a_1) At day 20, the heart tubes are formed from the heart cords. (*b*) By day 21, the medial migration of the heart tubes has brought them together within the forming pericardial cavity. (*c*) A single heart tube is completed at day 22 and is suspended in the pericardial cavity by the dorsal mesocardium.

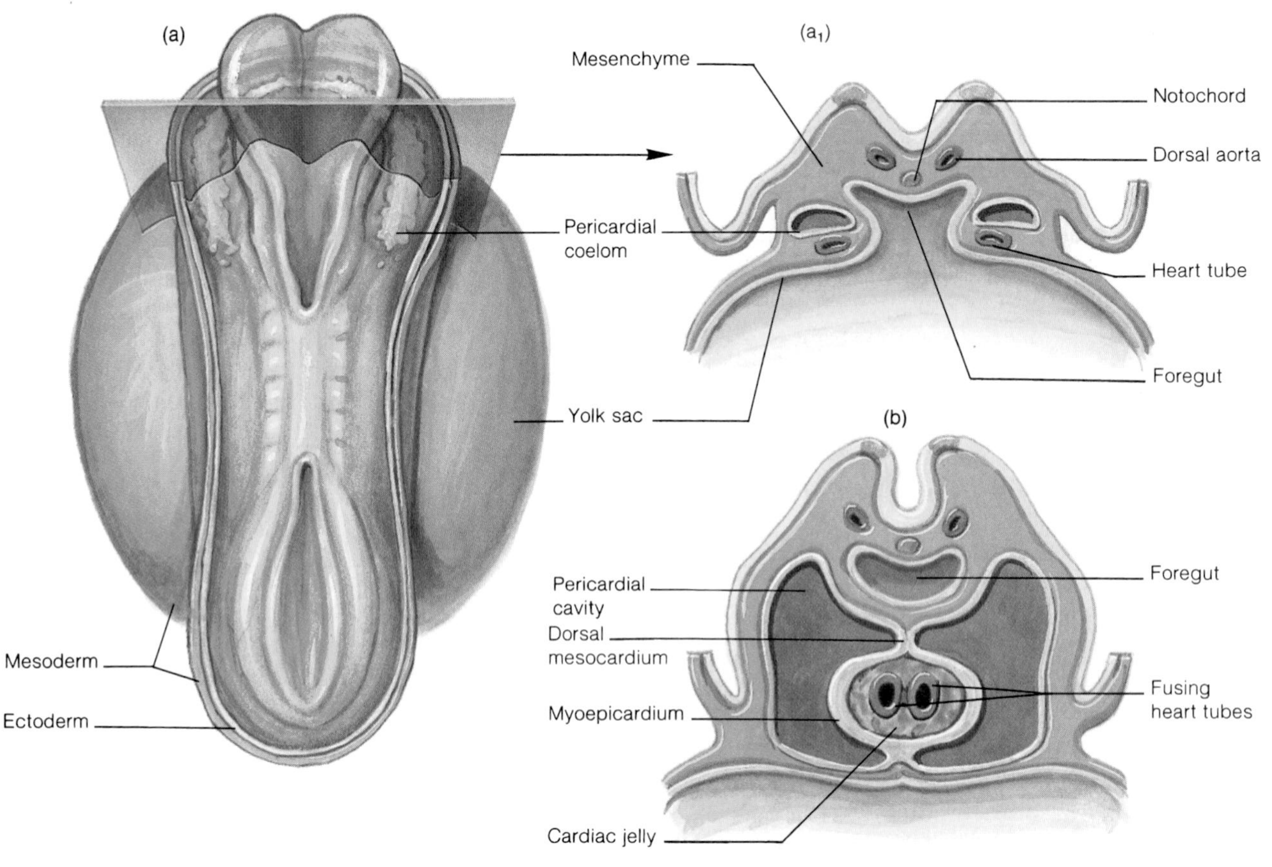

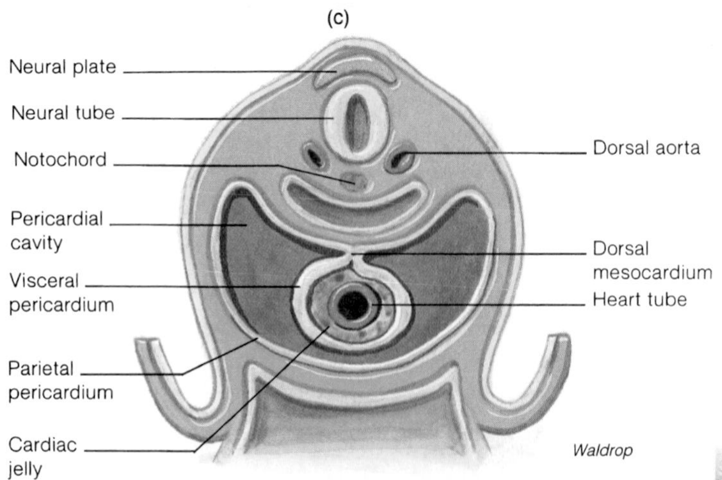

Explanation

The cardiovascular system is one of the first systems to form in the embryo, delivering nutrients to the mitotically active cells and disposing of waste products through its association with the maternal blood vessels in the placenta. Blood is formed and begins circulating through the vessels by the pumping action of the heart approximately 25 days after conception.

Throughout pregnancy, the fetus is dependent on the mother's circulatory system for exchange of nutrients, gases, and wastes. To accommodate this arrangement, some unique structures of the fetal circulatory system develop. The discussion that follows focuses on the development of the fetal heart and associated major vessels. The circulation of blood through the fetal cardiovascular system has already been discussed under the heading "Fetal Circulation."

Development of the Heart

The development of the heart from two separate segments of mesoderm requires only 6 to 7 days. Heart development is first apparent at day 18 or 19 in the *cardiogenic* (*kar″de-o-jen′ik*) *area* of the mesoderm layer (exhibit I). A small paired mass of specialized cells called **heart cords** form here. Shortly after, a hollow center develops in each heart cord, and each structure is then referred to as a **heart tube.** The heart tubes begin to migrate toward each other on day 21 and soon fuse to form a single median heart tube. During this time, the heart tube undergoes dilations and constrictions, so that when fusion is completed during the fourth week, five distinct regions of the heart can be identified. These are the **truncus arteriosus, bulbus cordis, ventricle, atrium,** and **sinus venosus** (exhibit II).

After the fusion of the heart tubes and the formation of distinct dilations, the heart begins to pump blood. Partitioning of the heart chambers begins during the middle of the fourth week and is complete by the end of the fifth week. During this crucial time, many congenital heart problems develop.

Major changes occur in each of the five primitive dilations of the developing heart during the week-and-a-half embryonic period beginning in the middle of the fourth week. The truncus arteriosus differentiates to form a partition between the aorta and the pulmonary trunk. The bulbus cordis is incorporated in the formation of the walls of the ventricles. The sinus venosus forms the **coronary sinus** and a portion of the wall of the right atrium. The ventricle is divided into the right and left chambers by the growth of the **interventricular septum.** The atrium is partially partitioned into right and left chambers by the **septum secundum.** An opening between the two atria called the **foramen ovale** persists throughout fetal development. This opening is covered by a flexible valve that permits blood to pass from the right to the left side of the heart.

The development of the heart is summarized in table 16.8.

Congenital defects of the cardiac septa are a relatively common form of birth abnormality. Approximately 0.7% of live births and 2.7% of stillbirths show cardiac abnormalities. Ventricular septal defects are the most common of the cardiac defects. An infant with a congenital cardiac defect may suffer from inadequate oxygenation of blood and thus be termed a "blue baby."

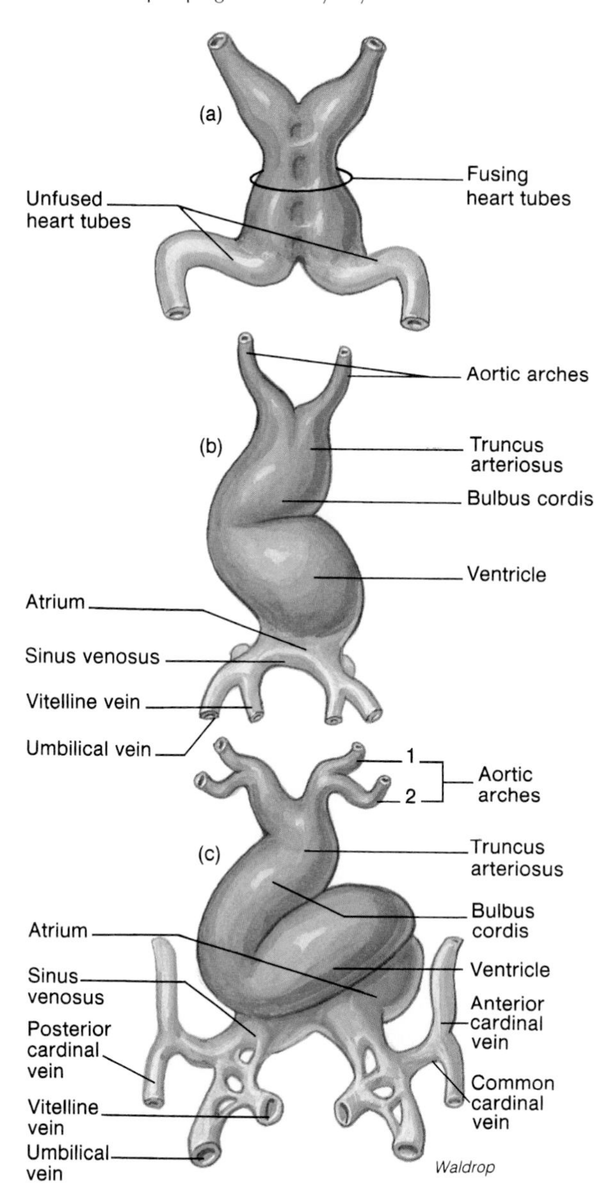

Exhibit II Formation of the heart chambers. (*a*) The heart tubes fuse during days 21 and 22. (*b*) The developmental chambers are formed during day 23. (*c*) Differential growth causes folding between the chambers during day 24, and vessels are developed to transport blood to and from the heart. The embryonic heart generally has begun rhythmic contractions and pumping of blood by day 25.

cardiogenic: Gk. *kardia*, heart; *genesis*, be born (origin)

Continued

DEVELOPMENTAL EXPOSITION

Continued

Table 16.8

Structural Changes in the Development of the Heart

Primitive Cardiac Dilation	Developmental Fate
Truncus arteriosus	Differentiates to form the aorticopulmonary septum, which partitions the ascending aorta and the pulmonary trunk
Bulbus cordis	Incorporated into the walls of the ventricles
Sinus venosus	Differentiates to form the coronary sinus and a portion of the wall of the right atrium
Ventricle	Divided into right and left chambers by growth of the interventricular septum
Atrium	Partially partitioned into right and left chambers by the septum secundum. The opening in the partition, called the foramen ovale, persists throughout development.

Table 16.9

Derivatives of the Aortic Arches

Aortic Arch	Derivative
First pair	No derivative
Second pair	No derivative
Third pair	Proximal portions form the common carotid arteries. Distal portions form the external carotid arteries.
Fourth pair	Right arch forms the base of the right subclavian artery. Left arch forms a portion of the aortic arch.
Fifth pair	No derivative
Sixth pair	Right arch: proximal portion persists as proximal part of the right pulmonary artery; distal portion degenerates. Left arch: proximal portion forms the proximal part of the left pulmonary artery; distal part persists as the ductus arteriosus.

Development of the Major Arteries

The formation of the major arteries occurs simultaneously with the development of the heart. The most complex and fascinating vascular formation is the development of the **aortic arches** associated with the pharyngeal pouches and branchial arches in the neck region (exhibit III). These aortic arches arise from the truncus arteriosus. Although six pairs of aortic arches develop, they are not all present at the same time, and none of them persists in entirety through fetal development.

The transformation of the six aortic arches into the basic adult arterial arrangement occurs between the sixth and eighth weeks of embryonic development. The first and second pairs of aortic arches disappear before the formation of the sixth. The third pair of aortic arches form the common carotid arteries and the external carotid arteries. The right fourth aortic arch forms the base of the right subclavian artery, and the left fourth arch contributes to part of the aortic arch. The first, second, and fifth pairs of aortic arches have no derivatives and soon atrophy. The right sixth aortic arch forms the proximal portion of the right pulmonary artery. The left sixth arch forms the proximal portion of the left pulmonary artery, and the distal portion of this arch persists as an embryonic shunt between the pulmonary trunk and the aorta called the **ductus arteriosus.**

The derivatives of the aortic arches are summarized in table 16.9.

Although the embryonic vessels that develop from the truncus arteriosus are called aortic arches, they should not be confused with the adult aortic arch; that is, the major systemic artery leaving the heart. Only the left fourth arch participates in the formation of the adult aortic arch.

An examination of the persisting aortic arches in the adults of different classes of vertebrates reveals interesting evolutionary relationships. Fish have six aortic arches persisting in the gill region. Reptiles have only one pair, a branch to the left and one to the right. All birds have a single right aortic arch, and all mammals have a single left aortic arch.

Exhibit III Formation of the aortic arch and major arteries of the thoracic region. (*a*) A schematic lateral view of the pharyngeal pouches and aortic arches within the neck region of a 28-day embryo. (Note that by this time arch five is no longer present.) (*b*) The aortic arches at 6 weeks. (Note that by this time arches one and two have largely disappeared.) (*c*) The aortic arches at 7 weeks. (*d*) The arterial configuration at 8 weeks.

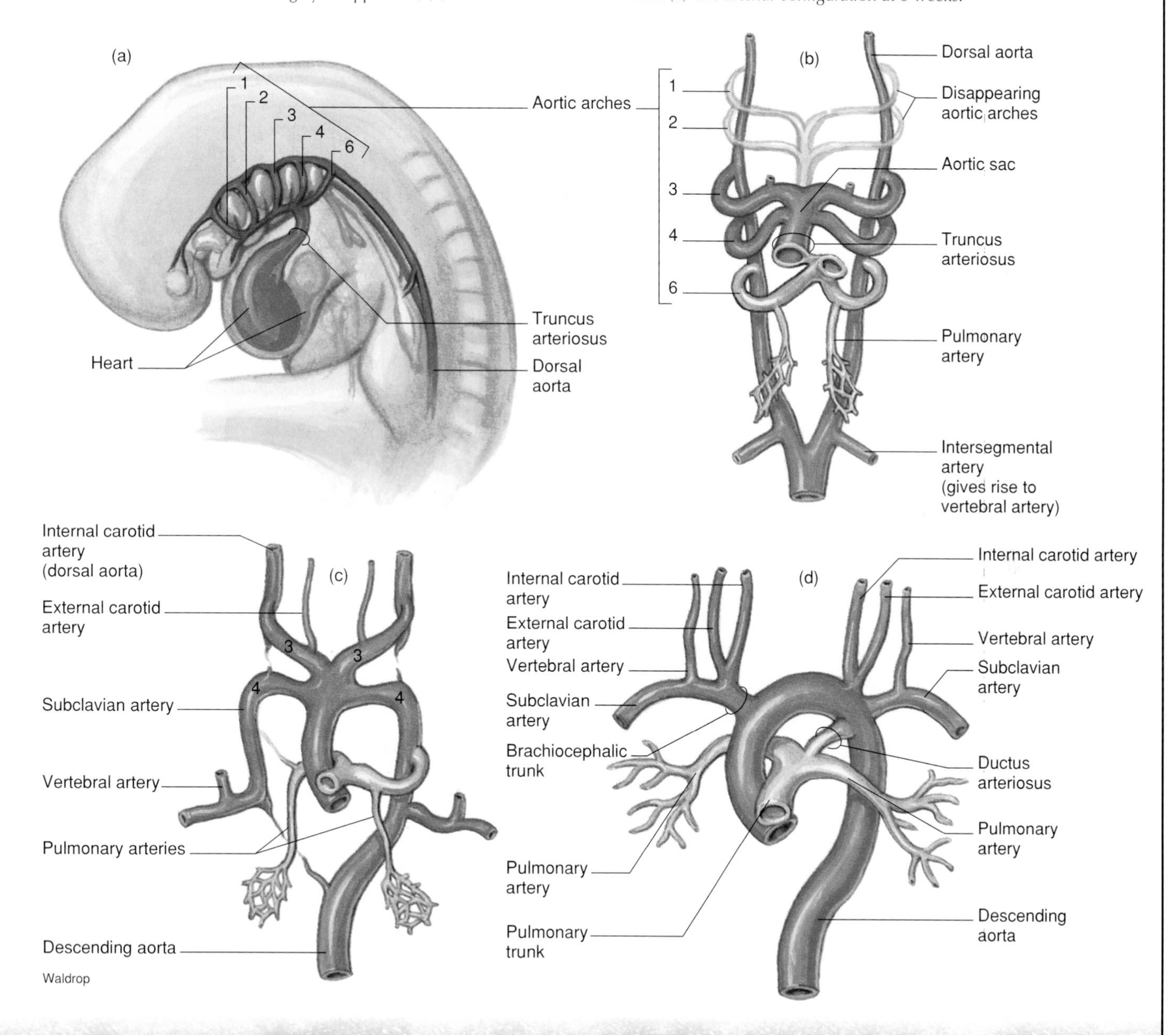

An abnormal tachycardia occurs if the heart rate increases when a person is at rest. This may result from abnormally fast pacing by the atria due to drugs or to the development of abnormally fast *ectopic pacemakers*—cells located outside the SA node that assume a pacemaker function. This abnormal atrial tachycardia thus differs from normal "sinus" (SA node) tachycardia. *Ventricular tachycardia* results when abnormally fast ectopic pacemakers in the ventricles cause them to beat rapidly and independently of the atria (fig. 16.46b). This is very dangerous because it can quickly degenerate to a lethal condition known as *ventricular fibrillation* (*fib″rĭ-la′shun*).

Ventricular Fibrillation

Fibrillation is caused by a continuous recycling of electrical waves through the myocardium. Because the myocardium enters a refractory period simultaneously at all regions, this recycling is normally prevented. If some cells emerge from their refractory periods before others, however, electrical waves can be continuously regenerated and conducted. The recycling of electrical waves along continuously changing pathways produces uncoordinated contraction and an impotent pumping action. These effects can be the result of damage to the myocardium.

Fibrillation can sometimes be stopped by a strong electric shock delivered to the chest—a procedure called **electrical defibrillation.** The electric shock depolarizes all the myocardial cells at the same time, causing them to enter a refractory state. The conduction of random recirculating impulses thus stops, and the SA node can begin to stimulate contraction in a normal fashion. Although this does not correct the initial problem that caused the abnormal electrical patterns, it can keep a person alive long enough to take other corrective measures.

Blood Disorders

Since blood is the functional component of the circulatory system, and since the circulatory system works in such close association with other body systems, blood analysis is perhaps the most informative part of a physical exam. Peripheral arterial pulsations, usually obtained at the radial artery, provide information about blood flow. Capillary filling, following **blanching,** is an indicator of peripheral arterial circulation and is generally tested at the nail bed. This is done by firmly pressing the thumbnail against the patient's toenail or fingernail and then quickly releasing the pressure. If the pinkish color returns quickly to the whitened (blanched) area, circulation is considered normal. Lack of peripheral coloration indicates vascular insufficiency.

Certain cardiovascular and blood abnormalities are expressed through the skin. **Cyanosis** (*si-ă-no′sis*) is characterized by a bluish coloration of the skin resulting from decreased oxygen concentration. **Anemia** is characterized by a pallor of the skin because of a deficiency of erythrocytes or hemoglobin. **Jaundice** is a condition in which the skin is yellowed because of excessive bile pigment (bilirubin) in the blood. **Edema** (*ĕ-de′mă*) is an excessive accumulation of interstitial (tissue) fluid, causing a swelling of a portion of the body. **Erythema** (*er″ĭ-the′mă*) is a redness of the skin usually caused by an infection, inflammation, toxic reaction, sunburn, or a lesion. None of these conditions are themselves diseases, but each is symptomatic of problems that may involve the cardiovascular system.

Blood analysis is an essential part of any thorough physical examination. Blood cell counts are used to determine the percentage of formed elements in the blood. An excess of red blood cells, called **polycythemia** (*pol″e-si-the′me-ă*) (more than 6 million/mm^3), may indicate certain bone diseases. Excessive leukocyte production, called **leukocytosis,** is generally diagnostic of infections or diseases within the body. **Leukopenia** is a decrease in leukocyte count. The disease **leukemia** causes the unrestricted reproduction of immature leukocytes, which depresses erythrocyte and platelet formation and causes anemia and a tendency to bleed. Coagulation time, blood sedimentation rates, prothrombin time, and various serum analyses are other blood tests that provide specific information about body function or dysfunction.

Several blood diseases are distinguished by their rate of occurrence. **Sickle-cell anemia** is an autosomal recessive disease that occurs almost exclusively in blacks. Although about 10% of American blacks have the sickle-cell trait, fortunately fewer than 1% have sickle-cell disease. The distorted shape of the diseased cells reduces their capacity to transport oxygen, resulting in an abnormally high destruction of erythrocytes. With the decrease in erythrocytes, the patient becomes anemic. **Mononucleosis** is an infectious disease that is transmitted by a virus in saliva, and is therefore commonly called the "kissing disease." Mononucleosis is characterized by atypical lymphocytes. It affects primarily adolescents, causing fever, sore throat, enlarged lymph glands, and fatigue.

Heart Diseases

Heart diseases can be classified as congenital or acquired. **Congenital heart problems** result from abnormalities in embryonic development and may be attributed to heredity, nutritional problems (poor diet) of the pregnant mother, or viral infections. Congenital heart diseases occur in approximately 3 of every 100 births and account for about 50% of early childhood deaths. Many congenital heart defects can be corrected surgically, however, and others are not of a serious nature.

Heart murmurs can be congenital or acquired. Generally, they are of no clinical significance; nearly 10% of all people have heart murmurs, ranging from slight to severe. A **septal defect** is the most common type of congenital heart problem (fig. 16.46). An **atrial septal defect,** or **patent foramen ovale,** is a failure of the fetal foramen ovale to close at the time of birth. A **ventricular septal defect** is caused by an abnormal development of the interventricular septum. This condition may interfere with closure of the atrioventricular valves and may be indicated by cyanosis and abnormal heart sounds. **Pulmonary stenosis** is a narrowing of the opening into the pulmonary trunk from the right ventricle. It may lead to a pulmonary embolism and is usually recognized by extreme lung congestion.

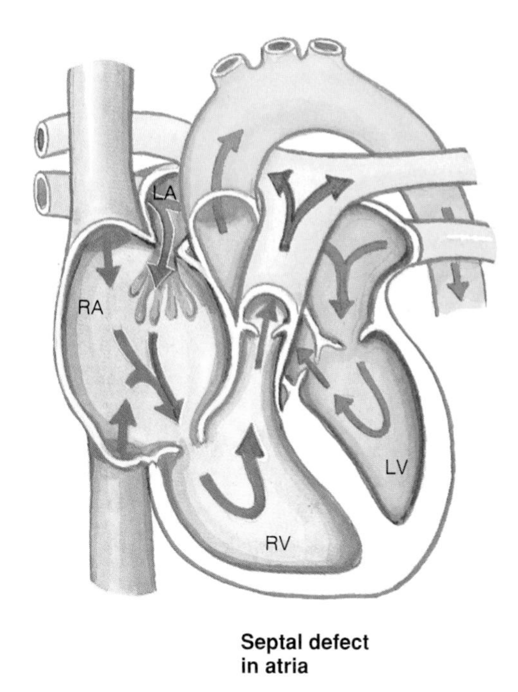

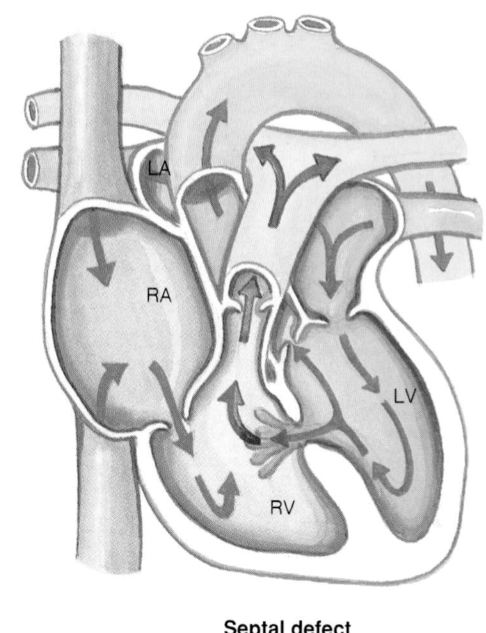

Figure 16.46

Abnormal patterns of blood flow as a result of septal defects.

The **tetralogy of Fallot** is a combination of four defects in a newborn that immediately causes a cyanotic condition. The four characteristics of this condition are (1) a ventricular septal defect, (2) an overriding aorta, (3) pulmonary stenosis, and (4) right ventricular hypertrophy (fig. 16.47). The pulmonary stenosis obstructs blood flow to the lungs and causes hypertrophy of the right ventricle. In an overriding aorta, the ascending portion arises midway between the right and left ventricles. Open-heart surgery is necessary to correct tetralogy of Fallot, and the overall mortality rate is about 5%.

Acquired heart disease may develop suddenly or gradually. Heart attacks are included in this category and are the leading cause of death in the United States. It is estimated that one in five individuals over the age of 60 will succumb to a heart attack. The immediate cause of a heart attack is generally one of the following: inadequate coronary blood supply, an anatomical disorder, or conduction disturbances.

Other types of acquired heart diseases affect the layers of the heart. **Bacterial endocarditis** is a disease of the lining of the heart, especially the cusps of the valves. It is caused by infectious organisms that enter the bloodstream. **Myocardial disease** is an inflammation of the heart muscle followed by cardiac enlargement and congestive heart failure. **Pericarditis** causes an inflammation of the pericardium—the covering membrane of the heart. Its distinctive feature is pericardial friction rub, a transitory scratchy sound heard during auscultation.

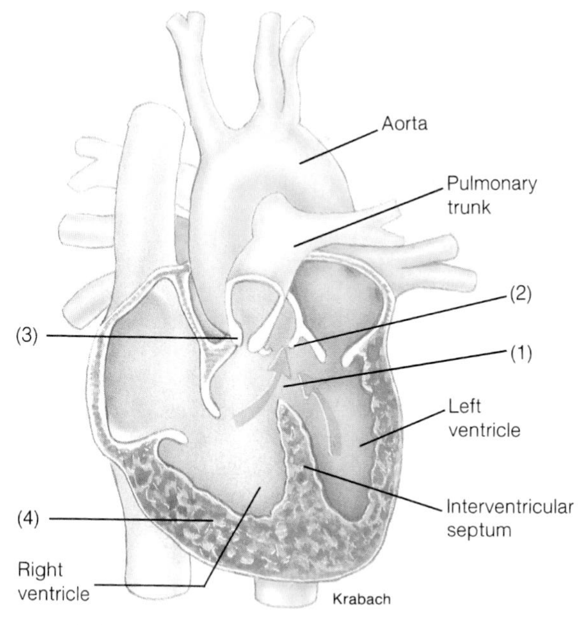

Figure 16.47

The tetralogy of Fallot. The four defects of this anomaly are (1) a ventricular septal defect, (2) an overriding aorta, (3) pulmonary stenosis (constriction), and (4) right ventricular hypertrophy (enlargement).

A tissue is said to be **ischemic** (*ĭ-ske'mik*) when it receives an inadequate supply of oxygen because of an inadequate blood flow. The most common cause of myocardial ischemia is atherosclerosis of the coronary arteries. The adequacy of blood flow is relative—it depends on the metabolic requirements of the tissue

tetralogy of Fallot: from Étienne-Louis A. Fallot, French physician, 1850–1911

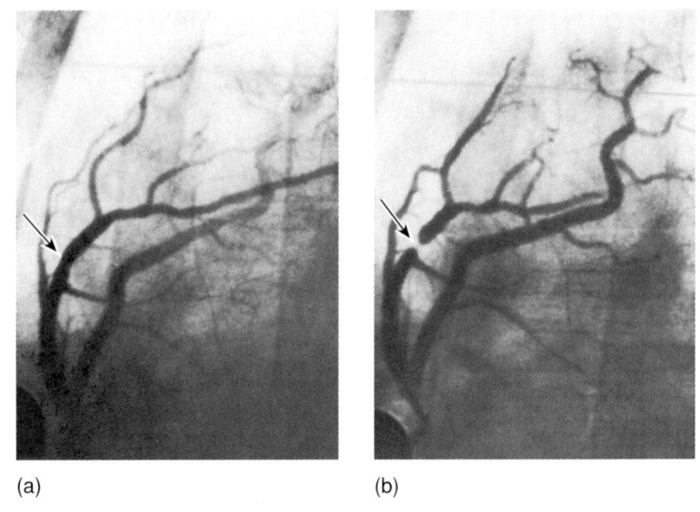

Figure 16.48

An arteriogram of the left coronary artery in a patient (*a*) when the ECG was normal and (*b*) when the ECG showed evidence of myocardial ischemia. Notice that a coronary artery spasm (see arrow in [*b*]) appears to accompany the ischemia.

for oxygen. An obstruction in a coronary artery, for example, may allow sufficient blood flow at rest but may produce ischemia when the heart is stressed by exercise or emotional factors (fig. 16.48). In patients with this condition, angioplasty or coronary artery bypass surgery may be performed (fig. 16.49).

Myocardial ischemia is associated with increased concentrations of blood lactic acid produced by anaerobic respiration of the ischemic tissue. This condition often causes substernal pain, which also may be referred to the left shoulder and arm, as well as to other areas. This referred pain is called **angina pectoris** (*an-ji'nă pek'tor'is*). People with angina frequently take nitroglycerin or related drugs that help to relieve the ischemia and pain. These drugs are effective because they stimulate vasodilation, which improves circulation to the heart and decreases the work that the heart must perform to eject blood into the arteries.

Myocardial cells are adapted to respire aerobically and cannot respire anaerobically for more than a few minutes. If ischemia and anaerobic respiration continue for more than a few minutes, *necrosis* (cellular death) may occur in the areas most deprived of oxygen. A sudden, irreversible injury of this kind is called a **myocardial infarction,** or **MI.**

Myocardial ischemia may be detected by characteristic changes in the ECG (fig. 16.50). The diagnosis of myocardial infarction is aided by examining the blood concentration of various enzymes that are released from the damaged cells.

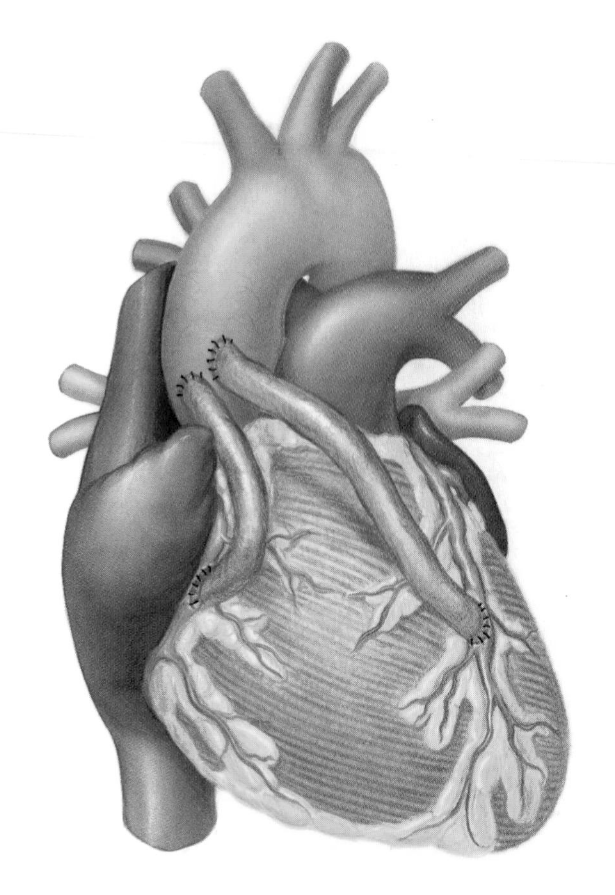

Figure 16.49

A double coronary artery bypass surgery. Several vessels may be used in the autotransplant, including the internal thoracic artery and the great saphenous vein.

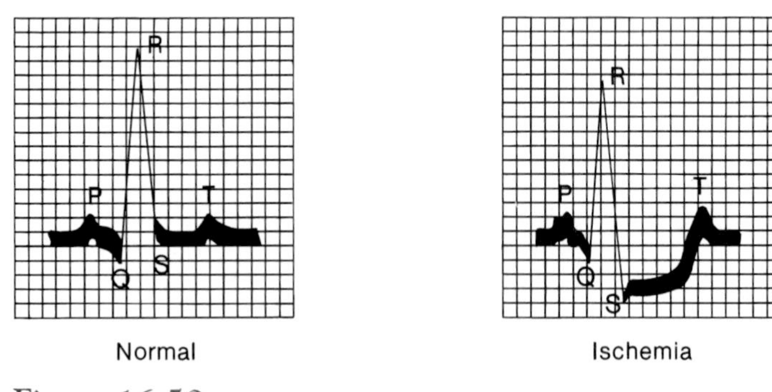

Figure 16.50

Depression of the S-T segment of the electrocardiogram as a result of myocardial ischemia.

Vascular Disorders

Hypertension, or high blood pressure, is the most common type of vascular disorder. In hypertension, the resting systolic blood pressure exceeds 140 mmHg. An estimated 22 million adult Americans are afflicted by hypertension. About 15% of the cases are the result of other body problems, such as kidney diseases, adrenal hypersecretion, or arteriosclerosis, and are diagnosed as secondary hypertension. Primary hypertension is more common and cannot be attributed to any particular body dysfunction. If hypertension is not controlled by diet, exercise, or drugs that reduce the blood pressure, it can damage various vital body organs, such as the heart or kidneys.

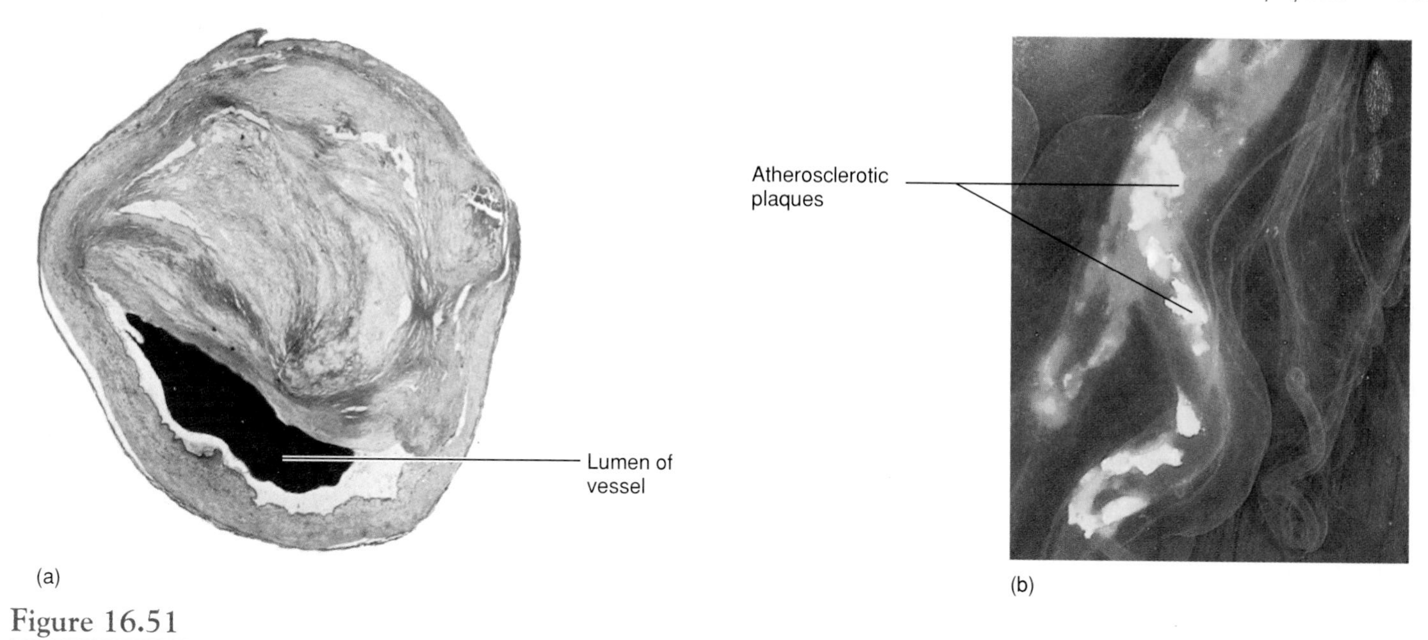

Figure 16.51

Atherosclerosis. (*a*) The lumen of a coronary artery is almost completely blocked by an atheroma. (*b*) A close-up view of the cleared left anterior descending coronary artery containing calcified atherosclerotic plaques. The heart is that of an 85-year-old female.

Arteriosclerosis (*ar-te″re-o-sklĕ-ro′sis*), or hardening of the arteries, is a generalized degenerative vascular disorder that results in loss of elasticity and thickening of the arteries. **Atherosclerosis** (*ath″er-o-sklĕ-ro′sis*) is a type of arteriosclerosis in which plaque material, called **atheroma** (*ath″er-o′-mă*), forms on the tunica intima, narrowing the lumina of the arteries and prohibiting the normal flow of blood (fig. 16.51). In addition, an atheroma often creates a rough surface that can initiate the formation of a blood clot called a **thrombus.** An **embolism,** or **embolus,** is a detached thrombus that travels through the bloodstream and lodges so as to obstruct or occlude a blood vessel. An embolism lodged in a coronary artery is called a *coronary embolism;* in a vessel of the lung it is a *pulmonary embolism;* and in the brain it is a *cerebral embolism,* which could cause a **stroke.**

The causes of atherosclerosis are not well understood, but the disease seems to be associated with improper diet, smoking, hypertension, obesity, lack of exercise, and heredity.

Aneurysms, coarctations, and varicose veins are all types of vascular disfigurations. An **aneurysm** (*an′yŭ-riz-em*) is an expansion or bulging of the heart, aorta, or any other artery (fig. 16.52). Aneurysms are caused by weakening of the tunicas and may rupture or lead to embolisms. A **coarctation** is a constriction of a segment of a vessel, usually the aorta, and is frequently caused by tightening of remnant of the ductus arteriosus around the vessel. **Varicose veins** are weakened veins that become stretched and swollen. They are most common in the legs because the force of gravity tends to weaken the valves and overload the veins. Varicose veins can also occur in the rectum, in which case they are called **hemorrhoids.** *Vein stripping* is the surgical removal of superficial weakened veins. **Phlebitis** (*flĕ-bi′tis*) is inflammation of a vein. It may develop as a result of trauma or as an aftermath of surgery. Frequently, however, it appears for no apparent reason. Phlebitis interferes with normal venous circulation.

arteriosclerosis: Gk. *arterio,* artery; *skleros,* hard
atheroma: Gk. *athere,* mush; *oma,* tumor
thrombus: Gk. *thrombos,* a clot
embolism: Gk. *embolos,* a plug

Disorders of the Lymphatic System

Infections of the body are generally accompanied by a swelling and tenderness of lymph nodes near the infection. An inflammation of lymph nodes is referred to as **lymphadenitis** (*lĭm-fad″en-i′tis*). In prolonged lymphadenitis, an **abscess** usually forms in the nodal tissue. An abscess is a localized pocket of pus formed by tissue destruction. If an infection is not contained by localized lymph nodes, **lymphangitis** may ensue. In this condition, red streaks can be seen through the skin extending proximally from the infected area. Lymphangitis is potentially dangerous because the uncontained infection may cause **septicemia** (*sep-tĭ-se′me-ă*) (blood poisoning).

The term **lymphoma** is used to describe primary malignancies within lymphoid tissues. Lymphomas are generally classified as **Hodgkin's disease lymphomas** or **non-Hodgkin's lymphomas.** Hodgkin's disease manifests itself as swollen lymph nodes in the neck and then progresses to involve the spleen, liver, and bone marrow. The prognosis for Hodgkin's disease is good if it is detected early. Non-Hodgkin's lymphomas include an array of specific and more obscure lymphatic cancers.

The lymphatic system is also frequently infected by metastasizing carcinomas. Fragmented cells from the original tumor may enter the lymphatic ducts with the lymph and travel to the lymph

septicemia: Gk. *septikos,* septic; *haima,* blood
Hodgkin's disease: from Thomas Hodgkin, English physician, 1798–1866

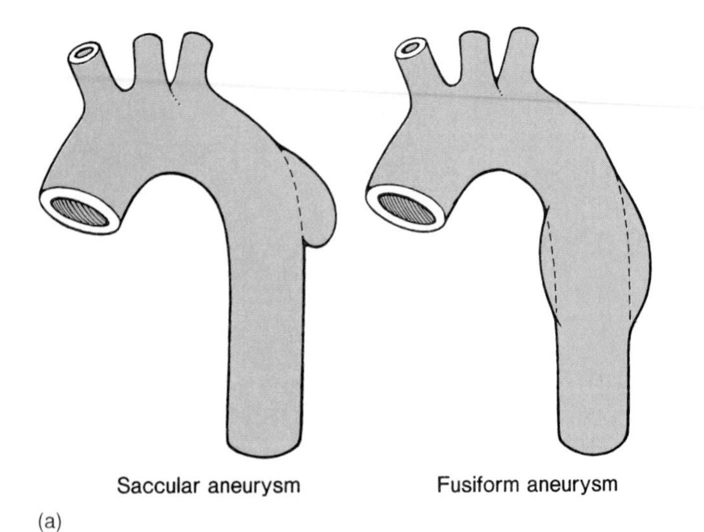

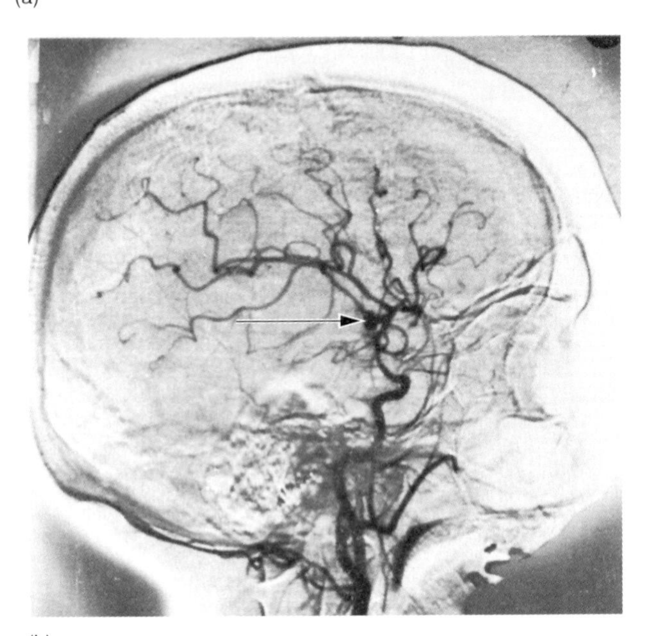

Figure 16.52

An aneurysm is an outpouching of the vascular wall. (*a*) A diagram of two types of aortic aneurysms and (*b*) an angiograph of a saccular aneurysm (see arrow) of the middle cerebral artery.

nodes, where they may cause secondary cancerous growths. Breast cancer will typically do this. The surgical treatment involves the removal of the infected nodes, along with some of the healthy nodes downstream to ensure that the cancer is eliminated.

Trauma to the Circulatory System

Hemorrhage and shock are two clinical considerations that directly involve the circulatory system. Knowledge of how to treat a victim experiencing these conditions is of paramount importance in administering first aid. Techniques to control bleeding and to administer first aid to victims experiencing shock are presented in the next two sections. Also review the important arterial pressure points presented in figure 16.31.

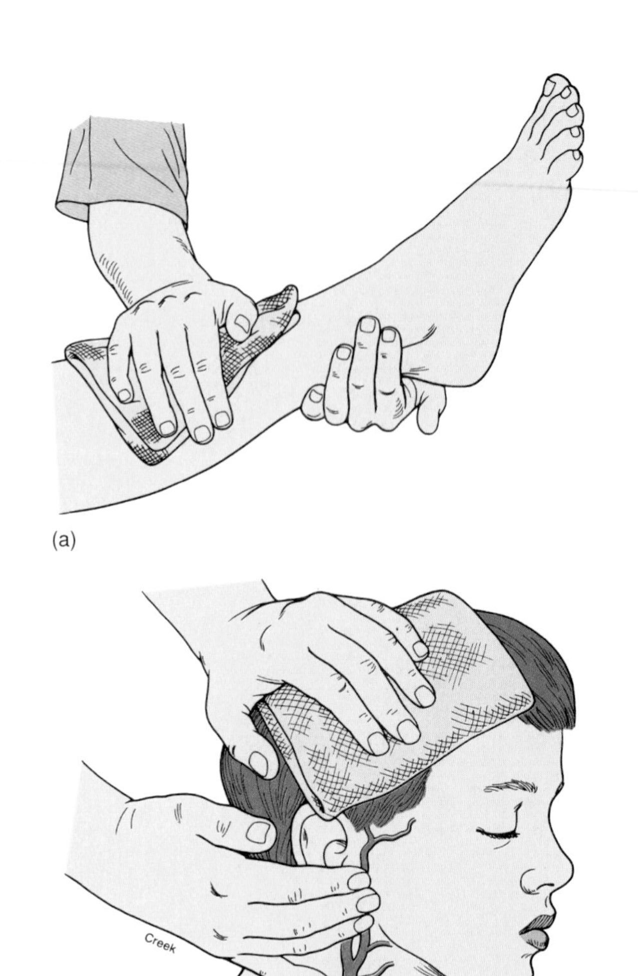

Figure 16.53

Control of bleeding. (*a*) Direct pressure and (*b*) direct pressure and compression at an arterial pressure point.

Control of Bleeding

Because serious bleeding is life threatening, the principal first-aid concern is to stop the loss of blood. The following are recommended steps in treating a victim who is hemorrhaging.

1. To reduce the chance that the victim will faint, lay the person down on a blanket (if available) and slightly elevate his or her legs. If possible, elevate the site of bleeding above the level of the trunk. To minimize the chance of shock, cover the victim with a blanket.
2. Without causing further trauma, carefully remove any dirt or debris from the wound. *Do not remove any impaling objects*. This should be done at the hospital by trained personnel.
3. Apply direct pressure to the wound with a sterile bandage, clean cloth, or an article of clothing (fig. 16.53).
4. Maintain direct pressure until the bleeding stops. Dress the wound with clean bandages or cloth lightly bound in place.

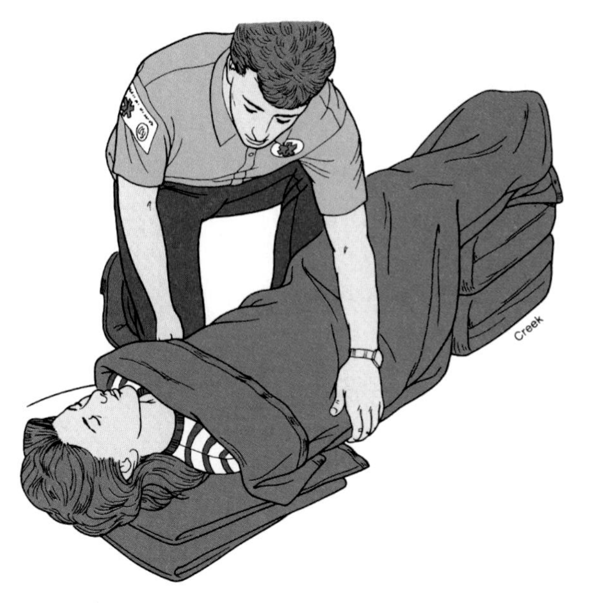

Figure 16.54

Treatment of a victim experiencing shock.

5. If the bleeding does not stop and continues to seep through the dressing, *do not remove the dressing*. Rather, place additional absorbent material on top of it and continue to apply direct pressure.
6. If direct pressure does not stop the bleeding, the pressure point to the wound site may need to be compressed. In the case of a severe wound to the hand, for example, compress the brachial artery against the humerus. This should be done while pressure continues to be applied to the wound itself.
7. Once the bleeding has stopped, leave the bandage in place and immobilize the injured body part. Get the victim to the hospital or medical treatment center at once.

Recognizing and Treating Victims of Shock

Shock is the medical condition that occurs when body tissues do not receive enough oxygen-carrying blood. It is often linked with crushing injuries, heat stroke, heart attacks, poisoning, severe burns, and other life-threatening conditions. Symptoms of patients experiencing shock include the following.

1. **Skin.** The skin is pale or gray, cool, and clammy.
2. **Pulse.** The heartbeat is weak and rapid. Blood pressure is reduced, frequently to below measurable values.
3. **Respiration.** The respiratory rate is hurried, shallow, and irregular.
4. **Eyes.** The eyes are staring and lusterless, possibly with dilated pupils.
5. **State of being.** The victim may be conscious or unconscious. If conscious, he or she is likely to feel faint, weak, and confused. Frequently, the victim is anxious and excited.

Most trauma victims will experience some degree of shock, especially if there has been considerable blood loss. Immediate first-aid treatment for shock is essential and includes the following steps.

1. Get the victim to lie down (fig. 16.54). Lay the person on his or her back with the feet elevated. This position maintains blood flow to the brain and may relieve faintness and mental confusion. Keep movement to a minimum. If the victim has sustained an injury in which raising the legs causes additional pain, leave the person flat on his or her back.
2. Keep the victim warm and comfortable. If the weather is cold, place a blanket under and over the person. If the weather is hot, position the person in the shade on top of a blanket. Loosen tight collars, belts, or other restrictive clothing. *Do not give the person anything to drink*, even if he or she complains of thirst.
3. Take precautions for internal bleeding or vomiting. If the victim has blood coming from the mouth, or if there is indication that the victim may vomit, position the person on his or her side to prevent choking or inhaling the blood or vomitus.
4. Treat injuries appropriately. If the victim is bleeding, treat accordingly (see arterial pressure points [fig. 16.31] and control of bleeding [fig. 16.53]). Immobilize fractures and sprains. Always be alert to the possibility of spinal injuries and take the necessary precautions.
5. See that hospital care is provided as soon as possible.

Clinical Case Study Answer

Mural thrombus (a blood clot adherent to the inner surface of one of the heart's chambers) is a fairly common complication of myocardial infarction. Once a thrombus forms within the heart, a piece may break off and travel throughout the body. This is the most likely cause of the symptoms in the right leg of our patient. Because the embolus traveled and lodged in the systemic circulation (as opposed to the pulmonary circulation), the mural thrombus was probably located in the left side of the heart. The embolus traveled to a point in the femoral artery and lodged there, thus occluding blood flow to the popliteal artery and its distal branches. The route of travel was as follows: left side of heart → ascending aorta → descending thoracic aorta → abdominal aorta → right common iliac artery → right external iliac artery → femoral artery.

The standard treatment for this problem is emergency surgery to extract the clot from the leg. Anticoagulation (blood-thinning) therapy is then continued or instituted.

Internal Affairs

■ How the *Circulatory System* Works with Other Body Systems
□ How Other Systems Work with the *Circulatory System*

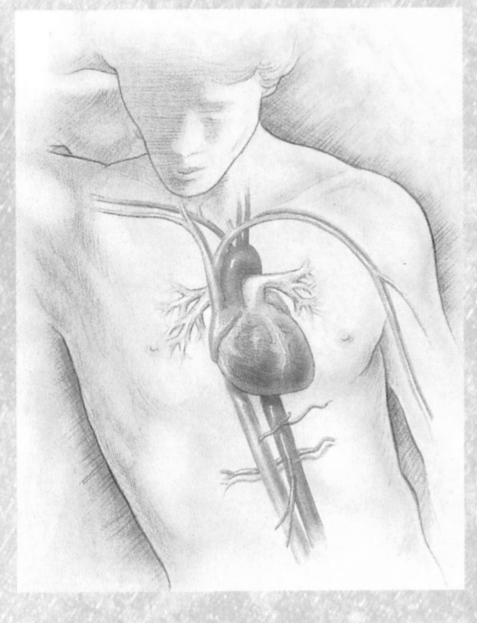

Integumentary System

■
- Transports O_2 and nutrients; eliminates CO_2 and metabolic wastes
- Clotting mechanism heals breaks in the skin surface

□
- Provides site for thermoregulation
- Serves as a blood reservoir

Skeletal System

■
- Transports O_2 and nutrients; eliminates CO_2 and metabolic wastes
- Delivers erythropoietin to bone marrow

□
- Protects the heart and thoracic vessels
- Provides a site (bone marrow) for hemopoiesis
- Serves as a reservoir of calcium needed for cardiac muscle contraction

Muscular System

■
- Transports O_2 and nutrients; eliminates CO_2 and metabolic wastes, including lactic acid
- Removes heat generated by exercise

□
- Muscle contraction assists blood movement

Nervous System

■
- Transports O_2 and nutrients; eliminates CO_2 and metabolic wastes
- Certain capillaries help to form the blood-brain barrier
- Cerebrospinal fluid is produced from and returned to the blood

□
- Provides autonomic regulation of cardiac rate and force
- Maintains blood pressure and controls blood distribution

Endocrine System

■
- Transports O_2, hormones, and nutrients; eliminates CO_2 and metabolic wastes

□
- Various hormones regulate blood pressure
- Epinephrine elevates the heart rate and increases contractile force
- Erythropoietin regulates the production of red blood cells
- Estrogen maintains vascular health in women

Lymphatic System

■
- Transports O_2 and nutrients; eliminates CO_2 and metabolic wastes
- Clotting response helps to restrict the spread of pathogens
- White blood cells and many plasma proteins are involved in immunity

□
- Maintains balanced amount of interstitial fluid within body tissues
- Spleen serves as a reservoir of red blood cells

Respiratory System

■
- Transports O_2 and nutrients; eliminates CO_2 and metabolic wastes

□
- Provides O_2 to bloodstream and eliminates CO_2 at pulmonary alveoli of lungs
- Serves as a site of exchange for blood gases
- Thoracic pump aids venous return

Digestive System

■
- Transports O_2 and nutrients; eliminates CO_2 and metabolic wastes

□
- Provides nutrients for blood formation, including iron and B vitamins
- Absorbs water and ions needed to maintain blood volume and pressure

Urinary System

■
- Transports O_2 and nutrients; eliminates CO_2 and metabolic wastes
- Blood pressure maintains kidney function

□
- Eliminates metabolic wastes from blood
- Helps to regulate blood volume and pressure
- Kidneys play a major role in erythropoietin production

Reproductive System

■
- Transports O_2, sex hormones, and nutrients; eliminates CO_2 and metabolic waste
- Local vasodilation causes physical changes during sexual arousal

□
- Estrogen maintains vascular health in women

Chapter Summary

Functions and Major Components of the Circulatory System (pp. 521–522)

1. The circulatory system transports oxygen and nutritive molecules to the tissue cells and carbon dioxide and other wastes away from tissue cells; it also carries hormones and other regulatory molecules to their target organs.
2. Leukocytes and their products help to protect the body from infection, and platelets function in blood clotting.
3. The components of the circulatory system are the heart, blood vessels, and blood, which constitute the cardiovascular system, and the lymphatic vessels and lymphoid tissue and organs of the lymphatic system.

Blood (pp. 523–529)

1. Blood, a highly specialized connective tissue, consists of formed elements (erythrocytes, leukocytes, and platelets) suspended in a watery fluid called plasma.
2. Erythrocytes are disc-shaped cells that lack nuclei but contain hemoglobin. There are approximately 4 million to 6 million erythrocytes per cubic millimeter of blood, and they are functional for about 120 days.
3. Leukocytes have nuclei and are classified as granular (eosinophils, basophils, and neutrophils) or agranular (monocytes and lymphocytes). Leukocytes defend the body against infections by microorganisms.
4. Platelets, or thrombocytes, are cytoplasmic fragments that assist in the formation of clots to prevent blood loss.
5. Erythrocytes are formed through a process called erythropoiesis; leukocytes are formed through leukopoiesis.
6. Prenatal hemopoietic centers are the yolk sac, liver, and spleen. In the adult, bone marrow and lymphoid tissues perform this function.

Heart (pp. 529–538)

1. The heart is enclosed within a pericardial sac. The wall of the heart consists of the epicardium, myocardium, and endocardium.
   (a) The right atrium receives blood from the superior and inferior venae cavae, and the right ventricle pumps blood through the pulmonary trunk into the pulmonary arteries.
   (b) The left atrium receives blood from the pulmonary veins, and the left ventricle pumps blood into the ascending aorta.
   (c) The heart contains right and left atrioventricular valves (the tricuspid and bicuspid valves, respectively); a pulmonary semilunar valve; and an aortic (semilunar) valve.
2. The two principal circulatory divisions are the pulmonary and the systemic; in addition, the coronary system serves the heart.
   (a) The pulmonary circulation includes the vessels that carry blood from the right ventricle through the lungs, and from there to the left atrium.
   (b) The systemic circulation includes all other arteries, capillaries, and veins in the body. These vessels carry blood from the left ventricle through the body and return blood to the right atrium.
   (c) The myocardium of the heart is served by right and left coronary arteries that branch from the ascending portion of the aorta. The coronary sinus collects and empties the blood into the right atrium.
3. Contraction of the atria and ventricles is produced by action potentials that originate in the sinoatrial (SA) node.
   (a) These electrical waves spread over the atria and then enter the atrioventricular (AV) node.
   (b) From here, the impulses are conducted by the atrioventricular bundle and conduction myofibers into the ventricular walls.
4. During contraction of the ventricles, the intraventricular pressure rises and causes the AV valves to close; during relaxation, the pulmonary and aortic valves close because the pressure is greater in the arteries than in the ventricles.
5. Closing of the AV valves causes the first sound (lub); closing of the pulmonary and aortic valves causes the second sound (dub). Heart murmurs are commonly caused by abnormal valves or by septal defects.
6. A recording of the pattern of electrical conduction is called an electrocardiogram (ECG or EKG).

Blood Vessels (pp. 538–542)

1. Arteries and veins have a tunica externa, tunica media, and tunica interna.
   (a) Arteries have thicker muscle layers in proportion to their diameters than do veins because arteries must withstand a higher blood pressure.
   (b) Veins have venous valves that direct blood to the heart when the veins are compressed by the skeletal muscle pumps.
2. Capillaries are composed of endothelial cells only. They are the basic functional units of the circulatory system.

Principal Arteries of the Body (pp. 542–554)

1. Three arteries arise from the aortic arch: the brachiocephalic trunk, the left common carotid artery, and the left subclavian artery. The brachiocephalic trunk divides into the right common carotid artery and the right subclavian artery.
2. The head and neck receive an arterial supply from branches of the internal and external carotid arteries and the vertebral arteries.
   (a) The brain receives blood from the paired internal carotid arteries and the paired vertebral arteries, which form the cerebral arterial circle surrounding the pituitary gland.
   (b) The external carotid artery gives off numerous branches that supply the head and neck.
3. The upper extremity is served by the subclavian artery and its derivatives.
   (a) The subclavian artery becomes first the axillary artery and then the brachial artery as it enters the arm.
   (b) The brachial artery bifurcates to form the radial and ulnar arteries, which supply blood to the forearm and hand.
4. The abdominal portion of the aorta has the following branches: the inferior phrenic, celiac trunk, superior mesenteric, renal, suprarenal, testicular (or ovarian), and inferior mesenteric arteries.
5. The common iliac arteries divide into the internal and external iliac arteries, which supply branches to the pelvis and lower extremities.

Principal Veins of the Body (pp. 554–560)

1. Blood from the head and neck is drained by the external and internal jugular veins; blood from the brain is drained by the internal jugular veins.
2. The upper extremity is drained by both superficial and deep veins.
3. In the thorax, the superior vena cava is formed by the union of the two brachiocephalic veins and also collects blood from the azygos system of veins.
4. The lower extremity is drained by both superficial and deep veins. At the level of the fifth lumbar vertebra, the right and left common iliac veins unite to form the inferior vena cava.

5. Blood from capillaries in the GI tract is drained via the hepatic portal vein to the liver.
   (a) This venous blood then passes through hepatic sinusoids and is drained from the liver in the hepatic veins.
   (b) The pattern of circulation characterized by two capillary beds in a series is called a portal system.

Fetal Circulation (pp. 560–562)

1. Structural adaptations in the fetal cardiovascular system reflect the fact that oxygen and nutrients are obtained from the placenta rather than from the fetal lungs and GI tract.
2. Fully oxygenated blood is carried only in the umbilical vein, which drains the placenta. This blood is carried via the ductus venosus to the inferior vena cava of the fetus.
3. Partially oxygenated blood is shunted from the right to the left atrium via the foramen ovale and from the pulmonary trunk to the aorta via the ductus arteriosus.

Lymphatic System (pp. 562–566)

1. The lymphatic system returns excess interstitial fluid to the venous system and helps to protect the body from disease; it also transports fats from the small intestine to the blood.
2. Lymphatic capillaries drain interstitial fluid, which is formed from blood plasma; when this fluid enters lymphatic capillaries, it is called lymph.
3. Lymph is returned to the venous system via two large lymph ducts—the thoracic duct and the right lymphatic duct.
4. Lymph filters through lymph nodes, which contain phagocytic cells and lymphatic nodules that produce lymphocytes.
5. Lymphoid organs include the lymph nodes, tonsils, spleen, and thymus.

Review Activities

Objective Questions

1. Which of the following is *not* a formed element of blood?
   (a) a leukocyte (c) a fibrinogen
   (b) an eosinophil (d) a platelet
2. An elevated white blood cell count is referred to as
   (a) leukocytosis. (c) leukopoiesis.
   (b) polycythemia. (d) leukemia.
3. Which of the following vessels transport(s) oxygen-poor blood?
   (a) the aorta
   (b) the pulmonary arteries
   (c) the renal arteries
   (d) the coronary arteries
4. Blood from the coronary circulation directly enters
   (a) the inferior vena cava.
   (b) the superior vena cava.
   (c) the right atrium.
   (d) the left atrium.
5. The first heart sound ("lub") is produced by the closing of
   (a) the aortic valve.
   (b) the pulmonary valve.
   (c) the right atrioventricular valve.
   (d) the left atrioventricular valve.
   (e) both a and b.
   (f) both c and d.
6. The QRS complex of an ECG is produced by
   (a) depolarization of the atria.
   (b) repolarization of the atria.
   (c) depolarization of the ventricles.
   (d) repolarization of the ventricles.
7. Which of the following does *not* arise from the aortic arch?
   (a) the brachiocephalic trunk
   (b) the left coronary artery
   (c) the left common carotid artery
   (d) the left subclavian artery
8. Which of the following does *not* supply blood to the brain?
   (a) the external carotid artery
   (b) the internal carotid artery
   (c) the vertebral artery
   (d) the basilar artery
9. The maxillary and superficial temporal arteries arise from
   (a) the external carotid artery.
   (b) the internal carotid artery.
   (c) the vertebral artery.
   (d) the facial artery.
10. Which of the following organs is (are) served by a portal vein?
    (a) the liver (c) both a and b
    (b) the pituitary gland (d) neither a nor b
11. The heart derives from mesoderm and is complete in embryonic form by
    (a) the third week.
    (b) the fifth week.
    (c) the tenth week.
    (d) the seventeenth week.
12. Which of the following fetal blood vessels carries fully oxygenated blood?
    (a) the ductus arteriosus
    (b) the ductus venosus
    (c) the umbilical vein
    (d) the aorta

Essay Questions

1. What are the functions of the circulatory system? Identify the body systems that work closely with the circulatory system in maintaining homeostasis.
2. Explain how the development of the aortic arches contributes to the formation of the major vessels associated with the heart.
3. Distinguish between granulocytes and agranulocytes. What are the functions of leukocytes? How do they differ from erythrocytes and thrombocytes?
4. Describe how the heart is compartmentalized within the thoracic cavity. What is the function of the pericardium?
5. Describe the cardiac cycle. Why is the sinoatrial node called the pacemaker of the heart?
6. Diagram and label a normal electrocardiogram. What is the significance of the deflection waves of an ECG?
7. Compare the structure and function of arteries, capillaries, and veins.
8. What is the difference between blood pressure and pulse pressure? What cardiac event determines the systolic blood pressure? What accounts for the diastolic blood pressure?
9. Name the vessels of the thoracic and shoulder regions that are not symmetrical (do not have a counterpart on the opposite side of the body).
10. Trace the path to the brain of a glucose injection into the median cubital vein of the arm. List in sequence all the blood vessels and chambers of the heart through which it passes.
11. What is significant about the hepatic portal system? What do we mean when we say that the liver has two blood supplies?
12. List the functions of the lymphatic system and describe the relationship between the lymphatic system and the circulatory system.

13. Briefly describe the development of the heart.
14. What are five fetal circulatory structures that cease to function in a newborn?
15. Distinguish between congenital and acquired heart diseases and describe several kinds of each.

Critical-Thinking Questions

1. What symptoms would a patient display with a low red blood cell count? A low white blood cell count? A low platelet count?
2. Examine figure 16.8c and predict the structures that might be harmed in the event of damage to the interventricular septum or ventricular wall.
3. The walls of the ventricles are thicker than those of the atria, and the wall of the left ventricle is the thickest of all. How do these structural differences relate to differences in function?
4. An endurance-trained athlete will typically have a lower resting cardiac rate and a greater stroke volume than a person who is out of shape. Explain why these adaptations are beneficial.
5. Some passenger planes are now equipped with defibrillators for use as emergency life-saving devices. Why is it critical that ventricular fibrillation receive attention within minutes? Is atrial fibrillation less urgent? Explain.
6. A hospitalized 45-year-old man developed a thrombus (blood clot) in his lower thigh following severe trauma to his knee. The patient's physician explained that although the clot was near the great saphenous vein, the main concern was the occurrence of a pulmonary embolism. Explain the physician's reasoning and list, in sequence, the vessels the clot would have to pass through to cause a heart problem.
7. Your sister's new baby was diagnosed as having a patent foramen ovale. What does this mean and why is the baby cyanotic?

SEVENTEEN

Respiratory System

Clinical Case Study 583

Introduction to the Respiratory System 583
- Physical Requirements of the Respiratory System 583
- Functions of the Respiratory System 583
- Basic Structure of the Respiratory System 583

Conducting Passages 585
- Nose 585
- Paranasal Sinuses 587
- Pharynx 587
- Larynx 588
- Trachea 590
- Bronchial Tree 591

Pulmonary Alveoli, Lungs, and Pleurae 592
- Pulmonary Alveoli 592
- Lungs 593
- Pleurae 595

Mechanics of Breathing 598
- Inspiration 598
- Expiration 599
- Respiratory Volumes and Capacities 600
- Nonrespiratory Air Movements 601

Regulation of Breathing 603

Developmental Exposition: The Respiratory System 604

Clinical Considerations 606
- Developmental Problems of the Respiratory System 606
- Trauma or Injury 607
- Common Respiratory Disorders 608
- Disorders of Respiratory Control 608

Internal Affairs 610

Clinical Case Study Answer 611

Chapter Summary 612

Review Activities 612

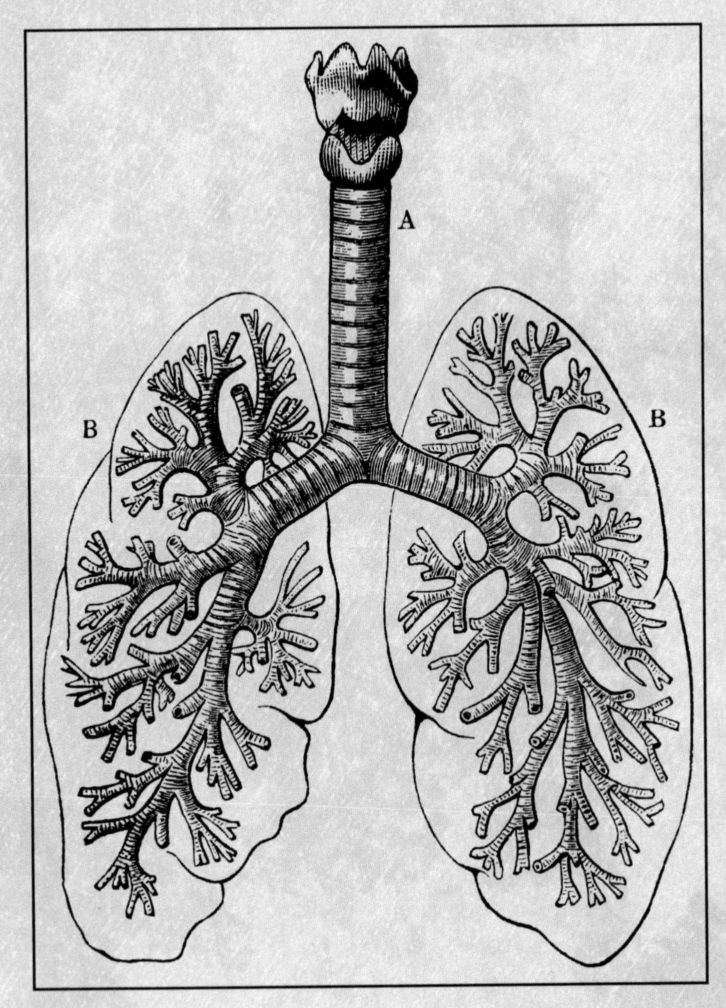

This nineteenth-century schematic of the larynx, trachea, and lungs—complete with branching airways—demonstrates a high level of understanding of respiratory system anatomy. Actually, the right lung is thicker and broader than the left; it is also shorter because the diaphragm rises on the right side to accommodate the mass of the liver.

Clinical Case Study

A 37-year-old male was brought to the emergency room after having been stabbed above the left clavicle with an ice pick. The patient's main complaint was pain in the left side of his chest. Initial evaluation revealed a minute puncture wound superior to the left clavicle, just lateral to the sternocleidomastoid muscle. The vital signs were normal except for a moderately high respiratory rate. The chest radiograph showed the left lung to be surrounded by blood and air, a condition known as *hemopneumothorax* (*he″mo-noo″mo-thor′aks*), and collapsed to half its normal size.

Explain how the air got inside the thoracic cavity (assuming it did not enter through the puncture wound). What is the term for the space where the air and blood are located? What membranes define this space?

Introduction to the Respiratory System

The respiratory system can be divided structurally into upper and lower divisions, and functionally into a conducting division and a respiratory division. The principal functions of the respiratory system are gaseous exchange, sound production, and assistance in abdominal compression.

Objective 1 Describe the functions associated with the term *respiration*.

Objective 2 Identify the organs of the respiratory system and describe their locations.

Objective 3 List the functions of the respiratory system.

The term *respiration* refers to three separate but related functions: (1) **ventilation** (breathing); (2) **gas exchange,** which occurs between the air and blood in the lungs and between the blood and other tissues of the body; and (3) **oxygen utilization** by the tissues in the energy-liberating reactions of cell respiration. Ventilation and the exchange of gases (oxygen and carbon dioxide) between the air and blood are collectively called *external respiration*. Gas exchange between the blood and other tissues are collectively known as *internal respiration*.

A relaxed adult breathes an average of 15 times a minute, ventilating approximately 6 liters of air during this period. This amounts to over 8,000 liters in a 24-hour period. Strenuous exercise increases the demand for oxygen and increases the respiratory rate fifteenfold to twentyfold, so that about 100 liters of air are breathed each minute. If breathing stops, a person will lose consciousness after 4 or 5 minutes. Brain damage may occur after 7 to 8 minutes, and the person will die after 10 minutes. Knowledge of the structure and function of the respiratory system is therefore of the utmost importance in a clinical setting.

respiration: L. *re*, back; *spirare*, to breathe

Physical Requirements of the Respiratory System

The respiratory system includes those organs and structures that function together to bring gases in contact with the blood of the circulatory system. In order to be effective, the respiratory system must comply with certain physical requirements.

1. The surface for gas exchange must be located deep within the body so that incoming air will be sufficiently warmed, moistened, and cleansed of airborne particles before coming in contact with it.
2. The membrane must be thin-walled and selectively permeable so that diffusion can occur easily.
3. The membrane must be kept moist so that oxygen and carbon dioxide can be dissolved in water to facilitate diffusion.
4. The system must have an extensive capillary network.
5. The system must include an effective ventilation mechanism to constantly replenish the air.

The respiratory system adequately meets all of these requirements, thus ensuring that all of the trillions of cells of the body will be able to carry on the metabolic processes necessary to maintain life.

Functions of the Respiratory System

The four basic functions of the respiratory system, not all of which are associated with breathing, are as follows:

1. It provides oxygen to the bloodstream and removes carbon dioxide.
2. It enables sound production or vocalization as expired air passes over the vocal folds.
3. It assists in abdominal compression during micturition (urination), defecation (passing of feces), and parturition (childbirth). The abdominal muscles become more effective during a deep breath when the air is held in the lungs by closing the glottis and fixing the diaphragm. This same technique is used when lifting a heavy object, in which case the diaphragm indirectly assists the back muscles.
4. It enables protective and reflexive nonbreathing air movements, as in coughing and sneezing, to keep the air passageways clean.

Basic Structure of the Respiratory System

The major passages and structures of the respiratory system are the *nasal cavity*, *pharynx*, *larynx* and *trachea*, and the *bronchi*, *bronchioles*, and *pulmonary alveoli* within the *lungs* (fig. 17.1).

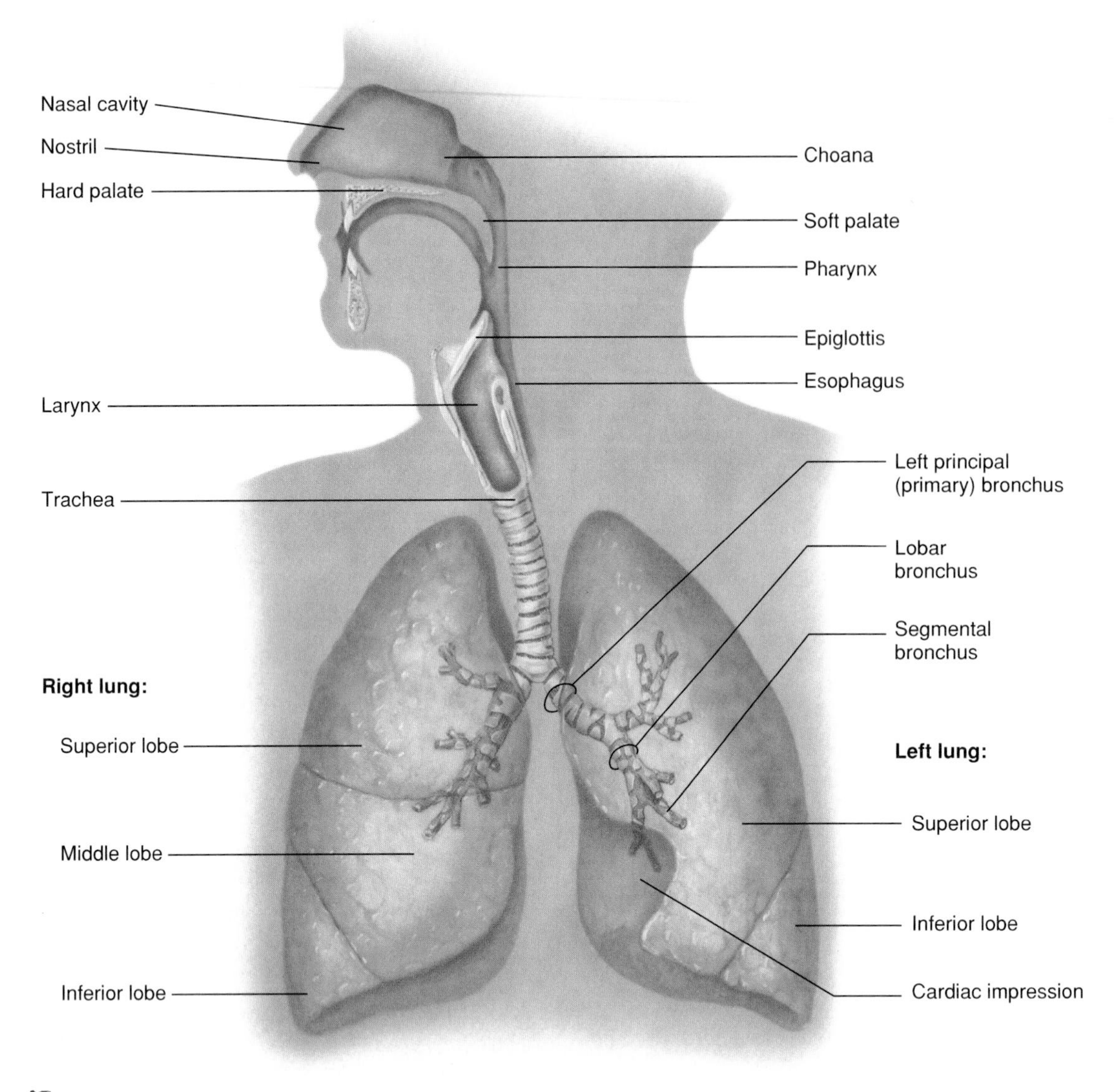

Figure 17.1

The basic anatomy of the respiratory system.

The structures of the **upper respiratory system** include the nose, pharynx, and associated structures; the **lower respiratory system** includes the larynx, trachea, bronchial tree, pulmonary alveoli, and lungs.

Based on general function, the respiratory system is frequently divided into a conducting division and a respiratory division. The **conducting division** includes all of the cavities and structures that transport gases to and from the pulmonary alveoli. The **respiratory division** consists of the pulmonary alveoli, which are the functional units of the respiratory system where gas exchange between the air and the blood occur.

Early Greek and Roman scientists placed great emphasis on the invisible material that was breathed in. They knew nothing about oxygen or the role of the blood in transporting this vital substance to cells. For that matter, they knew nothing about microscopic structures like cells because the microscope had not yet been invented. They did know, however, that respiration was essential for life. Early Greeks referred to air as an intangible, divine spirit called *pneuma.* In Latin, the term for breath, *spiritus,* meant life-force.

1. Define the terms *external respiration* and *internal respiration.*
2. What are the physical requirements of the respiratory system? What are its basic functions?
3. List in order the major passages and structures through which inspired air would pass from the nostrils to the pulmonary alveoli of the lungs.

Conducting Passages

Air is conducted through the oral and nasal cavities to the pharynx, and then through the larynx to the trachea and bronchial tree. These structures deliver warmed and humidified air to the respiratory division within the lungs.

Objective 4 List the types of epithelial tissue that characterize each region of the respiratory tract and comment on the significance of the special attributes of each type.

Objective 5 Identify the boundaries of the nasal cavity and discuss the relationship of the paranasal sinuses to the rest of the respiratory system.

Objective 6 Describe the three regions of the pharynx and identify the structures located in each.

Objective 7 Discuss the role of the laryngeal region in digestion and respiration.

Objective 8 Identify the anatomical features of the larynx associated with sound production and respiration.

The conducting passages serve to transport air to the respiratory structures of the lungs. The passageways are lined with various types of epithelia that cleanse, warm, and humidify the air. The majority of the conducting passages are held permanently open by muscle or a bony or cartilaginous framework.

Nose

The **nose** includes an external portion that protrudes from the face and an internal **nasal cavity** for the passage of air. The external portion of the nose is covered with skin and supported by paired **nasal bones,** which form the bridge, and pliable cartilage, which forms the distal portions (fig. 17.2). The **septal cartilage** forms the anterior portion of the **nasal septum,** and the paired **lateral cartilages** and **alar cartilages** form the framework around the **nostrils.**

The **vomer** and the perpendicular plate of the **ethmoid bone** (see fig. 6.17), together with the septal cartilage, constitute the supporting framework of the nasal septum, which divides the nasal cavity into two lateral halves. Each half is referred to as a **nasal fossa.** The **nasal vestibule** is the anterior expanded portion of the nasal fossa (fig. 17.3). Each nasal fossa opens anteriorly through the **nostril** (*naris*), and communicates posteriorly with the **nasopharynx** through the **choana** (*ko-a'nă*). The roof of the nasal cavity is formed anteriorly by the frontal bone and paired nasal bones, medially by the cribriform plate of the ethmoid bone, and posteriorly by the sphenoid bone (see figs. 6.17 and 6.20). The palatine and maxillary bones form the floor of the cavity. On the lateral walls of the nasal cavity are three bony projections, the **superior, middle,** and **inferior nasal conchae** (*kong'ke*), or **turbinates** (see fig. 6.26). Air passages between the conchae are referred to as **nasal meatuses** (*me-a'tus-es*) (fig. 17.3). The anterior openings of the nasal cavity are lined with stratified squamous epithelium, whereas the conchae are lined with pseudostratified ciliated columnar epithelium (figs. 17.4 and 17.5). Mucus-secreting goblet cells are present in great abundance throughout both regions.

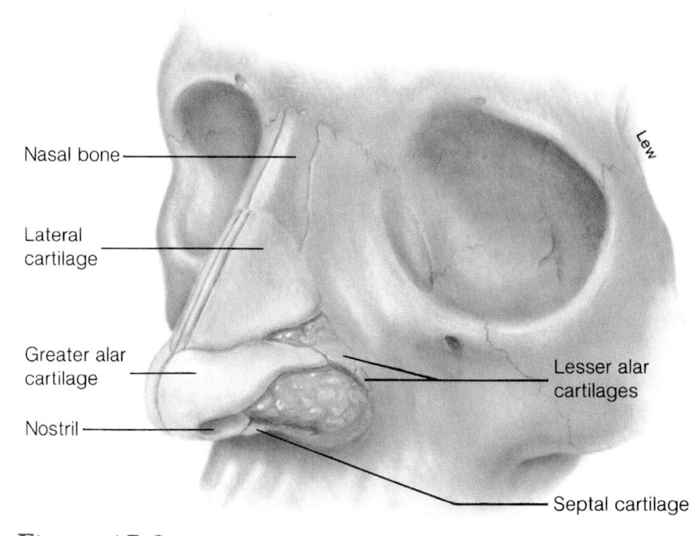

Figure 17.2

The supporting framework of the nose.

The three functions of the nasal cavity and its contents are as follows:

1. The nasal epithelium covering the conchae serves to warm, moisten, and cleanse the inspired air. The nasal epithelium is highly vascular and covers an extensive surface area. This is important for warming the air but unfortunately also makes humans susceptible to nosebleeds. Nasal hairs called **vibrissae** (*vi-bris'e*), which often extend from the nostrils, filter macroparticles that might otherwise be inhaled. Dust, pollen, smoke, and other fine particles are trapped along the moist mucous membrane lining the nasal cavity.
2. Olfactory epithelium in the upper medial portion of the nasal cavity is concerned with the sense of smell.
3. The nasal cavity affects the voice by functioning as a resonating chamber.

---

nose: O.E. *nosu*, nose
nostril: O.E. *nosu*, nose; *thyrel*, hole
choana: Gk. *choane*, funnel
concha: L. *choncha*, mussel shell
vibrissa: L. *vibrare*, to vibrate

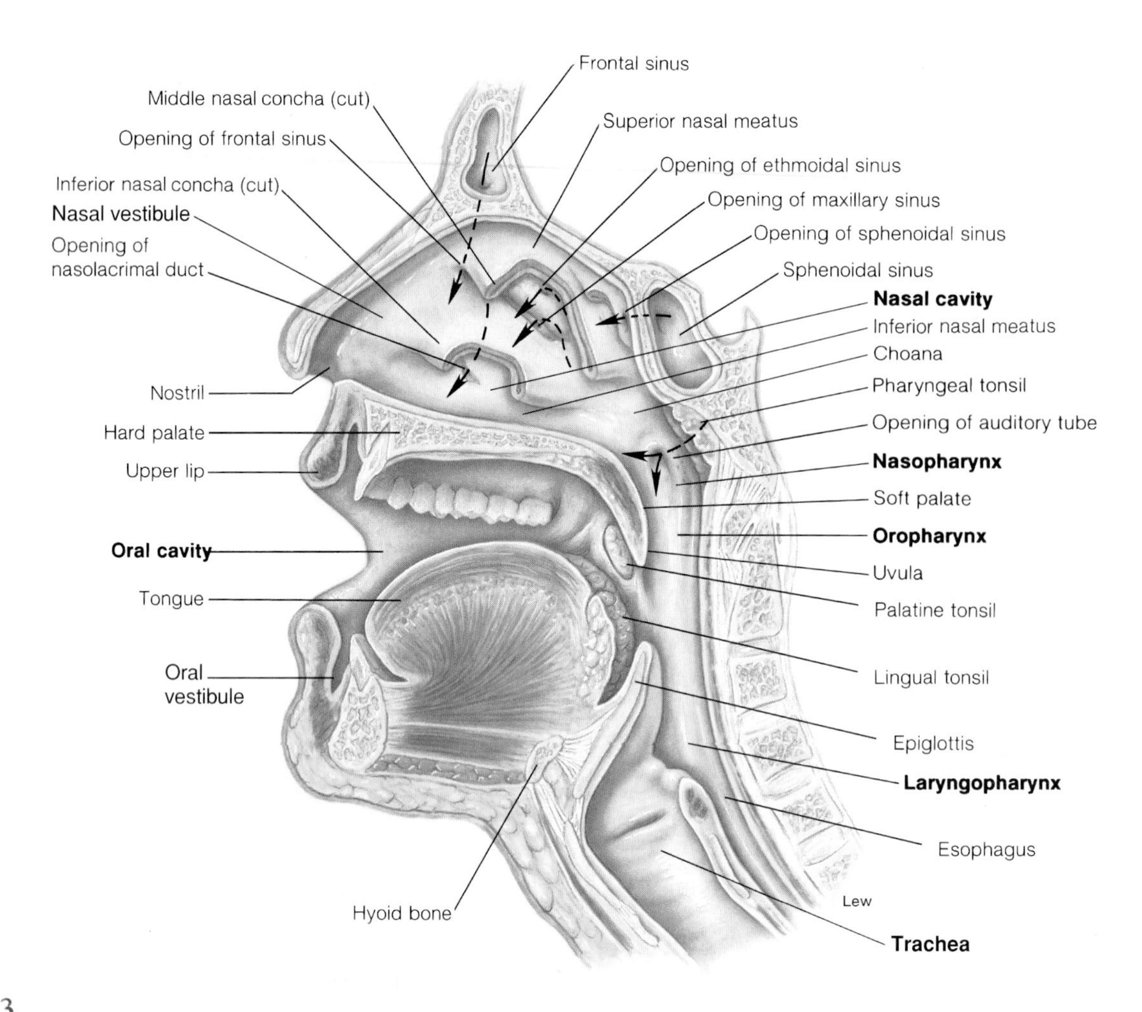

Figure 17.3

A sagittal section of the head showing the structures of the upper respiratory tract. There are several openings into the nasal cavity, including the openings of the various paranasal sinuses, those of the nasolacrimal ducts that drain from the eyes, and those of the auditory tubes that drain from the middle-ear chambers.

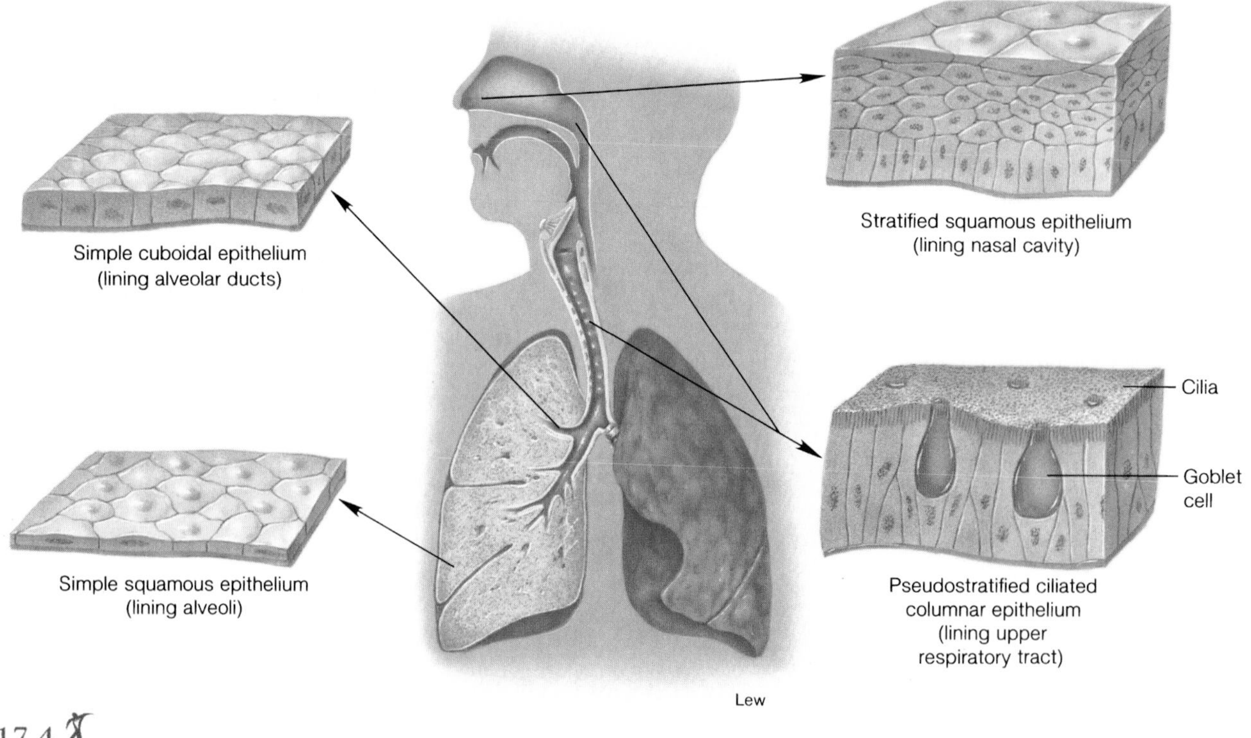

Figure 17.4

The various types of epithelial tissue found throughout the respiratory system.

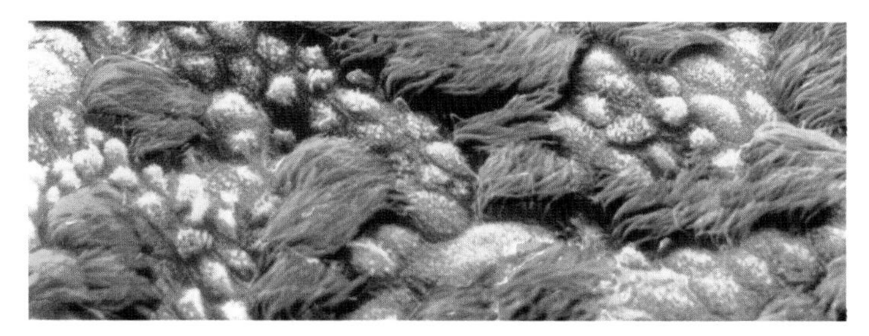

Figure 17.5

A color-enhanced scanning electron micrograph of a bronchial wall showing cilia. In the trachea and bronchi, there are about 300 cilia per cell. The cilia move mucus-dust particles toward the pharynx, where they can either be swallowed or expectorated.

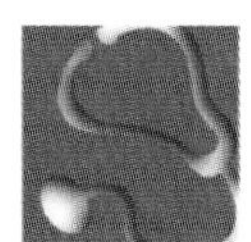

There are several drainage openings into the nasal cavity (see fig. 17.3). The paranasal ducts (discussed below) drain mucus from the paranasal sinuses, and the nasolacrimal ducts drain tears from the eyes (see fig. 15.11). An excessive secretion of tears causes the nose to run as the tears drain into the nasal cavity. The auditory tube from the middle-ear cavity enters the upper respiratory tract posterior to the nasal cavity in the nasopharynx. With all these accessory connections, it is no wonder that infections can spread so easily from one chamber to another throughout the facial area. To avoid causing damage or spreading infections to other areas, one must be careful not to blow the nose too forcefully.

Paranasal Sinuses

Paired air spaces in certain bones of the skull are called **paranasal sinuses.** These sinuses are named according to the bones in which they are found; thus, there are the **maxillary, frontal, sphenoidal,** and **ethmoidal sinuses** (fig. 17.6). Each sinus communicates via drainage ducts within the nasal cavity on its own side (see fig. 17.3). Paranasal sinuses may help to warm and moisten the inspired air. These sinuses are responsible for some sound resonance, but most importantly, they function to decrease the weight of the skull while providing structural strength.

You can observe your own paranasal sinuses. Face a mirror in a darkened room and shine a small flashlight into your face. The frontal sinuses will be illuminated by directing the light just below the eyebrow. The maxillary sinuses are illuminated by shining the light into the oral cavity and closing your mouth around the flashlight.

Pharynx

The **pharynx** (*far'ingks*) is a funnel-shaped passageway, approximately 13 cm (5 in.) long, that connects the nasal and oral cavities to the larynx of the respiratory system and the esophagus of the digestive system. The supporting walls of the pharynx are composed of skeletal muscle, and the lumen is lined with a mucous membrane. Within the pharynx are several paired lymphoid organs called **tonsils.** Commonly referred to as the "throat" or "gullet," the pharynx has both respiratory and digestive functions. It also provides a resonating chamber for certain speech sounds. The pharynx is divided on the basis of location and function into three regions (see fig. 17.3).

1. The **nasopharynx** serves only as a passageway for air, since it is located above the point of food entry into the body (the mouth). It is the uppermost portion of the pharynx, positioned directly behind the nasal cavity and above the soft palate. A pendulous **uvula** (*yoo'vyŭ-lă*) hangs from the middle lower portion of the soft palate. The paired **auditory** (eustachian) **tubes** connect the nasopharynx with the middle-ear cavities. The **pharyngeal tonsils,** or **adenoids,** are situated in the posterior wall of the nasal cavity.

   During the act of swallowing, the soft palate and uvula are elevated to block the nasal cavity and prevent food from entering. Occasionally a person may suddenly exhale air (as with a laugh) while in the process of swallowing fluid. If this occurs before the uvula effectively blocks the nasopharynx, fluid will be discharged through the nasal cavity.
2. The **oropharynx** is the middle portion of the pharynx between the soft palate and the level of the hyoid bone. Both swallowed food and fluid and inhaled air pass through it. The base of the tongue forms the anterior wall of the oropharynx. Paired **palatine tonsils** are located on the posterior lateral wall, and the **lingual tonsils** are found on the base of the tongue.
3. The **laryngopharynx** (*lă-ring"go-far'ingks*) is the lowermost portion of the pharynx. It extends inferiorly from the level of the hyoid bone to the larynx and opens into the esophagus and larynx. It is at the lower laryngopharynx that the respiratory and digestive systems become distinct. Swallowed food and fluid are directed into the esophagus, whereas inhaled air is directed anteriorly into the larynx.

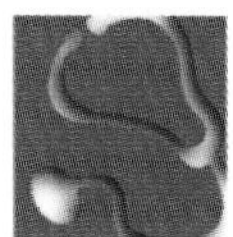

During a routine physical examination, the physician will commonly depress the patient's tongue and examine the condition of the palatine tonsils. Tonsils are lymphoid organs and tend to become swollen and inflamed after persistent infections. If after periods of infection, the tonsils become so large as to obstruct breathing, they may have to be surgically removed. The removal of the palatine tonsils is called a *tonsillectomy,* whereas the removal of the pharyngeal tonsils is called an *adenoidectomy* (*ad"ĕ-noid-ek'tŏ-me*).

sinus: L. *sinus*, bend or curve
pharynx: L. *pharynx*, throat

tonsil: L. *toles*, goiter or swelling
uvula: L. *uvula*, small grape
adenoid: Gk. *adenoeides*, glandlike

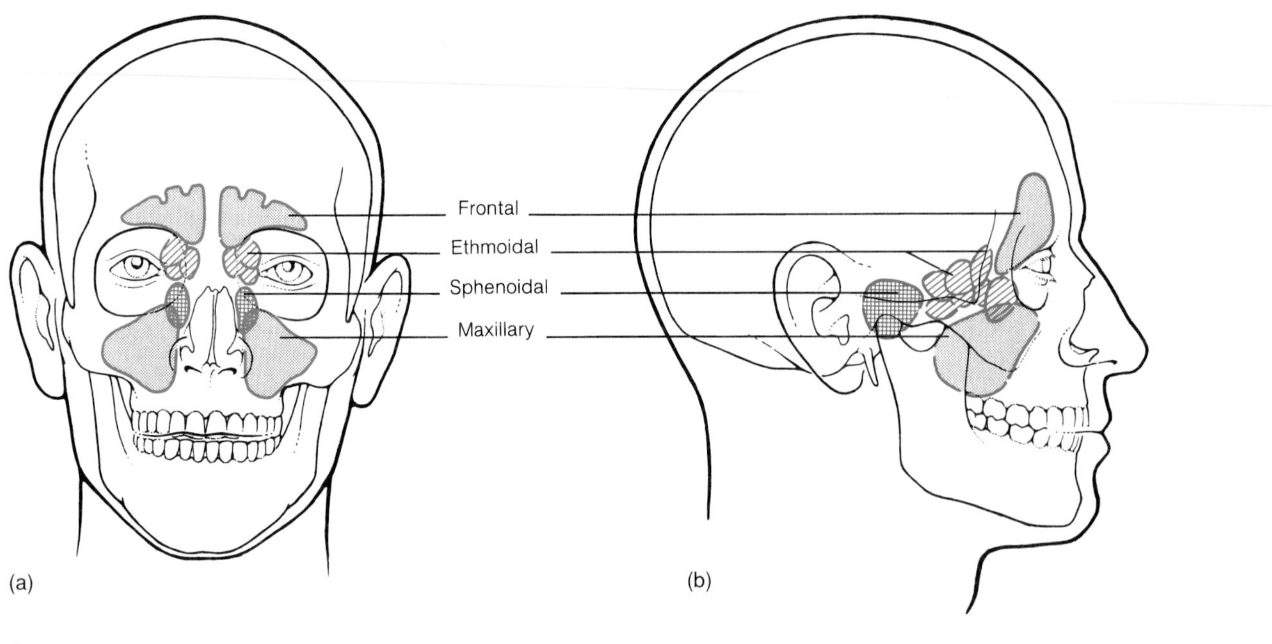

Figure 17.6

The paranasal sinuses.

Larynx

The **larynx** (*lar'ingks*), or "voice box," is a continuation of the conducting division that connects the laryngopharynx with the trachea. It is positioned in the anterior midline of the neck at the level of the fourth through sixth cervical vertebrae. The larynx has two functions. Its primary function is to prevent food or fluid from entering the trachea and lungs during swallowing and to permit passage of air while breathing. A secondary role is to produce sound.

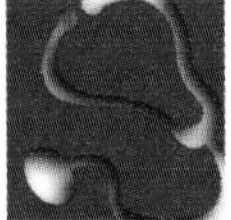

Laryngitis is the inflammation of the mucosal epithelium of the larynx and vocal folds, which causes a hoarseness of a person's voice or an inability to speak above a whisper. Laryngitis may result from overuse of the voice, inhalation of an irritating chemical, or a bacterial or viral infection. Mild cases are temporary and seldom of major concern.

The larynx is shaped like a triangular box (fig. 17.7). It is composed of a framework involving nine cartilages: three are large unpaired structures, and six are smaller and paired. The largest of the unpaired cartilages is the anterior **thyroid cartilage.** The **laryngeal prominence** of the thyroid cartilage is commonly called the "Adam's apple." It is an anterior vertical ridge along the larynx that can be palpated on the midline of the neck. The thyroid cartilage is typically larger and more prominent in males than in females because of the effect of male sex hormones on the development of the larynx during puberty.

larynx: Gk. *larynx*, upper windpipe
thyroid: Gk. *thyreos*, shieldlike

The spoon-shaped **epiglottis** (*ep"ĭ-glot'is*) has a framework of hyaline cartilage, referred to as the *epiglottic cartilage*. The epiglottis is located behind the root of the tongue where it aids in closing the **glottis,** or laryngeal opening, during swallowing.

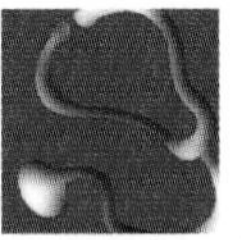

The entire larynx elevates during swallowing to close the glottis against the epiglottis. This movement can be noted by cupping the fingers lightly over the larynx and then swallowing. If the glottis is not closed as it should be during swallowing, food may become lodged within the glottis. In this case, the *abdominal thrust* (Heimlich) *maneuver* can be used to prevent suffocation. (See "Clinical Considerations" for how to perform this maneuver.)

The lower end of the larynx is formed by the ring-shaped **cricoid** (*kri'koid*) **cartilage.** This third unpaired cartilage connects the thyroid cartilage above and the trachea below. The paired **arytenoid** (*ar"ĭ-te'noid*) **cartilages,** located above the cricoid and behind the thyroid, are the posterior attachments of the *vocal folds*. The other paired **cuneiform cartilages** and **corniculate** (*kor-nik'yŭ-lāt*) **cartilages** are small accessory cartilages that are closely associated with the arytenoid cartilages (fig. 17.8).

Two pairs of strong connective tissue bands are stretched across the upper opening of the larynx from the thyroid cartilage anteriorly to the paired arytenoid cartilages posteriorly. These are the **vocal folds** (*true vocal cords*) and the **ventricular folds**

cricoid: Gk. *krikos*, ring; *eidos*, form
arytenoid: Gk. *arytaina*, ladle- or cup-shaped
cuneiform: L. *cuneus*, wedge-shaped
corniculate: L. *corniculum*, diminutive of *cornu*, horn

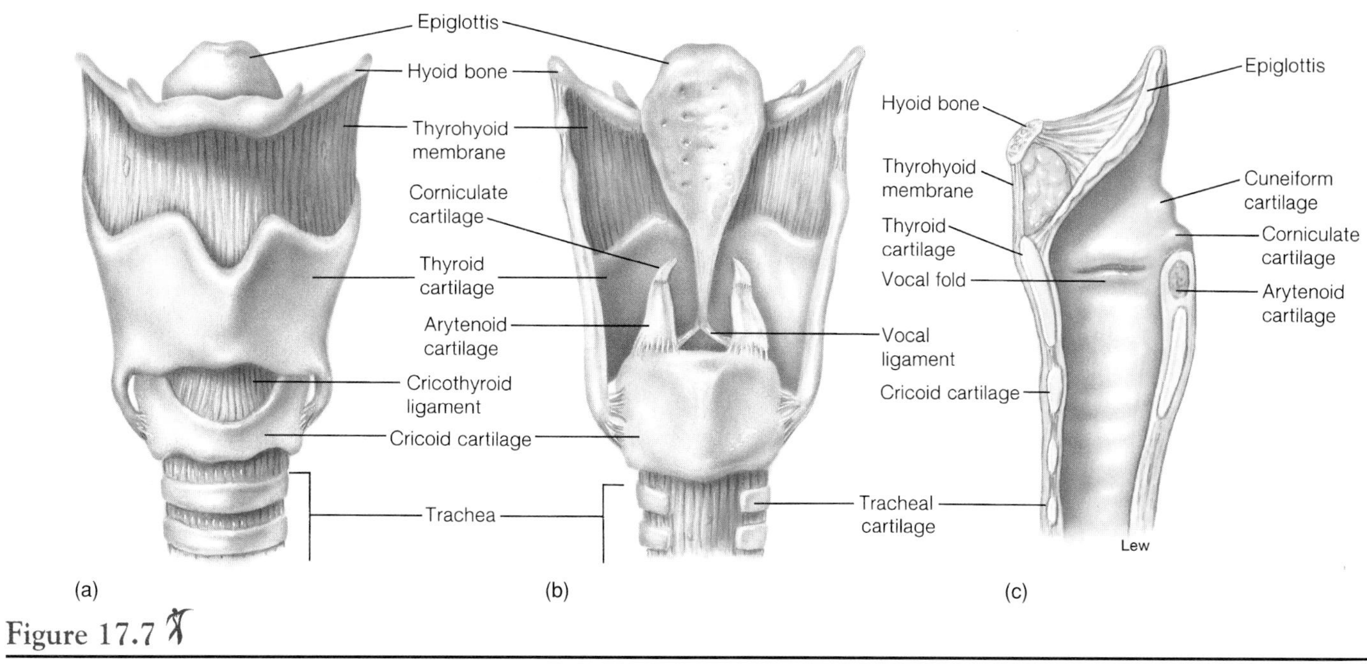

Figure 17.7

The structure of the larynx. (*a*) An anterior view, (*b*) a posterior view, and (*c*) a sagittal view.

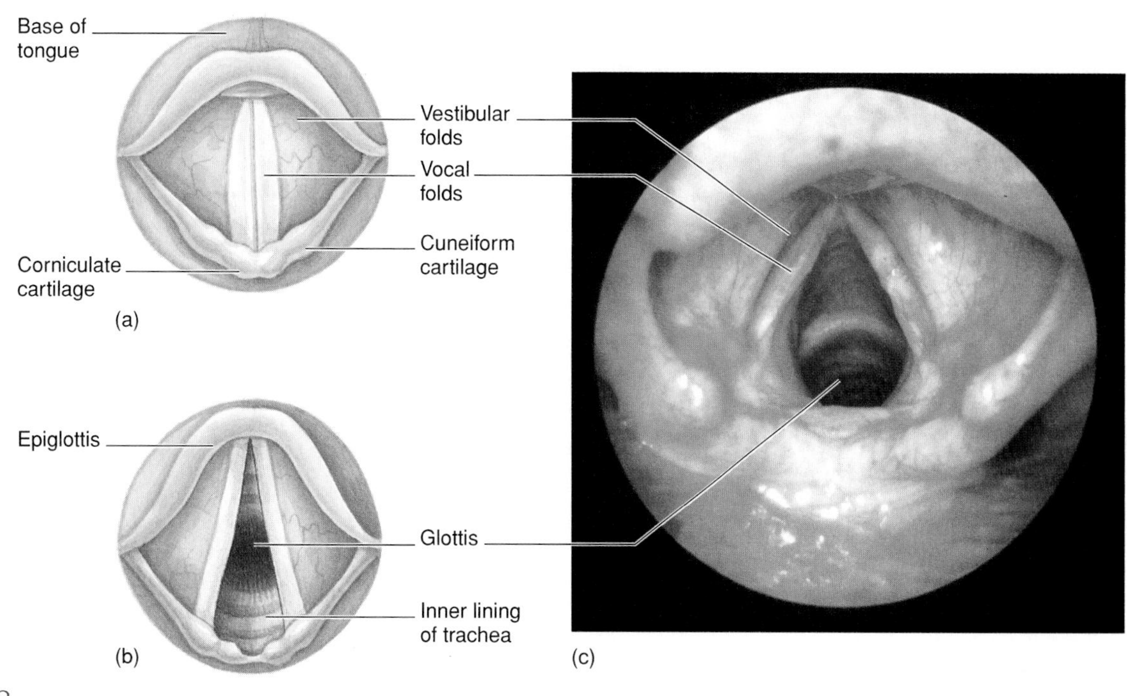

Figure 17.8

A superior view of the vocal folds (vocal cords). In (*a*) the vocal folds are taut; in (*b*) they are relaxed and the glottis is opened. (*c*) A photograph through a laryngoscope showing the glottis, the vestibular folds, and the vocal folds.

(*false vocal cords*) (fig. 17.8). The ventricular folds support the vocal folds, and are not used in sound production. The vocal folds vibrate to produce sound. Stratified squamous epithelium lines the vocal folds, whereas the rest of the larynx is lined with pseudostratified ciliated columnar epithelium. This is an important anatomical modification considering the tremendous vibratory action of the vocal folds in the production of sound.

The **laryngeal muscles** are extremely important in closing the glottis during swallowing and in speech. There are two groups of laryngeal muscles: **extrinsic muscles,** responsible for elevating the larynx during swallowing, and **intrinsic muscles** that, when contracted, change the length, position, and tension of the vocal folds. Various pitches are produced as air passes over the altered vocal folds. If the vocal folds are taut, vibration is

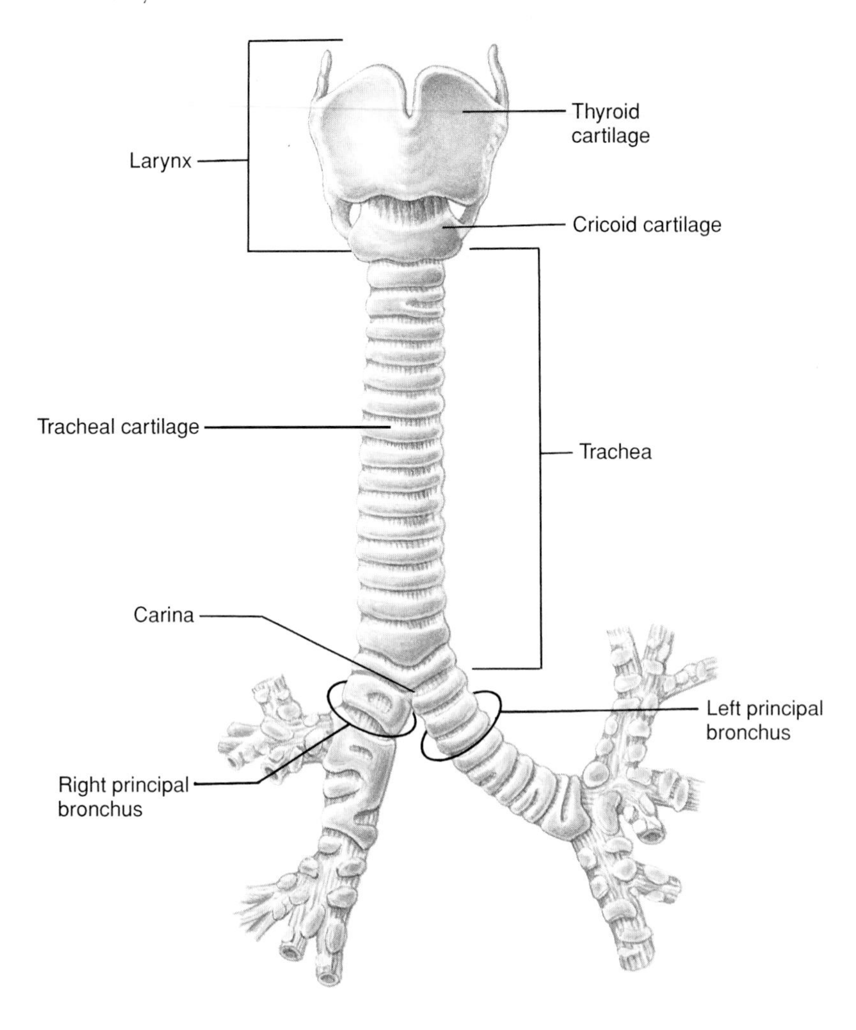

Figure 17.9

An anterior view of the larynx, trachea, and bronchi.

more rapid and causes a higher pitch. Less tension on the vocal folds produces lower sounds. Mature males generally have thicker and longer vocal folds than females; therefore, the vocal folds of males vibrate more slowly and produce lower pitches. The loudness of vocal sound is determined by the force of the air passed over the vocal folds and the amount of vibration. The vocal folds do not vibrate when a person whispers.

Sounds originate in the larynx, but other structures are necessary to convert sound into recognizable speech. Vowel sounds, for example, are produced by constriction of the walls of the pharynx. The pharynx, paranasal sinuses, and oral and nasal cavities act as resonating chambers. The final enunciation of words is accomplished through movements of the lips and tongue.

Trachea

The **trachea** (*tra'ke-ă*), commonly called the "windpipe," is a rigid tube, approximately 12 cm (4 in.) long and 2.5 cm (1 in.) in diameter, connecting the larynx to the principal (primary) bronchi (fig. 17.9). It is positioned anterior to the esophagus as it extends into the thoracic cavity. A series of 16 to 20 C-shaped hyaline cartilages form the supporting walls of the trachea (fig. 17.10). These tracheal cartilages ensure that the airway will always remain open. The open part of each of these cartilages faces the esophagus and permits the esophagus to expand slightly into the trachea during swallowing. The mucosa (surface lining the lumen) consists of pseudostratified ciliated columnar epithelium containing numerous mucus-secreting **goblet cells** (see figs. 17.4 and 17.5). It provides the same protection against dust and other particles as the membrane lining the nasal cavity and larynx. Medial to the lungs, the trachea splits to form the right and left principal bronchi. This junction is reinforced by the **carina** (*kă-ri'nă*), a keel-like cartilage plate (see fig. 17.9).

trachea: L. *trachia*, rough air vessel

carina: L. *carina*, keel

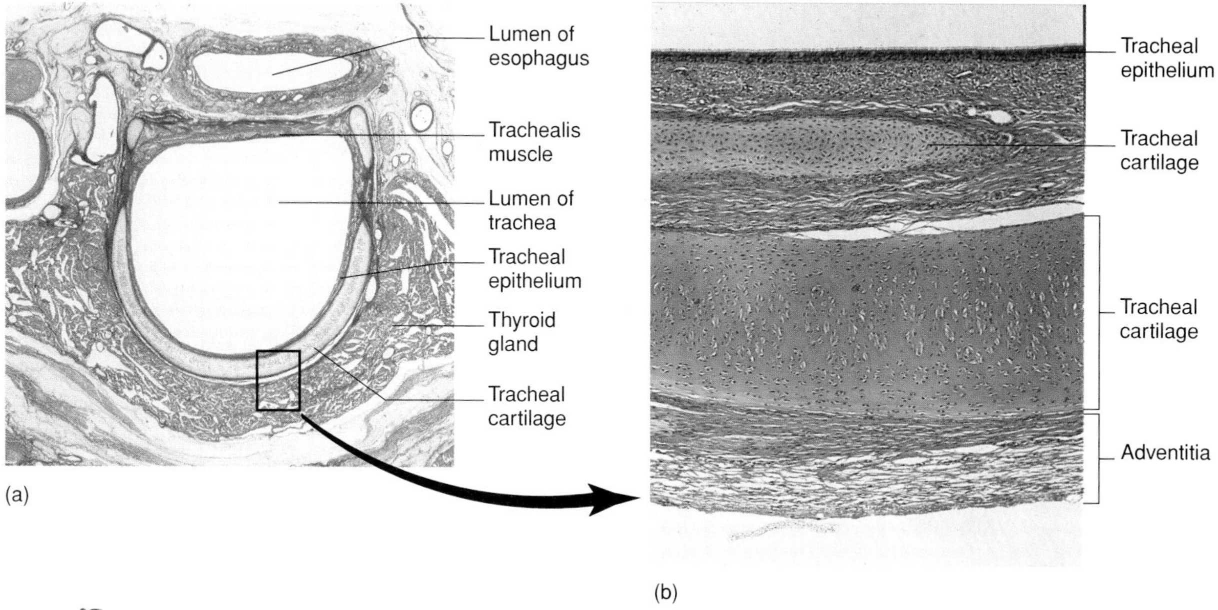

Figure 17.10

The histology of the trachea. (*a*) A photomicrograph showing the relationship of the trachea to the esophagus (3×) and (*b*) a photomicrograph of tracheal cartilage (63×).

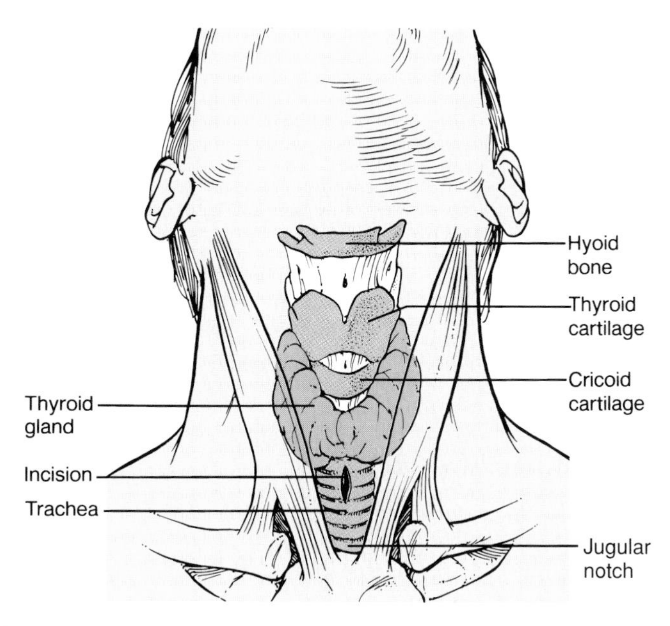

Figure 17.11

The site for a tracheostomy.

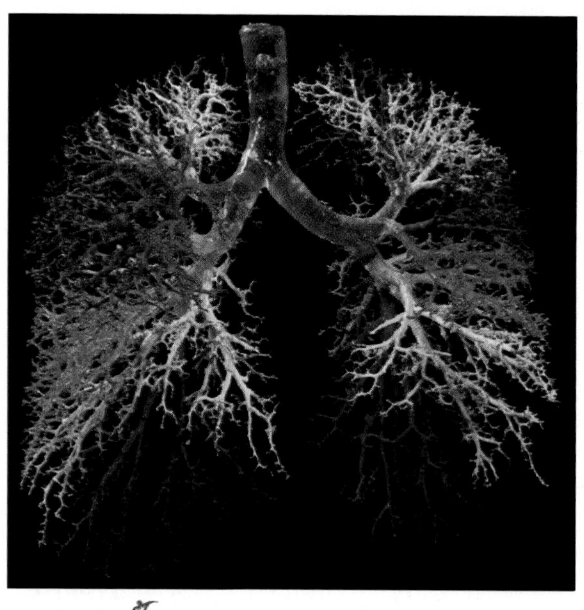

Figure 17.12

A photograph of a plastic cast of the conducting airways from the trachea to the terminal bronchioles.

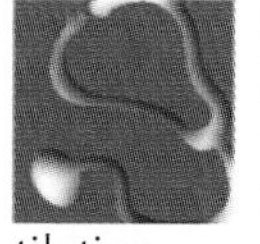

If the trachea becomes occluded through inflammation, excessive secretion, trauma, or aspiration of a foreign object, it may be necessary to create an emergency opening into this tube so that ventilation can still occur. A *tracheotomy* is the procedure of surgically opening the trachea, and a *tracheostomy* involves inserting a tube into the trachea to permit breathing and to keep the passageway open (fig. 17.11). A tracheotomy should be performed only by a competent physician as there is a great risk of cutting a recurrent laryngeal nerve or the common carotid artery.

Bronchial Tree

The **bronchial tree** is so named because it is composed of a series of respiratory tubes that branch into progressively narrower tubes as they extend into the lung (fig. 17.12). The trachea bifurcates into **right** and **left principal** (primary) **bronchi** at the level of the

bronchus: L. *bronchus*, windpipe

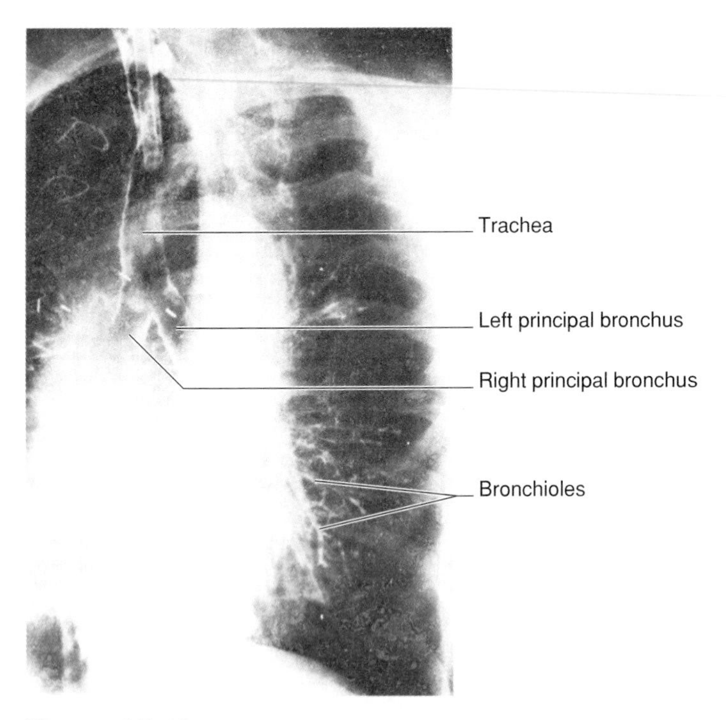

Figure 17.13

An anteroposterior bronchogram.

sternal angle behind the manubrium. Each principal bronchus has hyaline cartilage rings within its wall surrounding the lumen to keep it open as it extends into the lung. Because of the more vertical position of the right principal bronchus, foreign particles are more likely to lodge here than in the left principal bronchus.

The principal bronchus divides deeper in the lungs to form **lobar** (secondary) **bronchi** and **segmental** (tertiary) **bronchi** (see fig. 17.1). The bronchial tree continues to branch into even smaller tubules called **bronchioles** (*brong'ke-ōlz*). There is little cartilage in the bronchioles. The thick smooth muscle that encircles their lumina can constrict or dilate these airways. Bronchioles provide the greatest resistance to air flow in the conducting passages—a function analogous to that of arterioles in the circulatory system. A simple cuboidal epithelium lines the bronchioles rather than the pseudostratified columnar epithelium that lines the bronchi (see fig. 17.4). Numerous **terminal bronchioles** connect to **respiratory bronchioles** that lead into **alveolar ducts,** and then into **alveolar sacs** (see fig. 17.14). The conduction portion of the respiratory system ends at the terminal bronchioles, and the respiratory portion begins at the respiratory bronchioles.

Asthma is an infectious or allergenic condition that involves the bronchi. During an asthma attack, there is a spasm of the smooth muscles in the respiratory bronchioles. Because of an absence of cartilage at this level, the air passageways constrict.

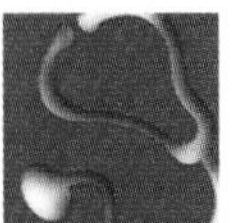

A fluoroscopic examination of the bronchi using a radiopaque medium for contrast is called *bronchography*. This technique enables the physician to visualize the bronchial tree on a bronchogram (fig. 17.13).

1. List in order the types of epithelia through which inspired air would pass in traveling through the nasal cavity to the alveolar sacs of the lungs. What is the function of each of these epithelia?
2. What are the functions of the nasal cavity?
3. Identify the structures that make up the nasal septum.
4. Describe the location of the nasopharynx and list the structures within this cavity.
5. List the paired and unpaired cartilages of the larynx and describe the functions of the larynx.
6. Describe the structure of the conducting airways from the trachea to the terminal bronchioles.

Pulmonary Alveoli, Lungs, and Pleurae

Pulmonary alveoli are the functional units of the lungs, where gas exchange occurs. Right and left lungs are separately contained in pleural membranes.

Objective 9 Describe the structure and function of the pulmonary alveoli.

Objective 10 Describe the surface anatomy of the lungs in relation to the thorax.

Objective 11 Discuss the structural arrangement of the thoracic serous membranes and explain the functions of these membranes.

Pulmonary Alveoli

The alveolar ducts open into **pulmonary alveoli** (*al-ve'ŏ-li*) as outpouchings along their length. Alveolar sacs are clusters of pulmonary alveoli (fig. 17.14). The alveolar ducts, pulmonary alveoli, and alveolar sacs make up the *respiratory division* of the lungs. Gas exchange occurs across the walls of the tiny pulmonary alveoli; hence, these minute expansions (0.25–0.50 mm in diameter) are the functional units of the respiratory system. The vast number of these structures (about 350 million per lung) provides a very large surface area (60–80 square meters, or 760 square feet) for the diffusion of gases. The diffusion rate is further increased by the fact that the wall of each pulmonary alveolus is only one cell layer thick, so that the total air-blood barrier is only one pulmonary alveolar cell with its basement membrane and one blood capillary cell across, or about 2 micrometers. This is an average distance because type II alveolar cells are thicker than type I alveolar cells

alveolus: L. diminutive of *alveus*, cavity

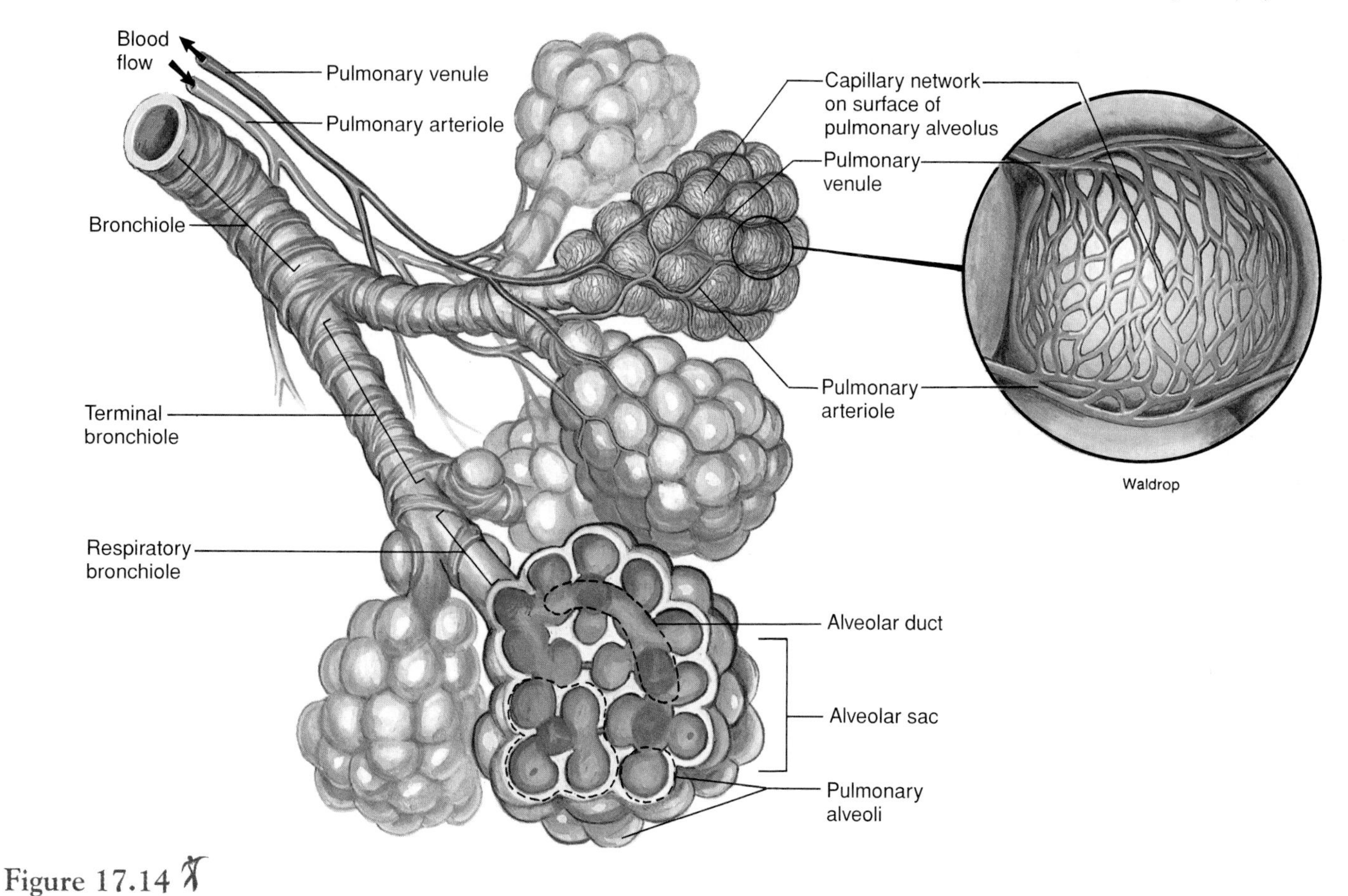

Figure 17.14

The respiratory division of the respiratory system. The respiratory tubes end in pulmonary alveoli, each of which is surrounded by an extensive pulmonary capillary network.

(fig. 17.15). Type I alveolar cells permit diffusion, and type II alveolar cells (septal cells) secrete a substance called *surfactant* that reduces the tendency for pulmonary alveoli to collapse.

Pulmonary alveoli are polyhedral in shape and are usually clustered together, like the units of a honeycomb, in the alveolar sacs at the ends of the alveolar ducts (fig. 17.16). Although the distance between each alveolar duct and its terminal pulmonary alveoli is only about 0.5 mm, these units together compose most of the mass of the lungs.

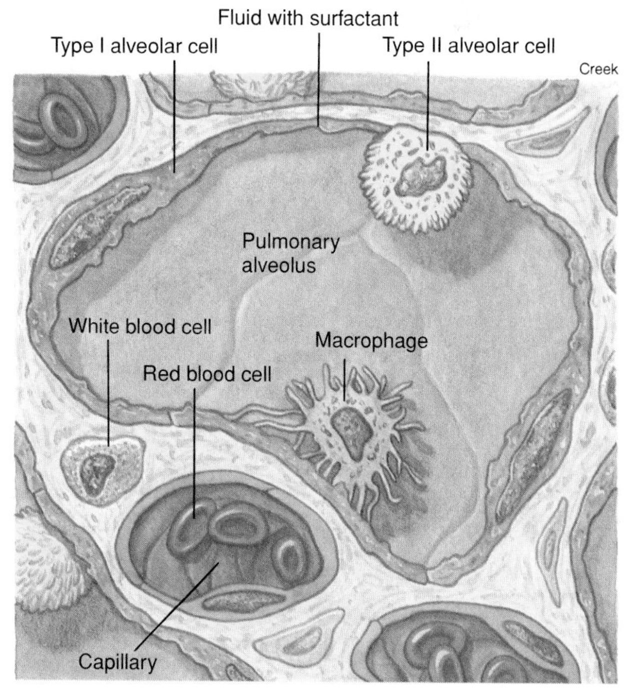

Figure 17.15

The relationship between a pulmonary alveolus and a pulmonary capillary.

Lungs

The large, spongy **lungs** are paired organs within the thoracic cavity (fig. 17.17). Each lung extends from the diaphragm to a point just above the clavicle, and its surfaces are bordered by the ribs to the front and back. The lungs are separated from one another by the heart and other structures of the **mediastinum** (*me″de-ă-sti′num*) which is the area between the lungs. All structures of the respiratory system beyond the principal bronchi, including the bronchial tree and the pulmonary alveoli, are contained within the lungs.

mediastinum: L. *mediastinus*, intermediate

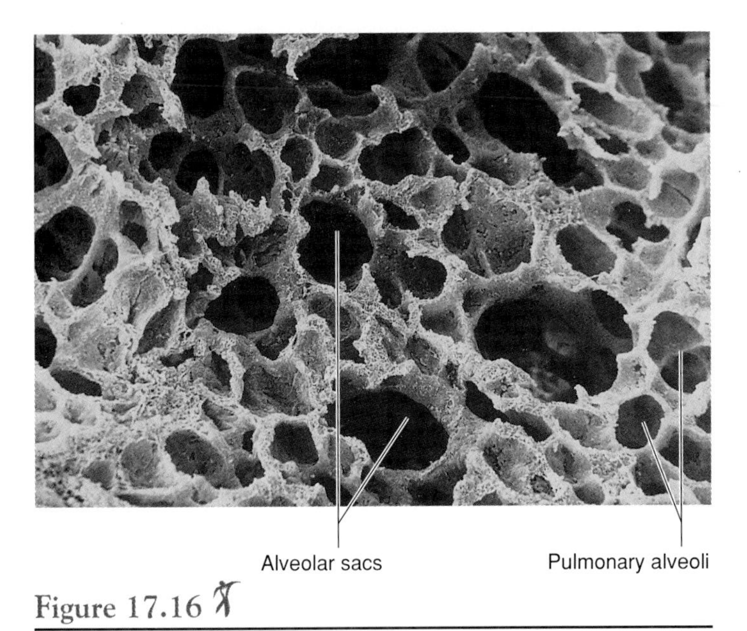

Figure 17.16

A scanning electron micrograph of lung tissue showing alveolar sacs and pulmonary alveoli.

Each lung has four surfaces that match the contour of the thoracic cavity. The **mediastinal** (medial) **surface** of the lung is slightly concave and contains a vertical slit, the **hilum** through which pulmonary vessels, nerves, and bronchi pass (fig. 17.18). The inferior surface of the lung, called the **base of the lung,** is concave as it fits over the convex dome of the diaphragm. The superior surface, called the **apex** (cupola) **of the lung,** extends above the level of the clavicle. Finally, the broad, rounded surface in contact with the membranes covering the ribs is called the **costal surface of the lung.**

Although the right and left lungs are basically similar, they are not identical. The left lung is somewhat smaller than the right and has a **cardiac impression** on its medial surface to accommodate the heart. The left lung is subdivided into a **superior lobe** and an **inferior lobe** by a single fissure. The right lung is subdivided by two fissures into three lobes: **superior, middle,** and **inferior lobes**

hilum: L. *hilum*, a trifle (little significance)
cupola: L. *cupula*, diminutive of *cupa*, dome or tub

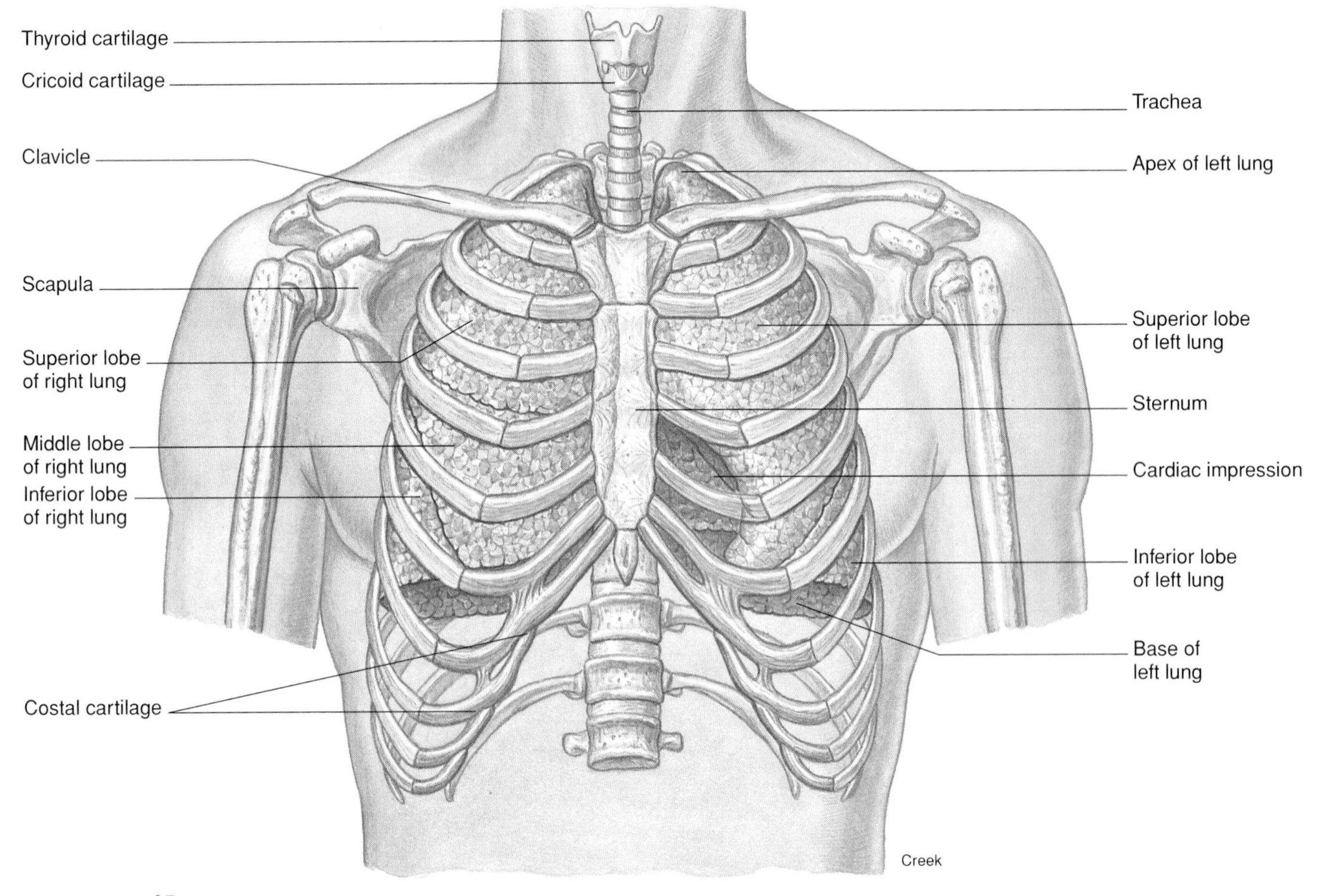

Figure 17.17

The position of the lungs within the rib cage.

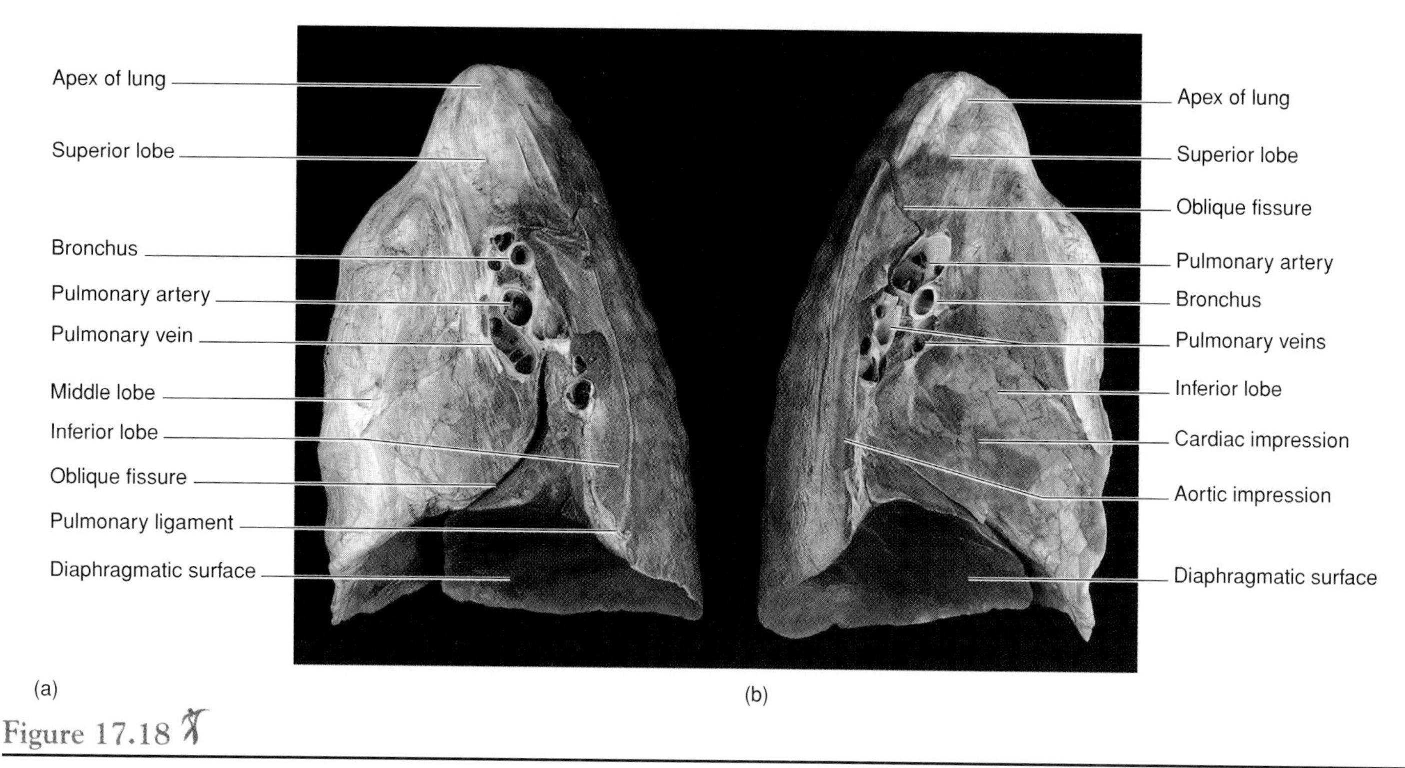

Figure 17.18

Lungs from a cadaver. (*a*) A medial view of the right lung and (*b*) a medial view of the left lung.

(see figs. 17.1, 17.17, and 17.18). Each lobe of the lung is divided into many small lobules, which in turn contain the pulmonary alveoli. Lobular divisions of the lungs make up specific **bronchial segments.** The right lung contains 10 bronchial segments and the left lung contains 8 (fig. 17.19).

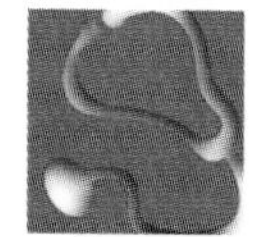

The lungs of a newborn are pink but may become discolored in an adult as a result of smoking or air pollution. Smoking not only discolors the lungs, it may also cause deterioration of the pulmonary alveoli. *Emphysema* (*em"fĭ-se'ma*) and *lung cancer* (see figs. 17.32 and 17.33) are diseases that are linked to smoking. If a person moves to a less polluted environment or gives up smoking, the lungs will get pinker and function more efficiently, unless they have been permanently damaged by disease.

Pleurae

Pleurae (*ploor'e*) are serous membranes surrounding the lungs and lining the thoracic cavity (figs. 17.20 and 17.21). The **visceral pleura** adheres to the outer surface of the lung and extends into each of the interlobar fissures. The **parietal pleura** lines the thoracic walls and the thoracic surface of the diaphragm. A continuation of the parietal pleura and between the lungs forms the boundary of the mediastinum. Between the visceral and parietal pleurae is the slitlike **pleural cavity.** It contains a lubricating fluid that allows the membranes to slide past each other easily during respiration. An inferiorly extending reflection of the pleural layers around the roots of each lung is called the **pulmonary ligament.** The pulmonary ligaments help to support the lungs.

The moistened serous membranes of the visceral and parietal pleurae are normally flush against each other like two wet pieces of glass, and therefore the lungs are stuck to the thoracic wall. The pleural cavity (intrapleural space) between the two moistened membranes contains only a thin layer of fluid secreted by the serous membranes. The pleural cavity in a healthy, living person is thus potential rather than real; it can become real only in abnormal situations when air enters the intrapleural space. Since the lungs normally remain in contact with the thoracic wall, they get larger and smaller along with the thoracic cavity during respiratory movements.

The thoracic cavity has four distinct compartments: a pleural cavity surrounds each lung; the pericardial cavity surrounds the heart; and the mediastinum contains the esophagus, thoracic duct, major vessels, various nerves, and portions of the respiratory tract. This *compartmentalization* has protective value in that infections are usually confined to one compartment. Also, damage to one organ usually will not involve another. For example, *pleurisy*, an inflamed pleura, is generally confined to one side; and a penetrating injury to the thoracic cavity, such as a knife wound, may cause one lung to collapse but not the other.

---

pleura: Gk. *pleura*, side or rib

pulmonary: Gk. *pleumon*, lung

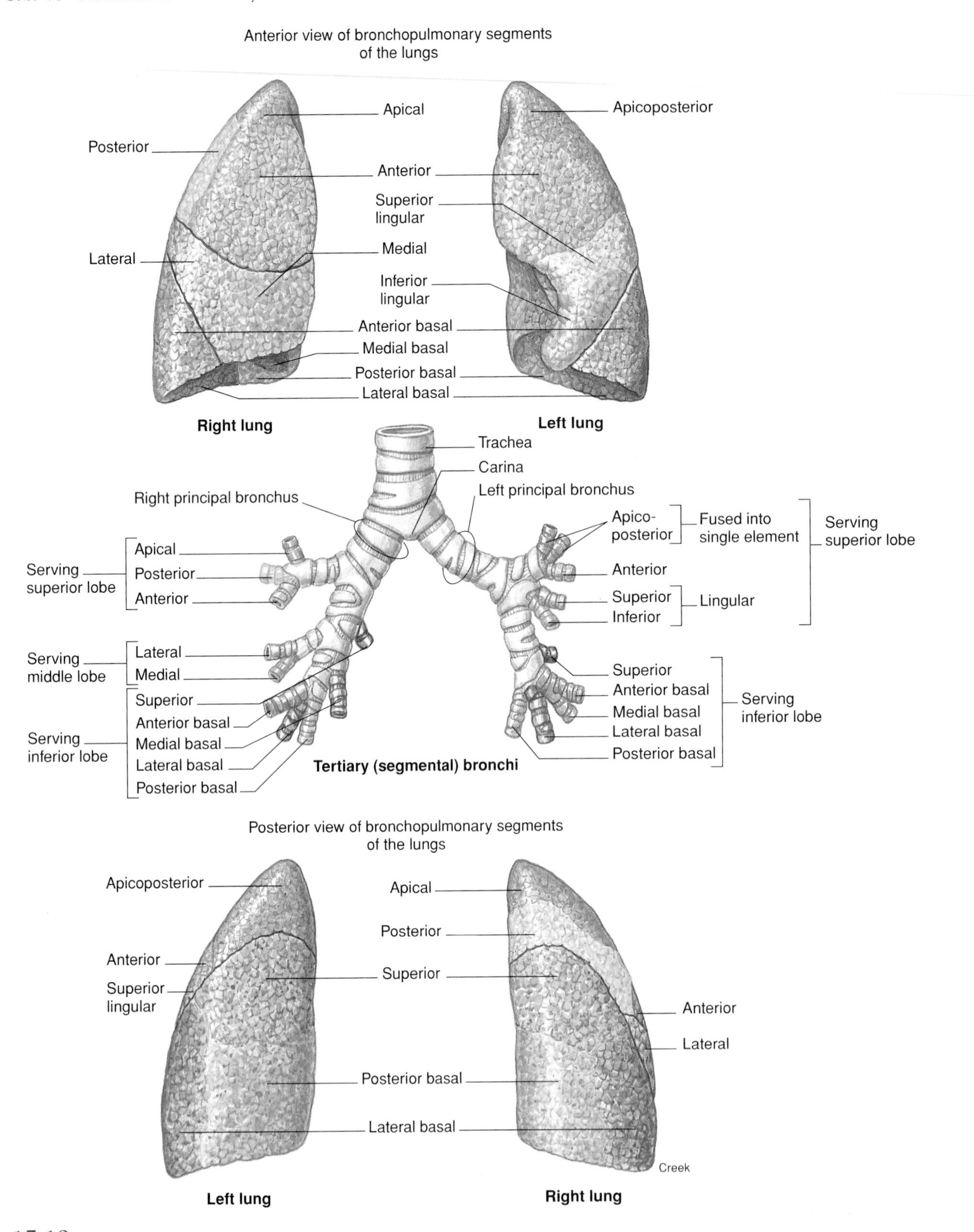

Figure 17.19

Lobes, lobules, and bronchopulmonary segments of the lungs.

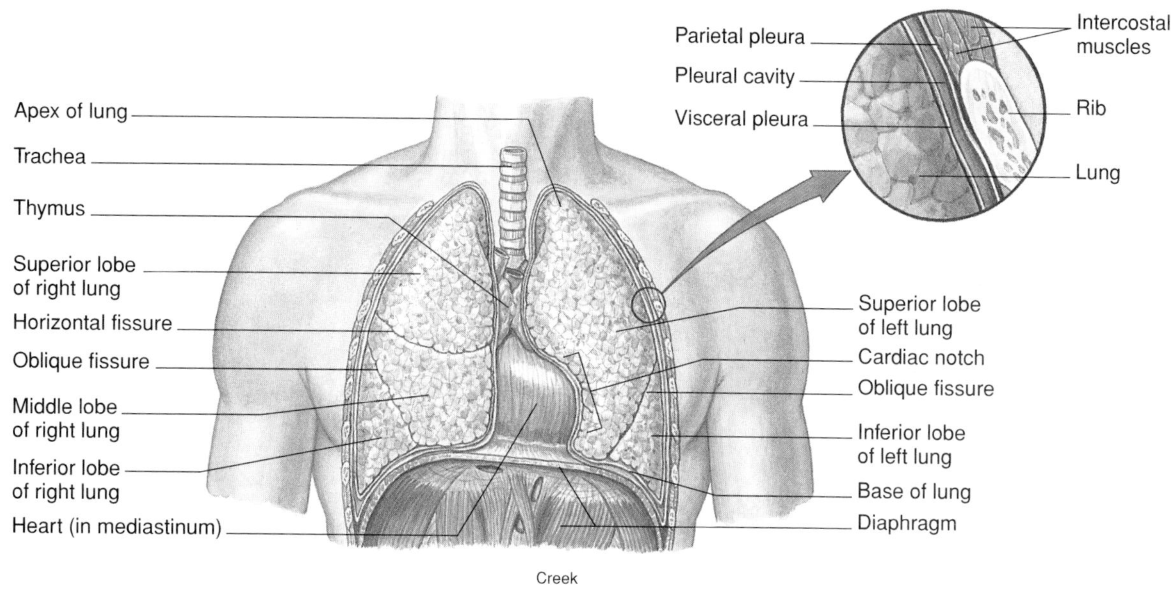

Figure 17.20

The position of the lungs and associated pleurae.

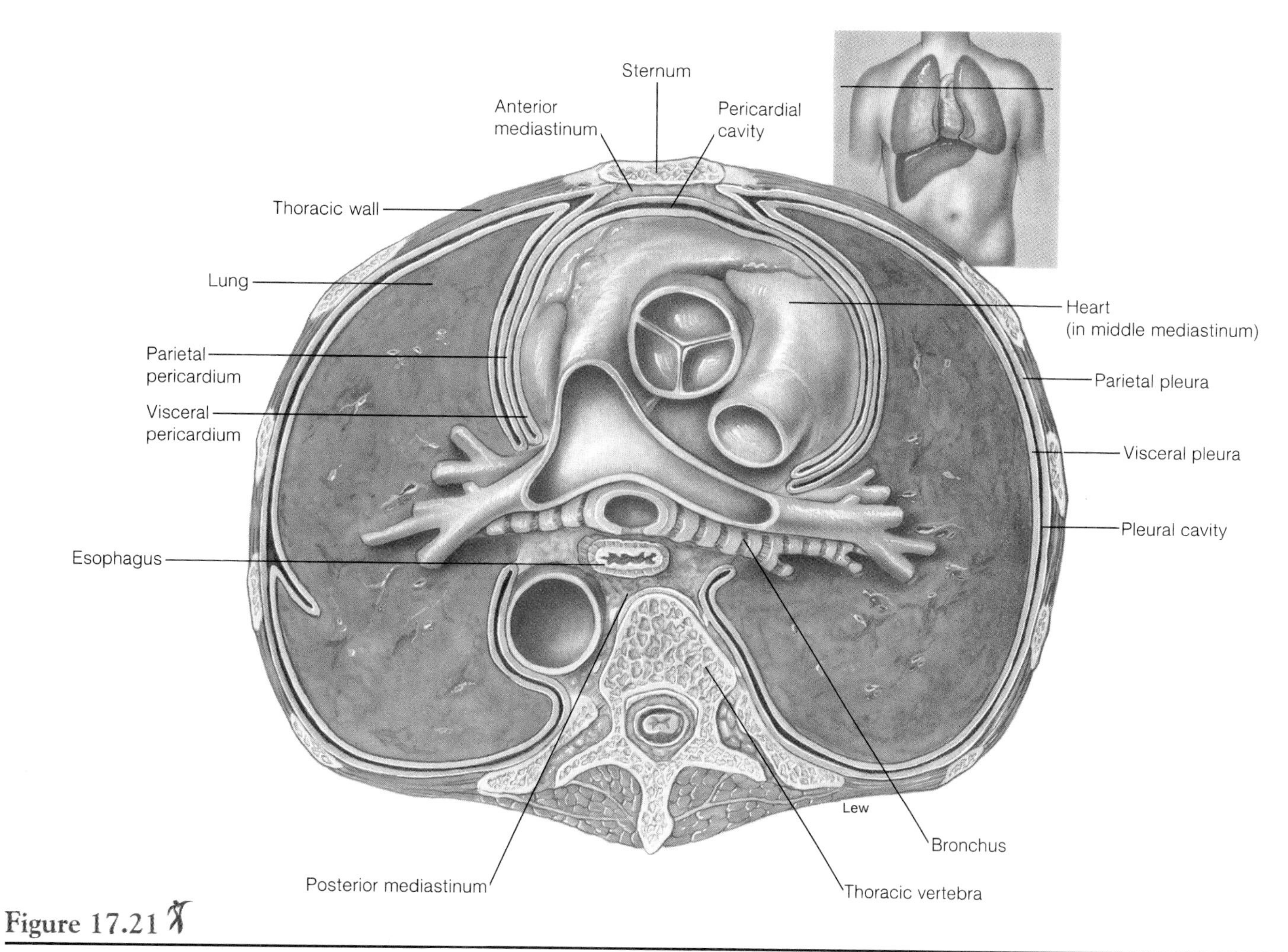

Figure 17.21

A cross section of the thoracic cavity showing the mediastinum and pleural membranes.

Table 17.1
Major Structures of the Respiratory System

Structure	Description	Function
Nose	Primary passageway for air entering the respiratory system; consists of jutting external portion and internal nasal cavity	Warms, moistens, and filters inhaled air as it is conducted to the pharynx
Paranasal sinuses	Air spaces in the maxillary, frontal, sphenoid, and ethmoid bones	Produce mucus; provide sound resonance; lighten the skull
Pharynx	Chamber connecting oral and nasal cavities to the larynx	Serves as passageway for air entering the larynx and for food entering the esophagus
Larynx	Voice box; short passageway that connects the pharynx to the trachea	Serves as passageway for air; produces sound; prevents foreign materials from entering the trachea
Trachea	Flexible tubular connection between the larynx and bronchial tree	Serves as passageway for air; pseudostratified ciliated columnar epithelium cleanses the air
Bronchial tree	Bronchi and branching bronchioles in the lung; tubular connection between the trachea and pulmonary alveoli	Serves as passageway for air; continued cleansing of air
Pulmonary alveoli	Microscopic membranous air sacs within the lungs	Functional units of respiration; site of gaseous exchange between the respiratory and circulatory systems
Lungs	Major organs of the respiratory system; located in the thoracic cavity and surrounded by pleural cavities	Contain bronchial trees, pulmonary alveoli, and associated pulmonary vessels
Pleurae	Serous membranes covering the lungs and lining the thoracic cavity	Compartmentalize, protect, and lubricate the lungs

A summary of the major structures of the respiratory system is presented in table 17.1.

1. Describe the structure of the respiratory division of the lungs and explain how this structure aids in gas exchange.
2. Compare the structure of the right and left lungs.
3. Describe the location of the mediastinum and list the organs it contains.
4. Describe the arrangement of the visceral pleura, parietal pleura, and pleural cavity. Comment on the functional significance of the compartmentalization of the thoracic cavity.

Mechanics of Breathing

Normal quiet inspiration is achieved by muscle contraction, and quiet expiration results from muscle relaxation and elastic recoil. A deeper inspiration and expiration can be forced by contractions of the accessory respiratory muscles. The amount of air inspired and expired can be measured to test pulmonary function.

Objective 12 Identify and describe the actions of the muscles involved in both quiet and forced inspiration.

Objective 13 Describe how quiet expiration occurs and identify and describe the actions of the muscles involved in forced expiration.

Objective 14 List the various lung volumes and capacities and describe how pulmonary function tests help in the diagnosis of lung disorders.

Breathing, or **pulmonary ventilation,** requires that the thorax be flexible in order to function as a bellows during the ventilation cycle. Breathing consists of two phases, *inspiration* and *expiration*. Inspiration (inhaling) and expiration (exhaling) are accomplished by alternately increasing and decreasing the volume of the thoracic cavity (fig. 17.22). Breathing in takes place when the air pressure within the lungs is lower than the atmospheric pressure; breathing out takes place when the air pressure within the lungs is greater than the atmospheric pressure.

Pressure gradients change as the size of the thoracic cavity changes. Not only must the thorax be flexible, it also must be sufficiently rigid to protect the vital organs it contains. In addition, it must provide extensive attachment surfaces for many short, powerful muscles. These requirements are met through the structure and composition of the rib cage. The rib cage is pliable because the ribs are separated from one another and because most ribs (upper 10 of the 12 pairs) are attached to the sternum by resilient costal cartilage. The vertebral attachment likewise allows for considerable mobility. The structure of the rib cage and associated cartilage provides continuous elastic tension, so that an expanded thorax will return passively to its resting position when relaxed.

Inspiration

The overall size of the thoracic cavity increases with inspiration (fig. 17.23). During relaxed inspiration, the muscles of importance

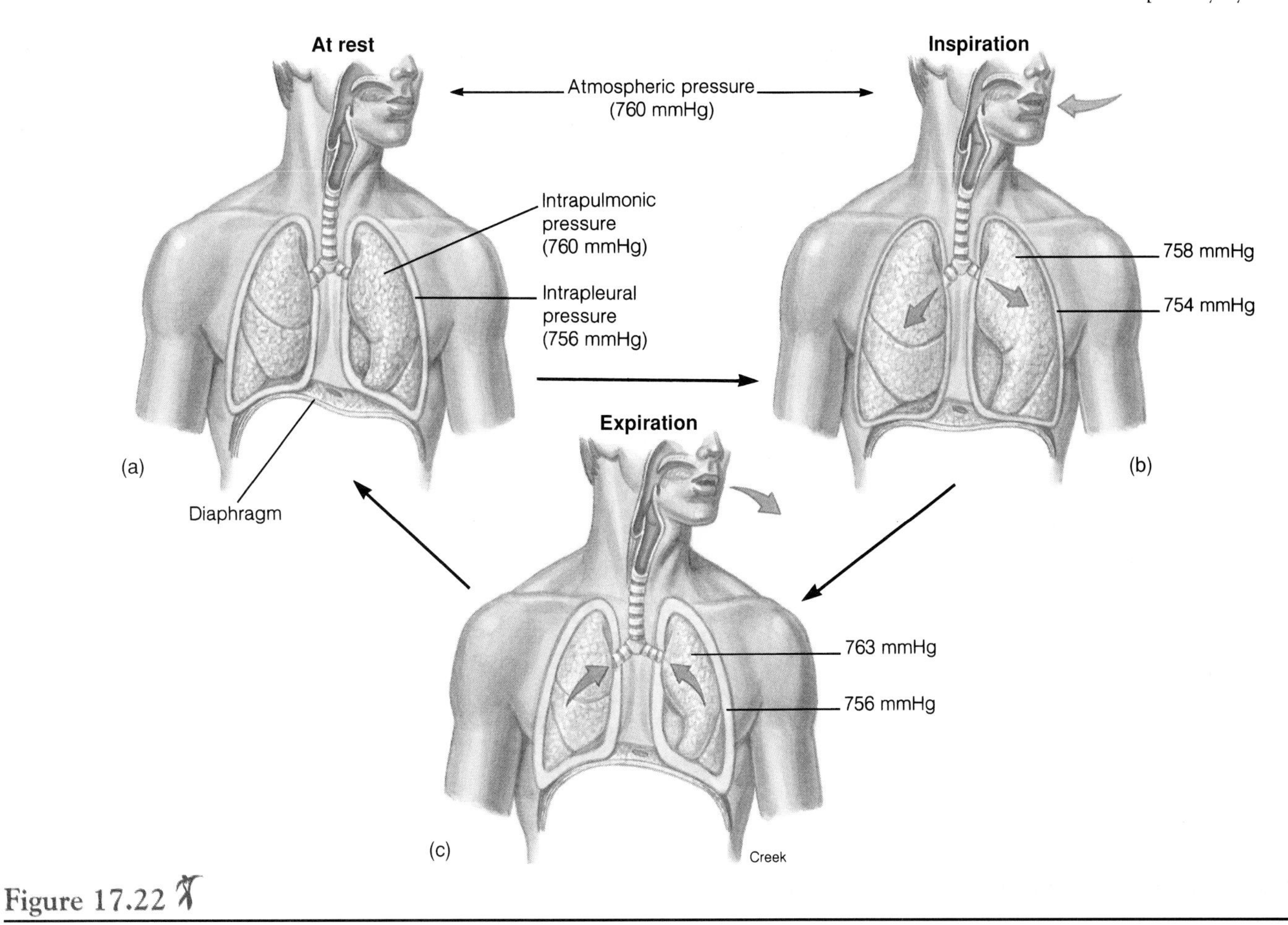

Figure 17.22

The mechanics of pulmonary ventilation. At rest (*a*), the atmospheric pressure at sea level and within the plural cavities is 760 mmHg. During inspiration (*b*), the diaphragm contracts, causing a decrease in intrapleural pressure and consequent inflation of the lungs. During expiration, the diaphragm recoils, causing an increase in intrapleural pressure and consequent deflation of the lungs.

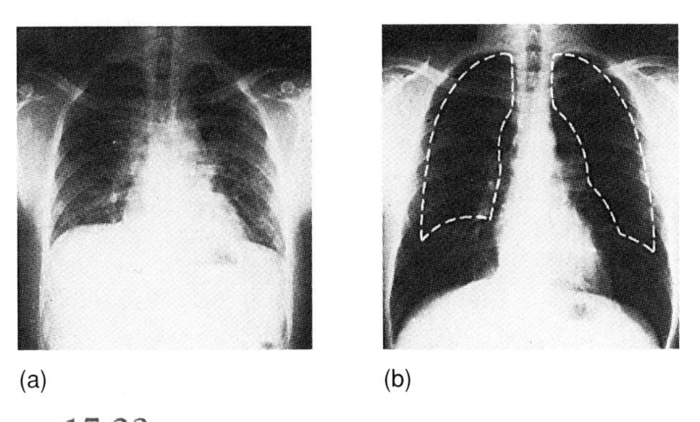

Figure 17.23

A change in lung volume, as shown by radiographs, during expiration (*a*) and inspiration (*b*). The increase in lung volume during full inspiration is shown by comparison with the lung volume in full expiration (dashed lines).

are the *diaphragm*, the *external intercostal muscles*, and the interchondral portion of the *internal intercostal muscles* (fig. 17.24). Contraction of the dome-shaped diaphragm causes it to flatten, lowering its dome. This increases the vertical dimension of the thoracic cavity. A simultaneous contraction of the external intercostal muscles and interchondral portion of the internal intercostal muscles increases the diameter of the thorax.

The *scalenes* and *sternocleidomastoid muscles* are involved in deep inspiration or forced breathing. When these muscles are contracted, the ribs are elevated. At the same time, the upper rib cage is stabilized so that the intercostal muscles become more effective.

The expanded thoracic cavity decreases the air pressure within the pleural cavities to below that of the atmosphere. It is this pressure difference that causes the lungs to become inflated.

Expiration

For the most part, expiration is a passive process that occurs as the muscles of inspiration are relaxed and the rib cage returns to its original position. The lungs recoil during expiration as elastic fibers within the lung tissue shorten and the pulmonary alveoli draw together. Lowering of the surface tension in the pulmonary alveoli, which brings on recoiling, is due to a lipoprotein substance called *surfactant* produced by type II alveolar cells. Surfactant is extremely important not only in causing the pulmonary alveoli to recoil during expiration, but also in equalizing the overall surface tension as the pulmonary alveoli expand and contract. A deficiency in

diaphragm: Gk. *dia*, across; *phragma*, fence

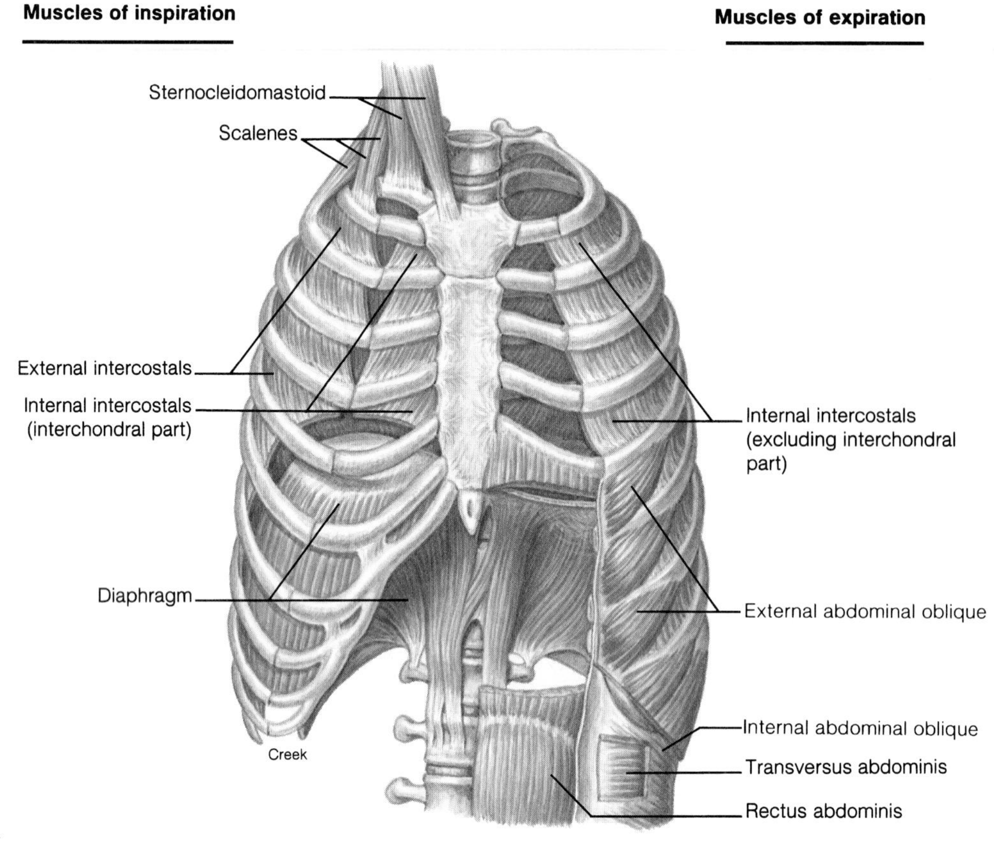

Figure 17.24

The muscles of respiration. The principal muscles of inspiration are shown on the right side of the trunk and the principal muscles of forced expiration are shown on the left side. For the most part, expiration is passive.

surfactant in premature infants can cause *respiratory distress syndrome* (*RDS*) or, as it is commonly called, *hyaline membrane disease*.

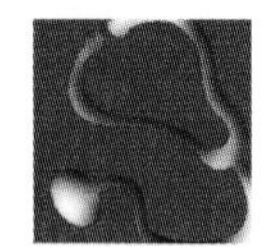

Even under normal conditions, the first breath of life is a difficult one because the newborn must overcome large surface tension forces in order to inflate its partially collapsed pulmonary alveoli. The transpulmonary pressure required for the first breath is 15 to 20 times that required for subsequent breaths, and an infant with *respiratory distress syndrome* must duplicate this effort with every breath. Fortunately, many babies with this condition can be saved by mechanical ventilators that keep them alive long enough for their lungs to mature and manufacture sufficient surfactant.

During forced expiration, such as coughing or sneezing, contraction of the interosseous portion of the internal intercostal muscles causes the rib cage to be depressed. The *abdominal muscles* may also aid expiration because, when contracted, they force abdominal organs up against the diaphragm and further decrease the volume of the thorax. Thus, intrapulmonary pressure can rise to 20 or 30 mmHg above the atmospheric pressure.

The events that occur during inspiration and expiration are summarized in table 17.2.

Respiratory Volumes and Capacities

The respiratory system is somewhat inefficient because the air enters and exits at the same place, through either the nose or the mouth. Consequently, there is an incomplete exchange of gas during each ventilatory cycle, and approximately five-sixths of the air present in the lungs still remains when the next inspiration begins.

The amount of air breathed in a given time and the degree of difficulty in breathing are important indicators of a person's respiratory status. The amount of air exchanged during pulmonary ventilation varies from person to person according to age, gender, activity level, general health, and individual differences. Respiratory volumes are measured with a *spirometer* (fig. 17.25). Any ventilatory abnormalities can then be compared to what is accepted as normal. The normal adult respiratory volumes and capacities are presented in table 17.3 and figure 17.26.

Table 17.2

Pulmonary Ventilation: Events of Inspiration and Expiration

Nerve Stimulus	Event
Inspiration	
Phrenic nerves	The diaphragm contracts, moving inferiorly, which measures the volume of the thorax. The diaphragm is the principal muscle involved in quiet inspiration.
Intercostal nerves	Contraction of the external intercostal muscles and the interchondral portion of the internal intercostal muscles elevates the ribs, thus increasing the capacity of the thoracic cavity.
Accessory, cervical, and thoracic nerves	Forced inspiration is accomplished through contraction of the scalenes and sternocleidomastoid muscles, which increases the dimension of the thoracic cavity anteroposteriorly. Pulmonary ventilation during forced inspiration usually occurs through the mouth rather than through the nose.
	As the dimension of the thoracic cavity increases, the pressure within the pleural cavities decreases; the lungs inflate because the atmospheric pressure is greater than the intraplural pressure.
Expiration	
	Nerve stimuli to the inspiratory muscles cease and the muscles relax.
	The rib cage and lungs recoil as air is forced out of the lungs because of the increased pressure.
Intercostal and lower spinal nerves	Forced expiration occurs when the interosseus portion of the internal intercostal and abdominal muscles are contracted.

*Some of the events during inspiration and expiration may occur simultaneously.

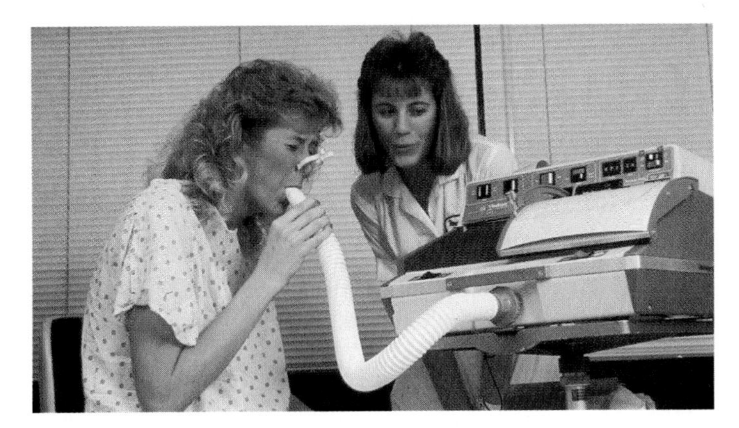

Figure 17.25

A spirometer. With the exception of the residual volume, which is measured using special techniques, this instrument can determine respiratory air volumes.

Table 17.3

Respiratory Volumes and Capacities of Healthy Adult Males

Volume	Quantity of Air	Description
Tidal volume (TV)	500 cc	Volume moved in or out of the lungs during quiet breathing
Inspiratory reserve volume (IRV)	3,000 cc	Volume that can be inhaled during forced breathing in addition to tidal volume
Expiratory reserve volume (ERV)	1,000 cc	Volume that can be exhaled during forced breathing in addition to tidal volume
Vital capacity (VC)	4,500 cc	Maximum amount of air that can be exhaled after taking the deepest breath possible: VC = TV + IRV + ERV
Residual volume (RV)	1,500 cc	Volume that cannot be exhaled
Total lung capacity (TLC)	6,000 cc	Total volume of air that the lungs can hold: TLC = VC + RV

From John W. Hole, Jr., *Human Anatomy and Physiology*, 6th ed.

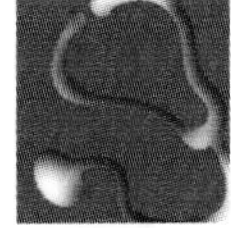

People with pulmonary disorders frequently complain of *dyspnea* (*disp'ne-ă*), a subjective feeling of shortness of breath. Dyspnea may occur even when ventilation is normal, however, and may not occur even during exercise, when the total volume of air movement is very high. Some of the terms used to describe ventilation are defined in table 17.4.

Nonrespiratory Air Movements

Air movements through the respiratory system that are not associated with pulmonary ventilation are termed *nonrespiratory movements*. Such movements accompany emotional displays such as laughing, sighing, crying, or yawning, or they may function to expel foreign matter from the respiratory tract, as in coughing and sneezing. Nonrespiratory movements are generally reflexive. Some of them, however, can be voluntarily initiated. These types of air movements and the reflexive mechanisms involved are summarized in table 17.5.

dyspnea: Gk. *dys*, bad; *pnoe*, breathing

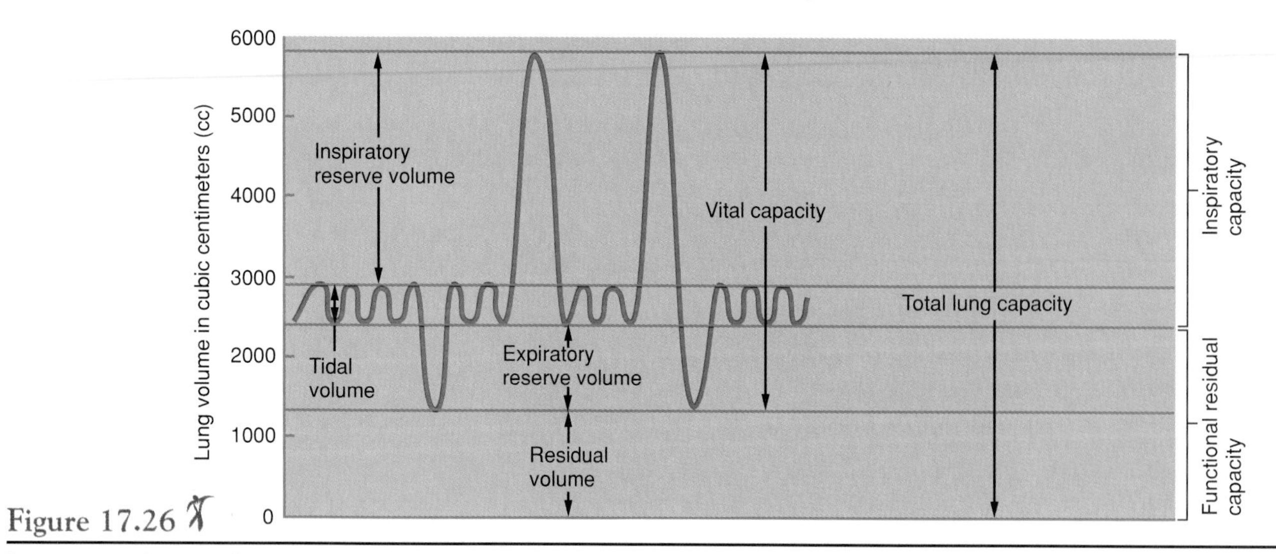

Figure 17.26

Respiratory volumes and capacities.

Table 17.4
Ventilation Terminology

Term	Definition
Air spaces	Alveolar ducts, alveolar sacs, and pulmonary alveoli
Airways	Structures that conduct air from the mouth and nose to the respiratory bronchioles
Alveolar ventilation	Removal and replacement of gas in pulmonary alveoli; equal to the tidal volume minus the volume of dead space times the breathing rate
Anatomical dead space	Volume of the conducting airways to the zone where gas exchange occurs
Apnea	Cessation of breathing
Dyspnea	Unpleasant subjective feeling of difficult or labored breathing
Eupnea	Normal, comfortable breathing at rest
Hyperventilation	Abnormally rapid, deep breathing; results in abnormally low alveolar CO_2
Hypoventilation	Abnormally slow, shallow breathing; results in abnormally high alveolar CO_2
Orthopnea	Inability to breathe comfortably while lying down
Physiological dead space	Combination of anatomical dead space and underventilated pulmonary alveoli that do not contribute normally to blood-gas exchange
Pneumothorax	Presence of gas in the pleural cavity (the space between the visceral and parietal pleural membranes) that may cause lung collapse

Table 17.5
Nonrespiratory Air Movements

Air Movement	Mechanism	Comments
Coughing	Deep inspiration followed by a closure of the glottis. The forceful expiration that results abruptly opens the glottis, sending a blast of air through the upper respiratory tract.	Reflexive or voluntary. Stimulus may be foreign material irritating the larynx or trachea.
Sneezing	Similar to a cough, except that the forceful expired air is directed primarily through the nasal cavity. The eyelids close reflexively during a sneeze.	Reflexive response to irritating stimulus of the nasal mucosa. Sneezing clears the upper respiratory passages.
Sighing	Deep, prolonged inspiration followed by a rapid, forceful expiration.	Reflexive or voluntary, usually in response to boredom or sadness.
Yawning	Deep inspiration through a widely opened mouth. The inspired air is usually held for a short period before sudden expiration.	Usually reflexive in response to drowsiness, fatigue, or boredom, but exact stimulus-receptor cause is unknown.
Laughing	Deep inspiration followed by a rapid convulsive expiration. Air movements are accompanied by expressive facial distortions.	Reflexive; may be voluntary to express emotions.
Crying	Similar to laughing, but the glottis remains open during entire expiration and different facial muscles are involved.	Somewhat reflexive but under voluntary control.
Hiccuping	Spasmodic contraction of the diaphragm while the glottis is closed, producing a sharp inspiratory sound.	Reflexive; serves no known function.

1. Describe the actions of the diaphragm and intercostal muscles during relaxed inspiration.
2. Describe how forced inspiration and forced expiration are produced.
3. Define the terms *tidal volume* and *vital capacity*.
4. Indicate the respiratory volumes being used during a sneeze, a deep inspiration prior to jumping into a swimming pool, maximum ventilation while running, and quiet breathing while sleeping.

Regulation of Breathing

The rhythm of breathing is controlled by centers in the brain stem. These centers are influenced by higher brain function and regulated by sensory input that makes breathing responsive to the changing respiratory needs of the body.

Objective 15 Describe the functions of the pneumotaxic, apneustic, and rhythmicity centers in the brain stem.

Objective 16 Identify the chemoreceptors and describe their pathway of innervation.

Pulmonary ventilation is primarily an involuntary, rhythmic action so effective that it continues to function even when a person is unconscious. In order for the neural control center of respiration to function effectively, it must possess monitoring, stimulating, and inhibiting properties so that the body can respond appropriately to increased or decreased metabolic needs. In addition, the center must be connected to the cerebrum to receive the voluntary impulses from a person who wants to change the rate of respiration.

The three **respiratory centers** of the brain are the rhythmicity, apneustic, and pneumotaxic areas (see fig. 11.29). The **rhythmicity area,** located in the medulla oblongata, contains two aggregations of nerve cell bodies that form the **inspiratory** and **expiratory portions.** Nerve impulses from the inspiratory portion travel through the phrenic and intercostal nerves and stimulate the diaphragm and intercostal muscles. Impulses from the expiratory portion stimulate the muscles of expiration. These two portions act reciprocally; that is, when one is stimulated, the other is inhibited. Stretch receptors in the visceral pleura of the lungs provide feedback through the vagus nerves to stimulate the expiratory portion.

The **apneustic** (*ap-noo'stik*) and **pneumotaxic** (*noo"mŏ-tak'sik*) **areas** are located in the pons. These areas influence the activity of the rhythmicity area. The apneustic center promotes inspiration and the pneumotaxic center inhibits the activity of inspiratory neurons.

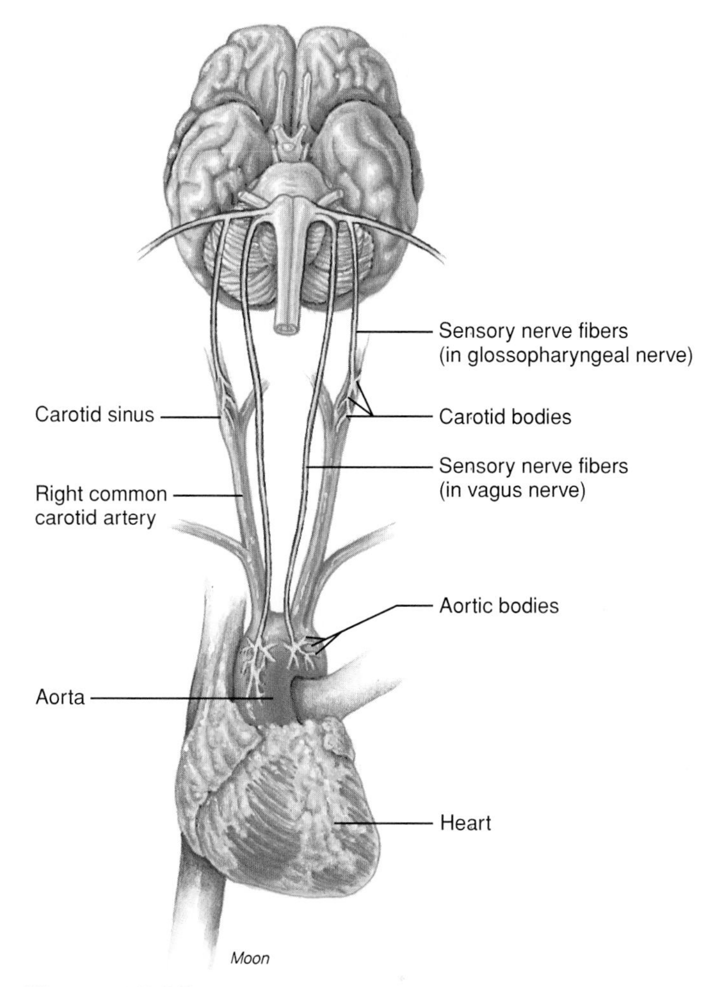

Figure 17.27

The peripheral chemoreceptors (aortic and carotid bodies) regulate the brain stem respiratory centers by means of sensory nerve stimulation.

The brain stem respiratory centers produce rhythmic breathing even in the absence of other neural input. This intrinsic respiratory pattern, however, is modified by input from higher brain centers and by input from receptors sensitive to the chemical composition of the blood. The influence of higher brain centers is evidenced by the fact that you can voluntarily hypoventilate (as in breath holding) or hyperventilate. The inability to hold your breath for more than a short period is due to reflex breathing in response to input from the chemoreceptors.

Two groups of chemoreceptors respond to changes in blood chemistry. These are the *central chemoreceptors* in the medulla oblongata and the *peripheral chemoreceptors*. The peripheral chemoreceptors include the **aortic bodies,** located in the aortic arch, and the **carotid bodies,** located at the junctions of the internal and external carotid arteries (fig. 17.27). These peripheral chemoreceptors control breathing indirectly via sensory neurons to the medulla oblongata. The aortic bodies send sensory information to the medulla oblongata in the vagus nerves; the carotid bodies stimulate sensory fibers in the glossopharyngeal nerve.

Developmental Exposition

The Respiratory System

Explanation

The development of the respiratory system is initiated early in embryonic development and involves both ectoderm and endoderm. Although all of the structures of the respiratory system develop simultaneously, we will consider the upper and lower systems separately because of the different germ layers involved.

Development of the Upper Respiratory System

Cephalization (*sef″ă-lĭ-za′shun*) is the evolutionary tendency toward structural and functional differentiation of the cephalic, or head, end of an organism from the rest of the body. In humans, cephalization is apparent early in development. One of the initial events is the formation of the nasal cavity at 3½ to 4 weeks of embryonic life. A region of thickened ectoderm called the **olfactory** (nasal) **placode** (*plak′ōd*) appears on the front inferior part of the head (exhibit I). The placode invaginates to form the **olfactory pit,** which extends posteriorly to connect with the **foregut.** The foregut, derived of endoderm, later develops into the pharynx.

The mouth, or oral cavity, develops at the same time as the nasal cavity, and for a short time there is a thin **oronasal membrane** separating the two cavities. This membrane ruptures during the seventh week, and a single large **oronasal cavity** forms. Shortly thereafter, tissue plates of mesoderm begin to grow horizontally across the cavity. At approximately the same time, a vertical plate develops inferiorly from the roof of the nasal cavity. These plates have completed their formation by 3 months of development. The vertical plate forms the nasal septum, and the horizontal plates form the hard palate.

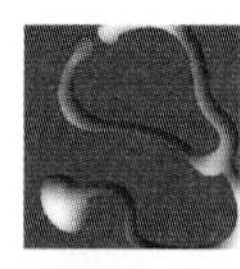

A *cleft palate* forms when the horizontal plates fail to meet in the midline. This condition can be corrected surgically (see fig. 17.28). The more immediate and serious problem facing an infant with a cleft palate is that it may be unable to create enough suction to nurse properly.

Development of the Lower Respiratory System

The lower respiratory system begins as a diverticulum, or outpouching, from the ventral surface of endoderm along the lower pharyngeal region (exhibit II). This diverticulum, which forms during the fourth week of development, is referred to as the **laryngotracheal** (*lă-ring″go-tra′ke-al*) **bud.** As the bud grows, the proximal portion forms the trachea and the distal portion bifurcates (splits) into a right and left principal bronchus.

The buds continue to elongate and split until the entire tubular network within the lower respiratory tract is formed (exhibit II). As the terminal portion forms air sacs, called **pulmonary alveoli,** at about 8 weeks of development, the supporting lung tissue begins to form. The complete structure of the lungs, however, is not fully developed until about 26 weeks of fetal development. Premature infants born prior to this time therefore require special artificial respiratory equipment to live.

---

cephalization: Gk. *kephale,* head

Exhibit I The development of the upper respiratory system. (*a*) An anterior view of the developing head of an embryo at 4 weeks showing the position of a transverse cut depicted in (a_1), (a_2), and (a_3). (a_2) Development at 5 weeks and (a_3) at 5½ weeks. (*b*) An anterior view of the developing head of an embryo at 6 weeks showing the position of a sagittal cut depicted in (b_1), (b_2), (b_3), and (b_4) at 14 weeks.

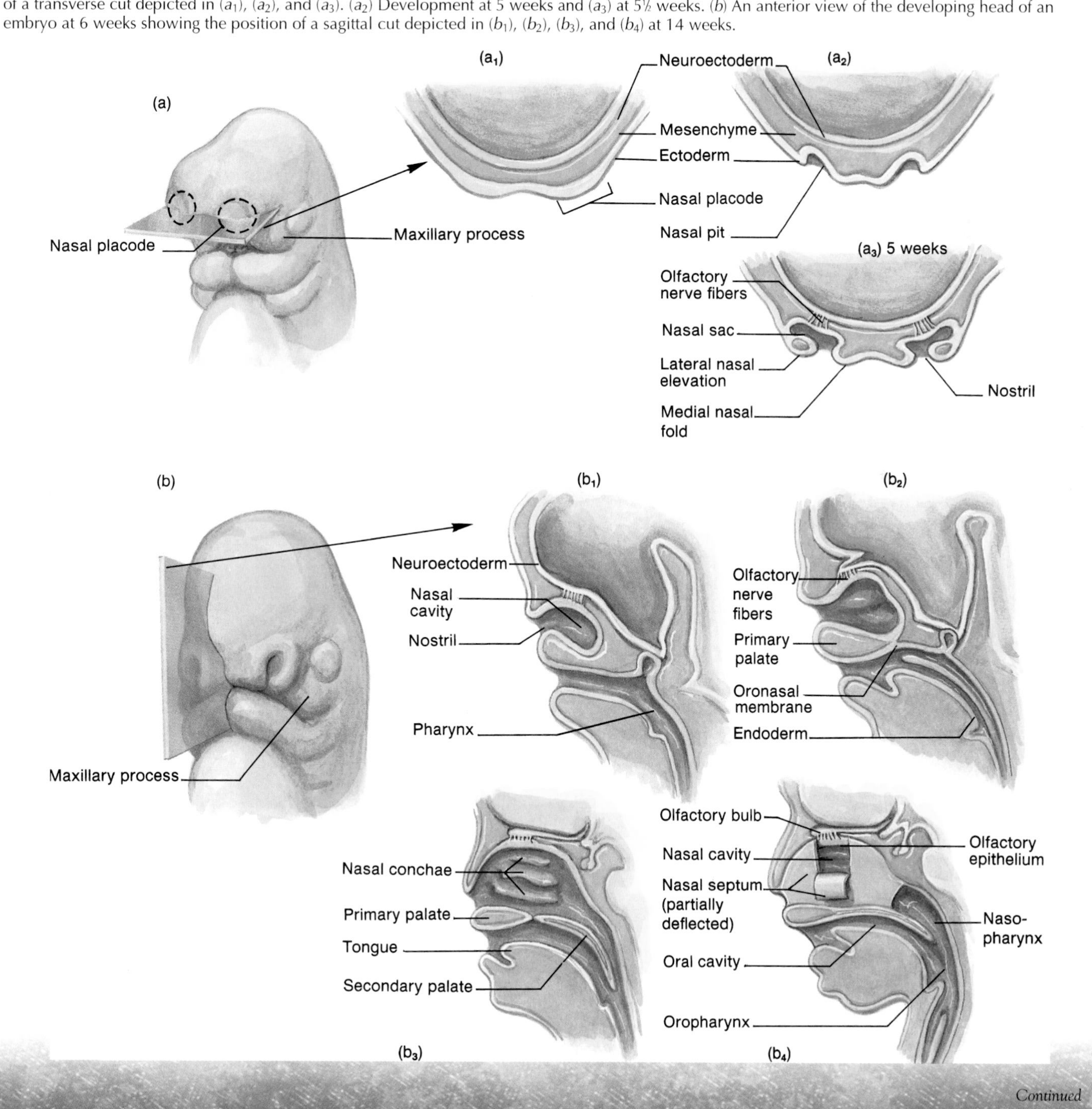

Continued

DEVELOPMENTAL EXPOSITION

Continued

Exhibit II The development of the lower respiratory tract. Developments in (*a*) and (*b*) occur during the first month. Those in (*c*), (*d*), and (*e*) occur during the second month.

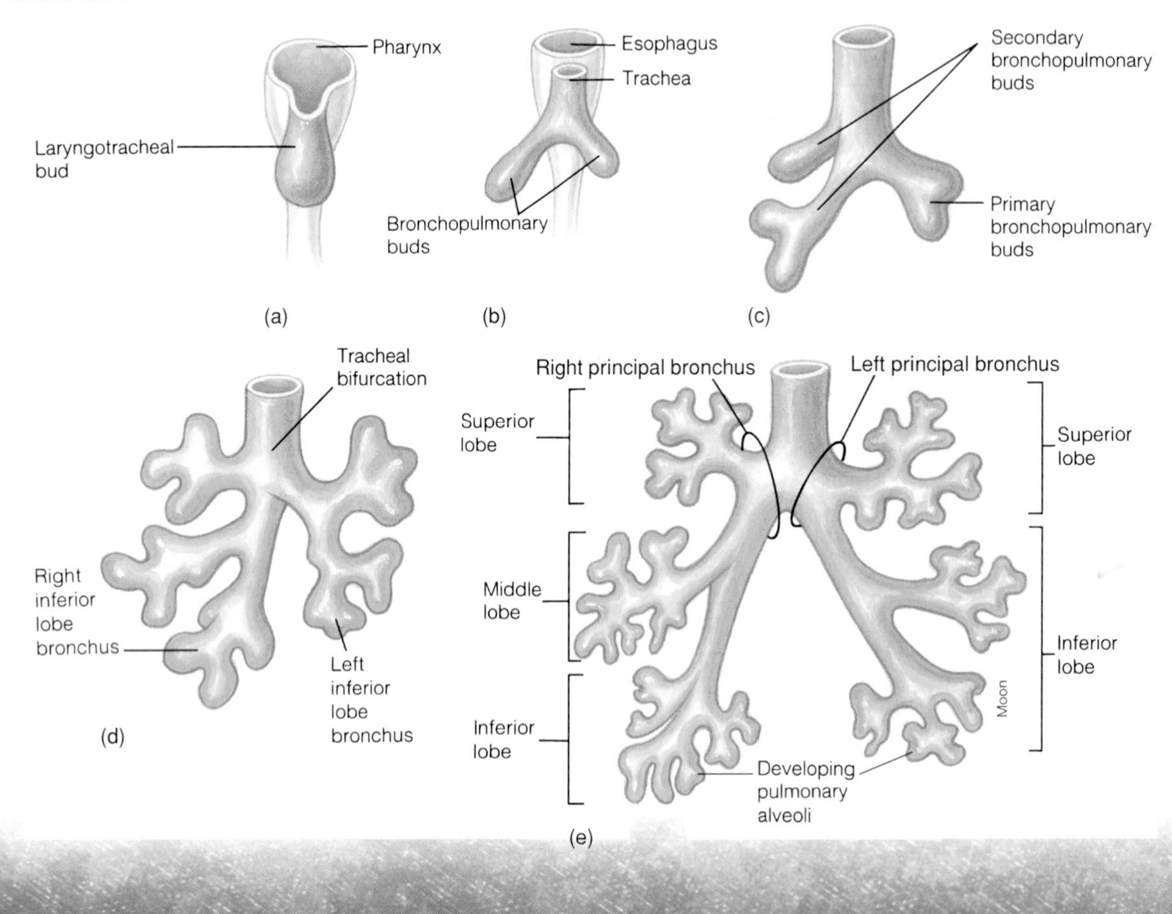

1. State where in the brain the three respiratory areas are located. Which of these three areas is responsible for autonomic rhythmic breathing?
2. Describe the locations of the peripheral chemoreceptors and identify the two paired cranial nerves that carry sensory impulses from these sites to the respiratory centers within the brain stem.

Clinical Considerations

The respiratory system is particularly vulnerable to infectious diseases simply because many pathogens are airborne, humans are highly social, and the warm, moistened environment along the respiratory tract allows pathogens to thrive. Injury and trauma are also frequent problems. Protruding noses are subject to fractures; the large, spongy lungs are easily penetrated by broken ribs; and portions of the respiratory tract may become occluded, since they also have a digestive function.

Developmental Problems of the Respiratory System

Birth defects, inherited disorders, and premature births commonly cause problems in the respiratory system of infants. A **cleft palate** is a developmental deformity of the hard palate of the mouth. An opening persists between the oral and nasal cavities that makes it difficult, if not impossible, for the infant to nurse. A cleft palate

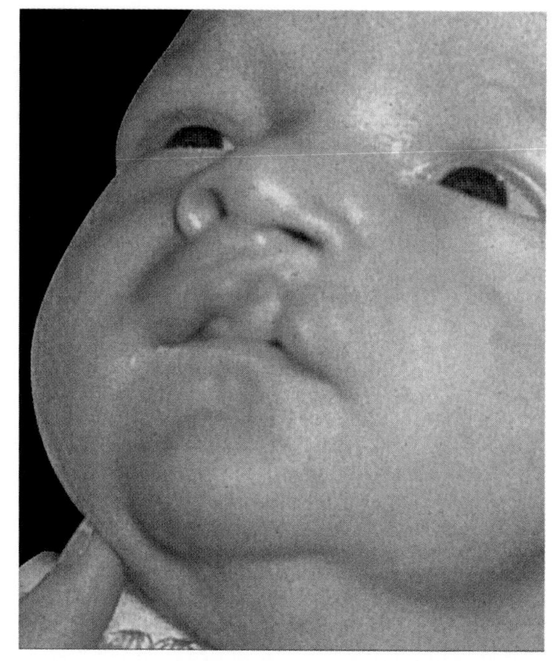

(a)

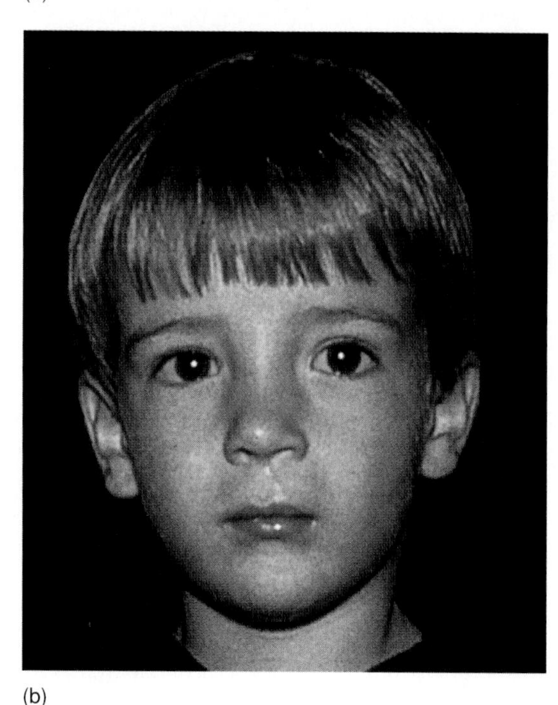

(b)

Figure 17.28

(*a*) An infant with a unilateral cleft lip and palate. (*b*) The same child following corrective surgery.

may be hereditary or a complication of some disease (e.g., German measles) contracted by the mother during pregnancy. A **cleft lip** is a genetically based developmental disorder in which the two sides of the upper lip fail to fuse. Cleft palates and cleft lips can be treated very effectively with cosmetic surgery (fig. 17.28).

As mentioned earlier, **hyaline membrane disease** is a fairly common respiratory disorder that accounts for about one-third of neonatal deaths. This condition results from the deficient production of surfactant. **Cystic fibrosis** is a genetic disorder that affects the respiratory system, as well as other systems of the body, and accounts for approximately one childhood death in twenty. The effect of this disease on the respiratory system is usually a persistent inflammation and infection of the respiratory tract.

Pulmonary alveoli are not developed sufficiently to sustain life until after week 20 of gestation. Thus, extrauterine life prior to that time is extremely difficult even with life-supporting devices.

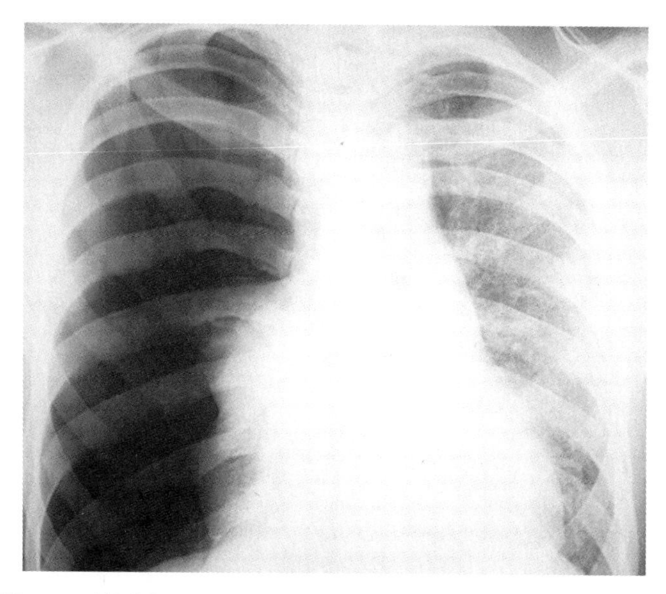

Figure 17.29

A pneumothorax of the right lung. The right side of the thorax appears uniformly dark because it is filled with air; the spaces between the ribs are also greater than those on the left, since the ribs are released from the elastic tension of the lungs. The left lung appears denser (less dark) because of shunting of blood from the right to the left lung.

Trauma or Injury

Humans are especially susceptible to **epistaxes** (*ep″ĭ-stak′sēz*) (nosebleeds) because the prominent nose can be easily bumped and because the nasal mucosa has extensive vascularity for warming the inspired air. Epistaxes may also be caused by high blood pressure or diseases such as leukemia.

When air enters the pleural cavity surrounding either lung, the condition is referred to as a **pneumothorax** (fig. 17.29). A pneumothorax can result from an external injury, such as a stabbing, bullet wound, or penetrating fractured rib, or it can be the result of internal conditions. A severely diseased lung, as in emphysema, can create a pneumothorax as the wall of the lung deteriorates along with the visceral pleura and permits air to enter the pleural cavity.

Choking on a foreign object is a common serious trauma to the respiratory system. More than eight Americans choke to death each day on food lodged in their trachea. A simple procedure termed the **abdominal thrust** (Heimlich) **maneuver** can save the life of a person who is choking. The abdominal thrust maneuver is performed as follows:

A. If the victim is standing or sitting:
   1. Stand behind the victim or the victim's chair and wrap your arms around his or her waist.
   2. Grasp your fist with your other hand and place the fist against the victim's abdomen, slightly above the navel and below the rib cage.
   3. Press your fist into the victim's abdomen with a quick upward thrust.
   4. Repeat several times if necessary.
B. If the victim is lying down:
   1. Position the victim on his or her back.
   2. Face the victim, and kneel on his or her hips.
   3. With one of your hands on top of the other, place the heel of your bottom hand on the abdomen, slightly above the navel and below the rib cage.
   4. Press into the victim's abdomen with a quick upward thrust.
   5. Repeat several times if necessary.

If you are alone and choking, use whatever is available to apply force just below your diaphragm. Press into a table or a sink, or use your own fist.

People saved from drowning and victims of shock frequently experience apnea (cessation of breathing) and will soon die if not revived by artificial respiration. The accepted treatment for reviving a person who has stopped breathing is illustrated in figure 17.30.

Common Respiratory Disorders

A cough is the most common symptom of respiratory disorders. Acute problems may be accompanied by dyspnea or wheezing. Respiratory or circulatory problems may cause **cyanosis** (*si′ă-no′sis*), a blue discoloration of the skin resulting from blood with a low oxygen content.

Although the **common cold** is the most widespread of all respiratory diseases, there is still no cure for this ailment—only medications that offer symptomatic relief. Colds occur repeatedly because acquired immunity for one virus does not protect against other viruses. Cold viruses generally incite acute inflammation in the respiratory mucosa, causing flow of mucus, sometimes accompanied by fever and/or headache.

Nearly all of the structures and regions of the respiratory passageways can become infected and inflamed. **Influenza** is a viral disease that causes an inflammatory condition of the upper respiratory tract. Influenza can be epidemic, but fortunately vaccines are available. **Sinusitis** is an inflammation of the paranasal sinuses. Sinusitis can be quite painful if the drainage ducts from the sinuses into the nasal cavity become blocked. **Tonsillitis** may involve any or all of the tonsils and frequently follows other lingering diseases of the oral or pharyngeal regions. Chronic tonsillitis sometimes requires a tonsillectomy. **Laryngitis** is inflammation of the larynx, which often produces a hoarse voice and limits the ability to talk. **Tracheobronchitis** and **bronchitis** are infections of the regions for which they are named. Severe inflammation in these areas can cause smaller respiratory tubules to collapse, blocking the passage of air.

Diseases of the lungs are likewise common and usually serious. **Pneumonia** is an acute infection and inflammation of lung tissue accompanied by exudation (accumulation of fluid). It is usually caused by bacteria, most commonly by the pneumococcus bacterium. Viral pneumonia is caused by a number of different viruses. **Tuberculosis** is an inflammatory disease of the lungs contracted by inhaling air sneezed or coughed by someone who is carrying active tuberculosis bacteria. Tuberculosis softens lung tissue, which eventually becomes ulcerated. **Asthma** is a disease that affects people who are allergic to certain inhaled antigens. It causes a swelling and blocking of lower respiratory tubes, often accompanied by the formation of mucus plugs. **Pleurisy** (*ploor′ĭ-se*) is an inflammation of the pleura and is usually secondary to some other respiratory disease. Inspiration may become painful, and fluid may collect within the pleural cavity. **Emphysema** (*em″fĭ-se′mă*) is a disease that causes the breakdown of the pulmonary alveoli, thus increasing the size of air spaces and decreasing the surface area (fig. 17.31). It is a frequent cause of death among heavy cigarette smokers.

Cancer in the respiratory system is known to be caused by the repeated inhalation of irritating substances, such as cigarette smoke. Cancers of the lip, larynx, and lungs (fig. 17.32) are especially common in smokers over the age of 50.

Disorders of Respiratory Control

A variety of disease processes can result in cessation of breathing during sleep, or *sleep apnea*. **Sudden infant death syndrome (SIDS)** is an especially tragic form of sleep apnea that claims the lives of about 10,000 babies annually in the United States. Victims of this condition are apparently healthy 2-to-5-month-old babies who die in their sleep without apparent reason—hence, the layperson's term, "crib death." These deaths seem to be caused by failure of the respiratory control mechanisms in the brain stem and/or by failure of the carotid bodies to be stimulated by reduced arterial oxygen.

Abnormal breathing patterns often appear prior to death from brain damage or heart disease. The most common of these abnormal patterns is **Cheyne–Stokes** (*chān′stōks′*) **breathing,** in which the depth of breathing progressively increases and then progressively decreases. These cycles of increasing and decreasing tidal volumes may be followed by periods of apnea of varying

---

cyanosis: Gk. *kyanosis*, dark-blue color
influenza: L. *influentia*, a flowing in

tuberculosis: L. *tuberculum*, diminutive of *tuber*, swelling
asthma: Gk. *asthma*, panting
emphysema: Gk. *emphysan*, blow up, inflate
Cheyne–Stokes breathing: from John Cheyne, British physician, 1777–1836, and William Stokes, Irish physician, 1804–78.

Mouth-to-Mouth Method

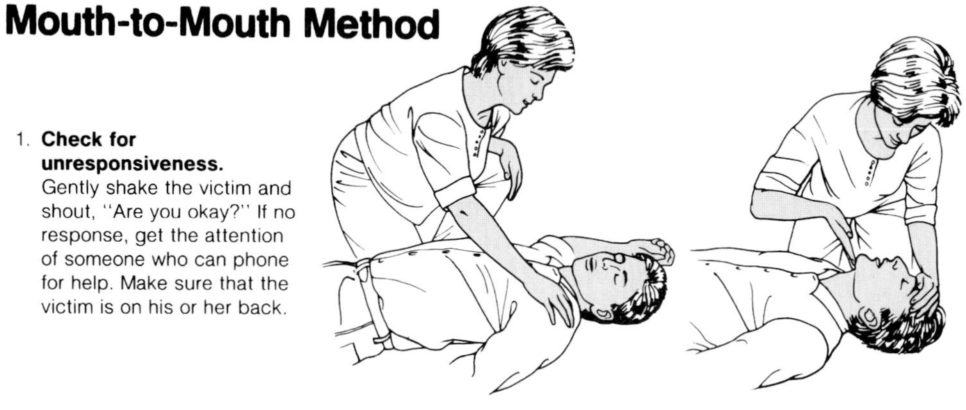

1. **Check for unresponsiveness.** Gently shake the victim and shout, "Are you okay?" If no response, get the attention of someone who can phone for help. Make sure that the victim is on his or her back.

2. **Open the airway.** Tilt the victim's head back by pushing on his or her forehead with your hand and lifting the chin with your fingers under his or her jaw. This will open the airway by moving the tongue away from the back of the victim's throat.

3. **Check for breathing.** Put your ear close to the victim's face to listen and feel for any return of air. At the same time, look to see if there is chest movement. Check for breathing for about 5 seconds.

4. **If no breathing, give two full breaths.** While maintaining the victim in the head-tilt position, pinch his or her nose to close off the nasal passageway. Take a deep breath, then seal your mouth around the victim's mouth and give two full breaths. (After the first breath, raise your head slightly to inhale quickly and then give the second breath.)

5. **Check for pulse.** While maintaining head tilt, feel for a carotid pulse for 5 to 10 seconds on the side of the victim's neck.

6. **Continue rescue breathing.** With the victim in the head-tilt position and his or her nostrils pinched, give one breath every 5 seconds. Observe for signs of breathing between breaths. For an infant, give one gentle puff every 3 seconds.

7. **Recheck for pulse.** Feel for a carotid pulse at 1-minute intervals. If the victim has a pulse but is not breathing, continue rescue breathing.

Mouth-to-Nose Method

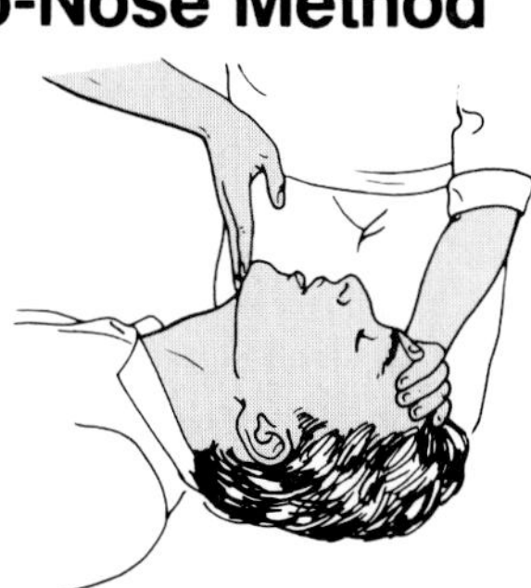

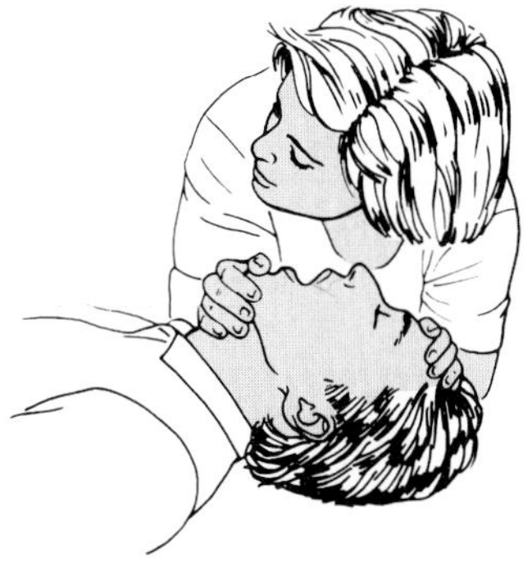

1. **Open the airway.** Place the victim in the head-tilt position as described above.

2. **Blow into the victim's nose.** Using the same sequence described above, blow into the victim's nose while holding his or her mouth closed.

3. **Feel and observe for breathing.** With the victim's mouth held open, detect for breathing between giving forced breaths.

Figure 17.30

Artificial respiration.

Internal Affairs

■ How the *Respiratory System* Works with Other Body Systems
□ How Other Systems Work with the *Respiratory System*

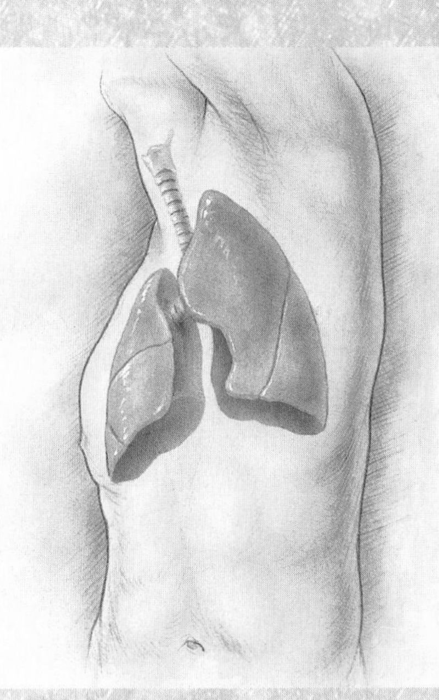

Integumentary System

- Provides O_2 and eliminates CO_2

- Nasal hairs and mucus prevent the entry of dust and other foreign material into respiratory tract

Skeletal System

- Provides O_2 and eliminates CO_2

- Rib cage provides a protective enclosure for the lungs
- Movement of the ribs assists in respiration

Muscular System

- Provides O_2 and eliminates CO_2

- Respiratory muscles ventilate the lungs, control position of the larynx during swallowing, and control tension on the vocal cords during speech production
- Muscle activity stimulates pulmonary ventilation by generating CO_2

Nervous System

- Provides O_2 and eliminates CO_2

- Monitors respiratory volume and blood gas levels
- Brain stem centers control the rate and depth of breathing
- Phrenic, intercostal, and other nerves innervate respiratory muscles

Endocrine System

- Provides O_2 and eliminates CO_2

- Epinephrine and norepinephrine dilate the bronchioles and stimulate ventilation

Circulatory System

- Provides O_2 and eliminates CO_2

- Blood serves as the transport medium for respiratory gases

Lymphatic System

- Provides O_2 and eliminates CO_2
- Tonsils in the pharynx house immune cells

- Prevents edema (retention of interstitial fluid)
- Protects against infection

Digestive System

- Provides O_2 and eliminates CO_2
- Contraction of respiratory muscles can aid defecation

- Provides nutrients for growth and maintenance of the respiratory system

Urinary System

- Provides O_2 and eliminates CO_2
- Expiratory effort against a closed glottis (Valsalva's maneuver) helps to empty the urinary bladder

- Disposes of metabolic wastes from respiratory organs
- Participates with lungs in regulation of blood pH

Reproductive System

- Provides O_2 and eliminates CO_2
- Valsalva's maneuver aids in childbirth

- Changes in respiratory rate and depth occur during sexual arousal

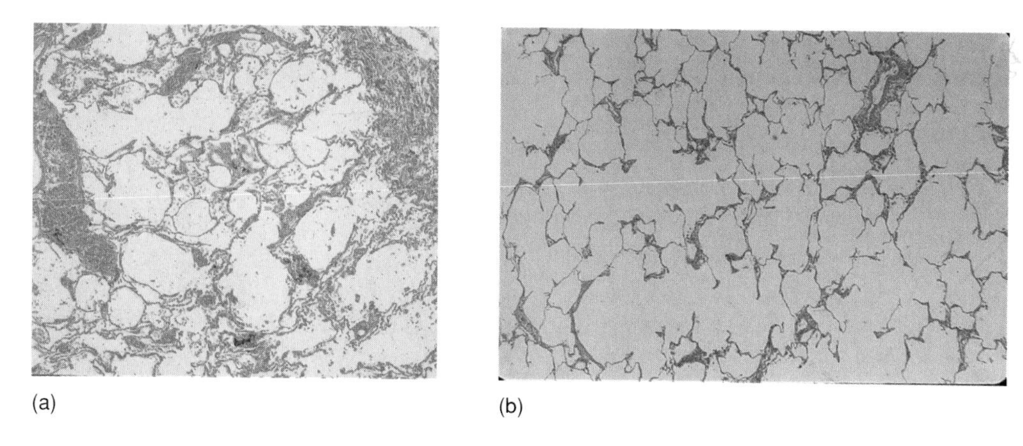

Figure 17.31

Photomicrographs of tissue (*a*) from a normal lung and (*b*) from the lung of a person with emphysema. In emphysema, lung tissue is destroyed, resulting in fewer and larger pulmonary alveoli.

durations. Cheyne–Stokes breathing may be caused by neurological damage or by insufficient oxygen delivery to the brain. The latter may result from heart disease or from a brain tumor that diverts a large part of the vascular supply from the respiratory centers.

Clinical Case Study Answer

The ice pick traversed the parietal pleura, the visceral pleura, and then entered the airway, at least at the alveolar–terminal bronchiole level, but possibly at larger airways. This allowed inspired air to escape from the airway into the pleural cavity. The air would have taken the following route: pharynx → larynx → trachea → left principal bronchus → lobar bronchus (of left upper lobe) → apical segmental (tertiary) bronchus → bronchioles through laceration into pleural space. This condition is treated by inserting a tube (tube thoracostomy) into the pleural cavity to allow suction evacuation of air and blood, which results in reexpansion of the lung. The laceration usually seals over within a day or two. Persistent bleeding may necessitate thoracotomy (incision of the chest wall) for repair.

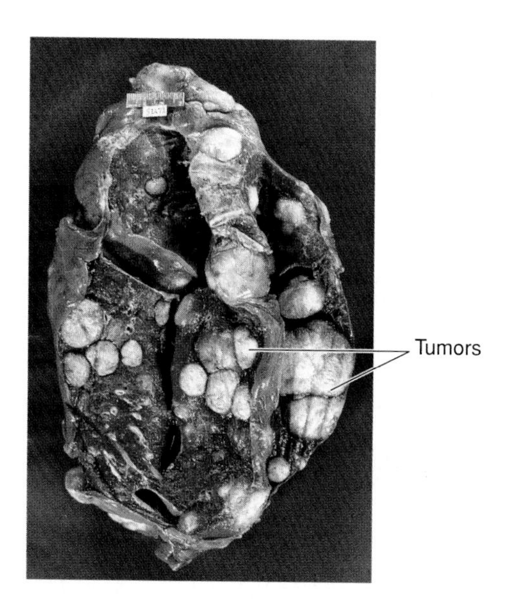

Figure 17.32

A cancerous lung. For people who regularly smoke a pack of cigarettes a day, the risk of developing lung cancer is 20 times greater than for people who have never smoked.

Chapter Summary

Introduction to the Respiratory System (pp. 583–584)

1. Respiration refers not only to ventilation (breathing), but also to the exchange of gases between the atmosphere, the blood, and individual cells. Within cells, the metabolic reactions that release energy are called cellular respiration.
2. In order for the respiratory system to function, the respiratory membranes must be moist, thin-walled, highly vascular, and differentially permeable.
3. The functions of the respiratory system include gaseous exchange, sound production, assistance in abdominal compression, and reflexive coughing and sneezing, and immune response.

Conducting Passages (pp. 585–592)

1. The nose is supported by nasal bones and cartilages.
2. The nasal epithelium warms, moistens, and cleanses the inspired air.
3. Olfactory epithelium is associated with the sense of smell, and the nasal cavity acts as a resonating chamber for the voice.
4. The paranasal sinuses are found in the maxillary, frontal, sphenoid, and ethmoid bones.
   (a) These sinuses lighten the skull and are lined with mucus-secreting goblet cells.
   (b) Sinusitis is an inflammation of one or more of the paranasal sinuses.
5. The pharynx is a funnel-shaped passageway that connects the oral and nasal cavities with the esophagus and larynx.
   (a) The nasopharynx, connected by the auditory tubes to the middle-ear cavities, contains the pharyngeal tonsils, or adenoids.
   (b) The oropharynx is the middle portion, extending from the soft palate to the level of the hyoid bone; it contains the palatine and lingual tonsils.
   (c) The laryngopharynx extends from the hyoid bone to the larynx and esophagus.
6. The larynx contains a number of cartilages that keep the passageway to the trachea open during breathing and closes the respiratory passageway during swallowing.
   (a) The epiglottis is a spoon-shaped structure that aids in closing the laryngeal opening, or glottis, during swallowing.
   (b) The vocal folds in the larynx are controlled by intrinsic muscles and are used in sound production.
7. The trachea is a rigid tube, supported by incomplete rings of hyaline cartilage, that leads from the larynx to the bronchial tree.
8. The bronchial tree includes a principal bronchus, which divides to produce lobar bronchi, segmental bronchi, and bronchioles; the conducting division ends with the respiratory bronchioles, which connect to the pulmonary alveoli.

Pulmonary Alveoli, Lungs, and Pleurae (pp. 592–598)

1. Pulmonary alveoli are the functional units of the lungs, where gas exchange occurs; they are small, thin-walled air sacs.
2. The right and left lungs are separated by the mediastinum. Each lung is divided into lobes and lobules.
   (a) The right lung is subdivided by two fissures into superior, middle, and inferior lobes.
   (b) The left lung is subdivided into a superior lobe and an inferior lobe by a single fissure.
3. The lungs are covered by visceral pleura, and the thoracic cavity is lined by a parietal pleura.
   (a) The potential space between these two pleural membranes is called the pleural cavity.
   (b) The pleural membranes compartmentalize each lung and exclude the structures located in the mediastinum.

Mechanics of Breathing (pp. 598–603)

1. Quiet (unforced) inspiration is due to contraction of the diaphragm and certain intercostal muscles. Forced inspiration is aided by the scalenes and the pectoralis minor and sternocleidomastoid muscles.
2. Quiet expiration is produced by relaxation of the respiratory muscles and elastic recoil of the lungs and thorax. Forced expiration is aided by certain intercostal muscles and the abdominal muscles.
3. Among the air volumes exchanged in ventilation are tidal, inspiratory reserve, and expiratory reserve volumes.
4. Nonrespiratory air movements are associated with coughing, sneezing, sighing, yawning, laughing, crying, and hiccuping.

Regulation of Breathing (pp. 603–606)

1. Ventilation is directly controlled by the rhythmicity center in the medulla oblongata, which in turn is influenced by the pneumotaxic and apneustic centers in the pons.
2. These brain stem areas are affected by higher brain function and by sensory input from chemoreceptors.
3. Central chemoreceptors are located in the medulla oblongata; peripheral chemoreceptors are located in the aortic and carotid bodies.

Review Activities

Objective Questions

1. Which of the following is a *false* statement?
   (a) The term *respiration* can be used in reference to ventilation (breathing) or oxygen utilization by body cells.
   (b) The incoming (inhaled) air that contacts the pulmonary alveoli is unchanged from that which surrounds the body.
   (c) As a resonating chamber, the nasal cavity is important in sound production.
   (d) It is only through the walls of the pulmonary alveoli that gaseous exchange occurs.
2. Which is *not* a component of the nasal septum?
   (a) the palatine bone
   (b) the vomer
   (c) the ethmoid bone
   (d) septal cartilage
3. An adenoidectomy is the removal of
   (a) the uvula.
   (b) the pharyngeal tonsils.
   (c) the palatine tonsils.
   (d) the lingual tonsils.
4. Which is *not* a paranasal sinus?
   (a) the palatine sinus
   (b) the ethmoidal sinus
   (c) the sphenoidal sinus
   (d) the frontal sinus
   (e) the maxillary sinus
5. Which of the following is *not* characteristic of the left lung?
   (a) a cardiac impression
   (b) a superior lobe
   (c) a single fissure
   (d) an inferior lobe
   (e) a middle lobe
6. The epithelial lining of the wall of the thoracic cavity is called
   (a) the parietal pleura.
   (b) the pleural peritoneum.
   (c) the mediastinal pleura.
   (d) the visceral pleura.
   (e) the costal pleura.

7. Which muscle group combination permits inspiration?
   (a) diaphragm, abdominal complex
   (b) internal intercostals, diaphragm
   (c) external intercostals, internal intercostals
   (d) external intercostals, diaphragm, internal intercostals (interchondral part)
8. The vocal cords are attached to
   (a) the cricoid and thyroid cartilages.
   (b) the cuneiform and cricoid cartilages.
   (c) the corniculate and thyroid cartilages.
   (d) the arytenoid and thyroid cartilages.
9. The maximum amount of air that can be expired after a maximum inspiration is
   (a) the tidal volume.
   (b) the forced expiratory volume.
   (c) the vital capacity.
   (d) the maximum expiratory flow rate.
   (e) the residual volume.
10. The rhythmic control of breathing is produced by the activity of inspiratory and expiratory neurons in
    (a) the medulla oblongata.
    (b) the apneustic center of the pons.
    (c) the pneumotaxic center of the pons.
    (d) the cerebral cortex.
    (e) the hypothalamus.

Essay Questions

1. Why is warming, moistening, and filtering the air so important for a healthy, functioning respiratory system?
2. Explain the structural and functional differences between the three regions of the pharynx.
3. What is a bronchial tree? Why are pulmonary alveoli rather than bronchial trees considered the functional units of the respiratory system?
4. Define each of the following terms relating to the structure of the lung: *base, hilum, apex, costal surface, fissure, lobe, lobule, pulmonary ligament,* and *bronchial segment.*
5. Diagram the location of the lungs in the thoracic cavity with respect to the heart. Identify the various thoracic serous membranes.
6. List the kinds of epithelial tissues found within the respiratory system and describe the location of each.
7. What protective devices of the respiratory system guard against pollutants, the spread of infections, and dual collapse of the lungs?
8. What are the functions of the larynx? List some of the clinical conditions that could involve this organ.
9. What are the advantages of compartmentalization of the thoracic organs?
10. Explain the sequence of pulmonary ventilation. Discuss the mechanisms of inspiration and expiration. How are air pressures related to ventilation?
11. Distinguish between tidal volume, vital capacity, and total lung capacity.
12. What is meant by a rhythmicity respiratory area? How are the apneustic and pneumotaxic areas related to the rhythmicity area?

Critical-Thinking Questions

1. Identify two places in the respiratory system where large surface areas of capillary networks are found. What is the function of each of these areas and why are the capillary networks important in the treatment of certain clinical conditions?
2. On what principal is the abdominal thrust maneuver based? When would mouth-to-mouth resuscitation rather than the abdominal thrust maneuver be used to revive a person?
3. The nature of the sounds produced by percussing (tapping) a patient's chest can tell a physician a great deal about the condition of the organs within the thoracic cavity. Healthy, air-filled lungs resonate, or sound hollow. How do you think the lungs of a person with emphysema would sound in comparison to healthy lungs? What kind of sound would be produced by a collapsed lung, or one that was partially filled with fluid?
4. Why do premature infants often require respiratory assistance (a mechanical ventilator) to keep their lungs inflated?
5. Nicotine from cigarette smoke causes mucus to build up and paralyzes the cilia that line the respiratory tract. How do these changes affect the lungs?

EIGHTEEN

Digestive System

Clinical Case Study 615

Introduction to the Digestive System 615

Serous Membranes and Tunics of the Gastrointestinal Tract 616
- Serous Membranes 616
- Layers of the Gastrointestinal Tract 617
- Innervation of the Gastrointestinal Tract 620

Mouth, Pharynx, and Associated Structures 620
- Cheeks, Lips, and Palate 620
- Tongue 622
- Teeth 622
- Salivary Glands 625
- Pharynx 625

Esophagus and Stomach 628
- Esophagus 628
- Swallowing Mechanisms 628
- Stomach 628

Small Intestine 632
- Regions of the Small Intestine 633
- Structural Modifications of the Small Intestine 634
- Mechanical Activities of the Small Intestine 635

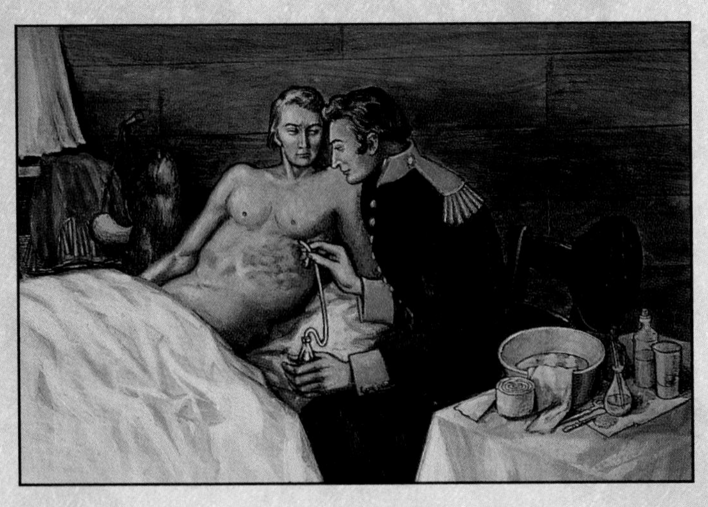

In 1822, when Canadian fur trapper Alexis St. Martin was accidentally shot at close range, the blast left a gaping hole in his side. The army doctor who saved his life studied St. Martin for 8 years, withdrawing gastric juice and lowering food samples into his stomach at regular intervals to monitor the digestive process.

Large Intestine 636
- Regions and Structures of the Large Intestine 636
- Mechanical Activities of the Large Intestine 638

Liver, Gallbladder, and Pancreas 639
- Liver 640
- Gallbladder 642
- Pancreas 643

Developmental Exposition: The Digestive System 644

Clinical Considerations 648
- Developmental Problems of the Digestive System 648
- Pathogens and Poisons 648
- Clinical Problems of the Teeth and Salivary Glands 649
- Disorders of the Liver 649
- Disorders of the GI Tract 650

Internal Affairs 651

Clinical Case Study Answer 652

Important Clinical Terminology 652

Chapter Summary 652

Review Activities 653

Clinical Case Study

A 25-year-old male construction worker was admitted to the emergency room after suffering a blow to the upper abdomen from a swinging beam. Initial assessment was significant for marked tenderness in the right upper quadrant of the abdomen and for vital signs and examination findings consistent with mild hemorrhagic shock. Intravenous fluids were administered, causing stabilization of vital signs. A chest radiograph revealed no abnormalities. Peritoneal lavage (see chapter 2, Clinical Case Study) was likewise negative for blood. There were no externally detectable signs of hemorrhage. A CT scan demonstrated a significant hematoma (collection of blood) deep within the substance of the liver, as well as a notable amount of blood in the small intestine. The decision was made to operate. Initial exploration revealed no trauma to the stomach or small intestine.

What is the likely source of the bleeding? Explain anatomically how the blood found its way into the small intestine, noting each step of its path. Given that the hepatic arterial system and hepatic venous system are possible sources of the bleeding, is there another system of blood vessels relative to the liver that could also be a source of hemorrhage?

Hint: As you study the digestive system, pay close attention to the location of each accessory digestive organ and note how each connects to the lumen of the gastrointestinal (GI) tract.

Introduction to the Digestive System

The organs of the digestive system are specialized for the digestion and absorption of food. The digestive system consists of a tubular gastrointestinal tract and accessory digestive organs.

Objective 1 Describe the activities of the digestive system and distinguish between digestion and absorption.

Objective 2 Identify the major structures and regions of the digestive system.

Objective 3 Define the terms *viscera* and *gut.*

Food is necessary to sustain life. It provides the essential nutrients the body cannot produce for itself. The food is utilized at the cellular level, where nutrients are required for chemical reactions involving synthesis of enzymes, cellular division and growth, repair, and the production of heat energy. Most of the food we eat, however, is not suitable for cellular utilization until it is mechanically and chemically reduced to forms that can be absorbed through the intestinal wall and transported to the cells by the blood. Ingested food is not technically inside the body until it is absorbed; and, in fact, a large portion of this food remains undigested and passes through the body as waste material.

The principal function of the digestive system is to prepare food for cellular utilization. This involves the following functional activities:

1. **Ingestion**—the taking of food into the mouth
2. **Mastication**—chewing movements to pulverize food and mix it with saliva
3. **Deglutition**—the swallowing of food to move it from the mouth to the pharynx and into the esophagus
4. **Digestion**—the mechanical and chemical breakdown of food material to prepare it for absorption
5. **Absorption**—the passage of molecules of food through the mucous membrane of the small intestine and into the blood or lymph for distribution to cells
6. **Peristalsis**—rhythmic, wavelike intestinal contractions that move food through the gastrointestinal tract
7. **Defecation**—the discharge of indigestible wastes, called *feces*, from the gastrointestinal tract

Anatomically and functionally, the digestive system can be divided into a tubular **gastrointestinal tract** (GI tract), or *digestive tract*, and **accessory digestive organs.** The GI tract, which extends from the mouth to the anus, is a continuous tube approximately 9 m (30 ft) long. It traverses the thoracic cavity and enters the abdominal cavity at the level of the diaphragm.

The organs of the GI tract include the **oral cavity, pharynx, esophagus, stomach, small intestine,** and **large intestine** (fig. 18.1). The accessory digestive organs include the **teeth, tongue, salivary glands, liver, gallbladder,** and **pancreas.** The term **viscera** is frequently used to refer to the abdominal organs of digestion, but actually viscera can be any of the organs (lungs, stomach, spleen, etc.) of the thoracic and abdominal cavities. **Gut** is an anatomical term that generally refers to the developing stomach and intestines in the embryo (see Developmental Exposition, p. 644).

It usually takes about 24 to 48 hours for food to travel the length of the GI tract. Food ingested through the mouth passes in assembly-line fashion through tract, where complex molecules are progressively broken down. Each region of the GI tract has specific functions in preparing food for utilization (table 18.1).

Although there is an abundance of food in the United States, so many people are malnourished that eating patterns have become a critical public health concern. Obesity is a major health problem. Grossly overweight people are at greater risk for cardiovascular disease, hypertension, osteoarthritis, and diabetes mellitus. People with good nutritional habits are better able to withstand trauma, are less likely to get sick, and are usually less seriously ill when they do become sick.

ingestion: L. *ingerere*, carry in
mastication: Gk. *mastichan*, gnash the teeth
deglutition: L. *deglutire*, swallow down
peristalsis: Gk. *peri*, around; *stellein*, compress
defecation: L. *de*, from, away; *faecare*, cleanse

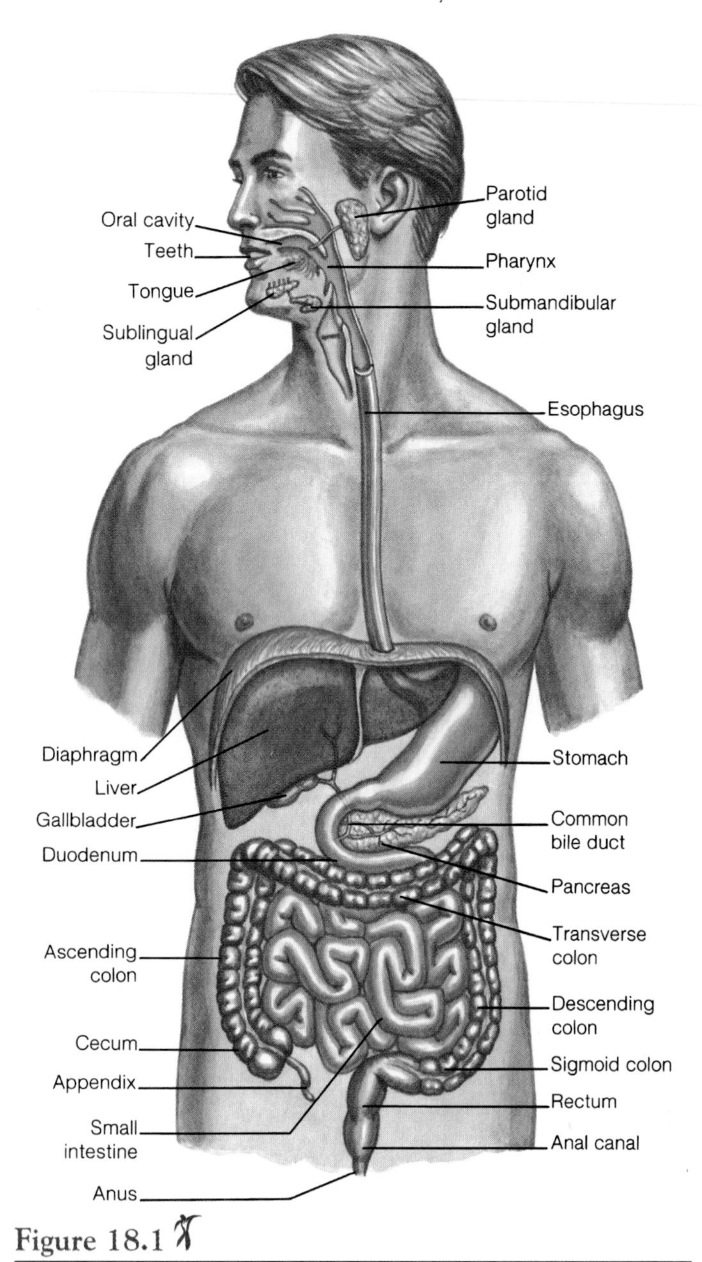

Figure 18.1

The digestive system.

1. Which functional activities of the digestive system break down food? Which functional activities move the food through the GI tract? Where does absorption take place?
2. List in order the regions of the GI tract through which ingested food passes from the mouth to the anus.
3. List organs of the GI tract and the accessory digestive organs.
4. Write a sentence in which the term *gut* is used correctly.

Table 18.1

The GI Tract: Regions and Basic Functions

Region	Function
Oral cavity	Ingests food; receives saliva; grinds food and mixes it with saliva (mastication); initiates digestion of carbohydrates; forms and swallows soft mass of chewed food called bolus (deglutition)
Pharynx	Receives bolus from oral cavity; autonomically continues deglutition of bolus to esophagus
Esophagus	Transports bolus to stomach by peristalsis; lower esophageal sphincter restricts backflow of food
Stomach	Receives bolus from esophagus; churns bolus with gastric juice; initiates digestion of proteins; carries out limited absorption; moves mixture of partly digested food and secretions (chyme) into duodenum and prohibits backflow of chyme; regurgitates when necessary
Small intestine	Receives chyme from stomach and secretions from liver and pancreas; chemically and mechanically breaks down chyme; absorbs nutrients; transports wastes through peristalsis to large intestine; prohibits backflow of intestinal wastes from large intestine
Large intestine	Receives undigested wastes from small intestine; absorbs water and electrolytes; forms, stores, and expels feces when activated by a defecation reflex

Serous Membranes and Tunics of the Gastrointestinal Tract

Protective and lubricating serous membranes line the abdominal cavity and cover the visceral organs. Specialized serous membranes support the GI tract and provide a structure through which nerves and vessels pass. The wall of the GI tract is composed of four tunics.

Objective 4 Describe the arrangement of the serous membranes within the abdominal cavity.

Objective 5 Describe the generalized structure of the four tunics that form the wall of the GI tract.

Serous Membranes

Most of the digestive viscera are positioned within the abdominopelvic cavity. These organs are supported and covered by serous membranes that line the cavities of the trunk and cover the organs within these cavities. Serous membranes are composed of simple squamous epithelium, portions of which are reinforced with connective tissue. Serous membranes secrete a lubricating serous fluid that continuously moistens the associated organs. The *parietal portion* of the serous membrane lines the body wall, and a *visceral portion* covers the internal organs. As

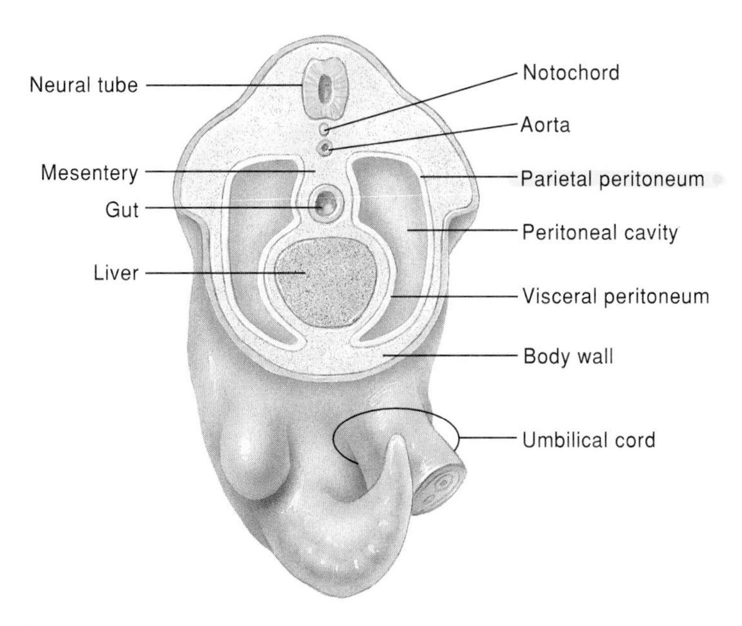

Figure 18.2

A diagram of the abdominal serous membranes.

described in the previous chapter, the serous membranes associated with the lungs are called pleurae (see fig. 17.21). The serous membranes of the abdominal cavity are called **peritoneal membranes,** or **peritoneum** (*per″ĭ-tŏ-ne′um*).

The ***parietal peritoneum*** lines the wall of the abdominal cavity (fig. 18.2). Along the posterior abdominal cavity, the parietal peritoneum comes together to form a double-layered peritoneal fold called the **mesentery** (*mes′en-ter″e*). The mesentery supports the GI tract, at the same time allowing the small intestine freedom for peristaltic movement. It also provides a structure for the passage of intestinal nerves and vessels. The **mesocolon** is a specific portion of the mesentery that supports the large intestine (fig. 18.3c,d).

The peritoneal covering continues around the intestinal viscera as the **visceral peritoneum.** The **peritoneal cavity** is the space between the parietal and visceral portions of the peritoneum. Certain abdominal organs lie posterior to the parietal peritoneum, and are therefore said to be *retroperitoneal*. Retroperitoneal organs include most of the pancreas, the kidneys, the adrenal glands, a portion of the duodenum, and the abdominal aorta.

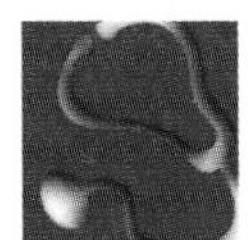

Peritonitis is a bacterial inflammation of the peritoneum. It may be caused by trauma, rupture of a visceral organ, an ectopic pregnancy, or postoperative complications. Peritonitis is usually extremely painful and serious. Treatment usually involves the injection of massive doses of antibiotics, and perhaps peritoneal intubation (insertion of a tube) to permit drainage.

Extensions of the parietal peritoneum serve to suspend or anchor numerous organs within the peritoneal cavity (fig. 18.3). The **falciform** (*fal′sĭ-form*) **ligament,** a serous membrane reinforced with connective tissue, attaches the liver to the diaphragm and anterior abdominal wall. The **greater omentum** (*o-men′tum*) (fig. 18.3*a*) extends from the greater curvature of the stomach to the transverse colon, forming an apronlike structure over most of the small intestine. Functions of the greater omentum include storing fat, cushioning visceral organs, supporting lymph nodes, and protecting against the spread of infections. In cases of localized inflammation, such as appendicitis, the greater omentum may actually compartmentalize the inflamed area, sealing it off from the rest of the peritoneal cavity. The **lesser omentum** (fig. 18.3*b*) passes from the lesser curvature of the stomach and the upper duodenum to the inferior surface of the liver.

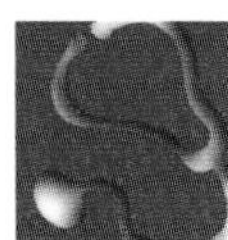

The peritoneal cavity provides a warm, moist, normally aseptic environment for the abdominal viscera. In a male, the peritoneal cavity is totally closed off from the outside body environment. In a female, however, it is not isolated from the outside, which presents the potential for contamination through the entry of microorganisms. A fairly common gynecological condition is *pelvic inflammatory disease (PID),* which results from the entry of pathogens into the peritoneal cavity at the sites of the open-ended uterine (fallopian) tubes.

Layers of the Gastrointestinal Tract

The GI tract from the esophagus to the anal canal is composed of four layers, or **tunics.** Each tunic contains a dominant tissue type that performs specific functions in the digestive process. The four tunics of the GI tract, from the inside out, are the *mucosa, submucosa, muscularis,* and *serosa* (fig. 18.4*a*).

Mucosa

The mucosa, which lines the lumen of the GI tract, is both an absorptive and a secretory layer. It consists of a simple columnar epithelium supported by the **lamina propria** (*lam′ĭ-nă pro′pre-ă*) (fig. 18.4**b**), a thin layer of connective tissue. The lamina propria contains numerous lymph nodules, which are important in protecting against disease. External to the lamina propria are thin layers of smooth muscle called the **muscularis mucosae.** This is the muscle layer responsible for the numerous folds, called *plicae circulares* (see fig. 18.20) that greatly increase the absorptive surface area of the small intestine portion of the GI tract. Specialized **goblet cells** in the mucosa throughout most of the GI tract secrete mucus.

Submucosa

The relatively thick submucosa is a highly vascular layer of connective tissue serving the mucosa. Absorbed molecules that pass through the columnar epithelial cells of the mucosa enter into blood vessels or lymph ductules of the submucosa. In addition to

peritoneum: Gk. *peritonaion*, stretched over
mesentery: Gk. *mesos*, middle; *enteron*, intestine

omentum: L. *omentum*, apron
tunica: L. *tunica*, covering or coat

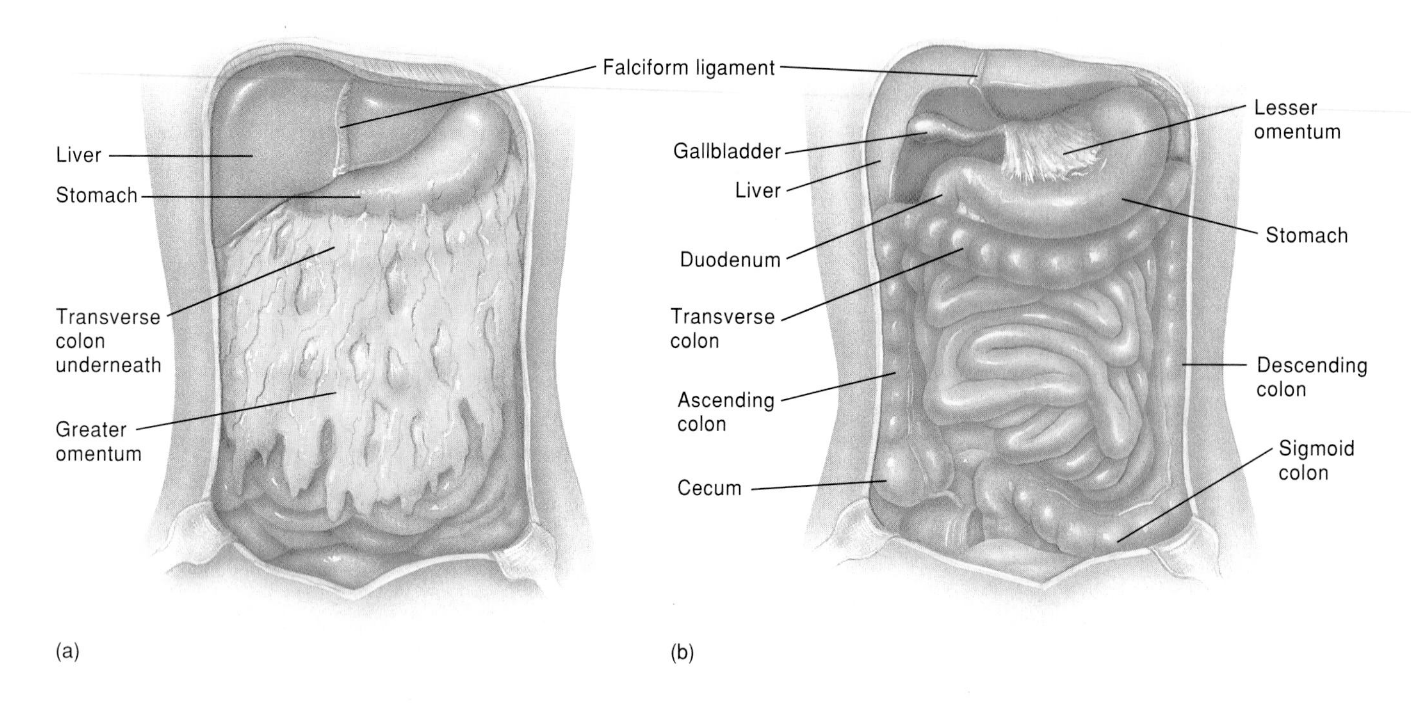

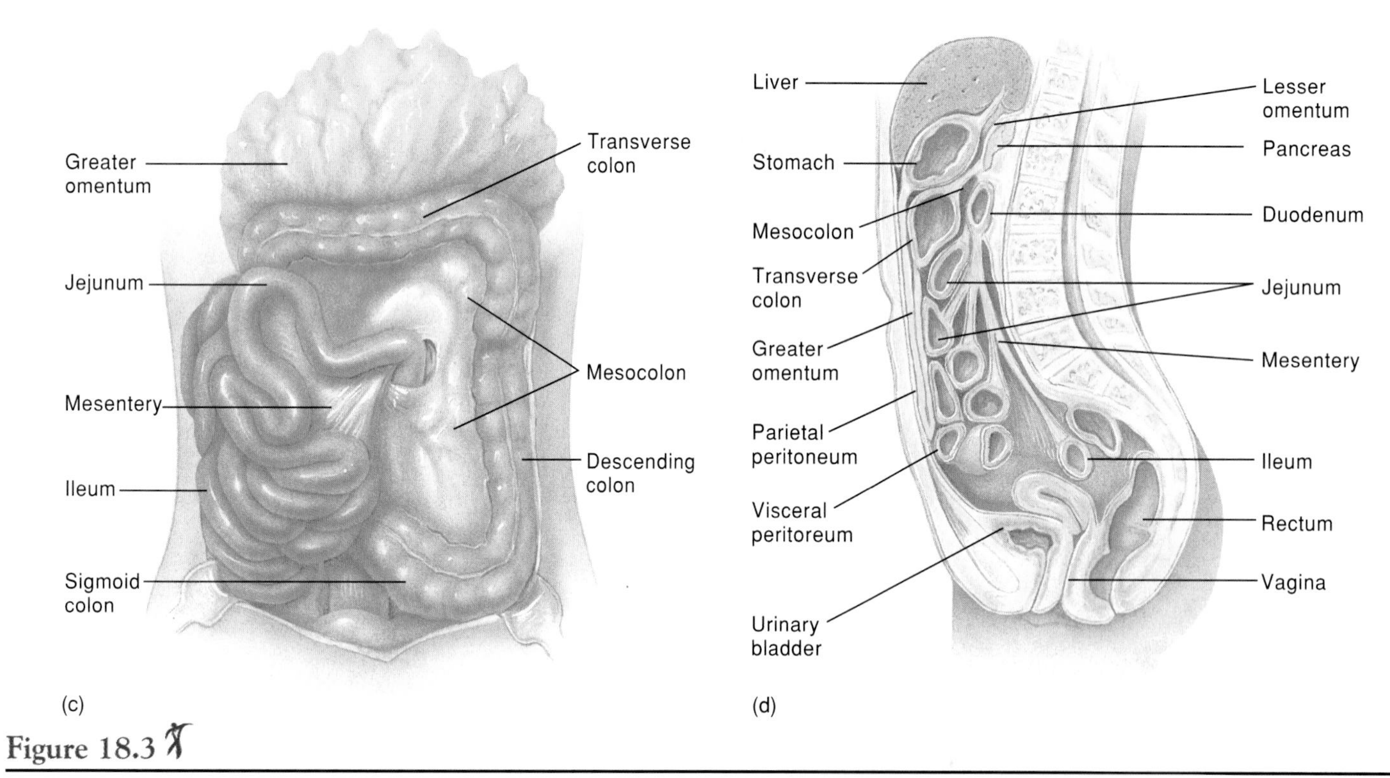

Figure 18.3

The structural arrangement of the abdominal organs and peritoneal membranes. (*a*) The greater omentum, (*b*) the lesser omentum with the liver lifted, (*c*) the mesentery with the greater omentum lifted, and (*d*) the relationship of the peritoneal membranes to the visceral organs as shown in a sagittal view.

blood vessels, the submucosa contains glands and nerve plexuses. The *submucosal plexus* (*Meissner's plexus*) (fig. 18.4*b*) provides autonomic innervation to the muscularis mucosae.

Tunica Muscularis

The tunica muscularis is responsible for segmental contractions and peristaltic movement through the GI tract. This tunic has an inner circular and an outer longitudinal layer of smooth muscle. Contractions of these layers move the food through the tract and physically pulverize and churn the food with digestive enzymes.

Meissner's plexus: from Georg Meissner, German histologist, 1829–1905

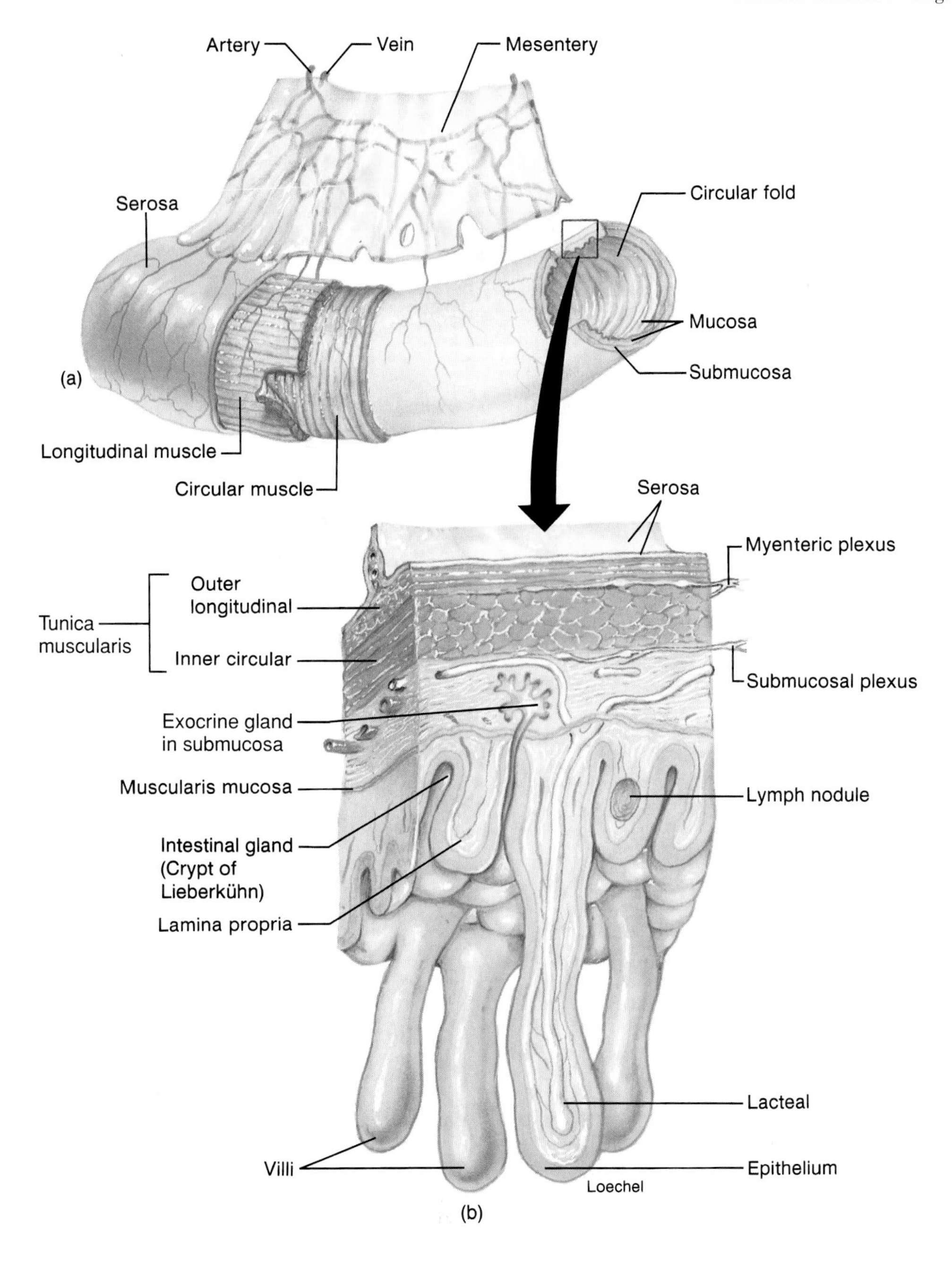

Figure 18.4

The tunics (layers) of the GI tract. (*a*) A section of the small intestine with each of the four tunics exposed and (*b*) a section showing the detailed structure and relative thickness of each tunic. (Note the location of the exocrine gland and the innervation of the small intestine.)

The *myenteric plexus* (*Auerbach's plexus*), located between the two muscle layers, provides the major nerve supply to the GI tract. It includes neurons and ganglia from both the sympathetic and parasympathetic divisions of the ANS.

Serosa

The outer serosa completes the wall of the GI tract. It is a binding and protective layer consisting of loose connective tissue covered with a layer of simple squamous epithelium and subjacent connective tissue. The simple squamous epithelium is actually the visceral peritoneum.

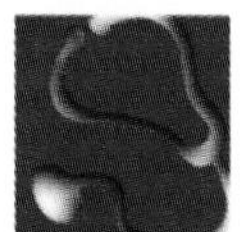

The body has several defense mechanisms to protect against ingested material that may be harmful if absorbed. The acidic environment of the stomach and the lymphatic system kill many harmful bacteria. A mucous lining throughout the GI tract serves as a protective layer. Vomiting, and in certain cases diarrhea, are reactions to substances that irritate the GI tract. Vomiting is a reflexive response to many toxic chemicals; thus, even though unpleasant, it can be beneficial.

Auerbach's plexus: from Leopold Auerbach, German anatomist, 1828–97

Innervation of the Gastrointestinal Tract

The GI tract is innervated by the sympathetic and parasympathetic divisions of the autonomic nervous system. The vagus nerves are the source of parasympathetic activity in the esophagus, stomach, pancreas, gallbladder, small intestine, and upper portion of the large intestine. The lower portion of the large intestine receives parasympathetic innervation from spinal nerves in the sacral region. The submucosal plexus and myenteric plexus are the sites where preganglionic neurons synapse with postganglionic neurons that innervate the smooth muscle of the GI tract. Stimulation of the parasympathetic neurons increases peristalsis and the secretions of the GI tract.

Postganglionic sympathetic fibers pass through the submucosal and myenteric plexuses and innervate the GI tract. The effects of sympathetic nerve stimulation are antagonistic to those of parasympathetic nerve stimulation. Sympathetic impulses inhibit peristalsis, reduce secretions, and constrict muscle sphincters along the GI tract.

1. Describe the position of the peritoneal membranes. What are the functions of the mesentery and greater omentum? Which organs are retroperitoneal?
2. List the four tunics of the GI tract and identify their major tissue types. What are the functions of these four tunics?
3. Compare the effects of parasympathetic and sympathetic inputs to the GI tract.

Mouth, Pharynx, and Associated Structures

Ingested food is changed into a bolus by the mechanical action of teeth and by the chemical activity of saliva. The bolus is swallowed in the process of deglutition.

Objective 6 Describe the anatomy of the oral cavity.

Objective 7 Contrast the deciduous and permanent dentitions and describe the structure of a typical tooth.

Objective 8 Describe the location and histological structures of the salivary glands and list the functions of saliva.

The functions of the *mouth* and associated structures are to form a receptacle for food, to initiate digestion through mastication, to swallow food, and to form words in speech. The mouth can also assist the respiratory system in breathing air.

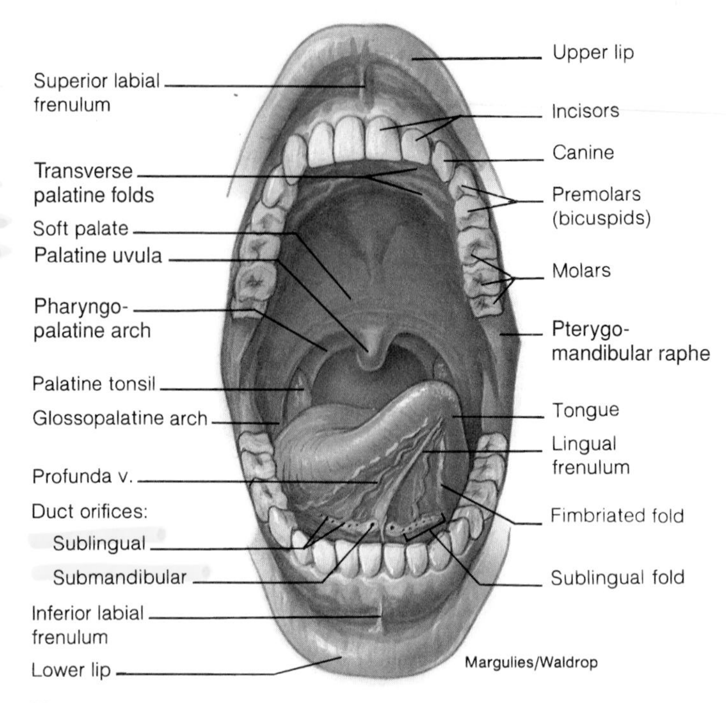

Figure 18.5

The superficial structures of the oral cavity.

The *pharynx,* which is posterior to the mouth, serves as a common passageway for both the respiratory and digestive systems. Both the mouth and pharynx are lined with nonkeratinized stratified squamous epithelium, which is constantly moistened by the secretion of saliva.

The **mouth,** also known as the **oral cavity** (fig. 18.5), is formed by the *cheeks, lips, hard palate* and *soft palate.* The **vestibule** of the oral cavity is the depression between the cheeks and lips externally and the gums and teeth internally (fig. 18.6). The opening of the oral cavity is referred to as the **oral orifice,** and the opening between the oral cavity and the pharynx is called the **fauces** (*faw'sēz*).

Cheeks, Lips, and Palate

The **cheeks** form the lateral walls of the oral cavity. They consist of outer layers of skin, subcutaneous fat, facial muscles that assist in manipulating food in the oral cavity, and inner linings of moistened stratified squamous epithelium. The anterior portion of the cheeks terminates in the superior and inferior lips that surround the oral orifice.

The **lips** are fleshy, highly mobile organs whose principal function in humans is associated with speech. Lips also serve for suckling, manipulating food, and keeping food between the upper and lower teeth. Each lip is attached from its inner surface to the gum by a midline fold of mucous membrane called the **labial**

pharynx: L. *pharynx,* throat
fauces: L. *fauces,* throat

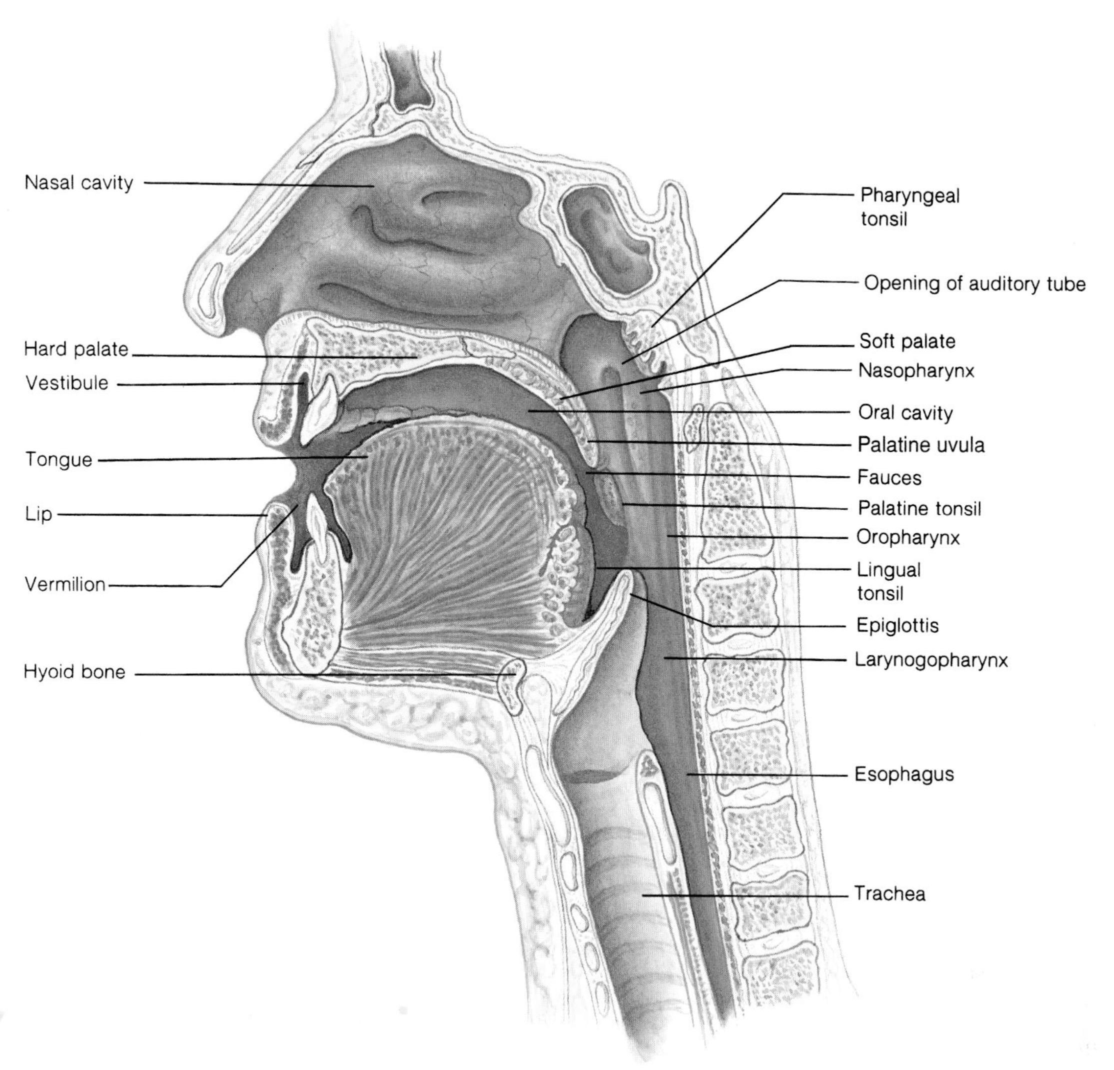

Figure 18.6

A sagittal section of the facial region showing the oral cavity, nasal cavity, and pharynx.

frenulum (*fren'yŭ-lum*) (fig. 18.5). The lips are formed from the orbicularis oris muscle and associated connective tissue, and are covered with soft, pliable skin. Between the outer skin and the mucous membrane of the oral cavity is a transition zone called the **vermilion.** Lips are red to reddish brown because of blood vessels close to the surface. The numerous sensory receptors in the lips aid in determining the temperature and texture of food.

Suckling is innate to newborns. Their lips are well formed for this activity and even contain blisterlike milk pads that aid in suckling. The wide nostrils and receding lower jaw of infants also facilitate suckling.

The **palate,** which forms the roof of the oral cavity, consists of the bony hard palate anteriorly and the soft palate posteriorly (figs. 18.5 and 18.6). The **hard palate,** formed by the palatine processes of the maxillae and the horizontal plates of the palatine bones, is covered with a mucous membrane. **Transverse palatine folds,** or *palatal rugae* (*roo'je*), are located along the mucous membrane of the hard palate. These structures serve as friction ridges against which the tongue is placed during swallowing. The **soft palate** is a muscular arch covered with mucous membrane and is continuous with the hard palate anteriorly. Suspended from the middle lower border of the soft palate is a cone-shaped projection called the **palatine uvula** (*yoo'vyŭ-lŭ*). During swallowing, the soft palate and palatine uvula are drawn upward, closing the nasopharynx and preventing food and fluid from entering the nasal cavity.

Two muscular folds extend downward from both sides of the base of the palatine uvula (figs. 18.5 and 18.6). The anterior fold is called the **glossopalatine arch,** and the posterior fold is the **pharyngopalatine arch.** Between these two arches is the **palatine tonsil.**

vermilion: O.E. *vermeylion*, red-colored

L. *uvula*, small grapes

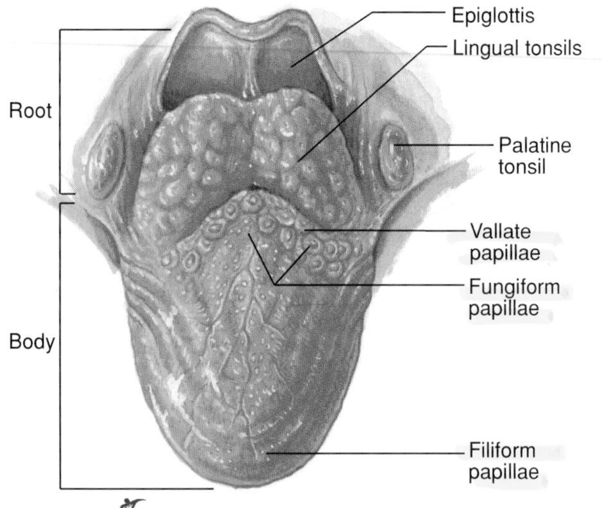

Figure 18.7

The surface of the tongue.

Tongue

As a digestive organ, the **tongue** functions to move food around in the mouth during mastication and to assist in swallowing food. It is also essential in producing speech. The tongue is a mass of skeletal muscle covered with a mucous membrane. Extrinsic tongue muscles (those that insert upon the tongue) move the tongue from side to side and in and out. Only the anterior two-thirds of the tongue lies in the oral cavity; the remaining one-third lies in the pharynx (fig. 18.6) and is attached to the hyoid bone. Rounded masses of **lingual tonsils** are located on the superior surface of the base of the tongue (fig. 18.7). The inferior surface of the tongue is connected along the midline anteriorly to the floor of the mouth by the vertically positioned **lingual frenulum** (see fig. 18.5).

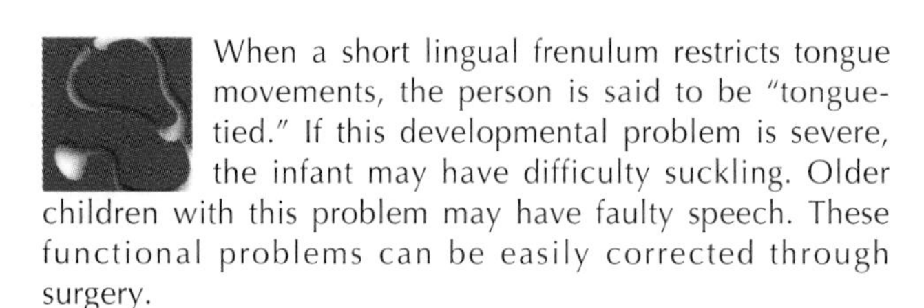

When a short lingual frenulum restricts tongue movements, the person is said to be "tongue-tied." If this developmental problem is severe, the infant may have difficulty suckling. Older children with this problem may have faulty speech. These functional problems can be easily corrected through surgery.

On the surface of the tongue are numerous small elevations called **papillae** (*pă-pil'ē*). The papillae give the tongue a distinct roughened surface that aids the handling of food. As described in chapter 15, some of them also contain taste buds that respond to sweet, salty, sour, and bitter chemical stimuli. Three types of papillae are present on the surface of the tongue: **filiform, fungiform,** and **vallate** (fig. 18.7). Filiform papillae are sensitive to touch, have tapered tips, and are by far the most numerous. These papillae lack taste buds and are not involved in the perception of taste. The larger, rounded fungiform papillae are scattered among the filiform type. The few vallate papillae are arranged in a V shape on the posterior surface of the tongue (see fig. 15.6).

Teeth

Humans and other mammals have *heterodont dentition*. This means that they have various types of **teeth** (fig. 18.8) that are adapted to handle food in particular ways. The four pairs (upper and lower jaws) of anteriormost teeth are the **incisors** (*in-si'sorz*). The chisel-shaped incisors are adapted for cutting and shearing food. The two pairs of cone-shaped **canines** (*cuspids*) are located at the anterior corners of the mouth; they are adapted for holding and tearing. Incisors and canines are further characterized by a single root on each tooth. Located behind the canines are the **premolars** (*bicuspids*), and **molars.** These teeth have two or three roots and somewhat rounded, irregular surfaces called **dental cusps** for crushing and grinding food.

Humans are *diphyodont* (*di-fi'ŏ-dont*); that is, normally two sets of teeth develop in a person's lifetime. Twenty **deciduous** (milk) **teeth** begin to erupt at about 6 months of age (fig. 18.9 and tables 18.2 and 18.3), beginning with the incisors. All of the deciduous teeth normally erupt by the age of 2½. Thirty-two **permanent teeth** replace the deciduous teeth in a predictable sequence. This process begins at about age 6 and continues until about age 17. The *third molars* ("wisdom teeth") are the last to erupt. There may not be room in the jaw to accommodate the wisdom teeth, however, in which case they may grow sideways and become impacted, or emerge only partially. If they do erupt at all, it is usually between the ages of 17 and 25. Presumably, a person has acquired some wisdom by then—hence, the popular name for the third molars.

A **dental formula** is a graphic representation of the types, number, and position of teeth in the oral cavity. Such a formula can be written for each species of mammal with heterodontia. Following are the deciduous and permanent dental formulae for humans:

Formula for deciduous dentition:

$$\mathbf{I}\ 2/2,\ \mathbf{C}\ 1/1,\ \mathbf{DM}\ 2/2 = 10 \times 2 = 20 \text{ teeth}$$

Formula for permanent dentition:

$$\mathbf{I}\ 2/2,\ \mathbf{C}\ 1/1,\ \mathbf{P}\ 2/2,\ \mathbf{M}\ 3/3 = 16 \times 2 = 32 \text{ teeth}$$

where **I** = incisor; **C** = canine; **P** = premolar; **DM** = deciduous molar; **M** = molar.

papilla: L. *papula*, little nipple
filiform: L. *filum*, thread; *forma*, form
fungiform: L. *fungus*, fungus; *forma*, form
vallate: L. *vallatus*, surrounded with a rampart

heterodont: Gk. *heteros*, other; *odous*, tooth
incisor: L. *incidere*, to cut
canine: L. *canis*, dog
molar: L. *mola*, millstone
deciduous: L. *deciduus*, to fall away

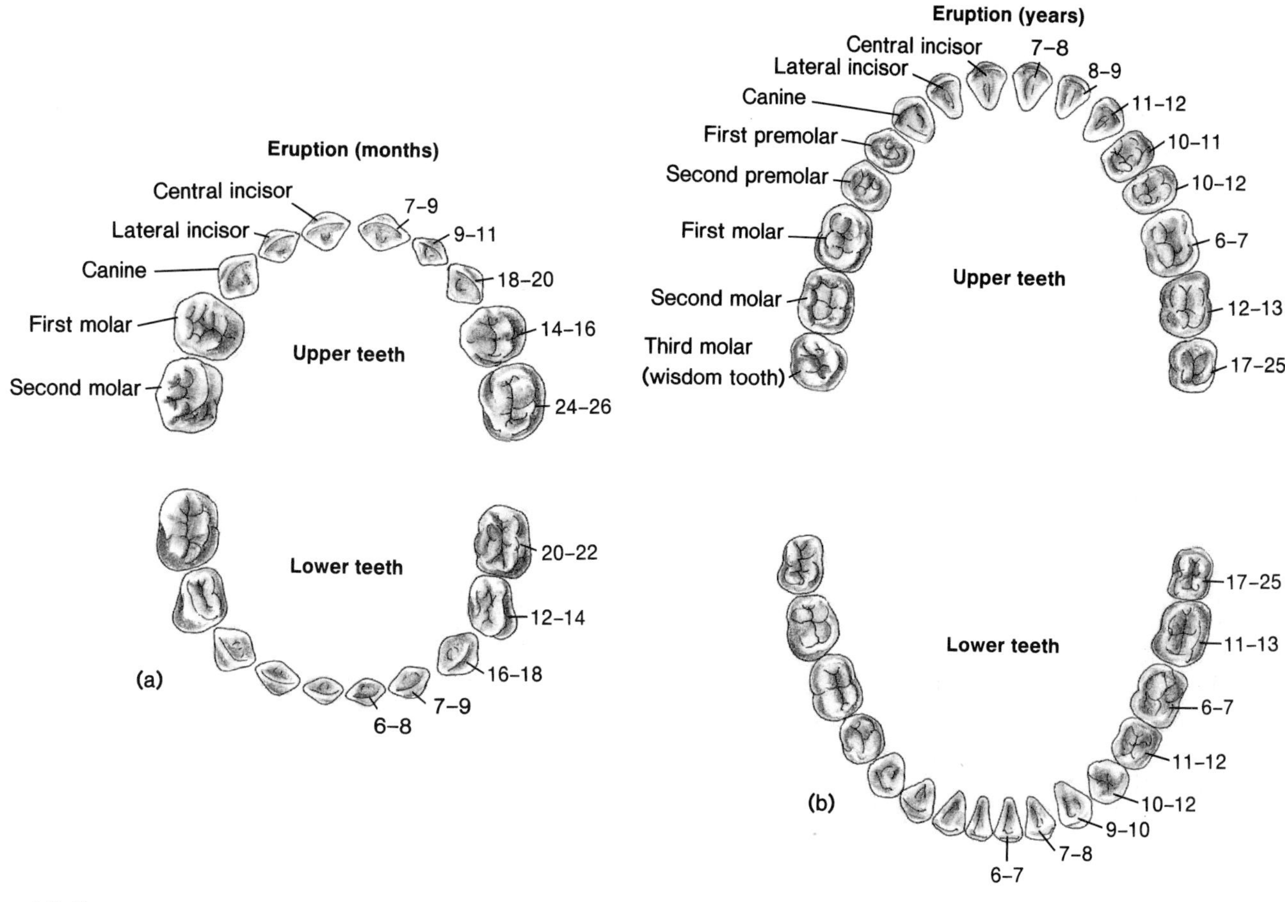

Figure 18.8

Dentitions and the sequence of eruptions. (*a*) Deciduous teeth and (*b*) permanent teeth.

Table 18.2

Eruption Sequence and Loss of Deciduous Teeth

Type of Tooth	Average Age of Eruption: Lower	Average Age of Eruption: Upper	Average Age of Loss
Central incisors	6–8 mos	7–9 mos	7 yrs
Lateral incisors	7–9 mos	9–11 mos	8 yrs
First molars	12–14 mos	14–16 mos	10 yrs
Canines (cuspids)	16–18 mos	18–20 mos	10 yrs
Second molars	20–22 mos	24–26 mos	11–12 yrs

Table 18.3

Eruption Sequence of Permanent Teeth

Type of Tooth	Average Age of Eruption: Lower	Average Age of Eruption: Upper
First molars	6–7 yrs	6–7 yrs
Central incisors	6–7 yrs	7–8 yrs
Lateral incisors	7–8 yrs	8–9 yrs
Canines (cuspids)	9–10 yrs	11–12 yrs
First premolars (bicuspids)	10–12 yrs	10–11 yrs
Second premolars (bicuspids)	11–12 yrs	10–12 yrs
Second molars	11–13 yrs	12–13 yrs
Third molars (wisdom)	17–25 yrs	17–25 yrs

The dental cusps of the upper and lower premolars and molars occlude for chewing food, whereas the upper incisors normally form an overbite with the incisors of the lower jaw. An overbite of the upper incisors creates a shearing action as these teeth slide past one another. Masticated food is mixed with saliva, which initiates chemical digestion and aids swallowing. The soft, flexible mass of food that is swallowed is called a *bolus* (*bo'lus*).

bolus: Gk. *bolos*, lump

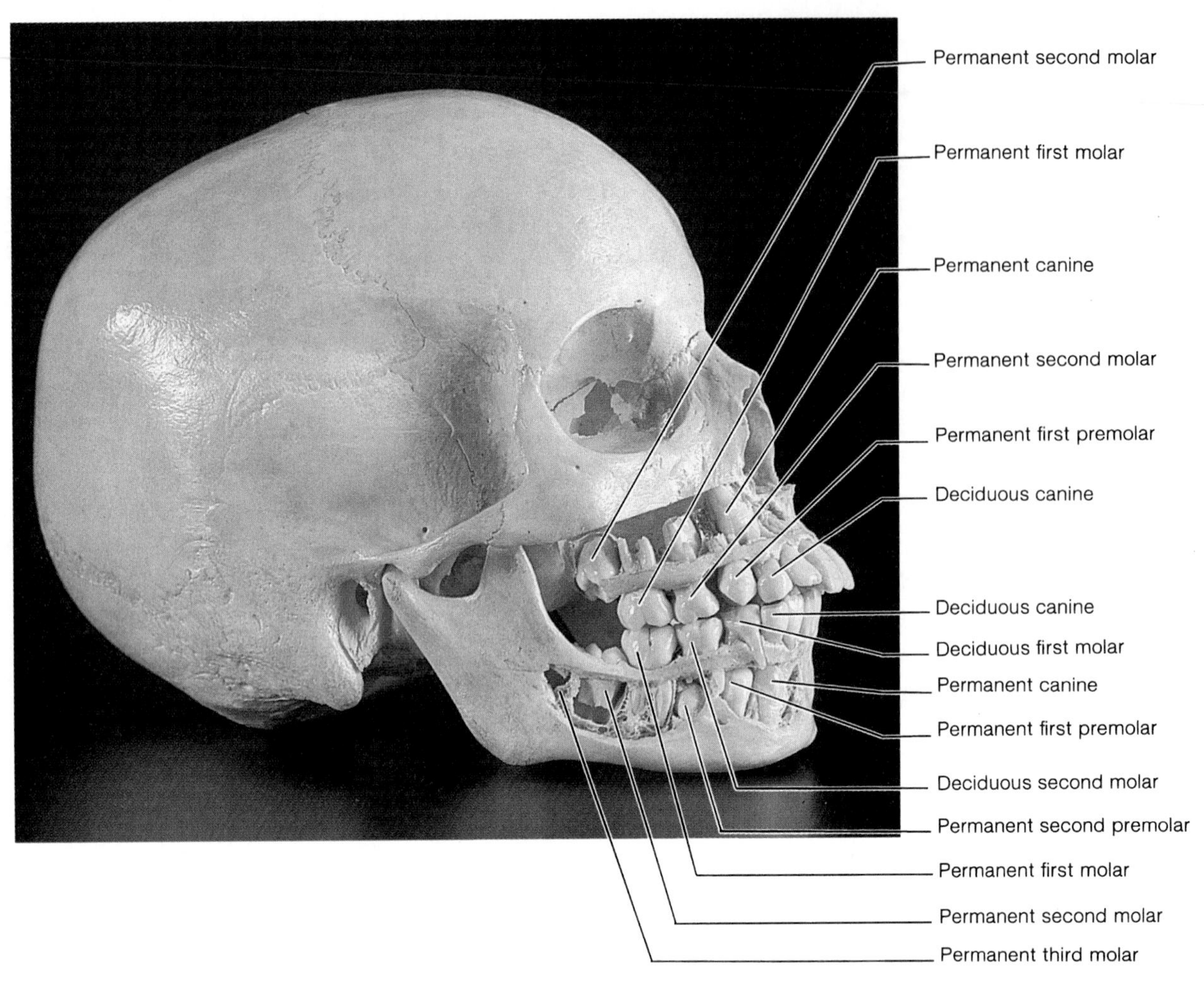

Figure 18.9

A skull showing the eruption of teeth in a youth about 10 years old.

A tooth consists of an exposed **crown,** which is supported by a **neck** that is anchored firmly into the jaw by one or more **roots** (fig. 18.10). The roots of teeth fit into sockets, called **dental alveoli,** in the alveolar processes of the mandible and maxillae. Each socket is lined with a connective tissue periosteum, specifically called the **periodontal membrane.** The root of a tooth is covered with a bonelike material called the **cementum;** fibers in the periodontal membrane insert into the cementum and fasten the tooth in its dental alveolus. The **gingiva** (*jin-ji'vă*) (*gum*) is the mucous membrane surrounding the alveolar processes in the oral cavity.

The bulk of a tooth consists of **dentin,** a substance similar to bone but harder. Covering the dentin on the outside and forming the crown is a tough, durable layer of **enamel.** Enamel is composed primarily of calcium phosphate and is the hardest substance in the body. The central region of the tooth contains the **pulp cavity.** The pulp cavity contains the **pulp,** which is composed of connective tissue with blood vessels, lymph vessels, and nerves. A **root canal,** continuous with the pulp cavity, opens to the connective tissue surrounding the root through an **apical foramen.** The tooth receives nourishment through vessels traversing the apical foramen. Proper nourishment is particularly important during embryonic development. The diet of the mother should contain an abundance of calcium and vitamin D during pregnancy to ensure the proper development of the baby's teeth.

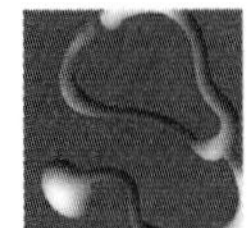

Even though enamel is the hardest substance in the body, bacterial activity may result in *dental caries* (*kar'ēz*), or *tooth decay.* Refluxed stomach acids also destroy tooth enamel and constant vomiting, as in the eating disorder *bulimia nervosa,* contributes to the development of dental caries. Cavities in the teeth must be artificially filled because new enamel is not produced after a tooth erupts. The rate of tooth decay decreases after age 35, but then problems with the gums may develop. *Periodontal diseases* result from plaque or tartar buildup at the gum line. This buildup pulls the gum away from the teeth, allowing bacterial infections to develop.

gingiva: L. *gingiva,* gum
dentin: L. *dens,* tooth

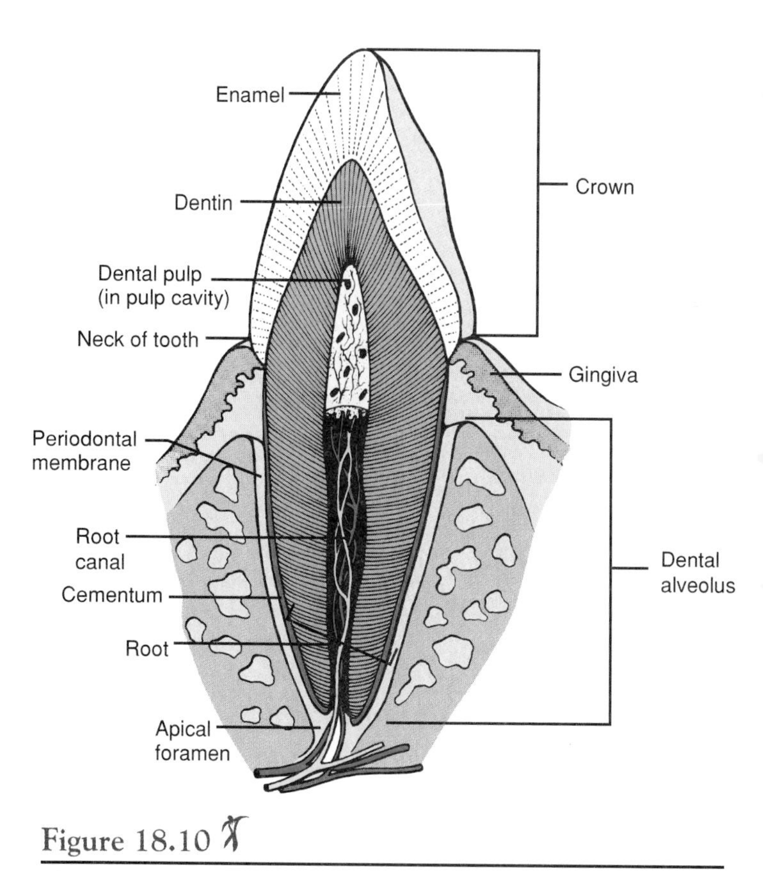

Figure 18.10

The structure of a tooth shown in a vertical section through one of the canines.

Salivary Glands

The **salivary glands** are accessory digestive glands that produce a fluid secretion called *saliva*. Saliva functions as a solvent in cleansing the teeth and dissolving food molecules so that they can be tasted. Saliva also contains starch-digesting enzymes and lubricating mucus, which aids swallowing. Saliva is secreted continuously, but usually only in sufficient amounts to keep the mucous membranes of the oral cavity moist. The amount of saliva secreted daily ranges from 1.0 to 1.5 L.

Numerous minor salivary glands are located in the mucous membranes of the palatal region of the oral cavity. However, three pairs of salivary glands that lie outside the oral cavity produce most of the saliva, which is transported to the oral cavity via **salivary ducts.** The three major pairs of extrinsic salivary glands are the *parotid, submandibular,* and *sublingual glands* (fig. 18.11).

The **parotid** (*pă-rot'id*) **gland** is the largest of the salivary glands. It is positioned below and in front of the auricle of the ear, between the skin and the masseter muscle. Saliva produced in the parotid gland drains through the **parotid** (Stensen's) **duct.** The parotid duct parallels the zygomatic arch across the masseter muscle, pierces the buccinator muscle, and empties into the oral cavity opposite the second upper molar. It is the parotid gland that becomes infected and swollen with the mumps.

The **submandibular gland** lies inferior to the body of the mandible, about midway along the inner side of the jaw. This gland is covered by the more superficial mylohyoid muscle. Saliva produced in the submandibular gland drains through the **submandibular** (Wharton's) **duct** and empties into the floor of the mouth on the lateral side of the lingual frenulum.

The **sublingual gland** lies under the mucous membrane of the floor of the mouth. Each sublingual gland contains several small ducts (Rivinus' ducts) that empty into the floor of the mouth in an area posterior to the papilla of the submandibular duct.

Two types of secretory cells, **serous** and **mucous cells,** are found in all salivary glands in various proportions (fig. 18.12). Serous cells produce a watery fluid containing digestive enzymes; mucous cells secrete a thick, stringy mucus. Cuboidal epithelial cells line the lumina of the salivary ducts.

The salivary glands are innervated by both divisions of the autonomic nervous system. Sympathetic impulses stimulate the secretion of small amounts of viscous saliva. Parasympathetic stimulation causes the secretion of large volumes of watery saliva. Physiological responses of this type occur whenever a person sees, smells, tastes, or even thinks about desirable food. Information about the salivary glands is summarized in table 18.4.

Pharynx

The funnel-shaped **pharynx** (*far'ingks*) is a passageway approximately 13 cm (5 in.) long connecting the oral and nasal cavities to the esophagus and larynx. The pharynx has both digestive and respiratory functions. The supporting walls of the pharynx are composed of skeletal muscle, and the lumen is lined with a mucous membrane containing stratified squamous epithelium. The pharynx is divided into three regions: the *nasopharynx*, posterior to the nasal cavity; the *oropharynx*, posterior to the oral cavity; and the *laryngopharynx*, at the level of the larynx (see fig. 17.3).

The external *circular layer* of pharyngeal muscles, called **constrictors** (fig. 18.13), compresses the lumen of the pharynx involuntarily during swallowing. The **superior constrictor muscle** attaches to bony processes of the skull and mandible and encircles the upper portion of the pharynx. The **middle constrictor muscle** arises from the hyoid bone and stylohyoid ligament and encircles the middle portion of the pharynx. The **inferior constrictor muscle** arises from the cartilages of the larynx and encircles the lower portion of the pharynx. During breathing, the lower portion of the inferior constrictor muscle is contracted, preventing air from entering the esophagus.

---

parotid: Gk. *para*, beside; *otos*, ear
Stensen's duct: from Nicholaus Stensen, Danish anatomist, 1638–86

Wharton's duct: from Thomas Wharton, English physician, 1614–73
Rivinus' ducts: from August Quirinus Rivinus, German anatomist, 1652–1723

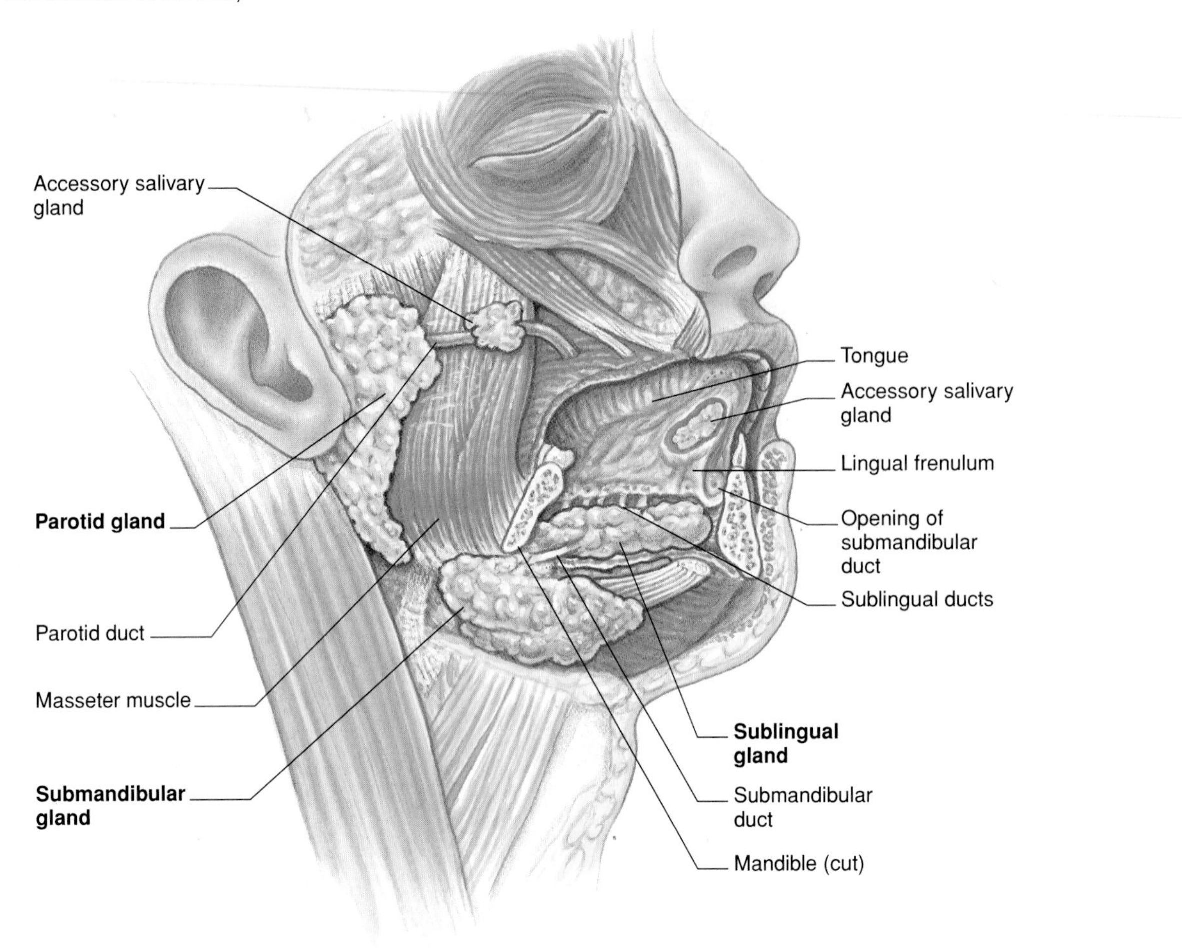

Figure 18.11

The salivary glands.

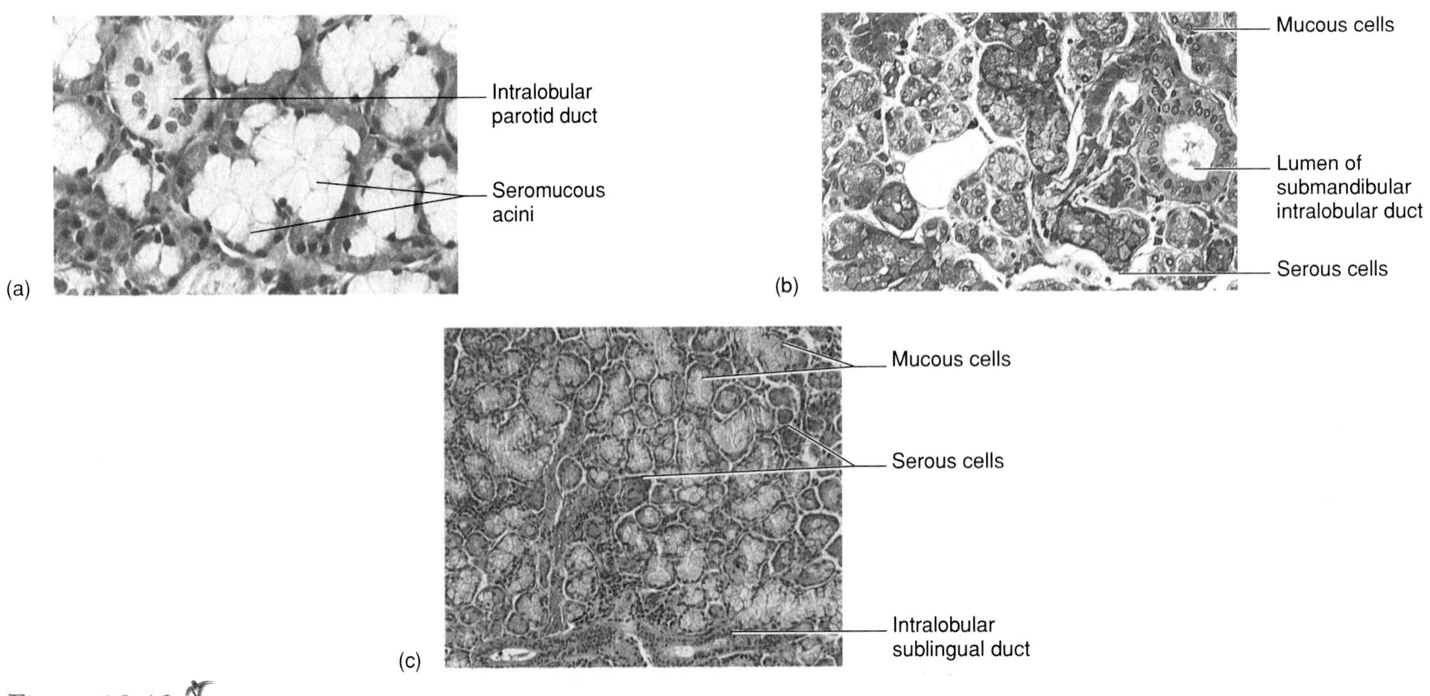

Figure 18.12

The histology of the salivary glands. (*a*) The parotid gland, (*b*) the submandibular gland, and (*c*) the sublingual gland.

Table 18.4
Major Salivary Glands

Gland	Location	Duct	Entry into Oral Cavity	Type of Secretion
Parotid gland	Anterior and inferior to auricle; subcutaneous over masseter muscle	Parotid (Stensen's) duct	Lateral to upper second molar	Watery serous fluid, salts, and enzyme
Submandibular gland	Inferior to the base of the tongue	Submandibular (Wharton's) duct	Papilla lateral to lingual frenulum	Watery serous fluid with some mucus
Sublingual gland	Anterior to submandibular gland under the tongue	Several small ducts (Rivinus' ducts)	Ducts along the base of the tongue	Mostly thick, stringy mucus; salts; and enzyme (salivary amylase)

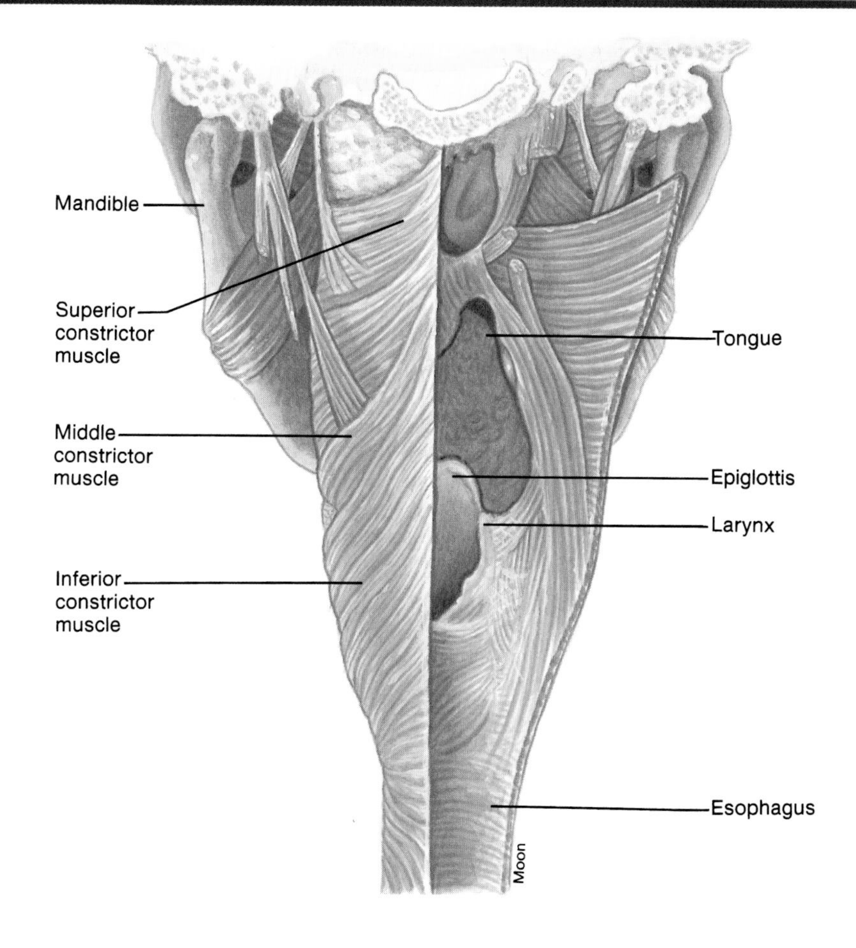

Figure 18.13

A posterior view of the constrictor muscles of the pharynx. The right side has been cut away to illustrate the interior structures in the pharynx.

The motor and most of the sensory innervation to the pharynx is via the pharyngeal plexus, situated chiefly on the middle constrictor muscle. It is formed by the pharyngeal branches of the glossopharyngeal and vagus nerves, together with a deep sympathetic branch from the superior cervical ganglion.

The pharynx is served by the ascending pharyngeal and inferior thyroid arteries, both of which are branches of the external carotid arteries. Venous return is via the internal jugular veins.

1. Define the terms *heterodont* and *diphyodont*. Which of the four kinds of teeth is not found in deciduous dentition?
2. Where are the enamel, dentin, cementum, and pulp of a tooth located? Explain how a tooth is anchored into its socket.
3. Describe the location of the parotid and submandibular ducts and state where they empty into the oral cavity.

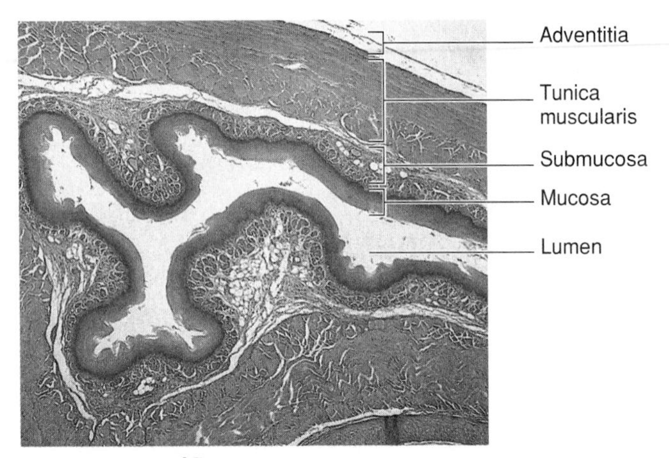

Figure 18.14

The histology of the esophagus.

The **lower esophageal** (gastroesophageal) **sphincter** is a slight thickening of the circular muscle fibers at the junction of the esophagus and the stomach. After food or fluid pass into the stomach, this sphincter constricts to prevent the stomach contents from regurgitating into the esophagus. There is a normal tendency for this to occur because the thoracic pressure is lower than the abdominal pressure as a result of the air-filled lungs.

The lower esophageal sphincter is not a well-defined sphincter muscle comparable to others located elsewhere along the GI tract, and it does at times permit the acidic contents of the stomach to enter the esophagus. This can create a burning sensation commonly called *heartburn,* although the heart is not involved. In infants under a year of age, the lower esophageal sphincter may function erratically, causing them to "spit up" following meals. Certain mammals, such as rodents, have a true lower esophageal sphincter and cannot regurgitate, which is why poison grains are effective in killing mice and rats.

Esophagus and Stomach

A bolus of food is passed from the esophagus to the stomach, where it is churned and mixed with gastric secretions. The chyme thus produced is sent past the pyloric sphincter of the stomach to the duodenum.

Objective 9 Describe the steps in deglutition.

Objective 10 Describe the location, gross structure, and functions of the stomach.

Objective 11 Describe the histological structure of the esophagus and stomach. List the cell types in the gastric mucosa, along with their secretory products.

Esophagus

The **esophagus** is that portion of the GI tract that connects the pharynx to the stomach (see figs. 18.1 and 18.15). It is a collapsible muscular tube, approximately 25 cm (10 in.) long, originating at the larynx and lying posterior to the trachea.

The esophagus is located within the mediastinum of the thorax and passes through the diaphragm just above the opening into the stomach. The opening through the diaphragm is called the **esophageal hiatus** (*ĕ-sof″ă-je′al hi-a′tus*). The esophagus is lined with a nonkeratinized stratified squamous epithelium (fig. 18.14); its walls contain either skeletal or smooth muscle, depending on the location. The upper third of the esophagus contains skeletal muscle; the middle third, a combination of skeletal and smooth muscle; and the terminal portion, smooth muscle only.

esophagus: Gk. *oisein*, to carry; *phagema*, food

Swallowing Mechanisms

Swallowing, or **deglutition** (*de″gloo-tish′un*), is the complex mechanical and physiological act of moving food or fluid from the oral cavity to the stomach. For descriptive purposes, deglutition is divided into three stages.

The first deglutitory stage is voluntary and follows mastication, if food is involved. During this stage, the mouth is closed and breathing is temporarily interrupted. A bolus is formed as the tongue is elevated against the palate through contraction of the mylohyoid and styloglossus muscles and the intrinsic muscles of the tongue.

The second stage of deglutition is the passage of the bolus through the pharynx. The events of this stage are involuntary and are elicited by stimulation of sensory receptors located at the opening of the oropharynx. Pressure of the tongue against the transverse palatine folds (palatal rugae) seals off the nasopharynx from the oral cavity, creates a pressure, and forces the bolus into the oropharynx. The soft palate and pendulant palatine uvula are elevated to close the nasopharynx as the bolus passes. The hyoid bone and the larynx are also elevated. Elevation of the larynx against the epiglottis seals the glottis so that food or fluid is less likely to enter the trachea. Sequential contraction of the constrictor muscles of the pharynx moves the bolus through the pharynx to the esophagus. This stage is completed in just a second or less.

The third stage, the entry and passage of food through the esophagus, is also involuntary. The bolus is moved through the esophagus by peristalsis (fig. 18.15). In the case of fluids, the entire process of deglutition takes place in slightly more than a second; for a typical bolus, the time frame is 5 to 8 seconds.

Stomach

The **stomach**—the most distensible part of the GI tract—is located in the upper left abdominal quadrant, immediately below the diaphragm. Typically J-shaped when empty, the stomach is

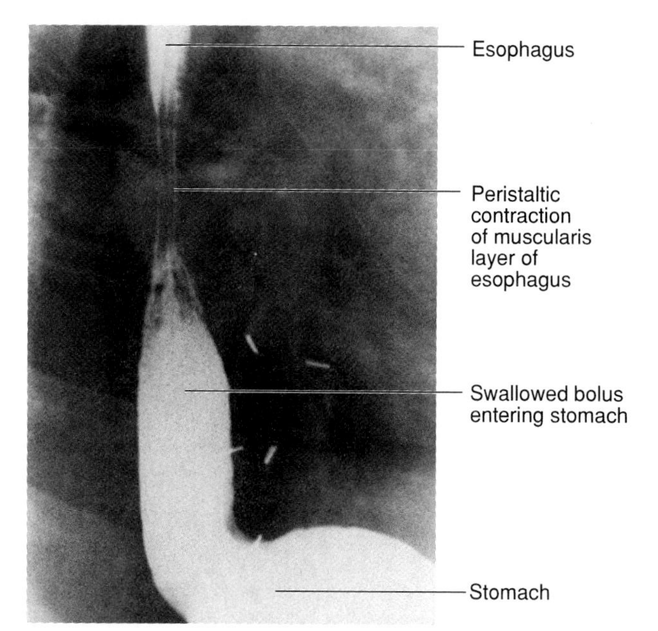

Figure 18.15

An anteroposterior radiograph of the esophagus showing peristaltic contraction and movement of a bolus into the stomach.

continuous with the esophagus superiorly and empties into the duodenal portion of the small intestine inferiorly (fig. 18.16). In the stomach, which serves as a "holding organ" for ingested food, the food is mechanically churned with gastric secretions to form a pasty material called **chyme** (*kīm*). Once formed, chyme is moved from the stomach to the small intestine.

The stomach is divided into four regions: the cardia, fundus, body, and pylorus (fig. 18.17). The **cardia** is the narrow upper region immediately below the lower esophageal sphincter. The **fundus** is the dome-shaped portion to the left of and in direct contact with the diaphragm. The **body** is the large central portion, and the **pylorus** is the funnel-shaped terminal portion. The **pyloric sphincter** is the modified circular muscle at the end of the pylorus, where it joins the small intestine. *Pylorus* is a Greek word meaning "gatekeeper," and this junction is just that, regulating the movement of chyme into the small intestine and prohibiting backflow.

The stomach has two surfaces and two borders. The broadly rounded surfaces are referred to as the **anterior** and **posterior surfaces.** The medial concave border is the **lesser curvature** (fig. 18.17), and the lateral convex border is the **greater curvature.** The lesser omentum extends between the lesser curvature and the liver, and the greater omentum is attached to the greater curvature.

The wall of the stomach is composed of the same four tunics found in other regions of the GI tract, with two principal modifications: (1) an extra *oblique muscle layer* is present in the muscularis,

chyme: L. *chymus*, juice
cardia: Gk. *kardia*, heart (upper portion, nearer the heart)
fundus: L. *fundus*, bottom
pylorus: Gk. *pyloros*, gatekeeper

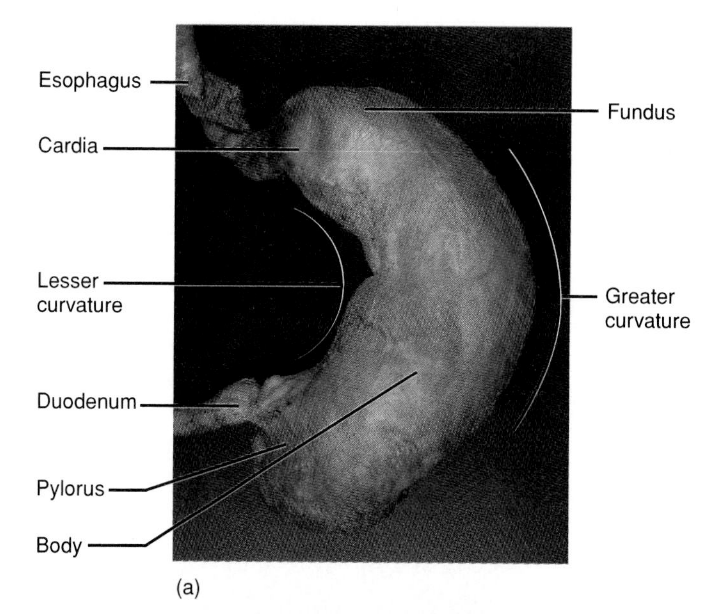

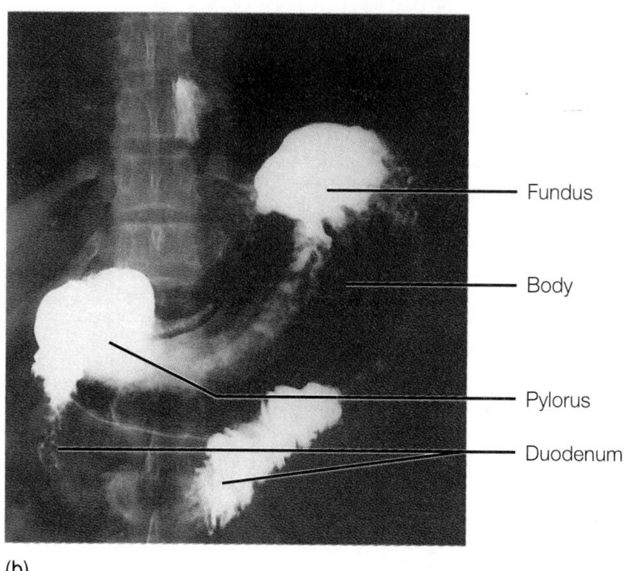

Figure 18.16

The stomach. (*a*) As seen from a cadaver, the stomach is a J-shaped organ. (*b*) A radiograph of the stomach is of clinical value for detecting ulcers, constrictions, or ingested objects. A patient swallows radiopaque barium, which coats the lining of the stomach and duodenum. These structures and any abnormalities will then show up clearly in the radiograph.

and (2) the mucosa is thrown into numerous longitudinal folds, called **gastric folds** or *gastric rugae*, which permit stomach distension. The mucosa is further characterized by the presence of microscopic **gastric pits** and **gastric glands** (figs. 18.18 and 18.19).

There are five types of cells in the gastric glands that secrete specific products.

1. **Goblet cells** secrete protective mucus.
2. **Parietal cells** secrete hydrochloric acid (HCl).
3. **Principal cells** (chief cells) secrete pepsinogen, an inactive form of the protein-digesting enzyme pepsin.

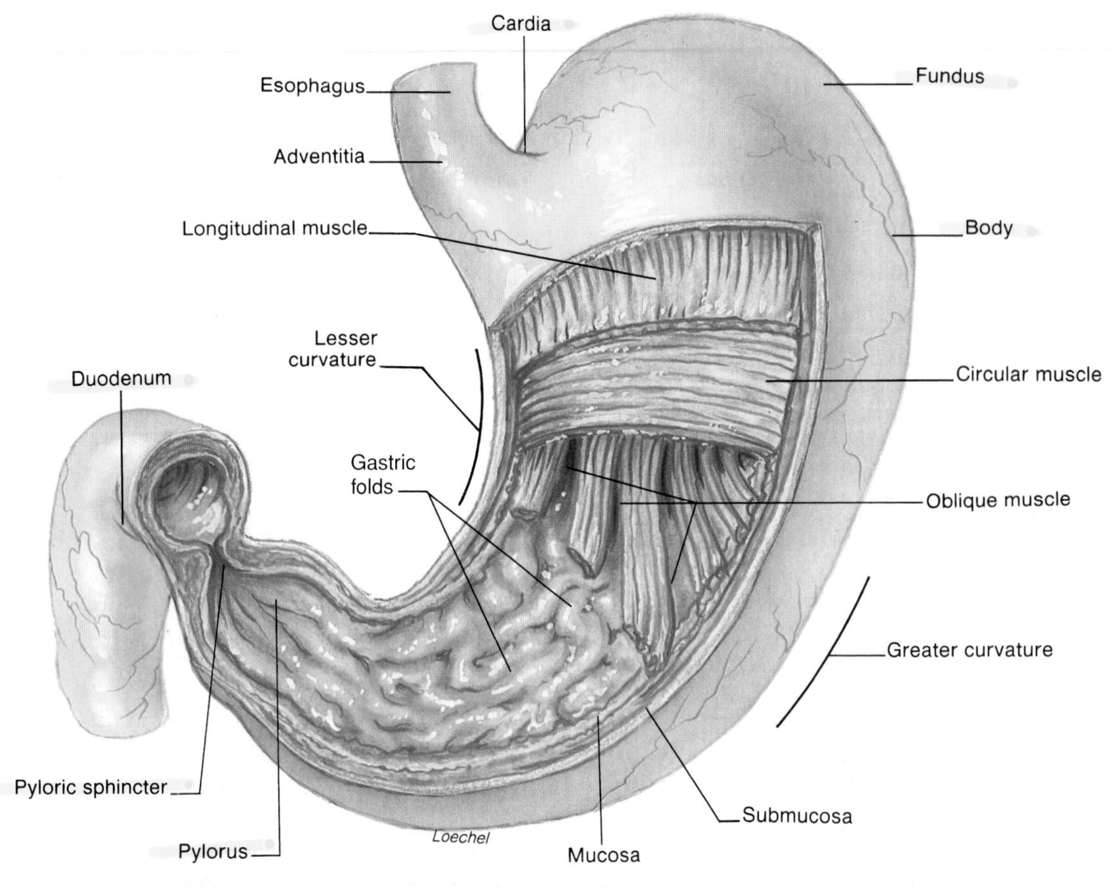

Figure 18.17

The major regions and structures of the stomach.

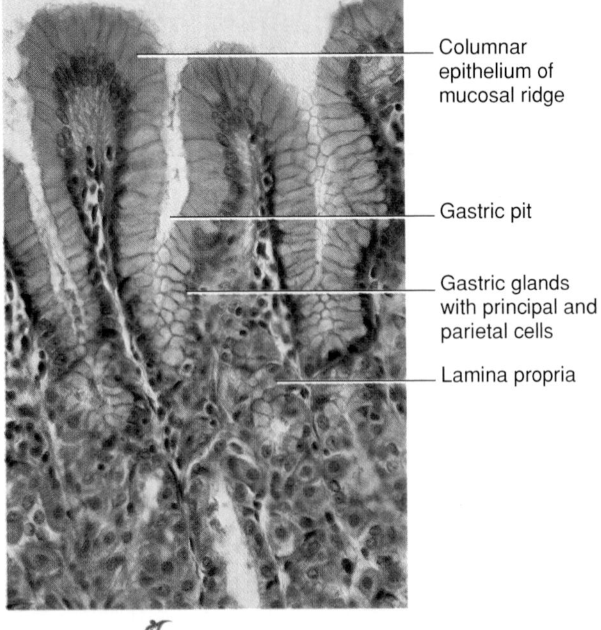

Figure 18.18

The histology of the mucosa of the stomach.

4. **Argentaffin** (*ar-jent'ă-fin*) **cells** secrete serotonin, histamine, and autocrine regulators.
5. **Endocrine cells** (G cells) secrete the hormone gastrin into the blood.

In addition to these products, the gastric mucosa (probably the parietal cells) secretes *intrinsic factor*, a polypeptide that is required for absorption of vitamin B_{12} in the small intestine.

The stomach is sensitive to emotional stress. Mucus, secreted by mucous cells of the stomach, is important in preventing hydrochloric acid and the digestive enzyme pepsin from eroding the stomach wall. *Peptic ulcers* may be caused by an increase in cellular secretion or by insufficient secretions of protective mucus. Another protective feature is the rapid mitotic activity of the columnar epithelium of the stomach. The entire lining of the stomach is usually replaced every few days. Nevertheless, in the United States approximately 10% of the male population and 4% of the female population develop peptic ulcers.

argentaffin: L. *argentum*, silver; *affinis*, attraction (become colored with silver stain)

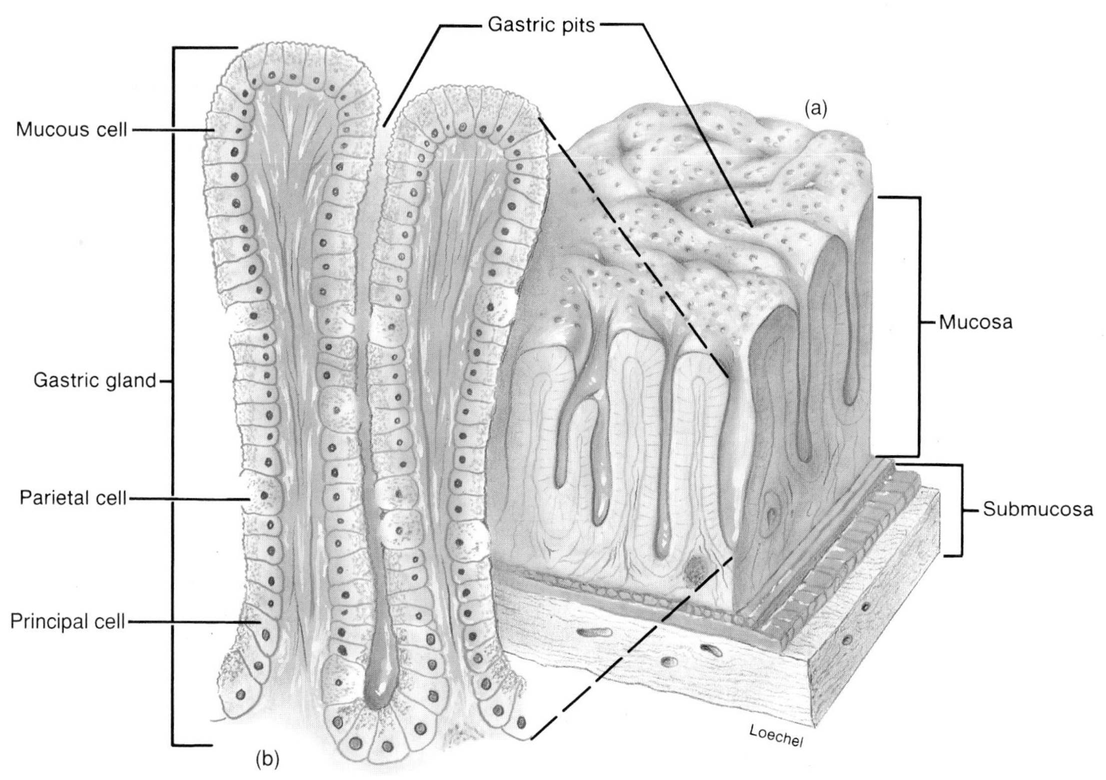

Figure 18.19

Gastric pits and gastric glands of the stomach mucosa. (*a*) Gastric pits are the openings of the gastric glands. (*b*) Gastric glands consist of mucous cells, principal cells, and parietal cells, each type producing a specific secretion.

Table 18.5
Phases of Gastric Secretion

Phase	Response
Cephalic phase	Sight, taste, small or mental stimuli evoke parasympathetic response via vagus nerves; 50–150 ml of gastric juice is secreted
Gastric phase	Food in the stomach stretches the mucosa, and chemical breakdown of protein stimulates the release of gastrin; gastrin stimulates the production of 600–750 ml of gastric juice
Intestinal phase	Chyme entering the duodenum stimulates intestinal cells to release intestinal gastrin; intestinal gastrin stimulates the production of additional small quantities of gastric juice

Regulation of gastric activity is autonomic. The sympathetic neurons arise from the celiac plexus, and the parasympathetic neurons arise from the vagus nerves. Parasympathetic neurons synapse in the myenteric plexus between the muscular layers and in the submucosal plexus in the submucosa. Parasympathetic impulses promote gastric activity, the phases of which are presented in table 18.5.

Vomiting is the reflexive response of emptying the stomach through the esophagus, pharynx, and oral cavity. This action is controlled by the **vomiting center** of the medulla oblongata. Stimuli within the GI tract, especially the duodenum, may activate the vomiting center, as may nauseating odors or sights, motion sickness, or body stress. Various drugs called *emetics* can also stimulate a vomiting reflex. The mechanics of vomiting are as follows: (1) strong, sustained contractions of the upper small intestine, followed by a contraction of the pyloric sphincter; (2) relaxation of the lower esophageal sphincter and contraction of the pyloric portion of the stomach; (3) a shallow inspiration and closure of the glottis; and (4) compression of the stomach against the liver by contraction of the diaphragm and the abdominal muscles. This reflexive sequence causes a forceful ejection of vomit. The feeling of *nausea* is caused by stimuli in the vomiting center and may or may not cause vomiting.

The only function of the stomach that appears to be essential for life is the secretion of *intrinsic factor.* This polypeptide is needed for the intestinal absorption of vitamin B_{12}, which is required for maturation of red blood cells in the bone marrow. Following surgical removal of the stomach (gastrectomy), a patient has to receive vitamin B_{12} (together with intrinsic factor) orally or through injections, so that he or she will not develop *pernicious anemia.*

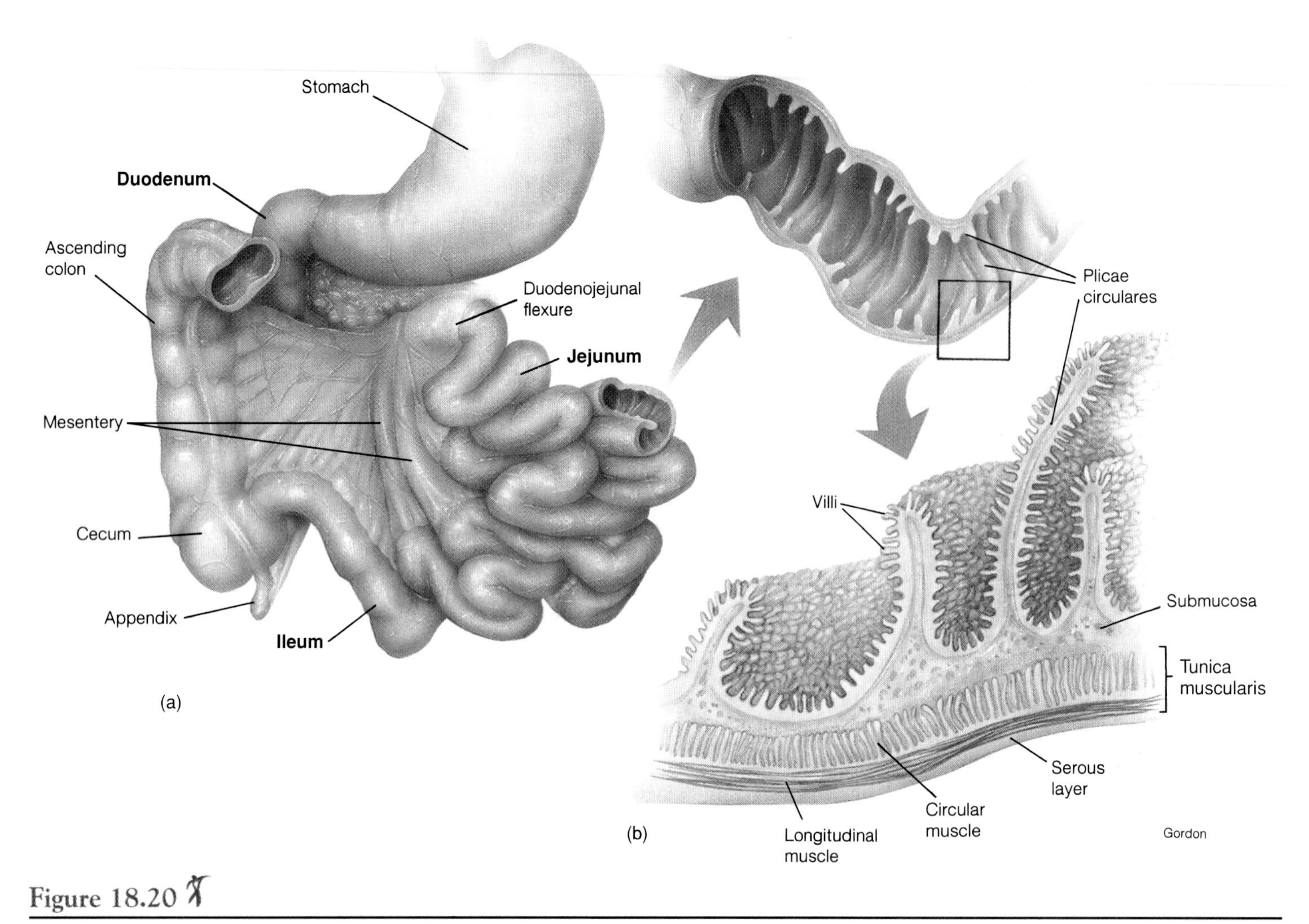

Figure 18.20

The small intestine in relation to the stomach and a part of the large intestine. (*a*) The regions and mesenteric attachment. (*b*) A section of the intestinal wall showing the mucosa and submucosa folded into structures called plicae circulares. (The regions of the small intestine are labeled in boldface type.)

1. Describe the three stages of deglutition with reference to the structures involved.
2. Describe the structure and function of the lower esophageal sphincter.
3. List the functions of the stomach. What is the function of the gastric folds?
4. Describe the modifications of the stomach that aid in mechanical and chemical digestion.

Small Intestine

The small intestine, consisting of the duodenum, jejunum, and ileum, is the site where digestion is completed and nutrients are absorbed. The surface area of the intestinal wall is increased by plicae circulares, villi, and microvilli.

Objective 12 Describe the location and regions of the small intestine and the way in which it is supported.

Objective 13 List the functions of the small intestine and describe the structural adaptations through which these functions are accomplished.

Objective 14 Describe the movements that occur within the small intestine.

The **small intestine** is that portion of the GI tract between the pyloric sphincter of the stomach and the ileocecal valve that opens into the large intestine. It is positioned in the central and lower portions of the abdominal cavity and is supported, except for the first portion, by the **mesentery** (fig. 18.20). The fan-shaped mesentery permits movement of the small intestine but leaves little chance for it to become twisted or kinked. Enclosed within the mesentery are blood vessels, nerves, and lymphatic vessels that supply the intestinal wall.

The small intestine is approximately 3 m (12 ft) long and 2.4 cm (1 in.) wide in a living person, but it will measure nearly

Table 18.6
Digestive Enzymes

Enzyme	Source	Digestive Action
Salivary enzyme		
Amylase	Salivary glands	Begins carbohydrate digestion by converting starch and glycogen to disaccharides
Gastric juice enzyme		
Pepsin	Gastric glands	Begins the digestion of nearly all types of proteins
Intestinal juice enzymes		
Peptidase	Intestinal glands	Converts proteins into amino acids
Sucrase	Intestinal glands	Converts disaccharides into monosaccharides
Maltase		
Lactase		
Lipase	Intestinal glands	Converts fats into fatty acids and glycerol
Amylase	Intestinal glands	Converts starch and glycogen into disaccharides
Nuclease	Intestinal glands	Converts nucleic acids into nucleotides
Enterokinase	Intestinal glands	Activates trypsin
Pancreatic juice enzymes		
Amylase	Pancreas	Converts starch and glycogen into disaccharides
Lipase	Pancreas	Converts fats into fatty acids and glycerol
Peptidases	Pancreas	Convert proteins or partially digested proteins into amino acids
Trypsin		
Chymotrypsin		
Carboxypeptidase		
Nuclease	Pancreas	Converts nucleic acids into nucleotides

twice this length in a cadaver when the muscular wall is relaxed. It is called the "small" intestine because of its relatively small diameter compared to that of the large intestine. The small intestine is the body's major digestive organ and the primary site of nutrient absorption. Its digestive enzymes, along with those of the salivary glands, gastric glands, and pancreas, are summarized in table 18.6.

The small intestine is innervated by the superior mesenteric plexus. The branches of the plexus contain sensory fibers, postganglionic sympathetic fibers, and preganglionic parasympathetic fibers. The arterial blood supply to the small intestine is through the superior mesenteric artery and branches from the celiac trunk and the inferior mesenteric artery. The venous drainage is through the superior mesenteric vein. This vein unites with the splenic vein to form the hepatic portal vein, which carries nutrient-rich blood to the liver (see fig. 16.37).

Regions of the Small Intestine

Based on function and histological structure, the small intestine is divided into three regions.

1. The **duodenum** (*doo"ŏ-de'num, doo-od'ĕ-num*) is a relatively fixed, C-shaped tube, measuring approximately 25 cm (10 in.) from the pyloric sphincter of the stomach to the **duodenojejunal** (*doo-od"ĕ-no"jĕ-joo'nal*) **flexure.** Except for a short portion near the stomach, the duodenum is retroperitoneal. Its concave surface faces to the left, where it receives bile secretions from the liver and gallbladder through the **common bile duct** and pancreatic secretions through the **pancreatic duct** of the pancreas (fig. 18.21). These two ducts unite to form a common entry into the duodenum called the **hepatopancreatic ampulla** (ampulla of Vater), which pierces the duodenal wall and drains into the duodenum from an elevation called the **duodenal papilla.** It is here that bile and pancreatic juice enter the small intestine. The duodenal papilla can be opened or closed by the action of the **sphincter of ampulla** (of Oddi). The duodenum differs histologically from the rest of the small intestine by the presence of **duodenal** (Brunner's) **glands** in the submucosa (fig. 18.22). These compound tubuloalveolar glands secrete mucus and are most numerous near the superior end of the duodenum.

duodenum: L. *duodeni*, twelve each (length of twelve fingers' breadth)

ampulla of Vater: from Abraham Vater, German anatomist, 1684–1751
sphincter of Oddi: from Ruggero Oddi, Italian physician, 1864–1913
Brunner's glands: from Johann C. Brunner, Swiss anatomist, 1653–1727

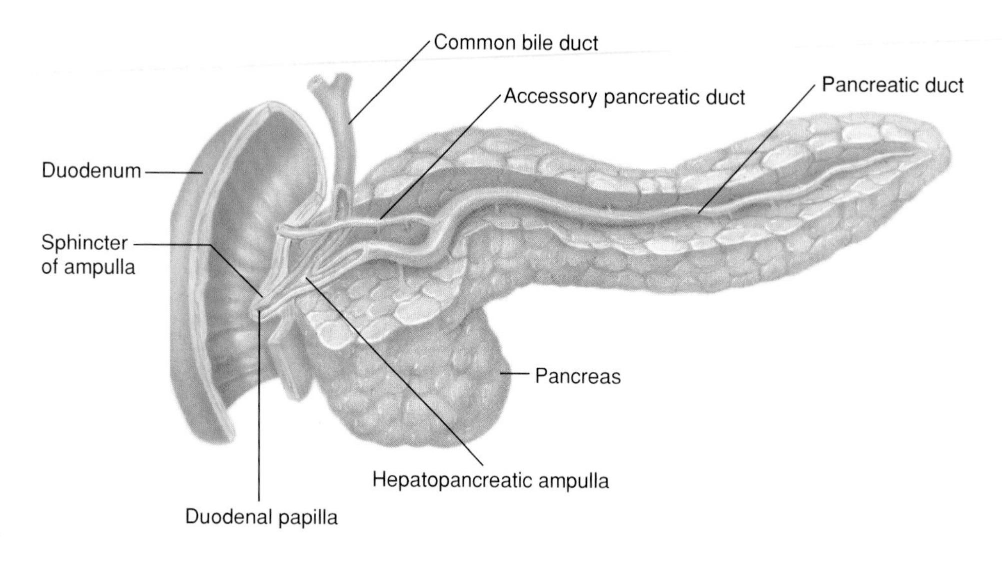

Figure 18.21

The duodenum and associated structures.

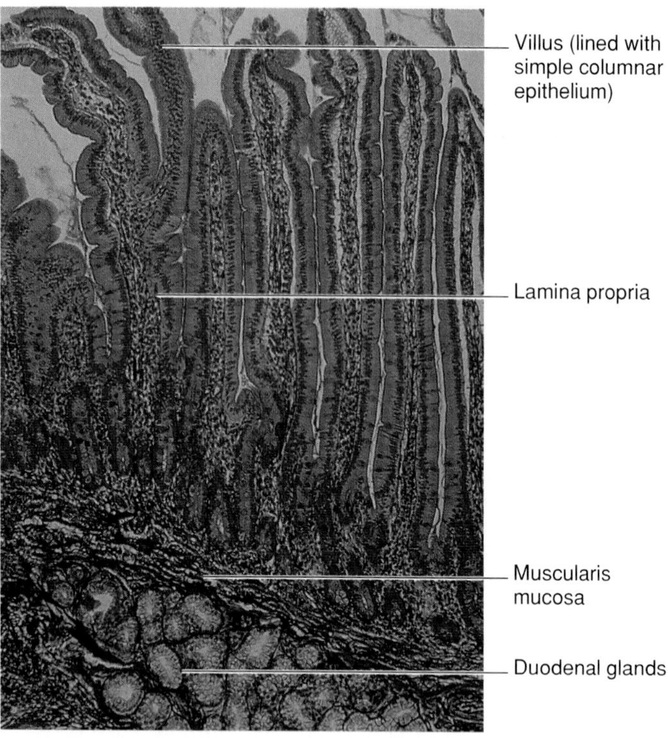

Figure 18.22

The histology of the duodenum.

2. The **jejunum** (*jĕ-joo'num*), which extends from the duodenum to the ilium, is approximately 1 m (3 ft) long. It has a slightly larger lumen and more internal folds than does the ileum, but its histological structure is similar to that of the ileum.
3. The **ileum** (*il'e-um*)—not to be confused with the ilium of the os coxae—makes up the remaining 2 m (6–7 ft) of the small intestine. The terminal portion of the ileum empties into the medial side of the cecum through the **ileocecal valve.** Lymph nodules, called **mesenteric** (Peyer's) **patches,** are abundant in the walls of the ileum.

Structural Modifications of the Small Intestine

The products of digestion are absorbed across the epithelial lining of the intestinal mucosa. Absorption occurs primarily in the jejunum, although some also occurs in the duodenum and ileum. Absorption occurs at a rapid rate as a result of three specializations that increase the intestinal surface area.

1. The **plicae** (*pli'se*) **circulares** are large macroscopic folds of mucosa (see fig. 18.20).
2. The **villi** (*vil'e*) are fingerlike macroscopic folds of the mucosa that project into the lumen of the small intestine (fig. 18.23).
3. The **microvilli** are microscopic projections formed by the folding of each epithelial cell membrane. In a light microscope, the microvilli display a somewhat vague brush border on the edges of the columnar epithelium (fig. 18.24). The terms **brush border** and **microvilli** are often used interchangeably in describing the small intestine.

The villi are covered with columnar epithelial cells, among which are interspersed the mucus-secreting goblet cells. The lamina propria, which forms the connective tissue core of each villus, contains numerous lymphocytes, blood capillaries, and a lymphatic vessel called the **lacteal** (*lak'te-al*) (fig. 18.23).

Peyer's patches: from Johann K. Peyer, Swiss anatomist, 1653–1712
plica: L. *plicatus*, folded
villus: L. *villosus*, shaggy
lacteal: L. *lacteus*, milk

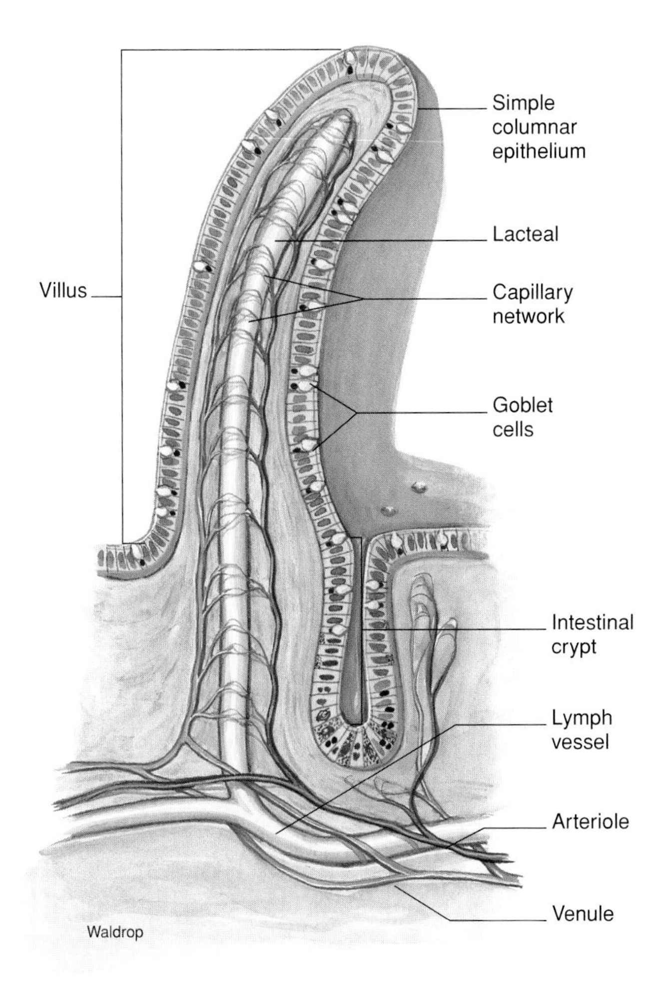

Figure 18.23

The structure of an intestinal villus and intestinal crypt.

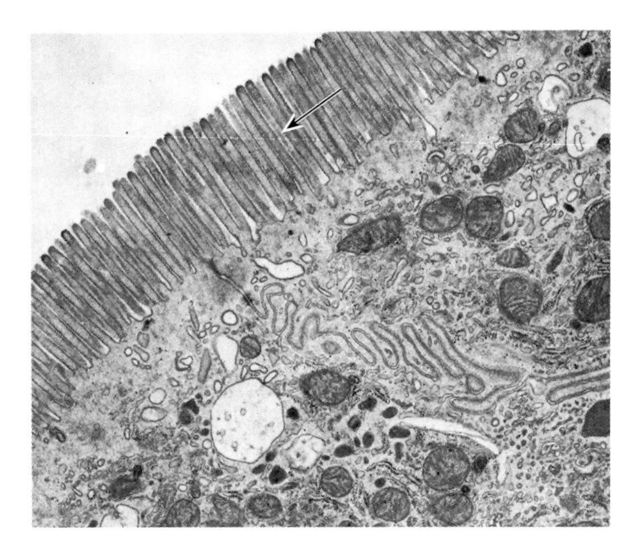

Figure 18.24

An electron photomicrograph of microvilli (arrow) at the exposed surface of a columnar epithelial cell in the small intestine.

Absorbed monosaccharides and amino acids enter the blood capillaries; absorbed fatty acids enter the lacteals.

Epithelial cells at the tips of the villi are continuously shed and are replaced by cells that are pushed up from the bases of the villi. The epithelium at the base of the villi invaginates downward at various points to form narrow pouches that open through pores into the intestinal lumen. These structures are called the **intestinal crypts** (crypts of Lieberkühn) (see fig. 18.23).

Mechanical Activities of the Small Intestine

Contractions of the longitudinal and circular muscles of the small intestine produce three distinct types of movement: *rhythmic segmentation*, *pendular movements*, and *peristalsis*.

Rhythmic segmentations are local contractions of the circular muscular layer. They occur at the rate of about 12 to 16 per minute in regions containing chyme. Rhythmic segmentations churn the chyme with digestive juices and bring it into contact with the mucosa. During these contractions, the vigorous motion of the villi stirs the chyme and facilitates absorption.

Pendular movements primarily occur in the longitudinal muscle layer. In this motion, a constrictive wave moves along a segment of the small intestine and then reverses and moves in the opposite direction, moving the chyme back and forth. Pendular movements also mix the chyme but do not seem to have a particular frequency.

Peristalsis (*per″ĭ-stal′sis*) is responsible for the propulsive movement of the chyme through the small intestine. These wavelike contractions are usually weak and relatively short, occurring at a frequency of about 15 to 18 per minute. Chyme requires 3 to 10 hours to travel the length of the small intestine. Both muscle layers are involved in peristalsis.

The sounds of digestive peristalsis can be easily heard through a stethoscope placed at various abdominal locations. These sounds can be detected even through clothing. The sounds, mostly clicks and gurgles, occur at a frequency of 5 to 30 per minute.

1. Describe the small intestine. Where is it located? How is it subdivided? How is it supported?
2. What are the primary functions of the small intestine?
3. List three structural modifications of the small intestine that increase its absorptive surface area.
4. Describe the movements of the small intestine. Which movements are produced by the circular layer of the tunica muscularis?
5. Which region of the small intestine is the longest? Which is the shortest? How long does it take a portion of chyme to move through the small intestine?

crypts of Lieberkühn: from Johann N. Lieberkühn, German anatomist, 1711–56

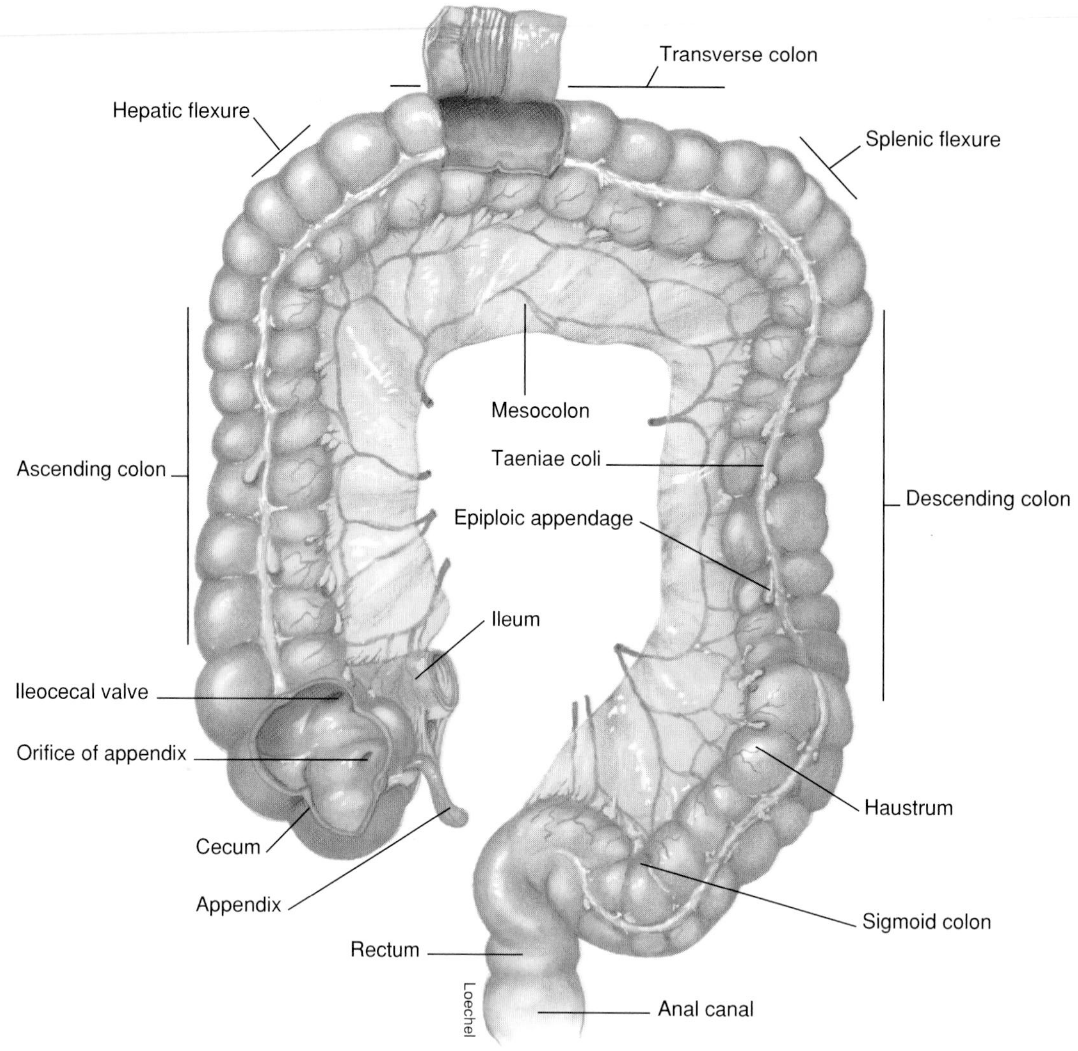

Figure 18.25

The large intestine.

Large Intestine

The large intestine receives undigested food from the small intestine, absorbs water and electrolytes from chyme, and passes feces out of the GI tract.

Objective 15 Identify the regions of the large intestine and describe its gross and histological structure.

Objective 16 Describe the functions of the large intestine and explain how defecation is accomplished.

The **large intestine** averages 1.5 m (5 ft) in length and 6.5 cm (2.5 in.) in diameter. It is called the "large" intestine because of its relatively large diameter compared to that of the small intestine. The large intestine begins at the end of the ileum in the lower right quadrant of the abdomen. From there, it leads superiorly on the right side to a point just below the liver; it then crosses to the left, descends into the pelvis, and terminates at the anus. A specialized portion of the mesentery, the **mesocolon,** supports the transverse portion of the large intestine along the posterior abdominal wall.

The large intestine has little or no digestive function, but it does absorb water and electrolytes from the remaining chyme. In addition, the large intestine functions to form, store, and expel *feces* from the body.

Regions and Structures of the Large Intestine

The large intestine is structurally divided into the cecum, colon, rectum, and anal canal (figs. 18.25 and 18.26). The **cecum** (*se'kum*) is a dilated pouch positioned slightly below the ileocecal valve. The **ileocecal valve** is a fold of mucous membrane at the junction of the small intestine and large intestine that prohibits the backflow of chyme. A fingerlike projection

cecum: L. *caecum*, blind pouch

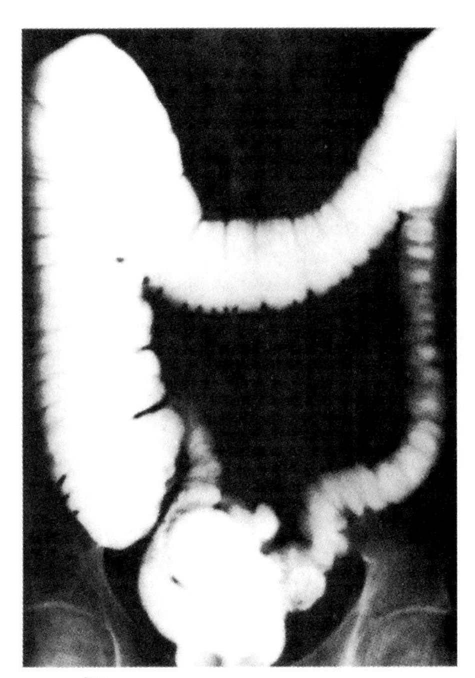

Figure 18.26

A radiograph after a barium enema showing the regions, flexures, and the haustra of the large intestine.

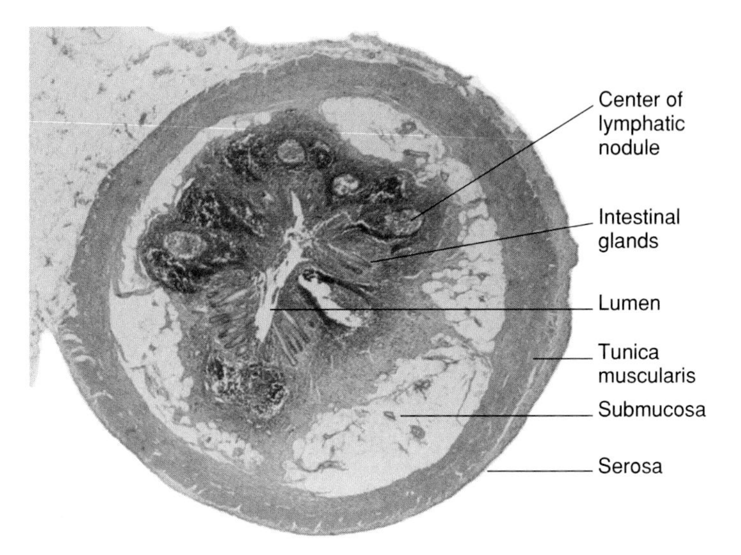

Figure 18.27

The histology of the appendix.

called the **appendix** is attached to the inferior medial margin of the cecum. The 8 cm (3 in.) appendix contains an abundance of lymphatic tissue (fig. 18.27) that may serve to resist infection. Although the appendix serves no digestive function, it is thought to be a vestigial remnant of an organ that was functional in human ancestors.

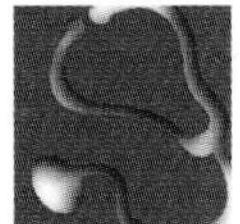

A common disorder of the large intestine is inflammation of the appendix, or *appendicitis*. Wastes that accumulate in the appendix cannot be moved easily by peristalsis, since the appendix has only one opening. Although the symptoms of appendicitis are quite variable, they often include a high white blood cell count, localized pain in the lower right quadrant, and loss of appetite. Vomiting may or may not occur. Rupture of the appendix (a "burst appendix") spreads infectious material throughout the peritoneal cavity, resulting in *peritonitis*.

The superior portion of the cecum is continuous with the colon, which consists of ascending, transverse, descending, and sigmoid portions (fig. 18.25). The **ascending colon** extends superiorly from the cecum along the right abdominal wall to the inferior surface of the liver. Here the colon bends sharply to the left at the **hepatic flexure** (right colic flexure) and continues across the upper abdominal cavity as the **transverse colon.** At the left abdominal wall, another right-angle bend called the **splenic flexure** (left colic flexure) marks the beginning of the **descending colon.** From the splenic flexure, the descending colon extends inferiorly along the left abdominal wall to the pelvic region. The colon then angles medially from the brim of the pelvis to form an S-shaped bend, known as the **sigmoid colon.**

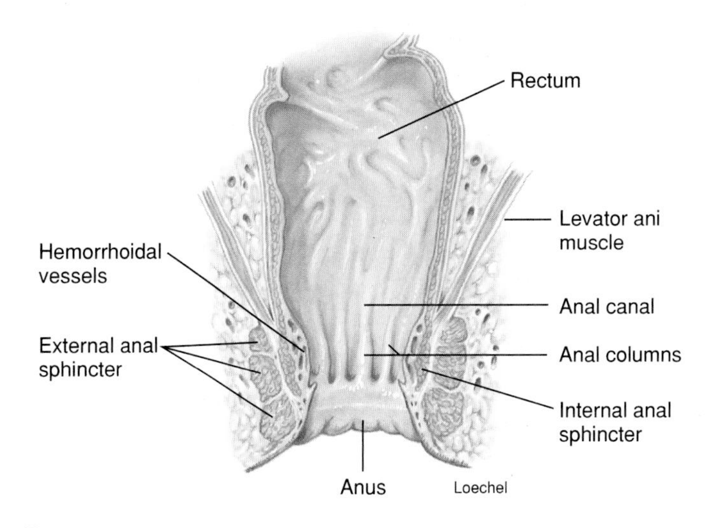

Figure 18.28

The anal canal.

The terminal 20 cm (7.5 in.) of the GI tract is the **rectum,** and the last 2 to 3 cm of the rectum is referred to as the **anal canal** (fig. 18.28). The rectum lies anterior to the sacrum, where it is firmly attached by peritoneum. The **anus** is the external opening of the anal canal. Two sphincter muscles guard the anal opening: the **internal anal sphincter,** which is composed of smooth muscle fibers, and the **external anal sphincter,** composed

appendix: L. *appendix*, attachment
colon: Gk. *kolon*, member of the whole

sigmoid: Gk. *sigmoeides*, shaped like a sigma, Σ
rectum: L. *rectum*, straight tube
anus: L. *anus*, ring

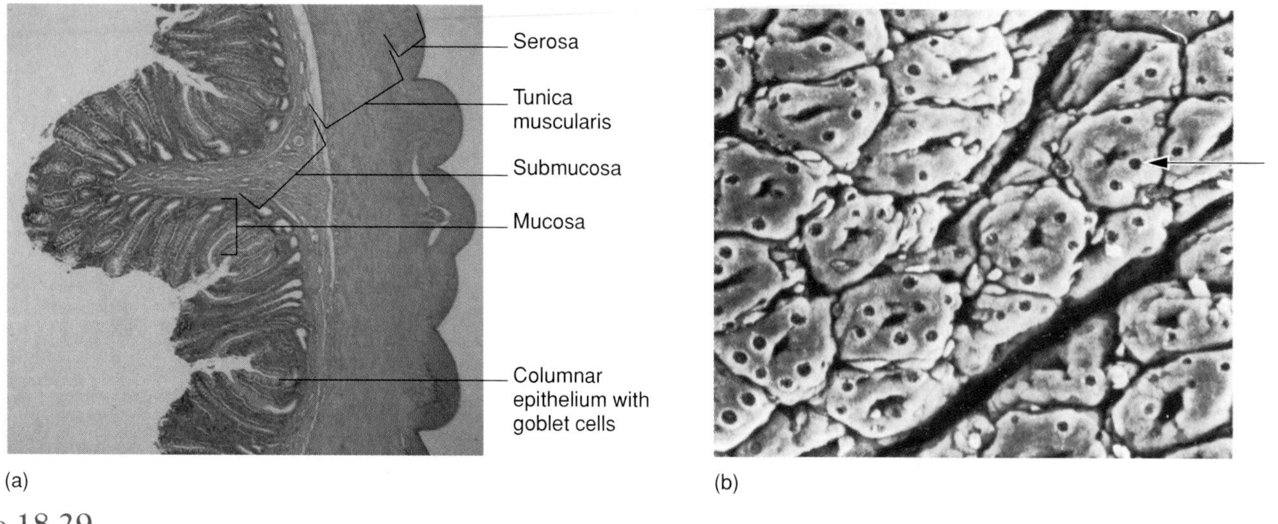

Figure 18.29

The histology of the large intestine. (*a*) A photomicrograph of the tunics. (*b*) A scanning electron photomicrograph of the mucosa. The arrow indicates the opening of a goblet cell into the intestinal lumen.

(*From R. G. Kessel and R. H. Kardon.* Tissues and Organs: A Text-Atlas of Scanning Electron Microscopy, © *1979 W. H. Freeman and Company.*)

of skeletal muscle. The mucous membrane of the anal canal is arranged in highly vascular, longitudinal folds called **anal columns.**

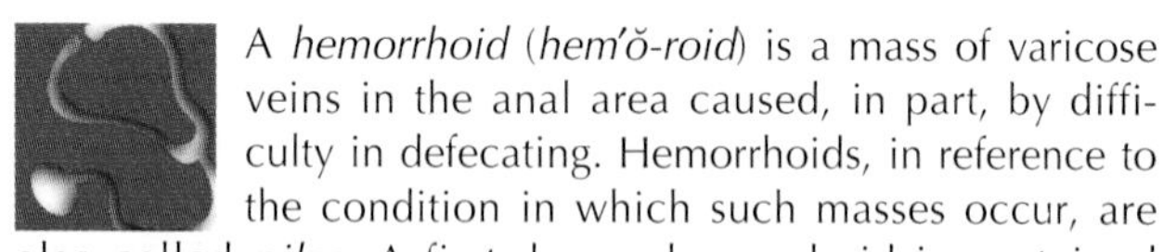

A *hemorrhoid* (*hem'ŏ-roid*) is a mass of varicose veins in the anal area caused, in part, by difficulty in defecating. Hemorrhoids, in reference to the condition in which such masses occur, are also called *piles.* A first-degree hemorrhoid is contained within the anal canal. A second-degree hemorrhoid prolapses, or extends outward during defecation. A third-degree hemorrhoid remains prolapsed through the anal orifice. Rubber band constriction is a common medical treatment for a prolapsed hemorrhoid. In this technique, a rubber band is tied around the hemorrhoid, constricting its blood supply, so that the tissue dries and falls off. In a relatively new treatment, infrared photocoagulation, a high-energy light beam coagulates the hemorrhoid.

Although the large intestine consists of the same tunics as the small intestine, there are some structural differences. The large intestine lacks villi but does have numerous goblet cells in the mucosal layer (fig. 18.29). The longitudinal muscle layer of the muscularis forms three distinct muscle bands called **taeniae coli** (*te'ne-e ko'li*) that run the length of the large intestine. A series of bulges in the walls of the large intestine form sacculations, or **haustra** (*haws'tra*), along its entire length (see figs. 18.25 and 18.26). Finally, the large intestine has small but numerous fat-filled pouches called **epiploic** (*ep-ĭ-plo'ik*) **appendages** (see fig. 18.25) that are attached superficially to the taeniae coli.

taenia: L. *tainia*, a ribbon
haustrum: L. *haustrum*, bucket or scoop
epiploic: Gk. *epiplein*, to float on

The sympathetic innervation of the large intestine arises from superior and inferior mesenteric plexuses, as well as from the celiac plexus. The parasympathetic innervation arises from the paired pelvic splanchnic and vagus nerves. Sensory fibers from the large intestine respond to bowel pressure and signal the need to defecate. Blood is supplied to the large intestine by branches from the superior mesenteric and inferior mesenteric arteries. Venous blood is returned through the superior and inferior mesenteric veins, which in turn drain into the hepatic portal vein that enters the liver.

Mechanical Activities of the Large Intestine

Chyme enters the large intestine through the ileocecal valve. About 15 ml of pasty material enters the cecum with each rhythmic opening of the valve. The ingestion of food intensifies peristalsis of the ileum and increases the frequency with which the ileocecal valve opens; this is called the **gastroileal reflex.** Material entering the large intestine accumulates in the cecum and ascending colon.

Three types of movements occur throughout the large intestine: peristalsis, haustral churning, and mass movement. **Peristaltic movements** of the colon are similar to those of the small intestine, although they are usually more sluggish in the colon. In **haustral churning,** a relaxed haustrum fills with food residues until a point of distension is reached that stimulates the muscularis to contract. Besides moving the material to the next haustrum, this contraction churns the material and exposes it to the mucosa, where water and electrolytes are absorbed. As a result of water absorption, the material becomes solid or semisolid, and is now known as *feces* (*fe'sēz*). **Mass movement** is a very strong peristaltic wave, involving the action of the taeniae coli, which moves the fecal material toward the rectum. Mass movements generally occur only two or three times a day, usually during or shortly after a meal. This response to eating,

Table 18.7
Mechanical Activity in the GI Tract

Region	Type of Motility	Frequency	Stimulus	Result
Oral cavity	Mastication	Variable	Initiated voluntarily; proceeds reflexively	Pulverization: mixing with saliva
Oral cavity and pharynx	Deglutition	Maximum of 20 per min	Initiated voluntarily; reflexively controlled by swallowing center	Clears oral cavity of food
Esophagus	Peristalsis	Depends on frequency of swallowing	Initiated by swallowing	Movement through the esophagus
Stomach	Receptive relaxation	Matches frequency of swallowing	Unknown	Permits filling of stomach
	Tonic contraction	15–20 per min	Autonomic plexuses	Mixing and churning
	Peristalsis	1–2 per min	Autonomic plexuses	Evacuation of stomach
	Hunger contractions	3 per min	Low blood sugar level	Feeding
Small intestine	Peristalsis	15–18 per min	Autonomic plexuses	Transfer through intestine
	Rhythmic segmentation	12–16 per min	Autonomic plexuses	Mixing
	Pendular movements	Variable	Autonomic plexuses	Mixing
Large intestine	Peristalsis	3–12 per min	Autonomic plexuses	Transport
	Mass movements	2–3 per day	Stretch	Fills sigmoid colon
	Haustral churning	3–12 per min	Autonomic plexuses	Mixing
	Defecation	Variable: 1 per day to 3 per week	Reflex triggered by rectal distension	Defecation

called the **gastrocolic reflex,** can best be observed in infants who have a bowel movement during or shortly after feeding.

As material passes through the large intestine, Na^+, K^+, and water are absorbed. It has been estimated that an average volume of 850 ml of water per day is absorbed across the mucosa of the colon. The fecal material that is left then passes to the rectum, leading to an increase in rectal pressure and the urge to defecate. If the urge to defecate is denied, feces are prevented from entering the anal canal by the internal anal sphincter. In this case, the feces remain in the rectum and may even back up into the sigmoid colon.

The **defecation reflex** normally occurs when the rectal pressure rises to a particular level that is determined largely by habit. At this point, the internal anal sphincter relaxes to admit feces into the anal canal.

During the act of defecation, the longitudinal rectal muscles contract to increase rectal pressure, and the internal and external anal sphincter muscles relax. Excretion is aided by contractions of abdominal and pelvic skeletal muscles, which raise the intraabdominal pressure and help push the feces from the rectum through the anal canal and out the anus.

The various mechanical activities of the GI tract are summarized in table 18.7.

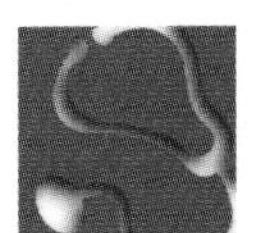

Constipation occurs when fecal material accumulates because of longer than normal periods between defecations. The slower rate of elimination allows more time for water absorption, so that the waste products become harder. Although uncomfortable and sometimes painful, this condition is usually not serious. *Diarrhea* occurs when waste material passes too quickly through the colon, so that insufficient time is allowed for water absorption. Excessive diarrhea can result in dangerous levels of dehydration and electrolyte imbalance, particularly in infants because of their small body size.

1. Identify the four principal regions of the large intestine and describe the functions of the colon.
2. Describe the haustra and the taeniae coli and explain their role in the movements of the large intestine.
3. Describe the location of the rectum, anal canal, and anal sphincter muscles and explain how defecation is accomplished.

Liver, Gallbladder, and Pancreas

The liver, consisting of four lobes, processes nutrients and secretes bile, which is stored and concentrated in the gallbladder prior to discharge into the duodenum. The pancreas, consisting of endocrine (islet) cells and exocrine (acini) cells, secretes important hormones into the blood and essential digestive enzymes into the duodenum.

Objective 17 Describe the location, structure, and functions of the liver.

Objective 18 Describe the location of the gallbladder and trace the flow of bile through the systems of ducts into the duodenum.

Objective 19 Describe the location, structure, and functions of the pancreas.

Three accessory digestive organs in the abdominal cavity aid in the chemical breakdown of food. These are the *liver*, *gallbladder*, and *pancreas*. The liver and pancreas function as exocrine glands

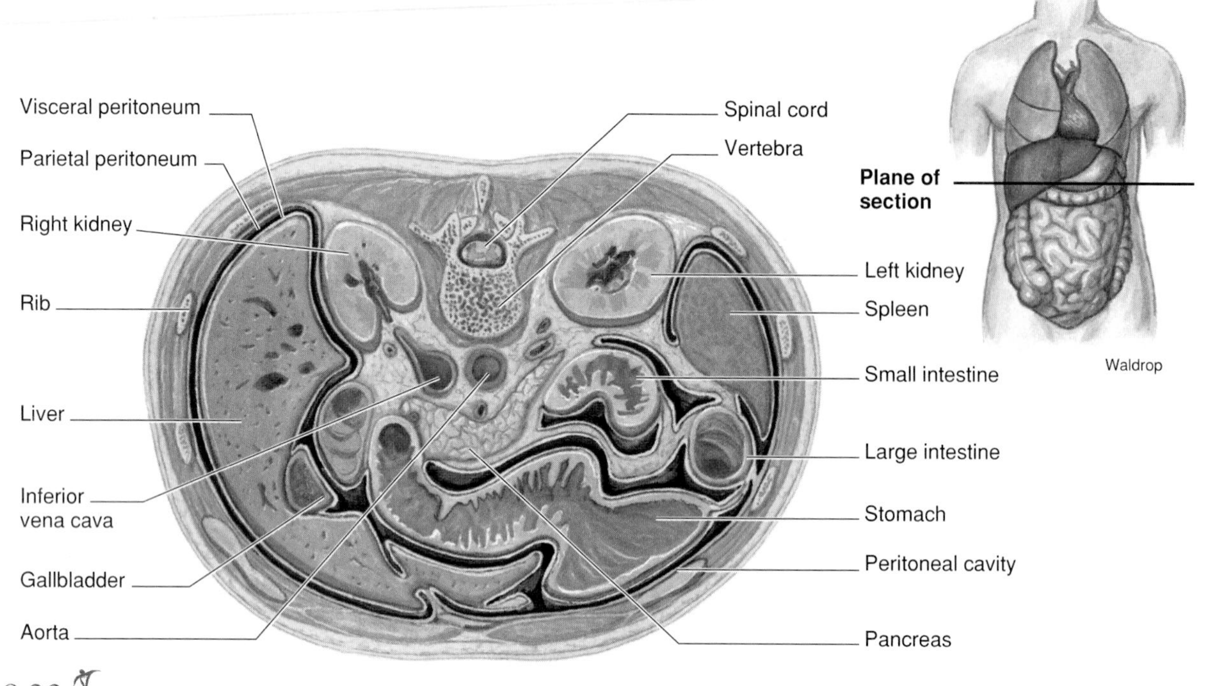

Figure 18.30

A cross section of the abdomen showing the relative position of the liver to other abdominal organs.

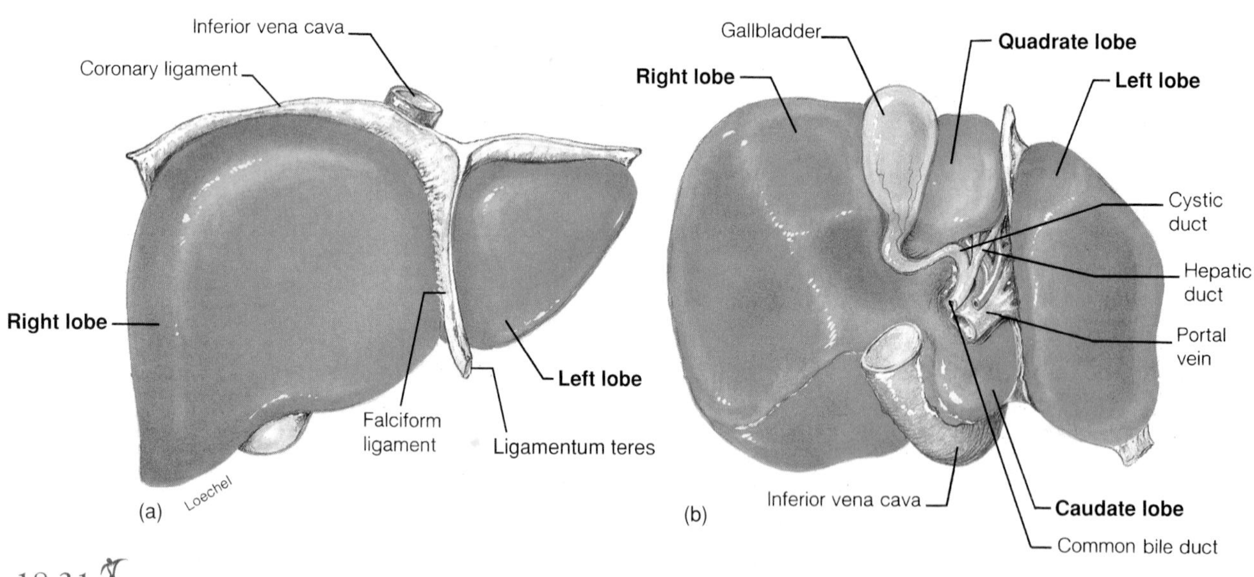

Figure 18.31

The liver and gallbladder. (*a*) An anterior view and (*b*) an inferior view. (The lobes of the liver are labeled in boldface type.)

in this process because their secretions are transported to the lumen of the GI tract via ducts.

Liver

The **liver** is the largest internal organ of the body, weighing about 1.3 kg (3.5–4.0 lbs) in an adult. It is positioned immediately beneath the diaphragm in the epigastric and right hypochondriac regions (see fig. 2.15 and table 2.4) of the abdomen (fig. 18.30). Its reddish brown color is due to its great vascularity.

The liver has four lobes and two supporting ligaments. Anteriorly, the **right lobe** is separated from the smaller **left lobe** by the **falciform ligament** (fig. 18.31). Inferiorly, the **caudate** (*kaw′dāt*) **lobe** is positioned near the inferior vena cava, and the

falciform: L. *falcis*, sickle; forma, form

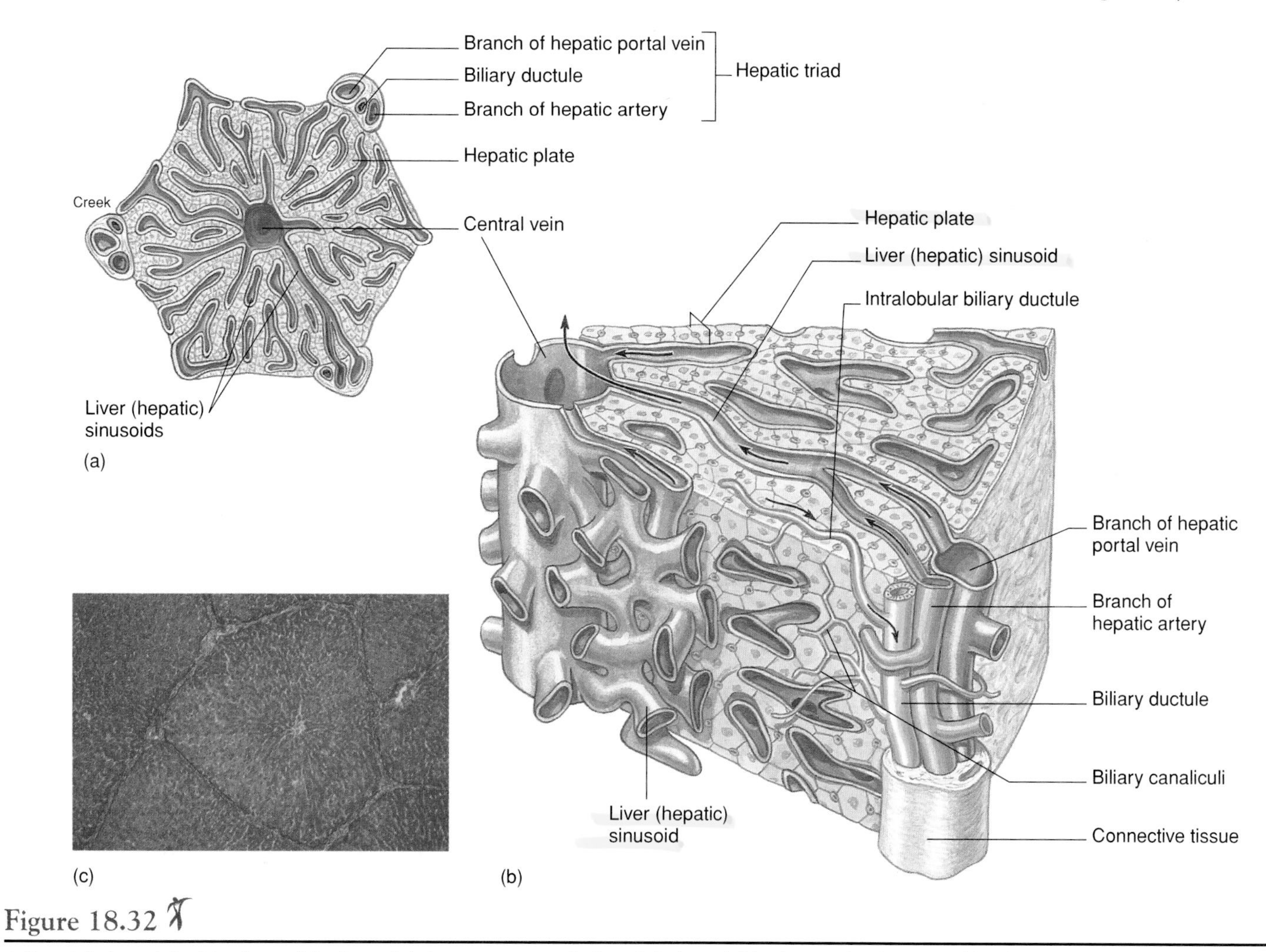

Figure 18.32

A liver lobule and the histology of the liver. (*a*) A cross section of a liver lobule and (*b*) a longitudinal section. Blood enters a liver lobule through the vessels in a hepatic triad, passes through hepatic sinusoids, and leaves the lobule through a central vein. The central veins converge to form hepatic veins that transport venous blood from the liver. (*c*) A photomicrograph of a liver lobule in cross section.

quadrate lobe is adjacent to the gallbladder. The falciform ligament attaches the liver to the anterior abdominal wall and the diaphragm. The **ligamentum teres** (round ligament) extends from the falciform ligament to the umbilicus. This ligament is the remnant of the umbilical vein of the fetus.

Although the liver is the largest internal organ, it is, in a sense, only one to two cells thick. This is because the liver cells, or **hepatocytes,** form **hepatic plates** that are one to two cells thick and separated from each other by large capillary spaces called **liver** (hepatic) **sinusoids** (fig. 18.32). The sinusoids are lined with phagocytic **Kupffer** (*koop'fer*) **cells,** but the large intercellular gaps between adjacent Kupffer cells make these sinusoids more highly permeable than other capillaries. The plate structure of the liver and the high permeability of the sinusoids allow each hepatocyte to be in direct contact with the blood.

The hepatic plates are arranged to form functional units called **liver lobules** (fig. 18.33). In the middle of each lobule is a **central vein,** and at the periphery of each lobule are branches of the hepatic portal vein and of the hepatic artery, which open into the spaces *between* hepatic plates. Portal venous blood, containing molecules absorbed in the GI tract, thus mixes with arterial blood as the blood flows within the liver sinusoids from the periphery of the lobule to the central vein (fig. 18.33). The central veins of different liver lobules converge to form the hepatic vein, which carries blood from the liver to the inferior vena cava.

The liver lobules have numerous functions, including synthesis, storage, and release of vitamins; synthesis, storage, and release of glycogen; synthesis of blood proteins; phagocytosis of old red blood cells and certain bacteria; removal of toxic substances; and production of bile. Bile is stored in the gallbladder and is eventually secreted into the duodenum for the emulsification (breaking down into smaller particles) and absorption of fats.

Bile is produced by the hepatocytes and secreted into tiny channels called **bile canaliculi** (*kan"ă-lik'yŭ-li*) located *within* each hepatic plate (fig. 18.33). These bile canaliculi are drained at the periphery of each lobule by **bile ducts,** which in turn drain

hepatic: Gk. *hepatos*, liver
Kupffer cells: from Karl Wilhelm von Kupffer, Bavarian anatomist and embryologist, 1829–1902

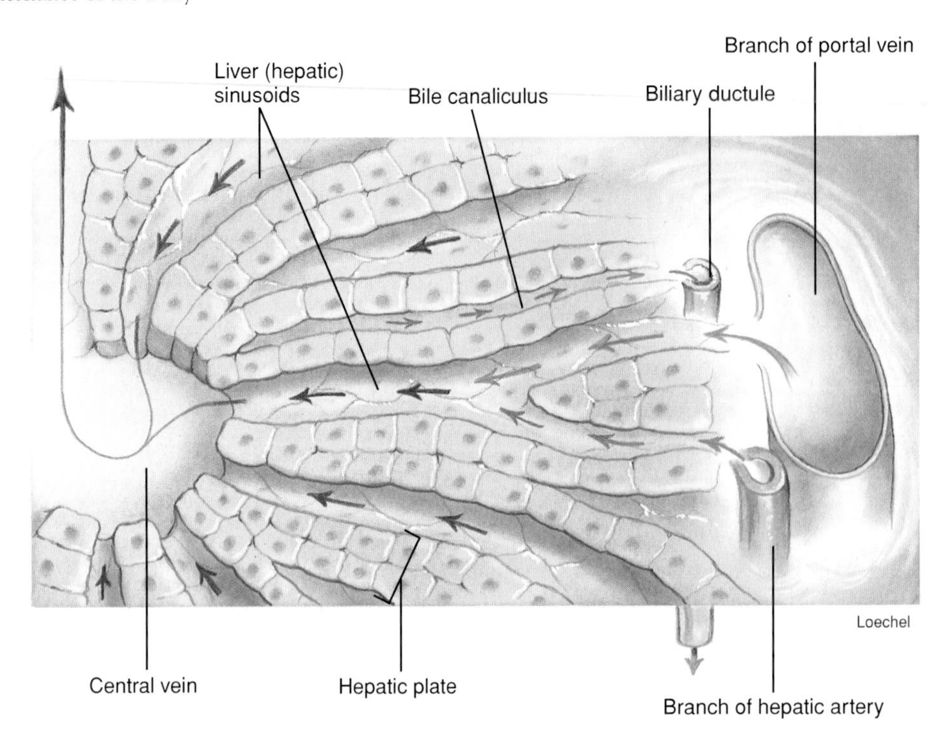

Figure 18.33

The flow of blood and bile in a liver lobule. Blood flows within sinusoids from branches of the hepatic portal vein to the central vein (from the periphery to the center of a lobule). Bile flows within hepatic plates from the center of a lobule to biliary ductules at the periphery.

into **hepatic ducts** that carry bile away from the liver. Since blood travels in the sinusoids and bile travels in the opposite direction within the hepatic plates, blood and bile do not mix in the liver lobules.

The liver receives parasympathetic innervation from the vagus nerves and sympathetic innervation from thoracolumbar nerves through the celiac ganglia.

Gallbladder

The **gallbladder** is a saclike organ attached to the inferior surface of the liver (figs. 18.31 and 18.34). This organ stores and concentrates bile. A sphincter valve at the neck of the gallbladder allows a storage capacity of about 35 to 50 ml. The inner mucosal layer of the gallbladder is thrown into folds similar to the gastric folds of the stomach. When the gallbladder fills with bile, it expands to the size and shape of a small pear. Bile is a yellowish green fluid containing bile salts, bilirubin (a product resulting from the breakdown of blood), cholesterol, and other compounds. Contraction of the muscularis ejects bile from the gallbladder.

Bile is continuously produced by the liver and drains through the hepatic ducts and bile duct to the duodenum. When the small intestine is empty of food, the **sphincter of ampulla** (see fig. 18.21) constricts, and bile is forced up the cystic duct to the gallbladder for storage.

The gallbladder is supplied with blood from the cystic artery, which branches from the right hepatic artery. Venous

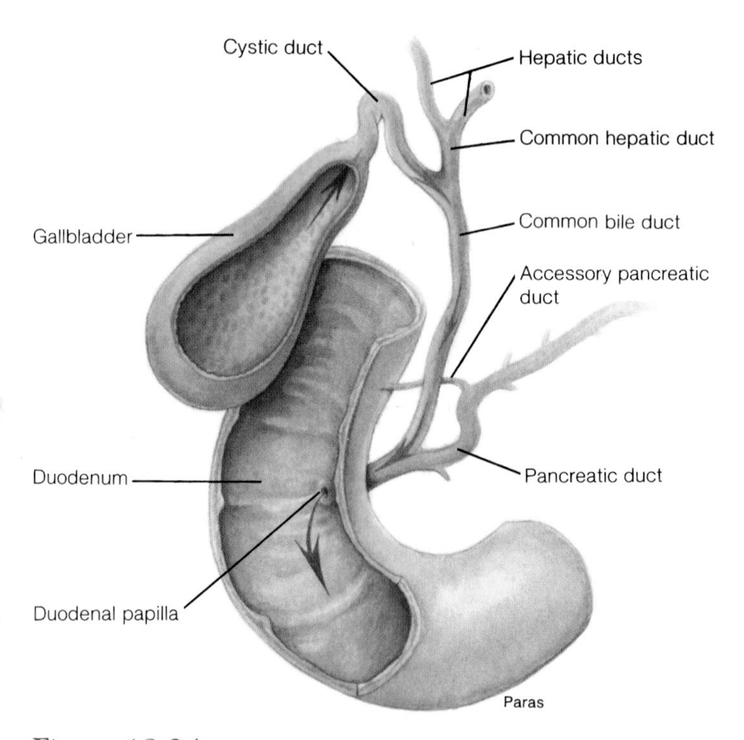

Figure 18.34

The pancreatic duct joins the bile duct to empty its secretions through the duodenal papilla into the duodenum. The release of bile and pancreatic juice into the duodenum is controlled by the sphincter of ampulla.

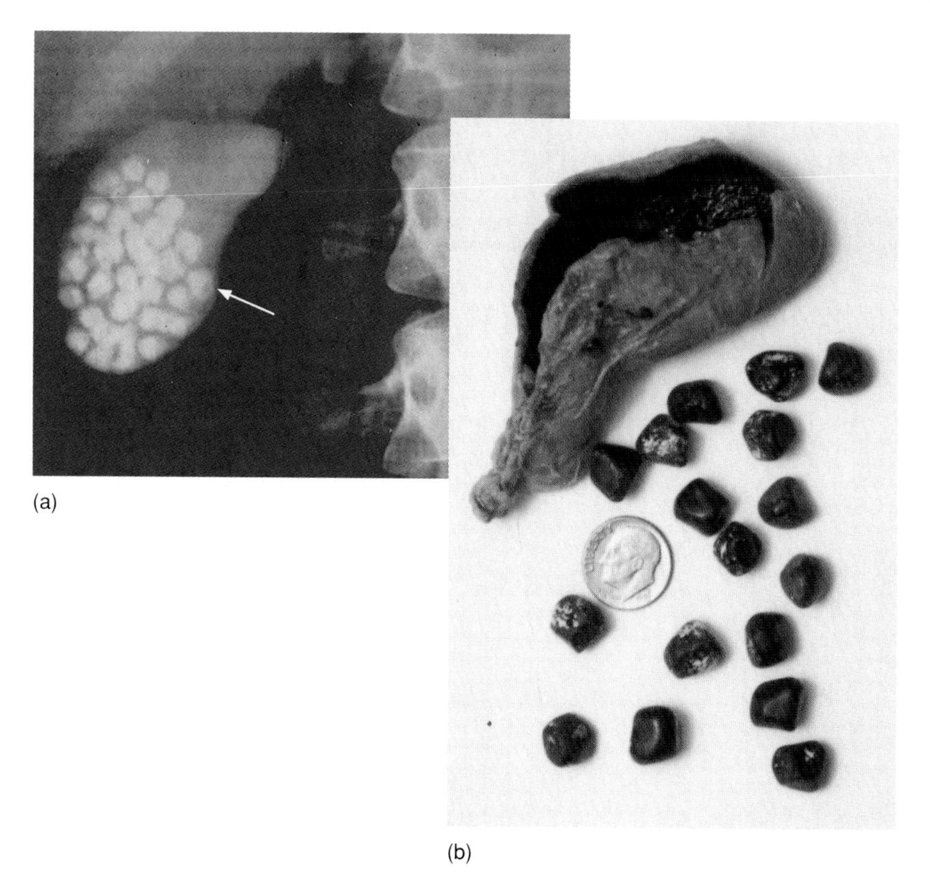

Figure 18.35

(*a*) A radiograph of a gallbladder that contains gallstones (biliary calculi). (*b*) Following surgical removal of the gallbladder (cholecystectomy), it has been cut open to reveal its gallstones. (Note their size relative to that of a dime.)

blood is returned through the cystic vein, which empties into the hepatic portal vein. Autonomic innervation of the gallbladder is similar to that of the liver; both receive parasympathetic innervation from the vagus nerves and sympathetic innervation from thoracolumbar nerves through the celiac ganglia.

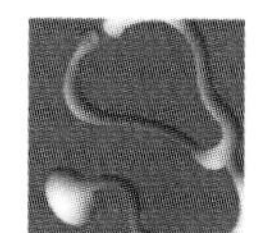

A common clinical problem of the gallbladder is the development of *gallstones.* Bile is composed of various salts, pigments, and cholesterols that become concentrated as water is removed. Cholesterols normally remain in solution, but under certain conditions they precipitate to form solid crystals. Large crystals may block the bile duct and have to be surgically removed. The radiograph in figure 18.35 shows gallstones in position, and the photograph shows removed gallstones.

Pancreas

The soft, lobulated **pancreas** is known as a *mixed gland* because it has both exocrine and endocrine functions. The endocrine function is performed by clusters of cells called the **pancreatic islets** (islets of Langerhans). The islet cells secrete the hormones *insulin* and *glucagon* into the blood. As an exocrine gland, the pancreas secretes *pancreatic juice* through the pancreatic duct (fig. 18.36), which empties into the duodenum.

The pancreas is positioned horizontally along the posterior abdominal wall, adjacent to the greater curvature of the stomach. It measures about 12.5 cm (6 in.) in length and is approximately 2.5 cm (1 in.) thick. It has an expanded **head,** positioned near the duodenum; a centrally located **body;** and a tapering **tail,** positioned near the spleen. All but a portion of the head is retroperitoneal. Within the lobules of the pancreas are the exocrine secretory units, called **pancreatic acini** (*as'ĭ-ni*), and the endocrine secretory units, called **pancreatic islet cells.** Each acinus consists of a single layer of epithelial acinar cells surrounding a lumen into which the constituents of pancreatic juice are secreted.

The pancreas is innervated by branches of the celiac plexus. The glandular portion of the pancreas receives parasympathetic innervation, whereas the pancreatic blood

pancreas: Gk. *pan,* all; *kreas,* flesh

islets of Langerhans: from Paul Langerhans, German anatomist, 1847–88
acinus: L. *acinus,* grape

DEVELOPMENTAL EXPOSITION

The Digestive System

Exhibit I A sagittal section of a 5-week-old embryo showing the development of the digestive system and its association with the extraembryonic membranes and organs.

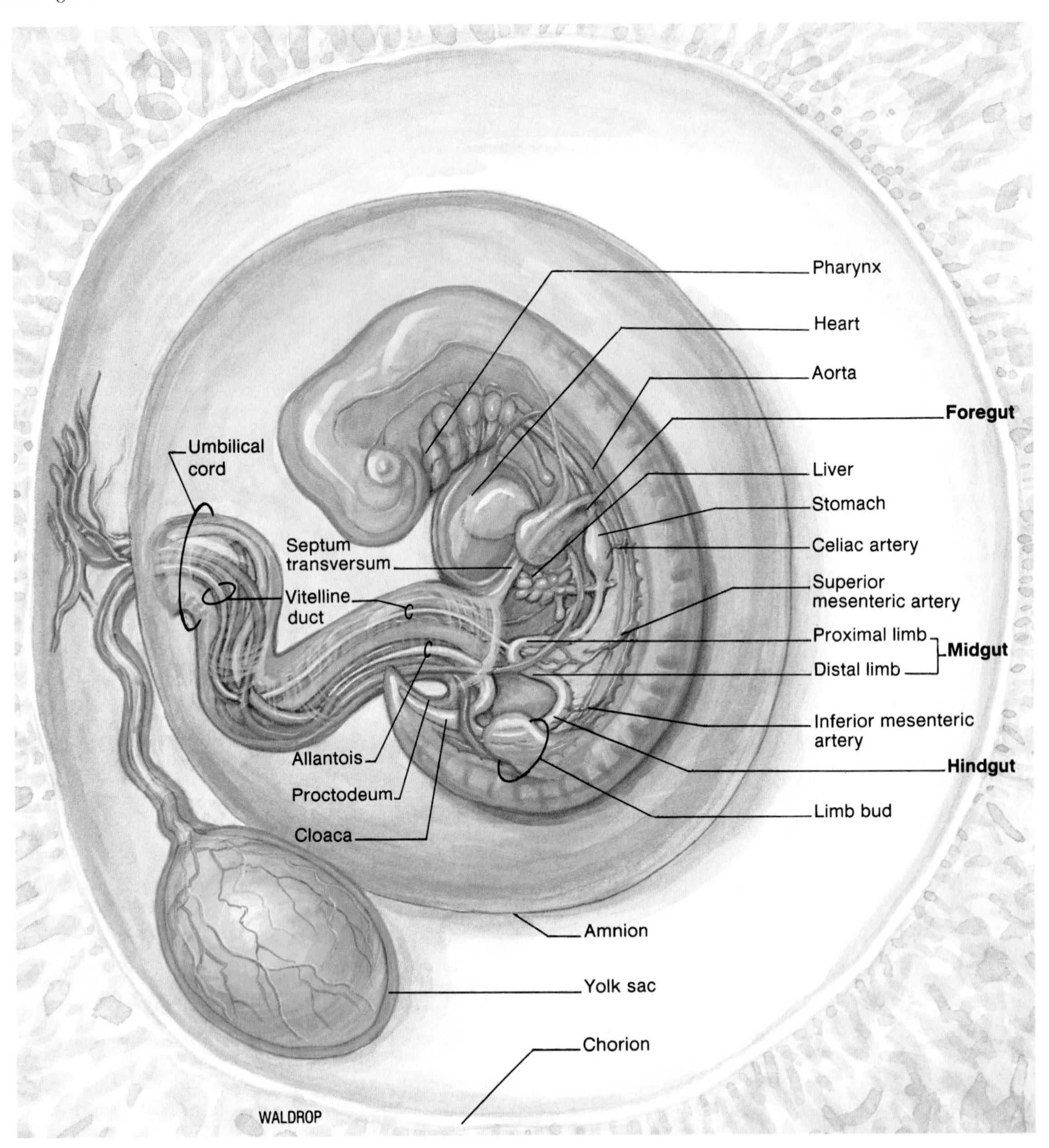

Explanation

The entire digestive system develops from modifications of an elongated tubular structure called the **primitive gut.** These modifications are initiated during the fourth week of embryonic development. The primitive gut is composed solely of endoderm and for descriptive purposes can be divided into three regions: the *foregut, midgut,* and *hindgut* (exhibit I).

Foregut

The **stomodeum** (*sto″mŏ-de′um*), or **oral pit,** is not part of the foregut but an invagination of ectoderm that breaks through a thin **oral membrane** to

stomodeum: Gk. *stoma*, mouth; *hodaios*, on the way to

Exhibit II Progressive stages of development of the foregut to form the stomach, duodenum, liver, gallbladder, and pancreas: (*a*) 4 weeks, (*b*) 5 weeks, (*c*) 6 weeks, and (*d*) 7 weeks.

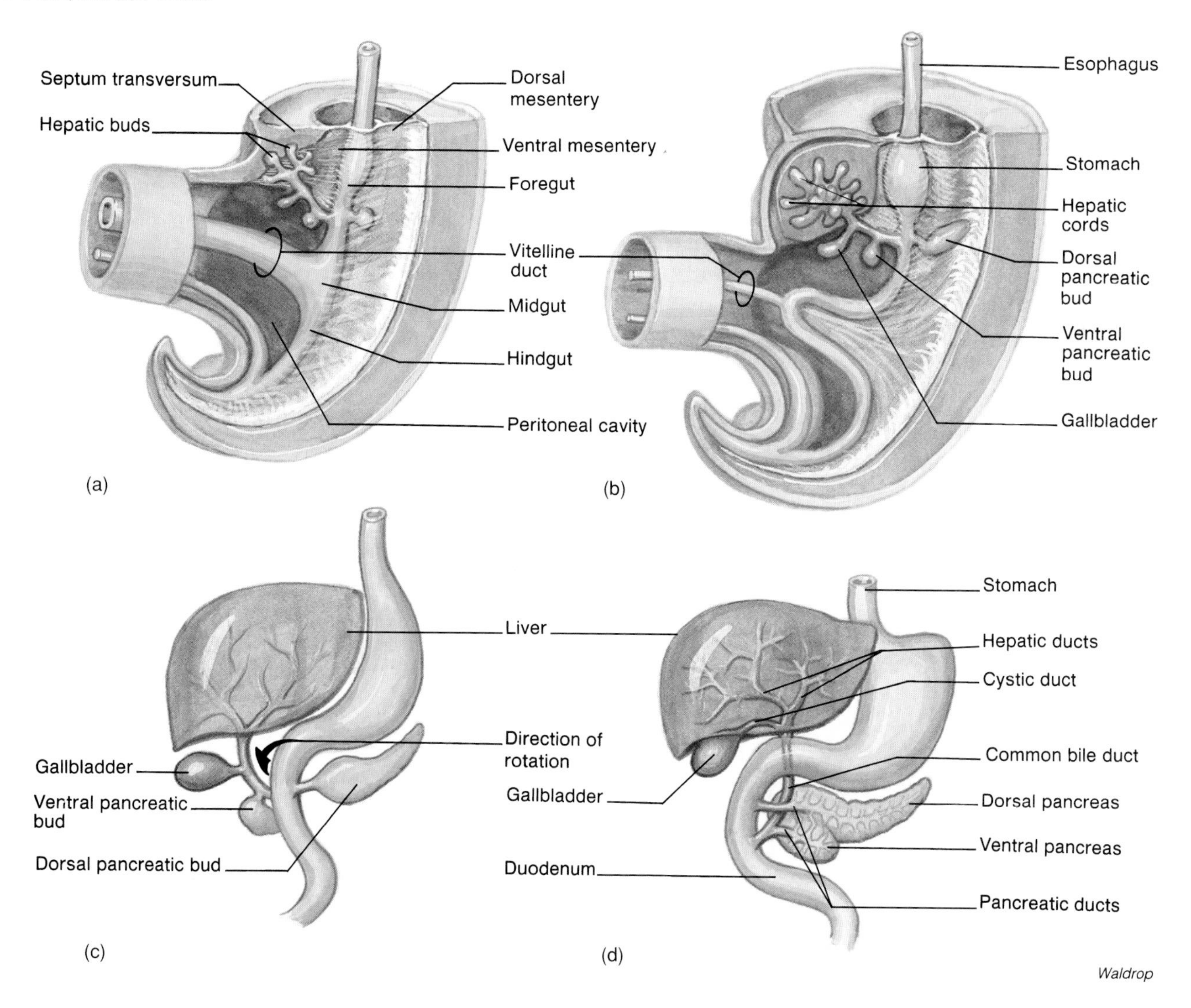

become continuous with the foregut and form part of the oral cavity, or mouth. Structures in the mouth, therefore, are ectodermal in origin. The esophagus, pharynx, stomach, a portion of the duodenum, the pancreas, liver, and gallbladder are the organs that develop from the foregut (exhibit II). Along the GI tract, only the inside epithelial lining of the lumen is derived from the endoderm of the primitive gut. The vascular portion and smooth muscle layers are formed from mesoderm that develops from the surrounding splanchnic mesenchyme.

The stomach first appears as an elongated dilation of the foregut. The dorsal border of the stomach undergoes more rapid growth than the ventral border, forming a distinct curvature. The caudal portion of the foregut and the cranial portion of the midgut form the duodenum. The liver and pancreas arise from the wall of the duodenum as small **hepatic** and **pancreatic buds,** respectively. The hepatic bud experiences incredible growth to form the gallbladder, associated ducts, and the various lobes of the liver (exhibit II). By the sixth week, the liver is carrying out hemopoiesis (the formation of blood cells). By the ninth week, the liver has developed to the point where it represents 10% of the total weight of the fetus.

The pancreas develops from dorsal and ventral pancreatic buds of endodermal cells. As the duodenum grows, it rotates clockwise, and the two pancreatic buds fuse (exhibit II).

Midgut

During the fourth week of the embryonic stage (exhibit I), the midgut is continuous with the yolk sac. By the fifth week, the midgut has formed a

Continued

Developmental Exposition

Continued

Exhibit III Progressive stages of the development of the midgut to form the distal portion of the small intestine and the proximal portion of the large intestine: (*a*) 5 weeks, (*b*) 6 weeks, (*c*) 7 weeks, (*d*) 10 weeks, and (*e*) 18 weeks.

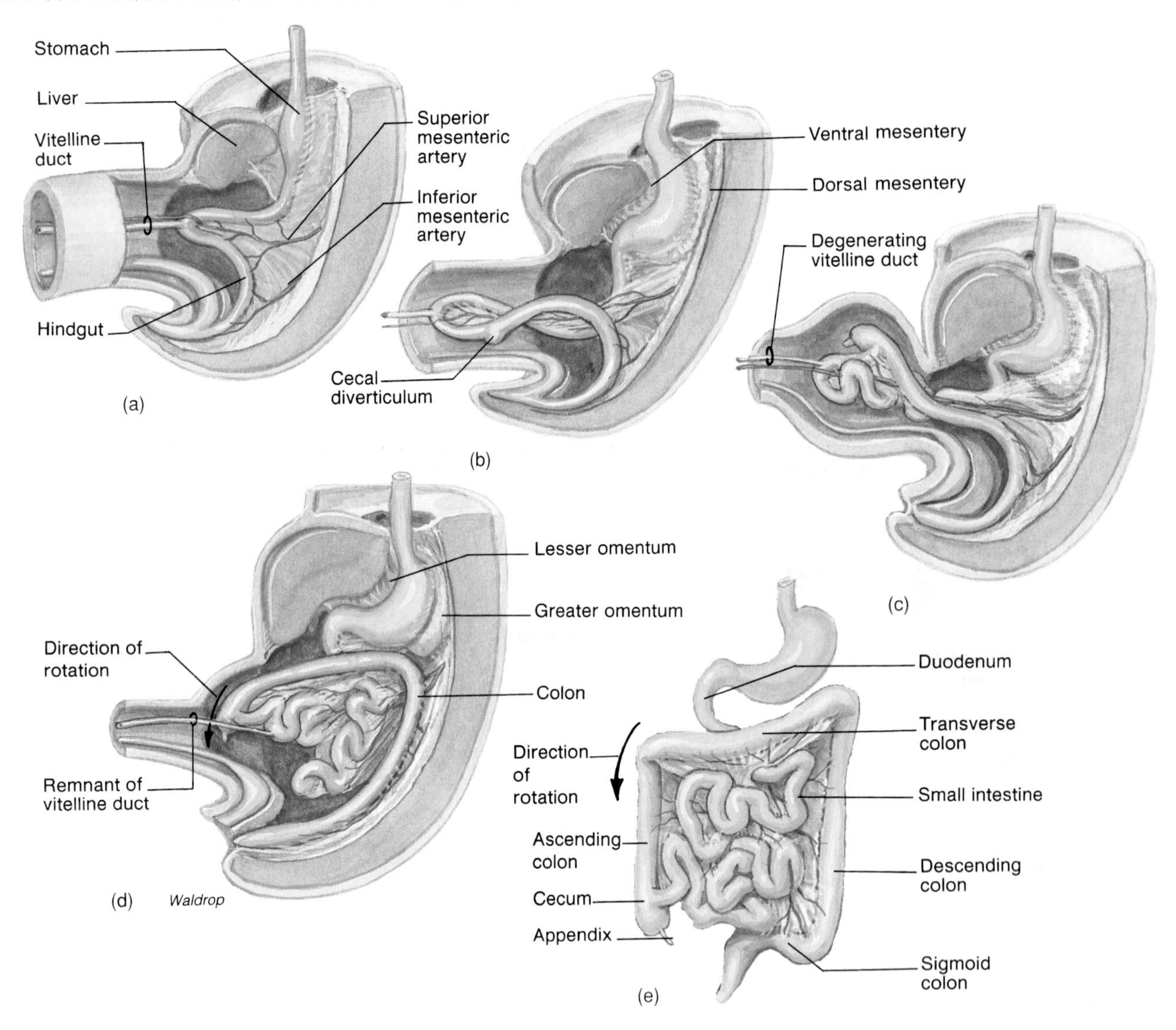

ventral U-shaped **midgut loop,** which projects into the umbilical cord (exhibit III). As development continues, the anterior limb of the midgut loop coils to form most of the small intestine. The posterior limb of the midgut loop expands to form the large intestine and a portion of the small intestine. A **cecal diverticulum** appears during the fifth week.

During the tenth week, the intestines are drawn up into the abdominal cavity, and further differentiation and rotation occur. The cecal diverticulum continues to develop, forming the cecum and appendix. The remainder of the midgut gives rise to the ascending colon and hepatic flexure (exhibit III).

Hindgut

The hindgut extends from the midgut to the **cloacal** (*klo-a'kal*) **membrane** (exhibit IV). The **proctodeum** (*prok"tŏ-de'um*), or **anal pit,** is a depression in the anal region formed from an invagination of ectoderm that contributes to the cloacal membrane. The **allantois** (*ă-lan'to-is*), which receives urinary wastes from the fetus, connects to the hindgut at a region called the

proctodeum: Gk. *proktos,* anus; *hodaios,* on the way to

Exhibit IV The progressive development of the hindgut illustrating the developmental separation of the digestive system from the urogenital system. (*a*) An anterolateral view of an embryo at 4 weeks showing the position of a sagittal cut depicted in (a_1), (*b*), and (*c*). (a_1) At 4 weeks, the hindgut, cloaca, and allantois are connected. (*b*) At 6 weeks, the connections between the gut and extraembryonic structures are greatly diminished. (*c*) By 7 weeks, structural and functional separation between the digestive system and the urogenital system is almost complete.

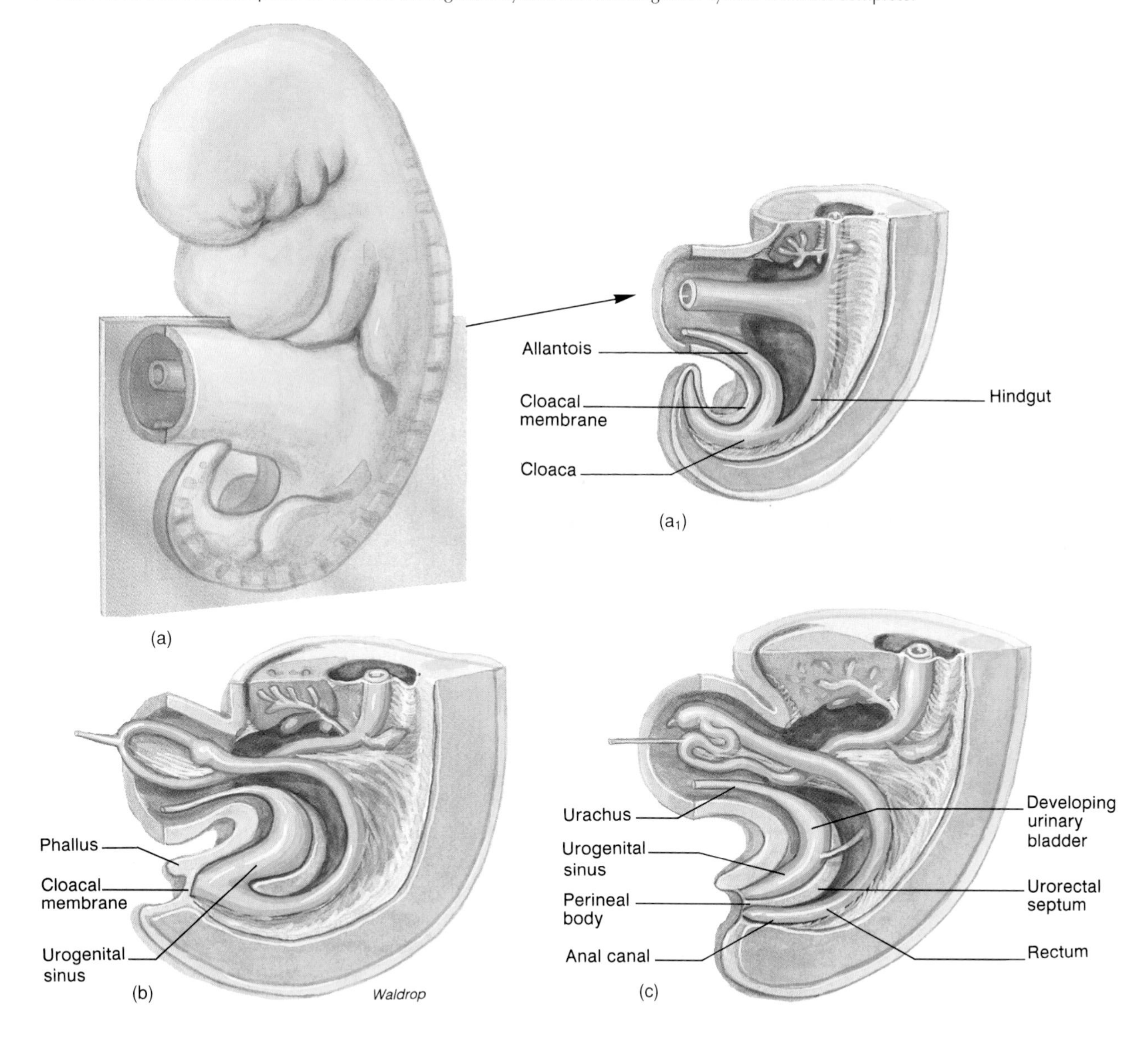

cloaca, as seen in exhibit IV. A band of mesenchymal cells called the **urorectal septum** grows caudally between the fourth and seventh week until a complete partition separates the cloaca into a dorsal **anal canal** and a ventral **urogenital sinus.** With the completion of the urorectal septum, the cloacal membrane is divided into an anterior **urogenital membrane** and a posterior **anal membrane.** Toward the end of the seventh week, the anal membrane perforates and forms the anal opening, which is lined with ectodermal cells. About this time, the urogenital membrane ruptures to provide further development of the urinary and reproductive systems.

cloaca: L. *cloaca*, sewer

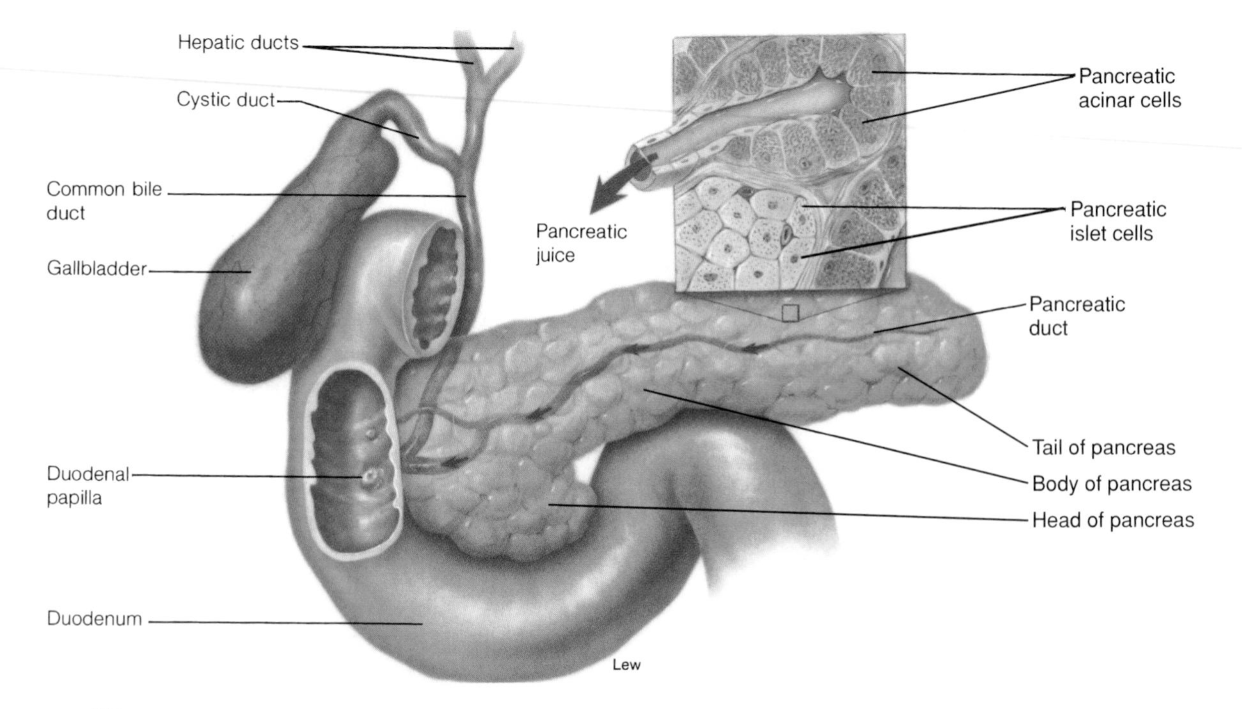

Figure 18.36

The pancreas is both an exocrine and an endocrine gland. Pancreatic juice—the exocrine product—is secreted by acinar cells into the pancreatic duct. Scattered islands of cells, called pancreatic islets (islets of Langerhans), secrete the hormones insulin and glucagon into the blood.

vessels receive sympathetic innervation. The pancreas is supplied with blood by the pancreatic branch of the splenic artery, which arises from the celiac artery, and by the pancreatoduodenal branches, which arise from the superior mesenteric artery. Venous blood is returned through the splenic and superior mesenteric veins into the hepatic portal vein.

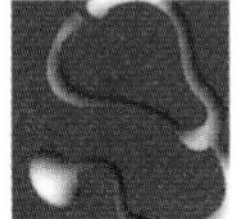

Pancreatic cancer has the worst prognosis of all types of cancer. This is probably because of the spongy, vascular nature of this organ and its vital exocrine and endocrine functions. Pancreatic surgery is a problem because the soft, spongy tissue is difficult to suture.

1. Describe the liver. Where is it located? List the lobes of the liver and the supporting ligaments.
2. List the principal functions of the liver.
3. Describe the structure of liver lobules and trace the flow of blood and bile in the lobules.
4. Explain how the liver receives a double blood supply.
5. Explain how the gallbladder fills with bile secretions and how bile and pancreatic secretions enter the duodenum.
6. Briefly state the exocrine and endocrine functions of the pancreas. What are the various cellular secretory units of the pancreas?

Clinical Considerations

Developmental Problems of the Digestive System

Most of the congenital disorders of the digestive system develop during the fourth or fifth week of embryonic life. A **cleft palate,** as described in chapter 17, is a congenital opening between the oral and nasal cavities; therefore, it involves both the digestive and respiratory systems (see fig. 17.29). **Esophageal atresia** (*ă-tre′ze-ă*), or failure to develop the normal structure of the esophageal-stomach area, is another disorder of the upper GI tract that requires surgery to correct. **Pyloric stenosis** is a common abnormality in which the pyloric sphincter muscle is hypertrophied, reducing the size of the lumen. This condition affects approximately 1 in 200 newborn males and 1 in 1,000 newborn females. Stenoses, atresias, and malrotations of various portions of the GI tract may occur as the gut develops. Umbilical problems involving the GI tract are fairly common, as is some form of *imperforate anus*, which occurs in about 1 in 5,000 births.

Pathogens and Poisons

The GI tract presents a hospitable environment for an array of parasitic helminths (worms) and microorganisms (fig. 18.37). Many of these are beneficial, but some bacteria and protozoa

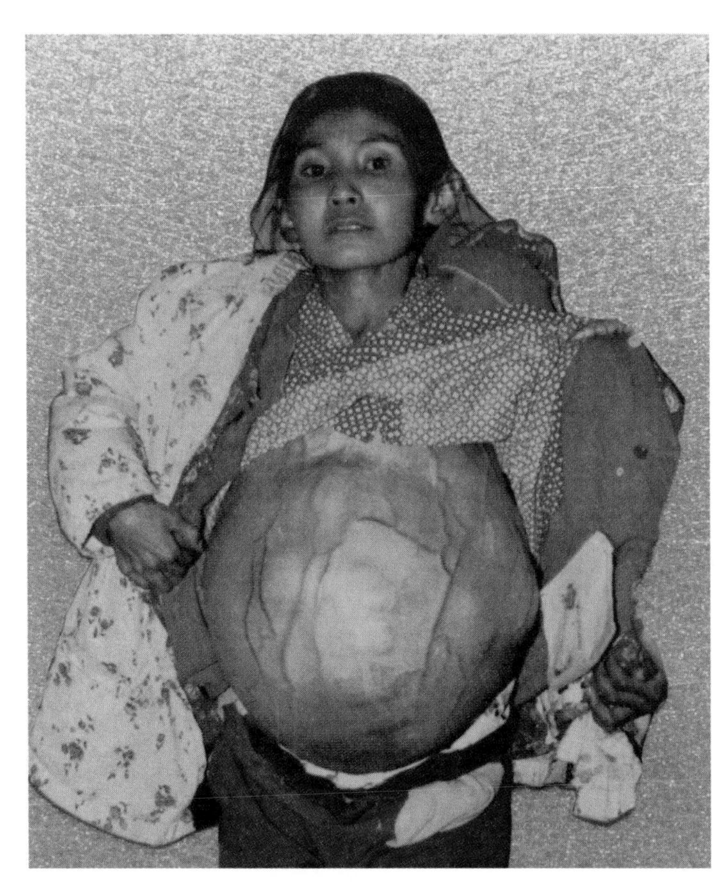

Figure 18.37

The gastrointestinal tract and the accessory digestive organs are susceptible to an array of pathogenic (disease-causing) agents. This woman is afflicted with hydatid cysts in her liver as a result of ingesting an egg from the tapeworm *Echinococcus granulosus*. If untreated, conditions such as this are eventually fatal.

can cause diseases. Only a few examples of the pathogenic microorganisms will be discussed here.

Dysentery (*dis'en-ter"e*) is an inflammation of the intestinal mucosa, characterized by frequent loose stools containing mucus, pus, and blood. The most common dysentery is **amoebic dysentery,** which is caused by the protozoan *Entamoeba histolytica*. Cysts from this organism are ingested in contaminated food, and after the protective coat is removed by HCl in the stomach, the vegetative form invades the mucosal walls of the ileum and colon.

Food poisoning is caused by consuming the toxins produced by pathogenic bacteria. *Salmonella* is a bacterium that commonly infects food. **Botulism,** the most serious type of food poisoning, is caused by ingesting food contaminated with the toxin produced by the bacterium *Clostridium botulinum*. This organism is widely distributed in nature, and the spores it produces are frequently found on food being processed by canning. For this reason, food must be heated to 120°C (248°F) before it is canned. It is the toxins produced by the bacterium growing in the food that are pathogenic, rather than the organisms themselves. The poison is a neurotoxin that is readily absorbed into the blood, at which point it can affect the nervous system.

dysentery: Gk. *dys*, bad; *entera*, intestine

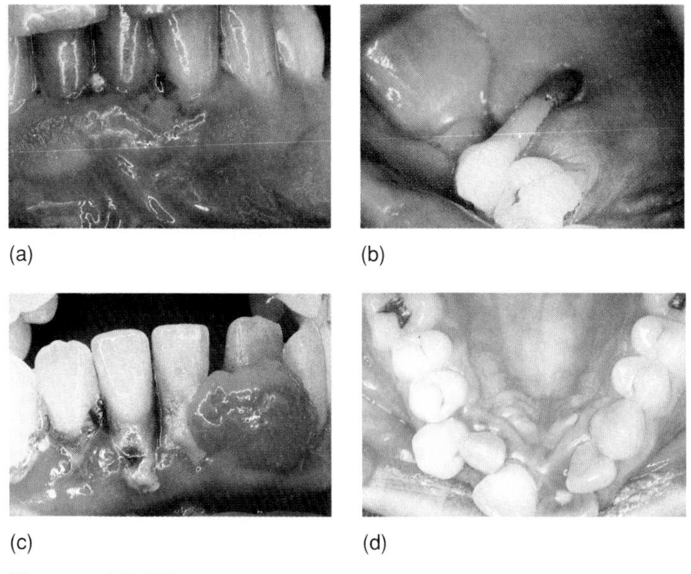

Figure 18.38

Clinical problems of the teeth. (*a*) Trench mouth and dental caries, (*b*) severe alveolar bone destruction from periodontitis, (*c*) pyogenic granuloma and dental caries, and (*d*) malposition of teeth.

Clinical Problems of the Teeth and Salivary Glands

Dental caries, or tooth decay, is the gradual decalcification of tooth enamel (fig. 18.38) and underlying dentin, caused by the acid products of bacteria. These bacteria thrive between teeth, where food particles accumulate, and form part of the thin layer of bacteria, proteins, and other debris called *plaque* that covers teeth. The development of dental caries can be reduced by brushing at least once a day and by flossing between teeth at regular intervals.

People over the age of 35 are particularly susceptible to **periodontal disease,** which involves inflammation and deterioration of the gingiva, dental alveoli, periodontal membranes, and the cementum that covers the roots of teeth. Some of the symptoms are loosening of the teeth, bad breath, bleeding gums when brushing, and some edema. Periodontal disease may result from impacted plaque, cigarette smoking, crooked teeth, or poor diet. It accounts for 80% to 90% of tooth loss in adults.

Mumps is a viral disease of the parotid glands, and in advanced stages it may involve the pancreas and testes. In children, mumps is generally not serious, but in adults it may cause deafness and destroy the pancreatic islet tissue or testicular cells.

Disorders of the Liver

The liver is a remarkable organ that has the ability to regenerate even if up to 80% has been removed. The most serious diseases of

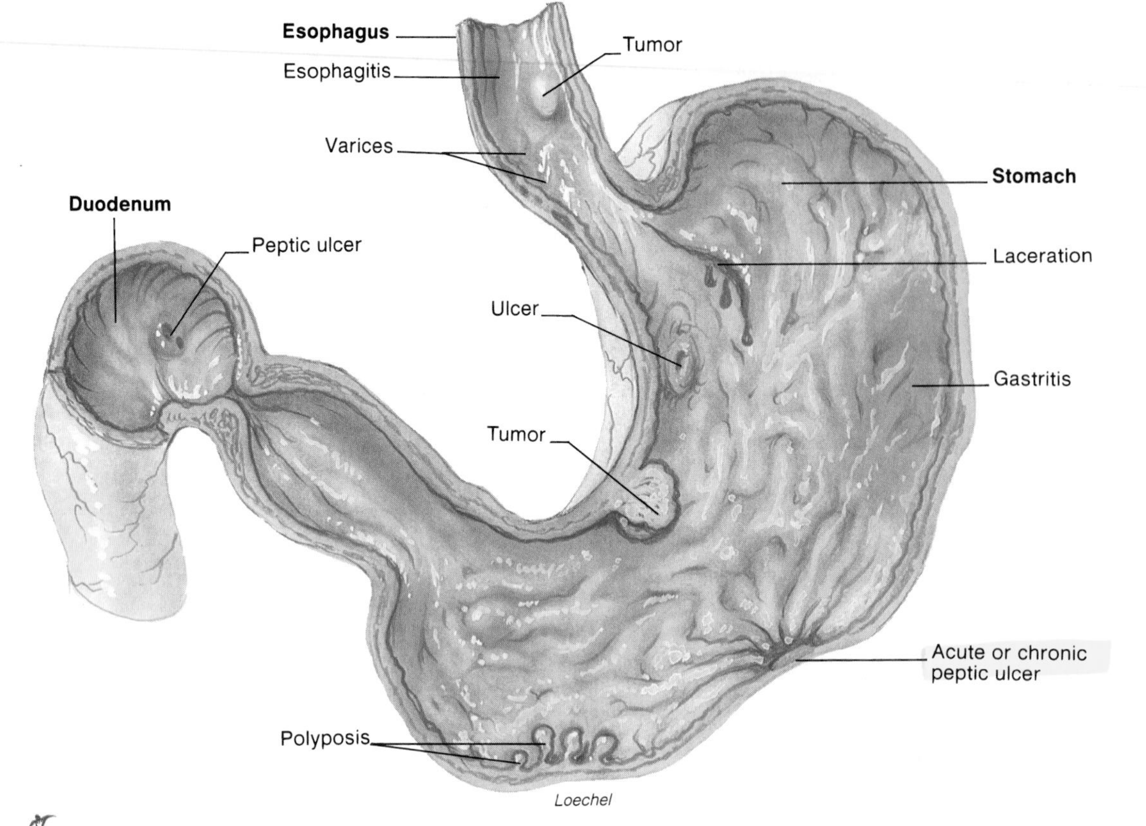

Figure 18.39

Common sites of upper GI disorders.

the liver (hepatitis, cirrhosis, and hepatomas) affect the liver throughout, so that it cannot repair itself. **Hepatitis** is inflammation of the liver. Certain chemicals may cause hepatitis, but generally it is caused by infectious viral agents. **Hepatitis A** (infectious hepatitis) is a viral disease transmitted through contaminated foods and liquids. **Hepatitis B** (serum hepatitis) is also caused by a virus and is transmitted in blood plasma during transfusions or by improperly sterilized needles and syringes. Other types of viral hepatitis are designated as hepatitis C, D, E, and G.

In **cirrhosis** (sĭ-ro′sis), the liver becomes infused with fibrous tissue. This causes the liver tissue to break down and become filled with fat. Eventually, all functions of the liver are compromised. Cirrhosis is most often the result of long-term alcohol abuse, but it can also result from malnutrition, hepatitis, or other infections.

Jaundice is a yellow staining of the tissues produced by high blood concentrations of either free or conjugated bilirubin. Since free bilirubin is derived from hemoglobin, abnormally high concentrations of this pigment may result from an unusually high rate of red blood cell destruction. This can occur, for example, as a result of Rh disease (erythroblastosis fetalis) in an Rh positive baby born to a sensitized Rh negative mother. Jaundice may also occur in healthy infants, since excess red blood cells are normally destroyed at about the time of birth. This condition is called *physiological jaundice of the newborn* and is not indicative of disease. Premature infants may also develop jaundice due to inadequate amounts of hepatic enzymes necessary to conjugate bilirubin and excrete it in the bile. In adults, jaundice is commonly exhibited when the excretion of bile is blocked by gallstones.

Hepatomas (*hep″ă-to′maz*) are tumors (usually malignant) that originate in or secondarily invade the liver. Those that originate in the liver (primary hepatomas) are relatively rare, but those that metastasize to the liver from other organs (secondary hepatomas) are common. Carcinoma of the liver is usually fatal.

Disorders of the GI Tract

Peptic ulcers are erosions of the mucous membranes of the stomach (fig. 18.39) or duodenum produced by the action of HCl. Prolonged exposure to agents that weaken the mucosal lining of the stomach, such as alcohol and aspirin, and abnormally high secretions of HCl increase the likelihood of developing peptic ulcers. Chronic stress can impair mucosal defense mechanisms, thereby increasing mucosal susceptibility to the damaging effects of HCl. A relatively recent finding is that most people who have peptic ulcers are infected with a bacterium known as *Helicobacter pylori*, which resides in the GI tract. Clinical trials have demonstrated that antibiotics that eliminate this bacterium appear to

cirrhosis: Gk. *kirrhos*, yellow orange
jaundice: L. *galbus*, yellow

Internal Affairs

■ How the *Digestive System* Works with Other Body Systems
□ How Other Systems Work with the *Digestive System*

Integumentary System

- Provides lipids that help to insulate dermal and subcutaneous tissues

- Covers and protects the body
- Helps to synthesize vitamin D needed for calcium and phosphorus absorption in the small intestine

Skeletal System

- Absorbs calcium and phosphate ions for formation and maintenance of bone tissue
- Provides lipids for storage in yellow marrow

- Supports and protects certain digestive organs
- Stores calcium and phosphate ions
- Teeth and mandible important for mastication

Muscular System

- Provides nutrients for muscle contraction and maintenance
- Liver metabolizes lactic acid from active muscles

- Supports and protects certain digestive organs
- Assists the processing of food—preparing, chewing, swallowing, and intestinal motility
- Controls voluntary defecation

Nervous System

- Provides nutrients for neural function and maintenance

- Provides autonomic innervation to the GI tract
- Control over skeletal muscles regulates ingestion and defecation
- Hypothalamus contains centers for hunger, thirst, and satiation

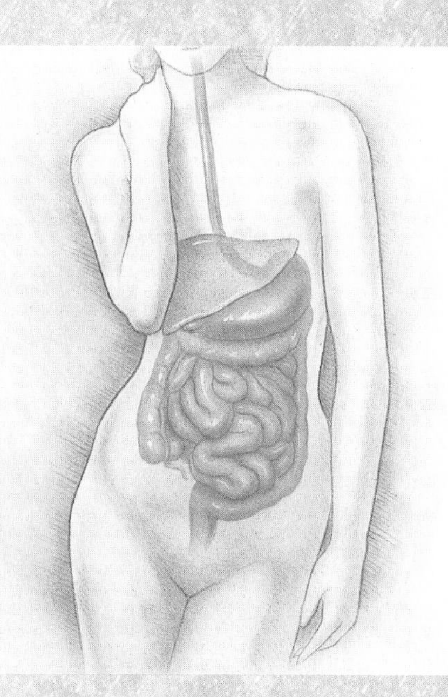

Endocrine System

- Provides nutrients for endocrine function and maintenance
- Liver deactivates hormones
- Pancreas contains hormone-producing cells

- Hormones regulate digestive activity

Circulatory System

- Provides nutrients for circulatory function and maintenance, including iron needed for hemoglobin synthesis
- Liver produces clotting proteins
- Liver excretes heme (as conjugated bilirubin)

- Transports O_2 and CO_2, nutrients, and fluids to and from the digestive organs

Lymphatic System

- Provides nutrients for lymphatic function and maintenance
- Acids and enzymes secreted by the GI tract provide nonspecific defense against microbes

- Maintains a balanced amount of intersitial fluid within the digestive organs
- Protects the GI tract against infection

Respiratory System

- Provides nutrients for function and maintenance of the respiratory organs
- Pressure of digestive organs against the diaphragm aids expiration

- Provides O_2 and eliminates CO_2
- Valsalva's maneuver aids defecation

Urinary System

- Provides nutrients for function and maintenance of urinary organs

- Eliminates metabolic wastes
- Helps to maintain the body's pH, ion, and water balance
- Kidneys transform vitamin D to its active form

Reproductive System

- Provides nutrients for function and maintenance of reproductive organs
- Supports the development of an embryo and fetus

- Influences metabolic rates through effects of steroids
- Developing fetus may crowd digestive organs; heartburn and constipation are common during pregnancy

help in the treatment of the peptic ulcers. It is now thought that *H. pylori* does not itself cause the ulcer, but rather contributes to the weakening of the mucosal barriers to gastric acid damage.

Enteritis, an inflammation of the mucosa of the small intestine, is frequently referred to as intestinal flu. Causes of enteritis include bacterial or viral infections, irritating foods or fluids (including alcohol), and emotional stress. The symptoms include abdominal pain, nausea, and diarrhea. *Diarrhea* is the passage of watery, unformed stools. This condition is symptomatic of inflammation, stress, and many other body dysfunctions.

A **hernia** is a protrusion of a portion of a visceral organ, usually the small intestine, through a weakened portion of the abdominal wall. Inguinal, femoral, umbilical, and hiatal hernias are the most common types. With a **hiatal hernia,** a portion of the stomach pushes superiorly through the esophageal hiatus in the diaphragm and protrudes into the thorax. The potential dangers of a hernia are strangulation of the blood supply followed by gangrene, blockage of chyme, or rupture—each of which can threaten life.

Diverticulosis (*di″ver-tik″yŭ-li′tis*) is a condition in which the intestinal wall weakens and an outpouching (diverticulum) occurs. Studies suggest that suppressing the passage of flatus (intestinal gas) may contribute to diverticulosis, especially in the sigmoid colon. **Diverticulitis,** or inflammation of a diverticulum, can develop if fecal material becomes impacted in these pockets.

Peritonitis is inflammation of the peritoneum lining the abdominal cavity and covering the viscera. The causes of peritonitis include bacterial contamination of the abdominal cavity through accidental or surgical wounds in the abdominal wall or perforation of the intestinal wall (as with an ulcer or a ruptured appendix).

Clinical Case Study Answer

The likely source of bleeding in such a case is from a rupture of the internal portion of the liver involving significant blood vessels, either of the hepatic arterial or venous circulation, but also possibly of the portal venous circulation. Trauma to the hepatic plates (see fig. 18.33) may allow blood to enter bile canaliculi. From there, the course of the blood is as follows: large hepatic ducts → common hepatic duct → common bile duct through duodenal papilla → duodenum of the small intestine.

In some cases, this type of bleeding can be stopped by introducing radiographic-guided catheters into the arterial system to block the blood vessels (angiographic embolization). However, surgery is sometimes required.

Important Clinical Terminology

chilitis (***ki-li′tis***) Inflammation of the lips.

colitis (***kŏ-li′tis***) Inflammation of the colon.

colostomy (***kŏ-los′tŏ-me***) The formation of an abdominal exit from the GI tract by bringing a loop of the colon through the abdominal wall to its outside surface. If the rectum is removed because of cancer, the colostomy provides a permanent outlet for the feces.

cystic fibrosis An inherited disease of the exocrine glands, particularly the pancreas. Pancreatic secretions are too thick to drain easily, causing the ducts to become inflamed and promoting connective tissue formation that occludes drainage from the ducts still further.

gingivitis (***jin-jĭ-vi′tis***) Inflammation of the gums. It may result from improper hygiene, poorly fitted dentures, improper diet, or certain infections.

halitosis (***hal-ĭ-to′sis***) Offensive breath odor. It may result from dental caries, certain diseases, eating particular foods, or smoking.

heartburn A burning sensation of the esophagus and stomach. It may result from the regurgitation of gastric juice into the lower portion of the esophagus.

hemorrhoids (***hem′ŏ-roidz***) Varicose veins of the rectum and anus.

nausea Gastric discomfort and sensations of illness with a tendency to vomit. This feeling is symptomatic of motion sickness and other diseases, and may occur during pregnancy.

pyorrhea (***pi″ŏ-re′ă***) The discharge of pus at the base of the teeth at the gum line.

regurgitation (vomiting) The forceful expulsion of gastric contents into the mouth. Nausea and vomiting are common symptoms of almost any dysfunction of the digestive system.

trench mouth A contagious bacterial infection that causes inflammation, ulceration, and painful swelling of the floor of the mouth. It is generally contracted through direct contact by kissing an infected person. Trench mouth can be treated with penicillin and other medications.

vagotomy (***va-got′ŏ-me***) The surgical removal of a section of the vagus nerve where it enters the stomach in order to eliminate nerve impulses that stimulate gastric secretion. This procedure may be used to treat chronic ulcers.

Chapter Summary

Introduction to the Digestive System (pp. 615–616)

1. The digestive system mechanically and chemically breaks down food to forms that can be absorbed through the intestinal wall and transported by the blood and lymph for use at the cellular level.
2. The digestive system consists of a gastrointestinal (GI) tract and accessory digestive organs.

Serous Membranes and Tunics of the Gastrointestinal Tract (pp. 616–620)

1. Peritoneal membranes line the abdominal wall and cover the visceral organs. The GI tract is supported by a double layer of peritoneum called the mesentery.
   (a) The lesser omentum and greater omentum are folds of peritoneum that extend from the stomach.
   (b) Retroperitoneal organs are positioned behind the parietal peritoneum.
2. The layers (tunics) of the abdominal GI tract are, from the inside outward, the mucosa, submucosa, tunica muscularis, and serosa.
   (a) The mucosa consists of a simple columnar epithelium, a thin layer of connective tissue called the lamina propria, and thin layers of smooth muscle called the muscularis mucosae.
   (b) The submucosa is composed of connective tissue; the tunica muscularis consists of layers of smooth muscle; and the serosa is composed of connective tissue covered with the visceral peritoneum.
   (c) The submucosa contains the submucosal plexus, and the tunica muscularis contains the myenteric plexus of autonomic nerves.

Mouth, Pharynx, and Associated Structures (pp. 620-627)

1. The oral cavity is formed by the cheeks, lips, and hard palate and soft palate. The tongue and teeth are contained in the oral cavity.
   (a) Lingual tonsils and papillae with taste buds are located on the tongue.
   (b) Structures of the palate include palatal folds, a cone-shaped projection called the palatine uvula, and palatine tonsils.
2. The incisors and canines have one root each; the bicuspids and molars have two or three roots.
   (a) Humans are diphyodont; they have deciduous and permanent sets of teeth.
   (b) The roots of teeth fit into sockets called dental alveoli that are lined with a periodontal membrane. Fibers in the periodontal membrane insert into the cementum covering the roots, firmly anchoring the teeth in the sockets.
   (c) Enamel forms the outer layer of the tooth crown; beneath the enamel is dentin.
   (d) The interior of a tooth contains a pulp cavity, which is continuous through the apical foramen of the root with the connective tissue around the tooth.
3. The major salivary glands are the parotid glands, the submandibular glands, and the sublingual glands.
4. The muscular pharynx is a passageway connecting the oral and nasal cavities to the esophagus and larynx.

Esophagus and Stomach (pp. 628-632)

1. Swallowing (deglutition) occurs in three phases and involves structures of the oral cavity, pharynx, and esophagus.
2. Peristaltic waves of contraction push food through the lower esophageal sphincter into the stomach.
3. The stomach consists of a cardia, fundus, body, and pylorus. It displays greater and lesser curvatures, and contains a pyloric sphincter at its junction with the duodenum.
   (a) The mucosa of the stomach is thrown into distensible gastric folds; gastric pits and gastric glands are present in the mucosa.
   (b) The parietal cells of the gastric glands secrete HCl, and the principal cells secrete pepsinogen.

Small Intestine (pp. 632-635)

1. Regions of the small intestine include the duodenum, jejunum, and ileum; the bile duct and pancreatic duct empty into the duodenum.
2. Fingerlike extensions of mucosa, called villi, project into the lumen, and at the bases of the villi the mucosa forms intestinal glands.
   (a) New epithelial cells are formed in the intestinal crypts.
   (b) The membrane of intestinal epithelial cells is folded to form microvilli; this brush border of the mucosa increases the absorptive surface area.
3. Movements of the small intestine include rhythmic segmentation, pendular movement, and peristalsis.

Large Intestine (pp. 636-639)

1. The large intestine absorbs water and electrolytes from the chyme and passes fecal material out of the body through the rectum and anal canal.
2. The large intestine is divided into the cecum, colon, rectum, and anal canal.
   (a) The appendix is attached to the inferior medial margin of the cecum.
   (b) The colon consists of ascending, transverse, descending, and sigmoid portions.
   (c) Haustra are bulges in the walls of the large intestine.
3. Movements of the large intestine include peristalsis, haustral churning, and mass movement.

Liver, Gallbladder, and Pancreas (pp. 639-648)

1. The liver is divided into right, left, quadrate, and caudate lobes. Each lobe contains liver lobules, the functional units of the liver.
   (a) Liver lobules consist of plates of hepatic cells separated by modified capillaries called sinusoids.
   (b) Blood flows from the periphery of each lobule, where branches of the hepatic artery and hepatic portal vein empty, through the sinusoids and out the central vein.
   (c) Bile flows within the hepatic plates, in bile canaliculi, to the biliary ductules at the periphery of each lobule.
2. The gallbladder stores and concentrates the bile; it releases the bile through the cystic duct and common bile duct into the duodenum.
3. The pancreas is both an exocrine and an endocrine gland.
   (a) The endocrine portion, consisting of the pancreatic islets, secretes the hormones insulin and glucagon.
   (b) The exocrine acini of the pancreas produce pancreatic juice, which contains various digestive enzymes.

Review Activities

Objective Questions

1. Viscera are the only body organs that are
   (a) concerned with digestion.
   (b) located in the abdominal cavity.
   (c) covered with peritoneal membranes.
   (d) located within the thoracic and abdominal cavities.
2. Which of the following types of teeth are found in the permanent but not in the deciduous dentition?
   (a) incisors (c) premolars
   (b) canines (d) molars
3. The double layer of peritoneum that supports the GI tract is called
   (a) the visceral peritoneum.
   (b) the mesentery.
   (c) the greater omentum.
   (d) the lesser omentum.
4. Which of the following tissue layers in the small intestine contains the lacteals?
   (a) the submucosa
   (b) the muscularis mucosae
   (c) the lamina propria
   (d) the tunica muscularis
5. Which of the following organs is *not* considered a part of the digestive system?
   (a) the pancreas (c) the tongue
   (b) the spleen (d) the gallbladder
6. The numerous small elevations on the surface of the tongue that support taste buds and aid in handling food are called
   (a) cilia. (c) villi.
   (b) rugae. (d) papillae.
7. Most digestion occurs in
   (a) the mouth.
   (b) the stomach.
   (c) the small intestine.
   (d) the large intestine.
8. Stenosis (constriction) of the sphincter of ampulla (of Oddi) would interfere with
   (a) transport of bile and pancreatic juice.
   (b) secretion of mucus.
   (c) passage of chyme into the small intestine.
   (d) peristalsis.
9. The first organ to receive the blood-borne products of digestion is
   (a) the liver. (c) the heart.
   (b) the pancreas. (d) the brain.
10. Which of the following statements about hepatic portal blood is *true?*
   (a) It contains absorbed fat.
   (b) It contains ingested proteins.
   (c) It is mixed with bile in the liver.
   (d) It is mixed with blood from the hepatic artery in the liver.

Essay Questions

1. Define *digestion*. Differentiate between the mechanical and chemical aspects of digestion.
2. Distinguish between the gastrointestinal tract, viscera, accessory digestive organs, and gut.
3. List the specific portions or structures of the digestive system formed by each of the three embryonic germ layers.
4. Define *serous membrane*. How are the serous membranes of the abdominal cavity classified and what are their functions?
5. Describe the structures of the four tunics in the wall of the GI tract.
6. Why are there two autonomic innervations to the GI tract? Identify the specific sites of autonomic stimulation in the tunic layers.
7. Define the terms *dental formula*, *diphyodont*, *deciduous teeth*, *permanent teeth*, and *wisdom teeth*.
8. Outline the stages of deglutition. What biomechanical roles do the tongue, hard palate and soft palate, pharynx, and hyoid bone perform in deglutition?
9. How does the stomach protect itself from the damaging effects of HCl?
10. Describe the kinds of movements in the small intestine and explain what they accomplish.
11. Diagram a villus and explain why villi are considered the functional units of the digestive system.
12. What are the regions of the large intestine? In what portion of the abdominal cavity and pelvic cavity is each region located?
13. Describe the location and gross structure of the liver. Draw a labeled diagram of a liver lobule.
14. Describe how the gallbladder is filled with and emptied of bile fluid. What is the function of bile?
15. List the functions of the large intestine. What are the biomechanical movements of the large intestine that make these functions possible?
16. Define *cirrhosis* and explain why this condition is so devastating to the liver. What are some of the causes of cirrhosis?

Critical-Thinking Questions

1. Technically, ingested food is not in the body. Neither are feces excreted from within the body (except bile residue). Explain these statements. Why would this information be important to a drug company interested in preparing a new oral medication?
2. The deciduous (milk) teeth don't matter because they fall out anyway. Do you agree or disagree with this statement? Explain.
3. Which surgery do you think would have the most profound effect on digestion: (a) removal of the stomach (gastrectomy), (b) removal of the pancreas (pancreatectomy), or (c) removal of the gallbladder (cholecystectomy)? Explain your reasoning.
4. Describe the adaptations of the GI tract that make it more efficient by either increasing the surface area for absorption or increasing the time of contact between food particles and digestive enzymes.
5. During surgery to determine the cause of an intestinal obstruction, why might the surgeon elect to remove a healthy appendix?
6. Explain why a ruptured appendix may result in peritonitis, while an inflamed kidney (nephritis) generally does not result in peritonitis.

NINETEEN

Urinary System

Clinical Case Study 656

Introduction to the Urinary System 656

Kidneys 657

- Position and Appearance of the Kidneys 657
- Gross Structure of the Kidney 658
- Microscopic Structure of the Kidney 658

Ureters, Urinary Bladder, and Urethra 664

- Ureters 664
- Urinary Bladder 665
- Urethra 666
- Micturition 667

Clinical Considerations 669

- Developmental Problems of the Urinary Organs 669

Developmental Exposition: The Urinary System 670

- Symptoms and Diagnosis of Urinary Disorders 672
- Infections of the Urinary Organs 673
- Trauma to the Urinary Organs and Functional Impairments 673

Internal Affairs 674

Clinical Case Study Answer 675

Chapter Summary 676

Review Activities 676

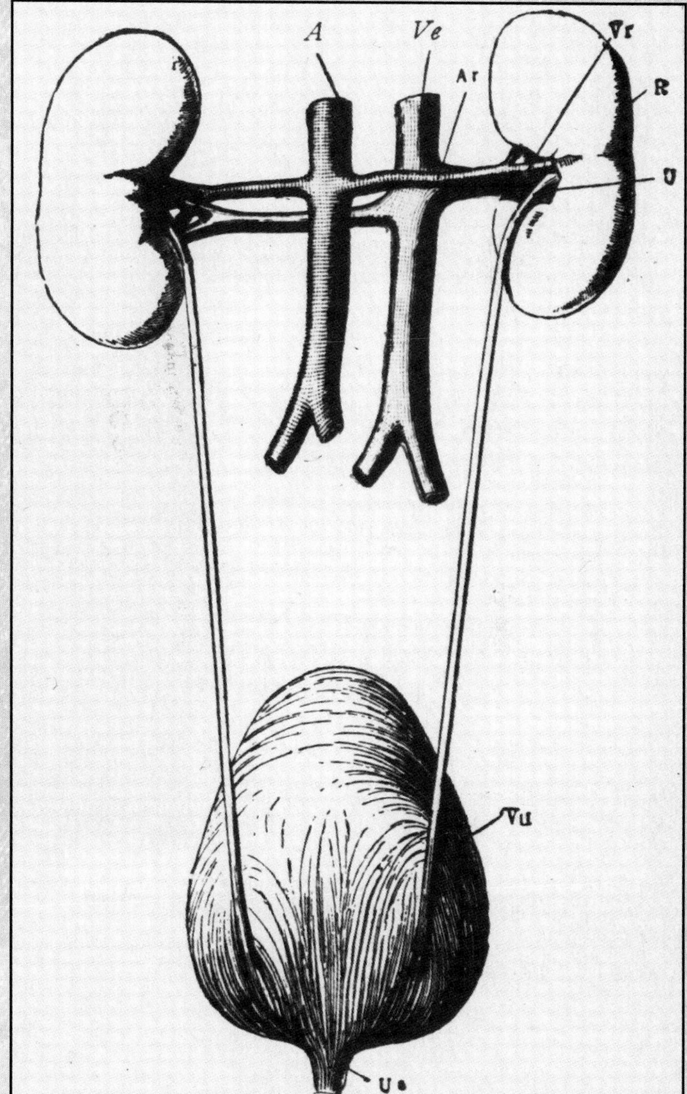

FIG. 156.—The Kidneys, Bladder, and their Vessels. Viewed from behind.

R, right kidney; U, ureter; A, aorta; Ar, right renal artery; Ve, vena cava inferior; Vr, right renal vein; Vu, bladder; Ua, commencement of urethra.

Although crude in its execution, this early medical illustration is anatomically accurate. The organs of the urinary system are shown in posterior view, with the renal arteries extending from the abdominal aorta to each kidney, and the renal veins extending from the kidneys to the inferior vena cava.

Clinical Case Study

A 17-year-old male was involved in a knife fight in which he sustained two stab wounds to the anterior abdomen. He was brought to the emergency room where he complained of mild abdominal pain and an urgent need to urinate. Although neither of the stab wounds was bleeding externally, examination by the surgeon revealed signs of moderate hemorrhagic shock. One wound was 3 cm below the right costal arch at the midclavicular line; the other was just medial to the right anterior superior iliac spine. The surgeon immediately ordered preparations for an emergency exploratory laparotomy. She noted that the patient's urinary bladder was quite full and chided the intern for not having placed a urinary catheter. Placement of the catheter yielded a brisk flow of bright red blood.

How would you explain the phenomenon of hematuria (blood in the urine) in this case? Which of the two stab wounds is most likely associated with the hematuria? Regarding the blood draining into the catheter, trace and explain its course of flow. Begin at a point in the abdominal aorta, and end the course with drainage into the catheter. Assuming that the surgeon would be prompted to remove the kidney in order to quickly control hemorrhage, what possible anatomical variant should she keep in mind?

Hints: Study the positions of the kidneys within the abdominal cavity and note the location of the supportive and protective serous membrane. Carefully examine the specific location of urine production and the route of urine passage through the urinary system.

Introduction to the Urinary System

The urinary system maintains the composition and properties of the body fluid that establishes the internal environment of the body cells. The end product of the urinary system is urine, which is voided from the body during micturition.

Objective 1 List the functions of the urinary system.

Objective 2 Identify the arteries that transport blood to the urinary system for filtration.

The urinary system, along with the respiratory, digestive, and integumentary systems, excretes substances from the body. For this reason, these systems are occasionally referred to as excretory systems. In the process of cellular metabolism, nutrients taken in by the digestive system and oxygen from inhaled air are used to synthesize a variety of substances while providing energy needed for body maintenance. Metabolic processes, however, produce cellular wastes that must be eliminated if homeostasis is to be maintained. Just as the essential nutrients are transported to the cells by the blood, the cellular wastes are removed through the circulatory system to the appropriate excretory system. Carbon dioxide is eliminated through the respiratory system; excessive water, salts, nitrogenous wastes, and even excessive metabolic heat are removed through the integumentary system; and various digestive wastes are eliminated through the digestive system.

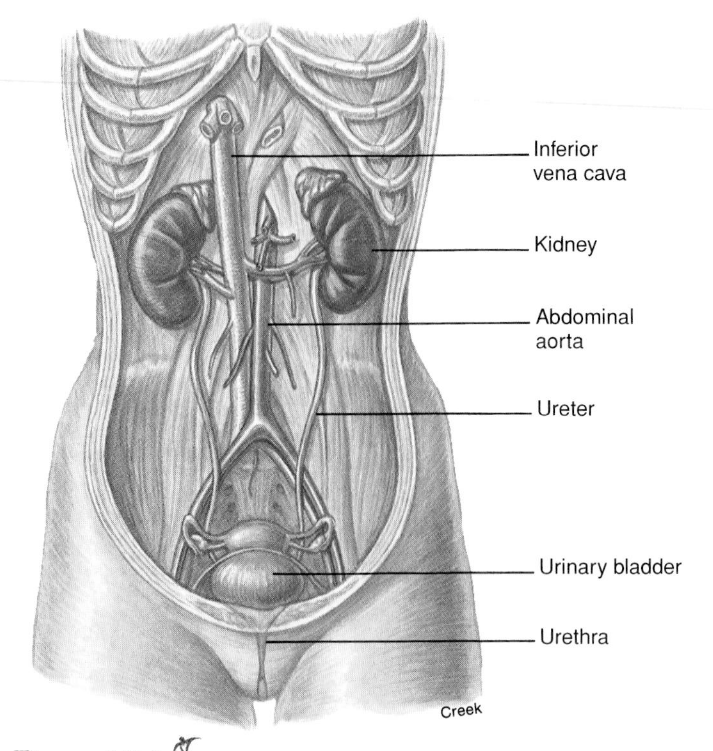

Figure 19.1

The organs of the urinary system are the two kidneys, two ureters, urinary bladder, and urethra.

The urinary system is the principal system responsible for water and electrolyte balance. Electrolytes are compounds that separate into ions when dissolved in water. Electrolyte balance is achieved when the number of electrolytes entering the body equals the number leaving. Hydrogen ions, for example, are maintained in precise concentration so that an acid-base, or pH, balance exists in the body.

A second major function of the urinary system is the excretion of toxic nitrogenous compounds—specifically, urea and creatinine. Other functions of the urinary system include the elimination of toxic wastes that may result from bacterial action and the removal of various drugs that have been taken into the body. All of these functions are accomplished through the formation of *urine* by the kidneys.

The urinary system consists of two *kidneys*, two *ureters*, the *urinary bladder*, and the *urethra* (fig. 19.1). Tubules in the kidneys are intertwined with vascular networks of the circulatory system to enable the production of urine. After the urine is formed, it is moved through the ureters to the urinary bladder for storage. *Micturition*, or voiding of urine from the urinary bladder, occurs through the urethra.

Blood to be processed by a kidney enters through the large *renal artery*. After the filtration process (see chapter 3), it exits through the *renal vein*. The importance of filtration of the blood is demonstrated by the fact that during normal resting conditions the kidneys receive approximately 20%–25% of the entire cardiac output. Every minute, the kidneys process approximately 1,200 ml of blood.

1. Drawing on your knowledge of the functions of the urinary system, list the basic substances that compose normal urine.
2. Explain the role of the renal vessels in maintaining homeostasis. Approximately how much blood is processed in the kidneys each minute?

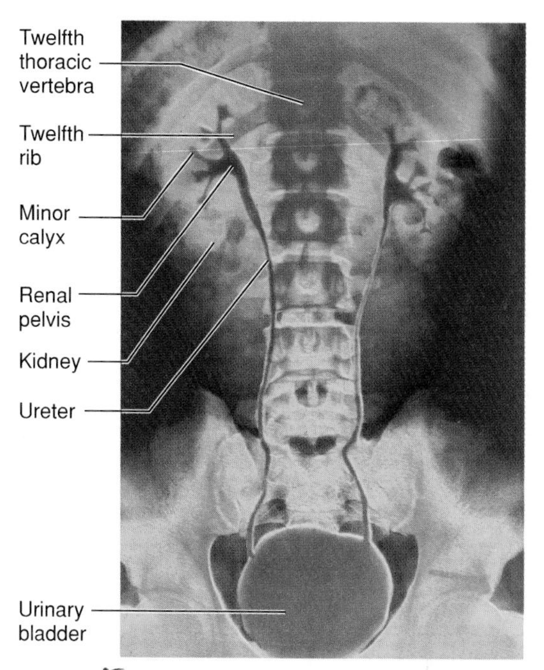

Figure 19.2

A color-enhanced radiograph of the calyces and renal pelvises of the kidneys, the ureters, and the urinary bladder. (Note the position of the kidneys relative to the vertebral column and ribs.)

Kidneys

The kidney consists of an outer renal cortex and an inner renal medulla that contains the renal pyramids. Urine is formed as a filtrate from the blood at the nephrons and collects in the calyces and renal pelvis before flowing from the kidney via the ureter.

Objective 3 Describe the gross structure of the kidney.

Objective 4 Describe structure of a nephron and explain how its components are oriented within the kidney.

Objective 5 Describe the position of cortical and juxtamedullary nephrons with respect to the gross structure of the kidney.

Position and Appearance of the Kidneys

The reddish brown **kidneys** are positioned against the posterior wall of the abdominal cavity between the levels of the twelfth thoracic and the third lumbar vertebrae (fig. 19.2). The right kidney is usually 1.5–2.0 cm lower than the left because of the large area occupied by the liver on the right side.

The kidneys are *retroperitoneal*, which means that they are located behind the parietal peritoneum (fig. 19.3). Each adult kidney is a lima-bean-shaped organ about 11.25 cm (4 in.) long, 5.5–7.7 cm (2–3 in.) wide, and 2.5 cm (1 in.) thick. The lateral border of each kidney is convex, whereas the medial border is strongly concave (figs. 19.2 and 19.3). The **hilum** (*hi'lum*) of the kidney is the depression along the medial border through which the **renal artery** enters and the **renal vein** and **ureter** (*yoo're'ter*) exit. The hilum is also the site for drainage of lymph vessels and innervation of the kidney.

Each kidney is embedded in a fatty fibrous pouch consisting of three layers. The **renal capsule** (fibrous capsule), the innermost layer, is a strong, transparent fibrous attachment to the surface of the kidney. The renal capsule protects the kidney from trauma and the spread of infections. Surrounding the renal capsule is a firm protective mass of fatty tissue called the **renal adipose capsule** (see fig. 19.3). The outermost layer, the **renal fascia,** is composed of dense irregular connective tissue. It is a supportive layer that anchors the kidney to the peritoneum and the abdominal wall.

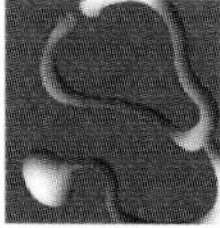

Although the kidneys are firmly supported by the renal adipose capsule, renal fascia, and even the renal vessels, under certain conditions these structures may give in to the force of gravity and the kidneys may drop a bit in position. This condition is called *renal ptosis* (*to'sis*) and generally occurs in extremely thin elderly people, who have insufficient amounts of supportive fat in the adipose capsular layer. It also may affect victims of *anorexia nervosa,* who suffer from extreme weight loss. The potential danger of renal ptosis is that the ureter may kink, blocking the flow of urine from the affected kidney.

hilum: L. *hilum*, a trifle

ptosis: Gk. *ptosis*, a falling

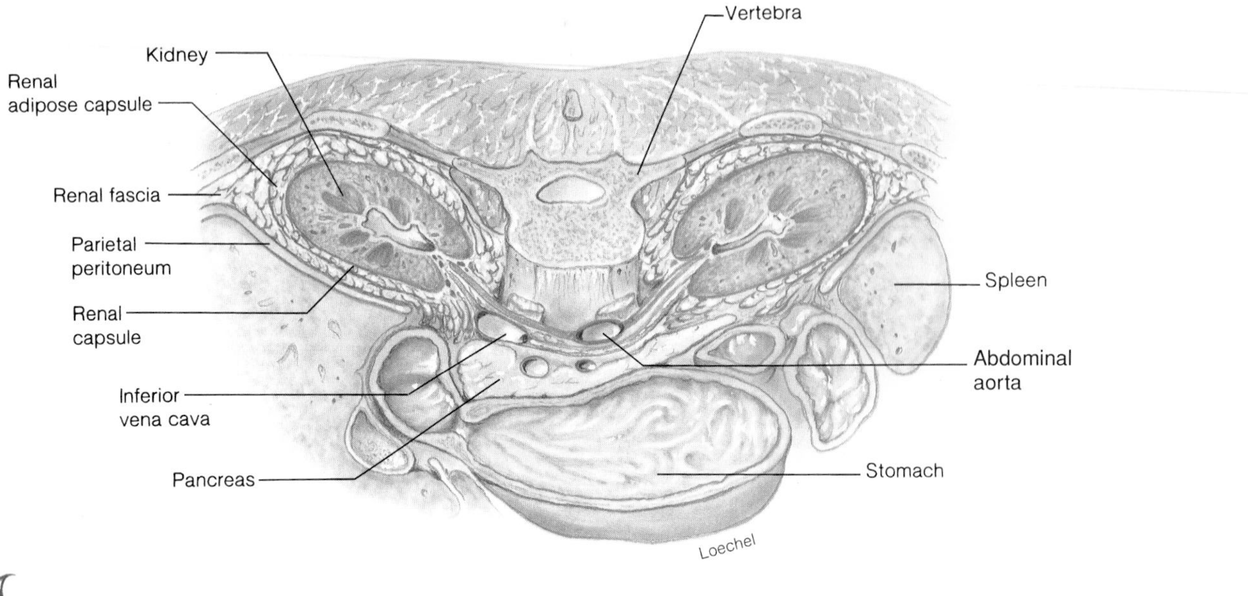

Figure 19.3

The position of the kidneys as seen in cross section through the upper abdomen. The kidneys are embedded in adipose tissue behind the parietal peritoneum.

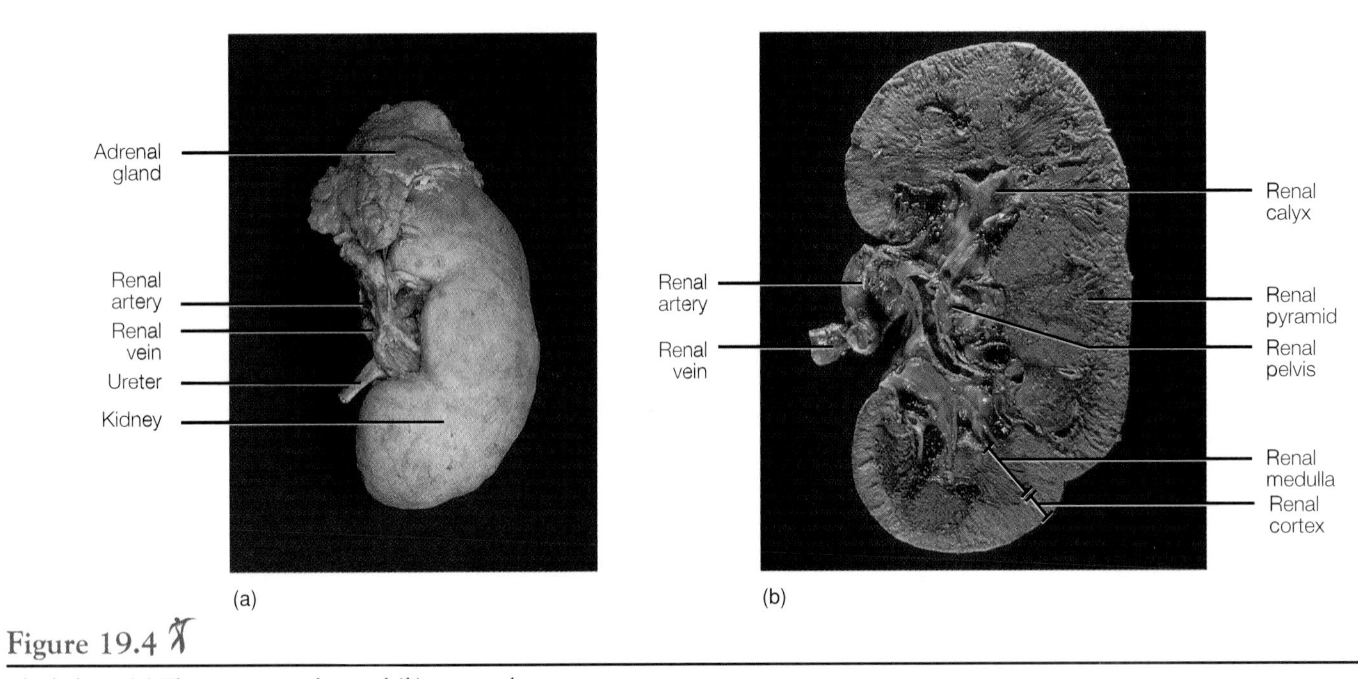

Figure 19.4

The kidney. (*a*) The anterior surface and (*b*) a coronal section.

Gross Structure of the Kidney

A coronal section of the kidney shows two distinct regions and a major cavity (figs. 19.4*b* and 19.5). The outer **renal cortex,** in contact with the renal capsule, is reddish brown and granular in appearance because of its many capillaries. The deeper **renal medulla** is darker, and the presence of microscopic tubules and blood vessels gives it a striped appearance. The renal medulla is composed of 8 to 15 conical **renal pyramids.** Portions of the renal cortex extend between the renal pyramids to form the **renal columns.** The apexes of the renal pyramids are known as the *renal papillae* (*pă-pil'e*). These nipplelike projections are directed toward the inner region of the kidney.

The cavity of the kidney collects and transports urine from the kidney to the ureter. It is divided into several portions. The papilla of a renal pyramid projects into a small depression called the **minor calyx** (*ka'liks*—in the plural, *calyces*). Several minor calyces unite to form a **major calyx.** In turn, the major calyces join to form the funnel-shaped **renal pelvis.** The renal pelvis collects urine from the calyces and transports it to the ureter.

Microscopic Structure of the Kidney

The **nephron** (*nef'ron*) is the functional unit of the kidney that is responsible for the formation of urine. Each kidney contains more

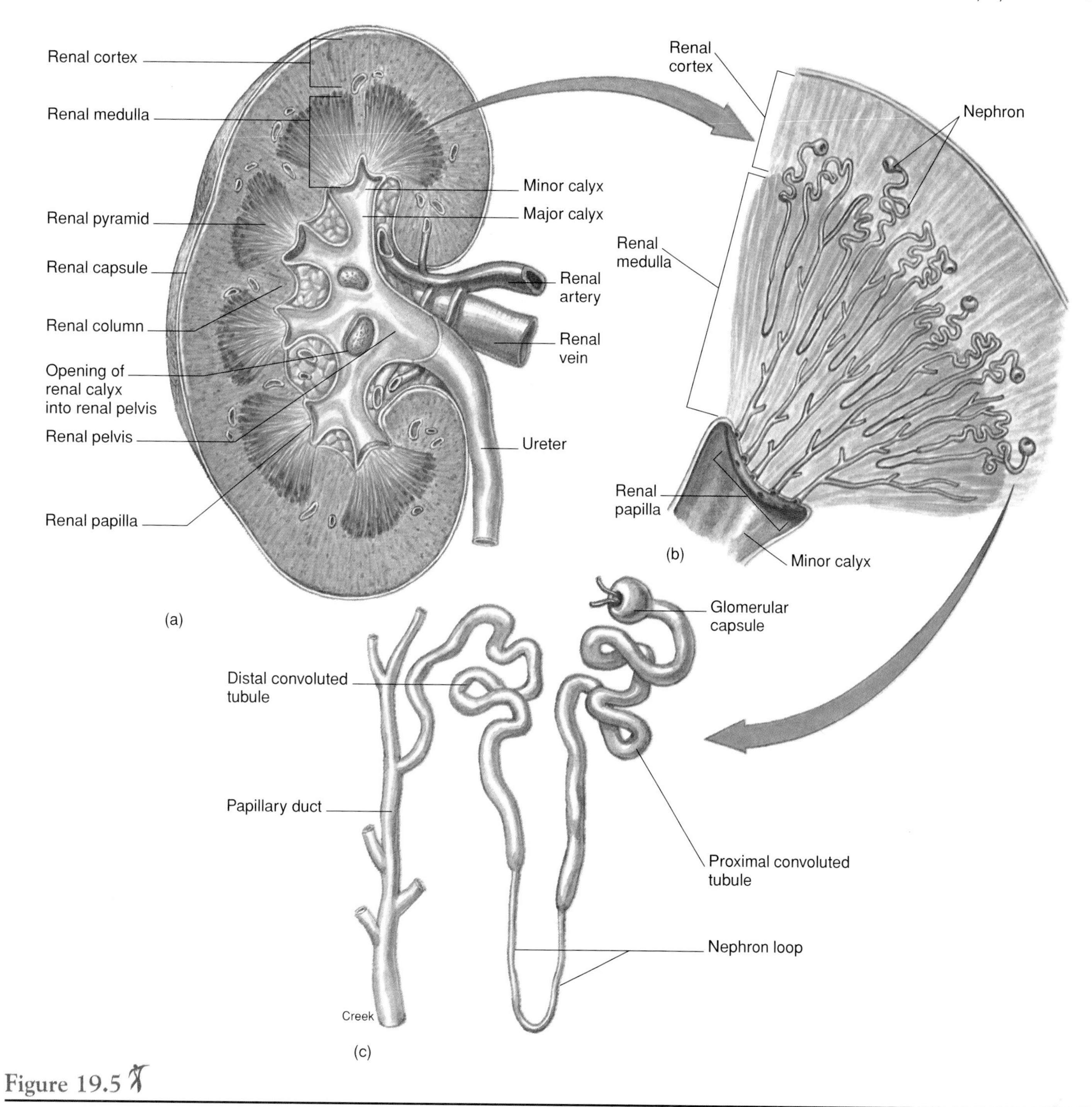

Figure 19.5

The internal structures of a kidney. (*a*) A coronal section showing the structure of the renal cortex, renal medulla, and renal pelvis. (*b*) A diagrammatic magnification of a renal pyramid to depict the renal tubules. (*c*) A diagram of a single nephron and a papillary duct.

than a million nephrons. A nephron consists of a **urinary tubule** and associated small blood vessels. Fluid formed by capillary filtration enters the urinary tubule and is subsequently modified by transport processes. The resulting fluid that leaves the tubule is urine.

Renal Blood Vessels

The kidneys have an extensive circulatory network to allow for the continuous cleansing and modification of large volumes of blood (fig. 19.6). Arterial blood enters the kidney at the hilum through the **renal artery,** which divides into **interlobar** (*in"ter-lo'bar*) **arteries** that pass between the renal pyramids through the renal columns. **Arcuate** (*ar'kyoo-āt*) **arteries** branch from the interlobar arteries at the boundary of the renal cortex and renal medulla. Small **interlobular arteries** radiate from the arcuate

arcuate: L. *arcuare*, to bend

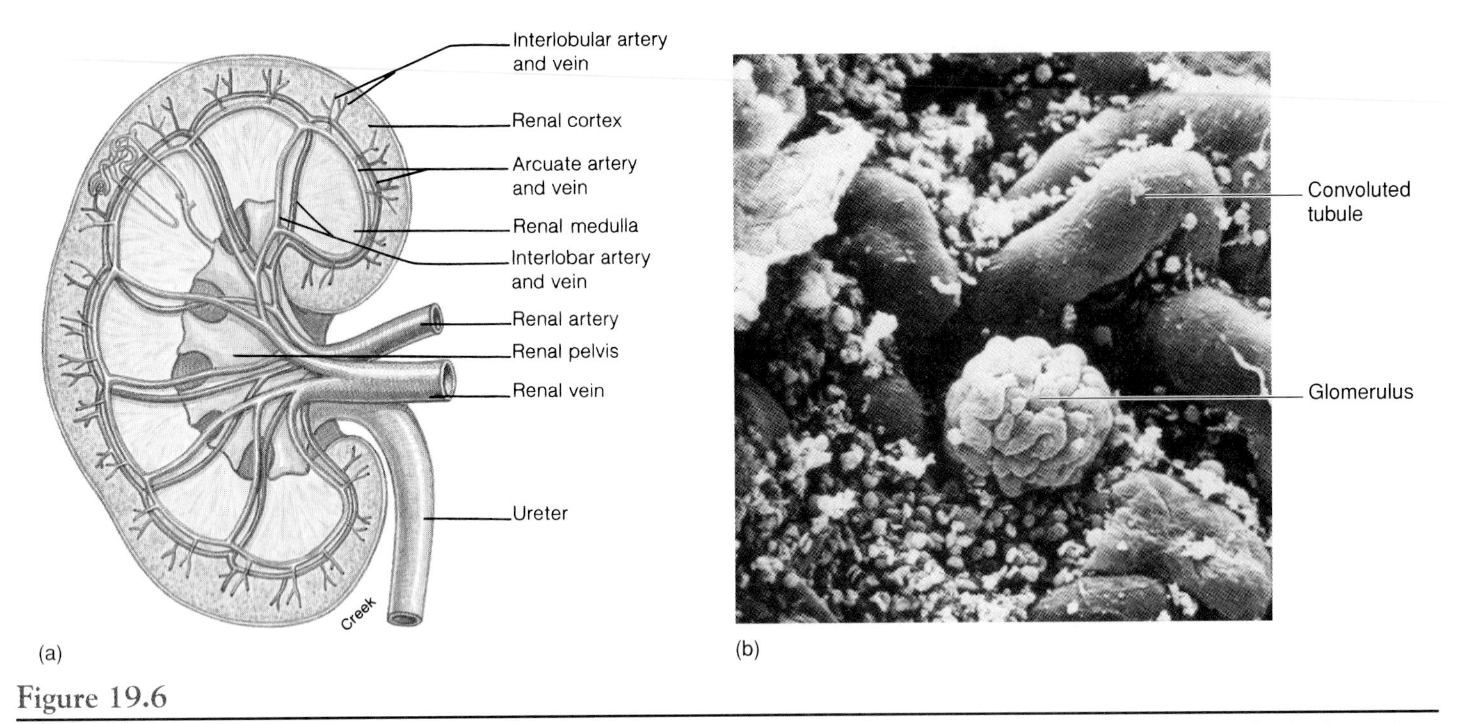

Figure 19.6

The principal arteries and veins of a kidney. (*a*) A coronal view of the pattern of renal vessels and (*b*) a scanning electron micrograph of a glomerulus.

arteries and project into the renal cortex. Microscopic **afferent glomerular arterioles** arise from branches of the interlobular arteries. The afferent glomerular arterioles transport the blood into ball-shaped capillary networks, the **glomeruli** (*glo-mer'yŭ-li*), which produce a blood filtrate that enters the urinary tubules. The blood remaining in the glomerulis leaves through **efferent glomerular arterioles.** This blood vessel arrangement is unique because blood usually flows out of a capillary bed into venules rather than into other arterioles. From the efferent glomerular arterioles, the blood enters either the **peritubular capillaries** surrounding the convoluted tubules or the **vasa recta** surrounding the ascending and descending tubules (fig. 19.7). From these capillary networks, the blood is drained into veins that parallel the course of the arteries in the kidney. These are the **interlobular veins, arcuate veins,** and **interlobar veins.** The interlobar veins descend between the renal pyramids, converge, and then leave the kidney as a single **renal vein** that empties into the inferior vena cava.

In summary, vessels are at a capillary level between the arterioles in each glomerulus of the kidney. The blood pressure at the glomerulus is strong enough to force water and dissolved wastes from the blood into the urinary tubular portion of the nephron. Thus, this capillary network produces the filtrate. A secondary capillary network of peritubular capillaries and vasa recta surrounds various tubular portions of the nephron. This capillary bed, however, is adapted for absorption rather than filtrate formation. It reabsorbs water and other substances that should not be excreted with the urine, reclaiming most of the filtrate produced in the glomerulus.

Although the kidneys are generally well protected in that they are encapsulated retroperitoneally, they may be injured by a hard blow to the lumbar region. Such an injury can produce blood in the urine, since the highly vascular kidneys are particularly susceptible to hemorrhage.

Nephron

The tubular nephron consists of a *glomerular capsule, proximal convoluted tubule, descending limb of the nephron loop (loop of Henle), ascending limb of the nephron loop,* and *distal convoluted tubule* (fig. 19.7).

The **glomerular** (Bowman's) **capsule** surrounds the glomerulus. The glomerular capsule and its associated glomerulus are located in the renal cortex of the kidney and together constitute the **renal corpuscle** (fig. 19.8). The glomerular capsule contains an inner visceral layer of epithelium, in contact with the glomerular capillaries and an outer parietal layer. The space between these two layers, called the **capsular space,** is where the glomerular filtrate collects.

The glomerular epithelium contains tiny pores called **fenestrae** (*fĕ-nes'tre*) that permit the filtrate to pass from the blood into the glomerular capsular space (fig. 19.8). Although the fenestrae are large, they are still small enough to prevent the passage of blood cells, platelets, and most plasma proteins into the filtrate. The inner layer of the glomerular capsule is composed of unique cells called **podocytes** (*pod'ŏ-sīts*) with numerous cytoplasmic extensions known as **pedicels** (*ped'ĭ-selz*). Pedicels interdigitate, like the fingers of clasped hands, as they wrap around the glomerular capillaries

glomerulus: L. diminutive of *glomus*, ball

Bowman's capsule: from Sir William Bowman, English anatomist, 1816–92

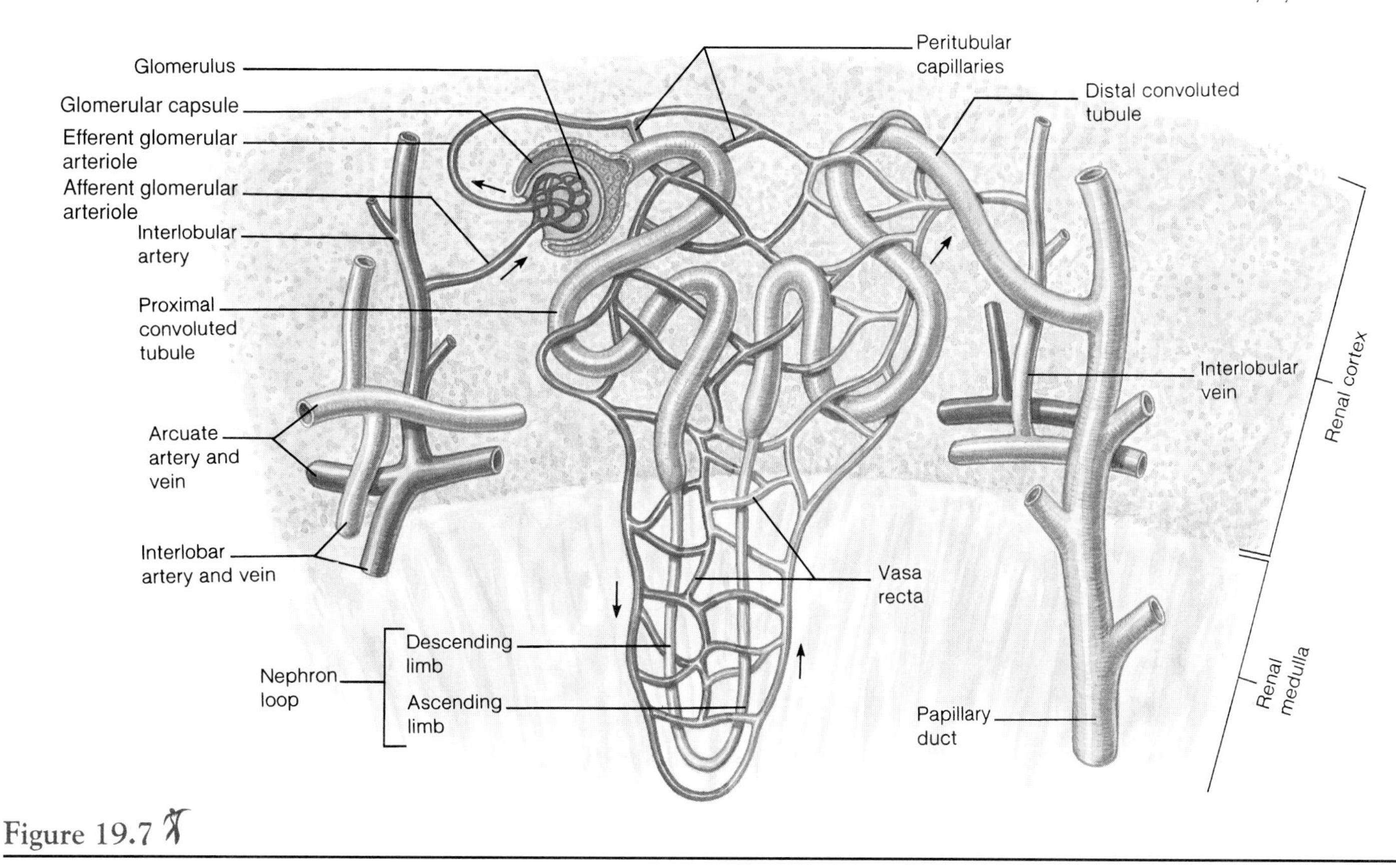

Figure 19.7

A simplified illustration of blood flow from a glomerulus to an efferent glomerular arteriole, to the peritubular capillaries, and to the venous drainage of the kidney.

(figs. 19.8 and 19.9). The narrow **slit pores** between adjacent pedicels provide the passageways through which filtered molecules must pass to enter the interior of the glomerular capsule.

Filtrate in the glomerular capsule passes into the lumen of the **proximal convoluted tubule.** The wall of the proximal convoluted tubule consists of a single layer of cuboidal cells containing millions of microvilli; these serve to increase the surface area for reabsorption. In the process of reabsorption, salt, water, and other molecules needed by the body are transported from the lumen through the tubular cells and into the surrounding peritubular capillaries.

The glomerulus, glomerular capsule, and proximal convoluted tubule are located in the renal cortex. Fluid passes from the proximal convoluted tubule to the **nephron loop** (loop of Henle). This fluid is carried into the renal medulla in the **descending limb** of the nephron loop and returns to the renal cortex in the **ascending limb** of the loop. Back in the renal cortex, the tubule becomes coiled again, and is called the **distal convoluted tubule.** The distal convoluted tubule is shorter than the proximal convoluted tubule and has fewer microvilli. It is the last segment of the nephron, and terminates as it empties into a **papillary duct** (collecting duct). Passing through the **renal papilla,** the papillary ducts empty fluid into the **minor calyx,** which in turn passes into the **major calyx.** Once within the calyces, this fluid is known as *urine*. From the major calyces, urine collects in the **renal pelvis** before it passes from the kidney into the ureter.

Two principal types of nephrons are classified according to their positions in the kidney and the lengths of their nephron loops. Nephrons that have their glomerali in the inner one-third of the cortex—called **juxtamedullary** (*juk″stă-med′yŭ-ler-e*) **nephrons**—have longer loops than the **cortical nephrons** that have their glomeruli in the outer two-thirds of the renal cortex (fig. 19.10).

The kidneys have an autonomic nerve supply derived from the renal plexus of the tenth, eleventh, and twelfth thoracic nerves. Sympathetic stimulation of the renal plexus produces a vasomotor vascular network response in the kidney. This response determines the circulation of the blood by regulating the diameters of arterioles.

The functions of the nephron are summarized in table 19.1.

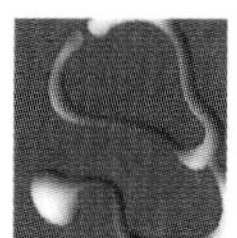

Urine from a healthy individual is virtually bacteria-free but easily becomes contaminated after voiding because its organic components serve as a nutrient source for contaminating microbes. The breakdown of urine by bacterial action produces ammonia.

A urinalysis is a common procedure in any routine physical examination. The appearance and pH of the urine are noted, as well as the presence of such abnormal constituents as albumin, blood, glucose, and acetone. Abnormal urine is symptomatic of a variety of diseases or problems in the urinary system.

---

loop of Henle: from Friedrich Gustav Jacob Henle, German physician, anatomist, and pathologist, 1809–85

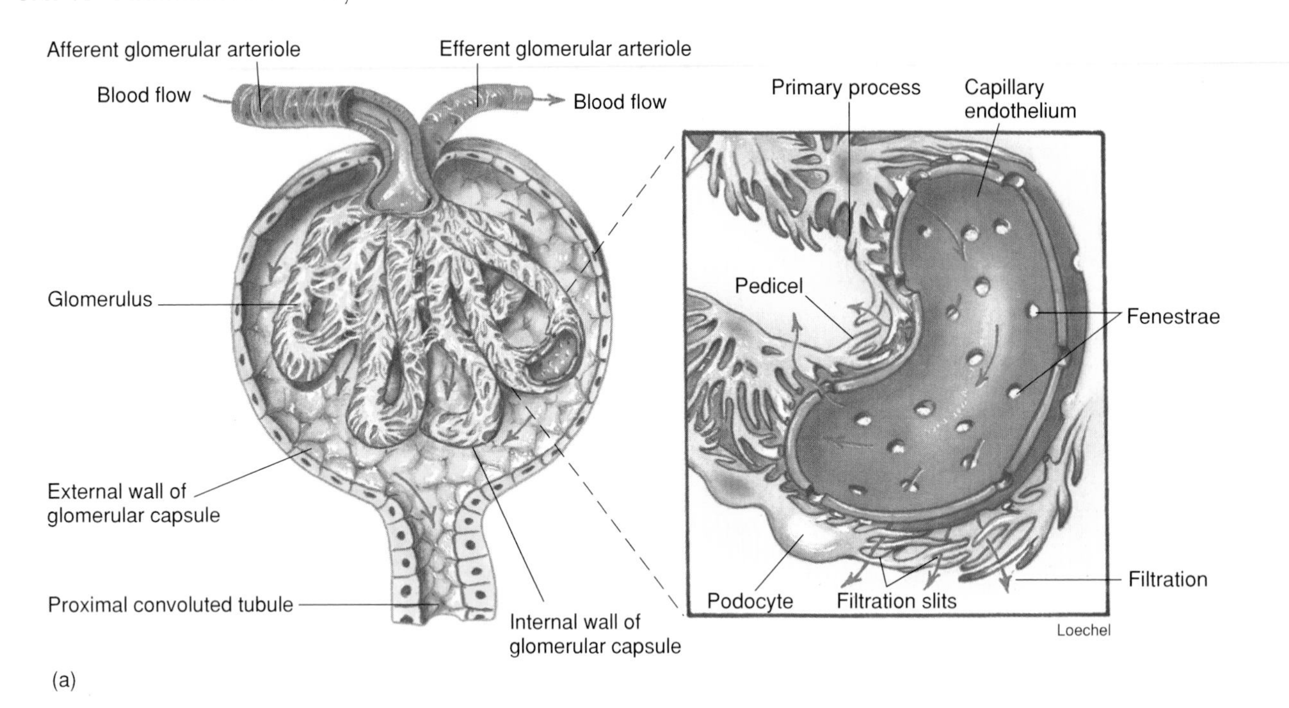

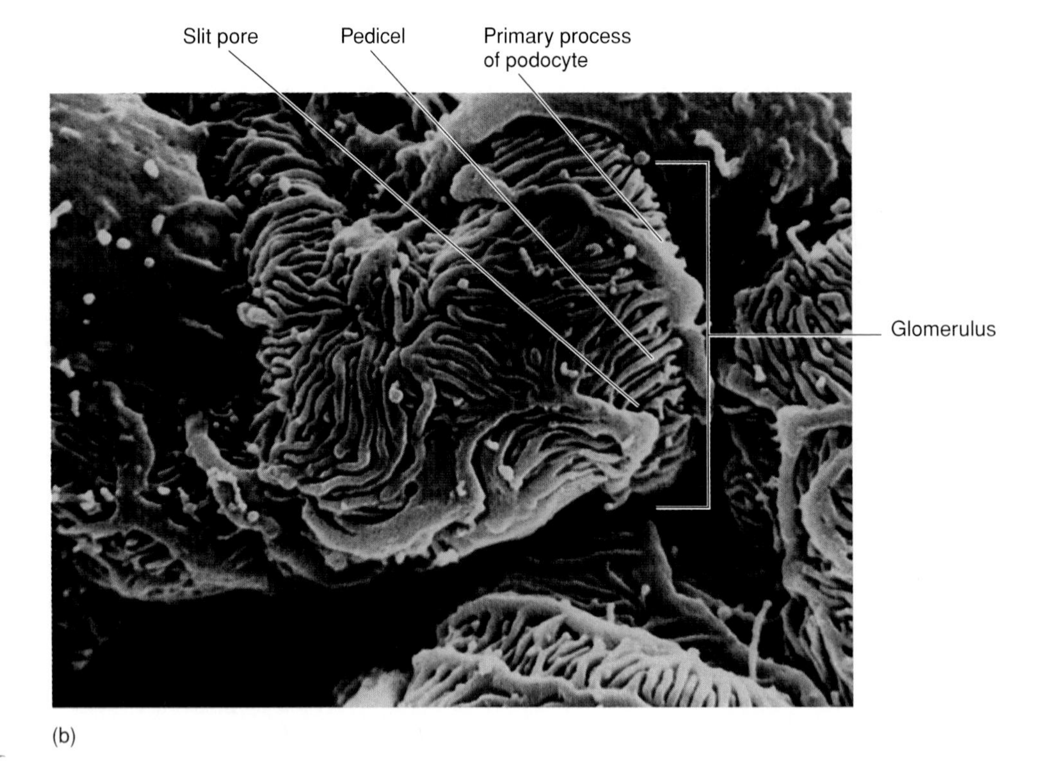

Figure 19.8

The structure of a glomerulus. (*a*) A renal corpuscle is composed of a glomerulus and a glomerular (Bowman's) capsule. Note that the diameter of the efferent glomerular arteriole carrying blood away from the glomerulus is smaller than that of the afferent glomerular arteriole transporting blood into the glomerulus. This is one of the contributing factors in the maintenance of high blood pressure within the glomerulus. The first step of urine formation is the filtration through the glomerular membrane into the glomerular space of the glomerular capsule. (*b*) A scanning electron micrograph of a glomerulus (8,000×).

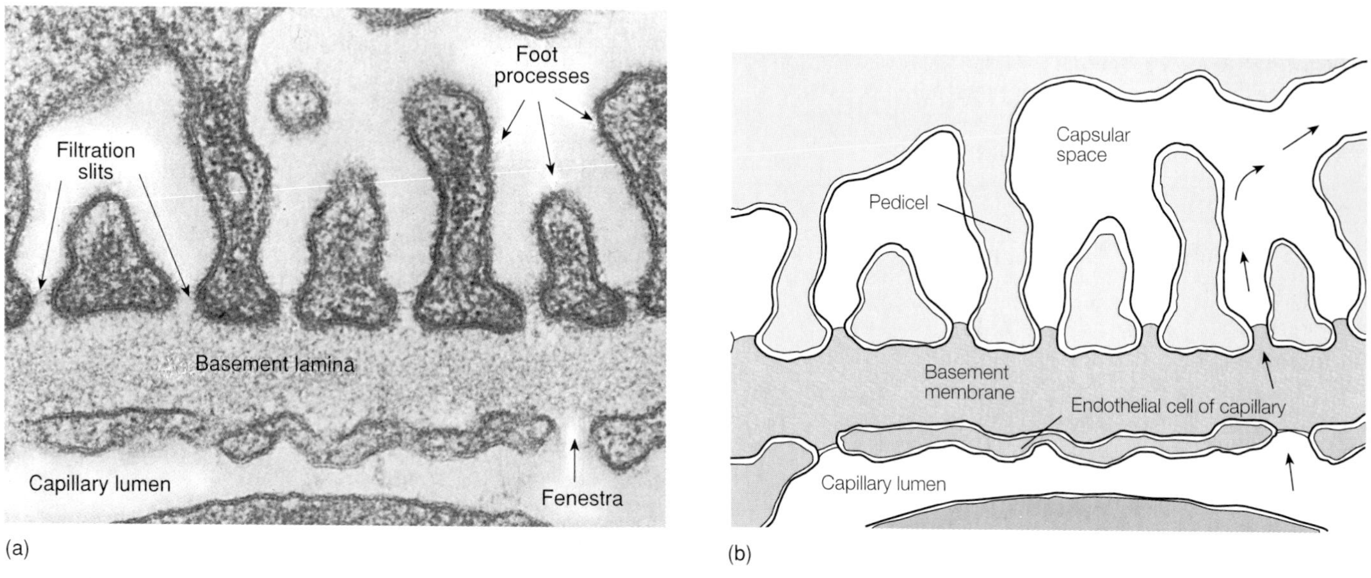

Figure 19.9

Glomerular filtration. (*a*) A transmission electron micrograph of a glomerular capillary and the glomerular membranes, and (*b*) accompanying line art. Substances in the blood are filtered through capillary fenestrae. The filtrate then passes across the basement membrane and through slit pores between the pedicels and enters the capsular space. From here, the filtrate is transported to the lumen of the proximal convoluted tubule (see fig. 19.8).

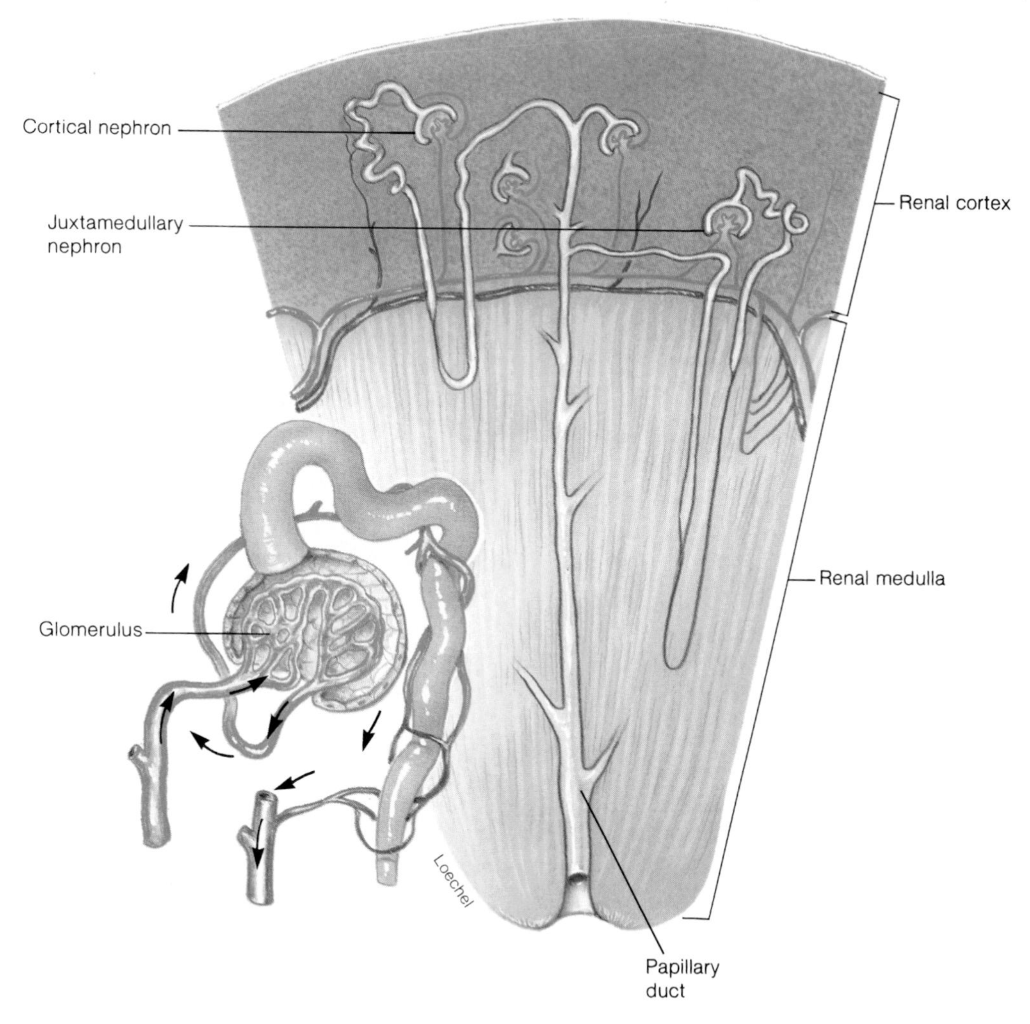

Figure 19.10

Cortical nephrons are located almost exclusively within the renal cortex. Juxtamedullary nephrons are located, for the most part, within the outer portion of the renal medulla. (Note the flow of blood through the glomerulus indicated with arrows.)

Table 19.1

Functions of the Nephron in Urine Formation

Structure	Function
Renal Corpuscle	
Glomerulus	Filtration of water and dissolved substances from blood plasma
Glomerular capsule	Receives glomerular filtrate
Nephron	
Proximal convoluted tubule	Reabsorption of water by osmosis Reabsorption of glucose, amino acids, creatine, lactic acid, citric acid, uric acid, ascorbic acid, phosphate ions, sulfate ions, calcium ions, potassium ions, and sodium ions by active transport Reabsorption of proteins by pinocytosis Reabsorption of chloride ions and other negatively charged ions by electrochemical attraction Active secretion of such substances as penicillin, histamine, and hydrogen ions
Descending limb of the nephron loop	Reabsorption of water by osmosis
Ascending limb of the nephron loop	Reabsorption of chloride ions by active transport and passive reabsorption of sodium ions
Distal convoluted tubule	Reabsorption of water by osmosis Reabsorption of sodium ions by active transport Active secretion of hydrogen ions Passive secretion of potassium ions by electrochemical attraction
Papillary duct	Passive reabsorption of water by osmosis

1. Describe the general appearance of the renal cortex and renal medulla.
2. Trace the course of blood through the kidney from the renal artery to the renal vein.
3. Trace the course of tubular fluid from the glomerular capsules to the ureter.
4. Draw a diagram of a nephron and label the renal cortex and renal medulla. Also label the structures within each region.

Ureters, Urinary Bladder, and Urethra

Urine is channeled from the kidneys to the urinary bladder by the ureters and expelled from the body through the urethra. The mucosa of the urinary bladder permits distension, and the muscles of the urinary bladder and urethra function in the control of micturition.

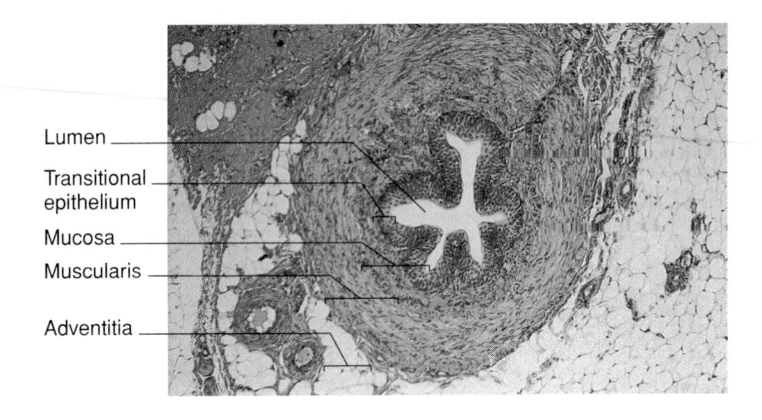

Objective 6 Describe the location, structure, and function of the ureters.

Objective 7 Describe the gross and histological structure and the innervation of the urinary bladder.

Objective 8 Describe the micturition reflex.

Objective 9 Compare and contrast the structure of the male urethra with that of the female.

Ureters

The **ureters** (*yoo-re'terz*), like the kidneys, are retroperitoneal. These tubular organs, each about 25 cm (10 in.) long, begin at the renal pelvis and course inferiorly to enter the urinary bladder at the posterolateral angles of its base. The thickest portion of a ureter, near where it enters the urinary bladder, is approximately 1.0 cm (0.4 in.) in diameter.

The wall of the ureter consists of three layers, or tunics. The inner **mucosa** is continuous with the linings of the renal tubules and the urinary bladder. The mucosa consists of transitional epithelium (fig. 19.11). The cells of this layer secrete a mucus that coats the walls of the ureter with a protective film.

The middle layer of the ureter is called the **muscularis.** It consists of inner longitudinal and outer circular layers of smooth muscle fibers. In addition, the proximal one-third of the ureter contains another longitudinal layer to the outside of the circular layer. Muscular peristaltic waves move the urine through the ureter. The peristaltic waves are initiated by the presence of urine in the renal pelvis, and their frequency is determined by the volume of urine. The waves, which occur from once every few seconds to once every few minutes, force urine through the ureter and cause it to spurt into the urinary bladder.

The outer layer of the ureter is called the **adventitia** (*ad"ven-tish'ă*). The adventitia is composed of loose connective tissue that covers and protects the underlying layers. In addition, extensions of the connective tissue anchor the ureter in place.

Figure 19.12

A renal stone (kidney stone) placed next to a dime for size comparison. Factors contributing to renal stone formation may include the ingestion of excessive mineral salts, a decrease in water intake, and overactivity of the parathyroid glands. A renal stone generally consists of calcium oxalate, calcium phosphate, and uric acid crystals.

The arterial supply of the ureter comes from several sources. Branches from the renal artery serve the superior portion. The testicular (or ovarian) artery (also called gonadal artery) supplies the middle portion, and the superior vesicular artery serves the pelvic region. The venous return is through corresponding veins.

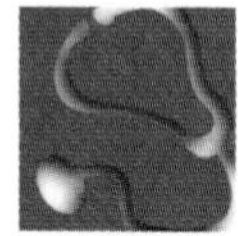

A *urinary stone* (calculus) may develop in any organ of the urinary system. A *renal stone* ("kidney stone") is one that forms in a kidney (fig. 19.12). A renal stone may obstruct the ureter and greatly increase the frequency of peristaltic waves in an attempt to pass through. The pain from a lodged urinary stone is extreme and extends throughout the pelvic area. A lodged urinary stone also causes a sympathetic ureterorenal reflex that results in constriction of renal arterioles, thus reducing the production of urine in the kidney on the affected side.

Urinary Bladder

The **urinary bladder** is a storage sac for urine. It is located just posterior to the symphysis pubis, anterior to the rectum (fig. 19.13). In females, the urinary bladder is in contact with the uterus and vagina. In males, the prostate is positioned below the urinary bladder.

The shape of the urinary bladder is determined by the volume of urine it contains. An empty urinary bladder is pyramidal; as it fills, it becomes ovoid and bulges upward into the abdominal cavity. The **median umbilical ligament,** a fibrous remnant of the embryonic urachus (see Developmental Exposition, p. 670), extends from the anterior and superior border of the urinary bladder toward the umbilicus. The base of the urinary bladder receives the ureters, and the urethra exits at the inferior angle, or apex. The region surrounding the urethral opening is known as the **neck** of the urinary bladder.

The wall of the urinary bladder consists of four layers. The **mucosa** (fig. 19.14), the innermost layer, is composed of transitional epithelium that becomes thinner as the urinary bladder distends and the cells are stretched. Further distension is permitted by folds of the mucosa, called **rugae** (*roo'je*), which can be seen when the urinary bladder is empty. Fleshy flaps of mucosa, located where the ureters pierce the urinary bladder, act as valves to prevent a reverse flow of urine toward the kidneys as the urinary bladder fills. A triangular area known as the **trigone** (*tri'gōn*) is formed on the mucosa between the two uretal openings and the single urethral opening (fig. 19.15). The internal trigone lacks rugae; it is therefore smooth in appearance and remains relatively fixed in position as the urinary bladder changes shape during distension and contraction.

The second layer of the urinary bladder, the **submucosa,** functions to support the mucosa. The **muscularis** consists of three interlaced smooth muscle layers collectively called the **detrusor** (*de-troo'sor*) **muscle.** At the neck of the urinary bladder, the detrusor muscle is modified to form the superior (called the internal sphincter) of two muscular sphincters surrounding the urethra. The outer covering of the urinary bladder is the **adventitia.** It appears only on the superior surface of the urinary bladder and is actually a continuation of the parietal peritoneum.

The arterial supply to the urinary bladder comes from the superior and inferior vesicular arteries, which arise from the internal iliac arteries. Blood draining the urinary bladder enters a vesicular venous plexus and then empties into the internal iliac veins.

The autonomic nerves serving the urinary bladder are derived from pelvic plexuses. Sympathetic innervation arises from the last thoracic and first and second lumbar spinal nerves to serve the trigone, urethral openings, and blood vessels of the urinary bladder. Parasympathetic innervation arises from the second, third, and fourth sacral nerves to serve the detrusor muscle. The sensory receptors of the urinary bladder respond to distension and relay impulses to the central nervous system via the pelvic spinal nerves.

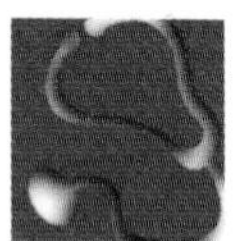

The urinary bladder becomes infected easily, and because a woman's urethra is so much shorter than a man's, women are particularly susceptible to urinary bladder infections. A urinary bladder infection, called *cystitis,* may easily ascend from the urinary bladder to the ureters, since the mucous linings are continuous. An infection that involves the renal pelvis is called *pyelitis;* if it continues into the nephrons, it is known as *pyelonephritis.* To reduce the risk of these infections, a young girl should be taught to wipe her anal region in a posterior direction, away from the urethral orifice, after a bowel movement.

calculus: L. *calculus*, small stone

trigone: L. *trigonum*, triangle
detrusor: L. *detrudere*, thrust or forced down

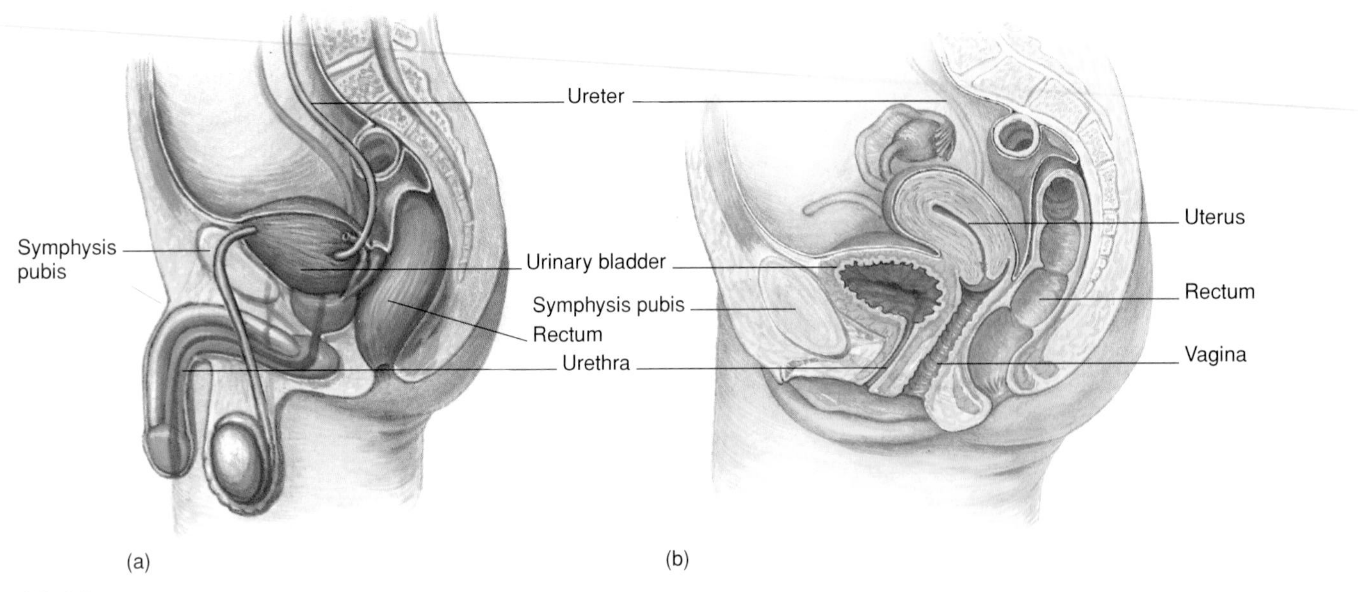

Figure 19.13

The position of the urinary bladder within the pelvic cavity. In the male (*a*), the urinary bladder is located between the symphysis pubis and the rectum. In the female (*b*), the urinary bladder is located between the symphysis pubis and the uterus and upper portion of the vagina. In a female, the volume capacity of the urinary bladder is diminished during the last trimester of pregnancy, when the greatly enlarged uterus exerts constant pressure on the urinary bladder.

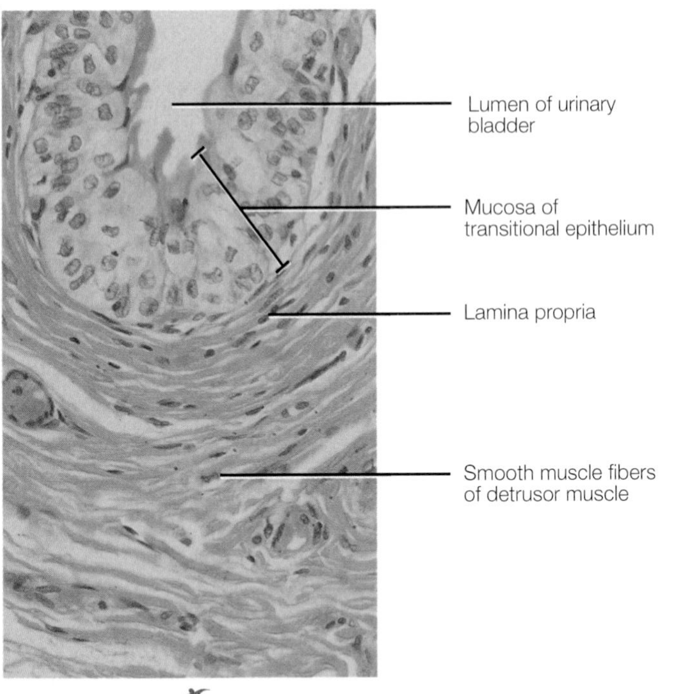

Figure 19.14

The histology of the urinary bladder (50×).

Urethra

The tubular **urethra** (*yoo-re'thră*) (fig. 19.15) conveys urine from the urinary bladder to the outside of the body. The urethral wall has an inside lining of mucous membrane surrounded by a relatively thick layer of smooth muscle, the fibers of which are directed longitudinally. Specialized **urethral glands,** embedded in the urethral wall, secrete protective mucus into the urethral canal.

Two muscular sphincters surround the urethra. The involuntary smooth muscle sphincter, the superior of the two, is the **internal urethral sphincter,** which is formed from the detrusor muscle of the urinary bladder. The lower sphincter, composed of voluntary skeletal muscle fibers, is called the **external urethral sphincter** (fig. 19.15).

The urethra of the female is a straight tube, about 4 cm (1.5 in.) long, that empties urine through the **urethral orifice** into the vestibule between the labia minora. The urethral orifice is positioned between the clitoris and vaginal orifice (see fig. 21.15). The female urethra has a single function: to transport urine to the exterior.

The urethra of the male serves both the urinary and reproductive systems. It is about 20 cm (8 in.) long, and is S-shaped because of the shape of the penis (see fig. 19.13). Three portions can be identified in the male urethra: the prostatic part, the membranous part, and the spongy part (fig. 19.15).

The **prostatic part of the urethra** is the proximal portion, about 2.5 cm long, that passes through the **prostate** located near the neck of the urinary bladder. The portion of the urethra receives drainage from small ducts of the prostate and two **ejaculatory ducts** of the reproductive system.

The **membranous part of the urethra** is the short portion (0.5 cm) that passes through the urogenital diaphragm and proximal portion of the penis. The external urethral sphincter muscle is located in this portion.

The **spongy part of the urethra** is the longest portion (15 cm), extending from the outer edge of the urogenital diaphragm to the external urethral orifice on the glans penis. This

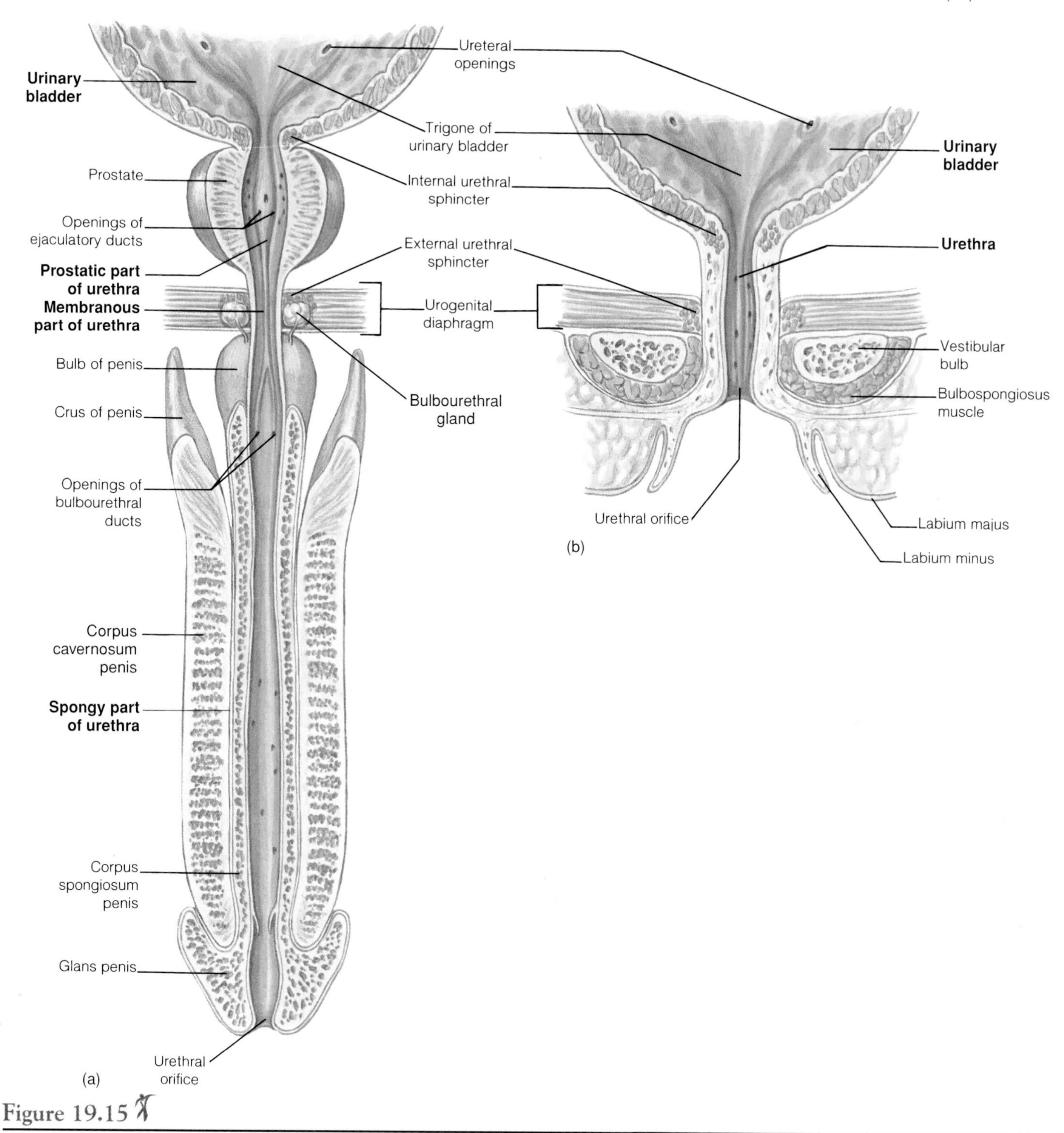

Figure 19.15

The urinary bladder and urethra. (*a*) The urethra of a male transports both urine and seminal fluid. It consists of a prostatic part that passes through the prostate, a membranous part that passes through the urogenital diaphragm, and a spongy part that passes through the penis. (*b*) The considerably shorter urethra of a female transports only urine.

portion is surrounded by erectile tissue as it passes through the corpus spongiosum of the penis. The ducts of the *bulbourethral glands* (Cowper's glands) of the reproductive system attach to the spongy part of the urethra near the urogenital diaphragm.

Cowper's glands: from William Cowper, English surgeon, 1666–1709

Micturition

Micturition (*mik″tŭ-rish′un*), commonly called urination, is a reflex action that expels urine from the urinary bladder. It is a complex function that requires a stimulus from the urinary bladder and a combination of involuntary and voluntary nerve impulses to the appropriate muscular structures of the urinary bladder and urethra.

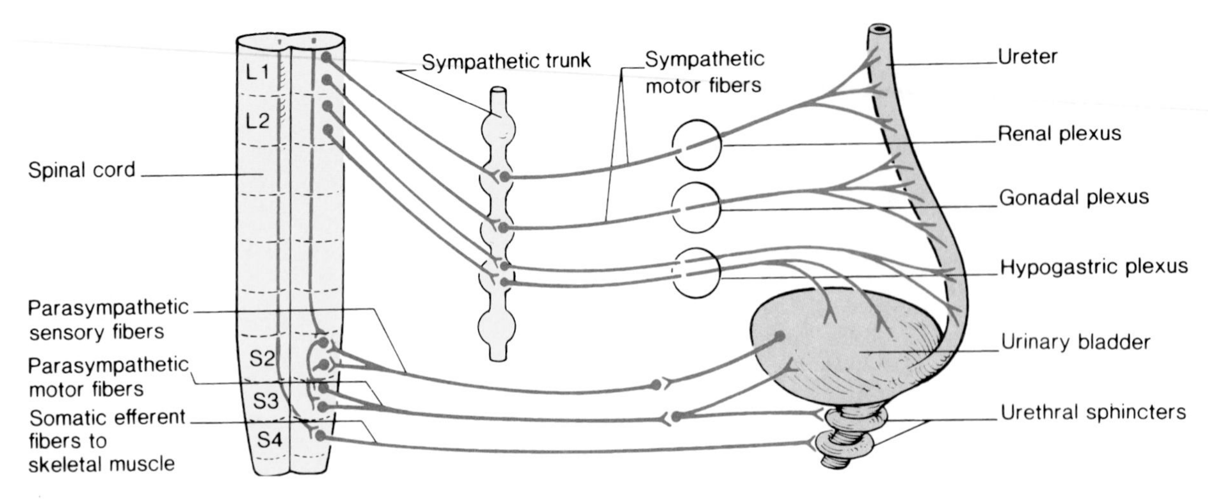

Figure 19.16

Innervation of the ureter, urinary bladder, and urethra.

In young children, micturition is a simple reflex action that occurs when the urinary bladder becomes sufficiently distended. Voluntary control of micturition is normally developed by the time a child is 2 or 3 years old. Voluntary control requires the development of an inhibitory ability by the cerebral cortex and maturation of various portions of the spinal cord.

The volume of urine produced by an adult averages about 1,200 ml a day, but it can range between 600 and 2,500 ml. The average capacity of the urinary bladder is 700 to 800 ml. A volume of 200 to 300 ml will distend the urinary bladder enough to stimulate stretch receptors and trigger the micturition reflex, creating a desire to urinate.

The micturition reflex center is located in the second, third, and fourth sacral segments of the spinal cord. Following stimulation of this center by impulses arising from stretch receptors in the urinary bladder, parasympathetic nerves that stimulate the detrusor muscle and the internal urethral sphincter are activated. Stimulation of these muscles causes a rhythmic contraction of the urinary bladder wall and a relaxation of the internal urethral sphincter. At this point, a sensation of urgency is perceived in the brain, but there is still voluntary control over the external urethral sphincter. At the appropriate time, the conscious activity of the brain activates the motor nerve fibers (S4) to the external urethral sphincter via the pudendal nerve (S2, S3, and S4), causing the sphincter to relax and urination to occur.

The innervation of the ureter, urinary bladder, and urethra is shown in figure 19.16, and the events of micturition are summarized in table 19.2.

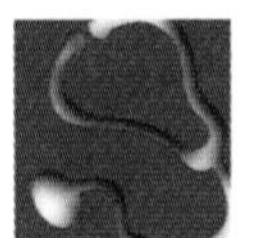

Urinary retention, or the inability to void, may occur postoperatively, especially following surgery of the rectum, colon, or internal reproductive organs. The difficulty may be due to nervous tension, the effects of anesthetics, or pain and edema at the site of the operation. If urine is retained beyond 6 to 8 hours, *catheterization* may become necessary. In this procedure, a catheter (tube) is passed through the urethra into the urinary bladder so that urine can flow freely.

Table 19.2

Events of Micturition

1. The urinary bladder becomes distended as it fills with urine.
2. Stretch receptors in the urinary bladder wall are stimulated, and impulses are sent to the micturition center in the spinal cord.
3. Parasympathetic nerve impulses travel to the detrusor muscle and the internal urethral sphincter.
4. The detrusor muscle contracts rhythmically, and the internal urethral sphincter relaxes.
5. The need to urinate is sensed as urgent.
6. Urination is prevented by voluntary contraction of the external urethral sphincter and by inhibition of the micturition reflex by impulses from the midbrain and cerebral cortex.
7. Following the decision to urinate, the external urethral sphincter is relaxed, and the micturition reflex is facilitated by impulses from the pons and the hypothalamus.
8. The detrusor muscle contracts, and urine is expelled through the urethra.
9. Neurons of the micturition reflex center are inactivated, the detrusor muscle relaxes, and the urinary bladder begins to fill with urine.

1. Describe the location and the structure of the ureters and indicate the function of the muscularis layer.
2. Describe the structure of the urinary bladder. What structural modifications permit the organ to be distended?
3. Compare the male urinary system with that of the female.
4. Explain the mechanisms by which micturition is controlled, with reference to the structures involved.

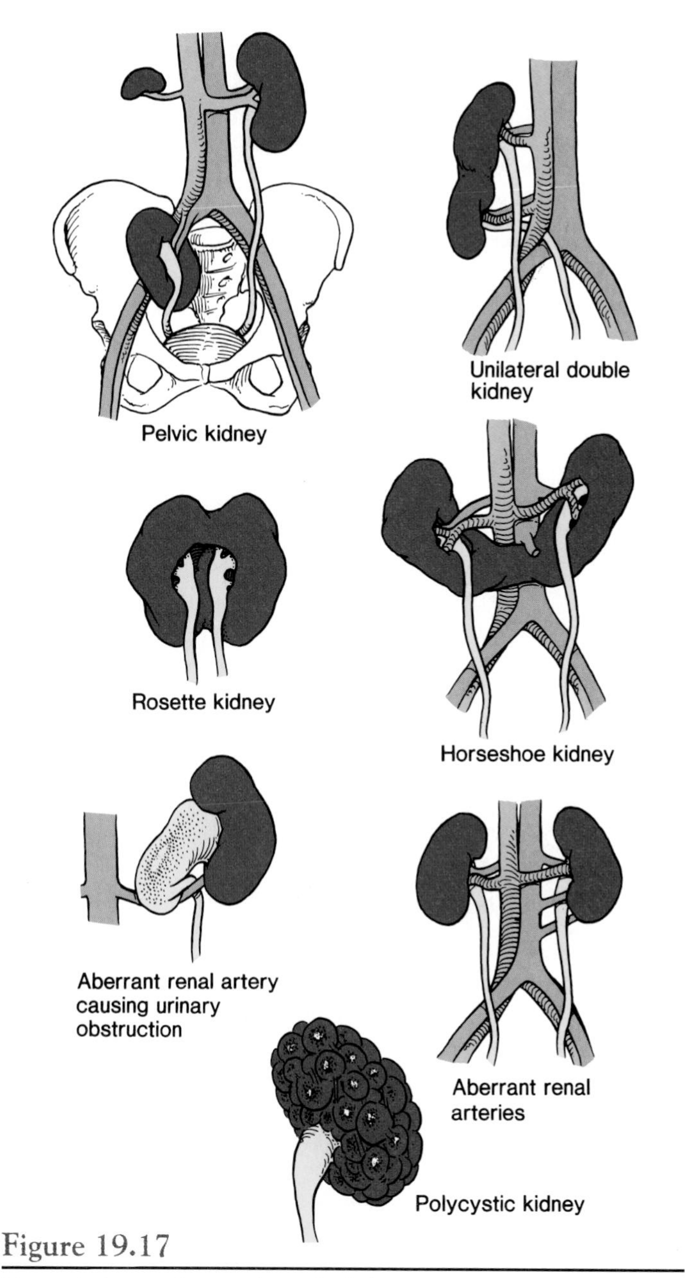

Figure 19.17

Congenital anomalies involving the kidneys.

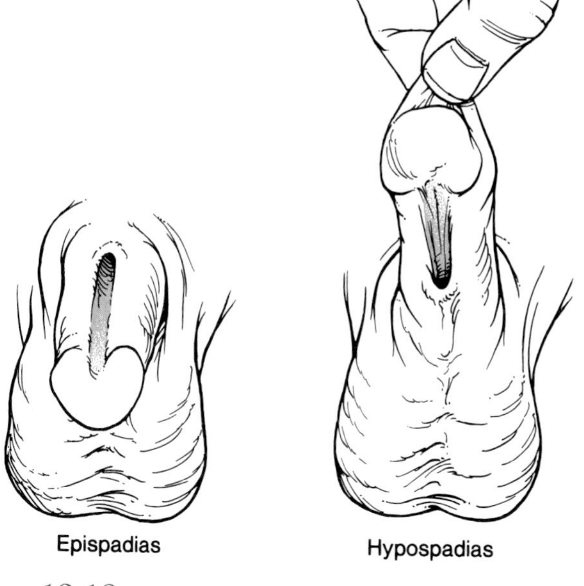

Figure 19.18

Epispadias and hypospadias of the male urethra.

Clinical Considerations

Urology (*yoo-rol'ă-je*) is the medical specialty concerned with dysfunctions of the urinary system. Urinary dysfunctions can be congenital or acquired; they may result from physical trauma or from conditions that secondarily involve the urinary organs.

Developmental Problems of the Urinary Organs

Abnormalities of the organs of the urinary system occur in about 12% of newborn babies. These deviations range from insignificant anomalies to those that are incompatible with life.

Kidneys

Common malformations of the kidneys are illustrated in figure 19.17. One common deformity is **renal agenesis** (*ă-jen'ĕ-sis*), the unilateral absence of a kidney as a result of failure of a uretic bud to develop. **Renal ectopia** is the condition in which one or both kidneys are in an abnormal position. Generally, this condition occurs when a kidney remains in the pelvic area. **Horseshoe kidney** refers to a fusion of the kidneys across the midline. The incidence of this asymptomatic condition is about 1 in 600.

Ureters

Duplication of the ureters and the associated renal pelvis is a frequent anomaly of the urinary tract. Unilateral duplication occurs in 1 in 200 births, whereas bilateral duplication occurs in 1 in 1,200 births. Occasionally there is partial duplication, which increases the propensity for urinary infections. A completely duplicated ureter frequently opens into areas other than the urinary bladder requiring surgical correction.

Urinary Bladder

Protrusion of the posterior wall of the urinary bladder, called **exstrophy of the urinary bladder,** occurs when the wall of the perineum fails to close. Associated with this condition, which occurs in about 1 in 40,000 births, are defective urethral sphincter muscles.

Urethra

The most common anomaly of the urethra is **hypospadias** (*hi"po-spād'e-as*), a condition in which the urethra of the male opens on the underside of the penis instead of at the tip of the glans penis (fig. 19.18). There is a similar defect in the female in which the

DEVELOPMENTAL EXPOSITION

The Urinary System

Exhibit I The embryonic development of the kidney. (*a*) An anterolateral view of an embryo at 5 weeks showing the position of a transverse cut depicted in (a_1). During the sixth week (b and b_1), the kidney is forming in the pelvic region and the mesonephric duct and gonadal ridge are prominent. The kidney begins migrating during the seventh week (b_2), and the urinary bladder and gonad are formed.

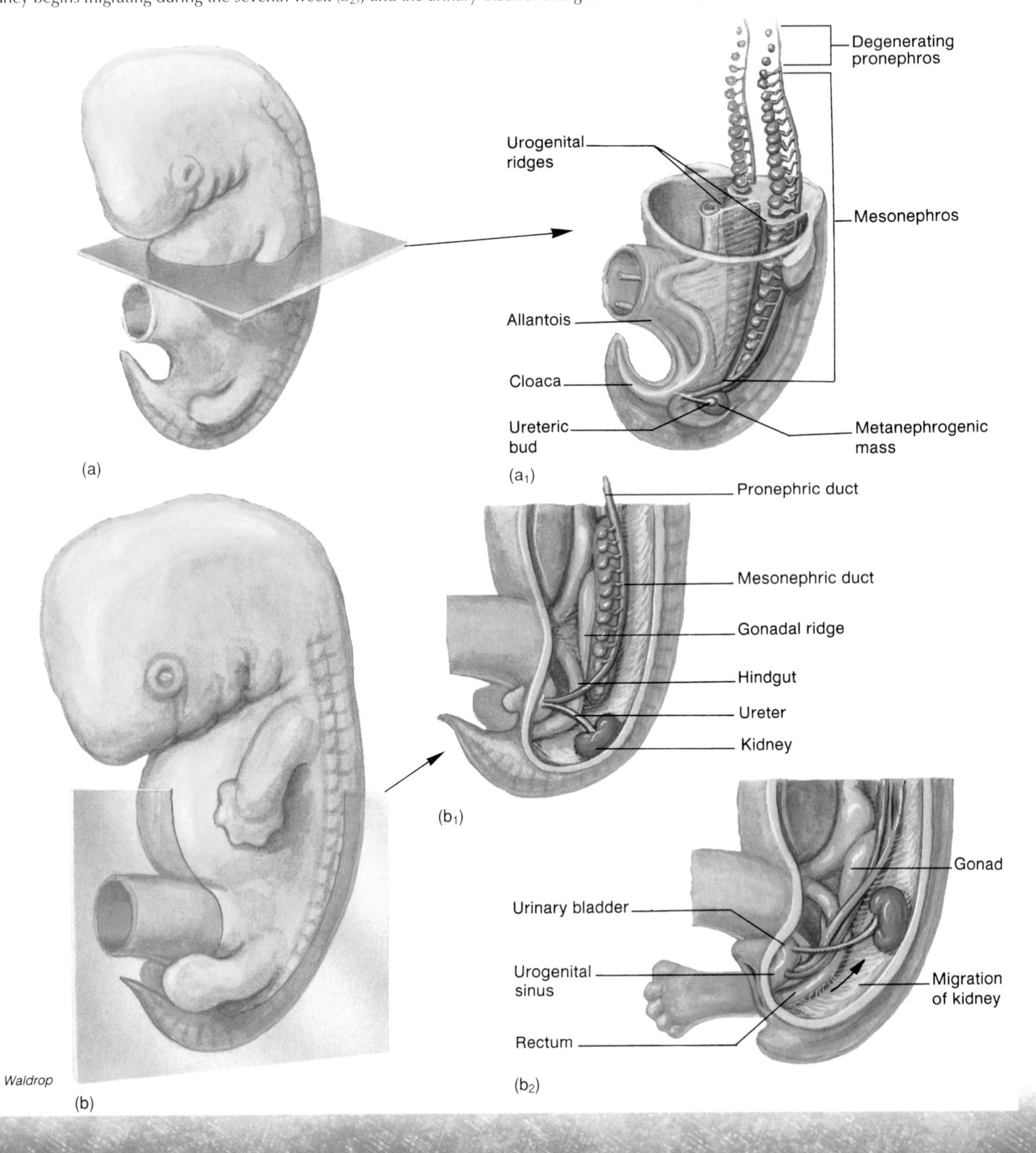

Explanation

The urinary and reproductive systems originate from a specialized elevation of mesodermal tissue called the **urogenital ridge.** The two systems share common structures for part of the developmental period, but by the time of birth two separate systems have formed. The separation in the male is not totally complete, however, since the urethra serves to transport both urine and semen. The development of both systems is initiated during the embryonic stage, but the development of the urinary system starts and ends sooner than that of the reproductive system.

Three successive types of kidneys develop in the human embryo: the *pronephros, mesonephros,* and *metanephros* (exhibit I). The third type, or metanephric kidney, persists as the permanent kidney.

The **pronephros** (*pro-nef'ros*) develops during the fourth week after conception and persists only through the sixth week. Of the three kidneys, it is the most superior in position on the urogenital ridge and is connected to the embryonic **cloaca** by the **pronephric duct.** Although the pronephros is nonfunctional and degenerates in humans, most of its duct is used by the mesonephric kidney (exhibit I), and a portion of it is important in the formation of the metanephros.

The **mesonephros** (*mez"ŏ-nef'ros*) develops toward the end of the fourth week as the pronephros becomes vestigial. The mesonephros forms from an intermediate portion of the urogenital ridge and functions throughout the embryonic period of development.

Although the **metanephros** (*met"ă-nef'ros*) begins its formation during the fifth week, it does not become functional until immediately before the start of the fetal stage of development, at the end of the eighth week. The paired metanephric kidneys continue to form urine throughout fetal development. The urine is expelled through the urinary system into the amniotic fluid.

The metanephros develops from two mesodermal sources (exhibit II). The glomerular part of the kidney forms from a specialized caudal portion of the urogenital ridge called the **metanephrogenic mass.** The tubular drainage portion of the kidneys forms as a diverticulum that emerges from the wall of the mesonephric duct near the cloaca. This diverticulum, known as the **ureteric** (*yoo"rĕ-ter'ik*) **bud,** expands into the metanephrogenic mass to form the drainage pathway for urine. The stalk of the ureteric bud develops into the ureter, whereas the expanded terminal portion forms the renal pelvis, calyces, and papillary

Exhibit II The development of the matanephric kidney (*a*) at 4 weeks, (*b*) 5 weeks, (*c*) 7 weeks, and (*d*) at birth. (d_1) A magnified view of the arrangement of papillary ducts within a papilla.

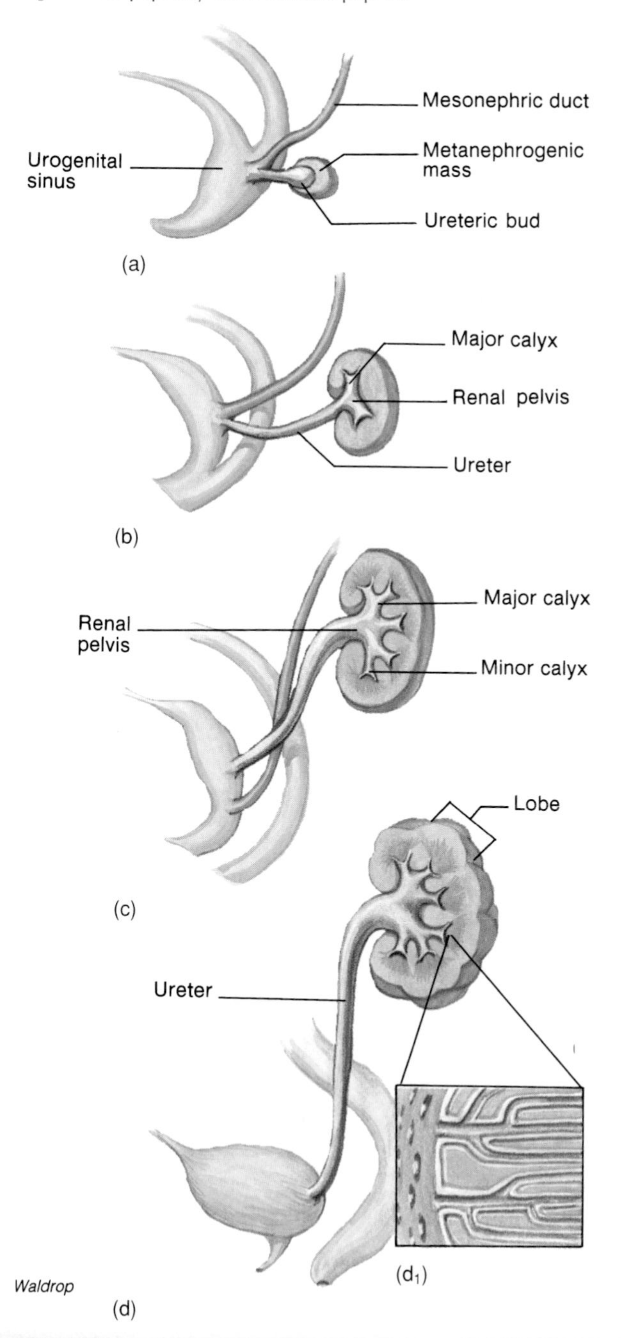

Continued

DEVELOPMENTAL EXPOSITION

Continued

ducts. A combination of the ureteric bud and metanephrogenic mass forms the other tubular channels within the kidney.

Once the metanephric kidneys are formed, they begin to migrate from the pelvis to the upper posterior portion of the abdomen. The renal blood supply develops as the kidneys become positioned in the posterior body wall.

The development of the kidneys illustrates the concept that ontogeny (embryonic development) recapitulates phylogeny (evolution). This means that the development of the three kidney types follows an evolutionary pattern. The larvae of a few of the more primitive vertebrates have functional pronephric kidneys. Adult fishes and amphibians have mesonephric kidneys, whereas adult reptiles, birds, and mammals have metanephric kidneys.

The urinary bladder develops from the urogenital sinus, which is connected to the embryonic umbilical cord by the fetal membrane called the allantois (exhibit I). By the twelfth week, the two ureters are emptying into the urinary bladder, the urethra is draining, and the connection of the urinary bladder to the allantois has been reduced to a supporting structure called the **urachus** (*yoo'ră-kus*).

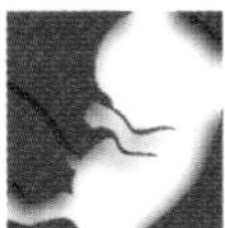

Occasionally a *patent urachus* is present in a newborn and is discovered when urine is passed through the umbilicus, especially if there is a urethral obstruction. Usually, however, it remains undiscovered until an enlarged prostate in an elderly male obstructs the urethra and forces urine through the patent urachus and out the umbilicus (navel).

urethra opens into the vagina. **Epispadias** is a failure of closure on the dorsum of the penis. Hypospadias and epispadias can be corrected surgically.

Symptoms and Diagnosis of Urinary Disorders

Normal micturition is painless. **Dysuria** (*dis-yur'e-ă*), or painful urination, is a sign of a urinary tract infection or obstruction of the urethra—as in an enlarged prostate in a male. **Hematuria** means blood in the urine and is usually associated with trauma. **Bacteriuria** means bacteria in the urine, and **pyuria** is the term for pus in the urine, which may result from a prolonged infection. **Oliguria** is a scanty output of urine, whereas **polyuria** is an excessive output. Low blood pressure and kidney failure are two causes of oliguria. **Uremia** (*yoo-re'me-ă*) is a condition in which substances ordinarily excreted in the urine accumulate in the blood. **Enuresis** (*en"yŭ-re'sis*), or **incontinence,** is the inability to control micturition. It may be caused by psychological factors or by structural impairment.

The palpation and inspection of urinary organs is an important aspect of physical assessment. The right kidney is palpable in the supine position; the left kidney usually is not. The distended urinary bladder is palpable along the superior pelvic rim.

The urinary system may be examined using radiographic techniques. An **intravenous pyelogram** (*pi'ĕ-lŏ-gram*) **(IVP)** permits radiographic examination of the kidneys following the injection of radiopaque dye. In this procedure, the dye that has been injected intravenously is excreted by the kidneys so that the renal pelvises and the outlines of the ureters and urinary bladder can be observed in a radiograph.

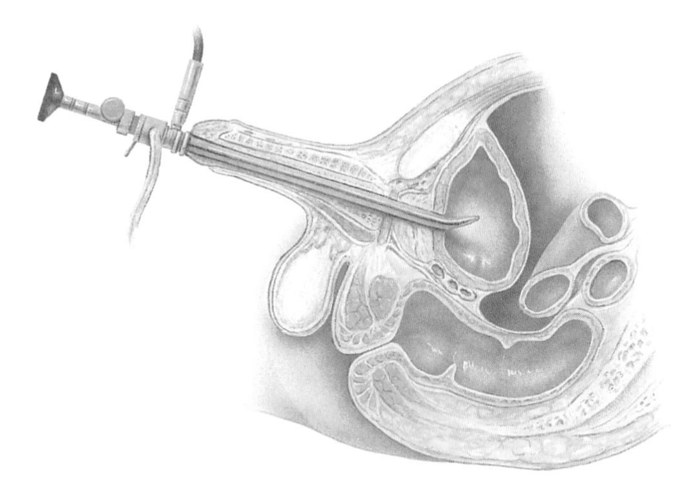

Figure 19.19

Cystoscopic examination of a male.

Cystoscopy (*sĭ-stos'kŏ-pe*) is the inspection of the inside of the urinary bladder using an instrument called a cystoscope (fig. 19.19). By means of this technique, tissue samples can be obtained, as well as urine samples from each kidney prior to mixing in the urinary bladder. Once the cystoscope is in the urinary bladder, the ureters and pelvis can be viewed through urethral catheterization and inspected for obstructions.

A **renal biopsy** is a diagnostic test for evaluating certain types and stages of kidney diseases. The biopsy is performed either through a skin puncture (closed biopsy) or through a surgical incision (open biopsy).

Urinalysis is a simple but important laboratory aspect of a physical examination. The voided urine specimen is tested for color, specific gravity, chemical composition, and for the presence of microscopic bacteria, crystals, and casts. *Casts* are accumulations of proteins that leaked through the glomeruli and were pushed through the tubules, like toothpaste through a tube.

Infections of the Urinary Organs

Urinary tract infections (UTIs) are a significant cause of illness and are also a major factor in the development of chronic renal failure. Females are more susceptible to urinary tract infections than are males because the urethra is shorter in females and the urethral and anal openings are closer together. The incidence of infection increases directly with sexual activity and aging in both sexes.

Infections of the urinary tract are named according to the infected organ. An infection of the urethra is called **urethritis** (*yoo″re-thri′tis*) and involvement of the urinary bladder is **cystitis** (*sis-ti′tis*). Cystitis is frequently a secondary infection from some other part of the urinary tract. **Nephritis** is inflammation of the kidney tissue. **Glomerulonephritis** (*glo-mer″yŭ-lo-nĕ-fri′tis*) is inflammation of the glomeruli. Glomerulonephritis may occur following an upper respiratory tract infection because antibodies produced against streptococci bacteria can produce an autoimmune inflammation in the glomeruli. This inflammation may permanently change the glomeruli and figure significantly in the development of chronic renal disease and renal failure.

Any interference with the normal flow of urine, such as from a renal stone or an enlarged prostate in a male, causes stagnation of urine in the renal pelvis and may lead to pyelitis. **Pyelitis** is an inflammation of the renal pelvis and its calyces. **Pyelonephritis** is inflammation involving the renal pelvis, the calyces, and the tubules of the nephron within one or both kidneys. Bacterial invasion from the blood or from the lower urinary tract is another cause of both pyelitis and pyelonephritis.

Trauma to the Urinary Organs and Functional Impairments

Trauma

A sharp blow to a lumbar region of the back may cause a contusion or rupture of a kidney. Symptoms of kidney trauma include hematuria and pain in the upper abdominal quadrant and flank on the injured side.

Pelvic fractures from accidents may result in perforation of the urinary bladder and urethral tearing. On a long automobile trip, it is advisable to stop to urinate at regular intervals because an attached seat belt over the region of a full urinary bladder can cause rupture of the urinary bladder in even a relatively minor accident. Urethral injuries are more common in men than in women because of the position of the urethra in the penis. In a straddle injury, for example, a man walking along a raised beam may slip and compress his urethra and penis between the hard surface and his pubic arch, rupturing the urethra.

Obstruction

The urinary system can become obstructed anywhere along the tract. Calculi (stones) are the most common cause, but blockage can also come from trauma, strictures, tumors or cysts, spasms or kinks of the ureters, or congenital anomalies. If not corrected, an obstruction causes urine to collect behind the blockage and generate pressure that may cause permanent functional and anatomic damage to one or both kidneys. As a result of pressure buildup in a ureter, a distended ureter, or **hydroureter,** develops. Dilation in the renal pelvis is called **hydronephrosis.**

Urinary stones (calculi) are generally the result of infections or metabolic disorders that cause the excretion of large amounts of organic and inorganic substances (see fig. 19.12). As the urine becomes concentrated, these substances may crystalize and form granules in the renal calyces. The granules then serve as cores for further precipitation and development of larger calculi. This becomes dangerous when a calculus grows large enough to cause an obstruction. The calculus also causes intense pain when it passes through the urinary tract.

Renal Failure and Hemodialysis

Renal output of 50 to 60 cc of urine per hour is considered normal. If the output drops to less than 30 cc per hour, it may indicate renal failure—the loss of the kidney's ability to maintain fluid and electrolyte balance and to excrete waste products. Renal failure can be either acute or chronic. **Acute renal failure** is the sudden loss of kidney function caused by shock and hemorrhage, thrombosis, or other physical trauma to the kidneys. The kidneys may sustain a 90% loss of their nephrons through tissue death and still continue to function without apparent difficulty. If a patient suffering acute renal failure is stabilized, the nephrons have an excellent capacity to regenerate.

A person with **chronic renal failure** cannot sustain life independently. Chronic renal failure is the end result of kidney disease in which the kidney tissue is progressively destroyed. As renal tissue continues to deteriorate, the options for sustaining life are *hemodialysis* (*he″mo-di-al′ĭ-sis*) or *kidney transplantation*.

Hemodialysis equipment is designed to filter the wastes from the blood of a patient who has chronic renal failure. The patient's blood is pumped through a tube from the radial artery and passes through a machine, where it is cleansed and then returned to the

Internal Affairs

How the *Urinary System* Works with Other Body Systems

How Other Systems Work with the *Urinary System*

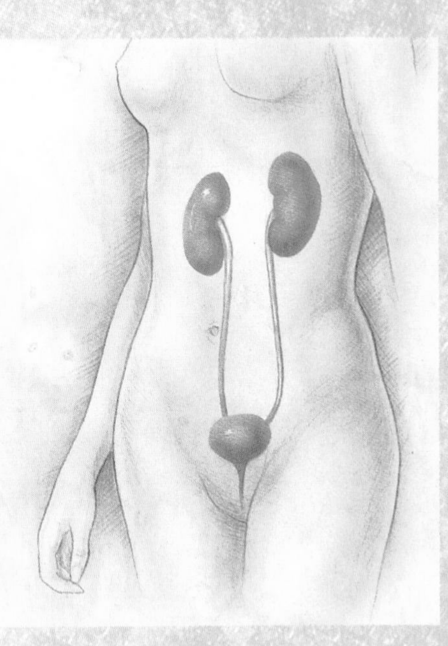

Integumentary System

- Eliminates metabolic wastes
- Maintains normal acid-base (pH), fluid, and electrolyte levels

- Covers the body and protects it from excessive fluid loss
- Provides site for evaporative water loss
- Skin plays role in vitamin D synthesis, along with the kidneys

Skeletal System

- Eliminates metabolic wastes
- Maintains normal acid-base (pH), fluid, and electrolyte levels
- Kidneys activate vitamin D, enabling absorption of the calcium and phosphorus needed for bone growth and maintenance

- Supports and protects certain organs of the urinary system
- Stores calcium and phosphate ions

Muscular System

- Eliminates metabolic wastes
- Maintains normal acid-base (pH), fluid, and electrolyte levels

- Supports and protects certain organs of the urinary system
- Assists the storage and voiding of urine

Nervous System

- Eliminates metabolic wastes
- Maintains normal acid-base (pH), fluid, and electrolyte levels

- Provides autonomic innervation to the urinary system
- Provides motor control of micturition

Endocrine System

- Eliminates metabolic wastes
- Maintains normal acid-base (pH), fluid, and electrolyte levels
- Kidneys produce the hormone erythropoietin

- Hormones help to regulate renal reabsorption of water and electrolytes

Circulatory System

- Eliminates metabolic wastes
- Maintains normal acid-base (pH), fluid, and electrolyte levels
- Regulates blood volume, pressure, and composition

- Transports O_2 and CO_2, nutrients, and fluids to and from the organs of the urinary system
- Blood pressure is vital for glomerular filtration

Lymphatic System

- Eliminates metabolic wastes
- Maintains normal acid-base (pH), fluid, and electrolyte levels
- Acidity of urine provides nonspecific defense against urinary tract infection

- Maintains a balanced amount of interstitial fluid within the organs of the urinary system
- Protects the urinary tract against infection

Respiratory System

- Eliminates metabolic wastes
- Maintains normal acid-base (pH), fluid, and electrolyte levels

- Provides O_2 and eliminates CO_2
- Kidneys may be damaged by inhaled toxic fumes

Digestive System

- Eliminates metabolic wastes
- Maintains normal acid-base (pH), fluid, and electrolyte levels
- Kidneys activate vitamin D, which is needed for the intestinal absorption of calcium and phosphorus

- Provides nutrients for growth, maintenance, and repair of the urinary system
- Liver, along with the kidneys, activates vitamin D
- Liver metabolizes blood-borne hormones to forms that can be excreted in urine

Reproductive System

- Eliminate metabolic wastes
- Maintains normal acid-base (pH), fluid, and electrolyte levels
- Urethra serves as common passageway for urine and sperm cells in males

- Provides organ (penis) in male for passage of urethra
- Enlarged prostate can cause urine retention and kidney damage in males
- Gravid uterus compresses urinary bladder and increases micturition frequency in females

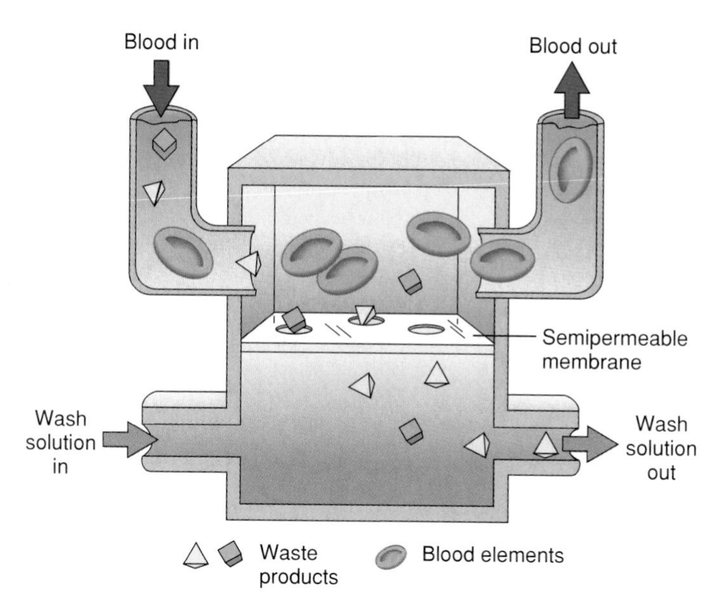

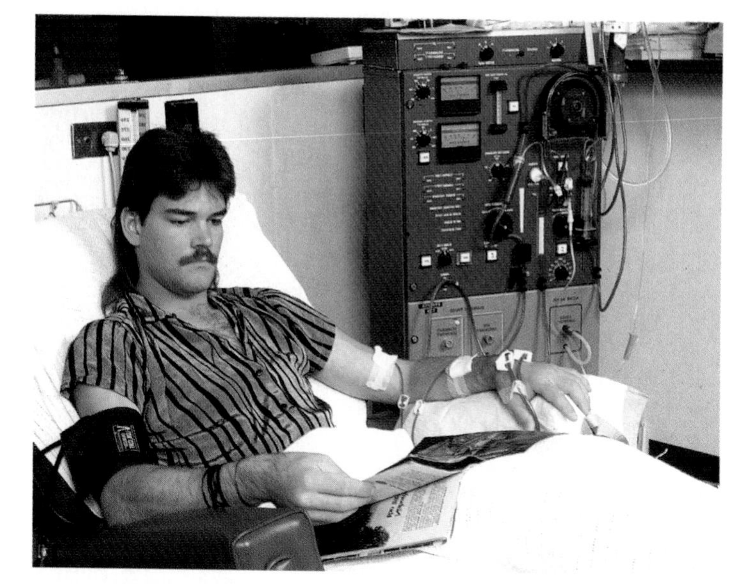

Figure 19.20

The hemodialysis process.

body through a vein (fig. 19.20). The cleaning process involves pumping the blood past a semipermeable cellophane membrane that separates the blood from an isotonic solution containing molecules needed by the body (such as glucose). Waste products diffuse out of the blood through the membrane, while glucose and other molecules needed by the body remain in the blood.

More recent hemodialysis techniques include the use of the patient's own peritoneal membranes for filtering. Dialysis fluid is introduced into the peritoneal cavity, and then, after a period of time, discarded after wastes have accumulated. This procedure, called *continuous ambulatory peritoneal dialysis (CAPD)*, can be performed several times a day by the patients themselves on an outpatient basis.

Urinary Incontinence

The inability to voluntarily retain urine in the urinary bladder is known as *urinary incontinence*. It has a number of causes and may be temporary or permanent. Emotional stress is a cause of temporary incontinence in adults. Permanent incontinence may result from neurological trauma, various urinary diseases, tissue damage within the urinary bladder or urethra, or weakness of the pelvic floor muscles. Remarkable advances have been made in treating permanent urinary incontinence through the implantation of an artificial urethral sphincter.

Clinical Case Study Answer

The hematuria experienced by the patient is probably the result of laceration caused by the upper abdominal knife wound. The lower right quadrant stab most likely did not violate the urinary tract. The course of blood seen in the catheter begins and proceeds as follows: abdominal aorta → right renal artery → smaller parenchymal artery → through the lacerated vessel(s) → into the lacerated urinary collecting system, either at the calyx or the renal pelvis or proximal right ureter → urinary bladder → into catheter. During the operation, the surgeon should keep in mind that in 2% to 4% of the population only one kidney is present. If, therefore, she is prompted to remove the damaged kidney, she should first confirm the presence of a second functioning kidney. If a second kidney is not present, every effort should be made to correct the problem without performing a nephrectomy, which would consign the patient to chronic hemodialysis or kidney transplant.

Chapter Summary

Introduction to the Urinary System (pp. 656–657)

1. The urinary system consists of two kidneys, two ureters, the urinary bladder, and the urethra.
2. The urinary system maintains the composition and properties of body fluid, which establishes the extracellular environment. The end product of the urinary system is urine, which is voided from the body during micturition.

Kidneys (pp. 657-664)

1. The kidneys are retroperitoneal, embedded in a renal adipose capsule.
2. Each kidney is contained by a renal capsule and divided into an outer renal cortex and an inner renal medulla.
   (a) The renal medulla is composed of renal pyramids separated by renal columns.
   (b) The renal papillae empty urine into the minor calyces and then into the major calyces, which drain into the renal pelvis. From there, urine flows through the ureter.
3. Each kidney contains more than a million microscopic functional units called nephrons.
   (a) Filtration occurs in the glomerulus, which receives blood from afferent glomerular arterioles.
   (b) Glomerular blood is drained by efferent glomerular arterioles that deliver blood to peritubular capillaries surrounding the nephron tubules.
   (c) The glomerular capsule and distal convoluted tubules are located in the renal cortex.
   (d) The nephron loop is located in the renal medulla.
   (e) Filtrate from the distal convoluted tubule is drained into papillary ducts that extend through the renal medulla to empty urine into the calyces.

Ureters, Urinary Bladder, and Urethra (pp. 664-668)

1. Urine is channeled from the kidneys to the urinary bladder by the ureters and expelled from the urinary bladder through the urethra. The detrusor muscle of the urinary bladder and the sphincter muscles of the urethra are used in the control of micturition.
   (a) Each ureter contains three layers: the mucosa, muscularis, and adventitia.
   (b) The lumen of the urinary bladder is lined by transitional epithelium, which is folded into rugae. These structures enhance the ability of the urinary bladder to distend.
   (c) The urethra has an internal sphincter of smooth muscle and an external sphincter of skeletal muscle.
   (d) The male urethra conducts urine during urination and seminal fluid during ejaculation. The female urethra is much shorter than that of a male and conducts only urine.
   (e) The male urethra is composed of prostatic, membranous, and spongy portions.
2. Micturition is controlled by reflex centers in the second, third, and fourth sacral segments of the spinal cord.

Review Activities

Objective Questions

1. Which of the following statements about the renal pyramids is *false?*
   (a) They are located in the renal medulla.
   (b) They contain glomeruli.
   (c) They contain papillary ducts.
   (d) They are supported by renal columns.
2. Renal vessels and the ureter attach at the concave medial border of the kidney called
   (a) the renal pelvis. (c) the calyx.
   (b) the urachus. (d) the hilum.
3. The renal medulla of the kidney contains
   (a) glomerular capsules.
   (b) glomeruli.
   (c) renal pyramids.
   (d) adipose capsules.
4. Urine flowing from the papillary ducts enters directly into
   (a) the renal calyces.
   (b) the ureter.
   (c) the renal pelvis.
   (d) the distal convoluted tubules.
5. Which of the following statements concerning the kidneys is *false?*
   (a) They are retroperitoneal.
   (b) They each contain 8 to 15 renal pyramids.
   (c) They each have two distinct regions—the renal cortex and renal medulla.
   (d) They are positioned between the third and fifth lumbar vertebrae.
6. A renal stone (calculus), would most likely cause stagnation of urine in which portion of the urinary system?
   (a) the urinary bladder
   (b) the renal column
   (c) the ureter
   (d) the renal pelvis
   (e) the urethra
7. Distention of the urinary bladder is possible because of the presence of
   (a) rugae.
   (b) the trigone.
   (c) the adventitia.
   (d) the transitional epithelium.
   (e) both a and d.
8. The detrusor muscle is located in
   (a) the kidneys.
   (b) the ureters.
   (c) the urinary bladder.
   (d) the urethra.
9. The internal urethral sphincter is innervated by
   (a) sympathetic neurons.
   (b) parasympathetic neurons.
   (c) somatic motor neurons.
   (d) all of the above.
10. Which of the following statements about metanephric kidneys is *true?*
    (a) They become functional at the end of the eighth week.
    (b) They are active throughout fetal development.
    (c) They are the third pair of kidneys to develop.
    (d) All of the above are true.

Essay Questions

1. Describe the location of the kidneys in relation to the abdominal cavity and the peritoneal membranes.
2. Diagram the kidney structures that can be identified in a coronal section.
3. Describe how the kidney is supported against the posterior abdominal wall. How is this support related to the condition called renal ptosis?
4. Trace a drop of blood from an interlobular artery through a glomerulus and into an interlobular vein. List in order all the vessels through which the blood passes. How do structural differences in afferent and efferent glomerular arterioles ensure the high blood pressure needed for filtrate formation?
5. In a male, trace the path of urine from the site of filtration at the renal corpuscle to the outside of the body. List in order all the structures through which the urine passes.
6. What is a nephron? Describe the two types of nephrons found in a kidney. Why are nephrons considered the functional units of the urinary system?
7. Describe the mechanism involved in the passage of urine from the renal pelvis to the urinary bladder.
8. Describe the urinary bladder with regard to position, histological structure, blood supply, and innervation.

9. Compare and contrast the urethra in the male and female.
10. What is the micturition reflex? Discuss the physiological and functional events of a voluntary micturition response.
11. What is a metanephrogenic mass? A ureteric bud? Discuss the sequential development of these embryonic structures to form the urinary system. How can a greater knowledge of this development process lead to a better understanding of congenital abnormalities?
12. Briefly describe the purpose of cytoscopy and urinalysis.
13. List four common congenital malformations of the urinary system. Which of these require surgical correction?
14. Define *dysuria, hematuria, bacteriuria, pyuria, oliguria, polyuria, uremia,* and *enuresis*.

Critical-Thinking Questions

1. Why is it more accurate to refer to the kidneys and associated structures as the urinary system rather than the excretory system?
2. Treatment with sulfa medications such as Gantrisin (sulfisoxazole) and broad-spectrum antibiotics such as tetracycline or ampicillin usually clear up the symptoms of cystitis very quickly. What is the danger of discontinuing the prescribed medication as soon as the symptoms are gone?
3. The neighborhood day-care center won't accept children who are still in diapers. You've tried to toilet train your 15-month-old boy, but you haven't made any progress at all. Should you persist in your efforts, or would it be better to wait? Explain.
4. What functions of a real kidney does an artificial kidney (dialysis machine) fail to duplicate?
5. Your friend's baby is due next month, and she is constantly running to the bathroom to urinate. Can you explain why?
6. Explain why a male should be particularly concerned if he has difficulty voiding urine.

TWENTY

Male Reproductive System

Clinical Case Study 679

Introduction to the Male Reproductive System 679
- Categories of Reproductive Structures 679

Perineum and Scrotum 681
- Perineum 681
- Scrotum 681

Testes 683
- Structure of the Testes 683
- Endocrine Functions of the Testes 684
- Spermatogenesis 685
- Structure of Spermatozoa 686

Spermatic Ducts, Accessory Reproductive Glands, and the Urethra 688
- Spermatic Ducts 688
- Accessory Reproductive Glands 689
- Urethra 690

Penis 691

Mechanisms of Erection, Emission, and Ejaculation 693
- Erection of the Penis 693
- Emission and Ejaculation of Semen 693

Clinical Considerations 695
- Developmental Problems of the Reproductive System 695

Developmental Exposition: The Reproductive System 696
- Sexual Dysfunction 699
- Diseases of the Reproductive System 700

Clinical Case Study Answer 702

Internal Affairs 703

Chapter Summary 704

Review Activities 704

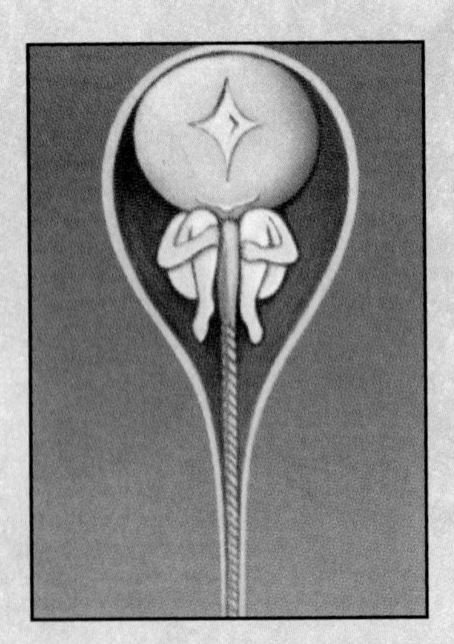

When sperm cells were first discovered and described by Antoni Van Leeuwenhoek in 1677, microscopists imagined seeing tiny men ("homunculi") curled up inside each spermatozoon. These preformed microscopic entities were thought to contain all the characteristics necessary to develop into human beings—a theory that dominated medical thinking throughout the eighteenth century.

Clinical Case Study

During a routine physical exam, a 27-year-old man mentioned to his family doctor that he and his wife had been unable to conceive a child after nearly 2 years of trying. He added that his wife had taken the initiative of having a thorough gynecological evaluation in an attempt to find out what was causing the problem. Her test findings revealed no physical conditions that could be linked to infertility. Upon palpating the patient's testes, the doctor found nothing unusual. When he examined the scrotal sac above the testes, however, the doctor appeared perplexed. He informed his patient that two tubular structures, one for each testis, appeared to be absent, and that they probably had been missing since birth. During a follow-up visit, the doctor told the patient that examination of his ejaculate revealed azoospermia (no viable sperm).

Explain how the result of the semenalysis relates to the patient's physical exam findings. What are the missing structures? Does it seem peculiar that the patient is capable of producing an ejaculate? Why or why not?

Introduction to the Male Reproductive System

The organs of the male and female reproductive systems are adapted to produce and allow the union of gametes that contain specific genes. A random combination of the genes during sexual reproduction results in the propagation of individuals with genetic differences.

Objective 1 Explain why sexual reproduction is biologically advantageous.

Objective 2 List the functions of the male reproductive system and compare them with those of the female.

Objective 3 Distinguish between primary and secondary sex organs.

Unlike other body systems, the reproductive system is not essential for the survival of the individual; it is, however, required for the survival of the species. It is through reproduction that additional individuals of a species population are produced and the genetic code passed from one generation to the next. This can be accomplished by either asexual or sexual reproduction. But sexual reproduction, in which genes from two individuals are combined in random ways with each new generation, offers the overwhelming advantage of introducing great variability into a population. This variability of genetic constitution helps to ensure that some members of a population will survive changes in the environment over evolutionary time.

The reproductive system is unique in two other respects. First, the fact that it does not become functional until it is "turned on" at puberty by the actions of sex hormones sets the reproductive system apart. By contrast, all of the other body systems are functional at birth, or shortly thereafter. Second, while the other organ systems of the body exhibit slight sexual differences, no other system approaches the level of dissimilarity of the reproductive system. Because sexual reproduction requires the production of two types of **gametes** (*gam'ēts*), or sex cells, the species has a male and female form, each with its own unique reproductive system. The male and female reproductive systems complement each other in their common purpose of producing offspring.

The functions of the male reproductive system are to produce the male gametes, **spermatozoa** (*sper-mat"ŏ-zo'ă*), and to transfer them to the female through the process of *coitus* (*sexual intercourse*) or *copulation*. The female not only produces her own gametes (called **oocytes** [*o'ŏ-sīts*], or **ova**) and receives the sperm from the male, but her reproductive organs are specialized to provide sites for fertilization, implantation of the developing embryonic mass (the blastocyst), pregnancy, and delivery of a baby. The more complex reproductive system of the female also provides a means for nourishing the baby through the secretion of milk from the mammary glands.

In this chapter we will consider the anatomy of the male reproductive system; the female reproductive system is the focus of chapter 21.

Categories of Reproductive Structures

The structures of the male reproductive system can be categorized on a functional basis as follows:

1. **Primary sex organs.** The primary sex organs are called *gonads;* specifically, the *testes* in the male. Gonads produce the gametes, or spermatozoa, and produce and secrete sex hormones. The secretion of male sex hormones, called *androgens*, at the appropriate times and in sufficient quantities causes the development of secondary sex organs and the expression of secondary sex characteristics.
2. **Secondary sex organs.** Secondary sex organs are those structures that are essential in caring for and transporting spermatozoa. The three categories of secondary sex organs are the sperm-transporting ducts, the accessory glands, and the copulatory organ. The ducts that transport sperm include the *epididymides*, *ductus deferentia*, *ejaculatory ducts*, and *urethra*. The male accessory reproductive glands are the *seminal vesicles*, the *prostate*, and the *bulbourethral glands*. The *penis*, which contains erectile tissue, is the copulatory organ. The *scrotum* is a pouch of skin that encloses and protects the testes.
3. **Secondary sex characteristics.** Secondary sex characteristics are features that are not essential for the reproductive process but are generally considered sexual attractants. In the male, they include body physique, body hair, and voice pitch.

The organs of the male reproductive system are shown in figure 20.1 and their functions are summarized in table 20.1.

gamete: Gk. *gameta*, husband or wife
spermatozoa: Gk. *sperma*, seed; *zoon*, animal
androgen: Gk. *andros*, male producing

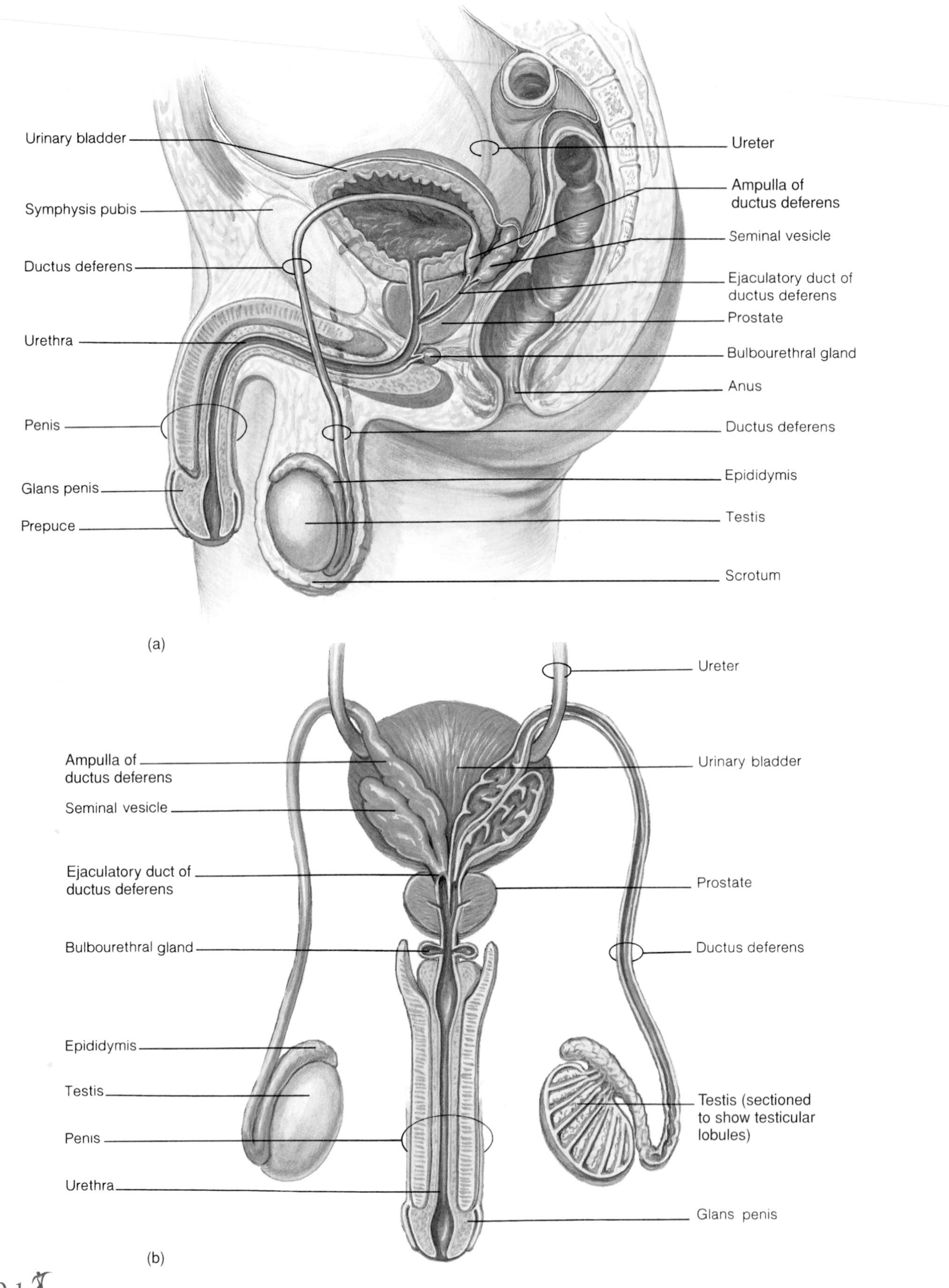

Figure 20.1

Organs of the male reproductive system. (*a*) a sagittal view and (*b*) a posterior view.

Table 20.1
Reproductive Organs of the Male

Organ(s)	Function(s)
Testes	
Seminiferous tubules	Produce spermatozoa
Interstitial cells	Secrete male sex hormones
Epididymides	Site of sperm maturation; store and convey spermatozoa to the ductus deferentia
Ductus deferentia	Store spermatozoa; convey spermatozoa to the ejaculatory ducts
Ejaculatory ducts	Receive spermatozoa and additives to produce seminal fluid
Seminal vesicles	Secrete alkaline fluid containing nutrients and prostaglandins
Prostate	Secretes alkaline fluid that enhances motility of spermatozoa and helps to neutralize the acidic environment of the vagina
Bulbourethral glands	Secrete fluid that lubricates the urethra and end of the penis
Scrotum	Encloses and protects the testes
Penis	Conveys urine and seminal fluid to outside of body; copulatory organ

1. What is the primary importance of sexual reproduction?
2. What are the functions of the male reproductive system?
3. List the organs of the male reproductive system and indicate whether they are primary or secondary sex organs.
4. With respect to the other body systems, discuss the latent (delayed) development of the reproductive system.

Perineum and Scrotum

The perineum is the specific portion of the pelvic region that contains the external genitalia and the anal opening. The scrotum, a pouch that supports the testes, is divided into two internal compartments by a connective tissue septum.

Objective 4 Describe the location of the perineum and distinguish between the urogenital and anal triangles.

Objective 5 Discuss the structure and function of the scrotum.

Perineum

The **perineum** (*per″ĭ-ne′um*) is a diamond-shaped region between the symphysis pubis and the coccyx (fig. 20.2). It is the muscular region of the outlet of the pelvis (see fig. 9.24). The perineum is divided into a *urogenital triangle* in front and an *anal triangle* in back. In the male perineum, the penis and scrotum are attached at the anterior portion of the urogenital triangle and the anus is located within the posterior portion of the anal triangle.

Scrotum

The saclike **scrotum** (*skro′tum*) is suspended immediately behind the base of the penis (see figs. 20.1 and 20.2). The functions of the scrotum are to support and protect the testes and to regulate their position relative to the pelvic region of the body. The soft, textured skin of the scrotum is covered with sparse hair in mature males and is darker in color than most of the other skin of the body. It also contains numerous sebaceous glands.

The external appearance of the scrotum varies at different times in the same individual as a result of the contraction and relaxation of the scrotal muscles. The **dartos** (*dar′tos*) is a layer of smooth muscle fibers in the subcutaneous tissue of the scrotum. The **cremaster** (*krĕ-mas′ter*) is a small band of skeletal muscle extending through the *spermatic cord*, a tube of fascia that also encloses the ductus deferens and testicular vessels and nerves (fig. 20.3). Both the dartos and the cremaster involuntarily contract in response to low temperatures to move the testes closer to the heat of the body in the pelvic region. The cremaster muscle is a continuation of the internal abdominal oblique muscle of the abdominal wall, which is derived as the testes descend into the scrotum. Because it is a skeletal muscle, it can be contracted voluntarily as well. When the scrotal muscles are contracted, the scrotum appears tightly wrinkled as the testes are pulled closer to the warmth of the body wall. High temperatures cause the dartos and cremaster muscles to relax and the testes to be suspended lower in the relaxed scrotum. The temperature of the testes is maintained at about 35°C (95°F), or about 3.6°F below normal body temperature, by the contraction or relaxation of the scrotal muscles. This temperature is optimal for the production and storage of spermatozoa.

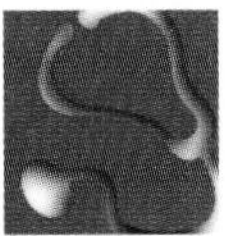

Although uncommon, *male infertility* may result from an excessively high temperature of the testes over an extended period of time. Frequent hot baths or saunas may destroy sperm cells to the extent that the sperm count will be too low to enable fertilization.

The scrotum is subdivided into two longitudinal compartments by a fibrous **scrotal septum** (fig. 20.3). The purpose of the scrotal septum is to compartmentalize each testis so that infection of one will generally not affect the other. In addition, the left

dartos: Gk. *dartos*, skinned or flayed
cremaster: Gk. *cremaster*, a suspender, to hang
septum: L. *septum*, a partition

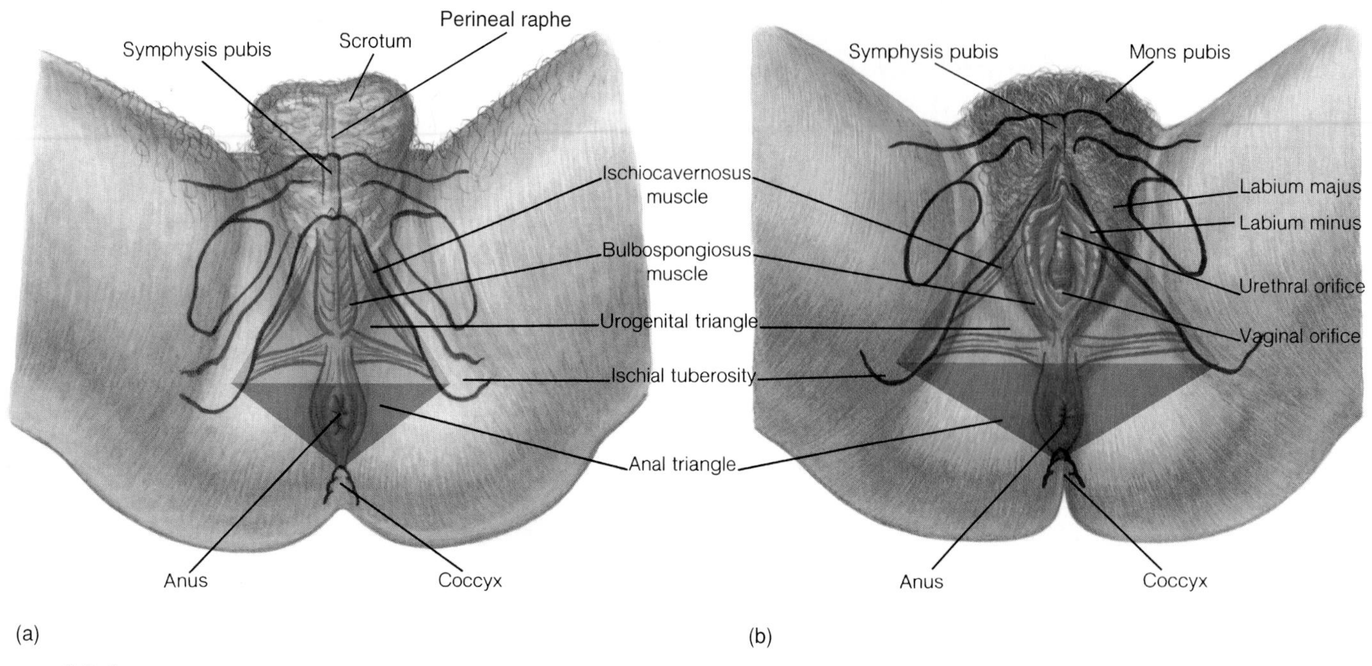

Figure 20.2

A superficial view of (*a*) the male perineum and (*b*) the female perineum (see chapter 21). The perineum is the region between the symphysis pubis and coccyx. It is divided into urogenital and anal triangles, indicated in red and purple respectively.

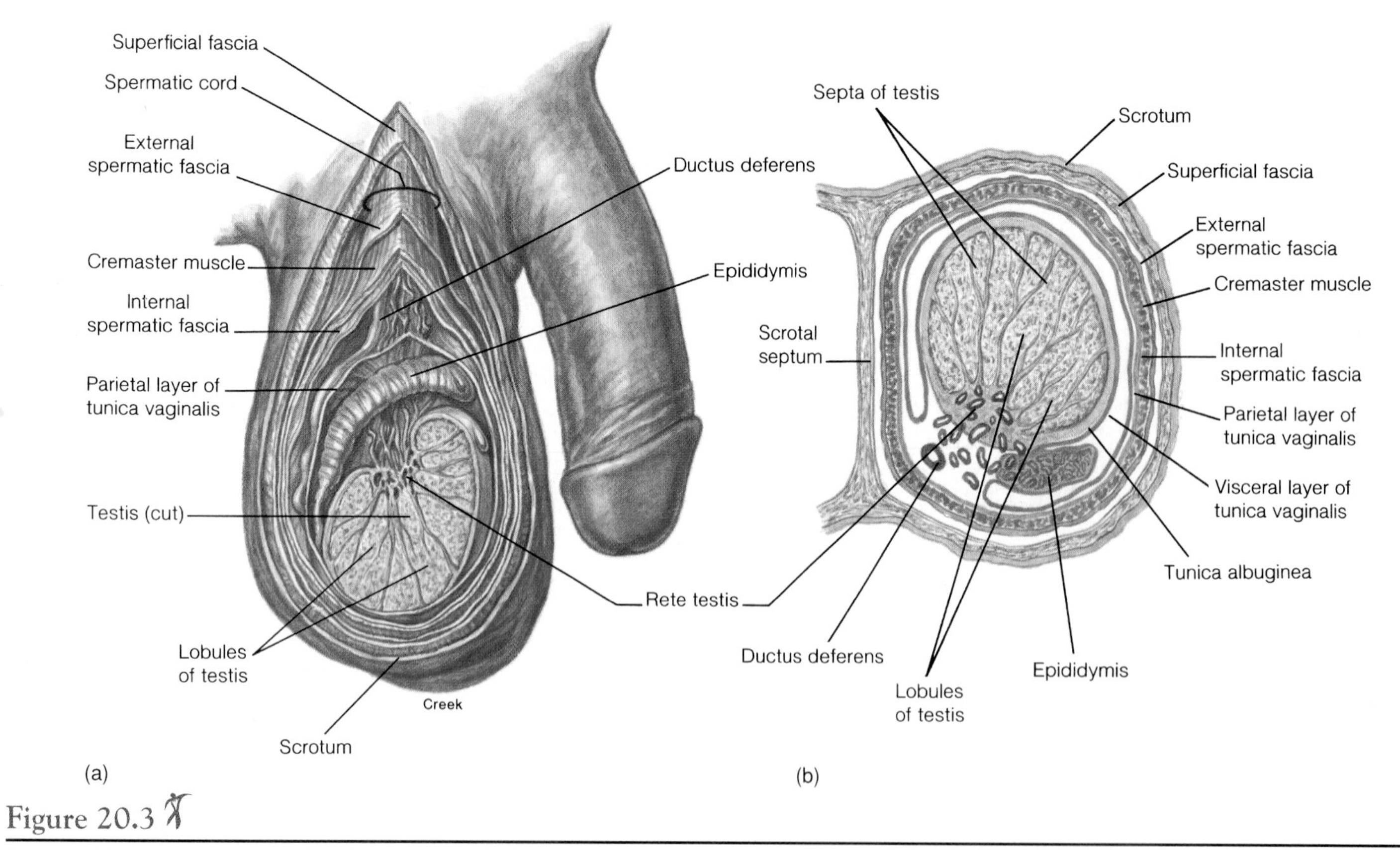

Figure 20.3

Structural features of a testis within the scrotum. (*a*) a longitudinal view and (*b*) a transverse view.

testis is generally suspended lower in the scrotum than the right so that the two are not as likely to be compressed forcefully together. The site of the scrotal septum is apparent on the surface of the scrotum along a median longitudinal ridge called the **perineal raphe** (*ra'fe*). The perineal raphe extends forward to the undersurface of the penis and backward to the anal opening (see fig. 20.2).

The blood supply and innervation of the scrotum are extensive. The arteries that serve the scrotum are the internal pudendal branch of the internal iliac artery, the external pudendal branch of the femoral artery, and the cremasteric branch of the inferior epigastric artery. The venous drainage follows a pattern similar to that of the arteries. The scrotal nerves are primarily sensory; they include the pudendal nerves, ilioinguinal nerves, and posterior cutaneous nerves of the thigh.

1. Describe the location of the perineum.
2. Explain how the temperature of the testes is maintained. Why is it important to maintain a particular testicular temperature?
3. Describe the innervation and the blood supply to the testes.

Testes

Located within the scrotum, the testes produce spermatozoa and androgens. Androgens regulate spermatogenesis and the development and functioning of the secondary sex organs.

Objective 6 Describe the location, structure, and functions of the testes.

Objective 7 Describe the effects of hormones on the growth and development of the male reproductive organs and explain how hormones are related to the development of male secondary sex characteristics.

Objective 8 List the events of spermatogenesis and distinguish between spermatogenesis and spermiogenesis.

Objective 9 Diagram the structure of a sperm cell and explain the function of each of its parts.

Structure of the Testes

The **testes** (*tes'tēz*—singular, *testis*) are paired, whitish, ovoid organs, each about 4 cm (1.5 in.) long and 2.5 cm (1 in.) in diameter. Each testis weighs between 10 and 14 g. Two tissue layers, or tunics, cover the testes. The outer **tunica vaginalis** is a thin serous sac derived from the peritoneum during the descent of the testes. The **tunica albuginea** (*al"byoo-jin'e-ă*) is a tough fibrous membrane that directly encapsulates each testis (fig. 20.3). Fibrous inward extensions of the tunica albuginea partition the testis into 250 to 300 wedge-shaped **testicular lobules.**

Each lobule of the testis contains tightly convoluted **seminiferous tubules** (fig. 20.4) that may exceed 70 cm (28 in.) in length if uncoiled. The seminiferous tubules are the functional units of the testis because it is here that *spermatogenesis*, the production of spermatozoa, occurs. Spermatozoa are produced at the rate of thousands per second—more than 100 million per day—throughout the life of a healthy, sexually mature male.

Various stages of meiosis can be observed in a section of a seminiferous tubule (see fig. 20.7*b*). The process begins as specialized germinal cells, called **spermatogonia** (*sper-mat"ŏ-go'ne-ă*), undergo meiosis to produce, in order of advancing maturity, the *primary spermatocytes*, *secondary spermatocytes*, and *spermatids* (see fig. 20.6). Forming the walls of the seminiferous tubules are **sustentacular** (Sertoli) **cells** (also called *nurse cells*) that produce and secrete nutrients for the developing spermatozoa embedded between them (see fig. 20.7). The spermatozoa are formed, but not fully mature, by the time they reach the lumen of a seminiferous tubule.

Between the seminiferous tubules are specialized endocrine cells called **interstitial cells** (cells of Leydig). The function of these cells is to produce and secrete the male sex hormones. The testes are thus considered mixed exocrine and endocrine glands because they produce both spermatozoa and androgens.

Once the spermatozoa are produced, they move through the seminiferous tubules and enter a tubular network called the **rete** (*re'te*) **testis** for further maturation (fig. 20.4). Cilia are located on some of the cells of the rete testis. The spermatozoa are transported out of the testis and into the epididymis through a series of **efferent ductules.**

In all, it takes from 8 to 10 weeks for a spermatogonium to become a mature spermatozoon. In the spermatic ducts, spermatozoa can remain fertile for several months, in a state of suspended animation. If they are not ejaculated, they eventually degenerate and are absorbed into the blood.

The testes receive blood through the testicular arteries, which in turn arise from the abdominal aorta immediately below the origin of the renal arteries. The testicular veins drain the testes. The testicular vein of the right side enters directly into the inferior vena cava, whereas the testicular vein of the left side empties into the left renal vein (see fig. 16.32).

pudendal: L. *pudeo*, to feel ashamed

tunica: L. *tunica*, a coat
vaginalis: L. *vagina*, a sheath
albuginea: L. *albus*, white
spermatogonia: Gk. *sperma*, seed; *gone*, generation
Sertoli cells: from Enrico Sertoli, Italian histologist, 1842–1910
cells of Leydig: from Franz von Leydig, German anatomist, 1821–1908
efferent ductules: L. *efferre*, to bring out; *ducere*, to lead

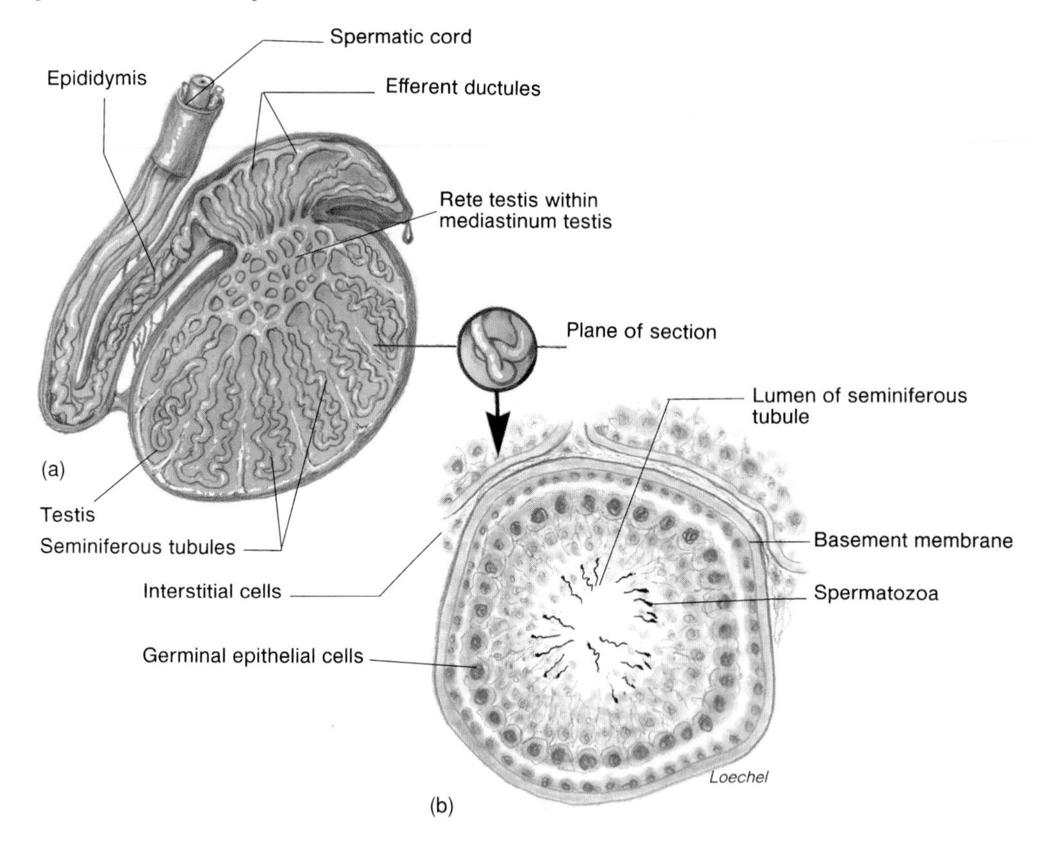

Figure 20.4

A diagram of the seminiferous tubules. (*a*) A sagittal section of a testis and (*b*) a transverse section of a seminiferous tubule.

Testicular nerves innervate the testes with both motor and sensory neurons arising from the tenth thoracic segment of the spinal cord. Motor innervation is primarily through sympathetic neurons, but there is limited parasympathetic stimulation as well.

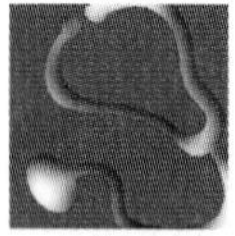

The most common cause of male infertility is a condition called *varicocele* (*var′ĭ-ko-sēl*). Varicocele occurs when one or both of the testicular veins draining from the testes becomes swollen, resulting in poor circulation in the testes. A varicocele generally occurs on the left side because the left testicular vein drains into the renal vein. The blood pressure is higher here than it is in the inferior vena cava, into which the right testicular vein empties.

Endocrine Functions of the Testes

Testosterone (*tes-tos′tĕ-rōn*) is by far the major androgen secreted by the adult testis. Androgens are sometimes called *anabolic steroids* because they stimulate the growth of muscles and other structures (table 20.2). Increased testosterone secretion during puberty is also required for the growth of the accessory sex organs, primarily the seminal vesicles and prostate. The removal of androgens by castration results in atrophy of these organs.

Androgens stimulate growth of the larynx (resulting in a deepening of the voice), hemoglobin synthesis (males have higher hemoglobin levels than females), and bone growth. The

Table 20.2
Actions of Androgens in the Male

Category	Action
Sex determination	Growth and development of the mesonephric ducts into the epididymides, ductus deferentia, seminal vesicles, and ejaculatory ducts
	Development of the urogenital sinus and tubercle into the prostate
	Development of the male external genitalia
Spermatogenesis	At puberty: Meiotic division and early maturation of spermatids
	After puberty: Maintenance of spermatogenesis
Secondary sex characteristics	Growth and maintenance of the accessory sex organs
	Growth of the penis
	Growth of facial and axillary hair
	Body growth
Anabolic effects	Protein synthesis and muscle growth
	Growth of bones
	Growth of other organs (including the larynx)
	Erythropoiesis (red blood cell formation)

Table 20.3

Stages of Meiosis

Stage	Events	Stage	Events
First Meiotic Division		**Second Meiotic Division**	
Prophase I	Chromosomes appear double-stranded. The two strands, called chromatids, contain identical DNA and are joined together by a structure known as a centromere. Homologous chromosomes pair up side by side.	Prophase II	Chromosomes appear, each containing two chromatids.
Metaphase I	Homologous chromosome pairs line up at equator; spindle apparatus is completed.	Metaphase II	Chromosomes line up single file along equator as spindle formation is completed.
Anaphase I	Homologous chromosomes separate; the two members of a homologous pair move to opposite poles.	Anaphase II	Centromeres split and chromatids move to opposite poles.
Telophase I	Cytoplasm divides to produce two haploid cells.	Telophase II	Cytoplasm divides to produce two haploid cells from each of the haploid cells formed at telophase I.

effect of androgens on bone growth is self-limiting, however, because androgens are ultimately responsible for the conversion of cartilage to bone in the epiphyseal plates, thus sealing the plates and preventing further lengthening of the bones.

Spermatogenesis

Spermatogenesis (*sper-mat″ŏ-jen′ĕ-sis*) is the sequence of events in the seminiferous tubules of the testes that leads to the production of spermatozoa. The germ cells that migrate from the yolk sac to the testes during early embryonic development become stem cells, or **spermatogonia,** within the outer region of the seminiferous tubules. Spermatogonia are diploid ($2n$) cells (with 46 chromosomes) that give rise to mature haploid ($1n$) gametes by a process of reductive cell division called *meiosis* (*mi-o′sis*).

Meiosis occurs within the testes of males who have gone through puberty and involves two nuclear divisions, as summarized in table 20.3. During the first part of this process, the DNA duplicates (prophase I) and homologous chromosomes are separated (during anaphase I), producing two daughter cells (at telophase I). Since each daughter cell contains only one of each homologous pair of chromosomes, the cells formed at the end of this *first meiotic division* contain 23 chromosomes each and are haploid (fig. 20.5). Each of the 23 chromosomes at this stage, however, consists of two strands (called *chromatids*) of identical DNA. During the *second meiotic division*, these duplicate chromatids are separated (at anaphase II), producing daughter cells (at telophase II). Meiosis of one diploid spermatogonia cell therefore produces four haploid daughter cells.

Actually, only about 1,500 stem cells migrate from the yolk sac to the embryonic testes. In order to produce many millions of spermatozoa throughout adult life, these spermatogonia duplicate themselves by mitotic division, and only one of the two cells—now called a **primary spermatocyte**—undergoes meiotic division (fig. 20.6). In this way, spermatogenesis can occur continuously without exhausting the number of spermatogonia.

When a diploid primary spermatocyte completes the first meiotic division (at telophase I), the two haploid daughter cells thus produced are called **secondary spermatocytes.** At the end of the second meiotic division, each of the two secondary spermatocytes produces two haploid **spermatids** (*sper′mă-tidz*). One primary spermatocyte therefore produces four spermatids.

The sequence of events in spermatogenesis is reflected in the wall of the seminiferous tubule. The epithelial wall of the tubule, called the **germinal epithelium,** is indeed composed of germ cells in different stages of spermatogenesis. The spermatogonia and primary spermatocytes are located toward the outer side of the tubule, whereas the spermatids and mature spermatozoa are located on the side facing the lumen (fig. 20.7).

At the end of the second meiotic division, the four spermatids produced by the meiosis of two secondary spermatocytes are interconnected—their cytoplasm does not completely pinch off at the end of each division. The development of these interconnected spermatids into separate mature spermatozoa (a process called *spermiogenesis*) requires the participation of another type of cell in the tubules—the sustentacular (nurse) cells, or Sertoli cells.

Sustentacular (*sus″ten-tak′yŭ-lar*) **cells** are the only nongerminal cell type in the tubules. Connected by tight junctions, they form a continuous layer around the circumference of each tubule. In this way, the sustentacular cells form a *blood-testis barrier*. The molecules from the blood must pass through the cytoplasm of sustentacular cells before entering germinal cells.

In the process of *spermiogenesis* (the conversion of spermatids to spermatozoa), most of the spermatid cytoplasm is eliminated.

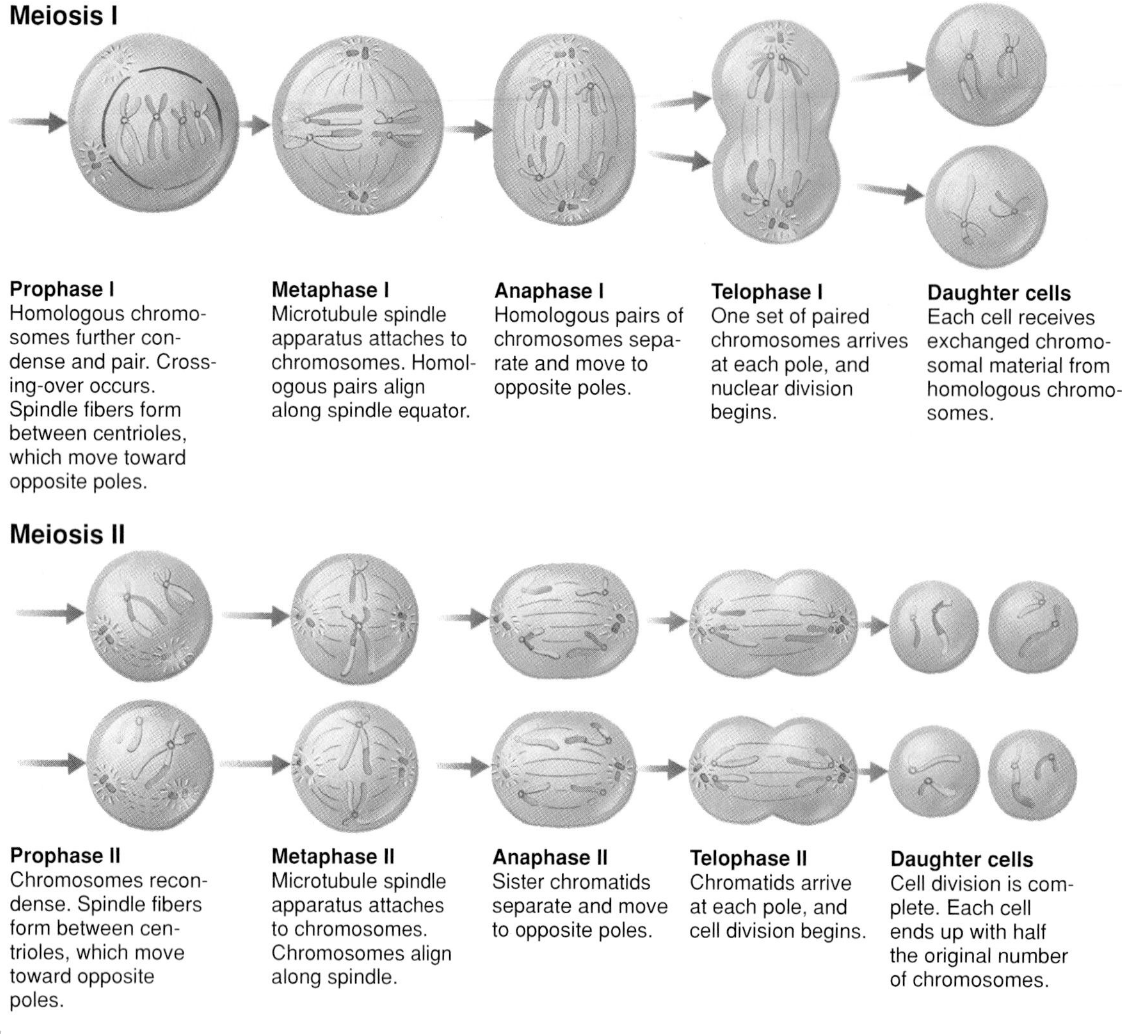

Meiosis I
Prophase I
Homologous chromosomes further condense and pair. Crossing-over occurs. Spindle fibers form between centrioles, which move toward opposite poles.
Metaphase I
Microtubule spindle apparatus attaches to chromosomes. Homologous pairs align along spindle equator.
Anaphase I
Homologous pairs of chromosomes separate and move to opposite poles.
Telophase I
One set of paired chromosomes arrives at each pole, and nuclear division begins.
Daughter cells
Each cell receives exchanged chromosomal material from homologous chromosomes.
Meiosis II
Prophase II
Chromosomes recondense. Spindle fibers form between centrioles, which move toward opposite poles.
Metaphase II
Microtubule spindle apparatus attaches to chromosomes. Chromosomes align along spindle.
Anaphase II
Sister chromatids separate and move to opposite poles.
Telophase II
Chromatids arrive at each pole, and cell division begins.
Daughter cells
Cell division is complete. Each cell ends up with half the original number of chromosomes.

Figure 20.5

Meiosis, or reduction division. In the first meiotic division, the homologous chromosomes of a diploid parent cell are separated into two haploid daughter cells. Each of these chromosomes contains duplicate strands, or chromatids. In the second meiotic division, these chromatids are distributed to two new haploid daughter cells.

This occurs through phagocytosis by the sustentacular cells of the "residual bodies" of cytoplasm from the spermatids. It is believed that phagocytosis of the residual bodies may transmit informational molecules from the germ cells to the sustentacular cells. The sustentacular cells, in turn, may provide molecules needed by the germ cells. It is known, for example, that the X chromosome of the germ cells is inactive during meiosis. Since this chromosome contains the genes needed to produce many essential molecules, it is believed that these molecules are provided by the sustentacular cells while meiosis is taking place.

Structure of Spermatozoa

A mature sperm cell, or **spermatozoon** (*sper-mat"ŏ-zo'on*) is a microscopic, tadpole-shaped structure approximately 0.06 mm long (figs. 20.7c and 20.8). It consists of an oval **head** and an elongated **flagellum.** The head of a spermatozoon contains a nucleus with 23 chromosomes. The tip of the head, called the *acrosome* (*ak'rŏ-sōm*), contains enzymes that help the spermatozoon penetrate the ovum. The flagellum contains numerous mitochondria spiraled around a filamentous core. The mitochondria provide the energy necessary for locomotion. The flagellum propels the spermatozoon with a lashing movement. The maximum unassisted rate of spermatozoa movement is about 3 mm per hour.

Recent findings indicate that ejaculated spermatozoa (plural of spermatozoon) can survive up to 5 days at body temperature—longer than previously thought. In the average fertile male, up to 20% of these sperm cells are defective, however, and are of no value. It is not uncommon for them to have enlarged heads, dwarfed and misshapen heads, two flagella, or a flagellum that is bent. Spermatozoa such as these are unable to propel themselves adequately.

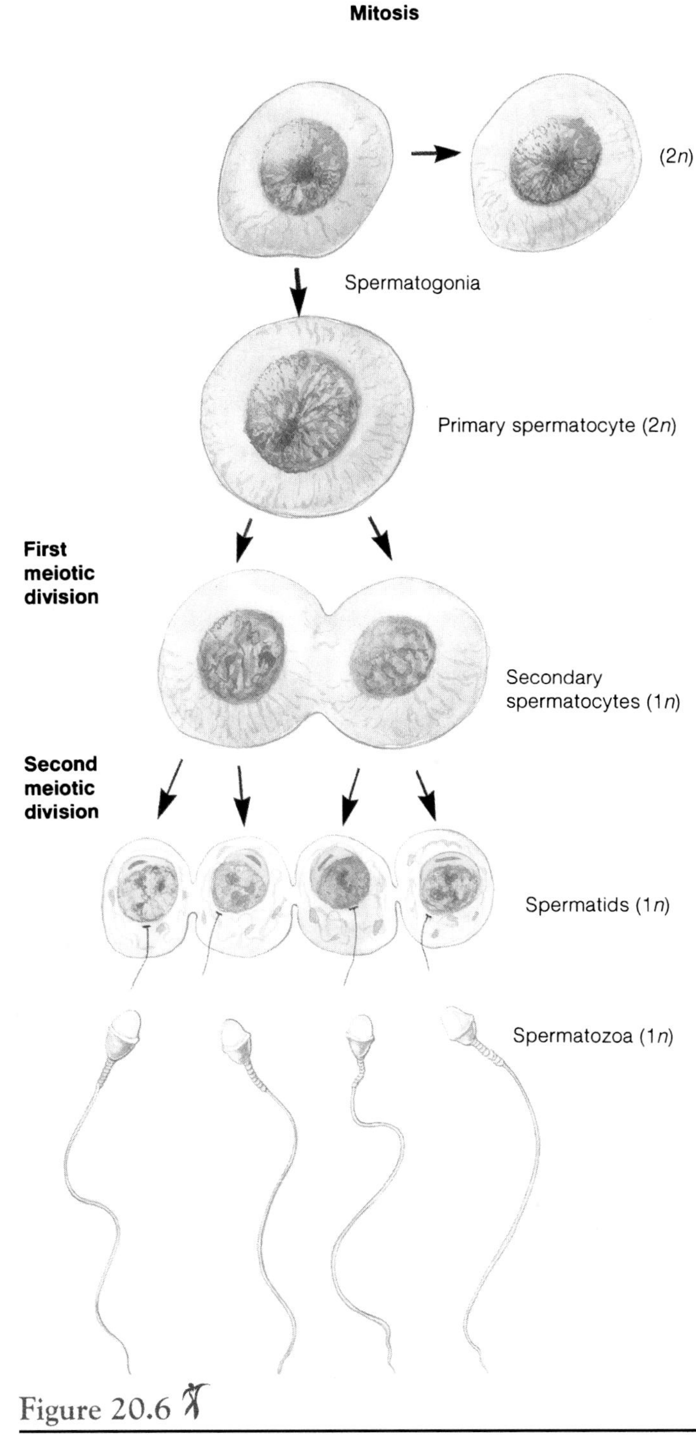

Figure 20.6

The process of spermatogenesis. Spermatogonia undergo mitotic division to replace themselves and produce a daughter cell that will undergo meiotic division. This cell is called a primary spermatocyte. Upon completion of the first meiotic division, the daughter cells are called secondary spermatocytes. Each of them complete a second meiotic division to form spermatids. Notice that the four spermatids produced by the meiosis of a primary spermatocyte are interconnected. Each spermatid forms a mature spermatozoon.

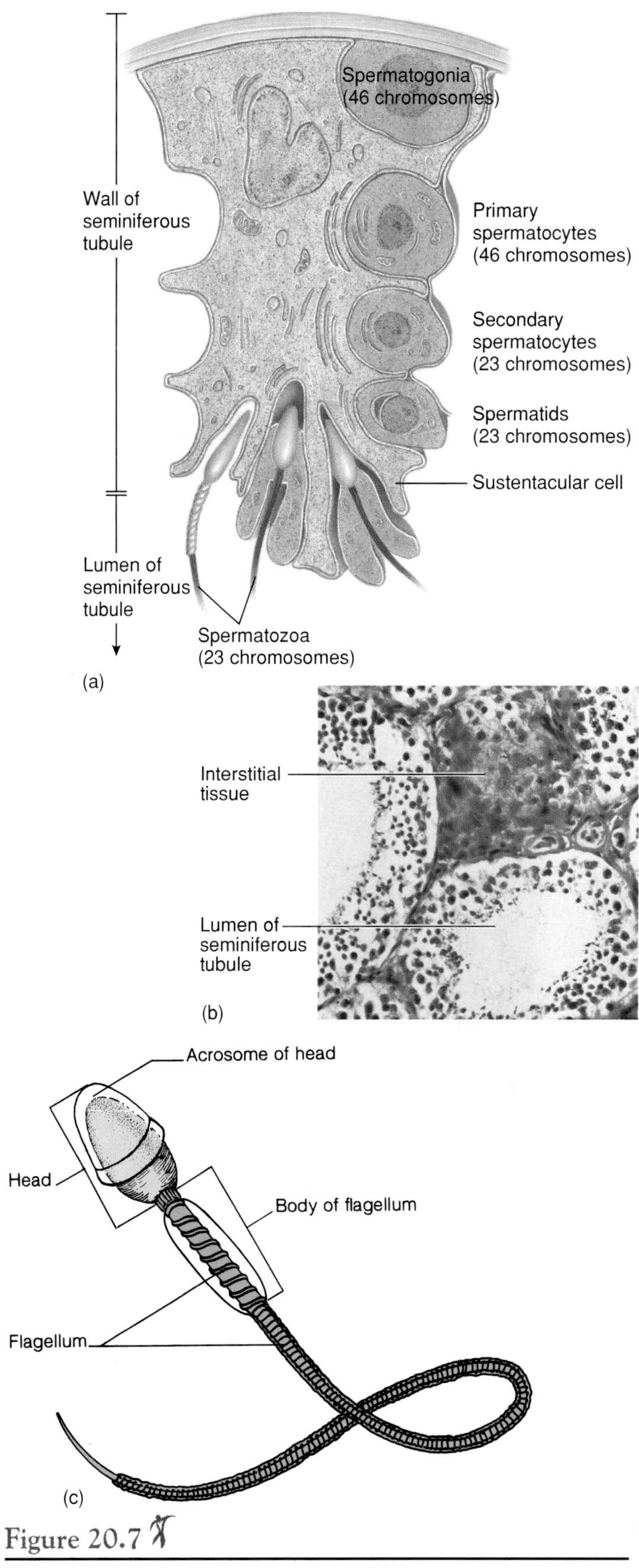

Figure 20.7

Seminiferous tubules and a spermatozoon. (*a*) The stages of spermatogenesis within the germinal epithelium of a seminiferous tubule, in which the relationship between sustentacular cells and developing spermatozoa is shown. (*b*) A photomicrograph of a seminiferous tubule and surrounding interstitial tissue, and (*c*) a diagram of a spermatozoon.

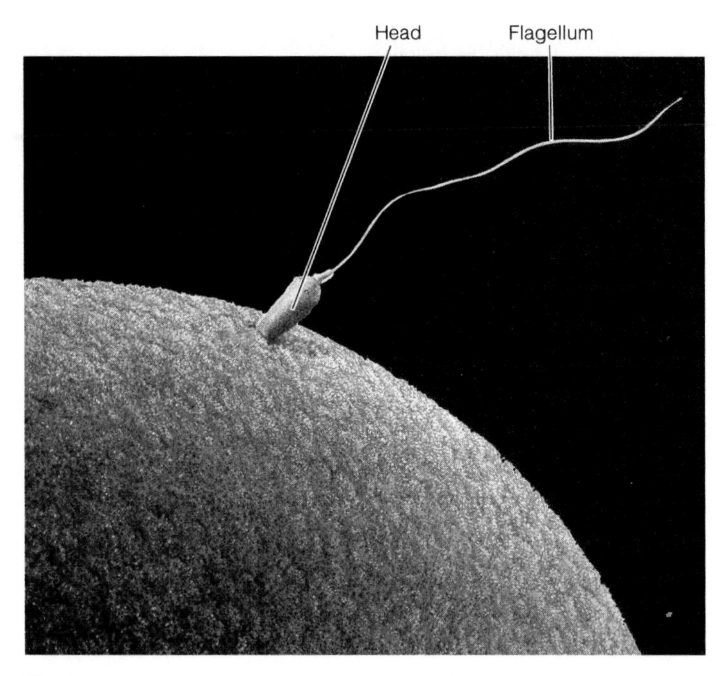

Figure 20.8

A scanning electron micrograph of a spermatozoon in contact with an egg.

1. Describe the location and structure of the testes.
2. What are the functions of the seminiferous tubules, germinal epithelial cells, and interstitial cells?
3. Discuss the effect of testosterone on the production of sperm cells and the development of secondary sex characteristics.
4. Diagram a spermatozoon and its adjacent sustentacular cell and explain the functions of sustentacular cells in the seminiferous tubules.

Spermatic Ducts, Accessory Reproductive Glands, and the Urethra

The spermatic ducts store spermatozoa and transport them from the testes to the outside of the body by way of the urethra. The accessory reproductive glands provide additives to the spermatozoa to form semen, which is discharged from the erect penis during ejaculation.

Objective 10 Describe the location and structure of each segment of the male duct system.

Objective 11 Describe the structure and contents of the spermatic cord.

Objective 12 Describe the location, structure, and functions of the ejaculatory ducts, seminal vesicles, prostate, and bulbourethral glands.

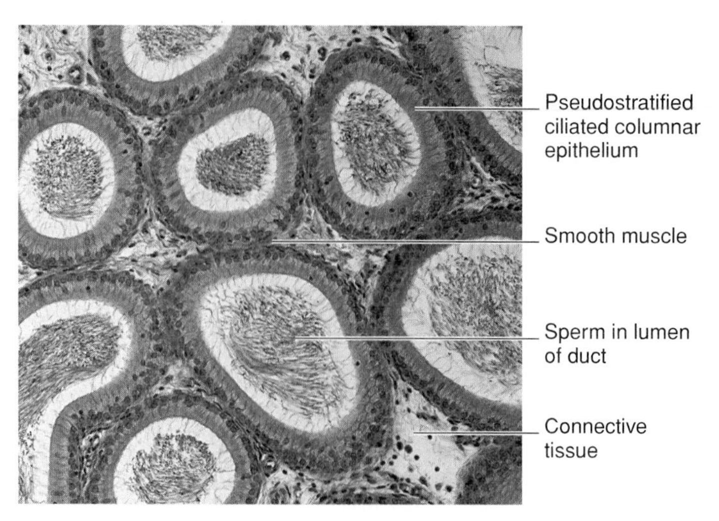

Figure 20.9

The histology of the epididymis showing sperm in the lumina (50×).

Spermatic Ducts

The spermatic ducts store spermatozoa and transport them from the testes to the urethra. The accessory reproductive glands provide additives to the spermatozoa in the formation of semen.

Epididymis

The **epididymis** (*ep″ĭ-did′ĭ-mus*—plural, *epididymides*) is an elongated organ attached to the posterior surface of the testis (see figs. 20.1 and 20.3). If it were uncoiled, it would measure 5.5 m (about 17 ft). The highly coiled, tubular *tail portion* contains spermatozoa in their final stages of maturation (fig. 20.9). The upper expanded portion is the *head*, and the tapering middle section is the *body*. The tail is continuous with the beginning portion of the ductus deferens; both store spermatozoa to be discharged during ejaculation. The time required to produce mature spermatozoa—from meiosis in the seminiferous tubules to storage in the ductus deferens—is approximately 2 months.

Ductus Deferens

The **ductus deferens** (*duk′tus def′er-enz*—plural, *ductus deferentia*) is a fibromuscular tube about 45 cm (18 in.) long and 2.5 mm thick (see fig. 20.1) that conveys spermatozoa from the epididymis to the ejaculatory duct. Also called the *vas deferens* (plural, *vasa deferentia*), it exits the scrotum as it ascends along the posterior border of the testis. From here, it penetrates the inguinal canal, enters the pelvic cavity, and passes to the side of the urinary bladder on the medial side of the ureter. The **ampulla** (*am-pool′ă*) of the ductus deferens is the terminal portion that joins the ejaculatory duct.

The histological structure of the ductus deferens includes a layer of pseudostratified ciliated columnar epithelium in contact

deferens: L. *deferens*, conducting away
ampulla: L. *ampulla*, a two-handled bottle

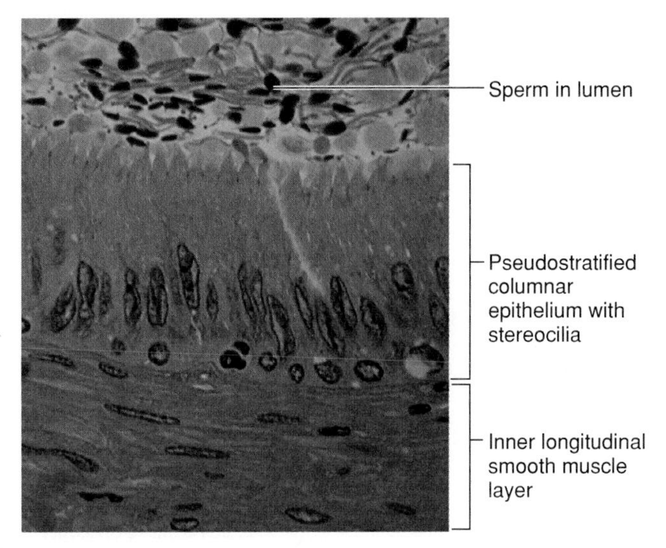

Figure 20.10

The histology of the ductus deferens (250×).

with the tubular lumen and surrounded by three layers of tightly packed smooth muscle (fig. 20.10). Sympathetic nerves from the pelvic plexus serve the ductus deferens. Stimulation through these nerves causes peristaltic contractions of the muscular layer, which propel the stored spermatozoa toward the ejaculatory duct.

Much of the ductus deferens is located within a structure known as the **spermatic cord** (see figs. 20.3 and 20.14). The spermatic cord extends from the testis to the inguinal canal and consists of the ductus deferens, the testicular artery and venous plexus, nerves, the cremaster muscle, lymph vessels, and connective tissue. The portion of the spermatic cord that passes anterior to the pubic bone can be easily palpated. The **inguinal** (*ing'gwĭ-nal*) **canal** is a passageway for the spermatic cord through the abdominal wall; it is a potentially weak area and a common site for development of a hernia.

Ejaculatory Duct

The ejaculatory (*ĕ-jak'yoo-lă-tor-e*) duct is about 2 cm (1 in.) long and is formed by the union of the ampulla of the ductus deferens and the duct of the seminal vesicle. The ejaculatory duct then pierces the capsule of the prostate on its posterior surface and continues through this gland (fig. 20.1). Both ejaculatory ducts receive secretions from the seminal vesicles and then eject the spermatozoa with its additives into the prostatic urethra to be mixed with secretions from the prostate. The urethra, discussed shortly, serves as a passageway for both semen and urine. It is the terminal duct of the male duct system.

Accessory Reproductive Glands

The accessory reproductive glands of the male include the *seminal vesicles*, the *prostate*, and the *bulbourethral glands* (see fig. 20.1). The contents of the seminal vesicles and prostate are mixed with the spermatozoa during ejaculation to form *semen* (*seminal fluid*).

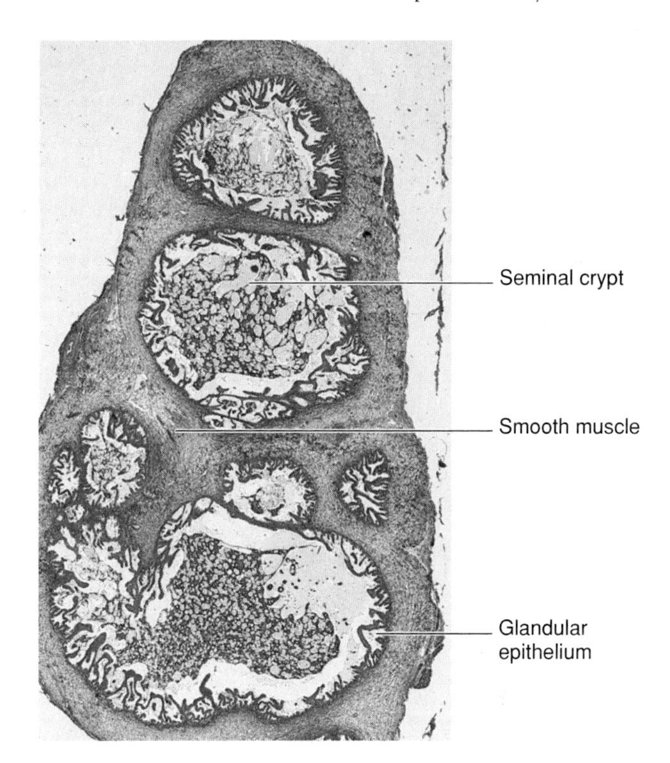

Figure 20.11

The histology of the seminal vesicle (10×).

The fluid from the bulbourethral glands is released in response to sexual stimulation prior to ejaculation.

Seminal Vesicles

The paired seminal vesicles, each about 5 cm (2 in.) long, are convoluted club-shaped glands lying at the base of the urinary bladder, in front of the rectum. They secrete a sticky, slightly alkaline, yellowish substance that contributes to the motility and viability of spermatozoa. The secretion from the seminal vesicles contains a variety of nutrients, including fructose, that provide an energy source for the spermatozoa. It also contains citric acid, coagulation proteins, and prostaglandins. The discharge from the seminal vesicles makes up about 60% of the volume of semen.

Histologically, the seminal vesicle has an extensively coiled mucosal layer (fig. 20.11). This layer partitions the lumen into numerous intercommunicating spaces that are lined by pseudostratified columnar and cuboidal secretory epithelia (referred to as glandular epithelium).

Blood is supplied to the seminal vesicles by branches from the middle rectal arteries. The seminal vesicles are innervated by both sympathetic and parasympathetic neurons. Sympathetic stimulation causes the contents of the seminal vesicles to empty into the ejaculatory ducts on their respective sides.

Prostate

The firm prostate (*pros'tāt*) is the size and shape of a chestnut. It is about 4 cm (1.6 in.) across and 3 cm (1.2 in.) thick and lies immediately below the urinary bladder, where it surrounds the

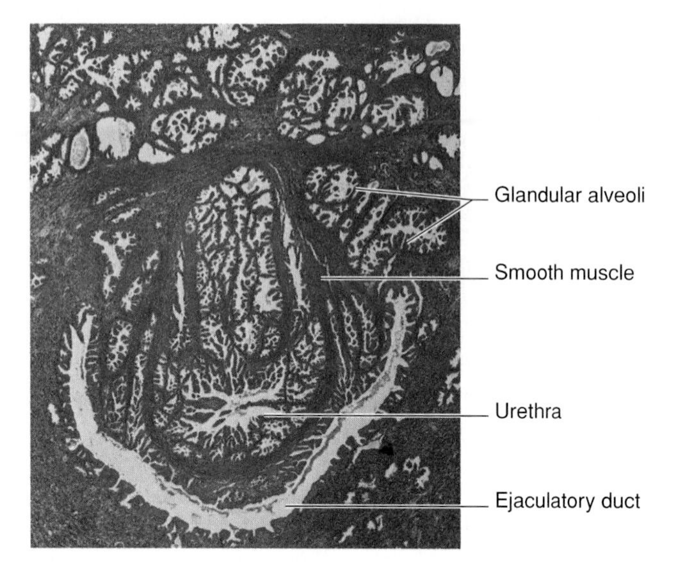

Figure 20.12

The histology of the prostate (50×).

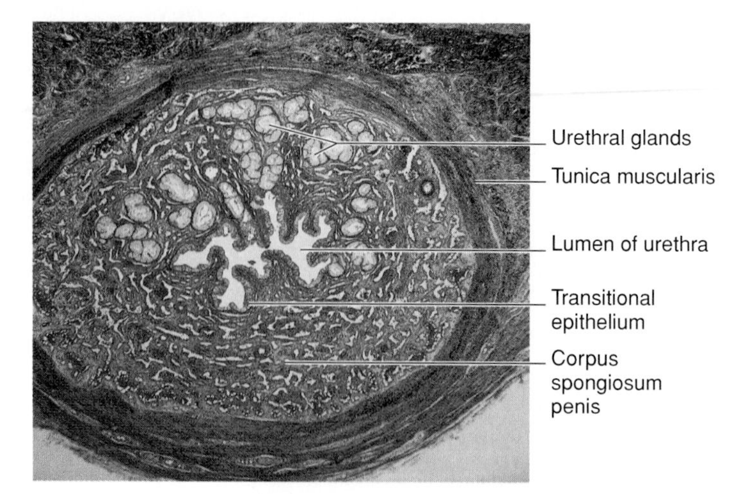

Figure 20.13

The histology of the urethra (10×).

beginning portion of the urethra (see fig. 20.1). The prostate is enclosed by a fibrous capsule and divided into lobules formed by the urethra and the ejaculatory ducts that extend through the gland. The ducts from the lobules open into the urethra. Extensive bands of smooth muscle course throughout the prostate to form a meshwork that supports the glandular tissue (fig. 20.12). Contraction of the smooth muscle expels the contents from the gland and provides part of the propulsive force needed to ejaculate the semen. The thin, milky-colored prostatic secretion assists sperm cell motility as a liquefying agent, and its alkalinity protects the sperm in their passage through the acidic environment of the female vagina. The prostate also secretes the enzyme *acid phosphatase*, which is often measured clinically to assess prostate function. The discharge from the prostate makes up about 40% of the volume of the semen.

Blood is supplied to the prostate from branches of the middle rectal and inferior visceral arteries. The venous return forms the prostatic venous plexus, along with blood draining from the penis. The prostatic venous plexus drains into the internal iliac veins. The prostate has both sympathetic and parasympathetic innervation arising from the pelvic plexuses.

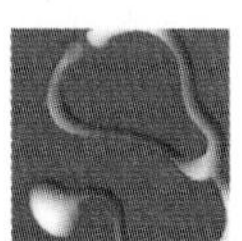

A routine physical examination of the male includes rectal palpation of the prostate. Enlargement or overgrowth of the glandular substance of the prostate, called *benign prostatic hypertrophy (BPH),* is relatively common in older men. It constricts the urethra, causing difficult urination. This condition may require surgery. In a *transurethral prostatic resection,* excess prostatic tissue is removed by use of a device inserted through the urethra.

Bulbourethral Glands

The paired, pea-sized, bulbourethral (Cowper's) glands are located below the prostate (see fig. 20.1). Each bulbourethral gland is about 1 cm in diameter and drains by a 2.5-cm (1-in.) duct into the urethra. Upon sexual arousal and prior to ejaculation, the bulbourethral glands are stimulated to secrete a mucoid substance that coats the lining of the urethra to neutralize the pH of the urine residue. It also lubricates the tip of the penis in preparation for coitus.

Urethra

The **urethra** of the male serves as a common tube for both the urinary and reproductive systems. However, urine and semen cannot pass through simultaneously because the nervous reflex during ejaculation automatically inhibits micturition. The male urethra is about 20 cm (8 in.) long, and S-shaped due to the shape of the penis. As described in chapter 19 (see fig. 19.15), it is divided into three regions, which are briefly reviewed here.

1. The **prostatic part of the urethra** is the 2.5-cm proximal portion that passes through the prostate. The prostatic urethra receives drainage from the small ducts of the prostate and the two ejaculatory ducts.
2. The **membranous part of the urethra** is the 0.5-cm portion that passes through the urogenital diaphragm. The external urethral sphincter muscle is located in this region.
3. The **spongy part of the urethra** is the longest portion, extending from the outer edge of the urogenital diaphragm to the external urethral orifice on the glans penis. About 15 cm long, this portion is surrounded by erectile tissue as it passes through the corpus spongiosum of the penis. The paired ducts from the bulbourethral glands attach to the spongy part of the urethra near the urogenital diaphragm.

The wall of the urethra has an inside lining of mucous membrane, composed of transitional epithelium (fig. 20.13) and surrounded by a relatively thick layer of smooth muscle tissue called the *tunica muscularis*. Specialized **urethral glands,** embedded in the urethral wall, secrete mucus into the lumen of the urethra.

1. List in order the structures through which spermatozoa pass as they travel from the testes to the urethra.
2. Differentiate between the ductus deferens and the spermatic cord.
3. Describe the structure and location of each of the accessory reproductive glands.
4. Why is the position of the prostate of clinical importance?
5. Describe the three portions of the male urethra.

Penis

The penis, containing the spongy urethra and covered with loose-fitting skin, is specialized with three columns of erectile tissue to become engorged with blood for insertion into the vagina during coitus.

Objective 13 Describe the structure and function of the penis.

The **penis** (*pe'nis*) when distended, serves as the copulatory organ of the male reproductive system. The penis and scrotum, which are suspended from the perineum, constitute the external genitalia of the male. Under the influence of sexual stimulation, the penis becomes engorged with blood. This engorgement results from the filling of intricate blood sinuses, or spaces, in the erectile tissue of the penis.

The penis is divided into a proximal attached *root*, which is attached to the pubic arch; an elongated tubular *body*, or *shaft;* and a distal cone-shaped *glans penis* (fig. 20.14). The **root of the penis** expands posteriorly to form the **bulb of the penis** and the **crus** (*krus*) **of the penis.** The bulb is positioned in the urogenital triangle of the perineum, where it is attached to the undersurface of the urogenital diaphragm and enveloped by the bulbocavernosus muscle (see fig. 9.24). The crus, in turn, attaches the root of the penis to the pubic arch (ischiopubic ramus) and to the perineal membrane. The crus, which is superior to the bulb, is enveloped by the ischiocavernosus muscle.

The **body of the penis** is composed of three cylindrical columns of erectile tissue, which are bound together by fibrous tissue and covered with skin (fig. 20.14). The paired dorsally positioned masses are named the **corpora cavernosa** (*kor'por-ă kav"er-no'să*) **penis.** The fibrous tissue between the two corpora forms a **septum penis.** The **corpus spongiosum** (*spon"je-o'sum*) **penis** is ventral to the other two and surrounds the spongy urethra. The penis is flaccid when the spongelike tissue is not engorged with blood. In the sexually aroused male, it becomes firm and erect as the columns of erectile tissue fill with blood (fig. 20.15).

crus: L. *crus*, leg, resembling a leg
cavernosa: L. *cavus*, hollow

Trauma to the penis, testes, and scrotum is common because of their pendent (hanging) position. Because the penis and testes are extremely sensitive to pain, a male will respond reflexively to protect the groin area. Urethral injuries are more common in men than in women because of the position of the urethra in the penis. "Straddle injuries" in which the urethra and penis are forcefully compressed between a hard surface and the pubic arch, may rupture the urethra.

The **glans penis** is the cone-shaped terminal portion of the penis, which is formed from the expanded corpus spongiosum. The opening of the urethra at the tip of the glans penis is called the **urethral orifice.** The **corona glandis** is the prominent posterior ridge of the glans penis. On the undersurface of the glans penis, a vertical fold of tissue called the **frenulum** (*fren'yŭ-lum*) attaches the skin covering the penis to the glans penis.

The skin covering the penis is hairless, lacks fat cells, and is generally more darkly pigmented than the rest of the body skin. The skin of the body of the penis is loosely attached and is continuous over the glans penis as a protective retractable sheath called the **prepuce** (*pre'pyoos*), or **foreskin.** The prepuce is commonly removed from an infant on the third or fourth day after birth, or on the eighth day as part of a Jewish religious rite. This procedure is called a *circumcision*.

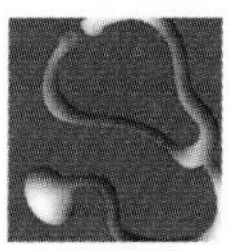

A *circumcision* is generally performed for hygienic purposes because the glans penis is easier to clean if exposed. A sebaceous secretion from the glans penis, called *smegma* (*smeg'ma*), will accumulate along the border of the corona glandis if good hygiene is not practiced. Smegma can foster bacteria that may cause infections, and therefore should be removed through washing. Cleaning the glans penis of an uncircumcised male requires retraction of the prepuce. Occasionally, a child is born with a prepuce that is too tight to permit retraction. This condition is called *phimosis* (*fĭ-mo'sis*) and necessitates circumcision.

The penis is supplied with blood on each side through the superficial external pudendal branch of the femoral artery and the internal pudendal branch of the internal iliac artery. The venous return is through a superficial median dorsal vein, which empties into the great saphenous vein in the thigh, and through the deep median vein, which empties into the prostatic plexus.

The penis has many sensory tactile receptors, especially in the glans penis, making it a highly sensitive organ. In addition, the penis has extensive motor innervation from both sympathetic and parasympathetic fibers.

glans: L. *glans*, acorn
corona: L. *corona*, garland, crown
frenulum: L. *frenulum*, diminutive of *frenum*, a bridle
prepuce: L. *prae*, before; *putium*, penis
phimosis: Gk. *phimosis*, a muzzling

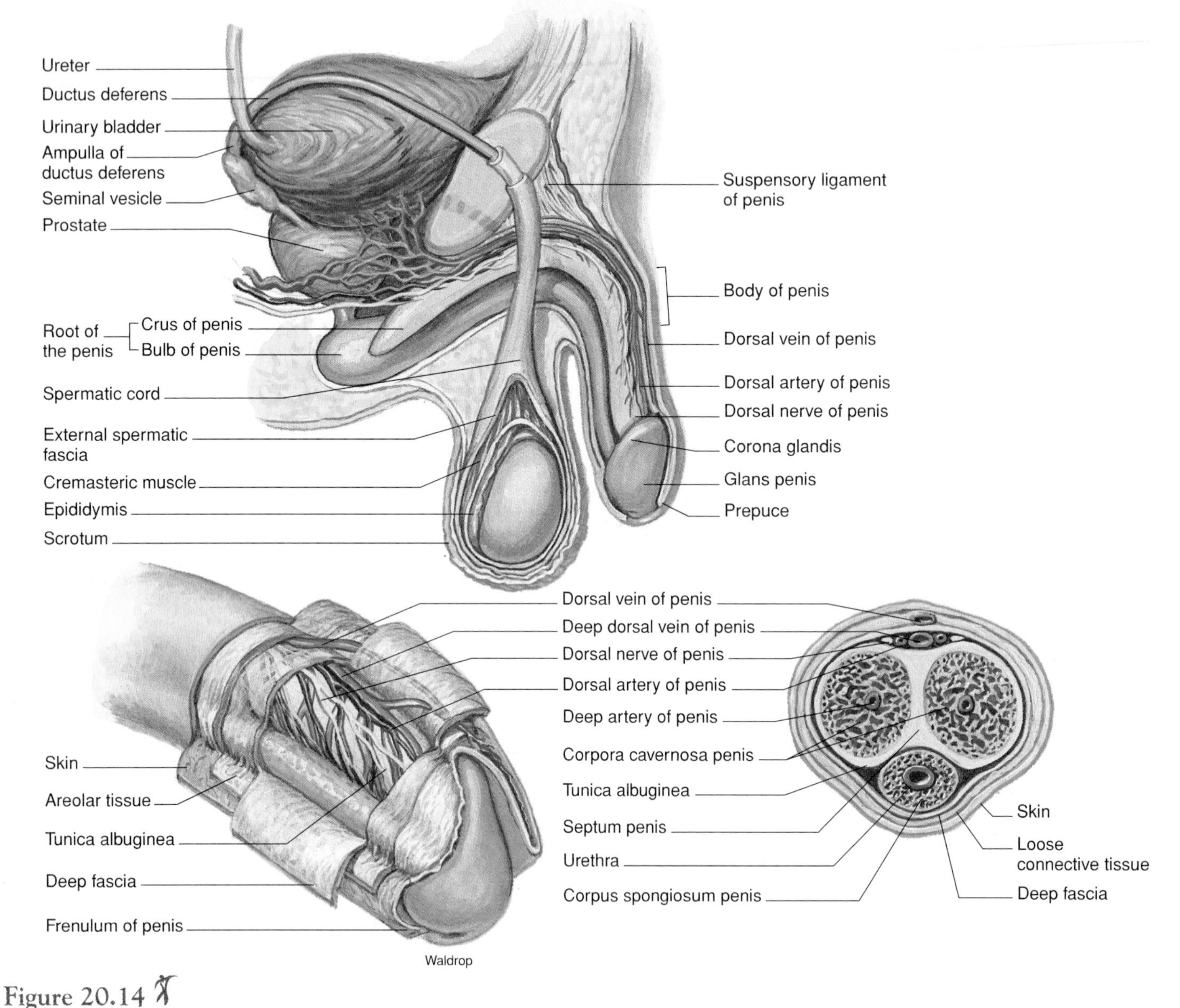

Figure 20.14

The structure of the penis showing the attachment, blood and nerve supply, and the arrangement of the erectile tissue.

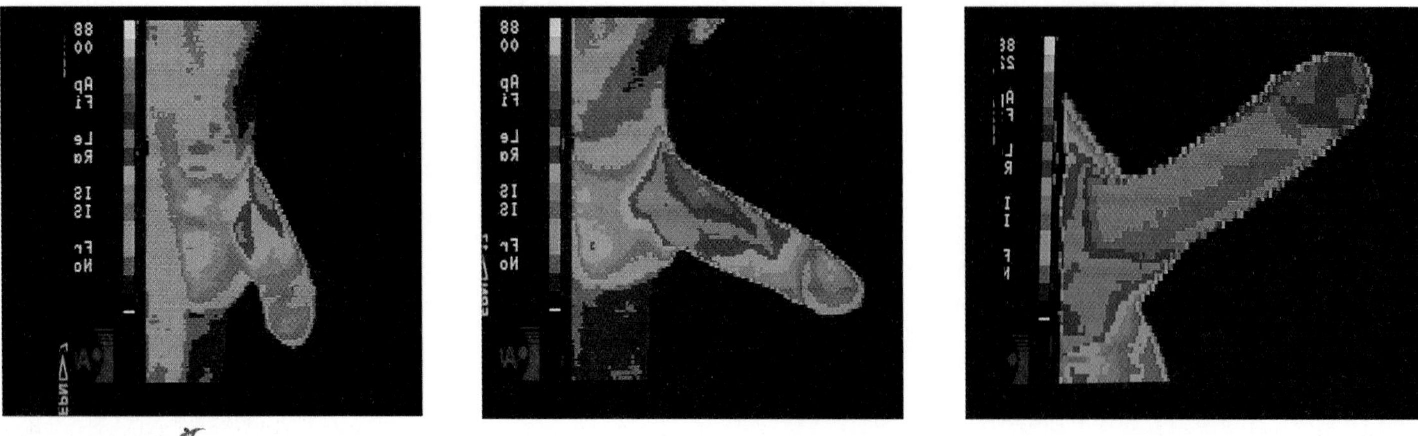

Figure 20.15

Thermograms of the penis showing differential heat radiation during erection. Red, yellow, and purple indicate the warmest portions; blue and green indicate the coolest.

1. Describe the position of the penis relative to the scrotum and accessory glands.
2. Describe the external structure of the penis and the internal arrangement of the erectile tissue.
3. Define *circumcision* and explain why this procedure is commonly performed.

Mechanisms of Erection, Emission, and Ejaculation

Erection of the penis results from parasympathetic impulses that cause vasodilation of arteries within the penis and a decrease in venous drainage. Emission and ejaculation are stimulated by sympathetic impulses, which result in the forceful expulsion of semen from the penis.

Objective 14 Distinguish between erection, emission, and ejaculation.

Objective 15 Describe the events that result in erection of the penis.

Objective 16 Explain the physiological process of ejaculation.

Objective 17 Describe the characteristic properties of semen.

Erection, emission, and ejaculation are a series of interrelated events by which semen from the male is deposited into the female vagina during coitus (sexual intercourse). *Erection* usually occurs as a male becomes sexually aroused and the erectile tissue of the penis becomes engorged with blood. *Emission* is the movement of spermatozoa from the epididymides to the ejaculatory ducts. *Ejaculation* is the forceful expulsion of the ejaculate, or *semen* (*se'men*), from the ejaculatory ducts and urethra of the penis. Emission and ejaculation do not necessarily follow erection of the penis; they occur only if there is sufficient stimulation of the sensory receptors in the penis to elicit the ejaculatory response.

Erection of the Penis

Erection of the penis depends on the volume of blood that enters the arteries of the penis as compared to the volume that exits through venous drainage (fig. 20.15). Normally, constant sympathetic stimuli to the arterioles of the penis maintain a partial constriction of smooth muscles within the arteriole walls, so that there is an even flow of blood throughout the penis. During sexual excitement, however, parasympathetic impulses cause marked vasodilation within the arterioles of the penis, resulting in more blood entering than venous blood draining. Also, during parasympathetic stimulation there is inhibition of sympathetic impulses to arterioles of the penis. At the same time, there may be slight vasoconstriction of the dorsal vein of the penis and an increase in cardiac output. These combined events cause the spongy tissue of the corpora cavernosa and the corpus spongiosum to become distended with blood and the penis to become turgid. In this condition, the penis can be inserted into the vagina of the female and function as a copulatory organ to discharge semen.

Erection is controlled by two portions of the central nervous system—the hypothalamus in the brain and the sacral portion of the spinal cord. The hypothalamus controls conscious sexual thoughts that originate in the cerebral cortex. Nerve impulses from the hypothalamus elicit parasympathetic responses from the sacral region that cause vasodilation of the arterioles within the penis. Conscious thought is not required for an erection, however, and stimulation of the penis can cause an erection because of a reflex response in the spinal cord. This reflexive action enables an erection in a sleeping male or in an infant—perhaps from the stimulus of a diaper. The mechanism of erection of the penis is summarized in figure 20.16.

The penis of many mammals contain a bone called an *os penis,* or *baculum,* of a highly variable shape. The human male, having no such bone, relies exclusively on blood-filled erectile tissue to give rigidity to the penis. On average, the erect penis in an adult is 15 cm (6 in.) long and 4 cm (1.5 in.) in diameter.

Ejaculation is the expulsion of semen through the urethra of the penis. In contrast to erection, ejaculation is a response involving the sympathetic innervation of the accessory reproductive organs. Ejaculation is preceded by continued sexual stimulation, usually through activated tactile receptors in the glans penis and the skin of the body. Rhythmic friction of these structures during coitus causes sensory impulses to be transmitted to the thalamus and cerebral cortex. The first sympathetic response, which occurs prior to ejaculation, is the discharge from the bulbourethral glands. The fluid from these glands is usually discharged before penetration into the vagina and serves to lubricate the urethra and the glans penis.

Emission and Ejaculation of Semen

Emission

Continued sexual stimulation following erection of the penis causes emission (fig. 20.17). Emission is the movement of sperm cells from the epididymides to the ejaculatory ducts and the secretions of the accessory glands into the ejaculatory ducts and urethra in the formation of semen. Emission occurs as sympathetic impulses from the pelvic plexus cause a rhythmic contraction of the smooth muscle of the epididymides, ductus deferentia, ejaculatory ducts, seminal vesicles, and prostate.

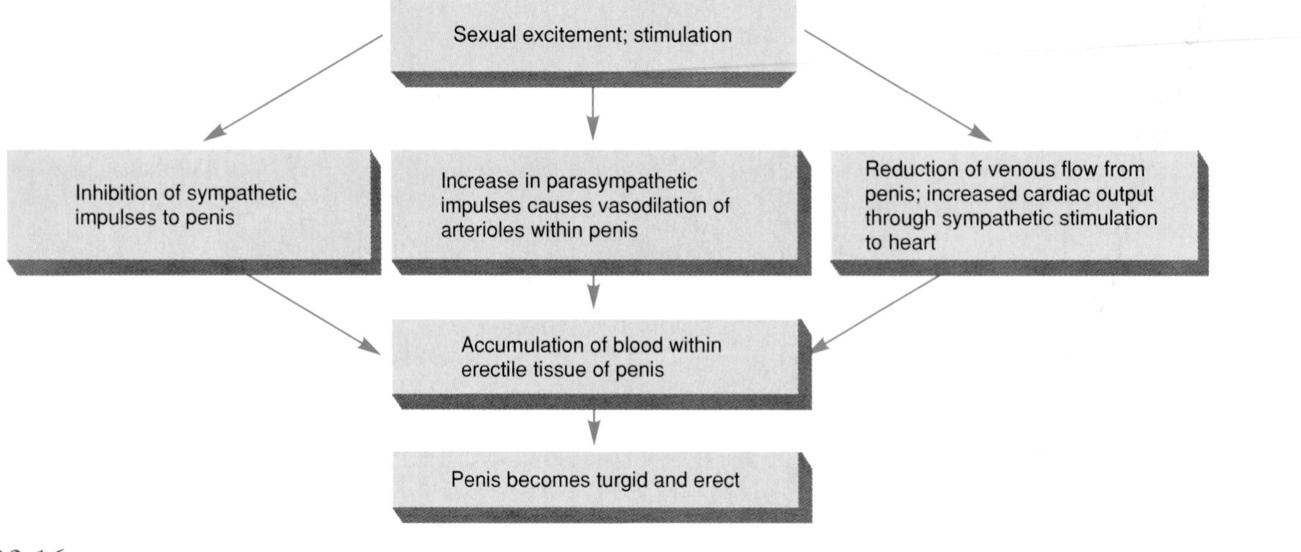

Figure 20.16

The mechanism of erection of the penis.

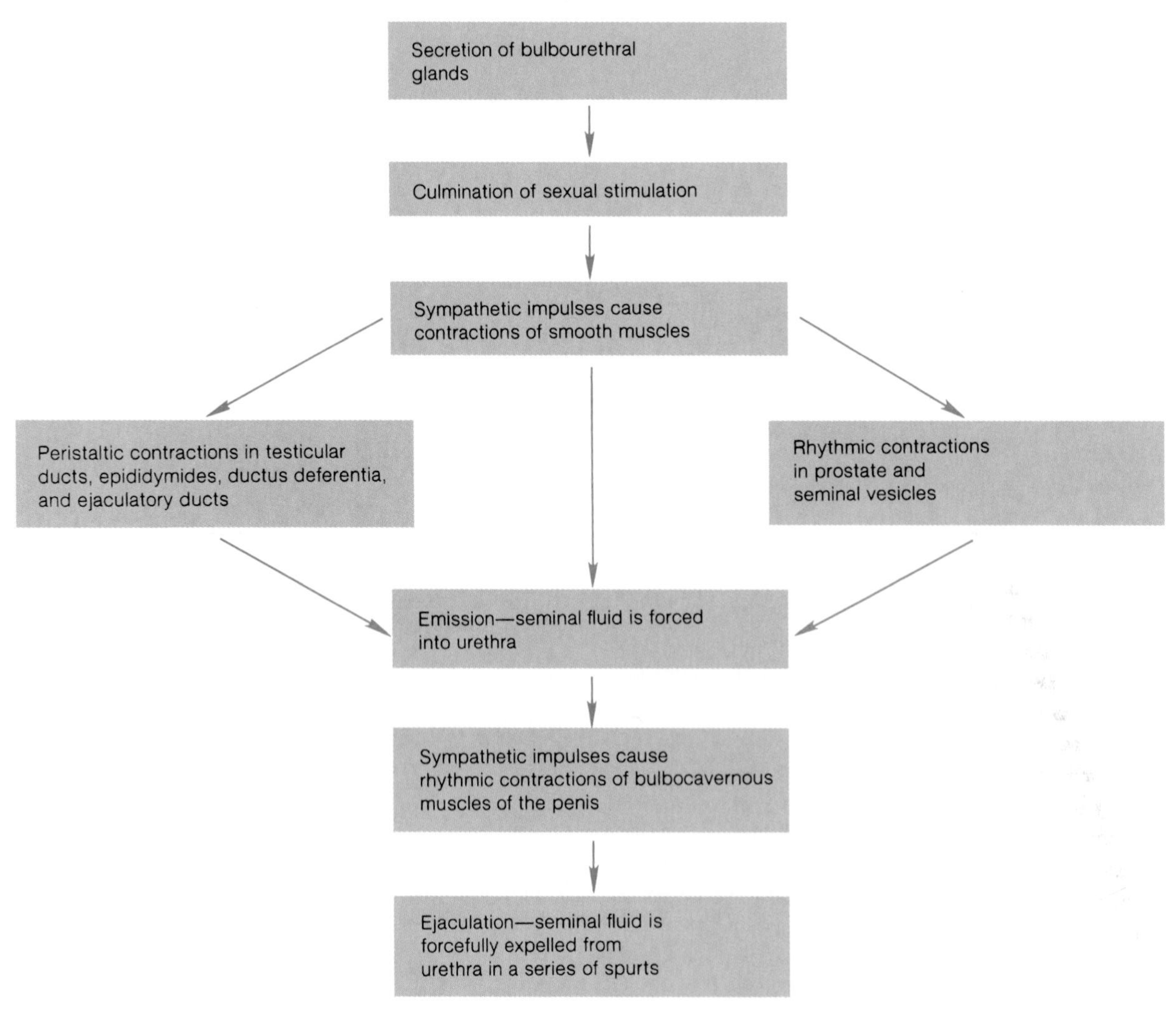

Figure 20.17

The mechanism of emission and ejaculation.

Table 20.4

The Clinical Examination of Semen

Characteristic	Reference Value
Volume of ejaculate	1.5–5.0 ml
Sperm count	40–250 million/ml
Sperm motility	
Percentage of motile forms:	
1 hour after ejaculation	70% or more
3 hours after ejaculation	60% or more
Leukocyte count	0–2,000/ml
pH	7.2–7.8
Fructose concentration	150–600 mg/100 ml

Modified from Glasser, L., "Seminal Fluid and Subfertility," *Diagnostic Medicine*, July/August 1981, p. 28. By permission.

Ejaculation

Ejaculation immediately follows emission and is accompanied by *orgasm*—the climax of the sex act. Ejaculation occurs in a series of spurts of semen from the urethra. This takes place as parasympathetic impulses traveling through the pudendal nerves stimulate the bulbospongiosus muscles (see fig. 20.2) at the base of the penis and cause them to contract rhythmically. In addition, sympathetic stimulation of the smooth muscles in the urethral wall causes them to contract, helping to eject the semen.

Sexual function in the male thus requires the synergistic action (rather than antagonistic action) of the parasympathetic and sympathetic divisions of the ANS. The mechanism of emission and ejaculation is summarized in figure 20.17.

Immediately following ejaculation or a cessation of sexual stimulus, sympathetic impulses cause vasoconstriction of the arterioles within the penis, reducing the inflow of blood. At the same time, cardiac output returns to normal, as does venous return of blood from the penis. With the normal flow of blood through the penis, it returns to its flaccid condition.

Adolescent males may experience erection of the penis and spontaneous emission and ejaculation of semen during sleep. These *nocturnal emissions,* sometimes called "wet dreams," are triggered by psychic stimuli associated with dreaming. They are thought to be caused by changes in hormonal concentrations that accompany adolescent development.

Semen

Semen, also called *seminal fluid,* is the substance discharged during ejaculation (table 20.4). Generally, between 1.5 and 5.0 ml of semen are ejected during ejaculation. The bulk of the fluid (about 60%) is produced by the seminal vesicles, and the rest (about 40%) is contributed by the prostate. Spermatozoa account for less than 1% of the volume. There are usually between 60 and 150 million sperm cells per milliliter of ejaculate. In the condition of *oligospermia* (*ol″ĭ-go-sper′me-ă*), the male ejaculates fewer than 10 million sperm cells per milliliter and is likely to have fertility problems.

Human semen can be frozen and stored in sperm banks for future artificial insemination. In this procedure, the semen is diluted with 10% glycerol, monosaccharide, and distilled water buffer, and frozen in liquid nitrogen. The freezing process destroys defective and abnormal sperm. For some unknown reason, however, not all human sperm is suitable for freezing.

1. Define the terms *erection*, *emission*, and *ejaculation*.
2. Explain the statement that male sexual function is an autonomic synergistic action.
3. Use a flowchart to explain the physiological and physical events of erection, emission, and ejaculation.
4. Describe the components of a normal ejaculate.

Clinical Considerations

Sexual dysfunction is a broad area of medical concern that includes developmental and psychogenic problems as well as conditions resulting from various diseases. Psychogenic problems of the reproductive system are extremely complex and beyond the scope of this book. Only a few of the principal developmental conditions, functional disorders, and diseases that affect the physical structure and function of the male reproductive system will be discussed here.

Developmental Problems of the Male Reproductive System

The reproductive organs of both sexes develop from similar embryonic tissue that follows a consistent pattern of formation well into the fetal period. Because an embryo has the potential to differentiate into a male or a female, developmental errors can result in various degrees of intermediate sex, or **hermaphroditism** (*her-maf′rŏ-dĭ-tiz″em*). A person with undifferentiated or ambiguous external genitalia is called a **hermaphrodite.**

True hermaphroditism—in which both male and female gonadal tissues are present in the body—is a rare anomaly. True hermaphrodites usually have a 46 XX chromosome constitution. **Male pseudohermaphroditism** occurs more frequently and generally results from hormonal influences during early fetal development. This condition is caused either by inadequate secretion of androgenic hormones or by the delayed development of the reproductive

hermaphrodite: Gk. (mythology) *Hermaphroditos*, son of Hermes (Mercury)

DEVELOPMENTAL EXPOSITION

The Reproductive System

Explanation

Sex Determination

Sexual identity is initiated at the moment of conception, when the *genetic sex* of the zygote (fertilized egg) is determined. The ovum is fertilized by a spermatozoon containing either an X or a Y sex chromosome. If the spermatozoon contains an X chromosome, it will pair with the X chromosome of the ovum and a female child will develop. A spermatozoon carrying a Y chromosome results in an XY combination, and a male child will develop.

Genetic sex determines whether the gonads will be testes or ovaries. If testes develop, they will produce and secrete male sex hormones during late embryonic and early fetal development and cause the secondary sex organs of the male to develop.

Embryonic Development

The male and female reproductive systems follow a similar pattern of development, with sexual distinction resulting from the influence of hormones. A significant fact of embryonic development is that the sexual organs for both male and female are derived from the same developmental tissues and are considered *homologous structures.*

The first sign of development of either the male or the female reproductive organs occurs during the fifth week as the medial aspect of each mesonephros (see chapter 19) enlarges to form the **gonadal ridge** (exhibit I). The gonadal ridge continues to grow behind the developing peritoneal membrane. By the sixth week, stringlike masses called **primary sex cords** form within the enlarging gonadal ridge. The primary sex cords in the male will eventually mature to become the seminiferous tubules. In the female, the primary sex cords will contribute to nurturing tissue of developing ova. Each gonad develops near a **mesonephric duct** and a **paramesonephric duct.** In the male embryo, each testis connects through a series of tubules to the mesonephric duct. During further development, the connecting tubules become the *seminiferous tubules,* and the mesonephric duct becomes the *efferent ductules, epididymis, ductus deferens, ejaculatory duct,* and *seminal vesicle.* The paramesonephric duct in the male degenerates without contributing any functional structures to the reproductive system. In the female embryo, the mesonephric duct degenerates, and the paramesonephric duct contributes greatly to structures of the female reproductive system. The distal ends of the paired paramesonephric ducts fuse to form the *vagina* and *uterus.* The proximal unfused portions become the *uterine tubes.*

Externally, by the sixth week a swelling called the **genital tubercle** is apparent anterior to the small embryonic tail (future coccyx). The mesonephric and paramesonephric ducts open to the outside through the genital tubercle. The genital tubercle consists of a *glans,* a *urethral groove,* paired *urethral folds,* and paired *labioscrotal swellings* (see exhibit II). As the glans portion of the genital tubercle enlarges, it becomes known as the *phallus.* Early in fetal development (tenth through twelfth week), sexual distinction of the external genitalia becomes apparent. In the male, the phallus enlarges and develops into the *glans* of the penis. The urethral folds fuse around the urethra to form the *body* of the penis. The urethra opens at the end of the glans as the *urethral orifice.* The labioscrotal swellings fuse to form the *scrotum,* into which the testes will descend. In the female, the phallus gives rise to the *clitoris,* the urethral folds remain separated as the *labia minora,* and the labioscrotal swellings become the *labia majora.* The urethral groove is retained as a longitudinal cleft known as the *vestibule.*

Descent of the Testes

The descent of the testes from the site of development begins between the sixth and tenth week. Descent into the scrotal sac, however, does not occur until about week 28, when paired inguinal canals form in the abdominal wall to provide openings from the pelvic cavity to the scrotal sac. The process by which a testis descends is not well understood, but it seems to be associated with the shortening and differential growth of the *gubernaculum,* which is attached to the testis and extends through the inguinal canal to the wall of the scrotum (exhibit III). As the testis descends, it passes to the side of the urinary bladder and anterior to the symphysis pubis. It carries with it the ductus deferens, the testicular vessels and nerve, a portion of the internal abdominal oblique muscle, and lymph vessels. All of these structures remain attached to the testis and form what is known as the **spermatic cord** (see fig. 20.14). By the time the testis has taken its position in the scrotal sac, the gubernaculum is no more than a remnant of scarlike tissue.

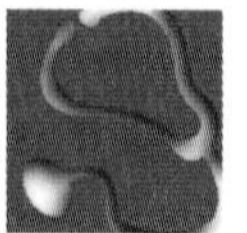

During the physical examination of a neonatal male, a physician will palpate the scrotum to determine if the testes are in position. If one or both are not in the scrotal sac, it may be possible to induce descent by administering certain hormones. If this procedure does not work, surgery is necessary. The surgery is generally performed before the age of 5. Failure to correct the situation may result in sterility and possibly the development of a tumorous testis.

homologous: Gk. *homos,* the same

gubernaculum: L. *gubernaculum,* helm

Exhibit I Differentiation of the male and female gonads and genital ducts. (*a*) An embryo at 6 weeks showing the positions of a transverse cut depicted in (a_1), (*b*), and (*c*). (a_1) At 6 weeks, the developing gonads (primary sex cords) are still indifferent. By 4 months, the gonads have differentiated into male (*b*) or female (*c*). The oogonia have formed within the ovaries (c_1) by 6 months.

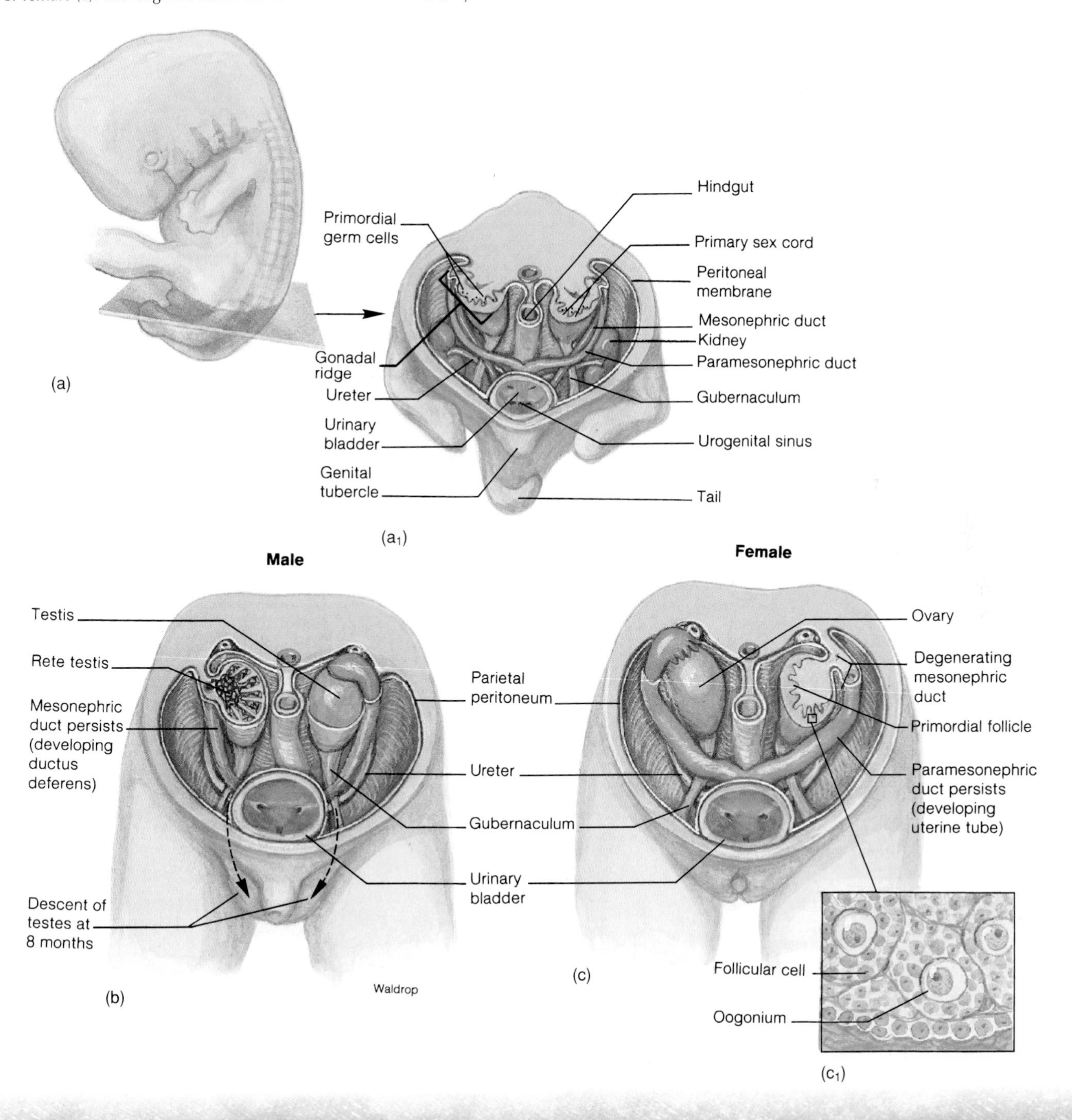

Continued

DEVELOPMENTAL EXPOSITION

Continued

Exhibit II (*a*) Differentiation of the external genitalia in the male and female. (a_1) A sagittal view. At 6 weeks, the urethral fold and labioscrotal swelling are differentiated from the genital tubercle. (*b*) At 8 weeks, a distinct phallus is present during the indifferent stage. By the twelfth week, the genitalia have become distinctly male (*c*) or female (*d*), being derived from homologous structures. (*e,f*) At 16 weeks, the genitalia are formed. (*g,h*) Photographs at week 10 of male and female genitalia, respectively.

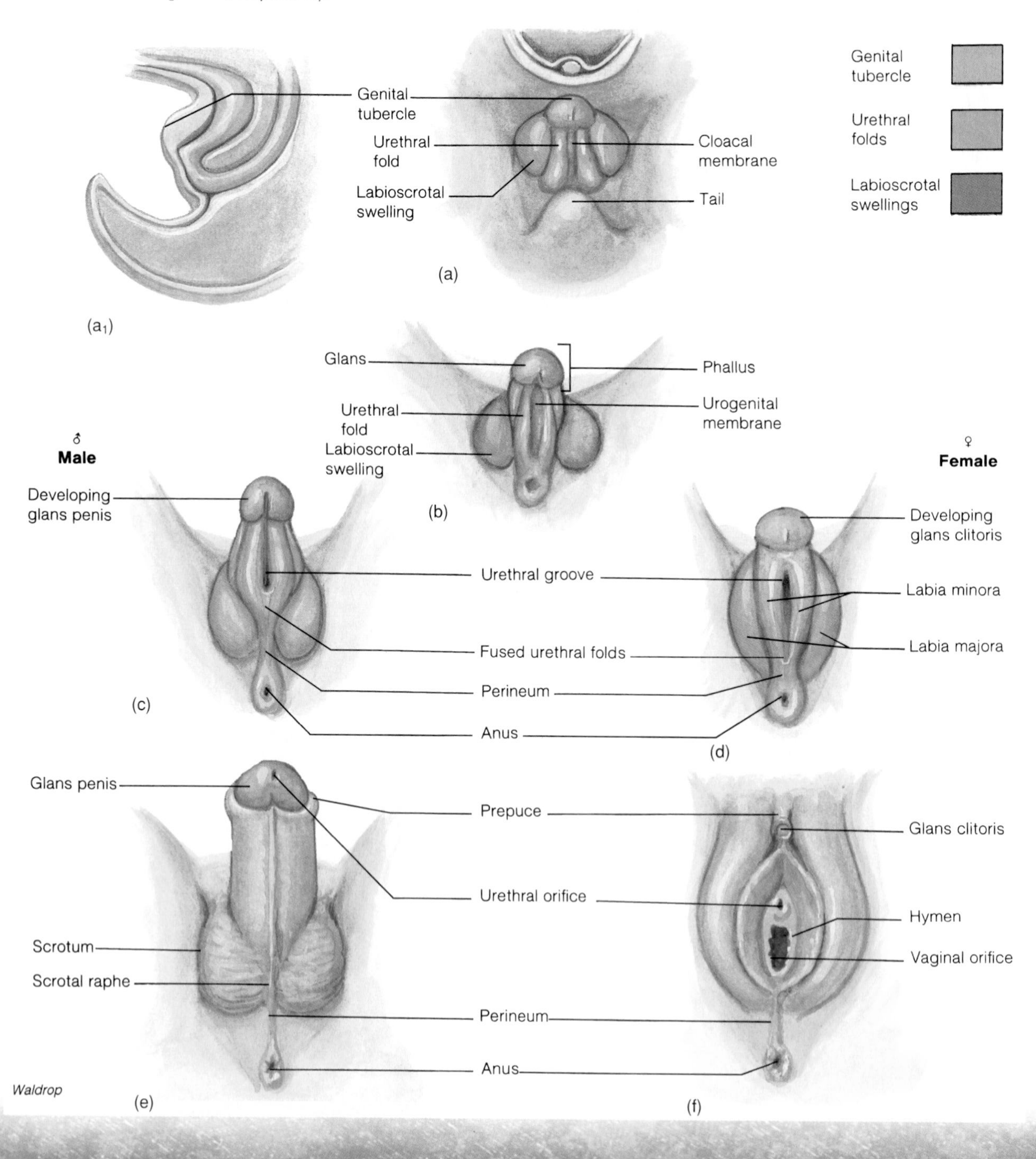

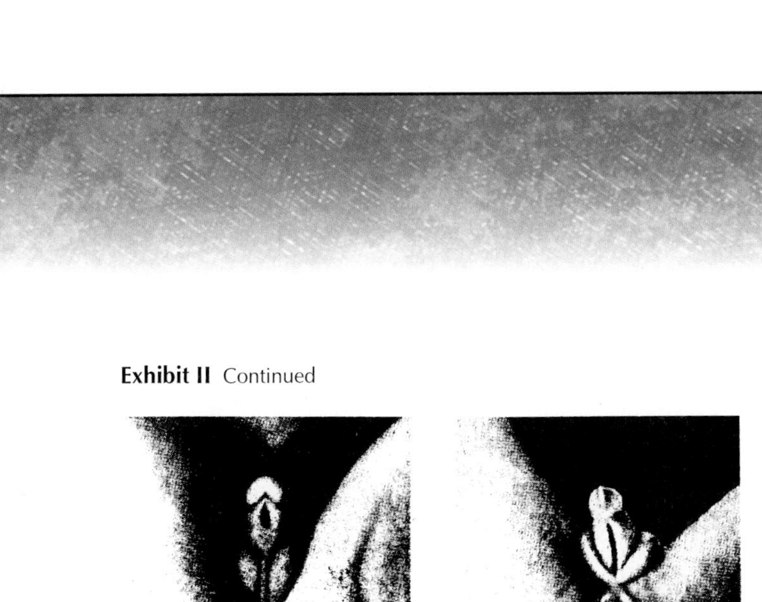

Exhibit II Continued

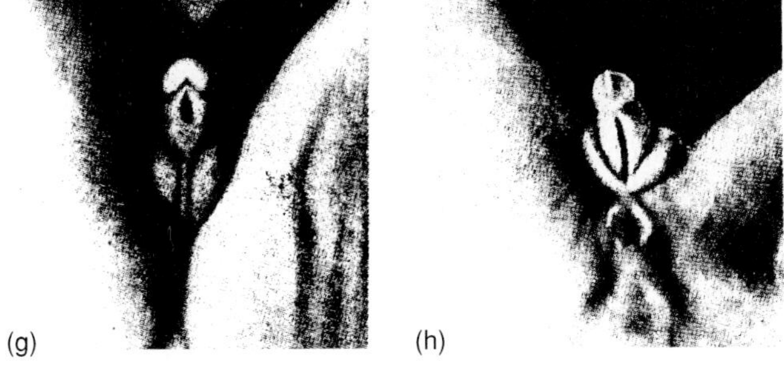

(g) (h)

Exhibit III The descent of the testes. (*a*) At 10 weeks, (*b*) at 18 weeks, and (*c*) at 28 weeks. During development, each testis descends through an inguinal canal in front of the symphysis pubis and enters the scrotum.

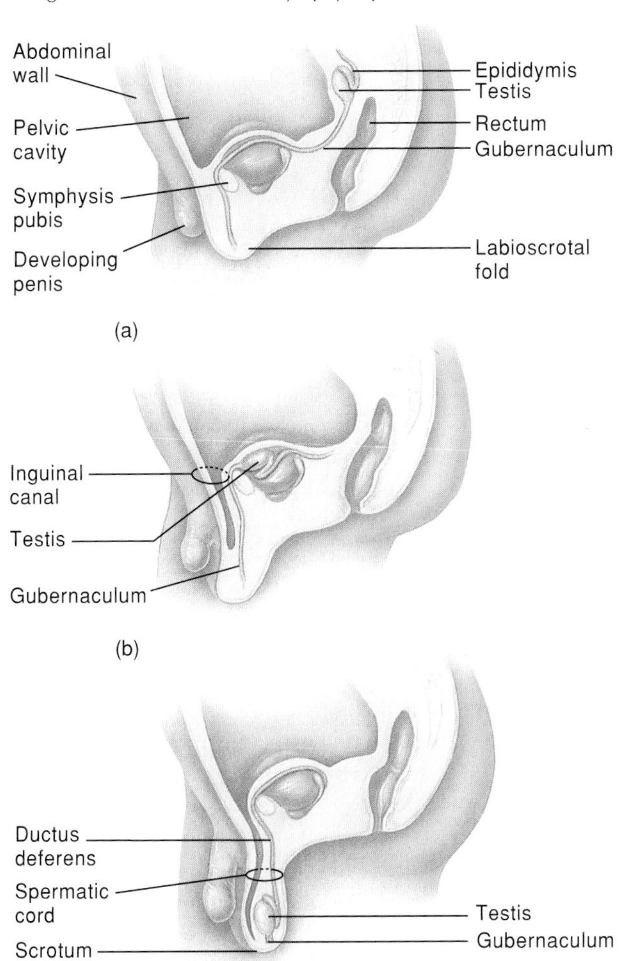

organs after the period of tissue sensitivity has passed. These individuals have a 46 XY chromosome constitution and male gonads, but the genitalia are intersexual and variable.

The treatment of hermaphroditism varies, depending on the extent of ambiguity of the reproductive organs. Although people with this condition are sterile, they may engage in normal sexual relations following hormonal therapy and plastic surgery.

Chromosomal anomalies result from the improper separation of the chromosomes during meiosis and are usually expressed in deviations of the reproductive organs. The two most frequent chromosomal anomalies cause Turner's syndrome and Klinefelter's syndrome. **Turner's syndrome** occurs when only one X chromosome is present. About 97% of embryos lacking an X chromosome die; the remaining 3% survive and appear to be females, but their gonads, if present, are rudimentary and do not mature at puberty. A person with **Klinefelter's syndrome** has an XXY chromosome constitution, develops breasts and male genitalia, but has underdeveloped seminiferous tubules and is generally mentally retarded.

A more common developmental problem than genetic abnormalities, and fortunately less serious, is cryptorchidism. **Cryptorchidism** (*krip-tor'kĭ-diz"em*) means "hidden testis" and is characterized by the failure of one or both testes to descend into the scrotum. A cryptorchid testis is usually located along the path of descent but can be anywhere in the pelvic cavity (fig. 20.18). It occurs in about 3% of male infants and should be treated before the infant has reached the age of 5 to reduce the likelihood of infertility or other complications.

Sexual Dysfunction

Functional disorders of the male reproductive system include impotence, infertility, and sterility. **Impotence** (*im'pŏ-tens*) is the inability of a sexually mature male to achieve and maintain penile erection and/or the inability to achieve ejaculation. The causes of impotence may be physical, involving, for example, abnormalities of the penis, vascular irregularities, neurological disorders, or certain diseases. Generally, however, the cause of impotence is psychological, and the patient requires skilled counseling by a sex therapist.

Infertility is the inability of the sperm to fertilize the ovum and may involve the male or female, or both. The most common cause of male infertility is inadequate production of viable sperm. This may be due to alcoholism, dietary deficiencies, local injury, variococele, excessive heat, hormonal imbalance, or excessive exposure to radiation. Many of the causes of infertility can be treated through proper nutrition, gonadotropic hormone treatment, or microsurgery. If these treatments are not successful, it may be possible to concentrate the spermatozoa obtained through

---

Turner's syndrome: from Henry H. Turner, American endocrinologist, 1892–1970

Klinefelter's syndrome: from Harry F. Klinefelter Jr., American physician, b. 1912

cryptorchidism: Gk. *crypto*, hidden; *orchis*, testis

impotence: L. *im*, not; *potens*, potent

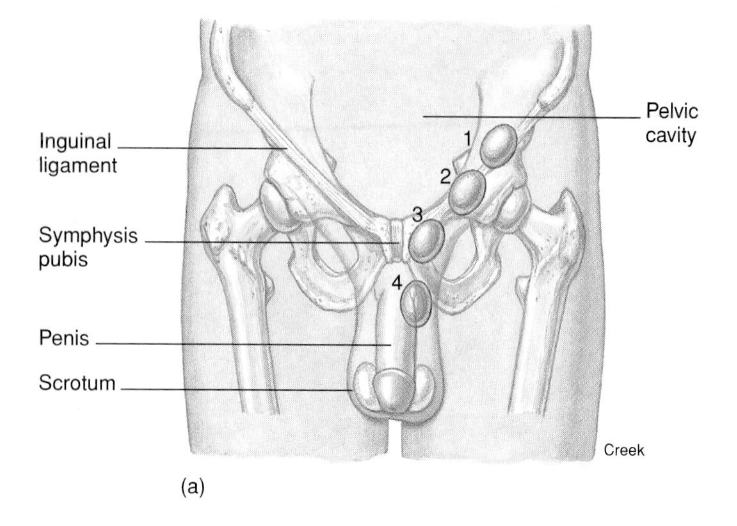

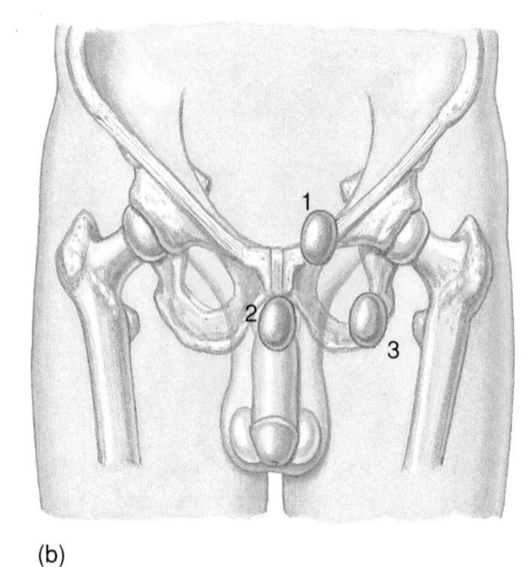

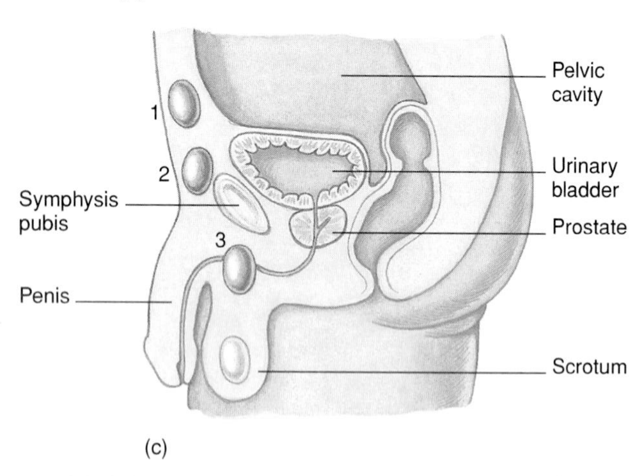

Figure 20.18

Cryptorchidism. (*a*) Incomplete descent of a testis may involve a region (1) in the pelvic cavity, (2) in the inguinal canal, (3) at the superficial inguinal ring, or (4) in the upper scrotum. (*b*) An ectopic testis may be (1) in the superficial fascia of the anterior pelvic wall, (2) at the root of the penis, or (3) in the perineum, in the thigh alongside the femoral vessels.

masturbation (in males, self-stimulation to the point of ejaculation) and use this concentrate to artificially inseminate the female.

Sterility is similar to infertility, except that it is a permanent condition. Sterility may be genetically caused, or it may be the result of degenerative changes in the seminiferous tubules (for example, mumps in a mature male may secondarily infect the testes and cause irreversible tissue damage).

Voluntary sterilization of the male in a procedure called a **vasectomy** (*vă-sek'to-me*) is a common technique of birth control. In this procedure, a small section of each ductus deferens near the epididymis is surgically removed, and the cut ends of the ducts are tied (fig. 20.19). A vasectomy prevents transport of spermatozoa but does not directly affect the secretion of androgens, the sex drive, or ejaculation. Since sperm cells make up less than 1% of an ejaculate, even the volume is not noticeably affected.

According to the National Center for Health Statistics, there are currently 5.3 million childless infertile couples in the United States. In 40% of the cases, it is the woman's infertility that prevents conception, in 40% it is the man's. In the remaining 20%, either both partners have some abnormality, or the cause of the problem is unknown. Of those couples who seek help for infertility, only 21% are eventually successful in producing a child.

Diseases of the Reproductive System

Sexually Transmitted Diseases

Sexually transmitted diseases (STDs) are contagious diseases that affect the reproductive systems of both the male and the female (table 20.5). They are transmitted during sexual activity, and their frequency of occurrence in the United States is regarded by health authorities as epidemic. Commonly called *venereal diseases*, STDs have not been eradicated mainly because humans cannot develop immunity to them and because increased sexual activity increases the likelihood of infection and reinfection. Furthermore, many of the causative organisms can mutate so fast that available drug treatments are no longer effective.

Gonorrhea (*gon"ŏ-re'ă*), commonly called clap, is caused by the bacterium gonococcus, or *Neisseria gonorrhoeae*. Males with this disease suffer inflammation of the urethra, accompanied by painful urination and frequently the discharge of pus. In females, the condition is usually asymptomatic, and therefore many women may be unsuspecting carriers of the disease. Advanced stages of gonorrhea in females may infect the uterus and the uterine tubes. A pregnant woman with gonorrhea who is not treated may transmit the bacteria to the eyes of her newborn during its passage through the birth canal, possibly causing blindness.

Syphilis (*sif'ĭ-lis*) is caused by the bacterium *Treponema pallidum*. Syphilis is less common than gonorrhea but is the more serious of the two diseases. During the *primary stage* of

sterility: L. *sterilis*, barren
vasectomy: L. *vas*, vessel; Gk. *ektome*, excision
venereal: L. (mythology) from *Venus*, the goddess of love
gonorrhea: L. *gonos*, seed; *rhoia*, a flow

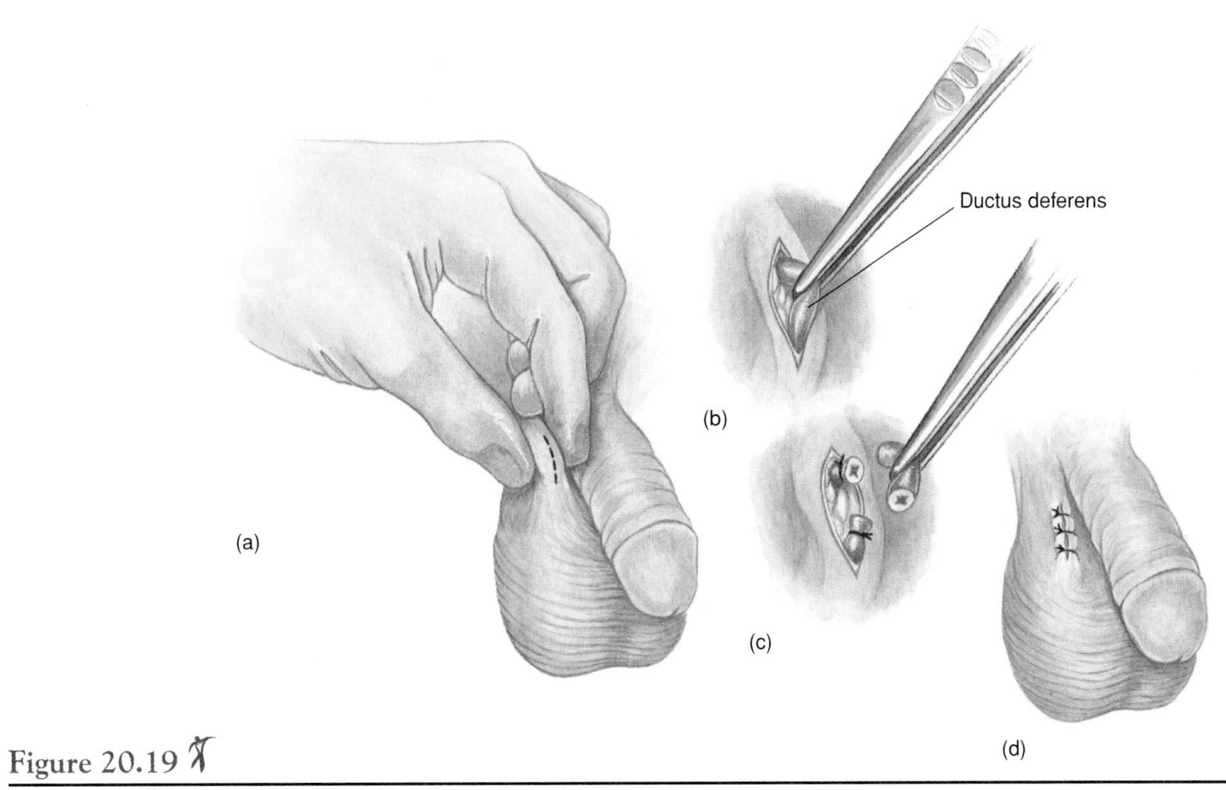

Figure 20.19

A simplified illustration of a vasectomy, in which a segment of the ductus deferens is tied and surgically removed through an incision in the scrotum. The procedure is then repeated on the opposite side.

Table 20.5
Sexually Transmitted Diseases

Name	Organism	Resulting Condition	Treatment
Gonorrhea	*Gonococcus* (bacterium)	Adult: sterility due to scarring of epididymides and testicular ductules (in rare cases, blood poisoning); newborn: blindness	Penicillin injections; tetracycline tablets; eye drops (silver nitrate or penicillin) in newborns as preventative
Syphilis	*Treponema pallidum* (bacterium)	Adult: gummy tumors, cardiovascular neurosyphilis; newborn: congenital syphilis (abnormalities, blindness)	Penicillin injections; tetracycline tablets
Chancroid (soft chancre)	*Hemophilus ducreyi* (bacterium)	Chancres, buboes	Tetracycline; sulfa drugs
Urethritis in males	Various microorganisms	Clear discharge	Tetracycline
Vaginitis in females	*Trichomonas* (protozoan)	Frothy white or yellow discharge	Metronidazole
	Candida albicans (yeast)	Thick, white, curdy discharge (moniliasis)	Nystatin
Acquired immune deficiency syndrome (AIDS)	Human immunodeficiency virus (HIV)	Early symptoms include extreme fatigue, weight loss, fever, diarrhea; extreme susceptibility to pneumonia, rare infections, and cancer	Azidothymidine (AZT), dideoxyinosine (ddll), and dideoxyctidine (ddC); new drugs being developed, including protease inhibitors
Chlamydia	*Chlamydia trachomatis* (bacterium)	Whitish discharge from penis or vagina; pain during urination	Tetracycline and sulfonamides
Venereal warts	Virus	Warts	Podophyllin; cautery, cryosurgery, or laser treatment
Genital herpes	Herpes simplex virus	Sores	Palliative treatment
Crabs	Arthropod	Itching	Gamma benzene hexachloride

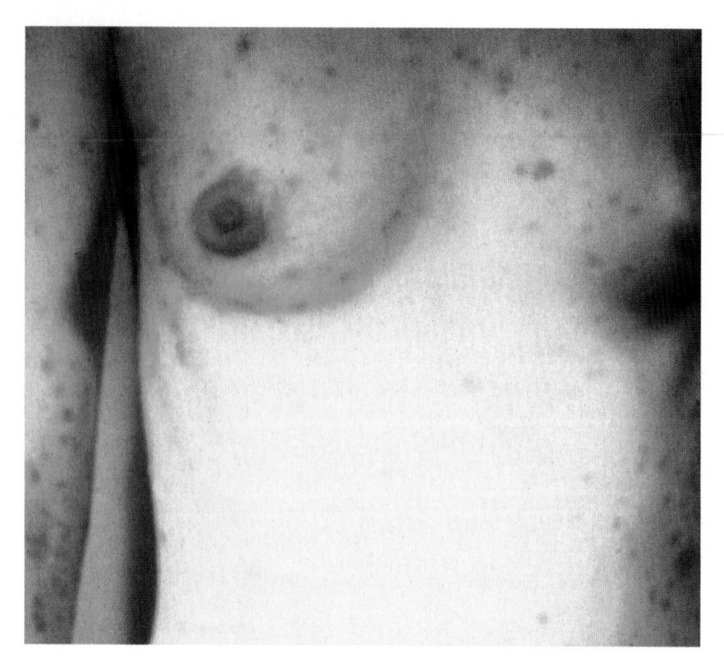

Figure 20.20

The secondary stages of syphilis as expressed by lesions of the skin.

Clinical Case Study Answer

The tubular structures that are apparently absent in our patient are the ductus deferentia. This condition, known as *congenital bilateral absence of the ductus deferentia*, prevents the transport of spermatozoa from the testes to the ejaculatory ducts. This explains the absence of spermatozoa in the patient's ejaculate. His ejaculate consists only of the secretions from the seminal vesicles (which in many cases are also absent or nonfunctional in this deformity) and the prostate. Until recently, this patient's condition would have categorically prevented him from becoming a father. Microsurgical extraction of spermatozoa from the epididymides is now possible, however, and has allowed many afflicted men to father children.

syphilis, a lesion called a *chancre* develops at the point where contact was made with a similar sore on an infected person. The chancre persists for 10 days to 3 months before the disease enters the *secondary stage*, which is characterized by lesions or a skin rash (fig. 20.20), accompanied by fever. This stage lasts from 2 weeks to 6 months, and the symptoms disappear of their own accord. The *tertiary stage* of untreated syphilis may occur 10 to 20 years following the primary infection. This stage is characterized by degenerative changes in various systems of the body that may lead to blindness, insanity, and death.

Acquired immune deficiency syndrome (AIDS) is a viral disease that is transmitted primarily through intimate sexual contact and through drug abuse (by sharing contaminated syringe needles). Additional information about this fatal disease, for which there is no known cure, is presented in table 20.5.

Disorders of the Prostate

The prostate is subject to several disorders, many of which are common in older men. The four most frequent prostatic problems are *acute prostatitis*, *chronic prostatitis*, *benign prostatic hypertrophy*, and *carcinoma of the prostate*.

Acute prostatitis is common in sexually active young men through infections acquired from a gonococcus bacterium. The symptoms of acute prostatitis are a swollen and tender prostate, painful urination, and in extreme conditions, pus dripping from the penis. It is treated with penicillin, bed rest, and increased fluid intake.

Chronic prostatitis is one of the most common afflictions of middle-aged and elderly men. The symptoms of this condition range from irritation and slight difficulty in urinating to extreme pain and urine blockage, which commonly causes secondary renal infections. In this disease, several kinds of infectious microorganisms are believed to be harbored in the prostate and are responsible for inflammations elsewhere in the body, such as in the nerves (neuritis), the joints (arthritis), the muscles (myositis), and the iris (iritis).

Benign prostatic hypertrophy (BPH) is an enlarged prostate from unknown causes. This condition occurs in nearly one-third of all males over the age of 60, and is characterized by painful and difficult urination. If the urinary bladder is not emptied completely, cystitis eventually occurs. People with cystitis may become incontinent and dribble urine continuously. Benign prostatic hypertrophy may be treated by surgical removal of portions of the gland through transurethral curetting (cutting and removal of a small section) or removal of the entire prostate, called *prostatectomy* (*pros″tă-tek′tŏ-me*).

Prostatic carcinoma, cancer of the prostate, is the second leading cause of death from cancer in males in the United States. It, too, is common in males over 60 and accounts for about 25,000 deaths annually. The metastases of this cancer to the spinal column and brain are generally what kills the patient. Advanced prostatic carcinoma is treated by removal of the prostate and frequently by removal of the testes, or orchiectomy (*or″ke-ek′tŏ-me*). An orchiectomy inhibits metastases by eliminating testosterone secretion.

Disorders of the Testes and Scrotum

An infection in the testes is called **orchitis** (*or-ki′tis*). Orchitis may develop as the result of a primary bacterial infection or as a secondary complication of mumps contracted after puberty. If orchitis from mumps involves both testes, it usually causes sterility.

A **hydrocele** (*hi′drŏ-sēl*) is a benign fluid mass within the tunica vaginalis surrounding the testis that causes swelling of the scrotum. It is a frequent minor disorder in male infants, as well as in adults. The cause is unknown.

chancre: Fr. *chancre*, indirectly from L. *cancer*, a crab

Internal Affairs

■ How the *Reproductive System* Works with Other Body Systems

□ How Other Systems Work with the *Reproductive System*

Integumentary System

- Sex hormones affect distribution of body hair and deposition of subcutaneous fat

- Covers and protects reproductive organs
- Provides sites for sexual sensory receptors
- Mammary glands represent specialized integumentary structures

Skeletal System

- Sex hormones stimulate bone growth and maintenance

- Pelvis provides protection and support for reproductive organs

Muscular System

- Sex hormones stimulate muscle growth and maintenance

- Certain pelvic muscles aid coitus; involuntary action of smooth muscles aids movement of gametes; abdominal muscles aid childbirth; dartos and cremaster muscles help to maintain proper temperature of testes

Nervous System

- Sex hormones influence brain development and sexual behavior

- Hypothalamus initiates gonadotropin function and lactation
- Sensory and autonomic nervous systems are involved in sexual arousal and orgasm

Endocrine System

- Certain hormones (gonadotropins and GnRH) regulate the function of gonads
- Gonads and placenta constitute part of the endocrine system

- Sex hormones aid development and maintenance of sex organs; influence sex drives and functions

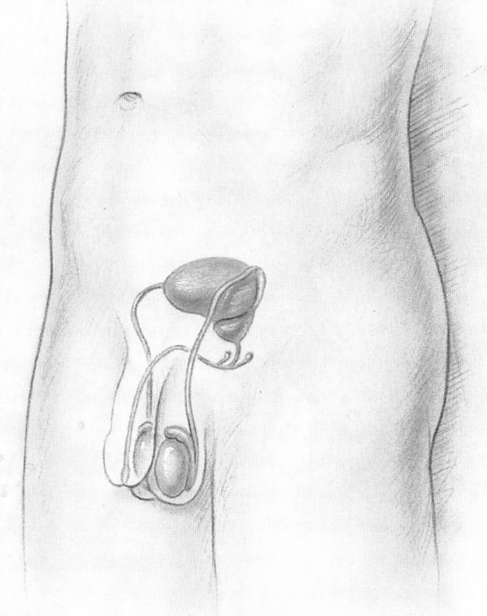

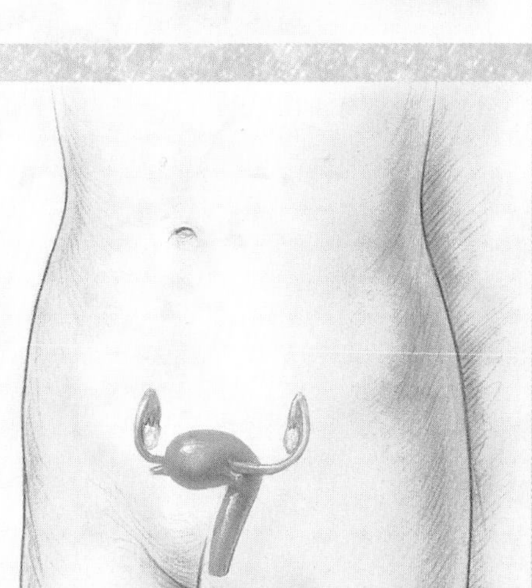

Circulatory System

- Estrogens maintain healthy blood vessels (lower cholesterol levels)
- Androgens stimulate erythropoiesis
- Pregnancy stimulates increase in blood volume and cardiac output; may contribute to varicose veins

- Transports O_2, sex hormones, and nutrients; eliminates CO_2 and metabolic wastes
- Vasodilation is necessary for erection

Lymphatic System

- Blood-testis barrier isolates and protects sperm cells from the immune system

- Maintains a balanced amount of interstitial fluid within reproductive tissues

Respiratory System

- Sexual arousal increases pulmonary ventilation
- Pregnancy interferes with the descent of the diaphragm during inspiration

- Provides O_2 to reproductive organs and eliminates CO_2
- Vital capacity and respiratory rate increase during pregnancy

Digestive System

- Developing fetus crowds digestive organs, which may cause heartburn and constipation

- Provides nutrients for organ function
- Provides nutrients for embryonic and fetal development in pregnant woman

Urinary System

- Pregnancy crowds the urinary bladder, causing more frequent micturition
- Prostatic hyperplasia may impede urine flow

- Eliminates metabolic wastes
- Male urethra transports semen during ejaculation

Chapter Summary

Introduction to the Male Reproductive System (pp. 679–681)

1. The common purpose of the male and female reproductive systems is to produce offspring.
2. The functions of the male reproductive system are to produce spermatozoa, secrete androgens, and transfer spermatozoa to the reproductive system of the female.
3. Features of the male reproductive system include the primary sex organs (the testes), secondary sex organs (those that are essential for reproduction), and secondary sex characteristics (sexual attractants, expressed after puberty).

Perineum and Scrotum (pp. 681–683)

1. The saclike scrotum, located in the urogenital portion of the perineum, supports and protects the testes and regulates their position relative to the pelvic region of the body.
2. Each testis is contained within its own scrotal compartment and is separated from the other by the scrotal septum.

Testes (pp. 683–688)

1. The testes are partitioned into wedge-shaped lobules; the lobules are composed of seminiferous tubules, which produce sperm cells, and of interstitial tissue, which produces androgens.
2. Spermatogenesis occurs by meiotic division of the cells that line the seminiferous tubules.
   (a) At the end of the first meiotic division, two secondary spermatocytes have been produced.
   (b) At the end of the second meiotic division, four haploid spermatids have been produced.
3. The conversion of spermatids to spermatozoa is called spermiogenesis.
4. A sperm consists of a head and a flagellum and matures in the epididymides prior to ejaculation.
   (a) The acrosome of the head contains digestive enzymes for penetrating an ovum.
   (b) The flagellum provides undulating movement of about 3 mm per hour.

Spermatic Ducts, Accessory Reproductive Glands, and the Urethra (pp. 688–691)

1. The epididymides and the ductus deferentia are the components of the spermatic ducts.
   (a) The highly coiled epididymides are the tubular structures on the testes where spermatozoa mature and are stored.
   (b) The ductus deferentia convey spermatozoa from the epididymides to the ejaculatory ducts during emission. Each ductus deferens forms a component of the spermatic cord.
2. The seminal vesicles and prostate provide additives to the spermatozoa in the formation of semen.
   (a) The seminal vesicles are located posterior to the base of the urinary bladder; they secrete about 60% of the additive fluid of semen.
   (b) The prostate surrounds the urethra just below the urinary bladder; it secretes about 40% of the additive fluid of semen.
   (c) Spermatozoa constitute less than 1% of the volume of an ejaculate.
   (d) The small bulbourethral glands secrete fluid that serves as a lubricant for the erect penis in preparation for coitus.
3. The male urethra, which serves both the urinary and reproductive systems, is divided into prostatic, membranous, and spongy portions.

Penis (pp. 691–693)

1. The penis is specialized to become erect for insertion into the vagina during coitus.
2. The body of the penis consists of three columns of erectile tissue, the spongy urethra, and associated vessels and nerves.
3. The root of the penis is attached to the pubic arch and urogenital diaphragm.
4. The glans penis is the terminal end, which is covered with the prepuce in an uncircumcised male.

Mechanisms of Erection, Emission, and Ejaculation (pp. 693–695)

1. Erection of the penis occurs as the erectile tissue becomes engorged with blood. Emission is the movement of the spermatozoa from the epididymides to the ejaculatory ducts, and ejaculation is the forceful expulsion of semen from the ejaculatory ducts and urethra of the penis.
2. Parasympathetic stimuli to arteries in the penis cause the erectile tissue to engorge with blood as arteriole flow increases and venous drainage decreases.
3. Ejaculation is the result of sympathetic reflexes in the smooth muscles of the male reproductive organs.

Review Activities

Objective Questions

1. The perineum is
   (a) a membranous covering over the testis.
   (b) a ligamentous attachment of the penis to the symphysis pubis.
   (c) a diamond-shaped region between the symphysis pubis and the coccyx.
   (d) a fibrous sheath supporting the scrotum.
2. The dartos is a layer of smooth muscle fibers found within
   (a) the scrotum. (c) the epididymis.
   (b) the penis. (d) the prostate.
3. Which statement regarding the interstitial cells (cells of Leydig) is *true?*
   (a) They nourish spermatids.
   (b) They produce testosterone.
   (c) They produce spermatozoa.
   (d) They secrete alkaline fluid.
   (e) Both b and d are true.
4. The bulb and crus of the penis are located at
   (a) the glans penis.
   (b) the corona glandis.
   (c) the body of the penis.
   (d) the prepuce.
   (e) the root of the penis.
5. Which of the following is *not* a spermatic duct?
   (a) the epididymis
   (b) the spermatic cord
   (c) the ejaculatory duct
   (d) the ductus deferens
6. Spermatozoa are stored prior to emission and ejaculation in
   (a) the epididymides.
   (b) the seminal vesicles.
   (c) the spongy urethra.
   (d) the prostate.

7. Urethral glands function to
   (a) secrete mucus.
   (b) produce nutrients.
   (c) secrete hormones.
   (d) regulate spermatozoa production.
8. Which statement is *false* regarding erection of the penis?
   (a) It is a parasympathetic response.
   (b) It may be both a voluntary and involuntary response.
   (c) It has to be followed by emission and ejaculation.
   (d) It is controlled by the hypothalamus of the brain and sacral portion of the spinal cord.
9. An embryo with the genotype XY develops male accessory sex organs because of
   (a) androgens.
   (b) estrogens.
   (c) the absence of androgens.
   (d) the absence of estrogens.
10. Which of the following does *not* arise from the embryonic mesonephric duct?
    (a) the epididymis
    (b) the ductus deferens
    (c) the seminal vesicle
    (d) the prostate

Essay Questions

1. What are the functions of the male reproductive system? How do these functions compare with those of the female reproductive system?
2. Discuss how and when the genetic sex of a zygote is determined.
3. Describe the location and structure of the scrotum. Explain how the scrotal muscles regulate the position of the testes in the scrotum. Why is this important?
4. Describe the internal structure of a testis. Discuss the function of the sustentacular cells, interstitial cells, seminiferous tubules, rete testis, and efferent ductules.
5. List the structures that constitute the spermatic cord. Where is the inguinal canal? Why is the inguinal canal clinically important?
6. Diagram a spermatozoon and describe the function of each of its principal parts.
7. Describe the position, histological structure, and functions of the epididymis, ductus deferens, and ejaculatory duct.
8. Compare the seminal vesicles and the prostate in terms of location, structure, and function.
9. Describe the structure of the penis and explain the mechanisms that result in erection, emission, and ejaculation.
10. Define *semen*. How much semen is ejected during ejaculation? What does it consist of and what are its properties?
11. Comment on the significance of the indifferent stage of development with respect to developmental abnormalities of the reproductive organs.
12. Define *perineum*. Contrast the appearance of the perineum of an embryo during the indifferent stage of genital development with that of an adult male.
13. Explain what is meant by homologous structures. List the structures of the male reproductive system that form from the phallus, urethral folds, labioscrotal swellings, and the gonadal ridge.
14. Distinguish between impotence, infertility, and sterility.
15. Define *hydrocele*, *orchitis*, and *orchiectomy*. What conditions would warrant an orchiectomy?

Critical-Thinking Questions

1. Using a diagram, describe the stages of spermatogenesis. Explain why a postpubescent male can produce billions of sperm cells throughout his life without using up all of the spermatogonia.
2. The reproductive system has been described as a "nonessential" body system because it is not necessary for the survival of the individual. Upon maturation, however, the reproductive system does produce hormones that maintain adult features as well as bone structure. These hormones also influence behavior, blood composition, and metabolism. Discuss your thoughts as to whether or not the reproductive system is nonessential.
3. As a safeguard against having more children, a 40-year-old man is considering a vasectomy. This procedure involves tying and surgically removing a short section of each ductus deferens within the spermatic cord (see fig. 20.19). Will the vasectomy affect his sexual performance? Will his sexual desire be altered? What will be the components and volume of his ejaculate? Will he be sterile or infertile? What happens to spermatozoa that cannot be ejaculated?
4. If it takes only one sperm cell to fertilize an egg, why are such huge quantities expelled during ejaculation?
5. A number of pathogenic (disease-causing) organisms are transmitted through sexual activity. From the standpoint of the pathogen, why does coitus provide an ideal means of propagation?

TWENTY-ONE

Female Reproductive System

Clinical Case Study 707

Introduction to the Female Reproductive System 707

Structure and Function of the Ovaries 709
- Position and Structure of the Ovaries 709
- Ovarian Cycle 710

Secondary Sex Organs 714
- Uterine Tubes 714
- Uterus 715
- Vagina 717
- Vulva 718
- Mechanism of Erection and Orgasm 719

Mammary Glands 719
- Structure of the Breast and the Mammary Glands 720

Ovulation and Menstruation 722

Clinical Considerations 723
- Diagnostic Procedures 723

Developmental Exposition: The Female Reproductive System 724
- Developmental Problems of the Female Reproductive System 726
- Problems Involving the Ovaries and Uterine Tubes 726
- Problems Involving the Uterus 728
- Diseases of the Vagina and Vulva 729
- Diseases of the Breasts and Mammary Glands 729
- Methods of Contraception 729

Clinical Case Study Answer 732

Chapter Summary 732

Review Activities 733

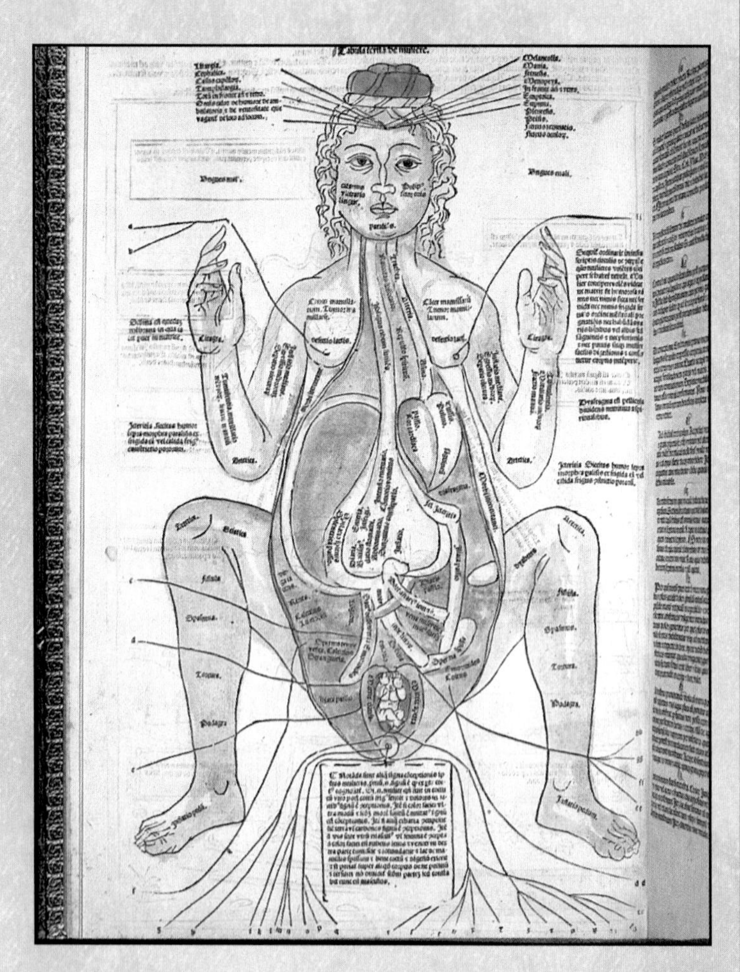

This illustration, after a woodcut from Fasciculus Medicinae *(1493) by Johannes de Ketham, shows the anatomy of a pregnant woman. The picture is annotated with a litany of familiar complaints, presumably those associated with pregnancy.*

Clinical Case Study

A 28-year-old female was brought to the emergency room following a 4-day history of moderate right-sided pelvic pain. On the morning of the fifth day the pain had become more severe, prompting her to seek medical attention. She complained of weakness and light-headedness, and stated that she hadn't had a period for about 8 weeks. A urine pregnancy test was positive. The consulting gynecologist said that a ruptured ectopic pregnancy was likely. He ordered a blood test, the results of which suggested that the patient had suffered a slight amount of hemorrhage. A *culdocentesis* (needle sampling of the peritoneal cavity via the posterior vaginal wall) was positive for pooled blood. The patient was prepared for surgery.

What is an ectopic pregnancy? Where is it most likely to occur? Briefly explain the sequence of events leading up to the rupture of the ectopic pregnancy beginning with ovulation. Differentiate normal from abnormal events. Explain how blood from a ruptured ectopic pregnancy can be aspirated through the vagina.

Hints: Carefully study the position of the uterus and uterine tubes with respect to the ovaries within the peritoneal cavity. Read about ectopic pregnancies in the clinical sections of this chapter and refer to figure 22.34 in chapter 22.

Introduction to the Female Reproductive System

The female reproductive system produces ova, secretes sex hormones, receives spermatozoa from the male, and provides sites for fertilization of an ovum and implantation of the blastocyst. Parturition follows gestation, and secretion from the mammary glands provides nourishment for the baby.

Objective 1 Explain the functional differences between the male and female reproductive systems.

Objective 2 Define *ovulation, menstruation,* and *menopause.*

Objective 3 Identify the primary and secondary sex organs in a female, and describe the female secondary sex characteristics.

The reproductive systems of the male and female have some basic similarities and some specialized differences. The two systems are similar in that (1) most of the reproductive organs of both sexes develop from similar embryonic tissues and are therefore *homologous;* (2) both systems have gonads that produce gametes and sex hormones; and (3) both systems experience latent development of the reproductive organs, which mature and become functional during puberty as a result of the influence of sex hormones secreted by the gonads.

The differences between the reproductive systems of the female and male are based on the specific functions of each in sexual reproduction and on the cyclic events that are characteristic of the female. The reproductive organs of a sexually mature, healthy male continuously produce male gametes, or spermatozoa, and transfer them to the female during *coitus* (*ko'ĭ-tus*) (*sexual intercourse*). A male does not produce any spermatozoa until puberty (at about age 13), but is then capable of producing viable spermatozoa throughout his life if he remains healthy. The gametes, or ova, of a female are completely formed, but not totally matured, during fetal development of the ovaries. The ova are generally discharged, or *ovulated,* one at a time in a cyclic pattern throughout the reproductive period of the female, which extends from puberty to menopause. *Menstruation* (*men"stroo-a'shun*) is the discharge of *menses* (blood and solid tissue) from the uterus at the end of each menstrual cycle. *Menopause* is the period marked by the termination of ovulation and menstruation. The reproductive period in females generally extends from about age 12 to about age 50. The cyclic reproductive pattern of ovulation and the age span of fertility are determined by hormones.

The functions of the female reproductive system are (1) to produce ova; (2) to secrete sex hormones; (3) to receive the spermatozoa from the male during coitus; (4) to provide sites for fertilization, implantation of the blastocyst (see chapter 22), and embryonic and fetal development; (5) to facilitate *parturition,* or delivery of the baby; and (6) to provide nourishment for the baby through the secretion of milk from the mammary glands in the breasts.

The organs of the female reproductive system, like those of the male, are categorized on a functional basis as follows:

1. **Primary sex organs.** The primary sex organs are called *gonads,* and in the female are known more specifically as the *ovaries.* Ovaries produce the gametes (ova), or eggs, and produce and secrete sex steroid hormones. Secretion of the female sex hormones at puberty contributes to the development of secondary sex characteristics and causes cyclic changes in the secondary sex organs that are required for reproductive function.
2. **Secondary sex organs.** Secondary sex organs (fig. 21.1 and table 21.1) are those structures that are essential for successful fertilization of the ovum, implantation of the blastocyst, development of the embryo and fetus, and parturition. The secondary sex organs include the *vagina,* which receives the penis and ejaculated semen during coitus and through which the baby passes during delivery; the external genitalia, which protect the vaginal orifice (opening); the *uterine* (fallopian) *tubes,* through which ovulated eggs are transported toward the uterus and where fertilization takes place; and the *uterus* (womb), where implantation and development occur. The

coitus: L. *coitio,* a coming together
menopause: Gk. *men,* month; *pausis,* cessation
ovaries: L. *ovum,* egg
vagina: L. *vagina,* sheath or scabbard

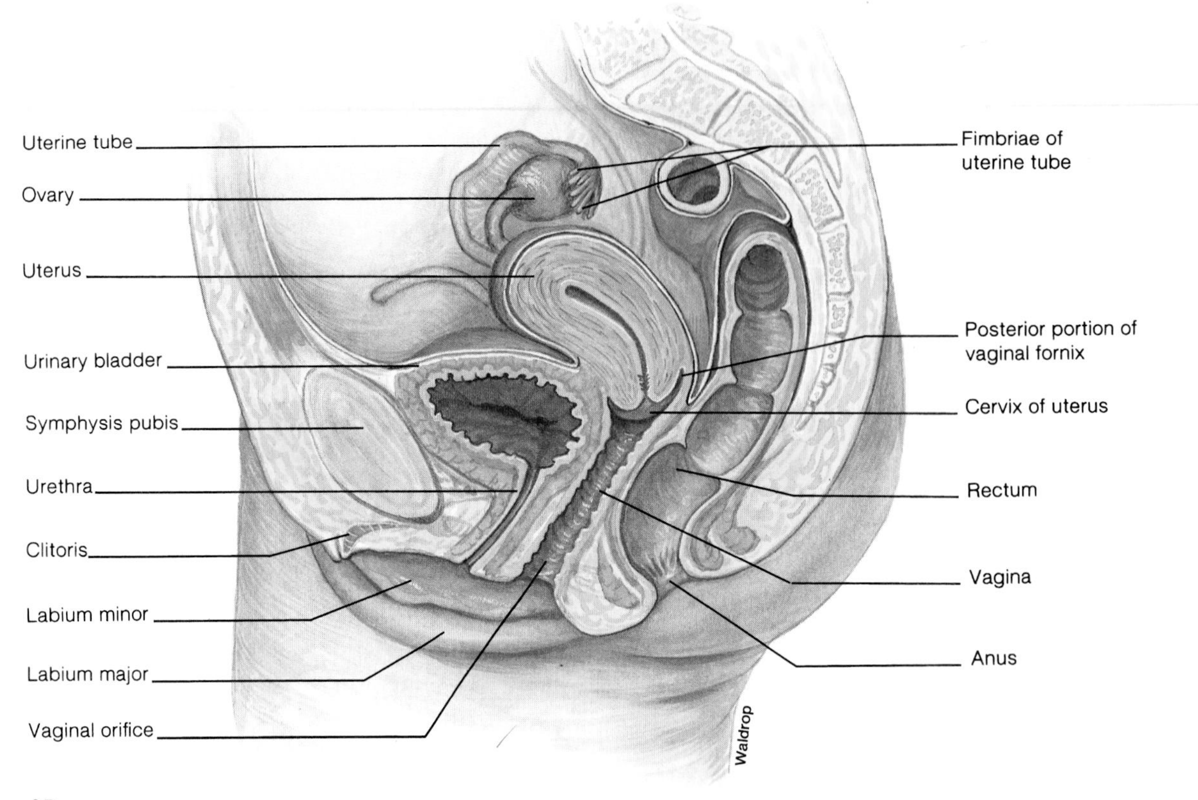

Figure 21.1

Organs of the female reproductive system seen in sagittal section.

Table 21.1
Reproductive Organs of the Female

Organ	Function/Description
Ovaries	Produce female gametes (oocytes) and female sex hormones
Uterine tubes	Convey oocytes toward uterus; site of fertilization; convey developing embryo to uterus
Uterus	Site of implantation; protects and sustains embryo and fetus during pregnancy; plays active role in parturition (childbirth)
Vagina	Conveys uterine secretions to outside of body; receives erect penis and semen during coitus and ejaculation; serves as passageway for fetus during parturition
Labia majora	Form margins of pudendal cleft; enclose and protect labia minora
Labia minora	Form margins of vestibule; protect openings of vagina and urethra
Clitoris	Glans of the clitoris is richly supplied with sensory nerve endings associated with feeling of pleasure during sexual stimulation
Pudendal cleft	Cleft between labia majora within which labia minora and clitoris are located
Vaginal vestibule	Cleft between labia minora within which vaginal and urethral openings are located
Vestibular glands	Secrete fluid that moistens and lubricates the vestibule and vaginal opening during coitus
Mammary glands	Produce and secrete milk for nourishment of an infant

muscular walls of the uterus play an active role in parturition. *Mammary glands* are also considered secondary sex organs because the milk they secrete after parturition provides nourishment to the child.

3. **Secondary sex characteristics.** Secondary sex characteristics are features that are not essential for the reproductive process but are generally considered to be sexual attractants. Distribution of fat to the breasts, abdomen, mons pubis, and hips; body hair pattern; and broad pelvis are examples of female secondary sex characteristics. Although the breasts contain the mammary glands, large breasts are not essential for nursing the young. In fact, all female mammals have mammary glands, but only human females have protruding breasts that function as a sexual attractant.

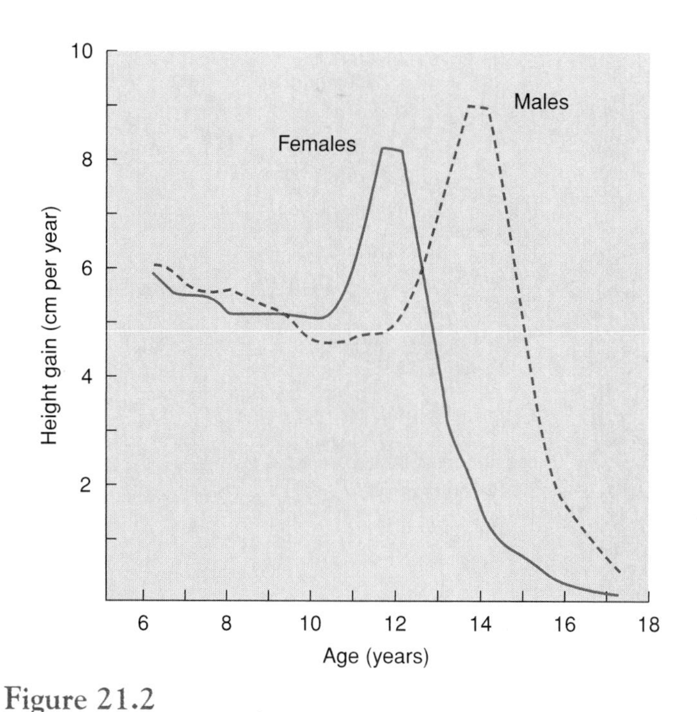

Figure 21.2

A comparison of male and female rates of growth in height.

The onset of puberty in females generally occurs between the ages of 12 and 14 (average, 12.6 years) in accordance with the nutritional condition, genetic background, and even sexual exposure of the individual. Generally, girls attain puberty 6 months to 1 year earlier than boys, accompanied by an earlier growth spurt (fig. 21.2). Puberty in girls is heralded by the onset of menstruation, or *menarche* (*mĕ-nar′ke*). Puberty results from the increased secretion of gonadotropic hormones from the anterior pituitary, which stimulates the ovaries to establish their ovarian cycles and sex steroid secretion.

The average age of menarche is later (age 15) in girls who are very active physically than in the general population. This appears to be due to a requirement for a minimum percentage of body fat for menstruation to begin, and may represent a mechanism favored by natural selection to ensure the ability to successfully complete a pregnancy and nurse the baby.

1. List the functions of the female reproductive system.
2. Explain the process by which puberty occurs. Define *menstruation* and *ovulation*. What is the usual span of a woman's reproductive years? Define *menses* and *menopause*.
3. Distinguish between the primary sex organs, secondary sex organs, and secondary sex characteristics.

Structure and Function of the Ovaries

The ovary contains a large number of follicles, each of which encloses an ovum. Some of these follicles mature during the ovarian cycle, and the ova they contain progress to the secondary oocyte stage of meiosis. During ovulation, the largest follicle ruptures and releases its secondary oocyte. The ruptured follicle becomes a corpus luteum and regresses to become a corpus albicans. These cyclic changes in follicular development are accompanied by changes in hormone levels.

Objective 4 Describe the position of the ovaries and the ligaments supporting the ovaries and genital ducts.

Objective 5 Describe the structural changes in the ovaries that lead to and follow ovulation.

Objective 6 Describe oogenesis and explain why meiosis of one primary oocyte results in the formation of only one mature ovum.

Objective 7 Discuss the hormonal secretions of the ovaries during an ovarian cycle.

Position and Structure of the Ovaries

Ovaries (*o′vă-rēz*) are the paired primary sex organs of the female that produce *gametes*, or *ova*, and the sex hormones, *estrogens* and *progesterone*. The ovaries of a sexually mature female are solid, ovoid structures about 3.5 cm (1.4 in.) long, 2 cm (0.8 in.) wide, and 1 cm (0.4 in.) thick. The color and texture of the ovaries vary according to the age and reproductive stage of the female. The ovaries of a young girl are smooth and pinkish. Following puberty, the ovaries are pinkish-gray and have an irregular surface because of the scarring caused by ovulation. On the medial portion of each ovary is a **hilum** (*hi′lum*), which is the point of entrance for ovarian vessels and nerves. The lateral portion of the ovary is positioned near the open end of the uterine tube (fig. 21.3).

The paired ovaries are positioned in the upper pelvic cavity, one on each lateral side of the uterus. Each ovary is situated in a shallow depression of the posterior body wall, the **ovarian fossa,** and secured by several membranous attachments. The principal supporting membrane of the female reproductive tract is the **broad ligament.** The broad ligament is the parietal peritoneum that supports the uterine tubes and uterus. The **mesovarium** (*mes″ŏ-va′re-um*) is a specialized posterior extension of the broad ligament that attaches to an ovary. Each ovary is additionally supported by an **ovarian ligament,** which is anchored to the uterus, and a **suspensory ligament,** which is attached to the pelvic wall (fig. 21.3).

Each ovary consists of four layers. The **superficial epithelium** is the thin outermost layer composed of cuboidal epithelial cells (see fig. 21.6). A collagenous connective tissue layer called the **tunica albuginea** (*al″byoo-jin′e-ă*) is located immediately below the superficial epithelium. The principal substance of the ovary is divided into an outer **ovarian cortex** and a vascular inner **ovarian medulla,** although the boundary between these layers is not distinct. The **stroma**—the material of the

stroma: Gk. *stroma*, a couch or bed

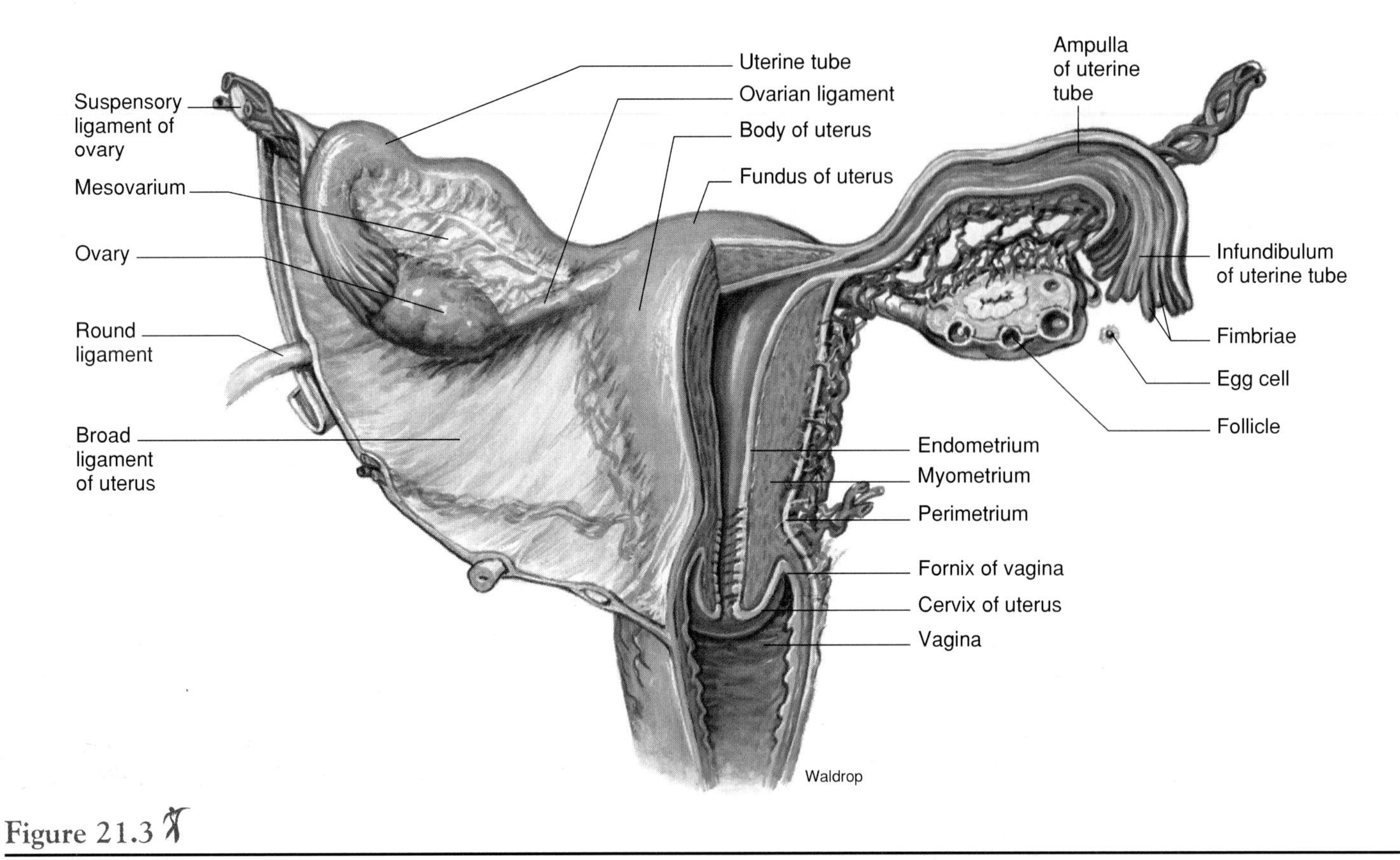

Figure 21.3

An anterior view of the internal female reproductive organs showing their positional relationships.

ovary in which follicles and blood vessels are embedded—lies in both the cortical and medullary layers.

Blood Supply and Innervation

Blood is supplied by ovarian arteries that arise from the lateral sides of the abdominal aorta, just below the origin of the renal arteries. An additional supply comes from the ovarian branches of the uterine arteries. Venous return is through the ovarian veins. The right ovarian vein empties into the inferior vena cava, whereas the left ovarian vein drains into the left renal vein.

The ovaries have both sympathetic and parasympathetic innervation from the ovarian plexus. Innervation to the ovaries, however, is only to the vascular networks and not to the follicular substance within the stroma. All of the vessels and nerves to an ovary enter by way of the hilum, which is supported by the ovarian ligament.

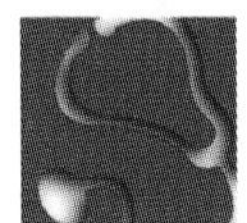

Normal, healthy ovaries usually cannot be palpated either by vaginal or abdominal examination. If the ovaries become swollen or displaced, however, they are palpable through the vagina. There are many types of nonmalignant tumors of the ovaries, most of which cause swelling and some localized tenderness. The ovaries atrophy during menopause, and ovarian enlargement in postmenopausal women is usually cause for concern.

Ovarian Cycle

The germ cells that migrate into the ovaries during early embryonic development multiply, so that by about 5 months of gestation the ovaries contain approximately 6 to 7 million primordial oocytes, called *oogonia* (*o″ŏ-go′-ne-ă*). The production of new oogonia stops at this point and never resumes. Toward the end of gestation, the oogonia begin meiosis, at which time they are called **primary oocytes** (*o′ŏ-sītz*). Meiosis is arrested at prophase I of the first meiotic division, and therefore the primary oocytes are still diploid (have 46 chromosomes). Although the ovaries of a newborn girl contain about 2 million oocytes, this number declines to 300,000 to 400,000 by the time she enters puberty. On average, 400 oocytes are ovulated during a woman's reproductive lifetime.

Primary oocytes that are not stimulated to complete the first meiotic division are contained within tiny follicles called **primordial follicles.** In response to stimulation by gonadotropic hormones, some of these oocytes and follicles get larger, and the follicular cells divide to produce the **follicular epithelium** that surrounds the oocyte and fills the follicle. A follicle at this stage in development is called a **primary follicle.**

Some primary follicles are stimulated to grow still bigger and develop a fluid-filled cavity called an *antrum* (*an′trum*). At this point, they are called **secondary follicles** (fig. 21.4). The follicular

oogonium: Gk. *oion*, egg; *gonos*, generation

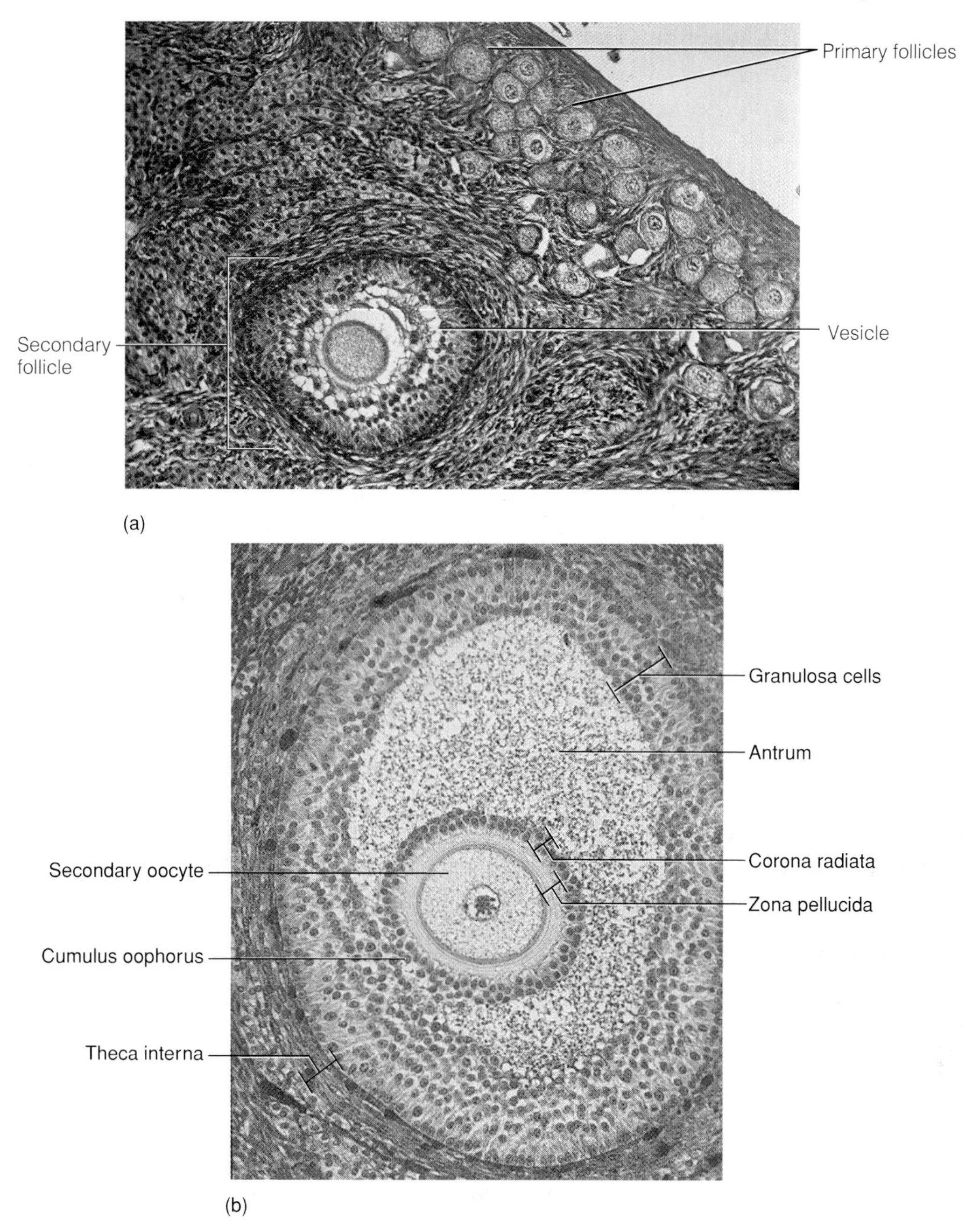

Figure 21.4

Photomicrographs of (*a*) primordial and primary follicles and (*b*) a mature vesicular ovarian follicle.

epithelium of secondary follicles forms a ring around the circumference of the follicle, called the *corona radiata*, and a mound that supports the oocyte. The mound is called the *cumulus oophorus* (*o-of'ŏ-rus*). Between the oocyte and the corona radiata is a thin gel-like layer of proteins and polysaccharides called the *zona pellucida* (*pĕ-loo'sĭ-dă*). Under stimulation of follicle-stimulating hormone (FSH) from the anterior pituitary, the follicular cells secrete increasing amounts of estrogen as the follicles grow. Interestingly, the follicular cells produce estrogen from its precursor testosterone, which is supplied by a layer of cells immediately outside the follicle called the *theca interna* (fig. 21.4*b*).

As the follicle develops, the primary oocyte completes its first meiotic division. This does not form two complete cells, however, because only one cell—the **secondary oocyte**—gets almost all of the cytoplasm. The other cell formed at this time becomes a small *polar body* (fig. 21.5), which eventually fragments and disappears. The secondary oocyte enters the second

corona radiata: Gk. *korone*, crown; L. *radiata*, radiate
cumulus oophorous: L. *cumulus*, a mound; Gk. *oophoros*, egg-bearing
zona pellucida: L. *zone*, girdle; L. *pellis*, skin

theca interna: Gk. *theke*, a box; L. *internus*, interior

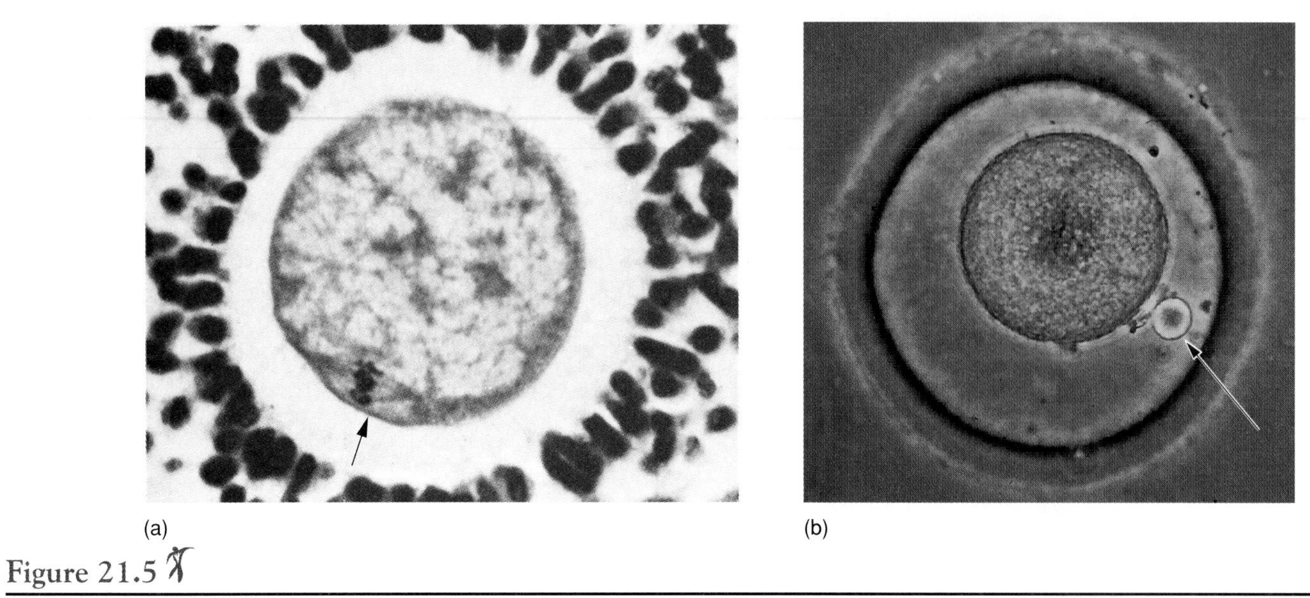

Figure 21.5

(*a*) A primary oocyte at metaphase I of meiosis. (Note the alignment of chromosomes [arrow]). (*b*) A mammalian secondary oocyte formed at the end of the first meiotic division and the first polar body (arrow).

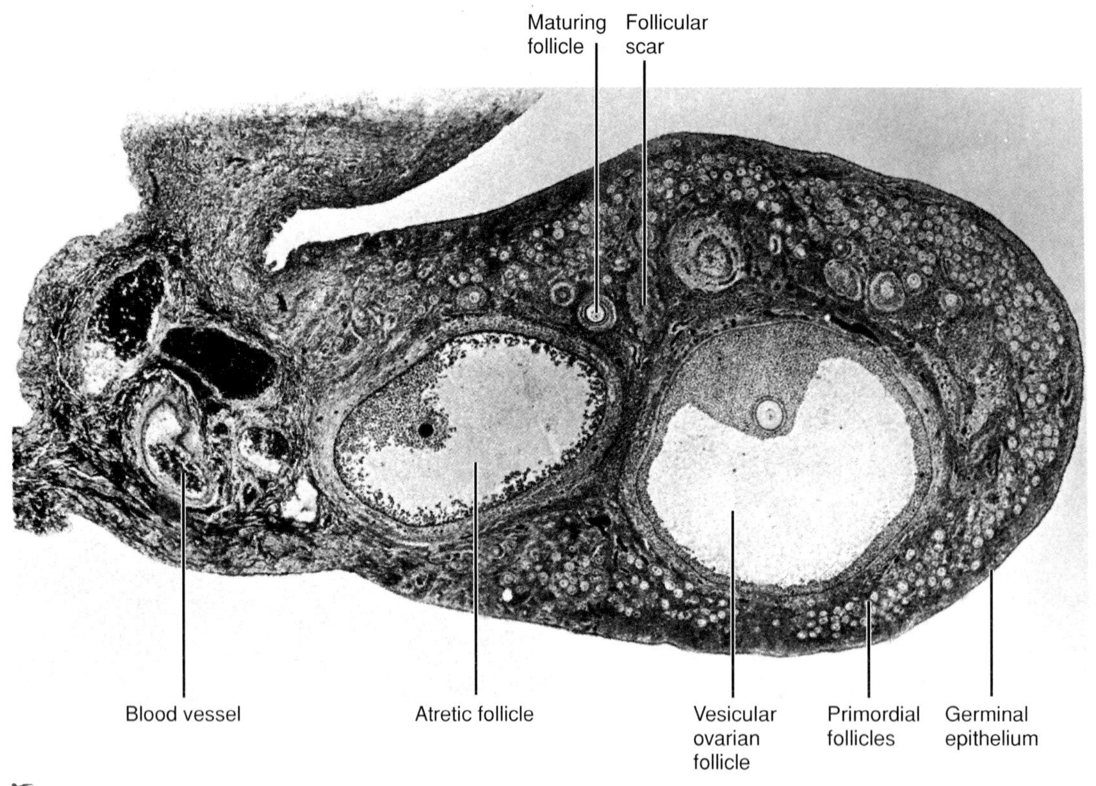

Figure 21.6

A vesicular ovarian follicle within the ovary of a monkey.

meiotic division, but meiosis is arrested at metaphase II and is never completed unless fertilization occurs.

Ovulation

By about the tenth to the fourteenth day following day 1 of a menstrual period, usually just one follicle has matured fully to become a **vesicular ovarian** (graafian) **follicle** (fig. 21.6). Other secondary follicles regress during that menstrual cycle and become *atretic*

---

graafian follicle: from Regnier de Graaf, Dutch anatomist and physician, 1641–73

atretic: Gk. *atretos*, not perforated

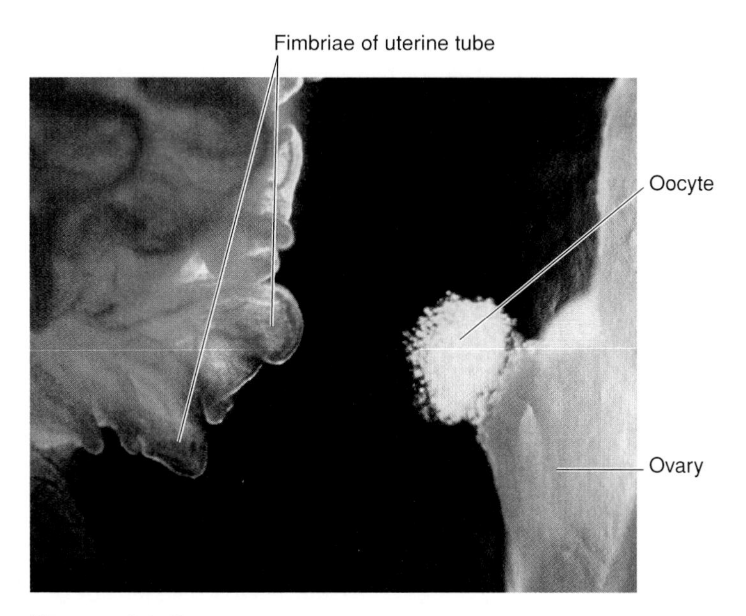

Figure 21.7

The release of the secondary oocyte from a human ovary. An ovulated oocyte is free in the peritoneal cavity until it is swept into the lumen of the uterine tube.

(*ă-tret′ik*). The vesicular ovarian follicle is so large that it forms a bulge on the surface of the ovary. Under proper hormonal stimulation (a sudden burst of luteinizing hormone from the anterior pituitary, triggered by a peak level of estrogen), this follicle will rupture—much like the popping of a blister—and extrude its secondary oocyte into the peritoneal cavity near the opening of the uterine tube in the process of *ovulation* (*ov-yŭ-la′shun*) (fig. 21.7).

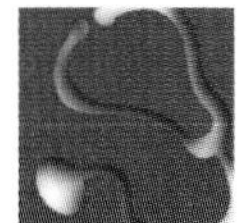

Most women are quite unaware that one follicle, approximately 2.5 cm (1 in.) in diameter, has ruptured and released its solitary egg. A substantial number of women, however (approximately 30%), experience a sharp, cramplike pain at the time of ovulation that may be confused with appendicitis.

The released secondary oocyte is surrounded by the zona pellucida and corona radiata. If it is not fertilized, it disintegrates in a couple of days. If a spermatozoon passes through the corona radiata and zona pellucida and enters the cytoplasm of the secondary oocyte, the oocyte completes the second meiotic division, becoming a mature ovum. In this process, the cytoplasm is again not divided equally; most remains in the *zygote* (fertilized egg), leaving another polar body, which like the first, disintegrates (fig. 21.8).

Changes continue in the ovary following ovulation. The empty follicle, under the influence of luteinizing hormone, undergoes structural and biochemical changes to become a **corpus luteum** (*loo′te-um*). Unlike the ovarian follicles, which secrete only estrogen, the corpus luteum secretes two sex steroid hormones: estrogen and progesterone. Toward the end of a

corpus luteum: L. *corpus*, body; *luteum*, yellow

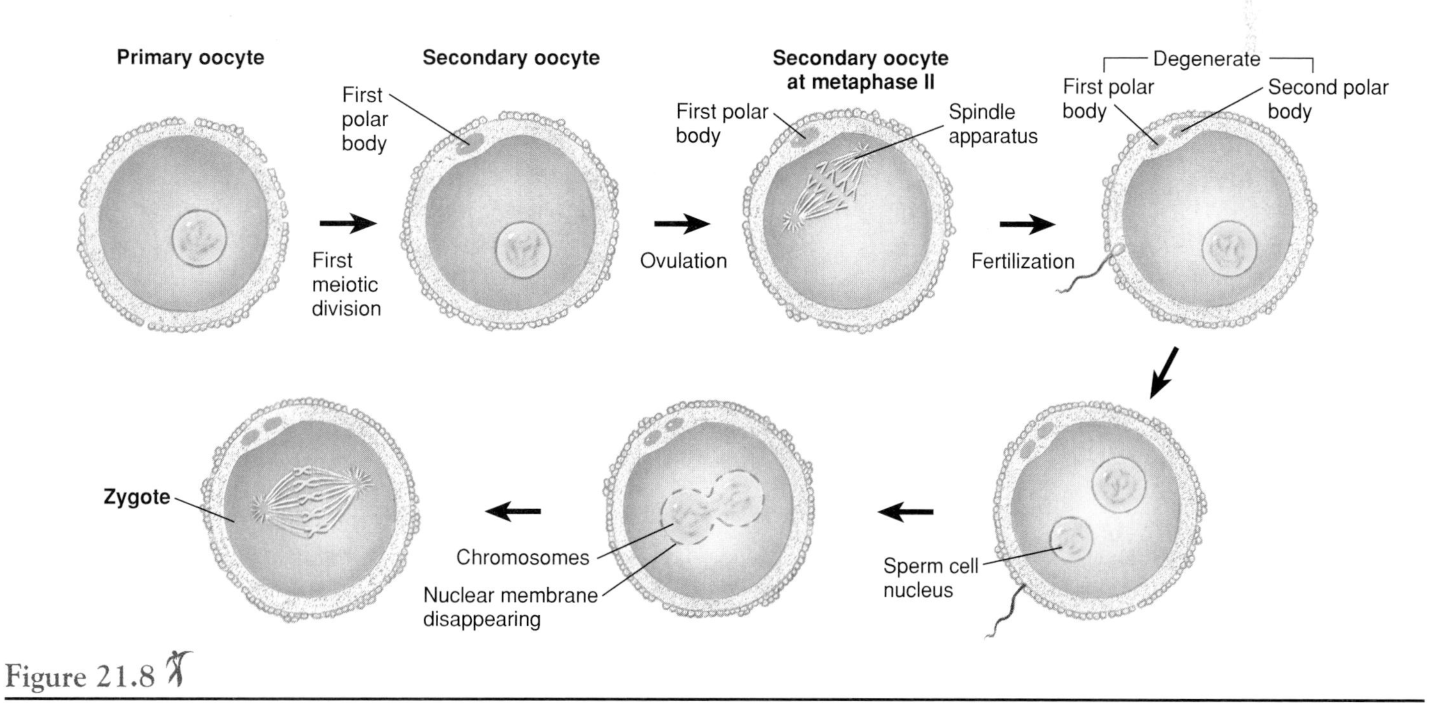

Figure 21.8

A schematic diagram of the process of oogenesis. During meiosis, each primary oocyte produces a single haploid gamete. If the secondary oocyte is fertilized, it forms a second polar body and its nucleus fuses with that of the sperm cell to become a zygote.

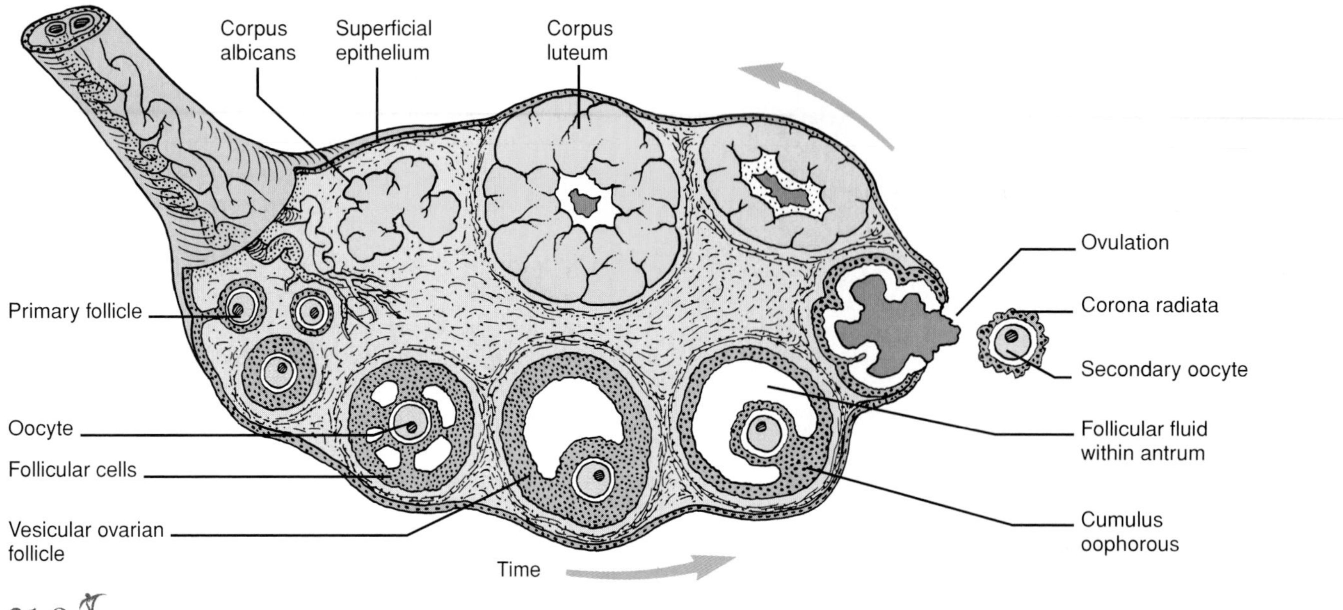

Figure 21.9

A schematic diagram of an ovary showing the various stages of ovum and follicle development.

nonfertile cycle, the corpus luteum regresses and is changed into a nonfunctional **corpus albicans** (*al'bĭ-kans*). These cyclic changes in the ovary are summarized in figure 21.9.

1. Describe the position of the ovaries relative to the uterine tubes and describe the position and functions of the broad ligament and mesovarium.
2. Compare the structure of a primordial follicle, primary follicle, secondary follicle, and vesicular ovarian follicle.
3. Define *ovulation* and describe the changes that occur in the ovary following ovulation in a nonfertile cycle.
4. Describe oogenesis and explain why only one mature ovum is produced by this process.
5. Compare the hormonal secretions of the vesicular ovarian follicle with those of a corpus luteum.

Secondary Sex Organs

The uterine tube conducts the zygote to the uterus, where implantation in the endometrium of the uterine wall typically occurs. The muscular layer of the uterus wall, or myometrium, is functional in labor and delivery. Sperm cells enter the female reproductive tract through the vagina, which also serves as the birth canal during parturition.

Objective 8 Describe how an ovum is moved through a uterine tube to the uterus.

Objective 9 Describe the structure and function of each of the three layers of the uterine wall.

Objective 10 Describe the structure and functions of the vulva and vagina.

Objective 11 Discuss the changes that occur in the female reproductive system during sexual excitement and coitus.

The secondary sex organs described in this section include the *uterine tubes*, *uterus*, *vagina*, and the *external genitalia* (*vulva*). The *mammary glands*, which are also considered secondary sex organs, are described in a separate section.

Uterine Tubes

The paired **uterine tubes,** also known as the *fallopian* (*fă-lo'pe-an*) *tubes*, or *oviducts*, transport oocytes from the ovaries to the uterus. Each uterine tube is approximately 10 cm (4 in.) long and 0.7 cm (0.3 in.) in diameter and is positioned between the folds of the broad ligament of the uterus (see fig. 21.3). The funnel-shaped, open-ended portion of the uterine tube, the **infundibulum** (see fig. 21.3), lies close to the ovary but is not attached. A number of fringed, fingerlike processes called **fimbriae** (*fim'bre-e*) project from the margins of the infundibulum over the lateral

albicans: L. *albicare*, to whiten

fallopian tubes: from Gabriele Fallopius, Italian anatomist, 1523–62
oviduct: L. *ovum*, egg; *ductus*, a leading

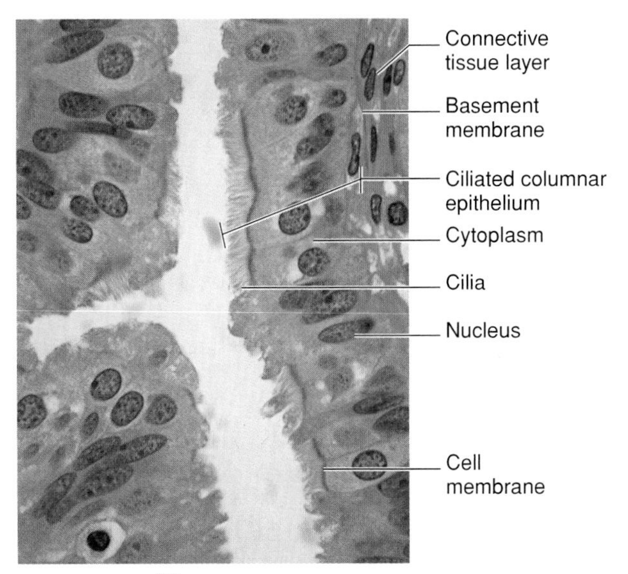

Figure 21.10

The histology of the uterine tube.

surface of the ovary. Wavelike movements of the fimbriae sweep an ovulated oocyte into the lumen of the uterine tube. From the infundibulum, the uterine tube extends medially and inferiorly to open into the cavity of the uterus. The *ampulla of the uterine tube* is its longest and widest portion (see fig. 21.3).

The wall of the uterine tube consists of three histological layers. The internal **mucosa** lines the lumen and is composed of ciliated columnar epithelium (fig. 21.10) that is drawn into numerous folds. The **muscularis** is the middle layer, composed of a thick inner circular layer of smooth muscle and a thin outer longitudinal layer of smooth muscle. Peristaltic contractions of the muscularis and ciliary action of the mucosa move the oocyte through the lumen of the uterine tube. The outer **serous layer** of the uterine tube is part of the visceral peritoneum.

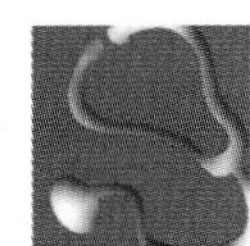

The term *salpinx* is occasionally used to refer to the uterine tubes. It is a Greek word meaning "trumpet" or "tube," and is the root of such clinical terms as *salpingitis (sal″pin-ji′tis)*, or inflammation of the uterine tubes; *salpingography* (radiography of the uterine tubes); and *salpingolysis* (the breaking up of adhesions of the uterine tube to correct female infertility).

The oocyte takes 4 to 5 days to move through the uterine tube. If enough viable sperm are ejaculated into the vagina during coitus, and if there is an oocyte in the uterine tube, fertilization will occur within hours after discharge of the semen. The zygote will move toward the uterus where implantation occurs. If the developing embryo (called a *blastocyst*) implants into the uterine tube instead of the uterus, the pregnancy is termed an *ectopic (ek-top′ik) pregnancy*, meaning an implantation of a blastocyst in a site other than the uterus (see fig. 22.34).

ectopic: Gk. *ex*, out; *topos*, place

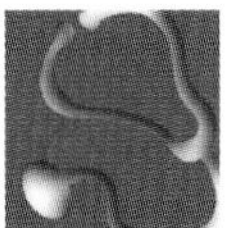

Since the infundibulum of the uterine tube is unattached, it provides a pathway for pathogens to enter the peritoneal cavity. The mucosa of the uterine tube is continuous with that of the uterus and vagina, and it is possible for infectious agents to enter the vagina and cause infections that may ultimately spread to the peritoneal linings, resulting in *pelvic inflammatory disease (PID)*. There is no opening into the peritoneal cavity other than through the uterine tubes. The peritoneal cavity of a male is totally sealed from external contamination.

The uterine tubes are supplied with blood through the ovarian and uterine arteries. Venous drainage is through uterine veins that parallel the arteries. Both the uterine artery and vein can be observed in the broad ligament that supports the uterine tube (see fig. 21.3).

The uterine tubes have both sympathetic and parasympathetic innervation from the hypogastric plexus and pelvic splanchnic nerves. The nerve supply to the uterine tubes regulates the activity of the smooth muscles and blood vessels.

Uterus

The **uterus** (*yoo′ter-us*) receives the blastocyst that develops from a fertilized oocyte and provides a site for implantation. Prenatal development continues within the uterus until gestation is completed, at which time the uterus plays an active role in the delivery of the baby.

Structure of the Uterus

The uterus is a hollow, thick-walled, muscular organ with the shape of an inverted pear. It is located near the floor of the pelvic cavity, anterior to the rectum and posterosuperior to the urinary bladder. Although the shape and position of the uterus changes dramatically during pregnancy (fig. 21.11), in its nonpregnant state it is about 7 cm (2.8 in.) long, 5 cm (2 in.) wide (through its broadest region), and 2.5 cm (1 in.) in diameter. The anatomical regions of the uterus include the uppermost dome-shaped portion superior to the entrance of the uterine tubes, called the **fundus of the uterus;** the enlarged main portion, called the **body of the uterus;** and the inferior constricted portion opening into the vagina, called the **cervix of the uterus** (fig. 21.12). The cervix projects posteriorly and inferiorly, joining the vagina at nearly a right angle.

The **uterine cavity** is the space within the fundus and body regions of the uterus. The narrow **cervical canal** extends through the cervix and opens into the lumen of the vagina (fig. 21.12). The junction of the uterine cavity with the cervical canal is called the **isthmus of uterus,** and the opening of the cervical canal into the vagina is called the **uterine ostium.**

fundris: L. *fundus*, bottom
cervix: L. *cervix*, neck

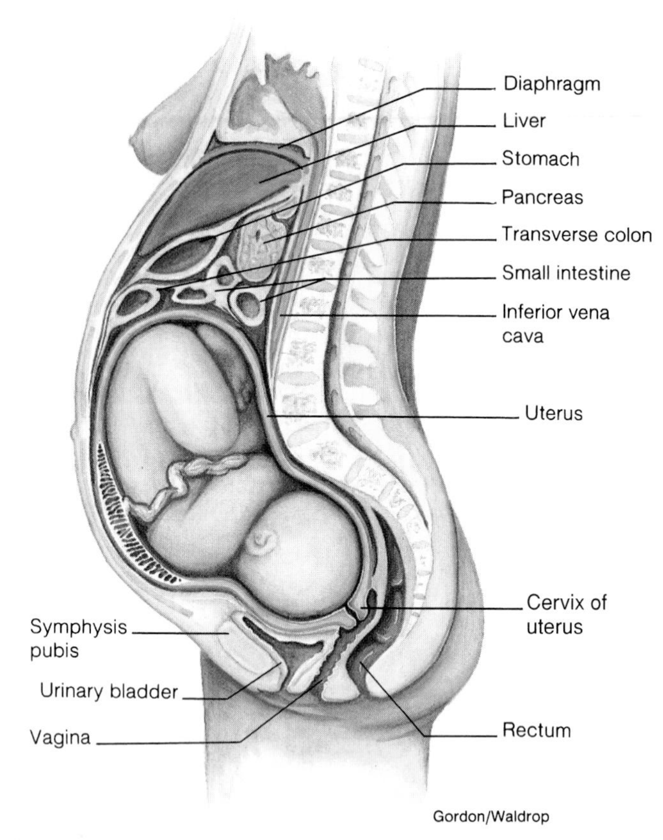

Figure 21.11

The size and position of the uterus in a full-term pregnant woman in sagittal section.

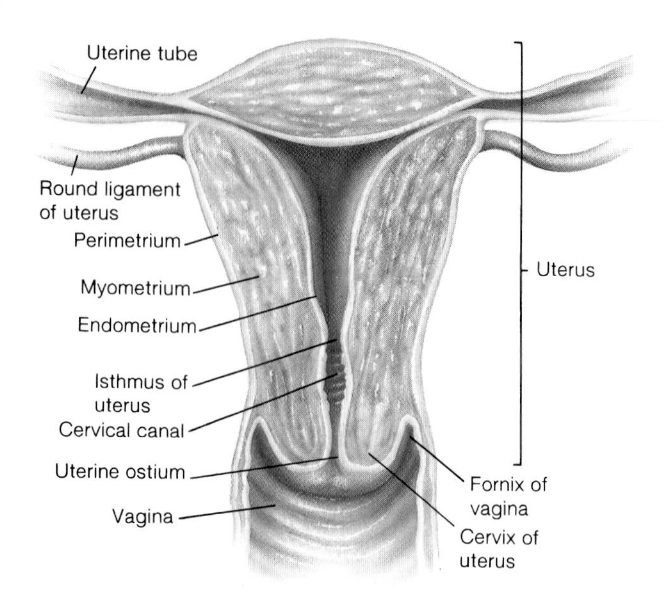

Figure 21.12

Layers of the uterine wall.

Support of the Uterus

The uterus is maintained in position by the muscles of the pelvic floor and by ligaments that extend to it from the pelvic girdle or body wall. The pelvic diaphragm, especially the levator ani muscle (see fig. 9.24), provides the principal muscular support to the vagina and uterus. The ligaments that support the uterus undergo marked hypertrophy during pregnancy and regress in size following parturition. They atrophy after menopause, which may contribute to a condition called *uterine prolapse*, or downward displacement of the uterus.

Four paired ligaments support the uterus in position within the pelvic cavity. The paired **broad ligaments** are folds of the peritoneum that extend from the pelvic walls and floor to the lateral walls of the uterus (see fig. 21.3). The ovaries and uterine tubes are also supported by the broad ligaments. The paired **rectouterine** (*rek″to-yoo′ter-in*) **folds** (not illustrated), which are also continuations of peritoneum, curve along the lateral pelvic wall on both sides of the rectum to connect the uterus to the sacrum. The **cardinal** (lateral cervical) **ligaments** (not illustrated) are fibrous bands within the broad ligament that extend laterally from the cervix and vagina across the pelvic floor, where they attach to the wall of the pelvis. The cardinal ligaments contain some smooth muscle as well as vessels and nerves that serve the cervix and vagina. The fourth paired ligaments are the **round ligaments** (see fig. 21.3). The round ligaments are actually continuations of the ovarian ligaments that support the ovaries. Each round ligament extends from the lateral border of the uterus just below the point where the uterine tube attaches to the lateral pelvic wall. Similar to the course taken by the ductus deferens in the male, each round ligament continues through the inguinal canal of the abdominal wall, where it attaches to the deep tissues of the labium majus.

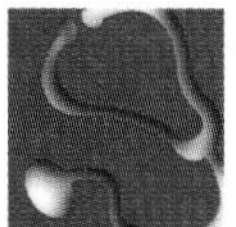

Although the uterus has extensive support, considerable movement is possible. The uterus tilts slightly posteriorly as the urinary bladder fills, and it moves anteriorly during defecation. In some women, the uterus may become displaced and interfere with the normal progress of pregnancy. A posterior tilting of the uterus is called *retroflexion,* whereas an anterior tilting is called *anteflexion.*

Uterine Wall

The wall of the uterus is composed of three layers: the perimetrium, myometrium, and endometrium (fig. 21.12). The **perimetrium,** the outermost serosal layer, consists of the thin visceral peritoneum. The lateral portion of the perimetrium is continuous with the broad ligament. A shallow pouch called the **vesicouterine** (*ves″ĭ-ko-yoo′ter-in*) **pouch** is formed as the peritoneum is reflected over the urinary bladder. The **rectouterine pouch** (pouch of Douglas) is formed as the peritoneum is reflected onto the rectum. The rectouterine pouch is the lowest point in the pelvic cavity and provides a site for surgical entry into the peritoneal cavity.

perimetrium: Gk. *peri*, around; *metra*, uterus
vesicouterine: L. *vesico*, bladder; *uterus*, womb
pouch of Douglas: from James Douglas, English anatomist and physician, 1675–1742

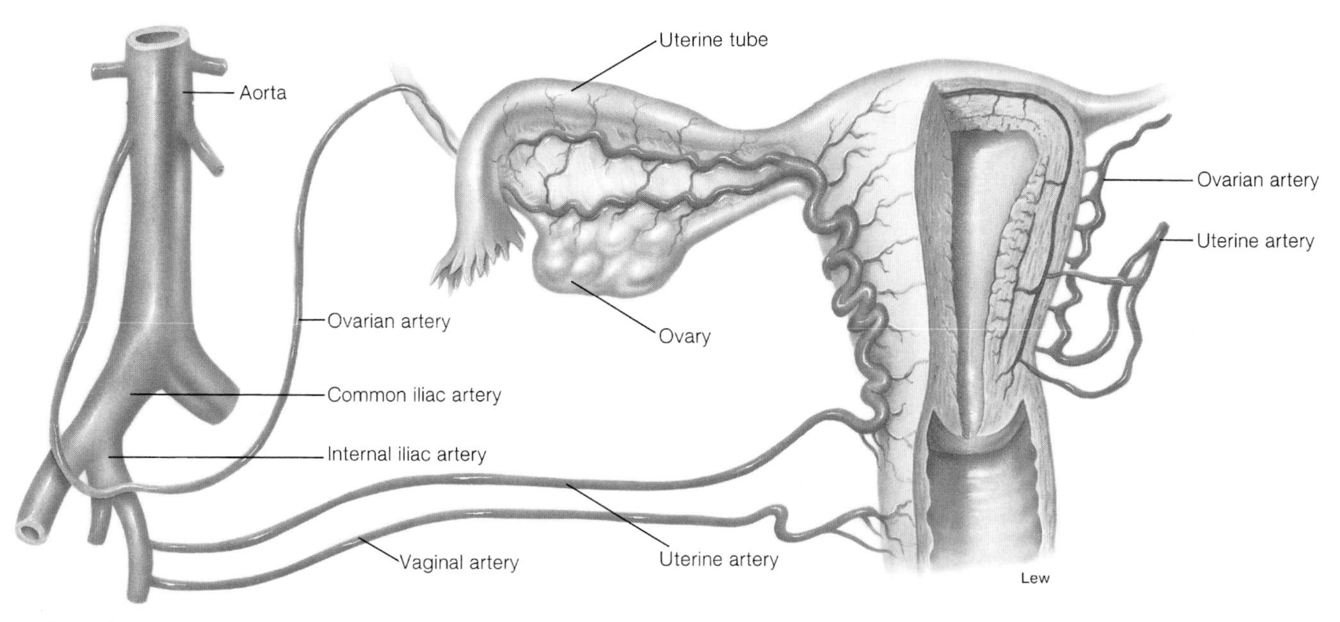

Figure 21.13

Arteries serving the internal female reproductive organs.

The thick **myometrium** is composed of three thick poorly defined layers of smooth muscle, arranged in longitudinal, circular, and spiral patterns. The myometrium is thickest in the fundus and thinnest in the cervix. During parturition, the muscles of this layer are stimulated to contract forcefully.

The **endometrium,** the inner mucosal lining of the uterus, is composed of two distinct layers. The superficial **stratum functionale,** composed of columnar epithelium and containing secretory glands, is shed as *menses* during menstruation and built up again under the stimulation of ovarian steroid hormones. The deeper **stratum basale** is highly vascular and serves to regenerate the stratum functionale after each menstruation.

The extent to which the uterus enlarges during pregnancy is nothing short of remarkable. From a fist-sized organ within the pelvis, it grows to occupy the bulk of the abdominal cavity, becoming about 16 times heavier than it was before conception. After parturition (childbirth), the uterus rapidly shrinks, but it may remain somewhat enlarged until menopause, at which time there is marked atrophy.

Uterine Blood Supply and Innervation

The uterus is supplied with blood through the uterine arteries, which arise from the internal iliac arteries, and by the uterine branches of the ovarian arteries (fig. 21.13). Each of these paired vessels anastomose on the upper lateral margin of the uterus. The blood from the uterus returns through uterine veins that parallel the pattern of the arteries.

The uterus has both sympathetic and parasympathetic innervation from the pelvic and hypogastric plexuses. Both autonomic innervations serve the arteries of the uterus, whereas the smooth muscle of the myometrium receives only sympathetic innervation.

Vagina

The **vagina** (*vă-ji'nă*) is the tubular, fibromuscular organ that receives sperm through the urethra of the erect penis during coitus. It also serves as the birth canal during parturition and provides for the passage of menses to the outside of the body. The vagina is about 9 cm (3.6 in.) long and extends from the cervix of the uterus to the vestibule. It is situated between the urinary bladder and the rectum, and is continuous with the cervical canal of the uterus. The cervix attaches to the vagina at a nearly 90 degree angle. The deep recess surrounding the protrusion of the cervix into the vagina is called the **fornix** (see fig. 21.3).

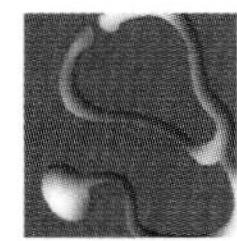

The fornix is of clinical importance since it permits palpation of the cervix during a gynecological examination. Occasionally, the deep posterior portion of the fornix provides surgical access to the pelvic cavity through the vagina. In addition, the fornix is important in the placement of two birth control devices—the cervical cap and the diaphragm (see fig. 21.26).

The exterior opening of the vagina, at its lower end, is called the **vaginal orifice.** A thin fold of mucous membrane called the **hymen** (*hi'men*) may partially cover the vaginal orifice.

The vaginal wall is composed of three layers: an inner mucosal layer, a middle muscularis layer, and an outer fibrous layer.

myometrium: Gk. *mys*, muscle; *metra*, uterus
endometrium: Gk. *endon*, within; *metra*, uterus
menses: L. *menses*, plural of *mensis*, monthly

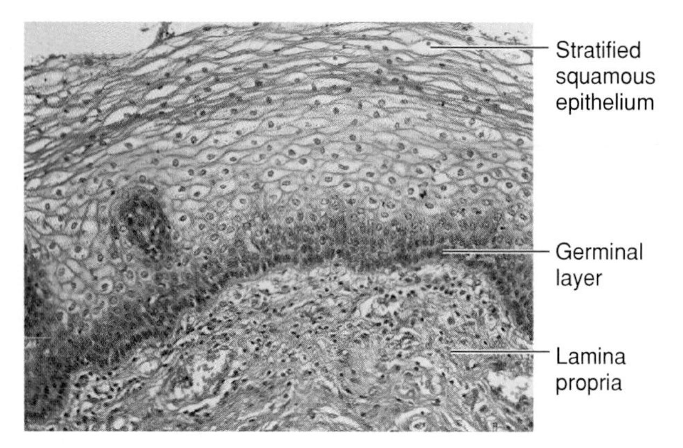

Figure 21.14

The histology of the vagina.

The **mucosal layer** (fig. 21.14) consists of nonkeratinized stratified squamous epithelium that forms a series of transverse folds called **vaginal rugae** (*roo'je*). The vaginal rugae permit considerable distension of the vagina for penetration of the erect penis. They also provide friction ridges for stimulation of the penis during coitus. The mucosal layer contains few glands; the acidic mucus that is present in the vagina comes primarily from glands within the uterus. The acidic environment of the vagina retards microbial growth. The additives within semen, however, temporarily neutralize the acidity of the vagina to assist the survival of the spermatozoa deposited within the vagina.

The **muscularis layer** consists of longitudinal and circular bands of smooth muscle interlaced with distensible connective tissue. The distension of this layer is especially important during parturition. Skeletal muscle strands near the vaginal orifice, including the levator ani muscle, partially constrict this opening.

The **fibrous layer** covers the vagina and attaches it to surrounding pelvic organs. This layer consists of dense regular connective tissue interlaced with strands of elastic fibers.

The blood supply to the highly vascular vagina comes primarily from the vaginal branches of the internal iliac artery. Blood is also supplied to the vagina through branches of the uterine, middle rectal, and internal pudendal arteries. Blood draining from the vagina returns through vaginal veins that parallel the course of the arteries.

The vagina has sympathetic innervation from the hypogastric plexus and parasympathetic innervation from the second and third sacral nerves. Sensory innervation is through the pudendal plexus and is especially well developed near the vaginal orifice.

Vulva

The external genitalia of the female are referred to collectively as the **vulva** (*vul'vă*) (fig. 21.15). The structures of the vulva surround the vaginal orifice and include the *mons pubis, labia majora, labia minora, clitoris, vaginal vestibule, vestibular bulbs*, and *vestibular glands*.

vulva: L. *volvere*, to roll, wrapper

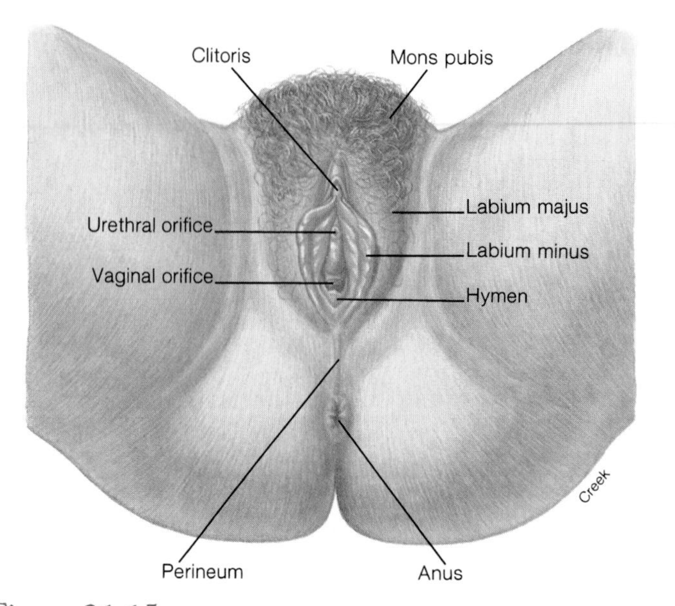

Figure 21.15

The external female genitalia.

The **mons pubis** is the subcutaneous pad of adipose connective tissue covering the symphysis pubis. At puberty, the mons pubis becomes covered with coarse pubic hair in a somewhat triangular pattern, usually with a horizontal upper border. The elevated and padded mons pubis cushions the symphysis pubis and vulva during coitus.

The **labia majora** (*la'be-ă mă-jor'ă*—singular *labium majus*) are two thickened longitudinal folds of skin that contain loose connective tissue and adipose tissue, as well as some smooth muscle. After puberty, their lateral surfaces are covered by pubic hair. The labia majora are continuous anteriorly with the mons pubis. They are separated longitudinally by the **pudendal** (*pyoo-den'dal*) **cleft** and converge again posteriorly on the **perineum** (*per"ĭ-ne'um*). The labia majora contain numerous sebaceous and sweat glands. They are homologous to the scrotum of the male and function to enclose and protect the other organs of the vulva.

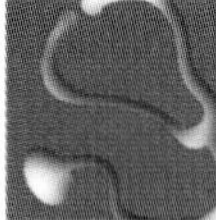

An *episiotomy* (*ĕ-pe'ze-ot"ŏ-me*) is a surgical incision, for obstetrical purposes, of the vaginal orifice that extends into the perineum. An episiotomy may be done during parturition to facilitate delivery and accommodate the head of an emerging fetus when laceration seems imminent. After delivery the cut is sutured.

Medial to the labia majora are two smaller longitudinal folds called the **labia minora** (singular, *labium minus*). The labia minora are hairless but do contain sebaceous glands. Anteriorly, the labia minora unite to form the **prepuce** (*pre'pyoos*), a hoodlike fold that partly covers the clitoris. The labia minora further protect the vaginal and urethral openings.

The **clitoris** (*klit'or-is, kli-tor'is*) is a small rounded projection at the upper portion of the pudendal cleft, at the anterior junction

mons pubis: L. *mons*, mountain; *pubis*, genital area

of the labia minora. The clitoris corresponds in structure and origin to the penis in the male; it is, however, much smaller and without a urethra. Although most of the clitoris is embedded in the tissues of the vulva, it does have an exposed **glans clitoris** of erectile tissue that is richly innervated with sensory endings. The clitoris is about 2 cm (0.8 in.) long and 0.5 cm (0.2 in.) in diameter. The unexposed portion of the clitoris is composed of two columns of erectile tissue called the **corpora cavernosa** that diverge posteriorly to form the **crura** and attach to the sides of the pubic arch.

The **vaginal vestibule** is the longitudinal cleft enclosed by the labia minora. The openings for the urethra and vagina are located in the vaginal vestibule. The external opening of the urethra is about 2.5 cm (1 in.) behind the glans of the clitoris and immediately in front of the vaginal orifice. The vaginal orifice is lubricated during sexual excitement by secretions from paired **major** and **minor vestibular glands** (Bartholin's glands) located within the wall of the region immediately inside the vaginal orifice. The ducts from these glands open into the vestibule near the lateral margins of the vaginal orifice. Bodies of vascular erectile tissue, called **vestibular bulbs,** are located immediately below the skin forming the lateral walls of the vestibule. The vestibular bulbs are separated from each other by the vagina and urethra, and extend from the level of the vaginal orifice to the clitoris.

The vulva is highly vascular and is supplied with arterial blood from internal pudendal branches of the internal iliac arteries and external pudendal branches from the femoral arteries. Extensive vascular networks are found within most of the organs of the vulva. The venous return is through vessels that correspond in name and position to the arteries.

During pregnancy, the vulva becomes swollen and bluish—especially the labia minora—due to increased vascularity and venous congestion. This discoloration is an important indicator of pregnancy. It appears at about the eighth to the twelfth week and becomes more apparent as pregnancy progresses.

The vulva has both sympathetic and parasympathetic innervation, as well as extensive somatic neurons that respond to sensory stimulation. Parasympathetic stimulation causes a response similar to that of the male: dilation of the arterioles of the genital erectile tissue and constriction of the venous return.

Mechanism of Erection and Orgasm

The homologous structures of the male and female reproductive systems respond to sexual stimulation in a similar fashion. The erectile tissues of a female, like those of a male, become engorged with blood and swollen during sexual arousal. During sexual excitement, the hypothalamus of the brain sends parasympathetic nerve impulses through the sacral segments of the spinal cord, which cause dilation of the arteries serving the clitoris and vestibular bulbs. This increased blood flow causes the erectile tissues to swell. In addition, the erectile tissues in the areola of the breasts become engorged.

Simultaneous with the erection of the clitoris and vestibular bulbs, the vagina expands and elongates to accommodate the erect penis of the male, and parasympathetic impulses cause the vestibular glands to secrete mucus near the vaginal orifice. The vestibular secretion moistens and lubricates the tissues of the vestibule, thus facilitating the penetration of the erect penis into the vagina. Mucus continues to be secreted during coitus so that the male and female genitalia do not become irritated, as they would if the vagina became dry.

The position of the sensitive clitoris usually allows it to be stimulated during coitus. If stimulation of the clitoris is of sufficient intensity and duration, a woman will usually experience a culmination of pleasurable psychological and physiological release called *orgasm*.

Associated with orgasm is a rhythmic contraction of the muscles of the perineum and the muscular walls of the uterus and uterine tubes. These reflexive muscular actions are thought to aid the movement of spermatozoa through the female reproductive tract toward the upper end of a uterine tube, where an ovum might be located.

Following orgasm or completion of the sexual act, sympathetic impulses cause a reduction in arterial flow to the erectile tissues, and their size diminishes to that prior to sexual stimulation.

1. Describe the structure and position of the uterine tubes and explain how an ovum is transported through these tubes to the uterus.
2. Describe the histological structure of the uterine wall and explain why the endometrium is subdivided into a stratum functionale and a stratum basale.
3. Describe the structures of the vagina and the vulva. What changes do these structures undergo during sexual excitement and coitus?

Mammary Glands

Mammary glands are modified sweat glands composed of secretory alveoli and ducts. The glands develop in the female breasts at puberty and function in lactation.

Objective 12 Distinguish between the mammary glands and the breast and describe the structure of the mammary glands.

In structure, the **mammary glands,** located in the **breasts,** are modified sweat glands and are a part of the integumentary system.

vestibule: L. *vestibule*, an entrance, court
Bartholin's glands: from Casper Bartholin Jr., Danish anatomist, 1655–1738

orgasm: Gk. *orgasmos*, to swell; to become excited

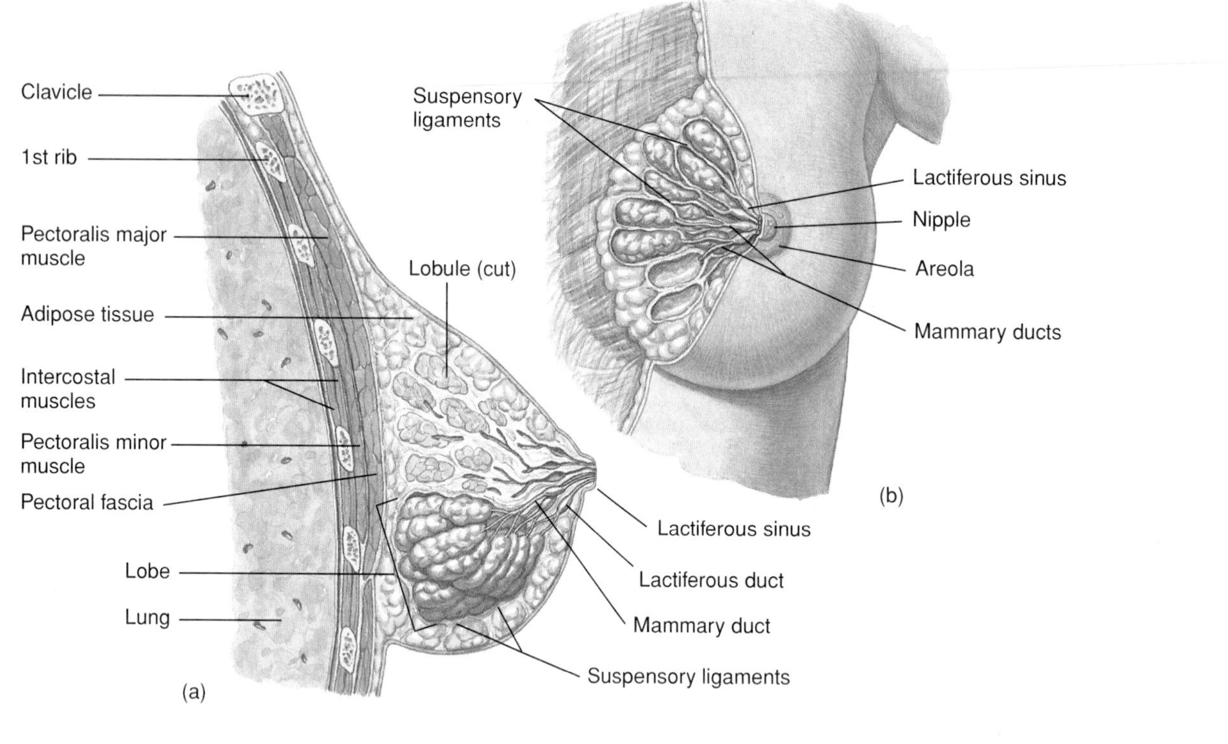

Figure 21.16

The structure of the breast and mammary glands. (*a*) A sagittal section and (*b*) an anterior view partially sectioned.

In function, however, these glands are associated with the reproductive system because they secrete milk for the nourishment of the young. The size and shape of the breasts vary considerably from person to person in accordance with genetic differences, age, percentage of body fat, or pregnancy. At puberty, estrogen from the ovaries stimulates growth of the mammary glands and the deposition of adipose tissue within the breasts. Mammary glands hypertrophy in pregnant and lactating women and usually atrophy somewhat after menopause.

Structure of the Breast and the Mammary Glands

Each breast is positioned over ribs 2 through 6 and overlies the pectoralis major muscle, the pectoralis minor muscle, and portions of the serratus anterior and external abdominal oblique muscles (see figs. 9.22 and 21.16). The medial boundary of the breast overlies the lateral margin of the sternum, and the lateral margin of the breast follows the anterior border of the axilla. The *axillary process of the breast* extends upward and laterally toward the axilla, where it comes into close relationship with the axillary vessels. This region of the breast is clinically significant because of the high incidence of breast cancer within the lymphatic drainage of the axillary process.

Each mammary gland is composed of 15 to 20 **lobes,** each with its own drainage pathway to the outside. The lobes are separated by varying amounts of adipose tissue. The amount of adipose tissue determines the size and shape of the breast but has nothing to do with the ability of a woman to nurse. Each lobe is subdivided into **lobules** that contain the glandular **mammary alveoli** (fig. 21.17). The mammary alveoli are the structures that produce the milk of a lactating female. **Suspensory ligaments** between the lobules extend from the skin to the deep fascia overlying the pectoralis major muscle and support the breasts. The clustered mammary alveoli secrete milk into a series of **mammary ducts** that converge to form **lactiferous** (*lak-tif'er-us*) **ducts** (fig. 21.16). The lumen of each lactiferous duct expands near the nipple to form a **lactiferous sinus.** Milk is stored in the lactiferous sinuses before draining at the tip of the nipple.

The **nipple** is a cylindrical projection from the breast that contains some erectile tissue. A circular pigmented **areola** (*ă-re-o'lă*) surrounds the nipple. The surface of the areola may appear bumpy because of the sebaceous **areolar glands** close to the surface. The secretions of these glands keep the nipple pliable. The color of the areola and nipple varies with the complexion of the woman. During pregnancy, the areola becomes darker in most women, and enlarges somewhat, presumably to become more conspicuous to a nursing infant.

Blood is supplied to the mammary gland through the perforating branches of the internal thoracic artery, which enter the breast through the second, third, and fourth intercostal spaces just lateral to the sternum, and through the more superficial mammary artery, which branches from the lateral thoracic artery. Venous return is through a series of veins that parallel the pattern of the arteries. A superficial venous plexus may be apparent through the skin of the breast, especially during pregnancy and lactation.

The breast is innervated primarily through sensory somatic neurons that are derived from the anterior and lateral cutaneous branches of the fourth, fifth, and sixth thoracic nerves. Sensory

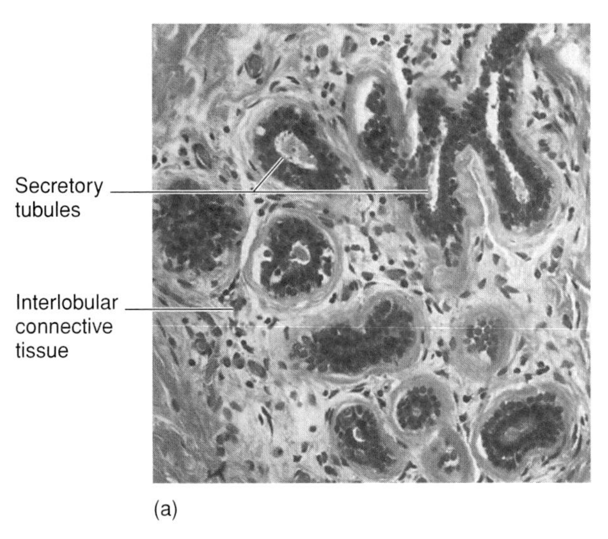

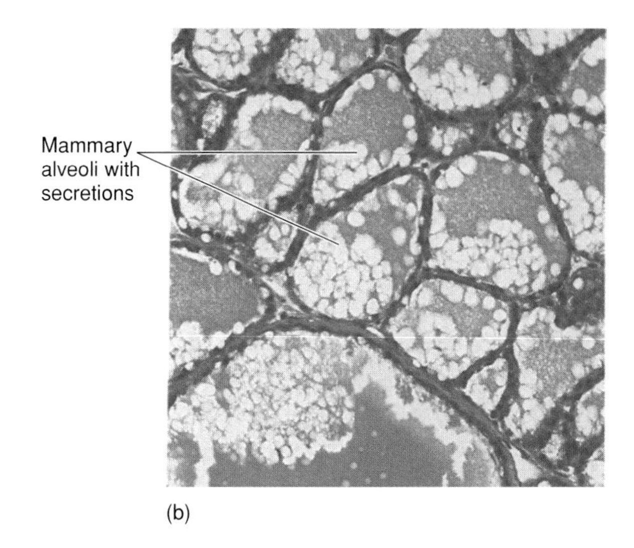

Figure 21.17

The histology of the mammary gland. (*a*) Nonlactating (63×) and (*b*) lactating (63×).

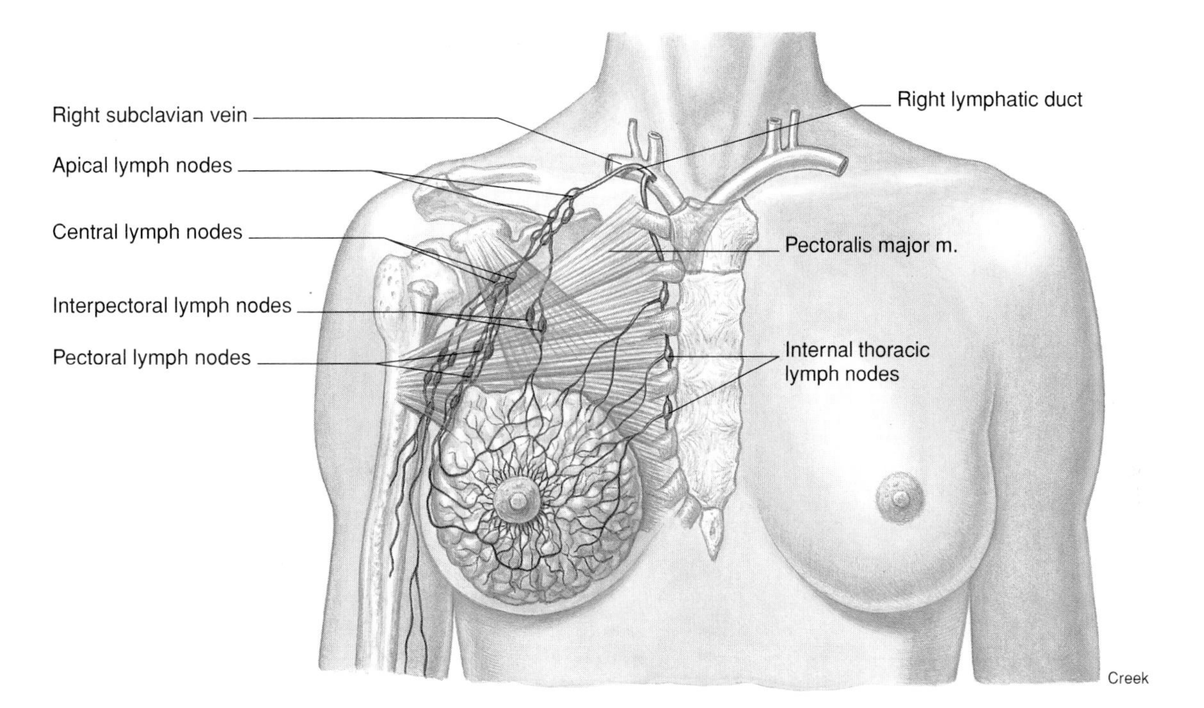

Figure 21.18

Lymphatic drainage of the mammary gland.

nerve endings in the nipple and areola are especially important in stimulating the release of milk from the mammary glands to a suckling infant (see chapter 14).

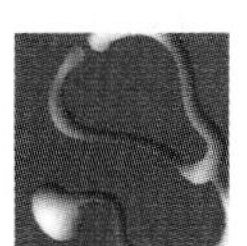

Lymphatic drainage and the location of lymph nodes within the breast are of considerable clinical importance because of the frequency of *breast cancer* and the high incidence of metastases. About 75% of the lymph drains through the axillary process of the breast into the pectoral lymph nodes (fig. 21.18). Some 20% of the lymph passes toward the sternum to the internal thoracic lymph nodes. The remaining 5% of the lymph is subcutaneous and follows the lymph drainage pathway in the skin toward the back, where it reaches the intercostal nodes near the neck of the ribs.

1. Describe the structure of the breasts and mammary glands and explain why variations in breast size do not affect the ability to lactate.
2. List in order the parts of the mammary glands through which milk passes during lactation.

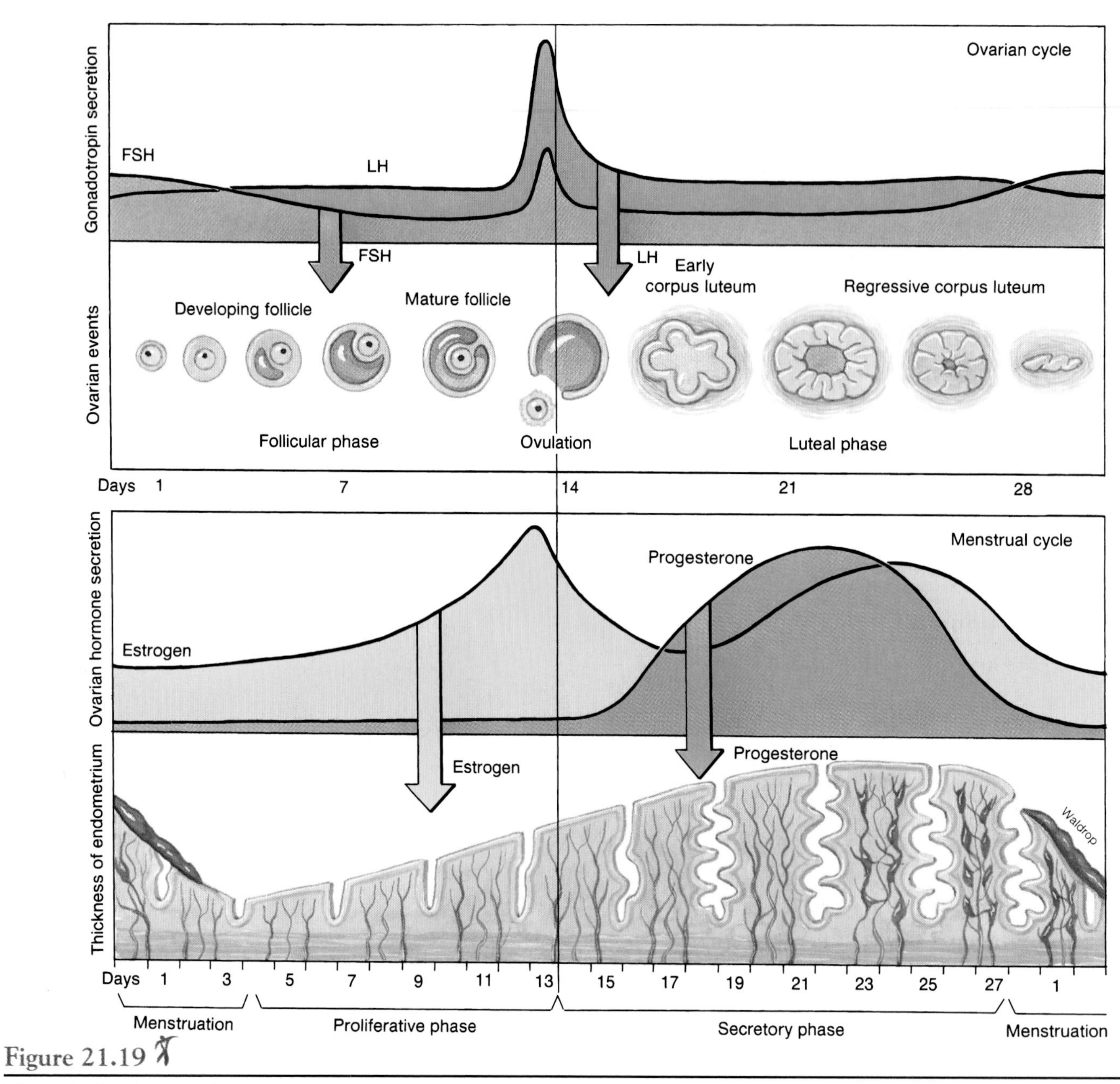

Figure 21.19
The cycle of ovulation and menstruation.

Ovulation and Menstruation

Ovulation and menstruation are reproductive cyclic events that are regulated by follicle-stimulating hormone (FSH) and luteinizing hormone (LH), secreted by the anterior pituitary, and by estrogen and progesterone, secreted by structures of the ovaries.

Objective 13 Describe the hormonal changes that result in ovulation and menstruation.

Objective 14 Describe the structural changes that occur in the endometrium during a menstrual cycle and explain how these changes are controlled by hormones.

Both ovulation and menstruation are reproductive functions of sexually mature females and are largely regulated by hormones from the anterior pituitary and the ovaries. Both occur approximately every 28 days, as menstruation follows ovulation and is regulated by the hormonal activity of the ovaries.

At ovulation, the wall of a follicle ruptures releasing a secondary oocyte that passes into the uterine tube. Ovulation typically occurs from alternate ovaries. If fertilization occurs, mitotic divisions are initiated and the blastocyst implants on the uterine wall (see chapter 22).

If the egg is not fertilized, the menstrual cycle is initiated usually 14 days after ovulation. The cycle is divided into three phases (fig. 21.19):

1. **Menstrual phase.** The menstrual phase, or *menstruation (men″stroo′a-shun)*, is characterized by a bloody discharge

of endometrial tissue from the uterus during the first 3 to 5 days of the cycle. The menstrual flow passes from the uterine cavity to the cervical canal and through the vagina to the outside of the body.

2. **Proliferative phase.** During the proliferative phase, days 5 to 14 of the cycle, endometrial tissue regrows.
3. **Secretory phase.** The secretory phase, lasting from day 14 to day 28, is characterized by an increase in glandular secretions and blood to the endometrium in preparation for nourishing a blastocyst. The last 2 or 3 days of the secretory phase may be characterized by cramping and external spotting of blood. This period of the cycle is sometimes referred to as the *premenstrual phase* (not shown in figure 21.19) and involves the initial breakdown of the endometrial lining.

The controlling center for ovulation and menstruation is the hypothalamus. On a regular cycle, the hypothalamus releases *gonadotropin-releasing hormone (GnRH)*, which in turn stimulates the anterior pituitary gland to release *follicle-stimulating hormone (FSH)* and *luteinizing hormone (LH)* at the appropriate time. The FSH stimulates the maturation of a follicle within an ovary. During the middle of the menstrual cycle, the anterior pituitary releases a large quantity of LH, in what is called an *LH surge*, and an increased amount of FSH. This surge in LH causes the mature follicle to swell rapidly and rupture. Ovulation occurs as the oocyte is discharged, along with its follicular fluid, toward the uterine tube.

Although there are several different female sex hormones, they all belong to two major groups that are referred to as *estrogen* and *progesterone*. The principal source of estrogen (in a nonpregnant female) is the ovaries. Estrogen, as associated with the menstrual cycle, causes the endometrium to thicken. It also plays an important role in the development and maintenance of the secondary sex organs and secondary sex characteristics. Progesterone is also secreted by the ovaries (in a nonpregnant female) and helps estrogen maintain the endometrium.

The principal events of ovulation and menstruation are outlined in table 21.2.

Table 21.2

Principal Events Surrounding Ovulation and Menstruation

1. The hypothalamus releases GnRH, which stimulates the anterior pituitary.
2. The anterior pituitary releases small amounts of FSH and LH.
3. FSH stimulates the maturation of a follicle.
4. Follicular cells produce and secrete estrogen.
   (a) Estrogen maintains secondary sexual traits.
   (b) Estrogen causes the uterine lining to thicken.
5. Toward the middle of the cycle, a surge in the release of LH from the anterior pituitary causes ovulation.
6. Follicular cells become corpus luteum cells, which secrete estrogen and progesterone.
   (a) Estrogen continues to stimulate the development of the uterine wall.
   (b) Progesterone stimulates the uterine lining to become more glandular and vascular.
   (c) Estrogen and progesterone inhibit the secretion of FSH and LH from the anterior pituitary.
7. If the oocyte is not fertilized, the corpus luteum degenerates.
8. As concentrations of estrogen and progesterone decline, blood vessels in the uterine lining constrict.
9. The uterine lining disintegrates and sloughs away as menstrual flow.
10. The anterior pituitary, which is no longer inhibited, again secretes FSH and LH.
11. The cycle is repeated.

1. Define *ovulation* and *menstruation*.
2. Explain the role of GnRH in regulating female reproductive functions.
3. Diagram the relative thickness of the endometrium during the three phases of the menstrual cycle.
4. Summarize the hormonal changes that regulate ovulation and menstruation.

Clinical Considerations

Females are more prone to dysfunctions and diseases of the reproductive organs than are males because of cyclic changes in reproductive events, problems associated with pregnancy, and the susceptibility of the female breasts to infections and neoplasms. The termination of reproductive capabilities at menopause can also cause complications as a result of hormonal changes. *Gynecology (gi″nĕ-kol′ŏ-je)* is the specialty of medicine concerned with dysfunction and diseases of the female reproductive system. *Obstetrics* is the specialty dealing with pregnancy and childbirth. Frequently a physician will specialize in both obstetrics and gynecology (OBGYN).

A comprehensive discussion of the numerous clinical aspects of the female reproductive system is beyond the scope of this text; thus, only the most important conditions will be addressed in the following sections. Coverage includes diagnostic procedures, developmental abnormalities, problems involving the ovaries and uterine tubes, problems involving the uterus, diseases of the vagina and vulva, and diseases of the breasts and mammary glands. In addition, the more popular methods of birth control are described.

Diagnostic Procedures

A gynecological, or pelvic, examination is generally included in a thorough physical examination of an adult female, especially prior to marriage, during pregnancy, or if problems involving the reproductive organs are suspected. In a gynecological examination, the physician inspects the vulva for irritations, lesions, or abnormal vaginal discharge and palpates the vulva and internal

DEVELOPMENTAL EXPOSITION

The Female Reproductive System

Explanation

Although genetic sex is determined at fertilization, both sexes develop similarly through the indifferent stage of the eighth week. The gonads of both sexes develop from the **gonadal ridges** medial to the mesonephros. **Primary sex cords** form within the gonadal ridges during the sixth week. The **genital tubercle** also develops during the sixth week as an external swelling cephalic to the cloacal membranes.

The ovaries develop more slowly than do the testes. Ovarian development begins at about the tenth week when **primordial follicles** begin to form within the ovarian medulla. Each of the primordial follicles consists of an **oogonium** (*o″ŏ-go′ne-um*) surrounded by a layer of **follicular cells.** Mitosis of the oogonia occurs during fetal development, so that thousands of germ cells are formed. Unlike the male reproductive system, in which spermatogonia are formed by mitosis throughout life, all oogonia are formed prenatally and their number continuously decreases after birth.

The **genital tract** includes the uterus and the uterine tubes. These organs develop from a pair of embryonic tubes called the **paramesonephric** (müllerian) **ducts.** The paramesonephric ducts form to the sides of the mesonephric ducts that give rise to the kidneys. As the mesonephric ducts regress, the paramesonephric ducts develop into the female genital tract (exhibit I). The lower

primordial: L. *prima*, first; *ordior*, to begin
follicle: L. diminutive of *follis*, bag
müllerian ducts: from Johannes P. Müller, German physician, 1801–58

Exhibit I The development of the female genital tract. (*a*) A lateral view and (*b*) an anterior view.

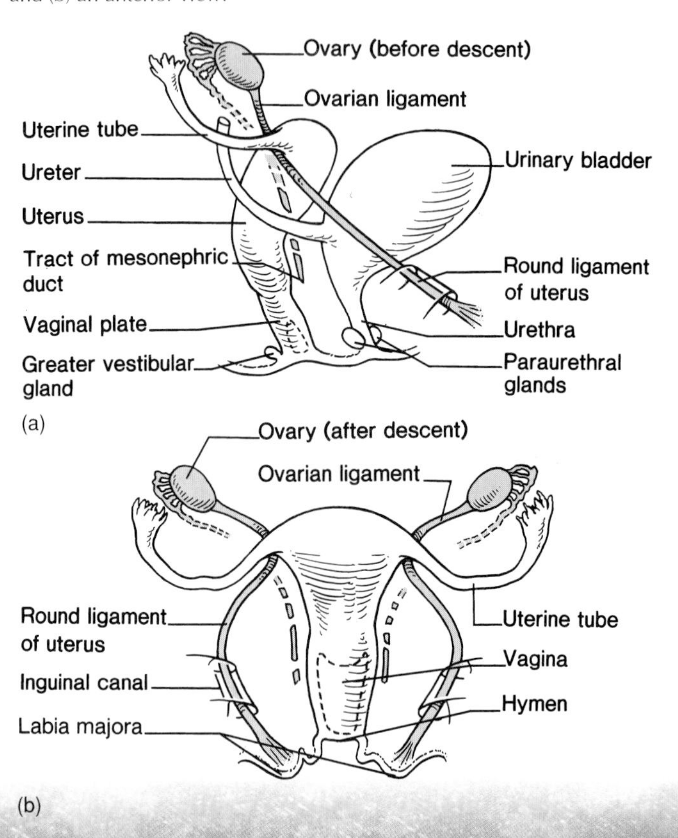

organs. Most of the internal organs can be palpated through the vagina, especially if they are enlarged or tender. Inserting a lubricated *speculum* into the vagina allows visual examination of the cervix and vaginal walls. A speculum is an instrument for opening or distending a body opening to permit visual inspection.

In special cases, it may be necessary to examine the cavities of the uterus and uterine tubes by *hysterosalpingography* (*his″ter-o-sal″ping-gog′ră-fe*) (fig. 21.20). This technique involves injecting a radiopaque dye into the reproductive tract. The patency of the uterine tubes, irregular pregnancies, and various types of tumors may be detected using this technique. A *laparoscopy* (*lap″ă-ros′kŏ-pe*) permits visualization of the internal reproductive organs. The entrance for the laparoscope may be through the umbilicus, through a small incision in the lower abdominal wall, or through the posterior fornix of the vagina into the rectouterine pouch. Although a laparoscope is used primarily in diagnosis, it can be used when performing a **tubal ligation** (fig. 21.21), a method of sterilizing a female by tying off the uterine tubes.

One diagnostic procedure that should be routinely performed by a woman is a *breast self-examination (BSE)*. The importance of a BSE is not to prevent diseases of the breast but to detect any problems before they become life-threatening. One in nine women will develop breast cancer during her lifetime. Early detection of breast cancer and follow-up medical treatment minimizes the necessary surgical treatment and improves the patient's prognosis. Breast cancer is curable if it is caught early.

A woman should examine her breasts monthly. If she has not yet reached menopause, the ideal time for a BSE is 1 week after her period ends because the breasts are less likely to be swollen and tender at that time. A woman no longer menstruating should just pick any day of the month and do a BSE on that

portions of the ducts fuse to form the uterus. The upper portions remain unfused and give rise to the uterine tubes.

The epithelial lining of the vagina develops from the endoderm of the urogenital sinus. The formation of a thin membrane called the **hymen** separates the lumen of the vagina from the urethral sinus. The hymen is usually perforated during later fetal development.

The external genitalia of both sexes appear the same during the indifferent stage of the eighth week (see page 698). A prominent **phallus** (*fal'us*) forms from the genital tubercle, and a **urethral groove** forms on the ventral side of the phallus. Paired **urethral folds** surround the urethral groove on the lateral sides. In the male embryo, these indifferent structures become masculinized by testosterone secreted by the testes. In the female embryo, in the absence of testosterone, feminization occurs.

In the process of feminization, growth of the phallus is inhibited, and the relatively small **clitoris** is formed. The urethral folds remain unfused to form an inner **labia minora,** and the labioscrotal swellings remain unfused to form the prominent **labia majora** (table 21.3). The external genitalia of a female are completely formed by the end of the twelfth week.

hymen: Gk. (mythology) Hymen was god of marriage; *hymen*, thin skin or membrane

Table 21.3
Summary of Homologous Structures

Indifferent Stage	Male	Female
Gonads	Testes	Ovaries
Urethral groove	Membranous urethra	Vaginal vestibule
Genital tubercle	Glans penis	Clitoris
Urethral folds	Spongy urethra	Labia minora
Labioscrotal swellings	Scrotum	Labia majora
	Bulbourethral glands	Vestibular glands

The structure of the hymen is characterized by tremendous individual variation. The hymen of a baby girl may be absent, or it may partially (or occasionally completely) cover the vaginal orifice, in which case it is called an *imperforate hymen.* An imperforate hymen is usually not detected until the first menstruation (menarche), when the discharge cannot be expelled. If the hymen is present, it may be ruptured during childhood in the course of normal exercise. On the other hand, a hymen may be so elastic that it persists even after coitus. Therefore, the presence of a hymen is not a reliable sign of virginity.

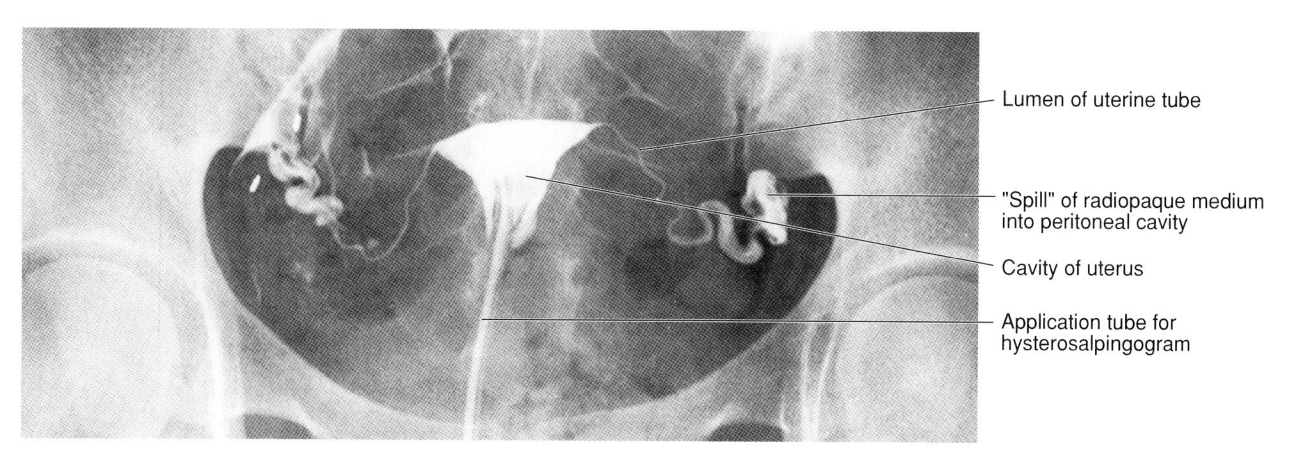

Figure 21.20

A contrast radiograph of the uterine cavity and lumina of the uterine tubes (hysterosalpingogram).

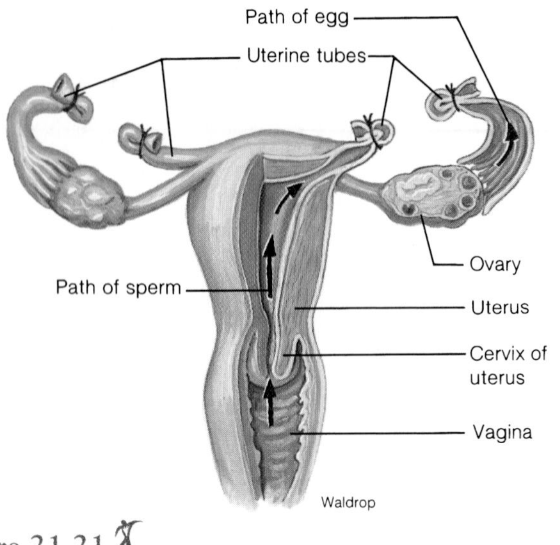

Figure 21.21

A tubal ligation involves the removal of a portion of each uterine tube. In actual practice, cautery, clips, or rings are used for tubal closure more often than ligation with suture thread.

same day on a monthly basis. Visual inspection and palpation (fig. 21.22) are equally important in doing a BSE. The steps involved in this procedure are the following:

1. **Observation in front of a mirror.** Inspect the breasts with the arms at the sides. Next, raise the arms high overhead. Look for any changes in the contour of each breast—a swelling, dimpling of skin or changes in the nipple. The left and right breast will not match exactly—few women's breasts do. Finally, squeeze the nipple of each breast gently between the thumb and index finger to check for discharge. Any discharge from the nipple should be reported to a physician.
2. **Palpation during bathing.** Examine the breasts during a bath or shower when the hands will glide easily over wet skin. With the fingers flat, move gently over every part of each breast. Use the right hand to examine the left breast, and the left hand for the right breast. Palpate for any lump, hard knot, or thickening. If the breasts are normally fibrous or lumpy (fibrocystic tissue), the locations of these lumps should be noted and checked each month for changes in size and location.
3. **Palpation while lying down.** To examine the right breast, put a pillow or folded towel under the right shoulder. Place the right hand behind the head—this distributes the breast tissue more evenly on the rib cage. With the fingers of the left hand held flat, press gently in small circular motions around an imaginary clock face. Begin at outermost top of the right breast for 12 o'clock, then move to 1 o'clock, and so on around the circle back to 12 o'clock. A ridge of firm tissue in the lower curve of each breast is normal. Then move in an inch, toward the nipple, and keep circling to examine every part of the breast, including the nipple. Also examine the armpit carefully for enlarged lymph nodes. Repeat the procedure on the left breast.

Another important diagnostic procedure is a *Papanicolaou (Pap) smear.* The Pap smear permits a microscopic examination of cells covering the tip of the cervix. Samples of cells are obtained by gently scraping the surface of the cervix with a specially designed wooden spatula. Women should have periodic Pap smears for the early detection of cervical cancer.

Developmental Problems of the Female Reproductive System

Many of the developmental problems of the female reproductive system also occur in the reproductive system of the male and were discussed in the previous chapter. Hermaphroditism and irregularities of the sex chromosomes, for example, are developmental conditions that cause a person to develop both male and female characteristics.

Other developmental abnormalities of the female reproductive system may occur during the formation of the uterus and vagina. Failure of the paramesonephric ducts to fuse normally can result in a **double uterus,** a **bicornuate** (*bi-kor′nyoo-āt*) **uterus,** or a **unicornuate uterus.** These types of uterine abnormalities and others are diagrammed in figure 21.23. In about 1 in 4,000 females, the vagina is absent. This is usually accompanied by an absence of the uterus as well.

Problems Involving the Ovaries and Uterine Tubes

Nonmalignant **ovarian cysts** are lined by cuboidal epithelium and filled with a serous albuminous fluid. These abnormal growths often can be palpated during a gynecological examination and may require surgical removal if they exceed about 4 cm in diameter. They are generally removed as a precaution because it is impossible to determine by palpation whether the mass is malignant or benign.

Ovarian tumors, which occur most often in women over the age of 60, can grow to be massive. Ovarian tumors as heavy as 5 kg (14 lb) are not uncommon, and some weighing as much as 110 kg (300 lb) have been reported. Some ovarian tumors produce estrogen and thus cause feminization in elderly women, including the resumption of menstrual periods. The prognosis for women with ovarian tumors varies depending on the type of tumor, whether or not it is malignant, and if it is, the stage of the cancer.

Two frequent problems involving the uterine tubes are salpingitis and ectopic pregnancies. **Salpingitis** is an inflammation of one or both uterine tubes. Infection of the uterine tubes is generally caused by a sexually transmitted disease, although secondary bacterial infections from the vagina may also cause salpingitis. Salpingitis may cause sterility if the uterine tubes become occluded.

Pap smear: from George N. Papanicolaou, American anatomist and physician, 1883–1962

bicornuate: L. *bi,* two; *cornu,* horn

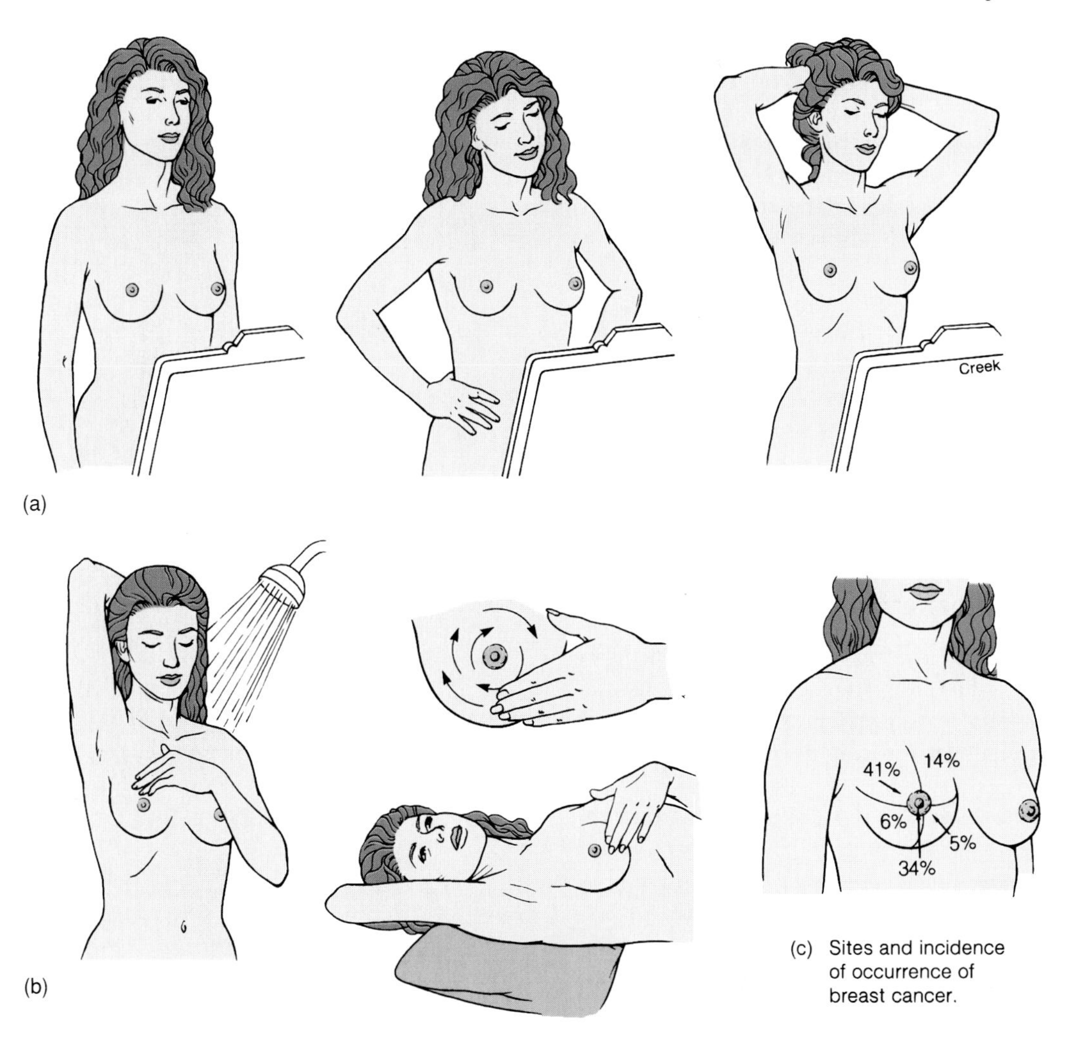

Figure 21.22

The technique of conducting a breast self-examination (BSE) involves (*a*) visual inspection and (*b*) palpation. The sites and incidence of occurrence of breast cancer are shown in (*c*).

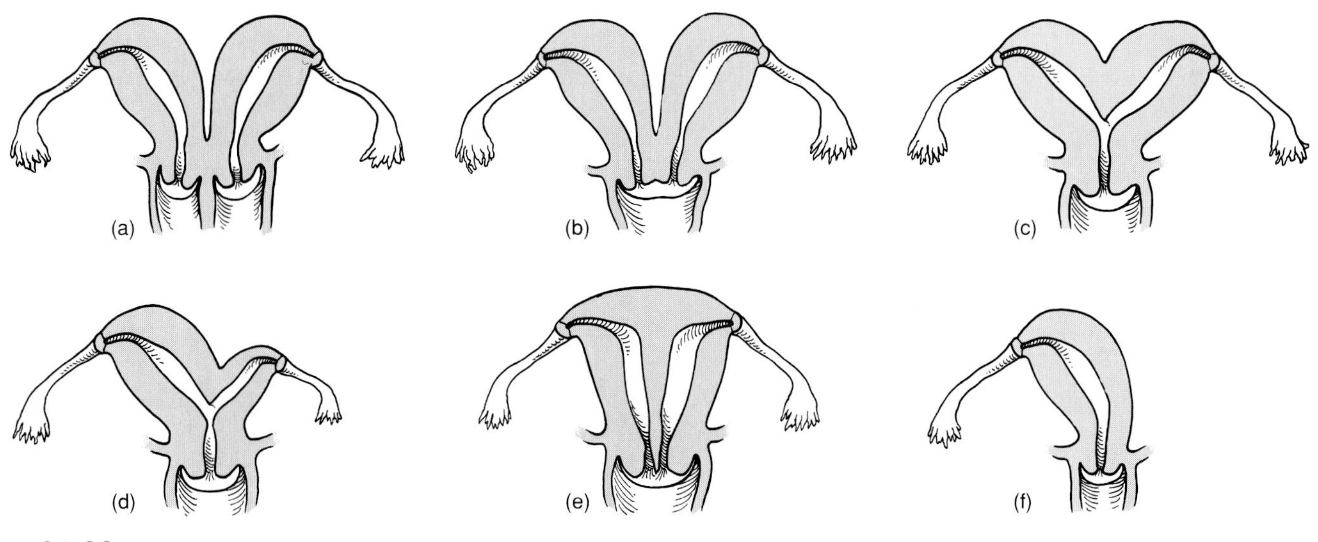

Figure 21.23

Congenital uterine abnormalities. (*a*) A double uterus and double vagina, (*b*) a double uterus with a single vagina, (*c*) a bicornuate uterus, (*d*) a bicornuate uterus with a rudimentary left horn, (*e*) a septate uterus, and (*f*) a unicornuate uterus.

Ectopic pregnancy results from implantation of the blastocyst in a location other than the body or fundus of the uterus. The most frequent ectopic site is in the uterine tube, where an implanted blastocyst causes what is commonly called a **tubal pregnancy.** One danger of a tubal pregnancy is the enlargement, rupture, and subsequent hemorrhage of the uterine tube where implantation has occurred. A tubal pregnancy is frequently treated by removing the affected tube.

Infertility, or the inability to conceive, is a clinical problem that may involve the male or female reproductive system. Based on the number of people who seek help for this problem, it is estimated that 10% to 15% of couples have impaired fertility. Generally, when a male is infertile, it is because of inadequate sperm counts. Female infertility is frequently caused by an obstruction of the uterine tubes or abnormal ovulation.

Problems Involving the Uterus

Abnormal menstruations are among the most common disorders of the female reproductive system. Abnormal menstruations may be directly related to problems of the reproductive organs and pituitary gland or associated with emotional and psychological stress.

Amenorrhea (*a-men″ŏ-re-ă*) is the absence of menstruation and may be categorized as normal, primary, or secondary. *Normal amenorrhea* follows menopause, occurs during pregnancy, and in some women may occur during lactation. *Primary amenorrhea* is the failure to have menstruated by the age when menstruation normally begins. Primary amenorrhea is generally accompanied by lack of development of the secondary sex characteristics. Endocrine disorders may cause primary amenorrhea and abnormal development of the ovaries or uterus.

Secondary amenorrhea is the cessation of menstruation in women who previously have had normal menstrual periods and who are not pregnant and have not gone through menopause. Various endocrine disturbances and psychological factors may cause secondary amenorrhea. It is not uncommon, for example, for young women who are in the process of making major changes or adjustments in their lives to miss menstrual periods. Secondary amenorrhea is also frequent in women athletes during periods of intense training. A low percentage of body fat may be a contributing factor. Sickness, fatigue, poor nutrition, or emotional stress also may cause secondary amenorrhea.

Dysmenorrhea is painful or difficult menstruation accompanied by severe menstrual cramps. The causes of dysmenorrhea are not totally understood but may include endocrine disturbances (inadequate progesterone levels), a faulty position of the uterus, emotional stress, or some type of obstruction that prohibits menstrual discharge.

Abnormal uterine bleeding includes **menorrhagia** (*men″ŏ-ra′je-ă*), or excessive bleeding during the menstrual period, and **metrorrhagia,** or spotting between menstrual periods. Other types of abnormal uterine bleeding are menstruations of excessive duration, too-frequent menstruations, and postmenopausal bleeding. These abnormalities may be caused by hormonal irregularities, emotional factors, or various diseases and physical conditions.

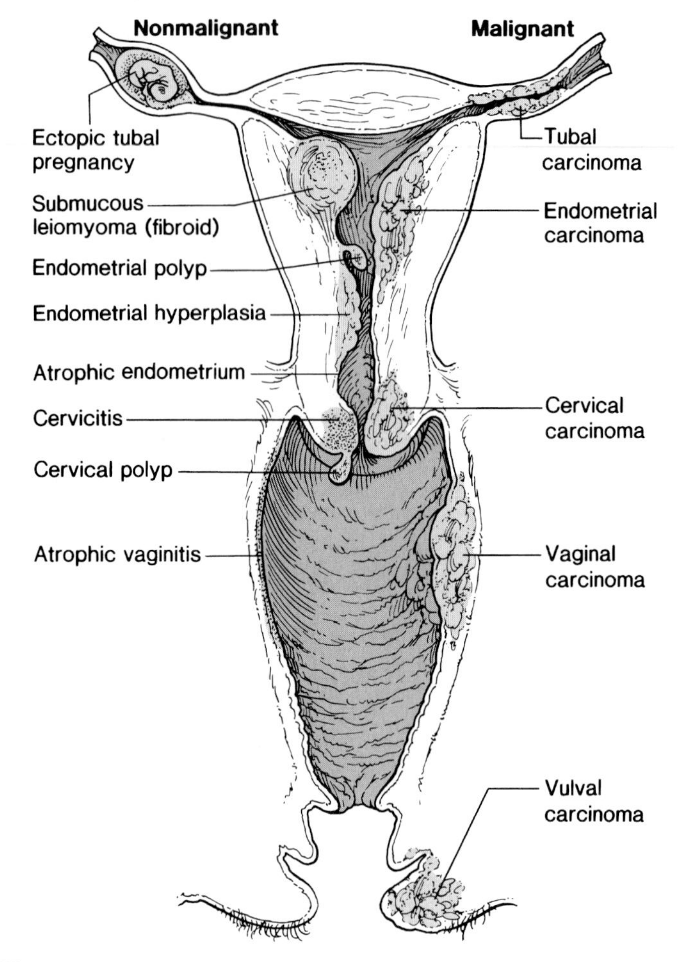

Figure 21.24

Sites of various conditions and diseases of the female reproductive tract, all of which could cause an abnormal discharge of blood.

Uterine neoplasms are an extremely common problem of the female reproductive tract. They include cysts, polyps, and smooth muscle tumors (leiomyomas), and most of them are benign. Any of these conditions may provoke irregular menstruations and may cause infertility if the neoplasms are massive.

Cancer of the uterus is the most common malignancy of the female reproductive tract. The most common site of uterine cancer is the cervix (fig. 21.24). Cervical cancer, which is second only to cancer of the breast in frequency of occurrence, is a disease of relatively young women (ages 30 through 50), especially those who have had frequent sexual intercourse with multiple partners during their teens and onward. If detected early through regular Pap smears, the disease can be cured before it metastasizes. The treatment of cervical cancer depends on the stage of the malignancy and the age and health of the woman. In the case of women for whom fertility is not an issue, a **hysterectomy** (*his″tĕ-rek′tŏ-me*) (surgical removal of the uterus) is usually performed.

---

neoplasm: Gk. *neos*, new; *plasma*, something formed

Endometriosis (*en″do-me″tre-o′sis*) is a condition characterized by the presence of endometrial tissues at sites other than the inner lining of the uterus. Frequent sites of ectopic endometrial cells are on the ovaries, on the outer layer of the uterus, on the abdominal wall, and on the urinary bladder. Although it is not certain how endometrial cells become established outside the uterus, it is speculated that some discharged endometrial tissue might be flushed backward from the uterus and through the uterine tubes during menstruation. Women with endometriosis will bleed internally with each menstrual period because the ectopic endometrial cells are stimulated along with the normal endometrium by ovarian hormones. The most common symptoms of endometriosis are extreme dysmenorrhea and a feeling of fullness during each menstrual period. Endometriosis can cause infertility. It is treated by suppressing the endometrial tissues with oral contraceptive pills or by surgery. An oophorectomy, or removal of the ovaries, may be necessary in extreme cases.

Uterine displacements are relatively common in elderly women. When uterine displacements occur in younger women, they are important because they may cause dysmenorrhea, infertility, or problems with childbirth. **Retroversion,** or **retroflexion,** is a displacement backward; **anteversion,** or **anteflexion,** is displacement forward. **Prolapse of the uterus** is a marked downward displacement of the uterus into the vagina.

An **abortion** is defined as the termination of a pregnancy before the twenty-eighth week of gestation. A **spontaneous abortion,** or **miscarriage,** occurs without mechanical aid or medicinal intervention and may occur in as many as 10% of all pregnancies. Spontaneous abortions usually occur when there is abnormal development of the fetus or disease of the maternal reproductive system. An **induced abortion** is removal of the fetus from the uterus by mechanical means or drugs. Induced abortions are a major issue of controversy because of questions regarding individual rights (those of the mother and those of the fetus), the definition of life, and morality.

Diseases of the Vagina and Vulva

Pelvic inflammatory disease (PID) is a general term for inflammation of the female reproductive organs within the pelvis. The infection may be confined to a single organ, or it may involve all of the internal reproductive organs. The pathogens generally enter through the vagina during coitus, induced abortion, or childbirth. Inflammation of the ovaries is called **oophoritis** (*o″of-ŏ-ri′tis*) and inflammation of the uterine tube is **salpingitis** (*sal″pin-ji′tis*).

The vagina and vulva are generally resistant to infection because of the acidity of the vaginal secretions. Occasionally, however, localized infections and inflammations do occur. These are termed **vaginitis,** if confined to the vagina, or **vulvovaginitis,** if both the vagina and external genitalia are affected. The symptoms of vaginitis are a discharge of pus (*leukorrhea*) and itching (*pruritus*). The two most common organisms that cause vaginitis are the protozoan *Trichomonas vaginalis* and the fungus *Candida albicans*.

Diseases of the Breasts and Mammary Glands

The breasts and mammary glands of females are highly susceptible to infections, cysts, and tumors. Infections involving the mammary glands usually follow the development of a dry and cracked nipple during lactation. Bacteria enter the wound and establish an infection within the lobules of the gland. During an infection of the mammary gland, a blocked duct frequently causes a lobe to become engorged with milk. This localized swelling is usually accompanied by redness, pain, and an elevation of temperature. Administering specific antibiotics and applying heat are the usual treatments.

Nonmalignant cysts are the most frequent diseases of the breast. These masses are generally of two types, neither of which is life-threatening. **Dysplasia** (fibrocystic disease) is a broad condition involving several nonmalignant diseases of the breast. All dysplasias are benign neoplasms of various sizes that may become painful during or prior to menstruation. Most of the masses are small and remain undetected. Dysplasia affects nearly 50% of women over the age of 30 prior to menopause.

A **fibroadenoma** (*fi″bro-ad″ĕ-no′mă*) is a benign tumor of the breast that frequently occurs in women under the age of 35. Fibroadenomas are nontender rubbery masses that are easily moved about in the mammary tissue. A fibroadenoma can be excised in a physician's office under local anesthetic.

Carcinoma of the breast is the most common malignancy in women (see fig. 10.59). One in nine women will develop breast cancer and one-third of these will die from the disease. Breast cancer is the leading cause of death in women between 40 and 50 years of age. Men are also susceptible to breast cancer, but it is 100 times more frequent in women. Breast cancer in men is usually fatal.

The causes of breast cancer are not known, but women who are most susceptible are those who are over age 35, who have a family history of breast cancer, and who are nulliparous (never having given birth). The early detection of breast cancer is important because the prognosis worsens as the disease progresses.

Confirming suspected breast cancer generally requires *mammography* (fig. 21.25). If the mammogram suggests breast cancer, a biopsy (tissue sample) is obtained and the tumor assessed. If the tumor is found to be malignant, surgery is performed, the extent of which depends on the size of the tumor and whether metastasis has occurred. The surgical treatment for breast cancer is generally some degree of *mastectomy (mas-tek′tŏ-me)*. A *simple mastectomy* is removal of the entire breast but not the underlying lymph nodes. A *modified radical mastectomy* is the complete removal of the breast, the lymphatic drainage, and perhaps the pectoralis major muscle. A *radical mastectomy* is similar to a modified except that the pectoralis major muscle is always removed, as well as the pectoral lymph nodes and adjacent connective tissue.

Methods of Contraception

The *rhythm method* of birth control continues to be used by many people, but the popularity of this technique has declined

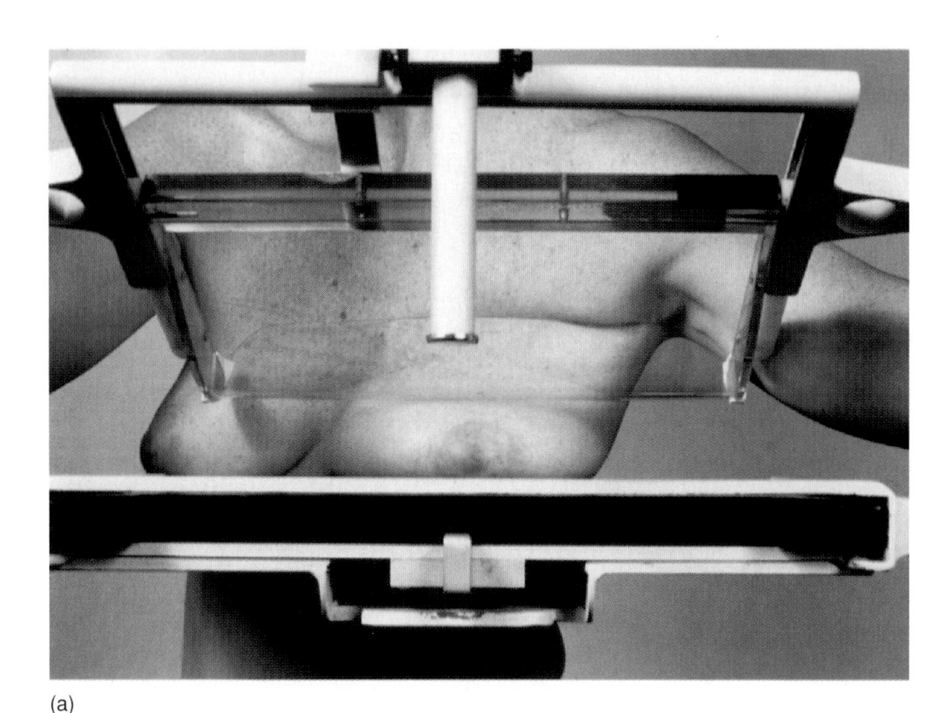

(a)

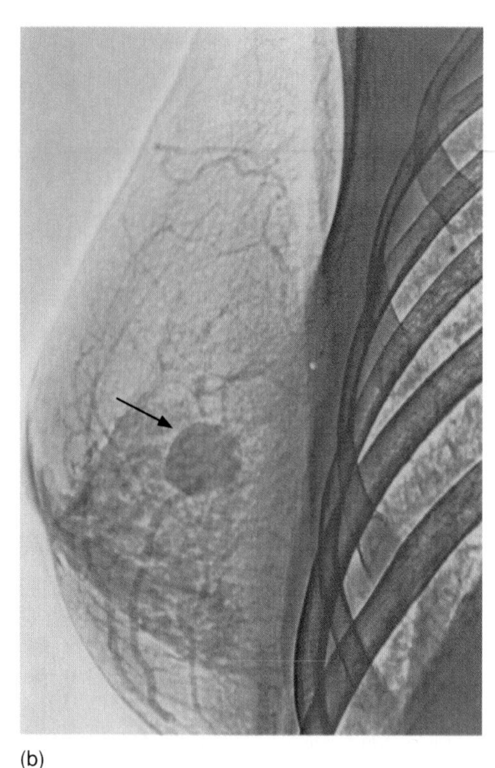

(b)

Figure 21.25

(*a*) In mammography, the breast is placed alternately on a metal plate and radiographed from above and from the side. (*b*) A mammogram of a patient with carcinoma of the breast. (Note the presence of a neoplasm indicated with an arrow.)

as more successful methods of contraception have been introduced (fig. 21.26). In the rhythm method, an attempt is made to predict the day of the woman's ovulation and restrict coitus to safe times of the cycle that allow no chance for fertilization to occur. The day of ovulation can be determined by a rise in basal body temperature or by a change in mucous discharge from the vagina. This technique has one of the highest failure rates of any of the widely used birth control methods, largely because women's cycles are often irregular.

More popular methods of birth control include *sterilization*, *oral contraceptives*, *intrauterine devices (IUDs)*, and *barrier methods*—including condoms for the male and female and diaphragms for the female. All of these techniques are effective, but they vary with respect to safety, side effects, and degree of efficacy.

Sterilization techniques include *vasectomy* for the male and *tubal ligation* for the female. In the latter technique (which accounts for over 60% of sterilization procedures performed in the United States), the uterine tubes are cut and tied. This is analogous to the procedure performed on the ductus deferentia in a vasectomy. It prevents fertilization of the ovulated ovum.

About 10 million women in the United States and 60 million women in the world are currently using **oral steroid contraceptives** ("the Pill"). These contraceptives usually consist of a synthetic estrogen combined with a synthetic progesterone in the form of pills that are taken once each day for 3 weeks after the last day of a menstrual period. This procedure causes an immediate increase in blood levels of ovarian steroids (from the pill), which is maintained for the normal duration of a monthly cycle. As a result of *negative feedback inhibition* of gonadotrophin secretion, *ovulation never occurs*. The entire cycle is like a false luteal phase, with high levels of progesterone and estrogen and low levels of gonadotrophins.

Since the contraceptive pills contain ovarian steroid hormones, the endometrium proliferates and becomes secretory, just as it does during a normal cycle. In order to prevent an abnormal growth of the endometrium, women stop taking the pill after 3 weeks. This causes estrogen and progesterone levels to fall, and permits menstruation to occur. The contraceptive pill is an extremely effective method of birth control, but it does have potentially serious side effects—including an increased incidence of thromboembolism and cardiovascular disorders. It has been pointed out, however, that the mortality risk associated with contraceptive pills is still much lower than the risk of death from complications of pregnancy—or from automobile accidents.

Another way in which to deliver hormonal contraceptives to a woman's body is by means of a **subdermal implant.** Implants are 2-in. rods filled with a hormonal contraceptive drug. They are implanted just under the skin, usually on the upper arm, through a tiny incision. The contraceptive hormone gradually leaches out through the walls of the rods and enters the bloodstream, preventing pregnancy for at least 5 years.

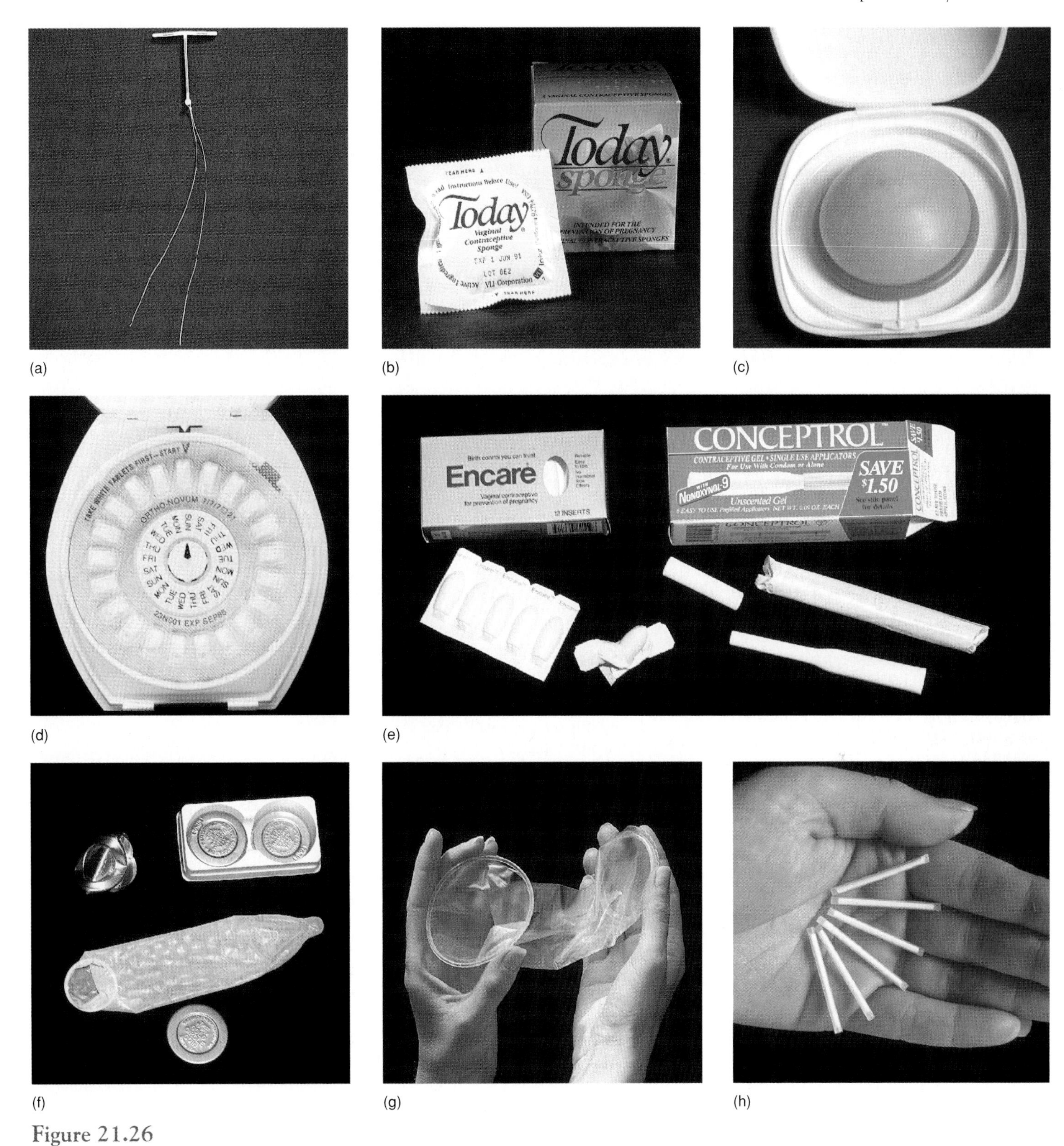

Figure 21.26

Various types of birth control devices. (*a*) IUD, (*b*) contraceptive sponge, (*c*) diaphragm, (*d*) birth control pills, (*e*) vaginal spermicides, (*f*) condom, (*g*) female condom, and (*h*) Norplant.

Intrauterine devices (IUDs) do not prevent ovulation, but instead prevent implantation of the blastocyst in the uterine wall in the event that fertilization occurs. The mechanisms by which their contraceptive effects are produced are not well understood but appear to involve their ability to cause inflammatory reactions in the uterus. Uterine perforations are the foremost complication associated with the use of IUDs. Because of the potential problems with IUDs, their use has diminished.

Barrier contraceptives—condoms, diaphragms, and cervical caps—are only slightly less effective than hormonal contraceptives or IUDs, but they do not have serious side effects. Barrier contraceptives are most effective when they are used in conjunction with spermicidal (sperm-killing) foams and gels. Many couples avoid them, however, because they detract from the spontaneity of sexual intercourse. Latex condoms offer an additional benefit; they provide some protection against sexually transmitted diseases, including AIDS.

Clinical Case Study Answer

An ectopic pregnancy is any pregnancy that implants outside of the uterine cavity. This is most likely to occur in the uterine tube. The events leading up to our patient's problem are as follows: An oocyte is extruded from the ovary during ovulation and received into the uterine tube. Soon after it enters the tube, the oocyte is fertilized, creating a zygote. Up to this point, the events are no different from those that occur in a normal pregnancy. In the case of tubal pregnancy, however, the transporting ability of the uterine tube fails, causing the conceptus to be retained in the tube. Implantation then occurs within tissues that are not well suited for that purpose. For example, the uterine tube does not expand well to accommodate a growing embryo, nor does it possess the necessary epithelium and glandular structures as does the endometrium of the uterus. The result is overexpansion and erosion of the tubal wall, leading to rupture and hemorrhage. Because the uterine tube is basically exposed to the peritoneal cavity, blood from the site of rupture can flow into and collect in the rectouterine pouch (pouch of Douglas or Douglas' cul-de-sac), which lies posterior to the vaginal fornix. There, it is easily aspirated by a needle placed through the vaginal wall.

Chapter Summary

Introduction to the Female Reproductive System (pp. 707–709)

1. The reproductive period of a female is the period between puberty (about age 12) and menopause (about age 50). In the course of this age span, cyclic ovulation and menstruation patterns occur in nonpregnant females.
2. The functions of the female reproductive system are to produce ova; secrete sex hormones; receive sperm from the male; provide sites for fertilization, implantation, and development of the embryo and fetus; facilitate parturition; and secrete milk from the mammary glands.
3. The female reproductive system consists of (a) primary sex organs—the ovaries; (b) secondary sex organs—those that are essential for sexual reproduction, characterized by latent development; and (c) secondary sex characteristics—features that are sexual attractants, expressed after puberty.

Structure and Function of the Ovaries (pp. 709–714)

1. The ovaries are supported by the mesovarium, which extends from the broad ligament, and by the ovarian and suspensory ligaments.
2. The ovarian follicles within the ovarian cortex undergo cyclic changes.
   (a) Primary oocytes, arrested at prophase I of meiosis, are contained within primordial follicles.
   (b) Upon stimulation by gonadotropic hormones, some of the primordial follicles enlarge to become primary follicles.
   (c) When a follicle develops a fluid-filled antrum, it is called a secondary follicle.
   (d) Generally only one follicle continues to grow to become a vesicular ovarian follicle.
   (e) The vesicular ovarian follicle contains a secondary oocyte, arrested at metaphase II of meiosis.
   (f) In the process of ovulation, the vesicular ovarian follicle ruptures and releases its secondary oocyte, which becomes a zygote upon fertilization.
   (g) After ovulation, the empty follicle becomes a corpus luteum.

Secondary Sex Organs (pp. 714–719)

1. The uterine tube, which conveys ova from the ovary to the uterus, provides a site for fertilization.
   (a) The open-ended portion of each uterine tube is expanded; its margin bears fimbriae that extend over the lateral surface of the ovary.
   (b) Movement of an ovum is aided by ciliated cells that line the lumen and by peristaltic contractions in the wall of the uterine tube.
2. The uterus is supported by the broad ligaments, uterosacral ligaments, cardinal ligaments, and round ligaments. The regions of uterus are the fundus, body, and cervix. The cervical canal opens into the vagina at the uterine ostium.
   (a) The endometrium consists of a stratum basale and a stratum functionale; the superficial stratum functionale is shed during menstruation.
   (b) The myometrium produces the muscular contractions needed for labor and parturition.
3. The vagina serves to receive the erect penis, to convey menses to the outside, and to transport the fetus during parturition.
   (a) The vaginal wall is composed of an inner mucosa, a middle muscularis, and an outer fibrous layer.
   (b) The vaginal orifice may be partially covered by a thin membranous hymen.
4. The external genitalia, or vulva, include the mons pubis, labia majora and minora, clitoris, vaginal vestibule, vestibular bulbs, and vestibular glands.

5. Impulses through parasympathetic nerves stimulate erectile tissues in the clitoris and vestibular bulbs; in orgasm, muscular contraction occurs in the perineum, uterus, and uterine tubes.

Mammary Glands (pp. 719–721)

1. Mammary glands, located within the breasts, are modified sweat glands.
   (a) Each mammary gland is composed of 15 to 20 lobes; the lobes are subdivided into lobules that contain mammary alveoli.
   (b) During lactation, the mammary alveoli secrete milk. The milk passes through mammary ducts, lactiferous ducts, and lactiferous sinuses and is discharged through the nipple.
2. The nipple is a cylindrical projection near the center of the breast, surrounded by the circular pigmented areola.

Ovulation and Menstruation (pp. 722–723)

1. Ovulation and menstruation are reproductive cyclic events that are regulated by hormones secreted by the hypothalamus, the anterior pituitary, and the ovaries.
2 The menstrual cycle is divided into menstrual, proliferative, and secretory phases.
3. The principal hormones that regulate ovulation and menstruation are estrogen, progesterone, follicle-stimulating hormone (FSH), and luteinizing hormone (LH).

Review Activities

Objective Questions

1. The cervix is a portion of
   (a) the vulva.
   (b) the vagina.
   (c) the uterus.
   (d) the uterine tubes.
2. Fertilization normally occurs in
   (a) the ovary.
   (b) the uterine tube.
   (c) the uterus.
   (d) the vagina.
3. The secretory phase of the endometrium corresponds to which of the following ovarian phases?
   (a) the follicular phase
   (b) ovulation
   (c) the luteal phase
   (d) the menstrual phase
4. Which of the following statements about oogenesis is *true?*
   (a) Oogonia, like spermatogonia, form continuously during postnatal life.
   (b) Primary oocytes are haploid.
   (c) Meiosis is completed prior to ovulation.
   (d) During ovulation, secondary oocytes are released from vesicular ovarian follicles.
5. The function of the mesovarium is
   (a) movement of the sperm to the ova.
   (b) nourishment of the ovarian walls.
   (c) muscular contraction of the uterus.
   (d) suspension of the ovary.
6. Which of the following layers is shed as menses?
   (a) the perimetrial layer
   (b) the fibrous layer
   (c) the functionalis layer
   (d) the menstrual layer
7. The transverse folds in the mucosal layer of the vagina are called
   (a) perineal folds. (c) fornices.
   (b) vaginal rugae. (d) labia gyri.
8. Which of the following is *not* a part of the vulva?
   (a) the mons pubis
   (b) the clitoris
   (c) the vaginal vestibule
   (d) the vagina
   (e) the labia minora
9. The paramesonephric (müllerian) ducts give rise to
   (a) the uterine tubes.
   (b) the uterus.
   (c) the pudendum.
   (d) both a and b.
   (e) both b and c.
10. In a female, the homologue of the male scrotum is/are
    (a) the labia majora.
    (b) the labia minora.
    (c) the clitoris.
    (d) the vestibule.

Essay Questions

1. Define *puberty*, *ovulation*, *menstruation*, and *menopause*.
2. Describe the follicular changes within the ovarian cortex during the events that precede and follow ovulation.
3. Define *oogenesis*. When is the process initiated and when is it completed?
4. Describe the gross and histologic structure of the uterus and comment on the significance of the two endometrial layers.
5. Identify the secondary sex organs and explain their functions.
6. Summarize the events of a menstrual cycle and explain the roles of estrogen and progesterone.
7. List the functions of the vagina and describe its structure.
8. Distinguish between the labia majora and labia minora, between the pudendal cleft and vaginal vestibule, and between the vestibular bulbs and vestibular glands.
9. List the events that cause an erection of the female genitalia, and explain the process of orgasm.
10. Describe the structure and position of the mammary glands in the breasts.
11. List the homologous reproductive organs of the male and female reproductive systems and note the undifferentiated structures from which they develop.
12. Define *gynecology*, *obstetrics*, *speculum*, *laparoscopy*, *breast self-examination*, and *Pap smear*.
13. Distinguish between normal, primary, and secondary amenorrhea. What causes each?
14. List the various kinds of uterine neoplasms. How is cancer of the uterus generally detected?
15. Distinguish between dysplasia, fibroadenoma, and carcinoma of the breast.

Critical-Thinking Questions

1. Why is the hormonal regulation of the reproductive system much more complex in females than it is in males?
2. Your 14-year-old daughter is serious about her gymnastics lessons and exercises rigorously for at least 2 hours every day. She has not yet had her first menstrual period and is beginning to get worried. How would you reassure her?
3. A wide variety of commercial douche preparations and scented aerosols can be purchased in drug stores and supermarkets as "feminine hygiene products." Why should women avoid using them?
4. Both sterility and impotence are sexual dysfunctions in men. Only one of these is associated with women, however. Explain.
5. "Contraceptive pills trick the brain into thinking you're pregnant." Explain what is meant by this statement.
6. With respect to homologous structures, explain why men have nipples? Why do mammary glands normally function only in females?
7. If a woman is on estrogen replacement therapy following menopause, why is it important to monitor her uterine lining?

TWENTY-TWO

Developmental Anatomy, Postnatal Growth, and Inheritance

Clinical Case Study 735
Fertilization 735
Preembryonic Period 737
- Cleavage and Formation of the Blastocyst 737
- Implantation 739
- Formation of the Germ Layers 740

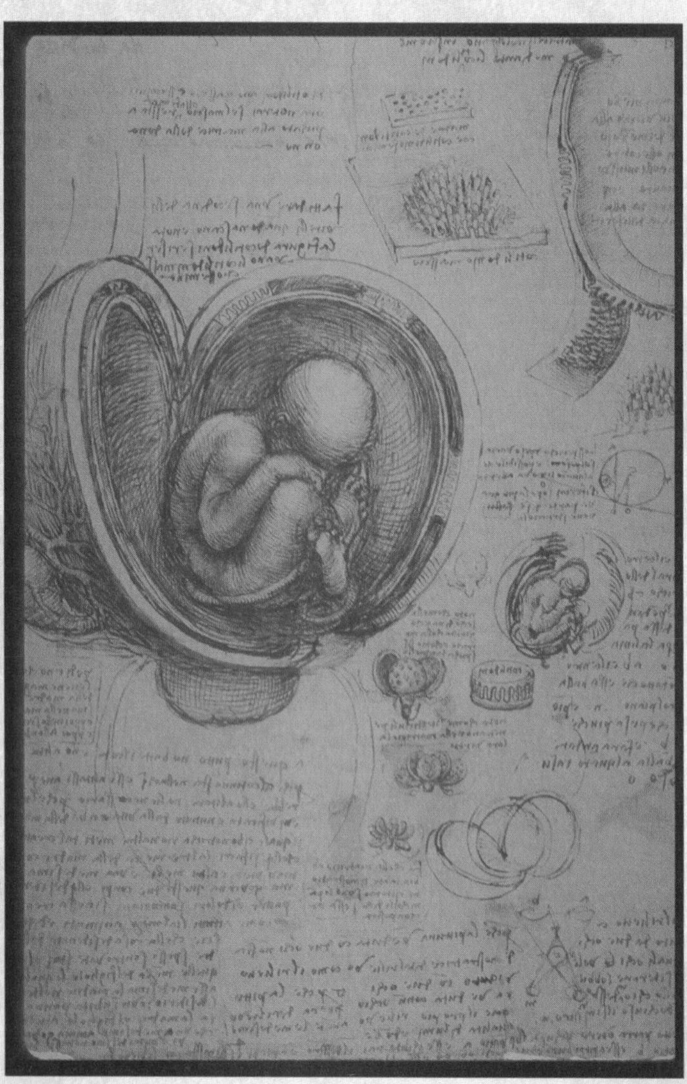

Among the drawings of Leonardo da Vinci are some famous studies of the fetus, annotated with the mirror script he invariably used. These superb renderings show a remarkable degree of accuracy, even including some quantitative calculations on the growth rate of the fetus in utero.

Embryonic Period 742
- Extraembryonic Membranes 742
- Placenta 745
- Umbilical Cord 746
- Structural Changes in the Embryo by Week 748

Fetal Period 752
- Structural Changes in the Fetus by Week 752

Labor and Parturition 755
Periods of Postnatal Growth 756
- Neonatal Period 756
- Infancy 757
- Childhood 757
- Adolescence 759
- Adulthood 760

Inheritance 762
Clinical Considerations 765
- Abnormal Implantation Sites 765
- In Vitro Fertilization and Artificial Implantation 766
- Multiple Pregnancy 766
- Fetal Monitoring 767
- Congenital Defects 769

Clinical Case Study Answer 769
Genetic Disorders of Clinical Importance 770
Chapter Summary 770
Review Activities 772

Clinical Case Study

A 27-year-old woman gave birth to twin boys, followed by an apparently single placenta. After examining the two infants, the pediatrician informed the mother that one of them had a cleft palate but that the other was normal. She added that cleft palate could be hereditary and asked if any family members had the problem. The mother said that she knew of none. Further examination of the placenta revealed two amnions and only one chorion.

Does the presence of two amnions and one chorion indicate monozygotic or dizygotic twins? How would you account for the fact that one baby has a cleft palate, while the other does not?

Hints: Read the section on multiple pregnancy at the end of the chapter and carefully study figure 22.40. Can any conclusions be drawn regarding genetic similarities between the two infants? Note that identical twins are always the same gender.

Fertilization

Upon fertilization of an ovum by a spermatozoon in the uterine tube, meiotic development is completed and a diploid zygote is formed.

Objective 1 Define *fertilization, capacitation,* and *morphogenesis.*

Objective 2 Describe the changes that occur in the spermatozoon and ovum prior to, during, and immediately following fertilization.

Fertilization refers to the penetration of an ovum (egg) by a spermatozoon (fig. 22.1), with the subsequent union of their genetic material. It is this event that determines a persons' biological inheritance. Fertilization cannot occur, however, unless certain conditions are met. First of all, an ovum must be present in the uterine tube—it can be there for at most 24 hours before it becomes incapable of undergoing fertilization. Second, large numbers of spermatozoa must be ejaculated to ensure fertilization. Although a recent study has shown that sperm cells can remain viable 5 days after ejaculation, this still leaves a "window of fertility" of only 6 days each month—the day of ovulation and the 5 days leading up to it.

As described in chapter 21, a woman usually ovulates one ovum a month, totaling about 400 during her reproductive years. Each ovulation releases an ovum that is actually a secondary oocyte arrested at metaphase of the second meiotic division. An ovum is surrounded by a thin layer of protein and polysaccharides, called the *zona pellucida,* and a layer of granulosa cells, called the *corona radiata* (fig. 22.1). These layers provide a protective shield around the ovum as it enters the uterine tube.

During coitus, a male ejaculates between 100 million and 500 million spermatozoa into the female's vagina. This tremendous number is needed because of the high fatality rate—only about 100 sperm cells survive to encounter the ovum in the uterine tube. In addition to deformed sperm cells—up to 20% in the average fertile male—another 25% will perish as soon as they contact the vagina's acidic environment. Still others are destroyed by the woman's immune cells, which recognize them as foreign cells. Even if they manage to reach the cervix, many sperm cells will get stuck there and never make it through the uterus.

If it finally encounters an ovum, the spermatozoon must penetrate the protective corona radiata and zona pellucida for fertilization to occur. To do this, the head of each sperm cell is capped by an organelle called an **acrosome** (*ak'rŏ-sōm*) (figs. 22.1 and 22.2). The acrosome contains a trypsinlike protein-digesting enzyme and *hyaluronidase* (*hi"ă-loo-ron'ĭ-dās*), which digests hyaluronic acid, an important constituent of connective tissue. When a spermatozoon meets an ovum in the uterine tube, an *acrosomal reaction* allows the spermatozoon to penetrate the corona radiata and the zona pellucida. A spermatozoon that comes along relatively late—after many others have undergone acrosomal reactions to expose the ovum membrane—is more likely to be the one that finally achieves penetration of the egg.

Experiments confirm that freshly ejaculated spermatozoa are infertile and must be in the female reproductive tract for at least 7 hours before they can fertilize an oocyte. Their membranes must become fragile enough to permit the release of the acrosomal enzymes—a process called *capacitation.* During in vitro fertilization, capacitation is induced artificially by treating the ejaculate with a solution of gamma globulin, free serum, follicular fluid, dextran, serum dialysate, and adrenal gland extract to chemically mimic the conditions of the female reproductive tract.

As soon as a spermatozoon penetrates the zona pellucida, a rapid chemical change in this layer prevents other spermatozoa from attaching to it. Therefore, only one spermatozoon is permitted to fertilize an oocyte. With the entry of a single spermatozoon through its cell membrane, the oocyte is stimulated to complete its second meiotic division (fig. 22.3). Like the first meiotic division, the second produces one cell that contains most of the cytoplasm and one polar body. The cell containing the cytoplasm is the mature ovum, and the second polar body, like the first, ultimately fragments and disintegrates.

At fertilization, the entire sperm cell enters the cytoplasm of the much larger ovum. Within 12 hours, the nuclear membrane in the ovum disappears, and the haploid number of chromosomes

zona pellucida: L. *zone,* a girdle; *pellis,* skin

corona radiata: Gk. *korone,* crown; *radiata,* radiate
acrosome: Gk. *akron,* extremity; *soma,* body
capacitation: L. *capacitas,* capable of
haploid: Gk. *haplous,* single; L. *ploideus,* multiple in form

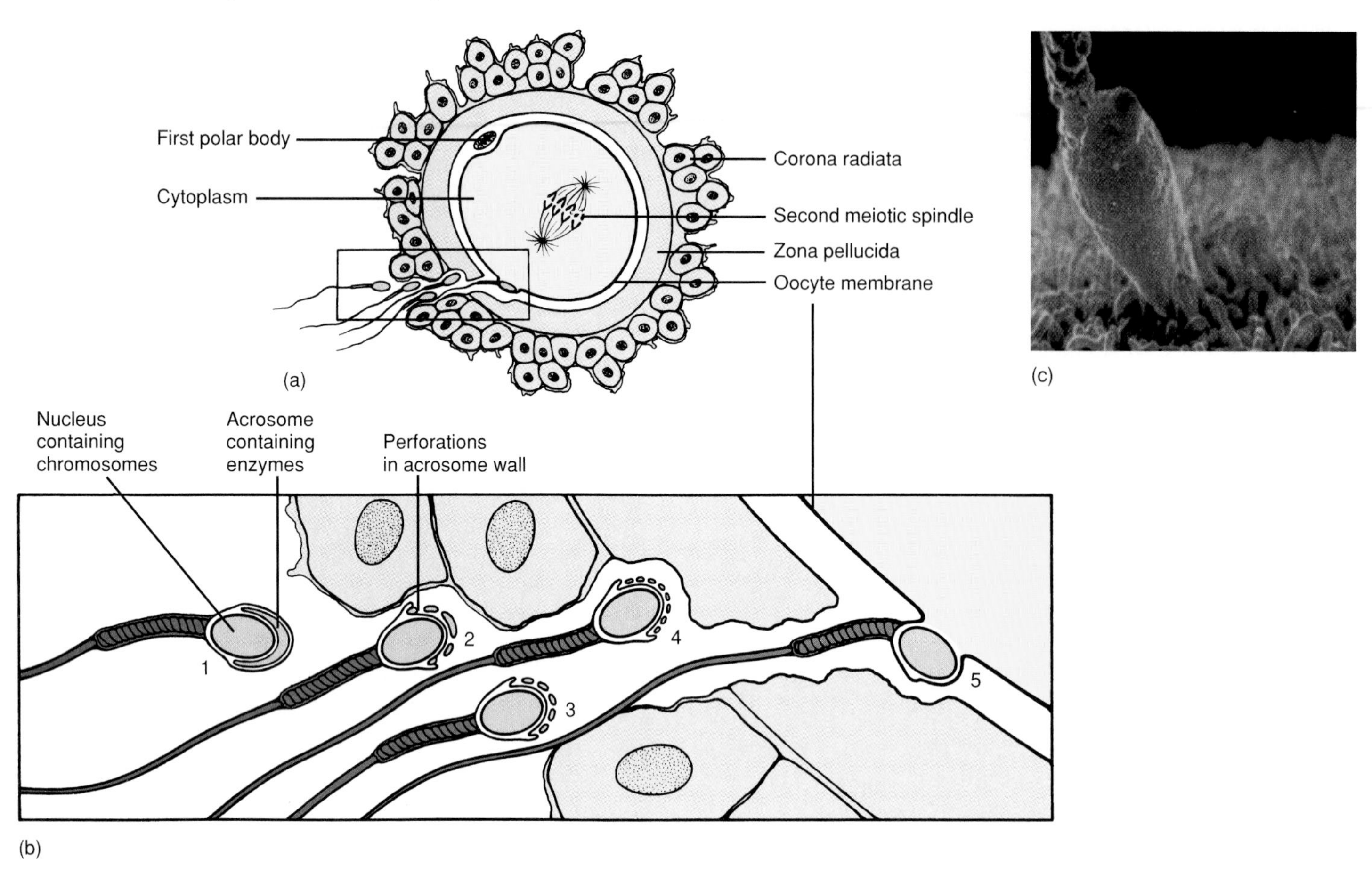

Figure 22.1

The process of fertilization. (*a,b*) As the head of the sperm encounters the corona radiata of the oocyte (2), digestive enzymes are released from the acrosome (3, 4), clearing a path to the oocyte membrane. When membrane of the sperm contacts the oocyte membrane (5), the membranes become continuous, and the nucleus and other contents of the sperm move into the egg cytoplasm of the oocyte. (*c*) A scanning electron micrograph of a sperm cell bound to the surface of an oocyte.

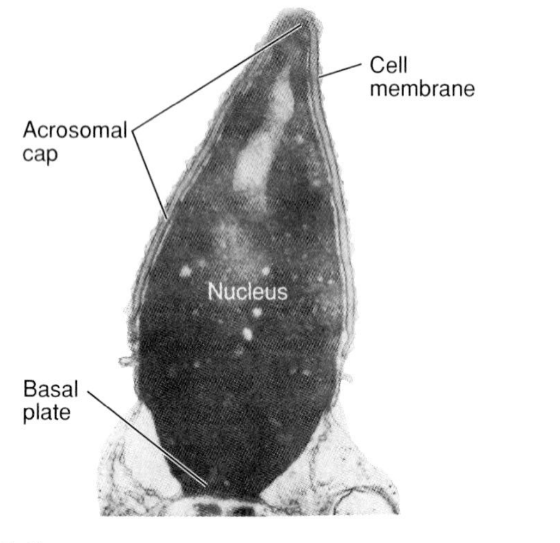

Figure 22.2

A transmission electron micrograph showing the head of a human spermatozoon with its nucleus and acrosome.

(23) in the ovum is joined by the haploid number of chromosomes from the spermatozoon. A fertilized egg, or **zygote** (*zi'gōt*), containing the diploid number of chromosomes (46) is thus formed.

Within hours after conception, the structure of the body begins to form from this single fertilized egg, culminating some 38 weeks later with the birth of a baby. The transformation involved in the growth and differentiation of cells and tissues is known as **morphogenesis** (*mor"fo-jen'ĭ-sis*), and it is through this awesome process that the organs and systems of the body are established in a functional relationship. Moreover, there are sensitive periods of morphogenesis for each organ and system, during which genetic or environmental factors may affect normal development.

Prenatal development can be divided into a *preembryonic period*, which is initiated by the fertilization of an ovum; an *embryonic*

zygote: Gk. *zygotos*, yolked, joined
diploid: Gk. *diplous*, double; L. *ploideus*, multiple in form
morphogenesis: Gk. *morphe*, form; *genesis*, beginning

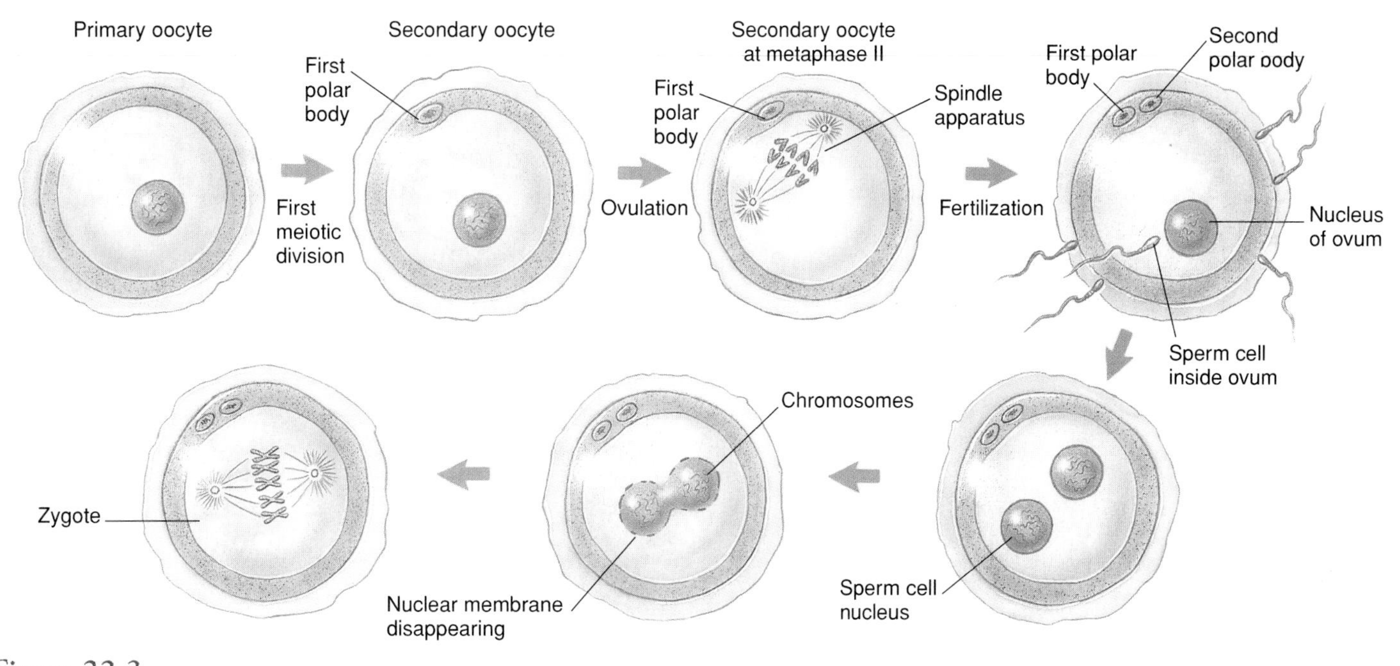

Figure 22.3

A secondary oocyte, arrested at metaphase II of meiosis, is released at ovulation. If this cell is fertilized, it will become a mature ovum, complete its second meiotic division, and produce a second polar body. The fertilized ovum is known as a zygote.

period, during which the body's organ systems are formed; and a *fetal period,* which culminates in parturition (childbirth).

1. Explain why capacitation and the acrosomal reaction of spermatozoa are necessary to accomplish fertilization of a secondary oocyte.
2. Discuss the changes that occur in a spermatozoon from the time of ejaculation to the time of fertilization. What changes occur in a secondary oocyte following ovulation to the time of fertilization?
3. Define *morphogenesis*. When does this process begin? What does it accomplish?

Preembryonic Period

The events of the 2-week preembryonic period include fertilization, transportation of the zygote through the uterine tube, mitotic divisions, implantation, and the formation of primordial embryonic tissue.

Objective 3 Describe the events of preembryonic development that result in the formation of the blastocyst.

Objective 4 Discuss the role of the trophoblast in the implantation and development of the placenta.

Objective 5 Explain how the primary germ layers develop and list the structures produced by each layer.

Objective 6 Define *gestation* and explain how the parturition date is determined.

Cleavage and Formation of the Blastocyst

Within 30 hours following fertilization, the zygote undergoes a mitotic division called **cleavage.** This first division results in the formation of two identical daughter cells called *blastomeres* (fig. 22.4). Additional cleavages occur as the structure passes down the uterine tube and enters the uterus on about the third day. It is now composed of a ball of 16 or more cells called a **morula** (*mor'yŭ-lă*). Although the morula has undergone several mitotic divisions, it is not much larger than the zygote because no additional nutrients necessary for growth have been entering the cells. The morula floats freely in the uterine cavity for about 3 days. During this time, the center of the morula fills with fluid passing in from the uterine cavity. As the fluid-filled space develops within the morula, two distinct groups of cells form, and the structure becomes known as a **blastocyst** (*blas'tŏ-sist*) (fig. 22.4). The hollow, fluid-filled center of the blastocyst is called the **blastocyst cavity.** The blastocyst is composed of an outer layer of cells, known as the **trophoblast,** and an inner aggregation of cells, called the **embryoblast** (*internal cell mass*) (see fig. 22.6). With further development, the trophoblast differentiates into a structure called the **chorion** (*kor'e-on*), which will later become a portion of the placenta. The embryoblast will become the embryo. A diagrammatic summary of the ovarian cycle, fertilization, and the morphogenic events of the first week is presented in figure 22.5.

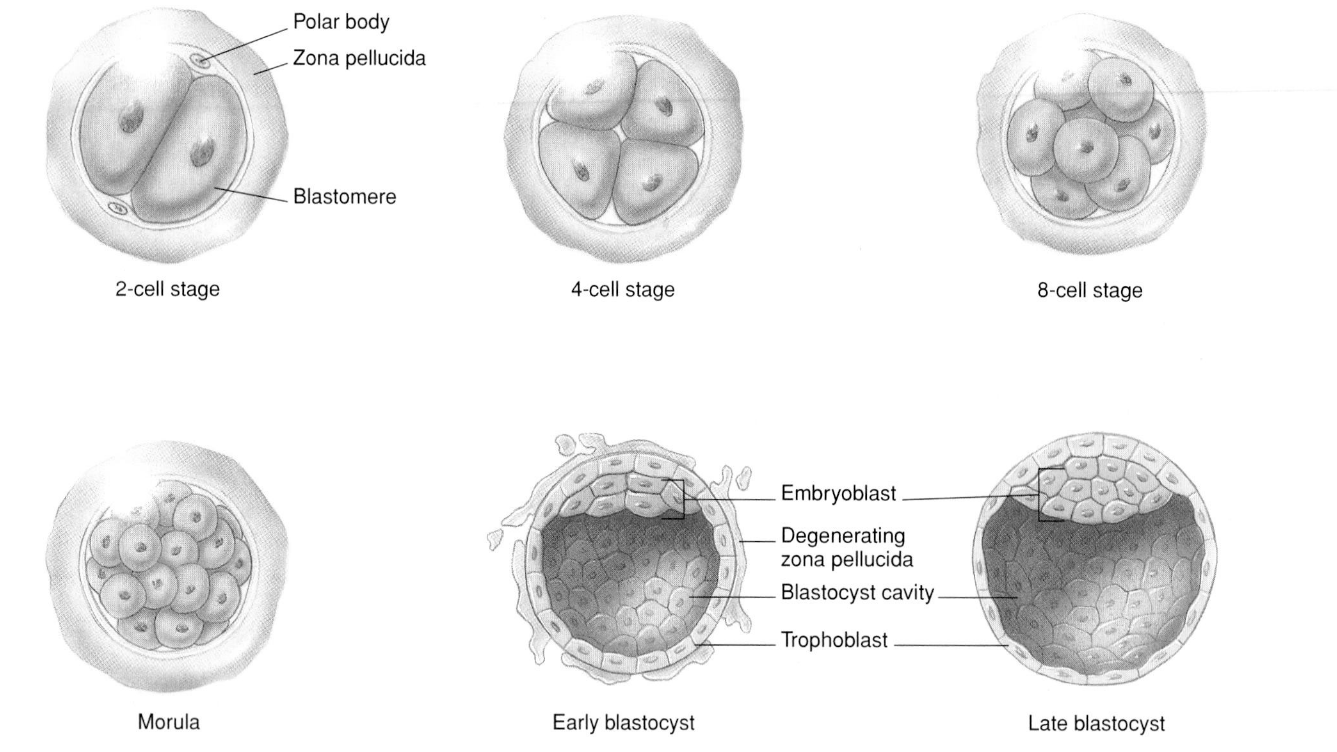

Figure 22.4

Sequential illustrations from the first cleavage of the zygote to the formation of the blastocyst. (Note the deterioration of the zona pellucida in the early blastocyst.)

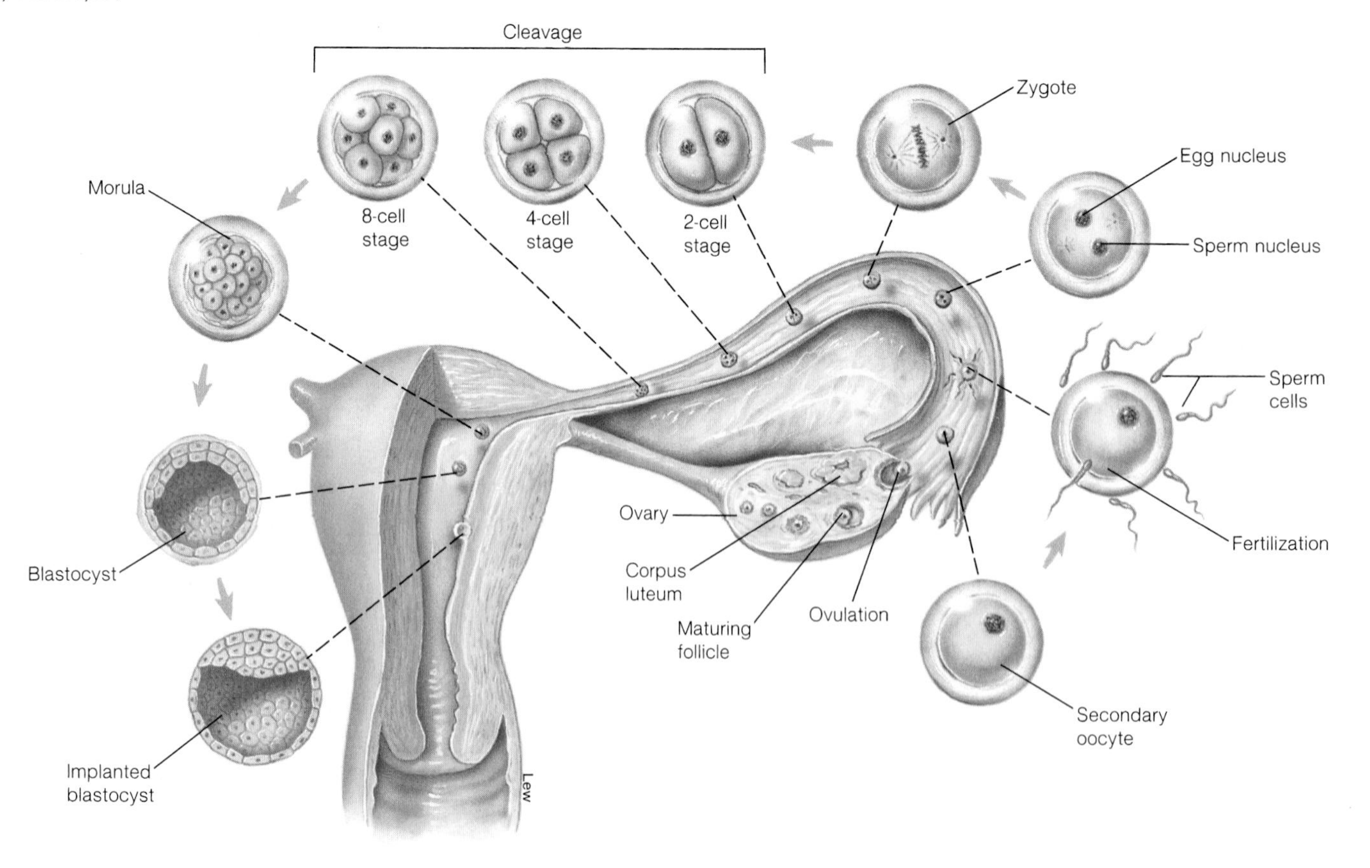

Figure 22.5

A diagram of the ovarian cycle, fertilization, and the events of the first week. Implantation of the blastocyst begins between the fifth and seventh day, and is generally completed by the tenth day.

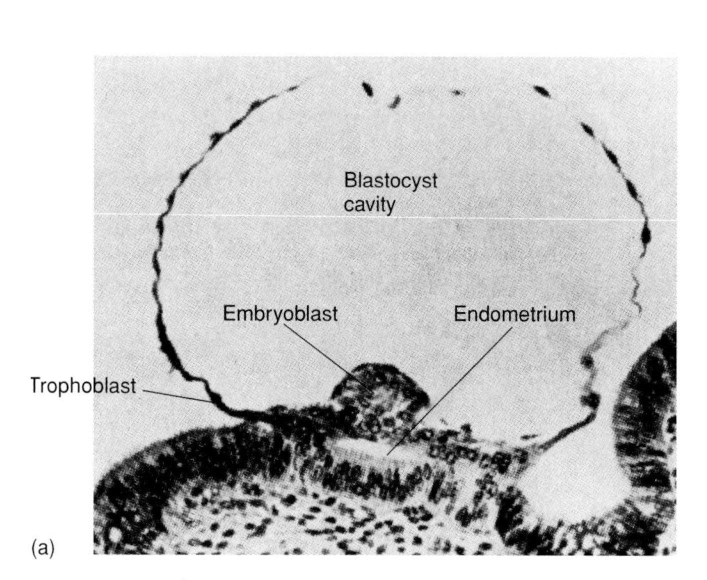

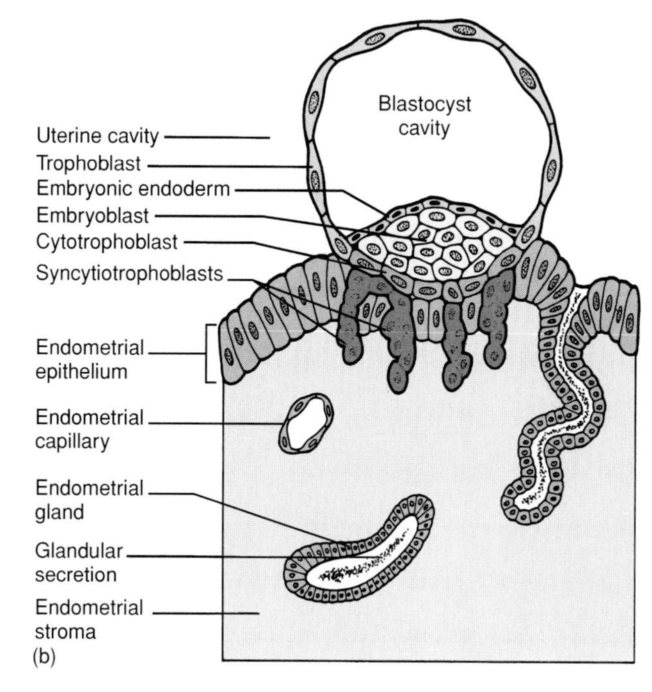

Figure 22.6

The blastocyst adheres to the endometrium on about the sixth day as seen in (*a*), a photomicrograph. By the seventh day (*b*), specialized syncytiotrophoblasts from the trophoblast have begun to invade the endometrium. The syncytiotrophoblasts secrete human chorionic gonadotropin (hCG) to sustain pregnancy and will eventually participate in the formation of the placenta for embryonic and fetal sustenance.

Implantation

The process of **implantation** begins between the fifth and seventh day following fertilization. This is the process by which the blastocyst embeds itself into the endometrium of the uterine wall (fig. 22.6*a*). Implantation is made possible by the secretion of *proteolytic enzymes* by the trophoblast, which digest a portion of the endometrium. The blastula sinks into the depression, and endometrial cells move back to cover the defect in the wall. At the same time, the part of the uterine wall below the implanting blastocyst thickens, and specialized cells of the trophoblast produce fingerlike projections, called **syncytiotrophoblasts** (*sin-sit″e-ŏ-trof′ŏ-blasts*), into the thickened area. The syncytiotrophoblasts arise from a specific portion of the trophoblast called the **cytotrophoblast** (fig. 22.6*b*), located next to the embryoblast.

The blastocyst saves itself from being aborted by secreting a hormone that indirectly prevents menstruation. Even before the sixth day when implantation begins, the syncytiotrophoblasts secrete **human chorionic gonadotropin** (*kor″e-on-ik go-nad-ŏ-tro′pin*) **(hCG).** This hormone is identical to LH in its effects, and therefore is able to maintain the corpus luteum past the time when it would otherwise regress. The secretion of estrogen and progesterone is maintained, and menstruation is normally prevented (fig. 22.7).

The secretion of hCG declines by the tenth week of pregnancy. Actually, this hormone is required only for the first 5 to 6 weeks of pregnancy because the placenta itself becomes an active steroid-secreting gland by this time. At the fifth to sixth week, the mother's corpus luteum begins to regress (even in the presence of hCG), but the placenta secretes more than sufficient amounts of steroids to maintain the endometrium and prevent menstruation.

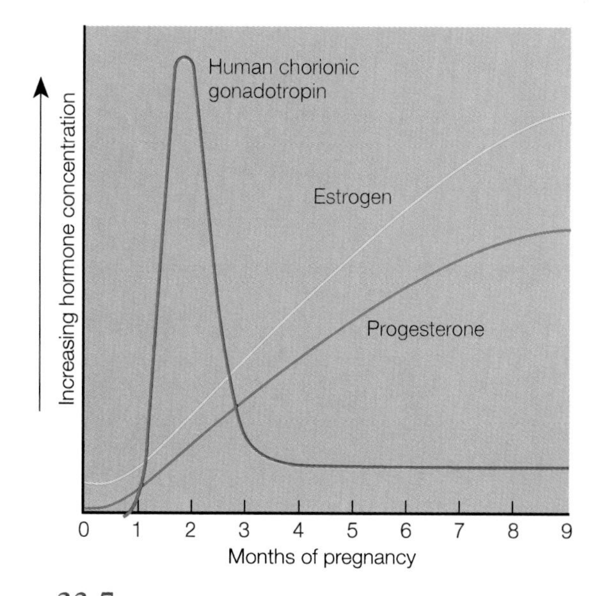

Figure 22.7

Human chorionic gonadotropin (hCG) is secreted by syncytiotrophoblasts during the first trimester of pregnancy. This pituitary-like hormone maintains the mother's corpus luteum for the first 5½ weeks of pregnancy. The placenta then assumes the role of estrogen and progesterone production, and the corpus luteum degenerates.

implantation: L. *im*, in; *planto*, to plant

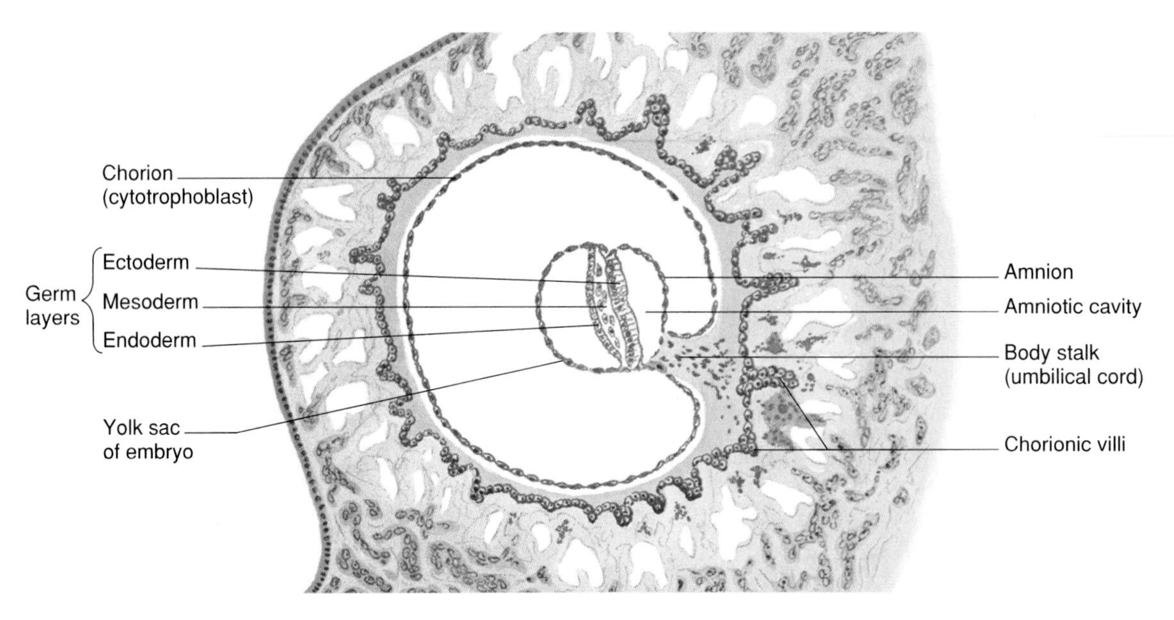

Figure 22.8

The completion of implantation occurs as the primary germ layers develop at the end of the second week.

Summary of Preembryonic Development

Morphogenic Stage	Time Period	Principal Events
Zygote	24 to 30 hours following ovulation	Egg is fertilized; zygote has 23 pairs of chromosomes (diploid) from haploid sperm and haploid egg and is genetically unique
Cleavage	30 hours to third day	Mitotic divisions produce increased number of cells
Morula	Third to fourth day	Solid ball-like structure forms, composed of 12 or more cells
Blastocyst	Fifth day to end of second week	Hollow ball-like structure forms, a single layer thick; embryoblast and trophoblast form; implantation occurs; embryonic disc forms, followed by primary germ layers

All *pregnancy tests* assay for the presence of hCG in the blood or urine because this hormone is secreted only by the blastocyst. Since there is no other source of hCG, the presence of this hormone confirms a pregnancy. Modern pregnancy tests detect the presence of hCG by use of antibodies against hCG or by the use of cellular receptor proteins for hCG.

Formation of the Germ Layers

As the blastocyst completes implantation during the second week of development, the embryoblast undergoes marked differentiation. A slitlike space called the **amniotic** (*am″ne-ot′ik*) **cavity** forms between the embryoblast and the trophoblast (fig. 22.8). The embryoblast flattens into the **embryonic disc** (see fig. 22.10), which consists of two layers: an upper **ectoderm,** which is closer to the amniotic cavity, and a lower **endoderm,** which borders the blastocyst cavity. A short time later, a third layer called the **mesoderm** forms between the endoderm and ectoderm. These three layers constitute the **primary germ layers** (fig. 22.8). Once they are formed, at the end of the second week, the preembryonic period is completed and the embryonic period begins.

The primary germ layers are especially significant because all of the cells and tissues of the body are derived from them. Ectodermal cells form the nervous system; the outer layer of skin (epidermis), including hair, nails, and skin glands; and portions of the sensory organs. Mesodermal cells form the skeleton, muscles, blood, reproductive organs, dermis of the skin, and connective tissue. Endodermal cells produce the lining of the GI tract, the digestive organs, the respiratory tract and lungs, and the urinary bladder and urethra.

The events of the preembryonic period are summarized in table 22.1. Refer to figure 22.9 for an illustration and a listing of the organs and body systems that derive from each of the primary germ layers.

ectoderm: Gk. *ecto,* outside; *derm,* skin
endoderm: Gk. *endo,* within; *derm,* skin

mesoderm: Gk. *meso,* middle; *derm,* skin

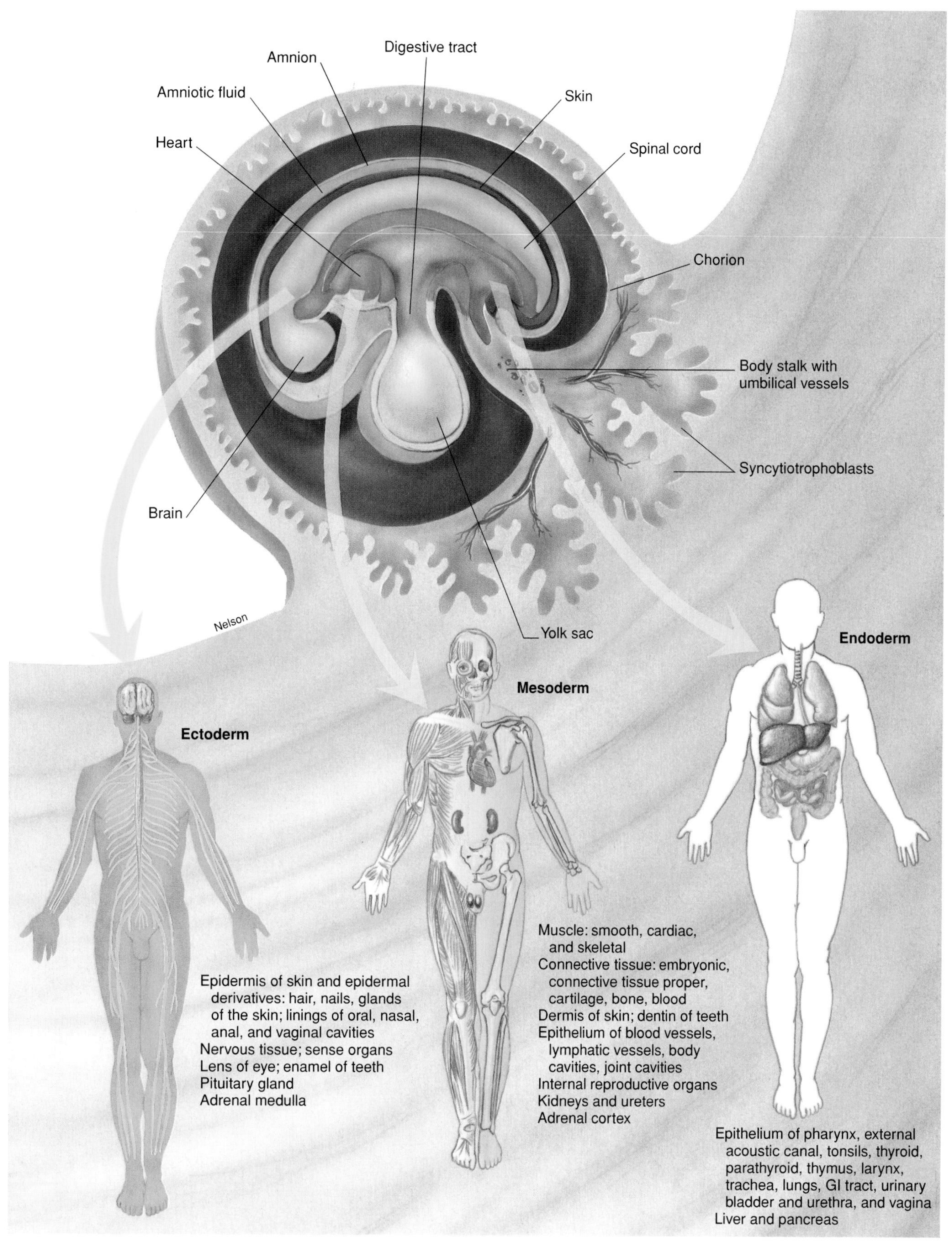

Amnion
Digestive tract
Amniotic fluid
Skin
Heart
Spinal cord
Chorion
Body stalk with umbilical vessels
Syncytiotrophoblasts
Brain
Nelson
Yolk sac
Endoderm
Mesoderm
Ectoderm
Muscle: smooth, cardiac, and skeletal
Connective tissue: embryonic, connective tissue proper, cartilage, bone, blood
Dermis of skin; dentin of teeth
Epithelium of blood vessels, lymphatic vessels, body cavities, joint cavities
Internal reproductive organs
Kidneys and ureters
Adrenal cortex
Epidermis of skin and epidermal derivatives: hair, nails, glands of the skin; linings of oral, nasal, anal, and vaginal cavities
Nervous tissue; sense organs
Lens of eye; enamel of teeth
Pituitary gland
Adrenal medulla
Epithelium of pharynx, external acoustic canal, tonsils, thyroid, parathyroid, thymus, larynx, trachea, lungs, GI tract, urinary bladder and urethra, and vagina
Liver and pancreas

Figure 22.9

The body systems and the primary germ layers from which they develop.

The period of prenatal development is referred to as *gestation*. Normal gestation for humans is 9 months. Knowing this and the pattern of menstruation make it possible to determine the baby's delivery date. In a typical reproductive cycle, a woman ovulates 14 days prior to the onset of the next menstruation and is fertile for approximately 20 to 24 hours following ovulation. Adding 9 months, or 38 weeks, to the time of ovulation gives one the estimated delivery date.

1. List the structural characteristics of a zygote, morula, and blastocyst. Approximately when do each of these stages of the preembryonic period of development occur?
2. Discuss the process of implantation and describe the physiological events that ensure pregnancy.
3. Describe the development of the placenta.
4. List the major structures that derive from each germ layer.
5. Define gestation. What is the average gestation time for humans? How is the parturition date determined?

Embryonic Period

The events of the 6-week embryonic period include the differentiation of the germ layers into specific body organs and the formation of the placenta, the umbilical cord, and the extraembryonic membranes. Through these morphogenic events, the needs of the embryo are met.

Objective 7 Define *embryo* and describe the major events of the embryonic period of development.

Objective 8 List the embryonic needs that must be met to avoid a spontaneous abortion.

Objective 9 Describe the structure and function of each of the extraembryonic membranes.

Objective 10 Describe the development and functions of the placenta and the umbilical cord.

During the embryonic period—from the beginning of the third week to the end of the eighth week—the developing organism is correctly called an **embryo.** It is at this period that all of the body tissues and organs form, as well as the placenta, umbilical cord, and extraembryonic membranes. The term *conceptus* refers to the embryo, or to the fetus later on, and all of the extraembryonic structures—the products of conception.

Embryology is the study of the sequential changes in an organism as the various tissues, organs, and systems develop. Chick embryos are frequently studied because of the easy access through the shell and their rapid development. Mice and pig embryos are also extensively studied as mammalian models. Genetic manipulation, induction of drugs, exposure to disease, radioactive tagging or dyeing of developing tissues, and X-ray treatments are some of the commonly conducted experiments that provide information that can be applied to human development and birth defects.

During the preembryonic period of cell division and differentiation, the developing structure is self-sustaining. The embryo, however, must derive sustenance from the mother. For morphogenesis to continue, certain immediate needs must be met. These needs include (1) formation of a vascular association between the uterus of the mother and the embryo so that nutrients and oxygen can be provided and metabolic wastes and carbon dioxide can be removed; (2) establishment of a constant, protective environment around the embryo that is conducive to development; (3) establishment of a structural foundation for embryonic morphogenesis along a longitudinal axis; (4) provision for structural support for the embryo, both internally and externally; and (5) coordination of the morphogenic events through genetic expression. If these needs are not met, a spontaneous abortion will generally occur.

The first and second of these needs are provided for by extraembryonic structures; the last three are provided for intraembryonically. The extraembryonic membranes, the placenta, and the umbilical cord will be considered separately, prior to a discussion of the development of the embryo.

Serious developmental defects usually cause the embryo to be naturally aborted. About 25% of early aborted embryos have chromosomal abnormalities. Other abortions may be caused by environmental factors, such as infectious agents or teratogenic drugs (drugs that cause birth defects). In addition, an implanting embryo is regarded as foreign tissue by the immune system of the mother, and is rejected and aborted unless maternal immune responses are suppressed.

Extraembryonic Membranes

At the same time that the internal organs of the embryo are being formed, a complex system of extraembryonic membranes is also developing (fig. 22.10). The **extraembryonic membranes** are the *amnion*, *yolk sac*, *allantois*, and *chorion*. These membranes are responsible for the protection, respiration, excretion, and nutrition of the embryo and subsequent fetus. At parturition, the placenta, umbilical cord, and extraembryonic membranes separate from the fetus and are expelled from the uterus as the *afterbirth*.

gestation: L. *gestatus*, to bear

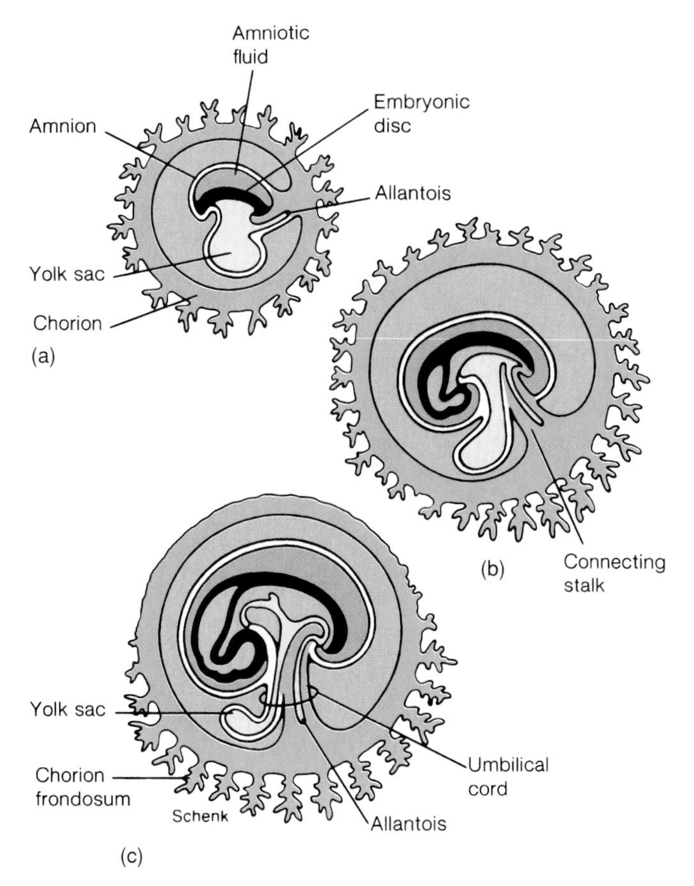

Figure 22.10

The formation of the extraembryonic membranes during a single week of rapid embryonic development. (*a*) At 3 weeks, (*b*) at 3½ weeks, and (*c*) at 4 weeks.

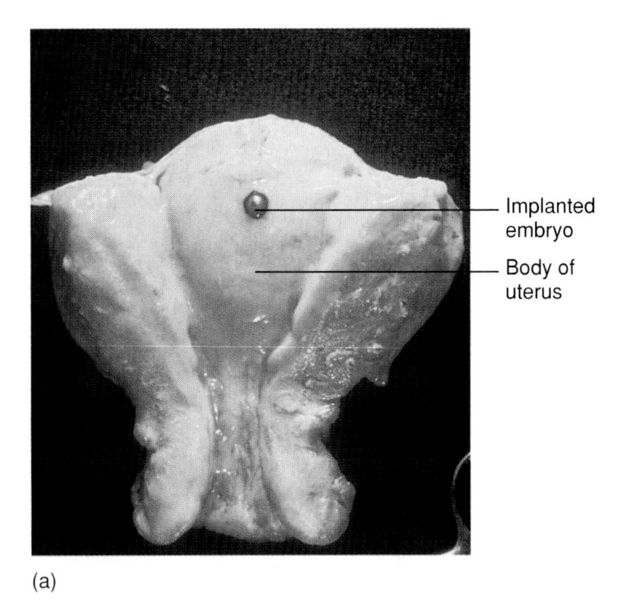

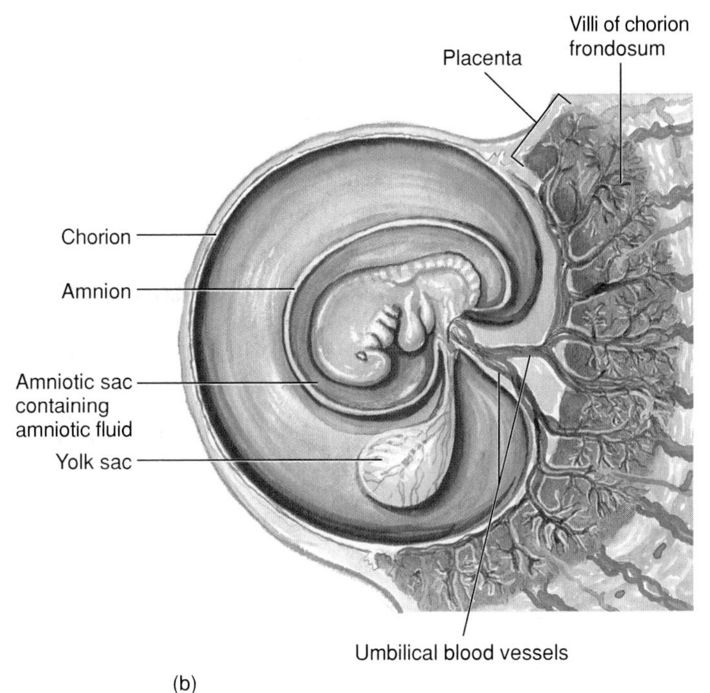

Figure 22.11

An implanted embryo at approximately 4½ weeks. (*a*) The interior of a uterus showing the implantation site. (*b*) The developing embryo, extraembryonic membranes, and placenta.

Amnion

The amnion (*am'ne-on*) is a thin membrane, derived from ectoderm and mesoderm. It loosely envelops the embryo, forming an **amniotic sac** that is filled with *amniotic fluid* (fig. 22.11*b*). In later fetal development, the amnion expands to come in contact with the chorion. The development of the amnion is initiated early in the embryonic period, at which time its margin is attached to the free edge of the embryonic disc (see fig. 22.10). As the amniotic sac enlarges during the late embryonic period (at about 8 weeks), the amnion gradually sheaths the developing umbilical cord with an epithelial covering (fig. 22.12).

The buoyant amniotic fluid performs four functions for the embryo and subsequent fetus.

1. It ensures symmetrical structural development and growth.
2. It cushions and protects by absorbing jolts that the mother may receive.
3. It helps to maintain consistent pressure and temperature.
4. It permits freedom of fetal movement, which is important for musculoskeletal development and blood flow.

Amniotic fluid is formed initially as an isotonic fluid absorbed from the maternal blood in the endometrium surrounding the developing embryo. Later, the volume is increased and the concentration changed by urine excreted from the fetus into the amniotic sac. Amniotic fluid also contains cells that are sloughed off from the fetus, placenta, and amniotic sac. Since all of these cells are derived from the same fertilized egg, all have the same genetic composition. Many genetic abnormalities can be detected by aspirating this fluid and examining the cells obtained, in a procedure called *amniocentesis (am″ne-o-sen-te′sis)*.

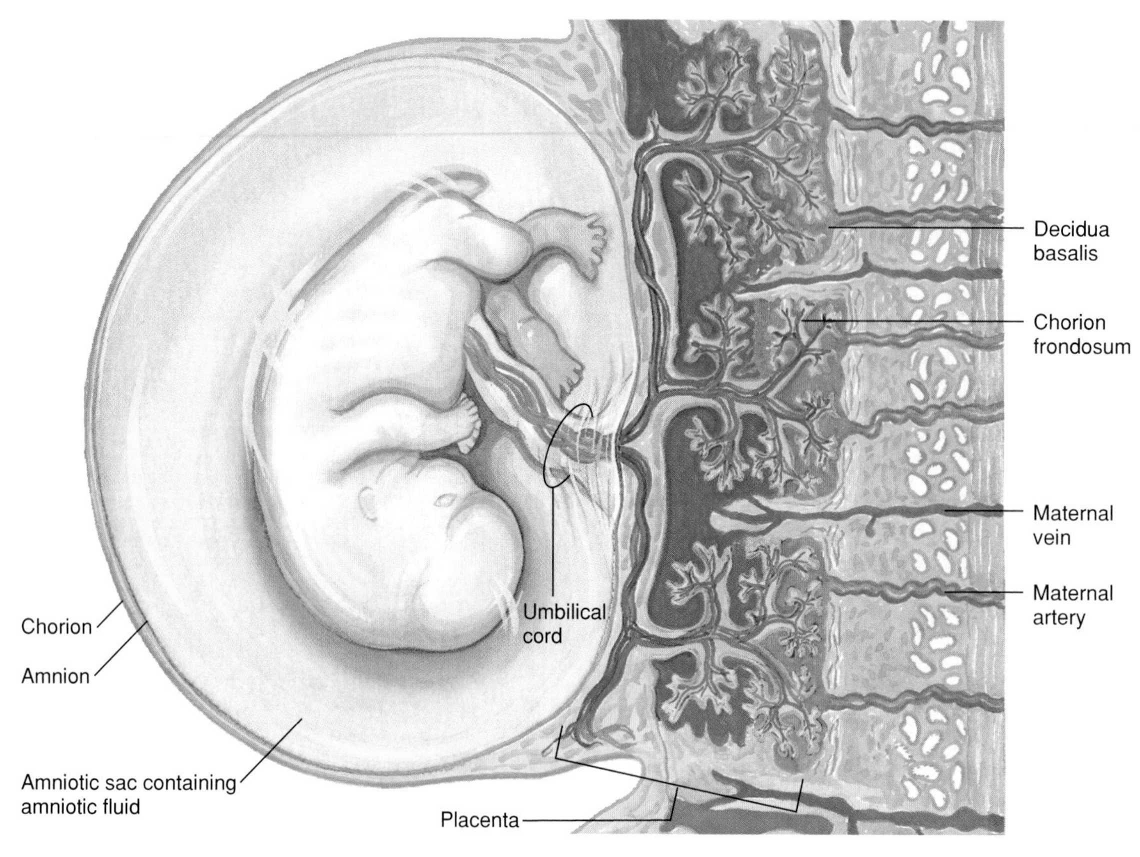

Figure 22.12

The embryo, extraembryonic membranes, and placenta at approximately 7 weeks of development. Blood from the embryo is carried to and from the chorion frondosum by the umbilical arteries and vein. The maternal tissue between the chorionic villi is known as the decidua basalis; this tissue, together with the villi, form the functioning placenta.

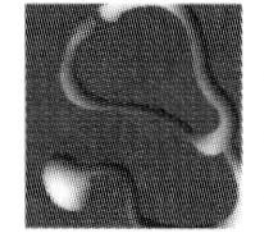

Amniocentesis (fig. 22.13) is usually performed at the fourteenth or fifteenth week of pregnancy, when the amniotic sac contains 175–225 ml of fluid. Genetic diseases, such as *Down syndrome,* or *trisomy 21,* (in which there are three instead of two number-21 chromosomes), can be detected by examining chromosomes. Diseases such as *Tay–Sachs disease,* in which there is a defective enzyme involved in formation of myelin sheaths, can be detected by biochemical techniques from these fetal cells.

Amniotic fluid is normally swallowed by the fetus and absorbed in the GI tract. The fluid enters the fetal blood, and the waste products it contains enter the maternal blood in the placenta. Prior to delivery, the amnion is naturally or surgically ruptured, and the amniotic fluid (bag of waters) is released.

As the fetus grows, the amount of amniotic fluid increases. It is also continually absorbed and renewed. For the near-term baby, almost 8 liters of fluid are completely replaced each day.

Yolk Sac

The yolk sac is established during the end of the second week as cells from the trophoblast form a thin *exocoelomic (ek″so-sĕ-lo′mik) membrane*. Unlike the yolk sac of many vertebrates, the human yolk sac contains no nutritive yolk, but it is still an essential structure during early embryonic development. Attached to the underside of the embryonic disc (see figs. 22.10 and 22.11), it produces blood for the embryo until the liver forms during the sixth week. A portion of the yolk sac is also involved in the formation of the primitive gut. In addition, primordial germ cells form in the wall of the yolk sac. During the fourth week, they migrate to the developing gonads, where they become the primitive germ cells (spermatogonia or oogonia).

The stalk of the yolk sac usually detaches from the gut by the sixth week. Following this, the yolk sac gradually shrinks as pregnancy advances. Eventually, it becomes very small and serves no additional function.

Allantois

The allantois forms during the third week as a small outpouching, or diverticulum, near the base of the yolk sac (see fig. 22.10). It remains small but is involved in the formation of blood cells and gives rise to the fetal umbilical arteries and vein. It also contributes to the development of the urinary bladder.

Down syndrome: from John L. H. Down, English physician, 1828–96

allantois: Gk. *allanto*, sausage; *iodos*, resemblance

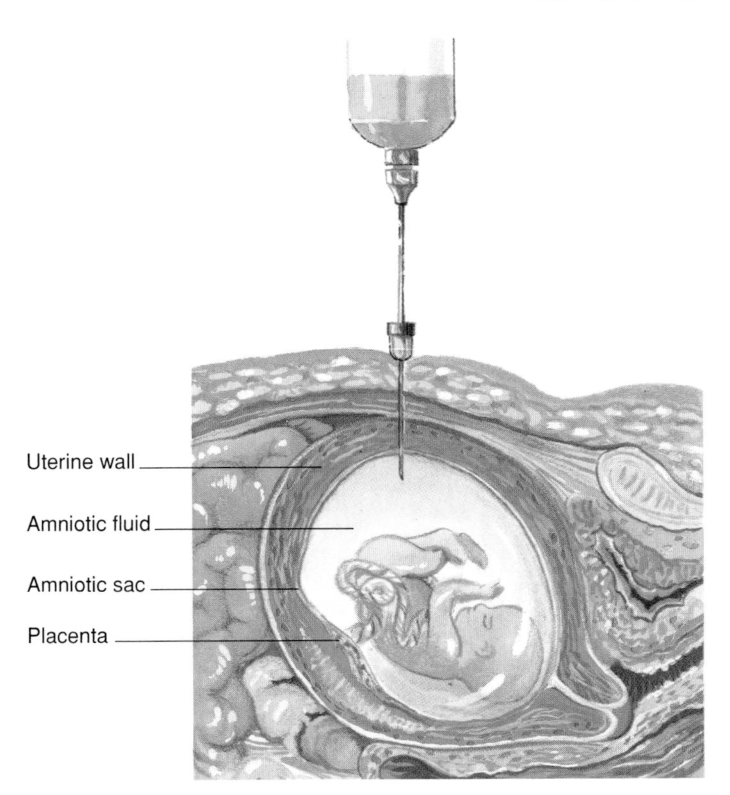

Figure 22.13

Amniocentesis. In this procedure, amniotic fluid containing suspended cells is withdrawn for examination. Various genetic diseases can be detected prenatally by this means.

The extraembryonic portion of the allantois degenerates during the second month. The intraembryonic portion involutes to form a thick urinary tube called the **urachus** (*yoo'ră-kus*). After birth, the urachus becomes a fibrous cord called the *median umbilical ligament* that attaches to the urinary bladder.

Chorion

The **chorion** (*kor'e-on*) is the outermost extraembryonic membrane. It contributes to the formation of the placenta as small fingerlike extensions, called *villi*, penetrate deeply into the uterine tissue (see fig. 22.10). Initially, the entire surface of the chorion is covered with villi. But those villi on the surface toward the uterine cavity gradually degenerate, producing a smooth, bare area known as the **smooth chorion.** As this occurs, the villi associated with the uterine wall rapidly increase in number and branch out. This portion of the chorion is known as the **villous chorion,** or **chorion frondosum** (*frondo'sum*). The villous chorion becomes highly vascularized with embryonic blood vessels, and as the embryonic heart begins to function, embryonic blood is pumped in close proximity to the uterine wall.

chorion: Gk. *chorion*, external fetal membrane
villous: L. *villus*, tuft of hair

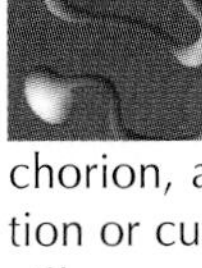

Chorionic villus biopsy is a technique used to detect genetic disorders much earlier than amniocentesis permits. In chorionic villus biopsy, a catheter is inserted through the cervix to the chorion, and a sample of chorionic villus is obtained by suction or cutting. Genetic tests can be performed directly on the villus sample, since this sample contains much larger numbers of fetal cells than does a sample of amniotic fluid. Chorionic villus biopsy can provide genetic information at 10 to 12 weeks' gestation.

Placenta

The **placenta** (*plă-sen'tă*) is a vascular structure by which an unborn child is attached to its mother's uterine wall and through which respiratory gas and metabolic exchange occurs (fig. 22.14). The placenta is formed in part from maternal tissue and in part from embryonic tissue. The embryonic portion of the placenta consists of the chorion frondosum, whereas the maternal portion is composed of the area of the uterine wall called the **decidua** (*dĕ-sid'yoo-ă*) **basalis** (see fig. 22.12), into which the villi penetrate. Blood does not flow directly between these two portions, but because their membranes are in close proximity, certain substances diffuse readily.

When fully formed, the placenta is a reddish brown oval disc with a diameter of 15 to 20 cm (8 in.) and a thickness of 2.5 cm (1 in.). It weighs between 500 and 600 gm, about one-sixth as much as the fetus.

Exchange of Molecules across the Placenta

The two **umbilical arteries** deliver fetal blood to vessels within the villi of the chorion frondosum of the placenta. This blood circulates within the villi and returns to the fetus via the **umbilical vein** (fig. 22.14). Maternal blood is delivered to and drained from the cavities within the decidua basalis, which are located between the chorionic villi. In this way, maternal and fetal blood are brought close together but do not normally mix within the placenta.

The placenta serves as a site for the exchange of gases and other molecules between the maternal and fetal blood. Oxygen diffuses from mother to fetus, and carbon dioxide diffuses in the opposite direction. Nutrient molecules and waste products likewise pass between maternal and fetal blood.

The placenta has a high metabolic rate. It utilizes about one-third of all the oxygen and glucose supplied by the maternal blood. In fact, the rate of protein synthesis is actually higher in the placenta than it is in the fetal liver. Like the liver, the placenta produces a great variety of enzymes that are capable of converting biologically active molecules (such as hormones and drugs) into less active, more water-soluble forms. In this way, potentially dangerous molecules in the maternal blood are often prevented from harming the fetus.

placenta: L. *placenta*, a flat cake

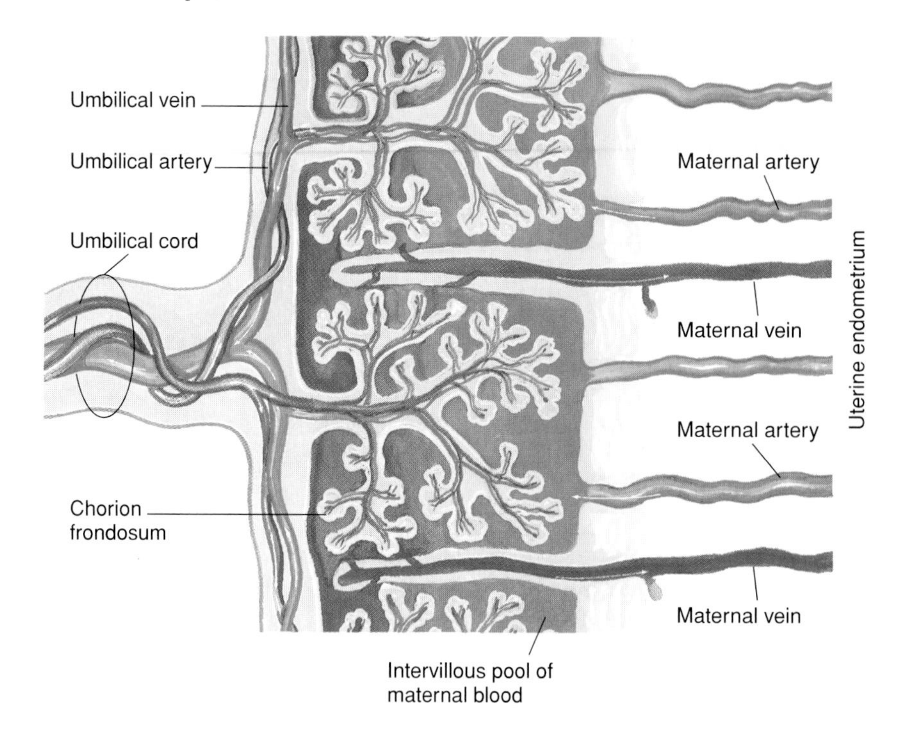

Figure 22.14

The circulation of blood within the placenta. Maternal blood is delivered to and drained from the spaces between the chorionic villi. Fetal blood is brought to blood vessels within the villi by branches of the umbilical artery and is drained by branches of the umbilical vein.

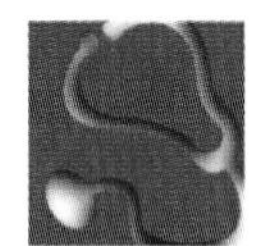

Some substances ingested by a pregnant woman are able to pass through the placenta readily, to the detriment of the fetus. These include nicotine, heroin, and certain antidepressant drugs. Excessive nicotine will stunt the growth of the fetus; heroin can lead to fetal drug addiction; and certain antidepressants can cause respiratory problems.

Although the placenta is an effective barrier against diseases of bacterial origin, rubella and other viruses, as well as certain blood-borne diseases such as syphilis, can diffuse through the placenta and affect the fetus. During parturition, small amounts of fetal blood may pass across the placenta to the mother. If the fetus is Rh positive and the mother Rh negative, the antigens of the fetal red blood cells elicit an antibody response in the mother. In a subsequent pregnancy, the maternal antibodies then cross the placenta and cause a breakdown of fetal red blood cells—a condition called *erythroblastosis fetalis.*

Endocrine Functions of the Placenta

The placenta functions as an endocrine gland in producing both glycoprotein and steroid hormones. The glycoprotein hormones have actions similar to those of some anterior pituitary hormones. They play a key role in maintaining pregnancy and ensuring that the fetus will receive optimal nourishment.

The placenta also converts androgens, secreted from the mother's adrenal glands, into estrogens to help protect the female embryo from becoming masculinized. In addition, the placenta secretes large amounts of *estriol,* a weak estrogen that helps to maintain the endometrium and stimulates the development of the mother's mammary glands, readying them for lactation.

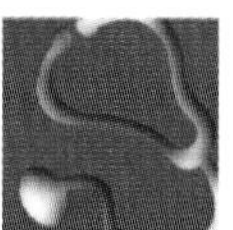

The production of estriol increases tenfold during pregnancy, so that by the third trimester estriol accounts for about 90% of the estrogens excreted in the mother's urine. Since almost all of this estriol comes from the placenta (rather than from maternal tissues), measurements of urinary estriol can be used clinically to assess the health of the placenta.

The hormones secreted by the placenta and the effects they exert on their target tissues are summarized in table 22.2.

Umbilical Cord

The **umbilical** (*um-bĭ'lĭ-kal*) **cord** forms as the yolk sac shrinks and the amnion expands to envelop the tissues on the underside of the embryo (fig. 22.15). The umbilical cord usually attaches near the center of the placenta. When fully formed, it is between 1 and 2 cm (0.5 and 1 in.) in diameter and approximately 55 cm (2 ft) long. On average, the umbilical cords of male fetuses are approximately 5 cm (2 in.) longer than those of female fetuses. The umbilical cord contains two **umbilical arteries,** which carry deoxygenated blood from the embryo toward the placenta, and one **umbilical vein,** which carries oxygenated blood from the placenta to the embryo. These vessels are surrounded by embryonic connective tissue called **mucoid connective tissue** (Wharton's jelly).

Wharton's jelly: from Thomas Wharton, English anatomist, 1614–73

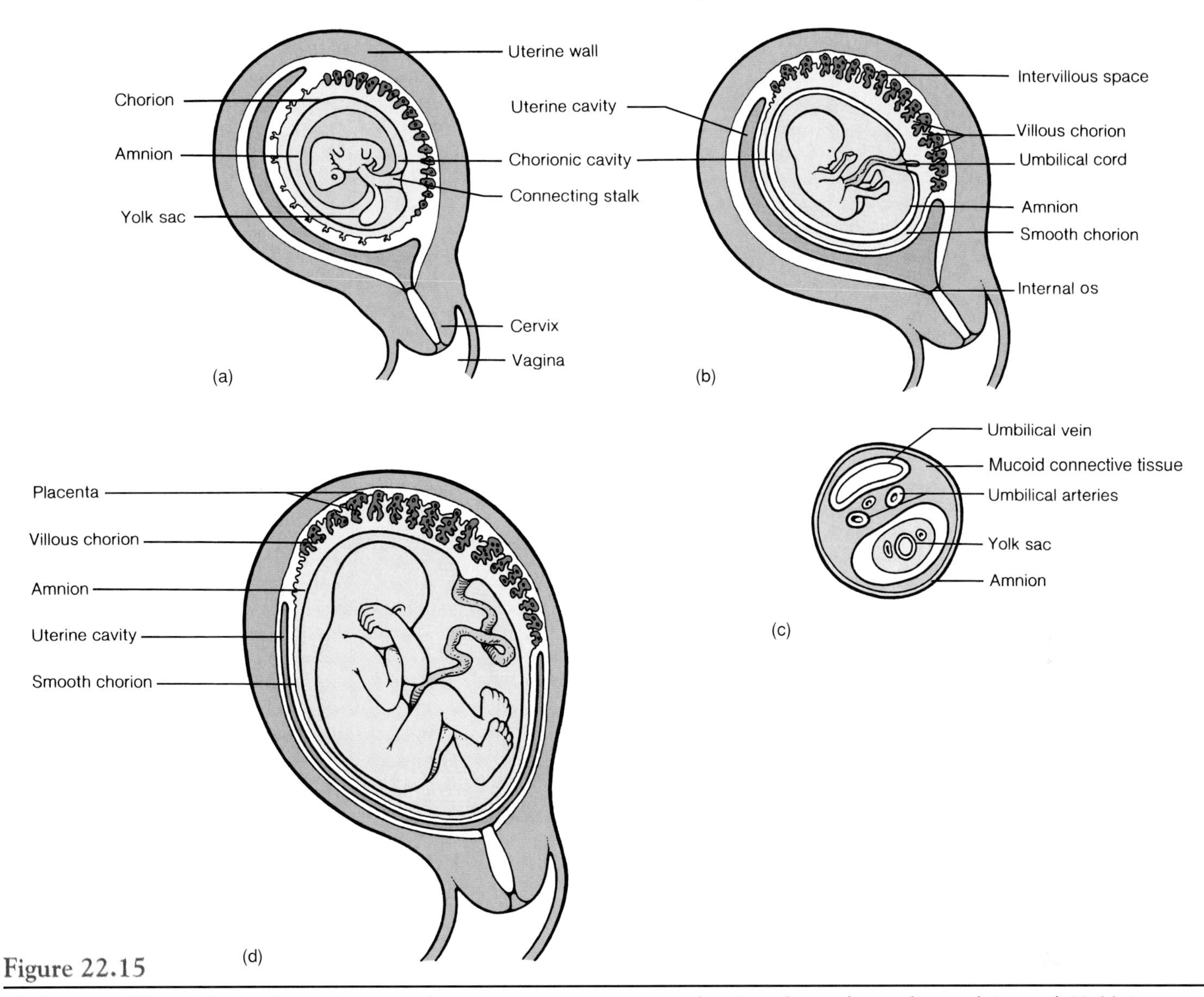

Figure 22.15

The formation of the umbilical cord and other extraembryonic structures as seen in sagittal sections of a gravid uterus from week 4 to week 22. (*a*) A connecting stalk forms as the developing amnion expands to surround the embryo, finally meeting ventrally. (*b*) The umbilical cord begins to take form as the amnion ensheathes the yolk sac. (*c*) A cross section of the umbilical cord showing the embryonic vessels, mucoid connective tissue, and the tubular connection to the yolk sac. (*d*) By week 22, the amnion and chorion have fused and the umbilical cord and placenta have become well-developed structures.

Table 22.2
Hormones Secreted by the Placenta

Hormones	Effects
Pituitary-like Hormones	
Chorionic gonadotropin (hCG)	Similar to LH; maintains mother's corpus luteum for first 5½ weeks of pregnancy; may be involved in suppressing immunological rejection of embryo; also exhibits TSH-like activity
Chorionic somatomammotropin (hCS)	Similar to prolactin and growth hormone; in the mother, hCS promotes increased fat breakdown and fatty acid release from adipose tissue and decreased glucose use by maternal tissues (diabetic-like effects)
Sex Steroids	
Progesterone	Helps maintain endometrium during pregnancy; helps suppress gonadotropin secretion; stimulates development of alveolar tissue in mammary glands
Estrogens	Help maintain endometrium during pregnancy; help suppress gonadotropin secretion; help stimulate mammary gland development; inhibit prolactin secretion; promote uterine sensitivity to oxytocin; stimulate ductule development in mammary glands

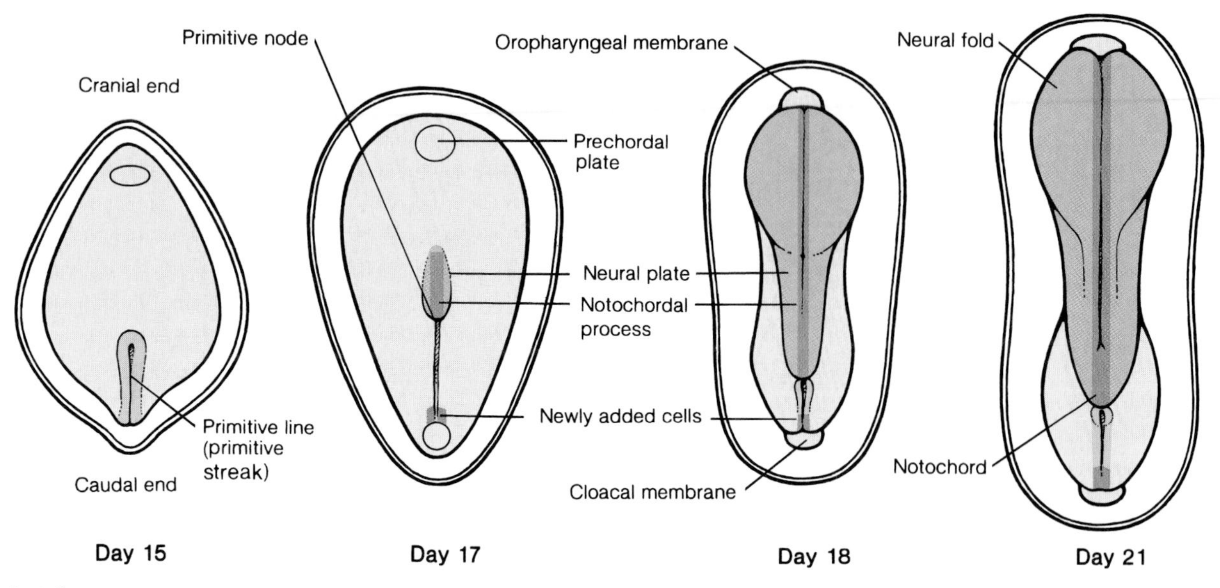

Figure 22.16

The appearance of the primitive line and primitive node along the embryonic disc. These progressive changes occur through the process of induction.

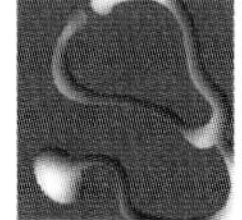

The umbilical cord has a helical, or screwlike, form that keeps it from kinking. The spiraling occurs because the umbilical vein grows faster and longer than the umbilical arteries. In about one-fifth of all deliveries, the cord is looped once around the baby's neck. If drawn tightly, the cord may cause death or serious perinatal problems.

Structural Changes in the Embryo by Week

Third Week

Early in the third week, a thick linear band called the **primitive line** (primitive streak), appears along the dorsal midline of the embryonic disc (figs. 22.10 and 22.16). Derived from mesodermal cells, the primitive line establishes a structural foundation for embryonic morphogenesis along a longitudinal axis. As the primitive line elongates, a prominent thickening called the **primitive node** appears at its cranial end (fig. 22.16). The primitive node later gives rise to the mesodermal structures of the head and to a rod of mesodermal cells called the **notochord.** The notochord forms a midline axis that is the basis of the embryonic skeleton. The primitive line also gives rise to loose embryonic connective tissue called **intraembryonic mesoderm** (mesenchyme). Mesenchyme differentiates into all the various kinds of connective tissue found in the adult. One of the earliest formed organs is the skin, which serves to support and maintain homeostasis within the embryo.

A tremendous amount of change and specialization occurs during the embryonic stage (figs. 22.17 and 22.18). The factors that cause precise sequential change from one cell or tissue type to another are not fully understood. It is known, however, that the potential for change is programmed into the genetics of each cell and that under conducive environmental conditions this change takes place. The process of developmental change is referred to as **induction.** Induction occurs when one tissue, called the *inductor tissue*, has a marked effect on an adjacent tissue, causing it to become *induced tissue*, and stimulating it to differentiate.

Fourth Week

During the fourth week of development, the embryo increases about 4 mm (0.16 in.) in length. A **connecting stalk,** which is later involved in the formation of the umbilical cord, is established from the body of the embryo to the developing placenta. By this time, the heart is already pumping blood to all parts of the embryo. The head and jaws are apparent, and the primordial tissue that will form the eyes, brain, spinal cord, lungs, and digestive organs has developed. The **superior** and **inferior limb buds** are recognizable as small swellings on the lateral body walls.

Fifth Week

Embryonic changes during the fifth week are not as extensive as those during the fourth week. The head enlarges, and the developing eyes, ears, and nasal pit become obvious. The appendages have formed from the limb buds, and paddle-shaped hand and foot plates develop.

induction: L. *inductus*, to lead in

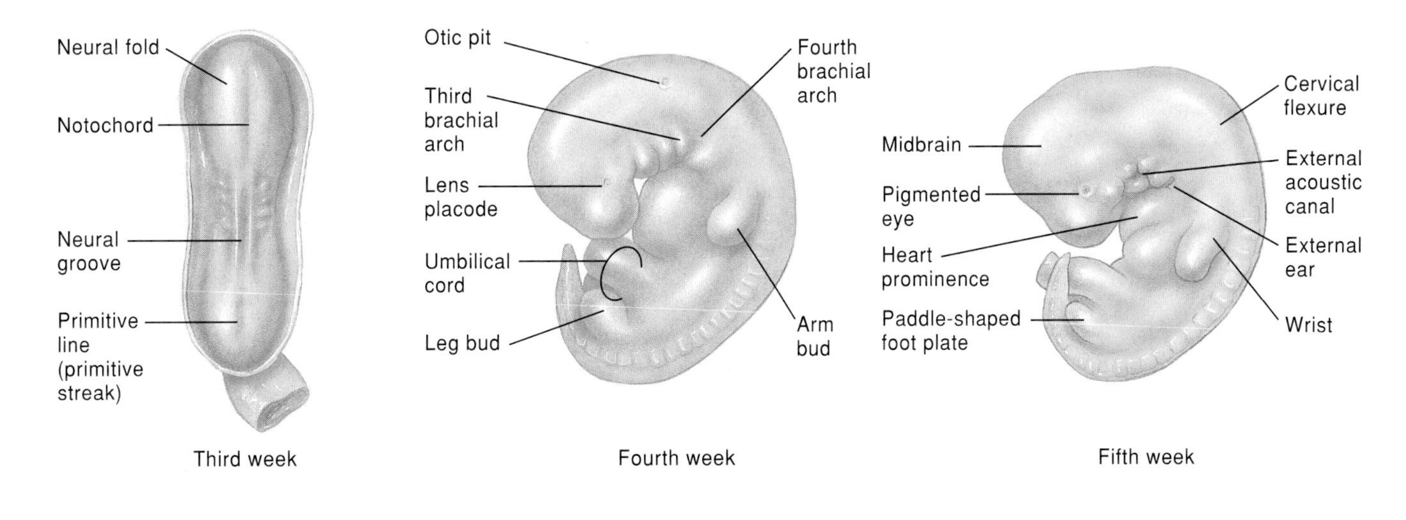

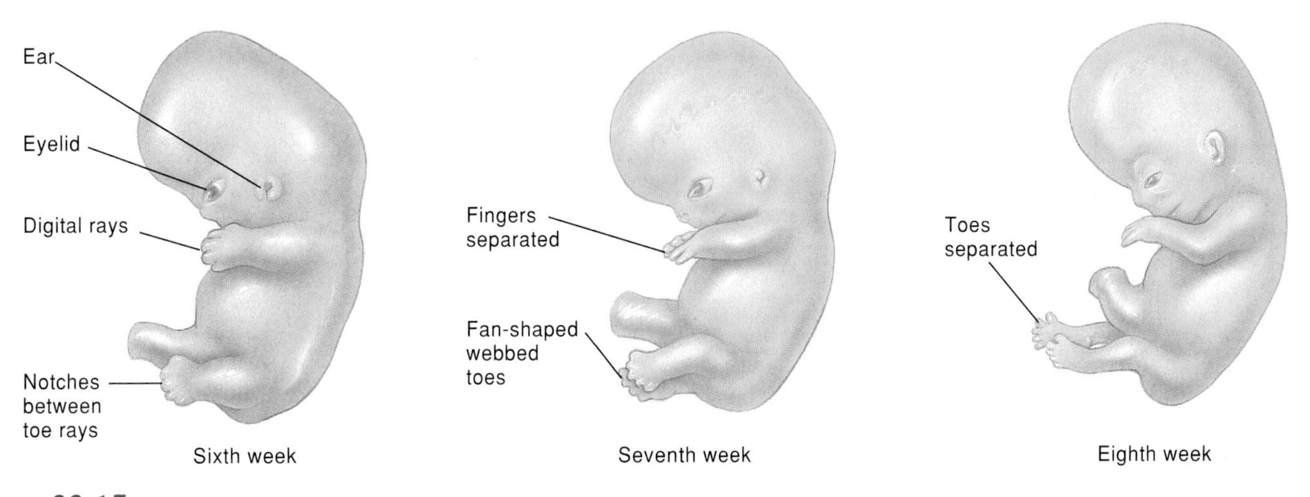

Figure 22.17

Structural changes in the embryo by weeks.

Sixth Week

During the sixth week, the embryo is 16–24 mm (0.64–0.96 in.) long. The head is larger than the trunk, and the brain is undergoing marked differentiation. This is the most vulnerable period of development for many organs. An interruption at this critical time can easily cause congenital damage. The limbs are lengthened and slightly flexed, and notches appear between the **digital rays** in the hand and foot plates. The gonads are beginning to produce hormones that will influence the development of the external genitalia.

Seventh and Eighth Weeks

During the last 2 weeks of the embryonic stage, the embryo, which is now 28–40 mm (1.12–1.6 in.) long, has distinct human characteristics (see figs. 22.17 and 22.18). The body organs are formed, and the nervous system is starting to coordinate body activity. The neck region is apparent, and the abdomen is less prominent. The eyes are well developed, but the lids are stuck together to protect against probing fingers during muscular movement. The nostrils are developed but plugged with mucus. The external genitalia are forming but are still undifferentiated. The body systems are developed by the end of the eighth week, and from this time on the embryo is called a **fetus.**

The most precarious time of prenatal development is the embryonic period—yet, well into this period many women are still unaware that they are pregnant. For this reason, a woman should abstain from taking certain drugs (including some antibiotics) if there is even a remote chance that she is pregnant or might become pregnant in the near future.

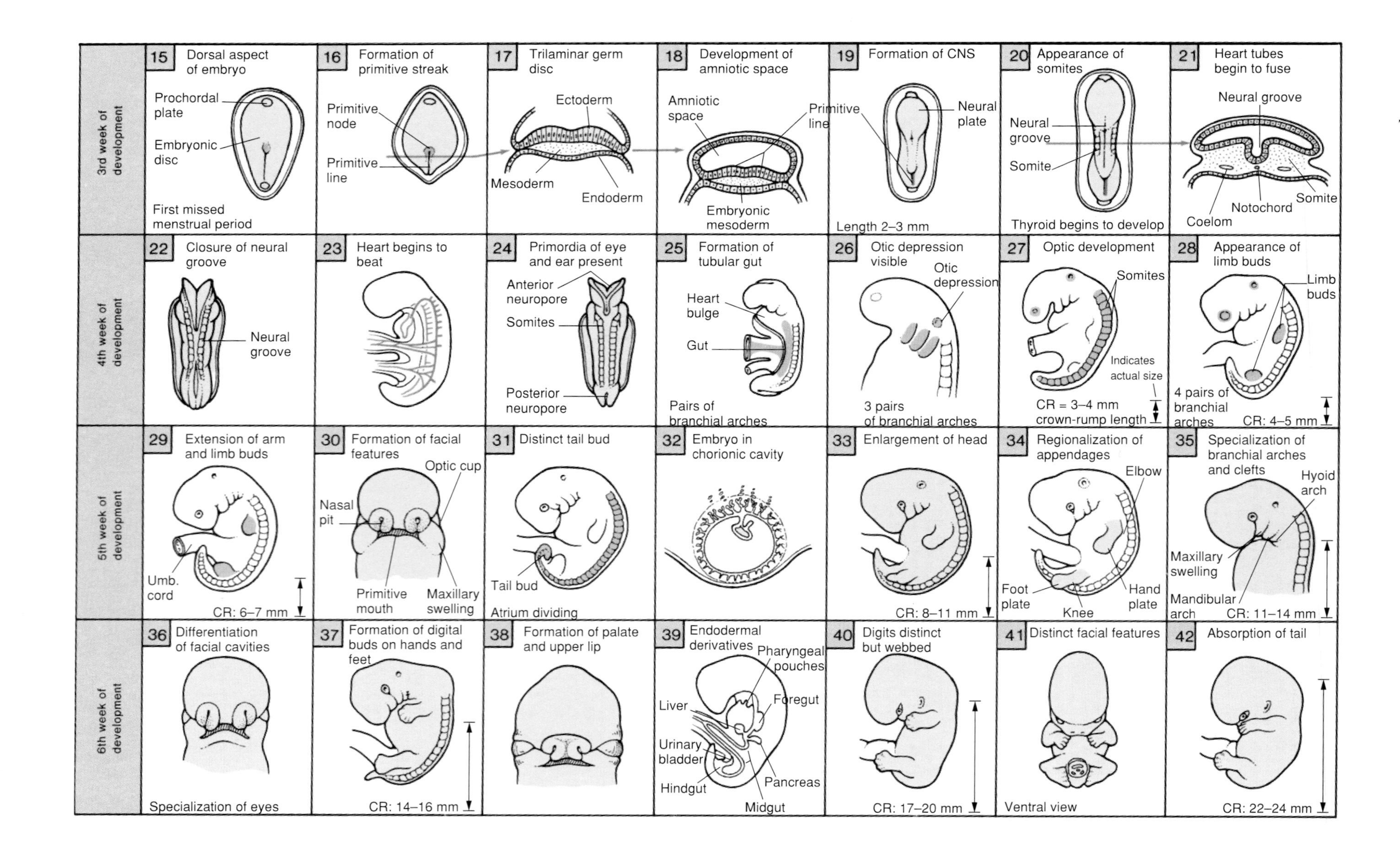
3rd week of development
15 Dorsal aspect of embryo
Prochordal plate
Embryonic disc
First missed menstrual period
16 Formation of primitive streak
Primitive node
Primitive line
17 Trilaminar germ disc
Ectoderm
Mesoderm
Endoderm
18 Development of amniotic space
Amniotic space
Primitive line
Embryonic mesoderm
19 Formation of CNS
Neural plate
Length 2–3 mm
20 Appearance of somites
Neural groove
Somite
Thyroid begins to develop
21 Heart tubes begin to fuse
Neural groove
Somite
Notochord
Coelom
4th week of development
22 Closure of neural groove
Neural groove
23 Heart begins to beat
24 Primordia of eye and ear present
Anterior neuropore
Somites
Posterior neuropore
25 Formation of tubular gut
Heart bulge
Gut
Pairs of branchial arches
26 Otic depression visible
Otic depression
3 pairs of branchial arches
27 Optic development
Somites
Indicates actual size
CR = 3–4 mm crown-rump length
28 Appearance of limb buds
Limb buds
4 pairs of branchial arches
CR: 4–5 mm
5th week of development
29 Extension of arm and limb buds
Umb. cord
CR: 6–7 mm
30 Formation of facial features
Optic cup
Nasal pit
Primitive mouth
Maxillary swelling
31 Distinct tail bud
Tail bud
Atrium dividing
32 Embryo in chorionic cavity
33 Enlargement of head
CR: 8–11 mm
34 Regionalization of appendages
Elbow
Foot plate
Knee
Hand plate
35 Specialization of branchial arches and clefts
Hyoid arch
Maxillary swelling
Mandibular arch
CR: 11–14 mm
6th week of development
36 Differentiation of facial cavities
Specialization of eyes
37 Formation of digital buds on hands and feet
CR: 14–16 mm
38 Formation of palate and upper lip
39 Endodermal derivatives
Pharyngeal pouches
Foregut
Liver
Urinary bladder
Hindgut
Pancreas
Midgut
40 Digits distinct but webbed
CR: 17–20 mm
41 Distinct facial features
Ventral view
42 Absorption of tail
CR: 22–24 mm

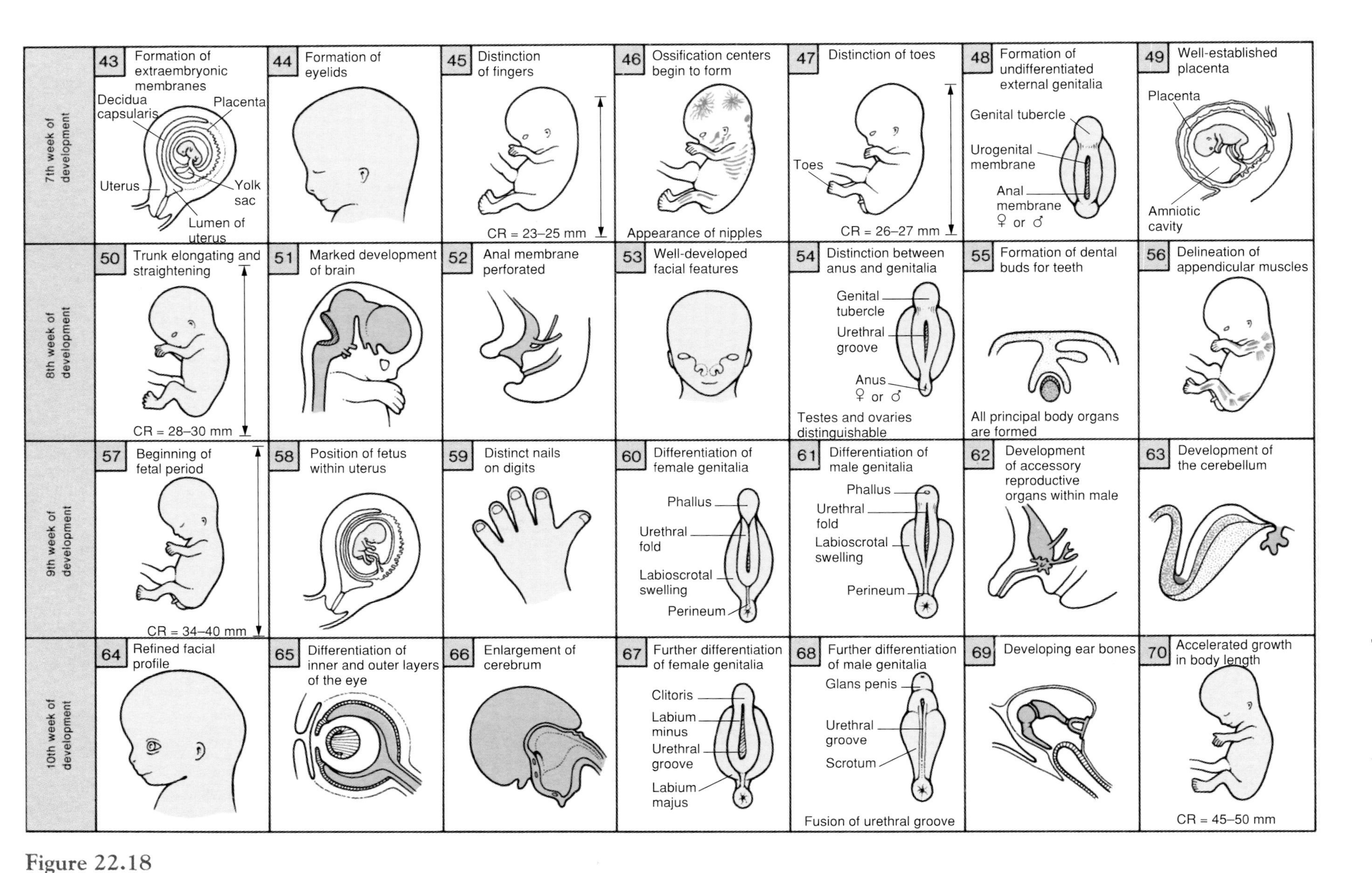

Figure 22.18

Major changes in embryonic and early fetal development from week 3 through week 10. (CR = crown-rump length, a straight-line measurement from the crown of the head to the developing ischium.)

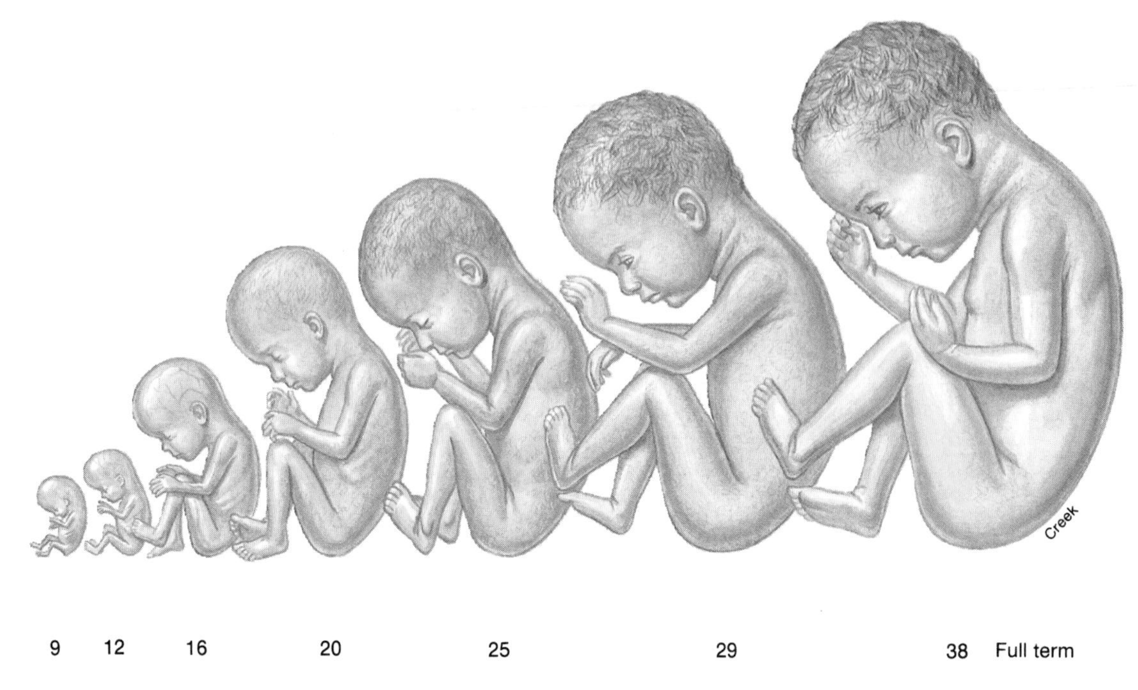

Figure 22.19

Changes in the external appearance of the fetus from week 9 through week 38.

1. Distinguish between embryo and fetus. Briefly summarize the structural changes that occur in an embryo between the fourth and eighth weeks of development.
2. Which of the five embryonic needs are met by the extraembryonic membranes?
3. Explain how each of the extraembryonic membranes is formed.
4. Identify the fetal and maternal components of the placenta and describe the blood circulation in these two components. How does gas exchange occur between mother and fetus?
5. What types of hormones are secreted by the placenta? Briefly summarize their effects.

Fetal Period

The fetal period, beginning at week 9 and culminating at birth, is characterized by tremendous growth and the specialization of body structures.

Objective 11 Define *fetus* and discuss the major events of the fetal period of development.

Objective 12 Describe the various techniques available for examining the fetus or monitoring fetal activity.

Since most of the tissues and organs of the body form during the embryonic period, the **fetus** is recognizable as a human being at 9 weeks. The fetus is far less vulnerable than the embryo to the deforming effects of viruses, drugs, and radiation. Tissue differentiation and organ development continue during the fetal stage, but to a lesser degree than before. For the most part, fetal development is primarily limited to body growth.

Structural Changes in the Fetus by Week

Changes in the external appearance of the fetus from the ninth through the thirty-eighth week are depicted in figure 22.19. A discussion of these structural changes follows.

Nine to Twelve Weeks

At the beginning of the ninth week, the head of the fetus is as large as the rest of the body. The eyes are widely spaced, and the ears are set low. Head growth slows during the next 3 weeks, while lengthening of the body accelerates (fig. 22.20). Ossification centers appear in most bones during the ninth week. Differentiation of the external genitalia becomes apparent at the end of the ninth week, but the genitalia are not developed to the point of sex determination until the twelfth week. By the end of the twelfth week, the fetus is 87 mm (3.5 in.) long and weighs about 45 g (1.6 oz). It can swallow, digest the fluid that passes through its system, and defecate and urinate into the amniotic fluid. Its nervous system and muscle coordination are developed to the point that it will withdraw its leg if tickled. The fetus begins inhaling through its nose but can take in only amniotic fluid.

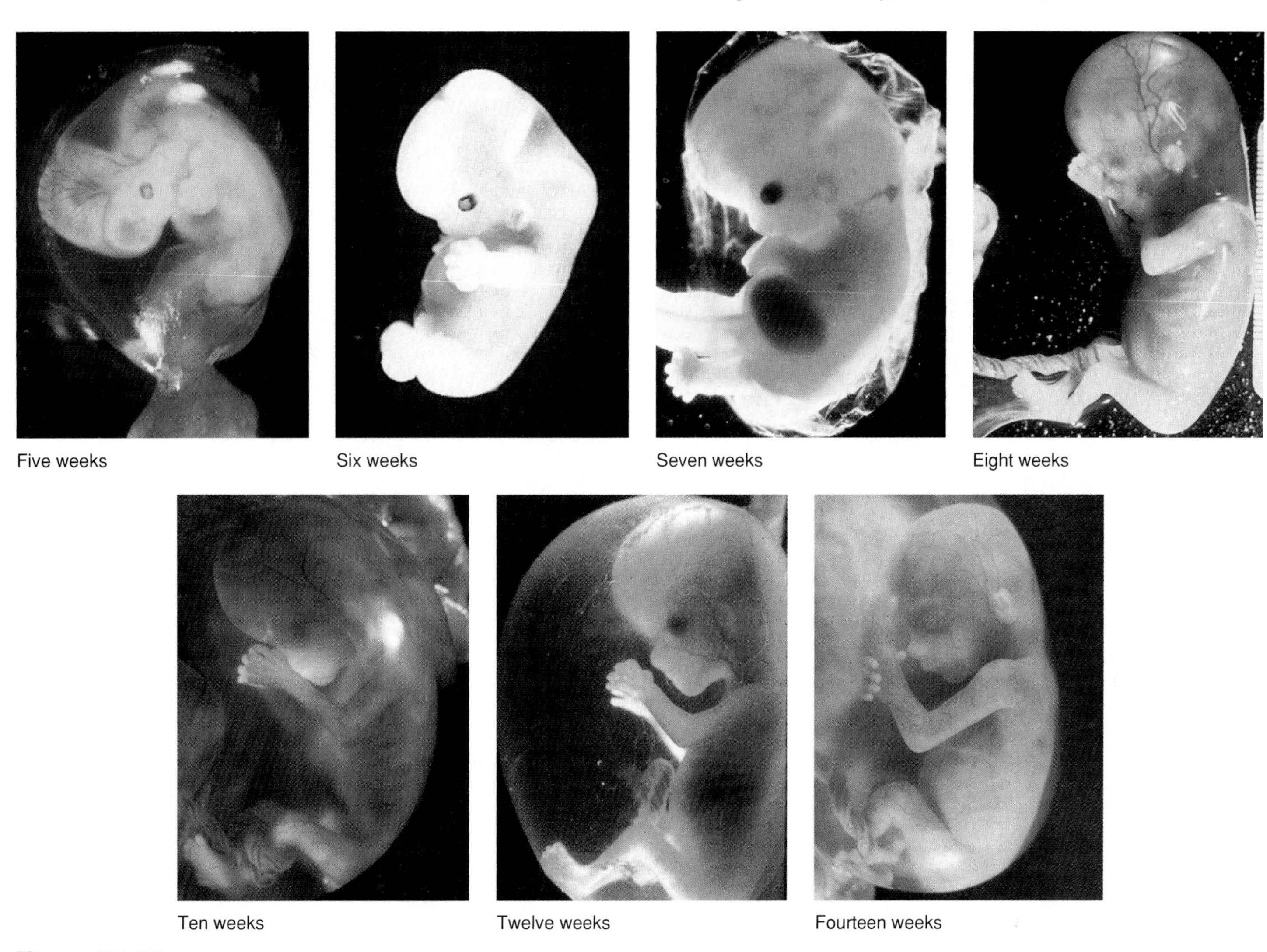

Figure 22.20

A photographic summary of embryonic and early fetal development.

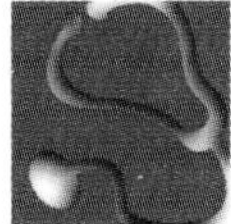

Major structural abnormalities, which may not be predictable from genetic analysis, can often be detected by *ultrasonography* (fig. 22.21). In this procedure, organs are bombarded with sound waves that reflect back in a certain pattern determined by tissue densities. For example, sound waves bouncing off amniotic fluid will produce an image much different from that produced by sound waves bouncing off the placenta or the mother's uterus. Ultrasonography is so sensitive that it can detect a fetal heartbeat several weeks before it can be heard with a stethoscope.

Thirteen to Sixteen Weeks

By the thirteenth week, the facial features of the fetus are well formed, and epidermal structures such as eyelashes, eyebrows, hair on the head, fingernails, and nipples begin to develop. The appendages lengthen, and by the sixteenth week the skeleton is sufficiently developed to show up clearly on radiographs. During the sixteenth week, the fetal heartbeat can be heard by applying a stethoscope to the mother's abdomen. By the end of the sixteenth week, the fetus is 140 mm long (5.5 in.) and weighs about 200 g (7 oz).

After the sixteenth week, fetal length can be determined from radiographs. The reported length of a fetus is generally derived from a straight-line measurement from the crown of the head to the developing ischium (crown-rump length). Measurements made on an embryo prior to the fetal stage, however, are not reported as crown-rump measurements but as total length.

Seventeen to Twenty Weeks

Between the seventeenth and twentieth weeks, the legs achieve their final relative proportions, and fetal movements, known as **quickening,** are commonly felt by the mother. The skin is covered with a thin, white, cheeselike material known as **vernix caseosa**

vernix caseosa: L. *vernix*, varnish; L. *caseus*, cheese

(a)

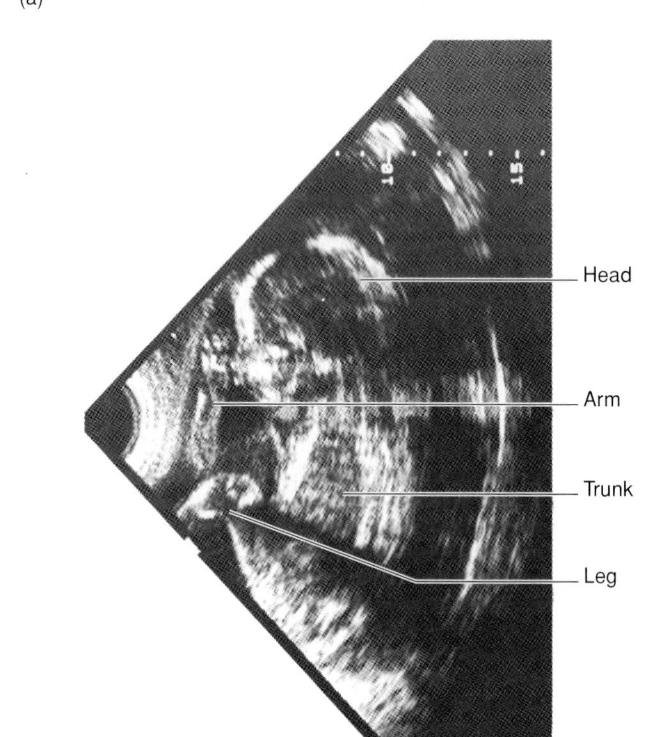

(b)

Figure 22.21

Ultrasonography. (*a*) Sound-wave vibrations are reflected from the internal tissues of a person's body. (*b*) Structures of the human fetus observed through an ultrasound scan.

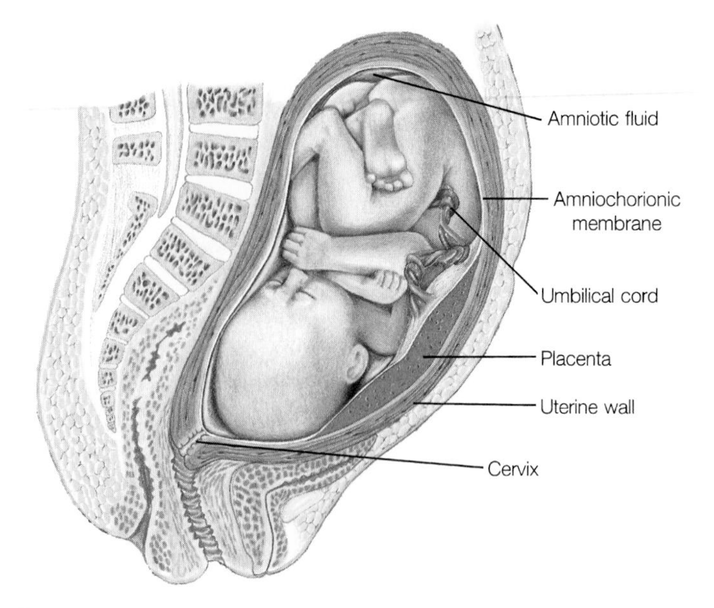

Figure 22.22

A fetus in vertex position. Toward the end of most pregnancies, the weight of the fetal head causes the body to rotate, positioning the head against the cervix of the uterus.

(*ka″se-o′să*). It consists of fatty secretions from the sebaceous glands and dead epidermal cells. The function of vernix caseosa is to protect the fetus while it is bathed in amniotic fluid. Twenty-week-old fetuses usually have fine, silklike fetal hair called **lanugo** (*lă-noo′go*) covering the skin. Lanugo is thought to hold the vernix caseosa in place on the skin and produce a ciliary-like motion that moves amniotic fluid. A 20-week-old fetus is about 190 mm (7.5 in.) long, and it weighs about 460 gm (16 oz). Because of cramped space, it develops a marked spinal flexure and is in what is commonly called the *fetal position,* with the head bent down, in contact with the flexed knees.

Twenty-One to Twenty-Five Weeks

Between the twenty-first and twenty-fifth weeks, the fetus increases its weight substantially to about 900 gm (32 oz). Body length increases only moderately (240 mm), however, so the weight is evenly proportioned. The skin is quite wrinkled and is translucent. Because the blood flowing in the capillaries is now visible, the skin appears pinkish.

Twenty-Six to Twenty-Nine Weeks

Toward the end of this period, the fetus will be about 275 mm (11 in.) long and will weigh about 1,300 gm (46 oz). A fetus might now survive if born prematurely, but the mortality rate is high. Its body metabolism cannot yet maintain a constant temperature, and its respiratory muscles have not matured enough to provide a regular respiratory rate. If, however, the premature infant is put in an incubator and a respirator is used to maintain its breathing, it may survive. The eyes open during this period, and the body is well covered with lanugo. If the fetus is a male, the testes should have begun descent into the scrotum (see exhibit III, chapter 20). As the time of birth approaches, the fetus rotates to a **vertex position** in which the head is directed toward the cervix (fig. 22.22). The head repositions toward the cervix because of the shape of the uterus and because the head is the heaviest part of the body.

lanugo: L. *lana,* wool

vertex: L. *vertex,* summit

Thirty to Thirty-Eight Weeks

At the end of 38 weeks, the fetus is considered full-term. It has reached a crown-rump length of 360 mm (14 in.) and weighs about 3,400 gm (7.5 lb). The average total length from crown to heel is 50 cm (20 in.). Most fetuses are plump with smooth skin because of the accumulation of subcutaneous fat. The skin is pinkish blue, even in fetuses of dark-skinned parents, because melanocytes do not produce melanin until the skin is exposed to sunlight. Lanugo is sparse and is generally found on the head and back. The chest is prominent, and the mammary area protrudes in both sexes. The external genitalia are somewhat swollen.

1. Explain why the ninth week is designated as the beginning of the fetal stage of development.
2. List the approximate fetal age at which each of the following occur: (a) first detection of fetal heartbeat, (b) presence of vernix caseosa and lanugo, and (c) fetal rotation into vertex position.
3. Compare ultrasound and radiographic techniques in determining fetal development and structure.

Labor and Parturition

Parturition, or childbirth, involves a sequence of events called labor. The uterine contractions of labor require the action of oxytocin, released by the posterior pituitary, and prostaglandins, produced in the uterus.

Objective 13 Describe the hormonal action that controls labor and parturition.

Objective 14 Describe the three stages of labor.

The time of prenatal development, or the time of pregnancy, is called **gestation.** In humans, the average gestation time is usually 266 days, or about 280 days from the beginning of the last menstrual period to **parturition** (*par″tyoo-rish′un*), or birth. Most fetuses are born within 10 to 15 days before or after the calculated delivery date. Parturition is accompanied by a sequence of physiological and physical events called **labor.**

The *onset of labor* is denoted by rhythmic and forceful contractions of the myometrial layer of the uterus. In *true labor*, the pains from uterine contractions occur at regular intervals and intensify as the interval between contractions shortens. A reliable indication of true labor is dilation of the cervix and a "show," or discharge, of blood-containing mucus that has accumulated in the cervical canal. In *false labor*, abdominal pain is experienced at irregular intervals, and cervical dilation and "show" are absent.

The uterine contractions of labor are stimulated by two agents: (1) **oxytocin** (*ok″sĭ-to′sin*), a polypeptide hormone produced in the hypothalamus and released from the posterior pituitary, and (2) **prostaglandins** (*pros″tă-glan′dinz*), a class of fatty acids produced within the uterus itself. Labor can indeed be induced artificially by injections of oxytocin or by the insertion of prostaglandins into the vagina as a suppository.

The hormone *relaxin,* produced by the corpus luteum, may also be involved in labor and parturition. Relaxin is known to soften the symphysis pubis in preparation for parturition and is thought to also soften the cervix in preparation for dilation. It may be, however, that relaxin does not affect the uterus, but rather that progesterone and estradiol may be responsible for this effect. Further research is necessary to understand the total physiological effect of these hormones.

As illustrated in figure 22.23, labor is divided into three stages:

1. **Dilation stage.** In this period, the cervix dilates to a diameter of approximately 10 cm. Contractions are regular during this stage, and the amniotic sac ("bag of waters") generally ruptures. If the amniotic sac does not rupture spontaneously, it is broken surgically. The dilation stage generally lasts from 8 to 24 hours.
2. **Expulsion stage.** This is the period of parturition, or actual childbirth. It consists of forceful uterine contractions and abdominal compressions to expel the fetus from the uterus and through the vagina. This stage may require 30 minutes in a first pregnancy but only a few minutes in subsequent pregnancies.
3. **Placental stage.** Generally within 10 to 15 minutes after parturition, the placenta is separated from the uterine wall and expelled as the *afterbirth*. Forceful uterine contractions characterize this stage, constricting uterine blood vessels to prevent hemorrhage. In a normal delivery, blood loss does not exceed 350 ml.

A *pudendal nerve block* may be administered during the early part of the expulsion stage to ease the trauma of delivery for the mother and to allow for an episiotomy. *Epidural anesthesia,* in which a local anesthetic is injected into the epidural space of the lumbar region of the spine, also may be used for these purposes.

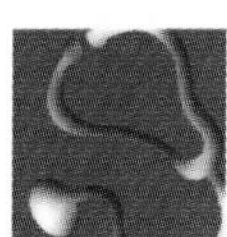

Five percent of newborns are born *breech.* In a breech birth, the fetus has not rotated and the buttocks are the presenting part. The principal concern of a breech birth is the increased time and difficulty of the expulsion stage of parturition. Attempts to rotate the fetus through the use of forceps may injure the infant. If an infant cannot be delivered breech, a *cesarean (sĕ-zar′e-an) section* must be performed. A cesarean section is delivery of the fetus through an incision made into the abdominal wall and the uterus.

gestation: L. *gestatus*, to bear

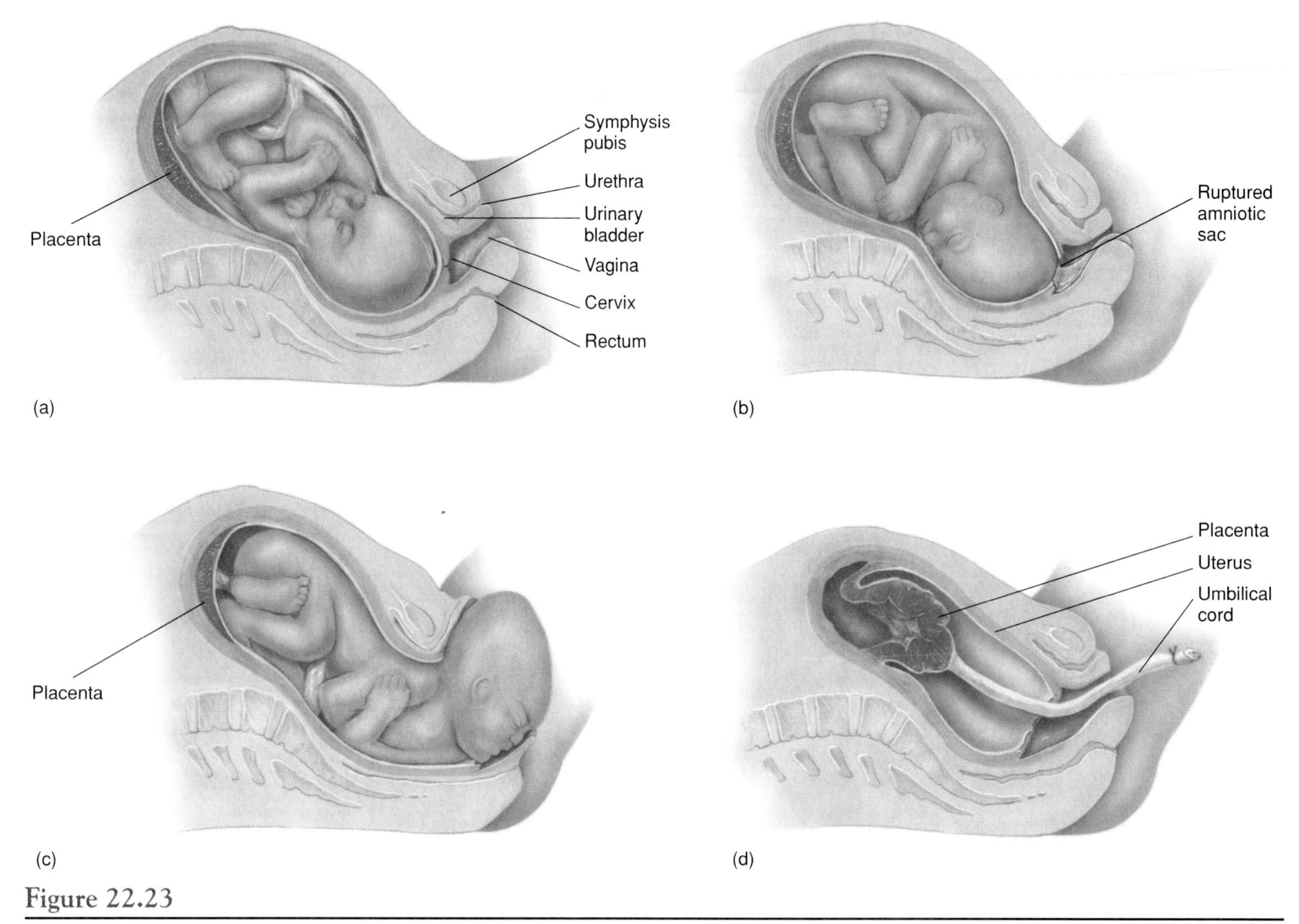

Figure 22.23

The stages of labor and parturition. (*a*) The position of the fetus prior to labor. (*b*) The ruptured amniotic sac during the early *dilation stage* of the cervix. (*c*) The *expulsion stage*, or period of parturition. (*d*) The *placental stage*, as the afterbirth is being expelled.

1. Distinguish between labor and parturition.
2. Explain the hormonal mechanisms responsible for labor and describe two techniques for inducing labor.
3. Describe the three stages of labor and state how long each stage lasts.

Periods of Postnatal Growth

The course of human life after birth is seen in terms of physical and physiological changes and the attainment of maturity in the neonatal period, infancy, childhood, adolescence, and adulthood.

Objective 15 Describe the growth and development that occurs during the neonatal period, infancy, childhood, and adolescence.

Objective 16 Define *puberty* and explain how its onset is determined in males and females.

Objective 17 Define the term *adulthood* and discuss sexual dimorphism in adult humans.

Neonatal Period

The **neonatal period** extends from birth to the end of the first month of extrauterine life. Although growth is rapid during this period, the most drastic changes are physiological. The body of a newborn must immediately adapt to major environmental changes, including thermal stress; rapid bacterial colonization of the skin, oral cavity, and GI tract; a barrage of sensory stimuli; and sudden demands on its body systems.

The most critical need of the newborn is the establishment of an adequate respiratory rate to ensure sufficient amounts of oxygen. The normal respiratory rate of a newborn is

neonatal: Gk. *neos*, new; *natus*, born

30 to 40 respirations per minute. An adequate heart rate is also imperative. The heart of a newborn appears to be large relative to the thoracic cavity (compared to the heart of an adult) and has a rapid rate that ranges from 120 to 160 beats per minute.

Most full-term newborn babies appear chubby because of the deposition of fat within adipose tissue during the last trimester of pregnancy. Dehydration is a serious threat because of the inability of the kidneys to excrete concentrated urine; large volumes of dilute urine are eliminated. Immunity is not well developed and is limited to that obtained from the mother through placental transfer. For this reason, newborns need to be guarded from exposure to infectious diseases.

Virtually all of the neurons of the nervous system are present in a newborn, but they are immature and the newborn has little coordination. Most infant behavior appears to be governed by lower cerebral centers and the spinal cord.

A newborn has many reflexes, some indicative of neuromuscular maturity and others essential for life itself. Four reflexes critical to survival are (1) the *suckling reflex*, triggered by anything that touches the newborn's lips; (2) the *rooting reflex*, which helps a baby find a nipple as it turns its head and starts to suckle in response to something that brushes its cheek; (3) the *crying reflex*, triggered by an empty stomach or other discomforts; and (4) the *breathing reflex*, apparent in a normal newborn even before the umbilical cord, with its supply of oxygen, is cut.

Babies born more than 3 weeks before the due date are generally considered *premature*, but because errors are commonly made in calculating the delivery date, prematurity is defined by neonatal body weight rather than due date. Newborns weighing less than 2,500 gm (5.5 lb) are considered premature. By this definition, approximately 8% of newborns in the United States are premature.

Postmature babies are those born 2 or more weeks after the due date. They frequently weigh less than they would have if they had been born at term because the placenta often becomes less efficient after a full-term pregnancy. Approximately 10% of newborns in the United States are postmature.

Infancy

The period of **infancy** follows the neonatal period and encompasses the first 2 years of life. Infancy is characterized by tremendous growth, increased coordination, and mental development.

A full-term child will generally double its birth weight by 5 months and triple it in a year. The formation of subcutaneous adipose tissue reaches its peak at about 9 months, causing the infant to appear chubby. Growth decelerates during the second year, during which time the infant gains only about 2.5 kg (5–6 lb). During the second year, the infant develops locomotor and manipulative control and gradually becomes more lean and muscular.

Body length increases during the first year by 25 to 30 cm (10 to 12 in.). There is an additional 12 cm (5 in.) of growth during the second year. The brain and circumference of the head also grow rapidly during the first year, but only moderately during the second. Head circumference increases by approximately 12 cm (5 in.) during the first year and only by an additional 2 cm during the second. The anterior fontanel (see fig. 6.13) gradually diminishes in size after 6 months and becomes effectively closed at any time from 9 to 18 months. It is the last of the fontanels to close. The brain is two-thirds of its adult size at the end of the first year and four-fifths of its adult size at the end of the second year.

By 2 years, most infants weigh approximately four times their birth weight and are 81 to 91 cm (32 to 36 in.) long. The body proportions of a 2-year-old are certainly not the same as those of an adult (fig. 22.24). Growth is a differential process, resulting in gradual changes in body proportions.

Deciduous teeth begin to erupt in most infants between 5 and 9 months. By the time they are a year old, most infants have 6 to 8 teeth. Eight more teeth erupt during the second year, making a total of 14 to 16, including the first deciduous molars and canine teeth.

The growth rates of children vary tremendously. Body lengths and weights are not always reliable indicators of normal growth and development. A more objective evaluation of a child's physical development is determined through radiographic analysis of skeletal ossification in the carpal region (fig. 22.25).

Childhood

Childhood is the period of growth and development extending from infancy to adolescence, at which time puberty begins. The duration of childhood varies because puberty begins at different ages for different people.

Childhood years are a period of relatively steady growth until preadolescence, which is characterized by a growth spurt. The average weight gain during childhood is about 3 to 3.5 kg (7 lb) per year. There is an average increase in height of 6 cm (2.5 in.) per year. The circumference of the head increases slightly—by about 3 to 4 cm (1.5 in.) during childhood—and by adolescence, the head and brain are virtually adult size.

The facial bones continue to develop during childhood (fig. 22.26). Especially significant is the enlargement of the sinuses. The first permanent teeth generally erupt during the seventh year, and then the deciduous teeth are shed approximately in the same sequence as they were acquired. Deciduous teeth are replaced at a rate of about four per year over the next 7 years.

Although there is an average rate of growth during childhood because of genetics, there is a wide range in what is considered normal growth. If, for example, the 8-year-olds who were among the tallest and heaviest 10% of their age group were to stop growing for a year while their classmates grew normally, they would still be taller than half of their contemporaries and heavier than three-quarters of them.

premature: L. *prae*, before; *maturus*, ripe
infancy: L. *in*, not; *fans*, speaking

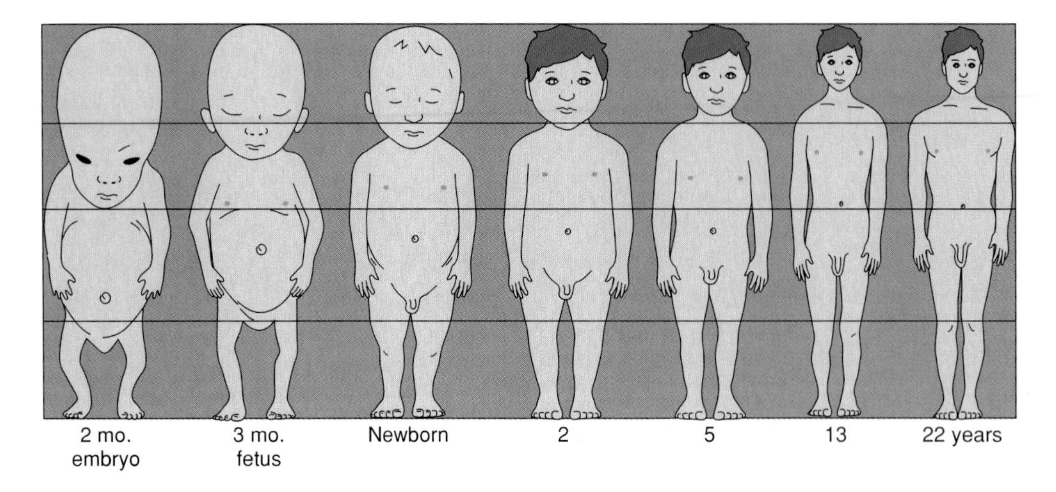

Figure 22.24

The relative proportions of the body from embryo to adulthood. The head of a newborn accounts for a quarter of the total body length, and the lower appendages account for about one-third. In an adult, the head constitutes about 13% of the total body length, whereas the lower appendages constitute approximately 50%.

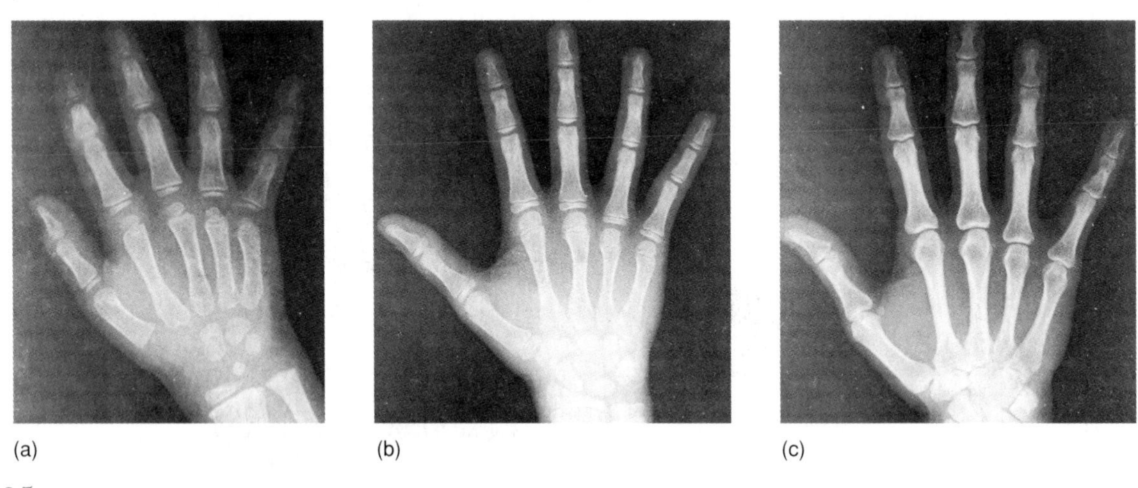

Figure 22.25

Radiographs of the right hand (*a*) of a child, (*b*) of an adolescent, and (*c*) of an adult.

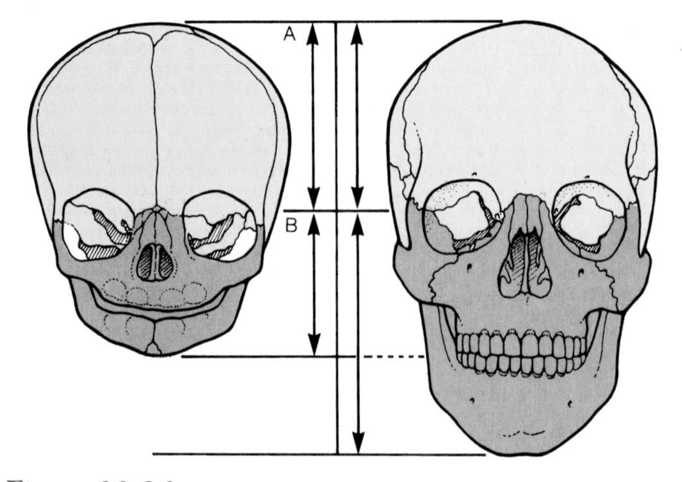

Figure 22.26

Growth of the skull. The height of the cranial vault (distance between planes A and B) is shown to be the same in both the infant and adult skulls. Growth of the skull occurs almost exclusively within the bones of the facial region.

During childhood, the average child becomes thinner and stronger each year as he or she grows taller. The average 10-year-old, for example, can throw a ball twice as far as the average 6-year-old. Visceral organs, particularly the heart and lungs, develop tremendously during this period, enabling a child to run faster and exercise longer.

Lymphoid tissue is at its peak of development during mid-childhood and generally exceeds the amount of such tissue in the normal adult. Children need the extra lymphoid tissue to combat childhood diseases, especially in countries where nutrition is poor and health care is minimal.

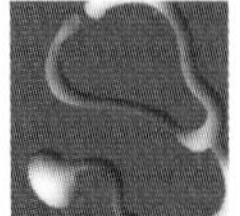

Childhood *obesity* can become a serious physical and psychological problem if not corrected. Overweight children usually exercise less and run a greater risk of contracting serious illnesses. Frequently, they are teased and rejected by classmates, which causes psychological stress. At least 5% of children in the United States can be classified as obese. Childhood obesity

Table 22.3

Sequence of Physical Development during Adolescence

Females	Age Span		Males
Growth spurt begins; breast buds and sparse pubic hair appear	10–11 yrs	11.5–13 yrs	Growth of testes and scrotum; sparse pubic hair appears; growth spurt begins; growth of penis begins
Appearance of straight, pigmented pubic hair; some deepening of voice; rapid growth of ovaries, uterus, and vagina; acidic vaginal secretion; menarche; further enlargement of breasts; kinky pubic hair; age of maximum growth	11–14 yrs	13–16 yrs	Appearance of straight, pigmented pubic hair; deepening of voice; maturation of penis, testes, scrotum, and accessory reproductive glands; ejaculation of semen; axillary hair; kinky pubic hair; sparse facial hair; age of maximum growth
Appearance of axillary hair; breasts of adult size and shape; culmination of physical growth	14–16 yrs	16–18 yrs	Increased body hair; marked vocal change; culmination of physical growth

and adult obesity usually go hand in hand. Obesity in adults is a major health problem considering that one of five adults is at least 30% over his or her ideal, healthy weight. A controlled diet and regular exercise are fundamental in correcting obesity.

Adolescence

Adolescence is the period of growth and development between childhood and adulthood. It begins around the age of 10 in girls and the age of 12 in boys. Adolescence is frequently said to end at 20 years of age, but it is not clearly delineated and varies with the developmental, physical, emotional, mental, or cultural criteria that define an adult.

Puberty (*pyoo'ber-te*) is the stage of early adolescence when the secondary sex characteristics become expressed and the sexual organs become functional. **Pubescence** (*pyoo-bes'ens*) refers to the continuum of physical changes during puberty, particularly in regard to body hair. Although puberty is under hormonal control, a complex interaction of other factors, including nutrition and socioeconomic forces, has a decisive influence on the onset and duration of puberty. Thus, for both sexes there is wide individual variation.

The end result of puberty is that additional **sexually dimorphic characteristics**—traits that distinguish the sexes—are apparent. The average adult male, for example, has a deeper voice and more body hair and is taller than the average adult female. Prior to puberty, male and female children have few major structural differences aside from the general appearance of the external genitalia.

Puberty actually begins before it is physically expressed. In most instances, significant amounts of sex hormones appear in the blood of females by the age of 10 and in males by the age of 11. Sexual changes are usually thought of as the only features of puberty, but major musculoskeletal changes take place as well. During late childhood, the body proportions of the sexes are similar, males being slightly taller. Under the influence of hormones, females experience a growth spurt in early adolescence that precedes that of males by nearly 2 years. During this time, females are temporarily taller. Once puberty begins in males, the heights of both sexes equalize, and at the culmination of puberty, males are approximately 10 cm (4 in.) taller than females on the average. By the time growth is completed at the end of adolescence, males are generally 13 cm (5 in.) taller than females.

Other dimorphic differences involving skeletal structures include a broadening of the pelvic girdle in females. The muscles of males become more massive and stronger than those of females. Females acquire a thicker subcutaneous layer of the skin during adolescence, which gives them more rounded contours.

Sexual maturation during adolescence includes not only the development of the reproductive organs but the appearance of secondary sex characteristics (see fig. 2.9). The sequence of sexual maturation and expression of secondary sex characteristics for both males and females is presented in table 22.3.

In females, the first physical indication of puberty is the appearance of **breast buds,** which are swellings of the breasts and slight enlargement and pigmentation of the areolar areas. Breast buds generally appear in healthy girls at about 11 years of age, but the age may range from 9 to 13 years. Approximately 3 years are required after the appearance of breast buds for maturation of the breasts. Pubic hair usually begins to appear shortly after the breast buds become apparent, but in about one-third of all girls, sparse pubic hair appears before the breast buds. Axillary hair appears a year or two after pubic hair.

The first menstrual period, referred to as **menarche** (*mĕ-nar'ke*), generally occurs at age 13 but may occur as early as the age of 9 or as late as 17. During puberty, the vaginal secretions change from alkaline to acidic.

The onset of puberty in males varies just as much as it does in females but generally lags behind by about 1½ years. The first indication of puberty in boys is growth of the testes and the appearance of sparse pubic hair at the age of 12 on the average. This is followed by growth of the penis, which continues for about 2 years, and the appearance of axillary hair. Vocal changes generally begin during early puberty but are not completed until midpuberty. Facial hair and chest hair (which may or may not be present) first

adolescence: L. *adolescere*, to grow up
puberty: L. *pubertas*, adult form
dimorphic: Gk. *di*, two; *morphe*, form

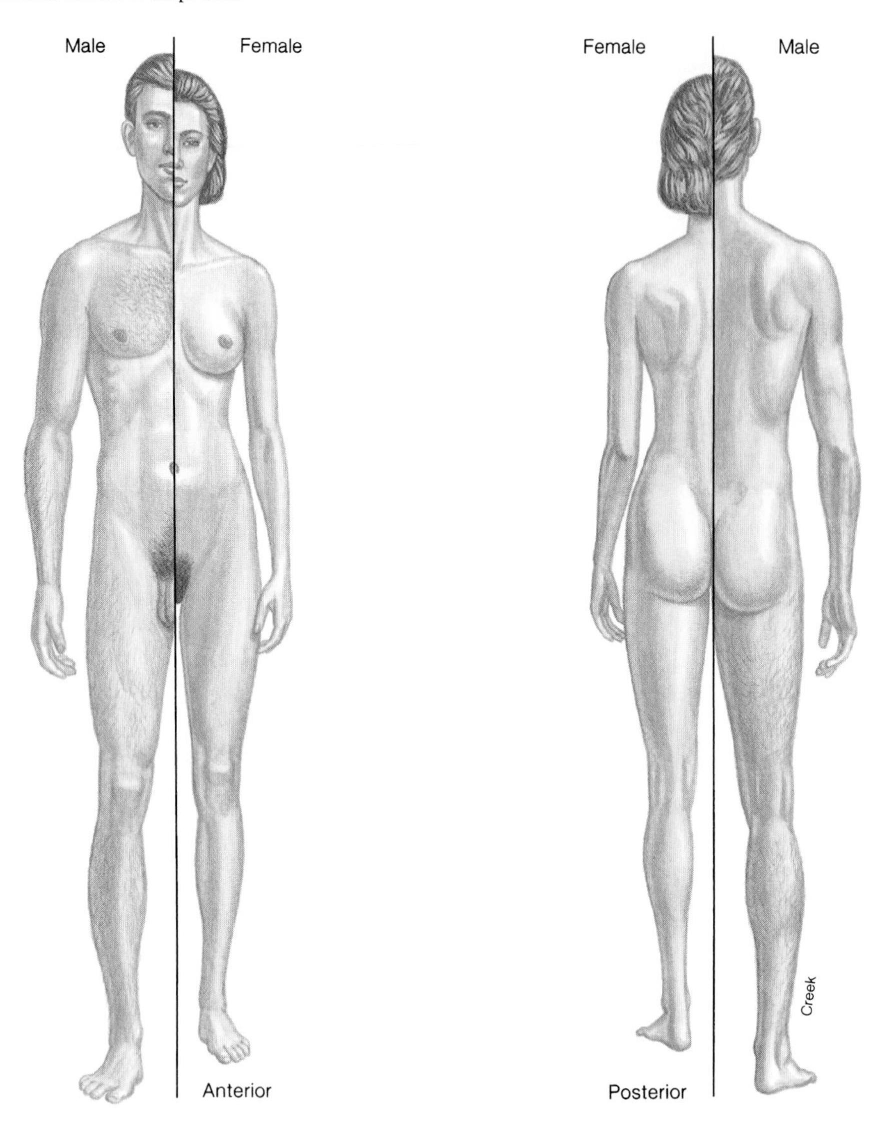

Figure 22.27

Relative differences in the physiques of an adult male and female. A male has proportionately longer appendages and a longer neck than a female.

appear toward the end of puberty. The mean age at which semen can be ejaculated is 13.7 years, but sufficient mature sperm for fertility are generally not produced until 14 to 16 years of age.

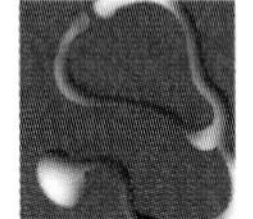

Acne is an inflammatory disease of the integument that frequently occurs during adolescence. The increase in hormonal activity responsible for the physical changes taking place during puberty also affects the activity of the sebaceous glands and promotes the formation of comedones and inflamed superficial pustules. Tension and emotional stress may also promote acne. Acne in teenagers may cause serious psychological problems related to the need for peer acceptance.

Adulthood

Adulthood is the final stage of human physical change. It is the period of life beyond adolescence. An adult has reached maximum physical stature as determined by genetic, nutritional, and environmental factors. Although skeletal maturity is reached in early adulthood, anatomical and physiological changes continue throughout adulthood and are part of the aging process.

Sexual dimorphism in human adults goes beyond the obvious anatomical differences. Males and females differ physiologically, metabolically, and behaviorally (psychologically and socially). Some of these differences manifest themselves prenatally and during childhood. Others are characteristics of adolescence and adulthood. It is uncertain to what extent specific dimorphisms of the sexes are genetically determined through hormonal action or influenced by environmental (including cultural) factors. It is also unclear how these governing factors are expressed in observed physical characteristics.

The shape of the adult body is determined primarily by the skeleton and attached muscles, and also by the subcutaneous connective tissue (especially adipose tissue) and extracellular body fluids. Although the body proportions of adult males and females vary widely between individuals (fig. 22.27), adult

Table 22.4
Body Composition of Average Adults

	Males				Females			
	Absolute		Relative (% Body Weight)		Absolute		Relative (% Body Weight)	
	Age 25	Age 65	Age 25	Age 65	Age 25	Age 65	Age 25	Age 65
Body weight	70.0 kg	70.0 kg			69.0 kg	60.0 kg		
Body fluids	41.0 liters	37.0 liters	58.9%	52.9%	30.8 liters	28.0 liters	51.3%	46.7%
Intracellular	24.0 liters	19.2 liters	34.3%	27.4%	16.6 liters	14.3 liters	27.7%	23.8%
Extracellular	17.2 liters	17.8 liters	24.6%	25.5%	14.2 liters	13.7 liters	23.6%	22.9%
Plasma volume	3,302 ml	2,940 ml	4.7%	4.2%	2,760 ml	2,462 ml	4.6%	4.1%
Lean body weight	56.3 kg	50.5 kg	80.4%	72.1%	42.0 kg	38.2 kg	70.2%	63.7%
Body fat	13.7 kg	19.5 kg	19.6%	27.9%	17.9 kg	21.8 kg	29.8%	36.3%
Skeletal weight	5.8 kg	5.7 kg	8.3%	8.1%	4.4 kg	4.2 kg	7.3%	7.0%

From K. H. Oleson, "Body Composition in Normal Adults" in *Human Body Composition*, Vol. 7:177–190. Copyright © 1965 Pergamon Press Ltd., Oxford, England. Reprinted by permission.

Table 22.5
Summary of Postnatal Periods

Period	Time Span	Physical and Behavioral Characteristics
Neonatal period	Birth to end of fourth week	Stabilizing of body systems necessary to carry on respiration, obtain nutrients, digest nutrients, excrete wastes, regulate body temperature, and circulate blood
Infancy	End of fourth week through second year	Tremendous growth; teeth begin to erupt; muscular and nervous systems develop so that motor activities are possible; verbal communication begins
Childhood	End of infancy to puberty	Consistent growth; deciduous teeth erupt and are replaced by permanent teeth; motor control improves; urinary bladder and bowel controls are established; intellect develops rapidly
Adolescence	End of puberty to adulthood	Maturing of reproductive system; growth spurts in skeletal and muscular systems; continued development of intellect and emotional maturity
Adulthood	End of adolescence to old age	Attainment of maximum physical stature and strength; anatomical and physiological degenerative changes begin
Senescence	Old age to death	Continuing of senescence; body becomes less able to cope with diseases and physical demands; death—usually from physical disturbances in the cardiovascular system or disease processes in vital organs

males generally have longer appendages than females, their shoulders are broader, and their pelvises are narrower. Males also have relatively longer necks.

The general body composition of males and females can also be compared. Mean data for body composition are summarized in table 22.4. Values for total body fluid and skeletal weight are lower for adult females than for adult males. Females, however, have a higher percentage of body fat. Other differences not shown in table 22.4 are that adult females have lower blood pressures, erythrocyte counts (hematocrits), basal metabolic rates, and respiratory rates than adult males. Females, however, have higher heart rates and oral temperatures.

Some of the physical and behavioral characteristics of each stage of human growth and development are summarized in table 22.5.

Physical anthropologists have long been interested in the body proportions of different racial groups. *Anthropometry* (*an″thrŏ-pom′ĭ-tre*) is the study of physical differences, particularly skeletal, between racial groups. Proportions in anthropometric studies are expressed in indices; that is, one measurement

anthropometry: Gk. *anthropos*, human; *metron*, measure

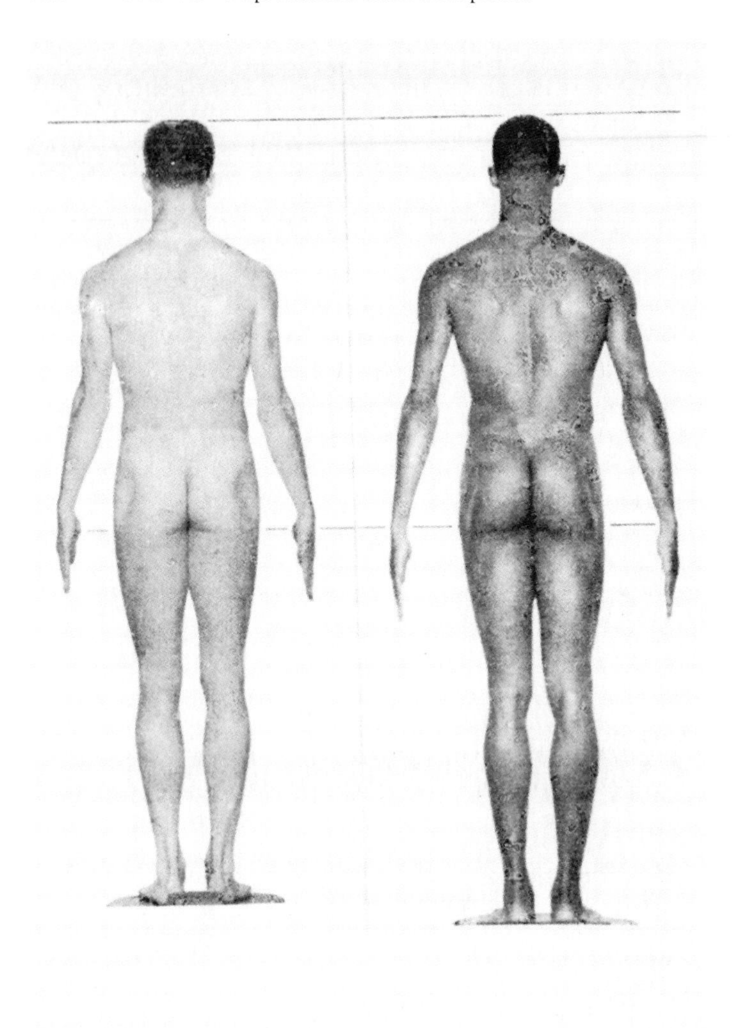

Figure 22.28

A comparison of Caucasian and African American physiques. The photographs of two Olympic 400-meter runners have been scaled so that both have the same sitting height.

reckoned as a percentage of another measurement. The cranial index, for example, is the breadth of the skull expressed as a percentage of its length.

A particularly standardized expression of racial differences are the indices of appendage lengths relative to sitting height. People of African ancestry, for example, have comparatively long appendages relative to sitting height; moreover, the forearm and leg are long relative to the brachium and thigh (fig. 22.28). Australian Aborigines have even longer legs proportionately than do Native Africans. People of African ancestry also have the narrowest pelvic girdle for a given shoulder width. People of Asian ancestry have relatively short appendage lengths to sitting heights. These differences provide distinct advantages and disadvantages in certain sports. Blacks have an advantage in many track events, particularly the sprints and high hurdles, whereas whites are generally adapted to distance running. Asians often excel in gymnastics and weight lifting.

1. Construct a table that lists the periods of postnatal growth from infancy through adolescence and indicate the events or characteristics of each period.
2. List four reflexes in a newborn that are critical for survival.
3. Define the terms *puberty*, *pubescence*, *sexual dimorphism*, and *menarche*. What is the average age of puberty and what brings it about?
4. Describe the physical characteristics of adulthood. Compare the body structure of an adult female to that of an adult male.

Inheritance

Inheritance is the acquisition of characteristics or qualities by transmission from parent to offspring. Hereditary information is transmitted by genes.

Objective 18 Define *genetics*.

Objective 19 Discuss the variables that account for a person's phenotype.

Objective 20 Explain how probability is involved in predicting inheritance and use a Punnett square to illustrate selected probabilities.

Genetics is the branch of biology that deals with inheritance. Genetics and inheritance are important in anatomy and physiology because of the numerous developmental and functional disorders that have a genetic basis. Knowledge of which disorders and diseases are inherited finds practical application in genetic counseling. The genetic inheritance of an individual begins with conception.

Each zygote inherits 23 chromosomes from the mother and 23 chromosomes from the father. This does not produce 46 different chromosomes; rather, it produces 23 pairs of *homologous chromosomes*. Each member of a homologous pair, with the important exception of the sex chromosomes, looks like the other and contains similar genes (such as those coding for eye color, height, and so on). These homologous pairs of chromosomes can be **karyotyped** (photographed or illustrated) and identified, as shown in figure 22.29. Each cell that contains 46 chromosomes (that is *diploid*) has two number-1 chromosomes, two number-2 chromosomes, and so on through chromosomes number 22. The first 22 pairs of chromosomes are called **autosomal** (*aw"tŏ-so'mal*) **chromosomes.** The twenty-third pair of chromosomes are the **sex chromosomes,** which may look different and may carry different genes. In a female these consist of two X chromosomes, whereas in a male there is one X chromosome and one Y chromosome.

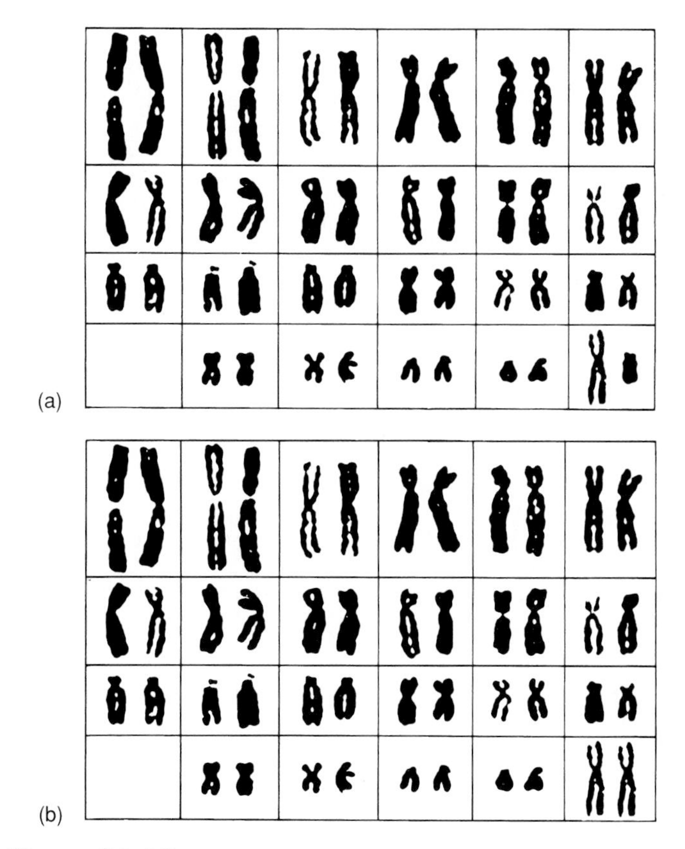

Figure 22.29

A karyotype of homologous pairs of chromosomes obtained from a human diploid cell. The first 22 pairs of chromosomes are called the autosomal chromosomes. The sex chromosomes are (*a*) XY for a male and (*b*) XX for a female.

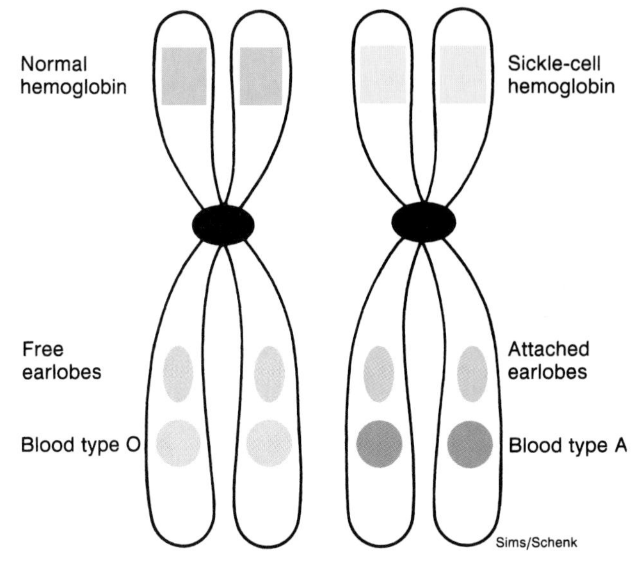

Figure 22.30

A pair of homologous chromosomes. Homologous chromosomes contain genes for the same characteristic at the same locus.

Genes and Alleles

A **gene** is the portion of the DNA of a chromosome that contains the information needed to synthesize a particular protein molecule. Although each diploid cell has a pair of genes for each characteristic, these genes may be present in variant forms. Those alternative forms of a gene that affect the same characteristic but that produce different expressions of that characteristic are called **alleles** (*ă-lēlz'*). One allele of each pair originates from the female parent and the other from the male. The shape of a person's ears, for example, is determined by the kind of allele received from each parent and how the alleles interact with one another. Alleles are always located on the same spot (called a **locus**) on homologous chromosomes (fig. 22.30).

For any particular pair of alleles in a person, the two alleles are either identical or not identical. If the alleles are identical, the person is said to be **homozygous** (*ho"mo-zi'gus*) for that particular characteristic. But if the two alleles are different, the person is **heterozygous** (*het"er-o-zi'gus*) for that particular trait.

Genotype and Phenotype

A person's DNA contains a catalog of genes known as the **genotype** (*jen'ŏ-tīp*) of that person. The expression of those genes results in certain observable characteristics referred to as the **phenotype** (*fe'nŏ-tīp*).

If the alleles for a particular trait are homozygous, the characteristic expresses itself in a specific manner (two alleles for attached earlobes, for example, results in a person with attached earlobes). If the alleles for a particular trait are heterozygous, however, the allele that expresses itself and the way in which the genes for that trait interact will determine the phenotype. The allele that expresses itself is called the **dominant allele,** the one that does not is the **recessive allele.** The various combinations of dominant and recessive alleles are responsible for a person's hereditary traits (table 22.6).

Table 22.6

Hereditary Traits in Humans Determined by Single Pairs of Dominant and Recessive Alleles

Dominant	Recessive	Dominant	Recessive
Free earlobes	Attached earlobes	Color vision	Color blindness
Dark brown hair	All other colors	Broad lips	Thin lips
Curly hair	Straight hair	Ability to roll tongue	Lack of this ability
Pattern baldness (♂♂)	Baldness (♀♀)	Arched feet	Flat feet
Pigmented skin	Albinism	A or B blood factor	O blood factor
Brown eyes	Blue or green eyes	Rh blood factor	No Rh blood factor

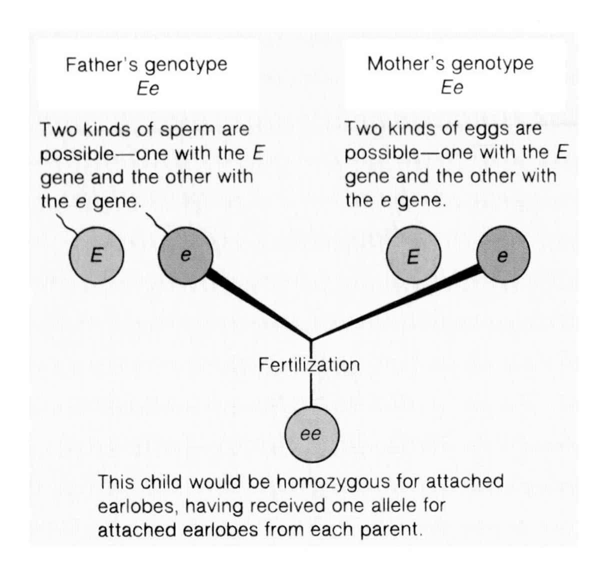

Figure 22.31

Inheritance of earlobe characteristics. Two parents with unattached (free) earlobes can have a child with attached earlobes.

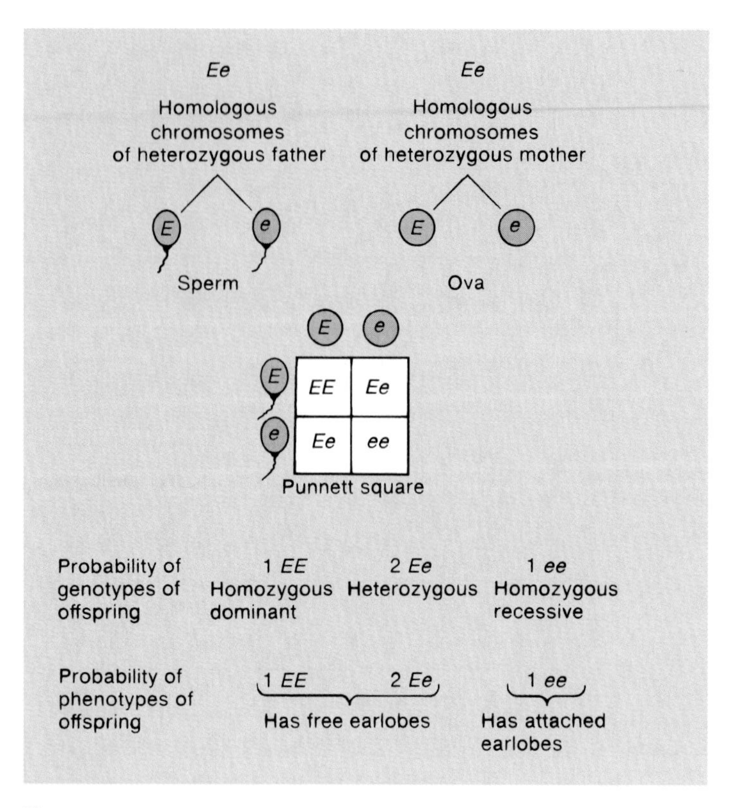

Figure 22.32

Use of a Punnett square to determine genotypes and phenotypes that could result from the mating of two heterozygous parents.

In describing genotypes, it is traditional to use letter symbols to refer to the alleles of an organism. The dominant alleles are symbolized by uppercase letters, and the recessive alleles are symbolized by lowercase. Thus, the *genotype* of a person who is homozygous for free earlobes as a result of a dominant allele is symbolized *EE*; a heterozygous pair is symbolized *Ee*. In both of these instances, the *phenotypes* of the individuals would be free earlobes resulting from the presence of a dominant allele in each genotype. A person who inherited two recessive alleles for earlobes would have the genotype *ee* and would have attached earlobes.

Thus, three genotypes are possible when gene pairing involves dominant and recessive alleles. They are *homozygous dominant (EE)*, *heterozygous (Ee)*, and *homozygous recessive (ee)*. Only two phenotypes are possible, however, since the dominant allele is expressed in both the homozygous dominant (*EE*) and the heterozygous (*Ee*) individuals. The recessive allele is expressed only in the homozygous recessive (*ee*) condition. Refer to figure 22.31 for an illustration of how a homozygous recessive trait may be expressed in a child of parents who are heterozygous.

Probability

A **Punnett** (*pun'et*) **square** is a convenient way to express the probabilities of allele combinations for a particular inheritable trait. In constructing a Punnett square, the male gametes (spermatozoa) carrying a particular trait are placed at the side of the chart, and the female gametes (ova) at the top, as in figure 22.32. The four spaces on the chart represent the possible combinations of male and female gametes that could form zygotes. The probability of an offspring having a particular genotype is 1 in 4 (.25) for homozygous dominant and homozygous recessive and 1 in 2 (.50) for heterozygous.

A genetic study in which a single characteristic (e.g., ear shape) is followed from parents to offspring is referred to as a **monohybrid cross.** A genetic study in which two characteristics are followed from parents to offspring is referred to as a **dihybrid cross** (fig. 22.33). The term *hybrid* refers to an offspring descended from parents who have different genotypes.

Sex-linked Inheritance

Certain inherited traits are located on a sex-determining chromosome, and are thus called **sex-linked** characteristics. The allele for *red-green color blindness*, for example, is determined by a recessive allele (designated c) found on the X chromosome but not on the Y chromosome. Normal color vision (designated C) dominates. The ability to discriminate red from green, therefore, depends entirely on the X chromosomes. The genotype possibilities are as follows:

X^CY	Normal male
X^cY	Color-blind male
X^CX^C	Normal female
X^CX^c	Normal female carrying the recessive allele
X^cX^c	Color-blind female

In order for a female to be red-green colorblind, she must have the recessive allele on both of her X chromosomes. Her

Punnett square: from Reginald Crundall Punnett, English geneticist, 1875–1967

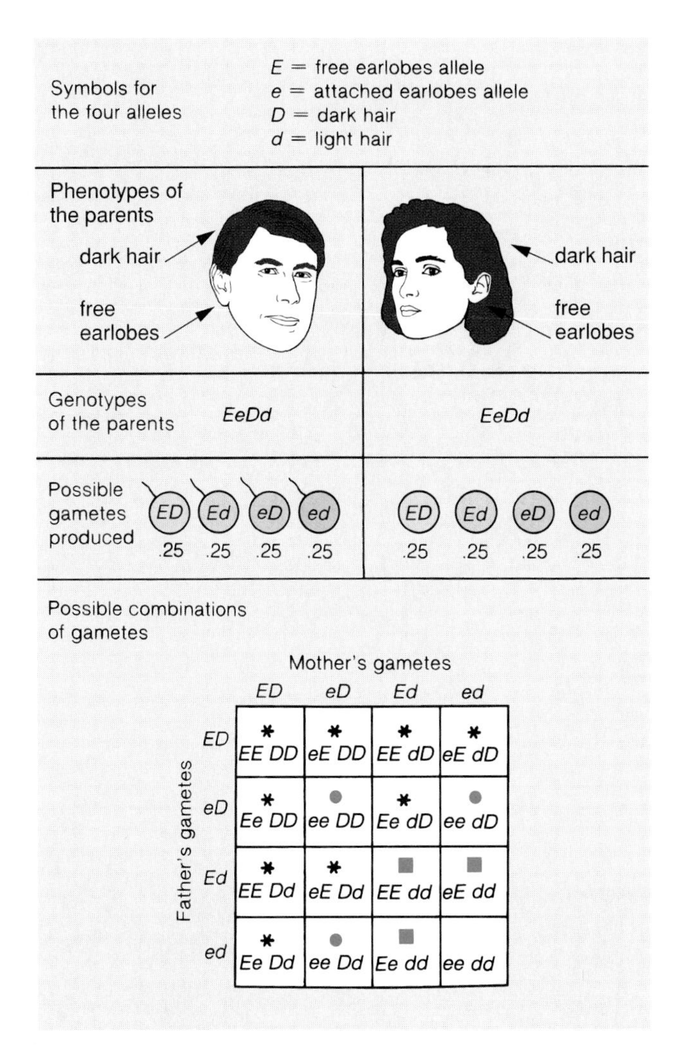

Figure 22.33

In a dihybrid cross, two pairs of traits are followed simultaneously. Any of the combinations of genes that have a *D* and an *E* (nine possibilities) will have free earlobes and dark hair. These are indicated with an asterisk (*). Three of the possible combinations have two alleles for attached earlobes (*ee*) and at least one allele for dark hair. They are indicated with a dot (•). Three of the combinations have free earlobes and light hair. These are indicated with a square (■). The remaining possibility has the genotype *eedd* for attached earlobes and light hair.

father would have to be red-green colorblind and her mother would have to be a carrier for this condition. A male with only one such allele on his X chromosome, however, will show the characteristic. Since a male receives his X chromosome from his mother, the inheritance of sex-linked characteristics usually passes from mother to son.

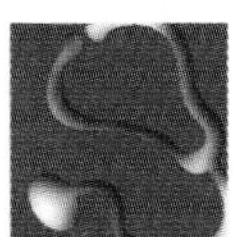

Hemophilia is a sex-linked condition caused by a recessive allele. The blood in a person with hemophilia fails to clot or clots very slowly after an injury. If *H* represents normal clotting and *h* represents abnormal clotting, then males with X^HY will be normal and males with X^hY will be hemophiliac. Females with X^hX^h will have the disorder.

1. Define *genetics, genotype, phenotype, allele, dominant, recessive, homozygous,* and *heterozygous*.
2. List several dominant and recessive traits inherited in humans. What are some variables that determine a person's phenotype?
3. Construct a Punnett square to show the possible genotypes for color blindness of an X^cY male and an X^CX^c female.

Clinical Considerations

Pregnancy and childbirth are natural events in human biology and generally progress smoothly without complications. Prenatal development is amazingly precise, and although traumatic, childbirth for most women in the world takes place without the aid of a physician. Occasionally, however, serious complications arise, and the knowledge of an obstetrician is required. The physician's knowledge of what constitutes normal development and what factors are responsible for congenital malformations ensures the embryo and fetus every possible chance to develop normally. Many of the clinical aspects of prenatal development involve what might be referred to as applied developmental biology.

In clinical terms, gestation is frequently divided into three phases, or **trimesters,** each lasting three calendar months. By the end of the **first trimester,** all of the major body systems are formed, the fetal heart can be detected, the external genitalia are developed, and the fetus is about the width of the palm of an adult's hand. During the **second trimester,** fetal quickening can be detected, epidermal features are formed, and the vital body systems are functioning. The fetus, however, still would be unlikely to survive if birth were to occur. At the end of the second trimester, fetal length is about equal to the length of an adult's hand. The fetus experiences a tremendous amount of growth and refinement in system functioning during the **third trimester.** A fetus of this age may survive if born prematurely, and of course, the chances of survival improve as the length of pregnancy approaches the natural delivery date.

Many clinical considerations are associated with prenatal development, and some of these relate directly to the female reproductive system. Other developmental problems are genetically related and will be mentioned only briefly. Of clinical concern for developmental anatomy are such topics as ectopic pregnancies, so-called test-tube babies, multiple pregnancy, fetal monitoring, and congenital defects.

Abnormal Implantation Sites

In an **ectopic** (*ek-top'ik*) **pregnancy** the blastocyst implants outside the uterus or in an abnormal site within the uterus

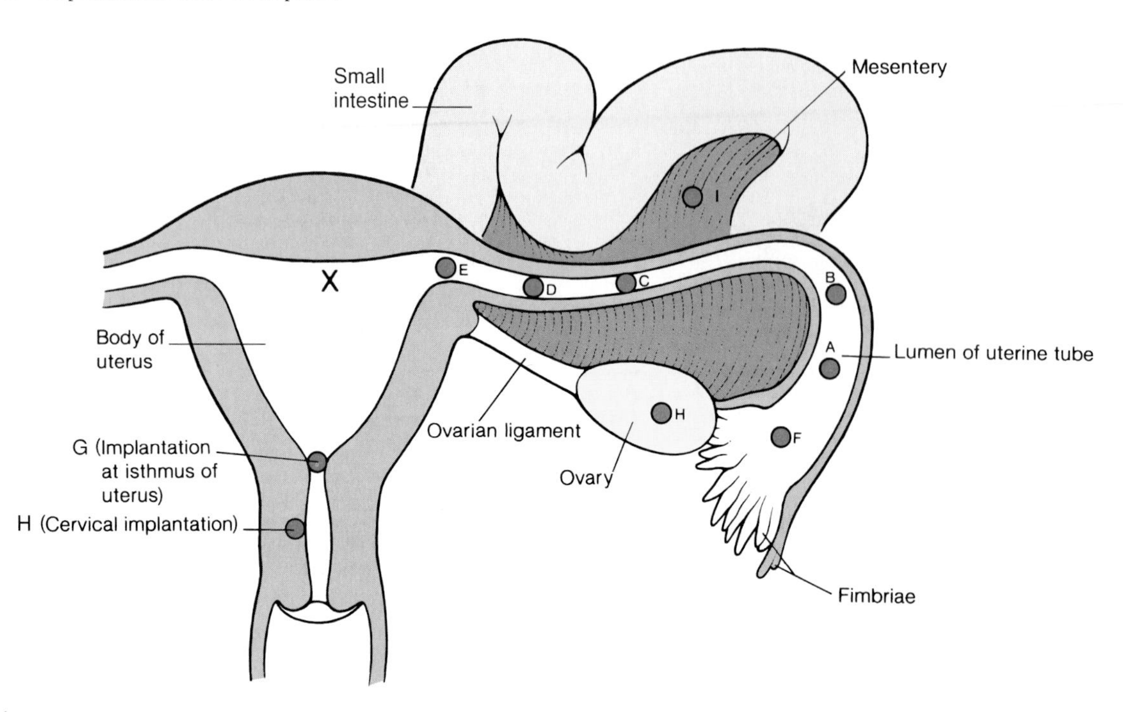

Figure 22.34

Sites of ectopic pregnancies. The normal implantation site is indicated by an X; the abnormal sites are indicated by letters in order of frequency of occurrence.

(fig. 22.34). About 95% of the time, the ectopic location is within the uterine tube and is referred to as a **tubal pregnancy.** Occasionally, implantation occurs near the cervix, where development of the placenta blocks the cervical opening. This condition, called **placenta previa** (*pre've-ă*) causes serious bleeding. Ectopic pregnancies will not develop normally in unfavorable locations, and the fetus seldom survives beyond the first trimester. Tubal pregnancies are terminated through medical intervention. If a tubal pregnancy is permitted to progress, the uterine tube generally ruptures, followed by hemorrhaging. Depending on the location and the stage of development (hence vascularity) of a tubal pregnancy, it may or may not be life-threatening to the woman.

In Vitro Fertilization and Artificial Implantation

Reproductive biologists have been able to fertilize a human oocyte in vitro (outside the body), culture it to the blastocyst stage, and then perform artificial implantation, leading to a full-term development and delivery. This is the so-called test-tube baby. To obtain the oocyte, a specialized laparoscope (fig. 22.35) is used to aspirate the preovulatory egg from a mature vesicular ovarian follicle. The oocyte is then placed in a suitable culture medium, where it is fertilized with spermatozoa. When the zygote reaches the blastocyst stage, it is introduced into the uterus for implantation and subsequent growth. In vitro fertilization with artificial implantation is a means of overcoming infertility problems due to damaged, blocked, or missing uterine tubes in females or low sperm counts in males.

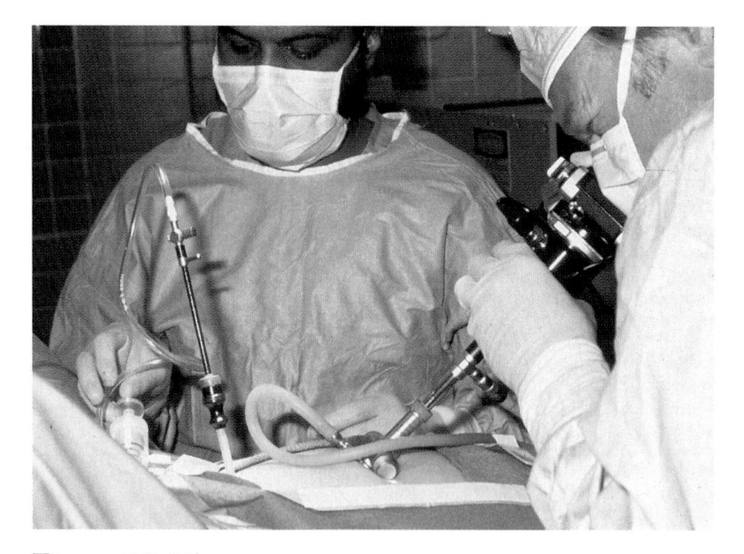

Figure 22.35

A laparoscope is used for various abdominal operations, including the extraction of a preovulatory ovum.

Multiple Pregnancy

Twins occur about once in 85 pregnancies. They can develop in two ways. **Dizygotic** (*di"zi-got'ik*) (fraternal) **twins** develop from two zygotes produced by the fertilization of two oocytes by two

previa: L. *previa*, appearing before or in front of

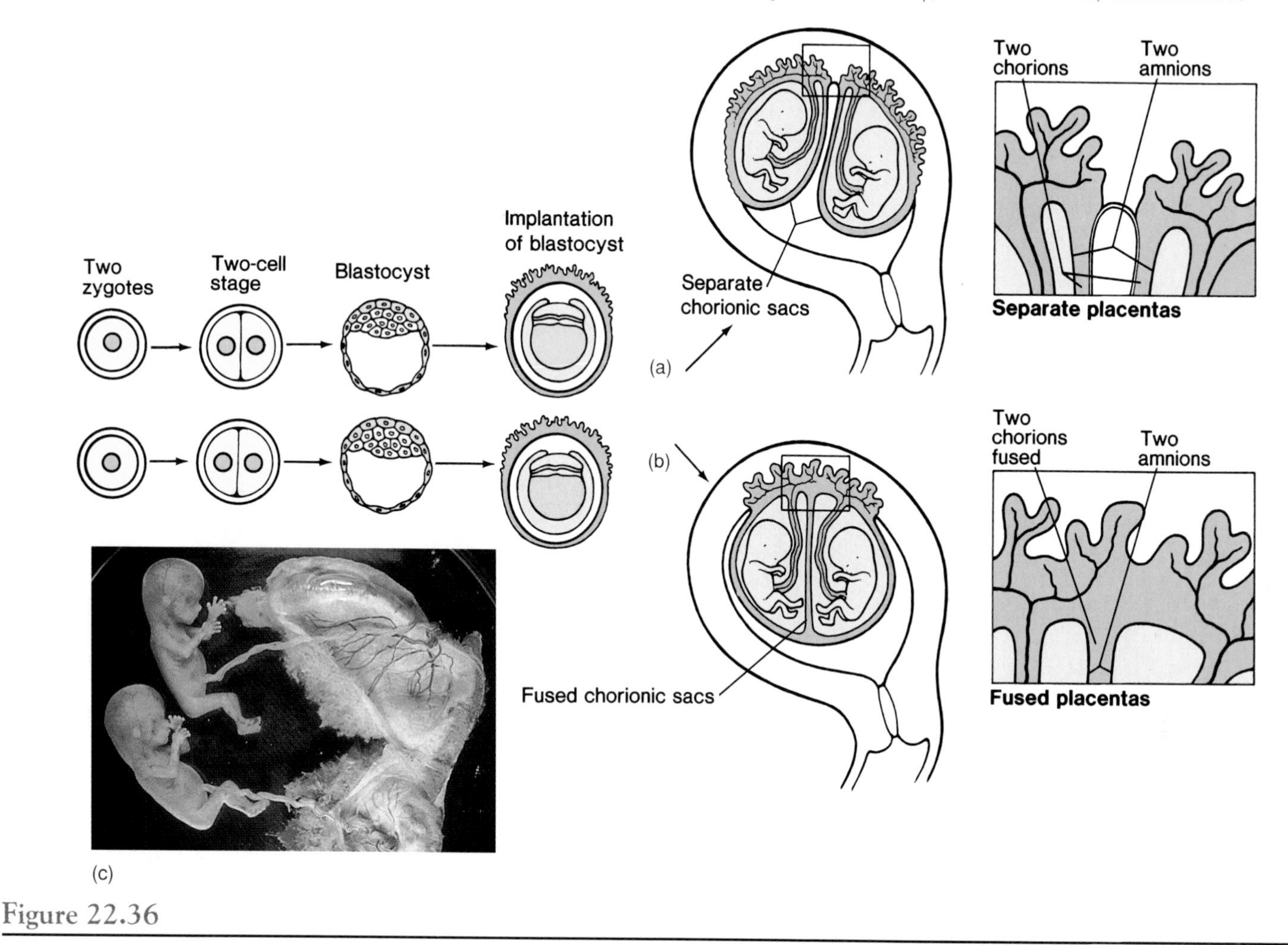

Figure 22.36

The formation of dizygotic twins. Twins of this type are fraternal rather than identical and may have (*a*) separate or (*b*) fused placentas. (*c*) A photograph of fraternal twins at 11 weeks.

spermatozoa in the same ovulatory cycle (fig. 22.36). **Monozygotic** (identical) **twins** form from a single zygote (fig. 22.37). Approximately one-third of twins are monozygotic.

Dizygotic twins may be of the same sex or different sexes and are not any more alike than brothers or sisters born at different times. Dizygotic twins always have two chorions and two amnions, but the chorions and the placentas may be fused.

Monozygotic twins are of the same sex and are genetically identical. Any physical differences in monozygotic twins are caused by environmental factors during morphogenic development (e.g., there might be a differential vascular supply that causes slight differences to be expressed). Monozygotic twinning is usually initiated toward the end of the first week when the embryoblast divides to form two embryonic primordia. Monozygotic twins have two amnions but only one chorion and a common placenta. If the embryoblast fails to completely divide, **conjoined twins** (Siamese twins) may form.

Triplets occur about once in 7,600 pregnancies and may be (1) all from the same ovum and identical, (2) two identical and the third from another ovum, or (3) three zygotes from three different ova. Similar combinations occur in quadruplets, quintuplets, and so on.

Fetal Monitoring

Obstetrics has benefitted greatly from advancements made in fetal monitoring in the last two decades. Before modern techniques became available, physicians could determine the welfare of the unborn child only by auscultation of the fetal heart and palpation of the fetus. Currently, several tests may be used to gain information about the fetus during any stage of development. Fetal conditions that can now be diagnosed and evaluated include genetic disorders, hypoxia, blood disorders, growth retardation, placental functioning, prematurity, postmaturity, and intrauterine infections. These tests also help to determine the advisability of an abortion.

Radiographs of the fetus were once commonly performed but were found harmful and have been replaced by other methods of evaluation that are safer and more informative. **Ultrasonography** employs a mechanical vibration of high frequency to produce a safe, high-resolution (sharp) image of fetal structure (fig. 22.38). Ultrasonic imaging is a reliable way to determine pregnancy as early as 6 weeks after ovulation. It can also be used to determine fetal weight, length, and position, as well as to diagnose multiple fetuses.

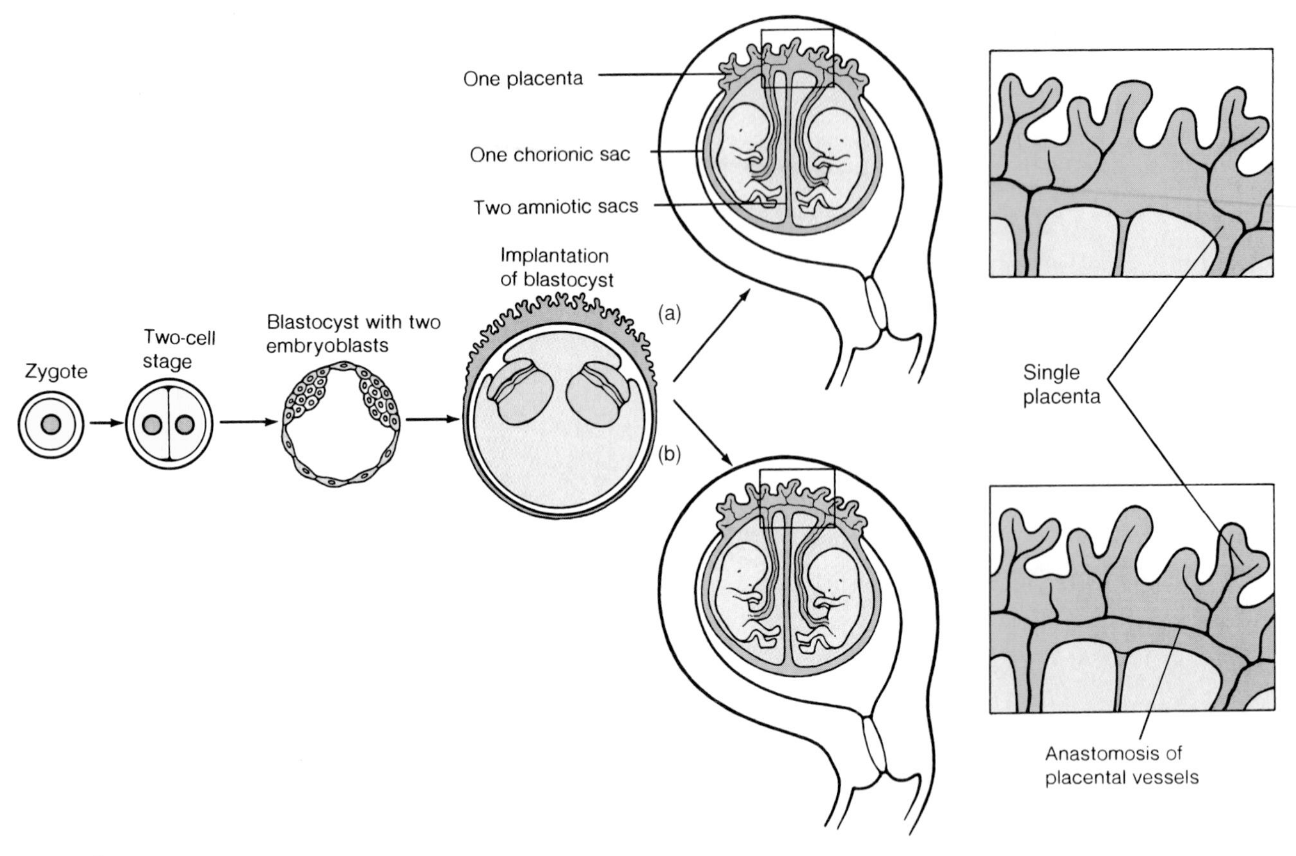

Figure 22.37

The formation of monozygotic twins. Twins of this type develop from a single zygote and are identical. Such twins have two amnions but only one chorion, and they share a common placenta.

Amniocentesis is a technique used to obtain a small sample of amniotic fluid so that it can be assessed genetically and biochemically (see fig. 22.13). A wide-bore needle is inserted into the amniotic sac with guidance by ultrasound, and 5–10 ml of amniotic fluid is withdrawn with a syringe. Amniocentesis is most often performed to determine fetal maturity, but it can also help to predict serious disorders such as *Down syndrome* or *Gaucher's disease* (a metabolic disorder).

Fetoscopy (fig. 22.39) goes beyond amniocentesis by allowing direct examination of the fetus. Using fetoscopy, physicians scan the uterus with pulsed sound waves to locate fetal structures, the umbilical cord, and the placenta. Skin samples are taken from the head of the fetus and blood samples extracted from the placenta. The principal advantage of fetoscopy is that external features of the fetus (such as fingers, eyes, ears, mouth, and genitals) can be carefully observed. Fetoscopy is also used to determine several diseases, including hemophilia, thalassemia, and sickle-cell anemia cases, 40% of which are missed by amniocentesis.

Most hospitals are now equipped with instruments that monitor fetal heart rate and uterine contractions during labor. These instruments can detect any complication that may arise during the delivery. The procedure is called *Electronic Monitoring of Fetal Heart Rate and Uterine Contractions (FHR-UC Monitoring)*. The extent of stress to the fetus from uterine contractions can be determined through FHR-UC monitoring (fig. 22.40). Long, arduous deliveries are taxing to both the mother and fetus. If the baby's health and vitality are presumed to be in danger because of a difficult delivery, the physician may decide to perform a cesarean section.

amniocentesis: Gk. *amnion*, lamb (fetal membrane); *kentesis*, puncture
fetoscopy: L. *fetus*, offspring; *skopein*, to view

Amniotic fluid
Placenta
Left cerebral hemisphere
Orbit of eye
Left hand
Uterine wall
Thorax

Figure 22.38

A color-enhanced ultrasonogram of a fetus during the third trimester. The left hand is raised, as if waving to the viewer.

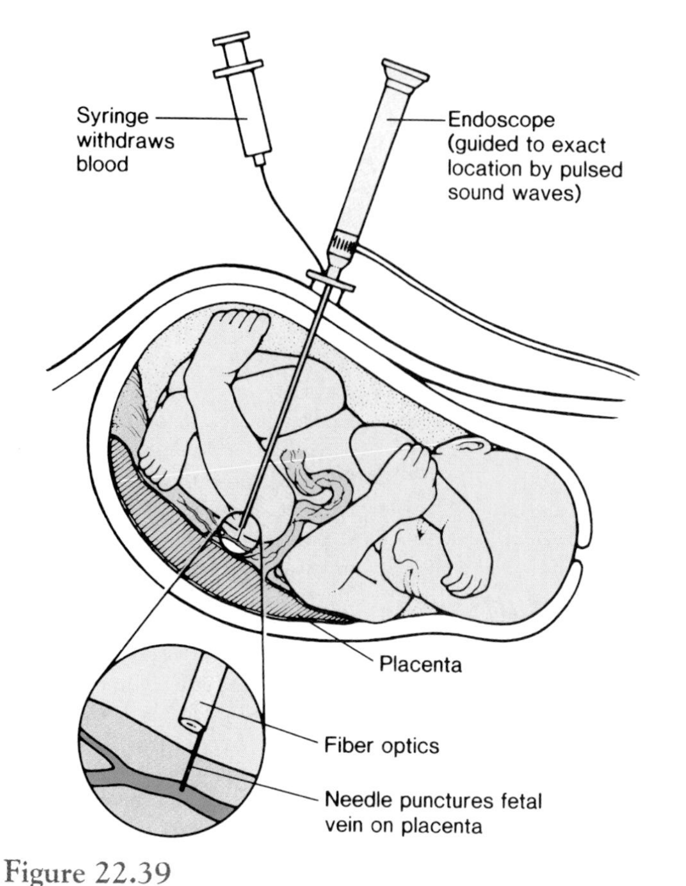

Figure 22.39

Fetoscopy.

Congenital Defects

Major developmental problems called **congenital malformations** occur in approximately 2% of all newborn infants. They may be caused by genetic inheritance, mutation (genetic change), or environmental factors. About 15% of neonatal deaths are attributed to congenital malformations. The branch of developmental biology concerned with abnormal development and congenital malformations is called *teratology* (*tar″ă-tol′ŏ-je*). Many congenital problems have been discussed in previous chapters, in connection with the body system in which they occur.

congenital: L. *congenitus*, born with
teratology: Gk. *teras*, monster; *logos*, study of

Clinical Case Study Answer

Two amnions and one chorion in all but very unusual cases prove the twins to be monozygotic. The two infants, therefore, are genetically identical. Thus, when considering genetic disorders such as cleft palate, one would expect a high degree of concordance (both twins of a monozygotic pair exhibit a particular anomaly). Many such defects however can be present in only one twin—a consequence of nongenetic factors, such as intrauterine environment. An example would be inadequate blood supply to only one twin, resulting in a defect in that twin but not in the other.

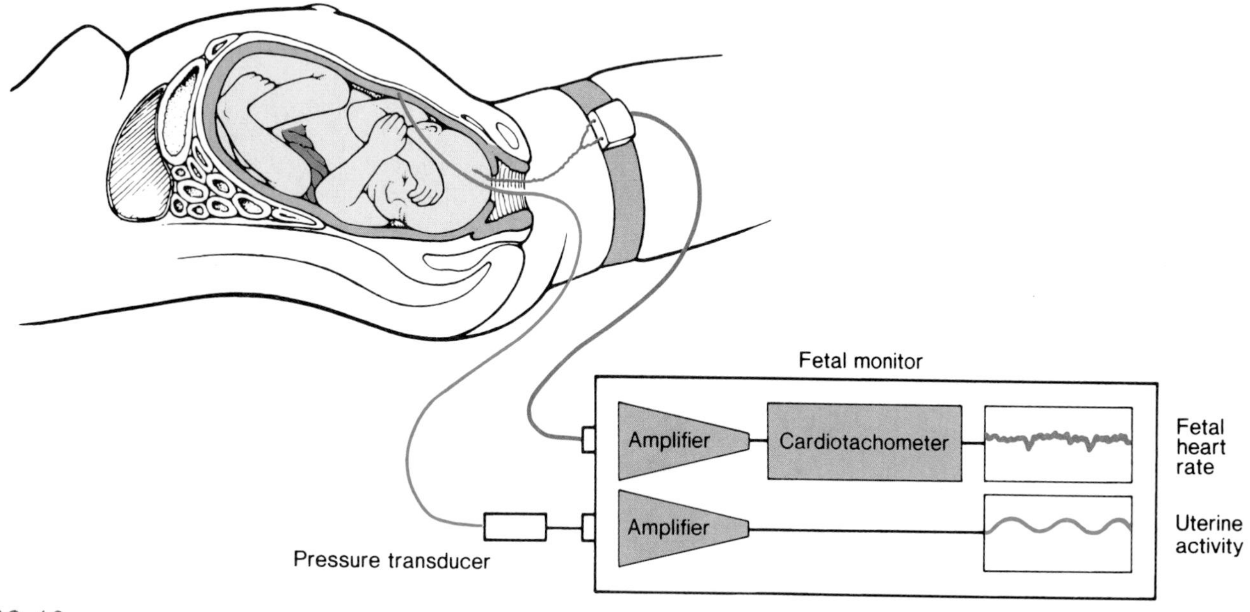

Figure 22.40

Monitoring the fetal heart rate and uterine contractions using an FHR-UC device.

Genetic Disorders of Clinical Importance

cystic fibrosis An autosomal recessive disorder characterized by the formation of thick mucus in the lungs and pancreas that interferes with normal breathing and digestion.

familial cretinism An autosomal recessive disorder characterized by a lack of thyroid secretion because of a defect in the iodine transport mechanism. Untreated children are dwarfed, sterile, and may be mentally retarded.

galactosemia (*gă-lak"tĕ-se'me-ă*) An autosomal recessive disorder characterized by an inability to metabolize galactose, a component of milk sugar. Patients with this disorder have cataracts, damaged livers, and mental retardation.

gout An autosomal dominant disorder characterized by an accumulation of uric acid in the blood and tissue resulting from an abnormal metabolism of purines.

hepatic porphyria (*por-fēr'e-ă*) An autosomal dominant disorder characterized by painful GI disorders and neurologic disturbances resulting from an abnormal metabolism of porphyrins.

hereditary hemochromatosis (*he"mŏ-kro"mă-to'sis*) A sex-influenced autosomal dominant disorder characterized by an accumulation of iron in the pancreas, liver, and heart, resulting in diabetes, cirrhosis, and heart failure.

hereditary leukomelanopathy (*loo"ko-mel"ă-nop'ă-the*) An autosomal recessive disorder characterized by decreased pigmentation in the skin, hair, and eyes and abnormal white blood cells. People with this condition are generally susceptible to infections and early deaths.

Huntington's chorea An autosomal dominant disorder characterized by uncontrolled twitching of skeletal muscles and the deterioration of mental capacities. A latent expression of this disorder allows the mutant gene to be passed to children before symptoms develop.

Marfan's syndrome An autosomal dominant disorder characterized by tremendous growth of the extremities, extreme looseness of the joints, dislocation of the lenses of the eye, and congenital cardiovascular defects.

phenylketonuria (*fen"il-kēt"n-oor'e-ă*) **(PKU)** An autosomal recessive disorder characterized by an inability to metabolize the amino acid phenylalanine. It is accompanied by brain and nerve damage and mental retardation. (Newborns are routinely tested for PKU; those that are affected are placed on a diet low in phenylalanine.)

pseudohypertrophic muscular dystrophy A sex-linked recessive disorder characterized by progressive muscle atrophy. It usually begins during childhood and causes death in adolescence.

retinitis pigmentosa A sex-linked recessive disorder characterized by progressive atrophy of the retina and eventual blindness.

Tay-Sachs disease An autosomal recessive disorder characterized by a deterioration of physical and mental abilities, early blindness, and early death. It has a disproportionately high incidence in Jews of Eastern European origin.

Huntington's chorea: from George Huntington, American physician, 1850–1916

Marfan's syndrome: from Antoine Bernard-Jean Marfan, French physician, 1858–1942

Tay-Sachs disease: from Warren Tay, English physician, 1843–1927, and Bernard Sachs, American neurologist, 1858–1944

Chapter Summary

Fertilization (pp. 735–737)

1. Upon fertilization of a secondary oocyte by a spermatozoon in the uterine tube, meiotic development is completed and a diploid zygote is formed.
2. Morphogenesis is the sequential formation of body structures during the prenatal period of human life. The prenatal period lasts 38 weeks and is divided into a preembryonic, an embryonic, and a fetal period.
3. A capacitated spermatozoon digests its way through the zona pellucida and corona radiata of the secondary oocyte to complete the fertilization process and formation of a zygote.

Preembryonic Period (pp. 737–742)

1. Cleavage of the zygote is initiated within 30 hours and continues until a morula forms; the morula enters the uterine cavity on about the third day.
2. A hollow, fluid-filled space forms within the morula, at which point it is called a blastocyst.
3. Implantation begins between the fifth and seventh day and is enabled by the secretion of enzymes that digest a portion of the endometrium.
   (a) During implantation, the trophoblast cells secrete human chorionic gonadotropin (hCG), which prevents the breakdown of the endometrium and menstruation.
   (b) The secretion of hCG declines by the tenth week as the developed placenta secretes steroids that maintain the endometrium.
4. The embryoblast of the implanted blastocyst flattens into the embryonic disc, from which the primary germ layers of the embryo develop.
   (a) Ectoderm gives rise to the nervous system, the epidermis of the skin and epidermal derivatives, and to portions of the sensory organs.
   (b) Mesoderm gives rise to bones, muscles, blood, reproductive organs, the dermis of the skin, and connective tissue.
   (c) Endoderm gives rise to linings of the GI tract, digestive organs, the respiratory tract and lungs, and the urinary bladder and urethra.

Embryonic Period (pp. 742–752)

1. The events of the 6-week embryonic period include the differentiation of the germ layers into specific body organs and the formation of the placenta, the umbilical cord, and the extraembryonic membranes. These events make it possible for morphogenesis to continue.
2. The extraembryonic membranes include the amnion, yolk sac, allantois, and chorion.
   (a) The amnion is a thin membrane surrounding the embryo. It contains amniotic fluid that cushions and protects the embryo.
   (b) The yolk sac produces blood for the embryo.
   (c) The allantois also produces blood for the embryo and gives rise to the umbilical arteries and vein.
   (d) The chorion participates in the formation of the placenta.
3. The placenta, formed from both maternal and embryonic tissue, has a transport role in providing for the metabolic needs of the fetus and in removing its wastes.
   (a) The placenta produces steroid and polypeptide hormones.
   (b) Nicotine, drugs, alcohol, and viruses can cross the placenta to the fetus.
4. The umbilical cord, containing two umbilical arteries and one umbilical vein, is formed as the amnion envelops the tissues on the underside of the embryo.
5. From the third to the eighth week, the structure of all the body organs, except the genitalia, becomes apparent.
   (a) During the third week, the primitive node forms from the primitive line, which later gives rise to the notochord and intraembryonic mesoderm.

(b) By the end of the fourth week, the heart is beating; the primordial tissues of the eyes, brain, spinal cord, lungs, and digestive organs are properly positioned; and the superior and inferior limb buds are recognizable.
(c) At the end of the fifth week, the sense organs are formed in the enlarged head and the appendages have developed with digital primordia evident.
(d) During the seventh and eighth weeks, the body organs are formed, except for the genitalia, and the embryo appears distinctly human.

Fetal Period (pp. 752–755)

1. A small amount of tissue differentiation and organ development occurs during the fetal period, but for the most part fetal development is primarily limited to body growth.
2. Between weeks 9 and 12, ossification centers appear, the genitalia are formed, and the digestive, urinary, respiratory, and muscle systems show functional activity.
3. Between weeks 13 and 16, facial features are formed and the fetal heartbeat can be detected with a stethoscope.
4. During the 17-to-20-week period, quickening can be felt by the mother, and vernix caseosa and lanugo cover the skin of the fetus.
5. During the 21-to-25-week period, substantial weight gain occurs and the fetal skin becomes wrinkled and pinkish.
6. Toward the end of the 26-to-29-week period, the eyes have opened, the gonads have descended in a male, and the fetus is developed to the extent that it might survive if born prematurely.
7. At 38 weeks, the fetus is full-term; the normal gestation is 266 days.

Labor and Parturition (pp. 755–756)

1. Labor and parturition are the culmination of gestation and require the action of oxytocin, secreted by the posterior pituitary, and prostaglandins, produced in the uterus.
2. Labor is divided into dilation, expulsion, and placental stages.

Periods of Postnatal Growth (pp. 756–762)

1. The course of human life after birth is seen in terms of physical and physiological changes and the attainment of maturity in the neonatal period, infancy, childhood, adolescence, and adulthood.
2. The neonatal period, extending from birth to the end of the fourth week, is characterized by major physiological changes.
   (a) The most critical need of the newborn is to establish adequate respiratory and heart rates. A normal respiratory rate is 30 to 40 respirations per minute, and a normal heart rate ranges from 120 to 160 beats per minute.
   (b) The four reflexes in the newborn critical to survival are the suckling reflex, the rooting reflex, the crying reflex, and the breathing reflex.
3. Infancy, extending from 4 weeks through the second year, is characterized by tremendous growth, increased coordination, and mental development.
   (a) By 2 years, most infants weigh about four times their birth weight and average between 32 and 36 inches in length.
   (b) Growth is a differential process resulting in gradual changes from infant to adult body proportions.
4. Childhood, extending from the end of infancy to adolescence, is characterized by steady growth until preadolescence, at which time there is a marked growth spurt.
   (a) During childhood, the average child becomes thinner and stronger each year as he or she grows taller.
   (b) The fact that disease and death are relatively rare during childhood may be due to the fact that lymphoid tissue is at its peak of development at this time; it is also present in greater amounts in children than in adults.
5. Adolescence is the period of growth and development between childhood and adulthood.
   (a) Puberty is the stage of early adolescence when the secondary sex characteristics are expressed and the sex organs become functional.
   (b) The end result of puberty is the structural expression of gender, or sexual dimorphism.
   (c) Menarche generally occurs in adolescent girls at the age of 13, but it may range from 9 to 17 years. At this time, vaginal secretions change from alkaline to acid.
   (d) The first physical indications of puberty are the appearance of breast buds in females and the growth of the testes and the appearance of sparse pubic hair in males.
   (e) Although semen may be ejaculated at age 13, sufficient mature spermatozoa for fertility are not produced until 14 to 16 years of age.
6. Adulthood, the final period of human physical change, is characterized by gradual senescence as a person ages.
   (a) Although skeletal maturity is reached in early adulthood, anatomical and physiological changes continue throughout adulthood and are part of the aging process.
   (b) Sexual dimorphism in human adults is evident anatomically, physiologically, metabolically, and behaviorally.
   (c) Male and female differences in the stature, proportions, and composition of the body may become more apparent with age.

Inheritance (pp. 762–765)

1. Inheritance is the passage of hereditary traits carried on the genes of chromosomes from one generation to another.
2. Each zygote contains 22 pairs of autosomal chromosomes and 1 pair of sex chromosomes—XX in a female and XY in a male.
3. A gene is the portion of a DNA molecule that contains information for the production of one kind of protein molecule. Alleles are different forms of genes that occupy corresponding positions on homologous chromosomes.
4. The combination of genes in an individual's cells constitutes his or her genotype; the observable expression of the genotype is the person's phenotype.
   (a) Dominant alleles are symbolized by uppercase letters and recessive alleles are symbolized by lowercase letters.
   (b) The three possible genotypes are homozygous dominant, heterozygous, and homozygous recessive.
5. A Punnett square is a convenient means for expressing probability.
   (a) The probability of a particular genotype is 1 in 4 (.25) for homozygous dominant and homozygous recessive, and 1 in 2 (.50) for heterozygous.
   (b) A single trait is studied in a monohybrid cross; two traits are studied in a dihybrid cross.
6. Sex-linked traits such as color blindness and hemophilia are carried on the sex-determining chromosome.

Review Activities

Objective Questions

1. The preembryonic period is completed when
   (a) the blastocyst implants.
   (b) the placenta forms.
   (c) the blastocyst reaches the uterus.
   (d) the primary germ layers form.
2. The yolk sac produces blood for the embryo until
   (a) the heart is functional.
   (b) the kidneys are functional.
   (c) the liver is functional.
   (d) the baby is delivered.
3. Which of the following is a function of the placenta?
   (a) production of steroids and hormones
   (b) diffusion of nutrients and oxygen
   (c) production of enzymes
   (d) all of the above apply
4. The decidua basalis is
   (a) a component of the umbilical cord.
   (b) the embryonic portion of the villous chorion.
   (c) the maternal portion of the placenta.
   (d) a vascular membrane derived from the trophoblast.
5. Which of the following could diffuse across the placenta?
   (a) nicotine (c) heroin
   (b) alcohol (d) all of the above
6. During which week following conception does the embryonic heart begin pumping blood?
   (a) the fourth week
   (b) the fifth week
   (c) the sixth week
   (d) the eighth week
7. Which of the following is the period of growth from birth to the end of the fourth week?
   (a) the neonatal period
   (b) the fetal period
   (c) infancy
   (d) the suckling period
8. The normal newborn heart rate is
   (a) 70–80 beats/min.
   (b) 120–160 beats/min.
   (c) 100–120 beats/min.
   (d) 180–200 beats/min.
9. The continuum of physical change in adolescence that regulates the growth of body hair is known as
   (a) puberty.
   (b) pubal progression.
   (c) pubescence.
   (d) dimorphism.
10. Which condition is *not* characteristic of the female as compared to the male?
    (a) lower blood pressure
    (b) higher basal metabolic rate
    (c) lower red blood cell count
    (d) faster heart rate
11. The first physical indication of puberty in females is generally
    (a) alkaline vaginal secretions.
    (b) a widening pelvis.
    (c) breast buds.
    (d) axillary hair.
12. Twins that develop from two zygotes resulting from the fertilization of two ova by two spermatozoa in the same ovulatory cycle are referred to as
    (a) monozygotic. (c) dizygotic.
    (b) conjoined. (d) identical.
13. Match the genotype descriptions in the left-hand column with the correct symbols in the right-hand column.
    homozygous recessive *Bb*
    heterozygous *bb*
    homozygous dominant *BB*
14. An allele that is *not* expressed in a heterozygous genotype is called
    (a) recessive. (c) genotypic.
    (b) dominant. (d) phenotypic.
15. If the genotypes of both parents are *Aa* and *Aa*, the offspring probably will be
    (a) ½ *AA* and ½ *aa*. (c) ¼ *AA*, ½ *Aa*, ¼ *aa*.
    (b) all *Aa*. (d) ¼ *AA* and ¼ *aa*.

Essay Questions

1. Describe the implantation of the blastocyst into the uterine wall and the involvement of the trophoblast in the formation of the placenta.
2. Explain how the primary germ layers form. What major structures does each germ layer give rise to?
3. Explain why development is so critical during the embryonic period and list the embryonic needs that must be met for morphogenesis to continue.
4. State the approximate time period (in weeks) for the following occurrences:
   (a) appearance of the arm and leg buds.
   (b) differentiation of the external genitalia.
   (c) perception of quickening by the mother.
   (d) functioning of the embryonic heart.
   (e) initiation of bone ossification.
   (f) appearance of lanugo and vernix caseosa.
   (g) survival of fetus if born prematurely.
   (h) formation of all major body organs completed.
5. Define the term *infancy* and discuss the growth and developmental events characteristic of this period of life.
6. Define the term *childhood* and discuss the growth and developmental events characteristic of this period of life.
7. Define the term *adolescence* and discuss the role of puberty in this period of life for both males and females.
8. Distinguish between puberty and pubescence.
9. Define the term *adulthood* and describe the characteristics of this period of life.
10. List some of the structural features of adults that exemplify sexual dimorphism.
11. Define *anthropometry* and give examples of anthropometric characteristics.
12. State the features of a genetic disorder that would lead one to believe that it was a form of sex-linked inheritance.

Critical-Thinking Questions

1. Write a short paragraph about pregnancy that includes the terms *ovum*, *blastocyst*, *implantation*, *embryo*, *fetus*, *gestation*, and *parturition*.
2. If a pregnant woman had a pack-a-day cigarette habit, what predictions could you make about the health of her baby? Justify your responses.
3. The sedative thalidomide was used by thousands of pregnant women in the 1960s to alleviate their morning sickness. This drug inhibited normal limb development and resulted in tragically deformed infants with flipperlike arms and legs. At what period of prenatal development did such abnormalities originate? What lessons were learned from the thalidomide tragedy?
4. Your friend just learned that she's pregnant and she wants to name the baby either Louis or Louise. How soon can she know for sure whether she is carrying a boy or a girl? Describe the technique by which the sex can be determined in vitro.
5. Cesarean sections must be performed when a baby cannot be delivered breech. Can you think of some other reasons for a C-section?
6. Hemophilia used to be called a royal disease because it plagued many male members of the royal families of Europe and Russia. Why are female hemophiliacs rare, as their absence from royal pedigrees shows?
7. In many states, laws prohibit consanguineous marriages—those between blood relatives such as siblings or first cousins. Why is it likely that a geneticist would endorse such restrictions?

Appendix A

Answers to Objective Questions with Explanations

Chapter 1

1. (a) The word *anatomy* is derived from the Greek prefix *ana*, meaning "up" and the suffix *tome*, "a cutting." In ancient times, the Greek word *anatomize* was more commonly used than the word *dissect*.
2. (a) In William Harvey's time, there were many misconceptions regarding the flow of blood. Using the scientific method, Harvey demonstrated the flow of blood and described the function of valves within veins.
3. (c) The concept of body humors was widely accepted by physicians in Greek and Roman times as an explanation for a person's disposition and general health. It was not until the Renaissance that this concept was gradually discarded in the light of scientific research.
4. (d) The Greeks held the body in great esteem and were eager to learn about its structure and function.
5. (a) Hippocrates is regarded as the father of medicine because of the sound principles of medical practice that he established. While the Hippocratic oath cannot be directly credited to him, it undoubtedly represents his ideals and principles.
6. (c) Phlegm was thought to be the body humor associated with the lungs. It was believed that too much phlegm, or not enough, could result in respiratory disorders.
7. (c) Andreas Vesalius completed *De Humani Corporis Fabrica* in A.D. 1543 when he was 28 years old. This beautifully illustrated book corrected many anatomical errors that had been handed down as fact and provided a visual guide for studying anatomy.
8. (d) The development of movable type allowed for widespread access to printed material, and was therefore important in ushering in the Renaissance.
9. (b) Aristotle thought that the heart was the seat of intelligence and that the warm blood from the heart was cooled by the fluids surrounding the brain.
10. (a) Wilhelm Konrad Roentgen discovered X rays in 1895. His achievement was recognized in 1901, when he was awarded the Nobel Prize.

Chapter 2

1. (b) Although chordates have a distinct head, thorax, and abdomen, so do many invertebrate organisms.
2. (a) Because we have prehensile hands, digits modified for grasping, and large, well-developed brains, we are classified as primates along with monkeys and great apes.
3. (b) The tissue level of body organization is intermediate between the cell and organ levels.
4. (a) The urinary system maintains blood homeostasis. Basically, the kidneys are filtering organs of the blood.
5. (b) The cubital fossa is located on the anterior surface of the upper extremity, at the junction of the brachium and antebrachium. It is the anterior surface of the elbow.
6. (c) An organ is a body structure composed of two or more tissue types.
7. (c) By definition, both the coronal and sagittal planes extend vertically through the body, but there is no vertical plane designation.
8. (b) The external genitalia are located within the perineum, which is the area between the coccyx and the symphysis pubis.
9. (b) Within the mediastinum are the principal (primary) bronchi, esophagus, and heart, along with major vessels that transport blood toward and away from these organs.
10. (b) The epigastric region of the abdomen is centrally located, medial to the right and left hypochondriac regions.
11. (c) The mesenteries are modified serous membranes that support portions of the small and large intestines within the abdominal cavity.
12. (c) A coronal section divides the body into front (anterior) and back (posterior) portions.
13. (c) The anatomical position provides a consistent frame of reference when describing structures of the body.
14. (d) Some listening can be accomplished without the aid of an instrument, but the technique of choice is to use a stethoscope.

Chapter 3

1. (c) Electrolytes are formed from the breakdown of inorganic compounds in water. All three kinds of electrolytes—acids, bases, and salts—are important for normal cellular function.
2. (d) The elements carbon, oxygen, nitrogen, and hydrogen compose over 95% of the body and are often linked to form molecules and compounds.
3. (a) The strong hydrolytic enzymes in lysosomes digest foreign molecules and worn and damaged cells.
4. (e) Ciliated cells promote movement of substances through the tubes or ductules of sections of the respiratory and reproductive systems. For example, mucus containing trapped particles is moved in the bronchioles and trachea, sperm are moved in the ductus deferentia and ova are moved in the uterine tubes.
5. (b) Osmosis is movement of water and solvent molecules through a semipermeable membrane as a result of concentration differences.
6. (c) Metaphase is the stage of mitosis immediately preceding the separation of the chromatids.
7. (d) Anaphase is the stage of mitosis immediately following metaphase; it precedes the formation of two identical cells.
8. (a) The Golgi complex synthesizes carbohydrate molecules in the production of glycoproteins.
9. (b) Compensatory hypertrophy is an increase in cell mass in response to greater demands, such as the enlargement of a muscle fiber as a result of exercise.
10. (b) Hyperplasia is a protective mechanism that ensures the availability of an adequate number of cells to perform a particular task.

Chapter 4

1. (b) The four principal types of tissues are epithelial tissues, connective tissues, muscular tissues, and nervous tissues. The integument, or skin, is an organ.
2. (c) Blood is a fluid tissue that flows through blood vessels.
3. (a) Many body structures derive from mesoderm, including cartilage, bone, and other connective tissue; smooth, cardiac, and skeletal muscle; and the dermis of the skin.
4. (b) Elastic and collagenous fibers are characteristic of certain kinds of connective tissues.
5. (d) The entire abdominal portion of the gastrointestinal (GI) tract contains simple columnar epithelium lining the lumen.
6. (a) Based on structural classification, the mammary glands are classified as compound acinar. Based on secretory classification, the mammary glands are classified as apocrine.
7. (c) Dense regular connective tissue is the principal component of tendons, which accounts for their great strength.
8. (a) Reticular tissue, which contains white blood cells, is phagocytic in cleansing body fluids.
9. (b) Cartilage tissue is slow to heal because it is avascular (without blood vessels). Cartilage generally derives its nutrients from surrounding fluids rather than from permeating blood vessels and capillaries.
10. (d) Like skeletal muscle, cardiac muscle is striated. As compared to skeletal muscle or smooth muscle tissues, however, cardiac muscle tissue is unique in containing intercalated discs and experiencing rhythmical involuntary contractions.

Chapter 5

1. (a) Ectoderm is the outermost of the three embryonic germ layers, and it is from this germ layer that hair, nails, integumentary glands, and the epidermis of the skin is formed.

2. (b) The appearance and general health of nails is an indicator of general body health and of certain dietary deficiencies, such as iron deficiency.
3. (b) The stratum lucidum is an epidermal layer that occurs only in the skin of the palms of the hands and the soles of the feet.
4. (a) The dermal papillae may contribute to surface features of the skin, such as print patterns, but does not contribute to skin coloration.
5. (c) Mitosis occurs in the stratum basale, and to a limited extent, in the stratum spinosum.
6. (a) The sebum from sebaceous glands is emptied into a hair follicle, where it is dissipated along the shaft of the hair to the surface of the skin.
7. (c) Lanugo, or fetal hair, is thought to be important in the development of the hair follicles.
8. (b) Melanoma is an aggressive malignant skin cancer that is life threatening if untreated.
9. (b) A second-degree burn is serious, but it generally does not require skin grafts.
10. (d) A comedo is a blackhead or whitehead resulting from a small, localized skin infection.

Chapter 6

1. (c) A bone, such as the femur, is considered to be an organ because it is composed of more than two tissues integrated to perform a particular function.
2. (a) Skin, rather than bone, is involved in the synthesis of vitamin D.
3. Matching: 1(c), 2(d), 3(e), 4(a), 5(b)
4. (c) The mandible articulates with the mandibular fossae of the temporal bones.
5. (a) The sella turcica of the sphenoid bone is located immediately superior to the sphenoidal sinus, where it supports the pituitary gland.
6. (d) Osteoclasts are important in releasing stored minerals within bone tissue and in the continuous remodeling of bone.
7. (a) Located in the squamous part of the temporal bone, the mandibular fossa is the depression for articulation of the condyloid process of the mandible.
8. (b) The crista galli is the portion of the ethmoid bone that contains numerous olfactory foramina for the passage of olfactory nerves from the nasal epithelium.
9. (d) Cervical vertebrae contain transverse foramina that permit the passage of the vertebral vessels to and from the brain.
10. (c) A person may gain some protection against osteoporosis by maintaining a healthy diet and a regular exercise program.

Chapter 7

1. (a) When in anatomical position, a person is erect and facing anteriorly. The subscapular fossa is found on the anterior side of the scapula. The supraspinatus and infraspinatus fossae are located on the posterior side.
2. (c) The sternal extremity of the clavicle articulates with the manubrium of the sternum. The acromial extremity of the clavicle articulates with the acromion of the scapula.
3. (d) The clavicle has a conoid tubercle near its acromial extremity.
4. (b) The bony prominence of the elbow is the olecranon of the ulna.
5. (b) Sesamoid bones are formed in tendons, and none of the carpal bones are sesamoid bones. Sesamoid bones are fairly common, however, at specific joints of the digits.
6. (d) Pelvimetry measures the dimension of the lesser pelvis in a pregnant woman to determine whether a cesarean delivery might be necessary.
7. (e) The linea aspera is a vertical ridge on the posterior surface of the body of the femur, where the posterior gluteal muscles of the hip attach.
8. (b) The intertrochanteric line is located on the anterior side of the femur between the greater trochanter and the lesser trochanter.
9. (d) Sex-related structural differences in the pelvis reflect modifications for childbirth. For example, the female pelvis as compared to that of the male has a wider pelvic outlet, a shallower and shorter symphysis pubis, and a wider pubic arch.
10. (a) Talipes is a congenital malformation in which the sole of the foot is twisted medially.

Chapter 8

1. (b) The structural classification of joints includes fibrous, cartilaginous, and synovial types.
2. (c) Synchondroses are cartilaginous joints that have hyaline cartilage between the bone segments.
3. (d) Syndesmoses are fibrous joints found in the antebrachium (forearm) and leg, where adjacent bones are held together by interosseous ligaments.
4. (d) Syndesmoses do not occur within the skull; rather, they are located only in the upper and lower extremities.
5. (c) The only saddle joint in the body is at the base of the thumb, where the trapezium articulates with the first metacarpal bone.
6. (d) Only the knee joints contain menisci.
7. (c) A pivotal joint is a synovial joint that permits rotational movement.
8. (b) The coxal (hip) joint has a wide range of movement, including hyperextension as the lower extremity is moved posteriorly beyond the vertical position of the body in anatomical position.
9. (b) The shoulder joint, with its relatively shallow socket and weak ligamentous support, is vulnerable to dislocation. In addition, we often place our arms in vulnerable positions as we engage in various activities.
10. (d) Rheumatoid arthritis is a chronic disease that frequently leaves the patient crippled. It occurs most commonly between the ages of 30 and 35.

Chapter 9

1. (b) The neuromuscular cleft is a slight gap within the neuromuscular junction where a motor nerve fiber and a skeletal muscle fiber meet.
2. (c) When there are many motor units present within a muscle, a person can be more selective at which ones are recruited, and thus have greater dexterity than when few motor units are present.
3. (e) Muscles are named on the basis of structural features, location, attachment, relative position, or function (action).
4. (d) When a nerve impulse reaches an axon terminal, a neurotransmitter chemical is released into the neuromuscular cleft at the neuromuscular junction.
5. (c) With respect to the muscles of the thigh, a single motor neuron serves a large number of muscle fibers and, therefore, has a low innervation ratio.
6. (a) The corrugator muscle is located beneath the medial portion of the eyebrow. When the muscle contracts, the eyebrow is drawn toward the midline.
7. (a) A description of a muscle's contraction is always made in reference to a person in anatomical position. In anatomical position, the shoulder joint is positioned vertically at 180°. When the pectoralis muscle is contracted while in this position, the angle of the shoulder joint is decreased; therefore, the joint is flexed.
8. (b) Although it is positioned on the anterior surface of the humerus, the biceps brachii muscle arises from the coracoid process of the scapula and from the tuberosity above the glenoid fossa of the scapula. Both heads insert on the radial tuberosity.
9. (c) Spanning both the hip and knee joints, the rectus femoris muscle can flex the hip joint and extend the knee joint.
10. (b) The tibialis posterior muscle is deep to the soleus muscle. Its tendinous insertion passes across the arches of the foot and inserts on the plantar surfaces of a number of foot bones. In this position, the tibialis posterior muscle supports the arches of the foot and plantar flexes and inverts the foot as it contracts.

Chapter 10

1. (e) The scalp is attached anteriorly to the supraorbital ridges. The eyebrows are attached to the scalp, just above the supraorbital ridges.
2. (b) The ala is a portion of the nose, lateral to the apex.
3. (d) The concha is a bony projection into the nasal cavity.
4. (d) The conjunctiva is a thin mucous membrane that covers the anterior surface of the eyeball and lines the undersurface of the eyelids.
5. (d) The cervix of the neck is the anterior portion. The cervical vertebrae can be palpated in the posterior portion, called the nucha.
6. (a) The carotid triangle is located on the lateral side of the neck and is bordered by the sternocleidomastoid, posterior digastric, and omohyoid muscles. Located within the carotid triangle are the common carotid artery, internal jugular vein, and vagus nerve.
7. (a) The ulnar nerve passes through the ulnar sulcus of the elbow. A fracture of the olecranon of the ulna often damages the ulnar nerve.
8. (b) The scapular muscles are usually obscure on an obese person.

9. (c) The axilla is the depression commonly known as the armpit. The anterior axillary fold is formed by the pectoralis major muscle and the posterior axillary fold consists of the latissimus dorsi muscle.
10. (a) The great saphenous vein and small saphenous vein are superficial veins of the leg that are frequently varicosed in elderly people.

Chapter 11

1. (d) The cerebellum is a structure within the metencephalon of the brain.
2. (a) The cerebral cortex is the outer portion of the cerebrum, and the cerebrum is a structure within the telencephalon of the brain.
3. (e) The medulla oblongata is a structure within the myelencephalon of the brain.
4. (c) As the nervous system matures in an infant, the neurolemmocytes wrap around the axons and some of the dendrites of neurons within the peripheral nervous system. Myelination is the process of forming myelin layers that protect neurons and aid conduction.
5. (d) Ganglia are collections of nerve cell bodies outside the CNS. Collections of nerve cell bodies within the CNS are called nuclei.
6. (a) A pseudounipolar neuron has a single process that divides into two, and its cell body is located in posterior root ganglia of the spinal and cranial nerves.
7. (a) Depolarization of an axon is produced by the movement of Na^+ into the axon and K^+ out of the axon. Once the impulse has completed its course to the axon terminal, repolarization of the axon is produced by the movement of K^+ into the axon and the movement of Na^+ out of the axon.
8. (a) The corpus collosum is composed of commissural fibers that connect the two cerebral hemispheres.
9. (d) The thalamus autonomically responds to pain by activating the sympathetic nervous system. It also relays pain sensations to the parietal lobes of the cerebrum for perception.
10. (b) The basal nuclei consist of cell bodies of motor neurons that regulate contraction of skeletal muscles. Basal metabolic rate is regulated, for the most part, in the hypothalamus and medulla oblongata.
11. (c) Located within the mesencephalon, the corpora quadrigemina is concerned with visual and hearing reflexes, the red nucleus is concerned with motor coordination and posture maintenance, and the substantia nigra is thought to inhibit forced involuntary movements.
12. (d) The fourth ventricle is located inferior to the cerebellum and contains cerebrospinal fluid.

Chapter 12

1. (a) Anatomically speaking, the PNS consists of all of the structures of the nervous system outside of the CNS. That means, then, that all nerves, sensory receptors, neurons, ganglia, and plexuses are part of the PNS.
2. (b) The oculomotor nerve innervates the medial rectus eye muscles that, when simultaneously contracted, cause the eyes to be directed medially.
3. (b) The oculomotor nerve innervates the levator palpebrae superioris muscle with motor fibers. This muscle elevates the upper eyelid when contracted.
4. (c) The vestibulocochlear nerve serves the vestibular organs of the inner ear with sensory fibers. These organs are associated with equilibrium and balance.
5. (a) Passing through the stylomastoid foramen, the facial nerve innervates the muscles of facial expression with motor fibers and the taste buds on the anterior two-thirds of the tongue with sensory fibers.
6. (b) The accessory nerve innervates several muscles that move the head and neck with motor fibers.
7. (c) The four spinal nerve plexuses are the cervical, brachial, lumbar, and sacral.
8. (d) Only the brachial plexus consists of roots, trunks, divisions, and cords. The nerves of the upper extremity arise from the cords.
9. (a) The median nerve arises from the brachial plexus.
10. (c) The knee-jerk reflex is an ipsilateral reflex because the receptor and effector organs are on the same side of the spinal cord.

Chapter 13

1. (d) Postganglionic neurons from the superior mesenteric ganglion innervate the small intestine and colon.
2. (d) Parasympathetic neurons within the oculomotor, facial, and glossopharyngeal nerves synapse in ganglia located in the head.
3. (c) Parasympathetic ganglia, also called terminal ganglia, synapse with the effector cells near or within the organs being served.
4. (c) Secreted from synaptic vesicles, the neurotransmitter chemical acetylcholine facilitates transmission across a synapse.
5. (d) Because the preganglionic neurons of the sympathetic division of the autonomic nervous system exit the vertebral column from the first thoracic to the second lumbar levels, the sympathetic division is also called the thoracolumbar division.
6. (c) Nerve impulses through the postganglionic sympathetic neurons in the heart release norepinephrine, which stimulates the heart to contract.
7. (b) The cooperative effect of sympathetic and parasympathetic stimulation is evident in the erection of the penis (parasympathetic response) and the ejaculation of semen (sympathetic response).
8. (d) Most blood vessels dilate in response to sympathetic stimulation. A few blood vessels (e.g., those serving the external genitalia) constrict in response to parasympathetic stimulation. Visceral blood vessels do not constrict in response to parasympathetic stimulation.
9. (e) Parasympathetic stimulation increases digestive activity, constricts pupils, and decreases the heart rate. Because atropine blocks parasympathetic stimulation, it results in decreased mucus secretion and GI tract movement and causes dilation of the pupils. It also results in an increased heart rate.
10. (c) The medulla oblongata is the structure of the brain that most directly controls the activity of the ANS. The medulla oblongata contains control centers for the circulatory, respiratory, urinary, reproductive, and digestive systems.

Chapter 14

1. (e) The adenohypophysis derives from the hypophyseal pouch in the roof of the oral cavity.
2. (b) The neurohypophysis derives from the neurohypophyseal bud of neuroectoderm within the developing brain.
3. (d) The adrenal medulla derives from neural crest ectoderm within the primitive coelomic cavity.
4. (c) The pancreas derives from an outpouching of the endoderm along the developing foregut.
5. (a) The thyroid gland derives from the thyroid diverticulum of endoderm within the developing pharynx.
6. (c) The sella turcica is a depression within the sphenoid bone that supports the pituitary gland.
7. (d) Secreted from the thyroid gland, thyroxine determines the basic metabolic rate of most organs and promotes the maturation of the brain.
8. (e) Activation of the adrenal medulla causes the secretion of epinephrine and norepinephrine. Both of these hormones prepare the body for greater physical performance—the fight-or-flight response.
9. (e) Insulin is secreted from beta cells within the pancreatic islets in response to a rise in blood glucose. Insulin stimulates the production of glycogen and fat.
10. (d) The adrenal medulla releases epinephrine upon receiving sympathetic nerve impulse stimulation.
11. (a) The thyroid gland releases thyroxine upon receiving TSH stimulation from the adenohypophysis of the pituitary gland.
12. (b) The adrenal cortex releases corticosteroids upon receiving ACTH stimulation from the adenohypophysis of the pituitary gland in response to stress.
13. (e) The adenohypophysis of the pituitary gland releases ACTH in response to receiving CRF stimulation from the hypothalamus.
14. (d) Both the adrenal cortex and the gonads secrete steroid hormones.
15. (a) In the case of an endemic goiter, growth of the thyroid is due to excessive TSH secretion, which results from low levels of thyroxine secretion.

Chapter 15

1. (d) In order for perception to occur, there must be a stimulus at a receptor site that causes a nerve impulse to be conducted to the cerebrum of the brain.
2. (b) Located within the dermis of the skin, in certain visceral organs, and near synovial joints, lamellated corpuscles respond to heavy pressures.

3. (d) Located within connective tissue capsules in synovial joints, joint kinesthetic receptors are stimulated by changes in position caused by movements at the joints.
4. (c) Because sensations of referred pain are consistent from one person to another, an understanding of this phenomenon is of immense clinical importance in diagnosing organ dysfunction.
5. (d) When an object is viewed at a distance of at least 20 feet by someone with normal vision, the suspensory ligaments of the eyes are taut and the lenses are flat because the ciliary muscles are relaxed.
6. (a) Composed of tightly bound elastic and collagenous fibers, the toughened sclera is avascular but does contain sensory receptors for pain.
7. (b) In dim light, sympathetic motor impulses cause the radially arranged smooth muscle fibers within the iris to contract, permitting the pupil to become larger.
8. (a) The semicircular canals contain endolymph and hair cells that respond to movements of the head and convey sensations to the brain that are important for maintaining equilibrium and balance.
9. (d) The vestibular window is at the footplate of the stapes and the cochlear window borders between the scala tympani and the middle-ear cavity. Both windows separate the middle ear from the cochlea within the inner ear.
10. (b) A concave lens corrects myopia (nearsightedness) by causing the light waves to focus deeper within the posterior cavity upon the fovea centralis.

Chapter 16

1. (c) Fibrinogen is a protein in blood plasma that aids in clotting.
2. (a) An excessive leukocyte count, called leukocytosis, is generally diagnostic of infections or diseases within the body.
3. (b) The pulmonary arteries transport deoxygenated blood from the right ventricle of the heart to the lungs.
4. (c) The right atrium receives venous blood from the superior and inferior vena cavae and the coronary sinus. The coronary sinus collects venous blood from coronary circulation prior to delivery into an opening of the right atrium.
5. (f) Closure of the atrioventricular valves causes the "lub" sound of the heart.
6. (c) The QRS complex represents the depolarization of the ventricles. During this interval, the ventricles of the heart are in systole and blood is being ejected from the heart.
7. (b) Arising from ascending aorta, the coronary arteries feed directly into the myocardium of the heart to ensure a rich blood supply to the cardiac muscles.
8. (a) The external carotid arteries supply blood to the entire head, excluding the brain. The paired internal carotid arteries and vertebral arteries (that unite to form the basilar artery) supply blood to the brain.
9. (a) Branching from the external carotid artery at the level of the mandibular condyle, the maxillary artery supplies blood to the teeth and gums of the upper jaw and the superficial temporal artery supplies blood to the parotid gland and to the superficial temporal region.
10. (a) The hepatic portal vein drains nutrient-rich blood into the liver, where the blood is processed by a venous portal system. An arteriole portal system provides the pituitary gland with blood.
11. (b) Although the heart begins pumping blood at 25 days following conception, its development is not complete until the end of the fifth week (at about 35 days).
12. (c) The umbilical vein receives oxygenated blood from the placenta and transports it to the fetal heart.

Chapter 17

1. (b) Inhaled air is cleansed, moistened, and warmed prior to its arrival at the pulmonary alveoli.
2. (a) The paired palatine bones support the nasal septum but are not part of its structure.
3. (b) Adenoid is the common name of the pharyngeal tonsil. An adenoidectomy is the removal of both pharyngeal tonsils.
4. (a) There are four paranasal sinuses, each named according to the bone in which it is located. Hence, we have ethmoidal, sphenoidal, frontal, and maxillary sinuses.
5. (e) Unlike the right lung that has three lobes, the left lung has only a superior lobe and an inferior lobe.
6. (a) The parietal pleura lines the wall of the thoracic cavity and the visceral pleura covers the lung. The space between the parietal pleura and the visceral pleura is referred to as the pleural cavity.
7. (d) When contracted, the muscles of inspiration increase the dimensions of the thoracic cavity, causing a decrease in the air pressure surrounding the lungs. Air flows through the respiratory tract, inflating the lungs.
8. (d) The vocal folds (cords) are attached between the arytenoid and thyroid cartilages on either lateral side of the glottis.
9. (c) The vital capacity is the greatest amount of air that can be exhaled following a maximal inhalation—an approximate volume of 4,500 cc of air.
10. (a) The nuclei for normal breathing are located within the medulla oblongata of the brain.

Chapter 18

1. (d) Viscera are the organs located within the trunk of the body. In the thoracic cavity, the viscera include the heart, lungs, and esophagus. In the abdominopelvic cavity, the viscera include the stomach, small and large intestines, liver, gallbladder, spleen, pancreas, kidneys, and adrenal glands, along with major vessels.
2. (c) The deciduous dentition includes 8 incisors, 4 canines, and 8 molars. The permanent dentition includes 8 incisors, 4 canines, 8 premolars, and 12 molars. The third molars are called wisdom teeth.
3. (b) It is through the supporting mesentery that vessels and nerves supply the abdominal viscera.
4. (c) Lacteals are lymph ductules found within the lamina propria of villi.
5. (b) Composed of lymphoid tissue, the spleen stores red blood cells and is an organ of the circulatory system.
6. (d) The papillae on the surface of the tongue provide a roughened surface for physically handling food. The papillae also support taste buds for responding to the chemical stimuli of various foods.
7. (c) Following the formation of chyme within the stomach, the food entering into the small intestine is ready for additional digestion and absorption. Bile and pancreatic juice enter the lumen of the duodenum to continue the chemical breakdown of food. Intestinal movements aid in its mechanical breakdown. The nutrients from digested food enter the bloodstream as absorption occurs within the small intestine.
8. (a) As the sphincter of ampulla opens, bile and pancreatic juice enter the duodenum. Stenosis of the sphincter of ampulla would prohibit the entry of these products.
9. (a) The hepatic portal vein transports absorbed nutrients within the bloodstream to the liver, where they are processed.
10. (d) Fats are absorbed into the lymphatic system. Proteins are broken down into amino acids before they are absorbed. Bile is secreted into the duodenum. The liver is served with blood from the hepatic portal vein and the hepatic artery; thus, it has a double blood supply, which becomes mixed at the capillary level.

Chapter 19

1. (b) The glomeruli are enclosed within the glomerular capsules in the renal cortex.
2. (d) The hilum of the kidney is the concave medial surface where the renal vessels enter and exit and where the ureter is located.
3. (c) Renal pyramids, containing the papillary ducts, are located within the renal medulla.
4. (a) The minor calyx receives urine directly from the papillary ducts. The urine then passes through the major calyx and into the renal pelvis.
5. (d) The kidneys are located high in the abdominal cavity between the levels of the twelfth thoracic and the third lumbar vertebrae.
6. (d) A renal calculus is most likely to cause a blockage of a ureter and backup of urine into the renal pelvis.
7. (e) Located only in the linings of the urinary bladder and ureters, transitional epithelium permits distension as urine is stored in the lumen of the urinary bladder.

8. (c) Collectively, the three layers of smooth muscle within the wall of the urinary bladder are referred to as the detrusor muscle. The detrusor muscle plays an active role in forcing urine from the urinary bladder during micturition.
9. (b) The internal urethral sphincter is actually a modified portion of the detrusor muscle and is innervated by parasympathetic neurons.
10. (d) Metanephros are the last type of developmental kidneys. They are continuously functioning throughout the fetal period, with the urine produced being expelled into the amniotic fluid.

Chapter 20

1. (c) The perineum is of functional and clinical importance because the external genitalia are located there.
2. (a) The dartos muscle is embedded within the wall of the scrotum. Along with the cremasteric muscle, it regulates the position of the testes within the scrotum through involuntary contraction in response to cold temperatures.
3. (b) Produced by interstitial cells, testosterone maintains male sexuality, including production of spermatozoa, activity of accessory sex organs, expression of secondary sex characteristics, and determination of sex drives.
4. (e) As components of the root of the penis, the bulb attaches to the undersurface of the urogenital diaphragm and the crus is the expanded proximal portion of the corpora cavernosa penis.
5. (b) The spermatic cord contains the ductus deferens, which is a spermatic duct. The spermatic cord, however, also contains spermatic vessels, a nerve, and the cremasteric muscle.
6. (a) The epididymides are long, flattened organs that store sperm in their last stage of maturation. Sperm are also stored in the ductus deferentia.
7. (a) Mucus secreted by the urethral glands lubricates the urethra and retards the entry of pathogens into the urinary bladder.
8. (c) Emission is movement of stored sperm from the epididymides and ductus deferentia to the ejaculatory ducts. Ejaculation is the forceful discharge of semen from the erect penis. Emission and ejaculation occur only if stimulation is sufficient, as in masturbation or coitus.
9. (a) The ovum from a female may be fertilized by a sperm cell containing either an X or a Y chromosome. If the sperm cell contains an X chromosome, it will pair with the X chromosome of the ovum and a female child will develop. A sperm cell carrying a Y chromosome results in an XY combination, and a male child will develop. The Y chromosome, therefore, is responsible for the subsequent production of androgens, which cause masculinization.
10. (d) The epididymides, ductus deferentia, and seminal vesicles derive from the mesonephric duct. The prostate arises from an endodermal outgrowth of the urogenital sinus.

Chapter 21

1. (c) The cervix of the uterus is the inferior constricted portion that opens into the vagina.
2. (b) The usual site of fertilization is within a uterine tube. From there, the fertilized egg continues to develop and enters into the cavity of the uterus about 3 days after conception.
3. (c) The secretory phase of the endometrium is characterized by an increase in glandular secretion and blood flow into the endometrium in preparation for implantation. During this time the ovary is in its luteal phase, characterized by a regressive corpus luteum.
4. (d) A secondary oocyte is discharged from an ovary during ovulation. If a spermatozoon passes through the corona radiata and zona pellucida and enters the cytoplasm of the secondary oocyte, the second meiotic division is completed and a mature ovum is formed.
5. (d) In addition to the mesovarium, each ovary is supported by an ovarian ligament and a suspensory ligament.
6. (c) The endometrium consists of a superficial stratum functionale layer that is shed as menses during menstruation and a deeper stratum basale layer that replenishes the stratum functionale layer after each menstruation.
7. (b) Vaginal rugae permit distension of the vagina during coitus and also during parturition.
8. (d) The vulva is the collective term for the female external genitalia. The vagina is an internal reproductive organ.
9. (d) The paramesonephric ducts give rise to the genital tract of the female reproductive system, which includes the uterus and the uterine tubes.
10. (a) Both the labia majora of the female and the scrotum of the male derive from the embryonic labioscrotal swellings.

Chapter 22

1. (d) The pre-embryonic period of developments lasts 14 days and is completed when the three primary germ layers have been formed and are in place to begin migration and differentiation.
2. (c) The yolk sac only produces blood for about a 2-week period (the third to the fifth week) prior to the formation of the liver. The liver then produces blood until the bone marrow is sufficiently developed to carry out the task.
3. (d) The placenta is an organ that serves many physiological functions, including the exchange of gases and other molecules between the maternal and fetal blood and the production of enzymes. In addition, it serves as a barrier against many harmful substances and as an endocrine gland in secreting steroid and glycoprotein hormones.
4. (c) The embryonic portion of the placenta is the chorion frondosum, and the maternal portion is the decidua basalis.
5. (d) Although the placenta is an effective barrier against diseases of bacterial origin, viruses and certain blood-borne diseases can diffuse through the vascular tissues. Furthermore, most drugs ingested by a pregnant woman can readily pass through the placenta, including nicotine, alcohol, and heroin.
6. (a) The embryonic heart begins pumping blood on about day 25, or during the fourth week of development.
7. (a) The neonatal period from birth to the end of the fourth week is characterized by major physiological changes, including the establishment of a stable heart rate and respiratory rate, and a consistent body temperature.
8. (b) A rapid heart rate of 120–160 beats/min. ensures an adequate oxygen supply to all of the cells and helps to maintain a constant body temperature.
9. (c) Pubescence accompanies puberty and refers to the continuum of physical changes that occur during this period of maturation, especially with regard to body hair.
10. (b) As compared to males, females have a lower basal metabolic rate. This may account, in part, for the longer life span of females.
11. (c) The average age at which breast buds appear in healthy girls in the United States is about 11 years, but the range is from 9 to 13 years.
12. (c) Dizygotic, or fraternal, twins may be of the same sex or of different sexes and are not any more alike than brothers or sisters born at different times.
13. homozygous recessive (bb)
heterozygous (Bb)
homozygous dominant (BB)
14. (a) A recessive allele is not expressed in a heterozygous genotype. This means that the particular recessive trait is not physically apparent.
15. (c) In a monohybrid cross, the probability of an offspring having a particular genotype is one in four for homozygous dominant and homozygous recessive and one in two for heterozygous dominant.

Appendix B

Laboratory Demonstrations in Anatomy

The study of human anatomy can be augmented by dissecting various vertebrate organs such as the brain, heart, eye, trachea, lung, and kidney.

Fresh organs obtained from a local slaughterhouse can be preserved by placing them in an embalming solution made according to the following formula:

Carbolic acid (melted crystals)	5 parts
Formalin (40%)	5 parts
Glycerin	5 parts
Water	85 parts

Carbolic acid is a disinfectant and helps maintain natural tissue color. Formalin is the main preservative. Glycerin prevents the anatomical structures from drying out too rapidly. Once an organ has been preserved, it can be stored for further, more detailed, dissection at a later date.

Using human cadavers is of obvious benefit in learning human anatomy. Cadavers have many advantages over anatomical charts, models, and illustrations when demonstrating size, position, and regional relationships within the human body. Furthermore, advanced anatomy students may be able to dissect the cadaver for use in an elementary human anatomy course.

Appropriate radiographs displayed on a viewing screen give an added dimension to the anatomy course. Outdated radiographs are usually available from doctors' offices, clinics, or hospitals.

Directions for preparing many interesting and valuable anatomical demonstrations can be found in the following books:

Hildebrand, M. 1968. *Anatomical preparations*. Berkeley: Univ. of California Press.

Tompsett, D. H. 1970. *Anatomical techniques*. 2d ed. Edinburgh: Livingstone.

Appendix C

Medical and Pharmacological Abbreviations

aa	of each
a.c.	before meals
A/G	albumin globulin ratio
ANS	autonomic nervous system
Bib.	drink
bid	twice a day
bihor	during two hours
B.M.R.	basal metabolic rate
bp	blood pressure
BUN	blood urea nitrogen
b.v.	vapor bath
c̄	with
caps.	capsule
c.b.c.	complete blood count
cc.	cubic centimeter(s)
cm.	centimeter(s)
CNS	central nervous system
Co., Comp.	compound
cr	tomorrow
CFS	cerebrospinal fluid
CT	computed tomography
CVP	central venous pressure
CXR	chest x-ray
d.	a day
D & C	dilatation and curettage
D.C.	discontinue
D, Det.	give
de d. in d.	from day to day
Dieb. secund	every second day
Dieb. tert.	every third day
dim.	one-half
d. in dup.	give twice as much
D. in p. aeq.	divide into equal parts
dr.	dram
D.T.D.	give of such doses
Dur. dolor.	while pain lasts
Dx	diagnosis
e	out of, with
ECG, EKG	electrocardiogram
EEG	electroencephalogram
e.m.p.	in the manner prescribed
feb	fever
GI	gastrointestinal
gm.	gram
gr.	grain
Grad.	gradually
gtt.	drop(s)
h.	hour
Hct	hematocrit
Hg.	mercury
Hgb	hemoglobin
hs	at bedtime
Hx	history
ibid.	in the same place
IM	intramuscular
incid.	cut
in d.	in a day
inj.	an injection
int. cib.	between meals
int. noct.	during the night
IV	intravenous
kg.	kilogram
Lat. dol.	to the painful side
M.	mix
man.	in the morning
mEq.	milliequivalent
mg.	milligram
ml.	milliliter
MRI	magnetic resonance imaging
Noct.	at night
Noct. maneq.	night and morning
N.P.O.	nothing by mouth
O.D.	right eye
o.d.	every day
Omn. hor.	every hour
Omn. man.	every morning
Omn. noct.	every night
O.S.	left eye

O.U.	both eyes
oz.	ounce
P	pulse
Part. aeq.	equal parts
pc	after meals
pCO_2	partial pressure of carbon dioxide
PNS	peripheral nervous system
PO	by mouth
pO_2	partial pressure of oxygen
p.p.a.	having first shaken the bottle
p.r.n.	as needed
pro. us. ext.	for external use
pt	patient
pt.	let it be continued
q.	each; every
qd	every day
qh	every hour
q. ___ h.	every ___ hours
qhs	every evening
qid	four times a day
qod	every other day
q.q.	also
q.s.	sufficient quantity
RBC	red blood cell
$\bar{s}$	without
Semih.	half an hour
Sig.	write, label
S.O.S.	if needed
sp. gr.	specific gravity
ss., $\overline{ss}$	one-half
s.s.s.	layer on layer
stat.	immediately
sum.	take
s.v.r.	alcohol
t.	three times
tab.	tablet
tid	three times a day
ung.	ointment
Ut. dict.	as directed
vic.	times
WBC	white blood cell

Appendix D

Units of Measurement and Their Equivalents

Apothecaries' Weights and Their Metric Equivalents

1 grain (gr) =
0.05 scruple (s)
0.017 dram (dr)
0.002 ounce (oz)
0.0002 pound (lb)
0.065 gram (g)
65. milligrams (mg)

1 scruple (s) =
20. grains (gr)
0.33 dram (dr)
0.042 ounce (oz)
0.004 pound (lb)
1.3 grams (g)
1,300. milligrams (mg)

1 dram (dr) =
60. grains (gr)
3. scruples (s)
0.13 ounce (oz)
0.010 pound (lb)
3.9 grams (g)
3,900. milligrams (mg)

1 ounce (oz) =
480. grains (gr)
24. scruples (s)
8. drams (dr)
0.08 pound (lb)
31.1 grams (g)
31,100. milligrams (mg)

1 pound (lb) =
5,760. grains (gr)
288. scruples (s)
96. drams (dr)
12. ounces (oz)
373. grams (g)
373,000. milligrams (mg)

Apothecaries' Volumes and Their Metric Equivalents

1 minim (min) =
0.017 fluid dram (fl dr)
0.002 fluid ounce (fl oz)
0.0001 pint (pt)
0.06 milliliter (ml)
0.06 cubic centimeter (cc)

1 fluid dram (fl dr) =
60. minims (min)
0.13 fluid ounce (fl oz)
0.008 pint (pt)
3.70 milliliters (ml)
3.70 cubic centimeters (cc)

1 fluid ounce (fl oz) =
480. minims (min)
8. fluid drams (fl dr)
0.06 pint (pt)
29.6 milliliters (ml)
29.6 cubic centimeters (cc)

1 pint (pt) =
7,680. minims (min)
128. fluid drams (fl dr)
16. fluid ounces (fl oz)
473. milliliters (ml)
473. cubic centimeters (cc)

Metric Weights and Their Apothecaries' Equivalents

1 gram (g) =
0.001 kilogram (kg)
1,000. milligrams (mg)
1,000,000. micrograms (μg)
5.4 grains (gr)
0.032 ounce (oz)

1 kilogram (kg) =
1,000. grams (g)
1,000,000. milligrams (mg)
1,000,000,000. micrograms (μg)
32. ounces (oz)
2.7 pounds (lb)

1 milligram (mg) =
0.000001 kilogram (kg)
0.001 gram (g)
1,000. micrograms (μg)
0.0154 grains (gr)
0.000032 ounce (oz)

Metric Volumes and Their Apothecaries' Equivalents

1 liter (l) =
1,000. milliliters (ml)
1,000. cubic centimeters (cc)
2.1 pints (pt)
270. fluid drams (fl dr)
34. fluid ounces (fl oz)

1 milliliter (ml) =
0.001 liter (l)
1. cubic centimeter (cc)
16.2 minims (min)
0.27 fluid dram (fl dr)
0.034 fluid ounce (fl oz)

Approximate Equivalents of Household Measures

1 teaspoon (tsp) =
4. milliliters (ml)
4. cubic centimeters (cc)
1. fluid dram (fl dr)

1 tablespoon (tbsp) =
15. milliliters (ml)
15. cubic centimeters (cc)
0.5 fluid ounce (fl oz)
3.7 teaspoons (tsp)

1 cup (c) =
240. milliliters (ml)
240. cubic centimeters (cc)
8. fluid ounces (fl oz)
0.5 pint (pt)
16. tablespoons (tbsp)

1 quart (qt) =
960. milliliters (ml)
960. cubic centimeters (cc)
2. pints (pt)
4. cups (c)
32. fluid ounces (fl oz)

Conversion of Units from One Form to Another

Refer to the preceding equivalency lists when converting one unit to another equivalent unit.

To convert a unit shown in bold type to one of the equivalent units listed immediately below it, multiply the first number (boldtype unit) by the appropriate equivalent unit listed below it.

Sample Problems:

1. Convert 320 grains into scruples (1 gr = 0.05 s).

$$320 \text{ gr} \times \frac{0.05 \text{ s}}{1 \text{ gr}} = 16.0 \text{ s}$$

2. Convert 320 grains into drams (1 gr = 0.017 dr).

$$320 \text{ gr} \times \frac{0.017 \text{ dr}}{1 \text{ gr}} = 5.44 \text{ dr}$$

3. Convert 320 grains into grams (1 gr = 0.065 g).

$$320 \text{ gr} \times \frac{0.065 \text{ g}}{1 \text{ gr}} = 20.8 \text{ g}$$

Body Temperatures in ° Fahrenheit and ° Celsius

° F	° C
95.0	35.0
95.2	35.1
95.4	35.2
95.6	35.3
95.8	35.4
96.0	35.5
96.2	35.7
96.4	35.8
96.6	35.9
96.8	36.0
97.0	36.1
97.2	36.2
97.4	36.3
97.6	36.4
97.8	36.6
98.0	36.7
98.2	36.8
98.4	36.9
98.6	37.0
98.8	37.1
99.0	37.2
99.2	37.3
99.4	37.4
99.6	37.6
99.8	37.7

° F	° C
100.0	37.8
100.2	37.9
100.4	38.0
100.6	38.1
100.8	38.2
101.0	38.3
101.2	38.4
101.4	38.6
101.6	38.7
101.8	38.8
102.0	38.9
102.2	39.0
102.4	39.1
102.6	39.2
102.8	39.3
103.0	39.4
103.2	39.6
103.4	39.7
103.6	39.8
103.8	39.9
104.0	40.0
104.2	40.1
104.4	40.2
104.6	40.3
104.8	40.4
105.0	40.6

To convert ° F to ° C:
Subtract 32 from ° F and multiply by 5/9.
_____ ° F – 32 × 5/9 = _____ ° C

To convert ° C to ° F:
Multiply ° C by 9/5 and 32
_____ ° C × 9/5 + 32 = _____ ° F

Appendix E

Related Web Sites

This appendix contains lists of web sites where useful information pertaining to most of the chapters can be found. If you have problems linking to any of the web sites listed, please visit the homepage for Van De Graaff's *Human Anatomy*, fifth edition, at www.mhhe.com/sciencemath/biology/vdghumananatomy/ for the most current web sites related to this textbook.

Chapter 1: History of Anatomy

http://hops.wharton.upenn.edu/~loren/Links/maloney/alchemy/quin_original.html
Featured is original text from an alchemy book in Middle English. If you have an IBM compatible, you'll be able to see something like the original phonetic pronunciations.

http://www.ea.pvt.k12.pa.us/medant/hyprtxts.htm
This page contains links to English translations of the writings of Hippocrates and several texts by Galen.

Chapter 2: Body Organization and Anatomical Nomenclature

http://www.med.wright.edu/SOM/academic/anatomy/beyond.html
Various imaging techniques are used to display the body, including radiographs and CT scans. Self-testing is featured.

http://www.npac.syr.edu/projects/vishuman/VisibleHuman.html
Pictures from the Visible Human Project. Various imaging techniques are used to view sections of a cadaver.

http://www.scar.rad.washington.edu/RadAnatomy.html
Radiographs from various perspectives.

http://www9.biostr.washington.edu/da.html
A human anatomy atlas illustrating numerous body systems.

http://www.largnet.uwo.ca/med/i-way.html
CT scans and ultrasounds of different parts of the body.

http://www.lifeart.com/
Free samples from the anatomy and physiology clipart packages are issued each month. You are invited to download the images.

Chapter 3: Cytology

http://the-tech.mit.edu/Chemicool/
A beautifully colored periodic table in which each element is linked to other tables that give detailed information about that element.

http://members.aol.com/jeff555555/table/ptable.html
Another periodic table with links. This one loads more quickly.

http://www.chemie.fu-berlin.de/chemistry/bio/amino-acids.html
This site includes a list of amino acids. Clicking on the name of an amino acid will bring up another page showing different models of it.

http://yip5.chem.wfu.edu/yip/organic/compgraf.html#viz
A linked movie at this site illustrates a rotating model of DNA. A model of an enzyme also can be viewed from a link on this site.

http://lenti.med.umn.edu/~mwd/cell_www/cell_intro.html
Some useful information about the chemical composition of cells.

http://esg-www.mit.edu:8001/esgbio/cb/org/animal.gif
A beautifully colored illustration of an animal cell, but with no links.

http://lenti.med.umn.edu/~mwd/cell_www/cell_intro.html
A good picture of the cell membrane and its constituents.

http://lenti.med.umn.edu/~mwd/cell_www/chapter2/membrane.html# SEMIPERM
Additional details about the chemicals that make up the cell membrane and a discussion of osmosis and diffusion.

http://lenti.med.umn.edu/~mwd/cell_www/chapter2/protein.html# MEMBRANE
Active transport proteins are discussed, and linked pictures illustrate their action.

http://lenti.med.umn.edu/~mwd/cell_www/chapter2/mitochondria.html
A picture and brief description of a mitochondrion. This site has links to codon information.

Chapter 5: Integumentary System

http://www.medic.mie-u.ac.jp/derma/anatomy.html
This site has a very clear color micrograph of a skin section with clickable areas for more details. The site includes links to other dermatology pages.

http://www-sci.lib.uci.edu/~martindale/Medical.html#Derm
Links to many good dermatology sites.

http://www.maui.net/~southsky/introto.html
A very useful page for skin cancer information. There are a number of links and detailed explanations.

http://www.pinch.com/skin/
This is a page with links to several skin disease pages, including a search link.

http://www.hkma.com.hk/std/menuskin.htm
Very detailed information about different types of eczema.

Chapter 6: Skeletal System: Introduction and the Axial Skeleton

http://weber.u.washington.edu/~dboone/key/subjects/arthritis/xxxxxx xx1_1.html
Links to sites on Paget's disease and other bone problems.

http://www.osteo.org/
Information on frequency of osteoporosis among different groups of people, prevention of the disease, and ways to slow its effects. There are additional links to other bone tissue disorders.

http://www.cs.brown.edu/people/oa/Bin/skeleton.html
Click on the bone and hear its name pronounced!

http://www.scar.rad.washington.edu/RadAnatomy.html
Radiographs from various perspectives.

http://anatomy.uams.edu/HTMLpages/anatomyhtml/pectorals.html
Lists of bones and muscles and an outline for major joints from the University of Arkansas.

Chapter 7: Skeletal System: The Appendicular Skeleton

http://www.cs.brown.edu/people/oa/Bin/skeleton.html
Click on the bone and hear its name pronounced!

http://www.scar.rad.washington.edu/RadAnatomy.html
Radiographs from various perspectives.

http://anatomy.uams.edu/HTMLpages/anatomyhtml/pectorals.html
Lists of bones and muscles and an outline for major joints from the University of Arkansas.

Chapter 8: Articulations

http://www.rad.upenn.edu/rundle/InteractiveKnee.html
Pictures, text, and links about this most complex of all joints.

http://www.meddean.luc.edu/develop/afang/KNEE.HTM
MRIs of the knee joint and associated structures, including muscles.

http://rpisun1.mda.uth.tmc.edu/se/anatomy/
This contains a clickable human. You can examine sections of the foot, knee, and elbow joints.

http://www.rad.washington.edu/Anatomy/TMJ/TMJ.html
Various types of images and accompanying text, and links concerning temporomandibular joint disorder.

http://weber.u.washington.edu/~dboone/key/subjects/arthritis/xxxxxx xx1_1.html
An arthritis site with links to numerous other disorders, including carpel tunnel syndrome and Paget's disease.

http://anatomy.uams.edu/HTMLpages/anatomyhtml/pectorals.html
Lists of bones and muscles and an outline for major joints from the University of Arkansas.

Chapter 9: Muscular System

http://rpiwww.mdacc.tmc.edu:80/se/anatomy/arm/
Viewers can click on any of several sections through the forearm to examine the positions of the muscles.

http://www.healthink.com/muscles.html

This site provides simple pictures and a brief summary of the functions and locations of certain skeletal muscles.

Chapter 11: Nervous Tissue and the Central Nervous System

http://neuron.duke.edu/

Go to this address to obtain a downloadable neuron simulation program. Complete documentation is available.

http://www.physiol.arizona.edu/CELL/Instruct/BodyElect.html

Very interesting material, pictures, and text from a University of Arizona course entitled "Body electric." There is information about excitable cells here.

http://www9.biostr.washington.edu/cgi-bin/imageform

Human brain atlases, including attractive graphics, a movie list, and three-dimensional reconstructions.

http://www.nucmed.buffalo.edu/nrlgy1.htm#fdg_normal_brainx31

You can see a PET scan of a normal brain, but there are also links to scans demonstrating pathological conditions.

http://www.coa.uky.edu/ADReview

This web page contains a great deal of useful information about Alzheimer's disease, with numerous links to other sites pertaining to the topic.

http://www.merck.com/!!s9OfD17kBs9OwL2Wwl/disease/preventable/hib/

Haemophilus B is a bacterium that can cause meningitis. This site offers a brief discussion of the mode of transmission and lists those at greatest risk.

http://www.loni.ucla.edu./humandata/human.html

This site displays brain sections using various imaging techniques.

http://uta.marymt.edu/~psychol/brain.html

You can click on the cerebral cortex, shown in vivid color, and get the names and brief descriptions of superficial structures.

http://www.geocities.com/HotSprings/1161/

This site features a multimedia of links providing information on polio and other diseases with similar symptoms.

Chapter 14: Endocrine System

http://www.niddk.nih.gov/DiabetesDocs.html

Information for patients with diabetes mellitus and conditions caused by diabetes.

http://home.ican.net/~thyroid/English/Guides.html

This site features multiple links to other websites concerning thyroid conditions. Pictures are available on some of the sites.

http://www.cs.sfu.ca/css/update/vol8/8.1-muscles.html

This site explains the effect of insulin-like growth factor (IGF) on the growth of muscle cells and relates this effect to aging. A clear drawing explains the roles of receptors and carrier molecules in this process.

Chapter 15: Sensory Organs

http://www.blindness.org/Latest_Research.html

This site features recent research news about macular degeneration, retinitis pigmentosa, and related disorders.

http://www.merck.com/!!s9OfD17kBs9OqO0OPa/disease/glaucoma/

Glaucoma and related diseases are discussed in layperson's terms.

http://www.tinnitus-pjj.com/

A good explanation of the cause of tinnitus, plus links to additional information and a bibliography.

http://www.adworks.com/dizzy/vestib.html

This website features numerous links that provide information about equilibrium disorders. Otoscopic exams are illustrated, and videos may be downloaded.

Chapter 16: Circulatory System

http://www.med.nagoya-u.ac.jp/pathy/Pictures/atlas.html

A photomicrographic atlas of blood cells and bone marrow. Includes very good pictures of various leukemias, thalassemia (a disorder affecting red blood cells), anemias, and masses of lymphocytes in the skin.

http://www.physiol.arizona.edu/CELL/Instruct/BodyElect.html

Very interesting material, pictures, and text from a University of Arizona course entitled "Body electric." There is information about excitable cells here.

http://www.merck.com/!!s9OfD17kBs9OwL2Wwl/disease/preventable/hepb/

Facts on hepatitis B symptoms, statistics, spread, and prevention.

http://cancer.med.upenn.edu/pdq.400113.html

A research report on leukemia.

http://sln2.fi.edu/biosci/heart.html

This website features a movie on open heart surgery (requires QTV) and other graphics. There are also activities related to EKGs and radiographs.

http://www.merck.com/!!s9OfD17kBs9Oma1w8T/disease/heart/

This site discusses risk factors for heart disease.

http://www-medlib.med.utah.edu/WebPath/ATHHTML/ATHIDX.html

Photomicrographs of atheroscleroses and thromboses and pictures of an infarcted heart are featured here.

http://www/vesalius.com/story/story.html

Heart illustrations and accompanying text appear at this site.

http://neurosurgery.mgh.harvard.edu/vaschome.htm#AVMs

This site provides information about strokes, aneurysms, and other neurological and vascular disorders. Links to other medical centers include diagrams and angiograms.

Chapter 17: Respiratory System

http://www.merck.com/!!s9OfD17kBs9OwL2Wwl/disease/preventable/pneu/

A brief explanation of pneumococcal pneumonia and a list of those at risk.

http://www9.biostr.washington.edu/cgi-bin/PageMaster?atlas:Thorax+ffpathIndex:Thoracic^Viscera+2

Beautiful three-dimensional color pictures of the lungs and other viscera.

Chapter 18: Digestive System

http://www.merck.com/!!s9OfD17kBs9OwL2Wwl/disease/preventable/hepa/

A simple explanation of what causes hepatitis A, and how to prevent this infection.

http://www.geocities.com/Paris/4664/introli.html

Lactose intolerance is discussed in layperson's terminology. A few links to other sources of information about this condition are included.

Chapter 19: Urinary System

http://www.gamewood.net/rnet/section/CME.htm

Featured here are a tutorial and links to other sites concerning disorders of the urinary system.

Chapter 20: Male Reproductive System

http://www.luc.edu/depts/biology/meiosis.htm

A good diagram of meiosis with a link to one of mitosis.

http://www.luc.edu/depts/biology/meiosis.html

A picture showing the steps of meiosis along with a brief explanation. Links to other sites are included.

http://www.merck.com/!!s9OfD17kBs9PBm0Hsz/disease/hiv/

Several aspects of HIV and AIDS are covered here, including an extensive presentation on the opportunistic infections associated with AIDS.

http://hawley-lab2.ucdavis.edu/Meiosis.html

A more technical discussion of meiosis.

Chapter 21: Female Reproductive System

http://cancer.med.upenn.edu/specialty/gyn_onc/ovarian/

This site provides information about ovarian cancer and technical data about specific types. Several links to other pages are included.

http://cancer.med.upenn.edu/classroom/colp/

Basically a tutorial on cervical cancer. It includes diagrams of normal tissue and progressive stages of cancer.

Chapter 22: Developmental Anatomy, Postnatal Growth, and Inheritance

http://www-medlib.med.utah.edu/WebPath/TUTORIAL/PRENATAL/PRENATAL.html

Information about prenatal diagnostic methods. Chromosomal abnormality images included.

http://www.ornl.gov/TechResources/Human_Genome/home.html

This is the homepage of the Human Genome Project. There are links to sites concerning specific genetic disorders.

http://www.informatik.uni-rostock.de/HUM-MOLGEN/index.html

Use this address instead of the above for the same information. This is what is called a mirror site.

http://www3.ncbi.nlm.nih.gov/htbin-post/Omim/dispmim?261600

This site contains information about PKU cause, clinical signs, genetics, population biology, and dietary recommendations.

http://www.menshealth.com/features/mensconf/indexg.html

A lively discussion of men's health issues.

Glossary

Key to Pronunciation

Most of the words in this glossary are followed by a phonetic spelling that serves as a guide to pronunciation. The phonetic spellings reflect standard scientific usage and can easily be interpreted following a few basic rules.

1. Any unmarked vowel that ends a syllable or that stands alone as a syllable has the long sound. For example, *ba, ma,* and *na* rhyme with *fay; be, de,* and *we* rhyme with *fee; bi, di,* and *pi* rhyme with *sigh; bo, do,* and *mo* rhyme with *go*. Any unmarked vowel that is followed by a consonant has the short sound (for example, the vowel sounds in *hat, met, pit, not,* and *but*).
2. If a long vowel appears in the middle of a syllable (followed by a consonant), it is marked with a macron (¯). Similarly, if a vowel stands alone or ends a syllable but should have short sound, it is marked with a breve (˘).
3. Syllables that are emphasized are indicated by stress marks. A single stress mark (′) indicates the primary emphasis; a secondary emphasis is indicated by a double stress mark (″).

A

abdomen (ab′-dŏ-men or ab-do′men) The region of the trunk between the diaphragm and pelvis.

abduction (ab-duk′shun) Movement of a body part away from the axis or midline of the body; movement of a digit away from the axis of the limb; the opposite of *adduction*.

accessory organs Organs that assist the functioning of other organs within a system.

acetabulum (as″ĕ-tab′yŭ-lum) The cup-shaped depression on the lateral surface of the hipbone (os coxae) with which the head of the femur articulates.

Achilles (ă-kil′ēz) **tendon** *See* tendo calcaneus.

actin (ak′tin) A protein in muscle fibers that, together with myosin, is responsible for contraction.

adduction (ă-duk′shun) Movement of a body part toward the axis or midline of the body; movement of a digit toward the axis of the limb; the opposite of *abduction*.

adipocyte (ad′ĭ-po-sīt) A fat cell found within adipose tissue.

adrenal cortex (ă-dre′nal kor′teks) The outer part of the adrenal gland that secretes steroid hormones.

adrenal glands Two small endocrine glands, one located above each kidney; also called the *suprarenal glands*.

adrenal medulla (mĕ-dul′ă) The inner part of the adrenal gland that secretes the hormones epinephrine and norepinephrine.

adventitia (ad″ven-tish′ă) The outermost epithelial layer of a visceral organ; also called *serosa*.

afferent (af′er-ent) Conveying or transmitting to.

afferent glomerular arteriole (ar-te′re-ōl) A vessel within the kidney that supplies blood to the glomerulus.

afferent neuron (noor′on) A sensory nerve cell that transmits an impulse toward the central nervous system.

agonist (ag′ŏ-nist) The prime mover muscle, which is directly engaged in the contraction that produces a desired movement.

alimentary (al″ĭ-men′tre) **canal** The tubular portion of the digestive tract; also called the *gastrointestinal (GI) tract*.

allantois (ă-lan′to-is) An extraembryonic membranous sac involved in the formation of blood cells. It gives rise to the fetal umbilical arteries and vein, and also contributes to the formation of the urinary bladder.

all-or-none principle The statement that muscle fibers of a motor unit contract to their maximum extent when exposed to a stimulus of threshold strength.

alveolar (al-ve′ŏ-lar) **sacs** A cluster of alveoli that share a common chamber or central atrium.

alveolus (al-ve′ŏ-lus) **1.** A *pulmonary alveolus* is an individual air capsule within the lung. The pulmonary alveoli are the basic functional units of respiration. **2.** A *dental alveolus* is a socket that secures a tooth (a tooth socket).

amniocentesis (am″ne-o-sen-te′sis) A procedure in which a sample of amniotic fluid is aspirated to examine cells for various genetic diseases.

amnion (am′ne-on) The innermost fetal membrane—a thin sac that holds the fetus suspended in amniotic fluid; commonly known as the "bag of waters."

ampulla (am-pool′ă) A saclike enlargement of a duct or tube.

ampulla of Vater (fa′ter) *See* hepatopancreatic ampulla.

anal (a′nal) **canal** The terminal tubular portion of the large intestine that opens through the anus of the GI tract.

anal glands Enlarged and modified sweat glands that empty into the anal opening.

anastomosis (ă-nas″tŏ-mo′sis) An anatomical convergence of blood vessels or nerves that forms a network.

anatomical position An erect body stance with the eyes directed forward, the arms at the sides, the palms of the hands facing forward, and the fingers pointing straight down.

anatomy The branch of science concerned with the structure of the body and the relationship of its parts.

antagonist (an-tag′ŏ-nist) **1.** A muscle that acts in opposition to another muscle. **2.** Any agent, such as a hormone or drug, that opposes the action of another.

antebrachium (an″te-bra′ke-um) The forearm.

anterior (ventral) Toward the front; the opposite of *posterior* (*dorsal*).

anterior root The anterior projection of the spinal cord composed of axons of motor, or efferent, fibers.

anus (a′nus) The terminal opening of the GI tract.

aorta (a-or′tă) The major systemic vessel of the arterial system of the body, emerging from the left ventricle.

aortic arch The superior left bend of the aorta between the ascending and descending portions; also called the *arch of aorta*.

apex (a′peks) The tip or pointed end of a conical structure.

apocrine (ap′ŏ-krin) **gland** A type of sweat gland that functions in evaporative cooling. It may respond during periods of emotional stress.

aponeurosis (ap″ŏ-noo-ro′sis) A fibrous or membranous sheetlike tendon.

appendix (ă-pen′diks) A short pouch that attaches to the cecum; also called the *vermiform appendix*.

aqueous (a′kwe-us) **humor** The watery fluid that fills the anterior and posterior chambers of the eye.

arachnoid (ă-rak′noid) The weblike middle covering (meninx) of the central nervous system.

arbor vitae (ar′bor vi′te) The branching arrangement of white matter within the cerebellum.

arm (brachium) That portion of the upper extremity extending from the shoulder to the elbow.

arrector pili (ă-rek′tor pi′li); plural, *arrectores pilorum* A bundle of smooth muscle cells attached to a hair follicle that, upon contraction, pulls the hair into a more vertical position, resulting in goose bumps.

arteriole (ar-te′re-ōl) A minute arterial branch.

artery (ar′tĕ-re) A blood vessel that carries blood away from the heart.

arthrology (ar-throl'ŏ-je) The scientific study of the structure and function of joints.

articular cartilage (ar-tik'yŭ-lar kar'tĭ-lij) A hyaline cartilaginous covering over the articulating surface of bones of synovial joints.

articulation (ar-tik'yŭ-la'shun) A joint.

arytenoid (ar"ĕ-te'noid) **cartilages** A pair of small cartilages on the superior aspect of the larynx.

ascending colon (ko'lon) That portion of the large intestine between the cecum and the hepatic (right colic) flexure.

association neuron (noor'on) A nerve cell located completely within the central nervous system. It conveys impulses in an arc from sensory to motor neurons; also called an *internuncial neuron* or *interneuron*.

atom The smallest particle of matter that characterizes an element.

atrioventricular (a"tre-o-ven-trik'yŭ-lar) **bundle** A group of specialized cardiac fibers that conducts impulses from the atrioventricular node to the ventricular muscles of the heart; also called the *bundle of His* or *AV bundle*.

atrioventricular node A microscopic aggregation of specialized cardiac fibers located in the interatrial septum of the heart; a component of the conduction system of the heart; also called the *AV node*.

atrioventricular valve A cardiac valve located between an atrium and a ventricle of the heart; also called an *AV valve*.

atrium (a-tre'um) Either of the two superior chambers of the heart that receive venous blood.

atrophy (at'rŏ-fe) A gradual wasting away or decrease in the size of a tissue or organ.

auditory Pertaining to the structures of the ear associated with hearing.

auditory tube A narrow canal that connects the middle-ear chamber to the pharynx; also called the *eustachian canal*.

auricle (or'ĭ-kul) **1.** The fleshy pinna of the ear. **2.** An ear-shaped appendage of each atrium of the heart.

autonomic nervous system (ANS) The sympathetic and parasympathetic portions of the nervous system that innervate cardiac muscle, smooth (visceral) muscles, and glands. The ANS functions largely without conscious control.

axilla (ak-sil'ă) The depressed hollow commonly called the "armpit."

axon (ak'son) The elongated process of a nerve cell that transmits an impulse away from the cell body of a neuron.

B

ball-and-socket joint The most freely movable of all the synovial joints (e.g., the shoulder and hip joints).

baroreceptor (bar"o-re-sep'tor) A cluster of neuroreceptors stimulated by pressure changes. Baroreceptors monitor blood pressure.

basal ganglion (gang'gle-on) A mass of nerve cell bodies located deep within a cerebral hemisphere of the brain.

basement membrane A thin sheet of extracellular substance to which the basal surfaces of membranous epithelial cells are attached; also called the *basal lamina*.

basophil(ba'sŏ-fil) A granular leukocyte that readily stains with basophilic dye.

belly The bulging, central part of a skeletal muscle.

benign (bĕ-nīn) Not malignant.

bicuspid (bi-kus'pid) **valve** The heart valve located between the left atrium and the left ventricle; the left AV valve; also called the *mitral valve*.

bifurcation (bi"fur-ka'shun) Forked; divided into two branches.

bile (bīl) A liver secretion that is stored and concentrated in the gallbladder and released through the common bile duct into the duodenum. It is essential for the absorption of fats.

bipennate (bi-pen'āt) Denoting muscles whose fibers are arranged on each side of a tendon like barbs on a feather shaft.

bipolar neuron (noor'on) A nerve cell with two processes, one at each end of the cell body.

blastocyst (blas'tŏ-sist) An early stage of embryonic development consisting of a hollow ball of cells with an embryoblast (inner cell mass) and an outer layer called the trophoblast.

blood The fluid connective tissue that circulates through the cardiovascular system to transport substances throughout the body.

blood-brain barrier A specialized mechanism that inhibits the passage of certain materials from the blood into brain tissue and cerebrospinal fluid.

bolus (bo'lus) A moistened mass of food that is swallowed from the oral cavity into the pharynx.

bone 1. The hard, calcified connective tissue forming the major portion of the skeleton. **2.** Any of the more than 200 anatomically distinct structures making up the skeleton.

bony labyrinth (lab'ĭ-rinth) A series of chambers within the petrous part of the temporal bone associated with the vestibular organs and the cochlea.

Bowman's (bo'manz) **capsule** *See* glomerular capsule.

brachial plexus (bra'ke-al plek'sus) A network of nerve fibers arising from spinal nerves C5–C8 and T1. Nerves arising from the brachial plexus supply the upper extremity.

brain The enlarged superior portion of the central nervous system located in the cranial cavity of the skull.

brain stem That portion of the brain consisting of the midbrain, pons, and medulla oblongata.

bronchial (brong'ke-al) **tree** The bronchi and their branching bronchioles.

bronchiole (brong'ke-ōl) A small division of a bronchus within the lung.

bronchus (brong'kus) A branch of the trachea that leads to a lung.

bulbourethral (bul"bo-yoo-re'thral) **glands** A pair of glands that secrete a viscous fluid into the male urethra during sexual excitement; also called *Cowper's glands*.

bundle of His (hiss) *See* atrioventricular bundle.

bursa (bur'sa) A saclike cavity filled with synovial fluid. Bursae are located between tendons and bones or at points of friction between moving structures.

buttock (but'ok) The fleshy mass on the posterior aspect of the lower trunk, formed primarily by the gluteal muscles; the rump.

C

calyx (ka'liks) The cup-shaped portion of the renal pelvis that encircles a renal papilla.

canaliculus (kan"ă-lik'yŭ-lus) A microscopic channel in bone tissue that connects lacunae.

canal of Schlemm (shlem) *See* scleral venous sinus.

cancellous (kan'sĕ-lus) **bone** Spongy bone; bone tissue with a latticelike structure.

capillary (kap'ĭ-lar"e) A microscopic blood vessel that connects an arteriole and a venule; the functional unit of the circulatory system.

carbohydrate Any of the group of organic molecules composed of carbon, hydrogen, and oxygen, including sugars and starches. A carbohydrate usually has the formula CH_2O.

carotid (kă-rot'id) **sinus** An expanded portion of the internal carotid artery located immediately above the point of branching from the external carotid artery. The carotid sinus contains baroreceptors that monitor blood pressure.

carpus (kar'pus) The proximal portion of the hand that contains the eight carpal bones.

cartilage (kar'tĭ-lij) A type of connective tissue with a solid elastic matrix.

cartilaginous (kar"tĭ-laj'ĭ-nus) A joint that lacks a joint cavity, permitting little movement between the bones held together by cartilage.

cauda equina (kaw'dă e-kwi'nă) The extension from the terminal portion of the spinal cord, where the roots of spinal nerves resemble a horse's tail.

caudal (kaw'dal) Referring to a position more toward the tail.

cecum (se'kum) The pouchlike portion of the large intestine to which the ileum of the small intestine is attached.

cell The structural and functional unit of an organism; the smallest structure capable of performing all of the functions necessary for life.

cementum (se-men'tum) Bonelike material that binds the root of a tooth to the periodontal membrane of the tooth socket (dental alveolus).

central canal The elongated longitudinal channel at the center of an osteon in bone tissue that contains branches of the nutrient vessels and a nerve; also called a *haversian canal*.

central nervous system (CNS) The brain and the spinal cord.

centrosome (sen'trŏ-sōm) A specialized zone of cytoplasm near the nucleus of a cell that contains a pair of centrioles.

cerebellar peduncle (ser"ĕ-bel'ar pĕ-dung'kul) An aggregation of nerve fibers connecting the cerebellum with the brain stem.

cerebellum (ser"ĕ-bel'um) That portion of the brain concerned with the coordination of skeletal muscle contraction. Part of the metencephalon, it consists of two hemispheres and a central vermis.

cerebral aqueduct (ser'ĕ-bral ak'wĕ-dukt) *See* mesencephalic aqueduct.

cerebral arterial (ar-te're-al) **circle** An arterial vessel on the inferior surface of the brain that encircles the pituitary gland. It provides alternate routes for blood to reach the brain should a carotid or vertebral artery become occluded; also called the *circle of Willis*.

cerebral peduncles (pĕ-dung′kulz) A paired bundle of nerve fibers along the inferior surface of the midbrain that conduct impulses between the pons and the cerebral hemispheres.

cerebrospinal (ser″ĕ-bro-spi′nal) **fluid** A fluid produced by the choroid plexus of the ventricles of the brain. It fills the ventricles and surrounds the central nervous system in association with the meninges.

cerebrum (ser′ĕ-brum or sĕ-re′brum) The largest portion of the brain, composed of right and left cerebral hemispheres.

ceruminous (sĕ-roo′mĭ-nus) **gland** A specialized integumentary gland that secretes cerumen, or earwax, into the external acoustic canal.

cervical (ser′vĭ-kal) Pertaining to the neck or a necklike portion of an organ.

cervical ganglion (gang′gle-on) A cluster of postganglionic sympathetic nerve cell bodies located in the neck, near the cervical vertebrae.

cervical plexus (plek′sus) A network of spinal nerves formed by the anterior branches of the first four cervical nerves.

cervix (ser′viks) **1.** The narrow necklike portion of an organ. **2.** The inferior end of the uterus that adjoins the vagina (cervix of the uterus).

chemoreceptor (ke″mo-re-sep′tor) A neuroreceptor that is stimulated by the presence of a chemical in solution.

chiasma (ki-as′mă) A crossing of nerve tracts from one side of the CNS to the other; also called a *chiasm*.

choanae (ko-a′ne) The two posterior openings from the nasal cavity into the nasopharynx; also called the *internal nares*.

chondrocranium (kon″dro-kra′ne-um) That portion of the skull that supports the brain. It is derived from endochondral bone.

chondrocyte (kon′dro-sīt) A cartilage cell.

chordae tendineae (kor′de ten-din′e-e) Tendinous bands that connect papillary muscles to the leaflets of the atrioventricular valves within the ventricles of the heart.

chorion (ko′re-on) An extraembryonic membrane that contributes to the formation of the placenta.

choroid (kor′oid) The vascular, pigmented middle layer of the wall of the eye.

choroid plexus A mass of vascular capillaries from which cerebrospinal fluid is secreted into the ventricles of the brain.

chromatophilic (kro″mă-to-fil′ik) **substances** Clumps of rough endoplasmic reticulum in the cell body of a neuron; also called *Nissl bodies*.

chromosomes (kro′mŏ-sōmz) Structures in the nucleus that contain the genes for genetic expression.

chyme (kīm) The mass of partially digested food that passes from the pylorus of the stomach into the duodenum of the small intestine.

cilia (sil′e-ă) Microscopic hairlike processes that move in a wavelike manner on the exposed surfaces of certain epithelial cells.

ciliary (sil′e-er″e) **body** A portion of the choroid layer of the eye that secretes aqueous humor. It contains the ciliary muscle that, when contracted, changes the shape of the lens.

circle of Willis *See* cerebral arterial circle.

circumduction (ser″kum-duk′shun) Circular movement of a body part in which a cone-shaped airspace is traced.

cleavage Rapid cell divisions of the zygote that occur approximately 30 hours after fertilization of the secondary oocyte.

clitoris (klit′or-is or kli′tor-is) A small erectile structure in the vulva of the female, homologous to the glans penis in the male.

coccygeal (kok-sij′e-al) Pertaining to the region of the coccyx; the caudal termination of the vertebral column.

cochlea (kok′le-ă) The spiral portion of the inner ear that contains the spiral organ (organ of Corti).

cochlear window A round, membrane-covered opening between the middle and inner ear, directly below the vestibular window; also called the *round window*.

coelom (se′lom) The abdominal cavity.

collateral A small branch of a blood vessel or nerve fiber.

colon (ko′lon) The first portion of the large intestine.

common bile duct A tube formed by the union of the hepatic duct and cystic duct that transports bile to the duodenum.

compact (dense) bone Tightly packed bone, superficial to spongy bone and covered by the periosteum.

conceptus (kon-sep′tus) The product of conception at any point between fertilization and birth. It includes the embryo or fetus, as well as the extraembryonic structures.

conduction myofibers Specialized cardiac muscle fibers that conduct electrical impulses from the AV bundle into the ventricular walls; also called *Purkinje fibers*.

condyle (kon′dīl) A rounded prominence at the end of a long bone, most often for articulation with another bone.

cone A type of photoreceptor cell in the retina of the eye that provides for color vision.

congenital (kon-jen′ĭ-tal) Present at the time of birth.

conjunctiva (con″jungk-ti′vă) The thin membrane covering the anterior surface of the eyeball and lining the eyelids.

connective tissue One of the four basic tissue types within the body. It is a binding and supporting tissue with abundant matrix.

conus medullaris (ko′nus med″yŭ-lar′is) The caudal tapering portion of the spinal cord.

convolution (con-vŏ-loo′shun) **1.** An elevation on the surface of a structure and an infolding of the tissue upon itself. **2.** One of the convex folds on the surface of the brain.

cornea (kor′ne-ă) The transparent convex anterior portion of the outer layer of the eyeball.

cornification (kor″nĭ-fĭ-ka′shun) The drying and flattening of the outer keratinized cells of the epidermis.

coronal (kor′ŏ-nal or kŏ-ro′nal) **plane** A plane that divides the body into anterior and posterior portions; also called a *frontal plane*.

coronary (kor′ŏ-nar″e) **circulation** The flow of blood in the arteries and veins in the wall of the heart; the functional blood supply of the heart.

coronary sinus A large venous channel on the posterior surface of the heart into which the cardiac veins drain.

corpora quadrigemina (kor′por-ă kwad″rĭ-jem′ĭ-nă) Four superior lobes of the midbrain concerned with visual and auditory functions.

corpus callosum (kor′pus kă-lo′sum) A large tract of white matter within the brain that connects the right and left cerebral hemispheres.

cortex (kor′teks) **1.** The outer layer of an internal organ or body structure, as of the kidney or adrenal gland. **2.** The convoluted layer of gray matter that covers the surface of each cerebral hemisphere.

costal (kos′tal) **cartilage** The cartilage that connects the ribs to the sternum.

cranial (kra′ne-al) Pertaining to the cranium.

cranial nerve One of 12 pairs of nerves that arise from the inferior surface of the brain.

cranium (kra′ne-um) The endochondral bones of the skull that enclose or support the brain and the organs of sight, hearing, and balance.

crest A thickened ridge of bone for the attachment of muscle.

cricoid (kri′koid) **cartilage** A ring-shaped cartilage that forms the inferior part of the larynx.

crista (kris′tă) A crest, such as the crista galli that extends superiorly from the cribriform plate.

cubital (kyoo′bĭ-tal) Pertaining to the elbow. The cubital fossa is the anterior aspect of the elbow.

cystic (sis′tik) **duct** The tube that transports bile from the gallbladder to the common bile duct.

cytology (si-tol′ŏ-je) The science dealing with the study of cells.

cytoplasm (si′to-plaz′em) In a cell, the material between the nucleus and the cell (plasma) membrane.

D

deciduous (dĕ-sij′oo-us) Pertaining to something cast off or shed in a particular sequence. Deciduous teeth are shed and replaced by permanent teeth in a predictable sequence.

decussation (de″kus-a′shun) A crossing of nerve fibers from one side of the CNS to the other.

defecation (def″ĕ-ka′shun) The elimination of feces from the rectum through the anal canal and out the anus.

deglutition (de″gloo-tish′un) The act of swallowing.

dendrite (den′drīt) A nerve cell process that transmits impulses toward the cell body of a neuron.

dentin (den′tin) The main substance of a tooth, covered by enamel over the crown of the tooth and by cementum on the root.

dentition (den-tish-un) The type, number, and arrangement of a set of teeth.

dermal papilla (pă-pil′ă) A projection of the dermis into the epidermis.

dermis (der′mis) The second, or deep, layer of skin beneath the epidermis.

descending colon That segment of the large intestine that descends on the left side from the level of the spleen to the level of the left iliac crest.

diaphragm (di′ă-fram) A sheetlike dome of muscle and connective tissue that separates the thoracic and abdominal cavities.

diaphysis (di-af′ĭ-sis) The body (shaft) of a long bone.

diastole (di-as′tŏ-le) The sequence of the cardiac cycle in which a heart chamber wall relaxes and the chamber fills with blood; especially ventricular relaxation.

diencephalon (di″en-sef′ă-lon) That part of the brain lying between the telencephalon and the mesencephalon. It includes the third ventricle, pineal gland, epithalamus, thalamus, hypothalamus, and pituitary gland.

digestion The process by which food is broken down mechanically and chemically into simple molecules that can be absorbed and used by body cells.

diploe (dip′lo-e) The spongy layer of bone positioned between the inner and outer layers of compact bone.

distal (dis′tal) Away from the midline or origin; the opposite of *proximal*.

dorsal (dor′sal) Pertaining to the back, or posterior, portion of a body part; the opposite of *ventral*.

dorsiflexion (dor″sĭ-flek′shun) Movement at the ankle or wrist as the dorsum of the foot or hand is elevated.

ductus arteriosus (duk′tus ar-te″re-o′sus) The blood vessel that connects the pulmonary trunk and the aorta in a fetus.

ductus deferens (def′er-enz); plural, *ductus deferentia* A tube that carries spermatozoa from the epididymis to the ejaculatory duct; also called the *vas deferens* or *seminal duct*.

ductus venosus (ven-o′sus) A fetal blood vessel that connects the umbilical vein and the inferior vena cava.

duodenum (doo″ŏ-de′num or doo-od′ĕ-num) The first portion of the small intestine that leads from the pylorus of the stomach to the jejunum.

dura mater (door′ă ma′ter) The outermost meninx.

E

eccrine (ek′rin) **gland** Any of the numerous sweat glands distributed over the body that function in thermoregulation.

ECG Electrocardiogram; EKG.

ectoderm (ek′tŏ-derm) The outermost of the three primary germ layers of an embryo.

edema (ĕ-de′mă) Abnormal accumulation of interstitial (tissue) fluid, causing the tissue to swell.

effector (ĕ-fek′tor) An organ, such as a gland or muscle, capable of responding to a stimulus.

efferent (ef′er-ent) Conveying away from the center of an organ or structure.

efferent glomerular arteriole (ar-te′re-ōl) A vessel within the kidney that conducts blood away from the glomerulus to the peritubular capillaries.

efferent ductules (duk′toolz) A series of coiled tubules that convey spermatozoa from the rete testis to the epididymis.

efferent neuron (noor′on) A motor nerve cell that conducts impulses from the central nervous system to effector organs, such as muscles or glands.

ejaculation (ĕ-jak″yoo-la′shun) The discharge of semen from the male urethra during climax.

ejaculatory (ĕ-jak′yoo-lă-tor-e) **duct** A tube that transports spermatozoa from the ductus deferens to the prostatic part of the urethra.

elastic fibers Protein strands found in certain connective tissue that have contractile properties.

elbow The synovial joint between the brachium and the antebrachium.

electrocardiogram (ĕ-lek″tro-kar′de-ŏ-gram) A recording of the electrical activity that accompanies the cardiac cycle; ECG or EKG.

electroencephalogram (ĕ-lek″tro-en-sef′ă-lŏ-gram) A recording of the brain-wave patterns or electrical impulses of the brain; EEG.

electromyogram (ĕ-lek″tro-mi′ŏ-gram) A recording of the electrical impulses or activity of a muscle; EMG.

embryology (em″bre-ol′ŏ-je) The study of prenatal development from conception through the eighth week in utero.

enamel (ĕ-nam′el) The outer dense substance covering the crown of a tooth.

endocardium (en″do-kar′de-um) The endothelial lining of the heart chambers and valves.

endochondral (en″do-kon′dral) **bone** Bones that develop as hyaline cartilage models first, and then are ossified.

endocrine (en′dŏ-krin) **gland** A ductless, hormone-producing gland of the endocrine system.

endoderm (en′dŏ-derm) The innermost of the three primary germ layers of a developing embryo.

endolymph (en′dŏ-limf) A fluid within the cochlear duct and membranous labyrinth of the inner ear that aids in the conduction of vibrations involved in hearing and maintaining equilibrium.

endometrium (en″do-me′tre-um) The inner lining of the uterus.

endomysium (en-do-mis′e-um) The connective tissue sheath surrounding individual skeletal muscle fibers, separating them from one another.

endoneurium (en″do-nyoo′re-um) The connective tissue sheath surrounding individual nerve fibers, separating them from one another within a peripheral nerve.

endoplasmic reticulum (en″do-plaz′mik rĕ-tik′yoo-lum) **(ER)** A system of interconnected tubules or channels running through the cytoplasm of a cell. Rough ER bears ribosomes on its membrane, while smooth ER does not.

endothelium (en″do-the′le-um) The layer of epithelial tissue that forms the thin inner lining of blood vessels and heart chambers.

eosinophil (e″ŏ-sin′ŏ-fil) A type of white blood cell characterized by the presence of cytoplasmic granules that become stained by acidic eosin dye. Eosinophils normally constitute from 2% to 4% of the white blood cells.

epicardium (ep″ĭ-kar′de-um) The thin outer layer of the heart; also called the *visceral pericardium*.

epicondyle (ep″ĭ-kon′dīl) A projection of bone above a condyle.

epidermis (ep″ĭ-der′mis) The outermost layer of the skin, composed of stratified squamous epithelium.

epididymis (ep″ĭ-did′i-mis); plural, epididymides (ep″ĭ-dĭ-dim′ĭdēz) A highly coiled tube located along the posterior border of the testis. It stores spermatozoa and transports them from the seminiferous tubules of the testis to the ductus deferens.

epidural (ep″ĭ-door′al) **space** A space between the spinal dura mater and the bone of the vertebral canal.

epiglottis (ep″ĭ-glot′is) A leaflike structure positioned on top of the larynx. It covers the glottis during swallowing.

epimysium (ep″ĭ-mis′e-um) The fibrous outer sheath of connective tissue surrounding a skeletal muscle.

epinephrine (ep″ĭ-nef′rin) A hormone secreted from the adrenal medulla whose actions are similar to those initiated by sympathetic nervous system stimulation; also called *adrenaline*.

epineurium (ep″ĭ-nyoo′re-um) The fibrous outer sheath of connective tissue surrounding a nerve.

epiphyseal (ep″ĭ-fiz′e-al) **plate** A layer of hyaline cartilage between the epiphysis and diaphysis of a long bone. It is responsible for the lengthwise growth of long bones.

epiphysis (ĕ-pif′ĭ-sis) The end segment of a long bone, separated from the diaphysis early in life by an epiphyseal plate, but later becoming part of the larger bone.

episiotomy (ĕ-pe″ze-ot′ŏ-me) An incision of the perineum at the end of the second stage of labor to facilitate delivery and avoid tearing of the perineum.

epithelial (ep″ĭ-the′le-al) **tissue** One of the four basic tissue types; the type of tissue that covers or lines all exposed body surfaces; also called *epithelium*.

eponychium (ep″ŏ-nik′e-um) The thin layer of stratum corneum of the epidermis of the skin that overlaps and protects the lunula of the nail.

erythrocyte (ĕ-rith′rŏ-sīt) A red blood cell.

esophagus (ĕ-sof′ă-gus) The tubular portion of the GI tract that leads from the pharynx to the stomach as it passes through the thoracic cavity.

estrogen (es′trŏ-jen) Any of several female sex hormones secreted from the ovarian (graafian) follicle.

etiology (e″te-ol′ŏ-je) The study of cause, especially of disease, including the origin and what pathogens, if any, are involved.

eustachian (yoo-sta′shun) **canal** *See* auditory tube.

eversion (e-ver′zhun) Movement of the foot in which the sole is turned outward; the opposite of *inversion*.

exocrine (ek′sŏ-krin) **gland** A gland that secretes its product to an epithelial surface, either directly or through ducts.

expiration The process of expelling air from the lungs through breathing out; also called *exhalation*.

extension Movement that increases the angle between parts of a joint; the opposite of *flexion*.

external (superficial) Located on or toward the surface.

external acoustic meatus (me-a′tus) An opening through the temporal bone that connects with the tympanum and the middle-ear chamber and through which sound vibrations pass; also called the *external acoustic meatus*.

external ear The outer portion of the ear, consisting of the auricle (pinna) and the external acoustic canal.

exteroceptor (ek″ster-o-sep′tor) A specialized sensory neuron located near the surface of the body that responds to stimuli from the external environment.

extraocular (ek″stră-ok′yŭ-lar) **muscles** The muscles that insert onto the sclera of the eye and act to change the position of the eye in its orbit (as opposed to the intraocular muscles, such as those of the iris and ciliary body within the eye).

extrinsic (eks-trin′sik) Pertaining to an outside or external origin.

F

face **1.** The anterior aspect of the head not supporting or covering the brain. **2.** The exposed surface of a structure.

facet (fas′et) A flattened, shallow articulating surface on a bone.

falciform (fal′sĭ-form) **ligament** The extension of parietal peritoneum that separates the right and left lobes of the liver.

fallopian (fă-lo′pe-an) **tube** *See* uterine tube.

false vocal cords *See* vestibular folds.

falx cerebelli (falks ser″ĕ-bel′i) A fold of dura mater anchored to the occipital bone. It projects inward between the cerebellar hemispheres.

falx cerebri (ser′ĕ-bri) A fold of dura mater anchored to the crista galli. It extends between the right and left cerebral hemispheres.

fascia (fash′e-ă) A tough sheet or band of fibrous connective tissue binding the skin to underlying muscles (superficial fascia) or supporting and separating muscles (deep fascia).

fasciculus (fă-sik′yŭ-lus) A small bundle of muscle or nerve fibers ensheathed in connective tissue.

fauces (faw′sēz) The passageway between the mouth and the oropharynx.

feces (fe′sēz) Material expelled from the GI tract by way of the anal canal during defecation; composed of food residue, bacteria, and secretions; also called *stool.*

fetus (fe′tus) A prenatal human after 8 weeks of development.

fibroblast (fi′bro-blast) An elongated connective tissue cell with cytoplasmic extensions that is capable of forming collagenous or elastic fibers.

fibrous joint A type of articulation bound by fibrous connective tissue that allows little or no movement (e.g., a syndesmosis).

filiform papillae (fil′ĭ-form pă-pil′e) Numerous small projections scattered irregularly over the entire surface of the tongue in which taste buds are absent.

filum terminale (fi′lum ter-mĭ-nal′e) A fibrous, threadlike continuation of the pia mater, that extends inferiorly from the terminal end of the spinal cord to the coccyx.

fimbriae (fim′bre-e) Fringelike extensions from the borders of the open end of the uterine tube.

fissure (fish′ur) A groove or narrow cleft that separates two parts, such as the cerebral hemispheres of the brain.

flagellum (flă-jel′um) A whiplike structure that provides motility for spermatozoa.

flexion Movement that decreases the angle between parts of a joint; the opposite of *extension*.

fontanel (fon″tă-nel′) A soft membranous gap between the incompletely formed cranial bones of a fetus or baby; commonly known as a "soft spot."

foot The terminal portion of the lower extremity, consisting of the tarsus, metatarsus, and digits.

foramen (fŏ-ra′men); plural, *foramina* (fŏ-ram′ĭ-nă) An opening in an anatomical structure, usually in a bone, for the passage of a blood vessel or nerve.

foramen ovale (o-val′e) An opening through the interatrial septum of the fetal heart.

forearm That portion of the upper extremity between the elbow and the wrist; also called the *antebrachium*.

formed elements The cellular portion of blood.

fornix (for′niks) **1.** A recess surrounding the cervix of the uterus where it protrudes into the vagina. **2.** A tract within the brain connecting the hippocampus with the mammillary bodies.

fossa (fos′ă) A depressed area, usually on a bone.

fourth ventricle (ven′trĭ-kul) A cavity within the brain, between the cerebellum and the medulla oblongata and pons, containing cerebrospinal fluid.

fovea centralis (fo′ve-ă sen-tra′lis) A depression on the macula lutea of the eye where only cones are located; the area of keenest vision.

frenulum (fren′yŭ-lum) A membranous structure that serves to anchor and limit the movement of a body part.

frontal **1.** Pertaining to the forehead or frontal bone. **2.** Pertaining to the frontal plane.

frontal plane *See* coronal plane.

fungiform papillae (fun′jĭ-form pă-pil′e) Flattened, mushroom-shaped projections interspersed among the filiform type on the surface of the tongue. Fungiform papillae contain taste buds.

G

gallbladder A pouchlike organ attached to the underside of the liver; serves as a storage reservoir for bile secreted by the liver.

gamete (gam′ēt) A haploid sex cell; either an egg cell or a sperm cell.

ganglion (gang′gle-on) An aggregation of nerve cell bodies located outside the central nervous system.

gastrointestinal tract (GI tract) A continuous tube through the anterior (ventral) body cavity that extends from the mouth to the anus; also called the *digestive tract*.

gene That portion of the DNA of a chromosome containing the information needed to synthesize a particular protein molecule.

gingiva (jin-ji′vă) The fleshy covering over the mandible and maxilla through which the teeth protrude within the mouth; also called the *gum*.

gland A cell, tissue, or organ that produces a specific substance or secretion for use in or for elimination from the body.

glans penis (glanz pe′nis) The cone-shaped terminal portion of the penis, formed from the expanded corpus spongiosum.

gliding joint A type of synovial joint in which the articular surfaces are flat, permitting only side-to-side and back-and-forth movements.

glomerular (glo-mer′yŭ-lar) **capsule** The double-walled proximal portion of a renal tubule that encloses the glomerulus of a nephron; also called *Bowman's capsule*.

glomerulus (glo-mer′yŭ-lus) The coiled tuft of capillaries surrounded by the glomerular capsule. It filters urine from the blood.

glottis (glot′is) A slitlike opening into the larynx, positioned between the vocal folds.

goblet cell A mucus-secreting unicellular gland associated with columnar epithelia; also called a *mucous cell*.

Golgi (gol′je) **complex** A network of stacked, flattened membranous sacs within the cytoplasm of a cell. It serves to concentrate and package proteins for secretion from the cell.

Golgi tendon organ *See* neurotendinous receptor.

gomphosis (gom-fo′sis) A fibrous joint between the root of a tooth and the periodontal ligament of the tooth socket.

gonad (go′nad) A reproductive organ—testis or ovary—that produces gametes and sex hormones.

gray matter The region of the central nervous system composed of nonmyelinated nerve tissue.

greater omentum (o-men′tum) A double-layered peritoneal membrane that originates on the greater curvature of the stomach. It hangs inferiorly, like an apron, over the contents of the abdominal cavity.

gross anatomy The branch of anatomy concerned with structures of the body that can be studied without a microscope.

gustation (gus-ta′shun) The sense of taste.

gut The GI tract or a portion thereof; generally used in reference to the embryonic digestive tube, consisting of the foregut, midgut, and hindgut.

gyrus (gi′rus) A convoluted elevation or ridge.

H

hair A threadlike appendage of the epidermis consisting of keratinized dead cells that have been pushed up from a dividing basal layer.

hair cells Specialized sensory cells, such as in the spiral organ, possessing numerous surface microvilli; receptors for the senses of hearing and equilibrium.

hair follicle A tubular depression in the dermis of the skin in which a hair develops.

hand The terminal portion of the upper extremity, consisting of the carpus, metacarpus, and digits.

hard palate (pal′it) The bony partition between the oral and nasal cavities, formed by the maxillae and palatine bones and lined by mucous membrane.

haustra (haws′tră) Sacculations, or pouches, of the colon.

haversian (hă-ver′zhun) **canal** *See* central canal.

haversian system *See* osteon.

head The uppermost portion of a human that contains the brain and major sense organs.

heart A four-chambered muscular pumping organ positioned in the thoracic cavity, slightly to the left of midline.

hemoglobin (he′mŏ-glo″bin) A substance in red blood cells consisting of the protein globin and the iron-containing red pigment heme. It functions in the transport of oxygen and carbon dioxide.

hemopoiesis (hem″ŏ-poi-e′sis) The production of red blood cells.

hepatic (hĕ-pat′ik) **duct** A duct formed from the union of several bile ducts that drain bile from the liver. It merges with the cystic duct from the gallbladder to form the common bile duct.

hepatic portal circulation The return of venous blood from the digestive organs and spleen through a capillary network within the liver before draining into the heart.

hepatopancreatic ampulla (hep″ă-to-pan″kre-at′ik am-pool′ă) A short ductule formed by the combined pancreatic and common bile ducts within the duodenal papilla; also called the *ampulla of Vater*.

hiatus (hi-a'tus) An opening or fissure; a foramen.

hilum (hi'lum) A concave or depressed area where vessels or nerves enter or exit an organ.

hinge joint A type of synovial joint in which the convex surface of one bone fits into the concave surface of another, confining movement to one plane (e.g., the knee and interphalangeal joints).

histology (his-tol'ŏ-je) Microscopic study of the structure and function of tissues.

horizontal (transverse) plane A directional plane that divides the body, an organ, or an appendage into superior and inferior or proximal and distal portions.

hormone (hor'mōn) A chemical substance produced in an endocrine gland and secreted into the bloodstream that acts on target cells to produce a specific effect.

hyaline (hi'ă-lin) **cartilage** A cartilage with a homogeneous matrix. It is the most common type, occurring at the articular ends of bones, in the trachea, and within the nose. Most of the bones in the body are formed from hyaline cartilage.

hymen (hi'men) A developmental remnant (vestige) of membranous tissue that partially covers the vaginal opening.

hyperextension Extension beyond the normal anatomical position of 180°.

hypertension Elevated or excessive blood pressure.

hypodermis (hi"pŏ-der'mis) Subcutaneous tissue that binds the dermis to underlying organs.

hyponychium (hi"pŏ-nik'e-um) A thickened supportive layer of the stratum corneum at the distal end of a digit, under the free edge of the nail.

hypothalamus (hi"po-thal'ă-mus) An important autonomic and neuroendocrine control center located below the thalamus within the diencephalon.

I

ileocecal (il'e-ŏ-se'kal) **valve** A modification of the mucosa at the junction of the small intestine and large intestine that forms a one-way passage and prevents the backflow of food materials.

ileum (il'e-um) The terminal portion of the small intestine between the jejunum and cecum.

incus (ing'kus) The middle of three auditory ossicles within the middle-ear chamber; commonly known as the "anvil."

inferior vena cava (ve'nă kă'vă) A large systemic vein that collects blood from the body regions inferior to the level of the heart and returns it to the right atrium.

infundibulum (in"fun-dib'yoo-lum) The stalk that attaches the pituitary gland to the hypothalamus of the brain.

ingestion The process of taking food or liquid into the body by way of the oral cavity.

inguinal (ing'gwĭ-nal) Pertaining to the groin region.

inguinal canal The passage in the abdominal wall through which a testis descends into the scrotum.

inner ear The innermost region of the ear, containing the cochlea and vestibular organs.

insertion The more movable attachment of a muscle, usually the more distal.

inspiration The act of breathing air into the pulmonary alveoli of the lungs; also called *inhalation*.

insula (in'sŭ-lă) A cerebral lobe lying deep to the lateral sulcus. It is covered by portions of the frontal, parietal, and temporal lobes.

integument (in-teg'yoo-ment) The skin; the largest organ of the body.

intercalated (in-ter'kă-lāt-ed) **disc** A thickened portion of the sarcolemma that extends across a cardiac muscle fiber and delimits the boundary between cells.

intercellular substance The matrix or material between cells that largely determines tissue types.

internal (deep) Toward the center, away from the surface of the body.

interstitial fluid Fluid between the cells; also called *tissue fluid*.

intervertebral (in"ter-ver'tĕ-bral) **disc** A pad of fibrocartilage positioned between the bodies of adjacent vertebrae.

intestinal gland A simple tubular digestive gland that opens onto the surface of the intestinal mucosa and secretes digestive enzymes; also called the *crypt of Lieberkühn*.

intramembranous (in"tră-mem'bră-nus) **ossification** A type of bone formation in which layers of bone are formed from mesenchymal cells without any cartilage model.

intrinsic (in-trin'zik) Situated in or pertaining to internal origin.

inversion Movement of the foot in which the sole is turned inward; the opposite of *eversion*.

iris (i'ris) The pigmented portion of the vascular tunic of the eye that surrounds the pupil and regulates its diameter.

islets of Langerhans (i'letz of lang'er-hanz) *See* pancreatic islets.

isthmus (is'mus) A narrow neck or portion of tissue connecting two structures.

J

jejunum (jĕ-joo'num) The middle portion of the small intestine, located between the duodenum and the ileum.

joint The point of juncture between two bones; an articulation.

joint capsule The fibrous tissue that encloses the joint cavity of a synovial joint.

K

keratin (ker'ă-tin) An insoluble protein present in the epidermis and in epidermal derivatives, such as hair and nails.

kidney (kid'ne) One of the paired organs of the urinary system that, among other functions, filters wastes from the blood in the formation of urine.

kinesiology (kĭ-ne"se-ol'ŏ-je) The study of body movement.

knee The region in the lower extremity between the thigh and the leg that contains a synovial hinge joint.

L

labial frenulum (la'be-al fren'yŭ-lum) A longitudinal fold of mucous membrane that attaches the lips to the gum along the midline of both the upper and lower lip.

labia majora (la'be-ă mă-jor'ă); singular, *labium majus* The portion of the female external genitalia consisting of two longitudinal folds of skin that extend downward and backward from the mons pubis.

labia minora (mĭ-nor'ă), singular, *labium minus* Two small folds of skin, devoid of hair and sweat glands, lying between the labia majora of the female external genitalia.

labyrinth (lab'ĭ-rinth) A complex system of interconnecting tubes within the inner ear that includes the semicircular ducts and the cochlear and vestibular labyrinths.

lacrimal canaliculus (lak'rĭ-mal kan"ă-lik'yŭ-lus) A drainage duct for tears at the medial corner of the eyelid. It conveys the tears medially into the nasolacrimal sac.

lacrimal gland A tear-secreting gland on the superior lateral portion of the eyeball, underneath the upper eyelid.

lactation (lak-ta'shun) The production and secretion of milk by the mammary glands.

lacteal (lak'te-al) A small lymphatic duct associated with a villus of the small intestine.

lacuna (lă-kyoo'nă) A small, hollow chamber that houses an osteocyte in mature bone tissue or a chondrocyte in cartilage tissue.

lambdoid suture (lam'doid soo'chur) The immovable joint in the skull between the parietal bones and the occipital bone.

lamella (lă-mel'ă) A concentric ring of matrix surrounding the central canal in an osteon of mature bone tissue.

lamellated corpuscle (lam'ĕ-la'ted kor'pus'l) A sensory receptor for pressure, found in tendons, around joints, and in visceral organs; also called a *pacinian corpuscle*.

lamina (lam'ĭ-nă) A thin plate of bone that extends superiorly from the body of a vertebra to form both sides of the arch of a vertebra.

lanugo (lă-noo'go) Short, silky fetal hair, which may be present for a short time on a premature infant.

large intestine The last major portion of the GI tract, consisting of the cecum, colon, rectum, and anal canal.

laryngopharynx (lă-ring"go-far'ingks) The lowermost portion of the pharynx, extending inferiorly from the level of the hyoid bone to the larynx.

larynx (lar'ingks) The short passageway between the pharynx and trachea that houses the vocal folds (cords); commonly known as the "voice box."

lateral Pertaining to the side; farther from the median plane.

lateral ventricle (ven'trĭ-kul) A cavity within the cerebral hemisphere of the brain that is filled with cerebrospinal fluid.

leg That portion of the lower extremity extending from the knee to the ankle.

lens **1.** A transparent refracting medium, usually made of glass or plastic. **2.** The transparent biconvex organ of the eye lying posterior to the pupil and iris. It focuses light rays entering through the pupil to form an image on the retina.

lesser omentum (o-men′tum) A peritoneal fold of tissue extending from the lesser curvature of the stomach to the liver.

leukocyte A white blood cell; variant spelling, *leucocyte*.

ligament (lig′ă-ment) A tough cord or fibrous band of connective tissue that binds bone to bone to strengthen a joint and provide flexibility. It also may support viscera.

limbic (lim′bik) **system** A group of deep brain structures encircling the brain stem, including the cingulate gyrus, the hypothalamus, the hippocampus, the amygdaloid nucleus, and various fiber tracts. The limbic system is associated with aspects of emotion and behavior and autonomic functions.

linea alba (lin′e-ă al′bă) A vertical fibrous band running down the center of the anterior abdominal wall.

lingual frenulum (ling′gwal fren′yŭ-lum) A longitudinal fold of mucous membrane that attaches the tongue to the floor of the oral cavity.

lipid (lip′id) Any of a group of organic molecules, including fats, phospholipids, and steroids, that are generally insoluble in water.

liver A large visceral organ lying inferior to the diaphragm in the right hypochondriac region. The liver detoxifies the blood and modifies the concentration of glucose, triglycerides, ketone bodies, and proteins in the blood plasma.

loop of Henle *See* nephron loop.

lower extremity A lower appendage, including the hip, thigh, knee, leg, and foot.

lumbar (lum′bar) Pertaining to the region of the loins; the part of the back between the thorax and pelvis.

lumbar plexus (plek′sus) A network of nerves formed by the anterior branches of spinal nerves L1 through L4.

lumen (loo′men) The space within a tubular structure through which a substance passes.

lung Either of the two major organs of respiration positioned within the thoracic cavity on either lateral side of the mediastinum.

lunula (loo′nyoo-lă) The crescent-shaped whitish area at the proximal portion of a nail.

luteinizing (loo′te-ĭ-ni″zing) **hormone (LH)** A hormone secreted by the adenohypophysis (anterior lobe) of the pituitary gland that stimulates ovulation and progesterone secretion by the corpus luteum, influences milk secretion by the mammary glands in females, and stimulates testosterone secretion by the testes in males.

lymph (limf) A clear, plasmalike fluid that flows through lymphatic vessels.

lymph node A small ovoid mass of reticular tissue located along the course of lymph vessels.

lymphocyte (lim′fo-sīt) A type of white blood cell characterized by a granular cytoplasm. Lymphocytes usually constitute about 20%–25% of the white cell count.

lymphoid tissue A type of connective tissue dominated by lymphocytes.

M

macrophage (mak′ro-fāj) A wandering phagocytic cell.

macula lutea (mak′yŭ-lă loo′te-ă) A yellowish depression in the retina of the eye that contains the fovea centralis, the area of keenest vision.

malignant Threatening to life; virulent. Of a tumor, cancerous, tending to metastasize.

malleus (mal′e-us) The largest and outermost of three auditory ossicles; attached to the tympanic membrane and articulating with the incus; commonly known as the "hammer."

mammary (mam′er-e) **gland** The gland of the female breast responsible for lactation and nourishment of the young.

marrow (mar′o) The soft connective tissue found within the inner cavity of certain bones that produces red blood cells.

mastication (mas″tĭ-ka′shun) The chewing of food.

matrix (ma′triks) The intercellular substance of a tissue.

meatus (me-a′tus) A passageway or opening into a structure.

mechanoreceptor (mek″ă-no-re-sep′tor) A sensory receptor that responds to a mechanical stimulus.

medial Toward or closer to the midline of the body.

mediastinum (me″de-ă-sti′num) **1.** A septum or cavity between two principal portions of an organ. **2.** The region in the center of the thorax separating the lungs; contains the heart and all of the thoracic viscera except the lungs.

medulla (mĕ-dul′ă) The innermost part; the central portion of such organs as the adrenal gland and the kidney.

medulla oblongata (ob″long-gă′tă) That portion of the brain stem between the spinal cord and the pons.

medullary (med′l-er″e) **(marrow) cavity** The hollow core of the diaphysis of a long bone in which marrow is found.

meiosis (mi-o′sis) A specialized type of cell division by which gametes, or haploid sex cells, are formed.

Meissner's corpuscle (mīs′nerz kor′pus′l) A sensory receptor found in the papillary layer of the dermis of the skin; responsible for fine, discriminative touch.

melanin (mel′ă-nin) A dark pigment found within the epidermis or epidermal derivatives of the skin.

melanocyte (mel′ă-no-sīt) A specialized melanin-producing cell found in the deepest layer of epidermis.

melanoma (mel″ă-no′mă) A dark malignant tumor of the skin; frequently forms in moles.

membranous (mem′bră-nus) **bone** Bone that forms from membranous connective tissue rather than from cartilage; also called *intramembranous bone*.

menarche (mĕ-nar′ke) The first menstrual discharge; occurs normally between the ages of 9 and 17.

meninges (mĕ-nin′jēz); singular, *meninx* (me′ningks) A group of three fibrous membranes covering the central nervous system, composed of the dura mater, arachnoid, and pia mater.

meniscus (mĕ-nis′kus); plural, *menisci* (mĕ-nis′ki or mĕ-nis′i) A wedge-shaped fibrocartilage in certain synovial joints.

menopause (men″ŏ-pawz) The period marked by the cessation of menstrual periods in the human female.

menstrual (men′stroo′al) **cycle** The rhythmic female reproductive cycle characterized by physical changes in the uterine lining.

menstruation (men″stroo-a′shun) The discharge of blood and tissue from the uterus at the end of menstrual cycle.

mesencephalic aqueduct (mez″en-sĕ-fal′ik ak′wĕ-dukt) The channel that connects the third and fourth ventricles of the brain; also called the *cerebral aqueduct* or the *aqueduct of Sylvius*.

mesencephalon (mez″en-sef′ă-lon) The midbrain, which contains the corpora quadrigemina, the cerebral peduncles, and specialized nuclei that help to control posture and movement.

mesenchyme (mez′en-kīm) An embryonic connective tissue that can migrate, and from which all connective tissues arise.

mesenteric (mes″en-ter′ik) **patches** Clusters of lymph nodes on the walls of the small intestine; also called *Peyer's patches*.

mesentery (mes′en-ter″e) A fold of peritoneal membrane that attaches an abdominal organ to the abdominal wall.

mesoderm (mes′ŏ-derm) The middle layer of the three primary germ layers of the developing embryo.

mesothelium (mes″ŏ-the′le-um) A simple squamous epithelial tissue that lines body cavities and covers visceral organs; also called *serosa*.

mesovarium (mes″ŏ-va′re-um) The peritoneal fold that attaches an ovary to the broad ligament of the uterus.

metabolism (mĕ-tab′ŏ-liz-em) The sum total of the chemical changes that occur within a cell.

metacarpus (met″ă-kar′pus) That region of the hand between the wrist and the digits, including the five bones that support the palm of the hand.

metarteriole (met″ar-te′re-ōl) A small blood vessel that emerges from an arteriole, passes through a capillary network, and empties into a venule.

metastasis (mĕ-tas′tă-sis) **1.** The spread of pathogens or cancerous cells from an original site to one or more sites elsewhere in the body. **2.** A secondary cancerous growth formed by transmission of cancerous cells from a primary growth located elsewhere in the body.

metatarsus (met″ă-tar′sus) The region of the foot between the ankle and the digits that includes five bones.

metencephalon (met″en-sef′ă-lon) The most superior portion of the hindbrain, containing the cerebellum and the pons.

microglia (mi-krog′le-ă) Small phagocytic cells found in the central nervous system.

microvilli (mi″kro-vil′i) Microscopic hairlike projections of the cell membranes of certain epithelial cells.

micturition (mik″tŭ-rish′un) The process of voiding urine; also called *urination*.

midbrain That portion of the brain between the pons and the forebrain.

middle ear The middle region of the ear, containing the three auditory ossicles.

midsagittal (mid-saj′ĭ-tal) **plane** A plane that divides the body or an organ into right and left halves; also called the *median plane*.

mitosis (mi-to'sis) The process of cell division that results in two identical daughter cells, containing the same number of chromosomes.

mitral (mi'tral) **valve** *See* bicuspid valve.

mixed nerve A nerve that contains both motor and sensory nerve fibers.

monocyte (mon'ŏ-sīt) A phagocytic type of white blood cell, normally constituting about 3%–8% of the white blood cell count.

mons pubis (monz pyoo'bis) A fatty tissue pad over the symphysis pubis in the female, covered by coarse pubic hair.

morphogenesis (mor"fo-jen'ĕ-sis) During prenatal development, the transformation involved in the growth and differentiation of cells and tissues.

morula (mor"yŭ-lă) An early stage of embryonic development characterized by a solid ball of cells.

motor area A region of the cerebral cortex from which motor impulses to muscles or glands originate.

motor nerve A nerve composed of motor nerve fibers.

motor neuron (noor'on) A nerve cell that conveys impulses away from the central nervous system to effector organs (muscles and glands). Motor neurons form the anterior roots of the spinal nerves.

motor unit A single motor neuron and the muscle fibers it innervates.

mucosa (myoo-ko'să) A mucous membrane that lines cavities and tracts opening to the exterior.

mucous (myoo'kus) **cell** A specialized unicellular gland that produces and secretes mucus; also called a *goblet cell*.

multipolar neuron (noor'on) A nerve cell with many processes originating from the cell body.

muscle A major type of tissue adapted to contract. The three kinds of muscle are cardiac, smooth, and skeletal.

muscularis (mus"kyŭ-lar'is) A muscular layer or tunic of an organ, composed of smooth muscle tissue.

myelencephalon (mi"ĕ-len-sef'ă-lon) The posterior portion of the hindbrain that contains the medulla oblongata.

myelin (mi'ĕ-lin) A lipoprotein material that forms a sheathlike covering around nerve fibers.

myeloid (mi'ĕ-loid) **tissue** The red bone marrow in which blood cells are produced.

myenteric plexus (mi"en-ter'ik plek'sus) A network of sympathetic and parasympathetic nerve fibers located in the muscular layer of the wall of the small intestine; also called the *plexus of Auerbach*.

myocardium (mi"o-kar'de-um) The middle layer of the heart wall, composed of cardiac muscle.

myofibril (mi"ŏ-fi'bril) A bundle of contractile fibers within muscle cells.

myofilament The filament that constitutes myofibrils. It is composed of either actin or myosin.

myogram (mi'ŏ-gram) A recording of electrical activity within a muscle.

myology (mi-ol'ŏ-je) The science or study of muscle structure and function.

myometrium (mi'o-me'tre-um) The layer or tunic of smooth muscle within the uterine wall.

myoneural (mi"o-noor'al) **junction** The site of contact between the axon of a motor neuron and a muscle fiber.

myopia (mi-o'pe-ă) A visual defect in which objects can be seen distinctly only when very close to the eyes; also called *nearsightedness*.

myosin (mi'ŏ-sin) A thick filament of protein that, together with actin, causes muscle contraction.

N

nail A hardened, keratinized plate that develops from the epidermis and forms a protective covering over the distal phalanges of fingers and toes.

nares (na'rēz) The openings into the nasal cavity; also called the *nostrils*.

nasal cavity A mucosa-lined space above the oral cavity, divided by a nasal septum. It is the first chamber of the respiratory system.

nasal concha (kong'kă); plural, *conchae* (kong'ke) A scroll-like bone extending medially from the lateral wall of the nasal cavity; also called a *turbinate*.

nasal septum A bony and cartilaginous partition that separates the nasal cavity into two portions.

nasopharynx (na"zo-far'ingks) The first or uppermost portion of the pharynx, positioned behind the nasal cavity and extending down to the soft palate.

neck **1.** Any constricted portion, such as the neck of an organ. **2.** The cervical region of the body between the head and thorax.

necrosis (nĕ-kro'sis) Cell death or tissue death as a result of disease or trauma.

neonatal (ne"o-na'tal) Concerning the period from birth to the end of 4 weeks.

nephron (nef'ron) The functional unit of the kidney, consisting of a glomerulus, glomerular capsule, convoluted tubules, and the nephron loop.

nephron loop The U-shaped part of the nephron, consisting of descending and ascending limbs; also called the *loop of Henle*.

nerve A bundle of nerve fibers outside the central nervous system.

neurilemma (noor"ĭ-lem'ă) A thin, membranous covering surrounding the myelin sheath of a nerve fiber.

neurofibril (noor'ŏ-fi'bril) One of many delicate threadlike structures within the cytoplasm of a cell body and the axon hillock of a neuron.

neurofibril node A gap in the myelin sheath of a nerve fiber; also called the *node of Ranvier*.

neuroglia (noo-rog'le-ă) Specialized supportive cells of the central nervous system; also called *glial cells* or *glia*.

neurohypophysis (noor"ŏ-hi-pof'ĭ-sis) The posterior lobe of the pituitary gland.

neurolemmocyte (noor"ŏ-lem'ŏ-sīt) A specialized neuroglial cell that surrounds an axon fiber of a peripheral neuron and forms the neurilemmal sheath; also called a *Schwann cell*.

neuron (noor'on) The structural and functional cell of the nervous system, composed of a cell body, dendrites, and an axon; also called a *nerve cell*.

neurotendinous (noor"ŏ-ten'din-us) **receptor** A proprioceptor found near the junction of tendons and muscles. It senses muscle tension and acts to prevent overuse of a muscle; also called a *Golgi tendon organ*.

neutrophil (noo'trŏ-fil) A type of phagocytic white blood cell, normally constituting about 60%–70% of the white cell count.

nipple A pigmented, cylindrical projection at the apex of the breast.

Nissl (nis'l) **bodies** *See* chromatophilic substances.

node of Ranvier (rahn-ve-a' or ran'vēr) *See* neurofibril node.

notochord (no'tŏ-kord) A flexible rod of tissue that extends the length of the back of an embryo.

nuclear membrane A double-walled membrane composed of protein and lipid molecules that surrounds the nucleus of a cell.

nucleoplasm (noo'kle-ŏ-plaz"em) The protoplasmic contents of the nucleus of a cell.

nucleus (noo'kle-us) The spherical or oval body within a cell that contains the genetic code.

nucleus pulposus (pul-po'sis) The soft, pulpy core of an intervertebral disc; a remnant of the notochord.

O

olfaction (ol-fak'shun) The sense of smell.

olfactory bulb A ganglion-like expansion of the olfactory tract, lying inferior to the frontal lobe of the cerebrum on either side of the crista galli of the ethmoid bone; receives the olfactory nerves from the nasal cavity.

olfactory tract The tract of axons that conveys impulses from the olfactory bulb to the olfactory portion of the cerebral cortex.

oligodendrocyte (ol"ĭ-go-den'drŏ-sīt) A type of neuroglial cell involved in the formation of the myelin of nerve fibers within the central nervous system.

oocyte (o"ŏ-sit) A developing egg cell.

oogenesis (o"ŏ-jen'ĕ-sis) The process of female gamete formation.

optic Pertaining to the eye.

optic chiasma (ki-az'mă) An X-shaped structure on the inferior aspect of the brain, anterior to the pituitary gland, where there is a partial crossing over of fibers in the optic nerves; also called the *optic chiasm*.

optic disc A small region of the retina where the fibers of the ganglion neurons exit from the eyeball to form the optic nerve; also called the *blind spot*.

optic tract A bundle of sensory axons between the optic chiasma and the thalamus that conveys visual impulses from the photoreceptors within the eye.

oral Pertaining to the mouth.

ora serrata (o'ră sĕ-ra'tă) The jagged peripheral margin of the retina.

organ A structure consisting of two or more tissues that performs a specific function.

organ of Corti (kor'te) *See* spiral organ.

organelle (or"gă-nel') A specialized cellular structure that performs a specific function for the cell as a whole.

organism Any individual life form; either unicellular or multicellular.

orifice (or'ĭ-fis) An opening into a body cavity or tube.

origin The place of muscle attachment—usually the more stationary point or proximal bone; opposite the insertion.

oropharynx (o″ro-far′ingks) The middle portion of the pharynx, located posterior to the oral cavity and extending from the soft palate to the level of the hyoid bone.

ossicle (os′ĭ-kul) One of the three bones of the middle ear; also called the *auditory ossicle*.

ossification (os″ĭ-fĭ-ka′shun) The process of bone tissue formation.

osteoblast (os′te-ŏ-blast) A bone-forming cell.

osteoclast (os′te-ŏ-klast) A cell that erodes or resorbs bone tissue.

osteocyte (os′te-ŏ-sīt) A mature bone cell.

osteology (os″te-ol′ŏ-je) The study of the structure and function of bone and the entire skeleton.

osteon (os′te-on) A group of osteocytes and concentric lamellae surrounding a central canal, constitutes the basic unit of structure in bone tissue; also called a *haversian system*.

oval window *See* vestibular window.

ovarian (o-va′re-an) **follicle** A developing ovum and its surrounding epithelial cells.

ovarian ligament A cordlike connective tissue that attaches the ovary to the uterus.

ovary (o′vă-re) The female gonad in which ova and certain sex hormones are produced.

oviduct (o′vĭ-dukt) The tube that transports ova from the ovary to the uterus; also called the *uterine tube* or *fallopian tube*.

ovulation (ov-yŭ-la′shun) The rupture of an ovarian follicle with the release of an ovum.

ovum (o′vum) The female reproductive cell or gamete; an egg cell.

P

pacinian corpuscle (pă-sin′e-an kor′pus′l) *See* lamellated corpuscle.

palate (pal′at) The roof of the oral cavity.

palatine (pal′ă-tin) Pertaining to the palate.

palmar (pal′mar) Pertaining to the palm of the hand.

palpebra (pal′pĕ-bră) An eyelid.

pancreas (pan′kre-us) A mixed organ in the abdominal cavity that secretes gastric juices into the GI tract and insulin and glucagon into the blood.

pancreatic (pan″kre-at′ik) **duct** A drainage tube that carries pancreatic juice from the pancreas into the duodenum of the hepatopancreatic ampulla.

pancreatic islet A cluster of endocrine gland cells within the pancreas that secretes insulin and glucagon; also called *islet of Langerhans*.

papillae (pă-pil′e) Small, nipplelike projections.

papillary (pap′ĭ-ler-e) **muscle** Muscular projections from the ventricular walls of the heart to which the chordae tendineae are attached.

paranasal sinus (par″ă-na′zal si′nus) An air chamber lined with a mucous membrane that communicates with the nasal cavity.

parasympathetic (par″ă-sim″pă-thet′ik) **division** Pertaining to the division of the autonomic nervous system concerned with activities that, in general, inhibit or oppose the physiological effects of the sympathetic nervous system.

parathyroid (par″ă-thi′roid) **gland** One of four small endocrine glands embedded in the posterior surface of the thyroid gland; secretes parathyroid hormone.

parietal (pă-ri′ĕ-tal) Pertaining to the wall of an organ or cavity.

parietal pleura (ploor′ă) The thin serous membrane attached to the thoracic walls of the pleural cavity.

parotid (pă-rot′id) **gland** One of the paired salivary glands located over the masseter muscle just anterior to the ear and connected to the oral cavity by the parotid duct.

parturition (par″tyoo-rish′un) The act of giving birth to young; childbirth.

pectoral (pek′tŏ-ral) Pertaining to the chest region.

pectoral girdle The portion of the skeleton that supports the upper extremities.

pedicle (ped′ĭ-kul) **1.** A constricted portion or stalk. **2.** The bony process that projects backward from the body of a vertebra, connecting with the lamina on each side.

pelvic (pel′vik) Pertaining to the pelvis.

pelvic girdle That portion of the skeleton to which the lower extremities are attached.

pelvis (pel′vis) A basinlike bony structure formed by the sacrum and ossa coxae.

penis (pe′nis) The male organ of copulation, used to introduce spermatozoa into the female vagina and through which urine passes during urination.

pennate (pen′āt) Pertaining to a skeletal muscle fiber arrangement in which the fibers are attached to tendinous slips in a featherlike pattern.

perforating canal A minute duct by means of which blood vessels and nerves from the periosteum penetrate into compact bone; also called *Volkmann's canal*.

pericardium (per″ĭ-kar′de-um) The protective serous membrane that surrounds the heart.

perichondrium (per″ĭ-kon′dre-um) A sheet of fibrous connective tissue that surrounds some kinds of cartilage.

perikaryon (per″ĭ-kar′e-on) The cell body of a neuron.

perilymph (per″ĭ-limf) A fluid of the inner ear that serves as a conducting medium for the vibrations involved in hearing and maintaining equilibrium.

perimysium (per″ĭ-mis′e-um) Fascia (connective tissue) surrounding a bundle (fascicle) of muscle fibers.

perineum (per″ĭ-ne′um) **1.** The pelvic floor and associated structures. **2.** The external region between the scrotum and anus in a male or between the vulva and anus in a female.

perineurium (per″ĭ-noor′e-um) Connective tissue surrounding a bundle (fascicle) of nerve fibers.

periodontal (per″e-ŏ-don′tal) **membrane** The fibrous connective tissue lining the dental alveoli.

periosteum The fibrous connective tissue covering the outer surface of bone.

peripheral (pĕ-rif′er-al) **nervous system (PNS)** The nerves and ganglia of the nervous system that lie outside the brain and spinal cord.

peristalsis (per″ĭ-stal′sis) Rhythmic contractions of smooth muscle in the walls of various tubular organs by which the contents are forced onward.

peritoneum (per″ĭ-tŏ-ne′um) The serous membrane that lines the abdominal cavity and covers the abdominal visceral organs.

Peyer's (pi′erz) **patches** *See* mesenteric patches.

phalanx (fa′langks); plural, *phalanges* (fă-lan′jēz) A bone of a digit (finger or toe).

pharynx (far′ingks) The organ of the digestive system and respiratory system located at the back of the oral and nasal cavities that extends to the larynx anteriorly and to the esophagus posteriorly; also called the *throat*.

phenotype (fe-nŏ-tīp) Observable features in an individual that result from expression of the genotype.

photoreceptor (fo″to-re-sep′tor) A sensory nerve ending capable of being stimulated by light.

physiology (fiz″e-ol′ŏ-je) The science that deals with the study of body functions.

pia mater (pi′ă ma′ter) The innermost meninx, in direct contact with the brain and spinal cord.

pineal (pin′e-al) **gland** A small cone-shaped gland located in the roof of the third ventricle.

pinna (pin′ă) The fleshy outer portion of the external ear; also called the *auricle*.

pituitary (pĭ-too′ĭ-tar-e) **gland** A small, pea-shaped endocrine gland situated in the sella turcica of the sphenoid bone and connected to the hypothalamus by the infundibulum; consists of anterior and posterior lobes; also called the *hypophysis*.

pivot joint A synovial joint in which the rounded head of one bone articulates with the depressed cup of another, permitting rotational movement.

placenta (plă-sen′tă) The organ of metabolic exchange between the mother and the fetus. It is expelled following birth.

plantar (plan′tar) Pertaining to the sole of the foot.

plasma (plaz′mă) The clear, yellowish fluid portion of blood in which cells are suspended.

platelets (plāt′letz) Small fragments of specific bone marrow cells that function in blood clotting; also called *thrombocytes*.

pleural (ploor′al) Pertaining to the serous membranes associated with the lungs.

pleural cavity The potential space between the visceral pleura and the parietal pleura.

pleural membranes Serous membranes that surround the lungs and provide protection and compartmentalization.

plexus (plek′sus) A network of nerve fibers, blood vessels, or lymphatics.

plexus of Auerbach (ow′er-bak) *See* myenteric plexus.

plexus of Meissner (mīs′ner) *See* submucosal plexus.

plicae circulares (pli′se ser-kyŭ-lar′ēz) Deep folds in the wall of the small intestine that increase the absorptive surface area.

pneumotaxic (noo″mŏ-tak′sik) **area** A portion of the respiratory control center in the pons that has an inhibitory effect on the inspiratory center in the medulla oblongata.

pons (ponz) The portion of the brain stem just above the medulla oblongata and anterior to the cerebellum.

popliteal (pop″lĭ-te′al or pop-lit′e-al) Pertaining to the concave region on the posterior aspect of the knee.

posterior Toward the back; also called *dorsal*.

posterior root An aggregation of sensory neuron fibers lying between a spinal nerve and the posterolateral aspect of the spinal cord; also called the *dorsal root* or *sensory root.*

posterior root ganglion (gang'gle-on) A cluster of cell bodies of sensory neurons located along the posterior root of a spinal nerve; also called a *sensory ganglion.*

postganglionic (pōst"gang-gle-on'ik) **neuron** The second neuron in an autonomic motor pathway. Its cell body is outside the central nervous system, and it terminates at an effector organ.

postnatal After birth.

preganglionic neuron The first neuron in an autonomic motor pathway. Its cell body is within the central nervous system, and it terminates on a postganglionic neuron.

pregnancy The condition in which a female is carrying a developing offspring within the body.

prenatal Existing or occurring before birth.

prepuce (pre'pyoos) A fold of loose retractable skin covering the glans of the penis or clitoris; also called the *foreskin.*

prime mover The muscle most directly responsible for a particular movement.

pronation (pro-na'shun) A rotational movement of the forearm that turns the palm of the hand posteriorly; the opposite of *supination.*

proprioceptor (pro"pre-o-sep'tor) A sensory nerve ending that responds to changes in tension in a muscle or tendon.

prostate (pros'tāt) A walnut-shaped gland surrounding the male urethra just below the urinary bladder. It secretes an additive to seminal fluid during ejaculation.

prosthesis (pros-the'sis) An artificial device to replace a diseased or worn body part.

protein Any of a group of large organic molecules made up of amino acid subunits linked by peptide bonds.

protraction (pro-trak'shun) Forward movement of a body part, such as the mandible, on a plane parallel with the ground; the opposite of *retraction.*

proximal (prok'sĭ-mal) Closer to the midline of the body or origin of an appendage; the opposite of *distal.*

pseudounipolar neuron (soo"do-yoo"nĭ-po'lar noor'on) A nerve cell in which only one process extends from the cell body; results from the fusion of two processes during embryonic development.

puberty (pyoo'ber-te) The period of development in which the reproductive organs become functional and the secondary sex characteristics are expressed.

pulmonary (pul'mŏ-ner"e) Pertaining to the lungs.

pulmonary circulation The circuit of blood flow between the heart and the lungs. Oxygen-poor blood from the right ventricle is oxygenated in the lungs and then returned to the left atrium of the heart.

pulp cavity A cavity within the center of a tooth that contains blood vessels, nerves, and lymphatics.

pupil The opening through the iris that permits light to enter the posterior cavity of the eyeball and be refracted by the lens.

Purkinje (pur-kin'je) **fibers** *See* conduction myofibers.

pyloric sphincter (pi-lor'ik sfingk'ter) A thick ring of smooth muscle encircling the opening between the stomach and the duodenum. It regulates the passage of food material into the small intestine and prevents backflow.

pyramid (pir'ă-mid) Any structure of the body with a pyramidal shape, including the renal pyramids in the kidney and the medullary pyramids on the inferior surface of the brain.

R

ramus (ra'mus) A branch of a bone, artery, or nerve.

raphe (ra'fe) A seamlike line or ridge between two similar parts of a body organ, as in the scrotum.

receptor (re-sep'tor) A sense organ or the specialized distal end of a sensory neuron that receives stimuli from the environment.

rectouterine (rek"to-yoo'ter-in) **pouch** A pocket formed by the deflection of the parietal peritoneum between the uterus and the rectum; also called the *pouch of Douglas* or *Douglas' cul-de-sac.*

rectum (rec'tum) That portion of the GI tract between the sigmoid colon and the anal canal.

red marrow (mar'o) A hematopoietic tissue found within the medullary cavity of certain bones.

red nucleus (noo'kle-us) An aggregation of gray matter of a reddish color located in the upper portion of the midbrain. It sends fibers to certain brain tracts and helps to coordinate muscular movements.

reflex A rapid involuntary response to a stimulus.

reflex arc The basic conduction pathway through the nervous system, consisting of a sensory neuron, association neuron, and a motor neuron.

regional anatomy The division of anatomy concerned with structural arrangement in specific areas of the body, such as the head, neck, thorax, or abdomen.

renal (re'nal) Pertaining to the kidney.

renal corpuscle (kor'pus'l) A glomerular capsule and its enclosed glomerulus; also called the *malpighian corpuscle.*

renal cortex The outer portion of the kidney, primarily vascular.

renal medulla (mĕ-dul'ă) The inner portion of the kidney, including the renal pyramids and renal columns.

renal pelvis The inner cavity of the kidney formed by the expanded ureter, into which the major calyces open.

renal pyramid Any of various pyramidal masses that are seen on longitudinal section of the kidney and that contain part of the nephron loops and the collecting tubules.

respiration The exchange of gases between the external environment and the cells of an organism.

respiratory center The structure or portion of the brain stem that regulates the depth and rate of breathing.

respiratory membrane A thin, moistened membrane within the lungs, composed of an alveolar portion and a capillary portion, through which gaseous exchange occurs.

rete testis (re'te tes'tis) The network of canals at the termination of the seminiferous tubules in the testis that is associated with the production of spermatozoa.

reticular (rĕ-tik'yoo-lar) **formation** A network of nervous tissue fibers in the brain stem that arouses the higher brain centers.

retina (ret'ĭ-nă) The inner layer of the eyeball that contains the photoreceptors.

retraction Backward movement of a body part, such as the mandible, on a plane parallel with the ground; the opposite of *protraction.*

retroperitoneal (ret"ro-per"ĭ-tŏ-ne'al) Positioned behind the parietal peritoneum.

rhythmicity (rith-mis'ĭ-te) **area** A portion of the respiratory control center located in the medulla oblongata. It controls inspiratory and expiratory phases.

ribosome (ri'bŏ-sōm) A cytoplasmic organelle composed of protein and RNA in which protein synthesis occurs.

right lymphatic (lim-fat'ik) **duct** A major vessel of the lymphatic system that drains lymph from the upper right portion of the body into the right subclavian vein.

rod A type of photoreceptor cell in the retina of the eye that is specialized for colorless, dim-light vision.

root canal The tubular extension of the pulp cavity into the root of a tooth. It contains vessels and nerves.

rotation Movement of a bone around its own longitudinal axis.

round window *See* cochlear window.

rugae (roo'je) Folds or ridges of the mucosa of an organ, such are those of the stomach or urinary bladder.

S

saccule (sak'yool) The saclike cavity in the membranous labyrinth within the vestibule of the inner ear that contains a vestibular organ for equilibrium.

sacral (sa'kral) Pertaining to the sacrum.

sacral plexus (plek'sus) A network of nerve fibers arising from spinal nerves L4 through S3. Nerves arising from the sacral plexus merge with those from the lumbar plexus to form the lumbosacral plexus that supplies the lower extremity.

saddle joint A synovial joint in which the articular surfaces of both bones are concave in one plane and convex, or saddle shaped, in the other plane, such as in the distal carpometacarpal joint of the thumb.

sagittal plane A vertical plane through the body that divides it into right and left sides.

salivary (sal'ĭ-ver-e) **gland** An accessory digestive gland that secretes saliva into the oral cavity.

sarcolemma (sar"kŏ-lem'ă) The cell membrane of a muscle fiber.

sarcomere (sar'kŏ-mēr) The portion of a skeletal muscle fiber between a pair of Z lines; the basic subunit of skeletal muscle contraction.

sarcoplasm (sar'kŏ-plaz"em) The cytoplasm within a muscle fiber.

scala tympani (ska'lă tim'pă-ne) The lower channel of the cochlea that is filled with perilymph.

scala vestibuli (vĕ-stib'yŭ-le) The upper channel of the cochlea that is filled with perilymph.

Schwann (shwahn) **cell** *See* neurolemmocyte.

sclera (skler′ă) The outer white layer of fibrous connective tissue that serves as a protective covering for the eyeball.

scleral venous sinus A circular venous drainage for the aqueous humor from the anterior chamber of the eye; located at the junction of the sclera and the cornea; also called the *canal of Schlemm*.

scrotum (skro′tum) The musculocutaneous sac that contains the testes and their accessory organs.

sebaceous (sĕ-ba′shus) **gland** An exocrine gland of the skin that secretes sebum.

sebum (se′bum) An oily waterproofing secretion of sebaceous glands.

semen (se′men) The thick, whitish secretion of the reproductive organs of the male, consisting of spermatozoa and secretions of the testes, seminal vesicles, prostate, and bulbourethral glands.

semicircular canals Tubular channels within the inner ear that contain the receptors for equilibrium; also called *semicircular canals*.

semilunar (sem″e-loo′nar) **valve** Crescent-shaped heart valves, positioned at the entrances to the aorta and the pulmonary trunk.

seminal vesicles (sem′ĭ-nal ves′ĭ-kulz) A pair of male accessory reproductive organs, lying posterior and inferior to the urinary bladder, that secrete additives to spermatozoa into the ejaculatory ducts.

seminiferous (sem″ĭ-nif′er-us) **tubules** Numerous small ducts in the testes, where spermatozoa are produced.

senescence (sĕ-nes′ens) The process of growing old; aging.

sensory area A region of the cerebral cortex that receives and interprets sensory nerve impulses.

sensory neuron (noor′on) A nerve cell that conducts an impulse from a receptor organ to the central nervous system; also called an *afferent neuron*.

septum (sep′tum) A membranous or fleshy wall dividing two cavities.

serous (se′rus) **membrane** An epithelial and connective tissue membrane that lines body cavities and covers visceral organs within these cavities; also called *serosa*.

Sertoli (ser-to′le) **cells** *See* sustentacular cells.

serum (se′rum) Blood plasma with the clotting elements removed.

sesamoid (ses′ă-moid) **bone** A membranous bone formed in a tendon in response to joint stress (e.g., the patella).

shoulder The region of the body where the humerus articulates with the scapula; also called *omo*.

sigmoid colon (sig′moid ko′lon) The S-shaped portion of the large intestine between the descending colon and the rectum.

sinoatrial (si″no-a′tre-al) **node** A mass of specialized cardiac tissue in the wall of the right atrium that initiates the cardiac cycle; the SA node; also called the *pacemaker*.

sinus (si′nus) A cavity or hollow space within a bone or other tissue.

sinusoid A small, blood-filled space in certain organs, such as the spleen or liver.

skeletal muscle A muscle that is connected at either or both extremities with a bone. It consists of elongated, multinucleated, striated skeletal muscle fibers that contract when stimulated by motor nerve impulses.

small intestine The portion of the GI tract between the stomach and the cecum whose function is the absorption of food nutrients; consists of the duodenum, jejunum, and ileum.

smooth muscle A specialized type of nonstriated muscle tissue, composed of spindle-shaped fibers with a single nucleus. It contracts in an involuntary, rhythmic fashion within the walls of visceral organs.

soft palate (pal′at) The fleshy posterior portion of the roof of the mouth, extending from the palatine bones to the uvula.

somatic (so-mat′ik) Pertaining to the nonvisceral parts of the body.

spermatic (sper-mat′ik) **cord** A structure of the male reproductive system that includes the ductus deferens, spermatic vessels, nerves, cremaster muscle, and connective tissue. The spermatic cord extends from a testis to the inguinal ring.

spermatogenesis (sper-mat″ŏ-jen′ĭ-sis) The production of male gametes, or spermatozoa.

spermatogonia (sper-mat″ŏ-go′ne-ă) Sperm stem cells within the seminiferous tubules of the testes; the progenitors of spermatocytes.

spermatozoon (sper-mat″ŏ-zo′on); plural, *spermatozoa* or, loosely, *sperm* A mature male sex cell, or gamete.

sphincter (sfingk′ter) A circular muscle that normally maintains constriction of a body opening or the lumen of a tubular structure and that relaxes as required by normal physiological functioning.

sphincter of ampulla (am-pool′ă) The muscular constriction at the opening of the common bile and pancreatic ducts; also called the *sphincter of Oddi*.

sphincter of Oddi (o′de) *See* sphincter of ampulla.

spinal cord The portion of the central nervous system that extends downward from the brain stem through the vertebral canal.

spinal ganglion (gang′gle-on) A cluster of nerve cell bodies on the posterior root of a spinal nerve.

spinal nerve One of the 31 pairs of nerves that arise from the spinal cord.

spinous (spi′nus) **process** A sharp projection of bone or a ridge of bone, such as on the scapula.

spiral organ The functional unit of hearing, consisting of a basilar membrane that supports receptor hair cells and a tectorial membrane located within the cochlea; also called the *organ of Corti*.

spleen (splēn) A large, blood-filled, glandular organ in the upper left quadrant of the abdomen. It is attached by mesenteries to the stomach.

spongy bone A type of bone with a latticelike structure; also called *cancellous bone*.

squamous (skwa′mus) Flat or scalelike.

stapes (sta′pēz) The innermost of the auditory ossicles of the ear that fits against the vestibular (oval) window of the inner ear; commonly known as the "stirrup."

statoconia (stat″ŏ-ko′ne-ă) Small, hardened particles of calcium carbonate in the saccule and utricle of the inner ear that are associated with the receptors of equilibrium; also called *otoliths*.

stomach A pouchlike digestive organ between the esophagus and the duodenum, lying just beneath the diaphragm.

stratified (strat′ĭ-fīd) Arranged in layers, or strata.

stratum basale (bă-să′le) The deepest epidermal layer, where mitotic activity occurs.

stratum corneum (kor′ne-um) The outer cornified layer of the epidermis of the skin.

stroma (stro′mă) A connective tissue framework in an organ, gland, or other tissue.

subarachnoid (sub″ă-rak′noid) **space** The space within the meninges between the arachnoid and the pia mater, where cerebrospinal fluid flows.

subdural (sub-door′al) **space** The narrow space between the dura mater and the arachnoid.

sublingual (sub-ling′gwal) **gland** One of the three pairs of salivary glands. It is located below the tongue and its duct opens into the floor of the mouth.

submandibular (sub″man-dib′yŭ-lar) **gland** One of the three pairs of salivary glands. It is located below the mandible and its duct opens to the side of the lingual frenulum.

submucosa (sub″myoo-ko′să) A layer of supportive connective tissue that underlies a mucous membrane.

submucosal plexus (plek′sus) A network of sympathetic and parasympathetic nerve fibers located in the submucosa of the small intestine; also called the *plexus of Meissner*.

sudoriferous (soo′dor-if′er-us) **gland** An exocrine gland that excretes perspiration, or sweat, onto the surface of the skin; also called a *sweat gland*.

sulcus (sul′kus) A shallow impression or groove.

superficial Toward or near the surface.

superficial fascia (fash′e-ă) A binding layer of connective tissue between the dermis of the skin and the underlying muscle.

superior Toward the upper part of a structure or toward the head; also called *cephalic*.

superior vena cava (ve′nă ka′vă) A large systemic vein that collects blood from regions of the body superior to the heart and returns it to the right atrium.

supination (soo″pĭ-na′shun) A rotational movement of the forearm that turns the palm of the hand anteriorly; the opposite of *pronation*.

surface anatomy The division of anatomy concerned with the form and markings of the surface of the body as they relate to the deeper structures.

surfactant (sur-fak′tant) A substance secreted by the pulmonary alveolar cells of the lung that reduces the surface tension and the tendency for the pulmonary alveoli to collapse after each expiration.

suspensory (sŭ-spen′sŏ-re) **ligament** **1.** A band of peritoneum that extends laterally from the surface of the ovary to the wall of the pelvic cavity. **2.** A ligament that supports an organ or a body part, such as that supporting the lens of the eye.

sustentacular (sus-ten-tak′yŭ-lar) Specialized cells within the testes that supply nutrients to developing spermatozoa; also called *Sertoli cells* or *nurse cells*.

sutural (soo′chur-al) **bone** A small bone positioned within a suture of certain cranial bones; also called a *wormian bone*.

suture (soo′chur) A type of fibrous joint found between bones of the skull.

sweat gland *See* sudoriferous gland.

sympathetic division Pertaining to the division of the autonomic nervous system concerned with activities that, in general, inhibit or oppose the physiological effects of the parasympathetic nervous system; also called the *thoracolumbar division*.

symphysis (sim′fĭ-sis) A type of joint characterized by a fibrocartilaginous pad between the articulating bones that provides slight movement.

symphysis pubis (pyoo′bis) The slightly movable cartilaginous joint between the two pubic bones of the pelvic girdle.

synapse (sin′aps) The functional junction between two neurons or between a neuron and an effector.

synaptic (sĭ-nap′tik) **cleft** The minute space that separates the axon terminal of one neuron from another neuron or muscle fiber. Neurotransmitter diffuses across the cleft to affect the postsynaptic cell.

synchondrosis (sin″kon-dro′sis) A cartilaginous joint in which the articulating bones are separated by hyaline cartilage.

syndesmosis (sin″des-mo′sis) A type of fibrous joint in which two bones are united by an interosseous ligament.

synergist (sin′er-jist) A muscle that assists the action of the prime mover.

synovial (sĭ-no′ve-al) **cavity** The space between the articulating bones of a synovial joint, filled with synovial fluid; also called a *joint cavity*.

synovial joint A freely movable joint in which a synovial cavity is present between the articulating bones; also called a *diarthrotic joint*.

synovial membrane The inner membrane of a synovial capsule that secretes synovial fluid into the joint cavity.

system A group of body organs that function together.

systemic (sis-tem′ik) Of, relating to, or affecting the organism as a whole.

systemic anatomy The division of anatomy concerned with the structure and function of the various systems of the body.

systemic circulation The circuit of blood flow from the left ventricle of the heart to the entire body and back to the heart via the right atrium; in contrast to the pulmonary circulation, which involves the lungs.

systole (sis′tŏ-le) The muscular contraction of a heart chamber during the cardiac cycle; ventricular contraction, unless otherwise specified.

systolic (sis-tol′ik) **pressure** Arterial blood pressure during the ventricular systolic phase of the cardiac cycle.

T

tactile (tak′til) Pertaining to the sense of touch.

taeniae coli (te′ne-e ko′li) The three longitudinal bands of muscle in the wall of the large intestine.

target organ A tissue or organ that is affected by a particular hormone.

tarsal gland An oil-secreting gland that opens on the exposed edge of each eyelid; also called the *meibomian gland*.

tarsus (tar′sus) Pertaining to the ankle; the proximal portion of the foot that contains the seven tarsal bones.

taste bud An organ containing the chemoreceptors associated with the sense of taste.

tectorial (tek′to′re-al) **membrane** A gelatinous membrane positioned over the hair cells of the spiral organ in the cochlea.

teeth Accessory structures of digestion adapted to cut, shred, crush, and grind food.

telencephalon (tel″en-sef′ă-lon) The anterior portion of the forebrain, constituting the cerebral hemispheres and related parts.

tendo calcaneus (ten′do kal-ka′ne-us) The tendon that attaches the calf muscles to the calcaneus; also called the *Achilles tendon*.

tendon A band of dense regular connective tissue that attaches muscle to bone.

tendon sheath A covering of synovial membrane surrounding certain tendons.

tentorium cerebelli (ten-to′re-um ser″ĕ-bel′i) An extension of dura mater separating the cerebellum from the basal surface of the occipital and temporal lobes of the cerebral cortex.

teratogen (tĕ-rat′ŏ-jen) Any agent or factor that causes a physical defect in a developing embryo or fetus.

testis (tes′tis) The primary reproductive organ of a male that produces spermatozoa and male sex hormones.

thalamus (thal′ă-mus) An oval mass of gray matter within the diencephalon that serves as a sensory relay center.

thigh The proximal portion of the lower extremity between the hip and the knee in which the femur is located.

third ventricle (ven′trĭ-kul) A narrow cavity between the right and left halves of the thalamus and between the lateral ventricles that contains cerebrospinal fluid.

thoracic (thor′ă-sik) Pertaining to the chest region.

thoracic duct A major lymphatic vessel of the body that drains lymph from the entire body, except for the upper right quadrant, and returns it to the left subclavian vein.

thorax (thor′aks) The chest.

thrombocytes (throm′bŏ-sīts) *See* platelets.

thymus (thi′mus) A bilobed lymphoid organ positioned in the upper mediastinum, posterior to the sternum and between the lungs.

thyroid cartilage The largest cartilage of the larynx. Its two broad processes join anteriorly to form the Adam's apple.

thyroid gland An endocrine gland located just below the larynx, in front of the trachea, consisting of two lobes connected by a narrow band of tissue called the isthmus.

tissue An aggregation of similar cells and their binding intercellular substance, joined to perform a specific function.

tongue A protrusible muscular organ on the floor of the oral cavity.

tonsil (ton′sil) A mass of lymphoid tissue embedded in the mucous membrane of the pharynx.

trabeculae (tră-bek′yŭ-le) **1.** Any of the supporting strands of connective tissue projecting into an organ and constituting part of its framework. **2.** Any of the fine spicules forming a network in spongy bone.

trachea (tra′ke-ă) The airway leading from the larynx to the bronchi, composed of cartilaginous rings and a ciliated mucosal lining of the lumen; commonly known as the "windpipe."

tract A bundle of nerve fibers within the central nervous system.

transection A cross-sectional cut.

transverse colon (ko′lon) That portion of the large intestine extending from right to left across the abdomen between the hepatic and splenic flexures.

transverse fissure (fish′ur) The prominent cleft that horizontally separates the cerebrum from the cerebellum.

tricuspid (tri-kus′pid) **valve** The heart valve located between the right atrium and the right ventricle; the right AV valve.

trigone (tri′gōn) A triangular area in the urinary bladder delimited by the openings of the ureters and the urethra.

trochanter (tro-kan′ter) A broad, prominent process on the proximolateral portion of the femur.

trochlea (trok′le-ă) A structure having the shape or function of a pulley, especially the part of the distal end of the humerus that articulates with the ulna.

trunk The thorax and abdomen together; also called *torso*.

tubercle (too′ber-kul) A small elevated process on a bone.

tuberosity (too″bĕ-ros′ĭ-te) An elevation or protuberance on a bone.

tunica albuginea (too′nĭ-ka al″byoo-jin′e-ă) A tough connective tissue sheath surrounding a structure (e.g., the capsule enclosing a testis).

tympanic (tim-pan′ik) **membrane** The membranous eardrum positioned between the outer and middle ear.

U

umbilical (um-bil′ĭ-kal) **cord** A flexible cordlike structure that connects the fetus with the placenta. It contains two umbilical arteries and one vein that transport nourishment to the fetus and remove its waste.

umbilicus (um-bil′ĭ-kus) The site where the umbilical cord was attached to the fetus; also called the *navel*.

upper extremity The appendage attached to the pectoral girdle, consisting of the shoulder, brachium, elbow, antebrachium, and hand.

ureter (yoo-re′ter) A tube that transports urine from the kidney to the urinary bladder.

urethra (yoo-re′thră) A tube that transports urine from the urinary bladder to the outside of the body.

urinary (yoo′rĭ-ner″e) **bladder** A distensible sac that stores urine; situated in the pelvic cavity, posterior to the symphysis pubis.

urogenital (yoo″ro-jen′ĭ-tal) **triangle** The region of the pelvic floor containing the external genitalia.

uterine (yoo′ter-in) **tube** The tube through which the ovum is transported to the uterus; the site of fertilization; also called the *oviduct* or *fallopian tube*.

uterus (yoo′ter-us) The hollow muscular organ in which a fertilized egg implants and develops into a fetus. It is located within the female pelvis between the urinary bladder and the rectum; also called the *womb*.

utricle (yoo′trĭ-kul) An enlarged portion of the membranous labyrinth, located within the vestibule of the inner ear.

uvula (yoo′vyŭ-lă) A fleshy, pendulous portion of the soft palate that blocks the nasopharynx during swallowing.

V

vacuole (vak′yoo-ōl) A small space or cavity within the cytoplasm of a cell.

vagina (vă-ji′nă) A tubular organ leading from the uterus to the vestibule of the female reproductive tract that receives the male penis during coitus.

vallate papillae (val′āt pă-pil′e) The largest of the papillae on the surface of the tongue. They are arranged in an inverted V-shape at the back of the tongue. Vallate papillae contain taste buds.

vasomotor (va-zo-mo′tor) **center** A cluster of nerve cell bodies in the medulla oblongata. It controls the diameter of blood vessels, and therefore has an important role in regulating blood pressure.

vein A blood vessel that conveys blood toward the heart.

vena cava (ve′nă ka′vă) One of two large vessels that return deoxygenated blood to the right atrium of the heart.

ventral (ven′tral) Toward the front or belly surface; the opposite of *dorsal*.

ventricle (ven′trĭ-kul) A cavity within an organ; especially those cavities in the brain that contain cerebrospinal fluid and those in the heart that contain blood to be pumped from the heart.

venule (ven′yool) A small vessel that carries venous blood from capillaries to a vein.

vermiform (ver′mĭ-form) **appendix** *See* appendix.

vermis (ver′mis) The narrow, middle lobular structure that separates the cerebellar hemispheres.

vertebral (ver′tĕ-bral) **canal** The tubelike cavity extending through the vertebral column that contains the spinal cord; also called the *spinal canal*.

vestibular (vĕ-stib′yŭ-lar) **folds** The supporting folds of tissue for the vocal folds within the larynx; also called *false vocal cords*.

vestibular window The oval opening in the bony wall between the middle and inner ear on which the footplate of the stapes rests; also called the *oval window*.

vestibule (ves′tĭ-byool) A space or region at the beginning of a canal, especially that of the nose, inner ear, and vagina.

villus (vil′us) A minute projection from the free surface of a mucous membrane, especially one of the vascular projections of the mucosal layer of the small intestine.

viscera (vis′er-ă) The organs within the abdominal or thoracic cavities.

visceral (vis′er-al) Pertaining to the membranous covering of the viscera.

visceral peritoneum (per″ĭ-tŏ-ne′um) A serous membrane that covers the surfaces of abdominal viscera.

visceral pleura (ploor′ă) A serous membrane that covers the surfaces of the lungs.

visceroceptor (vis″er-ŏ-sep′tor) A sensory receptor located within the visceral organs that responds to internal pain, pressure, stretch, and chemical changes.

vitreous (vit′re-us) **humor** The transparent gel that occupies the posterior cavity, located between the lens and retina of the eyeball.

vocal folds Folds of the mucous membrane in the larynx that produce sound as they are pulled taut and vibrated; also called *vocal cords*.

Volkmann's (folk′manz) **canal** *See* perforating canal.

vulva (vul′vă) The external genitalia of the female that surround the opening of the vagina; also called the *pudendum*.

white matter Bundles of myelinated axons located in the central nervous system.

wormian (wer′me-an) **bone** *See* sutural bone.

yellow marrow (mar′o) Specialized tissue within bone cavities in which lipids are stored.

Z

zygote (zi′gōt) A fertilized egg cell formed by the union of a spermatozoon and an ovulated secondary oocyte (ovum).

Credits

Photographs

Chapter 1

Opener: Corbis-Bettmann; **1.1:** © Mario Caprio/Visuals Unlimited; **1.3:** Published by permission of the Danish National Museum; **1.5:** The Trustees of the British Museum; **1.6: 1.7:** © John Watney/Science/Photo Researchers Inc.; Egyptian National Museum; **1.8:** Bibliotheque Nationale; **1.9, 1.10:** Fratelli Alinari; **1.11:** Courtesy of the New York Academy of Medicine Library; **1.12, 1.13:** Reproduced by gracious permissions of Her Majesty Queen Elizabeth, Royal Art Collection, Great Britain; **1.14:** From the Works of Andrea Vesalius of Brussels by J. Bade, C.M. Saunders and Charley P. O'Malley, pg. 1096, Dover Publications, Inc.; **1.15:** Johan Maurita Van Nassau Mawitsuia: The Hague; **1.16:** © Stock Montage; **1.17*a*:** Debi Stansbaugh; **1.17*b*:** © George Mosil/Visuals Unlimited; **1.17*c*:** © CNRI/SPL/Photo Researchers, Inc.; **1.15*b*:** From R.A. Robb, *Three-Dimensional Biomedical Imaging*, Vol. 1, 1985. Copyright CRC Press, Inc., Boca Raton, FL; **1.18*a*:** Courtesy of Kodak; **1.18*b*:** © Carroll H. Weiss/Camera M.D. Studios; **1.18*c*:** Courtesy of Utah Valley Regional Medical Center, Dept. of Radiation; **1.19*a*:** © Lester V. Bergman & Associates, Inc.; **1.19*b*:** From R.A. Robb, *Three-Dimensional Biomedical Imaging*, Vol. 1, 1985. Copyright CRC Press, Inc., Boca Raton, FL; **1.19*c*:** © Hank Morgan/Science Source/Photo Researchers, Inc.; **1.19*d*:** © Monte S. Buchsbaum, M.D.

Chapter 2

Opener: Corbis-Bettmann; **2.3*b*:** From C.D. Haagensen, *Diseases of the Breast*, 2/e, 1974, W.B. Saunders Company; **2.4:** © Dr. Sheril Burton; **2.7*a*:** © Joan L. Cohen/Photo Researchers, Inc.; **2.7*b*:** © John D. Cunningham/Visuals Unlimited; **2.7*c*:** © James L. Shaffer; **2.7*d*:** © L. Lee Rue III/Photo Researchers, Inc.; **2.7*e*:** © Cyril Toker/Photo Researchers, Inc.; **2.7*f*:** © Tom Hollyman/Photo Researchers, Inc.; **2.11*a*:** © Dr. Sheril Burton; **2.12*a*:** Kent M. Van De Graaff; **2.12*b*:** © A. Glauberman/Photo Researchers, Inc.; **2.12*c*:** © Martin M. Rotker/Photo Researchers, Inc.; **2.13:** © Dr. Sheril D. Burton; **2.15*a*, 2.16*a*:** Kent M. Van De Graaff

Chapter 3

Opener: A. Barrington Brown/Science Source/Photo Researchers, Inc.; **3.5*a*:** Keith R. Porter/Alpers D.H. and Seetharan D. *New England Journal of Medicine* 296/1977-1047; **3.5*b*:** © Per H. Kjeldsen, University of Michigan, Ann Arbor; **3.5*c*:** Visuals Unlimited; **3.6:** © BioPhoto Associates/Photo Researchers, Inc.; **3.7*a*:** © Keith R. Porter; **3.8*a*:** © David M. Phillips/Visuals Unlimited; **3.9*a*:** © Keith R. Porter; **3.10*a*:** © K.G. Murti/Visuals Unlimited; **3.11*a*:** © David M. Phillips/Visuals Unlimited; **3.13*a*:** From Joseph G. Gall, Microtubule fine structure. *Journal of Cell Biology* 31: (1966) pp. 639–643; **3.15:** © Don Fawcett/Photo Researchers, Inc.; **3.16*a*:** © Stephen L. Wolfe; **3.17:** © CNRI/Science Photo Library/Photo Researchers, Inc.; **3.21(all):** © Edwin A. Reschke; **3.22:** © Dr. Tony Brain/SPL/Photo Researchers, Inc.

Chapter 4

Opener: Bettmann Archive; **4.1*a*:** © Edwin A. Reschke; **4.1*b*:** Dr. Kerry L. Openshaw; **4.2*b*:** © Ray Simons/Photo Researchers, Inc; **4.3*b*:** © Edwin A. Reschke/Peter Arnold, Inc.; **4.4*b*, 4.5*b*, 4.7*b*, 4.8*b*, 4.9*b*:** Edwin A. Reschke; **4.10*b*:** © CNRI/SPL/Photo Researchers, Inc.; **4.13*b*:** © Biophoto/Science Source/Photo Researchers, Inc.; **4.2*b*:** © Ray Simons/Photo Researchers, Inc.; **4.14*b*:** Photo Researchers, Inc.; **4.15*b*:** © Edwin A. Reschke; **4.16*b*:** © Biology Media/Photo Researchers, Inc.; **4.17*b*:** © Edwin A. Reschke/Peter Arnold, Inc.; **4.18*b*:** Edwin A. Reschke; **4.19*b*:** © Edwin A. Reschke/Peter Arnold, Inc.; **4.20*b*:** © Edwin A. Reschke; **4.21*b*, 4.22*b*:** © Edwin A. Reschke/Peter Arnold, Inc.; **4.23*a–b*:** Kent M. Van De Graaff; **4.24*b*:** © Ed Reschke; **4.26*a–c*:** © Edwin A. Reschke; **4.27*b*:** © John D. Cunningham/Visuals Unlimited; **4.27*c*:** © 1984 Martin M. Rotker/Photo Researchers, Inc.

Chapter 5

Opener: Corbis-Bettmann; **5.2:** © Cabisco/Visuals Unlimited; **5.3:** © J. Burgess/Photo Researchers, Inc.; **5.4:** © Victor B. Eichler, Ph.D.; **5.7:** © Dr. Sheril D. Burton; **5.8:** © James M. Clayton; **5.9:** © Lester V. Bergman & Associates, Inc.; **5.10*a*:** World Health Organization; **5.10*b*:** George P. Bogumill, M.D.; **5.11*a*:** © Michael Abbey/Photo Researchers, Inc.; **5.11*b*:** Dr. Kerry L. Openshaw; **5.12*a*:** © Michael Abbey/Photo Researchers, Inc.; **5.12*b*:** Dr. Kerry L. Openshaw; **5.13*c*:** © John D. Cunningham/Visuals Unlimited; **5.15:** Dr. Kerry L. Openshaw; **5.18*a*:** © Zeva Oelbaum/Peter Arnold, Inc.; **5.18*b*:** © Dr. P. Marazzi/SPL/Photo Researchers, Inc.; **5.18*c*:** © James Stevenson/SPL/Photo Researchers, Inc.; **5.19*a*:** © Dorte Groning/Tierbild Okapia/Photo Researchers Inc.; **5.19*b*:** © James Stevenson/Photo Researchers Inc.; **5.19*c*:** © John Radcliff/SPL/Photo Researchers Inc.; **5.21*a–c*:** Courtesy of Drs. David A. Kappel, E. Phillips Polack, and Marjorie L. Bush, Plastic Surgery Inc.; **5.22:** © Dan McCoy/Rainbow; **5.25** © Art Wolfe/Tony Stone Images; **5.26:** (Child): From Science Year, The World Book Science Annual © 1973 Field Enterprise Education Corporation. By permission of World Book, Inc. (Woman): Black Star Publishing Co.; **5.27:** © Norman Lightfoot/Photo Researchers, Inc.

Chapter 6

6.6*a–b*: © BioPhoto Associates/Photo Researchers, Inc.; **6.8*a*:** From R.G. Kessel & K.H. Kardon, *Tissues and Organs: A Text Atlas of Scanning Electron Microscopy*; **6.8*b*:** © Victor B. Eichler, Ph.D.; **6.10:** © Ed Reschke; **6.11:** Courtesy of Utah Valley Regional Medical Center, Dept. of Radiology; **6.18:** Kent M. Van De Graaff; **6.22*a–b*, 6.34*a*, 6.36*a*:** Courtesy of Utah Valley Regional Medical Center, Dept. of Radiology; **EX 6.1*b*:** © Science VU/Visuals Unlimited; **6.41:** Blayne L. Hirshche; **6.42:** © CNRI/SPL/Photo Researchers, Inc.

Chapter 7

Opener: Corbis-Bettmann; **7.3:** Kent M. Van De Graaff; **7.8*b*:** © Dr. Sheril D. Burton; **7.10, 7.12, 7.17:** Courtesy of Utah Valley Regional Medical Center, Dept. of Radiology; **7.19*a*:** © Dr. Sheril D. Burton; **7.19*b*:** Courtesy of Utah Valley Regional Medical Center, Dept. of Radiology; **7.21*a*:** Kent M. Van De Graaff; **7.21*b*:** Courtesy Blayne L. Hirshche; **7.22:** Kent M. Van De Graaff; **7.24*e*:** Courtesy of Eastman Kodak Company

Chapter 8

Opener: © National Library of Medicine/SPL/Photo Researchers, Inc.; **8.5:** Courtesy of Utah Valley Regional Medical Center, Department of Radiology; **8.6:** © Paolo Koch/Photo Researchers, Inc.; **8.18*a–h*:** © Dr. Sheril D. Burton; **8.19*a*:** Kent M. Van De Graaff; **8.19*b*:** © Dr. Sheril D. Burton; **8.19*c*:** © Dr. Sheril D. Burton; **8.19*d*:** Kent M. Van De Graaff; **8.19*e–f*:** © Dr. Sheril D. Burton; **8.20*a–b*:** Kent M. Van De Graaff; **8.26, 8.28, 8.32, 8.34:** From *Clinical Anatomy Atlas*, R.A. Chase, et al., Mosby 1996 (Class Project, Stanford University); **8.35*a*:** Kent M. Van De Graaff; **8.35*b*:** © Lester V. Bergman & Associates, Inc.; **8.36*a*:** James Stevenson/Science Ph. Lib/Photo Research; **8.36*b*:** Richard Anderson BH-12-8; **8.37, 8.38*a–d*:** SIU, School of Medicine; **8.39:** Kent M. Van De Graaff

Chapter 9

Opener: © Mehan Kulyk/SPL/Photo Researcher, Inc.; **9.6*b*:** © Ed Reschke; **9.9*a–b*:** © Dr. H.E. Huxley; **9.9*c*:** From R.G. Kessel and R.H. Kardon: *Tissues and Organs: A Text-Atlas of Scanning Electron Microscopy*, W.H. Freeman and Company © 1979; **9.10*a*:** © Dr. H.E. Huxley; **9.11*a–b*:** Kent M. Van De Graaff; **9.12*b*:** #5P1267/Biophoto/Photo Researchers; **9.15*a–h*:** © Dr. Sheril D. Burton; **9.52:** © John Bird Photography

Chapter 10

Opener: Corbis-Bettmann; **10.1:** © Stock Montage; **10.4*a–b*, 10.5, 10.6*a–b*, 10.7, 10.8:** © Dr. Sheril D. Burton; **10.9*a*:** Kent M. Van De Graaff; **10.9*b*, 10.10*a–b*:** © Dr. Sheril D. Burton; **10.11, 10.12:** From S.D. Vidic and F.R. Suarez, *Photographic Atlas of the Human Body*, 1984 (pl 29, p. 38) St. Louis, C.V. Mosby Co.; **10.13:** © McGraw-Hill Higher Education/Karl Rubin, photographer; **10.14:** Kent M. Van De Graaff; **10.16, 10.17:** From S.D. Vidic and F.R. Suarez, *Photographic Atlas of the Human Body*, 1984 (pl 71, p. 62) St. Louis, C.V. Mosby Co.; **10.18*a–b*:** Kent M. Van De Graaff; **10.19, 10.20, 10.21, 10.22:** Kent M.

Van De Graaff; **10.23, 10.24:** From S.D. Vidic and F.R. Suarez, *Photographic Atlas of the Human Body*, 1984 (pl 115, p. 184) St. Louis, C.V. Mosby Co.; **10.25:** Kent M. Van De Graaff; **10.26:** McGraw-Hill Higher Education/Rubin; **10.27:** From S.D. Vidic and F.R. Suarez, *Photographic Atlas of the Human Body*, 1984 (pl. 39, p. 58) St. Louis, The C.V. Mosby Co.; **10.28:** From S.D. Vidic and F. R. Suarez, *Photographic Atlas of the Human Body*, 1984 (pl 40, p. 60) St. Louis, C.V. Mosby Co.; **10.29, 10.30:** McGraw-Hill Higher Education /Rubin; **10.31:** © Dr. Sheril Burton; **10.32:** McGraw-Hill Higher Education /Rubin; **10.33, 10.34, 10.35, 10.36, 10.37, 10.38, 10.39, 10.40, 10.41*a–b*:** Kent M. Van De Graaff; **10.42:** From S.D. Vidic and F.R. Suarez, *Photographic Atlas of the Human Body*, 1984 (pl 129, p. 210) St. Louis, C. V. Mosby Co.; **10.43:** S.D.Vidic, F.R. Suarez, 1984 *Photographic Atlas of the Human Body*, (pl. 130, 212) St. Louis, The C.V. Mosby Co.; **10.44:** Kent M. Van De Graaff; **10.45, 10.46:** © Dr. Sheril Burton; **10.47, 10.48, 10.49*a–c*, 10.50*a–d*:** Kent M. Van De Graaff; **10.51*a–b*:** Dr. Kenneth A. Johnson; **10.51*c*:** Lester V. Bergman & Assoc.; **10.52, 10.53:** From S. D. Vidic and F. R. Suarez, *Photographic Atlas of the Human Body*, 1984 (pl 206, p. 336) St. Louis, C. V. Mosby Co.; **10.54, 10.55, 10.56:** McGraw-Hill Higher Education/Rubin; **10.57:** © Wm. C. Brown Communications, Inc./Karl Rubin, photographer; **10.58:** McGraw-Hill Higher Education/Rubin

Chapter 11

Opener: Adam Hart-Davis/SPL/Photo Researchers, Inc.; **11.3:** © Ed Reschke; **11.6*b*:** Courtesy Dr. John Hubbard; **11.8*b*:** © Edwin A. Reschke; **11.17*a–d*:** Kent M. Van De Graaff; **11.18:** © Hank Morgan/Photo Researchers, Inc.; **11.23*a*:** © Martin Rotker; **11.38*a–b*:** © Kent M. Van De Graaff, Ph.D./Utah Valley Hospital, Department of Radiology; **11.41*b*:** © Per H. Kjeldsen, University of Michigan, Ann Arbor

Chapter 12

Opener: © Brooks Brown/Photo Researchers, Inc.; **12.2:** From R.G. Kessel and R.H. Kardon: *Tissues and Organs: A Text-Atlas of Scanning Electron Microscopy*, W.H. Freeman and Company © 1979; **12.27:** Kent Van De Graaff; **12.32*a–i*:** All: Photographed by Dr. Sheril D. Burton assisted by Dr. Douglas W. Hacking

Chapter 13

Opener: © Dr. Jeremy Burgess/SPL/Photo Researchers, Inc.

Chapter 14

Opener: © Mary Evans Picture Library/Photo Researchers, Inc.; **14.9:** © John Cunningham/ Visuals Unlimited; **14.13*b*:** © SIU/Nawrocki Stock Photo, Inc.; **14.14:** © Fred Hossler/Visuals Unlimited; **14.17:** © BioPhoto Associates/Photo Researchers, Inc.; **14.20:** © Edwin Reschke; **14.22:** © R. Calentine/Visuals Unlimited; **14.25*a*:** © Richard Hutchings/Photo Researchers Inc.; **14.25*b*:** a) dwarf - © Richard Hutchings/Photo Researchers, Inc. b) © Bettina Cirone/Photo Researchers, Inc. tallest woman; **14.25*c*, 14.26, 14.27:** © Lester V. Bergman & Associates, Inc.

Chapter 15

Opener: © Mary Evans Picture Library/Photo Researchers, Inc.; **15.8, 15.10, 15.13*a–i*:** © Dr. Sheril D. Burton; **15.14*a*:** © Carroll H. Weiss/Camera M.D. Studios, Inc.; **15.16*c*:** P.N. Farnsworth/University of Medicine & Dentistry, New Jersey Medical School; **15.19*b*:** © Per H. Kjeldsen, University of Michigan, Ann Arbor; **15.21:** Kent M. Van De Graaff; **15.21*b*:** ©A.L. Blum/Visuals Unlimited; **15.29, 15.30*a*:** Kent Van De Graaff; **15.30*b*:** © Southern Illinois University; **15.35, 15.38:** From R.G. Kessel and C.Y. Shih, *Scanning Electron Microscopy in Biology*, Copyright 1976, Springer-Verlag; **15.39*a*:** © Dean E. Hillman; **15.43:** Courtesy Dr. Stephen Clark

Chapter 16

Opener: Corbis-Bettmann; **16.4*b*:** © Bill Longcore/Science Service/ Photo Researchers #6H7313; **16.9:** © McGraw-Hill Higher Education, Inc./Karl Ruben, photographer; **16.17:** © Blair Seitz/Photo Researchers, Inc.; **16.21*b*:** From: Practische Intleedkunde from J. Dankmeyer, H.G. Lambers, and J.M.F. Landsmearr. Bohn, Scheltma and Holkema; **16.24*b*:** From *Practische Outleedkunde*, J. Dankmeyer, H.G Lambers, and J.M.F Landsmerr, Bohn, Scheltema & Holkkema; **16.40:** Ed Reschke; **16.42*b*:** Kent M. Van De Graaff; **16.45*a–b*:** Richard Menard; **16.48*a*:** Donald S. Bain from Hurst et al.: The Heart, 5/E © McGraw-Hill Book 6, 1982; **16.48*b*:** Donald S. Bain from Hurst et al.: The Heart, 5/E © McGraw-Hill Book 6, 1982; **16.51*a*:** American Lung Association; **16.51*b*:** © Lewis Lainey; **16.52*b*:** Courtesy of Utah Valley Regional Medical Center, Department of Radiology

Chapter 17

Opener: Corbis-Bettmann; **17.5:** © CNRI/SPL/Photo Researchers, Inc.; **17.8*c*:** © CNRI/Phototake, Inc.; **17.10*a*:** © John D. Cunningham/Visuals Unlimited; **17.10*b*:** © Edwin A. Reschke; **17.12:** John Watney Photo Library; **17.13:** Courtesy of Kodak; **17.16:** © David Phillips/Visuals Unlimited; **17.18*a–b*:** © McGraw-Hill Higher Education Group; **17.23*a*:** From J.H. Comroe, Jr., *Physiology of Respiration*, © 1974, Yearbook Medical Publishers, Inc., Chicago; **17.23*b*:** From J.H. Comroe, Jr., *Physiology of Respiration*, © 1974, Yearbook Medical Publishers, Inc., Chicago; **17.25:** © SIU/Photo Researchers, Inc.; **17.28*a–b*:** Kent M. Van De Graaff; **17.29:** Edward C. Vasquez R.T. C.R.T./Dept. of Radiologic Technology, Los Angeles City College; **17.31*a*:** © Martin M. Rotker; **17.31*b*:** © Martin M. Rotker; **17.32:** © Martin Rotker/Science Source/Photo Researchers, Inc.

Chapter 18

Opener: Corbis-Bettmann; **18.9:** Kent M. Van De Graaff; **18.12*a*:** © Biophoto Associates/Science Source/Photo Researchers, Inc.; **18.12*b*:** © Biophoto Associates/Photo Researchers, Inc.; **18.12*c*:** © BioPhoto Associates/Science Source/Photo Researchers, Inc.; **18.14:** © Edwin A. Reschke; **18.15:** Courtesy of Utah Valley Regional Medical Center, Dept. of Radiation; **18.16*a*:** Kent M. Van De Graaff; **18.16*b*:** Kent M. Van De Graaff; **18.18:** © Edwin A. Reschke; **18.22:** © Manfred Kage/Peter Arnold, Inc.; **18.24:** Keith R. Porter/Alpers D.H. and Seetharan D. *New England Journal of Medicine* 296/1977–1047; **18.26:** From W.A. Sodeman and T.M. Watson, *Pathologic Physiology*, 6/E, W.B. Saunders Company; **18.27:** © Lester U. Bergman and Associates; **18.29*a*:** © Edwin A. Reschke; **18.29*b*:** From R.G. Kessel and R.H. Kardon, *Tissues and Organs: A Text-Atlas of Scanning Electron Microscopy*, 1979, W.H. Freeman and Company; **18.32*c*:** © Victor B. Eichler, Ph.D.; **18.35*a*:** © Carroll Weiss/Camera M.D. Studios; **18.35*b*:** © Dr. Sheril D. Burton; **18.37, 18.38*a–c*:** Kent M. Van De Graaff; **18.38*d*:** John Van De Graaff, DDS & Larry S. Pierce, DDS both at Northwestern University

Chapter 19

Opener: Corbis-Bettmann; **19.2:** © SPL/Photo Researchers, Inc.; **19.4*a–b*:** Kent M. Van De Graaff; **19.6*b*:** © Biophoto Associates/Photo Researchers, Inc.; **19.8*b*:** © David M. Phillips/Visuals Unlimited; **19.9*a*:** © Daniel Friend from William Bloom and Don Fawcett, *Textbook of Histology*, 10th ed., W.B. Saunders, Co.; **19.11:** © Per H. Kjeldsen, University of Michigan, Ann Arbor; **19.12:** Kent M. Van De Graaff; **19.14:** © Bruce Iverson/Visuals Unlimited; **19.20*b*:** © SIU/Peter Arnold, Inc.

Chapter 20

Opener: Wm. C. Brown Publishers; **Exhibit 20.2*g–h*:** © Dr. Landrum Shettles; **20.7*b*:** © Biophoto Associates/Photo Researchers, Inc.; **20.8:** © Francis Leroy, Biocosmos/SPL/Photo Researchers, Inc.; **20.7, 20.10:** © Edwin A. Reschke; **20.11:** © Manfred Kage/Peter Arnold, Inc.; **20.12:** © Manfred Kage/Peter Arnold, Inc.; **20.13:** © Edwin A. Reschke; **20.15*a–c*:** © Lennart Nilsson/A Child is Born/Dell Publishing Company; **20.20:** Center For Disease Control/Atlanta

Chapter 21

Opener: © Jean-Loup Charmet/SPL/Photo Researchers, Inc.; **21.4*a–b*:** © Edwin A. Reschke; **21.5*a–b*:** From R.J. Balandau, *A Textbook Histology*, 10th ed. 1975 W.B. Saunders Co.; **21.6:** From W. Bloom and D.W. Fawcett, *A Textbook of Histology*, 10th ed., W.B. Saunders Co., 1975 © D.W. Fawcett; **21.7:** © Dr. Landrum Shettles; **21.10, 21.14:** © Edwin A. Reschke; **21.17*a–b*:** © Biophoto Associates/Photo Researchers, Inc.; **21.20:** Courtesy of Utah Valley Regional Medical Center, Dept. of Radiology; **21.25*a*:** SIU School of Medicine; **21.25*b*:** SIU School of Medicine; **21.26*a*:** McGraw-Hill Higher Education/Vincent Ho; **21.26*b–g*:** © McGraw-Hill Higher Education, Inc./Bob Coyle, photographer; **21.26*h*:** © Hank Morgan/Science Source/Photo Researchers, Inc.

Chapter 22

Opener: Planet Art; **22.1*c*:** © David Phillips/Visuals Unlimited; **22.2:** Luciano Zamboni, from Greep, Roy and Weiss, Leon: Histology, 3/E. © McGraw-Hill Book Company, 1973; **22.6*a*:** Courtesy of Ronan O'Rahilly, M.D., Carnegie Institute of Washington, Department of Embryology, Davis Division; **22.11*a*, 22.20*a*:** © Donald Yeager/Camera M.D. Studios; **22.20*b*:** © Dr. Landrum Shettles; **22.20*c*:** © Petit Format/Nestle/Science Source/Photo Researchers, Inc.; **22.20*d–f*:** © Donald Yeager/Camera M.D. Studios; **22.20*g*:** © Petit Format/Nestle/Science Source/Photo Researchers, Inc.; **22.21*a*:** © A. Tsiaras/Photo Researchers, Inc.; **22.21*b*:** Kent Van De Graaff; **22.25*a–c*:** Courtesy of Utah Valley Regional Medical Center, Department of Radiology; **22.28:** From J.M Tanner,

The Physique of the Olympic Athlete, 1964; **22.29a–b:** Babu, A. & Hirschhom K. A *Guide to Human Chromosomes Reflects* 3/e, White Plains, NY: March of Dimes Birth Defects Foundation, BD:OAS, 28(2), 1992. With permission of copyright holder; **22.35:** © Lester V. Bergman & Associates, Inc.; **22.36c:** © Dr. Landrum Shettles; **22.38:** Gregory Dellore, M.D. and Steven L. Clark, M.D.

Line Art

Chapter 1

1.2: From *A Short History of Anatomy and Physiology from the Greeks to Harvey* by C. Singer, 1957, Dover Publications, New York, NY. Reprinted by permission.

Chapter 9

9.51: Hollingsworth J.W., Hashizume A, Jablon S: Correlations between Tests of Aging in Hiroshima Subjects: An Attempt to Define "Physiologic Age." *Yale Journal of Biology and Medicine*, 38:11–26, 1965. Reprinted by permission.

Illustrators

Chris Creek: 2.3*a*, 2.5, 2.6, 3.14, 4.4*a*, 4.10*a*, 4.17, 5.13*a*, 6.12, Ex 6.1*a*, 8.22, 9.17, 10.3, 10.15, 11.46, 11.47, 15.16, 15.40, 15.41, 17.17, 17.19, 17.20, 18.32*a*,*b*, 20.18*a–c*, 21.18
Wayne Heim: 1.4, 3.20, 16.1
Rictor Lew: 22.3, 22.4
Peg Gerrity: Skeletal and vertebrae icons

Index

A

A bands, 234, 235, 237
Abdomen
  abdominal portion of aorta, 549–50, 551
  disease and trauma of, 327–29
  internal anatomy of, 304, 307, 309, 310
  peripheral sympathetic plexuses and ganglia, 426
  subdivision of into regions, 38, 39
  surface anatomy of, 301, 304, 305
  veins of, 558
Abdominal aorta, 428, 549–50, 551
Abdominal cavity, 41, 42
Abdominal muscle(s), 600. *See also* specific muscle
Abdominal oblique muscle, 248, 305
Abdominal reflex, 414, 415
Abdominal thrust maneuver, 588, 607–8
Abdominal wall, muscles of, 248–50
Abdominopelvic cavity, 41, 43
Abducens nerves, 392, 395, 399
Abduction movement, 201, 202, 205, 206
Abductor digiti minimi muscle, 265
Abductor pollicus longus muscle, 262, 263, 265
Abnormal implantation sites, of ova, 765–66
Aborigines, Australian, 762
Abortion, 729
Abscess, 575
Absorption, as function of digestive system, 615
Accessory digestive organs, 615
Accessory hemiazygos vein, 557
Accessory nerve(s), 393, 397, 399, 601
Accessory nipples, 25
Accessory olivary nuclei, 366
Accessory reproductive glands, 689–90
Accommodation, visual, 493
Acetabular labrum, 215
Acetabular notch, 179
Acetabulum, 178
Acetylcholine, (ACh) 429, 430
ACh. *See* Acetylcholine
Achilles reflex, 415
Achilles tendon, 320
Achondroplasia, 166
Acid(s), 49
Acid phosphatase, 690
Acinar glands, 85
Acne, 103, 113, 126, 293, 760
Acoustic nerve, 396
Acquired heart disease, 573
Acquired immune deficiency syndrome (AIDS), 701, 702, 732
Acromegaly, 161, 326, 446, 463, 464
Acromial extremity, 169
Acromion, 172
Acrosomal reaction, 735
Acrosome, 686, 735
ACTH. *See* Adrenocorticotropic hormone
Actin, 233, 234, 235
Active transport, 53
Acupuncture, 8
Acute pain, 475
Acute prostatitis, 702
Acute purulent otitis media, 516
Acute renal failure, 673
Acyclovir, 401
Adam's apple. *See* Thyroid cartilage
Addison, Thomas, 467
Addison's disease, 326, 465
Adduction movement, 201, 202, 205, 206
Adductor brevis muscle, 268, 269
Adductor longus muscle, 268, 269
Adductor magnus muscle, 268, 269
Adenine, 63
Adenohypophysis, 363, 445–46, 448, 460–61
Adenoidectomy, 587
Adenoids, 587
Adenyl cyclase, 443
ADH. *See* Antidiuretic hormone
Adipocyte(s), 90, 430
Adipose cells, 90
Adipose tissue, 90–91, 288, 289
Adolescence, growth of, 758, 759–60, 761
Adrenal cortex, 424, 441, 455, 462
Adrenal glands
  description of, 455–56
  development of, 462
  disorders of, 465, 467
  endocrine system and, 441
  structure and functions of, 424–25, 456, 458
Adrenaline. *See* Epinephrine
Adrenal medulla, 424, 441, 455, 462, 465
Adrenergic stimulation, of neurotransmitters, 429–30, 431
Adrenocorticotropic hormone (ACTH), 446, 448
Adrenogenital syndrome, 467
Adulthood, as stage of development, 758, 760–62
Adventitia, 664, 665
Aesculapius, 10
Afferent glomerular arterioles, 660
Afferent lymphatic vessels, 564
Afferent neuron(s), 342
African Americans, 762
Aging
  bone ossification and, 137, 138
  of cells, 69
  of intervertebral discs, 93
  of muscles, 283
  of nervous system, 383
  of skeletal system, 164
  of skin, 102, 124
  of vertebral column, 129, 158, 164
Aging atrophy, 67
Agranular leukocytes, 525
Agranulocytes, 526
AIDS. *See* Acquired immune deficiency syndrome
Air pollution, and respiratory system, 595
Air spaces, 602
Airways, 602
Akinesia, 377
Alar cartilage, 585
Alar nasal sulcus, 295
Alar plates, 378
Albinism, 106, 126
Albinus, Bernard, 226
Albumin, 528
Alcohol
  cerebellum and intoxication, 366
  cytological explanation of tolerance to, 56
  liver and processing of, 560, 650
Aldosterone, 456
Alendronate, 164
Alexander the Great, 10
Alexandria, and history of anatomy, 10–11
Alleles, 763
Allantois, 646–47, 744–45
Alligator, brain of, 26
Allogenic transplant, 528
Allografts, 99
All-or-none contraction, of heart, 422
Alopecia, 126
Alpha cells, of pancreas, 453
Alpha globulin, 528
Alpha waves, 356–57
Alveolar ducts, 592
Alveolar sacs, 592, 594
Alveolar ventilation, 602
Alveolus, 133
Alzheimer's disease, 385
Amblyopia, 494, 516
Amelia, 185, 329
Amenorrhea, 728
Amines, 442
Amino acid, 49
Amnesia, 380
Amniocentesis, 743–44, 745, 768
Amnion, 742, 743–44
Amniotic cavity, 740
Amniotic fluid, 743–44, 745
Amniotic sac, 743
Amoebic dysentery, 649
Ampulla
  of ductus deferens, 688
  of inner ear, 510
Ampulla of uterine tube, 715
Ampulla of Vater, 633
Amsler, Marc, 516
Amsler grid, 516
Amylase, 633
Anabolic steroids, 684
Anal canal, 637, 647
Anal membrane, 647
Anal pit, 646
Anal sphincter muscles, 250
Anal triangle, 681
Anaphase, 65, 66, 685, 686
Anatomical dead space, 602
Anatomical neck, of brachium, 172
Anatomical position, 34, 35
Anatomical snuffbox, 314
Anatomy
  characteristics of humans, 27–28
  definition of as science, 2
  history of, 2–20
  important contributors to, 5
  nomenclature of, 33–36
  regional and systematic approaches to, 30
Anatomy, Descriptive and Surgical (Gray), 6
Anatomy and Physiology of the Vertebrates (Owen), 6
Anconeus muscle, 258, 259
Androgens, 456, 679, 684–85
Anemia, 326, 524, 572
Anencephaly, 322, 380
Anesthetics and anesthesia
  bony landmarks of facial region and, 152
  brachial plexus, 403
  childbirth and, 755
  epidural block, 368
  general and local, 380
  trigeminal nerve and, 395
Aneurysm, 575, 576
Angina pectoris, 327, 574
Angiocardiography and angiocardiograms, 17, 567
Angiography, 17
Angora hair, 113
Angular gyrus, 359, 361
Angular movements, 201
Animal kingdom, 23
Animal Kingdom, The (Cuvier), 6
Ankle
  lower extremity and, 40
  muscles of, 273–78, 279
  talocrural joint, 217–18
Ankle reflex, 414, 415
Ankylosis, 224
Annular ligament, 211
Anorexia nervosa, 657
ANS. *See* Autonomic nervous system
Antagonistic effects, of sympathetic and parasympathetic innervation, 431
Antagonistic muscles, 231
Antebrachium, 39, 173–74, 313–14. *See also* Forearm
Anteflexion, of uterus, 716, 729
Anterior, as directional term, 36
Anterior body cavity, 41
Anterior cavity, of eye, 491
Anterior cerebral artery, 546
Anterior chamber, of eye, 491
Anterior commissure, 358
Anterior cervical triangle, of neck, 299

Anterior corneal epithelium, 487
Anterior corticospinal tracts, of spinal cord, 376
Anterior crest, of tibia, 180
Anterior crural muscles, 273–74
Anterior division, of lumbar plexus, 405
Anterior funiculi, of spinal cord, 374
Anterior horns, of spinal cord, 374, 378
Anterior humeral circumflex artery, 548
Anterior interventricular artery, 535
Anterior interventricular vein, 535
Anterior median fissure, of spinal cord, 373
Anterior pituitary hormones, 448, 449
Anterior ramus, of spinal nerve, 401
Anterior root, 400–401
Anterior surface, of stomach, 629
Anterior tibial artery, 553
Anterior tibial vein, 557
Anterior vagal trunk, 427–28
Anterior view, of body regions, 37
Anteversion, of uterus, 729
Anthropometry, 761–62
Antibodies, 50
Antidiuretic hormone (ADH), 447, 448
Antihelix, 294
Antitragus, 294
Anulus fibrosus, 164
Anus, 637
Anvil, 33
Anxiety, 385
Aorta
abdominal portion of, 549–50, 551
aortic arch, 542–44, 550
ascending portion of, 534, 550
segments and branches of, 550
sensory neurons and, 433
thoracic portion of, 548–49, 550
Aortic arch, 542–44, 550, 570, 571
Aortic area, 537
Aortic bodies, 603
Aortic (semilunar) valve, 532, 534
Apex
of heart, 529
of lung, 594
of nose, 294
of patella, 180
Aphasia(s), 359, 361
Apical ectodermal ridge, 185
Apical foramen, 624
Aplastic anemia, 524
Apnea, 602
Apneustic area, 603
Apocrine glands, 83, 84, 85
Apocrine sweat glands, 115
Aponeuroses, 229
Appendectomy, 305
Appendicitis, 637
Appendicular skeleton. *See also* Skeletal system
anterior and posterior view of, 170
clinical considerations, 186–88
development of, 185
muscles of, 228, 253–78
organization of, 129–31
pectoral girdle and upper extremity, 169, 172–74
pelvic girdle and lower extremity, 176–84, 186
Appendix, 305, 637
Appositional bone growth, 134
Aqueduct of Sylvius, 363
Aqueous humor, of eye, 491, 495
Arachnoid space, 368, 369
Arachnoid villi, 370
Arbor vitae, 364
Arches, of foot, 184
Architecture, of muscle fibers, 231
Arcuate arteries, 659
Arcuate fasciculus, 359
Arcuate popliteal ligament, 216
Arcuate veins, 660
Areola, 720
Areolar glands, 720
Argentaffin, 630
Aristotle, 5, 6, 7, 10
Arm, 172–73. *See also* Brachium; Upper extremity
Armadillo, elbow joint and extensor muscles of, 207
Arrectores pilorum, 110
Arrector pili muscles, 430
Arrhythmia, cardiac, 567, 572
Art, and anatomy, 2, 3
Arterial blood, 523
Arterial pressure points, 554
Arteriography, 567
Arterioles, 522
Arteriosclerosis, 69, 575
Artery(s). *See also* Blood vessels
abdominal portion of aorta, 549–50
aortic arch, 542–44
definition of, 522
development of, 570, 571
elastic connective tissue, 87, 89
female reproductive organs, 717
of neck and head, 544–47
of pelvis and lower extremity, 550–53
of shoulder and upper extremities, 547–48
structure and function of, 538
thoracic portion of aorta, 548–49
Arthralgia, 224
Arthritis, 221–22, 330
Arthrology, 192
Arthrometry, 224
Arthroncus, 224
Arthropathy, 224
Arthroplasty, 196, 224
Arthroscopy, 222
Arthrosis, 224
Arthrosteitis, 224
Articular cartilage, 134, 196, 220
Articular disc, 196, 220
Articulating surfaces, of bone, 133, 180
Articulations. *See* Joints
Artificial implantation, and pregnancy, 766
Artificial insemination, 695
Artificial ligaments, 196
Arytenoid cartilage, 588
Ascending colon, 637
Ascending limb, of nephron loop, 661, 664
Ascending lumbar veins, 557
Ascending pharyngeal artery, 546
Ascending portion of aorta, 534, 542–43, 550
Ascending tracts, of spinal cord, 372–73, 374–75, 376
Asian-Americans, 762
Aspirin, 348
Association fibers, 358
Association neuron(s), 342
Asthma, 592, 594, 608
Astigmatism, 515
Astrocytes, 340, 341, 342, 345
Astrocytomas, 382
Ataxia, 366
Atelectasis, 326
Atheroma, 575
Atherosclerosis, 575
Athlete's foot, 126
Atlantooccipital joint, 156
Atlas, 156
Atoms, 28, 29
ATP molecules, 443
Atresia, 511
Atretic secondary follicles, 712–13
Atrial septal defect, 572
Atrioventricular bundle, 535
Atrioventricular node (AV node), 535–36
Atrioventricular valves (AV valves), 529
Atrium, of heart, 529, 569, 570
Atropa belladonna, 430
Atrophy, of muscles, 67, 98, 282
Atropine, 430
Audiometry, 511
Auditory impairment, 517
Auditory nerve, 396
Auditory ossicles
axial skeleton and, 129, 131, 152, 153
body cavities and, 41
as characteristic of mammals, 25
development of, 513
middle ear and, 501, 502
Auditory (eustachian) tube, 23, 501, 513, 587
Auerbach, Leopold, 619
Auerbach's plexus, 619
Auricle(s)
of ear, 292, 293, 294, 499–500, 514
of heart, 529
Auricular abnormalities, 511
Auricular region, of head, 293, 294
Auricular surface, 158, 179
Auscultation, 34, 288, 327
Australoid race, 27, 762
Autograft, 121
Autographs, 99
Autonomic dysreflexia, 435–36
Autonomic motor fibers, 344
Autonomic motor nerve, 335
Autonomic nervous system (ANS). *See also* Nervous system
clinical case study, 420, 435
clinical considerations, 435–36
functions of, 429–33
higher brain centers and control of, 433–35
introduction to, 420–23
organization of nervous system, 334, 335
somatic motor system compared to, 421
structure of, 423–28, 429
Autopsy, and history of anatomy, 18
Autosomal chromosomes, 762
Autotransplant, 528
Avulsion fracture, 187
Axial muscles, 254–55, 256
Axial skeletion. *See also* Skeletal system
clinical considerations, 161–64
development of, 162–63
muscles of, 228, 241–53
organization of, 129, 131
rib cage, 159, 160
skull, 139–53
vertebral column, 153–59
Axilla, surface anatomy of, 311
Axillary artery, 548
Axillary branch, of brachial plexus, 404
Axillary fossa, 38
Axillary lymph nodes, 565
Axillary nerve, 403
Axillary process of breast, 720
Axillary region, 38
Axillary vein, 556
Axis, of vertebrae, 157
Axon(s), 97, 236–37, 338, 343
Axonal transport, 340
Axon terminals, 236, 347
Axoplasmic flow, 340
Aye-aye (primate), 26
Azygos vein, 557

B

Babinski, Joseph F., 414
Babinski's reflex, 377, 414, 415
Back, surface anatomy of, 301, 302
Bacteria. *See also* Infections
cellular dysfunction and, 68
food poisoning and, 649
Bacterial endocarditis, 573
Bacteriuria, 672
Baculum, 693
Baer, Karl Ernst von, 6
Balance, and sensory organs, 499–510
Balfour, Francis M., 6
Ball-and-socket joints, 199, 200
Baroreceptor(s), 472, 544
Barrier contraceptives, 731, 732
Bartholin, Casper, Jr., 719
Bartholin's glands, 15, 719
Basal cell carcinoma, 118
Basal metabolism rate (BMR), 463
Basal nuclei, 358
Basal plates, 378
Base
of heart, 529
of lung, 594
Base(s), 49
Basement membrane, 75
Basilar artery, 544
Basilar membrane, 504
Basilic vein, 556
Basle Nomina Anatomica (BNA), 16
Basophils, 525, 526
Ba spirit (ancient Egypt), 8
Bayliss, William M., 6
BBB. *See* Blood–brain barrier
Beaumont, William, 6
Bell, Charles, 396
Bell's palsy, 241
Belly, of muscle, 229
Benign neoplasms, 68
Benign prostatic hypertrophy (BPH), 690, 702
Beta cells, of pancreas, 453
Beta globulin, 528
Beta waves, 357
Biaxial joints, 199
Biceps brachii muscle, 257, 259
Biceps femoris muscle, 272
Biceps reflex, 414, 415
Bicornuate uterus, 726
Bicuspid area, 537
Bicuspid teeth, 622
Bicuspid valve, 531
Bile canaliculi, 641
Bile ducts, 641–42
Binocular vision, 482
Biofeedback, and autonomic nervous system, 420
Biology, anatomy and physiology as subdivisions of, 2
Biomechanics, of body movement, 203–8

Biopsy, 99
Bipedal locomotion, 28
Bipolar cells, in nasal mucosa, 392
Bipolar neuron, 342, 489–90
Birds, vision of, 482
Birth control, 458, 729–32. *See also* Contraception
Birth control pills, 730, 731
Bitemporal hemianopia, 446
Bitter taste, 481
Bladder. *See* Urinary bladder
Blanching, of blood, 572
Blastocoel, 75
Blastocyst, 74, 75, 715, 737, 738, 739, 740
Blastocyst cavity, 737
Bleeding, control of, 576–77. *See also* Hemorrhage
Blind spot, 494
Blister, 126
Blood. *See also* Red blood cell; White blood cell
  characteristics of, 523–24
  circulatory routes of, 534–35
  disorders of, 572
  formed elements of, 524–26
  hemopoiesis and, 527, 528
  plasma of, 528
  supply of to eyeball, 490–91
  supply of to skeletal muscle, 231–32
  vascular tissue, 94, 95
Blood-brain barrier (BBB), 342, 345, 372
Blood-cerebrospinal fluid barrier, 372
Blood clot, 521
Blood plasma, 94, 528
Blood platelets, 123
Blood pressure, 541–42, 567
Blood supply. *See* Blood vessels
Blood-testis barrier, 685
Blood vessels. *See also* Artery(s); Vein(s)
  autonomic nervous system and, 430
  as component of circulatory system, 522
  of ovaries, 710
  renal, 659–60
  structure and function of, 538–42
  of uterus, 717
Body
  of epididymis, 688
  of pancreas, 643
  of stomach, 629
  of uterus, 715
Body, human
  composition of average adult, 761
  compounds found in, 49
  elemental composition of, 48
  primary germ layers and systems of, 741
  skeletal muscles and support of, 227
Body cavities, 23, 40–43
Body membranes, 40–43, 80, 82
Body organization, 28–32
Body regions, 36–40
Body temperature
  endothermic animals and, 522
  hypothalamus and regulation of, 362
  variations in normal, 536
Boil, 126, 326
Bone(s). *See also* Skeletal system; specific bone
  of adult skeleton, 131
  development of, 162–63
  of foot, 183
  fractures of, 18, 135, 157, 159, 169, 174, 187–88, 193, 194
  growth of, 137–39
  prehistoric tools made of, 3
  structure of, 132–35
Bone cells, 135
Bone-lining cells, 135
Bone marrow, 134, 528
Bone scan, 163–64
Bone tissue, 94, 95, 135–36, 137
Bony callus, 188
Bony labyrinth, 502, 503, 513
Botulism, 649
Bowman, Sir William, 660
Bowman's capsule, 660, 662. *See also* Glomerular capsule
Brachial artery, 548
Brachial block, 403
Brachialis, 257, 259
Brachial plexus, 401, 403–4
Brachial vein, 556
Brachiocephalic trunk, 543–44
Brachiocephalic vein, 556
Brachioradialis muscles, 257, 259
Brachioradialis reflex, 414, 415
Brachium. *See also* Arm
  appendicular skeleton and, 172–73
  surface anatomy of, 311–12, 313, 315, 316
  upper extremity and, 39
Bradycardia, 567, 572
Bradykinin, 432
Braille, Louis, 474
Braille reading system, 474
Brain
  autonomic nervous system and, 433–35
  blood supply to, 544–47
  central nervous system and, 348–52
  cerebrum and, 352–61
  development of, 336–37
  diencephalon, 361–63
  mesencephalon, 363
  metencephalon, 364–66
  myelencephalon, 366–67
  planes of reference, 35
  taxonomic characteristics of human, 23, 25, 26, 28
  ventricles and cerebrospinal fluid, 370–72
Braincase, 292
Brain stem, 363
Brain waves, 356–57
Branched acinar glands, 85
Branched tubular glands, 85
Breast(s). *See also* Breast cancer; Breast self-examination
  diseases of, 729
  mammary glands and, 115, 311
  self-examination of, 311, 327
  structure of, 719, 720–21
  surface anatomy of, 304
Breast buds, 759
Breast cancer, 254, 327, 721, 729
Breast self-examination (BSE), 724, 726, 727
Breathing, mechanics and regulation of, 598–603
Breathing reflex, 757
Breech birth, 755
Bridge, of nose, 293, 295
Broad ligament, 709, 716
Broca, Pierre P., 359
Broca's area, 359
Bronchial arteries, 549
Bronchial segments, 595
Bronchial tree, 591–92, 598
Bronchioles, 592
Bronchitis, 608
Bronchography, 592
Bronchoscopy, 327
Brunner, Johann C., 633
Brunner's glands, 633. *See also* Duodenal glands
Brush border, 634
Buccal cavity, 140. *See also* Oral cavity
Buccinator muscle, 242
Bucy, Paul C., 497
Bulb, of penis, 691
Bulbar conjunctiva, 485
Bulbocavernosus muscle, 250
Bulboreticulospinal tract, 376
Bulbospongiosus muscle, 251
Bulbourethral glands, 667, 681, 689, 690
Bulbs of Krause, 473, 474
Bulbus cordis, 569, 570
Bulimia nervosa, 624
Bundle of His, 535. *See also* Atrioventricular bundle
Bunion, 198
Burn(s)
  clinical case study of, 102, 124
  integumentary system and types of, 119–20
  skin grafts and, 121
  trauma to cells and, 68
Bursa, 197–98
Bursitis, 219, 330
Butchering, of animals and anatomy in prehistoric times, 3
Buttock
  muscles of, 266, 267
  pelvic region and, 39
  surface anatomy of, 317–21

C

Cadavers, dissection of, 2, 4, 12, 13, 30. *See also* Dissection
Caffeine, 348
Calcaneofibular ligament, 217
Calcaneus, 184
Calcification, of bone, 137
Calcitonin, 450, 451–52
Calcitrol, 110
Calcium, in bone, 94, 132
Calculi, urinary, 673
Callus, 104, 109
Camera, analogues for eye structures, 482
Canaliculus
  of bone, 94
  of osteons, 136
Canal of Schlemm, 491
Cancer. *See also* Breast cancer; Lung cancer; Neoplasms; Skin cancer
  genetic factors in, 68
  lymph nodes and, 227, 565
  origin of term, 33
  pancreatic, 648
  respiratory system and, 608
  of uterus, 728
Candida albicans, 729
Canine teeth, 149, 622
Capacitation, and fertilization, 735
Capillary
  definition of, 522
  structure and function of, 538–40
Capillary fracture, 187
Capitate bone, 174
Capitulum, of humerus, 172–73
Capoid race, 27
Capsular space, in kidney, 660
Capsule. *See* Glomerular capsule; Joint capsule
Carbohydrates, 50, 53
Carboxypeptidase, 633
Carbuncle, 126
Carcinogens, 69
Carcinoma, of breast, 327, 729
Cardia, of stomach, 629
Cardiac arrhythmia, 567, 572
Cardiac catheterization, 534
Cardiac center, of medulla oblongata, 367
Cardiac cycle, 535
Cardiac muscle, 96, 97, 255
  autonomic nervous system and, 420, 422
Cardiac notch, 594
Cardiac output, 536
Cardiac veins, 535
Cardiac vessels, 530
Cardinal ligaments, 716
Cardiogenic area, and development of heart, 569
Cardiovascular assessment, 567, 572
Cardiovascular disease, 327, 556, 577. *See also* Heart; Myocardial infarction
Cardiovascular regulation, and hypothalamus, 362
Cardiovascular system, development of, 562, 568–71
Carina, 590
Carotene, 106
Carotid artery, 300, 544, 546. *See also* Common carotid artery
Carotid bodies, 544, 603
Carotid canal, 147
Carotid sheath, 300, 556
Carotid sinus, 544
Carpal bones, 129
Carpal tunnel syndrome, 330
Carpus, 40, 174
Carrier molecules, 52
Cartilage
  alar, 585
  articular, 134, 196, 220
  corniculate, 588
  costal, 91, 209
  cricoid, 298, 588
  cuneiform, 588
  elastic, 93
  epiglottic, 588
  fibrocartilage, 91, 92, 93
  hyaline, 91, 92, 93
  lateral, 585
  septil, 585
  thyroid, 297–98, 588
Cartilaginous joints, 192, 194–95, 200
Case studies, clinical
  appendicular skeleton, 169, 188
  autonomic nervous system, 420, 435
  axial skeleton, 129
  body cavities, 23, 43
  central nervous system, 334, 385
  circulatory system, 521, 577
  cranial nerves, 399
  development, 735, 769
  digestive system, 615, 652
  endocrine system, 439, 465
  female reproductive system, 707, 732
  integumentary system, 102, 124
  male reproductive system, 679, 702
  muscular system, 227, 283
  joints, 192, 222
  peripheral nervous system, 389, 414
  respiratory system, 583, 611

sensory organs, 471, 517
urinary system, 656, 675
Cat, vision of, 482
Catalase, 59
Cataracts, 511, 515
Catecholamines, 456
Catheterization
of heart, 567
of urethra, 668
Catholic Church, and history of anatomy, 12
Caucasian race, 27, 762
Cauda equina, 373
Caudate lobe, of liver, 640–41
Caudate nucleus, 358
Cave drawings, prehistoric and history of anatomy, 3
Cavities, of eye, 491. *See also* Body cavities; Oral cavity
CCK. *See* Cholecystokinin
Cecal diverticulum, 646
Cecum, 636
Celiac (solar) plexus, 424, 428
Celiac trunk, 549
Cell(s)
body organization, 28, 29
cell cycle, 62–67
cellular diversity, 47
chemistry of, 48–50
as functional units, 47
history of cell theory, 16
mitotic potential, 67
shapes of, 48
structure of, 50–62
trauma to, 67–68
Cell and Tissue (Hertwig), 6
Cell body, of neuron, 97, 338
Cell cycle, 62–67
Cell division, 64–67
Cell in development and heredity, The (Wilson), 6
Cell membrane, 50, 51, 52–54
Cell Pathology (Virchow), 6, 16
Cells of Leydig, 683
Cell theory, 16, 47
Cellular biology, definition of, 17
Cellular chemistry, 48–50
Cellular diversity, 47
Cellular hypertrophy, 67
Cellular structure, 50–62
Celsus, Cornelius, 5, 6, 11, 12
Cementum, 624
Central artery, of eye, 490–91
Central canal
of bone tissue, 94, 136
of spinal cord, 348, 374
Central chemoreceptors, 603
Central nervous system (CNS). *See also* Nervous system
brain structures and, 348–72
clinical considerations, 379–83, 385
organization of nervous system, 334, 335
sensory and motor fibers of, 345
spinal cord, 372–78
Central sulcus, 355
Central vein, 490–91, 641
Central vessels of retina, 499
Centriole(s), 59
Centromere, 63
Centrosome, 51, 59
Cephalic phase, of gastric secretion, 631
Cephalic vein, 556
Cephalization, 604
Cerebellar cortex, 364
Cerebellar peduncles, 364–66
Cerebellar vein, 556
Cerebellum, 364–66, 434–35
Cerebral angiogram, 379
Cerebral arterial circle, 544, 546
Cerebral artery, 546
Cerebral cortex, 355
Cerebral embolism, 575
Cerebral gyri, 355
Cerebral hypophysis, 363, 444
Cerebral palsy, 366, 382
Cerebral peduncle(s), 363
Cerebral sulci, 355
Cerebral thrombosis, 383
Cerebral vein, 556
Cerebrospinal fluid (CSF), 348, 370, 371, 372
Cerebrovascular accident (CVA), 383
Cerebrum
autonomic nervous system and, 434–35
central nervous system and, 352–61
functions of, 352
relative size of in vertebrates, 26
Cerumen, 118, 293
Ceruminous glands, 118, 500
Cervical canal, 715
Cervical cancer, 728
Cervical cap, 732
Cervical curve, of vertebral column, 154
Cervical enlargement, of spinal cord, 373
Cervical lymph nodes, 297, 298, 565
Cervical nerve(s), 601
Cervical plexus, 401, 402
Cervical region, 37
Cervical rib, 326
Cervical sympathetic ganglia, 426
Cervical triangles, of neck, 298, 299, 300
Cervical vertebra, 153, 155, 156–57
Cervix
of neck, 297
of uterus, 715
Cesarean section, 755
Chalazion, 516
Chambers
of eyeball, 491
of heart, 529–34
Chancre, 702
Chancroid, 701
Cheek(s), 620
Chemoreceptors
carotid bodies and, 544, 603
exteroceptors and, 472
olfactory sense and, 390, 472, 479
regulation of breathing by, 603
taste buds as, 396
Chest. *See* Thorax
Cheyne, John, 608
Cheyne-Stokes breathing, 608, 611
Chiasma, 390
Childbirth. *See also* Pregnancy
brachial plexus, 403
epidural block, 368
fetal skull and, 141
labor and parturition, 755–56
oxytocin and induction of labor, 447
pelvis and, 179–80
Childhood, as period of growth and development, 757–59
Chilitis, 652
Chimpanzee, brain of, 26
Chin, 294, 295
China, and history of anatomy, 8
Chlamydia, 701
Choana, 585
Cholecystokinin (CCK), 459
Cholera, 9
Cholinergic stimulation, of neurotransmitters, 429, 430, 431
Cholinesterase, 348
Chondritis, 224
Chondrocranium, 163
Chondrocyte(s), 91, 137
Chordae tendineae, 531
Chordata, 23–24
Chorda tympani branch, of facial nerve, 481
Chorea, 377
Chorion, 737, 745
Chorion frondosum, 745
Chorionic gonadotropin (hCG), 747
Chorionic somatomammotropin (hCS), 747
Chorionic villus biopsy, 745
Choroid, of sclera, 488, 491, 499
Choroid plexus, 363
Chromaffin cells, 455
Chromatin, 51, 52, 62
Chromatophilic substances, 338
Chromosomal abnormalities, 68, 742. *See also* Developmental disorders; Genetic disorders
Chromosomes. *See also* Genetic disorders; Genetics
autosomal, 762
diploid number of, 762, 763
homozygous, 763
structure and function of cells and, 62–63
X and Y, 762, 763
Chronic meningitis, 383
Chronic pain, 475
Chronic prostatitis, 702
Chronic renal failure, 673
Chyme, 629
Chymotrypsin, 633
Cigarette smoking
cancer and, 69, 611
cardiovascular effects of, 542
respiratory system and, 595, 608, 611
Cilia, 51, 61, 78
Ciliary arteries, of eye, 490
Ciliary body, 488, 491, 499
Ciliary ganglion, 425
Ciliary gland(s), 484
Ciliary muscle, of eye, 486, 488
Ciliary processes, 488
Ciliated cells, 61
Cineradiography, 17
Circle of Willis. *See* Cerebral arterial circle
Circular movements, 201–3
Circulatory routes, of blood, 534–35
Circulatory system
blood and, 523–28
clinical case study, 521, 577
clinical considerations, 566–67, 572–77
development of, 568–71
digestive system and, 651
fetal circulation, 560–62
functions and major components of, 32, 521–22
heart and, 529–37
integumentary system and, 125
interaction with other body systems, 578
lymphatic system and, 562–66
male reproductive system and, 703
muscular system and, 284
principal arteries of body, 542–53
principal veins of body, 554–60
respiratory system and, 610
skeletal system and, 165
urinary system and, 674
Circumcision, 691
Circumduction, 202–3
Circumflex artery, 535
Cirrhosis, of liver, 560, 650
Cisterna chyli, 564
Cisternal puncture, 379
Class, taxonomic, 25
Claustrum, 358
Clavicle
displacement of sternoclavicular joint and, 209
fracture of, 172
pectoral girdle and, 129, 169, 171
Clavicular notch, 159
Clavipectoral triangle, 311
Cleavage, and formation of blastocyst, 737, 738, 739, 740
Cleft lip, 149, 322, 607
Cleft palate
as congenital malformation, 161, 322
development of digestive system and, 648
development of respiratory system and, 604, 606–7
nursing and, 149, 604, 606
palatine processes and, 149
Clinical considerations. *See also* Clinical case studies; Disease
anatomical structure and function, 34
appendicular skeleton, 186–88
autonomic nervous system, 435–36
axial skeleton, 161–64
central nervous system, 379–83, 385
circulatory system, 566–67, 572–77
development and, 765–69
digestive system, 648–50, 652
endocrine system, 462–65, 467
female reproductive system, 723–32
heart sounds and, 537
histology, 98–99
integumentary system, 118–26
joints, 218–19, 221–22, 223
male reproductive system, 695, 699–702
muscular system, 278, 282–83
peripheral nervous system, 414, 415
respiratory system, 606–8, 609, 611
sensory organs, 510–11, 514–17
urinary system, 669, 672–73, 675
Clitoris, 708, 718–19, 725
Cloaca, 647
Cloacal membrane, 646
Closed fracture, 187
Clostridium botulinum, 649
Clostridium tetani, 241
Clubfoot, 186, 330
CNS. *See* Central nervous system
Coarctation, 575
Coccygeal cornua, 158
Coccygeal vertebra, 154
Coccygeus muscle, 250
Coccyx, 158
Cochlea
development of ear and, 513
inner ear and, 502, 503, 504
origin of term, 33
Cochlear duct, 504, 513
Cochlear nerve, 396
Cochlear nuclei, 396
Cochlear window, 501, 502
Codfish, brain of, 26
Coelom, 41
Coiled tubular glands, 85

Coitus, 679, 707
Cold, sensory receptors for, 474. *See also* Body temperature
Cold sore, 126
Cold viruses, 608
Colitis, 652
Collagen, 49, 69
Collagenous fibers, 86–87
Collateral branches, 338
Collateral ligaments
 of ankle, 217
 of elbow, 211, 214
 of knee, 216
Colles, Abraham, 187, 329
Colles' fracture, 174, 187, 329
Colloid, of thyroid follicular cells, 450
Colloid substances, 54
Colon, 637
Color, of lips and mucous membranes, 294
Color blindness, 496, 764–65
Colostomy, 652
Columnar cells, 54, 76
Coma, 367, 380–81
Comedo, 126
Comminuted fracture, 187
Commissural fibers, 358
Commissures, of eye, 484
Common bile duct, 633
Common carotid artery, 298, 543–44, 545
Common cold, 608
Common fibular nerve, 407, 409
Common hepatic artery, 549
Common iliac artery, 549
Common iliac vein, 558
Common nerve pathways, 475
Communication, and facial expression, 110–11
Compact bone tissue, 94, 133, 136
Comparative Embryology (Balfour), 6
Compartmentalization, of thoracic cavity, 595, 597
Compensatory hypertrophy, 67
Complementary effects, of sympathetic and parasympathetic innervation, 431
Complete fracture, 187
Compound fracture, 187
Compound gland(s), 82, 85
Compounds, chemical, 48–49
Computerized axial tomography (CT), 18, 19, 34, 379, 567
Conceptus, 742
Concha, 294. *See also* Nasal concha
Concussion, 380
Condoms, 731, 732
Conducting division, of respiratory system, 584
Conducting passages, of respiratory system, 585–92
Conduction, of nerve impulse, 471
Conduction aphasia, 359
Conduction deafness, 517
Conduction myofibers, 535
Conduction system, of heart, 535–36
Conductivity, of neurons, 347
Condylar process, 151
Condyle(s), 133, 180
Condyloid articulation, 199, 200
Condyloid canal, 147
Cone cells, 489–90
Cones, of eye, 54
Congenital bilateral absence of the ductus deferentia, 702
Congenital cataracts, 511
Congential deafness, 511
Congenital diaphragmatic (hiatal) hernia, 327
Congenital malformations. *See* Developmental disorders
Congenital megacolon, 328
Congenital syphilis, 326
Conjoined twins, 767
Conjugated proteins, 49
Conjunctiva, 294, 484–85, 499
Conjunctival sac, 485
Conjunctivitis, 516
Connecting stalk, 748
Connective tissue
 adipose tissue, 90–91
 blood or vascular tissue, 94, 95
 bone tissue, 94, 95
 cartilage, 91–93
 characteristics and classification of, 73, 86
 dense irregular, 88, 89
 dense regular, 88
 elastic, 88, 89
 embryonic, 86
 loose (areolar), 87
 reticular, 88
 skeletal muscles and, 229–30
 structure of, 86–87
Conoid tubercle, 169
Constipation, 639
Constriction, of pupil, 493–94
Continuous ambulatory peritoneal dialysis (CAPD), 675
Continuous capillaries, 539, 540
Contraception, methods of, 458, 729–32
Contraceptive sponge, 731, 732
Contractility, of muscle tissue, 227
Contrecoup fracture, 325
Conus medullaris, 373
Convergence, of eyeballs, 494
Convergent muscle, 231
Convolutions, of cerebrum, 355
Convulsion, of muscle, 285
Cooperative effects, of sympathetic and parasympathetic innervation, 431–32
Coracobrachialis muscle, 256
Coracohumeral ligament, 210
Coracoid process, of scapula, 169
Corn, 126
Cornea, 487, 491, 499
Corneal transplant, 487
Corniculate cartilage, 588
Cornification, of skin, 104
Corona glandis, 691
Coronal plane, 33–34, 35
Coronal suture, 141, 146
Corona radiata, 711, 735, 736
Coronary arteries, 535, 542–43
Coronary artery bypass surgery, 574
Coronary circulation, 535
Coronary embolism, 575
Coronary sinus, 530, 535, 569
Coronary sulcus, 530
Coronoid fossa, of humerus, 173
Coronoid process
 of mandible, 151
 of ulna, 174
Corpora cavernosa
 of clitoris, 719
 of penis, 691
Corpora quadrigemina, 363
Corpus albicans, 714
Corpus callosum, 353, 358
Corpuscles of touch, 473, 474
Corpus luteum, 713–14
Corpus spongiosum penis, 691
Corpus striatum, 358
Corrugator muscle, 242
Cortex layer, of brain, 348
Corti, Alfonso, 504
Cortical nephrons, 661, 663
Corticospinal tracts, of spinal cord, 375–76
Corticosteroids, 456, 458
Corticotropin-releasing hormone (CRH), 446, 447
Cortisol, 442, 456
Cortisone, 442
Costal angle, 159
Costal arch, 301
Costal cartilage, 91, 209
Costal groove, 159
Costal margin, 159
Costal notches, 159
Costal surface, of lung, 594
Costal tuberosity, of clavicle, 169
Costocervical trunk, 547
Costoclavicular ligament, 209
Coughing, 78, 602
Cowper, William, 667
Cowper's glands, 667, 690
Coxal (hip) joint, 214–15
Coxarthrosis, 224
Crabs, as sexually transmitted parasites, 701
Cramp, in muscle, 278, 282
Cranial bones, 139, 146–49
Cranial cavity, 139–40, 144
Cranial dura mater, 367
Cranial encephalocele, 322, 380
Cranial meningocele, 380
Cranial nerve(s), 334, 388, 389–99
Cranial region, 37
Cranial root, 397
Cranial sutures, 193, 200
Craniotomy, 166
Cranium, 37, 292
Cremasteric reflex, 414
Cremaster muscle, 681
Crest, of bone, 133
Cretinism, 463
CRH. *See* Corticotropin-releasing hormone
Cribriform foramina, 149
Cribriform plate, of ethmoid bone, 149
Crick, Francis H. C., 5, 6, 64
Cricoid cartilage, 298, 588
Crista, 57
Crista ampullaris, 507, 510
Crista galli, 149
Crossed extensor reflex, 412, 413
Cross-sectional plane, 34
Crown, of tooth, 624
Cruciate ligaments, 216
Crural muscles, 273–76, 277
Crural regions, 40
Crus, of penis, 691
Crusades, and history of anatomy, 12
Crutch paralysis, 403
Crying and crying response, 485, 602
Cryptorchidism, 328, 699, 700
Crypts of Lieberkühn, 635
CT. *See* Computerized tomography
Cubital fossa, 40, 312
Cubital lymph nodes, 565
Cubital region, 40
Cuboidal epithelial cells, 76
Cuboid bone, 184
Cumulus oophorus, 711
Cuneiform bone(s), 184
Cuneiform cartilage, 588
Cuneiform script, 5, 7
Cupula, 507, 508
Curettage, 99
Cushing, Harvey, 467
Cushing's disease, 465
Cushing's syndrome, 465
Cuspids, 622
Cutaneous absorption, 110
Cutaneous prepatellar bursa, 216
Cutaneous receptors, 110, 473
Cuticle, 112
Cuvier, Georges, 6
CVA. *See* Cerebrovascular accident
Cyanosis
 blood disorders and, 572
 cardiovascular and respiratory diseases and, 106, 291, 326, 572, 608
 color of mucous membranes and, 326
 neonate and, 291
Cyclic AMP (adenosine monophosphate), 443
Cyclopia, 511
Cycloplegic drug, 511
Cystic fibrosis, 112, 607, 652, 770
Cystic vein, 560
Cystitis, 665, 673
Cystoscopy, 672–73
Cytokinesis, 65
Cytology
 cell cycle, 62–67
 cellular chemistry, 48–50
 cellular structure, 50–62
 clinical considerations, 67–69
 definition of, 17
 introduction to, 47
Cytoplasm, 51, 54–61
Cytosine, 63
Cytotrophoblast, 739

D

Dandruff, 126
Dark Ages, and history of anatomy, 12
Dartos, 681
Darwin, Charles, 6
Daughter cells, 686, 687
David (sculpture by Michelangelo), 2, 3
da Vinci, Leonardo, 5, 13, 14, 734
Deadly nightshade plant, 430
Deafness, 517
Death, autopsy and cause of, 18
Decibels, of sound, 505
Decidua basalis, 745
Deciduous teeth, 622, 623
Decubitus ulcer(s), 108, 109, 126
Decussation, of spinal cord, 375
Deep brachial artery, 548
Deep circumflex iliac artery, 552
Deep digital flexor muscle, 261
Deep fascia, 230
Deep femoral artery, 552
Deep femoral vein, 557
Deep fibular nerve, 407
Deep infrapatellar bursa, 216
Deep motor branches, of cervical plexus, 402
Deep palmar arch, 548, 556
Deer, vision of, 482
Defecation, as function of digestive system, 615
Defecation reflex, 639
Defibrillation, electrical, 572
Definitive hair, 113

Deglutition, 615, 628
de Graaf, Regnier, 16, 712
De Humani Corporis Fabrica (Vesalius), 13, 14
Dehydration, 49
Delirium, 380
della Torre, Marcantonio, 13
Deltoid ligament, of ankle, 217
Deltoid muscle, 256
Deltoid region, 39
Deltoid tuberosity, 172
de'Luzzi, Mondino, 5, 6, 13, 14
Dendrites, 97, 338
Dendritic spinules, 338
Dendritic zone, 338
Dens, of vertebrae, 157
Dense irregular connective tissue, 88, 89, 91
Dense regular connective tissue, 88, 91
Dental alveolar process, 149
Dental alveoli, 624
Dental caries (cavities), 624, 649
Dental cusps, 622
Dental formula, 622
Dentate nuclei, 364
Dentin, 624
Dentistry, 293, 395
Dentoalveolar joint, 194
Deoxyribonucleic acid. *See* DNA
Depolarized nerve fiber, 347
Depressed fracture, 187
Depression
  in bone, 133
  central nervous system and mental, 385
  as type of movement, 203
Depressor anguli oris muscle, 242
Depressor labii inferioris muscle, 242
De re medicina (Celsus), 6, 11, 12
Dermabrasion, 126
Dermatitis, 103, 118, 119
Dermatoglyphics, 107
Dermatology, 126
Dermatome, 416
Dermis, 107
Descartes, René, 6, 419
Descending colon, 637
Descending limb, of nephron loop, 661, 664
Descending tracts, of spinal cord, 373, 374–75, 376
Descriptive terminology, 33, 34–36
Detached retina, 489, 515
Development, prenatal. *See also* Fetus; Growth
  adolescence and, 759
  of appendicular skeleton, 185
  of axial skeleton, 162–63
  of circulatory system, 568–71
  clinical case study, 735, 769
  clinical considerations, 765–69
  of digestive system, 644–47
  of ear, 512–14
  embryonic period of, 742–52
  of endocrine system, 460–62
  of eye, 498–99
  of female reproductive system, 724–25
  fertilization and, 735–37
  fetal period of, 752–55
  of integumentary system, 116–17
  labor and parturition, 755–56
  of male reproductive system, 696–99
  of muscular system, 280–81
  of nervous system, 336–37
  of peripheral nervous system, 416
  preembryonic period of, 737–42
  of respiratory system, 604–6
  of spinal cord, 378
  of synovial joints, 220
  of tissues, 74–75
  of urinary system, 670–72
Developmental disorders. *See also* Disease; Genetic disorders
  of abdominal region, 327–29
  of appendicular skeleton, 186
  of axial skeleton, 161
  of central nervous system, 380
  congenital malformations, 769
  of digestive system, 648
  of eyes and ears, 511
  of female reproductive system, 726, 727
  of head and neck, 322, 325
  of hip and lower extremity, 330
  of male reproductive system, 695, 699
  of respiratory system, 606–7
  of shoulder and upper extremity, 329–30
  of urinary organs, 669, 672
Diabetes insipidus, 447, 463
Diabetes mellitus, 454, 463, 465
Diagastric muscle, 246
Diagnostic procedures, for gynecological examination, 723–24, 726
Diapedesis, 525
Diaphragm, 25, 41, 248, 599
Diaphragm, as contraceptive device, 731, 732
Diaphragma sellae, 369
Diaphysis, of compact bone, 134
Diarrhea, 49, 329, 639
Diastole, 536
Diastolic pressure, 542
Diencephalon
  derivation and functions of, 352
  development of, 337
  epithalamus, 363
  hypothalamus, 362–63, 434
  pituitary gland, 363, 434
  section through, 354
  structure and function of, 361
  thalamus, 361–62
Diet. *See also* Nutrition; Nutritional disorders; Obesity
  fats, 91
  minerals, 132
  vitamins, 110, 132, 161, 631
Diffusion, through cell membranes, 53
Digastric muscle, 247, 394
Digestion, as function of digestive system, 615
Digestive enzymes, 633
Digestive system
  circulatory system and, 578
  clinical case study, 615, 652
  clinical considerations, 648–50, 652
  development of, 644–47
  endocrine system and, 466
  esophagus and stomach, 628–32
  functions of, 32
  integumentary system and, 125
  interaction with other body systems, 651
  large intestine, 636–39
  liver, gallbladder, and pancreas, 639–43, 648
  male reproductive system and, 703
  mouth, pharynx, and associated structures, 620–27
  muscular system and, 284
  nervous system and, 384
  respiratory system and, 610
  serous membranes and tunics, 616–20
  skeletal system and, 165
  small intestine, 632–35
  structure and functions of, 615, 616
  urinary system and, 674
Digit(s), 40. *See also* Fingers; Toes
Digital arteries, 548, 553
Digital rays, 185, 749
Dihybrid cross, 764
Dilation stage, of labor, 755, 756
Diphyodont dentition, 622
Diploe, 133, 134
Diploid number, 762, 763
Diplopia, 487, 494, 516
Directional terms, for human body, 34, 36
Disaccharides, 50
Discontinuous capillaries, 539, 540
Discourse on Method (Descartes), 6
Discover (journal), 19
Disease(s). *See also* Developmental disorders; Genetic disorders; Hormonal disorders; Infections; Nutritional disorders
  of abdomen, 328–29
  of blood, 572
  of breasts and mammary glands, 729
  cellular damage and, 68
  of digestive system, 648–49
  of ear, 516–17
  etiology of, 47
  of eye, 491, 516
  hair samples and, 112
  of head and neck, 326
  of heart, 572–74
  of hip and lower extremity, 330
  of joints, 221–22
  of liver, 649–50
  of male reproductive system, 700–702
  of muscles, 282–83
  of nervous system, 383, 385
  placenta as barrier against, 746
  of respiratory system, 608, 611
  of shoulder and upper extremity, 329–30
  of skin, 118
  tetanus, 241
  of thorax, 327
  tissue composition and, 98–99
  of vagina and vulva, 729
Disease atrophy, 67
Dislocated articular facet, 221
Dislocations, of joints, 211, 219
Displaced fracture, 187
Dissection, of human cadavers
  history of anatomy and, 12, 13
  regional approach to anatomy and, 30
  science of anatomy and, 2, 4
Distal, as directional term, 36
Distal convoluted tubule, 661, 664
Disuse atrophy, 67, 98
Diverticulitis, 652
Diverticulosis, 652
Dizygotic twins, 766–67
Dizziness, and vestibular organs, 508
DNA, 62, 63–64, 106, 763
Dominant allele, 763
Donor site, and skin graft, 121
Dopamine, 447
Dorsal hollow nerve cord, 23, 24
Dorsal interossei muscle, 265
Dorsal motor nucleus, 392, 396
Dorsal pedal artery, 553
Dorsiflexion movement, 201, 202, 206
Dorsum of foot, 40
Dorsum of hand, 40
Double helix, structure of DNA, 63, 64
Double uterus, 726, 727
Double vision, 487, 494
Douglas, James, 716
Down, John L. H., 744
Down syndrome, 68, 744, 768
Dreams and dreaming, 695
Drug abuse
  nervous system and, 380
  pregnancy and, 746
Drugs
  cellular dysfunction and, 68
  pregnancy and, 749
Dry skin, 103
DSR. *See* Dynamic spatial reconstructor
Duchenne, Guillaume G. A., 329
Ductus arteriosus, 561, 562, 570
Ductus deferens, 681, 688–89, 702
Ductus venosus, 561, 562
Duodenal glands, 633
Duodenal papilla, 633
Duodenojejunal flexure, 633
Duodenum, 633, 634
Duplication of ureters, 669
Dural sheath, 369
Dura matter, 367, 369
Dutrochet, René H., 16
Dwarfism, 161, 446, 463, 464
Dynamic spacial reconstructor (DSR), 18, 379
Dysentery, 649
Dyslexia, 382
Dysmenorrhea, 728
Dysplasia, 729
Dyspnea, 601, 602
Dysuria, 672

E

Ear(s). *See also* Hearing
  development of, 512–14
  disorders of, 511, 516–17
  inner ear, 502–5, 509, 510
  middle ear, 501
  outer ear, 499–500
  surface anatomy of, 293, 294
Eardrum, 500, 514
Earlobe(s), 293, 294, 500, 764
Eccrine sweat glands, 115
Ectoderm, 75, 740
Ectopic pacemaker, 572
Ectopic pregnancy, 715, 728, 732, 765–66
Eczema, 126
Edema, 87, 563, 572
EEG. *See* Electroencephalogram
Effector organ, 411
Efferent ductules, 683
Efferent lymphatic vessel, 564
Efferent neurons, 342. *See also* Motor neuron
Egg. *See* Ova
Egypt, and history of anatomy, 7–8
Ejaculation, of semen, 693–95
Ejaculatory ducts, 666, 681, 689
Elastic arteries, 538
Elastic cartilage, 93
Elastic connective tissue, 88, 89, 91

Elastic fibers, 86, 87
Elasticity, of muscle fibers, 227
Elastin, 69, 86
Elbow
bones of, 207, 211–12
muscles of, 257–58, 259
pelvic region and angle of, 318
ulnar nerve, 404
Elderly. *See also* Aging
integument of, 124
muscles and, 283
ossification of rib cage, 195
Electrical defibrillation, 572
Electrocardiogram, 536–37, 567
Electrocardiograph, 536
Electroencephalogram (EEG), 356–57, 358, 379–80
Electrolytes, 49, 362
Electromagnetic spectrum, 496
Electronic Monitoring of Fetal Heart Rate and Uterine Contractions (FHR-UC Monitoring), 768, 769
Electron microscope, 17, 73
Elements, 48
Elements of Physiology (Müller), 6
Elevation, as type of movement, 203
Embalming, Egyptian methods of, 7
Embedding, of tissue samples, 99
Embolism, 575
Embolus, 575, 577
Embryo. *See also* Development; Embryology
development of human, 74–75, 742
schematic diagram of chordate, 24
skeletal muscle and, 281
skull of, 163
structural changes in by week, 748–52
Embryoblast, 75, 737
Embryology, 10, 742
Embryonic connective tissue, 86
Embryonic Development and Induction (Spemann), 6
Embryonic disc, 740
Embryonic period, of prenatal development, 736–37, 742–52
Emission, of semen, 693–95
Emotions
crying response and, 485
higher brain centers and, 435
hypothalamus and, 363
Emphysema, 595, 608, 611
Enamel, of tooth, 624
Encephalitis, 382
Endemic goiter, 464
Endocarditis, 327, 529
Endocardium, 529
Endochondral bone formation, 137
Endochondral ossification, 91, 162
Endocrine cells, 630
Endocrine disorders, 463
Endocrine glands, 439
Endocrine system. *See also* Hormonal disorders; Hormones
adrenal glands, 455–58
circulatory system and, 578
clinical case study, 439, 465
clinical considerations, 462–65, 467
development of, 460–62
digestive system and, 651
functions of, 31
gonads, 458
hypothalamus and, 363
integumentary system and, 125
interactions with other body systems, 466
introduction to, 439–44
male reproductive system and, 703
muscular system and, 284
nervous system and, 384
pancreas, 453–54, 455
pineal gland, 458
pituitary gland and, 444–48, 449
placenta, 459, 462, 746, 747
respiratory system and, 610
skeletal system and, 165
testes and, 684–85
thymus, 459
thyroid and parathyroid glands, 450–53
urinary system and, 674
Endocrinology, 462
Endocytosis, 53
Endoderm, 75, 740
Endolymph, 502, 508, 510
Endometriosis, 729
Endometrium, 717
Endomysium, 229
Endoneurium, 344
Endoplasmic reticulum (ER), 51, 55–56
Endorphin(s), 349
Endosteum, 134
Endothelium, 76, 538
Endothermic animals, 522
Energy deficit, 68
Enkephalin(s), 349
Entamoeba histolytica, 649
Enteritis, 329, 652
Enterokinase, 633
Enuresis, 672
Enzymes, 50, 633
Eosinophil(s), 525, 526
Ependymal cells, 340
Ependymoma, 382
Epicanthic fold, 484
Epicardium, 529
Epicondyle(s), 133, 180, 181
Epicranius muscle, 242
Epidermis
coloration of, 106
development of, 116–17
glands in, 114–15, 118
hair, 111–13
nails, 113–14
names and characteristics of layers of, 103–6
surface patterns of, 106–7
Epididymis, 681, 688
Epidural anesthesia, 368, 755
Epidural space, 368, 369
Epiglottic cartilage, 588
Epiglottis, 588
Epilepsy, 381–82
Epimysium, 229
Epinephrine
adrenal glands and, 424–25, 431, 456
motor units of muscular system, 238
norepinephrine compared to, 457
therapeutic administration of, 442
Epineurium, 344
Epiphyseal plate
age and bone development, 139, 188
histological zones of, 137
linear bone growth and, 134, 188
photomicrograph of, 138
Epiphysiolysis, 166
Epiphysis, of spongy bone, 134
Epiploic appendage(s), 638
Epispadias, 669, 672
Epistaxes, 607
Epithalamus, 363
Epithelial tissue, 73, 74–85, 586
Epithelium
anterior corneal, 487
follicular, 710–11
germinal, 685
glandular, 82–85, 420
keratinized and nonkeratinized stratified squamous, 79, 80, 82
neuroepithelium, 76
pseudostratified ciliated columnar, 78, 79, 82
respiratory, 78
simple, 76–78, 82
simple ciliated columnar, 78, 82
simple columnar, 76–77, 82
simple cuboidal, 76, 82
simple squamous, 76, 82
stratified, 78–80, 82
stratified cuboidal, 80, 81, 82
superficial, 709
transitional, 80, 81, 82
Eponychium, 114
Equilibrium, mechanics of, 505–6
Erasistratus, 5, 11
Erb, Wilhelm H., 329
Erb-Duchenne palsy, 329
Erection, of penis, 692, 693, 694
Erector spinae muscle(s), 252, 253
Erythema, 106, 126, 572
Erythroblastosis fetalis, 746
Erythrocyte(s). *See also* Red blood cell
formed elements of blood and, 94, 95, 524, 525, 526
functions of, 521, 526
shape of cell, 48
Erythropoiesis, 528
Esophageal arteries, 549
Esophageal atresia, 648
Esophageal hiatus, 628
Esophagus, 298, 616, 628
Esotropia, 516
Estienne, Charles, 520
Estriol, 746
Estrogen
adrenal glands and, 456
classes of hormones and, 442
menstruation and, 723
ovaries and, 709, 723
placenta and pregnancy, 746, 747
Ethmoidal sinus, 149, 587
Ethmoid bone, 148–49, 392, 585
Ethnic groups, 27
Etiology, of disease, 47
Eupnea, 602
Eustachian tube, 23, 501, 513, 587
Eustachio, Bartolommeo E., 501
Eustachius (1524–74), 14
Eversion, 203
Examination, of tissue samples, 99
Exercise, benefits of for elderly, 283
Exocrine glands, 82, 84, 85, 439
Exocytosis, 53, 56
Exophthalmos, 464
Exotropia, 516
Experiments with Plant Hybrids (Mendel), 6
Expiration, and mechanics of breathing, 599–600, 601
Expiratory portions, 603
Expiratory reserve volume (ERV), 601
Expulsion stage, of labor, 755, 756
Exstrophy, of urinary bladder, 669
Extensibility, of muscle tissue, 227
Extension
of hand, 261–63
as type of movement, 201, 202, 205, 206
Extensor carpi radialis muscles, 261–62, 263
Extensor carpi ulnaris muscle, 262, 263
Extensor digiti minimi muscle, 262, 263
Extensor digitorium brevis muscle, 276
Extensor digitorum communis muscle, 262, 263
Extensor digitorum longus muscle, 273, 278
Extensor hallucis longus muscle, 273, 278
Extensor muscle(s), 207
Extensor pollicis brevis muscle, 262, 263
Extensor pollicis longus muscle, 262, 263
External, as directional term, 36
External acoustic canal, 293, 294, 500
External acoustic meatus, 147, 514
External anal sphincter, 637–38
External carotid artery, 544, 546–47
External hydrocephalus, 372
External iliac artery, 550
External iliac vein, 557–58
External intercostal muscles, 599
External jugular vein(s), 556
External occipital protuberance, 147
External otitis, 516
External respiration, 583
External urethral ostium, 691
External urethral sphincter, 666
Exteroceptors, 472
Extraembryonic membranes, 742–45
Extrapyramidal tracts, of spinal cord, 377
Extrinsic laryngeal muscles, 589
Extrinsic ocular muscles, 485–86, 497, 499
Extrinsic tongue muscles, 244
Eye. *See also* Eyeball; Ocular region
accessory structures of, 483–87
autonomic nervous system and, 430, 432
camera analogues for structure of, 482
development of, 498–99
disorders of, 511, 514–16
neural pathways and, 496–97
ocular muscles, 241, 244
shock and, 577
specialized cell membranes, 54, 55
stereoscopic vision, 28
Eyeball
accomodation of lens, 493
aqueous humor of, 495
blood supply to, 490–91
cavities and chambers of, 491
constriction of pupil, 493–94
convergence of, 494
fibrous tunic, 487
internal tunic, 488–90
refraction of light rays, 492–93, 495
retina, 488–90
transmission of light rays, 491
vascular tunic, 487–88
visual spectrum, 494, 496
Eyebrow(s), 293, 294, 483
Eyelash(es), 293, 294, 484
Eyelids, 293, 294, 483, 499
Eye movements, 497

F

Fabricus ab Aquapendente (Hieronymus), 6
Face, surface anatomy of, 292–94, 295
Facet, of bone, 133
Facial artery, 546

Facial bones, 139, 149–52
Facial expression, muscles of, 240, 241, 242, 293
Facial muscles, 25, 28
Facial nerves
  chorda tympani branch of, 481
  dysfunctions of, 399
  parasympathetic division of autonomic nervous system and, 428
  structure and function of, 392, 395–96
  taste and, 481, 482
Facial region, 37
Facilitated diffusion, 53
Fainting, 380
Falciform ligament, 617, 640
Fallopian tubes, 15, 707, 714–15. *See also* Uterine tubes
Fallopius, Gabriele (1523–62), 14, 714
Fallot, Étienne-Louis A., 573
False labor, 755
False ribs, 159
False vocal cords, 589
Falx cerebelli, 364, 369
Falx cerebri, 353, 369
Familial cretinism, 770
Family, taxonomic, 27
Fascia, 87, 229–30
Fasciculus, 229, 344
Fasciculus cuneatus tract, 376
Fasciculus gracilis tract, 376
Fasciculus Medicinae (de Ketham), 706
Fat. *See also* Adipose tissue; Diet
  carbohydrates and, 50
  health consequences of excessive, 91
Fauces, 620
Feces, 638
Feet. *See* Foot
Female. *See also* Female reproductive system
  adipose deposition, 289
  angle of elbow joint to pelvic region, 318
  body composition of average adult, 761
  development during adolescence, 759
  hypodermis and, 288
  pelvis of, 179–80
  sexual dimorphism in adults, 760–61
Female condom, 731, 732
Female reproductive system. *See also* Female; Female reproductive system
  clinical case study, 707, 732
  clinical considerations, 723–32
  development of, 697, 698, 724–25
  mammary glands, 719–21
  ovaries and, 709–14
  ovulation and menstruation, 722–23
  perineum, 682
  secondary sex organs, 714–19
  structure and functions of, 32, 707–9
Femoral artery, 552
Femoral circumflex arteries, 552
Femoral hernia, 285
Femoral nerve, 405, 406, 408
Femoral region, 40
Femoral triangle, 318–19, 552, 553
Femoral vein, 557
Femur
  bone growth and stress, 139
  compact bone tissue of, 136
  location and distinguishing features of, 186
  structure of lower extremity and, 131, 180, 186
Fenestrae, of glomerular epithelium, 660
Fenestrated capillaries, 539, 540
Fertilization, of ovum, 735–37
Fetal monitoring, 767–68
Fetal period, of prenatal development, 737, 752–55
Fetal position, 754
Fetoscopy, 768, 769
Fetus. *See also* Development; Infants; Neonates
  circulatory system of, 560–62
  structural changes in by week, 752–55
Fever blister, 126
FHR-UC Monitoring, 768, 769
Fibrillation, of muscles, 285
Fibrils, 51, 59, 60, 61
Fibrinogen, 123, 528
Fibroadenoma, 729
Fibroblast(s), 86, 123
Fibrocartilage, 91, 92, 93
Fibromyositis, 282–83
Fibrous joints, 192–94, 200
Fibrous layer, of vagina, 718
Fibrous pericardium, 529
Fibrous tunic, 487, 491
Fibula, 131, 181–82, 186
Fibular artery, 553
Fibular nerve, 407, 409, 410
Fibular notch, of tibia, 180–81
Fight or flight response, 429, 456, 457
Filiform papillae, 481, 622
Filtration, through cell membranes, 53, 54
Filum terminale, 373
Fimbriae, 714–15
Finger(s), 40, 258–63. *See also* Digits
Fingernails, 113–14, 291
Fingerprints, 106–7
First-degree burns, 119
First-degree frostbite, 121
First meiotic division, 685
First sound, of heart, 537
First trimester, of gestation, 765
Fissure, of bone, 133
Fissure of Rolando, 355
Fistula, 305
Fixation, of tissue samples, 99
Flaccid paralysis, 381
Flagella, 51, 61
Flat bones, 133
Flexion
  as type of movement, 201, 202, 205, 206
  of wrist, hand, and fingers, 260–61
Flexion creases and lines, in skin, 107
Flexor carpi radialis muscle, 260, 263
Flexor carpi ulnaris muscle, 261, 263
Flexor digiti minimi muscles, 265
Flexor digitorum longus muscle, 275, 278
Flexor digitorum profundus muscle, 263
Flexor digitorum superficialis muscle, 263
Flexor hallucis longus muscle, 275, 278
Flexor pollicus brevis muscle, 265
Flexor pollicus longus muscle, 261, 263
Flexor reflex, 412, 413
Floating ribs, 159
Fluoroscopic analysis, of heart, 567
Fold, of buttock, 318
Folia, 364
Follicle-stimulating hormone (FSH), 446, 448, 723
Follicular cells
  of ovaries, 724
  of thyroid gland, 450
Follicular epithelium, 710–11
Follicular hyperkeratosis, 103
Fontanel(s), 140–41, 291
Food poisoning, 649
Foot. *See also* Phalanges
  arteries of, 553
  bipedal locomotion and, 28
  bones of, 182–84, 186
  muscles of, 273–78, 279
  surface anatomy of, 320, 321
Foramen cecum, 462
Foramen lacerum, 148
Foramen magnum, 147
Foramen of Magendie, 371
Foramen of Monroe, 371
Foramen ovale
  ethmoid bone and, 148
  fetal heart and, 561, 562, 569
  maxillary nerve and, 392
Foramen rotundum, 148, 392
Foramen spinosum, 148
Foramina
  in cribiform plate of ethmoid bone, 392
  of Luschka, 372
  of skull, 141
Forearm
  bones of, 173–74
  muscles of, 257–63, 317
Foregut
  of digestive system, 644–45
  of respiratory system, 604
Foreskin, of penis, 691
Formed elements, of blood, 94, 524–26
Fornix, of vagina, 717
Fossa, of bone, 133
Fourth ventricle, of brain, 364, 366
Fovea, of thoracic vertebra, 157
Fovea capitis femoris, 180
Fovea centralis, 490
Fracture hematoma, 188
Fractures, of bone
  avulsion and, 187
  capillary, 187
  of cervical vertebrae, 157
  closed, 187
  Colles', 187
  comminuted, 187
  complete, 187
  compound, 187
  contrecoup, 325
  depressed, 187
  description of types, 187–88
  displaced, 187
  fissured, 187
  greenstick, 187
  of hip and lower extremity, 330
  impacted, 187
  of long bone and damage to epiphyseal plate, 135, 169, 194
  medical treatment of, 188
  nondisplaced, 187
  oblique, 187
  open, 187
  partial, 187
  Pott's, 187
  radiograph of healing, 18
  of ribs, 159
  of scaphoid, 174
  of shoulder and upper extremity, 329
  simple, 187
  of skull, 193, 325
  spiral, 187
  transverse, 187
Fraternal twins, 766–67
Freckles, 106
Free nerve endings, 473, 474
Frenulum, 620–21, 691
Frequency, of sound, 505
Frog, brain of, 26
Frontal bone, 146
Frontal lobe, of cerebrum, 355–56
Frontal plane, 34
Frontal sinuses, 146, 587
Frostbite, 68, 121
FSH. *See* Follicle-stimulating hormone
Functional murmur, of heart, 537
Functional proteins, 49–50
Functional syncytium, 422
Fundus
  of stomach, 629
  of uterus, 715
Fungiform papillae, 480, 622
Funiculi, of spinal cord, 374
Funnel chest, 326
Furuncle, 126

G

Gage, Phineas P., 435
Galactosemia, 770
Galen, Claudius, 5, 11–12, 13, 388, 419
Gallbladder, 18, 642–43
Gallstones, 18, 643
Gamete(s), 679, 709
Gamma globulin, 528
Ganglion, 335, 338, 390
Ganglion cells, of retina, 392
Ganglionic gliocytes, 340
Ganglionic neurons, 490
Gangrene, 98, 126
Garrod, Archibald, 6
Gas exchange, as function of respiration, 583
Gastrectomy, 631
Gastric folds, of stomach, 629
Gastric glands, 629, 631
Gastric inhibitory peptide (GIP), 459
Gastric phase, of gastric secretion, 631
Gastric pits, 629, 631
Gastric secretion, phases of, 631
Gastric vein(s), 560
Gastrin, 459
Gastrocnemius muscle, 274–75, 278
Gastrocolic reflex, 639
Gastroepiploic vein, 559
Gastroileal reflex, 638
Gastrointestinal (GI) tract
  autonomic nervous system and, 430
  disorders of, 650, 652
  endocrine system and, 459
  hypothalamus and activity of, 362
  innervation of, 620
  layers of, 617–19
  mechanical activity in, 639
  regions and functions of, 615, 616
Gaucher's disease, 768
Gender. *See* Female; Male; Secondary sex characteristics
Gene(s), 763. *See also* Chromosomes; Genetic disorders; Genetics
General anesthetic, 380
General conditions, and disease, 98
General senses, 472
Genetic counseling, 68

Genetic disorders
amniocentesis and, 744
cancer and, 69
chorionic villus biopsy, 745
cytology and, 68
mitochondria and, 58
Genetic Implications of the Structure of Deoxyribonucleic Acid (Crick & Watson), 6
Genetics. *See also* Genetic disorders
inheritance and, 762–65
medical, 68
Genetic sex, of zygote, 696
Geniculate ganglion, 392, 396
Geniculostriate system, 496
Genioglossus muscle, 244, 245
Geniohyoid muscle, 247
Genital herpes, 701
Genitalia. *See* Female reproductive system; Male reproductive system
Genital tract, 724
Genital tubercle, 696, 724, 725
Genitofemoral nerve, 406
Genotype, 763–64
German Anatomical Society, 16
Germinal epithelium, 685
Germinal (basal) layer, 116
Germ layers, 75, 740–42
Germplasm, The (Weismann), 6
Gestation, 742, 755. *See also* Childbirth; Pregnancy
GH. *See* Growth hormone
GH-IH. *See* Growth-hormone-inhibiting hormone
Gigantism, 161, 446, 463, 464
Gingiva, 624
Gingivitis, 652
GIP. *See also* Gastric inhibitory peptide
Glabellar reflex, 415
Glands
accessory reproductive, 689–90
acinar, 85
adrenal, 424–25, 441, 455–56, 458, 462, 465, 467
apocrine, 83, 84, 85, 115
areolar, 720
Bartholin's, 15, 719
branched acinar, 85
branched tubular, 85
Brunner's, 633
bulbourethral, 667, 681, 689, 690
ceruminous, 118, 500
coiled tubular, 85
Cowper's, 667, 690
duodenal, 633
eccrine sweat, 115
endocrine, 439
exocrine, 82, 84, 85, 439
gastric, 629, 631
holocrine, 83, 84, 85
mammary, 25, 115, 117, 249, 708, 719–21, 729
merocrine, 82–83, 84, 85
minor vestibular, 719
salivary, 625, 626, 627, 649
sebaceous, 114, 116, 293
simple, 82, 84, 85
of skin, 114–18
sudoriferous, 114–15
sweat, 105, 114–15, 116, 293
tarsal, 484
tubular, 82, 85
tubuloacinar, 82, 85
unicellular, 82, 85
urethral, 666, 690
vestibular, 708
Glandular epithelium, 82–85, 420
Glans clitoris, 719
Glans penis, 691
Glaucoma, 516
Glenohumeral joint, 210–22
Glenoid cavity, 169
Glenoid labrum, 210
Glial cells, 97, 338. *See also* Neuroglia
Gliding joints, 198, 200
Glioma, 382
Glisson, Francis, 16
Globins, 524
Globulin, 528
Globus pallidus, 358
Glomerular arterioles, 660, 661, 662
Glomerular capillary, 663
Glomerular capsule, 660, 662, 664
Glomerular filtration, 663
Glomerular membranes, 663
Glomerulonephritis, 673
Glomerulus, 660, 662, 663, 664
Glossopalatine arch, 621
Glossopharyngeal nerve
dysfunctions of, 399
gustatory pathway to brain and, 482
parasympathetic division of autonomic nervous system and, 428
structure and function of, 392, 396, 397
taste buds and, 481
Glottis, 588
Glucagon, 454, 455, 643
Glucocorticoid(s), 456
Gluteal arteries, 550, 552
Gluteal line(s), 179
Gluteal muscle(s), 266, 267
Gluteal nerve(s), 409
Gluteal region, 39, 317–18
Gluteal surface, 179
Gluteus maximus muscle, 266, 267
Gluteus medium muscle, 266, 267
Gluteus minimus muscle, 266, 267
Glycogen, 50
Glycoproteins, 56
GnRH. *See* Gonadotropin-releasing hormone
Goblet cells
ciliated cells and, 61
gastric glands and, 629
mucosa of gastrointestinal tract and, 617
simple epithelial tissue and, 77
trachea and, 590
unicellular glands and, 82, 83
Goiter, 452, 464
Golgi complex, 51, 56, 57
Golgi tendon organs, 476
Gomphoses, 194, 200
Gonad(s)
endocrine function of, 458
female reproductive system and, 441, 707
as homologous structure, 725
location of female and male, 441
male reproductive system and, 441, 679
Gonadal ridge, 696, 724
Gonadocorticoid(s), 456
Gonadotropin(s), 446
Gonadotropin-releasing hormone (GnRH), 447, 723
Gonorrhea, 700, 701
Goose, brain of, 26
Gorilla, hand of, 26
Gout, 770
Gouty arthritis, 222
Graafian follicles, 16, 712
Gracilis muscle, 268, 269
Graft bed, and skin graft, 121
Grand mal epilepsy, 381–82
Granular leukocytes, 525
Granulation tissue, 123
Granulocytes, 526
Grave robbing, and history of anatomy, 13
Graves, Robert James, 464
Graves' disease, 464, 465
Gray, Henry, 6
Gray commissure, of spinal cord, 374
Gray matter
of brain, 348
of spinal cord, 373–74
Gray rami communicantes, 423
Gray ramus, 401
Greater cornua, 152
Greater curvature, of stomach, 629
Greater omentum, 42, 617
Greater pelvis, 177–78
Greater splanchnic nerve, 424
Greater trochanter, 180
Greater tubercle, 172
Great saphenous vein, 557, 558
Greece, and history of anatomy, 9–11, 584
Greek language, and anatomical nomenclature, 2, 33
Greenstick fracture, 187
Gross anatomy, 17
Ground substance, 86
Growth. *See also* Development
of bone, 137–39
cell cycle and, 62
hormones and, 30, 446
periods of postnatal, 756–65
Growth hormone (GH), 446, 448, 463
Growth-hormone-inhibiting hormone (GH-IH), 446
Growth-inhibiting substance, 67
Guanine, 63
Gubernaculum, 696
Gums. *See* Gingiva
Gustatory cells, 480
Gustatory microvillus, 480
Gustatory pathway, 482
Gustatory receptors, 480, 481
Gustatory sense, 480–81
Gynecology, 723
Gyrus. *See* Cerebral gyri

H

Hair
as characteristic of mammals, 25, 111
development of, 116, 117
functions of, 112
kinds of, 113
Hair cells, 54, 55, 508, 509
Hair follicle, 112, 116
Hairline, 292
Halitosis, 652
Haller, Albrecht von, 6
Hallux, 184, 198
Hamate bone, 174
Hammer toe, 321
Hamstrings, 272, 273
Hand(s). *See also* Manus; Phalanges
arteries of, 548, 549
bones of, 139, 174, 175, 176, 199
development of, 758
muscles of, 258–65, 317
of primates, 25, 26
Hard palate, 149, 621
Hare lip. *See* Cleft lip
Harvey, William, 5, 6, 15
Haustra, 638
Haustral churning, 638
Havers, Clopton, 94, 136
Haversian canal, 94
Haversian systems, 136
H bands, 237
hCG. *See* Chorionic gonadotropin
hCS. *See* Chorionic somatomammotropin
Head
arteries of, 544–47
as body cavity, 41, 43
as body region, 37
of bone, 133
of epididymis, 688
of pancreas, 643
radiographic scans through, 19
surface anatomy of, 291–97
trauma and disease, 321–22, 325–26
upper respiratory tract, 586
veins of, 556
Headache, 380, 547
Health, and condition of fingernails, 114. *See also* Case studies; Clinical considerations; Diet; Disease; Genetic disorders; Nutritional deficiencies
Hearing. *See also* Ear
auditory impairment, 517
progressive loss of, 471
sensory organs and, 499–510
Heart. *See also* Cardiac muscle; Cardiovascular disease
autonomic nervous system and, 430
cardiovascular assessment, 567, 572
chambers and valves of, 529–34
characteristics of human, 25
circulatory routes and, 534–35
conduction system of, 535–36
congenital malformations of, 326
development of, 569–70
diseases of, 572–74
electrocardiograms of, 536–37
historical association of with emotions, 7
location and general description of, 529
as major component of circulatory system, 521–22
pericardial membranes, 42
serous membranes, 43
sounds of, 537
wall of, 529
Heart attack, 535. *See also* Cardiovascular disease
Heartbeat, variations in, 536
Heartburn, 628, 652
Heart cords, 569
Heart disease. *See* Cardiovascular disease
Heart murmurs, 537, 572
Heart sounds, 537
Heart tube, 569
Heat. *See also* Body temperature
muscles and production of, 227
sensory receptors for, 474
Heat exhaustion, 110
Heat stroke, 110
Heavy metals, in hair samples, 112
Heel, 40
Heimlich maneuver, 588, 607–8

Heliobacter pylori, 650, 652
Helicotrema, 504
Helix, of ear, 293, 294, 500
Hemarthrosis, 224
Hematocrit, 524
Hematuria, 672, 675
Heme, 524
Hemiazygos vein, 557
Hemiplegia, 381
Hemispheres, of brain, 353, 361
Hemivertebrae, 326
Hemocytoblasts, 528
Hemodialysis, 673, 675
Hemoglobin, 106, 524
Hemophilia, 765
Hemopneumothorax, 583
Hemopoiesis, 131–32, 527, 528
Hemorrhage, 553, 565, 576
Hemorrhoids, 575, 638, 652
Hemothorax, 326
Henle, Friedrich Gustav Jacob, 661
Heparin, 87
Hepatic bud, 645
Hepatic duct(s), 642
Hepatic flexure, of large intestine, 637
Hepatic plates, 641
Hepatic porphyria, 770
Hepatic portal system, veins of, 558–60. *See also* Liver
Hepatic portal vein, 559
Hepatic vein(s), 558–59
Hepatitis, 650
Hepatocytes, 641
Hepatoma(s), 650
Hepatopancreatic ampulla, 633
Hereditary hemochromatosis, 770
Hereditary leukomelanopathy, 770
Hereditary traits, and dominant and recessive alleles, 763
Hermaphroditism, 695
Hernias
- congenital diaphragmatic, 327
- definition of, 285, 652
- femoral, 285, 319
- hiatal, 285, 327, 652
- inguinal, 285
- umbilical, 285, 305

Herniated disc, 221, 408, 410
Herophilus, 5, 11
Herpes zoster, 401. *See also* Genital herpes
Hertwig, Oskar, 6
Heterodont dentition, 25
Heterografts, 99
Heterotransplant, 121–22
Heterozygous chromosomes, 763, 764
Hiatal hernia, 285, 327, 652
Hiccuping, 602
Hieronymus (Fabricus ab Aquapendente), 6
Hilum, 564, 594, 657, 709
Hindgut, 646–47
Hinge joints, 198, 200
Hip(s)
- bipedal locomotion, 28
- disease of and trauma to, 330
- joint, 214–15
- muscles of, 265–68

Hippocrates, 5, 9, 10
Hippocratic oath, 9, 10
Hirschsprung, Harold, 328
Hirschsprung's disease, 328
Hirsutism, 112
His, Wilhelm, Jr., 535
Histamine, 87
Histochemistry, 98
Histology
- classification of tissues, 73
- clinical considerations, 98–99
- of connective tissue, 85–95
- definition of, 17, 29, 73
- of ductus deferens, 689
- of duodenum, 634
- of epithelial tissue, 74–85
- of esophagus, 628
- of large intestine, 638
- of mucosa of stomach, 630
- of muscle tissue, 96–97
- of nervous tissue, 97, 98
- of prostate, 690
- of salivary glands, 626
- of seminal vesicle, 689
- of trachea, 591
- of urethra, 690
- of vagina, 718

Histopathology, 98
History, of anatomy
- China and Japan, 8–9
- Grecian period, 9–11
- Islam and classical Arabic period, 12
- Mesopotamia and Egypt, 5, 7–8
- Middle Ages, 12
- nineteenth century, 16
- prehistoric times, 2–4
- Renaissance, 12–14
- Roman era, 11–12
- seventeenth and eighteenth centuries, 14–16
- twentieth century, 16–19

History of Animals (Aristotle), 6, 10
Histotechnology, 98
Hives, 126
Hodgkin, Thomas, 575
Hodgkin's disease, 575
Holocrine glands, 83, 84, 85
Homeostasis, circulatory system and maintenance of, 522
Homer, 5, 9
Hominidae, 27
Homologous structures, of reproductive system, 696, 707, 725
Homo sapiens, 23, 27
Homotransplants, 99
Homozygous chromosomes, 763
Homozygous dominant alleles, 764
Homozygous recessive alleles, 764
Hooke, Robert, 16, 47
Horizontal plane, 34
Horizontal plates, of palatines, 150
Hormonal disorders, and axial skeleton, 161
Hormonal hypertrophy, 67
Hormone(s). *See also* Endocrine system
- actions of, 441–43
- adrenal cortex, 456
- classes of, 442
- control of secretion, 443, 444
- definition of, 439
- developmental changes and, 30
- pancreas, 455
- physiological effects of gastrointestinal, 459
- pituitary, 446–48, 449
- placenta and secretion of, 747
- proteins and, 50
- synthetic steroids, 238
- thyroid gland and, 451

Horse, brain of, 26
Horseshoe kidney, 669
Human chorionic gonadotropin (hCG), 462, 739, 740
Human Genome Project, 68
Humans, classification and characteristics of. *See also* Body
- anatomical characteristics of, 27–28
- class Mammalia and, 25
- family Hominidae, 27
- order Primates and, 25, 26
- phylum Chordata and, 23–24

Humeral circumflex arteries, 548
Humeral condyle, 172
Humeroradial joint, 211
Humeroulnar joint, 211
Humerus
- articular surface of, 213
- brachium and, 172–73
- growth of, 195
- muscles of, 254–57
- radiographs of, 171, 195
- structure of upper extremity and, 129

Humoral theory, of body organization, 9, 10
Hunger, hypothalamus and regulation of, 362
Huntington, George, 770
Huntington's chorea, 770
Hyaline cartilage, 91, 92, 93
Hyaline membrane disease, 600, 607
Hyaloid artery, 499
Hyaloid vein, 499
Hyaluronidase, 735
Hydrocele, 702
Hydrocephalism, 370
Hydrocephalus, 322, 372, 380
Hydrocortisone, 456
Hydrogen peroxide, 59
Hydrolysis, 49
Hydronephrosis, 673
Hydroregulation, and skin, 109
Hydroureter, 673
Hygeia, 10
Hymen, 717, 725
Hymenal tag, 291
Hyoglossus muscle, 245
Hyoid bone
- anterior view of, 153
- axial skeleton and, 131
- larynx and, 246, 297
- location and function of, 129, 152, 297

Hyperextension, 201, 205, 206
Hyperglycemia, 454
Hyperopia, 514–15
Hyperparathyroidism, 465
Hyperpigmentation, 103
Hyperplasia, 67
Hypertension, 67, 542, 567, 574
Hyperthermia, 68
Hyperthyroidism, 465
Hypertrophic zone, of epiphyseal plate, 137
Hypertrophy, of muscles, 67, 282
Hypodermis, 108–9, 288
Hypoglossal canal, 147, 393
Hypoglossal nerve, 393, 397–98, 399, 482
Hypoglossal nucleus, 366, 393
Hypoglycemia, 68, 465
Hyponychium, 114
Hypophyseal pouch, 461
Hypophysectomy, 446, 463
Hypopituitarism, 463
Hypospadias, 669 672
Hypotension, 327
Hypothalamo-hypophyseal portal system, 448, 450
Hypothalamo-hypophyseal tract, 447
Hypothalamus
- autonomic nervous system and, 433–34
- endocrine system and, 440–41
- structure and functions of, 352, 362–63

Hypothenar muscles, 263, 265
Hypothermia, 68, 110
Hypothyroidism, 326, 463
Hysterectomy, 728
Hysterosalpingography, 724, 725
H zone, 235

I

I bands, 234, 235, 237
ICSH. *See* Interstitial-cell-stimulating hormone
Identical twins, 767, 768, 769
Ileocecal valve, 634, 636
Ileum, 634
Iliac arteries, 550
Iliac crest, 178–79, 305, 318
Iliac fossa, 179
Iliac spine, 179
Iliac tuberosity, 179
Iliacus muscle, 265, 267
Iliac veins, 557–58
Iliad (Homer), 9
Iliocostalis muscle(s), 252, 253
Iliofemoral ligament(s), 214
Iliohypogastric nerve, 406
Ilioinguinal nerve, 406
Iliolumbar artery, 550
Iliopsoas muscle, 265
Ilium, 178–79
Immune system
- embryo and mother's, 742
- stress and, 458

Immunosuppressive drugs, 99
Impacted fracture, 187
Imperforate anus, 648
Imperforate hymen, 725
Impetigo, 126
Implantation, of fertilized ova, 739
Impotence, 699
Impulse conduction, in spinal cord, 372–73
Inborn Errors of Metabolism (Garrod), 6
Incisive foramen, 149
Incisors, 149
Incompetence, of heart valves, 326
Incontinence, urinary, 672, 675
Incus, 152, 153, 501
Indian subcontinent, human races of, 27
Induced abortion, 729
Induced tissue, 748
Induction, and process of developmental change, 748
Inductor tissue, 748
Infants. *See also* Development; Neonate
- brain of and oxygen deprivation, 349
- dehydration and, 49
- growth of, 757, 761
- surface anatomy of, 289–91

Infections. *See also* Disease
- infectious agents and, 68
- of urinary organs, 665, 673
- of uterus and vagina, 715

Inferior, as directional term, 36
Inferior angle, of scapula, 169
Inferior articular processes, 154
Inferior cerebellar peduncles, 366
Inferior colliculi, 352, 363

Inferior epigastric artery, 552
Inferior gemellus muscle, 266
Inferior gluteal artery, 550, 552
Inferior gluteal nerve, 409
Inferior lacrimal canaliculi, 485
Inferior limb buds, 748
Inferior lobe, of lung, 594–95
Inferior mesenteric artery, 550
Inferior mesenteric ganglia, 424
Inferior mesenteric vein, 559
Inferior nasal concha, 150–51, 585
Inferior oblique eye muscle, 394, 486
Inferior olivary nuclei, 366
Inferior orbital fissure, 149
Inferior phrenic artery, 549
Inferior phrenic vein, 558
Inferior rectus muscle, 394, 486
Inferior salivatory nucleus, 392
Inferior trunk, of brachial trunk, 401
Inferior vena cava, 530, 554, 558
Inferior vesicular artery, 550
Infertility
  female reproductive system and, 728
  male reproductive system and, 679, 681, 684, 699–700
Inflammation and inflammatory conditions, of skin, 118, 119
Influenza, 608
Infrahyoid muscle(s), 246
Infraorbital foramen, 149
Infraspinatus muscle, 256
Infraspinous fossa, 169
Infundibulum, 363, 446, 461, 714
Ingestion, as function of digestive system, 615
Ingrown toenail, 321
Inguinal canal, 689
Inguinal hernia, 285
Inguinal ligament, 310–11
Inguinal lymph nodes, 565
Inheritance, and genetics, 762–65
Injury. *See also* Fractures; Trauma
  to appendicular skeleton, 187–88
  to central nervous system, 380–81
  to knee, 216
  to respiratory system, 607–8
Inner ear
  ampullae within, 510
  cochlea, 502, 503, 504–5
  development of, 513
  maculae of, 509
  semicircular canals, 502
  vestibule, 502
Inner-ear cavity, 140
Innervation. *See also* Nervous system
  of gastrointestinal tract, 620
  of ovaries, 710
  of skin, 108
  of smooth muscle, 422
  of uterus, 717
Inorganic compounds, 49
Insertion, of muscles, 229
Inspection, as clinical procedure, 34
Inspiration, and mechanics of breathing, 598–99, 601
Inspiratory portions, 603
Inspiratory reserve volume (IRV), 601
Insula, of cerebrum, 356
Insulin, 454, 455, 465, 643
Insulin-dependent diabetes mellitus, 465
Insulin shock, 465
Integrative Action of the Nervous System, The (Sherrington), 6
Integument and integumentary system. *See also* Skin
  circulatory system and, 578
  clinical case study, 102
  clinical considerations, 118–26
  development of, 116–17
  digestive system and, 651
  endocrine system and, 466
  epidermal derivatives, 111–18
  functions of, 31, 109–11
  interaction with other body systems, 125
  layers of, 103–9
  male reproductive system and, 703
  muscular system and, 284
  nervous system and, 384
  respiratory system and, 610
  skeletal system and, 165
  structure and functions of, 102
  urinary system and, 674
Intention tremor, 377
Intercalated discs, 97, 422
Interclavicular ligament, 209
Intercondylar eminences, 180
Intercondylar fossa, 180
Intercostal muscle(s), 248, 599
Intercostal nerve(s), 601
Intercostal space(s), 159
Intercostal vein(s), 557
Interlobar arteries, of kidney, 659
Interlobar vein(s), of kidney, 660
Interlobular arteries, of kidney, 659–60
Interlobular vein(s), of kidney, 660
Intermediate muscles, of hand, 263, 265
Internal, as directional term, 36
Internal anal sphincter, 637–38
Internal anatomy
  of abdomen, 304, 307, 309, 310
  of shoulder and upper extremity, 314, 315, 316, 317
  of thorax, 305–7
Internal carotid artery, 544, 546
Internal cell mass, 737
Internal hydrocephalus, 372
Internal iliac artery, 550
Internal iliac vein, 558
Internal intercostal muscles, 599
Internal jugular vein(s), 556
Internal pudendal artery, 552
Internal respiration, 583
Internal thoracic artery, 547
Internal tunic, 488–90, 491
Internal urethral sphincter, 666
International Congress of Anatomists, 16
Interneuron(s), 342
Interphalangeal joint(s), 212–14
Interphase, 64, 65, 66
Interpretation, of nerve impulse, 471
Interracial marriage, 27
Intersegmental reflex arc, 412
Interstitial cells, testicular, 681, 683
Interstitial-cell-stimulating hormone (ICSH), 447
Interstitial fluid, 87, 522
Interstitial systems, of compact bone tissue, 136
Intertrochanteric crest, 180
Intertrochanteric line, 180
Intertubercular (bicipital) groove, 172
Interventricular foramen, 371
Interventricular septum, 529, 569
Interventricular sulci, 530
Intervertebral disc(s), 24, 93
Intervertebral foramina, 154
Intestinal crypts, 635
Intestinal phase, of gastric secretion, 631
Intracranial tumors, 382
Intraembryonic mesoderm, 748
Intramembranous ossification, 163
Intramuscular injection, 285, 318
Intraocular pressure, in eye, 491
Intrauterine devices (IUDs), 732
Intravenous pyelogram, 672
Intravertebral tumors, 382
Intrinsic factor, 630, 631
Intrinsic laryngeal muscles, 589
Intrinsic ocular muscles, 486, 497
Intrinsic tongue muscles, 244
Inversion, 203
In vitro fertilization, 766
Involuntary body movements, 97
Iodine, 452
Iodine-deficiency goiter, 452
Ionic charge, of cell membrane, 52
Ipsilateral reflex, 411, 412
Iris, 488, 491, 499
Iron-deficiency anemia, 524
Irregular bones, 133
Irritability
  of muscle tissue, 227
  of neurons, 347
Ischemia, 573–74
Ischial tuberosity, 179
Ischiocavernosus muscle, 250, 251
Ischiofemoral ligament, 214
Ischium, 179
Islam, and history of anatomy, 12
Islets of Langerhans, 453, 643. *See also* Pancreatic islets
Isographs, 99
Isometric contraction, 236, 238
Isotonic contraction, 236, 238
Isthmus
  of thyroid gland, 450
  of uterus, 715

J

Jacopo Berengario of Carpi, 12–13, 14
Janssen, Zacharius, 16
Japan, and history of anatomy, 8–9
Jaundice, 106, 291, 572, 650
Jejunum, 634
Joint(s). *See also specific joint*
  cartilaginous, 194–95
  classification of, 192
  clinical case study, 192, 222
  clinical considerations, 218–19, 221–22, 223
  coxal, 214–15
  elbow, 207, 211–12
  fibrous, 192–94
  glenohumeral, 210–11
  metacarpophalangeal and interphalangeal, 212–14
  principle articulations, 219
  sternoclavicular, 209
  synovial, 195–208
  talocrural, 217–18
  temporomandibular, 209
  tibiofemoral, 215–16
  types of, 200
Joint capsule, 89, 196, 220
Joint cavity, 220
Joint kinesthetic receptors, 476
Joint prostheses, 222, 223
Journals, and education in anatomy, 19
Jugular foramen, 147, 392, 393
Jugular notch, 159, 298
Jugular vein(s), 298, 300, 556
Juxtamedullary nephrons, 661, 663

K

Kaitai shinsho (Sugita), 6, 8
Karotype, 762, 763
Keratin, 49, 67, 79, 103
Keratinization, 103
Keratinized stratified squamous epithelium, 79, 80, 82
Keratinocytes, 103
Keratitis, 516
Keratoplasty, 487
Keratosis, 126
Ketham, Johannes de, 13, 706
Kidney(s)
  cell division in, 67
  congenital anomalies of, 669
  endocrine system and, 441
  gross structure of, 658
  microscopic structure of, 658–64
  position and appearance of, 657
Kidney stone, 665
Kidney transplantation, 673
Kinesiology, 192
Kinesiology of the Human Body under Normal and Pathological Conditions (Steindler), 6
Kinesthetic sense, 476
Kinocilium, 505–6
Klinefelter, Harry F., Jr., 699
Klinefelter's syndrome, 699
Klüver, Heinrich, 497
Klüver-Bucy syndrome, 497
Knee. *See also* Patella
  injury to, 192, 201, 222
  movement of, 201
  muscles of, 268, 270–73
  structure and functions of, 215–16
  structure of lower extremity and, 40
Knee-jerk reflex, 411, 412, 414, 415
Knee joint. *See* Tibiofemoral joint
Kölliker, Albrecht von, 6
Krause, Wilhelm J., 474
Kupffer, Karl Wilhelm von, 641
Kupffer cells, 641
Kyphosis, 221

L

Labia majora, 708, 718, 725
Labia minora, 708, 718, 725
Labioscrotal swellings, 725
Labor, and developmental anatomy, 755–56. *See also* Childbirth
Labyrinths, of ear, 502, 503
Lacrimal apparatus, 485
Lacrimal bone, of skull, 150
Lacrimal caruncle, 484
Lacrimal duct, 499
Lacrimal fluid, 485
Lacrimal gland, 485, 499
Lacrimal punctum, 485
Lacrimal sac, 485
Lacrimal secretions, 293
Lacrimal sulcus, 150
Lactase, 633
Lacteals, 562–63, 634
Lactic acid, and muscle fibers, 232, 278
Lactiferous duct(s), 720
Lactiferous sinus, 720
Lacuna, 91, 94, 136, 137
Lamarck, Jean, 16
Lambdoidal suture, 141, 147
Lamella, 94, 136

Lamellated corpuscle(s), 473, 474
Lamina, of vertebra, 154
Lamina propria, 617
Laminectomy, 156, 166
Langerhans, Paul, 453, 643
Language, and brain, 359, 361
Lanugo hair, 113, 291, 754
Laparoscopy, 724
Laparotomy, 43
Large intestine
basic functions of, 616
as digestive organ, 636–39
Laryngeal muscle(s), 589
Laryngeal prominence, 297, 588
Laryngitis, 588, 608
Laryngopharynx, 587
Laryngotracheal bud, 604
Larynx
muscles of, 246
respiratory system and, 588–90, 598
surface anatomy of, 297
Lateral, as directional term, 36
Lateral angle, of scapula, 169
Lateral apertures, of brain, 372
Lateral border, of scapula, 169
Lateral cartilage, 585
Lateral cervical ligaments, 716
Lateral commissure, of eye, 484
Lateral cord, of brachial plexus, 401
Lateral corticospinal tracts, of spinal cord, 376
Lateral crural muscles, 274
Lateral cutaneous femoral nerve, 406
Lateral epicondylitis, 173, 329
Lateral femoral circumflex artery, 552
Lateral funiculi, of spinal cord, 375
Lateral geniculate body, 496
Lateral horns, of spinal cord, 374, 378
Lateral malleolus, 181–82
Lateral rectus eye muscle, 395, 486
Lateral sacral arteries, 550
Lateral spinothalamic tract, 474, 477
Lateral sulcus, 355
Lateral ventricle, 353
Latin, and nomenclature in anatomy, 2, 23, 33. *See also* Rome
Latissimus dorsi muscle, 254–55, 256
Laughing, 602
Lavage fluid, 23, 43
Learning, and synapses between neurons within cerebrum, 355
Leather, and tanning process, 108
Leber's hereditary optic neuropathy, 58
Leeuwenhoek, Antoni van, 5, 15–16, 678
Left aortic arch, 25
Left atrioventricular (AV) valve, 531, 532
Left atrium, 531
Left common carotid artery, 543
Left common iliac artery, 549
Left coronary artery, 535, 542–43
Left gastric artery, 549
Left gastric vein, 560
Left gastroepiploic vein, 559
Left hemisphere, of brain, 353, 361
Left hepatic vein, 558
Left lobe, of liver, 640
Left ovarian vein, 558
Left principal bronchi, 591–92
Left pulmonary artery, 531
Left subclavian artery, 543, 547
Left subclavian vein, 556
Left suprarenal vein, 558
Left testicular vein, 558
Left ventricle, 531, 532, 533, 534
Leg. *See also* Lower extremity
fibula, 181–82
muscles of, 273–78, 279
structure of lower extremity and, 40
surface anatomy of, 319–21, 323, 324, 325
tibia, 180–81
Lens, of eye, 488, 491, 493
Lens capsule, 488, 499
Lens fovea, 499
Lens placode, 499
Lentiform nucleus, 358
Lesions, of skin, 118
Lesser curvature, of stomach, 629
Lesser omentum, 42, 617
Lesser pelvis, 177–78
Lesser sciatic notch, 179
Lesser splanchnic nerve, 424
Lesser trochanter, 180
Lesser tubercle, 172
Leukemia, 525, 572
Leukocyte. *See also* White blood cell
formed elements of blood and, 94, 95, 525, 526
protection against injury or infection as function of, 521
shape of cell, 48
Leukocytosis, 525, 572
Leukopenia, 572
Leukopoiesis, 528
Leukorrhea, 729
Levator ani muscle, 250
Levator labii superioris muscle, 242
Levator palpebrae superioris muscle, 244, 394
Levator scapulae muscle, 253–54, 256
Levers, and body movement, 204–5, 207, 208
Levodopa (L-dopa), 342, 378
Leydig, Franz von, 683
Leydig cells. *See* Interstitial cells
LH. *See* Luteinizing hormone
Lieberkühn, Johann N., 635
Life force, ancient Egyptian concept of, 8
Ligament(s), 88, 196, 716
Ligamentum capitis femoris, 214
Ligamentum denticulatum, 369
Ligamentum nuchae, 298
Ligamentum teres, 641
Light microscope, 73
Light rays, transmission and refraction of by eyeball, 492–93
Limb(s). *See* Arm; Leg; Lower extremity; Upper extremity
Limb bud(s), 185, 748
Limbic system, 434–35
Linea alba, 248, 250, 301
Linea albicans, 108
Linea aspera, 180
Linea semilunaris, 305
Lines of tension, in skin, 107
Lingual artery, 546
Lingual frenulum, 622
Lingual nerve, 482
Lingual tonsils, 587, 622
Linnaeus, Carolus, 6
Lipase, 633
Lip(s), 294, 295, 326, 620
Lipids, 50, 52
Liver. *See also* Hepatic portal system
alcohol and, 560
autonomic nervous system and, 430
cell division, 67
as digestive organ, 640–42
disorders of, 649–50
hyperplasia, 67
Mesopotamian depictions of, 7
Liver lobules, 641
Liver sinusoids, 641
Lobar bronchi, 592
Lobes
of cerebrum, 355–56
of liver, 640–41
of mammary gland, 720
Lobules, of mammary gland, 720
Lobus nervosa, 446, 461
Local anesthetic, 380
Local effect, of burns, 119
Long bones, 133–35, 138, 194
Longissimus capitis muscle, 246
Longissimus muscles, of vertebral column, 252, 253
Longitudinal arch, of foot, 184, 321
Longitudinal cerebral fissure, 353
Loop of Henle, 661
Loose connective tissue, 87, 91
Lordosis, 221
Lower esophageal sphincter, 628
Lower extremity
arteries of, 550–53
as body region, 40
bones of, 131, 176, 180–85, 186
disease of and trauma to, 330
surface anatomy of, 317–21
veins of, 557–58
Lower respiratory system, 584, 604, 606
Lower spinal nerve, 601
Lumbago, 283
Lumbar artery, 550
Lumbar curve, 154
Lumbar enlargement, of spinal cord, 373
Lumbar lymph nodes, 565
Lumbar plexus, 404–5, 406
Lumbar puncture, 379
Lumbar region, 39
Lumbar veins, 558
Lumbar vertebra, 153, 157
Lumbosacral plexus, 407
Lumbosacral trunk, 407
Lumbricales muscle, 265
Lumen, 54, 55
Lumina, of uterine tube, 725
Lunate bone, 174
Lung(s)
autonomic nervous system and, 430, 433
body cavities, 41
body membranes, 42–43
respiratory system and, 593–95
Lung cancer, 420, 435, 611
Lunula, 114
Luschka, Hubert, 372
Luteininzing hormone (LH), 446–47, 448, 723
Luxation, of joint, 219
Lymph, 522, 562–64
Lymphadenitis, 575
Lymphangitis, 575
Lymphatic capillaries, 562–63
Lymphatic nodules, 564
Lymphatic system
circulatory system and, 578
digestive system and, 651
disorders of, 575–76
endocrine system and, 466
functions of, 31, 562
integumentary system and, 125
lymph and lymphatic vessels, 562–64
male reproductive system and, 703
muscular system and, 284
nervous system and, 384
respiratory system and, 610
skeletal system and, 165
tonsils, spleen, and thymus as organs of, 565, 566
urinary system and, 674
Lymph ducts, 563
Lymph nodes
axillary region and, 38, 311
breast and, 721
cancer and removal of, 227
circulatory system and functions of, 522, 564–65, 566
Lymphoblasts, 528
Lymphocyte(s), 525, 526
Lymphoid tissue, 528
Lymphoma, 69, 575
Lymphatic vessels, 522
Lysosome(s), 51, 58–59
Lysozyme, 485

M

Macromolecule(s), 29, 52
Macrophagic cells, 103
Macula, 506, 509
Macular degeneration, 515–16
Magendie, Francois, 371
Magnetic resonance imaging (MRI), and history of anatomy, 18
Major calyx, 658, 661
Male. *See also* Male reproductive system
adipose tissue, 289
body composition of average adult, 761
development during adolescence, 759
pelvis of, 179–80
perineum and scrotum, 681–83
sexual dimorphism in adults, 760–61
Male pseudohermaphroditism, 695, 699
Male reproductive system. *See also* Reproductive system
clinical case study, 679, 702
clinical considerations, 695, 699–702
development of, 696–99
interaction with other body systems, 703
mechanisms of erection, emission, and ejaculation, 693–95
penis, 691–92, 693–95
spermatic ducts, accessory reproductive glands, and urethra, 688–90
structure and functions of, 32, 679, 680, 681
testes, 682, 683–87
Malignant melanoma, 118
Malignant neoplasms, 68
Malleolus, 182
Malleus, 33, 152, 153
Malnutrition, 615
Malpighi, Marcello, 5, 16
Maltase, 633
Mammalia, 25
Mammary alveoli, 720
Mammary ducts, 720
Mammary glands. *See also* Breast
breast self-examination and, 311
as characteristic of mammals, 25
diseases of, 729
function of, 115, 117, 708
pectoralis major muscle and, 249, 311
structure of, 719–21

Mammary region, 38
Mammary ridge, 25
Mammary tissue, 59
Mammography, 729, 730
Mandible, 151–52
Mandibular foramen, 151–52
Mandibular fossa, 147
Mandibular nerve, 392, 394, 395
Mandibular notch, 151
Manic-depressive psychosis, 381
Manubrium, 159
Manus. *See also* Hand
  bones of, 174, 175, 176
  structure of upper extremity and, 39, 40
Marfan, Antoine Bernard-Jean, 770
Marfan's syndrome, 770
Marrow, of bone, 134
Masseter muscles, 241, 243, 394
Mass movement, in large intestine, 638–39
Mast cells, 87
Mastectomy, 254, 729
Mastication
  as function of digestive system, 615
  muscles of, 241, 243
Mastoidal air cells, 501
Mastoid foramen, 147
Mastoiditis, 147, 501
Mastoid part, of temporal bone, 147
Mastoid process, 147, 283
Masturbation, 700
Matrix
  of blood, 528
  of tissue cells, 73
Maxilla, 149, 151
Maxillary nerve, 392, 394, 395
Maxillary sinus, 149, 587
Maxillary temporal artery, 546–47
McBurney's point, 305
Meatal plug, 514
Meatus
  external acoustic, 147, 514
  nasal, 585
Mechanical activities, in gastrointestinal tract, 635, 638–39
Mechanoreceptor(s), 472, 473
Meckel's diverticulum, 327
Medial, as directional term, 36
Medial border, of scapula, 169
Medial commissure, of eye, 484
Medial cord, of brachial plexus, 401
Medial epicondyles, 173
Medial femoral circumflex artery, 552
Medial lemniscus, 477
Medial malleolus, 180
Medial nerve, 406
Medial plantar artery, 553
Medial rectus muscle, 394, 486
Median aperture, of cerebrum, 371
Median branch, of brachial plexus, 404
Median cubital vein, 556–57
Median eminence, 448
Median furrow, of back, 301
Median nerve, of hand, 314
Median sacral crest, 158
Median umbilical ligament, 665, 745
Mediastinal surface, 594
Mediastinum, 41, 593, 597
Medical genetics, 68
Meditation, and autonomic nervous system, 420
Medulla, 112
Medulla oblongata, 366–67, 433
Medullary (marrow) cavity, 134
Megakaryoblasts, 528
Megakaryocytes, 525
Meiosis, 685, 712, 737
Meissner, Georg, 618
Meissner's corpuscles, 473, 474
Meissner's plexus, 618
Melancholy (term), 9
Melanin, 67, 103, 106, 112, 488
Melanoblasts, 116
Melanocyte(s), 103, 106, 116, 126
Melanocyte-stimulating hormone (MSH), 447, 448
Melanoma, 109, 118, 126
Melatonin, 458
Membranous ampulla, 502
Membranous epithelia, 75–76
Membranous labyrinth, 502, 503
Membranous part of urethra, 666, 690
Menarche, 709, 759
Mendel, Gregor, 6
Menes (circa 3400 B.C.), 5, 7
Ménière, Prosper, 517
Ménière's disease, 517
Meningeal branch, of spinal nerve, 401
Meningeal layer, of dura mater, 367, 369
Meningeal vein, 556
Meninges, 348, 367–69, 547
Meningioma, 382
Meningitis, 149, 369, 382
Meniscus, 93, 196, 216, 220
Menopause, 707, 710
Menorrhagia, 728
Menstrual phase, 722–23
Menstruation, 707, 709, 722–23
Mental foramen, 151–52
Mental illness, 381, 385
Mentalis muscle, 242
Mental retardation, 380
Merocrine glands, 82–83, 84, 85
Mesencephalic aqueduct, 353, 371
Mesencephalon, 337, 352, 363
Mesenchyme, 86, 116, 162, 748
Mesenteric ganglion, 424
Mesenteric lymph nodes, 565
Mesenteric patches, 634
Mesentery, 42, 617, 632
Mesocolon, 617, 636
Mesoderm, 75, 740
Mesonephric duct, 696
Mesonephros, 671
Mesothelium, 76
Mesovarium, 709
Metabolism, 47
Metacarpal arteries, 548
Metacarpal bone(s), 129
Metacarpophalangeal joint(s), 212–14
Metacarpus, 40, 174
Metanephrogenic mass, 671
Metanephros, 671
Metaphase, 65, 66, 685, 686, 712, 737
Metaplasia, 67
Metastasis, 68, 565
Metastatic neoplasms, 382
Metatarsal bone(s), 131, 186
Metatarsophalangeal joints, 184
Metatarsus, 40, 184
Metencephalon, 337, 352, 364–66
Metopic suture, 146
Metrorrhagia, 728
Michelangelo, 2, 3, 14
Microcephaly, 322, 325, 380
Microglia, 340, 341
Micromelia, 185, 329
Microscope and microscopy
  cytology and advances in, 47
  development of, 15–16, 47
Microscopic anatomy, and history of anatomy, 17
Microscopic Anatomy (Kölliker), 6
Microscopie Researches into Accordance in the Structure and Growth of Animals and Plants (Schwann), 6
Microtomography, 47
Microtubule(s), 51, 59, 60, 61, 338
Microvillus, 54, 55, 634
Micturition, 667–68
Midclavicular line, of back, 301
Middle Ages, and history of anatomy, 12
Middle cerebellar peduncles, 366
Middle cerebral artery, 546
Middle ear, 501, 513
Middle-ear cavity, 41, 140
Middle lobe, of lung, 594–95
Middle nasal conchae, 585
Middle rectal artery, 550
Middle sacral artery, 550
Middle trunk, of brachial plexus, 401
Middle vesicular artery, 550
Midgut, 645–46
Midgut loop, 646
Midsagittal plane, 34
Migraine, 380
Milia, 291
Mineralocorticoid(s), 456
Minerals, bone and storage of, 132
Minor calyx, 658, 661
Minor vestibular glands, 719
Miscarriage, 729
Mitochondrion, 51, 57–58
Mitosis, and cell cycle, 62, 65, 66
Mitral valve, 531
Mixed gland, 643
Mixed nerves, 344, 389
Modified radical mastectomy, 729
Molars, 149, 622
Molecule(s), 29, 48, 745–46
Mongolian spots, 291
Mongoloid race, 27
Monkey, ovary of, 712
Monoblasts, 528
Monocyte(s), 525, 526
Monohybrid cross, 764
Mononucleosis, 572
Monosaccharides, 50
Monosomy, 68
Monozygotic twins, 767, 768, 769
Monro, Alexander, Jr., 371
Mons pubis, 718
Morphogenesis, 736
Morula, 74, 737, 740
Motor end plates, 236–37
Motor facial nucleus, 392
Motor nerve(s), 344, 389
Motor nerve fiber, 335, 345
Motor (efferent) neuron, 232, 342, 411
Motor root, 394
Motor speech area, 359
Motor trigeminal nucleus, 392
Motor unit, 237–38, 239
Mounting, of tissue samples, 99
Mouth, 620. *See also* Oral cavity
Mouth-to-mouth and mouth-to-nose resuscitation, 609
Movements
  skeletal muscles and, 227
  synergistic muscles and, 231
  at synovial joints, 201–8
M phase, 65
MRI. *See* Magnetic resonance imaging
MSH. *See* Melanocyte-stimulating hormone
Mucocutaneous corpuscles, 474
Mucoid connective tissue, 746
Mucosa
  of gastrointestinal tract, 617
  nonkeratinized stratified squamous epithelium and, 79
  of ureter, 664
  of urinary bladder, 665
  of uterus, 715
Mucosal cell(s), of duodenum, 441
Mucosal layer, of vagina, 718
Mucous cell(s), 625
Mucous connective tissue, 86
Mucous membranes, 41, 82
Mucus, of stomach, 630
Müller, Johannes, 6, 16, 724
Müullerian ducts. *See* Paramesonephric ducts
Multicellular glands, 82, 84, 85
Multiple pregnancy, 735, 766–67
Multiple sclerosis, 383
Multipolar neurons, 342
Multiunit smooth muscle, 422, 423
Mummies and mummification, Egyptian, 7
Mumps, 649
Mural thrombus, 577
Murmur, of heart, 326
Muscarine, 430
Muscle(s). *See also* Muscular system; Skeletal muscle; *specific muscles*
  of appendicular skeleton, 253–78
  of axial skeleton, 241–53
  of eye, 486
  mitochondria and myoglobin, 57
  naming of, 239–41
  origin of term, 33
  of respiration, 600
  specialized cells, 67
Muscle atrophy, 98
Muscle contraction, 233–38
Muscle tissue, 73, 96–97
Muscular arteries, 538
Muscular dystrophy, 283, 330
Muscularis layer
  of ureter, 664
  of urinary bladder, 665
  of uterus, 715
  of vagina, 718
Muscularis mucosa, of gastrointestinal tract, 617
Muscular system. *See also* Muscles; Skeletal muscle
circulatory system and, 578
clinical case study, 227, 283
clinical considerations, 278, 282–83
development of, 280–81
digestive system and, 651
endocrine system and, 466
functions of, 31
integumentary system and, 125
interaction with other body systems, 284
introduction to, 227–28
male reproductive system and, 703
nervous system and, 384
respiratory system and, 610
skeletal system and, 165
urinary system and, 674
Musculocutaneous branch, of brachial plexus, 404
Musculotendinous (rotator) cuff, 211, 257
Mutations, genetic, 68. *See also* Genetic disorders
Myalgia, 285
Myasthenia gravis, 283
Myelencephalon, 337, 352, 366–67

Myelin, 340
Myelination, 340, 343
Myelin layer, 340, 342
Myeloblast(s), 528
Myeloid tissue, 528
Myeloma, 69
Myenteric plexus, 619
Mylohyoid muscle, 246, 247, 394
Myoblast(s), 280
Myocardial disease, 573
Myocardial infarction, 521, 574. *See also* Cardiovascular disease
Myocardial ischemia, 327, 574
Myocardium, 422, 529
Myofibril(s), 233, 234–35, 236
Myofilaments, 234, 235, 237
Myoglobin, 57
Myokymia, 285
Myology, 227
Myoma, 285
Myometrium, 717
Myopathy, 285
Myopia, 514
Myosin, 233, 234
Myotome(s), 280
Myotomy, 285
Myotonia, 285
Myringotomy, 516
Myxedema, 463–64

N

Nail(s), 113–14, 291
Naris, 294
Nasal bone(s), 150, 585
Nasal cavity. *See also* Nose
  body cavities and, 41
  bones enclosing, 150
  infections of, 149
  nose and, 585
  skull and, 139–40
Nasal concha, 149, 585
Nasal fossa, 140, 149, 585
Nasalis muscle, 242
Nasal meatuses, 585
Nasal region, of head, 292, 293–94, 295
Nasal septum, 140, 585
Nasal vestibule, 585
Nasolacrimal canal, 150
Nasolacrimal duct, 485
Nasopharynx, 585, 587
Natal cleft, 317
Nausea, 508, 652
Navel, 38, 301
Navicular bone, 184
Neck
  arteries of, 544–47
  body region, 37–38
  muscles of, 244–47, 249, 255
  surface anatomy of, 297–300
  of tooth, 624
  trauma and disease of, 321–22, 325–26
  of urinary bladder, 665
  veins of, 556
Necropsy, 12
Necrosis, 98
Negative feedback, and hormone secretion, 443, 444
Negroid race, 27
Neisseria gonorrhoeae, 700
Neonatal period, of growth, 756–57, 761
Neonate. *See also* Development; Infant
  brachial plexus, 403
  cardiovascular structures of, 562
  eye pigment, 488
  growth of, 756–57, 761
  jaundice in, 650
  respiratory distress syndrome, 600
  suckling and, 621
  surface anatomy of, 290
Neoplasm(s). *See also* Cancer; Tumors
  of bone, 162–64
  of central nervous system, 382
  genetic factors in, 68–69
  of muscle, 283
  of skin, 118
  of uterus, 728
Nephritis, 673
Nephron, 658–59, 660–61, 663, 664
Nephron loop, 661
Nerve(s). *See also* Nervous system
  classification of, 342, 344
  facial, 241
  skeletal muscles and, 231–32
  structure of, 346
Nerve blocks, 152, 380, 395
Nerve cell(s), 67. *See also* Neurons
Nerve fiber, 97
Nerve impulse, 347
Nerve pathways, 411
Nerve plexuses, and peripheral nervous system, 401–9
Nervous layer, of retina, 488, 489
Nervous system. *See also* Autonomic nervous system; Central nervous system; Innervation; Peripheral nervous system
  circulatory system and, 578
  development of, 336–37
  digestive system and, 651
  endocrine system and, 440, 466
  integumentary system and, 125
  interaction with other body systems, 384
  male reproductive system and, 703
  muscular system and, 284
  neurons and neuroglia, 338–46
  organization and functions of, 32, 334–38
  respiratory system and, 610
  skeletal system and, 165
  transmission of impulses, 346–48
  urinary system and, 674
Nervous tissue, 73, 97, 98
Neural crest, 337
Neural fold(s), 337
Neural groove, 337
Neural impulses, and endocrine glands, 443
Neural pathways
  for hearing, 505, 507, 508
  for somatic sensation, 477–78
  for vision, eye movements, and processing visual information, 496–97
Neural plate, 337
Neural tube, 337
Neurocranium, 163
Neuroendocrine reflex(es), 447
Neuroepithelium, 76
Neurofibril(s), 338
Neurofibril node(s), 340
Neuroglia, 97, 98, 337, 340–42
Neurohypophysis, 363, 445–46, 448, 460
Neurolemmal sheath, 340
Neurolemmocyte(s), 340, 343
Neurology and neurological assessment, 334, 379–80, 398
Neuromuscular cleft, 237
Neuromuscular junction, 236–37
Neuromuscular spindles, 476
Neuron(s)
  afferent neuron, 342
  association neuron, 342
  bipolar neuron, 342, 489–90
  classification of, 342, 346
  development of, 337
  efferent neuron, 342
  functions of nervous system and, 334, 339
  ganglionic neuron, 490
  interneuron, 342
  motor neuron, 232, 339, 342, 411
  multipolar neuron, 342
  nervous tissue and, 97, 98
  peripheral neuron, 343
  photomicrograph of, 340
  postganglionic neuron, 420
  postsynaptic neuron, 347
  preganglionic neuron, 420
  presynaptic neuron, 347
  principal components of, 338
  pseudounipolar neuron, 342
  regeneration of, 344
  sensory neuron, 232, 339, 342, 411, 433
  shape of cell, 47, 48
Neuropeptides, 349
Neuropore(s), 337
Neurosis, 381
Neurosyphilis, 383
Neurotendinous receptors, 476
Neurotransmitters
  autonomic nervous system and, 422, 429, 431
  diseases involving, 385
  synaptic transmission and, 346–48
Neutrophil(s), 525, 526
Nevus, 126
Newborn. *See* Development; Infant; Neonate
Nicotine, 542
Niferous tubules, 683
Nipple(s), 25, 720
Nissl, Franz, 338
Nissl body, 338
Nitrogenous bases, 63
NMR. *See* Nuclear magnetic resonance
Nociceptors, 472, 473, 474
Nocturnal emissions, 695
Nodes of Ranvier. *See* Neurofibril nodes
Nodose ganglion, 392
Nomenclature, of anatomy
  Greek and Latin languages, 2
  planes of reference and descriptive terminology, 33–36
  simplification and standardization of in twentieth century, 16
  system used in text, 16–17
Nomina Anatomica, 17
Nonarticulating prominences, of bone, 133
Nondisplaced fracture, 187
Non-Hodgkin's lymphoma, 575
Non-insulin-dependent diabetes mellitus, 465
Nonkeratinized stratified squamous epithelium, 79, 80, 82
Nonpigmented granular dendrocytes, 103
Nonrespiratory air movements, 601, 602
Norepinephrine
  adrenal medulla and secretion of, 425
  effects of, 456, 457
  neurotransmitters of autonomic nervous system and, 429, 431
Normal amenorrhea, 728
Norplant, 730, 731
Nose, as conducting passage of respiratory system, 585–87, 598. *See also* Nasal cavity
Nostril(s), 294, 295, 585
Notochord, 23, 24
Nucha, 297
Nuclear magnetic resonance (NMR), and history of anatomy, 18
Nuclear membrane, 51, 61
Nuclear pores, 61–62
Nuclease, 633
Nucleolus, 51, 62
Nucleolemma cisterna, 61
Nucleoplasm, 54
Nucleotides, 63
Nucleus
  of cell, 51–52, 61–62
  of neuron, 338, 348
Nucleus ambiguus, 366, 392, 393, 396
Nucleus cuneatus, 366
Nucleus gracilis, 366
Nucleus pulposus, 23, 24, 164
Nutrient foramen, 134, 172
Nutrition, clinical problems in, 615. *See also* Diet; Nutritional disorders; Obesity
Nutritional disorders
  anemia, 524
  axial skeleton and, 161
  conditions of skin indicating, 103
  goiter, 452, 464
  hair samples and, 112
  rickets, 94, 111, 132, 161
Nyoglossus muscle, 244

O

Obesity, 615, 758–59
OBGYN, 723
Oblique fracture, 187
Oblique muscle(s), 241, 244, 250, 394, 486
Oblique muscle layer, of stomach, 629
Oblique popliteal ligament, 216
Obstetrics, 723
Obstruction, of urinary system, 673
Obturator artery, 552
Obturator externus muscle, 266
Obturator foramen, 179
Obturator internus muscle, 266
Obturator nerve, 405, 406, 408
Occipital bone, of skull, 147
Occipital artery, 546
Occipital condyle(s), 147
Occipital lobe, of cerebrum, 356
Ocular muscle(s), 241, 244
Ocular region, of head, 292, 293, 294
Oculomotor nerve, 392, 393–94, 399, 428
Oculomotor nucleus, 392
Oddi, Ruggero, 633
Odontoid process, 157
Oily skin, 103
Olecranon, 173, 174
Olecranon bursa, 211
Olfactory bulb, 390, 391, 479

Olfactory hairs, 479
Olfactory nerve
    dysfunctions of, 399
    location and function of, 390, 391, 392, 393
    olfactory sense and, 479
Olfactory pit, 604
Olfactory placode, 604
Olfactory sense, 479
Olfactory tract, 390, 479
Oligodendrocyte(s), 340, 341
Oligodendroglioma, 382
Oligospermia, 695
Oliguria, 672
Olive, of medulla oblongata, 366
Omohyoid muscle, 246, 247
Omos, 39
Onset of labor, 755
On the Development of Animals (Beaumont), 6
On the Formation of Eggs and Birds (Hieronymus), 6
On the Formation of the Fetus (Hieronymus), 6
On the Generation of Animals (Aristotle), 6, 10
On the Motion of the Heart and Blood in Animals (Harvey), 15
On the Origin of Species (Darwin), 6
On the Parts of Animals (Aristotle), 6, 10
Oocyte(s), 679
Oogenesis, process of, 713
Oogonium, 710, 724
Oophoritis, 729
Openings, in bone, 133
Ophthalmic artery, 546
Ophthalmic nerve, 392, 394
Ophthalmic vein, 556
Ophthalmology, 293, 494, 511
Ophthalmoscope, 491, 511
Opponens digit minimi muscle, 265
Opponens pollicis muscle, 263, 265
Opposable grip, 28
Optic canal, 148, 392
Optic chiasma, 496
Optic cup, 499
Optic disc, 490, 494
Optic fissure, 499
Optic nerve
    development of, 499
    dysfunctions of, 399
    internal anatomy of eyeball and, 494
    location and functions of, 390, 392, 393, 394
Optic radiation, 496
Optic stalk, 499
Optic tract, 390, 496
Optic vesicle, 499
Optometry, 293, 511
Oral (buccal) cavity
    body cavities and, 41
    functions of, 616
    oral region and, 294
    skull and, 140
    superficial structures of, 296, 620
Oral membrane, 644–45
Oral orifice, 620
Oral pit, 644–45
Oral region, of head, 292, 295
Oral steroid contraceptives, 730
Ora serrata, 488–89
Orbicularis oculi muscle, 242
Orbit, of eye
    bones of, 145, 146
    eyeball and, 483
    skull and, 140
    structure of nasal cavity and, 41
    surface anatomy of cranium and, 292
Orchitis, 702
Order, taxonomic, 25–26
Organelle(s), 28, 29, 51, 54–61
Organic compounds, 49
Organism, 29
Organ of Corti, 396, 504–5. *See also* Spiral organ
Organs of Ruffini, 473, 474
Origin, of muscle, 229
Oronasal cavity, 604
Oronasal membrane, 604
Oropharynx, 587
Orthodontics, 139
Orthopedics, 166
Os coxa, 129, 186
Osmoreceptors, 362
Osmosis, 53, 54
Os penis, 693
Ossification
    bone growth and, 137, 138, 162
    of trepanated skulls, 4
Osteitis, 166
Osteoarthritis, 221–22
Osteoblast(s), 135, 137
Osteoblastoma, 166
Osteochondritis, 166
Osteoclast(s), 135
Osteocyte(s), 48, 94, 136
Osteogenic cancer, 69
Osteogenic cell(s), 135
Osteogenic sarcoma, 163
Osteoid, 88, 137
Osteoid osteomas, 163
Osteology, 129
Osteoma, 163
Osteomyelitis, 166, 326
Osteon(s), 136
Osteonecrosis, 166
Osteopathology, 166
Osteoporosis, 158, 164, 330
Osteosarcoma, 166
Osteotomy, 166
Otic capsule, 513
Otic fovea, 513
Otic ganglion, 427
Otic placode, 513
Otitis, 516
Otocyst, 513
Otorhinolaryngology, 293, 511
Otosclerosis, 516–17
Otoscope, 500, 501, 511
Outer ear, 499–500, 513
Ova (ovum), 47, 58, 709, 735–37
Ovarian arteries, 550
Ovarian cortex, 709
Ovarian cycle, 710–14
Ovarian cysts, 726
Ovarian (Graafian) follicles, 16, 712
Ovarian fossa, 709
Ovarian ligament, 709
Ovarian medulla, 709
Ovarian tumors, 726
Ovarian veins, 558
Ovary
    endocrine glands and, 441
    function of, 707, 708
    ovarian cycle, 710–14
    position and structure of, 709–10
    problems involving, 726, 727, 728
Oviduct(s), 714
Ovulation, 712–14, 722–23
Ovum. *See* Ova
Owen, Richard, 6
Oxygen utilization, as function of respiration, 583
Oxyphil cells, 453
Oxytocin, 447, 448, 755

P

Pacemaker activity, of single-unit smooth muscle, 422
Pacini, Filippo, 474
Paget, James, 161
Paget's disease, 161–62
Pain, and sensory receptors, 472, 474–76
Palate, 621. *See also* Cleft palate
Palatine bone, of skull, 150, 151
Palatine foramen, 150
Palatine process, of maxilla, 149
Palatine tonsil(s), 587, 621
Palatine uvula, 621
Palatoglossus muscle, 244, 245
Paleopathology, 4
Pallor, 291
Palmar (volar) arch(es), 548, 556
Palmar interossei muscle, 265
Palmaris longus muscle, 260, 263
Palmar ligament, 212, 214
Palmar region, 40
Palpation
    breast self-examination and, 726
    physical examination and, 34, 288, 327
Palpebra, 483
Palpebral conjunctiva, 484–85
Palpebral fissure, 484
Panacea, 10
Pancreas
    autonomic nervous system and, 430
    endocrine system and, 441, 453–54, 455, 462
    large intestine and, 643, 648
Pancreatic acini, 643
Pancreatic bud, 645
Pancreatic cancer, 648
Pancreatic cells, 56
Pancreatic duct, 633, 642
Pancreatic islets. *See also* Islets of Langerhans
    development of, 462
    disorders of, 465
    endocrine functions of, 441, 454
Pancreatic juice, 462, 643
Pancreatic vein, 559
Panhypopituitarism, 463
Papanicolaou, George N., 726
Papanicolaou (Pap) smear, 726
Papilla, 108, 622
Papillary duct, 661, 664
Papillary muscle(s), 531
Papilloma, 126
Pap smear, 726
Papule, 126
Parafollicular cells, of thyroid, 450
Parallel muscles, 231
Paralysis, 285, 381
Paralysis agitans. *See* Parkinson's disease
Paramesonephric duct, 696, 724
Paranasal sinus(es)
    functions of, 587
    infection of, 149
    position of in skull, 140, 146, 588
    respiratory system and, 598
Paranoia, 381
Paraplegia, 381
Parasympathetic division, of autonomic nervous system, 334, 420, 425–28, 429
Parathyroid gland, 297, 441, 452–53, 463–65
Parathyroid hormone (PTH), 453, 465
Paraventricular nuclei, 447
Parietal bone, of skull, 146
Parietal cell(s), of gastric glands, 629
Parietal lobe, 356
Parietal pericardium, 42, 529
Parietal peritoneum, 617
Parietal pleura, 42, 595
Parietal portion, of serous membrane, 616–17
Parkinson's disease, 342, 348, 378, 383, 385
Parotid duct, 16, 81, 625
Pars distalis, 446, 461
Pars intermedia, of adenohypophysis, 446, 461
Pars tuberalis, of adenohypophysis, 446, 461
Partial fracture, 187
Parturition, 707, 755–56
Patella, 131, 180, 186, 201
Patellar ligament, 215, 270
Patellar reflex, 415
Patellar region, 40
Patellar retinaculum, 216
Patellar surface, 180
Patellofemoral joint, 215
Patellofemoral stress syndrome, 201
Patent foramen ovale, 326, 572
Patent urachus, 327, 672
Pathogens, 68. *See also* Disease; Infections
Pathologic fracture, 187
Pathology, 73
Pavlov, Ivan, 6
PBI. *See* Protein-bound iodine test
Pectineus muscle, 268, 269
Pectoral girdle
    appendicular skeleton and, 129, 131
    muscles of, 253–54
    structure and functions of, 169, 172
Pectoralis major muscle, 254–55, 256
Pectoralis muscle(s), of pectoral girdle, 253–54
Pedicels, 660–61
Pedicle(s), 154
Pellagrous dermatitis, 103
Pelvic brim, 178
Pelvic cavity, 41, 666
Pelvic curve, of vertebral column, 154
Pelvic diaphragm, 250
Pelvic examination, 723–24, 726
Pelvic foramina, 158
Pelvic girdle
    appendicular skeleton and, 129, 131
    structure and function of, 176, 177–80, 186
Pelvic inflammatory disease (PID), 617, 715, 729
Pelvic inlet, 178
Pelvic outlet, muscles of, 250–51
Pelvic region, 38–39, 40
Pelvis
    arteries of, 550, 551, 552–53
    pelvic girdle and, 177–78
    surface anatomy of, 310–11
Pendular movements, of small intestine, 635
Penis
    autonomic nervous system and, 430

structure and functions of, 681, 691–92, 693–95
urethra and, 666–67
Pennate muscles, 231
Pepsin, 633
Peptic ulcer(s), 328–29, 630, 650, 652
Peptidases, 633
Perception, definition of, 471
Perceptive deafness, 517
Percussion, 34, 288
Percutaneous needle biopsy, 99
Perforating cutaneous nerve, 409
Perforating canal(s), 136
Perforating fibers, of bone, 134
Perforation, of tympanic membrane, 516
Pericapsulitis, 211
Pericardial arteries, 549
Pericardial cavity, 41, 42, 529
Pericardial fluid, 529
Pericardial membranes, 42
Pericarditis, 327, 529, 573
Perichondrium, 91, 220
Periderm, 116
Perikaryon, 97
Perilymph, 502, 513
Perilymphatic space, 513
Perimetrium, 716
Perimysium, 229
Perineal raphe, 683
Perineum
male reproductive system and, 681, 682
pelvic region and, 39, 40
surface anatomy of, 310–11
of vagina, 718
Perineurium, 344
Periodontal disease, 194, 624, 649
Periodontal membrane, 624
Periosteal bud, 137
Periosteal layer, of dura mater, 367
Periosteum, 94, 134
Peripheral chemoreceptors, 603
Peripheral ganglia, 424
Peripheral nervous system (PNS). *See also* Nervous system
clinical case study, 389
cranial nerves, 389–99
development of, 416
introduction to, 389
nerve plexuses, 401–9
organization of nervous system, 334, 335
reflex arc and reflexes, 410–15
sensory and motor fibers of, 345
spinal nerves, 400–401
Peripheral neuron, 343
Peristalsis, 615, 635
Peristaltic movements, 638
Peristaltic waves, 422
Peritoneal cavity, 23, 42, 617
Peritoneal lavage fluid, 23, 43
Peritoneal membranes, 42, 617
Peritoneum, 617
Peritonitis, 328, 617, 652
Peritubular capillaries, 660
Permanent teeth, 622, 623
Pernicious anemia, 326, 631
Peroneal artery, 553
Peroneus brevis muscle, 274, 278
Peroneus longus muscle, 274, 278
Peroneus tertius muscle, 278
Peroxisomes, 51, 59
Perpendicular plate, 148
Pes, bones of, 182–84, 186. *See also* Foot
PET. *See* Positron emission tomography
Petit mal epilepsy, 381
Petrosal ganglion, 392
Petrous part, of temporal bone, 147
Peyer, Johann K., 565, 634
Peyer's patches, 565, 634
pH, 49, 79
Phagocytosis, 53, 58, 59, 60, 88
Phalanges
of foot, 131, 184, 186
of hand, 129, 174
Phallus, 725
Phantom pain, 475–76
Pharyngeal pouches, 23, 24
Pharyngeal tonsils, 587
Pharyngopalatine arch, 621
Pharynx
as characteristic of chordates, 23
digestive system and, 620, 625
gastrointestinal tract and, 616
respiratory system and, 598
structure and functions of, 587, 625
Phasic receptors, 472
Phenobarbital, 56
Phenotype, 763–64
Phenylketonuria, 770
Pheochromocytomas, 465
Philtrum, 295
Phimosis, 691
Phlebitis, 575
Phlegm, 9
Phobia, 381
Phosphate group, 63
Phospholipid(s), 50, 52, 53
Photoreceptors
exteroceptors and, 472, 473
optic nerve and, 390
retina and, 493
specialized features of cell membranes and, 55
Photorefractive keratectomy, 515
Phrenic nerve, 401, 601
Phrenology, 333
Phylogeny, 23
Phylum, 23–24
Physical injury, to cells, 68
Physiological Elements (Haller), 6
Physiological jaundice, of newborn, 650
Physiology, 2
Pia mater, 369
PID. *See* Pelvic inflammatory disease
Pigeon breast, 326
Pigmented layer, of eye, 488, 489
Pineal gland, 363, 441, 458
Pinna, 292
Pinocytosis, 53, 59, 60
Piriformis muscle, 266
Pisiform bone, 174
Pitch, of sound, 505
Pituicytes, 446
Pituitary cachexia, 463, 464
Pituitary dwarfism, 463, 464
Pituitary gland
derivation and functions of, 352, 363
disorders of, 463
location of, 362
endocrine system and, 434, 440–41, 444–48, 449, 460–61
Pituitary-like hormones, 747
Pituitary stalk, 445
Pivot joint, 198, 199, 200
PKU. *See* Phenylketonuria
Placenta
as characteristic of mammals, 25
development and, 745–46
endocrine functions of, 441, 459, 462
fetal blood circulation and, 561
Placental stage, of labor, 755, 756
Placenta previa, 766
Planes of reference, 33–34
Plane suture, 193
Plantar arch, of foot, 553
Plantar artery, 553
Plantar flexion, 201, 202, 206
Plantaris muscle, 275, 278
Plantar nerves, 407
Plantar reflex, 414, 415
Plantar surface, 40
Plaque, dental, 649
Plasma membrane, 51, 52–54. *See also* Cell membrane
Platelets, blood, 525, 526. *See also* Thrombocytes
Plato, 10
Platysma muscle, 242
Pleura, 42–43, 595, 597, 598
Pleural cavity, 41, 42, 595
Pleural membranes, 597
Pleurisy, 41, 595, 608
Plexus, of peripheral nervous system, 390. *See also* Nerve plexuses
Plexus of Auerbach. *See* Myenteric plexus
Plexus of Meissner. *See* Submucosal plexus
Plicae circulares, 634
Pliny, 11
Plural word endings, guide to, 33
PMN. *See* Polymorphonuclear leukocytes
Pneumonia, 608
Pneumotaxic area, 603
Pneumothorax, 23, 326, 607
Podocyte(s), 660
Poison(s)
cellular dysfunction and, 68
digestive system and, 648–49
Polar body, 711
Polarized nerve fiber, 347
Poliomyelitis, 283, 330, 382
Pollex, 174
Polycythemia, 525, 572
Polydactyly, 186, 329
Polymorphonuclear (PMN) leukocytes, 525
Polysaccharides, 50
Polysynaptic reflex arc, 412
Polythelia, 25
Polyuria, 672
Pons, 364
Popliteal artery, 552
Popliteal bursa, 216
Popliteal fossa, 40, 216, 277, 319
Popliteal lymph nodes, 565
Popliteal vein, 557
Popliteus muscle, 275, 278
Portal system, 448, 546, 558. *See also* Hepatic portal system
Positron emission tomography (PET), 18, 349, 352
Postcentral gyrus, 356, 477
Posterior, as directional term, 36
Posterior auricular artery, 546
Posterior body cavity, 41
Posterior cerebral arteries, 544
Posterior cervical triangle, 299
Posterior chamber, of eye, 491
Posterior commissure, 363
Posterior communicating arteries, 544
Posterior cord, of brachial plexus, 401
Posterior crural muscles, 274–76, 277
Posterior cutaneous femoral nerve, 409
Posterior division, of lumbar plexus, 405
Posterior funiculi, of spinal cord, 375
Posterior horns, of spinal cord, 374, 378
Posterior humeral circumflex arteries, 548
Posterior intercostal arteries, 549
Posterior interventricular artery, 535
Posterior median sulcus, of spinal cord, 373
Posterior muscles, of neck, 244–46
Posterior pituitary, 447
Posterior ramus, of spinal nerve, 401
Posterior root, of spinal nerve, 400–401
Posterior sacral foramina, 158
Posterior surface, of stomach, 629
Posterior tibial artery, 553
Posterior tibial vein, 557
Posterior tubercle, 156
Posterior vagal trunk, 427–28
Posterior view, of body regions, 37
Postganglionic neuron, 420
Postmature infant, 757
Postnatal growth, periods of, 756–65
Postrotatory vestibular nystagmus, 508
Postsynaptic neuron, 347
Posture, and skeletal muscles, 227
Pott, Percivall, 187
Pott's fracture, 187
Pouch of Douglas, 716
Precapillary sphincter muscles, 538
Precentral gyrus, 355
Predation, and vision, 482
Preembryonic period, of prenatal development, 736, 737–42
Prefixes, glossary of, 33
Preganglionic neuron, 420
Pregnancy, and erector spinae muscles, 252. *See also* Childbirth; Female reproductive system; Labor; Parturition
Pregnancy tests, 740
Prehensile hands, 25, 26, 28
Prehistoric times, and knowledge of anatomy, 2–4
Premature infant, 757
Premature synostosis, 322, 325
Premenstrual phase, of menstruation, 723
Premolars, 149, 622
Prepuce
of clitoris, 718
of penis, 691
Presbyopia, 515
Pressure differences, and cell membranes, 54
Pressure receptors, 473–74
Presynaptic neuron, 347
Prey species, vision of, 482
Primary amenorrhea, 728
Primary follicle, 710
Primary germ layers, 75, 740
Primary neoplasms, 382
Primary oocytes, 710
Primary ossification center, 137
Primary sex cord(s), 696, 724
Primary sex organs
female, 707
male, 679
Primary spermatocyte, 685
Primates, 25–26
Primitive gut, 644
Primitive line, 748
Primitive node, 748
Primordial follicles, 710, 724
Principal cells
of parathyroid gland, 453
of stomach, 629

Principles of General Physiology (Bayliss), 6
P-R interval, 537
Probability, and inheritance, 764
Process, of bone, 133
Processing, of visual information, 497
Proctodeum, 646
Proerythroblasts, 528
Progesterone
adrenal cortex and secretion of, 456
classes of hormones and, 442
luteinizing hormone and, 446
ovaries and, 709
ovulation and menstruation, 723
placenta and secretion of, 747
Projection fibers, 358
Prolactin, 447, 448
Prolactin-inhibiting hormone (PIH), 447
Prolapse, of uterus, 729
Proliferation zone, of bone, 137
Proliferative phase, of menstruation, 723
Pronation, 202
Pronator quadratus muscle, 258, 263
Pronator teres muscle, 258, 263
Pronephric duct, 671
Pronephros, 671
Prophase, 65, 66, 685, 686
Proprioceptors, 366, 472, 476–77
Prosencephalon, 337, 352
Prostaglandin(s), 755
Prostate, 666, 681, 689–90
Prostatectomy, 702
Prostatic carcinoma, 702
Prostatic part of urethra, 666, 690, 702
Protection, as function of circulatory system, 521
Protein(s)
blood plasma and, 528
cell membrane and, 53
hormones and, 441, 442
structure and functions of, 49–50
Protein-bound iodine (PBI) test, 463
Proteolytic enzymes, 739
Protoplasm, 54
Protraction, 203
Proximal, as directional term, 36
Proximal convoluted tubule, 661, 663, 664
Proximal radioulnar joint, 211
Pruritus, 126, 729
Pseudohypertrophic muscular dystrophy, 770
Pseudostratified ciliated columnar epithelium, 78, 79, 82
Pseudounipolar neurons, 342
Psoas major muscle, 265, 266
Psoriasis, 126
Psychiatric disorders, and central nervous system, 381, 385
Psychomotor epilepsy, 381
Psychosis, 381
Psychosomatic illness, and hypothalamus, 363
Pterygoid muscles, 241, 243, 394
Pterygoid processes, 148
Pterygopalatine ganglion, 425
PTH. *See* Parathyroid hormone
Puberty, 30, 709, 759–60
Pubescence, 759
Pubic area, 38–39, 40
Pubic tubercle, 179
Pubis, 179
Pubofemoral ligament, 214
Pudendal cleft, 708
Pudendal nerve, 409
Pudendal omission cleft, 718
Pulmonary alveoli
development of, 604, 607
structure and functions of, 592–93, 594, 598
Pulmonary circulation, 534
Pulmonary embolism, 575
Pulmonary ligament, 595
Pulmonary plexuses, 427
Pulmonary semilunar valve, 531
Pulmonary stenosis, 572
Pulmonary trunk, 531
Pulmonary (semilunar) valve, 532
Pulmonary vein, 531
Pulmonary ventilation, 598, 601
Pulmonic area, 537
Pulp cavity, 624
Pulse pressure, 542, 554, 577
Punnett, Reginald Crundall, 764
Punnett square, 764
Pupil, of eye, 488, 493–94
Pupillary constrictor muscle, 486
Pupillary dilator muscle, 486
Pupillary reflex, 497
Purkinje, Johannes E. von, 535
Purkinje fibers, 535. *See also* Conduction myofibers
Pustule, 126
Putamen, 358
P wave, 537
Pyelitis, 665, 673
Pyelonephritis, 665, 673
Pyloric sphincter, 629
Pyloric stenosis, 327–28, 648
Pylorus, 629
Pyorrhea, 652
Pyramids, of medulla oblongata, 366
Pyuria, 672

Q

QRS complex, 537
Quadrate lobe, of liver, 641
Quadratus femoris muscle, 266, 271
Quadratus lumborum muscle, 252–53
Quadriceps femoris muscle(s), 268, 270, 318
Quadriplegia, 381
Quickening, 753

R

Races
characteristics of, 27
coloration of skin and, 106
differences in appendage lengths, 762
Radial artery, 548
Radial branch, of brachial plexus, 404
Radial keratotomy, 515
Radial nerve, 403, 414
Radial notch, of ulna 174
Radial recurrent artery, 548
Radial vein, 556
Radiation, cell injury by, 68
Radical mastectomy, 254, 729
Radiograms, of heart, 567
Radiographic anatomy, and history of anatomy, 17–19
Radiographs, 753
Radioimmunoassay (RIA) test, 463
Radiology, and history of anatomy, 17–19
Radius, 129, 173, 174, 193
Rami communicantes, 401
Ramus, 151, 179
Ranvier, Louis A., 340
RDS. *See* Respiratory distress syndrome
Reactant, 49
Reactive hypoglycemia, 465
Receptors. *See* Chemoreceptors; Mechanoreceptors; Pressure receptors; Proprioceptors; Sensory receptors; Tactile receptors
Recessive allele, 763
Recipient site, and skin graft, 121
Recti muscles, 486
Rectouterine folds, 716
Rectouterine pouch, 717
Rectum, 637
Rectus abdominis muscle(s), 248, 301
Rectus femoris muscle, 270, 271
Rectus muscle(s), 241, 244, 250, 394, 395
Recurrent laryngeal nerve, 397
Red blood cells, 47, 48. *See also* Erythrocyte
Red bone marrow, 134
Red-green color blindness, 496, 764–65
Red nucleus, 363
Red pulp, of spleen, 565
Referred pain, 475, 476
Reflex(es)
types of, 411–14
testing of, 34
Reflex arc, 410–15
Reflex integration, in spinal cord, 373
Refraction, of light rays by eyeball, 492–93, 495
Regeneration
of cut axon, 340
of neuron, 344
Regeneration tube, 340
Regional approach, to anatomy, 30
Regurgitation, 652
Relaxin, 755
Releasing factors, for hormones, 443
Religion, and history of anatomy, 12
Rembrandt, 15
Renaissance, and history of anatomy, 12–14
Renal adipose capsule, 657
Renal agenesis, 669
Renal arteries, 550, 657, 659
Renal biopsy, 673
Renal blood vessels, 659–60
Renal capsule, 657
Renal columns, 658
Renal corpuscle, 660, 664
Renal cortex, 658, 663
Renal ectopia, 669
Renal failure, 673, 675
Renal fascia, 657
Renal medulla, 658, 663
Renal papillae, 658, 661
Renal pelvis, 658, 661
Renal ptosis, 657
Renal pyramid(s), 658
Renal stone, 665
Renal vein, 558, 657, 660
Repolarization, of neurons, 347
Reproductive system. *See also* Female reproductive system; Male reproductive system
circulatory system and, 578
digestive system and, 651
endocrine system and, 466
functions of, 32
homologous structures of, 696, 707, 725
integumentary system and, 125
muscular system and, 284
nervous system and, 384
respiratory system and, 610
skeletal system and, 165
urinary system and, 674
Reserve zone, of bone, 137
Residual volume (RV), of lungs, 601
Resorption zone, of bone, 137
Respiration. *See also* Respiratory system
functions of, 583
muscles of, 248
shock and, 577
Respiratory bronchioles, 592
Respiratory center(s), of brain, 367, 603
Respiratory control, disorders of, 608, 611
Respiratory distress syndrome (RDS), 600
Respiratory division, of respiratory system, 584, 593
Respiratory epithelium, 78
Respiratory system
circulatory system and, 578
clinical case study, 583, 611
clinical considerations, 606–8, 609, 611
conducting passages of, 585–92
development of, 604–6
digestive system and, 651
elastic fibers in passageways, 87
endocrine system and, 466
function of, 32
integumentary system and, 125
interaction with other body systems, 610
male reproductive system and, 703
mechanics of breathing, 598–603
muscular system and, 284
nervous system and, 384
pseudostratified ciliated columnar epithelium, 78, 79
pulmonary alveoli, lungs, and pleurae, 592–98
regulation of breathing, 603
skeletal system and, 165
structure and functions of, 583–84
urinary system and, 674
Respiratory volumes and capacities, 600, 601
Resting potential, of nerve fiber, 347
Rete testis, 683
Reticular activating system (RAS), 367
Reticular connective tissue, 88, 90, 91
Reticular fibers, 86, 87
Reticular formation, of brain, 367
Reticulospinal tract(s), 376, 377
Retina
detachment of, 515
development of, 499
examination of with ophthalmoscope, 494
layers of, 493
structure and function of, 488–90, 491
Retinaculum, 229
Retinitis pigmentosa, 770
Retraction, 203
Retroflexion, of uterus, 716, 729

Retroperitoneal organs, 42, 617
Retroversion, of uterus, 729
Rhabdomyosarcoma, 283
Rh disease, 650
Rheumatic fever, 327
Rheumatism, 283
Rheumatoid arthritis, 59, 221
Rheumatology, 224
Rhombencephalon, 337, 352
Rhomboideus muscle(s), 254, 256
Rhythmicity area, of brain, 603
Rhythmic segmentations, of small intestine, 635
Rhythm method, of birth control, 729–30
RIA. *See* Radioimmunoassay test
Rib cage. *See also* Ribs
- axial skeleton and, 129
- bones of, 131
- dislocation of sternoclavicular joint, 209
- position of lungs within, 594
- structure and function of, 159, 160

Rib(s), 159, 160. *See also* Rib cage
Ribosome(s), 51, 56
Rickets, 94, 111, 132, 161
Right atrioventricular (AV) valve, 530–31, 532
Right atrium, 530
Right common carotid artery, 543–44
Right common iliac artery, 549
Right coronary arteries, 535
Right gastric vein, 560
Right gastroepiploic vein, 559
Right hemisphere, of brain, 353, 361
Right hepatic vein, 558–59
Right lobe, of liver, 640
Right lymphatic duct, 564
Right marginal artery, 535
Right ovarian vein, 558
Right principal bronchi, 591–92
Right pulmonary artery, 531
Right pulmonary trunk, 531
Right subclavian artery, 544, 545, 547
Right subclavian vein, 556
Right suprarenal vein, 558
Right testicular vein, 558
Right ventricle, 530–31
Rigor mortis, 99, 282
Risorius muscle, 242
Rivinus, August Quirinus, 625
Rivinus' ducts, 625
RNA, 56, 62
Rods, of eye, 54, 489–90
Roentgen, Wilhelm Konrad, 5, 17
Roentgenograph, 17
Rolando, Luigi, 355
Rome, and history of anatomy, 584. *See also* Latin
Root
- of penis, 691
- of tooth, 624

Root canal, 624
Root hair plexuses, 473, 474
Rotation movement, 202, 205, 206
Rotator cuff, 211, 257
Rough endoplasmic reticulum (rough ER), 55–56
Round ligaments, 716
Rubella virus, 329, 511
Rubrospinal tract, 376, 377
Ruffini, Angelo, 474
Rugae, 629, 665
Rule of nines, and treatment of burns, 120

S

Saccadic eye movements, 497
Saccular portion, 513
Saccule, 502, 505, 506–7
Sachs, Bernard, 383, 770
Sacral canal, 158
Sacral plexus, 405, 407–9
Sacral promontory, 158
Sacral region, 39
Sacral tuberosity, 158
Sacral vertebra, 153
Sacroiliac joint, 158, 179
Sacrum, 158
Sacs, of eye, 54
Saddle joint, 199, 200
Sagittal plane, 33–34, 35
Sagittal suture, 141, 146
St. Martin, Alexis, 614
Saliva, 625
Salivary duct(s), 625
Salivary gland(s), 625, 626, 627, 649
Salmonella bacterium, 649
Salpingitis, 726, 729
Salpingography, 715
Salpingolysis, 715
Salpinx, 715
Salts, 49
Salty taste, 481
Sanguine, use of term, 9
Saphenous nerve, 406
Saphenous veins, 558
Sarcolemma, 234
Sarcoma, 69
Sarcomere(s), 234, 235
Sarcoplasm, 234
Sarcoplasmic reticulum, 234
Sartorius muscle, 268–69, 271
Scala tympani, 504, 513
Scala vestibuli, 502, 503, 504, 513
Scalene muscles, 248, 599
Scaphoid bone, 174
Scapula
- pectoral girdle and, 129
- structure and function of, 169, 171, 172
- surface landmarks of back and, 301

Scapular muscle, 255–57
Scapular notch, 169
Scar tissue, 94, 123–24
Schizophrenia, 381, 385
Schleiden, Matthias, 5, 16, 47
Schlemm, Friedrich S., 491
Schwann, Theodor, 5, 6, 16, 47, 340
Schwann cells, 340, 343
Sciatica, 408
Sciatic nerve, 407, 408, 409
Science, and definition of anatomy, 2
Science (journal), 19
Science Digest (journal), 19
Scientific American (journal), 19
Sclera, 294, 487, 491, 499
Scleral venous sinus, 491
Sclerotome(s), 280
Scoliosis, 221
Scrotal septum, 681, 683
Scrotum
- development of, 696
- disorders of, 702
- structure and function of, 681, 682, 683

Sebaceous (oil) gland(s), 114, 116, 293
Seborrhea, 126
Seborrheic hyperkeratoses, 106
Sebum, 114
Secondary amenorrhea, 728
Secondary follicles, 710–11
Secondary oocyte, 711–12, 737
Secondary ossification center, 137
Secondary sex characteristics
- female, 708
- male, 679, 684

Secondary sex organs
- female, 707–8, 714–19
- male, 679

Secondary spermatocytes, 685
Second-degree burns, 102, 119
Second-degree frostbite, 121
Second-division nerve block, 152, 395
Second meiotic division, 685
Second sound, of heart, 537
Second trimester, of gestation, 765
Secretin, 459
Secretory phase, of menstruation, 723
Sectioning, of tissue samples, 99
Segmental bronchi, 592
Sella, 33
Sella turcica, 148
Semen, 695
Semicircular canals, 502, 507, 513
Semicircular ducts, 502, 505
Semilunar valve(s), 529
Semimembranous bursa, 216
Semimembranous muscle, 272
Seminal fluid, 695
Seminal vesicle(s), 681, 689
Seminiferous tubule(s), 681, 684, 687
Semispinalis capitis muscle, 245–46, 253
Semispinalis cervicis muscle, 253
Semispinalis thoracis muscle, 253
Semitendinosus muscle, 272
Senescence
- of nervous system, 383
- physical and behavioral characteristics of, 761

Senescence atrophy, 98
Sensation, somatic
- definition of, 471
- neural pathways for, 477–78

Senses. *See also* Gustatory sense; Hearing; Olfactory sense; Sensory organs; Visual sense
- classification of, 471–72
- definition of somatic, 472–73
- neural pathways for, 477–78
- proprioceptors, 476–77
- receptors for heat, cold, and pain, 474–76
- tactile and pressure receptors, 473–74

Sensory adaptation, 472
Sensory hair cells, 477
Sensory homunculus, 478
Sensory nerve(s), 344, 389
Sensory nerve fiber, 335, 345
Sensory neurons, 232, 342, 411, 433
Sensory organs
- classification of senses, 471–72
- clinical case studies, 471, 517
- clinical considerations, 510–11, 514–17
- gustatory sense, 480–81
- hearing and balance, 499–510
- olfactory sense, 479
- overview of sensory perception, 471
- somatic senses, 472–78
- specialized cell membranes, 54
- visual sense, 481–97

Sensory receptors, 108, 110, 344, 471, 474–76
Sensory root, 394
Septa, of cranial dura mater, 369
Septal defects, of heart, 572
Septicemia, 575
Septil cartilage, 585
Septum penis, 691
Septum secundum, 569
Serosa, of gastrointestinal tract, 619
Serotonin, 525
Serous cells, 625
Serous layer, of uterine tube, 715
Serous membranes, 41–43, 82, 616–17
Serous pericardium, 529
Serrate suture, 193
Serratus muscles, 253–54, 256
Sertoli, Enrico, 683
Sertoli cells, 683. *See also* Sustenacular cells
Serum, blood, 528
Sesamoid bone(s), 129
Sex. *See* Female; Male; Reproductive system
Sex chromosomes, 762, 763
Sex determination, 696
Sex-linked inheritance, 764–65
Sex steroids, 747
Sexual dimorphism, 759, 760–61
Sexual dysfunction, 699–700
Sexual intercourse, 679, 707
Sexually transmitted diseases (STDs), 700–702
Sexual response, and hypothalamus, 363
Shaft, of bone, 134
Shakespeare, William, 2
Sharpey, William, 134
Sharpey's fibers, 134
Sherrington, Charles, 6
Shin, 40
Shinbone. *See* Tibia
Shingles, 401
Shin splints, 285, 321, 330
Shock, treatment of, 576, 577
Shoes, and joint-related problems, 198, 203
Short bones, 133
Shoulder
- arteries of, 547–48
- disease of and trauma to, 329–30
- muscles of, 249, 254–57
- radiograph of, 171
- structure and function of, 210–11
- surface and regional anatomy of, 311–17
- upper extremity and, 39

Shoulder blade. *See* Scapula
Shoulder joint. *See* Glenohumeral joint
Siamese twins, 767
Sickle-cell anemia, 572
SIDS. *See* Sudden infant death syndrome
Sighing, 602
Sigmoid colon, 637
Sigmoid curvature, of spine, 28
Simple ciliated columnar epithelium, 78, 82
Simple columnar epithelium, 76–77, 82
Simple cuboidal epithelium, 76, 82
Simple diffusion, 53
Simple epithelium, 76–78, 82
Simple fracture, 187
Simple glands, 82, 84, 85
Simple mastectomy, 729
Simple squamous epithelium, 76, 82
Single unit smooth muscle, 422, 423
Singular word endings, guide to, 33
Sinoatrial node, 535–36

Sinus. *See also* Paranasal sinuses; Sinus venosus
of bone, 133
of lymph node, 564
Sinusitis, 608
Sinusoid(s), 540
Sinus venosus, 569, 570
Skeletal muscle
appendicular skeleton and, 228, 253–78
axial skeleton and, 228, 241–53
clinical case study, 227, 283
clinical considerations, 278, 282–83
development of, 280–81
eye and, 486
fibers and types of muscle contraction, 233–38
naming of, 239–41
respiratory system and, 610
shape of cells of, 48
structure of, 229–33
types of muscle tissue and, 96, 97
Skeletal muscle pump, 541
Skeletal system. *See also* Appendicular skeleton; Axial skeleton; Joints
bone growth, 137–39
bone tissue, 135–36, 137
circulatory system and, 578
digestive system and, 651
endocrine system and, 466
functions of, 31, 131–32
integumentary system and, 125
interaction with other body systems, 165
male reproductive system and, 703
muscular system and, 284
nervous system and, 384
organization of, 129
structure of bone, 132–35
urinary system and, 674
Skeleton. *See* Appendicular skeleton; Axial skeleton; Skeletal system
Skin. *See also* Integumentary system
coloration of, 106
conditions of indicating nutritional deficiencies or disease, 103
of neonate, 291
pH of, 79
shock and, 577
stratified squamous epithelium, 80
Skin cancer, 106, 109, 113, 118, 120
Skin grafts, 102, 120, 121–22
Skull
anterior and lateral views of, 142
auditory ossicles, 152, 153
axial skeleton and, 129
cranial bones and, 146–49
facial bones and, 149–52
fetal, 140
floor of cranial cavity, 144
frontal section of, 145
growth of, 758
hyoid bone, 152, 153
inferior view of, 143
inferolateral view of, 144
major foramina of, 141
nasal cavity and bones of, 150
sagittal view of, 143
structure of, 131, 139–41
Sleep, hypothalamic regulation of, 363
Sleep apnea, 608
Slipped disc, 24
Slit pores, of renal glomeruli, 661, 663
Small intestine
basic functions of, 616
as digestive organ, 632–35
endocrine system and, 441, 459
Small peroneus tertius, 274
Small saphenous vein, 558
Smegma, 691
Smooth chorion, 745
Smooth endoplasmic reticulum (smooth ER), 55–56
Smooth muscle(s)
autonomic nervous system and, 420, 422, 423
comparison of single-unit and multiunit, 423
eye and, 486
shape of cells of, 48
structure and function of, 96–97
Smooth pursuit movements, 497
Sneezing, 78, 602
Snellen, Herman, 511
Snellen's chart, 511
Soft corns, 321
Soft palate, 621
Soleus muscle, 274–75, 278
Solutes, 49
Solution, 49
Solvent, 49
Somatic death, 99
Somatic motor nerve, 335, 421
Somatic pain, 474–75
Somatic reflexes, 411–12
Somatic sensation, neural pathways for, 477–78
Somatic senses. *See* Senses; Sensory organs
Somatic sensory fibers, 344
Somatomammotrophin, 462
Somatostatin, 446
Somatotropin, 446, 448
Somesthetic senses, 473
Sound frequency, 505
Sounds, of heart, 537
Sound waves, 505
Sour taste, 481
Spallanzani, Lazzaro, 16
Spastic paralysis, 381
Special senses, 472
Spectrum, of light, 494, 496
Speculum, 724
Speech, as characteristic of humans, 28
Spemann, Hans, 6
Spermatic cord, 689, 696
Spermatic ducts, 688–89
Spermatids, 685
Spermatocytes, 683, 685
Spermatogenesis, 685–88
Spermatogonia, 683, 685
Spermatozoa, 48, 679, 686, 735
Spermicides, 731, 732
S phase, 64–65
Sphenoidal sinus, 148, 587
Sphenoid bone, of skull, 148
Sphenomandibular ligament, 209
Sphincteral muscles, 231
Sphincter of ampulla, 642
Sphincter of Oddi, 633
Sphygmomanometer, 541–42, 567
Spina bifida, 161, 380
Spinal column, 153
Spinal cord
central nervous system and, 372–78
as characteristic of chordates, 23, 24
development of, 378
injuries to, 381
meninges and, 368
vertebral column and, 153
Spinal ganglion, 401
Spinalis muscle(s), 252, 253
Spinal meningitis, 382
Spinal nerve(s)
peripheral nervous system and, 334, 400–401
pulmonary ventilation and, 601
scanning electron micrograph of, 391
structure of spinal cord and, 373
Spinal root, 397
Spindle fibers, 63
Spine, of bone, 133
Spinocerebellar tract(s), of spinal cord, 376
Spinothalamic tract(s), of spinal cord, 376
Spinous process, 154
Spiral fracture, 187
Spiral ganglion, 392, 396
Spiral organ, of inner ear
cochlear duct and, 504–5
cochlear nerve and, 396
cell membranes and, 54, 55
development of, 513
Spirometer, 601
Splanchnic nerve(s), 424
Spleen, 430, 565, 566
Splenectomy, 328, 565
Splenic artery, 549
Splenic vein, 559
Splenius capitis muscle, 245
Split brain, 497
Spondylitis, 224
Spongy bone, 94, 133, 136, 139
Spongy part of urethra, 666–67, 690
Spontaneous abortion, 729
Spoon nails, 103, 114
Sprain
of ankle, 218
of connective tissue, 88
as joint injury, 200, 218–19, 330
Sprengel, Otto G. K., 329
Sprengel's deformity, 329
Squamosal-dentary jaw articulation, 25
Squamous cell carcinoma, 118
Squamous epithelial cells, 48, 76
Squamous part, of temporal bone, 146–47
Squamous suture, 141, 193
Staining, of tissue samples, 99
Stapedius muscle, 501
Stapes, 33, 152, 153, 501
State of being, and diagnosis of shock, 577
Statoacoustic nerve, 396
Statoconia, 506, 509
Statoconial membrane, 506
Steindler, Arthur, 6
Stelluti, Francisco, 16
Stenosis, of heart valves, 326, 537
Stensen, Niels, 625
Stensen, Nicholaus, 16
Stensen's duct. *See* Parotid duct
Stereoscopic vision, 28
Sterility, and male reproductive system, 700. *See also* Infertility
Sterilization, and contraception, 730
Sternal angle, 159, 301
Sternal extremity, of clavicle, 169
Sternal region, 38
Sternoclavicular joint, 209
Sternoclavicular ligament, 209
Sternocleidomastoid muscle
accessory nerve and, 397
effects of removal of, 283
muscles of neck and, 244–45, 247
respiration and, 248, 599
surface anatomy of neck and, 298
Sternohyoid muscle, 246, 247
Sternothyroid muscle, 246
Sternum, 159
Steroid(s), 238, 441–43
Stethoscope, 567, 635
Stimulus, definition of, 471
Stomach
basic functions of, 616
as digestive organ, 628–32
endocrine system and, 441, 459
radiograph of with radiopaque contrast medium, 18
tissue types in, 29
Strabismus, 516
Strain. *See also* Sprain
of connective tissue, 88
of joint, 218
of muscles, 221, 278
Stratified cuboidal epithelium, 80, 81, 82
Stratified epithelium, 78–80, 82
Stratified squamous epithelium, 79, 80, 82
Stratum basale, 103, 105, 106, 717
Stratum corneum, 104, 105
Stratum functionale, 717
Stratum germinativum, 104
Stratum granulosum, 104, 105
Stratum lucidum, 104, 105
Stratum papillarosum, 108
Stratum reticularosum, 108
Stratum spinosum, 103–4, 105
Stress, and immune system, 458
Stress fractures, 330
Stretch receptors, 433
Stretch reflex, 411
Striate cortex, 496
Stroke. *See also* Cardiovascular disease
cerebellum and, 366
as cerebrovascular accident (CVA), 383
coronary embolisms and, 575
Stroma, 90, 709–10
Structural modifications, of small intestine, 634
Structural proteins, 49
Strychnine, 348
S-T segment, 537
Sty, 516
Styloglossus muscle, 244, 245
Stylohyoid muscle, 246, 247
Styloid process, 147, 174
Stylomandibular ligament, 209
Stylomastoid foramen, 147, 392
Subacromial bursa, 211
Subarachnoid space, 348, 368, 369, 370
Subclavian arteries, 543, 545, 547
Subclavian vein(s), 556
Subclavius muscle, 253–54, 256
Subcoracoid bursa, 211
Subcutaneous prepatellar bursa, 216
Subcutaneous tissue, 108–9, 288, 289
Subdeltoid bursa, 211
Subdermal implant, of hormonal contraceptive, 730
Subdural hemorrhage, 325
Sublingual gland, 625, 626, 627
Subluxation, of joint, 219
Submandibular duct, 16, 625
Submandibular ganglion, 425, 427
Submandibular gland, 625, 626, 627
Submucosa
of gastrointestinal tract, 617–18
of urinary bladder, 665
Submucosal plexus, 618
Subscapular fossa, 169

Subscapularis muscle, 256
Substantia nigra, 363
Suckling, and newborns, 621
Sucrase, 633
Suction lipectomy, 91
Sudden infant death syndrome (SIDS), 608
Sudoriferous glands, 114–15
Suffixes, glossary of, 33
Sugita, Genpaku, 5, 6, 8
Suicide, 68
Sulcus, of bone, 133. *See also* Cerebral sulci
Superficial cutaneous branches, of cervical plexus, 402
Superficial digital flexor muscle, 261
Superficial epithelium, 709
Superficial fibular nerve, 407
Superficial palmar arch, 548, 556
Superficial temporal artery, 546–47
Superior, as directional term, 36
Superior angle, of scapula, 169
Superior articular process(es), of vertebra, 154, 158
Superior articular surfaces, 156
Superior border, of scapula, 169
Superior cerebellar peduncles, 364
Superior colliculi, 363, 496, 497
Superior gemellus muscle, 266
Superior gluteal artery, 550, 552
Superior gluteal nerve, 409
Superior lacrimal canaliculi, 485
Superior limb buds, 748
Superior lobe, of lung, 594–95
Superior mesenteric artery, 549–50
Superior mesenteric ganglia, 424
Superior mesenteric vein, 559
Superior nasal conchae, 585
Superior nuchal line, 147, 292
Superior oblique muscle, 394, 486
Superior orbital fissure, 148, 392
Superior phrenic arteries, 549
Superior rectus muscle, 394, 486
Superior salivatory nucleus, 392
Superior thyroid artery, 546
Superior trunk, of brachial plexus, 401
Superior vena cava, 530, 554, 556
Superior vesicular artery, 550
Supination, 202
Supinator muscle, 258, 263
Supinator reflex, 414, 415
Supporting cells, 480
Suprahyoid muscle(s), 246
Supraoptic nucleus, 447
Supraorbital foramen, 146
Supraorbital margin, 146
Supraorbital ridges, 292
Suprapatellar bursa, 216
Suprarenal arteries, 550
Suprarenal glands. *See* Adrenal glands
Supraspinatus muscle, 256
Supraspinous fossa, 169
Surface anatomy
  clinical considerations, 321–22, 325–30
  definition of, 17
  of head, 291–97
  of infant, 289–91
  introduction to, 288–89
  of neck, 297–300
  of pelvis and perineum, 310–11
  of shoulder and upper extremity, 311–17
  of trunk, 300–309
Surface patterns, of skin, 106–7
Surfactant, and lungs, 599
Surgical neck, of brachium, 172
Surgical removal, of tissue, 99
Suspensatory ligament
  of breast, 720
  of eyeball, 488
  of uterus, 709
Sustentacular cells, 683, 685
Sutural bone(s), 129, 147
Suture(s), in skull, 193, 200
Swallowing mechanisms, 628
Sweat gland(s), 105, 114–15, 116, 293
Sweet taste, 481
Sylvius, Jacobus, 13–14, 363
Sylvius del la Boe, Franciscus, 355
Sympathetic division, of autonomic nervous system, 334, 420, 423–25
Sympathetic trunk ganglion, 401
Symphysis, 194, 195, 200
Symphysis pubis, 129, 301
Synapse(s), 347–48
Synaptic cleff, 347
Synaptic vesicle(s), 237, 347
Synchondrosis, 194–95, 200
Syncytial myotubes, 280
Syncytiotrophoblast(s), 739
Syndactyly, 186, 329
Syndesmoses, 193–94, 200
Synergistic muscles, 231
Synostosis, 193
Synovial fluid, 196, 220
Synovial joints, 192, 195–208, 220
Synovial membrane, 196, 220
Synovitis, 224, 330
Synthetic skin, 122
Syphilis, 326, 382–83, 700, 701, 702
Syringomyelia, 383
System(s), and body organization, 29–32
Systema naturae (Linnaeus), 6
Systematic approach, to anatomy, 30
Systemic circulation, 534–35
Systole, 536
Systolic pressure, 542

T

Tabes dorsalis, 383
Tachycardia, 567, 572
Tactile cells, 103
Tactile receptors, 472, 473–74
Taeniae coli, 638
Tail
  of epididymis, 688
  of pancreas, 643
Tail bone. *See* Coccyx
Talipes, 186, 330
Talocrural (ankle) joint, 217–18
Talofibular ligament(s), 217
Talus, 182, 184
Tarsal bone(s), 186
Tarsal gland(s), 484
Tarsal plate(s), 484
Tarsier (primate), 26
Tarsometatarsal joint(s), 184
Tarsus, 40, 182, 184
Taste, sense of, 480–81
Taste bud(s), 396, 480, 481
Taste pore, 480
Tatooing, 106
Taxon, 23
Taxonomy, and characteristics of humans, 23–27
Tay, Warren, 383, 770
Tay-Sachs disease, 383, 744, 770
T cells (thymus-dependent cells), 459
Tectal system, 496
Tectorial membrane, 505
Tectospinal tracts, of spinal cord, 376
Teeth
  characteristics of humans and, 28
  clinical problems of, 649
  as digestive organ, 622–24, 625
  gomphoses and, 194
  maxilla and, 149
  mineral storage in, 132
  orthodontics and, 139
Telencephalon, 337, 352
Telophase, 65, 66, 685, 686
Temporal arteries, 546–47
Temporal bone, of skull, 146–47
Temporalis muscle, 241, 243, 292, 394
Temporal lobe, of cerebrum, 356
Temporal process, 150
Temporal region, of head, 292
Temporomandibular joint, 147, 208, 209
Temporomandibular joint (TMS) syndrome, 209
Tendo calcaneous, 320
Tendon(s), 88, 229, 261. *See also* Tendon sheath
Tendonitis, 219
Tendon sheath, 197, 198, 229
Tennis elbow, 173, 212, 329
Tenosynovitis, 330
Tensor fasciae latae muscle, 266, 267
Tensor tympani muscle, 501
Tentorium cerebelli, 356, 364, 369
Teratology, 68, 769
Teres major and minor muscles, 256
Terminal bronchioles, 592
Terminal ganglia, 425
Tertullian, 11
Testes
  development of, 696, 699
  disorders of, 700, 702
  endocrine system and, 441
  function of, 681
  structure and functions of, 682, 683–87
Testicular artery(ies), 550
Testicular lobules, 683
Testicular vein(s), 558
Testosterone, 112, 442, 684
Tetanus, 241
Tetralogy of Fallot, 326, 573
Thalamus, 352, 360, 361–62
Thalidomide, 185, 329
Theca interna, 711
Thenar eminence, 314
Thenar muscles, 263, 265
Thermoreceptors, 472, 473, 474
Thermoregulation, and skin, 109–10
Theta waves, 357
Thigh
  bipedal locomotion and, 28
  lower extremity and, 40, 176
  muscles of, 265–73
  surface anatomy of, 318–19
Third-degree burns, 102, 119–20
Third-degree frostbite, 121
Third-division nerve block, 152, 395
Third molars, 622
Third trimester, of gestation, 765
Thoracic cavity, 41, 43, 595, 597
Thoracic curve, of vertebral column, 154
Thoracic duct, of lymphatic system, 564
Thoracic nerve(s), 601
Thoracic portion, of aorta, 548–49, 550
Thoracic region, 38, 326–27
Thoracic vertebra, 153, 156, 157
Thoracolumbar division, of autonomic nervous system, 423–25
Thoracotomy, 611
Thorax
  arteries of, 548–49, 550
  body regions, 38
  internal anatomy of, 305–7
  origin of term, 33
  serous membranes, 43
  surface anatomy of, 301, 303
  trauma to, 326–27
Thoroughfare channels, 538
Thrombocytes (platelets), 94, 526
Thrombus, 575, 577
Thumb, 314. *See also* Opposable grip
Thymine, 63
Thymosin, 459
Thymus, 441, 459, 565, 566
Thyrocalcitonin, 450
Thyrocervical trunk, 547
Thyroglossal duct, 462
Thyrohyoid muscle, 246, 247
Thyroid cartilage, 297–98, 588
Thyroid diverticulum, 461
Thyroid follicles, 450
Thyroid gland
  development of, 461–62
  disorders of, 463–65
  endocrine system and, 441, 450–52
  origins of term, 33
  surface anatomy of neck and, 297, 298
Thyroid scans, 463
Thyroid-stimulating hormone (TSH), 446, 448, 452
Thyrotropin, 448
Thyrotropin-releasing hormone (TRH), 446
Thyroxine, 442, 451, 452
Tibia, 131, 180–81, 186
Tibial arteries, 553
Tibialis anterior muscle, 273, 278
Tibialis posterior muscle, 275–76, 278
Tibial nerve, 407, 409, 410
Tibial tuberosity, 180
Tibial vein, 557
Tibiofemoral joint, 215–16
Tidal volume (TV), of lungs, 601
Tinnitus, 517
Tissue(s). *See also* Bone tissue; Muscle tissue; Vascular tissue
  analysis of and diagnosis of disease, 99
  body organization and, 28–29
  definition and classification of, 73
  disease and changes in composition of, 98–99
  prenatal development, 74–75
  transplantation of, 99
Tissue fluid, 522
Tissue-rejection reaction, 99
Toe(s), 40, 273–78, 279. *See also* Digits; Phalanges
Tongue
  as digestive organ, 622
  muscles of, 244
  nerves of, 396
Tonic receptors, 472
Tonometer, 511
Tonsillectomy, 565, 587
Tonsillitis, 587
Tonsils, 565, 566, 587
Tooth. *See* Teeth
Torso, 38. *See also* Trunk
Torticollis, 282

Total lung capacity (TLC), 601
Touch. *See* Senses; Tactile receptors
Toxic goiter, 464
Trabeculae, 136
Trabeculae carneae, 534
Trachea, 298, 590–91, 598
Tracheobronchitis, 608
Tracheostomy, 298, 591
Tracheotomy, 591
Trachoma, 516
Tragus, 293, 294
Transitional epithelium, 80, 81, 82
Transmission electron microscope (TEM), 55
Transplantation, of tissue, 99
Transportation, as function of circulatory system, 521
Transurethral prostatic resection, 690
Transverse acetabular ligament, 215
Transverse arch, of foot, 184
Transverse colon, 637
Transverse fissure, of cerebellum, 364
Transverse foramen, of cervical vertebra, 156
Transverse fracture, 187
Transverse humeral retinaculum, 210–11
Transverse ligament, of knee, 216
Transverse lines, 158
Transverse palatine folds, 621
Transverse plane, 33–34, 35, 39
Transverse process(es), of vertebra, 154
Transverse tubules, 234
Transversus abdominis muscle, 248, 250
Transversus perinei muscle, 250
Trapezium bone, 174
Trapezius muscle(s), 245, 253–54, 256, 298, 397
Trauma. *See also* Injury
  to abdomen, 328
  appendicular skeleton and, 187–88
  to brachial plexus, 403
  to cerebullum, 366
  to circulatory system, 576
  to facial nerve, 396
  to head and neck, 325
  hemmorhage and, 553
  to hip and lower extremity, 330
  to joints, 200, 218–19, 221
  to respiratory system, 607–8
  to shoulder and upper extremity, 329
  to sphenoid bone, 148
  to thorax, 326–27
  to urinary organs, 673
Traumatic fractures, 187
Trench mouth, 652
Trepanation, 3–4
Treponema pallidum, 700
TRH. *See* Thyrotropin-releasing hormone
Triangle of auscultation, 301
Triangles, of neck, 298–300
Triangular fossa, 294
Triceps brachii muscle, 258, 259
Triceps reflex, 414, 415
Triceps surae muscle, 274–75
Trichomonas vaginalis, 729
Trichosiderin, 112
Tricuspid area, 537
Tricuspid valve, 530–31
Trigeminal ganglion, 392, 394
Trigeminal nerve(s), 392, 394–95, 399
Trigone, 665
Triiodothyronine, 451–52
Trimesters, of gestation, 765
Triplets, and multiple pregnancy, 767
Triquetrum bone, 174
Trisomy, 68
Trochanter, of bone, 133, 180
Trochlea, 173, 241, 486
Trochlear nerve, 392, 394, 399
Trochlear notch, 174
Trochlear nucleus, 392
Trophic hormones, 446
Trophoblast, 75, 737
True ribs, 159
Truncus arteriosus, 569, 570
Trunk
  body cavities, 42
  body region, 38
  DSR scan through, 19
  surface anatomy of, 300–309
Trusor muscle, 665
Trypsin, 633
TSH. *See* Thyroid-stimulating hormone
Tubal ligation, 724, 730
Tubal pregnancy, 728, 732, 766
Tubercle, of bone, 133
Tuberculosis, 608
Tuberosity, of bone, 133
Tuberosity of calcaneus, 184
Tuberosity of radius, 174
Tubular glands, 82, 85
Tubules, 55
Tubuloacinar glands, 82, 85
Tulp, Dr. Nicholas, 15
Tumors, of ovaries, 710, 726. *See also* Cancer; Neoplasms
Tunic(s), of gastrointestinal tract, 617–19
Tunica albuginea, 683, 709
Tunica externa, 538
Tunica interna, 538
Tunica media, 538
Tunica muscularis, 618–19, 690
Tunica vaginalis, 683
Tuning fork tests, 511
Tunnel vision, 496
Turbinates, 585
Turner, Henry H., 699
Turner's syndrome, 68, 699
T wave, 537
Twins, 735, 766–67, 768
Tympanic cavity, 501, 502, 513
Tympanic membrane, 500, 516
Tympanic part, of temporal bone, 147
Tympanic recess, 501
Tympanum, 33
Type J receptors, 433
Tyrosine, 106

U

Ulcer(s), 328–29
Ulna, 129, 173, 174, 193
Ulnar artery, 548
Ulnar branch, of brachial plexus, 404
Ulnar nerve, 406
Ulnar notch, of radius, 174
Ulnar recurrent artery, 548
Ulnar vein, 556
Ultrasonography, 753, 754, 767
Ultrasound, of heart, 567
Ultrastructure, definition of, 17
Ultraviolet light, 494
Umbilical arteries, 561, 745, 746
Umbilical cord, 291, 561, 746, 747, 748
Umbilical hernia, 285, 305
Umbilical vein, 561, 562, 745, 746
Umbilicus, 38, 301, 305
Unicellular glands, 82, 85
Unicornuate uterus, 726
Upper arm birth palsy, 329
Upper extremity
  appendicular skeleton and, 129
  arteries of, 547–48
  as body region, 39
  bones of, 131
  disease of and trauma to, 329–30
  surface and regional anatomy of, 311–17
  veins of, 556
Upper leg, 40
Upper respiratory system, 584, 605
Urachus, 672, 745
Uremia, 672
Ureter(s), 81, 657, 664–65, 669
Ureteric bud, 671–72
Urethra, 666–67, 668, 669, 672, 690
Urethral folds, 725
Urethral glands, 666, 690
Urethral groove, 725
Urethral orifice, 666
Urethritis, 673, 701
Urinalysis, 463, 661, 673
Urinary bladder
  autonomic nerve stimulation and, 430
  disorders of, 669
  stratified squamous epithelium and, 80
  structure and function of, 665
  transitional epithelium and, 81
Urinary incontinence, 672, 675
Urinary retention, 668
Urinary stones, 673
Urinary system
  circulatory system and, 578
  clinical case study, 656, 675
  clinical considerations, 669, 672–73, 675
  development of, 670–72
  digestive system and, 651
  endocrine system and, 466
  functions of, 31
  integumentary system and, 125
  interaction with other body systems, 674
  kidneys, 657–64
  male reproductive system and, 703
  muscular system and, 284
  nervous system and, 384
  respiratory system and, 610
  skeletal system and, 165
  structure and functions of, 656–57
  transitional epithelium, 81
  ureters, urinary bladder, and urethra, 664–69
Urinary tract infections (UTIs), 673
Urinary tubule, 659
Urine, 661, 664. *See also* Micturition
Urogenital diaphragm muscle, 250
Urogenital membrane, 647
Urogenital ridge, 671
Urogenital sinus, 647
Urogenital triangle, 681
Urorectal septum, 647
Uterine artery, 550
Uterine cavity, 715, 725
Uterine displacements, 729
Uterine neoplasms, 728
Uterine ostium, 715
Uterine prolapse, 716
Uterine tube(s)
  anatomical nomenclature and, 17
  abnormalities of, 726, 727, 728
  contrast radiograph of, 725
  description and functions of, 708, 714–15
  epithelium of, 78
  as secondary sex organ, 707
Uterine wall, 716–17
Uterus
  autonomic nervous system and, 430
  blood supply and innervation of, 717
  function of, 707, 708
  lysosomes and, 59
  problems involving, 728–29
  structure of, 715
  support of, 716
  uterine wall, 716–17
Utricle, 502, 505, 506–7
Utricular portion, 513
Uvea, 491
Uvula, 33, 587

Vacuoles, 51, 59
Vagal trunk, 427–28
Vagina, 707, 708, 717–18, 729
Vaginal artery, 550
Vaginal orifice, 717
Vaginal rugae, 718
Vaginal spermicides, 731, 732
Vaginal vestibule, 708, 719
Vaginitis, 701, 729
Vagotomy, 652
Vagus nerve
  distribution of, 398
  dysfunctions of, 399
  parasympathetic division of autonomic system and, 428
  reflexes stimulated by input from sensory neurons in, 433
  structure and functions of, 392, 396–97
  structure of neck and, 300
Vagus nuclei, 366
Valgus stress, 192
Vallate (circumvallate) papillae, 480, 622
Valsalva, Antonio, 556
Valsalva's maneuver, 556
Valves, of heart, 529–34
Valvular auscultatory areas, 537
Valvular insufficiency, 537
Varicocele, 684
Varicose veins, 321, 541, 575
Vasa recta, 660
Vascular disorders, 574–75
Vascular processes, and blood-brain barrier, 342
Vascular shunts, 538
Vascular supply, of skin, 108
Vascular tissue, 94, 95
Vascular tunic, 487–88, 491
Vas deferens, 688
Vasectomy, 700, 701, 730
Vasomotor center, of medulla oblongata, 367
Vasopressin, 447
Vastus intermedius muscle, 270, 271
Vastus lateralis muscle, 270, 271
Vastus medialis muscle, 270, 271
Vater, Abraham, 633
Vein(s). *See also* Blood vessels
  of abdominal region, 558
  definition of, 522
  of head and neck, 556
  of hepatic portal system, 558–60

location of principal, 555
of lower extremity, 557–58
structure and function of, 538, 540–41, 554, 556
of thorax, 557
of upper extremity, 556–57
Vein stripping, 575
Venereal disease. *See* Sexually transmitted diseases
Venereal wart(s), 701
Venous blood, 523–24
Venous sinus thrombosis, 326
Venous valves, 541
Ventilation, and respiration, 583
Ventral spinothalamic tract, 477
Ventricle(s)
of brain, 348, 370–72
of heart, 529, 569, 570
Ventricular fibrillation, 327, 572
Ventricular folds, 588–89
Ventricular septal defect, 326, 572
Ventricular tachycardia, 572
Venule(s), 538
Vermilion, 621
Vermis, 33, 364
Vernix caseosa, 116, 291, 753–54
Vertebra, 28, 154–58
Vertebral arch, 154
Vertebral artery, 544, 547
Vertebral column
aging and compaction of, 93, 129, 164
bones of, 131
as characteristic of chordates, 24
injury to, 219, 221
muscles of, 251–53
structure and function of, 153–59
Vertebral foramen, 154
Vertebral region, 38
Vertex position, of fetus, 754
Vertigo, 508, 517
Vesalius, Andreas, 5, 6, 13–14, 128
Vesicouterine pouch, 716
Vesicular arteries, 550
Vesicular ovarian follicle, 712
Vestibular bulbs, 719
Vestibular ganglion, 392, 396
Vestibular glands, 708
Vestibular membrane, 504
Vestibular nerve, 396
Vestibular nuclei, 396
Vestibular organs, 396
Vestibular window, 501, 502
Vestibule
of inner ear, 502
of oral cavity, 620
Vestibulocochlear nerve
dysfunctions of, 399
neural pathway for hearing and sensory neurons of, 505, 508
structure and function of, 392, 396, 397
Vestibulospinal tract, 376, 377
Vibrissae, 585
Villi, of small intestine, 634
Villous chorion, 745
Virchow, Rudolf, 6, 16
Viruses, and cellular dysfunction, 68
Viscera, 41
Visceral effector organs, and autonomic nervous system, 422, 430
Visceral motor fibers, 344
Visceral organs, 41, 43
Visceral pain, 474–75
Visceral pericardium, 42
Visceral peritoneum, 42, 617
Visceral pleura, 42, 595
Visceral portion, of serous membrane, 616–17
Visceral reflexes, 411
Visceral senses, 472
Visceral sensory fibers, 344
Viscerocranium, 163
Visceroreceptors, 472
Visible light, 494
Vision. *See* Eye; Visual sense
Visual cortex, 393
Visual information, processing of, 497
Visual inspection, of surface anatomy, 288
Visual portion, of retina, 488
Visual sense. *See also* Eye
accessory structures of eye and, 483–87
basic structure of eye and, 481–82
function of eyeball and, 492–96
neural pathways and processing of information, 496–97
structure of eyeball and, 487–91
Visual spectrum, of light, 494, 496
Vital capacity (VC), of lungs, 601
Vitamins, 110, 132 , 161, 631
Vitiligo, 106
Vitreous body, 499
Vitreous humor, of eye, 491
Vocal cords, 588–89
Vocal structures, 28
Volkmann, Alfred, 136
Volkmann's canal, 136
Voluntary body movements, 97
Vomer, 151, 585
Vomiting, 619, 631, 652
Vomiting center, 631
Vulva, 718–19, 729
Vulvovaginitis, 729

W

Wakefulness center, in hypothalamus, 363
Wall, of heart, 529
Warts, 126, 701
Water
as component of body, 49
hypothalamus and regulation of, 362
as reactant or solvent, 49
Watson, James D., 5, 6, 64
Weismann, August, 6
Wernicke, Karl, 359
Wernicke's aphasia, 359
Wernicke's area, 359
Wharton, Thomas, 16, 86, 625, 746
Wharton's duct. *See* Submandibular duct
Wharton's jelly, 86, 746
Whiplash, 157, 325, 381
White blood cell(s), 47, 48, 58, 59. *See also* Leukocyte
White matter
of brain, 348, 355, 357–58, 359
of spinal cord, 373–74
White pulp, of spleen, 565
White rami communicantes, 423
White ramus, 401
William of Saliceto (1215–80), 12
Willis, Thomas, 16, 544
Wilson, Edmund B., 6
Wisdom teeth, 622
Withdrawal reflex, 412, 413
Work of the digestive glands, The (Pavlov), 6
Wounds. *See also* Injury; Trauma
to eyeball, 491
healing process and, 94, 123–24
Wrist, muscles of, 258–63
Wristdrop, 403
Wryneck, 282

X chromosome, 762, 763
Xenograph, 99, 122
Xiphisternal joint, 301
Xiphoid process, 159
Xiphos, 33
X-rays, 17, 68

Y

Yang, ancient Chinese concept of, 8
Yawning, 602
Y chromosome, 762, 763
Yellow bone marrow, 134
Yin, ancient Chinese concept of, 8
Yolk sac, 744

Z

Z lines, 234, 235, 237
Zona fasciculata, of adrenal cortex, 455
Zona glomerulosa, of adrenal cortex, 455
Zona pellucida, 711, 735
Zona reticularis, of adrenal cortex, 455
Zonular fibers, of eye, 488
Zygomatic arch, 146, 150
Zygomatic bone, 150
Zygomaticofacial foramen, 150
Zygomatic process, 146–47, 150
Zygomaticus muscle, 242
Zygote, 74, 713, 740

Prefixes and Suffixes in Anatomical and Medical Terminology

Element	Definition and Example
a-	absent, deficient, lack of: *atrophy*
ab-	off, away from: *abduct*
abdomin	relating to the abdomen: *abdominal*
-able	capable of: *viable*
ac-	toward, to: *actin*
acou-	hear: *acoustic*
ad-	toward, to: *adduct*
af-	movement toward a central point: *afferent artery*
alb-	white: *corpus albicans*
-algia	pain: *neuralgia*
ambi-	both: *ambidextrous*
angi-	pertaining to the vessels: *angiology*
ante-	before, in front of: *antebrachium*
anti-	against: *anticoagulant*
aque-	water: *aqueous*
arch-	beginning, origin: *archenteron*
arthr-	joint: *arthritis*
-asis	condition or state of: *homeostasis*
aud-	hearing, sound: *auditory*
auto-	self: *autolysis*
bi-	two: *bipedal*
bio-	life: *biopsy*
blast-	generative or germ bud: *blastocyst*
brachi-	arm: *brachialis*
brachy-	short: *brachydont*
brady-	slow: *bradycardia*
bucc-	cheek: *buccal cavity*
cac-	bad, ill: *cachexia*
calc-	stone: *calculus*
capit-	head: *capitis*
carcin-	cancer: *carcinogenic*
cardi-	heart: *cardiac*
cata-	lower, under, against: *catabolism*
caud-	tail: *cauda equina*
cephal-	head: *cephalic*
cerebro-	brain: *cerebrospinal fluid*
chol-	bile: *cholic*
chondr-	cartilage: *chondrocyte*
chrom-	color: *chromocyte*
-cide	destroy: *germicide*
circum-	around: *circumduct*
co-	together: *copulation*
coel-	hollow cavity: *coelom*
-coele	swelling, an enlarged space or cavity: *blastocoele*
con-	with, together: *congenital*
contra-	against, opposite: *contraception*
corn-	denoting hardness: *cornified*
corp-	body: *corpus*
crypt-	hidden: *cryptorchidism*
cyan-	blue: *cyanosis*
cyst-	sac or bladder: *cystoscope*
cyto-	cell: *cytology*
de-	down, from: *descent*
derm-	skin: *dermatology*
di-	two: *diarthrotic*
dipl-	double: *diploid*
dis-	apart, away from: *disarticulate*
duct-	lead, conduct: *ductus deferens*
dur-	hard: *dura mater*
dys-	bad, difficult, painful: *dysentery*
e-	out, from: *eccrine*
ec-	outside, outer, external: *ectoderm*

Element	Definition and Example
-ectomy	surgical removal: *tonsillectomy*
ede-	swelling: *edema*
-emia	pertaining to a condition of the blood: *lipemia*
en-	within: *endoderm*
enter-	intestine: *enteritis*
epi-	upon, over: *epidermis*
erythro-	red: *erythrocyte*
ex-	out of: *excise*
exo-	outside: *exocrine*
extra-	outside of, beyond, in addition: *extracellular*
fasci-	band: *fascia*
febr-	fever: *febrile*
-ferent	bear, carry: *efferent*
fiss-	split: *fissure*
for-	opening: *foramen*
-form	shape: *fusiform*
gastro-	relating to the stomach: *gastrointestinal*
-gen	an agent that produces or originates: *pathogen*
-genic	produced from, producing: *carcinogenic*
gloss-	tongue: *glossopharyngeal*
glyco-	sugar: *glycosuria*
-gram	a record, recording: *electroencephalogram*
gran-	grain, particle: *granulosa cells*
-graph	instrument for recording: *electrocardiograph*
gravi-	heavy: *gravid*
gyn-	female sex: *gynecology*
haplo-	simple or single: *haploid*
hem(at)-	blood: *hematology*
hemi-	half: *hemiplagia*
hepat-	liver: *hepatic portal*
hetero-	other, different: *heterosexual*
histo-	web, tissue: *histology*
holo-	whole, entire: *holocrine*
homo-	same, alike: *homologous*
hydro-	water: *hydrocoel*
hyper-	beyond, above, excessive: *hypertension*
hypo-	under, below: *hypoglycemia*
-ia	abnormal state or condition: *hypoglycemia*
-iatrics	medical specialties: *pediatrics*
idio-	self, separate, distinct: *idiopathic*
ilio-	ilium: *iliosacral*
infra-	beneath: *infraspinatus*
inter-	among, between: *interosseous*
intra-	inside, within: *intracellular*
-ion	process: *acromion*
-ism	condition or state: *dimorphism*
iso-	equal, like: *isotonic*
-itis	inflammation: *meningitis*
labi-	lip: *labium majus*
lacri-	tears: *lacrimal apparatus*
later-	side: *lateral*
leuk-	white: *leukocyte*
lip-	fat: *lipid*
-logy	science of: *morphology*
-lysis	solution, dissolve: *hemolysis*
macro-	large, great: *macrophage*
mal-	bad, abnormal, disorder: *malignant*
medi-	middle: *medial*
mega-	great, large: *megakaryocyte*
meso-	middle or moderate: *mesoderm*
meta-	after, beyond: *metatarsal*
micro-	small: *microtome*